KU-661-694

OXFORD MEDICAL PUBLICATIONS

OXFORD TEXTBOOK OF GERIATRIC MEDICINE

EDITORS

J. GRIMLEY EVANS

T. FRANKLIN WILLIAMS

OXFORD TEXTBOOK OF GERIATRIC MEDICINE

Edited by

J. GRIMLEY EVANS

Professor of Geriatric Medicine, University of Oxford

and

T. FRANKLIN WILLIAMS

Professor of Medicine, and Community and Preventive Medicine, University of Rochester, New York; formerly Director, National Institute on Aging, National Institutes of Health, Bethesda, Maryland, USA

Oxford New York Tokyo

OXFORD UNIVERSITY PRESS

1992

Oxford University Press, Walton Street, Oxford OX2 6DP
Oxford New York Toronto
Delhi Bombay Calcutta Madras Karachi
Petaling Jaya Singapore Hong Kong Tokyo
Nairobi Dar es Salaam Cape Town
Melbourne Auckland
and associated companies in
Berlin Ibadan

Oxford is a trade mark of Oxford University Press

Published in the United States
by Oxford University Press, New York

A catalogue record for this book is available from the British Library

Library of Congress Cataloging-in-Publication Data
Oxford textbook of geriatric medicine / edited by J. Grimley Evans and T. Franklin Williams.
(Oxford medical publications)
Includes bibliographical references.
1. Geriatrics. I. Evans, J. Grimley. II. Williams, T. Franklin, 1921– . III. Series.
[DNLM: 1. Geriatrics. WT 100 098]
RC952.094 1992 618.97—dc20 92-8669
ISBN 0–19–261590–4 (Hbk)

Set by BP Integraphics, Bath, Avon

Printed in Great Britain by Butler and Tanner Ltd, Frome, Somerset

Preface

There are few diseases or disabilities that are unique to late adult life and the authors and compilers of a textbook of geriatric medicine must avoid the danger of simply writing a textbook of general medicine. We therefore set out with the explicit intention of compiling a book as a companion volume to a standard textbook of internal medicine. Our authors were asked to start their chapters where the *Oxford Textbook of Medicine* leaves off, to tell our readers the *extra* things that doctors caring for elderly people need to know over and above what their training in general (internal) medicine has taught them.

Our authorship is international but with an inevitable predominance from the United Kingdom which has the longest experience in providing specialist services for an ageing population, and North America which has the greatest output of research in medical geratology (or gerontology to use the established but falsely coined term). We asked our authors to restrict themselves to some 30 references each, either in the form of a bibliography or separately cited items, but to follow whatever style of exposition best suited their disposition and subject matter. With a subject as young and as rapidly developing as geriatric medicine differences of opinion and controversies are inevitable and the reader must not be surprised to find differences of emphasis or even occasional contradictions among the various chapters of this book. In places, typically where practice differs between countries, we have inserted editorial comment in the text but have indicated such interpolations with 'curly' brackets. The authors of the chapters affected may agree or not with our comments and are in no way responsible for them.

We are grateful to our colleagues for their good-natured acceptance of these editorial tyrannies and for the erudition and hard work that have gone into their contributions to this volume. The hope of all of us is that it will contribute to the continuing improvement in the quality of the care and understanding that older people receive from their doctors.

J. Grimley Evans
T. Franklin Williams

Dose schedules are being continually revised and new side effects recognized. Oxford University Press makes no representation, express or implied, that the drug dosages in this book are correct. For these reasons the reader is strongly urged to consult the drug company's printed instructions before administering any of the drugs recommended in this book.

Contents

Contributors

MARY J. BAINES
Consultant Physician, St Christopher's Hospice, London

ARTHUR K. BALIN
Practitioner of Dermatology and Dermatologic Surgery; Scientific Director, Longevity Research Incorporated, Chester, Pennsylvania, USA

JOHN R. BARTLETT
Consultant Neurosurgeon, Brook General Hospital, London

R. N. BARTON
North Western Injury Research Centre, University of Manchester, UK

DAVID W. BENTLEY
Professor of Medicine and Head, Division of Geriatrics/Gerontology, State University of New York at Buffalo, USA

THOMAS P. BERESFORD
University of Michigan Alcohol Research Center, Ann Arbor, Michigan, USA

KLAUS BERGMANN
The Maudsley Hospital, London

MARC R. BLACKMAN
Associate Professor of Medicine, Johns Hopkins University School of Medicine; Clinical Chief, Division of Endocrinology and Metabolism, Francis Scott Key Medical Center; Guest Scientist, Gerontology Research Center, National Institute on Aging, NIH, Baltimore, Maryland, USA

DONALD L. BLIWISE
Director, Sleep and Aging Program, Stanford University Medical School, California, USA

FRANCOIS BOLLER
Director, Neuropsychologie de la Senescence Normale et Pathologique, Unite 324 INSERM, Centre Paul Broca, Paris, France

JACOB A. BRODY
Dean of the School of Public Health, University of Illinois at Chicago, USA

ELAINE M. BRODY
Philadelphia Geriatric Center, Pennsylvania, USA

G. A. BROE
Professor of Geriatric Medicine, University of Sydney; Consultant Neurologist, Concord and Royal Prince Alfred Hospitals, Sydney, Australia

A. J. BRON
Professor; Head of Department of Ophthalmology and Nuffield Laboratory of Ophthalmology, University of Oxford; Honorary Consultant, Oxford Eye Hospital, Oxford

EVAN CALKINS
Professor of Medicine and Family Medicine, State University of New York at Buffalo, USA

MIRA CANTRELL
Assistant Professor of Medicine, University of California in Los Angeles; Clinical Director, Academic Nursing Home Care Unit, VA Medical Center, West Los Angeles, California, USA

CHRISTINE K. CASSEL
Professor of Medicine and Public Policy, University of Chicago, Illinois, USA

HARVEY JAY COHEN
Director, Geriatric Research, Education, and Clinical Center (GRECC), Veterans Administration Medical Center; Director, Center for the Study of Aging and Human Development; Chief, Geriatrics Division, Duke University Medical Center, Durham, North Carolina, USA

KENNETH J. COLLINS
Member of Medical Research Council Staff and Honorary Clinical Lecturer, University College and Middlesex School of Medicine, University of London

L. ADRIENNE CUPPLES
Associate Professor of Public Health, Boston University School of Public Health, Massachusetts, USA

R. CURLESS
Research Fellow, Department of Medicine (Geriatrics), University of Newcastle upon Tyne, UK

PAUL J. DAVIS
Professor and Chairman, Department of Medicine, Albany Medical College; Physician-in-Chief, Albany Medical Center Hospital, Albany, New York, USA

WILLIAM C. DEMENT
Lowell W. and Josephine Q. Berry Professor of Psychiatry and Behavioral Sciences Director, Sleep Disorders Clinic and Research Center, Stanford University, California, USA

MARGARET F. DIMOND
University of Washington School of Nursing, Seattle, Washington, USA

SHAH EBRAHIM
Professor, Department of Health Care of the Elderly, Medical Colleges of The Royal London and St Bartholomew's Hospitals, London

ANN R. FALSEY
Instructor and Fellow, Infectious Diseases Unit, Department of Medicine, University of Rochester School of Medicine and Dentistry, New York, USA

ROY A. FOX
Professor and Head, Geriatric Medicine, Dalhousie University; Director, Centre for Health Care of the Elderly, Camp Hill Medical Centre, Halifax, Nova Scotia, Canada

SALLY FREELS
Assistant Professor of Biostatistics, University of Illinois at Chicago, USA

ROBERT P. FRIEDLAND
Clinical Director, Alzheimer Center, University Hospitals of Cleveland; Associate Professor, Department of Neurology, Psychiatry and Radiology, Case Western Reserve School of Medicine, Cleveland, Ohio, USA

DOROTHY BERLIN GAIL
Chief, Structure and Function Branch, Division of Lung Diseases, National Heart, Lung, and Blood Institute, National Institutes of Health, Bethesda, Maryland, USA

JEFFREY GARLAND
Consultant Psychologist, Rivendell Assessment Centre, Radcliffe Infirmary, Oxford

GARY GERSTENBLITH
Associate Professor of Medicine, Johns Hopkins University School of Medicine, Baltimore, Maryland, USA

ENOCH GORDIS
Director, National Institute on Alcohol Abuse and Alcoholism, Rockville, Maryland, USA

CAROLYN GREIG
Lecturer in Human and Applied Physiology, Academic Department of Geriatric Medicine and Human Performance Laboratory, Royal Free Hospital School of Medicine, London

J. GRIMLEY EVANS
Professor of Geriatric Medicine, University of Oxford

LUCIO GULLO
Professor, Department of Internal Medicine and Gastroenterology, St Orsola Hospital, Bologna, Italy

A. JULIANNA GULYA
Associate Professor of Otolaryngology – Head and Neck Surgery, Georgetown University, Washington DC, USA

NORTIN M. HADLER
Professor of Medicine and Microbiology/Immunology, University of North Carolina at Chapel Hill; Attending Rheumatologist, University of North Carolina Hospitals, USA

JEFFREY B. HALTER
Professor of Internal Medicine, Chief of the Division of Geriatric Medicine, Research Scientist and Medical Director of the Institute of Gerontology, and Director of the Geriatrics Center, University of Michigan and VA Medical Center, Ann Arbor, USA

RONALD C. HAMDY
Professor of Medicine; Cecile Cox Quillen Professor of Geriatric Medicine and Gerontology, Quillen College of Medicine, East Tennessee State University, Johnson City, USA

DAVID HAMERMAN
Professor of Medicine and Director, Resnick Gerontology Center, Albert Einstein College of Medicine, and Interinstitutional Programs in Aging, Einstein/Montefiore Medical Center, New York, USA

S. MITCHELL HARMAN
Chief, Endocrinology Section, Laboratory of Clinical Physiology, National Institute on Aging; Associate Professor of Medicine, Johns Hopkins University, Baltimore, Maryland, USA

CARL M. HARRIS
Associate Professor of Orthopedic Surgery, The University of Rochester, New York, USA

ARTHUR E. HELFAND
Professor and Chairman, Department of Community Health and Aging, Pennsylvania College of Podiatric Medicine; Adjunct Professor, Department of Orthopedic Surgery, Jefferson Medical College and Thomas Jefferson University Hospital, Pennsylvania, USA

A. S. HENDERSON
Director, NH and MRC Social Psychiatry Research Unit, The Australian National University, Canberra, Australia

WILLIAM H. HERMAN
Assistant Professor of Internal Medicine, University of Michigan Medical Center, Ann Arbor, USA

H. M. HODKINSON
Barlow Professor of Geriatric Medicine, University College and Middlesex School of Medicine, University of London

MICHAEL A. HORAN
Department of Geriatric Medicine, University of Manchester, UK

JAMES G. HOWE
Consultant Physician in Medicine for the Elderly, Airedale General Hospital, Keighley, West Yorkshire, UK

NIGEL M. HYMAN
Consultant Neurologist, Oxford Regional Health Authority, Oxford

O. F. W. JAMES
Professor of Geriatric Medicine, The Medical School, University of Newcastle upon Tyne, UK

RALPH H. JOHNSON
Hon. Consultant Physician and Neurologist, Oxford; Director of Postgraduate Medical Education, Oxford University

DEE JONES
Senior Research Fellow, Department of Geriatric Medicine, University of Wales College of Medicine, Cardiff

PAUL R. KATZ
Assistant Professor of Medicine, Division of Geriatrics, State University of New York at Buffalo, USA

KEVIN G. KINSELLA
Social Science Analyst, Center for International Research, U.S. Bureau of the Census, Washington DC, USA

THOMAS B. L. KIRKWOOD
Head, Laboratory of Mathematical Biology, National Institute for Medical Research, London

JOHN H. KLIPPEL
Clinical Director, National Institute of Arthritis and Musculoskeletal and Skin Diseases, Bethesda, Maryland, USA

EDWARD G. LAKATTA
Laboratory of Cardiovascular Science, Gerontology Research Center, National Institute on Aging, Baltimore, Maryland, USA

CLAUDE LENFANT
Director, National Heart, Lung, and Blood Institute, National Institutes of Health, Bethesda, Maryland, USA

RICHARD W. LINDSAY
Professor of Internal Medicine; Head, Division of Geriatrics, Department of Internal Medicine, University of Virginia, Charlottesville, USA

ZBIGNIEW J. LIPOWSKI
Professor Emeritus of Psychiatry, University of Toronto, Canada

DAVID A. LIPSCHITZ
Professor of Medicine; Director, Geriatric Research, Education, and Clinical Center (GRECC), John L. McClellan Veterans Hospital; Head, Division on Aging, University of Arkansas for Medical Sciences, Little Rock, USA

PETER S. LIPSKI
Senior Lecturer in Geriatric Medicine, Sydney University and Concord Hospital, Sydney, Australia

R. A. LITTLE
MRC Trauma Group and Scientific Director, North Western Injury Research Centre, University of Manchester, UK

DAVID M. MACFADYEN
Co-ordination with other Organizations, World Health Organization, Regional Office for Europe, Copenhagen, Denmark

W. J. MACLENNAN
Professor of Geriatric Medicine, Geriatric Medicine Unit, Department of Medicine of the Royal Infirmary of Edinburgh, UK

GEORGE M. MARTIN
School of Medicine, University of Washington, Seattle, USA

L. JOSEPH MELTON III
Head, Section of Clinical Epidemiology, Mayo Clinic: Professor of Epidemiology, Mayo Medical School, Rochester, Minnesota, USA

B. ROBERT MEYER
Professor of Clinical Pharmacology, North Shore University Hospital, Manhasset, New York, USA

TONI P. MILES
Assistant Professor of Epidemiology, School of Public Health, University of Illinois at Chicago, USA

LINDA A. MORROW
Instructor of Medicine, Harvard Medical School; Physician Scientist, Geriatric Research, Education, and Clinical Center, Brockton/West Roxbury Department of Veterans Affairs, Medical Center, Brockton, Massachusetts, USA

J. A. MUIR GRAY
Community Health Office, Radcliffe Infirmary, Oxford

ELAINE MURPHY
Professor of Psychogeriatrics, United Medical Schools, Guy's Hospital, London

GEORGE C. MYERS
Professor of Sociology and Director, Center for Demographic Studies, Duke University, Durham, North Carolina, USA

MICHAEL C. NEVITT
Assistant Professor, Department of Epidemiology and Biostatics and Department of Medicine, University of California, San Francisco, USA

E. A. NEWSHOLME
Reader in Cellular Nutrition, Department of Biochemistry, University of Oxford

DEAN C. NORMAN
Geriatric Research, Education, and Clinical Center, Veterans Administration Medical Center, Los Angeles, California, USA

MARK PARRY-BILLINGS
Postdoctoral Research Assistant, Cellular Nutrition Research Group, Department of Biochemistry, University of Oxford

RALPH A. PASCUALY
Medical Director, Sleep Disorders Center, Providence Medical Center, Seattle, Washington, USA

W. BRADFORD PATTERSON
Acting Director of Cancer Control, Dana Farber Cancer Institute, Boston, Massachusetts, USA

CAROL C. PILBEAM
Assistant Professor of Medicine, University of Connecticut Health Center, Farmington, USA

PATRICK RABBITT
Research Professor of Gerontology and Cognitive Psychology, University of Manchester, UK

PETER V. RABINS
Associate Professor of Psychiatry, Johns Hopkins University School of Medicine, Baltimore, Maryland, USA

LAWRENCE G. RAISZ
Professor of Medicine, University of Connecticut School of Medicine, Farmington, USA

N. W. READ
Professor of Human Nutrition, University of Sheffield, UK

MARCUS M. REIDENBERG
Professor of Pharmacology and Medicine, Cornell University Medical College, New York, USA

B. LAWRENCE RIGGS
Consultant, Division of Endocrinology and Metabolism; Purvis and Roberta Tabor Professor of Medical Research, The Mayo Clinic, Rochester, Minnesota, USA

N. A. ROBERTS
Senior Registrar, Department of Geriatric Medicine, Hope Hospital, Salford, UK

F. CLIFFORD ROSE
Director, Academic Unit of Neuroscience, Charing Cross Hospital, London

JOHN W. ROWE
President, Mount Sinai School of Medicine and Mount Sinai Hospital, New York City, USA

RICHARD W. SATTIN
Assistant Director for Science, Division of Injury Control, Centers for Disease Control, Atlanta, Georgia, USA

J. T. SCOTT
Consultant Physician, Charing Cross Hospital, London

ALLISON B. SEKULER
Department of Psychology, University of Toronto, Ontario, Canada

ROBERT SEKULER
Center for Complex Systems and Department of Psychology, Brandeis University, Waltham, Massachusetts, USA

D. GWYN SEYMOUR
Senior Lecturer, University Department of Geriatric Medicine, Cardiff Royal Infirmary, UK

H. G. M. SHETTY
Lecturer in Geriatric Medicine, University of Wales College of Medicine, Cardiff, UK

DIANE G. SNUSTAD
Assistant Professor, Division of Geriatrics, University of Virginia Health Sciences Center, Charlottesville, USA

RICHARD SUZMAN
National Institute on Aging, Bethesda, Maryland, USA

RAYMOND TALLIS
Professor of Geriatric Medicine, Hope Hospital Department of Geriatric Medicine, University of Manchester, UK

R. C. TAYLOR
Professor of Social Policy, University of Glasgow, UK

SUZANNE VAN H. SAUTER
Clinical Associate Professor of Medicine; Director, Rehabilitation Program Office, The University of Carolina at Chapel Hill, USA

DERICK T. WADE
Consultant in Neurological Rehabilitation, Rivermead Rehabilitation Centre, Oxford

ANGUS W. G. WALLS
Senior Lecturer in Restorative Dentistry, Dental School, University of Newcastle upon Tyne, UK

PHILIP A. WOLF
Professor of Neurology, Boston University School of Medicine and the Section of Preventive Medicine and Epidemiology, Evans Memorial Department of Clinical Research and Department of Medicine, University Hospital, Massachusetts, USA

LESLIE I. WOLFSON
Professor and Chairman, Department of Neurology, University of Connecticut School of Medicine, USA

PHILIP H. N. WOOD
Emeritus Director, Arthritis and Rheumatism Council Epidemiology Research Unit, University of Manchester; formerly Honorary Professor of Community Medicine, University of Manchester, UK

K. W. WOODHOUSE
Professor of Geriatric Medicine, University of Wales College of Medicine, Cardiff, UK

ARCHIE YOUNG
Professor of Geriatric Medicine, Royal Free Hospital School of Medicine, London

Introduction

J. Grimley Evans and T. Franklin Williams

The ageing of an organism is a progressive loss of adaptability as time passes. As we grow older we become less able to react adaptively to challenges from the external or internal environment. Examples of external challenges will include injury and infection and those from the internal environment will include arterial occlusion or deviant clones of cells. As homeostatic mechanisms become less sensitive, less accurate, slower, and less well sustained we sooner or later encounter a challenge that we are unable to deal with effectively and we die. The rise of death rates with age is the hallmark of senescence; in the human being this begins around the age of 13 and except for the deviance due to violent deaths in early adult life is broadly exponential thereafter (Fig. 1). If we did not age we would still die eventually from accident, disease, predation, or warfare but our risk of dying would be constant with age, or might even fall as natural selection weeded out less adaptable individuals from the population.

Loss of adaptability is also the key concept in medical practice among older people. They will be less able to adapt to minor errors in care (in drug dosage for example) and will need more help in recovering from disease or injury than will younger patients. The essence of good geriatric medicine is a scrupulous and comprehensive attention to detail.

The continuous and broadly exponential increase with age in vulnerability throughout adult life can be seen also in the use of health services and in the prevalence of chronic disease. There is no discontinuity in later life in any of these measures that could provide a biological justification for separating older people from the rest of the adult human race. The arbitrary definition of the state of being 'geriatric' beginning at age 65 or 75 is an historical accident and while it may have administrative convenience, it has no basis in biology or epidemiology, and may not always be in the best interests of older people. The processes that lead to disease and disability in old age are lifelong and can only be understood and modified through a lifelong perspective on ageing. The growth of geriatrics as a specialty in medicine must not be allowed to impede the development of this basis for research and prevention.

Medical students are still, too often, taught that in approaching the disability of an old person they should consider whether the problem is due to 'disease' or to 'normal ageing'. Unfortunately for this approach no one has produced a definition of 'normal ageing' that bears inspection, and 'disease' can be defined in many different ways. In effect students are being asked to separate the undefined from the indefinable. This makes neither for constructive thought nor good medicine, particularly if it is inferred that anything characterized as 'normal' is by definition not the proper concern of a doctor.

The question students should be taught to ask when faced with an elderly person with a disability is 'what can I do to improve the situation?'. If this is coupled with a knowledge of the range of therapeutic and prosthetic interventions available and how to assess their appropriate application the issue of whether the problem arises from 'normal' or disease' is seen to be irrelevant as well as meaningless.

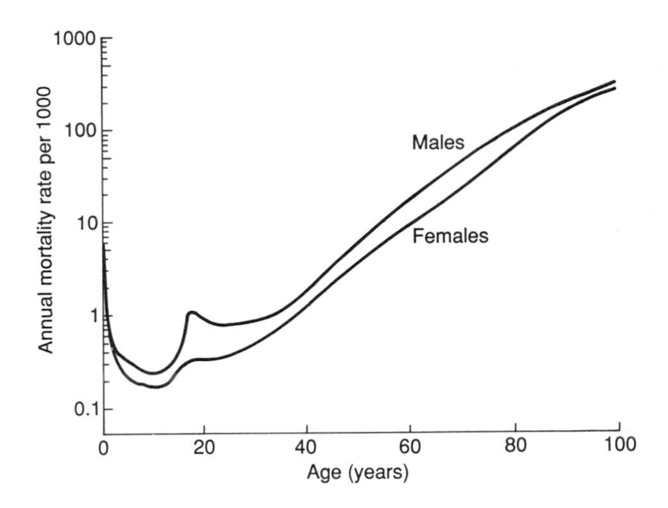

Fig. 1 Age- and sex-specific annual mortality rates, England and Wales, 1988.

It is important, however, to have some broad model of how age-associated changes come about, in order to structure thought about appropriate interventions. Table 1 sets out one such model based on an analysis of the origins of observed differences between the young and older members of a population. The first distinction is between those true ageing processes in which the old people have changed from what they were when young, from non-ageing factors in which the differences between young and old are not due to the old having changed during their lifetime. We may briefly consider each of the processes listed in Table 1.

Table 1 Differences between young and old

Non-ageing	Selective survival	
	Differential challenge	
	Cohort effects	
True ageing	Secondary (reactive)	
	Primary	Intrinsic
		Extrinsic

NON-AGEING

Selective survival

Those who reach extreme old age, say the ninth and tenth decades of life, are the survivors from their birth cohorts, selected for their greater resistance to whatever mortal challenges their generation had to face. It is therefore not surprising to find some evidence of selection for favourable genotypes among the very old compared with younger subjects (Takata *et al.* 1987). This process is sometimes characterized as an 'emergence of the biological elite'. More interesting would be the identification of psychological and behavioural factors linked to selective survival but it would be difficult to identify these without long-term prospective studies.

Cohort effects

In changing societies individuals born into successive generations are exposed to very different physical and social environments. The effects of these on individuals in their early years may persist throughout the rest of their lives. The habits of nutrition and food preferences acquired in early life will affect later susceptibility to vascular and other diseases, as will socially-determined habits such as smoking and alcohol consumption. Cultural changes will have a major effect on psychological capacity and functioning both through education and the more subtle influences of the media and government propaganda. The way in which cohort effects can be mistaken for age effects has been dramatically demonstrated in comparisons of cross-sectional and longitudinal studies of psychological functioning (Schaie and Strother 1968; Schaie 1989). Older people subjected to tests devised by young psychologists perform rather badly compared with the young but when followed up show much less decline with age than the cross-sectional comparisons would predict. The difference between the cross-sectional comparison and the longitudinal findings largely represent cohort effects.

Differential challenge

If ageing is loss of adaptability it can only be assessed by exposing individuals of different ages to the same challenge and measuring the effects. In many ways society is organized so that elder people are exposed to more severe challenges than are the young and the detrimental effects are too readily attributed to ageing alone. A classic example is hypothermia in the United Kingdom which is partly due to age-associated deterioration in the ability to maintain body temperature, but is also due to housing policies which put older people in poorer houses and colder environments (Fox *et al.* 1973). The quality of medical care which older people receive is often poorer than that offered to the young (Kane and Kane 1986) and the mentally impaired elderly patient may be particularly disadvantaged (Evans *et al.* 1980). However, because one expects older people, and especially mentally impaired older people, to do badly one may not notice that they are doing worse than they need because society has loaded the dice against them. This may reasonably be regarded as 'aggravated ageing'. As citizens as well as doctors we have a duty to be vigilant in countering such processes.

TRUE AGEING

This heading comprises those processes whereby individuals change as they grow older.

Secondary (reactive) ageing

One of the basic enigmas of ageing is why the body does not repair the ravages of age as well as it deals with some other insults such as minor wounds and infections. This question is addressed in Chapters 2.1 and 2.2. We need to be aware that the individual may make some adaptive responses to ageing because of the inveterate desire of doctors to 'normalize' anything that they encounter which seems 'abnormal'. If the abnormality is an adaptive response to some underlying impairment the patient could suffer from its 'correction'. At present, with the possible but ill-defined exception of high blood pressure in some individuals with impaired arterial supply to the hind brain, examples of secondary ageing in the physiological sphere have not been identified. We need, however, to be sensitive to their possible existence. Secondary ageing is more apparent in psychological functioning. Individuals may show a range of behavioural adaptations to age-associated impairments in memory and fluid intelligence. Obsessional behaviour is a well-recognized adaptation to unreliable memory for example.

At a species level the female menopause is probably an example of secondary adaptation to the age-associated decline in reproductive efficiency observed in all species. A menopause, defined as a genetically determined total cessation of reproductive capability at an age with low variance and approximately halfway through the maximum lifespan of the species, is almost exclusively a human and female characteristic. With a family social structure and a cumulative culture based on speech, there will on average come a time in a woman's life when in terms of getting her genes into succeeding generations (the measure of evolutionary success) she will be better engaged in contributing to the survival of her grandchildren, each containing 25 per cent of her genes, than in increasingly dangerous and unsuccessful attempts at producing more children of her own, even though each will contain 50 per cent of her genes. It is significant that a species of whale in which an important degree of grandmothering behaviour is seen also provides one of the very few well-substantiated examples of a menopausal phenomenon outside the human species (Marsh and Kasuya 1986). With his much lower biological investment in unsuccessful pregnancies the male will not be under similar evolutionary pressure to substitute grandparenting for reproduction. The disadvantage for him is that because of his larger potential reproductive capacity compared with the female he will be under less evolutionary pressure towards increased longevity (Clutton-Brock and Vincent 1991). This is presumably the explanation for a genetic component in the higher mortality rates of men in developed nations—as seen in Fig. 1. (There are, of course, other facts related to the female menopause which must be considered in the light of increasing life expectancy and the changes which follow the menopause. Today, on average, women in developed countries will live almost half their lives postmenopausally without the benefits, unless provided by therapy, of the oestrogens that helped sustain bones and minimize coronary heart disease but which themselves may increase the risks of cancers of uterus and breast.)

Primary ageing

Primary ageing is best viewed in terms of the traditional epidemiological model of the interaction between intrinsic (genetic and constitutional) and extrinsic (environmental and lifestyle) factors. The nature of the intrinsic determinants of ageing are difficult to define, and until recently the dissection of the intrinsic and extrinsic components of ageing was dependent on observations of different ageing patterns among populations living under differing environmental conditions. For example it has been recognized that although most adult cancers increase in incidence with age according to a power-law, epidemiological evidence suggests that 70 per cent or more are due not to intrinsic ageing but to the accumulation of exposure to environmental carcinogens (Doll and Peto 1981). The rise of blood pressure with age, once thought to be a 'normal' genetically determined change, is now acknowledged to be dependent on extrinsic factors interacting with genotype. The incidence of proximal femoral fracture is clearly under important environmental influence, although to what extent this influence is mediated through variation in the prevalence of osteoporosis or through the other determinants of fracture—falls and protective factors or responses in falling (Evans 1990)—remains unclear. It also seems that the high-tone hearing loss which is a universal and socially disabling feature of Western populations may be due almost as much to lifelong exposure to damaging levels of ambient noise as to intrinsically determined deterioration in auditory function (Goycoolea *et al.* 1986).

VARIABILITY IN AGEING

Because of the composite nature of ageing, genetic variability, and differences in exposure to extrinsic factors, individuals show wide variation in the pattern and rate of ageing. For reasons of brevity authors typically describe what happens on average with ageing in terms of movement of the mean value of some variable with age. Even where not made explicit, it needs to be borne constantly in mind that for virtually all such variables the variance increases with age so that whatever happens to the population mean, some individuals will show little or no age-associated change. A further consequence of this is the need to recognize the great heterogeneity of individuals in the older age groups. Referring to them collectively as 'the elderly' (something we have attempted to avoid in this volume) may inhibit appreciation that older people differ from each other more than do the young. Good science as well as good medicine requires us to respond to older people as individuals not as uniform members of an arbitrarily defined group.

OPTIMAL AGEING

The theory developed by Kirkwood (see Chapter 2.1) may provide a means whereby intrinsic ageing processes can be recognized directly rather than by a process of exclusion of extrinsic influences. Intrinsic influences on the rate and pattern of ageing will be sought in the control of damage detection and repair, particularly at an intracellular level. Energy-consuming proof-reading stages in informational transcription within the cell (Hopfield 1980) will be the particular target of study. In the long term we may look to the realistic possibility of modification of intrinsic ageing rates by biochemical or genetic manipulation. In the short and medium term, however, ageing processes will be more accessible through modification of extrinsic factors, by the creation of less damaging environments, and adoption of healthier lifestyles. Just as Doll and Peto (1981) identified the potential for prevention of cancer by combining the lowest incidence rates observed in a worldwide survey, we can use the epidemiological method to specify the target of *optimal ageing* by which to set targets for individuals and for populations. This will be a major conceptual advance on the traditional medical approach of setting standards on the basis of the population mean (often combined with standard deviations calculated with neither relevance nor statistical validity around it). The standards of optimal ageing should be set initially within populations, perhaps on a basis of the performance of the best 10 per cent of each age group. Neither the young nor the old should accept present ageing patterns as immutable; death is inevitable, disability may not be.

REFERENCES

Clutton-Brock, T.H., and Vincent, A.L.J. (1991). Sexual selection and the potential reproductive rates of males and females. *Nature*, **351**, 58–60.

Doll, R. and Peto, R. (1981). The causes of cancer: quantitative estimates of avoidable risks of cancer in the United States today. *Journal of the National Cancer Institute*, **66**, 1191–308.

Evans, J.G. (1990). The significance of osteoporosis. In *Osteoporosis 1990* (ed. R. Smith), pp. 1–8. Royal College of Physicians, London.

Evans, J.G., Prudham, D., and Wandless, I. (1980). A prospective study of fractured proximal femur: hospital differences. *Public Health* (London), **94**, 149–54.

Fox, R.H., Woodward, P.M., Exton-Smith, A.N., Green, M.F., Donnison, D.V., and Wilks, M.H. (1973). Body temperatures in the elderly: a national study of physiological, social and environmental conditions. *British Medical Journal*, **1**, 200–6.

Goycoolea, M.V., Goycoolea, H.G., Rodriguez, L.G., Martinez, G.C., and Vidal, R. (1986). Effect of life in industrialized societies on hearing in natives of Easter Island. *Laryngoscope*, **96**, 1391–6.

Hopfield, J.J. (1980). The energy relay: a proofreading scheme based on dynamic cooperativity and lacking all symptoms of kinetic proof reading in DNA replication and protein synthesis. *Proceedings of the National Academy of Sciences of the USA*, **77**, 5448–52.

Kane, R.A. and Kane, R.L. (1986). Self-care and health-care: inseparable but equal for the well-being of the old. In *Self-care and Health in Old Age* (eds. K. Dean, T. Hickey, and B.E. Holstein), pp. 251–83. Croom Helm, London.

Marsh, H. and Kasuya, T. (1986) Evidence for reproductive senescence in female cetaceans. *Report of the International Whaling Commission*, (special issue), **8**, 57–74.

Schaie, K.W. (1989). Perceptual speed in adulthood; cross-sectional and longitudinal studies. *Psychology and Aging*, **4**, 443–53.

Schaie, K.W. and Strother, C.R. (1968). A cross-sequential study of age changes in cognitive behavior. *Psychological Bulletin*, **70**, 671–80

Takata, H., Suzuki, Ishii, T., Sekiguchi, S. and Iri, H. (1987). Influence of major histocompatibility complex region genes on human longevity among Okinawan-Japanese centenarians and nonagenarians. *Lancet*, **ii**, 824–6.

SECTION 1
The ageing of populations and communities

1.1 Demography of older populations in developed countries

RICHARD SUZMAN, KEVIN G. KINSELLA, AND GEORGE C. MYERS

The ageing of populations is largely a phenomenon of the twentieth century, and stands as a testimony to the extension of our lives. At the same time, the growth of older populations poses a challenge to public policy. The world's more developed countries, which are the focus of this chapter, face a plethora of sociopolitical issues—ranging from provision of health care to social security to employment rights—that are directly linked to the changing age structure of their populations.* As we approach the year 2000, many of the debates around these issues will intensify and may well shift from the political to the ethical realm.

There is a growing awareness that the concept of 'elderly' is an inadequate generalization which obscures the heterogeneous nature of a population group that spans more than 40 years of life. 'The elderly' are at least as diverse as younger age groups in terms of personal and social resources, health, marital status, living arrangements, and social integration. To understand the dynamics of ageing, we need information on older populations from several interrelated perspectives: demographic, medical, social, and economic. This chapter seeks to lay out the demographic foundation upon which subsequent chapters can build.

A decade ago, demographic studies of older populations contained largely descriptive analyses of the distribution of those populations by age, sex, marital status, labour-force participation, living arrangements, and causes of death. The analytical tools used were predominantly those of ratios, the life table, and various decompositions. In recent years, the field has moved beyond an exclusively descriptive phase to probe the causes and consequences of changing age structures in populations. For example, biologically based models have been developed to elucidate and forecast the dynamics underlying the demographic transition in which life expectancy increases before the birth rate falls, and chronic diseases and disabilities become more prevalent. These models describe the transitions that occur as populations of individuals move from independent functioning to dependence, institutionalization, and death. Life tables measuring single, unilinear decrements in vital and functional status are giving way to tables that can describe the more complex, multiple, bidirectional changes. New theoretical models are being developed to investigate the deeper structural inter-relations between support ratios, intergenerational exchanges, population age structures, and the economics of pension, health, and other social insurance programmes.

* The 'developed country' category used in this chapter corresponds to the 'more developed' classification employed by the United Nations Statistical Office. Developed countries comprise all nations in Europe (including the Soviet Union, and excluding Turkey and Cyprus) and North America, plus Japan, Australia, and New Zealand.

POPULATION AGEING

The world's elderly population, 65 years of age and over, is currently growing at a rate of 2.5 per cent/year, considerably faster than the overall total population. In developed countries as a whole, the present older population numbers 146 million and will expand to 232 million by the year 2020. Sweden, with 18 per cent of its population aged 65 and over in 1990, has the highest proportion of elderly people of the major countries of the world. Other notably high proportions are found in Norway, the United Kingdom, Denmark, and what was the Federal Republic of Germany. Thirteen countries in Table 1 have elderly populations of 2 million or more today; by 2020, seven more will have reached this level.

In the simplest terms, population ageing refers to increasing proportions of older persons within an overall population age structure. Another way to think of population ageing is to consider a society's median age, the age that divides a population into numerically equal segments of younger and older persons. For example, the present median age in Portugal of 33 years indicates that the number of persons under the age 33 equals the number who have already celebrated their thirty-third birthday.

With the exception of the low median of 24 in Albania, today's median age in developed countries ranges from 29 in Ireland to nearly 39 in Sweden and West Germany (data for the reunified Germany are not readily available at the time of writing). The median will rise continually in every developed country during the next three decades, and is projected to exceed 47 years in Switzerland, Italy, and Germany by the year 2020 (Fig. 1). Unless birth rates rise unexpectedly in the coming years, these and many other societies face a future in which nearly half of their citizenry will be over the age of 50.

Within an elderly population, different age groups may grow at very different rates. An increasingly important feature of population ageing is the progressive ageing of the older population itself. The fastest growing age segment in many countries is the 'old old', defined here as persons aged 80 and over. This group currently constitutes 20 per cent of the overall elderly population in developed countries, and represents approximately 4 per cent of the total population in Scandinavia, France, (West) Germany, and Switzerland. Eight developed nations now have 'old old' populations in excess of 1 million, and six more nations will share this characteristic before the year 2020 (U.S. Bureau of the Census 1987).

Changes in the age structure of a population result from changes over time in fertility, mortality, and international migration (Myers 1990). Most societies historically have had high levels of both fertility and mortality. As prominent communicable diseases are eradicated and public health measures expanded, overall mortality levels decline and life expectancy at birth rises, while fertility tends to remain high. A large proportion of the initial improvement in mortality occurs among infants, meaning that more babies survive. Consequently, younger population age cohorts grow in size relative to older cohorts, and the population percentage of youths and young adults is relatively high. This is the situation today in many of the world's developing nations.

Populations begin to 'age' only when fertility falls and mortality rates continue to improve or remain at low levels. Countries that have both low fertility and low mortality have completed what demographers call 'the demographic transition', illustrated graphically in Fig. 2. In 1920, the United Kingdom's population age structure had the pyramidal shape common to societies with relatively high fertility and mortality. As early as 1920, however, one can see that fertility was falling significantly; the cohorts aged 0 to 4 and 5 to 9 years were noticeably smaller than those aged 10 to 14. By 1970, increases in life expectancy had helped to shift the centre of gravity of the population age structure upward, and persons aged 45 to 64 were a much greater share of the total. As the United Kingdom ages into the twenty-first century, the one-time pyramid will give way to a rectangular shape. By the year 2020, nearly half of the population will be aged 45 years or more, and the ranks of the old old will continue to swell.

The United Kingdom is typical of several developed countries of Europe which have had low fertility and mortality for decades. Germany, Sweden, and Hungary, for example, now have total fertility rates well below the natural replacement level of 2.1 children per woman. Successive small birth cohorts have contributed to the large proportions of elderly people in these societies. Thus, we see the importance of understanding past demographic trends, or disruptions thereof, when planning for the future. In spite of the relatively

Table 1 Elderly and 'old old' population in major developed countries, 1990 and 2020

	Elderly (65+ years) population (in thousands)		Elderly as percentage of total population		Old old (80+years) as percentage of total population	
	1990	2020	1990	2020	1990	2020
Western Europe						
Austria	1157	1595	15.1	21.3	3.6	5.2
Belgium	1470	2071	14.8	21.4	3.4	5.2
Denmark	801	1120	15.6	22.5	3.7	5.3
Finland	667	1149	13.4	22.9	2.9	5.7
France	7928	12119	14.1	20.2	3.8	5.2
Germany (FR)	9710	14993	15.5	22.5	3.9	7.2
Iceland	27	51	10.5	16.3	2.5	4.1
Ireland	400	620	11.4	16.2	2.1	3.8
Italy	8472	13078	14.7	23.3	3.1	7.4
Liechtenstein	3	6	10.5	20.6	2.2	4.0
Luxembourg	52	91	13.4	22.7	3.1	6.4
Netherlands	1928	3461	12.9	22.1	3.0	5.8
Norway	695	894	16.3	19.9	3.8	4.6
Sweden	1527	1975	17.9	22.9	4.4	6.3
Switzerland	986	1614	14.6	23.6	3.8	6.0
United Kingdom	8977	12108	15.7	20.4	3.6	5.9
Eastern Europe						
Bulgaria	1160	1744	13.0	19.2	2.3	4.5
Czechoslovakia	1849	3149	11.8	18.5	2.3	4.1
Germany (DR)	2141	3099	13.1	20.7	3.3	5.1
Hungary	1414	2186	13.4	21.0	2.6	5.3
Poland	3773	7243	10.0	17.4	2.0	3.6
Romania	2417	4588	10.4	17.7	1.7	5.0
USSR	27605	44864	9.6	13.1	2.3	3.2
Southern Europe						
Albania	197	529	6.0	11.3	1.5	3.0
Greece	1408	2237	14.0	22.6	3.2	7.6
Malta	37	79	10.4	19.6	1.9	4.6
Portugal	1361	2053	13.1	19.2	2.6	5.9
Spain	5246	8162	13.4	20.2	2.8	6.4
Yugoslavia	2250	4933	9.4	18.7	1.8	5.3
Other						
Australia	1897	3881	11.2	17.0	2.2	4.2
Canada	3053	6404	11.5	19.3	2.4	4.9
Japan	14655	31904	11.9	25.7	2.3	7.5
New Zealand	367	663	11.2	19.2	2.2	4.8
United States	31560	52067	12.6	17.7	1.9	2.9

Source. U.S. Bureau of the Census, International Data Base on Aging; and United Nations Department of International Economic and Social Affairs (1989). *Global estimates and projections of population by sex and age, the 1988 revision*, ST/ESA/SER. R/93, New York.

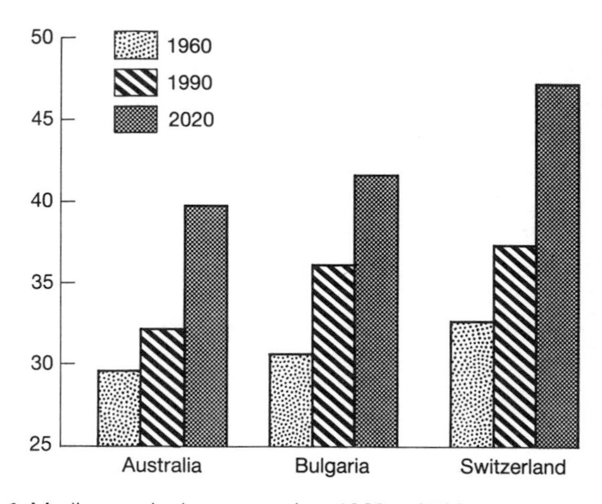

Fig. 1 Median age in three countries: 1960 to 2020.

high proportions of elderly people now observed in developed countries, there will be little change in most countries during the next 10 to 15 years, followed by accelerated growth in the early stages of the twenty-first century. The lack of change in the near future reflects low fertility during the Second World War and the preceding worldwide economic depression, while the subsequent growth stems from the 'baby boom' that characterized the postwar years in many Western nations. As the large postwar birth cohorts continue to advance in age, the proportions of elderly persons will begin to expand noticeably—in the United States, for example, the percentage of elderly people is expected to jump from 13 in the year 2000 to nearly 20 by 2020.

The general similarity in today's level of population ageing in developed countries, however, can mask important current and historical differences. Again, the timing and pace of fertility decline is usually the pre-eminent factor. Table 2 illustrates how rapidly the population of Japan is ageing compared with other developed nations; it will take fewer than three decades for Japan to double its proportion of elderly people from 7 to 14 per cent, compared to the 115 years this took for France. Moreover, by the year 2020, over a quarter of Japan's population will be 65 years old and over—the highest level for any country in the world.

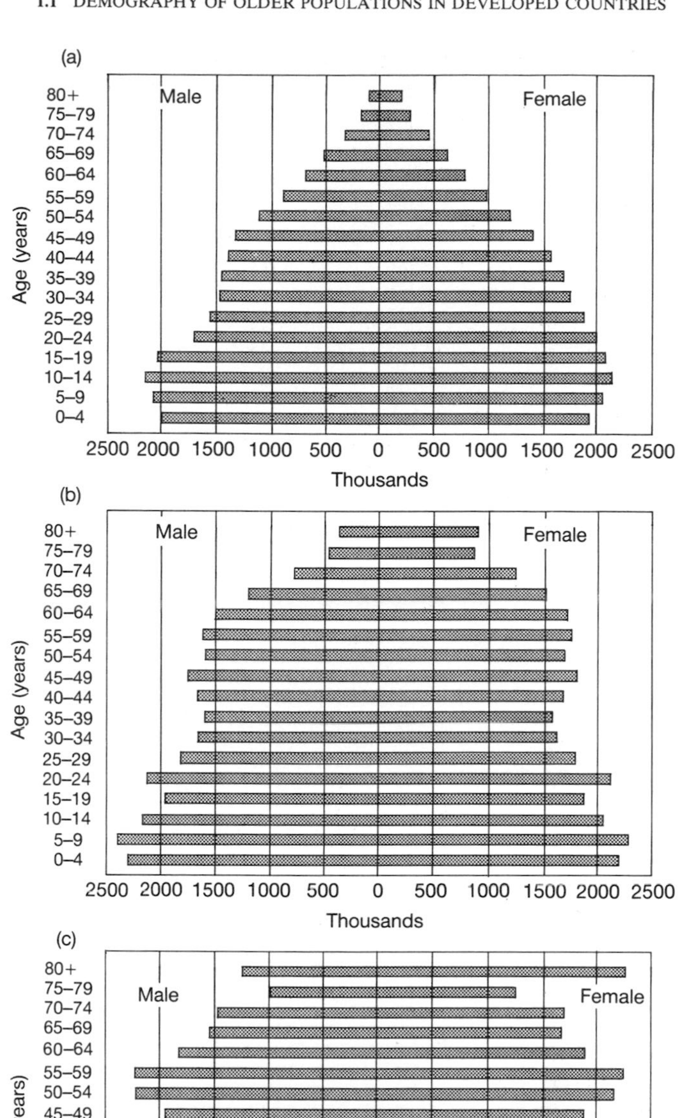

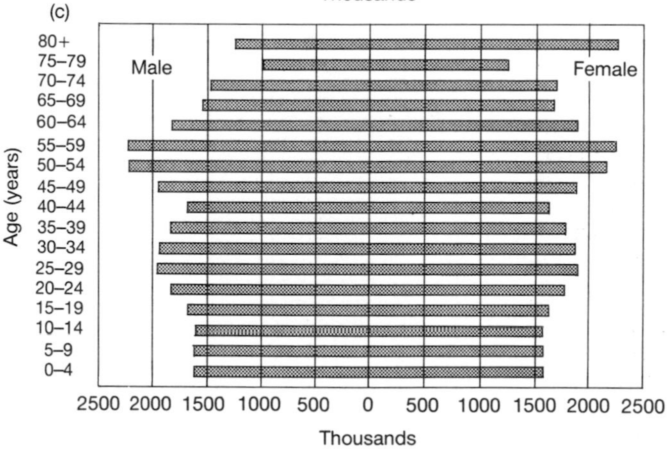

Fig. 2 Total population for the United Kingdom. (a) 1920; (b) 1970; (c) 2020.

LIFE EXPECTANCY

Industrialized countries generally have seen large gains in life expectancy at birth during the last half of the twentieth century. Average life expectancy at birth in Japan has now reached more than 79 years, the highest level in the world, and several European nations—Italy, Spain, Switzerland—have achieved levels of 78 years. While the recent pace of improvement has slowed somewhat, the level of increase that many countries continue to experience has confounded demographic predictions, and led to renewed research and debate over the biological limits to human life.

Women outlive men in every developed nation, with the female advantage exceeding 7 years in North America and several European nations (Fig. 3). In the countries of the Organization for Economic Co-operation and Development (**OECD**) as a whole, average life expectancy at birth increased by 8.5 years for females and by almost 6 years for males between 1950 and 1980. Although the sex difference has increased since the mid-1900s, there has been a stabilization or even a reversal of this trend in some countries in recent years (OECD 1988).

Given the low level of infant mortality that characterizes most developed countries, much of the current gain in life expectancy at birth is attributable to improvement in mortality among the elderly population. Owing in large part to the reduction of heart disease and stroke among middle-aged and older adults, gains in life expectancy at the age of 65 are now outpacing increases in life expectancy at birth. Men

Table 2 Speed of population ageing in selected countries

Country	Year in which the proportion of the population aged 65+ years reached or will reach: 7%	14%	Number of years required
Japan	1970	1996	26
United Kingdom	1930	1975	45
United States	1944	2010	66
Sweden	1890	1975	85
France	1865	1980	115

Source. Japan Ministry of Health and Welfare (1983) *Annual report on health and welfare for 1983: the trend of a new era and social security,* Tokyo; and Spencer (1989).

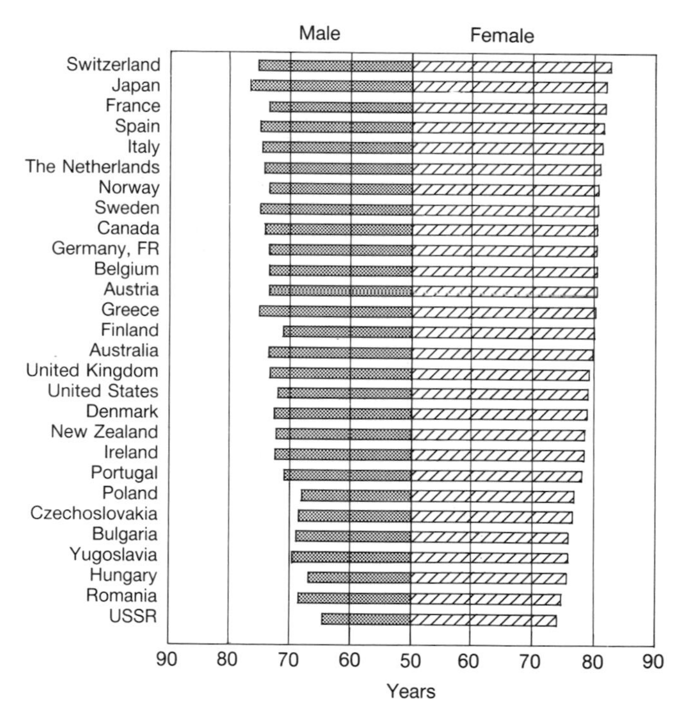

Fig. 3. Life expectancy at birth, 1990

who live to 65 typically can look forward to another 15 years of life, while women aged 65 can anticipate another 18 years.

The overall trend in improved life expectancy, however, is tempered by two factors. Some societies have not enjoyed linear progress in advancing life expectancy at old ages; the elderly population in Bulgaria, for example, lost ground between 1960 and 1980, even though life expectancy at birth increased. A similar pattern of decline or stagnation has been reported for some other Eastern European nations, and has been tentatively linked to increased smoking and consumption of alcohol. Secondly, research has noted that factors which contribute to reductions in mortality do not necessarily lead to a healthy extended life (see discussion below on health as active life expectancy).

IMPACT OF FUTURE MORTALITY RATES ON POPULATION SIZE

Although overall population ageing is determined primarily by birth rates, the pace at which death rates decline at advanced ages will play a significant role in determining future numbers, especially of the very old. The United States Bureau of the Census, which currently estimates the number of those aged 85 and over in the United States to be about 3.3 million (Fig. 4), has made several projections of the future

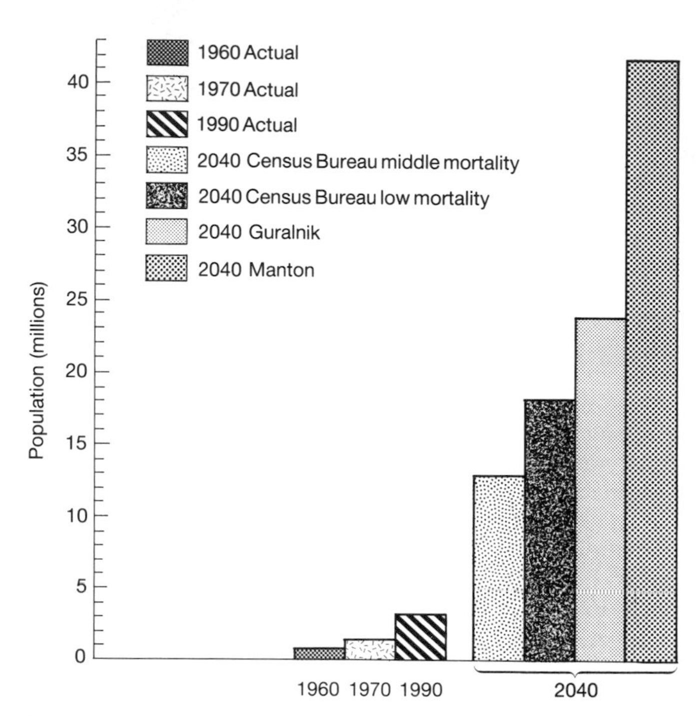

Fig. 4. Forecasts of the United States population (age 85 and over).

size of this age segment (Spencer 1989). The Bureau of the Census's middle-mortality series projections suggest that there will be 12.2 million people aged 85 and over in the year 2040, while their low-mortality series projection implies 17.9 million. As those who will be 85 years old and over in the year 2040 are already aged 35 and over, the differences in these projections result only from the assumptions about adult mortality rates, and are not affected by future birth or infant mortality rates. In the middle-mortality series, the Census Bureau assumes that life expectancy at birth will reach 81.2 years in 2080, while in the low-mortality (high life expectancy) series, life expectancy is assumed to reach 88.0 in 2080. In the low-mortality series, near-future mortality is assumed to decline at about the same rate as in the 1970s, but after 2010 the decline is much more gradual.

Alternative projections (see Fig. 4), using assumptions of a lower death rate, have produced even larger estimates of the future population of the United States aged 85+. Simply assuming that death rates will continue falling at about the recent 2 per cent rate of decline results in a projection of 23.5 million aged 85 and over in 2040 (Guralnik *et al.* 1988). Even more optimistic forecasts of future reductions in death rates have been made from mathematical simulations of potential reductions in known risk factors for chronic disease, morbidity, and mortality. Manton and Stallard (1991*b*) have used this method to generate an extreme 'upper bound' projection for the United States of over 40 million persons aged 85 and over in 2040. While such extreme projections are not necessarily the most likely, they do illustrate the potential impact of changes in adult mortality on the future size of the extremely old population, and underscore the uncertainty inherent in projections of the size and age composition of older populations.

ACTIVE LIFE EXPECTANCY

The extent to which a longer life will be a healthier one rather than one that contains more years of being chronically ill or disabled will have a potent impact on national health systems, most especially on the demand for long-term care. Compared with morbidity, disability is a more powerful determinant of the use of long-term care services. In order to conceptualize and assess the duration, dynamics, and aggregate levels of disability, a new approach has recently been developed in which overall life expectancy is partitioned into components (Fig. 5) such as active life expectancy (also

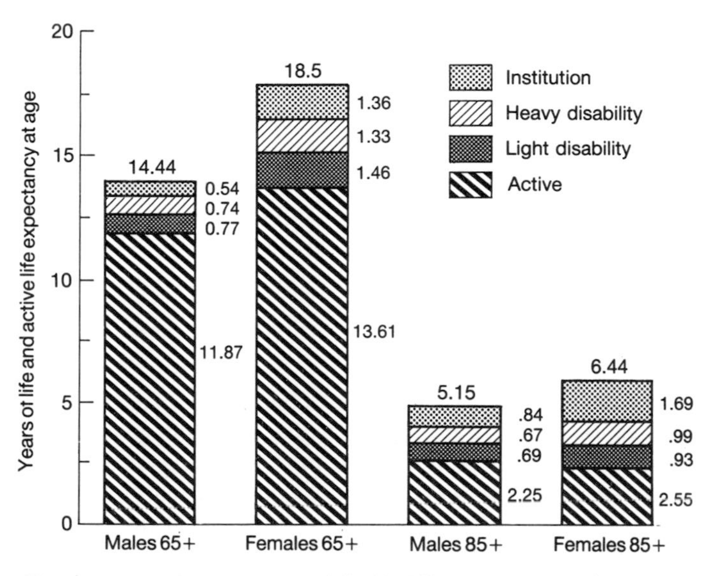

Fig. 5. United States active and disabled life expectancy estimates. (Source: National Long Term Care Survey 1982/1984, 1983 NCHS Life table.)

called health expectancy or disability-free life expectancy), lightly and heavily disabled life expectancy, and institutionalized life expectancy (Suzman *et al.* 1992). The definition of active life expectancy is not yet standardized, but generally refers to the expectation of life without a serious and chronic disability. While earlier computational approaches (e.g. Katz *et al.* 1983) could not capture the full dynamic nature of disability, more recent methodological advances (Manton 1989; Rogers *et al.* 1989; Manton and Stallard 1991*a*) make it possible to incorporate processes such as recovery and rehabilitation into the calculations. Similar attempts to develop time-weighted, quality-of-life measures have been hampered by the inherent subjectivity of the concept. Results of such calculations for the United States by Manton and Stallard (1991*a*) illustrate the sharp declines in the fraction of active life expectancy that occur between the ages of 65 and 85 years. The longer period of disability and institutionalization experienced by old females is also clearly apparent.

Indices of active life expectancy are increasingly being used to chart the progress and compare the efficacy of health and social systems. However, as international standardization of these measures is only now taking place (Robine 1989), and the longitudinal databases on functional status needed for the more sophisticated measurement are becoming available for the first time, geographical and time-series comparisons of cross-sectional estimates of active life expectancies, such as shown in Table 3, should be treated with caution. Methodological inconsistencies prevent any firm conclusion about whether or not active life expectancy has changed or remained constant over the last two decades, either in absolute or relative terms. So far, the available data tend to suggest that women can expect to spend more years in a disabled state then men, thus negating some of the benefit of their greater longevity.

Table 3 Life expectancy and active life expectancy at age 65, and percentage spent in an active state, for three countries.

	Life expectancy (in years)	Active life expectancy (in years)	Disabled life expectancy (in years)	Percentage spent in active state
Canada				
Men, 1978	14.4	8.2	6.2	56.9
Men, 1986	14.9	8.1	6.8	54.3
Women, 1978	18.7	9.9	8.8	52.9
Women, 1986	19.2	9.4	9.8	48.9
England/Wales				
Men, 1976	12.5	6.9	5.6	55.2
Men, 1985	13.4	7.7	5.7	57.4
Women, 1976	16.6	8.2	8.4	49.3
Women, 1985	17.5	8.9	8.6	50.8
United States				
Men, 1970	13.0	6.4	6.6	49.2
Men, 1980	14.2	6.6	7.6	46.4
Women, 1970	16.8	8.7	8.1	51.7
Women, 1980	18.4	8.9	9.5	48.3

Sources. Bebbington (1988, 1989); Crimmins *et al.* (1989); Wilkins and Adams (1983, 1989).

DIVERSITY OF THE OLDER POPULATION

The heterogeneity of the older population is manifested in the diverse demographic, social, and economic characteristics of old persons. These are brought about by behavioural features earlier in life, selective survival, and changes that occur in later life. The next sections will examine some important compositional characteristics of the older population, the factors that bring these features about, and their societal consequences.

URBAN AND RURAL AGED POPULATIONS

About 70 per cent of people aged 60 and older in developed countries currently live in urban areas. As the long-term trend of urbanization continues, the older population is expected to become even more concentrated. Although older women outnumber older men in every national population, the ratio of older men to older women is generally higher in rural than in urban areas. In the rural areas of several countries—e.g. Canada, Sweden, and Australia—older men actually outnumber older women. Conversely, elderly women are more likely than elderly men to live in urban areas. Preliminary evidence suggests that the sex difference in residential concentration is related to marital status. As elderly women are much more likely than elderly men to be widowed, it may be that urban residence provides widows the benefits of closer proximity to children and to social services.

The provision of health and other supportive services to ill and disabled older people in rural areas presents special challenges. Perhaps because of these difficulties, the percentage of older disabled people remaining in the community without being institutionalized is lower in predominantly rural areas than in urban areas.

EDUCATION AND INCOME

Education is an important determinant of economic attainment, health, and the ability to participate fully in modern societies. There are major differences in the educational attainment of older and younger people. While more than 90 per cent of persons aged 15 to 34 usually have completed a primary-school education, comparable rates for older people are often below 60 per cent. There also can be large differences among countries; while the population of people aged 65 and over in the United States has an unusually high proportion of primary-school graduates (75 per cent), fewer than half of Canadians aged 65 and over have completed this level of education. At higher educational levels, the gap between young and old widens even more, with high-school completion rates for elderly people often only 10 to 40 per cent of those for people aged 25–44. Completion of a post-secondary degree ranges from only 1 to 10 per cent among elderly people in developed countries.

Striking age-cohort differences also appear within older populations, and one of the most prominent differences between the young old and old old is the relatively low levels of attainment among the very old. As better educated cohorts reach old age and the oldest cohorts die, the average educational level of an elderly population increases thus narrowing the gap. This dynamic process of cohort improvement in education is one of the most important changes occurring within older populations, and is likely to have many ramifications for the life-styles, health status, and use of health care of tomorrow's older cohorts. Education is strongly associated with health. Those with higher education tend to live longer (Kitagawa *et al.* 1973). While the precise contributions of each of the many factors that account for this relationship have not been completely determined, it is clear that a well-educated and informed public is more receptive to public health campaigns and is better able to take a large share of responsibility for maintaining personal health.

The absence of adequate longitudinal studies has been a major obstacle to understanding economic status and the dynamics of impoverishment during old age. While widowhood appears to be an important correlate of poverty among older women, relatively little is known with certainty about the income and assets of older populations, their decisions about savings and consumption, and the role of health problems as a factor in impoverishment. However, current research is beginning to provide planners with a better basis for policy-making. Data from a comparable cross-sectional analysis (the Luxembourg Income Study) of six industrialized countries (the United States, Canada, the then Federal Republic of Germany, Norway, Sweden, and the United Kingdom) show that household income rises with age of the head of the household until retirement, and then declines. In the early 1980s, after-tax disposable income for households with a householder 75 years and over was, on average, only 78 per cent of the respective national means.

Retirement brings economic vulnerability as income from earnings declines and the ability to replace assets is often reduced. Industrialized societies have responded to the increased vulnerability of their elderly citizens with social insurance programmes with a wide variety of features. In almost all countries, social insurance benefits become more important with increasing age. However, the programmes differ significantly between countries in their impact on the economic security of older populations (de Jouvenel 1989). In the six-nation Luxembourg Income Study, the proportion of older people with incomes less than half the national median varied from 1 per cent in Sweden to 29 per cent in the United Kingdom. The percentage of elderly people aged 75 years and over with low incomes was much higher than the percentage for the 65–74 age group, except in Sweden.

SEX

Although more male than female babies are born each year, male mortality rates are usually higher than female rates at all ages. Thus, as a cohort ages, the proportion of females increases, usually producing a greater proportion of women by the age of 30 years. The same is true as a population ages, and this trend has become especially pronounced in developed countries. The large differences between female and male life expectancy and the lingering effects of the Second World War have resulted in elderly female/male ratios as high as 2 to 1 in parts of Europe (Germany, Austria). Among the old old (aged 80 and over), the proportion of females can reach 70 per cent. Hence it can be said that the social, economic, and health problems of 'the elderly' are in large part the problems of elderly women.

MARITAL STATUS

The marital status of older persons is a central feature of family structure that is closely related to living arrangements, support systems, survival, and economic and psychological well-being. Intact husband–wife families provide a continuity of the marital bond established through the life-course, and thus constitute a multiple support system for spouses in terms of emotional, financial, and social exchanges. Marital status plays a large part in determining the living arrangements of older persons, and directly influences the provision of care in coping with ill health and functional disability resulting from chronic disease.

Patterns of marital status are very different among elderly men and women. In a comparative study of 10 major industrialized countries (Myers 1991), the proportion of men aged 65 and over who were married (*circa* 1986) ranged from 67 per cent in Sweden to 76 per cent in Japan. The percentage of women who were married ranged from 28 in Hungary to 41 in Canada. The percentage of widowed men ranged from 14 to 21, while the corresponding percentages for women ranged from 44 (Sweden) to 61 (Hungary). Figure 6 illustrates the standard pattern for developed countries in which the sex difference in widow(er)hood increases with age. Among women in the oldest age categories (e.g. 80 years and over), three-quarters or more typically are widows. Several factors contribute to this sex difference: men marry women younger than themselves, women outlive men, and widows are less likely than widowers to remarry. Widowhood frequently results in major declines in the financial status of widows, and is often accompanied by changes in residence and living arrangements.

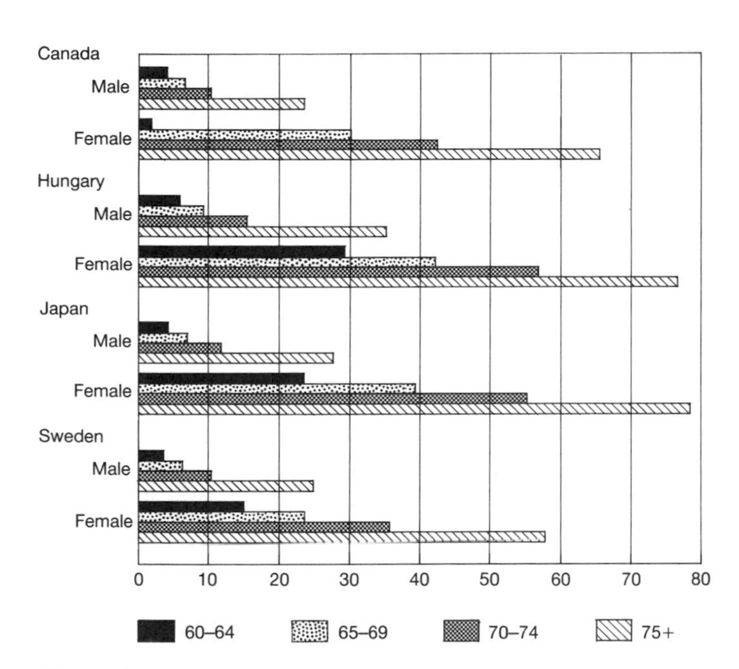

Fig. 6. Percentage widowed at older ages, circa 1986.

Changes in marital status over time have occurred slowly, but current trends for developed countries indicate that increasing proportions of the older population are married, declining proportions are widowed or never married, and the percentage of older people who are divorced or separated is small but steadily rising (Myers 1991). How long these trends in the proportions married and widowed will continue is an issue that is starting to be researched. As younger cohorts with higher divorce rates reach old age, the percentage of ever-divorced will increase, along with the potential complications implied by blended families, including the complex relationships and responsibilities operating between step-parents and stepchildren. Moreover, there are strong trends emerging among younger cohorts toward later ages at first marriage, increased cohabitation, and signs that larger numbers will never marry. Together with low levels of childbearing, these new trends, if they persist, imply that there will be generations of persons reaching the older ages in the next century without enduring marital ties and with fewer children to provide supportive services for older persons in need. For countries such as Sweden, such developments could be felt within the next 15 years. The time trajectory for other countries is more delayed, but similar developments are certainly likely, given the continuation of current patterns of family formation and dissolution.

LIVING ARRANGEMENTS

Living arrangements and residential quality are important aspects of the lives of older persons, and are determined by a host of factors including marital, financial, and health status, family size and structure, cultural traditions including kinship patterns, the value placed upon living independently, availability of social services and social support, and the physical features of the housing stock and the local community. In turn, living arrangements affect life satisfaction, health and, most importantly for those living in the community, the chances of institutionalization.

There are several intersections between living arrangements of older people and general social and economic policies. At the social level, these include broad policies that support the family as well as specific programmes that provide supportive services which permit older people who cannot live independently without personal assistance to remain in the community. Several economic policy issues also can affect living arrangements: increases in benefits that allow older people who wish to live independently of their children to do so; programmes that more easily allow the conversion of housing wealth into income; programmes that encourage the building of elder-friendly housing; and the discouragement of institutionalization through reimbursement disincentives. Data from New Zealand illustrate the interplay of an ageing population and changing social structure. A major change in New Zealand's household structure in recent years has been the growth of one-person households, which accounted for 20 per cent of all households in the 1986 census (up from 12 percent in 1971; New Zealand Department of Statistics 1990). Half of the one-person households in 1986 contained a person aged 65 or older, meaning that 10 per cent of all households in the country consisted of an elderly individual living alone.

Over the last three decades, most developed countries have experienced large increases in the numbers and proportions of elderly persons living alone (Table 4). The total for Canadian women, for instance, has more than doubled. Generally, between a fifth and a third of those aged 65 and over now live alone, with Japan having a very low proportion (9 per cent) while Sweden, the then Federal Republic of Germany, and Denmark, are close to 40 per cent (Kinsella 1990). Corresponding to widowhood patterns, two to three times more women than men live alone. Living alone increases with advancing age, but around the age of 80 or 85, the association drops or levels off, especially for women. Presumably this occurs as a result of health or economic factors that require

Table 4 Percentage of household population aged 65 years and over (unless noted) living alone, for selected countries (most recent date)

Europe	
Austria, 1980 (60+)	30.9
Belgium, 1981	31.9
Czechoslovakia, 1983 (60+)	32.4
Denmark, 1981	38.3
Finland, 1975	32.9
France, 1982	32.6
Germany (FR), 1982	38.9
Greece, 1981	14.7
Hungary, 1984	24.8
Ireland, 1981	20.1
Italy, 1981	25.0
Luxembourg, 1981	22.6
Malta, 1980 (60+)	10.5
Netherlands, 1982	31.3
Portugal, 1981	17.7
Spain, 1981	14.1
Sweden, 1982	40.0
United Kingdom, 1981	30.3
Other developed countries	
Australia, 1981	26.2
Canada, 1986	27.7
Japan, 1980	8.6
New Zealand, 1981	26.4
United States, 1987	30.4
Israel, 1983	26.1

Notes. Czechoslovakia—refers to urban areas; rural percentage is 24.5. Hungary—refers to pensioners and persons of retirement age. Sweden—refers to pensioners, with usual pension age being 65 years. United Kingdom—refers to men 65 years and over, women 60 years and over.
Source. Compiled at the U.S. Bureau of the Census from primary census and survey volumes, international compendia, and published research.

institutional caretaking, communal living, or sharing of housing costs.

INSTITUTIONALIZATION

Over the last several decades, the rising number of elderly individuals living alone has been accompanied by a similar increase in the number of older people in institutions. However, the overall proportions of older persons who are institutionalized has not risen markedly in most developed countries. Projections of rapid future growth of this institutionalized population and the attendant economic costs have occasioned intense policy debates about the medical, social, and economic dimensions of the problem. The debates have centred around issues such as the optimal balance between institutional and non-institutional care, the extent to which the family rather than the government should take responsibility for long-term care, and who should pay. Cross-national analyses permit us partially to disentangle the less plastic demographic imperatives from the more easily modifiable policy-determined factors. Cross-national comparisons of institutionalized populations are rendered hazardous by the absence of internationally consistent data. In this chapter, institutional living refers to receiving long-term care in either a medically oriented residential facility (e.g. nursing home) or a non-medical facility (e.g. home for the aged). Statistical differences between countries should be considered as indicative of orders of magnitude rather than as precise measurements of variance.

Comparative studies of institutional care (Doty 1985, 1991; supplemented by additional data from national sources) show that the proportions institutionalized for those aged 65 and over vary widely (Table 5), from less than 2 per cent in Hungary to nearly 11 per cent in the Netherlands. There is also considerable variation in the relative use of medical and non-medical facilities. Canada, (the Federal Republic of) Germany, Australia, and the United States rank high in the use of medical care, while the Netherlands, Switzerland, and Sweden show high rates of non-medical institutional care.

Table 5 Rates of institutional use for persons aged 65 years and over, for selected countries

Country	Year	Total	Medical facilities	Non-medical facilities
United States	1985	5.7	4.5	1.2
Canada	1980*	8.7	7.1	1.6
Belgium	1982	6.3	2.6	3.7
Denmark	1984*	7.0	N/A†	N/A
France	1982	6.3	5.3	1.0
Germany (FR)	1980	4.1	2.4	1.7
Great Britain	1981	4.0	N/A	N/A
Hungary	1984	1.9	N/A	N/A
Ireland	1982	3.6	N/A	N/A
Netherlands	1982/83	10.9	2.9	8.0
Sweden	1980	9.6	4.6	5.0
Switzerland	1982	8.9	2.8	6.1
Australia	1981	6.4	4.9	1.5
Japan	1981	3.9	3.1	0.8
Israel	1981	4.0	1.4	2.6

* = approximate date; †N/A = not available.

Notes. For reasons of definition, coverage, and temporal reference, rates of institutional usage may not be well suited to comparison across nations. Hence, differences between figures in this table should be taken as roughly indicative of different national propensities toward institutional use. Generally, a national usage rate reflects the percentage of all elderly persons resident in institutional settings at a particular point in time.

Great Britain: 1981 census data indicate 3 per cent of the elderly were 'usually resident' in communal establishments (mainly homes for the elderly, but also including hospitals, hostels, and hotels); another 1 per cent was present in such places (usually hospitals) though not usually resident.
Ireland: excludes elderly in general hospitals.

Sources. Doty (1985); UNDIESA (1985); Laquian (1988); Hugo (1988); United Kingdom Office of Population Censuses and Surveys and the Central Office of Information (1984); Hungarian Central Statistical Office (1986*b*).

Regardless of national setting, levels of institutionalization are strongly associated with increasing age (Table 6). In the United States, less than 2 per cent of the population aged 65–74 years resides in nursing homes; however, this figure rises to 6 per cent among those aged 75–84, and to 22 per cent among those aged 85 and older (Hing 1987). Comparable levels for very old populations include 31 per cent for those aged 85 and over in Canada; 49 per cent for those aged 86–90 and 76 per cent among those aged 91–96 in France; 41 per cent for those aged 90 and older in the Netherlands; 21–25 per cent for those aged 80 and older in Norway, Denmark, and Sweden; and 20 per cent for those aged 80 and older in Switzerland (Doty 1991). Levels are higher for women than for men, consistent with the fact that women

Table 6 Age-specific rates of institutional use by older persons, for selected countries, circa 1981

	Both sexes			
	Age 65–69	Age 70–74	Age 75+	Women Age 75+
Austria	1.6	2.4	7.3	8.8
Belgium	2.0	3.1	9.0	10.8
Canada	2.7	4.6	17.5	20.2
Denmark	1.6	2.8	13.4	15.3
France	2.2	2.9	9.1	10.7
Germany, DR	1.2	2.1	7.4	8.6
Italy	1.6	2.0	4.4	5.3
Japan	1.7	3.0	5.8	6.5
Luxembourg	2.7	4.4	11.6	14.1
Norway	0.9	2.1	11.0	12.7
Spain	1.2	1.8	3.7	4.2
Sweden	0.4	1.0	7.8	8.8
Switzerland	2.7	4.1	13.6	16.0
United Kingdom	1.2	1.9	7.8	9.2
United States	1.4	2.5	10.8	12.6

Sources: United Nations (1987): U.S. Congressional Budget Office (1988).

in institutions are far more likely than men to be widowed. Although the proportion of the older population that is institutionalized might be considered relatively small, the probability of an older person spending time in a nursing home is substantial. A set of projections for the United States estimated that 43 per cent of those who turned 65 in 1990 would enter a nursing home before they died, with 55 per cent of the nursing home users likely to have a total life-time use of at least 1 year, and 21 per cent of 5 or more years (Kemper and Murtaugh 1991).

The increase in the number of older people in institutions is a function of both the increasing number of older people and the rate of usage. That rate is itself a complex function of the age–sex–marital status composition of the older population, coupled with factors such as available beds, financial incentives and disincentives, and interfaces with the acute care and social insurance systems. The dynamics of age-associated transitions in functional status (in terms of the activities of daily living and instrumental activities of daily living) play a central role in the equation. Within the United States at least, certain data show that levels and mixes of functional capacity are highly predictive of future institutionalization. The usage rate will thus be influenced both by demographic factors such as increased numbers of very old widows, the availability of kin, and declining rural populations, and by policy decisions affecting living arrangements, nursing-home reimbursement, and bed capacity. A cross-national analysis (Doty 1991) that attempted to separate the demographic effect of variations in age–sex distributions from policy-determined factors concluded that the latter were most important. The supply of institutional beds and financing variables were found to be more important than home- and community-based care or support for family care-givers. Nevertheless, despite the considerable plasticity in rates of institutionalization, there does seem to be an irreducible base rate that is very strongly associated with being 80 years and older.

The high cost of appropriate nursing-home care has prompted many projections of the future size of institutionalized populations. Some are static component projections in which current rates of usage are multiplied by the expected future sizes of the affected age groups (although as we have seen, there is a high degree of uncertainty about the size of the future 'oldest old' population). Other projections have used more sophisticated models that include functional transition rates and changes in reimbursement policies. Most projections show a rapidly rising need for institutionalized long-term care. Thus, projections for The Netherlands suggest that by the year 2000, the numbers in old-age homes will increase by 50 per cent over those of the early 1980s, while the numbers in nursing homes will increase by 30 per cent (van der Wijst and van Poppel 1986). Projections for the United States (Manton and Stallard 1990) imply even greater increases (Fig. 7). As long-term care is so expensive, numerous governments are developing policies to limit nursing-home or old-age home admissions through changes in bed capacity, reimbursement, and increased supports for functionally disabled individuals in the community.

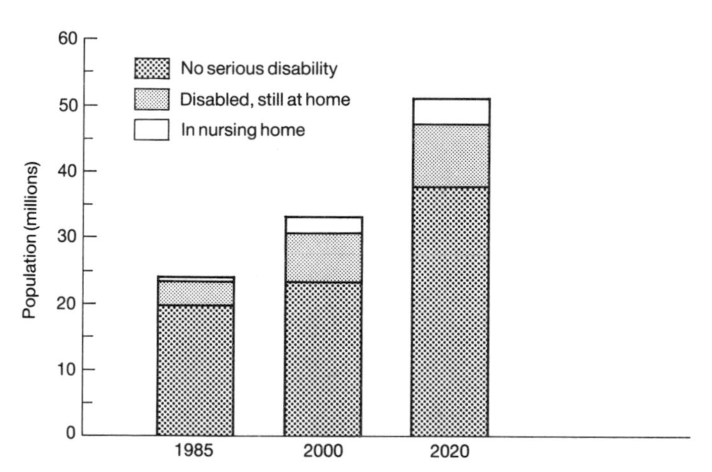

Fig. 7 Disability among older people: projections for the population 65 and older.

HEALTH EXPENDITURE

Between 1960 and 1987 the total public and private expenditure on health (all ages) as a fraction of gross domestic product (**GDP**) for the 23 OECD countries increased from an average of 3.8 to 7.4 per cent (Schieber and Poullier 1989). In 1987 the average *per capita* health spending for the OECD countries was $934 (US) ($1178 for the seven major industrialized countries within the OECD) (Table 7). At the low end were countries such as Turkey, Greece, Portugal, Spain, Ireland, New Zealand, and the United Kingdom, while the countries with high levels of expenditure were Norway, Switzerland, Sweden, Iceland, Canada, and the United States. Overall, there is a strong relationship between national wealth and health expenditures, with wealthier countries spending more *per capita* on health. The United States is, however, an outlier, as it spends a significantly larger fraction of its GDP on health (11.2 per cent in 1987) than would be predicted from the relationship that holds for other developed nations (Schieber and Poullier 1990).

Further, while health-care spending as a fraction of GDP

Table 7 Helath expenditures in OECD countries

	Per capita health spending in $US	Total health expenditures as a percentage of Gross Domestic Product[1]	
	1987	1960	1987
Australia	939	4.6	7.1
Belgium	879	4.6	7.2
Canada	1483	5.5	8.6
Denmark	792	3.6	6.0
France	1105	4.2	8.6
Germany	1093	4.7	8.2
Italy	841	3.3	6.9
Japan	915	2.9	6.8
Netherlands	1041	3.9	8.5
Sweden	1233	4.7	9.0
United Kingdom	758	3.9	6.1
United States	2051	5.2	11.2
Mean of 24 OECD Countries	934	3.8[2]	7.4[2]

[1]Schieber and Poullier (1989), based on OECD.
[2]Means are for 23 countries, excluding Turkey.

stabilized in the 1980s in most major European industrialized nations, it has continuously increased within the United States during the last decade. Population ageing by itself does not account for the major share of the increases in health expenditures in the industrialized nations. Other factors, such as the increases in the coverage of population eligible for public medical programmes in the 1960s and 1970s, and the considerable growth in real benefits were more important. Total *per capita* medical expenditures are greater for the elderly than non-elderly population and, within the older population, expenditures increase with age (Heller *et al.* 1986; OECD 1988). Within the United States, public and private personal-health expenditures were $3728 in 1987 for those aged 65–69 years and $9178 for those aged 85 and over, with hospital and especially nursing home expenses accounting for much of the age-group differences (Waldo *et al.* 1989).

Recent years have seen intense debate about the efficacy of health systems in nations that continue to age (Culyer 1990; Schieber and Poullier 1990). A general view is that the system in the United States is more expensive and less effective than most European health systems. But much of the trouble in trying to learn from and improve cross-national comparisons stems from uncertainty about the most appropriate measures of outcome. Differences in life expectancy at birth are perhaps the most basic standard, but is this still a good measure for industrialized nations? An extended set of indicators that measure age-specific health and functional status and expectancy may be required in order to assess health-system performance effectively.

LABOUR FORCE PARTICIPATION, RETIREMENT, AND PENSIONS

Policy-makers are confronted with several economic questions which arise in ageing societies that have changing balances of numbers in different age groups. What kinds of social support will a growing elderly population require, and how much will they cost? Will there be enough younger workers to provide an adequate tax base for this support? Will younger age groups be disadvantaged by an increasing emphasis on the needs of elderly people? How large a proportion of non-workers can a working population support? These questions have come to the fore partly as a result of changes in societies' economic behaviour.

Declines in retirement age, both statutory and actual, have been observed in many developed countries since 1960. The conjunction of trends such as weak economic growth, rising youth unemployment, and the entry of large numbers of women into the labour force, has spurred legislatures and businesses into lowering retirement ages and instituting various early retirement schemes. Concomitantly, rates of participation in the labour force at older ages have dropped, not only in response to these changes, but also because many of today's elderly people are able to afford to stop working. As life expectancy lengthens, individuals spend an increasingly smaller portion of their lives in remunerative work.

Declines in economic activity are especially notable among older men, who greatly outnumber older women in national labour forces. Nearly universal declines over time are seen among men aged 55 and over. In only a few developed nations (e.g. Poland, Japan, Norway) are 20 per cent or more of elderly men (65 and over) still considered to be economically active. In countries such as Hungary, Belgium, and France, the rate is 5 per cent or lower.

While paid work in the United States has virtually ceased for those over 75 years of age, the decline in unpaid productive activities such as housework, home maintenance, volunteer work, and help to people with chronic and acute problems is much less steep across age groups (Herzog *et al.* 1989). While a substantial fraction of those over the age of 75 work in these unpaid activities, there is little evidence to suggest on an aggregate level that retired people compensate by increasing their unpaid productive activity. Although the total agricultural labour force has shrunk rapidly in industrialized countries, the elderly people who remain economically active are more likely to be in agricultural than in other occupations. The latest census data show one-quarter to one-half of elderly workers in agriculture in countries as varied as Japan, Norway, Hungary, and Australia. The future occupational structure of elderly workers is likely to change markedly as more educated and affluent population cohorts reach older age.

The extent to which pensions are financed by contributions from employees, employers, and general tax revenues is, of course, different in each nation. However, population ageing has led developed countries to devote an increasing share of Gross Domestic Product to social security and public pensions. For example, between 1960 and 1981 the share within the United States increased by 76 per cent (to 7.4 per cent of the total), while in Norway and Sweden it more than doubled (to nearly 8 and 12 per cent, respectively). In somewhat different terms, pensions in the seven largest OECD economies increased as a fraction of social expenditures from 31 to 38 per cent between 1960 and 1975, while the cost increased by 40 per cent in real terms. After 1975, the rate of increase attributable to extended coverage and expanded real benefits decreased substantially, though governments still had to meet a 1.8 per cent annual rise in pension costs resulting from demographic changes.

Most industrial nations have financed their pensions on a pay-as-you-go basis. These schemes were popular after the Second World War because of their chain-letter distributional effects, with the earlier beneficiaries receiving far more in benefits than they had paid in contributions. As the schemes reached maturity, they were carried forward with expanded benefits by the strong economic growth of the 1950s and 1960s, combined with increasing numbers of workers. It has been argued that slowing economic growth rates, declining proportions of workers, and increasing proportions of retired people will result in workers paying into actuarially unfavourable plans, and that this will strain the implicit contracts between generations.

CONCLUSION

Concerns have been raised by the perception that over the last few decades, older citizens have fared better than children (Richman and Stagner 1986). This debate has occasioned a growing body of research into the nature of intergenerational exchanges and the economic impact of population ageing and changing support ratios. Several tentative conclusions have emerged. First, the future growth rate of the economy overshadows the impact of expected population ageing. Second, generations are interdependent, with future older cohorts dependent upon the productivity of the future labour force, which is in turn dependent upon the extent to which older generations are willing to invest in the educational capital of current cohorts of the young. Third, there are modifiable elements, and the potential effects of research on the health status and long-term care expenses of future generations of the very old are enormous (Schneider and Guralnik 1990; Suzman *et al.* 1992*b*). There are also a series of policy options available to slow population ageing, and reduce its fiscal impact (Martin 1991). These include pronatalist policies, encouraging immigration, reforming public pensions to ensure fiscal viability, and by decreasing early retirement and increasing working life expectancy at older ages. Finally, although many factors can influence the impact of population ageing in developed nations, there is an irreducible demographic imperative that will result in extensive changes to these societies.

References

Bebbington, A.C. (1988). The expectation of life without disability in England and Wales. *Social Science and Medicine*, **27**, 321–6.

Bebbington, A.C. (1989). Expectation of life without disability measured from the OPCS disability surveys. In *Proceedings of the first work-group meeting REVES*, International Research Network for Interpretation of Observed Values of Healthy Life Expectancy. Quebec (in press).

Crimmins, E.M., Saito, Y., and Ingegneri, D. (1989). Changes in life expectancy and disability-free life expectancy in the United States. In *Proceedings of the first work-group meeting REVES*, International Research Network for Interpretation of Observed Values of Healthy Life Expectancy. Quebec (in press).

Culyer, A.J. (1990). Cost containment in Europe. In *Health care systems in transition: the search for efficiency*. OECD, HCFA, ORD, DHHS, Baltimore.

de Jouvenel, H. (1989). *Europe's aging population. Trends and challenges to 2025*. Butterworth, Guildford, UK.

Doty, P. (1985). Long-term care provided within the framework of health care schemes. Paper presented at Twenty-First Meeting of the International Social Security Association Permanent Committee on Medical Care and Sickness Insurance, Brussels.

Doty, P. (1991). The oldest old and the use of institutional long-term care from an international perspective. In *The oldest old*, (ed. R. Suzman, D.P. Willis, and K.G. Manton). Oxford University Press, Oxford (in press).

Guralnik, J.M, Yanagishita, M., and Schneider, E.L. (1988). Projecting the older population of the United States: lessons from the past and prospects for the future. *Milbank Quarterly*, **66**, 283–308.

Heller, P.S., Hemming, R., and Kohnert, P.W. (1986). *Aging and social expenditure in the major industrial countries, 1980–2025*, Occasional paper 47, International Monetary Fund, Washington.

Herzog, A.R., Kahn, R.L., Morgan, J.N., Jackson, J.S. and Antonucci, T.C. (1989). Age differences in productive activities. *Journal of Gerontology: Social Sciences*, **44**, S129–38.

Hing, E. (1987). *Use of nursing homes by the elderly: preliminary data from the 1985 National Nursing Home Survey*, Advance Data from Vital and Health Statistics, No. 135, US DHHS Publn. (PHS) 87–1250. US PHS, Hyattsville.

Hugo, G. (1988). The changing urban situation in southeast Asia and Australia; some implications for the elderly. Paper prepared for the United Nations Conference on Aging Populations in the Context of Urbanization, Sendai, Japan, September 12–16.

Hungarian Central Statistical Office (1986). *Summary Report on the Situation of the Aged Population in Hungary*. Budapest.

Katz, S., Branch, L.G., Branson, M.H., Papsidero, J.A., Beck, J.C. and Greer, D.S. (1983). Active life expectancy. *New England Journal of Medicine*, **309**, 1218–24.

Kemper, P. and Murtaugh, C.M. (1991). Lifetime use of nursing home care. *New England Journal of Medicine*, **324**, 595–600.

Kinsella, K. (1990). *Living arrangements of the elderly and social policy: a cross-national perspective*, CIR Staff Paper (52). Center for International Research, U.S. Bureau of the Census, Washington.

Kitagawa, E.M. and Hauser, P.M. (1973). *Differential mortality in the United States*. Harvard Press, Cambridge.

Laquian, A. (1988). Changing age structure as a factor in planning urban development. Paper prepared for the United Nations International Conference on Aging Populations in the Context or Urbanization, Sendai, Japan, September 12–16.

Manton, K.G. (1989). Epidemiological, demographic, and social correlates of disability among the elderly. *Milbank Quarterly* (Suppl. 1 on *Disability policy: restoring socioeconomic independence*), **67**, 13–58.

Manton, K.G. and Stallard, E. (1991*a*). Cross sectional estimates of active life expectancy for the U.S. elderly and oldest old populations. *Journal of Gerontology: Social Sciences*, **46**, S170–82.

Manton, K.G. and Stallard, E. (1991*b*). Incorporating risk factors into projections of the size and health status of the U.S. elderly population. In *Forecasting and health of the oldest old*, (ed. K.G. Manton, B. Singer, and R. Suzman). New York (in preparation).

Martin, L. G. (1991). Population aging policies in East Asia and the United States. *Science*, **251**, 527–31.

Myers, G. (1990). Demography of aging. In *Handbook of aging and the social sciences*, (ed. R.H. Binstock and L.K. George), pp. 19–44. Academic Press, New York.

Myers, G. (1991). Demographic aging and family support for older persons. In *Family support to the elderly: the international experience*, (ed. Kemdig, Hashimoto, and Coppard).

Myers, G. (1991). Marital status dynamics at older ages. *Proceedings of the United Nations Conference on Aging Within the Context of the Family*, Kitakyusu, Japan. United Nations, New York (in preparation).

New Zealand Department of Statistics (1990). Changing structures of families and households in New Zealand and their consequences. Report to the United Nations Seminar on Demographic and Economic Consequences and Implications of Changing Population Age Structures, Ottawa.

Organization for Economic Cooperation and development (1988). *Ageing populations. The social policy implications*. OECD, Paris.
Richman, H. and Stagner, M. (1986). Children in an aging society: treasured resource or forgotten minority. *Daedalus* (Winter).
Robine, J.M. 1989. Estimating Disability-Free Life expectancy (DFLE) in the Western Countries in the Last Decade: How Can This New Indicator of Health Status Be Used. *World Health Statistics Q Rapport Trimestriel des Statistiques Sanitaires Mondiales* 42:141–150.
Rogers, R.G., Rogers, A., and Belanger, A. (1989). Active life among the elderly in the United States: multistate life-table estimates and population projections. *The Milbank Quarterly*, **67**, 370–411.
Schieber, G.J. and Poullier, J.P., (1989). International health care expenditure trends: 1987. *Health Affairs* (Fall), 169–77.
Schieber, G.J. and Poullier, J.P. (1990). Overview of international comparisons of health care expenditures. In *Health care systems in transition: the search for efficiency*. OECD, HCFA, ORD, DHHS, Baltimore.
Schneider, E.L. and Guralnik, J.M. (1990). The aging of America: impact on health care costs. *Journal of the American Medical Association*, **263**, 2335–40.
Spencer, G. (1989). *Projections of the population of the United States, by age, sex, and race: 1988 to 2080*, U.S. Bureau of the Census, Current Population Reports, series P-25, No. 1018. U.S. Government Printing Office, Washington.
Suzman, R., Manton, K.G., and Willis, D.P. (1992*a*). Introducing the oldest old. In *The oldest old*, (ed. R. Suzman, D.P. Willis, and K.G. Manton). Oxford University Press, Oxford (in press).
Suzman R., Harris, T., Hadley, E. and Weindruch, R. (1992*b*). The robust oldest old: optimistic perspectives for increasing healthy life expectancy. In *The oldest old*. Oxford University Press, Oxford (in press).
UNDIESA, United Nations Department of International Economic and Social Affairs (1985). *The world aging situation: strategies and policies*, ST/ESA/150. UNDIESA, New York.
United Kingdom Office of Population Censuses and Surveys and the Central Office of Information (1984). *Britain's Elderly Population.* 1981 Census Guide. London.
United Nations (1987). Demographic Yearbook, 39th issue. UN, New York.
U.S. Bureau of the Census (1987). *An aging world*, (ed. B.B. Torrey, K. Kinsella, and C.M. Taeuber), International Population Reports Series P-95, No. 78. U.S. Government Printing Office, Washington.
U.S. Congressional Budget Office (1988). *Changes in the Living Arrangements of the Elderly: 1960–2030*. U.S. Government Printing Office, Washington DC.
Van der Wijst, T. and van Poppel, F. (1986). Ageing of the population and its implications for the social and economic structure of West European society. Paper presented at Universite Catholique de Louvain Conference on Aging Populations and the Gray Revolution, Louvain-La-Neuve, Belgium.
Waldo, D.R., Sonnefeld, S.T., McKusick, D.R., and Arnett, R.H., III. (1989). Health expenditures by age group, 1977 and 1987. *Health Care Financing Review*, **10**, 111–20.
Wilkins, R. and Adams, O.B. (1983). *Healthfulness of life*, Montreal: The Institute for Research on Public Policy: 162.
Wilkins, R. and Adams, O. (1989). Health expectancy trends in Canada. In *Proceedings of the first work-group meeting REVES*, International Research Network for Interpretation of Observed Values of Healthy Life Expectancy. Quebec (in press).

1.2 Epidemiological issues in the developed world

JACOB A. BRODY, SALLY FREELS, AND TONI P. MILES

INTRODUCTION

As presented in the preceding section, an increase in the absolute number of persons aged 65 years and over is a phenomenon that is occurring in both developed and developing nations (Torrey *et al.* 1987; Kinsella 1988). Among developed countries there has also been a growth in that proportion of the entire population which is older. The global impact of an ageing population on society has implications beyond the demographic measures of absolute numbers of persons, population proportions, and rates of growth. Patterns of morbidity and mortality can be used to plan health-care resources to meet the needs of this population. By examining average expected remaining years at specific ages past 65, we can compare survival for older age cohorts among different countries. Marital status and rates of participation in the labour force provide insight into the social functioning of the older population. In addition, we seek answers to questions about patterns of disease and mortality. The period preceding death in ageing populations is one of declining health status caused by accumulating numbers of comorbid chronic diseases and conditions (Katz *et al.* 1983). Morbidity and the attendant need for long-term care will be a major economic concern of the twenty-first century (Brody 1985).

This chapter presents patterns of mortality and morbidity for the older adult population in selected developed countries. While useful for characterizing features of health status among populations of older persons, international data on health and vital statistics can be biased in many subtle ways. Population factors such as ethnicity, cohort life expectancy, and population density (urbanization) vary greatly among countries. The apparent risk factors for disease may be biased in countries with large immigrant populations. Emigration can artificially depress mortality rates if older persons who immigrated to an area return home to die. Comparable data for individual countries can be limited by the type of information required, adequacy and completeness of a country's recording system, and inability of international agencies to standardize information on health and vital status.

The type of information collected on a standard death certificate can have broad implications for research in ageing (Curb *et al.* 1987). Multiple pathological conditions can give rise to difficulty in analysing information on the cause of death for older persons. The underlying cause of death is usually reported when there is a need to attribute death to a single condition. It has been argued, however, that this practice underestimates the role of cardiovascular disease as a cause of death (Uemura 1988). Even among developed

Table 1 Mortality rates per 1000 by age and sex

		Men (age in years)			Women (age in years)		
Country	Year	65–74	75–79	80+	65–74	75–79	80+
United States	1983	38.0	68.9	128.5	20.1	38.8	92.8
Japan	1984	29.0	71.2	137.2	15.5	38.0	106.9
Germany (Federal Republic)	1984	46.0	79.2	152.6	23.3	48.9	121.4
Italy	1981	45.8	68.4	139.1	23.1	43.2	117.4
United Kingdom	1984	38.8	73.6	138.6	20.8	43.3	105.2
France	1983	41.5	80.8	154.5	19.3	41.6	123.5
Poland	1976	29.2	37.1	111.9	11.7	17.6	60.8
Canada	1984	34.9	70.4	132.8	17.6	36.5	98.7
Australia	1984	37.0	71.9	149.3	19.1	38.1	115.7
Hungary	1984	60.1	96.1	182.0	32.3	66.9	152.8
Belgium	1984	45.8	90.6	162.1	21.8	51.0	133.0
Greece	1983	35.7	72.4	132.3	21.6	49.8	128.6
Sweden	1984	34.6	77.0	146.9	17.4	40.2	113.6
Austria	1984	45.2	79.9	164.2	23.8	51.2	137.2
Denmark	1984	41.3	82.7	149.5	22.6	44.2	113.8
Norway	1984	37.7	77.6	151.3	17.5	40.3	116.1
Luxembourg	1984	46.4	91.3	176.3	25.1	54.1	143.1

Source: Demographic Yearbook, United Nations, 1985.

nations there is considerable variation in both diagnostic procedures as well as coding practices, which can lead to biased estimates of cause-specific mortality rates (Percy and Muir 1989). While recognizing these caveats, cross-national comparisons of survival and mortality data are not only warranted but essential to identify trends common to ageing populations.

Data on morbidity, comorbidities, and various functional measurements are also of great value. While data exist in all developed countries there are no international systematic convention of reporting (Brody 1989 *a, b*). We present some information from the United States in this chapter.

MORTALITY (TABLES 1–4)

Mortality risk was not uniformly distributed across the population aged 65 years and over in each listed country (Table 1). In all countries, there was a gradient of increasing risk by age. The change, by country, in risk of dying, varied. This is best illustrated by examining the risk of mortality for the older age groups relative to that of persons aged 65–74 years. For example, in Japan, where mortality rates for both men and women were among the lowest, the relative mortality risk for women aged 75–9 years was 2.5, and for women aged 80 years and older, 6.9. By contrast, a woman in the United States aged 75–9 years had a lower relative risk (1.9) as did those aged 80 years and older (4.6). The slower increase in mortality risk in the United States appears to be a function of both very low mortality rates among the oldest old as well as moderately elevated rates among the younger group.

In most countries, the proportionate contribution to all deaths made by men and women aged 65 years and over was well in excess of their representation in the population (Table 2). For women, a comparison of current (1981–4) mortality with data from 1976 reveals that all countries recorded an increase in proportion of deaths in persons over age 65. For men of 70 years and older a similar pattern was seen. Declines in the percentage of deaths for men occurred in France, Hungary, Austria, and Luxembourg, apparently confined to the 65–9 age group. Decreased premature mortality (deaths before 65 years) between 1976 and 1985, and ageing of the population over age 65 years were contributing factors to this increasing proportionate mortality. This phenomenon has accelerated through 1988 (Brody and Miles 1990).

Table 3 shows proportionate cause-specific mortality in four broad categories (heart disease, stroke, cancer, and all other causes) for persons aged 65 years and over in 13 countries. Heart disease was the leading cause of death in the United States and most other countries (28–43 per cent). Cancer was the second most frequent cause of death (10–24 per cent). Japan (22 per cent) and Greece (23 per cent) had the highest mortality rates for stroke. In France, where 'senility' is more widely used in death certification, 38 per cent of deaths were listed under all other causes, while the range among the rest of these countries was 22–33 per cent.

A closer inspection of mortality data for cancer and circulatory disease reveals differences between populations of older persons (Table 4). While ischaemic heart disease was a leading cause of death in Australia, England and Wales, Hungary, and the United States, among the Japanese, cerebrovascular disease caused the most deaths in both sexes. For cerebrovascular disease, mortality rates for women were higher than for men in Australia (7.6 versus 6.5 per 1000), England and Wales (8.6 versus 7.5 per 1000), and the United States (5.2 versus 4.8 per 1000). Mortality rates for all other circulatory diseases were slightly greater for women than for men in Australia (6.0 versus 5.8 per 1000) and England and Wales (7.2 versus 6.8 per 1000).

Lung and colon tumours were leading causes of death from cancer mortality for adults under 65 years. By contrast, among persons of 65 years and over, cancer outside the pulmonary and gastrointestinal sites causes death most frequently. This trend is observed in most industrialized countries. Mortality rates were highest in the 'all other' category for both sexes in the listed countries. Women in

Table 2 Deaths among persons 65 and over as a proportion (%) of all deaths by sex and age for selected years and selected countries

	Men (age in years)					Women (age in years)				
Country	65+	70+	75+	80+	85+	65+	70+	75+	80+	85+
United States										
1976	59.4	46.9	34.0	21.5	10.8	72.3	63.1	51.8	37.8	22.3
1983	63.4	51.4	37.4	23.7	12.6	76.3	67.3	55.7	41.8	26.7
Japan										
1976	61.7	49.5	34.6	19.1	7.6	72.2	62.6	49.6	32.9	16.7
1984	64.7	55.1	41.4	25.6	11.2	77.3	69.4	57.2	41.1	22.3
Germany, Federal Republic										
1976	70.0	55.6	36.8	20.2	9.1	81.7	71.3	56.0	36.7	17.8
1984	69.6	61.8	45.6	26.4	10.8	84.4	78.8	65.5	45.8	24.3
Italy										
1976	66.3	52.4	36.9	22.2	10.8	79.2	69.8	57.0	39.4	20.7
1981	69.1	56.0	39.4	23.4	10.8	82.4	73.8	60.6	43.3	23.6
United Kingdom										
1976	69.3	54.2	36.4	21.0	9.7	81.7	76.6	62.6	44.8	25.5
1984	72.1	60.0	42.5	24.3	10.3	83.4	75.7	63.1	46.0	27.0
France										
1976	66.4	53.9	38.3	22.4	11.1	83.3	76.1	64.8	48.1	28.6
1983	66.3	58.9	45.0	28.6	13.2	84.3	80.2	70.7	55.3	34.4
Poland										
1976	55.2	41.1	26.0	13.3	5.4	72.9	61.9	46.8	29.1	14.0
1981	55.7	44.5	29.0	15.0	5.8	74.4	65.3	50.6	32.7	15.5
Canada										
1976	59.0	47.4	34.6	22.5	12.1	71.2	62.3	51.3	37.8	23.0
1984	64.4	52.7	38.4	24.4	12.6	76.1	67.4	55.8	41.7	26.7
Australia										
1976	59.1	46.0	32.1	18.9	9.1	73.4	64.3	53.0	38.0	21.9
1984	64.0	51.9	36.2	21.3	9.9	77.5	68.8	56.4	42.0	25.4
Hungary										
1976	63.4	49.1	32.8	17.2	7.0	75.3	64.2	49.2	30.6	13.9
1984	57.9	49.6	33.6	18.4	7.1	75.8	69.1	53.4	34.9	16.9
Belgium										
1976	69.9	55.9	39.1	22.7	10.4	82.2	73.2	58.8	39.4	20.5
1984	69.3	60.1	43.9	26.4	11.8	83.0	77.6	65.8	47.9	26.1
Greece										
1976	68.1	55.7	40.8	26.2	14.6	79.2	70.1	57.4	39.8	24.2
1983	72.5	62.2	47.0	29.8	15.1	83.4	75.9	62.8	44.6	24.7
Sweden										
1976	73.6	61.0	45.2	28.4	14.0	82.9	74.6	61.8	44.3	24.8
1984	77.1	66.1	50.4	31.7	15.5	86.0	79.3	67.9	50.8	30.3
Austria										
1976	70.3	56.5	38.1	20.8	9.3	83.4	73.9	59.4	39.9	20.1
1984	67.1	59.6	44.0	25.8	10.4	84.2	78.8	66.0	47.0	24.9
Denmark										
1976	70.0	57.0	42.0	26.4	13.3	78.6	69.2	56.1	39.5	22.0
1984	72.1	60.5	44.8	28.0	14.3	81.1	72.9	60.9	45.3	27.4
Norway										
1976	70.7	58.5	43.3	27.4	13.4	82.3	74.1	61.8	44.0	24.4
1984	75.1	63.1	47.2	30.8	15.7	85.6	78.5	67.1	50.9	30.5
Luxembourg										
1976	66.9	53.0	36.5	20.2	8.9	80.4	70.1	53.3	33.8	17.4
1984	65.0	55.3	40.1	22.0	9.0	80.6	73.5	60.2	41.8	21.3

Source: Demographic Yearbook, United Nations, 1976 and 1985.

Table 3 Causes of death, persons aged 65 years and over, in selected countries

		Percentage of all deaths due to:			
Country	Year	Heart disease	Stroke	Cancer	All other
United States	1982	47.5	10.0	20.2	22.3
Japan	1984	24.4	22.1	21.0	32.5
Germany (Federal Republic)	1984	40.9	15.9	20.6	22.7
Italy	1981	37.6	16.1	19.3	27.0
United Kingdom	1983	38.1	13.4	20.2	28.3
France	1983	28.1	13.9	20.2	37.8
Canada	1984	41.6	9.9	23.2	25.3
Australia	1983	43.3	14.3	20.8	21.6
Belgium	1984	34.3	13.1	21.8	30.8
Greece	1983	28.2	22.9	16.2	32.7
Sweden	1984	42.2	11.7	19.7	26.4
Denmark	1984	36.6	10.7	23.8	28.9
Norway	1984	33.3	15.0	20.2	31.5

Source: Demographic Yearbook, United Nations, 1985.

Table 4 Mortality rates by sex and cause of death (per 1000)

	Neoplasms			Circulatory system		
	Malignant neoplasms of the trachea, bronchus, and lung	Malignant neoplasms of the stomach and colon	All other malignant neoplasms	Ischaemic heart diseases	Cerebrovascular diseases	All other circulatory diseases
Men 65 +						
United States 1982	4.4	2.0	7.5	20.0	4.8	8.6
Japan 1984	2.8	4.5	5.8	4.1	10.7	7.9
England/Wales 1982	6.1	2.6	7.5	21.1	7.5	6.8
Australia 1983	4.0	2.2	7.8	19.7	6.5	5.8
Hungary 1984	4.7	3.8	9.3	18.3	15.1	18.4
Women 65+						
United States 1982	1.1	1.4	5.3	13.9	5.2	7.2
Japan 1984	0.7	2.2	3.6	2.9	9.0	7.0
England/Wales 1982	1.3	1.8	6.0	12.9	8.6	7.2
Australia 1983	0.8	1.6	5.3	12.9	7.6	6.0
Hungary 1984	0.8	2.3	6.8	12.6	13.6	16.9

Source: International Data Base, Center for International Research, U.S. Bureau of the Census, 1987.

each country had higher mortality rates from stomach and colon cancer (1.4–2.3 per 1000) than from trachea, bronchus, and lung cancer (0.7–1.3 per 1000). Men in every country, except Japan, had higher mortality rates for trachea, bronchus, and lung cancer (2.8–6.1 per 1000) than for stomach and colon (2.0–4.5 per 1000). As women smokers age, these rates are changing.

MORBIDITY AND DISABILITY (TABLES 5, 6)

A range of health indicators are shown on Table 5 for the older United States' populations. This array was selected from (1984) data collected in a survey of measures of dependence in the physical functioning of the older, civilian, non-institutionalized population (National Health Survey 1989). Reversibility of specific conditions is not addressed, although it is clearly a factor. With exceptions all conditions increase sharply with age. Independent mobility was reported for 95 per cent of those aged 65–74 years and for 85 per cent of those over the age of 75; almost 80 per cent of those aged 75 and over could still do heavy housework. About half of each age group had arthritis and 40 per cent reported hypertension. Approximately 40 per cent of those 75 and over had impaired hearing and 10 per cent complained of daily incontinence.

With age the number of chronic diseases as well as the coexistence of chronic disease in the same person increases. A survey was conducted of the ageing, civilian, non-institutionalized population to ascertain the presence of comorbidities for nine chronic conditions—arthritis, hypertension, cataracts, heart disease, varicose veins, diabetes, cancer, osteoporosis or hip fracture, stroke, and Alzheimer's disease. Some chronic conditions are included in Table 5.

In Table 6 the percentage distribution of comorbidities is given. Only 10 per cent of men and 20 per cent of women over the age of 80 have none of the morbid conditions while more than half of the men and almost three-quarters of the women have two or more of these comorbidities.

The content of Tables 5 and 6 gives an indication of the types of information being collected throughout the developed world. These data are difficult to gather but are of great use in understanding pathology and functioning and their relationship to enlightened policy decisions in the

Table 5 Self-reported deficits in persons 65 years and over in the United States

	Rates per 1000		Number of people	
	Age: 65–74	75+	65–74	75+
Total population in thousands			16 288	10 145
Instrumental activities of daily living				
Preparing meals	18.0	71.2	293	722
Shopping for personal items	35.9	132.6	585	1345
Managing money	14.8	66.0	241	670
Using the telephone	8.5	37.0	138	375
Doing light housework	23.1	77.4	376	785
Doing heavy housework	105.8	227.9	1723	2312
Activities of daily living				
Bathing	35.2	106.5	573	1080
Dressing	29.3	66.0	477	670
Using the toilet	12.4	39.3	202	399
Getting in and out of bed or chair	17.7	46.6	288	473
Eating	6.3	17.5	103	178
Mobility status				
Independent	951.3	865.4	15 495	8779
Dependent getting outside; independent walking	8.7	39.5	142	401
Dependent walking; independent getting outside	10.6	12.6	173	128
Dependent getting outside and walking	19.2	62.5	313	634
Unknown	10.2	20.0	166	203
Continence status				
Continent	898.5	814.9	14 635	8267
Incontinent less than daily	39.2	73.2	638	743
Incontinent daily	47.5	89.1	774	904
Incontinent unknown frequency	5.5	10.6	90	108
Unknown	9.3	12.3	151	125
Reported impairments				
Visual impairment	73.2	138.5	1192	1405
Cataract	94.4	238.6	1538	2421
Hearing impairment	260.7	386.8	4246	3924
Deformity or orthopaedic impairment	165.2	171.3	2691	1738
Chronic conditions				
Ischaemic heart disease	137.1	133.0	2233	1349
Hypertension	393.6	397.1	6411	4029
Cerebrovascular disease	41.8	84.5	681	857
Emphysema	43.2	36.2	704	367
Chronic bronchitis	62.9	49.5	1025	502
Diabetes	93.6	85.5	1525	867
Arthritis	475.7	501.9	7748	5092

Source: Vital and Health Statistics, National Center for Health Statistics, Series 10, No. 107, March 1989.

Table 6 Percentage distribution of comorbid conditions by age and sex

	Men: number of chronic conditions			Women: number of chronic conditions		
Age (years)	0	1	2 or more	0	1	2 or more
60–69	30	35	35	23	32	45
70–79	22	31	47	14	25	61
80 +	19	28	53	10	20	70

Source: Vital and Health Statistics, National Center for Health Statistics Advance Data No. 17, May 1989.

future. We have not yet sorted out the key questions, methods, and data that would provide the necessary insights for any given country or for adequate international comparisons.

DISCUSSION

The percentage of deaths in people over the age of 65 years is related to the percentage of the entire population that is 65 and over. In 1988, in Sweden, with 17 per cent of the population aged 65 and over, only 18 per cent of all deaths occurred in people under age 65. In the United States, with 12 per cent of the population over age 65, only 23 per cent of deaths occurred before the age of 65.

Recently, we introduced two new terms—phase I and phase II mortality (Brody 1989*b*; Brody and Miles 1990). Phase I mortality is when only 20 per cent of deaths occur before the age of 65. It is obvious that this is already occurring. Once phase I occurs, gains in longevity will be realized to a lesser extent from reduction in the already low rate of premature mortality (death before the age of 65) and increasingly through gains in life expectancy after the age of 65. Phase II mortality is a period in which premature deaths reach a minimum and steady state. This should be about 14–15 per cent of deaths occurring before 65. Over time, we should be able to convert the phase II mortality figure from a percentage to a rate and make it analogous to the infant mortality rate. It seems unlikely that infant mortality will decline much below 5 per 1000 live births. A similar figure for phase II mortality would be of great use in monitoring comparative trends in countries and subpopulations, and also would be a key measure of the effectiveness of health practices.

With few exceptions, between 70 and 80 per cent of all deaths among those aged 65 and over are certified as being caused by heart disease, stroke, or cancer. There is variation over time and from country to country, but the strength and direction of this pattern is likely to persist. As increases in life expectancy are largely confined to life expectancy after 65 (Olshamsky *et al.* 1990), future gains in longevity will be related to further postponement or prevention of heart disease, stroke, and cancer. Morbidity from these diseases as well as from the numerous conditions that we have called age-dependent and non-fatal may not be affected by further postponement of fatal diseases. Instead, they are likely to accumulate in complex patterns of comorbidities and conditions that will detract from physical, mental, emotional, and social well-being.

Life expectancy is not greatly influenced by specific national policies for health care among developed nations (Brody and Schneider 1986; Brody, in press). Indeed the data presented above suggest considerable homogeneity of outcome in diverse health settings. Thus estimates of the impact on health care costs in the United States (Schneider and Guralnik 1990) can be generalized to the developed world; i.e. that health care costs will increase markedly unless we can achieve major improvements in the health of the growing numbers of older persons, especially the 'oldest old'.

It is commonly believed that the United States, unlike other advanced nations, does not provide systematic public insurance coverage for long-term care. There is, however, more mythology than reality in the relative completeness of long-term care coverage in other developed nations or its absence in the United States. Pamela Doty (1990), in her article 'Dispelling some myths' states: 'To draw useful lessons from other countries' experiences, however, we must first re-examine many of the preconceived notions we have about how their approaches to long-term care financing differ from ours'. She intelligently analyses the following four myths.

Myth 1. Other countries provide comprehensive coverage for long-term care along with acute-care benefits under their national health service or national health insurance programmes.

Myth 2. The United States is the only country that bases access to public coverage for long-term care on a means test or requires that individuals 'impoverish' themselves by first exhausting their capacity to pay privately for care.

Myth 3. Services' systems for long-term care in other advanced industrial countries are more integrated and 'rational', not fragmented like the system in the United States.

Myth 4. Other countries do a better job of preventing institutionalization through generous public funding of home and community-based alternatives.

The most difficult problem to work through is the issue of health care costs versus the preservation of a social system that provides a rational approach to healthy but progressively frail elderly people. The blend and interaction of social programmes and health care within a nation's economic and political fabric is something the entire developed world is addressing. Various approaches, some involving sweeping legislation, and others involving incentive-driven, stepwise pragmatism to induce change, are going on at all levels of all governments.

We are, at the same time, learning a great deal about the physiology of ageing as well as the necessary political and social options and consequences. We must inform each other and the developing world about those things that work better and under what circumstances. We must also try to quantify the reality that before death there is some pain, and try to reduce the pain to a minimum.

References

Aging in the eighties: the prevalence of comorbidity and its association with disability. In *Vital and Health Statistics*, No. 170. National Center for Health Statistics, Advance Data (1989), Hyattsville, MD.

Brody, J. (1985). Prospect for an ageing population. *Nature*, **315**: 463–6.

Brody, J.A. (1988). Changing health needs of the aging population. In *Research and the aging population*, Proceedings of the CIBA Foundation Symposium No. 134, pp. 208–15, Wiley, New York.

Brody, J.A. (1989a). Opportunities for international collaboration—comparisons of morbidity and mortality for chronic diseases in older persons. In *Proceedings, 1988 International Symposium on Data on Ageing*. National Center for Health Statistics, Hyattsville, MD.

Brody, J.A. (1989b). Toward quantifying the health of the elderly. (Editorial.) *American Journal of Public Health*, **79**, 685–6.

Brody, J.A. Aging: a vision from here to 2020. In *Health, Longevity and Vitality*, American Association of Retired Persons Monograph (in press).

Brody, J.A. and Miles, T.P. (1990). Mortality postponed and the unmasking of age-dependent non-fatal conditions. *Aging*, **2**, 283–9.

Brody, J.A. and Schneider, E.L. (1986). Disease and disorders of ageing: an hypothesis. *Journal of Chronic Diseases*, **39**, 871–6.

Curb, D., Miles, T., and White, L. (1987). Research implications of changes in 1989 revision of the U.S. Standard Certificate of Death in the Aging Population. In *Proceedings 'Data for an aging population issues in health, research and public now and into the 21st century'*. National Center for Health Statistics. DHHS Pub. In. No. (PHS) 88–1214. Hyattsville MD.

Doty, P. (1990). Dispelling some myths: a comparison of long-term care financing in the U.S. and other nations. *Generations*, **14**, 10–18.

Katz, S., Branch, L.G., Branson, M.H., Papsidero, J.A., Beck, J.C., and Greer, D.S. (1983). Active life expectancy. *New England Journal of Medicine*, **308**, 1218–24.

Kinsella, K. (1988). *Aging in the Third World*, International Population Reports Series, P-95, No. 79. U.S. Government Printing Office, Washington DC.

Physical functioning of the aged United States, 1984, Vital and Health Statistics Series 10: data from the National Health Survey, No. 167. National Center for Health Statistics, National Health Survey (1989).

Olshansky, S.J., Carnes, B., and Cassel, C. (1990). In search of Methuselah: estimating the upper limits of human longevity *Science*, **250**, 634–40.

Percy, C. and Muir, C. (1989). The international comparability of cancer mortality data: results of an international death certificate study. *American Journal of Epidemiology*, **129**, 934–46.

Schneider, E.L. and Guralnik, J.M. (1990). The aging of America: impact on health care costs. *Journal of the American Medical Association*, **363**, 2335–40.

Torrey, B.B., Kinsella, K., and Taeuber, C.M. (1987). *An ageing world*, International Population Reports Series, P-95, No. 78. U.S. Government Printing Office, Washington DC.

Uemura, K. (1988). International trends in cardiovascular diseases in the elderly. *European Heart Journal*, **9** (Suppl. D), 1–8.

1.3 The developing world

DAVID M. MACFADYEN

THE UNIVERSALITY OF DEVELOPMENT

If we accept the anthropologist's view that the most distinctive characteristic of man as a species is the human capacity to elaborate and transform culture (Cowgill 1986), then all cultures of the world are developing. Scholars can now draw upon a large number of studies of human cultures and they have observed that the definition of old age, in chronological terms, comes with transition to a modern society. Prior to such development, old age is associated with grandparenthood or succession to eldership. Anthropologists have also found support for the proposition that, as societies develop and modernize, the status of the aged person declines. These transitions in categorization and values occurred, historically, in developed societies. Consider the culture within which ageing policies first evolved in Europe. Around 1865, the process of demographic ageing slowly began within the French population. Simone de Beauvoir, in *La Vieillesse*, reminds us that from this time onward there were just too many old people for literature to pass over in silence (de Beauvoir 1970). Ageing persons, who previously were barely visible as a group, now featured more prominently in families, neighbourhoods and communities, and this was reflected in European literature of the 19th century. As the 20th century draws to a close, the same phenomenon is being experienced by the world as a whole. Every month, the net balance of the world population aged 55 or over increases by 1.2 million persons (Kane *et al.* 1989). Of this increase, 80 per cent occurs in the developing world. In effect, the human species will enter the 21st century with a sizeable new group of people and, almost everywhere in the world, it is the older age groups which are the fastest growing of all (see Fig. 1).

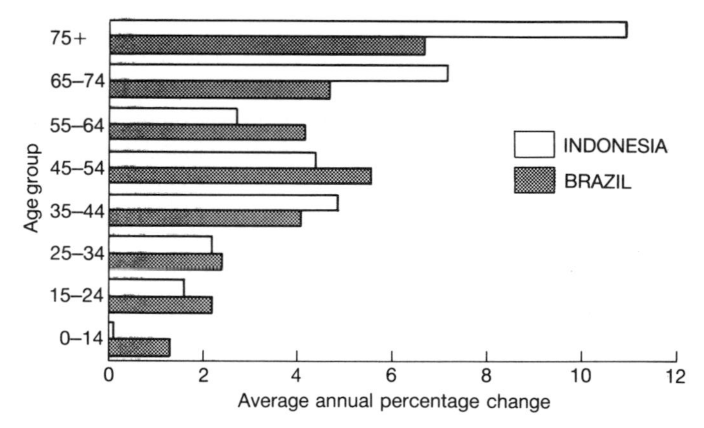

Fig. 1 Population change in Brazil and Indonesia by age group, 1988–2005.

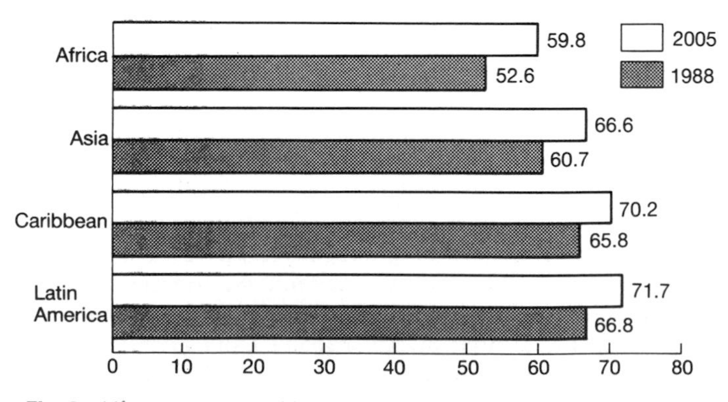

Fig. 2 Life expectancy at birth in years, developing regions, 1988–2005.

HEALTH PRIORITIES OF DEVELOPING COUNTRIES

The health services in most member states of the World Health Organization (**WHO**) are directed at improving the well-being of families through improved maternal and child health, the control of poverty-related diseases, and fertility regulation. Indeed, for the least developed countries, these are all-absorbing concerns. However, many developing countries have shaken free from the legacy of shortened life expectancy (Fig. 2) and have begun to see changes in the age composition of their population as a result of this, and of decline in fertility. These achievements of human survival are impressively documented in two publications of the United States Bureau of Census (Torrey *et al.* 1987; Kinsella 1988).

As in the industrialized world, attention is beginning to focus on shifts in survival at advanced age, in view of the consequences for health and social policies. In most developing countries, the population aged 75 years and older is growing faster than the older population in general. Future expansion of the oldest old will be especially pronounced in Asia.

Although geriatric medicine is not a priority issue for clinicians in most developing countries, there will be a growing need in coming years for all physicians to pay attention to the health issues associated with population ageing. A WHO-sponsored, cross-national study of the health and social aspects of ageing in four developing countries—the Republic of Korea, the Philippines, Fiji, and Malaysia (Andrews 1988)—compared and contrasted findings with those of a similar 11-country WHO study in Europe. In very broad terms, the physical and mental health and social patterns associated with ageing, as demonstrated by age-group and sex differences, were consistent throughout the four countries studied. Comparisons with European findings underlined the fundamental universality of age-related changes in biophysical, behavioural, and many social characteristics.

The greater importance of family support was evident in the four developing countries, with about three-quarters of those aged 60 and over living with children, often in extended families. The amount of adverse health-related behaviour and the prospect of changing patterns of illness point to a need to emphasize preventive health measures and measures directed to the maintenance of the physical and mental health of the ageing population.

CIRCUMSTANCES OF ELDERLY PEOPLE IN DEVELOPING COUNTRIES

Reports from the developing world express concern about the changing status of old people in society and within the family structure (Bose 1986). As in the developed world, governmental and voluntary social services are focused on the twin problems of income maintenance in old age and providing appropriate health care. Social actions in such countries are characterized (Little 1979) as being first:

> family care with few volunteers, some private old people's homes, lack of public funding, no domiciliary services, absence of training, followed by
>
> official services, some supervision of private homes, some public funding, an ageing component in training, and pilot domiciliary services.

This pattern generates a working model for developing services in the new ageing countries. Any service designed for elderly people should, at the outset, forge strong links between medical, social, and nursing professions as well as with voluntary organizations concerned with promoting the well-being of elders. The specific focus should be on:

- articulating any social services or institutional facilities with primary health care;
- using limited public funds to buttress voluntary effort, which is often religion-based;
- developing training for primary and referral-level care and pilot service programmes based upon an international inventory of good practices.

A step towards the last mentioned undertaking has been taken by Helpage International, which has reviewed studies and provided critiques of programmes for ageing in developing countries (Tout 1989). The text that follows is a further preliminary step in documenting approaches being followed by developing countries in organizing health care for older people. These draw mostly on sources from WHO, an organization which is searching for good practices that have relevance for countries burdened with a large number of health and economic problems.

HEALTH AND SOCIAL PROBLEMS OF THIRD-WORLD ELDERS

Few developing countries have information on the health and social conditions of elderly people, such as those collected from representative population samples in the Indian Ocean (Hamon and Catteau 1985). The WHO provided a guide as to how simpler data might be collected to plan service provision (WHO 1984). In 1985, 19 African countries responded to a United Nations' questionnaire on ageing that included questions on health and nutrition (United Nations 1985). In response to a question on whether the governments had identified health issues concerning the elderly as being of special policy concern, the following chronic diseases were mentioned: cardiovascular disease, hypertension, diabetes, diseases of digestive system, cancer, rheumatic diseases, and ophthalmic conditions. In addition, psychosocial problems were stressed in Kenya.

Information on specific causes of death among elderly people varies enormously by region and country, but reliable data are available from Caribbean and Latin American countries. On an aggregate basis, life expectancy is higher in these than in other developing regions (Fig. 2), and the Pan-American Health Organization has documented the changing pattern of disease in older age groups. Since at least the 1970s, cardiovascular disease has become the leading overall cause of death among persons aged 65 and over (Kinsella 1988). The increasing rates of cardiovascular deaths that have accompanied population ageing are best seen in Costa Rica and Uruguay, which approach (and at some ages exceed) those of the United States. Health care for older people in Latin America tends therefore to be organized as part of the services for the prevention and management of chronic diseases.

One country, Thailand, represents the Asian problem in microcosm. Survey data from 1981 (Chulalongkorn University 1985) reveal that the morbidity level of those aged 60 and over is significantly higher than the national average. The proportion of persons reported ill was 11.4 per cent of older people compared to 8 per cent for the population as a whole, and the older group experienced longer illnesses. The level of hospitalization for elderly people is higher than the national norm. Also of note is that nearly 40 per cent of elderly adults habitually smoke as compared with 29 per cent of the population aged 11 years and over. The Thai government is well aware that this necessitates long-term planning by all sectors. It established within its Five-Year National Plan of Action, a National Committee on Aging and the Department of Medical Sciences was given responsibility for subcommittees on Services Co-ordination and Research, plus Long-term Planning. Among the areas of responsibility of the latter is education, which is not limited to health workers but which extends to the 300 000 Buddhist priests, who are involved in distributing health information and in counselling (WHO 1989).

BUILDING ON THE INFRASTRUCTURE

Community care

African and Western Pacific countries emphasize the absolute necessity of building on to the services which already exist and of extending the role of the village or community health workers (WHO 1981) to provide comprehensive services covering health education, counselling, and total health management of the family, with the family.

At the present time, in a typical village of some 500 people there might be about 25 elders. An indication of the effectiveness of community care is obtained by evaluating their health status for, like other members of their families, these old people are prone to infections, parasitic infestations, anaemia, and accidental injury. In addition, they are likely to suffer from the chronic and degenerative conditions which require repeated or long-term care.

Home nursing

A Foundation was established in 1976 by the Singapore Ministry of Health as a registered charitable organization, subsidized by government and public donations. Although the Foundation employs 50 nurses to provide a home-nursing service, it is impossible for them to cover all the needs of all elders. The Singapore Trained Nurses Association therefore provided voluntary services as practical support to the Foundation's staff. Broadly, these services provide social programmes and meals, care for homebound elders and genitourinary nursing care to those elders who have need of such a service. All work is undertaken in close co-operation with governmental community and social workers. Singapore nurses have also moved into the work of the People's Association, lecturing to housewives on basic health and nursing care of elders, helping in senior citizens' clubs and in the community centres scattered throughout the island state (Skeet 1988).

Preventive services

But not all old people living in the community have obvious impairments or disabilities, and attention has to be given to the assistance required to help those who are well and independent to remain so. Regular 'check-ups' operated along the lines of 'under-fives' clinics are one way of ensuring this, and it has been suggested that community or family nurses specifically educated for this purpose would be able to take responsibility for such a service—aimed at maintaining mobility, social interaction, nutritional intake, eye care, and the wise use of drugs.

In Cuba, where each polyclinic serves a defined population, a nurse/physician health team is responsible for all members of some 120 families. The team is assisted by volunteers from the sector it serves. Activities in relation to elders include an annual visit to their homes and twice-a-year screening in the clinics for risk factors such as hypertension and hyperglycaemia. At-risk elders receive the same educational and screening schedules as younger members, but with additional control examinations for mental and social problems. This primary level of care therefore provides a continuity of health promotion and screening from year to year, giving frequent opportunities to redirect life-styles when necessary, and enabling people to help themselves to achieve a healthy and active old age.

Populations in transition to an affluent lifestyle, such as the island nation of Mauritius, would do well to take steps to avoid the diet-related epidemic of cardiovascular disease. In Mauritius, over the decade 1970 to 1980, when in many industrialized nations there were gains in life expectancy attributable to falling death rates from heart disease among older citizens, registered deaths from heart disease increased among older people, with an adverse effect on life expectancy (Fig. 3).

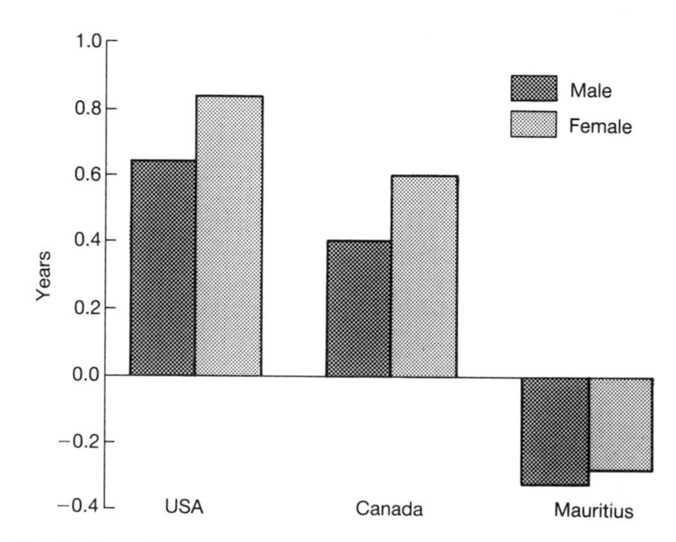

Fig. 3 Contribution, in years, of changes in mortality from heart disease at ages 65–84 to the increase in life expectancy at birth between 1970 and 1980.

Traditional medicine

The type of self-care undertaken by elders depends on their educational background and upon the environment. In rural areas, the elderly person is likely to resort to traditional practices. In some countries, Ministries of Health endeavour to limit the practice of traditional healers according to their capabilities, and to avoid commercialization of their practices—a particular problem of big cities.

Avoiding institutional care

Most developing countries have institutions to house elderly persons who are without shelter or food, although the numbers in institutional care are a tiny proportion of the older population. In some countries, special problems are presented by elders who left their own countries many years ago. Long-term care institutions in most developing countries are funded and run by the staff of voluntary organizations, who have little or no preparation in the care of elderly people. It is increasingly recognized that such institutions should be staffed by people with specialized knowledge of the needs of elders, and be supported by facilities for primary care and referral.

Urbanization and migration have been particularly acute in Latin America and the Caribbean and, in some situations, have resulted in many old people either living alone, being left alone for long periods, or being admitted to institutions. To help relieve the strain on the young women and others who are left to care for the two most vulnerable groups of people, i.e., the very young and the very old, some communities (e.g., in Peru) have established day centres or organized the formation of clubs. In both instances, these serve as educational and social meeting places. A guide to establishing and running these in developing countries is provided by the International Federation on Ageing (1984). Day-care placements are also being incorporated into old-age nursing home facilities in Cuba. With a view to reducing the need for more places in these homes, the ground floors of existing buildings are being adapted to house elders who need sheltered accommodation. The first 'Grandfather Centre' was established in Havana Province in 1970 and these now exist throughout the country. Essentially each centre is a converted house, situated in the middle of a town and used for day care of elders. Meals, including special diets for diabetics, are provided, and health-care activities are organized by the local polyclinic. City housing is often not suited for families with elderly members and urbanization has frequently resulted in segregation of older persons from persons of younger ages.

Continuity of care

Few countries have good medical records or follow-up schemes for the continuing care of old people when they leave hospital. To ensure appropriate and adequate family support for elders discharged from hospital (e.g., those who have suffered strokes), nurses in some countries now assess the family's capabilities and then supply relevant education. This includes guidance on rehabilitation training in order to improve the patient's performance of activities of daily living. This family training is given intermittently during the patient's hospital stay, with a period of intensive teaching after the discharge date has been agreed. Occasionally the family may be requested to stay overnight to learn the skills that are required by the patient at the end and beginning of each day. If necessary (e.g., when the patient has complex apparatus or appliances), the primary nurse visits after discharge.

Nevertheless, a large proportion of acute hospital beds are occupied by elderly patients, many of whom remain in hospital some time after the acute phase of their illnesses has passed, solely because they have unsatisfactory home conditions. In the majority of countries, the family and home to which a discharged elder is returning are rarely known by hospital staff and often are not suitably prepared to receive the old person when discharged. A system to enable easy referral between levels of care is vital for all age groups, but particularly so for old people. But all too frequently there is a lack of information on the community services available.

The advent of the family nurse practitioner in some Caribbean countries has helped to provide basic care in their own homes for old people with chronic illnesses or conditions. Nurses working in health centres in Jamaica have set aside 1 day each month for elderly ambulatory people. During that day, the entire staff of the centre caters to the social and educational as well as to the physical needs of their elderly patients. Other countries have scheduled special days for elders with specific conditions or diseases to attend clinics. Usually the cost of transport on these occasions is subsidized by the government (Skeet 1988).

Training programmes

It is now recognized that training in health care of the aged merits high priority and requires the immediate attention of all countries. A number in Asia/Oceania have already incorporated the subject of ageing in undergraduate and postgraduate medical, nursing, social science, and other curricula.

Courses vary widely in content, emphasis, and duration, but, by and large, they cover the following topics: the processes of ageing, including physiological and anatomical changes; ageing as a phase in the lifespan and not a disease; diseases of the elderly; mental health; social status; psychological effects of ageing; and methods of rehabilitation (WHO 1983).

In some developing countries of the Americas, subjects such as geriatrics and gerontology are now recommended for basic training and students are said to enjoy appropriate practical learning experiences with old people in hospitals, health centres, and nursing homes. Workshops are organized in some countries to train teachers.

Special concerns: nutrition

The nutritional status of old people in many countries has given rise to some concern. In several countries, government agencies, charitable organizations, and churches are frequently involved in supplying meals to old people in their own homes.

A STRATEGY TO MEET THE NEEDS OF FUTURE ELDERS

The successful planning of programmes and formulation of policies requires a national strategy. A template for building such policies is provided by the report of a WHO Expert Committee (WHO 1989). The process for developing such a strategy has been provided by a study *Perspectives on Aging in Belize* (Tout and Tout 1985). The two consultants who undertook the Belize study spent some months in that country, meeting representatives of government and voluntary organizations, travelling extensively, seeing the conditions under which many elders live, drawing upon existing information and case documentation, and then, finally, developing recommendations based on practices in similar countries. These recommendations not only take account of the needs of elderly people, but also consider these in the context of the country's economy, the governmental structure, the voluntary movement, and possibilities of international cofunding.

LESSONS FOR THE DEVELOPED WORLD

Current geographic data on the morbidity of older people are sparse. One of the tasks of the new WHO international research program on ageing (WHO 1987) is to gather comparable morbidity data. For example it is 20 years since Nordin, sponsored by WHO and the British Medical Research Council, visited nations representative of major land masses and ranges of climatic and dietary variables to evaluate calcium intake and the incidence of osteoporosis. Nordin's data support the observation that black subjects were less susceptible to bone loss. He found that osteoporosis with hip or vertebral fracture was infrequent in the elderly people of the islands of Jamaica, for example (Horowitz *et al.* 1989). All the world stands to gain by observing how geriatric research, services, and teaching evolve in developing regions.

References

Andrews, G.R. (1988). *Health and ageing in the developing world, Ciba Foundation Symposium*, **134**, 17–37.

Bose, A.B. (1986). Aging in India. *Zeitschrift für Gerontologie* **19**, 96–100.

Chulalongkorn University Institute of Population Studies (1985). *The Thai elderly population: a review of literature and existing data*. Population and Manpower Planning Division of the National Economic and Social Development Board, Bangkok.

Cowgill, D.O. (1986). *Aging around the world*. Wadsworth, Belmont, C.

de Beauvoir, S. (1970). *La vieillesse*. Éditions Gallimard, Paris.

Hamon, C. and Catteau, P. (1985). *Habitat et conditions de vie des personnes agées a la réunion*. DDASS, La Réunion.

Horowitz, A., Macfadyen, D.M., Munro, H., Scrimshaw, N.S., Steen, B., and Williams, T.F. (1989). *Nutrition in the elderly*. Oxford University Press.

International Federation on Ageing (1984). *An aging population: focus on day centres*. IFA, London.

Kane, R.L., Evans, J.G., and Macfadyen, D.M. (1990). *Improving the health of older people: a world view*. Oxford University Press.

Kinsella, K. (1988). *Aging in the Third World*. Bureau of Census, United States Government Printing Office, Washington.

Little, V. (1979). *In Reaching the aged: social services in forty-four countries* (ed. M. Teicher, D. Thursz, and J. Vigilante). Sage, California.

Skeet, M. (ed.) (1988). *The age of aging: implications for nursing*. World Health Organization, Copenhagen.

Torrey, B.B., Kinsella, K., and Taeuber, C.M. (1987). *An aging world*, International Population Report Series, No. 78, p. 95. United States Department of Commerce, Bureau of Census, Washington DC.

Tout, K. (1989). *Ageing in developing countries*. Oxford University Press.

Tout K. and Tout, J. (1985). *Perspectives on aging in Belize*. Help the Aged, London.

United Nations (1982). *International plan of action on aging: report of World Assembly on Aging Vienna*. United Nations, New York.

United Nations (1985). *Periodical on Aging*, Vol. 2, No. 1. UN, Vienna.

World Health Organization (1981). *Health care of the elderly, report on a working group (Manila 1981)*, (ICP/ADR/003). WHO Regional Office for the Western Pacific, Manila.

World Health Organization (1983). *Health care of the elderly: guide for teachers, asia and oceania region, outcome of workshop on education and training for care of the elderly (Singapore, 1983)*, (IRP/ADR 114). WHO, Copenhagen.

World Health Organization (1984). *Uses of epidemiology in the study of aging, report of a scientific group*, Technical report series, No. 706. WHO, Geneva.

World Health Organization (1987). *World health statistics annual*, pp. 11–13. WHO, Geneva.

World Health Organization (1989). *Health of Elderly People: report of a WHO expert committee*, Technical report series, No. 779. WHO, Geneva.

1.4 Social differences in an elderly population

R. C. TAYLOR

INTRODUCTION

Social scientists have made a major contribution to our understanding of human ageing in showing how it varies according to the location of the elderly person in the social structure. As a result, the elderly can no longer be considered or treated as a homogeneous group. This chapter explores the ways in which the elderly population is socially differentiated. It begins with an examination of the distribution of personal resources between age cohorts, the sexes, and social classes. This is followed by a more focused examination of the distribution of resources between a number of commonly identified subgroups—for example, those who live alone, the childless, and those recently discharged. It ends with an assessment of a risk group approach to geriatric case finding.

AGE, SEX, AND SOCIAL CLASS DIFFERENCES

Age

The distinction between the 'young' and the 'old' elderly is now commonplace and it assumes considerable importance in discussions of the demand for health and social services. We know that compared with those aged 65 to 74, those aged 75 years and over have a greater number of long-standing health problems, they experience more illness (General Household Survey 1971–8), and their use of services is twice, and in some instances four times as great (Grimley Evans 1981). We also know that the old elderly are more likely to live in poorer housing (Hunt 1978) and to have incomes below or on the margin of the state's standard of poverty (Townsend 1979). But in other respects we know that by comparison with their younger peers the old elderly are advantaged; they have larger families (Shanas *et al.* 1968), are more likely to have a child living in the neighbourhood (Abrams 1978, 1980), and are generally more satisfied with their lives (Savage *et al.* 1977).

Sex

The basis for many of the differences between elderly men and women lies in life expectancy. At age 65 a Scottish woman can expect another 15.7 years of life whereas a Scottish man can only expect another 11.7 years (Registrar General 1978). As a result of this greater longevity, elderly women are twice as likely to be widowed (Hunt 1978) and three times as likely to be living alone (Abrams 1978 1980). Of course, elderly women are also more likely to live alone

This chapter draws on work done with Graham Ford at the MRC Medical Sociology Unit in Aberdeen. Longitudinal analysis of the Aberdeen Styles of Ageing Study is proceeding with the assistance of Dr Martin Fischer at the MRC Medical Sociology Unit in Glasgow.

because they were less likely to marry in the first place, slightly over 10 per cent of all women over the age of 65 being spinsters whereas only 4 per cent of men over the age of 65 are bachelors (Hunt 1978). These demographic differences between men and women have important financial consequences. In his discussion of poverty among the elderly, Townsend (1979) shows that those never married and widowed have lower incomes and fewer assets than those currently married. There are also consequences for social support, men over the age of 65 being twice as likely as their female contemporaries to have a spouse, and slightly more likely to have a surviving adult child (Shanas *et al.* 1968).

Health constitutes another source of inequality between the sexes. Elderly women report more long-standing health problems, they experience more illness, and they report more consultations with their general practitioner. In fact, significant sex differentials are evident in all self-reports of health and illness (see General Household Survey 1971–8). Of course, all these self-reports are subject to cultural influences, including the expectation that men ought to be strong, healthy, and stoical. Thus, it has long been supposed that 'real' health differences may not be as great as they appear from self-reports. The Duke Longitudinal Study, based on physical examinations/physicians' ratings as well as self-reports, attempted to resolve the issue, but their findings are also rather ambiguous. It appears that there are health optimists and health pessimists and that the former tend to be male and the latter female; however, these tendencies were not statistically significant (see Maddox and Douglass 1973).

Elderly women also seem to have poorer psychological functioning than their male peers; they are more likely to suffer loneliness, anxiety, to have weaker self concepts (Atchley 1976), and lower levels of morale/life satisfaction (Abrams, 1978, 1980).

Social class

The best evidence for social class differences in the health of elderly people in the United Kingdom comes from the General Household Surveys. Taking the prevalence of long-standing illness as the best single indicator of ill health, most reports show that while the rates for those aged 65 and over show the same class pattern as is found in younger age groups, the gradient is less pronounced. The evidence from the Surveys and other sources (see review by Walker 1981) suggests that people from middle-class occupational backgrounds enjoy a health advantage in later life (sex and class differentials in morbidity are reviewed in DHSS 1980). Evidence for class differences in psychological functioning is rather sketchy. American researchers have consistently found a direct relationship between socioeconomic status and morale (see review by Larson 1978), but other dimensions have not received the same attention. The published record on the culture of poverty contains numerous studies suggesting working class—particularly lower working-class—disadvantage, but as most of these studies are descriptive the evidence is, at best, suggestive and inferential. The catalogue of working-class disadvantage is partly offset by their greater share of available family support. In Britain, Shanas (1968) and Abrams (1978, 1980), in studies 20 years apart, have confirmed that by comparison with their middle-class peers, working-class elderly people have more children and more siblings and are more likely to live with or near their close kin (see review by Allen 1979).

Combined effect of age, sex, and social class

While much is already known about the distribution of resources by age, sex, and social class, it is necessary to consider the effect of these variables simultaneously. The combination gives rise to eight age/sex/class subgroups: 'young' middle-class men; 'young' middle-class women; 'young' working-class men; 'young' working-class women; 'old' middle-class men; 'old' middle-class women; 'old' working-class men; 'old' working-class women. A range of detailed information is available in Taylor and Ford (1983*a*), and for present purposes it is sufficient to examine overall differences.

Table 1 shows cross-sectional data based on a random sample of the non-institutionalized elderly people in Aberdeen, Scotland, and shows subgroup deviations from the sample mean for a range of resource variables. It is immediately apparent that only one subgroup is consistently better off than the sample as a whole—the group of younger, working-class men. Younger middle-class men also emerge as an advantaged group, only falling below the sample mean in the availability of social support. The two groups of older women provide a sharp contrast, neither group ranking better than the sample as a whole on more than 2 out of the 12 resource variables. If we want to discriminate between these two groups to identify the one which is more disadvantaged, we have to examine each resource area in turn. For income, it is obvious that it is the working-class group which is massively disadvantaged. For social support, it is equally obvious that it is their middle-class peers who are most disadvantaged. As far as health is concerned there is little difference between the two groups. For the final resource area—psychological functioning—the middle-class group has a slight overall advantage. The limitations of this kind of approach notwithstanding, it is clear that the older working-class women constitute the single most disadvantaged group.

Longitudinal data from the same Aberdeen study provide confirmation of these conclusions. Table 2 shows the outcome at 5 years for each of the age/sex/class subgroups. If we discount the unreliable percentage of deaths among the older middle-class men, the class-based pattern of disadvantage is clear; working class men and women are more likely to have died over the 5-year period.

DIFFERENCES BETWEEN COMMONLY IDENTIFIED SUBGROUPS

A more focused approach to differences within the elderly population can be obtained from an analysis of the resource profiles of a number of commonly identified subgroups (see Taylor and Ford 1983*b*). The profile which will be presented for each group has its origins in the concept of 'personal resources', i.e., those resources that individuals draw upon when coping with difficulties. In the Aberdeen Styles of Ageing study, four categories of resource were used: income and savings, social support, health and psychological functioning. These were indexed by 19 key variables, which were standardized to have a mean of zero and a standard deviation of

Table 1 Age/sex/class subgroups: deviation from sample mean

	60–74 years				75+ years			
	Men	Women	Men	Women	Men	Women	Men	Women
Resource variables	MC	MC	WC	WC	MC	MC	WC	WC
Income	+16.61	+7.18	−2.41	−5.45	+2.78	−0.41	−5.44	−12.61
Currently married	+0.30	−0.10	+0.29	−0.01	+0.23	−0.35	+0.06	−0.35
Local children	−0.54	−0.39	+0.11	+0.45	−0.14	−0.63	+0.23	+0.22
Local siblings	−0.32	−0.27	+0.31	+0.39	−0.27	−0.48	−0.37	−0.33
Close friends	−0.07	+0.32	+0.32	−0.09	−0.01	+0.60	−0.45	+0.23
Chronic conditions	+0.74	+0.54	+0.48	−0.30	+1.36	−1.07	+0.15	−0.87
Symptoms	+0.70	+0.71	+0.56	−0.51	+1.07	−0.94	+0.19	−0.94
Functioning	+1.35	+0.93	+0.97	+0.52	+0.35	−2.89	−0.83	−2.17
Self-esteem	+0.73	+0.18	+0.39	−0.15	+0.92	−0.95	+0.06	−0.45
Self-competence	+1.62	−0.07	+0.19	+0.09	+0.31	−0.38	−0.15	−0.56
Morale	+0.23	−0.80	+1.02	−0.83	+2.54	+0.43	+0.89	−1.07
Health optimism	+0.76	+0.66	+0.06	−0.06	+0.20	−0.11	−0.90	−0.72

MC = middle class; WC = working class; + = better than; − = worse than.

Table 2 Age/sex/class subgroups: differences in survival over 5-year period

	60–74 years				75+ years			
	Men	Women	Men	Women	Men	Women	Men	Women
	MC	MC	WC	WC	MC	MC	WC	WC
Percentage dead	15	11	22	14	[51]*	28	40	45
Percentage too ill to interview	13	20	11	16	8	27	14	18
Percentage alive	72	69	67	71	41	45	46	37

* Small numbers; MC = middle class; WC = working class.

unity. In the following figures, the score for the group on each variable is represented as a bar on a bar graph; a bar above the line represents a variable on which the average score for the group is above that of the sample as a whole, while a bar under the line represents a below-average score. The bigger the bar, the more different the group from the sample as a whole. Those scores on which the departure of the group from the sample mean is statistically significant are represented in the figures by light stippling.

Those living alone

Accounting for about one-third of the elderly population, those living alone constitute the largest of the 10 groups considered here. In our sample of 619 persons, 216, or 35 per cent, lived alone. They are disproportionately old, 40.5 per cent being aged 75 and over (as opposed to 29.7 per cent in the sample as a whole) and disproportionately female (80.3 per cent, as opposed to 60.7 per cent in the sample as a whole).

For most of the measures of physical health, psychological functioning, and confidence, it can be seen (Fig. 1) that, while the group scores slightly worse than the population as a whole, the only departure which is statistically significant occurs in the number of chronic conditions experienced. It is in relation to various forms of social support that those living alone differ most. They have fewer intimates or confidantes available to them, and, viewing their profile as a whole, this is the most distinctive feature. However, despite having few confidantes, the group is not characterized by low levels of psychological functioning. One reason for this may be the compensatory effect of many friends, those living alone having significantly more friends than the rest of the population. The only other significant departure from the sample mean occurs in relation to household amenities, those living alone being less likely to have such household amenities as hot water, fitted bath, etc.

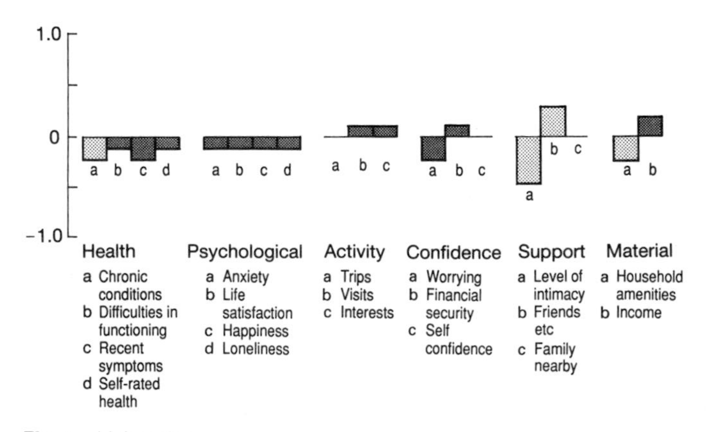

Fig. 1 Living alone.

Considering their profile as a whole, it can only be concluded that those living alone do not stand out as being significantly disadvantaged. Of course, it is a large and heterogeneous group, and encompassed within it are members of other groups—the single, the divorced/separated, and the recently widowed—whose profiles will be examined later.

The childless

Those without children constitute the second largest of the 10 groups. In the sample of 619 persons, 20 per cent (123) were childless. Compared with the rest of the elderly population, they are indistinguishable in terms of age and sex, but they are disproportionately middle class—55 per cent, as opposed to 35 per cent in the sample as a whole.

The most distinctive feature of their profile (Fig. 2) is found in the domain of social support. They have fewer intimates or confidantes and fewer family members living nearby, but this loss is partly compensated by the comparatively large number of friends. But being childless is not associated with significantly lower levels of health, psychological functioning, and confidence than those pertaining in the elderly population. It is possible that they are slightly more anxious and have rather less self-confidence, but they just fail to achieve statistical significance on both measures. As far as social activities are concerned, they are significantly less likely to receive visits and this is probably a direct consequence of being childless.

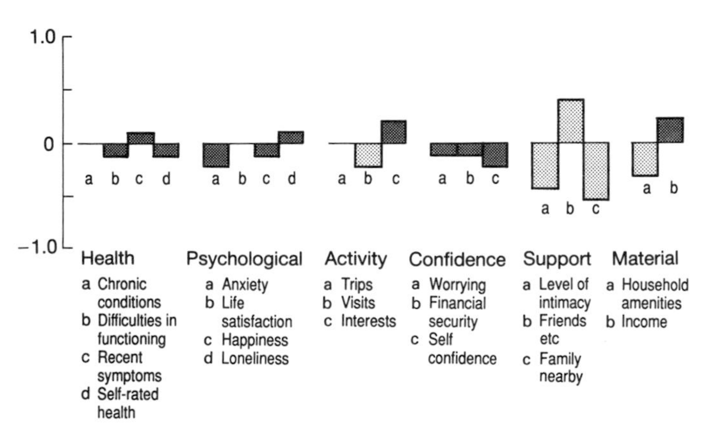

Fig. 2 Childless

The reason for their comparatively poor showing on household amenities is not immediately apparent. However, children constitute a major incentive for such home improvements as installing a bath or an inside toilet, and they also provide local authority tenants with housing points in any attempt they might make to secure better accommodation. Thus, in a city such as Aberdeen, in which over 60 per cent of the housing stock is owned by the local authority, over the years the childless seem to have been left behind in the poorer inner-city areas, and they have to live with the consequences in their old age.

The poor

Defining as poor those whose weekly income was below the Supplementary Benefit level identified a subgroup of 93, or 15 per cent, of the sample. They are indistinguishable from the rest of the elderly population in terms of age and sex, but they are disproportionately from working-class backgrounds—73.6 per cent, as opposed to 65.0 per cent in the sample as whole. Widows are also over-represented—45 per cent, as opposed to 32.6 per cent in the sample as a whole. The married are under-represented—37.3 per cent, as opposed to 52.2 per cent in the sample as a whole.

By comparison with the elderly population as a whole, they are clearly and massively deprived as far as income is concerned (Fig. 3), but as this is the measure on which they are defined, the observation is tautologous. On the related measure, household amenities, they are also significantly worse off than the sample as a whole, but this could also be seen as rather tautologous.

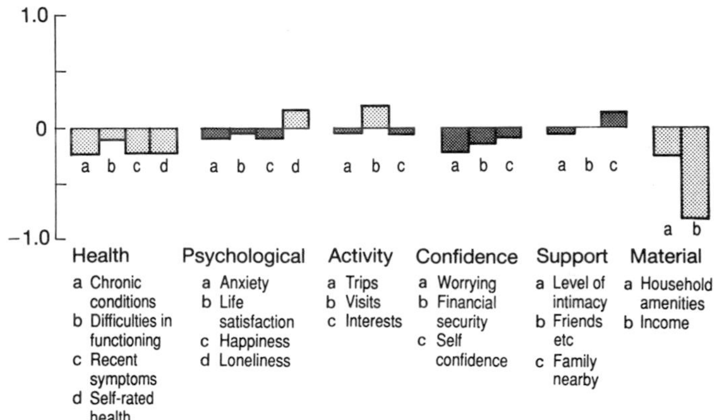

Fig. 3 Poor (below Supplementary Benefit).

Apart from being financially disadvantaged, the group is also rather comprehensively disadvantaged on health grounds, scoring significantly lower than the sample on all four health measures. This is an important finding, and additional analysis (not shown here) shows that the extent of health disadvantage is even greater among the poor who are widowed and divorced/separated. As regards their mental health, evidence of risk or disadvantage is less conclusive. The group scores worse than the sample as a whole on all confidence measures, but the differences are not statistically significant; a similar situation pertains in respect of scores on anxiety, life satisfaction, and happiness. The major exception—and this is statistically significant—occurs on the measure of loneliness. As a group, the poor are not as lonely as the rest of the elderly sample, and this finding is consistent with the rest of their profile. They receive significantly more visits than the sample as a whole, and they tend to have more family members living locally. These departures above the sample mean are clearly related: the poor are less likely to experience loneliness because they get more visits, and they probably get more visits because they tend to have more family members living locally.

On the basis of the data available, it is clear that the poor are not just disadvantaged in terms of their income and housing. Compared with the rest of the elderly population, they experience more illness and difficulties in functioning and have lower self-estimates of their health. However, it is important to note that this disadvantage is partially offset by their greater social contracts.

The very old

Almost 15 per cent of the sample (86 cases) were aged 80 and over. The group's composition is rather predictable, disproportionately female (74.1 per cent, as opposed to 60.7

per cent in the sample as a whole) and less likely to be currently married (19.5 per cent, as opposed to 52.2 per cent in the sample as a whole). They are also disproportionately from middle-class occupational backgrounds (40.4 per cent, as opposed to 35.0 per cent in the sample as a whole), but this result could have occurred by chance.

The profiles (Fig. 4) of the very old generally confirm what is already known about their health; they have a greater number of chronic conditions and difficulties in functioning, and they tend to report more symptoms and to have lower self-ratings of their overall health. As a consequence, they make fewer daily trips outside the house, but they do not seem to have fewer visits made to them.

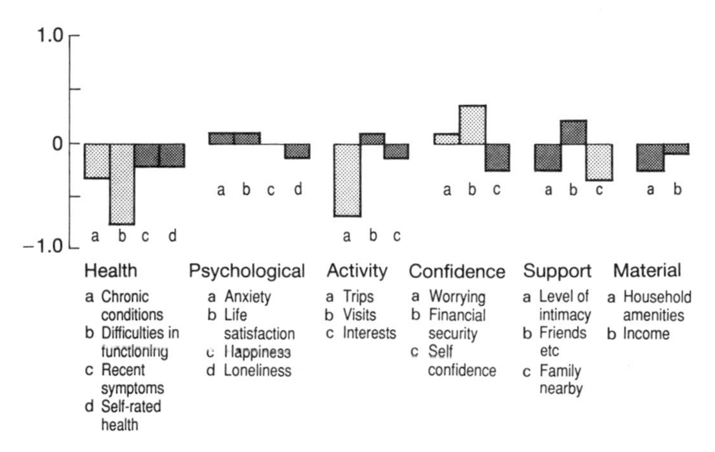

Fig. 4 Very old (80+ years).

On most measures of psychological functioning, the very old are either indistinguishable from the rest of the elderly population or are significantly better off. It is clear that they worry less and feel greater financial security. This latter finding is of particular interest in view of the objective evidence that over half of the very elderly (80 years and over) have incomes below or on the margin of the State's standard of poverty. A number of factors might account for this discrepancy: for example, the very old may have fewer needs; income may have a lower salience for them than, say, health; or it may be that, by their own time perspective, present levels of provision seem generous.

The only other respect in which the very old are indisputably disadvantaged in relation to the rest of the elderly is the number of family members living locally. This difference is almost entirely explained in terms of differential survival. The very old are, by definition, survivors. Many of their siblings, and in some cases even their adult children, are not.

The recently moved

Contained within the sample were 85 elderly people (13.7 per cent) who had moved home within the last 2 years. They are indistinguishable in terms of age, sex, and marital status, but they are disproportionately from working-class occupational backgrounds (75.8 per cent, as opposed to 65.0 per cent in the sample as a whole). It is immediately apparent that most deviations from the sample mean occur below the line, so that, as a group, they tend to be rather comprehensively disadvantaged (Fig. 5). This disadvantage is particularly noticeable, and statistically significant, on three measures—anxiety, loneliness, and worries—and containment within the psychological domain indicates the specific nature of the risks associated with residential mobility. It is worth observing that this psychological distress coincides with higher than average levels of intimacy and availability of family members.

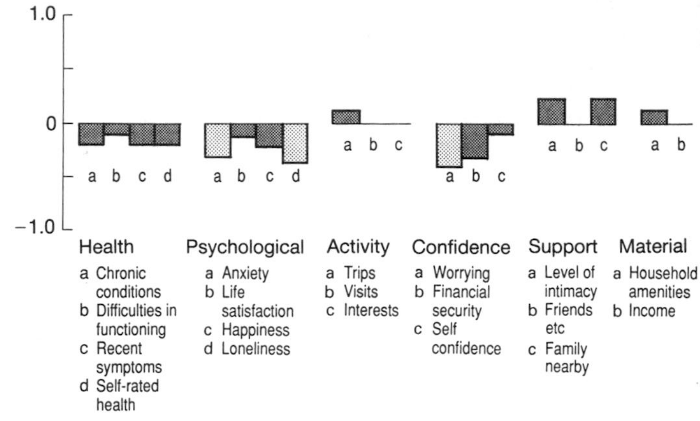

Fig. 5 Recently moved.

The recently discharged

The sample contained 83 cases (13 per cent) who had been discharged from hospital in the 2 years preceding the interview. As a group, they are indistinguishable from the rest of the population in terms of age, sex, marital status, and social class.

To define them as a risk group is partially tautological. They must all have been ill to enter hospital, and while some may have completely recovered after discharge, there will be many whose recovery is delayed, and some who never recover. But the frequency with which they are defined as a group at risk may also stem from more pragmatic considerations. Their entrances and exits from the hospital are documented, they are clearly identifiable, and they can (at least in theory) be routinely followed up as part of normal discharge procedure.

Their profile (Fig. 6) is dominated by poor health. They score significantly worse than the rest of the sample on all measures of physical health and also on our psychological measures of anxiety, loneliness, and general worrying. They also make significantly fewer trips outside the house, and have lower incomes. On the credit side, they report more confidantes and they tend to receive more visits and to have more family members living nearby. This is, therefore, a sense in which their physical and psychological condition mobilizes higher levels of personal support than are available to the rest of the population.

Of the groups so far examined, the recently discharged are clearly the most disadvantaged, and the profile merely underscores the need for follow-up services and support of various kinds.

The never-married

The sample contained 70 people (11.3 per cent) who had never married—46 spinsters and 24 bachelors. Apart from containing a greater proportion of women (76.0 per cent,

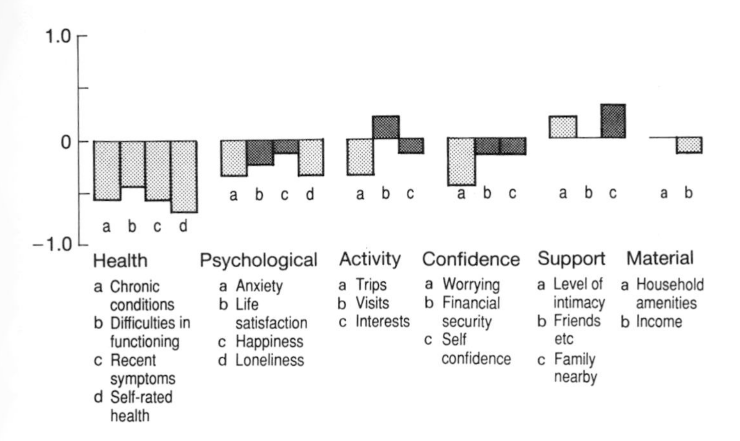

Fig. 6 Recently discharged.

as opposed to 60.7 per cent in the sample as a whole), the group is also disproportionately middle class (49.0 per cent, as opposed to 35.0 per cent). The most pronounced feature of their profile (Fig. 7) is found in the support domain. Compared with the rest of the elderly population, they have fewer confidantes and few family members living nearby. But, as already observed for other groups, there is clear evidence of compensation through friendship.

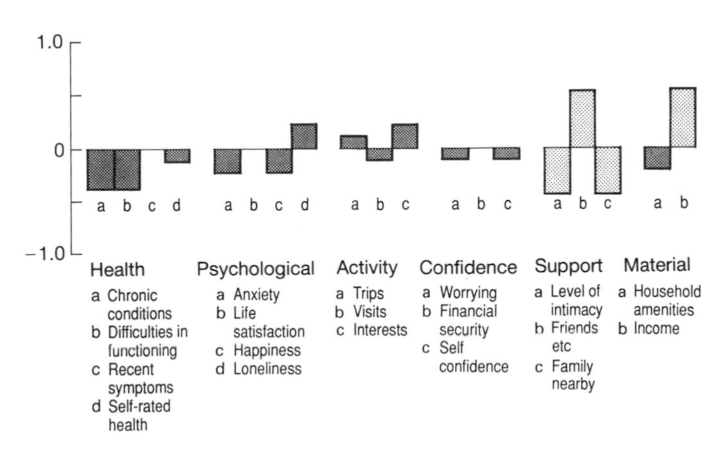

Fig. 7 Never married.

The recently widowed

The sample contained 202 widows and widowers, the majority of whom had been without their spouses for many years. Studies of widowhood have suggested that its psychological impact begins to tail off after about 2 years. Accordingly, 37 cases bereaved in the last 2 years were identified as a potential risk group. Their profile (Fig. 8) reveals little conclusive evidence of comprehensive decline. Indeed, the only respect in which the group departs significantly from the sample mean is in their income, and here it can be observed that the departure is in a favourable direction. While departures on other measures fail to reach statistical significance, they come closest on happiness, trips out, and worrying—all in the expected direction. But there are compensating tendencies, particularly in the confidence domain. On the basis of the available evidence, this group is not significantly disadvantaged by comparison with the rest of the elderly population.

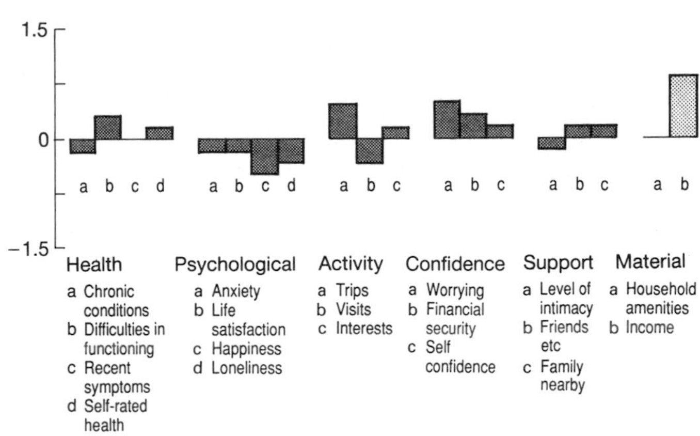

Fig. 8 Recently widowed.

The isolated

While 216 of the sample members lived alone and, by definition, had no current spouse, 54 were additionally disadvantaged in having no children or siblings living locally. They have been defined as socially isolated. This group contains a higher proportion of the older cohort than the sample as a whole (48.1 per cent, as opposed to 29.7 per cent). It also contains more women (78.1 per cent, as opposed to 60.7 per cent), and more of those from middle-class occupational backgrounds (45.9 per cent, as opposed to 35.0 per cent).

It is clear from the group's profile (Fig. 9) that the disadvantaged is fairly contained. With one exception, all significant departures from the sample mean occur in the support and material domains. Thus, apart from having poorer housing and better income the isolated are not significantly different from the rest of the elderly population in any respect other than those by which they have been defined.

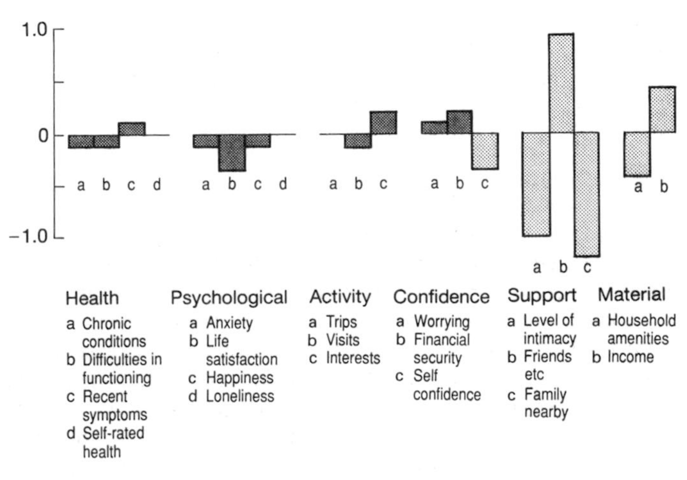

Fig. 9 Isolated (no spouses or near family).

Social Class V

The sample contained 51 men and women (8.2 per cent) whose previous occupations qualified them for membership of the Registrar General's Social Class V. On other structured variables—age, sex, and marital status—this group is indistinguishable from the sample as a whole.

It is clear from this group profile (Fig. 10) that members of Social Class V are only significantly disadvantaged in terms of income. On all other measures, apart from one, the group cleaves fairly closely to the sample mean. The exception is the number of family members living nearby. Whatever disadvantages members of Social Class V may experience in health or psychological functioning, and there is not conclusive evidence in the data, the family support available to them is significantly greater than that enjoyed by the rest of the elderly population.

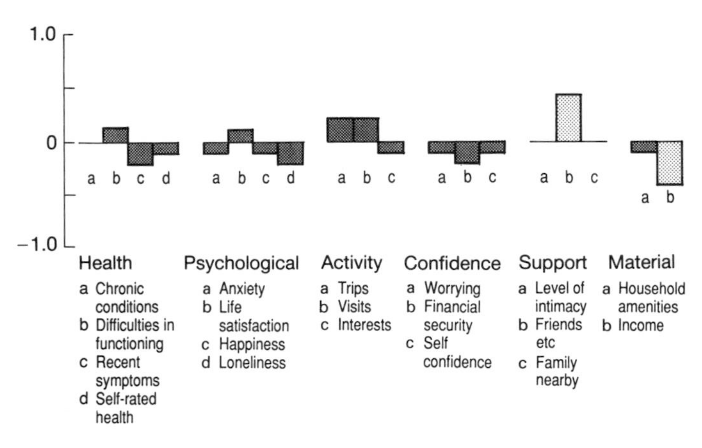

Fig. 10 Social Class V.

The divorced/separated

While the sample contained only 23 people (4 per cent) who were either divorced or separated, in new cohorts of elderly it is likely that this proportion will be doubled or even trebled. As might be expected, group members are younger than the sample as a whole, 15 per cent aged 75 and over, compared with 30 per cent ($p<0.05$). They are also more likely to be women, 78 per cent compared with 61 per cent, and middle class, 48 per cent compared with 35 per cent, but these differences could have occurred by chance.

On *a priori* grounds there are a number of reasons for considering the divorced/separated as a group at risk. By definition, they no longer have a spouse to rely on for support in illness, and the availability of other family members—particularly the potential set of in-laws—may have been reduced by the trauma of divorce proceedings. As far as their own health is concerned there is the stigma of divorce to be considered, a stigma which is likely to be rather marked in this age group.

In the overall profile (Fig. 11) there are considerable more departures below the sample mean, indicating that the group is disadvantaged in relation to the elderly population as a whole. The most significant disadvantages occur in the psychological and confidence domains. Compared with the rest of the population the divorced/separated experience less life satisfaction, less happiness, they are more lonely, and they worry more—both in general and in relation to their finance. On balance, their physical health is poorer than that of the population as a whole, they experience more chronic conditions and recent symptoms, but fewer difficulties in functioning.

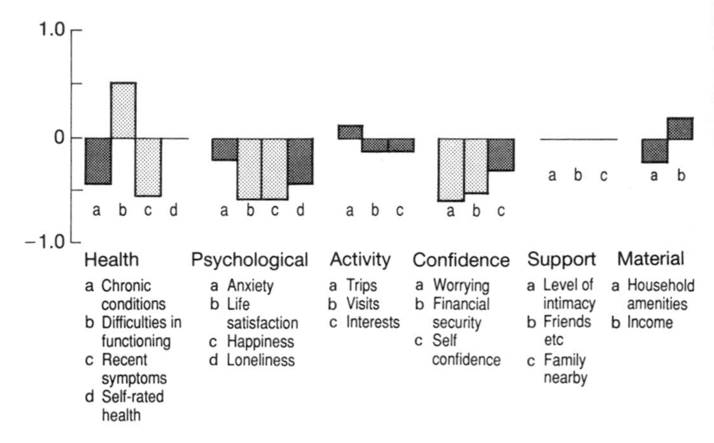

Fig. 11 Divorced/separated.

Contrary to expectations the social support available to them is neither greater nor less than that available to the sample as a whole. On the positive side, the only respect in which they are advantaged is that they experience fewer difficulties in functioning. The rather comprehensive disadvantage experienced by this group is all the more significant given that it is disproportionately a group of the young elderly. Clearly, this is a group to be closely watched in future cohorts of the elderly.

SUBGROUPS AND THE PREDICTION OF RISK

The profiles which have been presented show highly differentiated patterns of resources. This is their major function but inevitably questions are raised about their relevance for case-finding (Taylor and Ford 1983*c*)

There are good *a priori* grounds for suspecting that the groups analysed above offer only limited efficiency for case finding. First, they vary considerably in size. The largest, comprising those who live alone, accounts for about 30 per cent of the non-institutionalized elderly population; the smallest, the divorced/separated, accounts for less than 5 per cent. Groups defined on the basis of some threatening life-event are all fairly small. In the Aberdeen study the percentage of the non-institutionalized population experiencing various life-events over the course of a year was as follows: change of address, 7 per cent; discharged from hospital, 6 per cent; death of a spouse, 4 per cent; and retirement, 3 per cent. The implications for screening are obvious. Even if half of all group members are defined as 'cases', the small size of many groups mean that they will account for only a minute proportion of all cases in the population.

Secondly, the groups vary considerably in the extent and nature of their disadvantage. Risk profiles showed that it was only the very old and those recently discharged from hospital who were comprehensively disadvantaged. Groups such as the single and the childless were disadvantaged only

Table 3 Efficiency of case finding: cases in lower decile (i) as proportion of designated risk group (above diagonals); (ii) as proportion of total cases (below diagonal)

Risk group (proportion of total population)	Measures of well being (proportion of population in lowest decile)					
	Health (0.26)	Psych (0.33)	Confid (0.31)	Activ (0.37)	Support (0.53)	Material (0.21)
Very old (0.14)	0.46* / 0.24	0.33 / 0.14	0.22 / 0.10	0.51* / 0.19	0.56 / 0.15	0.24 / 0.16
Recent move (0.14)	0.40* / 0.21	0.46* / 0.19	0.45* / 0.20	0.39 / 0.14	0.50 / 0.13	0.17 / 0.11
Recent discharge (0.13)	0.51* / 0.26	0.43 / 0.17	0.38 / 0.16	0.36 / 0.13	0.41 / 0.10	0.26 / 0.16
Divorced/separated (0.04)	0.48 / 0.07*	0.45 / 0.05	0.49 / 0.06	0.50 / 0.05	0.57 / 0.004	0.16 / 0.03

Psych, psychological; Confid, confidence; Activ, activity (see Figures).
* Significant at 0.05 level.

in terms of social support, the poor in terms of material resources, recent movers only in terms of their mental health, and so on. The profiles also revealed important compensatory effects. In some groups, disadvantage in one domain was compensated by advantage in another. For example, while the poor were (by definition) disadvantaged in terms of material resources, they had more social support than their more affluent age peers. Similarly, while the single and childless were (again by definition) disadvantaged in terms of family support, they had more friends and confidantes.

The first test of case-finding efficiency was based on cross-sectional data in which caseness was defined by scores in the lowest decile of functioning (Taylor *et al.* 1983). Table 3 shows the proportions of those in the four most disadvantaged groups scoring in the lowest decline in each of the domains. Along the top of the table is shown the proportion of 'cases' in the population as a whole, down the left-hand side the four risk groups are listed by size, i.e. their proportion of the elderly population. The figures reported above and below the diagonals have to be related to both sets of marginal totals. Thus, in the very old group, 0.46 are in the lowest decile on one or more measures of health, 0.33 are in the lowest decile on one or more of psychological functioning, and so on. These figures have to be related to the proportions shown across the top, and it is evident that few are sufficiently higher or lower to be statistically significant. The figures below the diagonal indicate the proportion of all cases in the lowest decile accounted for by the designated groups. Taking the very old it can be seen that while 0.46 are in the lowest decile on one of the health measures, they account for only 0.24 of all cases in the lowest decile. This latter figure has to be related to the marginal total, in this case the proportion of the elderly population accounted for by the very old. Thus, while this group constitutes 0.14 of the total population, it accounts for 0.24 of all cases in the lowest decile on psychological functioning, and so on.

Overall, it will be clear that none of the groups is particularly efficient for case finding. Concentrating on health and psychological functioning, the domains of most relevance to health professionals, it can be seen that the proportion of all cases accounted for by any one group never rises above 0.26 (the recently discharged). Generally, the smaller the group, the smaller the proportion of cases accounted for. Thus despite the fact that around half of the divorced/separated are 'cases' on most of our measures, they never account for more than 0.7 of all cases. With larger groups the ratios are reversed. For example, only a third of the live alone group are 'cases', yet they account for almost half of all cases.

The second test of case-finding efficiency involves the use of longitudinal data from the Aberdeen study. Here caseness is defined in terms of outcome 5 years after the initial interview. Table 4 shows two rather unequivocal outcomes (death and extreme frailty) for the 10 subgroups. It is immediately obvious that for five groups—the childless, recently moved, never married, divorced/separated, and Social Class V—the likelihood of death/extreme frailty is lower than that of the sample as a whole. Only for one group—the very old—is the likelihood of death/frailty double that of the sample. Not surprisingly, the discriminant function of group membership is fairly low. Considering only those groups with the highest standardization coefficients—the very old (0.72) and those recently discharged (0.43)—knowledge of group membership correctly predicts 72 per cent of all deaths. This represents only a small improvement on the percentage (64) classified by chance.

Table 4 Efficiency of case finding: outcome at 5 years (percentages)

Groups	Outcome: Died (A)	Too ill to be interviewed (B)	A + B
Living alone	29.0	8.1	37.1
Childless	17.6	9.0	26.6
Poor	25.7	15.4	41.1
Very old	47.8	12.6	60.4
Recently moved	16.7	6.2	22.9
Recently discharged	36.3	13.9	50.2
Recently widowed	31.1	5.9	37.0
Never married	15.7	7.8	23.5
Isolated	34.2	9.5	43.7
Social Class V	18.9	7.4	26.3
Divorced/separated	12.2	6.5	18.7
SAMPLE	22.6	6.6	29.2

CONCLUSIONS

While the risk group approach to case finding does not reach acceptable levels of efficiency, knowledge of social differences between subgroups of the elderly population is important for planning and practice in the health and social services.

In relation to health service planning, four lessons emerge from the material reviewed in this chapter. First, comparison of the 'young' with the 'old' elderly reveals that it is only the latter group which is comprehensively disadvantaged. There is now a strong case for concentrating exclusively on those who are 75 years and over. Secondly, there is now sufficient evidence for the persistence of occupational/class-based inequalities to conclude that any formula for resource allocation should be weighted to discriminate in favour of areas with high concentration of elderly who were formerly in manual occupations. Thirdly, in relation to marital state, it now seems that the disadvantages experienced by the unmarried and widowed have been overestimated, whereas those experienced by the divorced/separated have been underestimated. The differences between the never married, particularly spinsters, and the divorced/separated are especially marked and should be reflected in future formulae for resource allocation. Fourthly, there is clear evidence of a need to distinguish between living alone and experiencing social isolation. At present the proportion of elderly living alone is often used as a proxy measure of risk; health service planners should attempt to develop more sensitive measures of social isolation.

In relation to geriatric practice, the single most important conclusion of the material reviewed in this chapter is that assessment must be comprehensive. There are two aspects to this. First, and most obviously, assessment must cover the full range of functioning—physical, social, and psychological. Secondly, given the recurrent finding that disadvantage in one domain is compensated by advantage in another, it is important that assessment pro formas are designed to detect the strengths as well as the weaknesses of resources.

References

Abrams, M. (1978 and 1980) *Beyond Three Score Years and Ten. First and Second Reports*. Age Concern, London.

Allen, G.A. (1979). *Sociology of friendship and kinship*. George Allen and Unwin, London.

Atchley, R.C. (1976). Selected social and psychological differences between men and women in later life. *Journal of Gerontology*, **31**, 204–11.

Department of Health and Social Security (Great Britain) (1980). *Inequalities in health*. DHSS, London.

General Household Survey (1971–8). *Reports*. HMSO, London.

Grimley Evans, J. (1981). Demographic implications for the planning of services in the United Kingdom. In *The provision of care for the elderly* (ed. J. Kinnaird, J. Brotherton, and J. Williamson), pp. 8–13. Churchill Livingstone, Edinburgh.

Hunt, A. (1978). *The elderly at home*, pp 41–57. HMSO, London.

Larson, R. (1978). Thirty years of research on the subjective well-being of older Americans. *Journal of Gerontology*, **33**, 109–25.

Maddox, G.L. and Douglass, E.B. (1973). Self assessment of health. *Journal of Health and Social Behaviour*, **14**, 87–93.

Registrar General (Scotland) (1978). *Annual report: part 2, population and vital statistics*, p.104. HMSO, Edinburgh.

Savage, R.D., Gaber, L. B., Britton, P. G., Bolton, N., and Cooper, A. (1977). *Personality and adjustment in the aged*. Academic Press, London.

Shanas, E., Townsend, P., Wedderburn, D., and Friis, H., eds. (1968). *Old people in three industrial societies*. Routledge and Kegan Paul, London.

Taylor, R. and Ford, G. (1983*a*). Inequalities in old age. *Ageing and Society*, **3**, 15–22.

Taylor, R. and Ford G. (1983*b*). The elderly at risk: a critical examination of commonly identified risk groups. *Journal of the Royal College of General Practitioners*, **33**, 699–705.

Taylor, R. and Ford, G. (1983*c*). Risk groups and selective case finding in an elderly population. *Social Science and Medicine*, **17**, 647–55.

Taylor, R., Ford, G., and Barber, J. M. (1983). *The elderly at risk*. Age Concern, London.

Townsend, P. (1979). *Poverty in the United Kingdom: a survey of household resources and standards of living,* pp. 784–822. Penguin, London.

Walker, A. (1981). Towards a political economy of old age. *Ageing and Society*, **1**, 73–94.

SECTION 2
Biological aspects of ageing

2.1 Biological origins of ageing

THOMAS B. L. KIRKWOOD

INTRODUCTION

The broad phylogenic distribution of ageing (see Comfort 1979; Finch 1990) indicates considerable antiquity of its origin. Why should we be interested in these distant origins when seeking to understand ageing in our own species today? Apart from a natural desire to know how our senescence fits into the biological scheme of things, there is a practical reason for asking why ageing occurs. The reason is that the way we answer this question influences the types of mechanisms we are likely to consider as potential causes of ageing. The history of science indicates the importance of the prevailing theoretical conception, or paradigm, in shaping and constraining the research effort (Kuhn 1970). If we are to arrive at a satisfactory understanding of the causes of ageing it is important that we select the right paradigm in which to work.

Three different views on the origins of ageing will be considered in this chapter. The first is that ageing is simply the inevitable price which a higher organism pays for complexity. In this view there is no requirement to account explicitly for an evolutionary origin of ageing, which is seen simply as a process of biological wear and tear. The two other views both propose evolutionary reasons why ageing occurs, but differ in the type of natural selection thought to have operated. The adaptive evolutionary view suggests that ageing itself is selectively advantageous. This leads to the idea that ageing is a programmed termination of life and that the life-cycle is effectively under genetic control from start to finish. The non-adaptive view suggests that ageing is deleterious, or at best selectively neutral, and that it has evolved as an indirect consequence of the forces shaping the life history. One particular version of the non-adaptive view, the 'disposable soma' theory, suggests that ageing results from natural selection tuning the life history so that fewer resources are invested in somatic maintenance than are necessary for indefinite survival. The disposable soma theory supports the stochastic, wear-and-tear concept of ageing, but does so on the basis that wear and tear follows from making the best use of the energy available to the organism rather than from being inevitable.

Before studying these views more closely we need to define ageing in a way that makes it possible to ask questions about its origins. This requires a definition which does not rely on specific aspects of senescence, so that the definition can be applied for comparative purposes in species where the hallmarks of senescence may not correspond with those well known to us in our own and other familiar species. The best definition for this purpose is a population-based definition that ageing is a progressive, generalized impairment of function resulting in a loss of adaptive response to stress and in a growing risk of age-related disease. Individuals may vary in the rate at which specific features of ageing develop, but the overall effect of these changes is summed up in the increase in the probability of dying, or age-specific death rate, within the population (Fig. 1). We take this pattern

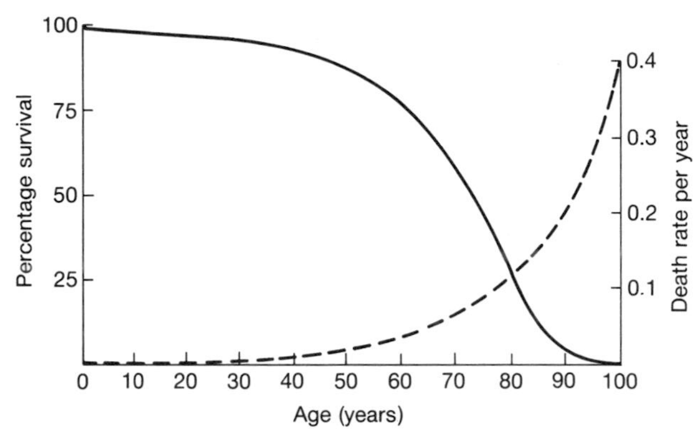

Fig. 1 Age-specific patterns of survival (continuous curve) and mortality (dashed curve) typical of a population in which ageing occurs. The example is of a human population with well-developed social and medical care (reproduced from Kirkwood and Holliday 1986 with permission).

of mortality as diagnostic of ageing when it is found in a species where death is not linked to some specific stage in the life-cycle, such as the rapid post-spawning death of Pacific salmon. In other words, ageing is the increasing tendency to failure of chronologically older individuals in a population where there is no obvious reason why, if ageing did not occur, the life-cycle of the individual should not extend indefinitely. The exclusion of species which undergo once-only, or semelparous, reproduction is important because in many of these species the commitment to begin the reproductive effort sets in train a specific sequence of physiological events that bring about the post-reproductive death of the adult. A discussion of the significance of the differences between post-reproductive death in semelparous species and ageing in repeatedly reproducing, or iteroparous, species may be found elsewhere (Kirkwood 1985).

WEAR-AND-TEAR THEORIES

There are many ways in which an organism can be damaged, ranging from a change in a single molecule to the loss of whole organs and structures. In a manner similar to the wear and tear of complex machines, it has been suggested that these intrinsic processes of biological deterioration present a fundamental barrier to the indefinite survival of higher organisms (Cutler 1978; Sacher 1978). This view finds support in the fact that many of the observable features of the ageing process do indeed resemble wear and tear, and also in the argument that, according to the second law of thermodynamics, ordered systems are intrinsically unstable and tend to give way to disorder.

The parallel with inanimate objects fails, however, to allow for the ability of living systems to repair themselves. The second law of thermodynamics tells us only that the degree of disorder, or entropy, increases in closed systems. Organisms are not closed; they take in nutrients, from which they extract energy. There is no fundamental reason why this flow of energy into the organism should not be used to maintain the level of entropy at a constant value. Indeed, it must be possible to maintain at least the germ-line in a steady state since, otherwise, species would fast become extinct. In fact, there are some species, for example sea anemones, where the powers of maintenance and renewal appear to be sufficient for the entire organism to survive indefinitely without visible deterioration (Comfort 1979).

To understand the scope and limitations of repair processes requires an analysis that takes account of the potential to evolve new repair mechanisms (Kirkwood 1981). We should not accept that because a particular kind of repair is beyond the capability of an organism in its present form it is necessarily impossible. The brain is an obvious example of a target in a higher organism that is vulnerable to irreversible damage because the death of neurones can disrupt the connectivity of its cell networks, and the information represented in these networks is then permanently lost. However, if natural selection had placed high enough importance on it, greater resilience to damage could have been incorporated within these networks, including perhaps the capacity to effect repair. An extreme example, not directly relevant to ageing, emphasizes the point that not all damage which is not repaired is necessarily irreparable. This is the regeneration of amputated limbs, which among vertebrates is restricted to certain small salamanders (Scadding 1977). While the proximate reason for this difference in regenerative ability between salamanders and other vertebrates depends upon the organization of cells, the ultimate reason depends upon natural selection and the evolutionary balance which must be struck between the costs and benefits of repair. In organisms which are very unlikely to survive during the time period required for a limb to regrow, the force of selection to acquire or retain limb regenerative ability will be minimal. It may be simply that the salamanders are near to a borderline beyond which the loss of a limb is a sufficiently serious threat to survival that it is not worthwhile to retain regenerative ability (Kirkwood 1981).

These considerations tell us that while wear and tear may play a role in ageing, it is not a sufficient explanation of why ageing occurs. This requires that we direct attention to the evolutionary theories.

ADAPTIVE EVOLUTIONARY THEORIES

The adaptive theories suggest that ageing confers some direct competitive advantage and that senescence is controlled in broadly the same way as development. The attraction of the adaptive theories is that they conform to the way we tend most easily to think about evolution, namely that new traits are produced to adapt the genotype in ways that are evidently fitter for survival. In view of the continuing appeal of these theories I shall examine in some detail why they are likely to be false.

One advantage suggested of ageing is that it prevents old and worn-out individuals from competing for resources with their progeny. This argument is plainly circular, as was pointed out by Medawar (1952), and need be considered no further. Another suggestion is that ageing helps to prevent overcrowding and so lessens the risk of severe depletion of resources (Wynne-Edwards 1962). It is conceivable that ageing could play such a role, but there are two very strong objections to this idea.

The first objection is that there is little evidence from natural populations that ageing is a significant contributor to mortality in the wild. Undoubtedly, an individual that lives long enough to experience senescence becomes more vulnerable to predators and other hazards. However, life tables for wild populations indicate that mortality during the early and middle periods of life is usually so great that few individuals survive long enough for ageing to have a measurable impact on the total death toll (see Fig. 2). This means that ageing

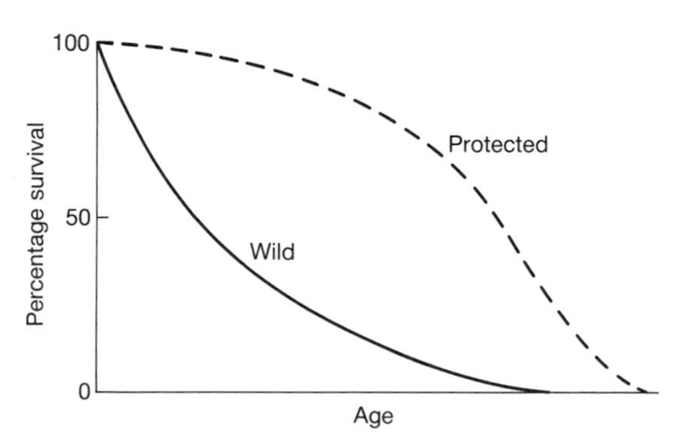

Fig. 2 The survival curve of a wild population tends to show little sign of age-related mortality, which may become apparent only when the population is transferred to a protected environment.

is in fact not needed to prevent overcrowding, nor is it easy to see how natural selection could have directly brought about the evolution of a trait which is so rarely seen under normal conditions. Some exceptions to the general rule that ageing is rare in the wild may possibly arise in the case of the larger animals. However, even in such species only a minority of individuals survive to old age, and in any case, these species evolved from smaller, more vulnerable ancestors in which it is reasonable to assume that ageing was already established.

A second objection to the idea that ageing evolved to prevent overcrowding is that the advantage, if it exists, is an advantage for the population rather than the individual. For the individual the best course, if other things are equal, will always be to live as long and reproduce as much as possible. This means that if ageing had evolved as a means to regulate population size, then a mutant in which the ageing process was inactivated would enjoy an advantage, and the mutant genotype would spread. For ageing to be stably maintained in the long term, it is necessary that 'group selection' for advantage to the population should outweigh the straightforward selection for advantage to the individual. The requirements for group selection to operate successfully against opposing selection at the level of the individual are very stringent (Maynard Smith 1976), and it is extremely unlikely that they apply to ageing. Briefly, it is necessary that the species is distributed among fairly isolated groups and that the introduction of a non-ageing mutant into a group should rapidly lead to the group's extinction.

Each of the above objections to the adaptive evolution of ageing is forceful in its own right. Taken together, the objections gain extra force because they complement each other. Where one objection applies less strongly, the other is intensified. For example, if ageing does generate significant mortality within a population so it is more plausible that it has some role to play, then the disadvantage of ageing to the individual is greater and the group selection argument is harder to sustain.

One final claim for the adaptive theories is that ageing is necessary for, or helps, evolution to occur (for a recent example, see Libertini 1988). The idea is that since evolution occurs through the operation of natural selection on successive generations, any process that accelerates the turnover of generations may result in a greater ability of a species to adapt to changes in its environment. The several weaknesses in this argument are as follows. Firstly, as I have discussed above, the generation time in the wild is determined for the most part not by ageing but by environmental hazards. Secondly, the argument assumes that the long-term advantage of adaptability outweighs the short-term disadvantage of reduced lifespan. This depends critically upon the assumed rate of change in the environment and the argument encounters the same complex difficulties that concern the evolution of sex and the optimal mutation rate (see Maynard Smith 1978). In fact, elevated recombination and/or mutation rates would be an alternative, more direct way to accelerate evolutionary change. Thirdly, for species which spread their reproduction over their lifetime, the critical factor in determining the rate of turnover of generations is not lifespan so much as the age at which individuals become reproductively mature. While there is force to the argument that species with long development times may be limited in their adaptability, the fact that such species tend also to have long lifespans does not establish that longevity itself poses a disadvantage. This would be true if a long lifespan necessitated slow reproductive maturation, but the causative link is much more likely to be the other way around.

NON-ADAPTIVE EVOLUTIONARY THEORIES

If ageing is not adaptive, then its evolution must be explained through the indirect action of natural selection. The non-adaptive theories of ageing are of two types: (i) ageing occurs because the power of natural selection declines during the lifespan; (ii) ageing occurs as the by-product of selection for some other trait.

Significance of the age-related decline in the force of natural selection

An important property of any life-cycle in which there is the opportunity for repeated reproduction between maturation and death is that the force of natural selection—that is, its ability to discriminate between alternative genotypes—weakens with age (Haldane 1941; Medawar 1952; Williams 1957; Hamilton 1966; Charlesworth 1980). The basic point is that because natural selection operates through the differential effects of genes on reproductive fitness, the force of natural selection must decline in proportion to the decline in the remaining fraction of the organism's total life-time expectation of reproduction. This is true whether or not the species exhibits ageing, since even if individuals do not age they are none the less susceptible to environmental mortality. If a gene effect is expressed early in life, it will influence the reproductive success of a larger proportion of individuals born bearing that gene than if it is expressed late, when many such individuals will already have died.

The attenuation in the force of natural selection with age means inevitably that there is only loose genetic control over the later portions of the lifespan. Indeed, it was suggested by Medawar (1952) that this might be sufficient in itself to account for the origin of ageing. Consider a species in which ageing does not initially occur. If a gene arises by mutation with an age-specific time of expression, and if the gene is a beneficial one, then natural selection can be expected to favour bringing forward its time of expression so that more individuals can benefit from it. Conversely, if the gene is harmful, selection would tend to defer its time of expression so its deleterious effects would be less damaging. Once a harmful gene has been so far delayed that it is expressed at an age when in the wild environment most individuals have died already, it is beyond the reach of any further opposing selection and can spread to fixation by random drift. Over many generations, Medawar (1952) suggested, there might thus accumulate a miscellaneous collection of late-acting, deleterious genes. In the normal environment these genes would only rarely have the opportunity to be expressed. In a protected environment, however, survivorship would be greater, and a significant fraction of individuals would experience the effects of these genes. The upshot would be that as a result of this process, ageing had appeared in the population.

That the declining force of natural selection with age is relevant to the origins of ageing is clear, but that an accumulation of late-acting deleterious mutations is a sufficient explanation for the origins of ageing is less certain. Experimental studies with Drosophila (Rose and Charlesworth 1980) have failed to confirm some predictions of this theory, while the theoretical objection has been raised that in the absence of ageing it is hard to see what would be the timing mechanism to determine 'lateness' in a life-cycle that potentially could last indefinitely (Kirkwood 1977).

A more plausible view, built on the same foundation as Medawar's, is the second type of non-adaptive theory, namely that ageing is a by-product of selection for other beneficial traits (Williams 1957). Williams' theory was developed using an argument similar to Medawar's except for the important difference that the genes in question were assumed to be pleiotropic, the same genes being responsible for both good effects early and bad effects late. Natural selection would then favour retention of the genes on the basis of their early benefits, but would defer as far as possible the time of expression of the deleterious effects to ages when survivorship in the wild environment would be low. The decline in the force of natural selection with age would ensure that even quite modest early benefits would outweigh severe harmful side-effects, provided the latter occurred late enough.

The pleiotropic genes theory avoids the problems faced by the late-acting deleterious genes theory and there is now some evidence from selection experiments in Drosophila and other insect species that the general trade-offs implied by the theory do exist (see Luckinbill *et al.* 1984; Rose 1984; Rose 1991). The pleiotropic genes theory does not specify, however, which particular genes are likely to have been responsible for the origins of ageing.

Ageing through optimizing the investment in maintenance

In this section, we describe a non-adaptive theory which is more specific about the nature of ageing processes. This is the 'disposable soma' theory (Kirkwood 1977, 1981; Kirkwood and Holliday 1979). The theory is named for its analogy with disposable goods, which are manufactured with limited investment in durability on the principle that they have a short expected duration of use. The term 'soma' is used in the sense introduced by Weismann (1891) to describe those parts of the body which are distinct from the 'germ-line' that produces the reproductive cells.

We consider how an organism capable of reproducing repeatedly during its life-cycle ought best to allocate energy among the different metabolic tasks it needs to perform. The organism may, in a sense, be viewed as an entity that transforms free energy from its environment into its progeny, and the law of natural selection asserts that those organisms (strictly, the genes that determine the phenotypes of the organisms) which are most efficient in this process are the ones most likely to survive (Townsend and Calow 1981). Part of the energy input must, however, be used for activities such as growth, foraging, defence, and maintenance. As all functions ultimately draw from the same total input of energy, there is an inevitable trade-off (direct or indirect) between the investment of energy in any one function and the investment in others. {Note that it is not necessary to assume that energy is in short supply, although for many species in their natural habitats this is the case. Even in populations with abundant energy supplies, those individual genotypes that best utilize the available energy will be the most successful.} As the following argument reveals, the optimum allocation of energy involves a smaller investment in somatic maintenance than would be required for the soma to last indefinitely.

Given the continual hazard of accidental death, to which no species is entirely immune, each individual soma can have only a finite expectation of life, even if it were not subject to ageing. When the soma dies, the resources invested in its maintenance are lost. Too low an investment in the prevention or repair of somatic damage is obviously unsatisfactory because then the soma may disintegrate before it can reproduce. However, too high an investment in maintenance is also wasteful because there is no advantage in maintaining the soma better than is necessary for it to survive its expected life-time in the wild environment in reasonably sound condition. In the latter case, the 'fitness' of the organism in terms of natural selection, that is, its ability to compete reproductively, would actually be enhanced by reducing the investment in somatic maintenance and channelling the extra energy into more rapid growth or greater reproductive output.

Fitness is therefore maximized at a level of investment in somatic maintenance which is less than would be required for indefinite somatic survival. The precise optimum investment in maintenance depends on the species' ecological niche. A species subject to high accidental mortality will do better not to invest heavily in each individual soma, but should concentrate instead on more rapid and prolific reproduction. A species which experiences low accidental mortality may profit by doing the reverse. The disposable soma theory thus not only explains why ageing occurs, but also suggests in broad terms how it is caused. As soon as the division between germ-line and soma evolved, the stage was set for the appearance of ageing.

GENETIC CONTROL OF LIFESPAN

As well as explaining the origins of ageing, the evolutionary theories need also to be able to account for the divergence of species' life-spans (see Table 1). This raises basic questions about the genetic control of ageing. What kinds of genes are involved? How many of them are there? How are they modified by natural selection to produce changes (usually increases) in life-span?

Table 1 Maximum recorded lifespans for selected mammals, birds, reptiles, and amphibians (from Kirkwood 1985)

	Scientific name	Common name	Maximum lifespan (years)
Primates	*Macaca mulatta*	Rhesus monkey	29
	Pan troglodytes	Chimpanzee	44
	Gorilla gorilla	Gorilla	39
	Homo sapiens	Man	115
Carnivores	*Felis catus*	Domestic cat	28
	Canis familiaris	Domestic dog	20
	Ursus arctos	Brown bear	36
Ungulates	*Ovis aries*	Sheep	20
	Sus scrofa	Swine	27
	Equus caballus	Horse	46
	Elephas maximus	Indian elephant	70
Rodents	*Mus musculus*	House mouse	3
	Rattus rattus	Black rat	5
	Sciurus carolinensis	Grey squirrel	15
	Hystrix brachyura	Porcupine	27
Bats	*Desmodus rotundus*	Vampire bat	13
	Pteropus giganteus	Indian fruit bat	17
Birds	*Streptopelia risoria*	Domestic dove	30
	Larus argentatus	Herring gull	41
	Aquila chrysaëtos	Golden eagle	46
	Bubo bubo	Eagle owl	68
Reptiles	*Eunectes murinus*	Anaconda	29
	Macroclemys temmincki	Snapping turtle	58+
	Alligator sinensis	Chinese alligator	52
	Testudo elephantopus	Galapagos tortoise	100+
Amphibians	*Xenopus laevis*	African clawed toad	15
	Bufo bufo	Common toad	36
	Cynops pyrrhogaster	Japanese newt	25

For the adaptive theories, if we suspend doubts about their plausibility, the control of life-span is straightforward, but the theories are uninformative about the nature and number of the genes involved. Any gene which has the effect of limiting lifespan might be considered, and there could be any number of them. A single death gene coupled to a suitable biological clock mechanism would suffice and would provide the simplest basis for modifying the life-span. However, there could be several genes.

For the non-adaptive theories, the kinds of genes proposed have been considered already in the previous section. The general nature of these genes is integral to the theories—late-acting deleterious genes in Medawar's theory; pleiotropic genes with early good effects and late bad effects in Williams' theory; and genes which regulate somatic maintenance in the disposable-soma theory. As regards numbers of genes, within each of the theories the same selection forces apply to any gene of the appropriate kind, so there are likely to be several, possibly many, genes involved. However, if there is a very large number of independent genes contributing to ageing, it is difficult to explain how the life-span can be altered, as modifying the expression of a single gene will do little to alter the rate of ageing, and multiple independent changes will be rare. This suggests either that there exists a relatively small number of principal genes responsible for ageing, or that the expression of the different genes is not independently regulated.

The evolution of increased life-span in the non-adaptive theories can be understood through the effects of reducing the risk of accidental death upon the rate of attenuation in the survival curve. An adaptation resulting in lower accidental mortality increases the force of selection at later ages. In Medawar's theory, this will then apply pressure to postpone further the expression of the late deleterious genes. In Williams' theory, the balance between the early beneficial and late harmful effects of the pleiotropic genes is shifted in favour of reducing the late harmful effects. In the disposable soma theory, the optimum investment in somatic maintenance is increased.

REPRODUCTIVE AGEING

So far I have not specifically considered ageing of the reproductive system, as distinct from other generalized aspects of the ageing process. Reproductive ageing is of particular interest because loss of reproductive function will accelerate the decline in the force of natural selection. The reverse is also true. In organisms that continue to grow indefinitely, and in which reproductive output increases with size, the decline in the force of natural selection is slowed. This may explain the considerable longevity of some species of fish.

The reason for leaving reproductive ageing until now is that a circular argument can arise if ageing of the reproductive system is not properly regarded as a feature that logically must follow the origin of ageing more generally. Weismann (1891), for example, originally suggested that ageing was necessary to rid a population of old and worn-out individuals that had produced their required quota of offspring and were of no further reproductive value (but see Kirkwood and Cremer (1982) for the later development of Weismann's views). Similar confusion can arise if the postreproductive death in semelparous species is not treated as distinct from ageing (see Introduction, and Kirkwood 1985).

After this preamble, it can be seen that although reproductive ageing takes its toll on reproductive function rather than on survival, there is nothing particularly special about it. Once the origin of ageing has been accounted for within an organism whose life history could otherwise extend indefinitely, reproductive ageing can be seen merely as a part of the generalized decline in function. In other words, if it does not matter that the organism does not survive indefinitely, then it does not matter that it does not reproduce indefinitely. For most species, reproductive ageing, like other aspects of the ageing process, is probably of little consequence in the wild. The special case of the human menopause is considered in the next section.

One final aspect of reproductive ageing which needs mention is the ageing of germ cells. Although the germ-line must, in a sense, be immortal, there is well-documented evidence of maternal and, to a lesser extent, paternal age effects in the frequency of genetic abnormalities (see, for example, Kram and Schneider (1978)). That age-related changes occur in germ cells is not particularly surprising. The increase in abnormal progeny, especially with maternal age, may reflect either less efficient screening for faults as a general consequence of ageing, or it may be due to the weakness of selection for late reproductive viability. Over a time-scale of generations, however, damage must not be allowed to accumulate in the germ-line, and this may be prevented by any of several mechanisms (Medvedev 1981).

EVOLUTIONARY ASPECTS OF HUMAN AGEING

Although this book is about geriatric medicine, the evolutionary aspects of human ageing have been left until now because (i) human evolution is comparatively recent, and (ii) in human populations certain features of ageing contrast markedly with those in the majority of other animal populations. To pay attention to these features too early could distort our understanding of how they must have arisen as modifications of the more general pattern.

First, our species is unique in the high frequency with which individuals survive to show clear signs of ageing, especially in the more affluent nations. This suggests a challenge to the idea that natural selection will usually operate so that ageing remains a potential rather than an actual phenomenon. The high incidence of ageing in modern human societies is doubtless due in part to the speed of recent social and cultural evolution, which is likely to have outstripped the potential for natural selection to modify our life history. Nevertheless, ageing is clearly seen, albeit less frequently, in more 'primitive' societies, and mention of it is found in the earliest human records. In terms of the disposable soma theory, it is conceivable that as accidental mortality was progressively reduced, under the influence of evolving human intellect and the associated trend to living in more protected social groups, there came a point where it is was no longer selectively worthwhile to increase the investment in somatic maintenance at the cost of further delaying growth and reducing reproduction (Kirkwood and Holliday 1986). Continuing selection pressure for further reduction in accidental mortality would have increased the average life-span while leaving the underlying rate of ageing unchanged. More individuals would therefore have begun to live long enough to age.

A second distinctive feature of human ageing is the clearly controlled shut-down of reproductive function that occurs in women at menopause. Only chimpanzees and macaques naturally exhibit similar changes and these are not so well defined. The menopause is sometimes cited as support for the adaptive theories, since it suggests that a strict genetic control of ageing may exist. A more plausible explanation of the menopause is that as soon as ageing became a significant feature of human or prehuman societies, long-lived females would have been faced with the serious risk of attempting to continue reproduction with an ageing soma. Child-bearing for humans is in any case complicated by the large size of the neonatal brain, and if reproduction were continued throughout the female life-span, the hazards associated with pregnancy and childbirth would have come to constitute a dominant cause of mortality in older women. It makes sense, therefore, to suppose that the menopause evolved as a means of removing older women from this risk and of preserving them for the important roles of rearing their later children, and possibly grandchildren, as well as sharing their valuable knowledge and experience with their kin group (see Medawar 1952). Seen in this way the menopause is not a primary feature of ageing but a secondary adaptation to lessen its deleterious effects.

CONCLUSIONS

The conclusions we can draw from studying the biological origins of ageing have broad implications for the way we perceive the ageing process. First, ageing needs to be explained in evolutionary terms, as it is not enough to regard it as just due to the inevitability of wear and tear. Secondly, the evolution of ageing as an adaptive process in its own right seems extremely unlikely. Thirdly, the non-adaptive theories in general offer the most plausible explanation for the evolution of ageing and longevity, and these theories make predictions that are amenable to experimental tests (Kirkwood and Rose 1991). Specifically, the disposable soma theory suggests that the efficiencies of somatic maintenance processes are crucial in determining longevity. Molecular studies using either a comparative approach or transgenic animals may serve to identify which of these processes play the most important roles.

A final word should be said about the way in which studies on the evolution of ageing throw light on the 'programming' of the lifespan. The point of issue between the adaptive and non-adaptive theories is not whether ageing is genetically controlled, as obviously it must be, but why and how this is arranged. This distinction is important as the theories influence the types of mechanism of ageing we are likely to consider as appropriate subjects for research.

References

Charlesworth, B. (1980). *Evolution in age-structured populations*. Cambridge University Press.

Comfort, A. (1979). *The biology of senescence*. (3rd edn). Churchill Livingstone, Edinburgh.

Cutler, R.G. (1978). Evolutionary biology of senescence. In *The biology of ageing*, (ed. J.A. Behnke, C.E. Finch and G.B. Moment). Plenum Press, New York.

Finch, C.E. (1990). *Longevity, Senescence and the Genome*. University of Chicago Press.

Haldane, J.B.S. (1941). *New paths in genetics*. Allen and Unwin, London.

Hamilton, W.D. (1966). The moulding of senescence by natural selection. *Journal of Theoretical Biology*, **12**, 12–45.

Kirkwood, T.B.L. (1977). Evolution of ageing. *Nature*, **270**, 301–4.

Kirkwood, T.B.L. (1981). Evolution of repair: survival versus reproduction. In *Physiological ecology: an evolutionary approach to resource use*, (ed. C.R. Townsend and P. Calow). Blackwell Scientific, Oxford.

Kirkwood, T.B.L. (1985). Comparative and evolutionary aspects of longevity. In *Handbook of the Biology of Ageing*, (2nd edn), (ed. C.E. Finch and E.L. Schneider.) Van Nostrand Reinhold, New York.

Kirkwood, T.B.L. and Cremer, T. (1982). Cytogerontology since 1881: a reappraisal of August Weismann and a review of modern progress. *Human Genetics*, **60**, 101–21.

Kirkwood, T.B.L. and Holliday R. (1979). The evolution of ageing and longevity. *Proceedings of the Royal Society, London* (B), **205**, 531–46.

Kirkwood, T.B.L. and Holliday R. (1986). Ageing as a consequence of natural selection. In *The biology of human ageing*, (ed. A.H. Bittles and K.J. Collins). Cambridge University Press.

Kirkwood, T.B.L. and Rose, M. R. (1991). Evolution of senescence: late survival sacrificed for reproduction. *Philosophical Transactions of the Royal Society, London* (B), **332**, 15–24.

Kram, D. and Schneider, E.L. (1978). Parental age effects: increased frequencies of genetically abnormal offspring. In *The Genetics of Aging* (ed. E.L. Schneider) Plenum Press, New York.

Kuhn, T.S. (1970). *The structure of scientific revolutions*, (2nd edn). Chicago University Press.

Libertini, G. (1988). An adaptive theory of the increasing mortality with increasing chronological age in populations in the wild. *Journal of Theoretical Biology*, **132**, 145–62.

Luckinbill, L.S., Arking, R., Clare, M.J., Cirocco, W.C., and Buck, S.A. (1984). Selection of delayed senescence in *Drosophila melanogaster*. *Evolution*, **38**, 996–1003.

Maynard Smith, J. (1976). Group selection. *Quarterly Review of Biology*, **51**, 277–83.

Maynard Smith, J. (1978). *The evolution of sex*. Cambridge University Press.

Medawar, P.B. (1952). *An unsolved problem of biology*. H.K. Lewis, London. (Reprinted in *The uniqueness of the individual*. Methuen, London, (1957).

Medvedev, Z.A. (1981). On the immortality of the germ line: genetic and biochemical mechanisms—a review. *Mechanisms of Ageing and Development*, **17**, 331–59.

Rose, M.R. (1984). Laboratory evolution of postponed senescence in *Drosophila melanogaster*. *Evolution*, **38**, 1004–10.

Rose, M.R. (1991). *Evolutionary Biology of Ageing*. Oxford University Press.

Rose, M.R. and Charlesworth, B. (1980). A test of evolutionary theories of senescence. *Nature*, **287**, 141–2.

Sacher, G.A. (1978). Evolution of longevity and survival characteristics in mammals. In *The genetics of ageing*, (ed. E.L. Schneider). Plenum Press, New York.

Scadding, S.R. (1977). Phylogenic distribution of limb regeneration capacity in adult Amphibia. *Journal of Experimental Zoology*, **202**, 57–68.

Townsend, C.R. and Calow, P. (1981). *Physiological ecology: an evolutionary approach to resource use*. Blackwell Scientific, Oxford.

Weismann, A. (1891). *Essays upon heredity and kindred biological problems*, vol. 1 (2nd edn). Clarendon Press, Oxford.

Williams, G.C. (1957). Pleiotropy, natural selection and the evolution of senescence. *Evolution*, **11**, 398–411.

Wynne-Edwards, V.C. (1962). *Animal dispersion in relation to social behaviour*. Oliver and Boyd, Edinburgh.

2.2 Biological mechanisms of ageing

GEORGE M. MARTIN

GENERAL CONSIDERATIONS

Some definitions

Social gerontologists and plant biologists often differentiate between the terms ageing and senescing (or senescence), using the former to describe all changes in structure and function of an organism from birth (or even fertilization) to death, while reserving the latter to describe events late in the life course that precede death. Most biogerontologists (including myself), however, judging by the way they conduct their experiments, use the terms ageing and senescence more or less synonymously to describe the structural and functional alterations that appear soon after an organism has completed its development, as defined by the emergence of sexual maturity, and (for most mammals) the cessation of major skeletal and organ growth. Some of these alterations are adaptive attempts to compensate for diminished function. Inevitably, however, there is a decline in the ability of the organism to maintain homeostasis and to mount a successful reaction to various types of injury. Thermodynamically, there is an inexorable increase in entropy or disorder of molecules and systems. In large populations of mature individuals, one observes an exponential decrease in the probability of survival as a function of a unit of chronological time (the Gompertz relationship). The ages at which this exponential decline appears and the slopes of the decline are powerfully determined by the constitutional genome, leading to vast differences in the maximum potential lifespan of different species; among mammalian species, the differences are of the order of thirtyfold. This is not to deny, however, the important potential of various types of environmental influence in modulating these events, either positively or negatively. To date, however, no agents have been proven to accelerate or decelerate an intrinsic rate of ageing, or specific aspects of ageing with the possible exception of dietary calories, which will be discussed below.

A major theoretical and practical difficulty in gerontology is the differentiation between intrinsic ageing and the impact of various environmental or life-style factors, or of various specific diseases. In certain cases, there may be a line of continuum between the two. For example, there is at present no hard evidence that any of the several histopathological and biochemical hallmarks of Alzheimer's disease are qualitatively distinct from what can be found in the brains of older people who, given the arbitrary threshold values for the sensitivities of psychometric tests, appear to be cognitively 'normal'. It may be that such depositions as β-amyloid and neurofibrillary tangles may result, in part, from age-related alterations in protein turnover or protein processing, although quantitatively these changes are far greater in persons with clinical evidence of Alzheimer's disease.

The term 'age-associated' is non-committal; it includes alterations that are simple functions of chronological time as well as those that are coupled to intrinsic, biological ageing. The latter might be diagnosed if it were shown that, among a group of closely related species, such as mammals, the ages of onset and the rates of progression of the phenotype of interest were inversely related to the maximum potential lifespan of the species. Thus, while more data are needed, most evidence to date is consistent with the view that several different types of age-associated neoplasms are related to intrinsic biological ageing and not simply to chronological time, as their prevalence rises substantially about midway through the lifespan (about 2, 4, 8, and 60 years, respectively, for house mouse, white-footed deer mouse, dog, and man, for example). Some workers would contest this interpretation (Peto *et al.* 1986). Unfortunately, current evolutionary theory would suggest that this simple approach of 'comparative gerontology' cannot be infallible.

Implications of evolutionary theory

The preceding chapter (Chapter 2.1) is of seminal importance in setting the stage for a further analysis of candidate mechanisms of ageing. Let us therefore briefly summarize the major conclusions of evolutionary biologists concerning the natural origins of ageing in age-structured iteroparous animals such as man and the vast majority of mammals. There is, first of all, a consensus supporting non-adaptive evolutionary theories; one would have to invoke group selection as a mechanism for an adaptive evolution of ageing and there is little support for the importance of group selection as opposed to selection for reproductive fitness at the level of the individual organism. Secondly, two central genetic phenomena appear to have some degree of support: (1) the accumulation of germ-line mutations that are expressed relatively late in the life course, when the force of natural selection is weak; and (2) the principle of antagonistic pleiotropic gene action, whereby selection for alleles that enhance reproductive fitness early in the lifespan exhibit negative effects post-reproductively. Perhaps the earliest speculation concerning a concrete example of the latter was by G. C. Williams (1957), who suggested that alleles acting during early phases of the lifespan to enhance calcium uptake, and thus providing sturdy bone structure, did so at the cost of subsequent calcium depositions in arterial walls.

This scenario has profound implications, as follows.

1. There is no need to invoke primary mechanisms of ageing based upon deterministic, 'programmed' gene actions which have as their direct aim, such outcomes as the 'turning on' of 'killer' genes or the 'turning off' of essential genes. Recent evidence for inactivation of the C-*fos* gene, which is essential for cellular proliferation, in senescent, non-proliferating human fibroblast cell cultures, has been interpreted as a 'terminal differentiation' of such cells (Seshardi *et al.* 1990). The *in-vivo* counterpart has not yet been demonstrated. (As a cautionary note, however, it is conceivable that some residue or caricature of such programmes that developed in hypothetical, ancestral,

semelparous species might still be expressed, to some degree, in certain iteroparous species or in particular individuals.)
2. There is no reason to believe, *a priori*, that the set of genetic loci of major importance in determining reproductive fitness in one species would be identical to those of another species, although there are likely to be degrees of overlap. Imagine, for example, the striking differences among species as regards loci selected because of behavioural patterns of importance in successful mating behaviour.
3. As numerous genetic loci are likely to be involved in the evolution of varying patterns of loci between species, one would expect plenty of opportunities for different patterns of ageing among individuals within a species, based upon genetic polymorphisms, mutations, and a complex matrix of genetic–environmental interactions.
4. Although, for any given putative mechanism of ageing, one could entertain a sizeable list of candidate gene loci that might modulate the rate of ageing, the rich variety of loci capable of influencing reproductive fitness or of accumulating mutations would strongly argue against the hypothesis of a single mechanism, process, or theory of ageing. This crucial question is by no means settled, however, as we shall see below.

A few major mechanisms or multiple, independent mechanisms of ageing?

While, as indicated above, a polygenic basis for lifespan or for the senescent phenotype does not necessarily obviate the proposal of just a few major mechanisms, or even a single major mechanism, the greater the number of genes shown to be playing a role, the greater the likelihood that the hypothesis of multiple independent mechanisms is correct. There are as yet no definite estimates in any organism regarding the genetic complexities involved, but a number of lines of evidence point to considerable complexity. In the fruit fly, *Drosophila melanogaster*, crosses between comparatively short-lived and long-lived strains (the latter serially selected for high fecundity among females late in the life course) clearly indicated a polygenic determination of lifespan, with genetic loci contributed by each of the three major chromosomes, although with a disproportionately high contribution from chromosome 3 and a relatively minor contribution (especially for males) from chromosome 2 (Luckinbill *et al.* 1988). In the roundworm, *Caenorhabditis elegans*, the experimental reassortment of genes from two strains having similar lifespans has produced numerous, recombinant, inbred lines with striking variations in both mean and maximum lifespans, consistent with a highly polygenic determination (Johnson 1987). On the other hand, there is also evidence that a single recessive mutation (age-1), with a pleiotropic effect expressed as a marked reduction of hermaphroditic self-fertility, can substantially enhance the maximum lifespan when compared with that of the parental, wild-type worms (Friedman and Johnson 1988).

In higher primates, estimates of the rate of change in cranial capacity of the hominid precursors of man (which can be statistically correlated with potential lifespan) have suggested that some 200 to 300 genetic loci might have been involved in the emergence of *Homo sapiens* and its unusually long lifespan (Cutler 1975). There are a number of uncertainties in estimates derived from comparisons of lifespans of different species, however, including the assumption that rates of change of amino acid substitutions in proteins are the appropriate 'molecular clock' for such projections. Other mechanisms, such as chromosomal rearrangements, may be of equal or greater significance; a single such rearrangement could potentially alter the regulation of expression of dozens or hundreds of genes. A study of progeroid mutations of man gave an estimated upper limit of several thousand genetic loci potentially capable of modulating particular aspects of the senescent phenotype, although it was suggested that only a small proportion of these might be of major significance to ageing (Martin 1978). Thus, a number of genetic approaches in several species point to a polygenic basis for ageing and hence would suggest caution in embracing a single, all-encompassing, mechanistic theory of ageing.

On the other hand, there is the striking evidence, from a wide variety of species (although the most reliable information is confined to rodents), that major (approximately half of) increases in the maximum lifespans of cohorts of experimental animals (generally consisting of somewhat less than a hundred to a few hundred individuals) can be consistently produced by the simple expediency of restricting the caloric intake to around 60 per cent of that of controls fed *ad libitum* (Weindruch and Walford 1988). The effect is most marked when the caloric restriction (in the face of an otherwise nutritionally sufficient diet) is commenced at weaning, but significant extensions of lifespan are also produced when adults are subjected to caloric restriction. In addition to lifespan, a surprising number (but not all) of the putative biomarkers of ageing so far examined appear to be influenced. There is also a striking retardation of certain diseases commonly observed in ageing rodents, notably chronic nephropathies and neoplasia. While some critics have suggested that the calorically restricted animals should in fact be considered the normal controls (i.e., those fed *ad libitum* may be 'overfed'), such an interpretation does not invalidate the observation of a dramatic effect of calories upon lifespan and a number of age-associated diseases. Thus, a detailed understanding of the molecular, cellular, and physiological effects of variations in caloric intake may lead us to a definition of a single, predominant mechanism of ageing. If so, then one could envisage interventions less daunting than the rigorous restriction of food. It should be emphasized, however, that there is as yet no good information on the effects of caloric restriction on putative biomarkers of ageing and on the lifespans of humans, although a modest pilot study in non-human primates by scientists at the United States National Institute on Aging is under way.

A SAMPLING OF CURRENT VIEWS ON PUTATIVE MECHANISMS OF AGEING

While numerous mechanistic theories of ageing have been proposed, it is fair to say that none of them has as yet been definitively established. It is difficult even to provide a satisfactory system of classifying the diverse ideas, many of which overlap extensively. They also vary substantially in the degree to which fundamental phenomena are invoked, as opposed to restatements of descriptive phenomenology.

However, the selected sample discussed below reflects a major segment of current thinking.

Ageing as a by-product of oxidative metabolism (free radical theory)

The free radical theory of ageing (Harman 1986) is one of the oldest and, in all likelihood, still the most popular, single mechanistic theory of ageing.

Chemical free radicals are atoms or groups of atoms with an unpaired electron. They are consequently highly reactive and capable of reacting with a variety of biologically important macromolecules, including DNA, protein, and lipid. Of special interest to the free radical theory of ageing are various oxy-radicals, mainly partially reduced products of oxygen, such as the superoxide radical. Many such substances are likely to have exceedingly short lifespans in biological tissues. The hydroxyl radical, for example, may react within a few molecular diameters of the site of formation, with a half-life of the order of a nanosecond. Nevertheless, a variety of lines of evidence indicates that cellular DNA can be readily attacked by the hydroxyl radical, potentially leading to mutagenesis and clastogenesis (chromosome breaks). Other compounds, however, may be capable of diffusing for substantial distances within a cell before reacting with a suitable substrate (Pryor 1986).

Fortunately, along with the evolution of aerobic organisms, there was developed the cytochrome system of respiration, ensuring that oxygen is largely reduced quadrivalently, thus minimizing the generation of highly reactive, partially reduced products. There is, however, a degree of 'leakiness' in the system, perhaps of the order of a few per cent. Other protective systems, such as the superoxide dismutases, catalase and glutathione peroxidase, provide a line of defence by enzymatically 'scavenging' partially reduced oxygen products. Moreover, damaged DNA may be repaired by sets of specific enzymes, and altered proteins and lipids may be degraded and replaced, although structural lesions—for example, the oxidation of amino acid side-chains and cysteine sulphydryl groups (Stadtman 1988)—may accumulate in proteins with intrinsically low rates of turnover, such as the lens crystallins and the collagen of connective tissues and blood vessels. The rates of development of such processes might be accelerated by oxidative attacks upon the system of proteases and ancillary proteins that are thought to recognize and preferentially degrade abnormal proteins. Given the large flux of oxygen, some deleterious consequences may occur, depending upon the balance between the rates of generation of the free radicals and the various defence mechanisms in particular tissues. Species-specific lifespans might therefore be attributable to variations in the constitutive baseline efficiencies of the several families of protective enzymes as well as to the rates and levels of induction subsequent to injury; all are under genetic control.

This, in essence, is the basis for the free radical theory of ageing; it might be regarded as a price we pay for an aerobic life-style. In terms of life-styles, the theory would predict that prolonged and stressful exercise might accelerate rather than retard ageing. Some support for this view comes from experiments in which vigorous exercise (to exhaustion) in rodents was shown to be associated with a three-to fourfold increase in the concentrations of free radicals, an increase in lipid peroxidation (as inferred from the generation of malondialdehyde), and biochemical evidence of mitochondrial damage (Davies *et al*. 1982). Such alterations are likely to lead to increased depositions of lipofuscin pigments ('ageing pigments'), which are believed to be the products of oxidative attack upon the lipoprotein constituents of cellular organelles. Lipofuscins are one of the few candidates for 'public' biological markers of ageing, as they accumulate in an amazing variety of ageing systems, ranging from fungi undergoing clonal senescence to mammalian myocardium, liver, skeletal muscles, testes, and neuronal subsets (although without any obvious correlation with cellular dysfunction). Moreover, the limited evidence available is consistent with the view that their rates of incorporation are inversely related to the maximum potential lifespans of mammalian species (Martin 1977, 1988).

The notion that ageing may be related, in part, to byproducts of oxidative metabolism can be reconciled with evolutionary theory, as one can imagine selection, for reproductive fitness, of alleles at many loci that serve to enhance oxygen flux in various tissues and to accelerate certain other metabolic processes capable of generating free radicals; ageing could emerge as a delayed, secondary, negative, pleiotropic effect. It obviously does not fit the findings on experimental caloric restriction, because, surprisingly, such restriction does not lead to a reduction in the metabolic rate (oxygen usage) per unit of lean body mass. It is conceivable, however, that caloric restriction increases the efficiency of the cytochrome system, thereby reducing the extent of univalent reduction of oxygen.

Ageing as a by-product of the flux of reducing sugars

A complex, non-enzymatic reaction between a variety of reducing sugars (the most relevant, in terms of concentration, being glucose) and the primary amino groups of proteins is one of a growing list of pathways to the production of post-translational alterations of proteins, one of the hallmarks of ageing in a variety of organisms. The resulting altered proteins, if they are long-lived, have complex, cross-linked end-products whose structures have not yet been fully elucidated. The initiating reaction is called 'glycation' ('non-enzymatic glycosylation'), and is followed by the formation of labile Schiff base derivatives of proteins, which slowly isomerize to more stable ketoamine adducts via the Amadori rearrangement (see Baynes and Monnier 1989 for an overview of the chemistry and biology).

This proposed molecular mechanism of ageing is supported by the clinical and pathological observations of what appears to be premature ageing (or aspects of ageing) in subjects with poorly controlled diabetes mellitus, particularly involving connective tissues and the vasculature. There is no correlation, however, between the concentration of blood glucose and the life-span of a species. This is not a fatal complication for the theory, as one may envisage a number of genetically controlled steps in the determination of the degree to which altered proteins emerge with ageing. One could also reconcile this proposition with evolutionary theories of ageing. It is supported by experiments with caloric restriction which have shown that restricted animals have significantly lower levels of glycated haemoglobin.

Some investigators have attempted to reconcile the oxidative injury and glycation hypotheses by pointing to the possibility of 'auto-oxidative glycation' of proteins. It is also apparent that the glycation idea can be reconciled with genomic instability theories of ageing, as there is evidence that DNA can act as a substrate for glycation and in fact may undergo mutagenesis as a consequence of such reactions.

Ageing as a decline in genomic stability

Many would look to DNA molecules as the ultimate macromolecular targets of ageing, because DNA specifies the information for all metabolic events, including the machinery for its own repair and the machinery for the removal of altered proteins. In the case of single-copy sequences specifying critical functions in obligate, postreplicative cells, such as neurones, two events (one affecting each of the alleles of the homologous chromosome pair) could result in cell death. In proliferating populations of cells, such events (for example, the homozygous loss of a tumour suppressor locus) could lead to the emergence of an age-related neoplasm.

A theory which provided the major impetus to the development of molecular gerontology, Leslie Orgel's protein synthesis error catastrophe theory (Orgel 1963, 1970) predicted an exponential rise in the number of point mutations in the somatic cells of ageing organisms towards the latter part of the lifespan. That proposal argued for the primacy, during ageing, of errors in genetic transcription and translation involving proteins that were themselves involved in protein synthesis. Depending upon the efficiencies of the proteases that scavenged the abnormal protein synthesizing machinery, essentially all proteins would be subject to synthetic errors, including all of the enzymes involved in DNA replication and repair, hence the inevitability of somatic mutation.

There is now a great deal of evidence arguing against the original form of the Orgel hypothesis. While abnormal proteins are definitely found in ageing organisms, they can be shown to be post-translational in origin rather than the results of errors in synthesis (Warner *et al.* 1987). Some role for somatic, genetic events in the genesis of various aspects of the senescent phenotype seems highly probable, however, given the substantial degree of genetic plasticity of mammalian somatic cells. The potential mechanisms are numerous for the case of that class of somatic, genetic events broadly classified as mutations. These include changes in gene dosage (chromosomal aneuploidy, tandem duplications, deletions, selective gene amplification, and shifts in ploidy); changes in the arrangements of genes (inversions and translocations); and modifications in the primary structure, or nucleotide base composition of genes (base substitution, depurinations and depyrimidinations, frameshifts, insertions via transposable elements, mitotic crossing over, gene conversion, and DNA-mediated transformation or transfection).

Much less is known about the molecular basis of changes in gene expression that are not based on alterations in nucleotide sequence or on gene dosage. These events are thought to underlie most developmental and physiological shifts in states of cellular differentiation, with some striking exceptions, such as the loss of the nucleus of mature red blood cells during mammalian erythropoiesis, and the rearrangement of immunoglobin and receptor loci of specialized lymphocyte lineages. Inappropriate (i.e. non-adaptive, or deleterious) shifts of gene expression are thought to occur during ageing. This idea has been termed 'dysdifferentiation' (Zs.-Nagy *et al.* 1988) or 'epimutation' (Holliday 1987). Its molecular basis is believed to involve changes in states of methylation of the cytosines of DNA and is thought to follow stochastic injurious events, including those resulting from free radicals, although some investigators have also suggested the possibility that changes in methylation at discrete domains of the genome could serve as a deterministic molecular clock.

One special form of change in gene expression involves a battery of genes on the X chromosome that is subject to facultative heterochromatization; this is the basis for the sex chromatin or Barr body. In mice, ageing can reactivate genes within such developmentally 'silenced' DNA, at least for the case of the HPRT gene (hypoxanthine phosphoribosyl transferase, the site of the Lesch–Nyhan mutation); this does not occur in humans, however (Migeon *et al.* 1988).

There is not enough information to decide which of the numerous mechanisms of genomic instability cited above are of the greatest significance to ageing, to what degree such changes are simply related to chronological age, as opposed to intrinsic biological ageing, and to what extent a given pattern is species specific. For the case of certain invertebrates, for example, there is strong evidence against a role for recessive mutations in the determination of lifespan (Maynard Smith 1965).

Arguments have been made, however, in support of the relative importance of large-scale (chromosomal type) mutations, at least for the case of proliferative populations of mammalian cells (Martin *et al.* 1985). It is of interest, for example, that the molecular pathology of the Werner syndrome, perhaps the most striking of the segmental progeroid syndromes of man (Martin 1978), involves a propensity to undergo relatively large deletions (Fukuchi *et al.* 1989). If this proposition is correct, it points to the special importance of protecting human populations from environmental clastogens (physical, chemical, and viral agents that produce chromosomal mutations). It will also be of considerable conceptual importance to evaluate the role of epimutation in ageing.

As pointed out in T. B. L. Kirkwood's 'disposable soma' concept of ageing, one could reconcile certain pathways of genomic instability to ageing with evolutionary theory, as the energetic investments in preventing and repairing varieties of DNA alteration would be considerable. Compromises in this area of quality control can be envisaged whereby relatively more 'energy' could be devoted to reproduction during the earlier phases of the life-span. For comparatively long-lived species, one would predict greater fidelity of the enzymes involved in DNA replication and DNA repair, with a decrease in the rate of somatic mutation; this is a basis for the intrinsic mutagenesis theory of ageing (Burnet 1974). Correlations, in mammalian species, between the efficiency of certain methods of DNA repair and potential lifespan lend some support to that proposal (Hart and Setlow 1974), but there is as yet no systematic evidence pointing to correlations between the maximum potential lifespan of various mammalian species and the rate of accumulation of point mutations.

There is as yet little information on the effects of caloric restriction on various aspects of genomic stability. Restriction does appear to retard the rate of decline in unscheduled DNA synthesis for some tissues of F344 rats (Weraarchakul *et al.* 1989).

Ageing as a decline in the rates of protein synthesis and turnover

About three-quarters of the large number of studies in various tissues of many organisms have indicated that ageing is associated with a decline in bulk protein synthesis (Richardson 1981). For a limited number of loci so far examined, this appears to be associated with a decline in the rate of gene transcription (see, for example, Richardson *et al.* 1987). There is evidence, in both fruit flies and mice, that the rate-limiting step could be the extent of peptide elongation, and that this decline is attributable to a deficiency in the amounts of the elongation factor EF-1α. The latter may be the result of decreased transcription, as a decline in the amount of mRNA for that factor precedes the decline in protein synthesis; moreover, the transfection of extra copies of that gene to fruit flies has resulted in an increased lifespan (Shepherd *et al.* 1989). To what extent these interesting findings can be generalized to other strains of fruit flies and to other species remains to be seen. It is conceivable that particular inbred lines may have different rate-limiting steps related to a decline in macromolecular synthesis. A key question is how such apparently specific transcriptional declines are brought about during the latter portions of the lifespan.

The body of knowledge linking ageing to a decline in the rates of protein turnover is more limited (Bienkowski and Baum 1983), but a number of studies (for example, Dubitsky *et al* 1985; Goldspink *et al.* 1985) do indicate such declines. An 'altered protein breakdown' theory of ageing has been proposed whereby compromised lysosomal pathways of protein degradation could result in the accumulation of abnormal proteins in aged cells, with the induction of cytosolic pathways that are presumed to lead to the excessive degradation of normal short-lived proteins, with deleterious metabolic effects (Dice and Goff 1987). Rothstein (1982) believes that the numerous post-translationally altered proteins in the tissues of aged animals are the outcome of subtle denaturations that follow from the prolonged 'dwell time' of proteins, the result of decline in the rates of both synthesis and degradation. It has yet to be demonstrated, however, that the accumulation of such altered proteins is functionally significant. One should recall that, for the case of heterozygotic carriers of numerous, recessive, inborn errors of metabolism, there is no discernible phenotype, despite the fact that there is typically a reduction by half in the concentrations of the affected enzymes (Pembrey 1987). It may be, however, that reductions by half in the concentrations of the functionally active molecules of numerous proteins, particularly if they are parts of some common metabolic pathways, could in fact contribute to the senescent phenotype. Immunological assays for the total populations of such molecules could be misleading, because it has been shown, for enzymes, that there is a decline during ageing in the proportion of enzymatically active molecules in the total of immunologically detectable molecules (Gershon and Gershon 1970).

The observation of a more or less systematic decline in protein synthesis and turnover and, especially, of a reproducible deficiency of a specific elongation factor does not obviously fit with the evolutionary theories discussed above. It is perhaps more consistent with deterministic theories of ageing or, at any rate, ideas that emphasize central regulatory pacemakers acting via neurohumoral factors.

Caloric restriction appears to ameliorate the age-associated decline in protein synthesis and turnover (Holehan and Merry 1986), thus suggesting that these phenomena are fundamental concomitants of ageing.

Ageing as the result of a neuroendocrine 'cascade'

According to this theory (Finch 1976), the peripheral physiological decrements of ageing could be the inevitable byproducts of the complex positive and negative feedback systems associated with the neuroendocrine controls of visceral function. The best studied subsystem has been the neuroendocrine-reproductive system of the female rodent, in which there is evidence of deleterious feedback effects of oestrogenic substances upon hypothalamic neurones; these effects can be attenuated by ovariectomy (Finch *et al.* 1984; Wise *et al.* 1989).

This model might be regarded as a special case of a more general 'systems analysis' or 'integrative physiology' view of ageing, although modern extensions of this approach introduce thermodynamic concepts of 'physiological noise', lethal fluctuations and catastrophes, non-linear mechanics, and bifurcation theory (Yates 1988), all of which are beyond the scope of this chapter.

One could reconcile these concepts with current evolutionary theory, as the undesirable late effects of feedback are presumably byproducts of systems that were selected in order to enhance reproductive fitness. Some gerontologists, however, regard neuroendocrine theories as examples of 'programmed' ageing in which the 'pacemaker' of ageing is in the central nervous system; such a view is indeed at variance with the evolutionary theory, although, as indicated earlier, it is conceivable that some evolutionary vestige of the type of 'programmed' ageing, as seen in semelparous organisms, might be operative in mammals.

There is not enough available information to evaluate this theory from the point of view of experimental caloric restriction.

Ageing as a decline in proliferative homeostasis

Ageing mammals characteristically have a striking multifocal array of tissue hyperplasia, often occurring side by side with regions of atrophy. Examples include the proliferation of arterial myointimal cells in atherosclerosis; adipocytes in regional obesity; chondrocytes, osteocytes, and synovial cells in osteoarthrosis; glial cells in regional neuronal atrophy; epidermal basal cells in verruca senilis; epidermal melanocytes in senile lentigo; epidermal squamous cells in senile keratosis; fibroblasts in interstitial fibrosis; fibromuscular stromal cells and glandular prostatic epithelium in benign prostatic hyperplasia; lymphocytes in ectopic lymphoid tissue; suppressor T cells in immunological deficiency; oral mucosal, squamous cells in leucoplakia; ovarian cortical stromal cells in ovarian stromal hyperplasia; pancreatic ductal epithelial cells in ductal hyperplasia and metaplasia; and sebaceous glandular epithelium in Fordyce disease (of oral mucosa) and in senile sebaceous hyperplasia (of skin) (Martin 1979).

The underlying mechanisms are unknown, but conceivably

could be related to alterations in cell–cell communication, qualitative and/or quantitative changes in cell receptors for various mitogens and cell-cycle inhibitors, or in the availability and quality of the effector agents themselves. While there is indeed a large published record pointing to changes in receptors in various types of ageing cells, very little is known about changes in cell–cell communications during ageing. It has been speculated, however, that the clonal senescence of somatic cells described below could underlie this inappropriate hyperplasia via a loss of normal feedback regulation among sets of related cell types (Martin 1979).

Whatever the underlying cellular and molecular mechanisms, one potentially important consequence of these alterations in the control of the mitotic cell cycle is the emergence of benign and malignant neoplasms, as prolonged cell proliferation could allow the expression of initiating events in oncogenesis via the opportunity for selection of secondary and tertiary alterations in the genome which further distance the cell from physiological controls.

Tissue atrophy, presumably due, in part, to the loss of the ability of potential precursor cells to replace effete cells, is the other side of the coin as regards the maintenance of proliferative homeostasis in ageing organisms. We have only limited information on this question *in vivo*, however, especially in humans. Some experiments with animals have indeed demonstrated a decrease in the baseline and induced proliferative behaviours of various cell types during ageing (Krauss 1981), but there are exceptions (Stemerman *et al.* 1982; Holt and Yeh 1989). Most of our information on the limited replicative potential of normal diploid somatic cells comes from experimental cell and tissue culture. Tissue culture has clearly established that the rate and extent of cellular outgrowth from explants declines as a function of donor age. The outcome of cell culture led to the famous 'Hayflick limit,' in which it was quantitatively established by Hayflick and Moorhead (1961) and numerous subsequent workers that mass cultures and individual clones of normal diploid cells from various animal and human tissues eventually cease to replicate unless there occur genetic alterations in the cells leading to a 'transformation' to unlimited growth. Many regard such culture systems as models for the study of cellular ageing and have described numerous biochemical and morphological changes associated with the *in-vitro* decline in growth potential.

The mechanisms underlying the gradual loss of proliferation are unknown. Hypotheses have ranged from that of an active, genetically programmed phenomenon (Smith *et al.* 1987), perhaps analogous to the terminal differentiation of stem-cell lineages (Martin *et al.* 1975; Seshardi *et al.* 1990), to stochastic losses of methyl groups in DNA (Holliday 1986). At least four different recessive genetic loci are involved in the escape from the limited lifespan (Pereira-Smith and Smith 1988). (It is possible that the locus for one such complementation group is on human chromosome 1 (Sugawara *et al.* 1990).)

Ageing as autoimmunity

This is one of the most venerable theories of ageing (Walford 1969) and is based upon a large body of evidence for a rise in the titres of autoantibodies in ageing animals and humans as well as an associated complex series of alterations in the immune system. Many simple, eukaryotic organisms with immune systems undergo biological ageing, however. While autoimmunity is likely to contribute to the senescent phenotype of mammals, it is unlikely to be the most fundamental underlying mechanism.

Ageing as the result of mechanical stress

For structures such as teeth, it is clear that mechanical wear and tear can contribute to the phenotype of ageing. Such effects, however, are likely to be more closely related to chronological time than to biological time.

CONCLUSIONS

We have seen how the concepts of evolutionary biology, as developed in the companion chapter (Chapter 2.1), lead to certain constraints regarding biological mechanisms of ageing. In age-structured populations that reproduce repeatedly, such as human beings and virtually all mammals, there is no theoretical basis for determinative, developmentally programmed mechanisms of senescence. There is no evidence of 'killer genes' designed to limit the lifespans of such organisms. A more plausible scenario is that ageing emerges as a byproduct of gene action, selected on the basis of an enhancement of reproductive fitness. Species-specific lifespans and the complex phenotype of senescence appear to be modulated by a large number of genes. Alleles at such loci, acting in concert with numerous environmental agents, could differentially influence numerous independent and pathogenetically overlapping biological mechanisms of ageing.

The above picture of ageing has important implications for the practice of geriatric medicine. It gives a scientific framework for the common clinical perception that there is enormous individual variation in the patterns of ageing and, consequently, underscores the importance of tailoring preventive medicine and management to the unique susceptibilities and strengths of the individual patient.

On the other hand, the outcome of experimental caloric restriction (yet to be confirmed in primates) argues for a major unification of apparently diverse mechanisms of ageing.

Which view is nearer the truth—the evolutionary/genetic view of numerous independent mechanisms—or the caloric restrictionist view of a universal, yet to be discovered, fundamental mechanism of ageing? Perhaps the answer is that there are indeed a few major pathways and numerous minor pathways, all of which are subject to both genetic and environmental controls.

References

Baynes, J.W. and Monnier, V.M. (ed.) (1989). *The Maillard reaction in aging, diabetes, and nutrition*. Liss, New York.

Bienkowski, R.S. and Baum, B. J. (1983). Measurement of intracellular protein degradation. In *Altered proteins and aging* (ed. R. C. Adelman and G. S. Roth), pp 55–80. CRC Press, Boca Raton, FL.

Burnet, M. (1974). *Intrinsic mutagenesis: a genetic approach to ageing*. Wiley, New York.

Cutler, R.G. (1975). Evolution of human longevity and the genetic complexity governing aging rate. *Proceedings of the National Academy of Sciences (USA)*, **72**, 4664–8.

Davies, K. J. A., Quintanilha, A. T., Brooks, G. A., and Packer, L.

(1982). Free radicals and tissue damage produced by exercise. *Biochemical and Biophysical Research Communications*, **107**, 1198–205.

Dice, J.F. and Goff, S.A. (1987). Error catastrophe and aging: Future directions of research. In *Modern biological theories of aging* (ed. H.R. Warner, R.N. Butler, R.L. Sprott, and E.L. Schneider). *Aging*, Vol. 31, pp. 155–68. Raven Press, New York.

Dubitsky, R., Bensch, K.G., and Fleming, J.E. (1985). Age-related changes in turnover and concentration of a subset of thorax polypeptides from *Drosophila melanogaster*. *Mechanisms of Ageing and Development*, **32**, 311–17.

Finch, C.E. (1976). The regulation of physiological changes during mammalian aging. *Quarterly Review of Biology*, **51**, 49–83.

Finch, C.E., Felicio, L.S., Mobbs, C.V., and Nelson, J.F. (1984). Ovarian and steroidal influences on neuroendocrine aging processes in female rodents. *Endocrine Reviews*, **5**, 467–97.

Friedman, D.B. and Johnson, T.E. (1988). A mutation in the *age*-1 gene in *Caenorhabditis elegans* lengthens life and reduces hermaphrodite fertility. *Genetics*, **118**, 75–86.

Fukuchi, K-I., Martin, G.M., and Monnat, R.J., Jr. (1989). The mutator phenotype of Werner syndrome is characterized by extensive deletions. *Proceedings of the National Academy of Sciences (USA)*, **86**, 5893–7.

Gershon, H. and Gershon, D. (1970). Detection of inactive enzyme molecules in ageing organisms. *Nature*, **227**, 1214–17.

Goldspink, D.F., Lewis, S.E., and Kelly, F.J. (1985). Protein turnover and cathepsin B activity in several individual tissues of foetal and senescent rats. *Comparative Biochemistry and Physiology (B)*, **82**, 849–53.

Harman, D. (1986). Free radical theory of aging: Role of free radicals in the origination and evolution of life, aging, and disease processes. In *Free radicals, aging, and degenerative diseases* (ed. J.E. Johnson, R. Walford, D. Harman, and J. Miquel), pp. 3–49. Liss, New York.

Hart, R.W. and Setlow, R.B. (1974). Correlation between deoxyribonucleic acid excision-repair and life-span in a number of mammalian species. *Proceedings of the National Academy of Sciences (USA)*, **71**, 2169–73.

Hayflick, L. and Moorhead, P.S. (1961). The serial cultivation of human diploid cell strains. *Experimental Cell Research*, **25**, 585–621.

Holehan, A.M. and Merry, B.J. (1986). The experimental manipulation of ageing by diet. *Biological Reviews of the Cambridge Philosophical Society*, **61**, 329–68.

Holliday, R. (1986). Strong effects of 5-azacytidine on the *in vitro* lifespan of human diploid fibroblasts. *Experimental Cell Research*, **166**, 543–52.

Holliday, R. (1987). The inheritance of epigenetic defects. *Science*, **238**, 163–70.

Holt, P.R. and Yeh, K.Y. (1989). Small intestinal crypt cell proliferation rates are increased in senescent rats. *Journals of Gerontology*, **44**, B9–14.

Johnson, T.E. (1987). Aging can be genetically dissected into component processes using long-lived lines of *Caenorhabditis elegans*. *Proceedings of the National Academy of Sciences (USA)*, **84**, 3777–81.

Krauss, S.W. (1981). DNA replication in aging. In *CRC handbook of biochemistry in aging*, (ed. J.R. Florini), pp. 3–8. CRC Press, Boca Raton, FL.

Luckinbill, L.S., Graves, J.L., Reed, A.H., and Koetsawang, S. (1988). Localizing genes that defer senescence in *Drosophila melanogaster*. *Heredity*, **60**, 367–74.

Martin, G.M. (1977). Cellular aging–postreplicative cells. A review (Part II). *American Journal of Pathology*, **89**, 513–30.

Martin, G.M. (1978). Genetic syndromes in man with potential relevance to the pathobiology of aging. *Birth Defects: Original Article Series*, **14**, 5–39.

Martin, G.M. (1979). Proliferative homeostasis and its age-related aberrations. *Mechanisms of Ageing and Development*, **9**, 385–91.

Martin, G.M. (1988). Constitutional genetic markers of aging. *Experimental Gerontology*, **23**, 257–67.

Martin, G.M. *et al.* (1975). Do hyperplastoid lines 'differentiate themselves to death'? *Advances in Experimental Medicine and Biology*, **53**, 67–90.

Martin, G.M., Fry, M., and Loeb, L.A. (1985). Somatic mutation and aging in mammalian cells. In *Molecular biology of aging: Gene stability and gene expression* (ed. R.S. Sohal, L.S. Birnbaum, and R.G. Cutler), pp. 7–21. Raven Press, New York.

Maynard Smith, J. (1965). Theories of aging. In *Topics in the biology of aging* (ed. P.L. Krohn), pp. 1–35. Wiley, New York.

Migeon, B.R., Axelman, J., and Beggs, A.H. (1988). Effect of ageing on reactivation of the human X-linked HPRT locus. *Nature*, **335**, 93–6.

Orgel, L.E. (1963). The maintenance of the accuracy of protein synthesis and its relevance to ageing. *Proceedings of the National Academy of Sciences (USA)*, **49**, 517–21.

Orgel, L. E. (1970). The maintenance of the accuracy of protein synthesis and its relevance to ageing: A correction. *Proceedings of the National Academy of Sciences (USA)*, **67**, 1476.

Pembrey, M.E. (1987). Genetic factors in disease. In *Oxford textbook of medicine* (2nd edn) (ed. D.J. Weatherall, J.G.G. Ledingham, and D. A. Warrell), pp. 4.1–4.47. Oxford University Press.

Pereira-Smith, O.M. and Smith, J.R. (1988). Genetic analysis of indefinite division in human cells: Identification of four complementation groups. *Proceedings of the National Academy of Sciences (USA)*, **85**, 6042–6.

Peto, R., Pavish, S.E., and Gray, R.G. (1986). There is no such thing as ageing and cancer is not related to it. In *Age-related factors in carcinogenesis*, IARC Scientific Publications No. 57 (ed. A. Likachev, V. Anisimov, and R. Montesano), pp. 43–53. IARC, Lyons.

Pryor, W.A. (1986). Oxy-radicals and related species: their formation, lifetimes, and reactions. *Annual Review of Physiology*, **48**, 657–67.

Richardson, A. (1981). The relationship between aging and protein synthesis. In *CRC handbook of biochemistry in aging* (ed. J. R. Florini), pp. 3–8. CRC Press, Boca Raton, FL.

Richardson, A. *et al.* (1987). Effect of age and dietary restriction on the expression of α_{2U}-globulin. *Journal of Biological Chemistry*, **262**, 12821–5.

Rothstein, M. (1982). *Biochemical approaches to aging*. Academic Press, New York.

Seshardi, T. and Campisi, J. (1990). Repression of c-*fos* transcription and an altered genetic program in senescent human fibroblasts. *Science*, **247**, 205–9.

Shepherd, J.C.W., Walldorf, U., Hug, P., and Gehring, W.J. (1989). Fruit flies with additional expression of the elongation factor EF-1α live longer. *Proceedings of the National Academy of Sciences (USA)*, **86**, 7520–1.

Smith, J.R., Spiering, A.L., and Pereira-Smith, O.M. (1987). Is cellular senescence genetically programmed? *Basic Life Sciences*, **42**, 283–94.

Stadtman, E.R. (1988). Protein modification in aging. *Journals of Gerontology*, **43**, B112–20.

Stemerman, M.B., Weinstein, R., Rowe, J.W., Maciag, T., Fuhro, R., and Gardner, R. (1982). Vascular smooth muscle cell growth kinetics in vivo in aged rats. *Proceedings of the National Academy of Sciences (USA)*, **79**, 3863–6.

Sugawara, O., Oshimara, M., Koi, M., Annab, L.A., and Barrett, J. C. (1990). Induction of cellular senescence in immortalized cells by human chromosome 1. *Science*, **247**, 707–10.

Walford, R.L. (1969). *The immunologic theory of aging*. Munksgaard, Copenhagen.

Warner, H.R., Butler, R.N., Sprott, R.L., and Schneider, E.L. (ed.) (1987). *Modern biological theories of aging*. *Aging*, Vol. 31. Raven Press, New York.

Weindruch, R. and Walford, R.L. (1988). *The retardation of aging and disease by dietary restriction*. Thomas, Springfield, IL.

Weraarchakul, N., Strong, R., Wood, W.G., and Richardson, A. (1989). The effect of aging and dietary restriction on DNA repair. *Experimental Cell Research*, **181**, 197–204.

Williams, G.C. (1957). Pleiotropy, natural selection, and the evolution of senescence. *Evolution*, **11**, 398–411.

Wise, P. M., Weiland, N.G., Scarbrough, K., Sortino, M.A., Cohen, I.R., and Larson, G.H. (1989). Changing hypothalamopituitary function: Its role in aging of the female reproductive system. *Hormone Research*, **31**, 39–44.

Yates, F.E. (1988). The dynamics of aging and time: How physical action implies social action. In *Emergent theories of aging* (ed. J. E. Birren, and V. L. Bengtson), pp. 90–117. Springer, New York.

Zs. Nagy, I., Cutler, R.G., and Semsei, I. (1988). Dysdifferentiation hypothesis of aging and cancer: A comparison with the membrane hypothesis of aging. *Annals of the New York Academy of Sciences*, **521**, 215–25.

SECTION 3
Infections

3.1.1 Immunology and ageing

ROY A. FOX

INTRODUCTION

Immune responses play an important role in defence against invading micro-organisms, the usual outcome being neutralization, killing, and a removal of the invader. As a result of such an interaction, immunity develops in the host, which means that future responses will be enhanced by being of greater intensity or by developing more rapidly. With advancing age, there is an increased incidence of infections that result in significant illness and death. Because of the known vulnerability of patients with immune deficiencies to various infections, the question of a link between impaired immunity and infections in elderly people has been raised.

Ageing has been considered to be the most common form of immune deficiency. The development of infection in an aged individual secondary to clinically identifiable immune deficiency is not common, and when it does occur is usually secondary to a specific disease state. The majority of infections occur in individuals without overt immunodeficiency. Much of the work on immune changes that has led to the concept of immune deficiency has involved apparently healthy subjects. In this case, the immune deficiency might be considered to be secondary to ageing. Is the immune deficiency described with normal ageing real? If so, is it of any clinical significance? These questions will be answered as far as possible in this section.

AGEING AND THE IMMUNE RESPONSE

Evidence has been accumulating for a number of years that there is deterioration in both humoral and cellular immunity with advancing age. Much of this work has been done in experimental animals, usually rodents, but a considerable amount of information has accumulated on changes in normal humans. Full details of all this work cannot be reviewed here but will be summarized. Most research on immune changes with age has been cross-sectional in design—for both humans and experimental animals. Work of this kind has revealed that the immune responses in the older age groups are less than in the younger. This is true for both man and experimental animals, and the experimental subjects have been assumed to be healthy and disease-free (normal). This is a difficult concept. Aged animal colonies are defined by the death of the animals, so that a colony that has lost 50 per cent of the animals from natural causes is considered to be old. The survivors have an increased chance of dying, and a significant number are likely to be suffering from a variety of diseases—some of which will prove fatal. The same is true for humans. The diseases that are present may not be detectable by normal clinical means. Thus any effect upon immune responses that might be considered to be a result of ageing may instead be a result of disease.

It could be argued that distinction is unimportant if the net effect is the same and there is a significant decline in immune function, with a concomitant increase in the incidence of infections. However, the issue is important if one is considering the possible contribution of waning immunity to ageing itself.

The immune response can be divided into individual components and each analysed separately. In response to infection various 'arms' of the response come into play, but for certain infections cellular immunity predominates—whilst for others it is the antibody response, humoral immunity. Both cellular and humoral immunity have been shown to deteriorate with age, although the most profound changes have been thought to be within the cellular immune reactions—those mediated by T lymphocytes.

Cell-mediated immunity

In intact humans, cell-mediated immunity is best measured by the delayed-hypersensitivity skin response. It has been known for a long time that skin reactivity to various antigens, for example tuberculin, declines with advancing age. It has also been suggested that this declining immunity contributes to declining health and increase susceptibility to disease—in a study of octogenarians a greater proportion of the anergic than of those with skin reactivity were dead within 2 years of follow-up (Roberts Thompson *et al.* 1974).

Cell-mediated immunity has been measured by means of contact sensitivity to the chemical sensitizing agent dinitrochlorobenzene. The response to this agent wanes with age (Grossman *et al.* 1975). Using a battery of five test antigens (tuberculin, candida, mumps, trichophyton, and varidase) it has been found that the expression of previously established delayed hypersensitivity declines with age. Similar changes are found in experimental animals, and this holds true for graft as well as tumour rejection. In summary, the evidence in man and experimental animals reveals a decline in cell-mediated immunity. The number and proportion of circulating T lymphocytes does not change with age.

These findings from *in-vivo* studies have been confirmed in the laboratory, although it is fair to say that the findings have not always been consistent. *In-vitro* lymphocyte transformation in response to the mitogens phytohaemagglutinin and concanavalin A declines with age. These responses are primarily dependent upon T lymphocytes. The pools of responsive cells appear to diminish, meaning that fewer cells are available, but the cells themselves have a limited reproductive capacity (Weksler *et al.* 1978). Not all studies confirm this (Kay 1979). Kinetic studies of the proliferative response of T cells from older humans point to a defect in the cell cycle in the transition from G0 to G1; the cells are arrested and few enter the S phase. The reasons for the arrest that explains the limited proliferative capacity are not clear; the receptors appear to be normal and the defect is likely to be subsequent to activation. The defect does not appear to be related to mobilization of intracellular calcium (Lustyik and O'Leary 1989), even though studies with high doses of phytohaemagglutin have shown an age-dependent decline in calcium concentration (Miller 1989, 1990). The markers remain the same so that this mosaic remains functional and cannot be picked up by counts of the total number of lymphocyte or of the subpopulation.

It appears that the functional deficits described for ageing T lymphocytes do not affect all cells but rather that there is a mosaic of responsive cells amid non-responsive cells (Miller 1989, 1990). With advancing age there is an increasing number of cells that appear normal but fail to respond to activating stimuli.

There has been extensive work on lymphokine production and responsiveness, and one of the major findings has been an age associated decline in the production of interleukin 2 (IL-2)—the T- and B-cell growth factor (Gillis *et al.* 1981; Nagel *et al.* 1988). Similar findings have been reported from work in experimental animals. It seems that there is defective production of IL-2 and that the cells themselves do not respond normally: addition of IL-2 to cultures of old cells will partially overcome the deficits (Miller 1984; Gottesman *et al.* 1985, Negoro *et al.* 1986). There are limited data on other factors and indeed some of it is conflicting. It would appear that the production of interleukin 1 (endogenous pyrogen) in man does not deteriorate with ageing (Jones *et al.* 1984; Rudd and Bannerjee 1989).

A subset of T lymphocytes is concerned with the control of immune responses (T-suppressor cells). The function of these cells appears to deteriorate with ageing and this is discussed later in the context of autoimmunity.

Table 1 Impaired cell-mediated immunity

In vivo	*In vitro*
Negative skin tests	Impaired lymphocyte transformation
Reduced contact sensitivity	Limited proliferative capacity
Prolonged graft survival	Reduced interleukin-2
Increased tumour incidence	Reduced T-helper and T-suppressor cells

In summary, cell-mediated immunity declines with ageing (Table 1). This part of the immune response is dependent upon T lymphocytes. The nature of the defect is becoming better understood, but the factors that lead to it are not known. Are the changes due to intrinsic changes within the cell, and irreversible, or are they secondary to external factors? This will be discussed in subsequent sections.

Humoral immunity

Most of the research in this area supports the conclusion that cell-mediated immunity is more affected by ageing than humoral immunity. However, changes in antibody production have been observed in both experimental animals and man. Circulating levels of natural antibodies and the primary antibody response in experimental animals decline with age. These changes are thought to be secondary to changes in T cells—an impairment of the T helper-cell response. Thus the most obvious change is seen within the humoral immune response to T-dependent antigens (Smith 1976). However, some of the changes described are probably due to intrinsic changes within the B cells. Antibody that is produced by ageing animals or man may be less effective and there is evidence that avidity declines (Goidl *et al.* 1976).

The functional capabilities of mature B cells in aged individuals appear to be similar to those of younger individuals. However, the number of mature, antigen-responsive B cells is reduced—at least in the mouse. B cells can be generated from the bone-marrow, stem-cell pool but the establishment of mature splenic B cells is clearly inhibited. In aged mice, the responsiveness of available B cells appears to be suppressed by anti-idiotypic antibodies. The B cells remain functional for less time in aged mice than in younger animals and are eliminated by the anti-idiotypic antibodies. Although this down-regulation appears to be the most likely explanation, there could be an intrinsic difference within the B cells of aged animals that accounts for the early loss (Zharhary and Klinman 1987). More recent work does not support the hypothesis of anti-idiotypic suppression but rather that there is a specific defect in the B cells or in the other factors that support B-cell generation (Zharhary 1988).

The changes described above do not have profound clinical implications. When the humoral immune response is examined in healthy individuals it is found that the antibody response to bacterial and viral vaccines does not decline with age. However, if one looks at aged individuals who are not healthy, then there is a decline in response. Fewer of the elderly subjects in many studies undergo seroconversion. When an antibody response is detected the rise in titre is delayed, the maximum attained is lower, and there is an earlier decline. It is concluded that ageing itself does not result in a significant impairment of humoral immune function (Table 2).

Table 2 Ageing and humoral immunity

Decline in natural antibodies
Vaccination intact
Anti-idiotypic suppression reduced mature B cells
Reduced antibody avidity

The lymphoid organs and stem cells

T and B lymphocytes have their origins within the bone marrow. Some studies show that bone-marrow stem cells, like other rapidly dividing cells, do not change significantly with age (Harrison 1975, 1983). However, with advancing age, fewer bone-marrow cells are capable of colonizing the spleen (Coggle and Proukakis 1970), or of seeding the thymus (Tyan 1977; Hirokawa *et al.* 1986), or of replicating (Ogden and Micklem 1976). This does not cause a functional deficit because the total number of nucleated cells in the bone marrow increases with age (Kay *et al.* 1979). One concludes that with the passage of time some functions of stem cells deteriorate, most particularly those concerned with self-repair or replication.

The decline in bone-marrow precursor cells is not restricted to the B cells and an anaemia of senescence due to declining haemopoietic ability has been reported (Lipschitz *et al.* 1981). In this situation there is also a slight degree of leucopenia.

The changes are slight and, as discussed in the previous section, do not become manifest as impaired antibody production. Stem cells from old animals retain the potential to develop fully but may be prevented from doing so. When old cells are taken from the old environment and transplanted into a young environment, and given time to recover, they can be shown to respond to various stimuli as efficiently as transplanted cells from younger animals (Harrison *et al.* 1977; Gutowski *et al.* 1984). This does not happen if old cells are placed in an old environment or if there is an insufficient period for recovery.

The bone-marrow cells from old mice retain the potential to generate T- and B-cell repertoires that are indistinguishable from those of young mice. These stem cells are characteristically more sensitive than younger cells and have a heightened sensitivity to the suppressive influences of the environment. The involuted thymus of old mice retains the ability to assist in the maturation of T cells, and there is no deterioration in the production of thymic hormones.

Mucosal immunity

In experimental animals the response to intraperitoneal injection of trinitrophenylated bovine gammaglobulin declines with age if measured in the spleen and peripheral lymph nodes, but remains elevated if assessed in the mesenteric nodes (Szewczuk *et al.* 1981). The work of Goidl and his coworkers has shown that with advancing age there is a loss of high-avidity, plaque-forming cells (Goidl *et al.* 1976). This has been confirmed by others and shown to occur at approximately 8 months of age in experimental mice, but that the same loss does not occur within the mesenteric nodes. In this site these high-avidity cells remain unchanged, at least to 24 months (Szewczuk and Campbell 1981). These findings indicate that the systemic and mucosal-associated immune systems do not have parallel, age-associated defects in T cell-dependent antibody formation.

The lack of impairment can also be observed in the secondary responses, with the response to T-independent antigens, and for all types this holds true for both intragastric and intraperitoneal immunization. If down-regulation is secondary to rising levels of auto anti-idiotypic antibodies, then these antibodies are not found in the mesenteric nodes.

Antibody responses in nasal secretions in response to influenza virus vaccine appear to be adequate in old age (Kluge and Waldman 1979). Other work has suggested that there is a small but significant decline in the levels of secretory IgA (Alford 1968).

The mucosal system appears to remain relatively free of age-associated defects. The same stem cells feed into the mucosal immune systems as into the systemic systems, which suggests that any intrinsic defect is essentially irrelevant to any deterioration in function that is found. It is of interest that the small intestine, which retains a fully functioning immune system during ageing, has a very low incidence of neoplasia and shows no increased incidence of cancer with advancing age.

Autoimmunity

The waning of humoral immunity has been described and the major change that is found appears to be in the thymic-dependent antibody response. This is explained by impaired T-helper cell function. It is thought that waning T-suppressor cell activity contributes to the rise with age in production of autoantibodies, a phenomenon that has been well described (Hooper *et al.* 1972).

Although there is conflicting evidence with regard to suppressor-cell activity, the majority of experimental results are consistent with this hypothesis. It is postulated that the B cells which are producing antibodies to self are not suppressed. Cohen and Ziff (1977) have suggested that there is increased stimulation of B cells to produce autoantibodies—by polyclonal activators. The nature of these activators has not been clearly defined, although evidence has been put forward that this is related to increased circulating levels of endotoxin (Horan and Fox 1984). The endotoxin that reaches the systemic circulation may act on a variety of target tissues, including the cells of the immune system. The result is polyclonal B-cell activation and autoantibody production, as well as adjuvanticity (Horan *et al.* 1984). Endotoxin is also cytotoxic in higher concentrations and may cause the release of sequestered antigens, resulting in further enhancement of autoantibody production (Habicht 1987). Autoantibody production increases because of the breakdown of self-tolerance, and the responsiveness to foreign antigens may be reduced by prior or prolonged exposure to endotoxin (Fujiwara and Kariyone 1981; Venkataraman and Scott 1979).

The increased autoantibody production is not paradoxical and appears to result from the loss of tolerance to self, the release of sequestered antigens—the B-cell activation by the polyclonal activators and the loss of suppressor-cell activity.

The immune theory of ageing states that changes in immune function are primary and that many or most other age changes are secondary to this, the age changes in other organs being secondary to autoimmune damage (Walford 1969; Burnet 1970). Most of the evidence cited in this review has supported the hypothesis that the major changes in immune function are secondary to changes elsewhere. It is conceivable that the altered autoimmunity that results from loss of suppressor activity and increased B-cell proliferation could result in organ damage. There is little evidence for this, although the incidence of diseases of unknown aetiology, such as hypothyroidism or pernicious anaemia, does increase with age. It is concluded that the immune changes do not contribute to the process of ageing.

Immediate hypersensitivity

Immediate hypersensitivity reactions are mediated by immunoglobulin E. IgE binds to mast cells by its heavy chain, and when the receptors bind with an allergen the configurational change triggers the mast cell to release its vasoactive substances, which account for the physiological changes of the allergic reaction. Little is known about this and ageing.

UNDERSTANDING IMPAIRED IMMUNITY

The observation that mucosal immunity remains intact in old age implies that there is nothing wrong with the stem cells. It appears to be more likely that any change that occurs with ageing is secondary to extrinsic effects. What is the explanation for this?

The abnormalities that have been described reveal that the major changes are of cell-mediated immunity or delayed hypersensitivity—the thymic-dependent immunity (Fig. 1).

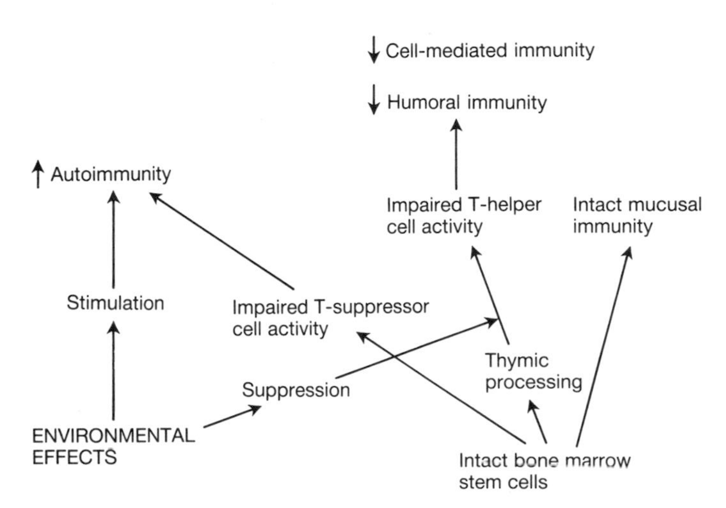

Fig. 1 Understanding the immune changes with ageing.

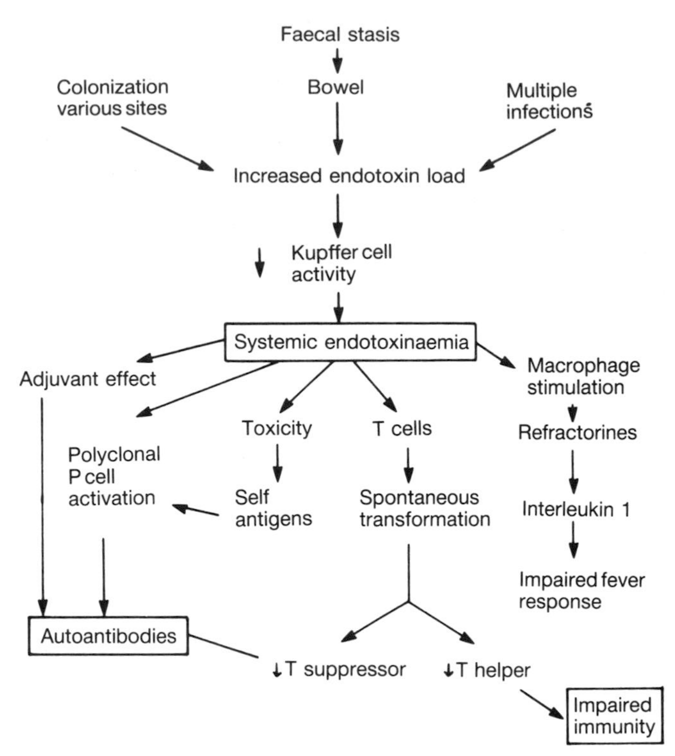

Fig. 2 Hypothesis for the role of endotoxin in impaired immunity.

There are changes within the humoral immune system but the most obvious are with thymic-dependent antibody responses. It has been suggested by many that these changes can best be explained by thymic involution; there is atrophy of the cortical tissue of the thymus as well as of the epithelial cells (Boyd 1932). This classical study, which is always cited, was done on a limited number of specimens obtained *post mortem* and it is quite possible that some of the changes were artefactual and related to the process of dying. Some recent reports reveal that even though the gland appears vestigial it retains full function (Chang and Gorczynski 1984*a*, *b*), as do the T cells (Miller 1984). The explanation has to be sought elsewhere, discussed in the previous section on autoimmunity, and it is suggested that most of the changes that occur can be related to the effects of increased circulating levels of endotoxin (Fig. 2). In this particular hypothesis it is suggested that there are age-associated changes in the liver Kupffer cells, which reduces their efficiency, and that there is an increased load of endotoxin from the colon and from other sites of bacterial colonization. The endotoxin has multiple effects.

There are other possibilities for an environmental effect, and the evidence from work already cited is that the internal environment of the aged differs from that of the young. It is of interest that the impaired cell-mediated immunity and the increased autoantibody production can be delayed, in the same way as other ageing effects, by calorie restriction. It has been known for more than 50 years that calorie restriction prolongs the life-span of rats (McCay *et al.* 1935); it will also delay the onset of autoimmunity and other immune changes (Fernandes 1984). These observations take us back to the effects of thymic involution, as this has been shown to be retarded by calorie restriction. Furthermore, exogenous thymic hormones can increase the numbers of circulating immature T cells in aged individuals and also enhance the immune responses (Pandolfi *et al.* 1983).

Another significant observation concerns the importance of nutrition. Elderly individuals with impaired immunity have important nutritional deficiencies and correction of these with supplements has resulted in enhancement of skin reactivity and *in-vitro* cellular immunity (Chandra *et al.* 1983).

INFECTION AND IMPAIRED IMMUNITY

Infections are a major cause of illness and death amongst elderly people. Respiratory infections are the fourth leading cause of death, and urinary tract infections afflict 15 per cent of elderly people.

There is no doubt that in children with congenital immunodeficiency states, and in adults with acquired immunodeficiency secondary to various diseases, the increased incidence of infections and of death from these infections can be linked directly to the immunodeficiency. There are no such clear links in elderly subjects. The important infections that occur amongst them are bacterial, but do not involve the type of organisms found in children with absent cell-mediated immunity. Zoster is a common problem in ageing individuals and it is thought that the infection occurs as a result of waning cellular immunity. There is some experimental evidence to support this; *in-vitro* cell-mediated immunity is reduced in patients with zoster when compared with controls, whilst antibody production is unchanged (Miller 1980). Few elderly patients develop widespread zoster infections and successful eradication is dependent upon cell-mediated immunity. Although waning immunity is linked to the cause, intact immunity prevents serious disease and promotes recovery—an apparent paradox.

The impaired immunity associated with ageing is not in itself enough to explain the increased incidence of infections. However, with ageing there is an increased incidence of various disease states, and increasing debility and decreasing reserve are likely to contribute to a decreased resistance to infection in a variety of ways.

IMPAIRED IMMUNITY AND OTHER DISEASES

Intact immune mechanisms, particularly T-dependent, are thought to be of great importance in the defence against cancer. It is postulated that as cells undergo malignant transformation and new antigens are expressed, an intact immune system will recognize the changes and eradicate the offending antigens. In the section on mucosal immunity it was pointed out that the immune system of the small intestine remains intact and that there is no increase in the incidence of cancer with age. This observation is consistent with the hypothesis that waning immunity contributes to increased cancer incidence.

'Natural killer' cells are thought to be important in defence against tumour cells and these have been extensively studied in relation to ageing. Various studies have shown that natural killer-cell activity may diminish with ageing, or remain the same, or increase (Solomon *et al.* 1988). It is well recognized that with ageing the response to stress is diminished and that there is less reserve. A recent observation with regard to natural killer-cell activity is of interest. Exercise has been shown to increase total leucocyte count, killer-lymphocyte activity, plasma interferon levels, interleukin-1 levels, and natural killer-cell activity. A study of the effect of exercise on natural killer-cell activity in older subjects found that ageing does not have a deleterious effect (Fiatarone *et al.* 1989); baseline and stimulated immunological functions were as great as or greater than in young subjects. This observation strengthens the argument that ageing itself has no major effect on immune function. However, the aged may well have impaired immunity for other reasons and this study leaves open the possibility that immobility and lack of exercise results in a failure periodically to boost the immune responses and particularly natural killer-cell activity, thus allowing the development of tumours and their new antigens.

Another area that needs examination is the immune response to oncogenic viruses. One study has shown that there is a loss of resistance to murine (Malony) sarcoma virus with advancing age and that this is correlated to an increased incidence of tumour growth (Pazmino and Yuhas 1973).

If impaired cellular immunity is linked to an increased incidence of cancer, then one might expect that boosting the immune response would lower the incidence. This appears to be the case, and in the experimental animal prolonged administration of thymic hormone reduces the incidence of spontaneous tumours (Anisimov *et al.* 1982). In experiments on increasing longevity by calorie restriction it was found that the incidence of tumours fell (Cohen 1979; Good *et al.* 1980; Weindruch 1989) and cellular immunity was maintained (Miller and Harrison 1985).

CONCLUSION

The immune system is complex and appears to become more complex with each passing year as more is learned. The immune system can be taken apart and studied in the test tube. Diseases and accidents of nature allow the system to be better understood and specific defects described. The immune system represents a series of complex multicellular homeostatic mechanisms. It is well recognized that it is within such interactive control mechanisms that one can first detect the effects of the wear and tear of ageing.

It seems likely that the waning immunity which results from intrinsic ageing itself may contribute in part to important and difficult clinical problems encountered in geriatric practice. Such changes become more important when they interact with other deleterious influences on immune function that are common in old age.

References

Alford, R.H. (1968). The effects of bronchopulmonary disease and aging on human nasal secretion IgA concentration. *Journal of Immunology*, **101**, 984–8.

Anisimov, V.N., Khavinson, V.K.H., and Morosov, V.G. (1982). Carcinogenesis and aging. *Mechanisms of Ageing and Development*, **19**, 245–58.

Boyd, E. (1932). The weight of the thymus gland in health and disease. *American Journal of Diseases of Children*, **43**, 1162–214.

Burnet, F.M. (1970). An immunological approach to ageing. *Lancet*, **ii**, 358–60.

Chandra, R.K., Joshi, P., and Au, B. (1983). Nutrition and immunocompetence of the elderly. Effect of short term nutritional supplementation on cell mediated immunity and lymphocyte subsets. *Nutrition Research*, **2**, 223–32.

Chang, M-P. and Gorczynski, R.M. (1984*a*). Peripheral (somatic) expansion of the murine cytoxic T lymphocyte repertoire. I Analysis of diversity in recognition repertoire of alloreactive T cells derived from the thymus and spleen of adult or aged DBA/2J mice. *Journal of Immunology*, **133**, 2375–80.

Chang, M-P. and Gorczynski, R.M. (1984*b*). Peripheral (somatic) expansion of murine cytotoxic lymphocyte repertoire. II Comparison of diversity in recognition repertoire of alloreactive T cells in spleen and thymus of young or aged DBA/2J mice transplanted with bone marrow cells from young or aged donors. *Journal of Immunology*, **133**, 2381–9.

Cohen, B.J. (1979). Dietary factors affecting rats used in ageing research. *Journal of Gerontology*, **34**, 803–7.

Cohen, P.L. and Ziff, M. (1977). Abnormal polyclonal B cell activators in NAB/NZW F1 mice. *Journal of Immunology*, **119**, 1534–7.

Coggle, J.E. and Proukakis, C. (1970). The effect of age on the bone marrow cellularity of the mouse. *Gerontologia*, **16**, 25–9.

Fernandes, G. (1984). Nutritional factors: Modulating effects on immune function and aging. *Pharmacological Review*, **36**, 123–9s.

Fiatarone, M.A., Morley, J.E., Bloom, E.T., Benton, D., Solomon, G.F., and Makinodan, T. (1989). The effect of exercise on natural killer cell activity in young and old subjects. *Journal of Gerontology*, **44**, M37–45.

Fujiwara, M. and Kariyone, A. (1981). Lipopolysaccharide-induced autoantibody II. Age related changed in plaque-forming cell response to bromelain-treated syngeneic erythrocytes. *International Archives of Allergy and Applied Immunology*, **66**, 161–72.

Gillis, S., Kozar, R., Durante, M., and Weksler, M.E. (1981). Immunological studies of aging. Decreased production of and response to T cell growth factor by lymphocytes from aged humans. *Journal of Clinical Investigation*, **67**, 937–43.

Goidl, E.A., Innes, J.B., and Weksler, M.E. (1976). Immunological studies of aging II. Loss of IgG and high avidity plaque-forming cells and increased suppressor cell activity in aging mice. *Journal of Experimental Medicine* **144**, 1037–48.

Good, R.A., West, A. and Fernandes, G. (1980). Nutritional modulation of immune responses. *Federation Proceedings*, **39**, 3098–104.

Gottesman, A., Walford, R.L., and Thorbeche, G.J. (1985). Proliferative and cytotoxic immune functions in aging mice: exogenous interleukin-2 rich supernatant only partially restores alloreactivity in vitro. *Mechanisms of Ageing and Development*, **31**, 103–13.

Grossman, J., Baum, J., Fusner, J., Condemi, J. and Gluckman, J. (1975). The effect of ageing and acute illness on delayed hypersensitivity. *Journal of Allergy and Clinical Immunology*, **55**, 268–75.

Gutowski, J.K., Innes, J., and Weksler, M.E. (1984). Induction of DNA synthesis in isolated nuclei by cytoplasmic factors. *Journal of Immunology*, **132**, 559–62.

Habicht, G.S. (1987). *Aging and acquired immunological tolerance in mice in aging and the immune response. Cellular and humoral aspects,* Immunology Series, vol. 31 (ed. E.A. Goidl, Marcel Dekker, New York).

Harrison, D.E. (1975). Normal functions of transplanted marrow cell lines from aged mice. *Journal of Gerontology*, **30**, 279–85.

Harrison, D.E. (1983). Long-term erythropoietic repopulating ability of old, young and fetal stem cells. *Journal of Experimental Medicine*, **157**, 1496–504.

Harrison, D.E., Astle, C.M. and Lerner, C. (1977). Stem cell lines from old immunodeficient donors give normal responses in young recipients. *Journal of Immunology*, **118**, 1223–7.

Hirokawa, K., Kubo, S., Utsayama, M., Kurashima, C., and Sado, T. (1986). Age related change in the potential of bone marrow cells to repopulate the thymus and splenic T cells in mice. *Cell Immunology*, **100**, 443–51.

Hooper, B., Whittingham, S., Matthews, J.D., MacKay, I.R., and Cudnow, D.H. (1972). Autoimmunity in a rural community. *Clinical and Experimental Immunology*, **12**, 79–87.

Horan, M.A. and Fox, R.A. (1984). Aging and the immune response—a unifying hypothesis. *Mechanisms of Ageing and Development*, **26**, 165–81.

Horan, M.A., Leahy, B.C., Fox, R.A., Streeton, T.B. and Hearney, M. (1984). Immunological abnormalities in patients with chronic bronchial suppuration, a possible relationship with endotoxaemia. *British Journal of Chest Diseases*, **78**, 66–74.

Jones, P.G., Kauffman, C.A., Bergman, A.G., Hayes, C.M., Kluger M.J., and Cannon, J.G. (1984). Fever in the elderly. Production of leucocyte pyrogen by monocytes from elderly persons. *Gerontology*, **30**, 182–7.

Kay, M.M.B. (1979). An overview of immune aging. *Mechanisms of Ageing and Development*, **9**, 39–59.

Kay, M.M.B., Mendoza, J., Diven, J., Denton, T., Union, N., and Lajines, M. (1979). Age related changes in the immune system of mice of medium and long-lived strains and hybrids.I Organ, cellular and activity changes. *Mechanisms of Ageing and Development,* **11**, 295–346.

Kluge, R.M and Waldman, R.H. (1979). Antibody to swine influenza virus in serum and nasal secretions of volunteers over the age of 55 years. *Journal of Infectious Diseases*, **140**, 635–6.

Lipschitz, D.A., Mitchell, C. and Thompson, C. (1981). The anemia of senescence. *American Journal of Hematology*. **11**, 47–54.

Lustyik, G. and O'Leary, J.J. (1989). Aging and the mobilization of intracellular calcium by phytohemagglutinin in human T cells. *Journal of Gerontology,* **44**, B30–6.

McCay, C., Crowell, M., and Maynard, L. (1935). The effect of retarded growth upon the length of life and upon ultimate size. *Journal of Nutrition*, **10**, 63–79.

Miller, A.E. (1980). Selective decline in cellular immune response to varicella zoster in the elderly. *Neurology*, **30**, 582–7.

Miller, R.A. (1984). Age associated decline in precursor frequency for different T cell mediated reactions with preservation of helper or cytotoxic effect per precursor cell. *Journal of Immunology*, **132**, 63–8.

Miller, R.A. (1989). The cell biology of aging: immunological models. *Journal of Gerontology*, **44,** B4–8.

Miller, R.A. (1990). Aging, and the immune response. In *Handbook of the biology of aging*, In (ed. E.R. Schneider and J.W. Rowe. pp. 157–71. Academic Press, San Diego.

Miller, R.A. and Harrison, D.E. (1985). Delayed reduction in T cell precursor frequencies accompanies diet-induced lifespan extension. *Journal of Immunology*, **134**, 1426–9.

Nagel, J.E. *et al.* (1988). Decreased proliferation, interleukin-2 synthesis and interleukin-2 expression are accompanied by decrease in RNA expression in phytohemagglutinin stimulated cells from elderly donors. *Journal of Clinical Investigation*, **81**, 1096–102.

Negoro, S. *et al.* (1986). Mechanisms of age-related decline in antigen-specific T-Cell proliferative response. IL-2 receptor expression and IL-2 induced proliferative response of purified TAC-positive T cells. *Mechanism of Ageing and Development*, **36**, 223–41.

Ogden, D.A., and Mickliem, H.S. (1976). The fate of serially transplanted bone marrow cell populations from young and old donors. *Transplantation*, **22**, 287–93.

Pandolfi, F. *et al.* (1983). T-dependent immunity in aged humans. II Clinical and immunological evaluation after three months of administering a thymic extract. *Thymus,* **5**, 235–40.

Pazmino, N.H. and Yuhas, J.M. (1973). Senescent loss of resistance to murine sarcoma (Malony) virus in the mouse. *Cancer Research*, **33**, 2668–72.

Roberts Thompson, I.C., Whittingham, S., Youngchaiyud, U., and Mackay, I.R. (1974). Ageing, immune response and mortality. *Lancet*, **ii**, 24–6.

Rudd, A.G. and Bannerjee, D.K. (1989). Interleukin 1 production by human monocytes in ageing and disease. *Age and Ageing*, **18**, 43–6.

Smith, A.M. (1976). The effects of age upon the immune response to type III pneumococcal polysaccharide and bacterial lipopolysaccharide in BALB/c, SJL/J and C3H mice. *Journal of Immunology*, **116**, 469–74.

Solomon, G.F., Fiatorone, M.A., Benton, D., Morley, J.E., Bloom, E., and Makinodan, T. (1988). Psychoimmunologic and endorphin function in the aged. *Annals of the New York Academy of Sciences*, **521**, 43–58.

Szewczuk, M., and Campbell, R.J. (1981). Differential effect of aging on the heterogeneity of the immune response to a T-dependent antigen in systemic and mucosal-associated lymphoid tissues. *Journal of Immunology*, **126**, 472–7.

Szewczuk, M.R., Campbell, R.J. and Jung, L.K. (1981). Lack of age associated immune dysfunction in mucosal-associated lymph nodes. *Journal of Immunology*, **126**, 2200–4.

Tyan, M.L. (1977). Age related decrease in mouse T cell progenitors. *Journal of Immunology*, **118**, 846–51.

Venkataraman, M. and Scott, D.W. (1979). B cell subsets responsive to fluorescein conjugated antigens. III Differential effect of *E. coli* lipopolysaccharide on T-dependent and T-independent response *in vivo*. *Immunology*, **38**, 519–27.

Walford, R.L. (1969). *The immunologic theory of aging*. Munksguaard, Copenhagen.

Weindruch, R. (1989). Dietary restriction, tumors, and aging in rodents. *Journal of Gerontology*, **44**, 62–71.

Weksler, M.E. (1978). The influence of immune function on lifespan. *Bulletin of the New York Academy of Medicine*, **54**, 964–9.

Weksler, M.E., Innes, J.B. and Goldstein, G. (1978). Immunological studies of aging. IV. The contribution of thymic involution to the immune deficiencies of aging mice and reversal with thymopoietin. *Journal of Experimental Medicine*, **148**, 996–1006.

Zharhary, D. and Klinman, N.R. (1987). The effects of aging on murine B-cell responsiveness. In *Aging and the immune response*, (ed. E.A. Goidl). Marcel Dekker, New York.

Zharhary, D. (1988). Age related changes in the capability of the marrow to generate B cells. *Journal of Immunology*, **141**, 1863–9.

3.1.2 Changes in non-immunological mechanisms of host defence in older persons

ROY A. FOX

When an individual is exposed to organisms with the potential for infection or clinical disease there are many factors that determine the outcome. In simple terms it will depend upon the number of organisms and their virulence, balanced against the resistance of the host. Minor changes in defence, resulting in reduced efficiency, might lead to infection if other circumstances lead to increased exposure to pathogens. Because of such changes the dose of inoculum required may be smaller. In this sense, many aged individuals can be considered to be predisposed to infection owing to a variety of changes in host defence. The changes in the immune system are considered separately; there are other important means of defence against infection, which will be discussed in this section.

MECHANISMS OF HOST DEFENCE

We live in an environment where micro-organisms are plentiful and the external surfaces, lined by epithelium, are constantly exposed to microbes. Colonization of different parts of the body is normal. This 'normal' population of micro-organisms is important in defence, for example within the bowel, the vagina, or on the skin surface, and prevents overgrowth of undesirable or pathogenic organisms. Changes in these populations may lead to infection of clinical significance.

The total population of organisms and thus of potential pathogens can be kept to a minimum by preventing attachment to epithelial surfaces, or if attached, by removal. Attachment can be prevented by prior occupation of binding sites by the commensal population already mentioned, or by agents that block binding.

Surfaces can be swept free of offending pathogens by a combination of voluntary and involuntary actions. For example, a voluntary action would be grooming or washing the skin. Involuntary mechanisms would include desquamation of the skin or the rapid flow of urine over the epithelium of the ureter or urethra washing organisms away. The next important defence is the natural barrier to infection present by an intact epithelium—the skin or the mucosa of the gastrointestinal, respiratory, or genitourinary tracts. Within these natural barriers and the tissues beneath them, other important mechanisms come into play, including the inflammatory responses mediated by a variety of chemical agents, mostly products of inflammatory cells, and the phagocytic cells themselves (Table 1).

Each of these forms of defence can be compromised in a number of ways and it is worthwhile to look at these in relationship to ageing and to diseases of the aged.

Table 1 Non-immunological mechanisms of defence

Commensal microbial population
Prevention of attachment—secretions
Mechanical removal
Involuntary—urine flow, mucus
voluntary—grooming, washing
Natural barrier
Intact epithelium, skin, mucous membrane
Inflammatory response
Phagocytic cells
Chemical mediators of inflammation

CHANGES WITH AGEING AND DISEASE

Ageing is associated with a decline in efficiency of natural defence mechanisms. It is not clear, to what extent these changes are extrinsically caused or due to intrinsic ageing processes. In considering them each important site of infection will be considered separately.

Respiratory system

Micro-organisms are inhaled into the respiratory tract and there is an aerodynamic filtration system in the upper airway. This natural barrier is lost by tracheostomy and this is the major reason why patients with tracheostomies are at increased risk of infection. A number of pathophysiological changes occur with advancing age that result in a reduction in the forced expiratory volume and vital capacity and an increase in dead space. Breathing becomes dependent upon the abdominal muscles because of the loss of rib-cage elasticity. Such changes may reduce the efficiency of the air filtration system and are likely to be more apparent in those who are less mobile and more functionally impaired. Stillness favours bacterial overgrowth, and may be an important factor in the increased incidence of chest infections in the elderly. Mobilization plays an important role in helping elderly patients in hospital to recover from chest infections, and patients that can be readily mobilized do better.

The mucus covering the membranes of the respiratory tract provides the next line of defence. Microbes trapped in mucus are prevented from attaching to epithelium and can be cleared by ciliary action and by coughing. There is no evidence that ageing reduces the production of mucus; in fact, prolonged exposure to environmental pathogens and many years of smoking may lead to the chronic mucus overproduction of chronic bronchitis. But the character of the secretions changes and the pathological changes of chronic lung disease lead to impaired mucociliary clearing. This is important because nocturnal aspiration of micro-organisms is normal throughout life. In aged individuals, clearance of aspirated organisms becomes less efficient and ageing and disease contribute to this. If the load of organisms is increased by oropharyngeal colonization, aspiration becomes a greater danger. In such predisposed individuals a chest infection may be precipitated by additional insults such as reduced mobility, drugs that reduce the cough reflex, or by dehydration and other factors making secretions more viscous. In a retrospective case-control study of nosocomial pneumonia in a geriatric assessment unit, we have found the use of nocturnal hypnotics to be significantly more frequent in the patients with nosocomial pneumonia than in the controls. We have also noticed that the patients with pneumonia are more debilitated, less mobile, score lower on the Barthel Index, and are more likely to be incontinent of urine (B. H. Clarke, K. MacPherson, and R. A. Fox, unpublished work).

If stroke or other neurological disorders affect the swallowing mechanism, aspiration of food or gastric contents may occur. In other situations the swallowing mechanism is intact but may be overwhelmed. For example, reflux of gastric contents through the lower oesophageal sphincter and into the mouth at night may lead to aspiration, and repeated reflux at night is a recognized cause of nocturnal wheezing and recurrent chest infections. It is of interest that for intubated patients in intensive care, the use of agents like H_2-antagonists, which reduce gastric acidity, also promotes colonization of the upper gastrointestinal tract and predisposes to aspiration pneumonia.

Inhaled bacteria that reach the alveoli are usually killed by pulmonary macrophages. Macrophage function can be impaired by a number of factors, such as smoking, alcohol ingestion, renal failure, and various drugs, but there is no evidence that intrinsic ageing has a deleterious effect. Chronic lung disease results in significant changes in the respiratory tract, and for most patients there are frequent exacerbations. The 'commensal population' is altered to include more potential pathogens, putting the patient at risk.

Urinary tract

Almost all urinary tract infections ascend from the perineal microbial population. With ageing the natural barrier to infection provided by a closed urethra becomes less efficient so that there is an increased likelihood of ascent from the perineum to the bladder. The female urethra is short, but at rest is normally closed. With advancing age there is mucosal atrophy, the epithelium becoming thinner, and there is loss of smooth muscle with replacement by connective tissue. This leads to laxity and reduced efficiency of this barrier. The same changes that predispose to urinary incontinence predispose to infection. A system that allows escape of urine also allows entrance of micro-organisms. Although the male urethra is longer and proves to be more efficient in early and middle adult life in preventing infection, deterioration in the structure reduces this efficiency in old age. In the eighth and ninth decades the incidence of urinary tract infection and of incontinence of urine in men approaches that of women.

If urine flow is reduced there is less efficient flushing away of contaminating organisms and an increased risk of infection. Reduced fluid intake is a real problem for patients in long-term care institutions. With advancing age there is often increasing trabeculation and the formation of diverticula in the bladder. These changes produce local stasis and, with other changes that increase the residual volume of urine and stasis, there is increased colonization and an increased likelihood of infection.

In men the most important cause of urinary stasis is outflow obstruction due to prostatic hypertrophy. In women, prolapse can lead to outflow obstruction as well as incontinence. In both sexes, neurological disease can lead to an atonic bladder with high residuals and infection. Constipation may also result in outflow obstruction or atony due to reflex relaxation. It is well recognized that in young women constipation can precipitate urinary tract infection and the same is probably true for the elderly. These conditions need to be recognized and managed in order to maintain a sterile urine (Table 2).

Table 2 Breakdown of natural barriers to infection of urinary tract

Age changes of urethra—mucosal atrophy, etc.
Reduced urinary flow
Dehydration
Stasis—obstruction
Instrumentation
Altered perineal microbial population
Ageing
Poor hygiene
Faecal incontinence
Atrophy

The frequency of urinary tract infections appears to be higher in patients with diabetes. This condition appears to alter the microbial population as well as reducing inflammatory responses.

Genital infections

Pathophysiological changes of ageing are particularly important in genital infections of women. Endocrine changes lead to epithelial thinning—reducing the efficiency of this natural barrier. At the same time, adherence of micro-organisms is more likely. The commensal microbial population of the vagina is a mixture of Gram-negative aerobes and anerobes. There is little evidence available on changes after the menopause but there appears to be a reduction of glycogen-splitting bacteria and so a reduction in acidity. This may lead to pathogenic, Gram-negative bacilli becoming more common in the vagina. Genitourinary secretions are reduced in postmenopausal women so that natural bactericidal activity is reduced.

The cyclical shedding of the endometrium is an important natural barrier against infection which ceases at the menopause (Horan *et al.* 1983).

Atrophic vaginitis and atrophy of the perineal skin results from postmenopausal endocrine changes in some women. Pruritus with scratching leads to further loss of the integrity of this natural barrier. This may result in colonization by pathogens and predispose to genital and urinary tract infection. A foreign body, for example a pessary inserted to correct uterine prolapse can lead to infection.

Gastrointestinal tract

A variety of natural barriers to infection of the gastrointestinal tract deteriorate with age. There is thinning of the mucosa and a reduction in mucus production. Gut motility may be reduced, predisposing to stasis and infection. Alteration of the normal flora may occur as a result of antibiotic administration.

Perhaps the most important barrier is gastric acidity, which kills many organisms. Between meals the stomach is normally free of bacteria. There are significant changes with age, and achlorhydria becomes progressively more common, being found in up to one-third of the people aged over 60. Acid production may be depleted by gastritis or pernicious anaemia. Chronic atrophic gastritis increases in frequency with age, and it has been shown that the degree of gastritis correlates with the achlorhydria. Treatment of other problems, such as acid-pepsin disease, with powerful antacids also reduces acid production. Loss of the natural acid barrier opens the gastrointestinal tract to infection. The importance of the acid barrier has been shown in outbreaks of enteric infection such as cholera, and also with bacterial dysentery in a geriatric hospital ward (Horan *et al.* 1984). In the latter case all patients who developed shigella infection had a reduction in acid brought about by the administration of drugs or disease. Another possible result of reduction in gastric acid is bacterial colonization of the upper intestine, resulting in malabsorption (Roberts *et al.* 1977).

The large bowel is normally heavily contaminated but this commensal population acts as a barrier to infection. The barrier breaks down when there is stasis or when there is a profound change in the microbial flora. Prolonged difficulty in defaecation with high intraluminal pressure leads to the development of diverticula. Blind sacs contain material that does not change or move out and may lead to local infection with the potential for systemic spread. The use of powerful broad-spectrum antibiotics may wipe out the natural commensal population, with bacterial overgrowth by harmful organisms—resulting in local and systemic infection.

Skin

Skin and soft tissue infections are not uncommon in elderly people. Aged skin appears dry and wrinkled. The epithelial layer thins and the support tissues alters and loses elasticity. There is a reduction in activity by the eccrine and apocrine sweat glands although the production of sebum does not change significantly. Minor trauma is more likely to cause damage and loss of integrity of the natural barrier. It is thought that bacterial colonization increases with ageing owing to changes in the state of hydration and in lipid content.

Senile pruritus is a common condition that threatens the integrity of the natural barrier of the skin through scratching. Pruritus needs to be treated to prevent scratching.

Any disease that results in immobility threatens the integrity of skin through pressure effects. Pressure sores provide a site for significant colonization and possible systemic spread. Patients with large pressure sores commonly die from overwhelming infection.

DRUGS

The natural barriers that have been discussed may be compromised by the use of drugs; indeed the therapeutic effect that is strived for may produce the unwanted breakdown of a natural barrier. The drugs with the most potential for damage to the natural barriers against infection are antibiotics. Antibiotics are widely prescribed for elderly patients, usually with good reason, but antibiotics can destroy the normal commensal populations of bacteria and cause superinfection. Perhaps the most common example would be the development of post-antibiotic diarrhoea, and infection by *Clostridium difficile*. Apart from such specific infections, the frequent use of antibiotics within a closed environment like a nursing home will lead to the development of a population of bacteria more likely to contain pathogenic organisms. Prescribing antibiotics to treat asymptomatic bacteruria in elderly people in institutions does not alter the overall prevalence of bacteruria but does increase the number of adverse drug reactions and promotes the emergence of resistant organisms (Nicolle *et al.* 1983, 1987). Antibiotic use should be carefully monitored in institutional settings and there needs to be more education of staff on this topic.

As already discussed, drugs that reduce the cough reflex or are sedating might increase the risk of pneumonia. It is wise practice to use such drugs sparingly, particularly if there are other factors that might be compromising the natural barriers against infection. Antacids, and particularly the potent H_2-antagonists, reduce gastric acidity and thus remove this natural barrier to infection. These drugs are now very widely prescribed.

CHANGES IN THE MICROBIAL POPULATIONS

Individuals admitted to hospital are exposed to a variety of new potential pathogens and elderly patients are at particular risk. Because of their underlying problems they are most likely to undergo invasive procedures that penetrate natural barriers, perhaps the most important being the use of intravascular devices. Nosocomial infection is preceded by successful colonization of the mucosa or skin, which provides a source of infection for the patients themselves and a reservoir for other patients (Phair and Reisberg 1984).

Infection is usually spread through hand carriage by health care personnel, and an important way of reducing infections would be to institute hand washing after contact with each patient. The rate of hand carriage is increased when personnel are dealing with excretions (Steere and Mulligan 1975), an extremely common occurrence with functionally impaired elderly patients. There should be a major effort in educating health care personnel in the dangers and causes of nosocomial infection.

With advancing age there is an increasing prevalence of bacterial colonization in the oropharynx and older individuals are more likely to have colonization of their oropharynx if they are functionally impaired. The less mobile and dependent individuals in nursing homes have a higher rate of colonization than fully mobile, independent individuals who are living at home. Colonization is likely in patients within the setting of the acute-care hospital. The organisms that predominate are usually Gram-negative bacilli and usually of one type (Valenti *et al.* 1978). These organisms pose a serious threat of pneumonia in patients at risk of aspiration and in whom clearance is ineffective.

The patient who has suffered cognitive decline presents a risk to others if excreting pathogenic material. For example, individuals with infectious diarrhoea may spread the material freely. The increased dependency of old age leads to the use of facilities where there is crowding—day care, day hospitals, and nursing homes. Within such settings there is potential for spread of infection (Stead 1981; Horan *et al.* 1984).

PHAGOCYTE AND NEUTROPHIL FUNCTION

Once infectious agents pass beyond the natural barriers the immunological mechanisms assume a greater role in host defence but non-specific defence mechanisms are also of importance. Perhaps the most important are the phagocytic cells and their mediators. There is no evidence of a decline in phagocytic function with age.

Although ageing itself does not appear to alter neutrophil function, some elderly individuals show impaired function which appears to result from specific disease or drugs. Macrophage function has been studied and there is no evidence of age-associated impairment; indeed, most of the experimental data indicate that macrophage function is intact in the elderly, and if anything is more efficient. This might compromise immunological reactions if antigen is destroyed more rapidly since larger doses of antigen will be required to initiate an immune response. In summary, however, there is no conclusive evidence that inflammatory reactions mediated by neutrophils and macrophages are impaired with ageing. It appears unlikely that impairment of this mechanism of defence contributes to the high risk of infection in elderly people.

CONCLUSION

Disease burden rather than chronological age is of most importance in predisposing to infection. The elderly who are experiencing decline in function represent a vulnerable population, particularly prone to further decline in function from the adverse effects of drugs and acute illnesses. Maintenance of integrity of the natural barriers and vigilance to reduce the population of pathogens that an elderly person is exposed to are likely to be useful strategies in reducing the risk of infection. At present these seem more useful strategies than trying to boost impaired immunity.

References

Horan, M.A., Puxty, J.A.H., Fox, R.A. (1983). Gynecologic sepsis as a cause of covert infection in old age. *Journal of American Geriatrics Society*, **31**, 213–15.

Horan, M.A., Gulati, R.S., Fox, R.A, Glew, E., Ganguli, L. and Kaeney, M. (1984). Outbreak of *Shigella sonnei* dysentery on a geriatric assessment ward. *Journal of Hospital Infection*, **5**, 210–12.

Nicolle, L.E., Bjornson, J., Harding, G.K.M., and MacDonnell, J.A. (1983). Bacteruria in elderly institutionalised men. *New England Journal of Medicine,* **309**, 1420–5.

Nicolle, L.E., Mayhew, J., and Bryan, L. (1987). Prospective randomised comparison of therapy and no therapy for asymptomatic bacteruria in institutionalised elderly woman. *American Journal of Medicine*, **83**, 27–33.

Phair, J.P., and Reisberg, B.E. (1984). Nosocomial infection. In *Immunology and infections in the elderly* (ed. R.A. Fox). Churchill Livingstone, Edinburgh.

Roberts, S.H., James, O.F.W., and Jarvis, E.H. (1977). Bacterial overgrowth syndrome without 'blind loop' a cause for malnutrition in the elderly. *Lancet*, **ii**, 1193–5.

Stead, W.W. (1981). Tuberculosis among elderly persons: an outbreak in a nursing home. *Annals of Internal Medicine*, **94**, 606–10.

Steere, A.C. and Mallison, G.T. (1975). Handwashing practices for the prevention of nosocomial infections. *Annals of Internal Medicine*, **83**, 683–90.

Valenti, W.M., Randall, G., Trudell, B.S., and Bentley, D.W. (1978). Factors predisposing to oropharyngeal colonisation with Gram-negative bacilli in the aged. *New England Journal of Medicine*, **298**, 1108–11.

3.2 Management of common clinical infectious problems

MIRA CANTRELL AND DEAN C. NORMAN

INTRODUCTION

Infectious diseases have a major impact on older persons. Many infections occur with greater frequency among them, and virtually all bacterial infections (e.g., pneumococcal pneumonia) and many viral infections (e.g., influenza) result in far higher rates of illness and death in the old than in the young (Yoshikawa 1983). (See the preceding chapter on frequency of infections and approaches to controlling them.) This chapter gives a brief review of the current management of several important infectious diseases. In order to understand the approach to specific infections, a review is first made of the unique clinical features of infections and of current antimicrobial therapy as it applies to older persons.

UNIQUE CLINICAL FEATURES OF INFECTIOUS DISEASES IN ELDERLY PEOPLE

In a significant proportion of elderly patients, particularly those who are very old and/or debilitated, infections may present in an atypical manner. This problem is compounded in nursing homes in the United States because most residents of these are frail, 80 years old or older, and suffer from multiple debilitating illnesses as well as cognitive impairment. Factors contributing to non-specific, blunted, or even absent classical signs and symptoms of infection in old age include: (1) under-reporting of illness; (2) differing causes than those in younger people; (3) coexisting diseases; and (4) altered response to illness.

UNDER-REPORTING OF ILLNESS

This is a common problem in elderly people and can be expected to be an especially important problem in those with cognitive impairment that results in poor communication.

DIFFERING CAUSES

These will often contribute to the severity of infections in this age group. A greater variety of pathogens is involved than in the younger age groups. Mixed infections involving aerobic or facultative anaerobic Gram-negative bacilli (hereafter referred to as Gram-negative bacilli) and obligate or strict anaerobic bacteria (hereafter referred to as anaerobes) are common. These variations in pathogens not only influence clinical presentation and prognosis but have important implications for the empirical choice of antimicrobial regimens. Empirical therapy will be discussed in the next section.

COEXISTING DISEASES

These are common in older patients and, not surprisingly, infection may be masked by them. For example, the diagnosis of septic arthritis in an elderly patient with chronic gouty arthritis may be overlooked because acute joint symptoms are mistakenly attributed to an exacerbation of gout.

ALTERED RESPONSE TO INFECTION

This is in part due to the above factors as well as changes with age. These include decreased immune response (see Chapter 3.1) and age-associated changes in chest-wall expansion and lung elasticity which contribute to a diminished cough reflex in elderly persons. There is often a disparity between the severity of an acute illness and the clinical presentation. Acute cholecystitis exemplifies this problem. A significant number of elderly patients with this infection may not have focal peritoneal signs on examination, and yet, despite a relatively 'benign' presentation, the risk of gangrene or perforation is high, particularly in the extremes of old age.

No discussion of the altered response to infectious diseases in older patients would be complete without a brief review of the significance and magnitude of fever, the cardinal sign of infection. Fever, or at least raised temperature, will occur in most infected elderly persons. But unlike children and young adults, in whom fever usually indicates a benign 'viral syndrome' or otitis media or pharyngitis, fever in elderly patients, especially in the very old, is typically associated with a serious infection, usually bacterial (Wasserman *et al.* 1989). Thus, whenever an elderly person presents with a fever, a careful search for bacterial infection should be made.

A small but significant number of elderly individuals with infection present with an absent or blunted fever response (Norman *et al.* 1985). This observation is supported by numerous reports of diminished fever responses in the old, with infections such as bacteraemia, pneumonia, tuberculosis, and infective endocarditis. The lack of a fever response to infection has other implications besides making early diagnosis difficult. It frequently portends a poor prognosis (Weinstein *et al.* 1983), which is explained in part by the evidence that fever itself is an important host defence mechanism (Norman *et al.* 1985).

The syndrome of fever of unknown origin {pyrexia of unknown origin in the United Kingdom} does occur in older persons. However, in one series, the cause of the erstwhile fever of unknown origin could often be determined, which is in contrast to events in younger individuals. Moreover, many of these cases in elderly patients were caused by treatable conditions such as intra-abdominal infection, bacterial endocarditis, and tuberculosis, and by connective tissue diseases such as temporal arteritis (Esposito and Gleckman 1978).

Acute infections may thus present atypically in older persons with minimal or subtle findings, often without signs pointing to specific involvement of a particular organ system. These altered presentations are summarized in Table 1.

Table 1 Non-classical presentation of infection in older persons

Any change of temperature in any direction from baseline
Any unexplained change in functional status or behaviour
Worsening cognition
Lethargy or agitation*
Anorexia or change in appetite
Falls
Incontinence
Focal neurological finding†
Tachypnoea‡

* Consider sepsis, central nervous system infection.
† Consider meningitis, brain abscess, endocarditis.
‡ Consider sepsis, pneumonia.

SPECIAL CONSIDERATIONS FOR ANTIMICROBIAL THERAPY IN ELDERLY PATIENTS

Mortality and morbidity

Rates of illness and death from infection are higher in elderly than in young persons. For example, bacterial pneumonia carries roughly five times the fatality rate for those over 70 years of age than for those less than 40 years old; acute cholecystitis for those over 70 has a fatality rate of 24 per cent compared to a negligible rate for those less than 40 (Yoshikawa and Norman 1987). A detailed discussion of factors contributing to higher rates of death and complications is beyond the scope of this chapter. Some of these factors are changes of ageing, which may be accelerated by lifestyle and chronic diseases, and by the presence of serious underlying illnesses.

The end result is diminished physiological reserve and compromised host defences. Older persons are also more likely to be in hospital and therefore exposed to virulent nosocomial pathogens, placing them at greater risk for developing bacteraemia and septic shock. Atypical presentations, coupled with difficulty in obtaining clinical specimens, may result in diagnostic delays and failure to start treatment promptly, which also contribute to greater illness and death.

Cost

The good management of institutions caring for older persons requires a policy for antibiotics to which both the management and practitioners subscribe, with regular review. The policy should include attention to the prevalence of particular micro-organisms in the institutions, to prevention of the emergence of resistant organisms, and to economic considerations of costs of drugs.

Pharmacological principles

In general, drugs with favourable pharmacokinetics and low toxicity such as β-lactam agents are preferred in most cases of bacterial infection in older patients. The greater effect of age on the pharmacokinetics of antibiotics is the common (but not universal) diminution in renal clearance. The decrement in age-associated renal function creates an advantage for some drugs that are relatively non-toxic, e.g., the newer β-lactams, which can be given in less frequent doses in older patients (Rho *et al.* 1991). Alternatively, for other agents that are potentially toxic, a decline in some function is a major disadvantage, e.g., aminoglycosides.

In view of these considerations, and the often diverse causes and mixed infections encountered, prompt empirical therapy, initially with broad-spectrum antimicrobial agents, is justified in the infected elderly patient. When the outcome of microbiological culture becomes known, treatment with agents effective across a narrower spectrum may begin, bearing in mind that older patients are more prone to drug toxicity (e.g., nephro- and ototoxicity from aminoglycosides, and ototoxicity from vancomycin).

SPECIFIC ANTIBIOTICS

This concise discussion will concentrate predominantly on newer drugs, as it is assumed that clinicians will be familiar with the older antibiotics.

β-Lactam drugs

This class of drugs is by far the most useful for treating infections in elderly patients because of antibacterial activity against many pathogens involved in infections of this age group, low toxicity, and favourable pharmacokinetics. A potential disadvantage is the high cost of some of these agents.

Penicillins

The newer penicillins include the acylampicillins (piperacillin, mezlocillin, azlocillin), which have largely replaced the carboxypenicillins (ticarcillin, carbenicillin). These drugs have a wider spectrum than penicillin G and provide antimicrobial activity not only against such important pathogens of older patients as *Streptococcus pneumoniae*, *Listeria monocytogenes*, non-enterococcal streptococci, and most anaerobes including the *Bacteroides fragilis* group, but also against such community-acquired Gram-negative bacilli as *Escherichia coli* and *Klebsiella* spp. (Drusano *et al.* 1984). Moreover, the spectrum is extended to *Enterobacter* spp. and some strains of *Serratia marcescens* and *Citrobacter* spp. None of these agents is effective against β-lactamase-producing strains of *Haemophilus influenzae* or staphylococci. Unfortunately, the role of these 'supercillins' in the treatment of infections in older patients is not yet clearly established.

Cephalosporins

These drugs are usually discussed in terms of their generation (first, second, and third), which is based on their antimicrobial spectrum.

First generation

The commonly used first-generation cephalosporins include cephalothin, cephapirin, cephradine, and cefazolin. These agents are active against important Gram-positive cocci such as *Staphylococcus aureus* and most streptococci including *Strep. pneumoniae*. Not all streptococci are susceptible, however, and a major deficiency in the spectrum of all cephalosporins is the lack of activity against the enterococcus (an important pathogen in urinary and intra-abdominal infections in older patients) and *L. monocytogenes* (an important meningopathogen).

First-generation cephalosporins have activity against most anaerobes except *B. fragilis*, and against certain Gram-negative bacilli such as *E. coli*, *Klebsiella* spp., and *Proteus mirabilis*. For elderly patients, cefazolin, which can be given twice or three times a day with good intramuscular absorption, is the most useful of these agents for treating infections such as pneumonia, cellulitis, urinary tract, and osteomyelitis. Cefazolin is also an ideal agent for prophylaxis against infection in certain 'clean' surgical procedures such as total hip replacement.

Second generation

These antibiotics have somewhat greater activity than the first-generation cephalosporins against Gram-negative bacilli, but at the expense of some loss of activity against Gram-positive bacteria. Cefoxitin and cefotetan are the only cephalosporins to have extended anaerobic coverage that includes the *B. fragilis* group, and are useful in treating 'mixed infections' such as diabetic ulceration of the foot or community-acquired aspiration pneumonia. Cefuroxime is unique in that it is the only first- or second-generation cephalosporin that can cross the blood–brain barrier as well as having reasonable activity against such respiratory pathogens as *Staph. aureus*, *Strep. pneumoniae*, β-lactamase producing strains of *H. influenzae* and *B. catarrhalis*, *Klebsiella* spp., and *E. coli*. These properties make cefuroxime an ideal agent for the empirical therapy of community-acquired pneumonia in older people. Moreover, an oral form of this drug (cefuroxime axetil) is now available in the United States. The other second-generation cephalosporins, which include cefamandole, ceforanide, cefonicid, and cefaclor, are less useful in the treatment of infected elderly patients.

Third generation

The currently available drugs include cefotaxime, ceftizoxime, ceftriaxone, moxalactam, cefoperazone, and ceftazidime. These important newer antibiotics are the most costly of the cephalosporins but offer the advantage of an expanded spectrum against several important, hospital-acquired, Gram-negative bacilli. Unfortunately, compared to first- or second-generation agents, these drugs have less activity against Gram-positive cocci (this is particularly true for moxalactam and ceftazidime). These antibiotics have variable activity against *B. fragilis*, with ceftazidime having poor activity. However, only ceftazidime has good activity against *Pseudomonas aeruginosa*. Most of these drugs penetrate well into the cerebrospinal fluid and have been a major advance in treating Gram-negative bacillary meningitis. Unfortunately, because they are not active against the important meningopathogen *L. monocytogenes*, they are not, by themselves, suitable for empirical therapy of meningitis in older patients. Table 2 summarizes our opinions on the most suitable third-generation cephalosporins for treatment of infections in older patients. Note that moxalactam is no longer recommended because of the increased risk of coagulopathy. With this exception, toxicity with any of the cephalosporins is infrequent and is similar to that of other β-lactam drugs.

Table 2 Recommended third-generation cephalosporins for infections in older patients

	Advantages
Cefoperazone	Favourable pharmacokinetics, i.e., twice-a-day dosing and dual mechanisms for clearance (hepatobiliary and renal)
Ceftriaxone	Favourable pharmacokinetics, i.e., once-a-day dosing unless patient seriously ill; good penetration into cerebrospinal fluid
Ceftazidime	Good activity against aerobic Gram-negative bacilli including *Pseudomonas aeruginosa*; good penetration into cerebrospinal fluid

β-Lactamase inhibitors

Two drugs (clavulanic acid, sulbactam) that irreversibly inhibit several bacterial β-lactamases have been introduced recently. When combined with β-lactam antibiotics that have a limited antimicrobial spectrum because of their β-lactamase sensitivity, these new drugs increase the antimicrobial activity. Amoxicillin has an antimicrobial spectrum similar to that of ampicillin (non-β-lactamase-producing strains of *H. influenzae*, streptococci including group D enterococci, most strains of *E. coli*, indole-negative *Pr. mirabilis*, and anaerobes excluding *B. fragilis*). When combined together as amoxicillin and clavulanic acid, the antimicrobial activity of amoxicillin is expanded to include β-lactamase-producing *Staph. aureus*, *H. influenzae*, *Klebsiella* spp., and anaerobes including *B. fragilis*. Furthermore, amoxicillin plus clavulanic acid is available as an oral preparation, making it an attractive broad-spectrum agent for ambulatory or non-institutional treatment of infections in older people. The other two drug combinations available, ticarcillin plus clavulanate and ampicillin plus sulbactam, have a similarly expanded spectrum of antibacterial activity, but only come in parenteral form in the United States. These appear to hold promise for empirical parenteral treatment of certain serious infections in elderly people.

Monobactams

These are β-lactam antibiotics that have only a single ring, in contrast to the more traditional two-ringed β-lactams, e.g., penicillins and cephalosporins. Aztreonam is the first of these newer antibiotics to become available commercially. Aztreonam has the low toxicity of a β-lactam drug but still has the anti-Gram-negative spectrum of the aminoglycosides, including activity against *Ps. aeruginosa*. However, aztreonam has no activity against anaerobes or Gram-positive organisms such as staphylococci or streptococci. This drug is effective in the treatment of septicaemia and urinary tract infection caused by Gram-negative bacilli. Thus, it has the potential to be an effective agent for the treatment of infections caused by Gram-negative bacilli in older patients without the risk of the oto- and nephrotoxicity associated with aminoglycosides.

Carbapenems

Imipenem is currently the only drug available from this newest class of antibiotics. It is marketed in combination with cilastatin, a renal dehydropeptidase inhibitor, which is needed to prevent breakdown of imipenem to a nephrotoxic metabolite. This drug is the most potent antimicrobial agent available for clinical use, with activity against virtually all important clinical isolates including *L. monocytogenes*, most Gram-negative bacilli, anaerobes including *B. fragilis*, and *Legionella* spp. Imipenem–cilastatin has demonstrable clinical efficacy for a wide variety of infections. Toxicity is similar to that of other β-lactams; however, it may be epileptigenic in elderly, debilitated patients, particularly if the dose is not adjusted for renal failure. As with other β-lactams, resistant bacterial strains have been reported, and thus routine or inappropriate use of imipenem–cilastatin for empirical therapy should be avoided. Rather, imipenem–cilastatin should be reserved for empirical therapy of life-threatening infections in older patients, under circumstances in which the cause is most likely Gram-negative bacilli with or without Gram-positive and anaerobic organisms, and the risk of potential aminoglycoside toxicity is high (e.g., in renal dysfunction and hypotension).

Fluoroquinolones

Norfloxacin and ciprofloxacin are currently the only quinolones that are commercially available, and both agents are marketed in oral preparations. However, several other fluoroquinolones, all of which are structurally derived from the parent compound nalidixic acid, may be released soon. These include pefloxacin, ofloxacin, amifloxacin, and enoxacin. The fluoroquinolones have a unique mechanism of antibacterial action; they inhibit an enzyme essential for bacterial replication (DNA-gyrase). Thus, cross-induction of β-lactamases does not occur as with β-lactam drugs. These drugs have remarkably high antimicrobial activity against Gram-negative bacilli, including *Ps. aeruginosa*, and relatively good activity is achieved against staphylococci (including methicillin-resistant strains), streptococci (including most strains of enterococci), and *H. influenzae*. However, fluoroquinolones have poor activity against anaerobes. Up to now, these drugs have been associated with few adverse reactions. Of the two currently available fluoroquinolones, norfloxacin should be used to treat urinary tract infections only, because blood concentrations are too low to treat other serious infections effectively. In contrast, ciprofloxacin achieves serum concentrations high enough to treat serious infections other than urinary tract infection. For example, it is effective in treating mild to moderately ill nursing-home residents with pneumonia (Peterson *et al.* 1988).

Aminoglycosides

Despite the availability of several newer antibiotics (described above), this class of drugs still has a role in the treatment of serious infections caused by Gram-negative bacilli in older patients. However, this use of aminoglycosides must be balanced against the increased potential with age for nephro- and ototoxicity. Nevertheless, these drugs, which include gentamicin, tobramycin, amikacin, and netilmicin, should be considered under the following conditions: (1) where the risk of death from the infection outweighs the risk of toxicity; (2) serious infections with *Ps. aeruginosa*; (3) infections caused by drug-resistant bacteria sensitive only to an aminoglycoside; and (4) a serious infection (e.g., infective endocarditis) best treated by a synergistic drug combination (one drug being an aminoglycoside) against a specific pathogen (e.g. enterococcus).

SPECIFIC INFECTIONS

In this section, a brief review of the management of specific infections in older persons is presented. For a more thorough review, consult specific references (for example, Yoshikawa and Norman 1987).

Pneumonia (see Section 12)

Pneumonia and influenza are the leading infectious cause of death for persons over the age of 65 (Yoshikawa and Norman 1987), with fatality rates several times higher than those of younger patients. The increased frequency and severity of this infection result from aspiration of virulent oropharyngeal pathogens in the setting of compromised host defences, i.e., diminished cough reflex, concomitant chronic medical problems that alter immune function, or diminished physiological reserve. Thus, one preventive measure to reduce the risk for pneumonia in older persons is to avoid circumstances that predispose or encourage aspiration, e.g., oversedation, alcohol intake, supine position.

The aetiology of pneumonia differs significantly between older and younger adults. *Strep. pneumoniae* and *Mycoplasma pneumoniae* are the leading pathogens in community-acquired pneumonia in young adults. Community-acquired pneumonia in older patients is usually bacterial, frequently with mixed organisms, and may be due to a wide variety of pathogens including *Strep. pneumoniae*, *H. influenzae*, and *Staph. aureus*, with a small number caused by Gram-negative bacilli (Verghese and Berk 1983). Pneumonia that occurs in long-term care facilities, particularly in the setting of recent or concurrent antimicrobial therapy, will have an even higher number of cases due to Gram-negative bacilli such as *K. pneumoniae* (Garb *et al.* 1978). On occasion, outbreaks of influenza A or B or respiratory syncytial virus may lead to several secondary cases of pneumonia in a chronic institutional setting. Nosocomial (hospital-acquired) pneumonia is typically caused by a mixed flora, with Gram-negative bacilli, anaerobes, *Staph. aureus* and *Strep. pneumoniae* being the most common isolates (Bartlett *et al.* 1986). Finally, two other pathogens are important aetiological agents of pneumonia in older people. These are *Legionella* spp., the incidence depending on geographic location, and *Branhamella catarrhalis*, an organism that produces β-lactamase, which is found mostly in elderly patients with chronic obstructive pulmonary diseases (Nicotra *et al.* 1986).

The clinical manifestations of pneumonia may differ between elderly patients and younger adults. Fever may be blunted or absent, and cough may be minimal. Tachycardia and tachypnoea are sensitive but non-specific findings. The physical examination will usually not find signs of consolidation and may be misleading in patients with chronic pulmonary diseases or congestive heart failure. The chest radiograph is a necessary complement to physical examination, and it may show patchy infiltrates rather than lobar consolidation. After coming to hospital, almost half of elderly patients will show radiological progression of pneumonia compared to only 10 per cent of younger patients; resolution of infiltrates will take longer for the elderly than for the younger persons with pneumonia (Andrews *et al.* 1987).

Optimal management of pneumonia will include a time in hospital for the majority of cases because of: (1) an atypical presentation, which may underestimate the seriousness of the infection; (2) coexistence of chronic diseases; (3) varied pathogens; and (4) the need to monitor response to therapy. Effective management of pneumonia may require: (1) stability of the patient's cardiopulmonary and haemodynamic functions; (2) 24-h clinical monitoring by trained nursing staff; (3) daily attendance by a physician; (4) intravenous therapy; and (5) full laboratory (including radiological) support.

All elderly patients with pneumonia should have sputum collected for Gram staining, a smear for acid-fast bacteria, and bacterial and mycobacterial cultures, as well as blood for culture. However, attempts to collect sputum in frail or ill elderly patients often meet with failure. Transtracheal aspiration, even in the hands of experienced clinicians, should be restricted to patients who fail on initial antibiotic therapy and are clinically deteriorating. Other routine laboratory tests including complete blood count, serum electrolytes, and renal function tests should also be made. In addition, often an arterial blood gas is needed and an electrocardiogram should be done. Diagnostic thoracentesis for culture, pH, and cytological examination of pleural fluid is indicated for co-operative patients without coagulopathy who have significant pleural effusions. More invasive diagnostic tests such as bronchoscopy or fine-needle aspiration may be indicated in selected cases (Yoshikawa and Norman 1987). Finally, most elderly patients with pneumonia should be skin tested for tuberculosis and coccidioidomycosis (in endemic areas only).

The approach to antimicrobial therapy for pneumonia in the older patient is derived from the principles discussed in the earlier sections. Thus, the often unrecognized severity of illness in this group justifies starting empirical therapy with broad-spectrum antibiotics. If and when the results of culture become known, a regimen with a narrower spectrum to reduce risk of superinfection and side-effects may be selected.

Table 3 summarizes our choice of empirical antimicrobial therapy for pneumonia in older patients. Alternatives to this may also be effective.

Table 3 Empirical antimicrobial therapy for bacterial pneumonia in older patients

Condition	Drug(s) of choice	Alternative choices
*Clinically stable**		
Good sputum for analysis	Treat according to Gram-stain results	
Poor sputum	Cefuroxime cefoxitin (or cefotetan)	Third-generation cephalosporin ampicillin–sulbactam ticarcillin–clavulanate ciprofloxacin†
*Clinically unstable** Nosocomial infection recently on antibiotics; or immunosuppressed	Third-generation cephalosporin [plus] aztreonam‡	Ticarcillin–clavulanate (or ampicillin–sulbactam) [plus] aztreonam‡

* Add erythromycin if legionnaire's disease suspected.
† If patient is mildly ill and can tolerate oral agent, ciprofloxacin plus or minus ampicillin (to ensure coverage of *Strep. pneumoniae*) may be used. Cefuroxime, axetil, or ampicillin–sulbactam may be other possible oral antibiotics.
‡ Or aminoglycoside if the risk of death from sepsis far outweighs the risk of aminoglycoside toxicity.

Tuberculosis

An increasing proportion of deaths from *Mycobacterium tuberculosis* infection and a disproportionately high number of active cases are among elderly persons. This association of age and tuberculosis is in part due to increased susceptibility from age-associated waning of cell-mediated immunity and from coexisting underlying illness in persons with previous exposure to or infection with *M. tuberculosis*. Although reactivation of prior (inactive) infection is the most common form of active tuberculosis in old age, outbreaks of primary infection have been reported in long-term care institutions where elderly patients have been exposed to an initially unrecognized index case (Stead *et al.* 1985).

The diagnosis of tuberculosis in older persons may be difficult because of their more subtle clinical presentations, atypical findings on chest radiographs, and problems with skin testing, which include decreased reactivity to tuberculin antigens. Tuberculous infection should be considered in elderly patients with unexplained weight loss, fever, recurrent pleural infusions, or any unexplained acute or subacute deterioration in functional status. Despite the lack of optimal sensitivity, the initial investigation in suspected cases should include skin testing with tuberculin antigen (purified protein derivative) and control dermal antigens to exclude cutaneous anergy. All older patients with pulmonary infiltrates should have multiple sputum specimens sent for Ziehl–Nielsen smears and mycobacterial cultures. Pleural biopsy should be attempted if a tuberculous effusion is suspected; bronchial biopsy and culture may be indicated in cases of suspected miliary tuberculosis and in other selected patients in whom the diagnosis is strongly suspected but the sputum is negative for mycobacteria. The investigation of extrapulmonary tuberculosis is beyond the scope of this chapter.

The treatment of choice for active pulmonary tuberculosis and extrapulmonary tuberculosis has been a combination of isoniazid (300 mg a day) and rifampicin (600 mg) for at least 9 months. More recently, trials have shown that a four-drug regimen (isoniazid, rifampicin, pyrazinamide, and streptomycin or ethambutol) for 2 months followed by isoniazid and rifampicin for 4 months (total of 6 months therapy) is as effective as the standard 9-month regimen (Grosset 1989).

Urinary tract infection

Whenever symptoms or laboratory evidence for urinary tract infection appear, a careful search for underlying causes must be made.

Uncomplicated community-acquired infection

The frequency of bacteriuria in those aged 65 and over ranges from 5 to 10 per cent for ambulatory men and 10 to 30 per cent for ambulatory women (Yoshikawa and Norman 1987). Symptomatic urinary tract infections, defined as the presence of significant bacteriuria (100000 or more bacteria per ml urine) and associated clinical manifestations, require specific antimicrobial treatment. Unfortunately, the typical clinical manifestations of urinary tract infection—fever, dysuria, frequency, urgency—may be subtle or even absent in older persons, who, instead, present with such symptoms as change in mental status, anorexia, weakness, or fatigue. The criteria for antimicrobial selection should be based on several factors: the bacteriological features of the urine, severity of infection, comorbidity, the presence of abnormalities of the genitourinary tract, and the characteristics of antimicrobial agents, including their pharmacokinetics and pharmacodynamics.

While the majority of urinary tract infections in older patients are caused by *E. coli*, a great variety of other Gram-negative organisms are also found: *Proteus* spp. (especially in elderly males) and *Klebsiella* spp., *Ps. aeruginosa*, *Serratia marcescens*, *Citrobacter* spp., *Providencia* spp., group D enterococci, and coagulase-negative staphylococci do cause these infections in older patients. These organisms are relatively infrequently isolated in healthy, ambulatory persons in the absence of genitourinary abnormalities, chronic bladder catheters, or recurrent urinary infections. Therefore, the antibiotics that are selected empirically for the initial treatment of community-acquired urinary tract infections should be those which provide broad Gram-negative coverage. To the extent that a centralized laboratory may be handling most cultures in an institution or community, it can provide information about the probabilities of different organisms in that locale.

The agents most commonly used today for treatment of uncomplicated, community-acquired urinary tract infections are trimethoprim–sulphamethoxazole, cephalosporins or ampicillin, and some of the newer fluoroquinolone and clavulanic acid combinations (Parsons 1988).

Trimethoprim–sulphamethoxazole and the cephalosporins provide good antimicrobial activity for most community-acquired uropathogens, i.e., *E. coli*, *Proteus* and *Klebsiella* spp. If enterococcal infection is suspected, ampicillin or amoxicillin is a good choice.

Norfloxacin and ciprofloxacin provide the advantages of broad coverage, less bacterial resistance, and twice-a-day administration. These drugs should be reserved for elderly patients with history of recurrent urinary infections, urinary tract infections caused by resistant pathogens, or intolerance to standard treatment. Magnesium- and aluminium-based antacids decrease the absorption of ciprofloxacin, and ciprofloxacin increases the blood level of theophylline so that careful monitoring of theophylline concentrations during such combination therapy is prudent. Amoxicillin combined with clavulanic acid is a relatively safe antibiotic and provides good microbiological coverage including of β-lactamase-producing bacteria.

The treatment of urinary tract infections is usually begun empirically pending the final results of culture and sensitivity testing. In elderly patients who appear septic or clinically ill, therapy should be started parenterally in hospital. Oral antibiotics may be started after clinical improvement. When the results of culture are known, treatment can be changed, if necessary. The therapeutic goal is to sterilize the urine and improve the patient's clinical status. Treatment is for 7 to 14 days, depending on the response. Although the value of single-dose and short-term therapy in the treatment of acute, uncomplicated urinary tract infections in young women has been well established, not enough is known about this form of therapy in older patients; studies show an unacceptably high rate of relapse with single-dose regimens (Rennenberg and Paerregaard 1984).

Complicated urinary tract infection

Complicated urinary tract infections may be defined as symptomatic infections associated with structural or functional abnormalities of the urinary tract (benign prostatic hypertrophy, stones, strictures, neurogenic bladder). This type of infection tends to relapse and is generally caused by a chronic focus of infection within the genitourinary tract. In addition, urinary tract infection associated with sepsis or renal involvement, or which persists, can be considered complicated.

Recurrent infections require prolonged treatment, often for as long as 12 to 14 weeks, with antibiotics that are effective against the offending pathogen. Norfloxacin, ciprofloxacin, and ofloxacin are all useful in the treatment of complicated or recurrent urinary tract infections for shorter periods of 4 to 6 weeks.

The use of antimicrobial agents in patients with persistent or intermittent asymptomatic bacteriuria is still open to debate. There is conflicting information on the effects of asymptomatic bacteriuria on illness and death in older persons. Some patients with asymptomatic bacteriuria have no discernible anatomical abnormalities of the urinary tract; the bacteriuria tends to be transient and often resolves without treatment. However, when it is chronic, asymptomatic bacteriuria is very difficult to eradicate permanently. Attempts to sterilize the urine permanently may lead to the emergence of resistant strains, a high rate of relapse, and significant adverse effects of medication (Lipsky 1989).

Antimicrobial chemoprophylaxis might be considered for a special subgroup of elderly patients, i.e., those who do not have major anatomical abnormalities of the genitourinary tract or long-term bladder catheters but who experience recurrent symptomatic infections or are at risk of developing serious complications as a result of these infections. After sterilization of the urine, follow-up with continuous low doses of trimethoprim–sulphamethoxazole (one-half of a tablet every evening) or trimethoprim (one-half of a tablet every evening) for 2 to 6 months will prevent bacteriuria in approximately 40 per cent of patients. In the majority, however, the bacteriuria will recur and persist, and prophylaxis may need to be extended indefinitely (Keys and Edson 1983).

Hospital- or institution-acquired urinary tract infection (see Chapter 3.3)

Urinary tract infections acquired in hospital are most often associated with indwelling bladder catheters. When there is the prospect of early removal of the catheter it is important to culture the urine in order to be able to treat symptomatic infections promptly. Patients with chronic indwelling urinary catheters almost always have bacteriuria, which is often polymicrobial; a quarter of these patients have group D enterococci. In such situations, antimicrobial treatment is reserved for patients who develop clinical signs of infection such as fever, chills, or changes in mental and/or functional status, inasmuch as repetitive treatment usually results in the emergence of resistant strains. The treatment of catheter-associated symptomatic urinary tract infection depends on the patients' clinical status. If they are clinically stable, empirical therapy with trimethoprim–sulphamethoxazole, fluoroquinolones, or a combination of ampicillin with a drug working in the Gram-negative spectrum (e.g., third-generation cephalosporins or aztreonam) is suggested. If they are clinically unstable or septic, a combination of ampicillin and an aminoglycoside (or third-generation cephalosporins) is a good initial choice until the results of culture and sensitivity testing become available. Although used by some clinicians, routine surveillance cultures in patients with chronic indwelling catheters are not likely to be useful because the bacterial flora changes rapidly (Breitenbucher 1984).

Pressure sores

Prevention of pressure sores should be among the cardinal goals of good care of patients who are at high risk because of immobility, malnutrition (often with hypoalbuminaemia), and/or skin soiling from incontinence. There should be specific plans for nursing care which emphasize regular turning and positioning to avoid prolonged pressure on any area, consideration of use of special mattresses, good skin care including immediate attention to any soiling, and restoration and maintenance of good nutrition (Allman 1989). {In the United Kingdom some centres make routine use of the Norton scale (Table 4) to assess risk of bedsores in elderly patients.}

Management of pressure sores must also be comprehensive

Table 4 The Norton Scale

Score	Physical condition	Mental state	Activity	Mobility	Incontinence
4	Good	Alert	Ambulant	Full	Not
3	Fair	Apathetic	Walks with help	Slightly limited	Occasionally
2	Poor	Confused	Chairbound	Very limited	Usually (urine)
1	Very bad	Stuporous	Bedfast	Immobile	Doubly

Score: ⩽14—risk; ⩽12—high risk.

to include systematic measures, positioning, local wound care, and patient education.

Systemic measures are the correcting or improving of conditions such as poor vascular perfusion, hypoalbuminaemia, incontinence, anaemia, and metabolic derangements. The treatment of associated medical or surgical conditions, especially those leading to immobility, is of the utmost importance. Nutritional requirements for wound healing significantly exceed the usual caloric intake. A diet rich in carbohydrates and protein should be provided, and hyperalimentation used, if deemed necessary. Additional dietary zinc and vitamin C may improve wound healing (Allman 1989).

The positioning of the patient is an integral part of the treatment, just as it is in prevention. The position should be changed every 2 or 3 h. Slight rotations of 20 to 30° are sufficient and can be accomplished with minimal physical effort. The position best suited for elderly patients is 30° oblique back, while 90° lateral should be avoided, as it may place too much pressure on the trochanteric area (Seiler and Stahelin 1986).

Local care of the wound varies according to the stage of involvement. Conservative treatment is used for stage I and II pressure sores, whereas stage III, IV, and closed pressure sores may require additional surgical intervention. Conservative treatment consists of cleansing and disinfection, debridement and control of infection.

So-called 'wet-to-dry' wound covering with normal saline four times a day is usually necessary to keep the area clean. Hydrogen peroxide, povidone–iodine, acetic acid, boric acid, and Dakin's solution (sodium hypochlorite) all have antiseptic properties. Unfortunately, most of these damage the surrounding healthy tissue, so that, if used, they should be highly diluted. In our experience, normal saline alone or in 50 per cent combination with hydrogen peroxide is beneficial for cleansing and disinfecting the sore. The wound should be loosely packed with mesh gauze soaked in solution and moistened again when removed because removing a dry dressing brings with it newly proliferated epithelium, which delays healing.

Various wound coverings—transparent adhesive dressing, hydrogel dressings, absorption (hydrocolloid) dressings—reduce the need for frequent changes of dressing. Some absorb the excess exudate and maintain a moist wound surface; most are permeable to oxygen and moisture, thereby minimizing maceration. They are easy to apply and cost effective, but must be changed frequently. Moreover, if the staff assume that the dressing is sufficient for healing and become less attentive in turning the patient, wounds may deteriorate. Debridement is necessary to clean the wound of dead tissue. This can be accomplished through changes of 'wet-to-dry' dressing, by enzymatic agents, or by cutting. Topical enzyme preparations liquefy the necrotic tissue but do not penetrate hard eschars and can cause local allergic and inflammatory reactions. They have no proven advantage over mechanical ('wet-to-dry') or surgical debridement. Surgical debridement with sharp dissection is quicker than other methods. Although it may cause transient bacteraemia, there are no reports of clinically apparent septic episodes after the procedure. Hydrotherapy with a whirlpool, Hubbard tanks, or jacuzzi is another useful method of debridement, with the added effect of providing thorough cleaning and wound disinfection (Revler and Cooney 1981).

Pressure sores are chronically contaminated wounds. Topical antibiotics may lower the bacterial counts, thereby facilitating healing, but they penetrate deeper wounds poorly. They may also cause localized tissue sensitivity, lead to the growth of resistant organisms, or cause a systemic reaction when applied over large areas (Seiler and Stahelin 1986). Topical antibiotics, if used at all (and local practices vary), should be used only for a short period of time, for example, preoperatively. The most commonly used agents are silver sulphadiazine, nitrofurazone, and metronidazole (Gomolin 1988).

Ultraviolet light and local oxygen may produce a local bactericidal effect. A placebo-controlled study of the effects of ultraviolet light on the healing time of superficial pressure sores showed improved healing in the experimental group (Wills *et al.* 1983). Similarly, hyperbaric oxygen is potentially bactericidal, although its usefulness in the treatment of pressure sores remains questionable.

Patients with sepsis, osteomyelitis, or cellulitis require systemic antibiotics directed against a mixed population of Gram-negative bacilli, anaerobic bacteria and Gram-positive cocci (Yoshikawa and Norman 1987). Effective treatment may be achieved with a combination of clindamicin, metronidazole, or chloramphenicol (if local practice permits its use) with an aminoglycoside, third-generation cephalosporin, or certain fluoroquinolones, as well as with ticarcillin–clavulanic acid and ampicillin–sulbactam.

An accurate assessment of the micro-organisms involved will aid in the management decision. The wound should be cultured from a punch biopsy using aerobic and anaerobic methods. A superficial swab culture of the purulent discharge is of little value. Blood cultures should be obtained in all cases where active infection is suspected, before beginning antimicrobial therapy.

Surgical treatment is indicated for primary closure, skin grafts (full thickness and split thickness), and skin flaps (cutaneous or musculocutaneous). Sometimes more extensive procedures, such as removal of bony prominences, amputations, and even hemicorporectomy, are needed for patients with multiple, non-healing ulcers, with severe contractures, or extensive osteomyelitis.

The treatment of pressure sores should be by a team approach. Achieving the best results requires a co-ordinated effort from the medical and nursing staff, aided by rehabilitation and dietary consultants. The formation of therapeutic teams leads to a significant decrease in the incidence and prevalence of pressure sores (Romm *et al.* 1982). Whenever possible, patients should be taught how to change position frequently in bed and in a wheelchair, and how to effect proper skin care. For all involved, continuous emphasis must be placed on proper implementation of preventive measures.

Herpes zoster

Herpes zoster is a painful, sometimes debilitating disease, seen principally among older persons, with an annual incidence of 10 per 1000 in persons older than 70 years. Epidemiological data indicate no significant sex difference in the age-adjusted incidence rates and no racial predilection. There are probably no significant seasonal variations, although an increase in incidence over time has been observed in recent years (Ragozzino *et al.* 1982). The severity of the disease and the incidence of complications increase with age. Postherpetic neuralgia occurs in up to three-quarters of patients over the age of 70, with zoster of the trigeminal distribution accounting for the pain of longest duration (Yoshikawa and Norman 1987). Ocular complications are predominantly seen in older persons, as is the generalized form of the disease.

The treatment of herpes zoster is directed toward ameliorating local pain and discomfort and toward preventing the complications. The management will vary according to the immunocompetence of the host.

Mild skin eruption with irritation may respond well to cool Burrow's solution applied as wet compresses several times a day. The efficacy of topical idoxuridine remains controversial, and its usefulness is limited because of the cumbersome method of application, cost, and occasional occurrence of contact dermatitis (Dawber 1974). The initial pain can be treated with non-narcotic or narcotic analgesics (aspirin, oxycodone, codeine) for a short time. (See also Section 17.)

Whether corticosteroids affect the duration of pain or prevent postherpetic neuralgia remains controversial. Studies done in non-immunosuppressed adults suggest that the early use of corticosteroids may reduce the severity of postherpetic neuralgia. It appears, however, that this additional pain relief can be demonstrated only during the early phase of infection and is not evident between 2 weeks and 6 months after the acute attack (Peterslund 1988). Other studies have suggested longer benefit of this treatment, including the observation that the effect of corticosteroids on pain was not detectable until 2 weeks into the treatment. If treatment with corticosteroids is chosen, the oral route is most commonly used with 50 to 60 mg prednisone a day initially, reduced weekly over 1 month.

Sklar *et al.* (1985) showed that adenosine monophosphate may be effective in treating and preventing postherpetic neuralgia; 82 per cent of patients over the age of 60 were pain-free after 28 days of treatment, and the treatment group had a shorter period of viral shedding and an earlier healing of skin lesions.

The analgesic effects of tricyclic antidepressant drugs have been used with moderate success for the treatment of postherpetic neuralgia. Amitriptyline in a dose of 65 to 75 mg/day produced a favourable response even in patients with long-standing symptoms (Max *et al.* 1988). The anticholinergic side-effects of amitriptyline in elderly patients can be significant, and careful monitoring of symptoms such as sedation, dizziness, dry mouth, and urinary retention is necessary.

Advances in antiviral drugs in recent years have established the usefulness of these agents in the treatment of acute herpes zoster in both immunocompromised and immunocompetent patients regardless of age. However, in elderly immunocompetent patients, postherpetic neuralgia rather than disseminated herpes zoster is the major complication. Thus, prevention of this neuralgia is the primary goal. It is not clear whether antiviral chemotherapy has a beneficial effect on postherpetic neuralgia.

Acyclovir has been used in the treatment of acute herpes zoster in the form of intravenous, intramuscular, and oral preparations. When used in previously healthy patients, it shortens the viral shedding, reduces early pain, and prevents the formation of new lesions, but it has only a modest if any effect on postherpetic neuralgia. When used in immunocompromised patients, it also significantly reduces the frequency of cutaneous dissemination and visceral complications (Nicholson 1984). The early treatment of acute herpes zoster in immunocompetent patients with oral acyclovir has been recommended (Peterslund 1988; Huff *et al.* 1988; Wood *et al.* 1988). Oral acyclovir can be given in doses of 800 mg five times a day for a 7 to 10 day course. It appears to be free of major adverse reactions, but some patients have reported nausea and vomiting. Nevertheless, in immunocompetent elderly patients with uncomplicated herpes zoster, the long-term benefit of acyclovir remains to be seen. Patients with ophthalmic zoster should be considered for therapy with oral acyclovir. This treatment reduces the incidence and the severity of keratitis and uveitis. Moreover, an effect can be demonstrated even in patients started on therapy as late as 7 days after the onset of cutaneous lesions (Cobo 1988).

Patients with disseminated zoster, zoster encephalitis, zoster pneumonitis, and those who are immunocompromised should initially receive intravenous acyclovir. Although both acyclovir and vidarabine have known efficacy in the treatment of zoster infection in an immunocompromised host, new data suggest that acyclovir should be considered the current treatment of choice because it is significantly more effective and less toxic (Shepp *et al.* 1988). The major side-effects of intravenous acyclovir are related to its propensity for nephrotoxicity; the dose should be adjusted according to the outcome of renal function tests.

The usefulness of interferon in the management of immunosuppressed patients with herpes zoster infection seems to be limited because of frequent adverse effects, although it has been shown to reduce the severity and dissemination of the disease and to lessen the visual involvement (Winston *et al.* 1988).

The prevention of herpes zoster infection may be achieved through the use of live attenuated vaccine. The vaccine enhances the immune response to varicella-zoster virus in older persons, but the clinical significance of this remains to be tested (Berger *et al.* 1984).

References

Allman, R. M. (1989). Pressure sores among the elderly. *New England Journal of Medicine*, **320**, 850–3.

Andrews, B. E. *et al.* (1987). Community acquired pneumonia in adults in British hospitals in 1982–1983. A survey of aetiology, mortality, prognostic factors and outcome. *Quarterly Journal of Medicine*, **62**, 195–220.

Bartlett, J. G., O'Keefe, P., Tally, F. P., Louie, T. J., and Gorbach, S. L. (1986). Bacteriology of hospital-acquired pneumonia. *Archives of Internal Medicine*, **146**, 868–71.

Berger, R., Luescher, D., and Just, M. (1984). Enhancement of varicella–zoster immune response in the elderly by boosting with varicella vaccine. *Journal of Infectious Diseases*, **149**, 647.

Breitenbucher, R. B. (1984). Bacterial changes in the urine samples of patients with long-term indwelling catheters. *Archives of Internal Medicine*, **144**, 1585–8.

Cobo, M. (1988). Reduction of the ocular complications of herpes zoster ophthalmicus by viral acyclovir. *American Journal of Medicine*, **85** (suppl. 2A), 90–3.

Dawber, R. (1974). Idoxuridine in herpes zoster: further evaluation of intermittent topical therapy. *British Medical Journal*, **ii**, 526–7.

Drusano, G. L., Schimpff, S. C., and Hewitt, W. L. (1984). The acylampicillins: mezlocillin, piperacillin, azlocillin. *Reviews of Infectious Diseases*, **6**, 13–21.

Esposito, A. L. and Gleckman, R. A. (1978). Fever of unknown origin in the elderly. *Journal of the American Geriatrics Society*, **26**, 498–505.

Garb, J. L., Brown, R. B., Garb, J. R., and Tuthill, R. W. (1978). Differences in etiology of pneumonias in nursing home and community patients. *Journal of the American Medical Association*, **240**, 2169–72.

Gomolin, I. H. (1988). Pressure sores in the elderly. II. Topical metronidazole therapy for anaerobically infected pressure sores. *Geriatric Medicine Today*, **7**, 93–9.

Grosset, J. H. (1989). Present status of chemotherapy for tuberculosis. *Reviews of Infectious Diseases*, **2** (suppl. 2), S347–52.

Huff, J. C. *et al.* (1988). Therapy of herpes zoster with oral acyclovir. *American Journal of Medicine*, **85** (suppl. 2A), 84–9.

Keys, T. F. and Edson, R. S. (1983). Antimicrobial agents in urinary tract infections. *Mayo Clinic Proceedings*, **58**, 165–8.

Lipsky, B. A. (1989). Urinary tract infections in men. *Annals of Internal Medicine*, **110**, 138–50.

Max, M. B., Schafer, S. C., Culnane, M., Smoller, B., Dubner, R., and Gracely, R. H. (1988). Amitriptyline, but not lorazepam, relieves postherpetic neuralgia. *Neurology*, **38**, 1427–32.

Nicholson, K. G. (1984). Antiviral therapy. *Lancet*, **ii**, 677–81.

Nicotra, B., Rivera, M., Luman, J. I., and Wallace, R. J., Jr. (1986). *Branhamella caterrhalis* as a lower respiratory pathogen in patients with chronic lung disease. *Archives of Internal Medicine*, **146**, 890–3.

Norman, D. C., Grahn, D., and Yoshikawa, T. T. (1985). Fever and aging. *Journal of the American Geriatrics Society*, **33**, 859–63.

Parsons, C. L. (1988). Protocol for treatment of typical urinary tract infections: criteria for antimicrobial selection. *Urology*, **32** (suppl.), 22–7.

Peterslund, N. A. (1988). Management of varicella zoster infection in immunocompetent host. *American Journal of Medicine*, **85** (suppl. 2A), 74–8.

Peterson, P. K. *et al.* (1988). Prospective study of lower respiratory tract infections in an extended-care nursing home program: potential role of oral ciprofloxacin. *American Journal of Medicine*, **85**, 164–71.

Ragozzino, M. W., Melton, L. J. III, Kurland, L. T., Chu, C. P., and Perry, H. O. (1982). Population-based study of herpes zoster and its sequelae. *Medicine*, **61**, 310–16.

Rennenberg, J. and Paerregaard, A. (1984). Single-day treatment with trimethoprim for asymptomatic bacteriuria in the elderly patient. *Journal of Urology*, **132**, 934–5.

Revler, J. B. and Cooney, T. H. (1981). The pressure sores: pathophysiology and principles of management. *Annals of Internal Medicine*, **94**, 661–6.

Rho, J., Jones, M., Woo, M., Castle, S. C., Smith, K., and Norman, D. C. (1991). Single dose pharmacokinetics of intravenous ampicillin plus sulbactam in healthy elderly compared to young adults. *Journal of Antimicrobial Therapy* (in press).

Romm, S., Tebbetts, J., Lynch, D., and White, R. (1982). Pressure sores: state of the art. *Texas Medicine*, **78**, 52–60.

Seiler, W. O. and Stahelin, H. B. (1986). Decubitus ulcers: treatment through five therapeutic principles. *Geriatrics*, **41**, 47–58.

Shepp, D. H., Dandliker, P. S., and Meyers, J. D. (1988). Current therapy of varicella–zoster virus infection in immunocompromised patients. *American Journal of Medicine*, **85** (suppl. 2A), 96–8.

Sklar, S. H., Blue, W. T., Alexander, E. J., and Bodian, C. A. (1985). Herpes zoster. The treatment and prevention of neuralgia with adenosine monophosphate. *Journal of the American Medical Association*, **253**, 1427–30.

Stead, W. W., Lofgren, J. P., Warren, E., and Thomas, C. (1985). Tuberculosis as an endemic and nosocomial infection among the elderly in nursing homes. *New England Journal of Medicine*, **312**, 1483–7.

Verghese, A. and Berk, S. L. (1983). Bacterial pneumonia in the elderly. *Medicine*, **62**, 271–85.

Wasserman, M., Levinstein, M., Keller, E., and Yoshikawa, T. T. (1989). Utility of fever, white blood cell, and differential count in predicting bacterial infections in the elderly. *Journal of American Geriatrics Society*, **37**, 534–43.

Weinstein, M. P., Murphy, J. R., Reller, R. B., and Lichtenstein, K. A. (1983). The clinical significance of positive blood cultures: a comprehensive analysis of 500 episodes of bacteremia and fungemia. II. Clinical observations with special reference to factors influencing prognosis. *Reviews of Infectious Diseases*, **5**, 54–70.

Wills, E. E., Anderson, T. W., Beattie, B. L., and Scott, A. (1983). A randomized placebo-controlled trial of ultraviolet light in the treatment of superficial pressure sores. *Journal of the American Geriatrics Society*, **31**, 131–3.

Winston, D. J. *et al.* (1988). Recombinant interferon alpha-2a for treatment of herpes zoster in immunosuppressed patients with cancer. *American Journal of Medicine*, **85** (suppl. 2A), 147–51.

Wood, M. J., Ogan, P. H., McKendrick, M. W., Care, C. D., McGill, J. I., and Webb, E. M. (1988). Efficacy of oral acyclovir treatment of acute herpes zoster. *American Journal of Medicine*, **85** (suppl. 2A), 79–83.

Yoshikawa, T. T. (1983). Geriatric infectious diseases: an emerging problem. *Journal of the American Geriatric Society*, **31**, 34–9.

Yoshikawa, T. T. and Norman, D. C. (ed.) (1987). Epidemiology of infectious diseases. In *Aging and clinical practice: infectious diseases. Diagnosis and treatment*, pp. 3–7. Igaku-Shoin Medical Publishers, New York.

3.3 Control of infections

ANN R. FALSEY AND DAVID W. BENTLEY

Infectious diseases are a significant cause of death and illness in older persons. The infections occur in part because of a depressed immune response associated with ageing (see Chapters 3.1.1 and 3.1.2) and with the changes in organ structure and function that accompany the multiple diseases and the medical and surgical procedures that older persons frequently experience. The control of infections in older persons is also greatly influenced by the setting, whether in the community, the acute hospital, or in long-term care facilities. In this chapter we discuss the major vaccines that should be considered for older persons, and then review the epidemiological risk factors, microbiology, and prevention of the more important infections that occur in older persons in various settings.

IMMUNIZATION OF OLDER PERSONS

Tetanus–diphtheria toxoid

Justification

Even in the developed countries of the world, where tetanus is uncommon, prevention can be justified in older persons because (1) 60 per cent of reported cases occur in persons of 60 years or older, with a high (60 per cent) case fatality rate; (2) approximately half of cases are related to lacerations as opposed to puncture wounds; and (3) the prevalence of a protective level of serum antitoxin (≥0.01 units/ml) is only 35 to 50 per cent in older persons residing in the community, and 30 to 50 per cent in elderly patients in long-term care facilities, with a lower proportion of females being protected (Bentley 1987). Although there is a similarly low prevalence of protective, antitoxin antibody levels against diphtheria in older persons, diphtheria is not as serious a problem for them as tetanus.

Immunizing agent

Because there is no natural immunity to the toxin (tetanospasm) of *Clostridium tetani*, active immunization is necessary to develop protective antibodies. The recommended immunizing agent for older persons at present is tetanus–diphtheria toxoid. Protective levels of antitoxin antibody occur in approximately 40 per cent of non-immune older subjects after the first dose, in 85 per cent after a second dose, and in all after the third dose of immunizing toxoid. The duration of protection is somewhat reduced in older patients: approximately a quarter of those immunized 8 years before have serum levels of antitoxin of less than 0.01 units/ml. Over 90 per cent, however, will respond to a booster immunization by producing protective levels of antibody. Adverse reactions are uncommon: the only contraindication to the toxoid is a history of neurological complications (febrile or non-febrile convulsions, encephalopathy, or focal neurological signs) or a severe hypersensitivity reaction to the previous dose.

Current recommendations

Immunization is recommended for all older persons who are unimmunized, inadequately immunized, or whose history of immunization is unknown. The routine immunizing schedule for older subjects requires a series of three doses (primary immunization); the booster immunization is administered every 10 years after the last dose, provided that the primary series has been completed. If the history of immunization is unknown, or if fewer than three doses were given, toxoid should be given after any wounds. The specific guidelines for tetanus prophylaxis in the management of wounds are the same as recommended for all adults.

Influenza virus vaccine

Justification

Although the rates of influenza-like illness in older persons residing in the community are relatively low during an influenza epidemic year (10 per cent), high rates of influenza-associated complications occur, including hospitalization (150–636/100000/year) and death (9–106/100000) (Barker and Mullooly 1980). During influenza outbreaks in long-term care facilities, where the overall vaccination rate is over half, the rates for vaccinees and non-vaccinees, respectively, are 21 to 33/100 for influenza-like illness, 3 to 7/100 for hospitalization, 4 to 10/100 for pneumonia, and 1 to 4/100 for death.

Immunizing agent

Vaccines of inactivated influenza virus have been the principal means for preventing influenza since the late 1940s. In general, the vaccine has contained both A and B type virus, usually the types isolated in the previous winter's influenza. Recent influenza vaccines comprise an inactive trivalent preparation containing H1N1 and H3N2 influenza A antigens and influenza B antigen. Owing to the frequently changing antigenic composition of the virus, annual vaccination with the current formulation is recommended.

Acute local reactions, primarily mild to moderate soreness around the vaccination site, occur in approximately one-third of elderly vaccinees and last 1 to 2 days. Systemic reactions, including fever with or without an influenza-like illness, occur in less than 1 per cent, begin 6 to 12 h after vaccination, persist for 1 to 2 days, and appear to be less severe in older persons. Recent improvements in the preparation and purification of the vaccine have led to these acceptable rates. Thus, the fear of adverse reactions often raised by patients (and especially by staff in long-term care facilities) is not warranted. The only contraindications to vaccination are a previous history of Guillain-Barré syndrome or anaphylactic hypersensitivity to eggs (Bentley 1984).

When influenza vaccine antigens closely matched with the

Dr Falsey was supported in part by NIH/NRSA Training Award in Immunology and Infectious Diseases (T32 AG00150).

epidemic strain are studied in placebo-controlled trials, their efficacy in reducing influenza infection in healthy older persons living in the community is as high as 96 per cent (Bentley 1987). Studies in nursing homes, however, have found considerably less efficacy in reducing uncomplicated, influenza-like illness (28–37 per cent) but impressive rates of efficacy in reducing complications, including admission to hospital (47 per cent), pneumonia (58 per cent), and death (76 per cent) (Patriarca *et al.* 1985).

Current recommendations

Prevention of influenza includes measures such as annual immunization with the current, inactivated, trivalent vaccine for high-risk older persons, especially those being treated for cardiopulmonary conditions. Patients in nursing homes and in other long-term care facilities are among the highest priority for targeted vaccination in the United States. Physicians, nurses, and other health care professionals who have extensive contact with these high-risk older persons should also receive influenza vaccine annually.

Amantadine (and rimantadine when licensed) is also recommended for prevention and treatment of influenza A virus infections in high-risk elderly patients in the United States. The dose is 100 mg, daily by mouth with reduction for renal insufficiency. Amantadine is currently recommended for prophylaxis before an outbreak: (1) to supplement protection afforded by vaccination (6–12 weeks), and (2) as the only prophylactic measure in patients for whom influenza vaccine is contraindicated (up to 12 weeks); and during an outbreak: (1) as an adjunct to late immunization (2 weeks), (2) as an adjunct to vaccination to provide additional protection to the highest-risk patients (2–3 weeks), and (3) as therapy in the treatment of uncomplicated influenza (5–7 days).

{Anti-influenzal prophylaxis, although available to individual patients in the community, and sometimes offered to all the residents of non-nursing institutions or elderly people, is not routinely used in long-term care/nursing facilities in the UK (Lennox *et al.* 1990). This is a controversial topic, but the geriatricians who do not advocate routine prophylaxis offer a twofold rationale. First, care objectives are set on an individual basis and for many patients in long-term facilities, most of whom are both mentally and physically severely disabled, prolongation of life is not a defined objective, although well-being, comfort, and autonomy are. Secondly, there are ethical uncertainties about the fact that few such patients are able to give informed consent to a procedure that is not aimed at immediate relief of distressing symptoms, or immediately life-saving, and may be largely for the benefit of people other than the patient. These considerations do not preclude the use of prophylaxis for individual patients for whom it is judged to be appropriate.}

Pneumococcal vaccine

Justification

Pneumococcal disease remains an important cause of illness and death in older persons. The estimated incidence of pneumococcal pneumonia is approximately 3/100000.year in persons living in the community and 15/100000.year in elderly patients in long-term care facilities. In addition, case fatality rates as high as 40 per cent for bacteraemia and 55 per cent for meningitis occur in older persons, despite the availability of potent antimicrobials such as penicillin. Moreover, even when penicillin is used in the first day or two of illness, there is a limited effect on the outcome of the disease among those 'destined to die' within the first 5 days of illness (Austrian and Gold 1964).

Immunizing agent

The current recommended immunizing agent is pneumococcal vaccine, polyvalent. The 'first generation' vaccine, introduced in the United States in 1978, contained purified capsular polysaccharides from 14 of the 83 different serotypes of *Streptococcus pneumoniae*. In July 1983, an expanded, 23-serotype vaccine was licensed, which included several more serotypes of *S. pneumoniae* that frequently cause pneumonia in long-term care facilities.

Vaccine-associated reactions occur within 24 h of injection in 10 to 15 per cent of elderly vaccinees, consist primarily of local discomfort, erythema, and induration, and last 1 to 2 days. Fever occurs in approximately 2 per cent and generally lasts less than 24 h. Severe local and systemic reactions with high fevers (103°F), headache, myalgias, and chills, have been reported in less than 1 per cent, usually in younger adults and persons revaccinated. The only contraindication to the vaccine is a history of allergy to one of the vaccine components, usually the phenol or thimerasol diluent, or a well-documented history of previous vaccination (see revaccination below) (Bentley 1984).

Randomized, controlled trials of the efficacy of pneumococcal vaccine in community-residing older adults have had less than satisfactory results. However, studies comparing the distribution of the vaccine serotypes of *S. pneumoniae* isolated from vaccinated and unvaccinated elderly persons with bacteraemia have shown an estimated efficacy of approximately 60 per cent for persons over 65 years of age, with or without chronic underlying diseases. A similar efficacy was found for bacteraemic patients over 55 years of age in recent case-control studies. Other studies, although uncontrolled, suggest that vaccination may be less effective in institutionalized elderly people (Bentley 1987). Most likely these conflicting results are due to the inclusion (or exclusion) of elderly patients who did not make or sustain 'protective' serum antibody levels. Nevertheless the data support the use of pneumococcal vaccine for certain, well-defined groups at risk (LaForce and Eickhoff 1988)

Current recommendations

Pneumococcal vaccine is recommended in the United States for all immunocompetent older persons who are at increased risk of pneumococcal disease or its complications, especially those with chronic illnesses (e.g., cardiovascular disease, pulmonary disease, diabetes mellitus, alcoholism, cirrhosis, or leakage of cerebrospinal fluid). In addition, it is recommended for immunocompromised older persons at increased risk for pneumococcal disease or its complications (e.g., those with anatomical or functional asplenia, including sickle-cell diseases, nephrotic syndrome, Hodgkin's disease, multiple myeloma, chronic renal failure, organ transplantation, or human immunodeficiency virus (**HIV**) infection, although such patients will probably have a suboptimal antibody response). Pneumococcal vaccine and influenza vaccine can be given at the same time, if different sites are used, without decreasing the antibody response of either vaccine or substantially increasing the side-effects (Centers for Disease Control 1989). The vaccine can be recommended for general use in long-term care facilities when epidemics or high endemic rates of vaccine-type pneumococcal pneumonia occur.

Because propensity for adverse reactions is correlated with elevated, prevaccination levels of antibody, and because further doses of pneumococcal vaccine provide a poor 'booster' response, the vaccine should be given only once during the lifetime of the older person. The optimal time for vaccination is unknown but probably is between 50 and 60 years of age. Revaccination with the 23-valent vaccine may be considered for older persons who received the 14-valent pneumococcal vaccine, provided that the individual patient is considered at highest risk for fatal pneumococcal infection (see recent recommendations for important details; Centers for Disease Control 1989).

NOSOCOMIAL INFECTIONS

Definitions

'Infection' is the replication of micro-organisms in the tissue of the host; 'disease' is the clinical expression of infection. 'Colonization' is the replication of micro-organisms on or in the host tissue without causing clinical disease. The term 'nosocomial' refers to colonization or disease not present or incubating when the patient entered the long-term care facility or hospital. Nosocomial infections usually occur 48 to 72 h after admission.

Prevalence/incidence

The frequency of nosocomial infections in adults admitted to general medical and surgical hospitals is approximately 5 per cent. Four major sites or types—urinary tract, surgical wound, lower respiratory tract, and bacteraemia—account for 90 per cent of all hospital-acquired infections. Estimates for site-specific rates per 100 admissions are: urinary tract, 2.4; surgical wound, 1.4; pneumonia, 0.6 and bacteraemia, 0.3 (Haley *et al.* 1981). Significant illness and death are attributable to nosocomial infections. In one study of 200 patients, there were 88 nosocomial infections and 63 patients died; infection was contributory in 52 of these 63 patients, most of whom were in their late 70s (Gross *et al.* 1980).

Risk factors

Several host factors increase the risk of nosocomial infection. Depressed immunity from steroids, cancer chemotherapy, malignancies, malnutrition, etc., occurs frequently. Often nosocomial infections result from the interruption of normal protective barriers by surgical procedures or 'medical' instrumentation, i.e., intravenous catheters, nasogastric tubes, indwelling urinary catheters, endotracheal tubes, and mechanical ventilation. The length of the preoperative stay and of surgery are also associated with increased rates of postoperative infections (Haley *et al.* 1981).

Increasing age alone is associated with higher rates of nosocomial infection. The decade-specific rates increase logarithmically after the age of 49 years, from one infection per 1000 discharges to 100 infections per 1000 discharges at 70 years. Although patients over 60 years of age account for less than a quarter of the total number of admissions to hospital, approximately 65 per cent of all nosocomial infections occur in this high-risk group (Gross *et al.* 1983). Site-specific infection rates increase with age for each of the four major sites (or types): fivefold for urinary tract infections and bacteraemia, threefold for pneumonia, and twofold for surgical wound infections (Haley *et al.* 1981).

Urinary tract infections

Epidemiology

Urinary tract infections account for approximately 30 to 50 per cent of all nosocomial infections (Turck and Stamm 1981). These or primary asymptomatic bacteriuria occur in 15 to 20 per cent of men and 25 to 50 per cent of women in long-term care facilities (Norman *et al.* 1987).

Risk factors

The majority of urinary tract infections are associated with instrumentation, either catheterization (80 per cent) or urological manipulation (20 per cent). Chronic indwelling urinary catheters (in place for 30 days or more) are associated with the highest rates (98 per cent) of bacteriuria; 77 per cent are polymicrobial. Lower rates of bacteriuria are found with intermittent catheterization and external condom drainage (Turck and Stamm 1981). Risk factors associated with the increased prevalence of infections in patients with indwelling catheters are more than 8 days of catheterization, interruption of the closed system, and catheterization by inexperienced individuals (Garibaldi *et al.* 1974). Other factors associated with increased risk of nosocomial urinary tract infections are female sex, debilitation, diabetes mellitus, and a serum creatinine of more than 2 mg/100 ml (177 μmol/l) (Platt *et al.* 1986).

Bacteriuria and urinary tract infections increase with age for both men and women. This increased risk can be demonstrated after only a single urethral catheterization. Age-related factors that promote the development of these infections in women include loss of pelvic support, changes in epithelium and increased rates of asymptomatic bacteriuria (Norman *et al.* 1987). Elderly males have decreased prostatic secretions and obstructive prostatic diseases. Elderly men and women experience greater rates of change in bladder function, incontinence, and increased residual volumes. The skin and mucous membranes of elderly patients are also more frequently colonized by Gram-negative bacilli. A patient in a long-term care facility is particularly at high risk, through greater debilitation, immobility and incontinence, and increased use of chronic urinary catheters.

Microbiology

Approximately 30 to 50 per cent of nosocomial urinary tract infections are caused by *Escherichia coli* with the remainder caused by *Proteus* and *Enterobacter* spp. With chronic indwelling urinary catheters, Providencia, Serratia, *Pseudomonas aeruginosa*, Enterococcus, and other resistant organisms are identified (Turck and Stamm 1981). The bacterial species causing such infections in long-term care facilities reflect the environmental flora, the patient's endogenous flora, antibiotic use, the frequent readmission of patients to hospitals, and care practices in nursing homes. These result in large numbers (65 per cent) of patients colonized or infected with antibiotic-resistant organisms (Gaynes *et al.* 1985).

Prevention

Prevention of urinary tract infection in both acute hospitals and long-term care facilities involves attention to the few correctable risk factors. General measures include the correction of reversible host factors, appropriate use of antibiotics, rigorous hand washing, and other infection control measures to reduce the risk of creating and spreading multiply resistant organisms. Minimizing the use of chronic indwelling urethral catheters and substituting intermittent catheterization, external condom catheters, and diapers are important measures in the long-term care facility.

{In the United Kingdom urinary tract infections in catheterized patients are not routinely treated unless symptoms or fever suggest bacteraemia or ascending infection. Some geriatricians advocate routine culture of the urine of catheterized patients in order to be able to choose appropriate antibiotics if indications for treatment appear. Others do not consider this cost-effective.}

Lower respiratory infections

Epidemiology

Pneumonia accounts for approximately 10 to 20 per cent of all nosocomial infections (Ungar and Bartlett 1983). Although it ranks third in frequency, behind urinary tract infections and surgical wound infections, it is the leading cause of death associated with nosocomial infection. The incidence of community-acquired pneumonia in older persons is 20 to 40 per 1000 persons per year. The incidence of nosocomial pneumonia is 8.6 per 1000 admissions per year and as high as 70 to 115 per 1000 per year in older persons in long-term care facilities. Death rates range from 10 to 40 per cent for bacterial agents causing minimum lung necrosis, e.g., *S. pneumoniae* and *Haemophilius influenzae*, to as high as 80 per cent for *Staphylococcus aureus* or Gram-negative bacilli that frequently cause necrosis of the lungs (Bentley 1986)

Risk factors

Nosocomial pneumonia occurs primarily as the result of silent aspiration of bacteria-laden oropharyngeal secretions; haematogenous spread and inhalation of contaminated aerosols are much less frequent. Many of the risk factors for nosocomial pneumonia involve alteration of the normal pulmonary clearance mechanisms, with increased risk of aspiration, i.e. decreased consciousness, nasogastric tubes, intubation, history of underlying lung disease, immunosuppression, prior antibiotic use, and prolonged stays in hospital. Surgical patients, especially after thoracoabdominal surgery, are more likely to develop pneumonia than medical patients (Celis *et al.* 1988). Admission to the medical intensive care unit frequently results in colonization (45 per cent) with aerobic Gram-negative bacilli and subsequent pneumonia (12 per cent) (Johanson *et al.* 1972).

Advanced age is associated with increased rates of, and a worse prognosis for, nosocomial pneumonia. Ageing-associated morphological and physiological changes that increase suceptibility to infection include loss of elastic tissue surrounding the alveoli, increasing anterioposterior diameter secondary to rib and vertebral decalcification, and weakening of the respiratory muscles. This leads to declining forced expiratory volume and forced vital capacity, and ventilation/perfusion mismatch (Dhar *et al.* 1976). Older patients are at increased risk for silent aspiration through reduced gag reflexes, an impaired cough mechanism, oesophageal dysmotility, and the reduced levels of consciousness frequently associated with cerebrovascular accidents. Older persons are also more commonly colonized by Gram-negative bacilli, with rates varying from approximately 20 per cent in those residing in the community to 40 per cent in patients in long-term care facilities, and to as high as 60 per cent in hospitalized elderly patients (Valenti *et al.* 1978).

Microbiology

Over 60 per cent of community-acquired pneumonia in older persons is due to *S. pneumoniae* and *H. influenzae* (non-typable). In the hospital setting, over 60 per cent of nosocomial pneumonias are due to Gram-negative bacilli. The pneumonia acquired in long-term care facilities may have a mixed flora (40 per cent), *S. pneumoniae* (35 per cent) or Gram-negative bacilli (15 per cent) (Bentley 1986).

Species of *Legionella* are a less common but well-recognized causes of nosocomial pneumonia in the elderly. The role of anaerobes is not well defined because of the difficulty in obtaining appropriate specimens. However, mixed infections secondary to aspiration may involve anaerobic organisms.

Prevention

There have been few studies of this, but probably most pneumonias in older persons are not preventable. General guidelines that may be helpful in reducing the occurrence of pneumonia include careful disinfection of respiratory devices, aseptic care of tracheostomies, and suctioning of ventilated patients. Respiratory therapy and early mobilization are important in postoperative care. Judicious use of antibiotics can limit the development of resistant pathogens. Simple techniques, such as raising the head of the bed, avoiding sedative drugs, alleviating abdominal distension, and providing dental care, are appropriate measures that may reduce the rates of pneumonia (Bentley 1986)

Skin and soft tissue infections

Postoperative wound infections

The occurrence of surgical wound infections ranges from 5 to 17 per cent, and these account for approximately 30 per cent of all nosocomial infections. The incidence of postoperative wound infections increases with increasing age. Infection rates as much as twofold higher than in younger persons have been found in patients over 65 years old (Haley *et al.* 1981).

General risk factors associated with postoperative wound infections include contaminated surgeries, retained foreign bodies, ischaemia, drains, delayed closure of wounds, and the length of the operation (Cruse 1986). Host factors associated with increased rates of infections include diabetes, obesity, and malnutrition. A number of age-specific changes in the skin appear to delay wound healing; these include a decrease in thickness of the epidermis and dermis, and loss of subcutaneous tissue. A relatively avascular dermis, loss of its resilience and elasticity, and a decrease by half in the turnover rate of the epidermis in elderly persons may also contribute to delayed wound healing (Gilchrest 1982).

The microbiology of postoperative wound infections is influenced by site of the surgery. Enterococci, enteric Gram-negative bacilli, and anaerobes are most common in gastrointestinal and gynaecological surgery. In orthopaedic and cardiothoracic surgical infections, staphylococci and streptococci predominate (Cruse 1986).

Prevention of postoperative wound infection is beyond the scope of this chapter. General considerations include making good the patient's condition preoperatively, especially by treating any underlying cardiopulmonary conditions, reducing the length of the operation, careful handling of tissues, appropriate use of prophylactic antibiotics, and strict attention to sterile technique.

Pressure sores and their related infections, including prevention and management, are discussed in Chapter 3.2. Contact isolation and wound precautions should be instituted in patients with multiply resistant, Gram-negative bacilli or *Staph. aureus*, depending on the degree and containment of purulent drainage.

Infections caused by herpes simplex and herpes zoster viruses may occur in elderly people. The incidence of zoster peaks at 50 to 80 years of age as a result of waning cellular immunity. Prevention of herpes simplex or zoster infection is not possible for the individual patient. Immunosuppressed patients and health care personnel not immune to varicella–zoster virus should avoid contact with patients with herpes zoster. Cutaneous lesions of herpes simplex or zoster remain infectious until dry and crusted. Precautions against spread of infected secretions are recommended for those individuals with localized disease. For patients with disseminated herpes infection, strict isolation is recommended because of possible airborne transmission (Valenti *et al*. 1981).

The varied course and management of herpes zoster are discussed in Chapter 3.2.

Bacteraemia

Primary bacteraemias account for approximately 6 per cent of hospital-acquired infections. Age-specific rates are 5 to 10 times greater in older than younger persons (Haley *et al*. 1981). The major portals of entry are percutaneous devices such as intravenous catheters, central lines, and Hickman or Broviac venous catheters. Approximately 25 000 patients in the United States develop device-related bacteraemia each year. Risk factors include granulocytopenia, immunosuppressive therapy, loss of skin integrity, severe underlying disease, and remote infection (Henderson 1985).

The incidence of bacteraemia in one long-term care facility was approximately 0.3 per 1000 patient-care days. Causative agents include aerobic Gram-negative bacilli (65 per cent), Gram-positive cocci (25 per cent), and mixed infections (10 per cent). *Staph. aureus*, *E. coli*, *Klebsiella* spp., Proteus, and Enterobacter are the most frequent pathogens. Portals of entry are urine (55 per cent), respiratory tract (10 per cent), and skin and soft tissue (15 per cent). Urinary tract-associated sepsis is the most common source of bacteraemia in older patients and is frequently caused by trauma, obstruction, or manipulation of the indwelling bladder catheter (Setia *et al*. 1984).

Prevention of bacteraemia is dependent on early detection and treatment of distant foci of infection, such as in the urinary or respiratory tract. Careful attention to sites of intravenous catheter is also imperative.

Long-term care facilities

Epidemiology

Infection is a significant cause of illness and death in these facilities, e.g., nursing homes. Prevalence rates range from 2 to 18 per cent, with an overall incidence of infection of approximately 4 to 10 cases per 1 000 patient-days per annum (Setia *et al*. 1985). In a 1-day study of seven urban nursing homes, the prevalence rate of nosocomial infections was 16.2 per cent (Garibaldi *et al*. 1981). The most frequent sites for infection were pressure sores (6 per cent), the conjunctiva (3.4 per cent), urinary tract (2.6 per cent), lower respiratory tract (2.6 per cent), upper respiratory tract (1.5 per cent), and gastroenteritis (1.3 per cent). Well-accepted criteria and definitions for infections acquired in such facilities have, however, not been defined; when they are, the above rates may require revision.

Differences from the hospital setting

Several risk factors predispose patients to these high rates of infection, including alterations in the immune response associated with advanced age, multiple underlying diseases, immobility, impaired cognitive functions, and urinary and faecal incontinence. In addition, these patients are exposed to longer periods of risk, with an average length of stay of 13 months. There are chronic indwelling bladder catheters, in place for more than 30 days, in 12 to 30 per cent of these patients (Setia *et al*. 1985).

The semiclosed environment promotes a higher risk for infection, with group activities, common washing and dining facilities, and crowding. These factors provide extensive opportunities for direct contact between patients, and between patients and staff. Patients are frequently transferred to and from hospitals, which provides an excellent opportunity for the exchange of multiply, resistant pathogens. These include aerobic Gram-negative bacilli, such as *Proteus, Providencia* and *Morganella* spp., as well as Enterococci and methicillin-resistant *Staph. aureus*.

Extensive and inappropriate use of antibiotics further contributes to the establishment of dangerous reservoirs of multiply resistant pathogens in these patients. Antibiotics affect the normal balance of the indigenous flora and select for resistant strains. In prevalence studies it has been found that, at any one time, 7 to 10 per cent of the patients in long-term care facilities are receiving systemic antibiotics (usually by mouth). Use of these antibiotics is inappropriate in approximately 40 per cent of cases (Zimmer *et al*. 1986). Sedatives and tranquillizers, which are frequently (up to 50 per cent of patients) used in such facilities may predispose to nosocomial pneumonia; anticholinergics can dry respiratory secretions and cause bladder dysfunction.

Most long-term care facilities in the United States have programmes for infection control and a designated infection control officer. Unfortunately, the officer, usually a nurse with other responsibilities, often has little formal training and on average spends less than 5 h a week on duties related to infection control. Many institutions have no specific policy for isolation procedures, urinary catheter care, pre-employment health screening, and employee sick leave. Procedures for screening residents on admission and for vaccination are also quite variable, with many residents not having had purified protein derivative, tetanus toxoid, and influenza and pneumococcal vaccines. Furthermore, direct patient care is most often by orderlies and aides to nurses who have little or no formal medical training (Crossley *et al.* 1985). These features, plus the rapid turnover of staff and high patient to staff ratios, all contribute to problems in educating the staff and add to the difficulties in controlling nosocomial infections in this setting.

Specific infections

Respiratory viral infections

These infections occur frequently in long-term care facilities and may be the cause of significant illness and death. Influenza is a highly contagious disease spread by respiratory droplets. Both influenza A and B viruses can cause nosocomial outbreaks among residents and staff, with infection rates of approximately 30 per cent in patients. Death rates are as high as 1 to 4 per 100 infected patients; it is mostly frail, debilitated patients who die, especially those with underlying cardiopulmonary disease. Several measures can be instituted when there are influenza outbreaks. These include closing the unit to new admissions, limiting patients' movements off the unit to essential medical activities, and discouraging personnel from 'floating' off or on the unit. Modified respiratory isolation, in which the patient remains room-bound for the first 5 days of the illness, is reasonable, but the delay in diagnosis may cause this measure to be ineffective once the patient is ill (Bentley 1986). Current recommendations on the use of amantadine prophylaxis and influenza vaccine have been discussed above.

Respiratory syncytial virus is a frequent respiratory pathogen for children under 2 years, but can also cause outbreaks in long-term care facilities and geriatric hospitals. Symptoms may mimic influenza and fatal pneumonias have been reported. Respiratory syncytial virus and influenza viruses can circulate together during the same outbreak of an influenza-like illness. The former is spread by large respiratory droplets and fomites, those requiring close contact with contaminated secretions. There is usually, therefore, a slow, steady spread of new cases as opposed to the more explosive outbreak associated with influenza infection. Important measures in the control of respiratory syncytial virus include many of those for influenza outbreaks, but with more attention to careful hand washing and the use of gowns and gloves when direct contact with secretion is unavoidable (Sorvillo *et al.* 1984).

Several outbreaks of parainfluenza virus infection have been reported. Symptoms include cough, fever, and pharyngitis. Attack rates range from 23 to 56 per cent, with low rates of pneumonia and death (Centers for Disease Control 1978).

Tuberculosis

Recently, several clusters or outbreaks of tuberculosis have been identified in long-term care facilities. Diagnosis may be delayed because of a low index of suspicion and concomitant confusing chronic conditions. Individuals with cavitary tuberculosis may infect large numbers of patients and staff, who then have a significant risk of developing active clinical disease (Stead 1981). The tubercle bacillus is spread by the airborne route; respiratory precautions are necessary in any patient with suspected active pulmonary tuberculosis. The two-step, purified protein derivative, skin test is recommended for all patients on admission, and every 1 to 2 years for patients and personnel with negative reactions (see also Chapter 3.2).

Infectious gastroenteritis

A number of agents may cause infectious gastroenteritis in patients in long-term care facilities. Viral gastroenteritis is the most common and is usually a self-limiting disease. Both Norwalk agent and rotavirus have been identified in outbreaks of diarrhoea (Marrie *et al.* 1982). Salmonellosis, from contaminated food or person-to-person spread, also occurs in these facilities (Schroeder *et al.* 1968). Other less common bacterial pathogens include Shigella, Yersinia, and Campylobacter. Although uncommon, *E. coli* (serotype-0157:H7) can cause outbreaks of haemorrhagic colitis and haemolytic uraemic syndrome with high death rates. Contaminated food is the usual vehicle of transmission (Ryan *et al.* 1986). Pseudomembranous colitis caused by a toxogenic strain of *Cl. difficile* also occurs, probably related to the widespread use of antibiotics. Enteric precautions should be instituted promptly whenever infectious gastroenteritis is suspected in order to reduce the spread of pathogens, including *Cl. difficile*.

Scabies

Scabies is a skin disease caused by the mite *Sarcoptes scabiei*, which results in a generalized, pruritic eruption. Outbreaks in long-term care facilities are caused primarily by direct contact but infection may be spread by contaminated objects such as the patient's clothes and bedding. Outbreaks of scabies in these facilities are frequent. Prevention of spread involves prompt recognition of suspected cases, scrapings of suspicious lesions to confirm the diagnosis, and treatment with lindane lotion or cream. In addition, the patient's immediate environment should be thoroughly cleaned, especially by laundering of clothes and bedding with heat drying.

Other

Conjunctivitis is frequently found in patients in long-term care facilities. Most infections are viral in origin or due to *Staph. aureus*. Nosocomial spread occurs, with clusters of cases among staff and patients following direct contact and poor hand washing practices.

Creutzfeldt–Jakob disease is a dementing illness caused

by a transmissible agent. It may be difficult to distinguish it from dementia of the Alzheimer type which is often present in patients in these facilities. The diagnosis should be suspected in a rapidly progressive dementia with extrapyramidal signs, cerebellar ataxia, and myoclonus. Such patients should be placed on blood and body fluids isolation until they can be carefully assessed by a neurological specialist. Some authorities feel that any demented patient should not be allowed to donate blood or organs (Gjadusek *et al.* 1977).

The role of the long-term care facility in caring for patients infected with HIV and afflicted with acquired immune deficiency syndrome (**AIDS**) is currently being defined. An increasing number of persons who are HIV-positive or have AIDS are living longer, with periods of relative stable, chronic disability, so the long-term care facility can be expected to play an increasing role in the continued care of selected patients. Effective implementation of universal precautions, plus the use of specific precautions for blood and body fluids will provide nursing staff with appropriate safeguards. In addition, they must become more knowledgeable and skilled in avoiding accidental wounds from needle sticks and sharp instruments (Bentley and Cheney 1989).

References

Austrian, R. and Gold, J. (1964). Pneumococcal bacteremia with especial reference to bacteremic pneumococcal pneumonia. *Annals of Internal Medicine*, **60**, 759–79.

Barker, W.H. and Mullooly, J. P. (1980). Impact of epidemic type A influenza in a defined adult population. *American Journal of Epidemiology*, **112**, 789–813.

Bentley, D. W. (1984). Immunization for the elderly. In *Immunology and infection* (ed. R. A. Fox), pp. 333–70. Churchill Livingstone, London.

Bentley, D.W. (1986). Infectious Diseases. In *Clinical geriatrics* (3rd edn) (ed. I. Rossman), pp. 438–71. Lippincott, Philadelphia.

Bentley, D.W. (1987). Immunizations in the elderly. *Bulletin of the New York Academy of Sciences*, **63**, 533–51.

Bentley, D.W. and Cheney, L. (1989). AIDS in the nursing home. In *Infections in the nursing home* (ed. A. Verghese and S. L. Berk), pp. 193–202. Karger, Zurich.

Celis, R., Torres, A., Gatell, J.M., Almela, M., Rodriguez-Roisin, R., and Agusti-Vidal, A. (1988). Nosocomial pneumonia: a multivariate analysis of risk and prognosis. *Chest*, **93**, 318–24.

Centers for Disease Control (1978). Parainfluenza outbreaks in extended care facilities—United States. *Morbidity and Mortality Weekly Report*, **27**, 475–6.

Centers for Disease Control (1989). Pneumococcal polysaccharide vaccine. *Morbidity and Mortality Weekly Report*, **38**, 64–76.

Crossley, K.B., Irvine, P., Kaszar, D.J., and Loewenson, R.B. (1985). Infection control practices in Minnesota nursing homes. *Journal of the American Medical Association*, **254**, 2918–21.

Cruse, P. (1986). Surgical infection: incisional wounds. In *Hospital infections* (2nd edn) (ed. J. V. Bennett, and P. S. Brachman), pp. 423–36. Little, Brown, Boston.

Dhar, S., Subramaniam, R.S. and Lenora, R.A.K. (1976). Aging and the respiratory system. *Medical Clinics of North America*, **60**, 1121–3.

Garibaldi, R., Burke, J.P., Dickman, M.L., and Smith, C.B. (1974). Factors predisposing to bacteriuria during indwelling urethral catheterization. *New England Journal of Medicine* **291**, 215–19.

Garibaldi, R.A., Brodine, S., and Matsumiya, S. (1981). Infections among patients in nursing homes: policies, prevalence, and problems. *New England Journal of Medicine*, **305**, 731–5.

Gaynes, R.P., Weinstein, R.A., Chamberlin, W., and Kabins, S.A. (1985). Antibiotic-resistant flora in nursing home patients admitted to the hospital. *Archives of Internal Medicine*, **145**, 1804–7.

Gilchrist, B. (1982). Age-associated changes in the skin. *Journal of the American Geriatrics Society*, **30**, 139–45.

Gjadusek, D.C. *et al.* (1977). Precautions in medical care of, and in handling materials from, patients with transmissible virus dementia (Creutzfeldt–Jakob disease). *New England Journal of Medicine*, **297**, 1253–8.

Gross, P.A., Neu, H.C., Aswapokee, P., van Antwerpen, C., and Aswapokee, N. (1980). Deaths from nosocomial infections: experience in a university hospital and a community hospital. *American Journal of Medicine*, **68**, 219–23.

Gross, P.A., Rapuano, C., Adrignolo, A., and Shaw, B. (1983). Nosocomial infections: decade-specific risk. *Infection Control*, **4**, 145–7.

Haley, R.W. *et al.* (1981). Nosocomial infections in U.S. Hospitals, 1975–1976: estimated frequency by selected characteristics of patients. *American Journal of Medicine*, **70**, 947–59.

Henderson, D.K. (1985). Bacteremia due to percutaneous intravascular devices. In *Principles and practice of infectious diseases* (2nd edn.) (ed. G. L. Mandell, R. G. Douglas, and J. E. Bennett), pp. 1612–20. Wiley, New York.

Johanson, W.G., Pierce, A.K., Sanford, J.P., and Thomas, G.D. (1972). Nosocomial respiratory infections with Gram-negative bacilli: the significance of colonization of the respiratory tract. *Annals of Internal Medicine*, **77**, 701–6.

LaForce, F.M. and Eickoff, T.C. (1988). Pneumococcal vaccine: an emerging consensus. *Annals of Internal Medicine*, **108**, 757–9.

Lennox, I.M., Macphee, A., McAlpine, C.H., Cameron, S.O., Leask, B.G.S., and Somerville, R.G. (1990). Use of influenza vaccine in long-stay geriatric units. *Age and Ageing*, **19**, 169–72.

Marrie, T.J., Lee, S.H.S., Faulkner, R.S., Ethier, J., and Young, C. H. (1982). Rotavirus infection in a geriatric population. *Archives of Internal Medicine*, **142**, 313–16.

Norman, D.C., Castle, S.C., and Cantrell, M. (1987). Infections in the nursing home. *Journal of the American Geriatrics Society*, **35**, 796–805.

Patriarca, P.A. *et al.* (1985). Efficacy of influenza vaccine in nursing homes: Reduction of illness and complications during an influenza A (H3N2) epidemic. *Journal of the American Medical Association*, **253**, 1136–9.

Platt, R., Polk, B.F., Murdock, B., and Rosner, B. (1986). Risk factors for nosocomial urinary tract infection. *American Journal of Epidemiology*, **124**, 977–85.

Ryan, C.A. *et al.* (1986). *Escherichia coli* 0157:H7 diarrhea in a nursing home: clinical epidemiological, and pathological findings. *Journal of Infectious Disease*, **154**, 631–8.

Schroeder, S.A., Aserkoff, B., and Brachman, P.S. (1968). Epidemic salmonellosis in hospitals and institutions: a five-year review. *New England Journal of Medicine*, **279**, 674–8.

Setia, U., Serventi, I., and Lorenz, P. (1984). Bacteremia in a long-term care facility: spectrum and mortality. *Archives of Internal Medicine*, **144**, 1633–5.

Setia, U., Serventi, I., and Lorenz, P. (1985). Nosocomial infections among patients in a long-term care facility: spectrum, prevalence, and risk factors. *American Journal of Infection Control*, **13**, 57–62.

Sorvillo, F.J., Iluic, S.F., Strassburg, M.A., Butsumyo, A., Shandera, W.X., and Fannin, S.L. (1984). An outbreak of respiratory syncytial virus pneumonia in a nursing home for the elderly. *Journal of Infection*, **9**, 252–6.

Stead, W.W. (1981). Tuberculosis among elderly persons: an outbreak in a nursing home. *Annals of Internal Medicine*, **94**, 606–10.

Turck, M. and Stamm, W. (1981). Nosocomial infection of the urinary tract. *American Journal of Medicine*, **70**, 651–4.

Ungar, B.L.P. and Bartlett, J.G. (1983). Nosocomial Pneumonia. In *Infections in the elderly* (ed. R.A. Gleckman and N. M. Gantz), pp. 91–116. Little, Brown, Boston.

Valenti, W.M., Trudell, R.G., and Bentley, D.W. (1978). Factors predisposing to oropharyngeal colonization with Gram-negative bacilli in the aged. *New England Journal of Medicine*, **298**, 1108–11.

Valenti, W.M. *et al.* (1981). Nosocomial viral infections: guidelines for prevention and control of respiratory viruses, herpes viruses, and hepatitis viruses. *Infection Control*, **1**, 165–177.

Zimmer, J.G., Bentley, D.W., Valenti, W.M. and Watson, N.M. (1986). Systemic antibiotic use in nursing homes: a quality assessment. *Journal of the American Geriatrics Society*, **34**, 704–10.

SECTION 4
Injuries in later life

4.1 Epidemiology and environmental aspects

RICHARD W. SATTIN AND MICHAEL C. NEVITT

BACKGROUND

Injuries are the fourth leading cause of death in the United States, exceeded only by heart disease, cancer, and cerebrovascular disease. Among older people, injuries are a major public-health concern, and they exact a tremendous toll in terms of death, illness, disability, and medical, economic, and emotional costs. Older people are at particularly high risk for some injuries, and, in otherwise equivalent circumstances, are more likely to sustain more severe injuries than a younger person. Moreover, older people often have a worse outcome from the same injury than younger people because of impaired tissue regeneration, decreased functional reserves, and poorer immunological function. After some events, such as a fall, a person can develop fear, a decrease in mobility and activity, or a change in the quality of life, even if not injured.

Despite this tremendous toll, injury in older people has not received much attention, until recently, from health professionals or the public as a preventable health problem (Committee on Trauma Research 1985; Sattin *et al.* 1988). At lease two misconceptions have contributed to this lack of attention. The first is that injuries are 'accidents', a word connoting randomness, occurrence without a pattern or predictability, and unanticipated damage that occurs quickly without the ability to prevent its effects. The word 'accident' hinders health professionals and the public from regarding injuries as preventable and should be avoided. The second misconception is that techniques used to study other diseases cannot be applied to injuries. As with other diseases, however, injuries can be understood as a problem in medical ecology, that is, as a relationship between a person (the host), an agent, and the environment.

THE ECOLOGY OF INJURY

Injury refers to harm or damage resulting from acute exposure to energy (mechanical, thermal, chemical, electrical, or radiational) or the absence of specific body needs, such as oxygen or heat (Baker *et al.* 1984). For example, too much thermal energy may damage cells, tissues, and other structures—resulting in a burn. Most injuries, however, result from excessive exposure to mechanical energy, such as during falls or car crashes.

The host–agent–environment model used to describe processes leading to communicable disease can also be used to describe processes leading to injury (Haddon and Baker 1981). The host is the injured person; the agent is the energy that damages tissues; and the environment includes all the conditions, circumstances, and influences surrounding and affecting the person. When components of the host, agent, and environment interact, injury may occur. To prevent this, measures are needed to alter the relationships between each of these components so as to interfere with the transmission of excessive energy.

The dose and duration of the dose of energy received, and the response of the organism to the transfer of energy, can determine the difference between an occurrence of injury or that of a chronic disease (Haddon and Baker 1981). A large energy load quickly transmitted usually results in an injury. If the same load is transmitted in smaller doses over time, the body can mobilize different responses, with the result being a chronic disease. For example, excessive ingestion of alcohol over many years might lead to cirrhosis of the liver, but acute alcohol poisoning—an injury—could lead to coma and death; exposure to the sun over many years might cause premature ageing or skin cancer, but acute overexposure—an injury—could result in a serious burn. Injury, therefore, can be considered a disease that has a short latency.

In injuries due to mechanical (impact) energy (the major cause of injury in older persons), the resistance of the body through inertial forces, the elasticity of the tissues, and the viscous tolerance of the organs play an important part (Committee on Trauma Research 1985). Inertial forces from excessive acceleration can lead to the tearing of an organ. An example is brain injury resulting from sudden acceleration of the skull—with the brain lagging behind. The elastic capacity of the tissues can be exceeded when the tensile or compressive strain exceeds the recoverable limit. Older people tend to have decreased elasticity of tissue. Fractures of the hip, ribs, and skull are examples of this mechanism. The viscous tolerance can be exceeded during high-speed impact, which can lead to contusion and possible rupture of an organ. Tissues and organ systems are sensitive to the rate of deceleration. For example, the heart may sustain damage from excessive motion rapidly applied to the sternum during a car crash. The same compression slowly applied would not necessarily damage the heart because it can tolerate gradual compression.

THE RESPONSE TO INJURY

Older people tend to have poorer responses to injury than younger people (Baker *et al.* 1984). With ageing, physiological changes occur in articular cartilage, bone, ligaments, and musculature (States 1985). These changes can lead to osteoporosis, arthritis, decreased muscle strength and mass, decreased joint flexibility, decreased elasticity and strength of collagen, and general discomfort and pain. People with these changes might respond more slowly during difficult or emergency events, or develop early and excessive fatigue, which might lead to an injury. In addition, the most effective energy absorber in the human body, the active musculature, is mostly dependent on muscle strength, which almost always decreases with age. During an injury, therefore, these changes in the musculoskeletal system can lead to a decreased ability to withstand the effects of mechanical energy.

The response to an injury among older people can be

influenced by underlying medical conditions and the use of alcohol (Committee on Trauma Research 1985; States 1985). Hypertension and atherosclerosis can increase the severity of cardiovascular injury. Osteoporosis decreases bone resistance to mechanical energy, which increases the risk of compression fracture from a given force and can result in fractures of the ribs and hip (Committee on Trauma Research 1985). Degenerative arthritis from prolonged use of steroids, congenital abnormalities, and decreased circulation, may also contribute to the occurrence of an injury (States 1985). The chronic use of alcohol interferes with tissue regeneration and immunological function, and so an otherwise equivalent injury may have a more severe outcome than for a non-drinker (Committee on Trauma Research 1985). The role of other drugs and other chronic diseases in the risk of injury needs further study.

APPROACHES TO INJURY PREVENTION

One way to identify options for controlling injury or to minimize the consequences of injuries is to use Haddon's matrix, which separates the injury 'event' into three distinct phases (Haddon 1970). Each phase of the Haddon matrix can also include information on the potential influence of the host, agent, and environment. Haddon's first is the 'pre-event phase', during which people are exposed to an energy source that can put them at risk of injury. (An example would be an older person who has taken a sedative or antidepressant and is approaching a floor area with deceptive visual patterns, or a stairway that is poorly lit or in disrepair.) The second is the 'event phase', during which the victim is acutely exposed to the energy source. (An example would be an older person falling on a hardwood floor.) The third is the 'post-event phase', during which the person is subject to factors that determine the final damage or disability stemming from the event. (An example would be an older person who has fallen and cannot get help quickly.) Although many diseases can be prevented by interventions in the first phase, many injuries or their serious consequences could best be prevented or ameliorated by interventions in the last two phases (Committee on Trauma Research 1985).

By following Haddon's matrix, various strategies for prevention can be used (Haddon and Baker 1981). These include preventing the marshalling of potentially injurious agents or reducing their amounts; preventing inappropriate release of the agent; modifying the release of the agent; separating the energy and the host in time or space or with physical barriers; modifying surfaces and basic structures; increasing resistance of the host to injury; beginning to counter the damage already done; and stabilizing, repairing, and rehabilitating the injured. Three general approaches can be used with each of these strategies: (i) persuading persons to alter their behaviour; (ii) requiring behavioural change by law or administrative rule; and (iii) providing automatic protection by correct and appropriate design of products and environments. Although each of these approaches has a role in prevention, the third tends to be the most effective.

AN OVERVIEW OF THE EPIDEMIOLOGY OF INJURY IN OLDER PERSONS

In 1986, there were more than 33000 deaths from injury among people aged 65 or older in the United States (unpublished data). By far the leading causes were falls, motor vehicles, and suicides—these will be discussed in detail later (Table 1). Injuries that result in death, however, represent

Table 1 Ten leading causes of injury death for persons aged 65 years or older, United States, 1986. (Source: National Center for Health Statistics, Detailed Mortality Tapes.)

Cause*	Number	Rate†
Falls	8313	28.5
Motor vehicles	6410	22.0
Suicide	6215	21.5
Surgical and medical procedures	1853	6.4
Aspiration–non-food	1412	4.8
Fire and flames	1383	4.7
Other and unspecified environmental and accidental (*sic*)	1328	4.6
Homicide	1295	4.4
Aspiration–food	956	3.3
Poisoning	763	2.6

* According to the International Classification of Diseases—9th Revision.
† Per 100 000 persons.

only a fraction of the impact on public health of injuries in the United States. Recent data show that more than 5.8 million injuries occur among older people, resulting in 21.3 injuries for every 100 persons per year (NCHS 1987). About 58.9 million days of restricted activity are associated with an acute injury among older people, and the number of bed days associated with acute injuries was 68.9 for every 100 older persons per year (NCHS 1987).

FALLS

Magnitude of the problem

Falls are the leading cause of death from injury in people 65 years or older. Approximately two-thirds of reported injury-related deaths in those aged 85 or older are due to falls (Baker *et al.* 1984). Of the 8313 fatal falls that occurred in the United States in 1986 among those aged 65 or older, 59 per cent were in the home or in a residential institution (unpublished data). The rate of fall-related deaths rises rapidly with increasing age for all race–sex groups aged 75 or older (Fig. 1). White males aged 85 or older have the highest death rates from falls.

Falls also lead to significant morbidity in older people. The rate of injury from non-fatal falls increases steadily by each 5-year age group for those aged 65 or older. Women have consistently higher rates than men for each 5-year age group. One specific type of injury, fracture of the proximal femur, increases exponentially by age in older persons, from 28.4 per 10000 persons aged 65 to 74 years to 251.4 per 10000 persons aged 85 years or more. For those aged 65 or older,

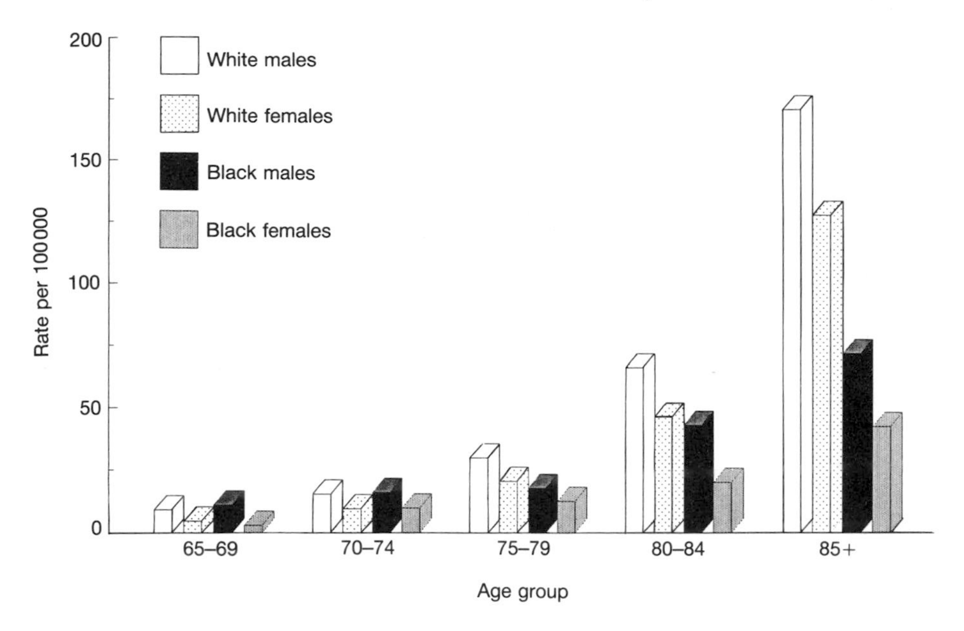

Fig. 1 Fall death rates, by age group, race, and sex (United States 1986).

the rate of hip fracture among white women is about twice that of white men, and white people have about twice the rate of hip fracture as all other races (Rodriguez *et al.* 1990).

Most falls in older people result in no or minor injury, with only a small percentage causing severe injury, such as a fracture (Kellogg International Work Group 1987). Estimates of the proportion of falls causing a fracture range from 4 to 6 per cent among people who can walk, with 1 per cent or less resulting in hip fractures (Tinetti *et al.* 1988; Nevitt *et al.* 1989). The incidence of falls among older people is considerably higher in long-term care facilities than in the community. The psychological trauma from falls may be severe and can result in loss of confidence in the ability to perform daily routine, restriction of activity, increased dependence, and decreased mobility (Kellogg International Work Group 1987). The results of immobility—deconditioning, muscle atrophy, and joint stiffness or contractures—can lead to more falls and further restrictions of mobility.

Aetiology

Falls among older people are the result of many pathophysiological and primary ageing processes, and of pharmacological, behavioural, and environmental factors. Some, such as syncopal falls, could be considered an acute symptom of a chronic or episodic cerebrovascular, cardiovascular, or neurological disorder. More commonly, falls are the result of decline in vision, gait, balance, sensory perception, strength, co-ordination, reflexes, and other aspects of neuromuscular function (Tinetti *et al.* 1988; Nevitt *et al.* 1989). Biological and functional variability within age groups, however, may be more important determinants of risk than age-dependent variations. Falls due to the effect of multiple medications are a potentially important problem, and antidepressant and sedative medications appear to increase the risk of falls. Falls due to these latter drug effects need to be distinguished from those due to the effect of the underlying diseases or the effect of drug–disease and drug–drug interactions.

About half of all injuries from falling occur in the home. Hazards in the home environment may increase the danger for falls or increase the chances of injury once a fall occurs. Such hazards include clutter, slippery floors, stairways in disrepair, unsecured mats and rugs, and lack of non-slip surfaces in baths. In addition, an older person may be influenced by more subtle environmental factors, such as poor lighting and visual and spatial design (Radebaugh *et al.* 1985). When a healthy, active older person falls, it often involves an unexpected hazard in the environment, one which could also cause a much younger person to fall. An older person with functional disabilities, however, might be less able to cope with a cluttered or poorly designed home, making it more likely that a fall and a serious injury will occur. Further work is needed to determine how environmental factors in the home interact with processes mentioned above in increasing or decreasing the risk of a fall or injury from it.

Approaches to prevention

The multifaceted, multifactorial nature of falls has prompted attempts to develop classifications. These focus on the circumstances of falls and are designed to provide information about the probable aetiological factors that will guide efforts at intervention. Unfortunately, work on classifying falls is still developmental and may be of limited value in understanding and preventing them because it is sometimes difficult or impossible to obtain valid information about the circumstances of a fall, and most falls have mixed causes.

The aetiology of falling is complex, and the significance of any individual fall is difficult to determine, for both health-care professionals and the person who falls. Though most falls in older people result in no injury or only minor injury, for a small proportion a fall or series of falls may signal serious acute illness, precipitous functional decline, or imminent death. For these reasons, clinicians must consider any fall seriously. A fall or near fall can provide the clinician with information about activities and circumstances that place a particular person at risk, including potential mismatches between a person's capabilities and his or her environment.

Efforts at prevention must strike a balance between protection from risk and the maintenance of mobility, functional activities, personal autonomy, and an acceptable quality of life. Reduction or adjustments in activity and mobility in response to a fall, however, are not enough to eliminate the risk of falling. Fear and excessive restrictions in activity may reduce risk of falling in the short term, but only increase the long-term risk by undermining self-confidence and physical conditioning. For the older person whose functional capacity is severely compromised, maintenance of even a minimum of independent mobility may entail substantial risk.

Unfortunately, we do not know much of specific interventions for specific risk factors in older people. The existing efforts at prevention and treatment are empirical, relying mostly on common sense and descriptive studies of the characteristics of falls and those who fall (Sattin *et al.* 1988). Efforts at prevention would benefit from an increased understanding of the mechanisms of injury and the use of the Haddon matrix. For example, the pre-event phase of injury might be affected by removing or altering energy sources that may increase a person's risk of falling or by altering pathophysiological conditions that would enable an older person to cope better. Proper design of stairways and lighting, better control over multiple drug prescriptions, exercise programmes designed for general muscle strengthening, and homes specifically designed for older people might be examples of this. Technological study to develop energy-absorbing flooring would be useful in the event phase of injury. The development of social networks could improve overall survival in older persons who do fall but cannot get help quickly.

MOTOR VEHICLE-RELATED INJURY

Magnitude of the problem

Motor vehicle-related injury is by far the leading cause of death for all ages in the United States, and for older people it is the leading cause up to the age of 75, when it is exceeded only by falls. Of the 6410 motor vehicle-related deaths among older people in the United States in 1986, about three-quarters occurred among motor-vehicle occupants and one-quarter among pedestrians.

For those aged 65 or older, death rates of motor-vehicle occupants per 100000 persons in the United States in 1986 were nearly twice as high for men (23.0) as for women (12.0) (Fig. 2). Black men have death rates similar to white men, but white women have consistently higher death rates than black women. The ratio of death rates for men compared with women, however, was much greater among black than white people for each 5-year age group over age 65. Death rates for older white men and white women increase steadily by age until the 80 to 84 years age category, after which the rates decrease. Death rates for older black men increase markedly between the 70 to 74 and 75 to 79 age category, after which the rates remain similar. For older black women, death rates remain about the same for each 5-year age group over age 65.

Although pedestrians accounted for 16 per cent of all motor vehicle-related deaths in 1986 for all ages, they accounted for 21 per cent for persons aged 65 to 74 and 28 per cent for those aged 75 or older (unpublished data). For persons aged 65 or older, men have twice the rate of

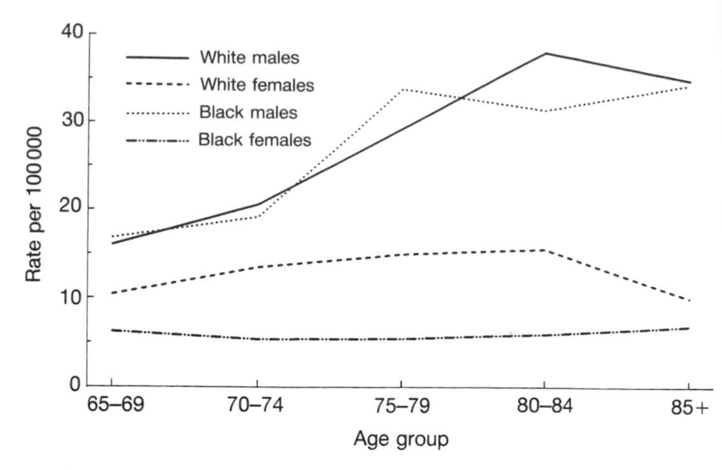

Fig. 2 Motor vehicle occupant death rates, by age group, race, and sex (United States 1986).

pedestrian deaths of women. Black men, however, have markedly higher pedestrian death rates for each 5-year age group for older people (Fig. 3). The death rate for white men is slightly higher than for white or black women until 75 to 79 years of age, beyond which the rate for white men becomes much higher.

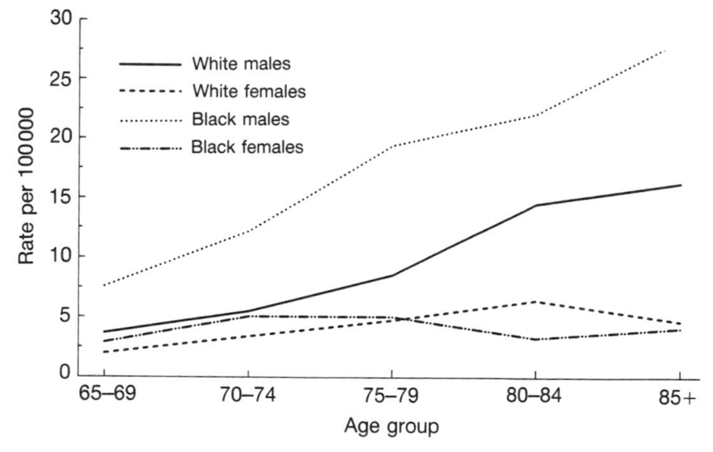

Fig. 3 Pedestrian death rates, by age group, race, and sex (United States 1986).

When exposure to risk is considered for deaths from motor vehicles, the death rate for drivers aged 75 or older is the same as that among those aged 20 to 24, but higher than in all other age groups (National Safety Council 1988). Based on the number of licensed drivers in the United States in 1987, statistics showed that drivers aged 75 or older had 61 fatal crashes per 100000 drivers compared with 36 fatal crashes for drivers of all ages. For total crashes (fatal and non-fatal), drivers aged 75 or older had 25 per 100 licensed drivers, compared with 20 for drivers of all ages. Per mile rather than by number of licensed drivers, drivers aged 75 or older are second only to drivers aged 15 to 19 years in the rate of crashes (TRB 1988). The absolute number of crashes for older drivers, however, is fewer than for younger drivers.

Older drivers have more traffic violations and crashes involving intersections than younger drivers, and about 29 per cent of pedestrian fatalities among older people occur at intersections compared with about 13 per cent among younger people (NHTSA 1988). Also, from a crash of otherwise equal severity, older people tend to have less tolerance of impact, require more time for healing, and have more complications after injury than younger people (States 1985; TRB 1988).

Aetiology

The car is a major means of mobility for older people in the United States, and the percentage of trips made by car by them is increasing. The highway transportation system, however, was designed without taking into account the capabilities and limitations of older people, whether drivers or pedestrians (TRB 1988). Older drivers are especially likely to have problems with giving right of way, observing stop signs and traffic signals, and manoeuvring cross-traffic turns. With older pedestrians, a decrease in visual, cognitive, and gait performance may make the task of avoiding a moving hazard more difficult, particularly at intersections. Vehicle design does not take into consideration what features might facilitate or impair the functioning of older drivers, and highway design has been developed primarily on the basis of performance measures obtained from young males (TRB 1988). Generally, older people need more light in order to see as well as younger people and have more difficulty with glare recovery. Older drivers can have results similar to younger drivers on standard tests of visual acuity, but still have to get much closer to highway signs at night in order to read them. This means they have less distance remaining in which to react to the information provided. (See also Chapter 18.14.2.)

The driving records of most older drivers are good, and older drivers tend to restrict their driving and alter their behaviour as they recognize their increasing limitations (TRB 1988). Compared with younger drivers, older drivers tend to decrease their alcohol intake as they age. Among older pedestrians who were fatally injured, the proportion with an elevated blood alcohol concentration was 10 per cent in 1986, by far the lowest of any adult age group (NHTSA 1988). Not all older people, however, are necessarily aware of their own particular driving or walking difficulties. For example, when asked to list manoeuvres that pose problems for them, older drivers ranked failure to give way at an intersection as the ninth out of 10, even though this is the major cause of crashes. Also, older people are more likely to be on medication and should receive more frequent advice from health professionals concerning its potentially adverse effects on driving or walking performance. {A study in the United Kingdom showed a link between road traffic accident and previous prescription of minor tranquillizers (Skegg *et al.* 1974).}

Approaches to prevention

Because motor vehicle-related and pedestrian injuries occur so suddenly, the tendency is to blame the injured as negligent (this tendency occurs in other injuries as well). An understanding of Haddon's matrix, however, can reveal important approaches to preventing these injuries. Improvements in the design of intersections (for both drivers and pedestrians), modifications to vehicles to increase crash protection, and improved visibility and legibility of road signs would increase the safety factor not only for older people but for all (TRB 1988).

In general, concerns about motor vehicle traffic have focused more on flow patterns and conservation of fuel than on pedestrian safety. For example, the traffic law in the United States allowing right turn on red lights—to reduce fuel consumption—appears to increase police-reported collision rates for older pedestrians in urban areas (CDC 1989). Environmental changes such as modification of stop-light signals to increase the pedestrian crossing time; roadway markings to emphasize pedestrian cross-walks, traffic lanes, and the direction of traffic flow; pedestrian signals on traffic islands; and larger speed-limit signs and increased police enforcement of the speed limit have been shown to reduce fatalities among older pedestrians (CDC 1989).

Relatively simple changes to the transport system can reduce the probability of injury from driving or walking for older people (Sattin *et al.* 1988; TRB 1988). Bus stops should be located beyond instead of before intersections so that drivers and pedestrians have a better view of the traffic. Better and more uniform delineation on highways (centre line and edge and lane lines, as well as guide signs and posts) would increase visibility and decrease confusion. Designing residential areas to limit through traffic severely and to reduce drastically the speeds of local traffic would decrease the chance of injury.

An important area of prevention involves screening and licensing of drivers (TRB 1988). Currently, all states require tests of static visual acuity for licence renewal. Other visual tests, such as contrast sensitivity, dynamic visual performance, and static acuity at low-level illumination, however, need to be further evaluated because they might help screening programmes to identify high-risk individuals. These programmes must consider the capabilities, regardless of age, of drivers with impairments and the demands placed on them by specified driving circumstances. Restriction of driving should be based on measures of driving performance and not on chronological age.

Older people who are restricted from driving need specific counselling on accepting the limitations and on alternative modes of transport. Physicians and other health-care professionals should counsel older people about the effects of alcohol, medication, and vision alteration on driving and walking performance, and particularly on problems at intersections. Further work is needed to determine the contribution of cognition, including dementia, and other chronic medical conditions on driving performance and the risk of motor vehicle crashes.

SUICIDE

Magnitude of the problem

Suicide among older people is a characteristic of all industrialized nations. In the United States, suicide is the third leading cause of death by injury among those aged 65 or more. The number of deaths reported on death certificates, however, may under-report by 10 to 15 per cent the true number of suicides (O'Carroll 1989; P. W. O'Carroll, personal communication). Personal bias, incomplete information, religious, social, financial, and family pressures, and failure to consider the possibility of suicide each contribute to this under-reporting.

The rate of suicide, which is higher in older than in younger people, increased by 21 per cent from 1980 (17.7 per 100000) to 1986 (21.5 per 100000) for persons aged 65 or more in the United States (unpublished data). The suicide rate for older men is more than six times (42.7 per 100000 persons) that for older women (7.1 per 100000 persons). White men have markedly higher suicide rates than all other race–sex groups, and these higher rates increase with increasing age (Fig. 4). White people of either sex are three times more likely to commit suicide than black people.

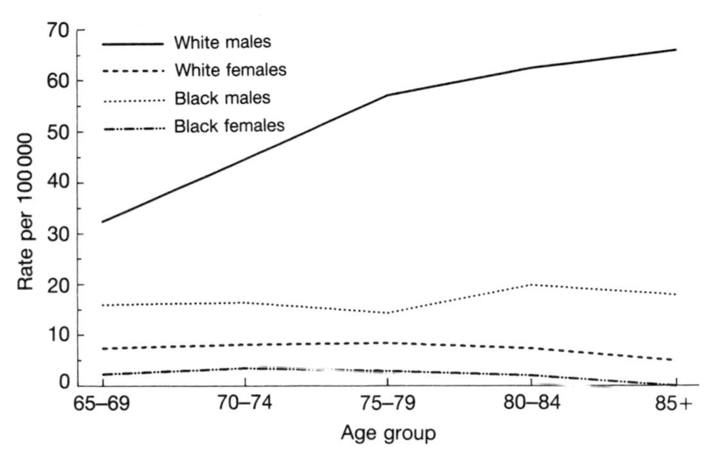

Fig. 4 Suicide death rates, by age group, race, and sex (United States 1986).

The first suicide attempt of an older person tends to be more serious and more likely to be successful than that of a younger person (O'Neal *et al.* 1956). Gestures or attempts as cries for help are not nearly as frequent in older as in younger age groups. About two-thirds of suicides in the United States among older people are from firearms, and another 13 per cent are from hanging, suffocation, or strangulation (unpublished data). Only 7 per cent of these deaths are from poisonings from liquids or solids. {In other countries with more rigid control of firearms, such as the United Kingdom, the majority of suicides are by poisoning.}

Aetiology

The incidence of clinical depressive illness preceding serious suicide attempts is quite high among older people (Batchelor and Napier 1953; O'Neal *et al.* 1956; Farbedrow and Shneidman 1970). Older people may be suddenly unable to cope with the biological, psychological, and social stresses of ageing or with physical illness. Physical illness is one of the most frequent precipitating factors for suicide in older people (Dorpat *et al.* 1968). Some potential diagnostic signs in someone contemplating suicide include hopelessness, alcoholism, withdrawal from social and personal relationships, bereavement (especially the feeling of isolation within the first year of a loss), preoccupation with the subject of death, and helplessness often stemming from placement in an institution. The feeling of exhaustion and failure, the fear of being a burden to others (financially and otherwise), and the loss of interest in pleasurable activities are all symptoms of high suicide potential, and their overt expression must be taken very seriously.

Efforts have been made to explain the high rate of suicides among older people, especially white men (Robins *et al.* 1977; Seiden 1981). One explanation is that the role of an older person in the modern (white) nuclear family is less purposeful than in the more traditionally oriented family in which social bonds are stronger and more firmly established. Another is that modern, youth-oriented cultures no longer esteem those who are older for their wisdom and experience, making older people feel estranged from society. Yet another explanation is that white men often have a greater sense of loss of power and influence as they age than do men from minorities who have never had these privileges. This relative deprivation may influence an older white man (who has other suicide risk factors) to commit suicide. Finally, the threat of his or her lifetime's savings being lost from the family because of the costs associated with chronic medical problems might influence an older person to commit suicide.

Approaches to prevention

Suicide can best be prevented by intervening at the pre-event phase. One approach would be to update physicians and other health professionals in the diagnosis, referral, and treatment of the susceptible older patient. {Studies in the United Kingdom have shown that a high proportion of elderly suicides have been in contact with their general practitioner in the 2 weeks before death.} Although not yet assessed for effectiveness in preventing suicide, suicide hot-lines, walk-in counselling, groups for suicidal individuals and their families, community-centred out-reach programmes are approaches now being tried to reduce youth suicide (Saltzman *et al.* 1988). Approaches like these, but with the focus on issues important to older people, might help to reduce their suicide rate. The predominant use of hand-guns in suicides among older people in the United States suggests that one approach to prevention might be limiting access to these weapons (Saltzman *et al.* 1988).

Social attitudes can play an important role in the prevention of suicide. Often societies place a premium on appearance, occupational status, power, and financial standing—attributes that almost always decline with age. People must learn to be realistic in their aspirations because the inability to achieve goals can lead to depression and other emotional problems. The general population needs to understand the capabilities and limitations of older people and their roles as valued members of the community. Underlying all of this is the need for greater acceptance of the ageing process and the inclusion of older people in the family structure. Although older people might have difficulty in coping with material adversity, physical pains, and illnesses, often it is more difficult for them to cope with feeling useless, unwanted, and superfluous. Communities that deal successfully with these social attitudes and recognize the value and purpose of their older population might have a positive impact on the prevention of suicide.

OTHER INJURIES

Though only the three most frequent causes of death by injury for older people are described in this chapter, similar principles of prevention apply to all other injuries. For example, fires are a potentially preventable cause of death among older people. Encouraging patients to stop smoking, promoting safety devices (e.g., smoke detectors, automatic sprinklers, clothing with flame-resistant properties, self-extinguishing cigarettes for persons who insist on smoking), and instituting such safety practices as fire escape drills are several approaches to reducing the number of deaths from residential fires among older people. Promoting fire-safety education and enforcing housing codes are additional approaches. Older people can also benefit from periodic checks of heaters and stoves; a public-health nurse could review a safety checklist of fire hazards during each visit to an older person's home. Finally, physicians and other health professionals could encourage older people to have a suitably placed smoke detector installed.

CONCLUSION

Injuries have serious consequences for older people, their families, and the health-care system. A multidisciplinary approach that includes people and disciplines not traditionally involved in public health could make safety for older people a priority. Developing such a focus for, and using, the principles of injury control would not only result in the prevention of many injuries and deaths among older people but would also increase the safety of society as a whole.

References

Baker, S.P., O'Neil, B., and Karpf, R.S. (1984). *The injury fact book*. Heath, Lexington, MA.

Batchelor, I.R.C. and Napier, M. (1953). Attempted suicide in old age. *British Medical Journal*, **ii**, 1186–90.

Centers for Disease Control (CDC) (1989). Queens boulevard pedestrian safety project—New York City. *Morbidity and Mortality Weekly Report*, **38**, 61–4.

Committee on Trauma Research, Commission on Life Sciences, National Research Council, and the Institute of Medicine (1985). *Injury in America*. National Academy Press, Washington.

Dorpat, T.L., Anderson, W.F., and Ripley, H.S. (1968). The relationship of physical illness to suicide. In *Suicidal behaviors* (ed. H.L.P. Resnik), pp. 209–19. Little Brown, Boston.

Farbedrow, N.L. and Shneidman, E.S. (1970). Suicide and age. In *The psychology of suicide* (ed. E.S. Shneidman, N.L. Farbedrow, and R.E. Litman), pp. 165–74. Science House, New York.

Haddon, W., Jr. (1970). On the escape of tigers: an ecologic note. *Technology Review*, **72**, 44–53.

Haddon, W., Jr. and Baker, S.P. (1981). Injury control. In *Preventive and community medicine* (ed. D. Clark and B. MacMahon), pp. 109–40. Little Brown, Boston.

Kellogg International Work Group on the Prevention of Falls by the Elderly (1987). The prevention of falls in later life. *Danish Medical Bulletin*, **34** (suppl. 4), 1–24.

National Safety Council (1988). *Accident facts*. National Safety Council, Chicago.

National Center for Health Statistics (NCHS) (1987). Current estimates from the National Health Interview Survey, United States, 1986. *Vital and health statistics*, Series 10, No. 164, pp. 16–52. US Government Printing Office, Washington.

National Highway Traffic Safety Administration (NHTSA) (1988). *Fatal accident reporting system 1986*. US Department of Transportation, Washington.

Nevitt, M.C., Cummings, S.R., Kidd, S., and Black, D. (1989). Risk factors for recurrent nonsyncopal falls: a prospective study. *Journal of the American Medical Association*, **261**, 2663–8.

O'Carroll, P.W. (1989). A consideration of the validity and reliability of suicide mortality data. *Suicide and Life-Threatening Behavior*, **19**, 1–16.

O'Neal, P., Robins, E., and Schmidt, E. (1956). A psychiatric study of attempted suicide in persons over sixty years of age. *Archives of Neurology and Psychiatry*, **75**, 275–84.

Radebaugh, T.S., Hadley, E., and Suzman, R. (ed.) 1985. Falls in the elderly: biologic and behavioral aspects. *Clinics in Geriatric Medicine*, **1**, 555–620.

Robins, L., West, P., and Murphy, G. (1977). The high rate of suicide in older white men. *Social Psychology*, **12**, 1–20.

Rodriguez, J.G., Sattin, R.W., and Waxweiler, R.J. (1989). The incidence of hip fractures, United States, 1970–1983. *American Journal of Preventive Medicine,* **5**, 175–81.

Saltzman, L. E., Levenson, A., and Smith, J.C. (1988). Suicides among persons 15–24 years of age, 1970–1984. *CDC Surveillance Summaries*, **37** (No. SS-1), 61–8.

Sattin, R.W., Nevitt, M.C., Waller, P.F., and Seiden, R.H. (1988). Injury prevention. In *Background papers from the Surgeon General's workshop on health promotion and aging* (ed. F.G. Abdellah and S.R. Moore), pp. D1–20. US Public Health Service, Washington.

Sattin, R.W., Rodriguez, J.G., DeVito, C.A., Lambert, D.A., and Stevens, J.A. (1991). The epidemiology of fall-related injuries among older persons. In *Reducing frailty and falls in older persons* (eds. R. Weindruch, E.C. Hadley, and M. Ory). Charles C. Thomas, Springfield, in press.

Seiden, R.H. (1981). Mellowing with age: factors influencing the nonwhite suicide rate. *International Journal of Aging and Human Development*, **13**, 265–84.

Skegg, D.C.G., Richards, S.M., and Doll, R. (1979). Minor tranquillizers and road accidents. *British Medical Journal*, **i**, 917–19.

States, J.D. (1985). Musculo-skeletal system impairment related to safety and comfort of drivers 55+. In *Drivers 55+: needs and problems of older drivers: survey results and recommendations*, Proceedings of the older driver colloquium, Orlando, FA (ed. J.L. Malfetti), pp. 63–76. AAA Foundation for Traffic Safety, Falls Church, VA.

Tinetti, M.E., Speechley, M., and Ginter, S.F. (1988). Risk factors for falls among elderly persons living in the community. *New England Journal of Medicine*, **319**, 1701–7.

Transportation Research Board, National Research Council (TRB) (1988). *Special report 218—transportation in an aging society*, Vol. 1. Transportation Research Board of the National Research Council, Washington.

4.2 Injury responses in old age

M. A. HORAN, N. A. ROBERTS, R. N. BARTON, AND R. A. LITTLE

VULNERABILITY TO INJURY

Both iatrogenic injuries (in the form of surgical operations) and accidental injuries are common among elderly people. It is widely held that they are more vulnerable to the consequences of injury than their younger counterparts. This is demonstrated by the observation that people over the age of 65 years account for only a quarter of surgical admissions but for about three-quarters of postoperative deaths, though this must also reflect the nature of the conditions for which the operations are performed. Similarly for accidental injuries: when death rates from these are plotted against age, there is a small peak in the 15 to 24 age group and a much larger peak in the group 75 to 84 years old. This pattern holds for injuries at all anatomical sites.

Head injuries

Head injuries are common in the general population and yet there is surprisingly little information on the assessment, prognosis, and continuing care of elderly patients injured in this way. Only 1 to 2 per cent of the total number of head injuries are in those aged over 80 years, but this simply reflects the relatively small number of people of this age in the population. The age-specific incidence of head injuries is high, reflecting the frequency of falls among elderly people. Domestic accidents cause the majority of head injuries in this age group and the remainder are largely accounted for by road traffic accidents.

The prognosis of head injury worsens sharply as old age is approached; after the age of 50 years in one study and after 60 years in another. Permanent cognitive impairment is also more frequently seen. Head injuries in old age have some unusual epidemiological features: intracranial mass lesions are common and epidural haematomas are rare. Recent findings suggest that the prognosis may not be as bleak as previously thought for those not deeply comatose and between the ages of 60 and 70 years. Aggressive treatment is appropriate in such patients. However, the outcome is poor for those over the age of 70 who are unconscious on admission, and aggressive therapy for this group of patients may be futile. Even those over the age of 70 years who have values of 13 to 15 on the Glasgow coma scale (what many would regard as minor head injuries) may well have serious intracranial damage (up to 64 per cent of such subjects in one study). This observation warrants serious attention and all elderly people coming to hospital with head injuries merit the most careful and detailed assessment.

Fractures of the upper femur

Most of the 'old age' peak in deaths from injury is explained by fractures of the upper femur. The annual occurrence of upper femoral fractures is 2.3 per 1000 men and 6.0 per 1000 women over the age of 65 years and the incidence rises exponentially with advancing age. Demographic changes indicate that hospital admissions with such fractures will increase markedly over the next two decades, and in the United Kingdom there is evidence that incidence rates are also increasing.

Elderly patients with upper femoral fractures often die. Most deaths occur within a month of injury but the death rate continues to be raised for considerably longer. No satisfactory explanations have yet been offered for these observations. Most studies have been short and the figures most often relate to deaths up to 6 months after the fracture (during which time it is estimated that approximately 30 per cent of such patients will have died). Most of the deaths within this time are attributable to cardiac, respiratory, and vascular complications (e.g., pulmonary embolism and stroke) and discriminant analysis has shown that these complications are major predictors of survival.

Upper femoral fractures are also associated with considerable morbidity but there have been few good studies examining outcomes other than technical ones (e.g., non-union of the fracture or late collapse of the femoral head). Only half of elderly patients with fractured femurs have returned home by 6 months and 10 per cent are still in hospital.

Age, an intracapsular fracture, and 'high dependency' before the fracture all appear to be associated with poor outcome in terms of walking ability at 6 months. A recent autopsy study suggests that frequently unrecognized complications at the operative site may be an important determinant of poor walking. The cardiorespiratory complications mentioned above as predictors of death may also be important determinants of the ultimate level of dependency, though other factors are also likely to be involved.

We will now go on to examine aspects of the host responses to injury but we must first discuss how injuries may be 'measured', a prerequisite for comparing responses to injury in different groups of patients.

Assessing the severity of injury

Great care should be taken when comparing injuries between different age groups. For example, because the bone is weaker, the femur of an old person can be fractured by a considerably smaller force than is required to produce a similar fracture in a young person, and the amount of soft tissue damage will also be less. However, severity of injury will be determined not only by the amount of tissue damage but also by the capacity to compensate for the loss of fluid from the circulation, which maintains blood flow to vital organs. There is evidence that such compensatory mechanisms may be impaired in old age.

There are several scoring systems for assessing the severity of injury but, because they were developed from data collected in young people, they cannot be applied indiscriminately to elderly people. For anatomically based systems such as the Injury Severity Score (**ISS**), reduced bone mass will tend to overestimate severity whereas reduced compensation will underestimate it. Physiologically based systems such as the Trauma Score and APACHE II take both of these factors into account, at least to some extent, thus ameliorating the problem. Some still consider it necessary to make an adjustment for age but this seems largely speculative and may not be appropriate for all applications. A comparison of deaths after injury with unadjusted ISS in patients aged over and under 50 years showed an increased fatality in the older age group at all ISS values (Fig. 1). The Geriatric Trauma Survival Score has recently been developed specifically for the prediction of survival after injury in old people. It incorporates the ISS together with an assessment of cardiac and septic complications. However, it is not well enough established for us to recommend its general use in clinical practice.

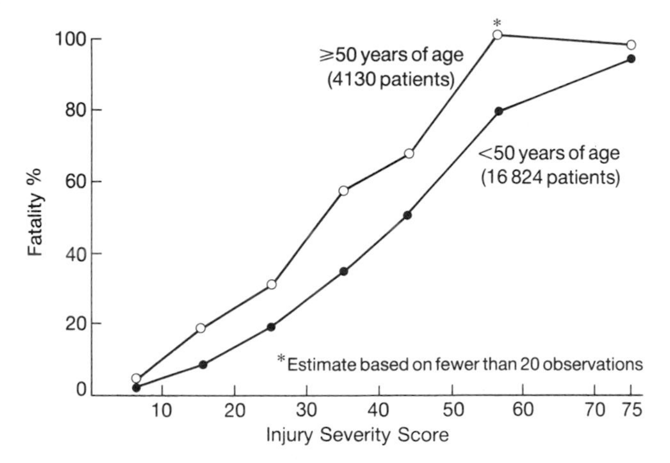

Fig. 1 Mortality rate *vs.* Injury Severity Score (ISS) in patients aged over or under 50 years with blunt injuries (reproduced with permission from Copes *et al.* 1988).

RESPONSES TO INJURY

Ebb phase

The first 24 to 48 h after accidental injury was termed the ebb phase by Cuthbertson, who thought that metabolic rate was decreased. In animal models of injury a striking fall in the thermoregulatory component of heat production does occur at this stage, but recent work has not shown this in man. Nevertheless, other aspects of thermoregulation are disordered: there is a fall in core temperature and measurement of 'preferred hand ambient temperature' shows that behavioural thermoregulation is impaired.

We have found no studies on the effects of ageing on thermoregulation after accidental injury, although there is evidence that heat loss during elective surgery causes metabolic demands that elderly patients may be unable to meet adequately. Ageing is known to impair several aspects of thermoregulation: the sensitivity to differences in environmental temperature, the ability to reduce heat loss, and the capacity to increase heat production. It is likely, therefore, that in old age the disturbances caused by trauma will be exaggerated. This will be particularly true in the undernourished, who are over-represented among injured elderly patients. It has been claimed that their thermoregulatory capacity is already impaired at the time of injury and that impaired balance induced by hypothermia may have caused their fall.

One of the most rapid effects of injury is to provoke well-known neuroendocrine responses. The activity of the sympathoadrenal system and hypothalamic–pituitary–adrenal (**HPA**) axis, and the secretion of a number of other hormones (growth hormone, prolactin, vasopressin, aldosterone, and glucagon), are increased. In some systems, such as the sympathoadrenal, the initial response increases with the severity of the injury and may not be seen at all with, for example, minor elective surgery. For other hormones, such as cortisol (Table 1), the relationship is more complex and it is the duration of the response, rather than its initial magnitude, that increases with injury severity.

There is evidence that, in general, ageing does not impair the ability of these neuroendocrine systems to respond to injury. Most thoroughly studied is the HPA axis. Shortly after accidental injury, plasma cortisol is similar in old and young patients for all severities of injury (Table 1). Several studies have shown that the same is true shortly after elective surgery. A few hours later, plasma cortisol is higher in elderly patients, and it is possible that this is due to decreased metabolic clearance of the hormone. However, as such age-associated differences are not seen when cortisol secretion is stimulated with exogenous adrenocorticotrophic hormone, it is likely that additional factors are involved.

Table 1 Plasma metabolites and hormones shortly after accidental injuries of differing severity

	Age (years)	Severity of injury			
		Minor	Moderate	Severe	Very severe
Cortisol	17–40	0.72	0.87	0.77*	0.73
μmol/l	>65	0.81	0.91	0.90*	0.66
Glucose	17–40	5.4	5.9	6.9	7.9
mmol/l	>65	6.0	6.2	7.3	8.7
Insulin	17–40	15	14	13	9
mu/l	>65	12	13	10	11
Lactate	17–40	1.5	2.4	2.9	3.9
mmol/l	>65	1.7	2.1	2.3	3.6
Free fatty acids	17–40	0.72	0.90	0.81	0.70
mmol/l	>65	0.81	0.86	0.81	0.75
Glycerol	17–40	0.06	0.09*	0.11	0.15
mmol/l	>65	0.09	0.20*	—	—

Severity of injury is based on Injury Severity Score (ranges 1–6, 8–12, 13–24 and 25–59). All blood samples were taken within 2h of injury. Values are means, either arithmetic (cortisol, insulin, free fatty acids) or geometric (glucose, lactate, glycerol); group sizes range from 5 to 98. The variance in most groups is large.

* Only pairs for which difference between ages is statistically significant ($p<0.05$ for cortisol, $p<0.01$ for glycerol by Wilcoxon rank-sum test); many differences between injury severity groups are significant.

The few studies of sympathoadrenal activity in elderly patients after trauma show that catecholamine concentrations rise at least as much as in younger patients. In one report, ageing was associated with an enhanced plasma noradrenaline response to cholecystectomy, in line with the general finding of increased noradrenaline concentrations, both basal and after other stimuli, but this difference was not found after either more minor elective surgery or severe accidental injury. Relevant studies of other hormones have been confined to isolated reports of patients undergoing elective surgery. One shows a reduced response of growth hormone to major surgery in elderly patients. In another study, neither plasma renin activity (which did not rise) nor plasma aldosterone (which rose transiently) differed between elderly and young patients after minor surgery. There are no relevant data for either glucagon or prolactin.

The influence of raised cortisol, aldosterone, catecholamine, and vasopressin concentrations on fluid and electrolyte balance has been investigated postoperatively but apparently not after accidental injury or specifically in elderly patients. After an operation there is a period of sodium and water retention, but its amount does not seem to be closely related to the changes in the concentrations of these hormones. For example, the vasopressin concentrations are higher than normally needed for maximal antidiuresis but this is not achieved. This discrepancy may be even more marked in elderly subjects owing to an age-associated decline in renal sensitivity to vasopressin. We must conclude that the factors responsible for determining the volume and concentration of urine soon after surgery are poorly understood and that what happens in elderly patients is unclear.

The neuroendocrine response is a major cause of the early metabolic changes after injury. The initial event is mobilization of the body fuels, glycogen, and triacylglycerol. Breakdown of liver glycogen increases the circulating concentration of glucose, uptake of which by insulin-dependent tissues (mainly skeletal muscle) is probably not stimulated to the expected extent. The lactate concentration also rises, owing partly to breakdown of muscle glycogen (which can be converted to glucose only via circulating lactate) and partly to hypoxia. Breakdown of triacylglycerol in adipose tissue raises the plasma glycerol and free fatty acid concentrations, although the latter is lower than expected after severe injury (Table 1) because the supply of albumin required for its removal from the fat depots is very susceptible to vasoconstriction. There are modest rises in the concentrations of ketone bodies, which are derived from free fatty acids in the liver.

Like the neuroendocrine response to injury, the early metabolic changes do not appear to be diminished in old age. Plasma glucose increases with severity of accidental injury in the same fashion as in younger patients, and there is no significant difference between age groups within the various categories of severity (Table 1). There is likewise no difference in plasma insulin, which fails to rise commensurately with glucose owing to adrenergic inhibition of insulin secretion (Table 1). Plasma lactate also increases similarly with injury severity in young and old patients (Table 1). This observation is of interest as lactate levels are often regarded as an index of hypoxia after injury. Either this is not correct or anatomically similar injuries do not cause more severe hypoxia in elderly patients at this early stage. Findings after elective surgery are mostly in keeping with these observations after accidental injury. Two studies of herniorrhaphy or cholecystectomy showed that old age had no effect on the plasma concentrations of glucose or insulin, and, while the increase in lactate after cholecystectomy was greater in elderly than in young patients, the difference was only transitory.

After injury the plasma concentrations of lipid metabolites are also no lower in elderly patients than in young ones. In samples taken within the first 2 h after minor or moderate accidental injury, plasma free fatty acids were similar in young and old while plasma glycerol was higher in elderly patients (Table 1); there was no difference in the concentration of the ketone body β-hydroxybutyrate. Once glycerol has been formed by lipolysis it cannot be reincorporated into triacylglycerol. Thus, it appears that in elderly patients both the gross rate of lipolysis and the rate of re-esterification of free fatty acids within adipose tissue are increased for reasons that are not clear. One report on the effects of ageing on lipid metabolites after elective surgery showed no difference in plasma glycerol, free fatty acids, or β-hydroxybutyrate between patients over and under 55 years old undergoing cholecystectomy. However, glycerol (and, to a lesser extent, free fatty acids) was higher in women than in men, and it is possible that such sex differences accounted partly for the higher glycerol concentrations after accidental injury in elderly patients, who were mostly women, their younger counterparts being mainly men.

Flow phase

The ebb phase is followed by the period of flow, a term coined by Cuthbertson to describe the increase in metabolic rate thought to characterize it. This concept has been amply borne out in numerous studies. The increase in metabolic rate is positively related to the severity of injury and there is much evidence that it reflects a further disturbance in thermoregulation, although the cause is not known. Burns and sepsis may cause disproportionate increases, the former because of evaporative heat loss and the latter because of pyrogens. The hypermetabolism is accompanied by a rise in nitrogen excretion due to net muscle protein catabolism. This is thought to be stimulated by various autacoids from cells and tissues involved in inflammation and repair, which are thus provided with essential substrates such as glucose and glutamine, but the detailed mechanisms remain speculative.

The effects of age on the hypermetabolism of the flow phase have apparently not been studied. Old people have an impaired ability to increase both cardiac output and respiratory gas exchange, which will reduce their maximal capacity for oxygen delivery to tissues. One would expect this to limit the increase in metabolic rate they can sustain after major injury. Whether the modest rises seen after less severe injuries are reduced commensurately, and whether this has adverse consequences, are not known.

Little is also known about protein metabolism in the elderly patient after trauma. Normal ageing is associated with a fall in protein turnover in the whole body (and particularly muscle) to which various factors contribute: decreased food intake, a lower skeletal muscle mass, and perhaps a reduced metabolic activity of the protein that remains. It is sometimes stated that the protein-catabolic response to injury is likewise decreased in old age, but there is little evidence for or against this view. One study showed that in elderly women with fractures of the femoral neck the values for nitrogen excretion were lower than some previously found for younger women after major elective surgery, but did not directly compare patients of different ages but similar degrees of injury. The elderly women were nearly all in negative nitrogen balance for over a week after injury, but this was improved by giving food supplements.

Injury also leads to disturbances in the metabolism of

plasma proteins, with increases in the concentrations of 'acute-phase reactants', and a fall in that of albumin. The effects of ageing on these changes have apparently not been reported. However, the acute-phase response is not specific to injury and several studies, using infection as the stimulus, have shown that the rise in at least one of these proteins, C-reactive protein, is well preserved in old age.

The flow phase also affects carbohydrate and fat metabolism. There is a shift from carbohydrate to fat as the preferred fuel, and this is not due solely to reduced food intake. The concentrations of the main substrates, glucose and free fatty acids, are inappropriately high and low, respectively. There is some evidence for increased metabolic clearance of free fatty acids from the circulation, at least after severe injury, whereas clearance of glucose is decreased because of insulin resistance, i.e., diminished ability of insulin to stimulate glucose uptake by insulin-dependent tissues, especially skeletal muscle. Little is known about the effects of ageing on any of these changes apart from one report showing exacerbation of the insulin resistance. After accidental injury, insulin concentrations were raised for 2 weeks, irrespective of age, for unknown reasons; however, in young patients, glucose soon fell to its normal value whereas in elderly patients it did not. Insulin resistance is known to develop with age and, possibly in conjunction with inhibition of insulin secretion, gives rise to a reduction in glucose tolerance in elderly people. After injury these changes are probably reinforced by immobility, which can cause insulin resistance even in healthy young subjects.

A further possible cause of insulin resistance in injured elderly patients is cortisol. After injuries of moderate severity in young people, plasma cortisol decreases to near normal over the first week, but in old people with fractures of the proximal femur raised cortisol levels may persist for 8 weeks or more. The cause of this phenomenon, which is not related to mobility, is not known, but it appears to reflect continuing stimulation at the hypothalamic or pituitary level rather than enhanced adrenal sensitivity to adrenocorticotrophic hormone or impaired cortisol clearance. In younger subjects, the only endocrine changes usually seen beyond the first week or so after injury, apart from the increase in insulin mentioned above, are falls in the tri-iodothyronine and, in men, testosterone concentrations. These hormones do not appear to have been studied in injured elderly people, but other evidence suggests that the decrease in both will be at least as great as in the young.

Fluid and electrolyte homoeostasis during the flow phase and beyond depends on many factors. These include the patient's level of consciousness, perception of thirst (which may be diminished), magnitude of insensible fluid losses (which will be high owing to the hypermetabolism of the flow phase), and the ability of the kidneys to cope with fluid deprivation or excess (which is often abnormal in elderly people).

One study has shown that after fluid restriction for 24 h in apparently healthy elderly and young men, the plasma osmolality and sodium and vasopressin concentrations were all significantly higher in the older men. However, they experienced less thirst and drank less fluid when its availability ceased to be restricted. This exaggerated vasopressin response to increased osmolality has been confirmed in other studies, which also suggest that it may be inappropriately prolonged, thus predisposing to hyponatraemia. However, one might predict that this will be offset by the decreased sensitivity of the aged kidney to vasopressin discussed earlier. In contrast, there is other evidence that the vasopressin response to pure volume depletion is reduced in old age (Section 7).

Ageing is also associated with a reduced ability to retain sodium when sodium intake is reduced, and considerable amounts may continue to be excreted in the urine. This is usually attributed to a state of relative hypoaldosteronism, secondary to reduced renin secretion, though enhanced production of atrial natriuretic peptide may also be important. Such a state predisposes to the development of salt and water depletion.

Both the enhanced vasopressin (and possibly also atrial natriuretic peptide) response and the reduced aldosterone response have implications for postoperative fluid therapy. Such therapy should be sufficient to prevent the development of water depletion. There seems little place for hypotonic fluids (including 5 per cent dextrose) and 0.9 per cent sodium chloride would be the preferred preparation.

CLINICAL MANAGEMENT

The present approach in the United Kingdom to the management of the elderly victim of trauma gives little cause for complacency. A report from the Royal College of Physicians of London (1989) states that 'hip fracture patients, almost all elderly and many frail, are in danger of receiving a second-class preoperative service'. The same report also considers deficiencies in peri- and postoperative management and goes on to propose guidelines for good practice. These guidelines are largely empirical and this reflects the dearth of knowledge about how elderly people respond to the stress of trauma. Notwithstanding, we shall use these guidelines as the basis for further discussion.

Pre- and perioperative management

The major considerations when a patient presents after trauma are to sustain life, to establish precise diagnoses, and to prevent complications. Very few elderly people present with multiple injuries, and fractures of the upper femur form most of the work-load. In this case, rehabilitation must commence at the time of presentation because the potential to do harm is great. The patient should be placed on an appropriate surface to distribute pressure evenly in an attempt to prevent decubitus ulcers. Further protection should be given for the heels. The patient should be kept warm (see above) and appropriate measures to control pain should be instituted straight away. Radiographs should be taken as soon as possible; if delays are anticipated, the patient should be transferred to the ward. No patient should remain in the accident and emergency department for more than an hour.

When the patient arrives on the ward he or she should be kept warm and placed on a pressure-distributing mattress (especially if traction is started). The ordinary King's Fund mattress is not adequate. A fuller medical history should be taken and examination done. The circumstances of the fall leading to the fracture should be determined and the nature of any other falls should be recorded. Details of the patient's drug treatment and domestic circumstances should also be noted or arrangements made to obtain them. The history and physical examination should be fairly detailed as coexisting diseases and nutritional status will influence further management. Blood should be taken to determine the haemoglobin and total white cell count as well as the plasma concentrations of sodium, potassium, urea, creatinine, glucose, and alkaline phosphatase. An elevated alkaline phosphatase concentration soon after fracture will not be due to the fracture and should prompt consideration of other explanations such as liver disease, osteomalacia, skeletal metastases, and Paget's disease of bone. Evidence of dehydration is common and usually requires treatment with isotonic salt solutions. Diabetes will almost always require treatment with insulin, at least in the short term.

Operative fixation of the fracture is the preferred treatment and this should take place within 24 h of admission unless there are overwhelming reasons why not. Some delay is permissible to improve the 'medical' condition of the patient, though this period should not be prolonged because, paradoxically, deterioration often occurs. During the operation, patients are usually well supervised but this may not be the case in the recovery area where not enough attention may be paid to pressure-sore prophylaxis and the prevention of heat losses.

Postoperative management

The major areas requiring attention are listed in Table 2. Prophylaxis against venous thrombosis is almost universally practised, usually by using low-dose subcutaneous heparin. The patient should be assessed frequently by both nursing and medical staff, paying particular attention to control of pain, prevention of decubitus ulcers and constipation, and the detection of complications such as infection and cardiac changes.

Table 2 Factors of special importance in the management of postoperative patients after upper femur fracture

Prophylaxis against deep vein thrombosis
Adequate analgesia
Decubitus ulcer prophylaxis
Prevention of constipation
Care with fluid balance
Nutritional support
Limited use of sedative medications
Repeated assessment for intercurrent illness and complications
Physical therapy while immobile
Early mobilization
Provision of properly designed furniture (e.g., chairs)
Early practice in activities of daily living
Consistency in the rehabilitation programme
Multidisciplinary discharge planning

Before and during the operation the patient is likely to have received enough intravenous fluid to compensate for blood loss and dehydration associated with the fracture. The aim of intravenous fluid therapy once the patient returns to the ward is to avoid dehydration, overhydration, and hyponatraemia until the patient is well enough to drink again, usually within 24 h. Great care is needed in the administration of intravenous fluids because of the pathophysiological changes described above. All patients on intravenous therapy should have their urea and electrolytes measured frequently and should be reviewed daily. It is very common to see patients who are hyponatraemic in the first few postoperative days. A number of factors are likely to be involved. In shocked patients the function of the sodium/potassium pump may be impaired, resulting in entry of sodium into cells from the extracellular compartment. The concentration of vasopressin is likely to be raised, resulting in water retention in excess of sodium, and the aged kidney has an impaired capacity to conserve sodium in the setting of reduced sodium intake as well as having an impaired capacity to excrete a water load. Many junior doctors are frightened of giving excessive sodium and it is common for patients to be given too much 5 per cent dextrose as a consequence. All these factors may contribute to the development of hyponatraemia with its associated hazards. We suggest that no more than 2 l of 0.9 per cent saline over the first 24 h is adequate in uncomplicated cases. Dextrose-containing solutions are usually best avoided.

Intravenous therapy should be discontinued once the patient is drinking normally, but oral intake and urine output need to be monitored. Fluid intake may be adequate but the intake of solutes (in the form of food) may not, again predisposing to hyponatraemia (especially in those already malnourished). Patients need an adequate intake of calories, protein, and minerals, both to supply nutritional needs and to prevent hyponatraemia. This can usually be achieved with liquid feeds in those with poor appetite and the help of the dietitians should be sought. Particularly in those already malnourished, it has been suggested that overnight nasogastric feeding through a fine-bore tube with free access to normal diet in the daytime may be advantageous. The hormone pattern during sleep is generally anabolic, with secretion of growth hormone and low cortisol concentrations; some animal experiments suggest increased protein synthesis during sleep.

Immobility coupled with protein catabolism and malnutrition leads to rapid wasting of the quadriceps femoris muscle and exercises should be started in the bed as soon as the patient is able. Early weight bearing and mobilization are also important to prevent quadriceps atrophy. Significant quadriceps wasting gives rise to difficulties in rehabilitation, with easy fatigue and knees 'giving way'. There is an urgent need to develop therapeutic interventions to maintain skeletal muscle bulk (particularly of the quadriceps femoris) in those unavoidably immobilized. (See also Chapter 4.4.)

Patients with diabetes mellitus require careful management and it should be recalled that the period of insulin resistance may persist for several weeks after injury, leading to varying demands for exogenous insulin. When patients are on a stable dose of insulin, continued vigilance is needed to prevent hypoglycaemia when the insulin resistance diminishes.

CONCLUSION

Injuries in old people are very demanding, in terms of both medical skill and use of resources. Many aspects of care could be considerably improved. The major change needed is one of attitude. Old people should not be left lying around accident and emergency departments developing decubitus ulcers while waiting for radiographs to be taken. The need for effective treatment and prophylaxis against complications is urgent and demanding of the very highest standards. The reader will now be aware of the large gap in our knowledge between an appreciation of the size and resource implications of the problem and our understanding of the underlying mechanisms that may predispose to a poor outcome. More basic and clinical research is needed and needed soon.

Bibliography

For a more complete bibliography the reader is referred to Horan *et al.* (1988).

Amacher, A.L. and Bybee, D.E. (1987). Toleration of head injury by the elderly. *Neurosurgery*, **20**, 954–8.

Barton, R.N. (1987). The neuroendocrinology of physical injury. *Baillière's Clinical Endocrinology and Metabolism*, **1**, 355–74.

Bastow, M.D., Rawlings, J., and Allison, S.P. (1983*a*). Undernutrition, hypothermia and injury in elderly women with fractured femur: an injury response to altered metabolism? *Lancet*, **i**, 143–6.

Bastow, M.D., Rawlings, J., and Allison, S.P. (1983*b*). Benefits of supplementary tube feeding after fractured neck of femur: a randomised controlled trial. *British Medical Journal*, **287**, 1589–92.

Blichert-Toft, M. (1975). Secretion of corticotrophin and somatotrophin by the senescent adenohypophysis in man. *Acta Endocrinologica*, **195**, suppl.

Copes, W.S., Lawnick, M., Champion, H.R., and Sacco, W.J. (1988). A comparison of Abbreviated Injury Scale 1980 and 1985 versions. *Journal of Trauma*, **28**, 78–86.

Cuthbertson, D.P. and Tilstone, W.J. (1969). Metabolism during the post-injury period. *Advances in Clinical Chemistry*, **12**, 1–55.

DeMaria, E.J., Kenney, P.R., Merriam, M.A., Casanova, L.A., and Gann, D.S. (1987). Survival after trauma in geriatric patients. *Annals of Surgery*, **206**, 738–43.

Evans, J.G., Prudham, D., and Wandless, I. (1979). A prospective study of fractured proximal femur: incidence and outcome. *Public Health*, **93**, 235–41.

Frayn, K.N. (1986). Hormonal control of metabolism in trauma and sepsis. *Clinical Endocrinology*, **24**, 577–99.

Håkanson, E., Rutberg, H., Jorfeldt, L., and Wiklund, L. (1984). Endocrine and metabolic responses after standardized moderate surgical trauma: influence of age and sex. *Clinical Physiology*, **4**, 461–73.

Horan, M.A., Barton, R.N., and Little, R.A. (1988). Ageing and the response to injury. In *Advanced Geriatric Medicine 7* (ed. J.G. Evans and F.I. Caird), pp. 101–35. Wright, London.

Le Quesne, L.P., Cochrane, J.P.S., and Fieldman, N.R. (1985). Fluid and electrolyte disturbances after trauma: the role of adrenocortical and pituitary hormones. *British Medical Bulletin*, **41**, 212–17.

McClure, J. and Goldsborough, S. (1986). Fractured neck of femur and contralateral intracerebral lesions. *Journal of Clinical Pathology*, **39**, 920–2.

Phillips, P.A. *et al.* (1984). Reduced thirst after water deprivation in healthy elderly men. *New England Journal of Medicine*, **311**, 753–9.

Royal College of Physicians of London (1989). Fractured neck of femur: prevention and management. Royal College of Physicians.

4.3 Temperature homeostasis and thermal stress

KENNETH J. COLLINS

There is substantial evidence that control of body temperature and the pattern of fever can alter with advancing years. This is likely to arise from structural and functional changes in the nervous system and a reduced blood supply to organs and tissues. In addition to the higher prevalence of degenerative diseases and joint disorders that lead to immobility in old age, there are decrements in cardiorespiratory function and an increase in the occurrence of mild confusional states; all of these encourage lack of response to thermal stress. Essential thermoregulatory effector organs such as skin blood vessels and sweat glands become less responsive, and central nervous mechanisms appear to lose some of their fine control and allow body temperature to oscillate between wider limits of internal temperature. The condition of thermoregulatory failure in many elderly people does not usually entail absence of response to ambient temperature change but there is a reduced thermoregulatory efficiency and a diminished potential to adapt to thermal stress.

Temperature homeostasis, like other homeostatic mechanisms, involves the maintenance of equilibrium in a system that displays inertia and adaptability. Older people react more slowly to the challenge of heat or cold, and diminished response and adaptability result in a wider 'hunting' pattern of adjustment and a lengthening of the period necessary to re-establish equilibrium. Most organs are affected by morphological and functional involution during ageing, with a gradual decline in performance. The effects of physical deconditioning or detraining are often part of the loss in physical capability, and similarly, disuse of thermoregulatory mechanisms may raise the threshold of response. Apart from these physiological changes, there are pathological causes of poor thermoregulation with age. Cerebrovascular disease may affect the function of the temperature-regulating centres of the brain and vasomotor control of heat loss. Hypothyroidism is more frequently found in old people and diminishes internal heat production, and many other pathological states affecting the endocrine, cardiovascular, and nervous systems may have a profound effect on temperature homeostasis. If it is the adaptive capacity rather than the equilibrium level of body temperature that is altered with age, then no changes in resting body temperature may be expected. Indeed, early studies found no evidence for progressive changes in body temperature with age. However, surveys of body temperatures in the elderly, using rectal, urine, and aural measurement, have revealed slightly lower core temperatures (by 0.2 to 0.3°C) than in younger people; unusually low body temperatures in about 10 per cent of an elderly population sample; or a bimodal distribution with one peak in the region of 37.0°C and another close to 35.5°C. The frequent inclusion of patients with disease or receiving medication, and the differences in environmental conditions surrounding each set of observations are of great importance in determining the body temperature observed, and it has therefore yet to be established that resting core temperature diminishes with age.

THERMOGENESIS

The metabolic rate is lower in older people whether measured as heat produced per kg body weight or as internal heat per square metre of body surface. The proportion of body mass made up of actively functioning cells is usually less and this results in an overall decrease in total heat production. When related to total body water, which also declines with age, or to lean tissue mass, basal oxygen consumption is constant. In addition, the more sedentary lifestyle of the older age group results in a smaller contribution to overall heat production from muscular activity.

Most of the evidence of shivering thermogenesis in cold environments suggests that shivering is reduced, or even absent, in elderly subjects. Investigations using convective cooling in air show that the ability to shiver is not lost even in those over 80 years of age, but changes occur in the characteristics of the shivering response. The high peaks of muscle contraction achieved by young people are usually not attained, and often a longer latent period is required to initiate maximum shivering. A contributory factor may be loss of motor power in the muscles with detraining. Age-associated changes in fast and slow motor units of muscle and lack of muscle bulk may help to account for these differences in shivering thermogenesis.

In the human adult, non-shivering thermogenesis from brown adipose tissue does not appear to play a major role in thermoregulation. Any contribution to internal heat production from this tissue in later years is likely to be small, for although a few cells of brown adipose tissue have been identified in mediastinal, perirenal and other regions in younger adults, they disappear almost entirely by the eighth decade of life.

CUTANEOUS VASOMOTOR RESPONSES

Blood flow to the skin plays an important part in thermoregulation and is the main defence against temperature stress in the zone of vasomotor control between shivering and sweating. A number of investigations have shown poor vasomotor responses in a proportion of old people when they are exposed to the cold. In longitudinal studies this proportion increases with age.

In most young adults it is possible to demonstrate spontaneous transient bursts of vasoconstrictor activity occurring three or four times a minute in a thermally neutral environment. This rhythm is apparently generated from the central nervous system because similar patterns can be found in neurograms recording sympathetic activity from peripheral nerves. With cold stimulation of the body, the rhythmic vasoconstrictor activity increases in frequency until it becomes continuous. In most elderly people this background vasomotor activity in neutral temperatures is absent and there is a longer latent period for vasoconstriction in the cold.

Although these findings indicate failure of autonomic vasomotor responses in some old people, it is difficult to dismiss the possible role of intrinsic arteriosclerotic changes with age. Measurements of the elasticity of peripheral blood vessels by pulse-contour analysis show significant decreases in compliance with age and loss of the dicrotic notch in pulse-wave forms. The effects of arterio- and atherosclerosis may therefore play an important part in the observed 'non-constrictor' responses.

The core-temperature threshold for vasodilatation may increase with age, although earlier investigations appear to demonstrate an exaggerated forearm blood flow in middle-aged men during exercise. Others have found that the blood flow: core temperature relationship is attenuated in older people. Comparison of young and old groups of healthy people exercising at the same absolute ($\dot{V}_{O_2}$) and relative (per cent $\dot{V}_{O_2}$ max.) intensity show a consistently lower, forearm blood-flow response in older men. During submaximal exercise in the heat, cutaneous blood flow is increased by a raised cardiac output as well as by selective vasoconstriction of splanchnic and renal vascular beds, so the finding of a lower cardiac output for a given $\dot{V}_{O_2}$ in older persons during light exercise may provide one explanation for the reduced peripheral blood flow in the older group.

SWEATING RESPONSES

Impairment of thermoregulation by diminished or absent sweating is a primary cause of heat exhaustion and heat stroke in hot conditions. Sudomotor dysfunction will increase thermal strain. The sweating response to thermal or neurochemical stimulation in people aged 70 years or more is usually found to be less than in young adults. There is a higher core-temperature threshold at which sweating is initiated in the old. Although endurance-exercise sweating in fit older people is reported to be unaffected by age, sweat production is lower in middle-aged people than in young during low-intensity, intermittent exercise. However, after rigorous heat acclimatization, fit older persons respond with higher core temperatures and lower sweat production than the young during exercise in dry heat.

THERMORECEPTION

Many of the senses are blunted in old age, with variable losses in vision, hearing, olfaction, and taste. At least part of this decline appears to be due to changes in the central nervous system itself. Less is known about age-associated changes in the function of thermoreceptors in the skin. Cutaneous cold receptors in apes are highly dependent for optimum function on a good oxygen supply. Reduced vascularity in the skin and a reduced number and sensitivity of peripheral nerves may affect the functioning of thermoreceptors in elderly people. Again, earlier studies demonstrated no systematic difference in thresholds of warm and cold sensation between young persons and those over 70 years of age. In tests on larger groups of people it has been found that many old persons are unable to match the high discrimination of the young. Psychophysiological factors may account for some of these age differences, for example in the degree of confidence with which individuals make their decisions. Signal detection analysis is a useful tool to detect such differences, and it has been shown that when temperature discrimination changes significantly in old people the criteria upon which old and young base their decisions may still be similar.

Reduced temperature sensation may place some old people at risk if they cannot readily detect changes in environmental temperature. One study of elderly people, aged between 74 and 86 years, showed that most responded to cold discomfort by an appropriate action to increase the temperature of their micro-environment. Some, however, experienced the feeling of cold only at unusually low temperatures and did not regulate the indoor climate promptly even when heating was available. Investigations of behavioural thermoregulation in controlled temperature conditions away from the home environment confirm that some elderly people lack the precision of ambient temperature control shown by young people.

THERMAL COMFORT

It is generally recognized that elderly persons prefer an ambient temperature for comfort that is slightly higher than that preferred by young people. This is consistent with a more sedentary lifestyle and pattern of activity at later ages. If, in controlled conditions, thermal comfort is measured when the extent of activity and amount of clothing are similar, there is little difference in preferred temperature between young and old people. A sedentary elderly person with hypothyroidism, however, may feel cold with a normal amount of indoor clothing even in a temperature of 24°C when the euthyroid elderly person would feel that 21°C was comfortable. The fall in basal metabolic rate between a 20-year-old and 65-year-old is approximately 5 W/m^2 of body surface area, and this might seem to imply a higher preferred comfort temperature in the elderly. However, insensible cutaneous water loss is about 5 W/m^2 less in the older person, and the two factors may therefore offset each other in the balance that underlies the sensation of thermal comfort.

CENTRAL NERVOUS CONTROL OF THERMOREGULATION

Changes in the hypothalamic threshold of sensitivity to temperature stimulation may be a primary cause of altered thermal homeostasis, both in 'setting' the deep body temperature and in functional responses to temperature stress. In animals there are reduced catecholamine concentrations in the preoptic, hypothalamic, temperature-sensitive area of the brain, and differences in the action of putative thermoregulatory neurotransmitters with increasing age.

Body temperature varies in a well-marked circadian rhythm that is partly intrinsic and party exogenous, falling to a minimum during sleep at night and rising during the day. The rhythm of body temperature appears to be synchronized with the sleep–wake cycle but there is an inherent difference in the way these two rhythmic changes are generated by 'pacemakers' in the brain. Where a person is deprived of 'cues' that time these rhythms, the sleep–wake cycle operates free and with a different periodicity in relation to the rhythm for body temperature. When this happens, for example if induced by constant illumination, body temperature becomes more labile and may fall to hypothermic levels when there is exposure to cold stress. There is some evidence that desynchronization of the different circadian rhythms can occur more frequently in old age and this may increase the risk of hypothermia.

Fever is a pathological process in which a regulated increase in core temperature occurs during infection. The anterior hypothalamic, preoptic area is the site in the brain sensitive to the effects of both exogenous and endogenous pyrogens. Old animals develop less of a fever in response to endotoxin than do young animals. There is a reduction or abolition of the second peak of fever and a large rise in plasma levels of adrenaline in old animals. It appears as if the older animals possess the drive to elevate their core temperature at the time of the second fever peak but cannot do so. The presence of fever in elderly people as a symptom of disease is often indeterminate; some elderly patients with pneumonia, for example, will develop very high temperatures while others may be afebrile. It is not unusual for pyrexia and leucocytosis to be absent in infections such as pyelonephritis, cholangitis, acute septicaemia, and pneumococcal bacteraemia commonly encountered in elderly patients. The response of the acute-phase protein affords a valuable, non-specific indicator of the severity of the infection in these circumstances.

COLD AND WINTER DEATHS

Statistics show that cold weather is associated with an increase in deaths, particularly among elderly people, and in Britain excess winter deaths double for every 9 years of advancing age over 40. In broad terms there is an immediate rise in mortality when mean day and night temperatures drop below freezing and the effects seem to persist over a period of 30 to 40 days after the end of the extremely cold period. A common measure of assessing this seasonal effect is through a seasonal or winter mortality ratio, calculated as the death rate for the January to March quarter divided by the average rate for the same calendar year. In the 1960s and 1970s Britain had the highest winter mortality ratios for temperate/cold countries in North America and Europe. Although differences now appear to be declining, there still remains a large discrepancy between Britain and countries such as Canada and Scandinavia, which have much colder winter climates. Bull *et al.* (1978) showed a strong relationship between environmental temperature and death rates for the years 1962 to 1967, and for people older than 60 years the slope of the regression of death rates on ambient temperature was steeper than for young adults. The length of time between the onset of a cold spell and the increase in mortality was 1 to 2 days for myocardial infarction, 3 to 4 days for strokes, and approximately 1 week for pneumonia and bronchitis. About half the excess winter deaths were certified as being caused by coronary or cerebral thrombosis and one of the underlying causes for this may be increased blood viscosity and haematocrit as a result of cooling. Keatinge *et al.* (1984) have shown increases in blood platelets, red cell counts, cholesterol, and arterial pressure with mild skin-surface cooling. These studies were carried out in young people; established arterial disease may be more susceptible to the potentially fatal effects of platelet aggregation and thrombosis formation in the cold. Another factor may be changes in reflex cardiovascular responses in elderly people. Arterial blood pressure increases more severely in cold conditions in older people with reduced baroreflex control (Collins *et al.* 1985).

Winter mortality statistics vary greatly in response to influenza epidemics, which nearly always affect the older age groups in particular. Some other causes of death including falls and femoral fracture also increase in the cold weather, perhaps contributed to by minor degrees of hypothermia and impaired neuromuscular co-ordination. This effect is likely to be more pronounced in old people suffering from undernutrition. In numerical terms, the average excess mortality in the winter months in Britain amounts to 40000, with only about 1 per cent being caused by certified hypothermia. None the less, hypothermia is an important condition from the gerontological standpoint as it represents an age-associated loss of adaptability of complex aetiology and pathogenesis.

HYPOTHERMIA

Although there is a continuum of physiological response as body temperature is reduced below normal, hypothermia is clinically defined as a state of subnormal body temperature when deep body temperature falls below 35.0°C. Even up to 25 years ago, hypothermia was thought to be a rare condition and there was no accurate information on its incidence. Few cases were recognized before admission to hospital and elderly people with hypothermia suffered a high fatality. Low environmental temperatures had a clear aetiological association, though social factors such as inadequate clothing, heating, and nutrition were also thought to be important. In a survey from the Royal College of Physicians (London) in the early months of 1965 it was then reported that hypothermia in hospital admissions was quite common, particularly affecting those under 1 year and over 75 years of age. In a further reported series of cases of hypothermia in Scotland, about half the 100 patients were found lying indoors on the floor after a fall due to an accident or illness, and one or more serious underlying disorder was recorded in 20 per cent of the cases (Maclean and Emslie-Smith 1977).

Most cases of hypothermia occur during the coldest winter months, but there is a generally higher risk in the elderly throughout the year, for the condition is not exclusively cold-induced. Hypothermia can occur when there is a marked deterioration in vital functions, especially when associated with malnutrition, existing illness, and infections. There is a high fatality in old people with the condition. Published series find this to be between 30 and 80 per cent of cases depending on the age of the patient, the severity and duration of the initial fall in deep body temperature, and the nature of any other intercurrent disease.

Predisposing conditions

Low body temperature may develop because of a number of extrinsic and intrinsic factors (Table 1). Old people with impaired thermoregulation, often living in the oldest dwellings that are difficult and expensive to heat in the winter, are most at risk. The vulnerable elderly in Britain are identified as those over 75 years of age, receiving supplementary benefit, and who are socially isolated. Ninety per cent of elderly people in England in 1976 were found to be living in their own or own family's accommodation, while only 8 per cent lived in sheltered housing or old people's homes. Thus a high proportion have at least some control over their own environment indoors and this proportion has not substantially altered in recent years. Cold exposure is an overriding cause of hypothermia and the number of cases rises measurably when the ambient temperature falls below 0°C.

Table 1 Factors contributing to hypothermia in elderly people

Extrinsic
Exposure to cold outdoor or indoor temperatures
Poor domestic heating and insulation
Insufficient warm clothing

Intrinsic
Physiological
- Low metabolic heat production
- Inadequate nutrition
- Immobility
- High surface area to mass ratio
- Impaired temperature regulation
- Blunted temperature perception

Clinical conditions leading to secondary hypothermia
- Endocrine: hypothyroidism, hypopituitarism, Addison's disease, diabetes mellitus
- Neurological: paraplegia, parkinsonism, Wernicke's encephalopathy, hypothalamic lesions
- Locomotor: osteoarthritis and other causes of immobility
- Mental: confusional states, dementia, depression
- Infection: bronchopneumonia, septicaemia
- Cardiovascular: myocardial infarction, pulmonary embolism, cerebral haemorrhage
- Drugs: psychotropics, hypnotics, tranquillizers, anti depressants, alcohol, hypoglycaemic agents, antithyroid agents, sympathetic and ganglion-blocking agents
- Other: erythematous skin disease, steatorrhoea, extensive Paget's disease of bone, extreme wasting in malignant disease and starvation, extensive burns

The greater tendency for hypothermia in old people can be partly accounted for by the physiological decline in thermoregulatory function revealed by cross-sectional and longitudinal studies. This has also been demonstrated in elderly survivors of hypothermic episodes in whom shivering was absent, the metabolic rate did not rise, and vasoconstriction was defective during moderate cooling. As a consequence, deep body temperature fell abnormally and progressively. These patients are at risk of developing further episodes of hypothermia precipitated by moderate cold exposure or by the use of even small amounts of drugs such as the phenothiazines.

Some old people admitted to hospital suffer from accidental hypothermia due to cold exposure and failing thermoregulation, but pathological conditions leading to secondary hypothermia are present in the majority. Diabetes mellitus is more common than hypothyroidism and hypopituitarism in causing hypothermia, which most often occurs in diabetic ketoacidosis when the arteriovenous difference in oxygen levels is abnormally low. Vasomotor dysfunction due to autonomic neuropathy can also contribute to impaired thermoregulation in elderly diabetic patients. In a number of neurological and locomotor disorders, immobility is a factor limiting the amount of heat generated and in parkinsonism there may also be autonomic dysfunction.

Clinical features

The clinical presentation of hypothermia is approximately related to core temperature. As moderate (core temperature 34–32°C) and severe (core temperature below 32°C) hypothermia develops, the patient becomes progressively more confused and disorientated, and this is followed by drowsiness and eventually coma. Impairment of consciousness becomes increasingly more likely as body temperature drops below 32°C. Some patients, however, may still be conscious at a rectal temperature of 27°C whereas others at 31°C are unconscious. Elderly people can sustain a chronic state of low body temperature for some days before reaching the stage of clinical hypothermia. Shivering becomes replaced by generalized muscular rigidity and vital signs become unrecordable. Tendon reflexes are slow and usually diminished, plantar responses are absent or extensor, and the light reflex sluggish. The patient usually develops a grey colour due to pallor and cyanosis, there is a distinctive firm, doughy consistency of the subcutaneous tissue, and a myxoedematous appearance. Normally warm regions of the abdomen, axillae, and groin feel cold. The diagnosis can be established by measurement of the rectal temperature using a low-reading clinical thermometer.

The heart rate slows in response to cold with sinus bradycardia or slow atrial fibrillation. The electrocardiogram usually shows some degree of heart block with an increase in PR interval in patients in sinus rhythm, and there is a delay in intraventricular conduction. The appearance of a J wave, shown by a characteristic deflection at the junction of the QRS and ST segment, is a typical sign. The size of the J wave varies from patient to patient. It is not related to the severity of the hypothermia and even in severe hypothermia it may be absent altogether. Respirations are very slow and shallow in severe hypothermia progressing to apnoea. Arterial P_{O_2} is low and the oxygen dissociation curve shifted to the left so that less oxygen is released to the tissues at a given P_{O_2}. Bronchopneumonia may develop without the usual clinical signs.

Gastric dilation is common and there is the risk of aspiration of gastric contents. Acute ulceration of the stomach can cause haematemesis. At autopsy, pancreatitis is often found but during life acute pancreatitis is frequently overlooked because few of the typical signs are present in the hypothermic patient.

Renal blood flow, glomerular filtration, and tubular function are impaired, oliguria is common, and acute tubular necrosis can occur. This may be due to ischaemia and to the direct effect of cold on the kidneys.

Haemoglobin and the haematocrit may be raised owing to a decrease in plasma volume. A moderate elevation of the white-cell count often occurs but a low or normal count does not exclude the possibility of an infection such as bronchopneumonia. Thrombocytopenia giving rise to bleeding is not uncommon, and this has been attributed to sequestration of platelets in the liver and spleen. Normal numbers of megakaryocytes can be found in the bone marrow at the time of the thrombocytopenia, and the platelet count rises again to normal as the hypothermic patient regains a normal body temperature. Multiple infarcts may occur in the myocardium, viscera, or pancreas.

The glucose concentration in hypothermia is often above 6 mmol/l and in the absence of diabetes mellitus, the blood sugar returns to normal as the patient's temperature rises. There is inhibition of glucose utilization at low body temperatures and fat then appears to be the main fuel for thermogenesis despite the apparent availability of glucose in the plasma. Raised serum levels of aspartate aminotransferase, hydroxybutyrate dehydrogenase, and creatine kinase are often present; very high levels of these muscle enzymes lead to the suspicion that the hypothermia is secondary to myxoedema. The levels fall to normal when body temperature rises, even before thyroid replacement is started. Serum levels of triiodothyronine, thyroxine, and thyroid-stimulating hormone can be significantly influenced by haematological changes induced by hypothermia, and caution is therefore necessary in interpretation. Raised cortisol levels are frequently found in hypothermia. There is evidence that cortisol utilization is poor and sometimes absent; the routine use of hydrocortisone in treatment has been discontinued.

Management

The primary aim of treatment is to restore body temperature to normal with the minimum of complications. Ideally, a rapid restoration of deep body temperature to avoid prolonged hypothermia would be most satisfactory. Although rapid, active, surface warming can be practised in fit young adults with accidental hypothermia, such a procedure is hazardous in the elderly and frequently leads to circulatory collapse. The initial after-drop in core temperature with rapid rewarming may precipitate hypotension and cardiac dysrhythmias. In severe hypothermia, however, active rewarming of even elderly patients has been shown to reduce fatality.

There is no standard treatment for the elderly victim of hypothermia; each patient should be considered as a separate problem for intensive care. This is because there is usually some underlying pathological condition which may at first be masked by low body temperature. Hypothermic patients require gentle handling for there is always a risk of cardiac dysrhythmias. Ventricular fibrillation is the most dangerous complication, occurring most frequently at and below deep body temperatures of 28°C.

Patients suffering from mild or moderate hypothermia with deep body temperature between 35 and 32°C should be transferred to a warm environment of 25 to 30°C for passive rewarming at the rate of about 0.5°C per hour. Those elderly who fail to rewarm at this rate are often suffering from hypothyroidism or are actually dying of their underlying disease. Fast rewarming carries the risk of hypotension so that the use of hot-water bottles or hot baths is contraindicated. Controlled oxygen administration through a Venturi mask is of benefit. Monitoring of rectal temperature, pulse, blood pressure, and the electrocardiogram are required; if there is a fall in blood pressure during treatment the patient should be temporarily cooled by lowering the room temperature. A few elderly hypothermic patients with a core temperature not lower than 33°C can be nursed successfully at home if conditions are favourable. Treatment then follows the line of slow rewarming in a room at 25°C or above. Bronchopneumonia frequently develops during or shortly after an episode of hypothermia in the elderly; there is therefore a place for prophylactic broad-spectrum antibiotic therapy.

Severe hypothermia with body temperature below 32°C may be treated in an intensive care unit with various methods of central warming, including warm gastric or colonic lavage, warm intravenous infusions of saline, or peritoneal dialysis with fluid warmed to between 38 and 43°C. Intermittent, positive-pressure ventilation is used to correct hypoxia and to re-expand collapsed alveoli. When there is severe depression of the respiratory centre, especially in hypothermia due to phenothiazines, anoxia may be the only drive to respiration. In these cases, oxygen therapy alone may produce apnoea and it is essential to institute artificial ventilation. Correction of dehydration and electrolyte disturbances is important, particularly in comatose patients. Monitoring of central venous pressure is essential to minimize the risk of pulmonary oedema following intravenous fluid replacement. Abnormal water retention, haemodilution, and very low serum electrolyte concentrations are also features of hypothermia associated with myxoedema and hypopituitarism.

The most favourable course from which the outcome for patients with core temperature below 32°C may be predicted during rewarming is one in which the patient rewarms spontaneously at about 0.5°C per hour without adverse haemodynamic effects. Less favourable is the course followed by patients whose spontaneous warming is complicated by one or more of the following conditions that demand specific treatment and intensive monitoring: unexpected hypotension, episodic sinus bradycardia or atrioventricular block, profound hypoglycaemia, persistent hypoxaemia, or intercurrent infection. A third clinical course carries an even worse prognosis, when spontaneous rewarming is complicated by acute pancreatitis, bronchopneumonia, or circulatory arrest. The worst prognosis is associated with failure to increase core temperature and clinical signs of shock not related to any obvious curable cause.

HYPERTHERMIA

Heat-related illnesses occur frequently during heat-waves in urbanized communities in temperate zones as well as in tropical and subtropical regions. The adverse effects of unexpected high environmental temperatures mainly affect the elderly, the very young, and those with chronic illness or debility. In one study in St Louis and Kansas City during the 1980 summer heat-wave, heat-stroke rates in persons aged 65 years or more were 12 to 13 times the rates in the remainder of the population. Peaks in daily mortality in nursing homes for the elderly generally lagged behind the corresponding peaks in maximum temperature, by one day. 'Heat-waves' have no precise meteorological definition and are usually determined by normal local temperature conditions. For example, in Greater London a heat-wave may consist of a period of 3 or more consecutive days when the maximum daily dry-bulb temperature exceeds 27°C (80°F). In New York City, maximum daily temperatures of 31°C (88°F) or more may be required to characterize heat-wave conditions. When heat-waves do occur, or when there are prolonged periods of unaccustomed hot weather, they are associated with significant increases in illness and death in the vulnerable groups.

In the United States about 4000 deaths each year have been attributed to the effects of heat-related illnesses, and 80 per cent of these deaths are of persons over 50 years of age. Other estimates put the average annual death rate directly associated with high ambient temperature at about 200. For the United Kingdom the annual total of deaths registered as due directly to excessive heat is usually only in single figures, but in a summer with a prolonged heat-wave, in 1976, 37 such deaths were recorded. There were 500 excess deaths in the same period of 1976, predominantly in those over 65 years of age, and there were increased hospital admissions of elderly patients suffering from strokes, transient ischaemic attacks, and subarachnoid haemorrhage. Mortality in the elderly population in the heat resembles that for cold winter conditions, that is with peaks in total cardiovascular, cerebrovascular, and respiratory deaths. To investigate the effects of temperature on mortality, it is necessary to study daily records because variations are obscured in monthly mortality statistics. This is probably because when a heat-wave leads to a sudden increase in mortality of those at greater risk, fewer die in the immediately succeeding weeks. With the predicted onset of global warming, hyperthermia in the elderly in countries such as the United Kingdom may begin to assume epidemiological and medical importance equal to that accorded to hypothermia.

Predisposing factors

Exposure to unaccustomed hot conditions, especially where there is high humidity, low air movement or a high level of insolation, commonly provides the physical elements leading to heat illness. Where there is a physiological demand for increased perfusion of the peripheral tissues, such as is needed for promoting heat loss by peripheral vasodilatation or the need to supply fluid to the sweat glands, diseases that severely restrict cardiac output will dispose to failure of thermoregulation. Renal disorders, which interfere with body fluid balance, will also impair thermoregulatory ability. The thermal strain is increased in obesity and in those who are physically unfit. Intrinsic factors predisposing to heat illness include dehydration, unacclimatization, old age, and particularly chronic disease involving disorders such as diabetes mellitus, cardiovascular disease, and gastrointestinal, respiratory, and febrile illnesses. Persons confined to bed or unable to care for themselves are at high risk from the effects of heat, as are alcoholics and those being treated with neuroleptic or anticholinergic drugs (Table 2).

Attention has been drawn to the relationship between fatality in elderly people in hot conditions and impairment of sweating. In old people, there is a marked reduction in sweating activity and an increase in the temperature threshold of sweating. The difference between the sudomotor response in young and old does not appear to be entirely due to different states of heat acclimatization or effects of drugs, although in general the elderly might be expected to be less heat-acclimatized. Sweat responses are smaller in healthy elderly than in young people when both are artificially acclimatized. Factors controlling water balance are also affected by age and elderly subjects have a reduction in osmoregulated vasopressin secretion and a decreased renal response to vasopressin. Elderly people are also less sensitive to increasing plasma

Table 2 Factors contributing to hyperthermia in elderly people

Extrinsic
Exposure to hot conditions especially with high humidity, high radiant heat or low air movement
Physical exertion in the heat
Excessive clothing in hot conditions
Intrinsic
Physiological
Lack of heat acclimatization
Dehydration
Low surface area to mass ratio in obese elderly people
Diminished sweating responses
Reduced renal efficiency
Clinical conditions leading to hyperthermia
Endocrine: hyperthyroidism, hyperpituitarism, diabetes mellitus
Neurological: autonomic dysfunction, cerebral haemorrhage (especially pontine or thalamic), head injury, cerebral tumour or abscess
Mental: confusional states, dementia
Infections: respiratory tract infections, gastrointestinal infections, septicaemia
Cardiovascular: ischaemic heart disease, congestive cardiac failure
Drugs: psychotropics, hypnotics, antidepressants, amphetamines, alcohol, anticholinergics
Other: skin disease causing inability to sweat

sodium and osmolality.

The elderly individual is more likely to be receiving many drugs at the same time, including some that cause or exacerbate heat illness. There are dangers in prescribing psychotropic drugs in hot conditions, particularly the phenothiazines, which interfere with thermoregulation, or anticholinergics that suppress sweating. Elderly patients are often treated long-term with diuretics, which are risk factors for dehydration and electrolyte disturbances. Many drugs contributing to the development of hyperthermia in the heat may also induce hypothermia in cold conditions.

Clinical features

Heat illnesses range from minor disorders such as syncope to life-threatening conditions including heat stroke. The initial physiological responses to heat may cause oedema of the feet and ankles due to peripheral vasodilatation, especially when combined with venous stasis in the legs with prolonged standing. Syncope can be precipitated in the heat by sudden postural changes and this is more likely in elderly people with postural hypotension. Skin disorders such as prickly heat occur in some individuals when sweat is allowed to accumulate on unventilated skin regions. Prickly heat causes considerable discomfort but the condition is not dangerous unless a large part of the surface area is affected and anhidrotic heat exhaustion develops. In urban temperate regions, even during heat-waves, it is unlikely that severe heat exhaustion due to water deficiency or salt deficiency will arise when it is possible to restrict physical activity in the heat and to have ready access to fluids. Heat-exhaustion syndromes and heat intolerance can, in the extreme, progress to heat stroke, which is characterized by hyperthermia with core temperature reaching 41°C or above, central nervous disturbances leading to convulsions and coma, and often a marked anhidrosis. For the majority of unacclimatized people, especially the physically unfit and the elderly, increased cardiovascular strain is the initial threat posed by environmental heat.

The manifestations of heat-stroke in elderly people are usually not typical of those in young adults. In old people, heat stroke often occurs in epidemic form and in the presence of predisposing disorders such as arteriosclerotic heart disease, congestive cardiac failure, diabetes mellitus, parkinsonism, and recent or old stroke. In young adults, heat-stroke is commonly exertional, arising in isolated cases, and is less frequently associated with an underlying chronic disorder. A history of prodromal illness may not be present but some patients complain of weakness, nausea, vomiting, dizziness, headache, breathlessness, anorexia, and a feeling of warmth. Salient clinical features are dehydration with anhidrosis in the majority of patients, coma with complete unresponsiveness to painful stimuli, and signs of pulmonary consolidation, often due to staphylococcal infection. In a series of 15 elderly patients with hyperpyrexia in whom serum sodium levels were estimated, six had concentrations above 150 mmol/l. Elderly patients have a diminished ability to combat dehydration because of depressed thirst sensation and failure of cardiovascular and renal mechanism for water conservation.

An initial haemodilution due to peripheral vasodilatation on exposure to heat is succeeded by an increased blood viscosity and coagulability during subsequent hyperthermia and dehydration. But there are also haemorrhagic manifestations in the form of petechial haemorrhages, ecchymoses, epistaxis, and haematemesis in the course of heat stroke. The haemorrhagic state has been attributed to increased capillary permeability, hypoprothrombinaemia, thrombocytopenia, and fibrinolysis. Disseminated intravascular clotting (DIC) has also been clearly demonstrated by histopathological studies in cases of heat-stroke. Shock syndrome is commonly associated with DIC, which worsens the condition by the dissemination of clots in the microcirculation.

Management

Patients with hyperthermia or heat-related illness should be quickly removed from the warm environment into the coolest, shaded area and further cooled by removing clothing, increasing air movement over the body surface, and sponging with tepid water. Circulatory shock occurs in many heat-stroke victims and intravenous fluids may be required to treat water and/or salt depletion. This should be done with caution for there is considerable danger of precipitating pulmonary oedema especially in heat-stroke patients with renal failure.

A heat-stroke emergency is best managed with a body cooling unit (Khogali and Hales 1983), which reduces core temperature rapidly by forced convective and evaporative cooling. It is imperative to employ any similar means of lowering core temperature, be it by sponging with tepid water or covering the patient with a thin, wetted sheet and creating a good air flow. Ice baths carry the risk of inducing vasoconstriction and shivering, which will increase heat load and reduce heat loss. Cooling should continue until the rectal temperature reaches 38.5°C; further aggressive cooling after this may produce hypothermia. In heat stroke in the elderly, the prognosis is very poor as with hypothermia, often due to the serious nature of underlying diseases. The hypothermic and antipyretic properties of phenothiazines and salicylates are of debatable value in the treatment of environmentally induced hyperthermia.

The protection of old people in hot environments involves restricting physical activity, maintaining a reasonable fluid intake, wearing light, loosely fitting clothing, and avoidance of extremes of heat stress with the appropriate use of air-conditioning or circulating or ventilating fans. In general, older people prefer and thrive better in a climate that is warm all the year round, provided they are equipped to deal with occasional extremes of temperature and humidity that might subject them to unacceptable levels of thermal stress. Fans decrease in cooling efficacy as ambient temperature rises to high levels above core temperature and they have sometimes been found to increase heat stress in very hot weather. Air-conditioned heat-wave shelters are of great potential benefit in urban areas experiencing frequent heat-waves.

Bibliography

Alderson, M.R. (1985). Season and mortality. *Health Trends* (HMSO) **17**, 87–96.

Allison, S.P. and Bastow, M.D. (1983). Undernutrition and femoral fracture. *Lancet*, **i**, 933–4.

Bastow, M.D., Rawlins, J., and Allison, S.P. (1983). Undernutrition, hypothermia and injury in elderly women with fractured femur: an injury response to altered metabolism. *Lancet*, **i**, 143–5.

Bittles, A.H. and Collins, K.J., eds. (1986). *The biology of human ageing*. Cambridge University Press.

Bull, G.M. and Morton, J. (1978). Environmental temperature and death rates. *Age and Ageing*, **7**, 210–24.

Collins, K.J. and Exton-Smith, A.N. (1983). Thermal homeostasis in old age. *Journal of the American Geriatrics Society*, **31**, 519–24.

Collins, K.J., Doré, C., Exton-Smith, A.N., Fox, R.H., MacDonald, I.C., and Woodward, P.M. (1977). Accidental hypothermia and impaired temperature homeostasis in the elderly. *British Medical Journal*, **1**, 353–6.

Collins, K.J., Easton, J.C., Belfield-Smith, H., Exton-Smith, A.N., and Pluck, R.A. (1985). Effects of age on body temperature and blood pressure in cold environments. *Clinical Science*, **69**, 465–70.

Ellis, F.P. (1972). Mortality from heat illness and heat-aggravated illness in the United States. *Environmental Research*, **5**, 1–58.

Foster, K.G., Ellis, F.P., Exton-Smith, A.N., and Weiner, J.S. (1976). Sweat responses in the aged. *Age and Ageing*, **5**, 91–101.

Fox, R.H., Woodward, P.M., Exton-Smith, A.N., Green, M.F., Donnison, D.V., and Wicks, M.H. (1973). Body temperatures in the elderly: a National study of physiological, social and environmental conditions. *British Medical Journal*, **1**, 200–206.

Keatinge, W.R., Coleshaw, S.R.K., Cotter, F., Mattock, M., Murphy, M., and Chelliah, R. (1984). Increase in platelet and red cell counts, blood viscosity, and arterial pressure during mild surface cooling: factors in mortality from coronary and cerebral thrombosis in winter. *British Medical Journal*, **289**, 1405–8.

Khogali, M. and Hales, J.R.S. (1983). *Heat stroke and temperature regulation*. Academic Press, New York.

Maclean, D. and Emslie-Smith, D. (1977). *Accidental hypothermia*. Blackwell Scientific, Oxford.

Pozos, R.S. and Wittmers, L.E. (1983). *The nature and treatment of hypothermia*. University of Minnesota Press.

4.4 Services for patients with proximal femoral fracture

J. GRIMLEY EVANS

Proximal femoral fracture is an important affliction of later life. In the United Kingdom and some other countries (Maggi *et al.* 1991) it is increasing in incidence for reasons that are not yet clear. If age-specific incidence rates continue to increase at their recent speed, the numbers of cases of proximal femoral fracture in the United Kingdom will increase by half in the next 10 years. This prediction may, however, prove too pessimistic. One series of British studies suggests that rates in men, having increased into the 1960s, stabilized in the 1970s although continuing to rise in women (Evans 1985). This pattern would be compatible with a long-term effect of cigarette smoking on the prevalence of osteoporosis and offers hope that rates in women will also stabilize in due course. However, even if incidence is not increasing, the growth in numbers of very elderly people in the population will lead to a 15 per cent increase in numbers of cases of proximal femoral fracture in the next decade. Already up to half of acute orthopaedic beds in some centres are occupied by patients suffering from this injury, so that even a 15 per cent increase in numbers could have a critical effect on orthopaedic services.

Clearly, prevention of osteoporosis and of the falls and impaired protective factors that lead to this type of fracture (Evans 1990) can provide the only radical approach to control, but there has been increasing concern in the last decade that management of elderly patients so afflicted could be improved. Certainly, results are not good; in an epidemiologically based study in Newcastle upon Tyne, England, there was a 6-month fatality of 40 per cent. This was inflated by a particularly high fatality among patients with pre-existing mental impairment at one hospital that was probably due to a low allocation of nurses to the orthopaedic ward (Evans *et al.* 1980). Later studies in the same city showed 6-month fatality rates of the order of 29 per cent, as has been found in other British epidemiological studies (Greatorex 1988). The British studies showed that only 50 per cent of patients had returned to their original abode by 6 months and 10 per cent were still in hospital. More than half of 6-month survivors suffered continuous or intermittent pain in the affected leg, and 17 per cent had additional problems such as wound infection or leg swelling. Improvements in operative technique, including better prevention of venous thrombosis, are needed but a significant part of the problem seems to lie with postoperative care and rehabilitation. This chapter considers some of the approaches that have been reported.

METHODOLOGICAL ISSUES

Efforts to evaluate the literature on different forms of management of proximal femoral fracture are frustrated by a widespread failure to standardize methodology. It is only possible to make plausible comparisons between different methods if the case material can be analysed to identify the results in *all* and *only* the cases of this fracture arising within a geographically defined population of elderly people and if every case is followed up to a fixed time from injury. The feasibility of this approach has been demonstrated in both descriptive studies (Evans *et al.* 1979*a*) and in the design of a randomized controlled trial (Hornby *et al.* 1989). Hospital-based series of 'consecutive cases' can be seriously misleading because studies have shown that patients admitted to different hospitals, even within the same town, can be widely different in health and lifestyle factors closely associated with clinical outcome in terms of survival and length of stay (Evans *et al.* 1980). Patients with proximal femoral fracture comprise two overlapping but contrasted populations. At one extreme are the 'superfit' elderly people who fracture their femur out of doors in the course of a busy life and generally do well after fracture. At the other extreme are the frail ill old people who fracture their hip in a fall indoors or in a nursing home, and for whom the fracture is merely one further event in a physical and mental decline leading inexorably to an impending death. The case-mix arriving at a hospital will vary with the characteristics of the catchment population and local referral patterns.

A second difficulty is to identify the baseline from which innovative management procedures are working. If things are very bad they can only get better and an innovation that seems to work well in such a situation might well be ineffective or even deleterious if introduced where baseline practice is better. Descriptive evaluations, comparison of patients from different residential districts, and 'before and after' studies are insufficiently rigorous to provide data that can justifiably be used for possibly expensive and potentially hazardous changes in practice. A third problem, even for randomized controlled trials of new procedures, is their sensitivity to possible 'negative Hawthorne' effects in their control groups. Where trials involve the early removal of a randomized sample of elderly patients with proximal femoral fractures from orthopaedic wards, there may be a feeling generated that the patients remaining are now in the wrong place and no longer the proper responsibility of the orthopaedic staff, and management may deteriorate. There is also an understandable desire on the part of orthopaedic staff for the study to succeed so that such removal will become a permanent feature of local practice. The design of such trials, therefore, should include an adequate establishment of the pretrial baseline to ensure that a 'positive' result indicates improvement in the intervention group rather than deterioration in the control group. A final problem in evaluation is that procedures and structures that work well in the hands of enthusiasts may be deleterious when deployed into general usage. Some form of long-term monitoring of care quality is needed.

OUTCOMES

A major preoccupation in the reported studies has been a reduction in the mean length of hospital stay of patients with these fractures. This reflects a proper concern with the patients' welfare, as it can usually be assumed that patients would rather be at home than in hospital. It also arises from a concern about the costs of care because length of stay has a marginal effect on hospital costs. It is necessary to bear in mind, however, that lowering hospital costs by reducing stay may merely transfer costs to community or other services. Studies need to be evaluated by an economist not a hospital accountant. Preoccupation with length of hospital stay should not distract researchers from what should be the outcome measures of central concern, namely survival, lifestyle, and functional status of patients at a fixed interval from injury. Most students suggest that 6 months is a suitable interval.

Robbins and Donaldson (1984) analysed the components of hospital stay in an English health district. Ten per cent of bed days were spent awaiting surgery, 3 per cent being got fit for surgery, 51 per cent recovering from surgery without complications, 1 per cent being treated for surgical complications, and 1 per cent for medical complications. Six per cent of bed days were accounted for by patients receiving conservative (non-operative) treatment for which there is now doubtful justification (Hornby *et al.* 1989). The remaining 28 per cent of bed days were spent awaiting discharge by patients who were medically and surgically fit for discharge. This analysis indicates that, in the United Kingdom at least, advances in the surgical management of fractures are unlikely to have the same effect on reducing length of stay as would more rapid access to operating-theatre time and more efficient discharge arrangements. This study was flawed, however, in its definition of 'medically and surgically fit for discharge', as patients were included in this category if they were awaiting 'geriatric assessment or bed' indicating that an unspecified proportion of them was still in need of hospital services. The importance of this point lies in the fact that it is easy enough to shorten length of stay by providing poorer care, and specifically by reducing rehabilitation. Studies by Fitzgerald *et al.* (1988) showed that a change from a billing to a fixed prepayment system of hospital reimbursement led to shorter hospital stays by elderly patients with proximal femoral fractures because they were discharged in a poorer functional state so that a higher proportion of them went to nursing homes rather than back into the community. There is apprehension that the recent changes in the funding structure of the United Kingdom health services may lead to similar pressure to curtail rehabilitation.

The question of what is a reasonable and desirable delay before operation on a patient with proximal femoral fracture has recently been debated in British publications. On the one hand it has been suggested that preoperative delay leads to deterioration in a patient's condition so that recovery is impaired. On the other hand, some patients might benefit from a period of preoperative resuscitation. There is also the consideration arising from a recent survey of general surgical care in the United Kingdom that emergency out-of-hours operations carried out by unsupervised junior staff carry a higher fatality rate than similar surgery by senior staff in the normal working day. While the issue remains controversial it seems that the best arrangement is for an elderly patient with a proximal femoral fracture to be operated on as soon as possible within a daily routine list, but that a delay introduced to rectify a defined medical problem is not associated with increased fatality (Harries and Eastwood 1991) (see also Chapter 4.2).

ORGANIZATION OF CARE

Good practice requires that reduction of length of hospital stay should only be regarded as a valid indicator of improved efficiency of care if it is associated with adequate measures of effectiveness in attaining optimal treatment outcomes. With this proviso, innovative approaches to improving care of patients with proximal femoral fracture can be classified broadly according to whether they are attempts to speed the discharge of the 'difficult' longer-stay patients, to expedite discharge of the 'good' short-stay group, or to produce an overall effect on patients of all types.

Services for the longer-stay patient

Several North American geriatric services based on the consultation team or evaluation unit are examples of triage of elderly patients into those who will do well anyway, those who will inevitably do badly, and those who will benefit from specialist geriatric intervention (see Chapter 23.2). There have been European experiments in triage of elderly patients with proximal femoral fracture to identify simply those who will not do well in ordinary orthopaedic care. An example of this approach is described by Thorngren *et al.* (1988). Using a locally derived prognostic index, patients identified as likely to present difficulty in rehabilitation and discharge are selected for intensive rehabilitation in a geriatric unit while those with a good prognosis are prepared for discharge in the orthopaedic ward. This approach is congruent with the general philosophy of geriatric medicine, which aims at early and preferably pre-emptive identification and management of elderly people's problems. Unfortunately the system has not been subjected to an epidemiologically based controlled trial. A less sophisticated proposal from Wallace *et al.* (1986) is to use a locally derived prognostic index to identify patients whose prognosis is so poor that they should be placed directly in permanent institutional care rather than offered rehabilitation. Again, there has been no experimental evaluation of the consequences of this approach. One wonders also if there is any group of patients other than the elderly in whom the identification of poor prognosis would lead to suggestions for withdrawal of care rather than for its intensification.

Speeding the shorter-stay patient

Sikorski *et al.* (1985) describe a service in which patients without significant medical problems other than a stabilizable fracture of the proximal femur and who lived near the hospital with a suitable home environment, were identified for early discharge. They were operated on by spinal anaesthetic without premedication and mobilized within hours of surgery while arrangements for direct discharge home were instituted. Forty-five of 50 patients were discharged within 5 days, representing a notable shift to the left of the cumulative discharge curve compared with a previous year before the introduction of the system. No economic evaluation of the system is presented, and its applicability clearly depends on the adequacy of housing and social support available to elderly people in the surrounding community.

A formalized attempt to shift part of hospital care into the community has been established at Peterborough and some other centres in England, and patients with proximal femoral fractures of good prognosis are among those eligible for accelerated discharge under the scheme. A comparison of patients entered in the scheme with patients from a different geographical area suggests that the enhanced community care reduces length of hospital stay without increase in overall costs (Pryor and Williams 1989). Unfortunately, the opportunity to introduce the scheme as part of a randomized evaluation was missed in its early stages so that its effectiveness and economic efficiency have never been rigorously established.

Approaches to improving management for all

These schemes comprise a variety of co-operative arrangements between geriatric and orthopaedic departments. (The coining of 'orthogeriatrics' to describe such activities is surely an etymological abomination.) Many are described but few evaluated. Some of the earlier forms of such 'co-operation' involved the rapid removal of elderly patients from orthopaedic wards to geriatric units soon after operation. This model raises the general organizational issue of whether patients should have to suffer the disturbance of interward or interhospital transfer to receive the care they need, or whether it would be more sensible for orthopaedic teams to be expected and trained to deal effectively with the patients coming to them for care. This can be achieved by the use of written protocols for the management and investigation of common problems in postoperative elderly patients, such as delirium and incontinence. One of the lessons learned by geriatricians in other contexts is that where clinicians do not have to take responsibility for the poor as well as the good results of their care they have little incentive to improve. However, one randomized trial in Scotland of the removal of elderly patients from the orthopaedic ward to a geriatric unit in another hospital apparently produced a reduction in length of stay and improved functional status (Kennie *et al.* 1988). More truly co-operative management of patients by geriatrician and orthopaedic surgeons has not produced striking results. In another Scottish study, Gilchrist *et al.* (1988) were unable to demonstrate any benefit from such joint management in a randomized trial against an orthopaedic rehabilitation ward without systematic geriatric input. This was in spite of a higher rate of identification of medical problems in the group of patients receiving geriatric assessment. One possible implication is that where benefits are found from geriatric involvement they are not due to management of otherwise unrecognized medical problems but rather have to do perhaps with more efficient manipulation of the machinery for hospital discharge and community care. There has been much speculation over the reasons for the failure of this approach compared with the success of Kennie *et al.* (1988). Too late an involvement by the geriatricians and that orthopaedic rather than geriatric staff retained overall responsibility for the patients have been among the suggestions made—by geriatricians. There may, however, be some fundamental difficulty with the model. Another controlled study of joint geriatric–orthopaedic management in Huddersfield, England, not only revealed no evidence of benefit but showed higher costs in the jointly managed group (Fordham *et al.* 1986).

The chief conclusion to be drawn from these results is that larger and better studies need to be done. It also seems desirable that such studies should be multicentre, as one suspects that whatever system is in place the results will be heavily determined by personal factors such as the enthusiasm of staff and the pattern of available rehabilitation and community resources. There may indeed be no universally applicable 'best buy', and clinicians and administrators must recognize the obligation to embody a formal evaluation by randomized controlled trial on an established baseline of any scheme they introduce. Anything less may lead to worse care of patients and waste of public funds.

However the care is organized there can be little doubt that the best results will be obtained if rehabilitation and discharge planning are begun as a positive, interdisciplinary programme immediately after operation. Muscular strength can be preserved and improved by active physiotherapy and enhanced meaningful activity, and with sensitive nursing regimens even on an orthopaedic ward it is possible for patients to resume their normal dress and their personal daily schedules and lifestyles in preparation for a return home.

References

Evans, J.G. (1985). Incidence of proximal femoral fracture. *Lancet*, **i**, 925–6.

Evans, J.G. (1990). The significance of osteoporosis. In *Osteoporosis 1990*. R. Smith (ed.), pp. 1–8. Royal College of Physicians, London.

Evans, J.G., Prudham, D. and Wandless, I. (1979*a*). A prospective study of fractured proximal femur; factors predisposing to survival. *Age and Ageing*, **8**, 246–50.

Evans, J.G., Prudham, D. and Wandless, I. (1979*b*). A prospective study of fractured proximal femur; incidence and outcome. *Public Health (London)*, **93**, 235–41.

Evans, J.G., Wandless, I. and Prudham, D. (1980). A prospective study of fractured proximal femur; hospital differences. *Public Health, (London)*, **94**, 149–54.

Fitzgerald, J.F., Moore, P.S. and Dittus, R.S. (1988). The care of elderly patients with hip fracture. Changes since implementation of the prospective payment system. *New England Journal of Medicine*, **319**, 1392–7.

Fordham, R., Thompson, R., Holmes, J. and Hodkinson, C. (1986). *A cost–benefit study of geriatric–orthopaedic management of patients with fractured neck of femur*, Discussion Paper 14. Centre for Health Economics, University of York.

Gilchrist, W.J., Newman, R.J., Hamblen, D.L. and Williams, B.O. (1988). Prospective randomised study of an orthopaedic geriatric inpatient service. *British Medical Journal*, **297**, 1116–18.

Greatorex, I.F. (1988). Proximal femoral fractures: an assessment of the outcome of health care in elderly people. *Community Medicine*, **10**, 203–10.

Harries, D.J. and Eastwood, H. (1991). Proximal femoral fractures in the elderly: does operative delay for medical reasons affect short-term outcome? *Age and Ageing*, **20**, 41–4.

Hornby, R., Evans, J.G. and Vardon, V. (1989). Operative or conservative treatment for trochanteric fractures of the femur. A randomised epidemiological trial in elderly patients. *Journal of Bone and Joint Surgery*, **71**-B, 619–23.

Kennie, D.C., Reid, J., Richardson, I.R., Kiamari, A.A. and Kelt, C. (1988). Effectiveness of geriatric rehabilitative care after fractures of the proximal femur in elderly women: a randomised clinical trial. *British Medical Journal*, **297**, 1083–6.

Maggi, S., Kelsey, J.L., Litvak, J. and Heyse, S.P. (1991). Incidence of hip fractures in the elderly: a cross-national analysis. *Osteoporosis International*, **1**, 232–41.

Pryor, G.A. and Williams, D.R.R. (91989). Rehabilitation after hip fractures. Home and hospital management compared. *Journal of Bone and Joint Surgery*, **71**-B, 471–4.

Robbins, J.A. and Donaldson, L.J. (1984). Analysing stages of care in hospital stay for fracture neck of femur. *Lancet*, **ii**, 1028–9.

Sikorski, M., Davis, N.J. and Senior, J. (1985). The rapid transit system for patients with fractures of proximal femur. *British Medical Journal*, **290**, 439–43.

Thorngren, M., Nilsson, L.T. and Thorngren, K-G (1988). Prognosis-determined rehabilitation of hip fractures. *Comprehensive Gerontology A*, **2**, 12–17.

Wallace, R.G.H., Lowry, J.H., McLeod, N.W. and Mollan, R.A.B. (1986). A simple grading system to guide the prognosis after hip fracture in the elderly. *British Medical Journal*, **293**, 665.

SECTION 5
Clinical pharmacology and ageing

5 Clinical pharmacology and ageing

B. ROBERT MEYER and MARCUS M. REIDENBERG

The passing of time is obvious; the change from infancy to old age is dramatic. It seems almost ridiculous to suggest that aging should *not* affect drug responses. Yet 'obvious facts' ... have a nasty habit of turning into anachronistic nonsense, so that it behooves us to examine critically the evidence on this relationship.

Lasagna (1956)

Clinical pharmacology is the scientific study of the effects of drugs in man. This discipline includes pharmacokinetics (the mathematical description of the absorption, distribution, metabolism, and excretion of drugs), pharmacodynamics (the description of drug effects and mechanisms of drug action), and toxicology (the description and analysis of the adverse effects of drugs and chemicals in the environment). Related to these areas of investigation is the scientific evaluation of newly developed drugs as they are studied in man.

A major goal of clinical pharmacology has been the elaboration of the fundamental principles underlying drug therapy. The aim of rational drug therapy is to use these principles to individualize drug therapy so as to achieve maximal therapeutic benefit while at the same time minimizing adverse effects.

The goal of improving therapeutics is nowhere more important than in older people. The elderly individual is more likely to be taking medication than a younger individual, more likely than a younger person to be taking multiple different medications, and more likely to suffer an adverse effect from that medication. Individuals over the age of 65 years constitute up to 18 per cent of the population in developed nations, but represent between 25 and 30 per cent of drug expenditures in those nations. In one study of ambulatory elderly individuals in the United Kingdom, approximately 87 per cent of the subjects were taking at least one prescription medication, the majority were taking at least two different prescription medications, and over one-third were taking three or more drugs.

The most frequently used classes of drugs by ambulatory elderly persons are cardiovascular drugs, analgesics (including non-steroidal drugs), gastrointestinal preparations including laxatives and antacids, and the sedative hypnotics (Table 1). Over-the-counter medications are widely used. As many as 70 per cent of individuals over the age of 65 use over-the-counter medications on a regular basis. Among old people in institutions, the pattern of drug utilization is somewhat different. While cardiovascular preparations and analgesics are frequently used (as in the ambulatory setting) the use of psychotropic drugs is substantially higher within institutions than in the community. Up to 25 per cent of patients in institutions may be receiving psychotropic drugs including such major tranquillizers as haloperidol and chlorpromazine as well as lesser drugs like the benzodiazepines or barbiturates. Patients are frequently (one-third of the time according to one study) on two or more psychotropic drugs simultaneously. Laxatives are also commonly used in institutions. In one government sponsored review of long-term care facilities in the United States a full 86 per cent of patients were receiving some sort of laxative. In another review at a skilled nursing facility 78 per cent of subjects were receiving laxative preparations.

ADVERSE DRUG REACTIONS

While the precise incidence of adverse reactions will vary according to the definition of an adverse effect and severity of the reactions which are deemed worthy of documentation, there is agreement that elderly individuals are more likely to suffer adverse effects from medications than the young (Table 2). In a study from the 1960s in Baltimore, Maryland, the incidence of adverse reactions was below 10 per cent in subjects in their twenties, but rose to 18 per cent in individuals in their seventies. Some more recent studies give lower rates, but a comparable trend is present.

Age associated increases in the incidence of adverse reactions have been particularly well described for certain groups of drugs. Some of these are listed in Table 3. The incidence of isoniazid hepatitis increases directly with age. According

Table 1 Most commonly used medications in the ambulatory elderly

Guttman (1977) (Washington DC) $n=447$		Freeman (1977) (Southampton UK) $n=941$		Chien *et al.* (1978) (Albany, New York) $n=244$	
Over-the-counter	(69%)	Cardiovascular	26%	Analgesics	(67%)
Cardiovascular	(61%)	Analgesics	17%	Over-the-counter	(40%)
Sedative-hypnotics	(17%)	Psychotropic	16%	Cardiovascular	(34%)
Antiarthritics	(12%)	Metabolic/endocrinologic	14%	Laxatives	(31%)
'Gastrointestinal' agents	(11%)	Antibiotics	9%	Antacids	(26%)
		Respiratory	8%	Antianxiety	(22%)
		Alimentary	5%		

Table 2 Frequency of adverse drug reactions in the young and the elderly

Location	Incidence in young	Incidence in old
Baltimore, Maryland	9.9% for 20–29 years 11.2% for 30–39 years	18.3% for 70–79 years 24.0% for 80 years
Belfast, Northern Ireland	3.0% for 20–29 years	21.3% for 70–79 years
Jerusalem, Israel	4.9% for 20–29 years	8.7% for 70–79 years

Table 3 Some drugs that are more likely to produce adverse effects in the elderly

Drug	Adverse effect
Benzodiazepines	Sedation, confusion
Nitrazepam, flurazepam, temazepam, diazepam, chlordiazepoxide	Ataxia
Non-steroidal anti-inflammatory drugs	Peptic ulcer disease Fluid retention
Benoxaprofen	Toxic hepatitis
Opiate analgesics	Sedation, confusion Constipation
Anticholinergics	Glaucoma, urinary retention
Antiarrhythmics	
Lidocaine	Confusion
Disopyramide	Urinary retention
Major tranquillizers	Malignant hyperthermia Tardive dyskinesias* Confusion, sedation
Diuretics	Dehydration, hyponatraemia Carbohydrate intolerance Orthostatic hypotension
Isoniazid	Hepatitis
Aminoglycosides	Renal and auditory injury

*These reactions are also more common in the young.

to data collected by the Centers for Disease Control, the incidence of isoniazid hepatitis (and not merely transaminase elevation) during the first 3 months of treatment rises from 0.2 per cent in individuals below 34 years of age to 2.2 per cent in those over 50 years of age. This increasing incidence may be a decisive factor in a decision for or against treatment with isoniazid.

Elderly people are more sensitive than the young to a variety of drugs acting on the central nervous system. This was first identified with morphine. When elderly individuals received a standard dose of morphine sulphate, they demonstrated greater analgesic effect than did younger individuals. Later studies showed a relation between age and the duration of pain relief (but not in the maximum intensity of pain relief) in individuals receiving graded doses of the drug.

Age-associated changes in sensitivity to the benzodiazepines are well established. In initial large surveillance studies on the use of benzodiazepines, the incidence of side-effects was generally low. Apart from isolated case reports, no particular comment on increased sensitivity in elderly people was made. Evans and Jarvis (1972) were the first to report the occurrence of a syndrome of immobility, ataxia, incontinence, and confusion in some of their patients maintained for long periods of time on ordinary therapeutic doses of nitrazepam. They noted that this syndrome was only observed in their elderly patients. Subsequently, the Boston Collaborative Drug Surveillance program (1973) reviewed records of patients receiving diazepam and chlordiazepoxide in ordinary therapeutic doses. The reported frequency of drowsiness and excess sedation rose from 4.4 per cent in subjects under 40 years of age, to 10.9 per cent in patients over the age of 70. Extension of these observations to patients taking flurazepam showed a similar trend. Adverse effects increased with age and dose of drug. Thirty-nine per cent of 41 subjects over 70 years of age who were taking large doses of the drug (30 mg or more daily) developed adverse effects (predominantly sedation). In contrast, only 2.7 per cent of 75 subjects who were under 40 years of age and taking the same dose developed adverse effects (Fig. 1).

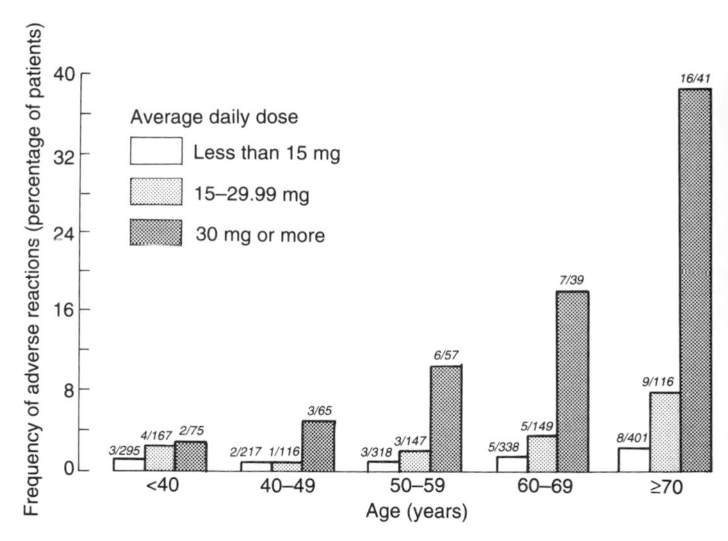

Fig. 1 The frequency of adverse reactions to flurazepam as a function of dose and age (reproduced from Greenblatt *et al.* 1977 with permission).

In addition to demonstrating exaggerated or more dramatic side-effects than may be seen in young persons, elderly people occasionally demonstrate the occurrence of side-effects not routinely associated with the use of a particular drug in a younger population. Non-steroidal anti-inflammatory drugs (NSAIDs) are among the most commonly prescribed medications in the world. They account for $2 billion in annual sales. Aged individuals represent a major portion of this market. The propensity of NSAIDs for producing gastrointestinal bleeding has been appreciated for many years. This effect has been attributed to local mucosal injury and gastric ulceration. Large studies with a significant percentage of patients under 60 years of age have usually failed to show an association between NSAID use and the occurrence of peptic ulcer disease. However, an increasing number of studies focusing on use of non-steroidal anti-inflammatory drugs (particularly aspirin) by elderly people have reported an association with peptic ulcer disease (Somerville *et al.* 1986). A recent case-control study reported by Griffin and colleagues (1988) of patients enrolled in the Tennessee Medicaid program suggested a substantially increased risk for fatal peptic ulcer disease or upper gastrointestinal haemorrhage in elderly subjects. They matched 122 patients who died from peptic ulcer disease or upper gastrointestinal haemorrhage to controls matched for age, sex, race, calendar year, and nursing home status. They then compared the rates of NSAID use in the 30 days prior to hospital admission (for the study cases) to NSAID use in their control population. The relative risk ratio for peptic ulcer disease in individuals over the age of 60 using NSAIDs in this study was between 4 and 5.

Hepatic injury has been extraordinarily rare in large populations of individuals taking non-steroidal drugs. Recent experience with benoxaprofen demonstrated that the administration of 'routine therapeutic doses' to elderly individuals produced significant accumulation of drug, and the occurrence of major hepatotoxic effects. This unexpected event led to the withdrawal of the drug from the market. The hepatotoxicity might have been prevented had dosage been lowered for elderly patients in proportion to their slowed elimination of the drug.

The full impact of the adverse effects of medication on elderly people is not generally realized. Adverse drug reactions are an important cause of hospital admission of elderly people (more so than in the young) and have a correspondingly greater effect upon the medical costs of this population and upon their functional status.

The occurrence of traumatic injury is one area where drug therapy may have an important but often unappreciated impact on elderly people. As a group, the elderly are second only to late adolescents and young adults in the frequency and severity of traumatic injury. Traumatic injury constitutes the sixth leading cause of death in elderly individuals. Between 30 and 40 per cent of individuals over the age of 65 will have a fall each year. Approximately 25 per cent of these falls will result in significant injury and one fall in 17 will result in a fracture.

Impairment of motor and cognitive skills persists for a period of time after the use of many psychoactive drugs. Castleden *et al.* (1977) assessed the effects of a single therapeutic dose of nitrazepam upon the performance of a simple test of psychomotor functioning in young and old individuals. Both young and old individuals showed impairment of performance which was not statistically significant for young subjects. For the older subjects, the effect was significantly different from the control measurement for a period of 36 h after a single dose of drug (Fig.2). {Temazepam is widely used in the United Kingdom on the assumption that it is a short-acting benzodiazepine. In fact the psychometric effects are more prolonged in elderly subjects than might be expected from the half-life of the parent compound in the plasma. This exemplifies the important point that drugs for the elderly should be assessed on the basis of the duration of their effects on the relevant end organs and not on their blood pharmacokinetics (Cook *et al.* 1983).}

Since elderly people are likely to suffer traumatic injury, take more medication than younger subjects, have a greater intensity of the CNS depressant effects of medication, and exhibit that effect for a longer period of time it is reasonable to suppose that they are at high risk for injury induced by the use of psychoactive drug use. Several studies have investigated this.

Investigators at Yale sought to identify the risk factors

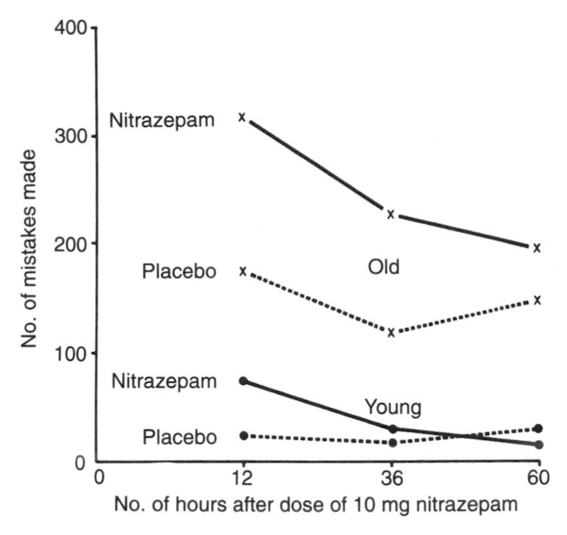

Fig. 2 Total number of mistakes made in psychomotor test by old and young age groups after receiving nitrazepam (10 mg) and placebo (reproduced from Castleden *et al.* 1977 with permission).

for falls in individuals over the age of 75 years who were living in the community (Tinetti *et al.* 1988). Their 1-year prospective study identified the most important risk factor for falls to be the use of sedative drugs (defined as benzodiazepines, phenothiazines, and antidepressants). The adjusted odds ratio for sedative drug use in individuals suffering falls was an impressive 28.3. Even when corrected for the presence of the next most important risk factor (pre-existent cognitive impairment) the adjusted odds ratio remained 2.5. The small number of subjects in the study prevented any conclusions concerning the relative risks associated with individual compounds, or the relationship of the falls to drug dose or duration of therapy.

Another large study noted an association between hip fractures and the use of long-acting benzodiazepines such as flurazepam and diazepam. These investigators found no association of fracture occurrence and the use of shorter acting compounds (Ray *et al.* 1987).

This and other evidence strongly supports a relationship between psychoactive drug use (particularly long-acting benzodiazepines and other sedatives) and traumatic injury in the elderly population.

COMPLIANCE

Compliance with a prescribed therapeutic regimen is a necessity for effective therapy. Virtually every review of this problem has found that under normal conditions patients frequently fail to follow a doctor's recommendations for therapy. The quoted frequency of non-compliance varies between 10 and 90 per cent. The higher incidence values reflect the compilation and reporting of a large number of trivial errors which are best disregarded. If non-compliance is defined as an alteration in therapy which may impair the likelihood of that therapy being successful, then non-compliance probably occurs between 10 and 50 per cent of the time.

Most studies of non-compliance have found that the more medications a patient takes, and the longer he or she is asked to take them, the more likely he or she will be to make important errors in administration, or decide to refuse to take the medication. Thus, it is individuals with multiple chronic illnesses taking multiple medications on a chronic basis that are least likely to comply with the full therapeutic regimen. This is often a description of the elderly patient.

Despite the preceding comments, the magnitude of the compliance problem among elderly patients remains a subject of controversy. In one study of compliance with simple antacid therapy, elderly individuals were actually more compliant than younger individuals. In a Swedish study, patients over the age of 65 who were taking fewer than four medications had non-compliance rates of 32 per cent. When asked to take four or more medications, elderly subjects had a non-compliance rate of 69 per cent. Younger individuals did better. When asked to take four or more medications, their non-compliance rate was 33 per cent. Other studies have also found increased numbers of errors in compliance by elderly patients. The greatest problems occur in those patients over the age of 75 years, who are taking multiple medications, and who have impaired functional status. In one study in a geriatric medicine unit, the incidence of non-compliance with therapy was 51 per cent by 10 days after hospital discharge.

Whether their compliance with medical therapy is similar to or somewhat poorer than that of younger individuals, it is clear that elderly patients may well have unique problems in achieving compliance with therapy. Cost of medication is a concern for individuals who are living on a limited and fixed income. The patient who is taking a variety of expensive medications over a prolonged period of time may find that the financial burden is overwhelming (Table 4). Impaired

Table 4 Typical retail costs for some commonly used drugs in the United States

Drug	Dose	Duration of therapy	Cost in 1990
Lanoxin®	0.25 mg daily	1 month	$3.00
Nifedipine	30 mg thrice daily	1 month	41.18
Enalapril	20 mg daily	1 month	39.30
Motrin®	600 mg thrice daily	1 month	6.60
Naprosyn®	375 mg daily	1 month	27.05
Ranitidine	150 mg twice daily	1 month	81.70
Haloperidol	2 mg twice daily	1 month	12.85
Norfloxacin	400 mg twice daily	2 weeks	60.40
Cephalexin	500 mg four times daily	2 weeks	26.45
Frusemide (furosemide)	40 mg daily	1 month	2.00
Temazepam	15 mg daily	1 month	2.40

vision may make it difficult for an individual to read prescription labels or to distinguish different tablets that are similar in appearance. Elderly individuals who were asked to identify paired samples of tablets of a variety of different colours failed to separate different tablets correctly up to 32 per cent of the time (Table 5). Sometimes even the mechanical process of picking up and holding a small pill may be problematic. Childproof containers may also be 'old-person-proof' for many patients. Impaired memory may make it difficult to remember complicated therapeutic regimens.

The physician will be doing his elderly patients a service by being aware of these special problems. Specific efforts to communicate with the patient, to simplify medication regimens, to recognize where relevant the issue of cost and discuss it with the patient, to direct the patient to obtain pill dispensing systems designed to maximize compliance, and to use easily opened containers with readable labels will all improve the patient's compliance. In some circumstances it may be better to let some conditions go untreated than exhaustively to treat each problem, and thereby create a complicated regimen which cannot be followed.

Table 5 Ageing and ability to discriminate coloured pills

Colour	Tablets	Percentage of correct tablet discrimination
Yellow	Aldomet 250 mg/inderal 80 mg	77.5
Yellow	Valium 5 mg/synthroid 0.1 mg	68.5
Yellow	Valium 5 mg/inderal 80 mg	85.5
Blue	Inderal 20 mg/hygroton 50 mg	75.0
Blue	Elavil 10 mg/apresoline 25 mg	90.0
Blue	Valium 10 mg/apresoline 50 mg	95.0
White	Valium 2 mg/Lanoxin 0.25 mg	75.0
White	Lanoxin 0.25 mg/Lasix 40 mg	55.0

(Adapted from Hurd and Blevins 1984.)

MECHANISMS OF ALTERED DRUG RESPONSE IN THE ELDERLY

There may be 'pharmacokinetic' or 'pharmacodynamic' reasons for the altered sensitivity of the elderly to drugs.

Pharmacokinetics

A variety of physiologic changes occur during the process of ageing. These include alterations in body composition, metabolic rate, hepatic blood flow and mass, and glomerular filtration rate. All of these have the potential to produce changes in how the body absorbs, distributes, metabolizes, and excretes foreign chemicals. The past decade has produced a large body of literature examining the potential impact of ageing on each of these processes. While these studies have identified a variety of changes in each of these important pharmacokinetic parameters, most of the abnormalities described are of relatively small magnitude. It is important to recognize that age-associated changes usually comprise only a small portion of the variation between individuals in pharmacokinetics. The effects of disease, genetics, concurrent use of other drugs, extent of alcohol consumption and smoking, and other factors usually have a greater impact than does age upon interindividual variation in drug disposition. As a result, while observed pharmacokinetic changes with ageing may be of interest to the pharmacokineticist, hepatologist, or gerontologist, most of these changes will have little or no direct impact upon clinical prescribing.

Absorption

A variety of age associated changes in gastric acidity, motility, small bowel surface area, and in splanchnic blood flow have been described. While from a theoretical point of view, the presence of these alterations might produce changes in absorption of compounds across the gut, this has not been substantiated by clinical investigations. Jori *et al.* (1972) demonstrated that the absorption of aminopyrine in subjects between 65 and 85 years of age was not significantly different from that in those aged 25 to 30 years. Similarly, no change in indomethacin, penicillin, phenylbutazone, ranitidine, or cimetidine absorption has been shown. Cusack *et al.* (1978) showed that while the rate of digoxin absorption was slower in older than in younger individuals, the extent of absorption was the same in the two groups.

Compounds such as iron, thiamine, and vitamin B_{12}, which undergo active transport across the intestinal mucosa, have a significant age-associated decline in absorption. However, the vast majority of drugs are absorbed by passive diffusion and not by active transport. The data would suggest that while the rate of passive absorption may be altered, the total amount of drug absorbed is the same in the aged individual as in the young. There are no data to indicate that the dosage of most drugs need be modified because of altered gastro-intestinal function in elderly people.

Distribution

Changes in body composition, including changes in body weight, an increase in body fat, a decrease in total body water and lean body mass, and alterations in serum protein concentrations have been seen in aged persons. Associated with these changes is an alteration in the volume of distribution for many drugs. Drug half-life is related to the volume of distribution by the equation

$$T_{1/2} = 0.693\left(\frac{\text{volume of distribution}}{\text{clearance}}\right)$$

Altered volume of distribution will therefore produce changes in drug half-life, and the duration of drug effect. For lipid-soluble drugs, the increase in body fat seen in older individuals would be predicted to produce an increase in volume of distribution. For water-soluble drugs, the decrease in total body water would be predicted to lead to a decrease in volume of distribution. To some extent, these predictions have been shown to hold true. The lipid soluble compounds chlordiazepoxide, diazepam, and thiopental have been shown to have increased volumes of distribution, and prolonged serum half-lives (Fig. 3).

Decreased total body water could lead to diminished volumes of distribution for water-soluble compounds. Such decreases in volume of distribution have been shown to occur with water-soluble compounds such as gentamicin, theophylline, and cimetidine. Digoxin is both water soluble and highly bound to muscle, and the relative decreases in muscle mass in elderly people may also contribute to increased serum concentrations after routine loading doses.

Alterations in serum protein concentrations in elderly subjects include changes in two principal binding proteins for drugs. Serums albumin concentrations decline slightly, while serum α_1-acid glycoprotein concentrations increase. These changes can produce alterations in distribution for a small number of highly protein bound compounds. Each case needs individual analysis to assess its clinical effects.

Phenytoin is significantly bound to albumin. The age associated decrease in albumin of 10 to 20 per cent may lead to an increase in the unbound fraction of phenytoin in the blood in elderly patients. This alteration in free drug concentration is followed by increased metabolic clearance (Fig. 4). Dosage is therefore not altered. However, since standard phenytoin assays measure 'total drug' and not 'free drug', the desired therapeutic concentration of phenytoin may be lower in elderly than in younger patients.

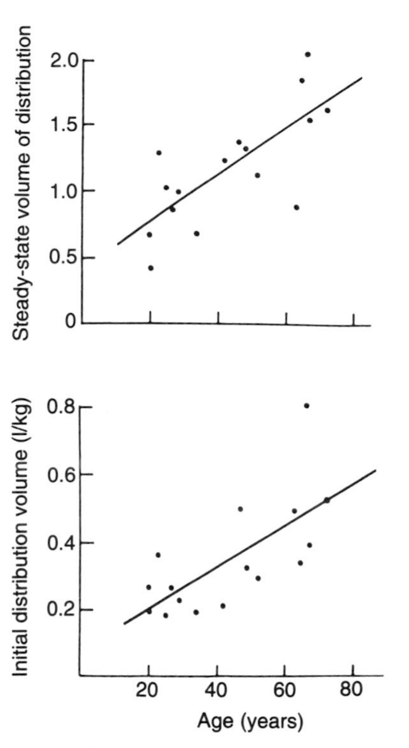

Fig. 3 Relationship of initial volume of distribution of diazepam (bottom panel) and steady state volume of distribution of diazepam (top panel) to age (reproduced from Klotz *et al.* 1975 with permission).

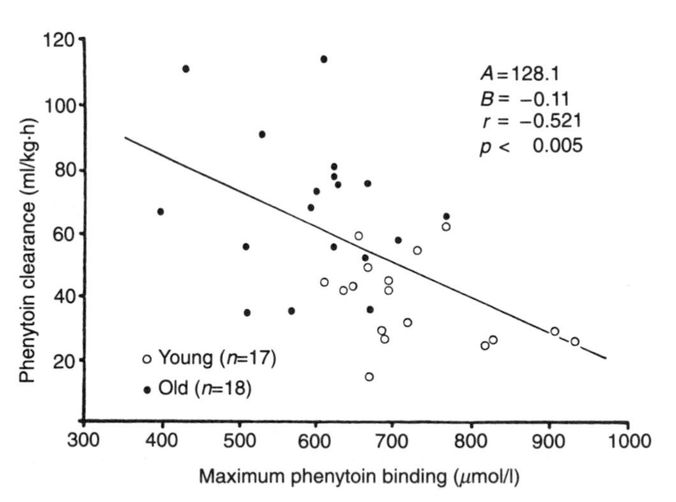

Fig. 4 Relationship of age to protein binding and clearance of phenytoin (reproduced from Hayes *et al.* 1975 with permission).

Lidocaine binds avidly to α_1-acid glycoprotein. An increase in total bound drug leads to an appearance of drug accumulation. This has erroneously led to recommendations for decreased dosing in elderly patients. However, since free drug concentrations are not substantially altered, no change in dose is indicated.

Metabolic clearance

Substantial alterations in liver mass and liver blood flow have been observed to occur in association with ageing. The effects of these changes are highly dependent upon the metabolic pathway and clearance rate of the drug being considered. A small number of compounds with extensive first-pass hepatic metabolism may have substantial alterations in bioavailability in the elderly. This appears to be due to diminished first-pass hepatic effect. Compounds implicated include propranolol and morphine.

Drugs that are metabolized by the liver may be characterized as either flow limited in their metabolism (high extraction ratio) or capacity limited (low extraction ratio). Investigations into the effects of age on hepatic drug metabolism has used antipyrine as an example of a capacity-limited drug, and indocyanine green as an example of a blood flow limited compound.

Studies of antipyrine clearance have shown an age-associated effect. The magnitude of the effect is small however, especially when placed in the context of the effects of tobacco. Approximately 3 per cent of the variability in dose to drug level seen with antipyrine could be attributed to age effect (Fig. 5). In contrast, the effect of smoking could account for 12 per cent of the population variability in antipyrine metabolism. Subsequent studies controlling for environmental factors such as tobacco have confirmed these findings. Thus, small decreases in hepatic clearance of antipyrine occur with ageing.

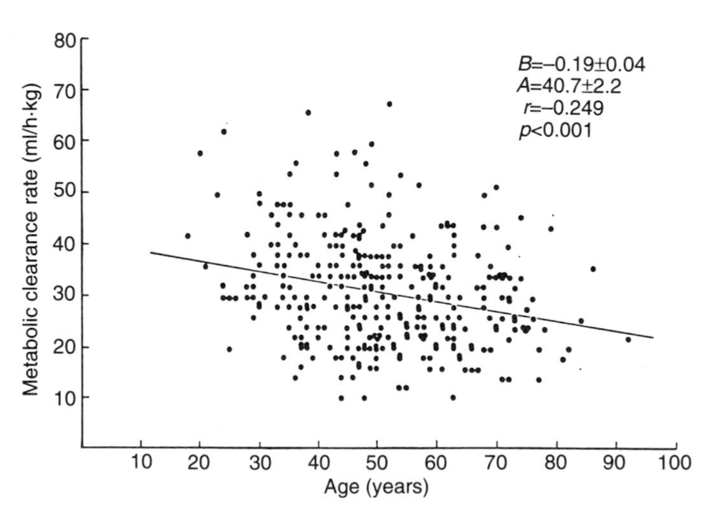

Fig. 5 Relationship of age to antipyrine clearance (reproduced from Vestal *et al.* 1975 with permission).

Indocyanine green has been used as a model of a compound with flow-limited metabolic clearance. Wood and Vestal (1979) reported that clearance of this compound was decreased by approximately 25 per cent in elderly compared with young subjects. A similar analysis of propranolol (which is also flow limited) suggested a similar reduction. This decline appeared to correlate with a decline in liver blood flow of comparable magnitude.

Hepatic drug metabolism may be grouped as Phase I (oxidation–reductions reactions) which involve the mixed-function oxidase system, and Phase II (conjugation reactions) which are independent of that system. Analysis of ageing effects on hepatic drug clearance of benzodiazepines has suggested that these phases of metabolism are affected differently by the ageing process (Table 6). Early studies of

Table 6 Relationship between pathway of metabolic clearance and age-associated changes in clearance for benzodiazepines

	Mixed-function (oxidase)	Conjugation (glucuronide)
Diazepam	Clearance decreased	—
Chlordiazepoxide	Clearance decreased	—
Lorazepam	—	No change
Oxazepam	—	No change
Nitrazepam	—	No change
Temazepam	—	No change

diazepam kinetics suggested that prolonged half-life (from 36 to 99 h) observed in elderly subjects predominantly reflected an increase in volume of distribution. Subsequent studies have also shown an age-associated decline in clearance. In one study the clearance of diazepam decreased by over one-third from 0.39 ± 0.1 ml/min.kg in young males to 0.24 ± 0.1 ml/min.kg in elderly men. Similar age-associated declines in metabolic clearance have also been reported with chlordiazepoxide. Both of these compounds are metabolized by the mixed-function oxidase system. In contrast, benzodiazepines which undergo conjugation such as lorazepam, temazepam, and oxazepam, do not appear to have alterations in metabolic clearance associated with ageing. In contrast to the cytochrome mediated reactions of the mixed-function oxidase system, the conjugation pathway is apparently not affected by ageing.

Even when studies of specific compounds have suggested the occurrence of potentially meaningful changes in metabolic clearance, the clinical impact of these changes may be vitiated by other compensatory factors. Theophylline kinetics has been a source of controversy and is a particularly relevant example. Investigations of theophylline clearance in elderly subjects have shown variable results, including increased, decreased, or unchanged clearance. Interpretation of these results has been complicated by the absence of definite information about dietary intake, protein-binding, and clearance of unbound theophylline. Recent data suggest that when dietary intake is controlled, unbound drug is measured, and clearance of free theophylline is calculated, the rates of both *N*-demethylation and *C*-oxidation are decreased by as much as 30 per cent. This magnitude of change would ordinarily be of potential clinical importance. However, the effects are offset by both a decrease in volume of distribution (theophylline is water-soluble) and a decrease in protein binding. The net effect of these changes is to produce no significant change in overall drug clearance. While moderate decreases in metabolic clearance of many drugs may well occur with ageing, the clinical impact of these changes is modest when compared to other factors such as the effects of disease, concomitant drug use, tobacco and alcohol use, and genetic constitution. Since for many commonly used drugs the average clearance in elderly people is about two-thirds that in young, the starting dose of many drugs for elderly patients should be only two-thirds that for the young.

Renal clearance

On average there is a decline in renal function with ageing, although longitudinal studies show that as many as one-third of people show no declines as they age (Lindeman *et al.* 1985). Serum creatinine concentrations reflect a balance between production of creatinine by muscle tissue, and clearance of creatinine by the kidney. Since creatinine production is decreased in the elderly associated with their decrease in muscle mass, 'normal' serum creatinine concentrations in these patients will not reflect the true decline in renal function. The Cockcroft and Gault equation has become a popular and effective technique for correcting for age-associated changes in creatinine production

$$\text{Creatinine clearance (in men)} = \frac{(140 - \text{age}) \times (\text{weight in kg})}{72 \times \text{serum creatinine (mg/dl)}}$$

Multiply results from this equation by 0.85 for women. Figure 6 shows a graphic representation of this equation for patients of varying ages and sizes with 'normal' or near normal serum creatinine concentrations. This equation is accurate for individuals without significant oedema, or severe cachexia. For drugs substantially excreted through the kidneys elderly patients need adjustment of dosage based upon glomerular filtration rate, just as do younger individuals with a similar degree of renal impairment (Table 7). There is no unique adjustment in dosing that is specifically related to age.

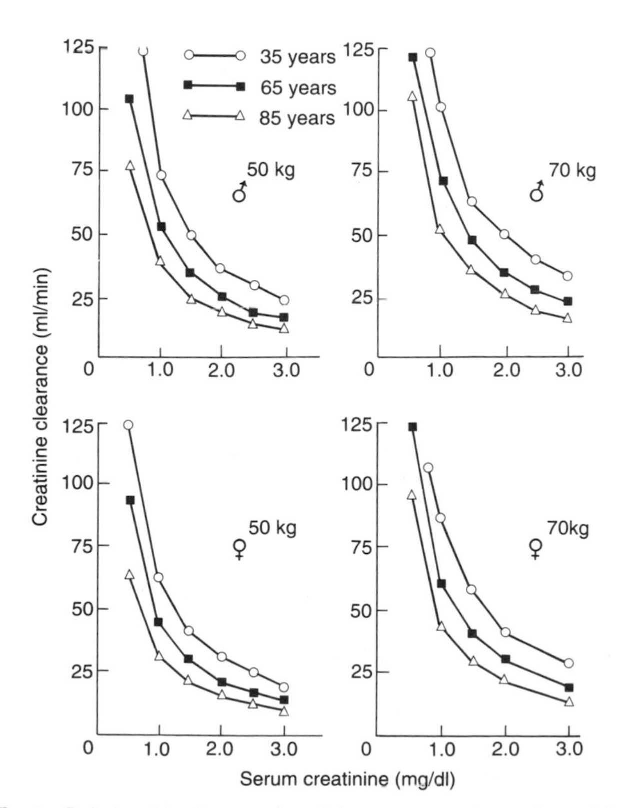

Fig. 6 Relationship of serum creatinine concentration and creatinine clearance to age and body weight for men and women.

The renal handling of drugs involves not only glomerular filtration, but also secretion and reabsorption by the renal tubules. Elderly people exhibit alterations in the renal tubular handling of water, salt, and acid. It would be reasonable to expect similar alterations in renal tubular handling of drugs. So far, however, there are few data available on the presence (or absence) of age-associated changes in the secretion of compounds by the organic base and acid secretory system of the renal tubules.

Table 7 Some drugs that require dosage adjustment with renal insufficiency

Digoxin
Acyclovir
Methotrexate
Vancomycin
Procainamide
Ethambutol
Most penicillin antibiotics
Aminoglycoside antibiotics
Cimetidine
Trimethoprim-sulphamethoxazole
Pentamidine
Lithium
Most cephalosporin antibiotics

Pharmacodynamics

In addition to alterations in pharmacokinetics, it is possible that elderly individuals have an altered tissue sensitivity to the effects of drugs. This subject has received less investigation than the area of pharmacokinetic change because it is much harder to study. Study design must eliminate the possibility of pharmacokinetic alterations before any definitive conclusions can be drawn concerning pharmacodynamic changes. Drug effects are often more difficult to define and measure than drug concentrations. In addition, given the important effects of concomitant drug use, disease, and other variables, care must be taken in the selection of an appropriate population for study.

Reviews of warfarin usage suggest that elderly subjects require lower doses of warfarin to achieve anticoagulation than do younger individuals. Since no substantive changes in warfarin pharmacokinetics can be demonstrated in elderly subjects, the sensitivity observed must reflect altered pharmacodynamic responsiveness. The observed increase in sensitivity may be related to the decreased rates of synthesis of clotting factors seen in elderly subjects. The observed effect of age is only demonstrable however when other variables are controlled. In a carefully selected population of individuals without concurrent diseases and confounding drug therapy, this relation is easily observed. However, in a consecutive series of patients in hospital receiving warfarin with extensive concurrent therapy and multiple disease processes, the contribution of age to individual variability in dose-response is negligible (Fig. 7).

Elderly subjects are generally more sensitive to the effects of psychoactive drugs such as morphine, diazepam, and nitrazepam on the central nervous system. This may reflect pharmacodynamic changes. Elderly individuals receiving nitrazepam have more difficulty in performing a simple test of psychomotor function than do younger individuals receiving the same dose of drug. Only minimal age-associated differences in nitrazepam pharmacokinetics have been identified, and the differences in psychomotor performance

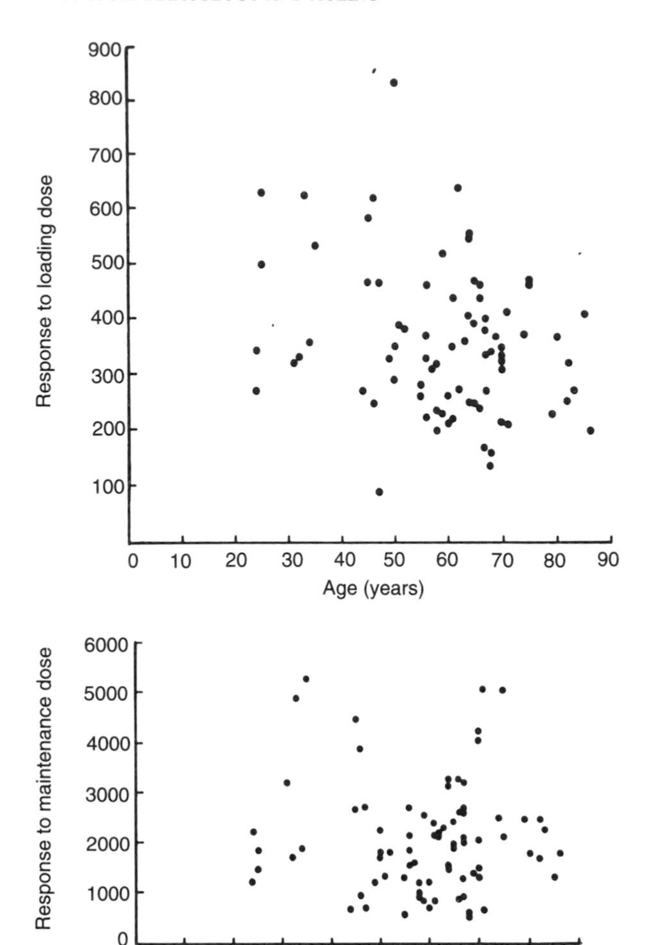

Fig. 7 Relative non-contribution of age to variability in warfarin dose requirement in a population of hospitalized patients. Top panel shows lack of correlation of age with loading dose-response index ($r = 0.22$ and $p > 0.05$). Bottom panel shows lack of correlation of age with maintenance dose-response index ($r = 0.06$) (reproduced from Jones *et al.* 1980 with permission).

observed have been attributed to pharmacodynamic changes. Nonetheless, the presence of changes in volume of distribution might raise questions about its relation to altered sensitivity. Other studies have assessed the acute effects of intravenous infusions of diazepam given to patients prior to cardioversion or upper gastrointestinal endoscopy. Elderly subjects require less diazepam and have lower serum drug concentrations to achieve comparable levels of sedation as younger people (Fig. 8). Since these studies are acute studies, they exclude the possible role of altered metabolic clearance of these drugs. Nonetheless, they fail to rule out the possibility of altered distribution of drug into the central nervous system.

Age-associated changes in sensitivity to drugs affecting the β-adrenergic system are also well established. Individuals between 50 and 65 years of age have less reduction of heart rate after propranolol than individuals between 25 and 30 years of age. This occurs despite the fact that serum propranolol levels are generally similar to or higher than those in the younger patient. This 'pharmacodynamic' insensitivity has been documented by the measurement of the effect of

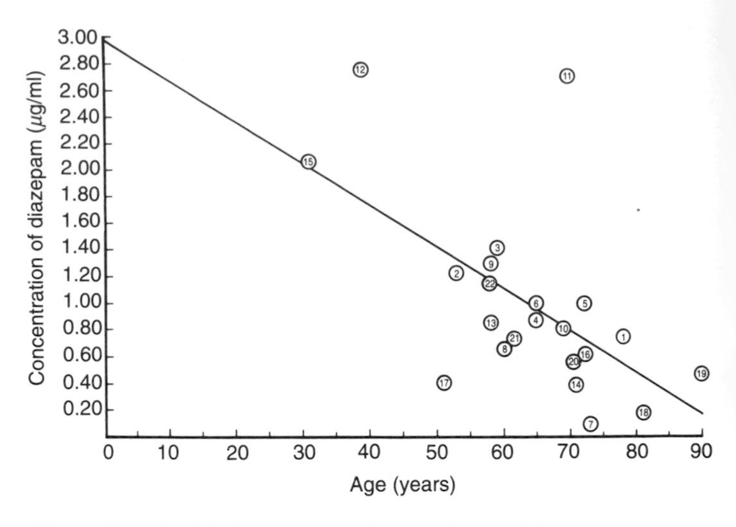

Fig. 8 Relationship between age of the patient and plasma concentration of diazepam which causes failure to respond to vocal, but not to painful stimuli (reproduced from Reidenberg *et al.* 1978 with permission).

propranolol on isoproterenol dose-response curves in young and elderly subjects (Fig. 9) This altered responsiveness

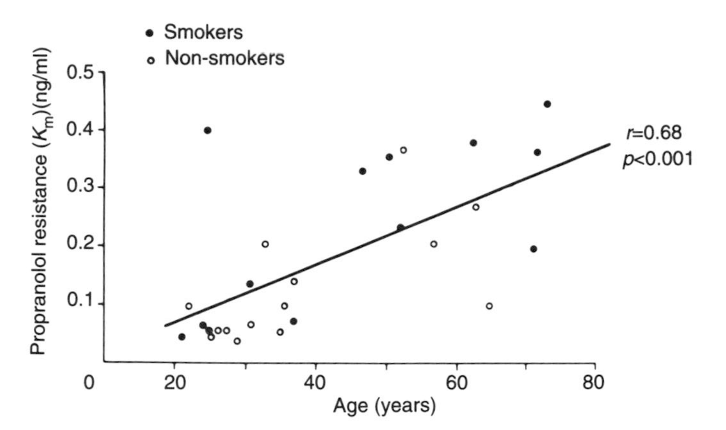

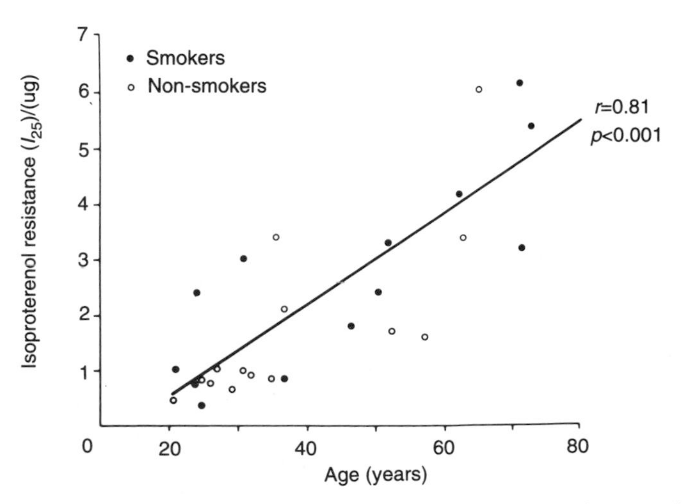

Fig. 9. Relationship between propranolol resistance and age (top panel) and relationship between isoproterenol resistance and age (bottom panel) in smokers and non-smokers (reproduced from Vestal *et al.* 1979 with permission).

appears confined to β_1 receptors. Measurements of isoproterenol-induced vasodilation and insulin release (β_2 effects) have failed to show significant changes associated with ageing. The mechanism for altered β_1 sensitivity remains obscure. Initial reports of decreases in the number or density of β-receptors on the lymphocytes of older individuals have not been confirmed. The observed alteration in sensitivity may then reflect alterations in postreceptor intracellular events. No alteration in alpha-adrenergic responsiveness with ageing has been seen.

'AGE EFFECTS' AND 'DISEASE EFFECTS'

Most of the studies leading to the conclusions discussed in the preceding paragraphs were studies of drug effects in 'healthy' older volunteers. The authors sought to exclude from their studies individuals with significant illness, hoping thereby to identify the effects of 'ageing' rather than 'disease' on drug pharmacokinetics and pharmacodynamics. (This approach is not able reliably to distinguish the effects of extrinsic and intrinsic ageing—see Section 1). The practising physician may find it more relevant to compare the 'sick' elderly patient to the 'sick' young patient. In many situations, the altered responsiveness of older individuals to the effects of drugs may not reflect the effects of the 'ageing' process so much as the effects of multiple disease processes superimposed upon ageing. Elderly people tend to take more medication because they have more illnesses. As a population, much of their altered responsiveness to medications may simply reflect the fact that they are, as a population, 'sicker' than younger individuals who are taking the same medication. Unfortunately, there is little quantitative information available on the interaction between severity of illness, age, and altered response to drug therapy.

DOSAGE ADJUSTMENTS FOR ELDERLY PATIENTS

As a rule, elderly individuals should begin drug treatment at a lower dose than younger patients. In the case of drugs with predominantly renal excretion, the adjustment of dose according to calculated or (preferably) measured creatinine clearance provides specific guidance on selection of an appropriate dose. When pharmacokinetic parameters such as hepatic clearance are considered, current knowledge provides substantially less precise recommendations on dose adjustment. The possibility of altered pharmacodynamic sensitivity also does not easily lead to precise recommendations on adjustment of dose.

Paediatricians have always been trained to adjust dosing according to weight. This has never been the practice in internal or geriatric medicine. As a result, elderly patients with low body weight often receive significantly higher doses of drug (on a mg/kg basis) than do younger patients. A study of prescriptions for flurazepam, cimetidine, or digoxin filled through the American Association of Retired Persons pharmacy service showed that the older patients tended to be smaller, and to receive higher doses of these drugs on a mg/kg basis than did younger patients. On a mg/kg basis, the smallest patients (<50 kg) received a dose of flurazepam that was 88 per cent higher than that given to the largest patients (>90 kg).

The development of dosing guidelines for elderly patients based upon weight is needed. Routine adjustment of dose according to patient weight might well lead to improved prescribing practices and better treatment in this population.

DRUG DEVELOPMENT AND ELDERLY PEOPLE

While elderly people are major consumers of drugs, until recently they have not participated in the clinical evaluation and development of new chemical entities. In fact, given their known sensitivity to many agents they were routinely excluded from these studies. As the importance of drugs in the geriatric population has been recognized, and as the potential importance of information on the pharmacokinetic and pharmacodynamic action of a drug in this population has been identified, this policy is changing. It can now be expected that in the United States a new drug that has a likelihood of significant use in the elderly will have been used in the elderly during phase III clinical trials.

Even as it becomes standard practice to include individuals over 65 years of age in clinical trials, it remains extremely important to collect data systematically on the experience of the elderly with new drugs. Few individuals over 75 years of age have been included in formal research protocols evaluating drug effects. Only with systematic collection of data can the information necessary for proper individualization of therapy for the very old be obtained.

General bibliography

Greenblatt, D.J., Sellers, E.M., and Shader, R.I. (1982). Drug disposition in old age. *New England Journal of Medicine*, **306**, 1081–8.

Lamy, P.P. (ed). (1990). Clinical pharmacology. *Clinics in Geriatric Medicine*, **6**(2).

Swift, C.G. (ed). (1987). *Clinical Pharmacology in the Elderly*, Marcel Dekker Inc, New York.

Vestal, R.E. (1982). Pharmacology and aging. *Journal of the American Geriatrics Society*, **30**, 191–200.

Vestal, R.E. (ed). (1984). *Drug Treatment in the Elderly*. ADIS Health Science Press, Sydney.

References

Boston Collaborative Drug Surveillance Program. (1973). Clinical depression of the central nervous system due to diazepam and chlordiazepoxide in relation to cigarette smoking and age. *New England Journal of Medicine*, **288**, 277–80.

Castleden, C.M., George, C.F., Marcer, D., and Hallett, C. (1977). Increased sensitivity to nitrazepam in old age. *British Medical Journal*, **1**, 10–12.

Chien, C.P., Townsend, E.J., and Ross-Townsend, A. (1978). Substance use and abuse among the community elderly: the medical aspect. *Addictive Diseases*, **3**, 351–72.

Cook, P.J., Huggett, A., Graham-Pole, R., Savage, I.T. and James I.M. (1983). Hypnotic accumulation and hangover in elderly inpatients: a controlled double-blind trial of temazepam and nitrazepam. *British Medical Journal*, **286**, 100-2.

Cusak, B., Horgan, J., Kelly, J.G., Lavan, J., Noel, J., and O'Malley, K. (1978). Pharmacokinetics of digoxin in the elderly. *British Journal of Clinical Pharmacology*, **6**, 439p–440p.

Evans, J.G. and Jarvis, E.H. (1972). Nitrazepam and the elderly. *British Medical Journal*, **4**, 487.

Faulkner, G., Prichard, P., Somerville, K., and Langman, M.J.S.

(1988). Aspirin and bleeding peptic ulcers in the elderly. *British Medical Journal*, **297**, 1311–3.

Freeman, G.K. (1977). Drug prescribing patterns in the elderly–a general practice study. In *Drugs and the elderly* (eds. J. Crooks and I.H. Stevenson), pp. 225–9. University Park Press, Baltimore.

Greenblatt, D.J., Allen, M.D., and Shader, R.I. (1977). Toxicity of high-dose flurazepam in the elderly. *Clinical and Pharmacological Therapeutics*, **21**, 355–61.

Griffin, M.R., Ray, W.A. and Schaffner, W. (1988). Nonsteroidal anti-inflammatory drug use and death from peptic ulcer in elderly persons. *Annals of Internal Medicine,* **109**, 359–63.

Guttman, D. (1978). Patterns of legal drug use by older Americans. *Addictive Diseases*, **3**, 337–56.

Hayes, M.J., Langman, M.J.S., and Short, A.H. (1976). Changes in drug metabolism with increasing age. I Warfarin binding and plasma proteins. *British Journal of Clinical Pharmacology* **2**, 69–72.

Hurd, P.D. and Blevins, J. (1984). Aging and the color of pills. *New England Journal of Medicine*, **310**, 202.

Jones, B.R., Baran, A. and Reidenberg, M.M. (1980). Evaluating patients' warfarin requirements. *Journal of the American Geriatrics Society*, **28**, 10–12.

Jori, A., DiSalle, E., and Quadri, A. (1972). Rate of aminopyrine disappearance from plasma in young and aged humans. *Pharmacology*, **8**, 273–9.

Klotz, U. Avant, G.R., Hoyumpa, A., Schenker, S., and Wilkinson, G.R. (1975). The effects of age and liver disease on the disposition and elimination of diazepam in adult man. *Journal of Clinical Investigation*, **55**, 347–59.

Lasagna, L. (1956). Drug effects as modified by aging. *Journal of Chronic Diseases*, **3**, 567–74.

Lindeman, R.D., Tobin, J., and Shock, N. (1985). Longitudinal studies on the rate of decline in renal function with age. *Journal of the American Geriatrics Society*, **33**, 278–85.

Ray W.A., Griffin, M.R., Schaffner, W., Baugh, D.K., and Melton, L.J. (1987). Psychotropic drug use and the risk of hip fracture. *New England Journal of Medicine*, **316**, 363–9.

Reidenberg, M.M. *et al.* (1978). Relationship between diazepam dose, plasma level, age and central nervous system depression. *Clinical Pharmacology and Therapeutics*, **23**, 371–4.

Somerville, K., Faulkner, G., and Langman, M.J.S. (1986). Non-steroidal anti-inflammatory drugs and bleeding peptic ulcer. *Lancet*, **i**, 462–4.

Tinetti, M.E., Speechley, M., and Ginter, S.F. (1988). Risk factors for falls among elderly persons living in the community. *New England Journal of Medicine*, **319**, 1701–7.

Vestal, R.E., Norris, A.H., Tobin, J.D., Cohen, B.M., Shock, N.W., and Andres, R. (1975). Antipyrine metabolism in man. Influence of age, alcohol, caffeine and smoking. *Clinical Pharmacology and Therapeutics*, **18**, 425–32.

Vestal, R.E., Wood, A.J.J., and Shand, D.G. (1979). Reduced beta-adrenoceptor sensitivity in the elderly. *Clinical Pharmacology and Therapeutics*, **26**, 181–6.

SECTION 6
Nutrition and ageing

6 Nutrition and ageing

DAVID A. LIPSCHITZ

INTRODUCTION

A complex interaction between an individual and his or her environment over time is an appropriate definition of human ageing. In relation to external variables that affect ageing perhaps none is more important than nutrition. There is evidence, in many animal models, that life-expectancy can be significantly extended by restricting food intake. Nutritional factors have been shown to contribute substantially to the aetiology of many diseases that occur in late life. With advancing age the risk of developing serious nutritional deficiencies also increases. This is due to age-associated reductions in total food intake combined with the presence of debilitating disease. Malnutrition increases functional dependency, morbidity, mortality, and use of health care resources. For a comprehensive review of nutrition and ageing the reader is referred to the recent World Health Organization book (Horwitz *et al.* 1989). This chapter will discuss the relevance of these findings and describe rational approaches to the diagnosis and management of nutritional problems in the elderly.

THE ROLE OF CALORIC RESTRICTION IN AGEING

Numerous animal studies have demonstrated that nutritional deprivation delays maturation and significantly prolongs life-expectancy. These investigations have shown that caloric restriction causes delays in virtually every biomarker of ageing (Masoro 1989, 1990). Classic studies have demonstrated delays in the appearance of the well-described age-associated declines in cell-mediated and humoral immunity. This effect has been suggested as the mechanism whereby dietary restriction results in the later appearance of neoplasms. Food restriction also leads to a marked reduction in the generation of free radicals which have been postulated to result in many of the declines in cellular function that occur with age. The mechanism by which food restriction results in prolongation of lifespan remains unknown. It appears that total calorie intake is a more important variable than is either total protein or fat intake. Caloric restriction results in the presence of leaner and more active animals who utilize energy very efficiently. Overall metabolic requirements are markedly reduced. The life-long diminution in metabolic activity has recently been suggested as important in prolonging life.

The importance of these observations in relation to humans remain unclear. Affluent societies consuming high caloric, high fat diets usually demonstrate the longest life expectancy. It must be emphasized however, that the shorter lifespan noted in the less affluent can be ascribed to pathologic malnutrition, defective sanitation, and the increased prevalence of communicable diseases. In first world societies, high fat, high calorie diets are associated with high prevalence rates of age-associated diseases, such as atherosclerosis, hypertension, and colon and breast cancer. For these reasons recent dietary recommendations have focused on the need for a prudent diet which not only restricts total and saturated dietary fats, but also avoids excessive calorie intake. These recommendations may be important in minimizing the role of nutrition risk in the common cancers that occur in the Western world.

ENERGY REQUIREMENTS

Ageing is often associated with a significant decrease in energy needs (McGandy *et al.* 1986; Munro *et al.* 1987). The major mechanism is a decrease in resting energy expenditure as a consequence of declines in muscle mass. Reduced thyroid function does not appear to contribute to the reduced energy needs of elderly people. Diminished energy needs also result from decreased physical activity which has been demonstrated longitudinally in men and confirmed in women. Decreased activity is the result primarily of coexisting diseases such as bone and joint disorders, loss of postural stability, and chronic diseases that may limit activity, such as angina pectoris or intermittent claudication. Reduced strength as a consequence of declines in muscle mass does, however contribute to reductions in mobility. It should be noted that a cardinal feature of ageing is a significant increase in variability. Thus the age-associated reduction in energy needs is not universal. In the study quoted above a full 25 per cent of the healthy male volunteers studied had no measurable reduction in energy requirements.

Total caloric (food) intake is determined primarily by energy needs. Thus a 30 per cent reduction in energy need will be accompanied by a 30 per cent reduction of food intake. This reduced caloric intake has been confirmed in both cross-sectional and longitudinal studies. As compared to younger subjects, individuals over the age of 70 consume on average a third less calories (Fig. 1). The importance of this effect relates to the fact that the average intake of all nutrients are reduced in parallel. Yet the requirement for virtually every other nutrient, with the exception of carbohydrates, do not decline significantly with age (Munro *et al.* 1984). As a consequence epidemiological studies of the dietary intake of healthy elderly individuals frequently reveal evidence of deficient intake (National Centre for Health Statistics 1974).

In contrast, biochemical assessments of nutritional status indicated that significant deficiencies of both macro and micronutrients (vitamin and minerals) are quite rare in ambulatory healthy elderly people (National Center for Health Statistics 1974; Yearick *et al.* 1980). This is explained by the fact that inadequate dietary intake of a nutrient is determined by the comparison of the actual intake with the recommended dietary allowance (RDA) for that nutrient. The recommended dietary allowance is generally much higher than an intake that would result in a nutritional deficiency. Neverthe-

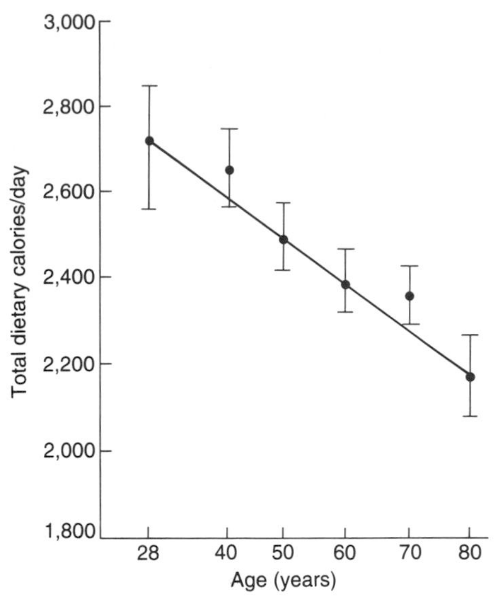

Fig. 1 Mean (± SEM) total caloric intake in a cohort of men ranging in age from 20 to 80. The data were obtained cross-sectionally (adapted from McGandy *et al.* 1966).

less decreased intake results in reduced reserve capacity. In the presence of disease with increased nutritional requirements or because of declining intake caused by anorexia, severe nutritional deficiencies are very common in elderly individuals with acute or chronic diseases in hospital or institutions.

PROTEIN REQUIREMENTS

On first principles it seems likely that, because of declines in muscle mass, ageing should result in decreased protein needs. At the time of writing the recommended dietary allowance for protein for younger subjects is 0.8 g/kg body weight. Studies of elderly people have shown that, even in healthy subjects, the requirements for protein is modestly increased. At the time of writing a protein intake of 1 g/kg body weight is recommended (Gersovitz *et al.* 1982). Most importantly the presence of acute or chronic diseases further increases protein requirements. In this circumstance protein intake in the older patient is frequently grossly inadequate. This is particularly important in wound healing and in decubitus ulcer, where inadequate protein intake adversely affects outcome.

Although ageing results in significant declines in muscle mass, protein synthetic and degradation rates are only minimally compromised. Visceral protein stores and turnover are generally unchanged with ageing so that no significant reductions are noted in serum albumin, retinol binding protein or prealbumin, which reflect visceral protein stores (Gersovitz *et al.* 1982).

FAT REQUIREMENTS

Ageing does not alter any of the specific requirements for any of the essential lipids. Advancing age is generally associated with an increase in the proportion of body weight as fat (Mitchell and Lipschitz 1982). This is the result of decreases in muscle mass accompanied by an increase in fat mass. Body fat stores increase until the seventh decade after which reductions in total weight and fat stores are frequently noted. Although obesity is not as common a problem in elderly as in younger individuals, approximately 20 per cent of subjects over the age of 65 are significantly overweight. Studies have shown that even in the very old obesity is associated with increased mortality. Furthermore there is evidence that the risks of atherosclerotic heart disease and stroke in elderly people can be reduced by consuming a diet low in saturated fats and cholesterol (McGandy 1988). These facts make dietary recommendations difficult. For older individuals a palatable acceptable diet is very important and drastic changes in dietary intake should, therefore be recommended only with caution and with careful clinical judgement. For individuals in their late 60s and early 70s who are generally healthy and ambulatory but significantly overweight, hypercholesterolaemic and perhaps hypertensive, an effort to reduce calories, fat, and sodium intake is warranted. In many circumstances drastic reductions in diet may not be beneficial. This particularly applies to institutionalized elderly people for whom medically prescribed diets are frequently not palatable, are not adequately consumed, and may result in weight loss. It must be noted that the value of serum or HDL cholesterol in the prediction of coronary artery disease is less for elderly than for younger subjects. For this reason the efficacy of aggressive dietary or pharmacologic attempts to lower cholesterol in subjects over the age of 70 is not yet clear.

WATER REQUIREMENTS

For older people fluid balance is extremely important because of their propensity to develop dehydration, and the ease at which overhydration can occur with compromised renal function or other disorders associated with fluid retention. As a general rule fluid intake should be 1 ml/kcal or 30 ml/kg body weight per day. Dehydration is extremely prevalent in hospitalized elderly patients and is a very common cause of an acute confusional state. This is primarily related to the well described age-associated decline in thirst drive. Studies have demonstrated a decreased ability of some elderly people to respond adaptively to fluid deprivation (Phillips *et al.* 1984). This becomes a particularly serious problem in frail elderly people who develop a minor pathological insult such as a respiratory or urinary tract infection resulting in fever, increased metabolism, and fluid loss. If fluid intake does not readily replace fluid lost, dehydration rapidly develops, leading to confusion, worsening dehydration, and the rapid development of a serious situation that may be life threatening, warrant admission to hospital, and necessitate a prolonged period of recuperation.

For the above reasons aggressive attempts at assuring adequate hydration of older persons is essential. Furthermore, this must commence soon after the development of a minor or major pathologic stress. Patients and their families must be educated to emphasize the importance of maintaining adequate fluid intake at all times and to monitor intake carefully if a minor illness develops or if fluid requirements are increased, as occurs during heat waves. In the older patient in hospital the possibility that confusion or delirium is caused by dehydration should be high on the differential diagnosis list. Physicians must ensure that their patients have adequate access to water. Furthermore total fluid intake should be carefully monitored by frequent weight and intake and output measurements.

MINERAL REQUIREMENTS

Numerous studies indicate that, for a wide variety of minerals and vitamins, intake is significantly lower than the recommended dietary allowance for a large fraction of ambulatory elderly people (Morley *et al.* 1986; Gary and Hunt 1986; Suter and Russel 1987)

Calcium

Of most importance is the evidence that life-long inadequate intake of calcium contributes to the high prevalence of osteoporosis in elderly people (see Chapter 14.1). It is generally recommended that calcium intake by elderly people be between 1.0 and 1.5 g/day.

Zinc

The prevalence of zinc deficiency is important because of the role that this mineral plays in food intake and in wound healing. In elderly subjects with chronic debilitating diseases modest zinc deficiency may contribute to anorexia. Although not clinically proven, there is also evidence that zinc supplementation aids in wound healing in general and in the healing of decubitus ulcers in particular. Zinc supplementation has also been shown to improve immune function and impede the rate of development of macular degeneration in elderly people.

Iron

In younger subjects iron deficiency is the most common cause of anaemia, and is the most common global deficiency leading to widespread morbidity and decreased work performance. Ageing is associated with a gradual increase in iron stores in both men and women. As a consequence iron deficiency is rare in elderly subjects and is invariably caused by pathological blood loss. It is important to emphasize that the anaemia of chronic disease, which is associated with iron-deficient erythropoiesis, including a low serum iron concentration and a reduced transferrin saturation, is frequently misdiagnosed as iron deficiency anaemia in elderly patients. This results in the inappropriate administration of oral iron and unnecessary invasive investigative procedures to identify the source of iron loss. The anaemia of chronic disease is associated with an impaired ability of the reticuloendothelial system to recirculate iron obtained from the breakdown of phagocytosed senescent red cells. Thus in the anaemia of chronic disease, iron stores are normal or increased whereas in iron deficiency iron stores are absent.

Selenium

There is evidence suggesting that selenium deficiency may contribute to age-associated declines in cellular function. The mineral may be involved in minimizing free radical accumulation as it is essential for the normal function of glutathione peroxidase. Significant selenium deficiency has been reported frequently in elderly people, although syndromes associated with selenium deficiency are very rare (cardiomyopathy, nail abnormalities, and myopathies). There is some evidence that selenium deficiency may contribute to a greater neoplastic risk and to decline in immune function.

Copper

Generally, ageing is associated with increases in serum copper concentrations, although the significance of this increase is unknown. Copper deficiency is very rare and has only been reported in total parenteral nutrition.

Chromium

Recent evidence has suggested an important role for chromium in carbohydrate metabolism. Studies have shown an age-associated decline in tissue chromium levels. It is possible that chromium deficiency may contribute to glucose intolerance in elderly people, although the therapeutic efficacy of chromium replacement is controversial.

VITAMIN REQUIREMENTS

Numerous studies have shown that dietary intake of many vitamins may be inadequate in elderly people (Morley *et al.* 1986; Garry and Hunt 1986; Suter and Russel 1987). These include an intake of 50 per cent or less for folic acid, thiamin, vitamin D, and vitamin E. In other studies intakes were shown to be less than 66 per cent of the recommended dietary allowance for most vitamins. It must be emphasized again that deficiencies identified on the basis of inadequate intake are invariably significantly higher than the prevalence of biochemical deficiency of most vitamins.

Water-soluble vitamins

Vitamin C

Numerous studies have indicated inadequate dietary intake of vitamin C in many elderly people. Others have shown a high prevalence of vitamin C supplementation. There is no evidence, however, that vitamin C deficiency is generally of clinical relevance in healthy elderly people or that replacement with megadoses of vitamin C is of value. In elderly subjects with chronic debilitating diseases there is some evidence that vitamin C supplementation improves the rate of wound and decubitus ulcer healing. There is little evidence that megadoses of vitamin C have any relevant side-effects, although false-negative faecal occult bloods have been reported, as have inaccuracies in serum and urine glucose determinations.

Thiamin

Clinically relevant deficiencies of the B vitamins are rare in elderly people. Thiamin deficiency is, however, common in elderly alcoholics and can be an important contributing factor in the development of disordered cognition, neuropathies, and perhaps cardiomyopathies. Relevant deficiencies of this vitamin are fairly common among elderly patients in institutions.

Folic acid

As with thiamin, folate deficiency in elderly subjects is predominantly found in alcoholics. It is also common in elderly subjects who are taking drugs which interfere with folate metabolism (trimethoprim, methotrexate, and phenytoin) or in disorders associated with increased folate needs (haemolytic anaemia and ineffective erythropoiesis). Folate deficiency may result in cognitive loss or significant depression and should always be included in the evaluation of an elderly subject with a memory disorder.

Vitamin B_{12}

Low serum vitamin B_{12} concentrations have been shown to occur in as many as 10 per cent of otherwise healthy elderly subjects. In many, a comprehensive evaluation indicates early pernicious anaemia, the most common cause of vitamin B_{12} deficiency, whereas in others no obvious cause can be identified. Vitamin B_{12} deficiency classically causes a severe megaloblastic anaemia, but not uncommonly, the non-haematological manifestations of vitamin B_{12} deficiency can occur in the absence of anaemia. These include gait disorders, sensory and motor deficits, and significant memory loss. The vitamin should be measured routinely in the evaluation of any elderly subject with disordered cognition or depression and replacement therapy should be given to subjects in whom low serum levels are found. The lower limit of normal varies in different laboratories but a value below 150 pg/ml is highly suspect and is an indication for replacement therapy.

Fat-soluble vitamins

Vitamin A

Recent evidence has suggested that vitamin A is one of the only nutrients in which requirements decrease with advancing age. Studies have shown that ageing is associated with an increase in absorption of vitamin A from the gastrointestinal tract accompanied by a reduction of hepatic uptake. These effects make elderly people susceptible to toxicity, if excessive amounts of the vitamin are consumed as a supplement. Side-effects of daily intakes in excess of 50 000 IU include headaches, lassitude, reduction in white cell counts, impaired hepatic function, and bone pain. The vitamin plays an important role in visual acuity but there is no evidence that vitamin A supplements improve age-associated declines in eyesight. Vitamin A and its precursor β-carotene have been suggested as exerting a protective effect against an array of neoplasms. Recent large-scale controlled trials have, however, failed to prove a beneficial effect of β-carotene on the development of skin cancers.

Vitamin D

In addition to its involvement in bone and neuromuscular metabolism (see Chapter 14.2) vitamin D also affects macrophage function in general and pulmonary macrophages in particular. This has led to the suggestion that vitamin D deficiency increases susceptibility to the development of pulmonary tuberculosis by compromising macrophage function. This has been suggested as contributing to the high prevalence of tuberculosis in nursing home patients in whom deficiencies are common and aggravated by diminished exposure to sunlight.

Vitamin E

Vitamin E (α-tocopherol) is abundant in the diet and deficiencies of the vitamin virtually never occur. It is important in the function of the enzyme glutathione peroxidase which is involved in free radical scavenging. The vitamin also affects the biophysical properties of the cell membrane reducing the age-associated increase in membrane microviscosity. It also influences immune function and recent evidence indicates that administration of the vitamin enhances immune function in elderly subjects and may reduce risk of infections. Despite these actions, which may improve age-associated declines in cellular function, there is no good evidence of a beneficial effect of vitamin E supplementation in subjects of any age.

Vitamin K

This vitamin is essential for the production of a number of factors involved in both the intrinsic and extrinsic clotting cascade. There is evidence that vitamin K administration is beneficial in elderly subjects who have unexplained prolongation of their prothrombin time. Although dietary intake is adequate, deficiencies can result from the administration of drugs that change with the vitamin's absorption or interfere with the intestinal bacterial flora.

NUTRITIONAL ASSESSMENT

Determining the nutritional status of elderly people involves an accurate assessment of body composition which is essential to define the presence of obesity and to detect individuals who are significantly underweight. Biochemical and haematological indices are needed to evaluate visceral protein stores and the functional impact of nutritional deficiencies (Mitchell and Lipschitz 1982). Nutritional assessment is made difficult by age-associated alterations in body composition and reductions in immunological function which mimic many of the alterations that occur as a consequence of malnutrition (Table 1).

Anthropometric methods

These include the determination of body composition and involve the comparison of a measured value against a reference standard of height and weight. The alterations in height, weight, postural changes, and mobility that occur with ageing combined with the lack of appropriate standards makes the assessment of body composition in the elderly difficult (Mitchell and Lipschitz 1982).

In both men and women height decreases by approximately 1 cm per decade after the age of 20. This is caused by vertebral bony loss, increased laxity of vertebral supportive ligaments, reductions in disc spaces, and alterations in posture. Historical estimations of height are also frequently inaccurate in elderly people and its measurement is difficult in bedfast patients or in those with significant postural abnormalities.

Table 1 Comparison of the effects of ageing and protein energy malnutrition

Ageing	Protein energy malnutrition
Immune function	
↓ T-cell number	↓ T-cell number
↓ T-helper cells	↓ T-helper cells
↑ T-suppressor cells	↑ T-suppressor cells
↓ Blastic response to mitogens	↓ Blastic response to mitogens
Anergy (10%)	Anergy (10%)
↓ B-cell function	B-cell function unaffected
↑ Autoantibody production	Autoantibody production not increased
Natural killer cells normal or ↑	↓ Natural killer cells
Anthropometric measurements	
↓ Lean body mass	↓ Lean body mass
↑ Fat stores	↓ Fat stores
Biochemical measurements	
Serum albumin unchanged	↓ Serum albumin
↓ Transferrin	↓ Transferrin
Haemoglobin unchanged	↓ Haemoglobin
↓ Granulocyte response to infection	↓ Granulocyte response to infection
Drug metabolism	
↓ Drug plasma clearance	↓ Drug plasma clearance
↓ Drug breakdown	↓ Drug breakdown
↓ Excretion	↓ Excretion

For this reason it has been suggested that alternatives to height should be used in the development of standards for body composition for the elderly. Options suggested include arm length and knee-height measurements.

In general, a gradual increase in weight occurs with advancing age, peaking in the early 40s in men and a decade later in women (Mitchell and Lipschitz 1982). After age 70 reductions in weight are not uncommon. Lean body mass decreases on average by approximately 6.0 per cent per decade after the age of 25 (Forbes and Reina 1970). By the age of 70 lean body mass has decreased on average by 5 kg in women and 12 kg in men. Thus, in elderly subjects, fat constitutes a greater percentage of total weight than it does in younger subjects (Fig. 2). Fat distribution also alters with ageing; truncal and intra-abdominal fat content increase while limb fat diminishes. Skinfold measurements are often employed to estimate fat and muscle stores. Although the triceps skinfold thickness is the most frequently obtained, multiple skinfold measurements are more reliable than single measurements. In elderly subjects subscapular and suprailiac skinfolds are the best predictors of fat stores in men, while the triceps skinfold and thigh measurements are of greater value in women. Total body water is also decreased in parallel to declines in lean body mass (Fig. 2).

Weight is still the most important measure of body composition and a history of recent weight loss is the single most important clue to the presence of a significant nutritional problem. Significant malnutrition can exist in elderly subjects who are overweight, and should be suspected in any subject in whom a significant degree of weight loss has occurred. In sick elderly individuals alterations in fluid balance may make interpretations of weight difficult. However, it is usually easy to identify those patients who are either significantly above or below their ideal body weight.

Although reference tables for nutritional assessment are

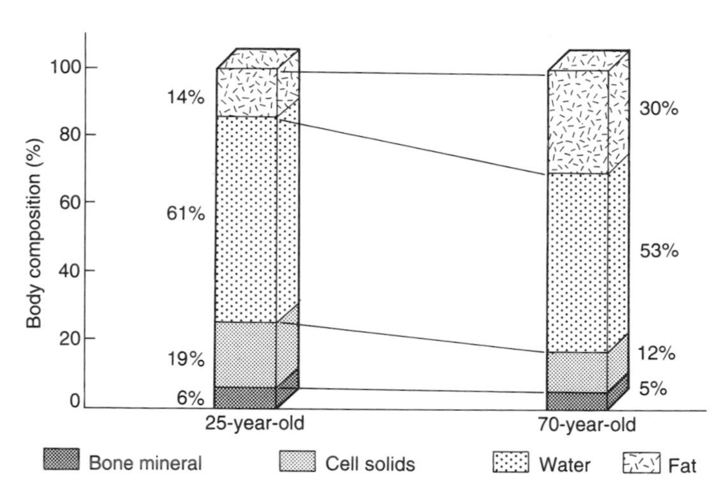

Fig. 2 Diagrammatic comparison of the effect of age on changes in body composition. Note the increased percentage of fat and the reduction of body water in the 70-year-old. This decrease reflects the age-related decline in lean body mass (adapted from Shock *et al.* 1964).

incomplete and frequently not representative of the entire population, some do exist and these should be employed when anthropometric determinations are obtained in the elderly (Masters and Lasser 1960; Forbes and Reina 1970).

Biochemical and haematological indices of nutritional status

Visceral protein stores are best assessed by the determination of the serum albumin; serum transferrin, prealbumin, and retinol-binding protein are also of value. Determination of serum transferrin in elderly subjects is complicated by the fact that it varies inversely with tissue iron stores. As a result of age-associated increases in tissue iron stores, older people may have transferrin values in a range that may be falsely interpreted as indicating visceral protein store depletion. In ambulatory healthy elderly subjects reduction in serum albumin or other measures of visceral protein status are very unusual.

Protein energy malnutrition (see below) is associated with the development of anaemia, accompanied by evidence of iron deficient erythropoiesis and declines in cell-mediated immune function. The nutrition-related alterations in immune and haematopoietic functions as a consequence of protein energy malnutrition have similarities to the declines in function which occur as a consequence of ageing (Table 1). Even in healthy elderly subjects anergy and lymphocytopenia have been reported in approximately 10 per cent. In addition an unexplained anaemia is not uncommon. For this reason ascribing declines in immune or haematological function to malnutrition in older individuals must be undertaken with appropriate clinical judgement.

MARASMUS

Marasmus is a clinical syndrome characterized by weight loss which is accompanied by marked depletion in both fat stores and muscle mass (McMahon and Bistrian 1990). Serum albumin is normal and visceral organ function remains largely intact. The disorder is caused by an inadequate intake of energy relative to needs. The classic cause is semistarvation, as occurs, for example, in anorexia nervosa. Similarly cachexia associated with cancer, chronic renal, pulmonary, and cardiac failure, and chronic infections present with the clinical features of marasmus.

While as many as 40 per cent of young men and women are overweight, only 10 per cent of subjects over the age of 70 are obese and 40 per cent are significantly underweight. Elderly individuals who fall below the 15th percentile of their ideal body weight can be diagnosed as having marasmus. Although the exact incidence is not clear, marasmus is common, particularly in chronically ill, hospitalized or institutionalized elderly people. In approximately 75 per cent of patients a cause will be identified (Table 2). The importance of depression and drugs, particularly digitalis, as potentially treatable causes of weight loss cannot be overemphasized. Poor dentition, and loss of taste and smell should also be considered in older patients. For institutionalized and dependent elderly people, inadequate access to food, failure to provide desired food choices, and decreased activity, which invariably occurs in recently institutionalized elderly patients, are common causes of weight loss. For about 25 per cent of elderly subjects an obvious cause cannot be identified. The aetiology of the weight loss in these individuals is unknown. Dietary intake data often indicate that energy and protein intake are adequate. It has been suggested that marasmus in the nursing home results largely from anorexia which may be reversible (Morley 1990).

Table 2 Common causes of weight loss in elderly people

Anorexia
Depression
Drugs e.g. digoxin
Diseases resulting in anorexia
Cancer
Chronic heart and pulmonary failure
Chronic infections
Polymyalgia rheumatica and other collagen vascular diseases
Single nutrient deficiencies
Vitamin A
Zinc
Environmental
Malabsorption
Intestinal ischaemia
Swallowing disorders
Neurological
Oesophageal candidiasis
Metabolic
Hyperthyroidism
Decreased activity
Inadequate access to food
Food preferences not met

As indicated above, weight loss has been shown to decline in subjects over the age of 70. It is important to emphasize, however, that at any age, significant recent weight loss of greater than 5 per cent of original weight should never be ascribed to 'normal ageing'. Nor can ageing explain the profound weight loss that characterizes marasmus. Marasmus is important because underweight older subjects have decreased reserve capacity and can develop serious nutritional problems both rapidly and with only minor stress.

PROTEIN ENERGY MALNUTRITION (PEM)

This disorder may be defined as a metabolic response to a pathological stress that is associated with a significant increase in the protein and energy needs required to maintain homeostasis (McMahon and Bistrian 1990). The pathophysiological changes that result in this disorder are illustrated in Fig. 3. The common conditions resulting in protein energy malnutrition are injury, burns, and infective or non-infective inflammation. In the acute setting, this response is both physiological and beneficial and assists in optimizing the body's response to injury. In young individuals the effects of this metabolic response become negative, and significant pathology results if, after a period of approximately 10 days, protein and energy needs are not met. Inadequate nutrient supply primarily affects organ systems with rapid cellular or protein turnover. The disorder is associated with marked depletion of visceral protein stores characterized by the presence of hypoalbuminaemia. Impairment of liver function contributes to the low serum albumin. Decreased clearance of drugs and toxins also occurs increasing the risk of toxicities and adverse drug reactions. The organ systems with the highest turnover of cells are the skin, immunohaematopoietic system, and gastrointestinal tract. Thus protein energy malnutrition is characterized by a dry skin and 'flaky paint' dermatitis. Impaired immune responses lead to compromised host defences increasing the risk of life-threatening infections. Malabsorption also develops as a result of impaired jejunal and ileal mucosal cell proliferation creating a vicious circle of malnutrition causing malabsorption and worsening malnutrition. As a result of disease and deficiencies of taste-related nutrients anorexia is usually present. The disorder is also referred to as hypoalbuminaemic malnutrition, and

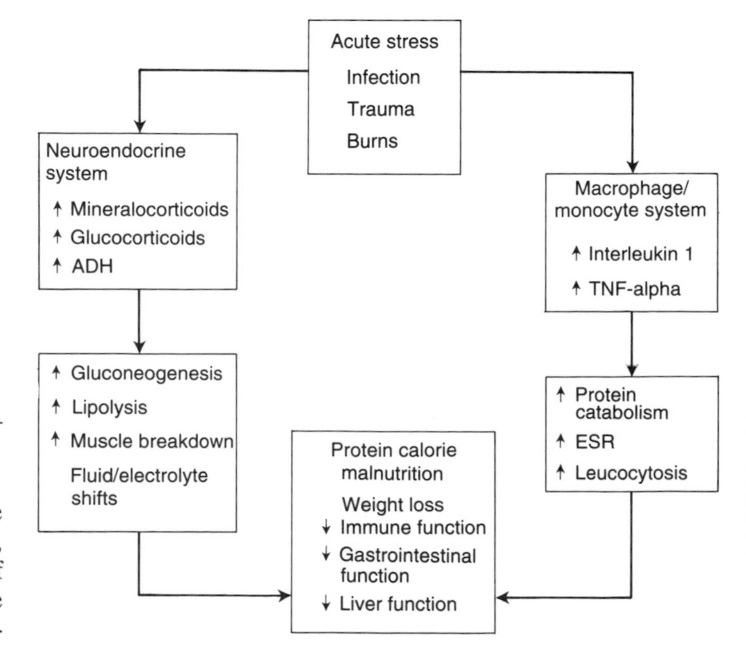

Fig. 3 The pathophysiological responses to stress that results in the development of protein energy malnutrition.

is usually diagnosed by the presence of a serum albumin level of less than 3.0 g/dl.

In older patients, relatively minor stress of short duration can result in protein energy malnutrition. Thus protein energy malnutrition is common in elderly patients who develop minor pulmonary and urinary infections is often found soon after an elective surgical procedure (Table 3). The problem in elderly patients is compounded by the ease with which these patients develop severe dehydration as a consequence of an age-associated decline in thirst drive. As already described, this can lead to the development of confusion, hypotension, and a vicious circle in which the overall condition of the patient can deteriorate very rapidly. Furthermore in contrast to younger subjects, the positive benefits of the disorder are limited to a very short period. If nutritional needs are not met within 2 to 3 days of the onset of the acute illness, the declines in immune, hepatic, and gastrointestinal function appear to contribute significantly to increased morbidity, mortality, and prolonged stays in hospital.

Table 3 Common causes of protein energy malnutrition in elderly people

Acute infections
Pulmonary
Urinary
Septicaemia
Decubitus ulcer
Acute cardiac or pulmonary failure
Trauma
Hip fracture

In the nursing home protein energy malnutrition should be suspected in any patient who develops an acute medical problem. It is also frequently seen in patients with chronic infections and in those with decubitus ulcers. Any patient presenting with confusion, lassitude, anorexia, decreased activity, or greater functional dependence may have developed an acute medical problem such as an infection which, if not treated, will result in the development of significant protein energy malnutrition.

Recent evidence has demonstrated that scant attention is paid to the nutritional status of acutely hospitalized elderly patients (Sullivan *et al.* 1989). In a prospective survey of these patients, a high prevalence of malnutrition was documented; 39 per cent had severe protein energy malnutrition and a further 33 per cent were less severely malnourished. In 17 per cent of patients adequate evaluation of their nutritional status was not possible because of a grossly inadequate clinical assessment and in no patient was a diagnosis of malnutrition recorded on the problem chart. In only 10 per cent was any form of intervention attempted and in no case was nutritional intervention adequate. In 20 patients enteral hyperalimentation was attempted; in three of these patients nasogastric feeding was never undertaken and in 80 per cent the tube was removed three or more times. Additional complications included diarrhoea, aspiration, and a fractured wrist in a patient who struggled in an attempt to remove restraints placed to prevent him from removing his nasogastric tube. These data indicate that in the institutions studied elderly patients are usually not screened appropriately for protein-energy malnutrition, the diagnosis is frequently missed or ignored, and nutrition support therapy is underutilized and often ineffectually managed with a high complication rate. {In geriatric practice in the United Kingdom, where physical restraints are proscribed, nasogastric and other forms of 'tube feeding' are much less often used than in the United States. The nutritional consequences of this policy have not been adequately evaluated.}

There is new evidence that proper evaluation of a patient's nutritional status may be very important. This view is based on a recent study which examined the impact of nutritional status on morbidity and mortality in a select population of geriatric rehabilitation patients (Sullivan *et al.* 1990). A prospective evaluation of 110 consecutive admissions to a geriatric rehabilitation hospital revealed a highly significant correlation between measures of nutritional status and risk of developing major infective and non-infective complications, as well as mortality. Most importantly, using discriminant function techniques, it was possible to demonstrate that the nutritional impact on outcome was independent of all the other non-nutritional variables known to modulate morbidity and mortality. The non-nutritional factors that were important in this regard were functional status, diagnosis, and the number of drugs prescribed. This information is important because it sets the stage for investigations that should be able to determine if aggressive nutritional rehabilitation of selected malnourished elderly patients decreases morbidity and mortality.

Management of protein energy malnutrition

Once a diagnosis of protein energy malnutrition has been made, clinical judgment is extremely important in deciding the appropriate time to commence nutritional support. In the acutely ill patient, attention should be directed first at correcting the major medical abnormalities. Thus, management of infections, control of blood pressure, and the restoration of metabolic, electrolyte, and fluid homeostasis must assume priority. During this period fluid and nutrient intake should be recorded so that an assessment of future needs can be made. Once the acute process has stabilized, daily calorie counts should be performed and the subjects should be encouraged by the staff to consume as much of their food as possible. If fluid overload is not a major concern, the use of polymeric dietary supplements between meals and in the late evening should be considered. The aim is to obtain a caloric intake of approximately 35 kcal per kg based upon an ideal rather than the actual body weight. It is our experience that by encouragement alone only 10 per cent of elderly subjects with protein energy malnutrition can consume sufficient food voluntarily to correct their nutritional deficiency. Thus, most subjects require a more aggressive form of nutritional intervention. Except for those patients with an abnormal gastrointestinal tract the most appropriate method of nutritional correction is enteral hyperalimentation through a small-bore nasogastric polyethylene catheter. These tubes are non-irritating and do not interfere with patient mobility or the ability to swallow food (but in most clinical settings require informed consent by the patient). It is extremely important that after the tube is passed, placement in the stomach be confirmed prior to commencing nutritional feedings. Infusions should begin with an undiluted, commercially available polymeric dietary supplement at a continuous rate of 25 ml/h. The supplement should contain no more than 1 kcal/ml as caloric-dense fluids are too viscous to pass through the tube with ease. The rate can gradually be increased so that after 48 h the total daily protein and calorie requirements of the patient are met by this route.

Enteral hyperalimentation has major side-effects (Table 4) of which the attending physician must be aware. One of the most commonly encountered is excessive fluid retention. When nutritional support begins, weight gain is invariably noted within the first 2 to 3 days. This almost certainly reflects fluid retention, as the weight gain is associated with significant reductions in the serum albumin and haemoglobin levels. The average increase in weight during this time in our patients was 1.3 kg, while the level of the serum albumin fell from a mean of 2.8 g/dl in patients prior to nutritional support to a value of 2.3 g/dl at day 3. Occasionally, and particularly in elderly subjects with inadequate renal function, excessive retention of fluid can result in peripheral oedema or even heart failure. When this occurs, diuretic therapy can correct the underlying problem or the use of calorie-dense supplements should be considered. Major alterations in circulating electrolytes have also been described. Hyponatraemia and hypocalcaemia occur frequently. In addition, hypophosphataemia and decreased serum magnesium levels can occur, worsening delirium. Hyperglycaemia and glycosuria are occasionally noted, and frank diabetic coma can develop. An additional problem seen occasionally is severe diarrhoea but the risk can be minimized if supplements are given by slow infusion. Bolus administration of dietary supplements through a nasogastric tube increases the risk of diarrhoea and, particularly in elderly patients, enhances the possibility of vomiting and aspiration pneumonia. Nutritional management requires a great deal of clinical skill, particularly when frail aged subjects are being supported. With suitable training and monitoring the side-effects of enteral hyperalimentation can be minimized and, when they occur, easily corrected.

Table 4 Risks of enteral feeding in elderly patients

Side-effects	Corrective actions
Mechanical risks	
Dysphagia	Use pliable non-irritating tubes
Pharyngitis	Feed patient with the upper body elevated
Oesophagitis	
Obstruction	
Pulmonary aspiration	
Gastrointestinal risks	
Nausea and vomiting	Formula dilution
Cramping	Continuous slow infusion
Malabsorption	Formula change
Diarrhoea	Antidiarrhoeal agents
	Lactose-free isotonic formulae
Metabolic risks	
Hyperglycaemia	Monitor carefully
Hyperosmolar dehydration	Appropriate infusion of fluids and electrolytes
Azotaemia	
Fluid and electrolyte disturbances	
Miscellaneous risks	
Intolerance of nasogastric tube	Consider feeding gastrostomy or jejunostomy
Tube frequently pulled out	Consider parenteral hyperalimentation
	Avoid restraints

The most difficult aspect of enteral hyperalimentation is the intolerance to the tube which is frequently seen in confused or unco-operative elderly subjects. As indicated above, these patients frequently remove the tube and (where ethically acceptable) restraints may be required to prevent this (Sullivan *et al.* 1989). It is our impression that this should be avoided if at all possible. For elderly subjects who are likely to require a prolonged period of enteral hyperalimentation a feeding jejunostomy or gastrostomy is strongly recommended. For the acutely ill older subject who will require short-term nutritional rehabilitation active attempts at voluntary feeding should be made. This is very labour-intensive and frequently unsuccessful (although the bedside help of a caring relative can sometimes make a significant contribution). A final alternative that may be of great value

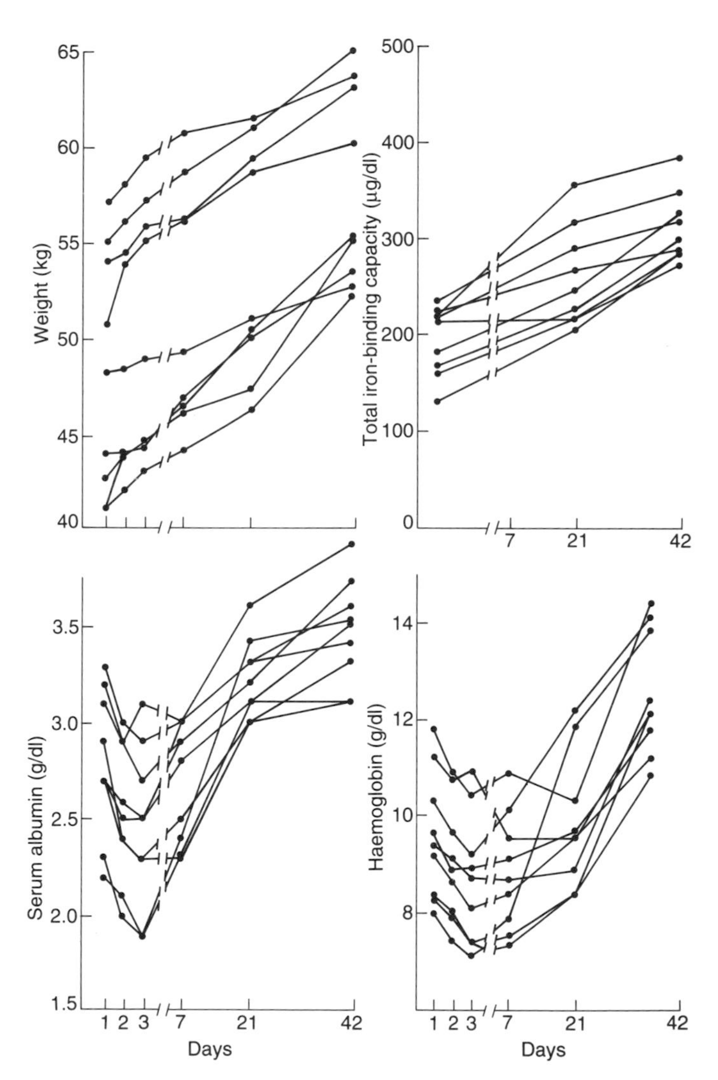

Fig. 4 Changes in weight, serum albumin, haemoglobin, and total iron-binding capacity in nine elderly subjects with severe protein energy malnutrition who received enteral hyperalimentation for 21 days, followed by a further 21 days of intensive nutritional rehabilitation (from Lipschitz and Mitchell 1982 with permission).

is the use of short-term parenteral hyperalimentation by which a patient can receive all his protein, carbohydrate, and fat needs via a peripheral line. It must be emphasized that older subjects do not tolerate prolonged periods of inadequate nutritional intake. The management of the confused and unco-operative older patient, who has an appropriate indication for nutritional intervention remains a difficult clinical problem requiring a qualified team of health professionals to obtain an adequate clinical outcome.

Although anecdotal evidence has demonstrated that aggressive nutritional intervention can result in weight gain and improved immune and haematologic functions, as well as the return of serum albumin, transferrin, and other indices of visceral protein stores to the normal range (Fig. 4), increases in muscle mass, as measured by anthropometric measurements, do not usually occur (Lipschitz and Mitchell 1982). Since a major goal of rehabilitation is to improve functional independence and improve strength, strategies aimed at improving muscle mass is particularly important. For this reason the recent observation that administration of recombinant growth hormone can improve muscle mass and performance in frail elderly subjects may prove particularly significant (Rudman *et al.* 1990). It may well be that this will become a useful tool as an adjunct to nutritional support in older patients receiving rehabilitation. Beneficial effects of appropriate exercise have also been reported (Fiatarone *et al.* 1990) (see Section 19). These studies emphasize the need for a comprehensive approach to the management of elderly malnourished patients. Aggressive nutritional intervention is only a part of a complete strategy aimed at restoring, in the appropriate patient, functional independence.

References

Fiatarone, M.A., Marks, E.C., Ryan, N.D., Meredith, C.N., Lipsitz, L.A., and Evans, W.J. (1990). High-intensity strength training in nonagenarians. *Journal of the American Medical Association*, **263**, 3029–34.

Frisancho, A.R. New standards of weight and body composition by frame size and height for assessment of nutritional status of adults and the elderly. *American Journal of Clinical Nutrition*, **40**, 808–19.

Forbes, G.B. and Reina, J.B. (1970) Adult lean body mass declines with age: some longitudinal observations. *Metabolism*, 19, 653–63.

Garry, P. J. and Hunt, W.C. (1986). Biochemical assessment of vitamin status in the elderly: effects of dietary and supplemental intakes. In *Nutrition and Aging* eds. M. Hutchinson and H.W. Munro, pp.117–137. Academic Press, New York.

Gersovitz, M., Motil, K., Munro, H.N., Scrimshaw, N.S., and Young, V.R. (1982). Human protein requirements: assessment of the adequacy of the current recommended dietary allowance for dietary protein in elderly men and women. *American Journal of Clinical Nutrition* **35**, 6–14.

Horwitz, A., *et al.* (eds). (1989). *Nutrition in the Elderly* Oxford University Press, Oxford.

Lipschitz, D.A. and Mitchell, C.O. (1982). The correctability of the nutritional, immune, and hematopoietic manifestations of protein calorie malnutrition in the elderly. *Journal of the American College of Nutrition*, **1**, 17–25.

McGandy, R.B. (1988). Atherogenesis and aging. In *Health Promotion and Disease Prevention in the Elderly* (eds. R. Chernoff and D.A. Lipschitz) pp. 67–75. Raven Press, New York.

McGandy, R.B., Barrows, C.H. Spanias, A., Meredith, A., Stone, J.L. and Norris, A.H. (1966). Nutrient intake and energy expenditure in men of different ages. *Journal of Gerontology*, **21**, 581–4.

McMahon, M.M. and Bistrian, B.R. (1990). The physiology of nutritional assessment and therapy in protein calorie malnutrition. *Disease-A-Month*, **7**, 375–417.

Masoro, E.J. (1989). Overview of the effects of food restriction. *Progress in Clinical and Biological Research*, **287**, 27–35.

Masoro, E.J. (1990) Physiology of ageing: nutritional aspects. Age and Ageing, **19** (Suppl. 1), S5–9.

Masters, A. and Lasser, R. (1960). Tables of average weight and height of Americans aged 65–94 years. *Journal of the American Medical Association* **172**, 658–62.

Mitchell, C.O. and Lipschitz, D.A. (1982). Detection of protein calorie malnutrition in the elderly. *American Journal of Clinical Nutrition* **35**, 398–406.

Morley, J.E. (1990). Nutrition and aging. In *Principles of Geriatric Medicine and Gerontology* (eds. W.R. Hazzard, R. Andres, E.L. Bierman and J.P. Blass) pp. 48–59. McGraw Hill, New York.

Morley, J.E., Silver, A.J., Fiatarone, M. and Mooradian, A.D. (1986). UCLA Grand Rounds: nutrition and the elderly. *Journal of the American Geriatrics Society* **34**, 823–32.

Munro, H.N., Suter, P.M., and Russel, R.M. (1987). Nutritional requirements of the elderly. *Annual Review of Nutrition* **7**, 23–49.

National Center for Health Statistics (1974). *First Health and Nutrition Examination Survey, United States, 1971–1982*. Health Service Administration, DHEW Publ. No. (HRA) 74–1219–1, Washington D.C.

Phillips, P.A., *et al.* (1984). Reduced thirst after water deprivation in healthy elderly men. *New England Journal of Medicine*, **311**, 753–6.

Rudman, D. *et al.* (1990). Effects of human growth hormone in men over 60 years old. *New England Journal of Medicine* **343**, 1–6.

Shock, N.W., *et al.* (1984). *Normal human aging: the Baltimore longitudinal study of aging*. U.S. Department of Health and Human Services. NIH Publ. No. 84:2450, Washington, D.C.

Sullivan, D.H., Moriarty, M.S., Chernoff, R., and Lipschitz, D.A. (1989). An analysis of the quality of care routinely provided to elderly hospitalized veterans. *Journal of Parenteral and Enteral Nutrition*, **13**, 249–54.

Sullivan, D.H., Patch, G.A., Walls, R.C. and Lipschitz, D.A. (1990). Impact of nutrition status on morbidity and mortality in a select population of geriatric rehabilitation patients. *American Journal of Clinical Nutrition*, **51**, 749–58.

Suter, P.M. and Russel, R.M. (1987). Vitamin requirements in the elderly. *American Journal of Clinical Nutrition* **45**, 501–12.

Yearick, E.S., Wang, M.S., and Pisias, S.J. (1980). Nutritional status of the elderly: dietary and biochemical findings. *Journal of Gerontology*, **35**, 663–71.

SECTION 7
Endocrine and metabolic disorders

7.1 Diabetes mellitus

LINDA A. MORROW, WILLIAM H. HERMAN, AND JEFFREY B. HALTER

DIAGNOSIS AND EPIDEMIOLOGY

Disorders of carbohydrate metabolism in the elderly population include glucose intolerance, which appears to be closely linked with ageing processes, and diabetes mellitus, a pathological condition that occurs with increased frequency among elderly individuals. Although the relationship between these entities is not clearly defined, the gradual impairment of tolerance to a glucose load with ageing may well contribute to the development of diabetes mellitus in those individuals who may be in some other way predisposed.

Table1 Diagnostic criteria for diabetes mellitus and impaired glucose tolerance[a]

		Oral glucose tolerance test	
Category	Fasting plasma glucose	1 h	2 h
Diabetes mellitus[b]	1. ≥7.8 mmol/l (2 occasions)		
	2. <7.8 mmol/l plus oral glucose tolerance test	≥11.1 mmol/l[c]	≥11.1 mmol/l
	3. Random hyperglycaemia plus classic diabetes symptoms		
Impaired glucose tolerance	<7.8 mmol/l	≥11.1 mmol/l[c]	7.8–11.1 mmol/l

[a]Criteria in non-pregnant adults are based on 75-g glucose load. All values are venous plasma glucose.
[b]Patients meeting any one of the three criteria have diabetes mellitus.
[c]One value ≥11.1 mmol/l at 30, 60, or 90 min meets the criterion.

Changes in the ability to regulate circulating glucose have been shown to occur with ageing by both cross-sectional and longitudinal studies. Both the fasting plasma glucose and plasma glucose levels 1 h after glucose ingestion increase with age. Changes in the fasting plasma glucose are small (average 0.06–0.11 mmol/l/decade), but 1 h post-glucose ingestion levels increase by 0.33–0.78 mmol/l/decade on average after the age of 30. In spite of these changes, the criteria recommended by the American Diabetes Association and the World Health Organization (**WHO**) for the diagnoses of impaired glucose tolerance and diabetes mellitus apply without modification to the elderly population (Table 1). These criteria are based upon the risk for the development of diabetes-related complications. Only symptomatic hyperglycaemia with an elevated random plasma glucose, overt fasting hyperglycaemia, or a markedly abnormal glucose tolerance test establish the diagnosis of diabetes mellitus in a person of any age.

The prevalence of diabetes mellitus increases with age. Multiple studies in such diverse populations as Polynesians, Finns, Pima Indians, Israelis, Maltese, and Americans (Table 2) have demonstrated a prevalence of diabetes mellitus that may approach 40 per cent in some populations of individuals aged 65 years and older. The figures from these studies are the result of screening populations; on an individual basis, many older adults are unaware that they have diabetes mellitus. Figure 1 emphasizes the dramatic disparity between diagnosed and undiagnosed diabetes mellitus in this age group in one study (NHANES II) made in the United States between 1976 and 1980. Nearly half of the individuals over the age of 65 with diabetes were previously undiagnosed.

The incidence of diabetes mellitus also increases with age. The incidence of overt diabetes in individuals over 65 is 0.5 to 1.0 per cent/year. Simultaneous analysis of incidence, prevalence, and mortality indicate an increase of almost 5 per cent/year in the number of people with diabetes mellitus, most of them in the elderly population.

Fewer data are available on impaired glucose tolerance and ageing, and differences in WHO and National Diabetes

Table 2 Prevalence of diabetes mellitus and impaired glucose tolerance among older adults in selected worldwide populations

Population	Age	Sex	Prevalence of diabetes mellitus (%)	Prevalence of impaired glucose tolerance (%)	Criteria
Finland	65–84	M	29.7	31.6	WHO
Israel	≥60	M and F	10.3	N/A	WHO and NDDG
Malta	65–74	M	19	N/A	
Polynesia					
Wallis Island	≥65	M	4.3	14.3	WHO
		F	10.0	17.6	
United States					
Pima Indians[a]	65–74	M	38	26	
		F	49	24	
NHANES II[b]	65–74	M	19.2/20.1	8.9/22.8	[a]NDDG/WHO
		F	16.5/17.4	9.4/22.7	

M=male; F=female; N/A, not assessed.
[a]Diabetes mellitus: 2-h postprandial plasma glucose ≥11.1 mmol/l. Impaired glucose tolerance: 2-h postprandial glucose 7.8–11.1 mmol/l.
[b]National Health and Nutrition Examination Survey II: figures are shown first with NDDG criteria, then WHO criteria.

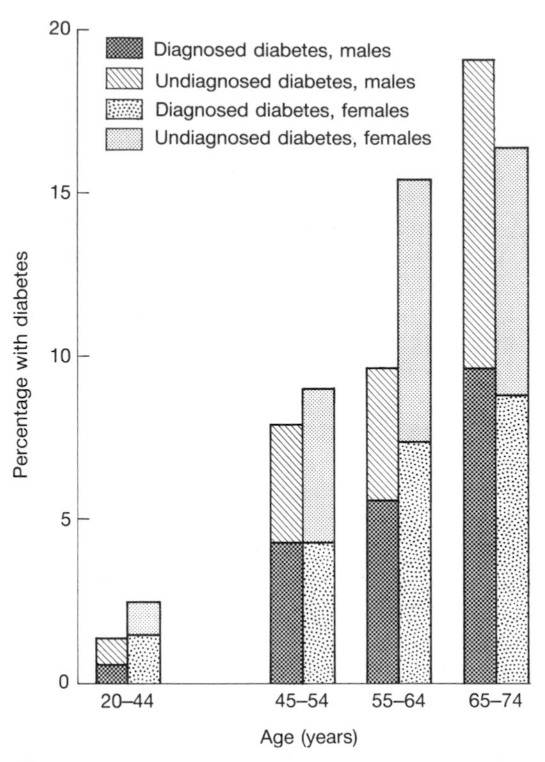

Fig. 1 Percentage of United States population aged 20 to 74 years with diabetes from NHANES II, 1976–1980 (reproduced from Harris *et al.* 1987 with permission).

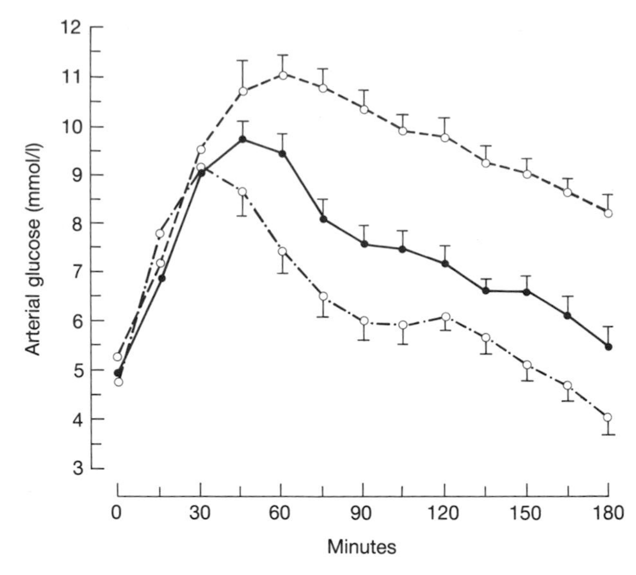

Fig. 2 Influence of age on oral glucose tolerance (100 g) in non-obese healthy men. *Dashed line*, 70–83 years (n=9); ●, 30–45 years (n=12); *stippled line*, 19–24 years (n=11) (reproduced from Jackson *et al.* 1982, with permission).

Data Group (USA) criteria are responsible for substantial variability in the reported prevalence of impaired glucose tolerance (see Table 2). However, in most studies, impaired glucose tolerance increases with age, and the prevalence, by WHO criteria, may be as high as 23 per cent for individuals over the age of 65. The significance of impaired glucose tolerance lies in its relationship to cardiovascular disease. Individuals with impaired glucose tolerance (but who do not meet the criteria for diabetes mellitus) have an increased mortality from cardiovascular disease, but no increase in microvascular complications. Recently noted age-associated increases in glycation of important proteins such as collagen and haemoglobin suggest other potentially harmful effects of impaired glucose tolerance.

PATHOGENESIS

Glucose intolerance of ageing

The glucose intolerance of ageing appears to be distinct from non-insulin dependent diabetes mellitus and could be the result of several changes that frequently accompany the ageing process. While gastrointestinal absorption of carbohydrates is probably delayed, and would actually improve glucose tolerance if unopposed, changes in absorption are more than counteracted by changes within the pancreas, liver, and peripheral tissues. These changes result in the well-documented decline in glucose tolerance to an oral glucose load demonstrated in Fig. 2.

The plasma immunoreactive insulin response to glucose ingestion is delayed and somewhat increased with ageing. However, high glucose levels in the older subjects make interpretation of these results difficult. Changes in pancreatic insulin secretion may occur as a result of age-associated changes in the structure and function of pancreatic β-cells, which have been found in experimental animals. Human studies suggest that elderly subjects have diminished β-cell sensitivity to glucose as compared to other stimulants of pancreatic insulin secretion. While suppression of hepatic glucose output by insulin may be somewhat delayed in non-diabetic elderly subjects, some investigators have found no changes as a result of ageing. Therefore this is probably of minor significance in accounting for alterations in glucose tolerance with ageing. The largest contributor may be that of the delay in insulin-mediated glucose uptake. Several investigators have attempted to characterize this abnormality. It appears that the number of insulin receptors does not change with ageing, but there is a primary postreceptor defect, perhaps related to a decrease in the number but not function of glucose transporters. This age-associated insulin resistance, coupled with impaired β-cell adaptation to insulin resistance, may largely account for the observed deterioration in glucose tolerance. Lifestyle factors such as physical fitness and adiposity may be important causes of some of these age-associated changes.

Diabetes mellitus

While insulin-dependent or type 1 diabetes mellitus can occur at any age, its relative incidence and prevalence are low compared to those of non-insulin dependent or type 2 diabetes mellitus, the predominant form of diabetes mellitus in people aged over 65. While few studies have been done with elderly patients, it is assumed that the pathogenesis of hyperglycaemia in elderly patients with non-insulin dependent diabetes is similar to that in younger individuals: accelerated hepatic production of glucose with inadequate insulin secretion for the level of hyperglycaemia, and peripheral tissue resistance to the actions of insulin to reduce hyperglycaemia. In some populations, diminished regulation of glycogen synthetase in muscle by insulin appears to be important in the insulin resistance of non-insulin dependent diabetes.

Certain predisposing factors (separate from the traditional risk factors of age, race, and family history) that may enhance or precipitate the development of diabetes mellitus may occur with increased frequency among elderly individuals. These include changes in body habitus and lifestyle, the presence of certain illnesses, and the use of some medications. With ageing there is an increase in adiposity at the expense of muscle tissue, a situation that may contribute to insulin resistance. The onset of illnesses that lead to decreased mobility, such as arthritis, congestive heart failure, or peripheral vascular disease, can lead to a sedentary lifestyle, which may also contribute to insulin resistance. Dietary changes, such as reduced carbohydrate intake or reduced total caloric intake, impair insulin secretion and can lead to insulin resistance. Acute illnesses associated with increases in stress hormones may impair insulin secretion and enhance insulin resistance, precipitating the onset of diabetes mellitus or even hyperosmolar hyperglycaemic non-ketotic coma. Medications used commonly by elderly individuals, such as those to treat hypertension and heart disease (diuretics, β-adrenergic antagonists), and chronic obstructive pulmonary disease (glucocorticoids), have actions that inhibit insulin secretion or enhance tissue resistance to the effects of insulin.

COMPLICATIONS OF DIABETES MELLITUS IN THE ELDERLY PATIENT

Acute complications

Elderly patients with diabetes mellitus are subject to two acute complications of metabolic control—hypoglycaemia and hyperglycaemia. Hypoglycaemia, while life-threatening in all age groups, may be particularly devastating for an older individual with little reserve capability. Similarly, acute hyperglycaemia may result in diabetic ketoacidosis or hyperglycaemic hyperosmolar non-ketotic coma, which have particularly high fatality rates among older people with diabetes.

Hypoglycaemia

Elderly patients with diabetes who are treated with either oral agents or insulin are at risk for the development of hypoglycaemia. This may be especially true when normoglycaemia is one of the goals of treatment. There are multiple factors that contribute to the increased susceptibility of elderly individuals to hypoglycaemia. Alterations in drug metabolism may prolong the duration of action and effectiveness of sulphonylureas. As both sulphonylureas and insulin are excreted by the kidneys, age-associated declines in renal function and diabetic nephropathy may lead to hypoglycaemia. Drug interactions may play a significant role, as elderly patients are the most frequent users of several drugs that enhance or prolong the action of sulphonylureas, such as salicylates, non-steroidal anti-inflammatory drugs, and β-adrenergic blocking agents.

Glucose counter-regulation is an important protective mechanism against hypoglycaemia, and is known to become impaired in individuals with both insulin-dependent and non-insulin dependent diabetes. Although no study has specifically addressed counter-regulation in older individuals with diabetes, some information is available about the components of the counter-regulatory mechanisms during ageing. While the activity of the sympathetic nervous system appears to be increased in ageing, with increased basal levels of noradrenaline secondary to increased production, end-organ sensitivity to β-adrenergic stimulation is impaired at many sites. This abnormality may result in the inability both to sense hypoglycaemia (reduced trembling and sweating) and to respond appropriately (impaired hepatic glycogenolysis). With long-standing diabetes mellitus, impairments in the functions of the autonomic nervous system secondary to diabetic neuropathy may result in decreased secretion of adrenaline. Thus, in older adults with diabetes mellitus, impairment of counter-regulation may result from changes in the autonomic nervous system secondary to diabetes mellitus, exacerbated by age-associated changes in the autonomic nervous system and the function of β-adrenergic receptors. Impairment in counter-regulation may also result from changes in glucagon secretion. While the effects of ageing on glucagon secretion are not yet clear, abnormalities in glucagon secretion in response to hypoglycaemia develop in younger people with both types of diabetes mellitus.

Other risk factors of hypoglycaemia in the older population are alterations that occur within the social setting such that hypoglycaemia might go undetected. An elderly, retired individual living alone, isolated from family and friends, could have an episode of hypoglycaemia unrecognized for hours or days. Such an episode may have serious sequelae.

Episodes of hypoglycaemia that do not respond to oral ingestion of carbohydrate are managed with the use of glucose, usually 50 per cent dextrose, intravenously followed by intravenous therapy with either 5 or 10 per cent dextrose for severe episodes that have resulted in coma. Hypoglycaemia resulting from oral hypoglycaemic agents requires extended monitoring and glucose therapy, especially with the longer acting agents such as glibenclamide (glyburide) or chlorpropamide with which hypoglycaemia may reappear within hours of successful initial treatment. However,the best treatment for hypoglycaemia is prevention through the use of reasonable treatment goals appropriate for each patient, patient education programmes, frequent monitoring of blood glucose and follow-up for patients using insulin, and well-structured support systems either through the family or the community.

Hyperglycaemia

Acute hyperglycaemic episodes are equally risky for the geriatric patient. Plasma glucose values above 11 mmol/l may be associated with symptoms that may significantly impair the functional abilities of older adults. These symptoms can include polyuria, which may lead to incontinence, dehydration, and falls, and visual impairment, which results from osmotic changes in the lens, impairing many skills necessary for independent living such as driving or administering medications.

The most extreme acute disorders of hyperglycaemia include diabetic ketoacidosis and hyperglycaemic hyperosmolar non-ketotic coma (**HHNC**). Both are frequently precipitated by underlying illnesses, and in both the death rate rises with age, to as high as 50 per cent in some studies of elderly patients. Diabetic ketoacidosis is relatively rare among elderly patients and treated similarly to that in younger patients. However, HHNC is most commonly seen among individuals over the age of 50 with non-insulin dependent diabetes and requires special consideration in the discussion of the elderly diabetic patient.

Factors that may contribute to the development of this syndrome among older adults are shown in Table 3. The high prevalence of diabetes mellitus (both known and undiagnosed) along with the impairment in glucose tolerance that accompanies the ageing process appear to place older individuals at special risk for HHNC. Several drugs used frequently with older patients have also been implicated. The mechanism by which thiazides exert their hyperglycaemic effect is unknown. Loop diuretics probably exacerbate the dehydration and volume depletion already present. β-Blockers may allow unopposed α-adrenergic inhibition of insulin secretion by catecholamines released during stress states. Glucocorticoids are known to impair tissue sensitivity to insulin, whereas anaesthetics are implicated by their effect to inhibit insulin release. Diphenylhydantoin is thought to increase glycogenolysis. Psychotropic agents can cloud consciousness and limit the patient's access to fluid.

Table 3 Factors associated with the increased frequency of hyperglycaemic hyperosmolar non-ketotic coma among older adults

Increased prevalence of diabetes mellitus
Impaired glucose tolerance
Drugs
Diuretics
Thiazides
Loop diuretics
β-Blockers
Glucocorticoids
Anaesthetics
Diphenylhydantoin
Psychotropic agents
Precipitating illnesses or events
Infections
Pneumonia
Influenza
Urinary tract infection
Septicaemia
Gangrene
Burns
Surgery
Cerebrovascular accident
Renal failure
Pancreatitis
Myocardial infarction
Impaired responses to stress hormones

Many illnesses and events have been associated with the development of HHNC,probably precipitating the syndrome by stimulating the secretion of catecholamines and other stress hormones. Impaired β-adrenergic responses to catecholamines in older humans would result in unopposed α-adrenergic inhibition of insulin secretion. This would tend to worsen the hyperglycaemia which results from the synergistic actions of adrenaline, glucagon, and cortisol. It is also possible that diminished β-adrenergic receptor mediated lipolysis might account for the low serum ketone levels which differentiate this syndrome from diabetic ketoacidosis.

The clinical picture of HHNC is frequently that of an elderly person with diet-controlled diabetes, living alone, who has the onset of an illness that may cause him or her to be bedridden. Impaired access to fluids, accompanied by the release of stress hormones (adrenaline, glucagon, and cortisol), and diminished insulin secretion results in worsening hyperglycaemia, dehydration, increasing osmolality, and eventually coma (the level of which is dependent upon the severity of the hyperosmolality). Upon admission to the hospital, the patient is often found to be severely dehydrated, with profound hypotension and shock. Plasma glucose values greater than 100 mmol/l have been reported. However, serum ketones are low or absent and the pH is usually above 7.3.

The initial treatment for HHNC, as in diabetic ketoacidosis is vigorous replacement of fluids and electrolytes. Appropriate management of these very ill patients may require invasive monitoring devices such as Swan–Ganz catheters, especially in elderly diabetic patients who may have underlying cardiac disease. Insulin is also required in the management of HHNC, and is best administered at low dose by intravenous drip. With hypotension there can be poor perfusion of muscle, and intramuscular injections of insulin may not be well or uniformly absorbed.

All patients who develop HHNC require an extensive search for the underlying cause of the metabolic decompensation. Precipitating events are shown in Table 3. Treatment of the cause is essential as the high fatality of this syndrome results largely from the underlying illnesses. These patients may develop recurrent bouts of HHNC if the underlying cause is not recognized and treated. Complications such as aspiration pneumonia, renal failure, arterial or deep venous thrombosis may cause considerable morbidity.

Chronic complications

Cerebrovascular disease

The incidence of stroke dramatically increases with age. However, even when rates of stroke are adjusted for age, systolic blood pressure, cigarette smoking, cholesterol, and left ventricular hypertrophy on ECG, the risk of stroke for patients between the ages of 45 and 74 with diabetes mellitus is 2.2 times that of patients without diabetes. The prevalence of stroke for those individuals over the age of 65 with diabetes is 11/100 compared to 5/100 for those without diabetes.

For individuals suffering stroke, the presence of diabetes mellitus increases the risk for greater resultant deficit and death. Five-year survival after a stroke for diabetic patients is only 20 per cent, compared with 40 per cent for non-diabetic patients. Additionally, the degree of glycaemia during the acute event is associated with the severity of the stroke. Studies are under way to determine if control of glycaemia during strokes will reduce the subsequent neurological deficit. Studies in animal models suggest that the use of glucose-containing solutions during acute cerebral ischaemic events may worsen the infarct, and it may be prudent to avoid these solutions in patients as well.

Patients with diabetes and hypertension should be actively treated for their hypertension. Those patients without known hypertension should be screened at each visit. Patients who use tobacco should be strongly encouraged to quit.

Eye disease

Diabetic eye disease is a major cause of illness among elderly people. In insulin-dependent diabetes, the prevalence of legal (registrable) blindness is most strongly associated with duration of diabetes. Among insulin-dependent diabetic subjects over 55 years of age, the prevalence of registrable blindness is about 12 per cent. Among patients with non-insulin-dependent diabetes, registrable blindness occurs earlier in the course of diabetes (about 1 per cent during the first 5 years of diabetes), and the prevalence increases with both duration of diabetes and age. Among patients with non-insulin dependent diabetes who are older than 65 years, the prevalence of registrable blindness is about 3 per cent.

When assigning causes of visual impairment, diabetic retinopathy is the sole or contributing cause of registrable blindness in about 86 per cent of insulin-dependent diabetic patients and in about 33 per cent of non-insulin dependent. In the latter group, cataract, glaucoma, and macular degeneration are more important contributors. Senile cataracts have been reported to occur in 12 per cent of the non-diabetic population, 19 per cent of diabetic patients, and up to 57 per cent of patients seen in diabetes clinics. Increasing age is the single greatest risk factor for cataract. In a recent population-based survey, cataract was found to be the most frequent cause of severe visual loss in subjects with non-insulin dependent diabetes. Among these patients, if older than 65 years, 96 per cent have nuclear sclerotic or posterior subcapsular cataract, or surgical aphakia.

Open-angle glaucoma is also an important cause of blindness in patients with non-insulin dependent diabetes. This glaucoma is uncommon in people less than 40 years of age but its prevalence increases with age. Glaucoma has been reported to occur in 1.8 per cent of the non-diabetic population, 2.4 per cent of diabetic patients, and in up to 4.8 per cent of patients seen in diabetes clinics. In a recent population-based survey, 7.5 per cent of patients with non-insulin dependent diabetes who were older than 65 reported histories of glaucoma, and intraocular hypertension was found in 10 per cent.

Trials have clearly demonstrated that focal photocoagulation of clinically significant diabetic macular oedema and panretinal laser photocoagulation for proliferative diabetic retinopathy and high-risk characteristics can significantly reduce the incidence of visual loss. It is also clear that cataract surgery, and medical and surgical treatment for open-angle glaucoma can reduce visual loss.

Despite compelling evidence that treatment may prevent visual loss or restore vision in people with diabetic eye disease, many do not receive regular ophthalmological examinations. A recent study showed that only 73 per cent of patients with insulin-dependent and 59 per cent with non-insulin dependent diabetes report ever having visited an ophthalmologist. Patients who are older at diagnosis and who have diabetes of shorter duration are less likely to have visited an ophthalmologist. A more recent study found that 5.2 per cent of patients with insulin-dependent and 3.9 per cent with non-insulin dependent diabetes attending a large diabetes outpatient clinic had undetected macular oedema or proliferative retinopathy requiring treatment.

All elderly subjects with diabetes should have annual ophthalmic examinations that include a history of visual symptoms, measurement of visual acuity and intraocular pressures, dilation of the pupils, and a thorough retinal examination. Patients with significant retinal disease or who have lost vision should still have regular eye examinations and should be referred for low-vision services including proper refraction, optical aids, and other techniques and devices.

Periodontal disease

Periodontal disease is a general term used to describe a group of localized infections affecting the tissues surrounding and supporting the teeth. The prevalence and severity of periodontal disease increases markedly with age. Up to one-third of adults aged 65 years and older have evidence of periodontal disease, and it has been estimated that, in non-insulin dependent diabetes, the prevalence of periodontal disease is almost three times greater than in non-diabetic individuals. In diabetes, the degree of glycaemic control appears to be associated with the severity of periodontal disease: individuals with poorly controlled diabetes have been found to experience significantly more periodontal disease than individuals with moderate or good control. In diabetes, altered microbial flora, impaired immunity, vascular changes, and abnormal collagen metabolism may contribute to the increased prevalence or severity of periodontal disease. Other factors associated with periodontal disease include smoking, vitamin C deficiency, dental restorations, and dental prostheses.

Health-care providers should recognize that periodontal infection may contribute to deterioration of glycaemic control and that periodontal disease, tooth loss, and poorly fitting dentures may limit food choices and contribute to inadequate nutrition in elderly subjects with diabetes. Effective self-care is essential to periodontal health, and diabetic patients should be encouraged to brush their teeth at least twice a day. The objective is to remove dental plaque in a gentle, systematic fashion on a regular basis. Regular professional dental care is another necessary component in the prevention and treatment of periodontal disease.

Diabetic kidney disease

Renal disease is a major cause of death and illness among diabetic patients. For diabetic patients older than 65 years, the renal mortality rate is 1.4 times that for non-diabetic subjects. In recent years, both the total number of patients with diabetic end-stage renal disease and the number older than 65 years have increased rapidly. In 1985, 28 per cent of people entering the Medicare end-stage renal disease (**ESRD**) Program had ESRD due to diabetes and 36 per cent of all new enrollees were older than 65 years. It is estimated that about half of all diabetic patients newly enrolled in end-stage renal disease programmes have non-insulin dependent diabetes.

In insulin-dependent diabetes, nephropathy follows a well characterized and predictable course: from the onset of diabetes, to the onset of albuminuria, to clinical diabetic nephropathy, to renal insufficiency or death. About 80 per cent of patients with insulin-dependent diabetes and microalbuminuria, defined as urinary albumin excretion above 15 μg/min, develop clinical diabetic nephropathy within 10 years. In insulin-dependent diabetes, the cumulative incidence of clinical diabetic nephropathy after 20 years of diabetes is about 40 per cent.

The course of diabetic nephropathy is less well characterized in non-insulin dependent diabetes. Only about 22 per cent of these patients who have microalbuminuria develop clinical nephropathy within 10 years. A number of studies have now confirmed that microalbuminuria is associated with increased cardiovascular disease and death. Cross-sectional studies of diabetic populations have demonstrated a lower prevalence of nephropathy in non-insulin dependent than in insulin-dependent diabetes, but this may reflect greater mortality among patients with non-insulin dependent diabetes and nephropathy. Recent longitudinal studies have demonstrated that the cumulative incidence of nephropathy in non-insulin dependent diabetes after 20 years of diabetes is 25 to 40 per cent, similar to that observed in the insulin-dependent type.

A host of risk factors including poor glycaemic control, hypertension, family history of hypertension, and cigarette smoking have been associated with a higher prevalence of nephropathy in insulin-dependent diabetes. Risk factors for nephropathy have been less clearly defined in non-insulin dependent disease. In a recent study, only age and initial level of glycaemia were associated with the development of persistent proteinuria in non-insulin dependent diabetes.

Studies in patients with both types of diabetes and diabetic nephropathy have shown that antihypertensive therapy can reduce, arrest, or reverse the rate of rise of albuminuria and can slow the rate of decline of kidney function. Recent studies in small numbers of patients with diabetic nephropathy have also suggested that improved glycaemic control and dietary protein restriction may slow the course of diabetic nephropathy.

Each year, all elderly subjects with diabetes should have a quantitative measurement of urinary albumin or protein excretion. Before a diagnosis of diabetic nephropathy is established, obstructive uropathy, infection, and other possible causes of renal disease should be excluded. In patients with diabetic nephropathy, hypertension should be aggressively treated and consideration should be given to dietary protein restriction. Nephrotoxic drugs and radiographic dyes should be avoided if at all possible. Older patients with diabetic nephropathy should be assessed and treated as necessary for cardiovascular risk factors.

Neuropathy

Elderly patients with diabetes who develop neuropathy may have no symptoms or may experience pain, sensory loss, weakness, and autonomic dysfunction. The prevalence of neuropathy appears to increase with age, independent of the duration of the diabetes, although epidemiologic data about diabetic neuropathy have been limited by the lack of clearly defined diagnostic criteria. Older patients with diabetes may present with any of the features of distal symmetrical polyneuropathy, focal neuropathy, or autonomic neuropathy that are found in younger patients.

Acute painful neuropathy may occur in elderly subjects in association with institution of insulin treatment or precipitous and severe weight loss. The syndrome is probably equivalent to diabetic neuropathic cachexia. Affected patients are usually in their 60s or 70s. The glucose intolerance is generally mild and men are affected more frequently than women. Pain in the distal extremities is severe and often worse at night. On examination there may be cutaneous hyperaesthesia and only minimal evidence of distal symmetric polyneuropathy. Patients appear depressed and have marked anorexia. The pain generally subsides in months and this is preceded by weight gain. Relapses are uncommon.

The onset and course of distal symmetrical polyneuropathy cannot be predicted for an individual patient, but increasing age appears to be an independent risk factor. Other risk factors include male sex, increased height, longer duration of diabetes, poor glucose control, hypertension, alcohol consumption, and cigarette smoking. The symptoms and signs of distal symmetrical polyneuropathy are similar whether the condition is associated with diabetes or a number of other conditions. The diagnosis of diabetic distal symmetrical polyneuropathy is therefore established only after other possible causes of neuropathy are excluded. If discovered, underlying secondary causes of distal symmetrical polyneuropathy should be treated.

Focal neuropathy includes a number of fairly characteristic focal and multifocal neuropathic syndromes including the cranial neuropathies, truncal neuropathies, mononeuropathies, radiculopathies, and plexopathies. Focal neuropathy is believed to occur after the acute occlusion of a blood vessel produces ischaemia in a nerve or group of nerves. The characteristics of focal diabetic neuropathy are sudden onset, an asymmetrical nature, and a self-limited course. Near total recovery generally occurs within 2 weeks to 18 months. Third cranial-nerve palsy, sometimes termed diabetic ophthalmoplegia, is the most common cranial mononeuropathy. It generally occurs in diabetic patients over 50 years of age. Truncal neuropathy occurs in patients over the age of 50 with long-standing though often mild diabetes. Pain and dysaesthesias are the heralding features. None of the focal neuropathies is unique to the diabetic patient, and other potential causes must be investigated.

Autonomic neuropathy may affect the psychomotor, pupillary, adrenomedullary, cardiovascular, gastrointestinal, and urogenital systems. Possible symptoms of autonomic neuropathy include heat exhaustion, poor night vision, unawareness of hypoglycaemia, orthostatic hypotension, painless myocardial ischaemia, gastroparesis, constipation, diarrhoea, faecal incontinence, bladder dysfunction, and sexual dysfunction. Dysfunction of the autonomic nervous system has been described early in the course of insulin-dependent and non-insulin dependent diabetes and appears to be most strongly associated with age and the presence of distal symmetrical polyneuropathy. Symptoms of autonomic neuropathy have been reported in approximately one-quarter of patients with non-insulin dependent diabetes attending a diabetes clinic, and tests of autonomic function have been reported to be abnormal in up to a half of such patients. Diabetic bladder dysfunction has been described in up to 15 per cent of diabetic patients.

The health-care provider should inquire about symptoms of diabetic neuropathy and perform a neurological examination including measurement of supine and standing blood pressures, and assessment of distal temperature, pin-prick, vibration, and position sense. Patients with evidence of neuropathy should have secondary causes excluded. Patients who have lost sensation in their feet should be instructed in foot care. Patients with large-fibre neuropathy should be prescribed proper footwear. Patients with orthostatic hypotension and those with sensory ataxia will need evaluation of their living situation and aids or assistance for ambulation to prevent falls.

Cardiovascular disease

Cardiovascular disease is the leading cause of illness and death in old age and diabetes. The annual risk for death from cardiovascular disease is two to three times greater for persons with diabetes than for persons without. It is estimated that 75 per cent of persons with onset of diabetes over the age of 60 die from cardiovascular disease.

Atherosclerosis causing cardiovascular disease in diabetic patients is biochemically and morphologically indistinguishable from atherosclerosis in non-diabetic subjects. In diabetes, however, atherosclerosis appears to be more extensive and severe and to progress more rapidly. Risk factors for atherosclerosis, including cigarette smoking, hypertension, and hypercholesterolaemia, are common in diabetic patients and have a similar relationship to cardiovascular disease in diabetic and non-diabetic subjects. They do not, however, explain the excess incidence of atherosclerosis in diabetic patients. Hyperglycaemia and hyperinsulinaemia may directly contribute to increased risk. Elevated glucose levels are known to have adverse effects on the function of endothelial cells and may contribute to atherogenesis by glycation of lipoproteins and other tissue proteins. Hyperinsulinaemia may promote atherogenesis by direct effects on the arterial wall or by effects on lipid metabolism, clotting factors, blood pressure, or the function of the sympathetic nervous system.

At each visit, the patient's blood pressure should be measured. At least once a year, patients should be asked about tobacco use and symptoms of vascular disease. Lipid levels should be measured annually and a baseline ECG should be obtained.

Although the importance of cardiovascular risk factors in old age has been recognized, trials of intervention have generally not been made in elderly people with or without diabetes. Nevertheless, it is reasonable to advise physical activity, appropriate diets, avoidance of cigarette smoking, and treatment of hypertension and hyperlipidaemia in elderly as well as young people with diabetes.

Amputation

Peripheral neuropathy, peripheral vascular disease, and infection contribute to the need for amputations. The development of peripheral neuropathy is usually the initiating condition. It contributes to the development of both foot deformities and insensitive feet. Foot deformities cause pressure points vulnerable to ulceration in insensitive feet. Peripheral vascular disease and infection can then lead to osteomyelitis and gangrene, which may ultimately necessitate amputation.

It is estimated that approximately half of the non-trauma amputations are for diabetic foot disease. Among persons over the age of 65, amputations are approximately 10 times more common among those with diabetes than among those without. Among individuals with diabetes the risk of amputation increases with age, male sex, cigarette smoking, and non-white race. For those with diabetes who have had an amputation, the risk of amputation of the contralateral leg is approximately 50 per cent after 4 years.

Prevention of amputation involves the prevention of peripheral neuropathy and peripheral vascular disease, and the prevention, early detection, and treatment of foot lesions. Several studies have demonstrated that improved foot care may reduce the frequency of amputations among those with diabetes. Programmes that have included education, increased clinical services, and comprehensive podiatry (chiropody) services have produced a 70 per cent reduction in amputations.

At each visit, the health-care provider should ask the patient to remove the shoes and socks and inspect the feet for lesions. Elderly diabetic patients with risk factors for amputation and their care-providers should be taught to examine the feet daily, to care for the feet, and to seek medical help if a lesion develops on the foot.

APPROACH TO MANAGEMENT

The treatment of diabetes mellitus in older patients uses the same four therapies as in younger diabetic patients (diet, exercise, oral agents, and insulin), but the approach to the older patient is somewhat different. For younger patients, a team approach for care of their diabetes has been recommended; for the older adult with diabetes a team approach often becomes imperative. The team approach takes contributions from many health-care providers including a nurse, physician, social worker, and dietitian, as well as the patient and the family. Team management of the older diabetic patient should begin at the time of diagnosis, not just when the patient requires insulin. All members of the team should participate to a varying extent in each phase of management—initial assessment, goal setting, treatment, and monitoring of therapy.

The means of monitoring the blood glucose at home should be made available to all patients with diabetes mellitus, whether they require insulin or not. If the patient is unable to do the monitoring, some other member of the patient's support system should be taught to do it. The frequency of monitoring depends on the patient's needs. For diet-controlled patients, or patients who are taking oral hypoglycaemic medications, once or twice a week may suffice. For insulin-treated patients, monitoring may be necessary several times each day, as it is with younger patients.

Initial assessment

The initial assessment of the older adult with diabetes should begin with a thorough history and physical examination. The purpose of this is threefold: (1) to evaluate the current extent of the disease; (2) to assess the risk factors for future complications of diabetes mellitus; and (3) to recognize other medical problems that might influence diabetes management, such as dementia or other chronic illnesses. The initial medical assessment should pay close attention to those end organs that may have already been damaged by long-standing undetected diabetes mellitus: the central nervous system (evidence of transient ischaemic attacks or cerebrovascular accidents), the eyes (glaucoma, cataracts, retinopathy), the cardiovascular system, including both the heart (myocardial ischaemia, infarction, or congestive heart failure) and the peripheral circulation (claudication, foot ulcers), the kidneys (proteinuria, impaired renal function), and the peripheral nervous system (signs and symptoms of both autonomic and peripheral neuropathy). Special attention should also be given to the blood pressure, level of blood lipids, and any history of tobacco use.

In addition to the medical history and physical examination, many elderly patients need a functional assessment that evaluates their capacity for self-care. This should discover what a patient actually does, not what they might or should be able to do. It should assess the patient's capability for the instrumental activities of daily living such as shopping, food preparation, using the telephone, self-administration of medications (both oral agents and insulin), and blood glucose monitoring, as well as the basic activities of daily living, which includes feeding, dressing, bathing, toileting, transferring, and ambulation. The functional assessment can be done by a physician, nurse, or trained assistant, and sometimes is best done in the patient's own home. The third portion of the initial assessment is that of a psychosocial evaluation, which considers the patient's support systems and assesses needs for community services such as meals-on-wheels, visiting nurses, or home health aids. While these items are also important for the younger individual with diabetes, the initial recognition and treatment of these issues in an elderly individual may mean the difference between independence and nursing home care.

Establishing treatment goals

Once the initial medical, functional, and social assessments have been completed, the next aspect of care for the older adult with diabetes is the establishment of appropriate treatment goals. For the care-givers of older diabetic patients, this may be an extremely difficult task as there is little scientific information available to guide decision-making. Some guidelines to help in this process are provided in Table 4. No studies exist which compare the efficacy and risks of tight diabetic control (fasting plasma glucose less than 8 mmol/l) for individuals above the age of 65. However, there are some proposed guidelines that may be acceptable for many patients.

For the 'younger' geriatric patient (65–75 years) with few other major medical problems, fairly tight control of blood glucose may be the most appropriate goal, as these patients have a life expectancy that will allow for the development of the chronic complications of diabetes mellitus. While the relationship between control of hyperglycaemia and macrovascular complications of diabetes mellitus has not been defined, several studies suggest that improved control of hyperglycaemia does reduce the risk of microangiopathic complications. To the extent that these studies can be extrapolated to elderly patients with diabetes, it seems prudent to aim for better control in those elderly patients who are otherwise generally healthy. For the older diabetic patient with multiple medical problems and reduced life expectancy, simple prevention of symptoms from hyperglycaemia is probably the appropriate therapeutic goal.

Table 4 Factors to consider for diabetes management in elderly patients

Patient's estimated remaining life expectancy
Patient preference and commitment
Belief of the primary-care provider: control vs. complications
Availability of support services
Economic issues
Coexisting health problems
Major psychiatric disorder
Major cognitive disorder
Major limitation of diabetes functional status
Complexity of medical regimen

After setting goals appropriate to the patient, the team should determine which treatment options may provide the desired results. Generally a mix of diet, exercise, and medications are needed, although some patients may be managed with diet alone. Each form of treatment has its own risks and benefits, which differ somewhat for older patients.

Therapeutic options

Diet

As in younger patients, diet is a major component of the treatment of diabetes mellitus. A weight-reducing diet may be appropriate for older patients who are overweight. However, it is unrealistic to expect any more success with weight-loss regimens than in younger patients. In fact, it may be more difficult for elderly individuals who are inactive to lose weight, as severe caloric restriction may be necessary. Current dietary recommendations are for a diet that is 50 to 60 per cent carbohydrate (most should be complex carbohydrates), no more than 30 per cent fat (less than 10 per cent as saturated fat), with the remainder as protein. High fibre content may also reduce postprandial glucose excursions. Diets should not be restricted to a level that interferes with good nutrition, and should take into account the patient's long-standing dietary pattern and food preferences. Some patients may have the most success with simplified dietary instruction that does not emphasize caloric restriction but rather limits high-fat foods and simple carbohydrates. Proper dietary counselling requires the services of a registered dietitian who can monitor the patient's eating habits, recommend a diet appropriate to the patient, and provide follow-up care to determine the efficacy of the dietary programme.

Exercise

The use of exercise as a means of treatment for the control of glycaemia continues to be controversial. Although no studies have specifically addressed the role of exercise in elderly diabetic patients, benefits demonstrated by studies of exercise programmes have included improvement in peripheral insulin sensitivity and a relative increase in the proportion of lean body mass. Exercise also lowers both serum triglycerides and low-density lipoproteins, increases high-density lipoproteins, and lowers both blood pressure and body fat, thus modulating risk factors for cardiovascular disease. These advantages, to the extent that they can be extrapolated to elderly patients with diabetes, suggest that a prudent exercise programme, such as walking for half an hour three times a week, may be of benefit.

However, exercise must be recommended cautiously as there may be substantial risks. Patients treated with insulin or oral hypoglycaemic agents may develop hypoglycaemia with vigorous exercise. Both coronary artery disease and peripheral vascular disease may be silent in the sedentary older adult with diabetes mellitus. Patients with peripheral neuropathy are at risk for developing skin breakdown and joint injury. No elderly patient should begin an exercise programme without a careful medical evaluation and an exercise tolerance test. However, most older adults with diabetes can perform some form of aerobic exercise and should be encouraged to do so.

Oral agents

When diet and exercise are not sufficient to reach therapeutic goals, then treatment with medication is necessary—either insulin or oral hypoglycaemic agents. Effective oral hypoglycaemic agents currently available include sulphonylureas and the biguanides, with sulphonylureas being the agents most commonly used.

Sulphonylureas act by stimulating insulin secretion, suppressing hepatic glucose production, and enhancing glucose uptake in peripheral tissues. There is no evidence to suggest that these agents act differently in older than in younger patients. Most older patients with non-insulin dependent diabetes are candidates for therapy with oral agents. The ease of administration (usually once a day) and the low incidence of side-effects make these agents attractive. However, there are risks for older patients associated with the use of sulphonylureas. The most significant risks are those of hypoglycaemia and hyponatraemia, both of which can cause changes in mental status and delay appropriate treatment. Other side-effects include allergic reactions, bone marrow toxicity, a disulfuram-like flushing reaction when chlorpropamide is combined with alcohol, hepatotoxicity with cholestatic jaundice, and gastrointestinal symptoms. The question of an increased incidence of cardiovascular deaths with the use of sulphonylureas has never been completely answered, but the ease of use of these drugs and low incidence of other side-effects would seem to outweigh this possible risk for most patients.

The agents currently available have been well outlined in *The Oxford Textbook of Medicine* and will not be repeated here. When choosing therapy for older patients, the most important points to consider are duration of action, site(s) of metabolism, drug interactions, and drug-specific side-effects. Drugs to be avoided are those with longer half-lives, those that interfere with water metabolism to cause hyponatraemia (especially when combined with thiazide diuretics), and, for those individuals with hepatic disease, the drugs that are extensively metabolized in the liver. When continuing therapy with sulphonylureas, the health-care provider should be certain that the agent and the dose are achieving the goals established for the patient. When the maximum dose of one sulphonylurea has not achieved the therapeutic goal, it is unlikely that changing to a different sulphonylurea will further improve the control of glycaemia. There is no currently defined role for the combined use of sulphonylureas and insulin. Added efficacy has been reported when sulphonylureas are combined with a biguanide.

The second group of oral hypoglycaemic agents available in some parts of the world are the biguanides. These act by impairing ion exchange across cell membranes, and have the effect of decreasing hepatic glucose production and increasing the use of glucose in peripheral tissue. They are used primarily in combination with sulphonylureas when a combination of sulphonylurea plus diet has failed in a patient who wishes to remain on oral therapy. Both the profile of side-effects (lactic acidosis, weakness, muscle pain, and gastrointestinal discomfort) and the primary route of excretion (renal) makes these drugs less than optimal choices for the treatment of elderly patients. Metformin is the only biguanide available in the United Kingdom and none of these agents is available for use in the United States.

Insulin

Insulin therapy is appropriate for any patient for whom treatment goals cannot be met by diet, exercise, and oral agents. This means that insulin can and should be used for patients whose goal is tight glycaemic control and for patients who require symptomatic relief not achieved by other means. The advantage of insulin therapy is the potential for excellent glycaemic control with adjustable dosing to cover changes in diet, exercise, and health status. The disadvantages of insulin therapy include the difficulties of administering it, the necessity for frequent blood glucose monitoring at home, and the potentially increased risk of hypoglycaemia.

Multiple skills are needed for patients to be independent in the administration and monitoring of insulin. Many of these skills, which are taken for granted in younger individuals with diabetes, must carefully be assessed in older diabetic patients. They include sufficient cognitive function to manage a complex regimen involving insulin and diet, adequate sight for reading labels, syringes, and reagent strips or reflectance meters, and fine motor control to draw up insulin and administer it. Conditions both related and unrelated to diabetes may cause impairments in any of these areas.

Deficiencies in self-care skills do not contraindicate the use of insulin therapy for those patients that require it. A week's supply of filled syringes can be left in the refrigerator by a family member, friend, or visiting nurse. Also available are prefilled cartridges that fit into pen-type syringes, with each click of the pen administering a preset dose of insulin. Jet guns have been marketed for patients who have difficulty with needle injections, and can reliably deliver insulin subcutaneously. Newer reflectance meters for monitoring blood glucose at home are simple to use and have large-print readouts for patients who are visually impaired. Also available are 'talking' meters. The successful approach to the older patient who newly requires insulin involves thoughtful anticipation of difficulties. Sometimes the best option is to admit the patient to hospital for initial diabetes control and education. Here, under careful observation, the necessary skills can be taught or reasonable alternatives developed. The decision to admit for insulin therapy must be made on an individual basis, dependent upon the patient, the patient's support system, and the community health services available to support the institution of insulin therapy as an outpatient.

Bibliography

Andres, R. (1971). Aging and diabetes. *Medical Clinics of North America*, **55**, 835–45.

Davidson, M.B. (1979). The effect of aging on carbohydrate metabolism:

a review of the English literature and a practical approach to the diagnosis of diabetes mellitus in the elderly. *Metabolism*, **28**, 688–705.

DeFronzo, R. (1981). Glucose intolerance and aging. *Diabetes Care*, **4**, 493–501.

Gerich, J.E. (1989). Oral hypoglycemic agents. *New England Journal of Medicine*, **321**, 1231–45.

Halter, J.B. and Christensen, N.J. (ed.) (1990). Diabetes mellitus in elderly people. *Diabetes Care*, **13** (suppl.2).

Harris, M.I., Hadden, W.C., Knowler, W.C., and Bennett, P.H. (1987). Prevalence of diabetes and impaired glucose tolerance and plasma glucose levels in U.S. population aged 20–74 yr. *Diabetes*, **36**, 530.

Herman, W.H. Teusch, S.M., and Geiss, L.S. (1985). Closing the gap: the problem of diabetes mellitus in the United States. *Diabetes Care*, **8**, 391–406.

Jackson, R.A. *et al.* (1982). Influence of ageing on glucose homeostasis. *Journal of Clinical Endocrinology and Metabolism*, **55**, 840–48.

Lipson, L.G. (ed). (1986). Diabetes mellitus in the elderly. *American Journal of Medicine*, **80** (Suppl. 5A).

Morley, J.E., Mooradian, A.D., Rosenthal, M.J., and Kaiser, F.E. (1987). Diabetes mellitus in elderly patients. Is it different? *American Journal of Medicine*, **83**, 533–44.

Morrow, L.A. and Halter, J.B. (1988). Carbohydrate metabolism in the elderly. In *The endocrinology of aging*, (ed. J.R. Sowers and J.V. Felicetta), p. 151. Raven Press, New York.

National Diabetes Data Group (1985). *Diabetes in America: diabetes data compiled 1984*, NIH publication 85–1468. National Institutes of Health, Bethesda, Maryland.

7.2 Disorders of the thyroid gland

PAUL J. DAVIS AND PAUL R. KATZ

Although the physiology of the pituitary–thyroid axis is altered in the course of normal ageing (Table 1), the ability of this axis to respond to stress is unchanged over the lifespan and circulating levels of thyroid hormone are maintained. Two age-dependent alterations in hormone physiology, however, are relevant to the diagnostic evaluation of patients suspected of having hyperthyroidism. First, a number of studies have shown that serum levels of tri-iodothyronine (**T_3**) decrease with age (Harman *et al.* 1984). It has been difficult to interpret these studies because of their failure to exclude the important effect of non-thyroidal illness on serum T_3 concentrations (Wartosfsky and Burman 1981), mediated by decreased conversion of circulating thyroxine (**T_4**) to T_3 in non-thyroidal tissues; the decline in T_3 in the most carefully studied, healthy aged subjects is very small (Harman *et al.* 1984). The serum total T_3 should not be used to screen subjects of any age for the presence of thyroid disease. In patients with thyrotoxicosis and an elevated serum T_4, the serum T_3 is sometimes normal or high-normal, rather than elevated, because of the severity of thyroid disease or the presence of coincident non-thyroidal illness (e.g. congestive heart failure or systemic infection) resulting in impaired peripheral conversion of T_4 to T_3. In patients with the syndrome of T_3-toxicosis—hyperthyroidism due exclusively to increased thyroidal secretion of T_3—the severity of the condition may blunt the rise in circulating levels of T_3, but these usually remain diagnostically elevated.

Second, the response of the pituitary to the systemic administration of thyrotropin-releasing hormone (**TRH**) may be sluggish in healthy elderly men (Ordene *et al.* 1983). Although this finding has been variable (Kabadi and Rosman 1987), the decreased response has been encountered in enough healthy aged men to view this as a significant impediment to diagnosis of disease of the pituitary–thyroid axis. A suppressed TRH response (manifested by failure of circulating levels of thyroid-stimulating hormone (thyrotropin, **TSH**) to rise after administration of TRH) is characteristic of hyperthyroidism, hypopituitarism, or ageing in men. Thus, a normal TRH test in elderly men is diagnostically useful, as it excludes hyperthyroidism; a suppressed response in this population is consistent with thyrotoxicosis (or pituitary disease), but is not diagnostic.

Table 1 Human ageing and thyroid hormone economy

Decreased average peripheral turnover (disposal rate) of thyroid hormone
Decreased average thyroidal secretion of thyroid hormone
Normal serum thyroxine (T_4) concentration
Normal or minimally reduced serum tri-iodothyronine (T_3) concentration
Normal free thyroid hormone concentration
Normal serum basal TSH concentration
Normal thyroid-gland response to exogenous TSH
Decreased pituitary TSH secretory response in men to exogenous thyrotropin-releasing hormone
Normal basal metabolic rate (corrected for metabolic cell mass)

THYROID DISEASES

Examination of the thyroid gland

Physical examination of the thyroid gland in an elderly subject offers several distinctive features. First, localization of the gland on palpation can be difficult; the dorsal kyphosis which is common in ageing can result in seclusion of the lower poles or more below the suprasternal notch. The thyroid so affected is described as *en plongeant*. On the other hand, visual examination under side-lighting of the pretracheal area, with the patient's neck slightly extended, can, in the elderly patient with symmetrical loss of strap-muscle mass, display the gland's anatomy very adequately. Second, the prevalence of carotid artery bruits increases with ageing and these can be misinterpreted as of thyroid origin. Careful stethoscopic exploration of the carotid artery and the poles of the thyroid with the paediatric bell will usually result in correct determination of the origin of the sound. Third, goitre is invariably found in young patients with hyperthyroidism and is also expected in the setting of Hashimoto's thyroiditis in younger subjects. Goitre should not be expected in elderly hyperthyroid patients, however (see below), and older subjects with Hashimoto's disease and attendant hypothyroidism usually lack goitre.

Goitre

Endemic goitre and hereditary goitre

Endemic goitre (iodide lack) and sporadic goitre, e.g. due to ingestion of dietary goitrogens, occur in younger subjects and do not appear *de novo* in elderly people. Pendred's syndrome (Friis *et al.* 1988) is the only hereditary defect in thyroid hormonogenesis that is consistent with normal somatic maturation and thus may be first detected in adult patients. This defect in metabolism of iodide is associated with goitre, deafness (which is usually mild), and eumetabolism or very mild hypothyroidism. It is rare for Pendred's syndrome to be recognized for the first time in an elderly patient.

Thyroiditis

Granulomatous ('subacute thyroiditis'; de Quervain's thyroiditis) and acute thyroiditis are rare in older patients. Acute thyroiditis is caused by direct spread of viral or, occasionally, bacterial infections from the tracheopharyngeal region. Hashimoto's thyroiditis (chronic lymphocytic autoimmune thyroiditis), in contrast, occurs with appreciable frequency in later life. Antithyroid antibody titres (antithyroglobulin and antimicrosomal) may be elevated in up to a quarter of apparently normal subjects over the age of 60 years. Patients over the age of 60 amount to 7 per cent of those with Hashimoto's disease confirmed by needle biopsy of the thyroid (Furszyfer *et al.* 1970). Autopsy studies have disclosed a similar prevalence. Because goitre may not be prominent in the older subject with lymphocytic thyroiditis, thyroid aspiration or biopsy are less likely to be carried out to confirm the presence of the disease; thus, the discrepancy in apparent incidence of thyroiditis (antithyroid antibody titre versus biopsy) cannot be rigorously evaluated.

As in younger patients, the significance of Hashimoto's disease in elderly people lies in its progression to hypothyroidism. Agoitrous hypothyroidism in middle-aged and elderly patients is attributable to antecedent lymphocytic thyroiditis. Hashimoto's thyroiditis with eumetabolism is not specifically treated. Appropriate management of hypothyroidism, regardless of its origin, is particularly important in the older subject and is discussed below.

Thyroid nodules

Autopsy studies have established that ageing is accompanied by increasing nodularity of several endocrine glands, e.g. thyroid and adrenal cortex. This tendency to develop nodules in the thyroid is expressed primarily at the subclinical level ('micronodularity'), but the prevalence of macronodules ('nodular goitre'), both multiple and solitary, increases with age (Rojeski and Gharib 1985). The significance of this finding is the risk that such nodules are cancerous (see next section). The risk of thyroidal cysts and sudden haemorrhage into a cyst—with acute appearance of an enlarging nodule—does not appear to increase with age. Nodularity sometimes represents localized thyroiditis, but the incidence of this does not rise with age.

Thyroid cancer

In terms of age distribution, number of cases of thyroid carcinoma peak in the fifth and sixth decades of life (Fig. 1). Case-fatality rates of thyroid cancer increase with age (Fig. 2), despite the fact that the histological appearances of the tumour are identical in young and old subjects. The thyroid cancer with the worst prognosis, regardless of age, is anaplastic; it occurs primarily in patients over age 50 years (Nel *et al.* 1985). Invasive Hurthle-cell cancer of the thyroid also has a peak incidence in the sixth decade (Watson *et al.* 1984). Medullary carcinoma of the thyroid originates in the calcitonin-producing cells of the gland and is usually encountered as a part of multiple endocrine adenomatosis syndrome (MEA, type IIa) in younger patients. However, sporadic cases of medullary carcinoma do occur and the poor prognosis of the sporadic disease is expressed primarily in the elderly patient (Samaan *et al.* 1988). Thus, regardless of histological type, thyroid cancer has a substantially worse prognosis in old patients. Consistent with this is the specific observation that papillary and follicular carcinomas in elderly people are less responsive to surgical and radioiodine therapy (Samaan *et al.* 1983). That the incidence of metastases from thyroid cancer is higher in younger patients but the prognosis better than in older subjects is a striking observation. Five-year survival in patients under the age of 40 years with a single differentiated distant metastasis is 92 per cent; in patients over 40 it is 38 per cent (Ruegemer *et al.* 1988); data for multiple pulmonary metastases from thyroid tumours are similar (Samaan *et al.* 1985). Local metastatic disease is readily susceptible to resection, with good results in young patients but poor results in the older subject.

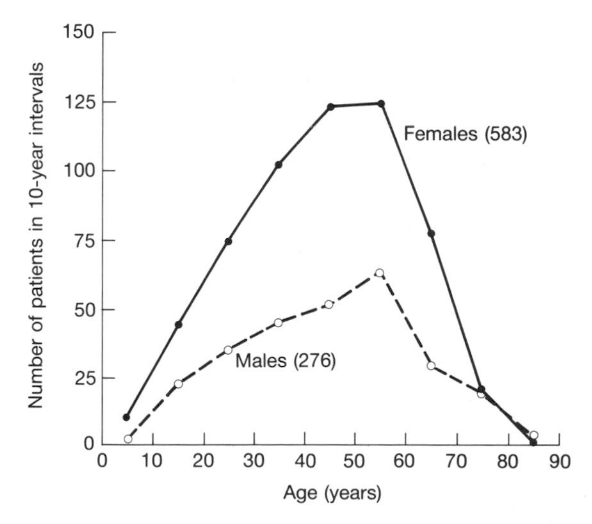

Fig. 1 Age distribution of patients with papillary carcinoma of the thyroid, treated at the Mayo Clinic, 1946–1970. (Reproduced from McConahey *et al.* (1986) with permission.)

Whether Hashimoto's thyroiditis is a risk factor for, or is protective against, thyroid cancer is not established. For example, one series reports an incidence of antithyroid antibodies in young patients with thyroid cancer that is four to five times that found in elderly patients with such tumours (Pacini *et al.* 1988), whereas data from the Mayo Clinic indicate that the absence of Hashimoto's thyroiditis increases the risk

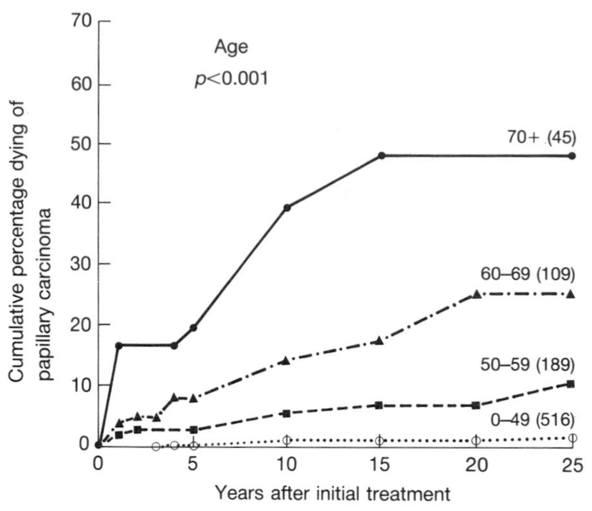

Fig. 2 Cumulative fatality, by age group, of papillary carcinoma of the thyroid evaluated at the Mayo Clinic. (Source as in Fig. 1.)

for papillary thyroid cancer in patients over the age of 50 years (McConahey *et al.* 1986). Such studies do not permit rigorous discrimination among such factors as histological typing, antibody type, and tumour behaviour. By 'tumour behaviour' is meant the relatively benign behaviour of metastatic thyroid cancer in young and its maleficent activity in old subjects. There may also be an increased risk of other tumours—such as myelo- and lymphoproliferative disorders, and thyroid lymphoma—in patients with Hashimoto's thyroiditis, but it is not clear that this is age-dependent (Holm *et al.* 1985).

Despite its worsening prognosis over the lifespan, thyroid cancer is treated in similar ways (surgery; radioiodine, where appropriate; chemotherapy of anaplastic and medullary carcinoma). It is generally agreed that suppression of endogenous secretion of TSH by administration of exogenous thyroid hormone (L-thyroxine)is indicated in elderly patients whose thyroid cancers have been approached surgically or with radioablation, as is the case in younger subjects. The reason for this is that TSH is thought to be a permissive factor in the emergence and/or progression of thyroid cancer. Suppression therapy appears to be less effective in elderly patients. The dose of L-thyroxine selected should be sufficient to reduce the serum concentration of endogenous TSH measured in a supersensitive radioimmunoassay to the lower limit of detectability.

HYPERTHYROIDISM

Aetiology

Epidemiological estimates of the incidence of hyperthyroidism are imprecise (Mogensen and Green 1980), but the disease is acknowledged to be common and, when classically expressed in any age group, readily recognized. Twenty per cent of hyperthyroid patients are over the age of 60 years. Subtle presentations of thyrotoxicosis occur in all age groups and in a quarter of affected aged patients (Davis and Davis 1974).

Thyrotoxicosis occurs in patients with diffuse thyroidal hyperplasia (diffuse toxic goitre, Graves' disease), nodular goitre, thyroiditis, or excessive administration of thyroid hormone. Graves' disease is an autoimmune disorder resulting from the production by lymphocytes of a TSH-receptor antibody (human thyroid-stimulating immunoglobulin, hTSIg) that is capable of inducing hyperactivity of the gland. A systematic, prospective study of the presence of hTSIg in the serum of elderly hyperthyroid patients has not been carried out. The incidence of diffuse enlargement on palpation or of diffuse radioiodide uptake on scanning normal-sized or non-palpable glands in older hyperthyroid patients is 25 to 50 per cent (Nordyke *et al.* 1988). Thus, it is expected that circulating levels of hTSIg will be raised in these patients. The mechanism of nodular toxic goitre remains controversial. Reports of levels of hTSIg and other immunoglobulins in this condition have been inconsistent (Grubeck-Loebenstein *et al.* 1985).

A relatively recently recognized form of hyperthyroidism in younger patients is painless thyroiditis with hypermetabolism. The pathophysiology of this syndrome involves autoimmune-mediated damage to the thyroid gland, with spillage of iodothyronines into the circulation. The condition is not associated with stimulating TSH-receptor antibodies, but with a Hashimoto's thyroiditis-like pattern of serum antibodies to thyroid antigens. Because hTSIg is absent and endogenous pituitary TSH is suppressed by increased levels of circulating thyroid hormone, thyroidal radioiodide uptake is very low, in contrast to hTSIg-mediated thyrotoxicosis (Nikolai *et al.* 1980). Hyperthyroidism associated with painless thyroiditis is a transient state of hypermetabolism, usually requiring minimal symptomatic therapy. The question of whether painless thyroiditis with hypermetabolism occurs with appreciable frequency in elderly people has not been specifically addressed in the published record. From available evidence, this syndrome is rare in older subjects.

Iodine-induced thyrotoxicosis (Jodbasedow) is another form of hyperthyroidism associated with little or no uptake of radioactive iodide by the thyroid gland. This usually but not invariably occurs with the introduction of iodide to geographical areas of decreased dietary iodide intake. The risk of iodine-induced thyrotoxicosis appears to increase with age, but the syndrome is rare.

The basis for monosystemic patterns of hyperthyroidism in elderly patients—e.g. clinical presentations limited to the cardiovascular system—is unclear, but raises the possibility that peripheral sensitivity to thyroid hormone may be altered organ-specifically. Studies in experimental animals indicate that iodothyronines have both direct and indirect (catecholamine-mediated, adrenergic receptor-dependent) effects on the myocardium (Ishac *et al.* 1983). A modest heightening of sensitivity of the myocardium to the action(s) of thyroid hormone could have important clinical effects with relatively small changes in circulating levels of iodothyronines and catecholamines. It has been suggested that older hypothyroid patients require lower doses of thyroid hormone replacement (100 to 125 μg) than younger hypothyroid subjects (150 μg or greater) (Davis *et al.* 1984) This is consistent with the concept of systemic increase in sensitivity to thyroid hormone in the aged, but dose-replacement data must be interpreted cautiously. Recent evidence indicates that, in the past, doses of thyroid hormone replacement in hypothyroid subjects of all ages have been excessive. The indicated replacement therapy for hypothyroidism is L-thyroxine. As biological potency may vary from one preparation to another, it is important to use a preparation from a reliable manufacturer.

A number of studies have shown that a small decrease in circulating levels of T_3 occurs in the course of apparently normal ageing; this observation is also consistent with the concept of increased tissue sensitivity over the human lifespan to this particularly active thyroid hormone analogue. The *in-vitro* response of human red-cell Ca^{2+}-ATPase activity to thyroid hormone diminishes with the age of healthy donors (Davis *et al.* 1987). This is a non-genomic (extranuclear) effect of thyroid hormone and cannot be extended, at this time, to important cell nucleus-mediated effects of iodothyronines. No categorical statement can yet be made about the sensitivity of peripheral tissue to thyroid hormone in the aged.

Despite studies emphasizing the importance of reducing the dose of thyroid hormone replacement in elderly patients, and despite the impression that the sensitivity of peripheral tissue to thyroid hormone may be altered with age, it is not clear that iatrogenic hyperthyroidism has occurred with greater frequency in elderly patients during the course of routine management of hypothyroidism or during suppression of endogenous TSH when treating thyroid nodules. The consequences of overtreatment of elderly patients with thyroid hormone, however, are more severe than in younger subjects; exacerbation of heart failure or angina pectoris are the most obvious results.

Clinical features

Diffuse toxic goitre and nodular goitre account for the majority of cases of hyperthyroidism in elderly subjects, with women predominating in a ratio of 5–10:1. In up to one-third of elderly hyperthyroid patients, the duration of symptoms attributable to thyroid overactivity may be prolonged, often exceeding 1 year. The majority of patients, however, come to medical attention within a relatively short time after onset of disease. Whether or not older patients with toxic nodular goitre experience a more indolent course than those with Graves' disease remains uncertain.

Many older hyperthyroid patients present with signs and symptoms indistinguishable from those experienced by their younger counterparts. Furthermore, the range of serum T_4 is similar in young and old thyrotoxic patients (Tibaldi *et al.* 1986). The presentations of hyperthyroidism due to Graves' disease and to toxic nodular goitre are similar in elderly patients; although malignant exophthalmos and thyrotoxic dermopathy are exclusive manifestations of Graves' disease, they are infrequently encountered in older hyperthyroid patients.

Unfortunately a large number of older patients suffer from a variety of acute and chronic diseases that often serve to distract the clinician's attention away from the possibility of thyroid disease, and so diagnostic delay is common.

As many of the symptoms reported by older thyrotoxic patients lack specificity, the clinician must have a high index of suspicion if he/she is to make the diagnosis early. Symptoms such as generalized weakness, shortness of breath, anxiety or nervousness, and palpitations are frequent in patients with hyperthyroidism, accounting for 4 of the 20 most commonly reported symptoms mentioned by a sample of euthyroid persons of 75 years and older during visits to the physician (White *et al.* 1986). Similarly, many older individuals are on multiple medications, some of which may mask symptoms of thyrotoxicosis and further contribute to a delay in diagnosis. Beta-adrenergic blocking drugs, for example are the sixth most frequently prescribed drug in elderly people and oppose many of the peripheral manifestations of thyroid hormone. Although thyrotoxicosis may be precipitated by the administration of iodinated radiographic dyes or iodine-containing drugs (e.g. amiodarone), this occurs in patients with multinodular glands that are relatively iodine-deficient. The likelihood of encountering drug-induced thyrotoxicosis in developed countries is extremely low because there is enough iodide in the diet.

In addition to non-specific symptoms, elderly hyperthyroid patients may manifest few, if any, of the classical symptoms and signs of thyrotoxicosis. Many thyrotoxic symptoms, such as weight loss, fatigue and dyspnoea, may be erroneously attributed to 'normal ageing', as may signs such as thinning and smoothness of the skin.

Certain of the presenting signs and symptoms of hyperthyroidism in elderly subjects deserve special mention (Table 2). Goitre, one of the classic signs of thyrotoxicosis, is absent in 25 to 50 per cent of all elderly patients with hyperthyroidism (Nordyke *et al.* 1988). In part, this may be secondary to kyphosis and age-associated changes in chest-wall diameter that lead to substernal displacement of the thyroid gland. Like exophthalmos and Graves' dermopathy (pretibial myxoedema), onycholysis and gynaecomastia are encountered infrequently in elderly people with hyperthyroidism.

As noted previously, symptoms of hyperthyroidism in elderly patients often suggest a 'monosystemic' rather than a constitutional or multisystemic illness; the pathogenesis of monosystemic hyperthyroidism is not well understood (Griffin and Solomon 1986). Predominantly, cardiac, gastrointestinal, or neuropsychiatric signs and symptoms are the usual clinical complexes encountered in patients with monosystemic disease.

Table 2 Frequency of signs and symptoms of hyperthyroidism (%)

	Young patients[a] (n=247)	Elderly patients[b] (n=85) (60–82 years)	Elderly patients[c] (n=25) (75–95 years)
Palpitation	89	63	36
Hyperhidrosis	91	38	—
Goitre	100	55	24
Tremor	97	69	8
Weight loss	85	39	44
Atrial fibrillation	10	39	32
Eye signs	71	57	12

[a]Williams (1946); [b]Davis and Davis (1974); [c]Tibaldi *et al.* (1986).

Cardiac findings

Although palpitation is commonly reported by elderly hyperthyroid subjects, it is often in the context of other major complaints. In fact, tachycardia is found in only half of elderly patients in contrast to virtually all young thyrotoxic subjects. Even though atrial fibrillation is common in elderly hyperthyroid patients and is encountered three to four times more often than in younger thyrotoxic subjects (Tibaldi *et al.* 1986), it is often associated with a normal or even low ventricular response. Presumably, this is on the basis of pre-existing disease of the myocardial conduction system. Symptoms of congestive heart failure are extremely common, occurring in up to 60 per cent of hyperthyroid elderly patients. In such patients, high-output cardiac failure and normal functioning heart valves may provide the clues to the presence of underlying hyperthyroidism. Refractory congestive heart failure and even acute pulmonary oedema are not uncommon. While hyperthyroidism may exacerbate angina, myocardial infarction in patients with thyrotoxicosis is rare.

Gastrointestinal findings

In contrast to younger patients, elderly thyrotoxic subjects usually do not complain of polyphagia. Anorexia, however, may occur in over 30 per cent of older patients. Surprisingly, constipation is encountered as frequently as diarrhoea and increased stool frequency, and may represent a pre-existing condition little affected by thyroid dysfunction. The symptom complex of weight loss, anorexia, and constipation occurs in up to 15 per cent of elderly hyperthyroid individuals, mimicking a malignancy of the gastrointestinal tract (Davis and Davis 1974).

Nutritional state and hyperthyroidism

Nutritional status has not been studied systematically in elderly thyrotoxic patients. In younger patients with hyperthyroidism, the plasma amino-acid profile is normal, despite the catabolic nature of the illness. Increased dietary intake in this younger group apparently compensates for hypermetabolism. The appreciable frequency of anorexia and decreased dietary intake in older individuals with thyroid hyperfunction encourages speculation that such patients may be inadequately nourished when thyrotoxic. Anaemia is uncommon in elderly hyperthyroid subjects. Trace-metal metabolism is disordered in hyperthyroidism, but it is not clear that the changes are clinically significant in any age group. Metabolic bone disease in hyperthyroidism in all age groups appears to be in the form of loss of bone mass (osteoporosis) in the axial skeleton (Seeman *et al.* 1982). Whether this loss significantly contributes to the progressively increased risk of vertebral crush fracture in ageing is not yet clear. Hypercalcaemia rarely occurs in elderly thyrotoxic patients but may occasionally be a presenting feature.

Neuromuscular findings

The neuromuscular manifestations of hyperthyroidism in elderly patients are varied and often suggest serious underlying non-endocrine illness. Tremor, although usually present in an elderly patient with thyrotoxicosis, is also common in euthyroid older people. The tremor most clearly resembles the physiological variety but, if it is coarse, may be confused with that of Parkinson's disease. Thyrotoxicosis may also be associated with weakness of proximal muscles, myalgia, and muscle cramps with normal or very low serum levels of creatine phosphokinase activity. Whereas myopathy is usually found in younger thyrotoxic patients on careful examination, such signs are found in only about 40 per cent of elderly patients (Davis and Davis 1974). Hyperthyroidism may occasionally present as weakness of the bulbar musculature with resulting dysphagia and ptosis. If associated with upper motor-neurone signs, these may be misinterpreted as due to cerebrovascular disease. Hyperactive deep-tendon reflexes are unusual in elderly patients with thyroid overactivity, and the characteristic rapid relaxation phase occurs in only a quarter of such patients (Davis and Davis 1974). Seizures may occasionally be the presenting manifestation of hyperthyroidism.

Psychiatric findings

In view of the high prevalence of psychiatric illness in the elderly population (10 per cent with depression; 5–10 per cent with dementia), the psychiatric findings in those with hyperthyroidism are noteworthy (White *et al.* 1986). Although there is no one distinctive pattern, emotional lability, nervousness, and a decreased attention span are common. Hyperthyroidism may also present as depression in elderly subjects. In such instances, this is often of the agitated variety and presents in an acute or subacute fashion and without historical precedent. At times, depression in hyperthyroidism may present in an 'apathetic form', in which patients manifest a retarded or stuporous depression. These patients may appear demented (Hall and Beresford 1988). As hyperkinetic features are present in only a quarter of elderly thyrotoxic patients, arriving at an accurate diagnosis in these older individuals is fraught with difficulty.

Laboratory findings

A diagnosis of thyrotoxicosis in an elderly patient is confirmed by the finding of an elevated serum T_4 in the presence of normal levels of serum thyroxine-binding globulin (TBG) and an additional, supportive test of thyroid function. Increases in TBG result in elevations of serum T_4 and are excluded by the routine use of the tri-iodothyronine (T_3) resin uptake or its derivative indexes. Occasionally the serum T_4 is normal in thyrotoxic patients in whom hypermetabolism is mediated exclusively by high circulating levels of T_3. So-called T_3-toxicosis affects a small minority (5 per cent) of thyrotoxic elderly patients. Conversely, the serum T_4 may be transiently elevated in the presence of acute illness, when in fact a true hyperthyroid state is not present. The frequency of isolated hyperthyroxinaemia in acute non-thyroidal illness may range as high as 15 per cent, although the exact frequency in elderly subjects has not been determined. Increases in T_4 concentration in the presence of non-thyroidal illness have been attributed to changes in binding-protein concentrations, in the affinity of TBG for T_4, or to a decline in the extrathyroidal metabolism of T_4 to T_3. (Borst *et al.* 1983). Differentiating authentic toxicosis from euthyroid ill patients with transient rises in serum T_4 can be difficult. Serum concentrations of free T_4 are modestly increased in 20 to 40 per cent of patients with non-thyroidal illness and do not discriminate between euthyroid hyperthyroxinaemia and thyrotoxicosis unless the concentrations are convincingly elevated (greater than twice normal).

The recent introduction of ultrasensitive immunoradiometric assays for TSH has simplified the diagnosis of hyperthyroidism. The finding of suppressed serum TSH in these assays greatly supports the diagnosis of hyperthyroidism, and the test is likely to be used to screen patients for both hyper- and hypothyroidism in the future. Undetectable levels are consistent not only with hyperthyroidism but also may be found in clinically euthyroid patients with autonomously functioning thyroid nodules and in some patients taking L-thyroxine (Ross 1987). The supersensitive TSH assay has obviated the need for diagnostic testing by TRH stimulation in most instances.

The thyroidal radioactive iodide-uptake test is a reliable confirmatory test in elderly patients with thyrotoxicosis. It is currently used primarily to determine the dose of radioactive iodide for therapy or to assess the possibility of the presence of painless thyroiditis. The iodide-uptake test is low in cases of painless thyroiditis with hyperthyroidism or may be normal in patients with toxic nodular or diffuse goitre if iodine loading has occurred previously by means of diet, medications, or radiographic dyes. Additional laboratory tests, such as the serum cholesterol, haemoglobin, calcium, and alkaline phosphatase, may be abnormal in hyperthyroidism but are not diagnostically useful.

Treatment

Several options are available both for the management of acute thyrotoxicosis and for long-term definitive therapy. Decisions should be highly individualized and based not only on the degree of thyrotoxicosis present at the time of diagnosis, but also on concomitant disease states and functional status that may affect the risks of surgery, pharmacological intervention, or thyroid ablation with radioactive iodide. Treatment is not, however, determined by the aetiology of the hyperthyroidism, nor is age alone a major determinant.

The mainstays of medical therapy are the thioamides propylthiouracil or methimazole (carbimazole). Although their onset of action is 2 to 4 weeks, they are administered early to block hormonogenesis and maintain steady-state in the later, non-acute phase. Propylthiouracil and methimazole do not prevent release of thyroid hormone already present in the gland, but propylthiouracil does inhibit peripheral conversion of T_4 to the more active T_3. Elderly patients on thioamides should be monitored closely for development of granulocytopenia, for they are more susceptible to marrow suppression than are younger patients. The rate of spontaneous remissions of hyperthyroidism during propylthiouracil therapy has not been well studied in elderly patients.

In elderly patients, ablative radioactive iodide is preferred over other options, particularly in those with other chronic illnesses and on multiple medications (Griffin and Solomon 1986; Levin 1987). Long-term therapy with propylthiouracil is used when the patient rejects therapeutic administration of radioactive iodide. In elderly patients with mild thyrotoxicosis, beta-adrenergic blockade may be used alone to control most of the symptoms and signs of hyperthyroidism before, during, or after the administration of ablative radioactive iodide.

The acute management of severe hyperthyroidism involves specific treatment of tachycardia, congestive heart failure, and fever, all of which reflect the effects of thyroid hormone on peripheral tissues. In addition, pharmacological therapy to inhibit synthesis and release of thyroid hormone is indicated. Rapid achievement of a euthyroid state is particularly desirable in elderly patients in whom atherosclerotic heart disease is prevalent and the risks of a prolonged hypermetabolic state significant.

In addition to its other roles in treating thyrotoxicosis, beta-adrenergic blockade is recommended in those with ventricular rates over 140 beats per minute (and perhaps lower), regardless of whether the underlying rhythm is sinus or atrial fibrillation. An obvious risk is present when tachycardia is accompanied by congestive heart failure. Beta-adrenergic blockade should then be used only with caution.

In rare cases of severe hyperthyroidism, inorganic iodide may be used to provide early inhibition of thyroid hormone production within the gland. However, one must recognize that the use of inorganic iodide precludes the use of therapeutic ^{131}I for several months. The onset of iodide action on hormone release and organification is within hours. Short-term use of corticosteroids is indicated in thyroid storm (fever, tachycardia, exaggerated findings of thyrotoxicosis). Among their many actions, corticosteroids block peripheral conversion of T_4 to T_3 but whether this is clinically important is unclear. Beta-blockade is the cornerstone of treatment of severe thyrotoxicosis, but inorganic iodide and corticosteroids are usually used as well.

Radioactive iodide therapy is effective and simple to administer, but several points must be kept in mind when treating older patients. It is desirable to control the patient's hyperthyroidism before radioiodide ablation of the thyroid. Thyroid storm has occurred in ^{131}I-treated thyrotoxic patients who were not protected with beta-blockers or who had not been given a thioamide to achieve the euthyroid state. Low-dose beta-blockade beginning 1 week before ablative ^{131}I is sufficient to prevent the occasional exacerbation of hyperthyroidism that can occur until 1 to 2 weeks after administration of the isotope. If thioamide therapy is to be used as pretreatment, it must be discontinued 3 days before the administration of radioactive iodine. Beta-adrenergic blockade has largely replaced thioamide in this role.

Overall, treatment of elderly patients with radioactive iodide ablation is effective and carries very low risk if the above-mentioned precautions are taken. More than 60 per cent of elderly hyperthyroid patients are rendered euthyroid or hypothyroid by a single dose of ^{131}I of 185–370 MBq. This susceptibility to the effect of radioiodine primarily reflects a patient subset with toxic diffuse goitre (Graves' disease). Control of toxic nodular goitre usually requires a larger dose of ^{131}I or multiple doses. Some clinicians induce hypothyroidism by intent in patients with diffuse or nodular toxic goitre through the use of a single large dose of radioiodine. This approach maximizes the therapeutic response to one dose of ^{131}I, but confers the requirement for lifelong thyroid hormone-replacement therapy. Follow-up is important but hypothyroidism after smaller doses of radioactive iodide appears to develop less frequently in elderly than in younger subjects, at least in short-term follow-up (Kendall-Taylor *et al.* 1984). Thyroidectomy is also an acceptable alternative for controlling hyperthyroidism in the older patient. Surgery is indicated particularly in patients with large goitres.

HYPOTHYROIDISM

Aetiology

Spontaneous hypothyroidism is a disease of elderly women. Seventy per cent of hypothyroid patients are more than 50 years of age and 80 per cent are women. This population of hypothyroid subjects is usually without both goitre and serum antithyroid antibodies. In contrast, Hashimoto's thyroiditis is a smaller, age-related mode of spontaneous hypothyroidism that occurs in adolescent girls and in women below middle-age who are goitrous, and is marked by the presence of circulating titres of antibodies specific for thyroid antigen(s), such as thyroid peroxidase ('microsomal antibody') or thyroglobulin. The goitre that occurs in these younger patients is a result of the increase in circulating TSH released by the pituitary in response to the lowered output of T_4 by the gland affected with thyroiditis. It is widely believed that spontaneous hypothyroidism in elderly subjects is also due to thyroiditis that developed earlier in life and that was associated with progressive destruction of the gland, leaving insufficient gland mass to respond to elevations in TSH. Destruction of the gland leaves no antigen late in life to continue to incite B-lymphocyte responses. Hypopituitary ('secondary') hypothyroidism occurs in less than 3 per cent of elderly hypothyroid patients.

Clinical findings

The clinical hallmarks of thyroid underfunction include lassitude, constipation, cold intolerance, and psychomotor retardation; they occur in both young and elderly hypothyroid patients. In the older subject, however, these symptoms, either singly or as a constellation, may be accepted by patient and physician as concomitants of 'normal ageing'. Previous medical histories of Hashimoto's thyroiditis, radioiodide or surgical treatment of hyperthyroidism, surgery for non-toxic goitre, or high-dose external irradiation of the neck, e.g. mantle irradiation for lymphoma, are obvious risk factors for hypothyroidism; difficult childbirth, particularly with intrapartum haemorrhage, may presage hypopituitary hypothyroidism several decades later. Altered behaviour patterns in an elderly person must raise the possibility of the presence of thyroid disease, both hypo- and hyperthyroidism. Certain pharmacological agents may promote hypothyroidism. These include lithium, carbutamide (an oral sulphonylurea available in continental Europe), and adjunctive interleukin therapy in the chemotherapy of certain types of cancer. In patients who are euthyroid but have histories of damage to the thyroid gland, e.g. Hashimoto's thyroiditis, excessive dietary iodide intake (as in the ingestion of kelp) may induce hypothyroidism.

Physical examination of the elderly hypothyroid patient may reveal the classical findings of psychomotor retardation, myxoedematous facies, bradycardia, and lengthened contraction and relaxation phases of the deep-tendon reflexes. All patients who present with core temperatures of 96.0°F (35.5°C) or lower should be screened for hypothyroidism. Interestingly, hypertension occurs in up to one-third of hypothyroid patients, and both supraventricular and ventricular tachyarrhythmias may occasionally be encountered. Hypoventilation, with attendant hypercarbia, may occur. Morbid obesity is rarely due to hypothyroidism. Although the accumulation of subcutaneous glycosaminoglycans and associated water (myxoedema) is non-pitting in quality, pitting oedema is seen in as many as 40 per cent of elderly hypothyroid subjects. Pitting oedema may represent concomitant disease—congestive heart failure of liver disease—or the increased vascular permeability of hypothyroidism. The exact nature of this defect remains unclear, but underlies in part the development of pleural and pericardial effusions and ascites in these patients. The accumulation of myxoedema may minimize skin wrinkling. Scalp hair is coarse in texture in younger hypothyroid patients, but can remain fine in elderly patients. {Strikingly well-preserved hair colour may be a feature of long-standing hypothyroidism in an elderly patient.} Unless euthyroid goitre (sporadic goitre) preceded the development of hypothyroidism, elderly subjects infrequently develop goitre with their hypothyroidism.

The cardiac examination may find brady- and tachyarrhythmias, muffled heart sounds—due to decreased myocardial contractility (infiltrative cardiomyopathy, reflecting myocardial accumulation of myxoedema) or to pericardial effusion—and the auscultatory features of asymmetrical septal hypertrophy. Pericardial effusion may be due to altered membrane permeability and include the accumulation of glycosaminoglycan; the effusion is rarely of haemodynamic significance. Reversible asymmetrical septal hypertrophy is an occasional complication of primary hypothyroidism (Santos *et al.* 1980).

The neuromuscular examination can be profoundly abnormal in hypothyroidism, although the most consistent findings in hypothyroid subjects of any age are delayed contraction and relaxation phases of the deep-tendon reflexes. Proximal, sometimes hypertrophic, skeletal myopathy may be seen. Cerebellar dysfunction and disordered cognition are manifestations of thyroid hypofunction more often encountered in the elderly than in younger subjects. Cerebellar signs include locomotor and truncal ataxia. Motor retardation is common and can be readily demonstrated by asking the affected patient to perform a simple motor task, such as buttoning a garment. Motor retardation usually improves with hormone replacement but the response of ataxic syndromes to thyroid hormone is unpredictable and cognitive problems frequently do not remit. Failure of therapy in this setting may reflect the concomitant presence of arteriosclerotic or other non-endocrine disease of the central nervous system that has progressed during a prolonged hypothyroid state.

The laboratory diagnosis of primary or thyroidal hypothyroidism in the aged patient, as in younger subjects, relies on the serum TSH assay. The serum T_4 by radioimmunoassay is low in both primary and secondary (hypopituitary) hypothyroidism; subsequent TSH assay distinguishes between the two syndromes. T_4 assay should always be carried out with a concomitant measurement of serum TBG. The latter can be measured directly by a relatively expensive specific radioimmunoassay or inferred from indirect assays of the saturation of hormone-binding sites on serum TBG (resin T_3 uptake test or other tests, which combine serum T_4 and TBG saturation measurements in 'index' tests). The serum T_3 concentration by radioimmunoassay is non-specifically lowered in non-thyroidal illness ('euthyroid sick syndrome' associated with impaired 'peripheral', i.e. extrathyroidal, conversion of T_4 to T_3), and this is not reliable in elderly subjects who may have accumulated several non-endocrine diseases of sufficient seriousness to reduce circulating T_3. The serum concentration of free T_4 is raised in non-thyroidal illness and thus may not be predictably low in the elderly patient with hypothyroidism and concomitant, but unrelated, systemic illness. Other laboratory tests, which do not directly measure hormone levels but can infer hypothyroidism, have included elevation of the serum total cholesterol, creatine phosphokinase, and lactic dehydrogenase activities, and decreased activity of angiotensin-converting enzyme. These do not provide reliable insights into metabolic state. The serum cholesterol content may be low in chronically ill, undernourished elderly hypothyroid subjects. Elevations of total creatine phosphokinase isoenzymes in serum are unpredictable in hypothyroidism; when present, they may suggest the concomitant presence of myocardial ischaemia. The MM creatine phosphokinase isoenzyme (of skeletal muscle origin) usually accounts for the rise in creatine phosphokinase in thyroid hypofunction, however.

Other laboratory tests that may be disordered in hypothyroidism, particularly in elderly patients, are the electrocardiogram, haematocrit/red-cell volume and haemoglobin concentration, serum carcinoembryonic antigen, serum sodium and arterial $P\text{CO}_2$. Disturbances of cardiac rhythm and rate are common in thyroid underfunction, as are non-specific changes in the ST–T wave on the electrocardiogram. Current of injury (localized ST-segment elevation) is not expected in hypothyroidism. Two-dimensional echocardiography can confirm the presence of hypothyroidism-associated asymmetrical septal hypertrophy.

Anaemia is common in hypothyroidism; frequently in women this is microcytic and hypochromic in character, reflecting iron deficiency that is unrelated to hypothyroidism and, in older women, had its origin in the premenopause. Macrocytic anaemia also occurs with appreciable frequency in hypothyroidism and may be a primary consequence of thyroid hypofunction perhaps due to alterations in the lipid content of the red-cell membrane or due to associated autoimmune pernicious anaemia. The latter is encountered in up to 10 per cent of patients with spontaneous (of presumed autoimmune origin) hypothyroidism. Circulating levels of carcinoembryonic antigen are mildly elevated in some patients with thyroid hypofunction, probably reflecting decreased metabolic clearance of the protein. This finding is of concern in patients with histories of colon carcinoma who then develop hypothyroidism independent of their bowel disease.

Hyponatraemia is frequent in hypothyroidism. It reflects decreased clearance of free water, which has been attributed to inappropriately normal or elevated circulating levels of arginine vasopressin in euvolaemic hypothyroid subjects, or to transcortical shunting of renal blood flow. Hypercarbia with or without significant respiratory acidosis can develop in hypothyroidism, reflecting central nervous system (medullary) insensitivity to CO_2 or, possibly, myopathy of respiratory musculature. Any elderly patient who presents with unexplained CO_2 retention should be screened for hypothyroidism.

Therapy

The treatment of hypothyroidism in an elderly patient can be difficult when significant disease of the coronary arteries is present. The frequency of heart disease in the aged general population is so great that management of thyroid hypofunction is usually tempered by the possibility of concomitant coronary arteriosclerosis. Whether hypothyroidism itself accelerates atherosclerosis remains a controversial issue. We believe that it is the hypertensive hypothyroid patient whose coronary artery disease is accelerated, but others view hypercholesterolaemia (for example, the secondary type IIa state seen in hypothyroidism or the mixed hyperlipidaemic type IIb state) as a mechanism of enhancement for vascular disease in elderly patients with thyroid underfunction.

Although it is possible to institute thyroid hormone replacement at full dosage levels (100–120 μg L-thyroxine daily) initially in stable hypothyroid subjects who have no evidence of heart disease, it is often impossible to exclude that disease and it is generally wise to institute a stepped hormone-replacement programme in elderly patients with moderate hypothyroidism. *Prima facie* evidence by history of coronary artery disease in newly diagnosed hypothyroid patients includes one or more of the following: angina pectoris, congestive heart failure, cardiomegaly, myocardial infarction, or cardiac tachyarrhythmia(s). A conservative, oral, stepped hormone replacement regimen is required when heart disease complicates thyroid hypofunction or when the medical history does not exclude the features of heart disease cited above. The regimen is 25 μg T_4 daily for 2 to 4 weeks with 25 μg increments (to 50 and 75 μg) at 2- to 4-week intervals thereafter. In patients with heart disease, 12.5-μg increments may be dictated by a changed pattern of angina pectoris or the supervention of cardiac arrhythmia or congestive heart failure. Full replacement therapy is achieved at 75–125 μg daily. That an appropriate (full replacement) dose of T_4 has been achieved is determined by serial measurements of the serum TSH. The dose should suppress the previously elevated serum TSH, as measured by a high-sensitivity method, into the normal (but not subnormal) range. {There is evidence that overtreatment with thyroxine has been common and one of the undesirable effects is an increased rate of postmenopausal bone loss (Stall *et al.* 1990).}

There is rarely, if ever, a role for replacement therapy with T_3 in hypothyroidism. We do not recommend combinations of T_4 and T_3 to treat hypothyroidism.

The finding of normal thyroid hormone concentrations in the presence of elevated thyrotropin levels has been termed subclinical hypothyroidism and occurs with increased frequency in elderly people. Prevalence estimates of subclinical hypothyroidism in community-dwelling elderly subjects range from 1 to 14 per cent (Drinka and Nolten 1988). Because only one double-blind, placebo-controlled trial addressing the suitability of replacement therapy for patients with this syndrome has been reported (Cooper *et al.* 1984), the management strategy is not established. The inevitability of later overt hypothyroidism is not clear in elderly patients with subclinical hypothyroidism, but some investigators advocate normalizing the TSH with cautious thyroxine replacement in the presence of high titres of thyroidal antibodies and marked elevation of TSH or in patients previously treated with ^{131}I. {In the United Kingdom current practice is usually simple follow-up with monitoring of thyroid function tests.}

Profound hypothyroidism ('myxoedema stupor', 'myxoedema coma') is a medical emergency marked by impaired cognition and hypothermia; hypotension may also be present. Therapy of such patients should be conducted in inpatient critical care areas and usually involves parenteral administration of T_4. In the previously diagnosed hypothyroid patient, altered sensorium, hypothermia, and hypotension are not specific for worsening hypothyroidism and may be due, for example, to septicaemia. On the other hand, patients who present with undiagnosed, mild hypothyroidism may lapse into myxoedema coma when non-thyroidal illness–sepsis or cardiovascular catastrophe–develops and results in enhanced turnover of the limited amount of available endogenous thyroid hormone.

References

Borst, G.C., Eil, C., and Burman, K.D. (1983). Euthyroid hyperthyroxinemia. *Annals of Internal Medicine*, **98**, 366–78.

Cooper, D.S., Halpern, R., Wood, L.C., Levin, A.A., and Ridgway, E.C. (1984). L-Thyroxine therapy in subclinical hypothyroidism. *Annals of Internal Medicine*, **101**, 18–24.

Davis, F.B., Lamantia, R.S., Spaulding, S.W., Wehmann, R.E., and Davis, P.J. (1984). Estimation of a physiologic replacement dose of levothyroxine in elderly patients with hypothyroidism. *Archives of Internal Medicine*, **144**, 1752–4.

Davis, P.J., and Davis, F.B. (1974). Hyperthyroidism in patients over age 60 years: clinical features in 85 patients. *Medicine*, **53**, 161–81.

Davis, P.J., Davis, F.B., Blas, S.D., Schoenl, M., and Edwards, L. (1987). Donor age-dependent decline in response of human red cell Ca^{2+}-ATPase activity to thyroid hormone *in vitro*. *Journal of Clinical Endocrinology and Metabolism*, **64**, 921–5.

Drinka, P.J. and Nolten, W.E. (1988). Review: subclinical hypothyroidism in the elderly: to treat or not to treat? *American Journal of Medical Science*, **295**, 125–8.

Friis, J., Johnsen, T., Feldt-Rasmussen, U., Bech, K., and Friis, T. (1988). Thyroid function in patients with Pendred's syndrome. *Journal of Endocrinological Investigation*, **11**, 97–101.

Furszyfer, J., Kurland, L.T., Woolner, L.B., Elveback, L.R., and McConahey, W.M. (1970). Hashimoto's thyroiditis in Olmstead County, Minnesota, 1935 through 1967. *Mayo Clinic Proceedings*, **45**, 586–96.

Griffin, M. and Solomon, D.H. (1986). Hyperthyroidism in the elderly. *Journal of the American Geriatrics Society*, **34**, 887–92.

Grubeck-Loebenstein, B. *et al.* (1985). Immunological features of non-immunogenic hyperthyroidism. *Journal of Clinical Endocrinology and Metabolism*, **60**, 150–5.

Hall, R.C., and Beresford, T.P. (1988). Psychiatric manifestations of physical illness. In *Psychiatry*, vol 2, (ed. J.O. Cavenar), pp 7–8. Lippincott, Philadelphia.

Harman, S.M., Wehmann, R.E., and Blackman, M.R. (1984). Pituitary–thyroid hormone economy in healthy aging men: basal indices of thyroid function and thyrotropin responses to constant infusions of thyrotropin-releasing hormone. *Journal of Clinical Endocrinology and Metabolism*, **58**, 320–6.

Holm, L-E., Blomgren, H., and Lowhagen, T. (1985). Cancer risks in patients with chronic lymphocytic thyroiditis. *New England Journal of Medicine*, **312**, 601–5.

Ishac, E.J.N., Pennefather, J.N., and Handberg, G.M. (1983). Effect of changes in thyroid state on atrial alpha- and beta- adrenoceptors, adenylate cyclase activity and catecholamine levels in the rat. *Journal of Cardiovascular Pharmacology*, **5**, 396–405.

Kabadi, U.M., and Rosman, P.M. (1987). Lack of influence of aging on hypothalmic–pituitary axis in normal subjects. *Journal of Clinical and Experimental Gerontology*, **9**, 189–99.

Kendall-Taylor, P., Keir, M.J., and Ross, W.M. (1984). Ablative radioiodine therapy for hyperthyroidism: long term follow up study. *British Medical Journal*, **289**, 361–3.

Levin, R.M. (1987). Thyrotoxicosis in the elderly. *Journal of the American Geriatrics Society*, **35**, 587–9.

McConahey, W.M., Hay, I.D., Woolner, L.B., van Heerden, J.A., and Taylor, W.F. (1986). Papillary thyroid cancer treated at the Mayo Clinic, 1946 through 1970: initial manifestations, pathologic findings, therapy, and outcome. *Mayo Clinic Proceedings*, **61**, 978–96.

Mogensen, E.F., and Green, A. (1980). The epidemiology of thyrotoxicosis in Denmark. *Acta Medica Scandinavica*, **208**, 183–6.

Nel, C.J.C. *et al.* (1985). Anaplastic carcinoma of the thyroid: a clinicopathologic study of 82 cases. *Mayo Clinic Proceedings*, **60**, 51–8.

Nikolai, T.F., Brosseau, J. Kettrick, M.A., Roberts, R., and Beltaos, E. (1980). Lymphocytic thyroiditis with spontaneously resolving hyperthyroidism (silent thyroiditis). *Archives of Internal Medicine*, **140**, 478–82.

Nordyke, R.A., Gilbert, F.I., Jr. and Harada, A.S. (1988). Graves' disease: influence of age on clinical findings. *Archives of Internal Medicine*, **148**, 626–31.

Ordene, K.W., Pan, C., Barzel, U.S., and Surks, M.I. (1983). Variable thyrotropin response to thyrotropin-releasing hormone after small decreases in plasma thyroid hormone concentrations in patients of advance age. *Metabolism*, **32**, 881–8.

Pacini, F. *et al.* (1988). Thyroid autoantibodies in thyroid cancer: Incidence and relationship with tumour outcome. *Acta Endocrinologica* (Copenhagen), **119**, 373–80.

Rojeski, M.T., and Gharib, H. (1985). Nodular thyroid disease: evaluation and management. *New England Journal of Medicine*, **313**, 428–36.

Ross, D.S. (1987). New sensitive immunoradiometric assays for thyrotropin. *Annals of Internal Medicine*, **104**, 718–20.

Ruegemer, J.J., Hay, I.D., Bergstralh, E.J., Ryan, J.J., Offord, K.P., and Gorman, C.A. (1988). Distant metastases in differentiated thyroid carcinoma: a multivariate analysis of prognostic variables. *Journal of Clinical Endocrinology and Metabolism*, **67**, 501–8.

Samaan, N.A. *et al.* (1983). Impact of therapy for differentiated carcinoma of the thyroid: an analysis of 706 cases. *Journal of Clinical Endocrinology and Metabolism*, **56**, 1131–8.

Samaan, N.A., Schultz, P. N., Haynie, T.P., and Ordonez, N.G. (1985). Pulmonary metastasis of differentiated thyroid carcinoma: treatment results in 101 patients. *Journal of Clinical Endocrinology and Metabolism*, **60**, 376–80.

Samaan, N.A., Schultz, P.N., and Kickey, R.C. (1988), Medullary thyroid carcinoma: prognosis of familial versus sporadic disease and the role of radiotherapy. *Journal of Clinical Endocrinology and Metabolism*, **67**, 801–5.

Santos, A.D., Miller, R.P., Matthew, P.K., Wallace, W.A., Cave, W.T. Jr., and Hinojosa, L. (1980). Echocardiographic characterization of the reversible cardiomyopathy of hypothyroidism. *American Journal of Medicine*, **68**, 675–82.

Seeman, E., Wahner, H.W., Offard, K.P., Kumar, R., Johnson, W.J., and Riggs, B.L. (1982). Differential effects of endocrine dysfunction on the axial and appendicular skeleton. *Journal of Clinical Investigation*, **69**, 1302–9.

Stall, G.M., Harris, S., Sokoll, L.J., and Dawson-Hughes, B. (1990). Accelerated bone loss in hypothyroid patients overtreated with L-thyroxine. *Annals of Internal Medicine*, **113**, 265–9.

Tibaldi, J.M., Barzel, U.S., Albin, J., and Surks, M. (1986). Thyrotoxicosis in the very old. *American Journal of Medicine*, **81**, 619–22.

Wartofsky, L. and Burman, K.D. (1981). Alterations in thyroid function in patients with systemic illness: the 'euthyroid sick syndrome'. *Endocrine Reviews*, **2**, 396–436.

Watson, R.G., Brennan, M.D., Goellner, J.R., van Heerden, J.A., McConahey, W.M., and Taylor, W.F. (1984). Invasive Hurthle cell carcinoma of the thyroid: natural history and management. *Mayo Clinic Proceedings*, **59**, 851–5.

White, L.R., Cartwright, W.S., Cornoni-Huntley, J., and Brock, D.B. (1986). Geriatric epidemiology. In *Annual review of gerontology and geriatrics*, vol. 6., (ed. C. Eisdorfer), pp. 215–311. Springer, New York.

Williams, R.H. (1946). Thiouracil treatment of thyrotoxicosis. I. The results of prolonged treatment. *Journal of Clinical Endocrinology*, **6**, 1–22.

7.3 The postmenopausal state

S. MITCHELL HARMAN AND MARC R. BLACKMAN

INTRODUCTION

The single most salient fact of reproductive function in ageing women is the menopause. This cessation of cyclic ovarian activity, which occurs at about 50 years of age, definitively divides a woman's mature life into reproductive and postreproductive phases and has profound physiological and psychological implications. Postmenopausal patients may have problems directly related to oestrogen deficiency such as atrophic vaginitis and/or dyspareunia, 'hot flashes' ('hot flushes' in the United Kingdom), and involution of breast and vulvar tissues. As (o-)estrogen replacement therapy (ERT; in the United Kingdom, hormone replacement therapy, HRT) is frequently discussed in the popular media, women may also consult the physician about the advisability of taking hormones for the prevention of long-term problems associated with lack of oestrogen, such as osteoporosis and coronary artery disease. Finally, women may present with symptoms related either to prolonged oestrogen deficiency (e.g. osteopenic bone fractures) or to use of HRT (e.g. vaginal bleeding due to endometrial hyperplasia or even carcinoma). It is therefore necessary for each physician involved in the care of older women to be well-informed about the menopause, its complications, and treatment. Many issues related to the menopause are at present controversial, and clinical research is being conducted in an effort to resolve uncertainties, but for the foreseeable future, doctors will have to act (or not act) on information that is incomplete. Thus, it is important for physicians to understand the physiology of the menopause, and the risks and benefits of menopausal hormone replacement.

PHYSIOLOGY OF THE MENOPAUSE

Ovarian structure and function

At birth, each human ovary contains all the approximately one million germ cells it will ever have. Each of these 'primary oocytes' nests inside a single layer of nurturing granulosa cells to form a structure known as the primordial follicle. In cycling women the monthly upsurge of follicle-stimulating hormone (**FSH**) recruits a cohort of one- to two-hundred ovarian follicles, which begin to grow. Of these recruits, only one will usually reach maturity. The rest undergo involution (atresia). Follicular growth occurs both by the enlargement of the oocyte and by mitotic increase in the number of surrounding granulosa cells. As follicles enlarge, they accumulate a surrounding outer layer of interstitial theca cells. Under the influence of luteinizing hormone, the theca cells secrete steroids, mainly the androgens—testosterone and androstenedione. The follicular granulosa cells, responding to FSH, convert these androgenic precursors to the oestrogens—oestradiol, the principal human oestrogen, and oestrone—by means of the enzyme aromatase. Thus, the ovarian follicle has dual functions, the production of mature germ cells and the secretion of active sex steroid hormones.

Events of the menstrual cycle

By convention, the days of the menstrual cycle are numbered beginning with the day of onset of menstrual flow. During the first 12 to 15 days (follicular phase), the single follicle destined for ovulation develops a lumen and becomes a mature 'Graafian' follicle. In the latter half of the follicular phase, this follicle is the main source of oestrogen. Besides their peripheral effects (see below), oestrogens also have 'feedback' regulatory activity at the pituitary and hypothalamus, where they first augment (positive feedback) and later inhibit (negative feedback) the secretion of luteinizing hormone and FSH. Under the influence of oestrogen the uterine endometrium proliferates to form a thick, multilayered lining with many straight glands. On or about the fourteenth day of the cycle, there is a high peak of secretion of luteinizing hormone, which stimulates ovulation. Next, the residual Graafian follicle is invaded by vascular connective tissue and theca cells and transforms into the corpus luteum, a body of densely packed epithelioid cells, which, stimulated by luteinizing hormone, secrete large quantities of the second major female sex hormone, progesterone, as well as oestradiol. Under the influence of progesterone, the hormone that dominates the latter half or 'luteal' phase of the cycle, the endometrium stops proliferating and becomes 'secretory' (saturated with fluid and glycogen). If implantation of a fertilized ovum fails to occur, the negative feedback effects of oestrogen and progesterone combined will inhibit secretion of luteinizing hormone, leading to involution of the corpus luteum. As oestrogen and progesterone levels fall, the thickened endometrial lining is shed down to the basal layer (menstruation). The cycle then begins again as, in the absence of suppression by sex steroid hormones, blood levels of FSH and luteinizing hormone rise and a new cohort of follicles is recruited.

Actions of oestrogens and androgens

Oestrogen action is not confined to the vagina, uterus, and hypothalamic–pituitary axis. Peripheral actions of oestrogen include growth and maintenance of breast tissue, maintenance of the mature female body habitus (probably by specific actions on subcutaneous fatty tissue of the thighs and hips), effects on plasma lipoprotein patterns (see below), and positive effects on calcium balance. The latter two actions have important implications with regard to risks of, respectively, atherosclerotic coronary-artery disease and osteoporosis. In addition, ovarian androgens, such as testosterone and androstenedione, play a role in fostering sex drive and libido, in developing and maintaining mature body (axillary and pubic) hair pattern, and in maintaining the sebaceous glands that keep skin and hair soft. The adrenal glands also contribute both oestrogens and androgens to the circulating steroid pool; however, the main adrenal oestrogen is oestrone, which is only about one-third as potent as oestradiol. About half the biologically active circulating androgen in women is of adrenal origin.

The menopausal transition and the menopause

Throughout the reproductive lifespan of a normal woman, atresia of recruited growing follicles, and also 'background' atresia of primary follicles, gradually deplete the ovaries of germ cells. By the time a woman is in her mid-40s, the number of follicles recruited at each cycle begins to decrease and the plasma oestrogen levels characteristic of the follicular phase become somewhat lower (Fig. 1). Cycle length and menstrual bleeding pattern may begin to vary, even in women with previously regular cycles. Frequently, the reduced follicular function in the early phase of the cycle leads to incomplete luteinization after ovulation with short or inadequate (low progesterone) luteal phase. This in turn results in a reduction in fertility. The decreased follicular feedback also causes gradual elevation of FSH in plasma (Fig. 1). As the process progresses, ovulation may fail from time to time, resulting in 'missed periods' with prolonged exposure to follicular-phase levels of oestrogen, unopposed by progesterone. This may lead to proliferative hyperplasia of the endometrium and irregular bleeding. Finally, as the number of available follicles drops below some critical number (probably a few hundred per ovary), cyclic bleeding stops altogether, levels of FSH and luteinizing hormone rise and remain elevated, and oestradiol and progesterone levels fall (Fig. 2). This transition phase typically takes from 1 to 2 years. Within

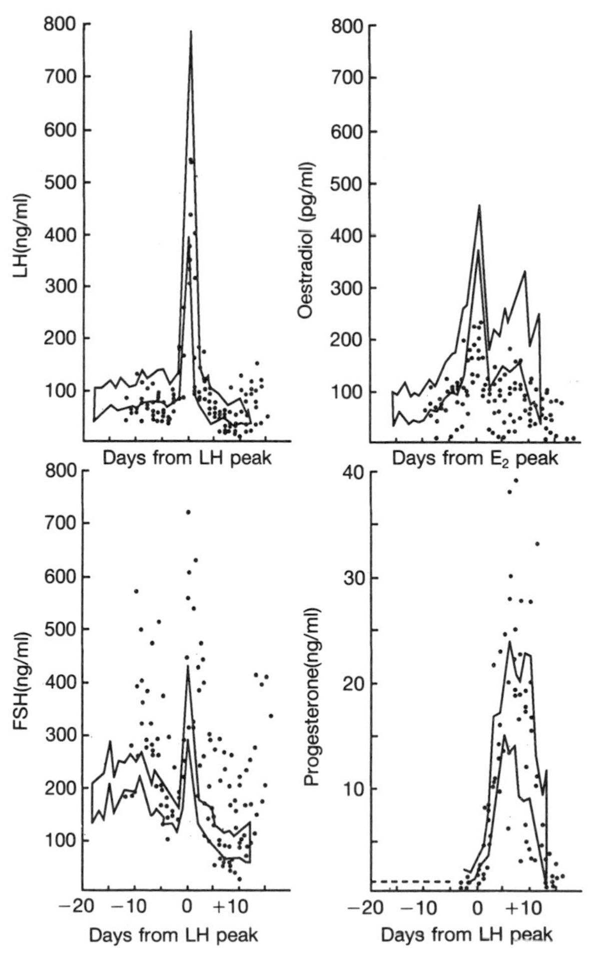

Fig. 1. Daily serum concentrations of luteinizing hormone (LH), follicular-stimulating hormone (FSH), oestradiol (E_2), and progesterone (P) during cycles of 8 women 46–56 years old are compared with the mean ±2 SEM in cycles of 10 women 18–30 years old (enclosed area). LH, FSH, and P are synchronized around the day of the LH peak and E_2 levels are synchronized around the day of the E_2 peak. Note the trend toward high FSH levels and low E_2 levels in the older cycling women. (Sherman, B. M., West, J. H., and Korenman, S. G. (1976). The menopausal transition: analysis of LH, FSH, estradiol, and progesterone concentrations during the menopausal cycles of older women. Reproduced from *Journal of Clinical Endocrinology and Metabolism*, **42**, 629–36, with permission.)

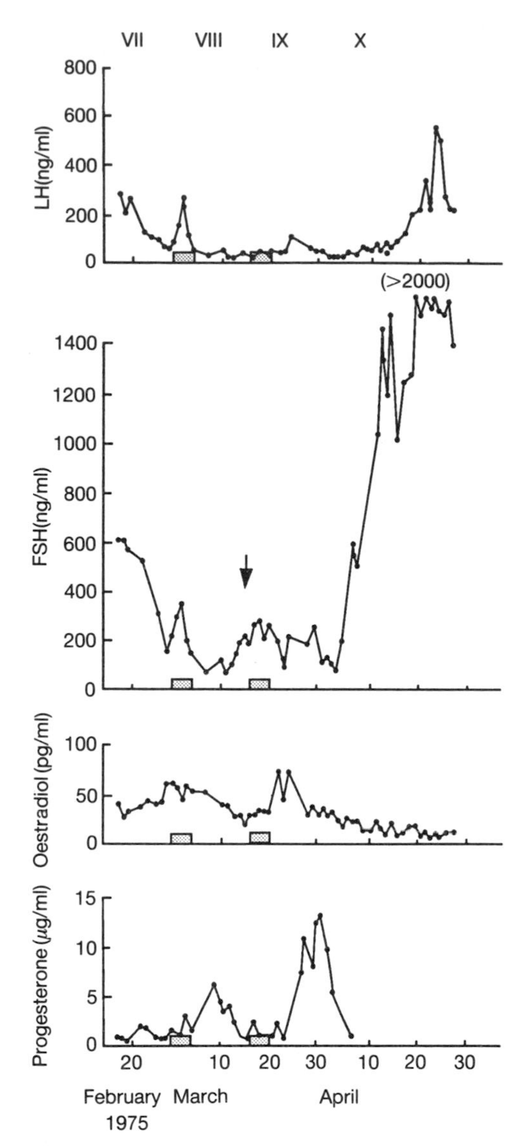

Fig. 2 Serum concentrations of LH, FSH, E_2, and P (abbreviations as in Fig. 1) are shown for cycles VII through X (of 10 cycles recorded) of a single 50-year-old woman during the menopausal transition. Date and year are the *x*-axis. Each shaded block above the *x*-axis indicates a period of vaginal bleeding. Note the decrease in oestradiol and progesterone and the steep rises in LH and FSH at the end of cycle X. (Sherman, B.M., West, J.H., and Korenman, S.G. (1976). The menopausal transition: analysis of LH, FSH, estradiol, and progesterone concentrations during the menopausal cycles of older women. Reproduced from *Journal of Clinical Endocrinology and Metabolism*, **42**, 629–36 with permission.)

Table 1 Plasma concentrations of sex steroids in normal women (ng/dl)

Steroid	Early follicular phase women 18–25 years (mean value, 15 cycles)	Postmenopausal women (51–65 years)	Ovariectomized women (51–62 years)
Oestrone	5.8± 1.6	4.9± 0.5	4.8± 0.5
Oestradiol	4.0± 0.3	2.0± 0.1	1.8± 0.4
Testosterone	44.0± 2.8	29.7± 4.0	12.0± 2.1
Dihydrotestosterone	33.0± 4.0	9.7± 2.1	<5
Androstenedione	184.0±16.0	99.0±13.0	64.0± 9.0
Dehydroepiandrosterone	550.0±43.0	197.0±43.0	126.0±36.0
Progesterone	31.0± 3.0	19.0± 3.0	18.0± 1.0

From Vermeulen, A. (1976). The hormonal activity of the postmenopausal ovary. *Journal of Clinical Endocrinology and Metabolism*, **42**, 247–53, with permission.

a year or two after the menopause, the ovary is shrunken and fibrotic and contains no remaining follicles. Residual steroid hormone secretion by epithelioid hilar cells and a few islands of interstitial (residual thecal) cells accounts for less than half of the circulating androgens and oestrogens in postmenopausal women.

Representative pre- and postmenopausal levels of sex steroid hormones and gonadotropins are shown in Table 1. After the menopause, ovarian oestradiol production is very low so that the dominant circulating oestrogen is the weaker oestrone. Ovarian androgen production is also reduced, and it has been reported that, with age, there is a relative decrease in adrenal androgen secretion as well. Thus, the postmenopausal woman is generally both oestrogen and androgen deficient relative to the younger adult female. The reduction of sex steroid hormone levels lead to derepression of the hypothalamic–pituitary axis with hypersecretion of luteinizing hormone and FSH.

THE MENOPAUSAL SYNDROME

At the time of the menopause, or shortly thereafter, many women go through a period of physical and psychological discomfort characterized as the 'menopausal syndrome'. This syndrome has a number of components, which may include vaginal inflammation with discharge, and burning pain or pruritus, increased susceptibility to urinary tract infections, dyspareunia, hot flushes, changes in libido (which may increase or decrease), and altered affect. The latter may range in severity from mild feelings of sadness or loss to severe depressive symptoms. Not all women will experience all components of the syndrome; in fact, many women pass through the menopause with little or no discomfort. Furthermore, it is not known why some women have a relatively easy time during the menopausal transition, while others experience severe or even disabling symptoms.

Atrophic vaginitis

The vaginal component of the menopausal syndrome is due to loss of oestrogen support of the vaginal lining epithelium. The oestrogen-stimulated vaginal lining is a thick, multilayered structure with deep papillae descending into the underlying vascular connective tissue, a layer of mitotically active basal cells, mid-level cuboidal cells with abundant stores of glycogen (eventually released into the vaginal lumen), and several layers of squamous cornified cells at the luminal surface. After the menopause, papillary depth is shallow, glycogen production is nearly absent, and the lining epithelium thins to only a few layers of cells, with little or no cornification at the surface. This atrophic lining is often infiltrated with leucocytes. The loss of glycogen leads to an increase in vaginal pH and an alteration of the normal vaginal flora from predominant acidophilic lactobacilli to a mixed flora of coliforms and other organisms more typical of an alkaline or neutral environment. The thinness of the lining epithelium and the loss of the cornified layer provide less barrier against trauma and make the vagina more susceptible to bacterial invasion. In this setting, it is not surprising that many women experience vaginitis. This atrophic vaginitis of menopausal women tends to be resistant to the usual topical or systemic antibacterial or antifungal agents used for vaginitis in younger women. Oestrogen therapy, however, is nearly always successful in ameliorating or eliminating the condition. Oestrogen may be applied locally as vaginal cream or suppositories, or given systemically as oral or transdermal preparations. Many physicians prescribe vaginal oestrogen under the impression that this will avoid exposing the patient to system oestrogen effects and the potential toxicity thereof. It has been shown, however, that systemic absorption from vaginal preparations produces biologically significant blood levels of oestrogens. Therefore, local oestrogen treatment may have little advantage over oral or transdermal therapy. The comparative effects of the latter two modes of therapy will be discussed below.

Urinary tract infections

The shortening of the urethra, relaxation of pelvic musculature, and altered vaginal flora characteristic of the oestrogen deficiency state also appear to enhance susceptibility of the postmenopausal female to urinary tract infections. These infections are similar to those in younger women with the following exceptions: first, they are more frequent and have more tendency to recur; second, they are more likely to produce urinary incontinence (see below); finally, because ageing is associated with an increase in susceptibility to systemic bacterial infection, probably related to alterations in lymphocyte function, there is a greater risk in elderly postmenopausal women that a urinary tract infection may develop into pyelonephritis and/or sepsis. For this reason, bladder infections in older women should be treated vigorously with appropriate antibacterial therapy. Initially, therapy may be empirical but should be modified, if necessary, after receiving the results of urine culture. Follow-up cultures should always be done within a few days of completing antibacterial therapy to assess sterilization of the urine. In the event that persistent bacteriuria or recurrent infections become a problem, the chronic daily use of a urinary sterilizing agent, such as hexamine (methenamine) hippurate, a combination sulpha agent, or an acidifying agent, such as ascorbic acid, should be considered (see Chapter 3.2).

Dyspareunia

Another major complaint associated with the menopausal syndrome is discomfort or pain during sexual intercourse. This problem may or may not occur in association with atrophic vaginitis, and may be of several types. Probably the most common is simple frictional discomfort due to a decrease in normal vaginal lubrication and thinning or inflammation of the vaginal lining. This type of pain is often relieved simply by use of an aqueous lubricant gel, but treatment with local or systemic oestrogen, and, in some cases, antibacterial or antifungal therapy, may be necessary to achieve long-term relief. The second type of dyspareunia is due to shortening and loss of elasticity of the vagina, which results in pain on deep penetration. This problem may be relieved by adjusting coital position (and vigour) to reduce depth of penetration. Although HRT helps prevent this problem if initiated at the onset of the menopause, its value once the condition is established is not certain. Finally, in the case of extreme vaginal atrophy, penetration may not be possible at all because of constriction of the introitus. This situation is most common where there has been a long period of abstinence and disuse atrophy, followed by an attempt to resume sexual activity. Physical methods, such as gradual dilation of the introitus with a graded series of dilators, can be helpful. The role of HRT once vaginal constriction has occurred is not clear. There is some evidence to support the concept that frequent, regular sexual intercourse in the postmenopausal period helps preserve vaginal elasticity as well as the lubrication response and thus may prevent the development of dyspareunia.

Hot flashes (hot flushes)

In the perimenopausal period, as oestrogen levels fall, many women report the onset of a characteristic complaint, which is commonly known as 'hot flashes' in the United States or 'hot flushes' in the United Kingdom. The symptoms vary somewhat, both in their frequency and severity as well as in the particular sensations experienced. Hot flushes usually begin with a sensation of chilliness or even a brief shudder. This is followed within a minute or so by the characteristic intense peripheral flush and a sensation of cutaneous heat, often accompanied by an outbreak of diaphoresis. Symptoms usually last several minutes and may occur as frequently as every 30 min, but typically are noticed eight to ten times per day, at irregular intervals. Although hot flushes occur in as many as 85 per cent of women around the time of the menopause, most women regard them only as a nuisance. In about 15 per cent of women, however, they may be so frequent and severe as to interfere with sleep and with normal daily activities.

At one time the accepted clinical teaching was that hot flushes were a psychological reaction to the subconscious perception of loss of femininity, attractiveness, fertility etc., engendered by the menopausal transition. The most common treatment was simple reassurance, unfortunately sometimes given in a patronizing manner. More recently, it has become apparent that there is a physiological basis for hot-flush symptoms in that measurable changes in body temperature and blood flow are temporally linked to characteristic hormonal events. In particular, it has been shown that in the perimenopausal period the characteristic secretory pulses of luteinizing hormone and FSH become augmented in amplitude as oestrogen feedback inhibition is reduced. Shortly before such an exaggerated pulse of luteinizing hormone begins, there is a brief rise in core body temperature of 1 to 2°C. it is during this rise that chilling may be experienced. This phase is followed by the onset of skin flushing and sweating, during which body temperature returns to normal. During the flush phase there is a decrease in skin resistance and an increase in cutaneous blood flow, which can be metered and recorded by appropriate monitors (Fig. 3). The coupling of physiological events and hormonal surges can be seen even during inter-

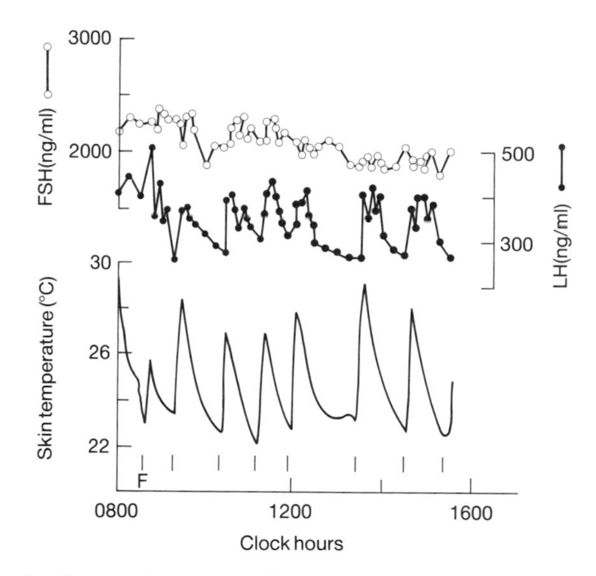

Fig. 3 Changes in cutaneous finger temperature and serum FSH and LH levels (abbreviations as in Fig. 1) in a postmenopausal woman during 8 h. Each tick above the *x*-axis represents onset of a subjective hot flush. Note the close temporal relationship among hot flushes, elevations of cutaneous temperature, and pulsatile LH release. (Tataryn, I. V., Meldrum, D. R., Lu, K. H., Frumar, A. M., and Judd, H. L. (1979). LH, FSH and skin temperature during the menopausal hot flash. Reproduced from *Journal of Clinical Endocrinology and Metabolism*, **49**, 152–3 with permission.)

vals when no subjective hot flush sensations are experienced, but nearly every subjective hot flush is accompanied by measurable physiological changes. It is suggested that the hypothalamic neural activity associated with pulsatile secretion of gonadotropin-releasing hormone, intensified by loss of negative feedback, spreads into adjacent thermoregulatory centres and alters their function temporarily. Hot flushes may or may not require therapy, depending on the degree to which they discomfort the individual patient. Fortunately, nearly all patients will obtain significant relief, and in the majority of cases complete alleviation of symptoms, from oestrogen therapy. Although other forms of therapy have been suggested to be effective at various times, including tranquillizers, alpha- and beta-adrenergic blocking agents, etc., only oestrogen replacement has been shown to be better than placebo in double-blind trials. The natural course of hot-flush symptoms is one of gradual resolution; however, this may require months to years. It is not uncommon for symptoms to recur when oestrogen therapy is discontinued.

Psychological symptoms

Although the hot flush *per se* is clearly related to altered neural and hormonal activity, the psychological response to this symptom, as well as to the other changes associated with the menopause, may range from little or none to profound alteration of affect and personality, depending on the woman. Symptoms vary from minor irritability and emotional lability to severe depression and withdrawal from usual activities. Sexuality is also commonly affected, with some women reporting an increase in libido, sometimes attendant on release from worries about conception, while other women have a reduction in sexual interest, usually associated with a perceived loss of attractiveness and femininity. There is some evidence that prior psychological instability predicts whether significant psychiatric disturbance will occur at the time of the menopause, but many women who have been apparently stable will also have some psychological symptoms associated with the 'change of life'. For the most part, these will be minor, self-limiting, and will resolve with sympathetic support from the physician and the family. In some cases a short course of a minor tranquillizer or an antidepressant may be justified. In more severe cases, psychiatric consultation is warranted. The value of oestrogen replacement for relief of psychological symptoms in the menopausal transition is controversial, but a number of studies have suggested that significant relief of psychological symptoms and an overall increase in sense of well-being and ability to concentrate may occur when oestrogen therapy is undertaken. Sleep disturbance is a common complaint during the perimenopausal phase. There is some evidence that waking during the night may be specifically linked to the physiological events underlying the hot flush phenomenon and consequently insomnia is often improved by oestrogen therapy.

OTHER PROBLEMS ASSOCIATED WITH OESTROGEN DEFICIENCY

Osteoporosis

The significance of oestrogen loss after the menopause in the pathogenesis of osteoporosis and the role of oestrogens in the prevention of postmenopausal bone loss are discussed in Chapter 14.1.

Coronary heart disease

Before the age of the menopause, women have a much lower rate of coronary heart disease than do men of comparable age. Specifically, rates of myocardial infarction, angina, and sudden cardiac death are significantly lower for women, even after allowing for known risk factors such as blood pressure and cigarette smoking. During the sixth to eight decades of life, rates in women become equal to or even slightly higher than those of age-matched men. Although this convergence is due in part to a slowing in the rate of rise of incidence with age in men, possibly as a consequence of changes in blood androgen and oestrogen levels after the age of 50, data from the Framingham study suggest that the level of risk of coronary heart disease is higher in postmenopausal women than in cycling women of the same age (Table 2). A number of non-randomized studies have found rates of coronary heart disease and cardiac death as well as all-cause mortality to be lower in women taking oestrogens long-term, although a few studies have shown no apparent effect or even higher mortality from coronary heart disease. The weight of evidence favours the hypothesis that women are protected from coronary atherosclerosis by their endogenous oestrogens and that such protection may be prolonged after the menopause by oestrogen therapy. If this is true, it may represent an even more compelling indication for HRT than does osteoporosis, as coronary heart disease is a far more common cause of death.

Table 2 Coronary heart disease incidence for women having a natural menopause (Framingham study: 24-year follow-up)

Age and menopausal status	Person–years	Coronary heart disease cases	(No./1000/year)
40–44 years old			
Premenopausal	4518	1	0.2
Menopausal	262	0	—
Postmenopausal	280	0	—
45–49 years old			
Premenopausal	3266	4	1.2
Menopausal	1068	1	0.9
Postmenopausal	1752	6	3.4
50–54 years old			
Premenopausal	600	1	1.7
Menopausal	884	6	6.8
Postmenopausal	5704	25	4.4

From Gordon, T., Kannel, W. B., Hjortland, M. C., and McNamara, P. M. (1978), Menopause and coronary heart disease, the Framingham study. *Annals of Internal Medicine*, **89**, 157–61, with permission.

It has been well established that plasma lipoprotein patterns predict risk of atherosclerosis. In particular, the greater the percentage of total cholesterol present in the high-density lipid (**HDL**) fraction and the less cholesterol in the low-density (**LDL**) fraction, the lower the risk of coronary heart disease. Cross-sectional measurements of HDL-cholesterol in normal men and women (Fig. 4(a)) have shown that mean levels are considerably higher in women than in men at every age and do not decrease in women at the age of the menopause, but rather continue to increase with age in women and, to a lesser extent, in men as well. In contrast, although LDL levels (Fig. 4(b)) are lower in young women than in age-matched men and increase gradually in both sexes through the 30s and early 40s, between 45 and 55 years, LDL-cholesterol increases by a relatively large increment and actually exceeds the LDL level in men after the age of 50. Thus, it would appear that HDL levels are higher in women, independent of endogenous oestrogen levels (and the menopause) and that LDL levels rise when oestrogen is deficient. Various investigations have established that sex steroid hormones have specific and highly significant effects on lipoproteins. In general, androgens appear to have deleterious effects, lowering HDL- and raising LDL-cholesterol, while oestrogens do the opposite.

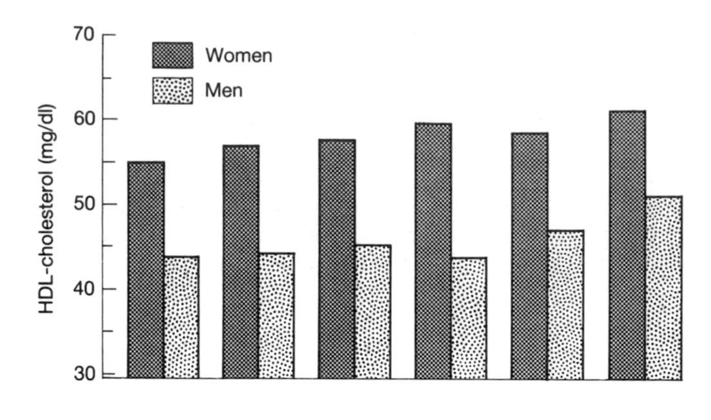

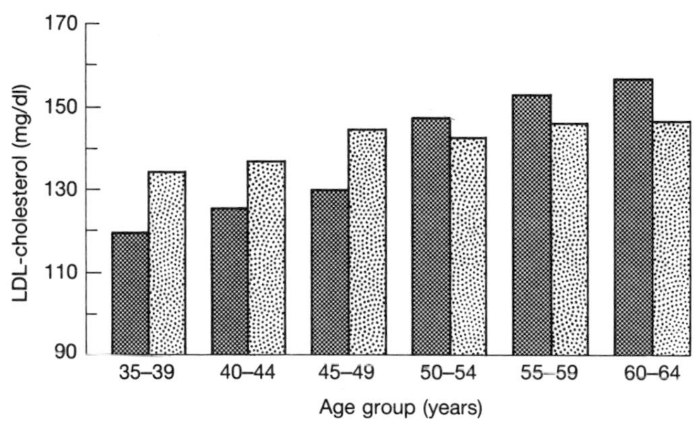

Fig. 4 Changes in plasma high-density (HDL) (upper panel) and low-density lipoprotein (LDL) (lower panel) cholesterol in women (black bars) and men (grey bars) with age in 5-year increments. Note that HDL-cholesterol is higher in women at all ages, with no significant change during the menopausal transition (ages 45–54). In contrast, LDL-cholesterol is lower in young women than in young men, but rises between ages 45 and 54 to exceed the male values (graphs constructed from tables in *The lipid research clinics* (1980) Vol. I, *The prevalence study*, pp. 70–9. National Institutes of Health, U.S. Dept of Health and Human Services, Bethesda, MD).

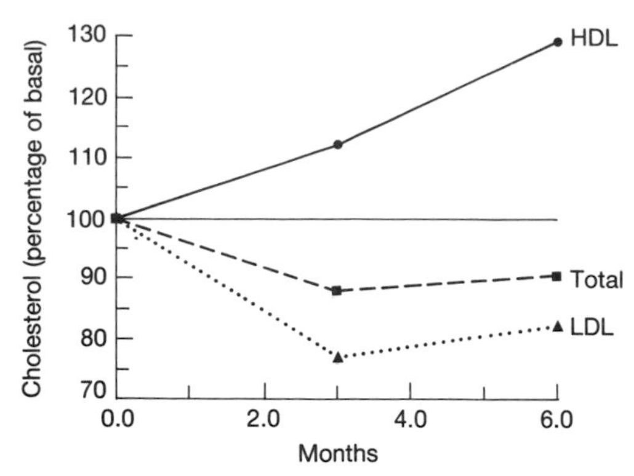

Fig. 5 Changes in total cholesterol (squares), HDL-cholesterol (circles) and LDL-cholesterol (triangles) (abbreviations as in Fig. 4) in 17 postmenopausal women are shown as the percentage of basal value after 3 and 6 months of treatment with 2 mg/day of oral oestradiol valerianate. Note the increase in HDL- and decreases in total and LDL-cholesterol levels (drawn from table in Tikkanen, M. J., Nikkila, E. A., and Vartainen, E. (1978). Natural oestrogens as an effective treatment for type II hyperlipoproteinaemia in postmenopausal women. *Lancet*, ii, 490–1).

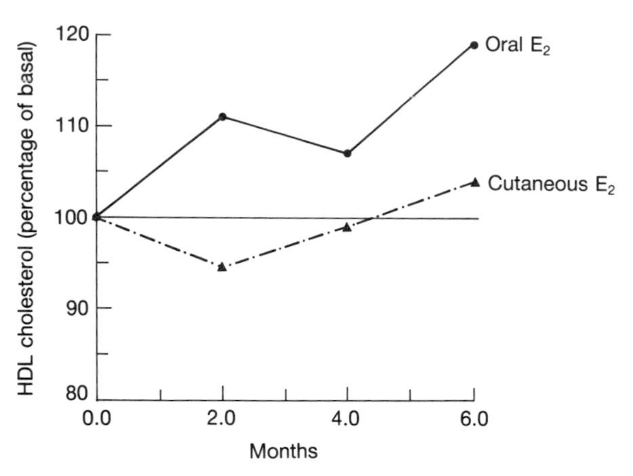

Fig. 6 Comparison of HDL-cholesterol levels, shown as the percentage of basal at 2-month intervals, in postmenopausal women treated with oral micronized versus cutaneous E_2 gel (abbreviations as in Figs. 1 and 4). Note significant rise in HDL with oral but not cutaneous E_2 (drawn from table in Fahraeus, L. and Wallentin, L. (1983). High-density lipoprotein subfractions during oral and cutaneous administration of 17β-oestradiol to menopausal women. *Journal of Clinical Endocrinology and Metabolism*, **56**, 797–801).

More recently it has become apparent that the route of steroid administration influences the pattern of lipoprotein effects observed. When oestrogens are given orally, there is a 10 to 30 per cent increase in HDL- and a significant reduction in LDL-cholesterol (Fig. 5). If oestrogens are given by a non-oral route (e.g. by injection or transdermally) this HDL effect is not observed (Fig. 6); however, prolonged (but not short-term) transdermal administration will lower LDL-cholesterol (Fig. 7). The differential effects of oral vs. parenteral oestrogens are thought to reflect the fact that portal absorption of orally administered steroid results in supraphysiological levels in the liver on first pass. The high portal levels of oral oestrogen significantly increase hepatocyte protein synthesis (see below). One protein affected is apo-AIC, the major apolipoprotein of the HDL subfraction. In addition, there may be oestrogen effects on hepatic lipoprotein receptors and on the enzyme hepatic lipoprotein lipase, all of which tend to increase HDL. {There is evidence however, that the passage of orally administered oestrogen through the liver may increase the production of various clotting factors and thus increase the risk of thromboembolic disease.}

To summarize, data suggest that use of non-oral exogenous oestrogens produces a lipoprotein pattern resembling that seen before the menopause, while oral oestrogen use has an added pharmacological effect to increase HDL beyond the high levels characteristic of women of any age. Because cycling women appear to be significantly protected from risk of coronary heart disease, it is reasonable to suppose that transdermal oestrogen confers a similar degree of protection; however, all currently available data showing decreased risk of coronary heart disease in oestrogen-replaced women are from studies of oral therapy. Whether the greater HDL level produced by oral therapy leads to added benefit is not known, nor will it be until long-term randomized studies comparing rates of coronary heart disease in oral and parenterally treated populations are completed.

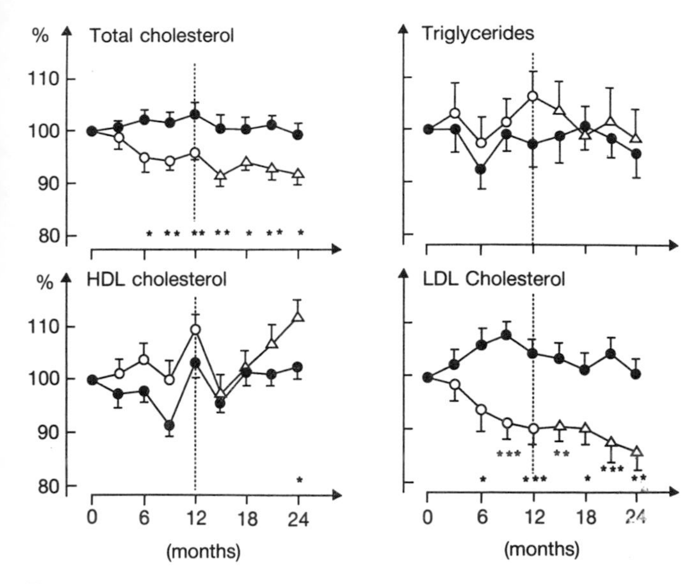

Fig. 7 Serum lipids and lipoproteins in postmenopausal women during 24 months of treatment with placebo (closed circles) or with percutaneous oestradiol (E_2) (open circles—first 12 months) or percutaneous E_2 with cyclic oral micronized progesterone (open triangles—second 12 months). Note that the significant decreases in total and LDL-cholesterol occur only after 6 months of therapy and continue after addition of progesterone while HDL-cholesterol is significantly greater in the treated group only at 24 months (abbreviations as in Figs. 1 and 4) (significance of difference from placebo group: *$p<0.05$, **$p<0.01$, ***$p<0.001$). (Reproduced from Jensen, J., Riis, B. J., Strom, V., Nilas, L., and Christiansen C. (1987). Long-term effects of percutaneous estrogens and oral progesterone on serum lipoproteins in postmenopausal women. *American Journal of Obstetrics and Gynecology*, **156**, 66–71, with permission.)

Urinary incontinence

Urinary incontinence is a distressing symptom at any age beyond infancy, resulting in embarrassment and unhappiness. When severe, it may lead to curtailment of normal social activities. The assessment of urinary incontinence and its differential diagnosis are discussed in Section 14. Vaginal and urethral atrophy due to oestrogen lack may contribute to the genesis of urgency incontinence. Whether oestrogen replacement therapy has a role in the treatment of incontinence beyond its value for relieving atrophic vaginitis is the subject of controversy and continued investigation, but some studies have shown positive effects of oestrogen in patients with stress incontinence. If symptoms are severe or resistant to medical therapy, urological, or gynaecological referral is appropriate, especially as cystometric studies may help in defining the diagnosis and surgical intervention (bladder resuspension, cystocele repair, etc.) is helpful in some cases.

Cosmetic and aesthetic considerations

As noted above, ovarian oestrogens and androgens have peripheral effects on skin and subcutaneous fat distribution that determine female habitus and secondary sexual characteristics. The most prominent change apparent at or shortly after the menopause is the loss of glandular breast tissue. Although, over a lifetime, gravity and obesity affect breast contour by gradually stretching the supportive ligaments, the atrophy caused by oestrogen deficiency results in relatively rapid loss of the rounded contour of the breast. The flattened atrophic postmenopausal breast is likely to be perceived as less feminine and attractive. In addition, there is withdrawal of subcutaneous fat from breast, hips, buttocks, and the labia majora, all of which may loosen and sag. The labia majora characteristically gape, exposing the labia minora. In postmenopausal women, pubic and axillary hair become sparse, grey, and lank. Scalp hair also thins and becomes drier and less lustrous owing to decreased oil secretion. In some cases, male-pattern type baldness may appear, probably related to decreased ratios of circulating oestrogen to androgen. The relative extent to which these changes are hormone dependent and hence preventable by oestrogen is not known. Clearly, oestrogen replacement, although it may alleviate or retard some of these changes, does not wholly or indefinitely prevent them. Despite this fact, many women feel that oestrogen replacement therapy provides significant benefit in terms of their body image and sexual attractiveness. Although this factor is often omitted from discussions of risks and benefits of oestrogen therapy, it is a major consideration for some patients and should not be overlooked.

OESTROGEN REPLACEMENT THERAPY

Type of oestrogen

Several types of oestrogen preparation are available and the number will almost certainly increase as replacement is more widely prescribed. The traditional and most commonly prescribed preparations, conjugated oestrogens, were originally extracted and purified from the urine of pregnant mares. These extracts are mixtures of oestrone, oestradiol, and reduced oestrogen metabolites, conjugated with sulphate and glucuronide. Currently there are also pure synthetic conjugated oestrogen preparations, such as oestrone sulphate. Conjugated oestrogens are relatively water soluble, compared to native steroid, and hence are well absorbed. Another type of oral oestrogen is micronized oestradiol, which depends for its bioavailability on the very small particle size, and hence large surface area for dissolution. Finally, there are synthetic oestrogens, such as ethinyl oestradiol, which are more potent than the natural compounds and have a prolonged half-life *in vivo*. Oestrogen is also available in transdermal patches and, in some countries skin creams and pellet implants, all of which supply 17-β oestradiol, the natural ovarian oestrogen.

Dose of oestrogen

The amount of oestrogen to be administered depends on the preparation used. The production rate of oestradiol in young women varies from 40 to 200 μg per day, depending on the phase of the menstrual cycle, but the average rate is 80–100 μg/day. Thus, the daily replacement dose of oestrogen should be equivalent to approximately 80 μg of oestradiol in its biological effect. Although normal blood levels of oestradiol are known to average about 100 pg/ml, oral oestrogens do not produce levels corresponding to the normal physiological pattern, and therefore estimates of dose must depend on clinical effects rather than plasma oestradiol measurements. In terms of such measures as vaginal cornification, relief from hot flushes, and urinary calcium excretion the physiological dose of conjugated oestrogen ranges from 0.625 to 1.25 mg/day, of micronized oestradiol from 1 to 2 mg/day, and of ethinyl oestradiol from 0.01 to 0.02 mg/day. Transdermal patches are available that deliver 50 or 100 μg/g/day of oestradiol. The higher-dose patch produces oestradiol blood levels in the physiological range and normalizes the clinical indices more completely than does the lower dose. Doses of oestrogen that give adequate clinical effects do not reduce the blood levels of FSH and luteinizing hormone to premenopausal levels. This may be in part because hormones other than oestrogens (inhibin, androgens, progesterone) normally act in concert with oestrogens to suppress gonadotropin secretion, and also in part because the sensitivity of pituitary feedback to sex steroids may be reduced with age.

Route of administration of oestrogen

As noted above, oral oestrogens are taken up from the intestine into the hepatic portal circulation and hence pass into the liver sinusoids, where they act upon and are altered by the hepatocytes. This results in conversion of a large proportion of administered oestradiol to oestrone and a smaller amount to oestrogen conjugates. Thus, with oral oestrogens, even when pure micronized oestradiol is given, the ratio of oestradiol to oestrone is much lower than in premenopausal women. As oestrone is only about one-third as potent as oestradiol, a greater amount of oestrogen must be given to produce the same physiological effects. The other consequence of this 'first pass' of high concentrations of oestrogen through the liver is the production of an unphysiological degree of oestrogen stimulation of hepatocyte function, including altered protein synthesis and altered bile composition. These effects have implications for alteration of blood pressure, risk of cholelithiasis (see below), synthesis of clotting factors, and lipoprotein pattern (see above). Thus, it has been said that it is not possible to attain truly physiological oestrogen replacement by the oral route. In contrast, transdermal or other parenteral routes of oestradiol replacement have been shown experimentally to approximate a physiological ratio of circulating oestradiol to oestrone and normal levels of hepatocyte function.

Use of progestogen

It is now common practice for menopausal women who receive oestrogen replacement therapy to be treated with a progestogen to antagonize the effects of oestrogens at the endometrium. This is because oestrogen, in the absence of opposing progesterone, causes hyperplasia of the endometrium and, over time, increases the risk of endometrial cancer by a factor of eight or more. Studies have shown that progestogens cause a loss of oestrogen receptors in endometrial epithelial cells and that women treated cyclically for at least 10 successive days with a progestogen have a rate of endometrial cancer as low as or lower than that of untreated women of comparable age. One study suggests that women so treated may have a lower incidence of breast cancer than does the untreated population, but this conclusion has not been confirmed. Thus, most experts recommend progestogen replacement only for women who have a uterus *in situ*. Although progestogen opposes oestrogen effects on the uterine epithelium and on plasma lipids (see below), it does not appear to do so at bone. In fact, some data suggest that progestogens may be synergistic with oestrogen in preventing bone loss. {Progestogens may reduce libido in some women and this may also need to be taken into account in prescribing for individual patients.}

Type of progestogen

Progestogens available for use include the 21-carbon compound, medroxyprogesterone acetate and a number of more potent orally active steroids derived from 19-nortestosterone and used in oral contraceptive preparations, including levonorgestrel, norethindrone, and ethynodiol. A number of studies have shown that oral 19-nortestosterone-derived progestogens act on the liver to increase LDL- and decrease HDL-cholesterol, opposing the beneficial effects of oestrogen. Medroxyprogesterone acetate has less of this 'androgenic' effect when given at its effective dose of 10 mg/day. The deleterious lipid effects of the 19-nortestosterone-derived steroids appear to be largely avoided if they are given in low (but still effective) doses (e.g. 250 μg/day for norethindrone or 100 μg/day for levonorgestrel) or if they are administered parenterally (as in an implant). Work is also proceeding on oral and transdermal preparations of progesterone, the natural human progestogen; results show no negative effects on plasma lipoproteins with such preparations.

Pattern of administration of progestogen

When a progestogen is given cyclically for 10 to 12 days each month and discontinued at the end of each cycle, there is a period of 2 to 5 days of vaginal withdrawal bleeding in most women. Although for some women this event is a reassuring proof of continued femininity, many perceive it as an unacceptable nuisance occurring at a time in life when they had anticipated freedom from this problem. For such women it is possible that a modified cyclic schedule in which progestogen is given for 12 to 14 days out of each 90 days may be more acceptable. Although such treatment prevents endometrial hyperplasia from building up and hence should lead to avoidance of carcinoma, there are no data on actual cancer risk in women so treated. Another possibility is to treat with a constant regimen of combined oestrogen and progestogen without cycling. Several studies have shown this mode to result in a quiescent endometrium after approximately 3 months of treatment; however, during the early phase of combined constant therapy, intermittent irregular bleeding may occur in a significant proportion of women. Again, as for 90-day interval progestogen treatment, there are no data about rates of endometrial carcinoma at present.

Duration of therapy

If the purpose of oestrogen treatment is relief of one or more of the acute manifestations of the menopausal syndrome, such as vaginitis or hot flushes, a few weeks to a few months of therapy may suffice. On the other hand, if therapy is given for long-term protection against osteoporosis or coronary heart disease, then many years of therapy will be required. Formerly, recommendations were for 5 or 10 years of treatment, but it has been shown that whenever oestrogen therapy is discontinued a period of rapid calcium loss ensues, similar to that occurring at the natural menopause. If the object of therapy is prevention of hip fractures, the rate of which peaks after age 70, then 10 years of treatment should delay the peak till approximately age 80. Thus, as the life expectancy in women increases past age 80, it may become important to continue oestrogen therapy for longer. Theoretically there is no objection to indefinite continuation, as long as the patient remains comfortable and has no side-effects. There is some evidence that oestrogen therapy remains as effective in women over 65 as in younger postmenopausal women, but it is possible that, with extreme old age, oestrogen effect may be attenuated. There is as yet little information on the merit (or lack of merit) of starting HRT in women who are already 10 years or more postmenopausal (i.e. aged 60 or older).

Selection of candidates for therapy

There is no consensus on who should or should not receive oestrogen therapy. Clearly, women with acute symptoms of the menopausal syndrome are candidates for short-term therapy, but controversy continues over which women should be given long-term treatment. Some experts in the field have recommended that any postmenopausal women without a clear contraindication should be considered a candidate for oestrogens, but most physicians adopt a more conservative approach. If HRT is considered as prophylaxis for osteoporosis, then it would be helpful to identify women at high risk for this problem at the time of the menopause. In general, risk factors for osteoporotic bone fractures include Caucasian race, fair skin, slender stature, sedentary lifestyle, family history of osteoporotic fractures, diabetes mellitus, and tobacco and alcohol use. Efforts have been made to devise specific tests or combinations of tests which would reliably predict risk of bone loss. (see Chapter 14.1) Thus, although there is hope that more objective indices of osteoporosis risk may be devised in the future, for the present physicians are left to decide which women they should treat based on subjective criteria and their perception of the overall risk–benefit ratio of therapy. It is our opinion that the potential beneficial effect of oestrogen therapy on risk of coronary heart disease and the elimination of excess risk of endometrial carcinoma by progestogen treatment weight the risk–benefit equation heavily in the direction of treatment, so that many more women should be given HRT than currently receive it. Absolute contraindications to oestrogen therapy include malignant disease, especially hormone-dependent tumours of breast or uterus, and melanoma. For reasons discussed below, the presence of liver disease or a history of venous thrombosis or thromboembolism contraindicate use of oral (but not necessarily parenteral) oestrogens. Relative contraindications include hypertension, gallbladder disease, and heart or renal failure.

PROBLEMS AND TOXICITY OF OESTROGEN REPLACEMENT

Minor side-effects

Women taking oral oestrogens may complain of nausea, sometimes with vomiting, similar to the gastrointestinal syndrome of pregnancy. Other typical symptoms are bloating, abdominal discomfort, or fluid retention. With the low doses used, these symptoms are usually mild, and may remit if therapy is continued for a few more weeks, or if doses are lowered temporarily. In studies with transdermal oestradiol the incidence of these symptoms appears to be lower, suggesting that some or all of them may be mediated by local effects of oestrogen on the upper gastrointestinal tract, first-pass effects of oestrogen on the liver, or by oestrogen metabolites produced during first pass through the liver. Women taking oestrogens may develop simple headaches or, on occasion, new onset, recurrence, or exacerbation of severe migraine headaches. These necessitate discontinuation of oestrogen. At the beginning of therapy, stimulation of breast tissue may lead to breast pain or tenderness. This usually subsides with continued therapy, but may require discontinuation of oestrogen in some patients, particularly those with a past history of painful polycystic breast disease. Finally, oestrogen stimulation of the uterus may cause some intermittent bleeding or spotting, even in women taking cyclic progestogen. This may be due to an inappropriate ratio of oestrogen to progestogen for the particular patient and may respond to lowering the oestrogen dose, or increasing the dose or number of days of treatment with progestogen. Bleeding may also be due to reactivation of a subendometrial uterine leiomyoma, to endometrial hyperplasia, or to cervical or endometrial cancer. Therefore, if bleeding fails to remit with adjustment of steroid therapy, patients should be evaluated by a gynaecologist.

Hepatic first-pass effects

It has long been recognized that prevalence of hypertension and incidence of thromboembolic disease (deep vein thrombophlebitis, pulmonary embolism, and thrombotic stroke) are increased in women taking oral contraceptives. These risks were shown to be related to the oestrogen rather than to the progestogen component, and, accordingly, the doses of oestrogen in oral contraceptives were decreased. More recently it has become apparent that these adverse effects of oral ocstrogen are mediated by its hepatic first-pass action on protein synthesis. A number of hepatic proteins increase in response to oestrogen. These include the hormone-binding globulins for sex steroids, cortisol, and thyroid hormones. Oestrogens appear to raise blood pressure by inducing hepatic synthesis of renin substrate (angiotensinogen), resulting in elevated levels of angiotensin. The particular factors responsible for increased intravascular clotting have not been identified, although a number of candidates have been suggested. Studies of women taking the lower doses of oral oestrogens typically given for postmenopausal therapy have not shown increases in clinical hypertension, thrombophlebitis, pulmonary embolus, or stroke, despite measurable increases in binding proteins and renin substrate. Furthermore, measurements of urinary excretion of fibrin breakdown products, an index of subclinical intravascular clot formation, do not increase. Nonetheless, non-oral oestrogen preparations, such as the oestrogen patch, have been devised and marketed based on the concept of avoiding first-pass hepatic effects. With these preparations there is no increase in any of the hepatic proteins measured and hence, presumably, an even lower risk of thromboembolism and hypertension.

There also appears to be an increase in clinical gallbladder disease in women taking oral oestrogens. This occurs both in women on oral contraceptives and those taking postmenopausal therapy. Whether this is due to increased formation of gallstones due to an oestrogen-induced alteration of bile chemistry or to some effect on the gallbladder itself, leading to symptomatic disease in the presence of pre-existing stones, is not clear. It is also not known whether cholelithiasis may be avoided by a non-oral route of oestrogen administration, although there is reason to presume that this may be so.

Hepatic-mediated effects of oestrogens and progestogens have been discussed above in the section on risk effects of oestrogen therapy on coronary heart disease.

Glucose tolerance

Oral contraceptive pills may be associated with decreased glucose tolerance and with onset of diabetes mellitus. It is not clear whether this effect is due to the oestrogen or the progestogen component. Although decreased glucose tolerance was at one time considered to be a potential side-effect of oestrogen therapy, recent studies have failed to show that this effect occurs with the lower doses of oestrogen used for postmenopausal treatment, demonstrating rather a neutral or slightly improved glucose tolerance in oestrogen-treated women.

Breast cancer

Many physicians hesitate to prescribe oestrogen from fear of inducing carcinoma of the breast. Although some breast cancers are oestrogen-responsive, the evidence is mixed and inconclusive as to whether postmenopausal HRT is associated with an increased risk of breast cancer, despite numerous studies of this question. The caveat to be considered with regard to most such studies is that their designs and patient populations have been such that a risk ratio of 2.0 or less could not be detected with confidence. On the other hand, the overall trend, when results of a large number of epidemiological studies of oestrogen and breast cancer risk are pooled, is neutral. A few studies have shown an increased risk, a few show a protective effect, and most studies give risk ratios near 1.0. In one large study, in which oestrogen appeared to increase the risk of breast cancer, this effect disappeared when women having a first-degree relative (mother or sister) with a history of breast cancer were subtracted from the population. Thus, it might be reasonable to recommend long-term treatment only to women without a strong family history of breast cancer. The role of progestogen in ameliorating any risk of breast cancer that oestrogen might produce is uncertain, despite a single study suggesting a protective effect. More research will be required on this question before conclusions can be stated with confidence.

OTHER PROBLEMS OF POSTMENOPAUSAL WOMEN

Vaginal bleeding

In the perimenopausal period, irregular vaginal bleeding may merely represent occasional ovarian reactivation due to the presence of a few residual follicles. On the other hand, vaginal bleeding by a woman of menopausal age, who has been without menses for 6 months or more and who is not taking oestrogens, strongly suggests disease of the reproductive system, such as infections, ulcerations, or tumours of uterus, cervix, or vagina. Thus, a woman manifesting postmenopausal bleeding should have an assessment of the vagina, cervix, and uterus, including endometrial biopsy. Most commonly, only endometrial hyperplasia will be found. This is due to uterine stimulation by unopposed oestrogen. The usual cause is overproduction of oestrogens by peripheral conversion of androgenic precursors, a process that is increased in obese women. Other possible causes include granulosa or theca cell tumours of the ovary, adrenal adenoma, and adrenal carcinoma. Thus, endocrine evaluation including measurement of plasma oestradiol, oestrone and testosterone, and urinary 17-ketosteroids should be undertaken to localize and define the source of steroid. Further measures would depend on the laboratory findings,but might include various imaging techniques to define ovarian or adrenal disease, laparoscopy, and surgical exploration.

Hirsutism and baldness

Some increase in, or darkening and coarsening of, facial and body hair is very common after the menopause. It is usually mild and confined to the moustache, chin, and sideburn areas. This is due to the physiological decrease in oestrogen/androgen ratio produced by the reduction of oestrogen secretion by the ovary. For the same reason, those women with a genetic susceptibility to male-pattern baldness may note gradual thinning of hair in the typical frontal and crown areas in their 60s and 70s. If more severe or rapidly progressive hirsutism occurs, with appearance of heavy beard growth or increased chest and abdominal hair, especially if accompanied by signs of virilization, such as deepening of voice, clitoromegaly, increased libido, or rapidly progressive hair loss, hyperandrogenism of a pathological type should be suspected. If the testosterone level is high (>100 ng/dl), ovarian disease is likely. This may consist of hyperplasia of thecal elements of hilus cells of the ovary, or may be due to benign or malignant tumours. In general, ovarian tumours (whether adenomas or carcinomas) tend to be associated with higher levels of testosterone (>300 ng/dl) while hyperplastic tissue secretes somewhat less. Tumours of the adrenal gland secrete large amounts of weak androgens (androstenedione, dihydroepiandrosterone, etc.) and can thus be detected by the increased excretion of urinary 17-ketosteroids or by measurements of the appropriate plasma steroids by radioimmunoassay.

Tumours

Like younger women, postmenopausal women are at risk for neoplasia of the reproductive organs. Certain of these tumours are more common after the menopause. The incidence of endometrial cancer peaks in the 60s, probably because the endometrium of younger women is subject to intermittent maturation and shedding due to cyclic exposure to progesterone. After the menopause, women maintain a level of tonic oestrogen secretion from the adrenal and ovary, which is unopposed by progesterone. This may be sufficient to induce endometrial hyperplasia, which can progress to endometrial cancer. Women who are obese are at greater risk for endometrial carcinoma because their oestrogen levels tend to be higher due to peripheral conversion of steroid precursors to oestrogen by adipose tissue. Vulvar carcinoma is also more common in elderly women. This malignancy of the junctional epithelium is often preceded by a premalignant condition characterized by areas of atrophy and metaplasia of the vulvar skin known as kraurosis vulvae. The relationship, if any, of this condition to hormone deficiency is unknown. Finally, it should be borne in mind that ovarian cancer is one of the most common malignancies of women and remains a major cause of death. Although the incidence of ovarian cancer peaks between 45 and 55 years, elderly women remain at risk for this disease.

Bibliography

Antunes, C.M. *et al.* (1979). Endometrial cancer and estrogen use: report of a large case-control study. *New England Journal of Medicine*, **300**, 9–13.

Bush, T.L. *et al.* (1983). Estrogen use and all-cause mortality. *Journal of the American Medical Association*, **249**, 903–6.

Chetkowski, R. J. *et al.* (1986). Biologic effects of transdermal estrodiol. *New England Journal of Medicine*, **314**, 1615–50.

DeFazio, J. and Speroff, L. (1985). Estrogen replacement therapy: current thinking and practice. *Geriatrics*, **40**, 32–48.

Gambrell, R.D., Jr. (1982). The menopause: benefits and risks of estrogen–progestogen replacement therapy. *Fertility and Sterility*, **37**, 457–74.

Jensen, J., Riis, B.J., Strom, V., Nilas, L., and Christiansen, C. (1987). Long-term effects of percutaneous estrogens and oral progesterone on serum lipoproteins in postmenopausal women. *American Journal of Obstetrics and Gynecology*, **156**, 66–71.

Quigley, M.E.T., Martin, P.L., Burnier, A.M., and Brooks, P. (1987). Estrogen therapy arrests bone loss in elderly women. *American Journal of Obstetrics and Gynecology*, **156**, 1516–23.

Samsioe, G. (ed). (1982). Management of the female climacteric: benefits and risks. *Acta Obstetrica et Gynecologica Scandinavica* (suppl. 106).

Sherman, B.M., West, J.H., and Korenman, S.G. (1976). The menopausal transition: analysis of LH, FSH, estradiol, and progesterone concentrations during the menopausal cycles of older women. *Journal of Clinical Endocrinology and Metabolism*, **42**, 629–36.

Solomon, D.H., Judd, H.L., Sier, H.C., Rubenstein, L.Z., and Morley, J.E. (1988). New issues in geriatric care. *Annals of Internal Medicine*, **108**, 718–32.

Stampfer, M.J., Willett, W.C., Colditz, G.A., Rosner, B., Speizer, F.E., and Hennekens, C.H. (1985). A prospective study of postmenopausal estrogen therapy and coronary heart disease. *New England Journal of Medicine*, **313**, 1044–9.

Tataryn, I.V., Meldrum, D.R., Lu, K.H., Frumar, A.M., and Judd, H.L. (1979). LH, FSH, and skin temperature during the menopausal hot flash. *Journal of Clinical Endocrinology and Metabolism*, **49**, 152–3.

Upton, V. (1982). The perimenopause: physiologic correlates and clinical management. *Journal of Reproductive Medicine*, **27**, 1–27.

Vermeulen, A. (1976). The hormonal activity of the postmenopausal ovary. *Journal of Clinical Endocrinology and Metabolism*, **42**, 247–53.

Weiss, N.S., Ure, C.L., Ballard, J.H., Williams, A.R., and Darling, J.R. (1980). Decreased risk of fractures of the hip and lower forearm with postmenopausal use of estrogen. *New England Journal of Medicine*, **303**, 1195–8.

Whitehead, M.I., Townsend, P.T., Pryse-Davies, J., Ryder, R.A., and King, R.J.B. (1981). Effects of estrogens and progestins on the biochemistry and morphology of the postmenopausal endometrium. *New England Journal of Medicine*, **305**, 1599–605.

7.4 The hypothalamic–pituitary axes

S. MITCHELL HARMAN AND MARC R. BLACKMAN

INTRODUCTION

At one time there was a widely prevalent belief that deficiencies of one or more hormones might provide a simple explanation for all or most of the physiological changes seen with ageing; however, no serious gerontologist would espouse a view so naive today. Although a more contemporary point of view would emphasize the complexity and multivariate character of the biological ageing process, sophisticated investigations have produced considerable evidence that: (a) characteristic alterations in hormone secretion and action do occur during ageing, both in man and in animals; and (b) certain metabolic and physiological changes associated with normal ageing may actually be secondary phenomena caused by alterations in hormone balance or hormone response.

An important consideration in the study of endocrinology and ageing is the interaction of disease states and hormones. Such interactions have been described for a wide variety of diseases and nearly all the major hormone axes. As ageing is associated with an increased susceptibility to (and prevalence of) various illnesses, studies of the effects of ageing on hormone balance may be confounded by the effects of coexisting chronic or acute disease on endocrine function. In addition, the complex interactions of diet, physical activity levels, sleep–wake cycles, body weight and composition, and medications, all of which may vary with the age of the population studied, must be accounted for, as all may affect measurements of hormonal function.

It should also be borne in mind that, although at its simplest, endocrine function is evaluated by direct measurements (usually by radioimmunoassay) of hormone concentrations in plasma or urine, there is a wide variety of complex factors (besides disease states, as mentioned above) that modulate and alter the final effects of the hormones measured. First, it is clear that many hormones are secreted rhythmically with characteristic diurnal and even circannual patterns. For many hormones (e.g. growth hormone), characteristic target-organ responses reflect an integrated exposure to many hours or days of varying hormone levels and may depend not only on the absolute amount but also on the pattern and timing of the secretion. Thus, hormone concentrations in single or even a few blood samples taken at short intervals may not provide an adequate index of overall hormone activity in the organism. For other hormones (e.g. adrenocorticotropin), levels vary rapidly in response to short-term stimuli, and the tissues response is immediate and short-lived. In these cases, measurements must be interpreted in the context of the relevant physiological states (e.g. stress, blood glucose level, etc.). Other factors that modulate the effects of hormones include their binding to plasma carrier proteins, in which case biological activity depends mainly on the free hormone fraction, and their chemical modifications (such as glycosylation), producing molecular heterogeneity and alterations in the ratio of biological action to hormone concentration as determined by immunoassay.

Finally, measurements of hormone concentration must be interpreted in relation to alterations of: (a) rates of secretion, which reflect not only the number of available secretory cells but also the modulation of their activity by other hormones, neurotransmitters, circulatory factors, and local between-cell (paracrine) and intracellular (autocrine) interactions; (b) rates of clearance from the plasma, as a high blood level of hormone may be due to a reduction in the rate of its metabolism rather than an increase in secretion; (c) altered target-tissue sensitivity, which may, in turn, reflect changes in the number or function of hormone receptors, alterations in the amount or activity of substances responsible for transduction of receptor signals (e.g. G proteins or phosphokinases), or changes in the cell's internal physiology that limit its capacity to respond to such signals (e.g. altered gene function at the level of transcription, translation, or postsynthetic processing).

It is likely that most changes observed in endocrine function with age will result from some combination of the factors reviewed above, rather than from a single cause. Moreover, it is typical of the physiology of ageing that basal activities are sustained at normal levels until very late in life. Early changes tend to reduce the maximum response of which the system is capable. Thus, effects of ageing may be revealed when the system is stressed so that reserve capacity is called on. All of the above principles must be considered in a critical review of experimental data pertaining to effects of ageing on endocrine function.

EFFECTS OF AGEING ON POSTERIOR PITUITARY FUNCTION

The posterior pituitary gland, or neurohypophysis, is an intrinsic part of the central nervous system, being an embryonic outgrowth of the basal hypothalamus., It is connected directly to the base of the brain by the pituitary stalk, with which it is continuous, and contains numerous axon terminals from cells located in the supraoptic and paraventricular nuclei of the hypothalamus. These axons terminate on a network of capillary sinusoids, rather than on other neurones, and contain a number of neurotransmitters, including the two peptides vasopressin and oxytocin.

Arginine vasopressin, an 8 amino-acid peptide, also known as antidiuretic hormone (**ADH**), is the major physiologically active product of the posterior pituitary gland. Although ADH is an arterial vasoconstrictor in high concentrations, its major action is to make the distal tubules and collecting ducts of the renal medulla permeable to water, allowing the passage of free water from their lumina back into the interstitial fluid and hence the circulation. When ADH concentrations increase, more water is reabsorbed from the glomerular filtrate, the volume of the urine decreases, and its concentration and osmolarity increase. In the absence of ADH activity, large volumes (>10 l/day) of dilute urine are excreted.

The secretion of ADH is normally controlled by two kinds of physiological input. In the usual state, ADH-secreting neurones are regulated by osmoreceptor cells in the central nervous system, which detect plasma osmolarity, determined mainly by sodium concentration. Solutes that are distributed freely across cell membranes (e.g. urea or glucose), although they may contribute to osmolarity as measured by freezing-point depression, etc., have little or no effect on the secretion of ADH. Increases in plasma osmolarity (i.e. high sodium concentrations) lead both to increased thirst (water intake) and to an increase in the rate of secretion of ADH (water conservation). A fall in plasma osmolarity, as when large quantities of water are consumed, decreases the secretion of ADH and its concentration falls rapidly to a low basal level. The 'set point' or level of plasma osmolarity at which this secretion begins to increase varies considerably among individuals and may change from time to time in a single individual depending on a number of factors, including the levels of thyroid and glucocorticoid hormones, illness, and stress. The other factor regulating secretion of ADH is blood pressure, as detected by baroreceptors at various sites. This influence is usually less important, but may over-ride osmolar regulation, as when hypotension (e.g. acute haemorrhage) leads to high levels of secretion despite normal osmolarity.

Clinically significant hyponatraemia is more common in older persons. This may be, in part, because they are more prone to illnesses or to taking medications (see below) that cause water retention or sodium loss at the kidney, but older persons also appear to have a greater likelihood of responding adversely to these influences. This tendency to hyponatraemia, however, does not appear to be due mainly to alterations with age in renal function. Although the maximal urinary-concentrating ability does tend to diminish with age, this decrease has recently been shown to be much smaller than previously reported; moreover, the renal response to exogenously administered ADH appears to be age-invariate. Rather, older people have been shown, on average, to have higher levels of secretion of ADH for a given osmolarity, which is to say, an alteration downward in their osmolar set point and higher 'basal' levels of plasma ADH. Nearly 75 per cent of patients with the syndrome of inappropriate ADH secretion (**SIADH**) are over 65 years of age.

Consistent with the observation of higher plasma levels of ADH in older persons, the hypothalamus and neurohypophysis show, not degeneration of neurones or axonal pathways, but rather changes typical of increased hormone synthesis. Similarly, there is an increased content of ADH in the hypothalamus with age. In addition, even when compared with young subjects having similar basal levels, older people have 2- to 2.5-fold greater increases in plasma ADH in response to increases in osmolarity induced by hypertonic saline infusion, and less suppression of ADH secretion after ingestion of ethanol. Thus, taken together with the finding of no decrease in distribution space or metabolic clearance rate of ADH, the available evidence points strongly to a tendency for augmented secretion of ADH with age in humans. An exception is that recumbency followed by standing up leads to a subnormal ADH response in older subjects. Given the brisk secretory response to osmolar stimuli, it is likely that the blunted volume–pressure regulation of ADH secretion with age is due to defects in the baroreceptors or afferent neural pathways, rather than in the neurohypophysis.

Clinical hyponatraemia may present in older people without obvious concomitants, but more commonly it is related to physiological stress, or to medication. It is essential that patients be differentiated into those with: (a) hypervolaemic hyponatraemia (e.g. fluid retention due to heart, liver, or kidney failure); (b) hypovolaemic hyponatraemia (in which increased secretion of ADH and water intake is an appropriate response to dehydration and salt loss due to diuretics, mineralocorticoid insufficiency, renal tubular salt-wasting, etc.); and (c) euvolaemic hyponatraemia (as in SIADH). The physician must determine whether cardiac or renal failure, stroke or other trauma to the central nervous system, medications (especially sulphonylreas or diuretics), or endocrine disease (hypothyroidism or adrenal failure) are present. Risk factors for SIADH also include surgery or other trauma, anaesthetics, and infections (e.g. pneumonia). Although ectopic secretion of ADH by neoplasms (especially small-cell carcinoma of the lung) must always be considered, it is not this form of excess of ADH to which older patients are particularly predisposed.

Symptoms of hyponatraemia include weakness, hyporeflexia, muscle cramps, lethargy, disorientation, and, in severe cases, coma and seizures. Mild euvolaemic hyponatraemia can be treated by removal of the offending agent (when possible) and moderate fluid restriction (usually less than 1.5 l/day of free water). In more severe cases (severe central nervous symptoms, plasma sodium <120 mEq/l), infusion of hypertonic saline may be necessary in the acute phase. This should almost always be done in consultation with an endocrine or renal specialist. The usual recommendation calls for the use of 3 per cent saline given at a rate of about 0.1 ml/kg.min until the plasma sodium reaches 125 mEq/l. In the elderly patient this should be done more slowly and stopped sooner than in younger patients in order to avoid excessive or overly rapid shifting of fluid out of the intracellular compartment (cerebral dehydration) and circulatory overload (congestive failure).

Patients with hyponatraemia require careful and frequent monitoring of fluid intake and output, cardiac status, and plasma sodium until the situation is completely rectified (sodium concentration of 130 mEq/l or more). In certain cases in which fluid restriction proves inadequate to treat the chronic hyponatraemia due to excessive secretion of ADH, oral demeclocycline, a tetracycline derivative that blocks the action of ADH on the renal tubule, may be useful. Again, close monitoring is essential, with attention to avoiding producing hyponatermia and its clinical consequences.

EFFECTS OF AGEING ON ANTERIOR PITUITARY FUNCTION

The anterior pituitary gland (adenohypophysis) is not a part of the central nervous system, but is derived from an outpouching (Rathke's pouch) of the epithelium of the dorsal pharynx that is sealed off from the digestive tract by formation of the basal cranial bones and comes to lie in the sella turcica just anterior to and in contact with the neurohypophysis. It consists of a number of cell types mixed in various proportions throughout the gland, each of which secretes one or more of a family of related protein hormones. These hormones exert a wide variety of effects throughout the body, regulating metabolism, growth, thyroid, adrenal, and reproductive function. The anterior pituitary has been referred to as the 'master gland' because several of its hormones, thyrotropin, adrenocorticotropin, and the gonadotropins, luteinizing hormone and follicle-stimulating hormone (**FSH**), stimulate hormone secretion by other endocrine glands (thyroid, adrenal cortex, and gonad). The other anterior pituitary hormones, growth hormone and prolactin, act on non-endocrine target tissues. Growth hormone exerts its effects directly at some sites, and indirectly at others by stimulating production of an insulin-like growth factor (**IGF-1**), also called somatomedin-C. Prolactin stimulates its target cells directly without intermediate factors.

Secretion of the anterior pituitary hormones is regulated by specific hypothalamic hormones or neurotransmitters, some of which stimulate and others of which inhibit. These factors, all but one of which are peptides, are products of neurosecretion and reach the anterior pituitary gland via a short network of portal vessels originating in the median eminence of the hypothalamus. The response of most kinds of anterior pituitary cells to hypothalamic input is further modulated by negative feedback inhibition from the products of their target cells, which reach the pituitary via the systemic circulation. In addition, these products may also influence production of the hypophysiotropic neurosecretory factors at the hypothalamic level.

In ageing humans the anterior pituitary gland is moderately decreased in size. It typically contains areas of fibrosis, local necrosis, and cyst formation. Cells may have extensive deposits of lipofuscin, and regional deposits of amyloid are also common. Recent immunocytochemical investigations have not revealed any prominent age associated alterations in the proportions of different types of pituitary secretory cells. The pituitary content of growth hormone, prolactin, and thyrotropin appears to be age-invariate, while luteinizing hormone and FSH are somewhat increased in older persons.

Growth hormone

Native growth hormone is a 191 amino-acid peptide secreted by the somatotropic cells of the pituitary gland. Although it is essential for normal growth and development in children, its role in adult life was long thought to be of little significance. Moreover, until recently, the expense and limited supply of human growth hormone from pituitary extracts restricted its use to children with deficiency and growth failure. Now that recombinant, synthetic human growth hormone is available it has become possible to assess its potential therapeutic role in adults of all ages. Recent evidence has made it increasingly apparent that growth hormone is an important anabolic hormone that reverses negative nitrogen balance, increases protein synthesis, causes positive calcium balance and bone growth, and stimulates lipolysis with concomitant increases in plasma free fatty acids and a reduction in the percentage of body fat mass. The fact that human ageing is typically characterized by loss of muscle and bone mass and an increase in the percentage of body fat suggests that the secretion and/or action of growth hormone may be reduced in older persons.

Most peripheral tissue actions of growth hormone are mediated by somatomedin-C, also known as IGF-1. Most of the circulating IGF-1 is generated in the liver, but it is also produced at other sites of growth-hormone action such as osteoblastic cells in bone, where it may act locally, In addition, circulating IGF-1 exerts feedback inhibitory effects on the secretion of growth hormone by pituitary somatotropic cells. Secretion of growth hormone is mainly modulated by two hypothalamic peptides released into the pituitary portal circulation. These are growth hormone-releasing hormone (**GHRH**), a 44 amino-acid peptide which stimulates release of growth hormone, and somatostatin, a peptide found in the hypothalamus as well as numerous other tissues, which inhibits the secretion of growth hormone (as well as the release of various other hormones).

The effect of ageing on the secretion of growth hormone in humans has been evaluated by a number of researchers. Some early studies reported unchanged baseline plasma levels of the hormone. However, it is normally secreted in rhythmic pulses, with highest frequency and amplitude associated with the onset (first 3–4 h) of sleep. Therefore, random single samples do not adequately characterize the daily dynamics of growth-hormone secretion. More recent studies, in which the hormone was measured in samples taken frequently over 24 h have shown a decrease in 24-h integrated concentrations of growth hormone, and a substantial decrease in amplitude of the spontaneous pulses of the hormone during sleep in elderly men and women. Despite the known alterations of sleep pattern in the aged, the decrease in the maximal secretory activity of growth hormone does not appear to be directly associated with the observed reduction in rapid eye-movement in sleep. Another major factor that modulates the secretion of growth hormone is the plasma oestrogen level. After puberty, secretion of growth hormone is significantly greater in women than men, and this difference disappears after the menopause, so that older persons of both sexes have similarly diminished levels of the hormone. Oral oestrogen treatment appears to augment spontaneous and exercise-induced secretion of growth hormone, but simultaneously reduces the blood levels of IGF-1. In contrast, physiological doses of oestradiol given transdermally to postmenopausal women do not alter basal plasma levels of growth hormone or IGF-1, and may actually decrease the secretory response of growth hormone to GHRH. The basis for the apparently discrepant effects of oral versus transdermal oestrogens may reside in the differential actions of the two routes on the hepatic modulation of IGF-1.

The responses of growth hormone in older persons to indirect secretagogues (exercise, L-dopa, arginine, and insulin-induced hypoglycaemia) have been variously reported as showing no change or a decrease with ageing. Some investigators have shown that there are growth-hormone responses to direct pituitary stimulation with GHRH, but that these are significantly reduced in apparently healthy older men and women. In another study, only a non-significant downward trend with age in the peak growth-hormone response was found in men. These discrepant findings could be explained by differences among the populations studied in a number of important physiological variables that modulate secretion of growth hormone. These include adiposity and lean body mass, caloric intake, psychological status (e.g. depression), and sex steroid hormone (especially oestrogen) levels.

Basal plasma levels of IGF-1 have been shown to decrease by 30 to 50 per cent on average in ageing men and women. Plasma levels of IGF-1 correlate well with the integrated spontaneous secretion of growth hormone and provide a good index of peripheral tissue exposure to that hormone. The ability of IGF-1 to respond to growth hormone, whether administered exogenously or derived endogenously in response to GHRH, is preserved with ageing. It is therefore likely that the observed age-associated decrease in IGF-1 reflects the decrease in circulating growth-hormone rather than some acquired tissue resistance to growth-hormone effect. Experiments have shown that cultured fibroblasts from elderly (and also progeric) human donors both bind and respond to IGF-1 similarly to those derived from young subjects, but that the synergism between glucocorticoid and IGF-1 in stimulating DNA synthesis by fibroblasts may be lost in cells from older donors. The physiological significance of this finding is not yet known.

Recent studies have shown that short-term (7 and 8 days) treatment with recombinant human growth hormone increases the circulating levels of IGF-1, improves nitrogen retention, and stimulates bone metabolism in elderly men and women. These findings have been corroborated in two longer-term (4 months and 6 months) studies of growth hormone deficient younger adults, in which treatment with recombinant human growth hormone produced increases in the plasma IGF-1, nitrogen retention, lean body mass, and basal metabolic rates and decreases in the percentage of body fat and the serum cholesterol. Similar findings, showing restoration of rates of protein synthesis with exogenous growth hormone, have been made in old rats. In contrast, others have observed that after 5 weeks of therapy the anabolic effects of exogenous growth hormone were lost in middle-aged obese men. The potential value of treatments (e.g. exogenous GHRH or growth hormone) that would restore growth hormone and IGF-1 in elderly patients to levels characteristic of those in younger people remains to be investigated, but such studies may have important implications for issues of gerontological relevance such as osteoporosis, healing of pressure sores or surgical wounds, and restoration of muscle strength, bone density, and lean body mass.

Prolactin

Although their absolute number depends on sex, age, and endocrine status (e.g. pregnancy), more than half of the secretory cells of the adult anterior pituitary are lactotropes, cells that secrete the 198 amino-acid peptide, prolactin. Despite this fact, prolactin, which stimulates the acinar tissue of the female (oestrogen-conditioned) breast to secrete milk, appears to subserve no normal physiological function in the non-lactating adult. Although prolactin has considerable structural homology with growth hormone, it has neither significant binding to growth-hormone receptors nor growth-promoting action. When secreted in excessive amounts, prolactin appears to be an 'antireproductive' hormone, suppressing sex steroid production, reducing sexual libido in men and women, and causing impotence in men. These antisexual actions do not appear to depend entirely on its suppression of androgen production, as exogenous androgen replacement usually fails to restore libido and potency until prolactin levels have been reduced. Regulation of prolactin is unlike that of the other anterior pituitary hormones in that the predominant central nervous system mediated effect on its secretions is inhibitory. Thus, disconnection of the anterior pituitary from the hypothalamus leads to increased rather than diminished secretion of prolactin. The hypothalamic inhibitory factor for prolactin is not a peptide (like the other hypothalamic factors), but has been shown to be dopamine, a catecholamine neurotransmitter, which is secreted into the pituitary portal system at the median eminence. Physiologically important stimulators of pituitary prolactin secretion include thyrotropin-releasing hormone (TRH) and oestrogens. The prolactin content of the human pituitary and the circulating basal or TRH-stimulated levels of prolactin are 30 to 50 per cent greater in women than men, independent of age.

Studies of the effects of ageing on prolactin secretion have produced varied and contradictory results and no consensus has emerged. Some investigators have reported a decrease in basal levels of prolactin at the menopause in women, but others have not. In older men, basal levels of prolactin have been reported to be unchanged or to increase. Studies of the diurnal secretory rhythm of prolactin have shown both no change and a loss of the normal nocturnal peak in older men. The secretory response of prolactin to an injection of TRH has been variously reported to increase, decrease, and remain unchanged with age, and sulpiride, a dopamine-receptor antagonist, has been found to produce similar increments of plasma prolactin in healthy young and old men.

Given the inconsistency of the experimental findings, it is most likely that alterations in prolactin secretion in humans with ageing are small and probably do not contribute to the observed decrease in sexual activity characteristic of the ageing male. None the less, it should be borne in mind that prolactin-secreting adenomata of the pituitary can occur at any age. Moreover, a number of pharmacological agents, including all the major and minor tranquillizers, some antihypertensives, and many antidepressants have been associated with elevations of prolactin, and elderly patients are more likely to be treated with multiple medicines. Therefore, older patients with reproductive or sexual complaints should have a detailed history of medication use, should be examined for galactorrhoea, and have their plasma prolactin measured. If a significantly elevated plasma prolactin is found, subsequent diagnosis and therapy should be directed at eliminating offending medications, if possible, or at detecting and treating a prolactin-secreting pituitary tumour.

The gonadotropins

The anterior pituitary secretes two gonadotropic hormones, luteinizing hormone and FSH. In the male, luteinizing hormone stimulates the interstitial (Leydig) cells of the testis to produce testosterone, and FSH initiates and maintains the function of the seminiferous tubules via its action on the Sertoli cells. In the female, luteinizing hormone elicits theca-cell production of androgenic steroids (mainly androstenedione), which are converted to oestrogens (mainly oestradiol) by the follicular granulosa cells. In cycling females, secretion of luteinizing hormone rises to a mid-cycle peak, which activates the ovulatory mechanism, while FSH induces granulosa-cell proliferation and follicular maturation. Both luteinizing hormone and FSH are large glycoprotein molecules consisting of two non-covalently bound subunits; the α subunit is common to both hormones (as well as to thyroid-stimulating hormone and chorionic gonadotropin), whereas the β subunit of each molecule is unique and confers both immunological and biological specificity. Central control of gonadotropin secretion is exercised by gonadotropin-releasing hormone (**GnRH**), a 10-amino-acid peptide secreted into the pituitary portal circulation at the median eminence by axonal terminations of neurosecretory cells located mainly in the arcuate nucleus of the anteromedial hypothalamus. Gonadal steroid hormones exert negative feedback control on both GnRH and gonadotropin secretion. An additional peptide factor called inhibin, secreted by Sertoli cells in the male and granulosa cells in the female, feeds back to suppress production of FSH.

Most studies of the effects of ageing on gonadal function in men have shown a gradual increase in both luteinizing hormone and FSH after the age of 50. The increase in luteinizing hormone does not appear to be accounted for by a decrease in its metabolic clearance, but no data exist as yet for FSH. Although a few men in their eighth and ninth decades may have very high blood levels of FSH and luteinizing hormone, approaching those seen in postmenopausal women, in most the elevations are more modest. The age-associated increase in FSH is disproportionately greater than that of luteinizing hormone, suggesting that the function of seminiferous tubules (and inhibin secretion) may be more prominently affected by ageing than is secretion of gonadal steroids. Consistent with this concept, decreased plasma levels of inhibin have been reported in elderly men, in parallel with the known decrease in sperm production. Histological examination of testes from aged men has shown varying degrees of tubular involution, hyalinization, and fibrosis, but this tends to be patchy and may be related to local vascular or autoimmune changes. Leydig cells have been reported to be normal in some studies and reduced in number in others. The total plasma testosterone in healthy men tends either to be unchanged or moderately decreased with age, depending on the population studied. Because ageing is associated with a significant increase in sex hormone-binding globulin (with a resultant increase in the fraction of testosterone bound), there is a decrease in plasma free testosterone that is disproportionate to the change in total testosterone but still of only modest magnitude in healthy men. Thus, some of the observed increase in luteinizing hormone (and FSH) in men may be due to a decrease in feedback inhibition by free gonadal steroid (i.e. partial failure of Leydig cells). This conclusion seems to be supported by data showing a diminished secretory reserve of Leydig cells, as demonstrated by a reduced response to exogenously administered gonadotropin. This could be due to altered Leydig-cell function, a decrease in Leydig-cell number, or both.

There are also data which suggest that the feedback sensitivity of the hypothalamic–pituitary axis to sex-steroid inhibition may be decreased with age in men. Such a decrease could explain in part the observed increase in plasma levels of FSH and luteinizing hormone despite only a modest change in plasma testosterone. This point is controversial, because other investigators, using a somewhat different experimental design, have found greater rather than lesser feedback sensitivity of steroids in elderly subjects. Another possible explanation of the increase in gonadotropins is that ageing results in secretion of gonadotropins with a decreased ratio of bioactivity to immunoactivity (**B/I** ratio). Evidence for an altered B/I ratio with age for luteinizing hormone has been found in some but not all studies. Similarly, greater heterogeneity of charge and size has been found in FSH extracted from pituitaries of older, as compared with younger men, and a decrease in the B/I ratio for FSH has been reported in one study of elderly men. Thus some findings suggest that the age-associated increases in FSH and luteinizing hormone measured by radioimmunoassay may be due in part to altered pituitary processing of the gonadotropin molecules so that a portion of the measured hormone is not actually bioactive. Finally, despite the increase in basal concentrations of gonadotropins, the pituitary secretory responses of luteinizing hormone to GnRH appear to be both delayed and diminished in elderly men, suggesting that there may be some age-associated decrease in pituitary gonadotropic secretory function. Whether such change is intrinsic to the pituitary gonadotropic cells, or represents a loss of prior conditioning of such cells due to altered hypothalamic function (e.g. chronic decrease in GnRH stimulation) has not been determined in humans.

In women, a rise in the plasma levels of luteinizing hormone and FSH far greater than that in ageing men marks the onset of the menopause, an event related mainly to failure of ovarian secretory as well as germinal function. This subject is covered more thoroughly in Chapter 7.3.

The alteration in gonadotropic function with age in men is quite subtle and generally has little clinical significance. The physician should be aware that elderly men complaining of symptoms of hypogonadism, especially impotence, may have moderately elevated luteinizing hormone and FSH levels with 'lower limit of normal' testosterone and free testosterone, without any obvious pathology of the reproductive system. On the other hand, patients with profound decrease in plasma testosterone (e.g. to levels less than 150 ng/dl) should be given the same diagnostic attention as would any case of suspected primary hypogonadism. It must be recalled, however, that serious chronic disease (such as cancer, renal or heart failure), especially if accompanied by malnutrition and debilitation, may be associated with profound hypogonadism. Hypogonadism induced by concomitant illness may be primary (with elevated gonadotropin levels), but more often will be found to be secondary (decreased plasma luteinizing hormone and FSH) or mixed (failure of gonadotropins to increase to compensate for testis failure). Treatment of elderly, mildly hypogonadal men with testosterone replacement is controversial and, at present, is not standard medical practice, first because there are no controlled studies that convincingly demonstrate beneficial effects, and second, because prostatic hyperplasia and/or cancer may be induced or exacerbated by androgen treatment.

The pituitary–adrenal axis

The adrenocorticotropic cells of the anterior pituitary gland synthesize a peptide hormone precursor molecule known as pro-opiomelanocortin. This molecule is subsequently processed by peptidases in the secretory cells and, depending on where the peptide chain is severed, may give rise to various combinations of adrenocorticotropic hormone, α and β melanocyte-stimulating hormone, and β-lipotropin, which by further cleavage can release β-endorphin. Pro-opiomelanocortin is secreted by a number of different cell types besides the pituitary corticotrope, and its postsynthetic processing varies from one location to another. In the anterior pituitary the main products are adrenocorticotropic hormone, a 39 amino-acid peptide, β-lipotropin and α-melanocyte-stimulating hormone. Adrenocorticotropic hormone is the main modulator of adrenocortical function, directly stimulating the synthesis of the adrenal cytochrome P-450 enzymes and dehydrogenases necessary for production of cortisol, the major adrenal glucocorticoid, from cholesterol.

Secretion of adrenocorticotropic hormone is regulated chiefly by corticotropin-releasing hormone, a 41 amino-acid hypothalamic peptide, which stimulates its release, and by glucocorticoids (both natural cortisol and exogenous compounds such as prednisone, dexamethasone, etc.), which inhibit its secretion at both the pituitary corticotropic and hypothalamic levels. Adrenocorticotropic hormone and cortisol are normally secreted rhythmically in response to discrete pulses of corticotropin-releasing hormone released by the hypothalamus into the pituitary portal circulation. These pulses increase to maximal amplitude in the early morning hours and become smaller and less frequent during the late morning and afternoon, leading to a highly reproducible diurnal secretion pattern for adrenocorticotropic hormone and cortisol, having is nadir in the afternoon and evening and peak between 4.00 and 8.00 a.m. Cortisol is one of the body's major 'stress hormones' so that any serious stress (traumatic, infectious, or psychological) over-rides the normal regulatory mechanisms and results in hours or even days of much-increased adrenocorticotropic hormone and cortisol secretion.

In the absence of adequate secretion of adrenocorticotropic hormone, secondary adrenal insufficiency develops. Because, without adequate glucocorticoid, organisms succumb to nearly any major stress, adrenocorticotropic hormone (unlike growth hormone, prolactin, or the gonadotropins) is a life-sustaining hormone. Primary (Addison's disease) and secondary glucocorticoid insufficiency are characterized by weight loss, anaemia, hypotension, hyponatraemia, weakness, and fatigue. Glucocorticoid excess is also potentially life-threatening. Excessive glucocorticoid produces Cushing's syndrome, which is associated with hypertension, glucose intolerance, centripetal obesity, negative nitrogen balance (loss of muscle mass and strength), loss of calcium from bone (development of osteoporosis), fragility of skin and blood vessels, poor healing of connective tissue, and altered immune function (increased susceptibility to bacterial infection). It should be noted that 'normal' ageing includes changes that, although of lesser degree, are reminiscent of those associated with glucocorticoid excess, such as loss of muscle, increased body fat, and decreased bone calcium. Therefore, the characterization of hypothalamic–pituitary–adrenal function with age is of considerable interest.

Early studies in which pituitary–adrenocortical function was deduced from measurements of random plasma cortisol levels and urinary excretion of 17-hydroxycorticoids, both with and without stimulation of the pituitary–adrenocortical axis (e.g. insulin–hypoglycaemia, metyrapone administration), did not detect systematic age-related alterations in glucocorticoid levels in plasma, although 24-h urinary excretion of glucocorticoid metabolites was generally reduced. The findings appeared to be explained by a decrease in metabolic clearance rate of cortisol with a compensatory reduction in secretion rate. Later studies employing sensitive radioimmunoassays for adrenocorticotropic hormone revealed little change with age in its responses to metyrapone, but suggested altered diurnal rhythmicity of spontaneous secretion, with a reduction of the overall nadir-to-peak excursion and a shifting of the peak to later morning.

Other investigators, however, have observed that older patients respond to various stresses (surgery, depression) with greater and more prolonged secretion of cortisol than is seen in similarly stressed younger patients. This alteration appears to be of greater duration and magnitude than can be explained by the relatively minor decrease in the metabolic clearance rate observed for cortisol. When dexamethasone was used to suppress secretion of adrenocorticotropic hormone and cortisol, there was less of an inhibitory response in healthy elderly subjects or older depressed patients. When measured in the evening (nadir) before and after administration of ovine corticotropin-releasing hormone, a non-significant but suggestive trend was found toward higher basal levels of adrenocorticotropic hormone and greater responses of this and cortisol with age, despite the fact that basal cortisol levels in the p.m. period were higher in the older men. Taken together, these data suggest that ageing may be associated with a tendency for the negative feedback action of glucocorticoids on secretion of adrenocorticotropic hormone to diminish. Further investigations, using more sensitive methods, should clarify whether there really is such a defect in feedback inhibition in humans and whether it may be associated with a subtle, but clinically significant, increase in 24-h integrated exposure to cortisol in older people.

The pituitary–thyroid axis

Pituitary thyroid-stimulating hormone (**TSH**), also known as thyrotropin, is, like the gonadotropins, a large heterodimeric glycoprotein. The TSH α-subunit is identical to that of luteinizing hormone and FSH, while specific bio- and immunoreactivity are conferred by the unique TSH-β-subunit. TSH is the major direct modulator of thyroid function, stimulating uptake and organification of iodine and production and secretion of thyroid hormone(s), as well as growth and increased vascularity of thyroid tissue. In the absence of TSH, thyroid hormone output becomes insufficient and clinical hypothyroidism occurs. As thyroid hormones are essential to normal metabolic activity, profound hypothyroidism results eventually in coma, circulatory and respiratory collapse, and death. Therefore TSH is, like adrenocorticotropic hormone, an essential life-sustaining hormone. Synthesis and secretion of TSH are stimulated by a cyclic hypothalamic tripeptide, TRH. Thyroid hormones feed back to inhibit basal and TRH-stimulated production of TSH. When hormone levels of thyroid are reduced by primary thyroid failure, basal levels of TSH are very high and the responses of TSH to TRH are greatly augmented. In the presence of excess thyroid hormone, blood levels of TSH are very low and the response of TSH to TRH is absent. New, highly sensitive assays for TSH can distinguish between normal and diminished basal plasma levels and hence provide highly reliable information about pituitary thyrotropic function and its modulation by thyroid hormones.

Early studies of the effects of ageing on secretion of TSH revealed normal or somewhat elevated levels in otherwise healthy men and women. Large, community-based studies sampling TSH in hundreds of subjects found significant elevations in approximately 3 per cent of older men and 8 per cent of older women, which were accompanied in some, but not all, by reduced levels of circulating thyroid hormones. These results suggest that undetected primary hypothyroidism is common in older people. It is well known that in early thyroid failure there may be a phase of pituitary compensation, as shown by increased secretion of TSH, so that, at first, circulating thyroid hormones remain within the normal range. This would account for the observation of subjects with high TSH but normal thyroid hormone levels. Another possible explanation for the increase in levels of TSH is an age-associated increase in heterotypic autoantibodies that cross-react with the anti-TSH antibodies in the immunoassays.

In studies of healthy ageing men, who had no evidence of hypothyroidism, an ultrasensitive assay for TSH has shown a modest but significant age-associated increase in basal levels and concomitant small decreases in free (but not total) thyroxine (T_4) and total and free tri-iodothyronine (T_3); hormone levels, however, were still within the normal range. The latter findings suggest that ageing may be associated with a subtle decrease in the secretion of thyroid hormone in the absence of identifiable thyroid disease. The secretory response of TSH to bolus intravenous administration of TRH has been variously reported to be reduced with age in men but not women, decreased in women but not men, and increased in both sexes. The reasons for these discrepant findings have not been fully resolved but may be related to confounding variables such as the presence of thyroid disease and concomitant non-endocrine illness in the study populations. Low-dose constant infusion of TRH produces a biphasic response in TSH, with early and late peaks. In one study, both early and late responses to constant infusion of TRH were of similar magnitude, timing, and duration in older and younger men. The expected augmentation of the TSH response, in the presence of the significantly lower levels of free thyroid hormone found in the older group, was not evident. Recently, 24-h monitoring of frequent blood samples has shown a 50 per cent reduction in spontaneous release of TSH in elderly men, despite normal levels of T_4 and slightly reduced T_3. These findings suggest that elderly men may have a subtle decrease in the basal secretion of TSH relative to the level of thyroid function.

It should be evident from the above that, despite the frequency of dry skin, cold intolerance, and a general slowing of body processes (and reduced basal metabolic rate), which occur in 'normal ageing', there is no evidence that ageing *per se* is normally a hypothyroid state. The major clinical significance of the physiological changes with age in pituitary–thyroid function relate to the interaction of thyroid function with systemic illness. Both severe acute (e.g. sepsis) and less severe chronic (e.g. renal or cardiac failure) illness may be associated with decreases in total and also in free plasma T_4. This so-called euthyroid sick syndrome is often difficult to differentiate from true hypothyroidism. One method of doing so is to demonstrate the augmented response of TSH to TRH expected in primary thyroid failure. This augmentation is not found in the euthyroid sick syndrome. Studies in severe illness have suggested that, in some patients, reduced secretion of TSH (the 'sick' thyrotrope) may even be the cause of the euthyroid sick syndrome. Unfortunately, in elderly subjects, and especially in elderly men, there is also frequently no augmentation of TSH secretion by low levels of T_4. Therefore, it is often difficult to distinguish between hypothyroidism and the euthyroid sick syndrome in the elderly patient.

Bibliography

Asnis, G.M., *et al.* (1981). Cortisol secretion in relation to age in major depression. *Psychosomatic Medicine*, **43**, 235–42.

Bellantoni, M., Harman, S.M., Cho, D., and Blackman, M. (1990). Transdermal estrogen replacement therapy decreases GH responsivity to GHRH in postmenopausal women of different ages. *Journal of Clinical Endocrinology and Metabolism*, **72**, 172–8.

Blackman, M.R. *et al.* (1986). Basal serum prolactin levels and prolactin responses to constant infusions of thyrotropin releasing hormone in healthy aging men. *Journal of Gerontology*, **41**, 699–705.

Blackman, M.R. (1987). Pituitary hormones and aging. *Clinics in Endocrinology and Metabolism*, **16**, 981–94.

Blichert-Toft, M. (1975). Secretion of corticotrophin and somatotrophin by the senescent adenohypophysis in man. Acta *Endocrinologica* (Copenhagen), **78** (suppl. 145), 15–154.

Burrows, G.N. *et al.* (1981). Microadenomas of the pituitary and abnormal sellar tomograms in an unselected autopsy series. *New England Journal of Medicine*, **304**, 156–8.

Conover, C.A. *et al.* (1985). Somatomedin-C-binding and action in fibroblasts from aged and progeric subjects. *Journal of Clinical Endocrinology and Metabolism*, **60**, 685–91.

Conover, C.A., Rosenfeld, R.G. and Hintz, R.L. (1985). Aging alters somatomedin-C–dexamethasone synergism in the stimulation of deoxyribonucleic acid synthesis and replication of cultured human fibroblasts. *Journal of Clinical Endocrinology and Metabolism*, **61**, 423–8.

Conover, C.A., Rosenfeld, R.G., and Hintz, R.L., (1987). Somatomedin C/insulin-like growth factor I binding and action in human fibroblasts aged in culture: impaired synergism with dexamethasone. *Journal of Gerontology*, **42**, 308–14.

Deslypere, J.P. and Vermeulen, A. (1984). Leydig cell function in normal men: effect of age, lifestyle, residence, diet and activity. *Journal of Clinical Endocrinology and Metabolism*, **59**, 955–62.

Goldstein, C.S., Braunstein, S. and Goldfarb, S. (1983). Idiopathic syndrome of inappropriate antidiuretic hormone secretion possibly related to advanced age. *Annals of Internal Medicine*, **99**, 185–8.

Gregerman, R.I. (1986). Mechanisms of age-related alterations of hormone secretion and action. An overview of 30 years of progress. *Experimental Gerontology*, **21**, 345–65.

Harman, S.M. and Tsitouras, P.D. (1980). Reproductive hormones in aging men. I. Measurement of sex steroids, basal luteinizing hormone, and Leydig cell response to human chorionic gonadotropin. *Journal of Clinical Endocrinology and Metabolism*, **51**, 35–40.

Harman, S.M. *et al.* (1982). Reproductive hormones in aging men: II. Basal pituitary gonadotropins and gonadotropin responses to luteinizing hormone-releasing hormone. *Journal of Clinical Endocrinology and Metabolism*, **54**, 547–51.

Harman, S.M., Wehmann, R.E. and Blackman, M.R. (1984). Pituitary–thyroid hormone economy in healthy aging men: basal indices of thyroid function and thyrotropin responses to constant infusions of thyrotropin releasing hormone. *Journal of Clinical Endocrinology and Metabolism*, **58**, 320–6.

Helderman, J.H. *et al.* (1978). The response of arginine vasopressin to ethanol and hypertonic saline in man: The impact of aging. *Journal of Gerontology*, **33**, 39–47.

Ho, K. Y. *et al.* (1987). Effects of sex and age on the 24-hour profile of growth hormone secretion in man: importance of endogenous estradiol concentrations. *Journal of Clinical Endocrinology and Metabolism*, **64**, 51–8.

Hossdorf, T. and Wagner, H. (1980). Prolaktin-Sekretion bei gesunden Männern und Frauen unterschiedlichen Alters. *Aktuel Gerontologie*, **10**, 119–26.

Johanson, A. J. and Blizzard, R.M. (1981). Low somatomedin-C levels in older men rise in response to growth hormone administration. *Johns Hopkins Medical Journal*, **149**, 115–17.

Kalk, W.J. *et al.* (1973). Growth hormone responses to insulin hypoglycemia in the elderly. *Journal of Gerontology*, **28**, 431–3.

Kirkland, J. *et al.* (1984). Plasma arginine vasopressin in dehydrated elderly patients. *Clinical Endocrinology*, **20**, 451–6.

Lang, I. *et al* (1987). Effects of sex and age on growth hormone response to growth hormone-releasing hormone in healthy individuals. *Journal of Clinical Endocrinology and Metabolism*, **65**, 535–40.

Marcus, R. *et al.* (1990). Effects of short-term administration of growth hormone to elderly people. *Journal of Clinical Endocrinology and Metabolism*, **70**, 519–27.

Marrama, P. *et al.*, (1984). Decrease in luteinizing hormone biological activity/immunoreactivity ratio in elderly men. *Maturitas*, **5**, 223–31.

Meites, J., Goya, R. and Takahashi, S. (1987). Why the neuroendocrine system is important in aging processes. *Experimental Gerontology*, **22**, 1–15.

Meyers, J.S., Matsumoto, A.M. and Bremner, W.J., (1986). Basal and clomiphene citrate (CC) stimulated secretion of luteinizing hormone (LH) and testosterone (T) in healthy young and elderly men. *Clinical Research*, **34**, 430A.

Pavlov, E.P. *et al.* (1986). Responses of growth hormone and somatomedin-C to GH-releasing hormone in healthy aging men. *Journal of Clinical Endocrinology and Metabolism*, **62**, 595–600.

Pavlov, E.P. *et al.* (1986). Responses of adrenocorticotropin, cortisol, and dehydroepiandrosterone to ovine corticotropin-releasing hormone in healthy aging men, *Journal of Clinical Endocrinology and Metabolism*, **62**, 767–72.

Rudman, D. *et al.* (1990). Effects of human growth hormone in men over 60 years old. *New England Journal of Medicine*, **323**, 1–6.

Tenover, J.S. *et al.* (1988). Decreased serum inhibin levels in normal elderly men: evidence for a decline in Sertoli cell function with aging. *Journal of Clinical Endocrinology and Metabolism*, **67**, 455–9.

Tenover, J.S., Dahl, K.D., Hsueh, A.J. and Lim, P. (1987). Serum bioactive and immunoreactive follicle-stimulating hormone levels and the response to clomiphene in healthy young and elderly men. *Journal of Clinical Endocrinology and Metabolism*, **64**, 1103–8.

van Coevorden, A., Laurent, E., Decoster, C. and Kerkhofs, M., (1989). Decreased basal and stimulated thyrotropin secretion in healthy elderly men. *Journal of Clinical Endocrinology and Metabolism*, **69**, 177–85.

Winters, S.J. and Troen, P. (1982). Episodic luteinizing hormone (LH) secretion and the response of LH and follicle-stimulating hormone to LH-releasing hormone in aged men. Evidence for coexistent primary testicular insufficiency and an impairment in gonadotropin secretion. *Journal of Clinical Endocrinology and Metabolism*, **55**, 560–5.

Winters, S.J., Sherins, R.J. and Troen, P. (1984). The gonadotropin-suppressive activity of androgen is increased in elderly men. *Metabolism*, **33**, 1052–9.

Zadik, Z. *et al.* (1985). The influence of age on the 24-hour integrated concentration of growth hormone in normal individuals. *Journal of Clinical Endocrinology and Metabolism*, **60**, 513–16.

7.5 Hypo- and hypercalcaemia

L. RAISZ AND CAROL C. PILBEAM

Although the most common disorders of mineral metabolism in older persons are the diseases of bone, such as osteoporosis, osteomalacia, and Paget's disease, which are discussed elsewhere in this volume, disorders of calcium regulation also occur with increasing frequency. This is particularly true of hyperparathyroidism, which occurs in many older individuals and which present special problems in diagnosis and management. While the general principles of calcium regulation as described in the *Oxford Textbook of Medicine* generally apply in older patients, there are changes in hormone synthesis and secretion and in organ response with ageing that affect both the relative importance and the efficacy of specific regulatory mechanisms.

THE DISTRIBUTION AND FUNCTION OF CALCIUM

The necessity for tight regulation of the extra- and intracellular calcium concentration is not diminished with ageing. However, the problem of maintaining these concentrations may be increased because of diminished intake and absorption of calcium, loss of skeletal stores, and accumulation of mineral in non-skeletal tissues, such as calcified vessels. While it is clear that the ageing organism must overcome obstacles in maintaining the extracellular calcium concentration, little is known about the effects of ageing on the problem of maintaining the intracellular concentration. As intracellular calcium is only 0.1 to 1 μmol, while the extracellular is over 1 mmol, a substantial concentration gradient must be maintained by active transport. This gradient is maintained by calcium extrusion and binding systems within the cell, which can rapidly take up or remove calcium. These include the sarcoplasmic reticulum and other microsomal structures as well as the cell membrane and mitochondria. Movement of calcium within the cell and into the cell through specific, gated, calcium channels regulates many cellular responses to hormones, neural mediators, and pharmacological agents. Cell damage is often associated with an impairment of this regulation and with intracellular accumulation of calcium. While some studies of cellular ageing suggest that impairment of intracellular calcium regulation is a marker for senescence, there is no evidence that this is a general phenomenon of ageing of the whole organism.

The tight regulation of extracellular calcium is probably necessary not only to enable the cell to maintain its intracellular concentration and distribution, but also because of the role of calcium in membrane function. Changes in extracellular calcium concentration result in changes in neuromuscular activity and affect the function of many organs, particularly the brain and kidney. To maintain these vital calcium concentrations, terrestrial organisms have developed a complex system of regulation in which there are large movements of calcium in and out of bone and across the gut and the kidney, with minimal changes in calcium ion concentration. The skeleton, containing 99 per cent of total body calcium, is a storehouse which the organism can draw upon to maintain the small but essential amounts of calcium in the extracellular fluid and cells during periods of calcium deprivation. However, the maintenance of extracellular and cellular calcium may occur at the expense of the structural integrity of the skeleton, particularly in older persons in whom calcium supplies and absorption are diminished.

The movement of calcium from bone to extracellular fluid, as well as the transport across intestine and kidney, are regulated largely by parathyroid hormone and 1,25-dihydroxyvitamin D (**1,25(OH)$_2$D**). The third calcium-regulating hormone, calcitonin, probably plays a much smaller role. The production of these hormones is regulated, directly or indirectly, by the ionized calcium concentration itself, providing a homeostatic feedback system which maintains the ionized calcium concentration with variations of less than 5 per cent in normal individuals.

Although it is the ionized calcium concentration that determines cell function and is actually regulated, in most clinical situations only the total plasma calcium concentration is measured. The normal total serum calcium ranges between 2.1 and 2.6 mmol (8.4 to 10.4 mg/100 ml). However, the ionized fraction is maintained over a narrower range of 1.2 to 1.3 mmol (4.8 to 5.2 mg/100 ml). Most of the non-ionized calcium is bound to serum proteins, particularly albumin, while a small amount is complexed to anions, such as phosphate, bicarbonate, and citrate. Because a decreased serum albumin is common in elderly individuals, particularly if there is some chronic illness, a low total serum calcium is frequently encountered in the presence of normal ionized calcium. Thus it is important either to measure the ionized calcium concentration directly or to estimate it from a simultaneous measure of serum albumin. A simple rough correction is to add back 0.2 mmol (0.8 mg/100 ml) of calcium for each 1 g/100 ml decrease in the serum albumin concentration below a normal of 4.0 g/100 ml. Thus a patient with a total calcium of 2.0 mmol and a serum albumin of 2.5 g/100 ml would have a deficit of approximately 0.3 mmol of protein-bound calcium (0.2 mmol × 1.5 g/100 ml albumin deficit), and the corrected total calcium of 2.3 mmol would be in the normal range. Conversely, a low albumin concentration may mask the small increase in ionized calcium concentration that occurs in mild hyperparathyroidism.

In interpreting any abnormality of serum calcium it is necessary to assess other ions that can either affect calcium binding or the response to changes in calcium concentration. For example, alkalosis results in an increase in neuromuscular irritability that can produce tetany, even without hypocalcaemia, while acidosis has a reverse effect and prevents tetany in some hypocalcaemic individuals. Potassium and magnesium levels can alter cellular responses to calcium, and magnesium can affect the secretion of and response to parathyroid hormone. Magnesium may also be able to displace calcium from its binding sites on albumin.

Measurement of the serum phosphate is important both

for diagnosis and for its possible effect on calcium distribution. It is important to realize that, unlike that of calcium, the plasma phosphate concentration can vary considerably with intake in the same individual, and has a diurnal rhythm. There are some data suggesting that the plasma phosphate increases with age, particularly in women. This may be due to increased release of phosphate from bone, as 88 per cent of total body phosphate is in the skeleton, or to age-related impairment of renal excretion of phosphate. However, some studies show a decrease in the serum phosphate in older individuals, possibly because the tendency for an increase is counterbalanced by an increase in secretion of parathyroid hormone and a resulting phosphaturia, or because phosphate intake and absorption are diminished.

CHANGES IN CALCIUM REGULATION IN ELDERLY PEOPLE

The three major organs for calcium regulation—bone, intestine, and kidney—all undergo age-associated changes which affect calcium transport and the response to calcium-regulating hormones. It is not clear how much of this change is due to ageing *per se* and how much to the accumulation of pathological events in elderly individuals. Some of the age-associated changes are due to changes in the production of other systemic hormones, particularly the reduction in sex hormones at the menopause in women, and the slower decrease in sex hormone production that occurs in some older men.

PARATHYROID HORMONE

The most important regulator of the serum calcium is parathyroid hormone. This single-chain, 84 amino-acid polypeptide is secreted in a pulsatile fashion and responds to changes in the serum calcium concentration. Secretion is diminished when the calcium concentration is elevated, but not blocked completely. The control is largely proportional, that is, as the serum calcium increases, the hormone secretion decreases. There is also a form of integral control, in that prolonged elevations or depressions in the serum calcium result in atrophy or hyperplasia of the parathyroid glands, respectively. When a hypercalcaemic state is reversed, there may be a period of transient hypoparathyroidism until the atrophied glands regenerate. More importantly, when a hypocalcaemic stimulus is removed, there is slow involution of the hyperplastic tissue. An increased serum calcium may not completely stop the secretion of parathyroid hormone from these hyperplastic glands and hypercalcaemia may result. In addition to being regulated by calcium, secretion of parathyroid hormone can be inhibited by $1,25(OH)_2D$ and stimulated, but probably only transiently, by catecholamines.

There is an age-associated increase in the serum concentration of both immunoreactive and biologically active parathyroid hormone. Although the reasons for this are not completely understood, there are several possible mechanisms. Synthesis and secretion of parathyroid hormone itself does not appear to be altered. The only morphological change is an increase in the number of oxyphil cells, which are rich in mitochondria and probably not major secretory cells. Early studies showing an increase in immunoreactive parathyroid hormone were at first considered to reflect an increase in the circulation of C-terminal fragments. These biologically inactive fragments are produced by the rapid degradation of parathyroid hormone, which has an extremely short half-life of only 2 to 3 min. There is cleavage and destruction of the first 32 to 34 amino acids, which are the biologically active portion of the molecule, while the C-terminal fragments continue to circulate and are excreted by the kidney. With the common (but not universal) decrease in renal function with age, the C-terminal fragments may accumulate. However, subsequent studies have shown that the increase in the hormone with age also obtains for the intact, 84 amino-acid, secreted form of the molecule, which is actually responsible for most biological activity on target organs. This has been measured both by new immunoassays and by highly sensitive cytochemical bioassays. An increase in the concentration of parathyroid hormone in the presence of a normal serum concentration of ionized calcium may indicate end-organ resistance to the hormone with ageing. However, the increase in parathyroid hormone may be due to a small decrease in ionized calcium within the normal range. The latter supposition is supported by the observation that moderate supplementation with vitamin D or calcium can lower the level of the hormone in older individuals.

The three major actions of parathyroid hormone which are responsible for maintaining the serum calcium concentration are increased bone resorption, increased renal tubular reabsorption of calcium, and an indirect increase in absorption of calcium from the gut, mediated by the increased renal synthesis of $1,25(OH)_2D$. The inhibition of renal tubular reabsorption of phosphate, which results in phosphaturia and a lowering of the serum phosphate, could also play a role in maintaining the serum calcium. Lowering the serum phosphate can increase bone resorption, decrease bone mineralization, and increase the synthesis of $1,25(OH)_2D$.

There is little evidence that the effects of parathyroid hormone on bone resorption and renal tubular reabsorption of calcium are altered with age. The phosphaturic effect appears to be increased. The reabsorption of phosphate is mediated by an increase in the production of cyclic AMP, which is, in turn, reflected in increased nephrogenous excretion of cyclic AMP. This excretion per unit glomerular filtrate actually increases with age.

The increased parathyroid hormone with age has a limited effect on the intestinal absorption of calcium, presumably because of an age-associated decrease in the ability of the hormone to stimulate 1α-hydroxylase in the kidney and increase production of $1,25(OH)_2D$ (Silverberg *et al.* 1989). This response appears to be further decreased in patients with osteoporosis. Not only is there a decrease in the synthesis of $1,25(OH)_2D$, but there may also be a decrease in the intestinal response to this hormone. Intestinal absorption of calcium is diminished with age in both man and experimental animals, and although it remains responsive to $1,25(OH)_2D$, the magnitude of the response is often decreased.

VITAMIN D

For further discussion of the effect of vitamin D, see Chapter 14.2. Despite the fact that it is really a prohormone, we continue to use the term vitamin D_3. Its formation in the skin by ultraviolet irradiation of 7-dehydrocholesterol does not require any external supply. However, vitamin D_3 (cholecalciferol) and vitamin D_2 (ergocalciferol) are present in the diet and are often given as supplements. The conversion of vitamin D in the liver to 25-hydroxyvitamin D (**25(OH)D**) provides the major circulating form of the hormone. The 25(OH)D is bound tightly to a specific, vitamin D-binding protein (Gc protein), which results in a half-life of approximately 8 days for this circulating precursor. The active form, $1,25(OH)_2D$, is produced in the kidney by 1α-hydroxylase. It may also be produced in some other tissues and cells, for example, the placenta and activated macrophages. The active, hormone form $1,25(OH)_2D$ is also bound to specific binding protein but has a much shorter half-life of less than 1 day. Serum values are substantially lower than for $25(OH)_2D$. Another metabolite of vitamin D, $24,25(OH)_2D$, is formed in the kidney and may have biological activity in growth and development, but is generally considered to be an inactive form in adults.

Although the major effect of $1,25(OH)_2D$ on mineral metabolism is to increase intestinal absorption of calcium and phosphate, it has many other effects on a variety of tissues. The vitamin D receptor, which belongs to the family of steroid hormone receptors and mediates the action of the hormone by nuclear binding and regulation of transcription, is present in most cell types in the body. The major effect of ageing on vitamin D metabolism is impaired production of $1,25(OH)_2D$ by the kidney, but there may also be decreased production of the prohormone in skin. The latter is due largely to decreased exposure to sun, although with ageing the capacity to produce the prohormone in skin is also diminished. The 25-hydroxylase in the liver is probably not a limiting factor in vitamin D activation with ageing. Although low levels of 25(OH)D are frequently encountered, the administration of vitamin D can rapidly restore these to normal. There may be some impairment of the absorption of vitamin D supplements in the form of either cholecalciferol or ergocalciferol, but the therapeutic doses used are usually large enough to overcome this.

$1,25(OH)_2D$ itself is an extremely potent hormone and doses of as little as $1\mu g$/day orally, in the form of the drug calcitriol, can produce toxic side-effects. Moreover, elderly individuals remain sensitive to the administration of this agent. Not only is there a brisk increase in calcium absorption if adequate doses are given but treated individuals may also show hypercalciuria and even hypercalcaemia, with consequent renal damage, in response to relatively small increases in dose.

CALCITONIN

Calcitonin is a potent inhibitor of bone resorption and is present in large amounts in the parafollicular or C cells of the thyroid gland, yet its existence was not suspected until 1960. The reason for this is probably that excesses and deficiencies of calcitonin are not associated with any abnormality of calcium regulation. Patients with medullary carcinoma of the thyroid, a tumour of the C cells that can produce enormous amounts of this hormone, do not show abnormal levels of serum calcium. Deficiency of calcitonin occurs in hypothyroidism, particularly when the thyroid is ablated surgically, yet these patients are able to regulate calcium normally.

Infusions of calcium in normal humans produce a transient increase in calcitonin secretion. Calcitonin in its monomeric, 32 amino-acid peptide form, may not be the only immunoreactive form in the plasma. Calcitonin is synthesized as a much larger precursor, and there is an alternative transcript for a hormone called 'calcitonin gene-related peptide', which is also produced by the C cells. This peptide is found in the nervous system and vessels and may have important regulatory functions in these tissues. Most studies show a lower rate of calcitonin secretion in women than in men, but there is controversy concerning the effect of age on secretion. An age-related decline has been reported, but not by all investigators. This discrepancy may be due to differences in assay systems.

The main importance of calcitonin in disorders of calcium regulation is as a therapeutic agent in hypercalcaemia. It is also useful as an inhibitor of bone resorption in Paget's disease and osteoporosis. Unfortunately, its effect in hypercalcaemia is short-lived. It is a potent and rapid inhibitor of bone resorption, but there appears to be an escape, which may represent down-regulation of calcitonin receptors or postreceptor desensitization. Thus, despite the continued administration of large doses of this hormone, the osteoclasts become resistant and hypercalcaemia recurs.

OTHER HORMONES THAT INFLUENCE CALCIUM METABOLISM

Although parathyroid hormone, $1,25(OH)_2D$, and calcitonin are the major calcium-regulating hormones, many other hormones can affect bone, kidney, or intestine and alter calcium regulation. Moreover, there are age-associated changes in the production of, or response to, these hormones. Growth hormone can stimulate bone growth, increase calcium absorption in the intestine, and increase tubular reabsorption of phosphate in the kidney. Both the activation of vitamin D to $1,25(OH)_2D$ in the kidney and the absorptive response in the intestine appear to be enhanced by growth hormone and decreased when it is deficient. The effect on bone growth is probably mediated by the stimulation of local production of insulin-like growth factor 1 (IGF-1, somatomedin C). With age there is a decrease in the secretion of growth hormone and in levels of somatomedin C in the blood. There may also be changes in the pattern of growth hormone secretion; sex- and age-associated differences in the frequency and magnitude of growth hormone pulses have been observed, which could influence the end-organ response. There is no evidence that either growth hormone or somatomedin C play an important part in calcium regulation, but they clearly have a role in bone formation, and thus influence mineral metabolism.

Thyroid hormones have a direct, stimulatory effect on bone resorption and can increase bone turnover. Hyperthyroidism is associated with a mild increase in the serum calcium concentration, and a decrease in levels of parathyroid hormone, which may be associated with some decrease in the production of $1,25(OH)_2D$ and in intestinal absorption of calcium. In hypothyroidism, the serum calcium usually remains normal; however, increases in the production of parathyroid hormone and 1,25$(OH)_2$D, and in intestinal absorption of calcium have been noted. Hypothyroidism may also bring about a partial deficiency of parathyroid hormone secretion. While there is no systematic change in the production of thyroid hormone with age, both hyper- and hypothyroidism become increasingly common in elderly people, particularly women.

Glucocorticoids have powerful effects on the skeleton, kidney, and intestine, and important interactions with parathyroid hormone and 1,25$(OH)_2$D. Their major direct action on the skeleton is to decrease bone formation. Glucocorticoids directly inhibit the intestinal absorption of calcium and oppose the action of vitamin D. They may also increase urinary excretion of calcium and phosphate. The inhibitory effect on intestinal absorption results in secondary hyperparathyroidism, which in turn increases bone resorption. As a result there is only a slight decrease in the serum calcium, which is well within the normal range. Hypercalcaemia can occur in adrenal insufficiency. Some of this may be an increase in total calcium through a haemoconcentration and an increase in the proportion bound to albumin, but increased renal tubular reabsorption of calcium and bone resorption may also be involved.

The effect of glucocorticoids on the intestine can be used in treatment of the hypercalcaemia of vitamin D intoxication, or for disorders that increase the synthesis of 1,25$(OH)_2$D. Glucocorticoids may also reverse hypercalcaemia in myeloma and some other malignancies. This may be due to inhibition of the effects of the osteoclast-activating factors, interleukin 1 and tumour necrosis factor, both of which can stimulate bone resorption in malignancy. Glucocorticoids do not decrease the plasma calcium in primary hyperparathyroidism, and this difference in response between patients with an excess of parathyroid hormone and other forms of hypercalcaemia has been used as a diagnostic aid.

Sex hormones play a crucial role in skeletal development during the adolescent growth spurt and are responsible for epiphyseal closure. However, they are also important in regulating calcium metabolism in the adult. Because the reduction in the output of ovarian hormone at the menopause is associated with accelerated loss of bone mass, most attention has been paid to the effects of oestrogens, but androgens also affect bone metabolism. In adults, loss of either oestrogen or androgen can result in an increase in the rates of bone resorption and formation, but the formation response is limited and bone loss occurs. At the menopause there is a slight rise in the serum calcium concentration within the normal range, accompanied by an increase in fasting urinary calcium and hydroxyproline excretion, presumably related to the increase in bone resorption. As this is not associated with an increase in levels of parathyroid hormone, it has been attributed to an increase in sensitivity of bone to that hormone or other resorption-stimulating hormones. Administration of oestrogen can prevent the changes that occur at the menopause. These changes may be responsible for the marked increase in the apparent incidence of primary hyperparathyroidism in postmenopausal women and the fact that oestrogens can lower the plasma calcium in mild hyperparathyroidism. While oestrogen can decrease serum calcium in hyperparathyroidism, it may increase serum calcium in carcinoma of the breast, presumably by an effect on tumour growth. Complementary responses have been seen with tamoxifen, an oestrogen inhibitor. Primary hyperparathyroidism is also being recognized with increasing frequency in older men. Here the role of sex hormones is less clear, although testosterone production does decrease with age in some men. The mechanisms of the sex hormone effects on bone have not been established, but recent evidence suggests that there are direct effects on bone cells and that there are sex hormone receptors in skeletal tissue.

LOCAL FACTORS INFLUENCING BONE METABOLISM

Bone turnover is regulated locally as well as systemically. There are now numerous agents that have been shown to act on bone resorption or formation and are produced either by bone cells themselves or by adjacent haematopoietic tissue. Among these are several products of leucocytes which were originally called osteoclast activating factors, but which are now known to be members of the cytokine family, including interleukin 1α and β, and tumour necrosis factor α and β. Interleukin 1 is probably produced by bone cells as well as haematopoietic cells and is a potent stimulator of bone resorption and inhibitor of bone formation. Tumour necrosis factor β (lymphotoxin) has been found in cultures of myeloma cells and could be responsible for the hypercalcaemia, osteolytic lesions, and severe osteoporosis seen in patients with multiple myeloma.

Prostaglandin E_2 and related products of arachidonic acid metabolism are local regulators of bone metabolism. Their effects are complex and biphasic, but the most prominent are stimulation of bone resorption and formation. Infusions of prostaglandin E_2 can increase the serum calcium concentration in experimental animals. There are cases in which this prostaglandin appears to mediate the hypercalcaemia of malignancy, but these are relatively uncommon.

There is evidence from experimental animals that oestrogen withdrawal can result in an increase in the production of prostaglandin E_2 in bone, and that there may also be changes in the production of interleukin 1 at the menopause. Changes in local production of factors with ageing have not been studied.

Epidermal growth factor and transforming growth factor α can stimulate bone resorption both directly and by increasing prostaglandin production in bone. Transforming growth factor β is also produced by tumours and has complex effects on bone, which include indirect stimulation of resorption by increasing prostaglandin synthesis. In fact, many factors that stimulate bone resorption also increase local production of prostaglandin E_2 in bone.

Recently, a parathyroid hormone-related peptide (**PTHrP**) which is probably responsible for most cases of humoral hypercalcaemia of malignancy has been identified. PTHrP has only eight of its first 13 amino acids in common with human parathyroid hormone. The rest of the molecule, which has been identified by its messenger RNA as being of 139 to 176 amino acids, differs almost completely from parathyroid hormone. The different lengths appear to be due to alternative gene splicing. The role of this peptide in calcium metabolism is not clear. A role in fetal calcium regulation and in calcium transport in the placenta and mammary glands has been suggested. Pathologically, it clearly is responsible for humoral hypercalcaemia of malignancy and can produce marked hypercalcaemia when injected in animals. Its actions on bone and kidney appear to be identical to those of parathyroid hormone, with a possible exception of the effect on 1α-hydroxylase in the kidney. In isolated kidney tissue, PTHrP can stimulate 1α-hydroxylase, but patients with humoral hypercalcaemia of malignancy usually do not have an increased concentration of 1,25$(OH)_2D$. Whether this is due to a difference in action *in vivo* or the presence of a second hormone which inhibits the formation of 1,25$(OH)_2D$ is not yet known.

HYPOCALCAEMIA

The frequency of hypocalcaemia depends on its definition. If we use a total serum calcium of less than 2.1 mmol, there will be a large number of patients who do not have true hypocalcaemia but have a normal ionized calcium and a low serum albumin concentration. This is particularly common in chronically ill, elderly patients. Deficiency of vitamin D, or abnormalities that impair absorption or synthetic pathways for the vitamin D hormone system, are common causes of hypocalcaemia in elderly people and are discussed in greater detail in Chapter 14.2.

Hypocalcaemia related to an impairment of the function of parathyroid hormone in maintaining the serum calcium is relatively uncommon but important to recognize because it can produce subtle and debilitating symptoms and is quite responsive to appropriate treatment. These parathyroid hormone-related forms of hypocalcaemia can be divided into three groups.

1. Primary parathyroid deficiencies due to a loss either of parathyroid gland tissue or an impairment of secretory function.
2. Relative parathyroid deficiencies, which are usually transient and due to a decrease in calcium supply that occurs so rapidly or is so severe that the parathyroid hormone response is delayed or inadequate.
3. Syndromes of resistance to parathyroid hormone in which secretion is high, but the end organs do not response adequately.

These divisions are not absolute and in many cases hypocalcaemia is due to multiple abnormalities.

Primary hypoparathyroidism

Permanent primary hypoparathyroidism is rare in elderly patients. Idiopathic hypoparathyroidism, an autoimmune disease, usually appears early in life. In older patients the most likely cause of hypoparathyroidism is surgical damage to the parathyroid glands, particularly after extensive thyroid surgery. Transient hypoparathyroidism is common after both thyroid and parathyroid surgery; it is presumably the result of damage to the normal parathyroid glands or their blood supply. In addition, prior hypercalcaemia can result in suppression of residual parathyroid gland function; this occurs in both hyperparathyroidism and hyperthyroidism. In the so-called hungry bone syndrome the marked increase in bone formation, which occurs in hyperparathyroidism and hyperthyroidism, results in the formation of unmineralized matrix that then takes up the mineral rapidly when the primary disease is cured surgically. This increases the demand on the residual parathyroid tissue. While severe permanent hypoparathyroidism is relatively rare after neck surgery, particularly when it is by an experienced endocrine surgeon, some studies have suggested that partial impairment of parathyroid gland function is more common. Such defects can be brought out by stressing the parathyroid gland, either by lowering ionized calcium with a chelating agent or by removing calcium with a diuretic. Whether or not such impairment is associated with any symptoms is not clear.

One relatively common form of hypocalcaemia in older people is that due to severe hypomagnesaemia, which is often associated with malnutrition and alcoholism. When the serum magnesium concentration falls well below normal, usually 0.4 mmol or less, secretion of parathyroid hormone is impaired. Thus, there is hypocalcaemia with paradoxically low circulating levels of the hormone. Infusion of magnesium results in a rapid release of hormone, but the serum calcium response may be somewhat slow. This is probably attributable to end-organ resistance, as magnesium-deficient bone appears to be less sensitive to resorptive stimuli. On the other hand, hypermagnesaemia, like hypercalcaemia, can also inhibit the secretion of parathyroid hormone. This could contribute to the hypocalcaemia seen in renal failure, and to the transient decreases in serum calcium that are occasionally found in patients treated with magnesium sulphate.

Relative hypoparathyroidism

Whenever the entry of calcium into extracellular fluid decreases or loss from extracellular fluid is accelerated, the serum calcium concentration decreases and parathyroid hormone secretion increases. Usually the parathyroid hormone response is rapid enough for the fall in serum calcium to be minimal, and the ionized calcium concentration remains within the normal range. However, under certain circumstances, clinically important hypocalcaemia may occur. Multiple mechanisms are usually involved. For example, when patients are treated with cytotoxic agents such as mithramycin, which markedly inhibit bone resorption, hypocalcaemia often occurs. Cytotoxic agents may also impair the function of the parathyroid glands or cause magnesium loss. Calcitonin rarely produces hypocalcaemia in normal individuals or in patients with hypercalcaemia of malignancy, but in patients with high bone turnover and intense osteoclastic resorption due to Paget's disease, transient hypocalcaemia can occur and may even be symptomatic.

Phosphate excess is a relatively common cause of hypocalcaemia. It can occur from rapid cell lysis in a pathological state such as rhabdomyolysis, or from treatment-induced lysis of cells in leukaemia and lymphoma. Phosphate retention occurs in acute and chronic renal failure, and if the level of phosphate is sufficiently high, the serum calcium concentration will be reduced. Excessive administration of phosphate, either intravenously, by enema or even occasionally orally, can produce hyperphosphataemia and a reduction in the serum calcium. The hypocalcaemia of phosphate excess is due to inhibition of bone resorption and increased mineralization of bone. However, phosphate can also cause deposition of calcium phosphate salts in soft tissues, including kidneys, blood vessels, and lungs, resulting in irreversible tissue damage. Chronic phosphate excess can also inhibit 1α-

hydroxylase in the kidney, and hence further reduce the calcium supply.

Hypocalcaemia is common in acute pancreatitis. It has been attributed to the sequestration of calcium as fatty acid salts in the pancreas and retroperitoneal space, and occasionally in the peritoneal cavity in the form of chylous ascites. In many cases of pancreatitis the total serum calcium is reduced because of hypoalbuminaemia but ionized calcium is normal.

Rapid entry of calcium into the skeleton in patients with osteoblastic metastases has been described as a cause of hypocalcaemia, but here again low serum albumin with low total but normal ionized calcium concentration is more common than true hypocalcaemia.

Resistance to parathyroid hormone

Inherited syndromes of parathyroid hormone resistance are usually identified in younger individuals but occasionally do not become clinically manifest until later in life. As parathyroid hormone is usually present in increased amounts, these cases are termed 'pseudohypoparathyroidism'.

In classical pseudohypoparathyroidism (type 1A), the response to parathyroid hormone is diminished because a protein that links the hormone's receptor to adenyl cyclase, the stimulatory guanyl nucleotide regulatory protein (G_s), is decreased. This decrease is present throughout the body and affects many other responses, but clinically hypocalcaemia is the most prominent. Other patients with pseudohypoparathyroidism do not have deficiency of the G_s protein but have other defects in cellular response to parathyroid hormone. Most of these patients can be identified by the presence of a high level of parathyroid hormone in the blood, and a low excretion of nephrogenous cyclic AMP and phosphate in response to the injection of human parathyroid hormone, which is now available for this diagnostic test. Pseudohypoparathyroidism is often associated with other abnormalities. Many of the patients with type 1A have a syndrome called Albright's hereditary osteodystrophy, with short metacarpals, short stature, and mental retardation. These patients are usually identified when young but occasionally missed. Older patients with hypocalcaemia are occasionally found to have resistance to parathyroid hormone, particularly if they have none of the phenotypic features of pseudohypoparathyroidism. These patients may have a defect that does not appear until their calcium regulatory system is stressed by age- or disease-related events.

Symptoms and signs of hypocalcaemia

The classical presentation of severe hypocalcaemia is tetany with carpopedal spasm. This usually does not occur until the ionized calcium concentration is below 0.8 mmol, unless there is accompanying alkalosis. Thus, latent tetany in cases of mild hypocalcaemia can be brought out by hyperventilation, producing respiratory alkalosis. Moreover, patients who have no defect in calcium regulation can also develop tetany if the alkalosis is sufficiently severe. Laryngeal spasm is an uncommon but potentially life-threatening complication of tetany.

Chvostek's sign, contraction of the facial muscles elicited by tapping on the facial nerve, is useful to test for latent tetany; however, this response can occur in normal individuals. Trousseau's sign is more specific: the arterial supply to the forearm is blocked by maintaining a tourniquet above systolic pressure 3 to 5 minutes; in hypocalcaemia the hand goes into flexion contracture. The rapidity of onset of contracture is proportional to the severity of hypocalcaemia and this test can therefore be used at the bedside to follow the response to therapy. The ECG can also be helpful because it shows prolongation of the QT interval, proportional to the degree of hypocalcaemia.

In chronic hypocalcaemia, regardless of cause, there is an increased incidence of cataracts and calcification of the basal ganglia. In children and occasional adults one can also look for polyglandular autoimmune failure, with associated adrenal insufficiency and chronic candidiasis. Many patients with chronic hypocalcaemia appear to have some defect in mental function or show neuropsychiatric abnormalities, which are quite non-specific.

Differential diagnosis

Because a low total serum calcium concentration may be due to decreased protein binding and does not necessarily indicate true hypocalcaemia, it is important to measure albumin or the ionized calcium concentration. For differential diagnosis it is useful to measure serum magnesium, potassium, phosphate, and bicarbonate. Measurements of renal function, vitamin D levels, and urinary excretion of calcium and phosphate are also helpful. Tests for malabsorption and pancreatic disease can be done when indicated. The distinction between primary hypoparathyroidism and pseudohypoparathyroidism can be made by accurate immunoradiometric assays of parathyroid hormone which measure the whole molecule of the hormone using two antibodies. However, if the level of hormone is high, it is worthwhile confirming the diagnosis by injection of the hormone and measurement of the renal response in terms of urine cyclic AMP and phosphate excretion. As pseudohypoparathyroidism is often familial, it is important to make this diagnosis and to identify other family members who might have minimal abnormalities but still require treatment.

As many episodes of hypocalcaemia in older patients occur after neck surgery, the diagnosis is often self-evident. It is possible to distinguish between transient and permanent damage to the parathyroid glands by following the levels of parathyroid hormone, but it may be more practical simply to withdraw therapy after the initial episode to determine whether the patient still requires treatment or has had restoration of parathyroid function. Even if there has been restoration of function, there is the possibility of secondary failure. Thus, patients who have had a bout of hypocalcaemia after thyroid or parathyroid surgery should be followed regularly to be certain that their serum calcium concentration is maintained in the normal range.

Treatment

Tetany can be reversed, regardless of its cause, by intravenous infusion of calcium, usually as calcium gluconate. While this is effective even in patients who have magnesium or potassium abnormalities, or alkalosis due to hyperventilation, it obviously is not an appropriate, long-term therapy. Some patients, particularly those who develop hypocalcaemia in the setting of osteomalacia or chronic renal failure, may require multiple injections or even prolonged intravenous infusion of calcium to restore the serum calcium to normal. An ampoule of calcium gluconate contains only 90 mg of calcium, but the skeleton contains 600 to 1000 g. Thus, a defect in mineralization can result in a deficit of several grams of calcium. Hypocalcaemia and tetany due to magnesium deficiency will respond only transiently to calcium and so magnesium must also be given. The response may be somewhat slow because of the end-organ resistance in magnesium-depleted patients.

As experience is gained with calcitriol, it is becoming considered the drug of choice in patients with hypocalcaemia lasting more than a few hours. Calcitriol can be begun in doses of 0.25 to 0.5 μg at 6- to 12-h intervals. This schedule is preferable because of its short half-life, although in chronic therapy daily administration is often effective. The serum calcium concentration should begin to rise within 24 h in patients given enough calcitriol. They should be carefully monitored so that hypercalcaemia can be avoided by reducing the dose when the serum calcium goes above the mid-normal range. If hypercalcaemia does occur, then the drug is stopped and the serum calcium usually falls within 24 h. Because tetany is such a disturbing symptom, especially if laryngeal spasms occur, we monitor the serum calcium concentration in patients after neck surgery and begin calcitriol early.

Calcitriol is also the most effective drug for the treatment of chronic hypoparathyroidism and pseudohypoparathyroidism, as well as renal failure. Although it is expensive and requires daily administration, it has several advantages over other metabolites. Calcitriol is the direct-acting, hormonal form of vitamin D and does not require further metabolic alteration. It is rapidly absorbed and cleared. Thus if hypercalcaemia or hypercalciuria develop, reduction of the dose will result in rapid reversal. While there is a greater margin of safety with vitamin D or with 25(OH)D, which is available as calcidiol, high doses are required because patients with hypoparathyroidism or pseudohypoparathyroidism cannot make $1,25(OH)_2D$ normally. The precursors may accumulate in the body so that when toxicity occurs, it is often prolonged. This may have to be treated not only by withdrawing the drug, but also by giving glucocorticoids to oppose the effect of vitamin D on the intestine. Dihydrotachysterol, a compound which can act directly on the $1,25\text{-}(OH)_2D$ receptor, has been used extensively in the past, but calcitriol appears to be more effective and reliable and is currently replacing this agent. For chronic therapy, a combination of a calcium supplement with calcitriol can reduce the required dose of the expensive hormonal medication. In some patients on calcitriol, excessive hypercalciuria can be controlled by the use of a thiazide diuretic and/or amiloride. Hypocalcaemia due to phosphate excess can be diminished by decreasing phosphate intake or using phosphate-binding compounds. Aluminium hydroxide gels have been used in the past, but may produce excessive deposition of aluminium in the bone and other tissues. Recent studies suggest that calcium carbonate in large doses can bind phosphate sufficiently to reduce its absorption.

HYPERCALCAEMIA

An increased concentration of serum calcium can occur over an extremely broad range and produce a variety of symptoms. Hence it is important not only to detect hypercalcaemia but to define its severity, which requires careful clinical assessment and may require measurement of ionized calcium. By definition an ionized calcium greater than 1.3 mmol is abnormal. This usually is accompanied by total serum calcium greater than 2.6 mmol, but if the albumin is low, the total serum calcium may be normal. False hypercalcaemia can occur when there is a high concentration of proteins which bind calcium. This is most commonly due to haemoconcentration, either because of diuretic therapy or after prolonged haemostasis during blood sampling. There are rare cases of myeloma in which atypical immunoglobulins bind calcium and produce a high total calcium with normal ionized calcium.

Multiple pathogenetic mechanisms can be responsible for hypercalcaemia. In most cases the primary abnormality is excessive bone resorption. This need not raise the serum calcium if renal excretion is increased enough or intestinal absorption is decreased, and most hypercalcaemic patients show an abnormality in these systems, as well as in bone. In malignancy the most common combination is increased bone resorption and decreased renal excretion. Increased intestinal absorption is an important factor in some cases of hyperparathyroidism and in vitamin D intoxication, sarcoidosis, and milk–alkali syndrome.

While it is useful to divide hypercalcaemia into mild and severe forms, there are patients who have a marked elevation of serum calcium with few symptoms and others who develop symptoms with only a modest increase. The rapidity with which calcium increases is probably an important factor. It is useful to know the duration of hypercalcaemia because cases with prolonged elevation of the serum calcium concentration are almost always due to primary hyperparathyroidism, although a few may be due to sarcoidosis or to familial benign hypocalciuric hypercalcaemia. In contrast, acute hypercalcaemia without prior elevation of serum calcium is most often due to malignancy. Regardless of the cause, a marked increase in the serum calcium can occur when the patient enters a vicious cycle of dehydration with further increase in the serum calcium concentration which, in turn, promotes fluid loss and decreases renal function. Elderly people are particularly prone to such hypercalcaemic crises because of age-related changes in kidney function that result in a delayed response to sodium deprivation and in decreased concentrating capacity. In addition, the sensation of thirst may be reduced in older patients, and they may have limited access to fluid. The problem is compounded by the frequent use of diuretics in the elderly patient and the fact that dehydration is often difficult to detect clinically.

Causes of hypercalcaemia

Primary hyperparathyroidism

An increase in parathyroid hormone secretion due to adenoma or polyglandular hyperplasia is the most common cause of hypercalcaemia. The incidence appears to increase with age, but this may be because mild hyperparathyroidism is often undiagnosed for many years. An increase in diagnosis at the menopause may be related to the loss of the opposing effect of oestrogen on parathyroid hormone-stimulated bone resorption. When this occurs, the serum calcium may rise from the high normal to abnormal. As more and more cases are being detected by routine screening for serum calcium concentration, the majority of patients are asymptomatic or have minimal symptoms which are difficult to ascribe to primary hyperparathyroidism. Severe and aggressive hyperparathyroidism is seen in a few cases, particularly in those with large adenomas, carcinomas, or with hyperplasia associated with other endocrine neoplasms. There are two multiple endocrine neoplasia syndromes in which the parathyroids may be involved. Hyperparathyroidism is associated with pancreatic and pituitary tumours in type 1, and with medullary carcinoma of the thyroid and phaeochromocytoma in type 2. It is important to take a careful personal and family history and to do laboratory studies to look for these other tumours, particularly because surgery on type-2 patients with an undetected phaeochromocytoma can cause a severe and possibly fatal hypertensive crisis.

In secondary hyperparathyroidism the parathyroid glands are hyperplastic because of prolonged stimulation by calcium deficiency. Patients with secondary hyperparathyroidism can become hypercalcaemic when the calcium deficiency is rapidly reversed. This occurs most often in those with chronic renal failure who are treated by phosphate depletion, become aluminium intoxicated, or have renal transplants. In such patients, hypercalcaemia persists because the hyperplasia of the glands persists, but this usually resolves slowly and does not require surgery. Transient hypercalcaemia can also occur after recovery from acute renal failure associated with rhabdomyolysis. This is probably due to the release of calcium previously deposited in the soft tissues back into the extracellular fluid.

Hypercalcaemia of malignancy

This type occurs quite frequently in elderly patients. It is often associated with metastatic involvement of bone and appears as a terminal event. There are several pathogenetic mechanisms. In patients who have hypercalcaemia without bone involvement, the so-called humoral hypercalcaemia of malignancy syndrome, a peptide, PTHrP, is released by the tumour and acts like parathyroid hormone to stimulate bone resorption and decrease renal excretion of calcium (see above). Haematological malignancies may also produce PTHrP, but in addition have been found to produce bone resorbing lymphokines, such as interleukin 1 and tumour necrosis factor. Some lymphomas may produce hypercalcaemia by converting 25(OH)D to 1,25$(OH)_2$D. Metastatic malignancy may produce hypercalcaemia by the same humoral mechanisms, but could also cause bone resorption by a direct effect of the tumour.

Vitamin D intoxication, sarcoidosis, and milk–alkali syndrome

These three cause hypercalcaemia largely by increased intestinal absorption of calcium, although stimulation of bone resorption may also play a role. The increase in intestinal absorption in sarcoidosis is probably due to increased synthesis of 1,25$(OH)_2$D by the abnormal macrophages. Other granulomatous disorders occasionally produce hypercalcaemia by a similar mechanism.

Familial benign hypocalciuric hypercalcaemia

This form is important to recognize because it should not be treated. Because elderly patients with primary hyperparathyroidism frequently do not show hypercalciuria, the differential diagnosis may be difficult. The mechanism of hypercalcaemia in these families is not clearly understood but appears to relate to an alteration in cation transport. Levels of parathyroid hormone are normal or minimally elevated, and urine calcium excretion is low.

Other forms of hypercalcaemia

Mild hypercalcaemia can occur in hyperthyroidism and occasionally in patients who are immobilized and have rapid bone turnover, for example, in elderly patients with extensive Paget's disease. However, hypercalcaemia with Paget's disease is more often due to coincidental primary hyperparathyroidism. Addison's disease is associated with hypercalcaemia, which is probably due to a combination of increased bone resorption and haemoconcentration. False hypercalcaemia can occur with dehydration, particularly in patients on thiazide or amiloride in whom urinary calcium excretion is also decreased.

Symptoms and signs of hypercalcaemia

Mild hypercalcaemia may be asymptomatic or associated with ill-defined, neuromuscular and psychiatric symptoms. In elderly patients, muscular weakness, irritability or depression, and mild gastrointestinal disturbances, including anorexia and constipation, are common. While they may be related to hypercalcaemia, these symptoms do not always improve after the hypercalcaemia is successfully treated. Chronic hypercalcaemia of whatever cause can produce nephrolithiasis, a decrease in bone mass, and impairment of renal function. Cystic bone lesions may occur in hyperparathyroidism and sarcoidosis. There is an association between hypertension and hypercalcaemia; the blood pressure of some, but certainly not all, hypertensive patients with hypercalcaemia will be improved after successful treatment therapy of the calcium imbalance.

Severe hypercalcaemia is associated with progressive dehydration through inhibition of renal tubular reabsorption of salt and water by the high calcium level, as well as the accompanying anorexia, nausea, and vomiting. When the total serum calcium concentration exceeds 3.5 mmol (14 mg/100 ml) or the ionized calcium concentration exceeds 1.7 mmol (6.8 mg/100 ml), mental confusion is common and further increases in the calcium concentration may result in coma and death and constitute a medical emergency.

Laboratory diagnosis

The diagnosis of primary hyperparathyroidism has been markedly improved with the advent of the highly specific, double antibody immunoassays which measure intact parathyroid hormone. Patients with primary hyperparathyroidism usually show only a modest increase in the serum calcium, although a hypercalcaemic crisis can occur. Phosphate concentration is low unless renal function is impaired. Renal calcium excretion varies widely and is often not increased in elderly patients. Unlike the earlier measurements of C-terminal fragments of parathyroid hormone, which were elevated with mild renal impairment, the assay of intact hormone is usually diagnostic in patients with mild to moderate renal failure and primary hyperparathyroidism. As the hypercalcaemia of malignancy due to PTHrP closely mimics primary hyperparathyroidism, the differential diagnosis depends largely on immunoassay and the presence of malignancy. Patients who have a malignancy and an elevated level of intact hormone probably have primary hyperparathyroidism in addition. In vitamin D intoxication the 25(OH)D value is helpful. Thyroid hormone levels will detect patients with hypercalcaemia due to thyrotoxicosis.

If the serum calcium has been mildly elevated over several years, this is strongly indicative of primary hyperparathyroidism, although it can occur occasionally in sarcoidosis. On the other hand, rapid onset of hypercalcaemia associated with anaemia, weight loss, a low serum albumin concentration, or an abnormal immunoglobulin, such as occurs in multiple myeloma, suggests that the patient has hypercalcaemia of malignancy. The measurement of $1,25(OH)_2D$ is not particularly helpful in differential diagnosis because it can be elevated in primary hyperparathyroidism, sarcoidosis, and occasional malignancies.

Treatment

Primary hyperparathyroidism

The treatment of primary hyperparathyroidism is surgical removal of the adenoma, or removal of three and a half glands in patients with multiglandular hyperplasia. Although not absolutely necessary, attempts to localize the parathyroids are helpful in elderly patients because this can shorten the time of surgical exploration. Indeed, some elderly patients can be explored under local anaesthesia if they have had successful preoperative localization of the adenoma. At present the thallium–technetium subtraction scan is the most useful of available tests, although magnetic resonance imaging and ultrasonography are also useful. Surgery is not necessarily indicated for all patients with primary hyperparathyroidism. Elderly asymptomatic patients may just be followed carefully if the serum calcium is not above 3.0 mmol total or 1.5 mmol ionized, but they must be warned to avoid dehydration. Oestrogen may be used to reduce the serum calcium in older women, but it is not known whether this will limit symptoms or prevent progression of the disease.

Hypercalcaemic crisis

Any patient with a total serum calcium concentration of more than 3 mmol (12 mg/100 ml) or an ionized calcium of more than 1.5 mmol (6.0 mg/100 ml) should be treated vigorously to lower the serum calcium. Patients with rapid onset of hypercalcaemia are usually dehydrated and therefore the first step is to rehydrate them with intravenous, isotonic sodium chloride. Because elderly patients may have difficulty in handling large fluid loads, the fluid balance should be carefully monitored, and in severe cases a central line should be placed to monitor venous pressure. Fluid overload can be treated with a diuretic. Frusemide (furosemide) is preferred because it increases urinary calcium as well as sodium excretion. While the calciuric effect of frusemide may help to reduce the serum calcium, it is also a potentially dangerous drug because if fluid replacement falls behind, the patient may become dehydrated again and the hypercalcaemia become worse. Potassium should be monitored during saline therapy, particularly in elderly patients who may have been potassium depleted by their prior illness, as both hypercalcaemia and its therapy can increase potassium excretion. Many patients have hypophosphataemia. If the serum inorganic phosphate concentration falls below 0.3 mmol (1 mg/100 ml), cautious replacement may be advised to prevent adverse effects of hypophosphataemia itself and because low phosphate can enhance bone resorption and make the hypercalcaemia worse. On the other hand, intravenous phosphate therapy must be used with great caution because it can increase the calcium phosphate product in the blood rapidly and result in soft tissue deposition of mineral with irreversible damage to the kidneys, vessels, and lungs.

If hypercalcaemia cannot be controlled by rehydration alone, then additional agents must be considered. Calcitonin can produce a transient decrease in the serum calcium, but this is usually short-lived, presumably due to escape from the inhibitory effects on osteoclasts. Prednisone therapy is indicated in patients with vitamin D intoxication, sarcoidosis, and haematopoietic malignancies. It may also be effective in some patients with solid tumours. It is not effective in hyperparathyroidism. In the past the most widely used drug in hypercalcaemia of malignancy was mithramycin, which will decrease the serum calcium concentration when given in doses of 15 to 25 μg/kg body weight intravenously over several hours. However, this drug is toxic to bone marrow, liver, and kidney and is being replaced by less toxic drugs. The bisphosphonates or diphosphonates, analogues of pyrophosphate which have a carbon linking the two phosphates so that they cannot be hydrolysed, are now used extensively in the treatment of hypercalcaemia. The agent currently available in the United States, disodium etidronate, is effective when given intravenously in doses of 7.5 mg/kg over 2 to 3 h daily for 3 to 6 days. Once the serum calcium has been reduced, it may be helpful to maintain the effect by giving oral etidronate, but this is not always necessary. New bisphosphonates are currently being tested or are already available in Europe. These are much more potent than etidronate and more effective in the treatment of hypercalcaemia of malignancy. {In the United Kingdom pamidronate (by slow intravenous infusion) is proving effective in the management of the hypercalcaemia of malignancy.} In occasional cases the serum calcium cannot be reduced by the measures described above. In these patients, dialysis or haemofiltration can be used to remove calcium rapidly and lower the serum concentration to a safer level. Diagnosis of the cause of hypercalcaemia should be made as rapidly as possible because specific treatment, such as removal of a parathyroid adenoma or surgical or chemotherapeutic attack on a malignancy, may be life-saving.

Bibliography

Forero, M.S. *et al.* (1987). Effect of age on circulating immunoreactive and bioactive parathyroid hormone levels in women. *Journal of Bone and Mineral Research*, **2**, 363–6.

Francis, R.M., Peacock, M., Storer, J.H., Davies, A.E.J., Brown, W.

B., and Nordin, N.E.C. (1983) Calcium malabsorption in the elderly: the effect of treatment with oral 25-hydroxyvitamin D_3. *Journal of Clinical Investigation*, **13**, 391–6.

Ireland, P. and Fordtran, J.S. (1973). Effect of dietary calcium and age on jejunal calcium absorption in humans studied by intestinal perfusion. *Journal of Clinical Investigation*, **52**, 2672–81.

Lindeman, R.D., Tobin, J.D., and Shock, N.W. (1985). Longitudinal studies on the rate of decline in renal function with age. *Journal of the American Geriatrics Society*, **33**, 278–85.

Silverberg, S.J., Shane, E., Luz de la Cruz, R.N., Segre, G.V., Clemens, T.L., and Bilezikian, J. P. (1989). Abnormalities in parathyroid hormone secretion and 1,25-dehydroxyvitamin D_3 formation in women with osteoporosis. *New England Journal of Medicine*, **320**, 277–81.

Sokoll, L.J. and Dawson-Hughes, B. (1989). Effect of menopause and aging on serum total and ionized calcium and protein concentrations. *Calcified Tissue International*, **44**, 181–5.

Tibblin, S., Pålsson, N., and Rydberg, J. (1983). Hyperparathyroidism in the elderly. *Annals of Surgery*, **197**, 135–8.

Tsai, K.S., Heath, H., Kumar, R., and Riggs, B.L. (1984). Impaired vitamin D metabolism with aging in women. *Journal of Clinical Investigation*, **73**, 1668–72.

Yendt, E.R., Cohanim, M., and Rosenberg, G.M. (1986). Reduced serum calcium and inorganic phosphate levels in normal elderly women. *Journal of Gerontology*, **41**, 325–30.

Young, G., Marcus, R., Minkoff, J.R., Kim, L.Y., and Segre, G.V. (1987). Age-related rise in parathyroid hormone in man: the use of intact and midmolecule antisera to distinguish hormone secretion from retention. *Journal of Bone Mineral Research*, **2**, 367–74.

SECTION 8
Gastroenterology

8.1 The ageing mouth

ANGUS W. G. WALLS

INTRODUCTION

The mouth has been described as a mirror of bodily well being, and is used as such by some health care workers, notably by experts in Chinese traditional medicine. It is not surprising that the structural and functional changes taking place in the remainder of the body have their parallels within the stomatognathic system. In addition, there are changes in association with increasing age that are unique to the mouth.

Superimposed upon the known changes in oral tissues, there is evidence of a dramatic improvement of oral health in the population of the United Kingdom, matching that which is occurring, or has occurred, in other developed countries. This improvement is characterized by many more people retaining some of their natural teeth into their advancing years. As yet this pattern of progress has had little impact upon the older age groups within the United Kingdom. However, it is anticipated that the proportion of the population 65 or older who have no natural teeth will fall significantly during the next 20 to 30 years (Fig. 1) (Todd and Lader 1991).

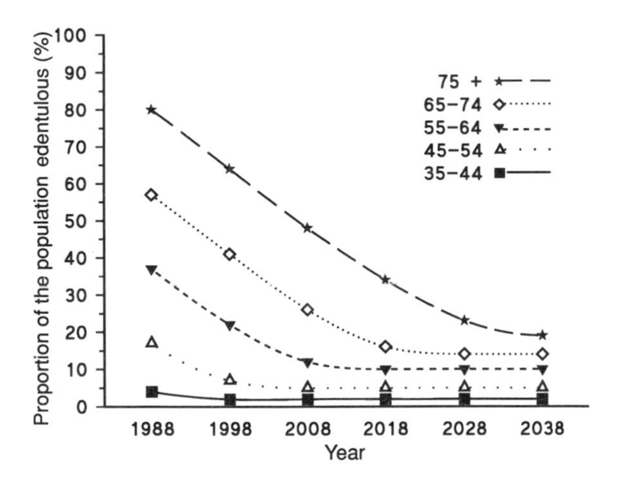

Fig. 1 Projected changes in the proportion of the population edentulous. The projections are based upon the results of the Adult Dental Health Surveys of 1968, 1978, and 1988, and of the General Household Survey (Todd and Lader 1991).

The reasons for these demographic changes are unclear, but they must include an elevation in the level of awareness of oral health in the population, and probably reflect an increasing reluctance to accept the replacement of natural teeth with dentures. These changes should be welcomed by all health care professionals, but they will result in a marked alteration in the needs of ageing patients in terms of oral health. It will be no longer sufficient to provide a denture bowl and cleanser for all. There will be an increasing need for regular supervised and/or assisted oral hygiene within long-stay care facilities, and for oral hygiene instruction within occupational therapy programmes designed to facilitate the return of patients into the community. Such activity will be costly in terms of personnel. In many older subjects oral hygiene instruction would have to be provided on a 1:1 basis. The resources required to satisfy this need are not available at the present time.

This chapter will be divided into two parts, the first dealing with age-associated changes within the oral tissues and the second with the pathological changes within the oral tissues which are more prevalent with increasing age. The latter will include some aspects of the oral manifestations of systemic disease processes and of systemic medication.

NORMAL AGE CHANGES

Soft tissues

The soft tissues of the mouth include the oral mucous membrane, the salivary glands, and the muscles of mastication. Age-associated changes in structure can be perceived in each tissue.

The oral mucous membranes

There are three broad categories of oral mucosa, masticatory mucosa (which covers the palatal vault and the attached gingiva), lining mucosa (which comprises the mobile lining tissues within the mouth including the cheeks, the soft palate, the floor of the mouth and the ventral surfaces of the tongue), and specialized mucosae (for example, that covering the lips and the dorsum of the tongue). Each of these mucosal types has a specific structure which is best able to satisfy the requirements for mobility, flexibility, or resistance to surface abrasion that are found in the various sites within the mouth (for example the lining mucosae are non-keratinized whilst the masticatory mucosae undergo parakeratinization). As with so many tissues the detection of age-associated changes in form is complicated by natural variation between individuals. In principle the changes are similar to those seen in skin, which are described in detail elsewhere within this book. However, as many of the changes in skin are attributable to exposure to actinic radiation, the oral mucosa is minimally affected, with the exception of the vermilion border of the lip.

There are a number of clinically recognizable changes in all of these tissues, including a loss of elasticity and of surface texture. There is an impression that the mucosal tissues are thinner, with increased prominence of underlying features (for example sebaceous glands). At a histological level the epithelial thickness remains constant with age, although there is some evidence of epithelial thinning on the lateral surface of the tongue (Scott *et al.* 1983). There is some alteration in the epithelial-connective tissue interface with shortening of the rete pegs, associated with a reduction in rete area and in alteration from a ridge form to a papillary architecture (Scott *et al.* 1983). There is no evidence of any alteration in the rate of cell renewal with age (Maidhof and Hornstein 1979).

It would seem likely that the perceived changes in mucosal form are a result of changes within the subepithelial connective tissues. Age changes within the lamina propria are poorly described although they are likely to mirror those seen within the dermis including an apparent increase in elastin with reduced elasticity, decreased quantities of collagen with a relative increase in insoluble collagen, and possible alterations within the proteoglycans of the ground substance resulting in altered levels of interstitial fluid.

Salivary glands

There are many references in the literature linking ageing to reduced salivary flow. Early studies on salivary function in older subjects tended to confirm these statements (Bertram 1967). The influence of medication and of systemic disease upon salivary flow were poorly understood, and these studies did not use carefully selected populations. The results of a number of studies of salivary function, where the population sample was selected to give healthy unmedicated individuals, have been reported recently. The consensus from these studies is that there is no detectable reduction in flow from the major salivary glands (Baum 1986). However, Pedersen *et al.* (1985) reported a reduction in flow from the submandibular gland and Gandara *et al.* (1985) reported a reduction in stimulated flow from the minor salivary glands with increasing age.

Morphologically there are significant changes in gland composition with age. These have been described most completely in the submandibular gland, where between 25 and 50 per cent of the acinar units may be lost between youth and old age (Scott 1987). This loss is accompanied by an increase in fibroadipose tissue and an increase in ductal volume relative to parenchyma. The changes within the parotid gland are less well documented. Andrew (1952) reported that in extreme old age up to 50 per cent of the acinar elements would be replaced by adipose tissue. More recently Drummond *et al.* (1987) have demonstrated a reduction in parotid gland density with age using computerized axial tomography. They attribute this finding to an increase in fibrofatty tissue within the gland.

The most profound morphological changes are found in the myriad of minor salivary glands within the mouth. Unlike the major glands age changes in the minor glands result in progressive fibrosis rather than an increase in adipose tissue. The fibrosis can result in complete obliteration of the secretory structures (Drummond and Chisholm 1984).

The reason for the apparent discrepancy between the results of flow measurement and those for gland morphology is unclear. There is a very wide range of normal flow rates for the adult population. This would make detecting small reductions in flow difficult in a cross-sectional study of older subjects. In addition it is presumed that there is a significant functional reserve within the salivary glands which is slowly 'degraded' as acinar tissue is lost with age. This would permit secretory levels to remain within the 'normal range' for glands with marked loss of secretory units. There is also some evidence of elevated sympathetic tone in older subjects, which would in turn lead to elevated salivary secretion resulting in apparently normal flow from glands with reduced secretory ability. This theory is supported by the findings of a recent study which demonstrated that depression of salivary secretion was more profound in a group of older subjects than in younger individuals following a single bolus dose of atropine (Rashid and Bateman 1990).

Whilst it has been demonstrated that there is no obvious relationship between salivary flow and ageing, there is some evidence to suggest that there is alteration in salivary composition with increasing age. The rate of protein synthesis reduces with increasing age (Kim 1987). This reduction does not result in any apparent qualitative alteration in secretory proteins. At the same time there are some changes in salivary composition, notably a reduction in sodium and potassium ion concentration and a reduction in total protein content (Chauncey *et al.* 1981). The effects of these alterations upon the physiological functions of saliva, which include lubrication, remineralization, and antimicrobial and buffering activity, as well as assisting with digestion, is unclear.

Muscle

Newton *et al.* (1987) have been able to demonstrate a reduction in the bulk of both the masseter and medial pterygoid muscles with increasing age. This is thought to be a reflection of a reduction in the number of functional motor units within these muscle groups, in a manner similar to the alterations in general muscle bulk. In addition it has been demonstrated that older subjects have less accurate control of their masticatory muscles (Yemm *et al.* 1985). These two effects are cumulative, and probably result in a reduced functional masticatory capacity, irrespective of the state of the dentition.

Hard tissues

The hard tissues of the mouth include the bony skeleton of the mandible and maxilla and the temporomandibular joint.

Bone

The hard tissues which comprise the mandible and maxilla can be divided into two broad areas, the basal bone and the alveolar process. This latter is a functional matrix of cancellous and cortical bone which provides anchorage and support for the teeth. Osteoporosis occurs within mandibular bone, and there is a reasonable level of agreement between the level of osteoporosis to be found in the radius and in the mandible of the same individual.

The alveolar process is lost (by gradual resorption) if the teeth are extracted and is destroyed by the inflammatory processes involved in chronic periodontal disease. The pattern of bone loss associated with tooth extraction varies markedly from individual to individual. It may be influenced by systemic factors and is also affected by the quantity of residual alveolar bone support at the time of extraction, the wearing of dentures, and by diet. In addition the most common pattern of tooth extraction (with loss of posterior teeth and retention of anterior teeth) produces a concave bony architecture in the posterior regions of the mandible. This will be subject to convex flexure as a result of masticatory function resulting in modelling resorption of this region.

Thus, the pattern of bone loss seen in an ageing individual

is a complex interaction of past disease, the effects of tooth loss, and ageing. At its extreme a combination of a mandibular resorption and osteoporosis can result in severe thinning of the mandible with an associated risk of pathological fracture.

Temporomandibular joint

There is an increased prevalence of arthritic change and damage in the internal meniscus of the joint with age (Carlson *et al.* 1979). However, it is likely that many of these changes are of functional origin rather than due to age *per se*.

The dentoalveolar complex

This comprises the supporting structures of the teeth (the alveolar bone and the periodontium), the mineralized components of the teeth, and the dental pulp. Age changes in alveolar bone have already been described above.

Periodontium

The periodontium comprises the supporting tissues of a tooth and includes the gingiva (gums), the fibrous attachment between the surface of the root and the alveolar bone of the jaw, the alveolar bone itself, and the cementum (the mineralized tissue on the surface of the roots of the tooth into which the fibrous attachment apparatus is inserted). It is a unique structure in that it both supports the tooth in function and maintains a 'tight junction' between the soft tissues of the mouth and surface of the tooth (which is of ectomesodermal origin) as it passes through the gingiva. There is no evidence of age-associated alterations within the gingival epithelium, or its underlying connective tissue. The structure of the fibrous periodontal ligament becomes more irregular with a decrease in the fibrous and cellular components (Severson *et al.* 1978). Cementum is deposited continually on the surface of the roots of teeth throughout life. The increase in cemental thickness is greatest at the apex of the root of the tooth.

Teeth

The teeth comprise two mineralized tissues. These are enamel, which is the most highly calcified tissue in the body and is acellular, and dentine, which has similar mineral density to bone and is a vital tissue with cell bodies (the odontoblasts) lying within the dental pulp.

By definition enamel cannot undergo any age changes, as there is no capacity for cellular modification in this tissue. There are however a number of acquired changes. These include alteration in the chemistry of the enamel surface as a result of ionic interchange with the oral environment, producing a fluoride rich surface layer with a coarser crystalline structure. This surface layer is probably more resistant to decay than subsurface enamel. Vertical fracture lines develop within the enamel, probably as a result of cyclical stressing of the enamel in function. These fracture lines are filled with protein from the saliva with consequent increase in the organic component of the enamel.

Conversely there are two well established age changes in dentine. Firstly, dentine is deposited throughout life on the inner aspect of the pulp chamber (secondary dentine formation), which results in a progressive diminution in size of the pulp chamber. The secondary dentine is identical to primary dentine (that present when the apex of the root of the tooth forms subsequent to its eruption into the mouth). Secondly, the degree of mineralization of the dentine increases with increased mineral levels, both within the bulk of the dentine and via centripetal deposition of mineral within the dentinal tubules.

Irregular tertiary dentine may also be found in older teeth. This forms as a direct response to a noxious stimulus, e.g. dental caries, or wear of the tooth substance bringing the pulp chamber into closer proximity to the oral environment.

These alterations in form have a twofold effect. The progressive increase in thickness and mineralization of the dentine will reduce its sensitivity to external stimuli and increase the width of this protective layer, preventing pulpal damage through tooth decay. In addition, the hypermineralized dentine has altered optical properties, with increased translucency in the root structure giving almost transparent roots in some instances. The combination of secondary and tertiary dentine found in the coronal area results in increased opacity and alteration in colour of the dentine which is reflected in an overall darkening of the coronal tissue with age.

The dental pulp

The dental pulp is the central core of vital soft tissue within each tooth. It provides nutrients for the cellular structure within dentine and contains the nerve cells which are responsible for the sensitivity of teeth when the dentine is exposed. The pulp is a loose fibrous connective tissue in youth with an irregular structure within the core of the pulp and a clearly defined neurovascular plexus immediately beneath the odontoblast layer. (The odontoblasts in dentine are analogous to osteoblasts in bone, although they are arranged in a palisade on the pulpal surface of the dentine with long cellular processes extending outward into tubules within the mineralized dentine.) With increasing age the pulpal volume decreases as a result of continuous deposition of dentine by the odontoblasts. This is associated with an apparent increase in fibrous tissue within the pulp (it is unclear whether this increase is absolute or relative as a result of decreased pulp volume), a reduction in pulpal vascularity, and degenerative changes within the nerve tissue which will ultimately result in denervation of the tooth. In addition to these phenomena there is an increase in pulpal mineralization in both focal (producing pulp stones), and diffuse forms. These alterations tend to render the teeth less sensitive to stimuli in older subjects. In addition some of the degenerative neuronal changes may be responsible for abnormal pain sensations in the oral region.

It is thought that many of these signs of ageing within the dental pulp are the result of functional stimuli, either of a mechanical or noxious nature, rather than age changes *per se*. There is very little evidence of alteration in pulpal tissue with age in teeth which have remained embedded within the bone of the jaws rather than erupting into function in the mouth.

Taste

Taste is the major oral special sense. The sensation of taste is a result of stimulation of specialized nerve endings in the mouth by chemicals in solution. Taste forms one part of the perception of flavour from food, the second element being formed by smell from the food passing up to the olfactory organ via the pharynx. Many of the nerve endings that are responsible for taste are grouped together in taste buds, found in the tongue, soft palate, and epiglottis. These receptors appear to be specific for each of the four basic taste modalities (sweet, sour, salt, and bitter) with certain zones within the mouth being responsible for perception of specific stimuli (Jenkins 1978). As yet, no morphological or biochemical differences have been elicited which account for these variations in sensitivity.

Early studies suggested a reduction in the number of taste buds with age but more recent, well-designed studies show no apparent variation in taste-bud number with age (Arvidson 1979). Despite the constant numbers of taste receptors in the older subject, there may be specific deterioration in taste perception as a result of variation in cell surface receptors or even alteration in salivary chemistry (see above).

The assessment of taste perception can be achieved by monitoring taste threshold or taste intensity. A number of well-designed studies have demonstrated age-associated changes in both of these variables (Bartoshuk *et al.* 1986; Weiffenbach *et al.* 1986). There is a trend for an increase in threshold stimulus and in perception of intensity with age. However, this trend does not apply to all taste modalities, for example the threshold intensity for salt and bitter stimuli appears to be affected by age whereas that for sweet and sour does not.

Notwithstanding these age-associated alterations in taste perception it must be remembered that texture, appearance and odour all have an influence upon the enjoyment of food. Individuals who complain that their food is not as appetizing now as it was 20 years ago may be aware of some variation in stimulus, but, their feelings may reflect simply the difference between home cooking and institutional food.

PATHOLOGICAL CHANGES IN THE OLDER SUBJECT

Pathological changes are found in all of the tissues described in the previous section. Some are of unknown aetiology, whilst others are associated with a systemic disease, or are an oral manifestation of systemic drug therapy.

Soft tissues

Oral mucosa

Oral mucosal diseases increase in prevalence with increasing age. The aetiological factors involved in this increase include age changes within the mucosal tissues, alterations in salivary flow and function (as a result of intercurrent disease or of systemic medication, see below), increased prevalence of systemic disease, and greater drug usage. There is a wide variety of oral mucosal diseases, some of which are of greater severity or are more common in older subjects. It is these two groups that will be described.

Oral cancer and precancerous conditions

Oral cancer registrations comprise some 1.2 per cent of all cancer registrations with an average of 2385 registrations annually in the United Kingdom (based on figures from 1980 to 1984) (Johnson and Warnakulasuriya 1991). The fatality rates for all oral neoplasms is 54 per cent, and this rises to 63 per cent for intraoral lesions (tumours of the lip and salivary glands are excluded). The mean age at registration of malignant disease for males has decreased over the last 20 years (Hindle and Nally 1991). The same study reported significant alterations in the patterns of oral cancer during this period with decreases in lip and salivary malignancies and a marked increase in those of the tongue and other intraoral sites, especially among females. As with the majority of malignant lesions, morbidity and mortality can be reduced by early detection and treatment of oral malignancy. Despite this there are several reports in the literature of significant delay in institution of appropriate management from the onset of symptoms. This delay may be attributable to the patient's reluctance to seek professional advice (Scully *et al.* 1986). Part of the problem is that fewer than 50 per cent of the population over the age of 45 attend for regular dental inspections, whilst 94 per cent of oral malignancies occur in this age group. This pattern of attendance is partly historical and partly associated with the high prevalence of edentulousness amongst older people at present. This prevalence is falling which may have a beneficial effect upon the detection of oral cancer at an early stage.

Early malignant lesions are often asymptomatic and present as either a white patch (leukoplakia), an exophytic growth with or without an ulcerated surface, a red patch (erythroplakia), or an ulcer (Fig. 2). The clinical features that should arouse suspicion of malignancy are persistent ulceration especially if it is pain free, and induration and/or fixation of the tissues. Any oral ulcer that has been present for more than 1 month should be actively investigated.

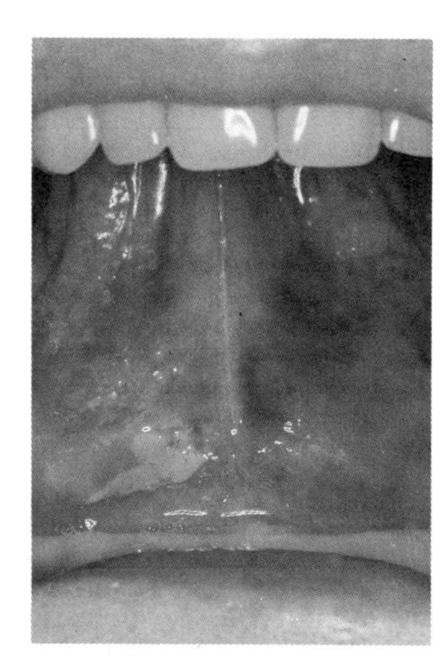

Fig. 2 An area of leukoplakia and erythroplakia in the floor of the mouth. On biopsy it was found that this lesion included elements of carcinoma-*in-situ*

Carcinoma of the lip is clearly visible and can be detected as a swelling or a crusted ulcerative lesion. Care should be taken not to confuse it with the healing lesion from a Herpes labialis infection which closely resembles labial carcinoma in appearance. Advanced oral cancers can be exophytic in nature with a warty necrotic or haemorrhagic surface or they can be ulcerative in appearance with rolled margins (Fig. 3).

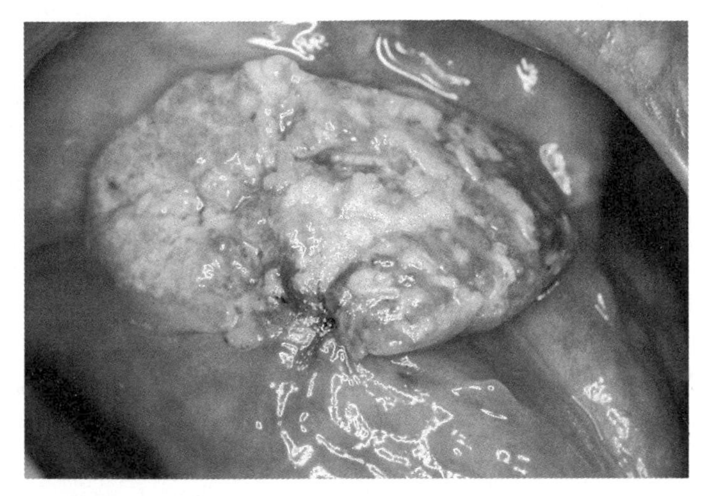

Fig. 3 A large, exophytic squamous cell carcinoma of the floor of the mouth.

Bony invasion is a common complication of carcinoma of the mucosa overlying the mandible or maxilla, and is a late complication of lesions of the floor of the mouth and the tongue. Metastatic spread to the lymphatic chain can occur at an early stage, although enlarged regional lymph nodes may also occur as a result of an aggressive inflammatory reaction to the lesion.

There are a number of accepted risk factors associated with oral squamous cell carcinoma.

Tobacco

Whether chewed or smoked, tobacco has been associated with oral malignancy (Moore 1971; Depue 1986). In addition there is an historical association between the use of a clay pipe and labial carcinoma. If the aetiological factor involved is chewing tobacco, the lesion tends to develop in proximity to the site where the tobacco quid is habitually held.

Pan

Pan is betel nut and lime wrapped in a betel leaf; other ingredients include tobacco and catchu. The chewing of pan is a widespread habit particularly in south-east Asia and India. The habit induces leukoplakia with a high risk of malignant change, possibly due to the reaction products of slaked lime and the alkaloids in betel nut.

Alcohol

It can be difficult to differentiate the effects of alcohol consumption from those of smoking. In studies where the effect of smoking has been controlled there is a positive relationship with alcohol consumption (Elwood *et al.* 1984). It may be significant that the rise in oral malignancy in the United Kingdom has occurred at a time when the alcohol consumption *per capita* has increased. Binnie and Wright (1986) have suggested that the association may be linked to liver dysfunction and nutritional deficiency rather than alcohol consumption *per se*.

Iron deficiency

Iron is an important factor in the maintenance of health of the oral epithelium (Rennie *et al.* 1984). It is well known that chronic iron deficiency leads to an increased risk of squamous cell carcinoma of the upper gastrointestinal tract, and it may be that a similar mechanism occurs in the mouth. It is possible that iron is a secondary factor in the development of cancer in the mouth as a result of the significant changes in the immune response seen in chronically deficient subjects (Joynson *et al.* 1972). Equally the thinned epithelium associated with iron deficiency may be more susceptible to carcinogenesis.

Infective agents

There are three infective agents that have been linked with oral carcinoma.

SYPHILIS

There is a long-standing association between syphilis and carcinoma of the tongue. There is no clear causative relationship and it may be that the medications used to treat syphilis before the development of antibiotics (i.e. arsenic and heavy metals), were more important instigators of the disease.

CANDIDIASIS

Some of the manifestations of oral candidiasis, specifically chronic hyperplastic candidiasis, are prone to undergo epithelial dysplasia (Cawson and Binnie 1980). However, the link between candida and malignant change is unclear, as other forms of chronic candidiasis do not carry the same risk.

VIRUSES

There is some evidence linking Herpes simplex virus type I (HSV-1) with oral carcinoma (Shillitoe 1982), possibly as a result of an interaction with tobacco smoke. There is also evidence that human papilloma virus is linked with both oral squamous cell papillomas and with oral carcinoma.

The nature of the interactions between these aetiological factors remains indistinct and has been summarized by Binnie and Wright (1986) (Fig. 4). There is no single aetiological agent whose removal would prevent the future development of oral cancer. Great care must be taken to monitor premalignant conditions.

There is one other form of squamous cell carcinoma which may have an oral presentation. Carcinoma of the maxillary antrum frequently spreads through the maxilla to present as an intraoral swelling or ulceration in the upper arch. An early sign of this problem in the edentulous subject is that the upper denture will become loose or mobile as a result of the altered contour of the palatal vault.

Details of the treatment of oral carcinoma are beyond the scope of this text. Treatment modalities commonly include radiotherapy and surgery either as individual items or in combination. Chemotherapy has only limited effectiveness against oral malignancies. The morbidity and mortality associated with this condition and with its management (especially if the management regimen includes extensive disfiguring surgery) is high. In addition the risks of the development of osteoradionecrosis, if operative procedures are undertaken on bony sites that have undergone radiotherapy, must be borne in mind when planning individual case man-

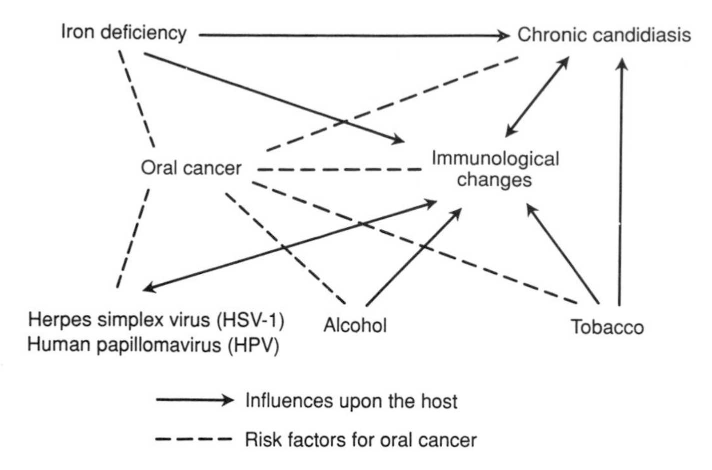

Fig. 4 A flow diagram illustrating the aetiological factors involved in oral cancer and their possible mechanisms of action (after Binnie and Wright 1986).

agement. Operative procedures would include the extraction of teeth: consequently patients who are to undergo radiotherapy are commonly rendered edentulous prior to treatment. All of these factors contribute to the morbidity associated with oral malignant disease.

Premalignant conditions of the oral mucosa

There are three conditions of the oral mucosa which may be premalignant.

Leukoplakia

This is simply defined as a white patch or plaque on the oral mucosa that cannot be removed by scraping and cannot be attributed to any diagnosable disease (Soames and Southam 1985). The risk of malignant transformation within such lesions is generally low, although there are certain high-risk sites (the ventrum of the tongue, the floor of the mouth, and the mucosa overlying the alveolar bone in the lower jaw adjacent to the tongue) where the risk of malignant change can be as high as 30 to 50 per cent (Fig. 2). Risk of malignancy may be higher if there are red areas within the white lesion.

Erythroplakia

These lesions have a bright red velvety appearance. They may arise in isolation or they may develop within pre-existing areas of leukoplakia. There is a high incidence of carcinoma-*in-situ* within such lesions which should be managed aggressively. It is not surprising to find that the same group of aetiological factors are associated with leukoplakia, erythroplakia, and squamous cell carcinoma (see above).

Lichen planus

The oral manifestations of this common dermatological disorder have a varied presentation, from the classical patterns of white papules arranged in linear reticular or annular patterns through dense patches of hyperkeratosis similar to those seen in leukoplakia to the erosive form where extensive shallow ulcers form as a result of sloughing of the mucosal surface (Fig. 5). This last form is often very painful and the oral lesions may be present for some time. Some 70 per cent of patients who present with skin lesions of lichen planus will also have oral lesions. There are also many subjects who exhibit the oral signs of this disease without any skin lesions. A small proportion of these oral lesions (less than 1 per cent according to Holmstrup and Pindborg 1979) will undergo malignant change. It is thought that the erosive forms are more likely to undergo such transformation, possibly as a result of epithelial atrophy and an increased susceptibility to carcinogens. The aetiology of lichen planus is unknown. It is likely that cell-mediated immune reactions are involved in the development of the lesions, but no circulating antibodies have been identified.

Lichenoid eruptions can occur in association with the use of a wide range of systemic medication including antimalarial agents, methyldopa, gold, frusemide, propranolol, and thiazide diuretics. The clinical and histological appearance of these lesions is identical to those of unknown aetiology, although they tend to regress on cessation of drug therapy.

Lupus erythematosus

The oral lesions of lupus erythematosus can be remarkably similar to those of lichen planus, with discrete areas of erythema/ulceration with a keratotic margin and some radiating striae.

Benign mucous membrane pemphigoid

Benign mucous membrane pemphigoid affects women more frequently than men and is rarely seen in patients under 50 years of age. The oral lesions are often the first sign of this serious condition. Intraoral bullae form following minor trauma. They are relatively tough as the plane of cleavage of the bulla is the basement membrane of the epithelium. Rupture will cause the formation of extensive ulcers, clearly demarcated from the surrounding mucosa. Such lesions persist for weeks and may heal with scar formation.

The most common oral lesions in dentate patients are of the attached gingiva, resulting in desquamative gingivitis. There may be some confusion in this circumstance between this condition and lichen planus. Biopsy will confirm the correct diagnosis. Steroids and other forms of immunosuppressive therapy may be required.

Denture-induced hyperplasia

Patients who have worn ill-fitting dentures for a number of years can develop fibrous hyperplastic masses at the periphery of the denture (see below).

Oral candidiasis

Oral infection with *Candida albicans* can occur both in acute and in chronic forms in the older patient.

Acute pseudomembranous candidiasis (thrush)

This occurs in debilitated subjects, commonly in those who are immunosuppressed and those with severe illness. It can occur in patients who have some remaining natural teeth as well as those who are edentulous. The presentation in adults is identical to that in younger patients, with the development of slightly raised white or cream plaques on the surface of the oral mucosa. The plaques have a curd-like consistency and can be stripped from the surface with ease leaving an erythematous and often bleeding mucosal bed.

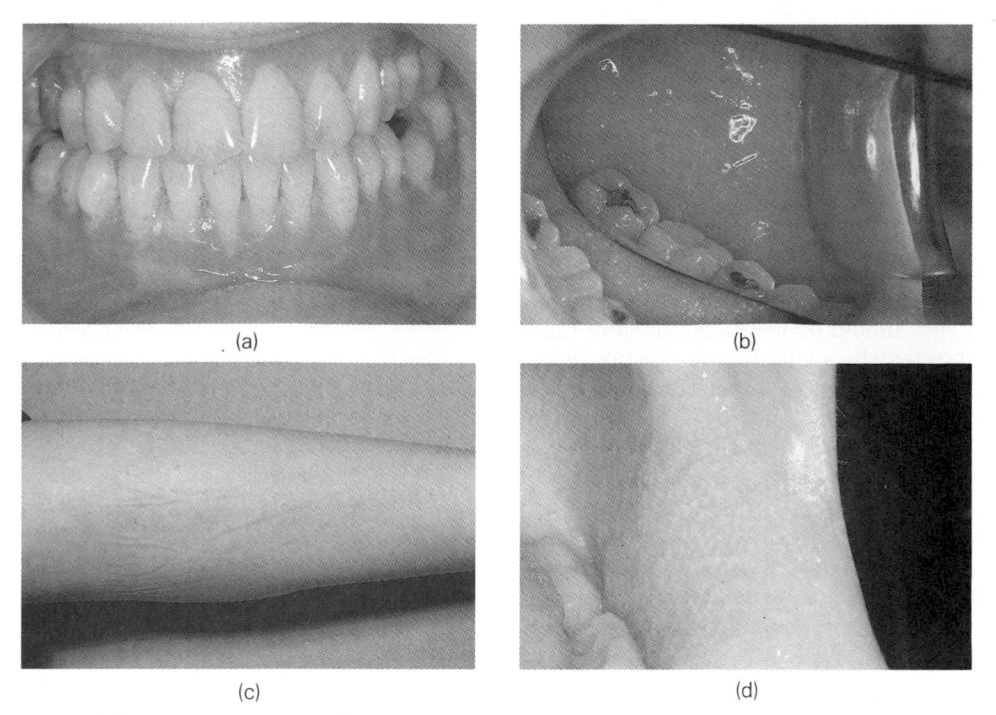

Fig. 5 (a) Erosive lichen planus of the attached gingiva with extensive ulceration of the buccal gingival mucosa. This condition can be mistaken for benign mucous membrane pemphigoid in the absence of other lesions of lichen planus. (b) Diffusive erosive lesions of lichen planus on the buccal mucosae of the same patient. (c) Typical skin lesions of lichen planus on the forearm of the same patient. (d) Classical striated lesions of lichen planus in the buccal mucosae.

Acute atrophic candidiasis

This commonly develops from superinfection of the oral environment with Candida as a result of the use of broad spectrum antibiotics.

Both forms of acute candidiasis respond well to topical antifungal therapy using oral lozenges (commonly nystatin or amphotericin B: miconazole oral gel is of value in patients with dentures). Systemic antifungals may be appropriate if the patient is also suffering from vulvovaginal candidiasis. Any underlying cause should be treated.

Chronic candidiasis

The most common form of chronic oral candidiasis in older age-groups is chronic atrophic candidiasis. This is associated with the wearing of some form of acrylic based denture and may be accompanied by angular cheilitis (Fig. 6).

The clinical presentation is of an erythematous area which corresponds to the outline of the overlying prosthesis. This condition occurs almost exclusively on the palatal mucosae. It is thought that the relative mobility of a lower denture prevents the development of an appropriate environment for candida overgrowth. Conversely the space between an upper denture and the mucosa is ideal, especially if the denture is worn continuously, is old or ill fitting, or where denture hygiene is poor.

Candida can often be cultured from the fitting surface of the denture where micropores and cracks in the surface of the acrylic provide an ideal culture zone. Angular cheilitis is a frequent complication of this condition with painful erythematous fissures at the angles of the mouth. Worn biting surfaces on prostheses, with an associated loss of height of the lower third of the face, or inadequate lip support from the labial surface of a denture, make angular cheilitis worse.

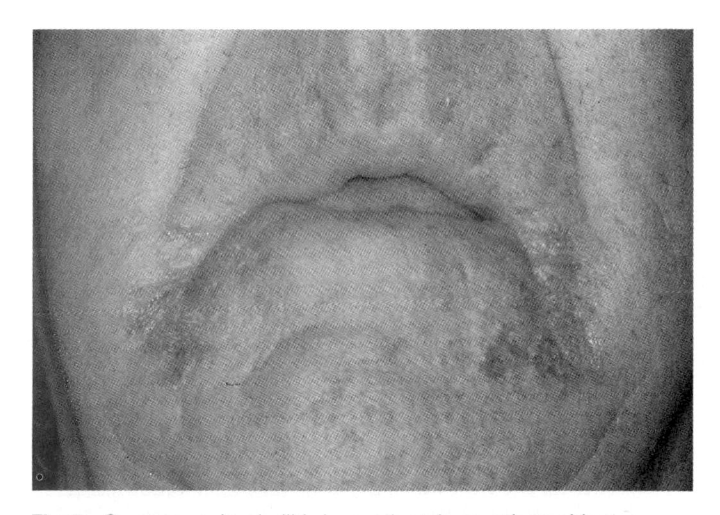

Fig. 6 Severe angular cheilitis in an edentulous patient with undiagnosed diabetes mellitus. Candida species could be cultured from swabs taken from angle lesions and from the fitting surface of the subject's upper denture.

Chronic candidiasis is associated with iron and folate deficiency. In addition the fissures of angular cheilitis are frequently infected with either *Staphylococcus aureus* or β-haemolytic streptococci. The presence of these commensals complicates treatment for this condition, necessitating the use of an appropriate topical antimicrobial agent.

Management of chronic atrophic candidiasis has three interlinked components.

Antifungal treatment

Antifungal lozenges can be sucked or miconazole oral gel can be applied to the fitting surface of the denture prior to insertion into the mouth. The latter drug can also be used topically on the lesions of angular cheilitis, as can other antifungal ointments.

Denture hygiene

Dentures should not be worn at night, and should be stored in an aqueous environment when not in the mouth. Patients with atrophic candidiasis should be instructed to store their dentures in 1 per cent sodium hypochlorite solution (available commercially for the sterilization of infant feeding bottles/nappies). In addition all loose debris should be removed from the surface of the dentures using a brush and soap, or a proprietary denture cleaner.

Correction of denture faults

Once the infection is brought under control any underlying denture faults should be corrected in an endeavour to prevent a recurrence. This may necessitate remaking the prosthesis.

Any underlying haematopoietic deficiency must be treated. If the lesions of angular cheilitis are infected with bacteria, then topical antimicrobials will be required. Chloramphenicol preparations designed for use in the eye and fusidic acid are useful in this circumstance, but carry a risk of contact sensitivity.

Salivary glands

Malignancy

A small proportion of oral neoplasms arise within the salivary glands. These include pleomorphic and monomorphic adenomas, mucoepidermoid tumours, and adenoid cystic carcinomas. The pleomorphic adenoma is by far the most common tumour (65 per cent of parotid and 55 per cent of minor salivary gland tumours). There is a higher prevalence of this tumour in women with the peak incidence during the fifth and sixth decades of life. Fortunately most pleomorphic adenomas are only locally invasive and respond well to surgical excision.

Xerostomia

Dry mouth is a common complaint amongst older subjects (Sreebny and Valdini 1987). However, there is little evidence of altered function in healthy unmedicated older subjects (this may be due, in part, to the wide variation in 'normal' salivary flow levels amongst adults). Conversely there are a number of well documented aetiological factors in xerostomia which increase in prevalence with increasing age (Table 1). Three of these aetiological factors will be discussed.

Table 1 Causes of xerostomia

Disease/condition		Examples
Drugs/medications		Anorectics, anticholinergics, antidepressants, antipsychotics, sedatives and hypnotics, antihistamines, antiparkinsonian agents, antihypertensive agents, and diuretics
Irradiation		Therapeutic radiation to the head and neck
Organic diseases	1.	Collagen vascular diseases: Sjögren's syndrome, rheumatoid arthritis, systemic lupus erythematosus, scleroderma
	2.	Hyposecretory conditions: primary biliary cirrhosis, atrophic gastritis, graft vs. host disease, pancreatic insufficiency, type V hyperlipoproteinaemia
	3.	Immunodeficiency disease: AIDS
	4.	Diabetes mellitus and diabetes insipidus
	5.	Hypertension
	6.	Cystic fibrosis
	7.	Neurological disease: Bell's palsy, tumours, surgery, trauma
	8.	Altered water balance: impaired water intake, dehydration through skin (fever, burns, excessive sweating), blood loss, emesis, diarrhoea, renal water loss, polyuria, osmotic diuresis, oedema
Gland malfunction		Aplasia, obstruction, trauma, infection
Psychogenic factors		Depressive illness, fear, excitement, anxiety
Decreased mastication		Animal and limited human studies have shown that intake limited to liquids or soft foods leads to decreased salivary flow. The contribution of this element to the feeling of mouth dryness is unknown

Adapted from Sreebny (1989).

DRUGS

Drugs can induce xerostomia in a number of ways, by affecting the central 'salivary centre', by influencing the autonomic control pathways peripherally, and by influencing water balance with relative dehydration and/or oedema. Sreebny and Swartz (1986) list over 400 drugs which have a 'xerostomic effect', many of which are available for sale by the pharmacist. There is little control of medication usage with these 'over the counter' products. This picture is complicated by multiple drug usage amongst older subjects. Levy *et al.* (1988) reported that an average of 3.2 medications 'with effects potentially important in dental cases' were consumed per person in an epidemiological study of subjects 65 or older. Seventy-four per cent of their sample were taking at least one drug with the potential to induce xerostomia.

Little is known about the mode of action of these medications, or why some pharmacological agents within the same family have a lesser effect than others. Even less is known about the potential for interaction between such drugs, although there is one report which specifically links tricyclic antidepressants and diuretics (Johnson *et al.* 1984). One disturbing finding is that uptake of drugs from the sublingual pouch was reduced in patients with xerostomia (Rasler 1986).

IRRADIATION

Mouth dryness can occur with relatively low radiation dosages (as little as 10 Gy). The severity of the xerostomia depends upon the dosage of radiation and tends to increase for some weeks after the cessation of radiotherapy owing to continued gland fibrosis (Mira *et al.* 1982). The reduction in salivary flow is accompanied by a decrease in bicarbonate and an increase in sodium, calcium, magnesium, and chloride ion and in protein concentrations within the saliva (Dreizen *et al.* 1976).

DISEASE

Autoimmune destruction of salivary glands is associated with the collagen vascular diseases. In its classical form this will comprise the triad of xerostomia, keratoconjunctivitis sicca, and rheumatoid arthritis which make up Sjögren's syndrome. Several serological abnormalities are present in such cases including rheumatoid factor. However, most subjects with xerostomia will have specific antibodies to salivary ductal epithelium. Xerostomia can occur in the absence of one of the other elements of the triad; indeed it has been reported that it is a complicating factor in up to 50 per cent of subjects with rheumatoid arthritis (Syrjanen and Syrjanen 1978).

Symptoms and signs

The symptoms and signs of this distressing condition are multiple (Table 2). The condition may be complicated by a superimposed candidiasis and angular cheilitis which makes the already thinned sensitive mucosa very sore. This can be a debilitating problem, affecting not only oral health and digestion, but also impairing speech with a marked effect upon the quality of life.

Table 2 Symptoms and signs of xerostomia

Oral symptoms	*Non-oral symptoms (associated with some conditions)*
Dryness of the mouth	Dryness of the throat
Altered taste	Dry or gritty eyes/blurred vision
Difficulty with speech, chewing dry food, swallowing.	Nasal dryness
	Dry skin
Recurrent oral/commissural candidal infection	Recurrent vaginal itching/ candidiasis
Burning/tingling tongue	Heartburn
Frequent ulceration of tongue, lips or cheeks.	Constipation
Difficulty in tolerating dentures	
Signs	
Tenderness of the glands if inflamed	
Angular cheilitis	
Dry lips	
Coating of the dorsum of the tongue	
Reduced keratinization of the lateral borders of the tongue	
Dry oral mucous membranes	
Difficulty in expressing saliva from gland orifices	
Root surface caries around many teeth	

Management

The objectives of management are threefold: (1) to identify and then to either remove or to ameliorate any aetiological factors; (2) to stimulate any residual salivary tissue to increase flow rates if possible; (3) effective replacement of saliva using artificial alternatives may be a final line of attack, together with preventive measures designed to reduce mucosal trauma and the risk of candida infection and to prevent the development of xerostomia induced dental caries.

Pharmacologically induced xerostomia is usually reversible. Whilst it may not be practicable to eliminate all potentially xerostomic agents the pattern of administration can be altered to give some symptomatic relief (by using frequent small doses of drugs, or by maximizing blood levels during the day with low levels at night). It may be possible to change the medication within a given group (for example substitution of protriptyline for amitriptyline) to reduce the xerostomic effects whilst maintaining the therapeutic benefits (Sreebny and Valdini 1987).

Salivary secretion can be enhanced using an acid stimulus within the mouth (commonly citric acid). Great care must be taken in dentate subjects lest the frequent use of such a stimulus exacerbates dental caries or erosive tooth surface loss. It has been suggested that citric acid buffered with bicarbonate may not cause demineralization. Chewing non-sugar containing chewing gum will also stimulate residual salivary flow. Pharmacological agents may also be used to stimulate salivary flow; these include pilocarpine and pyridostigmine (administered as pyridostigmine bromide syrup). Obviously these cholinergic agonists will have potential effects elsewhere in the body and should be used with caution (Greenspan and Daniels 1987).

There have been a wide variety of artificial salivas described based on glycerine, carboxymethyl-cellulose, and pig gastric mucin (Levine *et al.* 1987). Some preparations (e.g. glycerine and lemon mouthwash BP) have low pH and should be avoided in dentate subjects. Recently aerosol delivery salivary substitutes containing fluoride have become available. These have high pH and consequently can be used in both dentate and edentulous subjects (Joyston-Bechal and Kidd 1987). Salivary substitutes are not a universal panacea. Some patients find that the most practical solution is to take frequent sips of water, especially during meals.

Dentate subjects should also be given intensive preventive care to halt the development of caries. Such regimens include the use of high dose topical fluoride, fluoride mouth rinses, supersaturated calcium phosphate 'remineralizing' solutions and chlorhexidine gluconate mouth rinses (the latter is an antiplaque agent) (Johansen *et al.* 1987).

HARD TISSUES

Pathology of oral hard tissues is rare. Osteosarcomas may arise within the mandible or the maxilla, but they are unusual, as are bony metastases. Systemic disease which affects bony metabolism may also affect the jaws, e.g. Paget's disease of bone, and this in turn may influence the management of dental problems of an affected patient.

One common cause of anxiety amongst older subjects is the presence of benign bony exostoses which may occur in either the mandible or the maxilla. These can be single or multiple, and can be unilateral or bilateral in the mandible, where they always occur on its lingual aspect. In the maxilla there is usually a single torus in the midline of the palate. They usually give rise to anxiety when the mucosa overlying the bony swelling is traumatized during normal oral function. Such trauma results in a superficial ulcer which is painful. The subject then either looks for or feels for the painful area and finds a hard painful swelling. This is often mistaken for a malignant growth with all of the associated anxieties. Reassurance, after any appropriate investigations, is essential to avoid prolonged anxiety. A far better course is for all dentists to inform any patient who has a torus of its presence.

The aetiology of these tori is unclear. They can pose a significant problem during dental treatment if the area of the torus has to be covered by a denture. In such a circumstance the thin overlying mucosa can be traumatized easily giving chronic, painful ulceration. Such tori are often removed surgically to overcome this problem.

The dentoalveolar complex

The patterns of disease and the clinical problems that may arise from the dentoalveolar complex are primarily influenced by the presence or absence of teeth. The accepted image of the older subject is of an individual with no natural teeth, wearing complete dentures in both jaws. This pattern is in part historical and is undergoing rapid change. Traditionally, people have had their teeth removed, rather than seeking reparative care for dental problems. Indeed in some areas of the country the prophylactic removal of all natural teeth was regarded as prudent. This was undertaken either as a 21st birthday present or as a bride's father's gift to his daughter.

The level of dental awareness is now improving markedly and many more people are prepared to have regular routine dental care to maintain their natural dentition. This alteration in awareness is resulting in a dramatic change in the pattern of tooth retention with increasing age. The fall in the rates of edentulousness amongst the population aged 60 or more has been relatively small to date. If we look at cohort trends from the adult dental health surveys of 1968 through to 1988 (Fig. 1) we can see that many more older people will retain some of their natural dentition over the next 20 years. Superimposed upon these perceived national changes are underlying variations by region, by sex, and by social class (Figs. 7, 8), resulting in high rates of tooth retention in social classes I, II, and III who have greater expectations in terms of their continued function and quality of life and greater ability to pay dentists' fees. Dental pain or an inability to eat because teeth are lost at a stage in life where it is difficult for a person to adapt to wearing dentures (see below) is a potent cause of morbidity.

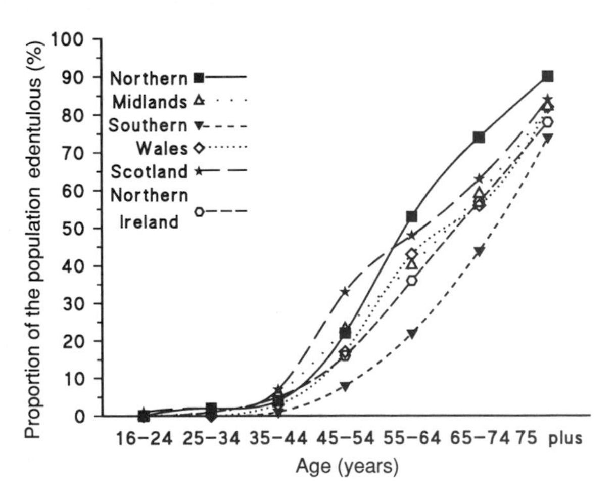

Fig. 7 Variation in the pattern of edentulousness by region within the United Kingdom. The lowest levels are found in the south of England, with the highest in the north of the country.

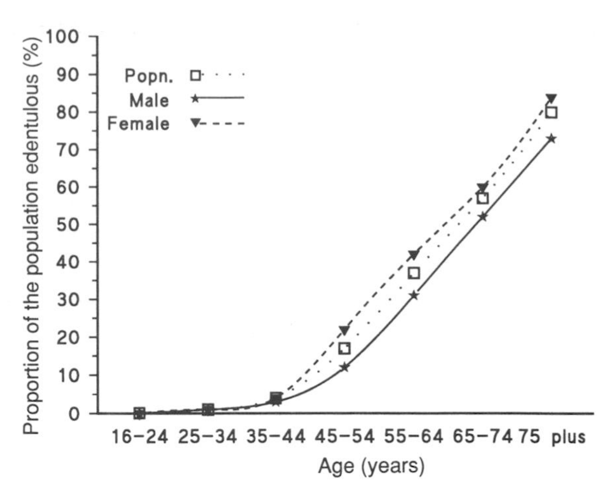

Fig. 8 Variation in the pattern of edentulousness by sex in the United Kingdom.

The nature of the dental care which is required by a partially dentate person is very different from that for an edentulous subject. There is a continued need for daily oral hygiene, which some older patients find difficult, if not impossible, and for regular checks of dental health including the provision of any necessary dental care. These tasks will place new burdens upon the carers, either at home or in long-stay accommodation, who may have to learn new skills to manage such patients. How many of us are capable of brushing somebody else's teeth?

Within the partially dentate group the patterns of dental disease vary from those found in younger adults. The disease processes which we need to consider are periodontal disease, dental caries, and tooth wear.

Periodontal disease

In epidemiological terms there is a progressive loss of attachment between teeth and their supporting alveolar bone with increasing age. The clinical manifestation of this loss is a gradual increase in tooth mobility as the bony support decreases. Ultimately the tooth may exfoliate, or develop an abscess and be removed by a dentist. This loss of attachment is not an age change, but is the result of a chronic disease state within the supporting structures of the tooth (Axelsson *et al.* 1991; Papapanou *et al.* 1991).

The aetiology of this disease is complex, and indeed the details are still unclear to the dental profession. In simplistic terms, if the surfaces of teeth are not cleaned regularly, bacterial deposits become firmly attached to tooth structure (dental plaque). As plaque matures the bacterial flora changes. Periodontal disease is produced by substances which are produced and released by such bacteria, especially those that lie in the crevice around the neck of the tooth between the gingiva and the tooth. The bacterial products either cause cellular damage themselves or stimulate the immune response producing inflammation. Some of the mediators of this inflammatory response (notably the T-cell mediated response) are thought to be responsible for much of the bony destruction seen in periodontal disease.

It is thought that disease is characterized by active phases of bony destruction, interspersed with inactive stages of defence/attempted repair. It is well established that individuals differ in their susceptibility to periodontal destruction.

By inference, those subjects who retain their natural teeth into older age have a degree of natural resistance to this disease. In addition the alterations in the immune response with ageing (notably the reduction in T-cell numbers and activity) result in an altered pattern of the disease (Holm-Pedersen *et al.* 1979, 1980).

Active phases of disease are characterized by a very florid inflammatory phase within the gingival tissues (Fig. 9), but with little bony destruction. Hence the rate of loss of support will tend to reduce and, consequently, it may be thought that periodontal disease is not a problem in older subjects. Whilst this may be so, the sequelae of past disease, namely gingival recession with exposure of root surface and tooth mobility, constitute a clinical problem.

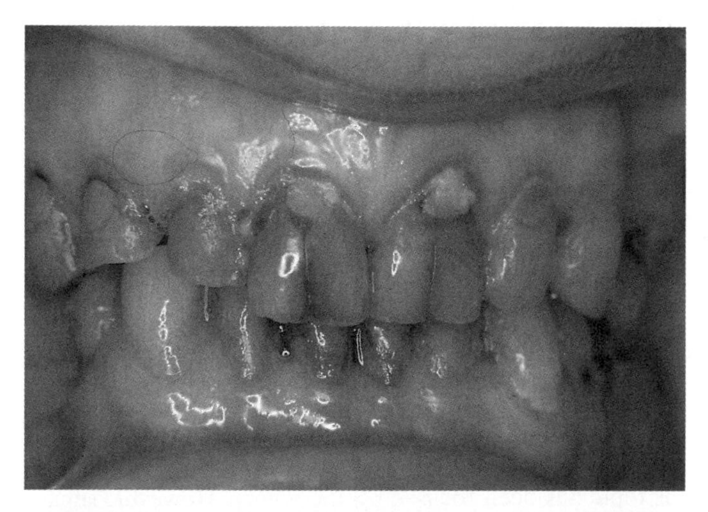

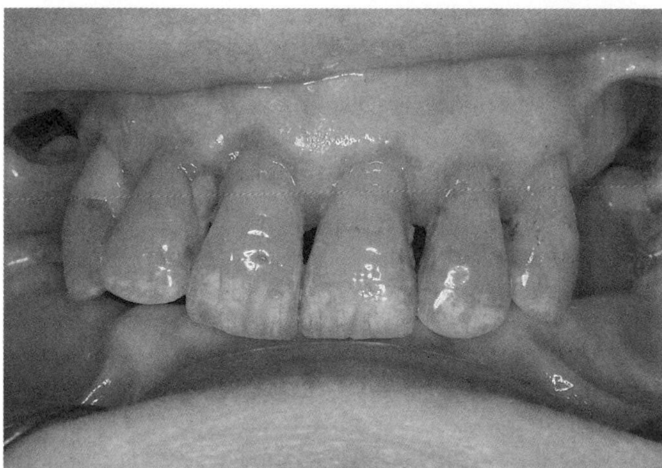

Fig. 9 Two examples of florid marginal gingival inflammation in subjects in their mid to late 70s. The severity of the marginal gingival inflammation is greater than would be seen in a younger subject for the same quantity of plaque.

In addition to these alterations in periodontal disease the gingiva may be involved in other disease processes. Desquamative gingivitis is very painful and is seen in benign mucous membrane pemphigoid, lichen planus, and in lichenoid drug reactions. Other gingival responses to drugs include gingival overgrowth. This has been described with phenytoin, cyclosporin, and the calcium channel blockers. Of this group, gingival overgrowth is most commonly associated with nifedipine although it is likely that all of the calcium channel blockers will exhibit this unwanted side-effect. The pathogenesis of drug induced gingival overgrowth is unclear. All the agents that can induce this problem can modify calcium metabolism within the cell. Mediation of the effect may be via altered function of the gingival fibroblasts and/or inhibited T-cell response with immunosuppression (Seymour and Heasman 1988; Seymour 1991; Seymour and Jacobs 1991).

Individual susceptibility to gingival overgrowth varies, as does the severity. Local factors (i.e. poor oral hygiene, defective or rough restorations, malalignment of the teeth), which encourage plaque formation, will make the overgrowth worse. The clinical manifestation is of fibrous gingival overgrowth which can obscure the crowns of the teeth. Management includes surgical removal of the excessive gingival tissue in association with appropriate oral hygiene instruction. It may be necessary to perform such surgical procedures repeatedly.

Management

Management strategies for periodontal disease in older subjects are very similar to those in younger patients. Achieving and maintaining a high standard of oral hygiene is vital to prevent any further progression of the disease. This goal is complicated in the older subjects by three factors;

Complex gingival architecture

Gingival recession as a result of ageing exposes large expanses of tooth surface into the mouth. The level of the gingiva may vary from tooth to tooth, and the gaps between the roots of teeth are wider than those between the crowns. In addition some premolar and nearly all molar teeth have either two or three roots. The exposure of the junction between these roots (the bi- or trifurcation) makes adequate cleansing of all tooth surfaces very difficult.

Alterations in manual dexterity

The alterations in fine motor skills associated with ageing are documented elsewhere. They have a detrimental effect upon the ability to complete oral hygiene, which may become an arduous task. If the problems of arthritic changes in the hands are superimposed on this then the burden is even greater. Specially modified toothbrushes with bulky handles will be required for such individuals. An electric brush may be of great benefit for older people for three reasons: first, the handle tends to be bulky as it contains both batteries and a motor; second, the brush head tends to be small so that access to difficult areas of the mouth is easier; and third, the mechanical brushing action removes much of the hard work from regular oral hygiene.

Reduced ability to learn new skills.

As a broad generalization people do not relearn their oral hygiene techniques as the gingival changes occur. It may be impossible to educate an older subject to clean all areas of their mouth with the necessary level of skill. In such a circumstance it may be necessary to compromise the ideal of retaining as many teeth as possible to prevent diseases and pain.

CHEMICAL PLAQUE CONTROL

Antimicrobial agents have a limited role in preventing periodontal disease. Metronidazole and tetracycline can be used to good effect during acute exacerbations but should not be given in the long term. The only topical antimicrobial which has any lasting demonstrable benefit is chlorhexidine gluconate (as a 0.1 or 0.2 per cent mouthwash). This can be used as an antiplaque agent with reasonable efficacy over a 6-month period. Resistant strains will develop with longer term use and chlorhexidine potentiates the staining of the natural teeth by oral fluids. There is a large volume of research underway to identify a surface active molecule which will compete with plaque for attachment to tooth substance and hence prevent plaque deposition. None is available as yet.

OTHER TREATMENT MODALITIES

Both surgical and non-surgical treatment options are available to attempt to arrest or reverse the damage produced by periodontal disease. The response of the gingival tissues to treatment in older subjects is very similar to that in the young, whichever treatment options are chosen.

Dental caries

Active caries of the crown of the tooth is largely a disease of children and younger adults. Older adults may require restorations in the coronal tissues, but this is commonly associated with failure of a previous restoration or decay around the margins of a restoration. Even in circumstances where the rate of development of caries on exposed root surfaces is high, the incidence of new coronal caries is low.

Caries of exposed root tissue can be a significant problem amongst older subjects. As has been described, there tends to be progressive loss of attachment of teeth to the supporting bone with age and this causes exposure of the root surface to the oral environment. It must be emphasized that this exposure is a consequence of disease rather than ageing *per se* (Fig. 10).

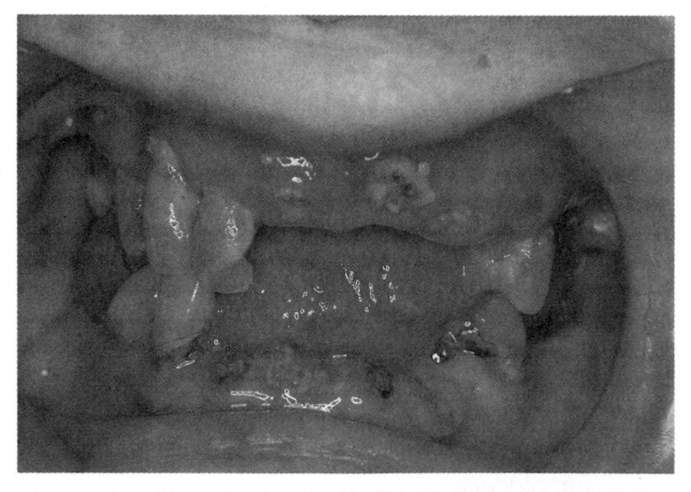

Fig. 10 Extensive coronal and root surface caries in a 74-year-old stroke patient. This woman had 12 intakes of sugar-containing foods daily and had not cleaned her teeth for 4 to 6 months.

Root surfaces comprise a combination of dentine and cementum. Both of these tissues have lower levels of mineralization than enamel and are more susceptible to surface attack, either in the form of demineralization and caries or of surface wear (see below). Interestingly, it would seem that the increased prevalence of root caries in older subjects is a manifestation of increased exposure of root surfaces into the mouth rather than an increased attack rate in older patients (Katz *et al.* 1985). This may not hold true for the very old (Donachie and Walls 1991).

Root surface decay tends to spread laterally around the necks of the teeth, within the superficial dentine, rather than centrally towards the dental pulp. It is characterized by the development of softened surface lesions which vary in colour from light tan to almost black. There may or may not be overt cavitation (loss of surface contour).

There are a number of important aetiological factors in this disease.

Xerostomia

Xerostomia, of whatever cause, will result in an increased incidence of root surface decay. This is probably a combined effect from reduction in salivary buffering/remineralization as a manifestation of reduced flow, and increased plaque deposition in such patients.

Diet

A diet comprising frequent snacks/drinks containing fermentable carbohydrate is just as likely to cause aggressive caries in older subjects as it is in the young. One added complication is sugar within medicines. The majority of the attention on this topic has been focused on the young. However, sugar-containing syrups/elixirs are just as damaging in the older adult, as are chewable or dispersible drugs which are acidic or which contain sugar. One particular culprit amongst this last group is chewable antacid preparations (Glenwright *et al.* 1988).

Oral hygiene

Not surprisingly, caries is associated with poor levels of oral hygiene. This can be a particular problem in subjects with dementia, or those with hemiparesis as a result of a stroke, where the provision of an adequate level of oral care is both difficult and time consuming for the carers.

Fluoride

Much of the positive benefit associated with fluoridation of water supplies is perceived in young children. However, a recent study from America (Hunt *et al.* 1989) has demonstrated that 'systemic' fluoridation has a benefit in terms of reducing the incidence of root caries, and in reducing the rate of development of new lesions in older populations with life-long exposure to fluoridated water as well as those who have lived in an area with fluoridated water for 30 to 40 years or more (i.e. subsequent to the complete formation of the teeth). It is likely that this benefit is a topical effect as tooth formation is completed during the third decade of life. The benefits are greater than those which accrue from topical fluoride application alone (e.g. the use of a fluoride-containing toothpaste (Jensen 1988)).

Management

Obviously an adequate level of oral hygiene is required but the problems are the same as those discussed above. There have been a number of studies of methods of preventing caries in xerostomic subjects using relatively intensive preventive regimens. These utilize a combination of professionally applied, high dose, topical fluoride, fluoride mouth rinses for daily use, chlorhexidine gluconate mouth rinse as an adjunct to fluorides and remineralizing mouth washes (supersaturated calcium phosphate solutions) (see Walls 1989 for review).

The restoration of deeper lesions has been revolutionized by the advent of tooth-coloured filling materials that can adhere to tooth tissues. The glass ionomer cements are particularly valuable as they also release fluoride into the local environment, helping to prevent secondary decay.

Tooth wear

Wear of the functional surfaces of teeth is an inevitable sequel to a lifetime's usage. The rate of the wear can be sufficiently rapid to cause a significant deficit in older patients. The deficit can be aesthetic, functional, as in an inability to eat (Fig. 11), or present as pain from the temporomandibular joints due to loss in vertical face height, or malpositioning of the mandible. There are a number of medical problems which can exacerbate tooth wear, especially erosive wear, during which the surface of the tooth is demineralized by acid and the softened surface is then worn away in function.

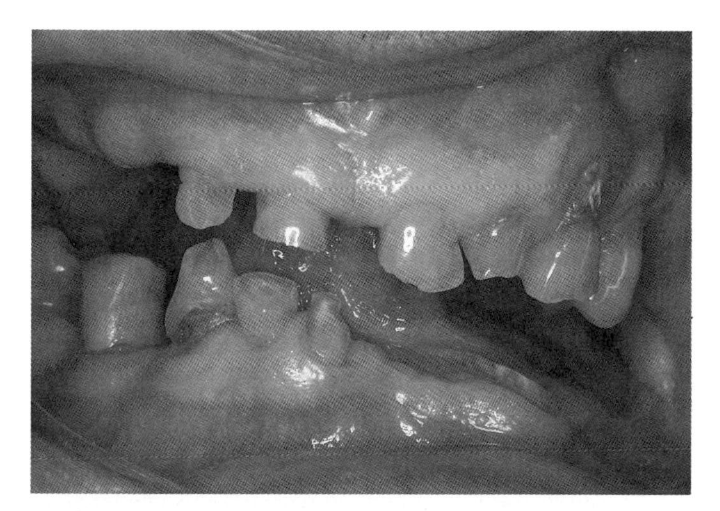

Fig. 11 Severe tooth wear of complex aetiology. This subject complained of both aesthetic and functional problems.

Gastric regurgitation in the form of frank vomiting or as a result of more insidious gastric reflux, can result in rapidly progressing tooth wear (Fig. 12). The chronic forms of regurgitation tend to manifest themselves in older subjects, with hiatus hernia as a common problem. A similar pattern of wear is seen in alcoholics and interestingly in subjects with long-standing diabetes mellitus (this may be a manifestation of gastric regurgitation from an autonomic neuropathy in association with altered salivary function).

There are two other commonly described modes of wear: attrition where tooth surface is worn away as a result of tooth to tooth contact in association with a parafunctional habit (either clenching or grinding), and abrasion where the tooth is worn by an extraneous abrasive; this is commonly a toothpaste/brush, but other objects can cause problems (Fig. 13).

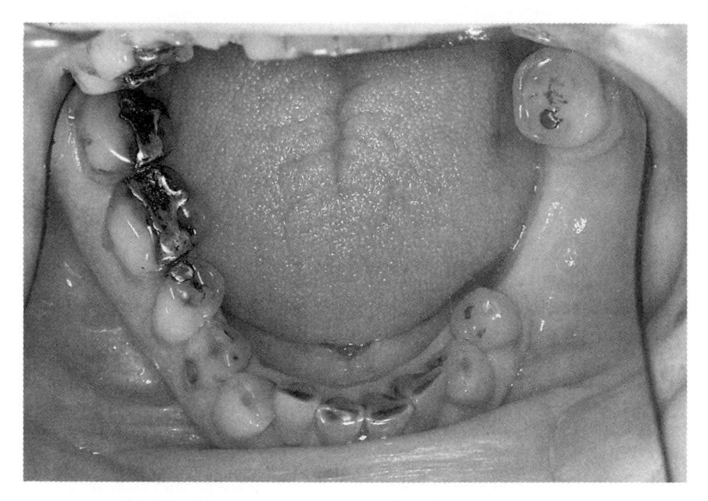

Fig. 12 Erosive tooth wear in an ageing alcoholic. Note the smooth tooth surfaces and that the metallic restorations are standing out of the surrounding tooth tissue.

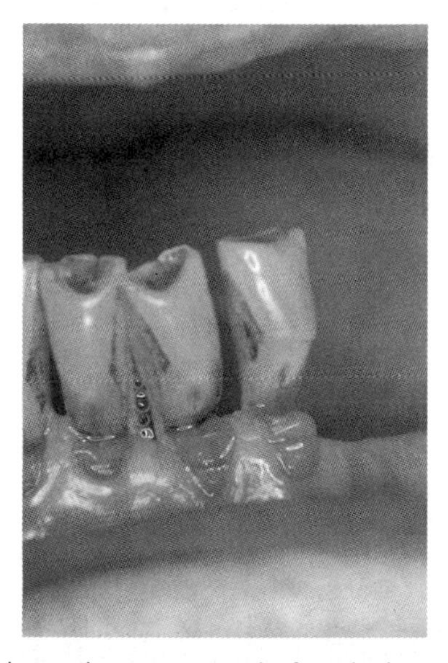

Fig. 13 Abrasive tooth wear as a result of overly vigorous use of a smoker's dentifrice and a scrubbing action with a toothbrush.

Management

The management of the worn dentition is a complex subject, but usually comprises detection and then prevention of any aetiological factors, followed by appropriate restorations. The restorative phase should be kept as simple as practicable in older subjects to try to ensure long-term success with decreasing ability on the part of the patient to cope with cleaning a complicated structure.

The edentulous person

Many older people have no remaining natural teeth and get by with poorly retained plastic prostheses. It has been said that it is one of the wonders of the modern world that people can function with full dentures.

It is a common misconception that once a person has a satisfactory 'set' of dentures then they are finished with the dentist for life. Nothing could be further from the truth. There are continued changes in the supporting structures of the denture, which involve progressive loss of underlying bone with flattening of the residual bony ridge (Fig. 14).

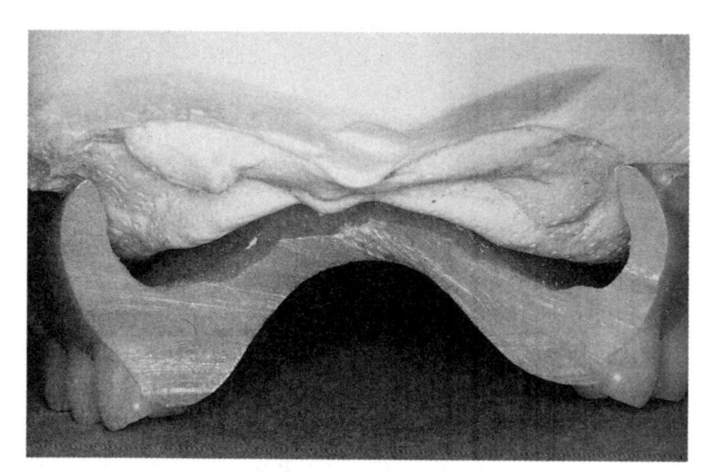

Fig. 14 A section through a cast of a patient's maxilla and his previous denture demonstrating a marked lack of adaptation of the denture base to the underlying tissues. This denture had been satisfactory until the patient underwent a course of chemotherapy and left the prosthesis out for some 4 weeks. His ability to control a poorly adapted prosthesis was lost during that time (illustration by courtesy of Mr I. D. Murray).

As a result the adaptation of a denture to the underlying mucosa deteriorates. Much of the positive retention for a denture is derived from its close adaptation to the supporting tissues. Hence it is surprising that many people can function perfectly well with poorly adapted prostheses. The explanation lies in methods used by individuals to stabilize dentures during function. These include active pressure from the muscles of the cheeks, tongue, and lips. The prostheses are effectively juggled by their owner during mastication.

The processes of adaptation and habituation to a prosthesis takes some time and are learned skills. If, for any reason, the practice of that skill is suspended, an older individual may have grave difficulty in relearning it. This is especially so if the dentures are intrinsically unstable because of poor fit, or if the subject has suffered some central impairment of neuromuscular control. Two causes of this 'suspension' of use of dentures are stroke, especially if the orofacial musculature is involved, and the use of cytotoxic drugs during the treatment of malignant disease. Rehabilitation of stroke victims is a complex process, but their ability to relearn the skills of denture wearing will be improved if the dentures are adjusted to improve the quality of fit. This can readily be achieved with simple bedside techniques in a temporary manner to overcome acute problems. Such temporary solutions can then be rendered permanent by remaking the prostheses.

Patients who are receiving, or who have received, cytotoxic drugs tend to have sore mouths, commonly a product of superinfection with normal oral commensals. In addition there will be an element of debilitation. These two factors often result in patients leaving their prostheses out for some time with the consequences already described for their usage. The lack of fit of dentures in this circumstance is commonly ascribed to weight loss which is a misconception. However, modifications in water balance can be an important factor in denture retention as a result of hydration/dehydration of the oral mucosae. This can be a particular problem in patients undergoing renal dialysis.

There are three common forms of oral pathology which are related to denture wear.

Atrophic candidiasis

This condition has already been described.

Ulceration

Alteration in the morphology of the alveolar ridge will result in overloading of the periphery of the denture as this soon becomes the only area in contact with the supporting tissues. This can in turn lead to two problems. Traumatic ulceration can ensue, especially where the overload occurs on an area with minimal soft tissue coverage. Such ulcers are painful and will not resolve unless the denture is left out of the mouth, or its margins are adjusted to prevent the local trauma.

Denture-induced hyperplasia (denture granuloma)

This is a proliferative fibrotic repair process which can result in the growth of large fibrous masses, usually in the buccal (cheek side) sulcus. Such masses are associated with the periphery of the denture and are common in an ageing population (Nordenram and Landt 1969). They are soft, mobile, and as a rule are not ulcerated. Obviously ulceration can occur as a result of trauma from the denture base, in which circumstance these benign overgrowths can be mistaken for oral cancer (Fig. 15). The hyperplastic lesions will tend to regress if the periphery of the denture is adjusted to prevent the trauma. Surgical removal may be required in some circumstances.

Treatment strategies for older patients should encourage minimal alteration to the surface shape of any prostheses to minimize the change perceived by the patient. Numerous denture duplication/copy techniques have been described to this end.

The transition to the edentulous state is a traumatic experience for all ages. This is especially so in older subjects with their reduced ability to adapt to change. Once again a strategy of gradual change should be adopted. Simple, all acrylic, part-dentures can be made to allow the patient to adapt to denture wear whilst retaining an element of support from any residual teeth. As these teeth are lost they can be added to the dentures one at a time, giving a phased transition from some to no teeth.

Miscellaneous

One subject that has not yet been covered is orofacial pain. There are a number of obvious causes of oral pain including oral ulceration and toothache, with or without abscess forma-

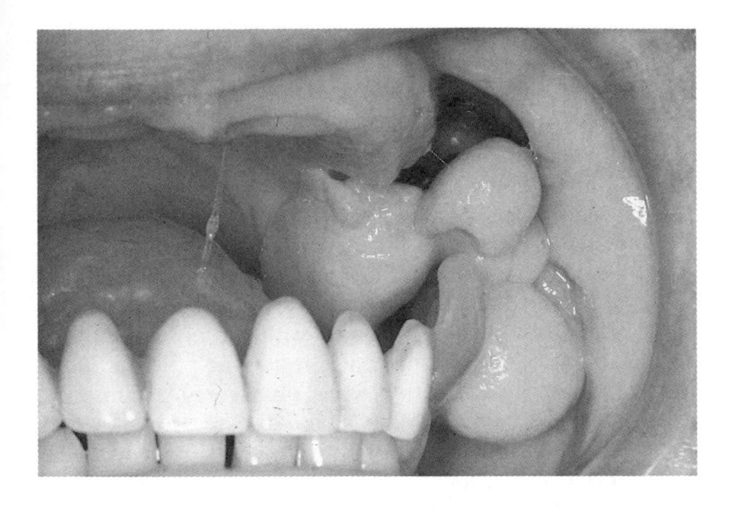

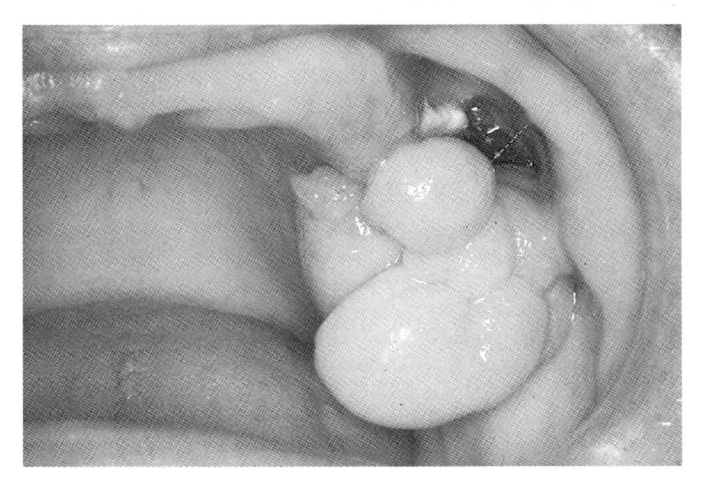

Fig. 15 Denture-induced hyperplasia. Extensive, exophytic, fibrous tissue resulting from chronic trauma to the denture bearing tissues from a poorly adapted denture (illustrations by courtesy of Mr I. D. Murray).

tion. (Toothache or abscess formation can occur in an apparently edentulous individual as a result of retained buried teeth or root remnants.) There are, however, others which are not quite so obvious, or for which the treatment is more complex.

Burning mouth

Painful or burning sensations in the mouth are an infrequent complaint amongst older subjects. Burning sensations can be associated with pathological changes including the lesions of erosive lichen planus and geographic tongue. However, such individuals are not suffering from burning mouth syndrome. This diagnosis should be reserved for patients who complain of painful burning sensations in the absence of any obvious pathological change. The burning sensation can be present on waking and persist during the day, or it can increase in severity gradually during the course of the day. The most common site for the burning sensations is the tongue, followed by the denture bearing areas and the lips. Burning mouth occurs with much greater frequency in women than in men.

There are a number of known aetiological factors including denture faults, parafunctional habits (including teeth clenching and tongue thrust), reduced salivary function, oral candidal infection, haematological disorders (both iron and vitamin B complex deficiencies), undiagnosed diabetes, drug reactions (Table 3), allergy (to foodstuffs or very rarely to acrylic resin), the climacteric, depressive illness, and cancerphobia.

Table 3 Aetiological factors in burning mouth syndrome (Lamey and Lamb 1988)

Denture problems
Oral parafunctional habits: e.g. teeth clenching or grinding, tongue thrust, etc.
Haematological disorders
Vitamin B complex deficiencies
Candidal infection
Reduced salivary gland function
The climacteric
Undiagnosed diabetes
Depression/anxiety states
Cancerphobia
Oesophageal reflux
Acrylic allergy

Strategies for effective care should be designed to identify and correct any aetiological factors. However, even when such strategies are adopted the chances of 'cure' for this condition are low, with only about one-third of patients undergoing complete remission of symptoms (Lamey and Lamb 1988). The presence of depressive illness or cancerphobia may take some time to elucidate, and careful counselling is required for such patients. Antidepressant medication may be required to alleviate symptoms. Lamey and Lamb (1988) used dothiepin to good effect in this role.

Facial pain

Pain associated with the facial musculature, commonly radiating from the region of the ear anteriorly across the face, can be a distressing complaint. Such discomfort may be associated with dysfunction of the temporomandibular joint and its associated musculature. The predisposing factors in this condition are stress, parafunctional habits, and derangement of the 'normal' relationship between the mandible and the maxilla with either a loss of vertical height or lateral/protrusive deviation of the mandible during closure. The symptoms in temporomandibular joint dysfunction syndrome can be associated with clicking or crepitus within the joint space during movement, especially in the presence of arthritic change to the condylar head. There is a considerable psychogenic component to the temporomandibular joint dysfunction syndrome.

A second condition, which also has a psychogenic component is atypical facial pain, where the discomfort is usually described as a continuous dull ache. The discomfort may be in either the upper or the lower jaw, and the sufferer blames their dentition/dentures. Unfortunately restorative treatment or extraction rarely resolve the problem, with the locus of pain moving to another site within the arch. The discomfort may be described as severe, but appetite and sleep are rarely affected.

Neuralgia

Idiopathic neuralgia of the trigeminal and (rarely) the glossopharyngeal nerves presents as severe bouts of sharp stabbing pain which is unilateral, is restricted to the distribution of the trigeminal nerve, is of sudden onset (and may be 'triggered' from an oral or perioral site ipsilaterally) and is characterized by pain-free periods between attacks. The aetiology remains unclear, but there is some association between trigeminal neuralgia and vascular abnormalities in the middle cranial fossa causing pressure on the nerve roots.

Trigeminal neuralgia commonly responds to oral carbamazepine, which must be given prophylactically for some time. Oral phenytoin can also be of some benefit. Injections of local anaesthetic can give local symptomatic relief, and cryosurgery or surgical decompression of the nerve roots may be attempted in severe cases. With any surgical approach there is the risk of permanent anaesthesia of the face and cornea, and occasionally anaesthesia dolorosa will develop. Glossopharyngeal neuralgia does not respond well to prophylactic regimens using either carbamazepine or phenytoin, and adequate pain relief can be difficult.

Finally, a condition where pain control can be problematical is post-herpetic neuralgia with continuous burning pain from dermatomes that have been affected by Herpes zoster. Treatment of the acute herpetic infection with steroids may reduce the likelihood of development of this complication to Herpes zoster infection (Eaglstein *et al.* 1970). Both antidepressant agents and chlorpromazine may be of some benefit if analgesics are unsuccessful (Diamond 1987). There is an elevated risk of suicide amongst such patients.

Cardiovascular causes of facial pain include migraine and migrainous neuralgia which are covered elsewhere. In addition the pain from angina pectoris or from a myocardial infarction can be referred to the mandibular area, usually over the point of the chin. This can obviously be differentiated from dental pain as it is triggered by such things as exercise and a cold environment rather than hot or cold intraoral stimuli or sweet foods.

CONCLUSIONS

There have been a number of studies which have demonstrated high levels of need for dental care amongst the population, and, at the same time, very low levels of uptake of care amongst older people, especially the edentulous portion of the population who only attend their dental practitioner when they experience problems with their dentures (Fiske *et al.* 1990). Equally, the standards of oral health amongst residents in long-stay care facilities are poor (Steele 1989). The impact of oral disease upon older people can be significant in terms of their physical, psychological, and social function. Greater care and awareness are required amongst both the population at large (in relation to themselves) and amongst carers (in relation to their charges) of the need for oral health and regular screening. Simple screening procedures can be undertaken on assessment by health care workers to assist in isolating dental problems (Regnard and Fitton 1989). Once a problem has been identified then appropriate care can be instituted, including seeking specialist advice if required.

Despite the obvious need for care there is a marked reluctance to undertake the tasks of oral hygiene for another person. Christensen (1988) observed 'Many trained nurses and other staff express an aversion towards dealing with people's oral care or cleaning dentures. (Paradoxically, they cope, without comment, with hygiene of other parts of the body, including the removal of faecal matter.) For some reason, cleaning the oral cavity is regarded as a particularly personal and intimate phenomenon, which may be why it is neglected.' This observation is nearly universal and the psychological barriers to caring for the mouths of others must be addressed, for the problems that we see now with the old edentulous population pale into insignificance beside the potential difficulties in store for the future when our ageing patients are at least partially dentate. Many of the disease processes that we see in the mouths of older subjects are similar to those of youth. However, the incidence, severity, and natural history of the diseases alter, presenting a challenge to all of the professionals who care for older patients.

References

Andrew, W. (1952). A comparison of age changes in man and of the rat. *Journal of Gerontology*, **7**, 178–90.

Arvidson K. (1980). Human taste: response and taste bud number in human fungiform papilae. *Scandinavian Journal of Dental Research*, **209**, 807–8.

Axelsson, P., Lindhe, J., and Nystrom, B. (1991). On the prevention of caries and periodontal disease. Results of a 15-year longitudinal study. *Journal of Clinical Periodontology*, **18**, 182–9.

Bartoshuk, L.M., *et al.* (1986). Taste and ageing. *Journal of Gerontology*, **41**, 51–7.

Baum, B.J. (1986). Evaluation of stimulated parotid saliva flow rate in different age groups. *Journal of Dental Research*, **60**, 1292–6.

Bertram, U. (1967). Xerostomia. *Acta Odontologica Scandinavica*, **25**, Suppl. 49.

Binnie, W.H. and Wright, J.M. (1986). Oral mucosal disease in the elderly. In *Dental Care for the Elderly*. (eds. B. Cohen and H. Thomson). pp. 42–4. Heinemann, London.

Carlsson G.E., Kopp, S., and Oberg, T. (1979). Arthritis and diseases of the temporomandibular joint. In *Temporomandibular Joint: Function and Dysfunction*. (eds. G.A. Zarb and G.E. Carlsson) Copenhagen, Munksgaard.

Cawson R.A. and Binnie, W.H. (1980). Candida leukoplakia and carcinoma: a possible relationship. In *Oral Premalignancy* (eds. I.C., Mackenzie E., Dabelsteen and C.A. Squier) pp. 59–66. University of Iowa Press. Iowa City.

Chauncey, H.H., *et al.* (1981). Parotid fluid composition in healthy aging males. *Advances in Physiological Science*, **28**, 323–8.

Christensen, J. (1988). Domiciliary care for the elderly patient. *Dental Update*, **8**, 284–90.

Depue, R.H. (1986). Rising mortality from cancer of the tongue in young white males. *New England Journal of Medicine*, **315**, 647.

Diamond, S. (1987). Post-herpetic neuralgia prevention and treatment. *Postgraduate Medicine*, **81**, 321–2.

Donachie, M.A. and Walls, A.W.G. (1991). Root caries experience and tooth retention in a group of ageing adults. *Journal of Dental Research*, **70**, 684 Abstr. 124.

Dreizen, S., *et al.* (1976). Radiation-induced xerostomia in cancer patients. *Cancer*, **38**, 273–8.

Drummond, J.R. and Chisholm, D.M. (1984). A qualitative and quantitative study of the human labial salivary glands. *Archives of Oral Biology*, **29**, 151–5.

Drummond, J.R., Newton, J.P., and Abel, R.W. (1987). Changes in human parotid gland density with age: a computed tomography study. *Journal of Dental Research*, **66**, 858.

Eaglstein, W.H., Katz, R., and Brown, J.A. (1970). The effects of early corticosteroid therapy on the skin eruption and pain of herpes zoster. *Journal of the American Medical Association*, **211**, 1681–3.

Elwood, J.M., Pearson, J.C.G. and Skippen D.H. (1984). Alcohol, smoking social and occupational factors in the aetiology of cancer of the oral cavity, pharynx and larynx. *International Journal of Cancer*, **34**, 603–12.

Fiske, J., Gelbier, S., and Watson, R.M. (1990). Barriers to care in an elderly population resident in an inner city area. *Journal of Dentistry*, **18**, 236–42.

Gandara, B.J., Izutsu, K.T., and Truelove, E.L. (1985). Age-related salivary flow rate changes in controls and patients with oral lichen planus. *Journal of Dental Research*, **64**, 1149–51.

Glenwright, H.D., Shaw, L., and Cook, C. (1988). Sugar-free prescriptions. *British Journal of Dentistry*, **164**, 6.

Greenspan, D. and Daniels, T.E. (1987). Effectiveness of pilocarpine in postradiation cancer. *Cancer*, **59**, 1123–5.

Hindle, I. and Nally, (1991). Oral cancer: a comparative study between 1962–67 and 1980–84 in England and Wales. *British Dental Journal*, **170**, 15–20.

Holm-Pedersen, P., Gaumer, H.R., and Folke, L.E (1979). Aberrant blastogenic response to LPS in experimental gingivitis of elderly subjects. *Scandinavian Journal of Dental Research*, **87**, 431–4.

Holm-Pedersen, P., Gawronski, T.H., and Folke, L.E. (1980). Composition and metabolic activity of dental plaque from young and elderly individuals. *Journal of Dental Research*, **59**, 771–6.

Holmstrup, P. and Pindborg, J.J. (1979). Erythroplakic lesions in relation to oral lichen planus. *Acta Dermatovenerologica*, **59** (Suppl. 85), 77–84.

Hunt, R.J., Eldredge, J.B., and Beck, J.D. (1989). Effect of residence in a fluoridated community on the incidence of coronal and root caries in an older adult population. *Journal of Public Health Dentistry*, **49**, 138–41.

Jenkins, G.N. (1978) Sensations arising from the mouth. In *The Physiology and Biochemistry of the Mouth*. pp. 542–58. Blackwell, London.

Jensen, M. (1988) The effect of a fluoridated dentifrice on root and coronal caries in an older adult population. *Journal of the American Dental Association*, **117**, 829–32.

Johansen, E., *et al.* (1987). Remineralization of carious lesions in elderly patients. *Gerodontics*, **3**, 47–50.

Johnson, G., Barenthin, I., and Westphal, P. (1984). Mouth dryness amongst patients in long term hospitals. *Gerodontology*, **3**, 197–203.

Johnson, N.W. and Warnakulasuriya, K.A.A.S. (1991). Oral cancer: is it more common than cervical? *British Dental Journal*, **170**, 171–2.

Joynson, D.H., *et al.* (1972). Defect of cell-mediated immunity in patients with iron-deficiency anaemia. *Lancet*, **ii**, 1058–9.

Joyston-Bechal, S. and Kidd, E.A.M. (1987). The effect of three commercially available saliva substitutes on enamel *in vitro*. *British Dental Journal*, **163**, 187–90.

Katz, R.V., Newitter, D.A., and Clive, J.M. (1985). Root caries prevalence in adult dental patients. *Journal of Dental Research*, **64**, 293 (Abstr. 1069).

Kim, S.A. (1987). Protein synthesis in salivary glands as related to aging. *Frontiers of Oral Physiology*, **6**, 96–110.

Lamey, P.J. and Lamb, A.B. (1988). Prospective study of aetiological factors in burning mouth syndrome. *British Medical Journal*, **296**, 1243–6.

Levine, M.J., *et al.* (1987). Artificial salivas: present and future. *Journal of Dental Research*, **66** (special issue), 693–8.

Levy, S.M., *et al.* (1988). Use of medications by a non-institutionalized elderly population. *Gerodontics*, **4**, 119–25.

Maidhof, R. and Hornstein, O.P. (1979). Autoradiographic study of some proliferative properties of human buccal mucosa. *Archives of Dermatological Research*, **265**, 165–72.

Mira, J.G., Fullerton, G.D., and Wescott, W.B. (1982). Correlation between initial salivary flow rate and radiation dose in the production of xerostomia. *Acta Radiologica et Oncologica*, **21**, 151–4.

Moore, C. (1971). Cigarette smoking and cancer of the mouth pharynx and larynx. *Journal of the American Medical Association*, **218**, 553–8.

Newton, J.P., *et al.* (1987). Changes in human masseter and medial pterygoid muscle with age: a study using computed tomography. *Gerodontics*, **3**, 151–4.

Nordenram, A. and Landt, H. (1969). Hyperplasia of the oral tissues in denture cases. *Acta Odontologica Scandinavica*, **27**, 481–91.

Papapanou, P.N., Lindhe, J., and Sterrett, J.D. (1991). Considerations on the contribution of ageing to loss of periodontal tissue support. *Journal of Clinical Periodontology*, **18**, 611–15.

Pedersen, W., *et al.* (1985). Age dependent decreases in human submandibular gland flow rates as measured under resting and post-stimulation conditions. *Journal of Dental Research*, **64**, 822–5.

Rashid, M.U. and Bateman, D.N. (1990). Effect of intravenous atropine on gastric emptying, paracetamol absorption, salivary flow and heart rate in young and fit elderly volunteers. *British Journal of Clinical Pharmacology*, **30**, 25–34.

Rasler, F.E. (1986). Ineffectiveness of sublingual nitroglycerine in patients with dry mucous membranes. *New England Journal of Medicine*, **314**, 181.

Regnard, C.F.B. Fitton, S. (1989). Mouth care: a flow diagram. *Palliative Medicine*, **3**, 76–9.

Rennie, J.S., MacDonald, D.G., and Dagg, H. (1984). Iron and the oral epithelium: a review. *Journal of the Royal Society of Medicine*, **77**, 602–7.

Scott, J. (1987). Structural age changes in salivary glands. *Frontiers of Oral Physiology*, **6**, 40–62.

Scott, J., Valentine, J.A., and St Hill, C.A. (1983). A quantitative histological analysis of the effects of age and sex on human lingual epithelium. *Journal de Biologie Buccale*, **11**, 305–15.

Scully, C., *et al.* (1986). Sources and patterns of referrals of oral cancer: role of the general practitioner. *British Medical Journal*, **293**, 599–601.

Severson, J.A., Moffett, B.C., and Kokich, V. (1978). A histological study of age changes in the adult human periodontal joint (ligament). *Journal of Periodontology*, **49**, 189–200.

Seymour, R.A. and Heasman, P.A. (1988). Drugs and the periodontium. *Journal of Clinical Periodontology*, **15**, 1–16.

Seymour, R.A. (1991). Calcium channel blockers and gingival overgrowth. *British Dental Journal*, **170**, 376–9.

Seymour, R.A. and Jacobs, D.J. (1991). Cyclosporin and the gingival tissues. *Journal of Clinical Periodontology*, in press

Shillitoe, E.J., Greenspan, D., and Greenspan, J. (1982). Neutralising antibody to Herpes simplex virus type 1 in patients with oral cancer. *Cancer*, **49**, 2315–20.

Soames, J.V. and Southam, J.C. (1985). *Oral Pathology*. pp. 127–30. Oxford University Press, Oxford.

Sreebny, L.M. (1989). Salivary flow in health and disease. *Compendium of Continuing Education in Dentistry*, **Suppl.13**, S461–S469.

Sreebny, L.M. and Swartz, S.S. (1986). A reference guide to drugs and dry mouth. *Gerodontology*, **5**, 75–99.

Sreebny, L.M. and Valdini, A. (1987). Xerostomia: a neglected symptom. *Archives of Internal Medicine*, **147**, 1333–7.

Steele, L.P. (1989). Oral status of older people. In *A Participative Approach to Oral Health*. pp. 6–8. Health Education Authority, London.

Syrjanen, S.M. and Syrjanen, K.T. (1984). Inflammatory cell infiltrate in labial salivary glands of patients with rheumatoid arthritis with special emphasis on tissue mast cells. *Scandinavian Journal of Dental Research*, **92**, 557–63.

Todd, J.E. and Lader, D. (1991). Total tooth loss—present and future. *Adult Dental Health 1988, United Kingdom*. pp. 9–18. HMSO, London.

Walls, A.W.G. (1989). Prevention in the ageing dentition. In *The Prevention of Dental Disease*. 2nd edn. (ed. J.J. Murray) pp. 303–26. Oxford University Press, Oxford.

Weiffenbach, J.M., Cowart B.J. and Baum, B.J. (1986). Taste intensity perception in ageing. *Journal of Gerontology*, **41**, 460–8.

Yemm, R., Newton, J.P., and Lewis, G.R. (1985). Age changes in human muscle performance. *Current Topics in Oral Biology*, 17–25.

8.2 The oesophagus

R. CURLESS AND O. F. W. JAMES

APPLIED ANATOMY

Pharynx

The two functions of the pharyngeal cavity are to direct the passage of food and liquid from the mouth to the oesophagus and to avert the aspiration of such contents into the larynx during swallowing. Pharyngeal anatomy may be considered under the following headings: (i) bony and cartilaginous structures and (ii) the striated muscles and their innervation. Alternatively, it may be considered as three intercommunicating spaces: the nasopharynx, oropharynx, and hypopharynx (Dodds *et al.* 1990).

Bony supporting structures

The roof of the mouth and the nasopharynx are bounded by the maxilla, the palatine, and the skull base superiorly. Laterally and anteriorly the margins are defined by the hinged mandible and hyoid bones which provide support for the base of the tongue and larynx. The posterior wall is formed by the cervical spine. The larynx contains four cartilaginous elements (epiglottis, cricoid, thyroid, and arytenoid) and is connected to the hyoid bone by the thyrohyoid membrane and muscles. It constitutes the anterior wall of the hypopharynx and upper end of the trachea.

Pharyngeal muscles and their innervation

Thirty-one paired striated muscles contribute to the oropharyngeal phases of swallowing. The action of the tongue is determined by its four intrinsic and four extrinsic muscles innovated by hypoglossal (XII) and ansa cervicalis (C1–C2) nerves respectively, except for the palatoglossus which is under vagal (X) innervation. The vagus also largely controls the soft palate and pharyngeal muscles together with the intrinsic laryngeal musculature via the recurrent laryngeal nerves. The major propulsive muscles of the pharynx are the three paired constrictors.

The oesophagus

Detailed descriptions of the anatomy of the oesophagus may be found elsewhere (Cuschieri 1986). There do not appear to be significant anatomical changes with normal ageing although there are few studies which directly address this question. Some understanding of normal oesophageal anatomy is required however to understand the pathological and functional features which present with advancing years.

The oesophagus is a hollow muscular tube arising as a continuation of the pharynx from the lower border of the cricoid cartilage and ending at the stomach. For clinical purposes measurements of oesophageal length are taken from the upper incisor teeth such that the oesophagus commences at 15 cm and the junction of the oesophagus and stomach is situated at about 40 cm. This is a crude estimate of the actual location of the gastro-oesophageal junction.

Four oesophageal segments have been proposed for the pathological classification and staging of malignancy with distances measured from the incisors. These are the cervical, upper thoracic, mid-thoracic, and lower thoracic segments.

There are several points of constriction in the course of the normal oesophagus which are of potential clinical importance in elderly people as sites for the lodging of food and tablets or, less frequently, foreign bodies. They are situated at the cricoid origin, the level of the aortic arch, the crossing of the left main bronchus, the left atrium and finally the passage through the diaphragmatic hiatus. Enlargement of the aortic arch or left atrium make these two sites the most common areas for lodgement in older populations.

Applied histology

While histological features of practical relevance to the endoscopist and pathologist have been reviewed (DeNardi 1991), there is virtually no information concerning possible histological changes in the oesophagus during normal ageing. The gross structure of the oesophagus is similar to that of the rest of the gastrointestinal tract, comprising four main layers, external, muscular, submucous, and mucous. The muscles of the oesophagus form an outer longitudinal and an inner circular layer. Superiorly, the posterior separation of the two bands of longitudinal muscle leaves an area of relative weakness (Killian's dehiscence). Distally, the longitudinal muscles of the oesophagus and stomach are continuous. The circular muscle divides in the region of the cardia to form the middle circular and inner oblique layers of the stomach. The proximal oesophagus comprises striated muscle traditionally regarded as extending as far as the first third of the organ. It is gradually replaced by smooth muscle fibres. Unlike skeletal muscle, these striated fibres are under visceral control. The mucosa lies in longitudinal folds in the resting state, and comprises three layers: epithelium, lamina propria, and muscularis mucosae. The epithelium is of non-keratinized stratified squamous type. One to two centimetres from the gastro-oesophageal junction a squamocolumnar junction is visible macroscopically as a serrated line (the Z line) where small projections of red gastric epithelium interdigitate with the paler squamous one. Thereafter, the distal 1 to 2 cm of the oesophagus is lined by columnar epithelium. This mucosal junction normally lies within the lower oesophageal sphincter.

The gastro-oesophageal area

The region between the oesophagus and stomach is the most complex and important area of the oesophagus. It contains the physiological lower oesophageal sphinter, the gastro-oesophageal junction and may occasionally give rise to oesophageal rings. Manometry demonstrates a physiological sphincter which at rest closes the oesophageal lumen and prevents reflux of gastric contents. There is, however, no distinct corresponding anatomical sphincter.

Blood supply

The inferior thyroid artery predominantly supplies the cervical oesophagus with the thoracic segment receiving branches of the bronchial arteries. The abdominal segment gains its supply from ascending tributaries of the left gastric and inferior phrenic arteries. Infarction is very rare owing to a widespread anastomotic network. Blood is drained from the proximal two-thirds of the oesophagus via the inferior thyroidal and azygos veins to the superior vena cava. The lower one-third and cardia drain via two routes: to the systemic circulation in branches of the azygos and left inferior phrenic veins and to the portal system via the left and short gastric veins. It is this portosystemic anastomosis which leads to the formation of varices in the presence of raised portal pressure.

Innervation

The extrinsic innervation of the oesophagus is from both the parasympathetic and sympathetic arms of the autonomic nervous system. The parasympathetic motor supply to the pharynx and oesophagus is mediated through the glossopharyngeal (IX) and vagal (X) cranial nerves. Branches of the recurrent laryngeal nerves supply the upper oesophagus, while the vagus itself innervates the body and distal portions. Sympathetic preganglionic fibres arise from cell bodies in the interomedial lateral columns of the fifth and sixth thoracic segments of the spinal cord to synapse in the cervical, thoracic and coeliac ganglia. The intrinsic innervation of the oesophagus arises in two ganglionic plexuses: the submucosal (Meissner's) plexus and the myenteric (Auerbach's) plexus.

Normal swallowing

Normal swallowing has recently been reviewed extensively elsewhere (Dodds *et al.* 1990). It may be considered as occurring in four orderly and highly co-ordinated phases: preparatory, oral, pharyngeal, and oesophageal. The first two are voluntary whereas the last two, once initiated, are entirely involuntary. This complex sequence of events requires fine neuromuscular control and involves bolus transport, the suspension of respiration, and protection of the airway. Overall co-ordination is thought to lie in paired medullary swallowing centres.

The oesophagus may be perceived as responsible for both the further transport of the ingested bolus from the pharynx to the stomach and also, via its two sphincter mechanisms, for the prevention of tracheal aspiration and gastro-oesophageal reflux. Detailed descriptions of oesophageal physiology may be found elsewhere (Christensen 1987).

At rest both oesophageal sphincters are tonically contracted. Unlike the remainder of the gastrointestinal tract the organ does not exhibit rhythmic interprandial contractions. The upper oesophageal sphincter, whose main component is the cricopharyngeus muscle, is an occluded region of 1 to 3 cm in length. The oesophageal body is flaccid with a small negative pressure which varies with respiration and cardiac contraction. The lower oesophageal sphincter is also in a state of tonic contraction. Physiologically a 2- to 4-cm region of occlusion may be detected by manometry. The basal pressure is thought to be 15 to 50 mmHg. An anatomically corresponding distinct muscular sphincter is not thought to exist. Lower oesophageal sphincter tone is probably maintained by a myogenic mechanism. Many gastrointestinal polypeptides can affect the lower oesophageal sphincter *in vitro* but no single substance has been definitively implicated *in vivo*.

Once the swallow is initiated a sequence of involuntary oesophageal relaxations and contractions occurs. Soon after the onset of pharyngeal contraction upper oesophageal sphincter tone is abolished by transient central inhibition. Pressure falls for 0.5 to 1 s whilst the bolus passes through the sphincter and this is followed by a short period of hypercontraction. A wave of primary peristaltic contraction originates just below the upper oesophageal sphincter proceeding down the body of the oesophagus. This primary peristaltic wave takes about 5 s to travel down the whole oesophageal body. The lower oesophageal sphincter starts to relax during this contraction and pressure within the sphincter falls to intragastric levels for 5 to 8 s; thus no barrier to the passage of the bolus is offered. Again there follows a brief hypercontraction of about 2 s before basal lower oesophageal sphincter tone returns.

Effect of age on swallowing

Only recently has quantitation of swallowing become possible. There are therefore few studies available to examine the effects of normal ageing on swallowing and most have included only small numbers of patients aged 65 or greater. Data on the very old (80 years or older) are even more sparse. Subject selection criteria have varied in different studies: the exclusion of pathology and concomitant medication use has not always been rigorous. The manometric methodology employed has also been subject to criticism. The effects of ageing may be divided into three types (Logemann 1990). Primary effects are those which may be ascribed to the ageing process itself. Secondary effects arise from those diseases seen in elderly persons. A third or tertiary effect has been suggested to incorporate those environmental and social and financial pressures which determine nutritional intake in the elderly.

Limited data available suggest that some primary ageing changes do occur: basal salivary volumes may fall although stimulated flow and salivary amylase activity have not consistently been shown to alter with age. Secretions from the glands of the vocal cords are also less abundant. Changes in the oropharyngeal phase of swallowing have been noted in persons aged 65 to 79. The bolus is held more posteriorly on the tongue. Swallowing is slower overall and additional tongue movements are noted. The initiation of the pharyngeal phase is about 0.5 s longer in old than young subjects. Once the pharyngeal phase has commenced the larynx may be elevated for longer. Limited data suggest that the pharyngeal peristaltic wave is up to 10 per cent slower. Overall pharyngeal transit time therefore may be longer.

The maximum resting pressure of the upper oesophageal sphincter has been found to decline with increasing age when measured in four radial orientations and there is delay in its relaxation. Data concerning the duration of opening of the upper oesophageal sphincter are conflicting but a greater proportion of swallows unaccompanied by its relaxation have been reported in older as compared to younger individuals. It is probable that the amplitude of contraction in both proximal and distal oesophageal body is reduced in normal individuals over age 80 but probably not in those under age 70. The proportion of disturbed peristaltic waves and tertiary contractions certainly increases with age and thus peristalsis may be less efficient with a consequent prolonged oesophageal transit time. Studies examining lower oesophageal sphincter relaxations and basal resting tone with respect to age have yielded conflicting results.

The meaning of these findings with respect to clinical practice is still unresolved. A recent radiological study of asymptomatic elderly people with a mean age of 83 found only 16 per cent had normal swallowing as defined in younger subjects (Ekberg and Feinberg 1991). This would suggest that our view of dysfunction derived from studies of younger populations may not apply to older cohorts. At present no data are available to suggest that the overall efficacy of swallowing in normal old age is impaired to a clinically important degree in terms of aspiration episodes. Until we have a clearer view of normal ageing, assessment of swallowing problems in the elderly will remain difficult. Secondary changes arising from pathological processes will be discussed later in this chapter.

SYMPTOMS

Despite the increasing sophistication of oesophageal investigation a thorough clinical history remains the cornerstone of diagnosis and the selection of appropriate tests. Our knowledge of the prevalence of oesophageal symptoms in the general population is not complete. The variety of definitions used by patients and doctors alike has hampered such studies. Nonetheless there is little doubt that symptoms normally attributable to the upper gastrointestinal tract are common. It is generally held that disease presentation in older people is different from that in younger adults. Whether this reflects physiological changes in symptom perception (for example modification of the pain threshold), the effect of concomitant chronic disease, a tendency for older people to attribute symptoms to 'old age', or the diligence with which doctors take histories in elderly patients is debatable.

In community studies, dyspepsia within the previous 6 months has been reported in over one-third of subjects although only one in four of these had consulted their family doctor. About half of those reporting dyspepsia had suffered from both upper abdominal discomfort/pain and heartburn (Jones and Lydeard 1989). Dysphagia has been reported in 10 per cent and cough on swallowing by up to 27 per cent of normal individuals although in only a small number (1–4 per cent) were such symptoms described as severe. In one American survey of elderly women the prevalence of dysphagia increased from 3.3 per cent in those aged between 65 and 69 to 13.5 per cent in those over aged 80, but this age-associated increase in symptoms has not been found elsewhere.

Within hospital and continuing care facilities the incidence of dysphagia is much higher. As many as one-third of old people in institutional care may experience some dysphagic symptoms. If those with difficulties in feeding are included the figures rise to over 50 per cent. It may thus be seen that a proper consideration of oral intake and swallowing difficulties should be a major concern for all those caring for elderly people. In this group of heavily dependent elderly people, presentation of pharyngo-oesophageal disorders may be atypical, not necessarily because of age itself but because of the background of chronic pathology. While the clear-cut onset of dysphagia following acute stroke is readily recognized, aspiration secondary to general debility, cognitive impairment, and depression are less readily identified although this is certainly a real phenomenon. Gastro-oesophageal reflux with consequent aspiration is more common in those spending prolonged periods in bed. Thus clinicians must be alert to the possibility that atypical or non-specific changes in dependent elderly patients may arise from pharyngo-oesophageal problems.

Dysphagia

Dysphagia is defined as difficulty in swallowing. It denotes a feeling that a bolus is impeded in its passage from mouth to stomach during the act of swallowing or within 15 s of deglutition. It is the main symptom of pharyngo-oesophageal disease and always requires investigation as to its cause. It should be distinguished from the feeling of a lump in the throat (globus) which is usually present between swallows and cleared by deglutition and also from the sensation that food is resting in the neck or chest that occurs typically 10 or more minutes after eating. 'Cortical inhibition' is a syndrome, encountered especially in elderly people, where food is repetitively chewed and then spat out. There is no obstruction and the patient can drink. The problem seems to be one of volition.

Dysphagia may conveniently be classified into oral, pharyngeal, and oesophageal on the basis of the phase of swallowing that is disturbed. The clinical presentation differs for each site. It is important to determine how long after the onset of swallowing dysphagia begins to occur. This is more reliable as an indicator of the site of the problem than the area of the chest to which the patient may point as being the site of hold-up. In particular, oesophageal problems are poorly localized. A short and progressive history suggests a carcinoma, whereas intermittent difficulties over a period of years are more typical of a benign stricture. However, in elderly patients symptom duration may not be particularly useful in making the distinction between benign and malignant causes of dysphagia. Webs or rings may cause well defined separate episodes of bolus impaction. The nature of the substances causing dysphagia should be established. Difficulties with both liquid and solid boluses from the outset make a motility disorder more likely. Dysphagia for solids only in the initial phase suggest a mechanical problem.

Oesophageal dysphagia usually described in terms of food sticking within the chest may be accompanied by mild discomfort to severe pain which persists until the bolus either passes through the blockage or is regurgitated. Patients will often adopt methods of relieving bolus impaction by taking liquid to 'wash it down'. In the elderly the enhanced role of non-oesophageal lesions such as gastric neoplasms, extrinsic compression, and peptic ulceration presenting as dysphagia must be emphasized.

Investigations

Examination

Careful physical examination, when combined with a detailed history, may enable a provisional diagnosis to be made in over 80 per cent of patients with dysphagia and will certainly guide the appropriate use of investigations.

Radiology

A chest radiograph should be performed since it may reveal obvious medistinal masses, cardiomegaly, retrosternal goitre, or a dilated oesophagus. There may be evidence of an aspiration pneumonia, oesophageal air/fluid level, or loss of the gastric air bubble in severe achalasia.

Disorders of the oropharynx require videofluoroscopy with the ability to store images for subsequent review and frame-by-frame analysis. An accurate and valid assessment of oropharyngeal dysfunction is possible in elderly patients if specific patterns are sought for. Single films may be useful in studying morphological changes but they are not helpful in the elucidation of motility disturbances. Assessment can be made in relation to differing bolus volumes and compositions as well as varying head and neck positions of the patient.

Radiological examination of the oesophagus combines several different examining techniques. The basic method is a full column examination with the patient prone drinking barium through a straw. A solid bolus such as a marshmallow sweet may be useful in detecting narrowings and dysmotility. Smaller lesions and mucosal abnormalities will require mucosal relief or double contrast techniques. A barium swallow will detect moderate or severe oesophagitis in up to 90 per cent of cases, but lesser grades are much less frequently demonstrated. The presence of an hiatus hernia may be an incidental finding especially in elderly people. Radiology should detect most lesions which cause symptomatic narrowing of the oesophageal lumen. Fluoroscopic observation is a sensitive qualitative test for motility disorders, particularly valuable in the evaluation of the pharyngo-oesophageal region.

In elderly patients upper gastrointestinal barium studies have the advantage of simplicity for the patient, lack of sedation, and safety, but may be complicated by frailty and immobility. Quality of these radiographs does decline with age especially in patients older than 75 years (Hawkins *et al.* 1991). When dysphagia is the predominant symptom radiology should always precede endoscopy.

Endoscopy

Endoscopy combines diagnosis by direct vision and the ability to biopsy the oesophageal mucosa together with the potential for treatment such as dilatation of strictures or injection of varices. Five per cent of all endoscopic examinations and 10 per cent of emergency upper gastrointestinal endoscopies are in patients over 80 years. It should be used as the first investigation of reflux disease or gastrointestinal bleeding and should always be performed when radiology has demonstrated an oesophageal stricture. It is of little value in the diagnosis of motility disorders. As a diagnostic procedure in elderly people upper gastrointestinal endoscopy would appear to be safe. In selected series up to 90 per cent of examinations reveal an abnormality and over 80 per cent lead to changes in management.

Intravenous sedation is used in most upper gastrointestinal endoscopies and may be combined with an opiate analgesic for some therapeutic procedures such as oesophageal dilatation. These drugs may lead to respiratory depression. Half of the complications and 60 per cent of deaths which arise from endoscopic procedures are cardiorespiratory. There is no large prospective study of complication rates from oesophageal endoscopic procedures alone but information from a large American database suggested that the number of cardiac complications, respiratory complications, and deaths were 0.4, 0.1, and 0.04 per cent respectively. There is no evidence that age *per se* is a major additional risk factor for such problems; rather, it is the concomitant presence of cardiorespiratory disease.

Arterial oxygen desaturation occurs frequently during upper gastrointestinal endoscopy as a result of drug-induced respiratory depression and obstruction of the airway by the endoscope itself. Fall in oxygen saturation below 90 per cent for greater than 1 min may occur in up to half of elderly patients undergoing this procedure. Preoxygenation and supplemental oxygen at flow rates between 2 and 4 l/min via nasal cannulae and mouth guards can largely abolish oxygen desaturation. The additional use of pulse oximeters and ECG monitoring is increasingly becoming a standard and desirable part of routine upper endoscopy, particularly in elderly patients. Comprehensive guidelines have been recently published by the British Society of Gastroenterology (1991).

Oesophageal manometry

This is the most direct method of assessing oesophageal motor function. While it allows measurement of the strength of muscular contractions it does not directly assess bolus propulsion. Manometry may therefore need to be combined with radiology or scintigraphy. Advantages of the technique are that it is of low risk and does not involve radiation exposure. Disadvantages include equipment cost, invasiveness and hence patient acceptability, and the need for considerable expertise in interpretation.

Oesophageal pressures may be detected by two methods: via continuous perfusion of side-hole catheters using low compliance infusion pumps, and via intraluminal transducers. The latter is the current method for recording pharyngeal pressures. Manometry of the oesophageal body records at up to eight sites, so that it is possible to construct a complete picture of the propagation of oesophageal pressure waves.

The main role of manometry is in the diagnosis of oesophageal motility disorders following radiological and endoscopic investigation. Sensitivity and specificity for the diagnosis of achalasia is high and manometry will assist in the characterization of the hypertensive oesophagus, diffuse oesophageal spasm, and other non-specific motor disorders. The clinical role of manometry in the investigation of high dysphagia is less clear and the yield in non-cardiac chest pain is low.

Ambulatory oesophageal pH monitoring

This technique quantitates the actual exposure time of distal oesophageal mucosa to gastric juice and has the highest sensitivity (over 85 per cent) and specificity (over 95 per cent) in the detection of gastro-oesophageal reflux disease. The development of ambulatory systems has allowed the test to be performed under physiological conditions within the normal environment of the patient.

A pH probe is placed 5 cm above the lower oesophageal sphincter and reflux is defined when the pH falls below 4. Patients record when they experience symptoms and when they retire to bed at night. A symptom index can be constructed, expressed as the percentage of symptoms associated with documented reflux over the total number of symptoms. A total 24-h pH score as well as the frequency and duration of individual episodes are recorded. Abnormal reflux is assessed by comparison with the ranges seen in healthy asymptomatic normal individuals.

Scintigraphy

The incorporation of a radioisotope (99m-technetium-sulphur colloid) either into a liquid or food bolus allows a gamma-camera and data processor to measure its passage from oesophagus to stomach. The gullet is divided into three segments and quantitation of transit through each segment together with total transit time can be computed. A qualitative assessment of the graphs may indicate peristaltic dysfunction.

While this has the advantages of low radiation exposure and non-invasiveness, the capital equipment costs are high. Only one or two swallows can be examined so intermittent episodes of difficulty in swallowing may be missed. There may also be technical problems due to distal oesophageal traces, including scatter from the gastric pool. The clinical role of scintigraphy is therefore not yet fully established.

DISEASE OF THE PHARYNX AND OESOPHAGUS

There are several methods by which disorders of the pharynx and oesophagus may be classified: these involve anatomical, pathological, clinical, and functional divisions. They are all to some extent arbitrary and many conditions may be placed in overlapping categories. Table 1 outlines one such classification based on a broad functional division into motility and mechanical problems and further subdivided into pharyngeal and oesophageal sites.

Table 1 Outline classification of diseases of the pharynx and oesophagus

Mechanical		Motility
Pharynx		
Intraluminal	Foreign bodies	Primary
Intrinsic	Inflammation	Cricopharyngeal achalasia
	Infection	
	Webs	Secondary
	Strictures	Upper motor neurone
	Neoplasia	Lower motor neurone
	Diverticula	Extrapyramidal
Extrinsic	Cervical vertebrae	Neuromuscular junction
	Retropharyngeal abscess	Muscles
	Thyroid	
	Vascular	
Oesophagus		
Intraluminal	Foreign bodies	Primary
	Drugs	Achalasia
Intrinsic	Infection	Diffuse oesophageal spasm
	Inflammation	Non-specific motility disorders
	Webs/rings	Reflux
	Strictures	Non-cardiac chest pain?
	Neoplasia	
	Rupture	
	Bleeding	
	Diverticula	
Extrinsic	Thoracic vertebrae	Secondary
	Vascular	Connective tissue diseases
	Mediastinal	Metabolic/toxic
		Chagas disease

Motility disorders of the pharynx

Motility disorders usually present with oropharyngeal dysphagia. The patient may complain of dysphagia localized to the suprasternal level, nasal or pharyngeal regurgitation, postdeglutition cough, or dysarthria. There may also be evidence of aspiration or associated neuromuscular disease.

Examination, radiology and possibly manometry should be performed. Endoscopy may be useful to exclude luminal lesions at the pharyngo-oesophageal level or more distally. Videofluoroscopic abnormalities can broadly be grouped as follows: (i) motility disturbances, (ii) bolus retention in pharyngeal recesses, (iii) pharyngeal stasis, and (iv) misdirected swallowing. Studies of selected patients with oropharyngeal dysphagia reveal functional abnormalities in about 80 per cent.

Cricopharyngeal achalasia

This term, meaning 'absence of slackening', is inappropriate since the problem appears to be one of inco-ordination with the rest of the pharyngeal phase of swallowing rather than inability of the cricopharyngeus muscle to relax. Incomplete opening of the upper oesophageal sphincter is usually secondary to neuromuscular disease and rarely due to a primary failure of cricopharyngeal relaxation. Consequently it is more common in elderly people. Videofluoroscopy may demonstrate incomplete or absent opening of the upper oesophageal sphincter but intraluminal manometry is the only method by which impaired relaxation may be reliably established.

Treatment for minor symptoms in the absence of aspiration and weight loss may be simple reassurance. Dilatation has been shown to give temporary relief only. Persistent symptoms may require cricopharyngeal sphincterotomy but the presence of concurrent gastro-oesophageal reflux, and hence the risk of aspiration, should first be excluded. Abolition or marked improvement in symptoms has been claimed in 85 per cent of those cases without underlying neurological disease. Success in the latter group is, however, very much lower.

Secondary motility disorders

Stroke (see Section 11)

Stroke disease forms the largest group of neurological conditions causing dysphagia. Acute difficulties with eating and drinking are very common in stroke victims with up to 45 per cent of patients having some clinical evidence of dysphagia. Older patients are significantly more likely to be afflicted. Impaired consciousness is associated with dysphagia in over two-thirds of these stroke victims and is the main determinant of early mortality. In most survivors dysphagia resolves within 2 weeks. However, even after correcting for conscious level, dysphagic patients have a higher risk of mortality at 6 months.

Contrary to previous beliefs, and despite bilateral upper motor neurone innervation of most of the cranial nerves involved with swallowing, unilateral hemisphere stroke does affect the swallowing mechanism. A large study of 357 conscious patients entering into a drug trial for stroke found nearly 30 per cent of single hemispheric strokes had swallowing impairment on the first poststroke day, and significantly more of these were over age 70. There was an inverse correlation between swallowing difficulty and functional outcome at 6 months. The prevalence of dysphagia may however have been underestimated since patients with the most severe dysphagia (therefore unable to take oral medication) were excluded. In patients with hemispheric strokes there may be variations in the type of dysfunction. Left-sided lesions tend to cause oral phase problems whereas those of the right interfere with the pharyngeal phase. Aspiration is often silent, the most common clinical association with aspiration being dysphonia.

Brain-stem strokes are thought particularly to affect cricopharyngeal relaxation. The classical lateral medullary syndrome (Wallenberg) results from occlusion of the posterior inferior cerebellar artery. Dysphagia is accompanied by ipsilateral ataxia, Horner's syndrome, lack of facial sensation, paresis of the palate, and incomplete vocal cord palsy. Clear-cut brain-stem syndromes such as this tend to be the exception rather than the rule. Many elderly patients will be found to have a variety of dysfunctions as a consequence of multiple vascular events.

Approaches to the acute management of dysphagic stroke patients vary from a 'wait and see policy' to parenteral therapy or nasogastric feeding. Definitive evidence to guide management is not yet available. Careful assessment by a speech therapist, usually on several occasions, may be very helpful in determining management and prognosis. In those with longer-term difficulties, a multidisciplinary approach is advocated. A bedside assessment of feeding abilities in addition to physical, cognitive, and behavioural function is made. A physiological appraisal with videofluoroscopy should follow.

Long-term management of dysphagia is problematic and raises important ethical issues. Prolonged nasogastric feeding is associated with mechanical, gastrointestinal, and metabolic complications. Misplacement of the feeding tube may result in pneumonia, abscess, or hydropneumothorax. Tube blockage, irritation and erosion of the posterior pharynx and oesophagus, stimulation of saliva production, and gastro-oesophageal reflux with aspiration all occur. Vomiting and diarrhoea may affect one-fifth of patients.

Occasionally pharyngostomy, gastrostomy, or jejunostomy are required and recent reports suggest that for the long-term management of severe poststroke dysphagia percutaneous endoscopic gastrostomy is now the treatment of choice in elderly patients (Finucane *et al.* 1991).

Parkinson's disease (see Chapter 18.12)

In Parkinson's disease not only are there difficulties in feeding due to tremor and rigidity but disorders of pharyngo-oesophageal function are also common. Drooling and oral phase dysfunction predominate. The degree of radiologically observed dysfunction correlates poorly with the severity of disease and reported symptoms. Oesophageal peristalsis may be reduced. Treatment with L-dopa may improve many of these symptoms. Persistent dysphagia in the absence of poorly controlled disease should prompt a search for other neurological and non-neurological disorders.

Motor neurone disease

Dysphagia and dysarthria are the initial symptoms in progressive bulbar palsy caused by motor neurone disease, although limb involvement is a more common mode of first presentation. Lesions above the medulla result in pseudobulbar palsy with spasticity of the bulbar musculature. There is no effective treatment and the dysphagia is usually steadily progressive.

Myasthenia gravis

In the elderly population myasthenia is usually a disease of men which contrasts with its female preponderance in the young. Dysphagia is an early symptom together with ocular, neck, and shoulder girdle muscle fatigue and is present in up to two-thirds of sufferers. Symptoms are worse in the evening and towards the end of a meal. The diagnosis and management of this disease are dealt with elsewhere.

Muscle disorders

Diseases of muscle may predominantly affect either striated or smooth muscle or both. Patients may present with either oropharyngeal or oesophageal dysphagia. Polymyositis and dermatomyositis are uncommon in elderly people but, where they occur, dysphagia is described in about half of those affected. Investigations reveal nasopharyngeal regurgitation, aspiration, and diminished pharyngo-oesophageal contractions. The degree of dysphagia parallels the severity of the muscular weakness. Both may be improved with steroids or other immunosuppressive drugs.

Two types of muscular dystrophy may produce oropharyngeal dysphagia in elderly people. Myotonic dystrophy is an autosomal dominant condition which usually presents before the age of 60 but occasionally later in life. Symptomatic swallowing difficulties are present in 50 per cent of patients. Oculopharyngeal muscular dystrophy is similarly a rare autosomal dominant condition but this typically presents after the age of 60. There is progressive bilateral ptosis and mild dysphagia. The prognosis is good and disability is usually small.

Other disorders that may involve muscle can lead to oropharyngeal dysphagia and include both hyper- and hypothyroidism, sarcoidosis, and mixed connective tissue disease. Systemic lupus erythematosus and scleroderma particularly involve the smooth muscle and lead to oesophageal dysphagia.

Oesophageal motor disorders

Primary

Several primary disorders of oesophageal motility may be found in patients with chest pain, dysphagia, and other oesophageal symptoms. Motility is disordered in the sense that it is statistically abnormal in comparison with healthy asymptomatic control populations in the absence of any other identifiable underlying disease. The nomenclature of these disorders has, until recently, been confusing and most of the data depicting normality have been obtained from young individuals. Normal ranges in the elderly population for various manometric parameters are far from well established. Extrapolation of 'younger' values to elderly patients when describing oesophageal motility and its abnormalities may well not be valid. Current concepts have recently been reviewed (McCord *et al.* 1991). Achalasia is the only disorder with a well established pathological basis.

Achalasia

The characteristics of achalasia are the absence of peristalsis throughout the oesophagus, incomplete relaxation, and raised basal tone of the lower oesophageal sphincter thought to be due to damage to the intrinsic and extrinsic innervation of both sphincter and oesophageal body.

Epidemiology

There is a paucity of large epidemiological studies of this disease. In England the incidence is probably 0.5 cases per hundred thousand population per year with a prevalence of between 7 and 13 patients per hundred thousand. There is probably an increase in incidence with age but no sex difference. The mean age at death is around 80 years. There appears to be geographical variation in the frequency of the disease. Mortality data suggest that it is particularly common in New Zealand, Ireland, and Sweden.

Pathophysiology

The main manometric findings in achalasia are lack of peristalsis in the oesophageal body (which may occasionally be limited to the distal two-thirds only), incomplete relaxation of the lower oesophageal sphincter and basal hypertonicity. Secondary abnormalities are supersensitivity to cholinergic agonists and gastrin. Histologically a decrease in or absence of the number of ganglion cells in the myenteric plexus in the region of the lower oesophageal sphincter is found. No primary defect in smooth muscle is recognized. Objects resembling Lewy bodies in the myenteric plexus have been described. Changes in intrinsic nerve fibres and vagal nerve abnormalities are reported. Whatever the primary lesion the result of these pathological changes is thought to be loss of the intrinsic inhibitory innervation of the lower oesophageal sphincter.

The aetiology is unknown. There are preliminary reports suggesting that certain HLA tissue antigens are more prevalent in achalasia than in control populations and rare family pedigrees have been described. However, evidence for a substantial genetic influence is lacking. Other hypotheses have centred on autoimmune reactions, nerve degenerations, and neurotropic infections. None has gained particular support.

Clinical features

The cardinal feature is dysphagia for both liquids and solids, which is intermittent initially but worsens with time. Patients acquire habits which allow them to finish meals more readily such as taking large quantities of water, the Valsalva manoeuvre, and backward arching. Chest pain is common but less frequent as a complaint in elderly patients. It is usually substernal and precipitated by meals, but is rarely severe. Burning pain or odynophagia suggests oesophagitis secondary to stasis or infection. Regurgitation of undigested food occurs when the oesophagus dilates and it is usual for pain to regress as a symptom at this advanced stage. Pulmonary complications secondary to aspiration occur in 10 per cent of patients. Loss of weight is usually present as a consequence of incomplete emptying of the oesophagus and also as a result of fear of eating. Failure to maintain weight is an indication for treatment. Rapid weight loss may suggest secondary achalasia due to the presence of a carcinoma.

Diagnosis

This is often delayed for several years in all age groups. A plain chest radiograph may reveal a widened mediastinum, air-fluid level, or absence of gastric air bubble. Barium radiology classically demonstrates defective or absent distal peristalsis in the oesophageal body. Dilatation which may become massive with time and evidence of residual food debris are also seen. The typical 'birds beak' is seen as the distal oesophageal lumen tapers off smoothly towards a poorly opening lower oesophageal sphincter. Sometimes numerous tertiary contractions are the major clue to the diagnosis. Detailed examination of the fundus is necessary to seek evidence of a secondary neoplastic process, especially in the elderly patient. Endoscopy is mandatory, not to make the diagnosis, but to exclude other diseases and reveal complications.

Manometry is the gold standard for the diagnosis of achalasia. The typical findings will confirm the diagnosis, although some patients with the disease have equivocal manometric recordings. Absent distal peristalsis and incomplete relaxation of the lower oesophageal sphincter are required for confident diagnosis. In subjects aged over 70 with achalasia the only manometric differences described compared with younger individuals are lower residual pressures at the lower oesophageal sphincter following swallowing.

Complications

Infection, stasis, and alcohol or drug injury can all lead to erosive oesophagitis in the distal part of the oesophagus and may increase the risk of complications following pneumatic dilatation. The association between achalasia and subsequent development of oesophageal carcinoma is controversial. An average prevalence of about 3 per cent is reported and figures are higher with prolonged follow-up. Tumours are usually sited in the midoesophagus and are histologically squamous in type. Effective treatment of the achalasia does not appear to abolish this risk. The role of screening endoscopy in patients with achalasia is yet to be established.

Treatment

The aim of treatment is to reduce pressure at the lower oesophageal sphincter and thereby improve oesophageal emptying while avoiding gastro-oesophageal reflux. The most effective non-surgical management is by pneumatic dilatation whereby a balloon is passed to the region of the lower oesophageal sphincter and inflated. It is thought that success results from the disruption of muscle fibres. A barium swallow using water soluble contrast is performed immediately after the dilatation to exclude perforation. Pneumatic dilatation is indicated in symptomatic patients but should be delayed in the presence of active oesophagitis and is contraindicated where there are epiphrenic diverticula or a prior history of perforation. Symptomatic improvement occurs in up to 90 per cent of patients treated and appears to be more likely in those with long-standing symptoms and in patients without very marked oesophageal dilatation. Repeat balloon dilatation can be performed but with a lower success rate.

There is a lack of prospective studies to guide clinicians to the most appropriate time for surgical intervention. The most common procedure is a modified Heller's myotomy. Symptomatic improvement is recorded in 70 to 95 per cent of patients with fatality of well under 1 per cent. The major late complication is that of reflux leading to oesophagitis and subsequent peptic stricture formation. Various antireflux procedures may therefore be performed with the basic myotomy.

At present most authorities would recommend that pneumatic dilatation be initially performed by an experienced clinician on one to three occasions in the majority of patients before operative myotomy is carried out.

Drug treatment has been tried but current results appear to offer only short-term benefit. In extremely frail old people with mild symptoms calcium antagonists or nitrates may act to reduce pressures at the lower oesophageal sphincter and may improve dysphagia for a few months.

Secondary achalasia

This term refers to the development of achalasia secondary to an underlying malignant disorder. Chagas' disease, chronic intestinal pseudo-obstruction, amyloidosis, and sarcoidosis may also give rise to achalasia-like syndromes. Secondary achalasia accounts for only about 4 per cent of all patients but, importantly, it occurs particularly in older age groups.

Diagnosis of secondary achalasia rests on a high degree of suspicion leading to meticulous endoscopic examination and biopsy/brushings. Treatment is that of the underlying neoplasm or other condition.

Diffuse oesophageal spasm

In addition to achalasia there is a series of primary motility disorders for which a pathological basis is not known. They are thought to occur more frequently with advancing age although epidemiological data are lacking. Additionally, similar motility patterns may be seen in some elderly individuals with no symptoms so that the relationship between manometric findings and symptoms is far from clear.

Initially based on radiographic findings of non-peristaltic (tertiary) contractions, diffuse oesophageal spasm is a syndrome characterized by symptoms of chest pain and dysphagia, variously known as the 'corkscrew' or 'rosary bead' oesophagus. Most patients have both chest pain and dysphagia, varying from mild and intermittent to severe and regular. In about half, the pain is related to meals but it may occur at other times and mimic angina in nature. Dysphagia for solid and liquids is not progressive, in contrast to achalasia.

The diagnosis is established by manometry where non-peristaltic contractions of the distal oesophagus following some but not all swallows are seen. Patients with diffuse oesophageal spasm are usually defined as those with non-peristaltic contractions following more than 10 per cent of swallows. Vigorous contraction waves and abnormalities of basal pressure and relaxation at the lower oesophageal sphincter are often found. Barium studies may show normal peristalsis in the proximal oesophagus but there is disruption distally where tertiary waves and lack of progression of the bolus may be demonstrated.

Diffuse oesophageal spasm is not usually a progressive condition and treatment involves reassurance as to its benign nature and avoidance of obvious precipitating factors. The use of smooth muscle relaxants such as nitrates and calcium antagonists have been anecdotally reported to have moderate success but controlled trials are lacking.

Presbyoesophagus

This term arose following the demonstration that 15 patients aged over 90 had motility patterns that were different from those of younger people. However, nearly all of these subjects had underlying neurological disorders or diabetes. Subsequent studies on patients allegedly without underlying disease provided conflicting findings about the prevalence of changes with age. Recent reports using modern manometric techniques suggest that only distal contractile amplitude and duration increase with age and that there is no age effect on lower oesophageal sphincter pressures. Until further studies with larger numbers of healthy individuals over age 80 are reported the term presbyoesophagus is of little practical use since it has no defined meaning. It should not be used as a substitute for the thorough search for such disorders as stricture, reflux, achalasia, and neoplasm in elderly patients.

Secondary motility disorders

These occur as part of a more generalized disease. Many affect the oropharyngeal phases of swallowing and have already been highlighted. Systemic lupus erythematosus and systemic sclerosis are thought to have a predilection for oesophageal smooth muscle.

Systemic sclerosis (scleroderma)

Although systemic sclerosis is the best studied of those connective tissue disorders which involve the gastrointestinal tract, it is not particularly a disease of elderly people. Patients suffer from both skin and visceral involvement with the oesophagus being the most commonly affected site. Atrophy of oesophageal smooth muscle with collagen replacement is seen histologically in 80 per cent of patients. There may be difficulties in mastication and oesophageal symptoms are present in roughly 50 per cent. Disordered motility is probably less likely to cause symptoms than those arising as a result of gastro-oesophageal reflux secondary to poor lower oesophageal sphincter tone and impaired clearance of refluxed material. Oesophagitis and strictures are often present and pulmonary complications secondary to aspiration occur.

Barium swallow may show dilatation and hiatus hernia or a distal stricture. In contrast to achalasia, however, the lower oesophageal sphincter eventually opens and reflux may be seen. Manometrically there will be low amplitude contractions and basal lower oesophageal sphincter pressure is low. Endoscopically, oesophagitis and strictures may be demonstrated.

Treatment is aimed primarily at aggressive management of reflux oesophagitis but this is much less effective than in simple reflux disease. Dilatation of structures may be necessary.

Other secondary motility disorders

Amyloidosis seen in elderly patients has occasionally been found to cause symptomatic oesophageal problems. It may present with either a clinical picture resembling achalasia or with inflammatory erosions. Treatment is symptomatic. Similarly, diabetes mellitus may be associated with diffuse motor disturbances throughout the digestive tract. Problems are less commonly encountered with the oesophagus than the stomach but manometric abnormalities have been reported in up to one-half of diabetic patients with peripheral neuropathy. Such patients are not usually symptomatic and oesophageal symptoms in an elderly diabetic should not be ascribed to the disease without proper investigation.

Gastro-oesophageal reflux disease

Intermittent episodes of reflux of gastric contents through the lower oesophageal sphincter is a normal physiological event. Should these episodes lead to symptoms or physical complications then the term gastro-oesophageal reflux disease is appropriate. Reflux disease is a consequence of the exposure of the lower oesophageal mucosa to gastric contents and acid is thought to be particularly relevant. The severity of reflux disease generally relates to the degree of oesophageal acid exposure. The physical complications of gastro-oesophageal reflux are thought to include oesophagitis, strictures, Barrett's metaplasia, bleeding, and pulmonary disease. Patients with those problems form a subset of the population with gastro-oesophageal reflux disease.

Hiatus hernia

An hiatus hernia is a protrusion of part of the stomach through the oesophageal hiatus of the diaphragm into the thoracic mediastinum. A uniform definition is lacking since surgeons, radiologists, and endoscopists use differing landmarks to identify the gastro-oesophageal junction. Four types are described.

1. Sliding hiatus hernia. This occurs when the oesophagus and gastro-oesophageal junction move easily through the hiatus, with the latter displaced into the thorax. They comprise the majority (90 per cent) of all hiatus hernia.
2. Para-oesophageal hiatus hernia. The gastro-oesophageal junction is normally placed but the gastric fundus and greater curvature protrude through the hiatus anteriorly to the oesophagus. Volvulus and strangulation may complicate this type more frequently.
3. Combination hiatus hernia. Both the gastro-oesophageal junction and gastric fundus are displaced.
4. Congenital short oesophagus. Present from birth, this is very rare and should be distinguished from Barrett's oesophagus and an hiatus hernia that has secondarily become fixed.

Hiatus hernia becomes more common with advancing age. Estimates of its prevalence vary from 50 to almost 100 per cent depending on the vigour used to search for it. Most sliding hiatus hernias are asymptomatic whereas para-oesophageal and mixed types may more usually give rise to mechanical effects. Chest tightness, dysphagia, bloating, dyspnoea, and satiety, exacerbated by meals and relieved by vomiting and belching, are described. Whereas para-oesophageal and combination types may warrant surgery the majority of people with a sliding hiatus hernia are improved or asymptomatic 10 years after its discovery regardless of age. Thus a sliding hiatus hernia in itself may almost be regarded as a non-pathological finding.

The exact relationship between gastro-oesophageal reflux disease and sliding hiatus hernia is unclear. Many of the symptoms formerly attributed to an hiatus hernia have come to be regarded as due to coexistent gastro-oesophageal reflux disease. Undoubtedly patients with hiatus hernia are over-represented in groups with the disease. The presence of an hiatus hernia appears to predispose to reflux symptoms and exacerbate gastro-oesophageal reflux disease (Petersen *et al.* 1991), perhaps by disturbing the action of the lower oesophageal sphincter.

Epidemiology

The epidemiology of gastro-oesophageal reflux disease may be regarded as guesswork. Definitions of the disease have been variable and the terms hiatus hernia and reflux used interchangeably. Mild symptoms are often ignored and may be regarded as normal by some individuals. The prevalence increases with age and has been estimated at about 5 per cent in 55 year olds of whom half have reflux oesophagitis. The age distribution of reflux oesophagitis suggest a peak in the age range 60 to 70 years; perhaps 25 per cent of patients are over age 75.

Complications are related to duration of exposure and therefore age-associated. Peptic strictures and Barrett's oesophagus peak between the ages of 50 and 70. Overall annual mortality rates from gastro-oesophageal reflux disease including postoperative mortality are thought to be as low as 0.16 per 100000. Thus despite the limitations in our knowledge gastro-oesophageal reflux disease appears to be a disease of older age and one with which the geriatrician may be frequently confronted.

Pathophysiology

The factors which contribute to excessive mucosal exposure to gastric contents are predominantly a high rate of reflux episodes and impaired clearance thereafter of the refluxed material (refluxate) back into the stomach. Additional factors may include the nature of the refluxate and the intrinsic mucosal protection mechanisms.

At rest, lower oesophageal sphincter pressure must be absent for reflux to occur. This may arise in two ways: by abnormal relaxation patterns or by lowered basal tone. Current evidence suggests that the majority of reflux episodes in both asymptomatic controls and those with gastro-oesophageal reflux disease without macroscopic oesophagitis occur as a result of transient lower oesophageal sphincter relaxation. In patients with severe oesophagitis it is estimated that defective basal tone of the lower oesophageal sphincter plays an important role in reflux episodes. The underlying mechanisms are unknown but it has been postulated to be a fault in the control of lower oesophageal sphincter pressure rather than an intrinsic muscle abnormality.

Once reflux has occurred, the development of oesophagitis depends upon the mucosal contact time of the refluxate, its nature, and the resistance of the mucosa. Contact time depends upon efficient volume clearance by gravity and oesophageal peristalsis as well as chemical clearing by saliva. The refluxate contains acid, pepsin, and bile salts. It is the acid, however, that is thought to be the major aggressive factor to oesophageal mucosa. Bile salts and pepsin perhaps have contributory roles in increasing mucosal permeability to hydrogen ions. It is still not known if the mucosal resistance mechanisms are also deranged in those with gastro-oesophageal reflux disease.

Clinical features

Most patients with gastro-oesophageal reflux disease have symptoms for between 1 and 3 years before presentation. It has recently been suggested that elderly patients present in a slightly different manner from younger patients. Typical symptoms in older people are regurgitation, respiratory problems, and vomiting rather than heartburn (Raiha *et al.* 1991), although roughly a third may have none of these. Furthermore, at least a third of elderly patients with gastro-oesophageal reflux disease suffer from other major conditions which are at least as important as the consequences of the gastro-oesophageal reflux disease itself. These have a substantial bearing upon management.

Unfortunately, as we have seen, the classical symptoms of heartburn and acid regurgitation clearly dominate at presentation in less than one-third of patients. Most complain of a variety of symptoms including epigastric pain, belching, retrosternal pain, nausea, waterbrash, bloating, and odynophagia. An association between gastro-oesophageal reflux disease and respiratory disease has been well established although the causal nature is debated. It is unclear whether direct aspiration of gastric contents into the respiratory tract or vagally mediated bronchoconstriction following exposure of oesophageal mucosa to acid is the predominant explanation. Symptoms such as onset of wheeze or cough should prompt a careful evaluation of the likelihood of reflux in an old person. There may be repeated episodes of bronchitis or aspiration pneumonia. Gastro-oesophageal reflux disease may also present with primary otolaryngological symptoms such as postnasal drip, hoarseness, and neck pain. Non-cardiac chest pain has been attributed to gastro-oesophageal reflux disease in up to 60 per cent of cases although these are not always 'pure' reflux episodes and associated oesophageal motor disorders may be involved. Since the likelihood that ischaemic heart disease, musculoskeletal disorders, and gastro-oesophageal reflux disease coexist increases with age, diagnosis and management is inherently more complicated in elderly patients.

Investigation

Endoscopy and biopsy

This should be the first line of investigation if uncomplicated gastro-oesophageal reflux disease is suspected. Diagnosis of moderate to severe oesophagitis causes little difficulty but milder degrees are subject to much interobserver variation. Thus the use of standard grading scales is to be encouraged. In up to 50 per cent of those with gastro-oesophageal reflux disease, endoscopy is macroscopically normal so biopsy should always be performed. Even histological criteria for oesophagitis are not uniform and the sensitivity of histological findings of inflammation as a marker of acid reflux is not high.

Barium radiology

In general this has little to offer in the diagnosis of gastro-oesophageal reflux disease. Most patients with acid reflux do not have oesophagitis, and thus the examination will be normal. Additionally, barium studies are insensitive to mild degrees of oesophagitis. The demonstration of unprovoked reflux or an hiatus hernia is not diagnostic of disease. Radiology is thus more useful in the detection of complications such as severe ulceration or stricture in patients with dysphagia.

Ambulatory oesophageal pH monitoring

This investigation is the current gold standard for the diagnosis of gastro-oesophageal reflux disease. Its main role is in the evaluation of those with typical symptoms but without endoscopic findings and those with atypical symptoms with or without endoscopic confirmation. It may also be needed to evaluate the results of treatment and is used for preoperative assessment. The total percentage of time that the oesophageal pH is less than 4 has good discriminant value and the symptom index allows correlation of symptoms with reflux events (Tytgat 1989).

Other ancillary investigations such as the Bernstein acid perfusion test, the acid reflux test, and radionuclide studies have now been largely replaced by ambulatory monitoring.

Treatment

Therapy is aimed towards the relief of symptoms and the prevention of complications. The degree of severity of the gastro-oesophageal reflux disease and the general condition of the patient must be carefully balanced in pursuing the appropriate management plan. While symptomatic relief is the aim in those without macroscopic oesophagitis, the prevention of complications assumes greater importance when disease can be documented endoscopically. In the elderly patient coexistent pathology and medication should be carefully taken into consideration.

Antireflux measures

Simple general measures may suffice. Postural advice about stooping and bending in addition to elevating the head of the bed to decrease nocturnal reflux should be given. Weight reduction, the wearing of loose rather than tight clothes, and the taking of small rather than very heavy meals are to be encouraged. Avoidance of eating within 3 to 4 h of retiring to bed may help. Smoking promotes reflux and impairs healing of the inflamed oesophagus as does alcohol. Elderly people are particularly likely to be prescribed drugs such as those with anticholinergic actions, nitrates, calcium antagonists, and theophyllines which decrease lower oesophageal sphincter pressure. These should be withdrawn if possible.

Antisecretory agents

The effectiveness of these drugs is directly related to their ability to suppress acid production and thus reduce the aggressiveness of the refluxate. Healing of oesophageal lesions requires greater acid suppression and for longer period than do peptic ulcers. H_2-receptor antagonists—ranitidine, cimetidine, famotidine, and nizatidine—have been studied extensively. Unfortunately many poorly designed trials with varying entry criteria, poorly documented severity of disease, and different dose regimens have been performed. Better studies have recently suggested that 50 to 70 per cent of patients will have complete or partial resolution of symptoms within 6 to 8 weeks but this does not correlate well with endoscopic healing. The subgroup of patients with poor response may require doses two or four times greater than conventionally accepted to be effective in overcoming acid production. Omeprazole, the first of a new class of proton pump inhibitors, produces prolonged and almost complete suppression of acid secretion. Ambulatory pH recordings show the virtual disappearance of acid reflux episodes. Eight weeks of treatment with omeprazole 40 mg at night achieves symptomatic relief and healing of oesophagitis in over 80 per cent of subjects. Direct comparison with H_2-receptor antagonists demonstrates the superiority of omeprazole and it is now the treatment of choice for severe or resistant oesophagitis.

Even after healing has been achieved with full remission of symptoms many patients with reflux oesophagitis relapse within 6 months if maintenance treatment is not continued. A substantial proportion of patients require the same maintainence dose as that necessary for healing. While omeprazole has been shown to prevent recurrence in a higher proportion of patients than H_2-receptor antagonists there is little experience as yet of its long-term use.

Prokinetic agents

Agents which increase lower oesophageal sphincter tone, augment the amplitude of oesophageal peristalsis, and promote gastric emptying have theoretical benefits in the treatment of gastro-oesophageal reflux disease. Disappointingly, clinical trials have shown only modest or no benefit with metoclopramide and domperidone. Cisapride is the newest agent and does not have antidopaminergic actions. It promotes postganglionic release of acetylcholine. It is potentially the most effective of these agents and early studies suggest its efficacy in moderate, but probably not severe, oesophagitis.

Mucosal coating drugs

Sucralfate is a glycosylated aluminium hydroxide salt which adheres to ulcerated surfaces and may therefore protect against acid and pepsin. Early clinical trials have suggested that it is moderately effective and further studies are awaited.

In summary, for severe oesophagitis omeprazole is the treatment of choice. Possibly combinations of cisapride and H_2-receptor antagonists may be satisfactory in patients with moderately severe disease and for prolonged maintenance.

Antireflux surgery

Less than 5 per cent of patients fail to respond to medical treatment. Indications for surgery in elderly patients should be extremely strict and the authors now believe that, with the introduction of omeprazole, very few patients indeed in this age group need to come to surgery. Very rarely a joint approach between physicians and specialist surgeons may still enhance management in patients with severe gastro-oesophageal reflux disease refractory to maximal medical therapy and who have stricture, ulceration, regurgitation, and pulmonary complications.

Surgery involves varying degrees of fundoplication which is thought to control reflux by imposing a non-relaxing high pressure zone around the lower oesophageal sphincter.

Benign strictures

About 80 per cent of benign strictures are thought to be peptic in origin and perhaps as many as one in ten of those with gastro-oesophageal reflux disease severe enough to consult a doctor will suffer from stricture formation. Other benign strictures include those that are postoperative (for example, myotomy for achalasia), sequelae to prolonged nasogastric intubation, due to scleroderma, and those secondary to ingestion of corrosive substances. Peptic strictures are typically a problem of patients in the seventh and eighth decades.

When oesophagitis becomes very severe ulcers appear in the mucosa and the inflammatory reaction may extend to the submucosa. The muscle layer is involved in the most advanced cases with secondary fibrosis. The oesophageal wall becomes thickened and shortened as strictures form. Most peptic strictures occur in the distal oesophagus and are 1 to 2 cm in length. Those that are more proximal are usually associated with Barrett's oesophagus at either the migrated squamocolumnar junction or within the columnar epithelium at the site of a chronic ulcer. Perhaps one third of patients with Barrett's oesophagus will progress to stricture formation.

The presenting symptom is dysphagia, initially intermittent but typically becoming progressive over a period of 1 to 2 years. There is often associated weight loss and there may be an antecedent history of heartburn and acid regurgitation but these symptoms frequently regress with the onset of dysphagia. In a quarter of patients no prior history is forthcoming, particularly in those with Barrett's changes.

In patients with dysphagia a barium swallow should be performed to indicate the location and length of the lesion. A narrow non-distensible area is seen and when benign the proximal oesophagus tapers symmetrically towards the upper end of the stricture. Unfortunately about a quarter of strictures have a misleading appearance on radiology, apparently benign strictures being malignant and vice versa. Endoscopy is therefore mandatory after radiology and allows biopsy and brushings at the site of the stricture. If the lesion is narrow an attempt to dilate the stricture sufficiently to allow samples to be acquired should be made. Correct histological diagnosis can be made in about 95 per cent of cases.

Therapy aims to relieve symptoms and prevent recurrence of stricture. In elderly patients the main-stays of treatment are medical suppression of acid production similar to that for reflux oesophagitis together with periodic dilation. Patients with associated Barrett's epithelium can be managed in a similar fashion. Adequate antireflux therapy reduces the need for repeated dilatation although as many as 75 per cent of elderly patients with strictures need dilatations at least at yearly intervals. Oesophageal dilatation under endoscopic control with guide-wire systems is now a safe and widely practised procedure. Detailed descriptions of the technical aspects can be found elsewhere (Tytgat 1989). The stricture is visualized with the endoscope; a guide-wire is then passed through the narrow lumen and dilators threaded along the wire. The major complication—perforation—occurs in less than 1 per cent of dilatations. It should be considered in patients complaining of chest pain or discomfort, dyspnoea or fever, and tachycardia following dilatation. Plain chest radiography can demonstrate a pneumomediastinum but may show no abnormality. A barium swallow with water soluble contrast medium should be performed if suspicion continues. Management of perforation should be carried out in conjunction with an experienced oesophageal surgeon. Often conservative management will be successful.

The indications for surgical treatment of benign peptic strictures are now very few indeed, particularly in patients over 75 years. Surgical mortality in such patients has been shown to be as high as 15 per cent in a recent series.

Barrett's oesophagus

The term Barrett's oesophagus or columnar lined oesophagus is now used to describe the condition where a variable length of squamous epithelium, usually at least 3 cm long, in the distal oesophagus is replaced by columnar epithelium. The shape of mucosal extensions may be in a discrete circular or non-circular band, in finger-like projections from the squamocolumnar junction, or as separate islands of columnar epithelium. The presence of columnar-lined oesophagus in older patients is almost always associated with gastro-oesophageal reflux disease which has led to destruction of squamous epithelium and subsequent healing by its replacement with more acid resistant columnar cells. Patients with columnar-lined oesophagus have been shown by means of pH studies to have greater acid exposure and poorer clearance than those with oesophagitis only or normal individuals.

The condition may affect any age group but the mean age has been reported at 64 years and there is a peak between 50 and 80. The male to female ratio is about 3:1. Among individuals with oesophageal symptoms the cumulative incidence of columnar-lined oesophagus is estimated at 4 to 5 per cent. It is of itself not necessarily symptomatic but most patients give a past history suggestive of gastro-oesophageal reflux disease. In 10 per cent the presentation is that of dysphagia secondary to inflammatory stricture. An ulcer may develop in the columnar epithelium in which case pain radiating to the back is predominant. Occasionally, this can perforate with life-threatening consequences.

Endoscopic appearances are usually distinctive with a sharp demarcation between pale squamous and velvety red columnar mucosa. There is absence of the folds of the intrathoracic stomach. Submucous vessels may be visible. Strictures and ulcers should be identified. Biopsies are essential to confirm the diagnosis and exclude dysplasia or carcinoma. These should be taken from the gastro-oesophageal junction and then at 1 to 2-cm intervals proximally even in the absence of visible lesions.

The treatment of columnar-lined oesophagus is effectively that of gastro-oesophageal reflux disease in that symptom relief, abolition of acid reflux, and where necessary dilatation of strictures are the mainstay, particularly in elderly patients.

Management of columnar-lined oesophagus is further complicated since it carries the risk of a dysplasia–carcinoma sequence with the prevalence of adenocarcinoma being 8 to 15 per cent, increasing with disease chronicity. It is thought that continued reflux on columnar cells initiates dysplasia leading to carcinoma-*in-situ* and ultimately frank carcinoma. Thus a further aim of therapy must be to minimize this risk of malignant change by aggressive treatment of acid reflux.

There is considerable debate about the effectiveness of screening for carcinoma-*in-situ* in the columnar-lined oesophagus (Atkinson 1989). Where high-grade dysplasia is detected surgery may be recommended but with elderly patients such a policy needs to be tempered by a general assessment of the life expectancy, fitness for surgery, and quality of life of the individual.

Non-cardiac chest pain

Recurrent 'angina-like chest pain' in the absence of demonstrable cardiac pathology is the subject of increasing attention. It is estimated that as many as one in three coronary angiograms in patients with anginal symptoms are normal. Despite the reasonable exclusion of heart disease many of these patients continue to suffer from recurrent disabling symptoms and hospital attendance. This diagnostic dilemma has recently attracted considerable attention in middle-aged individuals but there are little data directly applicable to the elderly population.

Evaluation of possible oesophageal chest pain should commence with the exclusion of cardiac disease. It is often impossible to discriminate between the two on the basis of clinical history alone. Age, risk factors, and family history will determine the invasiveness of investigations. A normal coronary angiogram indicates a favourable prognosis but some authors believe that many such patients may suffer from microvascular angina (syndrome X). Few subjects over 65 have been included in studies of this condition.

Musculoskeletal disease, peptic ulcer, and biliary and pancreatic disease, as well as depressive disorders should be considered as a cause of non-cardiac chest pain in elderly patients. When attention is turned towards the oesophagus recent reports suggest that about a third of patients have evidence of gastro-oesophageal reflux disease and a smaller proportion motility disorders (Richter *et al.* 1989). The treatment of these patients is unsatisfactory largely because definitive diagnosis is difficult. If gastro-oesophageal reflux disease is confirmed then medical therapy should be instituted.

Diverticula, webs, and rings

Pharyngo-oesophageal diverticulum (Zenker's)

This disorder of older people has a peak incidence in the seventh decade and a male proponderance. It usually presents chronically with upper dysphagia and regurgitation of undigested food after eating. Halitosis, coughing on swallowing, and the sensation of a lump in the throat may be troublesome. The two most serious complications are the rare development of squamous cell carcinoma within the sac and tracheal aspiration of pouch contents. It is likely to be an acquired disorder due to pulsion forces within the pharynx. The diverticulum emerges between the oblique heads of the inferior pharyngeal constrictor and transverse cricopharyngeus muscle.

Diagnosis is made most accurately with barium studies, lateral films demonstrating the diverticulum's posterior location. Endoscopy can be dangerous because of the risk of perforation.

Treatment of the diverticulum is surgical and is usually recommended at an earlier rather than later stage since the condition is progressive and risks of tracheal aspiration increase with time. In very elderly and debilitated patients a per oral endoscopic diathermy division of the wall between oesophagus and diverticulum may be carried out rather than more definitive surgery.

Midoesophagcal diverticula

These are of two sorts: pulsion and traction. Pulsion diverticula are similar to epiphrenic ones and arise due to raised intraluminal pressures associated with motility disorders. The more common type is the traction diverticulum which has a true muscular coat and is usually associated with mediastinal inflammation.

Usually asymptomatic, these diverticula can occasionally cause substernal pain and dysphagia. Complications include local abscess formation or fistula. Diagnosis with either barium studies or endoscopy is difficult because of the shallowness of the lumen. Treatment for symptomatic lesions is surgical.

Very rarely a condition of elderly patients occurs in which multiple small pouches arise from dilated submucosal glands lining the oesophageal wall. This is known as intramural pseudodiverticulosis. Radiologically these lesions resemble collar studs. Symptoms are usually of dysphagia and candidiasis is a complication in half of these patients.

Webs and rings

A web is a localized area of fibrosis covered with squamous epithelium occupying only part of the oesophageal wall. In the adult these may cause intermittent dysphagia or sudden bolus obstruction. A single dilatation usually suffices.

A particular postcricoid web is found in the Brown–Kelly–Paterson syndrome (also erroneously named Plummer Vinson syndrome). Usually a disorder of postmenopausal women, there may be sideropenic anaemia, intermittent dysphagia, glossitis, and koilonychia. The incidence of the disorder has fallen in recent years. Treatment comprises iron supplementation and dilatation. There is an enhanced risk of postcricoid carcinoma, so regular endoscopic surveillance at perhaps annual intervals is recommended.

A ring is covered by squamous epithelium on the upper surface and columnar on the lower. Schatzki's ring occurs at the gastro-oesophageal junction but the clinical significance and aetiology is unclear. It is not thought to be secondary to oesophagitis. Often an incidental finding at radiology these rings infrequently present with intermittent dysphagia to food. Treatment is by dilatation or merely by passage of the endoscope through the ring.

NON-PEPTIC STRICTURES

Non-malignant

All oesophageal strictures in old people should be regarded as malignant until proved otherwise. The vast majority of benign strictures will be secondary to reflux oesophagitis. Occasionally, the aetiology is secondary to collagen diseases, infections or very rarely to Crohn's disease. Other causes include strictures following radiotherapy and trauma (following instrumentation or impaction of a foreign body). Corrosive strictures may occur because of accidental or deliberate ingestion of acid or alkaline solutions but this is unusual in elderly people. Strictures secondary to medications are of greater importance. The emergency treatment of corrosive lesions is supportive and the use of high-dose steroids controversial. Oral feeding is withheld until the ability to swallow saliva can be confirmed. The physician should be vigilant against such complications as perforation and mediastinitis. Elective endoscopy, dilatation, gastrostomy, or reconstructive procedures may ensue.

External compression

The oesophagus may be compressed externally by surrounding structures giving rise to dysphagic symptoms. Dysphagia aortica is most common in elderly persons as a result of an impingeing thoracic aortic aneurysm or sclerotic descending aorta. Dilatation of the left atrium may be a further cause. Osteophytic projections from vertebrae are an uncommon cause of oesophageal swallowing difficulties. Mediastinal tumours should also be considered.

Oesophageal neoplasms

Carcinoma (see Section 10)

This is a disease of advancing age; most patients present over the age of 65, and the disease is responsible for 5200 deaths in England and Wales annually. It remains a dreaded disease because of the advanced stage of presentation with its poor prognosis and because of distressing symptoms associated with obstruction of the oesophagus. Median survival from the onset of dysphagia, the most common presenting symptom, is about 6 months. The majority (90 per cent) of carcinomas are squamous cell in origin. In the West, cancers of the lower third of the oesophagus are more common than those of the middle third, the smallest proportion being located in the upper third.

Epidemiology

The geographic variation in the incidence of this disease is great suggesting that the aetiology is related to environment. In the United Kingdom and among white North Americans annual incidence is about 5 per 100 000, in black North Americans 20, while rates in China and Iran are much higher at 109 and 262 respectively. Even within countries rates vary markedly. In China there is a seven hundredfold difference in mortality from squamous carcinoma of the oesophagus between the highest and lowest incidence areas.

Although no single cause has been implicated as responsible a number of risk factors are known. In China dietary nitrosamines in combination with vitamin and mineral deficiencies are likely to be important. The predominant carcinogenic agents in Western Europe and North America are thought to be alcohol and cigarette smoking. These independent risk factors are associated with 80 to 90 per cent of patients with oesophageal cancer.

The rare autosomal dominant condition tylosis is associated with a 95 per cent risk of distal oesophageal malignancy by the age of 65. Carcinoma probably occurs in about 5 per cent of individuals with achalasia after 20 years. In Brown–Kelly–Paterson syndrome (see above) there is ultimately a 10 per cent incidence of pharyngo-oesophageal malignancy. Patients with coeliac disease also have an enhanced risk of oesophageal cancer. The most important aetiological factor for adenocarcinoma is a Barrett's oesophagus.

Clinical presentation

Dysphagia is an almost universal symptom, usually initially for solids but, progressing over a few months to liquids and ultimately even saliva. The oesophageal lumen is usually at least 60 per cent stenosed with tumour before the onset of dysphagia and hence the disease is advanced. Weight loss is often rapid. There may be regurgitation, especially when recumbent at night with resultant cough and aspiration. The development of tracheo-oesophageal fistula may make pulmonary symptoms predominant. Odynophagia is present in half the patients.

Pain may result from local invasion of the spine, intercostal nerves, or aorta. Infiltration of surrounding structures may cause hoarseness or haemoptysis. Overt bleeding is unusual although exsanguinating haemorrhage can occur if the aorta is eroded.

Examination may reveal supraclavicular lymphadenopathy, a cervical mass, or tracheal deviation. Superior vena cava obstruction causes oedema of the arms with facial congestion. There may be Horner's syndrome. Evidence of metastatic spread to the liver should also be sought.

Investigations

The level of the cancer should be determined initially with barium studies, where luminal narrowing, irregularity, and rigidity will be seen. In the most advanced lesion there may be a typical 'apple core' appearance. Early lesions are not easily identified as malignant.

Endoscopy will normally provide a definitive diagnosis in 95 per cent of patients if multiple biopsy and brushings are combined. At endoscopy the distance of the lesions together with the biopsy positions should be carefully noted. Recently endoscopic ultrasound has shown some value in the detection of early lesions and in staging.

Patients with lesions in the upper third of the oesophagus should undergo bronchoscopy to determine the extent of tracheal or bronchial spread and to exclude a primary bronchial lesion. Further staging of the tumour requires CT scans of the chest and upper abdomen. Isotopic bone scans and even staging laparoscopy may be performed.

Treatment

The late presentation of the disease is the limiting factor in therapy. Early detection is the best hope of improving prognosis but cytological screening of asymptomatic individuals can at present only be justified in areas of extremely high risk such as parts of China.

There has been some improvement in the prognosis of patients with oesophageal carcinoma over the past decade both in respect of reduced postoperative mortality and of 5-year survival although this benefit is confined to specialist centres (Cuschieri 1991). Potentially curative resection is possible in only 20 per cent of patients and 5-year survival rates among this group are in the order of 20 to 40 per cent. The efficacy of radical radiotherapy is still highly debatable particularly in comparison with surgical treatment.

Surgical mortality increases with age for both elective and emergency procedures. Some authors have concluded that *en-bloc* resection after the age of 75 is unwise and that palliative resection should be advocated; others have indicated that in patients over age 80 radiotherapy may be the preferred option (Hishikawa *et al.* 1991).

For most patients treatment is palliative with the major aims of abolishing distressing dysphagia and ensuring adequate pain relief. Radiotherapy may offer partial or complete relief of dysphagia in up to two-thirds of patients but recurrence of symptoms occurs in over half of them within 6 months. Radiotherapy is of particular value in the patients with adenocarcinoma.

A further range of palliative procedures is available and while simple dilatation is ineffective, intubation, laser therapy, and intracavity irradiation all have their advocates (Brown 1991). There are few studies to indicate the most appropriate mode of treatment for an individual patient, and even fewer which have incorporated reliable and valid measures of quality of life in assessment of outcome. Intubation of the oesophagus with a prosthetic tube is widely used and may relieve swallowing difficulties for 6 months or occasionally longer. There are two techniques: traction tubes which require surgery to pull them through the oesophagus on a guide wire, and pulsion tubes inserted endoscopically. Endoscopic methods now predominate but even so carry mortality rates of 4 to 5 per cent in experienced hands. Complications include perforation, tube migration, aspiration, and obstruction of the tube. A special liquidized diet is almost always required so that palliation cannot be regarded as complete.

Nd-YAG laser therapy has recently become an increasingly attractive option for palliative treatment. Mortality rates and perforation are generally lower at about 1 and 2 to 5 per cent respectively. However several sessions may be required and duration of palliation is variable. Intra-oesophageal radiotherapy is also attracting increasing attention with the development of new technology making it simpler and safer. Efficacy data are limited but relief of dysphagia is considerably faster than after external beam irradiation. Direct comparisons of intubation and laser therapy are very few, indeed the procedures may be complementary in some patients. With further experience in the respective uses of these three techniques effective palliation should be increasingly possible for the majority of patients.

Other malignant oesophageal neoplasms

These include histological variants of both squamous and adenocarcinomas, usually with a bad prognosis. Small-cell carcinomas resembling oat-cell tumours of the lung may also be seen and can be associated with paraneoplastic syndromes. Rarely carcinoid tumours or sarcoma-like tumours are found.

Benign oesophageal tumours

Non-neoplastic tumours may arise from heterotopias, cysts, or granulomatous deposits. The most common benign tumour is the leiomyoma. Others include lipomas, neurofibromas, and giant cell tumours.

Oesophageal perforation

This is usually classified as spontaneous (Boerhaave's syndrome) or secondary. Rupture in Boerhaave's syndrome usually follows violent vomiting and is hence not truly spontaneous. It occasionally occurs after such activities as straining at stool or milder gastrointestinal upsets. The triad of presenting features—vomiting, lower chest pain, and subcutaneous emphysema—may be absent leading to fatal delays in diagnosis. Early surgery must be performed. This is in contrast to the disorder of submucosal intramural rupture of the oesophagus without perforation which may be managed conservatively with a good prognosis (Steadman *et al.* 1990).

Secondary causes of perforation are more prevalent in elderly people and include instrumental procedures, Barrett's ulcers, malignancy, anastomotic fistulae, and foreign bodies.

Infections

In frail elderly patients and in those who are immunocompromised infections of the oesophagus are a common but often overlooked problem which cause considerable morbidity and even mortality. Inadequate treatment may lead to local and systemic complications. Dysphagia, odynophagia, or enhanced awareness of the passage of food may be the presenting symptoms of an acute infectious oesophagitis. Equally, in the frail and debilitated patient, typical symptoms are often absent and while oral examination may reveal candidiasis it may be normal in many with oesophageal disease.

Candidiasis

Candida albicans is a normal commensal of the digestive tract, thought to be restrained by other intestinal flora, especially bacteria producing lactic acid. This fungus is the most prevalent infectious cause of oesophagitis and attacks may occur in old people in the absence of underlying disease. However it is most commonly associated with agents causing a breach of local mucosal defences such as mechanical obstructions, irradiation, or a defect in systemic immunity (for example malignancy, diabetes mellitus, or malnutrition). Most commonly, however, oesophageal candidiasis probably occurs following treatment with broad spectrum antibiotics.

Suspicion of infectious oesophagitis is the key to diagnosis, while a double contrast barium swallow may reveal a granular mucosa, ulceration, or nodularity. Endoscopy is the most useful procedure. Typically raised white adherent plaques with surrounding erythema and ulceration are seen. Brushings and culture aid diagnosis.

Treatment depends upon the severity of infection. Oral nystatin and/or miconazole or a related antifungal compound may suffice. Occasionally systemic treatment has been used.

Viral oesophagitis

Viral oesophagitis is predominantly caused by Herpes simplex virus (HSV) although cytomegalovirus (CMV) and Epstein-Barr virus (EBV) have been found in severely immunodeficient patients. These viral infections are a feature of immune impairment rather than age *per se*. They usually occur in a very frail old person with other severe, often malignant, disease.

Drugs and the oesophagus

This topic is of great importance to those caring for elderly patients. Up to a half of adverse drug events related to the oesophagus are reported in patients over 65. Drugs can affect the oesophagus in two ways: first, by direct damage to the mucosa and second by indirect effects following systemic absorption (Weinbeck *et al.* 1988).

The normal oesophagus can retain tablets for up to 20 min. Prolonged transit times are more likely in elderly people for the reasons discussed in earlier sections on motility disorders. Furthermore, extrinsic compression may also impede the passage of tablets. Drugs which have been particularly reported in connection with adverse oesophageal events include tetracyclines, potassium chloride, aspirin, and non-steroidal anti-inflammatory drugs.

Typically there may be sudden onset of retrosternal pain often at night. However, in older patients onset may be more vague presenting with stricture formation or bleeding. Early endoscopy may demonstrate single or multiple areas of erythema, friability, erosion or ulceration. Lesions are most commonly at the level of the aortic arch, the upper oesophageal sphincter or the gastro-oesophageal junction. The best form of management is prevention. If drugs known to cause oesophageal mucosal disease are required in a patient with a history of swallowing disorder then these should be taken in the upright position with adequate volumes of fluid. Liquid preparations may be substituted for tablets.

After systemic absorption some drugs may influence oesophageal function either by increasing or decreasing motility and lower oesophageal sphincter pressure. Symptoms may be exacerbated if there is pre-existing disease. Drugs causing motor stimulation are usually used therapeutically as in the case of prokinetic agents. These should be avoided in the presence of diffuse oesophageal spasm and achalasia. Those that inhibit motor activity and impair lower oesophageal sphincter tone tend to promote gastro-oesophageal reflux disease symptoms such as heartburn and acid regurgitation. These include anticholinergic agents, nitrates, calcium antagonists, β-blockers, benzodiazepines, and theophyllines.

References

Atkinson, M. (1989). Barrett's oesophagus—to screen or not to screen? *Gut*, **30**, 2–5.

Brown, S.G. (1991). Palliation of malignant dysphagia. *Gut*, **32**, 841–4.

British Society of Gastroenterology. (1991). Recomendations for standards of sedation and patient monitoring during gastrointestinal endoscopy. *Gut*, **32**, 823–7.

Christensen, J. (1988). Motor functions of the pharynx and oesophagus. In *Physiology of the gastrointestinal tract* (ed. L.R. Johnson). Raven Press, New York.

Cuschieri, A. (1986). Anatomy and physiology of the oesophagus. In Hennessy T.P.J., Cuschieri A. eds: *Surgery of the oesophagus* (eds T.P.J. Hennessy and A. Cushieri) pp. 1–16. Bailliere Tindall, England.

Cuschieri, A. (1991). Treatment of the carcinoma of the oesophagus. *Annals of the Royal College of Surgeons of England*, **73**, 1–3.

DeNardi, F.G. and Riddell, R.M. (1991). The normal esophagus. *American Journal of Surgery*, **15**, 293–309.

Dodds, W.J., Logemann, J.A., and Stewart, E.T. (1990). Radiologic assessment of abnormal oral and pharyngeal phases of swallowing. *American Journal of Roentgenology*, **154**, 965–74.

Ekberg, O. and Feinberg, M.J. (1991). Altered swallowing function in elderly patients without dysphagia:radiologic findings in 56 cases. *American Journal of Roentgenology*, **156**, 1181–4.

Finucane, P., Aslan, S.M., and Duncan, D. (1991). Percutaneous endoscopic gastrostomy in elderly patients. *Postgraduate Medical Journal*, **67**, 371–3.

Hawkins, S.P., Rowlands, P.C. and Shorvon, P.J. (1991). Barium meals in the elderly—a quality reassurance. *British Journal of Radiology*, **64**, 113–15.

Hishikawa, Y. *et al.* (1991). Radiotherapy for carcinoma of the oesophagus in patients aged eighty or older. *International Journal of Radiation Oncology, Biology, and Physics*, **20**, 685–8.

Jones, R. and Lydeard, S. (1989). Prevalence of symptoms of dyspepsia in the community. *British Medical Journal*, **298**, 30–2.

Logemann, J.A. (1990). Effects of aging on the swallowing mechanism. *Otolaryngology Clinics of North America*, **23**, 1045–56.

McCord, G.S., Staiano, A., and Clouse, R.E. (1991). Achalasia, diffuse spasm and non-specific motor disorders. *Clinical Gastroenterology*, **5**, 307–35.

Petersen, H., *et al.* (1991). Relationship between endoscopic hiatus hernia and gastroesophageal reflux symptoms. *Scandinavian Journal of Gastroenterology*, **26**, 921–6.

Raiha, I., Hietanen, E., and Sourander, L. (1991). Symptoms of gastroesophageal reflux disease in elderly people. *Age and Ageing*, **20**, 365–70.

Richter, J.E., Bradley, L.A., and Castell, M.D. (1989). Esophageal chest pain: current controversies in pathogenesis, diagnosis, and therapy. *Annals of Internal Medicine*, **110**, 66–78.

Steadman, C., *et al.* (1990). Spontaneous intramural rupture of the eosophagus. *Gut*, **31**, 845–9.

Tytgat, G.N.J. (1989). Dilatation therapy of benign oesophageal stricture. *World Journal of Surgery*, **13**, 142–8.

Weinbeck, M., Berges, W., and Lubke, H.J. (1988). Drug induced oesophageal lesions. *Clinical Gastroenterology*, **2**, 263–74.

8.3 Non-steroidal anti-inflammatory drugs and the gastrointestinal tract

H. G. M. SHETTY AND K. W. WOODHOUSE

Non-steroidal anti-inflammatory drugs (NSAIDs) are widely used, particularly by elderly people. It is estimated that around 23 million prescriptions of NSAIDs were issued in the United Kingdom in 1989; of these 12 million were to patients over 60 years of age (Anonymous 1990; Hazelman 1989). It has been recognized for over a century that NSAIDs cause gastrointestinal adverse effects. Gastric irritation caused by sodium salicylate was recognized as early as 1876 and led to the introduction of acetylsalicylic acid in 1899 (Hazelman 1989). In 1938, a gastroscopic study showed gastric haemorrhage or hyperaemia in 13 of 16 patients taking acetylsalicylic acid (Douthwaite and Lintott 1938). Since this time a large number of studies have reported an association between the use of NSAIDs and upper and lower gastrointestinal adverse effects.

NON-STEROIDAL ANTI-INFLAMMATORY DRUGS AND THE UPPER GASTROINTESTINAL TRACT

Non-steroidal anti-inflammatory drugs have been reported to cause dyspepsia, oesophageal, gastric, and duodenal ulcers, upper gastrointestinal bleeding, and perforation. There has been considerable debate about the role of NSAIDs in the causation of upper gastrointestinal damage, mainly because of the inadequacies of design of earlier studies (Kurata *et al.* 1982). More recent studies and meta-analyses have attempted to give a clear insight into this problem. Several recent studies have shown that the use of NSAIDs is associated with an increased risk of gastric ulceration, haematemesis and melaena, perforation, and death related to ulcers (Levy *et al.* 1988; Clinch *et al.* 1987; Duggan *et al.* 1986; Somerville *et al.* 1986; Griffin *et al.* 1988; Collier and Pain 1985; Jick *et al.* 1987). The relative risk of developing serious adverse gastrointestinal events with NSAID use has been reported to be approximately three times that in non-users (Gabriel *et al.* 1991). One study has shown that an excess of 0.44 episodes of gastrointestinal bleeding over the expected rate occurs for every 10000 person-months of NSAID treatment (Carson *et al.* 1987). It has been suggested that about 4000 ulcer deaths per year in the United Kingdom (about 90 per cent of total ulcer deaths) may be associated with NSAID use (Cockel 1987).

A number of studies have shown that elderly people are at greater risk of developing serious adverse gastrointestinal events with NSAID use than are younger patients. Perforated peptic ulcers have been found to be more commonly associated with consumption of NSAIDs in patients over 65 years of age (Collier and Pain 1985). A study which compared the trends in the frequency of admission for perforated peptic ulcer (United Kingdom from 1958 to 1982), prescription rates for NSAIDs, and changes in national smoking habits, detected an increased susceptibility of elderly women to peptic ulceration. This was suspected to be due to an increase in the use of NSAIDs rather than altered smoking habits; between 1967 and 1982 the annual prescription of NSAIDs for women over 65 years of age had increased threefold, and in the same period duodenal ulcer perforation rates had doubled in patients between 65 and 74 years and more than tripled in women aged 75 years or more (Walt *et al.* 1986). Similar patterns have been reported from Holland and Germany (Hagedoorn 1984; Sonnenberg *et al.* 1984). A recent study which included patients over 65 years of age showed that current users of non-aspirin NSAIDs were four times more likely than non-users to develop gastric/duodenal ulcer disease or upper gastrointestinal haemorrhage (Griffin *et al.* 1991). In another study, 56 per cent of elderly patients who had bleeding peptic ulcers on endoscopy were taking NSAIDs (Clinch *et al.* 1987). It has been estimated that, in the United Kingdom, about 2000 cases of gastrointestinal bleeding and 200 deaths per year occur in association with NSAID therapy in patients over 60 years of age (Somerville *et al.* 1986). It has been shown (i) that the relative risk of developing upper gastrointestinal ulceration or haemorrhage increases with increasing dose of NSAIDs (2.8 with low dose versus 8 with high dose) and (ii) the first 30 days of therapy are the period of greatest risk (Griffin *et al.* 1991). Other studies have reported that the risk is greatest during the first 3 months of NSAID treatment (Gabriel 1991). Some studies have suggested that elderly people taking NSAIDs are more likely to develop complicated, often silent ulcers and 4.7 times more likely to die from ulcer disease compared with non-users (Armstrong and Blower 1987; Griffin *et al.* 1988). However, it is thought that age-associated decline in general health and the presence of concurrent serious illnesses probably contribute to the increased risk of death in elderly people rather than the NSAID-induced disease *per se* (Langman 1989).

Some studies have suggested that certain NSAIDs are more dangerous than others. It has been shown that ibuprofen is associated with a lower risk of ulcer disease and bleeding (Somerville *et al.* 1986; Griffin *et al.* 1991), and there is a suggestion of increased risk with sulindac (Carson *et al.* 1987), meclofenamate (Griffin *et al.* 1991), and piroxicam (Gabriel *et al.* 1991). However, there is not yet sufficient information to prove that any given NSAID is safer or more dangerous than others.

A previous history of gastrointestinal events and one of concomitant steroid therapy have been identified as additional risk factors for NSAID-induced gastrointestinal adverse effects (Gabriel *et al.* 1991). It may also be that patients with rheumatoid arthritis are more prone to NSAID-induced gastric ulceration (Atwater *et al.* 1965; Collins *et al.* 1987). This has not been confirmed by other studies and a similar incidence has been demonstrated in patients with osteoarthrosis (Malone *et al.* 1986). No sex difference has been observed for the risk of developing gastrointestinal adverse effects with NSAIDs (Gabriel *et al.* 1991).

Mechanism of NSAID-induced upper gastrointestinal damage

The pathogenesis of NSAID-induced upper gastrointestinal injury has been investigated extensively. Several mechanisms are involved (Schoen and Vender 1989). The most important appears to be breakdown of the mucosal barrier by NSAIDs, allowing mucosal damage by the gastric acid. At low gastric pH, weakly acidic NSAIDs remain unionized; being lipid, soluble, they diffuse readily into the mucosal cells. At higher intracellular pH they dissociate into water-soluble ions and thereby become trapped intracellularly. The entrapped ions are thought to alter the cell membrane permeability, but the mechanism by which they do so is unclear. There is evidence that aspirin in high concentrations inhibits sodium and hydrogen ion transport across the mucosal cell, either by enzyme inhibition or uncoupling of oxidative phosphorylation, which in turn alters membrane permeability. Intraluminal hydrogen ions are thought to 'back diffuse' into the mucosal cells because of altered permeability, producing cellular damage. Increased hydrogen ion permeability has been observed with aspirin in normal volunteers (Baskin *et al.* 1976) and with indomethacin and fenoprofen in animal studies (Lin *et al.* 1975). Indomethacin has been reported to cause sustained back diffusion of hydrogen ions and bleeding at both neutral and acidic pH in canine gastric mucosa (Chvasta and Cooke 1972).

The need for the presence of acid for the production of mucosal damage is supported by the observation that achlorhydric patients are less susceptible to aspirin-induced gastric damage compared with normal persons, and at gastric pH between 6 and 7, the aspirin-induced gastric damage is considerably reduced (Schoen and Vender 1989).

NSAIDs produce a number of other effects which may result in the breakdown of the mucosal barrier (Schoen and Vender 1989). Aspirin, indomethacin, and phenylbutazone inhibit mucus secretion. Aspirin enhances pepsin-mediated proteolysis of mucus, reduces the viscosity of mucus, and increases its permeability to hydrogen ion. In rats it causes ulcers which produce mucus containing a low percentage of high molecular weight glycoprotein. Aspirin, indomethacin, and fenclofenac inhibit active bicarbonate secretion in the gastric mucosa. Normal amounts of intraluminal bicarbonate appear to protect against aspirin-induced erosions.

Prostaglandins are thought to have a 'cytoprotective' effect on gastric mucosa. NSAIDs inhibit prostaglandin synthesis and it is possible that the mucosal damage is a consequence of this. Significant inhibition of prostaglandin synthesis, which correlated with significant mucosal damage, has been demonstrated with aspirin, indomethacin, sulindac, and diclofenac (Rainsford and Willis 1982), although this has not been confirmed by several other studies (Redfern *et al.* 1987). Aspirin and indomethacin reduce mucosal blood flow. This observation supports the view that inhibition of prostaglandin synthesis due to NSAIDs causes mucosal ischaemia which in turn results in mucosal damage (Schoen and Vender 1989). Aspirin has been reported to decrease the rate of regeneration and migration of gastric epithelium, and increase epithelial cell loss (Schoen and Vender 1989). Whether these changes occur as a consequence of prostaglandin deficiency is unclear.

It is now generally believed that direct contact of NSAIDs with mucosa, resulting in permeability changes in mucosal cells, is the important step in the pathogenesis of gastric damage, and inhibition of prostaglandin synthesis may have additional effects (Rainsford and Willis 1982).

DISEASE IN THE SMALL AND LARGE BOWEL INDUCED BY NSAIDs

NSAID-induced lower gastrointestinal disease is being increasingly recognized, and it is possible that lack of recognition of this problem in the past accounts for the small number of reports compared with those of upper gastrointestinal involvement.

In a study that included 268 patients admitted with small or large bowel perforations and haemorrhage, it was noted that the patients were twice as likely to have taken NSAIDs as the matched controls (Langman *et al.* 1985). The expected incidence of lower bowel perforations has been estimated to be 10 per 100000 and that for lower bowel haemorrhages as 7 per 100000 (Langman *et al.* 1985). Elderly people taking NSAIDs appear to be at increased risk of developing lower gastrointestinal problems. Although mefenamic acid has been reported to cause colitis in a large number of patients, there is no clear evidence that lower gastrointestinal adverse effects are more common with any given NSAID.

Patients with NSAID-induced small bowel involvement may present with fatigue and weight loss. Multiple diaphragm-like strictures with a small central hole have been described on radiological and histopathological examination in such patients (Aabakken and Osnes 1989). In some patients ulcerations and haemorrhages with or without cicatricial narrowing have been seen. Crohn's disease-like distal ileitis has also been reported with NSAID therapy (Bjarnason *et al.* 1987). In low birth weight infants, indomethacin may produce haemorrhagic necrotizing enterocolitis with single or multiple isolated ulcerations in the distal ileum (Aabakken and Osnes 1989).

Mefenamic acid, flufenamic acid, naproxen, and phenylbutazone have been related to the development of ulcerative colitis in previously healthy individuals (Aabakken and Osnes 1989). Similarly, mefenamic acid and diclofenac may be related to the development of a Crohn's-disease-like syndrome in previously healthy subjects (Aabakken and Osnes 1989). Exacerbations of both ulcerative and Crohn's colitis have been described with a number of NSAIDs. In addition, naproxen and mefenamic acid may produce a non-specific colitis (Aabakken and Osnes 1989).

Perforation of the bowel may occur with a number of NSAIDs. Slow-release preparations and diverticulosis may increase the risk. Gastrocolic fistula formation has also been described (Aabakken and Osnes 1989).

Patients with colonic ulcers may present with haematochezia and elderly patients appear to be particularly at risk for this problem. Proctitis, rectovaginal fistula, and rectal bleeding have all been associated with the use of NSAID suppositories (Aabakken and Osnes 1989).

Pathogenesis of lower gastrointestinal damage

The exact mechanism of NSAID-induced lower gastrointestinal tract injury is unclear (Aabakken and Osnes 1989). It is thought to be due to loss of cytoprotective effect of prostaglandins since NSAIDs inhibit their synthesis. However, their cytoprotective effect in the lower gastrointestinal tract has not been clearly understood: they do stimulate mucus secretion and enhance mucosal blood flow in the intestine. Misoprostol, a prostaglandin analogue, has been shown to decrease the altered intestinal permeability associated with NSAIDs in man (Bjarnason *et al.* 1988).

PREVENTION AND TREATMENT OF NSAID-INDUCED GASTROINTESTINAL DAMAGE

As NSAID-induced gastrointestinal adverse effects can be potentially fatal, their use should be restricted to patients with clear indications for therapy. They should be avoided in elderly patients if possible. Elderly patients with osteoarthrosis should be treated with simple analgesics such as paracetamol (acetaminophen), and NSAIDs only considered for those who do not respond. Concurrent steroid therapy should be avoided if possible and other risk factors for peptic ulcer, such as smoking, should be avoided. NSAID therapy should be withdrawn if possible in patients with upper gastrointestinal ulceration or haemorrhage.

Drug therapy for prophylaxis and treatment of gastroduodenal injury due to NSAIDs is aimed at reducing gastric acid secretion and enhancing the 'cytoprotective' mechanisms.

Misoprostol, an ester of prostaglandin E_1, reduces gastric acid secretion, increases gastric mucus production, mucosal blood flow, and duodenal bicarbonate secretion (Dajani 1987*a*,*b*). It has been shown to reduce acute gastric injury due to aspirin, naproxen, ibuprofen, and tolmetin in the short term (Roth *et al.* 1989). Two double-blind, placebo-controlled studies have shown that misoprostol reduces the incidence of gastric ulcers and is helpful in healing gastroduodenal mucosal lesions due to NSAIDs. The first study included 420 patients who were on NSAIDs for osteoarthrosis. In patients receiving misoprostol 100 μg or 200 μg four times daily, the incidence of gastric ulcers over 3 months was 5.6 and 1.45 per cent respectively, compared with 21.7 per cent in the placebo group (Graham *et al.* 1988). In the second study, 239 patients with rheumatoid arthritis and NSAID-included gastroduodenal injury were treated with misoprostol or placebo for 8 weeks. Sixty-seven per cent of gastric ulcers healed in the misoprostol group compared with 26 per cent in the placebo group in spite of continued use of high doses of aspirin (Roth *et al.* 1989). The number of patients with duodenal ulcers was too small to provide information about the 'protective' effect of misoprostol in these studies, but the second study showed a trend in favour of misoprostol (Roth *et al.* 1989). Diarrhoea is a common adverse effect of misoprostol and appears to be dose-related. When used in doses of 200 μg four times daily it caused diarrhoea in 39 per cent of patients and necessitated withdrawal in 6 per cent in one study (Graham *et al.* 1988). The incidence of diarrhoea may be reduced by administering misoprostol after meals in doses of 200 μg or less. Postmenopausal bleeding, intermenstrual bleeding, and menorrhagia have also been reported with misoprostol.

H_2-antagonists have been used for both prophylaxis and treatment of NSAID-induced gastroduodenal injury. Although cimetidine has been shown to prevent gastric injury by aspirin in normal volunteers in the short term, there is no evidence that H_2-blockers protect against NSAID-induced gastric injury in the long term (McCarthy 1989; Hawkey 1990). However, they do appear to protect against NSAID-induced duodenal injury. A randomized, double-blind study in 263 patients with arthritis showed that ranitidine in doses of 150 mg twice daily protected against duodenal injury, but not gastric ulcerations. The incidence of duodenal ulcer was 1.5 per cent in the ranitidine group compared with 8 per cent in the placebo group at the 4th or 8th week, but the prevalence of gastric ulceration was 6 per cent in both groups at week 8 (Ehsanulla *et al.* 1988).

H_2-antagonists do heal NSAID-induced gastric and duodenal ulcers in 12 weeks or less. Duodenal ulcers appear to heal more easily with H_2-blockers, than do gastric ulcers. Continuation of NSAID therapy may not affect the healing of duodenal ulcers or small gastric ulcers, but may delay the healing and expose the patients to the risk of complications (McCarthy 1989).

At present there is insufficient information about treatment strategies for NSAID-induced lower gastrointestinal damage. The role of misoprostol is unclear. It is logical, as with upper gastrointestinal damage, to withdraw NSAID therapy if at all possible.

NSAIDs are used by a very large number of patients and universal prophylactic therapy with misoprostol or H_2-blockers, or both, is impracticable because of the enormous expense involved. The best way to reduce NSAID-induced gastrointestinal adverse effects is by restricting their use to patients with clear therapeutic indications and by avoiding them if possible in those with known risk factors. Prophylactic use of misoprostol is probably justified in the latter group of patients if they require NSAID therapy.

References

Aabakken, L. and Osnes, M. (1989). Non-steroidal anti-inflammatory drug-induced disease in the distal ileum and large bowel. *Scandinavian Journal of Gastroenterology*, **24** (Suppl. 163), 48.

Anonymous. (1990). Misoprostol for co-prescription with NSAIDs. *Drugs and Therapeutics Bulletin*, **28**, 25.

Armstrong, C.P. and Blower, A.L. (1987). Non-steroidal anti-inflammatory drugs and life threatening complications of peptic ulcerations. *Gut*, **28**, 527.

Atwater, E.C., Mongan, E.S., Wieche, D.R., and Jacox, R.F. (1965). Peptic ulcer and rheumatoid arthritis. A prospective study. *Archives of Internal Medicine*, **115**, 184.

Baskin, W., Ivey, K.J., Krause, W.J., Jeffrey, G.E., and Gemmell, R.T. (1976). Aspirin-induced ultrastructural changes in human gastric mucosa: correlation with potential difference. *Annals of Internal Medicine*, **85**, 299.

Bjarnason, I., *et al.* (1987). Non-steroidal anti-inflammatory drug-induced intestinal inflammation in humans. *Gastroenterology*, **93**, 480.

Bjarnason, I., Smethurst, P., Fenn, G.C., Lee, C.E., and Levi, A.J. (1988). Misoprostol reduced indomethacin-induced changes in human small intestinal permeability. *Gastroenterology International Supplement*, **1**, A884.

Carson, J.L., Strom, B.L., Soper, K.A., West, S. L., and Morse, M.L. (1987). The association of non-steroidal anti-inflammatory drugs with upper gastrointestinal tract bleeding. *Archives of Internal Medicine*, **147**, 85.

Chvasta, T.A. and Cooke, A.R. (1972). The effect of several ulcerogenic drugs on the canine gastric mucosal barrier. *Journal of Laboratory and Clinical Medicine*, **79**, 302.

Clinch, D., Banerjee, A.K., Levy, D.W., Ostick, G., and Faragher, E.B. (1987). Non-steroidal anti-inflammatory drugs and peptic ulceration. *Journal of the Royal College of Physicians of London*, **21**, 183.

Cockel, R. (1987). NSAIDs—Should every prescription carry a government health warning? *Gut*, **28**, 515.

Collier, D.S. and Pain, J.A. (1985). Non-steroidal anti-inflammatory drugs and peptic ulcer performation. *Gut*, **26**, 359.

Collins, A.J., Davies, J.St.J., and Dixon, A. (1986). Contrasting presentation and findings between patients with rheumatic complaints taking non-steroidal anti-inflammatory drugs and a general population referred for endoscopy. *British Journal of Rheumatology*, **25**, 50.

Dajani, E.Z. (1987*a*). Perspectives on gastric anti-secretory effects of misoprostol in man. *Prostaglandins*, **33** (Suppl.), 68.

Dajani, E.Z. (1987*b*). Mucosal protective activities of misoprostol in humans: an overview. In *Gastro-intestinal cytoprotection by prostaglandins: focus on misoprostol*. (ed. G. Bianchi Porro) pp. 21, 31. Cortina Int: Verona.

Douthwaite, A.H. and Lintott, G.A.M. (1938), Gastroscopic observation of the effect of aspirin and certain other substances on the stomach. *Lancet*, **ii**, 1222.

Duggan, J.M., Dobson, A.J., Johnson, H., and Fahey, P. (1986). Peptic ulcer and non-steroidal anti-inflammatory agents. *Gut*, **27**, 929.

Ehsanullah, R.S.B., Page, M.C., Tildesley, G., and Wood, J.R. (1988). Prevention of gastroduodenal damage induced by non-steroidal anti-inflammatory drugs: controlled trial of ranitidine. *British Medical Journal*, **297**, 1017.

Gabriel, S.E., Jaakkimainen, L., and Bombardier, C. (1991). Risk for serious gastrointestinal complications related to use of non-steroidal anti-inflammatory drugs. A meta-analysis. *Annals of Internal Medicine*, **115**, 787.

Graham, D.Y., Agrawal, N.M., and Roth, S.H. (1988). Prevention of NSAID-induced gastric ulcer with misoprostol: a multicentre, double-blind, placebo controlled trial. *Lancet*, **ii**, 1277.

Griffin, M.R., Ray, W.A., and Schaffner, W. (1988). Non-steroidal anti-inflammatory drug use and death from peptic ulcer disease in elderly persons. *Annals of Internal Medicine*, **109**, 359.

Griffin, M.R., Piper, J.M., Daugherty, J.R., Snowden, M., and Ray, W.A. (1991). Non-steroidal anti-inflammatory drug use and increased risk for peptic ulcer disease in elderly persons. *Annals of Internal Medicine*, **114**, 257.

Hagedoorn, D. (1984). Opmekelijke verschuivingen inhet epidemiologische patroon van het ulcus pepticum. *Nederlands Tijdschrift voor Geneeskunde*, **128**, 484.

Hawkey, C.J. (1990). Non-steroidal anti-inflammatory drugs and peptic ulcers. Facts and figures multiply, but do they add up? *British Medical Journal*, **300**, 278.

Hazelman, B.L. (1989) Incidence of gastropathy in destructive arthropathies. *Scandinavian Journal of Rheumatology*, **Suppl. 78**, 1.

Jick, S.S., Pevera, D.R., Walker, A.M., and Jick, H. (1987). Non-steroidal anti-inflammatory drugs and hospital admission for perforated peptic ulcer. *Lancet*, **ii**, 380.

Kurata, J.H., Elashoff, J.D., and Grossman, M.I. (1982). Inadequacy of the literature on the relationship between drugs, ulcers, and gastrointestinal bleeding. *Gastroenterology*, **82**, 373.

Langman, M.J.S. (1989). Epidemiologic evidence on the association between peptic ulceration and anti-inflammatory drug use. *Gastroenterology*, **96**, 640.

Langman, M.J.S., Morgan, L., and Worral, A. (1985). Use of anti-inflammatory drugs by patients admitted with small or large bowel perforations and haemorrhage. *British Medical Journal*, **290**, 347.

Levy, M., *et al.* (1988). Major upper gastrointestinal bleeding. Relation to the use of aspirin and other non-narcotic analgesics. *Archives of Internal Medicine*, **148**, 281.

Lin, T.M., Warrick, M.W., Evans, D.C., and Nash, J.F. (1975). Action of the anti-inflammatory agents, acetylsalycilic acid, indomethacin and fenoprofen on gastric mucosa of drugs. *Research Communications in Chemical Pathology and Pharmacology*, **11**, 1.

Malone, D.E., *et al.* (1986). Peptic ulcer in rheumatoid arthritis: intrinsic or related to drug therapy? *British Journal of Rheumatology*, **25**, 342.

McCarthy, D. (1989). Non-steroidal anti-inflammatory drug-induced ulcers: management by traditional therapies. *Gastroenterology*, **96**, 662.

Rainsford, K.D. and Willis, C. (1982). Relationship of gastric mucosal damage induced in pigs by anti-inflammatory drugs to their effects on prostaglandin production. *Digestive Diseases and Sciences*, **27**, 624.

Redfern, J.S., Lee, E., and Feldman, M. (1987). Effect of indomethacin on gastric mucosal prostaglandins in humans: correlation with mucosal damage. *Gastroenterology*, **92**, 969.

Roth, S. *et al.* (1989). Misoprostol heals gastroduodenal injury in patients with rheumatoid arthritis receiving aspirin. *Archives of Internal Medicine*, **149**, 775.

Schoen, R.T. and Vender, R.J. (1989). Mechanisms of non-steroidal anti-inflammatory drug-induced gastric damage. *American Journal of Medicine*, **86**, 449.

Somerville, K., Faulkner, G., and Langman, M. (1986). Non-steroidal anti-inflammatory drugs and bleeding peptic ulcer. *Lancet*, **i**, 462.

Sonnenberg, A., Muller, H., and Pace, F. (1984). Cohort analysis of peptic ulcer mortality in Europe. *Journal of Chronic Diseases*, **88**, 309.

Walt, R., Katschinski, B., Logan, R., Ashley, J., and Langman, M. (1986). Rising frequency of ulcer perforation in elderly people in the United Kingdom. *Lancet*, **i**, 489.

8.4 The stomach

O. F. W. JAMES

UPPER GASTROINTESTINAL HAEMORRHAGE

Acute upper gastrointestinal bleeding manifests itself either as haematemesis, melaena, or occasionally in elderly patients as rapid onset of breathlessness and symptoms of anaemia without definite evidence of a discrete haemorrhagic event. Meta-analysis of 44 different studies has confirmed that rising age leads to rising fatality from upper gastrointestinal bleeding (Morgan and Klamp 1988), and this risk may be as high as 17.5 per cent in individuals aged over 80 (Cooper *et al.* 1988). Death rates are higher among individuals who require surgery; more elderly patients require surgery and in a substantial proportion death is not directly related to gastrointestinal bleeding. In Britain the proportion of patients admitted to hospital with haematemesis and melaena who are over age 60 has recently been shown in one study to be 69.5 per cent (Holman *et al.* 1990). In this series of 430 patients from a District General Hospital 20 per cent were aged over 80. Fatality in this series was 0.8 per cent among individuals under age 60 rising to 9.5 per cent in those over 80. Overall mortality at 3.7 per cent was among the lowest reported by any group. Very similar figures in relation to age and mortality may be obtained worldwide (Branicki *et al.* 1990).

Important initial prognostic indicators appear to be the presence or absence of shock on admission, stigmata of recent or current bleeding (spurting vessel or active bleeding at the time of endoscopy, visible vessel or adherent clot in an ulcer base). The site of haemorrhage—lower oesophagus, stomach, or duodenum—and the cause of haemorrhage (apart from oesophageal varices) are probably not helpful in assessing prognosis.

NSAIDs and bleeding

The increasing trend towards upper gastrointestinal haemorrhage in an older group has been shown in elegant epidemiological studies to parallel the rise of use of non-steroidal anti-inflammatory drugs (NSAIDs) in several Western countries (Somerville *et al.* 1986; Bartle *et al.* 1986); to a lesser extent the same may be true of aspirin (Faulkner *et al.* 1988). In a Belgium case-controlled study the odds ratio of patients with upper gastrointestinal bleeding taking NSAIDs compared to case controls with no bleeding was 7.4, and there was also a significant difference in the use of aspirin (odds ratio 2.2). In this as in a number of other studies there was no significant difference in the use of other drugs—notably paracetamol and corticosteroids (Holvoet *et al.* 1991). Patients using NSAIDs are older and have higher fatality; use of NSAIDs is seen in patients bleeding from oesophageal lesions, gastric and duodenal erosions, and ulcers. Several studies have suggested that the risk of upper gastrointestinal haemorrhage attributable to the use of NSAIDs is as high as 30 per cent and use of aspirin 10 to 15 per cent. This epidemiological evidence of association between bleeding and use of anti-inflammatory drugs is part of a wider picture which connects peptic ulceration in general with anti-inflammatory drugs (see also Chapter 8.3).

Management of upper gastrointestinal bleeding

It is emphasized that modern management of upper gastrointestinal bleeding must be carried out with the closest possible collaboration between physicians and surgeons in an institution where high quality emergency endoscopy is readily available. The management of such patients consists of resuscitation, assessment of the site and severity of haemorrhage, and treatment. In patients with active bleeding and shock there is an urgent need for restoration of circulating blood volume and blood transfusion. Care must be taken in elderly individuals, who often have concomitant cardiovascular or pulmonary disease, that intravenous fluid replacement and transfusion is not overenthusiastic. In severely shocked patients the use of central venous pressure monitoring is recommended. Many elderly patients with upper gastrointestinal haemorrhage present with such symptoms as tiredness and breathlessness rather than with overt haemorrhage. In these as with all other patients in whom upper gastrointestinal haemorrhage is suspected it is vital to draw blood for measurement not only of haemoglobin and blood film indices but also for measurement of serum vitamin B_{12}, folate, and red cell folate, since once blood transfusion has been given accurate assessment of vitamin B_{12} and folate status becomes impossible for a period. Many elderly patients with severe anaemia in whom upper gastrointestinal bleeding is suspected turn out to have deficiencies of iron and of vitamin B_{12} or folate.

Unless extremely active bleeding continues it is best to resuscitate patients before endoscopy takes place. A planned endoscopy with consultation between physicians and surgeons and a consideration of various treatment options, usually carried out within 12 h of admission but not with undue haste except in the rare instances of continuing very severe haemorrhage or sudden severe rebleed, should be the aim of good management.

Drug treatment

Unfortunately no really convincing and reproducible evidence has been forthcoming to suggest that any drug treatment alone can successfully stop bleeding in the majority of patients regardless of site or cause (Walt 1990). In recent years well-planned controlled trials of treatment of acute upper gastrointestinal bleeding with the somatostatin analogue octreotide, the antifibrinolytic agent tranexamic acid, the H_2-antagonists cimetidine and ranitidine, and sucralfate have all shown no significant benefit although in each case small studies of selected patients or uncontrolled pilot trials have suggested possible efficacy.

Endoscopy therapy

Laser treatment

Laser photocoagulation using either an argon laser or the more expensive, more powerful Nd:YAG laser has been successfully used for endoscopic treatment of upper gastrointestinal bleeding for over 10 years. Meta-analysis of controlled trials of laser photocoagulation for bleeding ulcers versus no endoscopic treatment has shown significant results in favour of laser treatment for all endpoints at high levels of statistical significance (rebleeding, surgery, and death) (Swaine 1991). Unfortunately lasers, in particular the Nd: YAG laser, are extremely expensive and results of treatment while demonstrating benefit in high-risk patients do not by any means obviate all need for emergency surgery. It is suggested that laser treatment probably halves the need for emergency surgery with its attendant risks in these patients.

Thermal probes

These may be either monopolar or bipolar: recently, concerns about tissue damage caused by monopolar electrocoagulation have led to the more widespread use of bipolar heat probe units. Meta-analysis of well-conducted trials of thermal probe treatment for upper gastrointestinal bleeding suggests that a significant reduction in rebleeding and requirement for urgent surgery has been demonstrated but as yet no reduction in fatality.

Injection

The apparent efficacy of injection sclerotherapy for bleeding oesophageal varices led to the use of local injections around bleeding sites with adrenaline 1:10000 (epinephrine) or adrenaline plus sclerosant. Such treatment has the attraction of low cost. Several studies have now suggested a marked reduction in the need for emergency surgery following endoscopic injection treatment (which may be repeated after 24 h) from, for example, 41 per cent in a control group of high-risk patients to 15 per cent in those treated with adrenaline injections. Again no clear benefit in terms of fatality has yet been demonstrated.

These new endoscopic methods for treatment of upper gastrointestinal haemorrhage all have their proponents. In general all seem reasonably safe with very low incidence of severe complications. In experienced hands all three methods seem to reduce risk of rebleeding and the need for surgery. The likelihood is that all will be shown to reduce fatality. No studies addressed specifically to elderly high-risk patients have been carried out using any of these methods. It seems likely that because of cost and wide availability endoscopic injection of bleeding sites may become widely adopted over the next few years. Success will depend upon very careful monitoring of patients following treatment to make sure that no occult rebleeding is occurring.

All of these endoscopic methods for stopping bleeding are essentially buying time for treatment of underlying lesions. Thus medical treatment for peptic ulcer, gastritis, erosions, or other bleeding lesions should be started directly after the endoscopic treatment.

Surgery

Many surgeons still maintain that for actively bleeding ulcers or patients at high risk of rebleed, particularly elderly patients, early surgery is the best approach. In a typical series of well-managed patients this would mean that about 20 per cent of patients require surgery (Holman *et al.* 1990). Overall surgical fatality in a good centre is 10 to 15 per cent in patients over 60 but under 80 and 25 to 30 per cent in patients over age 80. These figures emphasize that only the sickest patients require surgery and that non-surgical methods for treating severe bleeding such as those outlined above may improve prognosis for this very high-risk group. It is the author's belief that an 80-year-old patient with acute upper gastrointestinal haemorrhage should choose a unit for hospital admission with wide experience of the management of large numbers of patients with gastrointestinal bleeding regardless of age, excellent endoscopic facilities, and close collaboration between physicians and surgeons.

Causes of upper gastrointestinal bleeding

Duodenal and gastric ulcers account for about half of non-variceal upper gastrointestinal bleeding episodes. Gastric erosions form a further quarter. Of the remaining quarter Mallory-Weiss tear, oesophagitis, and erosive duodenitis are the next most common causes followed by neoplasms (about 3 to 4 per cent of cases). Vascular ectasia or angiodysplasia is more commonly found in advanced age and may account for over 5 per cent of patients with upper gastrointestinal bleeding over age 65. It may be associated with aortic valve disease or hereditary haemorrhage telangiectasia. In most endoscopic series 10 to 15 per cent of patients have no diagnosis made and in a further 20 to 25 per cent there is more than one possible site for bleeding identified.

Recurrent severe upper gastrointestinal bleeding is comparatively rare and endoscopy after transfusion usually reveals the cause. The most difficult diagnoses to confirm are those associated with vascular abnormalities and those in patients with recurrent anaemia rather than overt sudden blood loss. Apart from the lesions listed above gastric erosions within large diaphragmatic hernias (Cameron erosions) may be difficult to diagnose. In elderly patients with obscure (i.e. routine upper gastrointestinal endoscopy and colonoscopy/barium enema negative), gastrointestinal bleeding the site is more commonly below the duodenum. In patients with marked gastrointestinal bleeding of obscure cause the sequence of investigations should probably be urgent endoscopy, colonoscopy, and abdominal angiography. The majority of those in whom an identifiable source of bleeding is not detected despite these investigations turn out to have vascular ectasia, often in the small bowel or colon. Such lesions increase in frequency with advancing age (Elta 1991).

PEPTIC ULCER

While peptic ulcer disease in general, and gastric ulcer in particular, have been declining, the prevalence, complications, and mortality among elderly people have been increasing. In the United States 80 per cent of the 10000 peptic ulcer-related deaths per year occur in people aged over 65 years and it has been estimated that current annual direct costs for patients with ulcer disease are 3 to 4 billion dollars (Gilinsky 1988; Isenberg *et al.* 1991). 'Geriatric' gastric ulcer now represents a major proportion of patients admitted to hospital with peptic ulcer disease.

Epidemiology

Since precise diagnosis of peptic ulcer disease is impossible without endoscopy or careful radiology the incidence and prevalence of peptic ulcer is hard to assess. However, complications such as perforation and haemorrhage may be studied more easily. It is thought that the annual prevalence of active gastric and duodenal ulcers in the United States is about 1.8 per cent of the population and while duodenal ulcer is more common in younger individuals there is an equal number of patients over 65 with duodenal and gastric ulcer being admitted to hospital both in England and the United States. In particular there has been a disproportionate increase in gastric ulcer patients admitted to hospital and mortality from gastric ulcer among women over 65 since about 1970.

It is thought that duodenal ulcer disease was rare until the beginning of the 20th century. Its incidence increased progressively until about 1950 in Westernized countries and has subsequently been declining. It has been suggested that there was a cohort of individuals born between roughly 1870 and 1910 predisposed to develop duodenal ulcer. This finding, confirmed in Britain, the United States, Europe, and Japan, suggests that one or more factors occurring during this time, possibly introduced in the early years of life was responsible for the rising incidence of the disease. Such factors as increasing industrialization, smoking, and dietary changes may be implicated. Although the incidence of ulcer disease and mortality rates for both duodenal and gastric ulcer decreased from 1970 there has been a substantial increase in hospital admissions of elderly patients, particularly women, with either ulcer haemorrhage or perforation during this period, and mortality has risen significantly in patients over age 75. This increase has been attributed to the steeply rising intake of NSAIDs in the growing population of very elderly people (Walt *et al.* 1986).

Aetiology

The history of our understanding of the aetiology and pathogenesis of gastritis, duodenitis, and peptic ulcer passes from the stage (roughly until 1970) when acid was thought to be the main culprit through the more recent ideas that the pathogenesis of these diseases occurred as a result of a balance between 'aggressive' factors (gastric acid and pepsin) with 'defensive' mucosal integrity. Most recently our ideas have been further modified and confused by the advent of *Helicobacter pylori* and the evidence that NSAIDs were a potent and important cause of gastric ulceration particularly in elderly people.

Gastritis and ulcer

Whether gastric ulcers are associated with concomitant use of NSAIDs, and regardless of the relationship between ulcers and *Helicobacter pylori*, most gastric ulcers are surrounded with an area of gastritis; this is usually widespread and may be superficial or atrophic. Most ulcers occur fairly close to the antral side of the junction between antral and oxyntic gland mucosae along the lesser curve of the stomach. With increasing age and possibly as a result of long-standing gastritis this junction between antral and oxyntic mucosa progresses higher and higher up the lesser curve so that the majority of non-NSAID associated gastric ulcers in elderly patients are found high on the lesser curve. In patients with gastric ulcer in whom there is no evidence of *H. pylori* infection it is possible that duodenogastric reflux of bile acids, lysolecithin, and pancreatic peptidases may damage the gastric mucosal barrier. The gastric mucosa, unlike other intestinal mucosae, is resistant to the diffusion of hydrogen ions from the gastric lumen and is protected by gastric mucus. It is possible that damage to this mucosal barrier—whether by NSAIDs, *H. pylori*, or other agents, such as ethanol or the contents of duodenogastric reflux, may predispose to ulcer formation. But, whereas it is well established that luminal agents (such as NSAIDs) and probably *H. pylori* are causally related to gastritis a definite causal relationship between *H. pylori* and a breach in the gastric mucosal barrier has not definitely been established. Conceivably in elderly patients impaired local gastric mucosal blood flow may also play a significant part in contributing towards ulceration.

Helicobacter pylori and the aetiology of gastritis and gastric ulcer

There have been descriptions of spirochaetes in the gastric mucosa for many years but it was not until the early 1980s that a firm relationship began to be established between the presence of a spirochaete, initially named *Campylobacter pylori*, and gastritis. The prevalence of gastritis in the normal population was known to increase with age so that up to 75 per cent of 'normal' people aged over 65 years showed some histological or endoscopic evidence. Recently direct and indirect studies have shown a close relationship between this increasing prevalence of gastritis and a parallel increase in infection with *H. pylori*. Over age 50 the prevalence of gastritis and serological evidence for infection with *H. pylori* levels off at about 50 per cent, the majority of infections being confirmed by positive histology.

Helicobacter pylori organisms usually lie in the gastric mucus layer just outside the epithelial cell surface; they rarely penetrate the intercellular junctions and occur infrequently within the cells themselves. They are virtually confined to 'gastric type' mucosa. They appear to be widely distributed in the antrum, body, and fundus of the stomach in most infected individuals but it is in the antrum where *H. pylori* is so closely associated with inflammation. Interestingly *H. pylori* appears to be less commonly associated with atrophic gastritis than with superficial gastritis and is seldom seen in the severe atrophic gastritis associated with pernicious anaemia. This has led to the hypothesis that the organisms need an acid milieu in order to flourish and may explain why, with advanced age, there is no increase in apparent prevalence of infection since the proportion of individuals with severe atrophic gastritis or pernicious anaemia and consequent hypochlorhydria or achlorhydria increases. Prolonged antral gastritis ultimately leads to metaplasia from oxyntic to antral type mucosa, hence the 'march' of gastric ulcers up the lesser curvature with advancing age.

NSAIDs, gastritis and gastric ulcer

Aspirin and the newer NSAIDs have been said by Morton Grossman to deliver a 'one-two' punch to the stomach—a direct topical damaging effect and a later systemic effect. Thus acute ingestion of aspirin or NSAIDs results in endoscopic evidence of superficial erosions, haemorrhage, and occasionally ulceration in almost all individuals. This occurs predominately in the antrum. The effect of chronic aspirin or NSAID ingestion is probably due to inhibition of the synthesis of prostaglandins (by inhibition of the key enzyme cyclo-oxygenase). Prostaglandins E and A stimulate gastric mucus and bicarbonate secretion and mucosal blood flow; their inhibition has the opposite effect.

While many patients with gastric ulcer, particularly those with gastritis, have reduced stimulated gastric acid secretion, almost all have some potential acid output. Unlike duodenal ulcer, gastric ulcer can develop with very small amounts of acid/pepsin suggesting that impairment of gastric mucosal defensive factors leading to increased sensitivity to luminal acid is the important feature in the aetiology. This might be seen as a 'final common pathway' of the gastritis and consequent mucosal barrier damage whether idiopathic, associated with *H. pylori*, or with use of NSAIDs.

Duodenal ulcer/duodenitis

Duodenitis is usually observed adjacent to duodenal ulcers but some patients have more widespread duodenal inflammation with or without the presence of ulcer. This more widespread duodenitis is usually associated with antral gastritis and such patients often have gastric ulcer either in the body of the stomach or in the immediate prepyloric area. Presumably the same underlying cause exists for these duodenal ulcer/duodenitis-associated gastric ulcers as for the duodenal inflammation/ulceration itself. In duodenitis the lamina propria is infiltrated with plasma cells, lymphocytes, and occasional eosinophils. Neutrophils invade the superficial epithelium. The villous epithelium may be atrophic and usually contains gastric metaplasia.

It is likely that duodenal ulcer disease, like gastric ulcer, may be the end result of a number of different insults and no completely unifying hypothesis as to its causation and pathophysiology may be possible. Nonetheless a number of specific abnormalities of physiology and predisposing causes can be identified which are important in understanding duodenal ulcer disease.

As a group patients with duodenal ulcer secrete more acid and pepsin both at rest and after stimulation than normal individuals. However, there is an enormous overlap between the two groups. In understanding duodenal ulcer disease it must be recognized that unlike most gastric ulcers many duodenal ulcers come and go spontaneously over a period of months in any given individual. In considering the aetiology we must examine a number of possible abnormalities seen in duodenal ulcer disease.

Gastric acid is produced in the parietal cells and patients with duodenal ulcer, as a group, have a greater number of parietal cells than normal subjects. Unfortunately there is a very large overlap between duodenal ulcer parietal cell mass and number and controls. Since parietal cell mass correlates with maximal stimulated gastric acid output patients with duodenal ulcer, as a group, do have higher maximal stimulated gastric acid but again there is a large overlap. Nocturnal acid secretion is also increased as is basal acid output in many patients with duodenal ulcer and in patients with duodenal ulcer as a whole compared with controls.

In addition to increased acid output a proportion (perhaps 40 per cent) of patients with duodenal ulcer have increased sensitivity to infused gastrin and increased basal and postprandial gastrin levels. Furthermore duodenal defence mechanisms are defective in patients with duodenal ulcer, as a group. In active duodenal ulcer disease bicarbonate production by the duodenal bulb is markedly reduced compared with normal individuals, both at rest and in response to acidification. The combination of increased gastric acid output and decreased duodenal bulb bicarbonate production would suggest that duodenal bulb pH would be lower. Really sensitive measurement is hard to achieve, but the likelihood is that in general this is true. The second 'defence mechanism' which is probably defective in patients with duodenal ulcer is that mucus gel is perhaps weaker and more easily permeated by pepsin in patients with duodenal ulcer.

As is the case with gastritis/gastric ulcer, relationships between infection with *H. pylori* on the one hand and NSAIDs on the other and duodenal ulcer appear to exist but exactly how the ulcers are produced and whether their production is mediated through the mechanisms of damage outlined above is unclear.

As far as elderly people are concerned one may say that such mucosa-damaging agents as smoking and high alcohol intake probably act by impairment of the duodenal mucosal barrier. The use of NSAIDs more commonly results in gastric than duodenal ulcers in elderly people, possibly because of the increasing proportion of the population with age who have impaired maximal gastric acid output. This may well account for the fact that duodenal ulcers are more common before age 60.

Although *H. pylori* infection, as identified by serology and histology, may be demonstrated in 80 per cent or more of patients with duodenal ulcer there is still no proof that *H. pylori* is causally related to duodenal ulcer. *H. pylori* appears to produce a factor which is cytopathic although its identity is still not clear. Interestingly, it has recently been suggested that infection with *H. pylori* is associated with elevated levels of gastrin (Levi *et al.* 1989). It is possible therefore to suggest that in young adult life individuals become infected; in some, perhaps in association with other factors (genetic or 'toxic' i.e. smoking or alcohol), antral gastritis with hypergastrinaemia and excess gastric acid output occurs. In early years this leads to duodenal ulcer; later, if gastritis persists and becomes atrophic, parietal cells are destroyed and gastric acid output decreases. Thus, in old age, gastric ulcer advancing up the stomach with the decrease in oxyntoxic cells is the more common pathology. At present much of the above is highly speculative.

Clinical features

The symptoms of peptic ulcer in elderly people are, as with so many other diseases, less specific than in younger individuals. Abdominal discomfort is less well localized, retrosternal pain associated with peptic ulcer rather than with oesophagitis or a cardiac cause is more common and loss of appetite with weight loss mimicking gastric carcinoma also occurs. A frequent presentation is with iron deficiency anaemia (Gilinsky 1988). Because of the absence of well-defined pain a higher proportion of elderly individuals with both gastric and duodenal ulcer present for the first time with complications, such as bleeding, perforation, or pyloric stenosis.

Giant ulcers

Giant ulcers of the stomach (⩾3 cm) or duodenum (⩾2 cm) are much more common in old age. These ulcers are usually chronic and frequently present with complications. Interestingly, giant gastric ulcers, despite an alarming appearance both at endoscopy or on radiology, are usually benign. They occasionally present with gastrocolic fistula.

Complications

Haemorrhage from peptic ulcer has been considered above.

The initial signs and symptoms of perforation in both duodenal and gastric ulcer are often absent in elderly patients and there is frequently delay both on the part of the patient in seeking medical advice and on the part of doctors in making the diagnosis. Up to one-half of perforated ulcers in elderly patients are probably acute, thus giving no previous history to help in the diagnosis.

Gastric outlet obstruction has dramatically decreased in frequency in the 15 years since the introduction of antisecretory drugs. Nonetheless, occasional patients still present with long-standing symptomatic disease, weight loss, and metabolic disturbances, together with vomiting and abdominal pain.

Treatment

Acute treatment for peptic ulcer will first be considered, followed by the question of long-term treatment strategies. While symptomatic treatment with antacids is still widely used and while, particularly in the United States, some physicians have returned to antacids in the treatment of peptic ulcer, the modern mainstays of treatment are histamine H_2-receptor antagonists, omeprazole, anti-*H. pylori* treatment involving tripotassium dicitratobismuthate, and 'cytoprotective' agents—notably sucralfate and prostaglandins.

H_2-receptor antagonists

The H_2-receptor antagonists, the most widely used of which are ranitidine, famotidine, and cimetidine, are widely perceived to be safe and effective drugs for the healing of both duodenal and gastric ulcers. They are also remarkably free of significant adverse effects although these can occur (see below). These drugs work largely by suppression of acid secretion, particularly nocturnal acid secretion. The conclusions which may be drawn from an enormous number of well-conducted trials in the use of these drugs for gastric and duodenal ulcers are as follows.

1. A night time dose is usually satisfactory in healing ulcers.
2. Between 60 and 95 per cent of peptic ulcers will be healed after 4 weeks of treatment at appropriate dose with any of the major H_2-receptor antagonists.
3. Larger ulcers may take longer to heal but by 8 weeks 90 to 95 per cent of peptic ulcers will have healed.
4. These drugs usually produce symptomatic relief rapidly, often before ulcer healing. Hence many patients use them as 'antacids'—taking them to relieve symptoms for which they would have previously taken antacids.
5. The speed and completeness of healing is related to dose. Conventional dosing suggests ranitidine 300 mg at night, cimetidine 800 mg at night or famotidine 40 mg at night (Pounder and Nwokolo 1990).

In a recent authoritative summary of the use of the H_2-receptor antagonists, Isenberg *et al.* (1991) concluded that there was no proven difference in the adverse reaction profiles. In particular there was little evidence that cimetidine has a greater propensity to produce confusion, depression, memory impairment, and hallucinations than the other drugs and age *per se* has not been found to be an independent risk factor for CNS toxicity. The newer H_2-receptor antagonists (famotidine) probably heal a higher proportion of ulcers more quickly, and possibly relieve symptoms more quickly. Ranitidine and famotidine suppress a great proportion of nocturnal acidity than does cimetidine at an equivalent dose.

Omeprazole

The proton pump inhibitor omeprazole heals peptic ulcers faster and in an even higher proportion than do the H_2-antagonists. The impression is that in patients with duodenal ulcer about 90 per cent are healed at 4 weeks with omeprazole 20 mg at night compared with 80 per cent of ulcers treated with ranitidine 300 mg at night. Pain is also resolved even more quickly with omeprazole. Secretory studies have shown that nocturnal acid secretion is more completely abolished with this drug. A single night time dose of omeprazole inhibits 90 to 99 per cent of gastric acid secretion over 24 h. At present there is no suggestion that this is not a safe drug for elderly patients, at least over a 1- to 2-month healing period.

Colloidal bismuth

Colloidal bismuth preparations—notably tripotassium dicitratobismuthate (bismuth chelate) were developed when it was found that these compounds formed a complex with mucus, creating a protective layer and augmenting mucus protection against acid–pepsin digestion; thus bismuth chelate was perceived as a cytoprotective. Only later was its action against *H. pylori* found to be of potential importance. Many studies have examined the possible role of bismuth chelate in healing peptic ulcers particularly in relation to removal of *H. pylori*. In summary, it seems likely that bismuth chelate given four times per day over a 2-month period has equivalent effectiveness in healing peptic ulcers as do conventional H_2-receptor antagonist treatments. The disadvantage of four times per day dosage, and its rather unpleasant taste causes potential compliance problems in elderly patients when compared with H_2-receptor antagonists or omeprazole. It is probable that bismuth chelate reduces the likelihood of ulcer relapse following cessation of antiulcer treatment when compared with the other agents. Whether this is due to its effect on *H. pylori* or in causing some prolonged improvement in the mucosal barrier is uncertain. Bismuth chelate appears to 'clear' *H. pylori* from the stomach during treatment, but does not eradicate the organism completely; 2 or 3 months after stopping treatment *H. pylori* organisms may again be detected in gastric biopsies. Complete eradication has been attempted with combinations of bismuth chelate and a wide variety of antibiotics. Almost all trials have shown a disappointingly low proportion of patients with eradication of *H. pylori* 1 or more months after a course of treatment. To emphasize the mystery of the relationship of *H. pylori* and peptic ulceration antibiotic treatment alone, which may sometimes clear the organism temporarily, has no marked effect on peptic ulceration. It is thought only a very few organisms need to remain resident in the stomach following a course of bismuth chelate and/or antibiotic treatment, for recolonization to occur. Triple treatment—bismuth chelate, metronidazole, and a second antibiotic, e.g. augmentin—may eradicate *H. pylori* and cure some ulcers.

Cytoprotective treatment

Sucralfate, a basic salt of sulphated sucrose, is particularly effective against acute ulcers caused by topical agents and stress. The mechanisms of its action are not clearly known but again it is thought to improve the mucosal barrier and retard back-diffusion of bile salts, pepsin, and acid from the lumen of the stomach. Clinical trials suggest that it is marginally less effective over a 1- to 2-month course than H_2-receptor antagonists or omeprazole. Because of its aluminium content it is not recommended for very long-term maintenance treatment. The place of sucralfate is probably in the treatment of stress ulcers and conceivably as an adjunct to other antiulcer treatment following acute bleeding episodes.

A number of synthetic prostaglandin E_1 and E_2 analogues have been used for ulcer treatment. Again these agents are thought to be cytoprotective and conceivably improve healing of ulcers through their effect on gastric blood flow. Unfortunately these agents are not as effective in ulcer healing as the other treatments discussed above and have the additional problem that they cause watery diarrhoea in 10 to 30 per cent of patients.

Maintenance treatment

Gastric ulcer

The relapse rate of gastric ulcer appears to be of the order of 50 per cent after 1 year (Wolsin *et al.* 1989). One of the main problems in determining maintenance treatment for both gastric and duodenal ulcers concerns the criteria on which ulcer 'relapse' is based. There is often a lack of correspondence between endoscopic appearances of ulcer (often with no symptoms) and symptomatic recurrence (often with no ulcer). Furthermore, in determining whether maintenance treatment for gastric ulcer should be given it should be borne in mind that several studies suggest that it is gastric ulcers associated with duodenal ulcers—more commonly seen in younger individuals—which have a tendency to relapse, rather than the isolated gastric ulcers, often high in the stomach, seen among elderly patients. The implication is that the former are part of the duodenal ulcer disease whereas the latter arise from different mechanisms. Other factors determining risk of relapse include heavy consumption of alcohol and cigarette smoking. To complicate the issue further the majority of elderly patients with gastric ulcer are likely to be using NSAIDs, if not continuously, at least intermittently. In the light of published long-term maintenance studies, and bearing cost and likelihood of compliance in mind, the author would recommend long-term maintenance treatment with a single night time dose of an H_2-receptor antagonist for elderly patients. There is some evidence that, if such a strategy were widely adopted, complications of gastric ulcer in elderly patients would be reduced.

Duodenal ulcer

Very similar considerations pertain to the long-term maintenance treatment of duodenal ulcer. Again no hard and fast rules can be made. In elderly patients presenting for the first time with duodenal ulcer—i.e. those who do not have an apparent long-term history of duodenal ulcer disease and who have no regular ingestion of NSAIDs—it might be reasonable merely to give a 2-month acute course of treatment for a duodenal ulcer and only carry out maintenance treatment if symptoms recur. Unfortunately, since many elderly patients present with the complications of the disease rather than merely symptoms of pain, this treatment strategy does have some risks. As far as cost is concerned several recent studies have suggested that maintenance treatment for peptic ulcer is probably no more expensive than offering no treatment and dealing with the complications (usually by admission to hospital) as they arise (Pounder and Nwokolo 1990).

NSAIDs and antiulcer treatment

Because of the accumulation of epidemiological evidence linking gastric ulcer and the complications of ulcers with NSAID use in elderly people there is now debate as to whether elderly patients taking regular NSAIDs, particularly if there is any history of abdominal symptoms, should take maintenance antiulcer treatment. It should however be stated that in patients with NSAID-related peptic ulcer, gastritis or duodenitis, the first question to ask is whether the NSAID is necessary. Is it being used for its anti-inflammatory activity rather than merely as an analgesic? If only analgesia is required then a change of drug, for example to paracetamol (acetaminophen), is logical. In patients developing well-defined peptic ulcer associated with NSAIDs the acute treatment is as outlined above for non-NSAID related ulcers. Probably omeprazole is now the first drug of choice for such ulcers since a higher proportion of patients have ulcers slow to heal with other agents. At present there is really no firm data on which to base recommendations as to whether antiulcer treatment should routinely be given with NSAIDs in elderly patients, and a number of clinical trials are in progress. At present, leading authorities suggest that in elderly patients with any prior ulcer history who need chronic treatment with NSAIDs concomitant maintenance antiulcer treatment, probably with a single night time dose of an H_2-receptor antagonist, should be given even in the absence of current symptoms suggestive of peptic ulcer. Although cytoprotective treatment with sucralfate or prostaglandins might appear more logical, compliance with these drugs is probably low.

Antacids

Antacids were the mainstay of ulcer treatment until the advent of the H_2-receptor antagonists. They still have a small part to play in symptom control although even now their mechanism of action is uncertain. Of the commonly used antacid preparations salts of calcium and aluminium are constipating, and those of magnesium tend to cause diarrhoea. Concern has been expressed at the possible association between long-term ingestion of aluminium salts and the pathogenesis of Alzheimer's disease. In elderly patients with renal insufficiency increase in serum calcium, magnesium, or aluminium may occur following significant ingestion of relevant antacids. In view of the efficacy and safety of other available antiulcer treatments routine use of antacids by elderly patients is not recommended, beyond very occasional use for non-specific symptoms.

Diagnosis

The diagnostic approach to symptoms suggestive of peptic ulcer is different in elderly compared with younger patients. Many authorities would now suggest that endoscopy is unnecessary in a patient under age 40 presenting with such symptoms as postprandial discomfort or night time discomfort relieved by food. Rather they would suggest that a therapeutic trial of antiulcer treatment is appropriate. In older patients, however, investigation of possible symptoms of peptic ulcer is mandatory because it is often hard to distinguish between the symptoms of ulcer and those of gastric carcinoma. Upper endoscopy is well tolerated by elderly patients and is the diagnostic examination of choice. While double contrast radiography, performed by an experienced radiologist, has an equivalent diagnostic yield, endoscopy carries the major advantage of allowing biopsy. All gastric ulcers regardless of appearance should be biopsied; biopsies should be taken at the edge of the ulcer and from a site adjacent to the ulcer. In a few instances histology of an ulcer may be equivocal or erroneous, even when multiple biopsies have been obtained. Accordingly, good practice dictates that repeat endoscopy should be undertaken for gastric ulcers to ensure that complete healing has occurred. This precaution is not necessary in the case of duodenal ulcers since duodenal malignancy is very rare. The presence or absence of *H. pylori* in biopsy specimens does not affect the type of antiulcer therapy used.

GASTRITIS

Gastritis—acute, chronic, focal, or diffuse inflammation of the stomach—has been constantly referred to in the preceding section on peptic ulceration. Nonetheless, since the major forms of gastritis are very widespread among elderly individuals, a brief section is devoted to descriptions and definitions of the major types of inflammatory lesion in the stomach even at the risk of some repetition.

The three major types of inflammatory disease of the stomach are acute erosive or haemorrhagic gastritis, and chronic non-erosive gastritis of two types: type A affects principally the fundal mucosa and type B affects principally the antral mucosa.

Acute erosive gastritis

Superficial ulceration or erosions of the stomach confined to the mucosal layer occur either because of very considerable external stress as in overwhelming multisystem diseases or following major trauma—so called stress ulcers—or as a result of ingestion of aspirin, other NSAIDs, large quantities of alcohol, or a number of other agents.

Erosive gastritis is often asymptomatic but patients may complain of some epigastric discomfort, nausea, or other dyspeptic symptoms. Erosive gastritis is a major cause of upper gastrointestinal bleeding and elderly patients may present either following major surgery or trauma on the one hand, or erosive gastritis may lead to haemorrhage in patients starting to take NSAIDs on the other. In some elderly people the haemorrhage can be occult and present with symptoms of profound anaemia rather than with haematemesis or malaena.

Endoscopy is the most sensitive method of diagnosis. The stomach is seen to contain multiple shallow ulcers looking something like superficial cigarette burns, with or without associated haemorrhage. The general appearance of the mucosa may be oedematous and friable, with red streaks and petechiae. It has been estimated that up to half of elderly patients taking NSAIDs for arthritic symptoms may have undiagnosed gastric erosions (Larkai *et al.* 1987).

The pathogenesis of this generalized acute gastric mucosal injury is still not clear. It is thought that the local action of prostaglandins within the mucosal layer to maintain mucosal integrity may be impaired and that the balance in favour of the maintenance of tight mucosal integrity may further be pushed towards local damage by the release of arachidonic acid metabolites and free radicals through such toxic agents as ethanol or NSAIDs. Local production of free radicals and focal impairment of mucosal blood flow may contribute to this multicentric damage. It should be noted that erosive gastritis may develop very quickly—within 24 h of a severe 'stress' such as major trauma, or very rapidly after starting treatment with an NSAID.

Treatment with sucralfate or prostaglandin E compounds like misoprostol may be considered but, as discussed above, H_2-receptor antagonist treatment or, in cases of severe haemorrhage, treatment with omeprazole are probably preferable.

Elderly patients admitted with acute upper gastrointestinal bleeding from NSAID- or aspirin-induced acute erosive gastritis in whom there are no focal major ulcers usually respond to intravenous H_2-receptor antagonist treatment. In very severe cases some authors suggest the addition of tranexamic acid and misoprostol.

Chronic non-erosive gastritis (types A and B)

Surveys in both the United Kingdom and the United States suggest that over half the population aged over 70 have some degree of chronic gastritis, and perhaps half of these have varying degrees of atrophy. In either case the endoscopic appearance of the gastric mucosa may be surprisingly normal, biopsy being necessary for diagnosis.

Type A involves the fundal (oxyntic) acid producing part of the stomach. Atrophy is confined to the fundus of the stomach, is usually autoimmune, not particularly related to infection with *H. pylori*, and is the usual cause of pernicious anaemia. Inflammation of the fundal mucosa leads to loss of parietal cells and ultimately inadequate secretion of intrinsic factor as well as acid with consequent malabsorption of vitamin B_{12}. The majority of these patients have antibodies against parietal cells and many against intrinsic factor itself. Because the antral mucosa is relatively well preserved, gastrin secretion is often high since there is a loss of feed-back inhibition of gastrin secretion by gastric acid.

In type B gastritis, accounting for 80 per cent of chronic gastritis seen in elderly people, the antrum is the main area of involvement but in severe prolonged cases there is also damage to the oxyntic mucosa leading to hypochlorhydria and ultimately achlorhydria. It is only recently becoming recognized that this form of gastritis not only leads to defective gastric acid production but also ultimately to inadequate secretion of intrinsic factor and hence vitamin B_{12} deficiency in the absence of autoantibodies to parietal cells and intrinsic factor in the serum. Arguments as to whether this type B gastritis is caused by *H. pylori* and, if so, whether this is the invariable cause have been discussed previously. The current consensus is that such is the case. Ultimately, if elderly patients are rendered almost or completely achlorhydric the gastric environment becomes inimical to *H. pylori* organisms and they disappear.

It should be emphasized that chronic gastritis is asymptomatic in the vast majority of elderly individuals although of course a number with type A gastritis develop pernicious anaemia. It is clear that acute infection with *H. pylori* may cause not only florid gastric inflammation but also symptoms of nausea, abdominal pain, and dyspepsia, although this is rare, certainly in old age. The relationship between chronic type B gastritis and peptic symptoms in the absence of any other definable lesion in the stomach is very hard to assess.

Other forms of gastritis

There are no other forms of inflammation of the stomach which are particularly confined to the elderly population. Almost all other forms of inflammation are found following gastric biopsy.

Granulomatous inflammation of the stomach is occasionally encountered and may be caused by Crohn's disease, sarcoidosis, and, more rarely, tuberculosis or fungal infections. In the last case these are usually found in terminally-ill elderly individuals.

A number of bacteria and viruses can occasionally cause gastric inflammation. Rarely α-haemolytic streptococci or other pyogenic bacteria cause a purulent inflammation of the submucosa, sometimes leading to perforation and peritonitis. This is a severe abdominal emergency which may be diagnosed by the appearance on plain abdominal radiographs of gas bubbles within the walls of the stomach. The appearances are similar to those of pneumatosis cystoides intestinalis and are very rarely also seen after endoscopic biopsy. This 'phlegmonous' gastritis carries a very grave prognosis with fatality of over 50 per cent in old age.

Hypertrophic gastropathy

This is also known as Menetrier's disease. It is a rare disease of unknown cause but is characterized by gross hypertrophy of the gastric folds produced by hyperplasia of gastric pit mucus cells. It is often accompanied by protein-losing enteropathy and is most commonly found in middle-aged to elderly men.

Patients may present with generalized upper abdominal symptoms, anaemia from blood loss, or with oedema as a result of hypoproteinaemia. Occasionally epigastric pain, vomiting, and diarrhoea may be very prominent. The diagnosis is confirmed by endoscopy and biopsy although the appearances on double contrast radiology are typical. At present no treatment has been found to be of proven benefit although occasionally patients with dyspeptic symptoms may benefit from conventional antiulcer medication. Occasionally in patients with very severe hypoproteinaemia resection of the worst affected portion of the stomach to reduce protein loss has been successful.

VASCULAR ECTASIAS

As with other vascular anomalies of the gut a variety of vascular ectatic disorders—notably angiodysplasia of the stomach are age-associated. It is unclear whether angiodysplasia should be regarded as an age-related degenerative process arising from chronic obstruction of submucosal veins or whether the lesions result from chronic mucosal ischaemia. The ectasias are dilated, distorted thin walled vessels lined by epithelium but by little or no smooth muscle. There is an association between aortic valve disease, particularly aortic stenosis and occult upper gastrointestinal bleeding, particularly from angiodysplasias (Clowse 1991).

Angiodysplasias of the stomach and duodenum may be seen in up to 3 per cent of upper endoscopies but their only important clinical manifestation is with bleeding. Coeliac or superior mesenteric angiography are usually necessary in cases of obscure gastrointestinal blood loss but offer little superiority to endoscopy in the diagnosis of bleeding from angiodysplastic lesions within sight of the endoscope. Treatment has been substantially improved with the advent of laser therapy, local injection, or bipolar heat coagulation.

Water melon stomach

This recently described vascular abnormality of the gastric antrum appears to be a discrete disorder unrecognized before the routine use of endoscopy. There is a striking appearance of parallel longitudinal red stripes in the gastric antrum. These stripes are on the tops of the longitudinal folds. The mean age of presentation is about 70 (Borsch 1987).

As with other ectasias presentation is with symptoms or signs of bleeding. The diagnosis is made both by the distinctive endoscopic appearance and by a typical histological finding of dilated vascular channels in the submucosal layer. The appearance is said to occur in only one in 10000 endoscopies but this may be because of previous lack of definition of this syndrome which was hitherto ascribed merely to 'gastritis'. Successful treatment of recurrent haemorrhage has recently been described using either endoscopic heat coagulation or laser treatment. Probably just over half of the cases of water melon stomach are associated either with collagen vascular disease or with cirrhosis and presumed portal hypertension.

GASTRIC TUMOURS (see Section 10)

Gastric carcinoma

Gastric cancer accounts for about 10 per cent of all cancers worldwide but the disease is far more common in the Far East than in Europe and the United States. There must be clues to the cause or causes of gastric cancer in these epidemiological findings. In Westernized countries there is a falling rate of gastric cancer. In the United States the incidence has dropped by 67 per cent between 1950 and 1979 and in England and Wales by 55 per cent. As with other gut cancers the incidence in most countries is age-associated.

Aetiology

There are two broad lines of thought regarding the causation of gastric cancer: these concern food and environmental toxins on the one hand and histological precursor lesions within the stomach on the other.

Dietary and environmental factors

The epidemiology of gastric cancer suggests that environmental, probably dietary, factors must play a part in carcinogenesis. The major candidates have included foods containing nitrite together with amides which form alkylating agents capable of mutagenesis within the stomach. Nitrite may be formed from nitrate and there is a striking correlation between *per capita* intake of nitrates and age-adjusted mortality for gastric cancer among many countries worldwide (Mirvish 1983). Models of gastric cancer in animals have been readily obtained using a number of nitrosamines. In specific areas of very high incidence certain foods—dried fish and pickles for example—appear to be associated with high incidence of cancer. Interestingly, although some case control studies have suggested that tobacco and alcohol are associated with increased risk of gastric cancer the fact is that over the past 40 years while tobacco consumption has increased in many countries and alcohol consumption has certainly not decreased there has been a striking decline in cancer rates (Boland and Scheiman 1991). Finally several recent studies have suggested that intake of some green vegetables, garlic, and some fruits are inversely related to cancer risk.

Atrophic gastritis

Atrophic gastritis is present in up to 90 per cent of patients with gastric cancer and this has led to the hypothesis that this is a premalignant condition. Furthermore, intestinal metaplasia may develop as a result of chronic atrophic gastritis. Again this development is age-associated (Sipponen *et al.* 1984). Epidemiologically the development of gastric cancer in a country of high risk may be related to the prevalence of intestinal metaplasia within that population. Several studies from Europe indicate that up to 10 per cent of individuals with chronic atrophic gastritis for 15 or more years develop gastric carcinoma.

Among patients with atrophic gastritis the development of gastric dysplasia almost certainly indicates increased cancer risk. Several recent studies have suggested that severe dysplastic change in the gastric mucosa indicates a markedly increased risk of the development of cancer although the exact quantitation of this risk is controversial (Craven 1991). Since pernicious anaemia is associated with atrophic gastritis one would expect elderly individuals with pernicious anaemia to be at increased cancer risk and this has always been widely stated. However, while there probably is some increase in cancer risk among these patients a recent careful 8-year endoscopic follow-up of 80 elderly patients with pernicious anaemia screened annually revealed only one case of early gastric cancer. There is no evidence that gastric ulcer is a predisposing cause to gastric cancer although partial gastrectomy, usually for peptic ulcer, appears to incur a two- to six-fold increase in subsequent development of cancer but not until 15 or more years following the original operation. Again, elderly patients operated upon for duodenal ulcer in their young adult life are at particular risk.

Because *N*-nitrosamine concentrations are highest in the achlorhydric stomach there may be an inter-relationship between histological precursor lesions seen in atrophic gastritis and increased dietary nitrosamine production in some individuals. A recent, extremely thorough, study has exonerated cimetidine, and by implication other H_2-receptor antagonists, from increased risk of gastric cancer (Colin-Jones *et al.* 1991).

Finally, recent interest has grown in the possible role of *Helicobacter pylori* in gastric carcinogenesis because of its causal relationship in the development of chronic atrophic gastritis. One recent prospective study in Britain suggested that as many as 35 to 55 per cent of cases of gastric cancer in the United Kingdom may be attributable to infection with *H. pylori* (Forman *et al.* 1991). Such is the high prevalence of evidence for past or present *H. pylori* infection in elderly individuals that a definitely causal association has, as yet, by no means been confirmed.

Presentation and diagnosis

The presenting features of stomach cancer are usually vague and non-specific. Since as many as 25 per cent of the normal elderly population complain of one or more 'abdominal' symptoms at any one time such symptoms as epigastric pain and bloating may pass unnoticed for a time. In areas of relatively low incidence such as most European countries or the United States this diagnostic difficulty has meant that the overwhelming majority of gastric cancers present at an advanced stage. In countries of high incidence, such as Japan, screening programmes—notably by endoscopy—have led to high rates of detection of early gastric cancers, curable by surgery. The analogy with breast screening in women in Westernized countries is reasonably close. In Western countries the question is whether endoscopic screening should be undertaken for such 'high risk' elderly groups as those with chronic atrophic gastritis, particularly in association with *H. pylori*, patients with pernicious anaemia, and patients who have undergone gastric surgery 15 or more years previously. Whether such surveillance is feasible or even desirable is as yet unanswered.

One study from Birmingham, England, has recently suggested that prompt endoscopy of all patients presenting for the first time with dyspepsia over the age of 40 increased the yield of potentially curable gastric cancers to over 60 per cent and the proportion of early gastric cancers rose from between 1 and 2 per cent to 26 per cent of the cancers detected (Hallissey *et al.* 1990). Further studies, specifically in elderly individuals, may provide answers to the cost–benefit relationship of such exercises among different populations.

Diagnosis

After history and examination, including palpation of lymph nodes, the diagnosis of gastric cancer should be made by endoscopy. While double contrast radiology of the stomach now provides high diagnostic accuracy in experienced centres endoscopy must be regarded as the initial investigation of choice giving the opportunity not only for direct visualization but for brush cytology and biopsy. In evaluation of all gastric ulcerated lesions multiple biopsies should be taken, preferably with brushings for cytology. Some authorities suggest that up to 10 biopsies of gastric ulcers or suspicious lesions must be taken to ensure a diagnostic rate of over 99.5 per cent. In practice four or five biopsies together with cytology confirm the diagnosis in over 96 per cent of patients.

Clinical course

As with a number of tumours prognosis of gastric cancer improves with age once surgical fatality associated with comorbidity has been taken into account. This is perhaps surprising since cancers of the upper third of the stomach have a worse prognosis and these are more common in elderly patients. Otherwise, prognosis is related to the degree to which the tumour has penetrated through the gastric wall. Lesions confined to the mucosa or submucosa—usually found on endoscopic screening—have an 80 per cent 5-year survival. The prognosis worsens if the tumour is found to have penetrated through the muscularis propria, the serosa, and into contiguous structures, or has metastasized.

Treatment

Treatment is almost exclusively surgical. The aim should be for total excision of the tumour and if this is confined to a single site then, provided the patient is generally fit enough, laparotomy should be carried out. Preoperative assessment by abdominal CT has been disappointing in detection of unexpected metastases but preoperative assessment by chest radiography, abdominal CT, and liver blood tests should be carried out to exclude obvious metastatic disease and give some idea of the extent of the tumour beyond that assessed by endoscopy. In general massive heroic upper abdominal resections are not recommended in elderly frail patients. Occasionally, since only 30 to 40 per cent of patients have resectable tumours palliative surgery may be necessary but it should be clear that such treatment should be kept to a minimum.

Unfortunately, no regime of chemotherapy has been shown to be of great benefit in any series of patients with inoperable gastric tumours. Such treatment is not recommended except in the context of controlled trials in experienced centres. Although radiotherapy has a few enthusiastic proponents there is little evidence of benefit in gastric cancer.

Beyond these measures general attention should be paid to nutrition and hydration. Otherwise the care of patients with gastric cancer may be regarded as similar to that of other patients with severe malignancy.

Gastric lymphoma

Although lymphomas comprise only about 5 per cent of all gastric malignancies the diagnosis is particularly important since these have a much better prognosis than adenocarcinomas. Often the gross appearance is indistinguishable. Over 40 per cent of gut lymphomas originate in the stomach.

The clinical history and presentation of these tumours is very similar to gastric carcinoma. Diagnosis is made by endoscopic biopsy and subsequent management is dictated by CT scan to assess the extent of lymph node involvement. If the lymphoma is confined to the stomach conventional treatment is with total gastrectomy. Subsequent treatment, or treatment of more widespread disease, is with combined chemotherapy. At present for elderly patients who are otherwise fit combination chemotherapy with cyclophosphamide, doxorubicin, vincristine, and prednisone (CHOP) is recommended. In very frail patients a modified approach may be indicated. Gastric lymphoma has no great predilection for advanced age.

Other malignancies

The other major malignant tumour of the stomach is a gastric carcinoid. These tumours comprise less than 1 per cent of gastric tumours and are extremely rare in old age.

POLYPS

The majority of gastric polyps are benign, hyperplastic polyps with no implications of premalignancy. They are present in up to 1 per cent of autopsy series and their occurrence increases with advancing age. It is not thought that they frequently cause symptoms nor can they be accounted as a cause for significant gastrointestinal blood loss. Such hyperplastic polyps account for about three-quarters of all gastric polyps. Less common are adenomatous polyps; these may be found particularly in patients with pernicious anaemia and may be considered as a premalignant lesion. Adenomatous polyps should be removed endoscopically and patients should subsequently be screened by endoscopy at annual intervals.

GASTRIC EMPTYING

Unlike the oesophagus, in which changes in motility with age are quite profound, alterations of gastric motility in elderly people are not so clear cut. Indeed there is probably little if any age-associated reduction in gastric emptying for solids.

There may be some modest reduction in gastric emptying for liquids in normal elderly individuals (Moore *et al.* 1983).

Far more important are the increase in use of medications causing delayed gastric emptying taken by elderly people and in diseases which also affect gastric motility. In particular, opiates, many antidepressants, anticholinergic agents, L-dopa, and calcium-channel blockers all cause delay in gastric emptying which may be significant.

Diabetic gastropathy

Although severe delay in gastric emptying attributable to autonomic neuropathy associated with diabetes has been said to be rare, recent studies using more sophisticated gastric emptying measurements have suggested that up to one-quarter of insulin-requiring diabetics have impaired gastric emptying of solid food. Symptoms of nausea and occasional vomiting are not infrequent in diabetics and it is often hard to establish a close link between these symptoms and measured abnormalities of gastric motility. Many older diabetics with poorly controlled type 2 diabetes have episodes of moderate hyperglycaemia and this in itself is associated with delayed gastric emptying of liquids. Treatment of elderly diabetics complaining of bloating, nausea, and vomiting, together with feelings of gastric fullness is often unrewarding; however, the use of prokinetic agents—particularly domperidone and cisapride on a regular basis—may improve these symptoms.

Effect of gastric surgery

Occasional elderly patients still experience bloating, fullness, and vomiting many years after previous gastric surgery. This is due to the effect of vagotomy and affects 1 to 5 per cent of patients postoperatively depending on the drainage procedure used with the vagotomy. Usually gastric emptying for liquids is far better than for solids in these patients and patients should be advised to eat frequent small liquid meals.

While prokinetic drugs are recommended for treatment our experience with these has been disappointing. Corrective surgery is occasionally necessary but once symptoms have been well established the results are often disappointing.

Investigation

Investigation of gastric motility disorders in elderly patients is seldom a high priority. However, in elderly individuals with symptoms of recurrent vomiting, nausea, and fullness, in whom upper endoscopy and barium radiology have shown no overt abnormality, studies of gastric emptying prior to treatment with a prokinetic agent (domperidone or cisapride) may be useful. Such investigations are not widely available and should be carried out in experienced centres. Conventional radiography is a poor method for assessment of gastric emptying. The best established methods are with scintigraphy in which a gamma-camera can be used to assess emptying of liquid or mixed meals incorporating a radiomarker, or real-time ultrasound in which a semiquantitative measurement of gastric emptying time, particularly of a liquid meal in the erect position, may be examined (Kupfer *et al.* 1985).

In general, symptoms of nausea, epigastric fullness, and vomiting should be thoroughly investigated in all elderly individuals. Except in those with any symptoms of dysphagia the first investigation should be upper endoscopy; in the absence of any detectable lesion radiology of the oesophagus and stomach should be next. If no clues are obtained from these investigations, clinical examination, and blood tests, and if possible adverse effects of medications or the effects of such concomitant diseases and diabetes or connective tissue disease have been excluded, then referral for gastric emptying studies is sensible.

References

Bartle, W.R., Gupta, A.K., and Lazor, J. (1986). Non steroidal anti inflammatory drugs and gastrointestinal bleeding: a case control study. *Archives of Internal Medicine*, **146**, 2365–7.

Boland, C.R. and Scheiman, J. (1991). Tumours of the stomach. In *Textbook of Gastroenterology* (eds. T. Yamada *et al.*) pp. 1353–79. J.B. Lippincott, Philadelphia.

Borsch, G. (1987). Diffuse gastric antral vascular ectasia: the water-melon stomach revisited. *American Journal of Gastroenterology*, **82**, 1333–8.

Branicki, F.J. *et al.* (1990). Bleeding peptic ulcer: a prospective evaluation of risk factors for rebleeding and mortality. *World Journal of Surgery*, **14**, 262–9.

Clowse, R.E. (1991). Vascular ectasias, tumors and malformations. In *Textbook of Gastroenterology* (eds. T. Yamada *et al.*) pp. 2172–88. J.B. Lippincott, Philadelphia.

Colin-Jones, D.G. *et al.* (1991). Post-cimetidine surveillance for up to ten years: incidence of carcinoma of the stomach and oesophagus. *Quarterly Journal of Medicine*, **78**, 13–19.

Cooper, B.T., Western, C.F.M. and Neumann, C.S. (1988). Acute upper gastrointestinal haemorrhage in patients aged 80 years or more. *Quarterly Journal of Medicine*, **58**, 765–74.

Craven, J.L. (1991). Gastric cancer. *Current Opinions in Gastroenterology*, **7**, 933–8.

Elta, G.H. (1991). Approach to the patient with gross gastrointestinal bleeding. In *Textbook of Gastroenterology* (eds. T. Yamada *et al.*) pp. 591–616. J. B. Lippincott, Philadelphia.

Faulkner, G., Prichard, P., Somerville, K., and Langman, M.J.S. (1988). Aspirin and bleeding peptic ulcers in the elderly. *British Medical Journal*, **297**, 1311–13.

Forman, D. *et al.* (1991). Associations between infection with *Helicobacter pylori* and risk of gastric cancer. *British Medical Journal*, **302**, 1302–5.

Gilinsky, N.H. (1988). Peptic ulcer disease in the elderly. *Scandinavian Journal of Gastroenterology*, **23** (suppl. 146), 191–200.

Hallissey, M.T. *et al.* (1990). Early detection of gastric cancer. *British Medical Journal*, **301**, 513–15.

Holman, R.A.E. *et al.* (1990). Value of a centralised approach in the management of haematemesis and malaena: experience in a district general hospital. *Gut*, **31**, 504–8.

Holvoet, J. *et al.* (1991). Relation of upper gastrointestinal bleeding to non steroidal anti inflammatory drugs and aspirin. A case control study. *Gut*, **32**, 730–4.

Isenberg, J.L., McQuaid, K.R., Laine, L. and Rubin, W. (1991). Acid-peptic disorders. In *Textbook of Gastroenterology* (eds. T. Yamada, *et al.*) pp. 1241–339. J.B. Lippincott, Philadelphia.

Kupfer, R.M., Heppell, M., Haggith, J.W. and Bateman, D.N. (1985). Gastric emptying and small bowel transit time in the elderly. *Journal of the American Geriatrics Society*, **33**, 340–3.

Larkai, E.N., Smith, J.L., Lidsky, M.D. and Graham, D.Y. (1987). Gastroduodenal mucosa and dyspeptic symptoms in arthritic patients during chronic non steroidal anti inflammatory drug therapy. *American Journal of Gastroenterology*, **82**, 1153–8.

Levi, S. *et al.* (1989). *Campylobacter pylori* and duodenal ulcers: the gastric link. *Lancet* **i**, 1187–9.

Mirvish, S.S. (1983). Intragastric nitrosamine formation and other theories. *Journal of the National Cancer Institute*, **71**, 629–41.

Moore, J.G., Tweedy, C., Christian, P.E. and Datz, F.L. (1983). Effect of age on gastric emptying of liquid-solid meals in man. *Digestive Diseases and Sciences*, **28**, 340–4.

Morgan, A.G. and Clamp, S.E. (1988). OMGE International upper gastrointestinal bleeding survey, 1978–1986. *Scandinavian Journal of Gastroenterology*, **23** (suppl. 144), 51–8.

Pounder, R.F. and Nwokolo, C.V. (1990). Duodenal ulceration. In *Recent Advances in Gastroenterology* (ed. R.E. Pounder), pp. 117–31. Churchill-Livingstone, Edinburgh.

Sipponen, P., Kekki, M., and Siurala, M. (1984). Age related trends of gastritis and intestinal metaplasia in gastric carcinoma patients and in controls. *British Journal of Cancer*, **49**, 521–30.

Somerville, K., Faulkner, G. and Langman, M. (1986). Non steroidal anti inflammatory drugs and bleeding peptic ulcer, *Lancet*, **i**, 462–4.

Swaine, P. (1991). Endoscopic treatment of peptic ulcer haemorrhage. *Clinical Gastroenterology*, **5**, 537–61.

Walt, R. *et al.* (1986). Rising frequency of ulcer perforation in elderly people in the United Kingdom. *Lancet*, **i**, 489–92.

Walt, R.P. (1990). Upper gastrointestinal bleeding. *Recent Advances in Gastroenterology* (ed. R. E. Pounder), pp. 101–16. Churchill-Livingstone, Edinburgh.

Wolsin, J.D. *et al.* (1989). Gastric ulcer recurrence. Follow-up of a double blind placebo controlled trial. *Journal of Clinical Gastroenterology*, **11**, 12–20.

8.5 Small intestine

P. S. LIPSKI AND O. F. W. JAMES

AGE-ASSOCIATED CHANGES IN MORPHOLOGY AND FUNCTION

Morphology

There are few studies on the structure and function of the normal ageing human small bowel. Postmortem studies have suggested age-associated atrophy of the muscularis mucosa and decrease in size and number of Peyer's patches. Although Webster and Leeming (1975) found that elderly people have broader and shorter small intestinal villi and calculated a reduced surface area for absorption many of the individuals whom they studied were malnourished and malabsorption was not specifically excluded. Nonetheless these findings were confirmed by Warren, Pepperman, and Montgomery (1978). However, Corrazza *et al.* (1986) in a comparison between young and over 60-year-old groups found no difference in surface area or mean enterocyte height between young and old. Lipski *et al.* (1992a) studied 39 subjects (age range 46 to 89) carefully selected to have no overt nutritional disease. There are no significant correlations between age and area of duodenal surface epithelium, area of crypts, area of lamina propia, heights of surface epithelium and villi, crypt depth, crypt to villus ratio, and number of intraepithelial lymphocytes. On balance it seems probable that as with so many other apparently age-associated variables age alone is not associated with marked changes in small bowel morphometry; rather it is other factors more commonly seen in age—for example undernutrition due to possible deficient dietary intake.

Absorptive function

Carbohydrate

There is no clinical evidence to suggest that significant malabsorption of carbohydrate occurs with advancing age. Indeed Beaumont *et al.* (1987) found no age associated deterioration in absorption of the test substances mannitol (passive absorption) and 3-*O*-methylglucose (active absorption). An apparent decline in D-xylose excretion with age may be accounted for completely by correcting for the age-associated decline in creatinine clearance (Arora *et al.* 1989). Feibusch and Holt (1982) have, however, demonstrated that there may be an impaired stress related absorption of carbohydrate in that, by increasing doses of carbohydrate in a test meal, elderly individuals had impaired absorptive capacity of high carbohydrate loads as reflected by assessment of breath hydrogen following the carbohydrate meal.

Fat

Although some early studies suggested that there was a decline in faecal fat with age, when carefully selected free living individuals are chosen and a well controlled fat intake is ensured, no change in faecal fat excretion is found between age 20 and 90 (Arora *et al.* 1989). Mylvaganam *et al.* (1989) found that fat absorption, as assessed by the ^{14}C-triolein breath test, was less in individuals over age 65 than in those below this age. Furthermore they suggested that there was a decline of 25 per cent in 8-h cumulative values of ^{14}C-triolein recovered in the breath between the ages of 65 and 87 years. Since this test assesses not only small intestinal fat absorption but also pancreatic function no firm conclusion as to the possible cause for this decline can be made. Again, however, by increasing dietary fat load given to elderly individuals malabsorption could be induced at levels of fat intake far above normal but which still produced no increased faecal fat excretion in young subjects (Werner and Hambraeus 1972).

Protein

Almost no studies of the integrity of protein absorption in old age have been carried out. Again Werner and Hambraeus (1972) demonstrated that faecal nitrogen increased in elderly individuals fed a very high protein load whereas younger individuals showed no such increase, suggesting a possible diminished intestinal or digestive reserve capacity for dietary protein with advanced age as for carbohydrate and fat.

Other nutrients

Calcium

It seems likely that calcium absorption does decline with age, probably starting after 60 years (Armbrecht 1988) although the studies which have suggested this decline were carried out before selection criteria for studies of ageing were as sophisticated as they have subsequently become. In addition elderly individuals show reduced ability to adapt to low calcium diets by increasing the efficiency of calcium absorption compared with younger people (Ireland and Fordtran 1973). Furthermore, achlorhydria, very commonly present in advanced age, leads to malabsorption of some calcium salts—for example calcium carbonate. Calcium carbonate is normally poorly absorbed but reacts with hydrochloric acid to form soluble calcium chloride, subsequently absorbed in the proximal small bowel. There is marked malabsorption of calcium carbonate in achlorhydric individuals (Recker 1985).

Iron

Although iron deficiency is common in old age this is usually due either to occult blood loss or deficient dietary iron intake. Again previous gastric surgery or achlorhydria may lead to decreased absorption of ferric iron since this is insoluble above pH 5, although ferrous iron and haem iron remain soluble and are absorbed normally (Marx 1979; Russell 1988).

Folate

The absorption of folate is normal in old age provided that gastric acid output is maintained. In the presence of atrophic gastritis, however, Russell *et al.* (1987) demonstrated malabsorption which could be reversed by the concomitant administration of hydrochloric acid. This group found that in normal elderly individuals the pH in the proximal small bowel was 6.7 versus 7.1 in those with atrophic gastritis and achlorhydria. This small difference in pH is critical, since folate absorption is optimal around pH 6.3 and negligible around pH 7. They suggest that other micronutrients, for example nicotinic acid, may be influenced in a similar way. These studies again demonstrate the extreme importance of selection in carrying out examinations of absorptive capacity and other function in relation to age. Because the proportion of individuals with atrophic gastritis and hence low gastric acid output increases with age the proportion of individuals with impaired absorption of iron, folate, vitamin B_{12} and other micronutrients must also rise and may well exceed 50 per cent among the general elderly population. However, this impaired absorption must not be seen as an age-related effect; rather it is the effect of age-associated disease or disorders.

Vitamin B_{12}

In carefully selected normal elderly individuals absorption of vitamin B_{12} by a whole body retention method shows no decline over a wide age range (McEvoy *et al.* 1982). Since vitamin B_{12} is usually associated with food protein and must be bound to intrinsic factor before absorption many elderly individuals have impaired absorption. Vitamin B_{12} must be digested away from dietary food protein by gastric acid and pepsin before linking with endogenous binders or intrinsic factor. Elderly individuals with atrophic gastritis lack both sufficient vitamin B_{12} and pepsin to perform this function in a complete fashion. Hence gastric atrophy, found in over 50 per cent of the over 70-year-old population, is associated with poor absorption of vitamin B_{12} (King *et al.* 1979; Russell 1988).

Malabsorption

Malabsorption often presents for the first time in old age. It may be caused by any of the diseases recognized in younger adults and there are no 'old age specific' diseases causing malabsorption. Nevertheless the most common causes of malabsorption are different among older individuals. Malabsorption is frequently silent and other diseases, more frequently seen with advanced age, may mask the intestinal problem. Significant small bowel disease is more likely to present as a complication rather than with gut symptoms themselves. Estimates of the prevalence of malabsorption in old age rely on only a few studies largely carried out on hospital subjects. Montgomery *et al.* (1978) suggested that at least 13 per cent of 'normal' elderly individuals had some malabsorption. McEvoy *et al.* (1983) suggested that about 5 per cent of individuals admitted to an acute geriatric assessment ward would have significant malnutrition of which just under one-half would have malabsorption as a cause. This was, however, a highly selected group.

Small bowel disease in elderly people does not usually present with pain unless Crohn's disease, mesenteric ischaemia, or neoplasm are present. Diarrhoea is not as common as in younger individuals and even when present is sometimes due to faecal impaction and overflow rather than steatorrhoea. Steatorrhoea is uncommon in the elderly, and fat malabsorption may be underestimated because of low dietary intake. When steatorrhoea and its symptoms are present they are more likely to be caused by pancreatic disease than small bowel disease. The most common symptoms are general malaise and such non-specific symptoms as 'off feet' as the combined effects of anaemia and weight loss lead to general debility. Anaemia, from deficiencies of iron, folate, vitamin B_{12}, or a combination of these is the most common laboratory finding – occurring in up to 100 per cent of patients in highly selected series (Clark 1972; Ryder 1963). The symptoms of bone pain due to osteomalacia are not infrequently found in malabsorption in old people, particularly as a late presentation of coeliac disease.

General debility may lead to poor dietary intake and indeed it is often extremely difficult to distinguish between malnutrition due to malabsorption or to inadequate dietary intake (McEvoy *et al.* 1983; Montgomery *et al.* 1986). It is therefore important that a careful dietary assessment and appropriate tests are made in the investigation of possible malabsorption in an old person.

Malabsorption is not of itself a diagnosis. Once the clinician has decided that malabsorption exists then appropriate investigations should be carried out to define the cause of the condition. Table 1 lists the main causes.

Investigation of malabsorption (Table 2)

Sensible investigation of malabsorption should take into account the range of differential diagnoses. As already indicated a dietary assessment in a patient with suspected malabsorption is vital to exclude poor diet alone as a cause for malnutrition. Basic blood tests should include haemoglobin

Table 1 The main causes of small bowel malabsorption in elderly people

Bacterial contamination of the small bowel
Anatomically normal small bowel
Small bowel diverticulosis
Postgastrectomy syndrome
Coeliac disease
Ischaemic bowel
Inflammatory bowel disease
Drugs
Radiation-induced enteritis
Motility disorders
Pseudo-obstruction
Diabetes
Scleroderma
Maldigestion
Postgastrectomy
Previous small bowel surgery
Enteric fistula (neoplastic or inflammatory)

and blood film, serum vitamin B_{12}, and folate and red cell folate (these are not infrequently abnormal even in the presence of a normal mean corpuscular volume (**MCV**) since mixed deficiency is common in elderly subjects). Serum albumin and bone chemistry should also be estimated in the routine blood test screen. If all of the above are normal significant malabsorption is unlikely but this does not exclude some specific small bowel diseases. Further nutritional assessment may include simple anthropometry—weight, height, Quetelet's index, (weight/height^2), triceps skinfold thickness (**TSF**), and mid-arm circumference (**MAC**). If symptoms and signs do not point to a specific cause for malabsorption then barium meal with careful follow through examination of the small bowel should be carried out. Specific radiological findings include diverticula, 'blind loops' from previous abdominal surgery, strictures, and fistulae. Unfortunately in patients with marked malabsorption, particularly in those with low serum albumin, the findings may be merely of flocculation of the barium or non-specifically thickened mucosal folds in the small bowel. In the absence of specific radiological findings endoscopy with endoscopic low duodenal biopsies and, if possible, aspiration of small bowel juice with aerobic and anaerobic cultures should be carried out.

Table 2 Order of investigations for malabsorption

	Result in malabsorption	Comment
Screening test		
Haemoglobin		
MCV	↑ or ↓	High MCV may be the only abnormality in coeliac disease
Blood film	Hypochromic, macrocytic	
	Hypochromic, microcytic	Also consider, gastrointestinal blood loss
Folate and RBC folate	↓	Exclude coeliac disease
Vitamin B_{12}	↓	Test for antibodies to exclude pernicious anaemia
Albumin	↓	
Calcium	↓	
Phosphate	↓	Exclude osteomalacia
Alkaline phosphatase	↑	
Triceps skinfold thickness	↓	Useful for monitoring improvement
Mid-arm circumference	↓	
Body mass index	↓	
Specific tests		
H_2 breath test	Normal or ↑	Early peak in bacterial contamination
^{14}C-glycocholic acid breath test	Normal or ↑	Abnormal in bacterial contamination
^{14}C-triolein breath test	Normal or ↓	Abnormal in small bowel malabsorption or pancreatic disease
Plain abdominal radiograph		Pancreatic calcification
Barium meal and follow-through		May identify small bowel diverticula and other pathology
Endoscopy/biopsy		Essential to exclude coeliac disease
Aspiration/culture of small bowel juice		A normal result does not exclude bacterial contamination

Specific tests of small bowel function

The value of absorption tests, particularly in elderly individuals, is highly questionable and with a few exceptions (see Table 2) such tests are unnecessary and frequently misleading in this patient group. They have been generally abandoned.

The D-xylose tolerance test was formerly used as an assessment of carbohydrate absorption. However, because of age-associated decline in renal function and the unreliability of urine collection the test has now fallen into disrepute.

The hydrogen breath test is based on the principle that hydrogen is produced in man only by the fermentation of carbohydrate in the gut by bacteria, usually the colonic flora. Hydrogen gas production is markedly increased when as little as 5 g carbohydrate is supplied to the faecal-type colonic bacteria (Levitt and Donaldson 1970). Hydrogen gas is readily diffusable and is excreted from the lungs; hence end-expiration breath samples may be collected and hydrogen measured using a portable electrochemical cell (hydrogen monitor). In fact the test is used in clinical practice not to test the integrity of carbohydrate absorption but rather to detect the presence of abnormal colonic-type bacteria in the small bowel. In addition in individuals without small bowel bacterial contamination the orocaecal transit time may be measured as the time from ingestion of lactulose or glucose to the time of appearance of hydrogen in the breath indicating that the sugar has reached the caecum and is undergoing digestion by the colonic bacteria (Rumessen *et al.* 1990). Unless there is decreased colonic transit time abnormal appearance of hydrogen in the breath within about 30 min of ingestion indicates small bowel contamination with faecal-type organisms (Corazza *et al.* 1990; King and Toskes 1986). In the investigation of possible bacterial contamination or of orocaecal transit the lactulose hydrogen breath test is cheap, safe, non-invasive, reproducible, and simple and easy for even frail elderly patients to manage. It has however only moderate specificity although considerable sensitivity (King and Toskes 1986).

Faecal fat collections are now virtually never used in elderly patients except in a research setting. The ^{14}C-triolein breath test is sensitive and specific for detecting fat malabsorption. The long-chain fatty acid is labelled with carbon-14 and added to a meal of known fat content. Normal absorption leads to metabolism and production of $^{14}CO_2$. Hourly samples of exhaled breath are measured for $^{14}CO_2$ activity for up to 8 h. The percentage of the administered dose may then be calculated. Low levels indicate fat malabsorption but do not distinguish between pancreatic disease or small bowel malabsorption (Mylvaganam *et al.* 1989).

Bacterial contamination of the small bowel

Bacterial contamination of the small bowel (**BCSB**) is the most common cause of occult malabsorption in elderly people. As in young people, the normal flora of the proximal small bowel in elderly people when fasting contains less than 10^4 organisms/ml including Gram-positive staphylococci, streptococci, lactobacilli, and a few fungi, principally derived from the oropharynx. Aerobic coliforms (*E. coli* or Pseudomonas) are occasionally present in small numbers but strict anaerobes are notably absent. Bacterial contamination of the small bowel is present when greater than 10^5 organisms/ml are present in small bowel juice. The population can usually be demonstrated to consist of colonic faecal-type flora which includes anaerobes such as Bacteroides and coliform species.

Pathogenesis of bacterial contamination of the small bowel

It is probable that several factors contribute to the development of BCSB (Table 3). Probably the most important factor in controlling the entry of faecal type organisms into the small bowel is the gastric acid barrier. If this is abolished (as when following gastric surgery, particularly if a 'blind loop' is caused) then circumstances exist for the passage of such organisms into the jejunum and subsequent proliferation with the development of BCSB. Even where gastric acid output is reasonable, if a 'sump' exists in the small bowel, as in multiple jejunal diverticula, then the normal 'housekeeping' mechanisms of gut immunology, bile acids, and peristaltic activity may not always stop a few organisms passing into the diverticula and proliferating, hence producing a BCSB.

Table 3 Conditions associated with bacterial contamination of the small bowel

Stagnation of intestinal contents
Afferent (blind) loop stasis—Bilroth II partial gastrectomy
Postoperative blind loops
Gastroenterostomy
Duodenal, jejunal diverticulosis
Obstruction caused by strictures, adhesions, or malignancy, e.g. in Crohn's disease
Excessive bacteria entering small bowel
Achlorhydria in otherwise anatomically normal gut
Fistula: gastrocolic, jeunocolic
Cholangitis
Disordered small intestinal motility
Diabetic autonomic neuropathy
Scleroderma
Amyloidosis
Chronic intestinal pseudo-obstruction
Immune deficiency
Tropical sprue

Clinical presentation and treatment

Among elderly individuals with no previous known malabsorption, presenting either with symptoms of malabsorption or with symptoms and signs of malnutrition the most common cause is BCSB. Over 50 per cent of such patients studied by Montgomery *et al.* (1986) and 70 per cent of the patients studied by McEvoy *et al.* (1983) were found to have BCSB. The presentation was non-specific with weight loss, 'off-feet', mental deterioration, chest infection, bone pain, and anorexia being common features, as well as diarrhoea. At least one and usually several nutritional deficiencies (macrocytic anaemia with abnormal vitamin B_{12} and/or folate, low serum albumin, low calcium, and raised serum alkaline phosphatase) occurred in all patients. Formal testing of such patients reveals that there is malabsorption of fat since many anaerobic bacteria are able to deconjugate and dehydroxylate conjugated bile acids sufficiently to impair micelle formation.

Treatment of BCSB is with antibiotics. McEvoy *et al.* (1983) found treatment with clindamycin together with full nutritional support only moderately successful. Recently among 16 patients with malabsorption associated with BCSB antibiotic treatment (amoxycillin and clavulanic acid) resulted in significant improvements in blood tests and weight gain in 13 of the 16 individuals. It is recommended that treatment which may be cyclical should be continued over many months (Haboubi and Montgomery 1992).

Is BCSB always harmful?

In 1984 Hellemans *et al.* (1984) suggested that up to one-half of a normal healthy ambulant elderly population might have BCSB on the basis of an abnormality of the glycocholic acid breath test. Subsequently Lipski *et al.* (1992) studied more than 100 fit elderly individuals free living in the community together with 73 elderly individuals from long-stay geriatric hospital wards but with no known active digestive disease and a control group of young fit ambulant individuals. About 20 per cent of both elderly groups versus 3 per cent of young subjects had abnormal glycocholic acid breath tests suggestive of possible BCSB. There was no association between abnormal breath tests and anthropometry, haematology, or biochemistry except for a marginally lower serum albumin in the fit community elderly (but no individual had a subnormal serum albumin). If an abnormal ^{14}C-glycocholic acid breath test does indicate BCSB then BCSB may be a concomitant of 'normal' ageing, not necessarily leading to ill health. In some, presumably a minority of individuals such bacterial contamination subsequently leads, for reasons that are unclear, to malabsorption that is reversible by appropriate antibiotics therapy. P.S. Lipski *et al.* (unpublished work) have also shown that in a number of elderly individuals undergoing upper gastrointestinal endoscopy for abdominal pain, upper small intestinal histology showed no difference between those with and without bacteria in their jejunal aspirate. This would again suggest that in many elderly individuals, perhaps in those with achlorhydria, bacteria may live in symbiosis with their host in the upper small intestine with no pathological effect. Further studies are needed in this important area.

Coeliac disease

This life-long disease may first present in later life (Price *et al.* 1977). Indeed it has been suggested that there is a bimodal distribution of presentation with a second peak over age 60, and in one British series over 20 per cent of patients presented at ages over 70 (Swinson and Levi 1980). The clinical presentation in old age usually occurs in one of three ways: (1) anaemia which turns out to be due to profound folate deficiency; (2) osteomalacia due to life-long malabsorption of vitamin D; (3) small bowel lymphoma which has superimposed itself upon long-standing coeliac disease. While diarrhoea is not a common feature in these patients a history may often be obtained of frequent loose stool during childhood and early adult life. Many of these patients have short stature. Other presenting features in old age may be hypoproteinaemia, impaired blood clotting, and peripheral neuropathy associated with multivitamin deficiency. We have recently seen such a patient presenting with coeliac disease aged 93. Finally, coeliac disease may come to light following previous gastric surgery for peptic ulcer (Brandt 1984).

Diagnosis and treatment

Clinical awareness of the possibility of coeliac disease in an elderly individual is of greatest importance. The diagnosis must be made by small bowel biopsy. No other investigations are necessary beyond general assessment of nutritional status and identification of specific deficiencies. The lower duodenal biopsy obtained with jumbo forceps at endoscopy is adequate for interpretation provided no other pathology (e.g. ulcer, severe duodenitis) is present. Examination reveals subtotal or total villous atrophy which is confirmed on microscopy. The mucosa is hypertrophied with stunted or absent villi and crypt cell production is greatly increased. The lamina propria is infiltrated by plasma cells, eosinophils, and mast cells. There are large numbers of intraepithelial lymphocytes. A counsel of perfection is to confirm the diagnosis by repeating the jejunal biopsy 3 or more months after complete gluten withdrawal. In an elderly person, however, provided that the clinical response is satisfactory this second biopsy may be omitted.

Treatment is by complete gluten withdrawal from the diet. Correction of nutritional deficiencies must also be carried out with particular attention to iron, folic acid, calcium, and vitamin D. Failure to respond to these measures must raise the possibility of a superimposed lymphoma in an elderly subject (Loughran *et al.* 1986).

Complications

The major complication is malignancy. While T-cell lymphomas have already been referred to, carcinomas of the gastrointestinal tract, especially in the pharynx and oesophagus, together with adenocarcinoma of the small bowel itself are all increased in patients with coeliac disease. Up to 10 per cent of patients with coeliac disease will have one or more of these malignancies (Holmes *et al.* 1976; Swinson *et al.* 1984). Occasionally single or multiple large benign ulcers, often associated with strictures and hard to distinguish from malignancy, may occur in older patients with coeliac disease, whether treated or untreated. Such ulcers should be removed by surgery since no other treatment appears effective (Hellier 1987).

Crohn's disease

A paradox exists in that, although Crohn's disease in advanced age seems uncommon, nonetheless several studies have shown that age-specific incidence rates are higher around age 70 than at any age after about 30. Thus, again, there appears to be a bimodal distribution of presentation age in this condition (Mayberry *et al.* 1979). Recently Stowe *et al.* (1990) showed that age-specific incidence rates for Crohn's disease in Rochester, New York were higher in the 7th, 8th, and 9th decades than in the 4th, 5th, and 6th. In other series the proportion of patients with Crohn's disease in whom a diagnosis was made over age 60 varies between 1 and 20 per cent (Fabricius *et al.* 1985).

Presentation

The symptoms seen in elderly patients are similar to those seen in the generality of patients with Crohn's disease. Fabricius *et al.* (1985) suggested that in those with distal ileal disease an acute exacerbation, often with obstructive symptoms or peritonitis, frequently precipitated the patient's first hospital admission and not infrequently required early laparotomy. In this series of 47 patients from a total of more than 600 Crohn's disease patients about one-half had predominantly colonic disease but distal colonic Crohn's disease was more common in the older than in the younger patients. Because of the acute presentation most patients had laparotomy and resection of their terminal ileal disease. Recurrence appeared less common among the older patients even in those followed for a long with a time than with younger individuals (Tchirkow *et al.* 1983). Mortality from Crohn's disease is lower in elderly patients presenting with Crohn's disease than in younger subjects.

Diagnosis and management

In all respects the diagnosis and management, together with the other associated features of the disease, are very similar in older patients to those in the generality of adults with Crohn's disease. Diarrhoea, weight loss, abdominal pain, and fever are typical as are symptoms which may be ascribed to enteric fistula and acute or subacute obstruction. Once the diagnosis has been considered the differential diagnosis is with ischaemic bowel disease, but this is usually much more acute. Occasionally, small intestinal ulceration, idiopathic, associated with coeliac disease, or due to enteric-coated potassium supplements, may mimic Crohn's disease (Holt 1985).

Since long-term steroid treatment or recurrent courses of steroids may be used in Crohn's disease, metabolic bone disease, particularly osteoporosis, is more common as long-term sufferers from the disease approach old age. Many such patients may have had malabsorption of calcium and vitamin D. Consideration should therefore be given, among other nutritional replacements, to careful calcium supplementation, possibly with cyclical bisphosphonate therapy (see Chapter 14.1).

SYSTEMIC DISEASES AFFECTING THE SMALL INTESTINE

Diabetes

Diabetic diarrhoea is characteristic of patients with long-standing insulin-dependent diabetes and has been thought to be associated with neuropathy and/or diabetic vascular disease. The diarrhoea is usually watery, preceded by abdominal discomfort, and, in elderly patients, may be complicated by faecal incontinence. It often occurs at night. The generalized nature of the autonomic neuropathy is manifest by features of delayed gastric emptying and plain radiography of the abdomen may show dilated segments of small intestine. Because of disordered motility bacterial contamination may complicate the picture. If the ^{14}C-glycocholic acid breath test is positive in an elderly diabetic with diarrhoea then broad spectrum antibiotic treatment is worth trying. In a few diabetics with disabling diarrhoea, treatment with clonidine and alpha-adrenergic agonists has resulted in significant decrease in stool volume (Fedorak *et al.* 1985). Weight loss due to diabetic diarrhoea is unusual despite the volume of stool passed; if weight loss is substantial a further cause should be sought (Brandt 1984).

Primary amyloidosis

This may present in old age with steatorrhoea and pseudo-obstruction although other systemic manifestations are almost always present. Uncontrollable diarrhoea and steatorrhoea are due to autonomic neuropathy caused by amyloid infiltration rather than direct involvement of the small bowel wall itself (French *et al.* 1965). The prognosis in generalized amyloidosis is very poor.

Rheumatoid arthritis

Small bowel involvement and malabsorption in patients with severe rheumatoid arthritis is well described. There may be secondary intestinal amyloidosis with a perivascular distribution in the lamina propria, lactase deficiency, and atrophic mucosal changes. The mechanism whereby these effects occur is not clear. Malabsorption of folic acid, iron, carbohydrate, and vitamins occurs. These disturbances may improve if the rheumatoid arthritis itself is improved by treatment (Pettersson *et al.* 1970).

INTESTINAL PSEUDO-OBSTRUCTION

In this condition, most common in old age, signs and symptoms of intestinal obstruction occur but there is no demonstrable obstructing lesion. The apparent obstruction may be isolated, in a short segment of intestine, or may be part of a more generalized process involving much of the gastrointestinal tract (Anuras *et al.* 1978). Transient acute or subacute pseudo-obstruction occurs in elderly patients who are suffering from one of a number of systemic diseases (Table 4). These include intra-abdominal sepsis and infections, chronic renal failure, collagen vascular diseases, and congestive cardiac failure.

Table 4 Conditions associated with chronic intestinal pseudo-obstruction

Abnormal gut motility
Diabetes mellitus
Myxoedema
Scleroderma and other collagen vascular diseases
Amyloidosis
Parkinson's disease
Sepsis
Heart failure
Gastrointestinal disease
Small bowel diverticula
Jejunoileal bypass
Drugs
Anticholinergic—antiparkinson, tricyclic antidepressant, phenothiazines
Beta-blockers
Narcotics

Although pseudo-obstruction more frequently affects the colon, the small intestine is not uncommonly involved. Symptoms include nausea, vomiting, colicky abdominal pain, abdominal distension, and diarrhoea. When the condition presents acutely for the first time the temptation is to assume that the 'obstruction' has a mechanical cause. Frequently however episodes regress spontaneously, each episode lasting for days or weeks. Occasionally in an elderly person the symptoms may become extremely persistent and debilitating (Golladay and Byrne 1981; Schuffler *et al.* 1981).

Primary intestinal pseudo-obstruction

This term refers to cases where no associated disease is found. In this condition, which is usually initially intermittent over months or even years, the syndrome usually progresses to a chronic fluctuant persistent obstruction in which distension and vomiting are prominent. It is extremely resistant to treatment. Radiographs often reveal abnormalities throughout the gastrointestinal tract, whether in primary pseudo-obstruction or in the pseudo-obstruction associated with other systemic disease.

The most common cause of transient pseudo-obstruction is electrolyte disturbance, particularly hypokalaemia but elevation and depression of serum calcium and magnesium may also result in paralysis of small intestinal peristaltic activity. Usually electrolyte imbalance results in simultaneous dilatation of small and large bowel and is reversed by correction of electrolyte disturbances.

SMALL INTESTINAL ISCHAEMIA

In the resting state over 20 per cent of the cardiac output is delivered to the gut, the major blood supply to the upper gut being from the coeliac trunk and the vessels arising from the superior mesenteric artery. Although the colon is more vulnerable to ischaemia because it lacks abundant interconnecting arcades, mesenteric ischaemia is not uncommon and is probably underdiagnosed in elderly individuals. It appears to be increasing in frequency and is most common in the 9th decade of life (Marston 1985).

Simple mesenteric embolus, usually a complication of rheumatic heart disease and arrhythmias, is probably extremely rare in the absence of other vascular disease. The distinction between embolic and non-embolic small intestinal ischaemia is therefore more complex since many individuals with extensive intestinal ischaemia are shown to have no obvious vascular occlusion at operation or autopsy. Multiple atheromatous lesions, often at the origins of the major visceral arteries, clearly predispose to ischaemia (Croft *et al.* 1981). It seems likely that a combination of low arterial blood flow, often in association with heart failure, myocardial infarction, shock, and hypotension from a variety of systemic causes, together with generalized atheromatous vascular disease may lead to such low perfusion of the splanchnic vessels that vasoconstriction can occur and ischaemia develop (Tytgatt and van Dongen 1984). More rarely, mesenteric venous occlusion may occur, particularly in hypercoagulable states or during intra-abdominal sepsis (Grendell and Ockner 1982).

Symptoms and diagnosis

The classic description of acute small bowel ischaemia is of the sudden onset of severe abdominal colicky pain followed by dull generalized abdominal pain. Nausea and sometimes vomiting are also quite common. Characteristically, there are few signs on abdominal examination during the early stage. Sudden bowel evacuation, sometimes with blood is more suggestive of colonic ischaemia. Often, unfortunately, the classical findings are not present. Frequently, major infarction presents only with mild cramping pains and the patient's general condition remains reasonable until subsequent sudden severe deterioration. Abdominal pain occurring in an individual with known heart disease, particularly cardiac arrhythmia, must raise the urgent suspicion of mesenteric ischaemia (Marston 1985).

Subsequently the patient presents with a moderately distended silent abdomen with some abdominal tenderness. As infarction develops the abdomen distends and the patient becomes frankly shocked. Other symptoms of advanced ischaemia include severe back pain and intense nausea with vomiting. Apart from the association with other cardiac and cardiovascular disease, digoxin, and diuretic therapy in an individual with abdominal pain, diagnosis is often difficult at the early stage. Fever, high leucocytosis, metabolic acidosis, and blood-tinged peritoneal fluid on needle paracentesis occur late in the condition. Plain radiography of the abdomen is by no means specific and shows gas-filled loops of small bowel with fluid levels. Jamieson *et al.* (1982) have suggested that in an individual with abdominal pain a marked rise in serum inorganic phosphate may correlate with the degree of ischaemic damage but this has not been confirmed.

Treatment

The features of treatment must include general resuscitation and correction of any possible exacerbating factors (reduction of excess use of digoxin or diuretics, improvement of heart failure). Antibiotic treatment should be given to combat bacterial invasion and toxaemia (this should be given despite negative blood cultures); the antibiotic of choice is probably a third generation cephalosporin. If hypotension persists despite correction of hypovolaemia and heart failure dopamine infusion may be necessary. The circulatory collapse associated with mesenteric infarction is an indication for use of careful monitoring of central venous pressure, preferably in an intensive therapy unit. Gastrointestinal decompression by nasogastric suction should be used in individuals with severe vomiting. If and when successful stabilization of the patient's condition is achieved surgery should be carried out to remove the infarcted gut.

Chronic intestinal ischaemia

Although the idea of ischaemic pain in the small bowel, occurring after a meal, analogous to angina pectoris, and arising from partial occlusion of one of the major small intestinal arteries leading to ischaemic pain in response to increased oxygen demand during digestion, is naturally most attractive this concept is controversial (Marston 1985). One reason for this is that stenosis or even complete occlusion of the main vessels of the gut is commonly found with no apparent ill effects in many old people at postmortem examination. Nonetheless a syndrome, in elderly patients, of abdominal pain coming on shortly after meals and where no other diagnosis (e.g. peptic ulcer or pancreatic disease) can be arrived at is now generally recognized. Typically the abdominal pain comes on 10 to 15 min after beginning a meal, reaches a plateau, and disappears after 1 to 2 h. The pain is cramping, occurs in the epigastrium, and may radiate to the back. The severity of the pain is related to the size of meal. Patients become afraid of the pain, reduce their food intake, and hence lose weight. On examination, no specific findings may be apparent although in about one-half of patients a systolic bruit may be heard in the upper abdomen. Unfortunately such bruits are frequently heard in elderly individuals with no abdominal symptoms.

Diagnosis

This is based on three aspects: (1) a typical and prolonged history, often in an individual with other more generalized vascular disease; (ii) the exclusion of other potentially treatable causes of similar symptoms, particularly peptic ulcer, hiatus hernia, pancreatic disease, upper abdominal cancer, and myocardial ischaemic pain referred to the abdomen; (iii) angiography with selective studies of the upper mesenteric vessels. Experience has taught that surgery will be unlikely to be of benefit unless at least two of the three major upper abdominal arteries show occlusive involvement.

Treatment

Medical treatment is unsatisfactory. In a carefully selected series of patients Marston *et al.* (1985) carried out arterial reconstruction in 22 of 63 patients who were referred for suspected intestinal angina; 15 patients showed some symptomatic improvement, and there were three operative deaths in this series. As Marston points out this represents the experience on a specialist unit over 18 years. The condition can thus be considered rare.

RADIATION-INDUCED ENTERITIS

The small bowel is relatively sensitive to radiation and, since pelvic radiotherapy, for example for gynaecological malignancy or tumours of the bladder is common, radiation damage to the small bowel, particularly the ileum is by no means rare. Damage is more likely in patients who have had previous pelvic surgery in whom adhesions fix the small bowel closer to the site of irradiation (Novak *et al.* 1979). In fact the majority of patients who develop radiation injury to the gut are women of average age around 50 years (Danielsson *et al.* 1991). It is suggested that radiation-induced enteritis (whether to the small or large bowel or both occurs in up to 16 per cent of patients undergoing pelvic radiotherapy (Danielsson *et al.* 1991). Occasionally the acute effects of pelvic radiotherapy lead to diarrhoea and transient malabsorption due to early mucosal damage to the small intestine. More commonly the symptoms arise months or even years after the radiotherapy, the lesions being due to fibrosis, vasculitis, and gross submucosal damage. There is venous and lymphatic ectasia and obliteration of arterioles and small arteries throughout the intestinal wall.

Patients may present with symptoms of intestinal obstruction, but malabsorption with diarrhoea or steatorrhoea is perhaps the most frequent presentation in isolated small bowel radiation enteritis and this may be followed by weight loss, hypoproteinaemia, and frank malnutrition. Radiation enteritis involving the colon is more likely to cause diarrhoea, pain, and haemorrhage.

The diagnosis is made by barium follow-through examination of the small bowel or, if the lesion involves the terminal ileum, by colonscopy and biopsy. Perhaps surprisingly the diagnosis may be overlooked and mistaken for other causes of pseudo-obstruction or even irritable bowel syndrome. Treatment is difficult; surgical excision of loops of affected bowel is often extremely difficult owing to intense fibrous adhesions between loops of bowel and poor healing of entero-anastomoses. Medical management hinges upon nutritional support with a high protein diet; vitamin B_{12} is the most common vitamin supplement needed because of the susceptibility of the terminal ileum to damage. Agents which slow intestinal transit, e.g. loperamide, may be marginally helpful but if symptoms of pseudo-obstruction rather than of diarrhoea occur this approach is contraindicated. Treatment of complicating bacterial contamination, occasionally due to terminal ileal damage, may improve nutrition. Bile salt induced diarrhoea may be strikingly improved by use of cholestyramine. High dose steroids have been shown to have no place in treatment. Unfortunately, in the relatively few elderly patients who present with this condition treatment is usually ineffective. In the majority the symptoms and nutritional deficiencies are unpleasant but manageable; in a substantial minority, however, once the condition has presented the course is progressive with the development of internal fistulae, severe malnutrition, and obstruction.

DRUG-INDUCED SMALL BOWEL DISEASE

Because elderly people consume higher quantities of drugs, and polypharmacy is more common than in younger individuals, the risks of adverse effects of drugs on the small intestine, as with many other organs, is high.

Ulceration

In the 1960s and 1970s small bowel ulceration caused by enteric-coated potassium tablets became not infrequent. Even now slow-release potassium tablets occasionally cause ulcerated strictures of the small bowel. It is though that local release of potassium within a short segment of small bowel caused spasm of local vessels with subsequent infarction, ulceration, and scarring with thickening of the bowel. Ultimately this leads to local stenoses (Boley *et al.* 1965). Slow release non-steroidal anti-inflammatory drugs may also be responsible for ulceration, perforation, or necrosis.

Motility and absorption

Drugs with anticholinergic or opioid properties may decrease small bowel motility and, as described above even cause pseudo-obstruction. Perhaps tricyclic antidepressants are particularly culpable in this respect since they are widely prescribed for elderly people. The other principal drug-related adverse effects are listed in Table 5.

Table 5 Drug-induced small bowel dysfunction

Drug	Mechanism
Tetracycline	Chelates iron
Anticonvulsants	Folate malabsorption by impairing mucosal folate conjugate
Biguanides	Malabsorption of fat, amino acid, carbohydrate, and vitamin B_{12} via direct mucosal damage
Irritant laxatives	Malabsorption of fat and carbohydrate and hypokalaemia—mucosal damage
Liquid paraffin	Dissolves fat-soluble vitamins and prevents their absorption
Cholestyramine	Binds bile salts—malabsorption of fat—soluble vitamins; also binds iron and vitamin B_{12}
Neomycin	Precipitates bile salts; damages mucosa; causes malabsorption of fat, carbohydrate, protein, iron, vitamin B_{12}, and other drugs
Colchicine	Mucosal injury—malabsorption of fat, carbohydrate, vitamin B_{12}
Aluminium hydroxide gels	Interfere with phosphate absorption
Ethanol	Mucosal injury—malabsorption of fat, fat-soluble vitamins, carbohydrate, vitamin B_{12}, folic acid

NEOPLASMS OF THE SMALL INTESTINE

Although the small intestinal mucosa has very rapid turnover and an enormous surface area, and although it is exposed to potential carcinogens in the diet the small intestine has a far lower rate of development of malignant tumours than the stomach or colon. Nonetheless, as with the majority of adult cancers, those found in the small intestine are age-associated and incidence is most frequent in later life. This is not true for benign tumours whose incidence and spectrum are unchanged throughout adult life.

Benign tumours

The four most common benign tumours of the small bowel are leiomyoma (smooth muscle tumours), adenomatous polyp, lipoma, and haemangioma. All of these tumours may present with obscure bleeding in an elderly person. Occasionally leiomyoma, adenomatous polyp, and lipoma may present as intussusception. Adenomas, particularly villous adenomas, may be found in the duodenum and may present therefore not only with obscure gastrointestinal bleeding but also with obstruction at the ampulla of Vater leading to obstructive jaundice or occasionally duodenal obstruction. Malignant change may occur in perhaps one-third of these villous tumours and this is apparently restricted to individuals over age 50.

Malignant tumours (see Section 10)

Carcinoid tumours

These are the most common malignant tumours of the small bowel. Their frequency rises steeply from the duodenum to the terminal ileum and the most common site is in the appendix. They may present with the mechanical effects of the tumours or with the effects of release of 5-hydroxytryptamine (5HT) from the tumours themselves, or more frequently from a large bulk of metastatic tumour in the liver. Carcinoid tumours may present with symptoms of intestinal obstruction, gastrointestinal haemorrhage, or occasionally intussusception. Rarely, in the duodenum, the tumours can present with symptoms of bile duct obstruction. More commonly tumours present with metastatic disease in the liver leading to hepatic enlargment, often with abdominal discomfort. The typical carcinoid syndrome of flushing, diarrhoea, and palpitations arises from the release of 5HT from the liver metastases. Long survival is frequently recorded even with metastatic carcinoid tumour. Small carcinoid tumours are not infrequently found incidentally at autopsy or at laparotomy for other unrelated causes.

Adenocarcinoma

Adenocarcinoma is much more common in the duodenum than elsewhere in the small intestine. These tumours present with epigastric pain, vomiting, and weight loss. Sometimes they may merely cause occult bleeding and anaemia; occasionally they present with upper small intestinal obstruction. Since they are most prevalent in the second part of the duodenum jaundice may develop in up to one-third of patients. Diagnosis is usually by upper gastrointestinal endoscopy and direct biopsy. Because of their site these tumours may occasionally be diagnosed when they are small and hence may be totally resected by partial pancreatoduodenectomy (Whipple's operation).

Carcinomas lower down in the small bowel are extremely rare in elderly people but may occasionally present either with small intestinal obstruction or perforation. Carcinomas may arise in association with areas of long-standing Crohn's disease in the small intestine. The mean interval between the diagnosis of Crohn's disease and the cancer is 18 years and the tumour is usually in the ileum. Whereas small bowel tumours typically arise at age over 60, and often over age 70, tumours arising in areas of Crohn's disease usually present much earlier in life (Hawker *et al*. 1982).

Leiomyosarcoma

This is a disease of middle age, rarely presenting over age 70. The tumours may be extremely large, presenting as a palpable mass in about one-half of patients. Diagnosis is now with abdominal ultrasound and angiography. Symptoms are similar to those of other malignant tumours in the small bowel except for mechanical effects due to their considerable size. Treatment is operative.

Lymphoma

These tumours may arise in relation to pre-existing coeliac disease or *de novo*. In the duodenum they may present with obstructive jaundice, and elsewhere with diarrhoea and occasionally pseudo-obstruction or true obstruction. Patients may develop secondary bacterial contamination proximal to the site of the lymphoma leading to steatorrhoea. Tumours can be large and are frequently palpable on abdominal examination. Barium follow-through examination may reveal segments of grossly abnormal small intestinal mucosa. This may be difficult to distinguish from an area of Crohn's disease or (rarely in Western countries) tuberculosis.

Even in very elderly individuals treatment with resection followed by chemotherapy may have remarkable results. We have seen an 80-year-old man alive and well 2 years after resection and chemotherapy for a massive ileal lymphoma.

OTHER CAUSES OF INTESTINAL OBSTRUCTION

Elderly people are particularly susceptible to gallstone ileus and intussusception.

Gallstone ileus

This is a mechanical intestinal obstruction caused by impaction of one or more gallstones within the lumen of the small intestine. It is extremely rare but occurs almost exclusively in individuals over age 60, usually over 70. There is a high preponderance of associated generalized degenerative diseases. Only about one-half of the patients have a previous history of possible gallbladder disease and previous jaundice attributable to biliary disease is reported by about 15 per cent. Symptoms of cramping abdominal pain and vomiting associated with intestinal obstruction are more prominent than any biliary symptoms. Perhaps because of delay in diagnosis and associated multisystem disease a toxic confusional state is a frequent presenting feature of these patients.

Diagnosis should be made (but is often missed) on the

basis of (a) air in the biliary tree on plain radiography, (b) visualization of the gallstone in the central abdomen (presumably in the small bowel), and (c) small intestinal obstruction. These stones are usually large, it is sometime mysterious how a gallstone of 3 cm diameter or more has escaped from the gallbladder via the cystic and common bile ducts, through the ampulla of Vater and into the small bowel; occasionally this is through an inflammatory fistula between the gallbladder and the adjacent small bowel.

Because these patients are usually very ill, surgery, following resuscitation, is kept to a minimum. Enterolithotomy and cholecystectomy should however be carried out together where possible.

References

Anuras, S., Crane, S.A., Faulk, D.L., and Hubel, K.A. (1978). Intestinal pseudo obstruction. *Gastroenterology*, **74**, 1318–24.

Armbrecht, H.J. (1988). Changes in the components of the intestinal calcium transport system with age. In *Aging in Liver and Gastro-Intestinal Tract* (ed. L. Bianchi, P.R. Holt and O.F.W. James) pp. 131–9. MTP, Lancaster.

Arora, S., *et al.* (1989). Effect of age on tests of intestinal and hepatic function in healthy humans. *Gastroenterology*, **96**, 1560–5.

Beaumont, D.M., Cobden, I., Sheldon, W.L., Laker, M.F., and James, O.F.W. (1987). Passive and active carbohydrate absorption by the ageing gut. *Age and Ageing*, **16**, 294–300.

Boley, S.J., Allen, A.C., Schultz, I., and Schwartz, S. (1965). Potassium induced lesions of the small bowel. *Journal of the American Medical Association*, 997–1000.

Brandt, L.J. (1984). The small intestine. In *Gastroinestinal Disorders of the Elderly* (ed L.J. Brandt) pp. 195–249. Raen Press, New York.

Clark, A.N.G. (1972). Deficiency states in duodenal diverticular disease. *Age and Ageing*, **1**, 14–23.

Corazza, G.R., Frazzoni, M., Galto, M.R.A. and Gasbarrini, G. (1986). Ageing and small bowel mucosa: a morphometric study. *Gerontology*, **32**, 60–5.

Corazza, G.R., *et al.* (1990). The diagnosis of small bowel bacterial overgrowth. *Gastroenterology*, **98**, 302–9.

Croft, R.J., Menon, G.P., and Marston, A. (1981). Does intestinal angina exist? A critical study of obstructed visceral arteries. *British Journal of Surgery*, **68**, 316–18.

Danielsson, A., Nyhlin, H., Persson, H., Stendahl, U. Stenling, R., and Suhr, O. (1991). Chronic diarrhoes after radiotherapy for gynaecological cancer: occurrence and aetiology. *Gut*, **32**, 1180–8.

Fabricius, P.J., Gyde, S.N., Shouler, P., Keighley, M.R.B., Alexander-Williams, J. and Allan, R.N. (1985). Crohn's disease in the elderly. *Gut*, **26**, 461–5.

Fedorak, R.N., Field, M. and Chang, E.B. (1985). Treatment of diabetic diarrhoea with cloridine. *Annals of Internal Medicine*, **102**, 197–9.

Feibusch, J.M. and Holt, P.R. (1982). Impaired absorptive capacity for carbohydrate in the ageing human. *Digestive Diseases and Sciences*, **27**, 1095–1100.

French, J.M., Hall, G., Parish, D.J., and Smith, W.T. (1965). Peripheral and autonomic nerve involvement in primary amyloidosis associated with uncontrollable diarrhoea and steatorrhoea. *American Journal of Medicine*, **39**, 277–84.

Golladay, E.S. and Byrne, W.J. (1981). Intestinal pseudo obstruction. *Surgery, Gynaecology and Obstetrics*, **153**, 257–73.

Grendell, J.H. and Ockner, R.K. (1982). Mesenteric venous thrombosis. *Gastroenterology*, **82**, 358–72.

Haboubi, N.Y. and Montgomery, R.D. (1992). Small bowel bacterial overgrowth in the elderly. Clinical significance and response to treatment. *Age and Ageing*, **21**, 13–16.

Hawker, P.C., Gyde, S.N., Thompson, H. and Allen, R.H. (1982). Adeno-carcinoma of the small intestine complicating Crohn's disease. *Gut*, **23**, 188–93.

Hellemans, J., *et al.* (1984). Positive $^{14}CO_2$ bile acid breath test in elderly people. *Age and Ageing*, **13**, 138–43.

Hellier, M.D. (1987). Coeliac disease. In *Diseases of the Gut and Pancreas*. (Eds. J.J., Misiewicz, R.E. Pounder, and C.W. Venables) pp. 619–39. Blackwell, Oxford.

Holmes, G.K.T., *et al.* (1976). Coeliac disease gluten free diet and malignancy. *Gut*, **17**, 612–19.

Holt, P.R. (1985). The small intestine. *Clinical Gastroenterology*, **14**, 689–724.

Ireland, P. and Fordtran, J.S. (1973). Effect of dietary calcium and age on jejunal calcium absorption in humans studied by intestinal perfusion. *Journal of Clinical Investigation*, **52**, 2672–8.

Jamieson, W.G, Marchuk, S., Rowsom, J., and Durhand, D. (1982). The early diagnosis of massive acute intestinal ischaemia. *British Journal of Surgery*, **69** (Suppl.), S52–3.

King, C.E., Leibach, J., and Toskes, P.P. (1979). Clinically significant vitamin B_{12} deficiency secondary to malabsorption of protein bound vitamin B_{12}. *Digestive Diseases and Sciences*, **24**, 397–402.

King, C.E. and Toskes, P.P. (1986). Comparison of the 1-gram [^{14}C]xylose, 10 gram lactulose-H_2, and 80-gram glucose-H_2 breath tests in patients with small intestinal bacterial overgrowth. *Gastroenterology*, **91**, 1447–51.

Levitt, M.D. and Donaldson, R.M. (1970). Use of respiratory hydrogen (H_2) excretion to detect carbohydrate malabsorption. *Journal of Laboratory and Clinical Medicine*, **75**, 937–45.

Lipski, P.S., Bennett, J.K., Kelly, P.J., and James, O.F.W. (1992*a*). Ageing and duodenal morphometry. *Journal of Clinical Pathology*, in press.

Lipski, P.S., Kelly, P.J., and James, O.F.W. (1992*b*). Bacterial contamination of the small bowel in elderly people: is it necessarily pathological? *Age and Ageing*, **21**, 5–12.

Loughran, T.P., Kadin, M.E., and Deeg, H.J. (1986). T-cell intestinal lymphoma associated with coeliac sprue. *Annals of Internal Medicine*, **104**, 44–7.

Marston, A. (1985). Ischaemia. *Clinical Gastroenterology*, **14**, 847–62.

Marston, A., Clarke, J.M.F., Garcia-Garcia, J., and Miller, A.L. (1985). Intestinal function and intestinal blood supply. *Gut*, **26**, 656–66.

Marx, J.J.M. (1979). Normal iron absorption and decreased red cell uptake in the aged. *Blood*, **53**, 204–11.

Mayberry, J., Rhodes, J., and Hughes, L.E. (1979). Incidence of Crohn's disease in Cardiff between 1934 and 1977. *Gut*, **20**, 602–8.

McEvoy, A., Dutton, J., and James, O.F.W. (1983). Bacterial contamination of the small intestine is an important cause of occult malabsorption in the elderly. *British Medical Journal*, **287**, 789–93.

McEvoy, A.W., Fenwick, J.D., Boddy, K., and James, O.F.W. (1982). Vitamin B_{12} absorption from the gut does not decline with age in normal elderly humans. *Age and Ageing*, **11**, 180–3.

Montgomery, R.D., *et al.* (1986). Cases of malabsorption in the elderly. *Age and Ageing*, **15**, 235–40.

Montgomery, R.D., *et al.* (1978). The ageing gut: a study of intestinal absorption in relation to nutrition in the elderly. *Quarterly Journal of Medicine*, **47**, 197–211.

Mylvaganam, K., Hudson, P.R., Herring, A. and Williams, C.P. (1989). ^{14}C-triolein breath test: an assessment in the elderly. *Gut*, **30**, 1082–6.

Novak, J.M., *et al.* (1979). Effects of radiation on the human gastroinestinal tract. *Journal of Clinical Gastroenterology*, **1**, 9–39.

Pettersson, T., Wegelius, O., and Skrifvars, B. (1970). Gastroinestinal disturbances in patients with severe rheumatoid arthritis. *Acta Medica Scandinavica*, **188**, 139–44.

Price, H.L., Gazzard, G.B., and Dawson, A.M. (1977). Steatorrhoea, in the elderly. *British Medical Journal*, **1**, 1582–4.

Recker, R.R. (1985). Calcium absorption in achlorhydria. *New England Journal of Medicine*, **313**, 70–3.

Rumessen, J.J., Hamberg, O., and Gudmand-Hoyer, E. (1990). Interval sampling of end-expiratory hydrogen (H_2) concentrations to quantify carbohydrate malabsorption by means of lactulose standards. *Gut*, **31**, 37–42.

Russell, R.M., *et al.* (1987). Folic acid malabsorption in atrophic gastritis. *Gastroenterology*, **91**, 1476–82.

Russell, R.M. (1988). Malabsorption and aging. In *Aging in the liver and Gastro-Intestinal Tract* (eds. L. Bianchi, P.R., Holt and O.F.W. James) pp. 297–307. MTP, Lancaster.

Ryder, J. B. (1963). Steatorrhoea in the elderly. *Gerontology Clinics*, **5**, 30–7.

Schuffler, M.D., *et al.* (1981). Chronic intestinal pseudo obstruction. *Medicine*, **60**, 173–96.

Stowe, S.P., Redmond, S.R., and Stormont, J.M. (1990). An epidemiological study of inflammatory bowel disease in Rochester, New York. *Gastroenterology*, **98**, 104–10.

Swinson, C. Slavin, A., Coles, E.C., and Booth, C.C. (1984). Coeliac disease and malignancy. *Lancet*, **i**, 111–14.

Swinson, C.M. and Levi, A.J. (1980). Is coeliac disease underdiagnosed? *British Medical Journal*, **281**, 1258–60.

Tchirkow, G. Lavery, I.C. and Fazio, V.W. (1983). Crohn's disease in the elderly. *Diseases of the Colon and Rectum*, **26**, 177–81.

Tytgat, G.N.J. and van Dongen, R.J.M. (1984). Ischaemic disease of the small and large intestine. In *Gastrointestinal Tract Disorders in the Elderly* (eds. J. Hellermans and G. Vantrappen) pp. 107–24. Churchill-Livingstone, Edinburgh.

Warren, P.M., Pepperman, M.A., and Montgomery, R.D. (1978). Age changes in small-intestinal mucosa. *Lancet*, **ii**, 849–50.

Webster, S.G.P. and Leeming, J.T. (1975). The appearance of the small bowel mucosa in old age. *Age and Ageing*, **4**, 168–74.

Werner, I. and Hambraeus, L. (1972). The digestive capacity of elderly people. *Nutrition in Old Age* (ed. L.A. Carlson) pp. 55–60. Almqvist and Wiksell, Uppsala.

8.6 Colonic disease in elderly people

N. W. READE

CONSTIPATION

Prevalence

Constipation is common in elderly people, particularly in those who are admitted to hospital. In a recent survey of 453 people living in Sheffield, we found that the prevalence of constipation was 12 per cent in those living in the community. This rose to 41 per cent of patients in acute geriatric wards and over 80 per cent of patients in long-stay geriatric wards. Most of the elderly constipated patients living in the community had had constipation all their lives, suggesting that constipation should not be regarded as a normal consequence of ageing. Instead, the abnormally increased prevalence in patients in hospital suggests an association with the immobility, illness, and mental changes that can occur with ageing.

Pathophysiology

Constipation is a symptom and can be a side-effect of many diseases, including metabolic disease such as hypothyroidism, colonic diseases, and neurological disorders. Constipation is also a side-effect of treatment with many drugs (Table 1), an important point in elderly patients who often have a variety of other diseases and can be the victims of polypharmacy. Despite the multiple and varied causes for constipation, no obvious pathological lesion can be found in the majority of constipated elderly patients.

Table 1 Drugs that may cause constipation in elderly patients

Opiate analgesics (including dextropropoxyphene)
Antacids (calcium and aluminium compounds)
Anticholinergic agents
Anticonvulsants
Antidepressants
Antiparkinsonism agents
Ganglion blockers
Diuretics
Iron
Antihypertensive agents
Psychotherapeutic drugs
Monoamine oxidase inhibitors
'Irritant' laxatives

A deficiency of dietary fibre?

The addition of bran to the diet increases stool weight, and improves stool consistency, reduces colonic transit time, and increases the frequency of defaecation. On average, constipated elderly patients living in the community take less fibre in their diet than elderly patients who are not constipated. It would, however, be a mistake to regard constipation as purely the result of a low fibre diet. Several studies have pointed out that constipated patients require more bran to achieve the same degree of stool weight than do normal subjects. Other studies in much younger subjects have indicated that personality may have a more important effect on bowel habit than fibre intake. Although an increase in fibre intake is important in the management of constipation in elderly patients, it is often unsuccessful by itself. These observations suggest that constipation is a motility disorder and bran may work because it functions as a mild laxative.

Faecal impaction

The majority of patients with constipation have faecal impaction or dyschezia. Masses of hard faeces accumulate in a commodious rectum. Anorectal function tests reveal no obvious sphincter spasm or obstruction in these patients. Anal pressures tend to be lower in patients with faecal impaction than they do in an age- and sex-matched control group. The internal anal sphincter relaxes normally to distension of the rectum with a balloon and there is no evidence that these patients block their own defaecation by contracting the external sphincter inappropriately. Patients with faecal impaction can expel simulated stools from the rectum more rapidly and with less effort than matched controls.

Not surprisingly, the rectal capacity is markedly increased in patients with faecal impaction so that volumes of 500 ml may be accommodated without difficulty. Rectal sensation is also considerably impaired. Thus, physiological data suggest that elderly patients with faecal impaction are unable to defaecate because they cannot detect the presence of faeces in the rectum until the mass becomes too large to expel. The propulsion of markers through the rest of the colon appears normal in these patients.

It is tempting to suggest that the persistent disregard of a call to stool in a confused, depressed, or immobile patient may lead to adaptive changes in rectal tone and sensitivity. A very similar condition can occur in children who withhold their stool and this can, in some cases, be treated with behavioural training. The success of behavioural training in some elderly patients suggests adaptation to faecal retention, but the combination of the hypercompliant and insensitive rectum with a weak anal sphincter and anal and perianal insensitivity, resembles the findings in patients with a lesion in the low spinal cord or cauda equina.

'High' faecal impaction may occur in elderly patients at the rectosigmoid junction and beyond the reach of digital rectal examination. Plain abdominal radiography may be needed to detect this condition. The differential diagnosis is usually rectosigmoid carcinoma.

Neuropathic causes of constipation in the elderly

Defaecation is thought to be triggered by a centre in the brain-stem that receives input both from the pelvic organs as well as the cortex. This centre in the pons then influences the function of a co-ordinating centre at the end of the spinal cord in the conus medullaris, which integrates the activity of visceral efferents that act on the intrinsic nervous system of the colon to cause colonic propulsion and internal anal sphincter relaxation with somatic efferents that inhibit contraction of the striated muscle of the pelvic floor. In theory, diseases of the nervous system can result in constipation by acting on several sites; the brain, the spinal pathway, the conus medullaris in the sacral cord, the cauda equina and pelvic parasympathetic nerves, and the intrinsic nerves of the colon.

Loss of neurones is a common accompaniment of ageing. It is not surprising, therefore, that neurological disease affecting the colon and anal sphincter is more common in elderly than in younger patients. Vascular and neoplastic disease of the brain and spinal cord may be particularly implicated in constipation that comes on rapidly in an elderly person.

Occult spinal disease or injury is probably as common a cause of disordered defaecation as of disturbed micturition. Degenerative disc disease is common in elderly people and its onset may be insidious. Patients with a neurological lesion involving the cauda equina or low spinal cord show distal colonic dilatation and hypomotility, decreased rectal tone, impairment of rectal sensation, and loss of rectocolonic reflexes. Their inability to defaecate is due to a combination of impaired colonic propulsion, impaired rectal sensation, and absent reflex contractions of the rectum.

Mechanical causes of constipation

Partial rectal prolapse

Women, particularly those who have had several children, tend to develop progressive weakness of the pelvic floor as they get older. This causes abnormal perineal descent on straining, which may obstruct defaecation by two mechanisms. First the rectum bulges into the perineum tightening the puborectalis and accentuating the anorectal angulation. Second, the distal segment of the pudendal nerve is stretched and compressed against the ischial spine. The resulting weakness of the external anal sphincter predisposes to rectal prolapse; the rectum herniates through a weak sphincter. The entry of the rectum into the anal canal produces an intense desire to defaecate.

Patients with a partial rectal prolapse often complain of a frequent urge to defaecate but an inability to pass stool, despite long periods of straining. Some may perceive a sensation of something coming down and may be able to deflect the prolapsing rectal wall by inserting a finger in the rectum.

Rectocele

The increasing weakness of connective tissue that is associated with ageing may cause the rectal contents to bulge into a weakened posterior vaginal wall, so that the force of defaecation is deflected anteriorly. Women with a rectocele notice that they are only able to pass a stool through the anus by inserting their fingers into the vagina and pressing backwards against the bulging rectal wall.

Haemorrhoids

Haemorrhoids are common in elderly subjects. Patients with non-prolapsed or first-degree haemorrhoids often complain of constipation and a feeling of obstructive defaecation. They say that they can feel the stool become obstructed half-way through a bulging anus. Manometric studies in patients with non-prolapsing haemorrhoids show abnormally high pressures, often associated with ultraslow waves, and during rectal distension, the pressure in the outer anal canal remains high. Recent evidence suggests that this pressure barrier is probably maintained by the abnormally high pressure in the anal cushions. Thus defaecation may be blocked by the hypertrophied anal cushions, a situation that is relieved after prolonged straining has caused the haemorrhoids to prolapse.

Examination

The following points are important.

1. Digital examination of the rectum and vagina is mandatory. Cancer of the pelvic organs is common in elderly people. Other important signs are the presence of large masses of faeces (faecal impaction), the total absence of faeces (colonic inertia), and the presence of a rectocele.
2. Absent anal and perianal sensation indicates a neurological lesion, but is common in patients with faecal impaction.
3. Excessive perineal descent on straining suggests the possibility of a mechanical obstruction of defaecation, but is common in older people.
4. Constipation can complicate many neurological disorders. Thus, it is important to examine the spine for deformity

and tenderness, and carry out a full neurological examination.

5. The abdomen should be examined for the presence of tenderness, masses, and gaseous or fluid distension.
6. Proctoscopy may reveal fissures, haemorrhoids, and ulceration or redness of the anterior rectal wall (solitary rectal ulcer or anterior mucosal prolapse).
7. Sigmoidoscopy may reveal neoplasms, strictures, diverticulosis, and colitis; it is not generally appreciated that patients with ulcerative colitis or pneumatosis coli are often severely constipated even though they may be passing blood and mucus from their inflamed mucosa.

Examination of the stool can be helpful, but is often overlooked. Small hard pellets suggest the irritable bowel syndrome and diverticular disease. Blood and mucus are found in neoplastic disease, but also in ulcerative colitis, pneumatosis coli, Crohn's disease, and haemorrhoids. Ribbon-like stools may suggest anal stenosis (which may follow surgery or trauma or Crohn's disease), but are also common in patients with the irritable bowel syndrome or haemorrhoids.

Diagnosis and investigation

A patient who presents with faecal impaction for the first time during admission to hospital with another condition probably requires no specific investigation for constipation. Investigations are required, however, in those patients with recent onset of chronic constipation, particularly when this cannot be explained by a change in their medication. Blood samples should be taken for thyroid function tests, calcium and urea, and electrolytes. It is important to remember that hypercalcaemia, hypokalaemia, and dehydration can all be associated with constipation. A barium enema and sigmoidoscopy is mandatory in such a patient to exclude neoplastic and inflammatory conditions and to identify diverticular disease, and megacolon, which may indicate an intrinsic or extrinsic neuropathy.

Defaecography and anorectal manometry may help to identify causes of obstructed defaecation, such as solitary rectal ulcer syndrome, non-prolapsing haemorrhoids, and abnormal descent of the pelvic floor.

Tests for neurological disorders may be indicated when constipation occurs rapidly or is associated with concomitant disorders of locomotion or micturition.

Measurement of gastric emptying and small bowel transit raise the possibility of a neuropathy involving the whole gut.

Treatment

Elderly patients admitted to acute geriatric wards with faecal impaction are often weak and unable to strain effectively. The impaction should be treated by manual evacuation and saline cathartics, followed by maintenance therapy with bulk laxatives (ispaghula) and if necessary irritant laxatives, and toilet training. The patients should be encouraged to sit on the toilet or a commode for 10 min twice a day. Excessive straining should be avoided and softening the stool with dioctyl or bulk laxatives can make evacuation easier. Bed pans are particularly stressful and tiring for elderly people, and whenever possible patients should use a commode.

Chronic administration of saline purgatives can occasionally cause dangerous fluid retention and should be avoided in elderly patients.

DIVERTICULAR DISEASE

Prevalence

Diverticular disease is probably a normal consequence of ageing. One-third of people over 60 and 90 per cent of people over the age of 90 have diverticulosis.

Pathogenesis

Histopathological studies show a shortening of the longitudinal muscle and thickening of the circular muscle of the distal colon. The disease is thought to result from excessive deposition of elastic tissue in the longitudinal muscle of the colon.

The normal taeniae coli act as suspension cables upon which the circular muscular arcs are suspended. This enables the circular muscle to contract against the taeniae in a very efficient way. A 17 per cent contraction of circular muscle reduces the luminal diameter by two-thirds. If the taenia shorten because of increase in elastic tissue they cause a disproportionate increase in the thickness of the circular muscle layer so that it folds up upon itself to produce the characteristic concertina-like appearance. Folds of circular muscle are much more closely apposed than normal and the bowel lumen can be readily occluded by minimal contraction of the circular muscle. Very high pressures develop in the short segments of colon between the circular muscle folds. The high pressures are thought to be responsible for pushing out pockets of mucosa (diverticula) through the gaps in the muscle that are the entry sites of blood vessels. The stagnant diverticula can readily become infected leading to diverticulitis, perforation, abscess formation, and frank haemorrhage from erosion of the large blood vessels that run in the necks of the diverticula. Repeated attacks of inflammation, particularly if they are associated with extramural abscess formation can ensheath the distal colon in a rigid case of fibrous tissue. This can lead to colonic obstruction due to impaired propulsion through the affected segment and fibrous stenosis.

Diverticular disease is uncommon in societies that consume a high fibre diet, and it has been suggested that such a diet may delay the onset of diverticular disease by several years. It is thought that elastogenesis is triggered by intermittent increases in wall tension. A diet that is low in residue is considered to cause intermittent distension. Bulky stools that cause more consistent distension of the colon induce less elastosis.

Presentation

Patients with diverticular disease often complain of constipation and abdominal cramping. These features resemble the symptoms of the irritable bowel syndrome and some investigators believe that irritable bowel syndrome precedes diverticulosis. It is likely that constipation is related to the moulding of colonic contents into hard faecal pellets by the abnormally powerful contractions in the sigmoid colon. Small, hard spheres (1 cm diameter) require more force and take longer to expel from the rectum than larger softer objects. Moreover, the rectum in the elderly patient, especially if he or she has features of the 'irritable bowel' is more sensitive than it is in the normal young person. Thus, the entry of small pellets is more likely to give rise to a desire to defaecate.

Diverticular disease can also present with acute emergencies such as profuse rectal bleeding, and acute left-sided abdominal pain, caused by diverticulitis. Fistulae can form between adjacent viscera. Vesicocolonic fistulae are particularly common. Small bowel obstruction is more commonly seen than large bowel obstruction and results from tethering of loops of small bowel that lie in the pelvis by adhesions surrounding pericolic inflammation or abscess.

Treatment

Diverticulosis and constipation is best managed by a combination of dietary fibre, bulk laxatives, and antispasmodics, but complications of diverticulitis, abscess, bleeding, and intestinal obstruction often need surgical intervention. Taenia myotomy is said to restore the affected colon to a normal length and prevent further symptoms.

FAECAL INCONTINENCE

Faecal incontinence is a common problem in elderly patients. Sphincter pressures are much lower, particularly in women, than they are in young people.

Faecal impaction with overflow incontinence

The most common cause of faecal incontinence is faecal impaction, even though sphincter pressures are no different in non-impacted elderly patients. The major cause of incontinence in impacted patients is the impairment of rectal sensitivity. When the rectum is distended, the internal sphincter often relaxes well before the patient is aware of any sensation of distension and is able to react with a compensatory contraction of the external anal sphincter. This problem is exacerbated by the very considerable impairment of anal and perianal sensation in impacted patients.

Management involves treatment of the impaction with laxatives and toilet training.

Pudendal neuropathy

Damage to the pudendal nerve increases with age, particularly in multiparous women. This is related to progressive weakness of the pelvic floor, leading to an abnormal degree of perineal descent which stretches the distal segment of the pudendal nerve and compresses it against the ischial spine. Damage to the nerve causes weakness of the external anal sphincter and also the external urethral sphincter. Thus elderly patients with weak pelvic floors often suffer from both faecal incontinence and stress incontinence of urine. If the patient is fit, does not have faecal impaction, and the sphincter pressures are not too low, then a surgical postanal repair of the pelvic floor often provides a good result. If this option is contraindicated, then judicious use of opiate-like antidiarrhoeal agents may harden the stools sufficiently to give the patients more control and confidence.

Other causes of faecal incontinence

The progressive weakness of the sphincters with age and the limited mobility of the patient often renders them at risk of incontinence if they have an attack of diarrhoea. Attention should be focused on the management of the diarrhoea in these cases. Vascular and neoplastic disease and spinal or root compression by degenerative disc disease are all common in an ageing population. These conditions may lead to incontinence by impairing rectal sensitivity and conscious control of the sphincter.

COLORECTAL NEOPLASMS

Benign neoplastic conditions

The most common form of colorectal tumour is the metaplastic polyp. These are found particularly in the rectum and increase in number with age. Metaplastic polyps are found in about 95 per cent of patients with carcinoma, and their incidence is raised in populations with a high risk of developing colorectal cancer.

Adenomas, the other benign tumour, are found particularly in the rectum and a great many of them produce symptoms of bleeding and passage of mucus. Malignancy rate varies from 10 to 40 per cent and differs with the pathology. Tubular adenomas have the lowest malignancy whereas the least common, the villous adenoma, has a malignancy rate of about 40 per cent. About 5 to 10 per cent of the population over 40 will have one or more adenomas and the risk rises with increasing age.

Carcinoma (see Section 10)

Carcinoma of the colon and rectum is the most common malignancy in old age (excluding prostatic carcinoma in man). Both ulcerative colitis and Crohn's disease are associated with a high risk of colorectal cancer and there are several studies that have indicated that the incidence of colonic carcinoma is higher in patients who have had cholecystectomy. This is probably related to increased entry of bile acids into the colon. Colorectal cancer is thought to be higher in populations that consume large amounts of meats and fat, but dietary fibre and possibly also carbohydrate may have a protective effect.

Clinical presentation

Although colonic cancer may classically present with rectal bleeding, abdominal pain and constipation, presentation is often insidious with increasing apathy and loss of weight. Anaemia and a change in bowel habit are important pointers to this condition.

Investigation

A combination of good quality double-contrast barium enema and colonoscopy should afford a diagnosis in 90 per cent of patients.

Treatment

Surgical excision is the only curative treatment. The outcome of surgical treatment is better than with cancer at many other sites.

INFLAMMATORY BOWEL DISEASE

Crohn's disease

Prevalence and pathogenesis

Crohn's disease is becoming more important in the elderly population, firstly because of the longer survival of patients with onset earlier in life and second because there is a small peak of incidence of new cases in the 6th and 7th decades. The older patients are more likely to be female, less likely to have ileal involvement, and particularly likely to have involvement of the left colon. This may lead to diagnostic confusion with diverticulitis since both may present with pain and diarrhoea, abdominal mass, rectal bleeding, and the development of internal fistulae. Clinical features favouring diagnosis of Crohn's disease include an anal lesion, a recto-vaginal fistula, and systemic complications including erythema nodosum, finger clubbing, arthritis, uveitis, and pyoderma gangrenosum. Radiographic evidence of involvement of the whole of the gastrointestinal tract, with deep fissures on radiographs and enteric fistulae strongly support the diagnosis. Colonic biopsy shows normal mucus content of epithelial cells in the presence of mucosal inflammation, and lymphoid aggegates in the mucosa and submucosa. Epithelioid granulomas only occur in about one-half of cases.

Treatment

This is similar to treatment in the young patient, and includes sulphasalazine, steroids, and surgery for complications. The role of elemental and exclusion diets in the long-term management of Crohn's disease looks very promising.

ULCERATIVE COLITIS

Prevalence

It is now well recognized that ulcerative colitis may present as a new disease in late life. A second peak at 60 to 70, restriction of the disease to the rectum, plus a higher incidence in females than males suggests that the aetiology may be different from the disease in the young. Diarrhoea is the most common presenting symptom in the late onset group: bleeding is a more common presentation in younger patients.

Presentation

The presentation of ulcerative colitis in old age requires full investigation since it may be very similar to that of diverticular disease or carcinoma. The inflammation is often confined only to the rectum but this may not always be the case. The disease often runs a mild and intermittent course with passage of loose stools and blood-stained mucus, tenesmus, and urgency.

Patients with active colitis often appear to retain colonic contents in the proximal colon; they have proximal colonic constipation and distal diarrhoea. Indeed, a less frequent bowel habit is often a useful signal that an attack of colitis is imminent.

Patients with extensive colitis lasting for 10 years or more have an increased risk of developing colorectal carcinoma. The risk increases with each subsequent decade of the disease by about 20 per cent. Because of the overlapping symptoms of carcinoma and colitis, regular colonscopy is important.

One condition that must be differentiated from ulcerative colitis in the elderly patient is antibiotic-associated colitis. Many elderly people receive broad spectrum antibiotics and this complication has a high fatality.

Treatment

Steroids and sulphasalazine are the mainstays of medical treatment of ulcerative colitis. Surgery is generally indicated for widespread disease that has been present for a long time, for complications, and when response to therapy has been poor.

COLONIC ANGIODYSPLASIA

Colonic angiodysplasia is a significant cause of intestinal bleeding in people aged 60 or more. Presentation may vary from acute massive haemorrhage to chronic anaemia. Colonoscopic appearance has been described as foot processes irradiating from a slightly raised centre of a cherry-red lesion. These lesions do not disappear on insufflation or aspiration of air from the colon. The cause of these vascular malformations is not known though they may result from long continued intermittent partial and low-grade obstruction of submucosal veins, that pierce the circular and longitudinal layers of muscle of the colon. This leads to dilatation of the submucosal veins and venules and the development of small arteriovenous communications.

The diagnosis can only be confirmed by angiography or colonoscopy and the treatment of the condition is either colonoscopic electrocoagulation or colonic resection.

ISCHAEMIC COLITIS

Pathogenesis and presentation

Because of its less well developed collateral circulation in the presence of pathogenic bacteria, bowel ischaemia is more likely to occur in the colon than in the small bowel. The clinical effects vary according to the magnitude and duration of the ischaemia. Gangrene of the colon presents as a fulminating abdominal catastrophe. The more common non-gangrenous form of the disease usually clears spontaneously but may lead to the formation of a fibrous stricture at the splenic flexure. It is this last condition that is often referred to as ischaemic colitis, primarily a condition affecting elderly people.

Typically the patient presents with acute pain in the left iliac fossa, fever, and a moderate amount of dark rectal bleeding. Examinations suggests a localized left-sided peritonitis, often confused with acute diverticulitis. Leucocytosis is found at an early stage. The acute episode usually subsides within a week. In about half the patients a stricture develops, which leads to a permanent abnormality on barium enema.

Diagnosis

The changes seen on barium enema are typical. 'Thumb printing' is the earliest change and consists of a series of blunt semiopaque projections into the bowel and seen in the area of the splenic flexure. These changes may disappear rapidly or persist for a few weeks or progress to the more mature changes of ulceration, narrowing of the bowel, and formation of a local stricture.

Treatment is conservative if possible but surgery may be needed if obstruction develops.

PNEUMATOSIS COLI

This uncommon condition also presents with the passage of blood-stained mucus and constipation. Endoscopic examination of the colon reveals that the epithelial surface is pushed up into mounds, which can be punctured to release gas. The mounds are produced by the formation of fermentation gases in the colonic submucosa. It is thought that the ischaemia of the colonic epithelium allows the invasion of anaerobic bacteria which then ferment endogenous protein and carbohydrate leading to the accumulation of gas.

Treatment with broad spectrum antibiotics helps but only for a comparatively short time. Longer-term resolution is afforded by raising the partial pressure of oxygen in the blood by getting the patient to breathe high concentrations of oxygen through a Venturi mask. Presumably the raised oxygen tension in the colonic mucosa discourages the growth of the fermentative bacteria at that site.

Bibliography

Bannister, J.J., Abouzekry, L.A., and Read, N.W. (1987). The effect of ageing on anorectal function. *Gut*, **28**, 353–7.
Brocklehurst, J.C. (1985). Colonic disease in the elderly. *Clinics in Gastroenterology*, **14**, 725–47.
Devroede, G. (1983). Conspitation, mechanisms and management. In *Gastrointestinal Disease* (eds. M. Sleisenger and J. S. Fordtran), 3rd edn. pp. 288–303. W. B. Saunders, Philadelphia.
Edwards, C.A., Tomlin, J., and Read, N.W. (1988). Fibre and constipation. *British Journal of Clinical Practice*, **42**, 26–32.
Read, N.W. and Timms, J. (1986). Defaecation and pathophysiology of constipation. *Clinics in Gastroenterology*, **13**, 937–65.
Read, N.W., Abouzekry, L., and Read, M.G. (1989). Anorectal function in elderly patients with faecal impaction. *Gastroenterology*, 959–66.
Whiteway, J. and Marson, B.C. (1985) Pathology of ageing—diverticular disease. *Clinics in Gastroenterology*, **14**, 829–46.

8.7 Diseases of the exocrine pancreas

LUCIO GULLO

Pancreatic diseases have been recognized with increasing frequency in the last 20 to 30 years; while some of this increase is certainly due to improved clinical knowledge and diagnostic techniques, the possibility of a real increase cannot be excluded. The elderly population is well represented among patients with pancreatic disorders: indeed, while chronic pancreatitis is a disease of the young, acute pancreatitis and pancreatic cancer tend to occur more frequently with advancing age.

In this chapter, I will describe inflammatory and neoplastic diseases of the exocrine pancreas as they are encountered in elderly patients. It should be noted that the clinical picture of these diseases as well as the diagnostic and therapeutic problems that they pose in older patients do not differ substantially from those in younger ones; however, some features are unique to elderly people, and these will be pointed out.

INFLAMMATORY DISEASES

Inflammatory diseases of the pancreas include acute and chronic pancreatitis. The definition of these diseases and their inter-relationship have been the object of considerable discussion in the past. It was long held that pancreatitis was a single disease entity, with chronic pancreatitis being considered a late stage of acute pancreatitis; now, however, there is sufficient evidence to establish that they are two distinct entities. Acute pancreatitis is an inflammatory condition that usually regresses when the cause is eliminated; if the cause persists, the disease may recur, but generally there is still no progression to chronicity. Chronic pancreatitis, on the other hand, is chronic *ab initio*, producing lesions which persist, and usually progress, even when the causal factor is eliminated.

Acute pancreatitis

Acute pancreatitis is an acute inflammatory disease which may be determined by various aetiological factors. Histologically, it is characterized by a wide spectrum of lesions in the pancreas and peripancreatic tissues, including oedema, fatty necrosis, parenchymatous necrosis, and haemorrhage. Clinically, abdominal pain and raised concentrations of pancreatic enzymes in the blood and urine are the most common features. In most patients (about 75 per cent), inflammation does not progress beyond the state of oedema and fatty necrosis, and the disease has a mild, uncomplicated course.

Incidence

Such data as are available, largely from studies done many years ago, indicate that the incidence in the general population of Western countries is about 10 to 20 cases per 100 000 per year. Because about half of the patients with acute pancreatitis are over 60 years of age, it may be estimated that the incidence in the elderly population is about 5 to 10 per 100 000 per year. The prevalence is not known.

Age and sex

Among elderly people, most cases are aged between 60 and 70 years in some studies (Fig. 1), and 70 to 80 years in others (Fig. 2). The slightly higher number of women (Fig. 2) may be partly due to a higher incidence of gallstone-associated pancreatitis in later life.

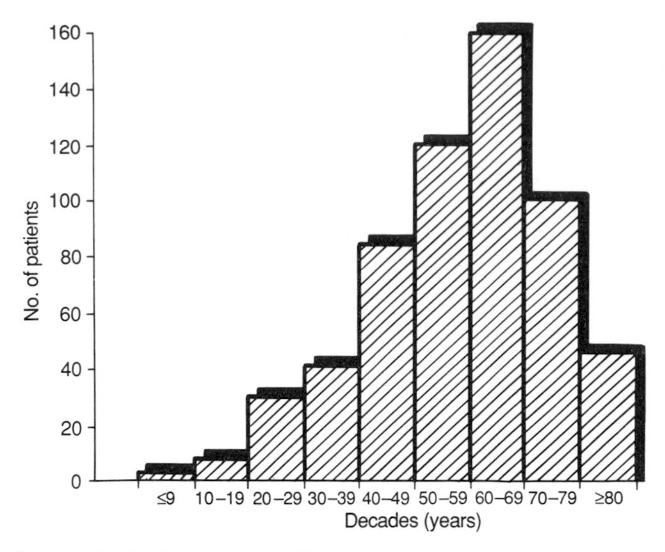

Fig. 1 Distribution by age of 590 patients with acute pancreatitis (constructed from data reported in Trapnell, J.E. and Duncan, E.H.L. (1975). Patterns of incidence in acute pancreatitis. *British Medical Journal*, **2**, 179–83).

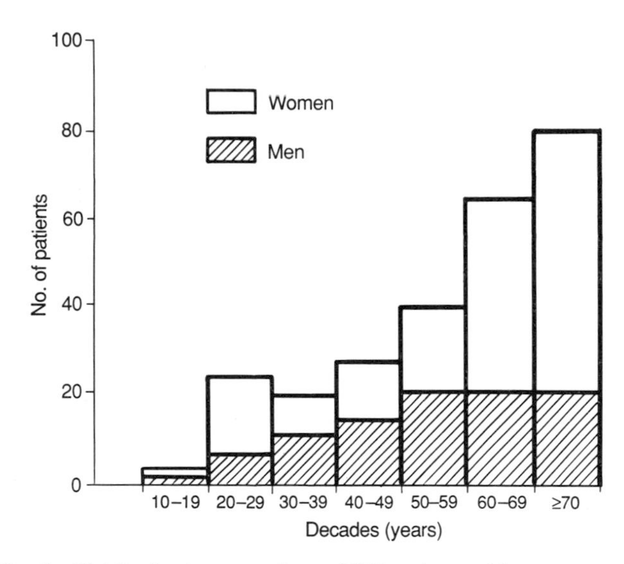

Fig. 2 Distribution by age and sex of 257 patients with acute pancreatitis (constructed from data reported in MRC. (1977). Death from acute pancreatitis. MRC multicentre trial of glucagon and aprotinin. *Lancet*, ii, 632–5).

Aetiology

Gallstones are the most frequent cause of acute pancreatitis in elderly people, present in up to 60 per cent of patients (Table 1). Several other factors or conditions may have an aetiological role, but these are less common, accounting collectively for about 15 per cent of the total (Table 1). Alcohol abuse is a frequent aetiological factor in patients under 50 years of age, but is rare in elderly patients. Postoperative pancreatitis, more common in later life, can complicate surgical procedures involving not only neighbouring organs (stomach and duodenum, biliary tract, and spleen), but also those more distant (especially the heart and aorta). Direct trauma to the pancreas and pancreatic ischaemia is thought to be responsible. In about 25 per cent of patients, no apparent cause is found.

Table 1 Main causes of acute pancreatitis in elderly people

Cause	%
Gallstones	60%
Surgery Alcohol abuse Drugs (frusemide, thiazides, azathioprine, sulphonamides, corticosteroids, oestrogens, tetracycline) Outflow obstruction (ampullary stenosis, pancreatic tumours, periampullary diverticula) Endoscopic retrograde cholangiopancreatography Endoscopic papillotomy Hypertriglyceridaemia Uraemia	15%
No apparent cause (idiopathic)	25%

Approximate percentages derived from the literature and our experience.

Pathogenesis

It is generally agreed that acute pancreatitis is the result of autodigestion of the pancreas by its own enzymes inappropriately activated within the gland. Trypsin is activated first, and, in turn, induces activation of the other proenzymes. Under normal conditions, autodigestion is prevented by several protective mechanisms: (1) synthesis of digestive enzymes as inactive zymogens; (2) segregation of these enzymes in vacuoles within the cellular cytoplasm; and (3) the presence of enzyme inhibitors in both pancreatic tissue and secretion. How these mechanisms are overcome and trypsin activated is unclear. In gallstone-associated pancreatitis, the triggering event is believed to be transient obstruction of the ampulla caused by a stone migrating from the biliary tree into the duodenum. This view is supported by the fact that stones are found in faeces of up to 95 per cent of patients with gallstone pancreatitis. The proposed mechanisms by which this event would precipitate acute pancreatitis are as follows: (1) reflux of bile into the pancreatic ductal system through a common channel; (2) reflux of duodenal contents into the pancreatic ducts because of a transient sphincter incompetence caused by the passage of the stone; and (3) a sudden increase in pressure within the pancreatic ducts. In each case, resulting alterations of the ductal mucosa would allow ductal contents to reach the parenchyma, with consequent injury to acinar cells and enzyme activation. Despite the large number of experimental studies, however, concrete evidence in support of these theories is still lacking.

Recently a new pathogenetic theory has emerged from studies performed in rats, in which acute pancreatitis was induced by a choline-deficient, ethionine-supplemented diet or by supramaximal pancreatic stimulation with the cholecystokinin analogue caerulein. In both these models, soon after induction of pancreatitis digestive enzyme secretion is blocked and the enzymes, which are normally separated from lysosomes, become colocalized with lysosomal hydrolases within large intracellular vacuoles. Here activation of trypsin by the lysosomal enzyme cathepsin B could initiate the cascade activation of the other pancreatic zymogens. The relevance of these findings to human acute pancreatitis is not yet clear; however, similar intracellular vacuoles have recently been described in human acute pancreatitis also.

However the trypsin activation occurs, this enzyme, in turn, also activates the other proenzymes, including phospholipase A and elastase, which have a major role in the development of parenchymal necrosis and haemorrhage. Trypsin also activates other cascade systems such as kallikrein-kinin, complement, coagulation, and fibrinolytic systems, leading to the release of vasoactive and cytotoxic substances, which cause vasodilatation, increased vascular permeability, leucocyte accumulation, and parenchymal damage. High concentrations of toxic oxygen radicals formed early in the course of the disease also contribute to acinar cell damage. Pancreatic enzymes and the other toxic substances may be released from the inflamed pancreas into the peripancreatic spaces, the peritoneal cavity, and systemic circulation; thus, they can damage not only pancreatic and peripancreatic tissues, but also organs distant from the gland.

The reasons why the course of acute pancreatitis can vary widely, from mild to severe or lethal, are unclear. An imbalance between activated proteases and their inhibitors in the pancreas and in systemic circulation (α_2-macroglobulin and α_1-antitrypsin) may play a significant role in determining the severity of the disease.

Clinical presentation

The clinical picture varies considerably with the severity of the disease. In milder forms, abdominal pain may be the only significant symptom and the disease can resolve in a few days; in the more severe forms, in addition to pain, various other symptoms and complications may occur, leading to a more serious and often fatal outcome. Between these two extremes, there is a wide range of intermediate forms.

Table 2 Main signs and symptoms of acute pancreatitis in elderly people

Pain	85–90%
Vomiting	80–90%
Anorexia	70–80%
Fever (usually less than 38.5°C)	80–90%
Abdominal tenderness	80–90%
Reduction of bowel sounds	60–70%
Jaundice	Less common
Tachycardia	Less common
Tachypnoea	Less common
Hypotension	Less common
Mental confusion	Less common
Body wall ecchymosis	Less common

Approximate percentages derived from the literature and our experience.

In the vast majority of patients (Table 2), abdominal pain is the primary symptom: usually it is sudden in onset, reaches maximal intensity within 30 to 60 min, and persists without relief for many hours (at least 10–12) or days. The pain is usually intense, often excruciatingly so; however, in some cases, especially in more elderly patients, it may be mild or the patient may complain of only a vague upper abdominal discomfort. The most frequent initial site of pain is the epigastrium (70–80 per cent), from which it can radiate to the right or left hypochondrium, throughout the abdomen or, more characteristically, into the back (T12–L2). Occasionally, pain originates in the right hypochondrium or in the lower abdomen. In about 5 to 15 per cent of cases pain may be absent; these painless forms often present with shock or coma, and are usually fatal.

Vomiting and varying degrees of nausea or anorexia are frequent (Table 2). There may be a low-grade fever; a higher temperature suggests the presence of septic complications. A mild, transient jaundice is seen in about 20 to 30 per cent of cases; it is usually due to oedema of the head of the pancreas with compression of the intrapancreatic portion of the common bile duct. A marked, persistent jaundice is due to stones obstructing the common bile duct, cholangitis or, more rarely, to compression by pseudocyst.

On physical examination, the patient may appear distressed and anxious, usually in proportion to the severity of the disease; some degree of mental confusion is fairly frequent, especially in more elderly patients. Dehydration may be a feature, especially when symptoms have been present for more than 24 h. Tachycardia, hypotension, and tachypnoea may also be present, mainly in severe forms. Though epigastric tenderness and some muscle guarding are common, true abdominal rigidity is unusual. Bowel sounds are usually diminished and sometimes absent. If a pseudocyst has formed, an epigastric mass may be palpable. Ecchymosis of the flanks (Grey-Turner's sign) or periumbilical area (Cullen's sign) are rare but ominous features, as they are indicative of severe haemorrhagic pancreatitis. On the whole, physical examination of the abdomen may be unremarkable, even in severe forms of the disease; indeed, a marked contrast between severity of symptoms and paucity of abdominal signs is characteristic of acute pancreatitis.

In the more severe forms, additional symptoms and signs due to local or systemic complications may be observed (Table 3).

Local complications

The most common local complication is pancreatic pseudocyst (30–40 per cent of cases), which usually develops about 1 to 6 weeks after onset of pancreatitis. This is a localized collection of fluid deriving from areas of liquefaction necrosis, containing tissue debris, pancreatic secretions, plasma protein, and blood. It arises within or adjacent to the pancreas and may extend into peripancreatic spaces, even far from the gland. Pseudocysts may be single or multiple, of varying dimensions, asymptomatic or symptomatic, causing persistent pain and raised serum pancreatic enzyme levels. Rupture, haemorrhage, infection, or compression of neighbouring viscera are possible complications. Spontaneous disappearance may occur in up to 40 to 50 per cent of cases.

Infection of necrotic tissue by Gram-negative bacteria from the intestinal tract is a serious and potentially lethal complication which occurs in about 50 to 60 per cent of patients with necrotizing pancreatitis, usually within 2 to 3 weeks of onset of the attack. It should be suspected when there is fever, leucocytosis, and clinical deterioration.

Pancreatic abscess is the most dreaded local complication, associated with high morbidity and fatality. Pain, fever, leucocytosis, ileus, and rapid deterioration of the patient are the main clinical manifestations. Without surgical intervention, fatality is almost 100 per cent.

Pancreatic ascites is a rare complication, usually due to

Table 3 Principal complications of acute pancreatitis

Local	Systemic
Pseudocysts	Cardiocirculatory
Infection of necrosis	Hypovolaemia
Abscess	Hyperdynamic circulatory state
Ascites	Hypotension
	Shock
	Non-specific electrocardiographic changes
	Myocardial infarction
	Pulmonary
	Pleural effusions
	Atelectasis
	Bronchopneumonia
	Pulmonary oedema
	Embolism of pulmonary arteries
	Hypoxaemia
	Acute respiratory distress syndrome
	Renal
	Oliguria
	Renal insufficiency
	Metabolic
	Hyperglycaemia
	Hypocalcaemia
	Hyperlipidaemia
	Abnormalities of the coagulation system
	Hypercoagulative state

disruption of secretory ducts or pseudocysts, with leakage of juice into the peritoneal cavity. It is characterized by high protein and pancreatic enzyme concentration in peritoneal fluid.

In rare cases, spreading peripancreatic necrosis may involve the small bowel or colon, and cause perforation with fistula formation. The spleen may also be involved by direct extension of necrosis or, secondarily, by splenic vein thrombosis, leading to splenic abscess, infarction, or haemorrhage.

Systemic complications

Cardiocirculatory

Hypovolaemia is an early and frequent complication. Plasma exudation from the pancreas and peripancreatic tissues and increased vascular permeability caused by excessive circulating vasoactive peptides are the main pathogenetic factors. Vomiting, nasogastric suction, and intestinal distension may be additional contributing factors. A loss of up to 30 to 40 per cent of the circulating plasma volume may occur early in necrotizing pancreatitis. Severe hypovolaemia and peripheral vasodilatation may result in shock. Fortunately, a better understanding of the pathophysiology of this disease and improved patient monitoring have considerably reduced its frequency.

In patients with necrotizing pancreatitis, a hyperdynamic circulatory state develops early in the course of the disease, characterized by decreased peripheral and pulmonary vascular resistances, increase in the shunt fraction of the pulmonary circulation, and a rise in the cardiac index. Vasoactive peptides released by the pancreas are believed to be responsible.

Non-specific electrocardiographic changes, severe myocardial ischaemia or infarction are not uncommon, especially in elderly patients.

Pulmonary

Pleural effusions (usually left-sided), atelectasis, bronchopneumonia, embolism of the pulmonary arteries, and pulmonary oedema can all be associated with acute pancreatitis. However, a significant number of patients (about 50 per cent, or even more among elderly patients) develop hypoxaemia in the absence of any of these pulmonary complications. This alteration is partly explained by arteriovenous shunting through the alveolar capillary bed, probably due to localized intravascular coagulation and/or the effects of vasoactive peptides. In some patients with severe, often fatal disease, pulmonary failure similar to that of the adult respiratory distress syndrome develops. Postmortem studies in these patients have shown changes similar to those seen in the so-called shock-lung syndrome. The pathogenesis of this serious complication is unclear; hypotheses include localized intravascular coagulation, increased vascular permeability, or a toxic action of phospholipase A and free fatty acids on pulmonary surfactant.

Renal

Oliguria is fairly common; usually it is due to hypovolaemia and regresses with adequate fluid replacement. Severe renal insufficiency is rare; in addition to hypovolaemia, localized intravascular coagulation with obstruction of glomerular capillaries probably plays a pathogenetic role.

Disturbances in carbohydrate metabolism

A mild and transient hyperglycaemia is common, whereas overt diabetes is rare. Hyperglycaemia is associated with low serum insulin levels and increased glucagon concentrations, indicating that the endocrine pancreas is also damaged.

Hypocalcaemia

This alteration develops in about one-third of patients, mainly those with necrotizing pancreatitis. It is primarily due to loss of serum albumin and the consequent fall in circulating protein-bound calcium. Sequestration of calcium in areas of fat necrosis plays a minor role.

Abnormalities of the coagulation system

Serum concentrations of factors V, VIII and fibrinogen rise together with those of other acute phase reactants. In patients with necrotizing pancreatitis, there is a tendency toward the development of a hypercoagulative state. Disseminated intravascular coagulation may occur, but is a rare event.

Other complications may occasionally be seen, including subcutaneous fat necrosis, arthritis and synovitis, fat necrosis in bone marrow and in the central nervous system, pancreatic encephalopathy, and sudden blindness due to retinal artery occlusion with aggregated granulocytes.

Fatality

There are few reports on case fatality rates for acute pancreatitis in elderly patients: a study published about 30 years ago reported a rate of 40 per cent and a more recent one reported 20 per cent. The latter percentage is about twice the current reported rate for all patients with acute pancreatitis. Severe multiorgan failure is the most frequent cause of death in elderly patients. Contrary to a common assumption, in a recent study of acute pancreatitis in patients over 70 years of age, coexistent medical illnesses did not significantly increase fatality.

Diagnosis

A detailed clinical history and careful physical examination are of fundamental importance for establishing a diagnosis as well as for the selection of a limited number of appropriate diagnostic procedures. Diagnostic investigations serve to: (1) confirm clinical diagnosis; (2) assess the severity of the disease; and (3) establish the cause. To these ends, various biochemical and imaging examinations are now available.

Laboratory tests

The determination of pancreatic enzymes in serum is the most useful test for confirming clinical diagnosis. At present, various enzymes can be measured, including amylase, pancreatic isoamylase, lipase, trypsin, and elastase. Serum concentrations of pancreatic enzymes usually rise within a few hours of the onset of the attack and then progressively decrease to normal within 2 to 7 days. Amylase is usually the first to return to normal levels, elastase the last. There is no relationship between the severity of pancreatitis and the degree of enzyme elevation; indeed, patients with mild oedematous pancreatitis may have very marked elevations, whereas those with extensive pancreatic necrosis may have normal or only slightly increased levels. The sensitivity of these various enzyme rises is high (95–100 per cent), especially during the first 24 to 48 h of the attack. With the exception of amylase, the specificity is also high (more than 90 per cent). In practice, the determination of only one enzyme is sufficient for confirming diagnosis. Amylase assay, for its simplicity and rapidity, is the most widely used. However, because various extrapancreatic diseases may cause hyperamylasaemia (Table 4), an alternative is lipase, which is more specific, and which can now be determined by very simple and rapid methods. The measurement of trypsin and elastase involves radioimmunoassay, which precludes routine usage. The determination of amylase in urine does not usually add further diagnostic information.

Enzyme assay is useful for diagnosis, but not for assessing the severity of the disease. To this end, other biochemical changes may be helpful; these include hyperglycaemia, increased levels of blood urea nitrogen (BUN), of lactic dehydrogenase, of bilirubin, alkaline phosphatase, and transaminases (the last three are especially helpful in gallstone-associated pancreatitis), and hypocalcaemia, hypoalbuminaemia, and hypoxaemia. Leucocytosis and low haematocrit may also be present. The frequency and degree of these changes correlate with the severity of acute pancreatitis, and are included in various objective systems for grading the severity of the disease (Tables 5 and 6).

In addition, there is an increase in serum acute phase proteins; among these C-reactive protein seems to be a good marker for necrotic pancreatitis (Fig. 3). However, because serum C-reactive protein rises 2 to 3 days after the onset of the attack, it cannot be used for early assessment of severity. In current studies, leucocyte elastase and interleukin 6 are showing promise as early markers for severity.

Hypertriglyceridaemia or hypercalcaemia may be a feature in the rare cases of acute pancreatitis due to abnormally high serum levels of triglycerides or hyperparathyroidism.

Table 4 Principal non-pancreatic causes of hyperamylasaemia

Salivary gland diseases
Parotitis
Mumps
Sialoadenitis
Tumours
Gastrointestinal disorders
Afferent loop syndrome
Peptic ulcer perforation
Intestinal obstruction
Mesenteric infarction
Acute appendicitis
Hepatobiliary diseases
Cirrhosis
Fulminant hepatitis
Acute cholecystitis
Acute and chronic renal insufficiency
Diabetic ketoacidosis
Lung and colon cancer
Gynaecological disorders
Ruptured ectopic pregnancy
Ovarian tumours and cysts
Salpingitis
Macroamylasaemia

Table 5 Clinical and laboratory values indicative of severe acute pancreatitis*

At admission:
Age over 55 years
White blood cell count over 16000/mm^3
Blood glucose over 11 mmol/l (200 mg/dl)
Serum LDH over 350 IU/l (normal: up to 225)
Serum aspartate transaminase over 250 Sigma Frankel Units per litre (normal: up to 40)
During initial 48 h:
Haematocrit fall greater than 10%
Blood urea nitrogen rise more than 5 mg/dl (plasma urea rise of 1.8 mmol/l)
Serum calcium below 2 mmol/l (8 mg/dl)
Arterial P_{O_2} below 8 kPa (60 mmHg)
Base deficit greater than 4 mEq/l
Estimated fluid sequestration more than 6 l

*Patients who display three or more of these values are predicted to have a severe attack. (From Ranson, J.H.C., *et al.* (1974). Objective early identification of severe acute pancreatitis. *American Journal of Gastroenterology*, **61**, 443–51.)

Imaging procedures

These may be very helpful for confirming the clinical diagnosis, assessing the nature and extent of structural damage, and also for excluding other pathological conditions.

Plain abdominal and chest radiographs furnish non-speci-

Table 6 Modified early prognostic signs for patients with acute pancreatitis associated with gallstones

At admission:
Age over 70 years
White blood cell count over 18 000/mm^3
Blood glucose over 12 mmol/l (220 mg/dl)
Serum lactate dehydrogenase (LDH) over 400 IU/l
Serum aspartate transaminase over 500 Sigma Frankel U per litre
During initial 48 h:
Haematocrit fall greater than 10%
Blood urea nitrogen rise more than 2 mg/dl (plasma urea rise of 0.7 mmol/l)
Serum calcium below 2 mmol/l (8 mg/dl)
Base deficit greater than 5 mEq/l
Estimated fluid sequestration more than 4 l

(From Ranson, J.H.C. (1979). The timing of biliary surgery in acute pancreatitis. *Annals of Surgery*, **189**, 654–62.)

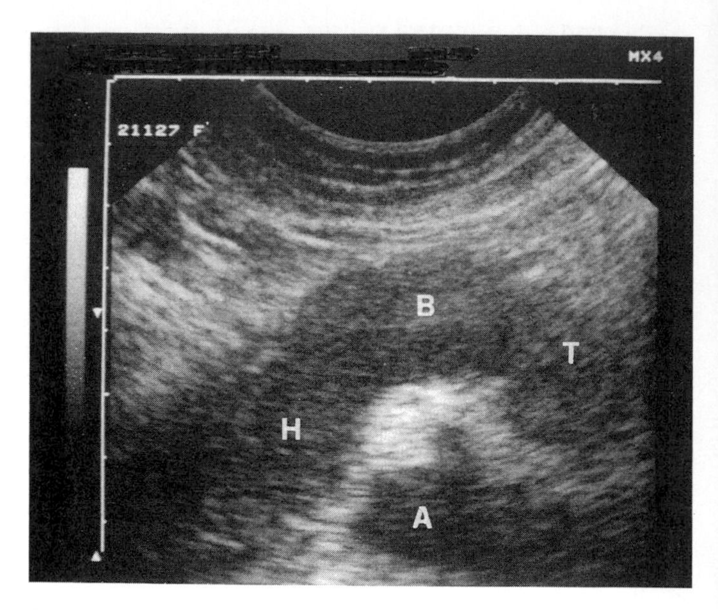

Fig. 4 Acute oedematous pancreatitis. Abdominal sonogram shows a diffusely enlarged and hypoechoic pancreas. Head (H), body (B), and tail (T) of the pancreas; aorta (A). (By courtesy of Dr P.L. Costa, Morgagni Hospital, Forlì.)

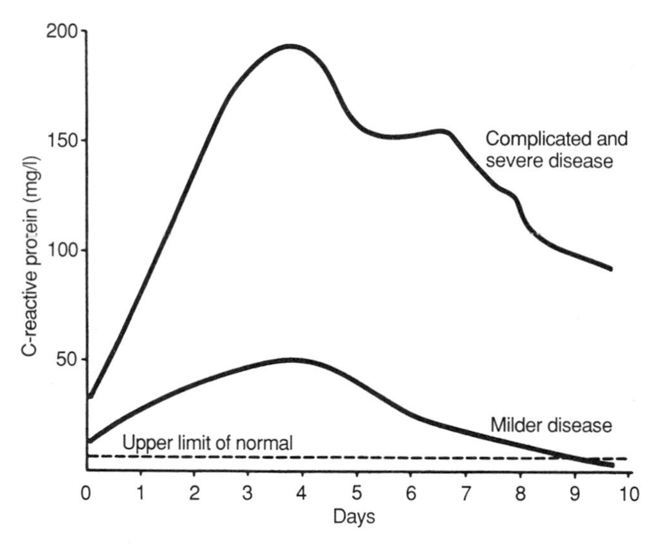

Fig. 3 C-reactive protein pattern in mild and in severe acute pancreatitis (reproduced with permission from Imrie, C.W. and Wilson, C. (1988). Systemic manifestations and the haematological and biochemical consequences of acute pancreatitis. In *Acute pancreatitis* (eds. G. Glazer and J.H.C. Ranson) pp. 227–50. Baillière Tindall, London).

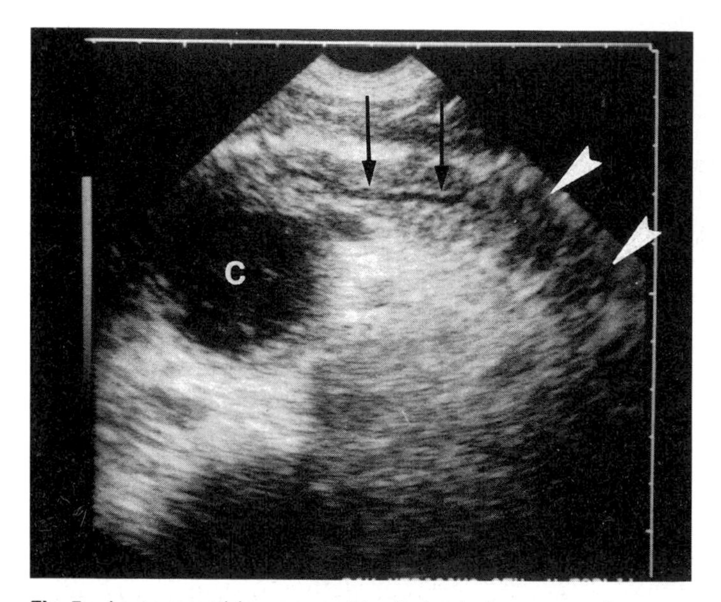

Fig. 5 Acute necrotizing pancreatitis. Abdominal sonogram shows a pseudocyst (C) of the pancreatic head and necrotic areas in the tail (white arrowheads); compression by the pseudocyst has caused a slight dilatation of the main pancreatic duct (arrows). (By courtesy of Dr P.L. Costa, Morgagni Hospital, Forlì.)

fic, but nonetheless useful, information. Abdominal films may demonstrate localized ileus, usually involving a jejunal loop ('sentinel loop'), gaseous distension of the transverse colon with collapse of the descending colon (colon cut-off sign), or a generalized ileus with gas/fluid levels in more severe forms of the disease. Chest radiography is helpful for demonstration of pleuropulmonary complications.

Ultrasonography and computed tomography are the most informative examinations. Abdominal ultrasound is a simple, non-invasive procedure which, in expert hands, has high sensitivity and specificity. It can give a rapid assessment of pancreatic changes and of possible complications such as pseudocysts (Figs. 4 and 5). In gallstone-associated pancreatitis, it can establish the aetiology quite readily. Unfortunately, the pancreas is not always well visualized by this technique because of the frequent presence of meteorism in patients with acute pancreatitis. If ultrasonography is not technically feasible, computed tomography may be indicated, remembering that this technique is unnecessary in patients with clinically obvious acute pancreatitis who have a mild uncomplicated course. Computed tomography with intravenous contrast enhancement is an essential examination for diagnosis of necrotizing pancreatitis, for establishing the site and extension of necrosis, and for detecting pancreatic and extra-pancreatic fluid collections (Figs. 6–9); in these conditions, its sensitivity and specificity are practically 100 per cent.

Until recently, endoscopic retrograde cholangiopancreato-

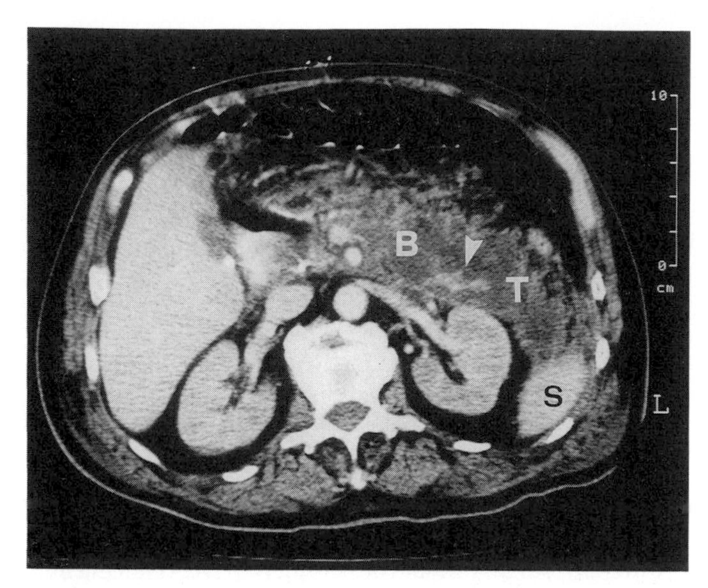

Fig. 6 Acute necrotic haemorrhagic pancreatitis. Contrast enhanced computed tomography showing a diffusely enlarged, non-homogenous pancreas, with hypo- and hyperdense areas and highly irregular, ill-defined anterior margin. The necrotic process extends into the left anterior pararenal space. Body (B) and tail (T) of the pancreas; spleen (S). Arrowhead indicates a hyperdense area due to haemorrhage. (By courtesy of Dr S. Papa, St Orsola Hospital, Bologna.)

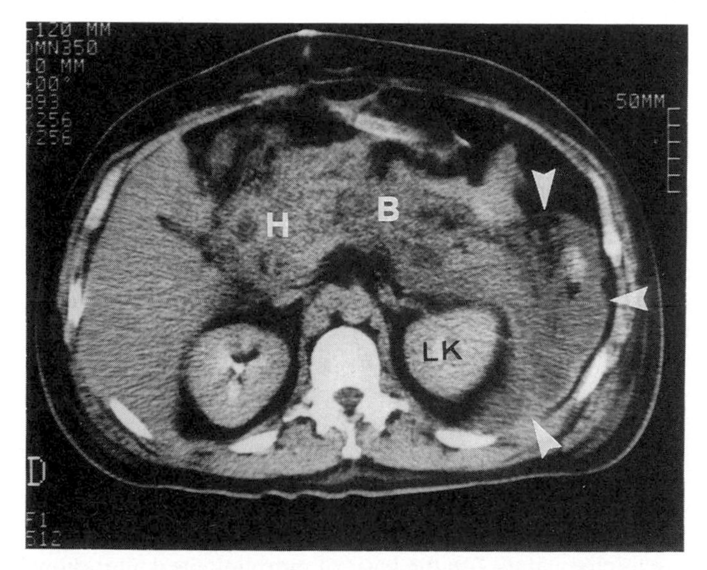

Fig. 7 Acute necrotizing pancreatitis. Contrast enhanced computed tomography. The pancreas is markedly enlarged, especially the head (H) and body (B); it is non-homogenous and margins are ill-defined. An extension of the necrotic process is evident in the left anterior and posterior pararenal spaces (arrowheads). LK, left kidney. (By courtesy of Dr P. Pisi, St Orsola Hospital, Bologna.)

graphy (**ERCP**) was considered to be contraindicated in acute pancreatitis, because of the risk of worsening pancreatic inflammation. This technique is currently performed in patients with biliary pancreatitis to clarify the aetiology in doubtful cases and to remove stones in the common bile duct.

The diagnostic role of magnetic resonance imaging is not

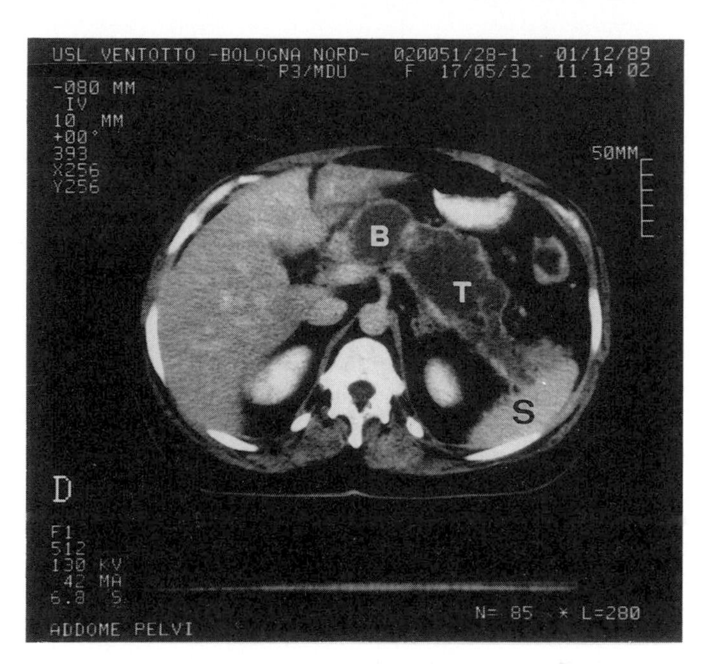

Fig. 8 Acute necrotizing pancreatitis. Contrast enhanced computed tomography. Liquefaction necrosis with pseudocyst formation in the body (B) and tail (T) of the pancreas. The hypodense areas are surrounded by a thin, enhancing rim of pancreatic tissue. S, spleen. (By courtesy of Dr S. Brusori, St Orsola Hospital, Bologna.)

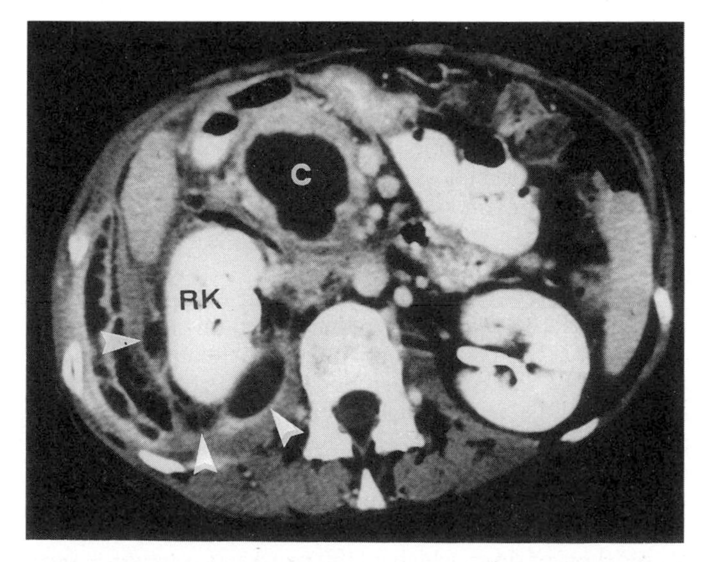

Fig. 9 Multiple fluid collections following acute pancreatitis. Contrast enhanced computed tomography shows a fluid collection (C) in the head of the pancreas surrounded by a weil-defined wall. Other fluid collections are also seen in the right posterior renal and lateral colic spaces (arrowheads). RK, right kidney.

well defined. At present, this technique adds no information to that provided by computed tomography.

In summary, clinical evaluation, serum amylase and/or lipase determination and ultrasonography are usually sufficient for the diagnosis of oedematous pancreatitis. For diagnosis of necrotizing pancreatitis, contrast-enhanced computed tomography should also be performed. In the vast majority of cases, these diagnostic procedures provide sufficient data for establishing a diagnosis and distinguishing oedematous from necrotizing pancreatitis.

In order to assess the severity of the disease, and especially to identify patients with an increased risk of life-threatening complications, various objective grading systems have been developed. The first and most widely known is that of Ranson (Tables 5 and 6). This or other modified versions are now widely employed because they enable a high proportion of severe cases to be correctly identified and appropriately managed.

Differential diagnosis

The differential diagnosis includes perforated peptic ulcer, mesenteric infarction, intestinal obstruction, and acute cholecystitis. In perforated ulcer there is a more abrupt onset of pain and more evidence of peritoneal irritation than in acute pancreatitis; moreover, the presence of free intraperitoneal air on plain abdominal films readily confirms the diagnosis. Acute mesenteric infarction should be suspected in the presence of severe abdominal pain and bloody stools, especially if the patient has a history of ischaemic cardiovascular disease. The analysis of peritoneal aspirate or lavage fluid is an important diagnostic method for differentiating acute pancreatitis from perforated peptic ulcer or mesenteric infarction. In patients with these last two conditions the aspirate has a foul odour and Gram-staining is usually positive. In intestinal obstruction, there is often a history of colicky pain; plain radiography of the abdomen reveals characteristic changes of mechanical obstruction (intestinal distension plus multiple gas/fluid levels). In acute cholecystitis, pain and tenderness are more intense in the right hypochondrium; ultrasonography can detect gallstones and may also be helpful in identifying inflammation of the gallbladder wall.

If any doubt remains, laparotomy is indicated, for both diagnosis and treatment.

Treatment

Therapy is largely supportive, aimed primarily at putting the pancreas to rest, controlling pain, and maintaining intravascular volume. Pancreatic secretion is minimized by withholding oral intake and by nasogastric aspiration. This latter measure prevents gastric contents from entering the duodenum, thus avoiding the release of pancreas-stimulating hormones from the duodenal mucosa; more importantly, it can also relieve vomiting and reduce gastrointestinal distension. H_2-blockers or drugs which inhibit pancreatic secretion, such as atropine, glucagon, calcitonin, and somatostatin, have been tried but not shown to be beneficial. Perhaps somatostatin, or its long-acting analogue octreotide, may reduce the incidence of local complications, but further studies are necessary before recommending their use.

Pain may be effectively relieved by non-steroidal anti-inflammatory drugs or, if severe, by meperidine or pentazocine. Coeliac plexus block or epidural anaesthaesia may rarely be required.

The maintenance of intravascular volume by infusing adequate amounts of fluids and electrolytes is one of the most important therapeutic measures. Any fluid deficit must be carefully monitored by charting pulse rate, blood pressure, urine output, haematocrit, and blood urea levels at regular intervals. Central venous pressure monitoring is indicated in severe attacks.

There is wide agreement that antibiotics should be administered only when infection develops. The protease inhibitor aprotinin (trasylol) has not proved effective.

Peritoneal lavage has been proposed to remove the potentially harmful substances present in pancreatic exudate, thus precluding their reabsorption into the circulation; while efficacy has been shown in some experimental studies benefit is not yet proven in humans.

Patients with oedematous pancreatitis can be treated on a general medical ward, but those with necrotizing pancreatitis should be admitted to an intensive care unit and, if possible, managed jointly with an experienced surgeon. In these patients, every effort must be made to prevent development of systemic complications but if they do occur, early recognition and vigorous treatment is of paramount importance. When the patient does not improve or worsens despite intensive therapy, surgery should be considered. Debridement and postoperative long-term drainage and lavage of the necrotic region and peritoneal cavity are measures which effectively reduce the incidence of complications and decrease fatality.

Local complications such as infected necrosis and abscess are also indications for surgery. Symptomatic pseudocysts can be treated by surgical drainage or, preferably, by ultrasound-guided percutaneous drainage. Asymptomatic pseudocysts usually do not require treatment; however, they should be periodically monitored by ultrasound, to detect any changes in size.

Once the acute episode has subsided, the patient can gradually resume a normal diet. Parenteral nutrition is indicated only when oral intake must be withheld for protracted periods of time. Once the patient's general condition permits, the cause of the pancreatitis is sought and, when possible, treated in order to prevent recurrence. Patients with gallstones should have a cholecystectomy during the same hospital admission; indeed, elderly patients generally tolerate elective surgery quite well.

Endoscopic sphincterotomy with extraction of stones in the common bile duct is a recently introduced technique which could be a valid alternative to surgery in selected, especially high risk, patients.

Chronic pancreatitis

Chronic pancreatitis is characterized by the persistence of pancreatic damage, even when the primary cause of the disease is eliminated. On the basis of morphological alterations of the pancreas, two forms may be distinguished: chronic obstructive and chronic calcifying pancreatitis. In the former, there is a dilatation of the ductal system and diffuse, uniform fibrosis and atrophy of the exocrine parenchyma proximal to an obstruction of the main pancreatic duct. This obstruction may be due to tumours, scarring after acute pancreatitis or trauma, or ampullary stenosis. Chronic calcifying pancreatitis is characterized by an irregular, lobular, spotty distribution of parenchymal fibrosis and atrophy, and by the presence in the excretory ducts of protein plugs, some of which may be calcified. Variable degrees of dilatation of the main pancreatic duct and its branches are also frequently seen. Usually caused by excessive alcohol consumption, this form of pancreatitis is by far the most frequent.

Incidence

Such data as are available suggest that, in the general population of Western countries, the incidence is about 5 to 10 cases per 100000 per year. Incidence and prevalence in the elderly population are unknown.

Age and sex

The clinical onset of chronic pancreatitis is usually between 30 and 40 years of age and onset after 60 years of age is rare (Fig. 10). Men are affected much more frequently than women, in a ratio of approximately 4:1.

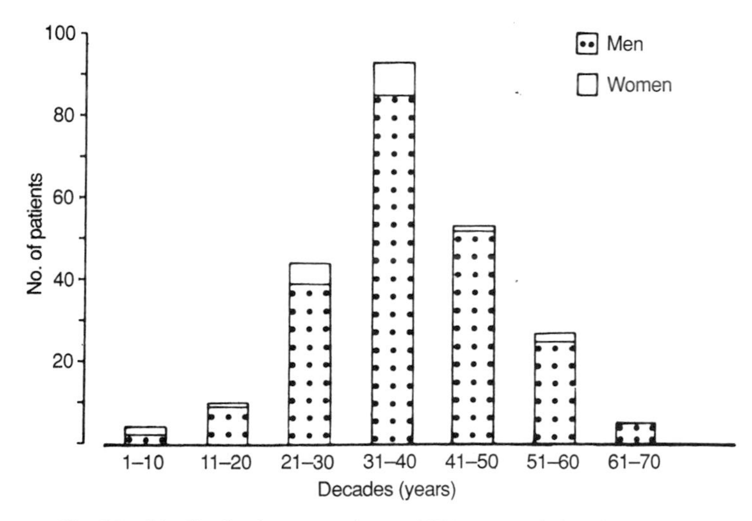

Fig. 10 Distribution by age and sex of 253 cases of chronic pancreatitis (reproduced with permission from Gullo, L., *et al.* (1977). Chronic pancreatitis in Italy. Aetiological, clinical and histological observations based on 253 cases. *Rendiconti Gastroenterologia*, **9**, 97–104).

Aetiology

In the general population of Western industrialized countries, about 70 per cent of cases of chronic pancreatitis are due to chronic excessive alcohol consumption. Heredity, biliary tract disease, obstruction of main pancreatic duct, hyperparathyroidism, hyperlipidaemia, and drugs are infrequent causes. About 20 per cent of cases are idiopathic (Table 7). Chronic pancreatitis which manifests itself for the first time in the elderly is almost always idiopathic (Table 7).

The mechanism by which alcohol causes chronic pancreatitis is unclear. Clinical and experimental studies suggest that it causes pancreatic secretory changes which result in the precipitation of pancreatic juice proteins in the ducts, with formation of plugs which later calcify to form stones. The resulting ductal obstruction leads to atrophy and fibrosis of the lobules drained by the obstructed ducts. The mechanisms that would be responsible for initiating protein precipitation and subsequent calcification are unknown, though a role has been suggested for the reduced biosynthesis of lithostatin, a pancreatic secretory protein which functions as a major stabilizer of calcium in pancreatic juice.

Table 7 Aetiology of chronic pancreatitis

	All age groups	Onset over 60 years of age
Alcohol abuse	70%	10% (alcohol abuse to drugs, combined)
Heredity Biliary tract disease Obstruction of the pancreatic duct Hyperparathyroidism Hyperlipoproteinaemia Drugs (corticosteroids)	10%	
No apparent cause (idiopathic)	20%	90%

Approximate percentages derived from the literature and our experience.

Clinical presentation

Elderly patients with chronic pancreatitis seen in clinical practice fall into two categories: (1) those with early or adult onset chronic pancreatitis, in whom the disease began in young or middle adulthood and who thus have an advanced pancreatitis, often with pancreatic calcification; and (2) those with late onset chronic pancreatitis, in whom the disease has first appeared when the patient is over 60 years of age.

Early or adult-onset chronic pancreatitis

Alcohol-induced pancreatitis is the most frequent form in this category. The natural history of alcoholic pancreatitis is characterized by an initial phase, lasting roughly 8 to 10 years, during which the main clinical manifestation is recurrent attacks of abdominal pain. The pain is generally intense, epigastric, and radiating to the right or left hypochondrium and to the back. These episodes last one or more days and tend to recur after intervals of unpredictable duration during which the patient feels well. In about half of the cases, attacks occur with increasing frequency and eventually necessitate surgical treatment; in the remaining half, they become less frequent and eventually cease as pancreatic insufficiency progresses. A second phase follows which is characterized by a lack of pain and the appearance of symptoms related to diffuse glandular destruction and resultant severe insufficiency. Thus, in elderly patients with this form of pancreatitis, the most frequent clinical manifestations of the disease are malabsorption and diabetes. These complications appear only late in the course of chronic pancreatitis, because more than 90 per cent of pancreatic function must be lost for malabsorption to become clinically apparent, and invasion of insular tissue by fibrosis usually occurs only in advanced stages of the disease.

Steatorrhoea (faecal fat excretion of more than 6 g/24 h) is observed in up to 60 to 70 per cent of patients (Table 8). It may be mild to moderate, detectable only by chemical analysis of the faeces, or severe, in which case the patient may pass loose, oily, foul-smelling stools. Deficiencies of the fat-soluble vitamins do not occur if malabsorption is correctly treated.

Diabetes of varying degrees is seen in about 50 to 70 per cent of patients; this is an insulin-dependent diabetes due to insular destruction by fibrosis. It has long been believed that diabetes secondary to chronic pancreatitis is almost

Table 8 Principal symptoms and complications of chronic pancreatitis in elderly patients

Steatorrhoea (faecal fat > 6 g/24 h)	60–70%
Diabetes	50–70%
Weight loss (usually mild)	60–80%
Pain	20–30%
Pancreatic calcification	60–70%
Pancreatic pseudocysts	20–30%
Obstruction of the common bile duct	Less common
Obstruction of the duodenum	
Splenic vein thrombosis	

Approximate percentages derived from the literature and our experience.

never complicated by vascular changes. However, we have recently demonstrated that the risk of retinopathy in this form of diabetes is similar to that of Type I diabetes, and that the frequency of this complication increases with the duration of diabetes (Fig. 11).

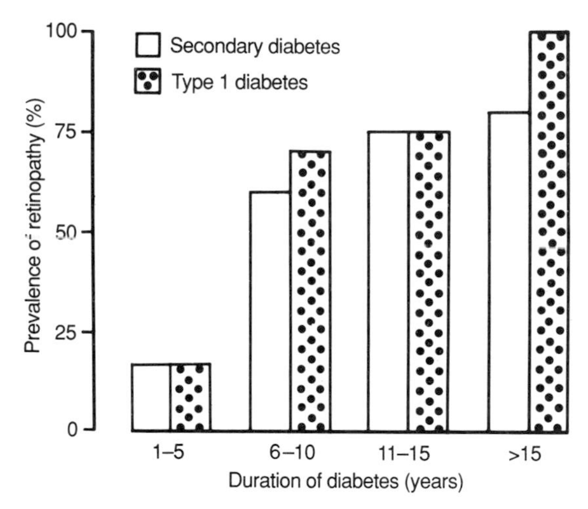

Fig. 11 Relationship between the prevalence of diabetic retinopathy and the duration of diabetes in 40 patients with chronic pancreatitis and secondary diabetes and in 40 Type 1 diabetics (reproduced with permission from Gullo, L., *et al.*, (1990). Diabetic retinopathy in chronic pancreatitis. *Gastroenterology*, **98**, 1577–81).

Steatorrhoea and diabetes may occur together or separately, and are frequently associated with weight loss (Table 8). If they are adequately treated, weight loss is mild; if not, the loss may be conspicuous and symptoms and signs of severe malnutrition may develop.

Late-onset chronic pancreatitis

This form of the disease develops in patients over 60 years of age, more commonly in men, the vast majority of whom have no history of alcohol abuse or other predisposing factors (Table 7). Other than age of onset, the main clinical distinction from chronic pancreatitis of the young is that, in the vast majority of cases, it is painless (Table 8) and when pain is present, it is often mild to moderate or atypical. The main clinical manifestations are steatorrhoea and diabetes with characteristics similar to those described in the preceding paragraph. The majority of these patients have pancreatic calcification. Arteritis of the lower limbs and/or coronary heart disease were reported in about 40 per cent of patients, suggesting that a vascular factor may have a role in its aetiology.

A very rare subtype of senile chronic pancreatitis was described by Sarles which he called primary inflammatory pancreatitis; it is distinguished by abundant foci of mononuclear cell infiltration in a diffusely atrophic and fibrotic pancreas, and by marked hypergammaglobulinaemia. The patients in whom this form was described were almost all women over 60 years of age. Clinically, steatorrhoea and weight loss were the symptoms most frequently observed; pain was absent or atypical. In one patient, undulating fever was present; in another, pancreatitis was associated with an aggressive chronic hepatitis. None had pancreatic calcification. The aetiology is unknown, though a viral or autoimmune origin was suggested.

For all forms of chronic pancreatitis in the elderly, physical examination is usually negative, apart from signs of malnutrition, if present. During painful periods, epigastric tenderness may be appreciated; if a pseudocyst is present, a mass may be palpated.

Complications

Pancreatic pseudocysts are a frequent complication (40–50 per cent) of alcoholic chronic pancreatitis, especially during the initial painful phase of the disease; they may also complicate late-onset chronic pancreatitis (Fig. 12) but the incidence in this form is unknown. Rarely, peripancreatic fibrosis may cause obstruction of the retropancreatic portion of the common bile duct or the duodenum (Fig. 13), or splenic vein thrombosis with consequent segmental portal hypertension.

Diagnosis

Chronic pancreatitis should be suspected in the presence of steatorrhoea, with or without a past history of recurrent attacks of upper abdominal pain, or in the case of diabetes in a patient with no family history of this disease and who is not overweight. It should also be considered when the patient has mild epigastric pain or vague abdominal discomfort of unclear origin, especially if associated with steatorrhoea and/or diabetes. Clinical suspicion is usually confirmed by pancreatic function tests and imaging techniques.

Tests of pancreatic function

Direct or indirect tests can be used. The direct tests consist of duodenal intubation and measurement in the duodenal aspirate of pancreatic bicarbonate and enzyme secretion after exogenous pancreatic stimulation with secretin and cholecystokinin. This is the most sensitive and specific means for diagnosing pancreatic insufficiency, valid even when the insufficiency is not severe. Unfortunately, the technique is invasive, time-consuming, and unpleasant for the patient, and thus is now little used in the clinic. The Lundh test, another method involving duodenal intubation, is based on measurement of enzyme activity after ingestion of a standard meal; it is less expensive but less sensitive than the former method.

The indirect or tubeless methods include oral, faecal, and blood tests. Two oral tests are in common use at present: the bentiromide or PABA test, based on the chymotrypsin-specific cleavage of *p*-aminobenzoic acid (PABA) from the

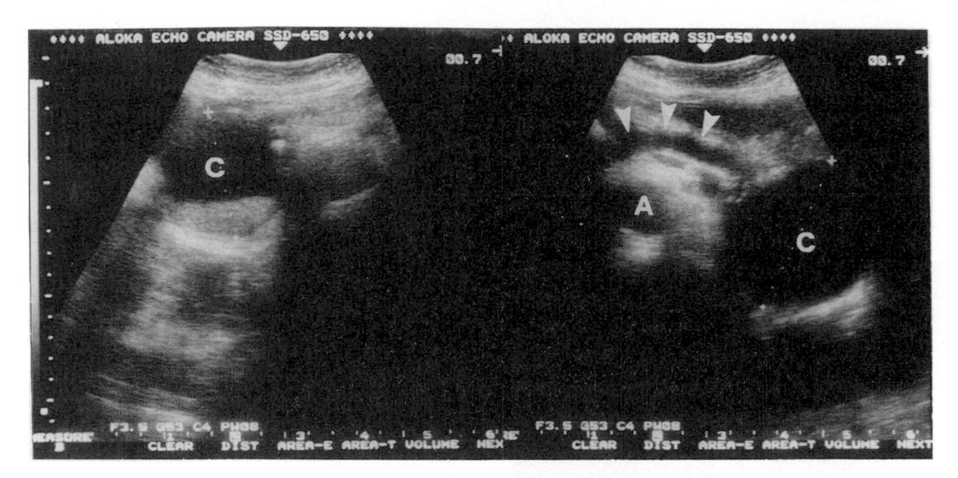

Fig. 12 Idiopathic chronic pancreatitis. Abdominal sonogram: the figure on the left shows a pseudocyst (C) in the pancreas head (5 cm); the figure on the right (same patient) shows a pseudocyst (C) in the pancreas tail (7 cm), and a dilated main pancreatic duct (arrowheads). A, aorta. (By courtesy of Dr S. Gaiani, St Orsola Hospital, Bologna.)

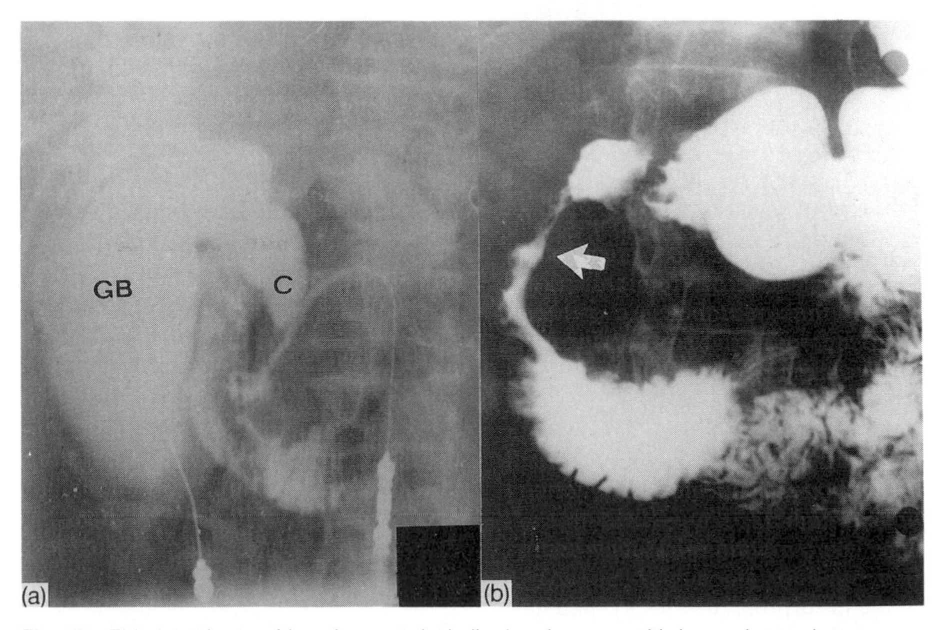

Fig. 13 Elderly patients with early onset alcoholic chronic pancreatitis in an advanced stage. (a) Peroperative transcholecystic cholangiography. Obstruction of the retropancreatic portion of the common bile duct (C) by peripancreatic fibrosis. GB, gallbladder. (b) barium meal and follow-through. Obstruction of the second portion of the duodenum (arrow) by peripancreatic fibrosis.

synthetic peptide *N*-benzoyl-L-tyrosil in the duodenum, and the pancreolauryl test based on the specific hydrolysis of fluorescein dilaurate by a pancreatic esterase. In both tests, the substrate is taken by mouth with breakfast; after hydrolysis in the duodenum, the marker (PABA or fluorescein) is absorbed and excreted in the urine. The amount of marker excreted in a given time is used as an index of pancreatic function.

The stool tests include faecal fat estimation and chymotrypsin determination. The amount of fat in faeces may be determined quantitatively using a reliable test such as the Van de Kamer chemical method. This test is useful mainly to assess the degree of steatorrhoea and to monitor treatment with pancreatic extracts. Chymotrypsin determination is a widely used procedure that can be done on a spot sample of stools.

The determination of serum trypsin concentration is the principal blood test in use. In patients with severe pancreatic insufficiency, serum trypsin concentrations are abnormally low. In milder forms of the disease, trypsin and the other serum pancreatic enzymes are normal, except during acute exacerbations when they may be transiently elevated.

The indirect tests are very simple to perform and particularly suitable for use with elderly patients. The sensitivity of these tests is only high when pancreatic insufficiency is severe.

In non-diabetic patients, an oral glucose tolerance test should be performed to assess endocrine pancreatic function.

Imaging procedures include plain abdominal radiography, which can show pancreatic calcification (present in about 60–70 per cent of cases), and ultrasonography, which, in addition to calcification, can demonstrate gross anatomic changes of the gland, dilatation of the pancreatic or biliary ducts, and pseudocysts (Figs. 12 and 14). In the vast majority

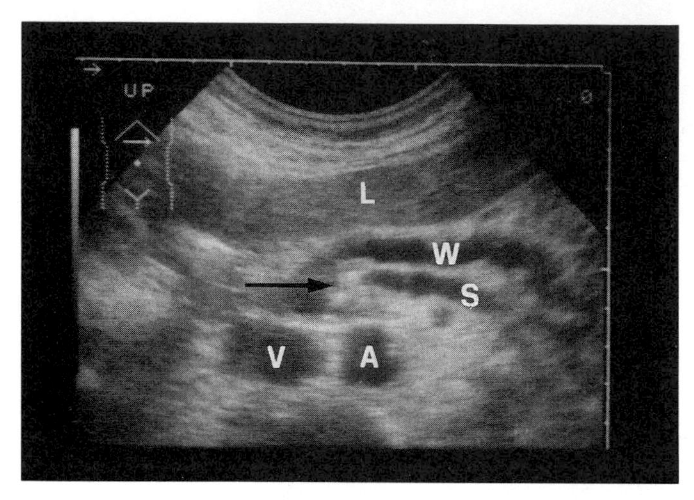

Fig. 14 Elderly patient with early onset alcoholic chronic pancreatitis in an advanced stage. Abdominal sonogram shows an atrophic pancreas with calcifications in the head (arrow) and marked, irregular dilatation of the main pancreatic duct (W). L, liver; V, inferior vena cava; A, aorta; S, splenic vein. (By courtesy of Dr P.L. Costa, Morgagni Hospital, Forlì.)

of cases, these imaging procedures, plus an indirect pancreatic function test (pancreolauryl test or faecal chymotrypsin) are sufficient to confirm the clinical diagnosis; when they are not, ERCP and/or computed tomography will usually help. When interpreting the ERCP images, however, it should be taken into account that, with ageing, a certain degree of dilatation of the pancreatic duct and its branches occurs, together with small cyst formation.

At times, especially in patients in whom steatorrhoea is the main symptom, differential diagnosis must be made with other diseases causing malabsorption in elderly patients, mainly bacterial contamination of the small bowel (see Chapter 8.5) and, more rarely, coeliac disease. The presence of calcifications or other morphological changes of the pancreas on plain abdominal radiography or ultrasonography, or the demonstration of pancreatic insufficiency usually resolve this diagnostic problem. When these examinations are inconclusive, the possibility of bacterial contamination may be pursued by the H_2 or ^{14}C-xylose breath test, or culture of intestinal aspirate. Coeliac disease may be diagnosed by xylose test, small bowel film series, and jejunal biopsy. In the rare cases in which there may be a suspicion of pancreatic cancer, careful clinical evaluation plus imaging techniques provide sufficient information to establish a diagnosis.

Treatment

Therapy is directed principally at pain relief, control of diabetes, and correction of malabsorption.

As stated above, pain is often absent or mild in the elderly patient and does not pose a particular therapeutic problem as it usually does in the young. Common analgesics are usually sufficient, or meperidine can be used when the pain is severe. In cases in which pain is resistant to analgesics, coeliac plexus block is the procedure of choice. Surgery should be avoided, if possible, because pain tends to disappear as pancreatic insufficiency worsens. Two surgical procedures are commonly performed for pain relief in chronic pancreatitis: lateral pancreaticojejunostomy when the pancreatic duct is dilated, and partial pancreatic resection (usually of the head) when it is not.

In alcohol-induced pancreatitis, avoidance of alcohol is recommended in the hope that this measure will prevent further painful attacks, though its efficacy is uncertain.

Malabsorption is treated by administering pancreatic extracts. There are several commercial preparations available which vary considerably in enzyme activity. It is important to choose those with the highest enzyme content and, preferably, those prepared in the form of enteric-coated microspheres that prevent enzyme inactivation by gastric acid. The dosage is adjusted on an individual basis, starting with 4 to 5 capsules or tablets per meal and increasing the number until satisfactory correction is achieved. If enteric-coated preparations are not used, and steatorrhoea does not improve, the addition of antacids or H_2-blockers may be indicated. Complete return to normal of faecal fat excretion is often difficult to achieve, especially when steatorrhoea is severe.

The control of diabetes is based on insulin administration and usually low doses are sufficient for control.

In order to maintain adequate caloric intake, dietary restrictions should be avoided, except in patients with marked steatorrhoea in whom a reduction in dietary lipids is indicated. Fat-soluble vitamins and other dietary supplements can be added to this regimen as required. It has been reported that about 40 per cent of patients with chronic pancreatitis develop subclinical vitamin B_{12} malabsorption and this can be corrected by oral pancreatic enzymes.

Symptomatic pseudocysts can be treated with non-surgical procedures such as ultrasound- or CT-guided percutaneous drainage. Some pseudocysts may respond to medical treatment with the long-acting somatostatin analogue octreotide. If these procedures are ineffective, surgical drainage may be necessary. Obstruction of the common bile duct or the duodenum by peripancreatic fibrosis requires surgical correction.

There is no reliable information on the course and nutritional consequences of chronic pancreatitis in elderly patients. When malabsorption and diabetes are recognized and adequately compensated, the course of this disease is usually uneventful and the patient's nutritional status is not significantly compromised. In some patients, severe malnutrition may develop, which may seriously affect well-being and quality of life.

CARCINOMA OF THE PANCREAS (see Section 10)

The principal tumours of the exocrine pancreas may be classified according to their postulated histogenesis into three groups: (1) duct cell tumours (ductal adenocarcinoma, mucinous cystadenocarcinoma, and giant cell carcinoma); (2) ductular-acinar tumours (serous cystoadenoma, papillary cystic tumour, pancreatoblastoma, ductuloacinar carcinoma, and acinar cell carcinoma); and (3) stem cell tumours (small cell carcinoma). Ductal adenocarcinoma is the most common, comprising about 85 to 90 per cent of all exocrine pancreatic tumours. It originates in the head of the pancreas in about 70 per cent of cases, in the body and tail in about 15 and 5 per cent, respectively, and at multiple sites in about 10 per cent.

The incidence of this carcinoma has increased considerably in the past 40 to 50 years. In the United States, age-adjusted annual mortality rates increased from 2.9 per 100 000 in 1920 to 9 per 100 000 in 1970. Pancreatic carcinoma is now the fifth or sixth leading cause of death from cancer in most Western countries. The reasons for this increased incidence are unknown.

Age and sex

Patients with pancreatic cancer are rarely under the age of 40 years, the model age being between 60 and 70 years (Fig. 15). There are more male patients than female, in a ratio of about 1.5 : 1.

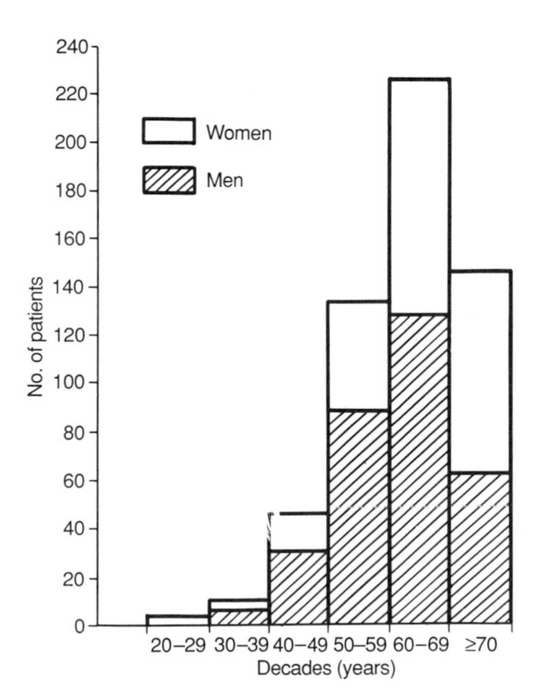

Fig. 15 Distribution by age and sex of 570 patients with pancreatic cancer (reproduced with permission from Gullo, L., *et al.* (1991). Aetiology of pancreatic cancer in Italy. A multicentre study, in press).

Aetiology and risk factors

The aetiology is unknown. Age is perhaps the most important known risk factor for pancreatic cancer; indeed, its incidence, which is about 10 per 100 000 annually in the general population, increases markedly with age, reaching up to 100 per 100 000 in the population over 80 years of age. Many attempts have been made to identify exogenous risk factors, but the results have been largely disappointing. Of the various factors studied, cigarette smoking has the highest associated risk but the strength of this association is very much lower than that observed between smoking and lung cancer. Coffee has been reported to be a risk factor in some studies but not in others. A diet rich in fat and protein, alcohol, and various chemical agents have also been incriminated as potential risk factors, but a significant association is not proven for any of them. Several studies indicate that diabetes may be a predisposing factor for pancreatic cancer. It has been suggested that chronic pancreatitis may also be a cause but convincing evidence is lacking.

Clinical presentation

Pancreatic cancer is asymptomatic in its early stages and becomes clinically apparent only when it has already reached an advanced stage. As a result, only 10 to 15 per cent of patients have resectable tumours at initial clinical presentation.

Jaundice is the most common presenting symptom of cancer of the head of the pancreas (Table 9). Owing to neoplastic infiltration of the retropancreatic portion of the common bile duct, jaundice usually progresses rapidly. It may be an isolated symptom, but is more frequently accompanied by pain. In the majority of cases, jaundice is not preceded by any disturbances; sometimes, however, it is preceded by a brief period (2–10 weeks) during which the patient may have had dyspeptic symptoms, mild upper abdominal pain, anorexia, or pruritus. Rarely, the jaundice is preceded by attacks of abdominal pain resembling those of acute pancreatitis.

Table 9. Main signs and symptoms of pancreatic cancer

Jaundice	60–80%
Pain	80–90%
Anorexia	80–90%
Weight loss	90–100%
Weakness	80–90%
Pruritus	20–30%
Diarrhoea	20–30%
Constipation	10–15%
Malnutrition	80–90%
Hepatomegaly	50–60%
Epigastric tenderness	80–90%
Palpable gallbladder	20–30%
Epigastric mass	40–50%
Ascites	5–10%

Approximate percentages derived from the literature and our experience.

In cancer of the body or tail of the pancreas, the most common presenting symptom is pain. With these locations of the tumour jaundice is uncommon and when present, is usually due to hepatic metastases or spread of the tumour to the pancreatic head.

Whatever the localization of the tumour, pain is generally very intense and persistent, giving the patient no respite. It is usually epigastric, frequently radiating to the right and/or left hypochondrium, but especially to the back. In the initial stages of carcinoma of the body, pain may be perceived exclusively in the back (T12–L2) and may thus be mistakenly treated as a dorsolumbar arthrosis. Characteristically, the pain improves somewhat on bending the trunk forward.

Anorexia, rapid weight loss, and weakness almost always accompany this disease (Table 9), though they may not be among the initial presenting symptoms. Other symptoms may appear which are due to impairment of exocrine and endocrine pancreatic function, including steatorrhoea and diabetes. While overt diabetes is detected only in about 15 to 20 per cent of patients, reduced glucose tolerance can be found in up to 60 per cent. A peripheral resistance to insulin may play a role in these disturbances of carbohydrate metabolism.

A mild-to-moderate anaemia is present in the majority of patients. Psychiatric disturbances and migratory thrombophlebitis have been reported in some studies, but these are rare and non-specific complications.

At physical examination (Table 9), signs of malnutrition and recent weight loss may be visible. Jaundice, when present, is usually readily evident and signs of scratching may be noted. Intense pain with deep palpation of the epigastrium is an almost constant finding to an experienced hand; in more advanced stages, a tumoural mass may be appreciated. Hepatomegaly is a frequent finding, usually due to metastases and/or cholestasis. The gallbladder can be palpated in about 20 to 30 per cent of patients with cancer of the head of the pancreas (Courvoisier's sign). Ascites, generally due to peritoneal invasion by the cancer, and peripheral oedema can be seen in patients with advanced disease.

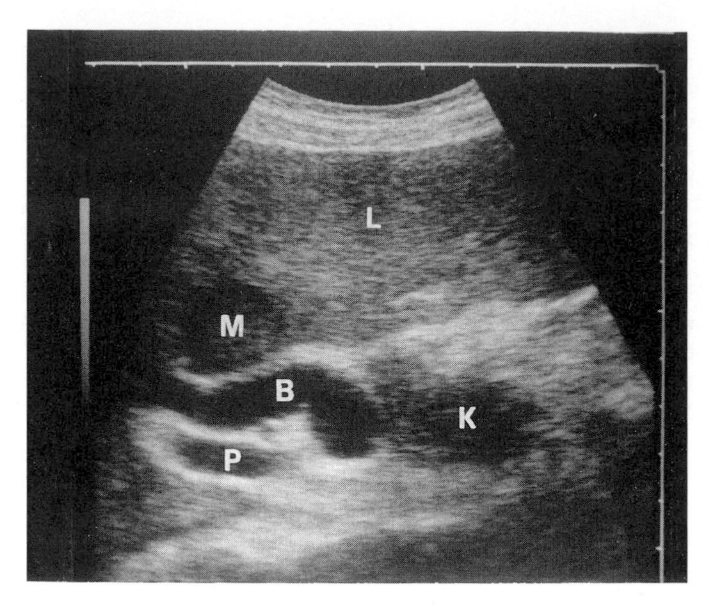

Fig. 17 Pancreatic carcinoma with metastasis. Abdominal sonogram shows a cancer (K) of the pancreas head with dilatation of the common bile duct (B) and hypoechoic liver metastasis (M). L, liver; P, portal vein. (By courtesy of Dr P.L. Costa, Morgagni Hospital, Forlì.)

Diagnosis

A detailed clinical history has a primary role. The sudden appearance of jaundice and/or pain with the above-described characteristics in an individual over 60 years of age, especially if associated with anorexia, or diabetes of rapid onset in a patient who is not obese and who has no family history of diabetes, should alert the physician to the possibility of pancreatic cancer. Once cancer is suspected, the diagnosis is generally confirmed by imaging techniques. Abdominal ultrasonography is the first examination performed; in expert hands, its sensitivity is high (about 80–85 per cent) (Figs. 16 and 17). If a suspicious mass is demonstrated, the next step should be ultrasound-guided percutaneous needle aspiration biopsy for histological confirmation, which has a diagnostic accuracy of more than 90 per cent. If doubt remains, or if aspiration biopsy cannot be performed, ERCP and/or computed tomography usually furnish sufficient information to establish a diagnosis, although some possibility remains that small tumours (2–3 cm or less) will be missed. As with other diseases, however, the overall diagnostic success rate depends, to a great extent, on the experience of the clinician and the expertise of imaging specialists.

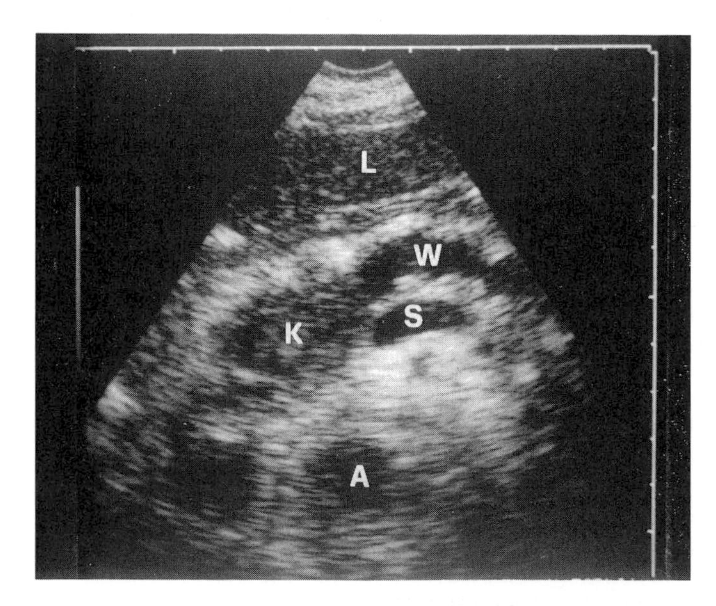

Fig. 16 Pancreatic carcinoma. Abdominal sonogram shows a small (3 cm) cancer (K) of the pancreas head, with marked dilatation of the main pancreatic duct (W). L, liver; S, splenic vein; A, aorta. (By courtesy of Dr P.L. Costa, Morgagni Hospital, Forlì.)

Endoscopic ultrasonography is a recently introduced technique which can permit an accurate assessment of pancreas morphology (Fig. 18). Its diagnostic value in pancreatic cancer has yet to be defined; especially promising results have been reported for diagnosis of endocrine tumours of the gland. The role of magnetic resonance imaging has yet to be defined.

Tests of pancreatic function are usually unnecessary; the determination of serum pancreatic enzymes is also of limited value because their behaviour is highly variable in this disease. Most of the serum tumour markers tested so far are non-specific and rise to abnormal levels only when the cancer is advanced. At present, the most widely used is CA 19–9, which has a sensitivity of about 80 to 85 per cent and a similar specificity. Carcinoembryonic antigen has a much lower diagnostic accuracy.

Varying degrees of elevation of serum bilirubin, alkaline phosphatase, γ-glutamyl transpeptidase, and transaminases are found in patients with cancer of the head of the pancreas with involvement of the common bile duct. In patients with cancer of the body and tail the finding of high levels of alkaline phosphatase and γ-glutamyl transpeptidase is generally indicative of hepatic metastases.

Once the diagnosis is made, the next task is to establish resectability of the cancer. To this end, computed tomography and selective angiography of the coeliac and superior

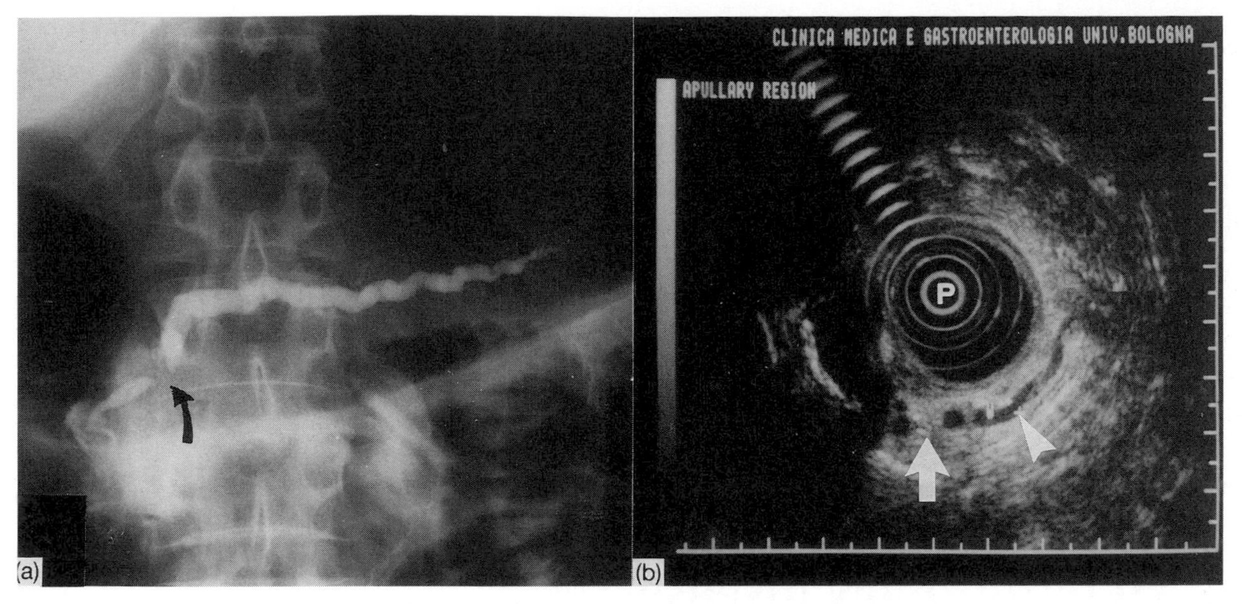

Fig. 18 Carcinoma of the head of the pancreas. (a) Endoscopic retrograde wirsungography showing marked stenosis (curved arrow) of a brief tract of the main pancreatic duct with proximal dilatation. (b) In the same patient, endoscopic ultrasonography demonstrates both the tumour (arrow) and resulting dilatation of the main pancreatic duct (arrowhead). P, probe. (By courtesy of Dr G.C. Caletti, St Orsola Hospital, Bologna.)

mesenteric arteries are particularly helpful for demonstrating local and distant diffusion of the tumour.

Differential diagnosis with chronic pancreatitis and other diseases causing obstructive jaundice can usually be accomplished by careful clinical evaluation and imaging techniques. Diagnostic laparotomy is rarely necessary.

Treatment

Stage I tumours should be resected if the patient's general condition permits. With more advanced cancer, palliative therapy only is indicated. Jaundice can be treated by surgical biliary bypass or by insertion of an endoprosthesis via a percutaneous transhepatic route, or endoscopically after sphincterotomy. These non-operative procedures have a lower fatality rate than surgery, but the possibility of cholangitis or blockage of the prostheses by biliary sludge are drawbacks to their use.

For non-resectable tumours which have not spread beyond the pancreas, the combination of radiation therapy and 5-fluorouracil (5-FU) can increase survival time in some patients. Intraoperative radiation of the tumour has been disappointing. Once metastasis has occurred, chemotherapy is largely ineffective.

Pain control is often difficult, especially in advanced stages of the disease. When narcotic analgesics are inadequate, coeliac plexus block or epidural anaesthesia can be effective. When present, diabetes is treated with insulin, steatorrhoea with oral pancreatic enzymes.

The course of pancreatic carcinoma is inexorable and most patients die within 5 to 6 months of diagnosis. Among those who have undergone resection of stage I tumours, the 1-year survival rate is about 20 per cent and 5-year, about 2 per cent. Some recent studies have reported a 5-year survival rate of 10 to 15 per cent, suggesting that improved diagnostic technology may be reducing the fatality rate slightly.

Bibliography

Acute pancreatitis

Acosta, J.M. and Ledesma, C.L. (1974). Gallstone migration as a cause of acute pancreatitis. *New England Journal of Medicine*, **290**, 484–7.

Beger, H.G., Bittner, R., Block, S., and Buchler, M. (1986). Bacterial contamination of pancreatic necrosis. A prospective clinical study. *Gastroenterology*, **91**, 433–8.

Glazer, G. and Ranson, J.H.C. (1988). *Acute pancreatitis. Experimental and clinical aspects of pathogenesis and management*. Baillière Tindall, London.

Imrie, C.W., *et al.* (1978). A single-centre double-blind trial of Trasylol therapy in primary acute pancreatitis. *British Journal of Surgery*, **65**, 337–41.

Neoptolemos, J.P., London, N.J., James, D., Carr-Locke, D.L., Bailey, I.A., and Fossard, D.P. (1988). Controlled trial of urgent endoscopic retrograde cholangiopancreatography and endoscopic sphincterotomy versus conservative treatment for acute pancreatitis due to gallstones. *Lancet*, **ii**, 979–83.

Park, J., Fromkes, J., and Cooperman M. (1986). Acute pancreatitis in elderly patients. *American Journal of Surgery*, **152**, 638–42.

Sarles, H., *et al.* (1989). Classifications of pancreatitis and definition of pancreatic diseases. *Digestion*, **43**, 234–6.

Steer, M.L. and Meldolesi, J. (1987). The cell biology of experimental pancreatitis. *New England Journal of Medicine*, **15**, 144–50.

Ventrucci, M., Pezzilli, R., Gullo, L., Platé, L., Sprovieri, G., and Barbara, L. (1989). Role of serum pancreatic enzyme assays in diagnosis of pancreatic disease. *Digestive Diseases and Sciences*, **34**, 39–45.

Chronic pancreatitis

Ammann, R. and Sulser, H. (1976). Die 'senile' chronische Pankreatitis—eine neue Nosologische einheit? *Schweizenische Medizinische Wochenschrift*, **106**, 429–37.

Gullo, L., Barbara, L., and Labò, G. (1988). Effect of cessation of alcohol use on the course of pancreatic dysfunction in alcoholic pancreatitis. *Gastroenterology*, **95**, 1063–8.

Gullo, L., Ventrucci, M., Naldoni, P., and Pezzilli, R. (1986). Aging and exocrine pancreatic function. *Journal of the American Geriatrics Society*, **34**, 790–2.

Gullo, L. and Barbara, L. (1991). Treatment of pancreatic pseudocysts with octreotide. *Lancet*, **338**, 540–1.

James, O., Agnew, J.E., and Bouchier, I.A.D. (1974). Chronic pancreatitis in England: a changing picture? *British Medical Journal*, **2**, 34–8.

Niederau, C. and Grendell, J.H. (1985). Diagnosis of chronic pancreatitis. *Gastroenterology*, **88**, 1973–95.

Sarles, H., *et al.* (1965). Observations on 205 confirmed cases of acute pancreatitis, recurring pancreatitis, and chronic pancreatitis. *Gut*, **6**, 545–58.

Sarles, H., Bernard, J.P., and Gullo, L. (1990). Pathogenesis of chronic pancreatitis. *Gut*, **31**, 629–32.

Carcinoma of the pancreas

Bornman, P.C., Harries-Jones, E.P., Tobias, R., Van Stiegmann, G., and Terblanche, J. (1986). Prospective controlled trial of transhepatic biliary endoprosthesis versus bypass surgery for incurable carcinoma of head of pancreas. *Lancet*, **i**, 69–71.

Fontham, E.T.H. and Correa, P. (1989). Epidemiology of pancreatic cancer. *Surgical Clinics of North America*, **69**, 551–67.

Gastrointestinal Tumour Study Group. (1979). A multi-institutional co-operative trial of radiation therapy alone and in combination with 5-fluorouracil for locally unresectable pancreatic carcinoma. *Annals of Surgery*, **189**, 205–8.

Ingber, S. and Jacobson, I.M. (1990). Biliary and pancreatic disease in the elderly. *Gastroenterology Clinics of North America*, **19**, 433–57.

Steinberg, W.M., *et al.* (1986). Comparison of the sensitivity and specificity of the CA 19–9 and carcinoembryonic antigen assays in detecting cancer of the pancreas. *Gastroenterology*, **90**, 343–9.

8.8 Hepatobiliary disease

K. W. WOODHOUSE AND O. F. W. JAMES

AGE-RELATED CHANGES IN HEPATIC STRUCTURE AND FUNCTION

Ageing and liver size

In early autopsy studies of liver size in relation to age findings ranged from an 18 per cent decline in liver size among men and 24 per cent decline among women between the ages of 20 and 80 to a fall in liver weight of almost 50 per cent between the 3rd and 10th decades. Because of the difficulties in interpreting autopsy studies non-invasive methods of determining liver size have been developed including computerized tomography, isotope scanning, and ultrasound. In the most extensive of these studies Wynne *et al.* (1989) calculated that average liver volume fell from 1474 ml at age 24 to 934 ml at age 91—a fall of 37 per cent. The changes were slightly more marked in women.

Liver blood flow

Although it was widely stated that liver blood flow falls with increasing age this assertion was only recently supported by the studies of Wynne *et al.* (1989) using indocyanine green who showed a decline in apparent liver blood flow of 35 per cent between the 3rd and 10th decades even allowing for changes in body weight. In addition liver perfusion (liver blood flow per unit volume of liver) also fell modestly by about 11 per cent.

Liver morphology

The microscopic appearance of the liver changes with ageing, becoming a darker brown colour owing to the accumulation of pigmented lipofuscin granules in liposomes and hepatocytes. Lipofuscins, a mixture of polymerized proteins and lipids, accumulate not only in the liver but elsewhere in the body. Ivy and Kitani have suggested that this may be due to reduced intracellular proteinolysis with age (Kitani 1990). Although the liver shrinks in advanced age individual hepatocytes tend to enlarge with nuclei showing increased DNA per nucleus and more polyploidy (Watanabe *et al.* 1978). Cells also have larger mitochondria and contain more protein much of which may be functionally inactive and represent the accumulation of 'junk'. In the absence of disease hepatocytes divide perhaps only two or three times during the human life-span (Popper 1986). The spaces between liver cells (space of Disse) enlarge in ageing and there is a corresponding increase in the supportive collagen matrix but the nature of the collagen appears unaltered (Grasedyck *et al.* 1980).

Liver function and ageing

In the absence of systemic or hepatic disease there are no important changes in conventional liver blood tests with ageing. Abnormal results in these standard liver function tests (serum bilirubin, serum transaminases, alkaline phosphatase, and serum proteins) may therefore be interpreted as a sign of possible liver disease just as in younger patients (Gambert *et al.* 1982). Minor transient abnormalities of liver function tests, particularly elevation in serum alkaline phosphatase, are not infrequently seen in elderly patients with acute infections or heart failure; these should revert to normal after treatment (Parker *et al.* 1986).

Decrements in some dynamic tests of hepatic function, particularly relating to the clearance of xenobiotic compounds such as drugs, have been noted in elderly subjects. Reductions attributable to a decline in liver function of between 0 and 50 per cent have been observed (Woodhouse and James 1990). It is likely that, after such factors as diet, smoking, nutrition, and gender have been taken into account, there is a broad decline in drug clearance of between 10 and 30 per cent between young adulthood and advanced old age. There seems little doubt that most of such changes are related to the fall in liver size, blood flow, and perfusion, since recent studies have shown no marked deterioration in the specific activities of several major drug metabolizing enzymes in man (Schmucker *et al.* 1990). This general modest decline in drug clearance, probably attributable to reduced liver blood flow and size rather than to changes in enzyme activity takes place both in drug oxidation and in conjugation (Woodhouse and James 1990). Similarly, galactose elimination has been shown to decline with normal ageing, and again this cytosolic function may well be reduced simply in relation to liver size (Marchesini *et al.* 1988). Few studies of liver synthetic function have been undertaken in humans. Marchesini *et al.* (1990) recently showed that there is a significant negative correlation between peak urea synthesis and age which may represent an example of the inability of elderly people to respond to stressful challenge although basal urea nitrogen synthesis is not an age dependent variable. Hepatic synthesis of cholesterol is reduced in old age, and there is reduction in total bile acid pool and possibly decreased synthesis of bile acids from cholesterol with corresponding increased biliary cholesterol secretion (Einarsson *et al.* 1985).

Investigations of liver disease

How to use liver function tests in elderly patients

All elderly patients with suspected liver disease should have routine tests of liver function, serum proteins, and prothrombin time/prothrombin ratio. If serum alkaline phosphatase is elevated serum-γ-glutamyl transpeptidase should be assessed to confirm cholestasis rather than possible bone disease. Blood count and blood film are also mandatory as a first stage in investigation of suspected liver disease. The haematology of liver disease is complex but there are no specific differences in features between old and young patients. Other blood tests will be discussed under specific liver diseases.

LIVER DISEASES

A detailed description of all liver disease which may occur in elderly people is inappropriate here. This chapter will describe areas in which hepatic disease presents special features or differences between old and young, and areas in which these differences may be important in diagnosis or therapy.

Viral hepatitis

Hepatitis A

Until recently the majority of old people in Westernized countries had acquired immunity to hepatitis A virus in early life. However, the proportion of non-immune individuals is now rising in Europe and North America and geriatricians may now expect to see an increasing number of sporadic cases of the disease. Death from hepatitis A, although uncommon, is age-related. For example in the United Kingdom the ratio of deaths to notifications rises steadily with age, being around 7 per 10000 in those aged 15 to 24 and 400 per 10000 in those aged over 65 (Forbes and Williams 1988). In addition, should fulminant hepatic failure develop, increased age is an adverse prognostic feature in hepatitis A as with all other forms of fulminant hepatic failure. The disease is said to be more protracted and possibly more cholestatic in the elderly. Hepatitis A vaccination is now available and screening of elderly people, particularly of travellers to less developed countries is indicated.

Hepatitis B

Post-transfusion hepatitis B is now very rare in European countries and North America but sporadic cases of acute hepatitis B do nonetheless occur. As with hepatitis A the disease is more cholestatic, with slower clearance of HBsAg, but the prognosis appears unaltered (Goodson *et al.* 1982, Laverdant *et al.* 1989). HBV vaccination produces unsatisfactory antibody responses in the elderly. Cook *et al.* (1987) suggest that this is due to lack of antibody-producing B cells, perhaps a facet of failure of the immune response in old age. Third-generation HBV vaccines with improved antibody responses in all ages may shortly be available. At present, however, it is still recommended that elderly travellers to the Far East, Africa, and Third World countries in general should receive prophylactic hepatitis B vaccination. The likelihood is that even without marked antibody response some protection against severe HBV infection will have been obtained.

Chronic HBV infection is a not infrequent cause of chronic liver disease in older patients, particularly in Mediterranean or Far Eastern countries. Such patients are usually HBsAg positive but are e antigen negative, often e antibody positive. This indicates lack of great infectivity and it is probable that the HBV genome has been incorporated into the hepatic nuclear material with the production of the HBsAg protein but with the persistence of little or no whole virus. The clinical manifestations, complications, and management of the cirrhosis which may arise from chronic HBV infection are similar to those of cirrhosis of other causes and will be considered below.

The major complication of chronic HBV infection in the elderly is the development of primary hepatocellular carcinoma. This is particularly common in men. In Europe and North America chronic alcohol consumption and coincident infection with hepatitis C virus (HCV) may further increase the likelihood of development of hepatocellular carcinoma but it is probably the length of time for which an individual has had cirrhosis which is the principal determinant (Melia *et al.* 1984).

In general the prognosis of chronic HBV infection relates to the severity of the underlying liver disease. Thus, in adults with chronic HBV infection and no other complicating factors, survival from the time of histological diagnosis is perhaps 55 per cent in those with cirrhosis and over 85 per cent in those without but no specific data on elderly patients are available. At present, treatment of elderly patients with cirrhotic HBV infection is largely symptomatic. There is no evidence for benefit of any specific antiviral treatments, and steroid treatment is strongly contraindicated since it may lead to persistent HBV infection and clinical deterioration. In general there are two objectives for treatment in precirrhotic patients—to prevent progression to cirrhosis with its complications and also to reduce or abolish infectivity, if possible by inducing an adequate immune response to the virus. Treatment with interferon will produce beneficial results in either or both of these therapeutic goals in over 50 per cent of selected patient groups (Thomas 1992). Very few elderly patients will present with precirrhotic chronic HBV infection with evidence of continuing active viral replication and infectivity (e antigen positive).

Hepatitis C

The virus responsible for the majority of post-transfusion non-A, non-B hepatitis in the United States, Japan, and much of Europe has recently been identified and antibody markers for HCV are now becoming available. Initially these were of doubtful sensitivity and specificity but new ELISA assays will help specific diagnosis and lead to epidemiological surveys. Only one serological survey of the prevalence of past or present HCV infection in an elderly community has so far been carried out. This Northern Italian study suggested that the prevalence of HCV infection among a large population of elderly individuals in residential care was similar to that of the younger general community in the same region (Chiaramonte 1992). Two other recent studies from France and Israel have suggested that over 70 per cent of cases of acute sporadic viral hepatitis in the elderly are now non-A, non-B in type (presumably HCV in the majority), blood transfusion being responsible for about half of these (Laverdant *et al.* 1989; Sonnenblick *et al.* 1988). Again cholestasis is a prominent clinical and biochemical feature and immediate and medium-term prognosis was reported to be better than had previously been thought with acute mortality of 4 and 10 per cent in these two studies. Progression to cirrhosis occurred in 20 per cent of patients within 5 years. Unlike acute hepatitis A and B, hepatitis C infection may quite often be followed by a carrier state. The proportion of these individuals who eventually become cirrhotic is unknown. Typically chronic hepatitis C runs an indolent, often asymptomatic, course with fluctuations in liver function tests, particularly transaminases. Diagnosis is based on liver biopsy, the specific anti-HCV antibody test, and, where possible, confirmation of a history of blood transfusion.

While the incidence and prevalence of HCV infection appears less common in Northern Europe than in Mediterranean countries, the United States, or Japan it is now suggested that as with HBV infection previous HCV infection may render an individual more susceptible to alcoholic liver disease. There is no well-accepted treatment for non-cirrhotic chronic HCV liver disease. As with HBV, long-term (6 months to 1 year) interferon treatment has shown some initially hopeful results in small controlled trials. At present it is not clear however whether this treatment merely suppresses HCV infection rather than eliminating the virus itself. Treatment with interferon is unpleasant and at present it is not recommended that elderly patients should undergo this therapy except after extremely careful assessment and in the context of a controlled trial in a major centre.

Hepatitis D (delta virus)

Delta virus is capable of infectivity only in the presence of hepatitis B virus, and is therefore termed a satellite virus. Infection may occur as a coinfection with HBV or the delta virus may infect a chronic HBV carrier (superinfection). The virus is common in Mediterranean countries and the Middle East but rare in the United Kingdom, Northern Europe, and the United States. Diagnosis is by demonstration of the anti-HDV antibody in serum or in liver biopsy using specific anti-HDV probes. HDV is responsible for apparently more severe attacks of HBV or relapse of previously quiescent chronic HBV liver disease (Rizzetto and Verme 1985). There are no studies of delta hepatitis specifically in aged individuals.

Other viral liver disease

Acute hepatic infections with other viruses—Epstein-Barr virus, cytomegalovirus, and herpes viruses—may very occasionally occur in elderly individuals, particularly in those who are immunocompromised by coexisting disease or by immunosuppressant treatment. Probably further specific hepatitis viruses still await discovery to extend the hepatitis virus alphabet. In particular, in the United Kingdom and Northern Europe more sensitive and specific serological tests for HCV are awaited to determine whether the substantial proportion of cases of sporadic presumed non-A, non-B viral hepatitis are due to HCV or to other hepatotropic viruses. The clinical cause of acute sporadic viral hepatitis in these instances is similar to that with the other forms of acute viral liver disease described above.

Bacterial infections

Pyogenic liver abscess

This is largely a disease of elderly people; over half of liver abscess patients in Western countries are over age 60. These patients present with non-specific symptoms—epigastric pain, weight loss, shortness of breath, and rigors—although in extreme old age fever and rigors have been reported as being less common (Sridharan *et al.* 1990). About 50 per cent of pyogenic liver abscesses now arise from biliary tract disease and ascending cholangitis rather than distant abdominal sepsis; no cause is found in many patients with liver abscess although there is some association with diabetes mellitus and metastatic cancer.

Diagnosis

Because of difficulty in diagnosis, one retrospective study (Sridharan *et al.* 1990) revealed misdiagnosis as malignancy in one-third of patients, the true diagnosis in these patients only being apparent at autopsy. Reactive pleural effusion, patchy shadowing on chest radiographs, or elevated right hemidiaphragm may be present in almost two-thirds of cases. Many patients have anaemia, and over 75 per cent with an elevated leucocyte count. Serum alkaline phosphatase is elevated in almost all patients but other liver function tests vary. Serum albumin is often low and globulin elevated. The diagnosis should be made by ultrasound examination. Ultrasound or CT scan reveals that abscesses are often loculated or multiple and sometimes débris in the middle of the abscess may make distinction from tumour difficult. In doubtful instances hepatic angiography may be helpful.

Treatment

Current treatment is now with percutaneous needle aspiration with subsequent percutaneous catheter drainage under ultrasound or CT scan control. A recent review of such treatment suggested that the success rate in 139 patients, many extremely elderly, was 90 per cent (Bertel *et al.* 1986). In most series of patients *Escherichia coli* is the most commonly isolated organism, although in about 50 per cent of abscesses anaerobic or mixed aerobic and anaerobic organisms are found. Antibiotic treatment should initially be via the intravenous (IV) route. This may be with ampicillin, aminoglycoside and metronidazole, or a third-generation cephalosporin with metronidazole. Treatment should be prolonged with perhaps 2 weeks of intravenous followed by up to 3 months of oral antibiotics. The progress of resolution should be followed by repeated ultrasound examinations. The drainage catheter may be removed from the liver after a few days. It is emphasized that in sick elderly patients with non-specific symptoms and abnormal liver function tests the index of suspicion of liver abscess, possibly from gallstone disease should be high and early ultrasound scanning is critically important.

Systemic bacterial infections

Abnormal tests of liver function may occur during systemic infections of many types. Furthermore non-hepatic infections in which no overt evidence of bacteraemia exists, such as diverticulitis, renal disease, soft tissue abscess, or endocarditis, may all be associated with abnormal tests. In one unselected series of largely elderly general hospital admissions non-hepatic bacterial infection accounted for almost 15 per cent of those with abnormal liver function tests (Parker *et al.* 1986). It is not clear why such hepatic abnormalities occur, although because of the portal circulation direct bacterial toxicity to the liver or endotoxinaemia may be contributory.

In an elderly patient with known infection and mild non-specific abnormalities of liver function no further hepatic investigations need be made unless these persist following resolution of the infective illness.

Drug-induced liver disease

Drug consumption rises progressively with age. Thus while 13 to 20 per cent of Westernized populations are over 65 this group accounts for 30 to 40 per cent of drug use. This rise continues throughout the later years and is particularly marked in elderly women. It is therefore hardly surprising that simply on a numerical basis hepatic adverse reactions are seen more frequently in elderly people. It is not yet clear whether they are at greater risk of developing hepatic adverse drug reactions when prescribing variables are taken into account. In the case of halothane there is no good evidence that elderly people are more likely to develop liver damage than are the young, but, once liver damage has occurred, hepatic failure and death from fulminant hepatic failure appear more common in older individuals than in the young (Neuberger and Williams 1981). This is probably due to such age-associated factors as decreased response to stress rather than any age difference in biotransformation of halothane itself. Benoxaprofen, formerly heavily marketed as a non-steroidal anti-inflammatory drug did show age-associated changes in its hepatotoxicity (see Section 5). Similarly there is some evidence that if liver damage does develop in relation to a few other drugs—notably alpha-methyl dopa and dantrolene—then the older the subject the more severe is likely to be the liver damage. Whether this is related to metabolite production, differences in immunological competence, or other factors is not known.

Alcoholic liver disease

Although alcoholic liver disease is perceived as being less common with advanced age a significant proportion of patients do present in later life. Indeed one study from Baltimore showed that the peak decade for presentation with cirrhosis was the 7th (Garagliano *et al.* 1979). In a British study 28 per cent of patients presented for the first time aged over 60 (Potter and James 1987), while in France a large retrospective study suggested that as many as 20 per cent of patients were over age 70 (Aron *et al.* 1979). Older patients have a higher proportion of symptoms and signs suggestive of severe liver disease—particularly abdominal swelling or jaundice—than do younger patients with alcoholic liver disease. Almost all patients presenting with alcoholic liver disease over age 70 had cirrhosis compared with only about 50 per cent of patients presenting under age 60 (Potter and James 1987). The vast majority of the elderly patients had been drinking in excess of 80 g alcohol per day for many years and there is little evidence that mean daily alcohol intake was less in elderly patients compared with younger ones at the time of presentation. Prognosis is related to age; 1- and 3-year fatality is 5 and 24 per cent respectively in those under age 60 at presentation but 34 and 54 per cent in those aged over 60. Three-year fatality in those presenting over 70 was 91 per cent.

National differences in presentation of alcoholic liver disease among elderly individuals may be related to referral patterns and it is unclear why old patients usually have more severe features and more advanced disease at presentation. Possibly family doctors are more reluctant to refer elderly individuals with minor features of alcoholic liver disease to hospital or perhaps older patients may present 'further down the line' than younger ones. Conceivably there is increased susceptibility of the older liver to the toxic effects of alcohol.

Investigation

Routine investigations should include liver function tests, serum albumin, blood film (noting raised mean corpuscular volume), clotting tests, and random blood or urine ethanol estimations. Unless marked ascites or impaired clotting or reduced platelets prevent it liver biopsy should be carried out since this has important prognostic value. No characteristic liver function test abnormalities are found in old people with alcoholic liver disease compared with the generality of patients. The clinical findings in older patients are also similar to those found in younger ones, although possibly signs of nutritional deficiency may be rather more marked.

Treatment

The most important treatment for alcoholic liver disease is abstinence; long-term prognosis in patients with the advanced disease is very dependent upon alcohol consumption.

Acute treatment

Acute alcohol withdrawal should be treated with chlordiazepoxide up to 80 mg daily although elderly patients may need less. In addition atenolol 50 mg daily is recommended for the first week in hospital to prevent the peripheral manifestations of alcohol withdrawal. Benzodiazepines should be withdrawn before hospital discharge. Unfortunately many elderly alcoholics are chronic abusers of benzodiazepines, and sudden simultaneous withdrawal of both alcohol and benzodiazepines is certainly not recommended.

In severe decompensated alcoholic liver disease there is some evidence that, in the absence of infection, corticosteroid treatment may be beneficial. Meta-analysis has suggested that treatment with, for example, 30 mg prednisolone daily for 2 weeks immediately after admission to hospital may be beneficial (Maddrey *et al.* 1986). Initially nutritional support with intravenous vitamins and minerals, particularly thiamine, and maintenance of blood sugar are very important.

Long-term treatment

Elderly individuals with established alcoholic liver disease should receive frequent long-term and careful follow-up. Abstinence should be encouraged, and complications of therapy—notably with diuretics—should be monitored. Elderly alcoholic cirrhotics have a high risk of development of primary hepatocellular cancer, and this should be watched for with repeated serum alpha-fetoprotein measurements, any rising trend being suggestive of the development of a small hepatocellular carcinoma.

Autoimmune liver disease

There are three main autoimmune liver diseases—autoimmune chronic active hepatitis, an attack on the liver cells themselves, primary biliary cirrhosis, in which the immune attack is on small intrahepatic bile ducts, and primary sclerosing cholangitis, in which the attack is upon larger intra- and extrahepatic bile ducts. The first two are by no means uncommon in elderly individuals in Europe and the United States.

Primary biliary cirrhosis

As many as 25 per cent of patients with primary biliary cirrhosis may present aged over 65, although the average age at presentation in a number of major series is between 55 and 60 years. Unlike alcoholic liver disease, primary biliary cirrhosis may present with a slightly milder and less symptomatic disease in older individuals although this is by no means universal. About one-third of patients have no symptoms of liver disease at the time of diagnosis; about one-half present with signs and symptoms of chronic liver disease and the remainder present with the complications of portal hypertension (Myszor and James 1990). In the United Kingdom and parts of Northern Europe the prevalence of primary biliary cirrhosis in women over age 50 may be as high as 1 in 1500.

Asymptomatic primary biliary cirrhosis

In older women this diagnosis is being increasingly made either because of the finding of a positive antimitochondrial antibody (AMA) in a patient with such symptoms as joint pains or suspected thyroid disease in whom an autoantibody profile has been requested, or because of the unexpected finding of persistently elevated serum alkaline phosphatase on biochemical screening, or enlarged liver on routine abdominal examination.

In women over age 50 with raised serum alkaline phosphatase and positive AMA the diagnosis of primary biliary cirrhosis is very probable but should be confirmed by liver biopsy. The prognosis for asymptomatic older patients with the disease appears virtually the same as for a normal age-matched control population until such a time as symptoms attributable to liver disease arise (Mitchison *et al.* 1990). In view of the excellent prognosis of asymptomatic individuals no intervention is recommended at present beyond regular review.

Symptomatic primary biliary cirrhosis

The major symptoms are pruritus, upper abdominal pain, and fatigue. The main signs are jaundice, some pigmentation, particularly over areas where scratching due to pruritus has occurred, and xanthomata. Typically liver function tests are cholestatic, and the AMA is almost invariably positive, at a titre greater than 1 in 40. Immunoglobulins may be elevated, particularly IgM. The diagnosis must be confirmed by liver biopsy, preferably under ultrasound control to exclude any extrahepatic biliary obstruction. Prognostic models have recently suggested that age is an independent adverse prognostic feature in primary biliary cirrhosis (Grambsch *et al.* 1989). Other important prognostic indicators are serum bilirubin, presence or absence of cirrhosis, serum albumin, and the presence or absence of complications of portal hypertension.

No universally accepted treatment for symptomatic primary biliary cirrhosis has been adopted but it has recently been suggested that the bile acid ursodeoxycholic acid may well be beneficial, particularly for symptomatic patients with precirrhotic liver histology (Poupon *et al.* 1991). Pruritus may be treated with cholestyramine. Some older women with primary biliary cirrhosis and prolonged cholestasis have accelerated bone thinning and this may perhaps be prevented by regular calcium supplements. Conceivably treatment with a bisphosphonate and calcium may prevent or even reverse this bone thinning.

Complications of portal hypertension

These occur in end-stage primary biliary cirrhosis, as in other forms of cirrhosis. Liver transplantation is now the standard treatment for patients with complications of cirrhosis but as yet almost no patients over age 65 have been offered this treatment in the United Kingdom. Other autoimmune disorders, particularly Sjögren's syndrome, rheumatoid arthritis, thyroid disease, mixed connective tissue disease or scleroderma, and fibrosing alveolitis are also associated with primary biliary cirrhosis.

Chronic active hepatitis

Autoimmune chronic active hepatitis, often associated with positive antinuclear and/or smooth muscle antibodies (AMA and SMA), is not as rare in advanced age as was once thought. The presenting symptoms are usually of fatigue, fluid retention, and jaundice; these may be insidious over a few weeks or months. The term chronic active hepatitis is really an histological description in which there is an inflammatory infiltrate, mainly of lymphocytes and plasma cells, expanding out from portal tracts into the liver lobules with associated piecemeal necrosis of liver cells. This may progress through fibrotic scarring and bridging between portal tracts and central veins to cirrhosis. In a number of old patients, particularly women, this disease may present with an accelerated and unremitting course. In a few individuals autoantibodies are all negative (as are viral markers and all markers of other possible liver diseases) but the disease is otherwise indistinguishable from classical autoimmune chronic active hepatitis.

Biochemical tests are variable but transaminases are usually markedly elevated as are serum immunoglobulins. The disease may be associated with arthralgia, haemolytic anaemia, or glomerulonephritis. A careful history to exclude a possible drug-related cause for such an illness must be taken. Physical signs of autoimmune chronic active hepatitis in older patients are as for other chronic liver diseases.

Treatment

Most patients respond to steroid treatment, usually together with azathioprine (initially 30 mg prednisolone daily, falling to a maintenance dose of 7.5 to 12.5 mg/daily). With this treatment the prognosis is reasonable with an 80 per cent 5-year survival. Treatment should almost always be maintained over many years since withdrawal may lead to serious relapse. Even in patients without autoantibodies a trial of steroid treatment is well worthwhile; if no response is obtained after 3 months then treatment can be withdrawn.

It is of interest that autoimmune chronic active hepatitis may actually be on the increase in older men and women and it has been suggested that in some instances the disease arises as a result of extreme persistence of the measles virus following a long previous primary infection (Robertson *et al.* 1987). Nonetheless, as with primary biliary cirrhosis, the reason for the onset of the immune attack is unknown.

Primary sclerosing cholangitis

This disease presents extremely rarely in patients over age 65, the mean age of onset being around 40 years. It is usually associated with ulcerative colitis and has a similar spectrum and clinical course to primary biliary cirrhosis. Very occasionally an elderly patient, usually a man with long-standing ulcerative colitis, may present with persisting abnormalities of liver function tests particularly raised serum alkaline phosphatase. The diagnosis rests on demonstration at endoscopic cholangiography of the typical irregular and beaded appearance of the intrahepatic biliary tree together with compatible liver histology and prolonged cholestatic abnormality of liver function tests. In such elderly patients the prognosis is usually good.

Cryptogenic liver disease

It has been suggested that there may be a specific form of 'senile cryptogenic cirrhosis' but this seems unlikely. Nonetheless in a major series from the Mayo Clinic 77 patients over age 70 from a total of about 33 500 autopsies were found to have a cryptogenic cirrhosis unsuspected in life (Ludvig and Bagenstoss 1970). More detailed investigation with new and future markers for past infection with hepatitis viruses together with greater histological definition of such diseases as α_1-antitrypsin deficiency and haemochromatosis have substantially reduced the proportion of individuals with cryptogenic liver disease. Nonetheless, patients at present designated as having cryptogenic cirrhosis have a poor prognosis and among elderly individuals often present with portal hypertension or even hepatocellular carcinoma.

Other parenchymal liver diseases

Haemochromatosis

In haemochromatosis there is increased hepatic uptake of iron from transferrin. At present the haemochromatosis gene has not yet been identified but it is known to be on the short arm of chromosome 6 close to the locus of histocompatibility antigen A3. Inheritance of haemochromatosis is autosomal recessive.

Although most patients with haemochromatosis, particularly men, present around age 40 a number of individuals with no previous known family history may present at age 70 or older. The presenting symptoms are then of chronic liver disease, usually with its complications and sometimes with hepatocellular cancer. Typically symptoms are of weakness, abdominal pain, lethargy, and jaundice, and about half of the patients also have diabetes. Investigation includes measurement of serum ferritin and iron saturation, liver function tests, and liver biopsy with histological evaluation of liver iron stores. Although serum ferritin may be elevated non-specifically in relation to hepatocellular damage, in a patient with serum ferritin over 1000 μ/l, iron saturation over 95 per cent, and HLA tissue typing A3, B7, or B14 the diagnosis is highly probable.

Treatment is by venesection. About 500 ml of blood should be removed at fortnightly intervals, and this should be carried out until serum ferritin returns to within the normal range. Despite frequent venesections anaemia very seldom occurs. In precirrhotic patients, once excess iron has been eliminated from the liver, the dangers of the complications of cirrhosis and other manifestations of haemochromatosis (diabetes, cardiomyopathy, testicular atrophy) are removed provided that continuing supervision of serum ferritin is maintained. Unfortunately the majority of elderly patients with haemochromatosis already have cirrhosis and while venesection reduces the likelihood of hepatocellular failure it does not prevent the development of hepatocellular carcinoma in up to 30 per cent of patients.

It is vital to screen the families of all individuals in whom a new diagnosis of primary haemochromatosis has been made. Serum ferritin estimations are a reliable indicator in men aged over 35. HLA tissue typing in first-degree relatives is also useful since only those relatives showing a similar HLA haplotype with the proband are at risk of development of the disease. Women very seldom develop haemochromatosis before the menopause because of menstrual blood loss and so may present for the first time in later life.

α_1-Antitrypsin deficiency

The normal protease inhibitor α-antitrypsin is synthesized in the liver. Its secretion is under the genetic control of a single autosomal dominant gene on the long arm of chromosome 14 and a number of phenotypic alleles have been identified. α_1-Antitrypsin deficiency may be suspected as a cause of chronic liver disease when periportal hepatocytes are seen to contain granules of a periodic acid-Schiff stain-positive diastase-resistant material. These granules are composed of abnormal α_1-antitrypsin whose transport out of the cell has been impaired. Individuals who are homozygous for the abnormal protein PiZZ very seldom reach old age since they develop either severe emphysema or cirrhosis in childhood.

At present controversy exists as to whether the possession of a Z heterozygous phenotype (present in about 3 per cent of Western Europeans) (Brind *et al.* 1990) predisposes to the development of cryptogenic chronic liver disease in elderly people. Some years ago it was suggested that such heterozygotes may have increased susceptibility to an indolent cirrhosis. More recently it seems possible that the presence of these α_1-antitrypsin granules in biopsies of elderly patients with otherwise cryptogenic cirrhosis may be no greater than in the population at large.

Liver cysts

These may be defined as liver-enclosed spaces which contain air or liquid. It is unclear what proportion are inherited or acquired. They are increasingly detected as a result of more widespread use of abdominal ultrasound.

Solitary liver cysts seldom cause symptoms; they are under low pressure and rarely grow at any speed. Very occasionally an elderly patient with upper abdominal pain of unknown cause may be found to have a large solitary liver cyst. It may be worth undertaking percutaneous needle aspiration of such a cyst under local anaesthetic on one occasion but such aspiration very rarely produces relief of symptoms. Polycystic disease of the liver is associated with polycystic disease of the kidneys and there is some overlap with the extremely rare condition of hepatic biliary cystic disease (Caroli's disease).

Occasionally polycystic disease may need to be distinguished from multiple hepatic metastases, since debris can sometimes be found in these totally benign cysts. Aspiration confirms the innocent nature of the lesions.

The liver in systemic disease

In addition to such conditions as heart failure there are a number of systemic diseases in which clinically relevant hepatic abnormalities may occur despite the fact that the liver is not the prime target of the illness.

Collagen vascular disease

Elderly patients with rheumatoid arthritis, polymyalgia rheumatica, and systemic lupus erythematosus (SLE) may all have abnormal tests of liver function. Histological examination may reveal granulomata, fibrosis, or other non-specific abnormalities. Curiously in none is a vasculitic lesion detected within the liver. Unless an additional separate liver disease is suspected it is therefore unnecessary to investigate mild abnormalities of liver function tests in these disorders.

Lymphoproliferative disorders

The liver is frequently involved in these diseases; it may be enlarged, and while focal abnormalities are not necessarily visible on imaging, non-specific lymphocytic infiltration with granuloma formation may be seen. In conjunction with a prolonged pyrexia of unknown origin and abnormality of liver function tests in the absence of other putative causes of liver disease, such findings are rather suggestive of lymphoma in an elderly patient. Sometimes, specific histological features of highly abnormal lymphocytes or Reed-Sternberg cells may provide a diagnosis. Occasionally the liver is massively enlarged in such conditions.

In the myeloproliferative disorders myelofibrosis and myelosclerosis the liver may be enlarged in conjunction with the spleen. Again liver function tests may be non-specifically and mildly abnormal. Occasionally, owing to massive splenic enlargement with consequent increased blood flow portal hypertension may arise in these conditions.

Heart failure

As has previously been stated, mild non-specific abnormalities of liver function tests are frequently seen in heart failure in the elderly. Occasionally if hypotension has occurred, particularly during myocardial infarction or after sudden severe heart failure, hepatic necrosis with a steep rise in serum transaminases may occur. If persistent right-sided heart failure exists in an elderly individual abnormalities in liver function tests may become irreversible and may ultimately be accompanied by decreased serum albumin and prolonged prothrombin time. Occasionally in such patients ascites may develop. Treatment should be directed at the underlying heart disease.

Complications of cirrhosis

Portal hypertension

The prognosis from upper gastrointestinal bleeding from all causes is worse in elderly patients than in young. However, several studies have now shown that when oesophageal varices bleed the immediate fatality and proportion of patients discharged from hospital is similar in elderly and younger individuals, at least in centres experienced in treating the complications of portal hypertension (Bullimore *et al.* 1989). We have recently shown that overall admission fatality for those over 65 years, presenting with oesophageal varices, was 26 per cent, not significantly different from the figure of 19 per cent in those under age 65 (K. W. Woodhouse and O. F. W. James, personal observations). It is not clear in an individual why varices bleed at any given moment but there appears to be a relationship between the portal pressure and also the size of the varices and their likelihood of bleeding. The higher the pressure, the larger the veins, and the more likely the subsequent haemorrhage (Rector and Reynolds 1985).

Management

The acute treatment of variceal haemorrhage is the same as for other upper gastrointestinal haemorrhage—resuscitation by transfusion, preceded by infusion of a plasma expander if necessary. In actively bleeding patients diagnosis of the source of bleeding requires emergency endoscopy, and subsequent treatment should be carried out with intensive care facilities. At present the acute management of variceal haemorrhage is controversial. However, a consensus is now emerging that in the absence of acute haemorrhage at the time of admission intravenous infusion with the somatostatin analogue octreotide to prevent earlier rebleeding is probably the treatment of choice (Burroughs *et al.* 1990). In patients in whom a site of current active bleeding is observed from oesophageal or upper gastric varices or who rebleed during somatostatin infusion balloon tamponade is highly effective. Great care should be used in the placement and management of such balloons and it is recommended that this treatment is only carried out in acknowledged centres.

Once emergency control of bleeding has taken place there are three therapeutic options in an elderly patient. The first, surgery—either oesophageal transection or gastric devascularization—is kept as a very last resort in elderly patients because of poor postoperative recovery. Over recent years it has been conventional to suggest that repeat endoscopic sclerotherapy to obliterate the oesophageal varices should be the treatment of choice (Westaby *et al.* 1985). While the incidence of rebleeding may be reduced by such treatment overall mortality has not been shown to be improved, and it is now being recognized that the eradication of oesophageal varices may lead to the development of gastric varices or portal gastropathy which are more difficult to treat. A third approach to long-term treatment is by using beta-blockers such as propranolol, possibly in combination with peripheral vasodilatation with long-term isosorbide (Lebrec 1992). Conceivably a combination of sclerotherapy and the lowering of variceal pressure with beta-blocker may provide the best long-term results. For an elderly patient a trial of propranolol treatment once the initial haemorrhage had been well controlled would seem reasonable in the first instance.

More recently long-term beta-blocker treatment has been suggested in the prevention of bleeding from varices in patients with cirrhosis and known oesophageal varices which have never bled. Several studies have now suggested that this may reduce both the likelihood of bleeding from the varices and mortality. Meta-analysis of these studies has confirmed that this therapy is indeed beneficial. Again such treatment requires long-term and careful follow-up of elderly patients and implies endoscopic confirmation of the presence of oesophageal varices in all patients found to have cirrhosis.

Cirrhotic ascites

Ascites may be conveniently classified into transudate or exudate depending upon its cause. Heart disease—constrictive pericarditis, right sided heart valve lesions or heart failure—and uncomplicated cirrhosis itself leads to a transudate with protein defined as being under about 15 g/l. Intraperitoneal carcinomatosis, severe intraperitoneal infections, and chylous ascites typically cause an exudative ascites with protein above about 25 g/l. Unfortunately there is considerable overlap between these but the distinction is still of some value.

It is now considered vital to take a diagnostic tap (with a venepuncture needle) of the ascitic fluid on hospital admission for all patients with cirrhotic ascites. It has become clear that this is particularly important since up to one-quarter of patients with cirrhotic ascites will have spontaneous bacterial peritonitis at the time of admission. A rapid diagnosis of spontaneous bacterial peritonitis can be made if the total polymorphonuclear white blood cell count is over 250/mm^3 (Reynolds, 1986). Fluid is also sent for protein content microscopy and culture. In patients with spontaneous bacterial peritonitis specific signs and symptoms of infection are often absent although abdominal pain, fever, and tenderness in a patient with ascites (particularly alcoholic) are highly suggestive. In the presence of a raised ascitic polymorphonuclear white blood cell count urgent antibiotic treatment should be started. Treatment with a third-generation cephalosporin, e.g. cephotaxime, is currently the treatment of choice. Spontaneous bacterial peritonitis carries a grave prognosis; over 75 per cent who have developed this complication of ascites are dead within 1 year of its first occurrence.

Treatment of ascites

In elderly individuals treatment of ascites need not necessarily be aimed at complete removal of all fluid, particularly as this may be accompanied by overenthusiastic use of diuretics with accompanying complications. In patients with moderate ascites sodium restriction (22 to 24 mmol/day) and bed rest underline sensible management. If serum sodium falls below 125 mmol/l the likelihood is that impaired water excretion has occurred in addition to sodium retention and fluid intake should be restricted to about 1 l/day.

Because hyperaldosteronism occurs in cirrhotic ascites the diuretic spironolactone which blocks tubular reabsorption of sodium by opposing the action of aldosterone is recommended as first-line treatment. This should start at 50 to 100 mg/daily and may be increased up to 600 mg/daily. In patients with peripheral oedema in addition to ascites a loop diuretic such as frusemide may be added. The progress of diuresis should be monitored by daily weighing together with frequent checks on electrolytes.

Contrary to previously held dogma large volume abdominal paracentesis, if accompanied by simultaneous intravenous infusions of albumin, is safe, rapid, and effective (Gines *et al.* 1987). Particularly in elderly subjects the advantage of abdominal paracentesis is that it relieves the patient of discomfort more rapidly than prolonged bed rest and diuretic treatment and may lessen the time spent in hospital. The procedure may be repeated on a number of occasions and may be followed by conventional treatment with diuretics. In order to minimize the risk of infection each paracentesis should take only 1 to 2 h, after which the paracentesis needle must be removed immediately. In several controlled or comparative studies between diuretic treatment and abdominal paracentesis the treatment complications and subsequent hospital admission for reaccumulation of ascites were similar in the two groups but total length of hospital stay was reduced in the abdominal paracentesis group. It seems sensible that in an elderly patient with recent onset of severe uncomfortable ascites paracentesis with simultaneous intravenous infusion of albumin should now be first-line treatment. Maintenance should still be with diuretics, dietary restriction of sodium, and attention to general nutrition.

Hepatocellular carcinoma (see Section 10)

In Europe and North America primary hepatocellular carcinoma is largely a disease of ageing like many other digestive tract cancers. Hepatocellular carcinoma is usually associated with cirrhosis regardless of the underlying cause and it is probable that the length of time during which an individual has had cirrhosis is the important determining factor. In these countries more than 50 per cent of patients with cirrhosis and hepatocellular carcinoma are over 60 and more than 40 per cent are over 70 at presentation (Cobden *et al.* 1986). The main underlying cause of the cirrhosis which has led to hepatocellular carcinoma varies from country to country. The combination of alcohol with previous hepatitis B or hepatitis C infection appears to be particularly important (Dazza *et al.* 1990).

Detection/presentation

First presentation of hepatocellular carcinoma in old age is often with the complications of cirrhosis—bleeding varices, ascites, or encephalopathy—and indeed, in an individual over age 70 presenting in this way, the index of suspicion that hepatocellular carcinoma underlies the sudden deterioration in the patient's condition should be high. In patients presenting primarily with the symptoms of the tumour itself then abdominal pain, weight loss, and symptoms of hepatic enlargement are typical. Bony metastases are not uncommon in elderly patients and thus bone pain may be a prominent feature.

Because prognosis depends to a great extent upon the size of tumour at first detection, and hence the treatment options available, screening of all elderly patients with known cirrhosis for the development of hepatocellular carcinoma is now recommended. At a minimum this should involve a 6-monthly check on serum alpha-fetoprotein but since this tumour marker is only elevated in about 70 per cent of patients with hepatocellular carcinoma in Western countries some authorities now suggest regular liver ultrasound examinations should also be carried out.

Diagnosis and treatment

Diagnosis may be made in a cirrhotic patient if serum alpha-fetoprotein is above 500 ng/m (normal range up to 50 ng/ml). In fact, if serum alpha-fetoprotein has been steady at under 50 ng/ml and rises above 100 ng/ml the index of suspicion that a small hepatocellular cancer has developed must be high. The size, extent, and number of tumours (these are frequently multicentric) must be ascertained by hepatic CT scan, preferably with angiography. Very small tumours are best detected by lipiodol angiography followed by subsequent CT scan. Ultrasound or CT-guided needle biopsy should be carried out.

Perhaps because of the underlying cirrhosis the overall prognosis for hepatocellular carcinoma is extremely poor with mean survival from presentation in many series being less than 6 months. At present no single treatment regime has shown unequivocal benefit. In patients with small single (or even multiple) tumours ultrasound-guided injection of ethanol appears moderately effective. Tumour resection of small single lesions is occasionally possible although again in the elderly patient major liver surgery appears to carry higher mortality and morbidity than in younger individuals (Fortner and Lincer 1990). Because the radiological agent lipiodol appears selectively retained by tumours within the liver efforts have been made to link various anticancer drugs (e.g. Adriamycin) to the lipiodol or to synthesize radioactive ^{131}I-lipiodol in order to give a local radiotherapy dose to the tumours. In small uncontrolled studies some slight benefit has been claimed for both of these treatments. Sometimes localized arterial embolization of feeder vessels to the tumours at angiography has also been carried out with some claims of improved survival in a few patients (Kasugai *et al.* 1989).

BILIARY DISEASE

Gallstones

The prevalence of gallstones increases with age, and in Western countries they are predominantly composed of cholesterol. In Europe the prevalence of gallstones/cholecystectomy in females aged 70 is about 30 per cent, and in males about 19 per cent (Jorgensen *et al.* 1990). By age 80 this figure may be as high as 40 per cent in women (Bateson and Bouchier 1975). There are two apparent reasons for this increase. First, cholesterol saturation of bile rises with age; this is due to increasing hepatic secretion of cholesterol in both men and women and to decreased bile acid production, as measured by an isotope dilution technique (Einarsson *et al.* 1985). Second, it is possible that the gallbladder is less sensitive to endogenous cholecystokinin with advancing age and this might lead to diminished gallbladder contraction following a meal (Poston *et al.* 1990). It has also been postulated that duodenal juxtapapillary diverticula whose prevalence steeply increases with age may be associated with biliary stasis, malfunction of the sphincter of Oddi, and reflux of duodenal contents into the common bile duct (Grace *et al.* 1990).

Gallbladder disease

Although gallstones in the gallbladder are regarded as a 'disease' we have seen that their prevalence in old age is so high that they may also be regarded as part of the ageing process. Further factors which increase the likelihood of gallstone formation include medications (clofibrate, oestrogens), ileal disease or resection, cirrhosis, and life-long obesity.

Episodic pain or biliary colic is the most frequent presenting symptom of gallbladder disease. The pain usually occurs 15 to 60 min after a meal and is worst in the right upper abdomen. Other less specific symptoms, such as bloating, belching, or a feeling of fullness, are of little diagnostic value. Of greater importance than occasional upper abdominal discomfort are cholecystitis and the complications of the passage of gallstones out of the gallbladder and into the biliary tree. Acute cholecystitis is manifest by fever and right upper abdominal pain and tenderness, with elevated white blood cell count. Blood cultures are frequently positive. While this may be a simple self-limiting illness the toxic effects of acute cholecystitis themselves and the risks of subsequent perforation or chronic biliary inflammation are so important that the disease should be treated urgently. Treatment should be with broad spectrum antibiotics (third generation cephalosporin is probably the first choice), together with rehydration. Urgent intervention is now generally favoured. Conventional treatment is with urgent cholecystectomy and in experienced hands mortality is little higher in the elderly than in young individuals (Hidalgo *et al.* 1989). In extremely ill and frail old people then cholecystostomy for decompression and drainage of the gallbladder may be preferable, particularly if perforation of the gallbladder with complicating peritonitis has already occurred. Subsequent operation may not now be needed to remove the gallbladder remnant (Kaufman *et al.* 1990), since endoscopists have turned their technical expertise to the problem of gallbladder disease. Percutaneous laparoscopic cholecystostomy is now routinely carried out in many centres, even with relief of obstruction of the cystic duct (Mirizzi's syndrome) (Vogelzang and Nemcek 1988). The ultimate is perhaps being approached in that percutaneous 'key-hole' endoscopic cholecystectomy in patients with cholecystitis is now being reported. Clearly these techniques require great expertise but laparoscopic cholecystectomy by an experienced surgeon should now be the treatment of choice for elderly patients whose gallstones are confined to the gallbladder.

With the risks of cholecystitis and escape of stones into the biliary tree the question arises as to what treatment to recommend in an elderly patient with occasional upper abdominal pain suggestive of biliary colic in whom investigation reveals gallstones in the gallbladder alone. There are four possible management strategies.

1. Masterly inactivity. If patients are extremely elderly, unwilling to undergo any further investigation or treatment, and have minimal symptoms then the physician may feel that further treatment is unwarranted despite the undoubted risks of morbidity and mortality associated with symptomatic gallstones.
2. Cholecystectomy may be carried out. As 'key-hole' cholecystectomy becomes more widespread then clearly this has become the operative treatment of choice (Cuschieri *et al.* 1989).
3. Lithotripsy. In this technique extracorporeal shock waves are directed at the gallstones. Patients suitable for lithotripsy should have predominantly cholesterol rich (radiolucent) stones, and a patent cystic duct demonstrated by ultrasound with stimulation of gallbladder contraction. This treatment is not suitable for acute cholecystitis. Adjuvant treatment is given with oral bile acid for up to 3 months after the procedure to dissolve the gallstone fragments. With new generation lithotripsy machines treatment is carried out under light sedation and may be on a day-case basis.
4. Gallstone dissolution treatment. There were great hopes that treatment with the bile acids chenodeoxycholic acid or ursodeoxycholic acid to dissolve gallstones would lead to a major reduction in the need for other intervention. Unfortunately these hopes have not been realized. While many patients have had gallstones successfully dissolved by oral bile acid therapy several drawbacks make it unrealistic in elderly patients. (i) Treatment is usually prolonged (1–2 years). (ii) Gallstones frequently reform after dissolution. (iii) Patients must have completely radiolucent gallstones, preferably small (<5 mm) in a functioning gallbladder. (iv) Even in these favourable circumstances successful complete dissolution of stones is as low as 20 per cent after 2 years of treatment. With the advent of other treatment options listed above and with the exception of its use as an adjuvant to lithotripsy the use of oral bile acid treatment for the dissolution of gallstones is now seldom recommended.

Acalculous cholecystitis

About 5 per cent of patients with acute cholecystitis do not have gallstones. This condition usually occurs in adults with some other major critical illness—particularly following severe trauma or major surgery. It is thought that there is a combination of bile stasis ('atony of the gallbladder') together with infection of the bile, sometimes with gas-forming organisms. The fatality of the condition is up to 50 per cent but this in part is because of the other associated illness and also because of the difficulty in making the diagnosis (Ullman *et al.* 1984). Ultrasound examination in an acutely ill patient with upper abdominal pain may show a markedly thickened gallbladder, sometimes with the presence of an adjacent fluid collection. Treatment is with very urgent cholecystectomy since delay almost invariably leads to perforation and death.

Carcinoma of the gallbladder (see Section 10)

Gallbladder carcinoma is age-associated and is most common at very advanced ages. Gallstones appear to increase the risk of gallbladder cancer but not to an enormous extent except where the gallbladder wall itself has calcified (Diehl 1983).

Occasionally small adenocarcinomas are detected in cholecystectomy specimens where the gallbladder has been removed for gallstones. In these circumstances the prognosis is good. Unfortunately, in the majority of instances, by the time the disease presents it has already spread beyond the confines of the gallbladder. Typical presenting symptoms and signs are upper abdominal pain, weight loss, a right upper abdominal mass, and jaundice. The diagnosis is often mistaken for that of some other form of biliary tract disease or liver disease. Gallbladder carcinomata are locally aggressive and overall 1-year survival is less than 10 per cent. Heroic surgery is seldom if ever indicated in elderly patients since operative fatality exceeds 20 per cent and even after such procedures 1-year survival is less than 10 per cent (Foster 1987). Treatment therefore must be palliative. Relief of biliary obstruction and symptomatic care are all that can at present be offered.

Diseases of the bile ducts

Choledocholithiasis

The presence of gallstones in the biliary tree (choledocholithiasis) presents a frequent challenge to physicians and surgeons caring for elderly patients. Biliary stones in elderly people may present with very non-specific symptoms, even suggesting a confusional state or dementia (really a chronic toxic confusional state) (Cobden *et al.* 1984), or with symptoms clearly referrable to the biliary tree such as obstructive jaundice and cholangitis. As with gallstones in general, the incidence of common bile duct stones increases with advancing age. Stones in the biliary tree are usually secondary to gallbladder disease; hitherto we have considered the gallbladder in isolation, but if stones exist in the biliary tree they must be considered in conjunction with the gallbladder. While most stones in the biliary tree arise in the gallbladder primary formation does occur in the bile ducts; these are usually brown pigment rather than cholesterol stones. They occur more frequently in circumstances of stasis—above biliary strictures for example.

Cholangitis

Cholangitis is an infective inflammation of the biliary tree; it is most commonly associated with the presence of gallstones and in many instances the likely source of the infection is from the duodenum although blood or lymph-borne infection certainly occurs not infrequently. The classical clinical presentation of cholangitis with intermittent fever and chills, jaundice, and abdominal pain (Charcot's triad) is present in perhaps 60 per cent of patients. However, jaundice is frequently absent. In a patient with fever, rigors, abdominal pain, and abnormal liver function tests cholangitis is the most likely diagnosis. The distinction from acute cholecystitis may be difficult; the greater the abnormality of the liver function tests the more likely is the biliary tree, rather than the gallbladder, to be the source of the problem.

Again treatment should be with urgent intravenous broad spectrum antibiotics (after blood cultures have been obtained). Patients are often frail and dehydrated and volume replacement should be carried out. Because of associated endotoxinaemia, renal impairment may occur and thus there should be careful monitoring of urine output, creatinine, and electrolytes.

Once antibiotic treatment has been instituted and the patient's general condition improved, endoscopic retrograde cholangiopancreatography (ERCP) with sphincterotomy and stone removal should be carried out. Even if there are stones remaining *in situ* in the gallbladder sphincterotomy and removal of the common duct stones show a good result in very elderly patients (Davidson *et al.* 1988). If large stones remain in the biliary tree then a nasobiliary cannula may be left in the common bile duct above the stone at the time of ERCP for subsequent introduction of a dissolution agent such as methyl tert-butyl ether (Brandon *et al.* 1988). Alternatively lithotripsy may subsequently be employed. Technological advances have meant that surgical exploration of the common bile duct with removal of stones for cholangitis has substantially reduced in frequency. Nonetheless this remains a viable option although associated morbidity and mortality may be a little higher.

Benign biliary strictures

Biliary strictures may arise as a result of gallstone disease or previous biliary tract surgery as well as in association with sclerosing cholangitis. If such a stricture is associated with gallstones and recurrent cholangitis or with prolonged 'obstructive' tests of liver function then after stone removal the stricture should be either dilated by endoscopic balloon dilatation or occasionally an endoprosthesis inserted to facilitate bile flow (Gallaher *et al.* 1985).

Cholangiocarcinoma

Bile duct carcinoma may arise at any point in the biliary tree. The usual presenting symptoms are jaundice and pruritus. Weakness and weight loss are usually also prominent. Epigastric discomfort and diarrhoea, related to steatorrhoea, may also occur. On examination the liver is usually smoothly enlarged and the patient jaundiced. Investigations reveal cholestatic liver function tests with negative markers for any possible chronic liver disease. Ultrasound examination usually reveals a dilated biliary tree down to the site of the obstruction, often at the hilum of the liver. If the lesion is high within the liver then percutaneous transhepatic cholangiography (PTC) rather than ERCP is the investigation of choice; in most instances both PTC and ERCP should be carried out to assess the extent of the lesion. If the obstruction is in the region of the common bile duct the differential diagnosis is with ampullary or pancreatic tumour, carcinoma of the gallbladder, obstruction by malignant glands, or benign gallstone related disease. Within the liver it is sometimes hard to distinguish bile duct tumours from hepatocellular carcinoma. Because of this wide differential diagnosis ultrasound guided biopsy or cytological aspiration is important. The tumour is often slow growing, distant metastases being relatively rare. It exerts its effects by its critical obstructive nature in the biliary tree. Even when completely untreated mean prognosis is more than 1 year.

Treatment

Occasionally in low bile-duct tumours surgical excision is possible although this is usually difficult in elderly patients. Palliative treatment may be carried out by introducing a stent through the tumour. If the tumour is low this can be carried out endoscopically. If the tumour is higher the stent is introduced percutaneously. After PTC an external draining catheter is left in place for up to 24 h. Following this decompression of the intrahepatic biliary tree the stent is introduced over the catheter into the liver and through the stricture. The catheter is removed leaving the internal stent in place (Dooley *et al.* 1984). Sometimes such stents block off but they can be replaced by either endoscopic or transhepatic routes. Mean survival in such patients may be more than 18 months and quality of life with reduced pruritus, improved weight, and well being is vastly improved.

References

Aron, E., Dupin, M., and Jobard, P. (1979). Les cirrhosis du troisieme age. *Annuales de Gastroenterologie et d'Hepatologie (Paris)*, **14**, 558–63.

Bateson, M.C. and Bouchier, I.A.D. (1975). Prevalence of gallstones in Dundee: a necropsy study. *British Medical Journal*, **4**, 427–30.

Bertel, C.R., Van Heerden, J.A., and Sheedy, P. F. (1986). Treatment of pyogenic hepatic abscess: surgical vs. percutaneous drainage. *Archives of Surgery*, **121**, 554–62.

Brandon, J.C., *et al.* (1988). Common bile duct calculi: updated experience with dissolution with methyl tertiary butylether. *Radiology*, **166**, 665–8.

Brind, A.M., Bassendine, M.F., Bennett, M.K., and James, O.F.W. (1990). Are alpha-1-antitrypsin granules in the liver always important? *Quarterly Journal of Medicine*, **76**, 699–710.

Bullimore, D.W., Miloszewski, K.J.A., and Losowsky, M.S. (1989). The prognosis of elderly subjects with oesophageal varices. *Age and Ageing*, **18**, 35–8.

Burroughs, A.K., *et al.* (1990). Randomised double-blind placebo controlled trial of somatostatin for variceal bleeding. *Gastroenterology*, **99**, 1388–95.

Chiaramonte, M., *et al.* (1992). In press.

Cobden, I., Bassendine, M.F. and James, O.F.W. (1986). Hepatocellular carcinoma in N.E. England. Importance of hepatitis B infection and ex-tropical military service. *Quarterly Journal of Medicine*, **60**, 855–63.

Cobden, I., *et al.* (1984). Gallstones presenting as mental and physical debility in the elderly. *Lancet*, **i**, 1062–4.

Cook, J.M., *et al.* (1987). Alterations in the human immune response to the hepatitis B vaccine among the elderly. *Cellular Immunology*, **109**, 89–96.

Cuschieri, A., El Ghany, A.A.B., and Holley, M.P. (1989). Successful chemical cholecystectomy: a laparoscopic guided technique. *Gut*, **30**, 1786–94.

Davidson, B.R., Neoptolemos, J.P., and Carr-Locke, D.L. (1988).

Endoscopic sphincterotomy for common bile calculi in patients with gallbladder *in situ* considered unfit for surgery. *Gut*, **29**, 114–20.

Dazza, M.C., *et al.* (1990). Hepatitis C virus antibody and hepatocellular carcinoma. *Lancet*, **i**, 1216.

Diehl, A.K. (1983). Gallstone size and the risk of gallbladder cancer. *Journal of the American Medical Association*, **250**, 2323–6.

Donaldson, R.M. (1982). Advice for the patient with 'silent' gallstones. *New England Journal of Medicine*, **307**, 815–17.

Dooley, J.S., Dick, R., and George, P. (1984). Percutaneous transhepatic endoprosthesis for bile duct obstruction: complications and results. *Gastroenterology*, **86**, 905–11.

Einarsson, K., Nilsell, K., Leijd, B., and Angelin, B. (1985). Influence of age on secretion of cholesterol and synthesis of bile acids by the liver. *New England Journal of Medicine*, **313**, 277–82.

Forbes, A. and Williams, R. (1988). Increasing age—an important adverse prognostic factor in hepatitis A virus infection. *Journal of the Royal College of Physicians London*, **22**, 237–9.

Fortner, J.G. and Lincer, R.M. (1990). Hepatic resection in the elderly. *Annals of Surgery*, **211**, 141–5.

Foster, J. (1987). Carcinoma of the gallbladder. In *Surgery of the Gallbladder and Bile Ducts* (eds. L.W. Way and C.A. Pellegrini), pp. 471–90. Saunders, Philadelphia.

Gallaher, D.J., *et al.* (1985). Non-operative management of benign postoperative biliary strictures. *Radiology*, **156**, 625–30.

Gambert, S.R., *et al.* (1982). Interpretation of laboratory results in the elderly. 1. A clinician's guide to hematologic and hepatorenal function tests. *Postgraduate Medicine*, **2**, 147–52.

Garagliano, C.F., Lillenfeld, A.M., and Mendelhof, A.I. (1979). Incidence rates of liver cirrhosis and related diseases in Baltimore and selected areas of the United States. *Journal of Chronic Diseases*, **32**, 543–54.

Gines, P., *et al.* (1987). Comparisons of paracentesis and diuretics in the treatment of cirrhotics with tense ascites. *Gastroenterology*, **93**, 234–40.

Goodson, J.D., *et al.* (1982). The clinical course of acute hepatitis in the elderly patient. *Archives of Internal Medicine*, **142**, 1485–8.

Grace, P.A., Poston, G.J., and Williamson, R.C.N. (1990). Biliary motility. *Gut*, **31**, 571–82.

Grambsch, P.M., *et al.* (1989). Extramural cross-validation of the Mayo PBC model. *Hepatology*, **10**, 846–50.

Grasedyck, K., *et al.* (1980). Aging of liver. Morphological and biochemical changes. *Mechanisms of Ageing and Development*, **14**, 435–42.

Hidalgo, L.A., *et al.* (1989). Influence of age on early surgical treatment of acute cholecystitis. *Surgery, Gynecology, Obstetrics*, **169**, 393–6.

Jorgensen, T., Kay, L., and Schultz-Larsen (1990). The epidemiology of gallstones in a 70 year old Danish population. *Scandinavian Journal of Gastroenterology*, **25**, 335–40.

Kasugai, H., *et al.* (1989). Treatment of hepatocellular carcinoma by transcatheter arterial embolisation combined with intraarterial infusion of a mixture of asplatus and ethiodized oil. *Gastroenterology*, **97**, 965–71.

Kaufman, M., *et al.* (1990). Cholecystostomy as a definitive operation. *Surgery, Gynecology, Obstetrics*, **170**, 533–7.

Kitani, K. (1990). Ageing and the liver. In *Progress in Liver Diseases*. (eds. H. Popper and F. Schaffner). Vol. IX, pp. 603–23. W. B. Saunders, Philadelphia.

Laverdant, C., *et al.* (1989). Les hepatites virales apres 60 ans. Aspects cliniques etiologiques et evolutifs. *Gastroenterologie Clinique et Biologique*, **13**, 499–504.

Lebrec, D. (1992). Pharmacological prevention of variceal bleeding and rebleeding. In *Therapy in Liver Diseases* (eds. J. Rodes and V. Arroyo) pp. 102–13. Doyma, Barcelona.

Maddrey, W., *et al.* (1986). Prednisolone therapy in patients with severe alcoholic hepatitis. *Hepatology*, **6**, 1202–8.

Marchesini, G., *et al.* (1988). Galactose elimination capacity and liver volume in ageing man. *Hepatology*, **8**, 1079–83.

Marchesini, G., *et al.* (1990). Synthesis of urea after a protein rich meal in normal man in relation to ageing. *Age and Ageing*, **19**, 4–10.

Melia, W.M., *et al.* (1984). Hepatocellular carcinoma in Great Britain; influence of age, sex, HBsAg status and aetiology of underlying cirrhosis. *Quarterly Journal of Medicine*, **53**, 391–400.

Mitchison, H.C., *et al.* (1990). Symptom development and prognosis in primary biliary cirrhosis. *Gastroenterology*, **99**, 778–84.

Myszor, M. and James, O.F.W. (1990). The epidemiology of primary biliary cirrhosis in North East England. *Quarterly Journal of Medicine*, **75**, 377–85.

Neuberger, J. and Williams, R. (1984). Halothane anaesthesia and liver damage. *British Medical Journal*, **289**, 1136–9.

Parker, S.G., James, O.F.W., and Young, E.T. (1986). Causes of raised serum alkaline phosphatase in elderly patients. *Modern Trends in Aging Research*, **147**, 153–7.

Popper, H. (1986). Aging and the Liver. In *Progress in Liver Diseases* (eds. H. Popper and F. Schaffner), Vol. VIII, pp. 659–83. Grune and Stratton, Orlando.

Poston, G.J. *et al.* (1990). Effect of age and sensitivity to cholecystokinin on gallstone formation in the guinea pig. *Gastroenterology*, **98**, 993–9.

Potter, J.R. and James, O.F.W. (1987). Clinical features and prognosis of alcoholic liver disease in respect of advancing age. *Gerontology*, **33**, 380–7.

Poupon, R.E., Balkau, B., Eschwege, E., Poupon, R., and the UDCA-PBC study group (1991). A multicenter, controlled trial of Ursodiol for the treatment of primary biliary cirrhosis. *New England Journal of Medicine*, **324**, 1548–54.

Rector, W.G. and Reynolds, T.B. (1985). Risk factors for haemorrhage from esophageal varices and acute gastric erosions. *Clinical Gastroenterology*, **14**, 139–44.

Reynolds, T.B. (1986). Rapid presumptive diagnosis of spontaneous bacterial peritonitis. *Gastroenterology*, **90**, 1294–5.

Rizzetto, M. and Verme, G. (1985). Delta hepatitis: present status. *Journal of Hepatology*, **i**: 187–9.

Robertson, D.A.F., *et al.* (1987). Persistent measles virus genome in autoimmune chronic active hepatitis. *Lancet*, **i**, 9–11.

Roll, J., *et al.* (1983). The prognostic importance of clinical and histological features in asymptomatic primary biliary cirrhosis. *New England Journal of Medicine*, **308**, 1–7.

Schmucker, D.L., Woodhouse, K.W., and Wang, R. (1990). Effect of age and gender on *in vitro* properties of human liver microsomal monooxygenases. *Clinical Pharmacology and Therapeutics*, **48**, 365–74.

Sonnenblick, M., Uren, R., and Tur-Kaspa, R. (1988). Non-a, non-B hepatitis in the aged. *Hepatology*, **8**, 1060 (abstr.).

Sridharan, G.V., Wilkinson, S.P., and Primrose, W.R. (1990). Pyogenic liver abscess in the elderly. *Age and Ageing*, **19**, 199–203.

Tauchi, H. and Sato, T. (1978). Hepatic cells of the aged. In *Liver and Aging* (ed. K. Kitani), pp. 3–19. Elsevier, Amsterdam.

Thomas, H.C. (1992). Management of chronic hepatitis B virus infection. In *Therapy in Liver Diseases* (eds. J. Rodes and V. Arroyo), pp. 242–7. Doyma, Barcelona.

Ullman, M., Hasselgren, P.O., and Tveit, E. (1984). Post traumatic and postoperative acute acalculous cholecystitis. *Acta Chirurgica Scandinavica*, **150**, 507–10.

Vogelzang, R.L. and Nemcek, A.A. Jr. (1988). Percutaneous cholecystectomy: diagnostic and therapeutic efficacy. *Radiology*, **168**, 29–34.

Watanabe, T., Shimada, H. and Tanaka, T. (1978). Human hepatocytes and aging. *Virchows Archiv B*, **27**, 307–16.

Westaby, D., MacDougall, B.R.D., and Williams, R. (1985). Improved survival following sclerotherapy for esophageal varices: final analysis of a controlled trial. *Hepatology*, **5**, 827–32.

Woodhouse, K.W. and James, O.F.W. (1990). Hepatic drug metabolism and ageing. *British Medical Bulletin*, **46**, 22–35.

Wynne, H.A., Cope, L., Mutch, E., Rawlins, M.D., Woodhouse, K.W., and James, O.F.W. (1989). The effect of age upon liver volume and applied liver blood flow in healthy man. *Hepatology*, **9**, 297–301.

SECTION 9
Cardiovascular disorders

9 Cardiovascular disorders

EDWARD G. LAKATTA AND GARY GERSTENBLITH

AGEING AND CARDIOVASCULAR STRUCTURE AND FUNCTION

Quantitative information on age-associated alterations in cardiovascular function is essential in attempting to differentiate those cardiovascular limitations of an elderly individual that relate to disease from those limitations that may fall within expected normal limits. However, interactions among lifestyle, disease, and ageing can have a substantial impact on cardiovascular function, and such interactions can alter the manifestations of 'pure' ageing effects on the cardiovascular system.

Occult disease can be easily overlooked and can cause severe functional impairments. This consideration is especially pertinent to investigation of the effect of age on cardiovascular function in man because coronary atherosclerosis, which increases exponentially in prevalence with age, is present in an occult form in at least as great a number of elderly persons as is the overt form of the diseaee (Lakatta 1985).

In addition to an increased prevalence of disease, changes in lifestyle occur with advancing age. These include changes in our habits of physical activity, eating, drinking, smoking, personality characteristics, etc. It has been well established in unselected populations that the average daily level of physical activity declines progressively with age (Lakatta 1985). Regular physical activity changes not only the function of the heart but also its size. Because the effect of physical conditioning can be so great, attempts to investigate to what extent a disease or an 'ageing process' alter cardiovascular function (particularly reserve function) must be controlled for the state of physical activity, or at least consider this in the interpretation of the results (Strandell 1964; Raven and Mitchell 1980).

Given the difficulty of quantifying and controlling of variables for lifestyle and detecting the presence of occult disease, it is not surprising that the literature contains widely differing perspectives of how cardiovascular function changes with age. Lakatta (1990) has provides an in-depth analysis of age-associated changes in cardiovascular function.

Cardiac output

Figure 1 summarizes the results of some studies that have measured cardiac output among individuals of a broad age range. The cardiac output (or cardiac index) at rest, as measured in different individuals (i.e. cross-sectional studies) of varying age, has been found to remain unchanged, to decrease substantially, or even to increase slightly with age. The differences probably stem from different criteria for selection in the various studies. The overall control of cardiac output results from a complex interaction of modulating influences (Fig. 2). Stroke volume is dependent on (1) the load before ejection (pre-load), measured as the end-diastolic volume; (2) the load during ejection of blood (after-load

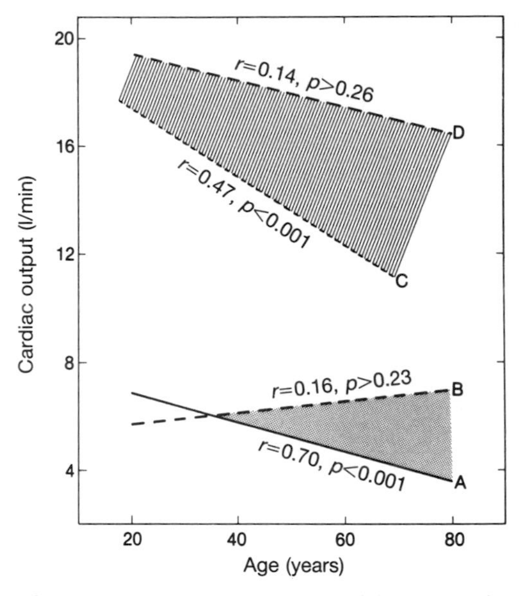

Fig. 1. Cardiac output measured at rest and during exercise at exhaustion in the upright position vs. age. Line A (least-squares linear regression) from Branfronbrener *et al.* (1965), lines B and D from Rodeheffer *et al.* (1984), line C from Julius *et al.* (1967).

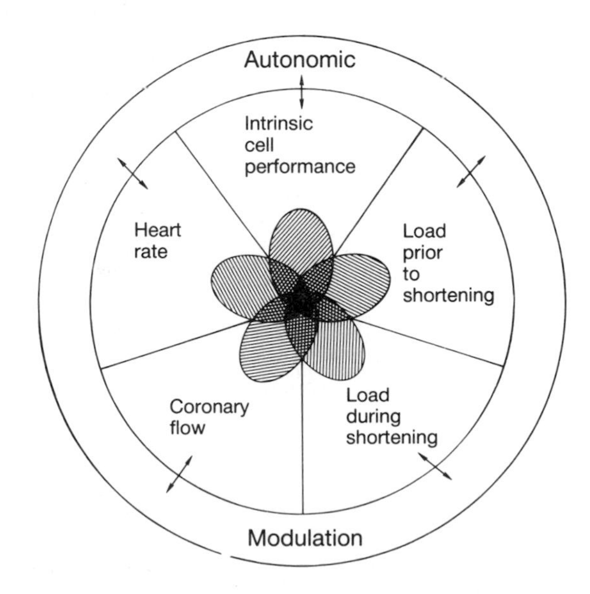

Fig. 2. Factors that regulate cardiovascular performance. The ovals have been drawn to overlap each other in order to indicate the interaction among these parameters. The bidirectional arrows also indicate that each factor is not only modulated by, but also in part determines, the autonomic tone. (Reproduced from Lakatta (1983), with permission.)

or impedance to the ejection of the blood from the ventricle); and (3) the inotropic or contractile state of the ventricular muscle, which depends on coronary flow. These determinants of stroke volume and the factors that determine heart rate govern the cardiac output. In some instances the changes in some of these variables that result from ageing or disease are compensatory and enhance overall cardiovascular function, whereas in other instances they may compromise function. In assessing the capacity of the intact cardiovascular system, it is often difficult to quantify the contribution of each factor in Fig. 2. However, in attempting to define the mechanisms that are operative during failure of the system, each of these regulatory factors must be studied, to the extent feasible, in isolation.

Resting heart rate is not greatly affected by age, although variation in sinus rate with respiration is diminished with advancing age (Lakatta 1990). The spontaneous variation in heart rate monitored over 24 h in men free from coronary artery disease also decreases with age. Resting left ventricular volume does not change significantly or may increase slightly with age in healthy adults (Gerstenblith *et al.* 1977; Rodeheffer *et al.* 1984). Longitudinal studies of chest radiography indicate that an increase in the cardiothoracic ratio from 0.405 to 0.427 occurs over a mean period of 12 years (Ensor *et al.* 1983). This is due to a slightly increased cardiac and decreased thoracic diameter; the latter may be due to decreased mobility of the rib cage.

Although the left ventricular volume at end-diastole is not diminished in healthy elderly individuals, many studies have observed that the rate at which the left ventricle fills with blood during early diastole is reduced by about half between 20 and 80 years of age. This may be due to age-associated alterations in passive, left ventricular stiffness but also probably reflects, in part, the age-associated, prolonged relaxation phase of cardiac muscle contraction. This reduction in filling rate is not large enough, however, to lead to a reduction in the end-diastolic filling volume at rest. Enhanced ventricular filling later in diastole in elderly subjects is an adaptive mechanism to maintain an adequate filling volume. This results largely from an enhanced atrial contribution to ventricular filling (Lakatta 1990).

After-load and impedance to ejection

The stroke volume is dependent on the end-diastolic and end-systolic volumes, the latter being governed in part by the after-load. Cardiac after-load or impedance to ejection at rest increases moderately with ageing in humans. In part this is due to an increase in arterial pressure. Most studies have shown that systolic pressure rises within the normal range with advancing age in men and women in economically developed societies (Kannel 1976).

Arterial stiffness increases with age owing to a diffuse process occurring within the vessel wall not explicable solely on the basis of atherosclerosis, and is manifest by an increase in pulse-wave velocity (Lakatta 1990). Owing to this increase in the pulse-wave velocity, reflected pulse waves from the periphery return back to the base of the aorta during the ejection period (O'Rourke 1982). This causes the systolic pressure to continue to rise into late systole in older individuals whereas in younger persons it peaks earlier, at the time of peak aortic flow. The attribution of the increase in vascular stiffness and the resulting increase in arterial pressure to 'normal' ageing can be seriously challenged by cross-cultural studies, which suggests that populations that consume more sodium chloride than others exhibit a more pronounced increase in arterial stiffness with age (Avolio *et al.* 1985).

Cardiac mass and vasculature

Moderate myocardial hypertrophy, which is a successful adaptation to maintain a normal heart volume and pump function in the presence of a modestly increased systolic arterial pressure, has been demonstrated to occur with ageing (Lakatta 1990), regardless of how subjects were screened for study (Fig. 3). This cardiac hypertrophy is due largely

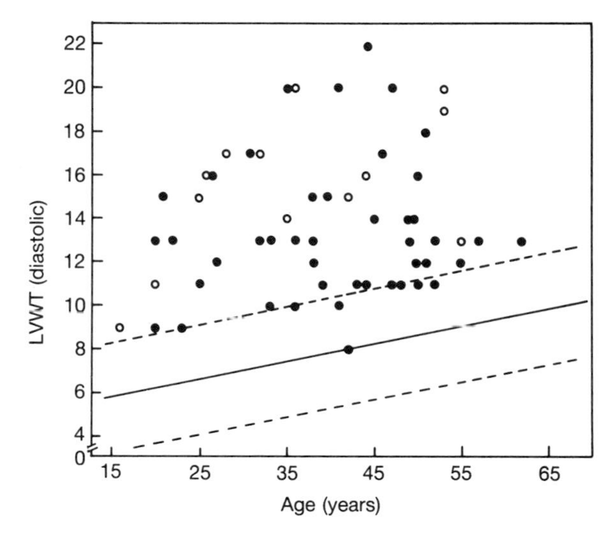

Fig. 3 Least-squares linear regression of left ventricular end-diastolic wall thickness (LVWT) on age (solid line = mean; dotted lines ±2 SD of the mean) in healthy men and women as measured by echocardiography. Circles indicate the LVWT in patients with severe hypertension or aortic valve disease. (Reproduced from Sjogren (1972), with permission.)

to an increase in the size of myocytes and may reflect an age-associated cardiac adaptation to the aforementioned changes in the arterial system. Specifically, the thickening of the left ventricular wall allows the end-systolic volume and ejection function at rest to remain normal in the face of an enhanced arterial impedance (Nichols *et al.* 1985). There are no data in normal humans about the effect of age on coronary flow or myofibre–capillary ratios. In the non-working isolated rat heart, coronary flow per gram of tissue under normoxic conditions is not altered with advanced age (Lakatta 1990). Earlier studies showed a decrease in the number of capillaries in the 26- and 27-month, isolated rat heart as compared with the 4-month specimen. However, it is difficult to interpret this as indicating that the senescent heart may be chronically ischaemic because it has been demonstrated that the capillary density is not fixed but can increase appropriately with the stimulation of chronic physical conditioning.

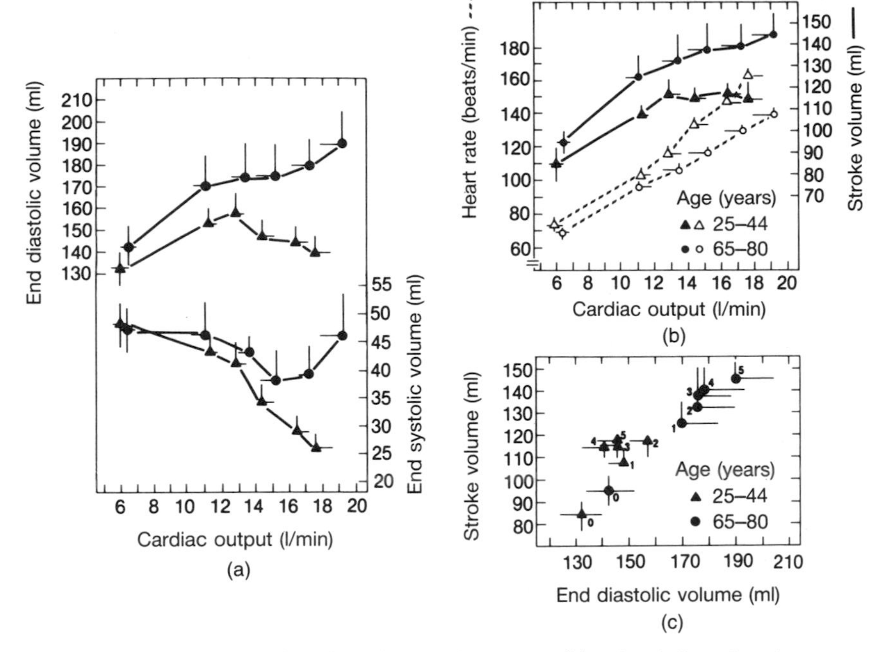

Fig. 4 The relationship of stroke volume and heart rate (b) and end-diastolic volume and end-systolic volume (a) to a given cardiac output at rest and during graded upright bicycle exercise in rigorously screened volunteer subjects. During vigorous exercise these older subjects have a diminution in heart rate but an increase in stroke volume compared with the younger subjects; this is not accomplished by a greater reduction in end-systolic volume but rather by an increase (as much as 30 per cent) in end-diastolic volume. This haemodynamic profile, redrawn in (c), is an example of Starling's law of the heart and resembles that observed during beta-adrenergic blockade. The numbers 0–5 indicate progressive exercise workloads from rest (workload 0). (Redrawn from Rodeheffer *et al.* 1984.)

Responses to postural stress

The haemodynamic response to postural stress is mediated in part by a change in cardiac volume. While earlier studies had suggested that ageing alters the haemodynamic response to a postural change (Granath *et al.* 1961, 1964), measurements of the effect of posture on absolute left ventricular volume have only recently been made in individuals who had been rigorously screened to exclude cadiovascular disease (Rodeheffer *et al.* 1986). Neither the supine or sitting cardiac output nor the average postural change in cardiac output, cardiac volume, or heart rate were age-related. The postural change in cardiac output among the different individuals was correlated (1) with a change in heart rate only in younger subjects, and (2) with a change in stroke volume in all age groups, but the slope of this relationship was greater in older than in younger subjects. The postural change in stroke volume was strongly correlated with a change in end-diastolic volume and this relationship did not vary with age. Thus, although the absolute postural variation in cardiac output among healthy subjects is not age-related, a given change in cardiac output with posture in an older individual depends more on a change in end-diastolic and stroke volume and less on a change in heart rate than in younger individuals. This indicates a greater reliance in older persons on the Frank–Starling mechanism than on heart rate for a given change in cardiac output in response to perturbations from the basal, supine state.

Exercise haemodynamics

While the increase in heart rate during exhaustive exercise is less in many elderly subjects than in younger ones, some elderly subjects can increase the stroke volume to compensate for the heart-rate deficit (Fig. 4). Thus, these elderly subjects use Starling's law of the heart as an adaptive mechanism to preserve cardiac output during exercise in the presence of a reduced heart rate. Elderly individuals who do not have this mechanism for increasing stroke volume and who also have a reduced heart rate compared with younger subjects do not increase cardiac output during severe exercise to the extent that younger individuals do (e.g. study C in Fig. 1). However, whether the reduced aerobic capacity of these individuals is exclusively limited by limitations of cardiac function is uncertain (see below).

The change in ejection fraction between rest and exercise is used clinically as a diagnostic test for the detection and quantification of the severity of cardiac disease, particularly ischaemic heart disease. Ejection fraction is thus of considerable clinical interest. In healthy individuals there is no age-associated change in ejection fraction at rest; at low levels of exercise, ejection fraction is also unchanged with age. At higher levels of exercise and at exhaustion an age-associated decrease in the change in ejection fraction occurs (Fig. 5), and is accompanied by less of a reduction in end-systolic volume in older individuals (see Fig. 4), reflecting an age-associated decrease in augmentation of myocardial contractility

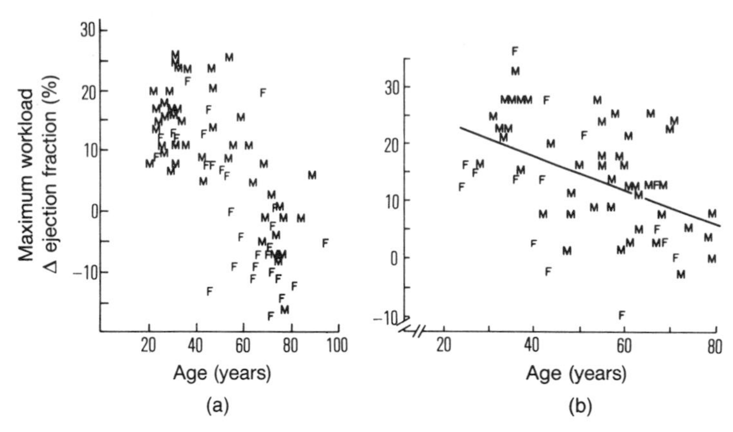

Fig. 5. (a) Effect of age on the change in left ventricular ejection fraction from the resting level to that at maximum voluntary exercise in apparently healthy subjects. However, during exercise a large number of these elderly volunteers exhibited left ventricular-wall motion abnormalities, possibly due to coronary artery disease and coronary insufficiency during the exercise stress. (Redrawn from Port *et al.* 1980.) (b) Effect of age on change in left ventricular ejection fraction from resting level to that at maximum voluntary exercise in subjects from the Baltimore Longitudinal Study of Aging. M, males; F, females. (Redrawn from Rodeheffer *et al.* 1984.)

or an increase in vascular loading. Even so, in most healthy subjects rigorously screened to exclude the presence of coronary artery disease, ejection fraction with exercise increases, although in older individuals this increase is smaller than in younger subjects (Fig. 5(b)). When a study population contains many elderly individuals with coronary artery disease, the ejection fraction during exercise does not increase, or it decreases below that at rest (Fig. 5(a)). This is a manifestation of the interaction between coronary artery disease and ageing, and must not be confused with an age-related change in cardiac pump function.

In a disease-free, cross-sectional population of normal human subjects, end-diastolic and systolic dimensions, ejection fraction, and the velocity of circumferential fibre shortening were found not to differ with age before or during a 30 mmHg increase in systolic pressure induced by phenylephrine (Yin *et al.* 1978). During the same pressor stress in the presence of beta-blockade (with propranolol), end-diastolic dimension increased in the older but did not change in the younger subjects. The ejection measurements of ventricular function, however, were unchanged. This study appears to show a small decrease in the contractility of the heart in normal old subjects, which is evident only after blockade of the sympathetic nervous system and in the face of a significant pressor stress. Under these conditions the heart of the older individual dilates more, thus utilizing the Frank–Starling mechanism, as is the case during postural changes and upright dynamic exercise.

Aerobic capacity

The extent of the decline in the maximum work capacity and $\dot{V}O_2$max and maximum cardiac output with advancing adult age varies with lifestyle, e.g. physical conditioning, and with presence of occult or clinical disease. The motivation to continue to exercise may decrease in sedentary elderly subjects and problems in the joints may limit maximum work capacity in some subjects.

Evidence to suggest that the central circulatory function limits the peak $\dot{V}O_2$ achieved during exercise in elderly individuals is indirect and unconvincing. The cardiac response for the work performed ($\dot{V}O_2$ achieved) in elderly subjects during exercise is as adequate as that in younger subjects (Strandell 1964; Julius *et al.* 1967). The lower peak cardiac output measured at exhaustion in elderly compared with younger individuals (e.g. in Fig. 2) could be the result of a lesser amount of work done by elderly subjects rather than the cause of a diminished capacity for work. Limitation of capacity could in part be due to non-cardiac, 'peripheral' factors that determine O_2 utilization, i.e. the difference in peak arteriovenous O_2 (Julius *et al.* 1967). One such factor is an age-associated reduction in muscle mass (Lakatta 1990). A decline in peak $\dot{V}O_2$ with age cannot be considered to be due to an age-related decline in central circulatory performance if an age difference in muscle mass or in the ability to shunt blood to exercising muscles cannot be excluded with certainty. This is not a trivial issue, given that a greater than tenfold increase in blood flow and O_2 utilization by muscle occurs during exercise. The impact of the normalization of peak $\dot{V}O_2$, on an index of muscle mass, the creatinine excretion, is shown in Fig. 6. Given these formidable obstacles to the interpretation of measurements of aerobic capacity in elderly subjects, the extent to which it declines due to age *per se* and the mechanisms of this decline need to be reassessed.

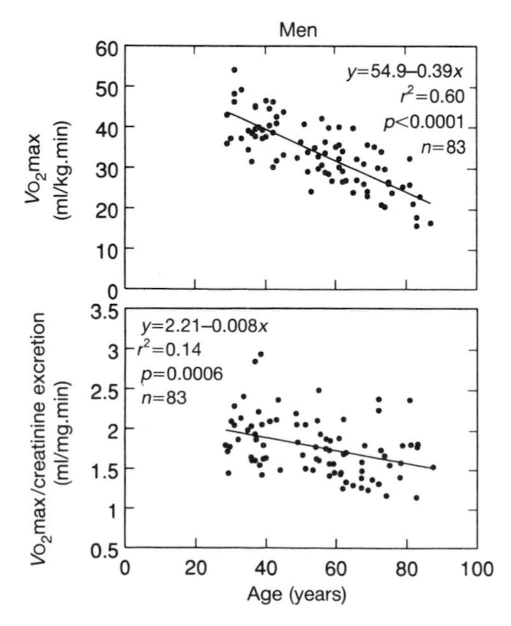

Fig. 6 Top panel: $\dot{V}O_2$max per kilogram body weight as a function of age in helathy, non-obese men. A strong negative relationship is apparent. Lower panel: $\dot{V}O_2$max normalized for urinary creatinine excretion in men as a function of age. Compared with the standard expression of $\dot{V}O_2$max per kilogram body weight, the age-associated decrease in $\dot{V}O_2$ normalized for muscle mass is markedly attenuated. (Reproduced from Fleg and Lakatta (1988), with permission.)

Cardiac muscle function

The intrinsic 'contractile state' or level of excitation–contraction coupling present in the myocardium itself (see Fig. 2) is difficult to ascertain in the intact circulatory system, given the interaction of the multiple modulators of cardiac function (Fig. 2). Thus, an understanding of the effect of age on intrinsic cardiac muscle performance has come from studies of isolated hearts or of cardiac muscle isolated from the hearts of animals. These have indicated that the capacity to develop force is not compromised in the senescent myocardium. However, characteristic age-associated changes in many aspects of excitation–contraction coupling mechanisms in cardiac muscle have been found (Lakatta 1987*a*). Specifically, in an isometric contraction, i.e. one in which the ends of the muscle are fixed, the transmembrane action potential (Fig. 7(a)), the myoplasmic [Ca^{2+}] (Ca_i) transient that initiates contraction (aequorin light (Fig. 7(c)), and the resultant contraction (Fig. 7(b)) are longer in heart muscle from senescent than from younger adult rats. It is noteworthy that the contraction time in humans, as determined non-invasively from measurement of time intervals of the cardiac cycle, is also prolonged. Prolonged systolic contractile activation may be a cause of the decrease in the observed early-diastolic filling rate discussed above. The rate at which the sarcoplasmic reticulum pumps Ca^{2+} decreases with ageing (Fig. 7(d)). In isotonic contractions, i.e. those in which one of the muscle ends is not fixed and in which macroscopic muscle shortening is allowed to occur, the speed and extent of shortening are less in cardiac muscle from senescent than from younger adult rats (Fig.7(f)). The myosin isozyme composition shifts from predominantly V_1 (rapid ATP hydrolytic isozyme) to predominantly V_3 (slower hydrolytic isozyme) with senescence (Fig. 7(e)). The altered pattern of mechanisms depicted in Fig. 7 allows the cardiac muscle of older hearts to generate force for longer after excitation. This enables the continual ejection of blood during late systole, an adaptation that is required in the face of enhanced vascular stiffness and early reflected waves. Thus, while each of the alterations in excitation–contraction coupling in Fig. 7 may individually be construed as a functional decline, consideration of the entire pattern of changes that occurs with ageing allows speculation that these and the associated cardiac hypertrophy may indeed be adaptive rather than decremental in nature.

Additional evidence in support of the notion that the age-associated changes depicted in Fig. 7 are adaptive stems from the observation that these do not occur independently of each other. Specifically, the pattern of changes in excitation–contraction coupling mechanisms that occurs with ageing can be mimicked in younger animals by chronic experimental hypertension (Lakatta 1987*b*). At least one of these changes in Fig. 7, the shift in the myosin isozyme type, is a change in genetic expression of protein synthesis, that is, a transcriptional change. Whether the changes in the various cell-membrane functions depicted in Fig. 7 are due to changes in the expression of proteins that determine or modulate ion channel or ionic carrier activities at the sarcolemma (prolonged action potential) or the sarcoplasmic reticulum pump (prolonged Ca^{2+} transient; decreases in the Ca^{2+} transport rate of sarcoplasmic reticulum) is at present unknown. However, it is tempting to speculate that as these changes do not occur independently of each other, they are all directed from within the genome. This would require a 'logic' within the genome that controls the simultaneous expression of multiple genes in order for cellular adaptation to occur. In the case of ageing and experimental hypertension, altered myocardial cell loading would be the macroscopic stimulus that signals this genetic 'logic' to alter protein expression. The subcellular signal that transduces the signal is not known but could be something like stretch or a change in ion gradients (Ca^{2+}, H^+) resulting from altered mechanical loading of cardiac cells.

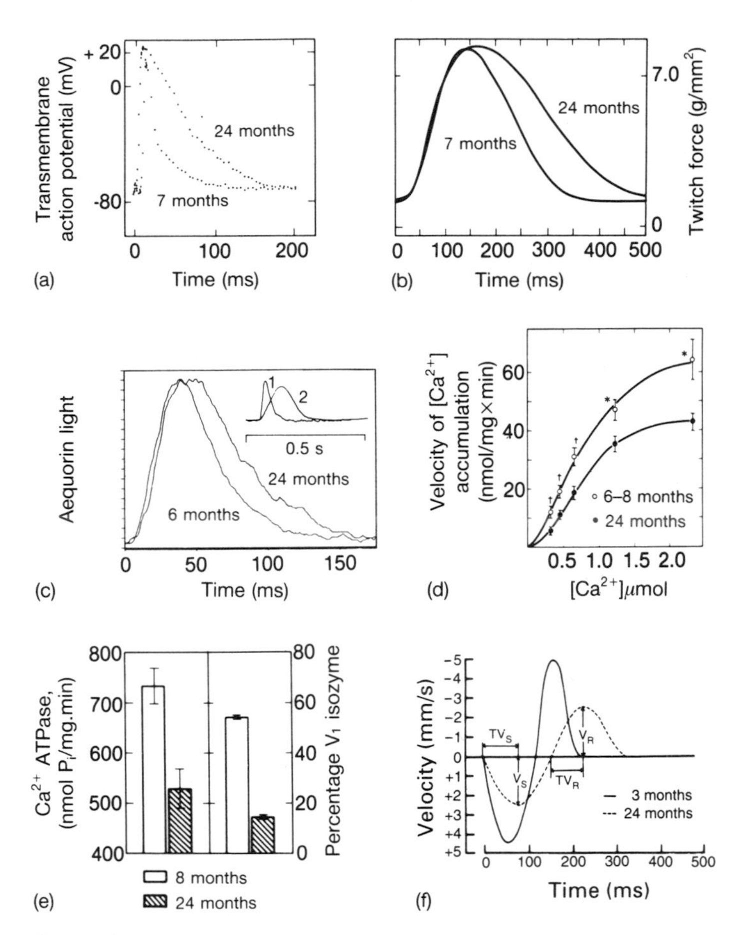

Fig. 7 Representative data depicting differences in various aspects of excitation–contraction coupling mechanisms measured between young adult (6 and 9 months) and senescent (24–26 months) rat hearts. (a) Transmembrane action potential (Wei *et al.* 1984); (b) isometric contraction (Wei *et al.* 1984); (c) cytosolic calcium transient measured by a change in the luminescence of aequorin which had been injected into several cells comprising the muscle preparation (Orchard and Lakatta 1985); (d) sarcoplasmic reticulum Ca^{2+} uptake rate (Froehlich *et al.* 1978); (e) Ca^{2+} stimulated myosin ATPase activity, and myosin isozyme composition (50 per cent) of the heterodimer (V_2) is included in the total percentage of V_1 (Effron *et al.* 1987); (f) velocity of shortening in isotonic muscle (Capasso *et al.*1983). (Reproduced from Weisfeldt *et al.* (1991), with permission.)

It is noteworthy that administration of thyroxine produces a pattern of change in these variables in the opposite direction to that of ageing and experimental hypertension, and can partially reverse age-associated changes. Short-term administration of thyroxine can reverse the age-associated changes in myosin isozyme expression (Effron *et al.* 1987).

Cardiovascular beta-adrenergic stimulation

The haemodynamic profile of elderly individuals shown in Fig. 4 is strikingly similar to that of younger subjects who exercise in the presence of beta-adrenergic blockade (Lakatta 1987*b*). These observations had led to the suggestion that perhaps the most marked changes in cardiovascular response to stress that occur with ageing in healthy subjects vigorously screened to exclude occult disease and highly motivated to perform exercise are due to a reduction in the efficacy of the beta-adrenergic modulation of cardiovascular function.

The average basal plasma level of noradrenaline has been found to increase with advancing age in many but not all studies. Virtually all studies have found that the average plasma level increases in response to stress, to a greater extent in elderly than in younger individuals (Lakatta 1990).

Beta-adrenergic stimulation modulates the heart rate, arterial tone, and myocardial contractility. Both the beta-adrenergic relaxation of arterial smooth muscle and the enhancement of myocardial performance (see below) facilitate ejection of blood from the heart. Deficits in either of these functions with ageing may be implicated in the alterations in the ventricular ejection pattern in some elderly individuals, as shown in Figs. 4 and 5.

There is abundant evidence to indicate that both arterial and venous dilatation (Lakatta 1990) to beta-adrenergic stimulation decline with age. Studies in isolated aortic muscle from young adult and senescent animals directly demonstrate a diminution in relaxation in response to beta-adrenergic agonists but not in response to non-adrenergic relaxants. During exercise a deficiency in arterial dilatation, in addition to structural changes that may occur within the large vessels with age, may contribute to a further increase in vascular impedance (Yin *et al.* 1981), over and above the increase that may already be present at rest. Some studies have indicated that alpha-adrenergic modulation of vascular function is not age-related (Lakatta 1987*b*).

The beta-adrenergic modulation of pacemaker cells in part accounts for the increase in heart rate during exercise. The effect of bolus infusions of beta-adrenergic agonists to increase heart rate diminishes with advancing age (see Lakatta 1987*b* for review). The maximum heart-rate response to infusion of isoproterenol is also diminished in senescent compared with younger adult beagles, and remains diminished even in the presence of full vagal blockade with atropine. In contrast, the maximum heart rate that could be elicited by external electrical pacing, which was far in excess of that elicited by isoprenaline (isoproterenol) infusion, is not age-related. Such infusions in intact rats have also produced a diminished increase in heart rate with age.

Summary

In summary, the resting heart rate is not age-related. There is little if any alteration in ventricular pre-load (diastolic volume), although the early rapid rate of filling is slowed. There is an increase in after-load to left ventricular ejection, which is due to arterial stiffening and is reflected in the age-associated increase in systolic blood pressure, but this is a modest, age-associated change in normal subjects and is compensated for in healthy individuals, in large part, by the age-associated, left ventricular hypertrophy. The net outcome of the age-associated changes in the architecture and contractile properties of the heart, in spite of changes in aortic distensibility, allows the aged heart to function normally at rest. Thus, the fraction of end-diastolic volume ejected with each beat (ejection fraction) does not decline with age. The velocity of circumferential shortening is also not age-related at rest. One of the major age-associated alterations in the cardiovascular response to exercise is a striking decrease in heart-rate response. Despite this decrease, overall cardiac output at any given work-load can be maintained in some healthy older individuals as a result of an augmentation of stroke volume above that seen in the young as exercise progresses. The main mechanism used to augment stroke volume in the older subject is the Frank–Starling mechanism. End-diastolic volume increases with exercise to a greater extent in some older subjects than in younger ones, leading to an increase in the volume of blood ejected from the left ventricle. The reduction in end-systolic volume and the increase in ejection fraction at peak exercise decrease with age and probably result from deficient myocardial performance or augmented after-load. This could be due in part to a deficiency in beta-adrenergic stimulation to enhance myocardial contractility or to reduced aortic impedance. Although there is a decrease in the maximum capacity for physical work, even among the healthy older subjects, this limitation in exercise is not solely due to limitations in the central circulation. Rather, the limitation in the aged subject may well be related to peripheral factors. It is most remarkable that cardiac output can be maintained at high levels at rest and during exercise in the community-dwelling, highly motivated, aged individual. Alterations in cardiac function that exceed the identified limits for ageing changes for healthy elderly individuals are most likely to be manifestations of the interaction between physical deconditioning and cardiovascular disease, which are, unfortunately, so prevalent within economically developed populations.

CARDIOVASCULAR DISEASE IN ELDERLY INDIVIDUALS

The proportion and number of the population in the United Kingdom and the United States that is 65 years of age or older is growing rapidly. In this age group, heart disease is the most frequent reason for admission to hospital and for death. The detection of disorders of the heart in elderly individuals depends in part on an understanding of how ageing alone affects the routine, clinically obtained measurements that are used to assess cardiac structure and function. The certainty of the answer depends on the population studied and the techniques used. It is important that the subjects examined are free of cardiovascular disease and that the methods used are uninfluenced by age-associated changes in other organ systems. The physiological changes described above provide an altered substrate upon which specific pathological conditions are imposed. The diagnosis and management of heart disease as well as the response to therapy are affected by these physiological changes.

Ischaemic heart disease

The chief cause of death in later years, as in middle age, is atherosclerotic disease. It has been estimated that 55 per cent of all deaths are due to coronary or cerebral ischaemia or infarction. Autopsy studies also indicate that there is a dramatic age-associated increase in the prevalence of significant disease, reaching as high as 50–60 per cent in men at the age of 60 years (Fig. 8) (Tejada *et al.* 1968; Elveback and Lie 1984), and a more recent study indicates that the prevalence of significant stenoses in individuals dying over the age of 60 has actually increased over the past 20 years (Elveback and Lie 1984).

The prevalence of symptomatic coronary disease is only 10 to 50 per cent of the true prevalence of disease. Stress testing, however, can significantly improve the sensitivity of identification of the population with disease. The sensitivity of exercise testing for the detection of coronary disease is increased with age, although the specificity declines somewhat (see Table 1) (Hlatky *et al.* 1984). The predictive accuracy of a negative test also decreases in older populations with a high prevalence of disease. Both sensitivity and specificity are improved by the addition of radionuclide stress testing and, with the use of both stress and resting criteria, true prevalence figures are approached (Fleg *et al.* 1990). The presence of both electrocardiographic and thallium scintigraphic evidence of ischaemia in asymptomatic individuals is associated, independent of conventional risk factors, with a 3.8-fold relative risk of subsequent coronary events (Fleg *et al.* 1990).

Symptoms of ischaemic heart disease in older persons do not differ greatly from those in younger persons with the exception that dyspnoea may be as common a presenting symptom as chest pain. If an older person's activities are limited by musculoskeletal, respiratory, or other problems, he or she may not be able to exercise enough to be aware of anginal symptoms. Pre-existing, age-associated changes in myocardial and pericardial compliance and diastolic relaxation may lead to more frequent and/or more rapidly appearing symptoms of heart failure, such as dyspnoea, than of chest pain in the setting of superimposed ischaemia.

Angina pectoris

Treatment of angina in elderly patients often begins with the detection of the reversible precipitating factors commonly found in this age group, including anaemia, congestive heart failure, 'masked' hyperthyroidism, and hypertension. Treatment of these may relieve angina. It should also be remembered that atherosclerosis is a progressive disease and that, although it has sometimes been asserted that reduction of risk factors is less important in the older patient, evidence now suggests that successful treatment of hypertension (European Working Party on High Blood Pressure in the Elderly 1985) and smoking cessation (Jajich *et al.* 1984) decrease fatalities from cardiovascular diseases in elderly people.

Nitrates, beta-blockers, and calcium-channel blockers are used to treat symptomatic ischaemia. The hypotensive effects of nitrates may be potentiated by altered cardiovascular reflexes, diminished plasma volume, incompetent venous valves, and other factors often present in older patients. Although the altered pharmacokinetics of propranolol may increase the plasma levels of the drug in older subjects, the physiological response has been reported to be diminished (Vestal *et al.* 1979). The choice of whether to use a beta-one-selective or non-selective, or a hydrophilic or lipophilic beta-blocker should be guided by associated medical conditions, particularly pulmonary, renal, or hepatic disease. It should be remembered that the ability of timolol and propranolol to decrease recurrent infarction and death in the postinfarction setting is at least as significant in older individuals as it is in younger patients (Beta Blocker Heart Attack Study Group 1981). Calcium-channel blockers have proved useful in the treatment of a wide range of patients who have rest angina. These agents dilate coronary and peripheral vessels and thus can potentially increase the myocardial oxygen supply and decrease the demand.

Although the National Heart, Lung and Blood Institute PTCA Registry initially reported increased in-hospital fatality associated with angioplasty and a lower success rate in patients over 65 years of age, more recent reports indicate low fatality (0.8 per cent), which does not differ from that in younger patients (Raizner *et al.* 1986). Angioplasty is now listed as an 'evolving indication' for treatment of angina in patients over 75 years of age (Bourassa *et al.* 1988). It is particularly useful in the older patient who has significant symptoms despite medical treatment and who is at increased risk from surgery because of other conditions including pulmonary disease, diabetes, and left ventricular dysfunction.

Myocardial infarction

The fatality of myocardial infarction in those over 70 years of age is significantly higher than that of younger individuals. This may be due not to an independent effect of age but to a greater severity and duration of disease, for when individuals of different ages are matched for clinical status and other variables that influence prognosis in coronary artery disease, the age difference in fatality is reduced or obliterated (Honey and Truelove 1957). The increased likelihood of heart failure, pulmonary oedema, and cardiogenic shock is probably due to preceding myocardial damage, larger infarctions, and/or poorer reserve in the remaining non-infarcted regions. The last could be related to changes in both intrinsic contractility and a diminished response to beta-adrenergic stimulation (see above). Cardiac rupture is also more common in older than younger patients and could be related to the increased prevalence of hypertension in older persons, a difference in the size of the infarcted region, or an age-related change in the inflammatory response to infarction.

Studies with catheterization demonstrate a high prevalence of coronary thrombosis in patients presenting within the first 6 h of transmural infarction. Several randomized studies have assessed the effectiveness of early thrombolysis on survival and left ventricular function in patients with known and suspected infarction and have shown, overall, that early thrombolysis decreases fatality and improves ventricular function. Some of these have reported analyses of subsets of patients of different ages (see Table 2) (GISSI 1986; AIMS Trial Study Group 1988; ISIS-2 1988; Wilcox *et al.* 1988). It is interesting that these studies also indicate, in the placebo groups, a several-fold increase in fatality in the subsets of older patients. The benefit of thrombolysis appears to increase with age up to 75 years, probably because of the age-related

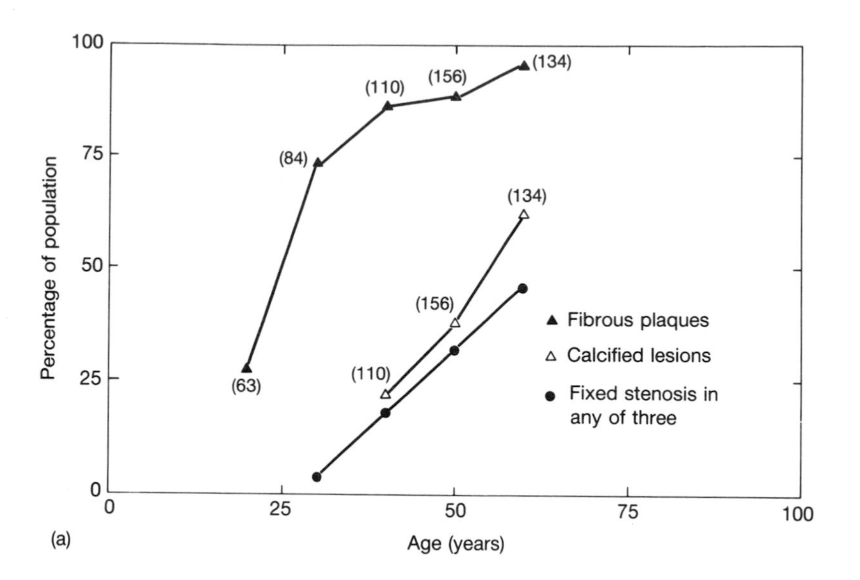

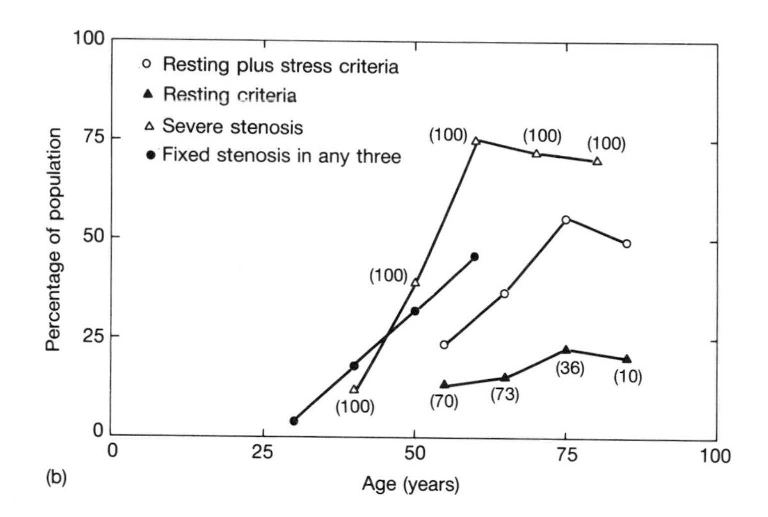

Fig. 8 (a) The effect of age on the prevalence of fibrous plaques, calcified lesions, and fixed stenosis greater than 50 per cent of vessel area in any major coronary artery in hearts from subjects who died of all causes: fibrous plaques; calcified lesions; fixed stenosis in any of three. Number of hearts examined is given by the number in parentheses adjacent to each symbol. Reconstructed from Tejada *et al.* (1968). (b) The effect of age on coronary artery stenosis in hearts from white males who died from all causes in Rochester, Minnesota, and New Orleans, and the prevalence of coronary artery disease in living participants in the Baltimore Longitudinal Study of Aging by resting criteria alone (history of angina or myocardial infarction; abnormal resting ECG, i.e., Minnesota codes 4:1, 25:1, or 26:1) and by resting plus stress criteria (ECG positive for ischaemia during maximum-exercise treadmill test, i.e., Minnesota code 11:1 or 18:1 or an abnormal thallium scan during maximum exercise but not at rest). The number of hearts in the Rochester study is given in parentheses: resting plus stress criteria; resting criteria; severe stenosis; fixed stenosis is any of three. Note the marked increase in the ability to detect the presence of coronary disease in living persons when stress criteria are employed in older participants in addition to the usual clinical epidemiological criteria. Reconstructed from White *et al.* (1950); Tejada *et al.* (1968); Elveback and Lie (1984).

Table 1 Effects of age on sensitivity and specificity of treadmill exercise test

Age (years)	Sensitivity (%)	Specificity (%)
<40	56	84
40–49	65	85
50–59	74	88
>60	84	70

Reproduced from Hlatky *et al.* (1984), with permission.

Table 2 Fatality (per cent) in age subsets (in years) after thrombolytic therapy for acute myocardial infarction

Study	Age	Treatment	Placebo
GISSI	⩽65	5.7	7.7
	65–74	16.6	18.1
	>75	28.9	33.1
ISIS-2	<60	4.2	5.8
	60–69	10.0	14.4
	>70	18.2	21.0
ASSET	⩽55	3.8	4.4
	56–65	6.5	7.9
	66–75	10.8	16.4
AIMS	<65	5.2	8.5
	65–70	12.2	30.2

GISSI: Thrombolytic agent—streptokinase. Fatality assessed at 21 days (GISSI 1986).
ASSET: Thrombolytic agent—tPA. Fatality assessed at 1 month (Wilcox *et al* 1988).
ISIS-2: Thrombolytic agent—streptokinase. Fatality assessed at 5 weeks (ISIS-2 1988).
AIMS: Thrombolytic agent—APSAC. Fatality assessed at one month (AIMS Trial Study Group 1988).

increase in fatality in the control groups. At greater ages, the benefit is less and the risk of haemorrhage may be increased. It should also be noted that older individuals have an increased prevalence of severe hypertension and history of cerebrovascular accidents, which are both contraindications for thrombolysis. For those without these associated conditions, age greater than 75 years is only a relative contraindication to thrombolytic therapy. In assessing the benefit/risk ratio in making the decision to use thrombolytics in these patients, therefore, it should be remembered that the benefit will vary with the size and site of the infarct and with time from onset of symptoms — older patients with large anterior infarcts who present within 3 h of onset will probably receive the most benefit. Otherwise, there are no specific changes in the choice of medical therapy for infarction that depend on the age of the patient, although the dose of some drugs may have to be decreased if renal or hepatic clearance is diminished.

Five to six per cent of elderly patients undergoing bypass surgery die early after the procedure (Gersh *et al.* 1983). This is somewhat more than the proportion of younger patients (2 to 3 per cent) but still very low and may be due in part to an increase in associated risk factors such as abnormal left ventricular function, other associated medical conditions such as diabetes, and more severe coronary artery disease. Independent predictors of perioperative and 5-year survival in patients 65 years or older from the CASS Registry are presented in Table 3. Age alone should not be considered a contraindication for cardiac surgery.

Table 3 Predictors of survival in patients undergoing coronary bypass surgery

Perioperative survival
1. Left main coronary stenosis ⩾70% in association with left dominant circulation
2. Left ventricular end-diastolic pressure
3. Current cigarette smoking
4. Pulmonary rales on auscultation
5. Number of associated medical diseases

Five-year survival (in perioperative survivors)
1. Number of associated medical diseases
2. Functional impairment due to congestive heart failure
3. Severity of abnormal left ventricular-wall motion (determined angiographically)
4. Left ventricular end-diastolic pressure

Congestive heart failure

The prevalence and incidence of heart failure increase so steeply with age that approximately 75 per cent of all ambulatory patients with heart failure are 60 years of age or older (Fig. 9) (McKee *at al.* 1971). Autopsy studies of elderly patients with heart failure reveal that the underlying changes are similar to those found in middle-aged individuals. These are ischaemic, hypertensive, and calcific degenerative valvular disease. In the absence of such changes, it is unlikely

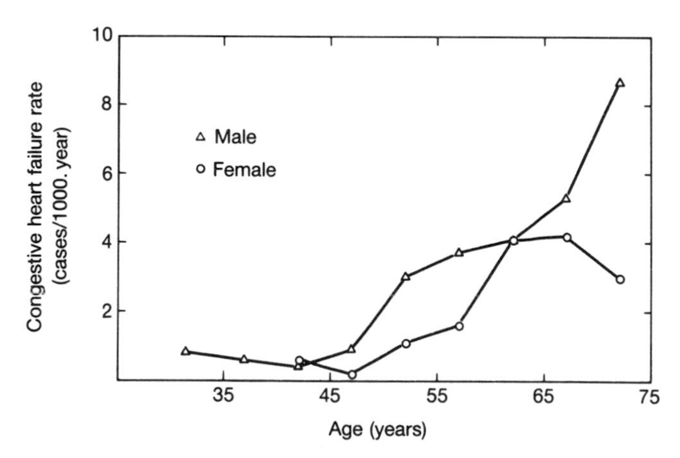

Fig. 9. Average annual incidence of congestive heart failure according to age and sex: Framingham Heart Study, 16-year follow-up results. A minimum of two major criteria and one minor criterion occurring concurrently were required for a diagnosis of congestive heart failure. Major criteria used were: paroxysmal nocturnal dyspnoea or orthopnoea; neck vein distension; pulmonary rales; cardiomegaly; acute pulmonary oedema; S_3 gallop; increased venous pressure ⩾16 cmH_2O; circulation time ⩾25 s; positive hepatojugular reflux. Minor criteria used were: ankle oedema; night cough; dyspnoea on exertion; hepatomegaly; pleural effusion; vital capacity one-third from maximum; tachycardia (rate of ⩾120 beats/min). A major or minor criterion was a weight loss of ⩾4.5 kg within 5 days of treatment. (Reconstructed from McKee *et al.* 1971.)

that ageing changes alone can cause failure. An interesting group of elderly patients with hypertensive hypertrophic cardiomyopathy has been described (Topol *et al.* 1985). Women made up more than three-quarters of the group and symptoms consisted primarily of dyspnoea and chest pain. The diagnosis is made by echocardiography, which shows exaggerated contractile function, small systolic and diastolic cavities, and prolonged and reduced early diastolic filling. Treatment with vasodilators is associated with clinical deterioration whereas patients treated with beta-blockers or calcium-channel blockers, which respectively extend the time course of and enhance diastolic relaxation, were improved.

Although the incidence and prevalence of congestive heart failure increase exponentially with age, the risk and benefit of the several types of treatment have not been thoroughly examined in older patients. Probably the most commonly used medication is digitalis. The pharmacokinetics of digoxin would be altered by the reduced volume of distribution and decline in creatinine clearance associated with age. It is important to remember that because muscle mass decreases with advanced age, creatinine clearance may be greatly diminished even though serum creatinine levels remain in the normal range. After oral or intravenous administration of the drug, the half-life is longer and serum levels higher in older individuals. Although the influence of age alone on the inotropic response to digitalis glycosides in humans has not been examined, the inotropic effect is diminished in senescence in both isolated rat myocardium and in intact dogs. This is apparently not due to an age difference in the relative inhibition of Na, K-ATPase. The serum level required to initiate toxic arrhythmias (specifically ventricular arrhythmias in the experimental beagle dog) is also not age-related. Also of interest are clinical reports that the drug can be withdrawn without clinical deterioration in some patients in sinus rhythm with chronic congestive heart failure, many of whom are elderly (Fleg *et al.* 1982). The magnitude of the response to digitalis is likely to be related to the cause of heart failure, its chronicity, and its severity as well as to the patient's age. In view of the high incidence of toxicity associated with digitalis in the elderly population, it may be prudent to begin treatment with vasodilators and/or diuretics in those elderly patients who are in sinus rhythm. The use of angiotensin-converting enzyme inhibitors may be particularly useful in this group, as enalapril has been shown to reduce fatality by 27 per cent in a group of individuals with severe congestive heart failure whose mean age was 70 years (Consensus Trial Study Group 1987).

Electrophysiology

Age-associated histological changes occur throughout most of the cardiac impulse and conduction system. Most hearts from elderly individuals show a decrease in the number of pacemaker cells in the sinus node. Although pronounced changes in the atrioventricular node have not been described, there are several age-related changes in the bundle of His. These include a loss of muscle cells and an increase in fibrous and adipose tissue as well as amyloid infiltration. There is also a decrease in the number of cells in the fascicle that connects the main bundle of His to the left bundle. Idiopathic bundle-branch fibrosis is the most common cause of chronic atrioventricular block in patients over 65 years of age. Although sinus bradycardia is frequently present in older persons, they can increase their heart rate in response to exercise or pharmacological interventions. The maximum heart rate achieved, however, decreases with age. Both in non-selected populations and in those free of cardiovascular disease (Fleg and Kennedy 1982) the prevalence of ectopic activity increases with age. In 98 elderly disease-free participants of the Baltimore Longitudinal Study of Aging Population, 24-h electrocardiographic recordings disclosed frequent (more than 100 during the monitoring period) supraventricular and ventricular premature beats, in 25 and 17 per cent, respectively, of the sample. Thirty-five per cent had multiform ventricular ectopic beats. The prevalence of ventricular tachycardia was 4 per cent, and couplets were present in an additional 11 per cent. The incidence of exercise-induced ventricular ectopy increases with age in otherwise healthy individuals as well, although short-term follow-up has not shown any subsequent increase in cardiac events.

Prolongation of the PR interval to 220 or 240 ms occurs frequently with advanced age and is not considered pathological. A leftward shift in the QRS axis also occurs with ageing in Westernized populations and may be related to fibrotic changes in the anterior superior division of the left bundle or in the myocardium, to mild left ventricular hypertrophy, or to a change in the spatial orientation of the heart in the chest. {This leftward axis shift was not seen, however, in a population with low prevalence of hypertension and coronary heart disease (Evans *et al.* 1982).}

The diagnosis of supraventricular and ventricular arrhythmias does not differ markedly in the older population (Table 4). It should be remembered that significant carotid stenoses and sensitivity to carotid sinus massage increase with age, and that increased atrial size and fibrosis may render the atria more prone to fibrillation. Previous studies have indi-

Table 4 Ventricular arrhythmias in 98 healthy subjects aged 60 to 85 years

Arrhythmias	No. subjects	Percentage of total
Ventricular		
Any	78	80
≥ 5 in any hour	76	78
≥ 30 in any hour	37	38
≥ 60 in any hour	12	12
≥100 in 24 h	17	17
Multiform	34	35
Ventricular	11	11
Ventricular tachycardia	4	4
R-on-T phenomena	1	1
Accelerated idioventricular rhythm	1	1
Supraventricular		
Any	86	88
Isolated ectopic beats	86	88
≥ 30 in any hour	22	22
≥100 in 24 h	25	26
Benign slow atrial tachycardia	27	28
Paroxysmal atrial tachycardia	13	13
Atrial flutter	1	1
Accelerated junctional rhythm	1	1

cated that 'lone' atrial fibrillation is associated with an increase in fatality from both stroke and cardiovascular diseases. Elderly patients with supraventricular arrhythmias should be evaluated for hyperthyroidism because cardiac signs and symptoms may be the presenting and major manifestations of this disorder.

The effect of ageing on the pharmacokinetics of digitalis, lidocaine, and quinidine has been investigated. As mentioned above, the steady-state plasma concentration of digoxin for any given dosage is increased in many older patients because of the reduced volume of distribution and decline in creatinine clearance. It is therefore recommended that, in general, the maintenance dose should be reduced to 0.125 mg daily or less in older patients. During the stress of an infarction, cardiac output and hepatic blood flow may be decreased more in older persons, and if so this would require a reduction in the dose of any lignocaine (lidocaine) administered. Mental confusion in an elderly patient in intensive care should not be automatically attributed to dementia because age-associated differences in cardiac performance and blood flow are likely to be exaggerated during stress.

The volume of distribution and serum levels of quinidine after acute intravenous injection are not age-related (Ochs *et al.* 1978). However, quinidine clearance is diminished, averaging 2.64 ml/min.kg in a group aged 60 to 69 years and 4.04 ml/min.kg in a younger group aged 23 to 29 years. The elimination of the drug in the same study was prolonged to 9.7 h in the older group compared with 7.3 h in the younger one. This may predispose to quinidine toxicity in the older population.

The experience with pacemaker therapy in elderly patients has been good, and the long-term survival of elderly patients with pacemakers is identical to that of an age-matched population without pacemakers (Fig. 10). The new models, which programme atrial and ventricular contractions, may be especially useful in this age group as diastolic filling and cardiac output may be more dependent on a properly timed atrial systole in an older, stiffer, left ventricle.

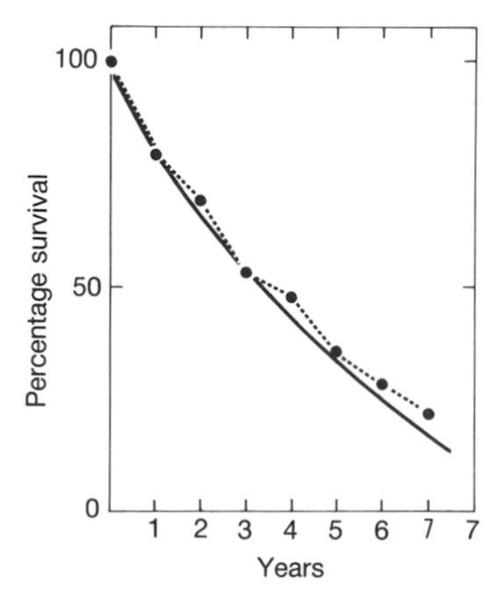

Fig. 10 Survival of patients over 80 years of age treated with a permanent pacemaker for complete heart block (dotted line) and that in an age-matched population (solid line). (Redrawn from Siddons (1974).)

Valvular heart disease

The most common cause of mitral valve disease in older persons is believed to be rheumatic disease. Of those with mitral disease of rheumatic origin, approximately two-thirds are said to have predominant incompetence and one-half to have aortic valvular disease as well. However, there is a major shift with increasing age in the underlying condition responsible for mitral regurgitation, from rheumatic to ischaemic heart disease, resulting in dysfunction of papillary muscles.

The physical signs of mitral stenosis are often obscured, and at times mitral stenosis is a surprise diagnosis in elderly women with chronic 'pulmonary' disease. There are no unusual findings on auscultation but, as in younger patients, the diastolic rumble and accentuated S1 may be less intense when cardiac output is diminished. Thus these signs may be obscured to the extent that associated disease may reduce cardiac output. The systolic murmur of mitral regurgitation may also be present but may be assigned less significance than a similar murmur found in a younger individual because of the increased prevalence of other systolic murmurs in older patients. The most important determinants of prognosis are the presence of failure and the atrial rhythm. Perhaps because of the stiffer left ventricle, which would increase dependence on atrial systole, the development of atrial fibrillation is an important unfavourable prognostic sign. Calcification of the mitral annulus, also usually asymptomatic, is most frequently diagnosed by the inverted C shape of calcium deposition present on the chest radiograph, or heavy calcification is seen in the mitral area on echocardiography. The usual murmur is related to mitral insufficiency, but stenosis may also occur. Complications include endocarditis and extension of the calcification into the bundle of His and peripheral bundle branches, resulting in conduction disturbances.

Significant aortic stenosis is not infrequent in older persons and the disease is often rapidly progressive. The most important signs of this are the extension of the murmur into late systole, a low pulse pressure, and a slowly rising carotid pulse. These signs may not be present, however, even in the setting of severe stenosis because of the age-associated increased stiffness of the central arteries. Electrocardiographic evidence of left ventricular hypertrophy is helpful. Although the echocardiographically determined thickness of the left ventricular wall increases with age, the change is minimal when compared with pathological states. Heavy calcification of the aortic valve on the echocardiogram also suggests significant obstruction. The main differentials are idiopathic hypertrophic subaortic stenosis, which can be diagnosed with echocardiography, and valvular sclerosis, which can be distinguished by the cardiac examination. Doppler echocardiography may also be particularly useful in establishing the severity of valvular stenosis.

Medical treatment of valvular disease in elderly patients is similar to that in younger patients, although the comments above about the use of digitalis and diuretics should be kept in mind. The fatality among those aged 65 years or greater undergoing aortic and mitral valve replacement is generally low. Because of poor long-term results with aortic valvuloplasty (Litvack *et al.* 1988), this should be considered as only palliative in patients with definite contraindications to surgery, which is the preferred treatment.

Hypertension

Hypertension is the most important reversible risk factor for myocardial infarction, congestive heart failure, stroke, and overall cardiovascular fatality in older persons. It is also known that treatment of hypertension decreases the risk of many of these outcomes in this age group (European Working Party on High Blood Pressure in the Elderly 1985). The diagnosis, evaluation, and appropriate therapy, therefore, is an important consideration in the medical care of older individuals.

Diagnosis

A careful definition of hypertension has assumed more importance with the realization that treatment of even mild elevations in diastolic pressure, i.e. in the range of 90 to 104 mmHg, results in a significant decrease in cardiovascular outcomes. The efficacy of therapy for isolated systolic hypertension has now been established (SHEP Cooperative Research Group 1991). The Joint National Committee on the Detection, Evaluation, and Therapy of Hypertension has recommended that treatment be considered in individuals over 60 years of age who have a systolic pressure of more than 160 mmHg in the presence of a diastolic pressure of less than 90. Two or more readings are required and should be obtained with the patient's arm at heart level and with a cuff that has a circumference of two-thirds of that of the patient's arm. The phenomenon of 'white coat' hypertension is more common in the younger age group but it might be useful to obtain readings either at home or at work to confirm the diagnosis.

The prevalence of 'pseudohypertension' may be more common in older persons. In this condition, the blood pressure reading is falsely elevated due to incompressibility of the brachial artery. The palpability of the brachial or radial artery even after inflation of the cuff and the presence of elevated readings without evidence of damage in target organs suggest this condition. Certainty requires measurement of intra-arterial pressure. It is also important to measure pressures when the individual is standing before beginning or increasing therapy because an orthostatic fall in pressure is more common in this age group and if the pressure falls to normal when the patient is upright, further therapy may result in compromise of cerebral flow. Postprandial falls in blood pressure also need to be borne in mind in the titration of treatment.

Evaluation

Evaluation of the older individual should focus in three areas. These are the presence or absence of damage in target organs, any reversible factors, and other risk factors for atherosclerosis. Renal impairment and left ventricular hypertrophy indicate the need for vigorous therapy and may also affect the choice of that therapy. The presence of ischaemic heart disease or other known risk factors would also make careful control of blood pressure more important and influence the therapeutic agent chosen. Reversible causes include hyperthyroidism, which commonly presents in older patients with cardiovascular manifestations, and stenosis of the renal artery due to atherosclerosis. Hypertension presenting in a patient over 55 years of age has a relatively increased likelihood of being due to renal atherosclerosis, and if this is present the hypertension may be curable with angioplasty.

Treatment

Non-pharmacological measures can be used to begin treatment in patients with isolated systolic and mild diastolic hypertension and as an adjunct to pharmacological agents in patients with moderate elevations of blood pressure. These include salt restriction, decreasing alcohol intake, and exercise. Weight reduction has been shown to lower pressure as well as left ventricular mass in young hypertensives, and may do so as well in older patients. Even if these measures are not successful in controlling pressure by themselves, they will decrease the amount of medication the patient requires and therefore the expense, inconvenience, and side-effects of that therapy. If pharmacological therapy is begun, it is important to do so with low doses and to increase the dosage slowly. Older individuals may be prone to hypotensive effects of agents because of pre-existing decreased plasma volume, diminished responsiveness of baroreceptors, incompetent venous valves, and alterations in the autoregulation of cerebral blood flow.

The choice of which agent to use should be based on several factors (see also Chapter 11.6). One is the pathophysiology, which in older individuals is more likely to be an increase in systemic vascular resistance. This would suggest that an agent which lowers vascular resistance would be particularly effective in an older population. Another consideration is the increasing likelihood of coexisting disease. In the presence of ischaemic heart disease, for example, it is often useful to choose an antihypertensive drug which also has anti-ischaemic properties, e.g. a calcium channel-blocker or beta-blocker. If patients have congestive heart failure, an angiotensin-converting enzyme inhibitor may favourably influence prognosis (Consensus Trial Study Group 1987). Other conditions, for example, pulmonary disease or gout, which are more common in older persons, may render them more prone to suffer adverse effects of some antihypertensives. Left ventricular hypertrophy has recently been demonstrated to increase the risk for the development of coronary disease in older individuals, which is independent of the risk associated with elevation of blood pressure *per se* (Levy *et al.* 1989). The choice of an antihypertensive that also has antihypertrophic properties, therefore, should be considered when left ventricular hypertrophy is present. Further considerations include the potential for adverse effects on lipid and glucose metabolism, and the cost and convenience of the therapy. Considerations of cost include not only that of the medication itself but also the number of visits to the doctor required and the expense of tests used to monitor for efficacy and any adverse effects. The goal of pharmacological therapy should be to lower systolic pressures to less than 160 mmHg and diastolic pressure to the 85 to 90 range.

References

AIMS Trial Study Group (1988). Effect of intravenous APSAC on mortality after acute myocardial infarction: preliminary report of a placebo-controlled clinical trial. *Lancet*, **i**, 545–9.

Avolio, A.P. *et al.* (1985). Effect of aging on arterial distensibility in populations with high and low prevalence of hypertension: comparison between urban and rural communities in china. *Circulation*, **71**, 202–10.

Beta-Blocker Heart Attack Study Group (1981). The beta-blocker heart attack trial. *Journal of the American Medical Association*, **246**, 3073–4.

Bourassa, M.G. *et al.* (1988). Report of the joint ISFC/WHO task force on coronary angioplasty. *Circulation*, **78**, 780–9.

Brandfronbrener, M., Landowne, M., and Shock, N.W. (1955). Changes in cardiac output with age. *Circulation*, **12**, 557–66.

Capasso, J.M., Malhotra, A., Remily, R.M., Scheuer, J., and Sonnenblick, E.H. (1983). Effects of age on mechanical and electrical performance of rat myocardium. *American Journal of Physiology*, **245**, H72–81.

Consensus Trial Study Group (1987). Effects of enalapril on mortality in severe congestive heart failure. *New England Journal of Medicine*, **316**, 1429–35.

Effron, M.B., Bhatnagar, G.M., Spurgeon, H.A., Ruano-Arroyo, G., and Lakatta, E.G. (1987). Changes in myosin isoenzymes, ATPase activity, and contraction duration in rat cardiac muscle with aging can be modulated by thyroxine. *Circulation Research*, **60**, 238–45.

Elveback, L. and Lie, J.T. (1984). Continued high incidence of coronary artery disease at autopsy in Olmstead County, Minnesota, 1950–1979. *Circulation*, **70**, 345–9.

Ensor, R.E., Fleg, J.L., Kim, Y.C., de Leon, E.F., and Goldman, S.M. (1983). Longitudinal chest X-ray changes in normal man. *Journal of Gerontology*, **38**, 307–14.

European Working Party on High Blood Pressure in the Elderly (1985). Mortality and morbidity results from the European Working Party on High Blood Pressure in the Elderly trial. *Lancet*, **1**, 1349–54.

Evans, J.G., Prior, I.A.M., Tunbridge, W.M.G. (1982). Age-associated change in QRS axis: intrinsic or extrinsic ageing? *Gerontology*, **28**, 132–7.

Fleg, J.L. and Kennedy, H.L. (1982). Cardiac arrhythmias in a healthy elderly population. Detection by 24-hour ambulatory electrocardiography. *Chest*, **81**, 302–7.

Fleg, J.L. and Lakatta, E.G. (1988). Role of muscle loss in the age-associated reduction in Vo_{2max}. *Journal of Applied Physiology*, **65**, 1147–51.

Fleg, J.L., Gottlieb, S.H., and Lakatta, E.G. (1982). Is digoxin really important in treatment of compensated heart failure? *American Journal of Medicine*, **73**, 244–50.

Fleg, J.L. *et al.* (1990). Prevalence and prognostic significance of exercise-induced silent myocardial ischemia detected by thallium scintigraphy and electrocardiography in asymptomatic volunteers. *Circulation*, **81**, 423–36.

Froehlich J.P., Lakatta, E G., Beard, E., Spurgeon, H.A., Wisfeldt, M.L., and Gerstenblith, G. (1978). Studies of sarcoplasmic reticulum function and contraction duration in young and aged rat myocardium. *Journal of Molecular and Cellular Cardiology*, **10**, 427–38.

Gersh, B.J. *et al.* (1983). Coronary arteriography and coronary artery bypass surgery: morbidity and mortality in patients ages 65 years or older. *Circulation*, **67**, 483–91.

Gerstenblith, G., Frederiksen, J., Yin, F.C.P., Fortuin, J.J., Lakatta, E.G., and Weisfeldt, M.L. (1977). Echocardiographic assessment of a normal adult aging population. *Circulation*, **56**, 273–8.

GISSI (Gruppo Italiano per lo studio della streptochinasi nell'infarto miocardico) (1986). Effectiveness of intravenous thrombolytic treatment in acute myocardial infarction. *Lancet*, **i**, 397–401.

Granath, A., Jonsson, B., and Strandell, T. (1961). Studies on the central circulation at rest and during exercise in the supine and sitting position. *Acta Medica Scandinavica*, **169**, 125–6.

Granath, A., Jonsson, B., and Strandell, T. (1964). Circulation in health old men studied by right heart catheterization at rest and during exercise in supine and sitting position. *Acta Medica Scandinavica*, **176**, 425–46.

Hlatky, M.A., Pryor, D.B., Harrell, F.E., Califf, R.M., Mark, D.B., and Rosati, R.A. (1984). Factors affecting sensitivity and specificity of exercise electrocardiography: multivariate analysis. *American Journal of Medicine*, **77**, 64–71.

Honey, G.E. and Truelove, S.C. (1957). Prognostic factors in myocardial infarction. *Lancet*, **i**, 1155–61.

ISIS-2 (Second International Study of Infarct Survival Collaborative Group) (1988). Randomised trial of intravenous streptokinase, oral aspirin, both, or neither among 17 187 cases of suspected acute myocardial infarction: ISIS-2. *Lancet*, **ii**, 349–60.

Jajich, C.L., Ostfeldt, A.M., and Freeman, D.H. (1984). Smoking and coronary heart disease mortality in the elderly. *Journal of the American Medical Association*, **252**, 2831–4.

Julius, S. Amery, A., Whitlock, L.S., and Conway, J. (1967). Influence of age on the hemodynamic response to exercise. *Circulation*, **36**, 222–30.

Kannel, W.B. (1976). Blood pressure and the development of cardiovascular disease in the aged. In *Cardiology in old age*, (ed. F.I. Caird, J.L.C. Dall, and R.D. Kennedy), pp. 143–75. Plenum, New York.

Lakatta, E.G. (1983). Determinants of cardiovascular performance: modification due to aging. *Journal of Chronic Disease*, **36**, 15–30.

Lakatta, E.G. (1985). Health, disease, and cardiovascular aging. In *Health in an older society* (ed. Institute of Medicine and National Research Council, Committee on an Aging Society), pp. 73–103. National Academy Press, Washington DC.

Lakatta, E.G. (1987*a*). Cardiac muscle changes in senescence. *Annual Review of Physiology*, **49**, 519–31.

Lakatta, E.G. (1987*b*). Catecholamines and cardiovascular function in aging. In *Endocrinology and metabolism clinics*, Vol. 16, *Endocrinology and aging*, (ed. B. Sacktor, pp. 877–91. W.B. Saunders, Philadelphia.

Lakatta, E.G. (1990). Heart and circulation. In *Handbook of the biology of aging*, (3rd edn), (ed. E.L. Schneider and J. Rowe), pp. 181–216. Academic Press, New York.

Levy, D., Garrison, R.J., Savage, D.D., Kannel, W.B., and Castelli, W. P. (1989). Left ventricular mass and incidence of coronary heart disease in an elderly cohort. *Annals of Internal Medicine*, **110**, 101–7.

Litvack, F., Jakubowski, A.T., Buchbinder, N.A., and Eigler, N. (1988). Lack of sustained clinical improvement in an elderly population after percutaneous aortic valvuloplasty. *American Journal of Cardiology*, **62**, 270–5.

McKee P.A., Castelli, W.P., McNamara, P.M., and Kannel, W.B. (1971). The natural history of congestive heart failure: the Framingham study. *New England Journal of Medicine*, **285**, 1441–50.

Nichols, W.W. *et al.* (1985). Effects of age on ventricular coupling. *American Journal of Cardiology*, **55**, 1179–84.

Ochs, H.R., Greenblatt, D.J., Woo, E., and Smith, T.W. (1978). Reduced quinidine clearance in elderly persons. *American Journal of Cardiology*, **42**, 481–5.

Orchard, C.H. and Lakatta, E.G. (1985). Intracellular calcium transients and developed tensions in rat heart muscle. A mechanism for the negative interval-strength relationship. *Journal of General Physiology*, **86**, 637–51.

O'Rourke, M.F. (1982). *Arterial function in health and disease*. Churchill Livingstone, New York.

Port, E., Cobb, F.R., Coleman, R.E., and Jones, R.H. (1980). Effect of age on the response of the left ventricular ejection fraction to exercise. *New England Journal of Medicine*, **303**, 1133–7.

Raizner, A.E. *et al.* (1986). Transluminal coronary angioplasty in the elderly. *American Journal of Cardiology*, **57**, 29–32.

Raven, P.B. and Mitchell, J. (1980). The effect of aging on the cardiovascular response to dynamic and static exercise. *Aging*, **12**, 269–96.

Rodeheffer, R.J., Gerstenblith, G., Becker, L.C., Fleg, J.L., Weisfeldt, M.L., and Lakatta, E.G. (1984). Exercise cardiac output is maintained with advancing age in healthy human subjects: cardiac dilatation and increased stroke volume compensate for a diminished heart rate. *Circulation*, **69**, 203–13.

Rodeheffer, R.J. *et al.* (1986). Postural changes in cardiac volumes in men in relation to adult age. *Experimental Gerontology*, **21**, 367–78.

SHEP Cooperative Research Group (1991). Prevention of stroke by antihypertensive drug treatment in older persons with isolated systolic hypertension. Final results of the systolic hypertension in the elderly program (SHEP). *Journal of the American Medical Association*, **265**, 2255–64.

Siddons, H. (1974). Deaths in long-term paced patients. *British Heart Journal*, **36**, 1201–9.

Sjogren, A.L. (1972). Left ventricular thickness in patients with circulatory overload of the left ventricle. *Annals of Clinical Research*, **4**, 310–8.

Strandell, T. (1964). Circulatory studies on healthy old men. With special reference to the limitation of the maximal physical working capacity. *Acta Medica Scandinavica*, **175** (suppl. 414), 2–44.

Tejada, C. *et al.* (1968). Distribution of coronary and aortic atherosclerosis by geographic location, rate, and sex. *Laboratory Investigation*, **18**, 509–26.

Topol, E.J., Traill, T.A., and Fortuin, N.J. (1985). Hypertensive hypertrophic cardiomyopathy of the elderly. *New England Journal of Medicine*, **312**, 277–83.

Vestal, R.E., Wood, A.J.J., and Shand, D.G. (1979). Reduced beta-adrenoreceptor sensitivity in the elderly. *Clinical Pharmacology and Therapeutics*, **26**, 181–6.

Wei, J.Y., Spurgeon, H.A., and Lakatta, E.G. (1984). Excitation–contraction in rat myocardium: alterations with adult aging. *American Journal of Physiology*, **246**, H784–91.

Weisfeldt, M.L., Lakatta, E.G., and Gerstenblith, G. (1991). Aging and cardiac disease. In *Heart disease. A textbook of cardiovascular medicine* (4th edn), (ed. E, Braunwald), in press. W.B. Saunders, Orlando.

White, N.K., Edwards, J.E., and Dry, T.J. (1950). The relationship of the degree of coronary atherosclerosis with age in men. *Circulation*, **1**, 645–61.

Wilcox, R.G., Olsson, C.G., Skene, A.M., von der Lippe, G., Jensen, G., and Hampton, J.R. 1988). Trial of tissue plasminogen activator for mortality reduction in acute myocardial infarction. *Lancet*, **ii**, 525–30.

Yin, F.C.P. *et al.* (1978). Age-associated decrease in ventricular response to hemodynamic stress during beta-adrenergic blockade. *British Heart Journal*, **40**, 1349–55.

Yin, F.C.P., Weisfeldt, M.L., and Milnor, (1981). Role of aortic input impedance in the decreased cardiovascular response to exercise with aging in dogs. *Journal of Clinical Investigation*, **68**, 28–38.

SECTION 10
Cancer in older people: an overview

10 Cancer in older people: an overview

W. BRADFORD PATTERSON

When someone survives to join the group we call 'older', cancer becomes a well-known experience, either from experience within the family, or among acquaintances. Cancer is infrequent before the age of 35 years, despite the public interest and sympathy for children and young adults who are occasionally struck. In later years, the overall incidence rises rapidly, in both men and women (Fig. 1). These age-associated increases are seen worldwide, although there are large variations in specific cancers in different countries.

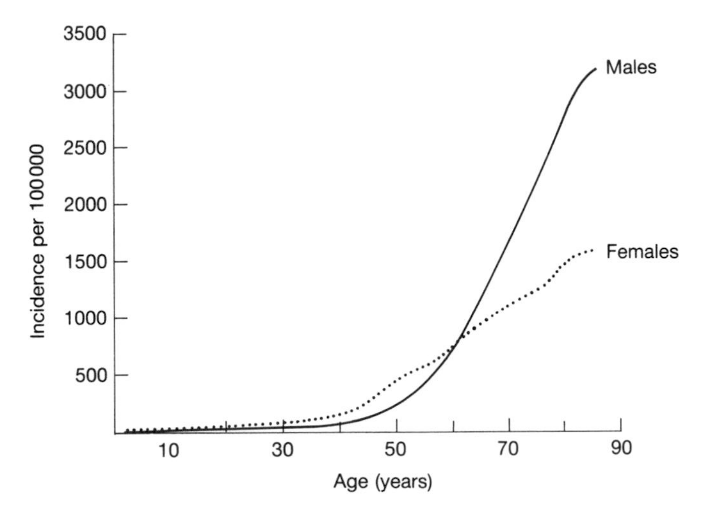

Fig. 1 Cancer incidence rates per 100 000 population, by age and sex, (from SEER incidence and mortality data, 1973–1977 (National Cancer Institute 1981). (Reproduced from Patterson, W.S., Yancik R., and Carbone, P. (1986) Malignant diseases. In *The practice of geriatrics.* (eds. E. Calkins, P.J. Davis and A.B. Ford). W.B. Saunders Co., Philadelphia, with permission.)

'Is cancer becoming a more common disease?' is a question asked by laymen and medically trained persons alike. In fact, not all types of cancer are increasing, although there is an overall absolute increased incidence. The apparent epidemic of cancer is created by several confluent trends.

1. The ageing of populations increases the population at risk, so that the numbers of incident and prevalent cases rise.
2. The incidence of smoking-related cancers, particularly lung cancer, has been rising for the past 60 years—as lung cancers in many 'advanced' countries make up a large portion of the total, this particular cancer has a great impact on total incidence, masking a decreased rate in some other cancers, such as stomach and cervix. The effect of increased cigarette smoking in Third World countries will not be seen in mortality statistics immediately.
3. As a topic open for public discussion, cancer has only been acceptable for the last few decades. In earlier years, a diagnosis of cancer was often kept secret and was certainly never listed in an obituary or in the media. In many cultures today, the prevention, treatment, costs, and pain of cancer are a common topic of conversation.

Thus, several readily identifiable reasons make cancer a more common and familiar disease than it was in the past.

This overview will selectively address topics that are germane to physicians in their diagnosis and treatment of cancer in older age groups. For those readers wishing a more detailed description of cancer management, many recent texts are available (Holland and Frei 1982; DeVita *et al.* 1985).

CAUSES OF CANCER IN OLDER PEOPLE

A major conceptual step forward in our understanding of cancer occurred when oncologists began to recognize and speak of it as a group of diseases, with a few characteristics in common but with many different causes, many different natural histories, and posing very different problems in management and prognosis.

Cancer has many causes and these are almost always multiple. It has been suggested that up to 70 per cent of cancers are due to environmental and lifestyle factors, which include tobacco use, high fat diets, alcohol abuse, excessive sun exposure, and specific carcinogens. This figure may be higher for those cancers appearing in older persons and is helpful in dispelling the notion that the causes of cancer are unknown. Understanding cancer causation becomes especially important when considering appropriate prevention.

The burgeoning field of cancer genetics (reviewed by Levine *et al.* 1989) deserves special attention although genetic predisposition in the Mendelian sense has until recently been considered a cause of cancer in only 15 to 20 per cent of cases. Rapid advances in molecular genetics are providing an increasing number of associations between genes and specific types of cancer. Investigations that identify these genes begin with chromosomal mapping and analysis of cancer cells taken from many different individuals with the same type of cancer. Chromosomal abnormalities are sought that are common to all or a significant proportion of these cancers. Knowledge about the clinical significance of the chromosomal abnormalities is still lagging, and many investigators are attempting to relate the specific genetic changes to the natural history of the cancer—for example, which changes predict for rapid progression versus an indolent course? Others are studying the presence of molecular markers that predict for susceptibility to specific chemotherapeutic agents. Research also involves the study of normal tissues, dysplastic but non-malignant tissues, and benign lesions such as colonic polyps, which are known to be associated with a higher incidence of malignancy in the same organ, in the search for genetic markers associated with the progression from benign to malignant. It is too soon to predict when these investigations will bear fruit for the clinician, but revolutionary changes are certain in our ability to identify those subjects who are at greater risk for particular cancers and to individualize treatment more effectively.

NATURAL HISTORY

The term cancer covers a broad spectrum of disease types, from those with sudden onset and rapidly lethal outcome to an indolent series of manageable lesions that require treatment but pose little threat to life in the main.

There is conflicting evidence as to whether the natural history of cancer varies with age. Doubling times of cancer cells are one index of clinical behaviour, and when doubling time is studied, there is such marked variation among cancers of the same type, e.g. ductal carcinoma of the breast, that differences based on patient age tend to become a minor variable. One can look at survival rates after diagnosis as an indicator of natural history, but this criterion ignores the effects of treatment itself, of treatment choices, and of individual tolerance to treatment, all of which can be major variables when assessing age effects, Table 1, taken from a review of many types of cancer, shows the marked variation in the effect of age on stage at diagnosis, which relates closely to survival rates. Clinicians should not consider a patient's age in predicting the rate of progression of a particular cancer. Other factors, including histological appearances, stage on presentation, and observation over even a limited period of time, are much better guides to prognosis than is age.

Table 1 Relationship of stage at diagnosis of common cancers to increasing patient age: studies of Holmes and Hearne and Goodwin *et al.* From Holmes (1989)

Site	Holmes and Hearne	Goodwin *et al.*
Breast	Increase	Increase
Cervix	Increase	Increase
Colon	—	Neutral
Colorectum	Neutral	—
Endometrium	Increase	Increase
Gallbladder/liver	—	Neutral
Kidney	Increase	Neutral
Lung	Decrease	Decrease
Melanoma	—	Increase
Ovary	Increase	Increase
Pancreas		Decrease
Prostate	—	Neutral
Rectum	—	Decrease
Stomach	—	Decrease
Thyroid	—	Increase
Urinary bladder	Increase	Increase

It is of interest that invasion and metastasis, hallmarks of the malignant process, are generally unaffected by age. Basal cell cancers of the skin erode locally, even if neglected, in young and old alike, while a small cancer of the lung, which is hardly visible on a chest radiograph, may first present clinically as a brain metastasis causing a seizure, also irrespective of the patient's age. The most logical explanation is that the cancer process has usually started years before, when a genetic event caused by some viral, hormonal, or environmental factor 'initiated' the cellular change. Much later, progression to a full-fledged cancer cell occurs due to the stimulus of a 'promoter' or 'promoters'. Thus, a breast cancer appearing in a 65-year-old woman can be regarded as the end result of a lengthy process, which may have started in her teenage years.

RESPONSES TO TREATMENT

Older persons respond to treatment as well as do younger patients. This concept is of great importance, having been well documented for a range of cancers and for the common modalities of treatment, which include surgery, radiation therapy, and chemo- or hormonal therapy. While there are exceptions, they are much less important than the general rule.

Surgeons have discovered that careful techniques of patient selection, preoperative assessment, safe anaesthesia with attention to tissue oxygenation, blood pressure, and meticulous postoperative care, allow cancer operations of all types to be carried out at any age with acceptably low rates of death and morbidity. These techniques are no different from those used with younger patients and for non-malignant conditions. Comorbid illness, rather than age itself, is the principal factor that makes a patient ineligible for a surgical procedure.

While radiotherapy may appear to be safer and less traumatic than a surgical operation for the treatment of cancer in an elderly patient, complications occur with both, and sometimes a 3-h operation is better tolerated than 6 weeks of daily visits for radiotherapy, with the attendant travel and continued stress of a challenging experience. Excellent outcomes for radiotherapy, whether curative or palliative, are not prevented by advancing age if attention is given to patient selection, technique, and to physical and psychosocial support during and after treatment (Brady and Markoe 1986).

Hormonal therapy is a particularly important modality for older patients with cancer of the breast, uterus, and prostate. While this treatment is palliative and not curative, the distinction is often blurred in the old and fragile person, when death from non-cancer causes is not a surprise. For example, the anti-oestrogen, tamoxifen, has been used as primary treatment in place of surgery for localized breast cancer in older women with no disadvantage apparent after 3 years of follow-up (Gazet *et al.* 1988). For endometrial cancer, progesterone has for many years played an important role in the management of metastatic disease. Metastatic cancer of the prostate in older men responds to treatment with diethylstilboestrol, while newer synthetic oestrogens, although more expensive, may avoid some of the fluid retention that poses a threat to the compromised cardiovascular systems common in older persons.

Before about 1980, the toxicity of chemotherapy discouraged its aggressive use for palliation in older patients. Also, some of the most common cancers occurring in people over 65—lung, colorectal, and prostate—have not responded well, or for very long, to any chemotherapeutic agents. Breast cancer has been a gratifying exception, often responding well to chemotherapy. During the past decade, improved techniques of chemotherapeutic administration and supportive care have made it possible to use a wide range of drugs that can yield worthwhile benefit in selected older patients, and many of these patients are very ready to accept some risk and discomfort for the chance of another year or more of life.

CANCER SCREENING, PREVENTION, AND EARLY DETECTION

Prevention

Older persons should be looked upon as excellent candidates for interventions designed to prevent cancer or to find it early (Warnecke 1989). Table 2 lists the cancers occurring in older

Table 2 Cancers that may be prevented

Cancer site	Prevention
Skin	Avoid excessive actinic exposure and use sun-screens
Lung, oral cavity, larynx, oesophagus, bladder	No cigarette smoking or tobacco chewing
Oral cavity, oesophagus, liver	Limit alcohol intake
Colon and rectum breast, endometrium	Reduce saturated fats in diet Increase fibre and antioxidant vitamins
Cervix	Periodic vaginal examination with cervical smear

people for which there is evidence that preventive measures are useful. Tobacco products, whether smoked or chewed, represent a major preventable cause of large numbers of cancer cases worldwide, primarily in people over 65. A reduction in smoking and chewing tobacco would lead to a substantial reduction in the incidence of and deaths from cancer of the oral cavity, larynx, lung, oesophagus, pancreas, and bladder. In the United States, cigarette consumption is falling, and the annual death rate in men from cancer of the lung appears to have peaked. Unfortunately, the export of American cigarettes to Asia and Africa is increasing, and consumption of tobacco products on these continents remains high or is rising.

Clinicians do have the power to influence their older patients to stop smoking. Research has demonstrated that when physicians discuss health risks with patients who smoke, and when information and help about cessation techniques are provided, a rate of stopping of 5 per cent or greater measured at 1 year can be anticipated. As it is now known that most smokers who stop are successful only after several tries, repeated attempts by physicians are indicated, and may make the cumulative cessation rate much higher. A review of many clinical trials of interventions for stopping smoking showed certain common features of the successful ones. These include the provision of information and personalized advice using a variety of formats and communication techniques, and reinforcement through repeated sessions (Kottke *et al.* 1988). Special techniques such as hypnotism are useful for some but not others. The use of a nicotine substitute by mouth to counteract the addictive potential of nicotine in cigarettes is probably most useful to those smokers whose withdrawal symptoms are so strong that they cannot tolerate even a few hours without a cigarette. For older persons, the knowledge that cancer risk can be lowered even after a long history of chronic smoking needs to be emphasized, to counteract the belief that the damage is already done.

The media and the public often focus on research that supports dietary modification and avoidance of sun exposure as keys to cancer prevention for adults. Without belittling their importance, a reduction of cancer incidence in humans by control of dietary factors is still unproven, and most of the skin cancers that relate to excessive exposure to ultraviolet sunlight, basal cell and squamous cancers, are non-lethal and easily treated. The development of melanoma appears to be related to sunburn and excess ultraviolent exposure during youth rather than to chronic exposure later in life.

Screening and early detection

The effect of early detection in reducing morbidity and death have been most obvious with cancer of the cervix and with melanoma. Widespread application of cervical cytology as a screening test in women of all ages has led to the earlier diagnosis of cervical cancer, often in the *in-situ* stage, when treatment is simpler and usually curative. Unfortunately, the percentage of older women who have never had a cervical smear remains high in populations such as the indigent and ethnic minorities. For this reason, efforts to ensure that all older women have at least one smear are of great importance.

Melanoma, although becoming more frequent, can be detected by careful examination of the skin by a practitioner who has learned the characteristic changes in pigmented skin lesions that signify malignancy. This kind of early detection works for adults of any age, and has been largely responsibile for the great improvement in control of melanomas, which used to be called the 'black death' but are now regularly cured in a large number of subjects.

Since mammography has been improved in sensitivity and made widely available, a reduction in deaths from breast cancer of up to 40 per cent now seems possible (Feig 1988). In the earliest trials, mortality appeared to be improved only in women at about menopausal age, but newer trials also show improvement for older women.

For cancer of the colon and rectum, early detection probably will reduce morbidity and mortality, but evidence is not as conclusive. For basal and squamous cancers of the skin and mucous membranes (oral, vaginal, and anal) finding smaller, asymptomatic cancers during a physical examination can lead to less extensive treatment with improved survival. Other common cancers, such as those of the lung and pancreas, do not merit aggressive attempts at early detection, either because trials have not shown such efforts to be effective, or because the treatment of lesions, even when 'early', has a low cure rate.

A distinction should be made between advising physicians to be alert and informed so that uncommon malignant lesions will be recognized on physical examination while still 'early', and advising against the routine use of expensive cancer tests, which will have a disappointing and almost useless pay-off. One is good, the other is not. A uniform policy on screening for cancer in asymptomatic older persons cannot be recommended.

The individual asking for a 'cancer check-up' deserves a thorough examination and appropriate tests. One set of recommendations, that of the American Cancer Society, is listed in Table 3. For a chronic smoker, a chest radiograph

Table 3 Summary of American Cancer Society recommendations for the early detection of cancer in asymptomatic people

Test or procedure	Population		
	Sex	Age (years)	Frequency
Sigmoidoscopy	M & F	50 and over	Every 3 to 5 years, based on advice of physician
Stool guaiac slide test	M & F	Over 50	Every Year
Digital rectal examination	M & F	Over 40	Every year
Cervical smear Pelvic examination	F	All women who are, or who have been, sexually active, or have reached age 18, should have an annual smear and pelvic examination. After a woman has had three or more consecutive satisfactory normal annual examinations, the smear test may be made less frequently at the discretion of her physician	
Endometrial tissue sample	F	At menopause, women at high risk*	At menopause
Breast self-examination	F	20 and over	Every month
Breast physical examination	F	20–40 Over 40	Every 3 years Every year
Mammography	F	35–39 40–49 50 and over	Baseline Every 1–2 years Every year
Chest radiograph			Not recommended
Sputum cytology			Not recommended
Health counselling and cancer check-up**	M & F M & F	Over 20 Over 40	Every 3 years Every year

*History of infertility, obesity, failure to ovulate, abnormal uterine bleeding, or oestrogen therapy.
**To include examination for cancers of the thyroid, testicles, prostate, ovaries, lymph nodes, oral region, and skin.

and perhaps sputum cytology should be done, despite the knowledge that finding a lung cancer may not lead to a better outcome than if one waited until symptoms developed. Not to include a radiograph when the patient asks for a cancer examination is hardly defensible. On the other hand, when physicians establish policies about what to do regularly for their patients who are over 65, annual chest radiographs cannot be defended as a screening test for lung cancer, as careful trials have not shown a reduction of mortality from lung cancer in groups so investigated.

A factor that strongly influences the usefulness of cancer screening in adults is the prevalence of different cancers in the populations under study. In Japan, where gastric cancer has a high prevalence relative to that in many other countries, extensive public screening campaigns, using fibreoptic endoscopy with mobile units, are justifiable as cost-effective; but in the United States, where the incidence of gastric cancer has been decreasing for 50 years, it is much less prevalent, and the cost of finding each cancer by such screening would be unacceptable.

In older persons, an occupational history may provide leads that influence cancer detection procedures. Workers exposed to asbestos who have also been smokers have an enormous risk of lung cancer as well as a lower rate of asbestos-related mesothelioma. Aniline dye workers may develop bladder cancer, as another example.

TREATMENT

Priorities

Clinicians are advised to put age low on their list of priority considerations in selecting cancer treatment. The overall effectiveness of cancer treatment does not diminish with increasing age, and old people may have a long life expectancy once rid of their cancer. At the top of a priority list are symptoms, ability to offer effective therapy, and rate of progression.

Cancer is sometimes a diagnosis that demands immediate, even emergency treatment, as when it causes severe pain, small-bowel obstruction, hypercalcaemia, early spinal-cord compression, or pathological fracture in a weight-bearing bone. To relieve severe pain, some treatment is indicated, and whether death is around the corner or a long remission is possible is irrelevant. Physicians and nurses have unfortunately tended to undermedicate patients with acute or chronic pain, because of fear of addiction or respiratory depression, but experience gained in the hospice movement and by careful investigations has shown these beliefs to be unwarranted. Good pain management can almost always provide relief that is rated by the patient as 'no pain' or 'some pain but tolerable' (Walsh 1987). It is of the greatest importance to diagnose correctly the source of pain. In an older person who has been taking intermittent codeine for moderate cancer pain, an increase in abdominal pain may be due only to a codeine-induced faecal impaction, or may relate to developing drug tolerance. The older patient may either under-report pain or magnify it owing to the ominous implications of a cancer diagnosis. Fortunately, old age does not of itself limit tolerance to pain-relieving drugs, and doses can be increased as tolerance develops. Medication should be provided by-the-clock to get ahead of the pain and then cut back when possible. In many institutions with expertise in cancer care, patients are provided with easy access to extra medication, to be used as needed, either by oral medication at the bedside or by sophisticated techniques such as patient-controlled analgesia (see Section 22).

When skeletal metastases soften bone, pain with or without neurological signs frequently precedes pathological fractures or vertebral collapse. If radiographs confirm an imminent or existing crisis, aggressive treatment to achieve bone stabilization is usually indicated, even when life expectancy is brief, as cord compression leads to para- or quadriplegia, and a fractured femur may mean permanent confinement to bed. Both radiotherapy and surgical bone stabilization can prevent these catastrophes and the consequent loss of self-sufficiency.

Just below these emergencies in a list of priorities are those requiring urgent treatment. Among these should be listed severe dysphagia and/or vomiting, early large-bowel obstruction, pulmonary insufficiency, and severe anaemia. Readily available treatments range from blood transfusions to laser endoscopy for opening a cancer-obstructed hollow viscus, making the difference between a patient who is wretched and one who can maintain some degree of control over the activities of daily living despite advancing cancer.

At the other extreme of a treatment priority scale are patients whose cancer may be lethal, but who are asymptomatic and whose disease may or may not be progressing. For such patients, unless the cancer is treatable with expectation of cure or useful remission, many of the diagnostic tests and treatments conventionally used may not be indicated. An example would be an octagenarian, in whom a faecal occult blood test led to a colonoscopic diagnosis of sigmoid cancer, but who on further study is found to have a massively enlarged, hard liver with already abnormal liver function tests. Ignoring for the moment psychological features and patient motivation, a case can be made for no sigmoid resection and no chemotherapy, but rather palliative management with a low-residue diet and generous laxatives to avoid possible obstruction. This patient has a short life-expectancy, is asymptomatic, and cancer-directed therapy will not lead to prolongation of life. It is not the risk of sigmoid resection in an 80-year-old that deters surgical treatment—these operations are routinely done with a low fatality rate. Rather it is the pending hepatic failure, which is a major cause of postoperative illness and death in any age group.

In the absence of major symptoms, other factors, such as the availability of effective treatment, the results of staging procedures, and the presence or absence of comorbid illness, take over as the dominant ones in deciding how to treat an elderly cancer patient.

Types of treatment

As stated earlier, all of the commonly used treatments in cancer can be used confidently in older patients. Tolerance to treatment is not reduced by age but by accompanying diseases that reduce the reserves of cardiovascular, hepatic, pulmonary, and renal function required when treatment imposes added stress on these vital systems.

Modern surgical techniques allow elderly persons to undergo operations that once carried a fearsome fatality rate with a risk not much increased over the risk in young adults. Complication rates are higher than in younger patients, mostly because of higher rates of comorbid conditions (Patterson 1989).

Radiotherapy, likewise, does not have to be curtailed just because of age. With the best available techniques and equipment, both curative and palliative treatment is highly effective.

Medical treatment of cancer with hormonal manipulation or chemotherapeutic drugs is the choice when cancer is not localized but widely disseminated. As most lethal cancers metastasize unless cured while localized, or unless death from other causes supervenes, one or more of the many drugs and hormones now available may have a role at some time during the natural history of a cancer in an elderly patient. Studies of chemotherapeutically treated patients on clinical trials have demonstrated that age over 70 is not associated with lower response rates or survival rates (Walsh *et al.* 1989). Only two of the commonly used drugs, methotrexate and methyl-CCNU, lead to more haematologic toxicity in older than younger patients. While cure is much less common with chemotherapy than with surgery or radiotherapy, because it is not used on curable patients in general, appropriate therapy can be highly successful (Table 4).

Table 4 Drug effectiveness in cancer

Degree of effectiveness	Type of cancer
Cure	Hodgkin's disease Histiocytic lymphoma Seminoma Breast (?)
Significant palliation with drugs	Breast Prostate Ovary Head and neck Small-cell carcinoma of the lung Lymphocytic lymphoma
Some effect	Colon and rectum Soft tissue sarcoma Cervix Thyroid Stomach Non-small cell carcinoma of the lung Glioma
Little or no effect	Pancreas Hepatoma Gallbladder and bile ducts Kidney Adrenal

The choice between cancer-directed therapy and symptomatic treatment

Knowing that cancer in older patients can be treated effectively and safely does not mean that state-of-the-art treatment which has cure or long-term remission as its objective is always appropriate. Table 5 lists factors for and against the selection of cancer-directed versus supportive-only therapy. Appropriate therapy varies from ignoring the cancer and treating only symptoms to using multimodal experimental therapy in a clinical trial.

When an elderly patient is fully rational, careful discussions

Table 5 Guidelines for decision-making in treatment of older cancer patients

	Cancer-directed therapy	Symptom-directed therapy
Objective	Cure, or long-term remission	Relief of symptoms
Effective therapy	One or more options	No good options
Patient expectations	Motivated	Indifferent
Comorbid illness	Absent or minor	Major
Resources needed	Hospital with full modern facilities, or cancer centre	Home care, hospice
Risks	Toxic side-effects, unwanted symptoms	May miss a chance of some life extension

of such pros and cons are desirable, with plenty of time allowed for exchange of ideas with family or others, and for additional questions that were not considered during the initial dialogue with the physician. All patients (and some physicians) need to understand that choosing a non-aggressive treatment option does not mean a grim, painful death whereas cancer-directed therapy offers the only opportunity for avoiding such an outcome. With either choice, careful control of symptoms can avoid the death process that is feared, and at least promise a death in which the family can say afterwards, 'Although sad, it was not as miserable as we had anticipated, and we know that he/she did not suffer unduly. We had the opportunity to say goodbye.' This is the outcome most desired by the survivors.

OBSERVATIONS ON SOME CANCERS COMMON IN OLDER PERSONS

Breast

As obesity is a risk factor for breast cancer and there is evidence that obese women also have a worse outcome after treatment, dietary modification, specifically a diet low in saturated fat, has been recommended as a preventive measure. This belief has yet to be confirmed by evidence from clinical trials, which have been started but which are costly and difficult to implement. It is also believed that long-term administration of tamoxifen may lower the risk of breast cancer for women at particularly high risk because of familial or other factors, and large clinical trials to test this hypothesis are under development in the United States and elsewhere. These opportunities for prevention, however, are probably not of as great importance to women over 65 as they are for younger women.

On the basis of several clinical trials mentioned earlier, mammography is now widely accepted as a means of lowering the death rate from breast cancer. Recent trials show benefit from mammographic screening in older women as well as younger. As the reduction in mortality may be as much as 30 per cent, the expense of this test is easy to justify, and annual mammography can be recommended for all older women whose cancer would be treated if discovered.

Breast self-examination is inexpensive, but is thought by some to create anxiety, while being unevenly practised and unproven as an effective technique for early detection of breast cancer. A breast cancer large enough to be felt (over 1 cm in diameter) cannot truly be called 'early'. However, on the basis of studies that link the practice of self-examination to the discovery of smaller lesions and to a better prognosis, regular self-examination should be encouraged and carefully taught to all those older women who are willing and interested.

The management of breast cancer in women of all ages has undergone dramatic changes in recent years. Radical surgery is much less favoured, as clinical trials have demonstrated to the satisfaction of most clinicians that survival is rarely altered by the kind or extent of the surgical procedure chosen to treat the primary tumour. In many patients the breast need not be sacrificed, and women are now much more involved in the selection of a treatment option. Multidisciplinary management should always be considered. Table 6 summarizes the therapeutic plans that may be selected in postmenopausal women. No single approach fits all. Recognition that breast-sparing procedures, i.e. segmental mastectomy or lumpectomy, are important for older as well as younger women has been slow to penetrate the male-dominated surgical profession. Now, instead of suggesting to older women that they will not miss the amputated breast (implying that they no longer need it as a sexual symbol), surgeons are beginning to realize that for women already suffering the myriad on-

Table 6 Options for managing primary breast cancer in older women

Stage*	Treatment option	Comment
I	Segmental mastectomy (lumpectomy) only	Trials underway in patients with small (less than 1 cm) lesions who recognize and accept risk of local recurrence
I	Simple (total) mastectomy with or without tamoxifen	In absence of palpable axillary nodes is expeditious and allows local anaesthesia
I, II	Segmental mastectomy and radiotherapy, with or without tamoxifen	One of the standard treatment options, with proven safety and efficacy
I, II	Modified radical mastectomy with or without tamoxifen and reconstruction	Same as above, for patients who prefer mastectomy or when segmental excision is not possible
III	Radiotherapy, with or without chemotherapy and deferred surgery	For extensive but regionally localized cancers
IV	Hormonal or chemotherapy cancer	Local treatment of primary is seldom indicated

*Definition of clinical staging system:

Stage I — Tumour 2 cm or less in diameter; axillary nodes not felt to contain tumour; no metastases.

Stage II — Tumour over 2 cm but less than 5 cm in diam. No 'grave signs', i.e. extension of tumour to skin or chest wall, oedema, or ulceration. Axillary nodes thought to contain tumour but mobile, not adherent.

Stage III — Small tumour with fixed axillary nodes; larger tumour, with 'grave signs', with or without axillary nodes.

Stage IV — Distant metastases present, with any size tumour and nodes.

slaughts of old age, loss of a breast may be a serious additional burden.

The use of adjuvant chemotherapy in the management of breast cancer confined to the breast and axillary nodes has not been shown to be effective in postmenopausal women, but adjuvant tamoxifen does have a role, particularly in Stage II cancer, and is at present being widely recommended (Henderson and Mouridsen 1988). For metastatic breast cancer in the older woman, the range of useful hormonal and chemotherapeutic manoeuvres is broad, and is well covered in current texts on breast cancer and oncology.

Colon and rectum

The incidence of these cancers rises in accordance with a power law with age in both sexes, and they share the lead spot in cancer incidence with lung cancer. Possibilities for prevention lie in the adoption of diets lower in saturated fats and higher in fibre and antioxidants, plus the removal of benign colorectal adenomas, which are precursors of cancer, through periodic screening. Because effective screening requires examination of the whole colon by air-contrast barium enema or colonoscopy, expense and patient acceptance are major considerations in any early detection programme. Present guidelines suggest the less effective but inexpensive rectal examination and faecal occult blood test annually, with radiographic examination or flexible sigmoidoscopy every 3 to 5 years for all older persons. Those who have a first-degree relative with a history of colorectal cancer have a significantly higher risk, and more frequent examinations may be justified.

Treatment of colorectal cancer has been fairly standard for many years, but two more recent developments have major implications for elderly people. First, more patients with rectal cancer can now be spared a permanent colostomy. A colostomy is burden enough at any age, but may be impossible to manage in a patient with failing eyesight, arthritis, or confusion. Surgeons are today more prone to do low intestinal anastomoses, in part owing to the convenience of mechanical stapling devices. Sphincter-saving procedures such as local excision (frequently in conjunction with radiotherapy) are also more widely accepted (Harms and Starling 1990). The second advance is the demonstration in clinical trials that adjuvant chemotherapy, sometimes also with radiotherapy, has a role after surgery in both colon and rectal cancer. Earlier trials were negative or equivocal, but recent large studies show a clear survival advantage to the treated group. These treatments are well described in recent publications (Mayer *et al.* 1989).

Lung cancer

This cancer, though largely preventable, is both the most common lethal cancer in the United States and causes the most deaths. Formerly much more common in men, it is now rapidly increasing in women, approaching or passing breast cancer incidence in many parts of the United States. Counselling older patients about smoking cessation is one of the few interventions available to physicians, as early detection has not been shown to alter mortality, and treatment is effective only for limited subsets of patients with lung cancer. Specifically, surgical resection is strongly indicated for peripheral lesions in patients of almost any age who have adequate pulmonary function and no evidence on computed tomographic examination of mediastinal node involvement. Five-year survival rates in these patients are relatively high (30 per cent or more) and rates of morbidity and mortality, even for older patients, are acceptably low.

Older patients with small-cell cancer of the lung should be considered for chemotherapy, as initial response rates are in the range of 60 to 80 per cent. However, patients and families may question the usefulness of this aggressive approach 6 months later, if the cancer reappears and runs a rapidly fatal course. For these, and most other patients with lung cancer, expert control of symptoms by the judicious use of radiotherapy and medications, plus psychosocial support by a medical care team, become the critically important interventions.

Prostate

Many facets of the natural history, diagnosis, and treatment of this cancer present dilemmas for the gerontologist. Almost entirely confined to older men, prostate cancer has a high incidence but a lower mortality, not because of good treatment but because many of this group die of other causes. The cancer is often found at autopsy, with death having occurred from non-malignant disease. As the cause or causes of prostate cancer are unclear, there are no guidelines for prevention. Early detection is sometimes possible by rectal examination, but only when the lesion develops under or very close to the prostatic capsule, and diagnosis is often made only when malignant glands are found by the pathologist after surgery for prostatic hypertrophy, or when a patient complains of bone pain and is found to have widespread skeletal metastases.

The treatment of localized prostatic cancer is controversial, and the patient's state of health may be given more weight than is justified. For example, should an 80-year-old man who is found at the time of transurethral prostatectomy to have a few microscopic foci of cancer be treated at all? Many will say no, reasoning that death from other causes 'seems' more likely than recurrence or metastases. No good data are available to answer this, nor is it likely that a clinical trial can be implemented ethically to test the question. If the cancer in the same patient is larger, or has been found by palpation, the controversy centres on whether radical perineal prostatectomy or radiation therapy offers the best chance of control at the least risk to the patient. The operation, when a nerve-sparing technique is used, can be done with relative safety and with less incontinence and impotence than occurred in past decades. Radiation therapy, on the other hand, avoids an operation and may provide as good local control as surgery. Neither modality prevents the eventual appearance of metastases in up to 30 per cent of patients.

For advanced but still regional prostatic cancer there is less debate, as operation is not indicated and radiotherapy can be important in preventing further local spread, with its pain and obstruction. In the presence of metastases, the decision to treat or not to treat depends on the presence or absence of symptoms. Controlled trials have shown no survival advantage when immediate hormonal therapy is used rather than delaying therapy until symptoms appear, which may be much later. Symptomatic metastases are either treated by orchidectomy or by administration of hormones.

SUMMARY

Cancer in older people should never be regarded as hopeless or inevitable. Rather, physicians should remember that great variability is a characteristic not only of the elderly, but of cancer. Thus, the range of preventive and therapeutic approaches needs to be broad, more so than in young adults, and participation by the patient, with his or her support system, gains in importance.

Prevention and early detection do not lose their usefulness when people pass the age of 65, although a decrease in frequency of cancer screening tests may be possible if a series of normal tests has been recorded in previous years. Also, older people do not lose their ability to resist cancer or to respond to treatment, despite the evidence for some deterioration in immune function as ageing occurs.

As general guides to treatment decisions, physicians are urged to keep in mind the following points:

1. Chronological age is of minor importance in selecting treatment, as compared with comorbid illness and patient choice. Average life expectancy for the 80-year-old is still about 8 years, and even at 90, 5 years more of life may be anticipated.
2. Symptomatic cancer deserves treatment at any age, but appropriate treatment may involve only symptom control, not cancer-directed therapy.
3. During the course of advanced cancer, therapeutic decisions that involve new interventions, or discontinuation of ongoing efforts, almost always require consultation among family and friends of the patient, and should entail repeated discussions rather than requiring an abrupt decision, which may be heart-rending.

References

Brady, L. W. and Markoe, A. M. (1986). Radiation therapy in the elderly patient. *Frontiers of Radiation Therapy and Oncology*, **20**, 80–92.

DeVita, V. T., Hellman, S., and Rosenberg, S. A. (1985). *Cancer: principles and practice of oncology*. Lippincott, Philadelphia.

Feig, S. A. (1988). Decreased breast cancer mortality through mammographic screening: results of clinical trials. *Radiology*, **167**, 659–65.

Gazet, J-C, *et al.* (1988). Prospective randomized trial of tamoxifen vs surgery in elderly patients with breast cancer. *Lancet* **i**, 679–81.

Harms, B. A. and Starling, J. R. (1990). Current status of sphincter preservation in rectal cancer. *Oncology*, **4**, 53–60.

Henderson, I. C. and Mouridsen, H., co-chairmen for the Early Breast Cancer Trialists Group (1988). Effects of adjuvant tamoxifen and of cytotoxic chemotherapy on mortality in early breast cancer. An overview of randomized trials among 28, 896 women. *New England Journal of Medicine*, **319**, 1681–92.

Holland, J. F. and Frei, E. F. III. (1982). *Cancer medicine* (2nd edn). Lee and Febiger, Philadelphia.

Holmes, F. F. (1989). Clinical evidence for a change in tumor aggressiveness with age. *Seminars in Oncology*, **16**, 34–40.

Kottke, T. E., Battista, R. N., DeFriese, G. H., and Brekke, M. L. (1988). Attributes of successful smoking cessation interventions in medical practice: a meta-analysis of 39 controlled trials. *Journal of the American Medical Association*, **5**, 2882–9.

Levine, E. G., King, R. A., and Bloomfield, C. D. (1989). The role of heredity in cancer. *Journal of Clinical Oncology*, **7**, 527–40.

Mayer, R. J., O'Connell, M. M., Tepper, J. E., and Wolmark, N. (1989). Status of adjuvant therapy for colorectal cancer. *Journal of the National Cancer Institute*, **81**, 1359–64.

National Cancer Institute (1981). SEER incidence and mortality date 1973–1977, Monograph 57. NIH Publication No. 81–2330. NCI, Washington DC.

Patterson, W. B. (1989). Surgical issues in geriatric oncology. *Seminars in Oncology*, **16**, 57–65.

Walsh, S. J., Begg, C. B., and Carbone, C. C. (1989). Cancer chemotherapy in the elderly. *Seminars in Oncology*, **16**, 66–75.

Walsh, T. D. (1987). Control of pain and other symptoms in advanced cancer. *Oncology*, **1**, 5–9.

Warnecke, R. B. (1989). The elderly as a target group for prevention and detection of cancer. In *Cancer in the elderly: approaches to early detection and treatment*, (ed. R. Yancik and J. Yates). Springer, New York.

SECTION 11
Stroke

11.1 Pathology of stroke

NIGEL M. HYMAN

The pathology of stroke is essentially the study of cerebral arterial disease. The two major pathological processes causing stroke are infarction, in which brain tissue is deprived of blood supply, and haemorrhage.

Roughly 85 per cent of acute strokes are due to occlusion of a cerebral artery causing cerebral infarction. The remaining 15 per cent are due either to primary intracerebral haemorrhage or haemorrhage secondary to aneurysmal rupture. Of the cases of cerebral infarction, 75 per cent occur in the territory of the middle cerebral artery, 15 per cent in the vertebrobasilar territory, and 10 per cent in the border zones between the territories of two major arteries. Cerebral infarction may be due to primary thrombosis in an artery or may be secondary to occlusion of the vessel by an embolus.

Atherosclerosis in the principal arterial disease causing occlusive stroke in elderly persons. The distribution of the atheroma reflects sites of turbulent blood flow and mechanical stresses on the vessel wall. Extracranially the three most important sites are the bifurcation of the common carotid and the origins of both the common carotid and vertebral arteries. Intracranially, important sites include the vertebral and basilar arteries, the carotid siphon, and the first parts of both the middle cerebral and posterior cerebral arteries.

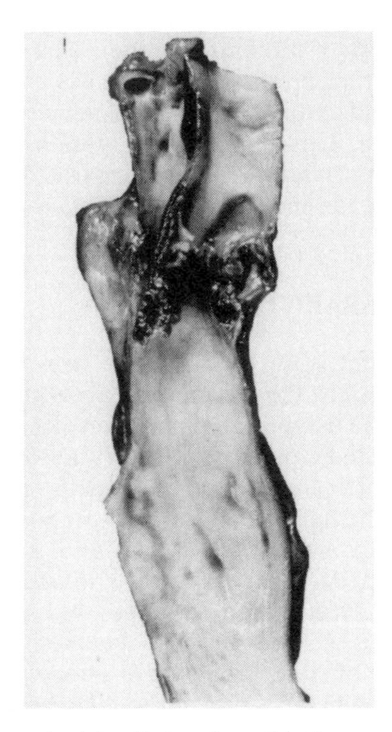

Fig. 1 Atheroma at origin of internal carotid artery causing severe stenosis.

THROMBOSIS OR EMBOLISM

The proportion of cerebral infarcts due to embolism is uncertain but is probably in the order of 50 per cent (Table 1). Difficulties in this differential diagnosis at a pathological level arise for several reasons. A recent embolus is softer than an *in-situ* thrombus and is usually relatively free from the vessel wall but an old lesion may be incorporated into the arterial wall and indistinguishable microscopically. Secondly, a source of embolism, for example from the heart, may be found in association with occlusion of the middle cerebral artery but this relationship could be coincidental rather than causal. About one-third of cerebral emboli arise from the carotid arteries, principally the bifurcation of the common carotid (Fig. 1). Cardiac sources include rheumatic valvular disease, prosthetic mitral valves, and both infective and non-infective valve vegetations. The latter, marantic endocarditis, is an underdiagnosed cause of stroke disease in patients with associated neoplasia. Mitral annulus calcification is common in elderly subjects but the precise association with stroke is unclear. Left ventricular mural thrombi are a source of embolic stroke, particularly complicating full-thickness myocardial infarction.

Table 1 Main causes of cerebral arterial embolism in elderly people

Sites of extracranial arterial atheroma: principally the common carotid bifurcation; the origins of the common carotid and vertebral arteries
Sites of intracranial arterial atheroma: principally the carotid siphon, vertebral arteries near junction with basilar artery and basilar artery itself
Cardiac origin
Left atrium: thrombi or myxoma
Mitral valve: vegetation in bacterial and marantic endocarditis; prosthesis; mitral annulus calcification
Left ventricle: mural thrombi from myocardial infarction
Aortic valve: sclerosis and calcification; prosthesis; vegetations in bacterial and marantic endocarditis
Trauma: direct trauma to neck; cervical fracture-dislocation

Although atheroma is by far the most common cause of thrombo-occlusive stroke in elderly people, there are other causes that remain important in view of their therapeutic implications (Table 2). Giant-cell arteritis can, rarely, involve the extracranial vertebral arteries to produce a completed brain-stem infarction. Polyarteritis nodosa characteristically causes multiple infarcts and haemorrhages in the cortex, white matter, basal ganglia, and brain-stem. Systemic lupus

Table 2 Main causes of cerebral arterial thrombosis in elderly people

Atheroma	
Vasculitis	Giant-cell (temporal) arteritis, polyarteritis nodosa, endarteritis obliterans due to tuberculosis, syphilis, etc.
Haematological disorders	Polycythaemia rubra vera, essential thrombocythaemia, hyperviscosity syndromes, e.g. myeloma

erythematosus is a rare disease in elderly people but can cause stroke, particularly in association with the lupus anticoagulant (Mueh *et al*. 1980). Endarteritis obliterans (a reaction of arteries to systemic bacterial infection) can complicate meningovascular syphilis and both pyogenic and tuberculous meningitis. Primary angiitis of the central nervous system (Calabrese and Mallek 1988) occurs in all age groups and can cause stroke, although headache and dementia are more common. The risk of cerebral infarction is increased in polycythaemia, whether primary or secondary, essential thrombocythaemia, and chronic myeloid leukaemia with very high white-cell counts. Thrombotic arterial lesions are described in hyperviscosity syndromes, particularly Waldenström's macroglobulinaemia and multiple myeloma.

CEREBRAL INFARCTION

A cerebral infarct is an area of brain in which the blood flow has fallen below the critical level necessary to maintain the viability of the tissue. Early infarction of a major vascular territory is reflected grossly at autopsy by a swollen, softened area of brain, which may be more easily detected by palpation than seen with the naked eye. It may be impossible to detect very recent infarction (less than 8 h) until a microscopical examination is made. The separation of infarcted tissue from adjacent relatively normal tissue takes 48 h, and by 3 to 4 days the lesion is well defined and the swelling is greatest. The boundaries of older infarcts are altogether crisper. A cavity 1 cm in diameter is thought to take 3 months to form and very large infarctions may never completely cavitate.

Microscopically the earliest changes include an increase in interstitial fluid within the white matter. Neurones develop ischaemic necrosis with nuclear pyknosis and there is necrosis of glial cells, capillaries, arterioles, and venules, swelling of axons and pallor of myelin. In pale infarcts there is little or no extravasation of red cells into the lesion but in haemorrhagic infarction the capillary and venular extravasation produces either discrete petechiae, confluent purpura, or frank haemorrhage. The distinction between pale and haemorrhagic infarction is probably unimportant in mild cases as nearly all infarcts show a few petechiae, particularly at the margins. Nevertheless, roughly 30 per cent of cerebral infarcts examined at autopsy are termed haemorrhagic. In those strokes thought to have an embolic basis, at least half are haemorrhagic (Fig. 2). The mechanism of haemorrhagic infarction is complex and may not be due to a single cause. The endothelium of the obstructed blood vessel becomes ischaemic, without vessel rupture, and as the embolus fragments and passes more distally blood seeps through the vessel wall into the adjacent brain substance. Another important factor, subsequent to the lysis of the embolus, is the sudden restoration of blood through the now patent vessel. The ischaemic capillaries rupture and there is secondary irrigation with consequent haemorrhage, particularly involving the cortex. Haemorrhagic infarction is very common in cardioembolic infarcts and less common in watershed infarcts and when infarction complicates vasospasm after subarachnoid haemorrhage.

Within the first week, after both pale and haemorrhagic infarction, capillary hyperplasia and microglial infiltration occur. Eventually the infarcted area shrinks and larger lesions show cavitation traversed by septa. As the lesion shrinks there may be compensatory enlargement of an adjacent ventricle and depression of the cortical surface.

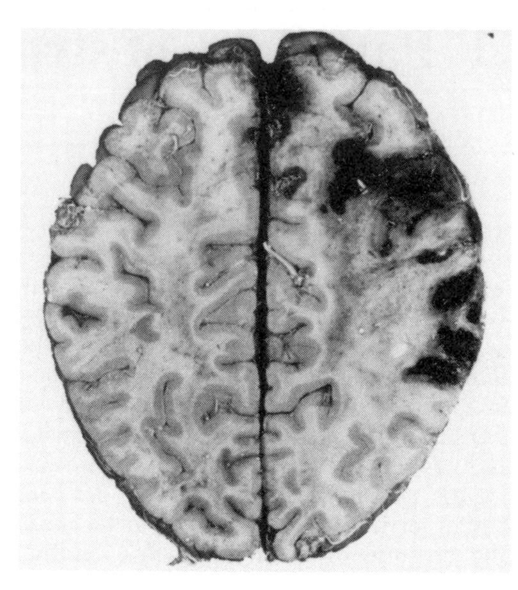

Fig. 2 Haemorrhagic infarction in distribution of both the anterior and middle cerebral arteries.

A gradient of blood flow exists after complete occlusion of an artery, and consequently there is absence of flow in the central zone of a vessel's territory and normal or increased flow at the periphery of the territory. It is suggested that this peripheral zone, which is ischaemic and swollen, has not reached a state of irreversible infarction. The 'ischaemic penumbra' hypothesis is attractive as a rationale for specific forms of treatment for acute stroke but remains controversial, largely because of variable results in animal experiments.

BRAIN SWELLING AFTER STROKE

Most deaths directly due to acute ischaemic strokes occur within the first 7 days after the onset and are secondary to a complication of cerebral oedema. The latter has two components; firstly, 'cytotoxic' oedema due to water entering the cells and producing astrocytic swelling; secondly, 'vasogenic' oedema at the periphery of the ischaemic lesion, which causes breakdown in the blood–brain barrier and allows extravasation of proteins into the extracellular space. The rise in oncotic pressure increases water entry into the ischaemic area, which in turn adds to the degree of swelling of the infarct. The oedema reaches a peak between 2 and 7 days after the onset and, in the case of major infarction in the territory of the middle cerebral artery, there is herniation of the ipsilateral cingulate gyrus beneath the free edge of the falx. More importantly there is downward displacement of the brain through the tentorial notch, with compression of the diencephalon and adjacent midbrain and subsequent brain-stem haemorrhages. Another complication of transtentorial herniation is compression of the posterior cerebral vessels with infarction of the ipsilateral occipital lobe (Fig. 3). Such lesions are usually haemorrhagic and may reflect venous rather than arterial occlusion.

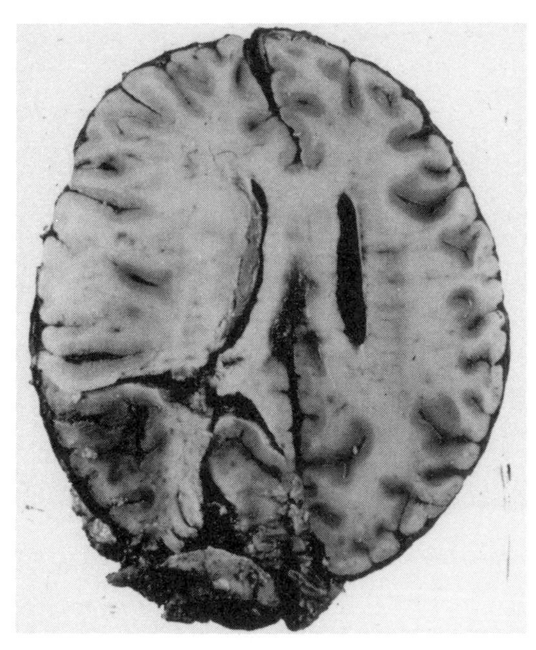

Fig. 3 Massive infarction of left hemisphere with haemorrhagic infarction of occipital lobe (see text).

TOPOGRAPHY OF CEREBRAL INFARCTIONS

Variations in individual anastomotic pathways, in association with the degree of atheromatous disease and the speed of occlusion, all contribute to the variation in size of the cerebral infarction, which typically occurs within the territory of the whole or part of a cerebral artery (Fig. 4). Anastomoses occur at the following sites.

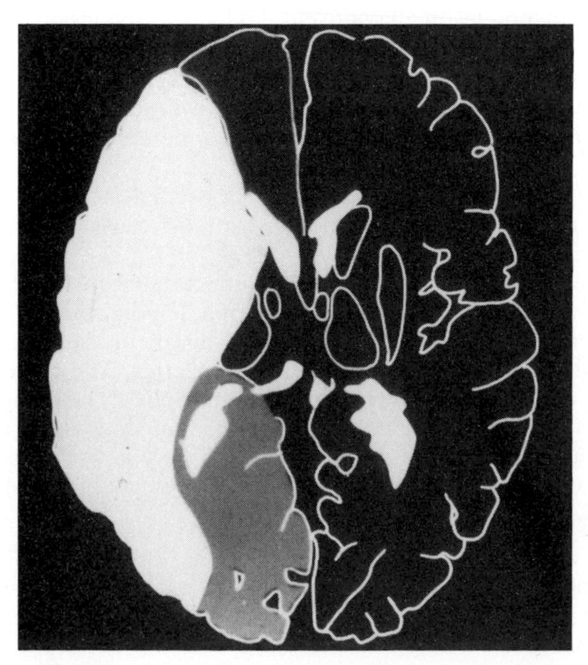

Fig. 4 Anterior, middle, and posterior cerebral arterial territories in horizontal plane.

1. Corticomeningeal anastomoses connect the cortical branches of the main cerebral arteries across their border zones.
2. Anastomoses are recorded in a quarter of autopsy cases within the circle of Willis.
3. Anastomoses occur between the external and internal carotid arteries via the ophthalmic artery and meningohypophyseal branch.
4. Anastomoses occur between the external carotid and the vertebral arteries via the occipital artery.

THE ANTERIOR CIRCULATION

The ophthalmic artery is the first branch of the intracranial internal carotid artery. Rather more distally the anterior choroidal artery leaves the parent artery and, in regard to the pathology of stroke, supplies to a varying degree the caudate nucleus, internal capsule, and optic tract. The anterior cerebral artery passes forward over the genu of the corpus callosum, supplying the orbital surface of the frontal lobe and the entire medial surface of the frontal and parietal lobes as far back as the parieto-occipital sulcus. The territory includes the motor and sensory cortex controlling the leg, the anterior four-fifths of the corpus callosum, and the anterior basal ganglia. A variable branch, the recurrent artery of Heubner, supplies the cranial nerve and arm section of the internal capsule. Isolated occlusion of the anterior cerebral artery is a rare course of stroke owing to the anastomosis with the anterior communicating artery. Emboli from the internal carotid artery are more likely to take a middle cerebral rather than an anterior cerebral route.

The territory of the middle cerebral artery is the most important for the pathology of stroke. It includes the lateral surface of the majority of the frontal and parietal lobes, the insula, superior and middle temporal gyri, and deep striatum. Infarction usually involves only part of this territory when due to occlusion of either the common or internal carotid artery (with vascular anastomoses at the base of the brain) or of the terminal middle cerebral branch (Fig. 5). Proximal, complete occlusion of the middle cerebral artery is frequently embolic and results in extensive infarction due to inadequate anastomoses. The lenticulostriate arteries are long, thin, penetrating vessels arising from the trunk of the middle cerebral artery. Primary thrombotic occlusion results in internal capsular infarction.

THE POSTERIOR CIRCULATION

The territory of the posterior cerebral artery includes the medial surfaces of the occipital lobe (Fig. 6), the inferomedial surface of the temporal lobe, the subthalamus, and thalamus. Infarction may be bilateral due to primary thrombosis of the basilar artery (Fig. 7) or ipsilateral secondary to embolic disease. Brain-stem infarcts vary in site and distribution, in part owing to the wide variation in vascular anatomy and anastomotic patterns. Infarction of the dorsolateral medulla is the most commonly encountered. It is more often due to thrombotic disease of the vertebral artery with secondary occlusion of the ostium of the posterior inferior cerebellar artery than to occlusion within that branch artery itself.

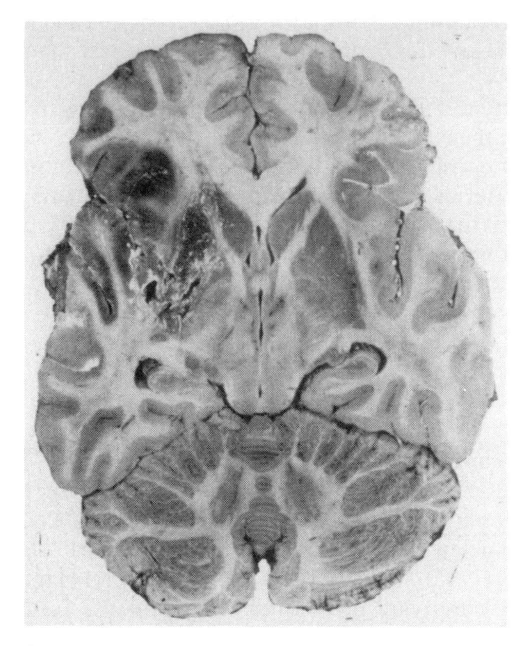

Fig. 5 Infarction in distribution of left middle cerebral artery.

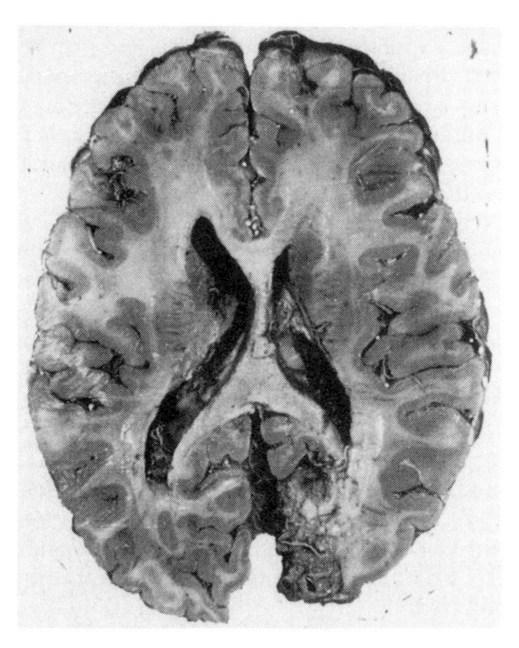

Fig. 6 Infarction of medial surface of right occipital lobe within posterior cerebral artery territory.

WATERSHED INFARCTION

Watershed infarctions (border zone lesions) are an important cause of stroke in elderly people and occur at the junction of two major arterial territories. The boundary between the middle and anterior cerebral arteries is particularly vulnerable but other potential watersheds are at the junction of the territories of the middle and posterior or the anterior and posterior cerebral arteries (see Fig. 4). These infarcts also occur between the territories of the cerebellar arteries or, more rarely, between the territories of the small arteries of the basal ganglia. The infarction is unilateral after, for example, occlusion of the internal carotid artery, or bilateral after generalized hypoperfusion as in systemic hypotension. Approximately 10 per cent of all cerebral infarctions are thought to be of the watershed type and it is unusual for them to be significantly haemorrhagic.

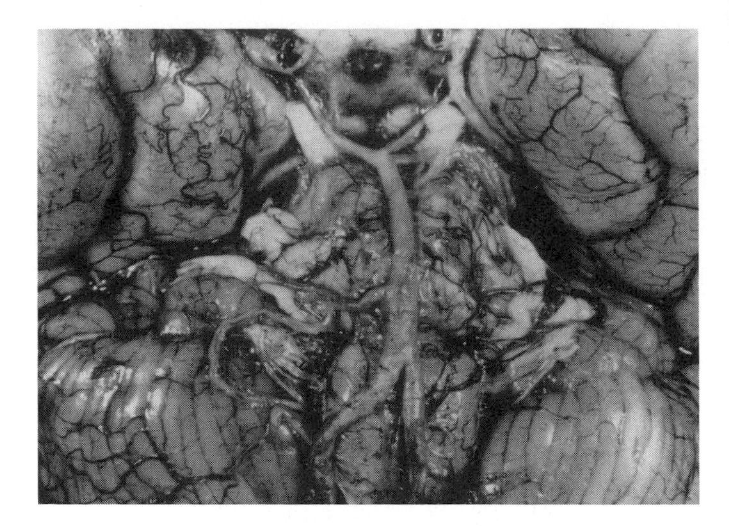

Fig. 7 Primary basilar artery thrombosis.

The pathogenesis is secondary to either systemic hypotension or microemboli. Hypotension is the most frequent cause, producing a failure of perfusion to the 'last fields of irrigation', which are the most distal branches of each major artery. The rate of reduction of blood flow is crucial as hypoperfusion of rapid onset and short duration after, for example, a cardiac arrest causes diffuse cortical cell death but not necessarily a border-zone infarct. If the hypotension is more gradual in onset and of longer duration, then characteristic watershed lesions occur. This does not entirely explain the asymmetry of the infarcts. The combination of stenosed vessels, either intra- or extracerebrally, and the speed of onset and duration of the hypotension all contribute to localize the lesions. The well-known association of this type of infarction with occlusion of the carotid artery at the bifurcation in the neck could be secondary to this cause. An alternative explanation is that showers of emboli from the thrombus, before actual occlusion, lodge in the watershed areas.

Other more unusual emboli, such as cholesterol crystals or those derived from tumours, can also cause watershed infarcts.

Certain patients with watershed infarcts have no detectable source of emboli or cause of systemic hypotension. These lesions are multiple and small, particularly at the junction of the anterior and middle cerebral arteries. The cortex is pale and scarred, known as 'granular atrophy'. The pathogenesis has been attributed to atheroma and sludging of flow at an arteriolar level. Other cases may be due to a primary disturbance of coagulation rather than a disease of the vessel wall.

SPECIFIC LESIONS

Lacunes

Lacunes consist of small, irregular softenings up to 1.5 cm in diameter that occur with greatest frequency in the deeper parts of the brain, particularly the basal grey nuclei and the anterior pons. *L'état lacunaire* consists of multiple lacunes and the clinical correlate is one of a progressive, step-wise disorder characterized by dementia, pseudobulbar palsy, and a shuffling gait. Solitary lacunes, in contrast, are a frequent incidental finding at autopsy (Fig. 8) but have been correlated

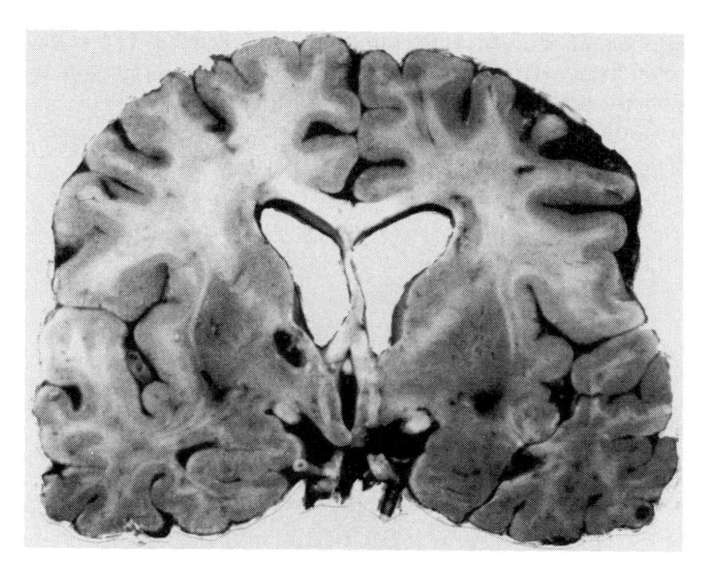

Fig. 8 Lacune within left internal capsule.

with clinical syndromes, particularly the pure motor stroke (Fisher 1982; Bamford and Warlow 1988). Pathological data are limited and there is doubt whether the underlying lesion is one of fibrinoid necrosis, lipohyalinosis, or microatheroma. Fibrinoid necrosis occurring in a setting of very high blood pressure seems unlikely. Lipohyalinosis as a cause of thrombotic occlusion occurs in non-malignant hypertension and consists of replacement of the muscle and elastic laminae by collagen, with an increase in subintimal hyaline material. The lesions, which stain readily for fat, include fatty macrophage replacement within thc disrupted walls. Microatheroma is the most likely cause of single lacunes but lipohyalinosis may represent an intermediate pathological stage between the fibrinoid necrosis of severe hypertension and the microatheroma of milder cases.

Binswanger's subacute arteriosclerotic encephalopathy

The most consistent pathological lesion in this disorder is of demyelination of white matter with loss of neurones and astrocytic gliosis. Affected patients are usually hypertensive and there is usually atheromatous involvement of both the basal cerebral arteries and the finer, penetrating, cortical vessels. Multiple lacunes in the basal nuclei are frequent, as well as lacunar infarction within the white matter adjacent to the more diffuse changes. The aetiology in these cases suggests disease of small vessels such as the lypohyalinosis implicated in lacunar development. Cerebral amyloid angiopathy can produce demyelination of white matter but with associated slit haemorrhages or frank lobar haematomas (Dubas *et al.* 1985). Chronic hypoperfusion and resulting ischaemia in the watershed area between the cortical medullary arteries and the long perforating arteries to the subcortical white matter has been advanced as a cause for the change in the white matter (Loizou *et al.* 1981).

The diagnosis of Bingswanger's encephalopathy has greatly increased since the introduction of computerized tomographic scanning and is based on the appearance of diffuse low-density lesions in the white matter, particularly in the periventricular area, centrum semiovale, and corona radiata distribution. Few of these cases have been confirmed by pathological examination and undoubtedly this appearance on a scan is of multiple pathologenesis and aetiology.

INTRACEREBRAL HAEMORRHAGE

Intracerebral haemorrhage is frequently a complication of hypertension, although an increasing number of cases in elderly subjects are recognized as secondary to amyloid angiopathy. Other causes include bleeding in association with berry aneurysms, mycotic aneurysms, arteriovenous malformations, arteritis, and neoplasms (particularly glioblastoma multiforme and melanoma). The frequency of deep ganglionic intracerebral haemorrhage is falling and this has been attributed to improved control of hypertension in the community. More superficial lobar haematomas, particularly in elderly, demented patients, may be secondary to amyloid angiopathy.

Ganglionic and thalamic haemorrhage

The most common site of massive haematoma is within the basal ganglia, particularly in the region of the putamen/claustrum (Fig. 9). After such a haemorrhage the brain rapidly

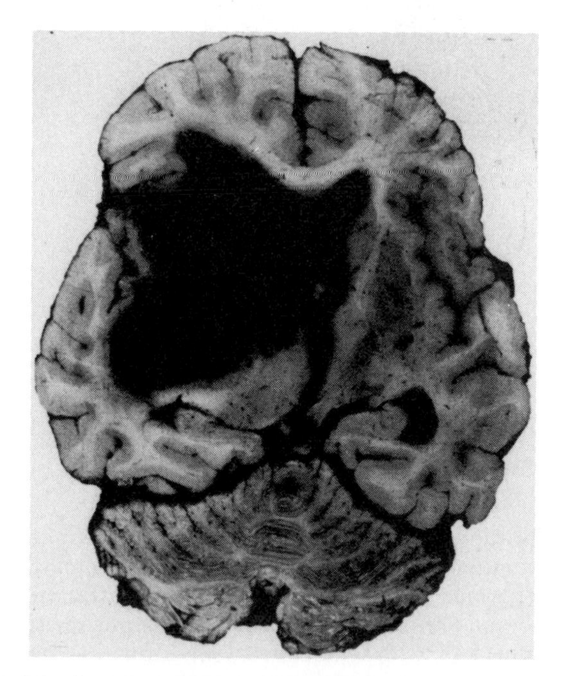

Fig. 9 Massive left ganglionic haemorrhage.

becomes swollen, particularly the affected hemisphere. If the brain is cut whilst still fresh, the clot, described as 'redcurrant jelly', is friable and easily dislodged, revealing a ragged cavity wall. The blood frequently ruptures into the ventricular system in fatal cases and may leak from the fourth ventricle into the subarachnoid space. Death results from downward displacement of the brain with associated brain-stem haemorrhages in a similar fashion to that described after massive cerebral infarction.

The perforating branches of the middle cerebral artery generally supply the area of the lesion but the aetiology of the haemorrhage remains controversial. Lypohyalinosis has been implicated by Fisher (1971), who identified a 'fibrin globe' with a ball of haemorrhage and a central core of platelet material filling the arterial lumen. Charcot and Bouchard in 1868 described miliary microaneurysms (resulting from 'periarteritis'), 250 to 300 μm in diameter, as a source of bleeding. A combination of both theories suggests that tiny leaks from lypohyalinotic vessels produce a false aneurysm which can subsequently rupture (Takebayashi and Kaneko 1983).

Cerebral amyloid angiopathy

This angiopathy is an increasingly recognized cause of intracerebral haemorrhage affecting both men and women equally among elderly demented patients (Vinters 1987). The pattern of haemorrhage is stereotyped causing, over time, multilobar peripheral intracerebral haematomas, particularly in the frontal and parietal regions, involving both the cortex and subcortical white matter (Fig. 10). Amyloid, demonstrated

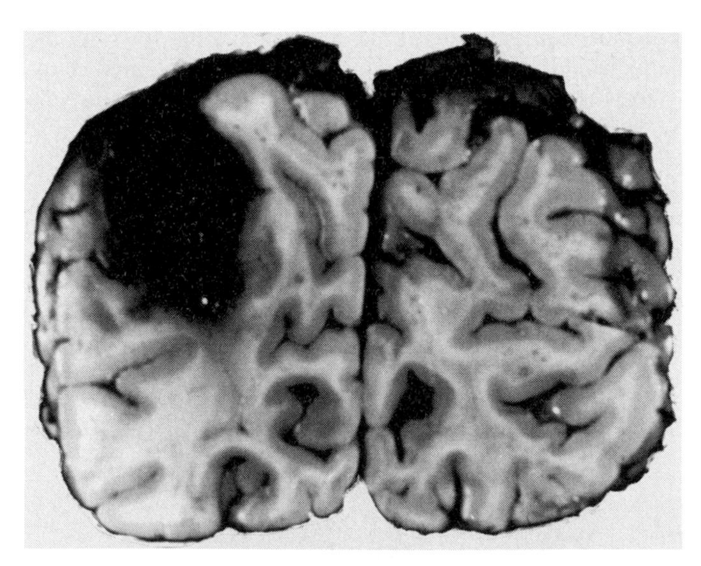

Fig. 10 Peripheral intracerebral haemorrhage due to amyloid angiopathy (by permission of Dr M. Esiri).

by application of the Congo red stain, is seen in the media and adventitia of vessels of the microcirculation, particularly in the leptomeninges and superficial cortex. Affected vessels may undergo fibrinoid degeneration; segmental dilatation of the vessel wall with microaneurysm formation has been reported, which is relevant to the complication of subsequent haemorrhage.

The condition is not associated with systemic amyloidosis. It occurs in association with Alzheimer's disease, Down's syndrome, and other rarer conditions but it remains unclear whether this is a causal relationship. The association with Alzheimer's disease is particularly well described and at least 40 per cent of those patients with haemorrhagic brain disease due to cerebral amyloid angiopathy show Alzheimer's disease at autopsy.

Primary brain-stem and cerebellar haemorrhage

Primary brain-stem haemorrhage usually arises in the pons and ruptures into the fourth ventricle, extending into the cerebellar peduncles rather than the medulla (Fig. 11). Cerebellar haemorrhage arises unilaterally, often near the dentate nucleus, and can produce both secondary brain-stem compression and hydrocephalus due to compression of the aqueduct.

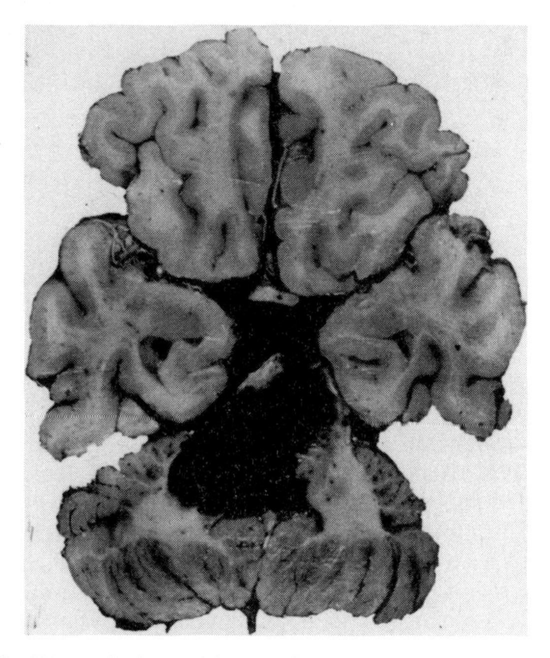

Fig. 11 Primary brain-stem haemorrhage.

Intracranial arterial aneurysms

The majority of intracranial aneurysms are of the so-called berry type consisting of thin-walled saccular dilatation; the larger ones may be multiloculated. They are considered to be acquired rather than congenital lesions and the most common size is that of a pea; unruptured aneurysms are described in about 2 per cent of all routine autopsies. Histologically the walls consist of a thin layer of connective tissue lined by endothelium and the normal muscular and elastic components are absent. The muscle of the media of the parent artery ends abruptly at the neck of the sac. Rupture occurs from the fundus, which represents the thinnest part of the wall of the aneurysm. The aetiology of the lesion remains unknown. Contributory factors include age, atheroma, and hypertension (Crompton 1966), although rupture can occur without their association. Aneurysms arise from the distal angle of points of arterial division and are situated chiefly on the vessels that form the circle of Willis; 90 per cent are sited in the carotid (anterior) territory and 10 per cent in the vertebrobasilar (posterior) territory. Some 15 per cent of aneurysms are multiple and are frequently symmetrical.

Blood from a ruptured aneurysm can enter directly into

the subarachnoid space to produce the classical features of a subarachnoid haemorrhage. Intracerebral passage of blood results in a stroke of acute onset and the site of the haematoma indicates the likely site of the aneurysm. For instance, a Sylvian haematoma is secondary to rupture of an aneurysm of the middle cerebral artery, and a haematoma in the medial aspect of the frontal lobe may well be secondary to rupture from an aneurysm of an anterior communicating artery. The majority of fatal cases result either from this type of intracerebral haemorrhage or from massive intraventricular haemorrhage. Another potential cause of 'stroke' is arterial spasm and subsequent infarction occurring between the third and seventh day after the ictus, generally but not exclusively in an artery close to the ruptured aneurysm (Crompton 1964). The spasm is not necessarily reversible; an inflammatory reaction occurs initially, with later thickening and fibrosis within the arterial wall, months or years afterwards (Hughes and Schianchi 1978).

Hydrocephalus occurs both acutely and as a late complication of subarachnoid haemorrhage. There is histological evidence of red cells blocking the arachnoid granulations in patients dying soon after the initial bleed.

Atherosclerotic fusiform aneurysms

Large fusiform aneurysms of the internal carotid or basilar artery are a complication of severe atherosclerosis in elderly patients. The walls of the artery are stretched owing to replacement of smooth muscle by fibrous tissue, and the lumen may be widened and tortuous (ectasia). The most common site is the supraclinoid segment of the internal carotid artery and the major complication is that of a compressive lesion within the cavernous sinus. The aneurysms rarely rupture but may thrombose and calcify.

Vascular lesions of venous origin

Cerebral venous thrombosis is a rare cause of stroke in elderly individuals. Frequently no obvious trigger is found but sagittal sinus thrombosis can complicate a 'hypercoagulable state' in association with systemic disorders such as dehydration, malnutrition, infections, and congestive cardiac failure, or as a non-metastatic complication of systemic carcinoma. Local sepsis, such as infection of the mastoid air-space, can produce a transverse sinus thrombosis, or a more remote lesion on the scalp or face can be complicated by a cavernous sinus thrombosis.

In those cases with recovery there may be partial recanalization of the sinus but in fatal cases there is characteristic symmetrical haemorrhagic infarction (sometimes with frank haematoma formation) involving both the cortex and white matter (Fig. 12).

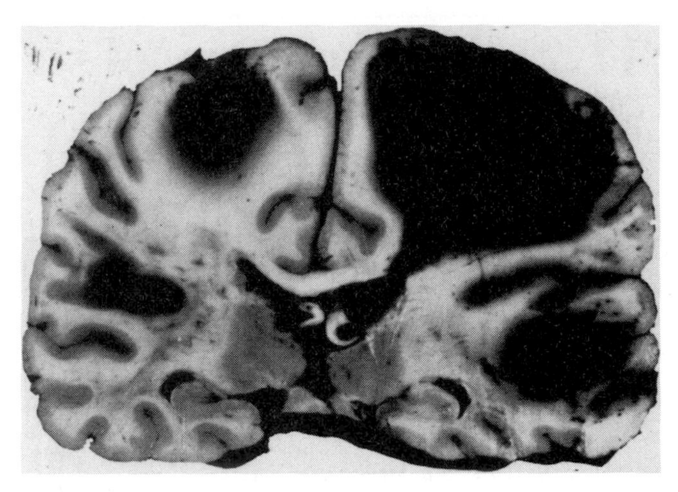

Fig. 12 Bilateral haemorrhage and haemorrhagic infarction due to venous sinus thrombosis.

References

Bamford, J.M. and Warlow, C.P. (1988). Evolution and testing of the lacunar hypothesis. *Stroke*, **19**, 1074–82.

Calabrese, L.H. and Mallek, J. A. (1987). Primary angiitis of the central nervous system. *Medicine*, **67**, 20–39.

Crompton, M. R. (1964). The pathogenesis of cerebral infarction following the rupture of cerebral berry aneurysms. *Brain*, **87**, 491–510.

Crompton, M. R. (1966). The pathogenesis of cerebral aneurysms. *Brain*, **89**, 797–814.

Dubas, F., Gray, F., Roullet, E., and Escourolle, R. (1985). Arteriopathic leukoencephalopathy. *Revue Neurologique*, **141**, 93–108.

Fisher, C. M. (1971). Pathological observations in hypertensive cerebral haemorrhage. *Journal of Neuropathology and Experimental Neurology*, **30**, 536–50.

Fisher, C. M. (1982). Lacunar strokes and infarcts: a review. *Neurology*, **32**, 871–76.

Hughes, J. T. and Schianchi, P. M. (1978). Cerebral artery spasm. A histological study at necropsy of the blood vessels in cases of subarachnoid haemorrhage. *Journal of Neurosurgery*, **48**, 515–25.

Loizou, L. A., Kendall, B. E., and Marshall, J. (1981). Subcortical arteriosclerotic encephalopathy: a clinical and radiological investigation. *Journal of Neurology, Neurosurgery, and Psychiatry*, **44**, 294–304.

Mueh, J. R., Herbst, K. D., and Rapaport, S. I. (1980). Thrombosis in patients with the lupus anticoagulant. *Annals of Internal Medicine*, **92**, 156–9.

Takebayashi, S. and Kaneko, M. (1983). Electron microscopic studies of ruptured arteries in hypertensive intracerebral haemorrhage. *Stroke*, **14**, 28–36.

Vinters, H. V. (1987). Cerebral amyloid angiopathy—a critical review. *Stroke*, **18**, 311–24.

11.2 Epidemiology of stroke

PHILIP A. WOLF AND L. ADRIENNE CUPPLES

INTRODUCTION

Cardiovascular disease is responsible for an increasing proportion of death and disability with advancing age, and accounts for almost half of deaths in persons above 65 years of age (Fig.1). A sizeable proportion of this illness and death

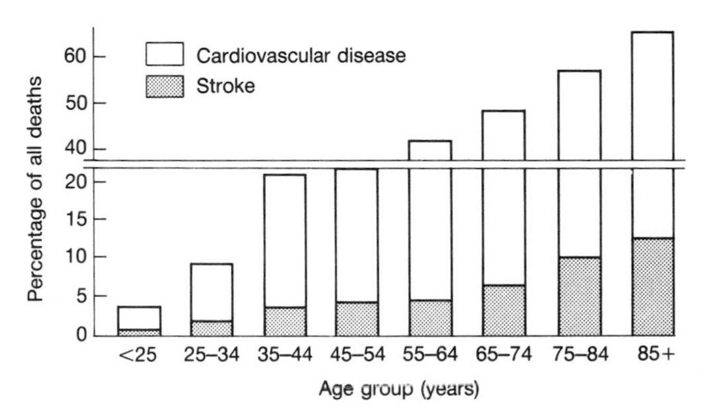

Fig. 1 Proportion of deaths due to cardiovascular disease and stroke by age, United States, 1983. Beginning at about 65 years of age, cardiovascular disease is responsible for almost 50 per cent of deaths. (Source: National Center for Health Statistics.)

is attributable to stroke, which accounts for nearly 20 per cent of all deaths from cardiovascular disease in older persons. Stroke is the third leading cause of death in the United States and is a major contributor to disability leading to the institutionalization of old people. In 1986, the American Heart Association estimated that 500 000 Americans had had a stroke, 147 800 had died of stroke, and remarked that many of the estimated 2 020 000 survivors of stroke required chronic care. Statistics for deaths cannot show the personal impact of stroke. To the functionally independent elderly person, stroke may exemplify a condition worse than death itself; robbing them of movement, sight, speech, and independence and thus heralding the end of useful life.

Although the development of medical and surgical treatment for impending stroke or stroke of recent onset should be pursued, it is likely that prevention is the key to reducing the impact of cerebrovascular disease. The development of atherosclerotic, cardiovascular disease, including stroke, has been related to a number of modifiable host and environmental factors. These risk factors have been identified and their impact assessed in recent years, chiefly through prospective epidemiological study. Furthermore, it has now been demonstrated that modification of key risk factors, notably control of hypertension, has substantially reduced the incidence of stroke. This review of the epidemiology of stroke in the geriatric population will rely principally on data collected in the Framingham study and on evidence, garnered from clinical trials, that modification of specific risk factors reduces the incidence or severity of stroke.

Assessment of risk factors in a cohort study offers special advantages for the investigation of stroke in the elderly because the data are collected systematically, often years before the appearance of clinical disease, allowing distinction between long-term and short-term effects. Prospective epidemiological study therefore provides the least distorted picture of the impact of risk factors on the development of cerebrovascular disease.

Stroke is chiefly a disease of older people, and hence a discussion of risk factors for stroke generally is pertinent to stroke in the geriatric age group, except for those relatively infrequent causes of premature stroke, such as pregnancy, oral contraceptives, migraine, and arteriovenous malformations.

THE OCCURRENCE OF STROKE

Data on the incidence of stroke are available from a number of populations around the world. The specific incidence rates depend on the age distribution of the population, whether recurrent as well as initial strokes are included, and on the completeness of case ascertainment. The occurrence of cardiovascular disease, including stroke and transient ischaemic attack (**TIA**), has been under study in the general population sample at Framingham, Massachusetts since 1950. The Framingham cohort of 5070 men and women, aged 30 to 62 years, and free of cardiovascular disease at entry into the study, has been followed for 32 years. This population consists of the respondents of a sample of two-thirds of the adults resident in the town in 1948, as well as their spouses and some volunteers. After the initial examination, in 1950, these 2252 men and 2818 women have been examined every 2 years; follow-up has been satisfactory, with 85 per cent presenting at each examination; only 7 per cent have been completely lost to follow-up. Sampling procedures, criteria, and methods of examination have been described elsewhere.

At each biennial examination, subjects were specifically and systematically asked about the sudden occurrence of speech difficulty, muscular weakness, numbness or tingling, visual defect, or other neurological symptoms including dizziness and double vision. Subjects with these symptoms of TIA or stroke were evaluated by the study neurologists to clarify the details and to determine if minimal criteria for these conditions were met. Detailed neurological evaluation at the time of the stroke or TIA has been carried out for more than 20 years. These clinic and hospital assessments by the study neurologists have helped to document the circumstances surrounding the event, facilitated classification according to type of stroke (particularly in the years before

Supported in part by grants 2-RO1-NS-17950–09 (Philip A. Wolf, National Institute of Neurological Disorders and Stroke), 1-RO1-AG-08122–01 (Philip A. Wolf, National Institute on Aging) and Contract NIH-NO1-HC-38038 (Philip A. Wolf, National Heart, Lung, and Blood Institute).

Table 1 Frequency of stroke by type, 32-year follow-up, the Framingham study

	Ages 35–64 years				Ages 65–94 years			
	Men		Women		Men		Women	
Stroke type	*N*	%	*N*	%	*N*	%	*N*	%
Atherosclerotic brain infarction	53	49.5	46	52.2	80	53.3	90	47.6
Cerebral embolism	12	11.2	13	14.8	32	21.3	47	24.9
Intracerebral haemorrhage	7	6.5	4	4.6	10	6.7	6	3.2
Subarachnoid haemorrhage	10	9.4	11	12.5	9	6.0	13	6.9
Other	2	1.9	6	6.8	4	2.7	1	0.5
Transient ischaemic attack only	23	21.5	8	9.1	15	10.0	32	16.9
TOTAL	107	100.0	88	100.0	150	100.0	189	100.0

computed tomography was available), and served to differentiate stroke and TIA from other neurological diseases. From all the available information it was often possible to determine if the stroke was haemorrhagic or infarctive in origin. One could often distinguish the subarachnoid haemorrhage of a ruptured berry aneurysm (or rarely, arteriovenous malformation) from intraparenchymatous haemorrhage, even in the era before the computed tomographic scan. These evaluations helped to separate cerebral infarction due to atherothrombosis of a large vessel from lacunar infarction, and from stroke secondary to embolism from the heart.

Frequency of stroke by type

Over the 32 years of follow-up in the Framingham study, stroke or TIA occurred in 534 persons; 257 in men and 277 in women (Table 1). There were few substantial differences in the relative frequency of specific types of stroke in two broad age groups, 35 to 64 years, and 65 to 94 years. Except for a higher proportion of cerebral embolism and somewhat less subarachnoid haemorrhage in the older group, the relative frequency of stroke by type was approximately the same in younger and elderly adults, and in men and women. In those aged 65 to 94 years, atherothrombotic brain infarction, which includes infarction resulting from atherothrombosis of a large vessel as well as lacunar infarction, was the most frequent manifestation and comprised about half of all strokes, regardless of age. TIAs alone, also a manifestation of occlusive, atherosclerotic, arterial disease, taken together with atherothrombotic brain infarction representing stroke resulting from atherothrombosis, account for approximately two-thirds of all symptomatic cerebrovascular disease. Intracranial haemorrhage was the mechanism in only 13 per cent of strokes: haemorrhage was intraparenchymatous as often as subarachnoid in men; in women it was twice as often subarachnoid (from a ruptured berry aneurysm) as intraparenchymatous.

Amyloid or congophilic angiopathy is responsible for a portion of intracerebral haemorrhages, particularly above the age of 70 years. As the incidence of 'hypertensive' intracerebral haemorrhage increases with age and amyloid angiopathy generally occurs after the age of 70, it is not easy to distinguish between these types. Advanced age, absence of a history of hypertension and presence of dementia of the Alzheimer type, lobar haemorrhage, and other features, such as multiple haemorrhages in different sites occurring simultaneously or serially, are thought to be characteristic of amyloid angiopathy. However, no epidemiological study of the incidence or relative frequency of amyloid haemorrhage has been made. In the absence of a 'gold standard' to facilitate diagnosis of haemorrhage due to amyloid angiopathy, epidemiological assessment will require the use of criteria for the diagnosis. Unfortunately, the use of these criteria to define the disease will necessarily preclude systematic study of these risk factors in this disease.

Incidence of stroke

The average, annual, age-specific incidence of stroke, including TIA, increased with age and doubled in each successive decade (Table 2). The incidence of myocardial infarction differed from that of its analogue in the cerebral arteries, atherothrombotic brain infarction, in several important ways (Table 3). In contrast to myocardial infarction, where the age-specific incidence in women lagged behind that in men by 20 years, the age-specific incidence of atherothrombotic brain infarction was similar in the two sexes. Furthermore, while myocardial infarction was several times more frequent in men than women, particularly at younger ages, the incidence of atherothrombotic brain infarction was only 35 per cent greater in men than women. It should be noted that these rates for the 32 years of follow-up were calculated by using a denominator that is different from that in some earlier publications; the 32-year follow-up population at risk consists of subjects free of all cardiovascular disease, not just stroke and TIA.

RISK FACTORS FOR STROKE

Atherogenic host factors

Deposition of atheromatous material in the arterial wall compromises the lumen and reduces the blood supply, which in the brain may result in cerebral infarction, the principal mechanism of stroke in two-thirds of cases. Knowledge of the importance of each risk factor and its net and joint effects may yield valuable clues to understanding the development of cerebrovascular disease and thereby lead to the prevention of stroke. As the pathological processes underlying stroke from atherothrombotic cerebrovascular disease differ from those which predispose to subarachnoid and intracerebral haemorrhage and from those which produce cerebral embo-

Table 2 Average annual incidence of stroke and transient ischaemic attacks per 1000, 32-year follow-up, Framingham study

	Men			Women		
Ages (years)	Person-years at risk*	No. of events	Rate	Person-years at risk*	No. of events	Rate
35–44	4516	—	—	5613	—	—
45–54	8032	33	2.1	10 033	24	1.2
55–64	9242	73	3.9	12 348	60	3.9
65–74	5158	90	8.7	7727	109	7.1
75–84	1482	58	19.6	2706	66	12.2
85–94	81	—	—	279	14	25.1
Age-adjusted rate†						
35–64 years			2.0			2.0
65–94 years			11.0			9.0
Male: female relative risk (35–84 years)				1.4 (1.2, 1.7)‡		

* Free of cardiovascular disease including stroke and transient ischaemic attack at examination 1.
† Age-adjusted by the direct method.
‡ Mantel–Haenszel relative risk, $p < 0.0001$ (95 per cent confidence limits).

Table 3 Average annual incidence of atherothrombotic brain infarction and myocardial infarction per 1000, 32-year follow-up, the Framingham study

	Atherothrombotic brain infarction		Myocardial infarction	
Age group (years)	Men*	Women*	Men*	Women*
45–54	1.0	0.7	5.3	1.2
55–64	2.0	1.3	9.4	3.0
65–74	4.7	3.7	13.7	5.3
75–84	10.5	5.4	20.2	8.7
Age-adjusted rate†	2.9	1.9	9.8	3.3
Male: female relative risk (95% confidence interval)	1.5 (1.2, 1.9)‡		3.0 (2.6, 3.5)‡	

* Free of cardiovascular disease including stroke and transient ischaemic attack at examination 1.
† Age-adjusted by the direct method.
‡ Mantel–Haenszel relative risk, $p < 0.001$.

lus, it seems reasonable to relate the levels of risk factors specifically to the occurrence of atherothrombotic brain infarction, the most frequent subtype of stroke.

Hypertension

Hypertension is the major precursor of stroke. For atherothrombotic brain infarction, the incidence of stroke was several times greater in definite hypertensives (blood pressure more than 160/95 mmHg) than normotensives (less than 140/90 mmHg), and even borderline hypertension carried an increased risk (Fig.2). This was true in both sexes and in all ages including those aged 75 to 84 years. There was no evidence to support the contention, with regard to stroke, that women tolerated hypertension any better than men or that the impact of hypertension waned above the age of 65.

Relative risk

The relative importance of each of the major risk factors may be deduced by comparing the size of the multivariate regression coefficients (Table 4). These coefficients were standardized for the varied units and scales, and take into account the contribution that other risk factors make to incidence of brain infarction (atherothrombotic brain infarction). The larger the standardized multivariate regression coefficient, the greater the impact on the incidence of stroke. Statistical significance indicates that the risk factor made a significant independent contribution to risk, even after age and other pertinent risk factors were taken into account.

For hypertension, the standardized multivariate regression coefficient was large, denoting a powerful and independent contribution of definite hypertension (blood pressure more than 160/95+ mmHg) to risk of atherothrombotic brain infarction in both age categories.

Systolic versus diastolic pressure

By comparing standardized coefficients (Table 4), it is apparent that those for systolic blood pressure are generally larger than for diastolic in men and women. Furthermore, by comparing the coefficients for systolic and diastolic pressure in each age group and by sex it is clear that systolic coefficients are generally larger than the diastolic, are similar in men and women, and do not decrease substantially with age. These findings contradict the notion that the effect of increased blood pressure on the incidence of stroke wanes with advancing age, or that systolic elevations are innocuous, particularly in the elderly. There is no evidence that women

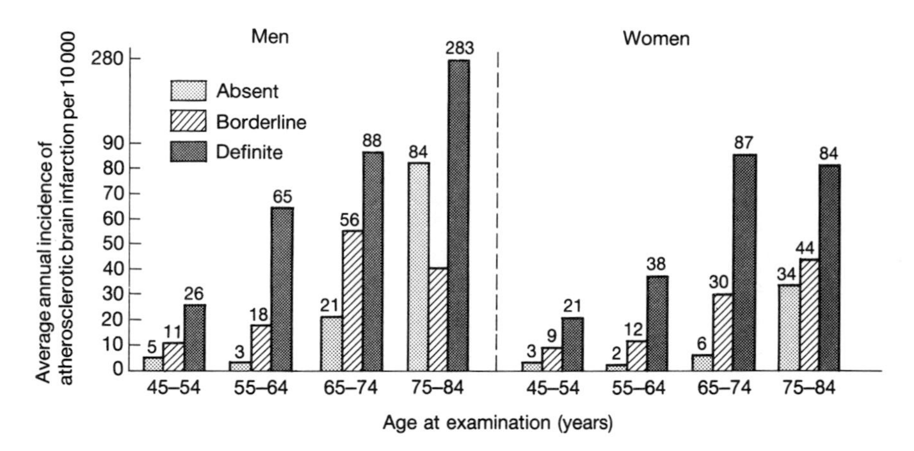

Fig. 2 Hypertension and incidence of atherosclerotic brain infarction, 32-year follow-up: the Framingham study.

Table 4 Comparison of risk factors for atherosclerotic brain infarction by means of standardized multivariate regression coefficients,* in men and women, ages 35 to 64 and 65 to 94, 32-year follow-up, the Framingham study

	Standardized multivariate regression coefficients			
	Men		Women	
Risk factor	35–64 years	65–94 years	35–64 years	65–94 years
Hypertension	0.837§	0.455§	0.840§	0.675§
Systolic BP	0.647§	0.421§	0.634§	0.564§
Diastolic BP	0.640§	0.184	0.663§	0.480§
Isolated systolic BP	−0.026	0.256‡	0.002	0.266‡
Left ventricular hypertrophy by ECG	0.128	0.204‡	0.104	0.202§
Cigarettes	0.391‡	0.076	0.003	0.179
Total serum cholesterol	0.122	−0.175	0.119	−0.231†
Metropolitan Relative Weight	0.158	−0.128	0.166	0.117
Glucose intolerance	0.030	0.149	0.112	0.288§
Haematocrit	0.350‡	0.087	0.060	−0.009

* Multivariate contains systolic BP, left ventricular hypertrophy by ECG, cigarette smoking, glucose intolerance, total serum cholesterol, relative weight and the other appropriate variable.
† $p<0.05$; ‡ $p<0.01$; § $p<0.001$.

tolerate hypertension better than men or that raised diastolic pressures are more important than systolic.

While the incidence of cerebrovascular disease was increased in hypertensives, it is apparent that the incidence of stroke generally, and atherothrombotic brain infarction specifically, was related to the level of the blood pressure.

Isolated systolic hypertension

With the disproportionate rise in systolic pressure that occurs with advancing age, isolated systolic hypertension becomes highly prevalent; above the age of 75, 18 per cent of men and 30 per cent of women have this condition. However, it is far from innocuous. Not only is systolic blood pressure as powerful a predictor of brain infarction as the diastolic, but isolated elevations of systolic pressure are also important. Even in the elderly (ages 65 to 84 years), there is at least a twofold increased risk of brain infarction among those with systolic pressures exceeding 160 mmHg, accompanied by diastolic pressures consistently below 95 mmHg (Table 4).

Because hypertension is the predominant contributor to the incidence of stroke, the importance of isolated systolic hypertension in the development of strokes was studied in the Framingham cohort, taking into account the degree of associated arterial rigidity. Arterial rigidity was estimated from pulse-wave recording using the degree of blunting of the diastolic notch as a measure of loss of elastic recoil. Although this is an imperfect measure of arterial rigidity, risk of cardiovascular disease was found to be related to the degree of blunting. Also, the prevalence of isolated systolic hypertension and pulse pressure were found to be related. All three—pulse pressure, systolic hypertension, and pulse-wave changes with blunting of the diastolic notch—increased with age.

Based on prospective data relating the future incidence of stroke to systolic pressure, diastolic pressure, and age and pulse-wave configuration, isolated systolic hypertension was found to be an independent risk factor for the development of stroke, taking associated arterial rigidity into account. Those with isolated systolic hypertension experienced stroke two to four times as often as normotensive persons. Although by itself diastolic pressure was related to the incidence of stroke, in the person with isolated systolic hypertension, the diastolic component added little to the risk assessment. In the men in this subgroup with systolic hypertension, the diastolic pressure was actually misleading. These findings strongly suggest that the increased risk of stroke associated with systolic hypertension is a direct result of the pressure and not merely a reflection of the underlying arterial rigidity.

Heart disease and impaired cardiac function

Although hypertension is the major risk factor for stroke of all types, at all levels of blood pressure those with evidence of impaired cardiac function, occult or overt, have a significantly increased risk of stroke. Cardiac abnormalities include prior coronary heart disease, cardiac failure, electrocardiographic abnormalities (particularly left ventricular hypertrophy), cardiac enlargement on radiographs, and atrial fibrillation. An increased incidence of stroke was found in the 2 months following acute cardiac infarction in the study based on a Rochester, Minnesota population between 1960 and 1979. The incidence of stroke was greater after transmural than subendocardial infarction. The increase in ischaemic stroke after myocardial infarction was presumed to result from embolization of intraventricular thrombus. In contrast, in a case-control, echocardiographic study, the frequency of stroke was greater in persons with left ventricular thrombi and the strokes were not restricted to the period immediately after infarction (MacMahon *et al.* 1989). In the Framingham study, where coronary heart disease was ascertained prospectively on biennial examination in a somewhat older population, all its manifestations, including uncomplicated angina pectoris, and clinically silent as well as overt myocardial infarctions, were significantly related to increased risk of stroke.

Left ventricular hypertrophy by ECG

Definite left ventricular hypertrophy by ECG, in part a reflection of the impact of prolonged or severe hypertension, increased in prevalence with age and blood pressure. The incidence of brain infarction was increased more than fourfold in persons with this electrocardiographic pattern (Table 4). This excess risk persisted even after the influence of other atherogenic precursors including blood pressure and age had been taken into account. Relating left ventricular mass, a more precise echocardiographic determination of an enlarged left ventricular wall, to cardiovascular disease including stroke has demonstrated a powerful relationship.

Atrial fibrillation

In association with rheumatic heart disease and mitral stenosis, atrial fibrillation is acknowledged to predispose to stroke. In recent years, chronic atrial fibrillation without valvular heart disease, previously considered to be innocuous, has been associated with a more than fivefold increase in the incidence of stroke. It is also the most prevalent cardiac arrhythmia in the elderly. In the Framingham study, the prevalence of atrial fibrillation more than doubled in successive decades and rose from 0.2 per 1000 for the ages 30 to 39 years to 39.0 per 1000 for the ages 80 to 89 years. Atrial fibrillation was particularly important in the elderly because the proportion of strokes associated with this arrhythmia increased steadily with age, rising from 6.7 per cent for 50- to 59-year-olds to 36.2 per cent for 80- to 89-year-olds (Fig. 3). Atrial fibrillation was present in one-fifth of all first strokes in the eighth decade of life and in approximately one-third of those in the ninth decade of life. The stroke was not a

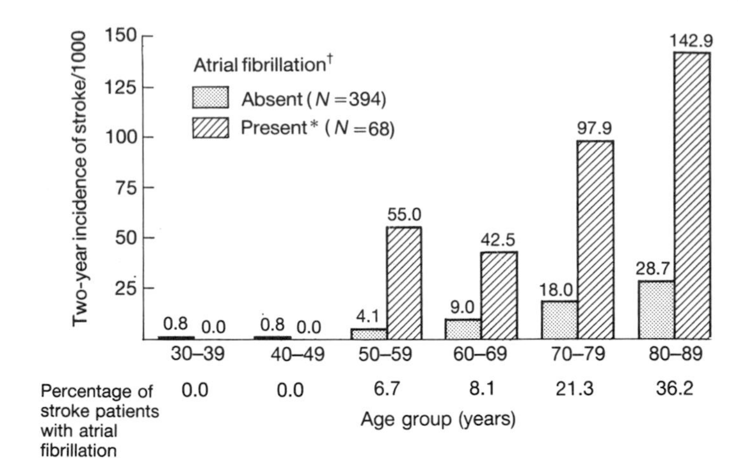

Fig. 3 Two-year stroke incidence in subjects with and without chronic non-rheumatic atrial fibrillation, men and women combined, 30-year follow-up; the Framingham study. * Significant increase with age for those with and without atrial fibrillation ($p < 0.001$). † Significant excess in those above age 50 years with atrial fibrillation ($p < 0.05$).

consequence of the often associated coronary heart disease or cardiac failure, as the relative risk of stroke rose with age for atrial fibrillation while the relative risk of stroke attributable to cardiac failure, coronary heart disease, and hypertension declined with age (Table 5). In addition to clinically evident stroke, a case-control study disclosed a greater number of silent cerebral infarcts in the patients with atrial fibrillation than in controls in sinus rhythm.

Although most persons with non-rheumatic atrial fibrillation do not sustain a stroke and the prevention of stroke with chronic warfarin anticoagulation is not innocuous, the findings of clinical trials strongly support the value of low-intensity coumadin therapy. Attempts to identify a subset of persons at particularly high risk of stroke have generally been unsuccessful. However, as 'low-intensity' anticoagulation with warfarin, with prolongation of the prothrombin time from 1.2 to 1.5 times that of the control, has significantly reduced the incidence of stroke, this is the preferred method of stroke prevention today. Prevention was accomplished with only minimal increase in significant bleeding in three separate, random-allocation, clinical trials. In two of these trials, bleeding did not occur despite prolongation of the prothrombin time to 'customary levels' of anticoagulation. However, as low-intensity treatment was equally effective, and the incidence of haemorrhagic complications quite low, this level of treatment seems indicated.

Blood lipids

The total serum cholesterol is significantly and independently related to the development of coronary heart disease in men and women, but the impact seems to diminish with advancing age, particularly in men. However, in persons over 55 years old, when the total serum cholesterol is separated into high-, low-, and very low-density lipoproteins, high-density lipoprotein-cholesterol is inversely, and low-density lipoprotein-cholesterol directly related to the development of coronary heart disease.

This association between total serum cholesterol, or of the lipoprotein-cholesterol fractions, and the occurrence of

Table 5 Estimated relative risk* of stroke for subjects with a stroke risk factor as compared to those without a stroke risk factor by 10-year age group (with 95 per cent confidence limits)

	Age group (years)			
Stroke risk factor	50–59	60–69	70–79	80–89
Atrial fibrillation	4.1‡ (1.5, 10.8)	2.6‡ (1.4, 4.9)	4.0‡ (2.6, 6.2)	4.8§ (2.5, 9.2)
Cardiac failure	4.1§ (1.8, 9.3)	2.5‡ (1.4, 4.4)	2.1‡ (1.3, 3.4)	1.5 (0.6, 3.4)
Coronary heart disease	2.9§ (1.8, 4.6)	1.7‡ (1.2, 2.5)	1.5† (1.1, 2.2)	1.1 (0.6, 2.1)
Hypertension	3.6§ (2.2, 5.7)	3.2§ (2.1, 4.7)	2.9§ (1.8, 4.7)	1.7 (0.7, 4.0)

* Each relative risk is adjusted for the other stroke risk factors.
Significant excess of strokes: † $p < 0.05$; ‡ $p < 0.01$; § $p < 0.001$.

stroke or brain infarction is less clear and consistent. Although a number of case-control studies have found increased odds for brain infarction and atherosclerosis of the carotid artery, links based on prospective data are inconclusive. Recently, a relationship between high levels of total serum cholesterol and ischaemic stroke, and low levels of serum cholesterol and intracerebral haemorrhage, was found after a 6-year follow-up of predominantly white males in the United States. These data, from the 350 977 screenees in the Multiple Risk Factor Intervention Trial, are based on a single measurement of total serum cholesterol at entry screening, and diagnosis on death certification of fatal stroke, but are none the less persuasive. In Framingham, the sole significant correlation between ischaemic stroke and blood lipids was in older women, and this was an inverse relationship: the higher the total serum cholesterol the lower the incidence of brain infarction. Six-year follow-up of the cohort after fractionation of blood lipids showed no significant positive association of low-density cholesterol with the incidence of brain infarction and no protective effect of high-density lipoprotein-cholesterol. The sole significant relationship of a lipoprotein-cholesterol fraction to stroke incidence was the negative association of low-density cholesterol to brain infarction in women; no relationship of triglyceride to risk of brain infarction was found.

Diabetes

Diabetic patients have increased susceptibility to atherosclerosis including stroke. Surveys of stroke and prospective studies confirm this increased risk of stroke. In the United States, in the period from 1976 to 1980, a medical history of stroke was 2.5 to 4 times more common in diabetics than in persons with normal glucose tolerance. In Framingham, peripheral arterial disease with intermittent claudication occurred more than four times as often in diabetics. The coronary and cerebral arteries are also affected but to a lesser extent. For atherothrombotic brain infarction, the impact of glucose intolerance, i.e., physician-diagnosed diabetes, glycosuria, or a blood sugar of more than 150 mg/100 ml, was greater in women than men and was significant as an independent contributor to incidence only in older women (Table 4). The age-adjusted, relative risk of atherothrombotic brain infarction was 2.5 in men and 3.6 in women with diabetes.

Obesity

Obese persons have higher blood pressures and glucose levels, as well as higher levels of atherogenic serum lipids. On that account alone, the increased prevalence of hypertension and impaired glucose tolerance associated with obesity could be expected to increase the incidence of stroke. However, when other relevant risk factors were taken into account, there was no residual independent increase in risk from overweight, here represented as the Metropolitan Relative Weight (Table 4). That should not be taken to imply that obesity does not exert an adverse influence on health but rather that this is probably mediated by blood pressure, impaired glucose tolerance, and other mechanisms. Recent studies have suggested that the pattern of obesity is also important, with abdominal deposition of fat being associated with disease.

Haematocrit

Data from Framingham have demonstrated the relationship of high-normal haematocrit to increased incidence of cerebral infarction. Confirmation of this relationship has come from an autopsy study of Japanese patients with stroke and from several clinical and radiological studies. In the 30-year follow-up data, an elevated haematocrit, within the normal range, and a generally not pathologically elevated red cell mass were significantly and independently associated with atherothrombotic brain infarction in men, even after the confounding effects of cigarette smoking, and hypertension had been taken into account (Table 4). The reason for the different impact in the two sexes is unclear.

Fibrinogen

Serum fibrinogen has been implicated in atherogenesis and in the formation of arterial thrombi. In addition, laboratory and clinical studies have shown that the serum fibrinogen, together with the haematocrit, have important effects on blood viscosity and cerebral blood flow. However, data from prospective epidemiological studies assessing the impact of fibrinogen on the incidence of stroke incidence are quite sparse. A recent prospective study showed that fibrinogen in concert with elevated systolic blood pressure was a potent risk factor for stroke in a sample of 54-year-old Swedish men followed for 13 years.

At the tenth biennial examination, the serum fibrinogen was measured in 1470 persons in the Framingham study, approximately half of the surviving cohort. Risk of stroke and total cardiovascular disease were increased in proportion to the level of fibrinogen; however, fibrinogen was also positively associated with most of the major risk factors for stroke, including age, hypertensive status, haematocrit, obesity, and diabetes. However, even when these risk factors were taken into account, fibrinogen exerted an independent impact on total cardiovascular disease ($p<0.01$), with a similar trend for stroke.

The increase in blood viscosity resulting from high-normal levels of haematocrit and fibrinogen may precipitate stroke in the territory of a critically stenotic, penetrating or major cerebral artery. Reduction of high-normal haematocrit by venesection has been shown to decrease blood viscosity and increase cerebral blood flow.

Race

The Japanese are known to have low rates of coronary heart disease and a high prevalence of stroke. A recent survey of six mainland Chinese cities disclosed a high incidence and prevalence of stroke in Chinese, similar to those of native Japanese in Japan. The disparity between the death rates for stroke and coronary heart disease, and presumably a similar disparity in the incidence rates, is usually attributed to the high prevalence of hypertension and the low levels of blood lipids in these racial groups. Cerebrovascular disease was the most frequently certified cause of death in Japan during the decades following the Second World War, and the most frequent mechanism of stroke was thought to be intraparenchymatous haemorrhage. In the 1980s, cancer has become the leading cause of death, with stroke falling to second place and coronary heart disease to third. In Japanese men in Hawaii and San Francisco, deaths attributed to stroke are also falling relative to the rates for those from coronary heart disease (and cancer).

In 1985, heart disease moved into second place as a cause of death in Japan, following cancer (156.4 per 100000) at 114.8 per 100000, and stroke fell to third with a death rate of 110.6 per 100000. Autopsy studies have shown that infarction, not haemorrhage, is the most frequent mechanism of stroke. There is also a difference in the site of the arterial lesion, with a predominance of intracranial disease in native Japanese which contrasts with the pattern in white Americans, where the extracranial arteries are the focus of most of the atherosclerotic occlusion. The Japanese migration to Hawaii and the western United States has created a natural experiment where a change in diet, consisting of an increase in protein and fats, has been associated with an increase in serum cholesterol and in the incidence of coronary heart disease. Recent clinical and angiographic studies in another racial group, black Americans, have suggested a predominance of intracranial and small vessel cerebral vascular disease, similar to the pattern found in the Japanese.

Family history of stroke

Although family history of stroke is perceived to be an important marker of increased risk, confirmation by epidemiological study has been lacking. Recently, in a cohort of Swedish men born in 1913, maternal history of death from stroke was found to be significantly related to the incidence of stroke. Other significant risk factors included hypertension, abdominal pattern of obesity, and fibrinogen level; however, maternal history of fatal stroke was independently related to stroke even after these variables had been taken into account.

Environmental factors

Cigarette smoking

Several large-scale prospective studies, including Framingham, have linked cigarette smoking to stroke. Even after other cardiovascular risk factors systolic blood pressure level, total serum cholesterol, left ventricular hypertrophy by ECG, obesity, and glucose intolerance, were taken into account, the incidence of stroke was 40 per cent higher in men and 60 per cent higher in women smokers. There was a similar significant relationship between cigarette smoking and the specific stroke subtype of atherothrombotic brain infarction. There was no evidence that the impact of smoking decreased with advancing age or of a lesser benefit of stopping smoking in the elderly.

In a cohort of female nurses, risk of stroke in light smokers (those smoking less than 15 cigarettes per day) was more than twice that of non-smokers. Heavy smokers (consuming more than 25 cigarettes per day) had a nearly fourfold increased risk of stroke when compared with non-smokers. This significantly increased risk of stroke persisted after other risk factors, including alcohol consumption, and use of oral contraceptives and postmenopausal oestrogen, had been taken into account. The way in which cigarette smoking predisposes to stroke may be discerned by examination of risk of stroke after stopping smoking. The incidence of stroke in smokers fell substantially within 2 years of quitting cigarettes and within 5 years risk of stroke fell to that of a non-smoker. This suggests that cigarette smoking acts by precipitating stroke in susceptible people rather than by promoting atherosclerosis.

Of considerable interest, the relative risk of subarachnoid haemorrhage also showed a dose–response relationship from 4-fold in light smokers to 9.8-fold in heavy smokers. The association between cigarette smoking and subarachnoid haemorrhage from aneurysm was found in men as well as women, both in Framingham and in New Zealand in case-control analyses. A similar relationship between cigarette smoking and stroke has been found in Hawaiian Japanese men after 10 years of follow-up in the Honolulu Heart Study, where cigarette smoking made a significant, independent contribution to risk of cerebral infarction and intracranial haemorrhage.

Alcohol

The impact of alcohol on the risk of stroke, like its influence on coronary heart disease, relates to the amount consumed. Non-drinkers and heavy users of alcohol have a higher incidence of stroke and coronary disease, while light or moderate consumers of alcohol have a lower risk of these diseases. This association between the level of alcohol consumption and the incidence of coronary heart disease may result from the increased high-density lipoprotein-cholesterol in moderate users of alcohol; heavy imbibers have more hypertension and higher levels of triglycerides. Total abstinence and heavy consumption of alcohol seem to have an adverse affect on the incidence of ischaemic stroke, while heavy drinking alone is related to haemorrhage stroke, both intraparenchymatous and subarachnoid. Some of this effect is mediated through higher blood pressures in heavy imbibers.

Data from the Framingham study also suggest an increased incidence of brain infarction and stroke with increased levels of alcohol use, but only in men. Other mechanisms by which the consumption of alcohol increases the risk of stroke include effects of the frequently associated cigarette smoking leading to haemoconcentration, increased haematocrit, and viscosity. Rebound thrombocytosis has been observed during abstinence. Acute alcohol intoxication provokes disturbances of cardiac rhythm, notably atrial fibrillation, and has been suspected as a precipitant of acute infarction in young people, and of subarachnoid haemorrhage.

Physical activity

Leisure- and work-associated, vigorous physical activity has also been linked to a lower incidence of coronary heart disease. More physically active longshoremen had lower rates of myocardial infarction but no reduction in the incidence of stroke. In a study of 17000 former Harvard students, those who were more physically active had about half the risk of fatal coronary heart disease and one-third the mortality rate of their least active fellow alumni. There have been no demonstrations of a reduction in the occurrence of stroke from vigorous, habitual, leisure-time physical activity. However, vigorous exercise exerts a beneficial influence on risk factors for atherosclerotic disease by promoting weight reduction, thereby lowering blood pressure, reducing the pulse rate, raising the high-density lipoprotein- and lowering the low-density lipoprotein-cholesterol, improving glucose tolerance, and by promoting a lifestyle conducive to favourable changes in detrimental habits such as cigarette smoking.

Signs of compromised cerebral circulation

The availability of surgical means to restore flow in a narrowed carotid artery stimulated interest in the early detection, by non-invasive means, of a compromised cerebral circulation before a stroke could occur. Two clinical indicators of a possibly compromised cerebral circulation are asymptomatic carotid bruit and TIA.

Asymptomatic carotid bruits

Although asymptomatic carotid bruit is associated with twice the incidence of stroke, the infarct may not be directly related to the underlying carotid stenosis. The carotid bruit merely serves as an indicator of generalized atherosclerosis, including the intracerebral vessels. Since the ninth biennial examination in 1966 in the Framingham study, carotid bruits have been routinely sought by auscultation. Over 8 years, bruits appeared in 171 subjects, 66 in men and 105 in women, all of whom had been asymptomatic and free of bruit. The prevalence of bruits increased with age, and was equal in men and women, rising from 3.5 per cent at 45 to 54 years, to 7.0 per cent at 67 to 79 years; the prevalence was higher in those with hypertension, coronary heart disease, and diabetes. TIAs appeared in eight and strokes in 20 of the subjects, a rate for stroke of more than twice that expected. More often than not, however, the cerebral infarction occurred in a different vascular territory from that of the carotid bruit, and often in the posterior circulation. Furthermore, not all the strokes were cerebral infarctions due to carotid stenosis. Ruptured aneurysm, atherosclerosis of the posterior circulation, emboli from the heart, and lacunar infarction were the mechanism of stroke in nearly half the cases. Interestingly, the incidence of myocardial infarction was also increased twofold in those with asymptomatic carotid bruit. General fatality was also increased: 1.7-fold in men and 1.9-fold in women, with 79 per cent of the deaths due to cardiovascular disease, including stroke. Carotid bruit is clearly an indicator of increased risk of stroke. However, in asymptomatic persons the bruit is chiefly a non-focal signal of advanced atherosclerotic disease and not necessarily an indicator of a local stenosis that will precede ipsilateral cerebral infarction in the territory of the affected carotid.

Transient ischaemic attacks

TIAs are reversible, focal neurological deficits, lasting from minutes to a day. Retrospective clinical data from patients who have sustained a stroke suggest that TIAs frequently precede the development of brain infarction. However, best estimates indicate that only 10 per cent of all strokes are actually preceded by TIAs. It is also estimated that atherothrombotic disease of the surgically accessible carotid artery accounts for less than 15 per cent of strokes.

In Framingham, 12 per cent of cerebral infarcts were preceded by a TIA. The average annual incidence of TIAs was similar in men and women and increased with age. Subjects developed neurological symptoms nearly four times as often as bona fide TIAs. Stroke developed in about 40 per cent of those with TIAs; half occurred within 3 months of the onset of the attacks and two-thirds within 6 months. In most cases, fewer than four attacks occurred before the stroke.

SURVIVAL AND RECURRENCE AFTER STROKE

Survival clearly depends on the type of stroke. In the Framingham cohort, 30-day case fatality rates ranged from 15 per cent for brain infarction and cerebral embolism, to 46 per cent in subarachnoid and 83 per cent in intracerebral haemorrhage; overall the rate was approximately 20 per cent. As expected, the incidence of death was clearly related to age; for brain infarction, the rates ranged from 8 per cent below 60 years of age, to 23 per cent above 70 years. Long-term survival was related to pre-existing cardiac disease and, to a lesser extent, hypertension. In the absence of antecedent cardiac failure, coronary heart disease, or hypertension, long-term survival after stroke generally, or brain infarction specifically, was close to that of the stroke-free cohort. The extent of survival was less in men. Stroke of all types tended to recur: the 5-year cumulative recurrence rates of brain infarction were 40 per cent for men and 20 per cent for women. Recurrent strokes were frequently of the same type as the first, and were more common when there had also been cardiac disease and hypertension before.

SECULAR TRENDS IN INCIDENCE AND MORTALITY

In the past 20 years a dramatic decline in death from stroke has occurred in most industrialized nations. In the United States the decline of one-half in deaths from stroke, a 5 per cent annual decrement, is an acceleration of the 1 per cent annual decline from 1915 to 1970. This accelerated decline supports the contention that modifiable environmental factors play a substantial role in the occurrence of stroke. The reduction in death from stroke has occurred in both sexes, in black and white subjects, in most Western and developed nations, and in all regions of the United States. Furthermore, death rates attributable to stroke have declined in the face of falling total death rates. In fact, the diminution in death from stroke has been a major contributor to the decline in total cardiovascular diseases, further substantiating that the decline in stroke is real and not an artefact of death certification or coding practices.

The accelerated decline in deaths from stroke has been attributed to realization of the importance of hypertension as the premier risk factor, irrespective of whether the mechanism was haemorrhagic or infarctive, and the demonstration that treatment will reduce the likelihood of stroke and of death from it. Controlled clinical trials of the treatment of hypertension have clearly demonstrated the efficacy of lowering blood pressure in reducing the incidence of stroke, by 40 per cent on average (Fig.4).

The decline in death rates from stroke may be a result of a decreasing incidence as well as from improved survival of patients with acute stroke. Evidence in support of the role of a declining incidence of stroke came from the community-based study of Rochester, Minnesota, where the incidence of stroke declined over time, coincident with improved control of hypertension. There has been evidence of declining rates of case fatality from acute stroke in the United States from 1970 to the present day. However, the availability of computerized tomography of the brain has improved diagnostic sensitivity by as much as a quarter and facilitated identification of many more of the milder strokes.

Thus, while the rates of case fatality from stroke may have fallen as a result of improved patient care, it is possible that in recent years, cerebral infarcts have been smaller and have produced milder deficits than before. It is also likely that milder strokes, which previously went undetected, are now routinely diagnosed by imaging. The widespread availability of a diagnostic test for stroke, particularly for stroke in the elderly, may be particularly important here, making detection of secular trends even more difficult.

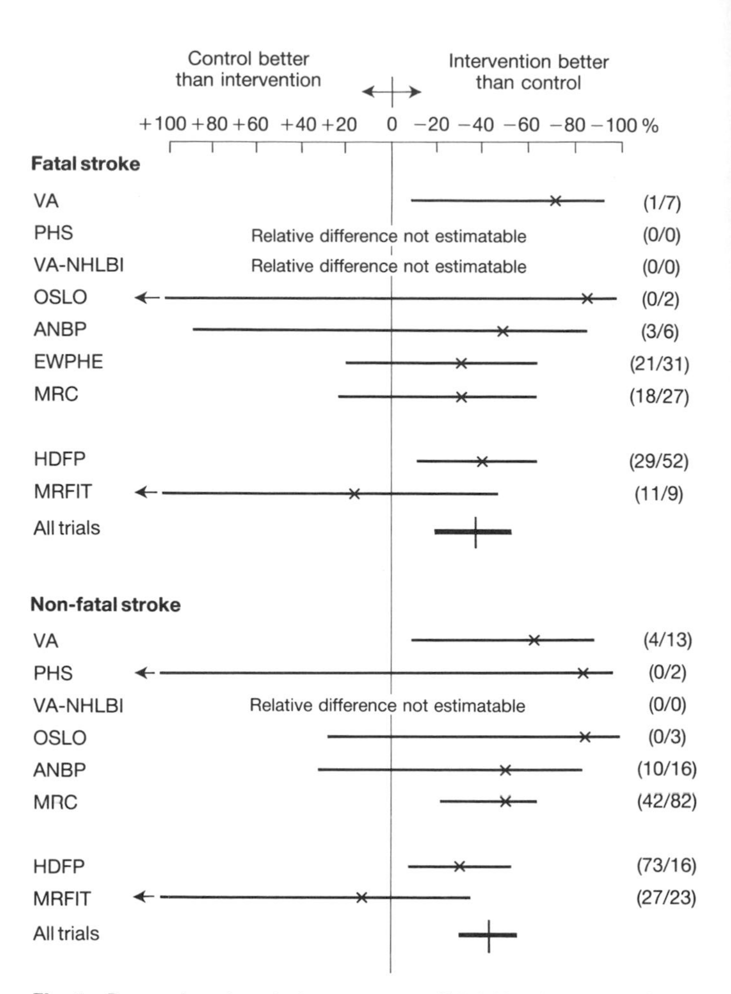

Fig. 4 Prevention of stroke by treatment of high blood pressure. Bar graph showing estimates (×) with approximate 95 per cent confidence intervals (−) of the relative difference in fatal and non-fatal stroke between intervention and control groups. The numbers in parentheses are the number of events (intervention/control). VA, Veterans Administration; PHS, United States Public Health Service Hospitals Cooperative Study; VA-NHLBI, VA National Heart, Lung, and Blood Institute Feasibility Study; OSLO, Oslo Study; ANBP, Australian National Blood Pressure Study; EE, European Working Party on Hypertension in the Elderly; MRC, (British) Medical Research Council; HDFP, Hypertension Detection and Follow-up Program; and MRFIT, Multiple Risk Factor Intervention Trial. (Source: *Hypertension* (1989) 13 (suppl. 1), 1–36, 1–44.)

IMPLICATIONS FOR PREVENTION

Starting with clinical observations of hypertensive patients, controlled clinical trials have consistently demonstrated the prevention of stroke by reduction of blood pressure. The initial Veterans Administration Cooperative Study clearly demonstrated the efficacy of such treatment in preventing stroke and in prolonging survival among treated, severe hypertensives whose pretreatment diastolic blood pressures were above 115 mmHg. Here the degree of benefit correlated with the levels of blood pressure before randomization and with the extent of control of blood pressure. This was followed by a trial of treatment of moderately severe hypertension (pretreatment diastolic blood pressure between 104 and 115 mmHg), which demonstrated a reduction in 75 per cent of treated patients. Since that time, control of moderately severe and severe hypertension has been consistently shown to reduce the occurrence of stroke (Fig. 4).

As 70 per cent of hypertensives have mild hypertension, with a diastolic blood pressure between 90 and 104 mmHg, the findings of a community-based trial, the Hypertension Detection and Follow-up Program, involving nearly 11 000 persons with high blood pressure including mild hypertension, were noteworthy. Overall, the 5-year death rate was significantly lower in the group given systematic and controlled antihypertensive treatment than in those given standard care. In the group with mild hypertension (entry diastolic blood pressure, 90–104 mmHg), the death rate was significantly reduced, by 20 per cent in those systematically treated. This reduction came mainly from a fall in deaths from cardiovascular disease, including a 45 per cent decline in stroke and a 46 per cent decrease in fatal acute myocardial infarction. Control of hypertension was followed by a commensurate reduction in deaths in the older age groups, in white men, and in black men and women.

The Systolic Hypertension in the Elderly Program, in a pilot study, demonstrated that systolic pressure could readily be reduced, usually with a thiazide diuretic alone; the reduction was well tolerated, and treatment cut hypertensive end-points including stroke by half. The sample was too small to show a statistically significant reduction in stroke and hypertensive end-points; a full-scale trial is under way.

The accelerating decline in the incidence of, and deaths from, cerebrovascular disease in recent years is a clear indication that stroke is not an inevitable consequence of ageing or genetic constitution. The decrease in deaths from stroke has occurred at all ages, including the eighth and ninth decades of life, providing evidence that control of hypertension and changes in environmental factors can effect changes in the occurrence of stroke. (See also Chapter 11.6).

Bibliography

Cupples, L.A. and D'Agostino, R.B. (1987). Some risk factors related to the annual incidence of cardiovascular disease and death using pooled repeated biennial measurements: Framingham Heart Study, 30-year follow-up. In *The Framingham study: an epidemiological investigation of cardiovascular disease*, Section 34, NIH publication No. 87–2703 (ed. W.B. Kannel, P.A. Wolf, and R.J. Garrison). National Heart, Lung and Blood Institute, Bethesda, MD.

Hypertension detection and follow-up program cooperative group (1979). Five-year findings of the hypertension detection and follow-up programs. I. Reduction in mortality of persons with high blood pressure, including mild hypertension. *Journal of the American Medical Association*, **242**, 2562–71.

Kannel, W.B. *et al.* (1970). Epidemiologic assessment of the role of blood pressure in stroke: The Framingham Study. *Journal of the American Medical Association*, **214**, 301–10.

Kannel, W.B. *et al.* (1981). Systolic blood pressure, arterial rigidity and risk of stroke: The Framingham Study. *Journal of the American Medical Association*, **245**, 1442–5.

Kurtzke, J.F. (1985). Epidemiology of cerebrovascular disease. In *Cerebrovascular survey report for the National Institute of Neurological and Communicative Disorders and Stroke* (ed. F.H. McDowell and L.R. Caplan), pp. 1–34.

MacMahon, S., Cutler, J.A. and Stamler, J. (1989). Antihypertensive drug treatment: potential, expected, and observed effects on stroke and on coronary heart disease. *Hypertension*, **13** (suppl. 1), I45–50.

Perry, H.M. *et al.* (1989). Morbidity and mortality in the systolic hypertension in the elderly program (SHEP) pilot study. *Stroke*, **20**, 4–13.

Reed, D.M. (1990). The paradox of high risk of stroke in populations with low risk of coronary heart disease. *American Journal of Epidemiology*, **131**, 579–88.

Whisnant, J.P. (1984). The decline of stroke. *Stroke*, **15**, 160–8.

Wolf, P.A. *et al.* (1987). In *The heart and stroke: exploring mutual cardiovascular and cerebrovascular issues* (ed. M. Furlan), pp. 331–55. Springer-Verlag, Berlin.

11.3 Diagnosis

SHAH EBRAHIM

THE STROKE SYNDROME

Stroke is a clinically defined syndrome of rapidly developing symptoms and signs of focal or global cerebral impairment lasting longer than, or leading to death within, 24 h and having a presumed vascular pathogenesis. The underlying causes of most focal and global neurological impairments of sudden onset are cerebral infarction, embolism, and haemorrhage, but differentiation of these vascular causes from other causes of neurological impairment is the main diagnostic task.

ACCURACY OF DIAGNOSIS

Strokes are so common that it is all too easy to fall into the trap of assuming that all global or focal neurological impairments are due to stroke, thus making false-positive diagnoses of stroke when in reality other diseases have caused the problem. A further and equally important error is to miss the symptoms and signs of a stroke and fail to make the diagnosis (i.e. a false-negative diagnosis) when in reality a stroke has caused the patient's illness. Table 1 summarizes some of the more common reasons for making both of these types of diagnostic error.

The assessment of patients who have had a previous stroke is particularly difficult as any intercurrent illness may make a residual hemiparesis worse. Consequently, the diagnosis of a new stroke or worsening of an old stroke will not be clear until several days have elapsed and the effects of treatment for the intercurrent illness have had a chance to have some impact. Grand mal fits, usually secondary to a previous stroke, are one of the most common causes of diagnostic confusion, and emphasize the need to obtain an accurate history of events leading up to the stroke. Monitoring progress over the first few days will also help in sorting out a residual postepileptic weakness (Todd's paresis) from a new stroke. A patient who is unconscious at the onset of the stroke but regains consciousness within a few hours and has a hemiparesis that recovers over the next few days is much more likely to have had a grand mal fit than a stroke.

An earlier stroke may be noted in the history and obvious signs of hemiplegia (or other focal impairments) ignored because a more acute illness (e.g. heart failure, gastrointestinal bleeding) appears to have caused the patient's presenting problem. In reality, the patient may turn out to have suffered a further stroke, thus 'previous stroke' appears in Table 1 as a cause of both false-positive and false-negative diagnostic errors.

Table 1 Causes of error in the diagnosis of stroke

False-negative diagnoses	False-positive diagnoses
Non-specific presentations	Mass lesions
Immobility	Primary tumours
Falls	Metastases
Incontinence of urine	Subdural haematoma
Social breakdown	Abscess
Impaired consciousness	Infection
Confusional states	Subacute bacterial endocarditis
Fractures	Meningitis
Neck of femur	Neurosyphilis
Humerus	Metabolic disturbances
After general anaesthetic	Hypoglycaemia
Previous stroke	Hyperosmolar states
Acute intoxication	Hypo/hypernatraemia
Alcohol	Previous stroke
Neuroleptics	Grand mal fit
	Intercurrent illness
	Acute hypotension
	Cranial arteritis
	Severe anaemia

A gradual onset of the stroke, although more common in patients with a mass lesion, does not have a very high predictive value as many patients with cerebral infarcts will also give such a history. A history or signs of head injury, an impaired or fluctuating level of consciousness disproportionate to the degree of hemiparesis, and headache should raise the suspicion of a subdural haematoma. General examination for signs of occult malignancy should not be omitted as cerebral metastases may present with a stroke syndrome. Postural hypotension of acute onset (e.g. due to gastrointestinal bleeding, silent myocardial infarction, dehydration, or drugs) may also mimic stroke.

Avoiding false-negative diagnoses also depends on obtaining a good history and a thorough examination of the patient. Careful examination of patients with fractures may reveal signs of an unsuspected hemiparesis, or more subtle signs such as nominal aphasia, agnosia, or apraxia. Some of the clinical signs that help in the diagnosis of stroke when there is a risk of making a false-negative diagnosis are shown in Table 2. Often it is necessary to re-examine the patient several times over a period of days, or even weeks, to confirm signs and monitor improvements in performance which increase the chances that an initial non-specific presentation was caused by a stroke.

The clinical distinction of stroke from non-vascular causes of a stroke syndrome is not completely accurate when compared with differential diagnosis at autopsy and/or by computed tomography (**CT**) or magnetic resonance imaging (**MRI**). False-negative clinical diagnoses (i.e. missing the diagnosis of stroke when it really is present) arise because the usual presentation of stroke may be masked by other conditions, leading to a low sensitivity. False-positive clinical diagnoses (i.e. diagnosing stroke when in fact the patient has other non-vascular disease) arise because other diseases masquerade as stroke, leading to a low specificity. Estimates of the diagnostic sensitivity of clinicians range from 67 to 100 per cent, and of specificity from 22 to 91 per cent. As cerebrovascular disease is so common, both types of error are easy to make if alternative causes for non-specific presentation of disease and the stroke syndrome are not considered. More experienced clinicians have been shown to be more accurate, making fewer false-positive diagnoses.

Table 2 Useful clinical signs of stroke

Perceptual impairments
Sensory inattention
Clock-face drawing
Visual inattention test
Tactile agnosia
Visual field loss
Motor impairment
Gait pattern
Drift of outstretched arms
Poor sitting balance
Flat nasolabial fold
False teeth difficult to keep in place
Cognitive impairment
Disorientation (time, place, person)
Memory loss
Reading, writing, arithmetic ability impaired
Speech impairment
Nominal dysphasia
Dysarthria
Tongue apraxia

Type of lesion: haemorrhage or infarction

The clinical differentiation of haemorrhage and infarction is not very accurate. Signs usually thought to be more likely in cerebral haemorrhage are initial loss of consciousness, headache, vomiting, neck stiffness, and bloodstained cerebrospinal fluid. Signs of meningeal irritation only occur if blood leaks into the subarachnoid space, so small haemorrhages may be clinically indistinguishable from thromboses. Clinical scoring methods improve the ability to diagnose cerebral haemorrhage but rely on signs of meningism together with absence of features of thromboembolic disease (history of hypertension, previous stroke, presence of heart disease). Such methods are more accurate than clinical impressions, but not accurate enough for diagnosis of individual patients, especially if anticoagulant treatment is contemplated.

Unfortunately, the validity of clinical scores has to be compared with that of CT scanning, which is not a very reliable method of detecting small haemorrhages or infarcts. Clinical scores may thus appear to be more accurate than they actually are, especially if normal CT scans are interpreted as indicating that a thromboembolic stroke has occurred.

LOCATION OF THE LESION

The site of the lesion will determine the symptoms and signs produced. Although hemisphere lesions are most common, and clinicians are familiar with speech and perceptual impairments that may occur in conjunction with hemiplegia, many patients with hemisphere damage may also have impairment of lower cranial nerve functions, in particular swallowing. Dysarthria, dysphagia, and gaze paresis in the first week after a stroke should not be taken as unequivocal evidence of a brain-stem lesion, and are probably more common with large hemisphere lesions. A weaker leg than arm, with associated urinary incontinence, is more common with thrombosis of the anterior cerebral artery. Isolated lesions of Wernicke's speech area may cause a fluent dysphasia, which can be mistaken for a confusional state or even psychotic illness.

Deep, small lesions in the region of the internal capsule can cause pure motor stroke without any cortical impairments, one of the lacunar syndromes. Lesions in the basal ganglia may be associated with 'thalamic' pain, hemiballismus, and hemichorea but are rarely seen. Lacunar syndromes are probably underdiagnosed in older patients, partly because of diagnostic 'laziness' and a tendency to include them with hemisphere lesions. It is not clear whether making the diagnosis of a lacunar syndrome is of any value in terms of prognosis as studies of its natural history have not yet been done.

Brain-stem lesions usually result in a severe stroke, because even a small lesion can cause a lot of damage. In addition to signs of hemiplegia (or quadriplegia), it is common to find ataxia (often leading to marked impairment of sitting balance), tinnitus, vertigo, persistent gaze paresis, dysarthria, and dysphagia. Often the patient is unconscious, with abnormal cyclical respiration. Even in younger people the diagnosis of cerebellar haematoma that presents in this manner, and may be surgically treated, is seldom made in life.

INVESTIGATIONS

To reduce diagnostic error, more investigations are often needed for older than younger patients, in part because of the more non-specific presentation of disease and because histories are often incomplete at the time of the acute event. In addition, many treatable, non-vascular causes of the stroke syndrome are more common in older patients (see Table 1). It is advisable to check the blood sugar, electrolytes, blood count, and ESR, syphilis serology, chest radiograph, and electrocardiogram of an elderly person presenting with a stroke syndrome. The use of other investigations will depend on the clinical findings and the outcome of these initial assessments. The utility of any investigation will depend on two factors: its accuracy (sensitivity and specificity) and the frequency of the disease to be detected by the investigation. The accuracy of an investigation is best measured by its likelihood ratio.

Likelihood ratios

A likelihood ratio can range from zero to infinity and is an indication of how useful a clinical sign or investigation is in increasing the chances the patient has the disease of interest. Taking a good history is usually regarded as the single most important part of diagnosis and this is reflected by the fact that likelihood ratios for information from the clinical history are high and, if symptoms or signs are present, lead to a large increase in the chances that a patient has the disease.

The likelihood ratio is a summary measure of accuracy, and is calculated as:

$$\text{Likelihood ratio} = \frac{\text{probability of a positive result in diseased people}}{\text{probability of a positive result in non-diseased people}}$$

which is equivalent to: sensitivity/(1 − specificity)

Its virtues are that it takes into account the ability of an investigation to pick up the diagnosis of interest, and can be simply multiplied by the pre-test (before use of the investigation) odds that the patient has the diagnosis of interest, to give the post-test (after use of the investigation) odds that the patient has the diagnosis. The post-test odds are the same as the predictive value of the investigation (i.e. the probability that a positive test result will be confirmed by the appropriate 'gold standard'). Clinicians usually use the probability of disease to measure frequency, but odds are closely related to probability. The relationship is given by:

$$P/(1 - P), \text{ where } P \text{ is the disease probability}$$

For example, a probability of 0.25 (or 25 per cent) is equivalent to odds of 1:3 or 0.33.

The likelihood ratios of clinical examinations range from around 4 to 25. Very useful investigations such as CT scanning for brain tumours or serial electrocardiograms for myocardial infarction have likelihood ratios that range from 30 to 100, whereas an exercise electrocardiogram for coronary artery disease has a likelihood ratio of 3.5 (using <1 mm depression of the ST segment as the criterion for abnormality).

The accuracy of clinical diagnosis of mass lesions (i.e. tumours, subdural haematomas, abscesses) from vascular causes of the stroke syndrome can be examined in this way. The pre-test (or examination) probability of a mass lesion is about 5 per cent, equivalent to odds of 0.053, in a typical series of patients presenting with signs suggestive of a stroke. The best reported sensitivity of clinical diagnosis is 100 per cent, with a specificity of 76 per cent, giving a likelihood ratio of (100/100 − 76), which is 4.2. The ability of clinical skill to increase the odds of the presence of a mass lesion is the product of the pre-examination odds and the likelihood ratio, 0.053×4.2, which is 0.22. These are the post-examination odds of a mass lesion, and can be converted back to a more easily understood probability:

$$\text{Posterior odds} / (1 + \text{posterior odds})$$

which in this case is 0.18, or 18 per cent. The clinician's skill improves the chances of a mass lesion being present, but is by no means infallible. It is important to realize that clinical symptoms and signs have a relatively low likelihood ratio for distinguishing mass lesions from vascular causes of the stroke syndrome, and as unsuspected mass lesions are relatively rare, even the most astute physician will, on average, be wrong four times out of five in his or her clinical hunch that the patient has a mass lesion. Consequently, clinicians who find that they are never wrong in their positive diagnoses of mass lesions are probably missing more cases than they are detecting.

CT scanning and MRI

Observing the patient's progress over time and a knowledge of the natural history of stroke are of value in deciding which patients should be scanned, because if patients with typical patterns of onset and recovery are excluded from further consideration, the remainder will be more likely to include those with mass lesions and thus the pre-test odds of a mass lesion will increase. The accuracy of diagnosis by CT scanning when compared with that of surgery or autopsy is very high for brain tumours, with reported likelihood ratios of 32 to 98. The postexamination odds after the clinician has seen the patient and suspects a mass lesion are 0.22 (see above). These can be used as the pre-test odds for the subsequent use of CT scanning, and multiplying them by the likelihood ratio of CT scanning for brain tumours gives a new post-test odds, after CT scanning and obtaining a scan demonstrating a tumour, of 7.1 to 21.8, equivalent to probabilities of 88 to 96 per cent that a mass lesion really is present. In practice the appearances of a tumour on a scan are usually taken as accurate enough for management but it is worth remembering that false-positive CT scans do occur, though rarely (hence the likelihood ratio, although high, is not equal to infinity), and if the natural history appears to be inconsistent with a diagnosis of cerebral tumour, a repeat scan should be done.

CT scanning is not nearly so good for detecting vascular strokes when compared with clinical diagnoses. Patients commonly present with clinical signs of a definite stroke that improves as expected but have normal CT scans. These scans are best considered as false negatives, although they are often thought to indicate an embolic or small thrombotic stroke. MRI has a higher sensitivity for very early infarcts. The CT scans of patients with impairment of consciousness or variable neurological signs can easily be overinterpreted, leading to false-positive diagnoses of lacunar stroke or small thromboses. It is worth reviewing the CT scans of patients whose clinical signs are inconsistent with the lesions seen with a neuroradiologist who has not seen the previous report. Moreover, the timing of the scan can be critical, as small haemorrhages may resolve quickly, leading to a normal scan or the appearances of an infarct; infarcts may not be apparent for a few days after the onset of symptoms, and if small may not show up at all or be easily missed. Patients who have an atypical clinical course, or features suggestive of a subdural haematoma, merit a scan.

CT and MRI scans do not contribute to the management of most patients with vascular strokes, although with the increasing evidence that antiplatelet agents are of benefit in secondary prevention of stroke, and the possibility that subcutaneous heparin may prevent deep vein thrombosis and pulmonary embolus after stroke, a case could be made for CT scanning early after onset to avoid giving such drugs to patients who have suffered cerebral haemorrhages. A policy of early CT scanning would be difficult and expensive to carry out, particularly in the United Kingdom. The risks of making a cerebral haemorrhage worse, provided that antiplatelet treatment is started several weeks after the acute event, are probably very small. The potential risk associated with subcutaneous heparin given during the first days of a stroke of unknown pathogenesis is not known.

Investigation of the carotid arteries

There is serious doubt about the benefits of carotid endarterectomy both in asymptomatic people and in patients who have had minor strokes or transient ischaemic attacks, and there is mounting evidence that the operation is widely, but often inappropriately, used in the United States. However, proponents of the operation maintain that, provided the perioperative risks of stroke and myocardial infarction are low (below 3–4 per cent), then the operation is associated with a reduction in subsequent risk of stroke. The majority of patients operated on are now over 65 years old in the United States, and operations on patients in their high 70s are now commonplace. Consequently, it is likely that in the future even more elderly patients will be investigated for their suitability for carotid endarterectomy.

The first prerequisite is that the patient is likely to have a lesion that might have been caused by embolus from the carotid arteries. An American audit of the use of carotid endarterectomy among patients over 65 years found that this was often not the case. The other major problems were that surgeons operated on patients at high operative risk, those with minimal carotid lesions, and those with a carotid lesion contralateral to the affected hemisphere. At present it is advisable for any patient offered carotid endarterectomy to obtain a second opinion.

Non-invasive assessments

The between-observer agreement on the presence of a significant cervical bruit is low, and so too is the accuracy of a cervical bruit in detecting significant stenosis, with both false positives and negatives. Consequently, in fit patients presenting with transient ischaemic attacks or with minor strokes of presumed embolic pathogenesis, and if safe carotid endarterectomy can be assured, investigation is appropriate.

Doppler ultrasonography has a likelihood ratio of around 9 for the presence of lesion, and of 0.11 for the absence of a lesion when compared with carotid angiography. Non-invasive investigations may be less accurate than suggested by the likelihood ratio because only those patients with apparent lesions tend to be investigated with the 'gold standard' of angiography and the investigations depend very much on skill. Given the uncertainty about the risks and benefits of carotid endarterectomy, and the risks of carotid angiography, it is wise to screen potential patients for carotid angiography with ultrasonography, provided that good quality studies are possible, as this will reduce the numbers of patients needing invasive investigation.

Carotid angiography

With digital subtraction techniques, carotid angiography is probably becoming more accurate but reported estimates (compared with operative findings) of sensitivity and specificity in detection of ulceration are only 73 and 63 per cent, a likelihood ratio of 1.97. Detection of stenosis is probably more accurate, although patients considered to have no ulceration or less than 50 per cent stenosis on angiography are very unlikely to be operated on, and consequently the correlation with operative findings in these patients cannot be made. A combination of a positive Doppler ultrasonogram and a positive carotid angiogram (i.e. 50 per cent or greater stenosis and/or ulceration) will increase the probability that an operable lesion is present to at least 90 per cent, thus reducing the chances of operating on patients with apparent rather than real lesions.

Carotid angiograms are subject to disagreement in their interpretation, even when broad groups (such as <10, 10–49,

50–99, 100 per cent stenosis) are used. Under these circumstances and with uncertainty about the benefits of surgery, together with evidence of widespread misuse of carotid endarterectomy, it is essential that angiograms are reviewed by a panel of surgeons and physicians before making a decision about operation.

Lumbar puncture

Lumbar puncture has two main uses: to diagnose meningitis, which in older patients may cause minimal signs of meningism, but may present with signs suggestive of a stroke; and to distinguish subarachnoid haemorrhage and cerebral haemorrhage from an infarction. This second indication has fallen into disuse since the advent of CT scanning, and because of the danger of tentorial herniation after lumbar puncture in patients with unsuspected mass lesions.

With increased interest in the use of antiplatelet agents after minor thrombotic strokes, the use of lumbar puncture to exclude cerebral haemorrhage might be helpful in centres without access to CT scanning. Lumbar puncture has a variable sensitivity for the detection of cerebral haemorrhage, ranging from 50 to 90 per cent, and specificity ranges from 95 to 100 per cent. This means that if a lumbar puncture shows no blood or xanthochromia, it is very likely that the CT scan will show either an infarct or be normal, but that when cerebral haemorrhage is present, it will often be missed by lumbar puncture. The likelihood ratio of lumbar puncture in diagnosing cerebral haemorrhage ranges from 14 to 25, which, given the pre-test probability of cerebral haemorrhage of about 15 per cent, has a useful effect of increasing the chances that a cerebral haemorrhage is present to about 70 to 80 per cent. As antiplatelet agents are relatively safe provided that they are not given to patients with a major cerebral haemorrhage, the risks of lumbar puncture to rule out cerebral haemorrhage may actually be overwhelmed by the benefits of treatment, but this approach requires further studies before it can be recommended.

The use of subcutaneous heparin after stroke to prevent deep vein thrombosis and pulmonary embolus is of particular concern. The small trials that suggested such treatment was useful in the first week of a stroke did not use CT scanning to rule out cerebral haemorrhage. Although no complications due to bleeding or worsening of neurological impairment occurred, the studies were too small to give adequate estimates of the risk of giving subcutaneous heparin to patients without first excluding those with haemorrhagic strokes. The use of lumbar puncture to exclude haemorrhage for this purpose should be avoided because, if anticoagulation (even with subcutaneous heparin) is given after lumbar puncture, there is a definite risk of producing a haematoma around the site of the puncture, which may lead to paraplegia.

CLINICAL DISAGREEMENT

Disagreement in clinical findings among clinicians is common, and many parts of the history and examination that may be taken for granted as reliable evidence with which to make a diagnosis are subject to high levels of disagreement, as shown in Table 3.

Table 3 Clinical agreement and disagreement on history and examination of stroke patients

Good agreement	Poor agreement
Side of weakness	History of previous stroke
Speech impairment	Mode of onset
Swallowing impairment	Stroke risk factors (e.g. hypertension)
Gaze paresis	Symptoms of meningism
Orientation*	Neck stiffness
Perceptual loss*	Conscious level
Activities of daily living*	Balance/gait
	Visual field loss
	Degree of weakness
	Mood
	Sensation (any modality)
	Plantar responses

* Only after training and use of standardized methods.

The causes of clinical disagreement are due to the examiner, the examined, and the examination. A tired doctor will make more mistakes, as will an inadequately trained one. Traditionally, neurological examination is left until the end of the examination, which may lead to a less thorough approach, particularly if the patient has particular difficulties, such as deafness or delirium. The medical 'shorthand' used to record notes is also a source of errors; a 'right CVA' may mean a right hemisphere lesion or a right hemiplegia. In general it is best to record observations and not inferences. A further problem is that the information most often recorded (plantar responses, reflexes, power, and tone) is seldom the most important for diagnosis, management, or prognosis. The examination environment is frequently inappropriate for the task of testing higher cognitive function, or enquiring about urinary continence or mood.

One solution is to attempt to standardize the collection of information, so that clinicians ask the same questions and record answers in the same way. Certainly this approach works when testing for perceptual and cognitive impairments. A second solution is to collect information appropriate to the immediate task. In the emergency room the need is to assess the urgency of the patient's condition and need for admission. The job of the admitting physician is to ensure that a diagnosis of stroke is reasonably likely, that other treatable causes of neurological impairment (e.g. hypoglycaemia, meningitis, mass lesions) have been excluded, and that information for immediate management has been obtained and recorded (e.g. swallowing, medication, need for pressure-sore prevention). Information needed subsequently for diagnosis, management, and prognosis will usually have several sources (such as friends, neighbours, old medical records), and will require a series of short, focused examinations of the patient in a quiet environment. Finally, it is vital that trainees and practitioners should check their examinations by repeating their key findings and observing others as they elicit the findings.

Bibliography

Allen, C.M.C. (1983). Clinical diagnosis of the acute stroke syndrome. *Quarterly Journal of Medicine*. (New series), **52**, 515–23.

Ebrahim, S. (1990). *Clinical Epidemiology of Stroke.* Oxford University Press, Oxford.

Ginsberg, M.D. and Cebul, R.D. (1983). Non-invasive diagnosis of caro-

tid artery disease. In *Cerebral vascular disease*, (ed. M.J.G. Harrison and M.L. Dyken), pp. 215–53. Butterworths, London.
Norris, J. W. and Hachinski, V.C. (1982). Misdiagnosis of stroke. *Lancet*. **i**. 328–31.
Sandercock, P., (1987). Asymptomatic carotid stenosis: spare the knife. *British Medical Journal*, **294**, 1368–9.
Sandercock, P., Molyneux, A., and Warlow, C.P. (1985). The value of CT scanning in patients with stroke: Oxfordshire Community Stroke Project. *British Medical Journal*, **292**, 193–6.
Shinar, D., *et al.* (1985). Interobserver reliability in the assessment of neurologic history and examination in the stroke data bank. *Archives of Neurology*, **42**, 557–65.
Shinar, D. *et al.* (1987). Reliability of the activities of daily living scale and its use in telephone interview. *Archives of Physical Medicine and Rehabilitation*, **68**, 723–8.
Sox, H.C., Blatt, M. A., Higgins, M.C., and Marton, K.I. (1988). *Medical decision making*, pp. 75–7; 345–6. Butterworths, London.
Toghi, H., *et al.* (1981). A comparison between the computed tomogram and neuropathological findings in cerebrovascular disease. *Journal of Neurology*, **224**, 211–20.

11.4 Care of the acute stroke patient

F. CLIFFORD ROSE

The three main elements of care of the patient with acute stroke are first, general care with respect to fluid balance; second, the prevention of complications and sequelae; and third, specific therapy.

GENERAL MANAGEMENT

General management should include ensuring fluid balance and nutrition, particularly where consciousness is impaired, with normal values of electrolytes, packed cell volume, glucose, blood gases, and pH. Patients with depressed consciousness should have frequent suctioning and oxygen; intubation and tracheostomy may prove necessary, especially if dysphagia persists. When intravenous or nasogastric feeding is required, the fluid and electrolyte balance must be maintained as patients with stroke get dehydrated rapidly and there may be inappropriate secretion of ADH or salt retention.

For incontinence, an external penile catheter may be required in men and an indwelling catheter in women not least for measurement of urinary output. {Many geriatric teams, particularly in the United States, place high priority on avoiding bladder catheterization and, if the patient is conscious, and his or her skin is not compromised by urinary leakage, prefer to try to manage incontinence by the use of pads and regular voiding.}

Bowel function should be carefully watched and rectal examination may be required to avoid constipation or faecal impaction. Loss of cortical function may cause faecal incontinence. {In geriatric practice, faecal incontinence is most commonly due to impaction, sometimes 'high impaction' at the rectosigmoid junction, and therefore not always detected by rectal examination.}

Cerebral oedema, which is maximal 4 days after a stroke but persists for a week or two, may be helped by raising the head.

Not all patients with an acute stroke are admitted to hospital, home conditions and the wishes of the patient's family and doctor, as well as age and severity, often being the determining factors. If the stroke is atypical or the diagnosis doubtful, if there is clinical deterioration or if the patient requires special nursing care because of confusion, incontinence, or dysphagia, then hospital admission is often essential.

Investigations will indicate if there is any underlying disease (e.g. polycythaemia, endocarditis, polyarteritis, myocardial infarction, or cardiac dysrhythmia) that needs treatment.

Ideally, every patient with stroke should have a computed tomographic (**CT**) scan to exclude non-vascular cerebral lesions (e.g. tumour) and to distinguish cerebral infarction from haemorrhage. This is particularly important in those that are deteriorating because emergency aspiration of a haematoma, whether extra- or intracerebral, can be life-saving. Angiography is indicated in a subarachnoid haemorrhage to determine the presence of an aneurysm (found in over half of cases) or angioma (about 5 per cent).

Hypertension should be treated only if it is persistent beyond the first day and is associated with such complications as congestive heart failure or hypertensive encephalopathy. The treatment of choice is a vasodilator drug but frusemide and other agents can also be given. Hypotensives must be used cautiously because reduction of cerebral perfusion pressure could encourage enlargement of an infarcted area owing to local breakdown of cerebral autoregulation in the region of damaged tissue. Pulmonary oedema may occur with large cerebral infarctions owing to intense adrenergic stimulation causing pulmonary hypertension and transudation.

PREVENTION OF COMPLICATIONS

Frequent turning is required to prevent decubitus ulcers, particularly in patients with sensory loss. Early physiotherapy is required to prevent joint contractions and the shoulder–hand syndrome. Early ambulation will help in the prevention of venous thrombosis, as will heparin (given subcutaneously, 5000 units every 8 h) if haemorrhagic stroke has been excluded.

Psychological problems are common in stroke, particularly depression and agitation, both of which can respond to pharmacological remedies as an adjunct to skilled and sensitive nursing care. {Early ambulation and a return as far as possible towards 'normal' daily routines of dress and socialization are thought to be psychologically and functionally beneficial as part of acute as well as of rehabilitative care.}

Over 10 per cent of patients with stroke have fits that may

require anticonvulsant medication. {For a high proportion of elderly patients who suffer fits in the acute phase of stroke, anticonvulsant medication can be withdrawn after a few weeks without recurrence. However, over 12 to 18 months after a stroke about 10 per cent of elderly patients develop epilepsy, perhaps due to cerebral scarring, and in these patients the condition is often persistent.}

SPECIFIC THERAPY

Rationale

The definitive treatment of an acute stroke is to maintain the oxygen and glucose supply. Maintaining oxygenation of ischaemic brain can be helped by reducing metabolic demand. The normal cerebral blood flow is over 50 ml/100 mg brain/min. Blood flow below 15 ml/100 mg brain/min will cause infarction if continued for more than a few minutes.

Dead neurones cannot be revived or replaced. The aim of acute treatment of stroke is to save threatened neurones, i.e. those that are still alive but perhaps not functioning normally. This concept of the 'ischaemic penumbra' has received supporting, but not conclusive, evidence from studies using positron emission tomography.

Even with reperfusion, there is experimental evidence from animal models and autopsies of human brain after cardiac arrest that neuronal damage can progress, presumably due to the cytotoxicity of damaged tissue.

The pharmacological remedies can be divided into:

1. treatment of blood viscosity by haemodilution;
2. antithrombotic agents;
3. thrombolytic agents;
4. vasoactive drugs;
5. anti-oedema agents;
6. metabolic activation.

Haemodilution

The chief factor responsible for the viscosity of whole blood is the haematocrit level and there is an inverse relationship between this and cerebral blood flow. If, by haemodilution, the haematocrit can be lowered to 33 per cent, cerebral blood flow and maximal oxygenation of tissues will occur under normal circumstances, but whether this applies to ischaemic tissue is uncertain. While haemodilution is widely used in some countries (e.g. Germany), its value is unproven.

Haemodilution can be either hypo-, iso-, or hypervolaemic. The Scandinavian isovolaemic study (Strand *et al.* 1984) initially indicated reduced fatality and improved functional recovery from acute stroke but a more recent study from Scandinavian community hospitals (Asplund 1987) did not confirm this benefit; the difference between these studies was unlikely to be due to delay in hospital admission or start of treatment, as evidenced by subgroup analysis (Scandinavian Stroke Study Group 1988). As this lack of benefit has also been confirmed by an Italian Study (Italian Acute Stroke Study Group 1988), isovolaemic haemodilution cannot be recommended.

Patients admitted to hospital with an acute stroke are often dehydrated, so giving a volume expander to produce hypervolaemic haemodilution seems logical, not least because the resulting increased cardiac output could increase cerebral blood flow (Grotta 1987). This was tried in 13 centres in the United States (Hemodilution in Stroke Study Group 1988), using pentastarch, a colloid solution of hydroxyethyl starch, which produced an expansion of plasma volume equal to 1.5 times the infused amount. The pentastarch was infused until the haematocrit was below 33 per cent and was continued for 3 days. Improvement was seen in those whose treatment started within 12 h of the stroke, but the trial was terminated because of deaths from cerebral oedema.

Antithrombotic agents

Anticoagulants are the main type of drug used clinically and then mainly for recurrent cerebral emboli (Cerebral Embolism Study Group 1983).

Cardiogenic embolus

The most common cause of cerebral embolus is cardiac dysrhythmia, chiefly atrial fibrillation, but also supraventricular tachycardia and ventricular ectopic beats. Embolism of thrombus either from a cardiac aneurysm or an akinetic left ventricular wall are the next common, followed by valvular lesions and prostheses. Anticoagulants are given not for acute treatment but for the prevention of recurrences, and consist of either heparin given parentally or oral vitamin K antagonists such as the coumarins, of which the most commonly used is warfarin.

{Opinions differ about the appropriate time to initiate anticoagulant therapy after cerebral embolism in view of the danger of creating a haemorrhagic infarct; 2 to 3 weeks seems to be common practice.}

Heparin

As this is not absorbed from the gastrointestinal tract, it has to be given parenterally, either as the sodium or calcium salt. When given intravenously, its anticoagulant effect is immediate but its half-life, varying in different individuals, is only about 1 h. Given subcutaneously, its effect lasts about 12 h.

Its mode of action is to prevent extension of thrombosis and the test used to measure this effect, and hence the dose necessary, is the activated partial thromboplastin time (**APTT**), the required therapeutic range being 1.5 to 2.0 times the normal control time.

In progressing thrombotic stroke, continuous intravenous therapy is required with an initial bolus of 5000 units, monitoring the APTT twice daily until the required level (Ockelford 1986).

Warfarin

This is the most widely used anticoagulant, having excellent bioavailability, with an average half-life of approximately 36 h. Nearly all the drug is bound to albumin and, as only the free fraction is therapeutically active, anything that affects the binding capacity of albumin, be it disease, diet, or other drugs, will have a profound effect on the anticoagulant activity.

Its effect takes at least 24 h, with a peak on the third or fourth day. The test used to measure its effect is the prothrombin time and a two- to threefold increase is required in Europe (where human brain thromboplastin is used) but only

1.25 to 1.5 times control in the United States (where rabbit brain thromboplastin is used).

The usual period for anticoagulants to be administered is not universally agreed but varies from 3 months (Ockelford 1986) to over a year, depending on the source of the cerebral emboli.

Thrombolytic agents

The two main types are proteolytic enzymes such as plasmin, and plasminogen activators such as urokinase and streptokinase. The latter convert plasminogen to plasmin, which hydrolyses fibrin and other proteins involved in clotting. Their value in the management of acute stroke has not yet been fully assessed.

Although this approach has been mooted for decades, progress has been slow because of the danger of cerebral haemorrhage seen with streptokinase. Urokinase has been successful (Abe 1986) in a double-blind trial when given orally. It has also been tried by intracarotid injection for occlusion of the middle cerebral artery (Mohr *et al.* 1988) but this was an uncontrolled study.

The advent of recombinant genetic techniques has revived interest because tissue plasminogen activator (**tPA**) can now be synthesized. Normally present in the vessel wall, tPA is not specific, its activity being limited to the fibrin of newly formed clot. Because of good experimental results (Papadopoulos *et al.* 1987) and its proven value in coronary thrombosis, it is now being tried in acute stroke.

The value of thrombolysis in acute stroke is still unproven but the indications are that if active intervention is required then patients need to be referred early, i.e. within 2h.

Vasoactive drugs

Vasodilators

These are likely to be ineffective because they dilate the normal vessels outside the infarcted area and 'steal' blood away from where it is most needed. They include carbon dioxide, papaverine, and cyclandelate.

Aminophylline

This has been used because experimentally it constricts the normal vessels and increases the blood supply to the infarct (Paulson 1974) but its clinical efficacy has not been demonstrated (Britton *et al.* 1980).

Naloxone

Anecdotally this is reputed to be rapidly effective (Jabaily and Davis 1984) but the dose required is at least five times that required for morphine overdosage (Baskin and Hosobuchi 1981). Its value has not been confirmed (Fallis *et al.* 1984).

Prostacyclin

This has been studied because it is an antiplatelet agent, as well as a vasodilator (Moncada 1983). Although a few open studies indicated improvements (Gryglewski *et al.* 1983; Miller *et al.* 1984), its efficacy has not been confirmed (Hsu *et al.* 1987).

Calcium channel blockers

After ischaemia, the low intracellular levels of calcium rise, partly due to an influx of calcium ions but also because of release of intracellular stores. Enzymes that degrade proteins (proteases) and phospholipids (phospholipases) are activated by this rise of intracellular calcium and produce leukotrienes and free radicals, which are cytotoxic. After a stroke, excitatory neurotransmitters such as glutamate are released, possibly due to the opening of calcium channels, so there is a rationale for giving stroke patients glutamate antagonists, or calcium blockers (Wong and Haley 1990), especially the dihydropyridines such as nimodipine (Gelmers *et al.* 1988) or nicardipine. Recent trials have been unpromising however.

Anti-oedema agents

Their effect is usually temporary but intravenous glycerol reduced mortality from acute stroke (Bayer *et al.* 1987). While steroids are effective in vasogenic oedema associated with cerebral tumour, trauma, and abscess, they do not seem to help the mixed anoxic and vasogenic oedema of cerebral infarction (Mulley *et al.* 1978; Norris and Hachinski 1986).

Metabolic activators

Although naftidrofuryl has been classed as a vasodilator, it is more likely to act as a metabolic activator (Steiner and Clifford Rose 1982) but it also has selective 5-hydroxytryptamine activity. Given intravenously, it reduced hospital stay (Steiner and Clifford Rose 1986*a*,*b*). Its effectiveness in acute stroke is still uncertain.

CONCLUSIONS

There is no consensus that any agent is effective in acute treatment. The clinical trial of naftidrofuryl is being repeated on a much larger, multinational scale, recruiting 500 patients, and glycerol trials are also being repeated. Steroids and haemodilution are now widely accepted as being unhelpful, whilst anticoagulants can be positively deleterious.

For these reasons the treatment of acute stroke is likely to be in general measures, whilst clinical trials on specific drugs are undertaken in stroke centres. New drug strategies are being investigated, for example, *N*-methyl-D-aspartase (NMDA) receptor antagonists and antioxidants, and trials are proceeding. Whatever their results, the declining incidence of stroke is likely to continue, perhaps aided by the promise of molecular biology in the prevention of atheroma.

References

Abe, T. (1986). Oral urokinase: absorption, mechanisms of fibrinolytic enhancement and clinical effect on cerebral thrombosis. *Folia Haematologica* (Leipzig), **113**, 122–36.

Asplund, K. (1987). Multicenter trial of hemodilution in acute ischemic stroke. 1. Results in the total patient population. *Stroke*, **18**, 691–9.

Baskin, D.S. and Hosobuchi, Y. (1981). Naloxone reversal of isochaemic neurological deficits in man. *Lancet*, **ii**, 272–5.

Bayer, A.J., Pathy, M.S.J., and Newcombe, R. (1987). Double-blind randomised trial of intravenous glycerol in acute stroke. *Lancet*, **i** 404–7.

Britton, M., de Faire, U., Helmers, C., Miah, K., and Rane, A. (1980). Lack of effect of theophylline on the outcome of acute cerebral infarction. *Acta Neurologica Scandinavica*, **62**, 116–23.

Cerebral Embolism Study Group (1983). Immediate anticoagulation of embolic stroke: a randomized trial. *Stroke*, **14**, 668–76.

Fallis, R.J., Fisher, M., and Lobo, R.A. (1984). A double blind trial of naloxone in the treatment of acute stroke. *Stroke*, **15**, 627–9.

Gelmers, H.J., Gorter, K., and de Weerdt, C. J. (1988). A controlled trial of nimodipine in acute ischaemic stroke. *New England Journal of Medicine*, **318**, 203–7.

Grotta, J.C. (1987). Current status of hemodilution in acute cerebral ischemia. *Stroke*, **18**, 689–90.

Gryglewski, R.J. *et al.* (1983). Treatment of ischaemic stroke with prostacyclin. *Stroke*, **14**, 197–202.

Hemodilution in Stroke Study Group (1989). Hypervolemic hemodilution treatment of acute stroke—the results of a randomized multicenter trial using pentastarch. *Stroke*, **20**.

Hsu, C.Y. *et al.* (1987). Intravenous prostacyclin in acute nonhemorrhagic stroke: a placebo-controlled double-blind trial. *Stroke*, **18**, 352–8.

Italian Acute Stroke Study Group (1988). Haemodilution in acute stroke: results of the Italian haemodilution trial. *Lancet*, **i**, 318–20.

Jabaily, J., and Davis, J.N. (1984). Naloxone administration to patients with acute stroke. *Stroke*, **15**, 36–9.

Miller, V.T., Coull, B.M., Yatsu, F.M., Shah, A.B., and Beamer, N.B. (1984). Prostacyclin infusion in cerebral infarction. *Neurology*, **34**, 1431–5.

Mohr, E., Tabuchi, M., Yoshida, T., and Yamadori, A. (1988). Intracarotid urokinase with thromboembolic occlusion of the middle cerebral artery. *Stroke*, **19**, 802–12.

Moncada, S. (1983). Biology and therapeutic potential of prostacyclin. *Stroke*, **14**, 157–68.

Mulley, G., Wilcox, R.G., and Mitchell, J.R.A. (1978). Dexamethasone in acute stroke. *British Medical Journal*, **2**, 994–6.

Norris, J.W., and Hachinski, V.C. (1986). High dose steroid treatment in cerebral infarction. *British Medical Journal*, **292**, 21–3.

Ockelford, P. (1986). Heparin 1986. Indications and effective use. *Drugs*, **31**, 81–92.

Papadopoulos, S.M., Chandler, W.F., Salamat, M.S., Topoi, E.J., and Sackellares, J.C. (1987). Recombinant human tissue-type plasminogen activator therapy in acute thromboembolic stroke. *Journal of Neurosurgery*, **67**, 394–8.

Paulson, O.B. (1974). Regional cerebral blood flow in cerebral infarction and in transient ishemic attacks. *Review of Electroencephalographic and Neurophysiological Clinics*, **4**, 210–16.

Scandinavian Stroke Study Group (1988). Multicenter trial of hemodilution in acute ischemic stroke: results of subgroup analyses. *Stroke*, **19**, 464–71.

Steiner, T.J. and Clifford Rose, F. (1982). Naftidrofuryl. In *Advances in stroke therapy*, (ed. F. Clifford Rose), pp. 71–7. Raven Press, New York.

Steiner, T.J. and Clifford Rose, F. (1986*a*). Randomized double-blind placebo-controlled clinical trial of naftidrofuryl in hemiparetic CT-proven acute cerebral hemisphere infarction. *Royal Society of Medicine, International Congress Symposium Series*, **99**, 85–98.

Steiner, T.J. and Clifford Rose, F. (1986*b*) Towards a model stroke trial. The single-centre naftidrofuryl study. *Neuroepidemiology*, **5**, 121–47.

Strand, T., Asplund, K., Eriksson, S., Hagg, E., Lithner, F., and Wester, P-O. (1984). A randomized controlled trial of hemodilution therapy in acute ischemic stroke. *Stroke*, **15**, 980–9.

Wong, M.C.W. and Haley, E.C. (1990). Calcium antagonists: stroke therapy coming of age. *Stroke*, **21**, 494–501.

11.5 Stroke rehabilitation

DERICK T. WADE

INTRODUCTION

This chapter introduces, briefly, a model of rehabilitation and discusses some general principles of stroke rehabilitation. The majority of the chapter then concentrates upon a series of specific problems. Where available, evidence is given but it must be realized that much rehabilitation, like much of medicine still lacks good evidence.

A MODEL OF REHABILITATION

The World Health Organization (1980) has published a detailed system for the classification of the consequences of any disease known as the *International Classification of Impairments, Disabilities, and Handicaps* (ICIDH). It is a good framework for discussing and understanding the management of neurological disability.

The most important concept this model introduces is that any illness can be considered at four levels: pathology, impairment, disability, and handicap. In practice these levels form a continuum, with many grey areas between them. It is important to retain the extremely useful underlying ideas, and not to get bogged down in sterile, semantic arguments.

Pathology

This refers to the damage or abnormal processes occurring within an organ or organ system. Synonymous terms are 'disease' and 'target disorder'. Pathology is the traditional focus of medical care. Doctors concentrate upon diagnosing the disease and then upon curing or reducing it if possible, and hospitals and medical systems are centred on this process. The crucial feature is that pathology is restricted to part of the whole body, usually an organ or system.

Impairments

These are the direct neurophysiological consequences of underlying pathology. They are usually referred to as 'symptoms and signs', and the ICIDH definition is '. . . any loss or abnormality of psychological, physiological or anatomical structure or function'. The ICIDH also notes that: 'Impairment represents exteriorisation of a pathological state'. In other words it is the immediate consequence of pathology as perceived by the person at the level of the whole body.

Disability

This refers to reduction in or loss of an ability. Most (but not all) impairments will disturb, to some extent, aspects of a patient's normal function. The functional loss, which manifests itself at the level of the person within his or her immediate environment, is termed disability. Examples include slowness of walking, needing help to dress, and being unable to cook a simple meal.

The ICIDH definition of disability is '... any restriction or lack (resulting from an impairment) of ability to perform an activity within the range considered normal for a human being'. The ICIDH also notes that: 'Disability represents objectification of an impairment, and as such represents disturbances at the level of the person.' In other words, disability is the personal nuisance caused by the pathology.

Handicap

This refers to the social and societal consequence of pathology, and in the patient's eyes it usually determines the severity of any illness. Handicap arises at the level of the patient's own social roles and activities. The most important distinguishing characteristic of handicap is that normality is judged with reference to the patient's own immediate social context, whereas normality for disability, impairment, and pathology is judged with reference to the population at large. Examples of handicap include the loss of a job, or divorce arising as a consequence of the disease.

The ICIDH definition for handicap is '... a disadvantage for a given individual, resulting from an impairment or a disability, that limits or prevents the fulfilment of a role that is normal (depending on age, sex, and social and cultural factors) for that individual'. The ICIDH also notes that: 'Handicap represents socialisation of an impairment or disability, and as such it reflects the consequences for the individual—cultural, social, economic and environmental—that stem from the presence of impairment and disability.' In other words, handicap is the freedom the patient has lost due to the pathology.

It is important to realize that handicap can arise directly from an impairment (or even from pathology), as well as from a disability. For example, a right hemianopia may cause no disability (indeed the patient may not notice it at all) but, once detected by a doctor, it might lead to loss of a driving licence and hence enormous handicap. In the case of handicap, the link with disability can be very slight and certainly environmental factors such as the legal framework, the family support, and the physical environment and financial support all have a major affect upon the final handicap.

Practical consequences arising from this model

Several useful ideas arise from this model

1. The focus of attention should pass from pathology to handicap as time passes.
2. The focus of attention should also pass from the patient to the environment as one progresses from pathology to handicap.
3. Treatment should be aimed as close to pathology as possible.
4. Assessments (measures) should only contain items relating to a single level.
5. The reference frame for judging normality changes (as mentioned above).

REHABILITATION: WHAT IS IT?

There are many definitions of rehabilitation, and this is one based upon the ICIDH model:

> Rehabilitation is a problem solving and educational process aimed at reducing the disability and handicap experienced by someone as a result of a disease, always within the limitations of available resources.

Put another way, it is acting upon pathology, impairment, or disability, or upon the intervening variables in order to reduce handicap. Its essence is the management of change, and the phrase 'the management of (neurological) disability' is perhaps a much better description of rehabilitation. Although the final goal is always to minimize handicap, it is in practice easiest and probably most effective to concentrate upon disability.

In managing a patient with disability, the following processes are important:

1. assessment—identification, analysis, and measurement of problems;
2. planning—analysing the problem(s) and setting goals;
3. treatment—intervention to reduce disability and handicap;
4. care—intervention to alleviate consequences of disability;
5. evaluation: checking on the effectiveness of any intervention (i.e.reassessment).

ASSESSMENT—THE KEY

Assessment in rehabilitation is no different from the approach used when considering any medical problem: it refers to the acquisition of information needed to come to some conclusion, be that a diagnosis, a prognosis, or a decision on intervention. The main purposes are:

1. detecting (diagnosing) problems, usually disabilities;
2. discovering the relevant contributing factors, usually impairments;
3. determining prognosis (with and without intervention);
4. establishing the role of treatment (any? what?);
5. evaluating change.

Assessments are tools to be used, not rituals to be completed. This chapter cannot discuss all the available measures and will simply give a classification of the main groups (Table 1) and introduce a few clinically useful assessments.

The overall level of physical disability is perhaps the most important aspect to assess, and certainly every doctor should be able to do this. The usual phrase used to describe suitable assessments is 'Activities of Daily Living' (**ADL**). The component items of the Barthel ADL index (Tables 2 and 3) illustrate the nature of ADL. Although there are very many ADL indices, the Barthel ADL index is a good diagnostic and measurement assessment for routine clinical use: it is valid, reliable, simple, quick, easily communicated, and relevant (Wade and Collin 1988).

Other useful formal measures of use are:

1. Motricity Index, for quantifying weaknesses (Wade and Langton-Hewer 1987*b*).
2. Rivermead Motor Assessment (RMA) (Lincoln and Leadbitter 1979);

Table 1 Neurological disability—a classification and some measures

Category	Measures
Subcategories	
Physical	
Overall	Barthel ADL index
Personal care	
Mobility	Gait speed, Functional Ambulation Categories
Dexterity	Nine Hole Peg Test
Communication	
Dysarthria	
Aphasia	Frenchay Aphasia Screening Test
Behaviour	
Intrapersonal	General Health Questionnaire
Interpersonal	
Safety	
Domestic	
Kitchen	Rivermead ADL Test
Housework	Rivermead ADL Test
Cognitive	
Orientation	Short Mental Test
Memory	Rivermead Behavioural Memory Test
Learning	
Problem solving	Any IQ test
Perception/spatial awareness	
Visual	
Reading	

3. Glasgow Coma Scale, especially the motor subsection, for quantifying coma (an important prognostic indicator) (Teasdale *et al.* 1979);
4. Frenchay Aphasia Screening Test (FAST), short, simple, non-specialist test for detecting and quantifying aphasia (Enderby *et al.* 1986);
5. Short Mental Test, for quantifying orientation (Hodkinson 1972);
6. Rivermead Behavioural Memory Test (RBMT) for quantifying memory loss (Wilson *et al.* 1989)
7. Behavioural Inattention Test (BIT) for detecting and quantifying neglect (Wilson *et al.* 1987)
8. Functional Ambulation Categories, to quantify need for help walking (Holden *et al.* 1974)
9. Gait speed over 10 m to quantify mobility (Wade *et al*, 1987);
10. Nine Hole Peg Test (NHPT) and Frenchay Arm Tests, to quantify manual dexterity (Heller *et al.* 1987);
11. Rivermead ADL assessment (Whiting and Lincoln 1980), domestic section for kitchen and domestic abilities;
12. Frenchay Activities Index, to assess 'social and leisure' function (Wade *et al.* 1985*a*);
13. Nottingham extended ADL scale (Nouri and Lincoln 1987);
14. General Health Questionnaire (Goldberg and Hillier 1979) to measure stress on relatives (probably best reserved for research or audit of a service, not as a routine clinical measure).

As stressed earlier, assessment is simply a process used to help plan further management. It should be carried out

Table 2 The Barthel ADL index

Bowels
- 0 = incontinent (or needs to be given enemata)
- 1 = occasional accident (once a week)
- 2 = continent

Bladder
- 0 = incontinent, or catheterized and unable to manage alone
- 1 = occasional accident (maximum once per 24 h)
- 2 = continent

Grooming
- 0 = needs help with personal care
- 1 = independent face/hair/teeth/shaving (implements provided)

Toilet use
- 0 = dependent
- 1 = needs some help, but can do something alone
- 2 = independent (on and off, dressing, wiping)

Feeding
- 0 = unable
- 1 = needs help cutting, spreading butter, etc.
- 2 = independent

Transfer (bed to chair and back)
- 0 = unable, no sitting balance
- 1 = major help (one or two people, physical), can sit
- 2 = minor help (verbal or physical)
- 3 = independent

Mobility
- 0 = immobile
- 1 = wheelchair independent, including corners
- 2 = walks with help of one person (verbal or physical)
- 3 = independent (but may use any aid, e.g. stick)

Dressing
- 0 = dependent
- 1 = needs help, but can do about half unaided
- 2 = independent (including buttons, zips, laces, etc.)

Stairs
- 0 = unable
- 1 = needs help (verbal, physical, carrying aid)
- 2 = independent

Bathing
- 0 = dependent
- 1 = independent (or in shower)

TOTAL
- 0–20

to sufficient depth to allow any necessary decision to be made, but no further. The doctor should encourage his colleagues to use a restricted range of measures so that all members of the team treating a patient can communicate with each other effectively.

PLANNING

It is in planning further action that the doctor may have his or her most important role. At all times, even immediately after the patient has had the stroke, it is important to be considering the future. Of course the doctor's immediate aims may be concerned with diagnosis and treatment of the pathology, although this often is simple, and can be carried out at home. Nonetheless it is still necessary to be considering the disability from the beginning.

Table 3 The Barthel ADL index guidelines

General
- The index should be used as a record of what a patient does, *not* as a record of what a patient could do.
- The main aim is to establish degree of independence from any help, physical or verbal, however minor and for whatever reason.
- The need for supervision renders the patient *not* independent.
- A patient's performance should be established using the best available evidence. Asking the patient, friends/relatives and nurses will be the usual sources, but direct observation and common sense are also important. However, direct testing is not needed.
- Usually the performance over the preceding 24–48 h is important, but occasionally longer periods will be relevant.
- Middle categories imply that the patient supplies over half of the effort.
- Use of aids to be independent is allowed.

Bowels (preceding week)
- If needs enema from nurse, then 'incontinent'.
- Occasional = once a week.

Bladder (preceding week)
- A catheterized patient who can completely manage the catheter alone is regarded as 'continent'.

Grooming (preceding 24–48 h)
- Refers to personal hygiene: doing teeth, fitting false teeth, doing hair, shaving, washing face. Implements can be provided by helper.

Toilet use
- Should be able to reach toilet/commode, undress sufficiently, clean self, dress, and leave.
- With help = can wipe self and do some of above.

Feeding
- Able to eat any normal food (not only soft food). Food cooked and served by others, but not cut up.
- Help: food cut up, patient feeds self.

Transfer
- From bed to chair and back
- Dependent: no sitting balance, unable to sit; two people to lift.
- Major help: one strong/skilled, or two normal people. Can sit up.
- Minor help: one person easily, *or* needs any supervision.

Mobility
- Refers to mobility about house or ward, indoors. May use aid.
- If in wheel chair, must negotiate corners/doors unaided.
- Help: by one, untrained person, including supervision and moral support.

Dressing
- Should be able to select and put on all clothes, which may be adapted.
- Half: help with buttons, zips etc. (check!), but can put on some garments alone.

Stairs
- Must carry any walking aid used to be independent.

Bathing
- Usually the most difficult activity.
- Must get in and out unsupervised, and wash self.
- Independent in shower = 'independent' if unsupervised and unaided.

Planning is best achieved by considering (and setting) goals. As with assessment, it is necessary to remember that goal setting is simply one method of planning, and it should not become an end in itself. Moreover, goals need to be set at various levels and with different degrees of specificity. Immediate goals can be specific, but distant goals are necessarily less certain. Some therapeutic goals are more detailed than others.

Failure to achieve a goal does not necessarily mean that intervention has failed: other explanations include setting unrealistic goals; the particular intervention was inappropriate (but another might work); the original analysis of the problem was faulty (e.g. another previously unrecognized impairment exists); or that more time is needed. However, if a goal is not achieved, it is important to analyse why and not simply to continue treatment in a possibly vain attempt.

The most important goal to discuss openly is that of the whole team treating the patient (doctor, nurse, occupational therapist, physiotherapist, and social worker, with perhaps psychologist and dietitian), so that each member can tailor their goals to the overall aim. Unfortunately it is this very goal which is often overlooked, and it should be the doctor's role to set the general course of rehabilitation as team leader.

In hospital practice much of this planning can be done easily and briefly, for example at a ward round. But it is important to discuss and make plans in full consultation with all others involved, including the patient and family where possible. If there are particular difficulties arising, usually with limited recovery, then it may be necessary to convene a case conference separate from the ward round so that more time can be spent discussing the problems and their possible solutions.

One way of adding focus is to concentrate upon planning discharge, with emphasis upon ensuring a smooth transfer to the community services. There is particular need to improve the process of discharge, given the evidence of how poor discharge planning is at present (Smith *et al.* 1981; Ebrahim *et al.* 1987).

Discharge of anyone with significant residual disability should be preceded by asking the following questions.

1. How much help is needed with personal ADL (use the Barthel)? And who will provide it?
2. How safe is the patient; how long can he be left alone? And how will the necessary supervision be given?
3. How will domestic ADL be provided (meals, housework, etc.)?
4. Has the house been assessed? Are all necessary alterations complete or in process?
5. Has all equipment needed been provided?
6. Are community carers aware of the patient and their role?
7. Has daytime occupation/relief been organized?
8. Have adequate follow-up arrangements been made?
9. Is planned inpatient relief care needed?

Lastly, anyone involved in making plans should remember this quotation from William Blake:

> 'He who would do good to another must do it in minute particulars. General good is the plea of the scoundrel, hypocrite, and flatterer.'

In other words, while it is important to have a general aim, the other vital component of effective and efficient rehabilitation is attention to detail. In military terms, planning has to be both strategic and tactical.

INTERVENTION

There is a large range of possible interventions, and to describe them all in detail would be impossible. Indeed, much depends upon the individual patient's circumstances and needs; and much depends upon the resources available locally. Even classifying the possible areas of intervention is difficult. Table 4 represents an attempt.

Table 4 Possible interventions in managing disability

None
Observe (only possible if disability not major)
Remove, reverse, or stop pathology
Not yet possible in stroke
Minimize impairments
Impairment-specific (symptomatic) drugs (e.g. baclofen)
Surgery (e.g. to contractures)
Orthoses (external equipment)
Prostheses (replacement of missing body parts/functions)
Prevent/avoid complications (e.g. shoulder pain)
Facilitate intrinsic recovery
Reduce 'learned non-use'
Minimize disability
Maximize use of recovering impairments
Teach/maximize adaptive techniques
Alter immediate, physical environment
Provide physical personal support
Counselling/support
Providing information
Minimize handicap
Alter general physical environment
Improve social environment
Provide money (allowances, etc.)
Provide transport
Arrange occupation, social stimulation
Help with any administrative formalities

When considering whether to take any action, and what action to take, the doctor needs to know:

1. the natural history of the problem;
2. ways of judging prognosis in the individual patient;
3. what treatments are available.

The next section discusses natural history and prognosis. Then problems arising at various stages of rehabilitation will be discussed.

NATURAL HISTORY AND PROGNOSIS

The natural history of many aspects of stroke has recently been studied in some detail, and can only be outlined briefly here. About 30 per cent of patients will die in the first 3 months, most of these dying primarily from the stroke itself within the first 3 weeks. Patients most likely to die are those with most severe strokes (Barer and Mitchell 1989), as evidenced by:

1. loss of consciousness;
2. gaze palsy;
3. complete paralysis;
4. urinary incontinence;
5. swallowing difficulties;
6. being 'unassessable' on most tests, etc.

Over the first week, about 20 per cent of patients will show deterioration in disability, presumably through further infarction which cannot be prevented. There may also be major fluctuations and rapid improvements in the first week. Consequently it is not always easy to give a certain prognosis at this stage.

Once the clinical state has settled, one can expect a period of rapid recovery over the first 2 to 3 weeks, followed by slower but continuing improvement over the next 5 months (Wade and Langton-Hewer 1987*a*). Whether much significant improvement occurs after 6 months is subject to debate, but it is certainly rare and limited in extent. In addition, the final (6-month) level of disability is closely related to the initial (first week) level (Wade and Langton-Hewer 1987*a*). These rules apply to:

1. overall physical disability (ADL);
2. walking (Wade *et al.* 1987);
3. manual dexterity (Heller *et al.* 1987);
4. aphasia;
5. memory and other cognitive functions (Wade *et al.* 1988).

Therefore the first prognostic indicator is simply the level of disability, global or specific, present at the end of the first week. Most traditional prognostic indices are simply markers of more severe strokes.

The main concern is to find indicators able to give more specific information, such as identifying patients likely to do better (or worse) than expected or identifying those likely to benefit from some specific treatment. Unfortunately, as we do not yet have firm proof of the effectiveness of any treatment (medical, surgical, physical, or any other), it is not possible to identify markers of likely response to treatment. Nonetheless, there are some clinically useful general rules.

Patients who are still incontinent of urine 2 to 3 days after their stroke are much more likely to die or, if they survive, to be immobile and in need of long-term institutional care. This has been confirmed repeatedly (Barer 1989), and applies whatever the underlying reason for the incontinence (e.g. coma, aphasia, immobility, pre-existing incontinence).

Patients who have no perceptible voluntary grip by 3 weeks are very unlikely to regain useful arm function (Heller *et al.* 1987). Aphasic patients who do not even recognize the meaning of common sounds are likely to be left with minimal communication abilities (Varney 1984).

THE FIRST WEEK

Rehabilitation should start immediately after the onset of stroke. The first need is to respond to the practical problems of care, usually feeding, excretion, and moving the patient. It is these problems that primarily determine whether a patient needs hospital admission, rather than any diagnostic or medical problems. The physical outcome is similar wherever care is given, at home or in hospital.

The second need is to counsel the patient and family. This

means giving factual explanations about stroke, the prognosis, and the management options. This should be supported with written information.

There is no evidence to support the traditional approach (by relatives at least) of resting the patient. From the onset the patient should be mobilized, and encouraged to be as independent as possible. This may speed recovery (Hamrin 1982), and should reduce the problems of deep vein thrombosis, depression, and dependency.

The carers (nurses, family) should be instructed on the most appropriate ways to handle the patient physically, and on communication where needed. This means there must be early involvement of therapists, which should minimize complications and might speed reduction of disability (Smith *et al.* 1982).

Although subcutaneous heparin may reduce radiologically demonstrated deep vein thromboses, there is in my view no evidence of any substantial clinical benefit to warrant the pain,inconvenience, and cost. Therefore a policy of early mobilization and (possibly) treating clinically obvious thromboses if they arise seems reasonable. {In the United States subcutaneous heparin is generally recommended.}

Dysphagia is common, and should be checked for by asking the patient to swallow a cup of water. If present, care should be taken in feeding. However, although patients with dysphagia are more likely to die (Gordon *et al.* 1987), there is no evidence that this arises because of the dysphagia and it is probably simply that dysphagia occurs with more severe strokes. As swallowing usually recovers quickly, it is rarely necessary to take any specific action.

THE FIRST 3 MONTHS

The majority of recovery occurs during the first 3 months (Skilbeck *et al.* 1983) and it is vital to monitor the change carefully, identifying and managing new problems as they arise, and ensuring that no avoidable complications are occurring. Of patients in hospital, the majority of survivors will have left by 3 months and most of those remaining will need long-term care. The doctor should particularly ensure that discharge plans are well made and executed, because discharge marks a major watershed in the patient's eyes.

Two frequent questions asked are: 'Does rehabilitation work?' and 'Does therapy work?' Given the all encompassing nature of rehabilitation (see Table 4), it is nearly impossible to answer the first question, and the second question is scarcely easier to answer, because therapists have many important roles including:

1. making a detailed assessment of any problems;
2. giving advice to the patient and any carers;
3. providing appropriate equipment, ensuring it is used correctly;
4. teaching adaptive techniques;
5. maximizing functional recovery.

However, some relevant issues will be discussed.

The first point to stress is how little therapy is actually received by patients, even in well-staffed units. It amounts to about 3.5 per cent of a patient's waking time, and is difficult to increase by much. There is, fortunately, some evidence suggesting that it is the early timing of a therapist's intervention and not the total time given which is important (Smith *et al.* 1982).

On the other hand, one randomized controlled study has found evidence of a dose–response relationship (Smith *et al.* 1981). In this study, patients discharged from hospital with continuing disability were asked to attend for (unspecified) outpatient therapy: 44 patients had weekly visits from a health visitor, 43 attended for outpatient therapy 3 half-days each week, and 46 attended for 4 full days of therapy a week. Attending for therapy led to some gains, more marked in the group being seen most, but whether the gains were clinically worthwhile is unclear.

In a second study, 95 patients were randomly allocated 1 week after stroke to receive either 'normal' or 'intensive' therapy most of which was given in a special stroke unit (Sivenius *et al.* 1985). Patients given intensive treatment made a faster recovery of motor power and had a faster recovery of function. This study needs to be replicated, preferably separating out the influence of more therapy from the influence of a specialized unit.

Speech therapy services have been investigated more thoroughly than all other rehabilitative services put together. Lincoln *et al.* (1984) could find no benefit arising from the service offered in Nottingham: volunteers probably help as much as trained therapists, provided that the patients are first assessed by trained therapists (Wertz *et al.* 1986), and supportive counselling may be as effective as specific therapy (Hartmann and Landau 1987).

About one-third of patients will be miserable, but not all these patients are clinically depressed. Many probably have emotionalism (also known as emotional lability) (House *et al.* 1989). Antidepressants can be useful but they should be used carefully in view of their high incidence of side-effects particularly in elderly patients.

Counselling, the provision of information, advice, and emotional support have received little investigation and studies have produced conflicting results (Towle *et al.* 1989). However, such support is a professional obligation of the physician and other carers.

Other studies have been carried out, and these have been reviewed (Wade and Langton-Hewer 1989). In general one can conclude that at present there is no incontrovertible evidence to support any particular general approach or specific treatment.

In the absence of firm evidence, and in the presence of so many possible ways of helping the patient, it is vital for the doctor to monitor critically the effect of any intervention he or she arranges, to try and assess its effectiveness. Obviously it may be difficult to prove benefit given the underlying 'natural' improvement, but at least close evaluation will reduce the chance of wasting resources that are not helping.

LONG TERM

The World Health Organization classification includes six types of handicap; physical independence; orientation; mobility; occupation; social integration; and economic self-sufficiency. Each should be considered.

About half of all survivors will be left with some physical disability, with many needing support at home and perhaps 5 to 10 per cent needing long-term institutional care. The doctor's role is to identify those patients who need help and to ensure that they receive whatever is available. In this way the 'physical independence' aspect of handicap and the stress upon the carer can be reduced.

Although about one-third of patients may suffer aphasia soon after stroke, relatively few will survive with continuing severe communication disability. This will lead to 'orientation handicap', and it is in practice difficult to alleviate. Groups run for patients with aphasia may help socially.

In most countries, little allowance is made for people with mobility disability and so they suffer considerable handicap. Public transport is often unusable, public places impassable, and pavements are often uneven and unsafe. Wherever possible (legally, financially, and in terms of safety), every effort should be made to achieve driving because loss of driving is strongly associated with 'depression'.

Ensuring adequate occupation and social integration is difficult in elderly people, even if they have not had a stroke. Many retired people give little thought to these aspects of their life, and public services rarely provide adequate facilities for the potentially huge demand. Instead, the voluntary sector is expected to provide help. The patient should be put in contact with day centres, self-support groups, and any other facilities available locally.

Few patients will have been working before their stroke, and economic self-sufficiency will be little affected by the stroke. However, the relative poverty of most patients does increase the difficulty of helping them to find a satisfying lifestyle after stroke, because the extra disability increase the cost of most activities.

ORGANIZATION

Improvements in the organization of stroke care services might be the best way of improving outcome after stroke. There is no definition of an ideal service, but some possible characteristics are shown in Table 5. No organization has been subjected to evaluation against these (or any other) criteria. There are relatively few trials of different patterns of care, or even of the components of any service. The studies have been reviewed (Wade 1989), and are summarized here.

Table 5 Characteristics of an ideal service

1. Accurate initial diagnosis, separating stroke from other causes of the stroke syndrome
2. Rapid, accurate identification of patients suitable for any specific treatment (when such becomes available)
3. Rapid provision of an adequate level of nursing support and care
4. Early (within 1 week) assessment to identify major disabilities and the underlying impairments:

with a view to:

5. Providing early intervention (rehabilitation) to:
   Mobilize patients rapidly and safely
   Avoid preventable complications
   Maximize use of any recovery
   Teach adaptive techniques where needed
   Ensure appropriate equipment is provided and used correctly
   Counsel, advise and inform the patient and family
6. Adequate monitoring of change in the patient, with:
   Forward planning of major transitions such as discharge home
   Follow-up, perhaps for at least 1 year
   A long-term contact point given for use after discharge
7. And, most important adaptability:
   Identifying *and* remedying deficiencies in the service
   Responding to new advances in management

Two conclusions can be drawn from studies on the effectiveness of stroke intensive care units (SICUs). First, some hospitals probably do offer a better standard of care than others—this is hardly surprising. Second, special units do not of themselves benefit patients in terms of reducing fatality or morbidity, but some hospitals probably do need to improve their standard of acute care. They might reduce the frequency of some complications (e.g. bronchopneumonia).

Stroke units are difficult to define, but generally include dedicated stroke wards, whether part of or separate from a general hospital. Some accept patients with acute stroke, whereas others do not, but their essence seems to be a concentration upon rehabilitation (the management of the disability, not acute medical and nursing care).

Conclusions are difficult to draw from the reported studies, in part because the special aspects of the unit are rarely defined in detail. Further, most units are set up and run by enthusiastic people and this, rather than the unit itself, may account for any positive effects. They may lead to more efficient use of rehabilitation staff, and patients might experience a slightly more rapid recovery with a greater level of eventual independence, especially in dressing, and less likelihood of needing long-term care. However, there is certainly no overwhelming evidence in favour of special stroke units, and it seems likely that general improvement in disability services would be of more help to more people.

However, the lessons learned from stroke units could be transferred to general hospital settings, and two studies have investigated stroke teams in hospital. Such stroke teams can be set up and run in British district hospitals within current medical resources.

Lastly, home care is an alternative approach. Many patients are already cared for at home, but providing additional resources (domiciliary therapy) neither decreases the use of hospital, nor improves the outcome for patients and their families (Wade *et al.* 1985*b*).

CONCLUSION

The management of a patient who has suffered a stroke is an exciting challenge, both intellectually and practically, and utilizes all the skills a doctor should have: diagnostic (logical deduction), leadership and management, decision making, counselling, etc. Stroke is a good model for all rehabilitation, and much written in this chapter can well be applied to most other disabling diseases. {For a United States perspective on stroke rehabilitation see Kelly (1990).}

References

Barer, D.H. (1989). Continence after stroke: useful predictor or goal of therapy? *Age and Ageing*, **18**, 183–91.

Barer, D.H. and Mitchell, J.R.A. (1989). Predicting the outcome of acute stroke: do multivariate models help? *Quarterly Journal of Medicine*, **70**, 27–39.

Ebrahim, S., Barer, D., and Nouri, F. (1987). An audit of follow-up services for stroke patients after discharge from hospital. *International Disability Studies*, **9**, 103–5.

Enderby, P.M., Wood, V.A., Wade, D.T., and Langton-Hewer, R. (1986). The Frenchay Aphasia Screening Test: a short, simple test for aphasia appropriate for non-specialists. *International Rehabilitation Medicine*, **8**, 166–70.

Goldberg, D.P. and Hillier, V.F. (1979). A scaled version of the General Health Questionnaire. *Psychological Medicine*, **9**, 139–45.

Gordon, C., Langton-Hewer, R., and Wade, D.T. (1987). Dysphagia in acute stroke. *British Medical Journal*, **295**, 411–14.

Hamrin, E. (1982). Early activation in stroke: does it make a difference? *Scandinavian Journal of Rehabilitation Medicine*, **14**, 101–9.

Hartmann, J., and Landau, W.M. (1987). Comparison of formal language therapy with supportive counselling for aphasia due to acute vascular accident. *Archives of Neurology*, **44**, 646–9.

Heller, A., Wade, D.T., Wood, V.A., Sunderland, A., Langton-Hewer, R., and Ward, E. (1987). Arm function after stroke: measurement and recovery over the first three months. *Journal of Neurology, Neurosurgery, and Psychiatry*, **50**, 714–19.

Hodkinson, H.M. (1972). Evaluation of a mental test score for assessment of mental impairment in the elderly. *Age and Ageing*, **1**, 233–8.

Holden, M.K., Gill, K.M., Magliozzi, M.R., Nathan, J., and Piehl-Baker, L. (1974). Clinical gait assessment in the neurologically impaired: reliability and meaningfulness. *Physical Therapy*, **64**, 35–40.

House, A., Dennis, M., Molyneux, A., Warlow, C., and Hawton, K. (1989). Emotionalism after stroke. *British Medical Journal*, **298**, 991–4.

Kelly, J. F. (1990). Stroke rehabilitation in elderly patients. In *Geriatric Rehabilitation* (eds. B. Kemp, K. Brummel-Smith, and J. W. Ramsdell), pp. 61–89. Little, Brown & Co., Boston.

Lincoln, N. and Leadbitter, D. (1979). Assessment of motor function in stroke patients. *Physiotherapy*, **65**, 48–51.

Lincoln, N.B. *et al.* (1984). Effectiveness of speech therapy for aphasic stroke patients: a randomised controlled trial. *Lancet*, **i**, 1197–200.

Nouri, F.M. and Lincoln, N.B. (1987). An extended activities of daily living scale for stroke patients. *Clinical Rehabilitation*, **1**, 301–5.

Sivenius, J., Pyorala, K., Heinonen, O.P., Salonen, J.T., and Riekkinen, P. (1985). The significance of intensity of rehabilitation of stroke—a controlled trial. *Stroke*, **16**, 928–31.

Skilbeck, C.E., Wade, D.T., Langton-Hewer, R., and Wood, V.A. (1983). Recovery after stroke. *Journal of Neurology, Neurosurgery and Psychiatry*, **46**, 5–8.

Smith, D.S. *et al.* (1981). Remedial therapy after stroke: a randomised controlled trial. *British Medical Journal*, **282**, 517–20.

Smith, M.E., Garraway, W.M., Smith, D.L., and Akhtar, A.J. (1982). Therapy impact upon functional outcome in a controlled trial of stroke rehabilitation. *Archives of Physical Medicine and Rehabilitation*, **63**, 21–4.

Teasdale, G., Murray, G., Parker, L., and Jennett, B. (1979). Adding up the Glasgow Coma Scale. *Acta Neurochirurgica*, **28**, (suppl.), 13–16.

Towle, D., Lincoln, N.B., and Mayfield, L.M. (1989). Service provision and functional independence in depressed stroke patients and the effect of social work intervention on these. *Journal of Neurology, Neurosurgery, and Psychiatry*, **52**, 519–22.

Varney, N.R. (1984). The prognostic significance of sound recognition in respective aphasia. *Archives of Neurology*, **41**, 181–2.

Wade, D.T. (1989). Organisation of stroke care services. *Clinical Rehabilitation*, **3**, 227–33.

Wade, D.T. and Collin, C. (1988). The Barthel ADL index: a standard measure of physical disability? *International Disability Studies*, **10**, 64–7.

Wade, D.T. and Langton-Hewer, R., (1987*a*). Functional abilities after stroke: measurement, natural history and prognosis. *Journal of Neurology, Neurosurgery, and Psychiatry*, **50**, 177–82.

Wade, D.T. and Langton-Hewer, R. (1987*b*). Motor loss and swallowing difficulty after stroke: frequency recovery and prognosis. *Acta Neurologica Scandinavica*, **76**, 50–4.

Wade, D.T. and Langton-Hewer, R. (1989). Stroke rehabilitation. In *Handbook of clinical neurology*, Vol 11(55) *Vascular diseases*, Part III, (ed. J.F. Toole), pp. 233–54. Elsevier, Amsterdam.

Wade, D.T., Legh-Smith, J., and Langton-Hewer, R. (1985*a*). Social Activities after stroke: measurement and natural history using the Frenchay Activities Index. *International Rehabilitation Medicine*, **7**, 176–81.

Wade, D.T., Langton-Hewer, R., Skilbeck, C.E., Bainton, D., and Burns-Cox, C. (1985*b*). Controlled trial of a home care service for acute stroke patients. *Lancet*, **i**, 323–6.

Wade, D.T., Wood, V.A., Heller, A., Maggs, J., and Langton-Hewer, R. (1987). Walking after stroke: measurement and recovery over the first three months. *Scandinavian Journal of Rehabilitation Medicine*, **19**, 25–30.

Wade, D.T., Wood, V.A., and Langton-Hewer, R. (1988). Recovery of cognitive function soon after stroke: a study of visual neglect, attention span and verbal recall. *Journal of Neurology, Neurosurgery and Psychiatry*, **51**, 10–13.

Wertz, R.T. *et al.* (1986). Comparison of clinic, home and deferred language treatment for aphasia: a Veterans Administration cooperative study. *Archives of Neurology*, **43**, 653–8.

Whiting, S. and Lincoln, N.B. (1980). An ADL assessment for stroke patients. *Occupational Therapy*, **43**, 44–6.

Wilson, B.A., Cockburn, J., and Halligan, P. (1987). Development of a behavioural test of visuospatial neglect. *Archives of Physical Medicine and Rehabilitation*, **68**, 98–102.

Wilson, B.A., Cockburn, J., Baddeley, A.D., and Hierns, R.W. (1989). The development and validation of a test battery for detecting and monitoring everyday memory problems. *Journal of Clinical and Experimental Neuropsychology* (in press).

World Health Organization (1980). *International Classification of Impairments. Disabilities and Handicaps*. WHO. Geneva.

NOTE

The Behavioural Inattention Test and the Rivermead Behavioural Memory Test are available from:

Thames Valley Test Company,
7–9 The Green
Flempton
Bury St Edmunds
Suffolk IP28 6E2
England.

The Frenchay Aphasia Screening Test and the General Health Questionnaire are available from:

NFER–Nelson
2 Oxford Road East, Windsor,
Berks SL4 1DF, England.

11.6 The prevention of stroke

J. GRIMLEY EVANS

Although more can be done for the victims of stroke than, sadly, commonly is done, the disagreeable fact remains that destroyed brain tissue cannot be replaced. Therapy and rehabilitation after a stroke can at best offer only limited benefits, and the hope for control of this horrible disease must lie in prevention.

Stroke is predominantly a disease of later life; 70 per cent of strokes happen to people aged over 65. Incidence rates of stroke increase as a power-law function of age and therefore produce a straight line when plotted as logarithms against the logarithm of age (Fig. 1). In contrast, mortality rates of stroke follow an exponential function of age and therefore will appear as a straight line if plotted as logarithms against untransformed age—a Gompertz plot. As Fig. 1 shows, these relationships lead to a convergence of incidence and mortality with age, which reflects the fact that the fatality of stroke also increases with age. That is to say, the older a stroke victim the more likely he or she is to die of the disease.

The significance of the power-law relation with age is that it is similar to the relation shown by most adult cancers, and it is known from epidemiological evidence that most adult cancers are caused by environmental (extrinsic) factors. This raises the hope that most strokes are also caused by environmental influences and therefore potentially preventable. More direct evidence for environmental causes of stroke lies in the variation in incidence between geographical regions and the decline in the mortality and incidence of stroke observed in many Western countries over the last four decades. How far this potential for control of the disease can be realized by the actions of clinicians rather than by public health programmes remains to be evaluated.

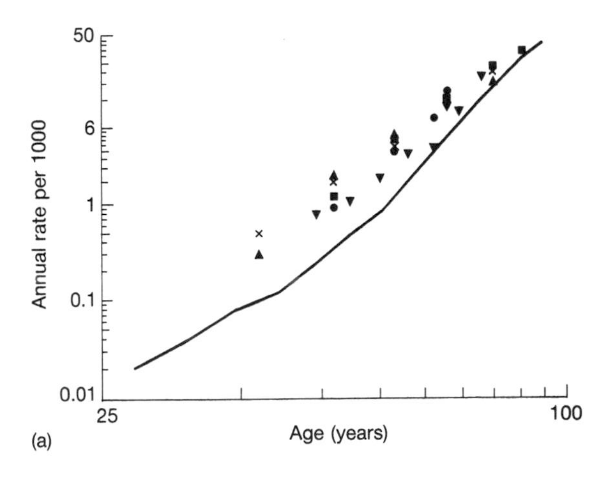

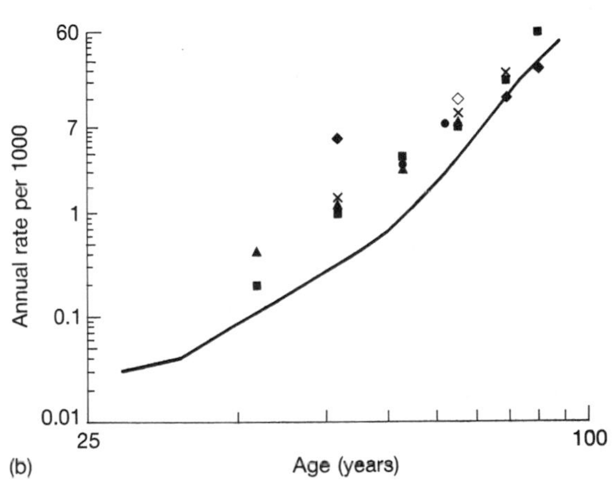

Fig. 1 Age-specific incidence rates (symbols) for stroke in seven studies of white populations: Middlesex County, CT, Framingham, MA, Los Angeles, CA, Rochester, MN, Denver, CO, and Kansas (Epidemiology Study Group 1972). The continuous line shows mortality rates from stroke for England and Wales 1981. (a) Men; (b) women.

THE PATIENT AT RISK

The epidemiology of stroke is reviewed in Chapter 11.2. Elderly patients at particular risk are those with high blood pressure, diabetes, peripheral vascular disease, previous transient ischaemic attacks or stroke, atrial fibrillation, and clinical or ECG evidence of coronary heart disease. (Table 1 shows the ECG features associated with a significantly increased risk of subsequent first stroke in an elderly British population). As these factors interact, attention to any of them may lower the risk of stroke in a patient with several factors.

BLOOD PRESSURE CONTROL

High blood pressure is common among elderly people but the assertion that blood pressure increases with age is an oversimplification. In cross-sectional data from population samples large enough to have statistically meaningful numbers in the later age groups, mean systolic pressure increases to a maximum in the first half of the eighth decade of life in men and the second half of the same decade in women, and then falls (Miall and Brennan 1981; Evans 1987). Diastolic pressure peaks in middle age and thereafter declines, so contributing to the age-associated increase in prevalence of isolated systolic hypertension. Although these declines in systolic and diastolic pressures in later life may partly reflect differential mortality of subjects with higher pressures, some older people do show a decline in blood pressure, sometimes but not always associated with a fall in body weight, and not necessarily associated with ill health or detectable cardiovascular disease. One of the consequences of these changes is that some elderly patients who have been on stable antihypertensive treatment for years may develop symptoms of hypotension if dosages are not revised.

Studies in many population groups have shown high blood pressure to be the dominant risk factor for stroke, but not all studies of elderly populations have found this. In Norway in the 1960s the relationship between blood pressure and

Table 1 Risk ratios for subsequent stroke associated with specifed electrocardiographic features (Minnesota codings in parentheses) in a British population of people aged 65 and over (Evans 1985).

ECG feature	Risk ratio		
	Men	Women	Persons
Q/QS waves (1.2,1.2,1.3)	—	—	163*
Left—axis deviation (2.1)	NS	NS	
High voltage LV compexes (3.1)	—	214***	
ST–J depression (4.1,4.2)	—	184**	210***
T-wave inversion (5.1,5.2)	308***	148*	
T-wave flattening (5.3)	—	147*	NS
Left bundle-branch block (7.1)	—	—	232***
Atrial fibrillation (8.3)	372***	—	291***
Multiple ventricular ectopics†	NS	NS	NS

Rates not computed for less than five events (—); NS = not statistically significant.

† Excluding subjects with atrial fibrillation.

*$p<0.05$; **$p<0.01$; ***$p<0.001$.

stroke mortality disappeared at ages over about 80 (Home and Waaler 1976). In a north-east English population in the late 1970s, blood pressure was not a risk factor for women aged 65 and over and was only a weak predictor of stroke in men of the same age group (Evans 1987). In a Finnish sample of people aged over 80, blood pressure showed an inverse relationship to stroke (Rajala *et al.* 1983). These variations presumably reflect differences in the interplay of the many epidemiological factors that affect people over long lifetimes (Evans 1984), and it is likely that as populations become less diverse culturally, old people will also show less variability. Until this has been shown to have happened, a prerequisite for introducing a screening programme for high blood pressure in order to prevent stroke in an elderly population should be evidence that high blood pressure is, in fact, an important risk factor for stroke in that particular population.

Measurement of blood pressure

The measurement of blood pressure of elderly subjects may present problems in both random and systematic errors. One form of error which has received particular attention is that of 'pseudohypertension' due to diminished compliance of the arterial wall, and the use of Osler's manoeuvre to assess the risk of being misled by a high sphygmomanometric measurement should be routine with older patients (Messerli *et al.* 1985). Osler's manoeuvre consists of blowing up the cuff to above systolic levels and palpating the brachial or radial artery. If the arterial wall can be felt as a rigid cord, the manoeuvre is positive and there is a risk of pseudohypertension being present. If the arterial wall cannot be felt, the manoeuvre is negative and pseudohypertension is unlikely.

In selected samples, under careful surveillance, reduction of high blood pressure can reduce the risk of stroke in elderly people (Staessen *et al.* 1988; SHEP 1991). This has been demonstrated at ages up to 80 (Amery *et al.* 1985) but there is a need for studies of larger numbers of individuals over that age. The frequency and severity of adverse effects of therapy in these studies have been acceptably low, but the subjects enrolled were highly motivated and higher rates of side-effects can be expected among less selected patients. Many of the side-effects of hypotensive medication may be overlooked unless specifically enquired for. However, the clinician and the patient should bear in mind that the relation between blood pressure and stroke is continuous. Any reduction in pressure will produce a corresponding measure of reduction in stroke risk. Although advocated as a possible ideal, there is therefore no need to 'normalize' the blood pressure to a diastolic below 90 or a systolic below 160 mmHg in order to produce at least some benefit. A more modest reduction of pressure may constitute a desirable compromise between benefit and adverse effects of treatment for some patients.

Impaired cerebral autoregulation is probably one mechanism underlying reports of disastrous initiation of hypotensive therapy in elderly patients. Antihypertensive therapy for elderly patients needs to be very gradual in effect to allow time for the autoregulatory range of the patient's cerebral circulation to adjust to the lower levels of systemic pressure.

Choice of antihypertensive therapy

For many years, thiazide diuretics dominated the management of hypertension in middle age and later life, but the recent availability of other forms of therapy calls for a reappraisal of this approach.

The pathophysiology of high blood pressure in old age shows some systematic differences from that in earlier adult life (Messerli *et al.* 1983), and the optimal physiological approach to treatment might therefore differ. In later life, hypertension is associated typically with a raised peripheral resistance, making diuretics, calcium-channel blockers, and vasodilators a rational choice of therapy. At younger ages, hypertension may have a hyperdynamic, high-output component to which β-blockers are a rational approach.

Diuretics have acquired a bad reputation for producing a high frequency of adverse effects among elderly patients, but this is partly due to inadequate care in their prescription. As far as their antihypertensive effects are concerned, thiazides show an early ceiling effect in their dose–response curve (McVeigh *et al.* 1988), whereas the response in terms of diuresis and metabolic changes continues to increase through higher doses. The incidence of thiazide-induced adverse effects in elderly patients may be unnecessarily high because when the blood pressure response to small doses of thiazide is inadequate, physicians increase the dose instead of adding a second-line drug.

A second reason for a high prevalence of diuretic-induced adverse effects among elderly people is that diuretics may be prescribed unnecessarily. A community survey in northern England (Evans 1985) revealed that among people aged 65 and over, 27 per cent of women and 18 per cent of men were taking diuretics regularly. It seems that in primary care, diuretics may be prescribed for elderly women who present with ankle swelling that is, in fact, due to stasis rather than to heart failure. Inevitably a number of these will run unnecessary risks of hypovolaemia or metabolic disturbance.

As monotherapy a thiazide diuretic is at least as effective a hypotensive agent as a calcium blocker, β-blocker, or prazosin (Kirkendall 1988). Diuretics appear to be as effective in the control of blood pressure in old patients as they are in younger patients. In some studies, calcium blockers have been found to be more effective in older patients than in young, but this has not always been seen. There is a more consistent trend for β-blockers to be less effective among elderly than among younger hypotensives. The lack of a hyperdynamic element to the hypertension of later life may be relevant to this. An alternative explanation is the greater proportion of elderly hypotensives who have low plasma renin levels compared with younger patients, but this interpretation is not universally accepted. Certainly there are no adequate grounds to justify the extra costs of 'renin-profiling' hypertensive elderly patients in the hope of identifying 'rational' antihypertensive therapy. The negative inotropic effects and production of fatigue by β-blockers are often a concern with older patients, as are the common presence of peripheral vascular disease and bronchospasm.

The main anxieties about diuretic therapy relate to their well-documented side-effects of hypokalaemia, hypomagnesaemia, hyperuricaemia, hyperglycaemia, and increase in levels of low-density lipoprotein cholesterol. Less common but potentially more dangerous problems include renal failure, hyponatraemia, and hypovolaemia with postural hypotension. The now well-recognized but long overlooked effect of thiazides in inducing impotence should not be ignored on the assumption that it is not relevant for elderly patients. For some elderly men, potency may be precarious and once abolished may not, as in younger men, be easily reinstated by stopping the thiazides. If prior enquiry of an elderly man indicates that potency is important but precarious, monotherapy with an angiotensin converting-enzyme (**ACE**) inhibitor may be the wisest choice, despite the need for especially careful induction and early monitoring of treatment (Croog *et al.* 1988).

Several studies have found that control of hypertension reduces the incidence of stroke but has no beneficial effect on coronary heart disease. One possible reason for this is that diuretic therapy induces by metabolic mechanisms as many 'extra' attacks of coronary heart disease as it prevents by lowering blood pressure. Diuretic therapy can induce cardiac ectopic activity, and it has also been known for many years that, in population studies, there is an association between the frequency of ventricular ectopic beats and subsequent cardiac death (Chiang *et al.* 1969). As the background incidence of ectopic beats increases with age and there is reason to suspect that the ageing heart, perhaps owing to an increased prevalence of ischaemic changes, will be more susceptible to serious dysrhythmic activity than will the younger heart, a drug-induced increase in ectopic activity is a worrying consideration in choosing therapy for an elderly hypertensive patient.

The diuretic-induced increase in ectopic activity is commonly attributed to reduced levels of potassium and magnesium in the body. There is evidence that potassium-sparing diuretics also conserve magnesium. In one study, elderly patients taking thiazides with potassium-sparing admixtures did not show the significantly low serum magnesium of those taking thiazides alone (Martin and Milligan 1987). The use of potassium-conserving combination therapy may be preferable for older patients than the use of potassium supplements. Older patients may not comply well with prescribed potassium supplements, and their self-chosen diet may be low in magnesium as well as potassium. Although British clinical tradition frowns on the prescription of combination tablets, with elderly patients the therapeutic benefits lost by pharmacological inflexibility may be more than regained by improved compliance.

In view of these concerns it is reassuring to note that in the trial conducted by the European Working Party on High Blood Pressure in the Elderly (**EWPHE**) (Amery *et al.* 1985), cardiac mortality was if anything reduced in the treated group, all of whom were receiving diuretic therapy. The diuretic therapy used in this trial comprised potassium-conserving combination tablets of hydrochlorothiazide 25 mg with triamterene 50 mg. Even this produced a statistically significant reduction in mean serum potassium levels in the treatment group. In view of the theoretical possibilities of cardiac arrhythmias, this is the only form of diuretic therapy for which therapeutic safety can be regarded as adequately established for elderly patients. Potassium supplements should not normally be given with a potassium-sparing diuretic and if an ACE inhibitor is required as a second-line drug, the potassium-conserving element of the diuretic should be withdrawn as there is a risk of hyperkalaemia.

The question remains whether some elderly hypertensive individuals may be at higher than average risk of diuretic-induced cardiac arrhythmia and therefore better treated with some other form of therapy. Two studies among mostly non-elderly subjects have produced evidence compatible with the hypothesis that diuretic therapy may increase mortality from coronary heart disease specifically among those subjects showing electrocardiographic abnormalities at induction (Hypertension Detection and Follow-up Program 1984; Multiple Risk Factor Intervention Trial 1985). For several methodological reasons, including small numbers, these findings cannot be regarded as established, and extrapolation of findings among younger adults to an older age group can be no more than speculative. Until better data are available, however, the prudent clinician might feel more inclined to use some first-line treatment other than diuretics in initiating antihypertensive therapy for an elderly patient whose electrocardiogram shows multiple ventricular ectopics or Q/QS features indicative of a previous myocardial infarct. An ACE inhibitor or a combination of a β-blocker with a calcium-channel blocker are possibilities. Apart from being often ineffective, β-blockers have a high frequency of side-effects in elderly patients, but this may be because they are often prescribed in unnecessarily high dosage. For atenolol, for example, 50 mg or even 25 mg daily may produce maximum hypotensive effect in an older patients compared with the 100 mg, which is usual dosage for younger adults.

Diuretics will precipitate some new cases of gout and diabetes and may cause deterioration in diabetic control among some elderly hypertensives. These effects will not be common in unselected primary-care populations but have to be considered in the context of the risk–benefit equation as a whole. On the basis of the findings of the EWPHE study, active treatment in 1000 patients over 1 year might be expected to prevent eleven fatal cardiac events, six fatal and eleven non-fatal strokes, and eight cases of severe heart failure at a cost of inducing seven cases of renal disease (one fatal), four cases of gout, and 71 cases of mild hypokalaemia (Fletcher and Bulpitt 1990). In situations where costs are strictly controlled, this level of incidence of gout probably does not justify routine screening for hyperuricaemia in patients being considered for antihypertensive therapy—not least because it is not established that this would prevent cases of clinical gout. Patients with diabetes or diabetic tendency would be better not treated with diuretics, and the first-line therapy of choice is likely to be an ACE inhibitor.

A serious side-effect of diuretic therapy that did not appear in the EWPHE study, but which is not uncommon as an emergency presentation in geriatric practice, is hyponatraemia. Possibly its absence in the EWPHE study reflects the safety of the hydrochlorothiazide/triamterene combination used in that trial. There is clinical evidence that some combination diuretics containing amiloride may be more likely to induce hyponatraemia. Elderly women under treatment for hypertension seem to comprise the main group at risk for hyponatraemia. The pathogenesis is obscure but may be related to interference with the thirst mechanism. Friedman *et al.* (1989) have shown that, in susceptible individuals, a single dose of a hydrochlorothiazide/amiloride combination induces a gain in body weight due to a polydipsia not shown by controls.

A beneficial side-effect of thiazides has recently been confirmed in good evidence that thiazide diuretics prescribed for the treatment of hypertension have a preventive effect on osteoporosis and proximal femoral fracture (Ray *et al.* 1989). Other things being equal, therefore, this antiosteoporotic effect makes thiazides preferable to other forms of treatment of hypertension or heart failure in middle-aged and elderly women.

In summary, therefore, there do not seem to be adequate grounds for prescribing anything other than a hydrochlorothiazide/triamterene compound as first-line treatment in the otherwise healthy elderly patient. ACE inhibitors should be used instead if the patient has potency problems, poorly controlled gout, or is diabetic. In view of the effects of thiazides on serum lipids, it is wise to give an elderly patient advice on a moderate cholesterol-lowing diet at the time of starting therapy. If the patient's ECG shows previous myocardial infarction or multiple ventricular ectopics, diuretics are best avoided and an ACE inhibitor or β-blocker preferred. Second-line treatment to thiazides may be an ACE inhibitor, methyldopa, or a calcium-channel blocker. If an ACE inhibitor is used as second-line treatment, the diuretic containing a potassium-sparing compound should be replaced by a simple thiazide such as bendrofluazide. Vasodilators such as hydralazine are often better given with both diuretic and β-blocking drugs as cotherapy. (See also Section 9.)

ATRIAL FIBRILLATION

As discussed in Chapter 11.2, atrial fibrillation is a risk factor for stroke, especially if combined with demonstrable valvular or ischaemic heart disease. Although paroxysmal atrial fibrillation in younger patients does not appear to increase risk of stroke, there is a strong clinical suspicion that an older patient who goes in and out of atrial fibrillation is at particularly high risk and this may underlie the high rate of stroke during the 6 months or so after atrial fibrillation has apparently become established. It has also been established that clinical stroke is only the visible part of the iceberg of brain damage in atrial fibrillation; with computed tomographic scanning of subjects without clinical evidence of stroke, brain infarcts are found more frequently in those with atrial fibrillation than in matched controls (Petersen *et al.* 1987).

There is now good evidence that, as primary prevention, anticoagulants reduce the risk of stroke in patients with atrial fibrillation (Petersen *et al.* 1989; Stroke Prevention in Atrial Fibrillation Study 1990; Boston Area Anticoagulation Trial for Atrial Fibrillation Investigators 1990). Inevitably this will be at the price of some haemorrhagic complications, but in terms of death and disability, carefully prescribed and supervised anticoagulant therapy confers more benefit than harm on elderly patients with atrial fibrillation. The margin of safety may be greater than the earlier trials suggested, as the Boston study showed that warfarin therapy is effective in low dose, with a target prothrombin-time ratio of 1.2 to 1.5. It has commonly been taught that anticoagulant therapy is hazardous in old people but this may not be the case provided that the greater pharmacodynamic effect of warfarin in older people is recognized and there is careful selection of patients and scrupulous clinical supervision (Wickramasinghe *et al.* 1988). Evidence that aspirin might be effective as an alternative to anticoagulants is not yet convincing for older patients.

In secondary prevention, anticoagulation appears to reduce the risk of recurrent stroke in patients with atrial fibrillation, but there is controversy about the time after a stroke that it is safe to introduce anticoagulant therapy. The risk of recurrence is highest in the early days but so is the risk of producing a haemorrhagic infarct. Most clinicians observe a period of 3 weeks before anticoagulation. It must be borne in mind that not all strokes in a patient with atrial fibrillation will necessarily be embolic, and computed tomography or magnetic resonance imaging are mandatory if anticoagulation is to be considered.

TRANSIENT ISCHAEMIC ATTACKS

Whether a transient ischaemic attack (TIA) represents a precursor of cerebrovascular disease or one of its minor manifestations is a debatable but unhelpful issue. Epidemiological data show that the risk of subsequent stroke is higher than average in people suffering a TIA, and although the increase in risk is highest in the year or so following an isolated TIA, some increase in risk may be demonstrable for a decade (Whisnant 1976). A clear-cut TIA with localizing neurological signs or symptoms in an elderly patient therefore needs urgent consideration of preventive intervention. The chief differential diagnosis in this situation is focal epilepsy with Todd's paresis, although transient cardiac arrhythmias may also need to be considered. Treatable causes of TIAs (Table 2) also need to be looked for in the individual case. The use of questionnaires aimed at detecting TIAs as a means of screening an elderly population for individuals at risk of stroke is much more problematical (Evans 1990) and more work needs to be done in this area.

A proportion of TIAs will occur in association with atherosclerotic plaques or stenosis in the carotid arteries. Provided that there is access to a specialist surgical centre with demonstrated good results in a large series of patients, there is now good evidence that for TIA patients (and those with minor stroke or retinal infarct) with carotid stenoses of 70 per cent or more, urgent endarterectomy will reduce the cumulative risk of stroke over the following 3 years (European Carotid Surgery Trialists' Collaborative Group 1991). Endarterec-

Table 2 Some treatable causes of transient ischaemic attacks

Giant-cell arteritis
Other arteritis
Embolism from the heart, e.g.
Atrial fibrillation
Infective endocarditis
Mural thrombus
Marantic endocarditis
Hyperviscosity syndromes, e.g.
Polycythaemia
Macrogobulinaemia

tomy is not indicated for minor stenoses of 0 to 29 per cent, while for intermediate degrees of 30 to 69 per cent the issue is uncertain. A carotid bruit may be found in approximately 12 per cent of elderly people. Such a bruit is a known but weak risk factor for stroke but this seems to be mostly because it demonstrates the presence of generalized vascular disease rather than identifying a specific emboligenic lesion. A high proportion of subsequent strokes in such subjects are not in the area of distribution of the affected carotid, and in one study the bruits disappeared in 60 per cent of subjects followed up for 3 years (Van Ruiswyk *et al.* 1990).

Despite somewhat defective data, the consensus is that aspirin therapy will reduce the risk of further TIA or completed stroke in subjects who have suffered a TIA (Sze *et al.* 1988). This will be at the expense of some haemorrhagic complications even if care is taken to exclude subjects with a recent history of peptic ulceration. Low doses of aspirin (150 mg daily) do not seem to be less effective than higher but are associated with lower risk of adverse effects.

As most elderly people in Western nations may be presumed to have some degree of vascular disease in later life, it has been postulated that perhaps aspirin might be of benefit in reducing the risk of stroke in healthy old people without demonstrable signs or history of vascular disease. Studies do not support this idea (Relman 1988). It seems that the general effect of aspirin is to reduce the risk of thromboembolic stroke but to increase the risk of haemorrhagic stroke. With some oversimplification, the data suggest that those subjects who are demonstrably at high risk of thromboembolism (i.e. have suffered a TIA or coronary heart disease) therefore benefit from aspirin therapy, particularly with regard to myocardial infarction, but old people at only average thromboembolic risk (average, that is, for Western populations) may lose more from the increased risk of cerebral haemorrhage than they gain from reduced risk of thromboembolism.

OTHER RISK FACTORS

Cigarette smoking

Cigarette smoking is a risk factor for stroke in young adults but the relative risk declines with age (Shinton and Beevers, 1989). None the less, there is convincing evidence that giving up smoking in later life confers benefit on the coronary arteries at least (Hermanson *et al.* 1988), and so it is good practice to advise older patients as well as younger to quit the habit.

Cold

There is a little-understood seasonal effect on stroke incidence, with higher rates at times of high wind-chill (Gill *et al.* 1988). This is one reason why it seems wise to advise elderly people to avoid exposure to cold, although there is no direct evidence that this will have any effect on the risk of stroke.

Obesity

There is no convincing evidence that control of obesity will have an effect on stroke incidence in later life. One study has suggested that it is obesity during early adult life rather than in old age that is important (Evans *et al.* 1980).

Alcohol

Reduction of high alcohol intake is wise general advice to an elderly patient but has not yet been shown to have a demonstrable effect in reducing stroke risk in samples of elderly populations.

Blood lipids

As noted above, it seems prudent to give elderly patients advice about modifying diet to offset the lipid-raising effects of thiazide diuretics, but a convincing major role for blood lipids in the aetiology of stroke in later life has not yet been sufficiently well demonstrated to justify dietary or pharmaceutical intervention as a means of preventing stroke in the general population (Khaw 1990). This situation may change as cohorts of elderly people survive from the ranks of middle-aged people exposed to lower lipid levels than have been current in economically advanced nations in recent decades. Geriatricians must be alert to the fact that, in preventive medicine for elderly people, the target may be moving, and possibly quite fast (Evans 1984).

References

Amery, A. *et al.* (1985). Mortality and morbidity results from the European Working Party on High Blood Pressure in the Elderly trial. *Lancet*, **i**, 1349–54.

Boston Area Anticoagulant Trial for Atrial fibrillation Investigators (1990). The effect of low-dose warfarin on the risk of stroke in patients with nonrheumatic atrial fibrillation. *New England Journal of Medicine*, **323**, 1505–11.

Chiang, B.N., Perlmann, L.V., Ostrander, L.D., and Epstein, F.H. (1989). Relationship of premature systoles to coronary heart disease and sudden death in the Tecumseh epidemiologic study. *Annals of Internal medicine*, **70**, 1158–66.

Croog, S.H., Levine, S., Sudilovsky, A., Baume, R.M. and Clive, J. (1988). Sexual symptoms in hypertensive patients. A clinical trial of antihypertensive medications. *Archives of Internal Medicines*, **148**, 788–94.

European Carotid Surgery Trialists' Collaborative Group (1991). MRC European Carotid Surgery Trial: interim results for symptomatic patients with severe (70–90%) or with mild (0–29%) carotid stenosis. *Lancet*, **337**, 1235–43.

Epidemiology Study Group (1972). 1. Epidemiology for stroke facilities planning. *Stroke*, **3**, 360–71.

Evans, J.G. (1984). Prevention of age-associated loss of autonomy: epidemiological approaches. *Journal of Chronic Diseases*, **37**, 353–63.

Evans, J.G. (1985). Risk factors for stroke in the elderly. MD dissertation. University of Cambridge.

Evans, J.G. (1987). Blood pressure and stroke in an elderly English population. *Journal of Epidemiology and Community Health*, **41**, 275–82.

Evans, J.G. (1988). The epidemiology of the dementias. In (ed. J.A. Brody and G.L. Maddox), *Epidemiology and aging: an international perspective*, pp. 36–53. Springer, New York.

Evans, J.G. (1990). Transient neurological dysfunction and risk of stroke in an elderly English population: the different significance of vertigo and non-rotatory dizziness. *Age and Ageing*, **19**, 43–9.

Evans, J.G., Prudham, D., and Wandless, I. (1980). Risk factors for stroke in the elderly. In *The aging brain: neurological and mental disturbances*, (eds. G. Barbagallo-Sangiorgi and A. N. Exton-Smith) pp. 113–26. Plenum, London.

Fletcher, A.E. and Bulpitt, C.J. (1990). Risk and benefit in the European Working Party on Hypertension in the Elderly Trial. *Age and Ageing*, Suppl. 2, P6.

Friedman, E., Hadei, M., Halkin, H., and Farfel. (1989). Thiazide induced hyponatremia. Reproducibility by single dose rechallenge and an analysis of pathogenesis. *Annals of Internal Medicine*, **110**, 24–30.

Gill, J.S., Davies, P., Ill, S.K. and Beevers, D.G. (1988). Wind-chill and the seasonal variation of cerebrovascular disease. *Journal of Clinical Epidemiology*, **41**, 225–30.

Hermanson, B. *et al.* (1988). Beneficial six-year outcome of smoking cessation in older men and women with coronary heart disease. Results from the CASS Registry. *New England Journal of Medicine*, **319**, 1365–9.

Holme, I. and Waaler, H.T. (1976). Five-year mortality in the City of Bergen, Norway, according to age, sex and blood pressure. *Acta Medica Scandinavica*, **200**, 229–39.

Hypertension Detection and Follow-Up Program Co-operative Research Group (1984). The effect of antihypertensive drug treatment on mortality in the presence of resting electrocardiographic abnormalities at baseline: the HDFP experience. *Circulation*, **70**, 996–1003.

Khaw, K.-T. (1990). Serum lipids in later life. *Age and Ageing*, **19**, 277–9.

Kirkendall, W.M. (1988). Comparative assessment of first-line agents for treatment of hypertension. *American Journal of Medicine*, **84** (suppl. 3B); 32–41.

McVeigh, G., Galloway, D., and Johnston, D. (1988). The case for low dose diuretics in hypertension: comparisons of low and conventional doses of cyclopenthiazide. *British Medical Journal*, **297**, 95–8.

Martin, B.J. and Milligan, K. (1987). Diuretic-associated hypomagnesemia in the elderly. *Archives of internal Medicine*, **147**, 1768–71.

Messerli, F.H., Sundgard-Riise, K., Ventura, H.O., Dunn, F.G., Glade, L.B., and Frohlich, E.D. (1983). Essential hypertension in the elderly: haemodynamics, intravascular volume plasma renin activity, and circulating catecholamine levels. *Lancet*, **ii**, 983–6.

Messerli, G.H., Ventura, H.O., and Amodeo, C. (1985). Osler's manouevre and pseudohypertension. *New England Journal of Medicine*, **312**, 1548–51.

Miall, W.E. and Brennan, P.J. (1981). Hypertension in the elderly. In *Hypertension in the young and old*, (ed. G. Onesti and K.E. Kimm), pp. 277–83. Grune and Stratton, New York.

Multiple Risk Factor Intervention Trial Research Group (1985). Baseline rest electrocardiographic abnormalities, antihypertensive treatment, and mortality in the Multiple Risk Factor Intervention Trial. *American Journal of Cardiology*, **55**, 1–15.

Petersen, P., Madsen, B., Brun, B., Pedersen, F., Gyldensted, C., and Boysen, G. (1987). Silent cerebral infarction in chronic atrial fibrillation. *Stroke*, **18**, 1098–100.

Petersen, P., Boysen, G., Gottfredsen, J., Andersen, E.D., and Andersen, B. (1989). Placebo-controlled, randomised trial of warfarin and aspirin for prevention of thromboembolic complications in chronic atrial fibrillation: the Copenhagen AFASAK study. *Lancet*, **i**, 175–9.

Rajala, S., Haavisto, M., Heikinheimo, R., and Mattila, K. (1983). Blood pressure and mortality in the very old. *Lancet*, **ii**, 520–1.

Ray, W.A., Griffin, M. R., Downey, W., and Melton, L.J. (1989). Long-term use of thiazide diuretics and risk of hip fracture. *Lancet*, **i**, 687–90.

Relman, A.S. (1988). Aspirin for the primary prevention of myocardial infarction. *New England Journal of Medicine*, **318**, 245–6.

SHEP Cooperative Research Group (1991). Prevention of stroke by antihypertensive drug treatment in older persons with isolated systolic hypertension. Final results of the Systolic Hypertension in the Elderly Program (SHEP). *Journal of the American Medical Association*, **265**, 2255–64.

Shinton, R. and Beevers. (1989). Meta-analysis of relation between cigarette smoking and stroke. *British Medical Journal*, **298**, 789–94.

Staessen, J., Fagard, R., Van Hoof, R., and Amery, A. (1988). Mortality in various intervention trials in elderly hypertensive patients: a review. *European Heart Journal*, **9**, 215–22.

Stroke Prevention in Atrial Fibrillation Study Group Investigators (1990). Preliminary report of the stroke prevention in atrial fibrillation study. *New England Journal of Medicine*, **322**, 863–8.

Sze, P.C., Reitman, D., Pincus, M.M., Sacks, H.S., and Chalmers, T.C. (1988). Antiplatelet agents in the secondary prevention of stroke. Meta-analysis of the randomized control trials. *Stroke*, **19**, 436–42.

Van Ruiswyk, J., Noble, H., and Sigmann, P. (1990). The natural history of carotid bruits in elderly persons. *Annals of Internal Medicine*, **112**, 340–3.

Whisnant, J.P. (1976). A population study of stroke and TIA: Rochester, Minnesota. In *Stroke*, (ed. F.J. Gillingham, C. Mawdsley, and A.E. Williams), pp. 21–39. Churchill Livingstone, Edinburgh.

Wickramasinghe, L.S.P., Basu, S.K. and Bansal, S.K. (1988). Long-term oral anticoagulant therapy in elderly patients. *Age and Ageing*, **18**, 388–96.

SECTION 12
The ageing respiratory system

12 The ageing respiratory system

DOROTHY BERLIN GAIL AND CLAUDE LENFANT

THE AGEING LUNG

The healthy respiratory system of a young adult is massively over-designed, such that even world-class marathon runners barely test its limits. Thus, although there are inevitable changes that take place over the decades, the healthy respiratory system of an octogenarian is still capable of supporting gas exchange at a level that is more than adequate to meet the usual demands. Even so, respiratory function declines with age, and it is important to understand the normal progression of these changes, and the implications of the loss in pulmonary reserve for the elderly person with lung disease. The changes in pulmonary structure and function that occur with normal ageing are the topic of this chapter.

Lung structure

Anatomical changes

Structural changes in the ageing lung are related to the gradual loss of elastic recoil (or increase in compliance) that occurs with advancing age. The lungs of elderly people are smaller than those of younger people and weigh about 20 per cent less in advanced old age (Krumpke *et al.* 1985). The large airways are for the most part unchanged, although some calcification of cartilage and an increase in diameter can occur, accounting for some of the increased anatomical dead space of the elderly person. The distal bronchioles (less than 2 mm in diameter) decrease in diameter, primarily because of decreased parenchymal support. In contrast, the respiratory bronchioles, alveolar ducts, and alveoli become progressively larger. Enlargement of the terminal air spaces with ageing may contribute to the decreased diffusion capacity of elderly people. Shortening and loss of alveolar septa, and thinning and disruption of alveolar walls result in a decreased alveolar surface area, and in less total surface area available for gas exchange. The internal surface area of the lung decreases from approximately 72 m^2 at 20 years of age to about 60 m^2 at 70 years. The number of alveoli per unit lung volume, however, is unchanged (Thurlbeck 1967).

Connective tissue changes

Age-associated changes in the connective tissue components of the lung (collagen, elastin, and proteoglycans) have been investigated in an attempt to seek biochemical correlates for these structural changes and for the functional changes in mechanical properties that occur with ageing. However, the information available is limited and somewhat confusing. Early studies of the collagen content of human lungs of various ages have provided conflicting results. Changes in the ratio of collagen types with age have not been examined in detail, but the limited information that is available suggests that ageing does not significantly alter the ratio of the major collagen types in the lung (type I/type III) (Krumpke *et al.* 1985). Ageing appears to be associated with changes in the cross-linking of collagen in skin and tendon, but there is no definitive information on age-related changes in cross-linking in the lung. There are fewer cross-links per collagen molecule in very old animals, and preliminary data on a small sample of human lungs suggest that non-reducible cross-links decline in later life (Reiser *et al.* 1987).

The proteoglycans of human lung parenchyma have been reported to decrease with advancing age (Krumpke *et al.* 1985). Most studies have found that the elastin content of the lung appears to increase with age, but interpretation of these studies is difficult (Pack and Millman 1988). Whether changes occur over time in the structure, distribution, or function of lung elastin, or in the balance of protease/antiprotease activity is not known. Thus, how biochemical changes in elastin relate to the decreasing compliance or the emphysematous changes observed in the ageing lung remains to be elucidated.

Methodological difficulties have precluded specific studies of how ageing might alter the expression or regulation of connective tissue components of the lung. The availability of specific probes for these complex molecules should allow more specific measurements in the future. The effects of air pollution, smoking, or infection and other lung disease on age-related alterations in the connective tissue structure and function also needs to be determined.

Chest wall and respiratory muscles

Ageing also results in changes in the chest wall and respiratory muscles. The compliance of the chest wall decreases with advancing age as a result of calcification of chondral cartilages and of skeletal deformities such as kyphoscoliosis. In addition, the shape of the rib cage undergoes a gradual change due to loss of lung recoil with ageing; as the functional residual capacity of the lung rises, the rib cage expands. The change in compliance of the chest wall increases the work of breathing. Thus, the contribution of the abdominal muscles and diaphragm to respiration is greater in older subjects. The function of respiratory muscle weakens somewhat with ageing, as maximal inspiratory and expiratory pressures in men and women, and maximal inspiratory pressure in women decrease with increasing age after the age of 55 years.

Pulmonary function

Measurements of lung compliance and pulmonary function

In general, the process of normal ageing results in a decline of lung function characterized by decreases in both vital capacity and measures of small airways function such as forced expiratory volume in one second (**FEV$_1$**) and maximal voluntary ventilation. Because of decreased elastic recoil, the lungs of elderly people hold a greater volume of air at a given transpulmonary pressure than do those of younger people. The lung pressure–volume curve is therefore shifted to the left in older people, although lung compliance, defined as the slope of the pressure–volume relationship, is unchanged. The compliance of the respiratory system as a whole, however, is decreased in elderly people. In fact, a 60-year-old individual does an estimated 20 per cent more work at a given level of ventilation than one who is 20 years of age (Pack and Millman 1988).

Because of the loss of elastic recoil and other age-associated changes in the airways, the closing volume (the lung volume during expiration at which the small airways begin to close) increases with increasing age. In people over the age of 65, small airways (at the level of the terminal bronchioles) generally close in dependent parts of the lung during tidal breathing. This increase in closing volume contributes, at least in part, to the reduced arterial oxygen tension found in elderly people. The loss of elastic recoil and the increased closing volume cause the residual volume to increase progressively in elderly persons. There is little change in the total lung capacity with ageing. The vital capacity (which represents the difference between total lung capacity and the residual volume) therefore gradually decreases with increasing age (Fig. 1). Maximal voluntary ventilation decreases by approximately 30 per cent between 30 and 70 years of age, primarily because of the decrease in compliance of the respiratory system and diminished strength in the respiratory muscles.

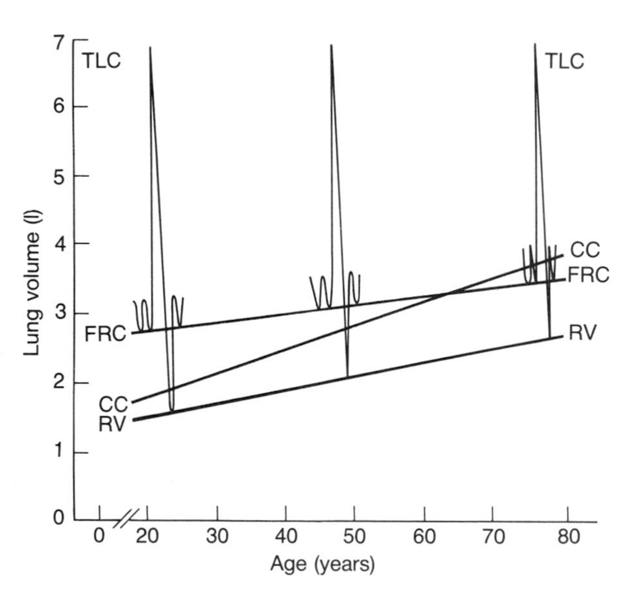

Fig. 1 Changes in lung volumes with age. CC = closing capacity; FRC = functional residual capacity; RV = residual volume; TLC = total lung capacity (from Peterson and Fishman 1982).

All measures of expiratory flow are decreased in older people as a result of the loss of elastic recoil pressure and the decreased strength of respiratory muscles. But differences in the maximal expiratory flow are significant only at lower lung volumes (below 30 per cent of the expired vital capacity). In general, the rate of fall in FEV_1 increases with age and is related to body size (Krumpke *et al.* 1985; Mahler *et al.* 1986). In a longitudinal study of 466 healthy non-smokers, the decline in FEV_1 started at about the age of 35 years in both sexes, but the rate of decline was not linear, and accelerated in later years to a decline of about 20 ml per year in persons of 65 years (Burrows *et al.* 1986). This decline in FEV_1 was smaller and occurred at a later age than suggested by earlier cross-sectional analyses. It should be noted that there is wide variability in measurements of FEV_1 in older persons. In fact, the intrasubject variability in FEV_1 greatly exceeds the expected decline in FEV_1 (Burrows *et al.* 1986). Smoking accelerates the decline in pulmonary function, including that of functional vital capacity and FEV_1 (Mahler *et al.* 1986). Even asymptomatic smokers have a significantly greater decrease in FEV_1 with age than non-smokers.

Gas exchange

The increase in closing volume with increasing age leads to areas of ventilation–perfusion inequalities, an increased alveolar–arterial difference in oxygen tension, and reduced arterial oxygen tension in elderly people. In contrast, the arterial carbon dioxide tension is unaffected by ageing. At 70 years of age, the average arterial Po_2 is about 75 mmHg (Sorbini *et al.* 1968) (Fig. 2). In spite of this reduction in arterial oxygen tension, arterial oxygen saturation is not significantly different in elderly people because of the shape of the dissociation curve for oxyhaemoglobin. However, when the Po_2 falls below 60 mmHg, the arterial oxygen saturation begins to fall more dramatically. In elderly persons with reduced cardiac output, oxygen delivery to the tissues is further compromised.

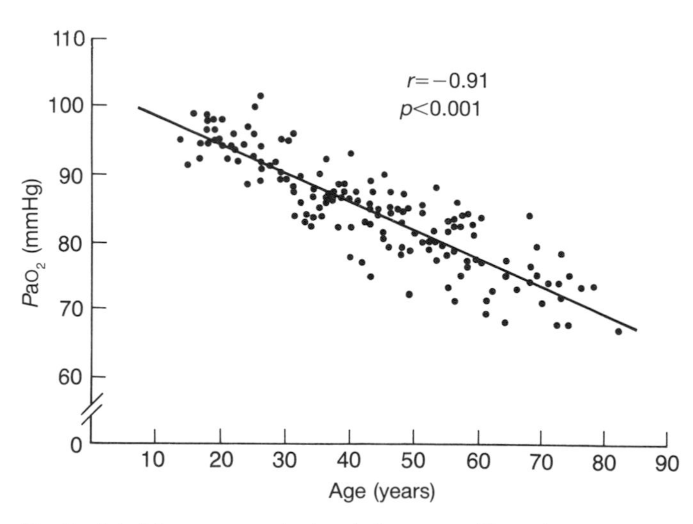

Fig. 2 Arterial oxygen tension in relation to age. The points represent duplicate estimations of Pao_2 in 152 normal supine adult subjects; the regression line is shown (from Sorbini *et al.* 1968).

A reduction in the measured diffusing capacity of the lung occurs with advancing age, primarily as a result of enlargement of terminal air spaces and the decrease in alveolar surface area. Loss of pulmonary capillary volume and ventilation/perfusion inequalities may also play a role.

Pulmonary circulation

Little is known about the effects of ageing on the pulmonary circulation. The compliance of the pulmonary arterial vasculature decreases with advancing age, but the changes are relatively small. Cor pulmonale and right heart failure are common sequelae of chronic obstructive pulmonary disease, yet information is lacking on how age *per se* affects these conditions. Pulmonary embolism in elderly people is discussed in the section on lung diseases below.

Control of ventilation

Several aspects of the ventilatory control system are affected by ageing. Breathing patterns in elderly subjects during both wakefulness and sleep are different from those of younger adults. But how these changes in ventilation occur and their clinical significance are not yet understood.

The ventilatory responses to both hypoxia and hypercapnia are reduced in elderly individuals (Fig.3); the hypoxic response of elderly men is about half that of younger men

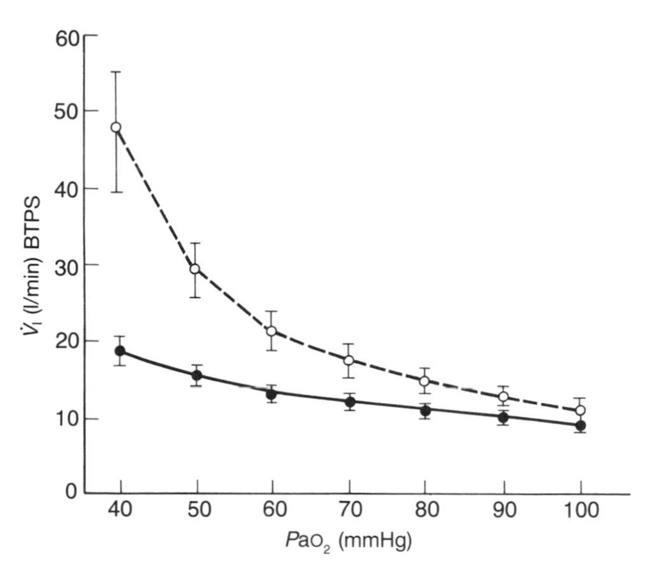

Fig. 3 Ventilatory response to isocapnic progressive hypoxia in eight young normal men (broken line) and eight normal men, aged 64 to 73 years (solid line); values are means ± SEM (from Kronenberg and Drage 1973).

(Kronenberg and Drage 1973). The mechanism of this change in ventilatory responsiveness with increasing age is not completely understood. It is believed to be due primarily to alterations in the central nervous mechanisms that control efferent outflow to the respiratory muscles (Pack and Millman 1988). Diminished perception of ventilatory stimuli (hypoxia or hypercapnia) and of respiratory mechanical loads may also play a role. Related work has demonstrated that under hypoxic conditions, the response to hypercapnia is lower in elderly than younger subjects. This suggests that ageing is also associated with reductions in ventilatory responsiveness to combined hypercapnia and hypoxia. Whatever the mechanism for these reductions in ventilatory responsiveness, they place the elderly person at increased risk of hypoxaemia with the added stress of acute or chronic respiratory illness.

The reduced ventilatory response in older subjects may also contribute to the increased frequency of Cheyne–Stokes breathing, which has increased prevalence in the elderly population. Compromised blood flow to the respiratory centres in the medulla is thought to contribute to this breathing disorder, which is frequently seen in patients with congestive heart failure.

Response to exercise

The changes in lung function that occur with normal ageing do not cause dyspnoea at rest or during normal activity. However, the capacity for strenuous exercise is greatly reduced. Limitations in pulmonary function, less efficient gas exchange, and decreased maximal cardiac output result in a decreased maximal oxygen uptake and anaerobic threshold during exercise with advancing age (Pack and Millman 1988).

Older people have an increased ventilatory response to carbon dioxide production ($\Delta\dot{V}_E/\dot{V}_{CO_2}$) during exercise as compared with younger people (Brischetto *et al.* 1984). The increase in ventilation during exercise compensates for the increased physiological dead space that occurs with ageing. Yet, as discussed above, elderly persons also have a decreased ventilatory response to carbon dioxide. The mechanism of these changes and their physiological or clinical significance are not known. However, they suggest that older individuals have an altered responsiveness to respiratory stresses.

Respiratory disturbances during sleep

There has been growing awareness in recent years that disturbances of breathing during sleep constitute a common problem among elderly people. During sleep, their pattern of ventilation is much more irregular than that of younger adults. The frequency of periodic breathing, apnoea, and episodes of oxygen desaturation during sleep increases with advancing age. The apnoea tends to occur predominantly in men, and primarily during light sleep (stage 1 or 2 sleep). It may be central or obstructive, and appears to be an extension of the periodic breathing observed in both young and old adults (Pack and Millman 1988). The mechanisms for these irregularities in breathing in older people during sleep are not known. Elderly subjects who develop apnoea during sleep appear to have larger periodic oscillations of ventilation while awake than do age-matched controls without apnoea (Pack and Millman 1986). This suggests that apnoea during sleep in some older subjects might be related to alterations in respiratory control present during wakefulness (for further discussion of sleep apnoea see Chapter 18.8).

Lung cell functions

In contrast to the amount of information that has been gathered concerning the effects of ageing on lung physiology, little is known about how the lung's cellular and metabolic activities are affected by advancing age. The past few decades have been a time of active research in lung cell biology and metabolism, yet practically nothing is known about how or if important activities such as pulmonary endothelial function or surfactant metabolism are altered with ageing. The sketchy information that is available suggests that alterations in lung cell functions with advancing age could affect the lung's response to various injuries including mechanical ventilation and oxidant stress. Much more needs to be done systematically to investigate these age-associated changes in lung cell function before any implications can be made about their possible significance to pulmonary diseases in older people.

Lung defence functions

Several aspects of the pulmonary defence systems have been shown to decline with advancing age. Older individuals appear to have slowed clearance of particles from the airways, although the mechanisms responsible are not completely understood (Krumpke *et al.* 1985). In contrast, the antibacterial or phagocytic functions of alveolar macrophages do not appear to be altered in old animals. Yet the function of alveolar macrophages is impaired by conditions such as smoking, malnutrition, and prolonged exposure to environmental pollutants. How or if these factors might interact to impair the function of alveolar macrophages and increase susceptibility to lower respiratory tract infections in the elderly is not known.

Older persons appear to be more sensitive than the general population to the effects of pollutants. Significant increases in fatality have been observed during episodes of severe pollution. How or if altered antioxidant activity in lung cells might contribute to this increased risk is unknown. Indeed, virtually nothing is known about the regulation or specific activity of pulmonary antioxidant enzymes, lung cell bioenergetics, or pulmonary repair mechanisms after exposure of older animals or humans to oxidant stress. Age-associated differences in the pulmonary response to ozone have been demonstrated in rats: total triphosphonucleotides are only slightly decreased in adult animals exposed to ozone, but are markedly decreased in aged animals; adult animals maintain a greater proportion of their available triphosphonucleotides in the reduced form (NADPH) compared to aged animals. These findings suggest that aged animals may not be able to maintain pulmonary reducing equivalents as efficiently as adult animals in the face of an oxidant insult (Montgomery *et al.* 1987). These preliminary investigations can only hint at the possible perturbations in pulmonary antioxidant defences in older human lungs.

A study of the drug-metabolizing activity of lung tissue obtained from rats of various ages found that, as in liver and kidney, a major decline in activity occurs between young adulthood and middle age rather than from middle age to senescence. Overall, there is a general declining trend in pulmonary drug-metabolizing ability in older animals. Certainly, more detailed information is needed concerning the decreased ability to metabolize various compounds with advancing age.

Active transport mechanisms

The ability to regulate the clearance of solutes and water from the air spaces is diminished in the lungs of aged rats (Goodman 1988). The rate of active sodium reabsorption could be stimulated by a beta-agonist (terbutaline) in young but not in aged rats. This implies that the mechanisms for maintaining alveolar fluid clearance may be altered in older animals. Further studies on the effects of ageing on active transport mechanisms in lung are needed. Whether the differences observed in rats also occur in humans is not known.

Airway cell reactivity

Beta-adrenergic responsiveness declines with age. Beta-adrenergic agonist affinity and both basal and hormonally stimulated adenylate cyclase activity are decreased in the lungs of aged rats (Scarpace and Abrass 1983), and receptor density may decrease with age as well. These findings suggest that the catecholamine pathway is somehow altered in senescent lungs. Similar changes have been found in ageing rat myocardium and human lymphocytes. Again, more studies are needed to determine if these changes occur in human lung. A study of alpha-adrenergic and cholinergic responsiveness found no significant change in airway reactivity with age in healthy adults (Davis and Byard 1988).

Lung surfactant

Surfactant production in aged animals has not been studied in great detail. The lecithin content and rates of synthesis in lung appear to be similar in adult and old rats. But a decrease in the number and volume density of lamellar bodies in type II alveolar cells in old monkeys has been reported, suggesting that surfactant synthesis or secretion may be impaired in these aged animals (Shimura *et al.* 1986). Whether surfactant reserves are relatively depleted in older persons is not known. A decreased reserve could cause elderly individuals to be more susceptible to surfactant depletion or dysfunction during mechanical ventilation, malnutrition, or pulmonary oedema, and would suggest that surfactant replacement may be a valuable adjunct to the treatment of pneumonia or severe respiratory distress in elderly patients. This is pure speculation, and further knowledge must await future studies on the regulation of pulmonary surfactant in elderly subjects.

LUNG DISEASES IN ELDERLY PEOPLE

Introduction and the magnitude of the problem

Respiratory diseases are a source of considerable burden in the older population. They account for approximately 200 000 deaths per year in people over the age of 65, and cause about 14 per cent of all deaths in the older United States population. In 1987, chronic obstructive pulmonary disease (**COPD**) and allied conditions, and pneumonia were the fourth and fifth leading causes of death, respectively, in people over the age of 65 years. The death rates from lung cancer, pneumonia, and COPD increase dramatically with increasing age (Fig. 4). As shown in Table 1, lung cancer, COPD, and pneumonia are the major causes of respiratory death in the elderly population, and the majority of deaths from these respiratory diseases occur in people who are 65 years of age and older.

The prevalence of COPD and allied conditions, including asthma, is approximately 14 per cent of people 65 years of age and older, an estimated 4 million people in the United States in 1986. Respiratory diseases accounted for over 8.5 million hospital days in elderly people in 1987 (Table 2). In this section the epidemiology and clinical features of some of the most prevalent respiratory diseases and conditions in the older population will be reviewed.

Obstructive pulmonary diseases

Chronic obstructive pulmonary disease

Epidemiology

Diseases causing chronic airway obstruction are a major source of illness and death in the elderly population. COPD

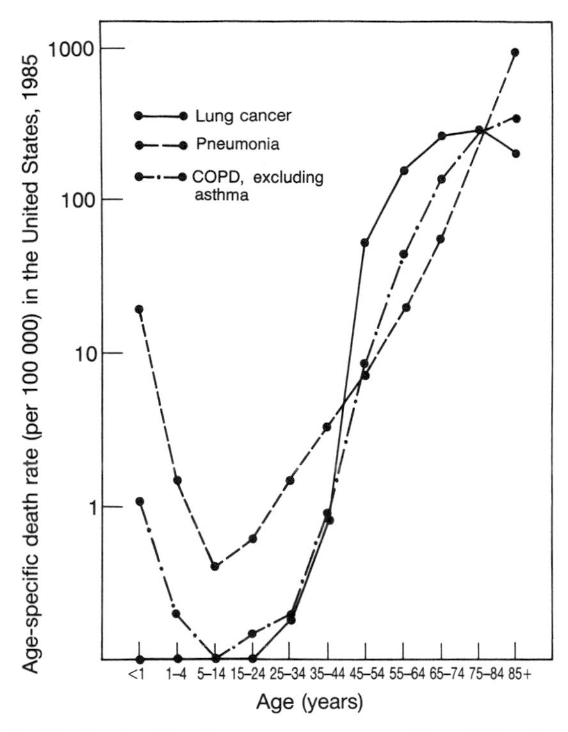

Fig. 4 Death rates from lung cancer, pneumonia, and chronic obstructive lung disease increase dramatically with advancing age (National Center for Health Statistics).

(emphysema and bronchitis) affected approximately 3.3 million people 65 years of age and older in the United States in 1987. The economic costs of COPD, excluding asthma, were $4.7 billion in the United States in 1986. Associated indirect costs of lost earnings and premature death were an additional $6 billion (National Heart Lung and Blood Institute estimates).

There is a striking increase in death rates from COPD with increasing age (Fig. 4). Fatality rates for COPD are more than twice as high in men than women at the ages of 65 to 74 years, and more than three times as high at 75 to 84 years and older. The age-adjusted death rate for COPD increased by 71 per cent between 1966 and 1986. During the same period, death rates for coronary heart disease decreased by 45 per cent. Death rates from COPD are still increasing in the United States, particularly among women and elderly people.

Early intervention

The major clinical features, natural history, and management of COPD have been addressed in several excellent reviews (Anthonisen 1988) and will not be discussed here. One aspect of treatment that is of particular interest concerns the question of whether early intervention can effectively prevent or slow down the deterioration of lung function in patients with COPD. This issue is being looked at from several different points of view. The effect of stopping smoking on the decline of lung function in patients with COPD, particularly elderly patients, is unclear. Stopping early in the course of the disease appears to be beneficial in slowing the decline of FEV_1 (Camilli *et al.* 1987). Bronchodilator therapy has also been associated with a slower decline in lung function (Anthonisen *et al.* 1986). The effects of both these interventions on lung function and respiratory illness in patients with early airways obstruction are being tested in the Lung Health Study, a large multicentre study supported by the National Heart, Lung, and Blood Institute.

Respiratory muscle training and rest

Another potential intervention and an active area for research relates to the role of respiratory muscle training and rest in altering the course of COPD. The effects of lung overexpansion on decreasing muscle efficiency, chronic undernutrition, and disuse atrophy of respiratory muscles, cause the patient with COPD to be vulnerable to acute fatigue of the respiratory muscles. This fatigue is thought to complicate acute respiratory failure due to COPD, but its prevalence in patients with COPD is unknown. Rest (i.e. mechanical ventilation) and training of respiratory muscles are part of the intensive efforts at rehabilitation that are sometimes required to restore a patient to maximal functional status. The role of chronic respiratory muscle fatigue in affecting the course of COPD is also uncertain. Whether respiratory muscle training is more efficacious than other forms of exercise in improving quality of life and providing patients with an increased reserve of respiratory muscle function remains to be determined.

Asthma

Asthma in older as in younger patients is characterized by hyper-reactive airways and diffuse airway obstruction with intermittent symptoms of chest tightness, wheezing, dyspnoea, and cough. In older patients, however, asthma often coexists with other medical conditions, most commonly COPD and cardiac disease, that complicates diagnosis and management. Patients should be educated to improve their compliance with treatment regimens, but compliance may be difficult to achieve for certain patients because of physical or other limitations.

Epidemiology

The prevalence of asthma in people over 65 years of age is essentially the same as in people under 65 years of age, i.e. approximately 40 cases per 1000 persons. Although asthma causes relatively few deaths among the elderly, it accounts for an estimated 700000 hospital days each year in the United States (see Tables 1 and 2). Approximately 1 million people in the United States over the age of 65 have asthma. Although asthma usually begins in childhood or early adult life, it can occur for the first time in the elderly individual. The incidence of asthma in later life is not known. Some reports suggest that approximately one-third of elderly asthmatics first develop asthma late in life (Braman and Davis 1986).

Epidemiological studies have detected an increasing number of deaths from asthma over the past 10 years, particularly in patients 75 years of age or older. However, the contribution of asthma to fatality is not clear because of confounding effects of multiple medical problems.

Aetiology

The aetiology of asthma is unknown. Airway hyper-responsiveness in asthma is thought to result from an imbalance

Table 1 Deaths from respiratory diseases among people aged 65 years and older; United States, 1985

Cause of death	ICD-9 code	Deaths*	Per cent of total deaths (all ages)
Tuberculosis	010–012	825	60
Lung cancer	160–165	77 032	61
Pulmonary embolism	415.0–415.9	7389	71
Pneumonia	480–486	56 949	87
Influenza	487	1880	92
Bronchitis, chronic and unspecified	490–491	2972	82
Emphysema	492	11 344	80
Asthma	493	2053	53
Other COPD and allied conditions	494–496	44 265	83

* National Center for Health Statistics.
COPD, Chronic obstructive pulmonary disease.

Table 2 Number of hospital days for patients 65 years of age and older, United States, 1987*

First-listed diagnosis	ICD-9 code	Number of days	Number of days per 100 000
Pneumonia	480–486	4 440 000	14 880
Influenza	487	46 000	150
COPD	490–492, 494–6	1 707 000	5720
Asthma	493	703 000	2360
Lung cancer	162	1 415 000	4740
Pulmonary embolism	415.1	272 000	910
Tuberculosis	010–012	76 000	250

* National Center for Health Statistics Health Interview Survey, US 1987 (provisional figures).
COPD, chronic obstructive pulmonary disease.

between adrenergic and cholinergic control of bronchial, smooth muscle tone. In the older patient, asthma is most likely to be precipitated by a respiratory viral infection, perhaps as a result of damage by the infectious agent to the airway epithelium, resulting in exposed nerve endings that function as irritant receptors and induce bronchospasm. Bronchoconstriction can be induced by a wide range of triggers including chemicals and dusts, exercise in cold air, gastro-oesophageal reflux, and aspirin and other non-steroidal anti-inflammatory agents. Atopy is usually not an important factor in asthma in older adults.

Clinical features and diagnosis

Intermittent wheezing, shortness of breath, chest tightness, and cough are typical symptoms of asthma. Paroxysmal nocturnal dyspnoea is a common complaint. Attacks usually involve production of thick, mucoid sputum. Elderly patients with intrinsic (non-allergic) asthma may present with continuous wheezing rather than the episodic wheezing of younger patients. On the other hand, wheezing may be absent entirely, and cough and dyspnoea may be the only symptoms of the disease.

Careful differential diagnosis is a major part of the proper management of the elderly patient with symptoms that suggest asthma (Table 3). There is considerable diagnostic overlap between asthma and COPD in older patients who are smokers or exsmokers. Chronic bronchitis, emphysema, and left heart failure can cause episodic nocturnal dyspnoea which mimics asthma. Wheezing in elderly patients is associated with these relatively common conditions as well as pulmonary embolism, pulmonary aspiration, and upper airway obstruction. Chest tightness may be misconstrued by the patient or the physician, or both, as angina pectoris.

Table 3 Differential diagnosis of asthma in the elderly person (from Bardana 1984)

Acute onset
Lower respiratory tract infection
Acute bronchitis
Pneumonia
Upper respiratory tract infection
Sinusitis
Congestive heart failure
Aspiration
Pulmonary embolism
Pneumothorax
Hyperventilation
Chronic onset
Emphysema
Chronic bronchitis
Occupational pulmonary disease
Tracheal or bronchial obstruction
Bronchogenic carcinoma
Tracheomalacia

One feature that distinguishes asthma from chronic bronchitis and emphysema is that the airway obstruction in asthmatics is reversible, either spontaneously or after inhalation of bronchodilators. However, reversible airways obstruction is not a consistent finding on pulmonary function testing in the elderly asthmatic (Braman and Davis 1986), and many patients have both a reversible and irreversible element to their airflow obstruction. Bronchial hyper-reactivity can be assessed in patients with normal spirometry by administration of bronchoconstricting agents. A decrease in FEV_1 of 20 per cent or more after inhalation of small doses of histamine or methacholine strongly suggests the diagnosis of asthma.

Treatment

Elderly asthmatics will often require understanding and supportive care. Careful evaluation of their medical status and individualized attention to pharmacological treatments are essential to avoid the exacerbation of coexisting medical conditions and to reduce the toxic effects of prescribed treatments.

Continuous pharmacotherapy with bronchodilators and corticosteroids is often required to control symptoms of asthma in the elderly patient. The selective use of aerosolized beta-2-adrenergic receptor agonists that cause bronchodilatation without exacerbation of arrhythmias or tremors is especially important. Beta-blocking agents used to treat hypertension, angina, or glaucoma may exacerbate an asthma attack by blunting the response to bronchodilators. Methylxanthines are potent bronchodilators that are most useful when aerosol sympathomimetics are not sufficient. In the elderly, very slow drug metabolism caused by heart or liver failure, and some antibiotics (especially erythromycin) can increase the potential toxicity of these drugs.

Corticosteroids are an important adjunct in the management of bronchial asthma. Use of inhaled corticosteroids is the safest form of chronic steroid therapy. Long-term exposure to systemic corticosteroids in the elderly patient can present special problems of bone demineralization and fractures, diabetes mellitus, activation of previous tuberculosis, accentuation of hypertension, and cataracts.

Cancer of the lung

The economic, medical, and social burdens of lung cancer in elderly people are enormous, and are likely to continue for some time. In 1985, 61 per cent of deaths attributed to lung cancer were of people 65 years of age or older (see Table 1). Prevention of lung cancer through control of cigarette smoking is the most important tool for eliminating deaths from lung cancer in the future. Until this can be accomplished, more effective detection and treatment of lung cancer in older patients is urgently needed.

Epidemiology

Lung cancer is now the most common cause of cancer death in the United States for both men and women; in 1985, it caused about 77 000 deaths in people 65 years of age or older (see Table 1). Men 65 years of age and older are three times more likely to develop lung cancer than men 45 to 64 years of age (O'Rourke and Crawford 1987). The incidence of this disease increases dramatically with increasing age, most probably because of the cumulative impact of a lifetime exposure to cigarette smoke. A decline in the incidence of lung cancer occurs in people who are 75 years of age or older, possibly because of a decrease in smoking prevalence due to cohort effects or selective survival. For smokers, the risk of lung cancer continues to increase with increasing age.

Aetiology

Approximately 80 to 90 per cent of bronchogenic carcinoma may be directly linked to cigarette smoking, with the risk of developing lung cancer increasing with total exposure to cigarette smoke. Data obtained from smoking histories taken in 1978 to 1980 suggest that the greatest lifetime cumulative exposure to cigarette smoking probably occurred in men currently in their seventh and eighth decades of life, and women now in their fifth and sixth decades (Harris 1983).

The risk for smokers of developing lung cancer approaches that of non-smokers approximately 10 to 20 years after smoking is stopped. Recent evidence suggests that people over the age of 65 years experience a decline in risk for lung cancer on stopping smoking similar to that of people under 65 years of age, and that for persons 65 years of age and older, the number of cigarettes smoked per day is more important in determining the risk of developing lung cancer than the number of years the individual has been smoking (Pathak *et al.* 1986). It is therefore most worthwhile strongly to encourage and support the elderly individual in his or her efforts to stop smoking.

Other factors that may contribute to risk for lung cancer include occupational exposures, particularly when combined with cigarette smoking, passive smoking, and family history.

Clinical features and natural history

Lung cancer is usually symptomatic at the time of diagnosis; only about 10 per cent of patients with lung cancer will be asymptomatic when first diagnosed. Early identification of the symptoms of lung cancer can be difficult in older patients. Table 4 lists the common symptoms in such patients that may be mistaken as signs of other illnesses and conditions. Lung carcinoma should always be suspected in older individuals, particularly smokers, who present with these symptoms. The range and frequency of lung cancer symptoms experienced by elderly patients are similar to those of younger patients, although older patients have been reported to experience more dyspnoea and less chest pain than younger patients (DeMaria and Cohen 1987).

Table 4 Lung cancer symptoms that may be confused with non-cancer symptoms in the elderly patient (from O'Rourke and Crawford 1989)

Lung cancer symptom	Comorbid disease or 'ageing explanation'
Cough	Chronic bronchitis
Dyspnoea	Emphysema, old age
Fever (postobstructive pneumonia)	Cold, flu
Weight loss	Depression, inactivity
Bone pain (bone metastases)	Arthritis, old age
Altered mental status (brain metastases, hypercalcaemia)	Dementia, old age

From: O'Rourke and Crawford, 1987.

The course of lung cancer is usually rapid and fatal. The average 5-year survival rate for all patients with lung cancer is 13 per cent, and for patients with localized tumours it increases to 33 per cent (Silverberg and Lubera 1988). White males over the age of 75 have a survival rate of 3 per cent for all stages of lung cancer, and 7 per cent for localized disease at 5 years after diagnosis (DeMaria and Cohen 1987).

In general, older patients receive less aggressive therapies or no definitive treatment at all for their lung cancers. The proportion of patients receiving treatment for lung cancer at the local stage was recently reported in one study to be 90 per cent for those aged 55 years or younger, and only 50 per cent for those of 75 to 84 years (Samet *et al.* 1986). This may explain, at least in part, why the survival rate for lung cancer decreases with increasing age (Fig. 5).

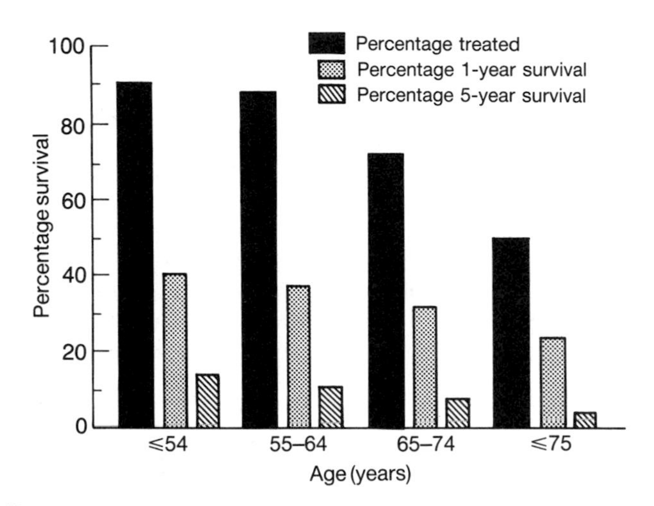

Fig. 5 A comparison of the trends of decreased treatment and survival from lung cancer with increasing age in the United States (from O'Rourke and Crawford 1987).

The frequency of metastases, and the stage or extent of disease are decreased in elderly patients with lung carcinoma (Filderman *et al.* 1986; DeMaria and Cohen 1987; Teeter *et al.* 1987). This is surprising, as at the time of diagnosis many cancers are more advanced in older patients than in the general population. Elderly patients with lung cancer, in contrast, experience more local lung cancer than younger patients. This trend is illustrated by a study of over 9000, histologically confirmed cases of lung carcinoma in which the proportion of localized lung cancer increased from 25 per cent in the 40- to 49-year age group to 42 per cent in those aged 80 or more (Teeter *et al.* 1987). These findings are supported by a study of almost 23000 patients which demonstrated an increase of local-stage disease with increasing age for all four lung cancer subtypes: squamous cell and small cell carcinoma, adenocarcinoma, and large cell carcinoma (O'Rourke *et al.* 1987), and by studies that found more patients over the age of 70 suitable for management by lobectomy than by pneumonectomy (LeRoux 1968).

The rise in the proportion of squamous cell carcinoma with increasing age may help to explain the increased prevalence of local-stage disease in older patients because this is the subtype most likely to be slow growing. It appears that older patients are less likely than younger to have their cancers confirmed histologically (Teeter *et al.* 1987). The relatively lower levels of metastatic disease in older patients might also, then, be a result of less aggressive staging procedures. Different smoking habits, and more frequent medical examinations (and earlier detection) in the older population may also play a part.

Treatment

The three main treatments for lung cancer are surgery, chemotherapy, and radiotherapy. There is considerable experience with surgery for this disease in older patients (Ginsberg *et al.* 1983; O'Rourke and Crawford 1987). Surgical resection is the treatment of choice for clinical stage I and II carcinoma (except small cell cancers; see below) for all age groups, although patients over 70 have a decreased survival when compared with younger patients. The Lung Cancer Study Group analysed 2200 patients who had resections for lung cancer (Ginsberg *et al.* 1983). The postoperative death rate for patients 70 years or older was 7.1 per cent (pneumonectomy, 5.9 per cent; lobectomy, 7.3 per cent). Patients under the age of 60 years had a 30-day death rate of 1.3 per cent, as compared with 4.1 per cent for patients 60 to 69 years, 7.1 per cent for patients 70 years or older, and 8.1 per cent for patients 80 or older. In other studies of patients over 70 years of age who had surgical resections the 5-year survival rates range from 11 to 46 per cent (DeMaria and Cohen 1987; O'Rourke and Crawford 1987).

Among patients with resectable, stage I lung cancer (except small cell carcinoma), age has very little effect on the recurrence rates after surgery, and is a less important determinant of long-term survival than are the size of the tumour, the extent of nodal involvement, the cell type, the presence of infection in the postoperative period, and the ability to carry out normal activities before surgery. Thus, age alone is not a contraindication to surgery although older patients will often have more coexisting medical conditions, including cardiovascular impairment and COPD, that will contribute to a higher risk of complications and the likelihood of a prolonged recovery period. The decision to proceed with surgery in an older patient requires careful evaluation of the patient's medical status, including cardiopulmonary function, tumour characteristics, psychological function, and personal preference.

Radiation therapy of carcinoma (except small cell cancer) can offer some benefit to older patients who are too compromised for surgical treatment, and can provide palliation and relief of symptoms. However, with radiotherapy as with surgery, the patient's overall medical condition, especially the pulmonary functional reserve, is the major limiting factor in determining whether or not the treatment can be given in curative doses. The most common complication of radiation therapy is pneumonitis. Although age *per se* does not influence the incidence of this, elderly patients may experience more severe pneumonitis than younger patients. This may be due to the greater number of pre-existing lung conditions in older patients, most notably emphysema, bronchiectasis, and fibrosis.

Chemotherapy, with or without radiotherapy, is the major option for treatment of small cell carcinoma of the lung. Toxic reactions, including leucopenia, febrile episodes, and cardiotoxicity are more frequent in older patients. However, the response rate to chemotherapy for this tumour is similar to that of younger patients. Older patients therefore can and should benefit from the recent advances in treatment of small cell carcinoma which have improved the prognosis of this disease. Recent progress in understanding the cell biology of lung cancers, particularly small cell carcinomas, may well lead to the development of new therapeutic interventions in the future.

Infectious diseases of the lung

(See also Chapter 3.2.)

Tuberculosis

Epidemiology

The incidence of pulmonary tuberculosis has been declining in the United States since the beginning of this century. Yet the percentage of new cases of tuberculosis in patients over the age of 65 is increasing, and elderly people are the largest group in this country with active tuberculosis.

Tuberculosis remains dormant for decades after the primary infection and then reactivates when the host response is impaired. Reactivation of pulmonary tuberculosis is the most common form of this infection in older individuals, but they can develop primary tuberculosis, especially in institutional settings such as nursing homes. The case rate of tuberculosis is estimated to be almost five times greater for nursing home residents than for elderly people living at home (Stead and To 1987).

Aetiology

The factors determining the risk of pulmonary tuberculosis for elderly persons are not well understood. A decrease in cell-mediated immunity is known to increase the risk of reactivation of dormant disease. Cancer and conditions that interfere with the immune system may increase their susceptibility to recrudescence of dormant tuberculous infection or the acquisition of new infection. The risk of pulmonary tuberculosis is increased by poor nutrition, alcohol abuse, diabetes, renal dialysis, and corticosteroids and other immunosuppressive drugs.

Clinical features and diagnosis

Reactivation tuberculosis can be difficult to diagnose in elderly patients because they often present with few or vague respiratory symptoms, atypical findings on chest radiographs, and coexisting illnesses, such as chronic bronchitis, which can obscure the symptoms of tuberculosis. The classic symptoms of fever, weight loss, night sweats, sputum production, and haemoptysis are found more often in young than older patients. Apical cavitation, which is typical of tuberculosis in the general population, is also not as common in elderly patients.

There is uncertainty about the reliability and interpretation of tuberculin skin testing in the older population (Stead and To 1987). A waning of tuberculin sensitivity appears to be part of the ageing process, and anergy has been reported to be increased to varying degrees in the elderly population. The older individual with tuberculosis may show no reaction when tested with tuberculin the first time, yet show a significant reaction if tested a second or even a third time after a period of a week or more. A negative tuberculin skin test in an elderly patient with pulmonary symptoms or unexplained fever should not preclude a sputum examination and culture for acid-fast bacilli (Stead *et al.* 1985).

Mycobacterium tuberculosis must be cultured from sputum or other clinical specimens to make a definitive diagnosis. At times, a presumptive diagnosis of tuberculosis, warranting a therapeutic trial of chemotherapy, may be made on the basis of clinical or radiographic findings in the absence of a positive culture. For example, when tuberculosis is suspected, but evaluations of all clinical specimens are negative, a diagnostic trial of drug therapy is indicated if there are no contraindications.

Treatment

Almost all cases of tuberculosis can be managed by chemotherapy, and the infectious phase can be brought under control in a few weeks. Treatment of the elderly patient with tuberculosis must take into account the possibility of adverse reactions to the drug, compliance, and the existence of underlying disease. The short course of chemotherapy currently recommended by the American Thoracic Society and the Centers for Disease Control employs isoniazid, rifampicin, and pyrazinamide for 2 months and isoniazid and rifampicin for 4 months. Nine-month regimens of isoniazid and rifampicin are equally effective. The risk of drug toxicity increases with age, and patients must be monitored for signs of toxic side-effects, especially hepatitis. The response to chemotherapy in patients with positive identification of *M. tuberculosis* in sputum is best evaluated by repeated examinations of sputum. Prophylactic therapy with isoniazid can be useful in patients of advanced age at increased risk of tuberculosis infection, but elderly patients are at high risk of developing isoniazid-induced hepatitis and other toxic effects of the drug, and prophylaxis should be selective.

Pneumonia

Epidemiology

Common bacterial pneumonias frequently complicate the treatment of patients with COPD and other chronic conditions. In 1987, infections of the lower respiratory tract (pneumonia and influenza) were the most common cause of deaths from infection, and the fifth overall cause of death in elderly people, accounting for approximately 61 000 deaths in people 65 years of age or older in the United States (National Center for Health Statistics, provisional figures). In 1987, pneumonia was responsible for an estimated 4.4 million hospital days for elderly patients in the United States (see Table 2).

Elderly residents of chronic care institutions have a reported incidence of pneumonia that is two to three times greater than among elderly people living in the community. At any one time, as many as 2 per cent of nursing home residents have a lower respiratory tract infection, and when elderly patients enter a hospital, they have a threefold greater incidence of nosocomial pneumonia than younger patients (Niederman and Fein 1986). The elderly also have an increased incidence of complications from pneumonia, such as bacteraemia or meningitis, and a higher death rate from lower respiratory infections than younger adults, especially in the presence of underlying illness. Death rates vary greatly, ranging from 10 to 80 per cent, and depend, in part, on the pathogen involved (Bentley 1984). The most common pathogens causing pneumonia in elderly patients are shown in Table 5.

Table 5 Bacterial pathogens causing pneumonia in older people in descending order of frequency (from Niederman and Fein 1986)

Community-acquired infection	Hospital-acquired infection
Streptococcus pneumoniae	*Klebsiella pneumoniae*
Enteric Gram-negative bacilli	Other enteric Gram-negative bacilli
Legionella pneumophila	*Legionella pneumoniae*
Haemophilus influenzae	*Streptococcus aureus*
Aspiration, including anaerobes	*Staphylococcus aureus*
Staphylococcus aureus	Aspiration, including anaerobes
	Haemophilus influenzae

Aetiology, pathogenesis, and clinical features

Pneumonia results most commonly from aspiration of upper airway pathogens. Colonization of the oropharynx with Gram-negative bacilli occurs more frequently in elderly subjects, particularly in those in hospital. Colonization can lead to more frequent aspiration and serious Gram-negative pneumonias. Ageing *per se* alters antibacterial defences, mucociliary transport, and bacterial adherence. In addition, a weaker cough, endotracheal intubation, drugs such as morphine, atropine, corticosteroids, and sedatives, and coexisting illnesses and malnutrition can further compromise host defences and lead to increased risk of pneumonia in elderly patients (Niederman and Fein 1986).

The diagnosis of pneumonia is sometimes easily missed in older patients because symptoms can be obscured by exacerbation of coexisting diseases or reduced by corticosteroids or other anti-inflammatory agents. Confusion is an important clinical sign. Fever and cough are usually present, but they may be reduced or absent in older patients. Approximately 20 per cent of elderly patients with community-acquired pneumonia are afebrile on admission. The respiratory rate is usually increased in the older patient with pneumonia, and tachypnoea can be present 24 to 48 h before the diagnosis is made. Incomplete consolidation and multiple-lobe penumonia are more common in elderly patients (Bentley 1984; Niederman and Fein 1986).

Cultures of respiratory tract secretions and blood should be made to provide positive identification of the pathogen but sputum samples are often hard to obtain and the results of these studies are not always diagnostic. Immunological staining tests and antibody techniques can be used to identify bacterial and viral agents. Respiratory and nutritional support, and fluid and electrolyte regulation are important components of the management of older patients with pneumonia. Recommendations for specific antibiotic regimens are given by Bentley (1984) and Niederman and Fein (1986). Prophylaxis with pneumococcal vaccines has been shown to be effective, but not in all cases (Niederman and Fein 1986) (see also Chapters 3.2 and 3.3).

Influenza

Elderly persons, especially those with chronic diseases, are at risk of serious complications from influenza virus infection and have a higher death rate than younger patients (Niederman and Fein 1986). Approximately 90 per cent of deaths from influenza occur in persons over 65 years of age (see Table 1). Vaccination is a prime factor in reducing death and illness from influenza, particularly in persons with cardiopulmonary disease and those who are residents of nursing homes. Amantadine should be considered for prophylaxis during epidemics of influenza A for unvaccinated elderly patients who are at increased risk or who have serious coexisting medical illnesses. Amantadine is also effective as a therapeutic agent that can reduce severity of influenza by about half if given early in the course of the disease.

{The use of anti-influenzal prophylaxis for demented patients in institutional care is less common in the United Kingdom than in the United States for legal, ethical, and cultural reasons; see Chapter 3.3.}

Pulmonary embolism

Pulmonary embolism is common in the older population because of the prevalence in the elderly of predisposing conditions including malignancy, cardiac disease, particularly atrial fibrillation and congestive heart failure, COPD, obesity, prolonged immobilization, and trauma, especially hip fractures, which carry a 40 to 70 per cent incidence of venous thromboembolism.

Epidemiology

Because antemortem diagnosis is difficult, largely owing to the variable and non-specific nature of the symptoms, and the presence of confounding underlying conditions, particularly cardiac or pulmonary disease, the exact incidence of pulmonary embolism in elderly people is unknown. A study of autopsy-confirmed pulmonary embolism in a nursing home population supports previous estimates that the cumulative incidence of pulmonary embolism in elderly patients is approximately 13 per cent (Taubman and Silverstone 1986).

Data from the PIOPED (Prospective Investigation of Pulmonary Embolism Diagnosis), multicentre, prospective clinical trial sponsored by the National Heart, Lung, and Blood Institute, suggest that pulmonary embolism has little effect on fatality, regardless of age, for this population (approximately 1000 patients who were referred for a ventilation/perfusion scan for evaluation of pulmonary embolism). In this study, the frequency of pulmonary embolism in patients under the age of 65 (34 per cent), and those 65 years or older (39 per cent) was not significantly different (PIOPED investigators, personal communication).

Clinical features

The respiratory and haemodynamic consequences of pulmonary embolism depend on the extent of the obstruction and on the presence of underlying cardiopulmonary disease. Many conditions common in the older population, for example, myocardial infarction, can be confused with pulmonary embolism, and vice versa. Pulmonary embolism should be suspected in any patient with COPD and cor pulmonale with sudden or progressively worsening dyspnoea. The diagnosis is supported by a reduction in the $Paco_2$ in a previously hypercarbic patient (Lippmann and Fein 1981). The sudden onset of arrhythmias in patients with pre-existing cardiac disease, sudden or progressive worsening of congestive heart failure, attacks of syncope or dizziness, particularly if dyspnoea is present, hyperventilation, and postoperative pneumonitis, should all raise suspicion of pulmonary embolism (Moser 1977).

There are no highly specific symptoms or clinical signs of pulmonary embolism. Dyspnoea is the most common presenting symptom. In patients with pre-existing cardiopulmonary disease, symptoms are more severe and pleuritic chest pain and haemoptysis are more common. Airway disease and chronic congestive heart failure predispose the patient to pulmonary infarction, and also, by mechanisms not yet understood, delay the resolution of an embolus (Moser 1977; Kelley and Fishman 1988). Tachycardia and tachypnoea are the most common clinical signs. Decreased breath sounds, râles, and abnormal heart sounds may also be present. In a retrospective study of patients over 65 years of age diagnosed as having a pulmonary embolus, 35 per cent had signs of deep vein thrombosis (Busby *et al.* 1988).

Diagnosis and treatment

The procedures available for diagnosis of pulmonary embolism have been discussed in depth (Kelley and Fishman 1988; Moser 1988). However, information specifically related to the elderly population is very limited. Ventilation–perfusion scans are well tolerated by elderly people and, in one study, confirmed the diagnosis in approximately half of patients with a suspected pulmonary embolus (Busby *et al.* 1988). However, interpreting a scan may be difficult in the older patient because of scarring, COPD, or panlobular emphysema. Pulmonary angiography is often required to confirm the presence of pulmonary embolism.

There is little information that specifically deals with prophylaxis of thromboembolism or treatment of pulmonary embolism in elderly patients. Anticoagulants are usually well tolerated by them (Busby *et al.* 1988), but the risk of complications is higher than in younger patients. The risk of bleeding with heparin therapy has been reported to be increased in women over the age of 60 years (Bell and Simon 1982). The use of thrombolytic agents is more controversial. They are recommended only for selected patients with a massive pulmonary embolism or severely compromised circulatory dynamics. Advanced age is a relatively minor contraindication to their use.

CONCLUSIONS

Although a great deal has been learned about the changes in pulmonary structure and function that occur with normal ageing, much more needs to be done. This is particularly true in regard of lung cellular functions and ageing. Virtually nothing is known about if or how these fundamental lung functions are altered with advancing age. Similarly, many important questions persist regarding the pathogenesis and treatment of lung diseases in the elderly. Thus, improved understanding of the ageing respiratory system in both health and disease remains an important priority for the future.

References

Anthonisen, N.R. (1988). Chronic obstructive pulmonary disease. *Canadian Medical Association Journal*, **138**, 503–10.

Anthonisen, N.R., Wright, E.C., and the IPPB Trial Group (1986). Bronchodilator response in chronic obstructive pulmonary disease. *American Review of Respiratory Disease*, **133**, 814–19.

Bardana, E.J., Jr. (1984). Treating the elderly asthmatic patient. *Comprehensive Therapy*, **10**, 15–23.

Bell, W.R. and Simon, T.L. (1982). Current status of pulmonary thromboembolic disease: Pathophysiology, diagnosis, prevention, and treatment. *American Heart Journal*, **103**, 239–62.

Bentley, D.W. (1984). Bacterial pneumonia in the elderly: clinical features, diagnosis, etiology, and treatment. *Gerontology*, **30**, 297–307.

Braman, S.S. and Davis, S.M. (1986). Wheezing in the elderly. *Geriatric Clinics of North America*, **2**, 269–83.

Brischetto, M.J., Millman, R.P., Peterson, D.D., Silage, D.A., and Pack, A.I. (1984). Effect of aging on ventilatory response to exercise and CO_2. *Journal of Applied Physiology: Respiratory, Environmental and Exercise Physiology*, **56**, 1143–50.

Burrows, B., Lebowitz, M.D., Camilli, A.E., and Knudson, R.J. (1986). Longitudinal changes in forced expiratory volume in one second in adults. *American Review of Respiratory Disease*, **133**, 974–80.

Busby, W., Bayer, A., and Pathy, J. (1988). Pulmonary embolism in the elderly. *Age and Ageing*, **17**, 205–9.

Camilli, A.E., Burrows, B., Knudson, R.J., Lyle, S.K., and Lebowitz, M. D. (1987). Longitudinal changes in forced expiratory volume in one second in adults. *American Review of Respiratory Disease*, **135**, 794–9.

Davis, P.B. and Byard, P.J. (1988). Relationships among airway reactivity, pupillary α-adrenergic and cholinergic responsiveness, and age. *Journal of Applied Physiology*, **65**, 200–4.

DeMaria, L.C., Jr. and Cohen, H.J. (1987). Characteristics of lung cancer in elderly patients. *Journal of Gerontology*, **42**, 540–5.

Filderman, A.E., Shaw, C., and Matthay, R.A. (1986). Lung cancer in the elderly. *Clinics in Geriatric Medicine*, **2**, 363–83.

Ginsberg, R.J. *et al.* (1983). Modern thirty-day operative mortality for surgical resections in lung cancer. *Journal of Thoracic and Cardiovascular Surgery*, **86**, 654–7.

Goodman, B.E. (1988). Characteristics and regulation of active transport in lungs from young and aged mammals. In *Membrane biophysics III: biological transport* (ed. M.A. Dinno and W.McD. Armstrong), pp. 263–74. Liss, New York.

Harris, J.E. (1983). Cigarette smoking among successive birth cohorts of men and women in the United States during 1900–1980. *Journal of the National Cancer Institute*, **71**, 473–9.

Kelley, M.A. and Fishman, A.P. (1988). Pulmonary thromboembolic disease. In *Pulmonary diseases and disorders*, Vol. 1 (2nd edn) (ed. A. P. Fishman), pp. 1059–86. McGraw-Hill, New York.

Kronenberg, R.S. and Drage, C.W. (1973). Attenuation of the ventilatory and heart rate responses to hypoxia and hypercapnia with aging in normal men. *Journal of Clinical Investigation*, **52**, 1812–19.

Krumpke, P. E., Knudson, R.J., Parsons, G., and Reiser, K. (1985). The aging respiratory system. *Clinics in Geriatric Medicine*, **1**, 143–75.

LeRoux, B.T. (1968). Influence of age on management of bronchial carcinoma. *Geriatrics*, **23**, 148–56.

Lippmann, M. and Fein, A. (1981). Pulmonary embolism in the patient with chronic obstructive pulmonary disease. *Chest*, **79**, 39–42.

Mahler, D.A., Rosiello, R.A., and Loke, J. (1986). The aging lung. *Geriatric Clinics of North America*, **2**, 215–25.

Montgomery, M.R., Raska-Emery, P., and Balis, J.U. (1987). Age-related difference in pulmonary response to ozone. *Biochimica et Biophysica Acta*, **890**, 271–4.

Moser, K. M. (1977). Pulmonary embolism. *American Review of Respiratory Disease*, **115**, 829–52.

Moser, K.M. (1988). Pulmonary embolism. In *Textbook of respiratory medicine* (ed. J.F. Murray and J.A. Nadel), pp. 1299–1327. Saunders, Philadelphia.

Niederman, M.S. and Fein, A.M. (1986). Pneumonia in the elderly. *Clinics in Geriatric Medicine*, **2**, 241–68.

O'Rourke, M.A. and Crawford, J. (1987). Lung cancer in the elderly. *Clinics in Geriatric Medicine*, **3**, 595–623.

O'Rourke, M.A., Feussner, J.R., Feigle, P., and Laszlo, J. (1987). Age trends of lung cancer stage at diagnosis. *Journal of the American Medical Association*, **258**, 921–6.

Pack, A.I. and Millman, R. (1986). Changes in control of ventilation, awake and asleep, in the elderly. *Journal of the American Geriatrics Society*, **34**, 533–44.

Pack, A.I. and Millman, R.P. (1988). The lungs in later life. In *Pulmonary diseases and disorders*, Vol. 1 (2nd edn) (ed. A. P. Fishman), pp. 79–88. McGraw-Hill, New York.

Pathak, D.R., Samet, J.M., Humble, C.G., and Skipper, B.J. (1986). Determinants of lung cancer risk in cigarette smokers in New Mexico. *Journal of the National Cancer Institute*, **76**, 597–604.

Peterson, D.D. and Fishman, A.P. (1982). The lungs in later life. In *Update: pulmonary diseases and disorders* (ed. A.P. Fishman), pp. 123–36. McGraw-Hill, New York.

Reiser, K.M., Hennessy, S.M., and Last, J.A. (1987). Analysis of age-associated changes in collagen crosslinking in the skin and lung in monkeys and rats. *Biochimica et Biophysica Acta*, **926**, 339–48.

Samet, J., Hunt, W.C., Key, C., Humble, C.G., and Goodwin, J.S. (1986). Choice of cancer therapies varies with age of patient. *Journal of the American Medical Association*, **255**, 3385–90.

Scarpace, P.J. and Abrass, I.B. (1983). Decreased beta-adrenergic agonist affinity and adenylate cyclase activity in senescent rat lung. *Journal of Gerontology*, **38**, 143–7.

Shimura, S., Boatman, E.S., and Martin, C.J. (1986). Effects of ageing on the alveolar pores of Kohn and on the cytoplasmic components of alveolar type II cells in monkey lungs. *Journal of Pathology*, **148**, 1–11.

Silverberg, E. and Lubera, J.A. (1988). Cancer statistics. *CA-A Cancer Journal for Clinicians*, **38**, 5–23.

Sorbini, C.A., Grassi, V., Solinas, E., and Muiesan, G. (1968). Arterial oxygen tension in relation to age in healthy subjects. *Respiration*, **25**, 3–13.

Stead, W.W. and To, T. (1987). The significance of the tuberculin skin test in elderly persons. *Annals of Internal Medicine*, **107**, 837–42.

Stead, W.W., Lofgren, J. P., Warren, E., and Thomas, C. (1985). Tuberculosis as an endemic and nosocomial infection among the elderly in nursing homes. *New England Journal of Medicine*, **312**, 1483–7.

Taubman, L.B. and Silverstone, F.A. (1986). Autopsy proven pulmonary embolism among the institutionalized elderly. *Journal of the American Geriatrics Society*, **34**, 752–6.

Teeter, S.M., Holmes, F.F., and McFarlane, M.J. (1987). Lung carcinoma in the elderly population. *Cancer*, **60**, 1331–6.

Thurlbeck, W.M. (1967). The internal surface area of nonemphysematous lungs. *American Review of Respiratory Disease*, **95**, 765–73.

SECTION 13
Joints and connnective tissue

13.1 Epidemiology of joint and connective tissue disorders

PHILIP H. N. WOOD

Musculoskeletal disorders in elderly people attract particular interest for two reasons. First, patterns of occurrence have distinct features, which is important clinically in regard to diagnosis, and especially differential diagnosis. Characteristics of distribution with age also command attention biologically, notably because of implications for aetiology and how their occurrence might be controlled. Second, less favourable host responses in elderly people determine that clinical management is rarely straightforward. Clinical objectives have to be formulated with a different emphasis, greater attention being paid to aspects of the individual's situation. Moreover, there are wide-ranging implications for the organization of supporting services.

Two clinical phenomena conspire to make differential diagnosis more difficult in elderly patients. The first is that associated characteristics which are strongly related to age assume a different significance; they tend to inject a greater component of 'noise', and their discriminatory value lessens. Secondly, biological features of the various rheumatic disorders lead to differences in their relative frequency of occurrence, so that the diagnostic spectrum is less familiar.

AGE-ASSOCIATED ATTRIBUTES

Variations in usage of the term arthrosis (or osteoarthritis) between specialists such as radiologists, pathologists, and clinicians confound appraisal of this condition. The clinical challenge is usually a patient presenting with pain that is presumed to arise from articular structures. In attempting to elucidate causes, it is customary to request a radiograph of the joint. When this is examined, any features observed have to be related to experience of the frequency and significance of such abnormalities. Two types of difficulty therefore arise.

Taking frequency first, abnormalities of at least mild degree may be seen in radiographs of joints in almost all individuals aged 65 years or more, being present in at least three joints in three-quarters of people in this age group (Lawrence *et al.* 1966). Even if attention is restricted to moderate or severe changes, such abnormalities may be observed in 64 per cent of elderly people, although changes of this severity are present in three or more joints in only about a third of people past retirement age. The other side of this coin, of course, is that, of individuals with radiographic features indicative of arthrosis in joints such as the knee and hip, 10 to 70 per cent (depending on chosen criteria) do not have symptoms related to these findings (Wood 1972; Felson 1988). The mere presence of radiographic changes is therefore of rather limited help.

The second difficulty concerns significance. Of the characteristics that may be identified, the most common are osteophytes. Undue importance tends to be attached to these: they appear to be a universally distributed, age-associated attribute, the pathological significance of which is therefore open to question (Wood 1971). This view is reinforced by the fact that there is little evidence to indicate that osteophytes *per se* do give rise to symptoms, the nature of Heberden's nodes, which may at times be symptomatic, being somewhat different. The significance of other features is also not straightforward. Even when narrowing of a joint is identified, it is still necessary to consider whether this is the immediate or only the underlying cause of symptoms. The therapeutic implications of these alternatives are important. As narrowing can hardly be reversed, other than by surgical replacement, the potential for intervention is severely limited if this change is the direct cause of problems. On the other hand, if it is antecedent the situation is potentially less disturbing because more proximate factors, such as consequential ligamentous strain (Kellgren 1939), may be amenable to correction or control in a way that is not possible with underlying loss of cartilage.

Analogous problems attend interpretation of articular chondrocalcinosis identified on radiographs. In those aged 70 years or more, slightly more than one-quarter may manifest such changes in the wrist, knee, or pubic symphysis (Ellman and Levin 1975), and yet only very rarely does this express itself as symptomatic pyrophosphate arthropathy. Laboratory evidence of a non-specific nature can give rise to similar difficulties. Erythrocyte sedimentation rates in excess of 20 mm/h (Westergren) are quite common in elderly people, and yet underlying disease cannot be detected in quite a proportion. In the past, rheumatoid factor has usually been encountered more frequently in older people, but non-rheumatological influences such as atmospheric pollution have been linked to this; certainly this could be relevant to the tendency for the frequency of positive tests to diminish over time. A number of autoantibodies show even more striking increases in frequency with age, so that the significance of such findings in older people is equally problematical. Finally, serum levels of uric acid also tend to be higher at older ages in both sexes, findings that are as likely to be related to diuretic therapy as they are to the presence of gout. The main lesson to be learned from all these observations is that, if a diagnosis is not established on clinical grounds, greater recourse may have to be made to specific investigations such as synovianalysis, arthroscopy, and biopsy.

THE RHEUMATOLOGICAL SPECTRUM

Diagnosis can be vulnerable to two influences, differences in the relative frequency of occurrence of the various rheumatic disorders with age, and possible variation in how specific individual conditions may express themselves in older people. Underlying these influences is the distinction to be made between incident cases (i.e. the frequency and characteristics of new onsets of particular disorders in older people) and the case mix encountered in everyday practice, where relative frequency of occurrence is determined by general biomedical characteristics of the individual conditions, including such aspects as differential survival of sufferers and the recruitment of incident cases.

Epidemiological data are scant, but available clinical reports, usually of small series, are generally compatible with what has been recorded in population surveys. However, such reports fail to reflect one very important aspect: the various forms of arthritis are responsible for 44 per cent of all physical disability in elderly people (Harris 1971); these conditions are the cause of being bedfast or housebound in 36 per cent of those so confined (Hunt 1978), and sufferers also form the hard core of attenders at many day centres. In such individuals the main condition is arthrosis in 72 per cent of those affected and rheumatoid arthritis in only 13 per cent, the principal limiting disability being located in the knee in half and elsewhere in the legs in a further quarter of those with articular disorders.

General properties associated with senescence, including changes in metabolism and other processes, and reduced life-expectancy, exert their influence on the pattern of rheumatic disorder in elderly people. For instance, the generally slow progress of a condition such as systemic sclerosis diagnosed for the first time in older persons can lead to an appearance of lesser severity, because those affected do not live long enough for the full ravages of the disease to become evident. The pattern observed is also determined by the emergence of new problems, relative frequencies being influenced by the preferential vulnerability of elderly persons to a disorder such as polymyalgia rheumatica. Age is one of the principal discriminators for this condition, and in persons over 70 years seen in consultative practice this disease is encountered almost half as frequently as rheumatoid arthritis; it also shows a female preponderance. The pattern in specialist clinics may be further confounded by difficulty in up to a quarter of patients 'on account of a surfeit of pathology with inadequate diagnostic precision for pinpointing the exact cause' (Grahame 1976).

With more classical rheumatic disorders, such as rheumatoid arthritis and gout, incident cases in older people fall on the tail of the distribution of onsets. In addition to forming a minority of all new cases of these conditions, those affected exhibit certain noteworthy differences in regard to what are customarily thought of as characteristic features. The female preponderance is less evident in rheumatoid arthritis starting in an elderly individual and, although most of the clinical features tend to be similar, the onset is more often abrupt, extra-articular manifestations are uncommon, and a relatively benign form of disease with a good prognosis is encountered more frequently, even in the presence of features usually associated with a more serious outlook, such as high titres of rheumatoid factor (Gibson and Grahame 1973). Gout first manifests itself after the age of 60 years in as many as a tenth of all persons affected by the disease, and almost a third of such presentations are in women; hereditary influences are less important, and the condition is more often secondary to therapy with thiazide diuretics or to myeloproliferative disorders.

ARTHROSIS

Some of the difficulties arising from how observations are regarded have already been mentioned, and how evidence is interpreted will be touched upon in the next section. However, the most common musculoskeletal problem to be encountered calls for further discussion. Similarity in the features observed at different sites of involvement with arthrosis has fostered the view that the condition is an almost universal phenomenon, to be explained by comparable mechanisms wherever it may be encountered. This in turn has also encouraged the notion of a degenerative process.

Such reasoning glosses over a number of aspects. One is that the concatenation of features is often treated as a whole, neglecting the fact that some of those included are fundamental whereas others are only associated or confounding. This has already been noted in regard to osteophytes, but it is also evident in insufficient attention being paid to the fact that what is observed to be similar structurally could nevertheless reflect just a common end-expression of a number of different processes. Regularity at what can be regarded as the micro level tends to deflect attention away from important locus-specific differences, in turn leading to rather sterile arguments about whether the chondrocyte or whatever is the originating source of trouble.

The most important clinical consideration is that significant discontinuities in the evidence get neglected. Structural abnormalities are relatively easy to study, and they have been productive of useful biomedical concepts. The concepts then tend to be linked to clinical problems on the basis of the observation that severe structural changes in a joint are often associated with symptoms. In turn this leads to a common clinical syllogism (Wood 1976): arthrotic changes cause symptoms; this patient has arthrotic changes; *ergo*, this patient's symptoms are due to the arthrotic changes. Although often a reasonable guide, the vulnerability of the conclusion is that the first premise does not state that all arthrotic changes always cause symptoms. In other words, available biomedical concepts have a limited applicability in the elucidation of an individual's subjective problems. This is because, as noted, there is appreciable discordance between clinical problems and evidence of underlying abnormalities such as may be revealed by radiographs. Moreover, structural changes are usually irreversible whereas symptoms in any one individual tend to be inconstant.

Although focusing on arthrosis, a closely similar reasoning applies to the understanding of back pain (Badley 1987). To pursue the implications in the present context, though, one would like to discuss the occurrence of arthrosis at each individual joint in turn, but unfortunately the evidence is insufficiently comprehensive to allow of this. All one can note is that the knee is the most commonly involved major joint, but the hip is undoubtedly the most disabling site. Heterogeneity occurs because three patterns of pathogenesis can be identified. Essentially normal cartilage may fail if subjected to abnormal or incongruous loading for long periods; damaged or defective cartilage can fail under normal conditions of loading; and articular cartilage can break up due to defective subchondral bone. Attractive though these ideas are, though, what constitute normal conditions of loading and how these may alter at different stages of life requires further clarification.

The prevalence of all forms of arthrosis increases with age, but unfortunately there is insufficient evidence to allow one to determine whether rates of change with age differ between the various joints—yet age patterns could indicate whether failure under normal or excessive loading was the more likely at any particular site. Overall the frequency of arthrosis appears to be similar in the two sexes, and men and women are more alike than different in their actual reporting of chronic joint symptoms. However, there are certain differences, though apparent sex differentials need to take account of treatment behaviour and disease state before conclusions can be drawn. Men not only seem to develop arthrosis more frequently in the wrists and hips, but they also often appear to be affected at younger ages—which has encouraged people to speculate about the role of trauma. On the other hand, arthrosis in the hands is encountered more frequently and with greater severity in women, and arthrosis of the knees and sacroiliac joints are also more often affected.

INTERPRETATION

Problems with interpretation in relation to certain features, such as radiographic changes, where it may be difficult to distinguish between the effects of disease and those of ageing, have already been noted. However, regard also needs to be paid to interpretation from a more fundamental biomedical point of view. The distribution of ages at onset of any disease usually follows a relatively simple, curvilinear pattern, and two aspects of such distributions pose interesting challenges.

The first is that observed distribution patterns may not be simple in form, either because there are suggestions of irregularity or bimodality, or because of departures from the relationship between the patterns in the two sexes in different parts of the distribution. Both features can be seen in rheumatoid arthritis, but it is change in sex patterns that is especially striking in elderly people as the female preponderance disappears, and only two classes of explanation can be offered for such findings. Different factors may be influencing disease development and expression, factors that operate differently between the sexes when compared with younger groups. Such a phenomenon has been observed with mortality from rheumatic fever in the past, the customary excess of women sufferers tending to disappear in wartime as male vulnerability is increased by herding together in close quarters, such as barrack rooms. The other type of explanation may be that the condition is heterogeneous, the disease called rheumatoid arthritis that is observed to develop in older people in fact being different and distinct from that affecting younger people. Unfortunately there have not been enough detailed clinical studies to provide evidence as to which of these two types of explanation might apply, although in fact the two are by no means mutually exclusive. As a result, the biological significance of the distribution in rheumatoid arthritis remains obscure.

The second challenge concerns the location of the distribution of ages at onset within the normal life-cycle. One of the more obvious examples is provided by the Hippocratic aphorism that women do not take the gout until the menses have ceased. In regard to human biography and bone and joint disease in the elderly, four patterns can be identified, as follows.

Concentration in the elderly population

Conditions such as polymyalgia rheumatica, Paget's disease of bone, and ankylosing hyperostosis do not show a gradient of ages at onset corresponding to what would be anticipated with ageing or degenerative processes (a sustained and progressive increase over a span of many years). The evidence is too scanty to permit much conjecture, but two comments are worth making. First, conventional possibilities, such as preferential exposure to particular agents, may have relevance to the incidence of fractures in elderly people but are difficult to conceive of in relation to microbial agents—although some old people may be congregated in residential accommodation, the majority are more notable for the solitariness of their existence. Secondly, a threshold concept has utility as a means of providing explanation for why progressive conditions such as osteoporosis tend to manifest themselves abruptly (e.g. vertebral collapse that suddenly impinges on nervous tissue). Progressive deterioration of the tissues may proceed to a qualitatively different state of swift collapse or vulnerability to some insult that formerly they were able to resist (a process perhaps in some way analogous to fatigue fracture in metal).

Increased frequency with age

The most obvious example is arthrosis, which conforms to the criterion of a sustained and progressive increase with age, but whether such a phenomenon should be equated with ageing or degeneration is another matter. It is undeniable that an arthrotic joint has degenerated from its formerly normal structure, but to generalize from this observation is tautologous and misleading—strictures that apply equally to cerebrovascular disease and many other supposedly degenerative disorders. The point being missed is that the term 'degenerative' suggests a universal process to which the tissues of the whole species are susceptible, whereas the phenomena of concern are non-universal, only a subpopulation being susceptible to such pathological changes at any given site. As already mentioned, the development of osteophytes appears to be a virtually universal age-associated trait, although the association is more likely to be some relatively non-specific concomitant of age, such as greater exposure to wear, rather than being due to ageing *per se*. Confusion is increased because, like osteophytosis, the disease osteoarthritis may well be related to wear, although in the latter case it seems that wear is accelerated and abnormal.

Lack of age gradient

Conditions appearing to occur with similar frequency in middle and late life include rheumatoid arthritis, gout, back troubles, and various forms of soft-tissue (non-articular) rheumatism. This is the most unsatisfactory pattern from the point of view of seeking generic explanations, for which there are a number of reasons. First, available data are rarely adequate to allow of certainty that any particular condition should not be assigned to one of the other three categories of age patterns. Secondly, confusion between prevalence and incident cases colours some of this uncertainty. Thirdly, given such limitations in the data, the influence of other distorting factors becomes potentially more important; for instance, there is evidence that older people make proportionately less use of health services in relation to their medically-defined 'needs', and lower rates of consultation cannot help but influence ascertainment (it is for this reason that backs and soft-tissue rheumatism have been discussed in this pattern category, even though morbidity data superficially suggest that their frequency is reduced in elderly people). Finally, there is the problem of heterogeneity, which makes it difficult to unravel what may be at the root of back pain, the relative contributions of different causes being obscured as conditions such as osteoporosis and osteomalacia become superimposed on pre-existing spondylotic changes. The same applies to soft-tissue rheumatism, where account has to be taken of changes such as the increased likelihood that shoulder pain might be due to an underlying lung cancer.

Reduced frequency with age

Although perhaps of least interest to clinical geriatrics, this pattern class is nevertheless instructive biologically. Here one might include ankylosing spondylitis, sprains and strains of joints and adjacent muscles, and lesions of the menisci in the knee, and the challenge is to interpret the tailing off of effects within a distribution, to which three types of explanation are of special relevance. First, coming most readily to mind, is a difference in risk in relation to age, such as older people being less exposed to some particular hazard—the converse of the situation applying to femoral neck fractures. The second type of explanation is that the susceptible subpopulation has been exhausted, either because all have been exposed at an earlier age or because vulnerability may be restricted to particular stages in the life-cycle, circumstances that may be relevant to the occurrence of ankylosing spondylitis. Thirdly, the explanation most commonly neglected, survivors to old age are a selected subpopulation, and this may operate in two ways. On the one hand the condition of interest may be strongly associated with some other disease which is itself lethal; this could apply to the link between rheumatoid arthritis and tuberculosis. More generally, the overall force of mortality experience earlier in life may have selectively removed persons who would otherwise have been susceptible to a particular condition. On the other hand, mortality associated with a specific disease obviously leads to attrition of those affected. Although most rheumatic disorders are generally regarded as having a low case-fatality rate, insurance companies nevertheless often load the life assurance ratings of persons affected by these conditions, and our own studies have certainly documented reduced survival in rheumatoid arthritis. At the moment such disease-specific life-expectancy data are too sparse to allow one to compute with any precision what effects they might have on patterns of occurrence at different ages. Similarly, evidence is too scanty to permit assignment of some rheumatic disorders, such as dermatomyositis, to one or other of the patterns just discussed.

BEHAVIOURAL INFLUENCES

Studies of public awareness have shown that although most respondents did not regard rheumatism as an inevitable accompaniment of growing old, older persons seemed more ready to accept arthritis as a normal part of the ageing process. It is therefore relevant to consider how such attitudes might be reflected in service utilization, and what detectable impact rheumatic disorders have on the life-style of those affected.

Health service utilization

Illness behaviour is certainly influenced by expectations, though whether corresponding modification of such behaviour is associated with differential attitudes on the part of elderly people in regard to specific classes of condition, such as disorders of the musculoskeletal system, has not been established. However, two considerations are relevant. First, at all ages the expectations of medical services for amelioration of rheumatic disorders tend to be unreasonably low. Secondly, superimposed on this there appears to be a lesser propensity for older people to make use of health services, despite their greater morbidity and the higher consultation rates these would lead one to expect. An important consequence is that measures of frequency derived from service contacts are likely to yield an underestimate of the true situation.

Primary care

Patterns of consultation with general practitioners reflect illness behaviour as much as they do changes in disease occurrence. Nevertheless such patterns can be instructive. Arthropathies such as rheumatoid arthritis, arthrosis, gout, and spondylitis show a marked increase in patient consulting rates with age. Rates for back troubles, including lumbago, sciatica, and disc lesions, and soft-tissue rheumatism (frozen shoulder, bursitis and tenosynovitis, and other non-articular rheumatism) are both lower in elderly people, though the margin is relatively small. In marked contrast, rates for sprains and strains are lower in elderly people than at any other period in life. Taken in conjunction with experience of fractures, it seems likely that age is associated with changes in the nature of trauma as well as in the body's response to such injury. Overall, musculoskeletal disorders make up about a quarter of the general practitioner's work for elderly patients—the three other big classes being circulatory and respiratory disorders, and symptoms and ill-defined conditions—and rheumatic conditions account for 33 per cent of all diseases and conditions present in patients appearing for consultation.

Although morbidity studies do not always allow one to subdivide the patterns of response to rheumatic disorders in general practice according to the age of patients, much can be learned from the overall experience. There is an impression that patients with rheumatic problems generally tend to be seen slightly less frequently on return visits than those with other conditions. Moreover, whereas more than one-third of all patients seen are referred for outside help, of rheumatic patients only a fifth are so referred; for each type of referral (specified investigation, specialist opinion or service, direct admission to hospital, and municipal authority services) the proportion of rheumatological patients referred was smaller than that for all diseases and conditions.

Admissions to hospital

It is instructive to compare what happens in elderly patients with the experience of those in all age groups. As with most classes of condition, the proportions of all admissions to hospital attributable to disorders of the musculoskeletal system remain relatively unchanged in the different age groups, at a level of between 4 and 5 per cent. This contrasts with circulatory disorders and neoplasms, which account for a proportionately greater share of admissions in elderly persons than they do at younger ages. Two-thirds of all admissions for disorders of the musculoskeletal system are to orthopaedic departments, but 18 per cent of admissions for arthrosis are to geriatric departments, and the latter deal with 11 per cent of admissions for rheumatoid arthritis.

Comparisons between elderly and very elderly patients reveal another important aspect. When admissions for broad categories of disease are related to quinquennia of age from 65 years onwards, virtually all show an increase in age-specific admission rates. However, three crude patterns can be detected. There is a very steep rise with respiratory disease, trauma, mental disorders, and 'senility'; a somewhat less marked increase with eye disorders, and neurological disease; and a more modest rise with other conditions. Disorders of the musculoskeletal system conform to the last pattern, admission rates tending to be fairly stable up to the age of 84 years but rising more sharply thereafter—which contrasts with neoplasms and eye disorders, these tending to fall off in the most elderly persons. As might be expected, there is a marked rise in mean durations of stay with age, the relevant figures for musculoskeletal disorders increasing from 25 days at 65 to 69 years to almost 70 days in those aged 90 or more and being somewhat longer in females throughout the age range. After mental disorders, neurological conditions, and peripheral vascular disease, arthritis and rheumatism lead the field for durations of stay.

Life-style

Although general awareness has become more enlightened, the false impression of old people being dependent and often having to live in institutions still needs to be challenged. Study of the disadvantages experienced by elderly people does reveal an increase, with progressive curtailment in activity. Most notable is the reduction in mobility, with more than a quarter of very elderly people being confined to their homes and a similar proportion being unable to have an all-over wash on their own. Being thus cut off from external resources at a time when dependence is increasing, the other noteworthy aspects are that two-fifths of very elderly people live alone and that their financial situation is so poor. What calls for special comment is a failure to benefit from adequate heating, which is a matter for therapeutic concern because ambient temperature exerts an appreciable influence on the severity with which pain and other rheumatic symptoms are experienced, and medication is a poor and more dangerous substitute for sufficient warmth.

A study of all elderly persons at home (Hunt 1978) allows the perspective to be extended. Having ascertained the minority who were confined to their houses, the remainder were asked about their health. Overall, slightly more than 40 per cent of the elderly respondents reported no disability, although by the age of 85 years or more this proportion had fallen to between a quarter and a fifth. Among those who did report illness or disability, arthritis and rheumatism was the most common problem, being responsible for difficulties in more than a third. Of the total sample, almost one in 20 were bedfast or housebound, the proportion rising to just over 1 per cent in the 65 to 69 quinquennium and to almost 21 per cent in those aged 85 years or more; the biggest single cause of these restrictions, accounting for 36 per cent of the total, was again the category of arthritis and rheumatism.

Restriction of mobility is one of the most serious health-related problems in elderly people, and, age group for age, severe loss is higher among women, who also tend to acknowledge arthritis and rheumatism more frequently. Of the arthritics in the survey, a greater proportion of the women were living alone, which presumably was related, among other things, to preferential mortality in men at younger ages. Despite this difference, though, the frequency of complaints about not being warm enough in bed was similar in the two sexes (about 10 per cent). Slightly greater proportions were not warm enough in their living rooms and kitchens. Highest on the list of 'official' visitors in the previous 6 months was an insurance representative (48 per cent of arthritic respondents), followed closely by the family doctor (40 per cent). The next most frequent were ministers of religion (19 per cent), home helps (16 per cent), and social-security officers and community nurses (both 11 per cent). With the exception of insurance and social-security visitors, contact rates tended to be higher for women than for men. No very clear pattern of provision for aids could be detected.

CONCLUSION

In this short compass I have endeavoured to indicate various perspectives on joint and connective tissue disorders in elderly subjects. Can these tell us anything about what might be anticipated in the future? In contemplating the likely health experience of future elderly populations, some aspects augur well and others less so. It seems likely that a large part of ill health in later life is extrinsically and societally induced by behaviours and attitudes. The present generation of elderly people grew up in times of great poverty and deprivation. Social conditions, for example housing, have in general improved as has medical care and some aspects of life-style. The social fabric has become more complex and potentially more isolating. Older people may face fewer and more restricted social roles, and frustration can be a major cause of illness. We cannot expect major reductions in the scale of problems for which some help from the health and social services will be needed. The strategic goal should be towards policies of enablement, seeking ways of maximizing the ability of old people to handle their health and other problems.

References

Badley, E.M. (1987). Epidemiological aspects of the ageing spine. In *The ageing spine*, (ed. D.W.L. Hukins and M.A. Nelson), pp. 1–17. Manchester University Press.

Ellman, M.H. and Levine, B. (1975). Chondrocalcinosis in elderly persons. *Arthritis and Rheumatism*, **18**, 43–7.

Felson, D.T. (1988). Epidemiology of hip and knee osteoarthritis. *Epidemiologic Reviews*, **10**, 1–28.

Gibson, T. and Grahame, R. (1973). Acute arthritis in the elderly. *Age and Ageing*, **2**, 3–13.

Grahame, R. (1976). The problems of diagnosis of arthritis in the elderly. *Proceedings of the Royal Society of Medicine*, **69**, 926–8.

Harris, A.I. (1971). *Handicapped and impaired in Great Britain*, part 1. Social Survey Division, Office of Population Censuses and Surveys. HMSO, London.

Hunt, A. (1978). *The elderly at home—a study of people aged 65 and over living in the community in England in 1976*. Social Survey Division, Office of Population Censuses and Surveys. HMSO, London.

Kellgren, J.H. (1939). Some painful joint conditions and their relation to osteoarthritis. *Clinical Science*, **4**, 193–201.

Lawrence, J.S., Bremner, J.M. and Bier, F. (1966). Osteoarthrosis—prevalence in the population and relationship between symptoms and X-ray changes. *Annals of the Rheumatic Diseases*, **25**, 1–24.

Wood, P.H.N. (1971). Rheumatic complaints. In *Epidemiology of non-communicable diseases*, (ed. E.D. Acheson). *British Medical Bulletin*, **27**, 82–8.

Wood, P.H.N. (1972). Radiology in the diagnosis of arthritis and rheumatism. *Transactions of the Society of Occupational Medicine*, **22**, 69–73.

Wood, P.H.N. (1976). Osteoarthrosis in the community. In *Osteoarthrosis*, (ed. V. Wright). *Clinics in Rheumatic Diseases*, **2**, 495–507.

13.2 Management of rheumatoid arthritis in the elderly patient

EVAN CALKINS

INTRODUCTION

Rheumatoid arthritis, which is a major concern for physicians and patients in early and mid life, assumes a different role among the rheumatic diseases in older persons. In this age group, osteoarthritis, osteoporosis, fibrositis and other poorly defined, localized inflammatory conditions, polymyalgia rheumatica, and musculoskeletal symptoms secondary to non-rheumatic diseases, such as metastatic carcinoma, assume a higher priority among primary-care providers and geriatricians. Although rheumatoid arthritis no longer occupies the centre stage among musculoskeletal diseases in this age group, the condition presents more than its share of challenges, both in differential diagnosis and in management, and many aspects require further study. Topics that have been the subject of totally conflicting views include whether or not the disease occurs with increased frequency among the elderly people; whether prognosis is better or worse than in younger individuals; the possibility that disease of abrupt onset may be a distinct subset rather than one end of a range of presentations; and the frequency of comorbid disease.

One example, which may explain a number of these conflicting observations, relates to the frequency of new-onset disease in older individuals. Although most rheumatologists are in agreement that the most frequent age of onset lies in mid life (35 to 55 years), one careful community-based epidemiological study (Linos *et al.* 1980) showed that the occurrence of new-onset disease increased progressively with advancing age, to 70 years or beyond, when it achieved an annual rate of more than twice that observed in persons aged 40 to 49 years.

It seems likely that a large number of elderly persons who experience symptoms and present signs that conform to the diagnostic criteria for rheumatoid arthritis do not seek medical care from a rheumatologist and are regarded, or regard themselves, as suffering from osteoarthritis, some other age-related inflammatory disorder, or simply the concomitants of old age. Clearly, concepts concerning the characteristics of the disorder or group of diseases and their management will vary depending on the population under consideration.

CLASSIFICATION AND DIAGNOSIS

Despite the diversity of views referred to above, the formulation shown in Table 1 provides a framework for discussion and further study. In elderly as with younger patients, the onset of rheumatoid arthritis is most commonly insidious, with the gradual development, over a number of months, of stiffness and swelling of the small joints of the hands, the metatarsophalangeal joints, wrists, and knees, and aching and limitation of motion of the neck, shoulders, and hips. Separating this syndrome from other forms of articular disease is both important and, at times, surprisingly difficult. Problems may arise in differentiating rheumatoid arthritis from the arthritis that occasionally accompanies the onset of a malignant disease (Chaun *et al.* 1984). More commonly, one is faced with differentiation from polymyalgia rheumatica (see Chapter 13.6), the frequency of which, among elderly patients, is probably greater than that of rheumatoid arthritis of new onset. In view of the benign course of the

Table 1 Clinical presentations of rheumatoid arthritis in the elderly

Insidious onset, seronegative, with objective findings in peripheral joints
Insidious onset, seronegative, involving primarily the neck and shoulders, and resembling polymyalgia rheumatica
Insidious onset, seropositive, with nodules and progressive destructive joint changes
Fulminating onset, seronegative, usually regarded as having an increased likelihood of remission
Sequelae of long-term rheumatoid arthritis, with onset in early or mid life

articular manifestations in polymyalgia rheumatica, the excellent clinical response to low-dose corticosteroid therapy, and the serious threat of concomitant cranial arthritis, this distinction is as important as it may be difficult (Healy, 1983, 1986). Even the distinction between rheumatoid arthritis and osteoarthritis, in the elderly patient, may present more difficulties than one would anticipate (MacFarlane and Dieppe 1983).

Occasionally, elderly rheumatoid patients will have, early in their disease, a positive test for rheumatoid factor and the appearance of nodules (Deal *et al.* 1985). Many elderly patients have rheumatoid factor, in low titre, even in the absence of rheumatic or other inflammatory disease. Nevertheless, a titre in the latex fixation test of 1/500 can be regarded as significant. In these patients, the disease usually follows an unremitting, destructive course, resembling that seen in younger persons with seropositive disease.

In approximately a quarter of patients with rheumatoid arthritis arising in later life, the condition has a fulminating onset accompanied by severe constitutional symptoms, including fever and inflammation of a number of joints, all developing over the course of days or weeks. This is one of the few syndromes of rheumatoid arthritis that may be accompanied by night sweats. Some have observed that patients with this presentation have a substantially greater chance of remission than those with disease of gradual onset but others refute this concept.

Approximately one-third of patients with rheumatoid arthritis occurring at any age continue to experience active disease over the course of many years, with progressive joint destruction and other chronic sequelae. The characteristics of long-standing rheumatoid arthritis in patients who grow old with their disease have not been well studied. Clinical experience indicates that some patients gradually experience a decline in activity of the rheumatoid process, leaving them to contend with the physical limitations secondary to the destructive process. In others, the disease remains active; they continue to develop new nodules and other constitutional manifestations, including both minor and major sequelae of vasculitis.

DIFFERENTIAL DIAGNOSIS

Musculoskeletal symptoms are not the sole province of the so-called rheumatic diseases, but may also reflect the presence of a wide range of other disorders, many of which occur with increased frequency among elderly people. Examples include Paget's disease of bone, sarcoidosis, hyper- and hypothyroidism, hyperparathyroidism, tuberculosis, metastatic carcinoma, multiple myeloma, and Hodgkin's disease. As the manifestations of many age-associated disorders resemble, at least in some respects, those of rheumatoid arthritis, it is extremely important to determine whether new symptoms in a patient with long-standing rheumatoid arthritis are due to an exacerbation of the rheumatoid process or to a concomitant, unrelated disorder.

In the conduct of this differential diagnosis, and also as a guide to therapy, it is helpful to make an assessment of the extent of disease activity. The most useful criteria, in elderly patients, are the duration of morning stiffness, the presence and extent of joint swelling and tenderness, the extent of elevation of the erythrocyte sedimentation rate, and the presence or absence of anaemia. Morning stiffness, lasting for a number of hours, is characteristic of active rheumatoid arthritis and also polymyalgia rheumatica. Although it may also be encountered in patients with disseminated lupus erythematosus, it is not seen in gout or most other rheumatic disorders. In osteoarthritis, morning stiffness frequently occurs, but rarely lasts for longer than 45 min. Both in rheumatoid arthritis and in polymyalgia rheumatica the duration of morning stiffness closely parallels the extent of disease activity.

While rheumatoid arthritis, especially in the elderly patient, may be accompanied by conspicuous constitutional manifestations, the hallmark is actual arthritis. Patients with active rheumatoid arthritis will also, almost without fail, have significant swelling and tenderness in the synovium of at least one joint.

In interpreting the erythrocyte sedimentation rate, one must bear in mind that many elderly persons will have moderately elevated values, in the neighbourhood of 30 mm/h, even in the absence of rheumatoid arthritis or other inflammatory disease. Values of 45 mm/h or higher are significant. The extent of anaemia is a valuable index of disease activity only if the patient clearly has the 'anaemia of chronic disease'. If the anaemia is due to coincidental iron deficiency it is not a valid reflection of active disease.

Obviously, the design of management is influenced significantly by the presence or absence of active disease. If the patient no longer has active disease, therapy can be directed specifically toward rehabilitation, together with the treatment of concomitant diseases. If the rheumatoid process is still active, however, the therapeutic programme must be modified accordingly, and some form of anti-inflammatory or antirheumatic medication should be maintained.

COMPREHENSIVE ASSESSMENT

In common with most chronic conditions in elderly people, the management of rheumatoid arthritis, especially if it involves significant limitation in functional capacity, is best carried out by a multidisciplinary team of professionals (Halstead 1976). Fundamental to this approach is a comprehensive assessment not only of the disease process or processes but also of the patient and his or her physical and socioeconomic environment (Table 2). Based on this, a therapeutic plan can be designed, immediate and long-term goals established, and criteria can be developed for determining the extent to which these goals are being achieved.

Assessment of the patient and his or her functional capacity is as important as assessment of the disease process itself because it provides a guide not only to efforts to prevent further decline in function but also in the design of necessary adaptive equipment and social support. A useful concept of functional classification by the World Health Organization (1980) differentiates impairments, disabilities, and handicaps. Impairments refer to the patient's organic dysfunctions—that is, anatomical, physiological, mental, and/or psychological defects. Disabilities refer to restrictions in the patient's manner or range of physical activities resulting from these impairments, and handicaps reflect the social disadvantages that an individual experiences in fulfilling roles that

Table 2 Comprehensive assessment of the elderly patient with rheumatoid arthritis

Assessment of the disease
Diagnosis: does the patient have rheumatoid arthritis?
Assessment of disease activity
Comorbid disease. Are recent onset symptoms due to other concomitant disease?
Do recent changes in symptomatology stem from sequelae of rheumatoid arthritis?
Drug inventory: can recent symptoms be attributed to inappropriate medication?
Impairments: anatomical or functional changes due to rheumatoid process
Assessment of the patient and environment
Disabilities: assessment of functional status
Handicaps: limitations in the ability to meet the demands or requirements of effective life
Psychological aspects and cognitive status—is the patient depressed?
Social support structure and economic status
Nutritional status
Adequacy of physical environment and assistive devices
Surgical risk
Therapeutic plan (establishment of long-term and short-term goals)
Rest and daily routine
Physical therapy
Pharmacotherapy
Psychological therapy and support; patient and family education
Addressing the social and physical environment
Surgical (orthopaedic) management

are regarded as normal, considering age, sex, and social and cultural factors. The manner in which these concepts can provide a framework for the comprehensive functional and psychosocial assessment of an individual patient have been well defined (Granger 1986).

Basic to this assessment is a complete examination of the joints, in which all joints, including the temporomandibular, dorsal and lumbar spine, and both large and small peripheral joints, are examined at rest and through their full range of motion. It is important, also, to assess the patient's ability to walk, rise from a chair, climb and descend stairs, feed, dress, and wash, and maintain continence (Liang and Jette 1981). In rheumatoid arthritis one is often impressed by the extent to which the patient has been able to retain these essential functions of daily living despite appalling deformities, such as hands with such severe ulnar deviation and subluxation that they look like claws.

In considering management goals, one must recognize that these differ markedly among individual patients (Spiegel *et al.* 1987). For some, they may include merely the ability to be maintained at home with support from the family and community. For others, the goal may be to continue to satisfy professional and social roles. Definition of the goal of management provides an important framework for programme design. For example, one does not plan extensive orthopaedic reconstruction of the hips or knees if massive deformities of the feet or other problems preclude eventual walking or independence.

Psychological factors, including chronic psychological stress, have a significant relationship with the symptoms and also the activity of the arthritic process and its ultimate course (Anderson *et al.* 1985). The patient who is angry and resentful or chronically depressed has a more difficult time coping with the disease than one who is able to accept its limitations yet still maintain strong motivation to enjoy life and to contribute to society. As the physician or other care provider gradually develops a relationship of trust and confidence, it often becomes clear that there are many instances in which the patient bitterly and repeatedly complains of some specific aspect of the disease, while the real problem relates to some personal conflict or frustration. A listening ear is one of the most useful instruments in both assessment and treatment. In many instances, once the 'real problem' has been identified and acknowledged, its significance becomes diminished.

Reasonably good cognitive capacity is essential to the patient's ability to achieve the goals of independence, mobility, and participation in community life. For those patients who have serious and irreversible cognitive limitations, the therapeutic goals must be scaled down substantially. The physician should also be aware of the possibility that cognitive limitations may, in fact, be secondary to medication and may be substantially reversed if the offending drug is stopped. Conversely, the presence of an added physical burden, such as active rheumatoid arthritis, in a person with established Alzheimer's disease may accentuate the cognitive defects or behavioural problems; considerable symptomatic improvement may be achieved through proper management of the rheumatoid process.

Many patients with long-standing rheumatoid arthritis become depressed. Although this may be a reaction to the restraints and discomforts caused by the disease, it may also be a side-effect of medication, especially the non-steroidal anti-inflammatory drugs (Goodwin and Regan 1982; Grigor *et al.* 1987). Patients who become depressed while receiving these agents should be given a trial with an alternative analgesic drug, before considering other methods of treatment of the depression.

Because of the limitations in functional capacity and the continuing need for psychological support, sound social support is a prerequisite for effective management of patients with advanced disease. Deficiencies in this assume a high priority in the overall care plan. In the absence of a built-in social support, a complex array of community services will sometimes prove effective; in other instances, long-term institutional care is required.

Patients with severe, long-standing rheumatoid arthritis frequently have a thin, emaciated appearance, and many elderly patients with this disorder suffer from severe protein-calorie malnutrition. This may be due, in part, to difficulties in chewing or swallowing because of arthritis of the temporomandibular joint, to chronic gastritis secondary to medication with non-steroidal anti-inflammatory agents, to difficulties in procuring or preparing food, or to manual problems with self-feeding. The patient may also suffer from coincidental dental disease and/or drug–nutrient interaction. If present, protein-calorie malnutrition contributes to muscle weakness and to diminished resistance to infection and stress. Maintenance of a good nutritional state is an important component of comprehensive care (Touger-Decker 1988).

One of the best ways to prevent a disability becoming a

handicap lies in the provision of an appropriate physical environment. This includes bathrooms and kitchens designed to facilitate self-care; avoidance of threats to safety such as loose rugs and poorly lit stairs; appropriate means of access from house to car or street; appropriately designed chairs and other furniture; and functional support such as canes, walkers, and wheelchairs. While interest on the part of the physician is extremely helpful in ensuring that these objectives are met, the professional knowledge and skills of an occupational therapist, who will make home visits, often provide the most effective way of addressing these problems.

MEDICAL MANAGEMENT

Despite the attention it has received in pharmaceutical advertising, pharmacotherapy is not the cornerstone of management, either of long-standing or new-onset rheumatoid arthritis in patients at any age. Instead, the use of drugs is only one component of a comprehensive programme. This concept, long held in the field of rheumatology, is especially important in the management of elderly persons, in whom the use of any drug has a higher likelihood of adverse side-effects than in younger individuals.

Although the general principles of comprehensive management of patients with rheumatoid arthritis, which pertain for younger people, are still appropriate in management of the elderly patient, there are important differences, as discussed next.

Rest

Rest is a dangerous form of treatment for elderly patients. All too often, those who undergo bed rest or a time in hospital for an acute, self-limiting disease or surgery 'lose their stride' and are never able to regain their former level of physical activity. As preservation of functional capacity is one of the major goals of therapy, bed rest must be viewed with great scepticism. By the same token, the constitutional symptoms accompanying rheumatoid arthritis are very real and it is not wise or practical for an older person to become fatigued. A useful approach is to spell out, in detail, a daily programme that will preserve limited periods of activity and opportunity for social interchange, yet allow enough rest to avoid fatigue.

For periods of bed rest, the patient's positioning is very important. The patient should avoid long periods in a hospital bed, with the mattress 'cranked up' in a chair-like position, knees flexed, head bent forward, and legs externally rotated, with feet pinned down by overly tight sheets. If possible, rest should be flat or nearly flat, with joints maintained in optimal position through use of appropriate supports.

Localized rest for a joint that is markedly inflamed, such as a wrist, is also important and can be effected with a lightweight splint. In patients who have begun to develop flexion deformities of the knee, a bivalve cast, with a hinge to allow gradual straightening, is a classical and still effective method of correction. During periods of acute joint inflammation, localized application of moist heat for periods of 20 to 30 min, two or three times per day, will usually lessen acute symptoms and muscle spasm.

Exercise

Exercises are undertaken for three reasons: preservation of range of motion of involved joints; avoidance of muscle contractures; and maintenance of muscle strength. In patients with rheumatoid arthritis, exercises should always be active. The role of the physiotherapist is not that of 'exerciser' but of 'coach'; responsibility for the conduct of the exercise must rest with the patient. A determination to get better on the patient's part is an essential element of a successful outcome. Although exercise programmes should be focused primarily on the most significantly involved joints, co-ordinated range-of-motion exercises for all major joints in the body should also be undertaken to ensure that contractures do not develop unexpectedly while attention is aimed on the site of most conspicuous involvement.

For those with minimal or moderate degrees of disease activity, aerobic exercises, the type often prescribed for elderly patients, provide social and recreational advantages and may enhance cardiac function but do not contribute to increased muscle strength. This must be achieved by active anaerobic exercises against resistance. Recent studies on the quadriceps muscle in patients with arthritis of the knee have shown that many elderly patients with either rheumatoid or osteoarthritis will show significant increases in quadriceps strength after exercises of this sort for periods of 20 min, three times a week, for 2 or 3 months (Fisher *et al.* 1991).

Drugs

The appropriate use of drugs in patients with rheumatoid arthritis is the subject of considerable divergence of view between individual physicians and between physicians in different countries. Although aspirin remains the 'gold standard' of first-line antirheumatic therapy, the frequency of adverse effects is substantially increased in elderly patients. Aspirin should usually be given to older patients in an enteric-coated or buffered form, and in a dose not exceeding 2.4 g/day. The principles governing use of non-steroidal antirheumatic agents are not substantially different in rheumatoid arthritis from those that apply in osteoarthritis. While these agents suppress the inflammatory manifestations of rheumatoid arthritis, including pain, there is no evidence that they contribute to the occurrence of a remission or alter the course of the disease.

What does one do, therefore, with an elderly patient with active rheumatoid arthritis if a well-designed programme of conservative management, together with judicious use of non-steroidal antirheumatic agents over the course of several weeks, is not accompanied by lessening of the evidence of disease activity? There is a wide range of views and practice among different physicians in this regard. As we do not yet have conclusive evidence about the merits of one course or another, one must conclude that any of a number of options may be pursued. The difference between success and failure depends more on the priority given to the many elements of comprehensive care and the meticulous attention to detail required by a programme of this sort than on the selection of a particular pharmacological regimen.

One approach to the patient who does not respond to a conservative programme, including non-steroidal antirheumatic agents, lies in the use of one of a number of agents that appear to have a suppressive action on the activity of the rheumatoid process itself, including hydrochloroquine, penicillamine, gold salts, and antimetabolic agents (see Kean *et al.* 1982, 1983). Although hydrochloroquine is an effective antirheumatic agent, the benefit appears to be limited to between 6 and 12 months. The drug may cause ophthalmological complications and, in my opinion, has no place in the treatment of elderly patients with rheumatoid arthritis. Penicillamine is an effective antirheumatic agent in elderly patients. Nevertheless, it is extremely toxic, equally or more so than gold. The fact that it can be administered orally in tablet form belies its toxicity. In one series, penicillamine therapy was accompanied by adverse events in 58 per cent of cases; these included rash, gastrointestinal symptoms, proteinuria, stomatitis, worsening of the arthritis, altered taste, leucopenia, thrombocytopenia, or myasthenia gravis. Careful follow-up of urinalysis and haematological data is essential. In my view, this drug must also be used with great caution, if at all, in the treatment of elderly patients with this disease.

For at least 50 years, courses of gold have been given to patients with rheumatoid arthritis. Although the drug is toxic, its dosage schedule, side-effects, and indications have been well studied and described. Gold therapy appears to be as effective in older as in younger individuals. Although toxic effects do not appear to be more frequent in elderly than in younger patients, they tend to be more serious, especially in those who are HLA-DR3 positive (Wooley *et al.* 1980). Whether the drug is given intramuscularly or orally, the patient should at first be seen personally by the physician every 2 weeks, and later at least every month, and a complete blood-cell count and urinalysis obtained. Many elderly patients find follow-up of this intensity difficult, especially in winter months. Furthermore, a risk that may be justified in a younger person, in whom a remission would produce benefits extending over many decades of life, may not be justified in an elderly person. Therefore, gold is infrequently used for elderly patients by most physicians in our group.

Methotrexate has recently been introduced as an antirheumatic agent (Tugwell 1989). The drug is given weekly in three divided doses, 2.5 mg each, at intervals of 12 h. If necessary, the dose may be doubled after an initial period of several weeks and then reduced as symptoms resolve. This agent, too, is very toxic. Liver function tests should be obtained before starting therapy and at intervals of 1 to 3 months so long as it is maintained. Complete blood-cell counts should be made initially and at monthly intervals. The drug is particularly effective in the treatment of psoriasis.

In my opinion, prednisone in a dose of 6 to 8 mg/day is the safest and most effective pharmacological agent in the treatment of elderly patients with continuously active rheumatoid arthritis. At first glance, the list of complications associated with corticosteroid therapy would suggest quite the opposite. When given in a large enough dose and long term, this drug not infrequently leads to hypertension, fluid retention, osteoporosis, and an increased frequency of diabetes mellitus. For most of these complications, however, the term 'large enough dose' provides the key. Experience has shown that, with the possible exception of osteoporosis, the occurrence of any of these complications is extremely rare if the dose is maintained in the range indicated above.

The effect of low-dose prednisone on the integrity of bones has been studied quite extensively (Dykman *et al.* 1985). The most important factor determining corticoid-induced osteopenia is the cumulative dose, the highest incidence of osteopenia occurring in patients who have received more than 30 g of prednisone. At a dosage of 7 mg/day this would require 11.5 years of continuous treatment. A programme of reasonable physical activity significantly lessens the loss of calcium associated with administration of corticosteroids. Thus, if low-dose prednisone therapy is started for an elderly person with rheumatoid arthritis with a view to enhancing physical functions, one must see to it that the patient actually increases his or her physical activity.

In short, the risk/benefit ratio for long-term systemic corticosteroid therapy appears to be favourable in elderly patients, in whom maintaining a relatively comfortable life and retaining physical functions and independence over the course of 8 or 10 years constitutes a highly attractive goal (Harris *et al.* 1983; Lockie *et al.* 1983; Masi 1983; Byron and Mowat 1985).

For any patient who has received long-term corticoid therapy, even in these low doses, it is wise to give additional corticosteroids during times of extreme stress or surgery.

Intra-articular injections of adrenal corticosteroids provide a helpful means of attenuating discomfort in individual joints, especially if there is demonstrable effusion. This procedure must be undertaken with care to avoid superimposed infection and is best reserved for occasional use. Indeed, accumulating evidence suggests that intra-articular administration of corticoids may be followed by destructive changes in joint cartilage. One should also be sensitive to the possibility that a single joint with disproportionate evidence of inflammation may be the site of a septic process.

The need for psychosocial support has been emphasized in the section on assessment. In most instances, this can be provided by a sensitive physician who is willing and able to take the time to listen. Education of the patient and family is extremely important, not only to ensure good compliance but because, even in the presence of a disease that is difficult to treat, knowledge about the condition almost always contributes to a patient's ability to respond in a more positive fashion. {Support groups may be provided by arthritis charities.}

ORTHOPAEDIC MANAGEMENT

Orthopaedic interventions are of two sorts: localized procedures undertaken early to prevent permanent functional loss, and major reconstructive procedures. One example of the former is the resuturing of ruptured tendons, primarily extensor tendons of the fingers and wrist. Failure to address this defect promptly results in serious limitation of function. Other localized procedures are release of carpal ligaments in patients with carpal tunnel syndrome and release of the ulnar nerve in an ulnar entrapment syndrome. Resection of the first metatarsal phalangeal joint in patients with severe symptomatic halux valgus and resection of the metatarsal heads in patients in whom these structures have become seriously prolapsed may provide considerable symptomatic benefit. In the rare patient who experiences subluxation of the atlanto-occipital joint, with neurological sequelae, emergency surgical repair may be required.

With increasing experience of arthroplasty of the hip, using the Charnley or other similar devices, there has been increasing recognition of the significant benefits of this procedure in relief of pain and restoration of function. With appropriate preoperative assessment (White 1985), these benefits are equally applicable in elderly and young patients and the procedure should not be discounted even in persons in their mid 80s who have become limited due to severe involvement of a single joint. It is important, however, to remind the patient that, to be effective, the procedure must be followed by many months of carefully guided physiotherapy (Bentley and Dowd 1986; Granger 1986). In addition, the chance of postoperative infection, a serious complication, appears to be increased. For the rest of the patient's life, the patient and his or her physician must be mindful of the possibility of the development of infection around the prosthesis from bacteraemia secondary to dental or urological procedures or localized infection elsewhere in the body, and appropriate prophylactic measures are essential (Poss *et al.* 1976).

Unfortunately, arthroplasty of the knee has proved even more complex than that of the hip. This is due in part to the increased incidence of postoperative phlebitis and, in part, because the postsurgical improvement in this highly complex joint, especially as regards function, may be less satisfactory than with the hip.

Surgical reconstruction of the hands may also be undertaken in older persons. However, many with severely deformed hands have become quite proficient in using them. Although certain localized deformities, such as rupture of tendons, provide clear-cut indications for surgery, it has yet to be determined whether more ambitious reconstruction, such as arthroplasty of metacarpal phalangeal joints, has a significant place in the management of elderly patients with this disease.

If patients with long-standing rheumatoid arthritis are to undergo general anaesthesia it is extremely important to take appropriate measures to avoid atlanto-occipital subluxation from careless movement of the anaesthetized patient's head and neck. In patients with advanced destructive rheumatoid arthritis, especially if neck symptoms are present, the preoperative evaluation should include radiograph of the cervical spine, with neck flexed and extended, to identify the possibility of early subluxation of this joint, and the patient should wear a protective neck collar during surgery. These patients also at risk for laryngeal obstruction due to cricolaryngeal arthritis. Laryngoscopy should be done preoperatively in these patients, especially if they have had hoarseness. If the condition is present, general anaesthesia should be avoided (see Section 21).

SUMMARY

Management of elderly patients with rheumatoid arthritis presents major challenges in diagnosis, differential diagnosis, functional and psychosocial assessment, and in the design and careful execution of a comprehensive programme of care. Reliance on the multidisciplinary-team approach, with comprehensive assessment as a common language, and meticulous attention to details of management provide important benefits to the patient and a constructive, optimistic atmosphere in which the appropriate professionals can do their best work.

References

Anderson, K.O., Bradley, L.A., Young, L.D., and McDaniel, L.K. (1985). Rheumatoid arthritis: review of psychological factors related to etiology, effect, and treatment. *Psychological Bulletin*, **98**, 358–87.

Byron, M.A. and Mowat, A.G. (1985). Corticosteroid prescribing in rheumatoid arthritis.—The fiction and the fact. *British Journal of Rheumatology*, **24**, 164–6.

Bentley, G. and Dowd, G.S.E. (1986). Surgical treatment of arthritis in the elderly. *Clinics in Rheumatic Diseases*, **12**, 291–327.

Chaun, H., Robinson, C.E., Sutherland, W.H., and Dunn, W.L. (1984). Polyarthritis associated with gastric carcinoma. *Canadian Medical Association Journal*, **131**, 909–11.

Deal, C.L., Meenan, R.F., Goldenberg, D.L., Anderson, J.J., Sack, B., Pastan, R.S., and Cohen, A.S. (1985). The clinical features of elderly-onset rheumatoid arthritis. A comparison with younger-onset disease of similar duration. *Arthritis and Rheumatism*, **28**, 987–94.

Dykman, R.T., Gluck, O.S., Murphy, W.A., Hahn, T.J., and Hahn, B.H. (1985). Evaluation of factors associated with glucocorticoid-induced osteopenia in patients with rheumatoid arthritis. *Arthritis and Rheumatism*, **28**, 361–8.

Fisher, N.M., Pendergast, D.R., Fresham, G.E., and Calkins, E. (1991). The effect of muscle rehabilitation in the muscular and functional performance of patients with osteoarthritis of the knees. *Archives of Physical Medicine and Rehabilitation*, in press.

Granger, C.V. (1986). Rehabilitation for the elderly. In *Practice of Geriatrics* (ed. E. Calkins, P.J. Davis, and A.B. Ford), p. 152. W.B. Saunders, Philadelphia.

Grigor, R.R., Spitz, P.W., and Furst, D.E. (1987). Salicylate toxicity in elderly patients with rheumatoid arthritis. *Journal of Rheumatology*, **14**, 60–6.

Goodwin, J.S. and Regan, M. (1982). Cognitive dysfunction associated with naproxen and ibuprofen in the elderly. *Arthritis and Rheumatology*, **25**, 1013–15.

Halstead, L.S. (1976). Team care in chronic illness: a critical review of the literature of the past 25 years. *Archives of Physical Medicine and Rehabilitation*, **57**, 507–11.

Harris, E.D., Jr., Emkey, R.D., Nichols, J.C., and Newberg, A. (1983). Low dose prednisone therapy in rheumatoid arthritis: A double blind study. *Journal of Rheumatology*, **10**, 713–21.

Healy, L.A. (1983). Polymyalgia rheumatica and the American Rheumatism Association criteria for rheumatoid arthritis. *Arthritis and Rheumatism*, **26**, 1417–18.

Healey, L.A. (1986). Rheumatoid arthritis in the elderly. *Clinics in Rheumatic Diseases*, **12**, 173–9.

Kean, W.F., Anastassiades, T.P., Dwork, I.L., Ford, P.M., Kelly, W.G., and Dok, C.M. (1982). Efficacy and toxicity of D-penicillamine for rheumatoid arthritis in the elderly. *Journal of the American Geriatrics Society*, **30**, 94–100.

Kean, W.F., Bellamy, N., and Brooks, P. M. (1983). Gold therapy in the elderly rheumatoid arthritis patient. *Arthritis and Rheumatism*, **26**, 705–11.

Liang, M.H. and Jette, A.M. (1981). Measuring functional ability in chronic arthritis. *Arthritis and Rheumatism*, **24**, 80–6.

Linos, A., Worthington, J.W., O'Fallon, W.M., and Kurland, L.T. (1980). The epidemiology of rheumatoid arthritis in Rochester, Minnesota: A study of incidence, prevalence and mortality. *American Journal of Epidemiology*, **111**, 87–98.

Lockie, L.M., Gomez, E., and Smith, D.M. (1983). Low dose adrenocorticosteroids in the management of elderly patients with rheumatoid arthritis: selected examples and summary of efficacy in the long term treatment of 97 patients. *Seminars in Arthritis and Rheumatism*, **8**, 373–81.

MacFarlane, D.G. and Dieppe, P.A. (1983). Pseudo-rheumatoid deformity in elderly osteoarthritic hands. *Journal of Rheumatology*, **10**, 489–90.

Masi, A.T. (1983). Low dose glucocorticoid therapy in rheumatoid arthritis (RA): Transitional or selected add-on therapy? *Journal of Rheumatology*, **10**, 675–8.

Pincus, T. and Callahan, L.F. (1986). Taking mortality in rheumatoid arthritis seriously—predictive markers, socioeconomic status and comorbidity. *The Journal of Rheumatology*, **13**, 841–5.

Poss, R., Ewald, F.C., Thomas, W.H., and Sledge, C.B. (1976). Complications of total hip-replacement arthroplasty in patients with rheumatoid arthritis. *Journal of Bone and Joint Surgery*, **58**, 1130–3.

Spiegel, J.S., Spiegel, T.M., and Ward, N.B. (1987). Are rehabilitation programs for rheumatoid arthritis patients effective? *Seminars in Arthritis and Rheumatism*, **16**, 260–70.

Touger-Decker, R. (1988). Nutritional considerations in rheumatoid arthritis. *Journal of the American Dietetic Association*, **88**, 327–31.

Tugwell, P. (1989). Methotrexate in rheumatoid arthritis. Feedback on American College of Physicians guidelines. *Annals of Internal Medicine*, **110**, 581–3.

White, R.H. (1985). Preoperative evaluation of patients with rheumatoid arthritis. *Seminars in Arthritis and Rheumatism*, **14**, 287–99.

Wooley, P.H., Griffin, J., Panayi, G.S., Batchelor, J.R., Welsh, K.I., and Gibson, T.J. (1980). HLA-DR antigens and toxic reaction to sodium aurothiomalate and D-penicillamine in patients with rheumatoid arthritis. *New England Journal of Medicine*, **303**, 300–2.

World Health Organization (1980). *International classification of impairments, disabilities and handicaps* (ICIDH). World Health Organization, Geneva.

13.3 Osteoarthritis

DAVID HAMERMAN

AGEING AND OSTEOARTHRITIS

There is a consistently documented association between age and the development of osteoarthritis (Peyron 1984; Davis 1988). The increase in the prevalence of osteoarthritis with age has been described as arithmetic until 50–55 years and geometric thereafter, particularly in women, and especially with osteoarthritis of multiple joints including the interphalangeal (Peyron 1984). In considering osteoarthritis of the peripheral joints, Peyron (1984) suggested an evolution that begins in the metatarsophalangeal joint after the age of 25, the interphalangeal and first carpometacarpal joints after 45, and later onset in the knee and then the hip. This is consistent with the variable involvement of joints in osteoarthritis (Cooke and Dwosh 1986), a critical point discussed later, for which there is no firm explanation. There is the almost universal presence of Heberden's nodes in the elderly population, and the lesser prevalence but important consequences of osteoarthritis of the knee and hip, which accounts for disability approaching 10 per cent of the population over 60, second only to cardiovascular diseases in the United States (Holbrook *et al.* 1984). Nevertheless, the few studies available of patients over 70 suggest that involvement of the knee or hip is neither inevitable nor universal. In an ambulatory population undergoing physical examination, the severity of complaints concerning the knee appeared to be unchanged after the seventh decade (Forman *et al.* 1983). About 70 per cent of the residents of a nursing home (mean age, 81 years) studied by radiographs had osteoarthritis of the knee yet only 45 per cent complained of knee pain (Gresham and Rathey 1975). In another study of 100 residents in a health-related facility (mean age, 86 years), 65 per cent had no symptoms from the knees, and 93 per cent none from the hips. Among the residents with complaints concerning the knee, 11 per cent had no physical findings, and only 24 per cent had physical and radiographic findings suggestive of osteoarthritis (Hamerman *et al.* 1988). While it is true that complaints of joint pain may diminish with age (Acheson 1983), Gilmet *et al.* (1981) found less than 2 per cent tibiofemoral osteoarthritis by radiographic examination in a group of 50 men and 50 women (mean age, 73 years) in an outpatient clinic who had no complaints referable to the knee.

FACTORS CONTRIBUTING TO THE DEVELOPMENT OF OSTEOARTHRITIS

Thus, while ageing is the single most important risk factor for the development of osteoarthritis, it is clearly only one of a number of contributing conditions, some of which are as follows (Davis 1988).

Sex

Women appear to have a higher prevalence and severity of osteoarthritis of the extremities—hand, knees, feet—while men have a higher prevalence and severity of osteoarthritis involving the spine, hips, and disc degeneration of the spine.

Race

Studies in South Africa have compared patterns of osteoarthritis between white and black population groups. Osteoarthritis of the hands, feet, and hips appears to be less prevalent in black than in white women; however, more black men had osteoarthritis of the hand than white men. However, even racial differences are influenced by associated conditions: the low incidence of osteoarthritis of the hip in one black population was found to parallel the virtual absence of dysplasia of the acetabulum or femoral head, while the high prevalence of osteoarthritis of the hip in another black African community was associated with acquired dysplasia of the hip.

Wear and tear

This is certainly the concept most widely associated with the development of osteoarthritis and would be in accord with its greater frequency in ageing. Repetitive mechanical impacts, such as occur in pneumatic-drill operators, dockers, cotton workers, dancers, and miners, may predispose to osteoarthritis, although Davis (1988) cites a number of studies where no such associations were found. Knee injuries, particularly with menisceal or ligamentous tears, seem especially likely to contribute to osteoarthritis. A high incidence of osteoarthritis of the knee joint (40 per cent) has been found within 15 years of meniscectomy.

Obesity

In the Framingham study cohort identified in 1948–1952, there was a strong and consistent association between being overweight or obese and having osteoarthritis of the knee 35 years later, especially in women (Felson *et al.* 1988). The association of obesity with osteoarthritis in other joints is less clear, especially in the hip, which is also weight-bearing. It is not known whether obesity is contributory or a consequence of decreased physical activity, nor is it clear whether the effects of obesity in apparently predisposing the knee joint to osteoarthritis are primarily mechanical or indirect through associated metabolic conditions (e.g. diabetes) that may affect the cartilage. Davis (1988) points out that the strong association between obesity and osteoarthritis of the knee is not reduced or eliminated after controlling for serum cholesterol, trigylcerides, serum uric acid, glucose intolerance, and blood pressure, thus not supporting the likelihood of a metabolic component in the association of obesity with osteoarthritis of the knee. Dietary factors related to osteoarthritis are poorly understood. Sokoloff (1985) made an interesting review of some associations with cereals and trace elements in endemic forms of osteoarthritis 'hardly known to the West' in remote parts of the world.

Genetic factors

In the development of Heberden's nodes, heredity appears to be an important factor, involving a single autosomal gene dominant in females and recessive in males, with long-delayed genetic expression peaking after the age of 70 (Davis 1988). The studies of Kellgren *et al.* (1963) suggest a similar inheritance pattern for primary generalized osteoarthritis with Heberden's nodes, one of the subsets of osteoarthritis. The non-nodal type appears to be polygenic, slightly more frequent in men, and less familial.

Conditions within the joint

In a sense, these aspects represent the final common pathway —the site where the disease expression of cartilage loss and attempts at repair are manifested as osteoarthritis. In addition to the multiple 'systemic' factors noted above that contribute to osteoarthritis are the many 'local' factors within the joint. Some of these appear to be clearly implicated pathogenetically, such as menisceal lesions in the knee, or a variety of congenital deformities in the hip. The role of other factors, particularly crystal deposition (calcium pyrophosphate dihydrate, hydroxyapatite, urates), so frequent in the elderly population (Bergstrom *et al.* 1986), is less certain and remains controversial. More advanced stages of osteoarthritis have been attributed by Reginato and Schumacher (1988) to deposits of calcium pyrophosphate; others have downplayed this association, regarding these crystals as 'opportunistic' (Dieppe and Watt 1985), and finding them in non-inflamed asymptomatic joints (Nuki 1984). On the other hand, destructive changes in the shoulders and knees have frequently been associated with the presence of hydroxyapatite crystals (Nuki 1984). Deposits of immunoglobulins and complement in joints were found in a high proportion of cases of nodal generalized osteoarthritis where there appeared to be an inflammatory component; such immunopathological findings were said to be much less frequent in 'secondary mechanical osteoarthritis' (Cooke and Dwosh 1986). It is not clear why certain joints, such as the ankle and shoulder, subjected to systemic influences and mechanical forces, are 'habitually spared' (Radin 1983) from the traditional expression of osteoarthritis. This is a question also raised by Dieppe (1984). Perhaps, in a disease that often evolves over decades, manifestations in these joints have not systematically been sought in the population over 80 years of age. Perhaps the complex interdependent factors such as shape (or geometry) of the articulating surfaces, the mechanical properties of the bone, cartilage, and periarticular tissues, and the integrity of the supportive tissues, as discussed by Bullough (1984), confer a degree of 'protection' uniquely on these joints.

An important association, particularly in relation to ageing, is that between osteoarthritis and osteoporosis. Peyron (1984) and Davis (1988) have summarized much of the data: diminished bone mass in the hip is associated with fracture of the femoral neck, whereas increased bone mass, not only in the femur but also in the second metacarpal, has a distinct positive correlation with osteoarthritis of the hip. The increase in bone density in the subchondral region may enhance the mechanical forces on the articular cartilage (Radin 1983) and promote its degeneration. Dequeker (1985) considered that osteoarthritis and postmenopausal osteoporosis represented two extremes: one (osteoarthritis) with above-normal bone mass, more body fat, more muscle strength, and fewer fractures; the other (osteoporosis) with osteopenia, a slender body, and little degenerative joint change.

OUTCOME

There is only limited information on the long-term outcome of osteoarthritis for the individual subject. Longitudinal studies would obviously be difficult to do because of the time needed and the rate of attrition. In one study (Hernborg and Nilsson 1977), patients seen between 1950 and 1958 and classified as having osteoarthritis of the knee by radiographic criteria were reviewed again in 1968. The final sample was small (87 joints in 71 patients from an original group of 2195), as 'few patients diagnosed as gonarthrosis before 1950 survived to 1968'. Of these, the symptoms in 15 improved, in 23 were unchanged, and in 49 were worse. Radiographic evidence of deterioration was actually more unfavourable than functional decline. By and large, disease of the medial compartment was observed, and this remained limited to that compartment.

In an analysis of patients first seen at a mean age of 37 years and followed for 5 to 11 years (mean 6.9) (Miller *et al.* 1973), changes in the medial compartment progressed in about half and most of these subjects had increased pain as well.

The Baltimore Longitudinal Study examined changes in the distal and proximal interphalangeal joints of the same individuals over time. In the distal joints, the worsening of osteoarthritis increased steadily with age and with time from the first visit. The changes in the proximal joints were less uniform (Plato and Norris 1979).

In summary, there are variables in assessing the outcome of osteoarthritis that relate to joints examined, the population surveyed, its age and sex, and the length of time of the study. While osteoarthritis of the distal interphalangeal joints seems to approach universality with advancing age, osteoarthritis of the hips and knees appears not to progress inevitably. Radiographic changes by no means always parallel clinical findings and symptoms (Hadler 1985), and patients may present for the first time with hip or knee complaints with no apparent radiographic changes in the involved joint. Attempting to determine the duration of symptoms at the time of the first visit is likely to be quite subjective; we do not know for how long the pathological process unfolds before the clinical manifestations are perceived in 'spontaneous' cases of osteoarthritis (i.e. those without trauma or preceding surgery), but it may be decades, and will certainly be highly variable from individual to individual. Clearly, we have no 'biomarkers' that reflect cartilage degeneration at the time of the initial visit to predict which patients will progress, although long-term studies on serum or joint fluid for constituents of cartilage matrix, such as proteoglycans (Hascall and Glant 1987) or keratan sulphate (Sweet *et al.* 1988) may have promise. Nor are there uniform criteria for measures of outcome, although many have been proposed, particularly in connection with clinical trials of non-steroidal anti-inflammatory drugs (**NSAID**). Bellamy (1986) has comprehensively reviewed the subject of evaluative indices.

One of the variables in outcome may relate to the individual's behaviour and lifestyle. Fries (1988) lists osteoarthritis among those 'morbid conditions which are seldom fatal and in which a very large number of very common problems may be prevented'. While preventive measures in these conditions 'do not extend life, the results are a pure gain in health'. Among the 'preventive' steps listed by Fries are reductions in obesity, injury, and inactivity. Yet in terms of life-style, the Framingham study suggested that smoking may exert a protective effect against the development of osteoarthritis (Felson *et al.* 1989). Whether this is a trade-off for smoking-related development of osteoporosis instead of osteoarthritis is unclear. Osteoarthritis in one individual may be manifest as the 'usual' organ decline with disability, while in another demonstrating 'optimal' ageing, cartilage changes may not evolve into osteoarthritis in a weight-bearing joint. Certainly, in any individual, the evolution of osteoarthritis will result from a combination of factors intrinsic within the joint and host, as noted above, combined with extrinsic factors, such as the life-style, over which we generally have more control.

One disquieting note relates to increased mortality in women with osteoarthritis. Women aged 55 to 74 with radiographic changes of osteoarthritis in the knee were said to be at increased risk for subsequent mortality compared to women with normal radiographs (Lawrence *et al.* 1987). In considering the practical and functional consequences of ageing in the context of the 'dilemma of the elderly rheumatic patient', Svanborg (1988) concluded that 'the role of early life-style and environmental factors as they relate to physical condition and joint function at older ages has to be explored more carefully'.

MANAGEMENT OF OSTEOARTHRITIS IN RELATION TO USE OF NSAIDs

Symptomatic improvement in osteoarthritis associated with the use of NSAIDs has to be weighed against a high frequency of drug-related illness among elderly patients.

The relationship of NSAID to osteoarthritis of the hip requires special comment. Doherty (1989) reviewed the condition of rapidly progressive hip disease characterized by cartilage and bone loss with few osteophytes, which was initially attributed to some of the NSAIDs, particularly indomethacin ('analgesic' or 'indomethacin hip'), and concluded that 'such rapidly destructive large-joint osteoarthritis is not specific to NSAID users'. However, in one study, patients with osteoarthritis of the hip on indomethacin had less articular cartilage, lost more joint space, and came to arthroplasty sooner than patients on azapropazone, a NSAID with less potent prostaglandin-inhibitory effects (Rashad *et al.* 1989). NSAIDs are likely to have multiple influences on joint components in osteoarthritis in multiple ways, but there seems little justification to support a role sometimes attributed to them as 'chondroprotective' (Doherty 1989).

A number of NSAIDs have been approved for use in osteoarthritis. In patients with little evidence of joint inflammation, Schlegel and Paulus (1986) recommend 'the lowest effective dose to minimize drug toxicity in this more susceptible older population', and suggest stopping the drug periodically. With evidence of joint inflammation, 'higher dosages are needed, and a NSAID is likely to be more effective than salicylates'. The range of adverse reactions, including gastric, hepatic, renal, haematopoietic, and cutaneous, are described in detail elsewhere (Morgan and Furst 1986; Huskisson *et al.* 1985). With reference to the range of interactions between NSAIDs and other medications, especially in the older patient (Furst 1988), they may (a) inhibit metabolism of warfarin, contributing to a prolonged anticoagulation time (particularly with phenylbutazone); (b) increase bleeding time (especially with acetylsalicyclic acid); (c) decrease lithium clearance; (d) increase free phenytoin, thereby depleting serum folate required for phenytoin clearance; (e) decrease the action of those diuretics acting on the proximal tubule, contributing to sodium retention and oedema; (f) decrease the hypotensive action of the β-blockers; (g) increase the hypoglycaemic effects of sulphonylurea compounds (particularly with phenylbutazone and aspirin).

{Recent evidence (Bradley *et al.* 1991) raises the possibility that simple analgesics such as paracetamol (acetaminophen) may be as effective as NSAIDs in the short-term treatment of osteoarthritis.}

CONCLUSION

The widespread numbers of musculoskeletal conditions among the ever-growing elderly population in the United States (Holbrook *et al.* 1984) and the United Kingdom (Taylor and Ford 1984) represent a 'major public health issue' (Hadler 1985), which may grow worse over the next several decades. Further research on clinical classifications, means of assessment, and 'biomarkers' is needed to dissect out that part in which the diagnosis of osteoarthritis can be accurately established and its course predicted with some reliability. A reappraisal of therapeutic approaches, emphasizing 'preventive measures' as discussed here, but started earlier in life, may lessen current dependency on drugs and surgery, with their attendant morbidity and costs. Community services must also receive more emphasis (Liang *et al.* 1988) for those homebound elderly people whose numbers are likely to grow, and for whom osteoarthritis may contribute to impairment of mobility.

References

Acheson, R.M. (1983). Editorial. Osteoarthrosis—the mystery crippler. *Journal of Rheumatology*, **10**, 180–3.

Bellamy, N. (1986). The clinical evaluation of osteoarthritis in the elderly. *Clinics in Rheumatic Diseases*, **12**, 131–54.

Bergstrom, G., Bjelle, A., Sorensen, L.B., Sundh, V., and Svanborg, A. (1986). Prevalence of rheumatoid arthritis, osteoarthritis, chondrocalanosis and gouty arthritis at age 79. *The Journal of Rheumatology*, **13**, 527–34.

Bradley, J.D., Brandt, K.D., Katz, B.P., Kalosinski, L.A., and Ryan, S.I. (1991). Comparison of an anti-inflammatory dose of ibuprofen, an analgesic dose of ibuprofen, and acetaminophen in the treatment of patients with osteoarthritis of the knee. *New England Journal of Medicine*, **325**, 87–91.

Bullough, P. (1984). Osteoarthritis: pathogenesis and aetiology. *British Journal of Rheumatology*, **23**, 166–9.

Cooke, T.D.V. and Dwosh, I.L. (1986). Clinical features of osteoarthritis in the elderly. *Clinics in the Rheumatic Diseases*, **12**, 155–72.

Davis, M.A. (1988). Epidemiology of osteoarthritis. *Clinics in Geriatric Medicine*, **4**, 241–56.

Dequeker, J. (1985). The relationship between osteoporosis and osteoarthritis. *Clinics in Rheumatic Diseases*, **11**, 271–96.

Dieppe, P. (1984). Editorial. Osteoarthritis: are we asking the wrong questions? *British Journal of Rheumatology*, **23**, 161–5.

Dieppe, P. and Watt, I. (1985). Crystal deposition in osteoarthritis: an opportunistic event? *Clinics in Rheumatic Diseases*, **11**, 367–92.

Doherty, M. (1989). 'Chondroprotection' by non-steroidal anti-inflammatory drugs. *Annals of Rheumatic Diseases*, **48**, 619–21.

Felson, D.T., Anderson, J.J., Naimark, A., Walker, A.M., and Meenan, R.F. (1988). Obesity and knee osteoarthritis. The Framingham Study. *Annals of Internal Medicine*, **109**, 18–24.

Felson, D.T., Anderson, J.J., Naimark, A., Hannan, M.T., Kannel, W.F., and Meenan, R.F. (1989). Does smoking protect against osteoarthritis? *Arthritis and Rheumatism*, **32**, 166–72.

Forman, M.D., Malamet, R., and Kaplan, D. (1983). A survey of osteoarthritis of the knee in the elderly. *Journal of Rheumatology*, **10**, 282–7.

Fries, J.F. (1988). Aging, illness and health policy: implications of the compression of morbidity. *Perspectives in Biology and Medicine*, **31**, 407–28.

Furst, D.E. (1988). Clinically important interactions of nonsteroidal antiinflammatory drugs with other medications. *Journal of Rheumatology*, **15** (Suppl. 17), 58–62.

Glimet, T., Masse, J.P., and Ryckerwaert, A. (1981). Frequency of painful knee in a population of 50 women and 50 men older than 65 years. In *Epidemiology of osteoarthritis*, (ed. J. G. Peyron), pp. 220–3, Ciba-Geigy, Paris.

Gresham, G.E. and Rathey, U.K. (1975). Osteoarthritis in knees of aged persons. Relationship between roentgenographic and clinical manifestations. *Journal of the American Medical Association*, **233**, 168–70.

Hadler, N.M. (1985). Osteoarthritis as a public health problem. *Clinics in Rheumatic Diseases*, **11**, 175–85.

Hamerman, D., Sherlock, L., Damus, K., and Habermann, E.T. (1988). Research in the teaching nursing home: an approach to assessing osteoarthritis of the hips and knees. *Einstein Quarterly Journal of Biology and Medicine*, **8**, 110–19.

Hascall, V.C. and Glant, T.T. (1987). Proteoglycan epitopes as potential markers of normal and pathological cartilage metabolism. *Arthritis and Rheumatism*, **30**, 586–8.

Hernborg, J.S. and Nilsson, B.E. (1977). The natural course of untreated osteoarthritis of the knee. *Clinical Orthopaedics and Related Research*, **123**, 130–7.

Holbrook, T.L., Grazier, K., Kelsey, J.L., and Stauffer, R.N. (1984). *The frequency of occurrence, impact and cost of musculoskeletal conditions in the United States*, pp. 1–180. American Academy of Orthopaedic Surgeons, Chicago.

Huskisson, E.C., Doyle, D.V., and Lanham, J.G. (1985). Drug treatment of osteoarthritis. *Clinics in the Rheumatic Diseases*, **11**, 421–31.

Kellgren, J.H., Lawrence, J.S., and Bier, F. (1963). Genetic factors in generalized osteoarthrosis. *Annals of the Rheumatic Diseases*, **22**, 237–55.

Lawrence, R.C., Everett, D.F., Coroni-Huntley, J., and Hochberg, M.C. (1987). Excess mortality and decreased survival in females with osteoarthritis (OA) of the knee. *Arthritis and Rheumatism*, **30** (suppl.), S130.

Liang, M.H., Partridge, A., Eaton, H., and Iverson, M.D. (1988). Rehabilitation management of homebound elderly with locomotor disability. *Clinics in Geriatric Medicine*, **4**, 431–40.

Miller, R., Kettelkamp, D.B., Lauberthal, K.N., Karagiorgos, A., and Smidt, G.L. (1973). Quantitative correlations in degenerative arthritis of the knee. *Journal of Bone and Joint Surgery*, **55A**, 956–62.

Morgan, J. and Furst, D.E. (1986). Implications of drug therapy in the elderly. *Clinics in the Rheumatic Diseases*, **12**, 227–44.

Nuki, G. (1984). Editorial. Apatite associated arthritis. *British Journal of Rheumatology*, **23**, 81–3.

Peyron, J.G. (1984). The epidemiology of osteoarthritis. In *Osteoarthritis, diagnosis and management*, (ed. R.W. Moskowitz, D.S. Howell, V.M. Goldberg, and H.J. Mankin), pp. 9–28. W.B. Saunders, Philadelphia.

Plato, C.C. and Norris, A.H. (1979). Osteoarthritis of the hand: longitudinal studies. *American Journal of Epidemiology*, **110**, 740–6.

Radin, E.L. (1983). The relationship between biological and mechanical factors in the etiology of osteoarthritis. *Journal of Rheumatology*, **10** (suppl. 9), 20–1.

Rashad, S., Hemingway, A., Rainsford, K., Revell, P., Low, F., and Walker, F. (1989). Effect of non-steroidal anti-inflammatory drugs on the course of osteoarthritis. *Lancet*, **ii**, 519–22.

Reginato, A.J. and Schumacher, H.R., Jr. (1988). Crystal-associated arthropathies. *Clinics in Geriatric Medicine*, **4**, 295–322.

Roth, S.H. (1988). Pharmacologic approaches to musculoskeletal disorders. *Clinics in Geriatric Medicine*, **4**, 441–62.

Schlegel, S.I. and Paulus, H.E. (1986). NSAIDs and analgesic therapy in the elderly. *Clinics in the Rheumatic Diseases*, **12**, 245–73.

Sokoloff, L. (1985). Endemic forms of osteoarthritis. *Clinics in the Rheumatic Diseases*, **11**, 187–202.

Svanborg, A. (1988). Practical and functional consequences of aging. In *The dilemma of the elderly rheumatic patient*, (ed. A. Calin), *Gerontology*, **34** (suppl. 1), 11–15.

Sweet, M.B.E. *et al.* (1988). Serum keratan sulfate levels in osteoarthritis patients. *Arthritis and Rheumatism*, **31**, 648–52.

Taylor, R. and Ford, G. (1984). The Chief Scientist reports: arthritis/rheumatism in an elderly population: prevalence and service use. *Health Bulletin*, **42**, 274–81.

13.4 Orthopaedic aspects of joint disease

CARL M. HARRIS

When considering musculoskeletal surgery for older patients, one is usually hoping to correct one of three general conditions. In the first of these, the patient is faced with a disabling but not life-threatening problem, which if successfully corrected might totally eliminate the disabling pain and limited mobility. An example of this would be an individual with a single osteoarthritic hip. A variant would be found in those with multiple joint involvement, such as in rheumatoid arthritis, where surgery, if successful, could be expected partially to resolve but not totally to eliminate the disability, even if multiple arthroplasties were to be considered. In this first condition, surgery can be elective with the patient in the best possible condition after being fully evaluated and fully informed of the risks.

The second common condition is characterized by trauma that is not only disabling but can be life threatening. Examples of this include a fractured hip in an elderly, debilitated individual or the multisystem trauma that often occurs in a motor vehicle accident. In this second condition, the urgency of treatment often means that the patient is in less than optimal general health.

The third general condition in which musculoskeletal surgery might be considered is a chronic disability that in itself is not life threatening, and which surgery, even if successful, will only partially improve. As examples the various tenotomies, tendon lengthenings, and joint fusions to treat spasticity after a cerebral vascular accident come to mind. Operative treatment is often delayed until a plateau of recovery has been achieved after a cerebrovascular accident. The interval between that accident and surgery is one of intensive rehabilitation, designed to prevent contractures and to promote relearning of motor, speech, and cognitive skills.

All the multiple illmesses that would fall into these three general categories share common concerns for the team of doctors, nurses, therapists, and social workers whose efforts must mesh for a satisfactory convalescence to be achieved. Some of the general concerns will be reviewed first, followed by brief discussions of special concerns as they relate to examples of specific illnesses in the three general categories.

MUSCULOSKELETAL FACTORS

The status of the musculoskeletal system should be closely evaluated. In elective surgery the range of motion of all major joints, the strength of various muscle groups, and some estimate of exercise tolerance should be documented. A standard method of reporting, e.g., that used for disability of the hip (Harris 1969) or knee (Install *et al.* 1976) is useful. Preoperatively, some thought should be given to planning the postoperative course. Many questions will arise such as: will the patient have the motor skills to use a walking frame, crutches, or other types of mechanical assistance? how much supervision will be required? is balance likely to be a problem? will there be a need for additional help to provide standby assistance with walking? will special nursing be required? Many of these questions can be answered before an elective operation and will be useful in planning for discharge. Obviously where there has been trauma to the musculoskeletal system, the treatment of this is the major reason for surgery, and usually the range of motion and strength cannot be well estimated. However, the use of an injury severity scale is helpful (see Chapter 4.2).

Evaluation of the musculoskeletal system must include adequate radiography and it is important to understand that the musculoskeletal history and physical examination should complement this examination. Paget's disease of bone can often be suspected on physical examination and confirmed by laboratory studies and radiographs. Spinal stenosis with concomitant problems in the legs may be suspected from the history and physical examination, and later confirmed by appropriate radiographs, computerized axial tomography, or magnetic resonance imaging.

A thorough examination of the jaw and neck should be made in anticipation of endotracheal intubation at the time of surgery. Any suspicion of abnormality should be confirmed by radiographs and, in certain instances, may be sufficient reason to recommend spinal or regional anaesthesia. It is prudent to obtain radiographs of the cervical spine in all patients with rheumatoid arthritis for whom endotracheal intubation is anticipated, because of the danger of cervical spine instability.

NEUROLOGICAL EVALUATION

A thorough neurological evaluation should be made. Weaknesses that may be subtle in the sedentary individual can become profound when new motor skills are being learned in the convalescent phase or when some assistance with walking is required. Problems of balance that may be central in origin or related to diabetic or alcoholic polyneuropathy may be exacerbated when other disabilities are superimposed. A carpal tunnel syndrome, which may be mild in the sedentary individual, can easily be exacerbated in an individual who is attempting to use crutches or a walking frame.

STATURE AND WEIGHT

This is an important aspect of evaluation and merits discussion from several points of view. Vaughan (1982), in discussing anaesthesia, has stated that the physiological risks of moderate obesity (15 to 25% overweight) are minimal. However, an individual who is more than 30 per cent overweight has an increased risk of illness and death. During the actual operation, technical difficulties may be encountered in securing intravenous lines, controlling the airway, placing percutaneous arterial lines, and achieving appropriate surgical positions for an obese person. Obesity is often associated with hypertension, altered serum lipids, and angina pectoris, as well as with increased total central and pulmonary blood volumes. There is usually an increased cardiac output, and sometimes an elevated left ventricular and diastolic pressure. Pulmonary consequences of obesity include a reduction in the compliance of the lung and chest wall, and an increase in work and energy costs of breathing. Closure of peripheral parts of the lung can occur, along with low ventilation/perfusion ratios, shunts, and systemic arterial hypoxaemia. Sugerman (1987) has pointed out that the respiratory insufficiencies and disorders of obesity can, in part, be corrected by weight loss.

Starting or establishing a programme of partial or non-weight bearing in the overweight patient postoperatively may be difficult. Many obese people have poorly developed motor skills, and they often lack enough strength in their upper body to manoeuvre themselves easily in these programmes. Nursing personnel and therapists are often exasperated by the difficulties in handling an overweight patient, and this irritation may have a negative effect on rapport with the patient.

If elective replacement arthroplasty is under consideration, remember there are more late failures due to loosening, wear, or breakage of the prosthesis in overweight individuals (Pellicci *et al.* 1979). If at all possible, try to get the patient to lose weight to near the ideal preoperatively. With this should go a programme to develop the motor skills and exercise tolerance.

CARDIOPULMONARY PROBLEMS

Cardiopulmonary aspects of obesity have been discussed above. Another major concern for which careful evaluation is required is coronary artery disease. In a study of approximately 35000 patients, Tarhan *et al.* (1972) found that the risk of perioperative myocardial infarction was approximately 6 per cent if the patient had had a prior myocardial infarction, compared with 0.13 per cent if they had not. If the infarction had occurred within the past 3 months, the risk of reinfarction was 37 per cent, diminishing to 16 per cent for infarctions occurring between 3 and 6 months beforehand. In addition, the fatality of perioperative infarction was 50 per cent. Steen *et al.* (1978) reported on a series of 73000 patients, of whom 587 had suffered a documented, prior myocardial infarction. Again the overall risk was 6 per cent, as in Tarhan's study, with a correspondingly high risk for patients within 3 months (27 per cent), diminishing to 11 per cent in the 3- to 6-month interval after the infarction.

Fowkes *et al.* (1982), in a study of 109000 operative procedures in which anaesthetics were used, found that perioperative fatality was increased 25-fold over that anticipated in the presence of congestive heart failure, 15-fold in chronic renal failure, 12-fold with ischaemic heart disease, and 8-fold with chronic obstructive pulmonary disease. Where there is concern and the procedure is elective, myocardial perfusion with thallium imaging substantially improves the ability to identify patients who are potentially at high risk (Strauss and Boucher 1986). Doppler ultrasonography and echocardiography are useful in diagnosing significant aortic stenosis. Goldman *et al.* (1977) have developed a multifactorial index of cardiac risk in non-cardiac surgical procedures and age is considered an independent risk factor. DelGuercio and Cohn (1980) advocated Swan–Ganz catheterization in all elderly patients undergoing elective major surgery, but this is not a widespread practice.

Remember that a spinal anaesthetic, or some type of regional anaesthetic, can be used for many procedures. This becomes valuable where there are potential problems with the airway or risk with endotracheal intubation. {In Europe, local or regional anaesthesia can be used in conjunction with a chlormethiazole infusion adjusted to provide varying appropriate levels of consciousness.}

Potassium imbalance is a potential source of cardiac irritability under anaesthesia. Acute hypokalaemia increases cardiac excitability, with an increased incidence of supraventricular tachycardia and premature ventricular contraction. Chronic severe hypokalaemia can lead to renal tubular dysfunction, and there may be enough impairment of muscle strength to necessitate mechanical ventilation. Hyperkalaemia may have a considerable effect on skeletal and heart muscles. Muscle weakness may be enough to impair breathing. Muscle relaxants, inhalational anaesthetics, and narcotics may potentiate dangerous hypoventilation. If the patient being considered for surgery has Parkinson's disease, or recently had a stroke (up to 6 months before), a major burn (up to 3 months), or some paralysis, succinylcholine may cause massive release of potassium. Within 5 days of many injuries, muscles become hypersensitive to succinylcholine. Hyperkalaemic patients will be more susceptible to drugs, for example, local anaesthetics, that decrease cardiac contractility. Aminoglycosides can also potentiate muscle relaxants used during surgery and lead to prolonged hypoventilation.

Trauma, sepsis, gastric aspiration, shock, and fat embolism are well-known causes of acute adult respiratory distress syndrome. This complication, still reported with a fatality rate as high as 60 per cent, is best managed by prevention. As will be discussed later, immediate fixation of fractured long bones, particularly where there has been multiple injury, decreases the risk of this complication and subsequent pulmonary dysfunction.

Good pulmonary toilet is imperative in the immediate postoperative phase. For many elective procedures, it may be wise to have a pulmonary therapist evaluate the patient preoperatively, to explain some of the procedures that will be encountered after surgery.

Often the greatest cardiopulmonary stress may not come at the time of the actual surgery but in convalescence, when the patient may be inadvertently pushed to the limit of a compromised exercise tolerance. If this is anticipated, one must be specific in outlining one's goals for the rehabilitative phase to physiotherapists.

GASTROINTESTINAL DISEASE

Most patients being considered for joint replacement will have received various anti-inflammatory medications. With many of these, the incidence of gastrointestinal distress and bleeding is high. As almost all patients with replacements in the leg, many with arthroplasties in the arm, and perhaps a majority of those with fractured long bones will be anticoagulated, the physician must be aware of any past or present problems with intestinal bleeding.

Most patients will have prophylactic antibiotics during or immediately after arthroplasty and major surgery to long bones. Any history of drug allergies or gastrointestinal sensitivities should be sought. Ulcerative colitis or Crohn's disease will pose special problems.

GENITOURINARY PROBLEMS

There is a high incidence of low-grade chronic infection in many elderly patients for replacement arthroplasty. This infection is a potential source of bacterial seeding. A second concern is that where there has been compromised renal function, postoperative electrolyte imbalance may occur. Renal failure can be induced by various medications including some prophylactic antibiotics and anticoagulants. The hypotension sometimes associated with major surgery can aggravate this. Prostatic hypertrophy, especially when associated with low-grade infection, is often best treated before elective arthroplasty because it is frequently associated with postoperative urinary retention.

BLOOD DISORDERS

Iron deficiency anaemia is frequently seen in older patients with diminished iron intake, malabsorption, or gastrointestinal blood loss. This should be recognized preoperatively, its causes appropriately diagnosed, and treated in the immediate preoperative and postoperative periods.

In recent years it has increasingly been the practice to have the patient donate his/her own blood for transfusion procedures (Thompson *et al.* 1987). Current practices commonly allow the patient to donate up to seven units of blood preoperatively, provided that anaemia or other problems do not occur. We allow this up to age 80, although there is no evidence that age confers additional hazards in autologous donations. In special circumstances the blood can be frozen for surgery that is planned for later than 35 days after donation, but it is recognized that certain blood factors are diminished by doing this. The use of autologous blood donation has greatly alleviated the concerns about blood-borne infection with human immunodeficiency virus and hepatitis virus.

PHERIPHERAL VASCULATURE

Mild peripheral arterial insufficiency, which may not be easily recognizable in the sedentary individual, can become a major problem in the postoperative phase if unusual demands are placed on the patient. If the history or physical examination suggest problems of claudication, carotid stenosis, or transient cerebral ischaemia, or if extensive arterial calcification is noted on the preoperative radiograph, Doppler studies or arteriography may clarify the problem. Of course, where there may be arterial injury, arteriography would be done as a matter of routine. In many instances of surgery on the limbs, a tourniquet will be used. This can be a source of concern if there is arteriosclerosis, and may warrant a sophisticated preoperative vascular evaluation.

In elective arthroplasty, particularly in the legs, the incidence of postoperative phlebothrombosis is high. It is assumed that the same risk is present with injuries, although it is obvious that good preoperative evaluations are not always made in such cases. Phlebothrombosis may occur in over half of the patients undergoing total hip arthroplasty. Of these, one in ten may suffer a pulmonary embolus if prophylactic measures are not carried out. These figures have justified the use of anticoagulants during the intraoperative and immediate postoperative phase (Consensus Conference 1986; Evarts 1987). Any history of phlebothrombosis increases these already high risks. Some surgeons feel that elective replacement arthroplasty should not be attempted within 1 year of a phlebothrombosis.

METABOLIC PROBLEMS

A great many elderly people will have chronic metabolic problems that influence musculoskeletal surgery. Osteoporosis and, in some populations, osteomalacia are quite common and are discussed in other sections. Three other commonly encountered endocrine problems are diabetes mellitus, thyroid disease, and adrenal suppression by exogenous steroids.

There is no one best way to manage diabetic patients in the perioperative period. It is generally felt that in the preoperative period, blood glucose levels of less than 250 mg per cent (14 mmol/l) should be achieved and that these should be maintained throughout the perioperative period. The regimen chosen should prevent wide swings in blood sugar as well as ketoacidosis. In the elective surgical patient, measurements of glycohaemoglobin Hba_{1c} may be useful, as elevations 3 to 6 per cent above normal may indicate poor control of blood glucose the preceding 4 to 6 weeks (Reynolds 1987).

Abnormalities of thyroid function can be extremely subtle and often difficult to diagnose in the presence of multiple systemic disease in an elderly patient. Remember that anaesthesia, surgery, infection, coincident illness, and fasting can all affect the manufacture, binding, peripheral conversion, hepatic release, and metabolism of thyroxine (T_4). Surgery has been reported to induce the shift in deiodinization of T_4 to tri-iodothyronine (T_3). The chief complication of hyperthyroidism in the surgical patient is thyroid storm, which has been reported to occur in anything up to 32 per cent of those with hyperthyroidism. Therefore, it is felt that elective surgery should not be carried out in thyrotoxic patients. If surgery is urgent, certain ways of treating thyrotoxic patients are available. These are the same as would be used if a patient with unrecognized or untreated hyperthyroidism developed thyroid storm postoperatively (Goldmann 1987).

Adrenal suppression can be caused by daily pharmacological doses of steroids (equivalent to more than 7.5 mg of prednisone daily) that suppress the hypothalamic–pituitary–adrenal axis. This suppression leads to decreased stimulation of the adrenal glands, leaving them unprepared for the stress of illness or surgery for a variable period of time. In common practice, if there is a history of steroid use for more than 2 to 3 weeks within the preceding year, it is considered prudent for the physician to give replacement steroids perioperatively at levels that would be approximately those of normal individuals subjected to the stress of surgery. This is about 100 mg of hydrocortisone acetate or hemisuccinate every 8 h. Should it be considered imprudent to use steroids, specific tests to measure adrenal function can be carried out preoperatively.

SKIN DISEASES

Various skin conditions can offer problems at the time of surgery. Individuals on long-term steroid therapy can develop quite fragile skin, and special handling may be required. For instance, removing the adhesive draping frequently used in surgery will take skin with it in many patients on long-term steroids. The presence of decubitus ulcers or severe stasis in the legs frequently causes surgery to be delayed until they can be cleared of all signs of infection. The presence of psoriatic patches may require modification of routine incisions.

RADIOGRAPHIC EVALUATION

Complete radiographic studies are essential and should include:

1. The involved joint and all bones involved in the joint complex. As an example, if hip arthroplasty is being considered, radiographs should include the opposite hip, the pelvis, and the entire femur. If knee arthroplasty is being considered, the radiographic evaluation should include the entire femur and tibia.
2. Weight-bearing films are essential in considering total knee arthroplasty and can be of considerable help in all replacements in the legs (Fig. 1). For example, an apparent discrepancy in leg length assumed to be caused by an arthritic hip may be due in part to a fixed pelvic obliquity. This is frequently more apparent in a weight-bearing view of the pelvis and would be a reason to obtain views of the thoracolumbar spine.
3. Appropriate ancillary films are important. Cervical spine films should be obtained in patients with rheumatoid arthritis undergoing endotracheal anaestheisa (see above). Shoulder, elbow, hand, and wrist films would also be appropriate in rheumatoid arthritic patients if prolonged use of ambulatory assistance is required.
4. Films of sufficient quality to estimate the size of the medullary canal are needed if intramedullary devices are to be used, as in the femoral component of a total hip replacement. These should also be good enough to give an estimate of osteopenia that may be present from disuse or disease.
5. In fracture treatment, a radiograph of good quality of the entire bone should be obtained.
6. Non-invasive imaging of the musculoskeletal system, particularly by magnetic resonance imaging, is becoming an important diagnostic tool. The spine, various fractures, and ischaemic necrosis of bone may best be evaluated in many instances by magnetic resonance imaging.

PREOPERATIVE PHYSICAL THERAPY AND OCCUPATIONAL THERAPY

In certain complex situations it is of considerable benefit to have patients evaluated preoperatively by the physiotherapist and occupational therapist. This evaluation can include some preoperative muscle training, as well as instruction in the use of the ambulatory assistance that may be required postoperatively. For handicapped patients, the occupational therapist often has suggestions that are helpful both preoperatively and postoperatively (Harris 1984).

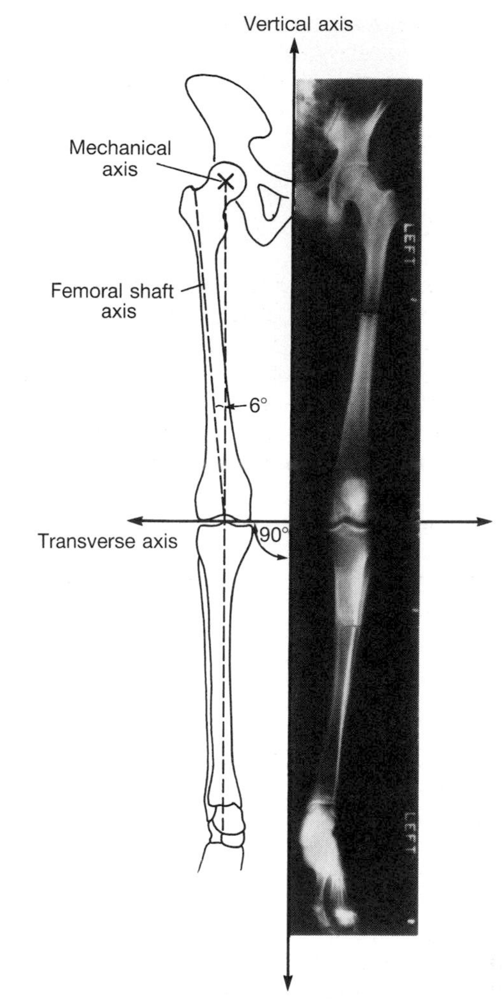

Fig. 1. Composite diagram of anatomical relationships in the lower extremity. Several factors are important: (a) in a normal standing configuration, a plumb line dropped from the hip will pass through the knee and bisect the talus; (b) axial lines drawn down the shaft of the femur and tibia subtend an angle of approximately 6°. The degree of varus (bowing) or valgus (knock-knee) is measured from this point.

TOTAL ARTHROPLASTIES

Patients for total arthroplasty of the hip or knee are excellent examples of the category of individuals faced with a disabling but non-life threatening complaint, which if successfully corrected might eliminate a painful and restrictive disability. The condition has diminished the quality of life. According to a 1982 report, an estimated 75000 total hip arthroplasties were performed in the United States annually. Approximately 60 per cent of these were in individuals over age 65, with another 25 per cent in individuals between the ages 55 and 64 (National Institute of Health Consensus Conference 1982). The number of total knee replacements done annually in the United States had earlier been estimated at 40000. Spokesmen for the prosthetic industry estimate that

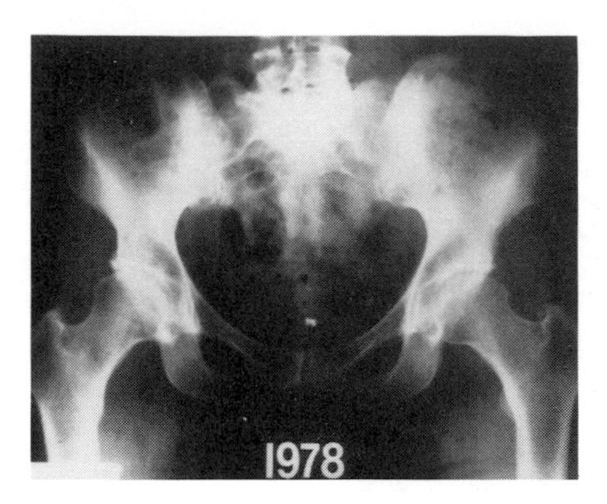

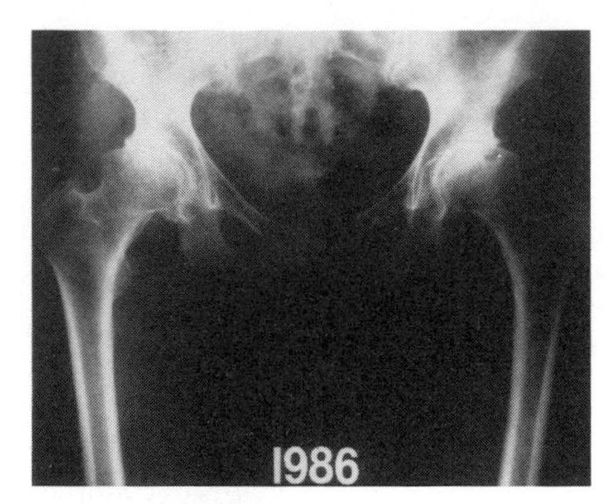

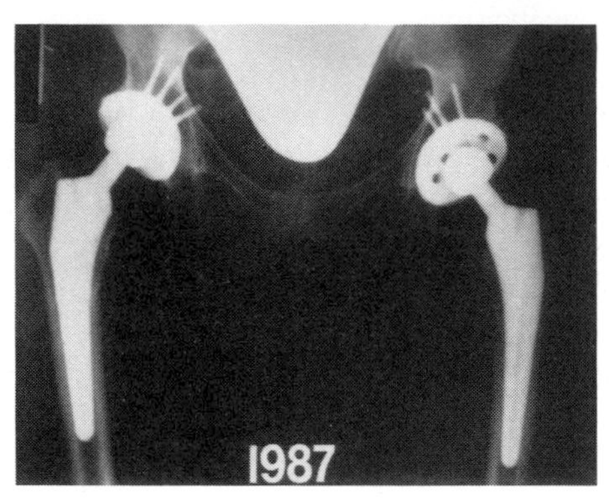

Fig. 2 Idiopathic osteoarthritis of the hip. Upper left (1978): relatively normal joint space with good femoral head contour. Upper right (1986): loss of joint space with femoral head flattening and osteophyte formation. Lower (1987): total hip arthroplasty using porous-coated uncemented components. Procedures done at age 61.

the number of joint arthroplasties done annually in the United States may have doubled since these figures were compiled.

The distinction between an anatomical abnormality and a functional disability is important (Figs. 2, 3). Many people, by various means of compensation, are able to accommodate to a potentially uncomfortable anatomical abnormality. It is not uncommon for people to tolerate a considerable disability when in a position of forced responsibility (a job, caring for a family member, or providing for a family). When these responsibilities are removed, the same anatomical abnormality may become the source of intolerable disability. Life circumstances may also change significantly so as to place more demands on an individual with some anatomical handicap. For example, a spouse who has been providing physical and emotional support may become disabled or die; a neighbour who in the past has provided a weekly shopping service moves; children who have been providing care for the household grow up and migrate. In instances such as these, pain and disability may increase with no apparent increase or change in the anatomical condition. Liang and Cullen (1984) have put it well: 'a patient's functional ability is determined by a complex interaction of physiological factors, psychosocial factors (coping skills, motivation, family or social support), socio-economic status, and environmental factors, e.g. transportation and adaptive devices'.

Another common observation in individuals with multiple joint abnormalities is that one joint is the source of most, if not all, complaints. Frequently, after successful arthroplasty of that joint, the other joints impose limitations on function that the patient did not anticipate. In the preoperative discussion one must consider these possibilities and tell the patient that multiple procedures may be needed to achieve satisfactory relief of pain and increase in function. As was pointed out by Burton *et al.* (1979), if arthroplasty does not fulfil the patient's expectations, they are less likely to rate the operation a success and will have a less positive attitude toward their general health.

Surgery is generally considered only when non-operative management fails to give the relief sought by the patient, and if the patient is willing to accept the risks. The risks of arthroplasty fall into three general categories:

1. The risks associated with undergoing major surgery.
2. The risk that various biomaterials will, with time and use, wear, break, or loosen.
3. The risks of failure not related to biomaterial failure. These include, but are not limited to, the immediate or late risk of infection, postoperative phlebothrombosis, and the risks associated with revision of a prosthesis in which there has been failure of a biomaterial with time.

The risks associated with major surgery have already been discussed in relation to the various systems. The risks associated with biomaterials are multiple and have given rise to some of the most intensive research in the field of musculoskeletal surgery today. Various types of non-corrosive metals, plastics, polymers, and ceramics are used as prosthetic materials that are introduced into bone and become composites with both cortical and cancellous bone in load transfer; this is particularly true in the legs. The various parts of this com-

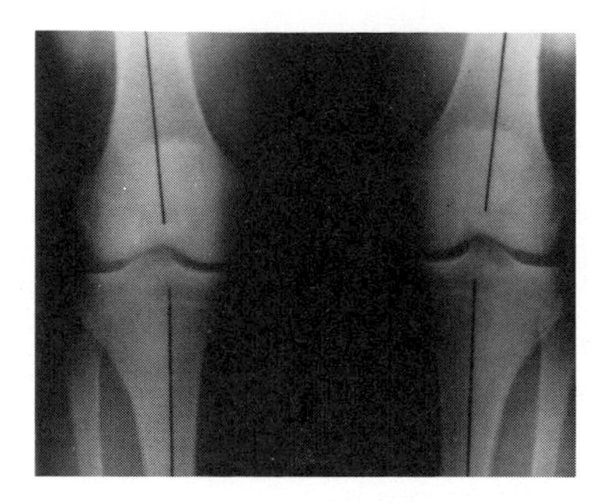

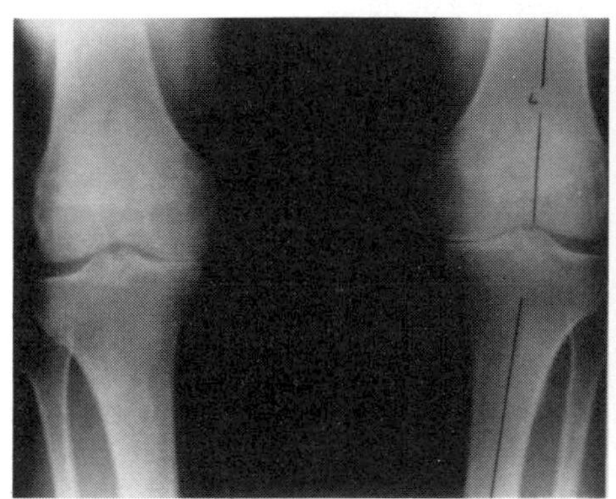

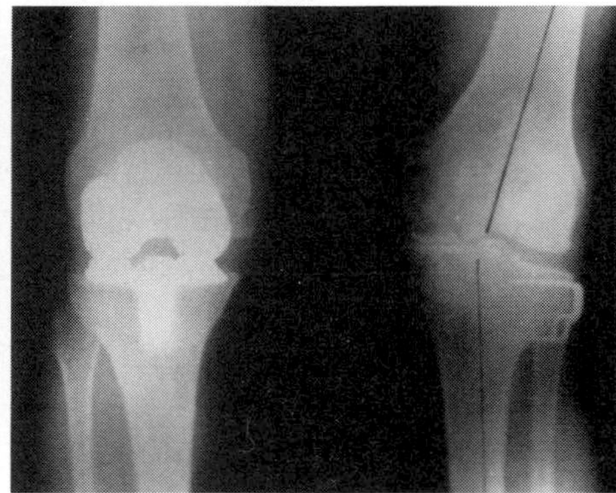

Fig. 3 A normal standing radiograph and pre- and postoperative views of high tibial osteotomy and total knee arthroplasty. (a) Normal weight-bearing AP radiograph of the knees showing a normal valgus femoral tibial axis and symmetrically thick cartilage space in the medial and lateral compartments of the knee. (b) osteoarthritis in a 62-year-old active male. A varus femoral tibial axis and loss of medial compartment joint space bilaterally. (c) Treatment of the patient shown in (b) by a high tibial osteotomy on the left by removing a medial based proximal tibial wedge (done at age 63) and a cemented total knee arthroplasty on the right (done at age 67). Both procedures remain satisfactory at age 76.

posite (plastics, metals, ceramics, cortical and cancellous bone) have different elastic moduli, which means that variable manifestations of strain (bending) arise when stresses (weight) are applied. One consequence is that shear forces develop at the interfaces between the various materials, and combinations of mechanical and biological forces cause loosening. Much research is focused on trying to find proper means of transferring the load and minimizing the various stresses between living tissue and the biomaterials, thus limiting the biological response that could lead to loosening or fracture of the prosthesis.

Charnley deserves the most credit for the development of total hip arthroplasty using ultra-high molecular-weight polyethylene (UHMWP), and while not the first to use polymethyl methacrylate, should also be given credit for development of the technique of fixation of prosthetic components using that material. He was also one of the first to experience significant difficulty with a biomaterial when he first attempted to use Teflon, which has a more favourable coefficient of friction than ultra-high molecular-weight polyethylene, for the acetabular component, and found it poorly tolerated. There was a rapid breakdown of Teflon and the development of large, non-caseating granulomas around the joint.

Several avenues of current research in arthroplasty are attempting to achieve better fixation and wear of the prosthesis, thus increasing the length of time between initial insertion and revision. One area of interest is to improve cementation by pressurizing the cement into the host cancellous bone. This technique has reduced the incidence of loosening at the bone–cement interface. Larger and better designed prostheses that are less likely to fracture because of metal fatigue have now almost universally replaced most of the smaller prostheses used in the 1960s and 1970s. The use of newer materials, including ceramics which have a coefficient of friction at the joint interface that is closer to that of the normal joint, is of considerable interest.

Increasingly porous coatings that allow bony ingrowth when applied to prosthetic devices are being used. The use of these prostheses eliminates the need for polymethyl methacrylate. The perceived advantages of this technique arise from the observation that the elastic modulus of the methacrylate is much at variance with that of cancellous bone and it is known that the greatest shear forces occur at the bone–cement interface. The porous surface greatly increases the area of the metal–bone interface. This distributes the load (body-weight stresses) over a much greater surface area than would be found in a smooth prosthesis, and thus the stress per unit area is much less. Successful use of porous-coated prostheses requires good, immediate fixation to prevent both micro- and macro-movements at the bone–prosthetic interface while bone is growing into the porosities. Bony ingrowth probably takes about as long as fracture healing and thus weight-bearing stresses need to be restricted for that time (Bobyn *et al.* 1980).

In following up prosthetic failures, a finding of interest is that polymethyl methacrylate, when fragmented, appears to induce osteolysis, which can accelerate bone loss around the prosthesis. There is some evidence that this process is stimulated by prostaglandin E (Goldring *et al.* 1983).

In the third category of risk for total arthroplasty, the risk of failure, the greatest short-term dangers are postoperative infection or deep venous thrombosis. As discussed earlier, the risks of phlebothrombosis are great enough for prophylactic, perioperative anticoagulation in the vast majority of instances.

To combat infection, multiple precautions are generally taken. If infection is suspected at preoperative evaluation, this is generally treated before arthroplasty. During surgery, special precautions are usually taken. These include special environments—various combinations of limited access to the operating suite, laminar air flow, high-efficiency particulate air filters, ultraviolet lights—and the use of topical antibiotic irrigation or other germicidal irrigations. In addition, special gowns or breath-exhaust systems for the operating-room personnel are used. The concern is to limit the introduction of bacteria into the wound at the time of surgery. To be transmitted as aerosols, bacteria must be on particles greater than 0.5 μm in diameter, and efforts are made to limit the number of these particles in the operating area or to sterilize them. Patients with known difficulties in wound healing, such as those on steroids, those with rheumatoid arthritis with vasculitis, or the diabetic patient with arterial insufficiency, have particular postoperative problems.

Myositis ossificans or heterotopic bone formation is a special postoperative problem in replacement arthroplasty of the hip. It may occur mildly in up to half of all cases, but it is extensive enough to become a disability in approximately 2 per cent. Patients most likely to develop myositis ossificans are those who have undergone previous hip surgery, or who have ankylosing spondylitis or disseminated skeletal hyperostosis. For high-risk individuals, various medications and postoperative radiation have been used with some success (Ayers *et al.* 1986).

When there is a late failure of a biomaterial, it is possible in most instances to carry out revision surgery. Remember that the risks faced at the original operation are not only still present but are increased because the patient is older and revision surgery is technically more difficult. Usually some bone stock has been lost as a result of the original procedure. Special prostheses are required and the risks of perioperative infection and myositis ossificans increase.

Late failure through infection is often difficult to treat. In certain instances, long-term antibiotics can be used, but in almost all circumstances, loosening of the prosthesis will occur. In most cases the prosthesis must be removed, with anticipation of it being reinserted at a later date after appropriate débridement and antibiotics. Occasionally the infection cannot be cured or controlled, even after extensive débridement and antibiotics. In many of these instances, the salvage procedure will be a joint fusion or a resectional arthroplasty. In either event, the disability will be worse than the patient had hoped for. In the hip, resectional arthroplasty is generally used, and walking will subsequently require an ischial weight-bearing brace or some type of aid (cane, crutch, or frame). In the knee, fusion will often be done. For most elderly patients, particularly those with other joint disabilities, these procedures leave a considerable handicap as the energy expenditure for walking is high.

STAGING PROSTHETIC INSERTIONS

Replacement arthroplasty, with rare exceptions, is elective. If it is known that multiple arthroplasties are to be done, some general principles are followed by many surgeons. First, if both a hip and a knee arthroplasty are anticipated, one generally does the hip first. The reason for this is that if the hip has a flexion contracture, there will frequently be a compensatory lordosis and compensatory knee flexion to allow an upright, weight-bearing stance. Correction of the flexion contracture in the hip will obviate the need for compensatory knee flexion on weight-bearing, thus making it more likely that knee arthroplasty with full extension will be successful at the second operation. Another reason for doing the hip before the knee is that, during hip surgery, the knee must frequently be flexed to near 90° with external rotation varus as stress is applied. Theoretically this could damage a knee arthroplasty. During knee arthroplasty, however, no extremes of position are required of the hip and no unusual stresses are applied.

Many surgeons doing replacement arthroplasties in the arm would prefer the patients not to be using crutches, walking frames, or canes. Therefore, it is preferable to do any surgery on the leg first.

TRAUMA

Treatment of trauma, to be effective, often has to be expedient; and for a variety of reasons patients will not be as in good health as most of those undergoing elective arthroplasty.

At all ages, trauma is a major source of musculoskeletal disability. For certain injuries, of which both hip and wrist fractures are examples, the incidence is skewed in favour of elderly people. The reason for this is most likely coincident osteoporosis. From 1970 to 1977, there were approximately six million fractures of all types in the United States (Holbrook *et al.* 1984). Their estimate of the incidence of hip fractures among the elderly population (average age, 65 years) was 88000 annually; Lewinnek *et al.* (1980) estimated a higher number (130000) annually for the same period.

Hip fractures have been the most extensively studied traumatic injury of elderly people, and numerous studies have emphasized the significant death, illness and expense associated with these fractures. Bauer *et al.* (1985) have stated that, in Sweden, hip fractures consume more hospital days than all abdominal operations for non-malignant conditions, and that when institutional convalescence is included in the cost, hip fractures are more expensive to individuals and society than any other surgical diagnosis. They also pointed out that in the United States, acute care for hip fractures cost more than one billion dollars annually. Miller (1978) reported on a group of 360 patients with hip fractures and estimated that the fatality rate of a population of a similar age without hip fractures was about 9 per cent per year; the death rate from all causes in the fracture group was 27 per cent in the year following injury, an increase of 19 per cent over that of the uninjured population. Hedlund *et al.* (1986) analysed 20538 intertrochanteric and femoral-neck fractures in Sweden from 1972 to 1981 and found that the rate of fractures in men doubled every 7.8 years after the age of 20 years, and every 5.6 years for women after the age of 30 years, suggesting that age-associated factors were the main risk {a cross-sectional analysis of hip fracture rates is provided by Maggi *et al.* (1991)}.

Ceder *et al.* (1980) examined indicators of rehabilitative potential, and the ability to return to a 'pre-fracture habitat', in patients with hip fractures in Ludon, Sweden. They found that the best predictive indicator for rehabilitation was the capacity to walk independently on visits outside the home before the fracture; a significant indicator of the potential for rehabilitation at home after a fracture was the ability to walk and carry out personal hygiene within the first 2 weeks.

Hip fractures are best treated if surgery is carried out as soon as the condition of the patient will permit. The choice of reduction with internal fixation or replacement arthroplasty will depend on the level of fracture and the personal bias of the surgeon. Efforts should be made to effect early independent walking and independence of activity in daily living. Evidence from Sweden suggests that the most effective early rehabilitation can be carried out in an acute hospital rather than a chronic-care institution.

It has been suggested that proximal femoral fractures can be treated without surgery in individuals who cannot walk (Winter 1987), although there is no unanimity of opinion that this is desirable (Hornby *et al.* 1989).

Trauma with multiple injuries or fractures appears to be less common with advancing age. In part, this may be because older people are less often exposed to high-velocity trauma in motor-vehicle accidents, or because many do not survive long enough to reach a hospital and thus are not included in most studies. The Transportation Research Board of the National Research Council for the United States (1988) has shown that 13 per cent of all automobile trauma occurs to people in the over-65 age group, which roughly equates with their proportion of the population. However, the same study finds that people aged over 65 are less likely to survive automobile trauma. Marx *et al.* (1986), in a study in Basle, Switzerland, found that in people aged over 65 years, 17 per cent of deaths were due to a single fracture, whereas death from this cause was highly unlikely in a younger group. The use of a scoring system for severity of injury is recommended for all individuals suffering from multiple trauma, as it allows one to compare data as well as providing prognostic information. However, Marx *et al.* felt that the scoring did not take into account the pre-existing handicaps that are sometimes found in older people.

As is true for younger patients, adult respiratory distress syndrome is a major problem among the old. For all ages, early operative intervention to stabilize fractures and minimize bleeding and the use of positive- and expiratory-pressure systems will lessen the chance of developing this syndrome (Pepe *et al.* 1984; Johnson *et al.* 1985).

CEREBROVASCULAR ACCIDENT

Cerebral thromboembolism is the most common cause of stroke and generally carries a better rehabilitative potential than cerebral bleeding. However, according to the studies of Clifford *et al.* (1985), approximately half of the survivors of an embolus will have some residual neurological problems.

After a stroke, there is usually a period of flaccid paralysis, after which there is enhancement in the tone in the adductors and flexors in both the arms and the legs. Failure of tone to recover in 6 weeks is generally considered a bad prognostic sign, as is failure to regain bladder and bowel control. Voluntary movement, when it occurs, usually starts proximally and extends distally. The extent of motor impairment depends upon the amount of damage to the cerebral cortex. Definitive surgical treatment is generally not begun until 6 months after the cerebrovascular accident so as to allow for spontaneous recovery. An effective programme of care is important in this interval to prevent contracture. A daily range of movements, splinting if necessary, limb positioning, and meticulous skin care are part of this programme.

Determining which groups of muscles are spastic or contracted frequently depends on a combination of skilled observation, examination under an anaesthesia, and often dynamic electromyography. The various procedures for muscle and tendon release, tendon lengthening, and tendon transfer all have as a goal a greater degree of independence, improvement of nursing care, and improvement in the quality of life. The potential for rehabilitation after a stroke is high. Waters (1978) pointed out that the number of patients with impairment in the arms and legs from stroke exceeds that of all other neurological diseases. At that time it was estimated that there were more than two million persons in the United States who had permanent neurological deficits after strokes. It was also pointed out that the average patient who survives a stroke beyond the first month has a life expectancy of more than 5 years. For details of more specific treatments of various deformities see Waters (1978) and Botte *et al.* (1988).

DISCHARGE PLANNING

Discharge from closely supervised convalescent care is generally more related to achieving performance goals than any other factor. For example, when an arthroplasty has been done in the leg, the patient generally remains in hospital until a level of independence is achieved that will allow unassisted transfer from a bed, independent walking with mechanical aids, use of a commode, and independent performance of an exercise regimen. After trauma, the extent of the injuries will obviously determine the level of independence before discharge from the acute hospital, and long-term convalescent care is often required.

As soon as early wound healing appears assured, and walking is achieved, it is possible for a patient with an arthroplasty of the leg to take a shower. This is often preferable to bathing because of the extreme hip and knee flexion required to get into a bath. When balance or weakness is a problem, a walking frame and stool should be taken into the shower so that the patient can sit down.

In early convalescence, dressing is frequently a problem. Various self-help devices can be used in accomplishing these tasks. Frequently, the patient cannot pull on long elastic stockings and some help is required with these. These hose are widely recommended to minimize peripheral oedema and phlebothrombosis postoperatively, and they are generally worn until there is no evidence of dependent oedema.

Driving is usually not recommended until muscle and joint recovery have reached the point where rapid movements can be carried out in an emergency. This would usually not be before the sixth postoperative week in the case of elective arthroplasty of the leg. In addition, when travelling either as a driver or passenger, use of a seatbelt is essential. Irreparable damage has been done to hip and knee arthroplasty when a patient has been thrown against the dashboard in a collision or sudden stop. Because the risk of phlebothrombosis remains for some time postoperatively, it is not wise for the patient to remain in a sitting position for extended periods. Our present recommendation is not to drive or remain in a sitting position for more than 90 min without a brief period of walking or change in position for at least 3 months postoperatively.

There is a slight risk of a late infection after arthroplasty. There is some controversy over the incidence of this complication, but it cannot be doubted that deep infections do occur around prosthetic devices when there has been a bacteraemia or septicaemia from a distant infection. It is generally recommended that any obvious infection be treated immediately in people who have had arthroplasty.

Whether individuals who have undergone joint arthroplasty need prophylactic antibiotic coverage for dental procedures is also somewhat controversial. Norden (1985) pointed out that there is little evidence to support the need for this practice. However, many physicians and dentists now follow a protocol similar to that recommended by the American Heart Association (1977) for individuals with heart-valve replacement undergoing dental procedures. This involves giving a bolus of 2 g of penicillin 1 h before the procedure and either 1 g 6 h after the procedure or 500 mg every 6 h for 48 h. A survey of United States dental schools found several other antibiotic regimens in use (Shay and Lloyd 1988).

References

American Heart Association (1977). Prevention of bacterial endocarditis (a committee report). *Journal of the American Dental Association*, **95**, 600.

Ayers, D., Evarts, C. M., and Parkinson, J. R. (1986). The prevention of heterotopic ossification in high-risk patients by low-dose radiation therapy after total hip arthroplasty. *Journal of Bone and Joint Surgery*, **68A**, 1423–30.

Bauer, G. C. H., Hansson, L. I., Lidgren, L., Stromqvist, B., and Thorngren, J. G. (1985). Comprehensive care of hip fractures. Scientific exhibit Las Vegas, American Academy of Orthopaedic Surgeons.

Bobyn, J. D., Pilliar, R. M., Cameron, H. U., and Weatherly, G. C. (1980). The optimum pore size for the fixation of porous-surfaced metal implants by the ingrowth of bone. *Clinical Orthopaedics and Related Research*, **150**, 236–70.

Botte, M. J., Waters, R. L., Keenan, M. E., Jordan, C., and Garland, D. E. (1988). Orthopaedic management of the stroke patient. Part II: treating deformities of the upper and lower extremities. *Orthopaedic Review*, **17**, 891–910.

Burton, K. E., Wright, V., and Richards, J. (1979). Patients' expectations in relation to outcome of total hip replacement surgery. *Annals of Rheumatologic Disease*, **38**, 471–4.

Ceder, L., Svensson, K., and Thorngren, K. G. (1980). Statistical prediction of rehabilitation in elderly patients with hip fractures. *Clinical Orthopaedics and Related Research*, **152**, 185–90.

Clifford, B., Waters, R. L., and Jordan, C. (1985). Stroke: the challenge. In *Orthopaedic Care of the Geriatric Patient* (ed. T. P. Sculco), pp. 323–45. C. V. Mosby, St Louis.

Consensus Conference (1986). Prevention of venous thrombosis and pulmonary embolism. *Journal of the American Medical Association*, **256**, 744–9.

DelGuercio, L. R. M. and Cohn, J. D. (1980). Monitoring operative risk in the elderly. *Journal of the American Medical Association*, **243**, 1350–5.

Evarts, C. M. (1987). Prevention of venous thromboembolism. *Clinical Orthopaedics and Related Research*, **222**, 98–104.

Fowkes, F., Lunn, J., Farow, J. C., Robertson, I. B., and Samuel, P. (1982). Epidemiology in anaesthesia III: mortality risk in patients with coexisting physical disease. *British Journal of Anaesthesiology*, **54**, 819–24.

Goldman, L. *et al.* (1977). Multifactorial index of cardiac risk in noncardiac surgical procedures. *New England Journal of Medicine*, **297**, 845–50.

Goldmann, D. R. (1987). Surgery in patients with endocrine dysfunction. *Medical Clinics of North America*, **71**, 499–509.

Goldring, S. R., Schiller, A. L., Roelke, M., Rourke, C. M., O'Neill, D. A., and Harris, W. H. (1983). The synovial-like membrane at the bone–cement interface in loose total hip replacements and its proposed role in bone lysis. *Journal of Bone and Joint Surgery*, **65A**, 575–84.

Harris, C. M. (1984). Joint replacement in the elderly. In *Rehabilitation in the aging* (ed. T. F. Williams), pp. 199–227. Raven Press, New York.

Harris, W. H. (1969). Traumatic arthritis of the hip after dislocation and acetabular fractures: treatment by mold arthroplasty: an end-result study using a new method of result evaluation. *Journal of Bone and Joint Surgery*, **51A**, 737–55.

Hedlund, R., Ahlbom, A., and Lindgren, V. (1986). Hip fracture incidence in Stockholm 1972–1981. *Acta Orthopaedica Scandinavica*, **57**, 30–4.

Holbrook, R. L., Grazier, K., Kelsey, J. L., and Stauffer, R. N. (1984). *The frequency of occurrence, impact and cost of selected musculoskeletal conditions in the United States*. American Academy of Orthopaedic Surgeons, Chicago.

Hornby, R., Evans, J. G., and Vardon, V. (1989). Operative or conservative treatment of trochanteric fractures of the femur: a randomised epidemiological trial. *Journal of Bone and Joint Surgery*, **71B**: 619–23.

Insall, J. N., Ranawat, C. S., Aglietti, P., and Shine, J. (1976). A comparison of four models of total knee-replacement prosthesis. *Journal of Bone and Joint Surgery*, **58A**, 745–73.

Johnson, K. D., Cadambi, A., and Seibert, G. B. (1985). Incidence of adult respiratory distress syndrome in patients with multiple musculoskeletal injuries: effect of early operative stabilization of fractures. *Journal of Trauma*, **25**, 375–84.

Lewinnek, G. E., Kelsey, J., White, A. A., and Kreiger, N. J. (1980). The significance and a comparative analysis of the epidemiology of hip fractures. *Clinical Orthopaedics and Related Research*, **180**, 35–43.

Liang, M. H. and Cullen, K. E. (1984). Evaluation of outcomes in total joint arthroplasty for rheumatoid arthritis. *Clinical Orthopaedics and Related Research*, **182**, 41–5.

Maggi, S., Kelsey, J.L., Litvak, J., and Heyse, S.P. (1991). Epidemio-

logy of hip fractures in the elderly: a cross-national analysis. *Osteoporosis International*, in press.
Marx, A. B., Campbell, R., and Harder, F. (1986). Polytrauma in the elderly. *World Journal of Surgery*, **10**, 330–5.
Miller, C. W. (1978). Survival and ambulation following hip fracture. *Journal of Bone and Joint Surgery*, **60A**, 930–4.
National Institute of Health Consensus Conference (1982). Total hip joint replacement in the United States. *Journal of the American Medical Association*, **248**, 1817–21.
Norden, C. W. (1985). Prevention of bone and joint infections. *American Journal of Medicine*, **78**, 229–32.
Pellicci, P. M., Salvati, E. A., and Robinson, H. J. (1979). Mechanical failures in total hip replacement requiring reoperation. *Journal of Bone and Joint Surgery*, **61A**, 28–36.
Pepe, P. E., Hudson, L. D., and Carrico, C. J. (1984). Early application of positive end-expiratory pressure in patients at risk for the adult respiratory-distress syndrome. *New England Journal of Medicine*, **311**, 281–6.
Reynolds, C. (1987). Management of the diabetic surgical patient. *Postgraduate Medicine*, **77**, 265–79.
Shay, K. and Lloyd, P. M. (1988). Dental schools' practices of prophylactic antibiotic coverage for patients with prosthetic joints. *Journal of Dental Education*, **52**, 567.
Steen, P. A., Tinker, J. H., and Tarhan, S. (1978). Myocardial reinfarction after anesthesia and surgery. *Journal of the American Medical Association*, **239**, 2566–70.
Strauss, H. W. and Boucher, C. A. (1986). Myocardial perfusion studies: lessons from a decade of clinical use. *Radiology*, **160**, 557–85.
Sugerman, H. J. (1987). Pulmonary function in morbid obesity. *Gastroenterology Clinics of North America*, **16**, 225–37.
Tarhan, S., Moifitt, A. E., Taylor, W. F., and Giuliani, E. R. (1972). Myocardial infarction after general anesthesia. *Journal of the American Medical Association*, **220**, 1451–4.
Thompson, J. D., Callaghan, J. J., Savory, C. G., Stanton, R. B., and Pierce, R. N. (1987). Prior deposition of autologous blood in elective orthopaedic surgery. *Journal of Bone and Joint Surgery*, **69A**, 320–4.
Transportation Research Board (1988). *Transportation in an aging society. Improving mobility and safety for older persons*. Special Report No. 218, *Vol.* 1: Committee Report and Recommendations. National Research Council, Washington, DC.
Vaughan, R. W. (1982). Definitions and risks of obesity. In *Anesthesia and the obese patient* (ed. B. R. Brown), pp. 1–7. F. A. Davis, Philadelphia.
Waters, R. L. (1978). Editorial comment: stroke rehabilitation. *Clinical Orthopaedics and Related Research*, **131**, 2.
Winter, W. G. (1987). Nonoperative treatment of proximal femoral fractures in the demented, nonambulatory patient. *Clinical Orthopaedics and Related Research*, **218**, 97–103.

13.5 Gout and other crystal arthropathies

J. T. SCOTT

Mineral deposition often occurs in joints, especially with advancing years. Metabolic disturbances and tissue damage are other predisposing factors. Deposits can occur in articular cartilage, synovial membrane, and periarticular structures.

Gout is the best known of these arthropathies, the crystal concerned being monosodium urate monohydrate. Calcium pyrophosphate dihydrate may be laid down in articular cartilage (chondrocalcinosis), while hydroxyapatite and other basic calcium phosphates are also deposited in relation to joints, characteristically in periarticular tissues.

GOUT

Gout is a disease with a strong familial tendency; it is seen predominantly in adult men, characterized by episodes of acute arthritis, and later also by chronic damage to joints and other structures. It is caused essentially by hyperuricaemia, an excess of urate in blood and tissues, which in some people (but not all) leads to the deposition of crystals of sodium urate in the joints and elsewhere. These crystalline deposits are now recognized as the cause of the acute gouty attack, and further accumulations of crystalline and amorphous urate form the tophi (Latin *tofus*, porous stone) which are a feature of the advanced, untreated disease.

With increasing knowledge of uric acid metabolism it has now become evident that there are many factors which can influence the development of hyperuricaemia and hence of gout. Gouty arthritis may therefore be regarded as the end result of a number of different biochemical processes. The term primary gout is used when hyperuricaemia is due principally to an inherited metabolic abnormality, and secondary gout when it is largely the result of an acquired disease or some environmental factor. In most patients with gout, however, a combination of inherited and environmental influences—particularly food (with regard to both its purine and calorie content), alcohol, and diuretic drugs—appears to be operating.

The pathology and clinical features of gout are described in standard texts, together with accounts of purine metabolism, causes of hyperuricaemia, factors influencing urate crystal deposition and the inflammatory response, and methods of treatment (Scott 1986; Nuki 1987).

Age and sex

Blood levels of uric acid are generally higher in men than in women because of differences in renal clearance (Fig. 1) and this is reflected in the predominance of gouty arthritis among men. After middle age, however, levels in women rise to approach those of men, and gout in elderly women becomes relatively less uncommon (Fig. 2).

A survey of 354 patients with gout (Grahame and Scott 1970) has shown that the peak age of onset lies in the fifth decade in men and in the sixth in females, 12 per cent of the total series experiencing their first attack of clinical gout after the age of 60 years. The female incidence after the age of 60 was 29 per cent compared to only 7 per cent below that age. Other features of gout starting in old age were a somewhat increased incidence of an underlying haematolo-

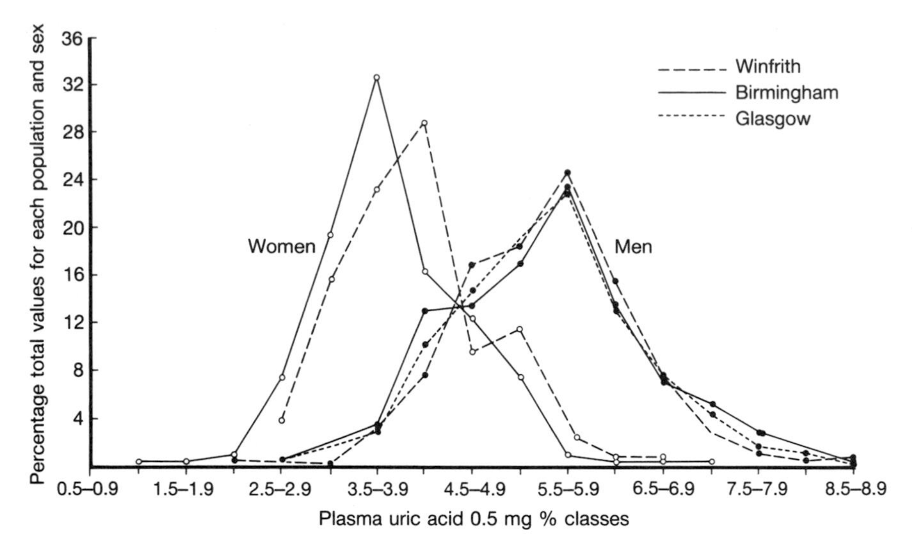

Fig. 1 Distribution curves of serum uric acid concentrations in three normal populations in the United Kingdom (from Sturge *et al.* 1977; reproduced by courtesy of the Editor of *Annals of the Rheumatic Diseases*).

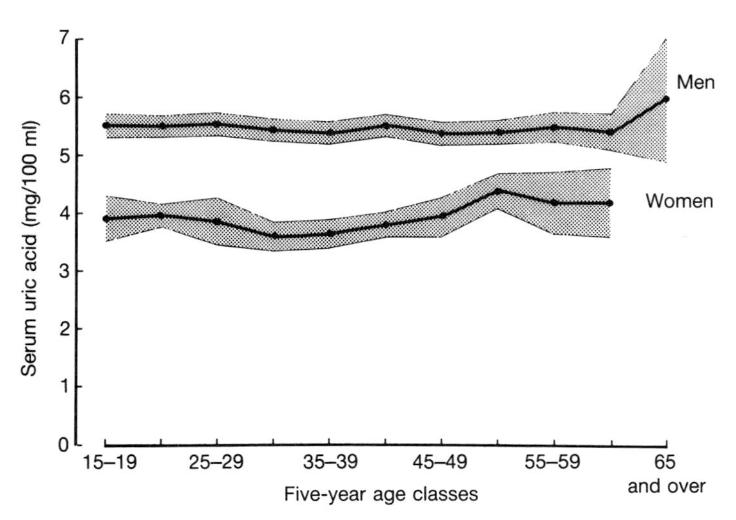

Fig. 2 Mean serum uric levels by 5-year age class from the combined populations shown in Fig. 1, totalling 1103 subjects. Stippled area is two SE on either side of mean. There is a significant rise in uric acid in women occurring at the fifth decade. The rise seen in old men does not reach significance because of the small number of subjects examined at this age. (Data from Sturge *et al.* 1977; reproduced by courtesy of the Editor of *Annals of the Rheumatic Diseases*.)

gical disorder, an absence of the predilection for higher social strata seen in the younger onset group, and a lower frequency of a positive family history.

Diuretics and gout

Most environmental factors contributing to hyperuricaemia and gout—namely purine intake, calorie consumption with consequent obesity, and regular alcohol ingestion—appear to be of no greater importance in the elderly than in younger patients with gout, perhaps less so. Special mention must, however, be made of diuretic-induced gout, which has been seen with increasing frequency over the past decade.

The hyperuricaemic action of thiazides and other diuretics, such as frusemide, ethacrynic acid, and chlorthalidone, is frequently encountered in clinical practice. Their mode of action is complex: whether due mainly to inhibition of tubular secretion of urate or to more avid resorption, it appears to be dependent upon contraction of extracellular volume (Manuel and Steele 1974).

A study of gout in general practice showed that in 10 per cent of 966 cases the condition was believed to be secondary, with induction by diuretics being the most frequent cause (Currie 1979). It has become apparent that this type of gout occurs especially among elderly women who have been taking diuretics for several years. In a study by MacFarlane and Dieppe (1985), 9 of 60 cases of gout were in women, all of whom had been taking long-term diuretics, whereas only one-third of the men with gout had done so. The men had a mean age of onset of gout of 46 years compared with 77 years for the women. There are similar data from other countries: for example, the striking increase in the prevalence of gout in Finland is partly attributable to diuretics (Isomaki *et al.* 1977).

A history of diuretic medication is therefore an important factor in assessment of the patient with gout. It must be remembered, however, that hyperuricaemia is a multifactorial condition and further investigation of an individual patient may reveal other contributory factors, such as inheritance (indicated by a positive family history), renal functional impairment, or occasionally a proliferative haematological disorder, as well as the environmental factors mentioned above.

Clinical features and diagnosis

Attacks of acute gout in the elderly may resemble those of younger people, usually occurring in a single joint, the metatarsophalangeal joint of the great toe being involved in 70 per cent of cases. The classical picture is one of a reddened, shiny, swollen, and exquisitely tender joint with distension of superficial veins, an appearance which is typical but which can be mimicked by infection or pyrophosphate arthropathy (see below).

Less typical modes of onset can be more deceptive. The

inflammatory episode is sometimes much less acute and a swollen joint such as a knee can easily be misdiagnosed as rheumatoid arthritis or osteoarthritis, especially if the condition is polyarticular.

A striking presentation, extremely rare in younger adults but not at all uncommon in the elderly, is where tophaceous deposits are formed in the absence of any history of acute attacks. This is found particularly in women with diuretic-induced gout where urate is deposited in pre-existing Heberden's nodes. The hard, swollen, distal interphalangeal joints of the fingers result from a combination of osteoarthritis and chronic tophaceous gout (Fig. 3). Of course, such lesions can also be the site of very painful, acute attacks.

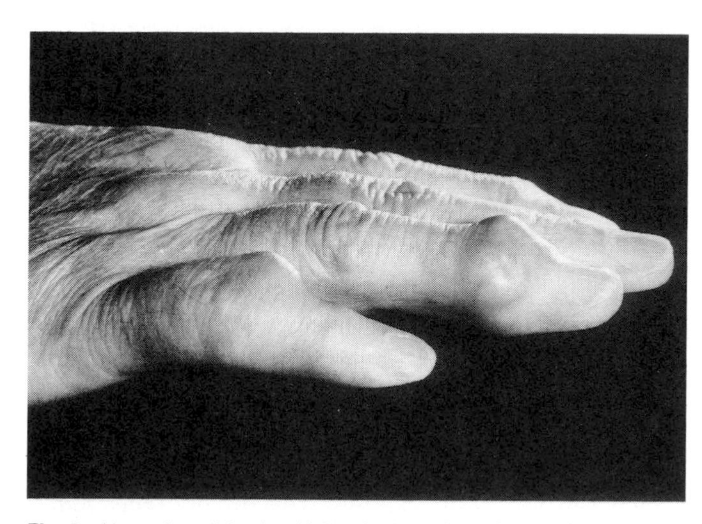

Fig. 3 Urate deposition in a Heberden's node at the distal interphalangeal joint in the right ring finger of an elderly woman. There is a similar lesion in the proximal interphalangeal joint of the little finger.

The diagnosis of gout is primarily clinical but must be checked by estimation of the serum uric acid, which is always raised (above 420 μmol/l, 7 mg%) unless it has been recently lowered by drug treatment. Absolute confirmation of gouty arthritis is provided by the demonstration of urate crystals in synovial fluid or from a tophus: crystals of sodium urate are needle-shaped, between 2 and 20 μm in length (Fig. 4), and show strong, negative birefringence by polarizing microscopy. Fluid is easily examined when an effusion is present in a large joint such as the knee, but a few drops obtained (with adequate local anaesthesia) from a small joint in a toe or finger can be enough to confirm the diagnosis.

Treatment

Acute gout

Treatment of acute gout is directed solely towards relief of inflammation as rapidly as possible, lowering the levels of uric acid (a process which may well prolong the episode) playing no part at this stage.

Colchicine is the time-honoured remedy, the usual dose being 1.0 mg followed by 0.5 mg every 2 h until the attack subsides. The most frequent toxic effect is diarrhoea, but this does not usually occur until a total dose of 5 to 6 mg has been taken. There is a considerable variation in individual sensitivity, however, and sometimes diarrhoea, nausea, or

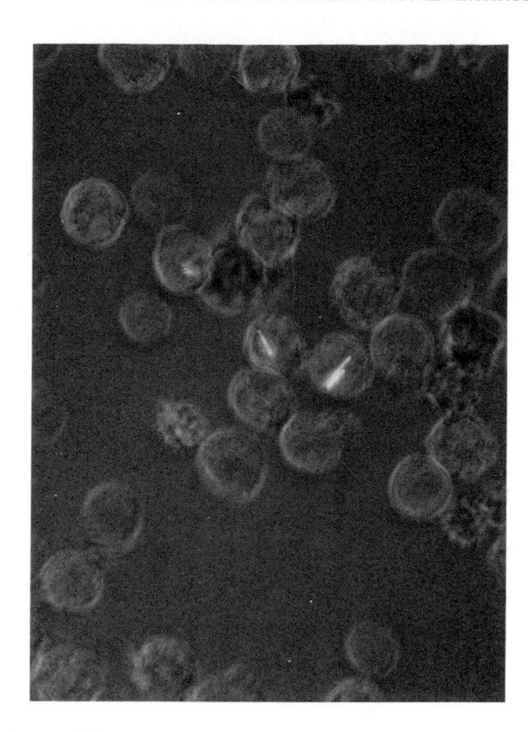

Fig. 4 Intracellular urate crystals seen in joint fluid by polarized light microscopy.

vomiting preclude the use of colchicine, such symptoms being especially troublesome in old people.

Phenylbutazone is highly effective in a dose of 200 mg, four times daily, until the attack is relieved. Because of the risk of marrow suppression (minuscule when the drug is given for only a few days), the drug has been withdrawn from prescription in the United Kingdom except for patients with ankylosing spondylitis, although some patients with gout express a strong individual preference for it.

Indomethacin in a dose of 50 mg, four time daily, is also useful, as is naproxen, 250 mg, three times daily; and indeed therapeutic effect has been shown with most of the other non-steroidal anti-inflammatory drugs such as fenoprofen, ketoprofen, fenbufen, and piroxicam. Azapropazone in a dose of 300 mg, three times daily, has both anti-inflammatory and uricosuric properties, although whether such a combination of effects is a significant therapeutic advance remains to be confirmed.

Non-steroidal anti-inflammatory drugs must be used with great care in elderly patients. Side-effects such as gastrointestinal intolerance and skin rashes are not infrequent, while fluid retention and a fall in glomerular filtration rate are also documented. Renal damage in patients with normal renal function is rare, but existing renal impairment can be aggravated, due to the effect of prostaglandin inhibition upon renal blood flow (Carmichael and Shankel 1985).

Long-term management

After just one or two attacks, with only a moderately elevated uric acid, there is little to be lost in awaiting the course of events. Regular prophylactic colchicine, given in a dose of 0.5 to 1.0 mg daily, can prevent recurrences of acute articular gout, but antihyperuricaemic agents are surely more satisfactory. They are certainly indicated in the presence of chronic changes in the joint or tophi; frequent acute attacks; evidence of renal damage; or gout accompanied by a considerably elevated serum uric acid—480 μmol/l or over.

The drug most commonly used in Britain to lower serum concentrations of uric acid is the xanthine oxidase inhibitor allopurinol. The final maintenance dose depends upon estimations of serum uric acid, but usually lies in the region of 300 mg daily. Old people tolerate the drug well enough, but severe reactions—exfoliative dermatitis, vasculitis, or bone marrow depression—have usually occurred in patients with reduced renal function, which must therefore be carefully assessed. Lower doses of allopurinol may be adequate (Cameron and Simmonds 1987).

Uricosuric drugs, such as probenecid, which lower the serum level of uric acid by increasing renal excretion, are generally less effective than allopurinol, and may produce renal deposition of urate. They should therefore be used only when there are adverse reactions to allopurinol.

PYROPHOSPHATE ARTHROPATHY (CHONDROCALCINOSIS)

Calcium pyrophosphate dihydrate is produced in large amounts but is normally rapidly hydrolysed to orthophosphate by abundant pyrophosphatase enzymes. Crystallization can occur in cartilage that has been damaged in one way or another, but other mechanisms are incompletely understood. Deposition takes place initially in the mid zone of hyaline cartilage around chrondocytes, and also in fibrocartilage. Shedding into joint fluid is associated with crystal-induced inflammation.

Apart from rare familial types and those with defined metabolic associations, the common sporadic cases are found exclusively in elderly people of both sexes (Hamilton 1986). Chondrocalcinosis is rare under the age of 50 years but has been described in nearly half of people over the age of 90.

Pyrophosphate is deposited in both fibrous and hyaline cartilage. The most common site is the meniscus of the knee (Fig. 5); also involved are the triangular ligament of the wrist joint, and the symphysis pubis. These radiological features are critical in diagnosis, together with the demonstration of pyrophosphate crystals when joint fluid is obtained for microscopic examination: the crystals are more pleomorphic than the fine needles of sodium urate and are positively birefringent under the polarizing microscope.

There are several patterns of joint involvement which may occur independently or, at different times, in association with one other:

1. *Asymptomatic chondrocalcinosis*. This has been demonstrated in various radiological and autopsy surveys of elderly people.
2. *Acute pyrophosphate arthropathy ('pseudogout')*. The affected joint becomes suddenly painful, warm, swollen, tender, and often red. The knee is the most common site, but other joints, especially the wrist, elbow, and ankle, are not infrequently involved, occasionally more than one simultaneously. Unlike urate gout, the great toe is only rarely affected. Attacks last for from a few days to several weeks and as with urate gout they may follow trauma, surgery, or acute illness. Fever, leucocytosis, and a raised sedimentation rate can occur, but there are usually no biochemical abnormalities. The acute attack can be treated by aspiration of joint fluid and injection of corticosteroid, together with administration of a non-steroidal anti-inflammatory drug. Colchicine is effective but less so than with urate gout.

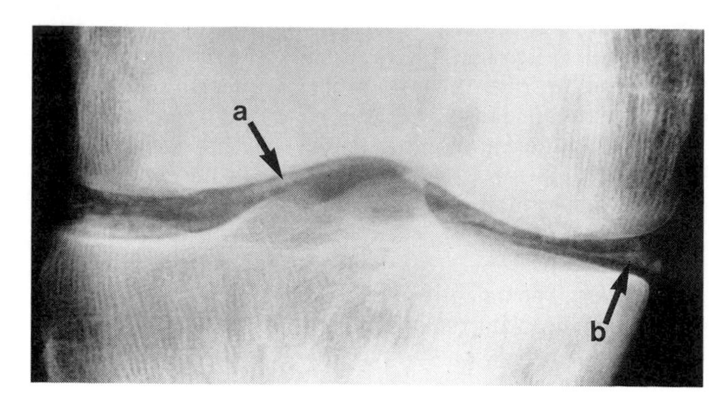

Fig. 5 Radiograph of knee joint showing chondrocalcinosis: deposition of calcium pyrophosphate is seen in (a) hyaline articular cartilage and (b) fibrocartilage of the meniscus.

3. *Osteoarthritis*. This is a common association, about half of patients with pyrophosphate arthropathy having some degree of osteoarthritis. This is seen most frequently in the knee joint but may involve sites that are not normally liable to osteoarthritis, such as the wrists, metacarpophalangeal joints, elbows, and shoulders. There is no doubt that pyrophosphate can lead to degenerative changes, but there can also be the opposite sequence of events, joints with advanced osteoarthritis becoming the site of crystal deposition.
4. *Chronic destructive arthropathy*. Pyrophosphate crystals can also produce a much more severe destructive arthritis, most commonly in the knee, hip, and shoulder joints.

Apart from situations where pyrophosphate is laid down in a joint which is already abnormal—e.g., in hypermobile joints, after meniscectomy, and in neuropathic joints—there are several conditions where an association with chondrocalcinosis has been demonstrated:

1. *Hyperparathyroidism*. The relation here is presumably with sustained hypercalcaemia. Pyrophosphate arthropathy is sometimes the presenting feature of hyperparathyroidism and a serum calcium is a necessary investigation in all patients with chondrocalcinosis, though it is nearly always found to be normal.
2. *Gout*. There is an association between these two forms of crystal arthritis, patients with gout showing a higher prevalence of chondrocalcinosis than controls. Urate crystals may enhance nucleation of calcium pyrophosphate.
3. *Haemochromatosis*. Acute and chronic forms of pyrophosphate arthropathy are found in about half of cases, superimposed on the characteristic degenerative arthritis, which affects first the small joints of the hands and later larger joints. It is likely that iron salts act as nucleating agents for crystal formation.
4. *Hypothyroidism*. There are several reports of this association.
5. *Hypophosphatasia*. These rare patients have very low

levels of alkaline phosphatase and an association with chondrocalcinosis is well documented. Inorganic pyrophosphate is a natural substrate for alkaline phosphatase.
6. *Hypomagnesaemia*. Pyrophosphate deposition has been described in some patients with low magnesium levels.

Associations have also been claimed with amyloidosis, diabetes, ochronosis, Wilson's disease, and Paget's disease of bone, the lack of controlled observations making evaluation of these difficult. It should again be stressed that although the possibility of such associations should be borne in mind when chondrocalcinosis is diagnosed, they are not found in the great majority of cases.

HYDROXYAPATITE AND JOINT DISEASE

Calcium apatite is commonly deposited in soft tissues related to joints, sometimes in well-defined sites such as the supraspinatus tendon or subdeltoid bursa. It is demonstrable by radiographs and is often asymptomatic, but may produce local pain, sometimes severe (acute calcific periarthritis). Local injection with lignocaine and corticosteroid is usually effective.

Apatite crystals have been found in osteoarthritic joints. They can be demonstrated by staining with alizarin red (Paul *et al.* 1983) but this merely confirms the calcific nature of the material, more specific identification depending upon sophisticated techniques such as scanning and transmission electron microscopy. Moreover, the extent to which their presence is causative or merely the result of damage to the joint is debatable. However, there does appear to be an apatite-associated arthropathy in which phagocytosis of crystals leads to release of proteases and collagenases. A striking example is the 'Milwaukee shoulder' (McCarty 1983), in which destruction of the glenohumeral joint is accompanied by a characteristic, upward subluxation of the humeral head.

References

Cameron, J.S. and Simmonds, H.A. (1987). Use and abuse of allopurinol. *British Medical Journal*, **294**, 1504–5.

Carmichael, J. and Shankel, S.W. (1985). Effects of non-steroidal anti-inflammatory drugs on prostaglandins and renal function. *American Journal of Medicine*, **78**, 992–1000.

Currie, W.J.C. (1979). Prevalence and incidence of the diagnosis of gout in Great Britain. *Annals of the Rheumatic Diseases*, **38**, 101–6.

Grahame, R. and Scott, J.T. (1970). Clinical survey of 354 patients with gout. *Annals of the Rheumatic Diseases*, **29**, 461–8.

Hamilton, E.B.D. (1986). Pyrophosphate arthropathy. In *Copeman's textbook of the rheumatic diseases* (ed. J.T. Scott), pp. 938–49. Churchill Livingstone, Edinburgh.

Isomaki, H., von Essen, R., and Ruutsalo, H.M. (1979). Gout, particularly diuretics-induced, is on the increase in Finland. *Scandinavian Journal of Rheumatology*, **6**, 213–16.

MacFarlane, D. G. and Dieppe, P. A. (1985). Diuretic-induced gout in elderly women. *British Journal of Rheumatology*, **24**, 155–7.

Manuel, M.A. and Steele, T.H. (1974). Changes in renal urate handling after prolonged thiazide treatment. *American Journal of Medicine*, **57**, 741–6.

McCarty, D.J. (1983). Crystals, joints and consternation. *Annals of the Rheumatic Diseases*, **42**, 243–53.

Nuki, G. (1987). Disorders of purine metabolism. In *Oxford textbook of medicine* (2nd edn), (ed D. J. Weatherall, J.G.G. Ledingham, and D.A. Warrell). Section 9, p. 123. Oxford University Press.

Paul, H., Reginato, A.J. and Schumacher, H.R. (1983). Alizarin red S-staining as a screening test to detect calcium compounds in synovial fluid. *Arthritis and Rheumatism*, **26**, 191–200.

Scott, J.T. (1986). Gout. In *Copeman's textbook of the rheumatic diseases* (6th edn), (ed. J.T. Scott), pp. 883–937. Churchill Livingstone, Edinburgh.

Sturge, R. A. *et al.* (1977). Serum uric acid in England and Scotland. *Annals of the Rheumatic Diseases*, **36**, 420–7.

13.6 Polymyalgia rheumatica and giant cell arteritis

JOHN H. KLIPPEL

Polymyalgia rheumatica and giant cell arteritis, diseases unique to later life, are among the more common of the inflammatory rheumatic syndromes. Although they are considered to be different clinical entities that probably share common pathogenetic features, the exact relationship between them has never clearly been defined.

EPIDEMIOLOGY

Prospective studies have revealed peak incidence rates of both diseases in the age group 70 to 79 years (Fig. 1; Boesen and Sorensen 1987). Polymyalgia rheumatica is substantially more common than giant cell arteritis. In all age categories, the diseases are two- to fourfold more frequent in women than men. The diseases appear to be more common in Caucasians, particularly those of Scandinavian background, and, perhaps, also in black subjects (Gonzalez *et al.* 1989).

POLYMYALGIA RHEUMATICA

Polymyalgia rheumatica is a systemic illness with both constitutional and musculoskeletal symptoms. There are no pathognomonic laboratory abnormalities and the diagnosis is entirely based on the clinical history, physical examination, and the findings of non-specific evidence of inflammation on laboratory testing.

Musculoskeletal manifestations

Symmetrical pain and stiffness of the neck, shoulders, low back, and pelvic girdle are the clinical hallmarks of polymyal-

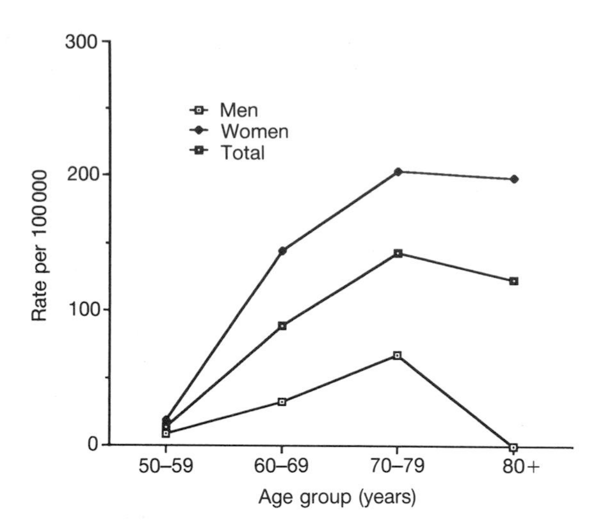

Fig. 1 The annual incidence of giant cell arteritis and polymyalgia rheumatica in Ribe County, Denmark, 1982–1983 (adapted from Boesen and Sorensen 1987, with permission).

gia rheumatica. In many patients, the onset is dramatic and acute such that the exact date, or even hour, of the first symptoms can be recalled. In others, the process is more insidious and evolves over weeks to months. The stiffness and pain are typically worse after periods of inactivity and particularly prominent in the morning. Movement during sleep often produces pain that awakens the patient and thus sleep disturbances are common. In severe cases, patients may be unable to get out of bed or dress themselves in the morning.

On physical examination, tenderness and limited range of motion of the neck, shoulders, and hips can be demonstrated. Muscle atrophy, joint contractures, and immobile, 'frozen' joints may develop. Because of pain, muscle strength is difficult to evaluate, although is generally normal. No evidence of muscle disease has been identified. Serum muscle enzymes, such as the creatine phosphokinase and aldolase, are not elevated. Similarly, no abnormalities are detected by electromyography or muscle biopsy.

Synovitis of both large and small, peripheral joints can be found in about one-third of patients, involvement of knees and wrists being the most common (Al-Hussaini and Swannell 1985). Compression of the median nerve and carpal tunnel syndrome from wrist synovitis have been reported. The extent of synovial involvement can be imaged with technetium-99 scans, in which increased uptake in axial and peripheral joints may be found. The synovitis is considered to be responsible for most of the musculoskeletal complaints.

Analysis of synovial fluid reveals mild inflammatory changes with leucocyte counts in the range of 300 to 5000/mm^3, mostly mononuclear cells. Standard radiographs of joints demonstrate only soft tissue swelling without loss of cartilage or bone erosions. On arthroscopy, vascular dilatation, engorgement, and perivascular oedema can be demonstrated. Histological examination of the synovium shows non-specific, mild inflammation with proliferation of synovial cells and scattered infiltration of lymphocytes and plasma cells (Chou and Schumacher 1984). The cellular infiltrate is neither organized into lymphoid follicles nor concentrated around synovial vessels. The synovial vasculature may be engorged, with slight perivascular fibrosis; there is no evidence of pannus, vasculitis, giant cells, or granuloma formation.

Constitutional symptoms

Indications of a systemic illness may be a prominent feature of polymyalgia rheumatica. These include high fevers, occasionally with shaking chills and night sweats (15 per cent), weight loss (20 per cent), anorexia (20 per cent), and malaise and asthenia (25 per cent) (Salvarani *et al.* 1987).

Laboratory abnormalities

Elevation of the erythrocyte sedimentation rate (**ESR**) is found in the great majority of patients (but not all) with polymyalgia rheumatica. Marked elevations of the Westergren ESR, often greater than 100 mm/h, are typical. Elevations of other acute phase serum proteins may be present (Andersson *et al.* 1988).

Other laboratory abnormalities may include a normochromic, normocytic anaemia, mild elevations of serum liver enzymes, particularly alkaline phosphatase, and thrombocytosis. Abnormalities of serum proteins include decreased albumin, and increases of fibrinogen, α_2-globulin and plasma von Willebrand factor (factor VIII antigen). Autoantibodies such as antinuclear antibody and rheumatoid factor are absent or found in low titres consistent with age norms.

Response to corticosteroids

Prompt and dramatic improvement in symptoms with low doses of prednisone (10–15 mg daily) is often used as a diagnostic test for polymyalgia rheumatica. The clinical response to corticosteroids must, however, be interpreted with great caution because many of the inflammatory, or even non-inflammatory, conditions that might be confused with polymyalgia rheumatica improve with corticosteroids.

Differential diagnosis

The differential diagnosis of polymyalgia rheumatica is extensive and includes giant cell arteritis, acute and chronic infectious processes, metabolic abnormalities, neoplastic diseases, and various other rheumatic conditions. For prognostic and therapeutic reasons, it is important to identify patients with coexistent polymyalgia and giant cell arteritis. Although opinions vary, there is probably little justification for routine biopsy of the temporal artery in patients with uncomplicated polymyalgia rheumatica. However, in those with suspected vasculitis based on symptoms (headache, visual disturbances, jaw 'claudication', etc.) or abnormal physical findings (vascular bruits or absent pulses), a temporal artery biopsy should be done. A negative finding, in general, is associated with a low risk for the subsequent development of giant cell arteritis (Hall *et al.* 1983).

The characteristic involvement of major organs in the connective tissue diseases such as systemic lupus erythematosus, polymyositis, and scleroderma, along with their high incidence of antinuclear antibodies, allow for easy differentiation between them and polymyalgia rheumatica. The two rheumatic conditions that provide the most difficulty in differential diagnosis are the chronic fatigue syndrome, or fibromyalgia, and, in particular, seronegative rheumatoid arthritis. The diffuse rather than girdle distribution of pain and tenderness, trigger points, absence of synovitis, normal ESR, and failure to respond to corticosteroids are helpful in identifying the patient with fibromyalgia (Buchwald *et al.* 1988). In patients with synovitis, distinguishing polymyalgia rheumatica from seronegative rheumatoid arthritis is much more problematical (Healey and Sheets 1988). Typically, these types of patients satisfy criteria for both diseases. Not surprisingly, a change in an original diagnosis of the polymyalgia to rheumatoid arthritis in patients with a disease course characterized by persistent synovitis is a fairly common occurrence (Palmer *et al.* 1986). In patients with early disease, findings more suggestive of rheumatoid arthritis include clinical involvement of the metatarsophalangeal joints, erosions of bone and cartilage, and an incomplete response to low doses of corticosteroids.

Treatment

Untreated, polymyalgia rheumatica is a chronic, self-limiting disease that carries a very small risk of progressing to giant cell arteritis. Low doses of corticosteroids, typically prednisone, 10 to 15 mg daily, are the standard treatment. Clinical improvement with low-dose prednisone is prompt and dramatic; most symptoms entirely resolve within several days. Administration of corticosteroids on alternate days is generally much less effective. The clinical response to non-steroidal anti-inflammatory drugs is less complete; however, these drugs may be useful in patients with very mild symptoms. The correction of laboratory abnormalities, particularly the elevated ESR, occurs much more slowly over a period of several weeks.

Gradual, slow reductions of the corticosteroid dose may be begun after 4 to 6 weeks of therapy. Recurrence of clinical symptoms during this reduction is an indication for increasing the dose. In a large series of patients with polymyalgia rheumatica, the median duration of prednisone therapy was approximately 3 years; one-third of the patients were able to discontinue prednisone by 16 months, but series vary in their outcome (Kyle and Hazelman 1990). The relapse rate with discontinuation of prednisone is reported to be around 20 per cent (Ayoub *et al.* 1985). Side-effects from corticosteroids are related to both peak and cumulative doses (Kyle and Hazelman 1989).

GIANT CELL ARTERITIS

Giant cell arteritis is a form of systemic vasculitis which typically involves medium to small arteries that originate from the thoracic aorta. Other terms used for this disease include cranial or temporal arteritis, or granulomatous arteritis.

The clinical symptoms are dependent on the anatomical distribution and extent of arterial involvement. Common presenting complaints include headache (75 per cent), jaw or tongue 'claudication' (51 per cent), scalp tenderness (47 per cent), and an assortment of visual symptoms, such as diplopia or blurred vision (12 per cent), amaurosis fugax (10 per cent), and, rarely, blindness (1 per cent) (Machado *et al.* 1988). The headache is often described as having a boring or lancinating quality; it may be localized to the temporal or occipital regions, or generalized. Approximately one-third of patients with giant cell arteritis have accompanying polymyalgia rheumatica. Deaths from the arteritis have been reported, secondary to stroke syndromes, myocardial infarctions, and dissections of the aorta (Save-Soderbergh *et al.* 1986). The prognosis may be worse in men than women (Delecoeuillerie *et al.* 1988).

The classical finding on physical examination is a prominent, beaded, tender, pulseless temporal artery. Tenderness or nodules may be detected along other arteries or in the scalp. Bruits may be found over large arteries such as the carotid, subclavian, axillary, or brachial; pulses may be absent in these. Aneurysms, dissections, and stenotic lesions of the aorta (Pereira *et al.* 1986) and involvement of the coronary arteries have been described. Although cases of involvement of arteries of the lung, abdominal aorta, and lower extremities have been reported, these are distinctly uncommon.

The neuro-ophthalmological manifestations typically result from involvement of the posterior ciliary arteries, branches of the ophthalmic artery (Mehler and Rabinowich 1988). These arteries provide the blood supply for the optic nerve and extraocular muscles. The ischaemia produced by involvement of these vessels leads to visual disturbances from either optic neuritis or from diplopia as a consequence of paresis of the extraocular muscles. Involvement of the central retinal artery by giant cell arteritis is distinctly unusual. Changes on fundoscopic examination are subtle, typically only pallor and oedema of the optic disc. Evidence of frank vasculitis with haemorrhages is very unusual. Optic atrophy is a relatively late finding.

Impairments of the central and peripheral nervous systems may develop secondary to giant cell arteritis. In the central nervous system these include transient ischaemic attacks, cerebrovascular accidents, seizures, vertigo (Caselli *et al.* 1988b), and various neuropsychiatric syndromes, such as changes in mental status, depression, acute disorientation, amnesia, and dementia (Pascuzzi *et al.* 1989). Involvement of the peripheral nervous system with both generalized and mono- or polyneuropathies is less common (Caselli *et al.* 1988a).

Diagnosis

The establishment of the diagnosis with certainty requires arterial biopsy, typically of the temporal artery. However, in patients with impending, severe vascular complications such as loss of sight or neurological impairment, treatment with corticosteroids should be started promptly and the biopsy postponed. Although the vascular lesion may be altered by corticosteroids, even after several days of corticosteroid treatment the characteristic appearances of the arteritis are still easily seen. Owing to the segmental, vascular distribution of the disease, a biopsy specimen of several centimetres in length is recommended to increase the chance of detecting lesional tissue (Vilasega *et al.* 1987). Angiography is generally not helpful, save for perhaps differentiating the arteritis from vascular disease secondary to atherosclerosis. The role of non-invasive vascular studies is uncertain. Doppler ultrasonography (Perruquet *et al.* 1986) and ocular pneumoplethysmography (Bosley *et al.* 1989) have been reported to be helpful in selected cases. Decreases in the amplitude of the ocular pulse, a measurement of diminished blood flow to the choroid from involvement of the posterior ciliary arteries, appears to correlate with abnormalities of the temporal artery.

Differences in the clinical manifestations and the populations affected generally help to differentiate giant cell arteritis from other forms of vasculitis. The other major form of vasculitis that primarily affects the vessels of the aortic arch is Takayasu's arteritis. However, this is predominantly a disease of young, oriental women. Similarly, whereas renal and intestinal involvement would be expected with polyarteritis and pulmonary involvement in hypersensitivity vasculitis, involvement of these organs is distinctly uncommon in giant cell arteritis.

Table 1 1989 criteria for the classification of giant cell (temporal) arteritis (Hunder *et al.* 1990)

Criterion	Definition
1. Age at onset ⩾50 years	Development of symptoms or findings beginning at or over the age of 50 years
2. New headache	New onset or type of localized pain in the head
3. Temporal artery abnormality	Tenderness to palpation of a temporal artery or decreased pulsation unrelated to arteriosclerosis of cervical arteries
4. Elevated ESR	ESR of 50 mm or more in 1 h by the Westergren method
5. Biopsy of artery	Biopsy specimen with artery showing vasculitis characterized by a predominance of mononuclear cell infiltration or granulomatous inflammation usually with multinucleated giant cells

For classification purposes, a patient shall be said to have a giant cell (temporal) arteritis if he/she has satisfied at least three of these five criteria. The presence of any three or more of the criteria yield a sensitivity of 93.5 per cent and a specificity of 91.2 per cent.

Recently a set of criteria for the classification of giant cell arteritis has been proposed (Hunder *et al.* 1990; Table 1). The presence of three, or more, of the criteria was found to be associated with a sensitivity of 93.5 per cent and a specificity of 91.2 per cent. Two of these criteria, increased age and elevation of the ESR, are important features in polymyalgia rheumatica. The failure to find an elevated ESR, however, does not preclude diagnosis. Cases of biopsy-documented giant cell arteritis with normal sedimentation rates have been described (Wong and Korn 1986).

Pathology

The characteristic lesions of giant cell arteritis occur in the media of the artery with an accumulation of inflammatory cells and focal necrosis (Chemnitz *et al.* 1987). Macrophages and granulomas containing multinucleated giant cells may be seen adjacent to the internal elastic membrane. The internal elastic membrane frequently becomes fragmented and calcified. Thickening of the intima, often with thrombosis, produces further narrowing or occlusion of the vascular lumen. Fibrinoid necrosis is uncommon. Patchy inflammation may be evident in the adventitia.

The predominant leucocytes that infiltrate the vessel wall are histiocytes and lymphocytes (mostly helper T cells). Plasma cells and eosinophils may be seen on occasion; however, polymorphonuclear leucocytes are infrequent. Immunofluorescent stains of involved vessels have revealed all classes of immunogloblins (IgG, M,A), fibronectin, and C1q. Intense accumulation of factor VIII may be seen along the endothelium, including the small blood vessels within the adventitia.

Treatment

Patients with giant cell arteritis require treatment with high doses of daily corticosteroids, typically 40 to 60 mg of prednisone. Guidelines for slow, gradual reduction of the dose, the need for increased doses for recurrences, the relapse rate with discontinuation, and the frequent prolonged courses of corticosteroids necessary are similar to those described for polymyalgia rheumatica (Salvarani *et al.* 1987; Kyle and Hazelman 1990). For patients who fail to respond to high-dose corticosteroids, or develop unacceptable side-effects, some success with cytotoxic or antimetabolite drugs has been described.

{STEROID-RESPONSIVE ANAEMIA OF THE ELDERLY

British geriatricians recognize this third syndrome as an additional variant to the cranial arteritis and polymyalgia rheumatica presentations of giant cell arteritis. Both cranial arteritis and polymyalgia rheumatica are typically associated with an anaemia. Steroid-responsive anaemia seems to be a manifestation of the disease in which the musculoskeletal and vasculitis symptoms are absent but the anaemia and, usually, some degree of constitutional disturbance predominate. The anaemia is usually mild (9 to 11 g/100 ml) and is most commonly normocytic and normochromic in type, but cases with microcytic and hypochromic anaemia have been described (Hume *et al.* 1973). The ESR is usually grossly raised. The patient typically complains of feeling tired, depressed, and inert; he or she is usually losing weight. Often the initial diagnosis is of occult malignancy, particularly if the anaemia is initially mistaken for an iron deficiency state. The response to steroid therapy is dramatic, and often the patient then becomes aware that he or she had in fact been experiencing minor polymyalgia rheumatica-like symptoms which had been so mild and so gradual in onset that they had not been recognized as abnormal. As is the case with other forms of giant cell arteritis, the syndrome may also present as a pyrexia of unknown origin.}

THE AETIOLOGY OF POLYMYALGIA RHEUMATICA AND GIANT CELL ARTHRITIS

The causes of these conditions are unknown. There is no satisfactory explanation as to why the diseases are seen exclusively in older populations. Similarly, it is unclear why the vascular involvement is largly confined to vessels of the thoracic aorta. It is tempting to speculate that an underlying vasculitis, mild and occult in polymyalgia and more advanced in arteritis, might be a common feature in the pathgenesis of the two diseases. However, even with careful study, abnormalities on temporal artery biopsy have been found in only 15 per cent or less of patients with uncomplicated polymyalgia rheumatica. Similarly, only about one-third of patients with biopsy-documented, often fulminating, giant cell arteritis have any of the clinical features associated with polymyalgia rheumatica.

The immunological reaction to components of the vascular wall in giant cell arteritis suggests a response to vascular injury that might develop from an environmental insult or even a primary autoimmune illness. No consistant abnormality in cellular or humoral immunity has been identified. Genetic factors are considered to be important in pathogenesis, with reports of disease in identical twins and first-degree relatives, and the predilection of the disease for white subjects, particularly those of Scandanavian background. Studies have revealed slight increases in the frequencies of HLA DR4 and Cw3 (Richardson *et al.* 1987; Cid *et al.* 1988).

References

Al-Hussaini, A.S. and Swannell, A.J. (1985). Peripheral joint involvement in polymyalgia rheumatica: a clinical study of 56 cases. *British Journal of Rheumatology*, **24**, 27–30.

Andersson, R., Malmvall, B.E., and Bengtsson, B.A. (1988). Acute phase reactants in the initial phase of giant cell arteritis. *Acta Medica Scandinavica*, **220**, 365–7.

Ayoub, W.T., Franklin, C.M., and Torretti, D. (1985). Polymyalgia rheumatica. Duration of therapy and long-term outcome. *American Journal of Medicine*, **79**, 309–15.

Boesen, P. and Sorensen, S.F. (1987). Giant cell arteritis, temporal arteritis, and polymyalgia rheumatica in a Danish county. A prospective investigation, 1982–1985. *Arthritis and Rheumatism*, **30**, 294–9.

Bosley, T.M., Savino, P.J., Eagle, R.C., Sandy, R., and Gee, W. (1989). Ocular pneumoplethysmography can help in the diagnosis of giant-cell arteritis. *Archives of Ophthalmology*, **107**, 379–81.

Buchwald, D., Sullivan, J.E., Leddy, S., and Komaroff, A.L. (1988). 'Chronic Epstein-Barr virus infection' syndrome and polymyalgia rheumatica. *Journal of Rheumatology*, **15**, 479–82.

Caselli, R.J., Daube, J.R., Hunder, G.G., and Whisnant, J.P. (1988a). Peripheral neuropathic syndromes in giant cell (temporal) arteritis. *Neurology*, **38**, 685–9.

Caselli, R.J., Hunder, G. G., and Whisnant, J.P. (1988b). Neurologic disease in biopsy-proven giant cell (temporal) arteritis. *Neurology*, **38**, 352–9.

Chemnitz, J., Christensen, B.C., Christoffersen, P., Garbarsch, C., Hansen, T.M., and Lorenzen, I. (1987). Giant-cell arteritis. Histological, immunohistochemical and electronmicroscopic studies. *Acta Pathologica, Microbiologica et Immunologica Scandinavica*, **95**, 251–62.

Cid, M.C. *et al.* (1988). Polymyalgia rheumatica: a syndrome associated with HLA-DR4 antigen. *Arthritis and Rheumatism*, **31**, 678–82.

Chou, C.T. and Schumacher, H.R., Jr. (1984). Clinical and pathologic studies of synovitis in polymyalgia rheumatica. *Arthritis and Rheumatism*, **27**, 1107–17.

Delecoeuillerie, G., Joly, P., Cohen-de-Lara, A., and Paolaggi, J.B. (1988). Polymyalgia rheumatica and temporal arteritis: a retrospective analysis of prognostic features and different corticosteroid regimens (11 year survey of 210 patients). *Annals of Rheumatic Diseases*, **47**, 733–9.

Gonzalez, E.B., Varner, W.T., Lisse, J.R., Daniels, J.C., and Hokanson, J.A. (1989). Giant-cell arteritis in the Southern United States. An 11-year retrospective study from the Texas Gulf Coast. *Archives of Internal Medicine*, **149**, 1561–5.

Hall, S., Persellin, S., Kurland, L., O'Brien, P.O., and Hunder, G. G. (1983). The therapeutic impact of temporal artery biopsy. *Lancet*, **ii**, 1217–20.

Healey, L.A. and Sheets, P.K. (1988). The relation of polymyalgia rheumatica to rheumatoid arthritis. *Journal of Rheumatology*, **15**, 750–2.

Hume, R., Dagg, J.H., and Goldberg, A. (1970). Refractory anaemia with dysproteinaemia: long-term therapy with low-dose steroids. *Blood*, **41**, 27–35.

Hunder, G.G. *et al.* (1990). The American College of Rheumatology 1990 criteria for the classification of giant cell arteritis. *Arthritis and Rheumatism*, **33**, 1122–8.

Kyle, V. and Hazelman, B.L. (1989). Treatment of polymyalgia rheumatica and giant cell arteritis. II. Relation between steroid dose and steroid side effects. *Annals of Rheumatic Diseases*, **48**, 662–6.

Kyle, V. and Hazelman, B.L. (1990). Stopping steroids in polymyalgia rheumatica and giant cell arteritis. *British Medical Journal*, **300**, 344–5.

Machado, E. B. *et al.* (1988). Trends in incidence and clinical presentation of temporal arteritis in Olmstead County, Minnesota, 1950–1985. *Arthritis and Rheumatism*, **31**, 745–9.

Mehler, M.F. and Rabinowich, L. (1988). The clinical neuro-ophthalmologic spectrum of temporal arteritis. *American Journal of Medicine*, **85**, 839–44.

Palmer, R.G., Prouse, P.J., and Gumpel, J.M. (1986). Occurrence of polymyalgia rheumatica in rheumatoid arthritis. *British Medical Journal*, **292**, 867.

Pascuzzi, R.M., Roos, K. L., and Davis, T.E. (1989). Mental status abnormalities in temporal arteritis: a treatable cause of dementia in the elderly. *Arthritis and Rheumatism*, **32**, 1308–11.

Pereira, M. and Kaine, J.L. (1986). Polymyalgia rheumatica and temporal arteritis: managing older patients. *Geriatrics*, **41**, 54–5, 59–60.

Perruquet, J.L., Davis, D.E., and Harrington, T.M. (1986). Aortic arch arteritis in the elderly. An important manifestation of giant cell arteritis. *Archives of Internal Medicine*, **146**, 289–91.

Richardson, J.E., Gladman, D.D., Fam, A., and Keystone, E.C. (1987). HLA-DR4 in giant cell arteritis: association with polymyalgia rheumatica syndrome. *Arthritis and Rheumatism*, **30**, 1293–7.

Salvarani, C. *et al.* (1987). Polymyalgia rheumatica and giant cell arteritis: a 5-year epidemiologic and clinical study in Reggio Emilia, Italy. *Clinical and Experimental Rheumatology*, **5**, 205–15.

Save-Soderbergh, J., Malmvall, B.E., Andersson, R., and Bengtsson, B. A. (1986). Giant cell arteritis as a cause of death. Report of nine cases. *Journal of the American Medical Association*, **255**, 493–6.

Vilasega, J., Gonzalez, A., Cid, M.C., Lopez-Vivancos, J., and Ortega, A. (1987). Clinical usefulness of temporal artery biopsy. *Annals of Rheumatic Diseases*, **26**, 282–5.

Wong, R.L. and Korn, J.H. (1986). Temporal arteritis without an elevated erythrocyte sedimentation rate. Case report and review of the literature. *American Journal of Medicine*, **80**, 959–64.

13.7 Connective tissue disorders

JOHN H. KLIPPEL

The term connective tissue disorders is used to describe a group of chronic inflammatory and degenerative rheumatic syndromes of unknown aetiology. The presence of cellular elements or products of the immune system at sites of disease suggests an immune-mediated pathogenesis for these syndromes. Autoantibodies, particularly antinuclear antibodies, develop in most patients. In several of the syndromes, relatively disease-specific antibodies have been identified.

Each of the clinical syndromes is characterized by primary involvement of individual target organs. However, both constitutional signs and symptoms and diffuse multisystem involvement may be seen. The connective tissue disorders pose particular problems in the elderly population. The clinical presentation and course of the disease may be substantially different from those described in younger patients. The effects of ageing on the immune system, the influences of endocrine changes, and an exposure to a different set of environmental agents are thought to contribute to the observed differences. The atypical clinical presentations, combined with the often prominent, non-specific constitutional symptoms, may cause long delays in diagnosis. In particular, differentiation from malignancy, chronic infections, or other rheumatic diseases is often difficult.

AUTOANTIBODIES AND THE CONNECTIVE TISSUE DISORDERS

Autoantibodies may be detected in the majority of patients with connective tissue disorders (Table 1). The profile of antibodies in individual syndromes is often helpful in diagnosis. Moreover, changes in antibody titre often correlate with fluctuations in inflammation ('disease activity') and are thus used to monitor disease and guide treatment. Several of the subtypes of antinuclear antibodies are highly disease specific. Examples include antibodies to double-stranded DNA and Sm in systemic lupus erythematosus, histadyl t-RNA synthetase (Jo-1) in inflammatory myositis, topoisomerase I (Scl-70) in diffuse scleroderma, and centromere proteins in limited scleroderma.

SYSTEMIC LUPUS ERYTHEMATOSUS

Systemic lupus erythematosus (**SLE**) is a chronic, relapsing and remitting, inflammatory syndrome that typically involves many organs. The most important pathological feature is the deposition of circulating antigen–antibody complexes along the vascular basement membranes of target organs. Organs commonly affected include the skin, joints, serosal surfaces, kidneys, heart, lungs, and central nervous system.

Lupus is the most common of the connective tissue disorders with an estimated prevalence of 500 patients per million. Women, particularly during the reproductive years, are at significantly increased risk; the female to male sex ratio is about 9 to 1. In addition, there appears to be an increase in the frequency of SLE in certain racial groups, particularly black Americans, Puerto Ricans, and persons of Asian background. In older patients, lupus is often of insidious onset with a relative sparing of involvement of major organs such as the kidneys and central nervous system (Maddison 1987; Ward and Polisson 1989).

Clinical

There is some similarity between the atypical lupus syndrome in elderly patients and Sjögren's syndrome with systemic disease manifestations (Bell 1988). Debilitating constitutional symptoms, such as fatigue, anorexia, weight loss, and fever, may be common and, in the absence of more classic clinical

Table 1 Autoantibodies in the connective tissue disorders

Autoantibody	Systemic lupus erythematosus	Primary Sjögren's syndrome	Inflammatory myopathies	Scleroderma
Antinuclear antibody (ANA)	++++	++	+++	+++
ANA subtype:				
DNA	+++	–	–	–
Histones	++	–	–	–
Sm	++	–	–	–
SS-A (Ro)	+	+++	+	+
SS-B (La)	+	+++	+	+
Jo-1	–	–	++	–
Scl-70	–	–	–	+(diffuse)
Centromere	–	–	–	+++(limited)
Rheumatoid factor	+	+++	+	+

++++ >75 per cent; +++ 75–50 per cent; ++ 49–25 per cent; + <25 per cent; – absent.

signs of lupus, cause a long delay in diagnosis (Baer and Pincus 1983).

Transient, peripheral, symmetrical polyarthritis combined with severe pain and stiffness in the limb girdle may mimic polymyalgia rheumatica (Hutton and Maddison 1986). The joint involvement is typically not chronic and rarely, if ever, produces cartilage loss, subchondral cystic changes, or bone erosions. The development of persistent synovitis in a single joint suggests a superimposed complication such as septic arthritis or osteonecrosis. Periarticular structures, particularly tendon sheaths, may be involved. Chronic inflammation of tendons and capsular structures may produce joint laxity with rheumatoid-like hand deformities (Fig. 1). Acute involvement of tendons may result in their rupture, most commonly Achilles or patellar tendons. A proximal inflammatory myopathy, often with a normal serum creatine phosphokinase, may cause weakness, pain, and muscle wasting.

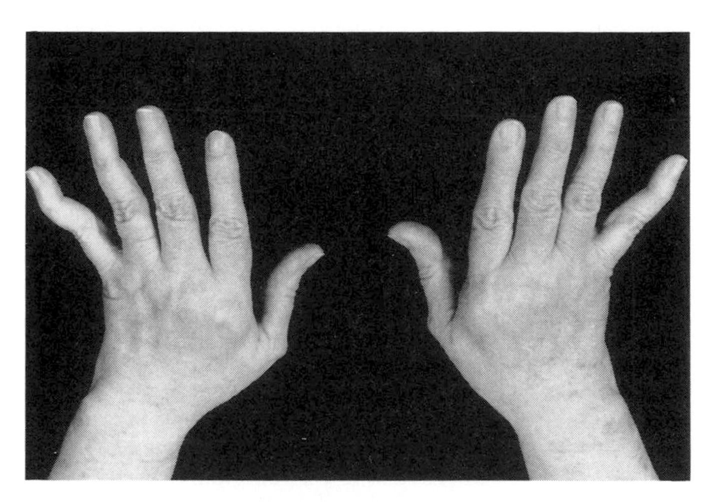

Fig. 1 Lupus arthropathy (Jaccoud's). Reversible, rheumatoid-like deformities of the hands may develop in the absence of cartilage or bone destruction on radiographs.

A wide variety of mucocutaneous lesions may be seen. The erythematous, maculopapular facial rash ('butterfly rash') and the atrophic, hyperkeratotic lesions of discoid lupus are most well known. However, a range of skin manifestations may be seen, including acute vasculitic rashes of the trunk and arms, psoriasis-like patches, bullae, urticarial eruptions, and angioedema. Many of the skin rashes are worsened by exposure to ultraviolet light. Superficial ulcerations of oral and genital mucosa are typically painless and often go undetected. Ulcerations of the nasal mucosa can lead to epistaxis and perforation of the septum. Diffuse or patchy alopecia in the absence of scarring of the scalp from discoid disease is generally entirely reversible. The regrown hair is often short, stubby, and brittle.

Sterile pleuritis, pericarditis, or peritonitis develop from inflammation of serosal surfaces. Fluid accumulation is usually modest, although on occasion massive ascites or pericardial tamponade may occur. Typically, the fluid has a low white-cell count, 3000 cells/mm^3 or less, predominantly made up of mononuclear cells. The fluid may contain reduced levels of complement and LE cells.

Pulmonary manifestations, particularly interstitial disease, are seen with increased frequency in older patients with SLE (Catoggio *et al.* 1984). Acute, transient basilar pneumonic infiltrates ('lupus pneumonitis') with non-productive cough, hypoxaemia, and dyspnoea must be distinguished from infectious causes. Rare serious pulmonary complications include alveolar haemorrhage with rapid obliteration of the lung fields and massive haemoptysis and pulmonary hypertension with cor pulmonale.

Non-infective vegetations may develop on the ventricular surfaces of valve leaflets (Libman–Sacks endocarditis). The vegetations are a potential nidus for superimposed bacterial infection, so antibiotic prophylaxis for dental or other surgical procedures is advised. An inflammatory myocarditis may lead to ventricular arrhythmias, conduction abnormalities, or intractable congestive heart failure. Ischaemic heart disease from coronary arteritis or, more commonly, unrelated atherosclerotic disease may be responsible for angina or myocardial infarction.

Gastrointestinal complaints are relatively infrequent, pain from peritonitis being perhaps the most common. Both acute and chronic pancreatitis have been described. Hepatitis, when found, is typically secondary to the use of salicylates or other non-steroidal drugs. Primary biliary cirrhosis has been reported to be increased in lupus. Vasculitis of mesenteric and intraabdominal organs may produce an acute abdomen that requires surgical exploration.

Clinical evidence of significant renal involvement is considered to be infrequent in older patients. Urine sediment, particularly red blood cells and red-cell casts, is indicative of active glomerular inflammation. Proteinuria, often in large amounts and resulting in the nephrotic syndrome, may be seen with lupus membranous nephropathy. Loss of renal function may be acute, similar to that of rapidly progressive nephritis, or, more typically, involves a slowly progressive rise of serum creatinine over the course of many months or years. Renal function is a poor guide to the actual severity of renal involvement because a substantial proportion of nephrons must be destroyed before functional impairment develops. In patients with persistent evidence of nephritis, renal biopsy is advised to determine the type and severity of renal disease. Hypercellularity may be confined to the mesangium or involve the glomerular capillaries in a focal segmental or diffuse (Fig. 2) distribution. Extensive glomerular sclerosis is generally associated with a poor outcome.

Various neurological and psychiatric manifestations may develop. Disturbances of mental function range from states of mild confusion with memory deficits and impairments of orientation and perception to major psychiatric disturbances of delirium, hypomania, or schizophrenia. Seizures are typically of the grand mal type, although petit mal, focal, and temporal lobe epilepsy have been described. There may be severe headaches with scotomata typical of the fortification spectra of migraines. Less common neurological disturbances include cranial neuropathies, transverse myelopathy, pseudotumour cerebri, chorea and hemiballismus, Parkinsonian-like tremors, and both sensory and motor peripheral neuropathies. Conventional tests of the central nervous system are frequently normal. The cerebrospinal fluid may show mild elevations of protein and IgG, oligoclonal bands on electrophoresis, and pleocytosis. Diffuse or focal changes may be found on electroencephalography. Arteriography rarely demonstrates evidence of vasculitis. The significance of

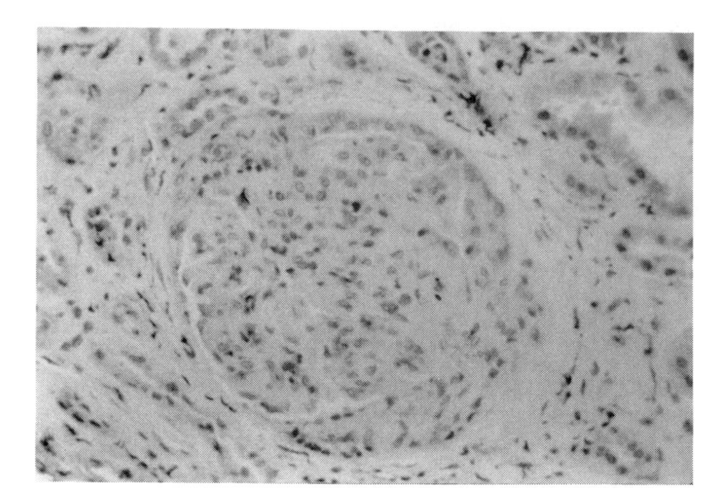

Fig. 2 Diffuse proliferative lupus glomerulonephritis. Advanced proliferation of mesangial cells, capillary endothelium, and cells of the capsule are evident (by courtesy of J. E. Balow).

recently reported abnormalities on nuclear magnetic resonance imaging remains to be determined.

Drug-induced lupus

Autoimmune syndromes induced by drugs are a very important consideration, particularly in the elderly population. Numerous drugs have been reported to be associated with the development of antinuclear antibodies (Table 2).

Table 2 Drugs associated with the development of antinuclear antibodies

Definite
Chlorpromazine, D-penicillamine, ethoxysuximide, hydralazine, isoniazid, methyldopa, phenytoin, practolol, procainamide, propylthiouracil, quinidine, sulphasalazine, trimethadione
Probable
Acebutolol, atenolol, captopril, carbamazepine, hydrazine, hydrochlorthiazide, labetalol, levodopa, lithium, mephytoin, methimazole, methylthiouracil, metoprolol, nitrofurantoin, oxprenolol, phenylbutazone, primidone, trimethadione

Although the majority of patients with drug-induced antinuclear antibodies are essentially asymptomatic, diseases that are clinically indistinguishable from connective tissue disorders, including SLE and polymyositis, may develop in the course of drug administration. Procainamide and hydralazine are by far the two most common drugs responsible for the drug-induced lupus syndrome. Prospective studies have shown that between half and three-quarters of patients treated with these agents develop antinuclear antibodies. However, only a fraction of these patients has clinical symptoms. The lupus syndrome that develops from drugs, particularly procainamide, is 'viral-like' with fever, myalgias, arthralgias, and with a high incidence of pleural and pulmonary symptoms. Although involvement of major organs such as the kidney (glomerulonephritis) has been described, it is distinctly uncommon. Characteristically, symptoms subside or completely resolve with discontinuation of the offending drug. However, treatment with anti-inflammatory drugs, including corticosteroids, may be indicated for acute, severe manifestations. Antinuclear antibodies, on the other hand, may remain present for months, or even years, after the drug has been discontinued.

Treatment

Anti-inflammatory agents are often useful in the treatment of minor lupus manifestations such as fatigue, fever, and joint symptoms. Enteric-coated aspirin in moderate doses is generally well tolerated and effective. Patients with active lupus seem more prone to the development of salicylate-induced hepatitis with increases in serum transaminase and transient impairments of renal function. These abnormalities are readily reversible upon stopping the salicylate. Non-salicylated, non-steroidal anti-inflammatory agents, such as ibuprofen, naproxen, tolmetin, and others, are similarly useful in management of mild lupus. There is no evidence, however, that any of these drugs is actually superior to salicylates. Moreover, they may produce adverse side-effects, including skin rashes and aseptic meningitis, that might be easily confused with manifestations of active lupus.

Adrenal corticosteroids should be reserved for patients who fail to respond to conservative drug therapy or those who have serious disease. Daily corticosteroids, occasionally needed in divided doses, are generally superior to alternate-day regimens. Manifestations such as fever, fatigue, polyarthritis, or serositis typically respond to treatment with low doses of prednisone, i.e., 10 mg to 20 mg daily. Acute severe or life-threatening manifestations such as pericarditis, myocarditis, acute glomerulonephritis or central nervous system disease are indications for higher doses, i.e., 60 mg prednisone daily, or more. Bolus intravenous methylprednisolone (1 g, or 15 mg/kg) is an alternative to conventional high-dose oral corticosteroids. The dose of oral corticosteroids should be kept constant until the inflammation is well under control. At that point, the dose should be cautiously reduced.

Antimalarial drugs have an important adjunctive role in the treatment of mild features of systemic disease, especially for the management of mucocutaneous manifestations. The daily dose of antimalarials should not exceed 250 mg chloroquine, 400 mg hydroxychloroquine, or 100 mg quinacrine. At these low doses the risks of retinal toxicity are extremely small. However, as a precaution, the eyes should be examined for disturbances in colour vision and retinal changes every 6 months during drug treatment. Additional complications of antimalarial drugs include gastrointestinal complaints, rashes, photosensitivity, and discoloration of the skin and hair.

Cytotoxic or antimetabolite drugs, such as azathioprine and cyclophosphamide, are regarded as experimental approaches to the treatment of lupus. They should be reserved for patients with serious forms who have failed conservative therapies, including high-dose corticosteroids. The drugs have been best studied in lupus nephritis in which controlled trials have demonstrated that they can retard the progression of chronic scarring within the kidney (Balow *et al.* 1984) and reduce the likelihood of end-stage renal failure (Austin *et al.* 1986). Other experimental approaches to the therapy of lupus include plasma exchange and total lymphoid irradiation.

SJÖGREN'S SYNDROME

Sjögren's syndrome is a chronic inflammatory disease characterized by lymphocytic infiltration of exocrine glands. The most prominent symptoms result from dysfunction of the major and minor salivary glands of labial, nasal, and hard-palate mucosa. However, diffuse involvement of glands of the upper and lower respiratory tracts, gastrointestinal tract, pancreas, vagina and skin may produce other complications. The syndrome may occur alone (primary Sjögren's) or in association with other rheumatic diseases, particularly rheumatoid arthritis, systemic lupus erythematosus, scleroderma, and inflammatory myopathies (secondary Sjögren's).

Recent surveys of elderly populations indicate a likely high prevalence of primary Sjögren's. In a study of 103 elderly white women, 40 complained of sicca (dryness) symptoms, and two patients (2 per cent) with definite primary Sjögren's syndrome were identified (Strickland *et al.* 1987). A similar study found focal lymphoid aggregates in minor salivary glands that were compatible with the appearances of Sjögren's syndrome in 8 of 62 (13 per cent) asymptomatic individuals (Drosos *et al.* 1988)

Clinical features

Dry eyes (xerophthalmia) and dry mouth (xerostomia) are the clinical hallmarks of Sjögren's syndrome. Eye symptoms are often insidious and generally involve complaints of a gritty, sandy, 'foreign body' sensation. The conjunctiva may become markedly injected. Accumulations of thick, ropey strands or crusted matter at the inner canthus may be noted on first arising in the morning. Ocular complications of advanced Sjögren's include corneal ulcerations, vascularization, and opacification.

The xerostomia causes a high frequency of periodontal disease with gingivitis and accelerated dental caries. Dysfunction of salivary and oesophageal glands also leads to difficulties with chewing and swallowing. Dry foods such as crackers or biscuits are particularly troublesome. Defects of both taste and smell may occur. Deep fissures and ulcerations of the tongue and buccal mucosa as well as angular stomatitis may develop.

Enlargement of parotid and other salivary glands may be episodic and associated with fever, tenderness, and erythema. Careful consideration must be given to other causes of salivary gland enlargement and dysfunction (Table 3). Secondary infection from obstruction of the gland ducts by inspissated matter must be considered where there are inflammatory changes of glands, particularly if unilateral. Rapidly growing, hard or nodular glands, on the other hand, should suggest malignant change.

Dryness of the oral pharynx, larynx, and tracheobronchial tree may cause hoarseness, recurrent otitis media and conduction deafness, or epistaxis. Involvement of the lower respiratory tract may be associated with recurrent bronchitis, non-productive cough, and an increased frequency of pulmonary infectious complications. Involvement of the glands of the vaginal mucosa leads to intense dryness with pruritus and dyspareunia.

In the majority of older patients, Sjögren's is confined to ocular and oral changes. However, serious multisystem disease may occur (Table 4). There are similarities between systemic Sjögren's syndrome and the other connective tissue disorders, particularly lupus (SLE) (Bell 1988).

Table 3 Causes of parotid or salivary gland enlargement or dysfunction

Anti-cholinergic drugs
Sedatives
Hypnotics
Narcotics
Phenothiazines
Atropine
Antihistamines
Antidepressants
Infiltrative disorders
Sarcoidosis
Amyloidosis
Haemochromatosis
Lymphoma
Leukaemia
Primary neoplasms of parotid
Systemic diseases
Cirrhosis
Diabetes mellitus
Hyperlipoproteinaemias
Obesity
Cushing's disease
Infection
Viral (Coxsackie, mumps, cytomegalic inclusion disease)
Bacterial (staphylococcal organisms typically secondary to obstruction of duct)
Fungal (actinomycosis or histoplasmosis)
Nutritional deficiency
Starvation
Vitamin deficiency (B_6, C, and A)

Table 4 Features of systemic Sjögren's syndrome

Gastrointestinal	Oesophageal stenosis, gastric hyposecretion, atrophic gastritis, pancreatitis, malabsorption
Pulmonary	Lymphocytic interstitial pneumonitis, pseudolymphoma, interstitial fibrosis
Renal	Tubular dysfunction, particularly type I renal tubular acidosis and hyposthenuria, and Fanconi's syndrome
Myopathy	Interstitial and perivascular fibrosis, inflammatory infiltrates, or both
Vasculitis	Cutaneous, mononeuritis multiplex, axonal neuropathy
Bone marrow	Thrombocytopenia
Nervous system	Cerebrovascular accidents, spinal cord syndromes, psychiatric or cognitive dysfunction, peripheral or cranial (trigeminal) neuropathies

Diagnosis

Biopsy of minor salivary glands is the most specific and definitive procedure for diagnosis; biopsy of major salivary glands is indicated only if there is a suspicion of malignancy. The condition is characterized by infiltration of the gland with multiple aggregates of lymphocytes, plasma cells, and occasionally macrophages. Destruction and atrophy of the acinar tissue and replacement with fat is an end result of the inflammation. The ductal cells may become hyperplastic so as to obliterate the lumen. However, the stroma and lobular architecture of the gland are generally preserved. Caution must be used in interpreting the histopathology of minor salivary glands in elderly individuals. Well-recognized changes associated with ageing include acinar atrophy, fibrosis, and ductal dilatation and hyperplasia; focal aggregates of lymphocytes and plasma cells compatible with the appearances of Sjögren's syndrome have also been described (DeWilde *et al.* 1986).

Measurements of the function and imaging of the anatomy of the major salivary glands are non-invasive procedures that may be helpful in monitoring the course of Sjögren's disease. The flow of saliva from the parotid gland may be directly measured. Structural abnormalities of acinar tissue and ductal atrophy can be demonstrated by injection of radiopaque dye into the parotid duct. The uptake, concentration, and excretion of $^{99}Tc^{m}$-pertechnate by the parotid glands may be measured by sequential scintophotographic techniques.

Several tests may be used to evaluate lacrimal function and ocular disease. A narrow strip of No. 41 filter paper placed in the lower lid can be used to measure the amount of wetness that occurs over 5 min (Schirmer's test). Normally, 15 mm or more of wetness develops; less than 5 mm is generally considered abnormal. On rose-bengal staining or slit-lamp examination of the eye, superficial erosions and punctate or filamentous keratitis (keratoconjunctivitis sicca) may be seen.

Sjögren's syndrome and lymphoma

An increased incidence of lymphomas, particularly of B-cell lineage, has been observed in patients with Sjögren's (Sheibani *et al.* 1988). The changes within the salivary and lymphoid tissues are thought to involve a continuum from benign lymphoid hyperplasia with polyclonal hypergammaglobulinaemia at one end to malignant proliferation at the other. An intermediate stage, termed pseudolymphoma, involves extraglandular extension of lymphoproliferation that is clinically and histologically benign. Clinical and laboratory findings associated with an increased risk of lymphoid malignancies include parotid enlargement, splenomegaly, lymphadenopathy and unexplained anaemia, hypogammaglobulinaemia, and a reduction in the titre of rheumatoid factor.

Treatment

The management of Sjögren's syndrome is largely symptomatic and consists of topical fluid replacement to prevent complications of dryness. Artificial tear preparations must be used on a regular basis. In addition to various eye drops, a slow-release capsule containing a polymer of hydroxyproline that is inserted beneath the inferior tarsal margin is available. Patients who wear contact lens must follow meticulous hygiene practices to avoid infection. Frequent small drinks are important; most patients soon learn to carry a bottle of water with them at all times and to keep a glass of water at their bedside. Sugarless tart candy or lemon-flavoured drops to stimulate saliva secretion benefit some patients. Drugs with anticholinergic properties should be avoided. Measures to prevent dental caries are of particular importance. Brushing and flossing for plaque control should be a part of the daily routine. In addition, topical stannous fluoride should be applied to the teeth nightly to promote remineralization and retard tooth damage. Saline nasal sprays and the use of humidifiers help to reduce nasal and oropharyngeal dryness. Lubricants relieve vulvar and vaginal pruritus and dyspareunia.

There is a very limited role for drugs in the management of patients with Sjögren's syndrome. Corticosteroids and cytotoxic/antimetabolite drugs should be reserved for those with life-threatening complications such as vasculitis or disease of the central nervous system. Topical corticosteroids to the eye accelerate corneal thinning and are contraindicated.

INFLAMMATORY MYOPATHIES

Several forms of chronic, idiopathic inflammatory and degenerative adult muscle syndromes are recognized (Plotz *et al.* 1989). Polymyositis and dermatomyositis, likely variants of a common pathological process, are the most common types. Inflammatory myopathies may be part of several connective tissue disorders including SLE, scleroderma, and Sjögren's syndrome ('overlap syndromes'). Inclusion-body myositis is a relatively recently recognized variant of inflammatory myopathy with distinctive clinical and, in particular, pathological features.

The peak incidence of adult inflammatory myopathies is in the fourth to sixth decades. More women than men are affected by most types of these myopathies except for inclusion-body myositis, which is characteristically a disease of men. There is no striking familial predisposition to any of the forms of inflammatory myopathy.

Clinical

The differential diagnosis of muscle pain and weakness is wide (Table 5). Of importance in the elderly population are electrolyte disturbances, endocrine abnormalities, particularly hypothyroidism, and drug-induced myopathic syndromes.

The clinical presentation of inflammatory myopathy may

Table 5 Causes of muscle weakness in elderly patients

Electrolyte disturbances	Hypokalaemia, hypocalcaemia, hypomagnesaemia, hypophosphataemia
Drugs and toxins	Alcohol, D-penicillamine, colchicine, clofibrate, others
Endocrine diseases	Hyperthyroidism, hypothyroidism, Cushing's or Addison's syndrome
Rheumatic diseases	Polymyalgia rheumatica, arthropathies of shoulders, hips, or knees, overlap syndromes
Neurological diseases	Myasthenia gravis, Guillain-Barré, lumbar/cervical cord syndromes
Infections	Bacterial (rickettsia, mycobacteria), Parasites (toxoplasma, trichinella) Viruses (Coxsackie, echovirus, influenza, human immunodeficiency virus, hepatitis B)
Infiltrative disorders	Amyloidosis

be either with acute pain and tenderness or, more typically, insidious weakness. Muscle involvement is very symmetrical in distribution. The proximal muscles are affected to a far greater extent than distal muscle groups; an exception is inclusion-body myositis in which distal muscle involvement is typical. Involvement of the legs commonly produces difficulty in getting out of a chair or the bath, or in climbing stairs, or falling while walking. Weakness in the arms typically causes difficulty with tasks that require hand movements above the shoulders, such as shaving, combing the hair, or reaching items from a high shelf. Pharyngeal muscles may be affected, leading to dysphasia or hoarseness. In advanced disease the accessory muscles of the neck may be involved such that the patient is unable to raise the head from the pillow. Muscles of the face, including extraocular muscles, are virtually never involved.

A diagnosis of dermatomyositis is used in patients who have an inflammatory myopathy in association with several distinct types of cutaneous lesions. Perhaps the most common are raised, violaceous scaly eruptions over the metacarpophalangeal and proximal interphalangeal joints (Gottron's papules). This same rash may be seen on the extensor surfaces of the knees and elbows, or the malleolar surfaces of the ankles. A somewhat different scaly hyperkeratotic rash may arise along the radial aspects and pads of the fingers. Dark pigmentation in the depths of the fissures gives the appearance of the calloused hands of a manual labourer. A bluish discoloration of the upper eyelid with periorbital oedema (heliotrope rash) may occur. {Oedema of the backs of the hands is a characteristic feature.} Erythematous rashes of the face and V-region of the neck may be similar to those of SLE. Involvement of the nasolabial folds, present in myositis and absent in lupus, may be a helpful point in differential diagnosis. Various non-specific changes of the periungual capillaries, including erythema, telangiectasia, and nailfold infarcts, may also occur.

Signs of systemic and multisystem illness may be prominent in inflammatory myopathies. These account for much of the illness and death and may confuse diagnostic efforts. High, spiking fevers, anorexia, and weight loss may initially suggest infection or malignancy. Gastrointestinal symptoms are common and stem from oesophageal dysfunction, reduced motility of the small bowel, and malabsorption. The development of cardiac or pulmonary involvement is the most serious of the systemic features of the inflammatory myopathies. Abnormalities of the cardiac conduction system or actual cardiomyopathy may lead to congestive heart failure, arrhythmias, or sudden death. Progressive dyspnoea from pulmonary involvement may develop from involvement of the diaphragmatic or intercostal musculature, or more commonly, from fibrosing alveolitis.

Diagnosis

Muscle inflammation results in increases in the serum levels of enzymes contained in muscle cells. Elevations of creatine phosphokinase and aldolase are commonly used to detect inflammatory muscle disease; elevations of various transferases and lactate dehydrogenase are less specific. The serum myoglobin and creatine may also be elevated; myoglobinuria may be detected as haemoglobin on urine dipstick testing. Elevations of acute-phase serum proteins lead to increases in the erythrocyte sedimentation rate. Antibodies against several t-RNA transferase enzymes are unique to patients with polymyositis/dermatomyositis. The most common of these antibodies, anti-Jo-1, occurs in a subset of myositis patients with pulmonary fibrosis, Raynaud's disease, and arthritis.

A number of abnormalities may be detected on electromyography. The electromyogram is perhaps most useful in differentiating inflammatory from neuropathic (denervation) myopathies. Findings considered to be characteristic of muscle inflammation include insertional irritability, repetitive, spontaneous discharges, and small-amplitude, short, polyphasic motor-unit potentials.

The most definitive procedure for diagnosing muscle inflammation is biopsy. Typically, this is taken from the quadriceps or biceps brachii muscles. The insertion of electromyographic needles may produce pathological changes in muscle, so the biopsy site should be as far as possible from where an electromyogram has been done. The basic histopathological features of the inflammatory myopathies are inflammatory cell infiltrates and evidence of muscle cell degeneration and regeneration. The cellular infiltrate is predominantly of lymphocytes, plasma cells, and macrophages; polymorphonuclear leucocytes are distinctly unusual. Both type I and type II muscle fibres are involved; affected individual muscle fibres vary in size.

The various types of inflammatory myopathies may have variations in their histopathological appearances. In polymyositis, the inflammatory reaction occurs primarily within fascicles. Necrotic fibres may be scattered or isolated, and perifascicular atrophy is not found. By contrast, in dermatomyositis, the inflammatory infiltrates are predominantly perivascular or in septa and perifascicular atrophy is rather characteristic. Necrosis and phagocytosis of muscle fibres tends to be more localized. In inclusion-body myositis, trichrome-stained frozen sections, or electron micrographs, may show myofibre vacuoles within the nucleus and cytoplasm. The vacuoles may be rimmed by basophilic, granular material. By electron microscopy, a characteristic membranous, whorl pattern is seen.

Myositis and cancer

Malignancies have been seen occasionally in association with inflammatory myopathies. Although typically they antecede the recognition of the muscle disease, there have been cases of simultaneous and late malignancies. Whether these are simply the random occurrence of two unrelated diseases or whether myositis or its treatment predispose to malignancy, or vice versa, is unclear. Most studies have failed to confirm an association (Manchul *et al.* 1985; Lakhanpal *et al.* 1986). In practical terms, malignancies in patients with myositis are inevitably apparent on careful physical examination combined with routine laboratory studies and chest radiographs. In the absence of clinical suspicion, there is no need for more exhaustive searches by imaging.

Treatment

High-dose corticosteroids are generally recognized as the standard drug treatment for patients with acute inflammatory myopathy. Therapy should be initiated with prednisone (or equivalent) at a dose of 1 mg/kg daily. Recent studies of high-dose, alternate-day therapy appear promising with less side-effects (Hoffman *et al.* 1983; Uchino *et al.* 1985). The response to corticosteroids is typically not dramatic, and weeks or more often months are required before convincing evidence of improvement is apparent. Once the disease process is under control, or unacceptable side-effects develop, the dose of corticosteroids should slowly be reduced. There is some indication that elderly patients require more therapy than younger patients (McKendry 1987). Patients who fail to respond to corticosteroids, or develop unacceptable side-effects, are candidates for alternative, experimental forms of therapy such as oral azathioprine (2–3 mg/kg daily) or low-dose oral (7.5–15 mg weekly) or intramuscular (0.5–0.8 mg/kg weekly) methotrexate.

SCLERODERMA

The terms scleroderma and systemic sclerosis are used interchangeably to describe a degenerative and inflammatory condition that ends in fibrosis. Skin thickening is most commonly confined to the distal extremities; however, involvement of proximal extremities, the face, and trunk may be seen with advanced disease. In addition, vasospasm and fixed, structural vascular disease with intimal hyperplasia account for involvement of the gastrointestinal tract, lungs, heart, and kidneys.

The slight predominance of scleroderma in women is less apparent in older age groups. The 10-year survival is estimated to be in the range of 60 to 70 per cent. Diffuse, rapidly progressive scleroderma with involvement of the lungs, heart, or kidneys is associated with a poor prognosis.

Clinical

Two clinical forms of scleroderma are recognized—diffuse and limited. Limited scleroderma is also referred to as the CREST (calcinosis, Raynaud's, esophageal dysphagia, sclerodactyly, and telangiectasia) syndrome. These types of scleroderma differ in modes of presentation, evolution of disease course, and autoantibodies (Lally *et al.* 1988).

Skin changes typically are first found in the hands and fingers, with diffuse, non-pitting swelling and puffiness. Over a period of several weeks to months the skin becomes thickened, hard, and shiny in appearance. Other skin changes include loss of folds and hair, dryness from loss of sebaceous glands, the development of punctate telangiectasia, and pigmentary changes. The skin involvement may be confined to the hands (sclerodactyly) or may spread centrally to affect the forearms, arms, face, anterior chest, and abdomen. Skin thinning, particularly over joints and malleolar surfaces, may lead to ulceration. Subcutaneous deposits of hydroxyapatite crystals form at sites of repeated trauma such as elbows, knees, and fingers. The subcutaneous masses may ulcerate to extrude white chalky material and become secondarily infected.

Cold-induced vasospasm (Raynaud's phenomenon) occurs in all patients with scleroderma. It may be the presenting manifestation and antedate the actual development of other features by many years. Structural abnormalities of nailfold capillaries may be evident grossly or on capillary microscopy. Ischaemia from severe and prolonged Raynaud's may lead to pain, pitting scars of the pulp or, on rare occasion, frank gangrene of the fingertips.

Polyarthralgias and non-specific, symmetrical polyarthritis may occur early. There is little actual bone or joint destruction except for osteolysis of the distal phalanx. Involvement of the synovial lining of tendon sheaths is particularly common, leading to crepitus, and palpable, coarse, leathery friction rubs over flexor and tendon sheaths of fingers, knees, and ankles. Fibrosis of the tendons and joint capsule combined with overlying changes in the skin may lead to flexion contractures.

Gastrointestinal symptoms are common, particularly in patients with severe Raynaud's. Hypomotility of the distal oesophagus leads to dysphagia, especially for solid foods. In addition, incompetence of the gastro-oesophageal sphincter may be associated with reflux, oesophagitis, and strictures. Metaplastic changes of the distal oesophagus are associated with an increased incidence of adenocarcinoma. Loss of smooth muscle of the proximal and distal intestine causes hypomotility with abdominal distension and bloating, and pseudo-obstruction. Weakening of the bowel wall leads to the formation of wide-mouthed diverticula of the large bowel, particularly of the transverse and descending colon, as well as of the jejunum and ileum. Malabsorption and diarrhoea may occur from incomplete mixing of bowel contents and bacterial overgrowth. The development of intraperitoneal air by dissection through the atrophic mucosa and muscular layers (pneumatosis intestinalis) is a rare complication of this bowel disease.

Abnormalities of pulmonary function, particularly a reduction in the diffusion capacity, can be demonstrated in most patients with scleroderma. The abnormalities worsen on exposure to cold, suggesting a Raynaud's-like, vasospastic process within the pulmonary vasculature. On physical examination and chest radiograph, bibasilar, interstitial disease with linear or nodular fibrosis is a frequent finding. Progression of the fibrosis with the development of respiratory insufficiency is one of the more serious consequences of scleroderma. In addition, patients with pulmonary fibrosis are at increased risk for the development of alveolar-cell carcinoma.

Involvement of the kidneys is the most serious complication of scleroderma. On urinalysis, microscopic haematuria and low-grade proteinuria may be found. Renal involvement is associated with the development of malignant hypertension. Prompt and aggressive medical treatment of the hypertension is essential to prevent rapidly progressive renal failure.

Treatment

The management of scleroderma is largely symptomatic. Wearing mittens and avoiding undue cold may minimize Raynaud's phenomenon; vasodilating drugs like prazosin or nifedipine relieve the symptoms and complications of vasospasm. Non-steroidal anti-inflammatory drugs and physiotherapy are helpful in the treatment of musculoskeletal complications, the latter particularly important for joint contractures. Conservative management of skin ulcerations and necrotic digits with occlusive, sterile dressings and immobilization promotes healing and prevents secondary infections. Broad-spectrum antibiotics such as tetracycline are helpful in patients with steatorrhoea or other features of intestinal malabsorption.

There is a very limited role for drugs in the primary management of scleroderma. Low-dose corticosteroids may be valuable in the management of acute inflammatory manifestations such as acute oedematous skin disease, acute pericarditis, or myositis. Colchicine, immunosuppressive agents, cyclosporin A and D-penicillamine have all been reported to be of benefit, although none has been rigorously studied.

References

Austin, H.A. *et al.* (1986) Therapy of lupus nephritis: controlled trial of prednisone and cytotoxic drugs. *New England Journal of Medicine*, **314**, 614–19.

Baer, A.N. and Pincus, T. (1983). Occult systemic lupus erythematosus in elderly men. *Journal of the American Medical Association*, **249**, 3350–2.

Balow, J.E. *et al.* (1984). Effect of treatment on the evolution of renal abnormalities in lupus nephritis. *New England Journal of Medicine*, **311**, 491–5.

Bell, D.A. (1988). SLE in the elderly— is it really SLE or systemic Sjögren's syndrome? *Journal of Rheumatology*, **15**, 723–4.

Catoggio, L.J., Skinner, R.P., Smith, G., and Maddison, P.J. (1984). Systemic lupus erythematosus in the elderly: clinical and serological characteristics. *Journal of Rheumatology*, **11**, 175–81.

De Wilde, P.C.M., Baak, J.P.A., van Houwelingen, J.C., Kater, L., and Slootweg, P.J. (1986). Morphometric study of histological changes in sublabial salivary glands due to aging process. *Journal of Clinical Pathology*, **39**, 406–17.

Drosos, A.A., Andonopoulos, A.P., Costopoulos, J.S., Papadimitriou, C.S., and Moutsopoulos, H.M. (1988). Prevalence of primary Sjogren's syndrome in an elderly population. *British Journal of Rheumatology*, **27**, 123–7.

Hashimoto, H., Tsuda, H., Hirano, T., Takasaki, Y., Matsumoto, T., and Hirose, S. (1987). Differences in clinical and immunological findings of systemic lupus erythematosus related to age. *Journal of Rheumatology*, **14**, 497–501.

Hoffman, G.S., Franck, W. A., Raddatz, D.A., and Stallones, L. (1983). Presentation, treatment, and prognosis of idiopathic inflammatory muscle disease in a rural hospital. *American Journal of Medicine*, **75**, 433–8.

Hutton, C.W. and Maddison, P.J. (1986). Systemic lupus erythematosus presenting as polymyalgia rheumatica in the elderly. *Annals of Rheumatic Diseases*, **45**, 641–4.

Lakhanpal, S., Bunch, T.W., Ilstrup, D.M., and Melton, L.J. (1986). Polymyositis–dermatomyositis and malignant lesions: does an association exist? *Mayo Clinic Proceedings*, **61**, 645–53.

Lally, E.V., Jimenez, S.A., and Kaplan, S.R. (1988). Progressive systemic sclerosis: mode of presentation, rapidly progressive disease course, and mortality based on an analysis of 91 patients. *Seminars in Arthritis and Rheumatism*, **18**, 1–13.

Maddison, P.J. (1987). Systemic lupus erythematosus in the elderly. *Journal of Rheumatology*, **14** (suppl. 13), 182–7.

Manchul, L.A. *et al.* (1985). The frequency of malignant neoplasms in patients with polymyositis-dermatomyositis. A controlled study. *Archives of Internal Medicine*, **145**, 1835–9.

McKendry, R.J. (1987). Influence of age at onset on the duration of treatment in idiopathic adult polymyositis and dermatomyositis. *Archives of Internal Medicine*, **147**, 1989–91.

Plotz, P.H., Dalakas, M., Leff, R.L., Love, L.A., Miller, F.W., and Cronin, M.E. (1989). Current concepts in the idiopathic inflammatory myopathies: polymyositis, dermatomyositis, and related disorders. *Annals of Internal Medicine*, **111**, 143–57.

Sheibani, K., Burke, J.S., Swartz, W.G., Nademanee, A., and Winberg, C.D. (1988). Monocytoid B-cell lymphoma. Clinicopathologic study of 21 cases of a unique type of low-grade lymphoma. *Cancer*, **62**, 1531–8.

Strickland, R.W., Tesar, J.T., Berne, B.H., Hobbs, B.R., Lewis, D.M., and Welton, R.C. (1987). The frequency of sicca syndrome in an elderly female population. *Journal of Rheumatology*, **14**, 766–71.

Uchino, M., Araki, S., Yoshida, O., Uekawa, K., and Nagata, J. (1985). High single-dose alternate day corticosteroid regimens in treatment of polymyositis. *Journal of Neurology*, **232**, 175–8.

Ward, M.M. and Polisson, R.P. (1989). A meta-analysis of the clinical manifestations of older-onset systemic lupus erythematosus. *Arthritis and Rheumatism*, **32**, 1226–32.

13.8 Back pain in elderly people

SUZANNE VAN H. SAUTER AND NORTIN M. HADLER

In a working-age population, back pain cannot be ignored. It is accompanied by significant costs for medical care and other treatments, as well as loss of productivity and disability. The impact of back pain in an older population has seldom been considered. Certainly, the risk of certain neoplastic or other diseases that may have back symptoms increases with age. None the less the frequency of acute back pain does not steadily increase in older populations (Cypress 1983; Reisbord and Greenland 1985). Rather, such complaints reflect a steady, substantial morbidity. Data from national surveys in the United States on visits to physicians by ambulatory patients aged 75 years and older show that back pain is the third most frequently mentioned symptom, and the most commonly mentioned musculoskeletal symptom.

Approximately 20 to 40 per cent of persons aged 65 and older report low back pain in the preceding year (Biering-Sorensen 1984; Lavsky-Shulan *et al.* 1985; Bergstrom *et al.* 1986) and the majority report that their first symptoms occurred before the age of 65. On the whole, the prevalence of back complaints declines slightly with age after 65 (Cypress 1983; Reisbord and Greenland 1985). Reported rates are higher for women after the sixth decade. Fifteen to 40 per cent of those with a history of back pain in the previous year report specific functional limitations as a direct result of their back complaints (Lavsky-Shulan *et al.* 1985). Five per cent claimed to have been in hospital at least once because of low back pain, and 1 per cent reported surgery to the lower back. Little is known about chronicity of symptoms, probability of recurrence, degree of functional impairment, and specific aetiopathology. Chronicity, severity, and disability may worsen with age, although the results of various studies are inconsistent (Frymoyer and Cats-Baril 1987).

Defining the cause of low back pain is difficult. In people aged 18 to 60 years the estimated probability of preceding, specific, treatable disease in those presenting with acute pain is 0.2 per cent (Liang and Komaroff 1982). Among elderly people, the proportion of back complaints with definable causes is certainly higher, but to what extent is uncertain. Many of the disorders associated with back pain are 'degenerative' in type and correlate with advancing age. Likewise, the incidence of other systemic diseases that may manifest as back pain increases with age. In a retrospective review of 259 consecutive patients aged over 50 years presenting for orthopaedic evaluation of low back or sciatic pain, there was a 7 per cent prevalence of malignancy, with no case of primary or secondary neoplasms in a comparison group aged under 50 years. Systemic diseases associated with back pain were more than twice as common in the older group of patients (Fernbach *et al.* 1976). Severe, acute, low back pain that causes the patient to writhe may be referred pain from a dissecting, expanding, or leaking aortic aneurysm. Unrelenting pain, pain that causes the patient to pace, or pain at rest, all suggest that there may be cancer or an infective process such as discitis, osteomyelitis, or epidural abscess. Evolving neurological deficits are most ominous. Possible causes include epidural abscess or haemorrhage, intra- or extradural tumour. Acute pain in patients with known osteoporosis or cancer might indicate pathological fracture (Frymoyer 1988). Thus uncommon causes of back pain must be excluded in an older population.

DEGENERATIVE DISEASES OF THE SPINE

Degenerative changes involve both major articulations of the spine, the facet joints and the intervertebral synchondroses. With advancing age these changes are universal. Structural disintegration of the disc begins as early as the third decade, with fibrillation of the nucleus pulposus, degeneration of cartilage in the end-plates, subchondral microfractures, and microscopic ruptures in collagenous tissue on both ventral and dorsal aspects of the annulus fibrosus. Biochemical changes with ageing include changes in the proteoglycans, decreased water content of the nucleus pulposus, and resultant thinning of the discs. Even though the majority of the population has radiographic evidence of degenerative disease of the spine by the age of 65, there is poor correlation with symptoms. Degenerative changes in the facet joints (osteoarthritis) and disc (spondylosis) are radiographic findings not diagnostic entities. In a radiographic survey of a random sample from a mixed urban–rural population aged 55 to 64 years, severity of osteoarthrosis was not associated with increasing lower back pain. The overall rate of complaints was approximately 50 per cent (Lawrence *et al.* 1966). Degenerative disease of the discs is more prevalent than lumbar osteoarthrosis; two-thirds of the men and nearly one-half of the women have radiographic evidence of disc degeneration. Less than one-third has osteophytosis involving facet joints of grade 2 or higher.

Diffuse idiopathic skeletal hyperostosis is another degenerative process that involves the enthesopathic attachments of the spine. Although patients with diffuse idiopathic skeletal hyperostosis may complain of diffuse lower back pain and stiffness, the symptoms are non-specific. Spinal mobility in such patients generally is little impaired and painless as compared with ankylosing spondylitis. Radiographically, spinal involvement is characterized by exuberant flowing osteophytes extending over four continuous vertebral bodies, relatively normal end-plates, and no osseous fusion of the sacroiliac joints. The last helps to distinguish this disorder from the spondyloarthropathies.

Spondylolisthesis is the forward subluxation of one vertebral body over the segment below it. In elderly people the most common cause is degenerative changes in the discs and facet joints, though late sequelae of trauma and congenital abnormalities must be considered. With forward movement of the vertebral body, the lamina moves anteriorly, causing narrowing of the spinal canal, and is one cause of spinal stenosis. Spondylolisthesis, like most other degenerative processes of the spine, is more often a radiographic finding than a clinical problem. Spondylolisthesis is found in about 5 per cent of men and 10 per cent of women. Back pain associated with spondylolisthesis is non-specific. Neurological symptoms, which may occur after irritation of the nerve roots or entrapment are uncommon, particularly at ages under 50 (most often of the L4 root).

Lumbar spinal stenosis

Spinal stenosis occurs with derangement of the structures that comprise the spinal canal: posterior bulging or herniation of the disc, hypertrophy/osteophytosis of bodies or facet joints, or anterior subluxation of the vertebral bodies themselves. Stenosis may occur centrally, compromising the cauda equina, or more laterally involving particular roots, or both. Spinal stenosis occurs more commonly in association with degenerative changes with advancing age, though there may be an underlying congenital or developmental abnormality. One or more levels may be involved. The mean age at diagnosis in one retrospective series was 63 years (Hall *et al.* 1985). Computed tomography (**CT**) and magnetic resonance imaging (**MRI**) demonstrate spinal stenosis as an incidental finding in many asymptomatic older patients (Hudgins 1983). Symptomatic spinal stenosis typically presents in an older man as mild to moderate, chronic, low back pain, along with pain, numbness, burning, or weakness in one or both buttocks, thighs, or legs, suggesting a sciatic radiculopathy. It is precipitated by standing or walking, and almost always alleviated by sitting or otherwise flexing forward at the waist. A few patients complain of persistent leg pain or discomfort at night. The discomfort of spinal stenosis may mimic the symptoms of vascular disease in being provoked by exercise and eased by rest—'neurogenic claudication'. Where present, distinguishing features include paraesthesiae or no pain relief by standing still, or exacerbating by prolonged standing. Low back pain that is relieved by forward flexion is suggestive of central spinal stenosis, where the cauda equina is compressed by a bulging or herniated disc anteriorly or redundant or hypertrophied ligamentum flavum posteriorly. With compression of the central canal, sphincter and sexual function may be affected (Hall 1983).

Physical findings are less sensitive than the history in suggesting the diagnosis. Measurable weakness in the lower extremities was found in only slightly more than one-third of patients (Hall *et al.* 1985). Other findings in a minority of patients included diminished or absent deep tendon flexes in the knee (18 per cent) or ankle (43 per cent), and a positive straight-leg raising test (10 per cent). Although decreased movement of the lumbar spine was seen in a majority of patients, this finding is also frequent in the asymptomatic elderly population (Einkauf *et al.* 1987). Other physical findings included increased pain on extension of the lumbar spine, weakness of the extensor hallucis longus, and sensory changes in a dermatomal distribution. {Occasionally, ankle jerks present at rest may be reduced or abolished by walking.} Plain films of the spine are not diagnostic for lumbar stenosis, but are indicated to exclude other conditions. CT and MRI allow detailed visualization of the bony margins of the spinal canal; myelography may still be done for preoperative evaluation (Frymoyer 1988).

Procedures to confirm a diagnosis of lumbar stenosis are warranted if there is progressive neurological impairment or if there is clinical suspicion of systemic disease, especially malignancy. Otherwise, a trial of conservative therapy is appropriate as up to 90 per cent improve without surgery (Swezey 1988). Pain and limitation in function are severe enough in only 10 per cent of documented cases for surgical decompression to be considered, though surgery was more frequently done in earlier series (Hall *et al.* 1985).

Herniated intervertebral disc

Extrusion of disc material into the spinal canal is responsible for only a small percentage of acute back complaints. Moreover, it is a frequent finding in asymptomatic individuals. Acute, symptomatic disc herniation is most frequent in persons under the age of 50, but does occur rarely in later years (Maistrelli *et al.* 1987). The approach in elderly patients is similar to that in the younger adult.

Symptoms from disc herniation may be acute, subacute, or minimal. The 'classic' patient can often pinpoint the moment of onset, with a sensation of 'tearing' or 'giving' in the low back. The pathoanatomical correlative of this event is the extrusion of disc material posteriorly or posterolaterally, causing distension of the posterior longitudinal ligament. In the acute state, pain is constant and exacerbated by movement, especially by activities that increase intradiscal pressure, such as sitting, flexing the spine, squatting, coughing, sneezing, or straining. Symptoms and physical findings correlate with the degree and the level of nerve-root compression. Sciatica, most commonly results from lesions at the levels of L4–5 or L5–S1. Less often there is herniation at the levels of L3–4 or L2–L3; older patients have increased risk for herniation at the higher levels. The most consistent physical finding in disc herniation is a positive straight-leg raising test.

The sensitivity of CT in diagnosing a herniated nucleus is approximately 92 per cent, but specificity is much more problematical (Schellinger 1984). The decision to use CT in the evaluation of suspected disc herniation depends on the level of diagnostic uncertainty or the progression of symptoms. Unless a cauda equina syndrome is present or there is reason to suspect systemic disease, confirmation of extruded disc material by CT will not alter conservative management in the acute phase.

A rare but potentially catastrophic complication, cauda equina syndrome, occurs in 1 to 16 per cent of cases of lumbar disc herniation (Kostuik *et al.* 1986). Defined by the triad of acute bladder or bowel sphincter dysfunction, saddle anaesthesia, and motor deficits in the lower extremities, cauda equina syndrome is the one unequivocal indication for surgery in back pain, and requires immediate referral to minimize permanent neurological sequelae. In the absence of these specific neurological findings, surgical intervention for herniated disc should not be entertained unless 2 months of conservative therapy have failed to provide relief of pain.

METABOLIC BONE DISEASE

Osteoporosis is the metabolic bone disorder of greatest clinical and economic significance in the elderly population (see Chapter 14.1). The major complications of osteoporosis are fractures of the vertebrae or hips. An estimated half-million new cases of vertebral compression fractures occur each year. In one series the majority of white women had at least one partial compression fracture of the spine by the age of 80. An unknown percentage of vertebral compression fractures are asymptomatic and found incidentally during radiographic screening for other reasons. Symptomatic compression fractures tend to present suddenly with moderate to severe pain, which occurs in association with minor trauma or effort such as lifting, coughing, sneezing, bending, or stretching. Point tenderness over the involved vertebra is common, as is paravertebral muscle spasm. The pain tends to be exacerbated by activity and relieved by immobility. Neurological problems are rare but can occur. As with other bone fractures, pain is a reflection of the disrupted periosteum. Such pain generally resolves in 1 to 3 months. So treatment is conservative.

The association of back pain with osteoporosis in the absence of radiographically evident compression fracture is problematical. It is widely suspected that subclinical fractures may be responsible for some episodes of acute backache. Crush fractures, in which the entire anteroposterior diameter of the vertebral body is reduced, are more specific for osteoporosis than wedge fractures, which involve the anterior portion, as wedge fractures can occur even when bone mass is normal (Jensen *et al.* 1982; Pitt 1983).

Paget's disease (osteitis deformans) occurs in 3 per cent or more of some radiographic or autopsy series of elderly subjects (see Chapter 14.3). The vertebrae are commonly involved, especially L2–4. Paget's disease is frequently asymptomatic, and may be only an incidental radiographic finding. Among symptomatic patients, back pain is common. Moreover, osteoarthritis, vertebral compression fractures, and spinal stenosis also are more common in these patients; therefore, to attribute back symptoms directly to pagetic involvement of the spine is a considerable diagnostic challenge.

Radiographically, pagetic lesions appear initially as osteolytic and later as osteoblastic lesions. The osteoblastic lesions produce a homogeneously increased density, a 'picture frame' appearance. With only one bone involved, biochemical tests are unlikely to be useful. In polyostotic disease, serum alkaline phosphatase and urinary hydroxyproline are elevated, but by this stage the diagnosis is usually evident by plain-film radiography. Treatment is not required for asymptomatic disease. Bone pain due to biochemically active

Paget's disease warrants a trial of treatment (see Chapter 14.3).

INFLAMMATORY DISEASE ASSOCIATED WITH LOW BACK PAIN

In the elderly population, late spinal complications of ankylosing spondylitis are important in a small percentage of patients with the disease who go on to develop the classic fused or bamboo spine. The patient with long-standing spondylitis who develops a sudden increase in back pain or spinal mobility should be promptly evaluated for possible spinal fracture (Hunter and Dubo 1978). Orthopaedic or neurosurgical consultation is indicated, and progressive neurological deficit is an indication for surgery. Spondylodiscitis is another rare complication of late ankylosing spondylitis. Associated with new onset of back pain, this destructive lesion involving the disc–bone margin or vertebral end-plate results from instability of an isolated spinal segment. Routine radiographs of the spine may not detect the lesion, but a bone scan will be abnormal with increased uptake of the radionuclide. Thus the new onset of back pain in the older spondylitic patient, especially if worsened by activity or movement, needs detailed evaluation.

Rheumatoid arthritis involving the hips may present as recurrent low back pain. The prevalence of lumbosacral subluxation and narrowing of disc spaces may be higher in rheumatoid patients than in others. Rheumatoid arthritis may uncommonly affect the lumbar spine but is reported to cause recurrent and more severe attacks of back pain. Histological changes characteristic of rheumatoid arthritis have been described in the synovium of lumbar facet joints from rheumatoid patients. Erosions of the surface of apophyseal joints, radiographically similar to rheumatoid erosions in other joints, have also been described.

BACK PAIN ASSOCIATED WITH SYSTEMIC, NON-RHEUMATIC DISEASE

Neoplastic disease

Metastatic lesions are the most common neoplasm of bone and 70 per cent of these occur in the axial skeleton. Primary tumours in the breast, lung, and prostate account for the majority of metastases to bone; thyroid, kidney, and adrenal gland contribute more rarely. Approximately half of the patients with one of the three main cancers will develop metastatic lesions of the skeleton. Back pain in the context of known or suspected malignancy should always provoke diagnostic intervention (Francis and Hutter 1963). Cancer is more prevalent in the older population and is a significant cause of back pain of new onset. Except when precipitated by acute collapse of an involved vertebral body, metastatic back pain tends to develop insidiously and progressively over weeks or even months. It may be local or radicular, and is classically worse at night and with recumbency, unlike most other types of back pain.

Although plain films of the spine are insensitive for detecting early metastatic lesions, they are the usual first step in evaluation. Demineralization of 30 to 50 per cent must occur before differences in bone density can be detected on routine radiography of the spine, unless the cortical margin is disrupted. Single isolated bony lesions are much less common than multiple metastases. Bone scans are a more sensitive screen in patients with the suspicion of malignant disease as the cause of back pain. The erythrocyte sedimentation rate is likewise highly sensitive, though even more non-specific for malignancy. Alkaline phosphatase, serum calcium, urinary hydroxyproline, or acid phosphatase may be abnormal, but have lower sensitivity and are not as useful as a means of general screening.

Infectious spondylitis

Pyogenic and tuberculous spondylitis are uncommon diseases but have a predilection for elderly subjects. Pyogenic osteomyelitis of the spine is rare, as are epidural abscesses, but both present as back pain. *Staphylococcus aureus* and Gram-negative bacilli are the most common organisms responsible for infection; the latter are probably related to urinary tract infections and genito-urinary manipulations {perhaps, more rarely, to cryptic cholecystitis}. Back pain is usually insidious and may be accompanied by fever, weight loss, night sweats, and an elevated sedimentation rate. Initial plain radiographs are not necessarily revealing though there may be erosion of subchondral end-plates. Progressive changes over 6 to 8 weeks include bony destruction and then reactive bone formation of end-plates adjacent to the infected disc. Multiple blood cultures and closed needle-biopsy of the infected disc are usually required to make the diagnosis.

Back pain developing weeks to several months after back surgery can be the result of infection in the disc space. Radiographs and a bone scan are difficult to interpret because of the surgery, so needle-biopsy aspiration of the disc space is required to make the diagnosis.

The spine is the most common site for skeletal tuberculosis. Lumbar segments, especially L1, are most frequently involved. The most common site of primary infection is the anterior portion of the disc spaces with secondary destruction of the vertebral bodies. Onset is usually insidious. Backache and constitutional symptoms may be present. Fever, leucocytosis, and a high sedimentation rate are frequently but not always present. Radiographic changes of osteopenia and erosion are not seen within the first 4 to 6 weeks of infection. A bone scan is highly sensitive for early discitis, but not specific, so biopsy and cultures are necessary (Gandy and Payne 1986).

HISTORY AND PHYSICAL EXAMINATION

A comprehensive history and general examination are essential in distinguishing acute from chronic, systemic from regional, and treatable from untreatable disorders (Svara and Hadler 1988).

Observe the patient entering the examination room. Abnormalities of gait and posture offer valuable insights into the level of functional impairment. Loss of lumbar lordosis, as well as mild to moderate degrees of kyphoscoliosis or pelvic tilt, are common in elderly people. These are all non-specific, and do not correlate with symptoms or disease. Decreased range of motion in the lumbar spine is not a useful discriminator in elderly patients with back pain. Studies of spinal mobility in normal subjects demonstrate a progressive decrease in mean range of motion in all planes, especially extension, with ageing (Einkauf *et al.* 1987). The number of abnormalities in physical findings increases with age in symptomatic as well as asymptomatic patients. Eighty per cent of those aged over 65 years will have one or more abnormality on examination of the back (Mayne 1977).

Findings on examination must be correlated with history. A couple of tender vertebral bodies with a history of nocturnal back pain should lead to considerations of metastatic cancer. Certainly constitutional symptoms, such as fever, weight loss, neurological complaints, all require a full evaluation.

The femoral stretch and passive straight-leg raising tests screen for upper lumbar (L2–4) and lower lumber or sacral nerve compression, respectively. In the femoral stretch test, with the patient prone and knees extended (straight), lift one foot at a time to flex the knee. Pain that radiates into the anterior thigh suggests nerve involvement. Absence or asymmetry of the patellar reflex (primarily L4) further helps to localize the lesion. Sciatic nerve (L4–S3) irritation can be elicited with the passive straight-leg raising test. The patient lies supine, one hip and knee flexed and the foot flat on the table, while the examiner lifts the contralateral leg with the knee in full extension. Pain that radiates distal to the knee constitutes a positive response. Symptoms that occur only beyond 60° of flexion, or that are felt only in the back or buttocks, are not specific for sciatic irritation. The ankle jerk (S1) may be compromised by sciatic radiculopathy; however, absence of this stretch flex is common in asymptomatic elderly persons. Therefore, this finding in isolation has no diagnostic significance (Svara 1988) {if bilateral}.

Sensory function also can be tested, although reliability is often problematical, especially in the presence of diabetes mellitus. Check pin-prick responses over the anterior lower extremities from L3 to S1 dermatomes. Reduced sensation between the upper buttocks (S3, 4, 5), or saddle anaesthesia, is a significant finding specific for cauda equina syndrome.

Careful assessment of hip joints is essential, as hip problems can mimic low back pain. Pain elicited on hip flexion or internal rotation suggests primary disease of the hip rather than a spinal source of pain. Referred back pain of visceral origin is also more common in elderly patients, so the abdomen must be examined for pulsation or mass. The association of metastatic spinal lesions with breast and prostate cancer necessitates careful attention to these aspects of the examination. Diminished sphincter tone and loss of perianal sensation suggest cauda equina compression. Presence of a palpable mass in the rectal vault or of occult blood in the stool requires further investigation of the gastrointestinal system. Finally, evaluate the lower extremities for arterial insufficiency: absent or asymmetrical distal pulses or mottling or decreased temperature in the feet suggest a possible vascular aetiology for low back or buttock pain (Svara 1988).

Radiographs should be obtained when the history and physical examination give reasons to suspect serious disease. The most common findings on films of the spine in elderly subjects are degenerative disc disease and facet-joint osteophytes, which do not correlate well with back pain or dictate specific therapy. Radiographic findings that do indicate the need for specific intervention include those of vertebral metastases and discitis. In one large retrospective review of 68 000 conventional examinations of the lumbosacral spine, clinically unsuspected positive findings were found in only 1 of 2500 (Nachemson 1976). The probability of identifying a treatable problem by radiography increases with careful selection of patients for study. Reasons for selection include fever, elevated sedimentation rate, a history of malignancy, and a history of trauma or risk factors for osteoporosis (Deyo 1986). When radiographs are indicated, anteroposterior and lateral lumbosacral spine films are sufficient in most cases. Although controversy persists, most evidence suggests that oblique films add little to the diagnostic sensitivity (Hall 1980; Rhea *et al.* 1980; Gehweiler *et al.* 1983).

Plain-film radiography is less sensitive than radionuclide imaging for detection of early destructive lesions, including spinal metastases and certain other tumours, inflammatory processes, and traumatic lesions (Schutte and Park 1983). The combination of normal screening laboratory studies and normal radiographs significantly reduces the probability of clinically important findings on the bone scan. In one retrospective review of 38 patients with back pain and normal radiographs who underwent bone scintigraphy, 31 had negative scans. The remaining seven, all of whom had laboratory abnormalities, fever, or both, had some systemic disease. Of those seven, all had a high sedimentation rate and at least one other laboratory abnormality (Schutte and Park 1983).

MANAGEMENT

Having eliminated, or significantly reduced, the probability of serious or systemic disease in the patient with back pain, what remains is the broad range of regional musculoskeletal disorders. Although the physician is reassured, the patient with the painful back may not be. The majority of patients who seek medical attention for low back pain are concerned about a serious illness (Williams and Hadler 1981). Assurances that the problem is not systemic, progressive, or destructive, and will 'most likely' resolve with time may be inadequate. The physician must listen for the unspoken fears and concerns of the patient, offer reasonable explanations, and discuss the expectation of spontaneous recovery. Loss of personal independence may be a great concern to the elderly patient with a backache who is anxious that he/she may have a progressive illness (Williams and Hadler 1983).

There is no single strategy for managing acute regional back pain in elderly subjects. Therefore, in the absence of evidence of systemic disease or neurological emergency, supportive care and close follow-up should be the guiding principles. Focusing on the patient's experience of back pain rather than attempting to fit it into the construct of a disease is a more palliative approach to the medical management of regional, low back pain (Svara 1988). For many patients, explanation and reassurance will allow them to tolerate a degree of discomfort, comfortable at least that they are doing themselves no harm and that there is no serious threat to their autonomy or functional status.

Standard medical recommendations for mechanical back pain include bed rest and analgesia. These are only appropriate in elderly people if certain qualifications are taken into account. Although there is no evidence that bed rest of any duration will speed healing and/or resolution of mechanical back pain, the rationale for prescribing a period recumbency is based on clinical experience that mechanical back pain is relieved by inactivity, and on biomechanical evidence that intradiscal pressure is lowest in recumbency. Two days of prescribed bed rest are as effective as seven for patients with acute mechanical back pain (Deyo *et al.* 1986). Patients should be educated as to postural modifications that will reduce biomechanical stress. Lying supine, sitting erect, or standing all exert less compressive force on the spine than being propped up on pillows or slouching in a chair, both of which should be avoided (Hadler 1984).

Analgesics, anti-inflammatory medications, sedatives--hypnotics, and alleged muscle relaxants are all used in low back pain. Aside from issues of compliance and cost, none of these has been demonstrated conclusively to be more effective than salicylate for younger adult patients (Hadler, 1985). For the elderly patient, there are additional caveats regarding pharmacological management. First, salicylates or non-steroidal anti-inflammatory medications should be used with added caution in all elderly patients, particularly those with a history of peptic ulcer, upper gastrointestinal bleeding, or renal insufficiency. Second, the sedative and other central nervous effects of benzodiazepines, tricyclic muscle relaxants, and narcotic analgesics may be enhanced in elderly subjects. Narcotics and sedatives should be used with care and under supervision and have been associated with falls and hip fractures. The patient with vertebral compression fracture may require narcotic analgesia during the acute phase. For many other patients, plain paracetamol (acetaminophen) may be adequate to alleviate symptoms pending the spontaneous resolution of the acute problem. Third, the administration of narcotic analgesics should be avoided, except in special circumstances, such as malignant disease that cannot be adequately palliated by other means. A clouded sensorium and obstipation is not a reasonable trade-off for a self-limiting backache.

PHYSICAL THERAPY

Although exercise programmes are widely used and advocated for prevention of recurrent acute low back pain, there is little scientific information about their effectiveness. Aerobic exercise may be effective in the returning to work of some persons chronically out of work because of low back pain. No studies have shown efficacy in an elderly population, nor have traction and manipulation been shown to be effective (Quinet and Hadler 1979).

OTHER THERAPIES

Chemonucleolysis is a marginal therapy at best and has no role in elderly patients. Epidural and intrafacet injections of steroids are no better than placebo. Rigorous evaluations of braces, corsets, and manipulation in elderly people are lacking but these appear to be of little benefit (Quinet and Hadler 1979).

SUMMARY

As a population ages, the cohort will have a higher incidence of serious disease causing severe backache. For anyone whose history and physical examination suggests that the cause is infectious, or malignant, or traumatic with neurological involvement, evaluation should be prompt and thorough. Even so, the majority of persons will experience a regional backache for which no specific cause can be identified. For these persons, education and conservative management appear to be the treatment of choice. Most people with backache will recover and can be reassured that the problem is painful but benign.

References

Bergstrom, G. *et al.* (1986). Joint disorders at age 70, 75 and 79 years—a cross-sectional comparison. *British Journal of Rheumatology*, **25**, 333–41.

Biering-Sørensen, F. (1984). A one-year prospective study of low back trouble in a general population. *Danish Medical Bulletin*, **31**, 362–75.

Cypress, B.K. (1983). Characteristics of physician visits for back symptoms: a national perspective. *American Journal of Public Health*, **73**, 389–95.

Deyo, R.A. (1986). Early diagnostic evaluation of low back pain. *Journal of General Internal Medicine*, **1**, 328–38.

Deyo, R.A., Diehl, A.K., and Rosenthal, M. (1986). How many days of bed rest for acute low back pain? A randomized clinical trial. *New England Journal of Medicine*, **315**, 1064–70.

Einkauf, D.K. *et al.* (1987). Changes in spinal mobility with increasing age in women. *Physical Therapy*, **67**, 370–5.

Fernbach, J.C., Langer, F., and Gross, A.E. (1976). The significance of low back pain in older adults. *Canadian Medical Association Journal*, **115**, 898–900.

Francis, K.C. and Hutter, R.V.P. (1963). Neoplasms of the spine in the aged. *Clinical Orthopaedics*, **26**, 54–66.

Frymoyer, J.W. (1988). Back pain and sciatica. *New England Journal of Medicine*, **318**, 291–300.

Frymoyer, J.W. and Cats-Baril, W. (1987). Predictors of low back pain disability, *Clinical Orthopaedics and Related Research*, **221**, 89–98.

Gandy, S. and Payne, R. (1986). Back pain in the elderly: updated diagnosis and management. *Geriatrics*, **41**, 59–62, 67–74.

Gehweiler, J.A., Jr. and Daffner, R.H. (1983). Low back pain: the controversy of radiologic evaluation. *American Journal of Roentgenology*, **140**, 109–12.

Hadler, N.M. (1984). Diagnosis and treatment of backache. In *Medical management of the regional musculoskeletal disorders*, (ed. N.M. Hadler), p. 3. Grune and Stratton, Orlando CA.

Hall, F.M. (1980). Back pain and the radiologist. *Radiology*, **137**, 861–3.

Hall, H. (1983). Examination of the patient with low back pain. *Bulletin of Rheumatic Diseases*, **33**, 1–8.

Hall, S. *et al.* (1985). Lumbar spinal stenosis. Clinical features, diagnostic procedures, and results of surgical treatment in 68 patients. *Annals of Internal Medicine*, **103**, 271–5.

Hudgins, W.R. (1983). Computer aided diagnosis of lumbar disk herniation. *Spine*, **8**, 604–15.

Hunter, T. and Dubo, H. (1978). Spinal fractures complicating ankylosing spondylitis. *Annals of Internal Medicine*, **88**, 546–9.

Jensen, G.F. *et al.* (1982). Epidemiology of postmenopausal spine and long bone fractures. A unifying approach to postmenopausal osteoporosis. *Clinical Orthopaedics*, **166**, 75–81.

Kostuik, J.P. *et al.* (1986). Cauda equina syndrome and lumbar disc herniation. *Journal of Bone and Joint Surgery* (A), **68**, 386–91.

Lavsky-Shulan, M. *et al.* (1985). Prevalence and functional correlates of low back pain in the elderly: the Iowa 65+ rural health study. *Journal of the American Geriatrics Society*, **33**, 23–8.

Lawrence, J.S. Bremner, J.M., and Bier, F. (1986). Osteo-arthrosis

prevalence in the population and relationship between symptoms and X-ray changes. *Annals of Rheumatic Diseases*, **25**, 1–24.
Liang, M. and Komaroff, A.L. (1982). Roentgenograms in primary care patients with acute low back pain: a cost-effectiveness analysis. *Archives of Internal Medicine*, **142**, 1108–12.
Maistrelli, G.L., Vaughan, P.A., Evans, D.C., and Barrington, T.W. (1987). Lumbar disc herniation in the elderly. *Spine*, **12**, 63–6.
Mayne, J.G. (1977). Symposium on arthritis in older persons. Section III. Osteoarthritis and back pain. Examination of the back in the geriatric patient. *Journal of the American Geriatrics Society*, **25**, 59–61.
Nachemson, A.L. (1976). The lumbar spine. An orthopaedic challenge. *Spine*, **1**, 59–71.
Pitt, M. (1983). Osteopenic bone disease. *Orthopedic Clinics of North America*, **14**, 65–80.
Quinet, R.J. and Hadler, N.M. (1979). Diagnosis and treatment of backache. *Seminars in Arthritis and Rheumatism*, **8**, 261–87.
Reisbord, L.S. and Greenland, S. (1985). Factors associated with self-reported back-pain prevalence. *Journal of Chronic Diseases*, **38**, 691–702.
Rhea, J.T., *et al.* (1980). The oblique view: an unnecessary component of the initial adult lumbar spine examination. *Radiology*, **134**, 45–7.
Schellinger, D. (1984). The low back pain syndrome: diagnostic impact of high-resolution CT. *Medical Clinics of North America*, **68**, 1631–46.
Schutte, H.E. and Park, W.M. (1983). The diagnostic value of bone scintigraphy in patients with low back pain. *Skeletal Radiology*, **10**, 1–4.
Svara, C.J. and Hadler, N.M. (1988). Back pain. *Clinics in Geriatric Medicine*, **4**, 395–410.
Swezey, R.L. (1988). Low back pain in the elderly. *Geriatrics*, **43**, 39–44.
Williams, M.E. and Hadler, N.M. (1981). Musculoskeletal components of decrepitude. *Seminars in Arthritis and Rheumatism*, **11**, 284–87.
Williams, M.E. and Hadler. N.M. (1983). Sounding board. The illness as the focus of geriatric medicine. *New England Journal of Medicine*, **308**, 1357–60.

13.9 Foot problems in older persons

ARTHUR E. HELFLAND

Foot problems are one of the most distressing and disabling afflictions associated with old age. A key factor in an older person's ability to remain socially active is the ability to walk effectively and comfortably.

CHANGES IN THE FOOT IN RELATION TO AGE

There are many factors that contribute to the development of foot problems in older persons. Of primary concern are age-associated changes and the presence of multiple chronic diseases. Other significant factors include the amount of walking, limitations in activity due to other causes, the length of any preceding time in hospital or other institutional care, the degree of social isolation, emotional adjustments to disease and life in general, and the effects of multiple medications. Optimal management of foot problems in the elderly patient requires a comprehensive team approach.

The skin is usually one of the first structures to show changes. There is usually a loss of hair below the knee joint and on the dorsum of the foot, with atrophy of the skin giving a parchment-like appearance. Brownish pigmentation is common and related to the deposition of haemosiderin. Hyperkeratosis may be present due to dysfunctional keratinization, as a residuum of pressure and atrophy of the subcutaneous soft tissue, and as a space replacement as the body adjusts to the changing stresses placed on the foot.

The toe-nails undergo degeneration and may have thickening and longitudinal ridging related to repetitive, small injuries, disease and nutritional impairment. Deformities of the toe-nails may become more pronounced and complicated by changes in the periungual nail folds, as onychophosis (hyperkeratosis) and tinea unguium (onychomycosis) are common and usually chronic in older persons. Onychomycosis is a constant focus of reinfection.

There is commonly progressive loss of muscle mass and atrophy of tissue due to decreased activity, which increases the susceptibility of the foot to injury; thus even minor trauma can result in a fracture or rupture of ligaments or tendons.

Many chronic diseases also produce degenerative changes in the foot. Examples include diabetes mellitus with neuropathy or angiopathy, rheumatoid arthritis, osteoarthritis, and various neuromuscular diseases.

Peripheral arterial insufficiency produces trophic changes, rest pain, intermittent claudication, coldness, and colour changes such as rubor and cyanosis. The presence of haemorrhage subungually or beneath hyperkeratotic tissue, particularly in the diabetic patient, demonstrates angiopathy, which can be an early finding in diabetic patients who are also developing retinal or renal disease. These changes in the feet predispose the patient to infection, necrosis, and tissue loss if care, which must include education, is not provided early and in a comprehensive and active manner.

The feet are fairly rigid structures that must carry heavy physical work-loads, both static and dynamic, throughout life. The foot itself is in the shape of a modified rectangle and bears static forces in a triangular pattern. The transmission of weight and force starts at heel strike, proceeds anteriorly along the lateral segment of the foot, medially across the metatarsal heads, to the first metatarsal segment for the push-off phase of the gait cycle. The varied activities of life produce many variations in both the structure and function of the foot, as the body adapts to the stresses placed upon it. Flat and hard floor surfaces force the foot to absorb shock, creating prolonged periods of micro-trauma, with the risk of inflammatory changes in bone and soft tissues.

Treatment should be directed towards eliminating the causes of trauma and redistributing the weight to non-painful areas of the foot. The primary goals of treatment are to relieve pain, to restore maximum function, and to maintain it.

Foot problems of a mechanical nature typically arise from the interaction between normal morphological variations, the capacity to adapt to stress, and the stressors acting on the foot itself. Morphological variations may be intrinsic to the foot itself, or extrinsic, such as arise from changes in the relationship of the legs, knees, thigh, and hip, and back to the foot. The common intrinsic changes include elements such as a hypermobile segment, pes cavus, atrophy of the interosseal muscles producing digiti flexus (hammer toes), and the development of hallux valgus, the so-called bunion deformity.

The foot must be considered as a total 'end-organ' of locomotion; changes in any part of the body that affect the foot are usually the result of a chain of events of a chronic and progressive nature. Once a link in the chain breaks, every effort must be made to prevent further damage and minimize the associated complications.

IDENTIFYING FOOT PROBLEMS

The management of foot problems in the elderly patient requires the early recognition of the aetiologic factors, the complaints and symptoms of the patients, physical signs, and the clinical manifestations of disease and degenerative changes, which may be local in origin or a complication of a systemic disease.

Degenerative joint diseases in the elderly foot, as a result of acute or repeated and chronic microtrauma, strain, obesity, and/or osteoporosis, may present with a variety of changes as listed in Table 1. These changes may result in pain, limited motion, and impaired walking. Gout may present as episodes of acute gouty arthritis and result in chronic manifestations of painful joints, stiffness, and joint deformity. Chronic rheumatoid arthritis may lead to the manifestations listed in Table 2. Early-morning stiffness, pain, fibrosis, ankylosis, contracture, deformity and impaired walking are characteristic effects in the foot.

Arterial insufficiency in the legs and feet may lead to complaints of fatigue, pain at rest, coldness, burning, colour changes, ulceration, cramps, oedema, claudication, and repeated infections. Primary physical findings include diminished or absent pedal pulses, with similar changes in the entire extremity, depending upon the location and extent of the occlusion. The hypertensive patient may have pulsations that are a false reflection of the vascular supply. The foot usually shows colour changes, i.e. pallor, rubor, or cyanosis; it is usually cool, with the skin dry and atrophic. Superficial infections are common and may be painful.

Table 1 Presentations of joint diseases in the elderly foot

Plantar fasciitis
Spur formation
Periostitis
Decalcification
Stress fractures
Tendonitis
Tenosynovitis
Residual deformities
Pes planus
Pes cavus
Hallux valgus
Digiti flexus (hammer toes)
Rotational digital deformities
Joint swelling

Table 2 Manifestations of chronic rheumatoid arthritis in the elderly foot

Hallux limitus
Hallux rigidus
Hallux valgus
Hallux abducto valgus
Cystic erosion
Sesamoid erosion
Metatarsophalangeal subluxation
Metatarsophalangeal dislocation
Interphalangeal subluxation
Interphalangeal dislocation
Digiti flexus (hammer toes)
Ankylosis (fused joints)
Phalangeal reabsorption
Talo-navicular arthritis
Extensor tenosynovitis
Rheumatoid nodules
Bowstring extensor tendons
Tension displacement
Ganglions
Rigid pronation
Subcalcaneal bursitis
Retrocalcaneal bursitis
Retroachillal bursitis
Calcaneal ossifying enthesopathy (spur)
Prolapsed metatarsal heads
Atrophy of the plantar fat pad
Digiti quinti varus
Tailor's bunion

The toe-nails may be discoloured, loose and thickened. Oedema may be present. Vascular blebs are common in the later stages of local occlusion. When necrosis and gangrene are present, the results of further diagnostic studies, in particular Doppler measurements, can guide the need for surgical intervention to revascularize the extremity. Preventive measures and early intervention are essential to avoid loss of tissue and threats to continued, independent functioning.

The elderly diabetic patient presents a special problem in relation to foot health. It has been projected that half to three-quarters of all amputations in the diabetic could be prevented by early intervention when disease is found, by improved health education, and by periodic, prophylactic evaluation. In addition to the risks of vascular insufficiency discussed above, diabetic neuropathy very commonly affects the feet. The patient typically presents with paraesthesiae, impaired sensation of pain and temperature, motor weakness, diminished or lost achilles and patellar reflexes, decreased vibration sense, a loss of proprioception, xerotic changes, anhidrosis, neurotrophic arthropapthy, atrophy, ulcers, and sometimes a difference in size between the feet. There is increased risk of infection, necrosis, and gangrene. There is often a loss of the plantar metatarsal fat-pad, which predisposes the patient to ulceration over any bony deformities.

Hyperkeratotic lesions form a focus for ulceration because of increased pressure on the soft tissues with an associated, localized avascularity from direct pressure and counter-pressure. Tendon contractures and claw toes (hammer toes) are common. When ulceration is present, the base is usually covered by keratosis, which retards and may prevent healing.

Radiographic findings in the foot of elderly diabetic

patients commonly include thin bone trabeculae, decalcification, joint changes, osteophytic formation, osteolysis, osteoporosis, and deformities.

General principles of management of foot problems in older diabetics include a decrease of local trauma by the use of orthotics, shoe modifications, and specialized foot wear, and efforts to maximize weight diffusion and dispersal. Physiotherapy and exercise can be used to improve the vascular supply to the foot, after adequate evaluation by Doppler and similar techniques. Ulcers usually require debridement and treatment with antiseptic compresses, and possibly topical antibiotics {although this is controversial, especially in institutional settings where it may enhance the spread of multiply resistant bacteria}. Radiographs should be obtained to detect infection of bones. Aggressive systemic antibiotic therapy and admission to hospital may be necessary to prevent amputation.

Asymptomatic elderly diabetic patients should be evaluated at least twice a year to identify problems at their earliest development. Patients with foot conditions requiring primary management should be seen every 4 to 8 weeks, depending upon the extent of complications. A multidisciplinary team approach is essential, including the primary physician, podiatrist (chiropodist), nurse, and, if needed, the vascular surgeon, social worker, pharmacist, and physiotherapist.

PRIMARY FOOT PROBLEMS AND THEIR MANAGEMENT

Changes in the nails, which commonly occur in older persons, may be the result of new or long-standing disease, injury, or functional abnormality.

Onychia is an inflammation involving the posterior nail wall and bed. It is usually precipitated by local trauma or pressure, or may be a complication of systemic diseases such as diabetes and an early sign of developing infection. Mild erythema, swelling, and pain are the most common findings. Treatment should be directed towards removing all pressure from the area, and tepid saline or povidone-iodine compresses used for 15 min, three times daily. Lambs' wool, tube foam, or modification to the shoe should also be considered to reduce pressure on the toe and nail. If the onychia is not treated early, paronychia may develop, with infection and abscess of the posterior nail wall. This may progress proximally and deeper structures may become involved; the potential for osteomyelitis, necrosis, and gangrene is obviously greater in the presence of diabetes and vascular insufficiency. Management includes establishing drainage, microbiological culture and appropriate choice of antibiotics, radiographs if there is suspicion of bony involvement, compresses, and early follow-up.

Deformities of the toe-nails are the result of repetitive minor injury, degenerative changes, or disease. For example, continued rubbing of the toe-nails against the shoe is enough to produce change. The initial thickening is termed onychauxis. Onychorrhexis with accentuation of normal ridging, trophic changes, and longitudinal striations may reflect a systemic disease and/or nutritional imbalance.

When debridement is not effected periodically, the nail elongates, continues to thicken, and is deformed by pressure from the shoe. Onychogryphosis or 'ram's horn nail' is usually complicated by fungal infection. The resultant disability and pain can prevent the older patient from wearing shoes. In addition, traumatic avulsion of the nail is more frequent with this condition. The exaggerated curvature of the nail may even penetrate the skin, with resultant infection and ulceration. Management should be directed towards periodic debridement of the nail both in length and thickness, with as little trauma as possible. The extent of onycholysis (freeing of the nail from the anterior edge) and onychoschezia (splitting) help to determine the amount of debridement. With the excess pressure of deformity, the nail grooves tend to become keratotic. When this occurs, debridement and the use of mild keratolytics and emollients, such as Keralyt gel and 10–20 per cent urea preparations, can be pursued at home. With onycholysis, subungual debris and keratosis develop, which increases discomfort and may generate pain. However, with degenerative changes, the sense of pain may be lost, which tends to defer care by the patient until some complicating condition occurs.

The most common non-bacterial infection of the toe-nails is onychomycosis, a chronic and communicable infection with Candida. It may cause distal subungual, white superficial, proximal subungual, or total dystrophic changes. In the superficial variety, the changes appear on the superior surface of the toe-nail and generally do not invade the deeper structures. In both the distal, proximal, and total dystrophic manifestations, the nail bed as well as the nail plate is infected. There is usually some degree of onycholysis and subungual keratosis.

The elderly patient usually presents with a chronic infection, involving one or more of the nail plates. The entire thickness of the plates is usually involved, with resultant hypertrophy and deformity. Pain is usually not a significant feature but can be caused by external pressure when the deformity becomes excessive. Avulsion of the nail, subungual haemorrhage, a foul, musty odour, and degeneration of the nail plate are common findings.

The condition is chronic and once the matrix of the nail is involved and hypertrophy and deformity have occurred the changes cannot be reversed, but the disease can be controlled by sustained treatment. Periodic debridement, aided by 20–40 per cent urea and the use of a topical fungicide in an alcoholic base to allow penetration provide the best approaches to management.

Ingrown toe-nails in older persons are usually the end result of deformity and inappropriate self-care. When the nail penetrates the skin, an abscess and infection result. If not managed early, periungual granulation tissue may form, which complicates treatment. In the early stage a segment of the nail can easily be removed with an English nail splitter and an onychotome and drainage established; povidone-iodine or saline compresses are then used for 15 min, three times a day, and antibiotics given as needed. Measures should be taken to prevent recurrence. Excision, fulgeration, desiccation, or caustics, such as silver nitrate (75 per cent), and astringents are used to reduce any granulation tissues present.

In all instances of ingrown toe-nails removal of the penetrating nail is primary. Partial excision of the nail plate and matrix can be achieved under regional anaesthesia by chemical cautery of the matrix area with phenol CP, for example.

With this procedure, postoperative management includes isopropyl alcohol compresses and topical steroid solutions, three times per day until healing is complete.

With ageing, we also find changes in the nail plate, which when viewed from distally, appears C shaped. When present, the pressure of the nail plate on the nail bed and folds produces onychophosis (hyperkeratosis in the nail folds) and discomfort, with complaints similar to an ingrown toe-nail. The condition may precipitate pressure ulceration and infection. When this condition is severe, early and total removal of the nail plate and matrix should be considered to avoid complications.

Subungual heloma or corn, when present, is usually associated with a subungual exostosis, spur, or hypertrophy of the tufted end of the distal phalanx. Initial treatment consists of debridement and protection of the involved toe, as well as the use of a shoe with a high toe-box. Surgical excision of the osseous deformity may be required if the condition cannot be managed conservatively.

In all cases of suspected bone changes, radiographs properly positioned to isolate the area of disease are appropriate. Bilateral weight and non-weight bearing studies are usually indicated.

A common problem in the elderly patient is dryness of the skin (xerosis), due to lack of hydration and lubrication. There is usually some evidence of dysfunctional keratinization that can be associated with xerosis. Fissures develop as a result of dryness, and when present on the heel, with associated stress, may lead to ulceration. Initial management includes the use of an emollient after hydration of the skin; 20 per cent urea is helpful as a mild and safe keratolytic. A plastic or styrofoam heel cup can be of assistance in minimizing trauma to the heel, thus reducing the potential for complications.

Pruritus is a common complaint of older people and is usually more severe in the colder weather. It is related to dryness, scalyness, decreased skin secretions, dysfunctional keratinization, environmental changes and defatting of the skin, which may be made worse by soaking the feet in hot water. Excoriations from scratching may be found on examination. Chronic tinea, allergic, neurogenic, and/or emotional dermatoses should be considered in the differential diagnosis and treated accordingly. Management consists of hydration, lubrication, protection, topical steroids if indicated, and judicious use of antihistamines in minimal doses to control the itching. If excoriations are infected, antibiotic therapy may be needed.

Hyperhidrosis and bromidrosis may occur. If local, measures should be undertaken to control the excessive perspiration and odour. Hydrogen peroxide, isopropyl alcohol, and astringents may be used topically. Neomycin powder will help control odour by reducing the bacterial decomposition of perspiration. Changes in the type of footwear and stockings/socks should be considered. Particular care should be provided in colder climates, as dampness can predispose to the vasospastic effects of cold.

Contact dermatitis may be the result of reactions to chemicals used in shoe construction, footwear fabrics, or socks/stockings. Limited and usually bilateral distribution of the skin lesions are helpful clinical findings in diagnosis. Skin tests can be used to identify the primary irritant. The general principles of management include removing the primary irritant, mild wet dressings, and the use of topical steroids.

Stasis dermatitis is usually due to venous insufficiency and chronic ulceration. It is more common in patients with dependent oedema. Management locally consists of elevation, mild wet dressings, topical steroids, antibiotics as indicated, and the supportive measures needed in the management of venous disease.

Pyodermas and superficial bacterial infections should be managed locally in a similar manner, and when present with tinea, treatment should be directed towards both conditions. Tinea pedis in the elderly patient is often an extension of onychomycosis, which serves as a focus of infection. With the chronic keratotic type common in elderly people it is more frequent in hot weather. Poor foot hygiene in many older patients and their inability to see their feet may motivate them to seek care only when the condition becomes severe. A wide variety of topical medications can usually control this condition. Solutions and/or creams (water washable or miscible) should be used if the patient is unable to remove an ointment base easily.

Other common dermatological manifestions that may be seen in elderly patients are those associated with atopic dermatitis, nummular eczema, neurodermatitis, and psoriasis. Simple or haemorrhagic bullae are due to trauma and friction from the shoe or related to systemic diseases such as diabetes mellitus. Management is directed towards eliminating pressure, with protection and drainage when appropriate. Supportive dressings and shoe modifications should be used as appropriate. Gait changes in older persons can magnify minor incompatibilities between foot and shoe, and result in local lesions.

Elderly patients who do not use footwear at home because they cannot bend expose their feet to foreign bodies and injury. For example, animal hairs will appear as keratotic plugs and require debridement and/or excision to relieve pain.

Common complaints of many older patients are the many forms of hyperkeratotic lesions, such as callus and corn, and their varieties, such as hard, soft, vascular, neurofibrous, seed, and subungual. The factors that help create these problems are associated with stress—compressive, tensile and/or shearing. The loss of soft tissue as a part of atrophy of the plantar fat pad increases pain and limits walking. Contractures, gait changes, deformities, loss of skin tone and elasticity, and the residuum of arthritis are all additional factors that need to be considered in management. Incongruity between the foot type (inflare, straight, or outflare) and the shoe is another factor to be considered.

Management and treatment should be directed towards the functional needs of the patients for daily living. Approaches to be considered include debridement, padding, emollients, modifications to and changes of shoes, orthoses, and surgical intervention. Materials to provide soft tissue replacement, weight dispersal, and weight diffusion are also indicated. It is also important to recognize that keratotic lesions of long standing are hyperplastic and hypertrophic, and that even when weight bearing is removed, they tend to persist. In a sense, hyperkeratotic lesions are a form of body protection to pressure and are symptoms of an abnormal state. If allowed to persist, enlarge, and condense, they become primary irritants. With pressure such as weight bearing and walking, they produce local avascularity, which can precipitate ulcerations. Pressure ulcers in the foot usually begin with subkeratotic haemorrhage. Once debrided and managed properly, they usually heal, but may be recurrent, unless adequate measures are taken to reduce the pressure to the localized areas of ulceration. Even with all measures, the problem may persist because of residual deformity and systemic diseases, such as diabetes mellitus. Thus management and monitoring are similar to any other chronic condition in older patients.

There is a variety of residual foot deformities that can be present in multiple combinations in older persons. Treatment can be non-surgical or surgical. Age itself should not be a determining factor in considering surgery. What is important is to determine what can be done to maintain quality of life for the patient. Consideration must also be given to the patient's ability to adapt to change, for to have an anatomically corrected joint in a patient who still cannot walk without pain defeats the object of treatment.

Conservative treatments include shoe-last changes, shoe modifications, orthoses, digital braces, physical medicine, exercises, and mild analgesics for pain. The residuum of these deformities can produce inflammatory changes such as periarthritis, bursitis, myositis, synovitis, neuritis, tendonitis, sesamoiditis, and plantar myofasciitis, for example, which need to be managed medically, physically, and mechanically to keep the patient walking and pain-free.

Fractures of the foot and toes may be the result of direct trauma and/or stress related to bone loss. Most uncomplicated and closed fractures that are in good position can be managed with a surgical shoe and supportive dressings, as a long as the joints distally and proximally are immobilized. Silicone moulds can be used for fractured digits and their use maintains position through healing, allowing the patients to maintain proper hygiene.

Shoe modifications that can be considered for older patients include mild calcaneal wedges to limit motion and alter gait; metatarsal bars to transfer weight; Thomas' heels to increase calcaneal support; long shoe counters to increase mid-foot support and control foot directions; heel flares to add stability; shank fillers or wedges to produce a total weight-bearing surface; steel plates to restrict motion; and rocker bars to prevent flexion and extension. Additional internal modifications include longitudinal arch pads, wedges, bars, lifts, and tongue or bite pads. The available orthoses include the rigid, semirigid, and flexible varieties, using materials such as plastic, leather, laminates, polyurethane, sponge or foam rubber, Kores, felt, latex, wood flour, Plastazote, Aliplast, and silicone to provide support, reduce pressure, and provide for weight diffusion and weight dispersion.

CONCLUDING REMARKS

Much of the ability to remain mobile in later life is related to foot health. Foot health education, such as developed by the United States Department of Health and Human Services (1970), the American Diabetes Association, and the American Podiatric Medical Association, are accessible to patients and professionals, and should be made available as part of education to elderly patients.

Bibliography

Helfand, A.E. (ed.) (1981). *Clinical podogeriatrics*. Williams and Wilkins, Baltimore.

Helfand, A.E. and Bruno, J. (ed.) (1984). Rehabilitation of the foot. In *Clinics in podiatry*, Vol. 1, No. 2. Saunders, Philadelphia.

US Department of Health and Human Services (1970). *Feet first*, NIH Publication No. 0-388-126. USGPO, Washington DC.

United States Government NIH Publication. (1982). *Final report of the 1981 White House conference on aging*, Vol. 3. Superintendent of Documents, Washington DC.

Williams, T.F., (ed.) (1984) *Rehabilitation in the aging*. Raven Press, New York.

Wilson, L.B., Simson, S.P., and Baxter, C. R. (ed.) (1984). *Handbook of geriatric emergency care*. University Park Press, Baltimore MD.

Witkowski, J.A. (ed.) (1983). Diseases of the lower extremities. In *Clinics in dermatology*, Vol. 1, No. 1. Lippincott, Philadelphia.

Yale, I. (ed.) (1980). *Podiatric medicine*, (2nd edn). Williams and Wilkins, Baltimore.

SECTION 14
Disorders of the skeleton

14.1 Involutional osteoporosis

B. LAWRENCE RIGGS AND L. JOSEPH MELTON III

GENERAL CONSIDERATIONS

By any standard, osteoporosis is one of the most important diseases encountered in geriatric practice. Everyone loses bone mass with ageing. During their life, women lose about 60 per cent of their trabecular bone and about 35 per cent of their cortical bone. Men are also affected but lose two-thirds of these amounts. Bone densities in the vertebrae and proximal femur are below the fracture threshold in half of the women at age 65 and in virtually all of them by age 80. As a consequence, fractures are common in the elderly population. It is estimated that 1.5 million fractures attributable to osteoporosis occur annually in the United States. The sites of these fractures are the vertebrae for 650000, the hip for 250000, the distal forearm (Colles' fracture) for 200000, and other skeletal sites for 400000. A third of the women over age 65 will have vertebral fractures, and the life-time risk of hip fracture in white women (15 per cent) is as great as the risks of breast, endometrial, and ovarian cancer combined. The life-time risk of hip fracture in men (5 per cent) is as great as the risk of prostate cancer.

The economic consequences of these fractures are staggering. The direct and indirect costs of osteoporosis in the United States have been estimated to be $7 to $10 billion annually. Much of this expense related to hip fracture. This catastrophic type of fracture is fatal in 12 to 20 per cent of cases. Half of the survivors are unable to walk unassisted, and a quarter are confined to long-term care in a nursing home.

{In the United Kingdom 25 to 30 per cent of victims of hip fracture are dead within 6 months and more than half of survivors suffer pain or leg swelling. Perhaps because of more readily available domiciliary services, fewer survivors go into nursing or residential homes than in the United States}.

PATHOPHYSIOLOGY

There are many causes of bone loss and fractures. In the elderly population, an increased propensity to fall and decreased ability to break the impact of a fall contribute to the risk, but the major cause of fractures due to osteoporosis is the increased fragility that results from bone loss. The bone that remains is relatively normal histologically but there is too little of it, so that fractures occur after minimal trauma. These factors are summarized in the model shown in Fig. 1 and are discussed below.

Supported by Research Grant AG-04875 from the National Institute on Aging, National Institutes of Health.

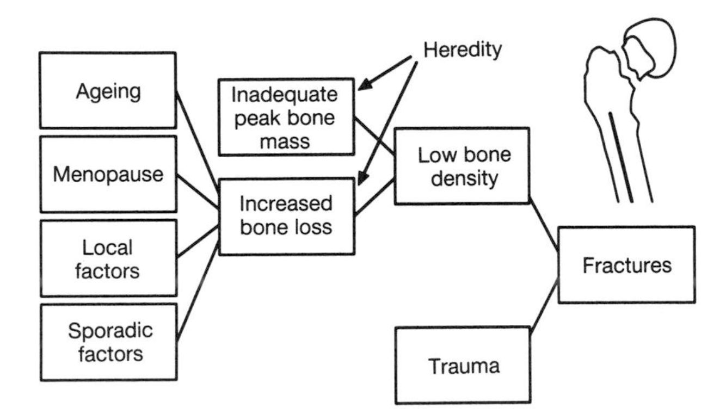

Fig. 1 Model of causes of bone loss and fracture (from Ciba Foundation Symposium with permission).

MAXIMAL BONE MASS

Maximal bone mass is reached at about the age of 25 to 30 years. After a period of stability at this level, the bone begins to lose mass and this loss continues throughout the remainder of life. The smaller the bone mass accumulated during skeletal growth and consolidation, the greater is the risk of fractures later in life as this bone loss ensues. Differences in initial bone density may explain, in part, the observed racial and sexual differences in the incidence of osteoporosis. White women have lighter skeletons than white men or black women; black men have the heaviest skeletons. This rank order corresponds to that for the occurrence of osteoporosis. Oriental women appear to be intermediate in both their initial skeletal density and their risk for osteoporosis. In all three races, however, women have lower peak bone mass than men do. Thus, women have an increased predisposition to osteoporosis, not only because of menopause-related bone loss (see below) but also because of lower initial bone density.

The factors that influence peak bone mass have not been well studied. Because of the larger requirement for calcium during bone growth and consolidation, it is reasonable to assume that inadequate calcium consumption during this period would decrease the accumulation of skeletal mass, and the limited data available support this idea. However, studies in twins and in osteoporotic and non-osteoporotic mother–daughter pairs have also shown that bone density has significant genetic determinants, which may explain the tendency to familial aggregation of osteoporotic patients.

BONE LOSS

Two distinct phases of bone loss can be recognized: a protracted, slow phase in both sexes that results in similar losses of both cortical and trabecular bone, and a transient accelerated phase after the menopause in women that results in a disproportionately greater loss of trabecular bone (Fig. 2).

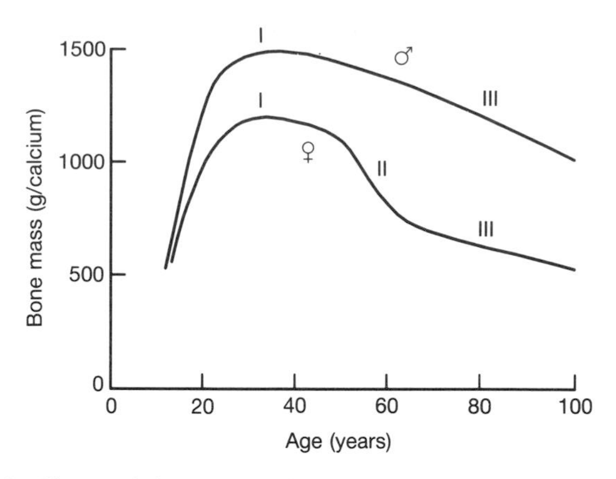

Fig. 2 Changes in bone mass with ageing in men and women, showing the phases (see text for details).

Cortical bone predominates in the shafts of the long bones whereas trabecular bone is concentrated in the vertebrae, pelvis, and other flat bones, and in the ends of long bones. Trabecular bone has a greater surface area than does cortical bone and is metabolically more active.

There is still some uncertainty about how much bone is lost over life with each of these phases. A reasonable estimate, however, is that the slow phase produces a loss of about 25 per cent from the cortical compartment and about 35 per cent from the trabecular compartment in both sexes. During the accelerated phase, postmenopausal women lose an additional 10 per cent from the cortical compartment and 25 per cent from the trabecular compartment.

Age-associated factors

Age is the most important empirical determinant of bone mass. Factors associated with ageing account for the slow phase of bone loss, which begins in cortical bone by the age of 40 and continues throughout life at a rate of about 0.6 per cent/year. In trabecular bone, it may begin even earlier and continues at a rate of about 0.7 per cent/year. This loss probably reflects the aggregate effects of several processes.

Although impaired osteoblast function could be caused by senescence, fracture healing is not delayed in older people. This suggests that ageing does not impair the response of osteoblasts to appropriate stimuli. More likely, the regulation of osteoblast activity is impaired by altered production of systemic or local growth factors. For example, circulating levels of growth hormone and of insulin-like growth factor I (**IGF-I**, somatomedin C), which mediates the effect of growth hormone on bone and cartilage, both decrease by almost one-half with ageing. Moreoever, bone cells synthesize and respond to a number of regulators of cell proliferation—especially IGF-I, IGF-II, and transforming growth factor—and the production of or response to one or more of these growth factors could decrease with ageing.

The dietary requirement for calcium is relatively high because of the obligatory faecal and urinary loss of 150 to 250 mg/day. When the amount of calcium absorbed from the diet is insufficient to offset these losses, calcium must be withdrawn from bone, which contains 99 per cent of the total body stores. However, in both sexes, active intestinal transport of calcium decreases with ageing, particularly after the age of 70. The main regulator of calcium absorption is the biologically active metabolite of vitamin D, 1,25-dihydroxyvitamin D ($1,25(OH)_2D$). There is evidence of age-associated decreases both in the responsiveness of the intestine to $1,25(OH)_2D$ and in the production of this metabolite.

The available data are conflicting, but the largest study of the effect of age on the serum $1,25(OH)_2D$ concentration showed that it increases until the age of 65 years and then decreases. This increase in the serum concentration of $1,25(OH)_2D$ at a time when calcium absorption decreases suggests that there is a primary impairment in intestinal responsiveness to the action of $1,25(OH)_2D$. In experimental animals, the concentrations of $1,25(OH)_2D$ receptors in both intestine and bone cells decrease with ageing.

Indirect studies in elderly women and direct studies in elderly rats have also shown impairment in the activity of the renal enzyme 25-hydroxyvitamin D (**25(OH)D**) 1α-hydroxylase, which is responsible for the conversion of 25(OH)D to $1,25(OH)_2D$. Thus, loss of enzymatic activity associated with the well-established loss of renal parenchyma with ageing probably decreases the rate of production of $1,25(OH)_2D$ in elderly people, especially in those whose glomerular filtration rate is less than 40 ml/min. This late-occurring defect aggravates the impairment in calcium absorption caused by intestinal resistance to the action of $1,25(OH)_2D$.

If the impairment in calcium absorption is physiologically significant, it should induce a compensatory secondary hyperparathyroidism. That the concentration of parathyroid hormone in the circulation increases with ageing has now been convincingly demonstrated by radioimmunoassay for specific NH_2 terminals, immunoradiometric assay for intact parathyroid hormone, and assay for biological activity. Increased secretion of parathyroid hormone with ageing would lead to increased bone turnover and, because of the age-related defect in osteoblast function, this would lead to increased bone loss.

Menopause

Women who have undergone oophorectomy in young adulthood have lower bone density in later life than non-oophorectomized women of the same age. Surgical menopause accelerates bone loss, and oestrogen replacement prevents or slows this loss in both the appendicular and the axial skeletons. Epidemiological studies have shown that postmenopausal administration of oestrogen decreases the occurrence of vertebral and hip fractures by about half. Thus, oestrogen deficiency at the menopause is an important cause of bone loss and subsequent fractures. Men do not undergo the equivalent of menopause, but gonadal function does decrease in some elderly men, and overt male hypogonadism is often associated with vertebral fractures.

The accelerated phase of postmenopausal bone loss in women decreases exponentially after the menopause and becomes asymptotic with the slow phase after 5 to 10 years. It is associated with a high rate of bone turnover— more osteoclasts are present and each of them creates a deeper resorption cavity. As assessed by studies of radiocalcium kinetics, there is an increase in bone accretion but an even greater increase in resorption. The primary effect of oestrogen is to decrease bone resorption. Although it was formerly believed that the action of oestrogen on bone was indirect, it now has been conclusively demonstrated that normal human bone cells contain sex steroid receptors and respond directly to treatment with these steroids.

Sporadic factors

Certain diseases, surgical procedures, and medications may be associated with the development of osteoporosis. Bone loss resulting from the presence of these factors is additive to the age-related, slow loss that occurs universally, and to the accelerated, postmenopausal phase. In about 20 per cent of women and about 40 per cent of men who present with vertebral or hip fractures, one or more of these sporadic factors can be identified (Table 1).

Table 1 Causes of secondary osteoporosis (modified from Smith 1988)

Immobility: general or local (e.g., in hemiplegic limb)
Endocrine
Hypogonadism
Hypercorticism—spontaneous or iatrogenic
Thyrotoxicosis—spontaneous or iatrogenic
Hypopituitarism
Chromosomal—Turner's syndrome
Other
Rheumatoid arthritis
Heparin
Cytotoxic agents
Vitamin C deficiency
Inherited
Osteogenesis imperfecta
Homocystinuria
With osteomalacia
Coeliac disease
Postgastrectomy
Renal osteodystrophy
Liver disease

In addition, a sedentary life-style, cigarette smoking, and excessive ethanol intake (more than two drinks daily), and a very low intake of calcium (<500 mg per day), may contribute to bone loss.

TRAUMA

Because the risk of fracture increases as the bone mass decreases with ageing, elderly persons with decreased bone density are uniquely subject to fractures from moderate trauma of the sort that rarely causes injury in young people. The most common cause of fractures among elderly persons is a 'simple' fall from a standing height or less, although a few hip fractures may be spontaneous and vertebral fractures frequently result from lifting or straining. The risk of falling increases with ageing, and at least a third of elderly persons living in the community fall once or more a year (rates are even higher among institutionalized persons). The pathophysiology of and risk factors for falling are discussed in Chapter 18.16.

It is important to note that the majority of falls do not result in specific injury, although fear of a subsequent fall may become disabling. In most studies, a fifth to a third of the documented falls led to a medically attended injury {only 5 per cent in studies in the United Kingdom}; but only about 1 per cent of these falls resulted in hip fracture. Thus, additional risk factors may be needed to account for the full range of outcomes. These factors are even less completely known that the risk factors for falling *per se*, but it is generally believed that elderly persons are less able to break the impact of a fall because they are weaker and have longer reaction times.

CLINICAL HETEROGENEITY

The available evidence suggests that there are at least two distinct syndromes of involutional osteoporosis that differ in clinical presentation and in their relationship to menopause and age.

Type I (postmenopausal) osteoporosis

This syndrome characteristically affects women within 15 to 20 years after the menopause and results from an exaggeration of the postmenopausal phase of accelerated bone loss. Much less commonly, a form of osteoporosis that is clinically similar to that in postmenopausal women occurs in men of the same age. Type I osteoporosis is characterized by a disproportionate loss of trabecular bone (Fig. 3), which results

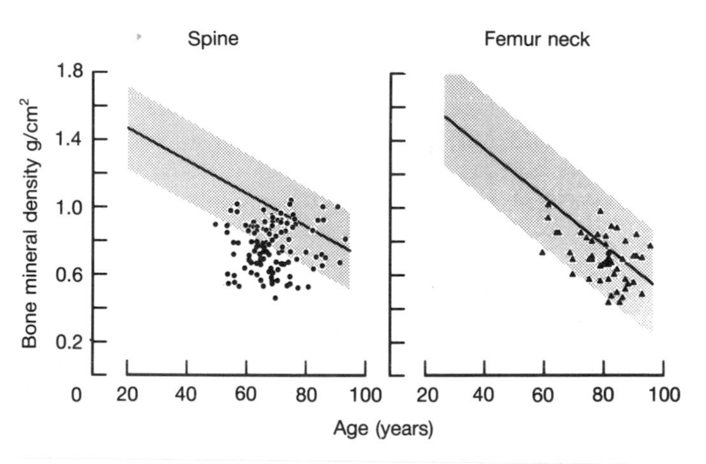

Fig. 3 Bone mineral density values for vertebrae (left) and femoral neck (right) plotted as function of age in 111 patients with vertebral fractures and 49 patients with hip fractures. The line represents the regression on age; stippled area represents the 90 per cent confidence limits for 166 normal women. (From Riggs and Melton 1986, with permission.)

in fractures of the vertebrae and distal forearm (Colles' fracture). The vertebral fractures usually are the 'crush' type associated with deformation and pain. Also, there is an increase in the incidence of ankle fractures and in the rate of tooth loss. The ankle and jaw bones contain large amounts of trabecular bone.

During the accelerated phase of bone loss, there is increased bone turnover. With its greater surface area, the trabecular bone has a rate of loss that is three times normal, but the rate of loss in cortical bone is only slightly greater than normal in patients with type I osteoporosis. The deeper resorption cavities lead to perforation of the trabecular plates and loss of structural trabeculae, which weakens the bones. The vertebrae are especially predisposed to acute collapse.

By the time of clinical presentation, however, tetracycline-labelled biopsies from the iliac crest may show bone turnover to be high, normal, or low. Thus, some osteoporotic patients may have reached a 'burned out' stage and will have little further loss of trabecular bone.

Only a relatively small proportion of postmenopausal women develop type I osteoporosis, although all postmenopausal women are relatively oestrogen deficient. The serum levels of sex steroids are similar in postmenopausal women with and without type I osteoporosis. Thus, one or more additional causal factors are present in postmenopausal osteoporotic women and interact with oestrogen deficiency to determine individual susceptibility. The possibilities include impaired coupling of formation to resorption, increased local production of interleukin 1 or another factor that increases bone resorption, prolongation of the phase of accelerated bone loss, low bone density at the inception of menopause, or some combination of these.

Type II (age-related) osteoporosis

This occurs in elderly men and women and results from continuation of the slow, age-associated phase of bone loss. It is manifested mainly by hip and vertebral fractures, although fractures of the proximal humerus, proximal tibia, and pelvis also are common. The trabecular thinning associated with the slow phase of the bone loss is responsible for gradual, usually painless, vertebral deformities. In contrast to the findings in type I osteoporosis, these deformities often are of the multiple wedge type, leading to dorsal kyphosis ('dowager's hump'). Bone densities in the proximal femur, vertebrae, and sites in the appendicular skeleton are usually in the lower part of the normal range adjusted for age and sex (Fig. 3). This suggests proportionate losses of cortical and trabecular bone and a rate of loss that is only slightly more than that in age-matched peers. The process causing type II osteoporosis, therefore, may affect virtually the entire population of ageing men and women and, as the slow phase of bone loss progresses, more and more of them will have bone densities below the fracture threshold.

The two most important age-associated factors are decreased osteoblast function and decreased calcium absorption leading to secondary hyperparathyroidism. The effects of all risk factors for bone loss encountered over a lifetime, however, are cumulative. Thus, the residual effects of menopausal bone loss many years before may help explain why elderly women have a two-fold greater incidence of hip fractures than elderly men, even though the rates of slow bone loss are similar.

DIAGNOSIS

General diagnostic procedures for metabolic bone disease and for osteoporosis are dealt with in the *Oxford Textbook of Medicine*.

The major diagnostic advance in recent years has been the development of practical methods for measuring bone density at the actual sites of the clinically important fractures. Three techniques are generally available—dual-photon absorptiometry (**DPA**), dual-energy X-ray absorptiometry (**DEXA**), and quantitative computed tomography (**QCT**) using single-energy scanning.

DPA uses transmission scanning with an isotope source (usually ^{153}Gd) that emits two energy peaks, thereby providing measurement of bone density that is independent of the thickness and composition of soft tissue. In the spine, the method has good reproducibility (2 to 3 per cent) and accuracy (4 per cent) with low radiation exposure ($<1\times10^{-5}$ Sv), but it measures the entire vertebra, which is only 70 per cent trabecular bone, and is sensitive to the effects of dystrophic calcification and vertebral compressions in the scanning area. When osteoarthritis of the spine or calcification of spinal ligaments and aorta are present, DPA may give misleading results for bone density of the lumbar spine in persons over the age of 70 years. DPA can also be used to assess bone density in the proximal femur.

DEXA is a major technological improvement. Its principle is similar to that of DPA but it uses an X-ray tube rather than an isotope to produce photons. The new technique has excellent reproducibility (1 to 2 per cent), low radiation exposure ($<3\times10^{-5}$ Sv), and a shorter scan time (5 to 10 min).

QCT can provide a measurement limited exclusively to the trabecular bone in the centre of the vertebral body. When it is carefully performed as a research procedure, a reproducibility of 2 to 3 per cent can be achieved but, for instruments in frequent use, the reproducibility may be as poor as 6 to 10 per cent. Accuracy is poor (12 to 30 per cent), mainly because of the confounding effect of marrow fat, and the radiation exposure is relatively high (220 to 1000×10^{-5} Sv depending on the quality of the scanner and the expertise of the operator). Nonetheless, QCT can provide useful information and, in contrast to DPA and DEXA, can be used to evaluate trabecular bone loss in the vertebrae of elderly patients. Methods such as single-photon absorptiometry that measure only the appendicular skeleton have limited usefulness in the evaluation of osteoporosis, a disease in which the major fractures occur in the axial skeleton.

TREATMENT

General measures

Acute back pain from vertebral collapse responds to analgesics, heat, and gentle massage to alleviate muscle spasm. Sometimes a brief period of bed rest is required. Chronic back pain is often caused by spinal deformity and thus is difficult to relieve completely. Instruction in posture and gait, and the institution of regular back-extension exercises to strengthen the paravertebral muscles usually are beneficial. Occasionally, an orthopaedic back brace is needed. All patients with osteoporosis should have a diet adequate in calcium, protein, and vitamins, should give up smoking and excessive alcohol consumption if possible, and should be reasonably active physically, avoid heavy lifting, and take precautions to prevent falls.

Drug therapy

The drugs currently approved by the United States Food and Drug Administration for the treatment of osteoporosis—calcium, oestrogen, and calcitonin—act by decreasing bone resorption. Calcium, which may act by decreasing secretion of parathyroid hormone, is safe, well-tolerated, and inexpensive. The effect of oestrogen on bone may be mediated by decreasing skeletal responsiveness to circulating parathyroid hormone. The mechanism of action of androgens and synthetic anabolic agents is probably similar to that of oestrogen, although some data suggest a weak stimulation of bone formation. Because of their masculinizing and hyperlipaemic effects, however, androgens and anabolic agents have a limited role in therapy.

Calcitonin acts directly on osteoclasts to decrease their activity. Calcium supplements must be given concurrently to prevent secondary hyperparathyroidism. The disadvantages of calcitonin include parenteral administration, relatively high cost, and the development of neutralizing antibodies in some patients (mainly in those receiving salmon calcitonin; these antibodies are less likely to occur when the more expensive human calcitonin is given). Clinical trials currently are evaluating calcitonin given by nasal insufflation. If this approach proves successful, calcitonin therapy will become more practical and will be better accepted by patients.

Although treatment with sodium fluoride for osteoporosis has been approved in some European countries, it is considered investigational in the United States and the United Kingdom. Fluoride stimulates bone formation, but calcium must be given concurrently to prevent impaired mineralization. Although this treatment increases vertebral bone mass, fluoridic bone is weaker than normal, and its efficacy in preventing fractures has not been convincingly demonstrated. {One recent study has suggested that fluoride therapy increases the incidence of fractures of long bones.}

Other promising regimens for osteoporosis that are currently being evaluated are several of the bisphosphonate drugs, the synthetic 1–34 analogue of parathyroid hormone, and phosphate–calcitonin combinations. {Cyclical intermittent therapy with oral etidronate (for example, 400mg daily for 2 weeks followed by 13 weeks without drug, repeated 10 times) has been reported to increase vertebral bone density and reduce vertebral fracture rate in two recent studies.}

Treatment of the individual patient

Therapy for patients with type I osteoporosis should be individualized. For those with mild disease, only calcium supplementation (1.0 to 1.5 g/day) need be used. For more severe disease, especially in women within 15 years after menopause, low-dose oestrogen therapy (such as cyclic doses of 0.625 mg of conjugated oestrogen per day, or 0.025 mg of ethynyl oestradiol per day) should be given. Because unopposed oestrogen therapy has a risk of endometrial hyperplasia, and therefore of carcinoma, concomitant progestogen therapy, 5 mg of medroxyprogesterone acetate per day, should be given during the last 14 days of each cycle. The side-effects of excessive hepatic production of coagulation factors, renin substrate, and bile cholesterol (which produce an increased risk of venous thrombosis, hypertension, and cholelithiasis, respectively) are due to exposure of the liver to a bolus of oestrogen in the first pass after oral administration. These problems can be decreased or eliminated by giving oestrogen via a transdermal patch (0.05 mg of 17-oestradiol per day) {or by subcutaneous implant}. If bone loss or fractures continue on hormone therapy, the dosages of both oestrogen and progesterone should be doubled.

Treatment with pharmacological dosages of vitamin D or active vitamin D metabolites should be reserved for patients with documented or suspected impairment in calcium absorption. Impaired calcium absorption can be inferred from a relatively low urinary calcium excretion (<75 mg/day), especially if this does not increase substantially during oral calcium supplementation.

Calcitonin is most likely to be effective in patients who have osteoporosis associated with high bone turnover. Without bone biopsy, this can be inferred from a serum phosphorus and urinary calcium that is in the upper portion of the normal range.

Patients with type II osteoporosis have already lost most of the bone that they ever will lose, and their bone density differs little from that of persons of the same age without fractures. The most important aspect of therapy is protection against falls. Potential household hazards such as freshly waxed floors, loose rugs, and raised edges of carpets tacked to the floors should be eliminated. The patient should be encouraged to wear simple Oxford-style shoes and to avoid using shoes with high heels. They should leave a light on at night so that they need not walk in the dark to the bathroom, and they should be extremely careful when walking on ice and snow.

Although the earlier idea that these elderly patients have uniformly low bone turnover appears to be incorrect, the efficacy of oestrogens or calcitonin in the treatment of type II osteoporosis has not been well documented. Drug treatment consists primarily of calcium (because the elderly have impaired intestinal absorption) and 1000 units of vitamin D per day (to correct any deficiencies that may be present). {Cyclical bisphosphonate therapy (see above) may also have a place in treatment.}

Prevention

In view of the magnitude of the problem, prevention is the only cost-effective approach. Indeed, if general steps are not taken now to retard bone loss, the already high costs of osteoporosis will double by the year 2025 because of continued ageing of the population.

Metabolic techniques for determining the level of calcium intake required to prevent negative calcium balance have given conflicting results, and population studies generally have not demonstrated a strong relationship between calcium intake and bone loss. A widely quoted study found that 1000 mg of calcium per day were required to maintain calcium balance in premenopausal women whereas 1500 mg/per day were required for postmenopausal women, but other investigators have found that balance could be achieved with smaller amounts. The recommended daily allowance has not been rigorously established for calcium, but currently it is set at 800 mg/day. This is near the mean dietary calcium intake for American men but in middle-aged and elderly women the intake is only 550 mg/day. Consequently, the dietary calcium level should be increased to at least the recommended daily allowance, 800 mg/day, for adults and to 1500 mg/day for adolescents and pregnant women. For those who have risk factors for osteoporosis, particularly if they are postmenopausal women, calcium supplements should be added.

While these recommendations seem prudent, their ultimate efficacy is uncertain. Three of six controlled trials of calcium supplementation showed significant short-term slowing of bone loss from the appendicular skeleton, but the reduction in bone loss was less than that achieved with oestrogen therapy. Two recent studies found that calcium supplementation at 1500 mg/day for women shortly after menopause had no effect on axial bone loss and only a weak effect on appendicular bone loss. Another study that measured rates of bone loss from the vertebrae and radius by serial measurements of bone density found no relationship between calcium intake and bone loss. By contrast, a prospective cohort study of a large number of elderly men and women living in a retirement community in the United States showed that those who had hip fractures within 12 years of initial evaluation had had a lower calcium intake according to the 24-h dietary recall interview made at the baseline examination.

Adequate intake of vitamin D is also important because there is an age-associated decrease in the ability of the skin to synthesize vitamin D and in the ability of the intestine to absorb it. Thus, housebound elderly persons are prone to vitamin D deficiency, particularly when they do not take supplementary vitamins and are consuming a diet with marginal vitamin D stores. In the United States, 10 per cent of an unselected group of elderly patients undergoing bone biopsy at the time of hip fracture were found to have subclinical osteomalacia. The problem appears to be even greater in the United Kingdom because of the higher latitude and the lack of vitamin D supplementation of dairy products.

Increased physical activity should be encouraged, and bone toxins, such as cigarettes and heavy alcohol consumption, should be eliminated. It is well established that skeletal stresses from weight-bearing and muscle contraction stimulate osteoblast function. Muscle mass and bone mass are directly related. Moreover, several recent prospective trials in postmenopausal women have shown that an experimental group enrolled in a regular exercise programme gained bone whereas sedentary controls lost it.

Because the postmenopausal phase of accelerated bone loss can be prevented by oestrogen replacement, the most effective way to decrease the incidence of future fractures is to give oestrogen replacement therapy to women at menopause. Women with premature surgical or natural menopause should have oestrogen replacement therapy at least until the usual age of menopause. At present, however, it probably is unwise to treat all perimenopausal women with oestrogen because of the side-effects and the high cost of follow-up. Thus, in order to select those perimenopausal women who would benefit most from treatment, measurements of the density of lumbar spine or hip should be made. Those women whose bone densities fall within the upper tertile are relatively protected against osteoporosis in the future and need not be treated with oestrogen. Those with densities in the lower tertile are at greater risk and, if there are no contraindications to oestrogen therapy, should be considered for treatment. Those with densities in the middle tertile should have the measurement repeated in 2 to 3 years and be considered for treatment then if there has been substantial additional bone loss. Persons with major risk factors for osteoporosis, such as those receiving glucocorticoid therapy, should have these measurements and be treated with an active drug if there is evidence of excessive bone loss. {Several studies have suggested that thiazide diuretics may increase bone density and reduce fracture rates. This knowledge may influence the choice of treatment of hypertension or mild heart failure in middle-aged and elderly women.}

CONCLUSIONS

Enormous strides have been made in recent years in understanding how bone loss develops, how bone turnover is regulated physiologically, and how changes in bone cell activity can be manipulated pharmacologically. Even greater progress is expected in the immediate future. Thus, there is every reason to be optimistic that this enormous public-health problem can begin to be brought under control within the coming decade.

Bibliography

Bikle, D.D., Genant, H.K., Cann, C., Recker, R.R., Halloran, B.P., and Strewler, G.J. (1985). Bone disease in alcohol abuse. *Annals of Internal Medicine*, **103**, 42–8.

Body, J., and Heath, H., III. (1983). Estimates of circulating monomeric calcitonin: physiological studies in normal and thyroidectomized man. *Journal of Clinical Endocrinology and Metabolism*, **57**, 897–903.

Consensus Conference (1984). Osteoporosis. *Journal of American Medical Association*, **252**, 799–802.

Cooper, C., Barker, D.J.P., and Wickham, C. (1988). Physical activity, muscle strength, and calcium intake in fracture of the proximal femur in Britain. *British Medical Journal*, **297**, 1443–6.

Cummings, S.R., Kelsey, J.L., Nevitt, M. C., and O'Dowd, K. (1985). Epidemiology of osteoporosis and osteoporotic fractures. *Epidemiological Review*, **7**, 178–208.

Cummings, S.R. *et al.* (1990). Appendicular density and age predict hip fracture in women. *Journal of the American Medical Association*, **263**, 655–8.

Eastell, R. and Riggs, B.L. (1987) New approaches to the treatment of osteoporosis. *Clinical Obstetrics and Gynecology*, **30**, 860–70.

Eastell, R., Delmas, P.D., Hodgson, S.F., Eriksen, E.F., Mann, K.G., and Riggs, B.L. (1988). Bone formation rate in older normal women: concurrent assessment with bone histomorphometry calcium kinetics, and biochemical markers. *Journal of Clinical Endocrinology and Metabolism*, **67**, 741–8.

Epstein, S., Bryce, G., Hinman, J.W., Miller, O.N., Riggs, B.L., and Johnston, C.C. (1986). The influence of age on bone mineral regulating hormones. *Bone*, **7**, 421–5.

Ettinger, B., Genant, H.K., and Cann, C.E. (1985). Long-term estrogen replacement therapy prevents bone loss and fractures. *Annals of Internal Medicine*, **102**, 319–24.

Forero, M.S. *et al.* (1987). Effect of age on circulating immunoreactive and bioactive parathyroid hormone levels in women. *Journal of Bone Mineral Research*, **2**, 363–6.

Heaney, R.P., Gallagher, J.C., Johnston, C.C., Neer, R., Parfitt, A.M., and Whedon, G.D. (1982). Calcium nutrition and bone health in the elderly. *American Journal of Clinical Nutrition*, **36**, 986–1013.

Holbrook, T.L., Barrett-Connor, E., and Wingard, D.L. (1988). Dietary calcium and risk of hip fracture: 14-year prospective population study. *Lancet*, **ii**, 1046–9.

Krolner, B., Toft, B., Pors Nielsen, S., and Tondevold, E. (1983). Physical exercise as prophylaxis against involutional vertebral bone loss: a controlled trial. *Clinical Science* **64**, 541–6.

Lindsay, R., Hart, D.M., Forrest, C., and Baird, C. (1980). Prevention of spinal osteoporosis in oophorectomized women. *Lancet* **ii**:;1151–4.

Mazess, R.B. (1983). The noninvasive measurement of skeletal mass. In *Bone and mineral research annual* 1, ed. W.A. Peck) pp. 223–279. Excerpta Medica, Amsterdam.

Meier, D.E., Orwoll, E.S. and Jones, J.M. (1984). Marked disparity between trabecular and cortical bone loss with age in healthy men: measurement by vertebral computed tomography and radial photon absorptiometry. *Annals of Internal Medicine*, **101**, 605–12.

Melton, L.J., III and Riggs, B.L. (1985). Risk factors for injury after a fall. *Clinical Geriatric Medicine*, **1**, 525–39.

Melton, L.J., III and Riggs, B.L. (ed.) (1988). *Osteoporosis: etiology, diagnosis, and management*, pp. 155–79. Raven Press, New York.

Peck, W.A. (1988). Research directions in osteoporosis. *American Journal of Medicine*, **84**, 275–82.

Raisz, L.G. (1988). Local and systemic factors in the pathogenesis of osteoporosis. *New England Journal of Medicine*, **318**, 818–28.

Reginster, J.Y. *et al.* (1987). 1-year controlled randomized trial of pre-

vention of early postmenopausal bone loss by intranasal calcitonin. *Lancet*, **ii**, 1481–3.
Riggs, B.L. and Melton, L.J. (1986). Medical progress series: involutional osteoporosis. *New England Journal of Medicine*, **314**, 1676–86.
Riggs, B.L., Seeman, E., Hodgson, S.F., Taves, D.R., and O'Fallon, W.M. (1982). Effect of the fluoride/calcium regimen on vertebral fracture occurrence in postmenopausal osteoporosis. *New England Journal of Medicine*, **306**, 446–50.
Riggs, B.L., Wahner, H.W., Melton, L.J., III, Richelson, L.S., Judd, H.L., and Offord, K.P. (1986). Rates of bone loss in the appendicular and axial skeletons of women: evidence of substantial vertebral bone loss before menopause. *Journal of Clinical Investigation*, **77**, 1487–91.
Riggs, B.L., Wahner, H.W., Melton, L.J., III, Richelson, L.S., Judd, H.L. and O'Fallon, W.M. (1987). Dietary calcium intake and rates of bone loss in women. *Journal of Clinical Investigation*, **80**, 979–82.
Riggs, B.L. *et al.* (1990). Effect of fluoride treatment on the fracture rate in postmenopausal women with osteoporosis. *New England Journal of Medicine*, **322**, 802–9.
Seeman, E., Wahner, H.W., Offord, K.P., Kumar, R., Johnson, W.J., and Riggs, B.L. (1982). Differential effects of endocrine dysfunction on the axial and the appendicular skeleton. *Journal of Clinical Investigation*, **69**, 1302–9.
Seeman, E., Melton, L.J., III., O'Fallon, W.M., and Riggs, B.L. (1983). Risk factors for spinal osteoporosis in men. *American Journal of Medicine*, **75**, 977–83.
Storm, T., Thamsborg, G., Steiniche, T., Genant, H.K., and Sørensen, O.H. (1990). Effect of intermittent cyclical etidronate therapy on bone mass and fracture rate in women with postmenopausal osteoporosis. *New England Journal of Medicine*, **322**, 1265–71.
Tinetti, M.E., Speechley, M., and Ginter, S.F. (1988). Risk factors for falls among elderly persons living in the community. *New England Journal of Medicine*, **319**, 1701–7.
Tsai, K., Heath, H., III, Kumar, R., and Riggs, B.L. (1984). Impaired vitamin D metabolism with aging in women: possible role in pathogenesis of senile osteoporosis. *Journal of Clinical Investigation*, **73**, 1668–72.
Watts, N.B. *et al.* (1990). Intermittent cyclical etidronate treatment of postmenopausal osteoporosis. *New England Journal of Medicine*, **323**, 73–9.
Weiss, N.S., Ure, C.L., Ballard, J.H., Williams, A.R., and Daling, J.R. (1980). Decreased risk of fractures of the hip and lower forearm with postmenopausal use of estrogen. *New England Journal of Medicine*, **303**, 1195–8.
Whyte, M.P., Bergfeld, M.A., Murphy, W.A., Avioli, L.V., and Teitelbaum, S.L. (1982). Postmenopausal osteoporosis: a heterogeneous disorder as assessed by histomorphometric analysis of iliac crest bone from untreated patients. *American Journal of Medicine*, **72**, 193–202.
Young, G., Marcus, R., Minkoff, J.R., Kim, L.Y., and Segre, G.V. (1987). Age-related rise in parathyroid hormone in man: the use of intact and midmolecule antisera to distinguish hormone secretion from retention. *Journal of Bone Mineral Research*, **2**, 367–74.

14.2 Osteomalacia

L. RAISZ AND CAROL C. PILBEAM

Osteomalacia is a metabolic bone disease characterized by impaired mineralization of newly formed bone matrix in mature lamellar bone. Although relatively uncommon compared with osteoporosis, it is important to recognize because treatment is often effective. Left untreated, osteomalacia can lead to weakened bones with fractures and skeletal deformities, and can cause bone pain and proximal muscle weakness. It is especially important to think of the diagnosis when treating elderly people who are at increased risk for both osteoporosis and osteomalacia.

Osteomalacia in elderly people usually results from decreased vitamin D availability and/or impaired conversion of vitamin D to its active form. A few cases are attributable to chronic hypophosphataemia and drugs that inhibit mineralization, such as bisphosphonates and fluoride. Two of the organs responsible for synthesis of vitamin D, skin and kidney, as well as the two major target systems, intestine and skeleton, can show substantial age-associated changes which affect vitamin D metabolism and action. In most elderly individuals, the changes are not great enough to impair mineralization, but they may be sufficient to contribute to the severity of age-associated bone loss and osteoporosis.

{Nutritional osteomalacia was not uncommon in northern Britain up to the early 1970s. Since then it has become rarer and this may be associated with a change in the diet of elderly people. During the 1970s in the United Kingdom older people showed a trend towards eating less butter and more margarines. By law in the United Kingdom margarines have to be fortified with vitamins A and D. More recently there is evidence that margarines may be displaced in the diets of elderly people by low-fat spreads, which are not subject to compulsory fortification with vitamin D.}

VITAMIN D METABOLISM AND ACTIONS

Many of the features of the vitamin D hormone system have now been elucidated. The major source of vitamin D_3 is the skin, where it is produced by the action of ultraviolet light on 7-dehydrocholesterol, although it can also be obtained from the diet. The belief that vitamin D_3 was a 'vitamin' and not a hormone derived in part from the circumstances of industrialized society in northern climates where exposure to the necessary ultraviolet irradiation can be so diminished that adequate amounts of vitamin D_3 are not formed in the skin and supplementation is necessary. Because there are only a few foods rich in vitamin D_3 (fatty fish, eggs, chicken liver), some countries fortify foods with vitamin D (milk in the United States, margarine in the United Kingdom). The dietary supplement and food additive used to be vitamin D_2, which is synthesized *in vitro* by ultraviolet photolysis of ergosterol, but it is now being replaced by synthesized vitamin D_3. Like other fat-soluble substances, both dietary vitamin D_2 and vitamin D_3, hereafter called vitamin D, are absorbed in the upper small intestine and enter the circulation primarily through lymphatic channels.

Vitamin D is bound to vitamin D-binding protein (**DBP**)

and carried to the liver where it is hydroxylated to 25-hydroxyvitamin D (**25(OH)D**), the major circulating metabolite. 25(OH)D circulates tightly bound to DBP at concentrations of 10–50 ng/ml (25–125 nmol) and has a relatively long half-life of about 15 days. Liver hydroxylation is not closely regulated and the circulating 25(OH)D level largely reflects the vitamin D reserve. 25(OH)D levels in healthy, ambulatory, elderly people are generally lower than in younger people, primarily owing to decreased exposure to sunlight and decreased dietary intake of vitamin D (Fig. 1). Although age-associated changes in liver metabolism occur, liver reserve is so large that vitamin D metabolism is usually unaffected. Some age-associated decline in the ability of the skin to make vitamin D_3 has been found, but increases in 25(OH)D levels on ultraviolet irradiation of elderly subjects are comparable to increases in the young. There may be some decrease in vitamin D absorption with age, but large doses of vitamin D seem to be equally well absorbed by both young and old.

The kidney is the site of the hydroxylation of 25(OH)D to 1,25-dihydroxyvitamin D (**1,25$(OH)_2$D**), the most active metabolite of vitamin D, which circulates at concentrations of 20–60 pg/ml (50–150 pmol) with a half-life of about 15 h. This activation step is stimulated by parathyroid hormone and low phosphate concentration, and inhibited by high calcium. With vitamin D deficiency, the stimulation by increased parathyroid hormone and decreased serum phosphate can maintain relatively high levels of 1,25$(OH)_2$D, and hence measurement of 1,25$(OH)_2$D is not a reliable indicator of substrate deficiency. The ability of parathyroid hormone to increase the renal 1α-hydroxylase appears to diminish with age, possibly owing to diminished renal mass, and this may account in part for the age-associated increase in parathyroid hormone. Some studies have also found a decrease in 1,25$(OH)_2$D levels with age, which may partially explain the well-documented decrease in calcium absorption in elderly people. Some of the decline in 1,25$(OH)_2$D levels after the menopause may be secondary to oestrogen withdrawal, as oestrogen replacement in postmenopausal women can increase both total and free levels of 1,25$(OH)_2$D.

The kidney also synthesizes 24,25$(OH)_2$D, and it is not clear if this is a biologically active compound or simply a waste product. There is some evidence that 24,25$(OH)_2$D can stimulate cartilage growth, but it does not appear to be necessary for mineralization. The vitamin D receptor has a high affinity for 1,25$(OH)_2$D but can bind 25(OH)D and 24,25$(OH)_2$D at much higher concentrations. The binding of 25(OH)D may be important in exogenous vitamin D intoxication when 25 (OH)D levels are very high but 1,25$(OH)_2$D levels are not.

The major actions of 1,25$(OH)_2$D in calcium homeostasis are to increase absorption of calcium and phosphate in the intestine and to increase bone resorption. It is generally agreed that the predominant role of vitamin D in bone mineralization is indirect, i.e., a consequence of the role that vitamin D plays in maintaining the extracellular calcium–phosphate product. For example, the impairment of mineralization seen in vitamin D deficiency in animal models can be reversed by returning the serum concentrations of calcium and phosphate to normal. However, 1,25$(OH)_2$D may also have direct effects on osteoblasts to enhance their differential function and to enable the cells to synthesize a matrix that can be mineralized. In addition, 1,25$(OH)_2$D can increase the production of the calcium-binding protein, osteocalcin, by bone cells.

Levels of 25(OH)D vary seasonally, with the nadir at the end of winter months, reflecting the importance of the endogenous photosynthesis of vitamin D. While short periods of low vitamin D stores may warrant treatment to avoid secondary hyperparathyroidism and increased bone turnover, a sustained period of low vitamin D is necessary to cause osteomalacia. The relationship of 25(OH)D levels to vitamin D-related bone disease is unknown, but values below 10 ng/ml (25 nmol) are considered to place individuals at high risk. Values below this level have been associated with increased levels of parathyroid hormone and alkaline phosphatase.

PATHOPHYSIOLOGY

After synthesis of matrix (osteoid) by osteoblasts, there is a period of 1 to 2 weeks during which this matrix matures before mineralization begins. In osteomalacia, mineralization is impaired and wide seams of osteoid accumulate. However, increased osteoid alone is not diagnostic, as this can occur in other conditions where bone matrix formation is rapid and mineralization is normal, such as hyperparathyroidism, hyperthyroidism, and Paget's disease. Definitive diagnosis requires a dynamic measure of the mineralization rate, which can be achieved by double labelling of the mineralization fronts with tetracycline or other calcium-binding compounds before bone biopsy. The absence of tetracycline labels or the finding of broad diffuse bands rather than a well-defined mineralization front are the diagnostic features of osteomalacia. In hypovitaminosis D, secondary hyperparathyroidism precedes the development of severe osteomalacia. Increased bone turnover is evident, with increased numbers of resorption lacunae and often a rather irregular trabecular-bone structure. These structural abnormalities, coupled with decreased mineralized bone, can lead to deformity and fracture. In advanced cases of osteomalacia, particularly in elderly subjects, the osteoblasts may become flattened and stop forming new matrix. While the severe case of osteomalacia is clearly recognizable, mild degrees of impairment of mineralization can occur in which there is only a small increase in the width of osteoid seams and a partial loss of mineralization fronts. Such mild or intermediate forms may be relatively common in elderly people and therapy may still be beneficial.

CAUSES OF OSTEOMALACIA

Vitamin D deficiency

While nutritional deficiency of vitamin D is rare, elderly people are a group at increased risk. As noted above, 25(OH)D levels fall with age (Fig. 1). Levels in elderly people vary not only with season but also with vitamin D intake, with the importance of diet depending on such things as season, form of dress, health status, and mobility. Latitude is also important because the appropriate spectral range of ultraviolet radiation for skin production of vitamin D is not available in northern latitudes during winter months. Studies in the United States and United Kingdom have found that up to 15 per cent of free-living elderly people have hypovitaminosis D. The lowest levels of 25(OH)D are seen among house-

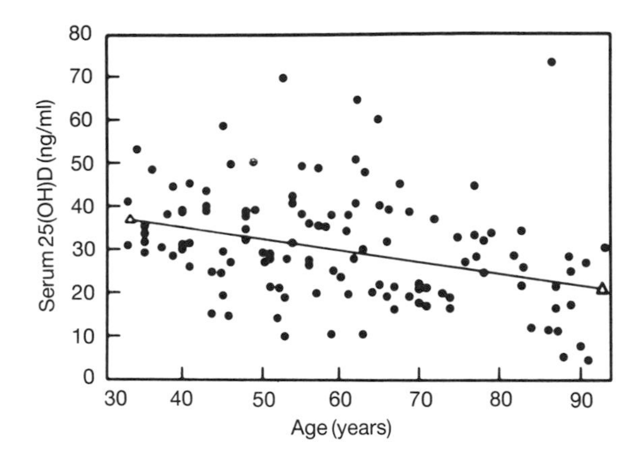

Fig. 1 Distribution of serum 25(OH)D levels by age for a sample of women from Rochester, Minnesota. (Reprinted with permission from Tsai, *et al.* 1987.)

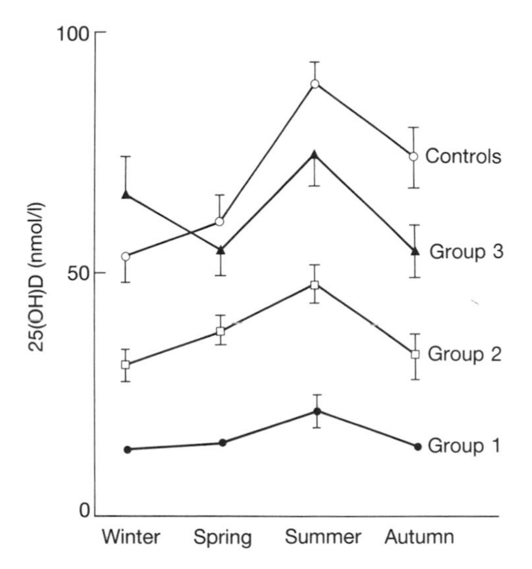

Fig. 2 Serum 25(OH)D levels in three groups of elderly people (65–80 years) and a young adult control group (30–50 years) in Finland. Group 1; long-stay (sick and immobile) geriatric patients; Group 2; residents at an old people's home; Group 3; healthy, ambulatory elderly living in their own homes. (Reprinted with permission from Lamberg-Allardt, 1984.)

bound and institutionalized elderly people (Fig. 2). As many as 25 per cent in the United States and 60 per cent in the United Kingdom of institutionalized elderly people who are not taking vitamin D supplements may be deficient. 400 i.u. (10 μg) daily of vitamin D, the equivalent of a single multivitamin tablet, is probably sufficient prophylaxis except for the most debilitated older patients, in whom plasma 25(OH)D levels should be determined and brought to a normal range by appropriate treatment.

Malabsorption

Patients with malabsorption due to gluten-sensitive enteropathy, gastric surgery, bowel resection, or intestinal bypass operations, as well as individuals with impaired fat absorption due to biliary or pancreatic disease, can develop osteomalacia. This occurs in part because of decreased absorption of exogenous vitamin D and increased loss of 25(OH)D from disruption of the enterohepatic circulation but also because of decreased calcium absorption. The calcium may be bound to fats in the intestinal lumen and thus unavailable, or the calcium transport systems in the intestine may be impaired and unresponsive to $1,25(OH)_2D$.

Renal disease

Although chronic renal disease produces impairment of 1α-hydroxylation, this does not lead to osteomalacia in the majority of patients because the calcium–phosphate product remains high as a result of the phosphate retention associated with decreased glomerular filtration. Patients with end-stage renal disease on maintenance dialysis generally have uraemic osteodystrophy, which is characterized by lesions that vary from predominant osteitis fibrosa, due to secondary hyperparathyroidism, to predominant osteomalacia. Osteomalacia is more likely to be the more prominent lesion in patients whose dialysates contain large concentrations of aluminium or who take large amounts of aluminium hydroxide gels. This is probably caused by both the depletion of phosphate, which is bound to the aluminium hydroxide gel, and the direct inhibitory effect of aluminium on mineralization. Osteomalacia due to renal tubular disorders is relatively rare in elderly people, although occasionally such patients with acidosis, hypophosphataemia, and osteomalacia are encountered.

Other causes

Patients taking anticonvulsants, such as phenytoin, carbamazepine, and phenobarbitone, can develop osteomalacia probably because of a combination of impaired metabolism of 25(OH)D and a direct inhibition of calcium absorption in the intestine. This problem can generally be overcome by a moderate increase in vitamin intake, and those at high risk, e.g., institutionalized elderly patients, should be given prophylactic doses of 800 i.u. daily (20 μg).

Tumour-induced osteomalacia may occur in elderly patients. The mechanism is not entirely clear, but certain tumours, usually sarcomas, may produce factors that either inhibit synthesis or increase degradation of $1,25(OH)_2D$.

Although one might expect to see osteomalacia in patients with chronic liver disease due to failure to synthesize 25(OH)D, this is not common. Most patients with chronic liver disease have osteoporosis and normal levels of 25(OH)D. Even if 25(OH)D levels are somewhat low, patients often fail to respond to vitamin D administration.

In economically advanced nations the congenital forms of rickets are not ordinarily encountered in elderly persons because they are diagnosed and treated much earlier. Some families with a mild defect in renal tubular transport of phosphate and in vitamin D activation have been encountered who develop osteomalacia only relatively late in life.

Osteomalacia can occur with some drugs that are used most frequently in the elderly. In particular, fluoride, used to treat osteoporosis, and bisphosphonates, used to treat Paget's disease and osteoporosis, can inhibit mineralization when taken chronically in large doses. It is recommended that patients be given prophylactic doses of 400–800 i.u. (10–20 μg) vitamin D, as well as an adequate intake of calcium, during treatment with these drugs.

MIXED OSTEOMALACIA IN ELDERLY PEOPLE

Not only are elderly individuals likely to have a combination of osteomalacia and osteoporosis, but the osteomalacia is likely to have a multifactorial aetiology. Poor intake of vitamin D, reduced exposure to sunlight, loss of renal 1α-hydroxylase, and impaired intestinal absorption of calcium can all contribute. Some patients may be phosphate-deficient because of decreased intake or the use of diuretics, laxatives, or antacids. Secondary hyperparathyroidism may be particularly marked in these elderly patients, not only because of their calcium deficiency, but also because of the lack of the normal feedback inhibition of $1{,}25(OH)_2D$ on secretion of parathyroid hormone.

CLINICAL FEATURES

Patients with osteomalacia often have generalized bone pain and tenderness. Skeletal deformities develop and there is an increased incidence of fracture. Weakness of the proximal muscles is common. This may be a consequence of secondary hyperparathyroidism or of the deficiencies of vitamin D, calcium, and phosphate. A depressed affect is also frequently present. In addition to true fractures, patients with osteomalacia can develop pseudofractures (Looser's zones) in which there is a loss of mineral, which appears as a discontinuity of the bone on radiographs. This may be associated with marked local tenderness. Bowing of the limbs and kyphosis can occur in elderly patients with osteomalacia. It is important to look for signs of an underlying disorder such as anaemia, steatorrhoea, and polyuria and polydipsia, and to check for previous gastrointestinal surgery.

One area of particular concern and controversy is the relationship between osteomalacia and fracture of the proximal femur. Low levels of 25(OH)D have been found in many patients with hip fracture (Fig. 3). Some studies have reported a high frequency of osteomalacia in patients with femoral neck fractures. However, the diagnosis of osteomalacia has been based on measurement of osteoid seams in samples obtained at surgery for hip fracture, without tetracycline labelling, and hence there is no definite evidence of impaired mineralization. In addition, several studies have suggested that there is a poor correlation between the blood levels of vitamin D and the presence or absence of increased osteoid on biopsies from in hip fracture patients. Nevertheless, a therapeutic trial of vitamin D would seem to be appropriate either when the 25(OH)D level is decreased or when there is histological evidence of osteomalacia.

{In studies from the United Kingdom it has not been clear if the histological osteomalacia found in patients with hip fracture has been causative or whether merely a consequence of the housebound and immobile state of a proportion of the patients before their fracture.}

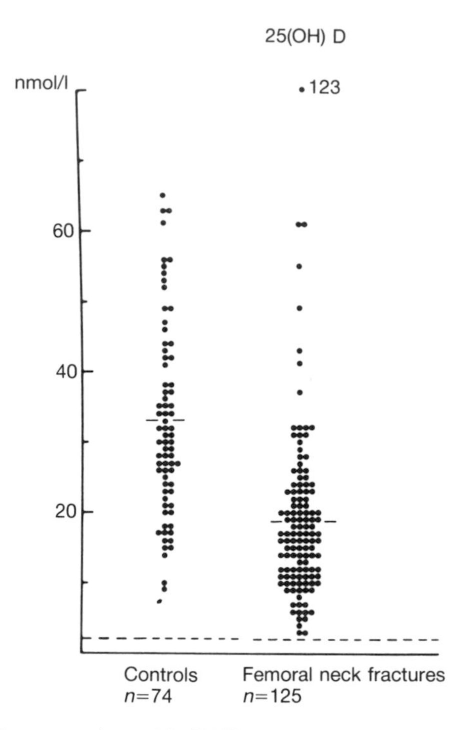

Fig. 3 Concentrations of 25(OH)D in patients with femoral neck fracture and control subjects of the same age. The dotted line marks the detection limits of the assay. (Reprinted with permission from Lips, *et al.* 1982.)

BIOCHEMICAL CHANGES

It is possible to have osteomalacia with a normal serum calcium, phosphate, alkaline phosphatase, and urine calcium. However, in most patients one or more of these values are abnormal. The combination of a mildly decreased serum calcium, a low serum phosphate, and a markedly low urine calcium with elevation of plasma alkaline phosphatase strongly suggests osteomalacia. {A plasma chloride at the upper end of the normal range may also be seen as a manifestation of the secondary hyperparathyroidism.} However, alkaline phosphatase values may be modestly elevated in a minority of normal older people.

Total plasma calcium can be low because of low albumin, and phosphate because of low intake, both of which are common in the older patient. Low urinary calcium is also relatively common. Thus a normal chemical pattern does not rule out, and an abnormal one does not establish, the diagnosis of osteomalacia. In patients with vitamin D deficiency, the most important measurement is the serum 25(OH)D. Serum $1{,}25(OH)_2D$ levels may not be low in patients with vitamin D deficiency but are low in patients with renal failure or renal tubular defects in $1{,}25(OH)_2D$ synthesis. Renal failure can be detected by measurement of blood urea nitrogen and creatinine or creatinine clearance; systemic acidosis detected by measurement of electrolytes. Acidosis due to a distal tubular defect should be considered if morning urine pH is constantly 6.5 or higher.

RADIOGRAPHIC CHANGES

In contrast to rickets in children, in which enlargement of the growth plate and cupping of the metaphyses are pathognomonic changes due to cartilage overgrowth in response to impaired mineralization, the radiographic changes of osteomalacia in adults are usually non-specific. The one exception is the pseudo-fracture or Looser zone. This localized area of demineralization can be seen in the long bones, the pubic rami of the pelvis, the ribs, and on the lateral margins of the scapulae. The vertebrae may show collapse or become biconcave (cod fish vertebrae). Coarse trabecular markings have been described in osteomalacic bone but are certainly not diagnostic. In renal osteodystrophy an alternation of dense and lucent areas in the vertebral body can produce the so-called 'rugger-jersey' spine.

BONE BIOPSY

The definitive diagnosis of osteomalacia can only be made by bone biopsy. In patients who have bone pain and a presentation that is atypical for osteoporosis, bone biopsy may be justified, especially if serum calcium phosphate, alkaline phosphatase, and urinary calcium are normal. On the other hand, a patient with a low 25(OH)D and one or more appropriate biochemical changes might be treated without bone biopsy, on the presumption that osteomalacia is responsible, at least in part, for the metabolic bone disease.

DIFFERENT DIAGNOSIS

In addition to assessing the component of osteoporosis that is likely to coexist with osteomalacia in elderly patients, it is important to consider other causes of proximal muscle weakness and bone pain. Rheumatological disorders, thyrotoxic myopathy, polymyositis, and occasionally lymphoma may mimic osteomalacia. Many patients are thought to have rheumatic disease and are treated with non-steroidal anti-inflammatory agents for long periods before the diagnosis of osteomalacia is made.

TREATMENT

Osteomalacia due to vitamin D deficiency will respond rapidly to vitamin D supplementation. There may be some justification for using 25(OH)D (calcidiol) because it is usually rapidly absorbed and one can measure the serum value after therapy to be certain that the desired level has been reached. {In the United Kingdom, alfacalcidol (1α(OH)D) and calcitriol (1,25$(OH)_2$D) are also in use.} On the other hand, these medications are much more expensive than vitamin D, and their use is probably only justified where there is some concern about absorption. 25(OH)D is absorbed in the intestine better than vitamin D itself. Indeed, in elderly patients with gastrointestinal disease, vitamin D may not be absorbed at all. Some patients have been given intramuscular vitamin D to overcome this problem. In patients with severe gastrointestinal disease, oral 25(OH)D may be reasonably well absorbed, but the blood level should still be checked. Patients with gluten-sensitive enteropathy will respond to a gluten-free diet. This diagnosis is sometimes difficult to make because steatorrhoea and other typical features are not present, and the diagnosis can only be made by jejunal biopsy.

Patients with impaired renal function and low 1,25$(OH)_2$D levels should be treated with 1,25$(OH)_2$D. Here, as the active hormonal form is being used, the margin of safety is narrower, and hypercalcaemia or hypercalciuria often develop when the dose is increased. This is particularly dangerous when serum phosphorus is high, because this will increase the likelihood of soft-tissue damage and worsening of renal function. Thus serum and urine calcium and phosphorus, and renal function, should be monitored in patients on calcitriol therapy.

{The half-lives of the hydroxylated vitamin D derivatives are shorter than that of vitamin D.} Vitamin D toxicity can be a problem when the prohormone is used, particularly because vitamin D is stored in soft tissue and can accumulate. Thus if toxicity occurs, its course may be quite prolonged. Vitamin D intoxication has been encountered in elderly patients taking doses equal to or greater than 50000i.u. a week. These large amounts may be necessary in patients with malabsorption, but 25(OH)D levels in the blood should be carefully monitored.

Bibliography

Dattani, J.T., Exton-Smith, A.N., and Stephen, J.M.L. (1984). Vitamin D status of the elderly in relation to age and exposure to sunlight. *Clinical Nutrition*, **38C**, 131–7.

Davies, M., Mawer, E.B., Hann, J.T., and Taylor, J.L. (1986). Seasonal changes in the biochemical indices of vitamin D deficiency in the elderly: a comparison of people in residential homes, long-stay wards and attending a day hospital. *Age and Ageing*, **15**, 77–83.

Hoikka, V., Alhava, E.M., Savolainen, K., and Parvianinen, M. (1982). Osteomalacia in fractures of the proximal femur. *Acta Orthopaedica Scandinavica* **53**, 255–60.

Lamberg-Allardt, C. (1984). Vitamin D intake, sunlight exposure and 25-hydroxyvitamin D levels in the elderly during one year. *Annals of Nutrition and Metabolism*, **28**, 144–50.

Lips, P. *et al.* (1982). Histomorphometric profile and vitamin D status in patients with femoral neck fracture. *Metabolic Bone Disease and Related Research*, **4**, 85–93.

Lips, P. *et al.* (1988). The effect of vitamin D supplementation on vitamin D status and parathyroid function in elderly subjects. *Journal of Clinical Endocrinology and Metabolism*, **67**, 644–50.

Lund, B. and Sorensen, O.H. (1979). Measurement of 25-hydroxyvitamin D in serum and its relation to sunshine, age and vitamin D intake in the Danish population. *Scandinavian Journal of Clinical and Laboratory Investigation*, **39**, 23–30.

Marel, G.M., McKenna, M.J., and Frame, B. (1986). Osteomalacia. In *Bone and mineral research, annual 4* (ed. W.A. Peck), pp. 335–412. Elsevier, Amsterdam.

Omdahl, J.L., Garry, P.J., Hunsaker, L.A., Hunt, W.C., and Goodwin, J.S. (1982). Nutritional status in a healthy elderly population: vitamin D. *American Journal of Clinical Nutrition*, **36**, 1225–33.

Parfitt, A.M. *et al.* (1982). Vitamin D and bone health in the elderly. *American Journal of Clinical Nutrition*, **36**, 1014–31.

Tsai, K.-S., Wahner, H.W., Offord, K.P., Melton, L.J., III, Kumar, R., and Riggs, B.L. (1987). Effect of aging on Vitamin D stores and bone density in women. *Calcified Tissue International*, **40**, 241–3.

Wicks, M., Garrett, R., Vernon-Roberts, B., and Fazzalari, N. (1982). Absence of metabolic bone disease in the proximal femur in patients with fracture of the femoral neck. *Journal of Bone and Joint Surgery* (Br), **64**, 319–22.

14.3 Paget's disease of bone

RONALD C. HAMDY

Symptomatic Paget's disease of bone predominantly affects the older population and has a peculiar geographical distribution. It is most common in England, especially in Lancashire; its frequency in Europe then decreases to the south, east, and north-east. The disease is rare in Scandinavian countries. In the United States, it is more common in the north than in the south. In Australia and New Zealand, it is more common among British descendants than among the rest of the population. The disease is, however, not limited to Caucasians and has been reported among the black population.

AETIOLOGY

The presence of inclusion bodies in the nuclei and cytoplasm of the osteoclasts suggests a slow viral infection, probably a paramyxovirus. Interest at present is centred on the measles and respiratory syncytial viruses. Retrospective epidemiological surveys, however, have not demonstrated a higher incidence of diseases caused by these viruses in patients than in controls. Also, the actual virus has not been identified, nor has it been possible to transmit the disease, so the viral aetiology is still hypothetical. Canine distemper virus has been suggested, on the basis of a case-control study, as a possible causative agent for Paget's disease, but other studies have been unable to confirm this.

The peculiar geographical distribution and tendency to familial aggregation suggest an autosomal-dominant, inherited disorder with a variable degree of penetration. The incidence among identical twins, however, is much less than would be expected and no evidence of major histocompatibility linkage has been demonstrated. It is therefore probable that Paget's disease is not an inherited disease and that genetic factors only increase the susceptibility to the putative viral infection.

The distribution of the skeletal lesions suggests that trauma may play a role in the development of the disease. Apart from the skull, weight-bearing bones are predominantly affected—the pelvis, femur, and tibia much more frequently so than the rest of the skeleton. The lumbar spine is more frequently affected than the thoracic spine, which is in turn more frequently affected than the cervical spine. Trauma appears to play an important role in the localization of the lesion, such as the billiard player or the treadle machine operator reported to have developing Paget's disease in constantly traumatized bones. On the other hand, many bones which are repeatedly traumatized, such as those of the feet and hands, are not commonly affected by Paget's disease, and trauma cannot explain the unusual geographical distribution. It is therefore probable that trauma only increases the susceptibility of particular osteoclasts to viral infections.

At present, therefore, although there are many hypotheses as to the aetiology of the disease, none has been reasonably proven beyond doubt, but the present evidence suggests a slow viral infection.

PATHOPHYSIOLOGY

Paget's disease is characterized by an excessive rate of bone turnover (Fig. 1). The osteoclasts are numerous, enlarged, multinucleated and very actively resorbing bone; their activity is paralleled by that of the osteoblasts and the rate of bone formation is also excessive. A large number of spindle-shaped osteoblasts can be seen lining the bone surface.

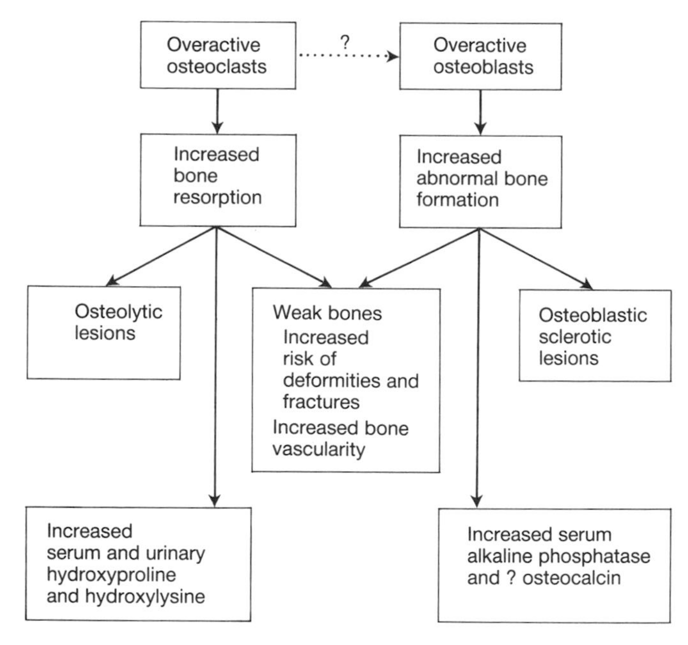

Fig. 1 Diagrammatic representation of the pathophysiology of Paget's disease of bone. (Adapted from Hamdy, R. C. (1981). *Paget's Disease of Bone*, p. 33. Praeger Publishers, Eastbourne, with permission.)

The newly formed bone is architecturally abnormal and, although larger than normal, it is mechanically weak. This results in deformities and fractures. Furthermore, the expanding bone may compress nerves or interfere with their blood supply. Finally, as a result of the rapid turnover, the affected bone is very vascular and may monopolize a large portion of the cardiac output. Various 'steal syndromes' have been reported.

The excessive rate of bone turnover increases the mobilization of calcium from the skeleton to the circulation and vice versa. This may lead to the development of extraosseous calcification, especially in the arteries and on the cardiac valves. There also appears to be an increased incidence of renal calculi.

CLINICAL MANIFESTATIONS

Paget's disease of the bone is by and large asymptomatic. Any symptoms and clinical manifestations vary according to the affected bones (Fig. 2).

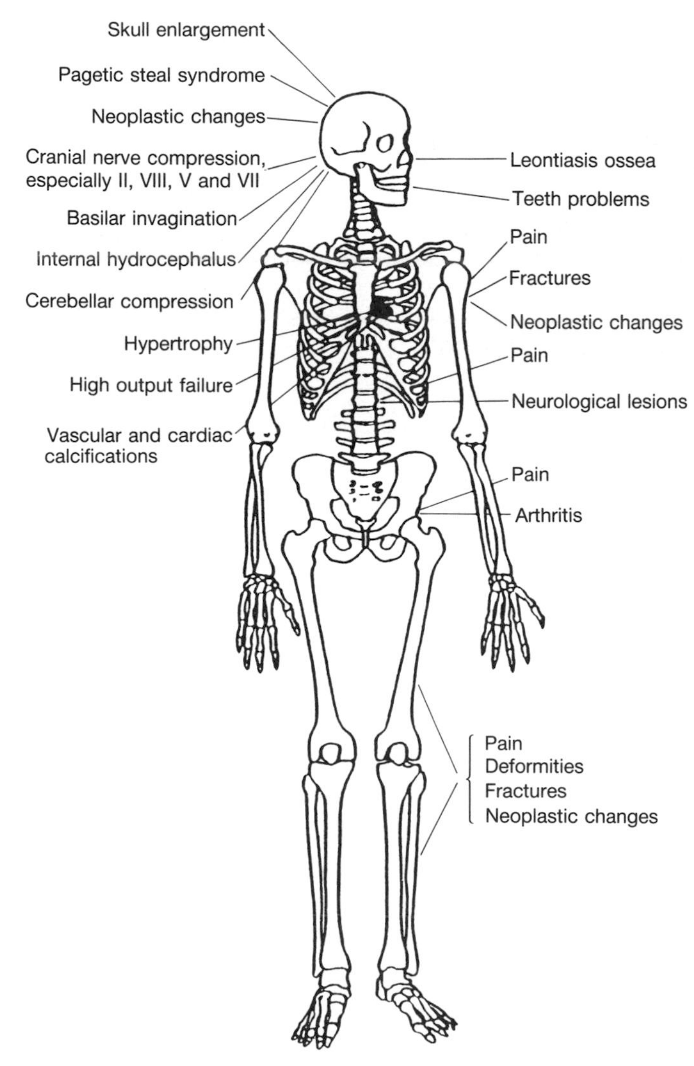

Fig. 2 Presentation and complications of Paget's disease of bone. (From Hamdy, R. C. (1981). *Paget's Disease of Bone*, p. 116. Praeger Publishers, Eastbourne, with permission.)

Pain

Only about 5 per cent of the population with Paget's disease complain of pain directly related to the pagetic process. Many more, however, may present with pain due to a complicating osteoarthritic process or nerve compression. The typical pagetic pain is usually constant, not precipitated by weight bearing or by exercise, and tends to be worse at night when the patient lies in bed and the limb is warm. Although typical osteoarthritic pain is precipitated by weight bearing and exertion and is relieved by rest, it is sometimes difficult to differentiate between these two types of pain clinically. Often a therepautic trial with a non-steroidal anti-inflammatory compound is justified.

Deformities

As the affected bones are mechanically weak, they cannot support the body weight and become bowed. These deformities may precipitate fissure fractures and often induce secondary osteoarthritic changes in neighbouring joints. The direction of bowing is in the line of least resistance. The tibia, for instance, bows anteriorly because in that direction it is only surrounded by skin and subcutaneous tissues whereas in all other directions it is surrounded by either strong muscles or the fibula. The femur, on the other hand, bows in the anterolateral direction because of all the muscle groups surrounding the femur, those in this part are the weakest. When the pelvis is affected, it often becomes triradiate, with the body weight pushing the upper part of the sacrum downwards into the pelvic cavity and the two femoral heads invaginating the pelvic cavity.

COMPLICATIONS

Paget's disease affecting the long bones

Fractures

Fractures of long bones complicate Paget's disease. Frequently these are preceded by fissure fractures and the patient often complains of pain well before the bone is completely broken. Fissure (or stress) fractures usually occur on the convex border of the bone and are perpendicular to its long axis.

The prognosis once a fracture develops is usually poor. The excessive vascularity of the bone may pose difficult problems of haemostasis to the operating surgeon, and if the lesion is not vascular, the brittle bone may make the insertion of a prosthesis difficult. Furthermore, even if the prosthesis is adequately inserted, the bone, being mechanically weak, is often unable to retain it so it is frequently displaced (Fig. 3). Finally, although callus formation may be good and indeed excessive, the newly formed bone may be affected by Paget's disease and therefore mechanically weak.

Neoplasia

Severe intractable pain, resistant to medication, should raise the suspicion of malignant neoplasia complicating Paget's disease. This occurs in less than 1 per cent of the patients. Bones of the upper limbs and the skull are more frequently affected than the rest of the skeleton. The prognosis in these instances is very poor, with less than half the patients surviving more than 12 months. This poor prognosis is due to the excessive vascularity of the lesion which facilitates the spread of the tumour, the often multicentric neoplastic changes with more than one centre simultaneously becoming malignant, and the usually older age of the patient which may interfere with the immune response to malignant changes.

Paget's disease affecting the skull

When the skull vault is affected, the head appears larger than normal. In most instances, the symptoms of Paget's disease of the skull are insidious and often erroneously attributed to the ageing process. Most, however, are reversible if detected early and adequately managed.

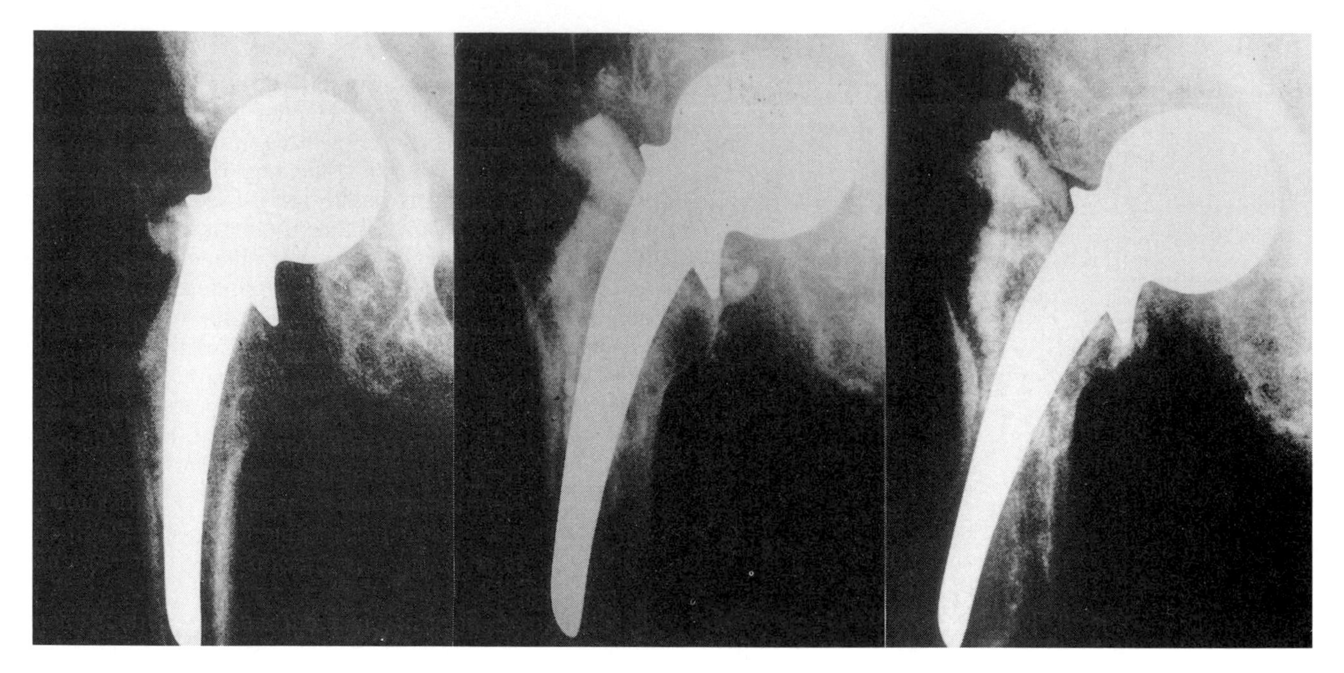

Fig. 3 The fate of a hip prosthesis inserted after a fracture-complicated Paget's disease of the femoral neck. The radiographs are taken at yearly intervals. Note the gradual displacement of the stem of the prosthesis as the shaft of the femur is not strong enough to keep the prosthesis in place. This complication may be prevented by using prostheses with longer stems or by administering active antipagetic therapy. (Reproduced from Hamdy, R. C. (1981). *Paget's Disease of Bone*, p. 53. Praeger Publishers, Eastbourne, with permission.)

Brain anoxia and impaired mental functions

The greatly increased vascularity of the pagetic bone may lead to relative ischaemia of the brain as a major part of the cardiac output is monopolized by it. Furthermore, anastomotic channels may develop between the territories of the internal and external carotid, thus further diverting blood from the brain to the highly vascular bone.

Platybasia and hydrocephalus

When the skull base is affected by Paget's disease, it also becomes mechanically weak and may become invaginated by the cervical vertebrae, giving rise to platybasia. This in turn may lead to distortion of the aqueduct of Sylvius or may obliterate the foramina of Luschka and Magendie and lead to hydrocephalus. Patients present with the characteristic triad of rigidity, impaired mental functions, and urinary incontinence.

Cranial nerve compression

Cranial nerves passing through narrow foramina in bone are particularly likely to become compressed; these include the IInd, Vth, VIIth, and VIIIth nerves. Patients may present with impaired visual acuity, deafness, atypical trigeminal neuralgia, and various degrees of hemifacial palsy or spasms. In addition to compression of the VIIIth nerve, deafness may be due to involvement of the ear ossicles by the pagetic process or to infiltration of the inner ear. With appropriate, timely treatment, the progression of hearing loss can be halted and even reversed.

Long tract involvement and cerebellar impairment

Various degrees of pyramidal signs and cerebellar impairment also may be present.

Paget's disease affecting the vertebrae

Most instances are asymptomatic and undetected until a radiographic study including the affected vertebrae is done for some unrelated reason. Many instances are detected when patients are examined for back pain. Entrapment neuropathies and spinal cord lesions may arise secondarily to compression of the spinal cord or nerve roots by the enlarging bone, or more frequently by interference with the local blood supply of the spinal cord.

Cardiovascular complications

Patients with extensive skeletal involvement tend to have a higher incidence of cardiomegaly, increased left ventricular mass, and larger left ventricular diastolic dimensions than control subjects. The incidence of hypertension, arrhythmias, and calcification of arteries and cardiac valves also tends to be higher in patients than in controls. The incidence of high-output heart failure is, however, low and may have been overemphasized in the past.

DIAGNOSIS

Clinical diagnosis

The clinical diagnosis of Paget's disease is simple when a bone like the tibia is affected because it becomes enlarged and bowed anteriorly. The skin on the affected side is often much warmer than on the non-affected side. Differences in temperature as high as 5°C between affected and non-affected limbs have been reported.

Paget's disease of the skull is also usually obvious if the vault is affected. The cranium may become irregular and have a corrugated appearance. The head becomes larger than normal and the face roughly triangular with the base of the

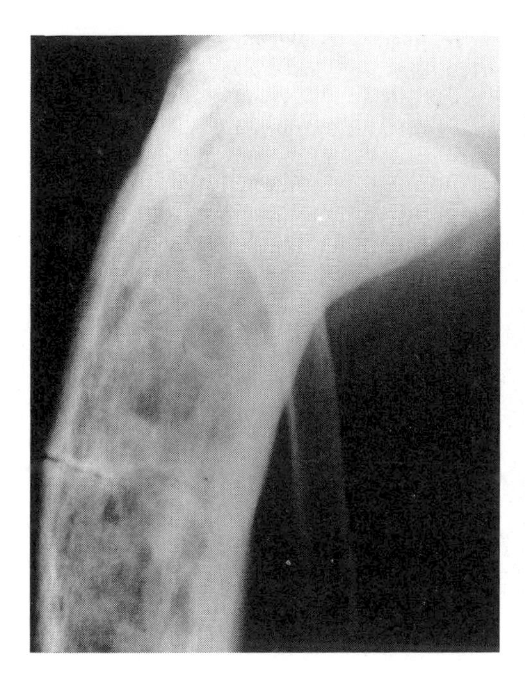

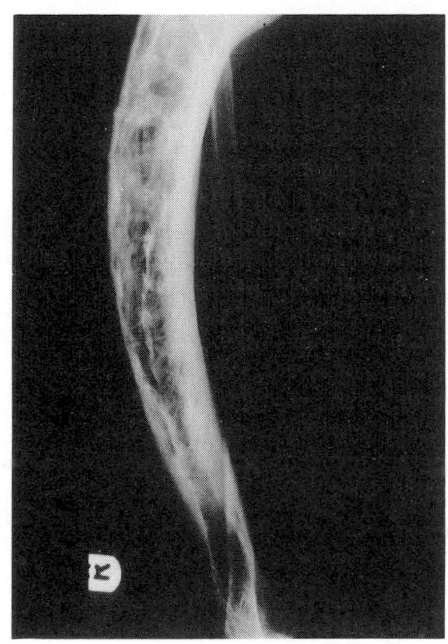

Fig. 4 Characteristic radiological features of Paget's disease of bone.

triangle on the forehead. If the mandible is affected also, then the face becomes rectangular. Not infrequently, Paget's disease is first detected by the patient's dentist, either because the patient's dentures are becoming too tight or because of other dental problems. Facial disfigurement is occasionally reported.

When the skull base is affected and complicated by platybasia, the neck appears shorter, often with the chin resting on the upper part of the chest. The patient may also present with pyramidal tract lesions, cerebellar impairment, or cranial nerve compression. Angioid streaks may be seen during fundoscopic examination. If, in addition, the patient is suffering from optic nerve compression, the fundus may appear congested and papilloedema may be obvious.

Most patients with Paget's disease, however, are not clinically diagnosed and the diagnosis is first made from radiological studies or biochemical blood tests done for unrelated diseases or as part of a routine 'screen'.

Radiological diagnosis

The radiological appearances of Paget's disease are characteristic. They do not, however, indicate the activity of the lesion and are slow to change as the lesion responds to treatment.

The characteristic features include areas of excessive bone resorption and irregular deposition. The cortex appears thickened and irregular (Fig. 4). In long bones, the disease usually starts at one end and progresses towards the other. It is very rare for the disease to start in the centre of the shaft. The edge of the progression is usually clearly defined as a V-shaped, translucent area at the very margin of the lesion. This is the area of excessive bone resorption and is closely followed by other areas of new bone formation and bone resorption. In late stages, the bone appears deformed. Incomplete, transverse, fissure fractures perpendicular to the long bone are sometimes present.

When the skull is affected, the diploic space is enlarged and if there is platybasia, the relationship between the anterior, middle, and posterior chambers of the skull is altered, with the skull base appearing flat or even invaginated by the cervical vertebrae (Fig. 5). Early cases of platybasia can

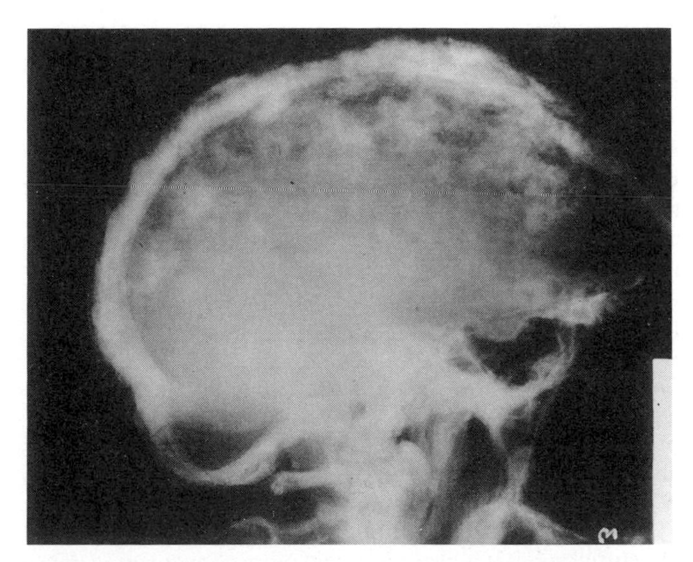

Fig. 5 Paget's disease of the skull with evidence of platybasia.

be diagnosed when the upper end of the odontoid process is above a line joining the posterior lip of the hard palate to the posterior end of the foramen magnum (Chamberlain's line), or 1 cm above a line joining the posterior lip of the hard palate to the lowermost end of the occipital bone (McGregor's line). Very rarely the area of excessive bone resorption is not followed by excessive formation. This condition is known as osteoporosis circumscripta and is usually localized to the calvarium.

When the pelvis is affected, it often becomes triradiate and is frequently complicated by osteoarthritis of the hip joints. When the osteoarthritic process complicates Paget's disease, the medial lip of the hip joint is more affected than the superior lip, which is usually more severely affected in degenerative osteoarthritis. In advanced cases, both the medial and superior lips are involved.

When the vertebrae are affected, the neural arches and processes are also involved. It is rare for the pagetic process to be limited only to the body of the vertebrae. The trabeculae are usually thickened and irregularly deposited.

Biochemical diagnosis

The excessive rate of bone resorption increases the urinary hydroxyproline excretion and the excessive rate of bone formation increases the serum alkaline phosphatase.

Biochemical measurements are often used to assess the activity of the lesion and its response to treatment. One of the first indices to change is the urinary hydroxyproline excretion rate, closely followed by the plasma alkaline phosphatase.

As the urinary hydroxyproline is affected by a number of factors including the ingestion of collagen, the ratio of hydroxyproline to creatinine excretion in the urine should be calculated. Often this ratio in a fasting urine specimen is used as an index of bone resorption because it is more convenient to obtain such a specimen than to collect urine over 24 h.

Various other biochemical ways have been suggested for monitoring the activity of Paget's disease. These include free plasma hydroxyproline, serum osteocalcin, and urinary hydroxylysine. The last is less affected by dietary intake than the urinary hydroxyproline level. The serum uric acid level also tends to be elevated.

Hypercalcaemia occurs if patients with Paget's disease are immobilized. Immobilization reduces the stimulus to new bone formation, but does not affect the rate of bone resorption, which continues at its previous rate, resulting in hypercalcaemia. This is particularly serious in older patients as it may precipitate hypercalciuria and renal calculi.

Skeletal scintigraphy

One of the main advantages of bone scintigraphy over radiology is that it gives a more precise picture of the activity of the lesion and is therefore useful in monitoring its progress and response to treatment. Scintigraphic uptake correlates with frequency of pain and biochemical abnormalities. Only rarely will a lesion evident on a radiograph not be visualized by scintigraphy. It is also easier and more convenient to determine the extent of the lesion and distribution of the disease throughout the skeleton by scintigraphy than by radiography. Scales are available to determine the activity of the lesion according to the scintigraphic appearance.

Uptake of technetium-99 reflects the bone vascularity; uptake of gallium-67 is more indicative of cellular activity and therefore may be a more useful way of assessing the degree of activity and response to treatment. When either isotope is given, pagetic lesions appear as areas of excessive uptake. The uptake is usually reduced when the lesion becomes malignant.

It must be emphasized, however, that many other bone diseases and in particular bone metastases may have increased uptake. Scintigraphic diagnosis should therefore be used to complement radiological diagnosis, but not to substitute for it.

TREATMENT

Most patients with Paget's disease do not need any specific treatment as most are asymptomatic. Furthermore, those who have pain often respond to simple analgesia or non-steroidal anti-inflammatory compounds. Whether or not to use specific antipagetic therapy depends on the affected bone and activity of the lesion. Guidelines are presented in Table 1.

The main indications for specific treatment include pain resistant to ordinary analgesia, fissure fractures, and neurological complications. Patients who are about to undergo orthopaedic surgery on an affected bone should also be treated appropriately as this will reduce the vascularity of the bone and is likely to improve the outcome of surgery. Surgical treatment is indicated for selected fractures, severe disabling arthritis, and gross deformities.

Several types of drug are now available to treat Paget's disease of bone, as follows.

Calcitonin

Calcitonin is secreted by the C-cells of the thyroid gland. Its main action is specifically to inhibit the osteoclasts. It also reduces the blood flow to bone and may have a centrally acting, analgesic effect. Synthetic human, salmon, porcine, and eel calcitonin are the main types commercially available. Except for the human type, they are all antigenic and induce the formation of antibodies. These, however, rarely block the action of calcitonin.

There is still no consensus of opinion as to the optimum dose of calcitonin. Although the majority of regimens vary between 50 units, three times a week, and 100 units daily, some centres suggest a much lower dose. The parenteral administration of calcitonin is usually followed by a rapid decline in urinary hydroxyproline and serum alkaline phosphatase. After the initial decline, however, these tend to reach a plateau well above the normal limits. Continuation of calcitonin in spite of this plateau has been shown, in some instances, to induce histological and even radiological improvement. When calcitonin is discontinued, the disease usually becomes active again after a period of quiescence of variable duration.

The main disadvantages of calcitonin include the need for parenteral administration and the high cost. A type for intranasal administration is now available. Side-effects of calcitonin include the development of flushing soon after its administration; this is particularly obvious with human calcitonin. Nausea is another complication, occurring usually 2 to 3 h after the dose and lasting for a similar amount of time. This effect is usually limited to the first 2 to 3 weeks of therapy and then tends to subside spontaneously. Rather than concomitantly prescribing antiemetics, the calcitonin can be given before the patient retires to bed at night.

Table 1 Summary of complications and management of Paget's disease (adapted from Hamdy (1981), p. 115, with permission of Praeger Publishers)

Site of lesion	Complications	Management: Active phase	Management: Quiescent phase
Long bones	Fractures Deformities Arthritis of neighbouring joints Malignancy	(A) No complications: observe progress; analgesics if painful Antipagetic drugs indicated if: (1) Risk of fracture substantial: Cortical thickness diminished; Transverse pseudofractures; Marked deformities (2) Pain unrelieved by analgesics (B) Fractures: Antipagetic drugs ± surgery (C) Malignancy: Suspected—biopsy Confirmed—? radical Surgery ± cytotoxic agents —? palliative	Observe
Skull base	Cranial nerve compression, especially II, VIII, V, and VII Basilar invagination: Involvement of —pyramidal tracts —lower cranial nerves —cerebellum Chronic brain failure: Obstruction of CSF, drainage, and hydrocephalus	Antipagetic drugs	Regular monitoring: Serum alkaline phosphatase and/or urinary hydroxyproline Fundus examination (papilloedema) Hearing and visual acuity Radiography of skull (for platybasia) Patient asked to report once symptoms manifest themselves
Vault	Brain ischaemia Malignancy	Observe Antipagetic drugs indicated for: intractable headaches, brain ischaemia, cosmetic reasons	Observe
Vertebrae:			
above L2	Paraplegia: Cord compression or ischaemia	Antipagetic drugs	As for skull base
below L2	Cauda equina lesion	As for skull vault	As for skull vault
Ribs, clavicles, scapulae, pelvis	Fractures and compression of adjacent structures	Observe Antipagetic therapy indicated if: (1) Intractable pain (2) Compression of neighbouring structures	Observe
Generalized	High-output heart failure	Antipagetic therapy if (1) Marked deformities (2) High-output failure	Observe

Bisphosphonates; etidronate disodium

Bisphosphonates inhibit the activity of Paget's disease and have several advantages over calcitonin. They can be given by mouth and are relatively inexpensive. Furthermore, the degree and duration of inhibition of the pagetic process is more significant and more sustained than with calcitonin. Resistance to second or third courses of treatment is also less likely to occur with bisphosphonates.

The main problem with bisphosphonates, however, is the induction of demineralization, fissure fractures, and even complete fractures in previously healthy bones, especially if large doses (e.g., 20 mg/kg body weight) are given. The recommended dose at present is 5 mg/kg body weight daily for not longer than 6 months. It is also recommended to have a drug-free interval of at least 90 days between courses of treatment. If larger doses are needed, they should be given for shorter periods if demineralization of healthy bones is to be avoided. As food may interfere with the absorption of bisphosphonates, they should be taken on an empty stomach and at least 2 h before the next meal.

These treatment 'courses' are usually followed by prolonged periods of quiescence which may last a few years. Bisphosphonates are sometimes given in conjunction with calcitonin.

Bisphosphonate derivatives

Many derivatives of the original bisphosphonates have been developed and these induce less demineralization in healthy bones. The degree and duration of inhibition of Paget's disease is very significant. A single infusion of 60 mg aminohydroxypropylidene bisphosphonate (AHPrBP, previously APD) can control the activity of the pagetic lesion for a period of at least 1 year in some patients, without significant side-effects.

Cytotoxic agents

Mithramycin has been used in the management of Paget's disease of the bone. Being a cytotoxic agent, it inhibits very active cells such as the pagetic osteoclasts. However, it also depresses the bone marrow and may induce hepatic and renal damage. Its use should therefore be limited to severe cases not responding to more conventional lines. Mithramycin is particularly useful in the management of hypercalcaemia. It should be restricted to short courses, preferably while the patient is in hospital.

CONCLUSIONS

Although Paget's disease of the bone is usually asymptomatic, clinicians should be aware of this condition as often patients may present with signs and symptoms that may not be directly associated with Paget's disease and may be erroneously attributed to the ageing process.

Relatively safe and effective drugs are currently available for the management of this condition. The ultimate prognosis for patients with Paget's disease of the bone is therefore good. Clinicians, however, must use their discretion to determine who, when, and how patients with Paget's disease are to be treated.

Bibliography

Arnalich, F. *et al.* (1984). Cardiac size and function in Paget's disease of bone. *International Journal of Cardiology*, **5**, 491–505.

Barker, D.J.P. (1984). The epidemiology of Paget's disease of bone. *British Medical Bulletin*, **40**, 396–400.

Breanndan Moore, S. and Hoffman, D.L. (1988). Absence of HLA linkage in a family with osteitis deformans (Paget's disease of bone). *Tissue Antigens*, **31**, 69–70.

Douglas, D.L. *et al.* (1981). Spinal cord dysfunction in Paget's disease of bone. Has medical treatment a vascular basis? *Journal of Bone and Joint Surgery*, **63B**, 495–503.

Haibach, H., Farrell, C., and Dittrich, F.J. (1985). Neoplasms arising in Paget's disease of bone: a study of 82 cases. *American Journal of Clinical Pathology*, **83**, 594–600.

Hamdy, R.C. (1981). *Paget's disease of bone*, Endocrinology and Metabolism Series, Vol. 1. Praeger Publishers Division, Greenwood Press, Westport, CT.

Lando, M., Hoover, L.A., and Finerman, G. (1988). Stabilization of hearing loss in Paget's disease with calcitonin and etidronate. *Archives of Otolaryngology, Head and Neck Surgery*, **114**, 891–4.

Levy, F., Muff, R., Dotti-Sigrist, S., Dambacher, M.A., and Fischer, J. A. (1988). Formation of neutralizing antibodies during intranasal synthetic salmon calcitonin treatment of Paget's disease. *Journal of Clinical Endocrinology and Metabolism*, **67**, 541–5.

Martin, B.J., Roberts, M.A., and Turner, J.W. (1985). Normal pressure hydrocephalus and Paget's disease of bone. *Gerontology*, **31**, 397–402.

Mills, B.G. *et al.* (1985). A viral antigen-bearing cell line derived from culture of Paget's bone cells. *Bone*, **6**, 257–68.

Smith, B.J. and Eveson, J.W. (1981). Paget's disease of bone with particular reference to dentistry. *Journal of Oral Pathology*, **10**, 233–47.

Stamp, T.C.B., Mackney, P.H., and Kelsey, C.R. (1986). Innocent pets and Paget's disease? *Lancet*, **ii**, 917.

Thiebaud, D., Jaeger, P., Gobelet, C., Jacquet, A.F. and Burckhardt, P. (1988). A single infusion of the bisphosphonate AHRPrBP (APD) as treatment of Paget's disease of bone. *American Journal of Medicine*, **85**, 207–12.

Vellenga, C.J., Bijvoet, A.L. and Pauwels, E.K. (1988). Bone scintigraphy and radiology in Paget's disease of bone: a review. *American Journal of Physiological Imaging*, **3**, 154–68.

SECTION 15
Nephrology and the genitourinary system

15 Nephrology and the genitourinary system

JOHN W. ROWE

EFFECTS OF AGEING ON KIDNEY STRUCTURE AND FUNCTION

In youth, renal capacity far exceeds the homeostatic need for waste excretion and regulation of salt and water balance. In old age, renal function, although often substantially diminished, still provides adequate regulation of these functions under normal circumstances. However, under stress or during treatment of extrarenal disorders, the age-related decline in function becomes clinically important.

The average combined weight of the kidneys falls from 270 g in young adulthood to approximately 180 g at 80 years of age. The total number of identifiable glomeruli decreases with age, roughly in accord with the changes in renal weight; the proportion of glomeruli that are hyalinized or sclerotic increases from 1 to 2 per cent in the third decade, to 12 per cent after the sixth decade. Distal tubular diverticula become much more common in older persons and may predispose them to pyelonephritis by harbouring micro-organisms, and may also be the origin of the simple retention cysts that are common in old age (McLachlan 1978).

Changes in the intrarenal vasculature with age—independent of hypertension or other renal diseases—are probably responsible for most age-related, clinically relevant changes in renal function. Normal ageing is associated with variable sclerotic changes in the walls of the large renal vessels. These changes do not encroach on the lumen and are augmented in the presence of hypertension. Smaller vessels appear to be spared, with fewer than 20 per cent of senescent kidneys from non-hypertensive persons displaying arteriolar changes.

The average renal plasma flow falls by approximately 10 per cent per decade from 600 ml/min in young adulthood to 300 ml/min by 80 years of age. There is a disproportionate loss of cortical flow with relative preservation of medullary flow (Hollenberg *et al.* 1974), accounting for the patchy cortical defects commonly seen in renal scans in healthy elderly adults (Friedman *et al.* 1972).

The major, clinically relevant defect of renal function arising from these histological and physiological changes is a progressive decline in the glomerular filtration rate (**GFR**), whether estimated by the clearance of inulin or creatinine. An age-adjusted, normative standard for creatinine clearance has been established. Average creatinine clearance is stable until the middle of the fourth decade, at which time a linear decrease of approximately 8 ml/min.1.73 m^2/decade begins (Rowe *et al.* 1976*a*).

The fall in creatinine clearance is not accompanied by an elevation in serum creatinine because muscle mass, from which creatinine is derived, falls with age at roughly the same rate as GFR. Thus, serum creatinine overestimates GFR in the elderly. Depression of GFR so severe as to result in elevation of serum creatinine to about 1.5 mg/100 ml (133 μmol/l) is rarely caused solely by normal ageing and indicates the presence of disease.

The change in renal function accompanying ageing is highly variable between individuals. For example, although it is possible to predict a deterioration in mean GFR with advancing age in groups of subjects, it is much less so for an individual. Despite the significant overall decrease in GFR with advancing age, some persons at the age of 80 years have a GFR typical of a 30-year-old (Rowe *et al.* 1976*b*). The age-related decline in GFR is greatest in persons with either high blood pressure or even with blood pressure in the upper range of normal (Lindeman *et al.* 1984).

The structural and functional changes in the kidney with age diminish the capacity to regulate the composition and volume of extracellular fluid. The aged kidney's response to sodium deficiency is blunted. In response to severe acute reduction in salt intake, aged patients are capable of reducing urinary sodium losses and reaching salt balance, but their response is delayed compared with that of younger adults (Epstein and Hollenberg, 1976). Thus, it is not unusual to find substantial amounts of sodium in the urine of elderly individuals on the second and third day after stopping oral intake. The resulting depletion of the extracellular volume is often of major clinical importance.

The salt-losing tendency of the senescent kidney is caused both by nephron loss, with increased osmotic load per nephron leading to mild osmotic diuresis, and the important, age-related alterations that occur in the renin–aldosterone system. Basal renin, whether estimated by plasma renin concentration or renin substrate, is diminished by 30 to 50 per cent in older persons. This basal difference between young and old is magnified by manoeuvres designed to augment renin secretion, such as salt restriction, diuretics, or upright posture. The lowered levels of renin are associated with 30 to 50 per cent reductions in the plasma concentration of aldosterone, which primarily reflect significant reductions in the rate of aldosterone secretion.

Just as older persons are more likely to develop depletion of the extracellular volume when salt deprived, volume expansion is also a commonly encountered problem. Because of the lower GFR, the senescent kidney is less able to excrete an acutely administered salt load than the younger kidney. During acute sodium loading the half-time for excretion in older persons is almost double that in younger ones. Geriatric patients, with or without pre-existing myocardial disease, are thus at risk for expansion of the extracellular fluid volume when faced with an acute salt load (from inappropriate intravenous fluids, dietary indiscretion, or, as commonly happens, after the administration of sodium-rich radiographic contrast agents such as those used in intravenous pyelography).

The clinical impact of the aged kidney's more limited ability to regulate salt balance under stress is compounded by similar abnormalities in water balance. The capacity of elderly individuals to conserve water and excrete a concentrated urine is blunted as compared with young adults. In response to water deprivation there is a modest (approximately 10 per cent) decline in maximum urine osmolality between 33 and 68 years of age. This may be secondary to age-related decreases in GFR and the mildly impaired renal response to vasopressin. An important clinical point is that a urine volume of 60 ml/h may represent the maximum capacity of an older patient to conserve water and therefore may under-represent the degree of depletion of extracellular volume (Shannon *et al.* 1984).

MANAGEMENT OF SALT AND WATER IMBALANCE IN OLDER PERSONS

General principles

The first step in proper management is detection of the disorder. Symptoms associated with hyponatraemia or hypernatraemia are non-specific and often overlooked. Lethargy, apathy, disorientation, anorexia, agitation, depressed sensorium, altered patterns of respiration, hypothermia, pathological reflexes, and seizures are all clinical symptoms of disorders of the volume and composition of the extracellular fluid. While the alterations in physiological regulation discussed above are insufficient in themselves to induce clinical abnormalities, when combined with diseases that limit renal perfusion, such as hypertension and congestive heart failure, these physiological changes increase the risks of azotaemia, and salt and water imbalance. Clinical estimates of vital signs and estimates of central and peripheral volume remain important features of the clinical examination. However, accurate clinical assessment of volume contraction is notoriously difficult in older patients owing to skin laxity and the high prevalence of orthostatic hypotension. The clinical approach requires awareness of the increased frequency of severe clinical disorders of this type in the elderly, and thus a low threshold to include estimates of blood urea nitrogen, creatinine, and electrolytes in all sick patients.

It is important to consider the influence of medications on renal function in older persons. The most common group of drugs, to cause hospital admission in elderly patients is the diuretics. As discussed in detail later, even low doses of these drugs can cause harm. Often secondary effects of other medications being taken concurrently worsen the clinical state (for example, sedatives taken when hyponatraemia has produced mild confusion). The evaluation of every older person must therefore include a careful review of all medications being taken, with appropriate adjustment of dosages.

The narrowed physiological reserve of older persons puts them at risk of adverse effects of treatment. Here, therapy for dehydration without careful monitoring can often lead to significant volume overload. Because there is great variation in reserve function among patients, therapy must include careful monitoring for the development of adverse side-effects.

Hypernatraemia

The incidence of severe hypernatraemia among older persons exceeds one case per hospital per month. Although the decline with age in water-conserving capacity is not so severe as to have clinical significance under conditions of free access to water, it may become important when fluid intake is limited (e.g., after surgery) or when fluid requirements increase (e.g., fever, intestinal losses). Under such conditions, elevation of the serum sodium to levels that result in impairment of mental function (higher than 160 mEq/l if dehydration is gradual; higher than 150 mEq/l if dehydration is rapid) are common in the geriatric age group.

When the renal capacity to conserve water is overwhelmed, thirst supervenes, stimulating the ingestion of water to reduce systemic hyperosmolality. Subjective measures of thirst and objective measures of drinking behaviour after water deprivation are decreased in older people and associated with impaired recovery from hyperosmolality (Phillips *et al.* 1984). Miller *et al.* (1982) described six elderly residents of long-term care centres with strokes that had not produced aphasia who had repeated hospitalizations with hypernatraemic dehydration. Their hypernatraemia was caused by hypodipsia or adipsia, as confirmed during studies with hypertonic saline infusions made after their recovery from the illness. All these patients required prescriptions of daily fluid intake to prevent dehydration. Patients likely to present with hypernatraemia thus include institutionalized, cognitively impaired persons who may have impaired thirst mechanisms and may be receiving sedatives and major tranquillizers likely to contribute to hypodipsia. Chronically ill older patients are at risk for protein–calorie malnutrition and sodium depletion, both of which further reduce medullary tonicity and renal water loss. The results of hypertonicity may be brain shrinkage, capillary haemorrhages, and permanent neurological injury if deficits are severe and prolonged. Rapid correction of hypertonicity to a serum osmolality of approximately 300 osmol/kg is suggested; this can be followed by a more gradual repletion of free water deficits over 36 to 48 h (Arieff and Guisado 1976; Mahowald and Himmelstein 1981). As most, if not all, hypernatraemic patients are also markedly volume depleted, initial therapy should often include intravenous infusions of half-normal saline until the blood pressure, level of consciousness, and urine flow become normal. This can then be followed by infusion of glucose and water solutions to further reduce serum sodium.

Hyponatraemia

The most commonly encountered electrolyte disturbance among elderly individuals is hyponatraemia. Eleven per cent of elderly patients in a 683-bed geriatric service were found to have serum sodium concentrations of less than 130 mEq/l. Sixty-one per cent of those individuals were symptomatic, albeit with non-specific complaints (Kleinfeld *et al.* 1979). Symptoms are determined by both the magnitude and rate of development of hyponatraemia. Anaesthesia and surgery both predispose older persons to hyponatraemic states, with the usual mechanisms being sustained vasopressin secretion plus the administration of hypotonic fluids. Chloropropamide-induced hyponatraemia secondary to prolonged stimulation of vasopressin release also occurs more commonly in older persons than in their younger counterparts. Other conditions with protean manifestations in older people, such as tuberculosis and hypothyroidism, are commonly associated with hyponatraemia and excess vasopressin release (Shannon *et al.* 1984).

Diuretics, especially thiazides, are important causes of impairment of free water excretion. In many hyponatraemic elderly patients admitted to hospital, diuretics are identified as the causative agent. In one large series, diuretic therapy was implicated in the development of severe hyponatraemia in 44 per cent of patients older than 65 years of age (Fichman *et al.* 1971).

The incidence of illness and death associated with hyponatraemia is significant. In patients with profound symptomatic hyponatraemia (Na^+, 105 ± 9 mEq/l), death and permanent neurological injury are not infrequent (Arieff *et al.* 1976; Ashraf *et al.* 1981).

The first step in treating elderly patients with hyponatraemia is to discontinue the medications or intravenous fluids that contribute to this condition. Standard therapeutic measures as discussed in the companion text, *The Oxford Textbook of Medicine*, such as fluid restriction or demeclocycline, can then be used. In view of the impressive extent of illness and death with severe hyponatraemia (Na less than 120 mEq/l), an aggressive approach, with intravenous administration of 500 ml of 5 per cent NaCl over a period of 12 to 18 h to increase the sodium concentration by 10 to 15 mEq/l, may be warranted.

For further discussion of the treatment of hyponatraemia see Berl (1990).

Acute renal failure

Acute renal failure is common in older patients because the major inciting events, including hypotension associated with marked volume depletion, major surgery, sepsis, and the injudicious use of antibiotics, are common in multiply-impaired elderly people, who are often especially vulnerable because of pre-existing, moderate renal insufficiency.

The management of acute renal failure in older people is a complex and demanding task worthy of the effort. The aged kidney retains the capacity to recover from acute ischaemic or toxic insults over the course of several weeks. In addition to the usual acute tubular necrosis, comprising 2 to 10 days of oliguria followed by a diuretic phase before recovery of function, 'non-oliguric' acute renal failure is being recognized with increased frequency. In these cases, renal function is transiently impaired, often for several days, after a hypotensive episode associated with surgery, sepsis, volume depletion, or, quite commonly, after the administration of nephrotoxic radiographic contrast agents. As the clinical hallmark of renal failure is generally a dramatic reduction in urine output, cases of non-oliguric acute renal failure may go unrecognized, resulting in inadvertent overdose of medications excreted predominantly via renal mechanisms, including digitalis preparations and aminoglycoside antibiotics such as gentamicin.

The management of elderly patients with full-blown, acute tubular necrosis is guided by the general principles used in younger patients, as discussed in the companion *Oxford Textbook of Medicine*, but some principles particularly relevant to elderly patients merit emphasis. The most important is the careful exclusion of urinary obstruction, particularly in men with prostatic hypertrophy or prostatic carcinoma, and in women with gynaecological malignancy. Prostatic hypertrophy and obstruction may be insidious and non-specific in its clinical course, and must be carefully evaluated.

The major causes of death during acute renal failure are acute pulmonary oedema, hyperkalaemia, and infection. Haemodialysis and peritoneal dialysis are effective in older people and substantially simplify management. One should not delay starting dialysis until a crisis supervenes in an elderly patient with acute renal failure.

Aside from starting dialysis, water and salt balance must be monitored carefully. Because of catabolism, the usual patient with acute renal failure will lose about one pound of body weight mass per day. Attempts to keep body weight constant will result in gradual expansion of the extracellular fluid volume and a consequent risk of pulmonary oedema. Similarly, overzealous fluid restriction may delay the recovery of renal function. In general, the administration of approximately 600 ml of fluid a day, in addition to insensible losses, provides adequate fluid balance.

Infection is a common and lethal complication of acute renal failure. Urinary infection secondary to unnecessary urinary catheterization is particularly common. Little is gained from placing a urinary catheter in an oliguric patient in whom volume status and serum levels of blood urea nitrogen, creatinine, and potassium are better guides to renal status than urinary output. Infection of intravenous lines is also common and these should be scrupulously monitored and discontinued when possible.

Chronic renal failure

The elderly patient with chronic renal insufficiency requires several special considerations. As has been discussed previously, serum creatinine generally fails to rise as high in elderly as in young persons, despite equivalent reductions in renal function, because muscle mass, the ultimate source of creatinine, falls with age. As serum creatinine underestimates the degree of renal failure, many debilitated, uraemic elderly patients will not be recognized as uraemic because their creatinine levels may be less than 10 mg/100 ml (885 μmol/l).

Chronic renal failure in older people often presents as decompensation of a previously impaired organ system. Examples include worsening of pre-existing heart failure through an inability to excrete salt and water, gastrointestinal bleeding in the presence of previous malignancy or ulcer, or mental confusion in a borderline demented patient who becomes increasingly azotaemic.

Diagnosis

Once the presence of chronic renal failure is established, the specific cause should be identified. Most chronic renal failure in older persons is due to chronic glomerulonephritis, hypertensive and atherosclerotic vascular disease, diabetes, or, in some cases, late-presenting polycystic kidney disease. As in acute renal failure, the most important diagnostic consideration is strict exclusion of potentially reversible causes, such as urinary tract obstruction, particularly in men with clinically 'silent' prostatism. Occlusion of renal arteries that may be reparable by balloon angioplasty or surgery, multiple myeloma, administration of nephrotoxic agents, or very commonly, chronic volume depletion, are all important, potentially treatable causes of chronic renal failure in older persons.

Management

While the management of chronic and end-stage renal failure is discussed in detail in the companion text, some major points merit emphasis. If no reversible component is identified, the patient should be followed closely so that the rate of loss of renal function can be judged accurately. Appropriate adjustments should be made in the dose schedules of all renally excreted medications, especially digoxin. Hypertension should be controlled carefully. As serum phosphate rises, phosphate-binding antacids should be given with meals to suppress parathyroid hormone and its resultant adverse effects on bone. As serum phosphate falls in response to treatment, serum calcium will generally rise toward the normal range. If hypocalcaemia persists after phosphate levels return to normal, it should be treated with preparations of vitamin D or its congeners in order to increase intestinal absorption of calcium.

Anaemia associated with chronic renal failure often requires more aggressive management in elderly patients because of coexisting cardiac disease. Recently, genetically engineered preparations of erythropoietin have become clinically available and are very useful in increasing red cell volume, obviating the need for frequent transfusion, especially in patients with severe coexistent cardiac disease.

Dietary management of elderly patients with chronic renal failure must be undertaken cautiously, with attention to adequate essential nutrients but often with little need to reduce protein intake or otherwise modify the patients' preferences. Many elderly patients ingest only 60 to 70 g of protein daily and 4 to 5 g of salt under normal conditions, and strict limitations of these dietary constituents is often unncessary.

Pruritus is a major problem in elderly uraemic patients, especially in the presence of coexisting xerosis. In addition to skin moisteners, ultraviolet treatments have been found effective and safe for this group. Administration of so-called antipruritic agents, such as antihistamines and ataractics, is rarely helpful because they act primarily by causing sedation and may produce adverse central nervous system effects in older people.

Chronic maintenance dialysis—generally haemodialysis but also chronic ambulatory peritoneal dialysis—remains the mainstay of treatment of elderly uraemic patients. Despite the preconceptions of many physicians and patients, most elderly patients do very well on dialysis, and the frequency of complications seems to be related more to the coexisting extrarenal disease than to age itself. Psychologically, elderly patients often adapt better to chronic dialysis than do their younger counterparts. Once it is clear that a patient will need dialysis at some time in the near future, early creation of an arteriovenous fistula for access is important, particularly in the elderly patient, because such fistulae often mature rather slowly. At present, renal transplantation is generally not considered in individuals over the age of 60 years.

SPECIFIC RENAL DISEASES

Nephrotic syndrome

The histological subtypes of nephrotic syndrome found in older individuals do not appear to differ substantially from those in younger populations (Fawcett *et al.* 1971). The most common cause of nephrosis in old age is membranous glomerulonephritis. As with younger adults, this histopathological change may take one of several possible clinical courses, including remission, stable moderate nephrotic syndrome, and an aggressive form with progressive loss of renal function moving to end-stage renal disease. In those individuals with the progressive form, which in some series is as many as half of all cases of membranous glomerulonephritis, treatment with corticosteroids every other day, or other forms of immunosuppressive therapy, appear to be of some value.

The second most common cause of nephrotic syndrome in older individuals is minimal change disease, which is as responsive to corticosteroid therapy in these patients as it is in younger adults. The prognosis in this disorder is generally favourable. A less common and much less responsive cause of nephrosis is amyloidosis. In all elderly individuals who present with heavy proteinuria, it is very important to exclude the presence of multiple myeloma, a disease with a progressively increasing prevalence with advancing age and which is often very responsive, at least initially, to immunosuppressive therapy.

Given the varied nature of the underlying lesion in nephrotic older persons and the substantial risk of adverse effects of immunosuppression in these patients, the renal biopsy remains an important procedure in their proper evaluation. In treatment it may be important to lean toward the use of immunosuppressive agents rather than corticosteroids to avoid the steroid-associated worsening of conditions such as glucose tolerance, hypertension, osteoporosis, muscular wasting, and cataracts that are common in older people.

Acute glomerulonephritis

Acute glomerulonephritis is often overlooked in older individuals. This is, in part, because the cardinal findings of oedema, hypertension, and a rise in blood urea nitrogen and creatinine levels are very common from other causes in this group, and often will not trigger a close examination of the urinary sediment. Another reason for imperfect recognition of glomerulonephritis in older persons is its altered presentation when compared with that in younger individuals. The presentation of glomerulonephritis is often non-specific, with nausea, malaise, and arthralgia. There is a striking predilection for initial symptoms of shortness of breath and cough, and the presence of bilateral pulmonary infiltrates on chest radiographs. The aetiology of these infiltrates is uncertain, but it does not appear to be simply volume overload as they occur very early on in the course, often before the development of oedema and hypertension.

It must be kept in mind that poststreptococcal glomerulonephritis occurs in elderly individuals as well as in children and carries a favourable prognosis. In other forms of glomerulonephritis, in which crescents are commonly found histologically, the prognosis is very poor and no therapies are very effective, although the experience in some centres suggests that high-dose immunosuppressive regimens have some value.

RENAL VASCULAR DISORDERS

Emboli

Emboli of renal arteries complicate acute myocardial infarction, chronic atrial fibrillation, subacute bacterial endocarditis, and aortic surgery or aortography. One specific subtype of renal embolism, cholesterol embolization, is especially important in elderly patients with diffuse atherosclerosis, and may occur spontaneously or as a complication of aortic surgery or angiography. A definitive diagnosis requires visualization of cholesterol crystals on renal biopsy (Smith *et al.* 1981).

The manifestations of renal emboli may vary from a clinically silent event to a full-blown syndrome of severe flank pain and tenderness, haematuria, hypertension, spiking fevers, marked reduction in renal function, and elevations of serum lactate dehydrogenase. Small emboli are very difficult to detect because renal scans may show focal perfusion defects in many, apparently normal, elderly patients due to an age-related reduction in the cortical vasculature. Major emboli are suggested by the finding of differential contrast excretion on pyelography and confirmed by renal scanning and aortography. For thrombotic emboli, surgery is generally not indicated, and anticoagulant therapy may be of benefit. In many cases when renal function is discernibly impaired, improvement occurs over a period of several days to weeks. The clinical course of cholesterol embolization is highly variable. Although most patients go on to progressive renal failure, some develop only moderate renal impairment and may regain renal function over time. No specific treatment is available.

Atherosclerotic and thrombotic occlusion of renal arteries

Occlusion of renal arteries frequently complicates severe aortic and renal atheroclerosis, especially in the setting of decreased renal blood flow caused by congestive heart failure or volume depletion. Such occlusion may be clinically silent. In patients in whom renal function was previously intact, the only manifestation of unilateral occlusion may be an increase in blood urea nitrogen and serum creatinine, and perhaps a modest increase in blood pressure. In cases with pre-existing renal impairment and azotaemia, renal arterial occlusion may precipitate congestive heart failure, marked hypertension, and the emergence of the uraemic syndrome.

The value of currently available diagnostic tests for detection of renovascular hypertension is influenced by age. The likelihood of false-positive findings for either the determination of stimulated plasma renin or saralasin infusion decreases markedly with age (Andersen *et al.* 1980). Thus, the older the patient, the more likely it is that a high stimulated renin or a positive saralasin infusion indicate significant renovascular hypertension. Patients with positive results from either of these studies will benefit from further evaluation, including timed intravenous pyelography, determination of the renal vein renin, and, ultimately, angiography. A careful evaluation for coexisting abdominal aortic aneurysm is required. Angiography should be conducted with the least amount of contrast possible in order to minimize the likelihood of a nephrotoxic reaction. When technically feasible, surgical revascularization should be considered. A substantial return of renal function can be obtained after prompt revascularization, and in some cases, recovery occurs even if surgery is delayed until several months after the vascular occlusion. Over the past several years, remarkable advances in angioplastic techniques have made available treatment of atherosclerotic occlusion of renal arteries to many patients who previously could not have been treated.

BENIGN PROSTATIC HYPERTROPHY

The distinction between age-dependent changes (i.e., normal ageing in the absence of disease) and age-related changes (those associated with pathological processes more common with advancing age) {'age-associated' is the preferred term in the United Kingdom} is especially difficult in the case of benign prostatic hypertrophy. This increases in prevalence dramatically with advancing age and is found in more than 90 per cent of men over the age of 80 years (Berry *et al.* 1984). The specific aetiology of this condition has not been fully elucidated, but hormonal effects are important because castration halts the hypertrophy of prostatic epithelium and reduces the size of the prostate. Recent studies suggest that alteration in the intraprostatic handling of testosterone and its metabolites may contribute to benign hypertrophy. A testosterone metabolite, dihydrotestosterone, is a growth factor for prostatic tissue; an intensive search is currently under way to develop pharmacological agents that will lower the levels of this metabolite and its effects (Wilson 1980). Should an effective and safe agent be developed, it promises to have widespread use and to decrease markedly the need for prostatic surgery.

Symptoms of benign prostatic hypertrophy can be divided into two general classes, obstructive and irritative. Obstructive syndromes include decreased force and calibre of stream, hesitancy, difficulty in starting the urinary stream, and, in the extreme, urinary retention. Irritative symptoms include frequency, nocturia, dysuria, urgency, and, in some cases, urge incontinence due to associated alterations in bladder compliance. When evaluating patients with symptoms suggestive of prostatism one should be mindful that these symptoms may also be due to other causes of urethral obstruction including prostatic cancer and strictures, and can also be mimicked by urinary tract infection.

Definitive determination of urethral obstruction is often difficult. Assessment of the prostatic size via rectal examination provides, at best, an indirect index of the role of the prostate in causing urinary symptoms. A more physiological measurement, determination of the urine flow rate, often requires sophisticated and expensive equipment. The overall size of the prostate, which is clearly related to the likelihood of obstruction, can be determined easily by echo sonography. In difficult or uncertain cases, urodynamic measurements and cystoscopy are indicated.

Although transurethral resection of the prostate has found widespread application, there has been no consensus on the specific indications for this procedure. Surgery is clearly indicated in individuals with large, postvoid residues of urine or impairments of renal function secondary to obstruction of the lower urinary tract. In the remainder, the results of surgery depend on the severity of symptoms. In a recent study of 318 men undergoing transurethral resection, most of whom were elderly, all those who suffered acute urinary retention improved postoperatively. In addition, over 90 per cent of patients who rated their preoperative symptoms as severe felt that their quality of life was improved after surgery. However, in patients who rated their symptoms as only mild or moderate, there was little discernible benefit from surgery, which is not without its risks. In this series, 24 per cent of patients suffered some postoperative complication, with 4 per cent having persistent incontinence and 5 per cent becoming impotent. Early surgery is thus not recommended in patients with relatively mild symptoms. These patients can be followed up closely for worsening of clinical status, urinary retention, or impairment in renal function (Fowler *et al.* 1988), with the patient's own perspective on the condition providing the physician with important guidance regarding the need for surgery.

URINARY INCONTINENCE

Urinary incontinence is a common, highly morbid, costly, and often neglected problem in elderly people. Defined as the involuntary loss of urine so severe as to cause social or hygienic consequences, urinary incontinence occurs in 15 to 30 per cent of older people living in the community and in as many as 50 per cent of residents of nursing homes (Diokno *et al.* 1986). While age-dependent changes in the lower urinary tract predispose older individuals to incontinence, the inability to control voiding always signals the presence of a pathological abnormality complicating normal, age-dependent changes. Urinary incontinence must not be accepted as an inevitable consequence of frailty. It must be rigorously detected and evaluated because many, if not most, older incontinent patients can be substantially improved if not cured. Incontinence is a symptom with multiple potential causes and a thorough evaluation, including a targeted history, a physical examination, and, in many cases, detailed laboratory testing, is necessary before a specific, underlying diagnosis can be made and a definitive course of therapy initiated (Resnick and Yalla 1985).

As in many geriatric disorders, detection of urinary incontinence is often difficult because of under-reporting. Many older individuals are embarrassed by their incontinence or under the misconception that it reflects a normal aspect of growing old and thus fail to report it to health providers. All detailed assessments of older persons should include specific questions regarding bladder function.

Urinary incontinence may be divided into two general forms, transient and fixed. Transient cases are those that develop as a complication of another illness or are associated with hospitalization and are generally self limited. The major causes of transient urinary incontinence include the following: delirium; drugs (especially agents that impair cognitive function or increase urine flow rate, such as loop diuretics); urinary tract infection; endocrine disorders, such as hyperglycaemia and hypercalcaemia; restricted mobility resulting in inability to reach the toilet; stool impaction; and depression with severe psychological regression. Generally, removal of the responsible cause(s) results in prompt return of continence.

There are four major forms of fixed incontinence: stress incontinence, urge incontinence, overflow incontinence, and a mixed form.

Stress incontinence is characterized by the involuntary loss of small amounts of urine in association with increases in abdominal pressure such as when coughing, sneezing, lifting heavy packages, or straining at stool. The most common underlying factor is laxity of the pelvic musculature, secondary to gynaecological surgery or multiple vaginal births. The tendency towards stress incontinence is aggravated by oestrogen deficiency. Stress incontinence is generally easy to diagnose from the history. Residual urinary volume, which is a cardinal aspect of a clinical evaluation of any incontinent elderly person, is low in uncomplicated cases of stress incontinence (McGuire *et al.* 1976; Mohr *et al.* 1983). In most cases, substantial improvement can be gained by a combination of regular pelvic-floor exercises and oestrogen treatment for those individuals who have complicating atrophic urethritis. Oestrogen can generally be applied via local creams or vaginal suppositories. For severe cases, a variety of relatively simple, urosurgical procedures has been developed to improve the function of the pelvic floor and decrease the likelihood of incontinence (Marshall *et al.* 1949).

Patients with uncomplicated urge continence are unable to inhibit urinary leakage when they feel the acute urge to void. In these cases, moderate amounts of urine are lost at each episode, which occur at intervals of several hours. The residual urine volume is generally low (less than 50 ml) and increases in abdominal pressure do not result in urinary leakage as occurs in stress continence. On detailed urodynamic testing, patients with urge incontinence generally display uninhibited bladder contractions. Underlying mechanisms associated with urge incontinence include lesions of the central nervous system such as stroke, demyelinating diseases, or degenerative diseases, all of which impair the central capacity to inhibit bladder contractions, and local irritating factors such as urinary infection or bladder tumours.

The most common therapeutic approach to urge incontinence is a general reduction in fluid intake, avoidance of medication that may increase the urinary flow rate, such as caffeine-containing fluids and diuretics, and, when necessary, the administration of specific bladder relaxants. The agents most commonly used as bladder relaxants include anticholinergic agents, the most frequent being oxybutynin, which decreases the strength of uninhibited contractions and the likelihood of urinary loss. All anticholinergic agents have, to variable degrees, systemic side-effects such as dry mouth, constipation, and confusion, and must be used with great caution in older patients. Other pharmacological agents that have found some use in the management of urge incontinence include direct smooth muscle relaxants such as flavoxate, and calcium channel blockers. As yet, no agent has been developed which is specific for inhibiting bladder contractions without the risk of systemic side-effects (Resnick *et al.* 1985).

Overflow incontinence is the result of urinary retention which progresses to the point at which the bladder can no longer expand and the pressure within the bladder exceeds the outlet pressure. Overflow incontinence has two basic causes, neurological abnormalities that impair bladder contraction and obstruction to the outflow tract, such as benign prostatic hypertrophy, tumours, or strictures. The clinical manifestations of overflow incontinence are generally distinctive, with the constant or near constant loss of small amounts of urine and the presence of a palpable bladder and a large, postvoid, residual urine. All patients with overflow incontinence require detailed urological evaluation to exclude a potentially reversible obstruction to the outflow tract.

Mixed urinary incontinence is from the simultaneous presence of more than one cause. In many patients a pre-existing abnormality, such as uninhibited contractions or impairment of the bladder outlet, which is not severe enough to itself result in incontinence, is complicated by a supervening, transient cause that overcomes the bladder capacity so that incontinence emerges. In such cases, treatment of the transient complicating event may result in restoration of continence. In other cases, two chronic abnormalities of the lower urinary tract occur simultaneously and treatment of either one may restore continence. In cases in which the clinical symptoms are confusing, a detailed urodynamic evaluation is often helpful in identifying the contributing pathophysiological abnormalities.

Behavioural techniques, such as pelvic floor exercises for stress incontinence, as well as biofeedback techniques and bladder training, may be very helpful adjuncts to the management of urinary incontinence (Hadley 1986).

Not all patients with urinary incontinence require urological consultation or detailed urodynamic evaluation. A reasonable approach includes a detailed history and physical examination with special attention to the pattern of incontinence and the presence of any neurological abnormality. As mentioned previously, a postvoid residual urine is an important aspect of the evaluation of all patients with incontinence. A detailed voiding chart, with the time of incontinence, amount of urine lost, presence of any symptoms, or associated events such as coughing, and straining at stool, is very helpful in discerning the likely dominant factors underlying incontinence. Based on this clinical information, the physician can generally design an initial therapeutic approach. If there is no or only slight clinical improvement, a detailed urodynamic evaluation is helpful.

Urinary incontinence does not reflect normal ageing and must be recognized as an important, often treatable, chronic disability in the older individual.

References

Andersen, G.H., Springer, J., Randall, P., Streeten, D.H., and Blakeman, N. (1980). Effect of age on diagnostic usefulness of stimulated plasma renal activity and salarasin test in detection of renovascular hypertension. *Lancet*, **ii**, 821–4.

Arieff, A.L. and Guisado, R. (1976). Effects on the central nervous system of hypernatremic and hyponatremic states. *Kidney International*, **10**, 104–16.

Arieff, A.L., Llach, F., and Massry, S. (1976). Neurologic manifestations and morbidity of hyponatremia: Correlation with brain water and electrolytes. *Medicine* (Baltimore), **55**, 121–9.

Ashraf, N., Locksley, R., and Arieff, A. (1981). Thiazide-induced hyponatremia associated with death or neurologic damage in outpatients. *American Journal of Medicine*, **72**, 43–8.

Berl, T. (1990). Treating hyponatremia: what is all the controversy about? *Annals of Internal Medicine*, **113**, 417–19.

Berry, S.J., Coffey, D.S., Walsh, P.C., and Ewing, L.L. (1984). The development of human benign prostatic hyperplasia with age. *Journal of Urology*, **131**, 474–9.

Diokno, A.C., Brock, B.M., Brown, M.B., and Herzog, A.R. (1986). Prevalence of urinary incontinence and other urological symptoms in the noninstitutionalized elderly. *Journal of Urology*, **136**, 1022–5.

Epstein, M. and Hollenberg, N.K. (1976). Age as a determinant of renal sodium conservation in normal men. *Journal Laboratory Clinical Medicine*, **87**, 411–17.

Fawcett, I.W., Hilton, P.J., Jones, N.F., and Wing, A. J. (1971) Nephrotic syndrome in the elderly. *British Medical Journal*, **2**, 387–8.

Fichman, M.P., Vorherr, H., Kleeman, G.R., and Telfer, N. (1971). Diuretic-induced hyponatremia. *Annals of Internal Medicine*, **75**, 853–63.

Fowler, F.J., Wennberg, J.E., Timothy, R.P., Barry, M.J., Mulley, A. G., and Handley, D. (1988). Symptoms, status and quality of life following prostatectomy. *Journal of the American Medical Association*, **259**, 3018–22.

Friedman, S.A., Raizner, A.E., Rosen, H., Solomon, N.A. and Sy, W. (1972). Functional defects in the aging kidney. *Annals of Internal Medicine*, **76**, 41–5.

Hadley, E. (1986). Bladder training and related therapies for urinary incontinence in elderly people. *Journal of the American Medical Association*, **256**, 372–9.

Hollenberg, N.K., Adams, D.F., Solomon, H.S., Rashad, A., Abrams, H.L., and Merrill, J.P. (1974). Senescence and the renal vasculature in normal man. *Circulatory Research*, **34**, 309–16.

Kleinfeld, J., Casimir, M., and Bona, S. (1979). Hyponatremia as observed in a chronic disease facility. *Journal of the American Geriatrics Society*, **27**, 156–61.

Lindeman, R.D., Tobin, J.D., and Shock, N.W. (1984). Association between blood pressure and the rate of decline in renal function with age. *Kidney International*, **26**, 861–8.

Mahowald, J.W. and Himmelstein, D.U. (1981). Hypernatremia in the elderly: Relation to infection and mortality. *Journal of the American Geriatrics Society*, **29**, 177–80.

Marshall, V.F., Marchetti, A.A., and Krantz, K.E. (1949). Correction of stress incontinence by simple vesicourethral suspension. *Surgical Obstetrics and Gynecology*, **88**, 509–18.

McGuire, E.J., Lytton, B., Pepe, V., and Kohorn, E.I. (1976). Stress urinary incontinence. *Obstetrics and Gynecology*, **47**, 255–6.

McLachlan, M.S.F. (1978). The aging kidney. *Lancet*, **ii**, 143–6.

Miller, P.D., Krebs, R.A., and Neal, B.S. (1982). Hypodipsia in geriatric patients. *Medicine* (Baltimore), **51**, 73–94.

Mohr, J.A., Roger, J., Brown, T.N., and Starkweather, G. (1983). Stress urinary incontinence: A simple and practical approach to diagnosis and treatment. *Journal of the American Geriatrics Society*, **31**, 476–8.

Phillips, P.A. *et al.* (1984). Reduced thirst after water deprivation in healthy elderly men. *New England Journal of Medicine*, **12**, 753–9.

Resnick, N.M. and Yalla, S.V. (1985). Management of urinary incontinence in the elderly. *New England Journal of Medicine*, **313**, 800–5.

Rowe, J.W., Andres, R., Tobin, J.D., Norris, A.H., and Shock, N.W. (1976a). Age-adjusted normal standards for creatinine clearance in man. *Annals of Internal Medicine*, **84**, 567–9.

Rowe, J.W., Andres, R., Tobin, J.D., Norris, A.H., and Shock, N.W. (1976b). The effect of age on creatinine clearance in man: A cross-sectional and longitudinal study. *Journal of Gerontology*, **31**, 155–63.

Shannon, R.P., Minaker, K.L., and Rowe, J.W. (1984). Aging and water balance in humans. *Seminars in Nephrology*, **4**, 346–53.

Smith, M.C., Ghose, M.K., and Henry, A.R. (1981). The clinical spectrum of renal cholesterol embolization. *American Journal of Medicine*, **71**, 174–80.

Wilson, J. D. (1980). The pathogenesis of benign prostatic hyperplasia. *American Journal of Medicine*, **68**, 745–56.

SECTION 16
Disorders of the blood

16 Disorders of the blood

HARVEY JAY COHEN

INTRODUCTION

Disorders of the blood are very common among elderly people. Debate continues to rage over what proportion of these changes are caused by age-associated alterations in the haemotopoietic system and what proportion by superimposition of the more frequently occurring acute and chronic illnesses of elderly people. The general approach to diagnosis is quite similar to that for younger individuals, although it may be modified by differing levels of expectation for certain processes and disorders. Although management of particular blood disorders may be in general similar to that in younger people, modifications may be required to accommodate the physiological alterations that occur with increasing age. This chapter will follow the overall format of the *Oxford Textbook of Medicine* and will not reiterate the information found there. Rather, it will concentrate on those areas in which there are issues of particular relevance to age-associated changes, and/or diseases that are particularly prominent in the older age group, and/or have manifestations and management issues that require a different approach in people of advanced age.

HISTORY AND PHYSICAL EXAMINATION

In the history and physical examination certain points should be borne in mind. The symptoms of haematological disorders may be subtle, sometimes chronic in onset, and rather non-specific. Therefore, it may be easy for the practitioner to assume that these are simply the changes of 'old age'. In fact, many older patients may themselves assume that changes such as increasing fatigue and weakness are occurring simply because they are 'getting older'. This tendency must be avoided by the physician and understood by the patient. Because of the increasing frequency of various disease processes with age, symptoms may also be masked by other manifestations of disease. Thus, secondary as well as primary haematological disturbances must be considered in the differential diagnosis for elderly individuals. Older patients may also present with manifestations that are rather atypical when compared with those of younger ones. Thus, the elderly person who presents with altered mental state, confusion, or delirium may be showing signs of a moderate anaemia that has finally reached the threshold at which function is compromised. In fact, functional alterations rather than specific and definable individual symptoms are often the presentation for elderly patients with disorders of the blood.

THE NORMAL BLOOD COUNT

In the absence of disease the stained blood-film components are unaltered by age. Whether certain components of the blood are quantitatively altered as a consequence of advancing age *per se*, or whether some of the changes that have been noted over the life course are those to which the increasing prevalence of other disease processes contribute remains controversial. This has been most disputed in defining the 'normal' range for haemoglobin and the haematocrit (O'Rourke and Cohen 1987; Zauber and Zauber 1987). This is of importance because defining the limit at which we begin to consider that an abnormal process such as anaemia is present will naturally affect the number of people regarded as having it. Of general population studies, a number have demonstrated that the mean haemoglobin declines slightly in both elderly men and women, with the mean and the great majority of the people remaining within the 'normal' range. However, most of these studies have included people in varying states of health, and are frequently flawed by being cross-sectional. When the lower limit of normal for haemoglobin is set at approximately 13 g/100 ml for men and 12 g/100 ml for women, there is an increasing prevalence with age of individuals who fall below this and who thus would be defined as anaemic. However, when such low values do occur, they by and large appear to be accompanied by processes that can explain them and are therefore not considered a normal concomitant of age. In many studies, as many as 20 to 25 per cent of people aged over 65 years will fall into the anaemic range by these criteria. For these reasons an appropriate cut-off point may be approximately 12 g/100 ml of haemoglobin for both men and women, indicating at least a moderate anaemia. Other measurements of red blood cells, such as mean corpuscular volume and mean corpuscular haemoglobin concentration, appear to be distributed normally in elderly subjects.

White blood-cell counts are largely unchanged, although longitudinal studies have shown a slight decrease in lymphocyte counts in the last 3 years of life, most likely concomitant with the occurrence of other diseases (Sparrow *et al.* 1980). Monocyte and platelet counts are unchanged. The erythrocyte sedimentation rate is raised above the usual normal range in 20 to 40 per cent of elderly people. However, most of these changes appear to be associated with various chronic or subacute illnesses or specific abnormalities of plasma proteins. When truly normal, healthy elderly people are assessed, the range of sedimentation rate is changed very little from that of normal young people (Crawford *et al.* 1987). Thus, significant deviations from this range should be assumed to be related to pathological processes.

BONE-MARROW ACTIVITY AND DISTRIBUTION

Histological studies have suggested that there might be a decline in marrow cellularity from 70 per cent in childhood to 30 per cent in the eighth decade of life (Hartsock *et al.* 1965). In animals, studies of radiolabelled iron distribution do not show marked changes with age in the pattern of marrow distribution (Boggs 1985).

HAEMATOPOIETIC STEM CELLS

Overall, for both pluripotential and committed stem cells, there appears to be little change in the basal level of function with advanced age. Thus, the levels of blast-forming units-erythroid (BFU-E) and granulocyte macrophage colony forming units (GM CFU) are little changed, although in some cases the latter appear to be decreased somewhat in older animals, perhaps more related to stressful conditions than age *per se*. However, there are age-associated deficiencies in the response of such stem cells to stimuli such as anaemia or infection. Thus, the basal function appears to be normal, but the homeostatic reserve appears to be somewhat diminished in response to stress (Cohen and Crawford 1986; Williams *et al.* 1986; O'Rourke and Cohen 1987; Resnitzky *et al.* 1987).

STEM-CELL DISORDERS

Acute leukaemia

Acute lymphocytic leukaemia (**ALL**) is the less common form in elderly adults. It is predominantly a disease of childhood, but has another peak of incidence in late life, though it is not clear whether these are all the same disease. With approaches to treatment that are similar to those designed for childhood ALL (prednisone – vincristine combinations as the base, sometimes adding methotrexate or *l*-asparaginase), this disorder appears to be much more responsive than acute myeloblastic leukaemia (**AML**) in the adult, but less responsive than the potentially curable forms of childhood ALL. However, these treatments can produce remissions in as many as 70 to 80 per cent of elderly patients, and this can frequently be achieved without the period of severe marrow aplasia required for the treatment of AML. This results in a lower rate of complications and more tolerable induction of remission. Over the long term, survival rates are poor, with as many as 80 per cent of elderly patients relapsing in the first year.

Acute myeloblastic leukaemia (AML)

This disorder is seen predominantly in elderly people, over half of all patients being over the age of 60 years at presentation (Freedman 1985). Presentation is often with non-specific symptoms, e.g., tiredness, weakness, anorexia, loss of weight. Frequently, the haematological findings are somewhat subtle, with peripheral blood cytopenia and few myeloblasts. These changes should be regarded with suspicion, and marrow examination of bone may reveal hypercellularity with replacement by immature myeloblasts. A further complicating factor is the high proportion of patients (up to one-third) who have had an antecedent preleukaemic or myelodysplastic phase. This might include cytopenia, sideroblastic anaemia, or other smouldering myelodysplastic syndromes. These patients may have a more indolent form of leukaemia that leads to a difficult decision when considering beginning treatment; watchful waiting and symptomatic treatment may be preferable to chemotherapy.

Standard chemotherapy involves a combination of agents, usually including cytosine arabinoside and anthracyclines such as daunorubicin. The goal is complete bone-marrow aplasia and normal cellular regrowth without leukaemic cells. In younger adults, there may be remission rates as high as 70 per cent, with improved overall survival. In some studies of elderly individuals, response rates as high as 50 per cent have been recorded, but most studies show lower rates than this (Champlin *et al.* 1989). Moreoever, there is a high incidence of death during induction, mostly because elderly patients cannot withstand the severe toxicity of the regimens, with resultant infection and bleeding. For those who do achieve remission, it may last as long as in the young, but the median survival is only in the general range of 1 year. Thus, the likelihood of achieving remission and a somewhat prolonged survival must be set against the high risk of early death, as well as the extensive comorbidity, and time required in hospital for most patients, even those that do not respond. Those with antecedent myelodysplastic syndromes have particularly poor responses. In one study, when this group of poor responders was excluded from the analysis, the remainder had similar responses to those of younger adults (Freedman 1985).

Because of these poor responses, suggested alternatives are lower doses of cytosine arabinoside, which may produce a similar period of response but with lower toxicity, or supportive care only (Copplestone *et al.* 1989; Powell *et al.* 1989). This decision should be made with regard to the physiological condition of the patient and the extent of and comorbidity as well as to their previous haematological state. It should include a frank discussion with patient and family of the likely short- and long-term outcomes.

Chronic granulocytic and lymphocytic leukaemia (CGL, CLL)

The approach to these disorders in the older patient is little different from that in the younger. CLL is chiefly a disease of old age: true cure or complete remission is rarely seen and the aim is to control symptoms; the decision to treat should be based on the current and expected quality of life. In general, treatment is not recommended for the asymptomatic stages where there is no evidence of impingement upon bone-marrow function. When treatment is required, a regimen of an alkylating agent plus prednisone is generally well tolerated by elderly individuals, especially as severe cytopenia and bone-marrow aplasia are not goals of therapy.

Polycythaemia and other myeloproliferative syndromes

These myeloproliferative disorders mostly occur in elderly individuals. Diagnosis demands a high index of suspicion because elderly patients with polycythaemia will often present with rather vague symptoms such as headaches, dizziness, or a thrombosis. Moreoever, because of the reduced cerebral blood flow that occurs with haematocrits of above 60 per cent, the elderly patient may present with atypical findings, such as altered mental status or confusion, especially if the haematological disorder is superimposed upon other conditions producing a marginally compensated cognitive state. In these cases, phlebotomy to lower the blood viscosity should first be done with due caution and not so quickly as to create haemodynamic changes that could further compromise the circulation. In the long run, however, phlebotomy alone is not enough to control the episodes of thromboembolism and bleeding, and specific therapy is required. Radioactive phosphorus (^{32}P) is effective and well tolerated. Hydroxyurea may be an excellent alternative for more rapid lowering of the blood counts; it may be especially useful in states such as primary thrombocytosis when rapid reduction of the platelet count is necessary. Myelosclerosis and myeloid metaplasia may be even more insidious in onset and overlap with the other myeloproliferative disorders in presentation. As these may be indolent in the older individual, and as chemotherapy may do more harm than good, symptomatic therapy is often the preferred approach.

Aplastic anaemia and other myelodysplastic syndromes

In elderly people, the predominant forms of aplastic anaemia are either idiopathic or presumed to be drug-induced. Elderly individuals may be particularly prone to the latter because they use more prescription and over-the-counter medicines than do younger persons. As bone-marrow transplantation is not at present in general use for patients over the age of 50, supportive care, and perhaps attempts at androgenic steroid treatment, are the only options. For the elderly patient with severe aplasia that does not spontaneously remit with drug withdrawal, the outlook is poor.

The myelodysplastic syndromes, including the spectrum of refractory anaemia, sideroblastic anaemia, and refractory anaemia with excess blast cells, are almost exclusively found in elderly adults (Gardner 1987). As mentioned above, they may present insidiously and be a precursor to subsequent myeloproliferative transformation. Supportive care is generally given. Such approaches as low-dose cytosine arabinoside have been used in efforts to achieve differentiation or maturation of cells in patients with myelodysplastic syndromes. However, despite their apparent attractiveness, they are associated with significant toxicity and must be used with caution, especially in elderly patients.

ANAEMIA

Iron metabolism and deficiency

Iron metabolism generally alters little with age. Although there may be a slight decrease in iron absorption, perhaps conditioned by increased gastric pH, this is rarely enough to create a deficiency state. In general, iron intake is adequate in elderly people. Some, such as those with very low caloric intake among nursing-home patients, those living alone or disabled, or those of very low socioeconomic status, may have a moderate decline in intake, but even this generally exceeds the minimum daily requirements, and dietary iron deficiency is extremely uncommon among elderly persons. Deficiency results almost exclusively (at least in developed nations) from blood loss, most commonly from the gastrointestinal tract, because the contribution from menstrual blood loss is absent in postmenopausal women.

The anaemia of iron deficiency is generally microcytic hypochromic, though in earlier phases, it may be normochromic. The diagnosis is generally made by laboratory investigation (O'Rourke and Cohen 1987; Babitz and Freedman 1988; Walsh 1989). Unfortunately bcause of the prevalence of other disorders in elderly people, serum iron and iron-binding capacity are not reliable indicators. There is a tendency for the serum iron to decrease somewhat with advancing age in the absence of iron deficiency, and many chronic illnesses can lower both serum iron and iron-binding capacity. Although the final diagnosis may be made by bone-marrow aspiration, stained for storage iron, one would like to avoid this procedure if possible in the elderly individual. Measurement of serum ferritin may allow one to do so. The normal range of serum ferritin tends to rise somewhat with increasing age, with the concentrations for older men and women converging. A low serum ferritin is diagnostic of iron deficiency and can be accepted without bone-marrow examination in some cases. A serum ferritin below 12 μg/l is a false/positive diagnosis in under 3 per cent of cases. A serum ferritin of under 45 μg/l in an elderly individual is highly predictive of iron deficiency, though liver disease and inflammatory disorders may sometimes raise the ferritin into this range. In elderly patients, the major differential diagnosis of iron deficiency is the anaemia of chronic disease. Though in its mild stages this is generally normochromic and normocytic, not infrequently it may be hyprochromic microcytic. In that case the serum ferritin, if low, is the best discriminator. Additional diagnostic information may be contributed by the red blood cell distribution width index, a measurement of anisocytosis made with automatic blood-cell counters. This develops early in iron deficiency and generally to a greater extent than in the anaemia of chronic disease. However, there is considerable overlap between the two.

Once the diagnosis of iron deficiency is made, the most important step in an elderly individual is to identify its cause. Because of the frequency of peptic ulceration, colon cancer, diverticulitis, and angiodysplasia, examination of the gastrointestinal tract should be thorough. A common complication is that elderly individuals may be taking various and numerous drugs that may cause gastrointestinal side-effects and bleeding. Nevertheless the cause can be identified in most elderly patients with iron deficiency anaemia. Treatment is generally by standard methods, but sometimes ferrous sulphate may be poorly tolerated because of increased gastric sensitivity. When this occurs, a paediatric liquid suspension may be considered as an alternative. If parenteral iron therapy is considered, intravenous administration may sometimes be preferable because of the difficulty of deep intramuscular injections in elderly individuals with decreased muscle mass.

Normocytic normochromic anaemias

The anaemic of chronic diseases, especially of those known to decrease marrow production (such as renal disease, malignant or inflammatory disorders, or infections to which elderly people are particularly prone) is the most common normocytic normochromic anaemia in this age group. The basic cause appears to be a reticuloendothelial cell blockade, sometimes complicated by decreased production of erythropoietin. Generally the anaemia is mild but can completely mimic that of iron deficiency. If the serum ferritin is markedly elevated one can be fairly secure in this diagnosis. Generally a serum ferritin about 50 μg/l is associated with this type of anaemia. The serum iron and iron-binding capacity are unreliable although, in the equivocal ranges of serum ferritin, very low iron saturation (under 10 per cent) is more suggestive of iron deficiency anaemia.

Megaloblastic anaemia

In elderly individuals, megaloblastic anaemias generally result from deficiency of folic acid or vitamin B_{12} (O'Rourke and Cohen 1987). As these disorders are usually correctable, diagnosis should be pursued vigorously. Folic acid deficiency most often results from nutritional deficiency, but other causes, such as malabsorption or increased utilization, can occur. General nutritional surveys show that the average daily folate intake of elderly persons is quite sufficient, and in the absence of illness or marked lack of eating, folate deficiency is not a widespread problem among elderly people living in the community in the United States. However, elderly people on low incomes, alcoholics, and elderly residents of institutions are more likely to have decreased intake and prevalence rates as high as 15 to 20 per cent have been reported in these groups. The serum folate may be helpful in diagnosis but as many as 10 per cent of apparently healthy aged individuals have folates below the normal limits. The red blood-cell folate is a much more reliable indicator of true tissue folate deficiency and can be used to substantiate the finding of an equivocal serum folate. In elderly individuals, folate deficiency has been associated with neuropsychiatric symptoms, but a direct causal relationship has not clearly been established.

The other major cause of macrocytic anaemia in elderly patients is deficiency of vitamin B_{12}. In contrast with folate deficiency, B_{12} deficiency is rarely produced by lack of dietary intake. Because of the large body stores of this vitamin, and the high content of vitamin B_{12} in the diet (at least in the United States) deficiency is rarely seen except in the strictest, life-long vegan or vegetarian. Other forms of malabsorption, such as arise after total gastrectomy or disease of the terminal ileum, may also produce vitamin B_{12} deficiency. But the most common form in old age is pernicious anaemia, due to impaired secretion of intrinsic factor by the gastric parietal cell, which accounts for two-thirds of cases of vitamin B_{12} deficiency. The lack of intrinsic factor seems to be secondary to chronic atrophic gastritis, a disorder of unknown aetiology that increases in frequency with age. The deficiency can produce a number of neurological changes including degenerative changes in peripheral nerves. Neuropsychiatric symptoms of delirium and confusion are a well-established feature of vitamin B_{12} deficiency, but a clear causal relationship between this deficiency and more fixed, true dementia has not been clearly shown. Moreover, descriptions of true dementia associated with vitamin B_{12} deficiency in the absence of any haematological abnormality are rare.

Measurements of vitamin B_{12} have been used as primary diagnostic tools in the assessment of anaemia, as well as in screening populations for vitamin B_{12} deficiency. There appears to be a progressive fall in the serum B_{12} with advancing age; this generally does not appear to reflect depleted tissue stores of B_{12}. Some of these low serum levels may be caused by alterations in transcobalamin, the B_{12}-binding protein, in elderly individuals. The diagnosis is established with the Schilling test, which can be done in any elderly individual capable of collecting 24-h urine samples. Whether there is true tissue deficiency of B_{12} may be approached by measurement of methylmalonic acid in serum or urine. This is frequently normal in borderline cases of low B_{12}, indicating that while true vitamin B_{12} deficiency may be prevalent in elderly persons, a low B_{12} is not pathognomonic for tissue deficiency, and may require further evaluation rather than immediate treatment (Nilsson-Ehle *et al.* 1989). Correct treatment in older as in younger adults should result in a brisk reticulocytosis with both B_{12} and folate replacement. A complicating feature may be the presence of combined deficiency, e.g., B_{12} and folate or one of these plus iron deficiency. If there is an initial response but not full recovery, a combined deficiency should be considered. In particular there may have been a limited amount of iron present in the bone marrow, that has been depleted during the initial rapid erythropoietic response to folate or B_{12} replacement. These individuals will also require iron replacement, and evaluation of the reason for reduced stores in the first place. {Treatment of severe pernicious anaemia should be carried out slowly with very small doses of vitamin B_{12} in elderly patients to reduce the risk of sudden death thought to be due to a massive shift of potassium from the extracellular fluid into red cells released from maturation arrest. Potassium supplements are commonly prescribed although not of proven efficacy.}

Disorders of the synthesis or function of haemoglobin

These disorders are largely those of younger individuals. However, it should be remembered that mild forms of thalassaemia may remain undetected and be found for the first time only when an elderly individual presents either for a totally unrelated disorder, or when the haemoglobin level falls further than the level at which it has been chronically maintained. In this case, investigations for microcytic anaemia must include the possibility of thalassaemia. It is important to distinguish this process from that of iron deficiency because inappropriate treatment with iron could result in overload. Likewise for the sickling disorders: sickle-cell anaemia usually results in death before old age, but those with sickle-cell trait appear to have normal survival, and when an elderly person with the trait becomes seriously ill for the first time in late life, he or she might be at risk for problems such as deoxygenation during anaesthesia or other stresses. Thus, if not previously done, appropriate investigations would be required in an elderly individual whose racial background predisposes to sickling.

Sideroblastic anaemias

These disorders are mostly seen in elderly individuals; the idiopathic forms have already been considered. The secondary form may frequently arise because of the number of the drugs that may be potentially responsible for it (such as isoniazid) and because of the prevalence of many systemic diseases that result in ring sideroblasts. These include infections, hypothyroidism, and rheumatological diseases.

Haemolytic anaemias

The inherited haemolytic anaemias are generally thought of as occurring in younger individuals, but these must be considered in older people as well. For example, patients with mild hereditary spherocytosis may well present for the first time with haemolysis under the stress of their first serious infectious disease in old age, after being in previous good health. Likewise, patients with enzyme deficiencies, such as of glucose 6-phosphate dehydrogenase, may not incur enough stress from either diseases or drugs to induce haemolytic manifestations until late in life. Therefore, in an elderly individual with an appropriate racial or ethnic background who presents with an acute haemolytic episode, enzyme deficiencies must be considered if no other apparent cause is found.

Immune haemolytic anaemia, especially the 'warm antibody', IgG-induced variety, is more common in elderly people. This is especially true of drug-induced immune haemolysis because older people use more drugs. Coombs-positive haemolytic anaemia frequently accompanies, and may be the first presentation of, chronic lymphocytic leukaemia, a disorder of older age, which contributes to its increased prevalence. Chronic cold-agglutinin disease, a disorder characterized by both haemagglutination and haemolysis caused by cold-reacting IgM antibodies, occurs almost exclusively in older individuals. This disorder should particularly be considered in differential diagnosis when the Coombs' test is positive but there is no IgG on the cell surface. In many individuals with cold-agglutinin disease, symptoms may be more related to the agglutination than to the haemolysis. The explanation for the increase in these types of antibodies with age is unknown. However, there is an increase in a wide variety of autoantibodies among elderly individuals (including rheumatoid factor and antithyroid antibodies) but frequently these do not result in specific disease. Non-immune forms of haemolytic anaemia, such as those related to infections, and both macro- and microintravascular haemolysis occur more often in old age because the primary conditions with which they are associated do so too.

LEUCOCYTES IN HEALTH AND DISEASE

Adequate neutrophil function is largely maintained throughout life. There is some evidence that, in elderly individuals, leucocytes may be less readily mobilized by stressful challenges such as bacterial invasion or steroids. This may be at the root of the clinical observation that infections induce a less vigorous granulocyte response in older people. However, accurate measurements of the maintenance of neutrophil function with age have been complicated by the differing populations studied. Some studies have shown that elderly individuals are less able to phagocytose and kill bacteria, and produce superoxide; other studies of narrowly defined, healthy populations have shown more 'normal' chemotaxis, bacterial killing, and oxidative metabolism (Corberand *et al.* 1986; Nagel *et al.* 1986; Udupa and Lipschitz 1987; Mege *et al.* 1988). This is further complicated by the fact that neutrophil function may be impaired secondarily by age-associated disorders such as diabetes, as well as by events more connected with nutritional deficiencies. Neutropenia is not an uncommon finding in elderly individuals, especially in relation to drug use. It may be either immune or non-immune, and may be a mild, chronic form or acute and severe.

Lymphocytes are functionally compromised with progressive age. The most dramatic aspect of this is the failure of lymphocytes from elderly individuals to respond appropriately to mitogenic stimuli (Cohen 1989). This appears to be predominantly a defect in the T-cell arm of the immune system related to cell-mediated immunity. Altered immunoglobin production by B lymphocytes, when seen, appears to be mainly a secondary outcome of altered T-cell regulation. There are no major quantitative or qualitative changes in other peripheral blood cells such as monocytes, eosinophils, and basophils.

LYMPHOPROLIFERATIVE DISORDERS

The lymphomas

Hodgkin's disease has a second peak of incidence in the elderly population. Because of its heterogeneous nature, it is not completely clear if the disease occurring in old age is the same as that in the young adult. Elderly individuals appear to present at a more advanced stage than their younger counterparts, and their histological type is more often lymphocyte depletion. In general, the responses to treatment and survival rates in older age groups are worse than those for the younger ones (Freedman 1985; Miescher and Jaffe 1988). Moreover, treatment-related complications may be more severe. In Hodgkin's disease treated with curative radiation, regeneration of bone marrow activity is considerably delayed in elderly patients, and substantial gastrointestinal toxicity may also occur. In some studies, many elderly patients were unable to undergo full staging of the disease because of difficulty in tolerating laparotomy. It is not clear, however, whether this is related more to co-morbidity than to age *per se*. Chemotherapy for Hodgkin's disease is poorly tolerated by elderly patients, with early toxicity and treatment-related deaths.

Non-Hodgkin's lymphomas are rarely localized in older people and tend to present as disseminated disease from the outset without obvious spread from node to node. It is particularly useful to define the histological type because this will help to define the therapeutic options (Miescher and Jaffe 1988). In general, the treatment is systemic chemotherapy. Histologically low-grade malignancies of favourable prognosis may be indolent; it is best to withhold treatment of these in elderly patients until the pace of the disease can be determined. In many instances, treatment will not be required for many years and when it is needed usually a single alkylating agent with or without vincristine and prednisone may produce good control and prolonged partial remissions.

On the other hand, the histologically high-grade lymphomas of unfavourable prognosis, especially when advanced, present a more aggressive picture. Treatment of these usually involves intensive multiagent chemotherapy (Miescher and Jaffe 1988). A large number of combinations have been devised; these commonly include cyclophosphamide, Adriamycin, prednisone, bleomycin, and methotrexate in varying numbers and combinations. Those regimens can produce cures in significant numbers of individuals and at least long-term remissions in many. However, the initial response and survival are often much worse in elderly patients (Solal-Celigny *et al.* 1987; Boyd *et al.* 1988; Jones *et al.* 1989; Tirelli 1989). This appears to be because full chemotherapeutic dosages cannot be given, owing to their excessive toxicity, and because treatment itself is too damaging. However, in other studies, although the same marked differences in survival were found, the deaths were due to other age-related diseases rather than to either toxicity or lymphoma (Vose *et al.* 1988). Most elderly patients, at least those over the age of 75, are not likely to be able to tolerate full doses of currently available regimens. Whether some remissions and better survival would come from lower dosage has not yet been established. Also, for those evaluating trials of different multiagent chemotherapies, knowledge of the age composition of the groups under treatment will be of great importance in estimating the equivalence of response.

Paraproteinaemia and myelomatosis

The incidence of monoclonal gammopathy is strikingly age-associated. The most frequent type, idiopathic monoclonal gammopathy (probably a better term than benign monoclonal gammopathy as, over time, this is not always benign), rises in prevalence from approximately 1 per cent in individuals over the age of 50 to as much as 14 per cent in those over the age of 90. This prevalence is some hundredfold greater than that of the related disease, multiple myloma. Upon detection of a monoclonal protein abnormality in the serum, the most pressing issue for determining treatment is whether active disease is present and, if not, what is the likelihood of its occurrence. The diagnosis of overt disease is made by standard clinical criteria in the elderly. Idiopathic monoclonal gammopathy (or monoclonal gammopathy of unknown significance) is defined by the absence of any clinical abnormalities other than the serum protein. There are intermediate forms, known as smouldering or indolent myeloma, that may be accompanied by a few clinical changes, but generally not severe anaemia or renal failure, and little in the way of bone lesions. In these instances, the patient should be followed over the course of months to see if there is any sign of progression. Frequently there will be none, and these patients may not require treatment for a number of years, if at all. There is no certain way of determining whether or not a particular monoclonal protein abnormality will become progressive. Transition to overt clinical disease occurs in approximately 2 per cent of these patients per year (among those who not not die from other causes). A high thymidine labelling index of plasma cells, or high levels of β_2-microglobulin are associated with transition to more aggressive disease, but these are not totally predictive of transition in the individual patient. Once the diagnosis of overt active multiple myeloma is made, treatment is generally required. Standard combination chemotherapy regimens have a response rate of approximately 50 to 60 per cent, with median survivals of approximately 3 years. Combinations of two and three agents appear to be well tolerated by elderly individuals, with response rates, survivals, and toxicity similar to those of younger people (Cohen 1988). There is no evidence that larger, more aggressive, drug combinations are more beneficial.

The hyperviscosity syndrome may be particularly troublesome in the elderly individual (Crawford *et al.* 1985). This direct effect of the abnormal serum protein is most prominent in Waldenström's macroglobulinaemia because of the intrinsically high viscosity of the IgM protein. It is, however, seen with IgG and IgA protein abnormalities as well. Hyperviscosity may be problematical because of an already precariously balanced circulatory system; the plasma volume may be expanded and congestive heart failure may occur in an elderly person with a reduced cardiovascular reserve. Careful plasmapheresis to reduce viscosity and plasma volume may be beneficial in the short term, but treatment with alkylating agents is generally required for long-term control. One must particularly avoid excessive blood transfusion in these patients; their haematocrit may appear low, particularly as a result of the increased plasma volume, but red blood-cell transfusion may further complicate the plasma viscosity and produce further cardiovascular collapse.

HAEMOSTASIS AND THROMBOSIS

There do not appear to be major alterations in clotting factors in elderly people, although mild deficiencies of congenital factors may first be noticed in older age. Certain non-specific factor inhibitors such as the lupus anticoagulant appear to occur more frequently in elderly persons, consistent with the general tendency towards an increase in autoimmune phenomena. Thrombocytopenia is not at all uncommon in elderly individuals because of the large number of disorders in which it may be a secondary feature, primary as well as secondary marrow disorders, and immunological causes. Although the general approach to treatment is similar across the age range one may be more reluctant to commit the elderly individual to chronic courses of corticosteroids, thus opting for earlier splenectomy when deemed appropriate. As with many of the conditions already described, the greater use of a variety of drugs by elderly people places them at particular risk for drug-induced thrombocytopenia.

The most common problem of haemostasis in elderly persons is 'senile purpura', which is presumably due to age-associated loss of subcutaneous fat and the collagenous support of small blood vessels. This needs to be distinguished from steroid-induced purpura or purpura secondary to vasculitis, which is less common. Secondary thrombocytosis associated with chronic inflammatory states, malignancies, and iron deficiency may occur but generally does not produce bleeding or thrombotic complications. Thrombocytosis due to the primary myeloproliferative disorders described earlier appears more likely to result in such complications.

Thrombosis-related events clearly have an increased incidence in advanced age. Venous thromboembolism is particularly prominent and arterial disease also increases. Thromboembolism is frequently related to stasis, a problem to which elderly people are particularly prone. Arterial disease may possibly be related to hypercoagulability. A number of studies have shown increased platelet aggregation with age (Kasjanovova and Balaz 1986; Vericel *et al.* 1988). There is also evidence of age-related hyperactivity of the haemostatic system due either to suppressed heparin–antithrombin III complex or excessive generation of factor Xa. These abnormalities could create a prethrombotic state, thus increasing susceptibility to subsequent thrombosis (Bauer *et al.* 1987). These conditions are managed as in younger individuals. However, anticoagulant therapy has been associated with increased bleeding complications in elderly people, perhaps due to altered vascular stability. Anticoagulation may also complicate multiple drug regimens, producing interactions with anti-inflammatory agents, antibiotics, and barbiturates. This simply means that anticoagulant therapy must be closely monitored by clinical and laboratory means.

BLOOD AND SYSTEMIC DISEASE

Because of the increasing prevalence of multiple diseases in elderly people, the haematopoietic system is particularly prone to secondary manifestations. Elderly individuals must be evaluated for age-associated changes in the haemotopoietic system that predispose to systemic changes, as well as for superimposed, age-associated diseases.

References

(References marked * are general reviews.)

*Babitz, L.E. and Freedman, M.L. (1988). Anaemia in the aged. *Comprehensive Therapy*, **14,** 55–64.

Bauer, K.A., Weiss, L.M., Sparrow, D., Vokonas, P.S., and Rosenberg, R.D. (1987). Aging-associated changes in indices of thrombin generation and protein C activation in humans. *Journal of Clinical Investigation* **80**, 1527–34.

Boggs, D.R. (1985). Hematopoiesis and aging. Mass and distribution of erythroid marrow in aged mice. *Experimental Hematology*, **13**, 1044–4.

Boyd, D.B. *et al.* (1988). COMBLAM III: infusional combination chemotherapy for diffuse large-cell lymphoma. *Journal of Clinical Oncology*, **6**, 425–33.

*Champlin, R.E., Gajewski, J. L., and Golde, D. W. (1989). Treatment of acute myelogenous leukemia in the elderly. *Seminars in Oncology: Cancer in the Elderly*, **16**, 51–6.

*Cohen, H.J. (1988). Monoclonal gammopathies and aging. *Hospital Practice*, **23**, 75–100.

*Cohen, H.J. (1989). Immune Regulation. In *Textbook of internal medicine*, Vol. 2, (ed. W.N. Kelley), pp. 2581–4. Lippincott, Philadelphia.

*Cohen, H.J. and Crawford, J. (1986). Hematologic problems. In *The practice of geriatrics*, (ed. E. Calkins, P. J. Davis, and A. B. Ford), pp. 519–31. Saunders, Philadelphia.

Connors, J.M. (1988) Infusions, age, and drug dosages: Learning about large-cell lymphoma (Editorial). *Journal of Clinical Oncology*, **6**, 407–8.

Copplestone, J.A. *et al.* (1989). Treatment of acute myeloid leukemia in the elderly: a clinical dilemma. *Hematological Oncology*, **7**, 53–9.

Corberand, J.X., Laharrague, P.F., and Fillola, G. (1986). Neutrophils of healthy aged humans are normal. *Mechanisms of Ageing and Development*, **36**, 57–63.

Crawford, J.C., Cox, E.B., and Cohen, H.J. (1985). Evaluation of hyperviscosity in monoclonal gammopathies. *American Journal of Medicine*, **279**, 13–22.

Crawford, J., Eye-Boland, M.K., and Cohen, H.J. (1987). The clinical utility of erythrocyte sedimentation rate, plasma viscosity and protein analysis. *American Journal of Medicine*, **82**, 239–46.

*Feedman, M.L. (ed.) (1985). Hematologic disorders. In *Clinics in geriatric medicine*, Vol. 4, pp. 699–924. Saunders, Philadelphia.

*Gardner, F.H. (1987). Refractory anemia in the elderly. *Advances in Internal Medicine*, **32**, 155–76.

Hartsock, R.J., Smith, E.B., and Petty, C.S. (1965). Normal variation with aging of the amount of hematopoietic tissue in the bone marrow from the anterior iliac crest. *American Journal of Clinical Pathology*, **43**, 326.

Jones, S.E., Miller, T.P., and Connors, J.M. (1989). Long-term follow-up and analysis for prognostic factors for patients with limited-stage diffuse large-cell lymphoma treated with initial chemotherapy with or without adjuvant radiotherapy. *Journal of Clinical Oncology*, **7**, 1186–91.

Kasjanovova, D. and Balaz, V. (1986). Age-related changes in human platelet function *in vitro*. *Mechanisms of Ageing and Development*, **37**, 175–82.

Mege, J.L. *et al.* (1988). Phagocytic cell function in aged subjects. *Neurobiology of Aging*, **9**, 217–20.

*Miescher, P.A. and Jaffe, E.R. (ed.) (1988). *Seminars in hematology: advances in chemotherapy for Hodgkin's and non-Hodgkin's lymphomas*, Vol. XXV, No. 2, Suppl. 2. Grune & Stratton, Philadelphia.

Nagel, J.E. *et al.* (1986). Age differences in phagocytosis by polymorphonuclear leukocytes measured by flow cytometry. *Journal of Leukocyte Biology*, **39**, 399–407.

Nilsson-Ehle, H. *et al.* (1989) Low serum cobalamin levels in a population study of 70- and 75-year-old subjects. *Digestive Diseases and Sciences*, **34**, 716–23.

*O'Rourke, M.A. and Cohen, H.J. (1987). Anemias. In *Geriatric medicine annual 1987*, (ed. R.J. Ham), pp. 237–66. Medical Economics, New York.

Powell, B.L. *et al.* (1989). Low-dose ara-C therapy for acute myelogenous leukemia in elderly patients. *Leukemia*, **3**, 23–8.

Resnitzky, P., Segal, M., Barak, Y., and Dassa, C. (1987). Granulopoiesis in aged people: Inverse correlation between bone marrow cellularity and myeloid progenitor cell numbers. *Gerontology*, **33**, 109–14.

Solal-Celigny, P. *et al.* (1987). Age as the main prognostic factor in adult aggressive non-Hodgkin's lymphoma. *American Journal of Medicine*, **83**, 1075–9.

Sparrow, D., Silbert, J.E., and Rowe, J.W. (1980). The influence of age on peripheral lymphocyte count in men: a cross-sectional and longitudinal study. *Journal of Gerontology*, **35**, 163.

Tirelli, U. (1989). Management of malignant lymphoma in the elderly: An EORTC retrospective evaluation. *Acta Oncologica*, **28**, 199–201.

Udupa, K.B. and Lipschitz, D.A. (1987). Effect of donor and culture age on the function of neutrophils harvested from long-term bone marrow culture. *Experimental Hematology*, **15**, 212–16.

Vericel, E., Croset, M., Sedivy P., et al. (1988). Platelets and aging. I. Aggregation, arachidonate metabolism and antioxidant status. *Thrombosis Research*, **49**, 331–42.

Vose, J.M. *et al.* (1988). The importance of age in survival of patients treated with chemotherapy for aggressive non-Hodgkin's lymphoma. *Journal of Clinical Oncology*, **6**, 1838–44.

*Walsh, J.R. (1989). Equivocal anemia in the elderly. *Journal of Family Practice*, **28**, 521–3.

Williams, L.H., Udupa, K.B., and Lipschitz, D.A. (1986). Evaluation of the effect of age on hematopoiesis in the C57BL/6 mouse. *Experimental Hematology*, **14**, 827–32.

Zauber, N.P. and Zauber, A.G. (1987). Hematologic data of healthy very old people. *Journal of American Medical Association*, **257**, 2181–4.

SECTION 17
Skin disease

17 Skin disease

ARTHUR K. BALIN

INTRODUCTION

In this chapter the physiological and biological basis for the increased susceptibility to skin disease in elderly people is reviewed and skin conditions that are particularly prevalent in older persons are discussed. Emphasis is given to those conditions the physician is likely to encounter during the course of daily practice. The reader is referred to recent dermatology texts for more detail regarding skin disease in older persons (Marks 1987, Balin *et al.* 1989; Newcomer and Young 1989).

The expression and treatment of cutaneous disease in older persons differ from those in younger adults. Anatomical changes in ageing skin result in altered physiological behaviour and susceptibility to disease. Decreased epidermal renewal and tissue repair accompany ageing. The rate of hair and nail growth typically declines, as does the quantity of eccrine, apocrine, and sebum secretion. There are alterations in immune surveillance and antigen presentation with ageing. The cutaneous vascular supply is decreased, leading to decreases in inflammatory response, in absorption, and in cutaneous clearance. Impaired thermal regulation, tactile sensitivity, and pain perception occur as one ages. These changes result in altered expression of cutaneous disease and indicate a need for specific modifications in treatment and prevention of cutaneous diseases in older persons.

The first section of this chapter summarizes the major changes that occur during the intrinsic ageing process of the skin to facilitate the recognition and treatment of skin disease in the older patient. The second section discusses the expression of skin disease in elderly people. The last section reviews selected specific clinical conditions prevalent in the elderly population.

CHANGES IN THE SKIN WITH AGE

Stratum corneum

There is little change in the number of layers of cells that compose the stratum corneum during ageing. The thickness of the stratum corneum and its resistance to the diffusion of water vapour are similar in young and old people. As a result the barrier function of the stratum corneum is preserved during ageing. The moisture content of the stratum corneum of aged skin is less than that in younger adults and as a consequence is somewhat more brittle. In addition, the corneocytes of aged skin have been found to become larger and less cohesive than those in younger skin.

Turnover of the stratum corneum reflects the renewal time of the epidermis and has been found to take longer in the aged individual.

Dry skin, and roughness of the skin with aberrant light scattering are consequences of these age-associated changes in the stratum corneum. The longer renewal time means that irritant and sensitizing substances that come into contact with the skin will remain longer and that substances, including medications, that are placed on the skin take longer to be shed. The treatment time needed to clear superficial fungal infections is increased because of the slower stratum corneum renewal.

Keratinocytes

The thickness of the epidermis between the rete ridges remains constant or decreases only slightly during ageing. There is, however, a pronounced effacement of the rete ridges. The basal keratinocytes with the highest degree of proliferative capacity and proliferative reserve are located at the bottom of the epidermal rete ridges and the effacement of these structures reflects the decreased proliferative reserve of the aged epidermis. The decrease in basal cells per unit area leads to a reduction in the reproductive compartment and in epidermal turnover. Basal cells from light-exposed regions of elderly donors show greater variability in their size, shape, and electron density than do similar regions from young donors. Basal cells from aged individuals have a paucity of microvilli and the basal cells are larger and cells in the spinous layer are smaller in elderly individuals. Certain epidermal functions are decreased in aged epidermis. For example, elderly skin is less responsive to the effect of ultraviolet irradiation in converting provitamin D_3 to previtamin D_3.

The decrease in the epidermal rete ridges leads to a reduced area of contact between the dermis and epidermis, resulting in an epidermis that separates from the underlying dermis more easily than in the younger individual. Simple trauma such as application and removal of a sticking plaster, or a tightly fitting shoe may peel off the epidermis in older persons.

The renewal of the epidermis, like the stratum corneum, is slower in persons older than 60 years and epidermal wound healing takes longer. As with the stratum corneum, substances that come into contact with the epidermis remain for a longer time before they are shed.

Cytoheterogeneity of the individual keratinocyte nuclei is observed in aged skin which may reflect some disordered regulation of proliferation and could contribute to the excess cutaneous growths such as seborrhoeic keratosis and skin tags that are nearly universal in older persons.

The epidermal response to photodamage differs from that seen in intrinsic ageing. The initial response to UV light is a hyperproliferative response to injury and a thickening of the epidermis. Late effects of severe UV irradiation injury result in marked epidermal atrophy.

The number of Langerhans cells is decreased in aged sun-protected skin and decreases even more in sun-damaged skin, leading to a decreased ability to sensitize with contact allergens. In addition, in experimental animals the UV light-induced decrease in Langerhans cells leads to improper antigen presentation resulting in the production of supressor T cells that impair tumour rejection.

Most of the studies that have measured the number of melanocytes as a function of age have only employed a small number of subjects and have not adequately controlled for the amount of light exposure the subject received before measurement. In aggregate, it does appear that there is some decline in the number of melanocytes with age. Probably more important, however, are the observations that the function of the remaining melanocytes is abnormal. One consequence of these changes is that UV exposure produces less effective pigment protection in older persons.

Greying of hair is a manifestation of loss of melanocyte function and is one of the earliest changes we associate with ageing. In an Australian study of 6000 males, some grey hair was found in 22 per cent of men 25 to 34 years, 61 per cent of men 35 to 44 years, 89 per cent of men 45 to 54 years, and 94 per cent of men older than 55 years. The observation that scattered individual hairs go grey independent of their neighbours illustrates the heterogeneity that is characteristic of ageing.

The dermis becomes thinner with age. In addition, it becomes less cellular and less vascular. Measurements of the total amount of dermal collagen reveal a decrease of about 1 per cent per year. The remaining collagen fibres thicken, become less soluble, have less capacity to swell and become more resistant to digestion by collagenase. Histologically, the collagen fibres appear to be deposited haphazardly in coarse rope-like bundles rather than in an orderly fashion as in younger skin.

Men have a thicker dermis than women. This may explain why female skin seems to deteriorate more readily with ageing. Thinner skin is more easily damaged by actinic damage and trauma.

Actinic damage and intrinsic ageing result in different changes in elastic tissue. In intrinsic ageing the fine subepidermal oxytalan fibres are lost, eventually contributing to superficial laxity, loss of resilience, and finely wrinkled appearance of the skin. These intrinsic degenerative changes in the elastic tissue begin at about age 30. As they progress, cystic spaces are seen by electron microscopy as the elastin matrix degenerates. The regression of the subepidermal elastic network permits old skin to be stretched over a large distance at low loads.

The changes of photoageing include a great increase in the elastotic material in the dermis. These changes are superimposed upon and eventually mask the intrinsic changes of ageing. As the actinic damage progresses, the elastic fibres become thicker, more numerous and tightly coiled. Histologically these tightly coiled elastic fibres appear fragmented.

The number of cells in the dermis decreases across the entire lifespan including the number of fibroblasts, macrophages, and mast cells. The fibroblast becomes a shrunken, narrower fibrocyte that contains decreased cytoplasm and there is a decrease in the turnover of the dermal matrix components.

The decrease in mast cells helps to explain the observation that it is harder to raise up wheals by histamine releasing drugs and the observation that urticaria is uncommon in older persons. Also, heparin, found in mast cells, stimulates capillary endothelial cell migration *in vitro* and its reduction may help contribute to the paucity of vasculature.

Microcirculation

Regression and disorganization of small vessels are a prominent feature of aged skin. As the rete ridges flatten, the capillary loops that were present in the dermal papillae disappear. In addition, the small vessels about the cutaneous appendages decrease. This is especially prominent in actinic damage. Vessels in intrinsically aged sun-protected skin become thinner and have a decrease (or absence) in the number of surrounding veil cells (Braverman *et al* 1986). The changes in the microvasculature of photodamaged skin differ from those found in intrinsically aged skin in that photodamaged skin manifests a marked thickening of the postcapillary venular walls.

The minimization of the cutaneous vasculature during ageing has profound clinical consequences. These include a decreased inflammatory response, decreased absorption, decreased clearance, decreased urticarial reactions, decreased sweating, delayed wound healing, impaired thermal regulation, easy bruisability, and a muted clinical presentation of many cutaneous diseases.

The superficial blood supply is particularly important for thermal regulation. The decreased vasculature can be observed in the pallor of aged skin. The temperature drop between the groin and the feet is greater. Older persons quickly experience coldness when the temperature falls and are more at risk for hypothermia and hyperthermia (see below under impaired sweating). Even a brief exposure to cold may lead to hypothermia. The hypothermia is due both to the inability to divert blood efficiently and a loss of insulating subcutaneous tissue. Younger persons vasoconstrict more, shiver more, and generate more metabolic heat.

Because of the decreased microvasculature, it takes longer to absorb substances applied to the skin and takes longer to clear substances injected into the skin. For example, Kligman has shown that it took twice as long for 65-year-old subjects to absorb radioactive testosterone rubbed on the skin or to resolve an intradermally injected saline wheal as it did for 30-year-old subjects (Balin and Kligman 1989).

Clinically, this decrease in clearance can prolong cases of contact dermatitis. Probably more important, however, is that many skin diseases are distinctive because of their pattern or degree of inflammation. We can be seriously hampered in the ability to diagnose disease in older patients unless we recognize that some of the cardinal signs of inflammation, including redness, heat, and swelling may be absent. Cellulitis, for example, can be much more difficult to recognize without these signs.

The decreased blood supply may also necessitate some modification in therapy. Fewer applications of a topical medication may be appropriate because of the decrease in clearance.

Cutaneous nerves

Cutaneous free nerve endings are little affected anatomically during ageing although tactile sensitivity is decreased. The number of Pacinian corpuscles decreases by about two-thirds from the age of 20 to 90. Meissner's corpuscles also decrease in number to a similar extent.

Physiological tests reveal less acuity in pain perception and the pain reaction threshold is increased; older persons are thus less capable of sensing danger and reacting appropriately. One consequence is that burns tend to be more serious and widespread.

Consequences

Overall, the dermal changes that occur during ageing have a number of disparate physiological consequences. These include: (1) skin is more easily damaged; (2) delayed wound healing; (3) decreased inflammatory response; (4) decreased protection from ultraviolet light; (5) decreased urticarial reaction; (6) wrinkling, sagging skin; (7) skin easily stretches under low loads; (8) loss of resilience; (9) diminished absorption; (10) altered thermal regulation; and (11) decreased sensitivity to pain and pressure.

Subcutaneous tissue

The subcutaneous tissue serves as a shock absorber and a high calorie storage depot. The subcutaneous tissue also modulates conductive heat loss. Generally, the proportion of the body that is fat increases until age 70 but there are great regional differences in the distribution of this fat. For example, the amount of subcutaneous fat is decreased on the face and dorsum of hands but increased around the abdomen and thighs. The subcutaneous tissue protects organisms from blunt and pressure-related trauma and serves as an insulator of heat loss. The loss of this protective padding results in an increase in problems of weight-bearing and pressure-prone surfaces, and other injuries, as well as the risk of hypothermia.

Eccrine sweat

There is a reduction in the overall number of sweat glands in aged skin along with a decrease in the functional capacity of the remaining glands. Recruitment of sweat by thermal stimuli takes longer and the density of actively secreting glands decreases with age. Impairment of evaporative heat loss due to attenuated dermal vasculature and decreased sweating leads to an increased risk of heat stroke during hot weather. There is a decrease in sweating in response to dry heat and to experimentally injected intradermal acetylcholine.

Apocrine sweat

There is a decrease in apocrine secretion with age that appears primarily due to the age-associated decrease in testosterone levels. The decreased apocrine secretion results in a decrease in body odour and the need for antiperspirants and deodorants is decreased.

Sebaceous glands

Sebaceous glands are also androgen-dependent. The size of the sebaceous glands increases with age, while the transit time of the individual maturing sebaceous cells is 4 to 6 days longer in the aged person. The sebaceous pore also gets larger with age. Despite the increased size of the sebaceous glands there is a decrease in sebum output by 40 to 50 per cent. Additionally, there is a decrease in the size the individual sebocyte, smaller cytoplasmic oil droplets within the sebocytes, decreased free cholesterol in sebum, and an increase in the squalene fraction of sebum. The proliferative activity of the sebaceous gland decreases with age. These changes lead to solar sebaceous hyperplasia which are huge sebaceous follicles. The decrease in sebum secretion may contribute to dry skin.

Sebaceous gland size increases with age in non-sun-exposed sebaceous glands such as the Fordyce spots within the mouth. However, sebaceous gland hyperplasia is worsened by chronic sun damage.

Hair

For scalp hair, the rate of hair growth declines and the diameter of the individual terminal hair decreases with age. There is an increased percentage of hairs in the telogen or resting stages of the cycle. Greying of hairs occurs because of a progressive loss of functional melanocytes from the hair follicle bulb. Hair growth, however, presents a paradox. Not all hair shows a decrease in growth with age. In women older than 65 there is an increase in hair on the lip and chin although the same women have a decrease in hair on the head, axillae, and pubis. Men lose scalp and beard hair but have an increase in the growth of hair over their ears, eyebrows, and nostrils. Understanding the mechanisms that are responsible for the androgen-dependent conversion of vellus hair to terminal hair and vice versa in different body regions at the same time may provide important insights into the processes of differentiation and ageing.

Nail growth

The rate of nail growth declines by an average of 35 per cent between the ages of 20 and 80 (Hamilton *et al.* 1955). Because of the decrease in nail growth rate treatment for fungal diseases of the nail should be prolonged in elderly patients. Nails become brittle and lustreless and longitudinal striations with ridging and beading form on the nail plate.

CLINICAL CONDITIONS PREVALENT IN ELDERLY PEOPLE

Several studies have tried to assess the prevalence of skin disease in aged subjects. However, the variables are so numerous that no two populations so far studied are similar. As a result, every statement about age-associated skin conditions must be guarded.

All surveys bear out the high incidence of skin abnormalities in older persons. Where sunlight is abundant there will, of course, be more patients with conditions such as solar lentigos, actinic keratoses, or solar comedones, which reflect cumulative exposure to radiation. High concordance is found for such common age-associated lesions as xerosis, angiomas, lax skin, and seborrhoeic keratoses.

The prevalence of less common diseases differs appreciably. For example, Tindall and Smith (1963) observed seborrhoeic dermatitis in 33 per cent and rosacea in 12 per cent. On the other hand, a Danish group found seborrhoeic dermatitis in only 7 per cent and rosacea in 0.2 per cent among 587 subjects in a municipal old peoples' home (Weismann *et al.* 1981). Eczematous conditions were not frequent in Tindall and Smith's series but were noted in 25 per cent in Droller's study of elderly people living at home (Droller

1953). In the course of 1 year in a chronic care facility in New York, Young found that over 65 per cent developed one skin disorder and 50 per cent had two (Young 1965). The Health and Nutrition Survey of a representative sample of all Americans aged 1 to 74 conducted by the National Center for Health Statistics in 1971–74 found that 56 per cent of those over 60 years and 66 per cent of those over 70 years had a skin condition that was serious enough to require medical attention (Johnson and Roberts 1977).

The common denominator in all studies is the frequency and multiplicity of skin abnormalities in aged people. Treatable skin conditions are worsened in the presence of the following: poor general health, emotional deterioration (chronic brain damage), and inactivity. In turn, these contribute to inability to provide proper daily skin care including cleaning and grooming: skin health is a function of skin care.

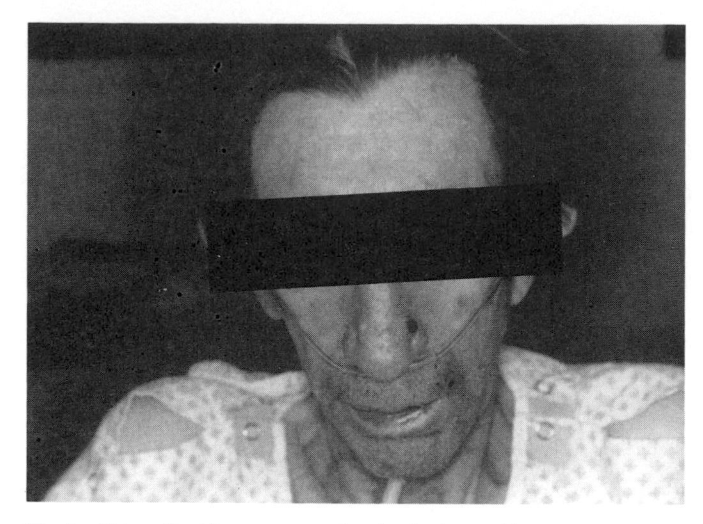

Fig. 1 This patient has severe seborrhoeic dermatitis characterized by yellow greasy scales with erythema of the involved skin. The eyebrows, nasolabial fold, and posterior auricular area are common locations for this disorder. The crusted lesion on his nose is a basal cell carcinoma.

THE EXPRESSION OF SKIN DISEASES IN ELDERLY PEOPLE

In older persons common skin disorders are frequently muted, and are so blurred and morphologically transformed that diagnosis may be delayed or missed altogether. Dermatitis, whatever its origin—irritation, allergy, stasis, microbial infection, drugs, and others—tends to behave differently in older than in younger persons. Unless quickly cleared, the dermatitis tends to become chronic, to spread widely (a process known as autoeczematization), and to respond sluggishly to treatment. Healing is slow and unpredictable. Thus, speedy diagnosis and treatment are exceedingly important to prevent chronicity, extension, and refractoriness.

The failure to react promptly to a toxic stimulus carries with it the danger of continued exposure to noxious agents. Redness appearing shortly after exposure warns a young person to desist. In an older person, because of a long latent period, applications of an irritating agent may continue until suddenly the tissue collapses, sometimes with ulceration. It is only in this sense that older persons may be characterized as reacting more vigorously to toxic agents. They should be cautioned regarding self-treatment with home remedies so abundantly at hand.

Some chronic diseases tend to regress in later years. Atopic dermatitis is rare. Plaque-type psoriasis, if not converted to a pustular eruption by overtreatment, usually declines. One might anticipate that hyperproliferative dermatoses would tend to fade as a result of age-associated declines in mitotic activity. Dandruff, a result of increased production of horny cells, disappears.

This is counterbalanced to some extent by the emergence of disorders which may achieve prevalence rates considerably higher than in earlier adult life. Seborrhoeic dermatitis (Fig. 1) of the scalp and face, in men particularly, is a striking example. Confinement to bed by severe illness, myocardial infarction for example, greatly aggravates seborrhoeic dermatitis which may generalize (Tager *et al*. 1964). Immobilization generally worsens chronic skin diseases. Rosacea (Fig. 2), starting in young adulthood, may become severe, culminating in such extreme forms as rhinophyma.

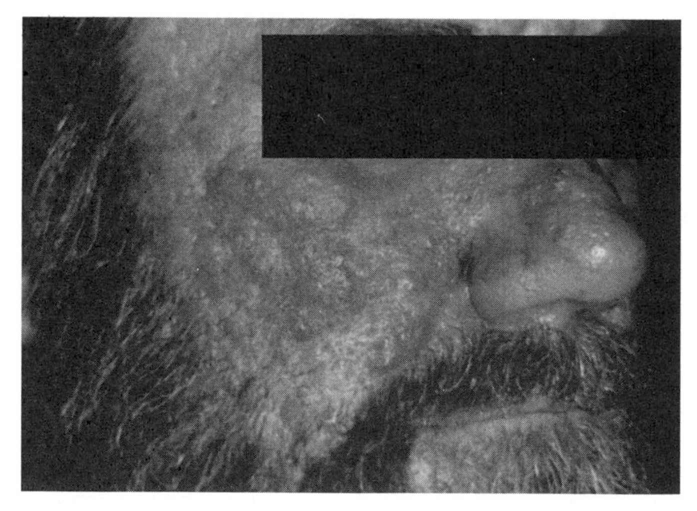

Fig. 2 Rosacea, formerly called acne rosacea, is characterized by erythematous papules, telangiectasia, and pustules on the central face. The condition can also involve the forehead and sides of the face.

Chronic photosensitivity reactions, especially those of allergic origin, reach their highest prevalence in older persons. These are visually disabling, maddeningly pruritic diseases which mainly localize on the face of older men. The two best known examples are allergic contact dermatitis due to airborne pollen (commonly ragweed in the United States) and photocontact allergy due to halogenated salicylanilides (optical bacteriostats) (Fig. 3). These may result in a severe photodermatitis which grotesquely thickens the skin. Such individuals are called persistent light-reactors and they are exquisitely sensitive to the entire ultraviolet spectrum; the allergen cannot always be identified. Actinically damaged skin with its attenuated blood supply is the substrate in which these photodermatoses develop. Sunlight worsens a number of skin disorders, such as rosacea.

Interdigital athlete's foot (Fig. 4) is common in old age often extending beyond the confines of the 5th interspace and is invariably accompanied by onychomycosis. Slower turnover of the horny layer, a depressed inflammatory

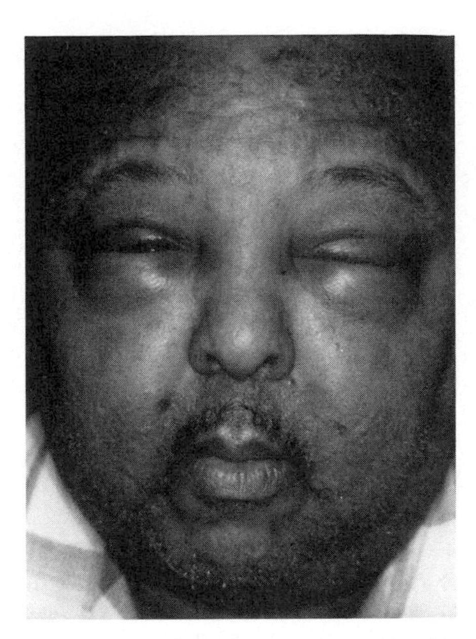

Fig. 3 This patient has an allergic contact dermatitis with erythema, oedema, and vesiculation involving the skin on the forehead and around the eyes.

response, and decreased cellular immunity contribute to chronicity.

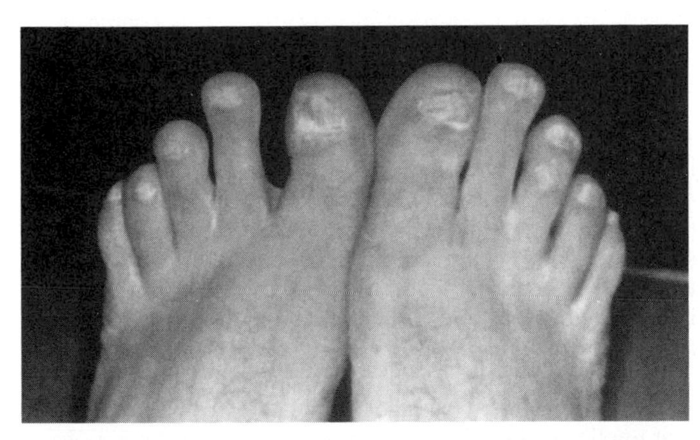

Fig. 4 Tinea pedis, which is caused by dermatophyte fungi, produces erythema, scaling, and fissuring of the skin of the feet. The infection can spread to involve the toenails leading to subungual hyperkeratosis and a yellowish discoloration to the nail.

The integument is at especially high risk of injury from burns, chemical irritation, and trauma owing to decreased sensory perception and slower reaction times. Particularly telling examples of chemical toxicity derive from the diminished capacity to mount a prompt inflammatory response. For example, a keratolytic solution of salicylic acid and propylene glycol, often used for dry skin, will, if applied twice daily to the face of a young person, incite scaling and redness in a few days. In an elderly person, the skin may remain silent for 2 to 3 weeks before suddenly exploding into a severe dermatitis from toxic overload. Household cleaning and disinfectant solutions, often used to stop itching, are a genuine hazard. The sensible, safe management of pruritus in older patients requires knowledge and experience.

Primary pyodermas due to *Staphylococcus aureus* and β-haemolytic streptococci may not call forth the customary signs of pain, heat, and redness. A furuncle may present as a cold abscess and cellulitis may show only an indolent swelling.

A high order of suspicion is indicated for every widespread eruption that cannot be readily identified. It is surprising how easy it is to miss a diagnosis of scabies which has only maddening pruritus as the signal feature, the lesions being otherwise unrecognizable. These cryptic cases can be the unsuspected source of major epidemics in institutions. Also, the expression of skin disease is modified by nutritional deficiencies. Scurvy is more frequent than often realized. Many old persons, particularly those living alone or with various disabilities, do not have an adequate diet. Patients with zinc deficiencies suffer exotic rashes which cannot be recognized clinically and which adversely affect cell-mediated immunity.

The expression of vitamin or mineral deficiencies are very easily overlooked, being regarded as part of the diverse cutaneous alterations which inevitably come with age. Thus, markedly xerotic skin due to iron deficiency with anaemia arouses no interest since almost all old people suffer from dry skin. Likewise, purpura is so familiar that diagnostic follicular haemorrhages in scurvy are not even seen. Perlèche from a lack of vitamin B will likely be put down to drooling of the corners of the mouth or diagnosed as moniliasis.

Finally, when infections are recognized, we should be on the lookout for exotic organisms, such as yeast-like fungi, unusual Gram-negative bacilli, and unfamiliar anaerobes. These should not be dismissed as contaminants.

SPECIFIC CLINICAL CONDITIONS PREVALENT IN ELDERLY PEOPLE

Pruritus

Pruritus, or itching, is the most common dermatological complaint of older people. 'Dry skin' is often credited with causing this incapacitating affliction, and it is true that xerosis is commonly observed. Seborrhoeic dermatitis is also very common in older persons, however, and there are numerous other specific conditions that can account for the itching symptoms. It is inappropriate to dismiss pruritus as banal.

The sensation of itching is detected by various types of nerve endings and transmitted by way of the sensory nerves, located below the dermal-epidermal junction, to the posterior nerve roots and the spinal cord. The 'bright and well-localized' sensation of spontaneous itch is transmitted by delta fibres of the A class of myelinated nerves, which are 10 μm in diameter and conduct at about 10 m/s. The unpleasant and poorly localized itch sensation is transmitted by C fibres of unmyelinated nerves which are 5.5 μm in diameter and conduct at about 1 m/s. Itch and pain are transmitted similarly, but not identically, by C fibres. Heat, for example, blocks itch but spares the sensation of pain. Pain can decrease itch perception. A great deal is unknown about the neurophysiology of itching. An ideal antipruritic agent, antihistamine, narcoleptic, or anti-inflammatory, has not been identified. One problem is the lack of a good model system in human beings in which to study the induction of pruritus and its control. Histamine, proteases, and other agents have been used to induce itching.

Clinically, a large number of conditions can cause pruritus.

These include: (1) xerosis; (2) infestations (e.g. pediculosis corporis, scabies, trichinosis, onchocerciasis); (3) metabolic or endocrine problems (hypo- or hyperthyroidism, diabetes mellitus); (4) malignant neoplasms (lymphoma, leukaemia); (5) chronic renal disease; (6) hepatobiliary disease (primary or secondary biliary cirrhosis); (7) drug ingestion (opiates, codeine, drug hypersensitivity, drugs that cause cholestasis as a side-effect such as erythromycin estolate, chlorpropamide, or chlorpromazine); (8) haematological disease (iron-deficiency anaemia, polycythaemia rubra, paraproteinaemia); (9) psychiatric problems (chronic depression and agitation, neurotic excoriations, delusions of parasitosis); (10) skin disease (many, including miliaria, folliculitis, irritant dermatitis from exogenous irritants such as chemicals, hairs, fibreglass, plant spicules).

Seborrhoeic dermatitis

Itching in older people may also be caused by inflammatory dermatoses. Seborrhoeic dermatitis is very common in elderly people, but its cause is unknown (Fig. 1). The prevalence of seborrhoeic dermatitis in parkinsonism, and at times of stress and fatigue, implicates neurological factors. Aetiologic roles for sebum, and yeast have been proposed. Seborrhoeic dermatitis does not develop unless the sebaceous glands are active, which probably accounts for its prevalence during infancy and in postpubertal individuals. The standard treatments for seborrhoeic dermatitis involve the use of tar shampoos and topical hydrocortisone. Recently, the use of topical ketoconazole has been shown to be effective.

Xerosis

There is much to be learned about xerosis, which is another cause of itching. Xerosis is due in part to decreased eccrine sweating, decreased sebum production, decreased water content of the stratum corneum, and decreased cohesion of corneocytes. Changes in production of sebum may contribute to the development of xerosis because the water-retaining property of the stratum corneum is reduced when the amount of sebum is reduced. The role played by the keratinization process in contributing to xerosis is also unclear. Keratinization is partially controlled by age-associated processes.

Cutaneous tumours of epidermal keratinocytes

A number of cutaneous tumours of epidermal keratinocytes are prevalent in older persons. Seborrhoeic keratoses (benign epithelial neoplasms) occur more frequently as individuals advance in age. They have been found in up to 88 per cent of persons over the age of 65; in one study about 50 per cent of those with seborrhoeic keratosis had 10 or more lesions (Tindall and Smith 1963). Seborrhoeic keratoses appear to be dominantly inherited, but they seldom appear before middle age. They are found on the skin in those areas of the body that are rich in sebaceous glands such as the trunk, face, and extremities. Seborrhoeic keratoses are sharply demarcated, brown, and slightly raised. They look as if they have been stuck on the skin surface. Most have a verrucous surface with a soft, friable consistency (Fig. 5).

The amount of melanin in a seborrhoeic keratosis is variable, and inexperienced observers have been known to mistake a seborrhoeic keratosis for a melanoma. If these lesions

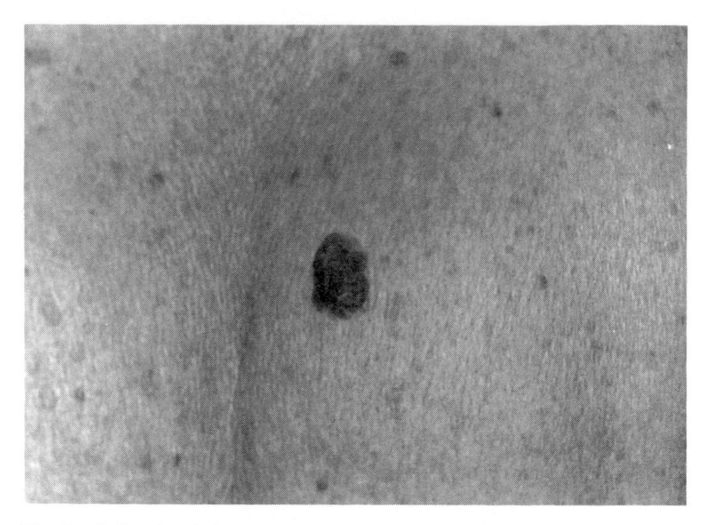

Fig. 5 Seborrhoeic keratoses can vary in colour from dark black to tan or flesh-coloured depending on the amount of melanin present. They are sharply demarcated, raised, and have a waxy, stuck-on appearance. Occasionally dark black lesions are mistaken for melanoma.

bleed, become irritated, or change in size, shape or colour, removal and pathological examination is indicated.

The sign of Leser-Trelat is the sudden appearance and rapid increase in size and number of seborrhoeic keratoses on skin that was previously blemish-free. This condition is associated with the development of an internal malignancy which is usually an adenocarcinoma (Curry and King 1980).

Several neoplastic conditions common in older persons are associated with environmental damage to the skin. These include actinic keratoses, Bowen's disease, squamous cell carcinoma, and basal cell carcinoma. The changes that most persons equate with ageing of the skin are due to chronic solar damage. Prolonged exposure to ultraviolet irradiation leads to cutaneous atrophy, alterations in pigmentation, wrinkling, dryness, telangiectasia, and solar elastosis. Wavelengths in the range 290 to 310 nm produce sunburn, and are also thought to be the irradiation mainly responsible for actinic damage to the skin. Some of the strongest evidence that implicates UV light as being important in the aetiology of epidermal tumours comes from epidemiological data correlating the incidence of tumours with degree of pigmentary protection. The individual principally at risk is light-skinned, is easily sunburned, and does not tan. Other strong epidemiological data correlate an increased incidence of skin tumours with decreasing latitude and increasing sun exposure.

Actinic keratoses

Actinic keratoses, or solar keratoses, are composed of clones of anaplastic keratinocytes confined to the epidermis and occurring commonly on sun-damaged skin of elderly individuals (Fig. 6). If left untreated, they may progress and invade through the basement membrane of the epidermal-dermal junction, thereby becoming invasive squamous cell carcinomas.

Actinic keratoses are extremely common in elderly individuals who have had extensive sun exposure. Actinic keratoses usually occur on skin damaged from sun exposure, such as the bald scalp, face, and forearms. They are more common

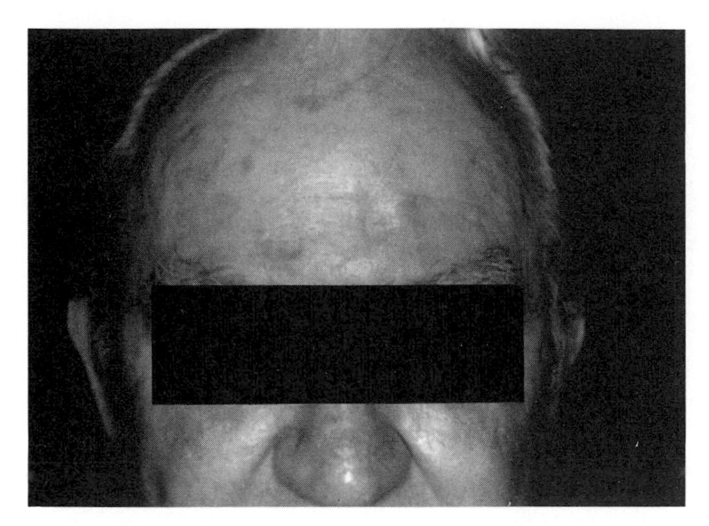

Fig. 6 Actinic keratoses on cheek and forehead. Each lesion occurs as a discrete, erythematous scaly papule with varying degree of induration. Actinic keratoses can occur on any part of the skin which has been chronically exposed to sunlight.

in fair skinned individuals, and are almost never seen in black subjects. These observations strongly suggest that chronic exposure to sunlight is an important aetiological factor. The carcinogenic property of sunlight resides mainly in the UVB range and can be adequately screened out by modern sunscreens.

Actinic keratoses occur as well as demarcated, scaly, rough papules on sun exposed skin surfaces. Colour varies from tan to red, but sometimes they are the same colour as the surrounding skin. As a result, some lesions are more easily palpated than seen. In some cases, known as pigmented actinic keratoses, increased amount of pigmentation renders the lesion a striking brown colour. Actinic keratoses are usually small, measuring from a few mm to 1 or 2 cm in size. Depending on the degree of prior sun exposure, a given patient may have one or a few lesions, or hundreds. There are often other signs of actinic damage in the surrounding skin, including wrinkling, dryness, and yellow discoloration from solar elastosis. Actinic keratoses can occur at the base of cutaneous horns.

Spreading pigmented actinic keratosis is an unusual variant of actinic keratosis. Clinically, it is characterized by large size (over 1 cm), brown pigmentation, and a tendency for centrifugal spread. Such lesions can mimic lentigo maligna in clinical appearance.

Histologically, actinic keratoses are well demarcated islands of abnormal keratinocytes with overlying parakeratosis. Their nuclei are large, irregular, and hyperchromatic, giving rise to a pleomorphic or atypical appearance. Changes of solar elastosis are invariably present in the underlying dermis.

Progression from carcinoma *in situ* to invasive squamous cell carcinoma

Progression of actinic keratosis to invasive squamous cell carcinoma occurs when buds of atypical keratinocytes extend deep into the dermis, leading to detached nests of abnormal cells capable of autonomous growth. Clinically, the lesion may become thicker, more indurated, and enlarged. Such signs, however, are not always present and are not substitutes for histological confirmation of dermal invasion.

Various studies indicate that, on the average, an individual with actinic keratoses would have a likelihood of 1 to 2 per cent per year or 10 to 20 per cent in 10 years of developing an invasive squamous cell carcinoma. Each actinic keratosis has a yearly incidence of progression to invasive squamous cell carcinoma of about 0.1 per cent (Marks *et al.* 1988).

In contrast to squamous cell carcinoma arising from burn scars, osteomyelitis sinuses, and chronic wounds, squamous cell carcinomas that originate from actinic keratoses metastasize infrequently. The rate of progression to metastasis of squamous cell carcinomas arising in actinic keratoses has been reported as ranging between 0.5 and 3 per cent depending on the series (Balin *et al.* 1988).

Treatment

A variety of therapeutic methods are available for the patient with actinic keratoses. Optimal choice depends on number of lesions, extent of involvement, and the patient's general state of health.

For the patient with a few lesions and little evidence of actinic damage, destructive techniques such as cryotherapy, curettage, electrodesiccation, chemical cauterization with phenol or trichloroacetic acid, or excisional surgery can be successfully employed. 5-Fluorouracil (5-FU) cream or solution (1 to 5 per cent) can be effectively employed topically to treat people with actinic keratoses. This treatment is particularly useful for patients with moderate actinic damage because it can uncover and treat subclinical lesions. 5-Fluorouracil can be used successfully to treat patients with widespread extensive actinic damage, although adequate treatment is usually prolonged (6 weeks or longer), uncomfortable, and unsightly for the patient. 5-Fluorouracil can be used for short periods (1 or 2 weeks) to identify the most advanced actinic keratoses, which can then be destroyed by cryotherapy. In selected patients with extensive involvement and numerous actinic keratoses, dermabrasion may be the treatment of choice. Dermabrasion is extremely effective in eradicating large numbers of actinic keratoses, particularly on the face and scalp. A number of experienced dermatologists believe that 5-fluorouracil is not as efficacious as dermabrasion in long-term prevention of recurrent dyskeratotic and malignant cutaneous disease (Field 1984).

Bowen's disease

Bowen's disease is another form of squamous cell carcinoma *in situ* (Fig. 7). It may occur anywhere on the skin but is more common on covered surfaces. Three factors have been implicated in the aetiology of Bowen's disease: exposure to UV irradiation, arsenic, and papova- and oncornaviruses.

Clinically, Bowen's disease appears as a slowly enlarging, erythematous patch of sharp but irregular outline showing little or no infiltration. Within the patch there are areas of crusting. Fifty-five per cent of patients have more than one lesion. Lesions of Bowen's disease can occur on the glans penis, vulva, or oral mucosa and in these locations are called erythroplasia of Queyrat.

Many cases of Bowen's disease develop in persons who ingested inorganic arsenic many years earlier. Some persons report that they received Fowler's solution, which was com-

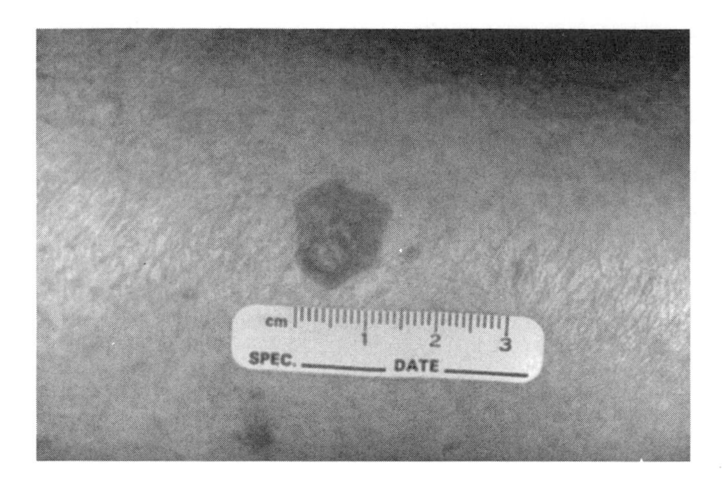

Fig. 7 Bowen's disease occurs more commonly on non-sun exposed areas but may occur anywhere on the skin. There is an erythematous lesion with a sharp but irregular outline. Scaling may be associated but these lesions are not usually indurated, in contrast to squamous cell carcinoma.

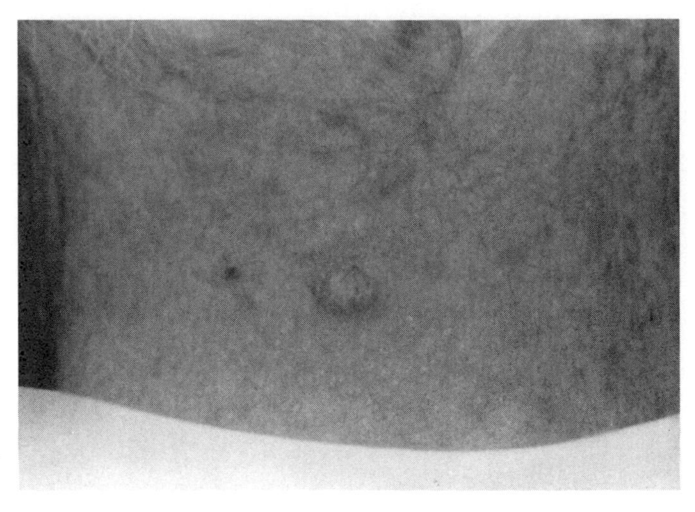

Fig. 8 This lesion on the neck shows the result of neglect of actinic damage resulting in invasive squamous cell carcinoma. The lesion is an erythematous nodule with scaling.

monly used to treat asthma and various other medical problems and which contained 1 per cent potassium arsenite. In other persons the source of arsenic is thought to have been well-water or insecticides. Arsenical keratoses of the palms and soles are verrucous, pale papules without surrounding inflammation. They occur in 40 per cent of patients who receive arsenic and histologically are analogous to Bowen's disease.

Fifteen to 30 per cent of patients with Bowen's disease on a non-sun-exposed site develop internal malignancies. Presumably this is due to exposure to arsenic (Graham and Helwig 1961).

The histological pattern seen in Bowen's disease is full-thickness epidermal dysplasia.

SQUAMOUS CELL CARCINOMA

In addition to actinic keratoses and Bowen's disease, conditions predisposing to squamous cell carcinoma include arsenic exposure, radiation exposure, scarring from a previous injury such as a burn or chronic leg ulcer, and exposure to heat (erythema ab igne). Squamous cell carcinoma may occur anywhere on the skin or mucous membranes, but it rarely arises from skin that appears normal. Clinically, there is commonly a shallow ulcer surrounded by a wide, elevated, and indurated border. Often the ulcer is covered by a crust that conceals a red, granular base. Occasionally raised verrucoid lesions without ulceration occurs.

Squamous cell carcinoma is a malignant, invasive carcinoma (Fig. 8). Histologically there are irregular masses of neoplastic epidermal cells proliferating downward and invading the dermis. Histological grading of squamous cell carcinoma depends on the percentage of keratinizing cells, percentage of atypical cells, number of mitotic figures, and depth of invasion. The incidence of metastasis varies from 0.5 to 3 per cent in squamous cell carcinomas arising in an actinic keratosis to 25 to 30 per cent in those arising in a chronic osteomyelitic sinus or in radiodermatitis.

BASAL CELL CARCINOMA

The most common skin cancer, and thus the most frequent cancer in the United States, is the basal cell carcinoma (Figs. 9, 10). This lesion is found most commonly on the head and

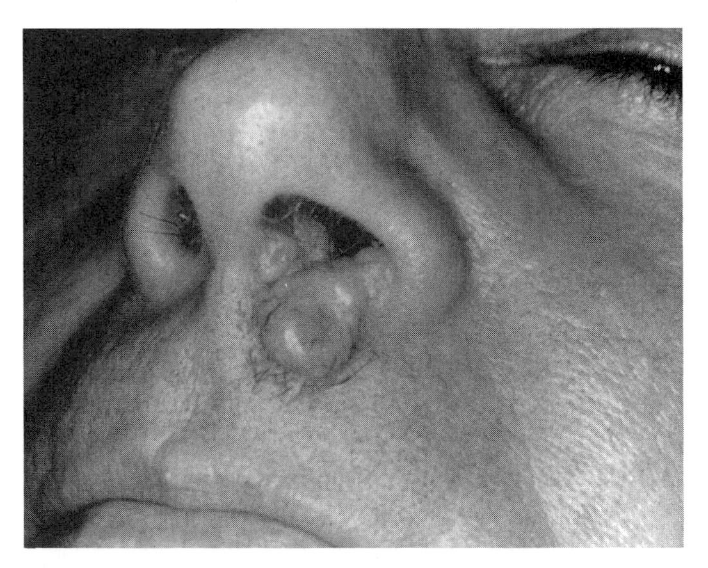

Fig. 9 Typical appearance of basal cell carcinoma with pearly raised border, telangiectasia, and central erosion.

neck of men, especially those who sunburn easily and have had chronic sun exposure. It is rare in darkly pigmented races. The distribution of lesions on the face, however, does not correlate well with the area of maximal exposure to light. These lesions are common on the eyelids, on the inner canthus of the eye, and behind the ear, and are not so common on the back of the hand or forearm. Basal cell epitheliomas generally occur on hair-bearing skin in adults, usually as single lesions. Predisposing features include chronic sun exposure, exposure to X-rays, burn scars, and xeroderma pigmentosum.

Basal cell epitheliomas rarely metastasize but can be locally

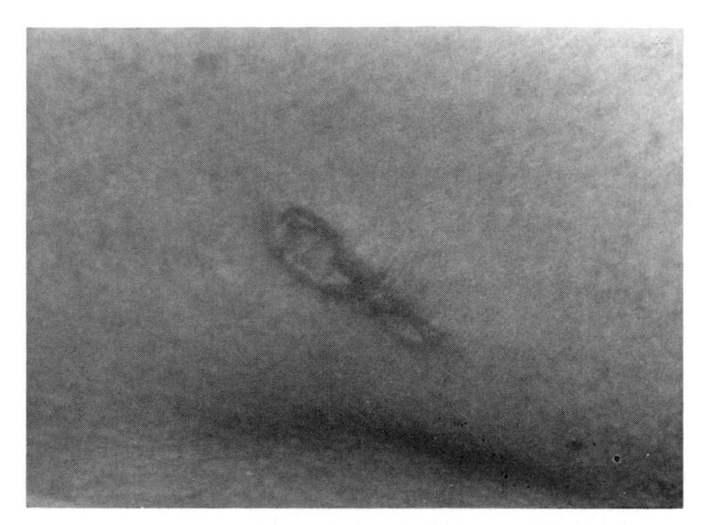

Fig. 10 Morphoeic or sclerotic basal cell carcinoma. The tumour infiltrates the dermis and extends well beyond the clinically apparent margin.

quite destructive. There are several different clinical types of basal cell epithelioma. Most common is the nodulo-ulcerative basal cell epithelioma, which begins as a small, waxy nodule with small telangiectatic vessels on the surface and a translucent, rolled border. The nodule increases in size and undergoes central ulceration. The typical lesion consists of a slowly enlarging ulcer surrounded by a pearly, rolled border. This represents the so-called rodent ulcer. Basal cell carcinomas can contain melanin pigment. The morpheaform, or fibrosing, basal cell epithelioma appears as an indurated, yellowish plaque with an ill-defined border. The overlying skin remains intact for a long time before ulceration develops.

Superficial, multifocal basal cell epitheliomas consist of one or several erythematous, scaling, only slightly infiltrated patches that slowly increase in size by peripheral extension. The patches are surrounded by a fine, thread-like, pearly border, and usually show small areas of superficial ulceration and crusting. The centre may show smooth, atrophic scarring. Superficial basal cell epitheliomas occur predominantly on the trunk.

TREATMENT OF SKIN CANCER

Bowen's disease, basal cell and squamous cell carcinoma can be treated by excisional surgery, curettage and electrodesiccation, radiotherapy, intralesional interferon injections, cryosurgery, and Mohs micrographic surgery. The treatment of each patient must be individualized, particularly in the elderly age group. However, surgical removal of these common cutaneous neoplasms provides the most direct and definitive therapy. Mohs micrographic surgery provides the optimum treatment of cutaneous cancers, particularly on the head and neck, since this technique provides the highest cure rate and conserves the most normal tissue. Mohs surgery involves the layered removal of the tumour with the entire undersurface of the excised tissue examined microscopically at the time of surgery. A map is made of any penetrating tumour roots and sequential layers of tissues are excised and examined until the borders of the excision are tumour free. It is advisable that recurrent basal cell and squamous cell cancers be removed by the Mohs micrographic surgical technique whenever possible to ensure complete removal of the cancer.

INFECTIONS AND INFESTATIONS

The decline in the immune system, the decrease in the rate of epidermal regeneration, and the decrease in the cutaneous vasculature, altered cutaneous pain perception, and any decrease in grooming and mobility and dexterity combine to make the elderly person more susceptible to infections and infestations. In addition, as reviewed above, the alteration in the cutaneous vasculature that accompanies ageing results in a decreased inflammatory response and a muted clinical expression of these disorders.

Fungal infections

Twenty to 80 per cent of people over 65 (depending on the population surveyed) have fungal infections of their skin or nails. The decreased rate of epidermal regeneration and the slower rate of nail growth contribute to this high prevalence. Tinea corporis may be diverse in its clinical presentation and is most often due to dermatophytes belonging to the genera Trichophyton, Microsporum, or Epidermophyton. Any red scaly rash deserves a potassium hydroxide (KOH) microscopical examination (and culture if there is a high index of suspicion) to exclude dermatophytes. Tinea cruris and Tinea pedis is most commonly due to *E. floccosum*, *T. rubrum*, and *T. mentagrophytes*. Clinically there is a well marginated, raised border that may be composed of multiple erythematous papules. In the groin, chronic scratching may cause lichen simplex chronicus and secondary bacterial infection may be present. Tinea pedis (Fig. 4) most often presents as fissuring, scaling, or maceration in the interdigital or subdigital area of the lateral toe webs. Onychomycosis includes all infections of the nail caused by any fungus, including non-dermatophytes and yeasts. This is particularly a problem in older persons where underlying nail disease allows the fungus to get under the nail. Distal subungual onychomycosis begins as a whitish or brownish-yellow discoloration at the edge of the nail or in the lateral nail fold. White superficial onychomycosis appears as white sharply outlined areas on the surface of the toenails and is most frequently caused by *T. mentagrophytes*. Proximal subungual onychomycosis is the least common form of onychomycosis and is due to the fungus invading the nail from the proximal nail-fold. Clinically a whitish to whitish-brown area is seen on the proximal part of the nail plate. In recent years, a number of effective topical antifungal agents have been developed, but usually they will not clear onychomycosis. Topical therapy will often confine the fungus to the toenails. Chemical removal of the nail with 40 per cent urea compounds or surgical avulsion of the nail combined with topical or systemic antifungal agents (griseofulvin or ketoconazole) may be employed to treat onychomycosis if clinically indicated.

Candidiasis

Candida albicans is a saprophytic yeast that colonizes the mucous membranes in 80 per cent of normal individuals. Candida infections are usually opportunistic and due to diminished host defences. Typically candida colonizes moist, macerated folds of skin. (Fig. 11) Obesity, occlusive clothing,

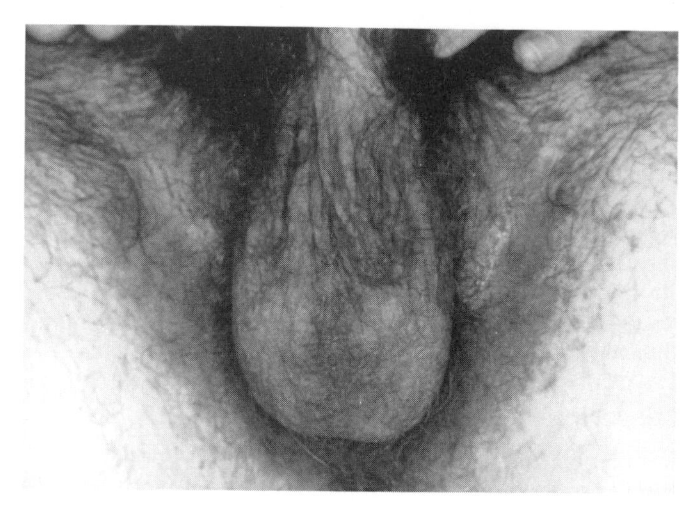

Fig. 11 Candidiasis. Erythema and satellite pustules in an area of moist macerated skin.

diabetes mellitus, and occupations that provide exposure to moist conditions predispose an individual to develop infection with *C. albicans*. Clinically there are pruritic, erythematous, macerated areas of skin in intertriginous areas with satellite pustules. The diagnosis is confirmed by potassium hydroxide examination and fungal culture. Treatment involves correction of the predisposing factors, topical therapy with any of a number of antifungal medications including clotrimazole, ketoconazole, amphotericin B, and, in refractory cases, systemic antifungal therapy with ketoconazole.

Erysipelas and cellulitis

Bacterial infection of the skin is common. It is important to note that the clinical presentation of bacterial infection may be altered in older persons. The cardinal signs of infection—erythema, swelling, warmth, and pain—may be absent owing to the diminished cutaneous vasculature in elderly subjects. In addition, pruritus leading to scratching can predispose to portals of entry for bacteria. The physician must be vigilant for cutaneous infection in older persons. Mupericin ointment, a relatively new topical antimicrobial agent that is particularly active against *Staphylococcus aureus*, has been useful for the long-term management of children with impaired cutaneous barriers and is useful on excoriations and surgical wounds in older persons.

Scabies

The mite *Sarcoptes scabiei*, var. *hominis* causes pruritic papules and vesicles with S-shaped or straight burrows. Most of the problems encountered with scabies are diagnostic rather than therapeutic. Crotamiton and lindane are effective treatments. Scabies may erupt in epidemics in nursing homes. *Sarcoptes scabiei* often inhabits the subungual region where the nail plates provide protection from effective scabicides. In epidemic situations it may be advisable to cut the nails as short as possible and brush the anti-infective agent along the fingertips and under the remaining edges of the nail.

Herpes zoster

Herpes zoster, commonly referred to as 'shingles' is caused by the varicella zoster virus, Herpes virus varicellae. Replication occurs within infected cell nuclei and intranuclear inclusions can be seen on stained preparations.

Chicken pox, or varicella, is caused by acute infection with this virus. Herpes zoster is due to reactivation of latent varicella zoster virus which was deposited in sensory dorsal root or cranial nerve ganglia during a prior acute infection. Herpes zoster can occur only in individuals previously infected with varicella and its incidence increases as a person ages with more than 60 per cent of the cases occuring in people over the age of 45. The highest incidence of disease occurs during the 5th to 7th decade and is approximately 5 to 10 cases per 1000 people per year in this age group. Men and women are equally affected; 10 to 20 per cent of all people will be affected by herpes zoster during their lifetime.

The mechanism for reactivation of the varicella zoster virus that results in herpes zoster is not fully known. Immunosuppression may contribute to reactivation by depression of cell-mediated immunity. Herpes zoster has been reported to develop from pressure on nerve roots due to tumour, after minor trauma to the skin or cornea, or after direct injury to nerves or the vertebrae. Onset of herpes zoster after trauma ranges from 1 day to 3 weeks following injury. Radiation therapy has been reported to activate latent virus in the ganglia of nerves supplying the treated dermatome.

Although specific mechanisms determining viral reactivation are unknown, some systemic diseases are associated with an increased incidence of herpes zoster. Herpes zoster occurs more frequently in patients with lymphoproliferative malignancies such as Hodgkin's disease, acute lymphocytic leukaemia, and chronic lymphocytic leukaemia than in the general population. Other disease states in which the incidence of zoster is higher include systemic lupus erythematosus, rheumatoid arthritis, and diabetes. Patients who have cancer or are immunocompromised due to immunosuppressive drug therapy, or exposure to X-rays are especially prone to developing herpes zoster. Whereas herpes zoster in cancer patients usually occurs in previously diagnosed advanced cancer or in those actively undergoing therapy, zoster in those with the human immunodeficiency virus (HIV) infection may actually be the presenting sign of the infection. There is not sufficient evidence, however, to suggest that otherwise healthy patients with zoster are at an increased risk of having or developing a malignancy. Recurrent herpes zoster occurs in approximately 2 per cent of healthy individuals, whereas immunosuppressed individuals have about a 5-fold increase in the frequency of recurrent herpes zoster. In 50 per cent of recurrent cases, lesions occur in the same dermatome as the first episode.

Clinical features

Clinically the first signs of localized herpes zoster are malaise, and pain and tenderness of the skin in the involved area. Some patients may experience burning pain, itching, shooting pains, or dysaesthesias. Pain may vary from mild to severe although children and young adults may experience no pain at all. Herpes zoster begins after this 3 to 4 day prodrome as grouped erythematous papules occurring unilaterally in the cutaneous distribution of a sensory nerve, or dermatome. (Fig. 12) Usually lesions appear posteriorly and progress to the anterior and peripheral distribution of the involved nerve. Vesicles or blisters develop in the area and new ones may continue to appear for a few days. In an uncomplicated case the vesicles become purulent owing to the accumulation of leucocytes and then dry up, form crusts, and begin to heal. In older patients, lesions may become necrotic and heal with scarring. Secondary bacterial infection can lead to delayed healing and scar formation at any age.

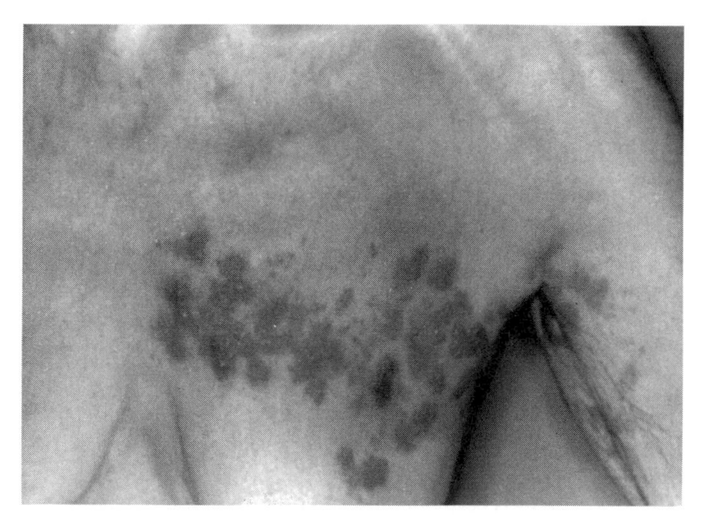

Fig. 12 Herpes zoster. Typical appearance of dermatomal distribution of herpes zoster. Grouped vesicles evolve into crusts and scabs on a brightly erythematous base.

Characteristically, lesions develop in a continuous or interrupted band in one or more dermatomes. Mucous membranes within the affected dermatomes are also involved. The affected dermatomes are usually hyperaesthetic. In about 15 per cent of cases the pain and skin lesions develop simultaneously. Lymph nodes draining the involved area may become swollen and tender. Occasionally, especially in immunocompromised patients, two or three adjacent dermatomes may be involved. Painful herpes zoster can be present without a rash developing. This has been called 'zoster sine herpete', or 'zoster sine eruptio'.

The dermatomes affected are located in the thoracic area in 50 per cent of cases, in the cervical area in 20 per cent, and in 15 and 10 per cent of cases in the trigeminal and lumbosacral areas, respectively. Herpes zoster of the ophthalmic branch of the trigeminal nerve occurs with an increased incidence in old age. When branches of the trigeminal nerve are affected, lesions erupt in the face, mouth, eye, or on the tongue. Lesions on the tip of the nose (Hutchinson's sign) indicate involvement of the nasociliary nerve and have been found to correlate with eye involvement which includes intraocular disease. Complications occur in about 50 per cent of patients with ophthalmic herpes zoster which is manifested by a red swollen conjunctiva and a superficial or deep keratitis. Argyll-Robertson pupils can result from involvement of the ciliary ganglia. Herpes zoster of the maxillary division of the trigeminal nerve produces vesicles on the uvula and tonsillar area while, when the mandibular division of the trigeminal nerve is involved, vesicles appear on the anterior part of the tongue, the floor of the mouth, and the buccal mucous membrane. The Ramsay Hunt syndrome, manifested by vesicular lesions of the pinna, auditory canal, and anterior two-thirds of the tongue, is associated with Bell's palsy, tinnitus, deafness, vertigo, decreased taste, decreased hearing, and meningitis. This syndrome is due to involvement of the sensory branch of the facial nerve and possibly the geniculate gangion. In orofacial herpes zoster toothache may be the presenting symptom.

Patients with herpes zoster may develop scattered lesions outside the involved dermatome. These lesions represent haematogenous dissemination probably due to circulating monocytes. The frequency of cutaneous dissemination reported in the literature is 25 per cent (Ragozzini *et al.* 1982). Dissemination occurs more frequently in elderly individuals and in immunosuppressed patients and reflects an inability of the host's immune system to keep the infection in check. Cutaneous dissemination develops in about 40 per cent of patients with Hodgkin's and non-Hodgkin's lymphoma. Evidence for dissemination occurs within 4 to 6 days after the appearance of the initial zoster lesions. The skin lesions which develop as a result of dissemination usually heal within 3 to 5 days. Dissemination may be accompanied by systemic signs of viraemia with fever, prostration, and lymphadenopathy.

Differential diagnosis

The diagnosis of herpes zoster can usually be made on clinical grounds. The presence of pain and unilateral grouped vesicles in a dermatomal distribution are most helpful. In general, lesions of herpes zoster do not cross the midline of the body. An exception to this rule, however, can be observed when, owing to local extension of virus through the skin rather than nerves, lesions cross the midline by 0.5 to 1.0 cm. Early lesions, and herpes zoster without a rash may create diagnostic dilemmas.

Distinguishing herpes simplex from herpes zoster can sometimes be difficult as recurrent herpes simplex occasionally produces grouped vesicles in a band-like pattern resembling a dermatome. Tzanck preparations from the base of blisters in both conditions show similar findings with multinucleated epithelial giant cells, nuclear margination of chromatin, and intranuclear inclusions. The Tzanck procedure can be helpful in distinguishing herpes simplex and zoster from other vesicular conditions such as allergic contact dermatitis and bullous impetigo. Histopathological examination by light microsopy also shows identical findings in the skin lesions of both herpes simplex and zoster. Helpful differential points include a history of recurrent lesions in the same area which is usual for herpes simplex but rare in herpes zoster, and the presence or absence of pain in the distribution of the rash. The pain associated with herpes zoster can be mistaken for visceral diseases such as myocardial ischaemia, renal colic, or cholecystitis. Very occasionally, herpes simplex and coxsackie virus infections can be a cause of dermatomal vesicular lesions. Diagnostic virology, including viral cultures and viral antibody titres, can be important in ensuring the proper diagnosis. Direct immunofluorescence of a cytologic smear of a vesicle with a monoclonal antibody to the varicella-zoster virus can provide an immediate definitive diagnosis.

Sequelae

Although herpes zoster usually resolves in non-immunocompromised hosts without sequelae, several important complications may occur especially in elderly subjects. Postherpetic neuralgia, one of the most intractable pain disorders, is the most common sequel of herpes zoster and is defined as pain persisting for 1 month or more after the rash has healed. Whereas postherpetic neuralgia lasting more than a year occurs in less than 10 per cent of patients younger than 50 years of age, it occurs in 35 to 50 per cent of patients over 60 years (Moragas and Kierland 1957). The cause of this higher incidence of postherpetic neuralgia in older patients might be explained by lowered cellular immunity to varicella zoster virus or to decreased ability to repair damaged tissue and nerve structures.

Ocular disease, including blindness, may result from involvement of the ophthalmic branch of the trigeminal nerve. Herpes zoster of the eyelids can lead to scarring with ectropion, lid retraction, and corneal damage. Conjunctivitis occurs frequently. Other ocular complications include episcleritis, dendritic keratopathy, and iritis. The syndrome of acute retinal necrosis has been attributed to reactivation of varicella zoster virus (Culbertson *et al.* 1986). This syndrome is characterized by retinal arteritis and retinitis with retinal detachment and has led to legal blindness in 64 per cent of patients diagnosed. Transient ocular muscle palsies develop in approximately 13 per cent of cases of ophthalmic herpes zoster and facial palsies occur in 7 per cent of cases. Several cases of contralateral hemiparesis following herpes zoster ophthalmicus due to either thrombosis (Eidelberg *et al.* 1986) or vasculitis have been reported in the literature.

Other neurological complications of herpes zoster include Bell's palsy which is rarely permanent, and leg weakness or foot drop resulting from motor nerve damage during lumbosacral zoster. Herpes zoster involving the sacral or anogenital area can produce abnormalities in bowel function and urination. Persistent encephalitis with progressive encephalopathy is seen in immunocompromised patients following herpes zoster.

Herpes zoster in immunosuppressed individuals is usually more severe, prolonged, and more likely to disseminate. Patients with cutaneous dissemination have a 5 to 10 per cent increased risk of serious complications such as encephalitis, hepatitis, and pneumonitis. Pneumonitis is the leading cause of death in such patients. With recent advances in antiviral therapy, disseminated herpes zoster is rarely fatal even in immunocompromised patients. Fatality rates have been estimated in the range 4 to 15 per cent in immunocompromised patients with disseminated disease (Strauss *et al.* 1988).

Treatment

Herpes zoster responds to intravenous acyclovir, and progression of the infection can be halted in immunosuppressed patients. Oral acyclovir is also useful, particularly if started early. While 200 mg five times daily is adequate therapy for herpes simplex, higher doses are more useful in herpes zoster. One useful treatment approach for herpes zoster that does not involve the ophthalmic nerve employs oral acylovir 800 mg five times daily for 7 to 10 days, then 200 to 400 mg five times daily for an additional 7 to 14 days depending upon the severity of the outbreak.

Rest and analgesics are sufficient for mild attacks. Topical application of acyclovir is useful and secondary bacterial infection can be treated with systemic and topical antibacterial preparations.

Corticosteroids given during the acute stage suppress the inflammatory changes and may reduce the incidence and duration of postherpetic neuralgia in otherwise healthy patients. A typical dose would be 40 mg of prednisone daily for 10 days, followed by a 3-week taper of the steroid. Dissemination of infection due to systemic steroids is a major risk in immunocompromised patients but is not usually a problem in healthy patients.

Postherpetic neuralgia can be treated with topical capsaicin (Bernstein *et al.* 1987). A topical anaesthetic mixture of lidocaine and prilocaine (EMLA cream) may be beneficial in some patients. Narcotic analgesics may be required but should be avoided if at all possible. Amitriptyline can be helpful for hyperaesthesia and burning pain and carbamazepine or other anticonvulsants for stabbing pain. In the elderly patient doses should be low and increased every few days as required. Transcutaneous electrical nerve stimulation can provide temporary relief of pain for some patients.

BULLOUS PEMPHIGOID

Bullous pemphigoid (Fig. 13) is a chronic, usually self-limited disease characterized by subepidermal blisters on the skin and mucous membranes (Lever 1979). In one study, 77 per cent of the patients were aged over 70. About one-third of patients will have involvement of the oral cavity. An urticarial or papular eruption or a non-specific dermatitis may be prodromal stages developing several months before the bullous eruption. The blisters are tense, variable in size, usually filled with a clear fluid, and develop on non-inflammatory or eryth-

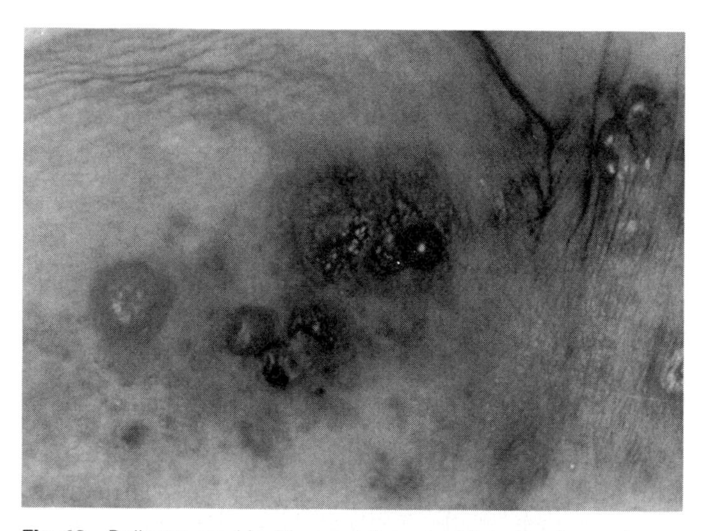

Fig. 13 Bullous pemphigoid, an autoimmune blistering disease in which IgG is deposited along the basement membrane zone of the epidermis, most commonly occurs in elderly persons and may result in death in severely debilitated individuals. Tense blisters arise on normal appearing skin or on an erythematous base. Secondary infection may develop after the blisters break.

ematous skin. The broken blisters usually heal quickly, and without scarring, if they do not become infected.

The cause of bullous pemphigoid is unknown. Exacerbations of bullous pemphigoid, as well as acute onset of the condition, have been reported to follow drug reaction, phototherapy, or other cutaneous injury. The component parts of the immunological reaction participating in the pathophysiology of bullous pemphigoid are gradually being identified, but there is no agreement as to what initiates these actions. Circulating basement membrane zone antibodies can be detected in the serum of patients, and deposits of immunoglobulin and complement can be identified in skin samples beneath the basal cells.

Uncomplicated bullous pemphigoid is frequently a self-limited disease and lasts from 3 to 6 years in a majority of patients. Fatality rates ranging from 10 to 30 per cent are reported from the disease and from complications of therapy. Systemic corticosteroid therapy is the predominant therapy for generalized bullous pemphigoid. Complications of high-dose, prolonged steroid therapy in persons of this age group, however, must be considered carefully and vigorously guarded against when treating bullous pemphigoid. Other therapies used to spare steroids may employ dapsone, or immunosuppressive agents such as azathioprine, cyclophosphamide, methotrexate, or cyclosporin. Cutaneous topical therapy to avoid infection and to provide a moist wound environment for re-epithelialization are important in the management of these patients.

STASIS DERMATITIS AND STASIS ULCERS

Venous insufficiency is responsible for stasis dermatitis and stasis ulcers (Fig. 14). The chronicity and refractory nature of this problem are well known. Stasis ulcers account for discouraging immobility and prolonged hospitalization of many otherwise healthy elderly people. Treatment is unsatisfactory. Most successful programmes are conservative and require the patient to elevate the legs, soak the ulcers, debride necrotic tissue, treat bacterial infections, manage the dermatitis with topical steroids and lubricants, and wait for the ulceration to re-epithelialize. Grafting the ulcers is sometimes required. Various alternatives to grafting have been alleged to speed the healing process, including application of amniotic membranes, gold foil, benzoyl peroxide, and sugar. Most of these modalities have been short-lived, without controlled studies demonstrating their efficacy. Recently, cultivation of epidermal cells and full thickness skin equivalents, as well as topical application of various cell-derived growth factors represent new and intriguing therapies.

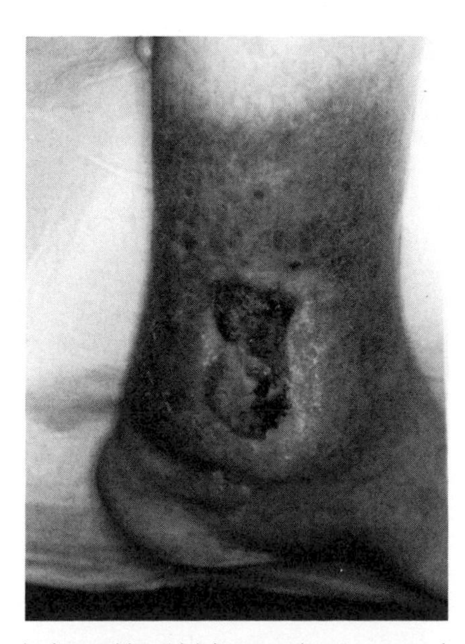

Fig. 14 Stasis dermatitis which is secondary to venous insufficiency. Leg ulceration due to severe stasis changes.

Another interesting development has been the discovery that patients with stasis dermatitis are more susceptible to allergic contact hypersensitivity to a variety of agents than are persons with other chronic dermatoses. Dermatitis and leg ulcers do not develop in all persons with varicose veins. Underlying immunological factors may be responsible for chronic stasis dermatitis and may have profound implications for new therapeutic approaches.

PHYSICAL APPEARANCE OF AGED SKIN

Although the appearance of aged skin is not normally considered as having medical significance, it does have great concern for our ageing population. There is increasing evidence that an individual's appearance is an important factor contributing to his or her self concept. Recent psychological studies have found that the physically attractive older person is more optimistic, more social, has a better personality, and feels he or she is in better health (Balin and Kligman 1989).

References

Balin, A.K. and Kligman, A.M. (eds). (1989). *Aging and the Skin*. Raven Press, New York.

Balin, A.K., Lin, A.N. and Pratt L. (1988). Actinic keratoses. *Journal of Cutaneous Aging and Cosmetic Dermatology*, **1**, 77–86.

Bernstein, J.E., *et al*. (1987). Treatment of chronic post-herpetic neuralgia with topical capsaicin. *Journal of the American Academy of Dermatology*, **17**, 93–6.

Braverman, I.M., Sibley, J., and Keh-Yen, A. (1986). A study of the veil cells around normal, diabetic and aged cutaneous microvessels. *Journal of Investigative Dermatology*, **86**, 57–62.

Culbertson, W.W., *et al*. (1986). Varicella Zoster virus is a cause of acute retinal necrosis syndrome. *Ophthalmology*, **93**, 559–69.

Curry, S.S. and King, L.E. (1980). The sign of Leser-Trelat. *Archives of Dermatology*, **116** 1059–60.

Droller, H. (1953). Dermatologic findings in a random sample of old persons. *Geriatrics*, **10**, 421–4.

Eidelberg, D. *et al*. (1986). Thrombotic cerebral vasculopathy associated with Herpes zoster. *Annals of Neurology*, **19**, 7–14.

Field, L.M. (1984). On the value of dermabrasion in the management of actinic keratoses. In *Controversies in Dermatology*. (ed. E. Epstein), pp. 96–102. W.B. Saunders, Philadelphia.

Graham, J.H. and Helwig, E.B. (1961) Bowen's disease and its relationship to systemic cancer. *Archives of Dermatology*, **83**, 738.

Hamilton, J.B., Terada, H., and Mestler, G.E. (1955). Studies of growth throughout the life span in Japanese: growth and size of nails and their relationship to age, sex, heredity, and other factors. *Journal of Gerontology*, **10**, 400–15.

Johnson, M.L.T. and Roberts, J. (1977). *Prevalence of dermatologic disease among persons 1–74 years of age*. Advance Data, No. 4, U.S. Dept. of Health Education and Welfare.

Lever, W.J. (1979). Pemphigus and pemphigoid. *Journal of the American Academy of Dermatology*, **1**, 2–31.

Marks, R. (1987). *Skin Disease in Old Age*. Lippincott, Philadelphia.

Marks, R., Rennie, G., and Selwood, T.S. (1988). Malignant transformation of solar keratoses to squamous cell carcinoma. *Lancet*, **i**, 795–7.

Moragas, J.M. and Kierland, R.R. (1957). The outcome of patients with Herpes zoster. *Archives of Dermatology*, **75**, 193–6.

Newcomer, V.D. and Young, E.M. (1989). *Geriatric Dermatology*. Igaku-Shoin, Tokyo.

Ragozzino, M.W., Melton, L.J., Kurland, L.T., Chu, C.P., and Perry, H.O. (1982). Population-based study of herpes zoster and its sequelae. *Medicine*, **61**, 310–16.

Strauss, S.E., *et al.* (1988). Varicella zoster virus infections. *Annals of Internal Medicine*, **108**, 221–35.

Tager, A., Berlin, C., and Schen, R.J. (1964). Seborrheic dermatitis in acute cardiac disease. *British Journal of Dermatology*, **76**, 367.

Tindall, J.P. and Smith, J.G. (1963). Skin lesions of the aged and their association with internal changes. *Journal of the American Medical Association*, **186**, 1039–42.

Weismann, K., Krakaver, R., and Wancher, B. (1981). Prevalence of skin disease in old age. *Acta Dermatovenerologica*, **66**, 352–3.

Young, A.W. (1965). Dermatogeriatric problems in the chronic disease hospital. *New York State Journal of Medicine*, **63**, 1748–52.

SECTION 18
Neurology and psychology: age-associated changes

18.1 The neurological examination of ageing patients

FRANCOIS BOLLER

Boller *et al.* (1987) have designed a form especially for recording the neurological and neuropsychological symptoms as well as the results of the neurological examination of elderly subjects with or without cognitive impairment. It consists of 102 items divided into history-taking, examination of mental status, and the rest of the neurological examination. The main domains examined are listed in Tables 1 and 2.

MENTAL STATUS EXAMINATION

It is now well established that major memory loss and intellectual decline in elderly subjects practically always constitute the symptoms of an abnormal condition. Standard measures of overall intellectual functioning, such as the Wechsler Adult Intelligence Scales (**WAIS**), generally show some age-associated decline, usually with a distinction between verbal abilities, which tend to be preserved, and non-verbal or performance abilities, which tend to deteriorate. This very stable pattern found in both sexes irrespective of socioeconomic status has been interpreted by Cattell (Horn and Cattell 1967) in terms of 'crystallized' and 'fluid' intelligence, the former representing the fruit of learning and experience. The tests included in the non-verbal part of the WAIS are timed and less familiar and therefore more demanding. Such factors do not account entirely for the differences, prompting some to suggest a greater vulnerability of the right cerebral hemisphere to ageing. The difference may, however, relate to the greater depth of processing (see Section 18.2) or greater redundancy of coding of early learning of language skills. In addition, it must be noted that many older persons—more than half of those in their 60s and 70s in one longitudinal study of a generally healthy population—perform as well on virtually all types of mental tests or those in their 20s.

Assessing intellectual abilities at the bedside or clinic should be based more on a general evaluation of the patient's behaviour (by observation and by history obtained from the family) than on a single informal test. For patients suspected of having dementia, the clinical evaluation may easily be supplemented by more formal but short assessments of mental status such as the mini-Mental State Examination (of Folstein *et al.* 1975). A score of less than 24 (out of a possible 30) should call for more in-depth evaluation, i.e. full psychometric testing through consultation with a neuropsychologist. On the whole, performance in the domains listed in Table 1, as tested in the clinic or at the bedside, should not show a clear-cut impairment with age. Formal neuropsychological tests are particularly useful when history and examination leave the clinician in doubt about the presence or absence of changes.

Research supported in part by the Institut National de la Santé et de la Recherche Medicale (INSERM), Paris, France.

Table 1 Principal domains to be assessed in the mental status examination

Overall intellectual abilities
General behaviour
Orientation for time, place, and person
Oral language
Reading (aloud and comprehension)
Writing
Right–left orientation
Calculations
Oral and limb praxis
Constructional abilities
Memory, recent and remote
Proverbs and similarities

Table 2 Principal domains to be assessed in the general neurological examination

Olfaction
Pupillary response (right, left)
Visual acuity (right, left)
Visual field
Extraocular movements
Facial nerve
Palatal elevation
Hearing (right, left)
Limb strength
Motor tone
Tremor, other abnormal movements
Deep tendon reflexes
Plantar (Babinski) sign
Release signs
Sensory examination (touch, pain, vibration)
Stereognosis and graphaesthesia
Cerebellar functions
Gait

'GENERAL' NEUROLOGICAL EXAMINATION

An impairment in sensory functions is an almost universal finding in the aged, even in the absence of disease. Olfaction, visual acuity, and hearing tend to be decreased and so is peripheral sensation, particularly vibration sense. Many of these impairments are due to changes occurring at the periphery (particularly the eye, the inner ear, and other receptors). Some of them, however, are probably due to changes in the central nervous system as shown, for example, by the delay in visual-evoked potential consistently found in elderly subjects. Many of these changes, for example loss of vibration sense and loss of ankle jerks, can be associated with a reduction in mean transmission velocity of nerves and an increase in the variance of velocity, which will produce most noticeable effects in the longer fibres. However, the prevalence of age-associated changes in neurological signs may have

Table 3 Prevalence (%) of abnormal signs in an industrial sample of subjects aged 50 to 93 years without overt neurological disorder (Jenkyn *et al.* 1985)

Age (years)	Nuchocephalic reflex	Glabellar tap	Snout reflex	Conjugate upward gaze	Conjugate downward gaze	Visual tracking	Paratonia
50–54	7	4	1	2	4	1	6
55–59	5	6	1	3	5	2	4
60–64	8	6	3	3	5	5	6
65–69	8	10	3	6	8	8	6
70–74	17	15	8	15	15	18	10
75–79	14	27	7	27	26	22	12
80–	29	37	26	29	34	32	21

been widely overestimated by the use of inappropriate clinical techniques. For example, Ipallomeni *et al.* (1984) showed that ankle jerks were present in 188 (94 per cent) of 200 consecutive patients in a geriatric department when tested by the plantar-strike method rather than the conventional tendon-strike method.

Changes also occur in the motor system, as shown at the level of the cranial nerves, by decreased ocular (especially upward gaze) and pupillary motility. Here again, these changes are in great part attributable to peripheral rather than central changes. On the other hand, strength is not normally affected.

The most striking neurological change is often a modification of posture and gait. Elderly persons are often found to have a stooped posture and a hesitant, small-stepped gait, together with some tremor and increased tonus of the extremities bearing, in fact, some resemblance to early parkinsonism. These changes may be due in large part to non-neurological factors such as arthritis or osteoporosis, but if they are not readily explicable call for more in-depth neurological evaluation.

NEUROLOGICAL EXAMINATION IN 'PRIMARY' DEMENTIAS

The neurological findings in patients with age-associated neurological diseases are described in other portions of this book. Here are described the salient characteristics of the neurological examination (excluding mental status findings) in patients presenting with dementia not related to either a systemic disease or to neurological diseases with easily detectable diagnostic features and therefore referred to sometimes as primary. These comprise essentially Alzheimer's disease, multi-infarct dementia, and occasionally less common disorders such as Pick's disease.

The neurological signs most commonly found in these patients are those known as primitive reflexes or release signs. Huff *et al.* (1987) found them in 55 per cent of patients with probable Alzheimer's disease, and Huff and Growdon (1986) found that they are related to the severity of the dementia. They consist of the glabellar sign and presence of snout, palmomental, and grasp reflexes. The exact pathophysiological basis of these signs (described in detail by Paulson (1977)) is unknown. They are thought to be related either to frontal-lobe lesions or to diffuse cortical atrophy, which result in the release of developmentally early reflexes from regulation by brain regions that develop later. They are found in normal infants and in some elderly subjects without evidence of neurological diseases—close to 10 per cent in the study of Huff *et al.* (1987). Therefore they are neither sensitive nor specific enough to serve as diagnostic markers for Alzheimer's disease. The age-specific prevalence of the snout reflex and glabellar tap, together with some other neurological features in an industrial sample of 2029 subjects aged 50 to 93 years (Jenkyn *et al.* 1985) are presented in Table 3.

Other abnormalities found in a significantly large percentage of patients with dementia include decreased olfaction and difficulty with stereognosis and graphaesthesia. There is also increased frequency of tremor (which may in turn produce abnormal response at the finger-to-nose and other 'cerebellar' tests of dexterity) as well as abnormal gait most commonly of the bradykinetic type found in extrapyramidal disorders. This could either represent disease of the frontal lobe or be due to concomitant Parkinson's disease known to occur more frequently in patients with Alzheimer's disease (Boller *et al.* 1980).

CONCLUSIONS

Even in 'normal' aged subjects, some changes can be found in the mental status examination and in the general neurological examination. What is the anatomical basis of these changes? The changes in posture and gait seen in normal elderly are most likely attributable to actual changes in the central nervous system. Neuropathological studies, particularly cell counts, tend to be rather controversial, but the age-associated cell loss found in the substantia nigra has been confirmed by several investigators, suggesting that the above clinical changes are related to an incipient Parkinson-like condition in older persons. The anatomical bases of other changes, however, particularly the rather subtle mental status changes that accompany ageing, remain to be determined.

References

Boller, F., Huff, F.J., Querriera, R., Kelsey, S., and Beyer, J. (1987). Recording neurologic symptoms and signs in Alzheimer's disease. *American Journal of Alzheimer's Care and Research*, **2**, 19–20.

Boller, F., Mizutani, T., Roessman, U., and Gambetti, P. (1980). Parkinson disease, dementia, and Alzheimer disease. Clinicopathological correlations. *Annals of Neurology*, **17**, 329–35.

Folstein, M.F., Folstein, S.E., and McHugh, P.R. (1975). Mini-mental state: a practical method for grading the cognitive state of patients for the clinician. *Journal of Psychiatric Research*, **12**, 189–98.

Horn, J.L. and Cattell, R.B. (1967). Age differences in fluid and crystallized intelligence. *Acta Psychologica*, 26, 107–29.

Huff, F.J., and Growdon, J. H. (1986) Neurologic abnormalities associ-

ated with severity of dementia in Alzheimer's disease. *Canadian Journal of Neurological Sciences*, **13**, 403–5.
Huff, F.J., Boller, F., Lucchelli, F., Querriera, R., Beyer, J., and Belle, S. (1987). The neurologic examination in patients with probable Alzheimer's disease. *Archives of Neurology*, **44**, 292–32.
Ipallomeni, M., Kenny, R.A., Flynn, M.D., Kraenzlin, M., and Pallis, C.A. (1984). The elderly and their ankle jerks. *Lancet*, **i**, 670–2.
Jenkyn, L.R. *et al.* (1985). Neurologic signs in senescence. *Archives of Neurology*, **42**, 1154–7.
Paulson, G.W., (1977). The neurologic examination in dementia. In *Dementia*, (2nd edn.), (ed. E. Wells), pp. 169–88. Davis, Philadelphia.
Schaie and Strother 1968.

18.2 Memory

PATRICK RABBITT

As for all other branches of cognitive psychology the main goal of cognitive gerontology is to use data on human performance to relate functional models for observed behaviour to events and changes in the brain and central nervous system. Lucid examples of the way in which this framework of description has been used to study human memory are found in accounts of personal experiments and theories such as those by Baddeley (1986), Tulving (1983), or Cermak and Craik (1979), and in tutorial texts such as Crowder (1976) and Wicklegren (1979). Excellent source-books now describe how functional models developed by cognitive psychologists have been used to guide and interpret neurophysiological investigations (e.g. a symposium by Squire and Butters (1984) and a tutorial text by Parkin (1987).

Most of these theories, experimental paradigms, and techniques of data analysis have been shaped by the assumption that the study of memory involves the identification of putatively independent 'functional modules'—and of mutually independent 'performance parameters' that are characteristic of these modules—which, albeit only in an optimistic future, may be mapped on to neurophysiological subsystems (Fodor 1983). Thus very similar databases are interpreted by Baddeley (1986) and Parkin (1987) to distinguish modular data-holding systems in a 'multistore model of memory', and by Wicklegren (1987) to identify operating characteristics of the memory system such as relative rates of trace acquisition and decay.

Inevitably this has generated a tradition that as soon as new 'models of memory systems' become fashionable they should be enlisted to guide comparisons between older and younger people. The expected outcomes of these comparisons are decisions as to whether age affects some of the hypothetical 'subsystems' or 'performance characteristics' that the new models specify, earlier or more severely than others. These goals are implicitly accepted in the best recent reviews of memory changes with age (e.g. Craik 1977; Klatzky 1988; Light and Burke 1988).

The assumption that ageing affects some systems, processes, or performance characteristics more than others agrees with the reports that older people very commonly give about their subjective experience of cognitive change. They seldom complain that all their cognitive faculties are deteriorating, but rather make specific complaints of loss of memory. Further, they complain of specific loss of a particular type of memory—for events in their recent pasts—with retained, or even enhanced, (cf. Ribot 1882) memory for remote events in childhood or early life. This subjective dissociation of perceived loss of efficiency in short and long-term memory is congenial to the distinctions between 'short-term', 'intermediate-term', and 'long-term' memory made by modular 'multistore' models such as that described by Parkin (1987).

The same assumptions reappear in the literature on applied clinical assessment of elderly people as a suggestion that normal ageing is accompanied by 'age-specific memory impairment' (cf. Crook *et al.* 1986; Crook and Larrabee 1988). This implies that different cognitive systems may 'age' at different rates both within and between individuals so that, in some people, the 'memory system' may be impaired more than other, putatively independent, cognitive structures.

These assumptions have had strong methodological consequences. In research on age changes in memory the standard procedure has been to compare very small (usually 20 or fewer) and very homogeneous (usually middle-class and well-educated) groups of older and younger people on experimental paradigms that have ingeniously been developed to obtain isolated performance indices supposed to be characteristic of particular functional processes. To do this, these paradigms deliberately seek to eliminate the effects of possible interactions between more than one of these processes and, moreover, by selection of subjects and choice of techniques of statistical comparison, these studies also seek to minimize the effects of variance between individuals. Obviously these are not the best possible procedures for investigating individual differences in performance nor for asking the unresolved questions as to whether neurophysiological ageing affects all cognitive systems to the same extent, whether it affects some systems more than others, and whether it affects interactions between series of independent systems as much as it affects the systems themselves. In short, use of this method assumes that we have already answered the basic questions about the nature of cognitive change with age that we use it to try to address.

These assumptions are inevitable in a neuropsychology which is founded on the technique of mapping tightly defined deficits on to focal lesions. It may be unhelpful in understanding changes that occur with normal ageing. As other chapters in this book make clear, 'normal ageing' involves diffuse cellular and neurochemical changes that affect the entire central nervous system, and there is as yet no clear evidence that it affects some particular structures or systems earlier, or more severely, than others. Further, it may be unhelpful to consider the 'ageing' of human memory solely as a consequence of degenerative, neurophysiological changes. Although people do suffer such changes (see Petit 1982), they also, throughout their lives, continue to acquire new information, to learn new cognitive skills, and to refresh data and skills that they have already mastered. Prolongation of life allows time to acquire a wider and more practised repertoire of skills. Recent studies show that extraordinarily high, but tightly 'domain-specific' skill may be attained by prolonged practice in particular, very specialized memory tasks. Thus Chase and Ericsson (1981), Ericsson *et al.* (1980), and Ericsson and Chase (1982) showed that while individuals may learn encoding techniques that allow them to increase their digit spans to 70 or above, the specificity of these mnemonic strategies prevents their transfer to other material so that their spans for letters remain unaffected at 7–9 items. Similarly Ericsson and Polsen (1988) found that the special mnemonic strategies that a restaurant waiter developed to allow him to remember orders for up to 20 different meals transferred to some, but not to other, task demands. Thus very marked differences between individuals in 'short-term' or 'long-term' memory efficiency may reflect their idiosyncratic acquisition and maintenance of specialized mnemonic skills rather than differences in the relative integrity of particular structures within their central nervous systems. The magnitude of this disparity between highly practised and unpractised memory skills is very much greater than those usually reported between the performances of groups of old and young subjects on tasks that are novel to both. It follows that, within individuals, disparities between levels of competence in practised and unpractised skills becomes increasingly marked in old age. For example, while scores on IQ tests that demand high working-memory capacity sharply decline with age, scores on vocabulary tests may remain unchanged until very late in life (Rabbitt 1986). Horn (1982) takes such data as evidence for a distinction between cognitive skills that require 'fluid intelligence' which are associated with efficiency of information processing and which decline with age, and those that depend on learned techniques or acquired information, which are relatively unaffected by ageing. The question as to whether some memory 'modules' 'age' faster than others may thus not provide useful information about differential rates of change in different neurophysiological 'modules' but rather only tell us which particular skills older individuals have acquired during their lifetimes, and which of these their current lives encourage them to maintain.

There are concrete illustrations as to how both changes in neural function and individual differences in the demands of daily living can jointly and independently determine differences in the relative ease with which recent and distant memories can be accessed. These illustrations are found when we experimentally examine Ribot's 'law' of memory ageing (Ribot 1982), based on the universal subjective impressions of all elderly people that they can remember remote much more vividly than recent life-events. This may readily be explained in terms of differential ageing of neurophysiological 'modules': for example by the assumption that changes in hippocampal and limbic structures, which serve immediate and short-term memory, proceed faster than in frontal or temporal cortex, which is also involved in much longer-term storage of information (Milner 1970; Warrington 1982). It is worth noting that here, as elsewhere, incommensurability of the performance indices which we can use to measure the relative efficiency of these hypothetically distinct memory systems should give us pause for thought. For example, just how are we to estimate the equivalent loss of information about specific autobiographical life-events in terms of a reduction in immediate memory span for digits? One possible way of circumventing this difficulty is to ask individuals to recall as many events as they can from particular periods of their lives and to compare the relative abundance, and the numbers, of details of the events that they can recall from different decades. Studies by Crovitz and Schiffman (1974), and Crovitz and Quinta-Holland (1976) suggest that while individuals of all ages recall many more recent than remote events, the ratio of distant to recent memories may be slightly greater for elderly groups.

Holland and Rabbitt (1990) asked individuals aged from 50 to 86 years to recall as many events as possible from the first, second, and third parts of their lives. Subjects also dated each event and rated the frequency with which they remembered having spontaneously recalled or rehearsed it during their recent daily lives. One subgroup of these people currently led active and independent lives; a second subgroup was made up of individuals who revealed no cognitive problems when objectively screened but, because of various physical disabilities, had been supported in institutional care for at least 2 years. The third subgroup lived in the same institutions but had been assessed as borderline cases of dementia on the Blessed dementia scale. Consistent with their parity on other cognitive measures, the community-resident and the cognitively intact, residential-care groups recalled similar total numbers of autobiographical events. Unsurprisingly the cognitively impaired group recalled very few life-events. Bearing this basic difference in mind, Fig. 1 gives the *proportions* of total numbers of life-events that members of each group recalled from successive decades of their lives.

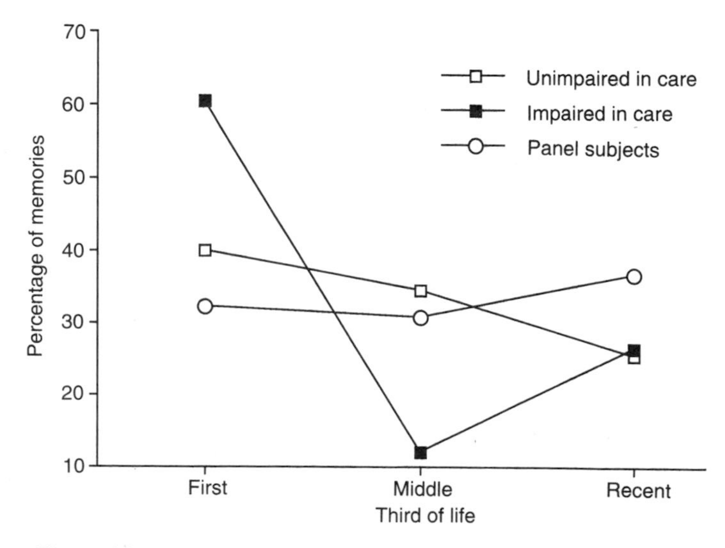

Fig. 1 Memories recalled from each third of life expressed as a percentage of the total recalled; institutionalized elderly people with and without cognitive impairment and independent volunteer ('panel') subjects.

Community residents showed the usual power relationship between abundance of events recalled and age decade (Rubin *et al.* 1986). Their ratings of the number of occasions on which they had spontaneously recalled each of these events during their recent everyday lives provided a sufficient explanation for this finding. All individuals reported that during the recent past they had spontaneously recalled and ruminated over recent events much more often than remote events. This is easily understandable if we consider the evolutionary advantages conferred by memory; i.e. what humans and other animals actually use their memories for. In order competently to manage an active life one must continually reference the recent past in order to anticipate and plan for the immediate future. The process of recalling and mulling over recent events is thus a staple of everyday mental life. In contrast memories of remote events are 'archival' in the sense that they much less frequently need to be consulted to anticipate and meet the demands of daily life.

In contrast, the few memories produced by the cognitively impaired group included very few recent but a high proportion of remote events. This might be taken as evidence that central nervous damage, which becomes progressively more common with age, affects brain systems involved in the encoding and storage of new information more radically than those involved in the retrieval of long-established memories. However, cognitively intact institutional residents, who recalled as many events in total as did active community dwellers, also recalled a relatively high proportion of remote as against recent events. These institutionalized individuals also reported very much more frequent spontaneous recall of, and intensive rumination over, the earlier memories that they recalled. This shift in balance of efficiency of spontaneous recall between recent and remote events is understandable in terms of social life in institutions in which there has long been little demand for active planning of personal life, and in which few novel or remarkable incidents occur. Recollection of interesting, though distant, autobiographical events gradually becomes the staple of solitary reverie and of conversational exchange.

Thus it may be very misleading to infer differential age changes in the efficiency of different 'memory systems' *only* from differences in the levels of single performance indices (such as quantitative estimates of the relative ease of retrieval of 'distant' and 'recent' memories) that have been shown by comparisons between small, highly selected groups of subjects. Not only the integrity of the 'biological' bases of memory but also the variety of social and practical demands that memory must serve can alter radically as age advances. Sustained practice may maintain particular memory databases or cognitive skills at a very high level without necessarily benefiting other, apparently very similar, types of performance.

For these reasons, and because it is doubtful whether the current successes of the fruitful and popular 'modular' approach can yet be better reviewed than they recently have been by Craik, Klatzky, and others, this chapter tries to describe what we know about memory changes in old age from a different, and much more general standpoint: that of questions recently raised by screenings of very large populations of ageing individuals.

Such screenings may seem very laborious and clumsy research tools, but they do allow us to approach questions that cannot be answered by comparing small groups of individuals on tightly controlled paradigms. For example, whether memory changes in old age are universal so that, to quote recent investigators, some 'benign, non-specific age related memory loss' is 'normal' in older groups (Crook *et al.* 1986, 1988) or whether memory loss is always the result of some specific pathology in the central nervous system and does not occur in fortunate (albeit rare) individuals who escape these changes until very late in life. To detect the presence, precisely specify the nature, and estimate the severity of pathological memory loss we need benchmark data as to what constitutes 'normal' memory function at any age. If we suppose that 'memory' is a general descriptive term for a variety of different activities carried out by a range of distinct cognitive or neuropsychological systems, we need to discover whether the rates of age-associated decrements in performance in putatively distinct tasks are correlated with each other or are independent. Data from such 'epidemiological' studies of memory are most appropriately analysed in terms of the models and techniques developed by psychometricians to interpret variance in performance among very large samples of individuals. Once basic data from population comparisons have been understood, the quite different models that neuropsychologists have developed in order to interpret differences in scores obtained by the same groups of subjects on different conditions of tightly controlled experimental paradigms may become useful. The point is not that either the 'psychometric' or the 'cognitive psychological' approach is superior, but that each answers questions that are complementary to those addressed by the other.

This population-centred approach raises three general questions, which examine rather than taking for granted, the assumptions of the 'modular' approach that we have discussed at such length: (1) when do changes first appear; (2) how fast do they then proceed; (3) do all cognitive systems change at the same rate, or do some change faster than, and independently of others? In particular is there any evidence that that changes in 'memory' can be considered as distinct from changes in other cognitive systems?

To learn precisely what evidence we would need in order to answer these questions we must consider the range of possible models for cognitive ageing. These are illustrated in Fig. 2. Figure 2(a) illustrates, with hypothetical data the invariable pattern obtained when average scores attained on any task by successive age groups are compared (i.e. 'cross-sectional' data on grip strength, IQ test score, learning rate for nonsense syllables, digit span, or virtually any other performance measure). A sharp improvement over the first 18 to 25 years is followed by a brief plateau, the duration of which varies between different tasks, and then by a continuous decline, which accelerates sharply in the eighth and ninth decades of life. Figure 2(b) illustrates an alternative model, which is suggested by some fragmentary longitudinal data from studies of individuals rather than of groups. The adult plateau may be extended until late in life, ending in an abrupt decline, possibly associated with life-concluding pathology (termed the 'terminal drop' by Jarvik 1983). Figure 2(c) illustrates the difficulty of distinguishing between the 'continuous decline' and 'terminal drop' models when only cross-sectional data are available. A necessary corollary of the 'terminal drop' model is that we would expect individuals

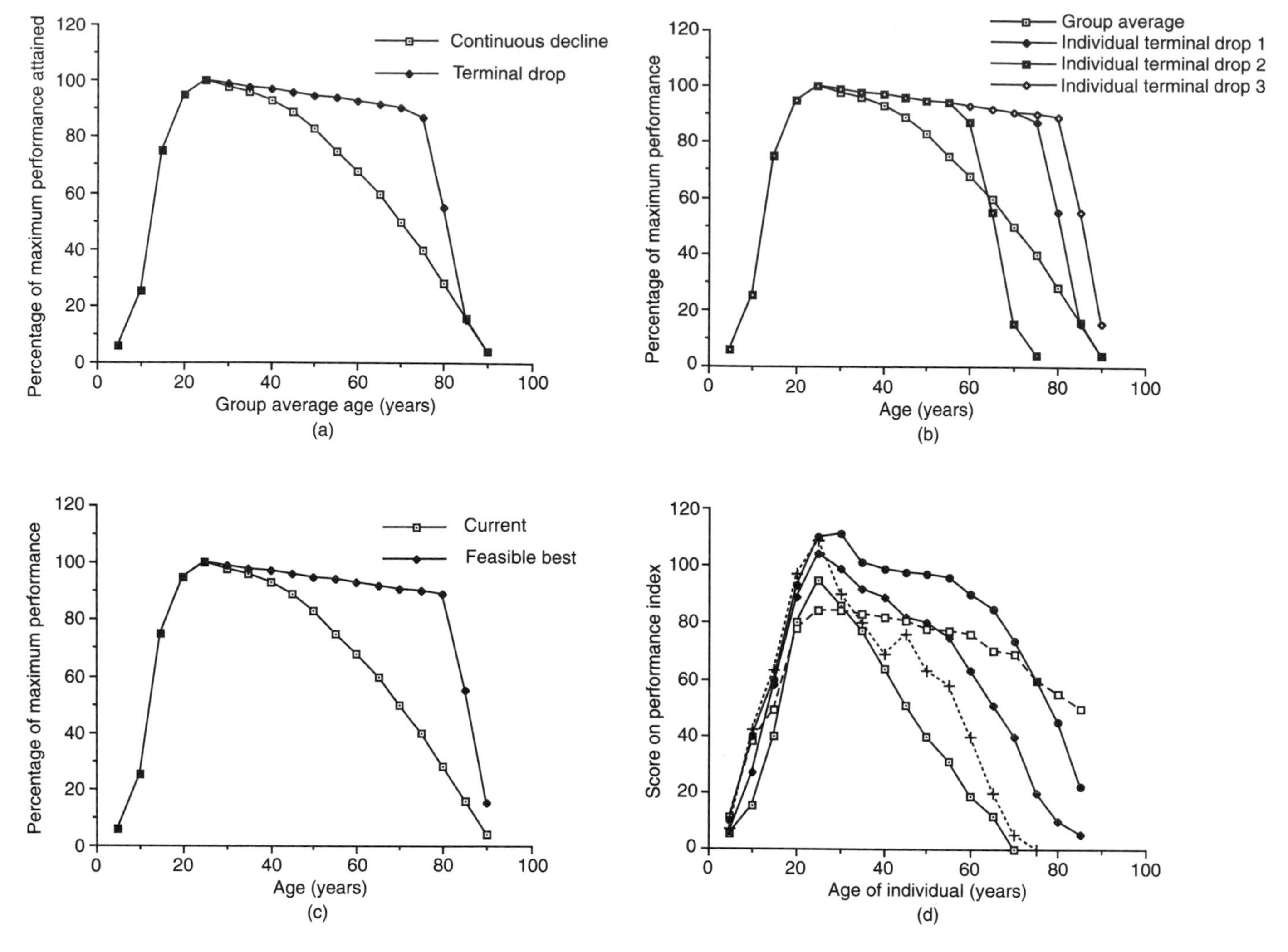

Fig. 2 (a) Theoretical longitudinal pattern of performance on 'continuous decline' and 'terminal drop' models. (b) Interindividual variation in age at which terminal drop occurs can produce group averages resembling continuous decline model. (c) Improvement in average population performance gained by delaying age at which terminal drop occurs. (d) Differences in trajectories of continuous decline will cause an age-associated increase in interindividual variation.

to suffer their declines at different ages. Consequently older samples would include increasing numbers of individuals who are currently undergoing their terminal declines and progressively fewer individuals who are still holding their performance plateaux. As Fig. 2(c) shows, typical cross-sectional data, i.e. means of performance scores from successive age groups, would thus continuously decline with increasing age. This makes the additional point that the variance of scores must increase in each successive age group in proportion to the decline in group mean. Figure 2(d) shows that increasing variance with decreasing mean scores across successive age groups is not a characteristic that is discriminatory between models since it would also occur if continuous declines begin at different ages and proceed at different rates.

It might seem that the question whether mean scores decline but between-individual variance increases with group mean age can readily be answered from the published record. However, the ubiquity of the 'modular' methodology of comparisons between small groups of individuals on an enormous variety of ingeniously diverse tasks has prevented this. A review was conducted of 15 experiments published between 1970 and 1989 in which paradigms were sufficiently comparable (intermediate-term recall of words and pictures), in which data were presented numerically rather than graphically, and in which both means and standard deviations of group scores were reported. Figure 3 plots group mean scores against the corresponding between-subject (within-age group) standard deviations in scores separately for younger and older groups in these 15 different comparisons.

At the lower end of the scoring range for subjects of all ages, variance must be attenuated because means are near zero (i.e. there are 'floor effects' in scores due to extreme task difficulty). Beyond this point within-group SDs must therefore increase with group means. The hypothesis that variance in performance between individuals increases with group mean age would be supported if the functions relating within group SDs to within group means were steeper across data points for older than for younger groups. However, Fig. 3 shows indistinguishable increases for older and younger subjects. Although the data were not analysed to make this point Fig. 3 casts an uncomfortable light on the methodology of at least two-thirds of these, in many cases very influential,

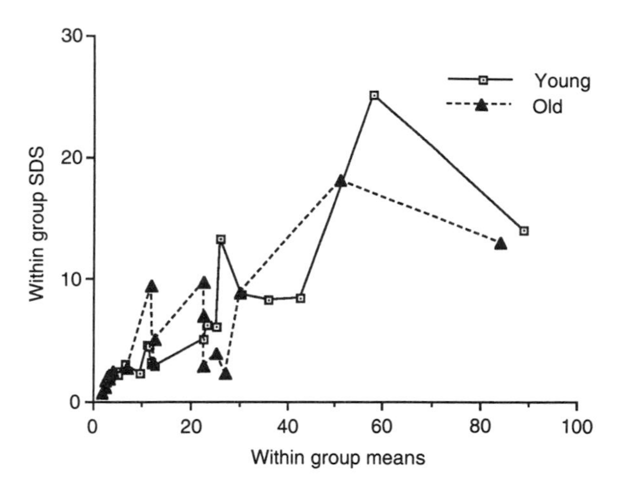

Fig. 3 Group means and standard deviations of scores in 15 published experiments comparing young and old subjects.

published comparisons. For both young and older subjects there is an inverted 'U' relationship between group means and SDs such that group mean scores of both less than 20 and more than 70 per cent correct are associated with reduced between-subject variance. If this sample correctly reflects the record this can only mean that in about 30 to 50 per cent of published studies, within-subject variance was curtailed either because the tasks on which groups were compared were so easy that one of the two (usually, of course, the young) had average scores approaching the maximum possible in the task, or were so difficult that one group (usually, of course, the older subjects) had average scores approaching zero. It follows that the group mean scores are also not trustworthy indices of the relative performance of the age-groups compared, and so only dubiously substantiate the theoretical points that these studies claim. Thus the chief use of Fig. 3 is as a caveat against unquestioning acceptance of two-thirds of the published data on age–memory comparisons, as a ready-reckoner that allows identification of useful as against potentially misleading data, and as a guide to appropriate choices of task difficulty in future studies (i.e. we can only draw useful conclusions when the average percentage correct scores for *both* young and old groups fall within the range 20–70 per cent).

Even if appropriate levels of task difficulty are used it is doubtful whether we can learn much about the effects of age on individual variability in memory performance from small-scale experiments that compare groups of 30 or fewer individuals, which are carried out using many different paradigms and for which older and younger subjects have been carefully selected to be closely comparable to each other on all criteria other than age (e.g. means and SDs of years of education, current verbal or estimated, young–adult IQ test scores, etc.). There is no good alternative to data on very large numbers of individuals distributed over a wide age range, who are all given the same tests. Among a number of different studies that are currently accumulating such data, P. Rabbitt, Bent and Abson (unpublished) have given five different memory tasks to 2100 individuals aged from 49 to 86 years who are active, self-maintaining, community residents of Newcastle upon Tyne and to 1600 residents of Greater Manchester, England.

Some methodological quirks, and outcomes, of such comparisons can be illustrated by comparing changes in means and standard deviations of age-group scores on two of these tasks: digit span, and free recall of a list of 30 words. Figures 4(a, b) are plots of means and SDs of percentage correct responses for successive, 5-year, age samples on these tasks. On the digit-span task, mean scores steadily decline, and within-group SDs steadily increase across successive age groups. This seems clear evidence that variability of performance between individuals increases with group mean age but it is possible that variance in younger age groups was curtailed because their scores approached the maximum possible on this task. In the free-recall task while group mean scores fall with age within group SDs remain approximately constant for all age groups.

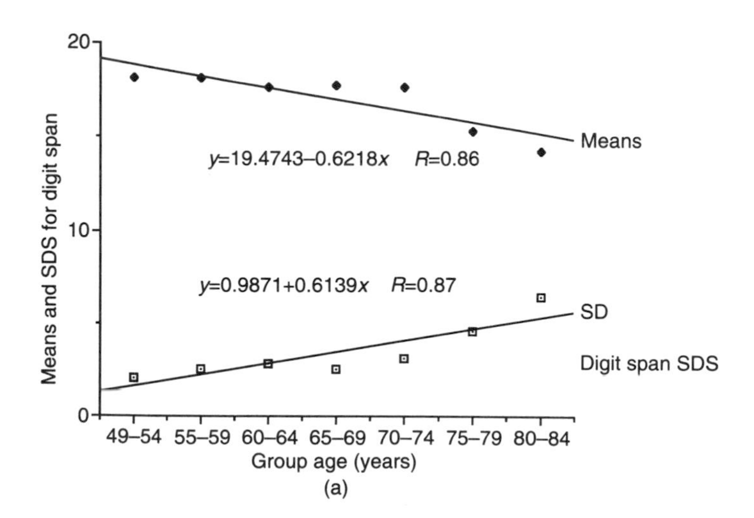

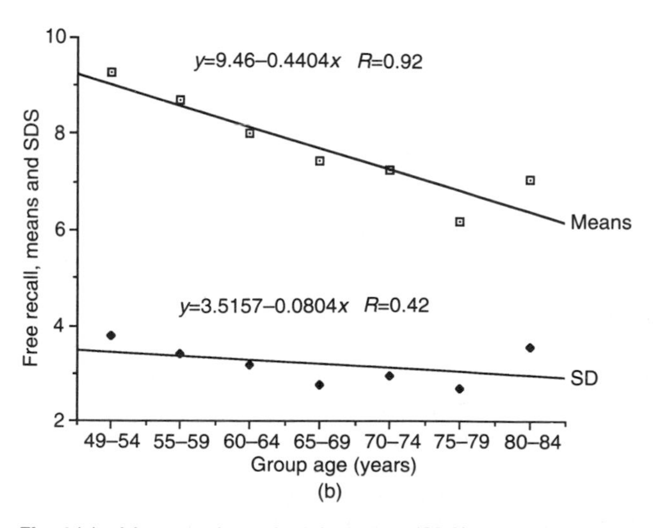

Fig. 4 (a) Means and standard deviations (SDS) of digit span by age for 2100 individuals. (b) Means and standard deviations (SDS) of free recall scores by age (2100 subjects).

Another possible reason for this discrepancy becomes evident when we examine the distributions of scores on these two tasks within successive decade samples. These are plotted in Fig. 5(a, b). Distributions for successive age groups are increasingly skewed towards the lower end. That is, while a few 70-year-olds still attain scores comparable with those

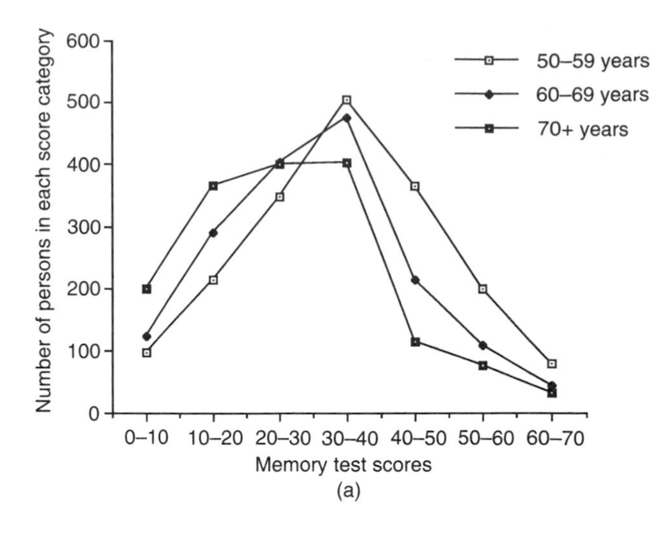

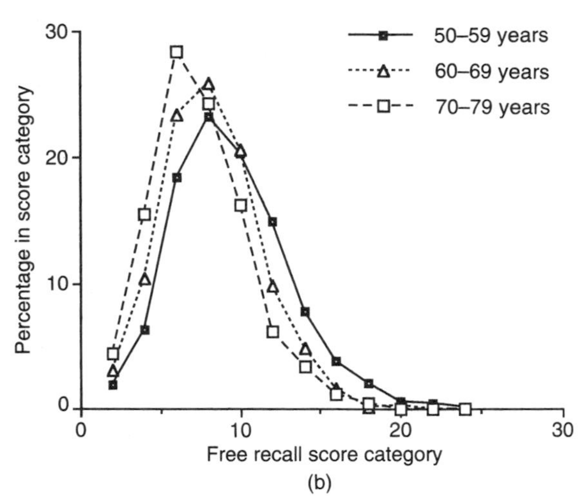

Fig. 5 (a) Distribution of memory test score by age decade (2100 subjects). (b) Percentage distribution of score for free recall of 30 items by age decade (2100 subjects).

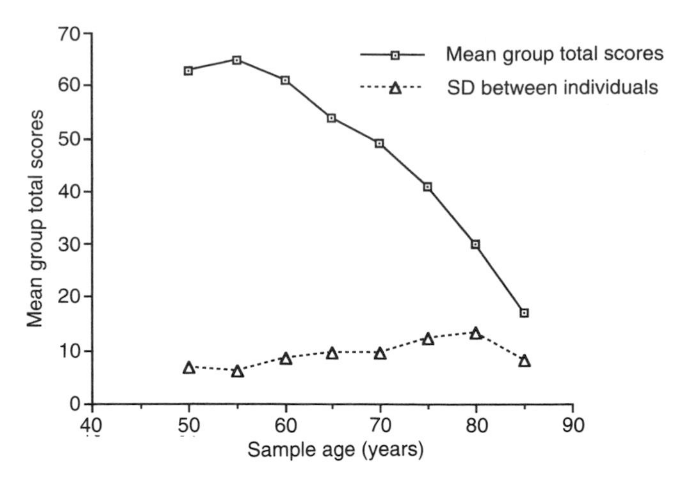

Fig. 6 Means and standard deviations (SD) of summed scores for three trials of free recall of 30 words by age group.

of the best 50- and 60-year-olds, most of them have drifted into the lower end of the scoring range. The difference is that in the case of the free-recall task the lowest end of the range of scores is near zero but in the digit span task it is as high as 60 per cent. This is a direct result of the scoring technique used. In a digit-span task, people almost always know when they fail to recall all digits in a presented list so that the task becomes distressing for the less able if they are permitted few or no successes. To avoid this, subjects were given 25 lists of digits, five of each length from 4 to 8 items. Scores are total numbers of lists of any length that were correctly reported, allowing even the least able individuals to have scores well above zero. In contrast, many older individuals found the free-recall task quite difficult and, as the distributions of scores in Fig. 5(b) suggest, within these groups the possible range of scores was curtailed by floor effects. To check this interpretation three groups of 100 individuals aged 50–59, 60–69, and 70–79 were given three trials at immediate free recall of different lists of 30 words. Means and SDs of their scores summed across all three lists are plotted in Fig. 6.

It seems that when tasks allow scoring over a very wide range so that even the least able individuals can attain non-zero scores and the most able subjects do not approach ceiling, and neither the upper nor the lower tails of distributions of scores are truncated, group mean scores decrease, and within-age-group SDs correspondingly increase with group mean age. One possible explanation for this is that individual rates of ageing are genetically determined and differences in age–performance trajectories become increasingly obvious late in life. Another is that as age advances, increasing numbers of individuals are likely to be affected by any or several of a variety of pathologies. It will be a long time before genetic differences in rates of ageing are understood, but the fact that pathologies which become increasingly common in old age do, indeed, contribute to age-associated increases in variability of memory is illustrated by a review of the accumulating literature on the effects of a variety of pathologies with advancing age (Holland and Rabbitt 1991). One clear illustration comes from a study of the cognitive effects of diabetes by P. Rabbitt, Metcalf and Bent (unpublished). Precisely the same memory tests administered to the Newcastle and Manchester populations described above were also given to a group of 168 active, community-resident diabetics, aged between 50 and 76 years. Within the very large, longitudinal screening sample from Manchester it was possible to match each individual diabetic in terms of age, sex, years of education, current socioeconomic status and score on the Mill Hill vocabulary test with a particular control subject who had no known pathology. Figure 7 shows distributions of scores on a test of immediate, free recall of 30 words and on cumulative learning of a list of 15 words across four successive trials for the diabetics and for their controls.

Compared to the distributions of scores for healthy controls, scores for diabetics are seen to be skewed towards the lower end of the range. Distributions of scores on other memory tasks and on performance IQ tests present similar pictures. Thus, with accumulating data on the 'cognitive epidemiology' of other age-associated conditions such as hypertension, the data confirm that increased variance in performance scores in older populations can result from the increase, with age, in the prevalence of a variety of pathologies—quite other than those such as strokes and dementias that specifically involve the central nervous system (Holland

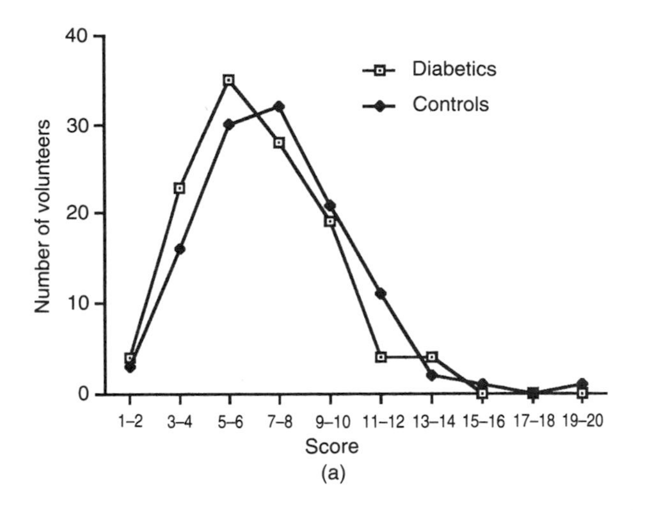

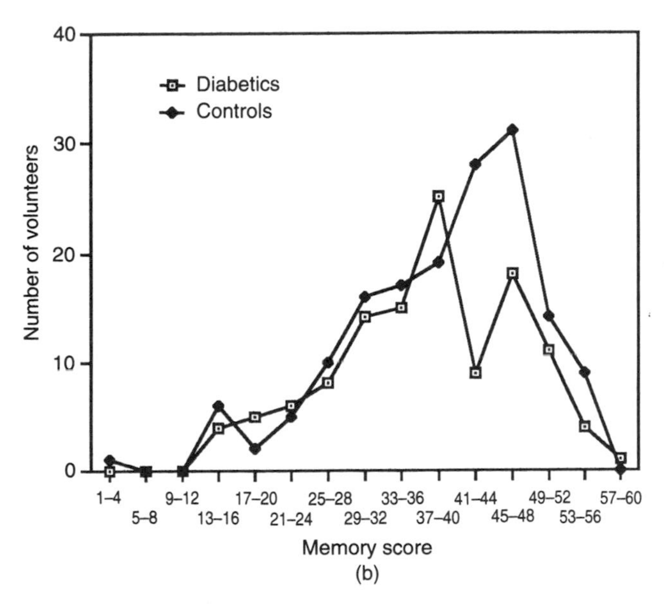

Fig. 7 (a) Distribution of visual free recall scores for diabetic and non-diabetic subjects. (b) Distribution of memory scores for diabetic and non-diabetic subjects.

and Rabbit 1991). Successive decade samples are characterized by increasing performance gaps between a dwindling number of healthy individuals and an increasing number of individuals who are affected by one or more different pathologies. There is, as yet, no clear evidence whether or not various pathologies produce distinct and characteristic changes in specific memory capabilities.

The next general question about the ageing of human memory is whether scores on all memory tasks change at the same rate or whether scores on some tasks change earlier, and more radically, than on others. An obvious approach is to examine relationships between scores on a wide variety of cognitive tests administered to very large numbers of individuals in successive decade samples. This has been done by P. Rabbitt, Bent and Abson (unpublished) for a sample of 2100 residents of Newcastle upon Tyne aged from 50 to 86 years who were each given performance and spatial IQ tests (the AH 4, parts 1 and 2; Heim 1966), two vocabulary tests (Mill Hill, parts A and B), four memory tests (cumulative learning, digit span, free recall, picture recognition). Table 1 gives the correlations between individuals' scores on each of these measures and their ages.

Table 1 Correlation coefficients between age and scores on various psychological tests

Task	Correlation with chronological age	Percentage of variance
AH4 (1)	−0.314	9.9
AH4 (2)	−0.39	15.0
Mill Hill (A)	−0.02	—
Mill Hill (B)	−0.06	—
Picture recognition	−0.22	4.8
Free recall	−0.23	5.3
Cumulative learning	−0.28	7.8
Digit span	−0.14	1.9

Correlations between memory tests and age are, as expected, negative and significant but are surprisingly low; age on its own does not account for more than 16 per cent of variance between subjects on any task. It is surprising that correlations between scores on memory tasks, while positive and significant, are, in general, also surprisingly low. The exception is the correlation of $r = 0.61$ between scores on two tasks that make very similar demands: i.e. free recall of a list of 30 words and cumulative learning over four successive trials of a list of 15 words. No other correlations between tasks show more than 17 per cent of common variance. These modest relationships are consistent with the hypothesis that memory skills may be intensely 'domain specific' so that, as they grow old, people may retain circumscribed islands of competence in some areas which contrast with greatly reduced efficiency in others. This may occur either because these different skills depend on independent subsystems of the central nervous system that may age at different rates, or because different skills are independently developed and maintained by practice in the face of different life-demands.

A third, striking point is that test scores on AH 4 IQ (part 1) correlate with chronological age much more strongly than do any other measures of performance. Indeed, when variance associated with IQ test scores is partialled out, the contribution of chronological age to individual variance in scores on all of the memory tests is very greatly reduced, and, in the case of free recall and digit span, drops below acceptable levels for significance with this large sample (the partial correlations with age adjusted for differences in IQ test scores are, for digit span, $r = -0.04$; for cumulative learning, $r = -0.022$; for free recall, $r = -0.17$ and with d' for picture recognition, $r = -0.22$). This does not, of course, mean that memory efficiency does not change with age but rather that nearly all the age changes which these tasks detect are picked up by the AH 4 (part 1) test.

These data raise two interesting possibilities. The first is that the four memory tasks make demands on a single, functional performance characteristic (share a single common factor), which, however, only modestly determines performance on any of them. This single performance characteristic is also well detected by the AH 4 IQ test, so that when individual differences in IQ test scores are partialled out there is little remaining common variance between tasks. A corollary of this hypothesis is that the single performance characteristic common to all four memory tasks is also the one most affected by age.

An alternative possibility is that the meagre intercorrelations between memory-task scores reflect the fact that each involves different, domain-specific, cognitive skills. This is very plausible, because these laboratory memory tasks have been developed to find unique performance indices that can be independently manipulated in laboratory experiments. In contrast, IQ tests have been developed to give the best possible predictions of performance in a wide range of complex everyday tasks and skills. Consequently the problems in any good IQ test will each require a wide range of different cognitive skills. Gross test scores will therefore tend to correlate significantly, albeit modestly, with scores on any of a wide range of laboratory tasks. In short, on this model, IQ tests pick up variance in individual performance in laboratory tasks because they incorporate tests of a very wide range of different cognitive skills rather than because they pick up any single 'master factor' crucial to most, or to all, tasks, which also happens to be the one most critically affected by advancing age.

To choose between these possibilities we would have to collect and compare scores on a much wider range of tasks. This is now in progress. However the present data can be analysed to throw light on a different practical question, central to the assessment of the cognitive status of elderly people: that is, whether we can define a syndrome of 'normal age-associated memory loss' distinct from the patterns of change associated with age-associated pathologies. As we have seen, chronological age did not emerge as a particularly good predictor of memory performance in this sample. A much better predictor was IQ test score. This raises the question as to whether the types of memory function that are picked up by IQ test scores are characteristically different from other patterns of changes in memory efficiency, which are not detected by tests such as the AH 4 but which do occur with increasing age. To examine this possibility memory test scores were transformed to 'z' scores for mutual comparability and summed to give an overall index of competence across all memory tests. Each individual's summed, memory test scores were then plotted against his or her IQ test score and the linear regression of memory test scores on age was computed. The overall correlation obtained between IQ test scores and summed memory test scores was $r = 0.64$. Using this function as a benchmark it was then possible to identify individuals whose memory test scores were 2 SDs greater or less than the empirically derived norms for their ages. These 48 individuals could then be examined in greater detail to discover whether any common factors might explain their atypically poor memories. This was done by Holland (*née* Winthorpe 1988). Her first finding was that the ratio of men to women in this 'memory impaired' sample was greater than in the volunteer population at large. Men were of average age and intelligence for the sample, but women were unusually elderly. Thus for women, but not for men, there is some limited support for the idea that the proportions of individuals who have memory test scores much lower than those normal for their current IQs increases with the age of the population we test—in other words, that for some individuals at least there may be an age-specific decrement in memory efficiency independent of a concomitant decrement in IQ test score.

It must be stressed that detailed examination of individuals with atypically poor memories revealed that they included a high proportion of 'frail elderly' who suffered from a variety of different pathologies (e.g. out of 48 individuals, 35 per cent had histories of unusually poor health: six had suffered from depression; one had been treated for other psychiatric symptoms; five suffered from severe arthritis; two were early cases of senile dementia; two had chronic bronchitis, one had suffered a mild stroke; one had suffered a head injury; and one suffered from cancer.)

The remaining 29 individuals reported no chronic illnesses and had good recent health histories. It is very unfortunate that neurophysiological records such as brain scans, were not available, because this would have allowed a direct check of the hypothesis that relatively faster changes in memory function were related to faster tissue loss in some areas than in others. Thus this analysis provides only weak and tentative support for the hypothesis that, as individuals grow older, some begin to experience changes in memory efficiency that are out of step with concurrent changes in other cognitive skills—such as those picked up by IQ tests. Even within a population of 1559 individuals, the number of these cases of memory impairment out of scale with IQ test attainment was too small (48) to reveal any pattern in the incidence of pathologies or of socioeconomic factors that apparently affected the strengths of associations between memory test scores and IQ test scores.

It is well-known that particular pathologies of the central nervous system (e.g. Korsakov's psychosis, or dementia of the Alzheimer type) are marked by a dissociation between measure of memory efficiency and scores on IQ tests. However, there is no evidence from the bulk of our sample that this dissociation is a *necessary* concomitant of 'normal' ageing. However, Holland's screening of the subset of individuals who had unusually poor memory-test scores for their levels of IQ suggests that it is at least worth while pursuing the possibilities that (a) there may be a small subgroup of individuals who, in the absence of any obvious pathology, show relatively greater change on memory test scores than on IQ test scores, and (b) that, at least among women, the incidence of individuals with this dissociation increases with age.

This returns us to the theoretical question as to precisely why IQ test scores so effectively predict performance on a variety of memory tasks; i.e. whether it is because they pick up individual variation in a single performance characteristic (in psychometric models, a 'single common factor') that is common to all tasks, or because they incorporate tests of a very wide range of different, independent ('modular') performance characteristics, each of which is critical to a different cognitive skill and so also to the laboratory tasks in which this skill is deployed. As appropriate data from large population screens are not yet available, the only alternative guidance comes from recent models of individual differences in IQ and/or changes in age, which have been derived from limited laboratory experiments on small samples of individuals in the tradition of a 'modular' cognitive psychology.

Salthouse (1986) follows the review by Birren *et al.* (1980) in concluding that the most sensitive index of general cognitive change in old age is slowing of information-processing rate. Eysenck (1986), Jensen (1982, 1985, 1986), and Vernon (1983) conclude, from a quite independent body of data gathered on young adults, that modest but ubiquitous negative correlations between individuals' scores on pencil-and-paper IQ tests and their choice reaction times or tachistoscopic recognition thresholds in simple laboratory tasks occur because a single factor, '*g*', which Spearman (1904) suggested was common to most IQ tests, can be reified in terms of individual differences in information-processing rates. Both Eysenck and Jensen further suggest that individual differences in information-processing rates are sensitive indices of individual differences in global neuronal efficiency in the central nervous system and so can provide a plausible neurophysiological, and perhaps ultimately a genetic (Jensen 1985), basis for individual differences in intelligence.

This comfortably agrees with Salthouse's idea that changes in information-processing rate result in correlated changes in all cognitive systems with advancing age and are reflected in the marked age decline in scores on performance IQ tests. There is evidence that this would also explain age changes in memory efficiency, as Waugh and Barr (1980, 1982) have shown that age differences in some memory tasks disappear if older individuals are allowed longer inspection time to compensate for their slower information-processing rates. A synthesis of the views of Eysenck and Jensen on the one hand, and of Salthouse on the other, would suggest that slowing of rate of information processing may be a single, common factor which affects performance on all memory tests with advancing age. Because information-processing rate is also the single factor, Spearman's '*g*', common to all performance IQ tests, these may be especially sensitive indices of age-associated slowing which affects all memory tasks. This would explain why IQ test scores predict memory task performance much better than does chronological age and, also, why peoples' performance on a given memory test may often be less well predicted by their performance on another memory test than by their IQ test scores. In terms of the contemporary literature on cognitive assessment of the aged, this would mean that 'age-specific memory impairment' would be a direct result of slowed information-processing rate, and so would not be distinguishable from differences in memory efficiency associated with differences in IQ test scores. In this case, as chronological age and IQ are, in effect, both indices of individual differences in information-processing rates, we would expect the relationship between memory test scores and IQ scores to remain unaltered by age. This is what the sampling of individuals whose test scores were 2 SD worse than their IQ-predicted norms does, indeed, suggest. The screen of the Newcastle sample could be analysed to provide a further test of this. Table 2 shows correlations between memory test scores and IQ test scores within each successive decade sample between 50 and 80 years. Correlations between AH 4 (part 1) and AH 4 (part 2) scores and memory test scores remain approximately the same across the three successive decade samples. This again suggests that loss in efficiency at the memory tests keeps pace with loss of efficiency on IQ tests between the ages of 50 and 80 years.

Table 2 Correlations between scores on AH 4 (1) and on four memory tasks within each of three successive age decades between 49 and 79 years

Group age	Picture recognition	Free recall	Cumulative learning	Digit span
49–59	0.41	0.45	0.49	0.15
60–69	0.33	0.43	0.44	0.33
70–79	0.40	0.36	0.46	0.45

These findings are consistent with the idea that, at least in healthy populations, age-associated memory losses which are due to global and diffuse changes in the central nervous system may affect all cognitive skills to approximately the same extent. These changes may be associated with a general decline in information-processing rate, and are distinct from the more specific impairments consequent on focal pathologies or lesions that involve particular neuroanatomical or neurochemical subsystems. This leaves us with the problem of how to find further, more sensitive tests of the hypothesis that changes in a single factor, information-processing rate, can indeed account for all differences in memory performance characteristics—whether these occur between individuals of the same age, or between different age groups.

All current models of memory specify functional relationships that would imply that slowing of information rate would lead to poorer subsequent recall. Notably the depth-of-processing theory, first proposed by Craik and Lockhart (1972) and elaborated by Craik and Tulving (1975), would predict that individuals with slower information-processing rates will achieve less adequate depth of processing (e.g. less elaborate encoding in terms of generating associations and mnemonics for the words that they have to remember). Depth-of-processing theory has been formally applied to account for age changes (Craik and Simon 1980, Craik 1982) and to memory impairment after ingestion of alcohol—which is also known to slow the rate of information processing (Craik 1977).

Another way in which slowed information-processing rate might be expected to impair recall is by reducing the rate at which items held in immediate memory can be rehearsed. This prediction is most direct in the 'working memory' model of Baddeley and Hitch (1974), which specifies that individual differences in the numbers of syllables which subjects can report back immediately after they hear them is limited by the maximum rate at which internal representations of these sounds can be circulated around the 'articulatory rehearsal loop' (Baddeley 1986).

Thus current models specify ways in which individual differences in information-processing rate, which may also markedly affect performance on pencil-and-paper IQ tests, can imply individual differences in the efficiency with which information is encoded, rehearsed, and subsequently recalled. What is at issue is whether all age differences in memory efficiency are directly, and solely, consequences of reduced information-processing rate or whether there are other performance parameters, unrelated to information-processing rate, that are unaffected by age and not detected by IQ test scores.

Goward and Rabbitt (described in Goward 1987) carried out a series of experiments directly to examine these points. In a first experiment, 72 individuals, 12 with high and 12 with low AH 4 test scores in each of the sixth, seventh, and eighth decades of life inspected lists of 32 concrete nouns that were presented, one at a time, on a computer-controlled monitor screen. They were required to identify each word and then to generate, and speak aloud, either one, two, or three associates to it. The time taken to generate associates was recorded and the next word was presented. Subjects then inspected a further list in which these 31 words appeared, one at a time, embedded in a list of 32 distractors. For all subjects the probability that a word would be recognized increased with the number of associates that had been produced to it. Older individuals, and those with lower AH 4 test scores took much longer to generate associates but, given parity in the number of associates generated, neither age nor IQ test score affected efficiency of recognition. These data are consistent with Waugh and Barr's (1980; 1982) findings and broadly support the hypothesis that differences in recognition efficiency due to age and IQ relate to individual differences in the speed with which elaborative encoding can be undertaken, and may disappear when subjects with slower information-processing rates are forced to attain the same levels of elaborative encoding as faster performers—and are allowed the extra time they need to do this.

A second experiment manipulated depth of processing by requiring a similarly selected group of 72 subjects to classify each word in lists of concrete nouns, presented to them one at a time, either 'deeply' in terms of the semantic category to which it belonged (e.g. living/non-living) or more 'shallowly' in terms of the presence or absence of a particular letter (e.g. S or E). Classification times and errors were recorded, and found to vary inversely with AH 4 test scores. Age *per se* had no effect. Subjects then inspected a list in which the 32 target words occurred one at a time, in random order, among 32 distractors to identify each word as being, or not being a member of the target list. As in many previous studies (e.g. Craik and Tulving 1975), for all age and IQ groups, recognition accuracy was greater for words that had been semantically classified than for those that had only been scanned to detect target letters. In both conditions classification times, and times taken to recognize items in the inspection list, reduced with IQ and increased with age. In contrast, recognition accuracy increased with group IQ and reduced with group age. A significant interaction between IQ and depth of processing indicated that semantic processing increased the recognition scores of subjects with lower test scores relatively more than those of subjects with higher test scores.

Up to this point these results are consistent with the idea that the well-documented age decline in efficiency of short-term memory may be due to reduced encoding efficiency caused by slowing of information processing rate, and further, that individuals' unadjusted scores on pencil-and-paper IQ tests predict their information-processing rates (i.e. reading tests both for initial classification and for subsequent recognitions of target words). However, more detailed inspection of these results suggests that some qualifications are necessary. In these experiments the effects of age on recognition accuracy were only partly removed by matching groups in terms of unadjusted IQ test scores. In other words it seems that IQ test scores detect some, but not all, age-associated changes that influence recognition accuracy. Further, the design of this experiment made it possible to ask whether IQ test scores predict recognition accuracy only in so far as they correlate with individual differences in information-processing rate, or whether IQ test scores pick up variance in recognition accuracy that are not detected by more direct measures of information-processing speed. This was done by using multiple regression analyses to compare the contributions of direct measures of information-processing speed (classification reading tests and recognition reading tests) and of IQ test scores, both as unique and as joint predictors of recognition accuracy. While both measures, on their own, gave modest significant predictions, direct measures of information-processing rate (i.e. reading tests) did not predict recognition accuracy when variance associated with difference in IQ test scores had been taken into consideration. In contrast, IQ test scores still predicted significant residual variance when that associated with reading tests had been partialled out. Thus we cannot be sure that AH 4 test scores predict recognition accuracy only in so far as they pick up individual differences in information-processing rate. They probably also detect individual differences in other performance parameters that also determine recognition efficiency.

In a third experiment Goward (1987) examined age and IQ test score differences in the context of Baddeley and Hitch's (1974) and Baddeley's (1986) specification for the operating characteristics of the articulatory rehearsal loop within their model of the working memory system. They suggest that the number of syllables that the loop can handle is determined by two performance parameters. The first is the rate at which information can be circulated within the loop; many experiments in different laboratories have shown that this is directly correlated with measures of information-processing speed, such as articulation rates for repetitive strings of syllable or maximum rates of reading aloud (see Baddeley 1986, pp. 75–107). A second performance parameter is the holding time of the loop, i.e. the maximum time for which an item can survive in the loop system without being refreshed by completion of a rehearsal cycle. This tight specification of the model allowed Goward to show that variations in syllable spans with age and IQ test scores are, indeed, mediated by individual differences in reading and articulation rates—and so by individual differences in information-processing speed. However, while indices of loop-holding times also varied markedly between individuals, their durations did not change either with age or with IQ test scores. Thus one performance characteristic that determines syllable span, the information-processing rate, varies with age and is modestly predicted by unadjusted IQ test scores. However, another—the rate at which an item held in the articulatory store decays unless it is refreshed by rehearsal—does not appear to vary with age and is not picked up by scores on the AH 4, pencil-and-paper IQ test.

To test this hint that the rates at which memory traces decay are unaffected by age and by IQ test scores, Rabbitt and Goward (1990) used a continuous recognition task developed by Shepherd and Teghtsoonian (1961). Subjects read a list of words, one at a time. Each, with equal probability, might either be novel or a repeat of a previously presented item. The numbers of presentations of other items intervening between successive repetitions of words varied, with equal probability, from 1 to 39. Individuals aged between 50 and 79 years, who had been scored on the AH 4 test, inspected each word and, as fast as possible, pressed one of two keys to categorize it as 'new' of 'old'. These responses were timed and recorded. As expected, the total numbers of words correctly recognized reduced with group age and increased with group mean IQ. The probability that a repeated word would be recognized as an 'old' item declined with the number of items and responses intervening between its first and subsequent presentations. This provided a measure of the rate at which a memory trace of a word declined with lapse of time and interference from other items. This measure was constant for all age and IQ groups. Once again, although age and IQ test scores partly overlapped in terms of their predictions of variance in overall recognition efficiency, they each independently accounted for a significant proportion of variance. However, when simple and direct measures of information-processing rate (recognition latencies) and IQ test scores were jointly evaluated as predictors of overall (percentage) correct scores, recognition times gave no independent prediction when variance associated with IQ test scores had been accounted for, but IQ test scores predicted significant residual variance in addition to that associated with recognition latencies. It seems that while IQ test scores do not pick up all the changes that reduce recognition-memory performance with increasing age, they do at least as well as direct measures of reaction time at picking up those factors associated with information-processing rate that contribute to recognition accuracy. In addition, they also detect other factors that significantly affect performance, but which are not detected by simple and direct laboratory indices of information-processing rate.

All this evidence comes from tasks in which relatively small amounts of information have to be remembered for relatively brief periods of time. In such tests of immediate memory we might expect individual differences in the speed and efficiency of encoding to outweigh other possible characteristics of the memory system such as the stability, and accessibility, of memory traces over long periods of time. Both performance data and lesion studies confirm that long- and short-term memory systems involve difference anatomical areas (Parkin 1987). Both folk wisdom and some formal laboratory comparisons suggest that while age markedly impairs immediate recall and learning of new information, data stored in long-term memory may remain accessible throughout a long lifetime. A recent illustration of this appears in P. Rabbitt, Abson, and Bent's (unpublished) study of 2100 Newcastle residents aged between 50 and 86 years. Figure 8 contrasts the increasing skew in distributions of AH 4 test scores for successive age-decade samples with the remarkable identity of distributions of Mill Hill vocabulary scores for all age groups.

These data replicate pioneering studies carried out some 40 to 60 years ago (e.g. Jones and Conrad 1933; Raven 1948; Jones 1952), which first drew attention to the relative preservation into old age of cognitive skills that depend on access to bodies of information stored in long-term memory. These bodies of data underlie Horn's (1982) distinction between vulnerability to age changes of dynamic systems that are required rapidly to process and briefly to store new information (whose degree of efficiency constitutes available 'fluid intelligence') and the contrasting robustness of systems involved in the long-term retention of information and its later access (which constitute available 'crystallized intelligence').

An obvious difficulty in comparing the relative extents of age decrements to long- and short-term memory systems is that the performance indices which we can derive for them are incommensurable, e.g. how can we compare percentage reductions in digit span or in free recall with loss of information about complex autobiographical events from long-term memory? How, indeed, can we assess what proportion

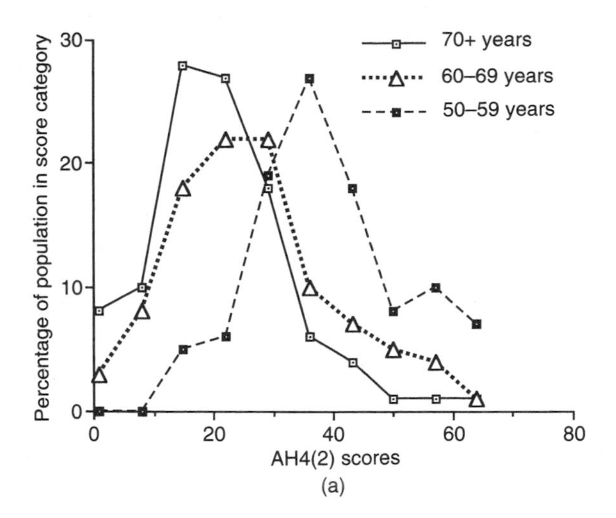

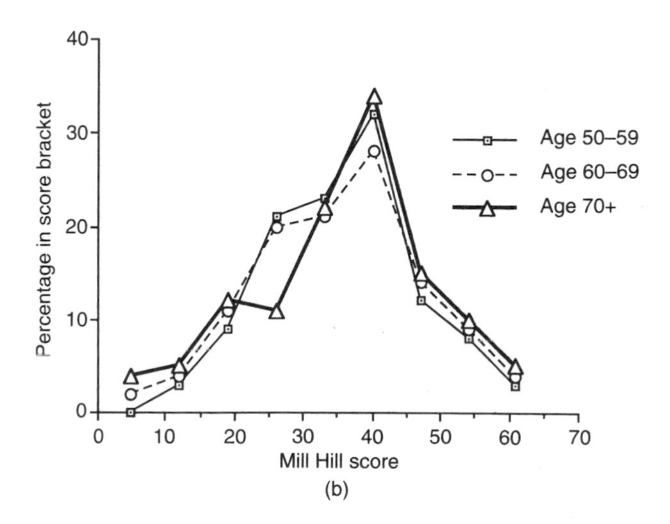

Fig. 8 (a) Percentage distribution of AH 4 (2) (intelligence) scores by age group (934 subjects). (b) Percentage distribution of Mill Hill (vocabulary) scores by age group (934 subjects).

of the total information once stored in long-term memory is no longer available?

One possibility is to compare accuracy and detail of memory for public events and personages. Such studies severely qualify the idea that age has little effect on the long-term memory system. As we have noted, Rabbitt and Winthorpe (1988) found that, like the young, older individuals accessed fewer remote than recent events, and that this difference was at least, partially explicable in terms of the relative frequency with which archival and recent memories are rehearsed, or accessed, in daily life. Given these similar patterns of chronological accessibility, younger people, and those with higher IQ test scores, recalled many more events than the older and less gifted. Thus current IQ test scores may predict ease of retrieval of information from long-term memory as well as rate of learning of new information, and efficiency of short-term memory. IQ effects were also found by Stuart-Hamilton *et al.* (1988), who gave 924 individuals aged from 50 to 86 years Stevens' (1979) 'famous names' test, which required them to recognize, among plausible distractors, names of people briefly but intensely featured in the media between 1920 and 1975. In all age groups, individuals with higher IQ test scores correctly identified more 'famous names'. In a second experiment on 106 volunteers aged from 18 to 36 and 50 to 82 years, Stuart-Hamilton *et al.* computerized the Stevens' test to measure the latency, as well as the accuracy, of name recognition. Because the older subjects had lived through periods when the earlier 'names' were 'famous', they recognized more 'famous names' in total. Recognition latencies, nevertheless, greatly increased with age. IQ test scores also predicted recognition latencies and the total numbers of names recognized.

Two further studies on recognition of photographs of public figures and of theme tunes of television serials, famous at various times over the last 50 years, help us to interpret these results (Maylor, unpublished). Older individuals recognized fewer people and tunes, and could recall less incidental information about those they did correctly identify. Age differences were particularly marked for people and tunes famous in the recent past. Individuals with higher current IQ test scores showed better recognition and recall of both recent and remote material.

Thus the general suggestion is that, for individuals of all ages, details of remote events become increasingly uncertain. Because the time-scales for remoteness are different for the old and for the young, current temporal distances of events are confounded with the ages when they were experienced. An event that occurred 30 years ago was experienced by individuals who are now aged 50 when they were aged 20, but by people now aged 70 when they were 40. Differences in recall between age groups might thus as easily reflect age changes in efficiency with which new information is encoded that have occurred between the ages of 20 and 40 years, as they might an increasing rate of loss of information from long-term memory, or increasing difficulties in accessing stored information in old age. Another factor that appears in the Stuart-Hamilton analysis of age increases in recognition latencies for 'famous names' is that, as people age, they take increasingly longer to access even such information as they can, eventually, reliably produce.

Holland (*née* Winthorpe 1988) used a more natural way of investigating individual differences in autobiographical recall by asking subjects aged 50 to 79 years to describe any events they wished from specified periods of their past lives in as much detail as possible. The 'experiment' took the form of a relaxed chat over coffee, with no time constraints, and subjects were encouraged to give as many, and as circumstantial, accounts of their experiences as they could recall. Nevertheless, older individuals and those with lower AH 4 test scores could only produce extremely impoverished accounts of their experiences in which a few, salient points were sparsely embellished with details. Younger individuals and those with higher IQs gave much more elaborate and detailed accounts. These age and IQ differences did not seem to be due to differences in social styles of communication because, during a subsequent session, Holland energetically and directly probed for missing or additional detail of accounts of experiences that she had previously recorded, but few or none were forthcoming.

Holland included these subjects among others who were also scored on recall of short stories and on measures of working-memory efficiency. All individuals with low IQ scores, especially those who had showed impoverished autobiographical recall, showed good memory for the salient points of the stories but very poor recall of details. In contrast, high scorers' IQ tests showed excellent recall of both salient points and details. An important dissociation appeared in the performance of individuals in their 70s; even those with high IQ test scores, comparable to those obtained by the more able 50- and 60-year-olds, showed impoverished recall of details. Over the whole population, measures of working-memory efficiency declined concomitantly with efficiency both at immediate recall of a story, at remote autobiographical recall, and with increasing age and lower IQ.

It is plausible that the mutual correlations between age, scores for performance IQ tests, working-memory efficiency, and amount of recall of details of stories occur because all these measures reflect individual differences in the rate of information processing. As we have seen from Goward's studies individuals who process information slowly will necessarily have poorer elaborative encoding and reduced working-memory capacity. A slower rate of information processing will mean that fewer propositions in text, whether salient or incidental, can be processed in unit time. Reduced working-memory capacity will compound this problem by reducing the maximum sizes of subunits of text within which information can be integrated. Individuals with a reduced rate of information processing will thus be wise to allocate their processing resources to preserve salient, rather than incidental, propositions: if they do not do this they will both lose the advantage of maintaining contextual links between the propositions they encode and will also only be able to report incoherent and disjunctive material. Thus the age-associated slowing of information-processing rate, which is sensitively reflected by scores in performance IQ tests, might explain qualitative as well as quantitative differences in the recall of complex material. This might also explain why elderly individuals can only give very impoverished accounts of recent, autobiographical events.

However, Goward's experiments do not explain Holland's findings that individuals who had lower IQ test scores, with concomitantly poor recall of text, and reduced working-memory capacity, also showed greatly impoverished recall of details from very remote, as well as from recent, incidents in their own lives. We may presume that events which they experienced as young adults were encoded at a time when their information-processing rates were much higher than they currently are. Consequently these individuals must, when young, have encoded and stored more information about their life-events than they are currently able to retrieve. This opens the possibility that the ease of retrieval of information from long-term storage may also be affected by slowing of the rate of information processing; if this is the case a distinction between 'fluid' and 'crystallized' cognitive abilities becomes much less clear-cut since a decline in the former would seem to entail a corresponding loss of access to the latter.

Bryan (*née* Core 1986) formally investigated this possibility by comparing groups of subjects in their 50s, 60s, and 70s who had identical scores on other vocabulary tests in terms of the times they took to identify pictures in the British Picture Vocabulary Scale (**BPVS**). In this test, pictures are graded from 'easy' (i.e. representations of objects or actions that are often named) to 'difficult' (i.e. those rarely named). As expected, subjects of all ages showed increases in identification latencies with item difficulty. Figure 9 shows naming latencies for groups of high and lower AH 4 test scores. High test scorers, as expected, have higher BPVS scores. However, more interestingly, even across the range of difficulty within which both high and low AH 4 scorers can correctly name all items, increases in latency with item difficulty are much steeper for the low than for the high AH 4 group.

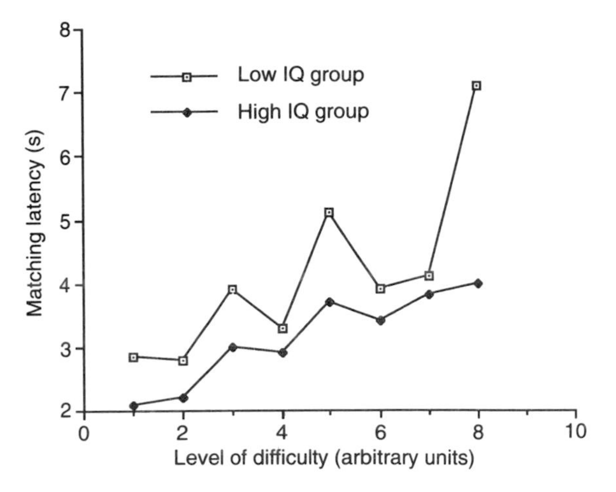

Fig. 9 British Picture Vocabulary Scale matching latencies for high and low intelligence groups (Core 1986).

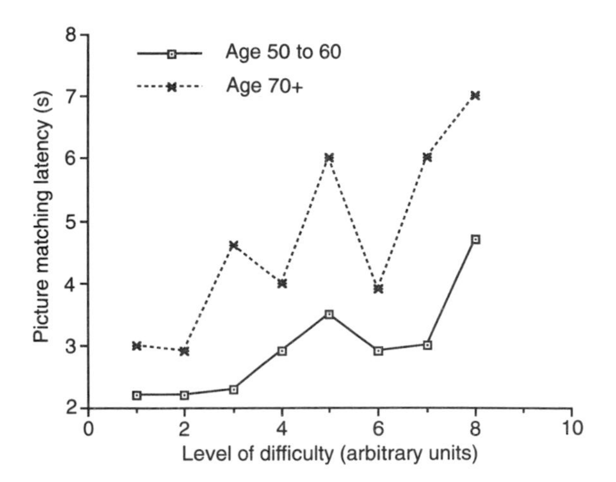

Fig. 10 British Picture Vocabulary Scale matching latencies for two age groups (Core 1986).

The same data are presented in Fig. 10, where name-retrieval latencies for older and younger subjects are compared. Again, older subjects are slower, and show a steeper increase in latency as a function of item difficulty, even over the range in which all subjects were correctly naming all pictures presented to them.

P. Rabbitt and T. J. Perfect (unpublished) have found precisely similar patterns of data in an experiment in which a computerized version of the Mill Hill vocabulary test allowed them to compare identification times for each individual Mill Hill test item and so compute the regressions of latencies across levels of item difficulty for age and IQ test scores. Perfect (1989) reports a very elegant series of experiments in which older and younger subjects with different levels of AH 4 test scores were compared on computerized 'trivial pursuit' tasks, in which they answered questions on some topics in which they were expert and on others in which they were only modestly informed. This allowed him to compare the effects on retrieval times from larger and smaller databases, between groups of individuals of different age and IQ test scores. Perfect found that recall latencies increased systematically with age, and reduced with scores for performance IQ tests and level of expertise (total amount known about a particular topic area). Once again, as in Core's studies, slowing of retrieval times with age and speeding with IQ test score, were much more marked on questions that individuals found difficult than on those they found easy.

When plotted as in Figs. 9 and 10, all these data show that age and IQ test scores slow retrieval times for infrequently recalled information ('rare' words or answers to 'difficult' trivial-pursuit questions) proportionately, as well as absolutely more than retrieval times for frequently recalled information ('common' words or answers to 'easy' trivial-pursuit questions). One possible explanation is that individual differences in age and IQ test scores affect one of two distinct functional systems (e.g. the 'long-term memory retrieval system', in which are held words or information that were last used a long time ago) more than another (e.g. an 'intermediate-term memory system' in which are held words or information last used a brief time ago). An alternative possibility is that interactions between age and IQ and in item difficulty retrieval are consequences of a simple scaling effect on decision times, such that the longer any decision takes, the more it will be delayed by slowing of the information-processing rate.

To test this, data in Fig. 10 were replotted to present response latencies for older individuals at each level of item difficulty as a function of response latencies for young individuals at the equivalent levels of item difficulty (Fig. 11(a)). Data in Fig. 9 were similarly replotted to show response latencies of low IQ groups as a function of response latencies of high IQ groups, again across equivalent levels of item difficulty. (Fig. 11(b)). Similarly data taken from Perfect (1989) are plotted to show percentiles of distributions of latencies for both 'know' and 'don't know' responses in a word-recognition task for young subjects plotted against those of old subjects (Fig. 11(c)).

All plots in Fig. 11 show very good fits to linear relationships. Perfect (1989) also presents evidence that percentile points for distributions of recall latencies for difficult and easy trivial-pursuit questions by older individuals are colinear with those obtained from much younger people. In short, in all Core's and Perfect's tasks the effects on memory-retrieval latencies of individual differences in either age or in IQ test scores can be expressed by a simple, multiplicative constant, irrespective of the difficulty of the retrieval process involved (i.e. irrespective of the relative ease, and so the probable frequency, with which the word, or the item of information, has previously been retrieved).

The next, more general question is whether these scaling effects of age and IQ are specific to the particular functional processes underlying information retrieval from long-term memory, or whether they are consequences of global differences in the rate of information processing that affect to the same degree all cognitive operations, even those in which memory is not at all involved.

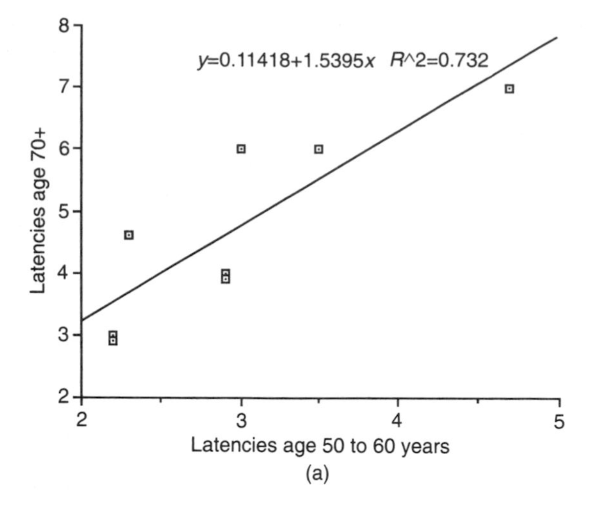

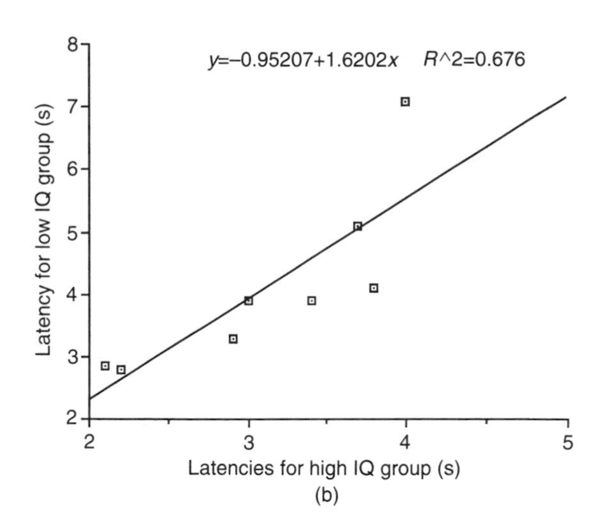

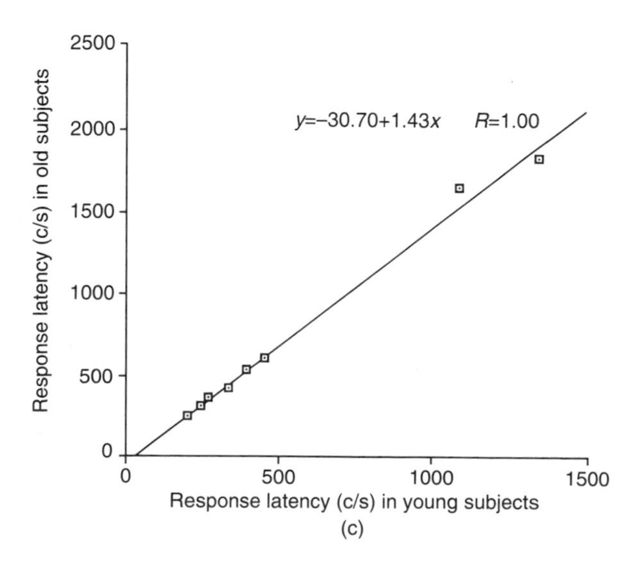

Fig. 11 (a) Association between latencies for British Picture Vocabulary Scale items of varying difficulty for two age groups. (b) Association between latencies for British Picture Vocabulary Scale items of varying difficulty for two intelligence groups. (c) Association between response latencies (percentile analysis) for younger and older subjects in word recognition task (Prentice 1989).

The same graphical analysis was applied to data gathered by me from a group of 168 individuals, 56 aged between 18 and 36 years and 112 aged from 68 to 76 years, all of whom had been screened on a computerized version of the Mill Hill vocabulary test, on three levels of difficulty of a choice–reaction time task in which they responded to the onset of each of either 2, or of 4, or of 8 lights by pressing the appropriate one of four keys, on a visual search task in which they searched for two symbols A and B on sheets of printed random letters and on a computerized version of the AH 4 (1) performance IQ test. The older group's mean response times per signal or item, and scanning times per letter of text were plotted against those of the young group. The resulting function was best described by a straight line (R = 0.93). It follows that, across all these diverse tasks, response latencies of older individuals can be generated by multiplying response latencies of younger individuals by the same multiplicative constant. The important point is that these scaling effects maintain equally well for tasks involving complex problem solving (which make demands on memory among other cognitive functions, e.g. AH 4 part 1), for retrieval of common or rare words from long-term memory (easy and difficult items on the Mill Hill A test), or for choice reaction time and visual search, in which demands on memory are minimal. Thus we can find no evidence that age differences in speed of retrieval of information from long-term memory reflect changes specific to any particular functional system that might be termed 'the long-term memory retrieval system'. They rather seem to be consequences of a diffuse, general slowing of information-processing rate, which affects, to much the same extent, all cognitive systems including those that do not involve either short- or long-term memory.

CONCLUSIONS

The dominant working assumption in cognitive psychology has been that all cognitive capabilities and, in particular, different memory capabilities can best be described as 'modular' in architecture (Fodor 1983; Minsky 1987). The goal of research has been to isolate these modules and to determine and measure their, hypothetically independent, performance characteristics by appropriate, very ingeniously circumscribed, laboratory experiments. Morale among investigators has been sustained by the hope of mapping descriptions of these functional modules on to the cognitive deficits observed in cases of highly localized brain damage. This conceptual framework has guided clever and useful work on the nature of memory changes in old age. The underlying assumption has been that age changes can be described in terms of differential effects on particular modular systems or performance characteristics. There have been insightful and complete reviews both of the conceptual framework underlying this approach (Klatzky 1988) and of the large bodies of data it has generated (Craik 1977; Light and Burke 1988).

The present review has attempted an alternative, and much more general, approach. Without questioning the obvious successes of 'modular' descriptions of the relationships between brain and behaviour, it has considered whether the diffuse changes in the central nervous system that occur even in healthy old people may not produce general, functional changes which simultaneously affect all cognitive systems, including all hypothetical subsystems in human memory. It has also raised the issue that old age is not solely a degenerative condition of the central nervous system. Long life offers more opportunity for the acquisition of useful information and skills. There is accumulating evidence both that prolonged practice can have more radical effects on levels of attainment of particular skills than can gross individual differences in attainment on intelligence tests (Sloboda *et al.* 1985) and that highly practised skills may be intensely domain-specific so that they represent isolated islands of competence which may not be predicted from general levels of performance on other tasks (Shiffrin and Schneider 1977; Schneider and Shiffrin 1977). Memory skills, in particular, seem to have this characteristic (Chase and Ericsson 1981; Ericsson *et al.* 1980; Ericsson and Chase 1982). The relevance of these issues for age comparisons was illustrated by evidence that patterns of relative accessibility of autobiographical memories from different parts of the life-span may be governed as much by social and personal uses to which these memories are put as by age-associated pathological changes in the central nervous system.

This more general framework requires discussion of evidence from batteries of tests given to very large and diverse populations of older individuals rather than of comparisons between small and highly selected groups of older and younger individuals. Data from these batteries suggest that levels of performance on particular laboratory memory tasks may indeed be 'domain-specific', in the sense that performance on one may only weakly predict performance on any other, unless tasks are virtually identical. Age changes in performance on these tests also seem to be modest. More surprisingly, such age changes as are detectable are well detected by a single, very brief, performance IQ test that assesses only a restricted range of activities (e.g. the AH 4 test; Heim 1966). Further studies showed that while mean levels of performance on most memory tasks decline steadily with age, variation in performance between individuals increases with group age. Preliminary data suggest that increasing spread of performance in older age groups may relate to increasing gaps between diminishing numbers of healthy individuals who show relatively little change in ability with age and growing numbers of individuals who suffer any of a variety of illnesses.

A series of laboratory studies suggested that individual differences in performance both on a variety of tests of immediate and short-term memory for simple material, and of memory for propositions in text, may all reflect a general change in information-processing rate with advancing age. Interestingly, individuals who showed unusually poor performance on such tasks also showed both qualitative and quantitative impairments in recall of events from their own lives.

The possible relationships between changes in speed of information processing and efficiency of retrieval of information from long-term memory were therefore examined. Integrity of data on long-term memory, as exemplified by unchanging scores on vocabulary tests, were found to contrast with a steady decline with age in IQ test scores and in the speed with which information reliably available in long-term memory can be accessed. More detailed analyses of such data provide no evidence that changes in efficiency of retrieval from long-term memory are determined by changes in the functional characteristics of any particular cognitive subsystem. These changes rather seem to be associated with general changes in the rate of information processing that affect all cognitive systems, even those not involving memory function. The fact that similar multiplicative constants serve to generate latencies for old from latencies for young subjects across a variety of different tasks, some involving memory and some not, suggests that all are affected by a common, global process. The strengths of correlations between scores on IQ tests and on memory tasks does not reduce with sample age from 50 to 80 years. This and other comparisons provide no evidence that changes in memory efficiency with age relate to changes in specific cognitive systems that 'age' at different rates; they are consistent with the hypothesis that memory changes reflect general changes in efficiency in all cognitive systems that are picked up by scores on a simple pencil-and-paper IQ test.

A caveat to this general picture is that in some tasks, the AH 4 IQ test picks up most but not all of the age-associated variance in performance between individuals. Further, while IQ tests predict all the individual variance in recall and recognition associated with simple and direct measures of information-processing rate, such as classification or recognition latencies, they also predict additional variance which these, very direct, measures of information-processing rate do not detect. To this extent we must regard as tentative the hypotheses that the AH 4 test picks up only a single and unique vector of change in cognitive efficiency that accompanies old age, and that the variance which it detects relates solely to individual difference in the rate of information processing.

These data do not, of course, question that useful descriptions of cognitive processes can be framed in terms of functional 'modules', nor that particular cognitive functions can be localized anatomically, nor the validity of a general approach that seeks to reify descriptions of hypothetical functional 'modules' by mapping them on to specific neuroanatomical, neurophysiological, or neurochemical subsystems within the central nervous system. However, these data do introduce the idea that conditions such as 'normal old age', and some of the many diverse pathologies associated with growing old, may produce global, 'non-specific' changes in the efficiency of the central nervous system that do not have distinct effects on specific systems or anatomical structures. So far, the best candidate description for these diffuse changes is a slowing in information-processing rate. This, as far as our data show, affects all cognitive systems. As far as we can tell the effects of this change in rate of information processing are not merely general, but quantitatively equal across all systems. Its effects on any particular task or hypothetical cognitive system may be described by a very similar and as simple, multiplicative constant as its effects on any other.

References

Baddeley, A.D. (1986). *Working memory*. Clarendon Press, Oxford.

Baddeley, A.D. and Hitch, G.J. (1974). Working memory. In *Recent advances in learning and motivation* (ed. G. Bower). Academic Press, New York.

Birren, J.E., Woods, A.M., and Williams, M.V. (1980). Behavioural slowing with age: causes, organisation and consequences. In *Aging in the 1980s: psychological issues* (ed.) L.W. Poon American Psychological Association, Washington DC.

Cermak, L.S. and Craik, F.I.M. (1979). *Levels of processing in human memory*. Erlbaum, Hillsdale NJ.

Chase, W.G. and Ericsson, K. A. (1981). Skilled memory. In *Cognitive skills and their acquisition* (ed.) J.R. Anderson. Erlbaum, Hillsdale, N.J.

Core, J. (1986). Comprehension and memory for everyday events by the elderly. Unpublished D.Phil thesis. University of Durham, England.

Craik, F.I.M. (1977). Age differences in human memory. In *Handbook of the psychology of aging* (ed. J.E. Birren and K.W. Schaie). Van Nostrand Reinhold, New York.

Craik, F.I.M. (1982). Selective changes in encoding as a function of reduced processing capacity. In *Coding and knowledge representation: process and structure in human memory* (ed. F. Klix, J. Hoffman and E. van der Meer). Elsevier, Amsterdam.

Craik, F.I.M. and Lockhart, R.S. (1972). Levels of processing: a framework for memory research. *Journal of Verbal Learning and Verbal Behaviour*, **11**, 671–84.

Craik, F.I.M. and Simon, E. (1980). Age differences in memory: the roles of attention and depth of processing. In *New directions in memory and aging* (eds. L.W. Poon *et al.*) Erlbaum, Hillsdale, N.J.

Craik, F.I.M. and Tulving, E. (1975). Depth of processing and the retention of words in episodic memory. *Journal of Experimental Psychology (General)*, **104**, 268–94.

Crook, T. and Larrabee, G.J. (1988). Age-associated memory impairment: Diagnostic criteria and treatment strategies. *Psychopharmacology Bulletin*, **24**, 509–14.

Crook, T., Bartus, R.T., Ferris, S.H., Whitehouse, P. Cohen, G.D., and Gershon, S. (1986). Age-associated memory impairment: proposed diagnostic criteria and measures of clinical change. Report of a National Institute of Mental Health Work Group. *Developmental Neuropsychology*, **2**, 261–76.

Crovitz, H.F. and Quinta-Holland, K. (1976). Proportion of episodic memories from early childhood by years of age. *Bulletin of the Psychonomic Society*, **7**, 61–2.

Crovitz, H.F. and Schiffman, H. (1974). Frequency of episodic memories as a function of their age. *Bulletin of the Psychonomic Society*, **4**, 517–18.

Crowder, R.G. (1976). *Principles of learning and memory*. Erlbaum, Hillsdale. N.J.

Ericsson K. A. and Chase, W.G. (1982). Exceptional memory. *American Scientist*, **70**, 607–15.

Ericsson, W.A., Chase, W.G., and Faloon, S. (1980). Acquisition of a memory skill. *Science*, **208**, 1181–2.

Eysenck, H.J. (1986). The theory of intelligence and the psychophysiology of cognitive. In *Advances in the psychology of human intelligence*, Vol. 3. (ed. R.J. Sternberg). Erlbaum, Hillsdale, N.J.

Fodor, J. (1983). *The modularity of mind: an essay on faculty psychology*. MIT Press, Cambridge, MA.

Goward, L.M. (1987). An investigation of factors contributing to scores on intelligence tests. Unpublished Ph.D. thesis. University of Manchester, England.

Heim, A.W. (1966). *The AH 4 test*. NFER-Nelson, Windsor.

Holland, C.A. and Rabbitt, P.M.A. (1991). The course and causes of cognitive change with advancing age. *Reviews in Clinical Gerontology*, **1**, 81–96.

Horn, J. (1982). The theory of fluid and crystalised intelligence in relation to concepts of cognitive psychology and aging in adulthood. In *Aging and cognitive processes* (ed. F.I.M. Craik and S. Trehub). Plenum Press, New York.

Jarvik, L.F. (1983). Age is in—is wit out? In *Aging of the brain* (ed. D. Samuel, S. Algeri, S. Gershon, V.E. Grimm, and G. Toffano). Raven Press, New York.

Jensen, A.R. (1982). Reaction times and psychometric 'g'. In *A model for intelligence* (ed. H.J. Eysenck). Springer-Verlag, Heidelberg.

Jensen, A.R. (1985). The study of the black-white difference on various psychometric tests: Spearman's hypothesis. *Behavioural and Brain Sciences*, **8**, 193–219.

Jensen, A.R. (1986) Intelligence: 'definition', measurement and future research. In *What is intelligence*? (ed. R.J. Sternberg and D.K. Detterman). Ablex, Norwood, N.J.

Jones, H.E. and Conrad, H.S. (1933). The growth and decline of intelligence: a study of a homogeneous group between the ages of ten and sixty. *Genetic Psychology Monographs*, **13**, 223–98.

Jones, H.E. (1955). Problems of aging in perceptual and intellective functions. In *Psychological aspects of aging*, pp. 135–39. American Psychological Association, Washington DC.

Klatzky, R.L. (1988). Theories of information processing and theories of aging. In *Language, Memory and Aging* (ed. L.L. Light and D.M. Burke). Cambridge University Press.

Light, L.L. and Burke, D.M. (ed.) (1988). *Language, memory and aging*. Cambridge University Press.

McCormack, P.D. (1979). Autobiographical memory in the aged. *Canadian Journal of Psychology*, **33**, 118–24.

Milner, B. (1970). Memory and the medial temporal regions of the brain. In *Biology of memory* (ed. K.H. Pribram and D.E. Broadbent). Academic Press, New York.

Minsky, M. (1987). *The society of mind*. Picador, London.

Parkin, A. (1987). *Memory and amnesia*. Blackwell, Oxford.

Perfect, T.J. (1989). Age, expertise and long term memory retrieval. Unpublished Ph.D. thesis. University of Manchester, England.

Petit, T.L. (1982). Neuroanatomical and clinical neuropsychological changes in aging and senile dementia. In *Aging and cognitive processes* (ed. F.I.M. Craik and S. Trehub). Plenum Press, New York.

Rabbitt, P.M.A. (1986). Memory Impairment in the elderly. In *Psychiatric disorders in the elderly* (ed. P.E. Bebbington and R. Jacoby). Mental Health Foundation, London.

Rabbitt, P.M.A. and Winthorpe, C. (1988). What do old people remember? The Galton paradigm reconsidered. In *Practical aspects of memory: current research and issues*, Vol. 1, *Memory in everyday life* (ed. M.M. Gruneberg, P.E. Morris, and R.N. Sykes). Wiley, Chichester.

Raven, J.C. (1948). The comparative assessment of adult intellectual ability. *British Journal of Psychology*, **39**, 12–9.

Ribot, T. (1882). *Diseases of memory*. Appleton, New York.

Rubin, D.C., Wetzler, S.E., and Nebes, R.D. (1986). Autobiographical memory across the lifespan. In *Autobiographical memory* (ed. D.C. Rubin). Cambridge University Press.

Schneider, W. and Shiffrin, R.M. (1987). Controlled and automatic human information processing. I. Detection, search and attention. *Psychological Review*, **84**, 1–66.

Shepherd, R.N. and Teghtsoonian, M. (1961). Retention of information under conditions approaching a steady state. *Journal of Experimental Psychology*, **62**, 302–9.

Shiffrin R.M. and Schneider, W. (1987). Controlled and automatic human information processing. II. Perceptual learning, automatic attending and a general theory. *Psychological Review*, **84**, 127–90.

Sloboda, J., Hermelin, B. and O'Connor, N. (1985). Exceptional musical memory. *Music Perception*, **3**, 155–70.

Spearman, C. (1904). General intelligence, objectively determined and measured. *American Journal of Psychology*, **15**, 201–93.

Stuart-Hamilton, I., Perfect, T., and Rabbitt, P.M.A. (1988). Remembering who was who. In *Practical aspects of memory: current research and issues*, Vol. 2, *Clinical and educational implications* (ed. M.M. Gruneberg, P.E. Morris, and R.N. Sykes). Wiley, Chichester.

Squire, L.R. and Butters, N. (1984). *Neuropsychology of memory*. Guildford Press, New York.

Tulving, E. (1983) *Elements of episodic memory*. Oxford University Press.

Vernon, P.A. (1983). Speed of information processing and general intelligence. *Intelligence*, **7**, 53–70.

Warrington, E.K. (1982). The double dissociation of short-and long-

term memory deficits. In *Human memory and Amnesia* (ed. L.S. Cermak). Erlbaum, Hillsdale NJ.
Waugh, N.C. and Barr, R.A. (1980). Memory and mental tempo. In *New directions in memory and aging* (ed. L.W. Poon, J. Fozard, L. Cermak, D. Arenberg, and L. Thompson). Erlbaum, Hillsdale, NJ.
Waugh, N.C. and Barr, R.A. (1982). Encoding deficits in aging. In *Aging and cognitive processes* (ed. F.I.M. Craik and S. Trehub). Plenum Press, New York.
Wicklegren, W.A. (1977). *Learning and memory*. Prentice Hall, Englewood Cliffs, NJ.
Winthorpe, C.A. (*née* Holland) (1988). Patterns of change in ageing memory. Unpublished Ph.D. thesis. University of Manchester, England.

18.3 Cognition

PETER V. RABINS

Cognition can be defined as the various thinking processes through which knowledge is gained, stored, manipulated, and expressed. Cognition can be broken down into a variety of functions including memory, language, praxis, visuospatial and perceptual function, and conceptualization/flexibility. This chapter will review the changes in these realms that may occur with healthy ageing (except for memory, which is discussed in Chapter 18.2), discuss how these changes can be differentiated from disease, and review how disorders in these realms can be elicited by the clinician.

DOES COGNITION CHANGE WITH HEALTHY AGEING?

Determining whether cognitive abilities change through the life-span has proved to be a complex task. Three complicating issues make this an especially difficult subject and effective techniques for overcoming the obstacles are still being developed.

One complication is the cohort effect, which ascribes differences in performance between groups to their living in different time periods and thus having been exposed to different environmental influences. The work of Schaie *et al.* (1973) demonstrates how this effect manifests itself in tests of fluid (newly learned) and crystallized (previously learned) intelligence. A crosssectional study carried out in 1958 found that fluid intelligence begins to decline at the age of 40 years while crystallized intelligence begins to decline at 60. However, when the individuals who were 40 years old in 1958 were retested 7 and 14 years later (i.e. at 47 and 54 years of age), their performance remained stable. This implies that the decline in fluid intelligence seen cross-sectionally was due to a time-period effect. That is, persons 40 years old in 1958 had had different experiences than those 54 years old in 1958. These different experiences rather than the ageing process account for the different findings of the cross-sectional and longitudinal studies. The cohort effect for intelligence testing is thought to result from older individuals having had less previous exposure to test taking, from the tests being less relevant to their lives, thereby eliciting less motivation to work hard at the task, and from the differential treatment by the examiners of subjects across the age span (for example, being less encouraging to elderly subjects during the first test sequence out of respect, deference, or lower expectation).

Secondly, studies of 'healthy ageing' can be confounded by the inclusion of subjects with early manifestations of illness. In the study of changes in intellect, the inclusion of subjects with early dementia is difficult to avoid because its earliest signs and symptoms can often be identified only in retrospect. To rule out this interference, investigators would have to retest subjects several years later; only those deemed 'healthy' at follow-up would be included in the group considered 'healthy' when first tested. Such a procedure would be confounded by any decline occurring in healthy individuals over several years; to include only those who do not decline would bias the study towards persons with exceptional, not usual, cognition.

A third issue confounding studies of the cognitive changes that occur with healthy ageing is the increasing prevalence of physical illnesses that affect the organs through which cognition is expressed. Arthritis, hearing deficits, and visual changes are three common conditions that interfere with the motor and perceptual abilities upon which testing depends; while many studies of the cognition of healthy elderly people have excluded individuals with deficits, not all studies report an effort to do so and some use screening procedures that would miss less overt deficits which could none the less subtly interfere with performance.

Difficulties can arise even when these interferences are taken into account. If a study of healthy ageing includes only 'super-healthy' individuals, it might underestimate the changes that commonly occur in a true random sample of ageing people. This type-2 error would be an important interference to guard against as it could lead to inappropriate diagnosis of, or investigation for, disease.

Such complexities have made it difficult to ascertain whether changes in cognitive function are to be expected in healthy elderly people, and if so, to determine their extent. Although one expert has concluded that 'the pattern of cognitive change in later life and its relation to brain functions has not been resolved' (Zarit *et al.* 1985), some generalizations can be made.

Speed of performance

Speed of performance is usually not regarded as a cognitive function on its own, yet it is operative in many aspects of cognition. Declines in speed occur in both motor performance and non-motor cognition through the age span beginning in the 40s and are the most clearly demonstrated age-associated changes. These declines in speed confound testing of specific cognitive functions because many cognitive tests are timed. Therefore, many studies of ageing effects of cognition use untimed versions of tests, even if the instrument was designed as a timed test.

Attention

Attention is the ability to concentrate on a task. Sustained attention, the ability to focus on a single task (e.g. digit span forward and backwards), and selective attention, the ability to ignore distracting or irrelevant information (e.g. crossing out only the letter 'f' in a long list of random letters) do not decline before the age of 80 years. On the other hand, divided attention, the ability to choose between two competing stimuli (for example, between different auditory stimuli being presented through ear phones to each ear) begins to decline in the fourth decade (Albert and Moss 1987).

Language

Vocabulary is among the cognitive functions least affected in ageing. Vocabulary remains stable through adult life into the 70s and is often used to determine premorbid intelligence because of its stability and because correlations between vocabulary and overall intelligence are high. However, the ability to name objects when shown either a picture or an object (confrontation naming) does begin to decline in the mid-70s. Characteristically, the older individual has difficulty in producing the actual name of the object; healthy individuals can usually correctly name the object when given a 'phonemic cue' (the first sound of the word). In healthy elderly people, paraphasic errors (see below) are uncommon and their presence suggests a disease impairing language. Naming deficits become prominent early in Alzheimer's disease.

Visuospatial ability

The ability to perceive objects and their relationship to one another depends on intact function of sensory organs. Declines in measures of the central processing of perception begin to occur in the seventh decade but some tests do not show decline until the eighth decade. Praxis, the ability to carry out complex physical activity (e.g. combing one's hair or imitating how one would comb hair) does not decline before the ninth decade.

Mental flexibility

The ability to change set, i.e. the ease with which an individual adapts to changing rules, declines in the eighth decade.

Abstraction ability

The ability to draw general rules from a series of individual pieces of data begins to decline in the seventh decade and difficulties as a group are clearly demonstrable in the eighth decade.

'Intelligence'

Intelligence tests were devised originally to predict school performance and their relevance to elderly people can be questioned. Intelligence refers to a general factor that reflects overall performances on measures of the abilities described above. The concept has been validated by finding high statistical correlations among performances on tests of different cognitive functions and by the stability of general measures through adulthood.

However, as might be suspected from the above discussion, some aspects of intelligence appear to be more susceptible to ageing effects than others. On the widely used Weschler Adult Intelligence Scale, tests that measure performance on tasks not previously learned, for example, arranging pictures in the proper sequence or duplicating a design with blocks, show clear declines by the early seventh decade. Performance on tests grouped together as verbal measures, for example, vocabulary and previously learned information, remain much more stable and do not decline until the eighth decade. A similar dissociation was noted by Cattell (1963), who found that changes in fluid intelligence begin to occur at the age of 60 years; changes in crystallized intelligence do not begin until 10 or 15 years later.

DISTINGUISHING HEALTHY AGE-ASSOCIATED CHANGES FROM DISEASE

For the clinician, the importance of delineating the expected, age-associated changes in cognition from those of disease lies in the early recognition and treatment of disease and the ability to reassure those who complain of changes unrelated to disease that they are not, or are unlikely to be, suffering from actual disease. Unfortunately, there is little information available that can guide the clinician in making the distinction in the earliest stages of dementia. Benign senescent forgetfulness and age-associated memory disorder (Crook *et al.* 1986) are two suggested terms that refer to non-pathological impairments, primarily to memory impairment. As memory impairment is often the earliest indicator of dementia, it can also be the harbinger of a progressive, global, intellectual decline.

Clinical experience suggests there are certain characteristics of memory difficulty that identify pathological memory impairment but there are few confirmatory data. Individuals with non-progressive, age-associated memory changes complain of memory loss during every-day activities such as producing the names of people or thinking of specific words. They report misplacing things. However, neither the patient nor other informants reports that important major events have been forgotten or that there are dramatic changes in function such as becoming lost while driving a familiar route or being unable to set a table or write a cheque.

Patients with early dementia often do not acknowledge impaired memory or other cognitive dysfunction. However, the presence of memory complaints cannot be used to exclude a dementing illness. Decline in non-memory functions such as language, praxis, and visuospatial perception suggests a disorder of cognition, as does dysfunction in every-day activities or a clear decline from previous level of function. Neuropsychological testing is helpful in unclear cases.

Complaints of memory loss and cognitive dysfunction can be symptoms of major or unipolar depression. It is common for depressives with secondary cognitive disorder to complain of memory loss and to appear, to the examiner, to be exaggerating their disability. When asked on direct questioning they will often acknowledge some depressive symptoms (Rabins *et al.* 1984), although they may relate their impairments to a 'physical' disorder. Patients with reversible cognitive disorder due to depression often express symptoms consonant with depression, such as self-blame, hopelessness, or hypochondriasis. They may have a history of depressive episodes earlier in life. This cognitive disorder often responds to direct treatment of the depression but up to half of individuals go on to manifest a progressive dementing illness.

POSSIBLE EXPLANATIONS FOR AGE-ASSOCIATED COGNITIVE CHANGES

As many myths exist about the ageing brain as about any other aspect of the human condition. The idea that millions of brain cells are being lost throughout the brain every day is widely held, although contradicted by recent evidence. Terry *et al.* (1987) have shown that large pyramidal cells (>90 μm in diameter) decline in number with age but smaller neurones decline minimally. These changes occur primarily in the frontal and temporal lobes. The arborization of brain dendrites declines with ageing, particularly in basilar, large pyramidal cells; conceivably this might lead to less interneuronal communication and thus explain declining speed of performance in cognitive function.

Brain volume declines with age and enlargement of ventricles and increased prominence of the sulci are seen as group phenomena among healthy elderly people; similar changes occur in patients with dementia. Although demented individuals as a group have more atrophy than age-matched controls, the overlap between normal and abnormal is so great that no measure of atrophy can reliably distinguish between a dementia and normal ageing in an individual. EEG frequency also declines in late life but remains above 8.5 Hz in healthy people. (See also Chapter 18.4.)

Neurochemical changes occur in the brain with age. Acetylcholine declines in the cortex and markers of dopamine, serotonin, and noradrenaline decline in pathways dependent on these neurotransmitters. The enzyme monoamine oxidase B, on the other hand, increases with age. As yet, there is no evidence linking these brain changes to declining intellectual function in healthy elderly people but the association is certainly plausible.

Vascular changes are universal in the brain of Western man and most individuals over the age of 75 have small numbers of neuritic plaques and/or neurofibrillary tangles, the hallmarks of Alzheimer's disease. The small number of these abnormal structures has suggested to many researchers that there is a qualitative as well as quantitative difference between healthy ageing, and dementia, but a minority of scientists believe that these structural and neurochemical changes (which could be aetiologically linked or have different causes) merely reflect a gradation from minor degrees of change in health to severe degrees of change in the diseased brain. Further research is necessary to clarify this issue.

The above paragraphs describe group changes. A significant minority of individuals show no cognitive change over time. In one study, one-third of 80 year olds performed in the same range as younger individuals (Benton *et al.* 1981). Other individuals are found to have widespread neuropathological abnormality at autopsy but to have expressed no evidence of dysfunction during life (Katzman *et al.* 1988).

ASSESSING COGNITION AT THE BEDSIDE AND IN THE CLINIC

An examination of mental status should be considered as part of the medical assessment of all older individuals. As with all aspects of clinical medicine, testing healthy individuals is necessary to understand and become skilled at the testing of the individual with disease.

History taking and behavioural observation

The clinician should suspect impairment in cognition if a patient repeats him- or herself frequently during the examination. Repetition is particularly of concern when it deals with issues on which a great deal of time has been spent. While patients vary in their ability to give a coherent sequence to their medical complaints, they should give a consistent history and, with the help of the examiner, be able to put events in the proper sequence. This ability is not affected by healthy ageing. Normal elderly individuals are fully alert. Patients who appear drowsy in the interview, who need to be reminded of a question several times (it is important to be sure that hearing and vision are intact), or who become easily distracted when speaking on a topic should be suspected of suffering from an impairment in level of consciousness (see Chapter 18.5).

Memory testing

Because memory is most sensitive to cognitive decline, deficits in this area can be picked up during the general interview as well as on specific testing. Cognitively normal individuals should know their birth date and the year in which important events such as marriage, birth of children, and retirement occurred. They should be able to give approximate dates of important events that have occurred during their life such as the Second World War. Orientation is a form of memory. Specific questions should be introduced with the statement that this is a routine part of the assessment. Patients rarely refuse to answer questions when asked in a supportive, direct manner. The patient should be asked date, day, month, season, and year. Missing the day of the week or exact date can be normal but missing the date by several weeks, not knowing the month, season or year is rare in healthy elderly people. {Five-minute recall of a name and address or of a small series of objects shown in pictures is also a useful clinical routine.}

Language assessment

The clinician should listen for spontaneous word-finding difficulty as well as for paraphasic errors. The latter are usually indicative of pathology in the language areas of the brain. Paraphasic errors take several forms and can consist of substituting one letter for another (for example, saying 'pet' instead of 'pen'), substituting one word for another (saying 'pen' instead of 'spoon') or mixing several words or letters together (saying 'porker' instead of 'quarter'). The ability to repeat is a sensitive test of language function provided that hearing is intact. Repetition of a phrase requires comprehension of the instruction and thus screens for both comprehension and expression abnormality, but it can be normal if there is a transcortical aphasic disorder. The ability to name objects (e.g. pen, watch, button) is tested by pointing to the object and asking the patient to name it. Loss of ability to name is not localizing but does indicate pathology; further testing should be done to determine if dysnomia reflects a specific language disorder or a generalized loss of vocabulary.

Serial seven subtractions

Individuals with a 7th-grade education or better should be able to count backwards from 100 by sevens. A person who has a 7th-grade education and is unable to do so may suffer impairment in attention, concentration, and/or mathematical calculation ability. {In addition to the effects of education, some population studies in the United Kingdom have shown that old men perform significantly better than old women in this particular test.}

Writing

For a person with intact motor strength, the ability to write depends on intact language and praxis skills. Normal individuals can spontaneously write a sentence when asked to do so. They should be encouraged to generate a sentence of their own. Figure 1 illustrates the progressive decline in language, including a paraphasic error ('miand' for 'mind'), of a patient with autopsy-confirmed Alzheimer's disease.

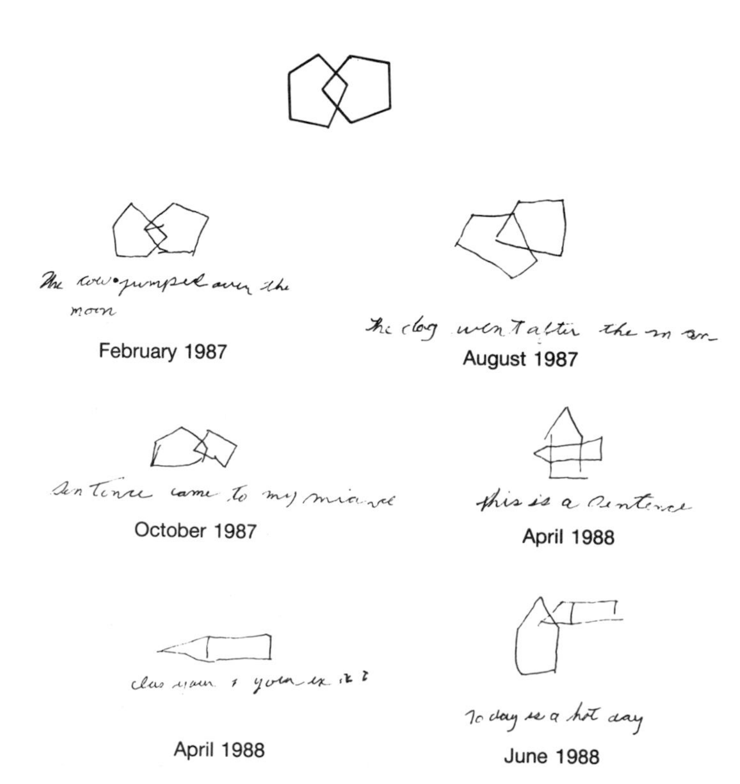

Fig. 1 Diagram copying and sentence writing in a 68-year-old man with probable Alzheimer's disease. The diagram at the top was copied by the patient.

Tests of construction and perception

The ability to copy a complex design such as interlocking pentagons (Fig. 1) remains intact in 80 per cent of individuals with adequate eyesight over the age of 80 (Farmer *et al.* 1987). Figure 1 illustrates the breakdown of this ability over time; the initial drawing demonstrates early but subtle dysfunction.

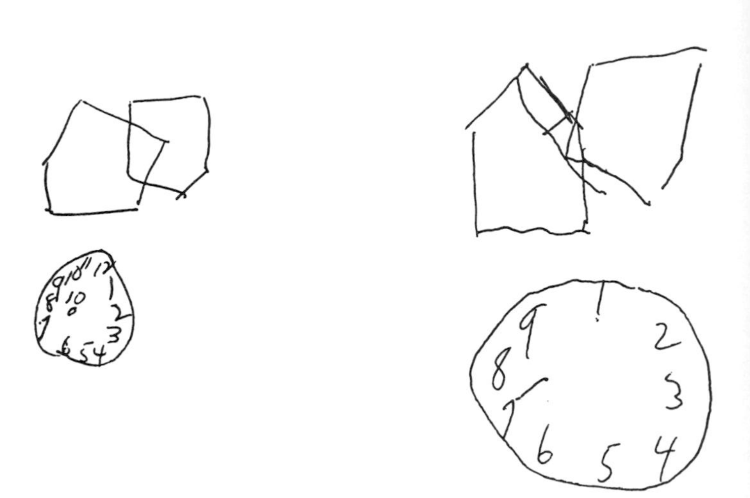

Fig. 2 Diagram copying and spontaneous clock drawings done 1 year apart by a patient with probable Alzheimer's disease. Interlocking pentagons were copied from the top drawing in Fig. 1. The left-hand drawing shows minimal abnormality in the diagram copying with clear abnormality in clock production. The right-hand drawing demonstrates deterioration in diagram copying while the clock drawing shows further deterioration in planning and visuospatial function.

Drawing a clock requires planning, visuospatial function, and praxis. The patient is first instructed to draw a clock with numbers. If this is done adequately he or she is then instructed to put the hands in for '10 minutes after 11'. Figure 2 demonstrates the testing of a mildly demented individual who was initially able to copy interlocking pentagons correctly but was unable to draw a clock correctly.

Mental flexibility and abstraction

Proverb interpretation is dependent upon cultural and educational experience. The examiner should first ask if the patient has heard the proverb before and consider it a test only when it has been known to the patient. The ability to identify an abstract category (for example, identifying an apple and orange as both being fruit; a table and chair as both being furniture) and to identify differences (for example, between a river and canal) is also culturally and educationally dependent and an inability to do so must be interpreted cautiously.

References

Albert, M.S. and Moss, M. (1987). *Geriatric Neuropsychology*. Guildford Press, New York.

Benton, A.L., Eslinger, B.J., and DaMasio, R. (1981). Normative observations on neuropsychological test performances in old age. *Journal of Clinical Neuropsychology*, **3**, 33–42.

Cattell, R.B. (1963) Theory of fluid and crystallized intelligence: a critical experiment. *Journal of Educational Psychology*, **45**, 1–22.

Crook, T. *et al.* (1986). Age-associated memory impairment: proposed diagnostic criteria and measures of clinical change—report of a National Institute of Mental Health work group. *Developmental Neuropsychology*, **2**, 261–76.

Farmer, M.E. *et al.* (1987). Neuropsychological test performance in Framingham: A descriptive study. *Psychological Reports*, **60**, 1023–40.

Katzman, R. *et al.* (1988). Clinical, pathological, and neurochemical changes in dementia: a subgroup with preserved mental status and numerous neocrotical plaques. *Annals of Neurology*, **23**, 138–44.

Rabins, P.V., Merchant, A., and Nestadt, G. (1984). Criteria for diagnosing reversible dementia caused by depression: validation by 2 year followup. *British Journal of Psychiatry*, **144**, 488–92.

Schaie, K.W., Labouvie, G.V., and Buech, B.U. (1973). Generational and cohort-specific differences in adult cognitive functioning: a fourteen-year study of indendent samples. *Developmental Psychology*, **9**, 151–66.

Terry, R.D., DeTeresa, R., and Hansen, L.A. (1987). Neocortical cell counts in normal human adult aging. *Annals of Neurology*, **21**, 530–9.

Zarit, S.H., Eiler, J., and Hassinger, M. (1985). Clinical assessment. In *Handbook of the Psychology of Aging* (ed. J.E. Birren and K.W. Schaie). Van Nostrand Reinhold, New York.

18.4 Dementia

ROBERT P. FRIEDLAND

INTRODUCTION

Dementia is an acquired impairment of intellectual and memory functioning caused by organic disease of the brain, which is not associated with disturbances in the level of consciousness, and which is of such severity that it interferes with social or occupational functioning. The term is not used in reference to individuals with mental retardation who have not acquired an adult level of intellectual development. Dementia refers to a clinical syndrome that has many causes. As a matter of definition, patients with dementia must have memory disturbances as well as defects in other mental abilities such as abstract thinking, judgement, personality, language, praxis, and visuospatial skills. The deficits must be of sufficient magnitude to interfere significantly with work or social activities (American Psychiatric Association 1980).

In the past, definitions of dementia have often included the requirement of a progressive and irreversible impairment. While most cases of dementia do show progressive decline and are not reversible, these characteristics are not relevant to the definition: dementia may be of sudden onset (e.g. after strokes or head injury), and many causes of dementia are completely reversible (e.g. subdural haematoma, drug toxicity, depression). Dementia may have onset before the age of 65 years (sometimes unfortunately referred to as 'presenile dementia'), or after the age of 65 years ('senile dementia'). It is inappropriate to use the terms 'dementia', 'senile dementia', or 'presenile dementia' as a final diagnosis in the individual patient, as they are merely symptomatic classifications, similar to the terms headache or seizure disorder. In the past, individuals with Alzheimer's disease were often referred to as having 'senility' or as being 'senile'. These two terms are now obsolete.

Furthermore, although there are on average some differences in the clinical course of early-onset versus later-onset dementia (for example, more rapid progression in early-onset cases), there is much overlap. Alzheimer's disease was originally used as a diagnosis only for cases under the age of 65 years. It is now recognized that the disease can affect people of any adult age, and is actually more common in the older age groups.

It is important to distinguish dementia from delirium. Delirium is a confusional state, usually of acute onset, characterized by disturbances of memory and orientation (often with confabulations), and usually accompanied by abnormal movements, hallucinations, illusions, and change in affect. In distinction from dementia, there is a reduced level of consciousness ('clouding') in delirium. Delirium usually fluctuates in intensity and is replaced by dementia if the underlying disorder is not resolved. Common causes of delirium include the encephalopathies caused by infectious diseases, toxic or nutritional factors, or systemic illness (see Chapter 18.5)

{Dementia may present acutely. It is not unusual for dementing people to restrict their daily activities to a regular and repetitive routine in a progressively circumscribed environment in order to reduce the anxiety produced by failing memory and reasoning powers in more demanding surroundings. The constant environment acts in effect as a prosthesis for failing memory so that such a person can exist in a state of 'compensated dementia'. This process of compensation may be reinforced, often unwittingly, by a spouse or other family members who may not recognize the severity of the subject's disability. If the environment of such a person is suddenly changed, for example by loss of a spouse, referral to hospital, or even by a well-intentioned but ill-advised holiday away from home with relatives or friends, the dementia may become acutely decompensated. The resulting sudden disorientation, anxiety, and disordered behaviour may be mistaken for a delirium. This condition is not uncommon in the casualty departments of acute hospitals and if not recognized and the patient returned rapidly to his or her familiar environment, the decompensation may prove irreversible or a superadded delirium (from drugs or dehydration) add to the difficulties of the situation and the sufferings of the patient (Evans 1982).}

The incidence of dementing illnesses increases markedly with advancing age. The most common cause of dementia in adults in the United States and Europe is Alzheimer's disease, which has an annual incidence of approximately 0.1 per cent for people between the ages of 60 and 65 years. This annual incidence increases to 1 per cent for people over the age of 65 years, and to over 2 per cent for people over the age of 80 years. In the United States, for healthy people between the ages of 75 and 85, new cases of Alzheimer's disease are as common as myocardial infarction and more common than stroke (Katzman *et al.* 1989). There is evidence to suggest that stroke is more common than Alzheimer's disease as a cause of dementia in Japan; the latest (unpublished) data actually indicate about the same prevalence of multi-infarct dementia in the United States and Japan, but considerably more Alzheimer's disease in the United States. The study by Evans *et al.* (1989) of the older population of East Boston, MA, concluded that 45 per cent of those aged 85 years or more suffered from Alzheimer's disease. Because of the rapid ageing of the world's population, and the declining incidence of heart disease and stroke, it is clear that dementing illnesses will become a more devastating problem in the future.

HEALTHY AGEING

Ageing in many healthy individuals is characterized by a modest slowing of motor, perceptual, and intellectual performance, as well as deterioration in memory functions. It is important to keep these changes in mind in the evaluation of older patients suspected of having dementia. However, not all cognitive functions decline at the same rate, and in many older patients, intellectual abilities, as revealed by tests of verbal performance, may be maintained until very late life. Memory may also be impaired with advancing age, particularly after the age of 70 years. Long-term or remote memory is less age dependent than short-term or recent memory. However, learning capacity itself is not impaired when tested in an untimed fashion. Changes in memory with ageing have been termed benign senescent forgetfulness, or age-associated memory impairment (**AAMI**). AAMI is most often of relatively mild severity, and does not significantly interfere with activities of daily living. In older subjects, life-events frequently cause depression, which may cause an increased concern on the part of the patients for deficits associated with AAMI. The items forgotten in AAMI are usually relatively unimportant (e.g. names of people recently met; location of the car in a car-park), while the items forgotten in dementia include those of greater significance (e.g. names of family members, presence of the car in a car-park). It may be difficult to distinguish AAMI from dementia, particularly in early disease, and in some cases the early stages of Alzheimer's disease are indistinguishable from AAMI. In order to conclude that the patient has dementia, definition requires that there be interference in social or occupational functioning. Dementia is never a 'normal' occurrence, no matter what age the patient has achieved.

In general, older people have a reduced capacity to deal with physiological and psychological stress. The behavioural changes accompanying healthy ageing are the result of a loss of neurones and neural connections. Of course, involutional changes are also often occurring, in varying degrees, in other organ systems during ageing in healthy people. As a result, older people are more sensitive than younger people to the deleterious effects of systemic illnesses and medications. It is also important to realize that the observed behaviour is produced by the interactions of the aged brain and the non-neural organs with the patient's environment. An older patient may be functioning well at home in familiar surroundings, but may have difficulty adapting to a new setting occasioned by life events (e.g. hip fracture, death of a spouse). It is important to consider these factors when evaluating the development of changes in mental functioning in older people that develop at the same time as changes in the patient's environment or general physical health.

DEMENTING ILLNESSES

Alzheimer's disease

Alzheimer's disease is responsible for 50 to 60 per cent of cases of dementia in adults. Memory loss is the most pronounced behavioural abnormality, and is usually the first symptom. Memory is impaired for recent events, with relative preservation of remote memory. In the early stages of the disease, memory impairment may be an isolated dysfunction, followed in time by the development of impairments of attention, language function (defective word finding with otherwise fluent speech), visuospatial abilities (drawing, route finding), praxis (purposeful movements), calculations, visual, auditory and olfactory perception, problem-solving ability, and judgement. Patients with Alzheimer's disease have difficulty shifting their mental set from one task to another. Depression, personality changes, apathy, and irritability are also common features of the disease. Language abilities and social skills my be remarkably preserved, even in the later stages, and patients with well established dementia may be able to maintain polite conversations with remarkable skill, and thus appear to be intact to the casual observer. Paranoid delusions, illusions, and hallucinations are seen in a minority of patients, usually in the later stages. Up to half of patients with Alzheimer's disease have limited awareness of their behavioural deficits. It should be emphasized that the behavioural features of this disease are highly variable from patient to patient: some patients may have preserved language function with impaired visuospatial abilities, while other patients at a similar stage in the overall disease process may show the reverse pattern of deficits. Motor function and urinary continence are usually not affected until late. This variability, or heterogeneity, in the behavioural manifestations of the disease is due to variations in the distribution of disease severity in brain regions (e.g. right-handed patients with severe involvement of the left temporal and parietal cortex will have relatively more marked language dysfunction). The disease is progressive, with survival, after onset, of 5 to 12 years duration.

In most cases of Alzheimer's disease the condition is of sporadic origin. However, in 10 to 15 per cent of cases there is familial inheritance, usually in an autosomal dominant pattern with complete penetrance. In some early onset familial cases a linkage to a marker on the long arm of chromosome 21 has been reported; there is also evidence for linkage to chromosome 19 in some families. The risk of developing Alzheimer's disease is approximately four times greater in the first-degree relatives of cases than in controls without a history of the disorder in first-degree relatives.

Alzheimer's disease is defined by its pathological hallmarks: neurofibrillary tangles (composed of paired, helical filaments), neuritic plaques, amyloid infiltration of vessel walls (amyloid or congophilic angiopathy), granulovascuolar degeneration, and Hirano bodies. There is also a loss of neurones, especially large pyramidal cells, and a marked loss of synaptic arborization, with gliosis. These abnormalities are most severe in the medial-basal temporal cortex (hippocampus and amygdala), basal forebrain (medial septal nucleus, nucleus of the diagonal band of Broca, and the nucleus basalis of Meynert) and in the posterior-lateral parietal and temporal cortices. The primary sensorimotor cortices are relatively spared in the disease, while the secondary association cortices and limbic systems are heavily involved.

The amyloid protein present in the cores of neuritic plaques and in the walls of cerebral and meningeal vessels (called A4 or beta protein) is composed of a 39–42 amino-acid chain whose sequence has been identified and found to be identical at the two sites. It is derived by proteolytic cleavage from a larger amyloid precursor protein (APP), which is coded for on the long arm of chromosome 21 at a site different from the site linked in some cases to familial Alzheimer's disease (Rumble *et al*. 1989). The precursor protein is found in many tissues of healthy animals and man, and appears to be an important cell-surface glycoprotein. Current work is focusing on the mechanism by which the normally present precursor protein is degraded to produce the amyloid protein that accumulates in the disease. Recently developed markers allow for the immunohistochemical identification of these proteins in brain, blood, and cerebrospinal fluid (Fig. 1) (Majocha *et al*. 1988). Neurofibrillary tangles have been reported to contain the microtubule-associated protein *tau*, as well as the amyloid A4 protein. Granulovacuolar degeneration contains accumulations of tubulin, and the protein actin is found in Hirano bodies. Qualitatively similar changes are seen, but to much less extent and with a more restricted distribution, in the brains of healthy older persons.

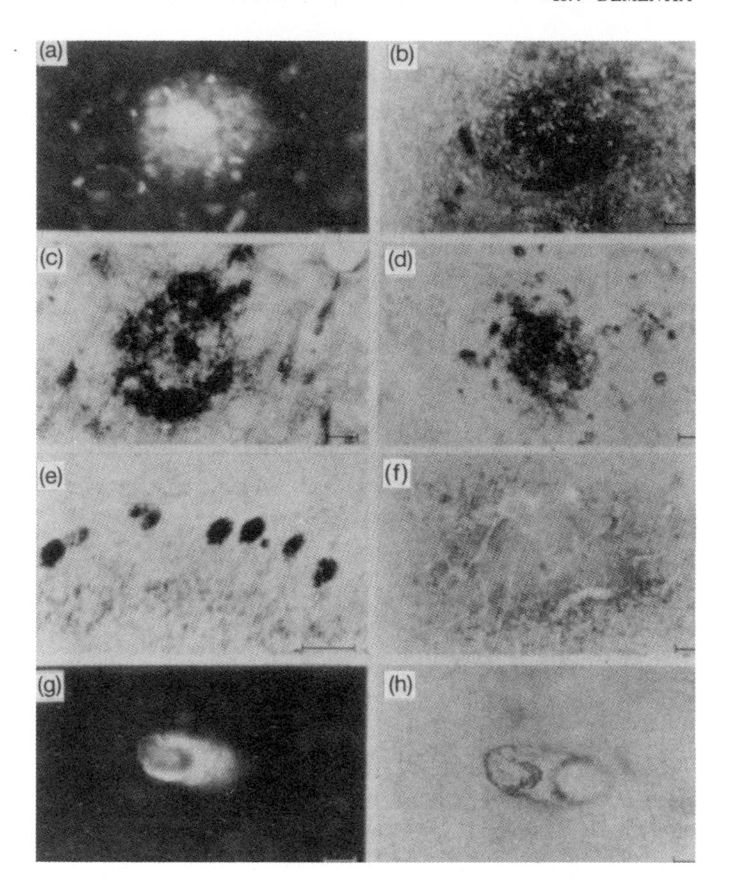

Fig. 1 Histochemical and immunological staining of amyloid deposits. (a) Prefrontal cortex from patient with Alzheimer's disease with thioflavin S and (b) with monoclonal antibody targeted against the Alzheimer amyloid (A4) protein; the same neuritic plaque is shown in both images. (Bar = 20 μm.) (c) Neuritic plaques from Alzheimer's disease in hippocampus and (d) prefronal cortex stained with anti-A4 antibody. (Bar = 20 μm.) (e) Neuritic plaques from Alzheimer's disease hippocampus stained with anti-A4 antibodies (Bar = 100 μm.) (f) Control hippocampus stained with anti-A4 antibody; no staining is seen. (Bar = 20 μm.) (g) Prefrontal cortex, Alzheimer's disease: blood vessel stained with thioflavin S and (h) with anti-A4 antibody; the same blood vessel is seen in both. (Bar = 20 μm.) (Reproduced with permission from Majocha *et al.* (1988).)

The most severe neurochemical deficiency in Alzheimer's disease is in the neurotransmitter acetylcholine. This deficiency is caused by the loss of basal forebrain neurones that send cholinergic projections throughout the cortex. Other neurotransmitter systems (and the localization of their cell bodies) that are also deficient in the disease include somatostatin (cortical interneurones), serotonin (dorsal raphe), noradrenaline (locus coeruleus), endogenous opioids (hypothalamus and brain-stem), corticotrophin-releasing factor (hypothalamus), and dopamine (substantia nigra pars compacta). It is believed that the severe memory disturbances found in Alzheimer's are an effect of the loss of cholinergic neurotransmission in the hippocampus and related limbic structures in the temporal lobe. The behavioural effects of the other neurotransmitter deficiencies are not well defined.

All individuals with Down's syndrome develop pathological evidence of Alzheimer's disease if they live past 35 years of age. Clinical evidence of dementia is evident in perhaps half of Down's subjects at this age. The neuropathological and neurochemical features of Alzheimer's are identical in persons with or without Down's syndrome. It is believed that triplication of the pre-A4 gene on chromosome 21 in Down's syndrome leads, by an as yet unknown mechanism, to overexpression of the gene and increased levels of pre-A4 and amyloid deposition (Rumble *et al* 1989).

There are a number of changes in organ systems other than the brain in Alzheimer's disease, including amyloid deposits in cultured fibroblasts from the skin of affected (but not from normal) persons, abnormal concentrations of several enzymes involved in glucose metabolism in white blood cells, and an almost universal weight loss that is apparently not accounted for by decreased intake or increased physical activity. These clues to the possibility of a more general disease process need further attention.

Multi-infarct dementia

The second most common cause of dementia in Western countries is multi-infarct dementia, which is responsible for up to 20 per cent of cases. In another 15 to 20 per cent of cases, multi-infarct dementia is found in coexistence with Alzheimer's disease. The infarctions that characterize the former may be superficial or deep, small or large, and many involve large, intermediate, or small vessels. Also, mixed cases of multi-infarct dementia are common. Strokes that are associated with this state are generally bilateral. Small infarctions, called lacunae (2–15 mm in diameter) may be distributed throughout the basal ganglia, internal capsule, and corona radiata and are associated with vascular lipohyalinosis, hypertension, and dementia (lacunar state). Clinical features associated with multi-infarct dementia include abrupt onset, history of strokes, focal symptoms and signs, stepwise deterioration, fluctuating courses, and history of hypertension. These symptoms and signs form the Hachinski ischaemia scale (Hachinski *et al.* 1975). Patients with Alzheimer's disease, without evidence of cerebral infarct, generally do not show these symptoms. One form of multi-infarct dementia with severe incomplete infarction of deep white matter is Binswanger's disease (subcortical arteriosclerotic encephalopathy). In Binswanger's disease, dementia is often accompanied by gait disturbance and pseudobulbar palsy (emotional incontinence, dysarthria, facial weakness).

Parkinson's disease

Parkinson's and Alzheimer's diseases have overlapping features. Dementia is found in 25 to 40 per cent of patients with Parkinson's disease. Other, less severe cognitive disturbances are found more commonly in the disease. Neuronal loss is seen in the dopaminergic substantia nigra pars compacta and ventral mesencephalic tegmentum, the cholinergic basal forebrain, the noradrenergic locus coeruleus, and the serotonergic dorsal raphe nucleus. Dementia in Parkinson's disease is associated with cell loss in these important projection systems, without intrinsic pathology in the cortex. In 15 to 30 per cent of cases of Parkinson's disease there may be the concurrent presence of Alzheimer's disease. Conversely, features of parkinsonism (bradykinesia, rigidity, masked facies) may also be found in 15 to 30 per cent of patients with Alzheimer's disease, usually in the later stages.

Creutzfeldt–Jakob disease

Creutzfeldt–Jakob disease is a rare form of dementing illness found throughout the world. It occurs with an incidence of approximately one in 1 million per year. It is characterized by a very rapid course, usually with death within 1 year of onset. Most patients have associated neurological deficits in addition to marked dementia. These additional features include weakness, spasticity, abnormal movements, myoclonus, seizures, and occasionally, fasciculations. Myoclonus is found in more than 80 per cent of patients and is often associated with characteristic abnormalities on the electroencephalographic tracing. Electroencephalography reveals periodic, sharp wave complexes appearing synchronously at regular intervals, often with burst suppression.

Creutzfeldt–Jakob disease is known to be caused by an unusual transmissible agent that is similar to the agent of a disease found in New Guinea called *kuru*, and also to scrapie, a disease of sheep and goats. These conditions may be transmitted only by direct contact with brain or other infected tissue. Five to 10 per cent of all cases of Creutzfeldt–Jakob disease are familial. The diagnosis of the disease is usually apparent because of the rapid course and abnormal EEG. Findings on computed tomography or magnetic resonance imaging are not particularly different from the abnormalities seen in Alzheimer's disease or healthy ageing. Several instances of iatrogenic transmission of the disease have been reported. These have occurred through dural grafts, corneal transplants, contaminated EEG depth electrodes, and therapeutic use of human growth hormone derived from human pituitary glands *post mortem*. There is no effective treatment for the condition, although the associated seizure disorder and myoclonic abnormalities may be amenable to symptomatic medication.

The treatable dementias

The treatable dementias (referred to often as 'chronic deliria' in the United Kingdom) are a large and multifaceted group of illnesses. Many of the causes of treatable dementia are clinically identical to Alzheimer's disease. It is estimated that 10 to 20 per cent of all cases of dementia are due to potentially treatable illnesses. Clarfeld (1988) has concluded from a review (chiefly of studies done in referral centres) that 11 per cent of patients with dementia had illnesses that partially or fully reversed with treatment. The true incidence of treatable dementia in the community may actually be higher. Prompt recognition and appropriate treatment for these conditions is of great importance, because delays in treatment can produce death or permanent disability. Conditions causing treatable dementia are listed in Table 1.

Cardiovascular disorders may cause dementia by causing chronic global impairment of cerebral blood flow (e.g. congestive heart failure, cardiomyopathy) or multiple episodes of acute impairment of the cerebral circulation (e.g. endocarditis, multi-infarct dementia, cardiogenic embolization). Disorders of any organ system can cause dementia by interfering in the metabolic homeostasis of the brain. Particularly important systemic disorders that may cause dementia include hypothyroidism, Cushing's disease, liver failure, uraemia, pulmonary insufficiency, anaemia, pellagra, hypoglycaemia, disorders of calcium metabolism (in particular, hyperparathyroidism), and vitamin B_{12} deficiency. Toxic states are the most common cause of treatable dementia, most often due to drug effects. The overall incidence of adverse drug reaction in patients in the older age groups is two to three times that found in young adults (Montamat *et al.* 1989). Older patients have numerous features that make them more sensitive than younger persons to the toxic effects of drugs on the central nervous system: polypharmacy, poor compliance, greater severity of disease, reduced blood volume, reduced lean body mass, diminished renal clearance, decreased cardiac output, lowered hepatic metabolism, poor nutritional status, and neuronal loss. For example, diazepam and nitrazepam have prolonged action in older patients because of an increased volume of distribution of these lipid-soluble agents {and possibly because of increased penetration into or diminished clearance from the brain}. Furthermore, older patients are sensitive to cognitive impairment induced by psychoactive and other drugs with anticholinergic properties.

Depression is a common illness among the older age groups. It is commonly associated with cognitive impairment in many cases, and mental impairment may be the initial complaint. Depression may be masked, without subjective recognition of depressed affect. Vegetative signs of depression (e.g. anorexia, weight loss, sleep disturbances), anhedonia, and past history of depressive illness may assist in recognition of the diagnosis. Depressed patients with cognitive impairment are usually quite vocal in their complaints concerning their deficits. Hypomanic patients may also per-

Table 1 The treatable* dementias

Cardiovascular disorders
Cardiac disorders
Multi-infarct dementia
Vasculitis
Subarachnoid haemorrhage (late effects)
Delayed effects of radiation
Vascular malformation
Systemic and metabolic disorders
Thyroid, parathyroid, pituitary, adrenal, liver, pulmonary, kidney, or blood disorders
Sarcoidosis
Porphyria
Systemic lupus erythematosus
Fluid and electrolyte abnormalities
Hypoglycaemia
Nutritional deficiency
Diabetes mellitus
Hyperlipidaemia
Toxic disorders
Drugs
Alcohol
Industrial agents
Pollutants
Heavy metals
Carbon monoxide
Affective disorders
Depression
Trauma
Seizure disorders
Infectious disorders
Meningitis
Encephalitis (including human immuno-deficiency virus infection and Lyme borreliosis)
Brain abscess
Syphilis
Whipple's disease
Neoplastic disorders
Direct and indirect effects of primary and metastatic tumors
Hydrocephalus
'Normal' pressure hydrocephalus
Obstructive hydrocephalus (acqueductal stenosis)
Genetic disease
Wilson's disease
Cerebrotendinous xanthomatosis
Hysteria
Sensory deprivation

*All of these disorders are considered to be treatable because there is the potential for slowing the rate of disease progression or reversing the behavioural deficit. For example, in multi-infarct dementia, control of the risk factors for stroke may slow the progress of the disease. While many of these disorders produce dementia in only a minority of patients, they are all capable of producing the dementia syndrome.

form poorly on psychometric tests because of distractability and restlessness. The dementia of depression has been referred to as 'pseudodementia' but this term is not appropriate, as the cognitive insufficiency of these patients is real, and not an artefact of their affective state. Depression with dementia is a treatable condition, with good response to antidepressant medication in many cases. However, Alzheimer's disease may also present with depression, and cases of depression with cognitive impairment at times go on to develop Alzheimer's disease. (For more discussion of depression, see Chapter 10.2.)

Continuing discussion of the table, alcohol can produce dementia by many mechanisms: head trauma, anoxia, ischaemic infarction, hepatic encephalopathy, malnutrition, and direct toxic effects. Neuronal loss in the Wernicke–Korsakoff syndrome may include the basal forebrain and locus coeruleus. Wernicke's encephalopathy is an important and often unrecognized cause of cognitive impairment. In older patients, chronic subdural haematoma may develop without a clear history of trauma. Dementia may be present in patients with subdural haematoma without marked disorders of consciousness or signs of increased intracranial pressure. Recurrent complex partial seizures can mimic dementia, and are suggested by a history of olfactory hallucinations or other episodic disturbances. Meningitis caused by fungi or the tubercle bacillus can also cause treatable dementia.

In the United States, human immunodeficiency virus infection (acquired immune-deficiency syndrome (AIDS)–dementia complex) is the most common cause of dementia among young adults in some urban areas but can also occur in older persons. AIDS–dementia complex may be clinically similar to Alzheimer's disease, and is suggested by a history of blood transfusion, drug addition, or sexual activity with partners infected with the virus. Chronic meningoencephalitis caused by infection with the tick-borne organism *Borrelia burgdorferi* (Lyme borreliosis) may also cause dementia, at times with pleocytosis of the cerebrospinal fluid and lesions of the deep cerebral white matter on magnetic resonance imaging. Dementia is also caused by the direct and indirect effects of neoplasms. Primary brain tumours (e.g. meningioma of the olfactory groove, colloid cyst of the third ventricle) can present with dementia. Dementia also occurs as a remote effect of systemic malignancy (e.g. hypercalcaemia, 'limbic encephalitis').

'Normal'-pressure hydrocephalus is another, uncommon cause of treatable dementia (Petereson *et al.* 1985). There is a disturbance of circulation of the cerebrospinal fluid with the hydrocephalus and normal fluid pressure on lumbar puncture but an episodically elevated pressure is found on monitoring of intracranial pressure. The disorder is suggested by the clinical triad of dementia, gait disturbance, and urinary incontinence. It may develop as a late complication of intracranial bleeding from trauma or rupture of an aneurysm, or as a complication of intracranial infection. It may also develop idiopathically. In normal-pressure hydrocephalus, placement of a shunt diverting the flow of cerebrospinal fluid from normal channels may cause resolution of the dementia.

CASE EVALUATION AND DIFFERENTIAL DIAGNOSIS

Patients with dementia require a comprehensive evaluation of the social, psychological, medical, psychiatric, and neurological factors that may be contributing to the observed decline. It is not possible to evaluate a case of dementia properly without viewing the patient in the context of his or her lifestyle and family environment. It is crucial to include the family in interviews with patients suspected of having dementia, as the loss of insight that is frequently seen in dementia may obscure important aspects of the clinical picture. The history should provide a view of the course, nature, and accompaniments of the cognitive deterioration. Of particular importance is information concerning hobbies and occupation, because of the opportunity to learn about the effects of the disease on the patient's performance and to uncover possible toxic exposures. The drug history is also crucial, as older patients are often taking many medications, sometimes with incorrect dosing. General physical examination is important for systemic illnesses that may affect cognitive function. The examination is designed to uncover signs and symptoms of systemic illness, and toxic or metabolic states. In particular, evidence of the adverse effects of drugs should be ascertained, including urinary retention, syncope, falls, postural hypotension, dyskinesias, extrapyramidal dysfunction, and sedation.

The neurological and psychiatric examinations focus on the patient's mental status and evidence of disturbance in awareness, orientation, insight, general behaviour, general information, memory, language function (spontaneous speech, repetition, comprehension, naming of object parts), praxis, visuospatial function, topographical orientation, problem-solving ability, judgement, calculations, and affect (including vegetative signs of depression). The presence of hallucinations, illusions, and delusional beliefs should be investigated. It is helpful to retain copies of the patient's actual performance (e.g. clock drawing) in order to facilitate comparisons at a later date. In this regard it is also crucial to do standardized tests for the quantitation of the cognitive deficit, such as the Mini-Mental State Examination (Folstein *et al.* 1975). Formal neuropsychological testing is often helpful in the evaluation of performance abilities. However, neuropsychological testing does not replace the mental status examination, which should be made by the physician.

The reminder of the neurological examination evaluates other possible evidence of involvement of the central or peripheral nervous system that may be of diagnostic importance (e.g. fundoscopy, ocular motility, other motor abilities, distal superficial and deep sensibility, Babinski signs, gait, carotid bruits). Laboratory tests needed in the evaluation of demented patients include complete blood count, erythrocyte sedimentation rate, serum electrolytes, vitamin B_{12} and folate, enzymes, syphilis serology, thyroid function tests, coagulation tests, urinalysis, chest radiograph, ECG, and computed tomography (without contrast) or magnetic resonance imaging of the brain.

Brain imaging is a crucial step in evaluating the presence of brain tumours or other space-occupying lesions. It has recently been found that in Alzheimer's disease, in contrast to findings in age-matched, normal subjects, repeat computed tomographic scans in 6 to 24 months show unequivocal evidence of increases in ventricular volume (i.e. further loss of brain tissue). Such re-examination may occasionally be useful in confirming a tentative diagnosis (Luxemberg *et al.* 1987). On magnetic resonance imaging, findings of severe, diffuse, deep lesions of white matter, with or without other evidence of cerebral infarction, are helpful in suggesting the diagnosis of multi-infarct dementia. Electroencephalography is also helpful in grading the severity of the encephalopathy. The tracing is often normal or only mildly impaired in Alzheimer's disease, while it is usually markedly disturbed in toxic or metabolic disorders, and usually has characteristic changes in Creutzfeldt–Jakob disease.

Tests on cerebrospinal fluid are indicated in the evaluation of patients who have atypical illness or are suspected of having infectious conditions. Lumbar puncture is not necessary in all cases.

The diagnostic process in dementia should focus on uncovering potentially treatable causes for the patient's illness. As Alzheimer's disease is currently a diagnosis of exclusion, it is not possible to rely on any specific abnormality to secure the diagnosis: it can only be concluded that a patient has the disease after the tests listed above have failed to demonstrate the presence of other illnesses that could be responsible for the dementing process. Cases with unusual features, such as incontinence in early disease, should raise suspicions that alternative disease processes are present. The term 'definite Alzheimer's disease' is reserved for those cases with histological evidence of the diagnosis provided by biopsy or autopsy (McKhann *et al.* 1984). 'Probable Alzheimer's disease' is used in those cases where there is no evidence of other illness that could account for the impairment, and in the presence of dementia with a typically progressive course. 'Possible Alzheimer's disease' is used when there are variations in the course, onset, or clinical features of the disease, or when a second systemic or brain disorder is present which could be responsible for the illness but is not considered to be the cause. The diagnostic accuracy in cases of probable Alzheimer's disease is in the region of 90 per cent. The diagnosis of multi-infarct dementia is suggested by a score on the Hachinski scale of 7 or more (Hachinski *et al.* 1975) and by the presence of severe lesions in white matter or multiple infarctions on computed tomography or magnetic resonance imaging. While the Hachinski ischaemia scale is helpful in patient evaluation, it should not be relied upon as an isolated measure to make the diagnosis of multi-infarct dementia.

Follow-up is often helpful in detecting the presence of dementing illness other than Alzheimer's disease. It should be understood that it is crucial to recognize the treatable dementing illnesses, even though the likelihood of their presence in an individual case of dementia is small. It is a grave error to miss the diagnosis of a potentially treatable process.

TREATMENT AND MANAGEMENT OF DEMENTIA

There is currently no treatment that can reverse the effects of Alzheimer's disease or multi-infarct dementia on memory or cognition. Work on treatment with choline, lecithin, physostigmine, and related compounds has failed to provide evidence for sustained significant therapeutic effects in patients with Alzheimer's disease. So-called nootropic agents with effects on cerebral blood flow and metabolism have also been found to be of no significant benefit to patients with Alzheimer's disease, multi-infarct dementia or related illnesses.

Other agents are being tested for possible therapeutic value, including drugs like tetrahydroaminoacridine (THA), which may increase effective concentrations of acetylcholine through blocking acetylcholinesterase, and acetylcarnitine, which may have multiple effects. Among theoretically promising developments is that of the potential use of nerve growth factor to stimulate compensation by remaining neurones for those damaged or lost in Alzheimer's disease or other degenerative conditions such as Parkinson's disease.

Despite the absence of curative drugs at present, there are many features of patients with dementing illnesses that can benefit from appropriate non-pharmacological as well as pharmacological interventions. The important aspects of care for these people, which family members as well as the professional staff of institutions must learn, are first to adapt the approach to the previous, established preferences and routines of the patient. A regular, familiar, daily routine is most important, together with learning what responses may ease the individual's tensions. Often snacks, or a stroll, or other diversions, can be quite effective.

At times the cautious use of drugs may be indicated to help treat otherwise uncontrollable agitation and hostility, depression, and sleep disturbances. In demented patients it is important to use psychoactive agents in small dosages, as these patients are relatively sensitive to the negative central nervous system effects of the drugs. In particular, drugs with anticholinergic activities should be used cautiously, in order to avoid negative effects upon primary memory and cognitive symptoms. Patients with Alzheimer's disease with depression may show a good response to treatment with antidepressants, with improved affect and general behaviour. Patients with multi-infarct dementia may be helped by the treatment of the risk factors for stroke, including cessation of smoking, and treatment of hypertension and hyperlipidaemias. Low-dose aspirin may also be of benefit in multi-infarct dementia and anticoagulants should be considered for patients in atrial fibrillation (see Chapter 11.6). In general, medical management should pay attention to the comorbidity (or excess morbidity) associated with dementia, as proper treatment of affective disturbances or systemic illness improve behaviour.

Management of dementia requires attention to both the patient and the family. Care-givers are subjected to severe stress and emotional hardships in dealing with the progressive loss of mental capacity in the loved one. Respite care can provide them with time to recover from the burdens of attending to the patient. Counselling for both patients and family can be helpful in increasing adaptation to illness. Attention must be paid to avoiding hazardous behaviours, planning for long-term care, maintaining daily activities, planning for legal matters (estate planning, securing durable power of attorney), and genetic issues. Support groups managed by local dementia associations may be very helpful to both patient and family.

Hazardous activities that may present a risk to the patient include car driving, walking out alone, cooking causing burns and fires, smoking, falls, occupational errors, incorrect compliance with pharmacotherapy and financial mismanagement. Taking some risks as long as they are not a danger to others, is fundamentally an individual's own choice. However, driving by patients with dementia should be discouraged.

Denial of illness may be a prominent feature of dementing illness and may be shared to varying degrees by the family. Medical professionals should be aware of the family unit's denial of disability, which may influence report of symptoms and response to guidance concerning everyday activities. Nursing assistance and social work counselling are critical elements of the care of the patient and family with dementing illness. The importance of autopsy in cases of dementia should be stressed, in view of the genetic factors at play in Alzheimer's disease, multi-infarct dementia, and related diseases. Reliable genetic information, provided by autopsy examination, may become valuable after future developments in our basic understanding of these illnesses.

References

American Psychiatric Association (1980). *Diagnostic and statistical manual of mental disorders*. APA, Washington DC.

Clarfield, A.M. (1988). The reversible dementias: do they reverse? *Annals of Internal Medicine*, **109**, 476–86.

Evans, D.A. *et al.* (1989). Prevalence of Alzheimer's disease in a community population of older persons. Higher than previously reported. *Journal of the American Medical Association*, **262**, 2551–6.

Evans, J.G. (1982). The psychiatric aspects of physical disease. In *The psychiatry of late life* (ed. R. Levy and F. Post), pp. 114–42. Blackwell, Oxford.

Fisher, C.M. (1982). Lacunar strokes and infarcts: a review. *Neurology*, **32**, 871–6.

Folstein, J.F., Folstein, S.E., and McHugh, P.R. (1975). 'Mini-mental State'—a practical method for grading the cognitive state of patients for the clinician. *Journal of Psychiatric Research*, **12**, 189–98.

Hachinski, V.C. *et al.* (1975). Cerebral blood flow in dementia. *Archives of Neurology*, **32**, 632–7.

Katzman, R. *et al.* (1989). Development of dementing illness in an 80 year old volunteer cohort. *Annals of Neurology*, **25**, 317–24.

Lishman, W.A. (1986) Alcoholic dementia: a hypothesis. *Lancet*, **i**, 1184–6.

Luxemberg, V.S., Haxby, J.V., Creasey, H., Sundaram, M., and Rapaport, S.I. (1987). Rate of enlargement in dementia of the Alzheimer type correlates with rate of neuropsychological deterioration. *Neurology*, **37**, 1135–40.

McKhann, G., Drachman, D., Folstein, M., Katzman, R., Price, D., and Stadlan, E. M. (1984). Clinical diagnosis of Alzheimer's disease: report of the NINCDS–ADRDA work group under the auspices of the Department of Health and Human Services Task Force on Alzheimer's disease. *Neurology*, **34**, 939–44.

Majocha, R.E., Benes, F.M., Reifel, J.L., Rodenrys, .M., and Marotta, C.A. (1988). Laminar specific distribution and infrastructural detail of amyloid in the Alzheimer's disease cortex visualized by computer-enhanced imaging of epitopes recognized by monoclonal antibodies. *Proceedings of the National Academy of Science (USA)*, **85**, 6182–6.

Montamat, S.C., Cusack, B.J., and Vestal, R.E. (1989) Management of drug therapy in the elderly. *New England Journal of Medicine*, **321**, 303–9.

Petereson, R.C., Mokri, B., and Laws, E.R. (1985). Surgical treatment of idiopathic hydroceplalus in elderly patients. *Neurology*, **35**, 307–11.

Rumble, B. *et al.* (1989). Amyloid A4 protein and its precursor in Down's syndrome and Alzheimer's disease. *New England Journal of Medicine*, **320**, 1446–52

18.5 Delirium and impaired consciousness

ZBIGNIEW J. LIPOWSKI

INTRODUCTION

Disturbances of consciousness range in severity from delirium through stupor to coma. The last two states refer to arousable and unarousable unresponsiveness, respectively (Plum and Posner 1980). Delirium has traditionally been viewed as a disorder of consciousness, one in which the latter is reduced or 'clouded' yet the patient is awake and able to respond to verbal stimuli. There is no generally accepted definition of consciousness. Hebb (Hebb 1972) defines it as the state of being normally awake and responsive, and of being capable to engage in complex thought processes to guide behaviour. In his view, insight, purpose, and immediate memory constitute the key features of consciousness. Psychiatrists and neurologists have usually regarded consciousness as a continuum of states of being aware of one's self and one's environment, and of being able to respond to external stimuli. This view may be considered to be a quantitative one, in the sense that one may be more or less conscious. A different way to define the concept of consciousness involves focusing on its contents, which may be altered by drugs, hypnosis, and other factors. This view may be regarded as a qualitative one, in that it implies the existence of qualitatively different experiential states occurring in the presence of either normal or reduced level of consciousness or awareness.

The term 'clouding of consciousness' has often been used since the late 19th century to refer to both the level and the contents of consciousness of a delirious patient. This vague and obsolete concept has recently been discarded from the definition of delirium (Diagnostic and Statistical Manual of Mental Disorders 1987). While the latter may still be regarded as a disorder of consciousness, it is best to define it in terms of its psychopathological features, which can allow its diagnosis in clinical practice.

DEFINITION

Delirium, also often referred to in the medical literature as 'acute confusional states', represents one of the organic mental syndromes (Diagnostic and Statistical Manual of Mental Disorders 1987). 'Organic' in this context implies that these syndromes constitute psychopathological manifestations of demonstrable cerebral dysfunction; which may be due to one or more precipitating organic factors (Lipowski 1983; Diagnostic and Statistical Manual of Mental Disorders 1987).

Delirium is currently defined as an acute organic mental syndrome, one characterized by global cognitive impairment, disturbances of attention, reduced level of consciousness, increased or reduced psychomotor activity, and disturbed sleep–wake cycle (Lipowski 1983; Diagnostic and Statistical Manual of Mental Disorders 1987; Lipowski 1990). Its onset is acute and its duration is brief, that is, usually less than 1 month. It may thus be regarded as a transient cognitive-attentional disorder.

The term 'delirium' stems from the Latin word 'delirare', which literally means to go out of the furrow ('lira', Latin for furrow), but whose figurative meaning is to be deranged, crazy, out of one's wits (Murray 1897). The word 'delirium' first appeared in the medical literature in the 1st century AD (Lipowski 1990). Until the 19th century it had been used by medical writers in two different ways: either as a synonym for insanity or, more specifically, as a designation for a transient, acute mental disorder associated with a variety of somatic, notably febrile, diseases (Lipowski 1990). In the 20th century, a variety of terms has been used more or less synonymously with delirium. 'Acute confusional states', 'acute brain syndrome', 'acute confusion', 'acute brain failure', and a host of other designations have been so used, resulting in a terminologic chaos that impedes communication. The Diagnostic and Statistical Manual of the American Psychiatric Association (1980) provided a welcome clarification by adopting the term 'delirium' and offering its definition and diagnostic criteria. The term 'acute confusional states', one widely used in the medical literature, remains the only acceptable synonym for 'delirium' {at least in the United States. In British usage 'acute confusional state' is a purely descriptive term for a clinical condition that may be due to delirium, decompensated dementia or functional mental disorders}. All the other related terms may be regarded as being either obsolete or idiosyncratic.

A term often used by both clinicians and patients in this context is that of 'confusion'. This ambiguous and overworked term has no generally agreed meaning and should not be used as if it referred to a diagnostic entity (Simpson 1984). It has been applied to both delirious and demented patients as well as to anybody having some difficulty in thinking with the usual clarity and coherence. Lishman (1987) rightly deplores the fact that this vague term has sometimes been applied as if it were the hallmark of organic mental states and used in their diagnostic classifications. If one wishes to state that a patient thinks incoherently, cannot grasp the situation, and is perplexed or disoriented, one should state this explicitly and avoid woolly statements such as 'the patient is confused'. Moreover, it is essential to establish in every case of 'confusion' whether a patient displaying the state suffers from delirium or dementia, or both, or from a non-organic mental disorder.

IMPORTANCE OF DELIRIUM IN GERIATRIC MEDICINE

Hodkinson (1976) asserts that, 'Acute mental confusion as a presenting symptom holds a central position in the medicine of old age. Its importance cannot be overemphasized, for acute confusion is a far more common herald of physical illness in an older person than are, for example, fever, pain or tachycardia.' Other geriatricians have referred to mental confusion as the 'very stuff of geriatric medicine' on account of its frequent occurrence, socially disruptive effects, and diagnostic significance as a cerebral manifestation of numerous diseases and toxic agents (Brocklehurst and Hanley 1976). In fact, delirium, as here defined, is one of the most common mental disorders among persons aged 65 years and older, and especially in the very old. Almost any physical illness or intoxication with even therapeutic doses of the commonly used medical drugs can give rise to this syndrome in later life. In an elderly patient delirium may be a presenting feature of pneumonia or myocardial infarction, for example, and failure to diagnose it early and to treat the underlying physical illness effectively may result in his or her death.

Delirium is important to the geriatrician not only because it is common and may constitute a presenting feature of potentially life-threatening physical illness or of drug intoxication. Its development in an elderly patient may result in serious, even lethal, complications (Lipowski 1983; Lipowski 1990). There is an ever-present risk of self-injury, as when an agitated and fearful delirious patient attempts to escape from an unfamiliar environment, such as a hospital ward, and falls and sustains a fracture or some other serious injury. Moreover, such mishaps may lead to subsequent litigation. Finally, the development of delirium in a patient in hospital tends to result in a longer stay and hence in a higher cost (Thomas *et al.* 1988). This economic aspect alone calls for familiarity with delirium and requires attempts to prevent it or at least to diagnose it early and institute appropriate investigations and treatment without delay.

INCIDENCE AND PREVALENCE

Few epidemiological studies on delirium have been carried out to date. A multicentre British study found that 35 per cent of elderly patients displayed features of this syndrome at some point during the index admission (Hodkinson 1980). Two studies of elderly patients admitted to general medical wards reported a prevalence of delirium of 16 per cent (Bergmann and Eastham 1974; Seymour *et al.* 1980). As Millard (1981) observes, however, 'In epidemiological surveys acute confusional states are less prevalent, but in the hospital service acute confusion is all too often equated with dementia.' A study of 2000 consecutive admissions to a medical department of patients aged 55 years and older found that 9.1 per cent showed moderate or severe dementia and 41.4 per cent of those demented were also delirious, while about 25 per cent of all delirious patients were demented (Erkinjuntti *et al.* 1986). These findings highlight the frequent association of these two syndromes among the elderly. Of 130 consecutive admissions to general medical units of patients aged 70 years and older 15 per cent had delirium on admission and an additional 11 per cent developed it during their stay in hospital (Francis *et al.* 1988). One may conclude that at least one in four elderly patients admitted to a general hospital will display delirium at some point during their stay, and this figure is liable to be higher among those aged 80 years and older. Postoperative delirium has been reported to occur in 10 to 15 per cent of elderly patients (Seymour 1986) and in about 50 per cent of those of this group who have undergone an operation for a femoral neck fracture (Berggren *et al.* 1987). All these figures indicate clearly that delirium is a common syndrome among the elderly in hospital.

AETIOLOGY

The term 'delirium' covers a set of psychopathological symptoms (i.e., a syndrome), caused by one or more organic factors that bring about widespread cerebral dysfunction. In an elderly patient, it is usually brought about by more than one organic factor (Liston 1982; Lipowski 1983; Levkoff *et al.*, 1986; Beresin 1988; Lipowski 1989; Lipowski 1990). Many such factors can cause this syndrome and may be classified into four major groups (Lipowski 1989; Lipowski 1990): (1) primary cerebral diseases; (2) systemic diseases affecting the brain secondarily; (3) exogenous toxic agents; and (4) withdrawal from certain substances of abuse, mainly alcohol and sedative–hypnotic drugs. The presence of one or more organic factors constitutes a necessary condition for delirium to occur. The syndrome is aetiologically non-specific, that is to say neither its occurrence nor clinical features give a clear indication of which particular aetiologic factor is involved in a given patient. A list of organic aetiologic factors of particular importance in later life is given in Table 1. One should stress again that in an elderly individual the syndrome is often of multifactorial origin.

Table 1 Common causes of delirium in the elderly

1. Intoxication with medical drugs: anticholinergics; sedative-hypnotics; diuretics; digoxin; cimetidine; antihypertensive and antiarrhythmic drugs; lithium; hypoglycaemic agents; L-dopa; non-steroidal anti-inflammatory agents; narcotics
2. Alcohol and sedative-hypnotic drug withdrawal (especially benzodiazepines)
3. Metabolic disorders: fluid and electrolyte imbalances; endocrine disorders; renal, hepatic, pulmonary failure; malnutrition; hypothermia; heat stroke
4. Cardiovascular disorders; myocardial infarction; congestive heart failure; pulmonary embolism; cardiac arrhythmias
5. Cerebrovascular disorders: stroke; transient ischaemic attacks; subdural haematoma; vasculitides; multi-infarct dementia; orthostatic hypotension
6. Infections: pneumonia; bacteraemia; septicaemia; urinary tract infection; meningitis
7. Neoplasms: intracranial, systemic
8. Trauma: head injury; burns; surgery
9. Epilepsy

Apart from the organic precipitating factors, one needs to take into account those factors which predispose a person to the development of delirium as well as those that facilitate its occurrence. The elderly, especially the very old, are more prone to it than are younger individuals. The predisposing factors include: (1) ageing of the brain; (2) cerebral damage and disease; (3) chronic systemic disease; (4) addiction to alcohol or sedative–hypnotic drugs, or both; (5) impairment of vision and hearing; and (6) impaired mechanisms of drug distribution and metabolism. Certain parts of the brain, notably the frontal lobes, the amygdala, putamen, thalamus, locus coeruleus, and the central cholinergic system, show selective cell loss as a result of the ageing processes (Creasey and Rapaport 1985). Senile plaques occur most often in the hippocampus, hippocampal gyrus, amygdala, and the middle cerebral cortical layers. Patients suffering from Alzheimer's disease show reduction of metabolic rates for glucose and oxygen in the parietal areas of the neocortex as well as right–left asymmetry (Friedland 1988). Moreover, there is loss of cholinergic neurons in the nucleus basalis of Meynert which project to most cortical areas and provide the major source of cholinergic input to the cerebral cortex. As a result, patients suffering from this disease have a cholinergic deficit, a factor that renders them prone to develop delirium in response to any factor that further reduces the availability of acetylcholine to the brain. On the whole, both the degenerative and the vascular diseases of the brain predispose the patient to delirium.

Other predisposing factors include high prevalence of chronic diseases as well as coexistence of multiple diseases and increased susceptibility of elderly people to the acute ones. Addiction to alcohol and sedative–hypnotic drugs is common in the elderly (Atkinson and Schuckit 1983). Also common is impairment of vision and hearing. Finally, age-related altered pharmacokinetics and pharmacodynamics of drugs make an elderly patient more susceptible to drug-induced delirium (Lipowski 1990).

Factors that facilitate the onset of delirium in an elderly person include psychosocial stress occasioned by bereavement or translocation, for example; sleep deprivation; sensory deprivation or overload; and immobilization (Lipowski 1983; Lipowski 1990).

COMMON PRECIPITATING FACTORS

Table 1 lists the more common causes of delirium in later life. Disorders most often causing the syndrome in an elderly patient include congestive heart failure, pneumonia, urinary tract infection, cancer, uraemia, malnutrition, hypokalaemia, dehydration and sodium depletion, and stroke (Liston 1982; Lipowski 1983; Levkoff *et al.* 1986; Beresin 1988; Lipowski 1990). Intoxication with commonly used medical drugs, especially those with anticholinergic effects, is probably the most common single cause of delirium in an elderly person. The syndrome is most likely to occur if several anticholinergic drugs are used concurrently, as is often the case (Blazer *et al.* 1983). Patients with Alzheimer's disease are particularly sensitive to the effects of these drugs (Sunderland *et al.* 1987). Excessive prescribing and polypharmacy as well as inadequate monitoring of drug intake by physicians are common problems, and are likely to contribute to the frequency of iatrogenic delirium in elderly people (Kroenke 1985).

PATHOGENESIS AND PATHOPHYSIOLOGY

These aspects of delirium are still poorly understood, no doubt as a result of its neglect by researchers. Two major pathogenetic hypotheses have emerged to date: (1) the neurochemical and (2) the psychosocial stress–cortisol excess hypotheses (Lipowski 1983). General reduction of cerebral oxidative metabolism has been proposed to account for the cognitive–attentional impairment and disturbances and the slowing of the background activity in the electroencephalogram that characterize delirium (Engel and Romano 1959). Consequently, any disease or toxic agent that leads to a reduction of the supply, uptake, or utilization of substrates for brain metabolism is potentially deliriogenic. Reduced cerebral metabolism caused by a factor such as hypoxia, for example, results in reduced synthesis of acetylcholine and possibly other neurotransmitters as well (Blass and Plum 1983). Acetylcholine is a central neurotransmitter necessary for normal cognitive functioning, attention, and sleep–wake cycle, and it is suggested by some investigators that its deficiency plays a key pathogenetic role in many, if not in all, cases of delirium (Blass and Plum 1983). Patients with Alzheimer's disease have an inherent cholinergic deficit and any factor, such as hypoxia or cholinergic blockade caused by the intake of an anticholinergic drug, is liable to reduce the brain acetylcholine levels even further and to precipitate delirium.

According to the second pathogenetic hypothesis, delirium in an elderly patient represents a reaction to acute stress of any origin, physical or psychosocial, one mediated by high plasma cortisol levels, or by increased sensitivity of the hypothalamus to their effects, or both (Kral 1975). This interesting hypothesis has still to be tested empirically. Cortisol does affect cerebral and mental functions, and its excess has been found to interfere with selective attention and hence with processing of information, that is, with the two psychological functions whose impairment constitutes an essential feature of delirium.

CLINICAL FEATURES, COURSE, AND OUTCOME

As no systematic studies of the clinical features of delirium in the elderly compared to those in younger patients have been carried out to date, it must be assumed that they do not differ in their key characteristics in the various age groups.

A global disturbance of cognition is a cardinal feature of delirium (Lipowski, 1983; Diagnostic and Statistical Manual of Mental Disorders 1987; Lipowski 1989; Lipowski 1990). The word 'global' in this context implies that the main cognitive functions, i.e., thinking, memory and perception, are all impaired or abnormal in some degree. Delirious patients display abnormalities of acquisition, processing, and retrieval of information. Their thinking is more or less incoherent, disorganized, and fragmented. Ability to think logically, grasp abstract concepts, solve problems, and guide action in a purposeful manner is invariably reduced to some extent. A delirious patient may exhibit delusions, usually of a persecutory nature, that are typically unsystematized, fleeting, and readily influenced by environmental stimuli; they are reported to occur in about 50 per cent of delirious elderly people (Simon and Cahan 1963).

Memory in delirium is impaired in its key aspects, that is, registration, retention, and recall. Characteristically, immediate recall is defective, probably as a result of shortened attention span. Both retrograde and anterograde amnesia of some degree of severity can be demonstrated. The patient has reduced ability to learn new material. Some degree of amnesia for the experience of delirium upon its resolution is the rule.

Finally, perception is disordered, in that the patient displays a reduced ability to discriminate and integrate percepts, and to relate incoming information meaningfully to previously acquired knowledge. He or she may have difficulty in telling apart percepts from images, dreams, and hallucinations. Illusions and hallucinations, most often visual or mixed visual and auditory ones, are common in delirium, but are neither invariably present nor necessary for the diagnosis of the syndrome. In many patients they occur only at night (Frieske and Wilson 1966).

A clinically important aspect of the cognitive impairment in delirium is the presence of disorientation of some degree of severity. The earliest to appear and the last to clear up is disorientation for time, that is, errors in giving the date, day of the week, and time of the day. In mild delirium this may be the only form of disorientation, while in more severe delirium, the patient is also disoriented for place, that is, unable to give a correct answer to the question where he or she is located. Moreover, patients may also be disoriented for person, that is unable to recognize and name correctly persons familiar to them. Typically, a delirious patient tends to misidentify an unfamiliar for a familiar place and person.

Attention is invariably disturbed, in that a delirious patient shows reduced ability to mobilize, sustain, and direct it selectively and at will. Alertness, that is, readiness to respond to external stimuli, may be either abnormally heightened or reduced. Some patients are mostly hyperalert but respond to stimuli indiscriminately and are distractible. Others are predominantly hypoalert and it is difficult to attract and maintain their attention. These attentional disturbances tend to fluctuate irregularly in severity in the course of a day and to be most marked at night. They render patients more or less accessible to anyone trying to make contact with them. Some authors maintain that delirium should be viewed as a global disorder of attention and that it reflects pathology of the anatomical substrates of selective and directed attention (Geschwind 1982). This view is based on the finding that acute confusional states occur frequently in right hemisphere infarctions and on the postulated dominance of that hemisphere for attention. This hypothesis remains untested.

A delirious patient not only demonstrates cognitive–attentional disturbances that are essential for the diagnosis of the syndrome, but also a disorder of psychomotor behaviour (Lipowski 1983; Diagnostic and Statistical Manual of Mental Disorders 1987; Lipowski 1990). This term refers to the observable verbal and non-verbal behaviour as well as to the reaction time. A delirious patient may be either hypoactive or hyperactive, or shift from hypo- to hyperactivity and vice versa. Hypoactive patients appear lethargic, respond slowly to verbal stimuli, and speak slowly and hesitantly; speech may be slurred. Their general motility is reduced. By contrast, a hyperactive delirious patient reacts to stimuli promptly, if indiscriminately, appears restless and agitated, and speaks freely and often loudly. In an extreme case, such a patient is in almost constant motion, tries to get out of bed, and screams. Some patients shift unpredictably from one to the other form of psychomotor activity. On the whole, elderly patients, notably those suffering from delirium in the course of one of the metabolic encephalopathies tend to be hypoactive and may be thought by the medical staff to be depressed rather than delirious. Patients suffering from alcohol or hypnotic–sedative drug withdrawal tend to be predominantly hyperactive.

The last essential feature of delirium involves disturbance of the sleep–wake cycle (Lipowski 1983; Diagnostic and Statistical Manual of Mental Disorders 1987; Lipowski 1990). The association between sleep pathology and delirium has been noted by countless medical writers since Hippocrates, but has been relatively neglected in the more recent literature (Lipowski 1983; Lipowski 1990). Night sleep is, as a rule, shortened and fragmented, and is often disturbed by vivid and frightening dreams, which may continue as hallucinations upon awakening. Patients may be unsure whether they are dreaming, hallucinating, or perceiving actual events. They are often restless and agitated during the night. During the day, many delirious patients are drowsy and tend to nap. In a few cases, total insomnia or reversal of the diurnal sleep–wake cycle may occur.

ASSOCIATED CLINICAL FEATURES

The clinical features discussed so far are considered to be essential for the diagnosis of delirium. In addition, a number of inconstant, or associated, clinical features may be observed. Emotions, ranging from apathy or depression to fear or rage, may occur and can give the false impression that the patient is suffering from a functional (non-organic) psychosis. Sympathetic nervous system hyperactivity is liable to accompany intense emotions, such as fear or rage, and be manifested by flushed face, tachycardia, dilation of the pupils, sweating, and elevated blood pressure. In metabolic encephalopathies (e.g. hepatic, renal) the patient may display asterixis, while in alcohol- or drug-withdrawal delirium a coarse tremor is usually present. Urinary and faecal incontinence often occur in the delirious elderly.

COURSE AND OUTCOME

Delirium has an acute onset, a matter of hours or a few days, and often manifests itself first during a sleepless night. During the day its severity tends to fluctuate, in that so-called lucid intervals occur unpredictably and have varying duration. On the whole, the severity of the symptoms tends to be highest at night. The duration of delirium varies. In a younger patient it seldom lasts longer than a week, but in an elderly one it may continue for a few weeks. The outcome is favourable in the majority of patients but in the elderly the syndrome is often a feature of a terminal illness, such as cancer, and is followed by death in about 15 to 30 per cent of patients (Lipowski 1983; Levkoff *et al.* 1986; Lipowski 1990). Thus in an elderly patient delirium should be viewed as a grave prognostic sign. In an unknown proportion of patients it is followed by dementia.

DIAGNOSIS AND DIFFERENTIAL DIAGNOSIS

Diagnosis of delirium involves two essential steps: first, recognition of the syndrome on the basis of its essential clinical features described above, and second, identification of its cause (or causes) as a precondition for effective treatment (Lipowski 1983; Lipowski 1990).

Clinical diagnosis of delirium is relatively easy in a patient who until recently had adequate intellectual function. Acute development in such a patient of global cognitive and attentional deficits and abnormalities that tend to fluctuate irregularly in severity during the day and reach their peak during the night is practically diagnostic. In addition, the presence of disturbed psychomotor behaviour and sleep–wake cycle

Table 2 Differential diagnosis of delirium and dementia

Feature	Delirium	Dementia
Onset	Acute, often at night	Insidious
Duration	Less than 1 month	More than 1 month
Course	Irregular fluctuations during daytime; worse at night	No diurnal fluctuations
Consciousness	Reduced	Normal
Attention	Always disordered; distractibility	Usually undisturbed
Orientation	Always impaired, at least for time	Variable
Memory	Recent and immediate impaired	Recent and remote impaired
Thinking	Always incoherent; may be oneiric	Impoverished
Perception	Hallucinations common	Hallucinations uncommon
Sleep–wake cycle	Always disturbed	Fragmented sleep
Physical illness or drug toxicity	Always present	Often absent

serves to support the diagnosis. The problem may be more difficult when the available history indicates the presence of dementia; this issue will be discussed below. The diagnosis needs to be confirmed by bedside testing of cognitive functions and attention. One of the commonly used clinical diagnostic scales, such as the Mini-Mental State Examination (Anthony *et al.* 1982), may be applied for this purpose. One needs to inquire about the patient's state of orientation in time and for place and person. Immediate memory and attention are tested by asking the patient to substract serially 7 or 3 from 100; to count from 20 backwards; and to recall three words and three objects after 5 min. The patient should be asked to state the circumstances of and reasons for his or her admission to hospital. Errors on these simple cognitive tests indicate the presence of cognitive impairment, while distractibility becomes apparent as one observes the patient's accessibility and performance.

Once the syndrome has been diagnosed on clinical grounds and with the aid of bedside cognitive testing, the search for its underlying cause (or causes) should begin without delay. All drugs taken by the patient, especially those with anticholinergic effects, are suspect even if given in therapeutic doses. They should be withheld or their doses be reduced. A physical, including a neurological, examination is mandatory, and should be supplemented by a battery of routine laboratory tests, including urinalysis, haemogram, blood chemistry, serology, ECG, and chest radiograph. The electroencephalogram (EEG) is usually the most reliable laboratory test for delirium, in that it typically shows some degree of slowing of the background rhythms, or focal abnormalities, or both (Engel and Romano 1959). Alcohol- and drug-withdrawal deliriums represent an exception to this rule: the EEG tends to show low-amplitude fast activity (Brenner 1985). In an elderly patient, however, the diagnostic value of the EEG is reduced, since in patients with dementia, especially that of Alzheimer's disease, it tends to be diffusely slow (Brenner *et al.* 1988). If either the EEG or the neurological examination indicates that a focal brain lesion may be present, further investigations with a CT scan and magnetic resonance imaging are indicated (Koponen *et al.* 1987).

Differential diagnosis of delirium is practically important if proper investigations and treatment are to be instituted (Lipowski 1983; Lipowski 1989; Lipowski 1990). Such diagnosis may be difficult in a patient whose history is unavailable and previous intellectual functioning is unknown: a history of intellectual decline over a period of months or years immediately suggests dementia. Delirium is often superimposed on the latter (Erkinjuntti *et al.* 1986). Both of these syndromes have in common global cognitive impairment and differential diagnosis may present difficulties (Hodkinson 1980). Table 2 summarizes the key points that may aid clinicians in making such a diagnosis.

Delirium also needs to be distinguished from a functional psychosis, such as schizophrenia, mania, and depression, in the course of which some elderly patients may display cognitive impairment, or 'pseudodelirium' (Lipowski 1983). Pseudodelirium, like pseudodementia (Wells 1979), may be suggested by inconsistencies in cognitive tests and the presence of marked depressive or manic features, as the case may be. Evidence of physical illness is usually lacking and the EEG is likely to be normal.

TREATMENT

It is important, whenever possible, to prevent the onset of delirium and to investigate and treat it before it becomes a disruptive and even life-threatening emergency. Prevention involves avoidance of polypharmacy in elderly patients and of prescribing more than one drug with anticholinergic activity. Moreover, drug intake should be closely monitored by physicians. Patients at risk for delirium need to be identified early and observed closely. They include the very old, the demented, the depressed, and those with impaired vision and hearing. Evidence of urinary tract infection, proteinuria, elevated white blood count, and low serum albumin on admission have been identified as markers for the development of delirium (Levkoff *et al.* 1988). Familiarity with the prodromal symptoms of the syndrome on the part of the medical and nursing staff is important. Such symptoms include insomnia, vivid dreams, anxiety, restlessness, and transient hallucinations. Monitoring of cognitive functions throughout the patient's stay in hospital should be mandatory.

Treatment of delirium involves two essential aspects: first, the underlying cause (or causes) needs to be removed or treated; and second, symptomatic–supportive measures need to be instituted (Lipowski 1983; Lipowski 1989; Lipowski 1990). Guidelines for treatment are summarized in Table 3. General supportive measures imply ensuring fluid and electrolyte balance, adequate nutrition, and vitamin supply. The patient is best cared for in a quiet, well-lit room. Good nursing care, supporting, reassuring, and orienting the patient, is crucial, and may help prevent the development of severe agitation and emotional distress (Williams *et al.* 1985).

Table 3 Treatment of delirium

A. Treat or remove underlying cause (or causes)
B. Symptomatic and supportive
   1. Ensure water and electrolyte balance, nutrition, vitamin supply (especially B-complex)
   2. Supportive and orienting nursing care
   3. Close monitoring of mental status and behaviour
   4. Ensure sleep with short-acting benzodiazepine (lorazepam) or hydroxyzine hydrochloride
   5. Place patient in quiet environment
   6. Treat agitation with high-potency neuroleptic (e.g. haloperidol)
   7. Avoid physical restraints
   8. Allow trusted family members to stay with patient

A markedly agitated and restless patient needs to be adequately sedated. Haloperidol may be regarded as relatively the safest and most effective drug for this purpose (Steinhart 1983; Lipowski 1989; Lipowski 1990). It is not likely to induce severe orthostatic hypotension, unlike the phenothiazine drugs, such as chlorpromazine, has low anticholinergic activity, and does not readily cause oversedation. It does have extrapyramidal side-effects, however. A low dose, such as 0.5 mg given orally or intramuscularly twice a day may suffice, but in severe agitation higher doses may be required. Benzodiazepines should be used in alcohol and drug withdrawal deliriums, and in hepatic encephalopathy. {In the United Kingdom chlormethiazole is widely regarded as the drug of choice for alcohol-withdrawal delirium.} In severe anticholinergic delirium, physostigmine salicylate, 1 to 2 mg, may be given slowly intravenously or intramuscularly and repeated after 15 min (Nilssen 1982). It needs to be given cautiously to avoid cardiac arrhythmia and seizures. It is contraindicated in patients with a history of heart disease, asthma, diabetes, peptic ulcer, and bladder or bowel obstruction.

References

Anthony, J.D., Le Resche, L., Niaz, U., Von Korff, M.R., and Folstein, M.R. (1982). Limits of the Mini-Mental State as a screening test for dementia and delirium among hospital patients. *Psychological Medicine*, **12**, 397–408.

Atkinson, J.H. and Schuckit, M.A. (1983). Geriatric alcohol and drug misuse and abuse. *Advances in Substance Abuse*, **3**, 195–237.

Beresin, E.V. (1988). Delirium in the elderly. *Journal of Geriatric Psychiatry and Neurology*, **1**, 127–43.

Berggren, D., *et al.* (1987). Postoperative confusion after anesthesia in elderly patients with femoral neck fractures. *Anesthesia and Analgesia*, **66**, 497–504.

Bergmann, K. and Eastham, E.J. (1974). Psychogeriatric ascertainment and assessment for treatment in an acute medical ward setting. *Age and Ageing*, **3**, 173–88.

Blass, J.P. and Plum, G. (1983). Metabolic encephalopathies in older adults. In *The Neurology of Aging*. (eds. R. Katzman and R.D. Terry), pp. 189–220. F.A. Davis Co., Philadelphia.

Blazer, D.G., Federspiel, C.F., Ray, W.A., and Schaffner, W. (1983). The risk of anticholinergic toxicity in the elderly: a study of prescribing practices in two populations. *Journal of Gerontology*, **38**, 31–5.

Brenner, R.P. (1985). The electroencephalogram in altered states of consciousness. *Neurologic Clinics*, **3**, 615–31.

Brenner, R.P., Reynolds, C.F., and Ulrich, R.F. (1988). Diagnostic efficacy of computerized spectral versus visual EEG analysis in elderly normal, demented and depressed subjects. *Electroencephalography and Clinical Neurophysiology*, **69**, 110–17.

Brocklehurst, J.C. and Hanley, T. (1976). *Geriatric Medicine for Students*, p. 59. Churchill Livingstone, Edinburgh.

Creasey, H. and Rapoport, S.I. (1985). The aging human brain. *Annals of Neurology*, **17**, 2–10.

Diagnostic and statistical manual of mental disorders. (1980). (3rd edn.). American Psychiatric Association, Washington, D.C.

Diagnostic and statistical manual of mental disorders. (1987). (3rd edn. revised). American Psychiatric Association, Washington, D.C.

Engel, G.L. and Romano, J. (1959). Delirium: a syndrome of cerebral insufficiency. *Journal of Chronic Diseases*, **9**, 260–77.

Erkinjuntti, T., Wikström, J., Palo, J., and Autio, L. (1986). Dementia among medical inpatients. *Archives of Internal Medicine*, **146**, 1923–6.

Francis, J., Kapoor, W.N., and Martin D. (1988). Delirium on medical services: prospective study of elderly admissions. *Gerontologist*, **28**, 44A.

Friedland, R.P. (Moderator). (1988). Alzheimer disease: clinical and biological heterogeneity. *Annals of Internal Medicine*, **109**, 298–311.

Frieske, D.A. and Wilson, W.P. (1966). Formal qualities of hallucinations: a comparative study of the visual hallucinations in patients with schizophrenic, organic and affective psychoses. In *Psychopathology of Schizophrenia*, (eds. P.H. Hoch and J. Zubin), pp. 49–62. Grune and Stratton, Inc., New York.

Geschwind, N. (1982). Disorders of attention: a frontier in neuropsychology. *Philosophical Transactions of the Royal Society of London, Biology*, **298**, 173–85.

Hebb, D.O. (1972). *Textbook of Psychology* (3rd edn.). W.B. Saunders Company, Philadelphia.

Hodkinson, H.M. (1976). *Common Symptoms of Disease in the Elderly*, p. 24. Blackwell Scientific Publications, Oxford.

Hodkinson, H.M. (1980). Confusional states in the elderly. In *The Aging Brain* (eds. G. Barbagallo-Sangiorgi and A.N. Exton-Smith), pp. 155–9. Plenum Press, New York.

Koponen, H., Hurri, L., Stenbäck, U., and Riekkinen, P.J. (1987). Acute confusional states in the elderly: a radiological evaluation. *Acta Psychiatrica Scandinavica*, **76**, 726–31.

Kral, V.A. (1975). Confusional states: description and management. In *Modern Perspective in Psychiatry of Old Age* (ed. J.G. Howells), pp. 356–62. Brunner/Mazel Inc., New York.

Kroenke, K. (1985). Polypharmacy. Causes, consequences, and cure. *American Journal of Medicine*, **79**, 149–52.

Levkoff, S.E., Besdine, R.W., and Wetle, T. (1986). Acute confusional states (delirium) in the hospitalized elderly. *Annual Review of Gerontology and Geriatrics*, **6**, 1–26.

Levkoff, S.E., Safran, C., Cleary, P.D., Gallop, J., and Phillips, R.S. (1988). Identification of factors associated with the diagnosis of delirium in elderly hospitalized patients. *Journal of the American Geriatrics Society*, **36**, 1099–104.

Lipowski, Z.J. (1983). Transient cognitive disorders (delirium, acute confusional states) in the elderly. *American Journal of Psychiatry*, **140**, 1426–36.

Lipowski, Z.J. (1989). Delirium in the elderly patient. *New England Journal of Medicine*, **320**, 578–82.

Lipowski, Z.J. *Delirium: (Acute Confusional States)*. Oxford University Press, New York, 1990.

Lishman, W.A. (1987). *Organic Psychiatry*, 2nd edn. Blackwell Scientific Publications, Oxford.

Liston, E.H. (1982). Delirium in the aged. *Psychiatric Clinics of North America*, **5**, 49–66.

Millard, P.H. (1981). Last scene of all. *British Medical Journal*, **283**, 1559–60.

Murray, J.A.H. (ed.) (1897). *A New English Dictionary on Historical Principles*. Vol. 3, p. 165. Clarendon Press, Oxford.

Nilsson, E. (1982). Physostigmine treatment in various drug-induced intoxications. *Annals of Clinical Research*, **14**, 165–72.

Plum, F. and Posner, J.B. (1980). *Diagnosis of Stupor and Coma* (3rd edn.). F.A. Davis Company, Philadelphia.

Seymour, G. (1986). *Medical Assessment of the Elderly Surgical Patient*. Croom Helm Ltd., London.

Seymour, D.G., Henschke P.J., Cape, R.D.T., and Campbell, A.J. (1980). Acute confusional states and dementia in the elderly: the role of dehydration/volume depletion, physical illness and age. *Age and Ageing*, **9**, 137–46.

Simon, A. and Cahan, R.B. (1963). The acute brain syndrome in geriatric patients. *Psychiatric Research Reports*, **16**, 8–21.

Simpson, C.J. (1984). Doctors' and nurses' use of the word confused. *British Journal of Psychiatry*, **142**, 441–3.

Steinhart, M.J. (1983). The use of haloperidol in geriatric patients with organic mental disorder. *Current Therapeutic Research*, **33**, 132–43.

Sunderland, T., *et al.* (1987). Anticholinergic sensitivity in patients with dementia of the Alzheimer type and age-matched controls. *Archives of General Psychiatry*, **44**, 418–26.

Thomas, R.I., Cameron, D.J., and Fahs, M.C. (1988). A prospective study of delirium and prolonged hospital stay. *Archives of General Psychiatry*, **45**, 937–40.

Wells, C.E. (1979). Psuedodementia. *American Journal of Psychiatry*, **136**, 895–900.

Williams, M.A., Campbell, E.G., Raynor, W.J., Mlynarczyk, S.M., and Ward, S.E. (1985). Reducing acute confusional states in elderly patients with hip fractures. *Research in Nursing and Health*, **8**, 329–37.

18.6 Headaches and facial pain

JAMES G. HOWE

Backache and headache are the most common symptoms that human beings suffer (Crook *et al.* 1984). While backache is common in elderly individuals, people seem to complain less about headaches as they grow older. Zeigler *et al.* (1977) found that only 1 per cent of the population aged over 65 years which they sampled had started to complain of severe headache in the previous year, and only 18 per cent of the men and 29 per cent of the women over 65 complained of disabling or severe headaches. There are three causes of head pain that become more common with increasing age: temporal arteritis (see Chapter 13.6), trigeminal neuralgia, and post-herpetic neuralgia. Elderly people are subject to pain caused by other pathological processes in the head and neck just like younger people, but migraine and tension headache are much less common.

In many studies the prevalence of both migraine and headache declines after middle age (Goldstein and Chen 1982); Leviton *et al.* (1974) reported a slightly higher risk of death before the age of 70 for migraine sufferers. Migraine does not disappear altogether, for Whitty and Hockaday (1968) reviewed a group of patients who had attended a hospital outpatient clinic up to 20 years before and found that half of them were still having migraine attacks after the age of 65.

Headache may be less common in elderly people but it is not unknown. When attempting a differential diagnosis of headache or facial pain, it is useful to have a simple mental picture of the pain-sensitive structures in the head and the sensory pathways, and to analyse the time-course of the pain. It is particularly helpful to imagine and review all the tissues and organs in the solid part of the head that would be sliced off by a coronal cut from the vertex to the chin, remembering that they are all innervated by the trigeminal nerve. This means that pain from supratentorial intracranial structures and the sinuses is referred to the front of the head. In humans, stimulation of the first cervical dorsal root refers pain to the frontal and orbital regions (Kerr 1961), but most disease of the upper cervical spine refers pain solely to the occiput and upper part of the neck (Edmeads 1979*a*).

Degenerative arthropathy in the neck is universal in older people, yet neck pain is not a common complaint in geriatric wards and clinics. When severe neck pain does occur in an elderly person, secondary carcinoma should be excluded (Edmeads 1979*a*).

Depressed patients may complain of headache, and patients with chronic headaches may become depressed (Romano and Turner 1985). In elderly people, depression is more likely to generate somatic complaints, being associated with an increase in hypochondriasis (Leading Article 1984), and the present generation of elderly people are much more likely to offer somatic complaints to their doctor than emotional problems.

Atypical facial pain is not a diagnosis that should be made in elderly patients. Frazier and Russell (1924) appear to have been the first to use this term to distinguish trigeminal neuralgia from other kinds of facial pain. Since then it has been used in the neurological literature to describe chronic facial pain in the middle-aged with no evidence of an organic cause, and psychological factors have usually been blamed (Lascelles 1966; Friedman 1969). In dental practice, similar patients are diagnosed as suffering from the temporomandibular joint pain–dysfunction syndrome (Feinmann and Harris 1984; Yusuf and Rothwell 1986). Problems like this do not seem to bring patients to geriatricians.

It is uncommon for rheumatoid or degenerative arthritis to cause pain in the temporomandibular joint (Cawson 1984). Pain in the jaw precipitated by talking or chewing can be caused by ischaemia in cranial arteritis, so-called 'jaw claudication'.

Headache is also less common in elderly patients with cerebral tumours. Godfrey and Caird (1984) found that only 10 per cent of their series of patients had headache and only

2.5 per cent had papilloedema. In contrast, up to 43 per cent of patients with subdural haematoma report headache (Cameron 1978) but it is rare in patients with ischaemic stroke (Edmeads 1979*b*). Facial or head pain is usually a prominent feature with tumours involving bones, sinuses, and meninges (Wasserstrom *et al.* 1978).

Paget's disease of the skull can cause headache, cranial nerve palsies, and ataxia and spasticity if there is cranio-vertebral distortion due to basilar invagination (Friedmann *et al.* 1971).

Carbon monoxide poisoning and carbon dioxide retention in chronic obstructive airways disease both cause headache. Old people with faulty gas appliances are at risk of carbon monoxide poisoning which, if not fatal, can cause headache and confusion (James 1984).

Painful diseases involving the eyes, teeth, and sinuses are usually obvious but in a confused or dysphasic patient who appears to have head pain, such possibilities should be systematically considered, as should parotitis. This is more likely to occur in a frail, dehydrated, old person on an anticholinergic drug such as an antidepressant.

INVESTIGATION

The erythrocyte sedimentation rate should be measured in all elderly patients with headache and plain radiographs of the skull, sinuses, or neck bones may be needed. Computed tomography should not be ruled out on grounds of age alone, as palliative treatment of tumours and hydrocephalus can be successful in elderly patients, subdural haematomas can be removed, infections cured, and Paget's disease treated (Cameron 1978; Godrey and Caird 1984).

SPECIFIC CONDITIONS

Temporal arteritis is dealt with in Chapter 13.6. Two other conditions that are particularly important in elderly persons will be discussed here.

Trigeminal neuralgia

This is a common cause of severe pain, which becomes more common in older age groups. Disease or damage to the trigeminal nerve is presumed to cause increased firing in the nerve as well as impaired efficiency of central inhibitory mechanisms and leads to paroxysmal bursts of neuronal activity which cause sudden, excruciating paroxysms of facial pain (Fromm *et al.* 1984). The lightning explosions of pain are often triggered by minimal cutaneous stimuli at particular places and patients will often be unable to eat, shave, wash their faces, or blow their noses because of it. The pain rarely attacks all three sensory divisions of the nerve and the first division is less often affected than the other two. The presence of sensory loss or motor weakness should arouse suspicion of a progressive structural lesion affecting the nerve or nucleus, as should signs of involvement of neighbouring cranial nerves. Many different lesions appear to be able to cause the chronic irritation that results in trigeminal neuralgia, including chronic oral and dental disease, tumours, plaques of multiple sclerosis, and, most commonly of all, tortuous blood vessels (Fromm *et al.* 1984). Progressive elongation and tortuosity of arterial loops with advancing age and atheroma seem to account for the increased prevalence of trigeminal neuralgia in old age.

Glossopharyngeal neuralgia is much less common. The pattern of pain is similar to trigeminal neuralgia but it is triggered from, and felt in, the throat, sometimes radiating to the ear. Involvement of vagal fibres may cause autonomic symptoms such as syncope or bradycardia (Rushton *et al.* 1981).

The pain of trigeminal neuralgia can be suppressed with anti-convulsant drugs such as carbamazepine, phenytoin, and clonazepam. We have found clobazam effective and easier to use in frail, elderly patients who are often intolerant of the other drugs because of ataxia. Baclofen has also been used, with some success (Fromm *et al.* 1980). In elderly patients, a low starting dose and gradual increases in dose are essential if dangerous ataxia is to be avoided with carbamazepine and phenytoin.

When drugs are ineffective, surgery should be considered and a number of options are open to the patient, depending on their fitness. Peripheral nerve injection, followed if necessary by avulsion of the nerve to denervate the trigger area, can be worthwhile. Where this is not successful, radiofrequency lesions of the nerve root or ganglion can be tried. This is a safe technique and has a relatively low risk of side-effects but the pain can recur (Tew 1979). Microvascular decompression of the trigeminal nerve is said to be effective in over 90 per cent of patients (Richards *et al.* 1983). However, this is a major operation, requiring a posterior fossa craniotomy. Discussion with a neurosurgeon should not be denied to elderly patients in severe pain, as simple procedures are often helpful.

Post-herpetic neuralgia

Half of those who reach the age of 85 can expect an attack of herpes zoster (Hope-Simpson 1965) and the number of patients still suffering pain 1 year after the acute attack rises with increasing age. More than 60 per cent of those aged over 70 with involvement of the trigeminal nerve, which is the dermatome most commonly affected, and nearly half of those over 70 who have shingles elsewhere on the body, report persisting pain 1 year after the acute attack (de Moragas and Kierland 1957). The severity and duration of the pain, which is felt in the dermatome where the vesicles appeared, is worse in elderly people. It can be agonizing and persistent, is usually described as 'burning', with increased painful and abnormal sensitivity to any stimulus in the affected area. Stabbing, neuralgic-type pain can also occur and sufferers quickly become demoralized and depressed. Portenoy *et al.* (1986) provide an authoritative and up-to-date account of the pathophysiology and management of post-herpetic neuralgia and recommend that pain continuing 2 months after the onset of the rash should be labelled post-herpetic neuralgia in order to compare therapeutic regimens.

Recently, oral and intravenous acyclovir have been shown to speed resolution and reduce the pain of the acute attack, but so far there is no evidence that this drug reduces the prevalence or severity of post-herpetic neuralgia (McKendrick *et al.* 1986). Acyclovir should definitely be used in all elderly patients with trigeminal herpes zoster and in elderly patients with complications of acute herpes zoster. It is not clear whether corticosteroids should be used at the time of the acute attack, but Portenoy *et al.* (1986) recommend it.

The scheme recommended by Portenoy *et al.* for the management of post-herpetic neuralgia can be followed at a day hospital. They recommend the simultaneous prescription of a tricyclic antidepressant, physiotherapy, and occupational therapy to increase activity, build confidence, and improve function, and transcutaneous electrical nerve stimulation (**TENS**). Small, portable stimulators are inexpensive, reliable, and useful in other pain syndromes but the response is variable and different electrode placements and stimulation protocols should be tried before abandoning it. If triggering of painful paroxysms is a feature, then anticonvulsants may be helpful. Portenoy *et al.* emphasize the importance of psychological, functional, and social approaches to management from the start. Putting the patient back in control of at least some aspect of his problem is helpful in post-herpetic neuralgia as in other chronic disorders.

Instruction in methods of relaxation and distraction and encouraging the patient to keep a diary to demonstrate increasing levels of physical and social activity are all helpful. An anaesthetic spray, or instructions on counter-irritation by rubbing the affected area, or loaning a TENS apparatus to use at home are all ways in which the patient can regain control over the symptoms. Neurolytic and surgical techniques should be avoided in post-herpetic neuralgia as they are generally unhelpful.

CONCLUSIONS

Headache may not be common in elderly people and diagnosis can be made more difficult by problems with communication and multiple pathology. Where pain develops rapidly and increases in severity, especially when associated with neurological signs, prompt and full investigation is needed. The pattern of pain in the more common conditions is usually characteristic and successful treatment is possible in most patients.

References

Cameron, M.M. (1978). Chronic subdural haematoma: A review of 114 cases. *Journal of Neurology, Neurosurgery and Psychiatry*, **41**, 834–9.

Cawson, R.A. (1984). Pain in the temporomandibular joint. *British Medical Journal*, **288**, 1857–8.

Crook, J., Rideout, E., and Browne, G. (1984). The prevalence of pain complaints in a general population. *Pain*, **18**, 299–314.

de Moragas, J.M. and Kierland, R.R. (1957). The outcome of patients with herpes zoster. *Archives of Dermatology*, **75**, 193–6.

Edmeads, J. (1979*a*). Headaches and head pains associated with diseases of the cervical spine. *Medical Clinics of North America*, **62**, 533–44.

Edmeads, J. (1979*b*). The headaches of ischaemic cerebrovascular disease. *Headache*, **19**, 345–9.

Feinmann, C. and Harris, M. (1984). Psychogenic facial pain. *British Dental Journal*, **156**, 165–8.

Frazier, C.H. and Russell, E.C. (1924). Neuralgia of the face. An analysis of seven hundred and fifty four cases with relation to pain and other sensory phenomena before and after operation. *Archives of Neurology and Psychiatry*, **11**, 557–68.

Friedman, A.P. (1969). Atypical facial pain. *Headache*, **9**, 27–30.

Friedmann, P., Sklaver, N., and Klawans, H.L. (1971). Neurological manifestations of Paget's disease of the skull. *Diseases of the Nervous System*, **32**, 809–17.

Fromm, G.H., Terrence, G.F., Chatta, A.S., and Glass, J.D. (1980). Baclofen in the treatment of refractory trigeminal neuralgia. *Archives of Neurology*, **37**, 768–71.

Fromm, G.H., Terrence, G.F., and Maroon, J.C. (1984). Trigeminal neuralgia: Current concepts regarding aetiology and pathogenesis. *Archives of Neurology*, **41**, 1204–7.

Godfrey, J.B. and Caird, F.I. (1984). Intracranial tumours in the elderly: Diagnosis and treatment. *Age and Ageing*, **13**, 152–8.

Goldstein, M. and Chen, T.C. (1982). The epidemiology of disabling headache. In *Headache, Advances in Neurology*, Vol. 33 (ed. M. Critchley, A.P. Friedman, S. Gorini, and F. Sicuteri). pp. 377–90, Raven Press, New York.

Hope-Simpson, R.E. (1965). The nature of herpes zoster: A long term study and a new hypothesis. *Proceedings of the Royal Society, London* (Biol), **58**, 9–20.

James, P.B. (1984) Carbon monoxide poisoning. *Lancet*, **ii**, 810.

Kerr, F.W.L. (1961). A mechanism to account for frontal headache in cases of posterior fossa tumour. *Journal of Neurosurgery*,**18**, 605–9.

Lascelles, R.G. (1966). Atypical facial pain and depression. *British Journal of Psychiatry*, **112**, 651–9.

Leading Article (1984). Headache and depression. *Lancet*, **i**, 495.

Leviton, A., Malvea, B., and Graham, J.R. (19740. Vascular disease, mortality and migraine in the parents of migraine patients. *Neurology*, **24**, 669–72.

McKendrick, M.W., McGill, J.I., White, J.E., and Wood, M.J. (1986). Oral acyclovir in acute herpes zoster. *British Medical Journal*, **293**, 1529–32.

Portenoy, R.K., Duma, D., and Foley, K.M. (1986). Acute herpetic and post-herpetic neuralgia: Clinical review and current management. *Annals of Neurology*, **20**, 651–64.

Richards, P., Shawdon, H., and Illingworth, R. (1983). Operative findings on microsurgical exploration of the cerebello-pontine angle in trigeminal neuralgia. *Journal of Neurology, Neurosurgery and Psychiatry*, **46**, 1098–101.

Romano, J.M. and Turner, J.A. (1985). Chronic pain and depression: Does the evidence support a relationship? *Psychological Bulletin*, **97**, 18–34.

Rushton, J.G., Stevens, J.C., and Miller, R.H. (1981). Glossopharyngeal (vagoglossopharyngeal) neuralgia: a study of 217 cases. *Archives of Neurology*, **38**, 201–5.

Tew, J.M. (1979). Treatment of trigeminal neuralgia. *Neurosurgery*, **4**, 93–4.

Wasserstrom, W.R., Glass, J.P., and Posner, J.B. (1978). Diagnosis and treatment of leptomeningeal metastases from solid tumours. *Cancer*, **49**, 759–72.

Whitty, C.W.M. and Hockaday, J.M. (1968). Migraine: a follow up study of 92 patients. *British Medical Journal*, **i**, 735–6.

Yusuf, H. and Rothwell, P.S. (1986). Temporo-mandibular joint pain–dysfunction in patients suffering from atypical facial pain. *British Dental Journal*, **161**, 208–12.

Zeigler, D.K., Hassanein, R.S., and Couch, J.R. (1977). Characteristics of life headache histories in a non-clinic population. *Neurology*, **27**, 265–9.

18.7 Dizziness

W.J. MACLENNAN

INTRODUCTION

Dizziness is a common symptom in old age, which confronts the clinician with major problems in diagnosis and management. Patients use the term to describe sensations that may include unsteadiness, light-headedness, anxiety, vertigo, or even syncope. Even the most detailed questioning may fail to elicit their exact nature. A further difficulty is that physical examination may reveal a whole series of abnormalities, and it is difficult to determine which of these is responsible for the symptom. Finally, the patient may present to a neurologist with little experience of geriatric medicine, to a geriatrician with little experience of otolaryngology, or to an otolaryngologist with little experience of cardiology. A catholic medical education, and effective rapport with specialist colleagues is thus a prerequisite for the effective management of this condition.

AETIOLOGY

Ageing

Ageing has an adverse effect on most of the end organs and neuronal pathways concerned with balance. An example is the decline in peripheral position sense. Degeneration of position-sense receptors in the synovial joints of the neck may also have an adverse effect on balance, but there is, as yet, no firm clinical evidence of this (Wyke 1979; Sharma and MacLennan 1988). A further but unsubstantiated possibility may be that ataxia is the result of damage to neuronal pathways within the midbrain. Experimental lesions of the posterior vermis of the cerebellum in animals, and damage to this site in humans produce a truncal ataxia, a picture that is found in many elderly people who are unsteady on their feet, but who have few other focal neurological signs.

Ageing is associated with a reduction in the number of hair-cell receptors within the vestibular apparatus, and a concomitant reduction in the number of fibres in the vestibular nerve. There is no clear relationship between these changes and clinical evidence of vestibular dysfunction or ataxia in old age.

Vision also has an important part to play in regulating balance, so that old people whose balance is already compromised by proprioceptive defects, midbrain damage, and, possibly, vestibular degeneration are at particular risk if they suffer from one of the range of eye defects common in old age (Sharma and MacLennan 1988).

Midbrain function is dependent upon an adequate blood supply and this may be compromised by postural hypotension. This is fairly common, even in healthy old people, but is not often associated with symptoms (see Chapter 18.10).

Disease in old age

A wide range of disorders common in old age may be responsible for dizziness. It is important to recognize, however, that in many cases even detailed investigation may fail to reveal a specific cause. Thus, a survey of 740 old people presenting to an otological clinic with dizziness revealed that only 21 per cent had diseases specifically associated with this (Belal and Glorig 1986). The remainder were disabled by one or more of the degenerative processes associated with ageing, and were labelled as suffering from 'presbyastasis'.

Table 1 lists possible causes for recurrent episodes of dizziness in elderly people. This is not comprehensive in that many less common disorders have been omitted. However, it forms a useful framework on which a history, physical examination, and investigation can be based. A single episode of dizziness may be the first manifestation of one of these disorders. Alternatively it may herald a brain-stem infarction or haemorrhage, or may indicate a sudden fall in the intracranial perfusion pressure resulting from haemorrhage, dehydration, or cardiovascular catastrophe.

Table 1 Causes of dizziness in old age

System	Disorders
Neurological	Vertebrobasilar insufficiency
	Carotid artery stenosis
	Subclavian steal
	Cerebellar ischaemia
	Drop attacks
	Epilepsy
	Anxiety neurosis
	Depression
Cardiovascular	Carotid sinus hypersensitivity
	Cardiac arrhythmias
	Aortic stenosis
	Postural hypotension
Metabolic	Hyperventilation
	Diabetes/hyperglycaemia
Otological	Positional nystagmus
	Vestibular neuronitis
	Infection
	Vascular damage
	Menière's syndrome
	Acoustic neuroma
	Drug side-effects
Other	Ocular
	Cervical spondylosis
	Cough syncope
	Anaemia
	Micturition syncope

ASSESSMENT

History

Assessment of dizziness is simplified if a systematic approach is taken to history taking, physical examination, and investigation. A variety of flow systems and algorithms has been devised (Luxon 1983, 1984; Fowler 1984). These are often compromised by the problems involved in getting a clear history of episodes from elderly patients.

Given that this may be a major problem, an attempt should be made to distinguish the rotational symptom of vertigo from other sensations. In the presence of the former, attention should be directed to physical examination and investigation of the inner ear and its neuronal connections. A history of the symptom being associated with neck movements, changes in the position of the head, or exposure to drugs may also provide useful information.

If there is no clear history of vertigo, then symptoms of imbalance in the lower limbs, light-headedness, black-outs, anxiety, or depression may point to the aetiology. Again, the association of any of these with a particular medication may be relevant.

Physical examination

A full examination of the cardiovascular and neurological systems is essential. In relation to the first of these, particular attention should be paid to identification of the cardiac rhythm, to palpation of the carotid arteries, and to auscultation over the heart valves and over the subclavian, carotid, and subclavian arteries. Lying and standing, systolic and diastolic blood pressures should be recorded, the latter being estimated 2 min after the patient has assumed an erect position. Neurological signs of particular relevance include nystagmus, intention tremor, past pointing, and abnormalities of the corticospinal tract. Adequate evaluation of position sense is difficult, but one way of quantifying this is to score out of 20 the ability of patients to identify correctly five movements each of the right and left halluces and ankles. Balance should be tested by asking the patient to stand with feet comfortably apart, first with eyes open and then with eyes closed. A more accurate evaluation of balance may be obtained by asking the patient to assume a series of positions ranging from standing with legs comfortably apart with eyes open, to standing with one foot in front of the other with eyes closed (Gabell and Simons 1982). Additional stress can be placed on balance by asking the patient to turn his or her head repeatedly to the right and left in sequence, and to rotate 360° around a position on the floor. The response of the righting reflexes to a firm push over the manubrium can also be evaluated.

Where appropriate, a more detailed examination should include testing for positional nystagmus. In summary, the patient is seated on a table, the head is turned 45° to one side, and then lowered to an angle 30° below the horizontal; the test is then repeated with the head turned to 45° in the opposite direction; and the test finally repeated with the head looking forwards. With each manoeuvre a careful evaluation is made of any eye movements (Venna 1986*a*).

Balance may also be compromised by locomotor problems so that the feet and ankles, hip, and knee joints should be examined. The range of movement within the cervical vertebrae should also be tested.

Investigations

Simple baseline investigations of dizziness should include haemoglobin, blood urea and electrolyte levels, a random blood glucose concentration, an ECG, and a chest radiograph. Although cervical spondylosis may cause dizziness or ataxia in a variety of ways, osteoarthritic changes are almost universal in the necks of old people so that radiography is of no diagnostic value (Adams *et al.* 1986). More sophisticated tests should be made only if there are clear symptoms and signs pointing to a treatable condition.

Cardiovascular investigations

Sensitivity of the carotid sinuses is tested by lightly massaging in turn the right and left common carotid arteries for 5 s each (Sugue *et al.* 1986). A sinus is diagnosed as hypersensitive if massage causes a cardioinhibitory pause of more than 3 s or a drop in blood pressure of more than 50 mmHg. The pulse rate is measured on an ECG. The blood pressure should be regularly monitored with a cuff and sphygmomanometer. Complications of sinus massage are rare, but emboli may be released from an atheromatous plaque. Even less commonly the manoeuvre may induce permanent asystole. It is essential, therefore, that the equipment and expertise for cardiopulmonary resuscitation are immediately available.

Ambulatory electrocardiographic monitoring should be run for at least 24 h, but periods of up to 7 days may be necessary if an arrhythmia is particularly elusive. There should be a system whereby the patient records symptoms to identify if these are temporally related to arrhythmic episodes, which are not uncommon in elderly people without dizziness.

Lesions of the subclavian, vertebral, and carotid arteries can be identified by retrograde transfemoral or brachial angiography. However, the techniques are invasive, and are associated with a significant though low incidence of serious morbidity. These are likely to be supplanted by non-invasive tests such as ultrasonography. When investigation of carotid artery stenosis by Doppler and real-time ultrasonography was compared with angiography, there was 95 per cent agreement between the tests for patients with more than 50 per cent stenosis (Lo *et al.* 1986). Sonography is also effective in diagnosing stenosis of vertebral and basilar arteries (Ringlestein *et al.* 1985). Again, however, it is only comparable with angiography in identifying stenosis of greater than 60 to 80 per cent.

Neurological investigations

The key investigation in more detailed evaluation of vestibular function is the caloric test (Hinchcliffe 1983). More detailed information can be obtained if the test is combined with electro-oculography.

Electrophysiological tests are of value in excluding the possibility of epilepsy (see Chapter 18.11). They may also be useful in monitoring the course of transient ischaemic attacks. An example is that brain-stem, auditory-evoked responses separate out different parts of the labyrinthine pathway and have been used to identify and monitor temporary neurological damage after vertebrobasilar, transient ischaemic attacks (Factor and Detinger 1987). The test is also useful in distinguishing an acoustic neuroma from Menière's disease.

Though computerized tomography is often of immense value in establishing neurological diagnosis, it is less sensitive in identifying lesions of the midbrain and cerebellum. The technique has identified cerebellar atrophy associated with ageing but there was no correlation between these changes and ataxia. It may be that truncal ataxia is associated with changes in the vermis and that computerized tomography is insufficiently sensitive to identify these. Magnetic resonance imaging may be of some value in this context. Certainly, this has already been used successfully to distinguish between small cerebellar infarcts and haemorrhages, and to identify small cerebellar infarcts within 24 h of their onset.

Even if a clinician takes a full history, makes a detailed examination, and organizes appropriate tests he or she may be left with no convincing explanation for dizziness, or with several minor abnormalities, any one of which or any combination could account for the symptom. The clinical reality of dizziness in old age is rarely as simple as algorithms or the following detailed account of single disorders would suggest. A considerable amount of frustration, compromise, and empiricism is inevitable in dealing with this difficult area.

VERTEBROBASILAR INSUFFICIENCY

This condition may present as recurrent, non-specific episodes of dizziness (Venna 1986*b*). Eventually, however, there are the additional symptoms of midbrain ischaemia, which include limb numbness or weakness, facial paraesthesiae, dysarthria, diplopia, and visual impairment. These may or may not be accompanied by clinical signs of permanent damage to the midbrain, cerebellum, or occipital cortex. The condition is usually associated with atheroma of the vertebral or basilar arteries (Ausman *et al.* 1985). Opinions differ, but many clinicians consider that lesions of the distal vertebral arteries or basilar arteries are much more likely to cause problems than stenosis at their origins. Stenosis is only likely to have a significant effect on blood flow if it occludes more than 60 per cent of the arterial lumen. An exception is where there also is stenosis of the carotid artery resulting in a 'steal' through the circle of Willis from the posterior part of the brain. The extent to which emboli cause vertebrobasilar symptoms is uncertain. Emboli from the stenotic ostia of the vertebral arteries have been documented but there is no information on their incidence.

Vertebrobasilar insufficiency is often overdiagnosed in elderly patients. An example might be one with dizziness and truncal ataxia associated with cerebellar degeneration who coincidentally has symptoms and signs of cervical spondylosis. The error may be of little importance if it leads to a policy of non-intervention, but may have serious consequences if angiography and reconstructive surgery are contemplated. A reasonable policy would be to reserve investigation for patients who give a clear-cut history of focal neurological symptoms in addition to dizziness.

Even if there is clinical evidence of vertebrobasilar investigation, complex and, more particularly, invasive investigations should be reserved for patients in whom surgical intervention would be of benefit. A review of patients with vertebrobasilar disease has shown that they had a 5-year survival rate of 78 per cent compared with one of 90 per cent for age- and sex-matched controls (Mouzfarrrig *et al.* 1986). The rate of stroke was 17 times greater in the first group. It is likely, however, that this considerable difference would be truncated in old age. The distinction would be even more blurred in patients with evidence of generalized vascular disease, dementia, or serious coincidental conditions.

Ultrasonography is establishing itself as the investigation of choice in the diagnosis of stenosis of vertebral or basilar arteries. Angiography remains necessary for identifying the anatomical site of a lesion once the diagnosis of stenosis has been made. This procedure is reported as having a death rate of 0.6 per cent, and a 7.1 per cent incidence of transient complications. These risks can be eliminated by using intravenous digital subtraction angiography, but this does not give such an effective definition of the blood vessels.

Anticoagulants have been used in the treatment of vertebrobasilar insufficiency, but no controlled, randomized studies on this have been completed. Retrospective studies suggest that anticoagulants may reduce the incidence of stroke, but that the benefit is limited to the first 6 months of treatment (Wishart *et al.* 1978). Until there is more convincing evidence of its efficacy, anticoagulant therapy should not be used in the routine management of elderly patients with disease of the vertebrobasilar arteries.

Antiplatelet agents are easier to administer and have fewer side-effects. Several studies of these in patients with transient ischaemic attacks suggest that subgroups with vertebrobasilar disease also derive benefit from treatment (Canadian Cooperative Study Group 1978; Webster and Lewin 1983). Interim findings from a study of transient ischaemic attacks involving all sites suggest that, if aspirin is used, the dose should be 300 mg daily (UK-TIA Study Group 1988).

Various surgical procedures have been used to improve the vertebrobasilar circulation including endarterectomy and the forming of anastomoses between the extra- and intracranial circulations. Benefits are uncertain.

CAROTID ARTERY STENOSIS

The diagnosis, investigation, and treatment of stenosis of the carotid artery are described in Section 9.

AORTIC STENOSIS

This condition may cause dizziness, syncope, or transient ischaemic attacks. Such episodes may follow exercise or a change in posture when the cardiac output is unable to change in response to an increased demand. They also may occur spontaneously as a result of a transient arrhythmia or atrioventricular block.

SUBCLAVIAN STEAL

This is a rare but clinically intriguing condition in which stenosis of the subclavian artery proximal to the origin of vertebral artery results in a diversion or 'steal' of blood from the opposite vertebral artery, or from the circle of Willis via the basilar artery. The classical features are dizziness or the symptoms of brain-stem anoxia precipitated by physical activity in the affected arm. Associated signs are bruit over the stenotic artery, and an absent or diminished radial pulse.

Stenosis of the subclavian artery is a fairly common finding in elderly patients, but is only associated with a haemodynamic diversion in around 1 in 20 cases. Even fewer go on to develop symptoms of transient brain-stem ischaemia or a full-blown stroke.

Both the stenosis and the 'steal' can be identified by continuous wave Doppler ultrasonography. As surgery has no effect upon survival in patients with stenosis of the subclavian artery, it should be reserved for those with symptoms of a 'steal'. It is of no value if the stenosis or 'steal' is asymptomatic.

The defect can be treated by a bypass operation in which the affected axillary artery is connected to the other one by a Dacron, reversed vein, or polytetrafluorethylene graft tunnelled across the sternum. The haemodynamic consequences of this are impressive, with 90 per cent of grafts remaining patent over 10 years. An alternative in subclavian stenosis is percutaneous transluminal angioplasty. If dilatation of the stenosis is successful, subsequent restenosis is rare. The procedure is ineffective if the artery is completely occluded, and this can only be corrected by a bypass operation.

CEREBELLAR ISCHAEMIA

A cerebellar infarction sometimes has a dramatic onset with vomiting, ataxia, vertigo, and headache associated with truncal ataxia, ipsilateral gaze palsy, and a lower motor neurone, facial palsy. Around 80 per cent of cases, however, present with a vague history of dizziness and vertigo, while around a quarter have no focal cerebellar signs. Lesions of the cerebellar vermis, in particular, produce an atypical picture, with vertigo, downbeat nystagmus, and truncal ataxia, but no signs of past pointing, intentional tremor, or dysdiadokokinesia.

Computerized tomography is useful in identifying a haemorrhage or an infarct of 7 to 10 days duration, but usually misses one of recent onset. Magnetic resonance imaging provides much better definitions of small cerebellar lesions, and is successful in picking up infarcts within the first 24 h.

It has been argued that some cases of recurrent vertigo could be a manifestation of recurrent transient ischaemic attacks involving the cerebellum. It would be extremely difficult to confirm this, but if the diagnosis was suspected it would seem reasonable to provide prophylactic treatment with 300 mg of aspirin daily (UK-TIA Study Group 1988).

DROP ATTACK

This condition is defined as being a fall associated with a sudden loss of strength and muscle tone in the legs. The sufferer may be investigated for dizziness because he or she has difficulty in describing the nature of the attacks, and agrees with the clinician when asked a leading question about the sensation of dizziness. More detailed questioning should elicit the information that the attack was not preceded by any sensations in the head, and that there was no loss of consciousness during attacks.

EPILEPSY

The aura preceding a fit of any kind may produce a sensation of vertigo but one of the manifestations of temporal lobe epilepsy may be vertigo. There are also cases in which diseases of the labyrinth or middle ear provoke attacks of vertigo which are associated with grand mal seizures (vestibulogenic epilepsy; see Chapter 18.11).

ANXIETY NEUROSIS AND DEPRESSION

Patients suffering from an anxiety neurosis may complain of dizziness along with a wide range of other symptoms. This is at its most florid in a panic attack. The patient may give a graphic account of a feeling of constriction in the chest or pit of the stomach, followed by palpitations, sweating, suffocation, light-headedness, and unsteadiness. These often engender a sensation of extreme terror, and a conviction of impending death. A detailed history and physical examination are necessary to differentiate these symptoms from those associated with serious cardiovascular disease. It is a matter of fine judgement as to whether intensive laboratory investigation will reassure a patient if the outcome is negative, or convince him or her that there must be something seriously wrong if the clinician is going to such lengths to find a disorder. Sometimes symptoms are the result of a metabolic derangement associated with hyperventilation (see below).

More recently recognized are symptoms associated with discontinuation of long-term treatment with tranquillizers, particularly benzodiazepines. These have been estimated as occurring in between 14 and 44 per cent of cases, and frequently persist for more than 6 months. Amongst a range of symptoms, patients experience an increase in sensory awareness, gastrointestinal disturbance, headache and insomnia, vertigo, and a sense of imbalance. The recommended approach is to withdraw the benzodiazepines gradually in steps equivalent to a dose of 2.5 mg of diazepam or less. This should be backed up with a high level of support, counselling, and even psychotherapy.

Elderly patients with depression frequently present with physical symptoms (Chapter 20.2). Detailed questioning may be required to tease out the depressive elements of an illness which is causing severe physical incapacity or associated with extreme anxiety. Antidepressant therapy may complicate the picture further by causing ataxia associated with sedation, or light-headedness associated with postural hypotension.

CAROTID SINUS HYPERSENSITIVITY

The healthy response to massage of the carotid sinus is slowing of the pulse rate and a marginal fall in the blood pressure. In some individuals the response is accentuated and associated with symptoms of dizziness or syncope.

In old age, however, there is an actual decline in the sensitivity of the carotid sinus and over the age of 60 only 25 per cent of men and 20 per cent of women show even a marginal reduction in pulse rate in response to sinus massage (Mankikar and Clark 1975). If there is hypersensitivity in old age, however, it is much more likely to cause symptoms than in youth. This may relate to the effects of fibrous tissue degeneration and atherosclerosis on the ability of the cerebral circulation to adjust to changes in blood flow.

Carotid sinus hypersensitivity presents with recurrent episodes of dizziness or syncope which may or may not be associated with manoeuvres such as sudden movements of the hand and neck, or wearing a tight collar. The diagnosis is confirmed by the test described earlier. Symptoms associated with asystole or a fall in blood pressure during the test are usually less severe than those occurring spontaneously. This is because the patient is usually investigated while recumbent, whereas attacks usually occur while standing. Nonetheless, if a patient does not experience any symptoms during a positive test, it is likely that the hypersensitivity is a coincidental finding, and unrelated to attacks of dizziness or syncope.

The management of the disorder is dependent upon whether or not the patient experiences recurrent attacks. In one review, two-thirds of patients with a single attack suffered no further symptoms (Sugue *et al.* 1986). In the remainder who suffer from episodes of asystole, treatment with an anticholinergic agent may be tried. An example is propantheline given in a dose of 15 to 30 mg, three times daily. This treatment is often only partially effective and fails to eliminate attacks completely. It should be reserved for patients in whom more aggressive therapy is inappropriate.

The most effective approach to recurrent episodes of asystole is to insert a cardiac pacemaker (Sugue *et al.* 1986). A simple, ventricular-demand pacemaker is usually effective in achieving permanent relief of symptoms.

The minority of patients with a hypotensive response to hypersensitivity requires a different approach to treatment. One is to use the sympathomimetic agent ephedrine in a dose of 15 to 30 mg, three times daily. This is of only limited efficacy and is likely to cause arrhythmias or infarction in patients with myocardial ischaemia. Treatment with a ventricular-demand pacemaker is ineffective and, indeed, often increases the frequency and severity of attacks. This paradox is thought to be due to the atria contracting against an asynchronous contraction of the ventricles. This increases the intra-atrial pressure to stimulate a stretch reflex, which in turn provokes an episode of hypotension. The problem is resolved by using atrioventricular segmental pacing so that atrial and ventricular contractions are co-ordinated.

Even more aggressive forms of treatment include transection of the glossopharyngeal nerve, resection of Hering's nerve, stripping of the carotid sinus plexus, or irradiation of the sinus. Such therapy may be appropriate in young patients with severe symptoms unrelieved by pacing, but is rarely relevant in the treatment of elderly people.

SICK SINUS SYNDROME

Attacks of dizziness or syncope associated with a persistent arrhythmia or complete heart block present no major diagnostic problems. Transient arrhythmias are much more difficult to identify.

These are frequently related to dysfunction of the sinoatrial node. This may be associated with abnormal impulses within the node, or impaired conduction from the node to the rest of the atrium. A wide range of cardiac disorders may cause this, but the more important in old age are ischaemia, chronic rheumatic heart disease, and local amyloidosis. Fibrotic infiltration associated with ageing may also interfere with nodal function.

The abnormality may be accentuated or precipitated by a variety of drugs that alter myocardial irritability or conductivity. The most important of these are the cardiac glycosides and β-adrenergic blocking agents, and potassium supplements and sparing agents associated with hyperkalaemia. Others such as methyldopa, cimetidine, and calcium-channel antagonists are less frequently implicated, but should be avoided in patients known to have dysfunction of the sinus node.

Sinus node dysfunction is associated with a variety of transient or persistent arrhythmias (Table 2). A sinus arrhythmia is physiological. Indeed one of the features of degeneration associated with ageing is a decline in beat-to-beat variation in the pulse rate. In sinus node dysfunction the arrhythmia may be accentuated to the extent that there are pauses of more than 2 s and the average heart rate is less than 40 beats per minute. A more common abnormality is a sinus bradycardia. This rarely causes symptoms unless the heart rate falls below 40 beats per minute. A bradycardia may also be the result of an ectopic atrial focus with a low rate.

Table 2 Abnormal rhythms associated with the sick sinus syndrome

Sinus arrhythmia	Sinus pauses
Sinus bradycardia	Bradycardia–tachycardia syndrome
Slow atrial rhythm	Atrial fibrillation

A more complex arrhythmia is associated with sinoatrial block or sinus arrest. This may produce wide variations in conduction ratios, or Wenckebach block. At a more advanced stage in the disorder the sinus pause is followed by a supraventricular escape rhythm. This produces intermittent episodes of bradycardia and tachycardia, in which the tachycardia may take the form of atrial flutter, atrial fibrillation, or ectopic atrial or junctional tachycardias. With further degeneration of the sinus node the rhythm progresses to that of permanent atrial fibrillation, a situation in which it is at least clear that an abnormal rhythm is present.

The initial investigation should be that of a standard, 12-lead ECG. This may show P-wave abnormalities, or record the heart during a period of arrhythmia. Frequently, however, it fails to elicit any diagnostic abnormalities.

It is then reasonable to make an ambulatory electrocardiographic recording. The major limitation of this technique is that even young healthy adults show a range of arrhythmias. Amongst the more common of these are a sinus bradycardia, first- and second-degree atrioventricular block, sinus pauses, and nodal escape rhythms. In old age, cardiac arrhythmias are even more common and several studies have found that they were only marginally more common in patients with attacks of dizziness than elderly, asymptomatic controls. Table 3 details the prevalence of arrhythmias in men and women aged over 80 years who had no clinical evidence of cardiovascular disease (Kantelip *et al.* 1986). As a consequence, it is extremely difficult to link an arrhythmia to symptoms, and a diagnosis is often made by taking a standard, 12-lead ECG during an attack of dizziness or syncope.

If patients with the sick sinus syndrome are subjected to carotid sinus massage a proportion will show evidence of sinus hypersensitivity. It is uncertain whether there is a direct relationship between the two conditions, or if they merely occur coincidentally. In terms of the practical management of patients, the issue is of little importance.

Table 3 Prevalence (in percentages) of arrhythmias recorded by 24-h tape in men and women over the age of 80 years (from Kantelip *et al.* 1986)

Supraventricular tachycardia	28
Nocturnal sinus pauses	12
Ventricular extrasystoles	32
Multifocal ventricular extrasystoles	18
Supraventricular extrasystoles	100
Couplets	8
Run of ventricular extrasystoles	2

The only effective treatment for sinus node dysfunction is to insert a cardiac pacemaker. Although there are theoretical advantages in using an atrial pacemaker, many patients have concurrent defects of atrioventricular conduction, so that the simplest approach is to fit a ventricular demand instrument. It is likely that in the future more effective treatment will involve the use of multiprogrammable, multinode, and dual chamber pacemakers.

There is evidence that hydralazine in a dose of 25 to 50 mg twice daily reduces the frequency of symptoms in patients with sinus node dysfunction. The effect of this is unpredictable, so that it should only be used in the rare situation where cardiac pacing is inappropriate.

POSTURAL HYPOTENSION

This potent and common cause of dizziness in elderly people is discussed in Chapter 18.10.

METABOLIC DISORDERS

Hyperventilation

One of the most elusive causes of dizziness is hyperventilation. Although the stereotype is that of a young, anxious woman, it also is encountered in elderly people.

Hyperventilation is usually due to anxiety. The sequence is that irrational stress leads to anxiety which leads to an increased respiratory rate which leads in turn to an increase in anxiety and so on (Brashear 1984). Several physical disorders, such as renal or liver failure, organic brain disease, diabetic ketoacidosis, pulmonary embolism and interstitial pulmonary fibrosis, may cause hyperventilation, but, in these circumstances, symptoms of the underlying disorder usually outweigh those of the hyperventilation.

The patient is usually unaware of a rapid respiratory rate, and classical symptoms include paraesthesiae, dizziness, unsteadiness and blurred vision, chest pain, and dry mouth, sweating, tremor and tachycardia (Lum 1982). The cause for all of these is a respiratory alkalosis. Chest pain may be due to a forceful heartbeat striking a hypersensitive precordium, or to aching from overexercised intercostal muscles. This is often associated with tender spots which can be palpated over the precordium. Dizziness, unsteadiness, and blurred vision may result from cerebral vasoconstriction, while paraesthesiae are a direct effect of alkalosis on peripheral nerves. Other symptoms are an effect of the increased release of catecholamines, and it can be difficult to establish whether these precede or follow the metabolic disorder. Deliberate overbreathing, however, induces symptoms of anxiety, and, if continued may cause symptoms of depersonalization and hallucinations.

Although symptoms are due to a respiratory alkalosis, the biochemical confirmation of this can be elusive, and a more practical approach is to make the diagnosis on the basis of a cluster of appropriate signs and symptoms, ensuring, of course, that other important disorders have been excluded.

Symptoms can often be improved by discussing the nature of the disorder with the patient. Explanation and reassurance help to break the vicious circle of anxiety and rapid respiration (Brashear 1984). More complex regimens involving behaviour therapy and biofeedback have been reported, but require a high level of co-operation and motivation, even if they are locally available. The traditional approach of correcting the alkalosis by asking patients to breathe in and out of a bag may be undignified, uncomfortable, or alarming, and is rarely effective.

If reassurance fails, treatment with a β-adrenergic blocking agent, for example, propranolol, 40 mg twice daily, can be tried. A more aggressive approach is to give a tricyclic antidepressant such as imipramine, starting off at a dose of 25 mg daily and progressively increasing this to a maximum of three times daily or until side-effects intervene. Benzodiazepines are best avoided in this chronic and refractory condition.

Hyperglycaemia/hypoglycaemia

Patients with diabetes mellitus are more likely to suffer from attacks of dizziness than are healthy people. The association is related to a variety of factors which include coincidental cerebrovascular disease, postural hypotension associated with an autonomic neuropathy, or dehydration precipitated by an overnight osmotic diuresis.

The effects of hypoglycaemia are well recognized. Less clear-cut is the relationship between insulin and dizziness. It may be that, in old age, insulin has a direct effect on autonomic function, provoking postural hypotension (Page and Watkins 1976). In practical terms, this means that if an elderly diabetic patient on insulin has recurrent bouts of dizziness, these are not necessarily due to episodes of hypoglycaemia.

OTOLOGICAL DISORDERS

Readers requiring a detailed account of the aetiology, diagnosis, and treatment of disorders affecting the inner ear are referred to general textbooks on the subject, and to a work on the problem as it relates to the elderly edited by Hinchcliffe (1983). The following sections only cover issues of particular relevance to the practical management of elderly patients presenting with dizziness.

Positional vertigo

In this condition, attacks of dizziness are provoked by changes in the position of the head. Attacks may be induced by turning in bed, rising from bed, bending and stretching, extending the neck, or reaching for a high shelf, but the history of this is often less specific so that a standard test for positional nystagmus and vertigo described earlier should be made on all elderly patients in whom another obvious cause for dizziness is not apparent. Despite claims that the condition is diagnostic of inner ear disease, conditions involving the brain-stem are often implicated (Baloh *et al.* 1987). Table 4 gives details of patients of all ages attending a neuro-otology clinic and shows that, while the majority suffered from inner ear disorders, around 1 in 20 had a vertebrobasilar disorder. It is likely that this proportion would have been substantially higher if attention had been focused on the elderly patient. Given the degree of diagnostic uncertainty, it is important that patients with benign postural nystagmus should have bithermal caloric testing and electronystagmography to exclude disorders of the brain-stem.

Table 4 Diagnoses in 240 patients with benign paroxysmal positional vertigo (Baloh *et al.* 1987)

Disorder	Number	%*
Idiopathic	118	49
Post-traumatic	43	18
Viral neurolabyrinthitis	37	15
Vertebrobasilar insufficiency	11	5
Menière's disease	5	2
Postsurgical (general)	5	2
Postsurgical (ear)	5	2
Ototoxicity	4	2
Others	12	5

* Percentages taken to nearest whole number.

In most patients with benign positional vertigo the condition is labelled as being idiopathic. The probable explanation is that there is degeneration of the otolithic membrane of the utricle so that otoconia are released (Schnuknecht 1969). These are deposited on the cupulae within the posterior duct. At this site, changes in the position of the head result in the otoconia stimulating sensors within the cupula, so inducing a sensation of vertigo. The condition is usually self-limiting, resolving after a period of several weeks or months as the offending particles are cleared from the endolymph. In a few patients the otoconia become firmly attached to the cupula so that the condition persists indefinitely.

A review of patients referred for investigation suggests that the peak incidence of the idiopathic form of the disorder is between 50 and 69 years in both men and women (McClure 1986). This might be an artefact, however, as very old people are less likely to be referred to specialist clinics.

Post-traumatic vertigo may be the result of a fall downstairs or backwards on ice in which an elderly person strikes the occiput. Shearing forces are transmitted to the temporal bone with dislodgement of otoconia within the utricle (Pearson and Barber 1973).

Viral neurolabyrinthitis may be distinguished from other vestibular disorders by its acute onset and its association with cochlear symptoms such as severe tinnitus. The more acute symptoms usually settle within a few days, but an unfortunate minority of patients is left with a persistent feeling of malaise and unsteadiness. Bacterial infection should be suspected if there is a long history of middle ear disease.

Though Menière's disease is a much less common cause of positional vertigo it also has a peak incidence between the ages of 50 and 69 years (McClure 1986; Baloh *et al.* 1987). Characteristic features are impairment of hearing associated with episodes of vertigo usually lasting several hours. The high prevalence of presbyacusis in elderly people limits the value of hearing loss as a discriminating sign in Menière's disease. The condition is occasionally associated with recurrent drop attacks, postulated to result from recurrent stimulation of the cupula of the affected ear.

The first essential in treatment of positional vertigo is an accurate diagnosis. The administration of labyrinthine sedatives to old people whose dizziness is unrelated to inner ear disease is likely to make their ataxia worse, and increase rather than diminish their risk of falls. They also are responsible for a high incidence of drug-induced parkinsonism and tardive dyskinesia.

The least harmful approach is to recruit the patient into a vestibular training programme in which he or she is trained to use visual and proprioceptive information to replace a loss of vestibular sensation (Table 5; Baloh 1983). Obvious limitations to the schedule are that it presupposes that vision and position sense are intact; that the patient is ambulant, supple, and fit; and that he is highly motivated and able to remember a complex sequence of instruction. It is of little value to the typical patient in a geriatric day hospital with poor vision and ataxia, crippled with arthritis, and suffering from a mild degree of dementia.

Table 5 Exercises for patients with positional vertigo (Oosterveld 1985)

Exercises in bed
Eye movements up and down; eye movements side to side; focus on moving finger—perform slowly at first and then more rapidly
Head movements backwards and forwards; then turning from side to side—perform slowly at first and then more rapidly—perform first with eyes open and then with eyes closed

Exercises sitting
Eye and head movements as above
Shoulder shrugging and circling
Bending to pick objects from the ground

Exercises standing
Eye, head, and body movements as above
Stand up from sitting, first with eyes open and then with eyes closed
Throw ball from hand to hand above head, and then under one knee
Stand up from sitting position, turning round each time

Exercises walking
Walk across room with eyes open and then closed
Walk up and down on slope with eyes open and then closed
Walk up and down stairs with eyes open and then closed
Walk while throwing and catching a ball
Play a ball game involving stooping, stretching, and aiming

Do three sessions per day, each lasting 5 min

If vertigo is clearly due to an otological disorder, if more conservative measures have been ineffective, and if the symptom is sufficiently severe to run a high risk of side-effects, then drug treatment should be considered (Oosterveld 1985).

Antihistamines are particularly effective as labyrinthine suppressants. They include cinnarizine, cyclizine, dimenhydrinate, meclozine, and promethazine. All have general, sedative side-effects, but cinnarizine probably has a more selective effect on the vestibular systems than the others. It should be prescribed in a dose of 30 mg, three times daily.

Phenothiazines have also been used in the treatment of vertigo. In addition to sedating patients they produce the whole gamut of anticholinergic effects. Even more worrying in old age, is the frequency with which they cause parkinsonism. Drugs in this group include prochlorperazine and thiethylperazine.

A further approach is to improve the circulation of the labyrinthine area with vasodilators. Drugs used for this purpose include nicotinic acid, cyclandelate, and piracetamine (Oosterveld 1985). More recently, betahistine has been recommended as the drug of choice in this situation. The standard maximum dose is 16 mg, three times daily, but additional benefit may be obtained by increasing this to 24 mg, three times daily.

Surgery has sometimes been successful in relieving symptoms where positional vertigo is severe and continuous (Møller *et al.* 1986). The problem is rare, however. A variety of operations have also been devised for the treatment of Menière's disease, but there is considerable doubt about their efficacy. Old people more likely to benefit from surgery are those with vertigo secondary to long-standing disease of the middle ear.

Acoustic neuroma

The unlikelihood of dizziness being caused by an acoustic neuroma is illustrated by the fact that it was diagnosed in only 1 out of 240 patients presenting with benign positional vertigo (Baloh *et al.* 1987). None the less, the consequences of missing the disorder are disastrous, so that any elderly patients with a suspicious history should be adequately investigated by audiometry, caloric testing, and electronystagmography. Unfortunately, a high prevalence of presbyacusis and other causes of dizziness in old age may make it difficult to screen out patients with characteristic symptoms. There is no easy answer to the problem.

Ototoxicity

Old people are at particular risk from the ototoxic effects of aminoglycosides. Large doses of aspirin may also cause deafness and ataxia, but the fact that this has largely been superseded by less toxic, non-steroidal anti-inflammatory agents means that this is now uncommon (Ballantyne 1970). An even rarer cause of inner ear dysfunction is electrolyte imbalance associated with a massive dose of loop diuretic.

MISCELLANEOUS CAUSES OF DIZZINESS

Elderly patients in whom visual defects are superimposed on an underlying proprioceptive defect are particularly ataxic. The problem may be accentuated if, after cataract surgery, their peripheral vision is distorted by thick convex lenses.

Cervical spondylosis

In cats, elimination of the sensory input from the articular capsules of the cervical vertebrae causes severe ataxia, suggesting that damage to these joints could have a catastrophic effect on balance in humans (McClure 1986). The obvious inference is that cervical spondylosis is an important cause of dizziness and ataxia in old age. A major obstacle to establishing the hypothesis is that cervical spondylosis is almost universal in old people so that there is no way of comparing balance in elderly patients with and without the disorder. One of the few practical issues is that, in this situation, immobilization of the neck with a cervical collar would be likely to accentuate rather than alleviate dizziness and ataxia.

Cough syncope

This condition is most commonly encountered in middle-aged men with chronic obstructive airways disease, who suffer from attacks of dizziness or syncope after severe bouts of coughing (De Mana *et al.* 1984). Electroencephalographic monitoring of these attacks suggests that they follow a rise in the pressure of the cerebrospinal fluid which interferes with the cerebral circulation. Treatment should be directed at achieving weight loss and alleviating respiratory symptoms with appropriate courses of antibiotics. One of the most useful measures is to persuade the sufferer to stop smoking.

Micturition syncope

Elderly patients who rise to pass urine during the night may suffer attacks of dizziness or syncope once voiding is complete (Kapoor *et al.* 1985). The condition often afflicts patients who already suffer from postural hypotension, and occurs at night when sympathetic activity is at its lowest ebb. Decompression of the bladder stimulates vagal activity, which further compresses an already marginal cerebral circulation.

There is little information on the effective treatment of the disorder, possibly because attacks are usually spasmodic and thus difficult to monitor. A reasonable approach might be to suppress vagal activity with an agent such as propantheline in a dose of 30 mg each evening, but controlled study of this would be difficult to organize.

Haematological abnormalities

Severe anaemia, polycythaemia, and disorders associated with hyperviscosity, such as multiple myeloma, may all be associated with dizziness or vertigo (Luxon 1984).

References

Adams, K.R.H., Yung, M.W., Lye, M., and Whitehouse, G.H. (1986). Are cervical spine radiographs of value in elderly patients with vertebrobasilar insufficiency? *Age and Ageing*, **15**, 57–9.

Ausman, J.I., Shrontz, C.E., Pearce, J.E., Diaz, F.G., and Crecelius, J.L. (1985). Vertebrobasilar insufficiency—a review. *Archives of Neurology*, **42**, 803–8.

Ballantyne, J. (1970). Iatrogenic deafness. *Journal of Laryngology*, **84**, 969–1000.

Baloh, R.W. (1983). The dizzy patient. Symptomatic treatment of vertigo. *Postgraduate Medicine*, **73**, 317–214.

Baloh, R.W., Honrubias, V., and Jacobson, K. (1987) . Benign positional vertigo: clinical and oculographic features in 240 cases. *Neurology*, **37**, 371–8.

Belal, A. and Glorig, A. (1986). Dysequilibrium of ageing (presbystasis). *Journal of Laryngology and Otology*, **100**, 1037–41.

Brashear, R.E. (1984). Hyperventilation syndrome: managing elderly patients. *Geriatrics*, **39**, 114–25.

Canadian Cooperative Study Group (1978). A randomised trial of aspirin and sulphinpyranzone in threatened stroke. *New England Journal of Medicine*, **299**, 53–9.

Clarke, A.N.G. (1987). Postural hypotension in the elderly. *British Medical Journal*, **295**, 683.

De Mana, A.A., Westmoreland, B.I., and Sharborough, I.W. (1984). EEG in cough syncope. *Neurology*, **34**, 371–4.

Factor, S.A. and Detinger, M.P. (1987). Early brainstem auditory evoked responses in vertebrobasilar transient ischaemic attacks. *Archives of Neurology*, **44**, 544–7.

Fowler, H.M.A. (1984). Dizziness and vertigo. *British Medical Journal*, **288**, 1739–43.

Gabell, A. and Simons M.A. (1982). Balance coding. *Physiotherapy*, **68**, 286–8.

Hinchcliffe, R. (ed.) (1983). *Hearing and balance in the elderly*. Edinburgh, Churchill Livingstone.

Kantelip, J.P., Sagee, E., and Duchene-Marullaz, P. (1986). Findings on ambulatory electrocardiographic monitoring in subjects older than 80 years. *American Journal of Cardiology*, **57**, 398–401.

Kapoor, W.N., Peterson, J.R., and Karpf, M. (1985). Micturition syncope. A reappraisal. *Journal of the American Medical Association*, **253**, 796–8.

Leading article (1988). Cerebellar stroke. *Lancet*, **i**, 1031–2.

Lo, L.Y., Ford, C.S., McKinney, W.M., and Toole, J.F. (1986). Asymptomatic bruit, carotid and vertebrobasilar transient ischaemic attacks—clinical and ultrasonic correlation. *Stroke*, **17**, 65–8.

Lum, L.C. (1982). Hyperventilation syndromes in medicine and psychiatry: a review. *Journal of the Royal Society of Medicine*, **80**, 229–321.

Luxon, L.M. (1983). Dizziness in the elderly. In *Hearing and balance in the elderly* (ed. R. Hinchcliffe), pp. 402–52. Edinburgh, Churchill Livingstone.

Luxon, L.M. (1984). 'A bit dizzy'. *British Journal of Hospital Medicine*, **32**, 315–21.

Mankikar, G.D. and Clark, A.N.G. (1975). Cardiac effects of carotid sinus massage in old age. *Age and Ageing*, **4**, 86–94.

McClure, J.A. (1986). Vertigo and imbalance in the elderly. *Journal of Otolaryngology*, **15**, 248–52.

Møller, M.B., Moller, A.R., Janetta, P.J., and Sekhar, L. (1986). Diagnosis and surgical treatment of disabling positional vertigo. *Journal of Neurosurgery*, **64**, 21–8.

Mouzfarrig, N.A., Little, J.R., Fourlan, A.J., Leatherman, J.R., and Williams, G.W. (1986). Basilar and distal vertebral artery stenosis: long term follow up. *Stroke*, **17**, 938–42.

Oosterveld, W.J. (1985). Vertigo—current concepts in management. *Drugs*, **30**, 275–83.

Page, M.M. and Watkins, P.J. (1976). Provocation of postural hypotension by insulin in diabetic autonomic neuropathy. *Diabetes*, **25**, 90–5.

Pearson, B.W. and Barber, H.O. (1973). Head injury. Some otoneurologic sequelae. *Archives of Otolaryngology*, **97**, 81–4.

Ringlestein, E.B., Zeimer, H., and Pock, K. (1985). Non-invasive diagnosis of intracranial lesions in the vertebrobasilar system. A comparison of Doppler sonographic and angiographic findings. *Stroke*, **16**, 848–55.

Schnuknecht, H.I. (1969). Cupulothisis. *Archives of Otolaryngology*, **90**, 113–26.

Sharma, J.C. and MacLennan, W.J. (1988). Causes of ataxia in patients attending a falls laboratory. *Age and Ageing*, **17**, 94–102.

Sugue, D.D., Gersh, B.J., Holmes, D.R., Wood, D.L., Osborn, M.J., and Hammill, S.C. (1986). Symptomatic 'isolated' carotid sinus hypersensitivity: natural history and results of treatment with anticholinergic drugs or pacemaker. *Journal of the American College of Cardiologists*, **7**, 158–62.

UK-TIA Study Group (1988). United Kingdom transient ischaemic attack (UK-TIA) aspirin trial: interim results. *British Medical Journal*, **296**, 316–20.

Venna, N. (1986*a*). Dizziness, falling, and fainting: differential diagnosis in the aged (part 1). *Geriatrics*, **41**, 30–6, 39, 42.

Venna, N. (1986*b*). Dizziness, falling and fainting: differential diagnosis in the aged (part II). *Geriatrics*, **41**, 31–3, 36–7, 41–5.

Webster, B.B. and Lewin, M. (1983). Anticoagulation in cerebral ischaemia. *Stroke*, **14**, 658–63.

Wishart, J.P., Cartlidge, N.E., and Elveback, L.R. (1978). Carotid and vertebro-basilar transient ischaemic attacks: Effects of anticoagulants, hypertension, and cardiac disorders on survival and stroke occurrence—a population study. *Annals of Neurology*, **3**, 107–15.

Wyke, B. (1979). Cervical articular contributions to posture and gait: their relation to senile disequilibrium. *Age and Ageing*, **8**, 251–9.

18.8 Sleep disorders

DONALD L. BLIWISE, RALPH A. PASCUALY, AND WILLIAM C. DEMENT

With rare exceptions, the prevalence of nearly all primary sleep disorders increases with age, and sleep disturbance is one of the most common problems encountered by the geriatrician. In this chapter we will summarize the current state of knowledge of the major sleep disorders in old age: sleep apnoea; restless legs (Ekbom's) syndrome/periodic leg movements; insomnia; sleep disturbance of dementia (sundowning); and narcolepsy. First, however, we will describe the clinical approach to evaluation of elderly patients with known or suspected sleep disorders.

TAKING THE HISTORY

It is useful to categorize a patient's sleep complaints into disorders of initiating or maintaining sleep (DIMS), disorders of excessive sleepiness (DOES), or parasomnia disorders, reflecting events or symptoms occurring during the sleep period (ASDC 1979).

Taking a thorough clinical history of a significant sleep-related complaint in an elderly patient is time consuming, and enough time should be allotted to allow a thorough medical and psychiatric review. Subjective reports of both difficulty in sleeping or daytime sleepiness are characteristically inaccurate and every effort should be made to have the spouse, bed-partner, or live-in companion present during some part of the consultation. This informant should be questioned about snoring, irregular breathing, excessive sleepiness, sleep habits, unusual nocturnal behaviours during sleep, changes in cognitive or emotional function associated with sleep disturbance, use of drugs and alcohol, and general psychosocial factors. It must be kept in mind, however, that partners who sleep in separate beds or bedrooms, who may be hearing impaired, who are taking sedating drugs, or who are unusually sound sleepers could be unreliable historians.

The clinical history should be organized as a review of the 24-h sleep/wake cycle, beginning with the night-time routine. Questions should be asked about variability in the daily cycle from day to day, weekday to weekend, home versus travel, and workday versus vacation time. The timing of evening meals and drinks, medications, use of caffeinated products, and the presence of environmental noise or discomfort should be established.

Establishing the onset of the complaint and any associated changes in health or psychosocial function that occurred at the time is particularly important. Patients who have complaints of DIMS and DOES may appear to have symptoms of relatively recent onset, but which, upon further questioning of an informant, may reveal a long-standing and chronic episodic problem that has become more aggravated in the recent past. The presence of a long-standing complaint should neither be discouraging nor imply a functional problem. Individuals with sleep disorders often see many physicians over many years before a definitive evaluation is made.

It cannot be overemphasized that the investigation of sleep complaints in older persons must include undiagnosed medical and psychiatric conditions. This is particularly true for the latter where elderly patients with a focus on a sleep complaint may be somatizing psychiatric difficulties in a manner similar to that of the patient with a chronic pain or other physical complaint. In addition, the role of suboptimal management or exacerbation of pre-existing conditions must be considered. The treatment of specific symptoms of known diseases may restore the normal sleep pattern or contribute to an overall improvement along with other interventions. Examples of this would involve the aggressive management of rheumatoid arthritis, congestive heart failure, Parkinson's disease, or chronic obstructive lung disease.

Evaluation of daytime somnolence

The geriatrician must understand that daytime sleepiness is an important medical symptom and should not be dismissed with superficial explanations. Common, erroneous diagnoses of clinically significant sleepiness include ageing, boredom, retirement, leisure, nap behaviour, poor night-time sleep habits, depression, or the sequelae of stroke. Patients often deny or minimize clinically significant sleepiness and misinterpret it as lethargy, fatigue, or lack of energy. In all cases of sleepiness, a corroborating history should be obtained. Only in the most obvious cases are direct questions such as 'Are you abnormally sleepy?' likely to provide clinically useful information. History taking should focus on the presence of drowsiness during specific day-to-day activities, as reported by the patients and observed by the informant. Certain clinical cues will identify the seriously sleepy patient. These individuals have drowsiness that is persistent from day to day. Others will see them fall asleep spontaneously in quiet situations such as reading, watching television, or travelling as a passenger. They may report or be observed to have difficulty maintaining wakefulness while driving for less than 1 h. These individuals may also take frequent naps during the day, which may be erroneously characterized as a positive habit. Obvious severe pathological sleepiness is indicated by a history of falling asleep while driving reasonable distances, while eating, while on the toilet, during conversation with others, or while waiting on the phone.

It should never be assumed that the presence of clinically severe, excessive sleepiness is caused by trouble sleeping at night (DIMS). In general, patients with severe complaints of being unable to sleep well at night do not have severe excessive sleepiness, although they may complain of fatigue or weariness. A key factor is whether the individual with DIMS is actually able to nap substantially during the daytime. Patients are seldom able to achieve sound sleep at any point during the 24-h day.

It is important to distinguish daytime sleepiness from hypersomnolence. Hypersomnolence refers to prolonged total sleep times during a 24-h period, with or without sleepiness. Patients may have persistent complaints of excessive sleepiness and drowsiness while maintaining a relatively normal total sleep time during a 24-h period. Conversely, prolonged total sleep times without daytime sleepiness when awake may suggest a primary psychiatric diagnosis and is not typical of primary sleep disorders or neurological conditions.

Evaluation of insomnia

The clinical approach to the evaluation of insomnia is made more difficult by the fact that insomnia is a vague term that is used to refer to a variety of sleep-related symptoms. There is even some evidence that an individual's use of the label 'insomnia' may be unrelated to the number of hours he or she typically sleeps at night. Classically, insomnia is broken down into difficulty with falling asleep at the beginning of the night (initial insomnia), difficulty in returning to sleep after nocturnal awakenings, usually during the first two-thirds of the night (middle insomnia), and awakening too early in the morning, usually during the last third of the night (early morning awakening). In practice, many patients have some overlap of these three categories of complaint. Insomnia may also refer to frequent brief awakenings, or complaints of broken, restless sleep, often accompanied by recall of dream fragments. On occasion, a patient may complain of insomnia although there is no subjective sleep disturbance or daytime consequence other than a belief that the sleep period is too short.

Some specific lines of inquiry are worthwhile for each type of insomniac complaint. All patients with *initial insomnia* should be questioned about specific discomfort in the legs or calves that interferes with sleep onset, characterized by restlessness in the legs and relieved by leg movement. A positive answer suggests that a thorough review of restless leg syndrome and periodic leg movements should be pursued. In the absence of such symptoms, it is important to take care to establish the presence and extent of excessive arousal associated with the process of going to sleep. Arousal refers to a psychophysiological process that results in high levels of mental activity or emotional distress, and may include anxiety and depression or other effects less easily characterized, such as demoralization, hopelessness, and despair. Enhanced somatic sensations may include increased pain, over-awareness of normal body sensations such as heartbeat, noise sensitivity, muscle tension, or vague, migrating, physical discomforts. The development of increased psychophysiological arousal may be secondary to environmental factors or poor sleep hygiene, as well as to conditioned responses that the patient is unaware of. Environmental factors may include noise, disturbing bed-partners, or lack of privacy. Sleep hygiene factors encompass poor bedtime habits, late-night eating and drinking, watching late-night, disturbing programmes on television, or personal activities such as arguments, paying bills, or planning for 'the future'. Arousal triggered by conditioning is more difficult to identify, but may be suspected if the sleep disturbance occurs only under certain circumstances and reliably disappears with a change in conditions. More commonly, the patient is conditioned to become aroused by the very process of attempting to sleep, and by repeated failure and apprehension about perceived consequences of sleep loss.

When evaluating a complaint of a middle insomnia, many specific factors associated with or causing awakenings need to be evaluated to establish their relative importance. Conclusions made about causal relationships between specific nocturnal events and nocturnal awakenings can be misleading (i.e., urinary frequency may be secondary to frequent awakenings rather than vice versa). In patients who snore, particular attention should be given to symptoms suggestive of sleep apnoea. In snorers, awakenings associated with snorting, gasping, headache, dyspnoea, choking, chest pain, acid reflux, or night sweats should trigger suspicions of possible sleep apnoea. Broken, fragmented, and restless sleep, with or without leg kicking being reported by the bed-partner or patient, may suggest periodic leg movements. To a lesser extent such movements may be present with sleep-related complaints including unexplained fatigue or drowsiness during the daytime, without the patient having any awareness of a nocturnal sleep disorder.

Early morning awakening has long been considered a hallmark of endogenous affective disorders, but less commonly may also be found in the anxious elderly patient. In practice it is often difficult to distinguish a complaint of middle insomnia from early morning awakening because the patient may remain awake and be unable to return to sleep after awakening at any point during the night. Nonetheless, in patients with this specific complaint, careful consideration should be given to the full range of aetiological factors including sleep apnoea syndrome, which may be substantially aggravated during REM (rapid eye movement) sleep in the early morning hours. Complaints of early morning awakening may also reflect progressively earlier bedtimes in an effort to attempt to achieve longer total sleep times. Some physiologists have explained this phenomenon as an advanced-phase syndrome (see below).

The duration of the insomnia will have practical, clinical, and therapeutic significance. An insomnia of a few weeks' duration suggests a lengthy medical differential diagnosis, while a complaint of chronic insomnia for many years without other associated symptoms may suggest a primary sleep disorder or a chronic psychiatric disorder. Organ system function during the nocturnal hours should be reviewed specifically, including gastrointestinal symptoms, muscular or skeletal discomforts, urinary urgency, cramps, breathing difficulties, palpitations, cough, temperature discomfort, nightmares, and anxiety. Episodes of confusion, falling out of bed, falls, and near falls at night should also be specifically sought. Finally, a thorough psychiatric review is essential, including any past history of depression, anxiety, suicidal behaviour, panic, phobia, and previous use of psychotropic drugs. The review of endogenous symptoms with depression easily follows from open-ended questions about family life, socioeconomic concerns, diet, and daily activity. Dementia may be associated with nocturnal sleep disturbance and daytime sleepiness; therefore, memory function should also be investigated. Although the sleep disturbances associated with dementing illness are typically parasomniac in nature (see below), the clinician evaluating insomnia in elderly people should always investigate changes in memory and intellectual function.

Perhaps most crucial in the evaluation of DIMS is a thorough review of current medications, over-the-counter drug use, alcohol, and illegal drugs. In history taking, attempt to establish previous treatments for sleep disturbance as carefully as possible, including drug dosages and regimens, durations, side-effects, and the patient's general response to them. Inquire specifically about rebound insomnia or a history suggestive of drug habituation or abuse. Attempt to corroborate a history of alcoholism, if suspected. Home remedies and treatments given by various professionals should be specifically asked about. Chronic use of sedative hypnotics may result in chronic sleep disturbance. Habituation leads some patients to increase the dose to restore the therapeutic effect; physical and psychological tolerance then develop. Continuity of sleep, particularly in the second half of the night, is disrupted because the drug rapidly loses its sedative effect in the tolerant individual. Sleep latency may be gradually prolonged and unpredictable so night-to-night variability ensues.

Many other medications can also interfere with sleep; these include aminophylline, sympathomimetics, bronchodilators, antimetabolites, chemotherapeutic agents, thyroid preparations, anticonvulsants, monoamine oxidase inhibitors, adrenocorticotropic hormone, methyldopa and antiparkinsonism drugs. Propranolol and quinidine, and other antiarrhythmics, may produce parasomniac-like disturbances of nightmares and night terrors. On occasion, patients develop insomnia in conjunction with substantial usage of opiates, particularly if their insomnia began in association with an injury. Finally, over-the-counter medications must not be overlooked in the development of insomnia in old age. Of particular interest to the geriatrician are tolerance to sleep aids containing scopolamine and antihistamine, antidiarrhoeal preparations containing belladonna alkaloids, cold and allergy preparations containing sympathomimetics such as pseudoephedrine and phenylpropanolamine, and excessive doses of analgesics containing salicylates. Scopolamine and belladonna alkaloid may produce an anticholinergic toxic psychosis initially manifested by agitation in presenting an acute organic brain syndrome. Sympathomimetics and the early stages of salicylism are associated with high-arousal insomnia.

Evaluation of parasomniac behaviours

Parasomniac behaviours include snoring and gasping for breath (suggestive of sleep apnoea), and nocturnal agitation, violence, wandering, or confusion (typical of some cases of dementia). In addition, episodes of sleep talking, shouting associated with limb movement, falling out of bed, or falling without clear recall of the incident, all require careful assessment to rule out drug reactions, cardiovascular disease, and diseases of the central nervous system, including epilepsy

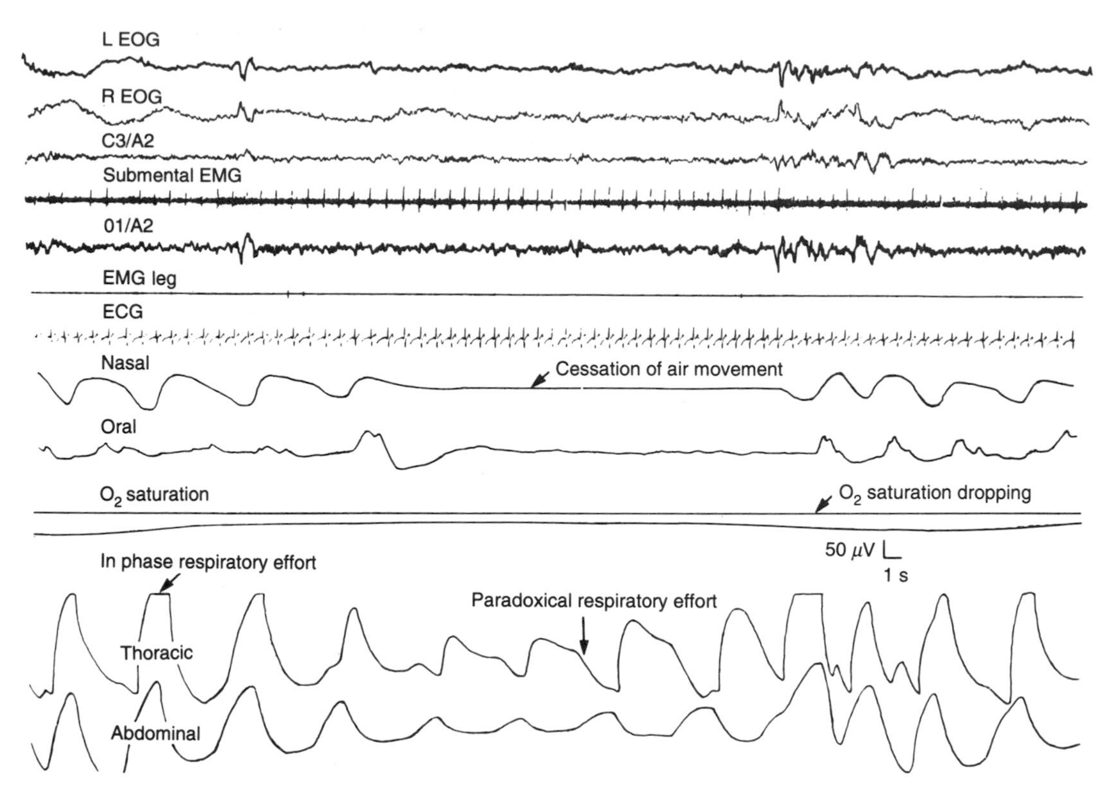

Fig. 1 Typical polysomnographic tracing of sleep apnoea. Notice out-of-phase thoracic and abdominal movements as patient attempts to resume breathing. Additionally, appreciable arousal occurs (top 6 channels) at the end of the apnoea (reproduced by permission from W. Mendelson, *Human sleep*, New York: Plenum, 1987).

and dementias. Night-time somnambulism of new onset should be diagnosed with particular care in the elderly patient. Although this condition is common and easily diagnosed on clinical grounds in children and young adults, a thorough review should be made before reaching this diagnosis in an older individual. If vivid recall of dreams is associated with vigorous physical activity (i.e. diving off the bed while dreaming of jumping out of a burning room), a disorder of REM sleep behaviour should be considered and polysomnographic studies should be made promptly, as these patients are at risk for serious injury. On the other hand, in the absence of such dream recall and in the presence of even mild levels of cognitive impairment, such behaviours may signify the so-called 'sundown syndrome'. Sundowning is a poorly understood disorder that may represent upheaval of the circadian sleep/wake system (see below).

Indications for referral to a sleep disorders centre

In general, disorders of excessive sleepiness and those involving parasomniac behaviours are more likely to benefit from a polysomnographic evaluation in an accredited sleep disorders centre. Referral of a patient with disorders of initiating or maintaining sleep for this evaluation depends largely on the extent to which a primary sleep disorder (periodic leg movement/restless legs syndrome), psychiatric disorders (major affective disorder), or uncorroborated sleep complaints may be involved. In one series of patients with disorders of initiating or maintaining sleep, about half received additional or altered diagnoses on the basis of polysomnography (Jacobs *et al.* 1988). This figure might be expected to be higher in elderly patients.

In the United States, both nocturnal polysomnography, consisting of a minimum of a 6-h physiological recording involving sleep stages, respiration, oxyhaemoglobin desaturation, electrocardiogram, and surface electromyogram of the anterior tibialis and the daytime multiple sleep latency test (**MSLT**), have been included under the proposed new Medicare guidelines. For hypersomnolence conditions, such as sleep apnoea and narcolepsy, both a nocturnal polysomnogram and the MSLT are required. For parasomniac behaviours involving sleep-related violent behaviour or seizures, up to two nights of recordings may be needed. Diagnostic testing may also be indicated in persistent insomnia (defined as insomnia four or more nights per week for at least 6 months), providing that the insomnia has been unresponsive to short-term pharmacological treatment and behavioural/sleep hygiene measures, and that a coexisting psychiatric cause has been ruled out.

SPECIFIC SLEEP DISORDERS

Sleep apnoea

Sleep apnoea and sleep-related respiratory disturbances are relatively common in the elderly population. In such conditions, respiration ceases (apnoea) or is greatly reduced (hypopnoea) for from 10 s up to as long as several minutes (Fig. 1). In recent years, the topic of sleep-related breathing disorders has seen burgeoning interest, largely due to the recognition that ventilation undergoes neurogenic, sleep-related alterations.

Without question the two cardinal symptoms of sleep

apnoea are snoring, particularly when intermittent and punctuated by snorting, and daytime fatigue and drowsiness. Snoring is a relatively common symptom and care must be taken to differentiate continuous noise emanating from the upper airway from discontinuous gasping. It is probably true that most individuals with sleep apnoea snore but not all individuals who snore have sleep apnoea. As mentioned above, daytime hypersomnolence can be a notoriously difficult symptom to gauge by history, and frequently the spouse or bed-partner can provide valuable corroboration. Other symptoms reported by some elderly individuals with sleep apnoea include headache (particularly upon awakening in the morning), mild dysphoria, automatic behaviour, irritability, nocturnal sweating, and dry mouth in the morning.

Epidemiology and risk factors

As early as the mid 1970s, epidemiological studies of self-reported snoring suggested that the prevalence of sleep apnoea probably increased with age. As suggested above, figures derived from such studies are likely to overestimate the prevalence of sleep apnoea although there is probably no reason to suspect the error is greater in older than in younger subjects. Thus, at the age of 40, habitual snoring is present in about 30 per cent of men and 20 per cent of women but, by 60, the figures rise to 50 and 40 per cent, respectively. Though different surveys have inquired about snoring in a variety of different ways, these figures are generally in accordance across studies. The age effects are independent from obesity. In every decade, the proportion of obese (i.e. body mass index (weight/height^2) >27) habitual snorers is approximately double that of non-obese habitual snorers. Smoking is also positively related to snoring, even apart from age and obesity, suggesting that upper airway inflammation and oedema may be involved. There is little information on snoring in the population aged over 80 years, but what findings there are suggest a decreased prevalence in both men and women. We do not know to what extent this reflects a survivorship phenomenon, snoring being associated with excess mortality, or merely a bias introduced by the fact that so many elderly persons live alone and do not know whether they snore or not.

In sleep disorders centres, the number of cases of sleep apnoea clearly increases with age, at least up to 70 years. Figure 2 shows data pooled from over 2500 clinic patients seen in 18 sleep clinics over 15 months. The major disadvantage of data such as these is that they may be biased in relation to health-care use or symptomatic status because they are not from a representative sample of the general population. Over the last 10 years, over 20 different research teams have confirmed that high prevalence of sleep apnoea in small samples of healthy, aged volunteers. Many of these individuals have been free of sleep/wake symptoms. Of these studies, the prevalence study from San Diego is by far the largest (about 400 cases) to use the most representative sampling of a socioeconomically diverse, ambulatory, elderly population, in this case over 65 years of age (Ancoli-Israel *et al.* 1987). Of their sample, 24 per cent (28 per cent of men, 19.5 per cent of women) met the criterion of an apnoea index of five events per hour. In two related but independent samples of elderly patients in hospital for medical reasons and of nursing-home residents, prevalence figures of 37 and 41 per cent, respectively, were obtained. These figures suggest that literally millions of aged individuals meet this definition of sleep apnoea, and, at the most general level, the prevalence of sleep apnoea is related to the health of the population under consideration. This may require a redefining of the sleep apnoea syndrome to include clinically important levels of illness.

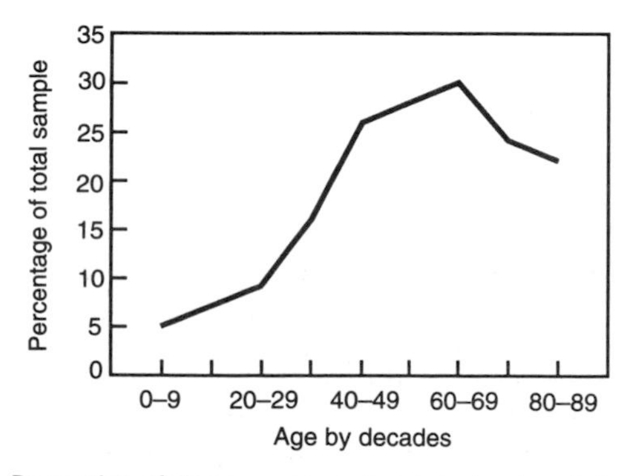

Fig. 2 Proportion of sleep apnoea cases seen by decade in sleep disorders clinics.

Although there is no question that ageing is a major risk factor for sleep apnoea, there are numerous other risk factors as well (see Fig. 3). The predominance in men, estimated at 30 to 1 in middle age, probably continues at a similar level in old age. Without question, the increase in average expected body weight through to the age of 70 plays a part in at least some of the age-related prevalence in sleep apnoea. In one study, the body mass index (weight/height^2) accounted for more variance in sleep apnoea than did age *per se* (Bliwise *et al.* 1987).

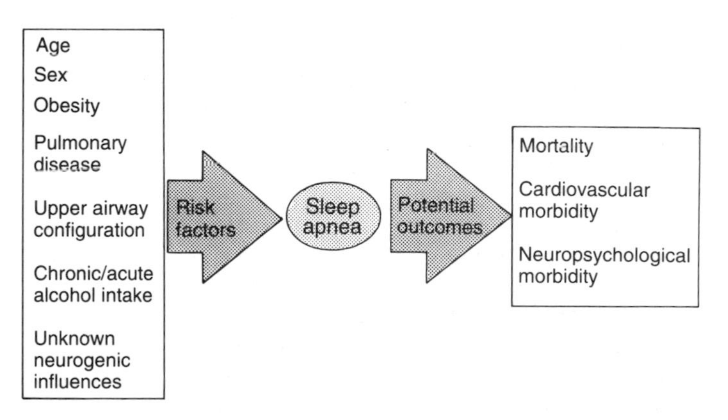

Fig. 3 Heuristic model showing risk factors for and potential outcomes of sleep apnoea.

Other risk factors for sleep apnoea in elderly persons include acute and chronic alcohol intake, though some have claimed the acute effects are detectable only in men (Block *et al.* 1986). Certain hypnotics such as flurazepam may predispose individuals to sleep apnoea, though such an effect has not been reported with hypnotics that have a shorter elimination half-life. Acute sleep deprivation may also worsen mild sleep apnoea; sleep deprivation decreases the output to the upper airway musculature and blunts the hypoxic and hypercapnic ventilatory responses. Body position may be a risk factor in many mild cases of apnoea, with the supine position rendering patients particularly vulnerable. Whether older individuals are particularly susceptible to positional effects is unclear, but it appears that such effects are most marked in non-REM sleep and in less obese individuals. A final major risk factor for sleep apnoea in older people is undoubtedly the structure of the upper airway. Several studies have found that the pharynx and glottis have smaller cross-sectional areas in individuals prone to sleep apnoea. The relative importance of each of these risk factors in the apnoea of old age is unclear, but undoubtedly many factors may operate in any one individual.

Mechanisms

The mechanisms responsible for sleep apnoea in any age group remain poorly understood and are clearly multifactorial. At one time, polysomnographically defined, central apnoeas (related to mixed and obstructive events) were thought to have the greater likelihood of involvement of neurogenic ('central') as opposed to upper airway ('obstructive') factors in the apnoea of old age. It is now clear that the vast majority of cases of sleep apnoea at all ages appear to be obstructive. Moreoever, it is now known that the function of upper airway musculature is appreciably altered during sleep, thus arguing for a considerable influence of the central nervous system. The genioglossus and stylopharangeus, for example, usually show phasic inspiratory activity immediately before diaphragmatic contraction. Presumably patients with sleep apnoea have diminished phasic oscillatory output or, possibly, suffer a lack of co-ordination between the diaphragm and upper airway muscles such that the upper airway is susceptible to collapse during negative inspiratory pressure. The pathophysiology is specific to sleep. Even in unaffected individuals, the transition from wakefulness to sleep is associated with reductions in muscle tone in the upper airway. In apnoeic patients, who are known to have narrower oropharyngeal lumina as shown by both computerized tomography and acoustic reflection, the patency of the upper airway may be even more precarious. Some studies have even shown correlations between the severity of apnoea and the area of pharynx. Because ageing is known to be positively related to changes in pharyngeal resistance, this might explain why elderly persons are susceptible to airway collapse during sleep, though the findings are stronger in men than in women (White *et al.* 1985).

Although anatomical factors are likely to predispose elderly individuals to sleep apnoea, factors related to functional state in the control of ventilation are undoubtedly involved because the obstruction of the upper airway occurs only during sleep. One plausible model is that, because of the mild hypoventilation that occurs during sleep (which results in a rise in Pa_{CO_2} of 3 to 9 mmHg), the gain of the respiratory control system is more likely to be overdriven (Dempsey and Skatrud 1986). This would result in an unmasking of the hypocapnic apnoea threshold. Episodes of periodic breathing, which frequently occur at the onset of non-REM sleep (stages 1 and 2 in normal persons), could accentuate the instability by lowering the apnoea threshold. Such periodic breathing is also known to be driven by natural (e.g., high altitude) and artificially induced hypoxaemic states. Because periodic breathing increases with age, the lack of stability in the system could then be further disrupted by the small decreases in P_{O_2} that are known to occur with ageing, which would place the older individual on the steeper portion of the oxyhaemoglobin desaturation curve.

There are, however, a number of reasons to suspect that the overdriven respiratory control may be an inaccurate explanation of sleep apnoea in old age. First, as the development of sleep apnoea in elderly people is likely to begin in their 40s and 50s, the small reductions in P_{O_2} found in old age are unlikely to be the hypoxic stimulus for the later development of apnoea. Secondly, there is evidence that respiratory control is more likely to be underdriven in old age. In the waking state, even apart from changes in the respiratory musculature and the loss of elastic recoil in the lung wall, the sensitivity of central chemoreceptors is blunted in elderly persons. Studies of mouth occlusion pressure show that hypercapnic and hypoxic responses are blunted as much as 50 to 60 per cent in older persons. Thirdly, during sleep, minute ventilation does not differ between young and old subjects, and CO_2 levels rise by similar amounts. If the unmasking of an hypocapnic apnoea threshold leads to the greater amount of sleep apnoea in older persons, relatively higher levels of CO_2 during sleep would be expected in them. Taken as a whole, these facts suggest that other mechanisms may be involved.

One alternative mechanism involves the observation that respiratory abnormalities in the sleep of older persons may occur in wakefulness as well (Pack *et al.* 1988). Elderly subjects with sleep apnoea had a 40- to 60-s oscillation of respiration when awake when compared with elderly persons without sleep apnoea. This suggests that the fundamental defect seen in older people with sleep apnoea may be an abnormality in the periodicity of the control of the arousal state. This is not at all inconsistent with the many polysomnographic studies that have shown frequent bouts of sleepiness during the daytime and frequent brief arousals during sleep in old age. If this hypothesis is correct, it would suggest that the elderly individuals who are most at risk for developing sleep apnoea are those who have a greater tendency for fragmentation of the sleep/wake rhythm. The implication would be that poor soundness of sleep predisposes aged individuals to sleep apnoea.

Death

Of all the outcomes potentially associated with sleep apnoea, none is more dramatic than the possibility that disturbed respiration during the night will lead to death during sleep. British and French teams have made inconclusive findings on very small cohorts (less than 20 subjects) of elderly subjects with sleep apnoea; the Stanford and San Diego groups have reported on cohorts of sizeable numbers of elderly individuals, in some cases studied for up to 10 years (Ancoli-Israel *et al.* 1988; Bliwise *et al.* 1988) A major feature of both of these studies is the relatively small number of individuals lost to follow-up (about 1 to 2 per cent). The Stanford study suggests that an individual with respiratory disturbance index of more than 10 events per hour was nearly three times as likely to die within 5 years, though these findings held in only a univariate model. More complex multivariate models could not show the effect, but because the cumulative mortality rate of the cohort of 200 had only reached about 10 per cent, the results may change as the study continues. In the San Diego study, cumulative mortality rates were somewhat higher (15 per cent for the independently living elderly subjects and 20 per cent for the elderly patients in acute hospitals), but the only cohort for which mortality data have been presented is the nursing-home sample where the cumulative mortality rate approached 50 per cent. In that sample, a combined product of the apnoea and hypopnoea indices strongly predicted mortality in women within 6 months of study. In both the Stanford and San Diego studies, sleep apnoea was associated with death during the nocturnal hours.

A second set of studies performed at Henry Ford, Montefiore, Gainesville, and Stanford University Hospitals have compared treated and untreated groups of referred patients with sleep apnoea. Two of the centres reported higher mortality in untreated (or conservatively treated) patients than among those aggressively treated with tracheostomy or nasal continuous positive airway pressure; however, two of the studies could not find differences in survivorship. The relevance of these studies to the older population may be limited. First, the mean age of the patients in these studies was approximately 50 years. Secondly, all of these studies used samples derived from clinical sources which are probably not at all representative of any aged population. Thirdly, the retrospective nature of the study designs dictates that assignment to treatment was not random. Finally, the number of cases lost or unspecified (in some cases approaching half) is simply too large to be considered a meaningful clinical trial.

In summary, studies of elderly cohorts to date are at least suggestive of an association between sleep apnoea and death but the retrospectively based clinical trials are simply too methodologically flawed. Obviously, controlled clinical trials with randomized assignment will be necessary to evaluate whether sleep apnoea is definitely associated not only with death but also with disease.

Cardiovascular disease

Premature ventricular contractions occur during apnoeic episodes, although there is some disagreement about their prevalence in apnoea patients, with estimates as low as 20 per cent and as high as 75 per cent. Perhaps the most well-recognized cardiovascular effect of sleep apnoea, however, is the cyclic brady-tachycardia which occurs during the apnoeic events. Although such R-R cycling does not meet traditional epidemiological definitions for the presence of cardiac disease, this pattern is virtually pathognomonic for the presence of sleep apnoea. In general, the longer the apnoea and the more severe the oxygen desaturations the more severe the bradycardia. After this, hypoxaemic stimulation of the carotid body presumably occurs, which then produces a vagal response. A similar mechanism is assumed to underlie the drop in cardiac output reported in a few studies. Sympathetic tone is enhanced as breathing resumes, and tachycardia, increased vascular resistance, and rises in systemic and, possibly, pulmonary pressure occur. Interestingly, apnoeic patients with sympathetic denervation may not show increases in systemic pressure during their apnoeas. Another effect of increased sympathetic tone may be increased blood viscosity. In a small series, approximately two-thirds of general hospital patients with unexplained polycythaemia were shown to have sleep apnoea (Kryger *et al.* 1982). The higher concentrations of plasma and urinary catecholamines in patients with sleep apnoea are also assumed to reflect this increase in sympathetic activation.

Right heart failure has often been attributed to sleep apnoea but newer evidence suggests this will be found in only about 1 in 10 apnoeic patients, and even then only in those who have waking hypercapnia and hypoxaemia. More common and more relevant for the elderly population, however, is left heart failure, because left-axis deviation is a particularly common finding among elderly people and left ventricular hypertrophy is strongly associated with morbidity in the Framingham studies. Surprisingly, no studies have linked left heart failure and sleep apnoea in an elderly population. Echocardiographic studies in middle-aged individuals with apnoea have shown a larger end-diastolic diameter of the left ventricle and higher pre-ejection period/left ventricular ejection ratios than in controls. Systemic hypertension has been associated with sleep apnoea in a number of studies, even when age, sex, and obesity were taken into account statistically, although the fact that most of the patients in these studies were taking antihypertensives and were on average about 50 years of age may limit the relevance of these findings for the elderly population.

Many of the population-based studies of snoring provide epidemiological evidence, albeit indirect, linking sleep apnoea and cardiovascular disease in older (75 years and above) age groups. Two separate Italian studies showed that associations between measured hypertension and habitual snoring may be stronger in individuals in their 70s than in younger persons. One Canadian study of reported hypertension suggested that these associations persist well into the ninth decade. Relationships between inferred sleep apnoea and cardiovascular disease are not limited to systemic hypertension. Finnish and English data suggest that habitual snorers are over 10 times as likely to have experienced a stroke than non-snorers, at least in men up to the age of 70, and habitual snoring was a risk factor for angina and ischaemic heart disease after hypertension, obesity, smoking status, and alcohol usage were controlled for statistically (relative risk = 2).

Neuropsychological dysfunction

The most profound neuropsychological outcome associated with sleep apnoea is daytime somnolence. Studies with the MSLT have shown that sleep apnoea in older individuals is related to short sleep latencies (i.e., ease of falling asleep) in this daytime test procedure. All higher order mental functions such as memory, abstract reasoning, psychomotor speed, and even simple attention, will incur deficits if the individual is unable to maintain an optimal level of arousal. Neuropsychological studies of both middle-aged and older non-demented patients with sleep apnoea have shown that it is associated with a host of neuropsychological functions including deficits in visuospatial memory, immediate and delayed verbally mediated contextual memory, psychomotor sequencing and manipulation, new verbal learning and facial recognition, and pure motor dexterity. There is some suggestion, however, that particularly among elderly persons in optimal health, such deficits may be far more modest if detectable at all. Some psychometric studies have also shown mild dysphoria, somatization, and hostility associated with sleep apnoea in old age. Whether any or all of these changes relate more strongly to the accumulated sleep deprivation caused by the repetitive brief arousals or to intermittent nocturnal hypoxaemia is still a matter of debate, with most of the evidence at present falling on the side of the hypoxaemia (Bliwise 1989*b*). Whatever the cause, several reports of treatment (but without a control group) have confirmed the apparent reversibility of these deficits.

A second issue in discussing neuropsychological function in sleep apnoea involves dementia. If sleep apnoea is either a cause for or an effect of dementia, the condition should occur more frequently in demented patients. The evidence on this topic is mixed at present. There is some suggestion that women (but not men) with Alzheimer's disease may have higher rates of disordered breathing during sleep than do age-matched controls. Yet if hypoxaemia lies behind most of the above mentioned neuropsychological deficits seen in sleep apnoea, we would expect marked desaturation in the sleep of demented patients, and this has not been found to date. Several other points deserve mention. First, there is some suggestion that, even if sleep apnoea and dementia were causally unrelated, some elderly people will have both conditions by chance alone because the two have a mutual, age-related prevalence. In such cases, sleep apnoea may make awakening from sleep in demented patients a particularly vulnerable period for confusion (Bliwise *et al.* 1989*a*). Secondly, studies of sleep in dementia are diagnostically imprecise. Given the relationships between snoring and hypertension and stroke mentioned above, sleep apnoea may be more likely related to multi-infarct dementia than Alzheimer's disease. Some studies have presented data consistent with this possibility though autopsy verification was lacking.

Treatments

Without question the treatment of choice for the vast majority of individuals with sleep apnoea is nasal continuous positive airway pressure. This presents a small (2.5 to 15 cmH_20) positive pressure through a mask attached to a blower which then the patient wears during sleep. Continuous positive airway pressure works essentially as a pneumatic splint by simply preventing airway occlusion during negative inspiratory pressure. It is clear that in nearly all cases reported, it is an effective treatment for sleep apnoea. Compliance rates vary between 65 and 83 per cent, depending on the patient population being considered (Nino-Murcia *et al.* 1989). The most common side-effects include dry nose and throat, and sore eyes (from mask leakage). Nasal obstruction prevents nightly usage in about a third of the cases. Most of the problems can be countered with humidification, frequent mask replacement, and nasal decongestants, respectively. Pretreatment counselling and contact with support staff are essential in overcoming initial reluctance to using the equipment and long-term problems of compliance. To date no adverse consequences (e.g., pneumothorax) have been reported with nasal continuous positive airway pressure, although there has been some concern for patients with congestive heart failure about the increased expiratory effect required to overcome the continuous positive flow. Others have claimed that simple increased expiratory resistance can even result in normalization of respiration during sleep, but the mechanism underlying such an effect remains unclear.

Of the other treatments for sleep apnoea, surgical options, including uvulopalatopharyngoplasty and maxillofacial restructuring have received the most attention. The efficacy of the former has been liberally estimated at 50 per cent, based on greater than 50 per cent improvement in the respiratory disturbance index. The maxillofacial approach often involves multiple operations, is a difficult process, and not well-suited to elderly people. Gross upper airway obstruction (e.g., deviated septum, tonsillar hypertrophy) might be viable possibilities for surgical correction in selected cases. Tracheostomy, the first described treatment for sleep apnoea, is now seldom used, but remains an effective and important option.

Pharmacological treatments, including medroxyprogesterone, protriptyline, and acetazolamide, all received some initially encouraging reports but later accounts were less optimistic. A few have suggested prosthetic devices that either keep the tongue from collapse or induce a mild prognathia during sleep, but compliance and efficacy are uncertain. For patients with obstructive pulmonary disease and sleep apnoea, nocturnal administration of oxygen may be helpful in raising the oxygen saturation, but is not by itself a significant treatment for sleep apnoea.

In mild cases of sleep apnoea, more conservative treatment options can be used. Avoidance of the supine position, for example, may be helpful in some cases. Given the relative importance of body weight, weight loss, even of the order of 4 to 9 kg may have some benefit. Alcohol before bed should be avoided and hypnotics are also likely to exacerbate a mild condition.

Apart from how, the issue of when to treat sleep apnoea in the aged remains a clinical decision based on careful polysomnographic evaluation (including the extent of oxygen desaturation), and assessment of cardiovascular condition and clinical symptoms. Nocturnal sleep disruption and associated daytime fatigue and somnolence are particularly salient issues and must not be overlooked. Elderly patients are acutely aware of any improvement in daytime alertness. Assessment with the daytime MSLT can be an important factor in directing treatment. Chronic fatigue in the aged often has an insidious effect, leading to decreased energy for social interactions and life activities. All such factors should be weighed in the decision to manage actively the sleep apnoea of old age.

Restless legs syndrome/periodic leg movements in sleep

Restless legs syndrome (Ekbom's syndrome) is characterized by paraesthesiae that are difficult for the patient to describe. These are usually in the calves and cause an irresistible urge to move the legs. The patient may use a variety of different words to describe the symptoms, but may mention sensations of restlessness, or creeping and crawling sensations, that are characteristically relieved by leg movement, standing, or walking. The relief of these symptoms by leg movement helps distinguish them from other symptoms related to peripheral neuropathy, vascular insufficiency, or musculoskeletal disease. The symptoms occur more frequently in the evening or in bed and usually disappear in most cases during the daytime hours. In more severe cases, the patient simply is unable to sleep and may develop severe emotional disturbance and dysfunction. The restless paraesthesiae may be

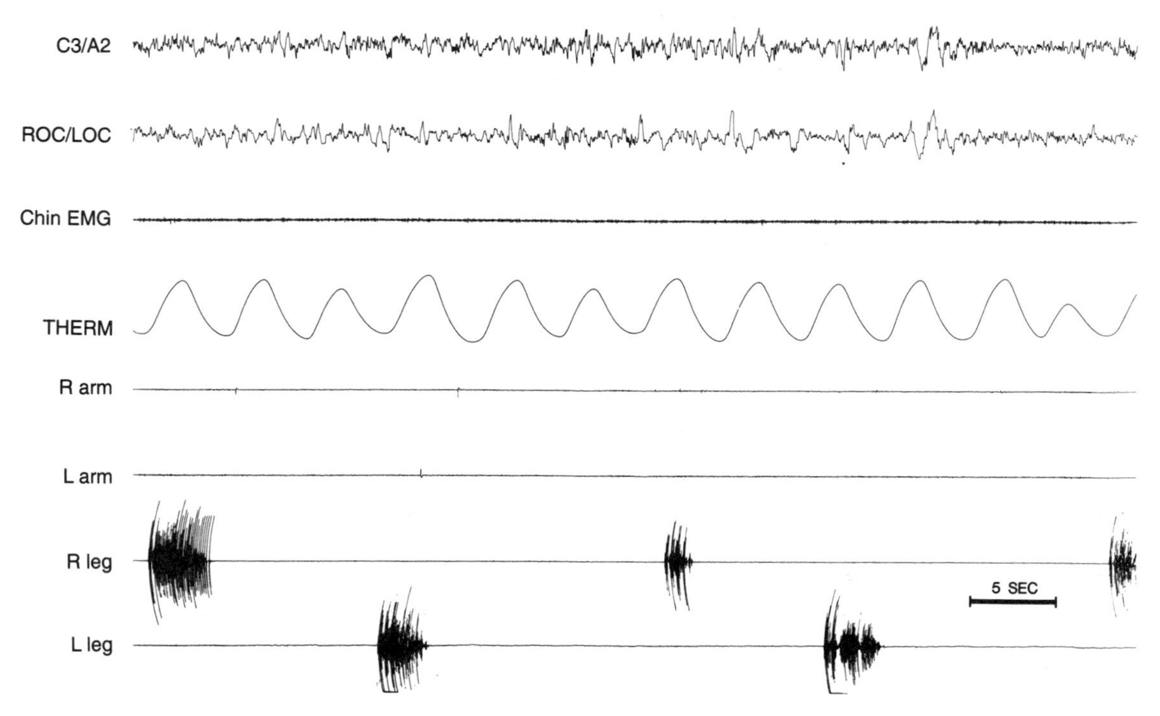

Fig. 4 Typical polysomnographic tracing of periodic leg movements in sleep. Notice in this case alternating bilateral movements confined to lower limbs.

aggravated by chronic sleep deprivation, apprehension about sleep loss, and expectation of further restlessness, creating a vicious circle. The prevalence of the symptoms of restless legs is unknown but apparently increases with age. The condition is clinically distinct from the more common problem of nocturnal leg cramps.

Restless legs syndrome clearly overlaps with a polysomnographically defined condition called periodic leg movements in sleep. Most patients with the syndrome also tend to have periodic leg movements but the converse is not necessary the case, i.e. many individuals with the leg movements are otherwise asymptomatic. A good measure of association between the two is not yet available. Periodic leg movements are involuntary movements usually limited to unilateral or bilateral extension of the big toe and can include flexion of the foot at the ankle and partial flexion of the knee and the hip. The patient is usually unaware of these movements, which may occur up to 600 or 700 times throughout the night. The substantial arousal in the central nervous system caused by these leg movements can result in complaints of restless, broken sleep, or, less typically, to daytime fatigue and drowsiness. Periodic leg movements have sometimes been referred to as 'nocturnal myoclonus'; however, the term is a misnomer to the extent that the condition is non-epileptiform and occurs in the absence of paroxysmal EEG activity. Figure 4 shows a sample polysomnographic tracing of periodic leg movements.

These movements have a clear age-related prevalence into old age and, unlike sleep apnoea, show no obvious drop in the oldest age groups studied to date. Prevalence rates for the over-50 population vary between 22 and 45 per cent, depending upon the type of sample and the way in which the periodic movements are defined. We stress, however, that many of these individuals are asymptomatic. In one study, 40 per cent of those insomniacs with periodic leg movements reported the experience of leg twitching, and only 18 per cent of those insomniacs without those movements experienced this symptom. However, 60 per cent of elderly insomniacs with periodic leg movements never experienced twitching (Bliwise *et al*. 1985).

The mechanisms underlying these movements remain obscure, although they appear confined to the lower limbs. This raises the possibility that motor neurone dysfunction may be involved, as larger nerve fibres could be more sensitive to damage, but there is little information to support this idea at present. Another possibility is some type of transient functional suppression of the pyramidal tract. A positive Babinski sign is found in a few patients with periodic leg movements. Some central dysfunction at the pontine level or higher is also likely, given the reported abnormalities in the glabellar reflex (Wechsler *et al* 1986). Vascular involvement has also been implicated in that symptoms like cold feet, which are related to poor circulation in the legs, have been associated with periodic leg movements.

Both the restless legs syndrome and periodic leg movements are notoriously refractory to most treatments. Levodopa/decarboxylase inhibitor combination (e.g. co-careldopa, in relatively low doses of 10–100 or 25–250 mg, one or two tablets at bedtime), may provide relief. Opiates have also been effective in relieving the paraesthesiae, although tolerance can be a problem. In milder cases, non-specific sedating agents such as chlormethiazole or lorazepam allow the patient to ignore the paraesthesiae by producing increased drowsiness. Some have favoured clonazepam, but this must be used in low doses, 0.5 mg to 2.0 mg, as it may give significant problems with confusion, disorientation, sedation, and behavioural change in elderly patients. Equivocal success has been claimed for a variety of other medications including baclofen, pentoxifylline, and antiepileptics. {In some patients restless legs respond to correction of iron deficiency.}

Insomnia

Risk factors, prevalence, aetiology

Numerous physiological factors operate in reducing the depth, quality, and length of nocturnal sleep in older persons. These normal, age-associated changes include: loss of deep (stages 3 and 4) sleep; an increased proportion of total sleep spent in stage 1 (transitional) sleep; and increased frequency of brief, transient arousals of less than 15 s. REM sleep may change only minimally with ageing. In addition, other factors such as increased sympathetic tone, higher autonomic arousal, and an elevated auditory arousal threshold (even within stage 2 sleep) may predispose older persons to poorer quality sleep. Some have suggested that the pattern for elderly persons to become sleepy earlier in the evening and awaken during the night may actually reflect an age-associated change in what is termed a 'phase-advance' of the circadian sleep–wake cycle. We stress that many of these changes are seen even among aged persons who sleep well. Because the prevalence of specific sleep disorders such as periodic leg movements and sleep apnoea also increases with age, and the many medical illnesses that disturb sleep are so common in aged populations, these factors combine to predispose the elderly person to insomnia. Nonetheless, it is important to appreciate that although most elderly persons probably experience sleep of poorer quality and duration than when younger, a smaller proportion of these individuals actually complain of insomnia. Insomnia is a symptom and must be evaluated as such, particularly in terms not only of its medical but also its psychosocial implications.

The prevalence of insomnia in the elderly population varies somewhat from survey to survey, depending upon the question asked. In a survey based in Houston, Texas (Karacan *et al.* 1983), about 28 per cent of the men over 65 and about 35 per cent of the women experienced difficulty in falling asleep sometimes, often, or always. These figures were 55 and 65 per cent for difficulty maintaining sleep and 25 and 30 per cent for early morning awakening, respectively. Some polysomnographic studies have suggested a more EEG-definable sleep disturbance in older men than in older women, but data generated from the surveys of the National Institute on Aging and National Center for Health Statistics (Cornoni-Huntley *et al.* 1986) corroborate the self-reported sex differences. These data also suggested that awakening during the night 'most of the time' afflicts about 25 to 30 per cent of elderly people. There were also some urban/rural differences, with urban elderly people having more trouble falling asleep and awakening too early than rural subjects, who appear to have more difficulty with nocturnal awakenings. In parallel with the age-associated prevalence of all these sleep complaints is the use of hypnotic medication, which also increases with age. Again, estimates vary depending upon the question asked. In the Houston survey, about 12 per cent of the men aged 65 or over used hypnotics at least sometimes, as did about 15 per cent of the women. A Finnish survey (Partinen *et al.* 1983) showed that about 7 per cent of men aged 60 and above and 11.5 per cent of women aged 60 and above used prescription sleeping pills on at least 10 days during the year preceding the survey.

Several large studies with psychological tests, such as the Minnesota Multiphasic Personality Inventory, have consistently shown, at least in younger persons, that chronic poor sleep is invariably related to anxiety, dysthymia, or other character (axis II personality) disorders (Kales and Kales 1984). Given all the aforementioned normal physiological changes associated with poor sleep in old age, the question is whether older persons with insomnia also show similar patterns of psychopathology. The studies to date suggest that, although physiological changes may predispose elderly people to poor sleep, the complaint of insomnia, even in old age, continues to be associated with anxiety and depression. None the less, the role of physical illness cannot be minimized. There is epidemiological evidence that bronchitis/bronchial asthma was associated with middle insomnia, diabetes with initial and middle insomnia, and musculoskeletal pain was primarily associated with initial insomnia (Gislason and Almqvist 1987).

Current evidence indicates that about 15 per cent of elderly patients with disorders of initiating or maintaining sleep can be diagnosed as suffering from psychophysiological insomnia. The term 'psychophysiological' refers to the process by which external stress and internal psychological conflicts produce tension and anxiety that is discharged through somatic channels. The somatization of this tension/anxiety results in physiological arousal, which continues into the sleep period. A psychological sketch of these patients drawn from several studies shows them to be introverted individuals whose anxieties are turned inward, with prominent, chronic low-grade depression and ruminative thinking. On psychological tests they appear as repressors and sensation avoiders, but are low on measures of overall psychopathology. They often have multiple tension-related complaints such as headaches, back pains, and palpitations. These patients often strike the geriatrician as worried, tense, and hypochondriacal. They may develop an anxious apprehension at bedtime that comes from the repeated failed attempts to achieve sleep and their anticipation of fatigue the following day. Conscious attempts to fall asleep only result in further arousal and loss of confidence in the normal physiological process of sleeping. Polysomnographic studies of these patients indicate that they take longer to fall asleep than insomniac patients with dysthymic disorders and also have more alphawave intrusions within sleep (Hauri and Fisher 1986).

Treatments

Treatments for insomnia can be resolved into pharmacological as opposed to non-pharmacological approaches. This does not necessarily mean that usage of medications obviates the practice of good sleep hygiene. In fact, in selected patients, it may be possible to substitute a programme (see below) of fixed sleep/wake schedule, avoidance of naps, and reduction of alcohol and caffeine once a hypnotic drug has induced a pattern of sound sleep. Before being prescribed sedative hypnotics, a patient should have completed a thorough history for medical and sleep disorders, for the exclusion of a history suggestive of sleep apnoea, and for other specific diagnoses including depression.

Sedative hypnotics are most effectively prescribed in patients who have insomnia of recent onset, characterized by difficulty in falling asleep or maintaining sleep with a previous history of normal sleep patterns. Patients with a history of chronic insomnia of many months or years are less likely to be cured of their complaint with sedative hypnotics, and more likely to develop problems with tolerance or dependence. Before prescription of a sedative hypnotic, patients should be thoroughly educated on the overall treatment plan of a time-limited trial. There should be emphasis on a plan to taper and discontinue these medications within a period of several weeks. Patients should receive written information outlining possible side-effects, drug interactions, decreased alertness at night, daytime hangover, and possible impairments of gait, memory, or cognitive function. They should also be given sleep diaries, which will allow an accurate daily record of their sleep behaviours, medications, and sleep complaints.

When considering pharmacological treatment, adverse effects on daytime function and night-time sleep enter equally into consideration. Initially, care should always be taken to identify those patients whose complaints of daytime problems, fatigue, and dysfunction are focused on their sleep when there are other physical and psychosocial contributors of equal or a greater importance. A sedative hypnotic may increase daytime sleepiness and a patient may be unaware of the extent of impairment. Use of the MSLT while under short-term dosage with flurazepam has demonstrated that subjects are demonstrably sleepier on the days after taking this medicine than at baseline. The improvement in daytime alertness that follows the use of a short-acting benzodiazepine suggests that it is the long half-life of many sedative hypnotics that results in daytime sleepiness (Carskadon *et al.* 1982). Night-time sedation also has adverse nocturnal effects by increasing the arousal threshold, impairing balance, and, with benzodiazepines, possibly producing anterograde amnesia for night-time events (Scharf *et al.* 1986). These effects may produce decreased ability to respond to emergencies, worsening of sleep apnoea, or an increased incidence of nocturnal falls.

The choice of initial medication should be determined by a variety of clinical factors. Patients whose main problem is an inability to fall asleep should begin on a drug with a short elimination half-life and relatively rapid absorption at the lowest dose. In the United Kingdom, chlormethiazole is preferred as the hypnotic of first choice for elderly patients. Low-dose antidepressants of the sedative type, such as doxepin, amitriptyline, and trazodone, as well as antihistamines (in some countries), are often used for sleep-onset insomnia. For the patient with primary insomnia, there is little to recommend selecting an antidepressant or an antihistamine to induce sleep in favour of what are clearly effective sedative hypnotics available on the market today. There is no evidence that properly used sedative hypnotics with careful follow up and education are more difficult to discontinue than low-dose tricyclic medications. In fact, antidepressants may have substantial problems for elderly patients including anticholinergic reactions, hypotension, excessive sedation, and cognitive impairment.

Individuals with problems maintaining sleep may have ongoing difficulties with short-acting sedative hypnotics that wear off after 4 to 6 h. In these patients, a drug with an intermediate half-life, such as temazepam (15 mg) may be preferable, with the caveat, for the elderly patient, that there may be substantial morning hangover and less obvious impairment during the following day. The longer acting benzodiazepines, such as flurazepam and nitrazepam, accumulate long-lived active metabolites, with the increased likelihood of excessive daytime sleepiness, adverse drug actions, and the possibility of interaction with diminished kidney function. These medications, however, may provide beneficial daytime sedation in patients with some level of anxiety related to their sleep disturbance. With careful monitoring, they have a place in selected patients who do not benefit or respond to the shorter-acting medications. {Flurazepam and nitrazepam are not recommended for elderly patients in the United Kingdom.}

There is little to recommend the use of antianxiety agents (alprazolam, diazepam) as sleeping pills unless a diagnosis of anxiety disorder has been made or sedative hypnotics have failed. In the absence of frank sundowning-like behaviour in demented patients (see below), the use of phenothiazines in the elderly as simple sedative hypnotics should be condemned. There is no evidence supporting such a use and development of tardive dyskinesia and significant postural hypotension are well-known complications of these medications.

The role of non-pharmacological substances to promote sleep probably deserves wider attention because these can allow some substitution for pill-taking behaviour and diminish apprehension over discontinuation of medication. Until recently the amino acid precursor of serotonin, L-tryptophan, was a widely used food additive with demonstrated hypnotic effects in the sleep laboratory. Documented cases of eosinophilia/myalgia syndrome linked to L-tryptophan resulted in its being withdrawn from production in late 1989. It remains to be seen whether other natural food supplements may be useful insomnia treatments, though some evidence suggests that herbal preparations containing valerian can improve sleep in both humans and animals.

Once treatment has been initiated, there may be demands for increases in dose and repeat prescriptions for chronic use. We stress that before the first prescription, patients should be clearly instructed that they are not to increase the dose without at least phone consultation. They should also be advised that if the medication becomes ineffective after an initial positive response, increases in dose will not be in their best interest, and that supervised discontinuation of the medication or an alternative treatment plan will be recommended. In general, a previously effective dose should never be increased. Patients should also be encouraged to use medication on an alternate-night or sporadic regimen as much as possible. This may diminish the development of tolerance, allow drug-free nights to normalize sleep patterns, and may encourage some fortitude in coping with sleep disruption. Alternate-night regimens are not desirable in patients whose insomnia is characterized by apprehension and rumination about possible sleep loss. In these patients, concern over the use of any medication may produce pill-taking dependence triggered by concerns about whether or not to take a pill.

Individuals should be seen regularly during the time of prescribing a sedative hypnotic, usually weekly or every other week, with sleep diaries to be carefully reviewed at each visit. Prescriptions should not be routinely repeated by phone, nor should patients be allowed to miss appointments because they say they wish only to continue on sedative hypnotics.

Physicians caring for elderly patients should rarely prescribe enough hypnotics for more than 2 weeks of therapy. Short-term prescriptions ensure close follow-up in patient compliance. Upon discontinuation of sedative hypnotics sleep may be disrupted for several nights and patients should be specifically told of this possibility. Clinically, it may be difficult to differentiate 'rebound insomnia' from re-emergence of the original complaint. Individuals who develop dependence on sedative hypnotics or clear recurrence of their symptoms upon tapering and withdrawal should be referred for further psychiatric assessment or sleep disorders consultation.

Although behavioural treatments (relaxation, autogenic training, biofeedback, self-hypnosis) have been used extensively to treat (primarily initial) insomnia in younger individuals, there have been very few attempts to treat elderly patients with these techniques. The available results are mildly optimistic, although elderly patients with chronic medical conditions were excluded in most studies so the target population may be limited. A newer approach, sleep restriction therapy (Spielman *et al.* 1987), has also met with some limited success in elderly patients. In this treatment, the usual rules of sleep hygiene (use bed only for sleep, maintain regular bedtime/wake-up hours, avoid alcohol and caffeine) are supplemented by the additional stipulations to avoid all daytime napping (even unintentional 'dozing' and 'resting of the eyes') and drastically curtail the time available in bed for sleep (often to as short as 5 or 6 h). As sleep efficiency improves, the individuals are allowed slightly more time in bed (e.g., an increment of 15 min) every 3 to 5 days. Essentially this treatment uses voluntary sleep deprivation, which has been shown to have consolidating effects on depth of sleep in elderly subjects (Carskadon and Dement 1985), to improve length and quality of sleep. Preliminary findings suggest that elderly persons derive some benefits from a 4-week sleep restriction programme with daily contact (Friedman *et al.* 1991).

Sleep/wake disturbance in dementia (sundowning)

The profound sleep/wake disturbance occurring in some demented patients represents one of the most difficult management problems in clinical geriatrics. In fact, one survey of care-givers showed this as the single most common reason for a demented family member finally to enter an institution. Such sleep disruption goes beyond the simple inability to sleep in that the demented patient becomes agitated, confused, sometimes aggressive, and disoriented in a temporally specific pattern. The typical time for patients to show such behaviours is at or near sunset and sometimes throughout the night. Such sundowning often has a daytime component of excessive napping and relative docility. Figure 5 shows an example of sundowning-like behaviour in a demented patient, as recorded with a self-contained wrist actigraph for measuring simple activity.

Exactly who sundowns, how long the behaviours actually last, and when in the course of the dementia such behaviours occur are all unanswered questions (Bliwise 1989*a*). There is even contention about what behaviours constitute sundowning. Some have argued that sundowning is essentially nocturnal delirium but because the definition of delirium includes reference to a disturbed sleep/wake cycle such a definition is circular. None the less, prevalence data on delirium in elderly patients may offer some cursory notion as to how common sundowning actually is. In admissions to hospital for acute care in general medical wards, 20 to 30 per cent of older patients are delirious on admission or become delirious in hospital (Gillick *et al.* 1982; Levkoff *et al.* 1986). Upon admission to a skilled nursing facility, 19 per cent of patients met the criteria of the American Psychiatric Association *Diagnostic and Statistical Manual of Mental Disorders* (3rd edn. revised) for delirium, though the prevalence on a given day or night tended to be much lower.

It is important to keep in mind that, be it called sundowning or nocturnal agitation, delirium in elderly patients may represent a large number of metabolic, toxic, and biochemical abnormalities. Acid–base or electrolyte imbalance, endocrinopathies, liver failure, uraemia, hypovitaminosis, or hypovolaemia can lead to a delirious state in elderly patients, which may then be exacerbated during the night. Medications are another major cause of delirium in these patients including antiarrhythmics, antidiarrhoeals, steroidal and non-steroidal anti-inflammatories, antihistamines, and analgesics. A particularly common problem occurs with the medication of patients with Parkinson's disease.

Although parkinsonian patients may have sleep problems directly related to their illness, the medications used to treat them (amantadine, anticholinergics, and dopaminergics) may actually worsen the problem. Psychosis is common in the long-term management of these patients, and is commonly preceded by the presence of vivid nocturnal disorientation, disturbed sleep, or increased motor activity after the onset of sleep. The drugs may induce nightmares and night terrors, along with mild toxic reactions presenting as nocturnal confusion and wandering. Terrifying hypnagogic and hypnapompic hallucinations in association with panic reactions may also be seen. Inadequate night-time medication may result in exacerbation of symptoms during the sleep period, which is best treated by changing the drug schedule.

Although there are many such causes of sundowning, demented patients often show behavioural disturbances in the absence of any such factors. The mechanisms underlying this are poorly understood. In as much as the clinical problem is typically manifested across 24 h (excessive daytime dozing and disruptive nocturnal agitation), dysfunction of the circadian timing system is implied. A relevant animal model could be selective deterioration of the suprachiasmatic nucleus of the hypothalamus, because primates with such lesions show loss of circadian rhythms. Because the neuronal degeneration characteristic of Alzheimer's disease frequently involves subcortical structures, such selective deterioration in at least some demented patients appears likely. Based on this hypothesis, other components of 24-h circadian physiology (body temperature, diurnal variation in prolactin, melatonin, luteinizing hormone) might be expected to show marked changes in dementia; some evidence indicates that this has not always been the case (Prinz *et al.* 1984), although certainly these studies have been biased to the extent that the most agitated patients were almost certainly excluded from consideration. There is some suggestion that hyperarousal in the hypothalamic–pituitary–adrenal axis in dementia (assessed by dexamethasone suppression testing) is related

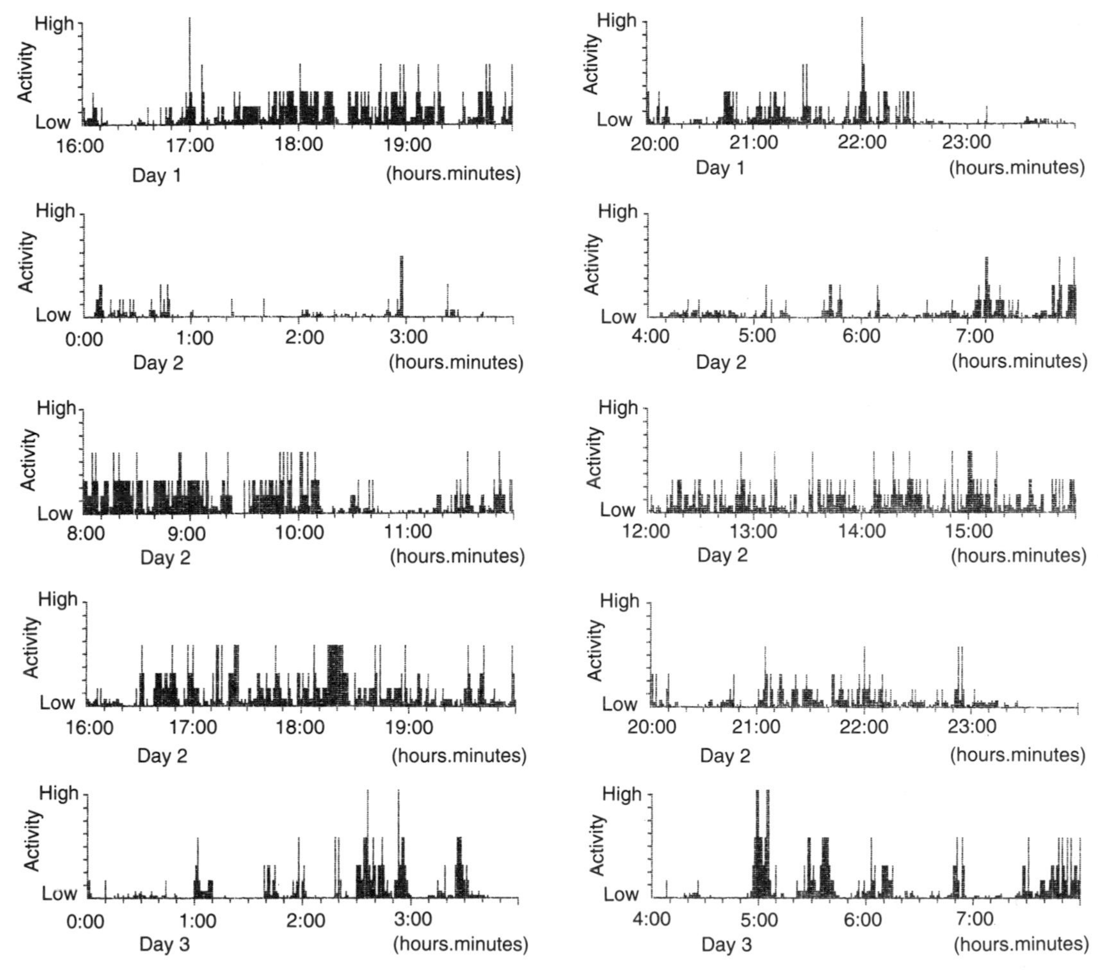

Fig. 5 Example of sundowning in a demented patient recorded with a self-contained wrist actigraph. Notice peaks of activity beginning at 1:00 a.m. on day 3 (last line).

to blunted diurnal variation in cortisol levels, but whether this portends sundowning is uncertain.

Treatments for the sundowning associated with dementia are often inadequate and the condition is extraordinarily difficult to control. The usual hypnotics are ineffective in most cases and antipsychotics are often used. Medications such as haloperidol (0.5–1.0 mg) or thioridazine (25–75 mg) are often used but side-effects (extrapyramidal and postural hypotension, respectively) are a problem. Some clinicians suggested that clonidine (0.4–1.2 mg) may be helpful. These medications tend to be overprescribed in institutional settings, and non-pharmacological methods, although not yet tested experimentally, might be worthy of consideration. These include active restriction of daytime sleep (Bliwise *et al.* 1990), avoidance of disruptive nocturnal bed-checks, and increased daytime light exposure (Campbell *et al.* 1988).

A condition described as REM-sleep behaviour disorder (Schenck *et al.* 1987) may also have some relevance for sundowning. In this, patients (typically elderly men) literally appear to act out their dreams by engaging in apparently complex, purposeful behaviours during REM sleep. At least a few of the patients in the original series were demented, but most were not. Clonazepam has been reported to be a successful treatment at dosages of 0.5 to 1.5 mg at bedtime.

Narcolepsy

As is the case for sleep apnoea, so the hallmark symptom of narcolepsy is persistent, profound daytime sleepiness. Patients with narcolepsy may also have symptoms of cataplexy (loss of muscle tone with strong emotion), sleep paralysis (an inability to move despite being conscious at the beginning or end of the sleep), and hypnagogic hallucinations. It is these extra symptoms that aid in the differential diagnosis from sleep apnoea. Narcolepsy must be diagnosed polysomnographically, preferably with a nocturnal and a further daytime test (MSLT or a related procedure). The daytime test allows documentation of the multiple onsets of REM sleep so characteristic of this condition.

Narcolepsy is one of the few sleep disorders whose prevalence shows no age-associated increase in elderly persons. Available evidence instead suggests that the highest incidence is in the second and third decades of life. Because the condition is life long and does not remit, its prevalence remains relatively constant (estimated at 0.01 per cent of the general population) into old age. The discovery that class II major histocompatability-complex antigens are present in such patients suggest a genetic component for the disorder, although the basis for association and mode of inheritance remain unclear (Juji *et al.* 1984; Guilleminault 1989).

Although more frequently encountered in younger

patients, particularly as the often bizarre and dramatic symptoms develop, the physician may encounter an elderly narcoleptic patient who has never been diagnosed and evaluated polysomnographically. Particularly among older patients who have received more limited exposure to medical care over the years, the diagnosis may have never been made. Because the tendency to fall asleep during the day increases with age even in normal persons, the aged narcoleptic patient may have even greater problems with daytime somnolence.

Treatments for narcolepsy include stimulant medication for sleepiness (e.g., pemoline, methylphenidate, amphetamine, and others) and tricyclic antidepressants for cataplexy (e.g., protriptyline) and REM-related phenomena. Frequently, concurrent usage of a tricyclic and stimulant can reduce the required dosage for the latter. For the elderly narcoleptic patient with ischaemic heart disease, dosages of such medications should be kept as low as possible. There is also some evidence that regularly scheduled naps can minimize the untoward effects of daytime somnolence.

References

Ancoli-Israel, S., Klauber, M.R., Kripke, D.F., and Parker, L. (1988). Increased risk of mortality with sleep apnea in nursing home patients: preliminary report. *Sleep Research*, **17**, 140.

Ancoli-Israel, S., Kripke, D.F., and Mason, W. (1987). Characteristics of obstructive and central sleep apnea in the elderly: an interim report. *Biological Psychiatry*, **22**, 747–50.

Association of Sleep Disorders Centers. (ASDC) (1979). Diagnostic classification of sleep and arousal disorders, first edition. *Sleep*, **2**, 1–137.

Bliwise, D.L. (1989*a*). Dementia. In *Principles and practice of sleep medicine* (ed. M. Kryger, T. Roth, and W.C. Dement), pp. 358–64. Saunders, Philadelphia.

Bliwise, D.L. (1989*b*). Neuropsychological function and sleep. *Geriatric Clinics of North America*, **5**, 381–94.

Bliwise, D.L., Petta, D., Seidel, W., and Dement, W.C. (1985). Periodic leg movements during sleep in the elderly. *Archives of Gerontology and Geriatrics*, **4**, 273–81.

Bliwise, D.L. *et al.* (1987). Risk factors for sleep disordered breathing in heterogeneous geriatric populations. *Journal of the American Geriatrics Society*, **35**, 132–41.

Bliwise, D.L., Bliwise, N.G., Partinen, M., Pursley, A.M. and Dement, W.C. (1988). Sleep apnea and mortality in an aged cohort. *American Journal of Public Health*, **78**, 544–7.

Bliwise, D.L., Yesavage, J.A., Tinklenberg, J., and Dement, W.C. (1989). Sleep apnea in Alzheimer's disease. *Neurobiology of Aging*, **10**, 343–6.

Bliwise, D.L., Bevier, W.C., Bliwise, N.G., Edgar, D.M., and Dement, W.C. (1990). Systematic 24-hour behavioral observations of sleep/wakefulness in a skilled care nursing facility. *Psychology and Aging*, **5**, 16–24.

Block, A.J., Hellard, D.W., and Slayton, P.C. (1986). Effects of alcohol ingestion on breathing and oxygenation during sleep: Analysis of the influence of age and sex. *The American Journal of Medicine*, **80**, 595–600.

Campbell, S.S., Kripke, D.F., Gillin, J.C., and Hrubovcak, J.C. (1988) Exposure to light in healthy elderly subjects and Alzheimer's patients. *Physiology and Behavior*, 42, 141–4.

Carskadon, M.A. and Dement, W.C. (1985). Sleep loss in elderly volunteers. *Sleep*, **8**, 207–21.

Carskadon, M.A., Seidel W.F., Greenblatt, D.J., and Dement, W.C. (1982). Daytime carryover of triazolam and flurazepam in elderly insomniacs. *Sleep*, **5**, 361–71.

Cornoni-Huntley, J., Brock, D.B., Ostfeld, A.M., Taylor, J.O., and Wallace, R.B. (ed.) (1986). *Established populations for epidemiological studies of the elderly*, NIH Publication No. 86–2443. National Institute on Aging, Washington DC.

Dempsey, J.A. and Skatrud, J.B. (1986) A sleep-induced apneic threshold and its consequences. *American Review of Respiratory Disease*, **133**, 1163–70.

Friedman, L., Bliwise, D.L., Yesavage, J.A., and Salom, S.R. (1991) A preliminary study comparing sleep restriction and relaxation treatments for insomnia in older adults. *Journal of Gerontology: Psychological Sciences*, **46**, P1–P8.

Gillick, M.R., Serrell, N.A., and Gillick, L.S. (1982). Adverse consequences of hospitalization in the elderly. *Social Science and Medicine*, **16**, 1033–8.

Gislason, T. and Almqvist, M. (1987). Somatic diseases and sleep complaints. *Acta Medica Scandinavica*, **221**, 475–81.

Guilleminault, C. (1989). Narcolepsy syndrome. In *Principles and practice of sleep medicine* (ed. M.H. Kryger, T. Roth, and W.C. Dement), pp. 338–46. Saunders, Philadelphia.

Hauri, P. and Fisher, J. (1986). Persistent psychophysiologic (learned) insomnia. *Sleep*, **9**, 38–53.

Jacobs, E.A., Reynolds, C.F., III., Kupfer, D.J., Lovin, P.A., and Ehrenpreis, A.B. (1988). The role of polysomnography in the differential diagnosis of chronic insomnia. *American Journal of Psychiatry*, **145**, 346–9.

Juji, T., Satake, M., Honda, T., and Doi, Y. (1984). HLA antigens in Japanese patients with narcolepsy. All patients were DR2 positive. *Tissue Antigens*, **24**, 316–19.

Kales, A. and Kales, J.D. (1984). *Evaluation and Treatment of Insomnia*. Oxford University Press, New York.

Karacan, I., Thornby, J.I., and Williams, R.L. (1983) Sleep disturbance: a community survey. In *Sleep/wake disorders: natural history, epidemiology, and long-term evolution* (ed. C. Guilleminault and E. Lugaresi), pp. 37–60. Raven Press, New York.

Kryger, M.H., Mezon, B.J., Acres, J.C., West, P., and Brownell, L. (1982). Diagnosis of sleep breathing disorders in a general hospital: experience and recommendations. *Archives of Internal Medicine*, **142**, 956–8.

Levkoff, S.E., Besdine, R.W., and Wetle, T. (1986). Acute confusional states (delirium) in the hospitalized elderly. *Annual Review of Gerontology and Geriatrics*, **6**, 1–26.

Nino-Murcia, G., McCann, C.C., Bliwise, D.L., Guilleminault, C., and Dement, W.C. (1989). Compliance and side effects in sleep apnea patients treated with nasal CPAP. *Western Journal of Medicine*, **150**, 165–9.

Pack, A.I., Silage, D.A., Millman, R.P., Knight, H., Shore, E.T., and Chung, D-C.C. (1988). Spectral analysis of ventilation in elderly subjects awake and asleep. *Journal of Applied Physiology*, **64**, 1257–68.

Partinen, M., Kaprio, J., Koskenvuo, M., and Langinvaino, H. (1983). Sleeping habits, sleep quality, and use of sleeping pills: a population study of 31,140 adults in Finland. In *Sleep/wake disorders: natural history, epidemiology and long-term evolution* (ed. C. Guilleminault and E. Lugaresi), pp. 29–35. Raven Press, New York.

Prinz, P.N. *et al.* (1984). Circadian temperature variation in healthy aged and in Alzheimer's disease. *Journal of Gerontology*, **39**, 30–5.

Scharf, M.B., Kauffman, R., Brown, L., Segal, J.J. and Hirschowitz, J. (1986). Morning amnestic effects of triazolam. *Hillside Journal of Clinical Psychiatry*, **8**, 38–45.

Schenck, C.H., Bundlie, S.R., Patterson, A.L., and Mahowald, M.V. (1987). Rapid eye movement sleep behavior disorder: a treatable parasomnia affecting older adults. *Journal of the American Medical Association*, **257**, 1786–9.

Spielman, A.J., Saskin, P., and Thorpy, M.J. (1987) Treatment of chronic insomnia by restriction of time in bed. *Sleep*, **10**, 45–56.

Wechsler, L.R., Stakes, J.W., Shahani, B.T., and Busis, N.A. (1986). Periodic leg movements of sleep (nocturnal myoclonus): an electrophysiological study. *Annals of Neurology*, **19**, 168–73.

White, D.P., Lombard, R.M., Cadieux, R.J., and Zwillich, C.W. (1985). Pharyngeal resistance in normal humans: influence of gender, age, and obesity. *Journal of Applied Physiology*, **58**, 365–71.

18.9 Subdural haematoma

JOHN R. BARTLETT

INTRODUCTION

The subdural space is a potential cavity or space between the dura and the arachnoid mater. Normally it contains a thin film of serous fluid. It does not communicate with the subarachnoid space. Blood entering the subdural space produces a subdural haematoma which, if small, neither increases the intracranial pressure nor interferes with the function of the brain, and may therefore be symptomless. Acute haemorrhage into the subdural space is most often due to injury. Severe injuries in which the forces provided sufficient energy to render the patient unconscious and fracture the skull generally produce a cortical laceration, contusion, and/or rupture of cerebral vessels. Brain damage is a prominent feature, so signs of disordered function are prominent. In contrast, a trivial injury may rupture one of the fragile cortical veins as it enters the sagittal sinus, allowing blood to enter the subdural space. In the upright posture the intracranial pressure is low and blood can enter the subdural space easily, which may explain why large chronic subdural haematomas are usually a sequel of trivial injury. There is no primary damage to the brain, so signs are minimal.

However, not all acute haematomas are due to severe trauma. A few are associated with haemorrhagic diatheses, notably anticoagulants, and rarely with intracranial haemorrhage due to aneurysm or tumour. Like chronic haematomas the effects tend to be those of a large, space-occupying lesion within the cranial cavity.

While it is customary to classify subdural haematomas as acute, subacute, and chronic, it is equally important to take account of the extent of the underlying primary brain damage. For the most part, whereas an acute haematoma is associated with severe primary brain damage, a chronic haematoma is not, and the two conditions tend to form distinct clinical entities.

NATURAL HISTORY AND PATHOLOGY

A subdural haematoma is a dynamic and not a static entity. An initiating factor, usually an injury, allows blood to enter the subdural space. The presence of blood in that space provokes an inflammatory reaction in the overlying dura, which leads to the gradual removal of the haematoma. This process is not always successful and an expanding, life-threatening haematoma forms. Certain factors, which have the common feature of allowing a large volume of blood to enter the subdural space, favour this development. They are conveniently divided into those which increase the potential size of the subdural space and those reducing the coagulability of the blood. Examples of the former include advancing age, which produces a gradual reduction in the size of the brain (Tomlinson 1979); cerebral atrophy (whether due to parenchymal disease or secondary to vascular insufficiency); and hydrocephalus (especially if there is a functioning diversion of cerebrospinal fluid). Bleeding eventually ceases as a result of coagulation and/or the counter-pressure of raised intracranial pressure.

The natural history of haematomas in the subdural space is conveniently divided into three stages—acute, the first week; subacute, the second and third weeks; and chronic, beginning at the fourth week—but the distinction is not absolute, each stage merging imperceptibly into the next. Resolution of the haematoma or death from cerebral compression may occur during any stage.

As soon as the healing process begins the haematoma ceases to be an homogeneous entity. The haematoma capsule, which is the result of the dural reaction, is characterized by capillary growth, formation of sinusoids resembling veins (but without the normal investing layers), and fibroblastic activity. As the haematoma gets older the number of capillaries is reduced and the fibroblasts mature. The haematoma is gradually absorbed and the inner endothelial layer adjacent to the pia-arachnoid comes to lie against the inner membrane, which is ultimately indistinguishable from normal dura. Not all cases resolve. In some patients the haematoma increases in size, producing signs due to distortion of the brain or raised intracranial pressure. Two theories have been put forward to explain the increase in size of chronic haematomas. Gardner (1932) postulated that the capsule acts as a semipermeable membrane and that as the haematoma is broken down into smaller particles, fluid is drawn in and increases the volume. There are several objections to this theory, on which the rational use of osmotic diuretics depends. Fresh erythrocytes are regularly found in the haematoma fluid. Ito *et al.* (1976), using red cells labelled with ^{51}Cr, demonstrated that daily fresh haemorrhage may amount to 10 per cent of the volume of the haematoma. Weir (1980) showed that the osmotic pressure of the haematoma fluid is the same as cerebrospinal fluid. Studies of the ultrastructure of the 'membrane' make it clear that the capsule is not a simple semipermeable membrane. Putnam and Cushing (1925) postulated that bleeding occurs intermittently from the outer membrane into the cavity. This theory explains why fresh red cells and albumin are found in fluid aspirated from chronic haematomas and also gives a convincing reason for the growth.

Clearly the development of the vascular membrane or capsule is the means by which blood is removed from the subdural space as well as source of material for those cases where the haematoma increases in size. What determines why some haematomas enlarge while others resolve is not clear but there is no doubt that absolute size is an important factor. Apfelbaum *et al.* (1974) pointed out that the larger the volume the smaller proportionately is the surface area. If the rate of absorption is directly related to the surface area of absorbing membrane, then larger haematomas will take longer to absorb. Recurrent haemorrhage could interrupt the absorption process at any time and lead to an increase in size of the haematoma. It is possible that haematomas must reach a critical size to become chronic and perhaps physical activity on the part of the patient favours recurrent minor haemorrhage, which might explain why the formation of chronic haematoma is rare in patients who have sustained more severe injuries (brain swelling reducing the potential size of the subdural space) with haemorrhage into the subdural space. Evacuation of haematomas through burr holes is not a complete process: the brain does not always expand and a subsequent computed tomographic (**CT**) scan may show little initial change. Why evacuation does not precipitate further haemorrhage from the fragile capillaries into the cavity is also unclear but removal of the fibrinolytic activity and fibrin degradation products may be important. None the less a minor procedure, a burr hole or twist-drill aspiration, seems to tip the balance back in favour of the healing process.

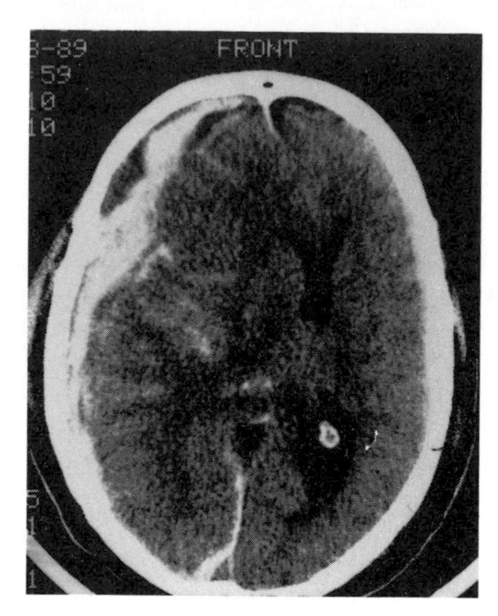

Fig. 1 A CT scan of patient who sustained a severe head injury complicated by extensive acute subdural haemorrhage. There is marked shift of the midline structures which is mainly due to the haemorrhagic contusion and swelling of the cerebral hemisphere. Acute subdural haemorrhage is usually a reflection of a severe primary injury to the brain and carries a grave prognosis.

THE RELATIONSHIP BETWEEN CLINICAL PRESENTATION, PATHOLOGY, AND OUTCOME

Acute subdural haematomas, due to head injury, are generally associated with contusion and swelling of the brain, which restricts their size (Fig. 1). Coma and focal signs are therefore prominent. Recovery of function is usually incomplete and fatality is high.

Acute extradural haemorrhage, which may be confused with subdural haemorrhage, is rare in elderly people because as age increases the dura becomes more adherent to the bony vault. On CT scan the haematoma is lenticular (Fig. 2). The distinction is important because extradural haemorrhage, if treated, carries a better prognosis.

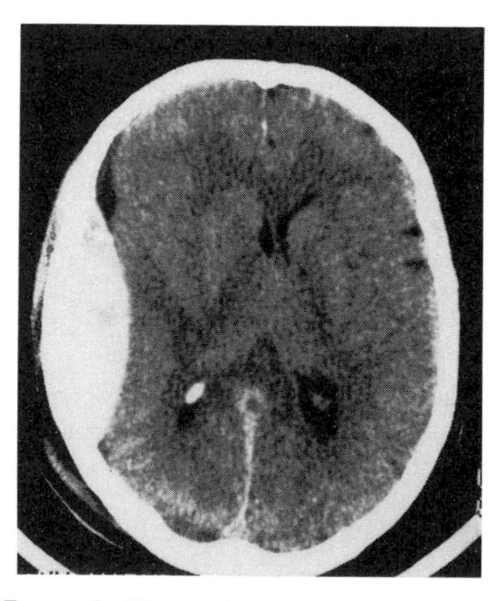

Fig. 2 A CT scan of a 63-year-old patient showing the typical lenticular appearance of an acute extradural haematoma. This patient developed a mild hemiparesis and increasing drowsiness but was able to carry out simple commands immediately before evacuation. She made a full recovery. Extradural haematoma carries a good prognosis, if treated early, and should not be confused with an acute subdural haematoma.

Contrasted are the chronic haematomas, which are amongst the largest space-occupying lesions encountered in clinical practice. It is well known that timely drainage often produces a dramatic recovery. Chronic haematomas exert their pressure over a large area of undamaged brain and the local forces on the brain are small. For these reasons, with well-directed management the prognosis is excellent and restoration of function is to be expected. Focal signs are usually minimal or may be completely absent. Symptoms and signs, confusion and intellectual deterioration often with headache, generally suggest a diffuse disorder of cerebral function. The rise in intracranial pressure responsible for these effects is rarely sufficient to produce papilloedema. Progressive reduction of consciousness with the development of defects of ocular movements and pupillary abnormalities indicate impaired upper brain-stem function. These signs are due to the brain-stem distortion produced by any large, space-occupying lesion within the cranial cavity. When present to a marked degree, these are signs of approaching death. They are an indication that evacuation of the haematoma is urgent; recovery of function is often incomplete when surgery is delayed to this extent.

CLINICAL DAIGNOSIS

It is not possible to establish or refute the diagnosis with certainty on clinical criteria alone.

Acute haematoma

The acute haematoma is generally associated with a clear history of head injury. Such patients are usually seen in accident and emergency departments where patients are assessed for the global effects of the injury. A group of British neurosurgeons have set out guidelines (Suggestions from a group of neurosurgeons 1984), applicable to adults, for hospital admission (Table 1); skull radiography (Table 2); and consultation with a neurosurgeon (Table 3). Incomplete recovery of consciousness, significant scalp trauma, focal signs, and the presence of a skull fracture are all associated with increased risk of an intracranial haematoma, which may be intracerebral, subdural, or extradural, singly or in combination. A full discussion of the management of trauma is outside the scope of this chapter.

Table 1 Indications for admission to a general hospital following head injury

1. Confusion or any other depression of the level of consciousness at the time of examination
2. Skull fracture
3. Neurological symptoms or signs
4. Difficulty in assessing the patient—for example, alcohol, epilepsy, or other medical condition
5. Lack of a responsible adult to supervise the patient; other social problems

Note. Brief amnesia after trauma with full recovery is not sufficient indication for admission. Relatives or friends of the patient should receive written advice about changes that would require the patient to be returned immediately to hospital.

Table 2 Guidelines for skull radiography after recent head injury

1. Loss of consciousness or amnesia at any time
2. Neurological symptoms or signs
3. Cerebrospinal fluid or blood from the nose or ear
4. Suspected penetrating injury
5. Scalp bruising or swelling

Table 3 Indications for consultation with a neurosurgeon in cases of recent head injury

1. Fractured skull with any of the following: confusion of worse impairment of consciousness, one or more epileptic fits, or any other neurological symptoms or signs
2. Coma continuing after resuscitation
3. Deterioration in the level of consciousness
4. Confusion or other neurological disturbances persisting for more than 8 h
5. Depressed fracture of the skull vault
6. Suspected fracture of the skull base (cerebrospinal fluid rhinorrhoea or otorrhoea, bilateral orbital haematoma, mastoid haematoma, or evidence of penetrating type of injury

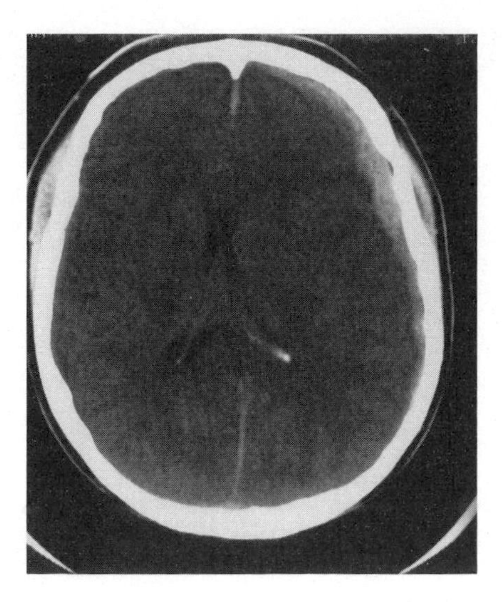

Fig. 3 A CT scan showing an extensive narrow rim of blood in the subdural space in a patient on anticoagulant therapy. The brain shows no sign of intrinsic damage or pre-existing atrophy. This haematoma produced a degree of life-threatening compression with ocular motor palsies and required emergency evacuation.

The sudden development of severe headache in a patient on anticoagulants should arouse the suspicion of an acute subdural haemorrhage (Fig. 3), especially when there is a paucity of focal neurological deficits referable to the cerebral hemispheres or cerebellum. Because of the coagulation defect, bleeding is usually profuse so features of raised intracranial pressure, drowsiness, and disorders of ocular movements (signs of brain-stem compression) are common. The dearth of focal signs generally differentiates subdural from intracerebral haematoma and the relative lack of neck stiffness and absence of a Kernig's sign differentiates the condition from primary subarachnoid haemorrhage. It must be emphasized that acute spontaneous haemorrhage into the subdural space with normal blood coagulation is very rare. However, if there is serious doubt about the diagnosis, the matter must be resolved by specialist investigation.

Chronic haematoma

In the absence of any history of injury, chronic haematoma may be very difficult to diagnose. Diffuse symptoms of a cerebral disorder (personality change and intellectual change) are common; focal signs may occur (such as a mild hemiparesis) but signs of focal tissue destruction (such as hemiplegia, hemianopia, and global aphasia) are rare; signs of brain-stem compression (altered consciousness, defective ocular movements, particularly upward gaze, which is difficult to assess in elderly patients, and pupillary inequality) are relatively common. In any patient with a short history of intellectual deterioration, defects in blood coagulation, or circumstances that are associated with repeated mild injury should alert the clinician to the possibility of a chronic subdural haematoma. Most patients present with a history of 6 to 8 weeks' duration; a history of symptoms over more than 3 months is rare.

Many patients have mild and persistent headache without any physical signs after head injury and only rarely is this due to a chronic haematoma. Even if a haematoma is demon-

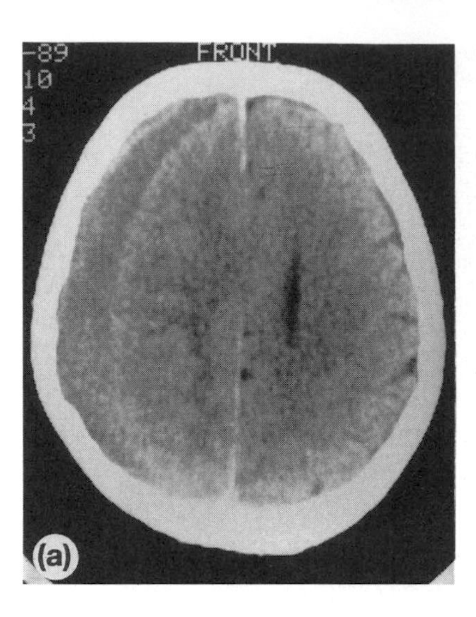

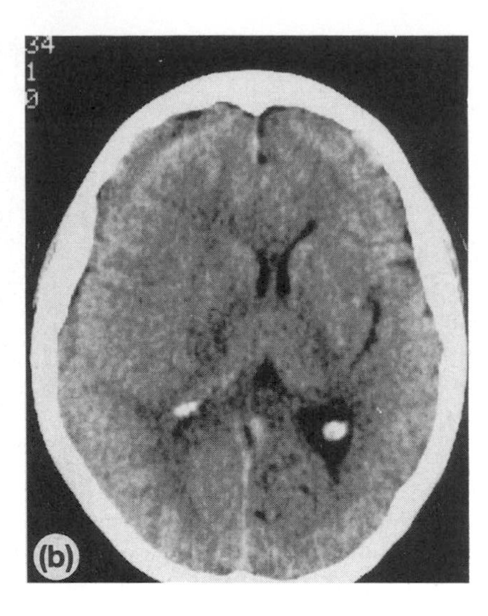

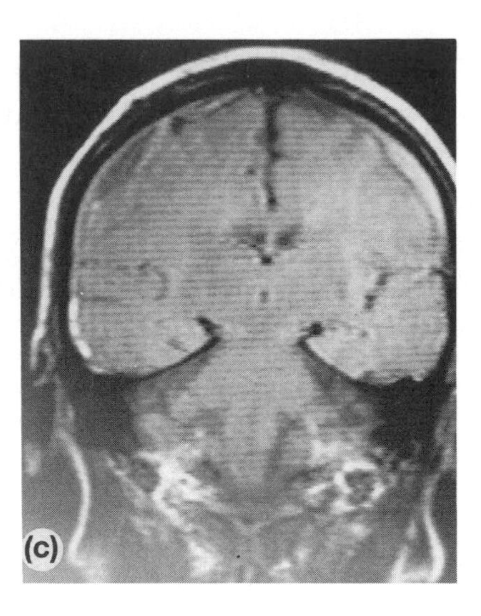

Fig. 4 CT scans showing a large almost isodense chronic subdural haematoma. (a) The thick and extensive crescentic extracerebral lesion is clearly demonstrated. (b) There is a marked shift of the midline structures and medial displacement of the choroid plexus. The cerebral hemisphere on the side of the haematoma is not swollen and has a normal density. (c) T_1 weighted magnetic resonance image showing bilateral subdural haematomas of differing ages.

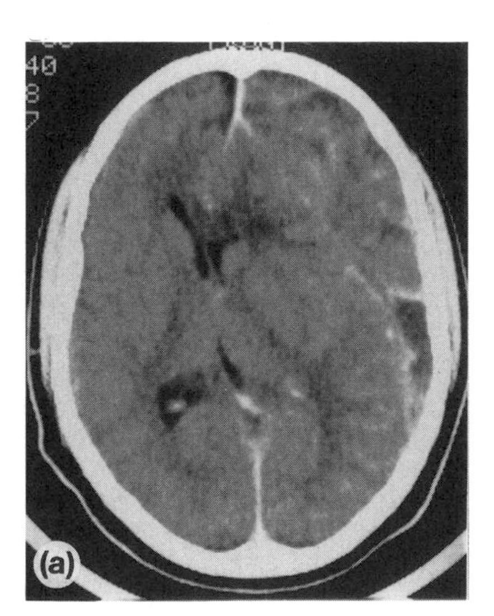

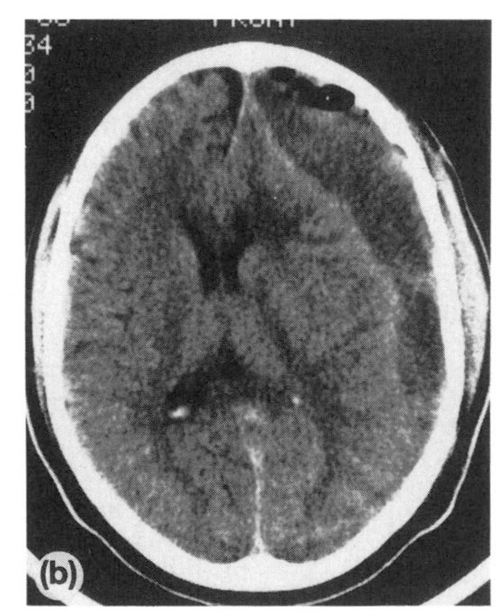

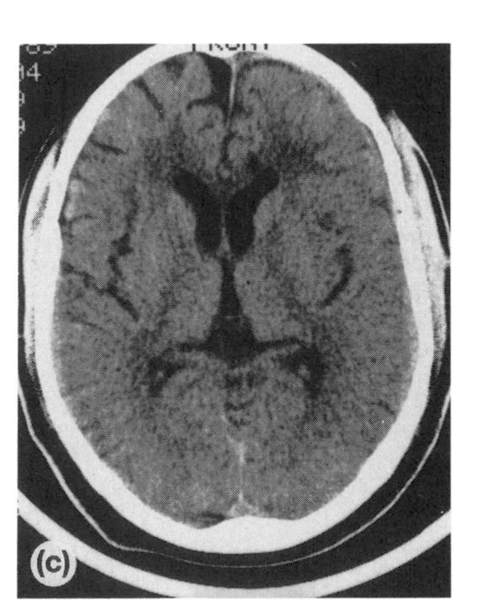

Fig. 5 CT scan of a loculated haematoma presenting with mild headache and, unusually for a chronic haematoma, papilloedema. (a) Before evacuation. (b) After an attempt at burr hole drainage which was repeated with equal lack of success. (c) Three months later, after several weeks of bedrest followed by restricted physical activity, showing complete resolution.

strated it is usually small, and left alone absorbs without recourse to surgery. The routine investigation of such patients with CT is not justified.

Differentiation of chronic subdural haematoma from diffuse cerebrovascular disease and intracranial tumour may be very difficult. If present, marked fluctuation in the level of consciousness is very characteristic of chronic haematoma and is possibly due to recurrent haemorrhage. When papilloedema is present, there is a clear indication for specialist investigation.

INVESTIGATION

The definitive investigation is now CT scanning or magnetic resonance imaging (**MRI**), (Fig. 4(a–c)). Before the introduction of these methods, diagnosis depended on angiography or burr-hole exploration of the subdural space. The plain skull radiograph may reveal a displaced pineal. Isotope imaging has proved disappointing for routine use and is not a satisfactory substitute for good clinical appraisal and CT or MRI if indicated. The acute, subacute, and chronic phases in the natural history of subdural haematoma correspond to the hyper-, iso-, and hypodense appearances on a CT scan. The three stages are equally distinct on MRI: the acute haematoma is isointense with brain on T_1- and hypointense on T_2-weighted images; during the subacute phase, lysis of the red cells and oxidation of deoxyhaemoglobin to methaemoglobin tends to shorten T_1 and lengthen T_2, which will increase the intensity on either T_1- or T_2-weighted images. The continued breakdown of methaemoglobin produces compounds that are not paramagnetic and have

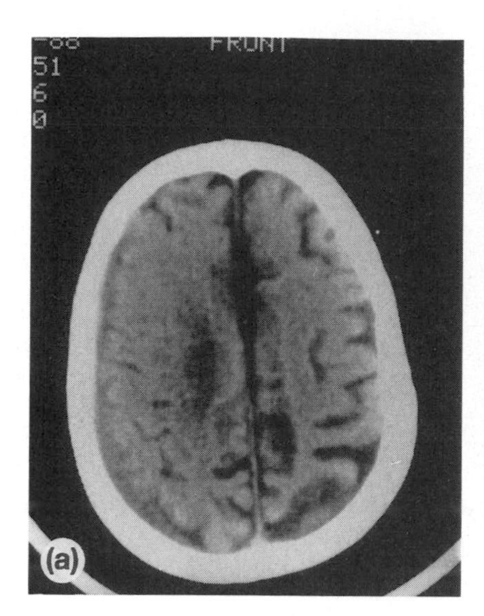

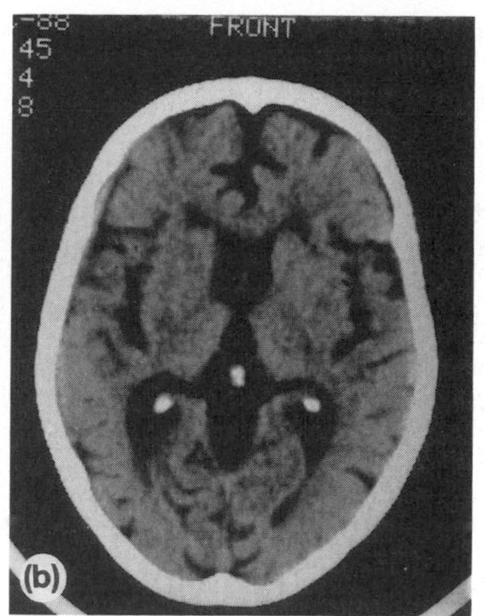

Fig. 6 (a) A CT scan showing a surface collection with a density a little greater than CSF and partial obliteration of the cortical sulci consistent with a chronic subdural haematoma or hygroma. (b) A lower section in the same patient shows severe cerebral atrophy and no midline shift. There is nothing to be gained from evacuation of lesions of this kind.

reduced T_1 values, and thus the signal intensity is reduced. Ultimately the fluid, if it is not fully absorbed, becomes a subdural hygroma behaving similarly to cerebrospinal fluid on MRI (and in this respect analogous to CT).

While CT has proved a very reliable means of making the diagnosis, it is still possible to miss isodense, bilateral, subdural haematomas and it may be necessary to resort to angiography or exploratory burr holes in an exceptional case.

MANAGEMENT

It is almost universally accepted that symptomatic progressive haematomas should be evacuated, whether acute or chronic. Treatment has two aims—first, to save life when this is threatened by compression of the midbrain; and second, to remove (sufficient of) the haematoma to tip the balance of the healing process back in favour of the patient. Acute haematomas are usually solid, and it is not possible to remove them without recourse to craniotomy. Chronic haematomas have a large liquid component, which is easily evacuated through one or more burr holes. Epilepsy is usually a reflection of injury to or underlying pathology in the brain; it is most usually associated with acute traumatic haematoma and is rare in the chronic variety.

Occasionally the cavity of the haematoma is loculated and a craniotomy is necessary; this is most likely to occur in the subacute case and is perhaps a reflection of patchy liquefaction. In long-standing haematomas, loculation is more likely, owing to further haemorrhage (Fig. 5). The importance of loculated haematomas lies in the difficulty in obtaining enough reduction in volume to relieve brain-stem compression and intracranial pressure without craniotomy. Fortunately such circumstances are rare.

The knowledge that many haematomas resolve spontaneously has blurred the distinction between acute and chronic haematomas and led to the exploration of conservative methods of management. Bender and Christoff (1974), who reported 100 cases, demonstrated unequivocally that all do not need removal. However, a chronic haematoma with progressive symptoms and signs may have catastrophic effects on the nervous system. Attempts to treat chronic haematoma with infusions of 20 per cent mannitol failed in a controlled trial (Gjerris and Schmidt, 1974), despite Suzuki and Takaku's (1970) report of success in an uncontrolled series.

A difficulty yet remains. There are patients whose symptoms are a consequence of the underlying conditions that predispose to the accumulation of fluid similar to cerebrospinal fluid in the subdural space. Such conditions might be demonstrated radiographically, for example wide cortical sulci and atrophy (Fig. 6). Evacuation of such collections does not restore brain function. It is unfortunate that small but insignificant haemorrhages occur into these collections, which are then branded as chronic subdural haematomas requiring surgery. The management of these problems is best resolved by frank discussion of the nature and significance of the findings with the family rather than ill-considered surgery.

Burr-hole evacuation remains the treatment of choice for chronic haematomas with progressive symptoms and signs. Solid, usually acute, haematomas and those chronic haematomas that rapidly and repeatedly reaccumulate require craniotomy. Small haematomas discovered simply because the means of diagnosis is readily available can be managed, in the absence of clinical deterioration, conservatively.

Bibliography

Bartlett, J.R. (1984). Should chronic subdural haematomas always be evacuated? In C. Warlow and J. Garfield (ed.), *Dilemmas in the management of the neurological patient*, pp. 215–22. Churchill Livingstone, Edinburgh.

Markwalder, T.M. (1981). Chronic subdural haematomas: a review. *Journal of Neurosurgery*, **54**, 637–45.

Guthkelch, A.N. (1982). The aetiology and evolution of chronic subdural haematomas. In J.M. Rice Edwards. *Topical reviews in neurosurgery*, Vol. 1 (ed.), pp. 122–33. Wright, Bristol.

References

Apfelbaum, R.I., Guthkelch, A.N., and Shulman, K. (1974). Experimental production of subdural haematomas. *Journal of Neurosurgery*, **40**, 336–46.

Bender, M.B. and Christoff, N. (1974). Nonsurgical treatment of subdural haematomas. *Archives of Neurology*, **31**, 73–9.

Gardner, W.J. (1932). Traumatic subdural haematoma with particular reference to the latent interval. *Archives of Neurology and Psychiatry*, **27**, 847–58.

Gjerris, F. and Schmidt, K. (1974). Chronic subdural haematoma—surgery or mannitol treatment. *Journal of Neurosurgery*, **40**, 639–42.

Ito, H., Yamamoto, S., Komai, T., and Mizukoshi, H., (1976). Role of hyperfibrinolysis in the etiology of chronic subdural haematoma. *Journal of Neurosurgery*, **45**, 26–31.

Putnam, T.J. and Cushing, H. (1925). Chronic subdural haematoma: its pathology its relation to pachymeningitis haemorrhagica and its surgical treatment. *Archives of Surgery*, **11**, 329–93.

Suggestions from a group of (British) neurosurgeons. (1984). Guidelines for initial management after head injury in adults. *British Medical Journal*, **288**, 983–5.

Suzuki, J. and Takaku, A. (1970). Nonsurgical treatment of chronic subdural haemotoma. *Journal of Neurosurgery*, **33**, 548–53.

Tomlinson, B.E. (1979). The aging brain. In *Recent advances in neuropathology*, (ed. W. Thomas Smith and J.B. Cavanagh), pp. 129–59. Churchill Livingstone, Edinburgh.

Weir, B. (1980). Oncotic pressure of subdural fluids. *Journal of Neurosurgery*, **53**, 512–15.

18.10 Orthostatic hypotension in elderly people

RALPH H. JOHNSON

INTRODUCTION

A medical student at the beginning of the nineteenth century was taught a straightforward and balanced conclusion about the subject of this chapter, which is difficult to better nearly two centuries later:

> 'A syncope is when the action of the heart, and along with it that of the arteries, is suddenly and very much lessened; whence the animal powers, the senses and voluntary motions, immediately cease. Various kinds of nervous diseases . . . every kind of debility . . . especially great loss of blood, and many kinds of poisons produce fainting. Whatever weakens the motion of the blood through the brain tends to produce fainting. The mere posture of the body may either bring on or keep off fainting, or remove it after it has already come on. This disorder may sometimes be of little consequence and easily removed; at others, very dangerous, not only as a symptom, but even in itself, as sometimes terminating in death.' (*Edinburgh Practice of Physic, Surgery and Midwifery* 1803.)

Later in the nineteenth century the method of measuring blood pressure was developed and it was only subsequent to that discovery that the findings obtained with a sphygmomanometer were related to clinical disorders already recognized. It was not until 1932 that Sir Thomas Lewis delineated the pathophysiological changes in a vasovagal faint, and, when orthostatic hypotension was first described by Bradbury and Eggleston, it was looked upon as a clinical curiosity. The introduction of hypotensive drugs, and also tranquillizers and antidepressants, many of which may precipitate orthostatic hypotension, led to widespread recognition of the need to check a patient's blood pressure both lying and standing.

A fall in blood pressure initially produces a feeling of light-headedness and dizziness, then clouding of consciousness and total loss of consciousness may supervene. The causes of this disorder may be transient, when the loss of consciousness is known as syncope or they may be persistent. Upright posture, micturition, and defaecation may all contribute to the development of syncope, and the mechanisms are discussed by Johnson *et al.* (1984). It is the purpose of this chapter to review the latter disorders. Although the rapid progression of symptoms through dizziness to syncope may lead to clinical suspicion of orthostatic hypotension, there are patients who suffer from this problem in whom symptoms are masked: an elderly person who gets out of bed slowly and sits in a chair, or who only takes a few steps, may spend part or even much of the day confused through cerebral hypoperfusion as a result of reduced blood pressure, and yet the diagnosis may only be made fortuitously. It is essential, therefore, to make lying and standing measurements of blood pressure a clinical routine in the assessment of elderly patients.

This review will consider the normal regulation of blood pressure in the elderly person and the maintenance of cerebral perfusion, and then review the disorders in which it is disturbed. In many patients the symptoms occur because of autonomic failure, either as a separate syndrome or secondary to well-recognized disorders. However, in some elderly patients, it appears that orthostatic hypotension does not result from autonomic failure but is probably due to degeneration affecting the vascular system; this situation, often called descriptively orthostatic hypotension of the elderly is next reviewed. In both groups of patients, glucose ingestion may further precipitate symptoms and this chapter comments on this problem and also reviews treatment.

BLOOD PRESSURE REGULATION IN NORMAL ELDERLY PEOPLE

Arterial blood pressure depends primarily on peripheral resistance and cardiac output. It is kept relatively constant by a number of mechanisms, particularly the baroreceptor reflexes. With increasing age, the usual range of arterial blood pressure gradually increases. Systolic blood pressure increases much more markedly than diastolic blood pressure. There is also a wider range of systolic blood pressure in the old compared with the young, although the range of diastolic pressure is much less. These changes may be due to alterations in the sensitivity of the baroreceptor reflex or in the structure of vessels themselves (Johnson 1976).

The chief baroreceptors are those in the carotid sinus, near the bifurcation of the carotid artery. Others occur in the arch of the aorta. These receptors pass afferent impulses to the brainstem via the glossopharyngeal and the vagal nerves, respectively. The efferent part of the reflex arc depends on parasympathetic fibres in the vagus nerves, activation of which causes slowing of the heart, and sympathetic fibres, both to the heart and to blood vessels. Increased sympathetic activity causes cardiac acceleration and peripheral vasodilatation. The baroreceptor reflexes are called into play when the circulation is stressed. Thus, when a person stands, there is an increase in heart rate, which is usually greater in the young than in the old. In elderly people, the increase may be only 10 to 15 beats/min, compared with 20 beats/min in younger people. The difference implies that baroreceptor activity diminishes with age (Bristow *et al.* 1969).

BARORECEPTOR REFLEX SENSITIVITY

Baroreceptor reflex sensitivity reduces with age, at least up to 65 years, and with raised blood pressure (Fig.1) (Bristow *et al.* 1969). It may be assessed by recording blood pressure

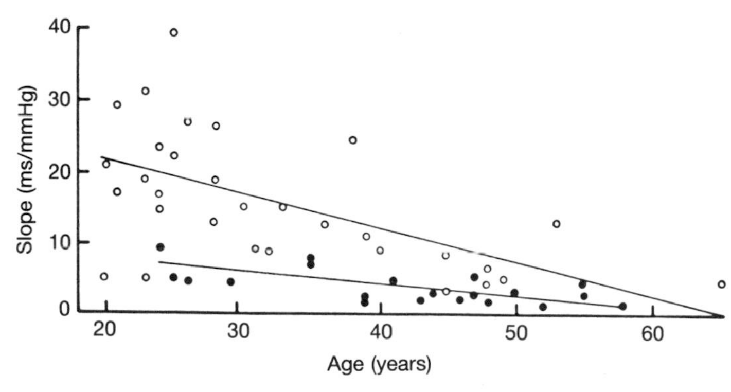

Fig. 1 Pulse interval related to change in blood pressure (ms/mmHg) as an index of baroreflex sensitivity related to age in normotensive subjects (○) and subjects with hypertension (mean arterial blood pressure above 100 mmHg) (●) (taken with permission from Bristow *et al.* 1969). Baroreceptor sensitivity is reduced with increasing age and with hypertension.

through an intra-arterial cannula, and heart rate with an electrocardiograph, during procedures which alter blood pressure. Interpretation from any single measurement must be cautious unless the abnormality is gross, as the correlation between techniques is poor. The techniques have been reviewed by Johnson *et al.* (1984); these may be physiological or pharmacological.

The most commonly used physiological procedure is Valsalva's manoeuvre. In this an acute rise of intrathoracic pressure is obtained by asking the subject to take a deep inspiration and then to maintain a constant attempt at expiration without expiring air. This is done by closing the glottis, usually keeping the nose and mouth closed at the same time. The subject may also be asked to blow into a manometer. Raised intrathoracic pressure is maintained for about 10 s at 40 mmHg. During and after this manoeuvre the blood pressure and heart rate are recorded. The observation of an 'overshoot' above the original blood pressure in the few seconds following the procedure is an indication of baroreceptor-mediated, reflex vasoconstriction. Absence of such an 'overshoot' and failure of the demonstration of any reflex vasoconstriction during the manoeuvre is evidence of either partial or complete baroreflex failure. The changes in blood pressure during the various parts of the manoeuvre have been reviewed (Johnson and Spalding 1974; Johnson *et al.* 1984). Of particular importance is not only absence of an 'overshoot' in systolic blood pressure during recovery from the manoeuvre, but also a lower heart rate during the manoeuvre than in the recovery phase, and a fall in mean blood pressure during the manoeuvre below 50 per cent of the resting mean.

Pharmacological techniques involve intravenous infusion of pressor agents such as angiotensin or phenylephrine. These drugs cause a transient rise in blood pressure and an associated bradycardia due to the consequent baroreflex activity. They have the advantage that no voluntary effort is required. Neck suction may be used, but digital stimulation of the carotid sinus has resulted in death, and neck suction may also have serious consequences. Systolic blood pressure may be plotted against heart rate or pulse (R–R) intervals, and the relationship of change in heart rate to the alteration in blood pressure is an index of baroreflex sensitivity. Decreased sensitivity with ageing (Gribbin *et al.* 1971) may be due to decreased distensibility because of rigidity of the arterial wall in the carotid sinus. This has been shown in postmortem specimens (Fig. 2) (Winson *et al.* 1974). It is also possible that there may be degenerative changes in the afferent nerves of the baroreflex arc and that these are age dependent.

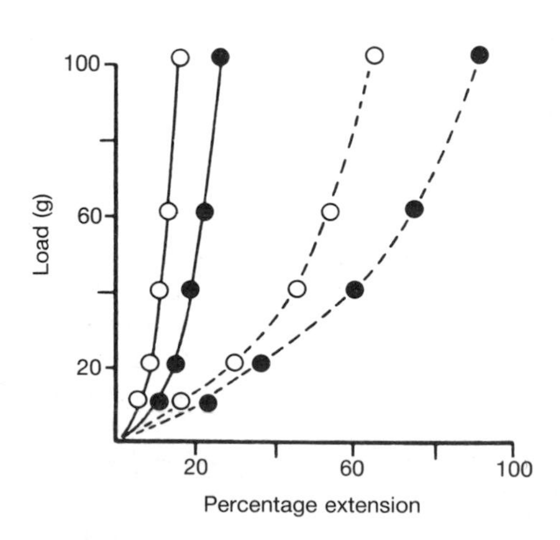

Fig. 2 Circumferential extensibility of the right carotid sinuses (○) and right common carotid arteries (●) in a 27-year-old man (– – –) and a 75-year-old woman (——) (taken with permission from Winson *et al.* 1974). The carotid sinus is more rigid than the common carotid artery in both subjects, and the vessels of the older subject are more rigid than those of the younger.

CEREBRAL BLOOD FLOW

Maintenance of an adequate blood flow to the brain depends both upon the adequacy of the major arterial systems (or the development of anastomoses between them), and also upon autoregulation of the small arteries and arterioles within the brain itself. The major arterial vessels are the two internal carotid arteries and the basilar artery and, at the base of the brain, they are joined together by the vessels of the circle of Willis through which a collateral circulation can develop if there is an occlusion of one of the major vessels.

Autoregulation implies the maintenance of a constant blood flow, despite changes in perfusion pressure, as a result of intrinsic mechanisms within the blood vessels themselves. There may also be involvement of vasomotor nerves. Maintenance of a constant cerebral blood flow occurs within a wide range of mean arterial pressure in normal subjects. In normotensive subjects this ranges from about 60mmHg to about 140mmHg. In chronic hypertensives this range is elevated, so that higher blood pressures are tolerated without an increase in cerebral blood flow (Strandgaard *et al.* 1973).

When the mean blood pressure falls below the lower limit of autoregulation, symptoms of cerebral ischaemia develop. These include dizziness, disorientation, and eventually loss of consciousness. Cerebral autoregulation is generally maintained in patients with orthostatic hypotension so that symptoms only develop when the mean blood pressure falls below the limits in which autoregulation can be maintained. This may be shifted downwards so that mean arterial pressures as low as 40 mmHg may be tolerated without symptoms (Bannister 1988). Although cerebral vessels are innervated by both adrenergic and cholinergic fibres, which may be derived from the superior cervical ganglia, no significant functional role for them has been defined in the regulation of cerebral blood flow. A physiological role of sympathetic fibres cannot, however, be discounted, although cerebral autoregulation is maintained even if sympathetic pathways are completely divided, as in transection of the cervical spinal cord (Nanda *et al.* 1976). In some elderly patients with orthostatic hypotension, however, failure of cerebral autoregulation has been demonstrated and this failure may contribute to the development of symptoms (Fig. 3) (Wollner *et al.* 1979). This finding implies that such patients may be at particular risk of brain damage with only minor falls of blood pressure. The cause of their failure of cerebral autoregulation is still not understood and it is uncertain whether such patients show evidence of autonomic failure. They may have altered blood volume control and abnormal control of plasma vasopressin release, or there may be degenerative changes in the blood vessels themselves.

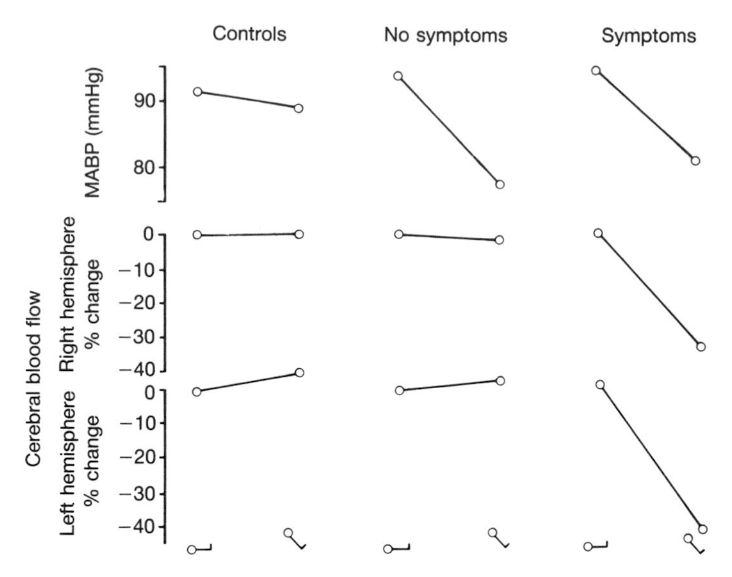

Fig. 3. Mean changes in mean arterial blood pressure (MABP, mmHg) and cerebral blood flow (percentage change) with change of posture (feet down) in elderly patients with orthostatic hypotension and controls. The patients with symptoms of cerebral ischaemia associated with the minor fall in systemic arterial pressure had bilateral or unilateral failure in cerebral autoregulation. These observations suggest that impaired autoregulation may put some elderly patients at risk of brain damage if they suffer only minor falls of blood pressure (data from Wollner *et al.* 1979; graph from Johnson *et al.* 1984, with permission).

HUMORAL CHANGES WITH CHANGE OF POSTURE

Change of posture causes a rise in adrenaline and noradrenaline. The concentrations of these hormones are higher in elderly patients at rest than in younger subjects, so that higher basal concentrations and higher concentrations during various stresses, including standing, are found in elderly patients (Young *et al.* 1980). Concentrations of renin and angiotensin are raised during baroreflex stimulation. Vasopressin release is dependent on afferent stimuli from baroreceptors (reviewed by Johnson *et al.* 1984). Baroreceptor dysfunction may therefore lead to abnormal control of plasma vasopressin release and, consequently, of blood volume. Preliminary observations of vasopressin release in elderly patients with orthostatic hypotension of non-neurological origin suggest that they have an exaggerated release of vasopressin in response to standing (Fig. 4) (Robinson *et al.* 1990).

PREVALENCE OF ORTHOSTATIC HYPOTENSION IN ELDERLY PEOPLE

Orthostatic hypotension has been recognized as a frequent condition in old people. The diagnosis depends upon the postural fall of blood pressure being greater than that found in a normal population, and this is usually defined as a fall of more than 20 mmHg in systolic blood pressure or of more than 10mmHg in diastolic blood pressure. In several series the proportion with a fall of this extent has ranged from 10 to 20 per cent of all people over the age of 65 years, although the majority of studies have been of patients in institutional care (Caird *et al.* 1973). In 5 per cent of a personal series the fall in systolic blood pressure was more than 40 mmHg (Johnson *et al.* 1965). The patients were not receiving any drugs with a hypotensive side-effect. The proportion is much lower in elderly subjects living in the community but, even so, in one series, the overall prevalence was 10.7

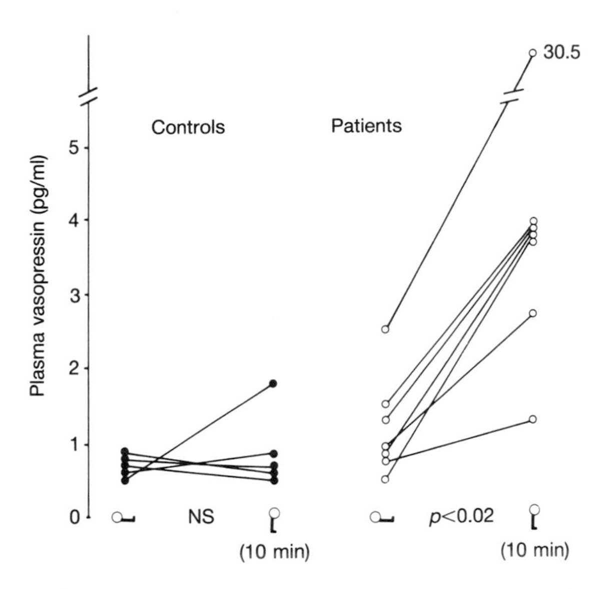

Fig. 4. Plasma vasopressin concentrations in subjects with (○) and without (●) orthostatic hypotension before and 10 min standing. Results are plotted as individual results and as group means ± SE (taken with permission from Robinson *et al.* 1990).

per cent. Increased incidence in institutions may sometimes be related to dehydration and consequent hypovolaemia. It has, however, been suggested that there is no increased incidence in relation to age. In that study, however, the patients were only supine for 5 min before they stood. It is, therefore, important to use standard conditions to determine whether the condition is, in fact, present. The patient should be lying supine for at least 20 min and should then stand of his or her own accord. The blood pressure should be taken 2 min and 5 min after standing.

ASSESSMENT OF ORTHOSTATIC HYPOTENSION

As noted above, it is essential to measure blood pressure both lying and standing in elderly patients. In order to be certain that any observation is not related to sudden standing or due to bed rest after the patient being restricted at home and gradually becoming immobile, it is usually essential to admit the patient to hospital and measure blood pressure lying and standing, 4-hourly, for several days. The extent of the problem will then be apparent, together with its relationship to meals. Careful clinical assessment to determine possible contributory factors, such as water intake and output, the presence of varicose veins, and the contribution of other features, such as cardiac arrhythmias, may then be assessed.

The diagnosis of autonomic failure depends upon careful assessment of autonomic function. This requires assessment of the activity of the sympathetic and parasympathetic systems separately. Sympathetic activity is likely to be disrupted if orthostatic hypotension of a major degree is diagnosed. Catecholamine changes and other tests may then be studied. Parasympathetic function depends upon careful assessment of changes in heart rate, not only on standing but with other provocations such as Valsalva's manoeuvre and deep breathing. The tests required have been described in detail by Johnson (1984), Johnson *et al.* (1984), and Bannister (1988).

CAUSES OF ORTHOSTATIC HYPOTENSION IN ELDERLY PEOPLE

Causes of orthostatic hypotension (Table 1) have been reviewed by Johnson and Spalding (1974), Johnson *et al.* (1984) and in Bannister (1988). Some earlier studies suggested that elderly patients might have orthostatic hypotension as a result of cerebrovascular disease (Johnson *et al.* 1965). However, part of the evidence depended upon the absence of a normal vasoconstrictor response to Valsalva's manoeuvre, and, as already described, it is now clear that this becomes blunted with ageing. Cerebrovascular disease does not appear to alter the lability of blood pressure. An important account by Caird *et al.* (1973) demonstrated the multifactorial nature of the condition. Among the contributory factors are immobility and environmental influences. Even in young patients, orthostatic hypotension may develop when they rise rapidly after long periods (days or weeks) in bed. The combination of a warm bed and a warm room may also contribute to orthostasis. It may at first become apparent after an elderly patient gets out of a hot bath, but many normal individuals of a much younger age have experienced black-outs if this is done too quickly. Other contributory causes in the elderly may include defaecation syncope and cough syncope.

Leaving these causes aside, there are three major groups which should be delineated and, even in these, other factors may contribute to the development of orthostatic hypotension. The one of major significance in terms of incidence is the effect of drugs. The second is the wide range of disorders, particularly neurological, in which it may be a symptom; the third is a group, of considerable significance in the elderly, in which the condition occurs without any definite neurological association and the causation still requires further study.

Drugs

Numerous drugs have orthostatic hypotension as a side-effect, as described in standard textbooks on therapeutics. These include antihypertensive agents, and even the more recently introduced agents, such as calcium channel blocking agents and angiotensin-converting enzyme inhibitors, may have this side-effect. Diuretics may also contribute to its development, particularly in hypertensive patients. Antidepressants are particularly prone to precipitate orthostatic hypotension by blocking cardiovascular reflexes.

Some drugs used in the management of Parkinson's disease, including levodopa and bromocriptine, may produce major degrees of orthostatic hypotension. The combination of a dopa-decarboxylase inhibitor with levodopa appears to reduce this side-effect. Tranquillizing agents may also precipitate or exacerbate orthostatic hypotension by blocking autonomic pathways.

A good drug history is therefore very important. Stopping the offending drug may result in rapid alleviation of symptoms.

Table 1 Neurological disorders in which patients of all ages may develop orthostatic hypotension and the parts of the reflex pathway affected

		Efferent pathway	
Afferent pathway IX or X nerves	Central (brainstem integration)	Spinal cord	Sympathetic chain postganglionic nerves
Acute polyneuropathy	Acute polyneuropathy (?)	Trauma	Acute polyneuropathy
Chronic alcoholism (?)	Acute alcoholism	Transverse myelitis	Pure autonomic neuropathy
Diabetes mellitus (?)	Brainstem lesions	Syringomyelia	Idiopathic orthostatic hypotension
Adie's syndrome	Familial dysautonomia (?)	Intramedullary tumours	Chronic polyneuropathy
Renal failure	Anorexia nervosa (?)	Extramedullary tumours	Chronic alcoholism
Haemodialysis	Drugs	Intermediolateral column degeneration: idiopathic orthostatic hypotension, multiple system atrophy, Parkinson's disease	Diabetes mellitus
			Tumours (non-metastatic complication)
			Rheumatoid arthritis (?)
			Acute intermittent porphyria
			Amyloidosis
			Pernicious anaemia
			Dopamine-β-hydroxylase deficiency
			Failure of catecholamine release, anorexia nervosa
			Drugs

Clinical disorders associated with orthostatic hypotension

In studies of pathophysiology, it is convenient to consider the part of the baroreflex arc affected. It has been shown that in some conditions the lesion may be on the afferent side of the arc, in some it may be central, affecting the brainstem or spinal cord, and in others the condition results from a peripheral neuropathy affecting sympathetic, and some times also parasympathetic, activity.

In detailed descriptions of orthostatic hypotension, acute autonomic neuropathy is usually described first but this is very rare. It may also occur as a result of acute lesions of the spinal cord or acute alcoholism. In both of these problems, however, the cause is obvious.

Among chronic neurological disorders, the following are of particular significance.

Idiopathic orthostatic hypotension, multiple system atrophy (Shy–Drager syndrome).

In these disorders the autonomic nervous system progessively fails. Patients are usually middle aged or elderly, and the system fails as a whole over years or decades, progressing through loss of sweating, impotence, sphincter disturbances, and orthostatic hypotension, of which the last problem is usually the most incapacitating. A distinction must be made between idiopathic orthostatic hypotension, in which autonomic failure occurs alone, and orthostatic hypotension with other non-autonomic neurological deficits. It has been suggested that these two conditions are a continuum but, in most patients, it appears that autonomic failure either occurs independently or in association with other degenerations which develop *pari passu*; the condition is then called multiple system atrophy, first described by Shy and Drager (1960). In both conditions there is neuronal loss in the intermediolateral columns, the site of sympathetic preganglionic cell bodies (Johnson *et al.* 1966).

In multiple system atrophy, lesions may occur in many other parts of the brainstem and spinal cord, including olivopontocerebellar atrophy and degeneration of a variety of tracts including the corticospinal, corticobulbar, extrapyramidal, and cerebellar. Clinical findings may therefore include parkinsonism, spasticity, and ataxia. There is overlap between many of these disorders in which degeneration of parts of the nervous system develops (Fig. 5). Certain combinations are more common than others and these give rise to well-recognized clinical states, but the separation is sometimes somewhat arbitary.

Other structures than the intermediolateral columns may be involved in causing autonomic failure, including the afferent or brainstem part of the circulatory reflex arc, the hypothalamus, or preganglionic vagal neurones. In idiopathic orthostatic hypotension, in which there is sympathetic failure alone, it has been suggested that the postganglionic sympathetic fibres or terminals are particularly affected. In both these conditions there is a reduction in circulating catecholamines. Although it has been suggested that patients with idiopathic orthostatic hypotension may have low levels, whereas those with multiple system atrophy, or at risk of developing it, usually have normal basal levels, this has not been substantiated (Johnson 1983).

Patients with these disorders have a marked increase in alpha-receptor concentration, related to the generally lower plasma concentrations of noradrenaline. Another problem is sleep apnoea, which appears to be a complication of lesions in the region of the respiratory centres of the brainstem (see Chapter 18.8).

In a few patients, familial orthostatic hypotension has been described; they also have other features of the Shy–Drager syndrome.

Parkinson's disease

As noted above, in patients with orthostatic hypotension associated with multiple system atrophy, Parkinsonian features may develop. Occasionally, they may precede the autonomic failure. There also appears to be an association of autonomic failure with idiopathic paralysis agitans. Distinction, however, between idiopathic Parkinson's disease and

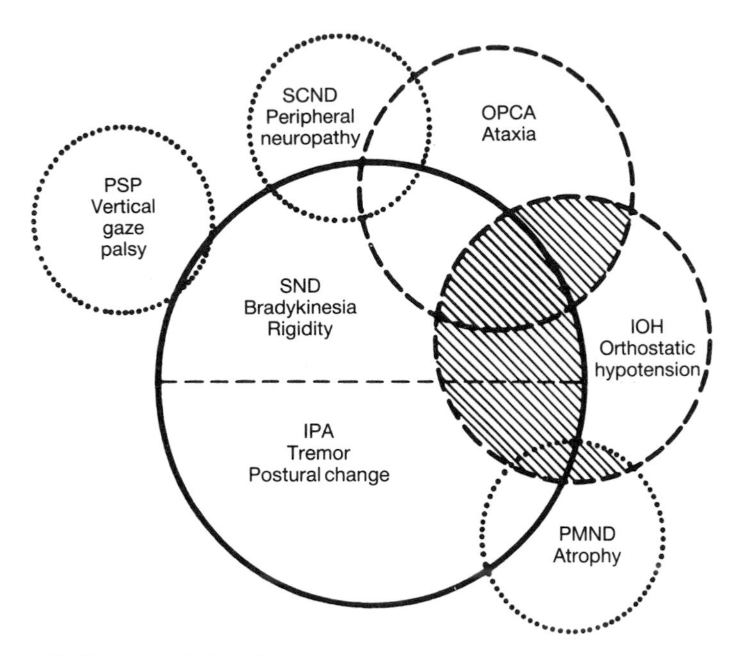

Fig. 5 Inter-relationships between idiopathic orthostatic hypotension and other neurological degenerative disorders. IPA denotes idiopathic Parkinson's disease; SND, striatonigral degeneration; PSP, progressive supranuclear palsy; IOH, idiopathic orthostatic hypotension; OPCA, olivopontocerebellar atrophy; PMND, parkinsonism with motor neurone disease; and SCND, spinocerebellar-nigral degeneration. The central solid circle represents the parkinsonism syndrome. The shaded area indicates the clinical findings in multiple system atrophy (Shy–Drager syndrome) in which there may be considerable variability in the neurological findings in association with orthostatic hypotension. (Adapted from Ropper and Hedley-White, 1983, by Johnson *et al.* 1984, and reproduced with permission.)

parkinsonism with multiple system atrophy depends on the finding of other disorders in association with the parkinsonian features.

In general, circulatory reflexes are intact in patients with Parkinson's disease, but there is frequently some moderate reduction in sympathetic activity.

Orthostatic hypotension is a common problem of drug therapy in Parkinson's disease and may develop in 10 per cent of patients treated with levodopa. Improvement occurs using a dopa-decarboxylase inhibitor. Fludrocortisone may be used as treatment in association with levodopa therapy.

Diabetes mellitus

Autonomic neuropathy, affecting both parasympathetic and sympathetic nerves, is particularly common in insulin-dependent diabetes of long duration. Impotence, bladder problems, nocturnal diarrhoea, and orthostatic hypotension are all commonly reported when the condition develops. Autonomic neuropathy may be a contributing factor to the development of diabetic ulcerations of the foot. It is possible that cold hypersensitivity of denervated blood vessels in the feet might make ischaemia more likely. Good insulation and keeping the feet warm, and also dry, are all of importance. Autonomic neuropathy may also contribute to the development of Charcot's joints, as discussed in the reviews already mentioned.

It is now clear that diabetics in whom autonomic neuropathy develops have a poor prognosis. The fatality rate is about 50 per cent in the period 30 months to 5 years after the development of autonomic neuropathy (Ewing *et al.* 1980). The cause is not always clearly related to autonomic neuropathy. A high proportion of sudden deaths occur, and some may be due to the development of sleep apnoea. Cardiorespiratory arrests appear to be a common explanation of sudden death. It is important to avoid the use of respiratory depressant drugs, and anaesthesia should be very carefully monitored in these patients.

As noted below, in the section on postprandial orthostatic hypotension, insulin may aggravate orthostatic hypotension and it is important to be aware of this possible complication, which occurs in those patients who have developed autonomic neuropathy.

Chronic alcoholism

Chronic alcoholics may have orthostatic hypotension during withdrawal from alcohol. It has been suggested that this is frequently related to vasodilatation during thiamine administration, but this remains unconfirmed. In general, orthostatic hypotension is uncommon in chronic alcoholics. Although there is pathophysiological evidence of parasympathetic degeneration in some patients, sympathetic function remains relatively unimpaired as far as vascular control is concerned (Johnson 1988; Johnson *et al.* 1989). An early sign is, however, a peripheral sympathetic neuropathy causing anhidrosis of the hands and feet.

Chronic brainstem and spinal cord lesions

Any space-occupying lesion involving these structures may affect sympathetic pathways and produce orthostatic hypotension. Tumours in the posterior fossa have been described with this complication. Spinal cord tumours may also produce it. It is a common complication in the later stages of syringomyelia but is rare in the early stages of this disorder. Syringobulbia appears to be more commonly related to its development. Syncope in syringobulbia may, however, result from hindbrain herniation as a result of change in size in the syrinx after coughing or straining: it is not necessarily a sign of loss of autonomic control.

Carcinoma

A variety of carcinomas may present with orthostatic hypotension due to autonomic neuropathy or have it as a complication. It has been reported with bronchial carcinoma (Park *et al.* 1972) and with carcinoma of the pancreas, and may be more common than is appreciated because many patients become debilitated and have problems in rising or appear to develop postural symptoms due to long periods in bed.

Other causes

Orthostatic hypotension due to autonomic failure may also occur in a wide range of other conditions including pernicious anaemia (Eisenhofer *et al.* 1982) and renal failure. Although tabes dorsalis is rarely seen now, orthostatic hypotension is a well-recognized association with it.

There is a variety of conditions which are more likely to develop at a younger age, including acute intermittent porphyria and anorexia nervosa. The differential diagnosis of the condition is wide ranging, as is shown in Table 1. This range of conditions should be borne in mind when the diagnosis is being considered.

Orthostatic hypotension without a definite neurological association

As already described, orthostatic hypotension is frequent in institutionalized elderly patients and may be an important cause of debility in them. In some, a specific disorder with which orthostatic hypotension is commonly associated may be diagnosed, as already described, but in a large proportion there is no evidence of such a disease. There are several disorders of a non-neurological kind that are frequently associated with its occurrence (Fig. 6) (Caird *et al.* 1973). Such

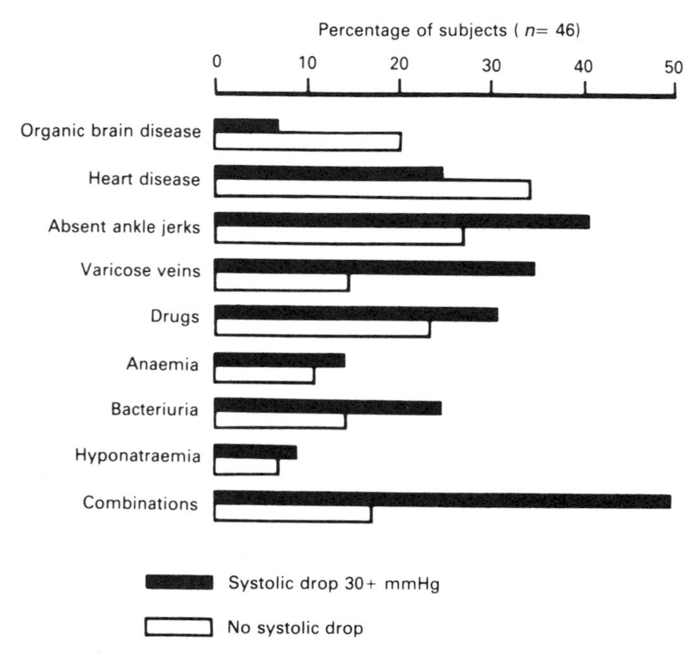

Fig. 6 A study of 496 patients, aged 65 years or more, living at home in which a comparison of the frequency of a series of factors was made between subjects with a drop in systolic pressure of 30 mmHg or more on standing (46 subjects) and a matched group without such a drop. No factor was statistically significant alone, but a combination of factors was significantly more frequent in those with orthostatic hypotension (taken with permission from Caird *et al.* 1973).

is the wide range of possible contributory factors that it is often unclear which is the most important and the following summary attempts to delineate some of the possible issues.

In a personal study, I suggested that the development of orthostatic hypotension in some elderly patients may be related to reduced baroreflex activity as the heart rate and blood pressure changes with Valsalva's manoeuvre were not marked. Efferent sympathetic activity was present and it was therefore suggested that the deficit might be in the baroreceptors themselves or their connections in afferent nerves or the brainstem (Johnson *et al.* 1965). However, ageing produces a depression of baroreflex responses without orthostatic hypotension necessarily being present (Johnson *et al.* 1969). A study of R–R intervals when elderly patients with orthostatic hypotension stood up has suggested that there may be a defective autonomic reflex (White 1980), but our own attempts to repeat those observations were not successful, even though we were examining elderly patients with marked orthostatic hypotension (Fig. 7) (Robinson *et al.* 1983).

We obtained further evidence for normal autonomic function in orthostatic hypotension in the elderly. A rapid rise in concentrations of noradrenaline takes place in patients with this disorder (Fig. 8). The resting concentrations were lower, even though they had a higher supine diastolic blood pressure. Further, if autonomic failure was the cause for orthostatic hypotension in these elderly patients, raised concentrations of alpha-receptors might be expected, but our observations show that receptor number is depressed rather than exaggerated (Fig. 9) (Robinson *et al.* 1990). These findings, in association with the observation that there was no significant difference in the heart rate responses on standing compared with those of normal elderly subjects, imply that autonomic failure was not present in this group of patients (Robinson *et al.* 1983). It is possible that they have increased rigidity of blood vessels, which would be in keeping with observations of elevated diastolic and systolic blood pressure (MacLennan *et al.* 1980). A positive correlation between postural changes in systolic blood pressure and systolic blood pressure on lying in such elderly patients has been reported. These findings suggested that a change of structure in the vascular tree may be more important than autonomic dysfunction. Loss of elasticity in arterial walls has been demonstrated. Postmortem specimens from the carotid artery are less distensible with increasing age (see Fig. 2) (Winson *et al.* 1974), and this may contribute to loss of baroreflex sensitivity (see above). Other mechanical factors have been implicated, such as varicose veins or thrombophlebitis (Caird *et al.* 1973).

Failure of sympathetically mediated vasoconstriction may occur due to adrenoreceptor dysfunction. There is evidence of decreased beta-adrenoreceptor activity as measured by cyclic AMP response to beta-stimulation in elderly subjects. This could be related to the generally higher plasma concentrations of noradrenaline found in the elderly. There may, however, be a primary disturbance of the cellular response to beta-receptor stimulation rather than alteration in beta-receptor number. It has been suggested that there could be a reduction in alpha-adrenoreceptor activity in relation to age, but further studies, *in vitro*, have failed to show this relationship.

Some patients with a minor degree of orthostatic hypotension have hyponatraemia, and potassium depletion has also been reported. The relationship of vasopressin release in such subjects has not been studied.

As noted in a previous section, some elderly patients have failure of cerebral autoregulation (Wollner *et al.* 1979) and the development of symptoms in this group may be particularly obvious if it occurs. The causation of dysautoregulation in these patients has not been explained.

POSTPRANDIAL REDUCTION IN BLOOD PRESSURE IN ASSOCIATED WITH ORTHOSTATIC HYPOTENSION

Cerebrovascular accidents are particularly prone to occur following meals and it has been suggested that this may be due to a hypotensive phase at that time. Postprandial hypotension

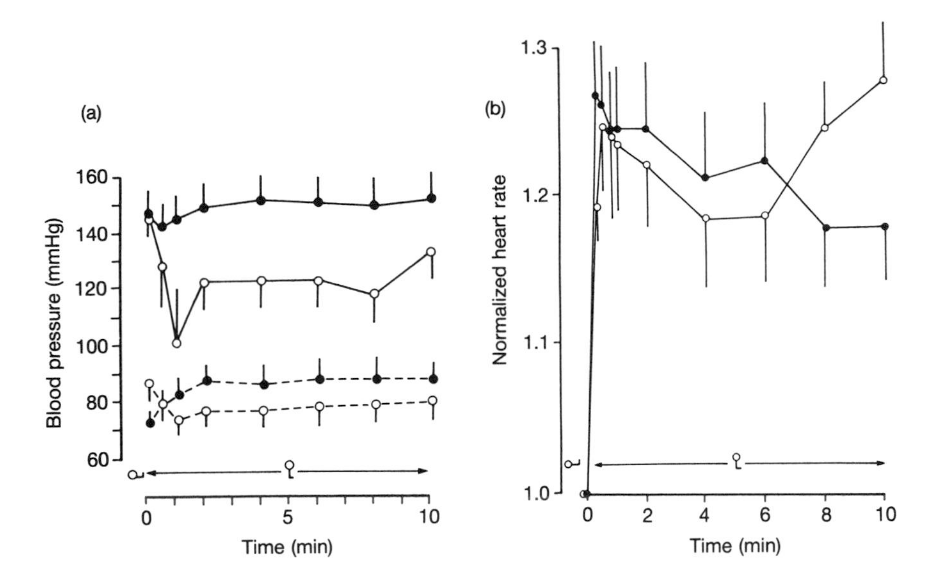

Fig. 7 (a) Systolic (——) and diastolic (– – –) blood pressure in subjects with (○) and without (●) orthostatic hypotension before and during standing; results are plotted as means ± SE. (b) Normalized heart rate (instantaneous heart rate derived from ECG R–R interval/preceding lying heart rate) in subjects with (○) and without (●) orthostatic hypotension during standing; results are plotted as means ± SE (taken with permission from Robinson *et al.* 1983).

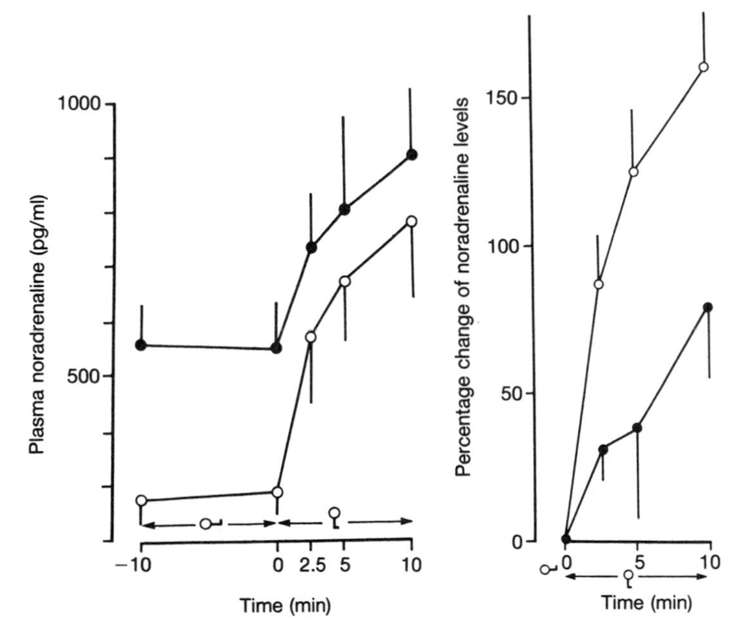

Fig. 8 (a) Plasma noradrenaline concentrations in subjects with (○) and without (●) orthostatic hypotension before and during standing; results are plotted as means ± SE. (b) Percentage changes of plasma noradrenaline concentrations in subjects with (○) and without (●) orthostatic hypotension during standing; results are plotted as means ± SE (taken with permission from Robinson *et al.* 1983).

occurs in patients with orthostatic hypotension due to autonomic failure, and this may be the explanation in some elderly patients. It has, however, been observed in elderly people with orthostatic hypotension in whom no evidence of autonomic failure is present (Robinson *et al.* 1985). In another series of institutionalized elderly patients there was a significant fall in blood pressure after a meal, whether or not the patients had a previous history of syncope (Lipsitz *et al.* 1983). Whether the patients also had evidence of orthostatic hypotension was not, however, defined. It seems clear from these observations in elderly subjects that the mechanisms involved in these patients are not primarily related to autonomic dysfunction.

In normal individuals, ingestion of food is not normally associated with any change in blood pressure and there is a marked increase in intestinal blood flow. Heart rate and cardiac output also increased and there is a fall in peripheral blood flow in the forearm, associated with a rise in vascular resistance in arm muscles. Overall, however, peripheral resistance falls and this compensates for the increased cardiac output (Mathias *et al.* 1988). These changes depend upon both sympathetic nervous activity and the release of vasoactive hormones.

In patients with autonomic failure, carefully studied when supine, a fall in blood pressure occurs within 10 to 15 min of ingestion of food (Bannister 1988). The fall may be so severe that dizziness and mental confusion may occur. In patients with autonomic failure the fall is not accompanied by changes in plasma adrenaline or noradrenaline, which normally increase to a small extent. There are no changes in plasma electrolytes or osmolality. Concentrations of gastrointestinal hormones increase both in patients with autonomic failure and in normal subjects. One hormone which has a vasodilatory effect, neurotensin, increases to a greater extent in patients with orthostatic hypotension and may therefore contribute to the development of hypotension (Mathias *et al.* 1988). Other hormones whose concentrations increased

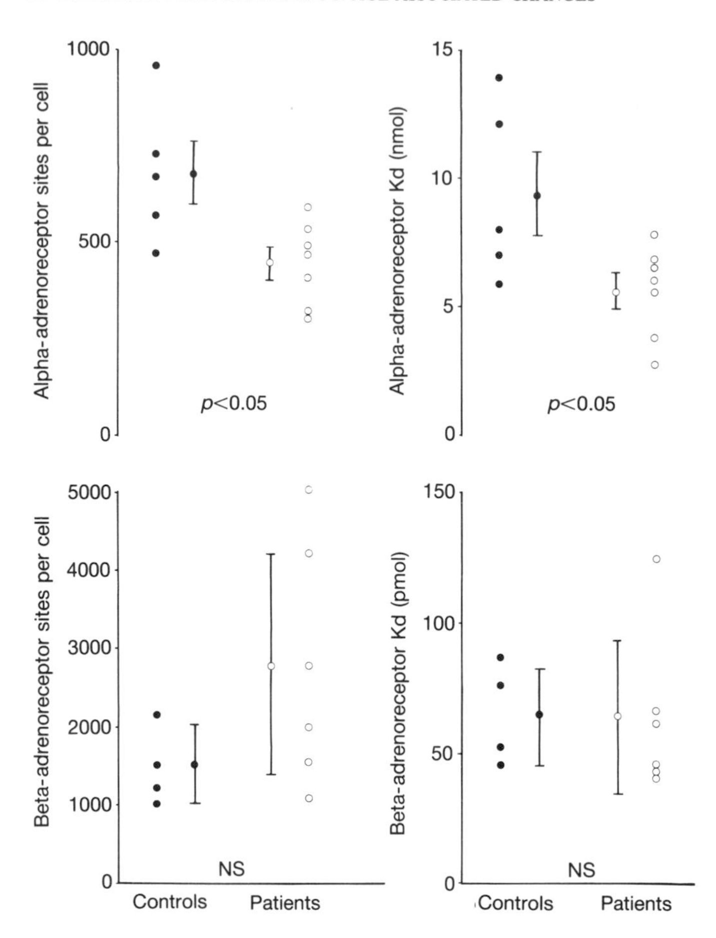

Fig. 9 Adrenoreceptor sites per cell and binding affinity, (kd) on isolated platelets (alpha-adrenoreceptor) and lymphocytes (beta-adrenoreceptor) in subjects with (○) and without (●) orthostatic hypotension. Results are plotted as individual results and as group means ± SE (taken with permission from Robinson *et al.* 1990).

similarly in both groups have vasodilatory effects, particularly vasoactive intestinal polypeptide, which may influence splanchnic vasodilatation. It could cause hypotension in patients with autonomic failure as there is no compensatory baroreflex activity leading to reflex vasoconstriction via sympathetic nerves as would occur in normal subjects.

Another hormone which may be important is insulin, for it has been shown that intravenous insulin can lower blood pressure substantially in patients with autonomic failure (Fig. 10), independently of changes in blood glucose. Insulin probably causes hypotension via splanchnic vasodilatation. Glucose, when compared with protein or fat produces the most marked development of hypotension in this group of patients, and this is probably because it is responsible for insulin release.

Another feature which may contribute to postprandial hypotension in patients with autonomic failure is the rapid rate of gastric emptying that occurs in many of the patients with this disorder. This is analogous to the 'dumping' syndrome after gastric surgery. Although some of the symptoms in that disorder, such as sweating, occur as a result of an increase in autonomic activity, the hypotension which also develops may be due to the release of vasoactive hormones.

Patients who are on drugs causing autonomic dysfunction may also show marked postprandial hypotension. This was

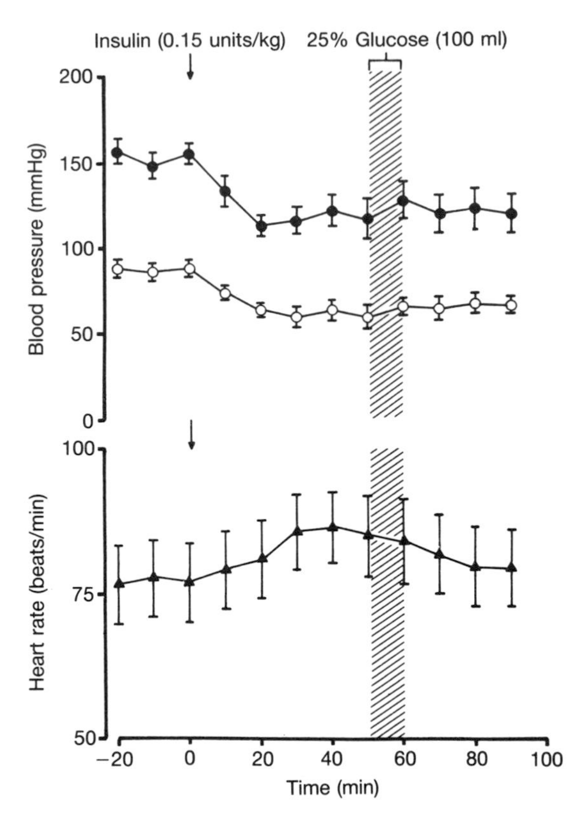

Fig. 10 Mean (SEM) systolic (●——●) and diastolic (○——○) blood pressure and heart rate in five patients with autonomic failure before and after intravenous insulin and after reversal of hypoglycaemia (taken with permission from Mathias *et al.* 1987).

first reported in hypertensive patients receiving ganglion blockers. A similar mechanism may therefore be occurring in those elderly patients who develop postprandial symptoms and are receiving drugs which cause some degree of autonomic blockade. Elderly diabetics, who may have some of the features of autonomic failure, may also develop hypotension, either as a result of insulin injections (Fig. 11) (Page and

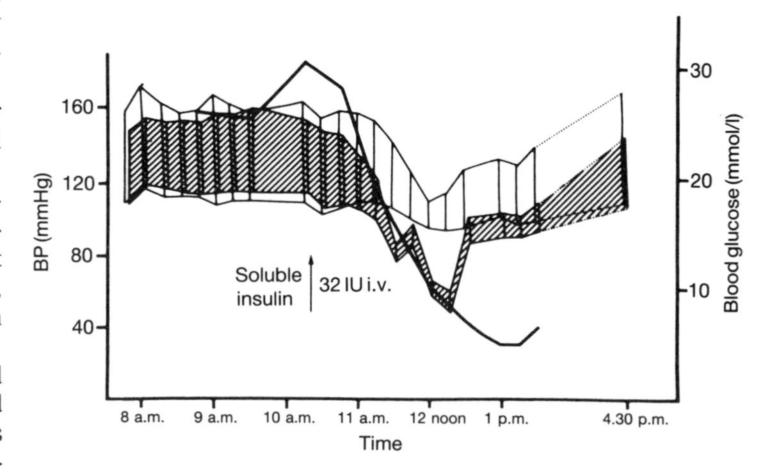

Fig. 11 The effect of intravenous insulin on supine (unhatched area) and standing (hatched area) blood pressure in a diabetic subject with autonomic neuropathy. The blood pressure was taken both lying and standing on each occasion that it was measured. Blood glucose is shown by the continuous line. Orthostatic hypotension developed after injection of insulin (taken with permission from Page and Watkins 1976).

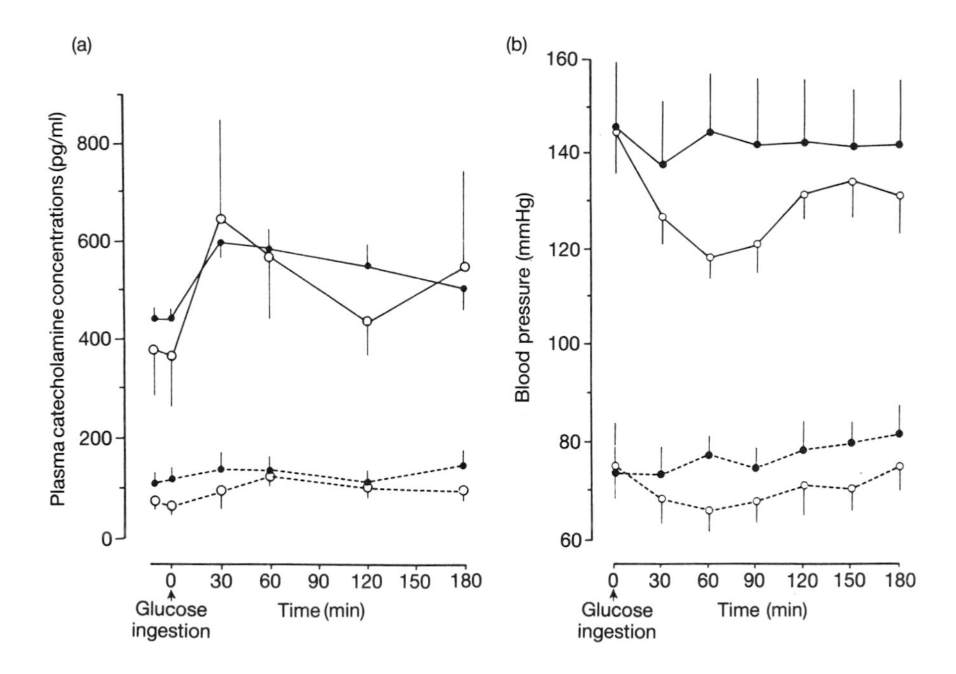

Fig. 12 (a) Systolic (——) and diastolic (– – –) blood pressures in subjects with (○) and without (●) orthostatic hypotension, before and after glucose ingestion; results are plotted as means ± SE. (b) Plasma noradrenaline (——) and adrenaline (– – –) in subjects with (○) and without (●) orthostatic hypotension, before and after glucose ingestion. Results are plotted as means ± SE (taken with permission from Robinson *et al.* 1985).

Watkins 1976), or after meals, when it is probably due to the release of other vasoactive hormones.

Elderly patients, as noted previously, may develop postprandial hypotension without any evidence of autonomic failure being present (Robinson *et al.* 1985). Postprandial hypotension is particularly likely to develop in those elderly subjects who have orthostatic hypotension of the type discussed in the previous section. In these subjects, catecholamine concentrations rise normally in response to glucose (Fig. 12). It is possible that, in this group of patients, the hypotension results from blunting of the baroreflex response, as has been observed in elderly patients after administration of glucose. The deterioration in baroreflex function may be due to the release of insulin after oral ingestion of glucose. There is, therefore, the possibility that a dual mechanism contributes to postprandial hypotension in elderly patients. First, there may be splanchnic vasodilatation from elevation of the concentrations of insulin and other vasoactive peptides, and, secondly, baroreflex activity may be depressed as a direct affect of insulin upon the baroreflex from the carotid sinus.

The management of this condition depends primarily upon its recognition. Advice may then be given about the frequency of meals and the quantity of food ingested at them. Alcohol should be avoided because it may contribute to orthostatic hypotension by causing peripheral vasodilatation. Maintenance of the fat and protein content of the diet should be achieved, as these have a smaller effect in producing postprandial hypotension than carbohydrate.

Table 2 Treatment of neurogenic orthostatic hypotension (from Johnson *et al.* 1984, with permission)

1. Mechanical supports:
   - Elastic stockings, abdominal, inflation suit
2. Blood volume expansion:
   - Response to raising head of bed
   - 9-α-Fluorohydrocortisone (fludrocortisone)
3. Vasoconstrictors:
   - Phenylephrine
   - Ephedrine
   - Midodrine
   - Noradrenaline
   - Tyramine with monoamine oxidase inhibition
   - Dihydroergotamine
   - Ergotamine
   - Yohimbine
4. Inhibition of prostaglandins:
   - Indomethacin
   - Flurbiprofen
   - Clonidine
5. β-agonist:
   - Prenalterol
6. β-blockers (inhibiting vasodilatation, reducing tachycardia during orthostatic hypotension):
   - Pindolol
   - Propranolol
7. Atrial pacemaker

Treatments 1 and 2 are used together. If they are inadequate, treatments 3, 4 and 5 may be tried successively. The importance of treatment 6 remains to be established by further use.

MANAGEMENT

Orthostatic hypotension may lead to rapid loss of consciousness with standing and the patient may fall and suffer complications, such as bruises, abrasions, fractures, or burns, if in the kitchen or in front of a fire. The consequences of orthostatic hypotension may also be serious in patients when sitting, perhaps with the additional contribution of the effect of food, as they may suffer persistent cerebral ischaemia with consequent confusion and disorientation and the possibility of developing a 'stroke'.

It has already been emphasized that the condition is frequently multifactorial and the assessment of contributory factors is therefore important (Caird *et al.* 1973). A good drug history may indicate an agent that is precipitating or exacerbating the problem. The possibility of a clearly defined disorder, such as diabetes mellitus, should also be considered. In many patients, however, no clear indication of a major cause is found. It is in this group that the descriptive diagnosis 'orthostatic hypotension in the elderly' is used. In patients with orthostatic hypotension, a number of procedures may help their control. These include pressure to the limbs and the abdomen with elastic stockings and a rubber roll-on. Patients need to be properly measured for these garments, which must be full length and firm. They may require help in dressing with them. Elastic has the disadvantage that it may be particularly unpleasant in hot weather. Tilting up the head of the bed at night has been shown over many years to cause some improvement in many patients with the disorder, probably by stimulating the renin–angiotensin system (Bannister 1988).

A wide range of drugs has been recommended but perhaps the most useful remains fludrocortisone, although it has the disadvantage that it increases both standing and supine blood pressure. A list of drugs is given in Table 2. Many have the additional complication that they require careful control of dose and will require introduction in hospital. A drug which is receiving some attention, at present, is xamoterol. This is a beta-1-adrenoreceptor partial agonist with considerable intrinsic sympathomimetic activity. Caffeine may also be useful in some patients. Although it may raise blood pressure, either by stimulting sympathetic nervous activity or by causing renin release, another possibility is that it blocks adenosine receptors. Coffee and other caffeine-containing foods may assist in the long-term management of postprandial hypotension. A somatostatin analogue has also been studied as it appears to block release of vasoactive peptides.

As one of the major problems is postprandial hypotension, avoidance of heavy meals, alcohol, and food with high glucose content may be of assistance.

The wide range of drugs implies that the condition is one which is very difficult to treat and the responses are frequently poor. Progressive elevation and encouragement to walk or take a few steps as often as possible prevent complete immobility developing. Moreover, because patients may respond by a variety of physiological mechanisms, including increased vascular tone and expansion of blood volume, together with the release of some vasoactive substances such as those concerned with the renin system, it may be necessary to try several therapeutic drugs in succession before the best possible management is achieved.

These patients are nevertheless likely to require very considerable care and assistance. It is frequently difficult for them to manage continuously at home. Support by Social Services, including Meals-on-Wheels and an attendance allowance for a helper, together with day care, may assist those patients still able to live in their own home. Continuing assessment and assistance are vital.

References

Bannister, R. (ed.) (1988). *Autonomic Failure*. Oxford University Press.

Bristow, J.D., Gribbin, B., Honour, A.J., Pickering, T.G., and Sleight, P. (1969). Diminished baroreflex sensitivity in high blood pressure and ageing man. *Journal of Physiology*, **202**, 45–46P.

Caird, F.I., Andrews, G.R., and Kennedy, R.D. (1973). Effect of posture on blood pressure in the elderly. *British Heart Journal*, **35**, 527–30.

Eisenhofer, G., Lambie, D.G., Johnson, R.H., Tan, E.T.H., and Whiteside, E.A. (1982). Deficient catecholamine release as the basis of orthostatic hypotension in pernicious anaemia. *Journal of Neurology, Neurosurgery and Psychiatry*, **45**, 1053–5.

Ewing, D.J., Campbell, I.W., and Clarke, B.F. (1980). The natural history of diabetic autonomic neuropathy. *Quarterly Journal of Medicine*, **49**, 95–108.

Gribbin, B., Pickering, T.G., Sleight, P., and Peto, R. (1971). Effect of age and high blood pressure on baroreflex sensitivity in man. *Circulation Research*, **29**, 424–31.

Johnson, R.H. (1976). Blood pressure and its regulation. In *Cardiology in old age* (ed. F.I. Caird, J.L.C. Dall, and R.D. Kennedy), pp. 101–26. Plenum, New York.

Johnson, R.H. (1983). Autonomic dysfunction in clinical disorders with particular reference to catecholamine release. *Journal of the Autonomic Nervous System*, **7**, 219–32.

Johnson, R.H. (1984). Clinical assessment of sympathetic function in man. *Methods and Findings in Experimental and Clinical Pharmacology*, **6**, 187–95.

Johnson, R.H. (1988). Autonomic failure in alcoholics. In *Autonomic failure* (ed. R. Bannister), pp. 690–714. Oxford University Press.

Johnson, R.H. and Spalding, J.M.K. (1974). *Disorders of the autonomic nervous system*. Blackwell, Oxford.

Johnson, R.H., Smith, A.C., Spalding, J.M.K., and Wollner, L. (1965). Effect of posture on blood pressure in elderly patients. *Lancet*, **i**, 731–3.

Johnson, R.H., Lee, G. de J., Oppenheimer, D.R., and Spalding, J.M.K. (1966). Autonomic failure with orthostatic hypotension due to intermediolateral column degeneration. *Quarterly Journal of Medicine*, **35**, 276–92.

Johnson, R.H., Smith, A.C., and Spalding, J.M.K. (1969). Blood pressure response to standing and to Valsalva's manoeuvre: independence of the two mechanisms in neurological diseases including cervical cord lesions. *Clinical Science*, **36**, 77–86.

Johnson, R.H., Lambie, D.G., and Spalding, J.M.K. (1984). *Neurocardiology: the inter-relationships between dysfunction in the nervous and cardiovascular systems*. Saunders, London.

Johnson, R.H., Lambie, D.G., and Eisenhofer, G. (1989). Effects of ethanol on autonomic function. In *The human metabolism of alcohol*, Vol. 3 (ed. R. Batt and K. Crow), pp. 189–203. CRC Press, Boca Raton, FA.

Lipsitz, L.A., Nyquist, R.P., Wei, J.Y., and Rowe, J.W. (1983). Postprandial reduction in blood pressure in the elderly. *New England Journal of Medicine*, **309**, 81–3.

MacLennan, W.J., Hall, M.R.P., and Timothy, J.I. (1980). Postural hypotension in old age: is it a disorder of the nervous system or of blood vessels? *Age and Ageing*, **9**, 25–32.

Mathias, C.J., da Costa, D.F., Fosbraey, P., Christensen, N.J., and Bannister, R. (1987). Hypotensive and sedative effects of insulin in autonomic failure. *British Medical Journal*, **295**, 161–3.

Mathias, C., da Costa, D., and Bannister, R. (1988). Postcibal hypotension in autonomic disorders. In *Autonomic failure* (ed. R. Bannister), pp. 367–80. Oxford University Press.

Nanda, R.N., Wyper, D.J., Johnson, R.H., and Harper, A.M. (1976). The effect of hypocapnia and change of blood pressure on cerebral blood flow in men with cervical spinal cord transection. *Journal of the Neurological Sciences*, **30**, 129–35.

Page, M. McB. and Watkins, P.J. (1976). Provocation of postural hypotension by insulin in diabetic autonomic neuropathy. *Diabetes*, **25**, 90–5.

Park, D.M., Johnson, R.H., Crean, G.P., and Robinson, J.F. (1972). Orthostatic hypotension with recovery after radiotherapy in a patient with bronchial carcinoma. *British Medical Journal*, **3**, 510–11.

Robinson, B.J., Johnson, R.H., Lambie, D.G., and Palmer, K.T. (1983). Do elderly patients with an excessive fall in blood pressure on standing have evidence of autonomic failure? *Clinical Science*, **64**, 587–91.

Robinson, B.J., Johnson, R.H., Lambie, D.G., and Palmer, K.T. (1985). Autonomic responses to glucose ingestion in elderly subjects with orthostatic hypotension. *Age and Ageing*, **14**, 168–73.

Robinson, B.J., Stowell, L.I., Johnson, R.H., and Palmer, K.T. (1990). Is orthostatic hypotension in the elderly due to autonomic failure? *Age and Ageing*, **19**, 288–96.

Ropper, A.H. and Hedley-White, E.T. (1983). Parkinsonism associated with other neurologic manifestations. *New England Journal of Medicine*, **308**, 1406–14.

Shy, G.M. and Drager, G.A. (1960). A neurological syndrome associated with orthostatic hypotension. *Archives of Neurology*, **2**, 511–27.

Strandgaard, S., Olesen, J., Skinhoj, E., and Lassen, N.A. (1973). Autoregulation of brain circulation in severe arterial hypertension. *British Medical Journal*, **1**, 507–10.

White, N.J. (1980). Heart-rate changes on standing in elderly patients with orthostatic hypotension. *Clinical Science*,**58**, 411–13.

Winson, M., Heath, D., and Smith, P. (1974). Extensibility of the human carotid sinus. *Cardiovascular Research*, **8**, 58–64.

Wollner, L., McCarthy, S.T., Soper, N.D.W., and Macy, D.J. (1979). Failure of cerebral autoregulation as a cause of brain dysfunction in the elderly. *British Medical Journal*, **1**, 1117–18.

Young, J.B., Rowe, J.W., Pallotta, J.A., Sparrow, D., and Landsberg, L. (1980). Enhanced plasma norepinephrine response to upright posture and oral glucose administration in elderly human subjects. *Metabolism*, **29**, 532–9.

18.11 Epilepsy

RAYMOND TALLIS

The term epilepsy should be used not to refer to a single seizure but to imply a continuing tendency to epileptic seizures. In epidemiological studies, the term is usually used where a patient has suffered from more than one unprovoked seizure of any type. Seizures are defined pathophysiologically as being due to paroxysmal discharges of cerebral activity, in which a critical mass of neurones fires synchronously.

EPIDEMIOLOGY

More than 5 per cent of the population have at least one afebrile seizure during their lives. On general grounds, one would expect that the proportion of the population with epileptic fits would rise with age, each age carrying the cumulative prevalence of earlier ones. Epilepsy in younger patients, however, tends to remit, a fact reflected in the overall prevalence rate for active epilepsy, which has been estimated at between 3 and 6 per 1000. The elderly population will include patients whose epilepsy had presented earlier in life; and with better treatment, and consequently longer survival of such patients, this contribution may be expected to increase.

The impressions that seizures are comparatively rare among elderly people and that epilepsy beginning for the first time in old age is uncommon are false. There is a steep rise in both the prevalence of epilepsy and in the incidence of new cases above the age of 50 years (Fig. 1). Hauser and Kurland (1975) found that the annual incidence of epileptic seizures rose from 12 per 100 000 in the 40 to 59 age range to 82 per 100 000 in those over 60. This is remarkably close

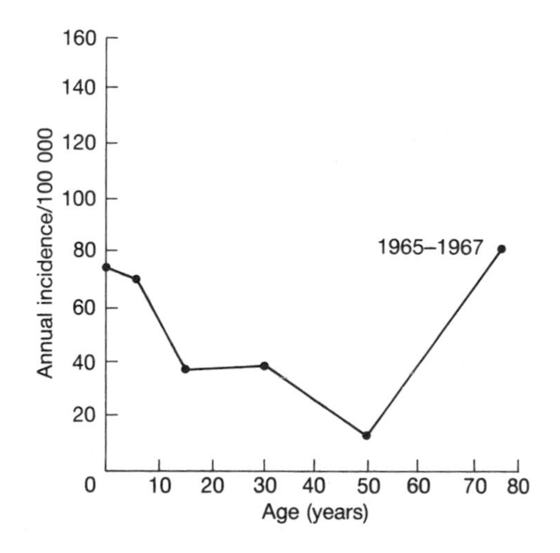

Fig. 1 The incidence of epilepsy at different ages (derived from the data of Hauser and Kurland 1975).

to the incidence observed in a recent Danish study which included all patients over 60 in a well-defined population and arrived at an incidence of definite epilepsy of 77 per 100 000. The National General Practice Survey of Epilepsy (**NGPSE**),* a prospective, community-based study, identified 1200 patients with newly diagnosed or suspected epilepsy in its 3-year study period (1984 and 1987). Twenty-four per cent of new cases of definite epilepsy were over the age of 60. The excess incidence of epilepsy in the elderly is illustrated in Fig. 2 derived from the NGPSE. Estimates of the prevalence of epilepsy are, for the reasons discussed by Hauser and Kurland in their classical study, unreliable. Nevertheless, they showed a rise in prevalence in older sub-

* I am grateful to Drs Yvonne Hart and Simon Shorvon who kindly allowed me to quote preliminary data from the NGPSE.

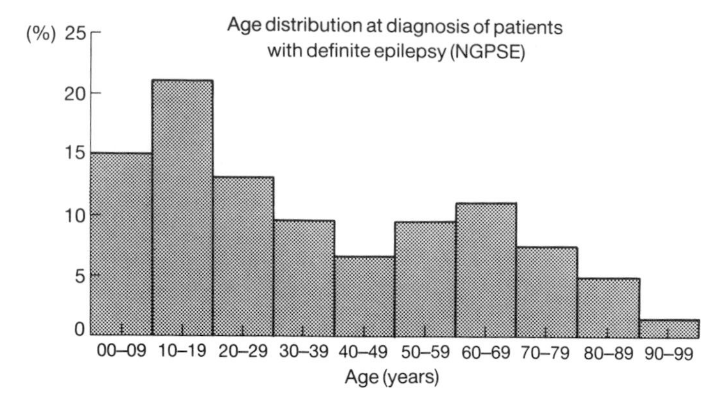

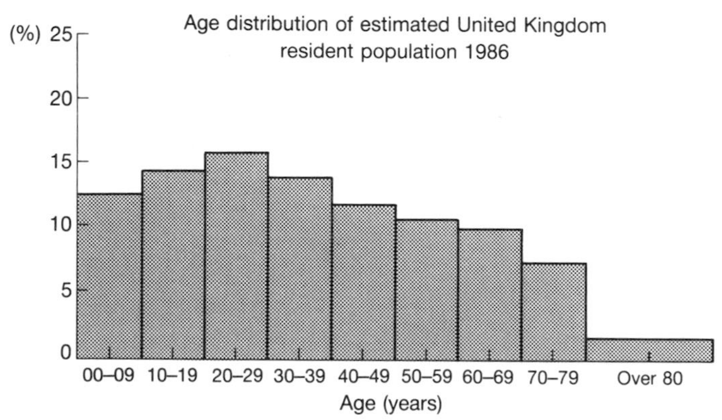

Fig. 2 Age distribution at diagnosis of patients with definite epilepsy (unpublished data by permission of National General Practice Survey of Epilepsy).

jects: from 7.3/1000 in the 40 to 59 age range to 10.2 for those over 60.

There are several reasons why the incidence and prevalence of epilepsy in elderly subjects may be underestimated. Minor seizures may not be reported; even if reported, they may not be recognized for what they are, getting lost in the pathologically rather 'noisy' situation of the biologically aged person; or, even if recognized, not referred to hospital and included in hospital-based or hospital-biased series. This last possibility was well demonstrated in the NGPSE: 20 per cent of elderly patients with seizures were not referred to hospital, compared with only 4 per cent of younger patients. This is despite the fact that the elderly group contributed a disproportionate number of the additional cases of 'possible' or 'probable' epilepsy noted in the series, confirming the increased diagnostic uncertainty in the aged. It would appear that general practitioners' hunger for diagnostic precision falls with the increasing age of the patient.

TYPES OF SEIZURES

The manifestations of epilepsy are complex and varied, and the methods of classifying seizures correspondingly complex. In 1969, the International League Against Epilepsy published its classification. The scheme was in part motivated by an attempt to match clinical phenomena with electroencephalographic features but was not entirely successful because clinical and electrical phenomena are not easily mapped on to one another. Since the 1969 classification, objective and sophisticated methods of studying seizures have become more widely available and in 1981 the International League suggested a revised classification. This correlates types of clinical seizures with ictal and interictal electroencephalographic features. Table 1 gives those parts of the classification relevant to elderly patients.

Table 1 Classification (suggested by the International League against Epilepsy) of epilepsy in elderly patients

I. Partial (focal, local)
A. Simple partial seizures (consciousness not impaired)
  1. With motor signs (focal motor with or without march; versive; postural; vocalization)
  2. With somatosensory or special-sensory symptoms (somatosensory; visual; auditory; olfactory; gustatory; vertiginous)
  3. With autonomic symptoms or signs (epigastric sensation, pallor, sweating, flushing, piloerection, pupillary dilatation)
  4. With disturbances of higher cerebral function (dysphasic; dysmnesic; cognitive; affective)
  5. Illusions, e.g. macropsia
  6. Structured hallucinations, e.g. music, scenes
B. Complex partial seizures (with impairment of consciousness; sometimes beginning with simple symptoms)
  1. Simple partial seizure followed by impairment of consciousness or automatisms
  2. Impairment of consciousness at the outset, with or without automatisms
C. Partial seizures evolving to secondarily generalized seizures (with convulsive manifestations)

II. Generalized seizures (convulsive or non-convulsive)
A.1. Absence seizures
  a. Impairment of consciousness only
  b. With mild clonic components
  c. With atonic components
  d. With automatisms
  e. With autonomic components
A.2. Atypical absences (changes in tone may be more pronounced and onset/cessation less abrupt than in A.1)
B. Myoclonic seizures
  Multiple or single myoclonic jerks
C. Clonic seizures
D. Tonic seizures
E. Tonic-clonic seizures
F. Atonic seizures

III. Unclassified epileptic seizures

Primary generalized seizures are those in which the first clinical events suggest involvement of both hemispheres from the outset. This is confirmed by EEG discharges during a fit, which are bilateral from the outset. Such seizures may be convulsive or non-convulsive. In the former case, motor manifestations are bilateral from the outset. In non-convulsive seizures, there is impairment or interruption of consciousness without motor manifestations. Impairment of consciousness may be the first event in a convulsive seizure. In partial seizures, the first changes suggest activation of neurones limited to part of one cerebral hemisphere. A partial seizure is classified as simple if consciousness is not impaired and as complex if it is impaired. Impairment of consciousness in a partial seizure usually implies bilateral spread of seizure activity. Generalization of electrical activity may also lead to generalized convulsions. The clinical correlative of this will be loss of consciousness and tonic–clonic features supervening on initially focal symptoms.

Primary generalized convulsive seizures correspond roughly to the classical 'grand mal' attack without a preceding aura. A grand mal attack preceded by an aura or other focal features corresponds to 'partial seizures evolving to secondary generalized seizures'. 'Minor' or 'focal' epilepsy covers simple and complex partial seizures. Most of the latter were previously classified as 'temporal lobe epilepsy' as, amongst the focal seizures, it is those originating in the temporal lobes that are most likely to be associated with disturbances of consciousness. Some temporal-lobe attacks may take the form of simple partial seizures with autonomic or psychic symptoms.

There is little satisfactory information about the relative proportion of different types of epilepsy in elderly people. Most hospital-based series have emanated from neuromedical centres which tend to attract more complex cases and in which elderly patients are under-represented. Community-based series, on the other hand, often lack sufficient clinical electroencephalographic evidence to ensure accurate classification. The study of Luhdorf *et al.* (1986) is both population-based and supported by electroencephalography. This 5-year study of incidence found that 51 per cent of elderly-onset patients had apparently primary tonic-clonic seizures, 42 per cent partial or secondary generalized seizures, and 6 per cent unclassifiable seizures. Of those with seemingly primary generalized seizures, however, 38 per cent had focal abnormalities on the EEG. In total, over 70 per cent of subjects had seizures that were either clinically or electrically partial or partial in origin. In the NGPSE, the proportions of different seizures for the overall population were partial or secondarily generalized, 52 per cent; generalized from the outset, 43 per cent; and unclassified, 7 per cent. In the elderly cases, the proportions were partial or secondarily generalized, 73 per cent; primary generalized, 17 per cent; and unclassified, 10 per cent. This confirms the tendency for fits in elderly patients to be of focal origin, reflecting their relationship to focal cerebral pathology. Except in those cases where fits are due to some systemic cause, such as a metabolic disturbance or a drug lowering the convulsive threshold, it is probably that elderly-onset seizures are, with very few exceptions, all focal in origin. Certainly, the harder one looks, the more one is likely to find a candidate focal cause.

DIAGNOSIS

The diagnosis of an elderly patient who presents with episodes suggestive of seizures may be considered under three headings.

1. Are the episodes seizures or not?
2. What type of seizures are they?
3. What is their underlying cause?

Are the episodes seizures or not?

The most common feature of epilepsy in the elderly patient, as at any other age, is transient impairment or loss of consciousness. Sometimes disturbances of consciousness may be forgotten and the patient will report only a fall. The differential diagnosis will then encompass the numerous other causes of falls in elderly people. A fall occurring in the absence of any obvious environmental cause and which cannot be confidently attributed to orthopaedic, cardiovascular, or non-epileptic neurological factors should raise the suspicion of a fit.

Where there is a story either from the patient or witness of loss of consciousness, the differential diagnosis will include other causes of transient loss of consciousness, in particular hypoglycaemic episodes and syncopal attacks due to temporary impairment of cerebral circulation. It must be remembered that hypoglycaemic episodes in elderly people may not be associated with the characteristic features that are due to excessive adrenergic activity and that many hypoglycaemic attacks occur at night. However, there will usually be an obvious precipitating cause, such as hypoglycaemic medication. It must not be forgotten that severe hypoglycaemia may itself precipitate epileptic seizures.

Differentiating syncopal attacks from seizures may be very difficult. In the absence of eye witness reports, a history will necessarily be incomplete. The following features are particularly helpful in discriminating fits from faints: a well-defined aura; clear progression from a tonic to a clonic phase; tongue-biting, incontinence, or focal neurological features during an attack; and stupor or prolonged confusion, headache, and transient neurological signs after the attack. Post-event confusion and headache are particularly useful pointers to a fit. The situation is complicated in elderly patients, however, because there may be coexistent conditions that predispose to syncope and it is well known that cerebral anoxia, as for example in carotid sinus syncope or cardiac arrhythmia, may itself cause convulsions.

There are, of course, numerous causes of syncope (transient loss of consciousness due to any cause of an acute decrease in cerebral blood flow) in elderly individuals. Features that may favour syncope over seizures include a history of pallor and sweating; gradual onset preceded by palpitations; an association with the assumption of the erect posture, voiding, or neck turning; and medication known to cause postural hypotension. Again, however, the situation is confusing because complex partial seizures affecting the temporal lobes may present with autonomic features. As there are many causes of syncope in an elderly person and as, moreover, many elderly patients may have features suggestive of cerebrovascular disease, it may prove impossible to determine whether or not transient cerebral symptoms are cardiac or cerebral in origin. Even ambulatory ECG and EEG may not permit a confident diagnosis. Non-specific abnormalities on an EEG or cardiac arrhythmias recorded on a 24-h tape unrelated to the symptoms may add to the confusion. In the case of many 'funny turns' it may be necessary, after a careful history, examination, and appropriate investigations, simply to wait and see. A therapeutic trial of anticonvulsants as a diagnostic test is not usually to be recommended as it will rarely produce a clear answer and will add the burden of possibly unnecessary drug treatment to the patient's troubles.

What type of seizures are they?

The most powerful diagnostic tool in this context is a careful history, preferably supplemented by information from a relative, carer, or other eyewitness. The presence of focal neurological signs and of focal changes on an EEG may provide supplementary evidence. Focal or secondary generalized seizures usually imply focal neurological damage, whereas primary generalized seizures may be idiopathic (rare in elderly subjects) or, more commonly, due to a metabolic or other systemic disturbance lowering the seizure threshold. Occasionally, however, such a disturbance may trigger off a discharge in a focus of pre-existing neurological damage. Although the type of fit is sometimes clear from the history, as when there is a story of a Jacksonian march or recurrent, stereotyped hallucinations, localization of the precise origin of seizure activity will often depend on investigations, in particular an EEG.

Besides the 'noisy' situation in which seizures may occur in an elderly person—inadequate history, concurrent pathology, etc.—there may be other special diagnostic difficulties. In all age groups, partial seizures, especially complex partial seizures with or without automatisms, may be labelled as non-specific confusional states or even, where there are affective or cognitive features or hallucinations, as manifestations of functional psychiatric illnesses. Patients with non-convulsive epileptic status may present with acute behavioural changes—withdrawal, mutism, delusional ideas, paranoia, and vivid hallucinations. Fluctuating mental impairment may easily be attributed to other causes of recurrent confusional states or even misread as part of a dementing process.

Because of age-associated changes in the brain and because epilepsy in elderly people usually takes place against the background of cerebral damage, postictal states may be very prolonged. At least 14 per cent of patients in one series suffered a confusional state lasting 24 h or more, and in some cases it could persist as long as a week. This may well cause diagnostic problems where the history is incomplete. A focal, postictal paresis (Todd's palsy) is also more frequent in elderly patients. This may lead to misdiagnosis of stroke, and such patients have been occasionally referred to stroke units. This is particularly likely to happen where fits occur against a background of known cerebrovascular disease, and a recurrent stroke may be incorrectly diagnosed.

What is the underlying cause?

Elderly-onset seizures will almost invariably be due to an underlying cause. This may require treatment in its own right and such treatment may lead to remission of seizures.

AETIOLOGY OF EPILEPSY IN THE ELDERLY PATIENT

Cerebrovascular disease

It is becoming increasingly clear that cerebrovascular disease is the main cause of epilepsy in elderly people. Most series indicate not only that vascular disease is the most common cause (30–50 per cent of cases) and but that it also accounts for an even higher proportion of those cases in which a cause is found (up to 75 per cent). The more carefully cerebrovascular disease is sought in epileptic patients, the more frequently it is found. One series compared the appearances of the computed tomography (**CT**) scan of patients who had late-onset epilepsy and no evidence of cerebral tumour with those of age- and sex-matched controls. There was an excess of ischaemic lesions in epileptic patients. In half of the epileptic patients who were found to have CT evidence of vascular disease, clinical examination was normal.

Epilepsy commonly follows stroke and may also precede it. In one series, epilepsy occurred in 8 per cent of hemiplegic stroke patients. Another examined the incidence of post-stroke epilepsy in patients with angiographically proven, occlusive disease of the carotid or middle cerebral artery and found that fits occurred in 17 per cent of the carotid cases and 11 per cent of the middle cerebral artery strokes in a 3-year follow-up. There was a predominance of partial motor attacks in these patients. Conversely, studies have shown an excess over controls of previous epilepsy in patients admitted to hospital with acute stroke, suggesting that in a proportion of elderly patients, epilepsy may be the earliest manifestation of cerebrovascular disease.

Other cerebral disorders

Clinicians are often concerned that very late-onset epilepsy may indicate a cerebral tumour. Most series indicate that this applies only to a minority of cases, ranging from 10 to 15 per cent in most series. High and lower figures have been reported, due to the different populations being studied (itself a reflection of different referral patterns), to the different extent to which patients are investigated and, related to this, the different proportion of cases in which no cause is found. In most cases, tumours are either metastatic or (inoperable) gliomas, though a few meningiomas are found. Until there is information on adequately documented, adequately investigated, and sufficiently large, population-based series, one cannot be certain what proportion of cases of very late-onset epilepsy is due to treatable and non-treatable tumours. All that one can say at present is that treatable tumour does not appear to have the aetiological importance attributed to it in traditional teaching. The relatively rare sensory epilepsy appears to be more often associated with tumours than other forms of epilepsy.

A variable proportion of seizures (about 2 to 14 per cent in different series) are attributed to non-vascular cerebral degeneration. Again, the data are insufficient and will remain so until large series with uniform access to CT scanning facilities are reported. Subdural haematoma is an important but remediable cause, especially as very elderly patients are prone to this condition because of cerebral atrophy. As it may occur after a relatively trivial injury, the diagnosis may be missed. Direct brain damage due to head injury is itself a relatively uncommon cause of elderly-onset epilepsy in most series. Seizures may occur during the course of severe cerebral infections (meningitis, encephalitis, or cerebral abscess) but under such circumstances should not strictly be called 'epilepsy'. After recovery from such infections, however, epilepsy may arise due to scarring.

Metabolic and toxic causes

Recent series have underlined the importance of toxic and metabolic causes, accounting for between 10 and 15 per cent of cases of elderly-onset seizures. Of the metabolic causes, uraemia seems to be most common. In one recent series of late-onset seizures, alcohol or alcohol withdrawal was the main cause in nearly a quarter of cases; moreover, alcohol misuse appeared to be the sole precipitating factor in 20 per cent of cases of status epilepticus. As alcohol abuse is becoming increasingly prevalent amongst elderly people (especially women), this will continue to be a major aetiological factor.

A wide range of drugs has been suspected of causing convulsions. It is often difficult to prove that a given drug caused convulsions in a particular case but in certain drugs the probability of a causal relationship seems to be high. Drug-induced seizures are particularly likely to occur when the drug is given in high dosage or parenterally or to patients with impaired drug handling. Aminophylline, which has a narrow therapeutic index, and whose disposition may be altered by cigarette smoking, is especially prone to cause generalized seizures. Psychotropic drugs, including tricyclic antidepressants and phenothiazines, are also particularly important. Benzodiazepine withdrawal may cause fits. Repeated hypoglycaemic episodes due to excessive insulin or oral hypoglycaemics may precipitate recurrent seizures.

Very occasionally one encounters a patient whose seizures may be idiopathic, presenting for medical help for the first time in old age. Such patients may have had a lifetime of untreated epilepsy. Without such a long history, it is difficult to sustain a diagnosis of idiopathic epilepsy.

Table 2 Causes of seizures in elderly patients

Cerebral disease
Cerebrovascular disease
Neoplasia—primary and secondary
Alzheimer's disease and other non-vascular causes of cerebral degeneration
Subdural haematoma
Post-traumatic
Infective or postinfective (meningitis, encephalitis, abscess)
Cardiovascular disease
Tachyarrhythmias
Stokes–Adams attacks
Metabolic
Renal failure
Hepatic failure
Hypoglycaemia
Hyperosmolar diabetes
Hypocalcaemia
Water and electrolyte disturbance
Myxoedema
Hypoxia
Hypercapnia
Toxic
Drugs and drug withdrawal
Alcohol and alcohol withdrawal
Idiopathic

INVESTIGATIONS

Investigations will be directed towards supporting the clinical diagnosis of epileptic seizures (and so differentiating episodes from other causes of 'funny turns' and other transient impairments), towards defining the type of epilepsy (although it can be safely assumed that the vast majority of cases of epilepsy in old age are focal in origin), and towards identifying remediable underlying causes.

General investigations

These will be determined by the history and by the findings on examination, as well as by consideration of likely causes. It is particularly important to rule out metabolic causes (see Table 2) as they are specially amenable to treatment. An estimate of γ-glutamyl transferase may be a useful marker of alcohol consumption. Control of blood glucose in diabetic patients on treatment should be reviewed, especially where the presenting problem is that of nocturnal seizures in a patient on oral hypoglycaemic agents. Carotid sinus massage, ambulatory ECG recordings, and other investigations may be indicated if the problem is thought to be that of cardiac arrhythmias rather than epilepsy or if it is thought that the patient is having fits secondary to such arrhythmias.

Lumbar puncture used to be carried out routinely in patients presenting with fits. In the absence of evidence of acute infection of the nervous system, there is no indication for this investigation. Moreover, lumbar puncture may be dangerous unless a space-occupying lesion has been ruled out. The absence of symptoms suggestive of raised intracranial pressure or of papilloedema on examination does not guarantee that the patient has not got a space-occupying lesion. The only indication for carrying out a lumbar puncture is the remote possibility of neurosyphilis, and this possibility will be even more remote if blood testing does not show abnormal serology. A chest radiograph may reveal a relevant primary neoplasm. While skull radiography may show evidence of raised intracranial pressure, intracranial calcification, or other evidence of an intracerebral neoplasm, it is rarely helpful.

Electroencephalography

Slavish reliance upon an EEG to make or to refute a diagnosis of epilepsy is potentially dangerous. A routine EEG may support the diagnosis of epilepsy, especially if clear-cut paroxysmal discharges are observed. The absence of such activity on a routine recording does not, however, rule out the diagnosis. It must be recalled that most recordings last for only 20 min and that ictal or diagnostic interictal activity occurs only intermittently. The range of normality increases with age, so that discriminating normal from abnormal becomes more difficult in an elderly patient. Non-specific abnormalities are more common in old age. It follows from this that a diagnosis of epilepsy should not be made on EEG findings alone. A focal abnormality on an EEG may support the clinical diagnosis of a clinical origin for fits and suggest a local neurological cause. In those fits where there is an inadequate history or where the focal phase is too brief to be observed clinically, its observation may suggest a focal origin for the first time and guide further investigation. Persistent gross abnormalities on an EEG would strongly support a focal structural lesion.

In summary, while the EEG may provide useful supporting evidence for the diagnosis of epilepsy, it should not over-rule the clinical diagnosis nor provide its sole basis. It should also be added that an EEG, in this age group as in any other, cannot alone determine the need for a treatment in a newly diagnosed case, establish the adequacy of treatment, or predict the safety of discontinuing therapy.

Neuroradiology

If all elderly patients who had seizures not explicable in terms of an acute illness, metabolic problems, or alcohol were referred for neuroradiological investigation, the pressures on such services would be overwhelming. There is no doubt that a CT scan is an extremely powerful tool for picking up structural lesions of the brain. Moreover, there is a good deal of evidence that the older the patient the greater the chance of a positive scan. As many as 60 per cent of very late-onset patients with epilepsy may show a structural lesion on a CT scan. These facts would argue for routine scanning, however, only if identification of such lesions were going to influence management in a positive way. The discovery of a space-occupying lesion amenable to neurosurgical removal would be an obvious example. However, in only a minority of patients with elderly-onset epilepsy is a neoplasm or subdural haematoma the cause and in only a small proportion of cases with tumour would neurosurgical intervention be appropriate. As already noted, in most series tumours are more likely to be gliomas or metastases than meningiomas. Even where a meningioma is diagnosed, neurosurgical treatment may not always be indicated. There is an impression that some meningiomas in old age may be relatively inert and there is no doubt that craniotomy is often tolerated poorly by elderly patients. Even so, it is sometimes useful to have a definitive diagnosis, though treatment for the underlying condition may not be available or considered inappropriate. Arguable indications for CT scanning are given in Table 3.

Table 3 Indications for computerized tomography or magnetic resonance imaging in elderly-onset epilepsy

Strong
Progressive neurological signs
Focal neurological signs not explicable by stroke
Suspicion of subdural haematoma as cause of fits
Epilepsy proving difficult to control
Symptoms/signs of raised intracranial pressure
Less strong
Focal epilepsy
Very active focus on EEG

With the advent of CT and magnetic resonance imaging, the role of isotope scanning is less clear. Unlike the CT scan, it has a low rate of detection for infratentorial tumours, but these less commonly give rise to epilepsy. Moreoever, a single scan cannot usually distinguish between a vascular and neoplastic lesion, and repeated scanning may be required. Isotope scanning is sometimes used as a 'screening' procedure to identify those patients who warrant referral for CT scanning. The requirement that an elderly person should 'earn' his or her CT or magnetic resonance scan by producing evidence of non-resolving focal lesions on an isotope scan (and possibly also an EEG) may not, however, represent the best option for the individual patient or, indeed, the most economical use of resources.

TREATMENT

General measures

Many elderly patients will remember a time when epilepsy was often associated with severe cerebral damage, either as cause or consequence of fits, when fits were often uncontrolled, and when epileptic patients were stigmatized. They may need, therefore, to be reassured that, in the vast majority of cases, fits do not indicate serious brain damage, that they are not going mad, and that the fits themselves can be controlled by medication.

Patients may want to know whether fits are brought on by any particular activity and whether, for this reason, they should lead restricted lives. The advice in this age group is the same as that given to any patients: avoid only those activities that would mean immediate danger if a fit occurred. A home visit by an occupational therapist to look for potential sources of danger—unguarded fires, etc.—may be helpful. In the case of frequent fits, especially where there is a warning aura, the patient may wish to be provided with an alarm system. Factors that are known to precipitate fits, such as inadequate sleep, excess alcohol, or sudden alcohol withdrawal, should be avoided. The patient should be warned that alcohol will increase the side-effects of medication and that other drugs may have a convulsant effect or interact with anticonvulsants.

The patient must be advised of the regulations regarding driving. In the United Kingdom, for example, anyone diagnosed as having epilepsy and holding a driving licence must notify the Driver and Vehicle Licensing Centre and stop driving until further directed by the Centre. The onus of responsibility to inform the Centre lies with the patient and not with the doctor. The regulations for ordinary vehicle licences, which came into force in 1982, require that a person suffering from epilepsy has to fulfil the following conditions before the licence can be restored:

1. freedom from any epileptic attack during the period of 2 years prior to the date when the licence is to have effect; or
2. attacks only while asleep during a period of at least 3 years prior to the date when the licence is to have effect.

In the case of a single fit, driving should be stopped pending specialist advice and the doctor should write for advice to the Medical Adviser at the Centre, without necessarily disclosing the patient's name. In these cases, it is usual for the licence to be suspended for 12 months.

Drug treatment

Though the published record on anticonvulsant drug therapy is enormous, it contains relative little specific reference to elderly patients. In those few drug trials from which elderly patients are not actually excluded, they are seriously under-represented. Most of what we think we know about anticonvulsant therapy in the ageing brain has been extrapolated from studies on younger patients, many of whom do not have the focal lesions that are typical in elderly epileptic patients and all of whom lack the age-associated changes seen in older people. It is to be hoped that this serious lack of information will be remedied shortly.

Starting treatment

Should patients who have a single major seizure be treated at once or should a recurrence be awaited? Many physicians do not treat a single major seizure, assuming that only a minority of patients will go on to have further seizures. Recent studies, however, have shown that the recurrence rate may be over 60 per cent by 1 year. Although age does not figure clearly as a predictive factor for recurrence in all studies that have addressed this question, one would expect the recurrence rate to be higher in the kind of elderly patients seen by geriatricians, among whom there is an especially high proportion of cases associated with cerebral, and in particular, cerebrovascular disease. Several other considerations may encourage early treatment. There have been suggestions that untreated fits may themselves predispose to more fits. Major seizures are more dangerous in elderly people, who are prone to injury, in particular fractures, and in whom postictal states may be prolonged. On the basis of these observations, there would appear to be a case for treating a single, clearly documented, tonic–clonic seizure that is not apparently related to alcohol or its withdrawal, drugs, infectious diseases of the nervous system, or a metabolic disturbance. Against this, however, there is the increased expectation of problems with anticonvulsant therapy in the aged. Much more work must be done before definite recommendations can be made about treating the first tonic–clonic seizure. Most physicians would treat two or more seizures.

There is even less information regarding the prognosis of untreated minor seizures. Treatment of a single minor episode is probably overzealous and it would seem to be reasonable to wait and see how frequent and how upsetting the episodes are before embarking on drug therapy.

The choice of medication

There is now considerable evidence that the vast majority of adult patients with either primary or secondary, generalized seizures or partial seizures can be controlled on a single drug. Phenytoin, carbamazepine, and sodium valproate may all be considered as first-line, broad-spectrum anticonvulsants. Monotherapy with any of these controls approximately four-fifths of patients in most series. Where monotherapy is unsuccessful, it is very often due to poor compliance or associated with widespread brain damage. Adding a second drug rarely contributes anything other than additional side-effects and it has been shown that in patients who are on more than one drug, withdrawal of the second or third drug may actually improve control. As well as increasing the probability of adverse effects, multiple therapy makes drug interactions more likely, makes it more difficult to evaluate the effect of individual drugs, and increases problems with compliance. There is some evidence that if monotherapy with one anticonvulsant gives unsatisfactory control, it is worthwhile trying monotherapy with another.

The general principle of monotherapy may have to be modified when we consider the place of the new generation of anticonvulsants such as progabide, vigabatrin, milacemide, and gabapentin. Their place is still very uncertain and there is negligible evidence about their use in elderly patients at present. They have been tried out only in resistant epilepsy, which is why they have been used only as add-on therapy, and we do not yet know whether they may be useful as monotherapy.

Reviews of clinical trials with the three main drugs, phenytoin, carbamazepine, and sodium valproate, indicate that they have approximately equal efficacy, and the choice of drug will therefore be influenced by considerations of toxicity and, to a lesser extent, cost. The toxicity of anticonvulsants has been investigated intensively, although relatively few studies have included significant numbers of elderly patients.

The gross neurological side-effects include ataxia, dysarthria, nystagmus, dizziness, unsteadiness, blurring and doubling of vision, reversible dyskinesias, and asterixis. Again, reviews indicate that although some toxic effects may occur more frequently with certain drugs, there is so much overlap that most side-effects cannot be attributed with certainty to any one drug. The effects are generally dose-related and in the general adult population can usually be avoided or minimized by careful titration of dosage. In elderly patients, whom one would expect to be more vulnerable to minor adverse effects, the differences between anticonvulsants may be more important. This particularly relates to those effects that impinge on cognitive function.

There is now a good deal of information about the impact of anticonvulsants on cognitive function in the general adult population. There is a consensus that polytherapy and excessive blood levels are associated with cognitive impairment. Of the commonly used, broad-spectrum anticonvulsants, maximum deficits are seen with phenytoin and lesser effects are observed with sodium valproate and carbamazepine. Phenytoin has adverse effects on memory, concentration, mental speed, and motor speed in volunteers, and there is a relationship between serum levels of phenytoin and memory impairment. There appear to be less marked cognitive effects with valproate and carbamazepine.

Of the non-neurological side-effects, osteomalacia may be particularly important because this is more likely to occur in patients with poor dietary intake of vitamin D and reduced exposure to sunlight, which already puts them at risk. Phenytoin, in particular, induces enzymes in the liver, and so accelerates metabolism of vitamin D. In one series there was biochemical evidence of osteomalacia in almost a fifth of patients on anticonvulsants for over 10 years and nearly half of these had histological evidence on bone biopsy. Recent studies have shown that sodium valproate, unlike phenytoin or carbamazepine, does not cause hypocalcaemia or reduced levels of vitamin D.

With the exception of reversible thrombocytopenia in patients receiving sodium valproate, blood dyscrasias are rare. Hepatic dysfunction may occur with all three drugs but it has been a matter of particular concern with valproate as it may occasionally lead to hepatic failure. How relevant this rare event is to the elderly population is uncertain, as most of the reported cases have been in young children on multiple therapy.

On the basis of relative toxicity, therefore, the choice of first-line, broad-spectrum anticonvulsant would seem to lie between carbamazepine and sodium valproate. Nevertheless, in the absence of work specifically directed at the elderly population, any firm recommendations must be regarded as premature. The case for properly conducted comparative trials in biologically aged, epileptic patients is stronger than the case for the routine first choice of any particular drug.

Other considerations may influence the choice of anticonvulsant. Phenytoin may be taken in a single daily dose, unlike sodium valproate and carbamazepine, which have to be taken twice or three times a day (the recently available, sustained-release formulation of carbamazepine may modify this). This is clearly an advantage in that it simplifies regimens, so improving compliance, and may be important in those patients who depend on others (e.g. a district nurse or carer) to help with their medication. Phenytoin also has the advantage that there is a predictable relationship between blood levels and efficacy, and between blood levels and side-effects. With carbamazepine the relationship between blood levels and efficacy is less predictable, although there is a strong relationship between levels and side-effects. In sodium valproate, the blood levels may predict neither efficacy nor all side-effects, though certain side-effects, such as tremor, may be dose-related. The saturation kinetics of phenytoin (see below), however, may make adjustment of the dosage difficult.

Dosages

The dosages recommended for the general adult population may be inappropriate for elderly patients. There is considerable evidence to support an age-associated increase in pharmacodynamic sensitivity to certain anticonvulsants. For example, a study of carbamazepine found a greater effect on body sway with one 400-mg dose, despite the absence of pharmacokinetic differences. Even more important than age-associated changes in pharmacodynamic sensitivity are the altered pharmacokinetics of anticonvulsants in older patients.

The concentration of anticonvulsants in the nervous system reflects the free or unbound concentration in the plasma rather than that bound to protein. As albumin concentrations tend to be lower in elderly people, higher free concentrations of certain drugs are to be expected. This has been demonstrated in the case of phenytoin, sodium valproate, and certain benzodiazepines. The differences are particularly marked with valproate. There is also reduced clearance of certain anticonvulsants. Single-dose studies have shown reduced clearance of valproate; in multiple-dose studies the maximum rate of phenytoin clearance and the clearance of unbound valproate are both reduced. In the case of phenytoin, there is an increased plasma half-time. This is due in part to reduced clearance but also to the fact that the volume of distribution for lipid-soluble drugs is increased in elderly individuals because of the increase of fatty tissue as a proportion of the total body mass. There is therefore a longer interval between initiation of a drug dosage and the attainment of a steady-state. This does not appear to apply to valproate.

The information just given should not lead to an exaggerated estimate of present knowledge of age-associated changes in pharmacodynamics or pharmacokinetics, or of its applicability to an individual patient. Such changes are often derived by comparing mean values for young and old groups. Differences within these groups may be at least as important as differences between them. In the case of phenytoin, for example, only 20 per cent of the interindividual variation noted in one series was attributable to age alone. In this context, as so often in clinical geriatrics, age is more important as a marker of unpredictable variability than of predictable change.

Other sources of unpredictability arise from concurrent diseases, particularly those that affect hepatic metabolism or that, for a variety of reasons, lead to a further reduction in albumin and hence protein binding. Renal impairment appears to be less important for most anticonvulsants. The multiple pathology associated with old age will often mean multiple medication; many drugs interact with anticonvulsants and they interact with one another. Interactions affecting anticonvulsants occupy over 10 per cent of Appendix 1 of the *British National Formulary*. Predicting plasma levels in a patient who is on more than two interacting drugs is even more difficult. Finally, there is the problem of compliance. Epileptic patients of all age groups comply poorly with their medication—which is not surprising in view of the chronicity of treatment, the purely prophylactic nature of the benefit, and the frequency of side-effects. There is little evidence that most elderly patients are much worse in this respect; nevertheless, poor or variable compliance will be another reason for the lack of predictable relationship between prescribed dose and plasma level, and between the doctor's action and the patient's response.

There is now sufficient evidence to suggest that the initial dose of phenytoin in an elderly person should not be more than 200 mg. It would seem reasonable to commence carbamazepine at 200 mg total daily dose and sodium valproate at 400 mg total daily dose. Except where fits are frequent and control is a matter of urgency, increases in dosage should be gradual. This is particularly applicable to phenytoin where, near the therapeutic range, an increment of as little as 25 mg may cause a marked rise in blood levels.

Anticonvulsant monitoring

The long-term management of epileptic patients has been enormously improved by the introduction of anticonvulsant monitoring. This is particularly useful where fits are not controlled by average doses of drugs, where there are doubts about compliance, where there are signs of intoxication, where there are odd neuropsychiatric syndromes, where there is a sudden loss of control of fits, where new interacting drugs are introduced, or where there are other diseases that may complicate treatment. The considerations discussed earlier, which indicate an increased unpredictability between prescribed dose and blood levels in the patient, make anticonvulsant monitoring particularly appropriate in the elderly.

It must be appreciated, however, that the most important part of monitoring the patient is not measurement of anticonvulsant levels but the use of information derived from history and examination. The patient or relative should keep a record of seizures; moreover, the patient should always be accompanied by a well-informed relative, neighbour, or carer to insure that as accurate as possible an account of events is obtained. Independent witnesses may also help the physician to pick up adverse effects, which may be subtle in the elderly and, if not actively looked for, missed. Some attempt should be made to assess compliance and this should always be discussed with the patient. Increasing the dose because of poor control due to variable compliance has been misinterpreted as implying insufficient dosage may lead to disaster. It is vital to emphasize the need to take medication consistently and indefinitely; some patients may have the impression that anticonvulsants need to be taken only when fits occur or as a 'course'. Finally, doctors should be aware that generic substitution may be associated with alteration in control and/or an increase in side-effects. This is particularly important in the case of phenytoin, where different preparations have markedly different bioavailability.

Anticonvulsant levels are most useful when they are used to answer a particular question or to resolve a particular uncertainty. Phenytoin is an especially appropriate drug for monitoring. First, its saturation kinetics means a non-linear relationship between dose and blood level: near the therapeutic range there will be a very steep dose–blood level curve. Secondly, it has a propensity to produce adverse neuropsychiatric effects, which may present non-specifically or be lost in the 'noise' of other neurological and other non-neurological pathology. Thirdly, interindividual variation in kinetics is more marked than with the other two broad-spectrum anticonvulsants. Fourthly, there is a close correlation, at least at the population level, between blood values of this drug and, on the one hand, efficacy and, on the other, side-effects. Finally, because of the long half-life of phenytoin, single samples taken at random give a good approximation of the steady-state level.

The place of anticonvulsant monitoring is less well-defined for valproate and carbamazepine. However, the dosage of carbamazepine is a poor predictor of serum concentration (though after the initial period of enzyme induction the relationship between dose and plasma concentration in an individual is relatively linear) and many of the side-effects do appear concentration-dependent. The relationship between the concentration of carbamazepine and the clinical response is complicated by the varying extent of metabolism to its active metabolite and individual pharmacodynamic variability. The values given for the therapeutic range should be interpreted with caution: seizure control may be achieved throughout a very wide range of concentrations. Moreover, a single measurement may be meaningless because of great variations of concentration during a dose interval. Both peak (3 to 4 h after a dose) and trough (just after the next dose) levels need to be measured. In the case of sodium valproate, there is little correlation between blood levels and pharmacological effect and, as there may be diurnal variation in drug clearance, repeated levels on the same dose may show wide variation. Only a few of the side-effects, such as tremor, are concentration-dependent. Monitoring, however, may help to rationalize treatment in patients on 'polypharmacy' and to identify the cause for failure of treatment when a patient is on an apparently adequate dose. Samples should be taken at a standard time in relation to doses.

Overdoing or over-interpreting levels of anticonvulsant may lead to mismanagement. As already indicated, levels are only a small part of the clinical picture and the results obtained from the laboratory must be interpreted in the light of the overall presentation. Therapeutic ranges defined in general adult populations may not apply to the elderly population and certainly will not necessarily apply to an individual elderly patient. Doses should not be adjusted in fit-free, non-toxic patients simply to bring the levels into the 'therapeutic range'. A patient with symptoms suggestive of intoxication should not be required to continue on the same dose of an anticonvulsant simply because the values from the laboratory fall within the notional therapeutic range. It must be remembered that laboratory results may be incorrect for all sorts of technical reasons, ranging from the time the specimen was taken, through the labelling of the specimen, to the method used in the measurement.

Can anticonvulsants be stopped in elderly patients?

The withdrawal of anticonvulsants is currently under intensive investigation. At present, specific information about its advisability in elderly patients is lacking but in view of the fact that late-onset epilepsy, partial seizures (which are, of course, more common in the elderly), and the presence of known cerebral pathology (also more common in elderly epileptic patients) are associated with an increased rate of relapse, one may have reluctantly to concede that withdrawal of therapy should not be attempted in most elderly patients who have had a good reason to be placed on anticonvulsants in the first instance. Whether patients in prolonged remission can be maintained on lower, even 'subtherapeutic' doses of anticonvulsants is something that will need clarifying. It is certainly worthwhile exploring and should give pause to a physician who, finding an elderly patient fit free on 100 mg of phenytoin daily, concludes that this patient does not require anticonvulsants and withdraws it. The result may be disastrous.

PROGNOSIS FOR THE ELDERLY EPILEPTIC PATIENT

The standard mortality ratio is increased in epileptic patients. One study, however, found that this increase was less marked in those diagnosed over 60 years of age (1.6 for the first 5 years, 1.8 for the next 5 years) than for those diagnosed in youth or middle age (2.5 for those aged 20–59). Mortality from heart disease was increased in epileptic subjects over the age of 65 in the Minnesota study, but the increase appeared to be confined to those in whom the seizures were symptomatic of cerebrovascular disease. Small studies of sudden, unexplained deaths in younger epileptic patients seemed to indicate that poor control is the most important factor. In aged epileptic patients, of course, there would be other causes of sudden death.

CONCLUSION

There is a good deal that we do not know about epilepsy in elderly people: epidemiological information is still only scanty; the role of cerebrovascular disease and the relative importance of other causes are still incompletely defined; the views of elderly patients themselves about epilepsy and the psychological impact of fits are largely unexplored; the prognosis of untreated epilepsy has not been properly addressed and the appropriate time to commence medication not determined; and the preferred choice of drug, the ideal dosage, and blood levels are not known. There is now more research than hitherto in this field and it is to be hoped that by the time the second edition of this textbook is published, many of these questions will at least have been addressed, even if they still have not been completely answered.

Bibliography

Brodie, M.J. and Hallworth, M.J. (1987). Therapeutic monitoring of carbamazepine. *Hospital Update*, **13**, 57–63.

Chadwick, D. (1987). Overuse of monitoring of blood concentrations of antiepileptic drugs. *British Medical Journal*, **294**, 723–4.

Chadwick, D. (1988). The modern treatment of epilepsy. *British Journal of Hospital Medicine*, **39**, 104–11.

Elwes, R.D., Johnson, A.L., Shorvon, S.D., and Reynolds, E.H. (1984). The prognosis of seizure control in newly diagnosed epilepsy. *New England Journal of Medicine*, **311**, 944–7.

Hauser, W.A. and Kurland, L.T. (1975). The epidemiology of epilepsy in Rochester, Minnesota 1935 through 1967. *Epilepsia*, **16**, 1–16.

Hopkins, A., Garmon, A., and Clarke, C. (1988). The first seizure in adult life: value of clinical features, electroencephalographs, and computerised tomographic screening in prediction of seizure recurrence. *Lancet*, **i**, 721–6.

Laidlaw, J., Richens, A., and Oxley, J. (1988). *A textbook of epilepsy*. Churchill Livingstone, London.

Leading article (1988). Sodium valproate. *Lancet* **ii**, 1229–31.

Luhdorf, K., Jensen, L.K., and Plesner, A. (1986). Etiology of seizures in the elderly. *Epilepsia*, **27**, 458–63.

Mawer, G. (1988). Specific pharmacokinetic and pharmacodynamic problems of anticonvulsant drugs in the elderly. In *Epilepsy and the elderly* (ed. R. C. Tallis). Royal Society of Medicine Services, London.

Reynolds, E.H., Shorvon, S.D., Galbraith, A.W., Chadwick, D. Dellaportas, C.I., and Vydelingum, L. (1981). Phenytoin monotherapy for epilepsy: a longterm prospective study, assisted by serum level monitoring, in previously untreated patients. *Epilepsia*, **22**, 475–88.

Tallis, R.C. (ed.) (1988). *Epilepsy and the elderly*. Royal Society of Medicine Services, London.

Trimble, M.R. (1987). Anticonvulsant drugs and cognitive function: a review of the literature. *Epilepsia*, **28** (suppl. 3), S37–45.

18.12 Parkinson's disease and related disorders

G. A. BROE

PARKINSON'S DISEASE AND AGEING

Parkinsonism, characterized by bradykinesia, rigidity, postural defects and, less often, a characteristic 3–8 c.p.s. tremor, is common among very elderly people and extremely common in those in hospitals and institutions, in whom it may be caused by neuroleptic drugs or form part of more widespread degenerative disease of the central nervous system. Idiopathic Parkinson's disease itself is closely associated with ageing, the prevalence rising from 8 per 100 000 under 50 years of age to about 2000 per 100 000 or 2 per cent of the general population at 70 years of age, and continuing to rise with advancing age.

Incidence rates also increase with age to reach a maximum at about 75 years. The decline in the rate of new cases after this age is most likely due to a decrease in case ascertainment related to problems in the diagnosis of Parkinson's disease in very elderly people and the failure of this group to seek expert medical care for their condition. These are the findings from community-based epidemiological studies (Schoenberg 1987). A number of autopsy studies show 10 times this prevalence of Lewy bodies in brains from the general elderly population, also rising with advancing age from 60 to 90 years. Gibb and Lees (1988) have presented a convincing argument that incidental Lewy-body disease corresponds to presymptomatic Parkinson's disease. It is clear that idiopathic or Lewy-body disease is age-associated, and its incidence rises exponentially with advancing age. An opposing concept, that Parkinson's disease is a disorder of late middle life, has been taught by neurologists for more than a century and has recently been revived by Koller *et al.* (1986), who describe a mean age of onset of 57 years in six combined series, with a peak incidence between the ages of 60 and 69 years and a decreased susceptibility at older ages. This concept is based on the use of clinical series rather than population samples.

The bias towards younger patients with this disease in most clinical series has had adverse effects on the management of the frailer and older population who comprise the client group for geriatric services, as a review of the history of drug therapy in Parkinson's disease will show (see below). Geriatricians are well aware from clinical experience that parkinsonism in older patients often requires a specific approach in terms of both diagnosis and management, particularly in the 'frail' elderly patient. This is not simply a factor of chronological age, as many elderly patients with this disease in their 70s or 80s have the same clinical and pathological features as younger patients and require the same clinical approach.

Four factors distinguish a significantly different, elderly parkinsonian population. First, late-onset Parkinson's disease commonly has clinical differences from idiopathic Parkinson's disease in younger age groups (Table 1); second, overt multiple disease processes are common in very elderly people, both within and outside the brain; third, multifactorial aetiology with summation of subclinical conditions is characteristic of many syndromes of ageing including parkinsonism; fourth, the most common cause of parkinsonism in elderly individuals is prescribed drugs. Each of these factors affecting the diagnosis and management of Parkinson's disease in late life will now be discussed in more detail.

Late-onset Parkinson's disease

Idiopathic Parkinson's disease is not a homogeneous entity, either clinically or pathologically. Even within the range of idiopathic Lewy-body disease responsive to levodopa there are clinical differences between early and later forms: a rapidly progressive and more aggressive form of the disease occurs in younger subjects with early dyskinesias and early fluctuations in response to levodopa; in contrast, later onset often means a more slowly progressive disorder with a stable long-term response to levodopa but an increased incidence of dementia.

In subjects of under 70 years of age the onset of the disease is usually unilateral, the most common initial feature is tremor, and while the symptoms and signs almost invariably

Table 1 Late-onset variant of idiopathic Parkinson's disease

	Classical idiopathic Parkinson's disease	Late-onset Parkinson's disease
Age (years)	40–90+	70–90+
Resting tremor	Asymmetrical, early and marked	Symmetrical, later and less marked
Rigidity	Asymmetrical	Symmetrical
Bradykinesia	Asymmetrical	Symmetrical
Gait/balance defect	Asymmetrical and late	Symmetrical and early
Fluctuation in response to levodopa	Early and marked	Late and mild
Dementia	Uncommon 8% at presentation <70 years; 20–30% during course	Common 40% at presentation 70+ years; >70% during course

become bilateral, they usually remain asymmetrical, involving one side significantly more than the other.

Late-onset Parkinson's disease is a variant of the idiopathic disease with increasing incidence over 70 years of age. Rigidity and bradykinesia are more bilateral and symmetrical from the onset, and tremor is less marked. The typical flexed posture, reduced arm-swing, and slowed gait are also bilateral and symmetrical. The disorder is often slowly progressive and benign in terms of the motor symptoms. These clinical features are responsive to low-dose levodopa; dyskinesias and fluctuation in response are uncommon, even after prolonged treatment.

Late-onset Parkinson's disease was previously termed senile parkinsonism (Broe 1987). The use of the term 'senile parkinsonism' by Jellinger (1987) for a disease of short duration with severe mental changes and severe, cortical, Alzheimer-like changes makes it inappropriate to continues its use for late-onset disease as defined here. While cognitive deficits and dementia are more common in this elderly group the dementia is often mild, slowly progressive, and does not have the cortical features of Alzheimer's disease. The cognitive defects are also partly dopa-responsive, at least in the early stages of therapy.

Classical idiopathic Parkinson's disease with asymmetrical onset and the asymmetrical progression of resting tremor, rigidity, and bradykinesia, and with a lower prevalence of dementia, also occurs at least as commonly as the late-onset type in very elderly persons. Further epidemiological studies are required to determine the proportion of classical and late-onset cases in the older age group.

The pathology of idiopathic Parkinson's disease is reviewed by Jellinger (1987) and will not be discussed in detail. The characteristics of late-onset disease in terms of its clinical picture, slow course, and dopa-responsiveness suggest that it is true, idiopathic, Lewy-body Parkinson's disease with a primary nigrostriatal dopaminergic deficiency due to loss of pigmented cells of the substantia nigra. The clinical features and the outcome of pathological studies suggest that the disease process may be more widespread in the central nervous system, both anatomically and biochemically, in the very elderly subjects who do not have classical idiopathic Parkinson's disease. In late-onset disease there may be more diffuse Lewy-body disease, as described by Yoshimura (1988) and others, to explain the altered clinical pattern (see Table 1) and higher prevalence of dementia. Alternatively, the clinical differences may be explained by more widespread changes that result from the summation of Lewy-body disease with other subclinical lesions in the ageing brain including Alzheimer's. The pathogenesis of late-onset Parkinson's disease remains speculative in the absence of longitudinal studies of ageing and its pathological correlates.

Multiple pathology in the ageing brain

Multiple disease processes both inside and outside the brain are characteristic of very elderly people and a natural consequence of their high prevalence of degenerative diseases.

Within a random sample of community-living elderly people (n=808), four disorders were responsible for neurological disability, contributing to almost half of the total disability (n=227). These disorders were senile dementia of the Alzheimer type (14 per cent); stroke including vascular dementia (18 per cent); Parkinson's disease (3 per cent); and senile gait disorder (13 per cent) (Akhtar *et al.* 1973). All four disorders increase in prevalence with advancing age (Table 2). It is therefore quite common in clinical populations of elderly patients to see multiple, overt brain diseases with various admixtures of Parkinson's disease, Alzheimer's disease, multiple infarcts, and age-associated ataxia.

Table 2 Prevalence (percentage) of neurological disorders in a random sample of community-living elderly people (Glasgow–Kilsyth Study)

Disorder	Age group (years)	
	65–74 (n = 488)	75+ (n = 320)
Dementia	4.3	14.1
Dementia of the Alzheimer type	2.5	10.9
Vascular dementia	1.4	2.5
Other dementias	0.4	0.6
Stroke	7.0	7.8
Parkinson's disease	1.6	1.6
Essential tremor	0.8	3.1
	(n = 227)	(n = 81)
Gait ataxia	4.4	32.1

Sources: Akhtar *et al* (1973); Broe *et al.* (1976).

Such complex patients often present with functional deficits in 'activities of daily living' or with falls, gait disorder, incontinence, acute confusion, and cognitive decline, rather than with classical neurological syndromes.

Multifactorial aetiology in the ageing brain

Within the very elderly population (75+ years) who do not have overt, multiple disease processes, it is likely that the common conditions outlined above may coexist in subclinical forms. In this age group, neurological syndromes such as parkinsonism may result from the product or summation of common subclinical states in the ageing brain. It is generally agreed that clinical parkinsonism does not occur until 80 per cent of nigral function is lost, and that a high degree of compensation through a variety of mechanisms maintains normal function until the 80 per cent threshold is passed. In an elderly population with subclinical pathology in other systems (e.g. Alzheimer changes in the cholinergic system, cell loss in the cerebellum, etc.), these homeostatic or compensatory mechanisms are likely to be impaired, and clinical parkinsonism may appear at a level of loss of nigral function that is usually subclinical, e.g. of the order of 50 to 60 per cent. Quinn (1986) has suggested a similar mechanism for the excess of dementia in the parkinsonian population, and Broe and Creasey (1989) have suggested summation of subclinical states to explain the clinical syndrome of senile gait disorder.

Drug-induced Parkinson's disease

Prescribed drugs are the most common cause of parkinsonism in older people. The susceptibility of elderly subjects to drug-induced parkinsonism presumably reflects the high prevalence of asymptomatic Parkinson's disease (10–15 per cent), and of loss of cells in the substantia nigra with advancing age, together with the loss of compensatory mechanisms in the brain due to subclinical disease in other systems.

The most potent dopamine-blocking drugs are the major tranquillizers, the fluorinated phenothiazines, and haloperidol used in the management of the major psychoses. However, the most common offending drugs for community-living elderly people are those of reputedly low risk: prochlorperazine used for vertigo, metoclopramide used for nausea, and thioridazine used for anxiety, confusion, and insomnia. In elderly subjects drug-induced parkinsonism occurs with low-risk drugs, and with low doses and short courses of high-risk drugs. It carries a high morbidity, producing falls and fractures, hypothermia, postural hypotension, dysphagia, and the complications of immobility including hypostatic pneumonia and pressure sores. Prevention is therefore vital and this means reducing the use of dopamine-blocking agents to an absolute minimum in elderly people. Prochlorperazine is of no value in the treatment of vertigo of any cause and should no longer be used for elderly patients; metoclopramide should not be used routinely as a postanaesthetic or postsurgical agent to prevent vomiting, and for elderly patients the alternative use of domperidone should be considered; phenothiazines should not be used for anxiety states or insomnia; and the use of phenothiazines and haloperidol to control delirium in elderly subjects should be seen as a medical strait-jacket with additional long-term complications and should be at low dose only, and when all other measures are ineffective.

Finally, the response to withdrawal of the dopamine-blocking agent, which is the appropriate therapy for drug-induced parkinsonism, may be very prolonged or incomplete in elderly individuals and may be followed, in a significant proportion, by prolonged or permanent tardive dyskinesia.

AETIOLOGY

The cause of idiopathic, Lewy-body, Parkinson's disease is unknown. It may well be a multifactorial disorder in view of its clinical variability, the lack of specificity of the Lewy body for this disease, and the many known causes of symptomatic parkinsonism.

The environmental hypothesis

The lack of family history in most cases, the lack of concordance in identical twins, cross-cultural differences in prevalence, and preliminary investigations that demonstrate an increased risk with a rural environment all suggest an environmental aetiological factor or factors. The major identified risk factor for idiopathic Parkinson's disease is, however, age. This is quite compatible with the environmental hypothesis in that chronological age may simply be an index of an extended time for the effect of environmental agents or toxins, either acting directly or through mechanisms such as free-radical formation. Brain ageing is not associated with generalized neuronal loss. It is associated with the loss of specific neuronal populations, particularly the pigmented cells of the substantia nigra and locus ceruleus, and with the accumulation of abnormal structures, including Lewy bodies, in the brains of those without definable neurological disease before death. It is unlikely that the progressive loss of specific neuronal populations, with or without the accumulation of abnormal structures, is due to a generalized biological ageing factor as has been suggested. It is a more parsimonious hypothesis that the age-associated loss of nigral cells and the reduction in biochemical markers of dopaminergic activity are due to extrinsic or environmental factors. These may be the same as, or additional to, the agent or agents responsible for Parkinson's disease.

Some specific environmental factors have been implicated in symptomatic parkinsonism. These include dietary features in the parkinsonism–dementia complex of Guam, repeated head trauma in the parkinsonism of the punch-drunk syndrome, and the recent observation that the toxin 1-methyl-4-phenyl-1,2,3,6-tetrahydropyridine (**MPTP**) can provoke parkinsonism in drug addicts and also produce subclinical nigrostriatal damage detectable with scanning by positron-emission tomography. In the case of the Guamanians and boxers, clinical parkinsonism may occur many years after the environmental insult.

These observations have led to the hypothesis that Parkinson's disease results from subclinical nigrostriatal damage early in life, followed by age-associated neural attrition (Calne and Peppard 1987). However, in the search for environmental agents causing the disease, we should be examining risk factors acting throughout the human lifespan including old age. Experimental evidence in animals suggests that the ageing brain is more susceptible to the toxic effects of MPTP and that the pathological changes in older, exposed animals more closely resemble the human disorder (Langston 1987; Ricaurte *et al.* 1988). Postencephalitic parkinsonism, presumed to be viral in origin, may also become clinically evident

Table 3 Late-onset Parkinson's disease—misdiagnoses

Diagnosis	Misdiagnosis
Essential tremor: presenting in any slowed or elderly person	Parkinson's disease
État lacunaire: presenting as gait disorder, slowing, and hypertonia	Parkinson's disease
Gait disorder: presenting with flexion, slowing,and mild gait disorder without hypertonia	Parkinson's disease
Dementia of the Alzheimer type: presenting with an extrapyramidal syndrome and gait disorder	Parkinson's disease
Parkinson's disease: presenting with 'stiffness' and slowing in the presence of age-associated joint disease	Arthritis
Parkinson's disease: presenting with psychomotor slowing, apathy, and inertia	Depression
Parkinson's disease: presenting with psychomotor slowing, apathy, and inertia (in the presence of abnormal tests of thyroid function	Hypothyroidism

some years after the original episode of encephalitis lethargica. In general, however, postencephalitic parkinsonism progresses very little, even with advanced ageing. Autopsy studies of elderly cases of the postencephalitic disease by Gibb and Lees (1987) showed no evidence of active neuronal death as seen in autopsies of age-matched, Lewy-body disease. This clinical and pathological evidence does not support the hypothesis that disease progression in idiopathic Parkinson's disease can be attributed to intrinsic ageing processes.

DIFFERENTIAL DIAGNOSIS OF LATE-ONSET PARKINSON'S DISEASE

Unlike classical Parkinson's disease, the late-onset disease is commonly misdiagnosed (Table 3): as arthritis because of complaints of early-morning stiffness and difficulty getting out of bed; as depression or hypothyroidism because of generalized psychomotor slowing; or simply as 'normal ageing'. Diagnosis is more difficult in the presence of multiple disease processes such as a previous stroke, osteoarthritic hips and knees, osteoporosis with crush fractures and flexed posture, true depressive illness, or of non-specific findings, on routine investigation, of mildly abnormal thyroid function tests or mildly reduced levels of vitamin B_{12}

Benign essential tremor

One of the most common diagnostic errors is to mistake benign essential tremor for Parkinson's disease in a slightly stooped and slowed, elderly person. Essential or heredofamilial tremor is very common in elderly people, with a prevalence of 1 per cent in the age group 65 to 74, and 3 per cent in those 75 and over (Broe *et al.* 1976). Except by the coincidence of two common age-associated diseases it is not related to parkinsonism. Essential tremor is clinically associated with lack of arm-swing, difficulty in manipulating fine objects, and cogwheeling due to the tremor but not with any other features of bradykinesia nor with true rigidity. In elderly subjects, essential tremor is often slow at 4 to 6 Hz. It may also be asymmetrical, pill-rolling in character, and superficially resemble a coarse, resting tremor. It is in fact a postural or action tremor and may be elicited by movement such as holding a cup or pen or by maintaining a posture with the hands outstretched. It is not inhibited by action as is parkinsonian tremor and, for this reason, coarse essential tremor is far more disabling when attempting to drink from a cup than is an equivalent resting tremor. Essential tremor disappears with the limb relaxed; however, it is often quite difficult for an elderly person to relax, particularly if there is associated disease such as arthritis, cervical spondylosis, or other causes of pain or hypertonia. Finally, essential tremor, unlike parkinsonian tremor, commonly involves the head, jaw, and pharynx. Despite the many differentiating features on careful examination, almost half the elderly patients with essential tremor are inappropriately given antiparkinsonian drugs, often for prolonged periods and often with toxic consequences.

État lacunaire

Multiple small strokes, particularly when located adjacent to the internal capsule, are commonly misdiagnosed as Parkinson's disease in elderly people. The syndrome of multiple lacunar infarcts or 'état lacunaire' is produced by bilateral corticospinal lesions. It mimics parkinsonism—with bilateral slowing, impaired fine movements, hypertonia, gait disorder, and minimal if any corticospinal weakness. However, careful examination usually shows asymmetrical hyper-reflexia and often extensor plantars. Although a history of small strokes may be difficult to elicit in an elderly, cognitively impaired patient, characteristic signs may be present, including pseudobulbar or reflex crying, brisk jaw-jerk, and a typical gait disorder diagnostic of this syndrome, marche-à-petit pas, in which the patient's feet stick to the floor with tiny, stuttering steps. This may be intermittent or only present on turning. There may also be a history of hypertension or diabetes, and often more diffuse vascular disease.

Other disorders with parkinsonism

Idiopathic Parkinson's disease needs to be differentiated from other disorders of basal ganglia in old age that may produce the parkinsonian syndrome, and from more diffuse central nervous disorders. The classic Steele-Richardson-Olzewski syndrome or progressive supranuclear palsy, in which parkinsonism is accompanied by a characteristic stare, loss of voluntary conjugate deviation of gaze, and pseudobulbar dysarthria, is encountered in elderly patients. However, progressive supranuclear palsy in old age more commonly presents as a slowly progressive, atypical parkinsonism with early unexplained falls and personality change involving a mixture of frontal impulsiveness, loss of affect and of concrete thinking, followed by the classical eye signs. Multisystem atrophy (olivopontocerebellar degeneration; Shy–Drager syndrome) is rare in elderly people. A degree of parkinsonism is not uncommon in senile dementia of the Alzheimer type and may occur in hypertensive end-vessel cerebrovascular disease or 'etat lacunaire'. In most situations where parkinsonism is part of a more diffuse degenerative disease of the central nervous system, it is unlikely to respond to levodopa, or early and serious neuropsychiatric toxicity is likely to occur. The appropriate diagnostic and therapeutic approach is to identify those complex elderly patients who have evidence of major motor involvement outside the extrapyramidal system (pyramidal or cerebellar signs); who have evidence of cognitive dysfunction atypical for Parkinson's disease; who have atypical parkinsonism particularly with marked axial rigidity, disproportionate dysarthria, and swallowing defects; who have hyper-extension with a tendency to fall backwards rather than the more typical flexed posture and tendency to fall forwards. Such patients should be given levodopa with great caution, if at all, and followed closely for toxic effects. In atypical cases the prognosis given to the relatives should also be guarded and community support organized in anticipation of a difficult time for carers.

Gait disorder

Finally, Parkinson's disease in elderly people needs to be distinguished from the age-associated, clinical deficits that have been commonly, but incorrectly attributed to 'normal' ageing in the sense of a generalized, biological, ageing factor. Senile gait disorder (Broe and Creasey 1989) is a syndrome that has been defined in cross-sectional, neuroepidemiological studies of community-living, elderly individuals. It affects about 4 per cent of the 'young' old (65–74 years) and about a quarter of the 75+ group, and occurs in the absence of a specific neurological disease, in particular stroke dementia or parkinsonism. The neurological aspect of senile gait disorder does include an extrapyramidal component, with slowing of gait and reduced step-length, but this is in the absence of rigidity and resting tremor. There is also a cerebellar component with a mildly widened gait base, and a cognitive component with mild decline in memory and impairment of adaptive abilities. Other neurological deficits that have in the past been attributed to 'normal' ageing are seen more commonly in senile gait disorder than in elderly individuals without disordered gait; these include flexed posture, action tremor, impaired upward gaze, and absent ankle jerks (Table 4).

Table 4 Gait disorder—a multifactorial syndrome

Flexed posture
Gait ataxia
Motor slowing
Mild cognitive impairment
Reduced upward gaze
Absent ankle jerks
Reduced distal vibration sense
Action tremor

Senile gait disorder is a multifactorial syndrome without an identified pathological basis. The conjecture that it represents the summation of subclinical, age-associated disease processes in the brain—including Lewy bodies, Alzheimer changes, and loss of Purkinje cells—remains a conjecture in the absence of autopsy studies related to its diagnosis in life, although the speculation is consistent with the known neuropathology of 'normal' ageing. Its clinical importance, in the context of this discussion of Parkinson's disease in elderly individuals, lies in the slender margin between therapeutic efficacy and toxicity in the use of antiparkinsonian drugs in this age group and the need to make as accurate a diagnosis as possible of idiopathic Parkinson's disease before using these drugs. Senile gait disorder has a significant extrapyramidal component in terms of flexion and motor slowing, yet it is clearly multifactorial in origin, with subtle but definite neurological deficits in multiple systems, and does not respond to levodopa therapy. If both slowing and rigidity are present, then late-onset Parkinson's disease should be considered and a trial of low-dose levodopa may be indicated.

ASSOCIATED CLINICAL FEATURES OF LATE-ONSET PARKINSON'S DISEASE

Other clinical features of Parkinson's disease more common in the elderly patient are postural hypotension, constipation, impairment of bladder function, difficulty in swallowing, sleep disorders, and the early development of a gait and balance disorder, all of which are poorly responsive to dopa, unlike the cardinal clinical features which respond well to low-dose levodopa. Postural hypotension causes more clinical problems in elderly than in younger patients, although it has the same pathogenesis and clinical features, and similar difficulties with drug management. Constipation with faecal impaction and spurious diarrhoea can be a serious problem in frail, elderly, parkinsonian patients and is a particular risk for the elderly patient in hospital or bedbound with any intercurrent illness. Impairment of bladder function, with frequency and incontinence, is more common and more difficult to evaluate in elderly parkinsonians. It is important to assess cognitive function as an index of frontal control of urinary continence in these patients before any consideration of urodynamic studies, medication, or surgery. A disorder of gait and balance unresponsive to dopa is not uncommon in younger parkinsonians late in the course of the disease after years of therapy. This disorder, in a milder form, appears early in late-onset disease and probably represents more widespread changes in the nervous system, including the loss of cerebellar Purkinje cells commonly seen in 'normal' ageing.

COMPLICATIONS OF LATE-ONSET PARKINSON'S DISEASE

Specific deficits of cognitive function including slowing of cognitive processing (bradyphrenia), reduced flexibility of cognitive processing, and mild memory impairment are common in elderly parkinsonians, even in the absence of dementia. These predispose this group to acute confusional state with all its potential complications (Table 5) including worsening of the parkinsonism, infection, dehydration, falls, incontinence, and pressure sores. Overt dementia is more

Table 5 Late-onset Parkinson's disease complications

Complication	Pathogenesis
Falls—resulting in: Loss of confidence; isolation; Colles fracture; fractured neck of femur	Progressing disease; inadequate therapy; antiparkinsonian drug toxicity; worsening gait disorder due to drugs, e.g. prochlorperazine, metoclopramide, other phenothiazines, tricyclics, etc.; postural hypotension due to drugs, e.g. dopa, cardiac drugs, hypotensives, diuretics, tricyclics, phenothiazines, etc.
Immobility—resulting in: Pressure sores; constipation; faecal impaction/spurious diarrhoea; hypostatic pneumonia; hypothermia	Progressing disease; inadequate therapy; drugs, e.g. prochlorperazine, metoclopramide, phenothiazines, sedatives, etc.; intercurrent infection, surgery, etc.
Neuropsychiatric disorders Delirium; visual hallucinations; paranoid states; incontinence	Progressing disease; intercurrent infection; antiparkinsonian drug toxicity; other drugs, e.g. beta-blockers, tricyclics, anticholinergics and antihistamines, sedatives, etc.

common in patients with late-onset Parkinson's disease than in younger parkinsonians and carries with it a high risk of complications in the patient admitted to hospital. Parkinson's disease in elderly people may be exacerbated during an intercurrent viral illness, chest or urinary tract infection, and may improve spontaneously after the infection resolves unless complications supervene. The coexistence of Parkinson's disease and dementia or confusion produces a particularly high risk for the development of pressure sores in elderly patients.

NEUROPSYCHOLOGY OF PARKINSON'S DISEASE

Cognitive impairment, behavioural changes, and mental disturbances are common and important in elderly patients with Parkinson's disease—as part of the disease process, as complications of drug therapy, and as part of the multiple pathological changes associated with advanced ageing.

It is much easier to establish the importance of disturbed mental function in elderly parkinsonians than to determine the causes or pathogenesis of the wide variety of deficits in cognition and behaviour that may occur in this disease. Major neuropsychological syndromes or mental disorders associated with ageing—dementia, delirium, hallucinations, and depression—are all more common in elderly parkinsonians than in the general elderly population and, with the exception of depression, they are much more common in the older than in the younger parkinsonian population. The relationship of these mental disorders to the milder cognitive deficits of idiopathic, Lewy-body, Parkinson's disease or to the summation of multiple disease processes in the ageing brain is clearly important but can only be speculated on in the absence of appropriate longitudinal and pathological studies.

In terms of management, it is important first, to know the prevalence of the major mental disorders in elderly parkinsonians and to determine the presence or absence of these disorders in the individual patient; second, to look for and treat other reversible causes rather than attribute the mental disorder to Parkinson's disease *per se*; third, to define the pathogenesis of dementia, delirium, hallucinations, and depression in Parkinson's disease not only in terms of scientific interest for theories of brain function but also in order to counsel the patient and carers and to plan long-term management, including drug therapy, behavioural strategies, and community support.

Delirium (see Chapter 18.5)

Delirium is the most common and serious toxic effect of antiparkinsonian drugs in elderly patients, reflecting both the high prevalence of cognitive impairment in this group and the high potential for neuropsychiatric toxicity in the drugs used. Delirium is a potentially reversible disorder, some 80 per cent of survivors returning to their previous level of function. It is both the most common and most neglected mental disorder of elderly patients in hospital. It carries a poor rate of accurate diagnosis by physicians and all too often a management programme—disastrous for the patient—aimed merely at reducing disruptive behaviour with restraints, bed rails, and phenothiazines. Prevention is therefore vital and in the elderly parkinsonian patient this means avoiding anticholinergics and bromocriptine; using low doses and small increments of levodopa; minimizing the use of sedatives, hypnotics, and general anaesthetics; avoiding unnecessary surgery and admissions to hospital; and minimizing all drug therapy.

Delirium is easily diagnosed in the restless, hallucinating patient and the drowsy, toxic patient who can be seen to have reduced attention span, poor concentration, or impaired consciousness. The diagnosis is easily missed in the apathetic, inert, elderly parkinsonian who is quietly confused and immobile, and in the paranoid, confused elderly patients who are misdiagnosed as having functional psychoses and given phenothiazines. Such drugs immobilize the patient but often worsen the confusion. Management consists of treatment of the (usually multiple) precipitating factors; attention to hydration; avoiding phenothiazines because of their anticholinergic and their dopa-blocking effect; and the prevention of complications, particularly bedsores. This essentially means good nursing in a proper environment where restraints, bed rails, phenothiazines, and sedatives can be avoided—a situation that is all too rarely available in our hospitals.

Elderly parkinsonian patients may also have recurrent hallucinations with variable degrees of insight and memory in the absence of other features of acute confusion. These are commonly drug-related and may occur with any antiparkinsonian agent, although most often anticholinergics. They may also occur in the absence of antiparkinsonian medication, usually at sunset and often in patients with impaired visual acuity. A wide variety of other monosymptomatic delusions or paranoid ideations may also be drug-related or, in the later stages of the disease, continue after withdrawal of medications. Clinical experience suggests these behavioural disorders are probably no more common in elderly than in late-stage, younger patients.

Depression

A large number of reviews have described a high prevalence (30 to 90 per cent) of depression in Parkinson's disease. This has commonly been attributed to the neurotransmitter disturbances and considered to be endogenous in origin. A recent, important review challenges this concept and considers the depression of idiopathic Parkinson's disease to be primarily reactive in type (Taylor *et al.* 1988).

Depression in elderly parkinsonian patients has not been specifically reviewed. Depressive illness itself is not uncommon in elderly people and clinical experience suggests it is less common in elderly parkinsonians than in younger patients. Perhaps this reflects a greater acceptance of disability by ageing people. Depression can present problems in diagnosis in elderly parkinsonians in three ways. First, long-term depressive patients may fail to have their developing parkinsonism recognized by their physician. Second and conversely, a depressive illness developing in a patient with long-standing, late-onset Parkinson's disease may be concealed by the already present apathy, inertia, and psychomotor retardation of the disease itself. Third, the parkinsonian features of apathy and inertia, together with slowing of cognitive processing and sleep disturbance, may lead to false diagnosis of depressive illness in elderly parkinsonians (see Table 3).

True depression is, however, a common enough management problem in this group to merit some specific comments, particularly in relation to drug therapy. The tricyclic antidepressants remain the mainstay of treatment for major depression in elderly people, despite their significant anticholinergic side-effects. Both peripheral (vision, bladder function) and central (delirium, gait disorder) anticholinergic side-effects are more common in elderly subjects, but delirium causes most problems. A careful trial of mianserin, with less anticholinergic effect, is reasonable in cognitively impaired patients with a clinical diagnosis of endogenous depression and Parkinson's disease treated by levodopa. If the depression is serious, life-threatening or unresponsive, electroconvulsive therapy should be considered.

Dementia (see also Chapter 18.4)

There has been a long-standing debate about dementia in Parkinson's disease, which reflects in part problems in defining and measuring cognitive functions and in defining 'clinically significant' dementia. In the older age group there have also been problems in defining idiopathic Parkinson's disease and separating this disorder from other syndromes, in particularly multi-infarct brain disease and Alzheimer's disease.

In a critical analysis of the methodological weaknesses inherent in cross-sectional studies based on clinical populations, and taking into account the problems in definition, Brown and Marsden (1984) have challenged the prevailing view that about one-third of patients with Parkinson's disease will become demented. Dementia as defined by criteria outlined in the American Psychiatric Association (1980) *Diagnostic and Statistical Manual of Mental Disorders* (**DSM III**) or 'clinically significant' dementia certainly occurs in a proportion of patients with Parkinson's disease but, at least in younger age groups, it is uncommon, usually mild, and occurs late in the course of the illness. Their estimate of 20 per cent for developing dementia is more reasonable than previous, higher estimates.

Clinicians who see a large number of elderly patients with idiopathic Parkinson's disease would, however, agree that a high proportion are demented on presentation and that dementia occurs frequently during the course of the illness. These clinical impressions are borne out in part by recent studies, using DSM-III criteria (Table 6), by Hietanen and Teravainen (1988), and by Reid *et al.* (1989) in a series of 100 cases comparing younger-onset with *de novo* later-onset patients.

Table 6 Dementia in Parkinson's disease (DSM III criteria)

	Younger onset	Later onset
Hietanen and Teravainen (1988) ($n = 108$)	2% (<60 years)	25% (>60 years)
Reid (personal communication) ($n = 100$)	8% (<70 years)	39% (>70 years)

The major determinants of dementia in Parkinson's disease are therefore late onset and advanced age rather than duration of disease (which was of equal or longer duration in the younger groups). Dementia, however, is much more common in elderly parkinsonians than in community-living elderly people, in whom the prevalence of 'clinically significant' dementia in a large number of studies ranges around 10 per cent. In terms of management, however, it is reasonable to adopt the same approach for the diagnosis and management of dementia in Parkinson's disease as for elderly patients in general, rather than to assume it is part of the disease process.

Several mechanisms may be at work in the pathogenesis of dementia in Parkinson's disease, with summation of these in the elderly group. First, a progression of the cognitive deficits related to loss of dopaminergic pathways involved in idiopathic Parkinson's disease, in particular those of the mesocorticolimbic system (Agid *et al.* 1987); second, a cholinergic deficit due to age-associated, Alzheimer-like changes in the hippocampus and cortex. As postulated by Quinn (1986), the excess of dementia in elderly parkinsonians could be explained by the summation of these two conditions rather than by a parkinsonian dementia or an Alzheimer dementia *per se*. Further contributions to the total neuropsychological deficit may come from loss in other neurotransmitter systems including cholinergic projections from the nucleus basalis of Meynert and the septal region, and noradrenergic projections from the locus ceruleus.

Alternatively it can be postulated that a subgroup of elderly patients with idiopathic Parkinson's disease have a slowly progressive dementia due to more widespread, Lewy-body disease rather than a summation of this change with those of Alzheimer type (Yoshimura 1988).

The nature of the dementia in Parkinson's disease of later life is as variable as its pathogenesis. Clinically, dementia appears less common in the elderly group with classical, idiopathic Parkinson's disease. It is more common in the late-onset group with bilateral, symmetrical signs (late-onset disease) but is clinically unlike Alzheimer dementia. The dementia in late-onset Parkinson's disease is slowly progressive like the disorder itself, with prominent slowing of cognitive processing, rigidity of thinking, difficulty in changing set, and perseverant thought processes, associated with mild or moderate memory impairment and absence of language disorder and apraxias. The actual frequency of this type of frontolimbic dementia in Parkinson's disease is, however, undetermined in epidemiological studies. As would be expected by the coincidence of two common disease processes, some very elderly, demented parkinsonian patients do manifest the severe amnesia, language disorder, and apraxias of Alzheimer's disease.

The management of elderly, demented parkinsonian patients is a common problem for geriatricians. All antiparkinsonian medication is accompanied in the 'frail' elderly patient by risk of neuropsychiatric toxicity, but the presence of mild or even moderate dementia does not contraindicate a trial of low-dose levodopa, with appropriate caution, if the physical indications are there for its use. Some cognitive deficits are even dopa-responsive, particularly inertia and slowed cognitive processing, but this response is invariably short-lived. Anticholinergics and bromocriptine are contraindicated for very elderly patients in the presence of dementia. Other aspects of management, including behavioural strategies, community services, and support for carers, depend on the nature and degree of the dementia, physical disability, and any associated disturbances of behaviour and continence.

Cognition and behaviour

In a recent comprehensive review, Brown and Marsden (1987) found no good evidence that patients with Parkinson's disease necessarily have widespread, 'clinically significant', cognitive impairment. Comparison of groups of patients with Parkinson's disease and controls reveals evidence of mild impairment in certain areas of cognitive function. There is, however, increasing evidence to confirm the clinical view that elderly parkinsonians show more evidence of widespread neuropsychological deficits as well as an increased prevalence of dementia compared with the general population of the same age. Both the deficits and the dementia are clinically important as they relate closely to complications of the disease, drug therapy, and general management.

In a brief description of cognition and behaviour in Parkinson's disease, four areas will be discussed in relation to ageing: (i) visuospatial function; (ii) memory function; (iii) speed of information-processing; (iv) behaviour.

Visuospatial function

Defects in visuospatial function have received more attention than any other area of cognitive impairment in Parkinson's disease. This contrasts with the recognized absence of impairment of parietal language function. The presence of any parietal disorder of visuospatial function in Parkinson's disease remains controversial and is not supported by recent neuropsychological research or clinical experience. In clinical terms, cognitively impaired patients with Parkinson's disease do not have the visuospatial defects and topographical disabilities of patients with Alzheimer's disease, and do not present the same management problems.

Memory function

Both verbal and visuospatial memory may be impaired in elderly parkinsonians on neuropsychological testing and in day-to-day events as described by patients and carers. The impairment is commonly mild and slowly progressive. The memory loss is more marked for free recall than for recognition tasks and to a certain extent responds to cues, indicating a 'frontal' memory disorder rather than the more severe and progressive, posterior hippocampal amnesia of Alzheimer's disease. This corresponds with its slower course in Parkinson's disease.

Speed of information-processing

One of the causes of apparent memory impairment in Parkinson's disease is the failure, on the part of health professionals, to allow time for the patient to process information. Slowing of information-processing is one of the most commonly reported cognitive impairments of Parkinson's disease in studies of choice reaction time. This slowing is clinically obvious and significant in the elderly group, who require a corresponding change in behaviour by their attendants and carers. Given more time, a patient with Parkinson's disease will answer the question; given more time, a patient with Alzheimer's disease will commonly forget the question.

Behaviour

There is increasing evidence in the neuropsychological record to support the clinical view that the cognitive impairments of Parkinson's disease primarily relate to 'frontal' processing of information in terms of both memory impairment and reduced speed, and possibly also the 'visuospatial' defect. The behavioural impairments of Parkinson's disease also fit a 'frontal' pattern. The disease in elderly people is characterized by a degree of apathy, inertia, and inflexibility of thought-processing, or difficulty in changing from one set of thought processes to another. This behavioural pattern is described with damage to the frontal lobes or frontal connections, particularly the frontolimbic system. It is defined by Tate (1987), in traumatic frontal-lobe damage, as a 'disorder of drive' with inertia, inflexibility, and adynamia, and is contrasted with a frontal 'disorder of control' characterized by disinhibition, impulsivity, and facetiousness. These frontal syndromes can be viewed as primary disorders of attention through loss of feedback control via loops linking the frontal lobes with subcortical structures.

The 'disorder of drive' characteristic of Parkinson's disease may result from an inability to shift attention from internal stimuli and a consequent failure to attend appropriately or rapidly to environmental stimuli—'the needle in the groove' syndrome. In Parkinson's disease this presumably relates to impairment of a complex loop connecting the basal frontal cortex with the basal ganglia (Brown and Marsden 1987).

An understanding of this behavioural syndrome in Parkinson's disease is likely to be of increasing importance in the management of elderly patients in several ways. First, the depression of Parkinson's disease may reflect loss of initiation and drive and/or inability to change set from perseverant depressive thoughts; second, paranoid ideation, not uncommon in elderly patients with Parkinson's disease, may reflect a perseverance of disordered perception; third, the predilection of patients with Parkinson's disease to develop toxic delirium may relate to impaired attention to the environment. Further research in these 'frontal' deficits is likely to lead to a better understanding of similar functional changes in so-called 'normal' ageing and of the relationship of these changes to possible dopaminergic defects in th mesocorticolimbic system.

ANTIPARKINSONIAN DRUGS

Introduction

The very elderly population of Parkinson's disease sufferers commonly has an altered response to drug therapy, reflecting both diffuse, age-associated changes in the central nervous system and altered drug metabolism. Drug trials are invariably carried out on fitter, younger populations. Regimens so established are then used for older populations with similar clinical syndromes but of multifactorial aetiology, and/or with multiple disease processes and altered drug metabolism. In the early days of levodopa, because of this lack of understanding of the special therapeutic needs of 'frail' elderly people, patients were all too frequently overdosed resulting in toxicity and treatment failure or, because of the consequent toxicity, denied the benefits of levodopa in low dose and treated with even more toxic alternatives particularly anticholinergics. Although levodopa, with or without dopa-decarboxylase inhibitors, is now the drug of choice for idiopathic Parkinson's disease at all ages, we are entering a new era of dopamine agonists and facing a push towards the earlier use of existing agonists in an attempt to prevent the late consequences of levodopa therapy. Before examining the present therapeutic alternatives, it is instructive to review the history of levodopa and Parkinson's disease over the past 30 years with particular reference to the older population.

Levodopa therapy: 1969–1989

The problems of age-associated variations in clinical features, in choice of dosage and side-effects of drugs, and in the clinical course of Parkinson's disease in elderly subjects have been given inadequate attention.

After the first demonstrations of a therapeutic effect of levodopa in humans by Birkmeyer and Hornykiewicz (1961), the main clinical series on which modern drug therapy of Parkinson's disease was based between 1969 and 1970 used high doses with rapid increments in dosage for predominantly younger populations (Table 7). Three of these studies, involving a higher proportion of elderly patients, found a negative correlation between old age and response to treatment. Early studies by Sacks *et al.* (1970, 1972) reported severe toxic confusional states in aged patients treated with levodopa. In all of these studies, high doses and rapid increments of levodopa were used and the need for low dosage and gradual increments to reduce toxicity in at-risk groups was not, in general, recognized.

A significantly different approach to levodopa therapy in elderly patients was brought out in a number of studies from geriatric units published between 1973 and 1974 (Table 8).

Table 7 Early levodopa trials

Trial	No. of subjects	Average age (years)	Average daily dose (g)
Barbeau (1969	86	60	4.8
Cotzias *et al.* (1969)	17	51	5.8
Godwin-Austen *et al.* (1969)	18	56	3.0–8.0[a,b]
Klawans and Garvin (1969)	105	65[a]	2.0–6.0[a,b]
Mawdsley (1970)	32	61	4.0–6.0[a]
Mones *et al.* (1970)	152	55[a]	3.0–4.0[a]
Peaston and Biachine (1970)	22	65	3.0[b]
Stellar *et al.* (1970)	91	60[a]	3.0–5.0[a,b]

[a] Estimated; [b] negative correlation between old age and response to levodopa.

Table 8 Early levodopa trials in elderly patients

Trial	No. of subjects	Average age (years)	Average daily dose (g)
Broe and Caird (1973)	16	76*	1.7
Grad *et al.* (1974)	15	76	1.9
Sutcliffe (1973)	50	70	1.0–2.0*
Vignalou and Beck (1973)	122	70–90	2.4

* Estimated.

These clearly showed the benefit of smaller increments and low dose in achieving a good therapeutic response and reducing toxicity; they showed a low incidence of serious nausea and vomiting on plain levodopa in low dosage and a low incidence of neuropsychiatric toxicity, even in very elderly and demented subjects. Broe and Caird (1973) demonstrated the serious neuropsychiatric toxicity caused by adding anticholinergics to levodopa therapy in this group. Vignalou and Beck (1973), in their series of 122 subjects aged 70 to 90 years, showed that an equivalent therapeutic response could be obtained with lower doses of levodopa, producing a much reduced incidence of dyskinesias. Low-dose levodopa was not, however, generally accepted for a further decade. Indeed the next therapeutic step was the introduction of combined therapy with peripheral dopa-decarboxylase inhibitors, which reduced systemic dopamine levels, nausea, and vomiting in younger subjects, and achieved even higher striatal concentrations of dopamine.

Geriatric units using low-dose, plain levodopa found that most elderly patients with late-onset Parkinson's disease had a stable, long-term response to therapy. European experience (Birkmeyer 1976) with lower maintenance doses of levodopa was similar. However, the continuation of high-dose levodopa therapy in the United States in particular, and the general acceptance of combination therapy with peripheral dopa-decarboxylase inhibitors to produce higher levels of striatal dopamine was accompanied by an increasing recognition in the 1970s of fluctuations in response as a complication of long-term levodopa. In an initial reaction that was potentially adverse as far as elderly patients were concerned, Calne Fahn and others, in a series of articles from

1976, advocated that levodopa therapy should be delayed and treatment started with anticholinergic drugs without taking into account their high toxicity in the aged (Editorial 1976; 1978). This was followed, over the decade from 1976, by a gradual and welcome tendency for neurologists to reduce levodopa dosage, and in the 1980s the recognition that anticholinergics are to be avoided in Parkinson's disease of later life.

There is a current move towards the introduction of bromocriptine or other dopamine agonists early in the course of Parkinson's disease in younger patients to reduce the incidence of fluctuations in response. In the elderly patient bromocriptine has greater neuropsychiatric toxicity and the long-term use of low-dose levodopa in elderly parkinsonians is not commonly accompanied by severe fluctuations in response. Levodopa used alone, commenced early, and given in low dose remains the drug of choice in idiopathic Parkinson's disease of late onset (Hely *et al.* 1989).

Absorption of levodopa in elderly patients

In a series of papers, Evans *et al.* (1980, 1981*a*, *b*) have shown that elderly subjects have a threefold increase in the absorption of oral levodopa compared with young controls. This increase occurred despite a much prolonged gastric emptying rate in elderly subjects. These studies suggest that gastric dopa-decarboxylase activity is reduced in elderly subjects and that small-bowel absorption is consequently increased; this results in higher plasma levels of levodopa crossing the blood–brain barrier and higher striatal levels of dopamine for an equivalent oral dose (Broe 1987). These changes in absorption in elderly subjects are in part responsible for their lower dosage requirements and may in part explain the more stable long-term therapeutic response to levodopa, which resembles the effect of controlled release preparations in younger patients.

Initiation

Levodopa with carbidopa and levodopa with benserazide have equivalent therapeutic effects in elderly people. The value of the peripheral dopa-decarboxylase inhibitors in these preparations is dubious in very elderly patients. However, initiation with levodopa 100 mg/carbidopa 10 mg, which is recommended as a starting dose, is equivalent to old-fashioned plain levodopa, because carbidopa at this dose is ineffective. Increments should not be made at less than weekly intervals in outpatients and a close watch kept for mental disturbances, which are the main toxic effects in very elderly patients, although rare with this low-dose regimen if the diagnosis is correct. The maximum dose is the same in the well-preserved, healthy, elderly patient with classical Parkinson's disease as in young patients, but doses above 250 mg levodopa/25 mg carbidopa, three times a day, would be rare. Patients with late-onset Parkinson's disease commonly have a satisfactory response at doses between 100 mg/10 mg, thrice daily, to 200 mg/20 mg, thrice daily—levels at which the carbidopa probably remains ineffective. The rare elderly patient who develops nausea at these doses should be changed to levodopa 100 mg with carbidopa 25 mg or benserazide 25 mg in a reduced total dose initially. Dyskinesias are rare in elderly patients on this low dosage. The early development of mental disturbances, most commonly delirium or hallucinations, on low dosage often means a wrong diagnosis, usually a parkinsonian presentation of Alzheimer's disease, or multi-infarct brain disease. Failure to respond to reasonable levels of levodopa also commonly means a wrong diagnosis or multiple disease processes in the brain.

Other drugs

Initiation of therapy with bromocriptine, or with a combination of low-dose levodopa and low-dose bromocriptine, to prevent future fluctuations in response is not an issue in the very elderly patient in whom late fluctuations are uncommon. Bromocriptine is in general less effective and, in the elderly patient, far more toxic than levodopa, producing a high rate of psychiatric complications. The addition of other antiparkinsonian agents to levodopa is rarely helpful in very elderly patients. Anticholinergics are the most toxic, inducing delirium, memory impairment, and behavioural disturbances. Amantadine carries a particularly high risk in elderly subjects, although it is not in itself a highly toxic (or effective) antiparkinsonian agent. Amantadine has a relatively long duration of action and is excreted unchanged in the urine. Elderly patients have frequent renal impairment, not necessarily reflected in their serum creatinine levels, and are at high risk of psychiatric toxicity in the usual dose of 200 mg per day.

Maintenance

Patients with late-onset Parkinson's disease, as outlined in earlier sections, respond well to low-dose levodopa initially and have a more stable, long-term response. The development of late fluctuations in response and an on-off effect is less common than in younger age groups but equally difficult to manage when it occurs. The management techniques for fluctuations in response are similar for all age groups, but the use of bromocriptine and newer dopamine agonists for severe on-off effects is limited in elderly patients by their neuropsychiatric toxicity. Selegeline, a selective monoamine-oxidase B inhibitor, is of some value in fluctuation in response in younger patients. Its use in elderly patients has not been evaluated; however, there is no reason to expect specific toxicity in the older group. The use of antiparkinsonian drugs in the presence of dementia has been discussed above.

OVERALL MANAGEMENT

Idiopathic Parkinson's disease in both younger and older patients is relatively easy to manage in the early stages of the uncomplicated disease; while the clinical features are dopa responsive this response is stable and the level of disability remains mild. At this early stage, overall management consists of preventing complications by using appropriate drugs, minimizing other medications, and avoiding unnecessary surgery.

Prevention is particularly important in elderly parkinsonian patients with moderately advanced disease and neuropsychological impairment—a group prone to serious complications (outlined above under 'Clinical Features' and 'Neuropsychology') including delirium, falls and fractures, chest and bladder infections, incontinence of urine, severe constipation, and pressure sores. Preventive measures include avoiding dopamine-blocking drugs, particularly those of reputedly 'low risk'—prochlorperazine, metoclopramide, and thioridazine, which is also strongly anticholinergic; avoiding all drugs with central anticholinergic action; minimizing the use of sedatives, hypnotics, anaesthetics, and surgery; carefully assessing and minimizing all drug therapy; avoiding immobilization.

The management of incontinence illustrates the complex mix of motor and cognitive impairments in Parkinson's disease. Frequency, urge incontinence, and nocturnal incontinence are all relatively common in elderly parkinsonians with moderately advanced disease. The majority who are investigated for bladder symptoms have a hyperactive detrusor and often a degree of cognitive deficit. These patients may benefit from intensive bladder retraining, if cognitive function is relatively intact, or simply require frequent toileting if more cognitively impaired to remain dry during the day. The diagnosis of true outflow-tract obstruction, requiring prostatic surgery and carrying a significant risk of postsurgical incontinence, is particularly difficult in elderly parkinsonians and requires careful clinical and urodynamic study before surgery. Staskin *et al.* (1988) have demonstrated that, while three-quarters of patients with Parkinson's disease have detrusor hyper-reflexia, incontinence after surgery correlates more with poor or absent voluntary sphincter control. The assessment of cognitive function as an index of frontal control of bladder function should therefore precede urodynamic studies in very elderly patients with Parkinson's disease and incontinence, and should certainly precede any consideration of peripheral anticholinergic drugs or surgery.

Management in more advanced parkinsonian patients is chiefly aimed at supporting independent functioning in the community for as long as possible and supporting carers, particularly where significant dementia or behavioural deficits predominate. Progressive assessment of functional disabilities in mobility, in instrumental activities of daily living (cooking, housework, shopping, budgeting) and in personal care (bathing, feeding, dressing, toileting) are of value in determining the patient's needs for therapy, the family's needs for community support (home care, delivered meals, community nurse, transport) and the carers' needs for respite. The most important determinant of stress on carers and their need for respite is disturbed behaviour, usually in the context of a progressive dementia.

Respite given by regular attendance at a day centre, or intermittent attendance at a day hospital for therapy, often produces functional improvement in the patient and relief of stress for the carer, so delaying any need for long-term institutional care.

References

Agid, Y., Javoy-Agid, F., and Ruberg, M. (1987). Biochemistry of neurotransmitters in Parkinson's disease. In *Movement Disorders 2* (ed. C.D. Marsden and S. Fahn), pp. 166–230. Butterworths, London.

Akhtar, A.J., Broe, G.A., Crombie, A., McLean, W.M.R., Andrews, G.R., and Caird, F.I. (1973). Disability and dependence in the elderly at home. *Age and Ageing*, **2**, 102–10.

American Psychiatric Association (1980). *Diagnostic and statistical manual of mental disorders* (3rd edn). APA, Washington.

Barbeau, A. (1969). L-dopa therapy in Parkinson's disease. *Canadian Medical Association Journal*, **101**, 791–800.

Birkmeyer, W. (1976). In *Advances in parkinsonism*, (ed. W. Birkmeyer and O. Hornykiewicz), p. 407. Editiones Roches, Basel.

Birkmeyer, W. and Hornykiewicz, O. (1961). Der 1-dioxyphenylalanin (l-dopa)—effeckt bei der Parkinson-akinese. *Wiener Klinische Wochenschrift*, **73**, 787–8.

Broe, G.A. (1987). Antiparkinsonian drugs. In *Clinical pharmacology in the elderly* (ed. C.G. Swift), pp. 473–509. Marcel Dekker, New York.

Broe, G.A. and Caird, F.I. (1973). Levodopa for parkinsonism in elderly and demented patients. *Medical Journal of Australia*, **1**, 630–5.

Broe, G.A. and Creasey, H. (1989). The neuroepidemiology of old age. In *The clinical neurology of old age* (ed. R. Tallis), pp. 51–65. Wiley, London.

Broe, G.A., Akhtar, A.J., Crombie, A., McLean, W.M.R., Andrews, G.R., and Caird, F.I. (1976). Neurological disorders in the elderly at home. *Journal of Neurology, Neurosurgery and Psychiatry*, **39**, 362–6.

Brown, R.G. and Marsden, C.D. (1984). How common is dementia in Parkinson's disease? *Lancet*, **ii**, 1262–5.

Brown, R.G. and Marsden, C.D. (1987). Neuropsychology and cognitive function in Parkinson's disease: an overview. In *Movement Disorders 2* (ed. C.D. Marsden and S. Fahn), pp. 99–123. Butterworths, London.

Calne, D.B. and Peppard, R.F. (1987). Aging of the nigrostriatal pathway in humans. *Canadian Journal of Neurological Sciences*, **14**, 424–7.

Cotzias, G.C., Papavasiliou, P.S., and Gellene, R. (1969). Modification of Parkinsonism—chronic treatment with l-dopa. *New England Journal of Medicine*, **280**, 337–45.

Editorial (1976). Alternatives to levodopa. *British Medical Journal*, **i**, 1169–70.

Editorial (1978). Drugs for Parkinson's disease. *Lancet*, **i**, 754–5.

Evans, M.A., Triggs, E.J., Broe, G.A., and Saines, N. (1980). Systemic availability of orally administered l-dopa in the elderly parkinsonian patient. *European Journal of Clinical Pharmacology*, **17**, 215–21.

Evans, M.A., Broe, G.A., Triggs, E.J., Cheung, M., Creasey, H., and Paull, P.D. (1981*a*). Gastric emptying rate and the systemic availability of levodopa in the elderly parkinsonian patient. *Neurology*, **31**, 1288–94.

Evans, M.A., Triggs, E.J., Cheung, M., Broe, G.A., and Creasey, H. (1981*b*). Gastric emptying rate in the elderly; implications for drug therapy. *Journal of the American Geriatrics Society*, **29**, 201–5.

Gibbs, W.R.G. and Lees, A.J. (1987). The progression of idiopathic Parkinson's disease is not explained by age-related changes. Clinical and pathological comparisons with post-encephalitic parkinsonian syndrome. *Acta Neuropathologica* (Berlin), **73**, 195–201.

Gibb, W.R.G. and Lees, A.J. (1988). The relevance of the Lewy body to the pathogenesis of idiopathic Parkinson's disease. *Journal of Neurology, Neurosurgery and Psychiatry*, **51**, 745–52.

Godwin-Austen, R.B., Tomlinson, E.B., Frears, C.C., and Kok, H.W.L. (1969). Effects of l-dopa in Parkinson's disease. *Lancet*, **ii**, 165–8.

Grad, B., Werner, J., Rosenberg, G., and Wener, S.W. (1974). Effects of levodopa therapy in patients with Parkinson's disease: statistical evidence for reduced tolerance to levodopa in the elderly. *Journal of the American Geriatrics Society*, **22**, 489–94.

Hely, M. *et al.* (1989). Sydney Multicentre Study of Parkinson's Disease—a report of the first 3 years. *Journal of Neurology, Neurosurgery and Psychiatry*, **52**, 328–52.

Hietanen, M. and Teravainen, H. (1988). The effect of age of disease onset in neuropsychological performance in Parkinson's disease. *Journal of Neurology, Neurosurgery and Psychiatry*, **51**, 244–9.

Jellinger, K. (1987). The pathology of parkinsonism. In *Movement Disorders 2* (ed. C.D. Marsden and S. Fahn), pp. 124–65. Butterworths, London.

Klawans, H.L. and Garvin, J.S. (1969). Treatment of parkinsonism with l-dopa. *Diseases of the Nervous System*, **30**, 737–46.
Koller, W. *et al.* (1986). Relationship of aging to Parkinson's disease. In *Advances in Neurology*, Vol. 45 (ed. M.D. Yahr and K.J. Bergman), pp. 317–21. Ravin Press, New York.
Langston, J.W. (1987). M.P.T.P. insights into the etiology of Parkinson's disease. *European Neurology*, **26**, 2–10.
Mawdsley, C. (1970). Treatment of parkinsonism with levodopa. *British Medical Journal*, **1**, 331–7.
Mones, R.J., Elizan, T.S., and Siegal, G.J. (1970). Evaluation of l-dopa therapy in Parkinson's disease. *New York State Journal of Medicine*, **70**, 2309–18.
Peaston, M.J.T. and Bianchine, J.R. (1970). Metabolic studies and clinical observations during l-dopa treatment of Parkinson's disease. *British Medical Journal*, **i**, 400–3.
Quinn, N.P. (1986). Dementia and Parkinson's disease—pathological and neurochemical considerations. *British Medical Bulletin*, **42**, 86–90.
Reid, W.G.J., *et al.* (1989). The neuropsychology of *de novo* patients with idiopathic Parkinson's disease: the effects of age of onset. *International Journal of Neuroscience*, **48**, 205–17.
Ricaurte, G.A., Langston, J.W., Irwin, I., DeLanney, L.E., and Forno, L.S. (1988). The neurotoxic effect of M.P.T.P. on the dopaminergic cells of the substantia nigra in mice is age-related. *Neuroscience Abstracts*, **11**, 631.
Sacks, O.W., Messeloff, C., Schwartz, W., Goldfarb, A., and Kohl, M. (1970). Effects of l-dopa in patients with dementia. *Lancet*, **i**, 1231.
Sacks, O.W., Kohl, M.S., Messeloff, C.R., and Schwartz, W.F. (1972). Effects of levodopa in parkinsonian patients with dementia. *Neurology*, **22**, 516–19.
Schoenberg, B.S. (1987). Epidemiology of movement disorders. In *Movement disorders 2* (ed. C.D. Marsden and S. Fahn), pp. 17–32. Butterworths, London.
Staskin, D.S., Vardi, Y., and Siroky, M.B. (1988). Post-prostatectomy continence in the parkinsonian patient: the significance of poor voluntary sphincter control. *Journal of Urology*, **140**, 117–18.
Stellar, S., Mandell, S., Waltz, J. M., and Cooper, I. S. (1970). L-dopa in the treatment of parkinsonism. *Journal of Neurosurgery*, **32**, 275–80.
Sutcliffe, R.L.G. (1973). L-dopa therapy in elderly patients with parkinsonism. *Age and Ageing*, **2**, 343–8.
Tate, R.L. (1987). Issues in the management of behaviour disturbance as a consequence of severe head injury. *Scandinavian Journal of Rehabilitation Medicine*, **19**, 13–18.
Taylor, A.E., Saint-Cyr, J.A., and Lang, A.E. (1988). Idiopathic Parkinson's disease: revised concepts of cognitive and affective status. *Canadian Journal of Neurological Sciences*, **15**, 106–13.
Vignalou, J. and Beck, J. (1973). La l-dopa chez 122 parkinsoniens de plus de 70 ans. *Gerontologia Clinica*, **15**, 50–64.
Yoshimura, M. (1988). Pathological basis for dementia in elderly patients with idiopathic Parkinson's disease. *European Neurology*, **28** (suppl. 1), 29–35.

18.13.1 The ageing eye

A. J. BRON

INTRODUCTION

The eye is supported and protected by the orbit and adnexal structures, the orbital fat, extraocular muscles, lids and lacrimal glands. None of these structures escapes the effects of ageing. Although there are distinct age-associated diseases of the eye which are genetically determined, it is likely that many more are due to the prolonged, cumulative action of physical and chemical influences. This includes, in particular the action of light and other forms of irradiation, as well as the products of metabolism. Some features of ageing are of little functional consequence; others are disease processes, which at their worst may lead to blindness. A list of age-associated diseases is given in Table 1.

ANATOMY AND PHYSIOLOGY OF THE EYE

It is important to understand some of the normal aspects of ocular anatomy and physiology before considering those changes brought about by ageing. The eyeball is protected by the action of the lids as in blinking or closure, and the ocular surface is continuously bathed by the tears, whose aqueous component is chiefly from the lacrimal gland. Eye movements are finely co-ordinated by the reciprocal contractions of the extraocular muscles to maintain alignment of the eyes in the same visual direction. The globe has a tough outer corneoscleral envelope, whose corneal portion is highly transparent. With the tears it forms the front refracting surface of the eye. The tear film confers a mirror-like optical surface to the cornea whose high refracting power lies in the difference in refractive index between air and tears. The optical homogeneity of the cornea is due to the packing of its constituent collagen fibrils, which are of extremely small diameter and organized in bundles within which the fibrils lie in parallel, and are separated by no more than a quarter the wavelength of the incident light. Back-scatter of light is reduced to a minimum by destructive interference and light transmission is almost 100 per cent. The collagen fibrils are embedded in a ground substance (proteoglycan) whose negative charge and mutual repulsion tend to cause corneal swelling. This is counteracted by the pumping action of the endothelial cells which line the posterior surface of the cornea and maintain it at a constant hydration, which is critical for corneal transparency.

The other refracting elements of the eye responsible for forming a sharp image of the visual world upon the retina are the aqueous humour, lens, and vitreous. The refractive power of the lens arises from the high protein content of its fibres (one-third of its wet weight), mainly due to the presence of its lens-specific proteins, the crystallins. This reduces the refractive jump between cytoplasm and fibre membrane, and optical homogeneity is further achieved by the small extracellular space between adjacent fibres.

The lens is suspended from the ciliary body in such a way

Table 1 Age-associated changes and diseases of the eye

	Site	Changes	Disease
Tears	Tear flow	+	Keratoconjunctivitis sicca
Lids			Entropion, ectropion, ptosis
Globe	Scleral rigidity	+	
Cornea	Corneal sensitivity	−	Hudson–Stähli line
			Spheroidal degeneration
	Endothelial density	−	Lipid arcus
			Pinguecula (pterygium)
			Cornea guttata
			Fuchs' dystrophy
Pupil	Miosis		
Chamber	Volume	−	Chronic open-angle glaucoma
Pressure		+?	Angle closure glaucoma
Aqueous flow		−	
Outflow resistance		+	
Lens	Size	+	Cataract
	Accommodation	−	
	Short-wave absorption	+	
	Scatter	+	
	Water-insoluble protein	+	
	Crystallin aggregates	+	
	Fluorescent chromophores	+	
	Ascorbate	−	
	Reduced glutathione	−	
	Post-translational modification	+	
Vitreous	Syneresis		
	Posterior detachment		
Retina	Rod disorder		Drusen
	Rod density	−	Age-related macular degeneration
	Cone disorder		Diabetic retinopathy
	Cone density	−	
	Retinal pigment epithelium disorder*		
	Retinal pigment epithelium density	−	
	Lipofuscin	+	
	Morphological change		
	Ganglion cell axons	−	
	Retinal vascularity	−	
	Choroidal vascularity	−	

that contraction allows the lens to take up a more curved shape. This act of accommodation, increasing its refractive power, achieves focusing. The iris diaphragm, lying anterior to the lens, is perforated by the pupil, which regulates the entry of light into the eye, miosis occurring in response to bright light or accommodation. The neuroretina is derived embryologically from the optic cup, the outer layer forming the retinal pigment epithelium and the inner the closely apposed photoreceptor layer of rods and cones. The blood supply of the retina is dual, with the choroid supplying the outer retina and the retinal vessels supplying its inner part. Light energy falling on the retina is captured by the rod and cone pigments and converted into a neural signal. This is partially processed by retinal neurones and then transmitted along the ganglion cell axons to the lateral geniculate body, where further processing occurs. Information then passes to the primary visual (striate) cortex and thence to the parastriate cortex, where elements of the visual message representing colour, form, luminance, motion, and stereopsis are segregated and independently transferred in parallel to multiple peristriate areas. Here furtherprocessing occurs, which results in the construction of a single visual percept.

The globular shape of the eye, essential for its optical functions, is the result of relatively high intraocular pressure (between 11 and 21 mmHg), which is maintained by the continuous secretion of aqueous humour by the ciliary gland. Pressure homeostasis is achieved by the removal of aqueous via the trabecular meshwork and outflow channels at the drainage angle of the eye.

CUMULATIVE INFLUENCES IN AGEING

Ageing and senescence result from the interaction of genetic and environmental factors. The internal milieu is determined in a major way genetically while the external milieu depends in part on a cell's own metabolism and in part on that of neighbouring cells. It will also reflect, in greater or lesser degree, its continuity with the wider extracellular compartment. These factors differ for the different tissues of the eye. Various blood–ocular barriers exist. The blood–aqueous barrier limits the entry of large molecules into the aqueous, while the outer blood–retinal barrier (at the level of the retinal pigment epithelium) limits entry for such molecules from the choroid into the retina. The tight junctions of the retinal capillary endothelial cells perform a similar function as the inner blood–retinal barrier. Anions are removed from the retina or aqueous by pumps located in the retina or ciliary body; selected molecules are secreted by the ciliary body into the aqueous, or by the retinal pigment epithelium into the retina. Those events affect the environments to which the different tissues are exposed over a lifetime.

The eye is exposed to the physical effects of sunlight during the daylight hours and to other sources of electromagnetic radiation at other times of day. The damage caused to the tissues of the eye by exposure to light depends on the amount of energy absorbed (Draper's Law), and this in turn depends on the spectral characteristics (wavelength composition) of the source (Fig. 1) The far ultraviolet and infrared wavelengths (**UV** and **IR B** and **C**) are almost completely absorbed by the cornea, sclera, and conjunctiva. Some UV B (295–315 nm) and the UV A (315–400 nm) passes through the cornea and is absorbed by the lens. Visible light (400–780 nm) and the IR A (700–1400 nm) are transmitted by the lens and impinges on the retina, where it may be absorbed by the photopigments and the melanin within the retinal pigment epithelium. Transmission of UV A and B, and IR A by the cornea allows these wavelengths to reach the pigmented tissues of the iris.

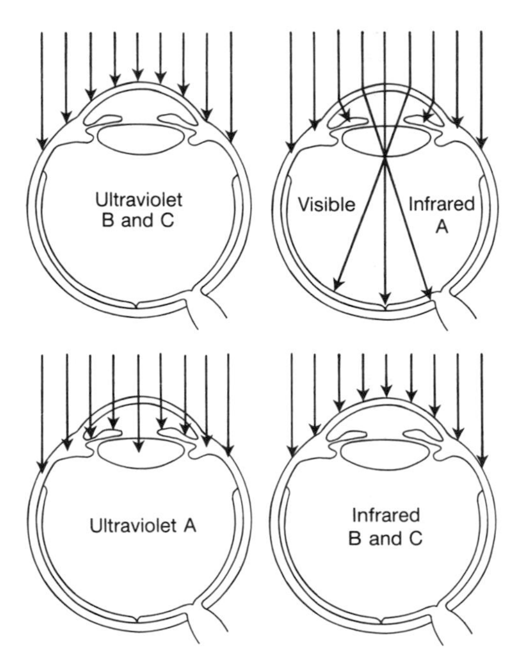

Fig. 1 Absorption of optical radiation. The wave bands are: UV B and C, 200–315 nm; UV A, 315–400 nm; visible and infrared (IR), 400–1400 nm; and IR B and C, 1400–10 000 nm. A narrow waveband in the UV B, 295–315, will pass through the cornea and may be a hazard to the lens. (From Marshall 1985, with permission.)

The absorption of photon energy by cellular chromophores generates a series of reactive oxygen species which are potentially damaging to the tissues. The most important of these are singlet oxygen and superoxide. Singlet oxygen causes peroxidation of the polyunsaturated fatty acids of cell membranes to alter membrane function. The free radical, superoxide, can damage biomolecules directly by transfer of its unpaired outer orbital electron or by formation of hydrogen peroxide, which itself can cause oxidative damage to cells. Reactive oxygen species are also generated by all aerobically metabolizing cells during the transfer of electrons along the cytochrome chain in the reduction of molecular oxygen to water. About 2 per cent of the energy leaks from the system within the mitochondria with the formation of superoxide, hydroxyl radical, and hydrogen peroxide. The damaging effects of these reactive species on the tissues of the eye, as in other organs, is combated by the presence of scavenger molecules. Vitamin E is present in cell membranes and combats lipid peroxidation. Superoxide dismutase scavenges superoxide, and catalase scavenges hydrogen peroxide. Vitamin C and reduced glutathione are powerful reducing agents which co-operate to protect the cell against oxidative stress. The maintenance of these agents within the cell by diet, transport, or synthetic processes is probably important in protecting the tissues from the cumulative damage which will be manifest as ageing.

Another cumulative event affecting the function of biomolecules, is post-translational modification of proteins. The process of glycation involves the formation of a Schiff-base compound between an aldose, such as glucose, and the free amino groups of a protein. Further chemical rearrangement may occur to form stable end-products. Haemoglobin A1C was the first such product identified but there are now numerous examples involving collagen, neuroprotein, fibrin, lens crystallins, and Na^+K^+-ATPase. Such modification can lead to conformational and functional changes in the proteins which, over a period of time, will be attributable to ageing. Ageing in this case results from the cumulative action of an endogenous factor which is operative in the normal subject and will be amplified in the diabetic.

Another example of post-translational modification of protein is the carbamylation of proteins by cyanate. Urea, which is ubiquitous in the body, can undergo hydrolysis to form hydrocyanic acid. Cyanate reacts with free amino groups of proteins to modify them in a manner similar to that which occurs with glycation. As in diabetes and glycation, it would be anticipated that this process of carbamylation would be amplified in uraemia. Case control studies have shown that plasma glucose and urea, even within the high normal range, are associated with an increased risk of senile cataract.

GROWTH OF THE EYE AND AGEING

Growth is an age-dependent phenomenon which is usually considered separately from ageing. The eyeball does not reach its full size at birth, and, for instance, increases in length in the first 5 years with a small increase up to the age of 12 years. This is paralleled by other changes in ocular dimensions.

The lens of the human eye, however, grows throughout life. The sagittal thickness is said to increase almost linearly with age, increasing at a rate of about 25 μm per year (Fig. 2) (Sparrow 1989). This relentless growth causes a progressive shallowing of the anterior chamber that in predisposed eyes (such as the small eye of the hypermetrope) can precipitate an attack of angle-closure glaucoma by narrowing the peripheral recess of the chamber. Thus growth can be a determinant of this age-related disorder.

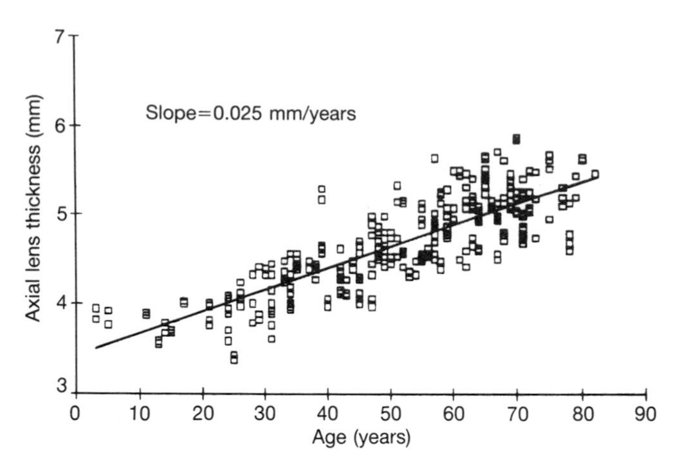

Fig. 2 Sagittal increase in lens thickness with age (from Sparrow 1989, with permission).

SOME ASPECTS OF SENESCENCE IN THE EYE

Tears

The dry eye of keratoconjunctivitis sicca is caused by a reduction in the aqueous component of the tears due to a loss of lacrimal function. Hypertonicity of tears causes damage to the ocular surface, particularly of the exposed part of the eye, causing irritation, pain, and even visual disability (Bron 1986). Tear formation falls with age, while the prevalence of keratoconjunctivitis sicca rises. With age too, there is an infiltration of the lacrimal gland by round cells and this is accompanied by acinar and ductal atrophy. This explains the loss of secretory function. A similar process occurs in Sjögren's syndrome and raises the question whether an autoimmune process is involved in both disorders (Abdel-Khalek *et al.* 1978; Damato *et al.* 1984).

Orbit

A relative loss of orbital fat occurs in old age, which leads to enophthalmos, and a deepening, particularly of the upper but also of the lower, tarsal fold. Lid skin thins and wrinkles, much as it does in other parts of the body. Muscular laxity and weakening of the inferior tarsal ligament may encourage the occurrence of entropion or ectropion, while a similar event affecting the levator and its aponeurosis may lead to a senile ptosis.

The globe and sclera

The corneoscleral coat becomes more rigid with age. This may be seen grossly during intraocular surgery, where there is less tendency for the eye to lose its shape when fluid is removed, and it is also recorded as a loss of compliance of the ocular coats in studies of the pressure–volume relationship of the enucleated eye. This change in the expressibility of blood from the choroidal vessels influences the interpretation of indentation (Schiotz) tonometry, so that a correction for 'scleral rigidity' must be made, which is age-related. The corneoscleral envelope also becomes more rigid with age, which explains why glaucoma occurring in the first few years of life is accompanied by globe enlargement (buphthalmos), while this does not occur in chronic open-angle glaucoma. This increased rigidity results in part from an increased cross-linking of collagen, because human collagen is less extractable in old age than in infancy, though the difference is small.

This observation does not apply however, to that part of the sclera (the cribriform plate) which transmits the fibres of the optic nerve. The elastic properties of this multilamellar structure are retained in adult life, and in the presence of raised ocular pressure it becomes bowed backwards to create the classical picture of glaucomatous optic cupping. Recently it has been demonstrated that the collagenous elements of the cribriform plate are different from those of the sclera proper, which may explain this difference in behaviour.

Scleral plaques are vertical, oval zones of increased scleral translucency lying just anterior to the insertions of the horizontal rectus. Their frequency increases from the fifth decade and they have no functional effect (Cogan and Kuwabara 1959; Norn 1983).

Cornea and conjunctiva

The fetal cornea is steeply curved and this curvature, which is greater in the vertical, accounts for the presence of the astigmatism 'with the rule', which is commonly found at birth. The radii of curvature increase in the late teens, i.e., the cornea flattens and the hypermetropia lessens (Weale 1982). Corneal diameter increases little after birth, while growth of the globe is almost complete by 3 to 5 years of age. Corneal sensitivity declines with age (Fig. 3) as does that of the conjunctiva and lid margin.

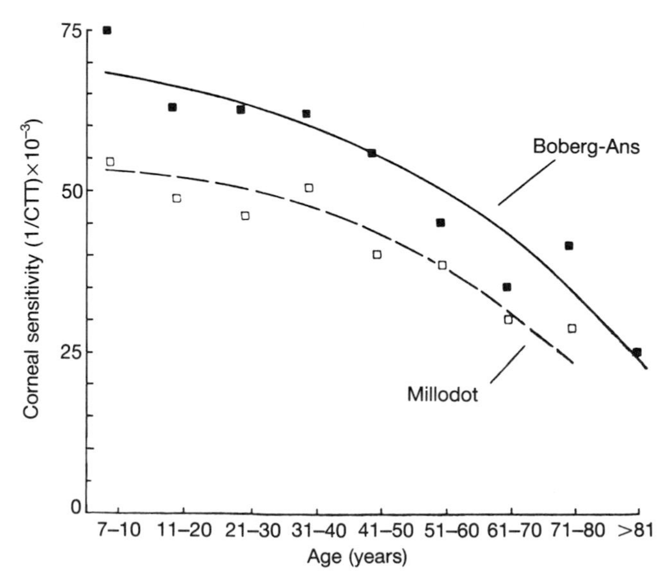

Fig. 3 Decline in corneal sensitivity with age (from Millodot 1977, with permission).

A number of degenerative conditions affect the cornea. The Hudson–Stähli line is a horizontal brown line in the corneal epithelium lying at the junction of its middle and lower third and due to the presence of iron in its basal cells. It has no effect on vision. It is present as early as 2 years of life, and becomes increasingly visible with age. Its frequency is 2 per cent at 10 years, 11 per cent at 20 years, 14 per cent at 30 years, and 40 per cent by 60 years (Norn 1968). The source of the iron is not established but its distribution is probably determined by the centripetal flow of epithelial cells over the cornea from the limbus.

Lipid arcus is a yellowish-white stromal deposit which appears in the peripheral cornea with increasing frequency and density from middle age. It is composed chiefly of cholesterol ester thought to be derived from the plasma lipoproteins. The prevalence of lipid arcus is said to be related to the plasma cholesterol level below the age of 50 years and its occurrence in youth (arcus juvenilis) is associated with hypercholesterolaemia, usually hyperbetalipoproteinaemia. Winder (1983), however, was unable to confirm a dose-related effect between the density of arcus and the duration of exposure to the raised plasma low-density lipoprotein.

Spheroidal degeneration is a yellow-brown, granular deposition in Bowman's layer and the superficial stroma of the cornea, which also affects the neighbouring interpalpebral conjunctiva. It occurs in specific geographic latitudes. Bietti *et al.* (1955) described it in the Dahlak Islands and Freedman (1973) called it Labrador keratopathy and recognized its relationship with exposure to ultraviolet radiation. The deposition in the corneal stroma extends horizontally across the cornea and, in its higher grades, it is a blinding condition. Ultrastructural studies show an electron-dense deposit in the stroma. Johnson and Overall (1978) have suggested that it is the product of ultraviolet irradiation of plasma proteins which have diffused into the cornea. Johnson (1981) and others have brought forward strong epidemiological evidence to suggest that spheroidal degeneration is caused by years of exposure to high levels of ambient light, probably in the wavelength 290–300 nm. Direct sunlight or light reflected from sand (as in the Dahlak islands) or snow (as in Eastern Canada) is thought to be responsible.

A similar aetiology has been adduced for pinguecula, a raised degenerative condition of the interpalpebral conjunctiva, and of pterygium, a fleshy conjunctival elevation, which grows on to the cornea. Various studies have shown that the prevalence of spheroidal degeneration and pinguecula not only rises with age, but is higher in males, probably because of a more outdoor life. The slope of prevalence is steepest at a latitude of 55–60° N, where ultraviolet exposure is at its highest. It falls off for both sexes on either side of this latitude (Johnson 1981).

Removal of water from the cornea by the corneal endothelium is the major active process responsible for its transparency. The newly developed technique of specular micrography of the living human corneal endothelium now provides a clear picture of the changes in this important monolayer of cells with age. The human endothelial cell undergoes only rare mitosis after birth, and even after injury the mitotic response is limited. In youth, the cornea is covered by approximately 500 000 cells arranged with beautiful hexagonal regularity. With age, there is a gradual fall in endothelial density and increasing variation in cell size and shape (polymegathism) (Fig. 4). This reflects the gradual loss of endothelial cells with time, and their replacement by a spreading enlargement of the remaining cells. This response is seen in exaggerated form after any endothelial injury and

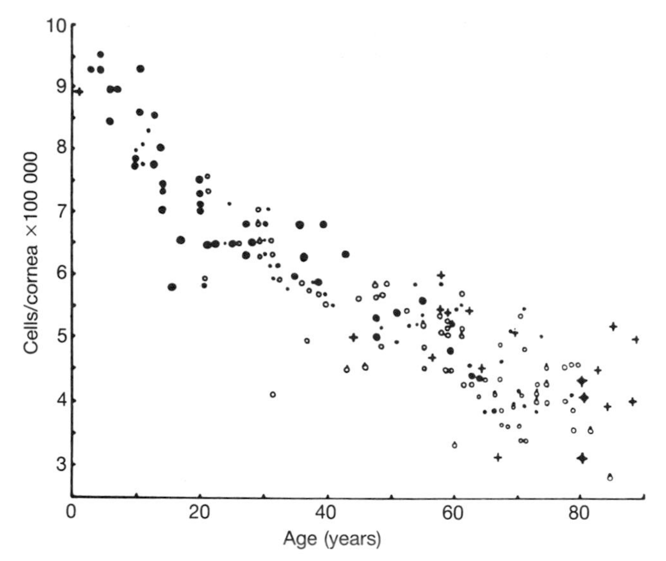

Fig. 4 Fall in endothelial density with age, in the living human eye. Small dots represent male subjects; large dots, female subjects; circles, female subjects with apparent cell-free areas; and circles with dot above, male subjects with cell-free areas. Eye bank eyes are represented by cross and cross with central dot for cell-free areas (from Laule *et al.* 1978, with permission).

is therefore a routine occurrence after cataract surgery and corneal grafting. Some reports have suggested that the loss of endothelial cells may continue after surgery, and it therefore follows that eyes which have undergone, say, cataract surgery, may be at risk of endothelial failure at some later date as their reserve of endothelial cells is slowly eroded. As the endothelium provides the major water pump for the cornea, its failure leads to corneal oedema and loss of transparency (Fuchs' dystrophy).

Because of the importance of the endothelium in maintaining corneal transparency, and the effects of age on endothelial cell density, researchers have been concerned to establish whether ageing adversely affects the outcome of certain kinds of surgery involving the cornea. Waltman and Cozean (1979) found that the endothelial cell loss associated with cataract surgery by phakoemulsification increased with age. Olsen (1980) found that the relatively greater loss of endothelial cells in the region of the wound was accentuated by age. Although it was anticipated that younger corneal donors would confer a higher endothelial density on corneal grafts, most studies but not all have not confirmed this. Linn *et al.* (1981) found a surprising, positive correlation between host age and donor cell count, which suggested that there are factors in the younger eye that support the survival of graft endothelial cells. In keeping with the above, most studies show no relationship between donor age and graft survival. However, Harbour and Stern (1983) found that graft success was related to donor age.

Cornea guttata and Fuchs' dystrophy

Corneal endothelial cells are concerned with the secretion of its basal lamina, Descemet's membrane. This process proceeds throughout life with thickness increasing with age. In old age, focal excrescences appear in the lamina, which are visible on specular microscopy as Descemet's warts. These are a common feature of the peripheral cornea

(Hassle–Henle's warts), and less so of the central cornea where they are encountered as 'cornea guttata'. The prevalence of cornea guttata increases with age (Lorenzetti *et al.* 1967) (Fig. 5). Cornea guttata reflects an impairment of endothelial cell function and is associated with a sixfold increase in endothelial permeability, despite a normal corneal thickness. Cornea guttata is associated with a slower recovery of corneal thickness to normal after cataract extraction and is regarded as a risk factor for corneal decompensation in this situation. Endothelial warts may be found in profusion in Fuchs' endothelial corneal dystrophy, where a generalized failure of endothelial pumping function leads to corneal oedema and visual loss. The condition may be inherited as a dominant, and both inherited and sporadic forms present in later life. There are also inherited endothelial dystrophies which occur in infancy and early adult life.

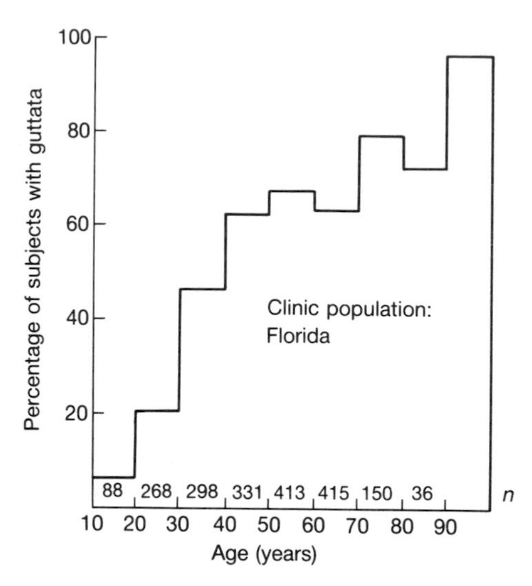

Fig. 5 Prevalence of cornea gutta (plotted from data in Lorenzetti *et al.* 1967).

Anterior chamber

The intraocular pressure lies between 11 and 21 mmHg (mean 16 mmHg) and results from the balance between aqueous flow and outflow resistance. Aqueous leaves the eye via the trabecular meshwork and it is thought that the major site of outflow resistance resides in that part which is closest to the canal of Schlemm, the endothelial meshwork. With age there is a two- to threefold increase in width of the cortical zone of the trabecular sheets, with some degenerative change and thinning of the trabecular core. The endothelial cells that line the spaces of the meshwork and come in direct contact with the aqueous are reduced in number but increased in area, like the endothelial cells of the cornea. They show degenerative cytoplasmic changes with an increase in pigmentary and other inclusions (Tripathi and Tripathi 1987). These sclerotic changes, and possibly changes in the amount of mucopolysaccharide ground substance that occupies the intertrabecular spaces, may account for the increase in outflow resistance in elderly people.

Most studies report a small but steady increase in intraocular pressure with age (Armaly 1965; Hollows and Graham 1966; Bankes *et al.* 1968; Schappert-Kimmijer 1971). However, no association between intraocular pressure and age was found in the Framingham study (Leibowitz *et al.* 1980), which also failed to find the higher mean ocular pressures in women shown by the other studies. Bengtsson (1981) found an increase with age, which he attributed to an increase in blood pressure in his sample. Aqueous flow falls with age, and it may be that differences in intraocular pressure between studies reflect differences in interaction between the effects of age on flow and resistance in different populations.

Chronic open-angle glaucoma (**COAG**) is a condition where damage is caused to the optic nerve head by the intraocular pressure, in the presence of open angles. A figure of 21 mmHg is traditionally taken as the cut-off point for the diagnosis of COAG (2 SD above mean ocular pressure), but because the diagnosis of COAG is often made on the basis of characteristic glaucomatous optic cupping and field loss at pressures below this level, there is a continuous relationship between ocular pressure and nerve damage, with the lower limit of pressure that causes damage not yet having been established. The frequency of COAG rises steeply from the age of 50 years (Fig. 6). In classical COAG, with ocular pressure raised significantly above 21 mmHg, there is an increase in outflow resistance. Structural changes that have been found in the trabecular meshwork in COAG appear to be an exaggeration of changes seen with ageing. A reduction in width of the trabecular spaces has been encountered, and even, but less commonly, a reduction in calibre or obliteration of the collector channels that drain the canal of Schlemm. However, these changes are less often seen in the early stages of COAG and therefore do not fully explain the increase in ocular pressure (Tripathi and Tripathi 1987).

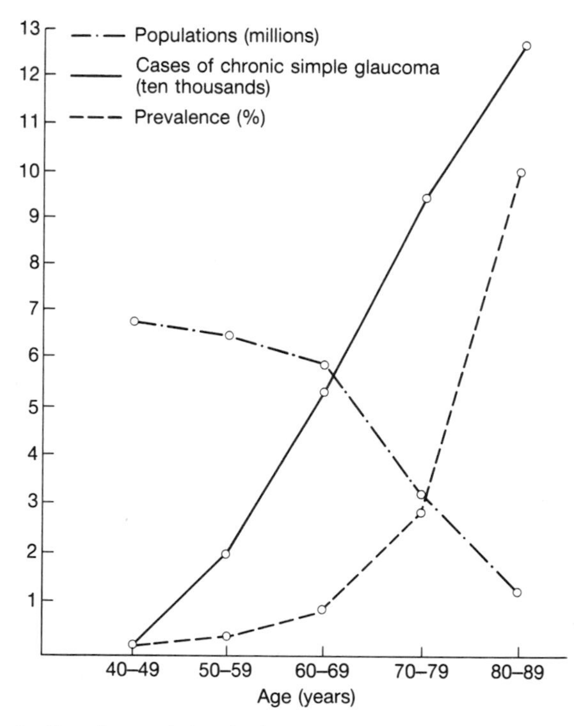

Fig. 6 Prevalence of chronic simple glaucoma (from Crick 1980, with permission).

As mentioned earlier, the depth of the anterior chamber decreases with age because of the growth of the lens. This exposes predisposed eyes with narrow chamber angles to the risk of angle-closure glaucoma in later life. The prevalence of angle-closure glaucoma varies widely between different populations for genetic reasons. Thus in Eskimos, who have very shallow anterior chambers, the prevalence is 2.9 per cent compared to 0.08 per cent in South Wales.

The crystalline lens

The transparency of the lens results from the optical homogeneity of its parts. The lens fibres resemble one another and are closely packed with a minimum of extracellular space. The high concentration of specific lens proteins (alpha, beta, and gamma crystallin) within the fibres minimizes scatter from the membrane–cytoplasm interface. It has been calculated that the short-order interactions of the crystallins confer optical properties similar to those of glass.

The ion–water balance of the lens essential for transparency is maintained in part by an outward sodium pump residing in the epithelium and superficial cortical fibres. Its effectiveness is ensured by a selective permeability of the epithelial and fibre membranes. The energy required for transport, growth, and synthetic processes is derived from active metabolism. Epithelial respiration is chiefly aerobic and supplies about 30 per cent of the energy requirements of the lens. The superficial cortical fibres supply the remainder by anaerobic glycolysis; a lens incubated anaerobically and supplied with glucose remains transparent.

The lens is equipped with a protective mechanism against oxidative stress. Scavenger molecules include vitamin E within the lens membranes, and glutathione, which is synthesized within the lens from precursor amino acids. Glutathione is important in maintaining protein thiols in the reduced state, such as the lens crystallin thiols, or the thiol that is close to the reactive site of Na^+K^+-ATPase. It also maintains ascorbate in the reduced state and scavenges peroxides and radiation-induced radicals. Vitamin C, already in high concentration in the aqueous (about 10–20 times the plasma level) is transported into the lens and achieves a high concentration here also. It is a powerful reducing agent that can co-operate with glutathione in scavenging carbon-centred radicals and will scavenge superoxide. Other compounds thought to perform similar roles in the lens include carotenoids, choline, taurine, and thioredoxin-T.

Growth of the lens is a complex event that is due essentially to the addition of generations of new fibres to the outer surface of the cortex. This is achieved by the mitosis of pre-equatorial epithelial cells that differentiate into fibres and elongate, behind the epithelium anteriorly and in front of the lens capsule posteriorly. Once established, the fibres pass through a cycle of terminal differentiation, in which nuclei and organelles are lost and the cells lose their metabolic capacity. Thus the fibres at the centre of the lens, the lens nucleus, are the oldest and are metabolically inert, while the superficial fibres of the cortex are the youngest and metabolically active.

In addition to the crystallins, the fibres contain cytoskeletal proteins such as actin, tubulin, and vimentin, and numerous enzyme proteins. Some cytoskeletal proteins disappear from the fibre shortly after it is formed, while the crystallins are retained throughout life and may be subjected to post-translational modification. Recently it has been demonstrated that as new fibres are added to the surface of the lens, so the fibres of the nucleus become compacted so that the rate of cortical addition is actually about twice that calculated from measurements of total lens thickness (25 μm/year; Sparrow 1989). Although increase in lens thickness is almost linear with age it is known that the rate of epithelial mitosis decreases, so that it may be that the rate of central compaction also decreases with age. The number of pre-equatorial cells available for mitosis does not appear to alter with age, despite the increased circumference of the lens that occurs with lens growth. Thus new fibres are wider rather than more numerous as the lens grows and it may be that reduced rate of epithelial mitosis can be compensated by a differential increase in the size of lens fibres.

Growth of the lens has certain implications for the remodelling of its component parts. Anterior capsular thickness increases with age while that of the posterior capsule remains almost unaltered. This observation is supported by turnover studies in the rat which suggest that it is the epithelium of the lens, which lies anteriorly, that is predominantly engaged in capsule synthesis. Capsular turnover must also include a remodelling process to permit growth of the lens, but this does not appear to have been studied. Capsular synthesis is slowed down in the ageing lens.

The lens fibres and epithelial cells are electrically coupled by gap junctions. Fibres are also connected by reciprocal interdigitations of their plasma membranes that are of 'ball and socket' type in the younger cortical fibres and of 'tongue and groove' type in the older, nuclear fibres. With age there is increasing complexity of interdigitation towards the centre of the lens, and there is uncertain integrity of the plasma membranes.

Remodelling of the growing lens must include the establishment of new gap junctions between new fibres and epithelial cells and with existing fibres, as well as new fibre interdigitations. The tips of the lens fibres meet and interdigitate anteriorly and posteriorly at seams that are termed the lens sutures. Their complexity increases with age, so that the fetal sutures are Y-shaped while those of the adult branch with increasing complexity to form a symmetrical, stellar pattern (Fig. 7). However, with age in adult lenses, the sub-branches are of unequal number anteriorly and posteriorly so that there is increasing disorder of the suture pattern (Williams and Kuszak 1986). Certain forms of cataract take their morphology from the suture pattern (Bron and Brown 1986).

With continued growth of the lens, its ability to recoil into

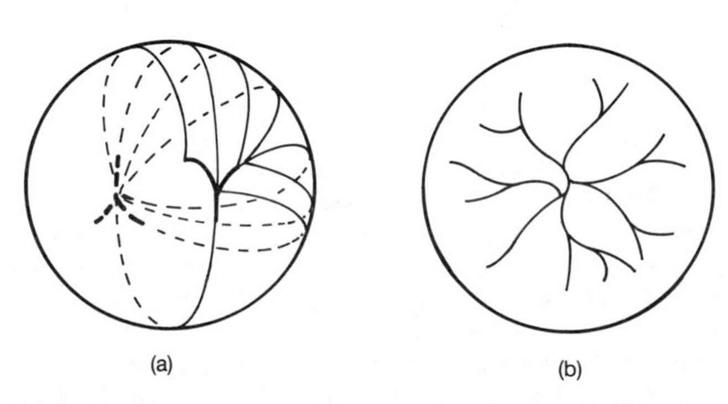

Fig. 7 The fibre suture pattern in human lens. (a) Infantile, (b) adult.

a more curved shape when zonular tension is relaxed during accommodation is reduced. Thus by the fifth decade there is little accommodative function left, and the emmetrope (who can see sharply in the distance when accommodation is completely relaxed) requires a positive spectacle addition for near vision. This is termed presbyopia. There is functional evidence that ciliary muscular activity is retained or even increased in the presbyopic eye, but this is a controversial area.

A number of changes in the chemistry of the lens occur with age, some of which are thought to predispose to cataract. A series of yellow chromophores accumulates in the lens, particularly in its nucleus, which absorb strongly at the shorter wavelengths of visible light (Lerman 1980). A chromophore whose concentration is more stable is 3-hydroxykynurenine, a fluorescent derivative of tryptophan, but other chromophores have been identified. A protective role has been ascribed to these pigments in screening the retina from the damaging effects of blue light (see below) (Fig. 8). Another soluble chromophore is P2-B, whose level is high at birth and falls with age as other, water-insoluble chromophores appear. The latter are highly bound to the lens crystallins and contribute to increasing coloration of the lens nucleus. They include anthranilic acid, an oxidation product of tryptophan, bityrosine, and the betacarbolines, as well as advanced products of the Maillard reaction (see below).

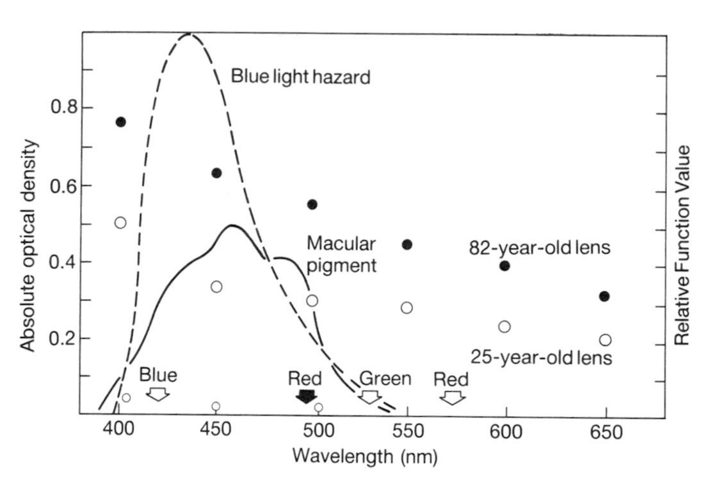

Fig. 8 Graph to show the spectral absorption characteristics of the macula lutea, and the lenses of a 25-year-old and an 82-year-old man, compared to the spectral distribution and relative magnitude of the 'blue-light hazard'. Absorption by the yellow macular pigment matches the spectral distribution of the 'blue-light hazard' almost perfectly. The optical density of the pigment of the ageing lens makes this an even better protective filter although the spectral match is less good. (From Marshall 1985, with permission.)

With age there is a fall in enzyme levels in the lens nucleus, and protein turnover there which is already low, decreases to negligible levels (Harding and Dilley 1976). Gamma crystallin, which is found predominantly in the nucleus, falls in concentration, and there is a fall in the water-soluble and increase in the water-insoluble crystallin fractions, which chiefly affects the nucleus. Both water-soluble and water-insoluble high molecular-weight crystallin aggregates appear, whose presence probably accounts for the increase in light scatter from the nucleus with age (Harding and Crabbe 1984).

There is increasing evidence of post-translational modification of the lens crystallins with age. A list of such changes is given in Table 2. Such changes alter the surface charge

Table 2 Post-translational changes with possible conformational effects on lens crystallins

Glycation	Loss of positive charge
Carbamylation	Loss of positive charge
Mixed disulphide formation with oxidized glutathione	Three additional charges
Phosphorylation	Extra positive charge
Proteolysis of crystallins	Loss of section of chain
Ultraviolet degradation of tryptophan	More polar residues
Deamidation	Extra negative charge
Methionine oxidation	Hydrophobic polar group
Aspartyl racemization	Changed direction of backbone

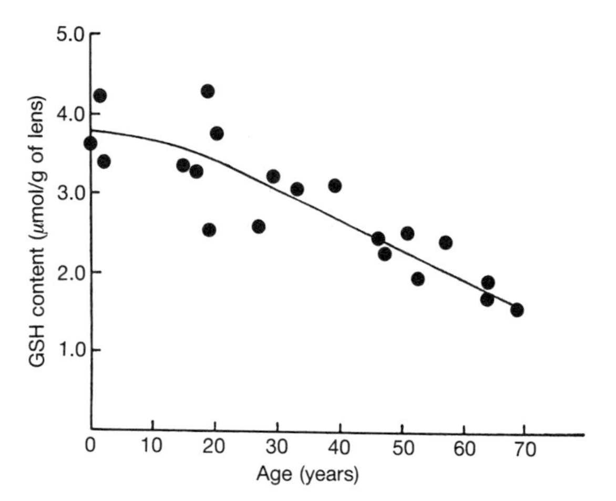

Fig. 9 Fall in lens free glutathione with age (from Harding 1970, with permission).

on these globular proteins and encourage unfolding of the molecule and the exposure of buried thiol groups. Oxidation of these groups in turn is thought to result in the formation of high molecular-weight, covalently cross-linked aggregates, responsible particularly for the increased scattering of light found in nuclear cataract. The fall in glutathione concentration which is found in older lenses, in the nucleus more than the cortex, would render the crystallins more susceptible to this kind of oxidative damage (Fig. 9).

Glycation of the crystallins occurs at the epsilon amino (side-chain amino) groups of their constituent lysine moieties, exposed at the surface of the proteins. Both alpha and beta crystallins are rich in lysine; beta is rich in thiols and alpha is less so. Both proteins will form adducts with glucose and other aldoses, with a loss of surface positive charge. This disturbs the tertiary structure and will permit the formation of disulphide-bonded aggregates. Gamma crystallin is in rich thiols but poor in lysine and therefore aggregation by this mechanism is less likely, although gamma crystallin can be incorporated within aggregates of other crystallins.

Glycation of crystallin amino groups results in a Schiff-base compound in which the covalent bond of the adduct is stabilized by an Amadori rearrangement. The material then undergoes further 'browning' (Maillard) reactions to produce brown products that are themselves able to cross-link proteins (advance glycosylation end products; **AGE**). An agent derived from AGE products and apparently involved in cross-linking, is furazoyl-furanyl-imidazole (**FFI**), which has the same fluorescence and absorption characteristics as a chromophore extracted from nuclear cataracts (Pongor *et al.* 1984) However, further studies have suggested that FFI may in fact be artefactually generated. Ascorbate, whose concentration in the lens is far higher than that of glucose, can also react in this way to form adducts with similar spectroscopic features.

Carbamylation of crystallins has been identified in lenses from Pakistan with advanced cataract and implies the formation of adducts between crystallin lysine and cyanate, a product of urea hydrolysis. This has not been found in clear lenses. While both glycation and carbamylation would be expected to induce protein unfolding and induce the risk of forming light-scattering, high molecular-weight aggregates, only glycation will lead to the formation of a brown chromophore of the kind described above. Carbamylation can therefore be expected to make a contribution to white nuclear cataract, but not to brunescent cataract.

Cataract

Cataract, or lens opacity is the term given to any light-scattering feature in the lens. Its location, size, or density determines whether it affects vision. By far the majority of cataracts are age-related or senile cataracts, whose prevalence rises steadily from the fifth decade (Table 3). Cataract is the first cause of self-reported visual disability (Kahn and Moorehead 1973), and the third cause of newly registered blindness in the United States. Similar prevalence rates are reported from Great Britain and New Zealand, while the frequency is greater in Egypt, Tunisia, and in the Punjab, where the age-adjusted prevalence was almost three times higher than in the Framingham study.

Table 3 Percentage prevalence of lens opacities; provisional data from NHANES (1971–1972). *Vision research. A national plan,* NIH Publication 83-2473

	Age group (years) (both sexes)	Lens opacities	Lens opacities causing visual loss
	1–5	0.4	0.1
	6–11	0.6	—
	12–17	1.3	0.2
	18–24	2.4	0.3
	25–34	2.8	0.2
	35–44	4.1	0.8
	45–54	12.2	2.6
	55–64	27.6	10.0
	65–74	57.6	28.5
Males	1–74	8.4	3.1
Females	1–74	10.3	3.6
All		9.4	3.4

Senile cataract has three distinct morphological forms, which may occur independently or in combination: cortical-spoke cataract affects groups of lens fibres within the superficial cortex; subcapsular cataract affects those fibres which have most recently formed; and nuclear cataract affects the oldest fibres, those of the nucleus, in the form of a diffuse, white, scattering opacity or brunescence. Subcapsular cataract occurs at an earlier age than the other forms. Nuclear cataract is the most common form encountered whether alone or in combination with other forms (Tables 4 and 5).

Table 4 Age-specific prevalence of cataract (%) (6/9 or worse)

Age group (years)	NHANES[1]	Framingham*	Melton Mowbray†
45–54	2.6		
52–64		4.5	
55–64	10.0		
65–74	28.5	18.0	
75–85		45.9	
76–84			41.6
85 and over			65.2

* Leske and Sperduto (1983); † Gibson *et al.* (1985).

Table 5 Age-specific prevalence of cataract type (%)*

Age group (years)	Cataract type: alone or in combination Nuclear	Cortical	Posterior Subcapsular
52–54	10.5	7.4	4.1
65–74	34.0	20.1	10.5
75–85	65.5	27.7	19.7
All	25.6	14.3	8.3
(Pure types)	13.3	6.5	2.1

* From Sperduto and Hiller (1984). Data from Framingham survey (1973–75).

Cataract is one of the major causes of blindness in the world. Of all the risk factors identified for cataract, age is the most important. In the developed world, most people have some degree of lens opacity by the age of 80 years. In the NHANES survey in the United States, 58 per cent of those aged 65 to 74 years had lens opacities, while 29 per cent had a visual defect. The frequency of each major form and some minor forms of cataract is age-related.

Of the other risk factors for cataract, diabetes is the most important. Ederer *et al.* (1981) found an excess risk of cataract in both Framingham and NHANES (National Health and Nutrition Examination Survey) studies. Data from the Framingham study showed a significant risk for cataract resulting in visual loss in the 65- to 74-year age group (relative risk 4.02); in the NHANES study there was also a significant association for cataract (relative risk 2.97) in this age group and also in the 65- to 74-year age group (relative risk 1.63). No such association was shown in this age group in the Framingham study. Both studies confirm an excess of cataract in diabetics in the 50- to 64-year age group which disappears in later life. This may reflect a higher mortality in diabetics with cataract. A recent case-control study in Oxford, involving 300 cataract patients and 609 controls, has shown an increased risk for cataract extraction in the age group 50 to 79 years (relative risk 6.2) with a threefold greater risk for women (relative risk 10.5). The role of glycation in this increased risk is not established. About 2 per cent of lens crystallins are glycated in non-diabetic and 3 per cent in diabetic lenses, and it is likely that other lens proteins are also glycated. Glycation inhibits $Na^{+}K^{+}$-ATPase and could influence ion–water balance in the lens, leading to swelling and ultimately to a breakdown of fibre structure. It may be relevant that the non-cataractous diabetic lens is about 10 years thicker than expected for age, an effect which could be caused by overhydration rather than increased growth.

It is accepted that the juvenile form of diabetic cataract results from an osmotic swelling of the lens due to an increased flux through the polyol pathway and an accumulation of sorbitol (van Heyningen 1977). This explanation is, however, inadequate for adult cataract where there is the accumulation of sorbitol is not enough to produce an osmotic effect. An alternative possibility is that there is enough flux through the pathway to consume reduced coenzyme (NADPH) required for synthetic processes such as the regeneration of reduced glutathione. According to this view the diabetic process would render the lens more susceptible to oxidative stresses, to which the lens is subjected in diabetic and non-diabetic alike. Although no decrease in the amount of vitamin E has been found in cataractous lenses, there is a fall in both glutathione and ascorbate (Bron and Brown 1987), affecting the nucleus more than the cortex, and lending support to the general view that cataract overall is due to the decreased ability of the lens to combat oxidative stress. There is a fall in lens glutathione with age (Fig. 9). It is likely none the less that the detailed manner in which cataract is brought about differs for different forms of cataract.

Recently, an age-related increase in the permeability of lens membranes has been demonstrated in clear lenses and proposed as an initiating factor in cataract. It may well explain in part the increase in lens thickness that occurs with age and is usually attributed to growth. Cortical cataract is associated with a progressive failure of cation pumps and an increased passive permeability of lens membranes (Gandolfi *et al.* 1986). It seems likely that the initial problem in cortical cataract is a focal or patchy disorder of ion–water regulation in groups of fibres, which results in the formation of water clefts and localized fibre breakdown. This process, initially confined to the wedge-shaped zones of spoke opacity, may in time proceed to more diffuse swelling of lens fibres and cortical opacification.

An associated event is proteolysis, with leakage of low molecular-weight crystallins into the aqueous and a fall in total protein content of the lens. This may be due to the activation of certain neutral proteinases by a rise in local calcium concentration. Calpain II, for instance, will degrade crystallins in experimental cortical cataract and a role has been proposed for the loss of a natural calpain inhibitor in cortical cataract. In its most extreme form, proteolysis is seen in Morgagnian cataract, where the lens cortex is completely liquefied and only the dense, brunescent, nuclear part of the lens remains, whose cross-linked crystallins are resistant to digestion.

In posterior subcapsular cataract an opacity appears under the posterior capsule as a result of the migration of abnormal fibre cells of epithelial origin from the lens equator. This is a disorder of epithelial mitosis, differentiation, and elongation, presumably implicating disordered gene regulation. Reduced production of lens fibres results in reduced lens growth. Recent studies in our laboratory have shown that senile posterior subcapsular cataract differs from that which occurs in renal transplant patients receiving systemic steroids, in that lens growth is not retarded in those patients and a zone of normal fibres inserts itself between the posterior capsule and the opacity.

Nuclear cataract is characterized by brunescence and increased light-scattering. This is accompanied by an accumulation of water-insoluble at the expense of water-soluble crystallins, mainly in the nucleus, and the appearance of water-soluble and -insoluble high molecular-weight aggregates. One of these is the so-called HM 3, a cross-linked aggregate of crystallins, some of which are disulphide cross-linked to membrane proteins and some, again, by non-disulphide bonds. The other aggregate is HM 4, which is a water- and urea-insoluble aggregate found in the darkest brunescent lenses containing crystallins cross-linked by non-disulphide covalent bonds. This fraction accounts for most of the chromophores cross-linked to crystallins (Table 6).

Table 6 Chromophores reported in the lens

3-hydroxykynurenine glucoside
Anthranilic acid
P_2-B
Bityrosine
β-Carbolines
Advanced glycation end-products

As the free radical scavengers, glutathione and ascorbate, are decreased in concentration in the nuclear cataractous lenses and there is an increased crystallin unfolding, the increased disulphide cross-linked aggregates are thought to result from the oxidation of exposed crystallin thiols.

A number of risk factors have been identified for cataract (Tables 7 and 8). As discussed in the Introduction, many of the effects attributed to ageing probably result from the prolonged action over many years of multiple environmental influences, some of which are exogenous, such as diet, radiation, smoking and drug action, and some of which are endogenous, such as the levels of plasma calcium, glucose, urea, and cholesterol. Many of these influences render the lens more susceptible to the effects of oxidative stress, and in any case there appears to be a gradual fall in the levels of

Table 7 Selected risk factors for cataract*†

Factor (+age (years) and sex)	Relative risk
Diabetes (All)	6.20
Age 50–69	7.90
70–79	5.20
Males	3.20
Females	10.50
Myopia (females)	2.30
Diarrhoea (70–79)	2.20
Spironolactone	2.30
Heavy beer drinking	2.08
Heavy smoking	1.97
Systemic steroids	1.79

* Van Heyningen and Harding (1986*a*, *b*); † Harding and van Heyningen (1988).

Table 8 Factors showing significant difference in cataract and control populations by univariate analysis (from Clayton *et al.* 1984)

Variable	Cataract patients	*p* value
Diabetes	More	<0.0001
Cardiovascular disease	More	0.0015
Psychiatric illness	More	0.0157
Physical injury	More	<0.0250
Major tranquillizers	More	<0.0250
Diuretics	More	0.0060
Alcohol: abstainers/heavy drinkers	More	<0.0001
Tobacco	More	0.0730
Topical eye drops	More	<0.0018
Serious bacterial illness	Less	<0.0001
Antibiotics	Less	0.0114
High blood pressure	More	<0.0001
Multiparity	More	0.0126
Glaucoma	More	0.0011
Uveitis	More	0.0150
Macular degeneration	Less	<0.0001
Blood:		
Urea	Higher	<0.0001
Glucose	Higher	<0.0001
Bilirubin	Higher	0.0109
Total CO_2	Lower	0.0269
Creatinine	Higher	0.0003
Calcium	Lower	<0.0001
Phosphate	Lower	0.0005
Cholesterol	Lower	<0.0001
Total protein	Lower	<0.0001
Albumin	Lower	<0.0001

those elements within the lens that will normally combat it. The high prevalence of cataract in the Punjab has been associated with a low current protein intake and with a past history of severe diarrhoea. Biochemical changes induced by diarrhoea might be responsible for cataract in some geographic areas.

Cataract may be caused experimentally by almost all regions of the electromagnetic spectrum. For some time the view has been held that certain geographic differences in cataract prevalence could be explained by different levels of exposure to sunlight. Data from the NHANES study were used to show a higher frequency of cataract in persons over 65 years, in locations with large amounts of sunlight. A greater frequency of cataract was found in Australian aborigines from areas with high levels of ultraviolet light as shown by residential radiation data, but no relationship between cataract and residence was found for non-aborigines. Although there have been further reports supporting a relationship between exposure to ultraviolet radiation and cataract, they have not always been able to exclude sociological and environmental factors that could influence interpretation. The effect, if it exists, may not be major.

The vitreous

The vitreous at birth is a relatively homogeneous, gel-like structure; it becomes increasingly more fluid and less homogeneous with age. The vitreous has a sparse arrangement of fine collagen fibrils but its structure is more condensed in the periphery, or cortex, than in its core. The cortex gains collagenous insertion at certain sites on the retinal surface: around the margin of the optic nerve head, around the macula, and along the retinal vessels. It also attaches along a broad band across the ora serrata, straddling the peripheral retina and the adjacent pars plana of the ciliary body. This attachment is referred to as the vitreous base. With age, the vitreous base narrows and moves posteriorly. Liquefaction of the vitreous gel (syneresis) occurs increasingly from middle age and may at any time be accompanied by an inward collapse of the gel, causing the cortex to separate from its peripheral attachments. This posterior vitreous detachment is heralded by a shower of vitreous floaters and sometimes a symptom of flashing lights. A small but significant number of vitreous detachments lead to the formation of a retinal hole and may therefore predispose to a retinal detachment (Sebag and Balazs 1985; Eisner 1987).

The retina

Anatomy

The function of the retina is to convert light into a neural signal. The neural elements involved in this process are, from behind forwards, the photoreceptors (rods and cones), the bipolar cells, and the ganglion cells whose axons are collected together as the optic nerve and terminate in the lateral geniculate body. Amacrine and horizontal cells lying at the level of the bipolar cells co-operate with them in the retinal processing of the neural signal. The photoreceptor cells are in intimate contact with the pigmented retinal epithelial cells, whose finger-like processes engage the stacked discs of their outer segments and remove them externally while they are regenerated centrally. This process of outer segment renewal provides an important protective mechanism for the photoreceptor and its efficiency differs from the rods and cones. The retinal pigment epithelium is involved in nutritional exchanges with the photoreceptors, for instance supplying retinol required for the manufacture of the visual pigments. These exchanges are controlled in part by the outer blood–retinal barrier, whose tight junctions seal the contiguous basal aspects of the retinal epithelial cells.

The cells of the neuroretina and retinal pigment epithelium do not normally divide after birth. As both the visual pigments and melanin absorb light, they are subject to the cumulative damaging effects of light upon the retina. This is mitigated for the outer segments in part by disc turnover, and for the pigment epithelium by scavenger molecules within the cells. The amount and character of the light falling on the retina is in turn modified by the action of the pupil and filtering by lens and luteal pigments, (see below). The inner segments of the photoreceptors and the other neural elements of the retina, including the ganglion cells, are displaced away from the fovea and thereby achieve a more perfect image formation upon the foveal cones. At the foveal pit so formed, the concavity creates a bright reflection that is seen with the ophthalmoscope in youth but disappears in old age.

The macula lutea, a zone 7 to 10° wide centred on the fovea, is so called because of its yellow colour, which is visible ophthalmoscopically. The colour is due to a carotinoid pigment, the xanthophil lutein, occupying chiefly the inner connecting fibres of the foveal cones, which radiate from the fovea as the fibre layer of Henle. It is found elsewhere in the retina between the inner and outer nuclear layers. The luteal pigment absorbs wavelengths between 430 and 490 nm with a peak at 465 nm. General confirmation of its spectral characteristics have been obtained by spectral absorption studies and colour matching. The pigment is absent at birth and accumulates from dietary sources. It is also absent in the albino eye. Accumulation has been prevented in macaque monkeys by dietary restriction. Its natural role may be one of reducing the effects of chromatic aberration, and also reducing glare and dazzle and increasing contrast. But its spectral absorption characteristics allow it to filter out the shorter wavelengths (blue and violet), which are potentially most damaging to the photoreceptors and particularly to the cones. This could explain the findings of Lawill *et al.* (1977), who studied the retinal damage from chronic light exposure in rhesus monkeys. They found that the central macular area consistently showed less damage than the area just surrounding the macula. Absorption by the yellow pigment of the lens is in the near-ultraviolet range, and overlaps that of the macular pigment (see Fig. 8).

The optic disc marks the site of exit of the retinal neurones from the eye through the scleral foramen. The depression in the disc, the optic cup, is surrounded by the neuroretinal rim, whose width and area reflect the number of neurones leaving the eye. There is a loss of neurones from the optic nerve with age. In chronic glaucoma the enlargement and deepening of the cup, and narrowing of the rim are due to death and loss of optic neurones.

Retinal damage from radiation

The young retina is exposed to visible light (400–780 nm) and to the near infrared (780–1400 nm). With ageing there is decreasing exposure at the short-wave (blue) end of the spectrum (300–400 nm) due to the accumulation of yellow chromophores in the lens. Thus there is a 75 per cent transmission at 10 years of age, and 20 per cent at 80 years (Lerman 1980). In this way the retina is thought to be increasingly protected with age from the damaging effects of the so-called blue-light hazard.

By the same token, when the lens is removed at cataract extraction, the lens filter is removed, and patients will be exposed for the remainder of their lives not only to the effects of light at the shorter wavelengths but also to that of UV A, which is normally screened out by the young and old lens alike. At the time of cataract surgery, patients are at special risk for retinal radiation exposure because, at this time, they are exposed to a prolonged period of intense light after the lens has been removed. Because of this immediate risk, it is current practice to insert a yellow filter in front of the microscope light-source after the extraction and, because of the longer-term risk, it is now customary in most units to insert tinted lens implants that filter out damaging ultraviolet wavelengths.

The evidence for the potential of visible light to damage retina was first produced in the studies of Noell *et al.* (1966), who showed that light could damage rod receptors in nocturnal animals (rats). More recently such damage has been demonstrated in diurnal species such as the monkey, and it is apparent that the cones, perhaps surprisingly in view of their functioning at higher ambient light levels, are more susceptible to light damage than the rods.

Sykes *et al.* (1981) exposed monkeys to a broad-spectrum fluorescent light-source for 12 h and demonstrated damage to cones with retinal irradiances between 195 and 361 W/cm^2 and rod damage at the higher irradiance of 361 to 615 W/cm^2. Damage was greater at the fovea, which in primates receives the highest retinal irradiance. Hollyfield and Basinger (1980) showed that damage is increased once the diurnal cycle of cone repair is exceeded. In earlier studies in monkeys, using repetitive monochromatic flashes, Harwerth and Sperling (1971) showed that cone damage was most readily produced by exposure to blue light.

The retinal pigment epithelium can also be damaged by visible light, using short suprathreshold exposures of relatively high irradiance. Both pigment epithelium and photoreceptors are preferentially affected by shorter wavelengths at the blue end of the spectrum, which has given rise to the concept of the 'blue-light hazard'. As has been noted above, this hazard is to some extent filtered out by the yellow macular pigment and, increasingly with age, by the yellow chromophores of the lens. The blue-light hazard has been recognized in the construction of international safety regulations concerning the prolonged viewing of laser light. It is now possible to calculate the relative retinal hazard of any light source on the basis of its total retinal irradiance and the spectral emission of the source (Sliney and Wolbarsht 1980).

Morphological age changes in the retina

Each rod outer segment is made up of a stack of discs, each of which is a membrane-limited structure rich in rhodopsin. The stack contains about 1000 discs per cell and these are turned over at the rate of 1 to 3 per hour, or 30 to 100 discs per day. New discs are formed centrally, and older discs are displaced outwards towards the retinal pigment epithelium. The entire stack of discs is turned over in about 2 weeks. Disc turnover by the pigment epithelium has a diurnal rhythm, with removal by day and renewal by night. There is a small loss of rods with age, but perhaps more important is an increase in outer segment membranes and a loss of the orderly packing of the discs, so that the stack takes on a slightly crumpled appearance. This has been attributed to a reduced phagocytic capacity of the pigment epithelial cells. The growth rate of these cells in culture declines with age.

The structure of the cone outer segments differs from that of the rods. The plasma membranes of the cone discs are not separate as in the rods but in continuity. The cone stack renews much more slowly, with renewal estimated at 9 to 12 months. This slower turnover may render the cones more susceptible to cumulative damage. Gartner and Hendkind (1981) found a net loss of cones with age, particularly at the fovea, while Curcio (1986) found cone densities in several older donor eyes that were well within the range for younger adults. Support for the phenomenon of cone loss in the primate comes from studies in the rhesus monkey, in which density drops by 20 per cent between 5 and 20 years, which, with a lifespan of 25 years would be equivalent in the human retina to a loss of 3 to 4 per cent per decade (Marshall 1987). There is an approximately linear loss of cone membranes in human eyes between birth and 40 years, and this accelerates thereafter. There is also a loss of membrane order, as in the rods. The loss of cones is associated with an increasing diameter of the remaining cones, whose average diameter and range of diameter increase with age (Marshall 1987).

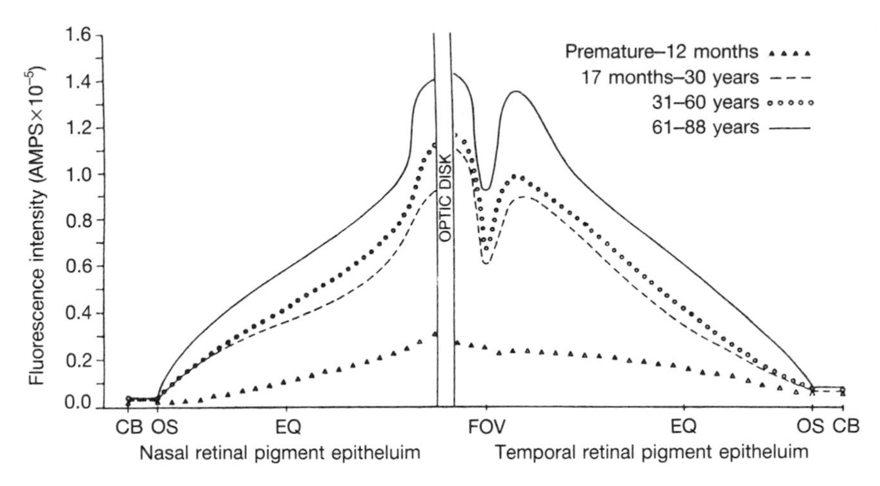

Fig. 10 The topographic distribution of lipofuscin in the pigment epithelium and its variation with age (from Wing *et al.* 1978, with permission).

This loss of photoreceptor order may explain the abnormal Stiles–Crawford effects seen in older adults because this effect is dependent on photoreceptor orientation (Smith *et al.* 1988). There is a fall in the visual pigment density of cones over the ages of 22 to 50 years, as measured by television spectroscopy, which is in keeping with the conceived loss of cone cells. However, van Norren and van Meel (1985) found this loss to be relatively minor.

The mature retinal pigment epithelial cell does not divide. With age, these cells die and, like the endothelial cells of the cornea, the deficiency in the cell monolayer is made up by a spreading and thinning out of existing cells. The pigment cells actively degrade the products of rod discs and, in early life, digestion is complete and no residue is left in the cell. With age the capacity of the cell to degrade disc material is reduced. The undigested products, probably the remnants of membrane lipid peroxidation, accumulate in phagosomal particles in the cell as the autofluorescent age pigment lipofuscin. Lipofuscin first appears in the pigmented epithelium at the age of 10 years, and by the age of 40 years occupies about 8 per cent of the cytoplasm. By 80 years, more than 20 per cent of the cell is occupied (Fig. 10).

It has been shown *in vitro* that division and metabolic activity in the pigment retinal epithelial cell can be inhibited by feeding with lipofuscin particles. It is of interest that these cells from the macula grow less well than those from the periphery, and that this differential increases with age. In keeping with this behavioural difference, the macular pigment epithelial cells contain more lipofuscin granules (Wing *et al.* 1978). This group also demonstrated that the distribution of lipofuscin parallels the density of the rods, perhaps emphasizing the preponderant origin of the granules from rod disc membranes. However, they are also derived from the cones and are present at the fovea.

The retinal pigment epithelium rests on a basal lamina which is fused to that of the choroid and is called Bruch's membrane. Bruch's membrane acts as a diffusion barrier to large particles passing in either direction between the choroid and retina. It thickens with age, and is probably an increasing barrier to diffusion of solutes. It undergoes degenerative changes involving its collagenous and elastic components and may calcify in a patchy manner. With time there is an accumulation of debris both within the membrane and anterior to it, which is interpreted as lipofuscin material that has been externalized from the pigmented epithelial cell. This focal debris becomes visible ophthalmoscopically as fine yellowish 'drusen', from about the age of 40 years (Sarks 1976), and their density increases with age.

Other structural changes in the retina with age include a displacement of nuclei from the outer nuclear layer into the adjacent outer plexiform and photoreceptor layers (Gartner and Hendkind 1981). Also, there is about a 50 per cent loss of ganglion cell axons from birth to 70 years of age, and this is reflected ophthalmoscopically by a loss of the foveal reflex (due to shallowing of the foveal pit,) and an enlargement of the optic cup over the age of 60 years due to thinning of the neuroretinal rim. Other degenerative changes at the optic nerve head with age are lipofucsin deposition, and the appearance of corpora amylacea arising from the degeneration of local astrocytes.

Ageing also brings about a reduction in blood supply to the retina, although the functional significance of this is unknown. There is a fall in capillary density in the retina, with a widening of the foveal avascular zone. There is a loss of peripheral capillaries as well as changes in the central retina. There is a systematic decrease in choroidal capillaries from the seventh to ninth decades.

Age-associated retinal disease

Debris from the retinal pigment epithelium collects on the inner surface of Bruch's membrane and is visible ophthalmoscopically as retinal drusen by the age of 40 years. Although drusen at low density have little discernible effect on vision, a substantial accumulation may have visual consequences. Drusen may separate the pigment epithelium from Bruch's membrane and this detachment may interfere with photoreceptor function. Occasionally there may be a rip in the detached sheet of pigmented epithelium which will similarly affect photoreceptor function in the region of the defect. Drusen changes at the fovea increase the risk of age-related macular degeneration (**ARMD**). ARMD is a condition of foveal and macular receptor loss which is a major cause of registered blindness. In the United Kingdom it accounts for

47 per cent of the registered blind (HMSO 1979), and for 63 per cent of cases of low vision (Robbins 1981). ARMD is often bilateral and there are genetic factors in its development. In the United States it is encountered in 4 per cent of the population between the ages of 65 and 74, in 17 per cent between the ages of 75 and 84, and in 22 per cent of those above the age of 85 years. The development of breaks in Bruch's membrane with age can allow the ingress of new vessels into the space below the pigmented epithelium, with the development of a more extensive affection of the macula termed disciform degeneration. In this, a neovascular membrane in the subepithelial space leads to an oedematous, exudative, and haemorrhagic condition which ultimately organizes and leads to the progressive destruction of the photoreceptors over a disc-shaped zone of the macula.

CHANGES IN VISION DUE TO AGEING

There are progressive changes in visual function from about the fifth decade of life onwards, and perhaps because of the well-documented senescent changes in retinal morphology cited above, many functional losses had, until recently, been attributed to retinal ageing. Weale (1982), however, emphasized that a number of past studies had failed to take into account the powerful effect of senile miosis in reducing retinal illuminance or the effects of increasing light scatter by the lens and the differential absorption of light by its accumulated yellow pigment. Thus undue weight may have been given to the role of retinal senescence. More recent studies have shown that even when optical factors are adequately considered, there is still an important neural component to the age-related loss of visual function. This has important implications for visual performance in old age.

In considering the changes in visual function that occur with ageing some attempt may be made to identify the role of pupillary, lens, or retinal factors (Weale 1982). Thus senile miosis contributes to the rise in absolute threshold to white light, to reduction in acuity to standard or Landolt C targets, to contraction of the visual fields, and to the fall in critical fusion frequency and the increased incremental threshold background luminance that occurs with age. It also contributes to the reduced size of 'a' and 'b' waves on the electroretinogram associated with reduced retinal illuminance during the test. Increased lens scatter is responsible for increased glare threshold and probably plays a part in increased glare recovery time. Relative changes in spectral response are the result of the filtering effect of the yellow lens pigment. Verriest (1972) found no significant difference in spectral sensitivity in young and old subjects who had undergone cataract extraction. Gunkel and Gouras (1963) studied the effect of age on scotopic visibility thresholds. They found that the relative reduction in spectral sensitivity in phakic eyes compared to aphakic eyes was greatest for short wavelengths (the violet end of the spectrum,) less for red, and least for white light. A similar explanation accounts for changes in trichromatic matching that occur with age. Senile miosis has also been invoked to account for the impaired colour-vision performance reported by Lakowski and Oliver (1974), but Barca and Vaccari (1977) do not believe that the increase in error score is accounted for by decreased luminance alone. A paradoxical rise in sensitivity to blue light found after the age of 50 years has been attributed to an age-related increase in lens fluorescence, so that the absorption of light at short wavelength is counteracted by emittance at a longer wavelength at the green end of the spectrum.

Increasing evidence points to a loss of retinal function with age. There is a slight rise in luminance threshold for longer wavelengths with age in aphakic as well as phakic subjects, which could have a neural basis, but senile miosis cannot be excluded as a cause. Jay *et al.* (1987) found an age-related fall in visual acuity after cataract extraction, with the average acuity at age 50 years being 6/5 and at 90 years, 6/12. Such studies in aphakics, however, do not usually address the question of whether the surgery itself might be responsible for a change in retinal function in an age-associated manner.

Many studies have reported a loss of contrast sensitivity with age that preferentially affects the higher spatial frequencies. This is in keeping with the change in modulation transfer function of the excised lens, which was found to affect preferentially the intermediate and high spatial frequencies. Higgins *et al.* (1988) found a greater loss of sensitivity at high spatial frequencies than could be explained by increased lens density. A loss of 0.3 log units occurred at 16 c.p.d between the third and seventh decade, which would imply a reduction in retinal illuminance of 0.6 log units if an increase in lens density were its sole explanation. This was outside the variation in lens density expected over this period (0.01–0.0.3 ± 0.15) as cited by Pokorny *et al.* (1987).

Absorption or scattering of light by the media affects high, more than low spatial frequencies; low spatial frequencies are relatively immune to optical factors (Campbell and Green 1965; Kelly 1972). Sloane *et al.* (1988) examined the relationship between contrast sensitivity and luminance in young (19–35 years) and old (68–79 years) subjects. Sensitivity decreased for both groups with decreasing luminance, but at low spatial frequencies (e.g., 0.5 c.p.d) the decrease is greater in old age, with the implication that it is due to neural rather than optical factors. They suggest that the use of relatively high luminances may explain those studies which found little or no loss with age at low spatial frequency. Support for the observation of losses of contrast sensitivity due to neural senescence also comes from studies using laser-generated interference gratings whose visibility is minimally affected by light absorption and scatter by the ocular media.

Temporal modulation of a grating stimulus modifies the contrast sensitivity response. At high luminance and low spatial frequency, a moderately fast rate of flicker enhances sensitivity, while it is reduced at low luminance. This effect has a neural basis. Sloane *et al.* (1988) have demonstrated, using an optimum flicker rate for enhancement of 7.5 Hz, that in older subjects the enhancement is less at high luminance and is reduced less at low luminance than in young subjects. Other workers, but not all, have reported deficits in temporal processing in the elderly. Some of the discrepancies between reports may reflect the differences in the effects of luminance on flicker enhancement at the lower spatial frequencies in relation to luminance levels used in particular studies.

A recently introduced psychophysical term that has an important bearing on visual performance is the 'useful field of vision' (**UFOV**), a measure of that portion of the visual field providing access to visual information usable in specified ways. It determines the ability to respond to peripheral targets while fixating centrally. The UFOV is reduced by foveal stimulation, a central task demand, and by distracting elements in the visual background, particularly when these are eccentric. Both the number of distracting stimuli and the degree of similarity between the distractors and the target impair performance.

The UFOV declines with age for extrafoveal acuity, letter identification, and for peripheral localization within a visually cluttered scene. For briefly presented targets, a secondary central task has a much greater effect on the peripheral performance of older than younger observers (Ball *et al.* 1987).

In keeping with such reports, it is recognized that the frequency of difficulty reported by older adults for everyday tasks involving peripheral vision, visual search, and cluttered visual scenes is about five times that encountered by younger adults, and may be predicted by measurement of the UFOV. It is envisaged that a constricted UFOV could affect performance by forcing an observer to make more fixations to scan the visual scene. Although fixational stability is not affected by age, there is an age-related increase in latency of ocular movements, which would place elderly people at a disadvantage in everyday situations involving visual search because of the imposed increase in processing time. The difficulty experienced by elderly people is greater when required to perform simultaneous central and peripheral visual tasks. This is consistent with the results of other studies that show an age-related increase in difficulty in the ability to ignore irrelevant detail or perform divided attention tasks. In general it is accepted that older individuals have a slower speed of perceptual processing.

Ball *et al.* (1987) found that elderly subjects could improve their performance of selected UFOV tasks with practice. Salthouse and Somberg (1982) found that age differences persisted, despite training, in a visual scanning task, while Walsh (1988) found less improvement in the old than the young in a visual search task.

The impact of declining visual functions on everyday performance is difficult to gauge but has attracted attention in the task of car driving. Most studies have found no link between age-related visual field loss, as measured by perimetry, and driving performance, although there is a relationship between accidents and severe binocular field loss. Nonetheless, older subjects in general are found to suffer more accidents and figure more often in cited traffic offences. Their accident profiles indicate that they have difficulty in processing information from the periphery (i.e., failure to heed signs, to give way, and to turn safely, and more frequent involvement in accidents at junctions). In the study by Avolio *et al.* (1986), a relationship was found between driving accident rates and performance in tasks involving visual clutter.

BLINDNESS

The many different definitions of blindness have made comparison between countries a difficult task (World Health Organization (WHO) 1966). A WHO study group proposed a definition of visual impairment to embrace many of those based on visual acuity. In the United Kingdom, for registration purposes, blindness implies a vision of less than 3/60 binocularly, and partial sight less than 6/60, in the presence of a full peripheral field. This is modified when the field is contracted. Category 1 of the WHO definition is a best vision of less than 6/18; category 5 is no light perception (WHO 1973).

In England in 1982 there were 109 700 adults on the blind register, with about 12 000 newly registered (about 1 in 3000 of the adult population). An estimated 115 000 were registered in England and Wales in 1980, with 132 000 projected for 1992 (Table 9) (Cullinan 1977). Each year, about half of the registered men and 63 to 65 per cent of the women are over 75 years, and the rise of about 20 per cent in registrations since 1958 is almost entirely explained by increased longevity, particularly for women. There are more men than women registered as blind up to the age of retirement, which may reflect the fiscal and retraining benefits of registration (Cullinan 1977, 1987). The age-specific prevalence rates for England and Wales are given in Table 9, while causes of blindness are given in Tables 10 and 11 (DHSS 1988).

Table 9 Age-specific prevalence rates among the registered blind in England and Wales, 1980 (from Cullinan 1987)

Age group (years)	Prevalence (per 1000)
0–4	0.09
5–15	0.23
16–64	0.85
65–74	4.48
75+	23.24

In 1980 there were 51 420 persons registered as partially sighted in England and Wales. It is well recognized that both categories of registration under-represent those who might qualify; by half for partial sight and 20 per cent for blindness. Cullinan (1977) conducted a national random survey of some 18 000 households in England and Wales and gave an estimated prevalence of visual disability (using the WHO category 1 definition: less than 6/18), of 5.2/1000. This represented two and a half times the then number of registered blind. Over half of these were over 75 years of age and three-quarters were in their retirement years. Eighteen per cent were ascribed to macular degeneration. Nearly half of those with visual disability had never attended an eye specialist, and in over half, visual performance could be improved by better home lighting. Table 12 gives prevalence figures for other parts of the world, emphasizing the wide geographic variation. The demography of blindness is considered by Goldstein (1974).

Macular degeneration, diabetic retinopathy, glaucoma, and cataract are the major causes of blindness in the United States (Kahn and Moorehead 1973) and England and Wales (DHSS 1988), and diabetic retinopathy is the most frequent cause of blindness in the adult population of working age. In diabetics, blindness results either from proliferative retinopathy, which is most common in type I diabetics, or maculopathy, which is most common in type II diabetics. Because type II diabetes is much more prevalent than type I it also is responsible for more cases of the proliferative form of the disease. Background and proliferative retinopathy and maculopathy increase with age in both forms of diabetes. This reflects that their prevalence is related to the duration of the diabetes. Insulin-dependent diabetics show a low frequency of retinopathy in the first 5 years after diagnosis and

Table 10 Causes of blindness in England among new registrations, year ended March 1981*

	All ages	0–15 years		16–64 years		65–74 years		75–84 years		85 years and over	
Condition	M & F	M	F	M	F	M	F	M	F	M	F
Diabetic retinopathy	7.9	0.9	1.9	17.1	22.3	13.1	16.5	3.1	4.7	1.3	0.9
Macular and posterior pole degeneration	37.0	1.8	13.3	9.3	8.8	26.4	29.9	43.8	45.9	50.7	52.5
Open-angle glaucoma	8.0	1.8	1.0	3.8	2.8	11.9	6.6	11.2	10.2	6.3	6.9
'Senile' cataract	5.8	—	1.9	0.3	2.8	2.2	6.8	4.9	7.3	4.4	10.1
Myopia	4.0	4.5	1.0	6.5	7.8	7.0	6.1	3.1	2.7	1.9	1.9
Optic atrophy	3.4	20.7	10.5	11.0	9.7	2.2	2.2	1.6	1.6	1.7	1.5

* From DHSS (1988): expressed as a percentage of the total number in each age/sex bracket.

Table 11 Causes of blindness in England among new registrations, year ended March 1981 (DHSS 1988)

Condition	65–74 years (%)	75 years and over (%)
Diabetic retinopathy	15.1	3.0
Macular and posterior pole degeneration	28.4	47.7
Open-angle glaucoma	8.8	9.2
Senile cataract	4.8	7.2
Myopia	6.5	2.5
Optic atrophy	2.2	1.6

Table 12 Geographic variation in prevalence rates for blindness

Country	Rate/1000
Netherlands (1959)	0.55
USA (NHANES) (1962–64)	1.50
England (1982)†	2.40
Iceland (1954)	3.00
Kenya (1960)	10.00
Zambia (1978)	22.00
India (rural)‡	35.00
Sudan (rural)‡	45.00
Egypt (rural)‡	48.00
Tunisia	70.00

* Cited by Cullinan 1987.
† DHSS 1988.
‡ Urban rates are less than half these.

a rising prevalence thereafter. About 60 to 80 per cent show retinopathy after 10 years of their disease, and 80 to 100 per cent after 20 years. In a large population-based study from Wisconsin, there was a 50 per cent prevalence of proliferative retinopathy after 15 years of the disease (Klein *et al.* 1984*a*). In non-insulin dependent diabetes about half have retinopathy 15 years after diagnosis, and about 20 per cent will have proliferative retinopathy (Klein *et al.* 1984*b*). Caird *et al.* (1968), in Oxford, estimated that 2 to 7 per cent of all diabetics would ultimately become blind, while a more recent population-based study in Rochester, Minnesota, found a cumulative incidence of bilateral blindness due to diabetic retinopathy of 8.2 per cent after 20 years (Dwyer *et al.* 1985). Klein *et al.* (1984*c*) found that 12 per cent of insulin-dependent diabetics were blind after 30 years of diabetes, compared to 5 per cent of non-insulin dependent diabetics. It is likely that in the future, effective prophylactic laser photocoagulation will reduce the incidence of blindness from this cause.

The annual incidence of macular degeneration responsible for a vision of 6/9 or worse was calculated by Podgor *et al.* (1983) from the Framingham study findings. The incidence was 0.5 per cent at 55 years and 6.5 per cent at 75 years. (Macular change without visual loss was 3 per cent at 55 years and nearly 6 per cent at 75 years.) For cataract, comparable figures are 1.2 per cent at 55 years, and 15 per cent at 75 years (with 10 per cent and 37 per cent respectively for lens change causing no visual defect).

The presence of sight-threatening retinopathy may have implications for overall longevity. Davis *et al.* (1979) found that the life expectancy of diabetics was no different from non-diabetics in the presence of minimal or no retinopathy, while those with moderate retinopathy had about 25 per cent excess mortality over the study period of 7 years. Those with proliferative retinopathy had more than a 50 per cent excess mortality over this period. This risk for death in diabetics with cataract is also said to be increased (Podgor *et al.* 1985).

References

Abdel-Khalek, L.M.R., Williamson, J., and Lee, W.R. (1978). Morphological changes in the human conjunctival epithelium II. In keratoconjunctivitis sicca. *British Journal of Ophthalmology*, **62**, 800–6.

Armaly, M.F. (1965). On the distribution of applanation pressure. I. Statistical features and the effect of age, sex and family history of glaucoma. *Archives of Ophthalmology*, **73**, 11–18.

Avolio, B.J., Kroeck, K.B., and Panek, P.E. (1986). Individual differences in information-processing ability as a predictor of motor vehicle accidents. *Human Factor*, **27**, 577–88.

Ball, K.K., Beard, B.L., Miller, R.L., and Roenker, D.L. (1987). Mapping the useful field of view as a function of age. *Gerontologist*, **27**, 166A.

Bankes, J.L.K., Perkins, E.S., Tsolakis, S., and Wright, J.E. (1968). Bedford glaucoma survey. *British Medical Journal*, **1**, 791–6.

Barca, L. and Vaccari, G. (1977). Illuminance dependence of 100 hue response for normal subjects of different ages. *Atti della Fond. G. Ronchi*, **32**, 412–18.

Bengtsson, B. (1981). Aspects of the epidemiology of chronic glaucoma. *Acta Ophthalmologica*, **148** (suppl.).

Bietti, G.B., Guerra, P., and Ferraris de Gaspare, P.F. (1955). La dystrophie corneene nodulaire en ceinture de pays tropicaux a sol aride. *Bulletin de la Societé Ophtalmologie Francaise*, **68**, 101–29.

Bron, A.J. (1986). Prospects for the dry eye. Duke Elder lecture. *Transactions of the Ophthalmological Society (UK)*, 1986, **104**, 801–26.

Bron, A.J. and Brown, N.A.P. (1986). Lens structure and forms of cataract. In *The lens: transparency and cataract* (ed. G. Duncan), pp. 3–11. Eurage, Rijwijk.

Bron, A.J. and Brown N.A.P. (1987). Perinuclear lens retrodots: a role for ascorbate in cataractogenesis. *British Journal of Ophthalmology*, **71**, 86–95.

Caird, F., Pirie, A., and Ramsell, T.G. (1968). *Diabetes and the eye*, pp.1–7, 110–121. Blackwell, Oxford.

Campbell, F.W. and Green D.G. (1965). Optical and retinal factors affecting visual resolution. *Journal of Physiology* (*London*), **181**, 576–93.

Clayton, R.M., Cuthbert, J., Seth, J., Phillips, C.I., Bartholomew, R.S., and Reid, J.M. (1984). Epidemiological and other studies in the assessment of factors contributing to cataractogenesis. In *Human cataract formation.*, Ciba Symposium 106, pp. 25–47.

Cogan, D.G. and Kuwabara, T. (1959). Focal senile translucency of the sclera. *Archives of Ophthalmology*, **62**, 604–10.

Crick, R.P. (1980). Computerised clinical data base for glaucoma. Ten years experience. *Research Clinicial Forums*, **2**, 29–39.

Cullinan, J. (1987). The epidemiology of blindness. In *Clinical Ophthalmology* (ed. S. Miller) pp. 157–78. Wright, London.

Cullinan, T.R. (1977). The epidemiology of visual disability. In *Studies of visually disabled people in the Community*, HSRU Report No. 28. University of Kent, Canterbury.

Cullinan, R.T. *et al.* (1979). Visual disability and home lighting. *Lancet*, **i**, 642–4.

Curcio, C.A. (1986). Aging and topography of human photoreceptors. *Journal of the Optical Society of America*, **A3**, 59.

Damato, B.E., Allan, D., Murray, S.B., and Lee, W.R. (1984). Senile atrophy of the human lacrimal glands: the contribution of chronic inflammatory disease. *British Journal of Ophthalmology*, **68**, 674–80.

Davis, M.D. *et al.* (1979). Prognosis for life in patients with diabetes: relation to severity of retinopathy. *Transactions of the American Ophthalmological Society*, **77**, 144–70.

Department of Health and Social Security (DHSS) (1988). *Causes of blindness and partial sight among adults in 1976/77 and 1980/81, England.* HMSO, London.

Dwyer, M.S., *et al.* (1985). Incidence of diabetic retinopathy and blindness: a population-based study in Rochester, Minnesota. *Diabetes Care*, **8**, 316–22.

Ederer, F., Hiller, R., and Taylor, H. (1981). Senile lens changes and diabetes in two population studies. *American Journal of Ophthalmology*, **91**, 381–95.

Eisner, G. (1987). Clinical anatomy of the vitreous. In *Biomedical foundations of ophthalmology*, Vol. 1 (ed. T.D. Duane and E.A. Jaeger). Harper and Row, Philadelphia.

Freedman, A. (1973). Labrador keratopathy and related disease. *Canadian Journal of Ophthalmology*, **8**, 286–90.

Gandolfi, S.A., Melli, E., Tomba, M.C., and Maraini, G. (1986). Membrane damage in human senile cataract. Evidence from radio tracer flux measurements. In *The lens: transparency and cataract* (ed. G. Duncan), pp. 97–102. Eurage, Rijwijk.

Gartner, S. and Hendkind, P. (1981). Aging and degeneration of the human macula. 1. Outer nuclear layer and photoreceptors. *British Journal of Ophthalmology*, **65**, 23–8.

Gibson, J.M., Rosenthal, A.R., and Lavery, J. (1985). A study of the prevalence of eye disease in the elderly in an English community. *Transactions of the Ophthalmological Society* (*UK*) **104**, 196–203.

Goldstein, H. (1974). Incidence, prevalence and causes of blindness. Statistics for the United States and selected countries in Asia and Europe. *Public Health Review*, **3**, 5–37.

Gunkel, R.D. and Gouras, P. (1963). Changes in scoptopic visibility thresholds with age. *Archives of Ophthalmology*, **69**, 4–9.

Harbour, R.C. and Stern, G.A. (1983). Variables in McCarey Kaufman corneal storage. *Journal of Ophthalmology*, **90**, 136–42.

Harding, J.J. (1970). Free and protein-bound glutathione. *Biochemical Journal*, **117**, 957–60.

Harding J.J. and Crabbe, M.J.C. (1984). The lens: development, proteins, metabolism and cataract. In *The eye*, Vol. 1b (3rd edn), (ed. H. Davson), pp.207–492. Academic Press, London.

Harding, J.J. and Dilley, K.J. (1976). Structural proteins of the mammalian lens: a review with emphasis on changes in development, ageing and cataract. *Experimental Eye Research*, **22**, 1–73.

Harding, J.J. and van Heyningen, R. (1988). Drugs, including alcohol, that act as risk factors for cataract, and possible protection against cataract by aspirin-like analgesics and cyclopenthiazide. *British Journal of Ophthalmology*, **72**, 809–14.

Harwerth, R.S. and Sperling, H.G. (1971). Prolonged colour blindness induced by intense spectral lights in Rhesus monkeys. *Science* (*NY*), **174**, 520–3.

Higgins, K.E., Jaffe, M.J., Caruso, R.C., and de Monasterio, F.M. (1988). Spatial contrast sensitivity: effects of age, test–retest and psychophysical method. *Journal of the Optical Society of America*, A, **12**, 2173–80.

HMSO (1979). *Blindness and partial sight in England 1969–1976*, Reports on Public Health and Medical Subjects No. 129. HMSO, London.

Hollows, F.C. and Graham, P.A. (1966). Intraocular pressure, glaucoma and glaucoma suspects in a defined population. *British Journal of Ophthalmology*, **50**, 570–86.

Hollyfield, J.G. and Basinger, S.F. (1980). RNA metabolism in the retina in relation to cyclic lighting. *Vision Research*, **20**, 1151–5.

Jay, J.L., Mammo, R.B., and Allan, D. (1987). Effect of age on visual acuity after cataract extraction. *British Journal of Ophthalmology*, **71**, 112–15.

Johnson, G.L. (1981). Aetiology of spheroidal degeneration of the cornea in Labrador. *British Journal of Ophthalmology*, **65**, 270–283.

Johnson, G.J. and Overall, M. (1978). Histology of spheroidal degeneration of the cornea in Labrador. *British Journal of Ophthalmology*, **62**, 53–61.

Kahn, H.A. and Moorhead, H.B. (1973). *Statistics on blindness in the model reporting area 1969–1970*, US Department of Health, Education and Welfare, Publication No. (NIH) 73–427. US Government Printing Office, Washington.

Kelly, D.H. (1972). Adapatation effects on spatio-temporal sine-wave thresholds. *Vision Research*, **12**, 89–101.

Klein, R. *et al.* (1984*a*). The Wisconsin epidemiologic study of diabetic retinopathy. II. Prevalence and risk of diabetic retinopathy when age at diagnosis is less than 30 years. *Archives of Ophthalmology*, **102**, 520–6.

Klein, R. *et al.* (1984*b*). The Wisconsin epidemiologic study of diabetic retinopathy. III. Prevalence and risk of diabetic retinopathy when age at diagnosis is 30 or more years. *Archives of Ophthalmology*, **102**, 527–32.

Klein, R., Klein, B.E.K., and Moss, S.E. (1984*c*). Visual impairment in diabetes. *Ophthalmology*, **91**, 1–9.

Lakowski, R. and Oliver, K. (1974). Effect of pupil diameter on colour vision test performance. *Modern Problems in Ophthalmology*, **13**, 307–11

Laule, A., Cable, M.K., Hoffman, C.E. and Hanna, C. (1978). Endothelial cell population changes of human cornea during life. *Archives of Ophthalmology*, **96**, 2031–5.

Lawill, T., Crockett, S., and Currier, G. (1977). Retinal damage secondary to chronic light exposure; thresholds and mechanisms. *Documenta Ophthalmologica*, **44**, 379–402.

Leibowitz, H.M. *et al.* (1980). The Framingham eye study monograph. *Surveys in Ophthalmology*, suppl. 24.

Lerman, S. (1980). *Radiant energy and the eye*, Functional Ophthalmology Series, Vol. XX. Macmillan, New York.

Leske and Sperduto, R. D. (1983). The epidemiology of senile cataract: a review. *American Journal of Epidemiology*, **118**, 152–65.

Linn, J.G. Jr., Stuart, J.C., and Warnicki, J.W. (1981). Endothelial morphology in long-term keratoconus corneal transplants. *Ophthalmology*, **88**, 761–9.

Lorenzetti, D.W.C., Uotila, M.H., Parikh, N., and Kaufman H.E.

(1967). Central corneal guttata. *American Journal of Ophthalmology*, **64**, 1155–8.

Marshall, J. (1985). Radiation and the aging eye. *Journal of Ophthalmology, Physiology and Optics*, **5**, 241–63.

Marshall, J. (1987). The aging retina. Physiology or pathology. *Eye*, **1**, 282–95.

Millodot, M. (1977). The influence of age in the sensitivity of the cornea. *Investigative Ophthalmology and Visual Science*, **16**, 240–2.

Noell, W.K., Walker, V.S., Kang, B.S., and Berman, S. (1966). Retinal damage by light in rats. *Investigative Ophthalmology*, **5**, 450–73.

Norn, M.S. (1968). Hudson-Stähli' line of the cornea. *Acta Ophthalmologica*, **46**, 106–18, 119–275.

Norn, M.S. (1983). *External Eye. Methods of examination*. Scriptor, Copenhagen.

Olsen, T. (1980). Cornea thickness and endotheliall damage after intracapsular cataract extraction. *Acta Opthalmologica*, **58**, 424–33.

Podgor, M.J., Leske, C., and Ederer, F. (1983). Incidence estimates for lens changes, macular changes, open-angle glaucoma and diabetic retinopathy. *American Journal of Epidemiology*, **118**, 206–12.

Podgor, M.J., Cassel, G.H., and Kannel, W.B. (1985). Lens changes and survival in a population based study. *New England Journal of Medicine*, **313**, 1438–44.

Pokorny, J., Smith, V.C., and Lutze, M. (1987). Aging of the human lens. *Applied Optics*, **26**, 1437–40.

Robbins, H.G. (1981). Low vision care for the over 80s. *Australian Journal of Optometry*, **64**, 243–51.

Salthouse, T.A. and Somberg, B.L. (1982). Skilled performance: the effects of adult age and experience on elementary processes. *Journal of Experimental Psychology General (Washington)*, **111**, 176–207.

Sarks, S.H. (1976). Aging and degeneration in the macular region: A clinico-pathological study. *British Journal of Ophthalmology*, **60**, 324–41.

Schappert-Kimmijer, J. (1971). A five-year follow up of subjects with intraocular pressure of 22–30 mmHg without anomalies of optic nerve and visual field typical for glaucoma at first investigation. *Ophthalmologica*, **162**, 289–95.

Sebag, J. and Balazs, E.A. (1985). Human vitreous fibres and vitreoretinal disease. *Transactions of the Ophthalmological Society (UK)*, **104**, 123–8.

Sliney, D. and Wolbarsht, M. (1980). *Safety with lasers and other Optical sources*. Plenum Press, New York.

Sloane, M.E., Owsley, C., and Alvarez, S.L. (1988). Aging, senile miosis and spatial contrast sensitivity at low luminance. *Vision Research*, **28**, 1235–46.

Smith, V.C., Pokorny, J., and Diddie, K.R. (1988). Colour matching and the Stiles–Crawford effect in observers with early age-related macular changes. *Journal of the Optical Society of America*, **A5**, 2113–21.

Sparrow, J.M. (1989). The lens in diabetes. D.Phil. thesis, University of Oxford.

Sperduto, R.D. and Hiller, R. (1984). *Ophthalmology*, **91**, 815–18.

Sykes, S.M., Robinson, W.G., Waxler, M., and Kuwabara, T. (1981). Damage to monkey retina by broad spectrum fluorescent light. *Investigative Ophthalmology and Visual Science*, **20**, 425–34.

Tripathi, R.C. and Tripathi, B.J. (1987). Functional anatomy of the anterior chamber angle. In *Biomedical Foundations of Ophthalmology* (ed. T.D. Duane and E.A. Jaeger), pp.1–88. Harper and Row, Philadelphia.

van Heyningen, R. (1977). The biochemistry of the lens: selected topics. In *Scientific Foundations of Ophthalmology* (ed. E.S. Perkins and D.W. Hill). Heinemann Medical, London.

van Heyningen, R. and Harding J.J. (1986*a*). Risk factors for cataract: diabetes, myopia and sex. In *Modern trends in aging research* (ed. Y. Courtois, B. Forette, and D. Knook). INSERM/John Libbey, Paris.

van Heyningen, R. and Harding, J.J. (1986b). A case-control study of cataract in Oxfordshire: some risk factors. *British Journal of Ophthalmology*, **72**, 804–8.

van Norren, P. and van Meel, G.J. (1985). Density of human cone photopigments as a function of age. *Investigative Ophthalmology and Visual Science*, **26**, 1014–16.

Verriest, G. (1972). The relative spectral luminous efficiency in different age groups of aphakic eyes. *Die Farbe*, **21**, 17–25.

Walsh, D.A. (1988). Aging and visual information processing: potential implications for everyday seeing. *Journal of the American Optical Association*, **59**, 301–6.

Waltman, S.R. and Cozean, C.H. (1979). The effect of phacoemulsification on the corneal endothelium. *Ophthalmic Surgery*, **10**, 31–3.

Weale, R.A. (1982). A biography of the eye: development, growth and age. H.K. Lewis, London.

World Health Organization (1966). *Blindness (information collected from various sources)*, Epidemiology Vital Statistics Report 19, pp. 437–511. WHO, Geneva.

World Health Organization (1973). *The prevention of blindness (Report of a WHO Study Group)*, WHO Technical Report Series, No. 518. WHO, Geneva.

Williams, R.D. and Kuszak, J.R. (1986). An asymmetrical lattice-work of sutures revealed by SEM in adult primate (human and baboon) lenses. *Investigative Ophthalmology and Visual Science*, **27** (suppl.) 268 (abstr. 2).

Winder, A.F. (1983). Relationship between corneal arcus and hyperlipidaemia is clarified by studies in familial hypercholesterolaemia. *British Journal of Ophthalmology*, **67**, 789–94.

Wing, G.L., Blanchard, G.C., and Weiter, J.L. (1978). The topography and age relationship of lipofuscin concentration in the retinal pigment epithelium. *Investigative Ophthalmology and Visual Science*, **17**, 601–7.

18.13.2 Visual perception and cognition

ROBERT SEKULER AND ALLISON B. SEKULER

The past two decades have seen substantial increases in the number of vision researchers who study the visual aspects of ageing and age-associated processes. This heightened concentration of effort has brought a remarkable growth in our understanding of many aspects of vision, including the ocular changes that accompany human ageing (Owsley and Sloane 1990) (see also Chapter 18.13.1). Although advances in understanding ocular structure are most welcome, they can only tell part of the story of the ageing visual system. This chapter tells another part of the story, describing recent efforts that have been made to understand the functional changes of ageing vision.

Several years ago, an editorial encouraged vision researchers to concentrate on vision and ageing, emphasizing the real-world dimensions of their nexus (Sekuler *et al.* 1983). That editorial did much to set the research agenda relevant to this chapter. The ultimate goal of the research programme reviewed here is a comprehensive understanding of vision under everyday conditions. Needless to say, the nexus of vision and ageing has real and important consequences, both for society and for individuals. For example, as a result of visual impairments among elderly people, each year in the United States more than 40 000 hip fractures are estimated to occur (Felson *et al.* 1989). By understanding precisely what aspects of visual impairment are responsible for such injuries, one can begin to devise methods for ameliorating the problem. In this context, it is important to understand how visual functions might be diminished by artificial means, including various medications commonly taken by older persons (Collins 1989), as well as how visual functions might be diminished by natural anatomical and physiological changes with ageing.

VISUAL CHANGES IN DAILY LIFE

Although laboratory research and clinical observations have provided much of our knowledge about vision and ageing, important insights can be derived from a new and surprisingly direct approach: simply asking older people about their visual problems. Consider, for example, the thoughtful introspections of a 70-year-old vision scientist (NRC 1987; p.7):

> It seems to me that I must concentrate harder now and that I require higher levels of illumination than I formerly did in order to have the same perceptual results. Just plain seeing in simplified situations, as in routine vision testing, seems as good as quick as ever—but perceiving the meaning of a complex, changing scene is definitely more difficult and slower. I see the parts almost as well as I ever did but the organizing of the perception as a whole seems to be more time-consuming and to require more attention.

Extending this emphasis on observers' self-reports to a great many respondents, Hakkinen (1984) combined extensive personal interviews with an array of ophthalmological tests in her study of several hundred elderly people. Most interesting were clear discrepancies between the results of clinical tests and subjects' self-assessment of their visual function under everyday conditions.

Carefully constructed and refined questionnaires have been used to advantage in a number of recent studies. One widely used questionnaire was developed by Kosnik *et al.* (1988), and administered to subjects aged from 18 to 95 years. Respondents rated the frequency with which they had difficulty performing 18 different visual tasks such as reading, recognizing objects, picking out a face in a crowd, seeing in dimly lit environments, and seeing moving objects. Respondents were also asked to rate the severity of their problems with each task.

Factor analysis revealed that older respondents had particular difficulties with regard to five visual factors, or clusters of tasks. The troublesome clusters were speed of visual processing, light sensitivity, dynamic vision, near vision, and visual search. For these clusters, the percentage of respondents reporting difficulty increased by a factor of 2 to 6 across the adult lifespan. Examining the growth of these self-reported difficulties revealed substantial diversity in rate, suggesting that various aspects of everyday vision 'age' differentially.

Versions of this questionnaire on ageing and vision have yielded a variety of other results (Kline *et al.* 1991; Kosnik *et al.* 1990).

One consistent finding bears emphasis: older respondents report a number of visual difficulties that would not be detected by standard clinical tools. For example, a great many older respondents complain that their visual processing is noticeably slower than it was formerly. This reduced speed of visual processing is probably independent of the increase in reaction time commonly associated with ageing (Wilkinson and Allison 1989). It would be valuable if future research attempted to verify these self-reported complaints by independent, objective methods.

EXTENT OF VISUAL FIELD

The visual field is that region of space in which objects are visible at once with steady fixation. The boundaries of the field determined by a target of given size and luminance define a single isopter. Studies with various techniques and instruments confirm that the isopters in older adults are constricted compared with those in younger observers (Drance *et al.* 1967). It had been assumed that much of this shrinkage of visual field was caused by optical factors, including senile miosis. However, a recent study casts doubt upon this assumption (Johnson *et al.* 1989), suggesting instead that the shrinkage is neural in origin.

Johnson and Keltner (1986) reported a prevalence of visual-field loss of 3 to 3.5 per cent for individuals between 16 and 60 years of age. The prevalence doubled for people aged 60 to 65, and nearly redoubled for individuals over 65.

It is important to appreciate that most individuals with field loss were unaware of this condition. Despite the lack of awareness, the shrinkage in the visual field is sufficiently great to be of consequence to everyday vision. In a study of 10 000 applicants for drivers' licences, Johnson and Keltner (1986) discovered that drivers with loss of visual field in both eyes were twice as likely to be involved in traffic accidents or violations as age-and sex-matched controls who had no field loss. Through demonstrating a connection between loss of visual field and driving performance, Johnson and Keltner suggested that diminished peripheral vision placed older drivers at greater risk of accident. This suggestion was amplified by Keltner and Johnson (1987).

THE EFFECTIVE FIELD OF VIEW

Recent work raises some questions about the usefulness, under everyday conditions, of data on observers' visual fields assessed by standard clinical perimetry. Much of this recent work has focused on the *effective* visual fields in elderly subjects. As an introduction to this question, again consider the introspections of the 70-year-old vision scientist (Morgan 1988):

> I have the impression that my working visual fields, in contrast to my clinically measured fields (which are normal), are somewhat reduced ... Objects coming from my right that I missed on casual observation sometimes suddenly appear in my field. I think I have to make a greater effort than before to perceive objects in the periphery of my field. If I give my full attention to detecting peripheral objects, as in visual field testing, my performance is excellent. But when my attention is divided, as in driving, I think that there has been a decrease in the size of my visual field.

As noted earlier, older observers commonly report having difficulty finding sought-after objects in scenes that are visually cluttered. These anecdotal reports are confirmed by various empirical studies using highly simplified visual displays (Cerella 1985; Scialfa and Kline 1988). Other studies made use of a laboratory analogue specifically designed to mimic the cluttered visual environments about which older observers complained. For example, Sekuler and Ball (1986) used a computer-driven display to determine how well a single, randomly positioned peripheral target could be localized in the presence of other peripheral, distractor stimuli. In some conditions the observer had to perform a concurrent task using foveal vision. Note that when neither a concurrent task nor peripheral distractors were present, this task resembled that which is used clinically to assess visual fields; but, with a concurrent foveal task as well as peripheral distractors, the task more nearly resembled those encountered in everyday life. The response required of observers, radial localization, was selected because it was relatively impervious to optical changes that accompany ageing: decreased retinal illumination and image degradation (Post and Leibowitz 1980). Older observers proved to be extremely susceptible to the effects of irrelevant stimulus 'distractors'. When the briefly presented target appeared in isolation, observers of all ages did equally well; however, when the target was surrounded by distractor stimuli, older observers alone were impaired in their performance. These results suggest that under everyday conditions, older observers experience a particularly large discrepancy between their useful field of vision and their visual fields measured via standard perimetry.

The disadvantage suffered by the older observers proved to be surprisingly amenable to amelioration. Some of Sekuler and Ball's older observers were given an opportunity to practise radial localization for several dozen additional trials during each of 5 days. Practice produced a steady improvement in radial localization and, hence, an expansion in the useful field of vision. After rest periods that ranged from 3 to 5 weeks, the older observers were again tested. Despite the lengthy interval, performance on the retention test did not differ from performance on the last day of practice, suggesting that virtually all the improvement had endured. The potency of the practice effect can be gauged from an unsolicited remark offered by one older observer. After just 2 days' practice, radial localization had already become so much easier for her that she insisted the researchers must have increased the duration of the target display.

In a further study, Ball *et al.* (1990) found virtually no relation between the size of an observer's useful field of vision, measured by means of radial localization in the presence of distractors, and the size of the observer's visual fields measured perimetrically. The study also found a high degree of variability among older individuals, both on reported problems related to visual search and in the size of useful field of vision. The size of the *useful* field of vision task, but not performance on standard perimetry, was well correlated with self-assessed difficulties in everyday visual search and speed of visual processing.

Ball *et al.* (1990) showed clearly that the concurrent task, which required the processing of foveal information, had the greatest deleterious impact on radial localization by older observers. This result suggested that older observers' peripheral vision in particular could be strongly affected by simultaneous demands on their attention.

SPATIAL VISION

Much effort has been devoted to age-associated changes in various aspects of spatial vision. Most of this work has focused on the ability to resolve fine spatial details, usually assessed in terms of visual acuity. Pitts (1982) collated the results of eight different, large-sample studies of visual acuity as a function of age. These data, collected by various researchers over a period of nearly a century, reveal a systematic trend: diminished acuity as a function of age, commencing at approximately 50 years. The data also show large differences among studies, both in the absolute levels of acuity at any one age and in the rate at which acuity declines with age. Unfortunately, the data are presented as arithmetic means for each cohort, with no indication of interobserver variation. Experience suggests that such variation would be appreciable, particularly for older cohorts (Owsley and Sekuler 1984).

Contrast sensitivity, introduced to the clinic within the past two decades, assesses the contrast needed to see targets of varying spatial structure. Typically, but not necessarily, the targets are gratings whose alternating dark and light bars have a sinusoidal luminance profile. Measurements of threshold contrast are reported for a range of spatial frequencies, from low (wide bars) to high (fine bars). More than a dozen studies, mostly with small samples and diverse methods, have not produced an entirely consistent picture (Crassini *et al.* 1988). However, there seems to be a strong and growing consensus about a number of points. At low spatial frequencies, contrast sensitivity does not vary with age (Owsley *et al.* 1983). At intermediate and high spatial frequencies (e.g. greater than 4 cycles/degree of visual angle), contrast sensitivity declines steadily with age, beginning as early as 30 years (Owsley *et al.* 1983; Higgins *et al.* 1988). Moreover, Sloane *et al.* (1988) showed that age-associated decline in contrast sensitivity is greater with targets of low space-averaged luminance.

Much energy and ingenuity have been expended to identify the sources of declining contrast sensitivity with age. Possibilities include optical factors (e.g. intraocular light scatter, reduced retinal illumination), photoreceptor properties (e.g. alterations in the regularity and spacing of the receptor mosaic, morphological changes that reduce the area of each receptor's effective aperture—Marshall (1987); Werner *et al.* (1990)), as well as post-receptor, neural factors. Prime among the last of these would be changes in the number of cells within the cerebral cortex and in the connectivity of those cells (Braak and Braak 1988). The number and variety of possible mechanisms demand a unified, sophisticated theoretical approach such as the development of an explicit, ideal-observer model for the ageing eye (Geisler 1989). Such a model would allow one to determine the extent to which each factor, jointly or separately, can limit spatial vision in the elderly subject.

Whatever its origins, age-associated reduction in contrast sensitivity may be implicated in some findings on reading deficits in older readers. Akutsu *et al.* (1991) examined the maximum achievable reading rates with various sizes of character. With characters whose size supports highest reading rates in young readers (average age, 21.6 years), older readers (average age, 68.7 years) do almost as well as their younger counterparts. However, with text composed of very small or very large characters, older observers' reading rate drops to about 70 per cent of their younger counterparts. This decrease, according to Akutsu *et al.*, is probably the result of the older observers' decreased contrast sensitivity. It is important to note that despite their somewhat poorer acuity (in Snellen notation, 6/6.8 vs. 6/5.9), the older readers achieved maximum reading speeds not significantly different from those achieved by the younger readers. So advancing age *per se* has little or no effect on maximum reading speed. At first, this finding seems to be inconsistent with older persons' self-assessment that their visual processing is slowed (see above). Of course the self-assessment may be valid but only for some set of tasks other than reading or for some measure of reading other than maximum speed. {Cohort effects may also be relevant here.}

Various studies have attempted to establish connections between either visual acuity or contrast sensitivity, and changes in performance on tasks representative of everyday vision, including the perception of faces, road signs, and objects (Owsley *et al.* 1981; Owsley and Sloane 1987). Generally, visual acuity tends to be poorly correlated with everyday tasks, while contrast sensitivity shows promising correlations. Such findings suggest that age-associated changes in contrast sensitivity could account for some portion of age-associated changes in the perception of certain real-world targets. The details of these results aside, it is instructive to consider why such attempts have succeeded or failed. Conditions under which acuity is typically assessed differ in important ways from the conditions under which perception must operate everyday. For one thing, both the contrast (better than 90 per cent) and illumination (comfortably photopic) of a typical acuity chart are high; in everyday life, contrast and illumination levels vary widely. For example, contrast of the typical real-life target is probably low to moderate, i.e., 20 to 30 per cent.

Additionally, targets on an acuity chart are isolated and static; in everyday life, objects important for vision are often aggregated and/or dynamic. Vision research confirms the dangers of extrapolating from conditions of high-contrast, well-illuminated, well-spaced, static targets to real-life targets with quite different characteristics. Among older people, in particular, low-contrast, poorly illuminated, aggregated or dynamic targets place older observers at a gross disadvantage (Adams *et al.* 1988).

COLOUR VISION

Advancing age brings a diminished ability to discriminate subtle differences in hue in the blue–green range (Knoblauch *et al.* 1987). This acquired colour deficit, particularly noticeable under conditions of reduced illumination, resembles the genetically determined tritan deficit that results from abnormally low sensitivity of the eye's short-wavelength system. In ageing, though, the functionally equivalent change has a different origin, and is apparently connected to changes in the density of ocular media and in part to changes in receptor sensitivity. Preliminary results suggest that the changes in colour vision are characteristic of a quite early stage of age-associated maculopathy, a stage earlier than changes in visual acuity (Applegate *et al.* 1987).

Recently, it has been suggested that some subtle changes in colour vision result from sensitivity changes in post-receptor processes (Werner *et al.* 1990). These changes include shifts in the spectral loci of unique hues, wavelengths that produce sensations of elemental blue, green, yellow, and red. The significance of such effects for everyday colour vision remains to be demonstrated.

PERCEPTION OF DYNAMIC DISPLAYS

Perhaps the most prominent functional decline with age is the visual system's ability to detect temporal change. According to Kline (1987) 'temporally contiguous visual stimuli that would be seen as separate by young observers are often seen as fused or "smeared" by older persons'. Kline and Schieber (1982) suggest that age-associated reduction in temporal resolution manifests itself in three ways: a reduced value for critical flicker fusion, alterations in the temporal extent of visual masking, and an increased duration of visual after-images. Other causes, not fully understood, are probably responsible for the well-documented decline in spatial acuity assessed with moving targets. This ability, dynamic visual acuity, may play an important part in everyday vision, particularly in the perception of moving targets.

A moving target, such as a car, is a complex perceptual object. Depending upon the circumstances, observers may be called upon to judge the moving object's spatial structure, speed, and direction, as well as other attributes. Little is currently known about possible age-associated changes in the principal responses to a moving object: judgments of speed or direction. The single study of speed perception as a function of age (Brown and Bowman 1987) showed that speed discrimination was not impaired with age. Speed discrimination thresholds measured at three different eccentricities, 0, 4 and 32°, were virtually identical for groups of young and old observers (mean ages of 21 and 67 years, respectively). The extent to which one may generalize from these results may be limited by the use of just one target size and contrast, and by the use of only one standard speed. Equally limited was the only one study that has compared direction discrimination in older and younger observers (Ball and Sekuler 1986). Testing with just a single target speed (10°/s), thresholds for discriminating small differences in target direction were measured relative to several different standard directions, e.g. upward, leftward, etc. The direction discrimination thresholds of older observers were as much as double those of younger observers, a result not explicable in terms of optical differences between the groups.

Practice can improve many perceptual abilities, and this could be of great significance as we grow older. Ball and Sekuler's study is interesting mainly for examining the effect of practice on performance. With repeated testing, direction discrimination improved in both older and younger observers, and at approximately the same rate. With practice, the older observers achieved levels of discrimination that were equivalent to those shown by the younger observers at the start of the study. Also, the older observers retained this gain over a 10-week rest period with no discernible loss and in the absence of further practice. Although Ball and Sekuler could not identify the mechanisms behind the measured improvement in older observers' performance, their results demonstrated clearly that, at least for some tasks, older observers might profit from perceptual training. Future research should determine whether this sort of training on a very simple task generalizes to more complex, everyday 'uses' of directional information.

A MODEL EVERYDAY TASK: DRIVING

For a variety of reasons, the connection between ageing and driving seems worth dwelling upon. First, driving, along with reading, is the most complicated, visually dominated task that most people perform. Second, driving is clearly an arena in which visual information must be co-ordinated with other information, including memories and cognitions. Third, in many industrialized countries, the self-esteem, mobility, and independence of older people depends on driving. This last fact becomes especially noteworthy because diminished vision plays a major role when older drivers decide to abandon their privilege of holding a driver's licence (Kosnik *et al.* 1990).

One study of driving and ageing drew upon 348 participants in the Baltimore Longitudinal Study of Ageing (Kline *et al.* 1991). The study used a questionnaire that collected information about a respondent's visual difficulties under everyday conditions, as well as information about the difficulty, if any, with each of 18 different visual aspects of driving. Examples included problems with oncoming headlights, seeing the instrument panel at night, and judging speed. Eight visual aspects of driving proved to be age-associated. These aspects included reading a street-sign rapidly, seeing past dirt or rain on the windscreen, reading dim instrument panels, being bothered by windscreen glare, and judging one's own speed. They also include difficulty with merging traffic, a sense that other vehicles are moving too quickly, and the unexpected appearance of vehicles in the periphery. The most striking feature of these data was the strong congruence between many problems reported by older drivers and the types of automobile accidents most likely to involve older drivers.

In an extraordinarily ambitious investigation of ageing and driving, Owsley *et al.* (1991) examined how accident frequency in older drivers related to variables including ocular health status, visual sensory function, visual attention (as measured by the useful field of vision), and mental status. The participants (average age, 70 years) were 53 drivers for whom official driving records and accident reports were available. The size of the useful field of vision and accident frequency were significantly correlated, as were mental status and accident frequency. To illustrate how number of accidents related to the useful field of vision, Owsley *et al.* compared the accident rates of the 27 participants who had the largest useful fields of vision against the accident rates of the 26 participants with smallest useful fields. Individuals with the more constricted useful fields suffered more than four times the number of accidents. The relationship between visual and driving abilities becomes even more dramatic when a more detailed analysis of accidents is made, dividing them into types such as improper backing up or collisions at intersections. The latter, the most common type of accident in the sample, resulted from the driver's failure to yield the right of way or to notice another vehicle in the intersection. In intersections, size of peripheral visual field or awareness of other vehicles would appear to be crucial. So it is no surprise that individuals with the smallest useful fields of vision were involved in intersection accidents almost 16 times more often than were individuals with the largest useful fields of vision.

LESSONS AND CAUTIONS

Throughout, this chapter has played a number of repeating themes. In an effort to highlight those themes, and in order to provide a stronger framework for future research, it is worth attempting to restate the three key themes here. First, vision is not a single, unitary function, but a collection of separable ones. This point has been made repeatedly for more than a century (see Sekuler and Owsley 1982). Recently, the existence of multiple maps in the visual cortex has been suggested as one possible physiological substrate for some of these separable functions (deYoe and van Essen 1988). Whatever the substrate, though, the diversity of visual function in everyday life requires the co-operation of several different visual subsystems. Secondly, as we must respect the diversity of visual functions, we must understand and appreciate the diversity of ageing individuals. After all, 'people do not lose their sight in age cohorts, marching lock step together in obedience to some chronological imperative; people lose their sight as individuals, each in her or his own way and according to a highly individual timetable' (Sekuler 1991). Thirdly, vision, except under highly controlled artificial conditions, depends not just upon the afferent pathway, but also upon contributions from elsewhere, including systems related to memory and cognition. William James (1892) put it aptly when he wrote: 'Whilst part of what we perceive comes through our senses from the object before us, another part (and it may be the larger part) always comes ... out of our own head.' James' point is especially important for older persons. Clearly, when there is diminished access to sensory input, supplementation by cognition becomes increasingly important. Therefore, when studying vision in elderly people, and in others with reduced sensory function, one must pay very careful attention to concomitant changes in cognition.

References

Adams, A.J., Wang, L.S., Wong, L., and Gould, B. (1988). Visual acuity changes with age: some new perspectives. *American Journal of Optometry and Physiological Optics*, **65**, 403–6.

Akutsu, H., Legge, G.E., Ross, J.A., and Schuebel, K.J. (1991). Psychophysics of reading. X. Effects of age-related changes in vision. *Journal of Gerontology: Psychological Sciences* (in press).

Applegate, R.A., Adams, A.J., Cavender, J.C., and Zisman, F. (1987). Early color vision changes in age-related maculopathy. *Applied Optics*, **26**, 1458–62.

Ball, K. and Sekuler, R. (1986). Improving visual perception in older observers. *Journal of Gerontology*, **41**, 176–82.

Ball, K., Owsley, C., and Beard, B. (1990). Clinical visual perimetry underestimates peripheral field problems in older adults. *Clinical Vision Science*, **5**, 113–25.

Braak, H. and Braak, E. (1988). Morphology of the human isocortex in young and aged individuals: qualitative and quantitative findings. *Interdisciplinary Topics in Gerontology*, **25**, 1–15.

Brown, B. and Bowman, K.J. (1987). Sensitivity to changes in size and velocity in young and elderly observers. *Perception*, **16**, 41–7.

Cerella, J. (1985). Age-related decline in extrafoveal letter perception. *Journal of Gerontology*, **40**, 727–36.

Collins, M. (1989). The onset of prolonged glare recovery with age. *Ophthalmic and Physiological Optics*, **9**, 368–71.

Crassini, B., Brown, B., and Bowman, K. (1988). Age-related changes in contrast sensitivity in central and peripheral retina. *Perception*, **17**, 315–32.

deYoe, E.A. and van Essen, D.C. (1988). Concurrent processing streams in monkey visual cortex. *Trends in Neurosciences*, **11**, 219–26.

Drance, S.M., Berry, V., and Hughes, A. (1967). Studies of the effects of age on the central and peripheral isopters of the visual field in normal subjects. *American Journal of Ophthalmology*, **63**, 1667–72.

Felson, D.T., Anderson, J.J., Hannan, M.T., Milton, R.C., Wilson, P.W.F., and Kiel, D.P. (1989). Impaired vision and hip fracture. *Journal of the American Geriatrics Society*, **37**, 495–500.

Geisler, W.S. (1989). Sequential ideal-observer analysis of visual discrimination. *Psychological Review*, **96**, 267–314.

Hakkinen, L. (1984). Vision in the elderly and its use in the social environment. *Scandinavian Journal of Social Medicine*, **35**, 5–60.

Higgins, K.E., Jaffe, M.J., Caruso, R.C., and deMonasterio, F.M. (1988). Spatial contrast sensitivity: effects of age, test-retest, and psychophysical method. *Journal of the Optical Society of America A*, **5**, 2173–80.

James, W. (1892). *Psychology: a briefer course*. Holt, New York.

Johnson, C. and Keltner, J. (1986). Incidence of visual field loss in 20,000 eyes and its relationship to driving performance. *Archives of Ophthalmology*, **101**, 371–5.

Johnson, C.A., Adams, A.J., and Lewis, R.A. (1989). Evidence for a neural basis of age-related visual field loss in normal observers. *Investigative Ophthalmology and Visual Science*, **30**, 2056–64.

Keltner, J.L. and Johnson, C.A. (1987). Visual function, driving safety, and the elderly. *Ophthalmology*, **94**, 1180–8.

Kline, D.A. (1987). Ageing and the spatiotemporal discrimination performance of the visual system. *Eye*, **1**, 323–9.

Kline, D.A. and Schieber, F. (1982). Visual persistence and temporal resolution. In *Aging and visual function* (ed. R. Sekuler, D. Kline, and K. Dismukes), pp. 231–44. Liss, New York.

Kline, D.A., Kline, T.J.B., Fozard, J.L., Kosnik, W., Schieber, F., and Sekuler, R. (1991). Vision, aging and driving: the problems of older drivers. *Journal of Gerontology: Psychological Sciences* (in press).

Knoblauch, K. *et al.* (1987). Age and illuminance effects in the Farnsworth-Munsell 100-hue test. *Applied Optics*, **26**,1441–8.

Kosnik, W., Winslow, L., Kline, D., Rasinski, K., and Sekuler, R. (1988). Age-related visual changes in daily life throughout adulthood. *Journal of Gerontology*, **43**, 63–70.

Kosnik, W., Sekuler, R., and Kline, D. (1990). Self-reported visual problems of older current and former drivers. *Human Factors*, **32**, 597–608.

Marshall, J. (1987). The ageing retina: physiology or pathology. *Eye*, **1**, 282–95.

Morgan, M. (1988). Vision though my ageing eyes. *Journal of the American Optometric Association*, **59**, 278–80.

NRC (National Research Council) (1987). *Work, ageing, and vision: report of a conference*. NRC, Washington DC.

Owsley, C.J. and Sekuler, R. (1984). Visual manifestations of biological ageing. *Experimental Ageing Research*, **9**, 253–5.

Owsley, C. and Sloane, M. E. (1987). Contrast sensitivity, acuity and the perception of 'real-world' targets. *British Journal of Ophthalmology*, **71**, 791–6.

Owsley, C. and Sloane, M.E. (1990). Vision and Aging. In *Handbook of neuropsychology*, Vol. 4 (ed. F. Boller and J. Grafman). Elsevier, Amsterdam.

Owsley, C., Sekuler, R., and Boldt, C. (1981). Aging and low contrast vision : face perception. *Investigative Ophthalmology and Vision Science*, **21**, 362–5.

Owsley, C.J., Sekuler, R., and Siemsen, D. (1983). Contrast sensitivity throughout adulthood. *Vision Research*, **23**, 689–99.

Owsley, C., Ball, K., Sloane, M.E., Roenker, D.L., and Bruni, J.R. (1991). Visual/cognitive correlates of vehicle accidents in older drivers. *Psychology and Aging* (in press).

Pitts, D.G. (1982). The effects of ageing on selected visual functions: dark adaptation, visual acuity, stereopsis and brightness contrast. In *Aging and visual function* (ed. R. Sekuler, D. Kline, and K. Dismukes). Liss, New York.

Post, R.B. and Leibowitz, H.W. (1980). Independence of radial localization from refractive error. *Journal of the Optical Society of America*, **70**, 1377–8.

Scialfa, C.T. and Kline, D.W. (1988). Effects of noise type and retinal eccentricity on age differences in identification and localization. *Journal of Gerontology: Psychological Sciences*, **43**, 91–9.

Sekuler, R. (1991). Why vision changes with age. *Geriatrics*, **46**, 96–100.

Sekuler, R. and Owsley, C.J. (1982). The spatial vision of older humans. In *Ageing and visual function* (ed. R. Sekuler, D. Kline, and K. Dismukes), pp. 185–202. Liss, New York.

Sekuler, R., and Ball, K. (1986). Visual localization: age and practice. *Journal of the Optical Society of America A*, **3**, 864–7.

Sekuler, R., Kline, D., Dismukes, K., and Adams, A.J. (1983). Some research needs in ageing and vision perception. *Vision Research*, **23**, 213–216.

Sloane, M.E., Owsley, C., and Jackson, C.A. (1988). Aging and luminance-adaptation effects on spatial contrast sensitivity. *Journal of the Optical Society of America A*, **5**, 2181–90.
Werner, J.S., Peterzell, D.H., and Scheetz, A.J. (1990). Light, vision, and ageing. *Optometry and Vision Science*, **67**, 214–29.
Wilkinson, R.T. and Allison, S. (1989). Age and simple reaction time: decade differences for 5,325 subjects. *Journal of Gerontology: Psychological Sciences*, **44**, 29–35.

18.14 Disorders of hearing

A. JULIANNA GULYA

INTRODUCTION

Hearing loss of ageing (presbycusis) afflicts 10 million elderly citizens in the United States (one-third of those 65 to 74 years of age and half of those 75 to 79 years), and can have a profound impact upon their lives (Vital and Health Statistics 1985). The embarrassment engendered by misunderstanding others encourages social withdrawal. Poor hearing breeds the suspicion that others are mumbling, or worse yet, a paranoia that others are conspiring to keep one from overhearing conversations. The victim enters a cycle of suspicion, isolation, loneliness, and depression—certainly an unhappy prospect for one's twilight years.

This chapter, using the foundation of knowledge provided in the *Oxford Textbook of Medicine*, specifically examines the histopathology of presbycusis as well as its varied clinical presentations, and concludes with suggestions for the evaluation and management of the individual patient.

HISTOPATHOLOGY

Presbycusis is the term used to describe the clinical manifestations of ageing of the auditory system. For a discussion of the biological mechanisms thought to be involved, see Chapters 2.1 and 2.2. Many factors combine in individually determined permutations to result in the bilaterally symmetrical loss of hearing in elderly people; in the examination of human material, it is exceedingly difficult to separate those histopathological changes of the auditory system associated with hearing loss that are solely attributable to intrinsic ageing from those due to noise exposure, toxins, disease processes, genetic influence, diet, vascular disorders, and climate, to mention just a few extrinsic factors. Men generally suffer a greater degree of hearing loss with ageing than do women (Matkin and Hodgson 1982).

Changes in the external and middle ear do not appear to contribute significantly to the sensorineural hearing loss of ageing (Etholm and Belal 1974). Thus more attention has been focused on the inner ear.

Those cochlear structures that have been particularly scrutinized for age-associated changes are the organ of Corti, including its sensory cells, the first-order neurones, and their afferent dendrites, the stria vascularis, and the basilar membrane (Fig. 1). Clinical correlation of audiometric configuration with specific patterns of histopathological alteration has been relatively successful, and Schuknecht (1974) has defined four 'types' of presbycusis: sensory, neural, metabolic, and mechanical (or cochlear conductive).

Sensory presbycusis

This is defined audiometrically by a bilaterally symmetrical, abruptly dropping, pure-tone threshold curve with excellent speech discrimination scores (Fig. 2). Its hearing loss is generally noticed by those in middle age, although the histopathological correlative, hair-cell loss in the basal cochlea, may begin as early as infancy (Fig. 3).

The age-associated degeneration of the organ of Corti progresses very slowly, only affecting a few millimetres of the basal cochlea even in the very aged (Schuknecht 1974). The outer hair cells, especially those in the third row, are most severely affected, but the degeneration can encompass the first row of outer hair cells, extend to the inner hair cells, and eventually culminate in the disappearance of the entire organ of Corti. A secondary neuronal degeneration occurs as well.

Ultrastructural examination suggests that lipofuscin accumulation and the formation of giant cilia precede cellular loss (Soucek *et al.* 1987). The amount of intracellular lipofuscin found in the cochlear hair cells, Hensen's cells, Claudius' cells, the pillar cells, and spiral ganglion cells has been positively correlated with increased individual age (Ishii *et al.* 1967), tendency to autolysis (Gleeson and Felix 1987), and the extent of hearing loss (Raafat *et al.* 1987).

Neural presbycusis

The age of onset of neural presbycusis appears to be determined primarily by genetic factors. The condition audiometrically is a loss of speech discrimination out of proportion to the loss of pure-tone thresholds (Schuknecht 1974) (Fig. 4). Speech discrimination, in comparison to pure-tone perception, apparently requires a greater proportion of surviving neurones to maintain a higher level of signal integration and transmission; this observation accounts for the phenomenon

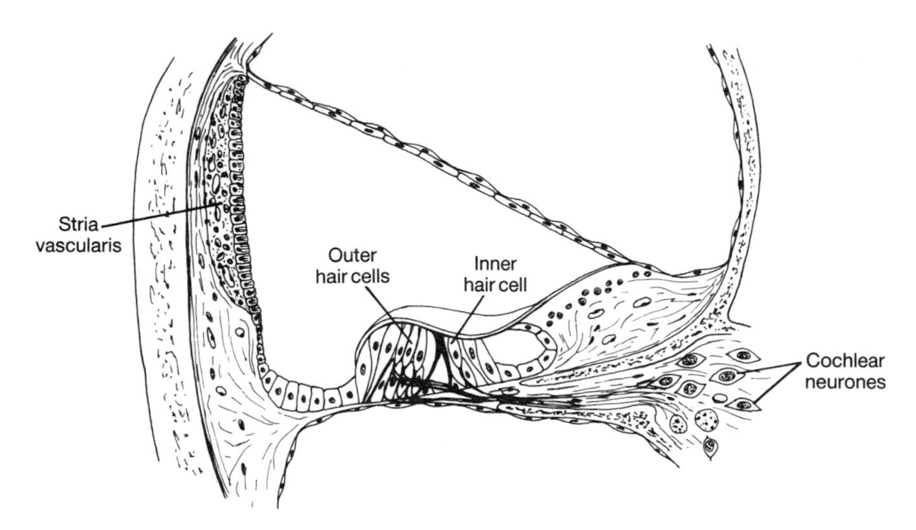

Fig. 1 Normal cochlear structures.

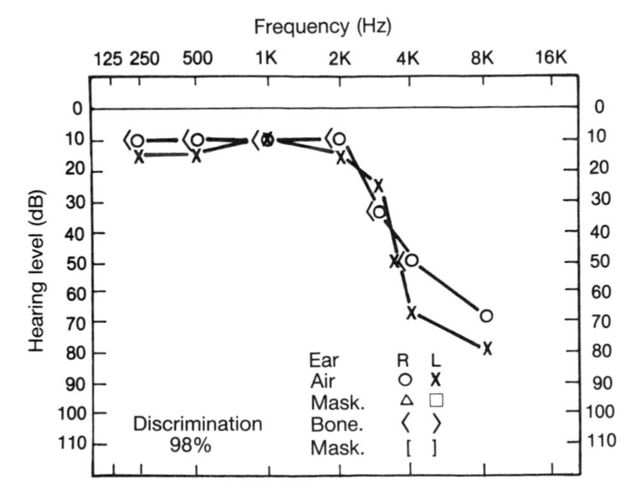

Fig. 2 Sensory presbycusis, defined audiometrically.

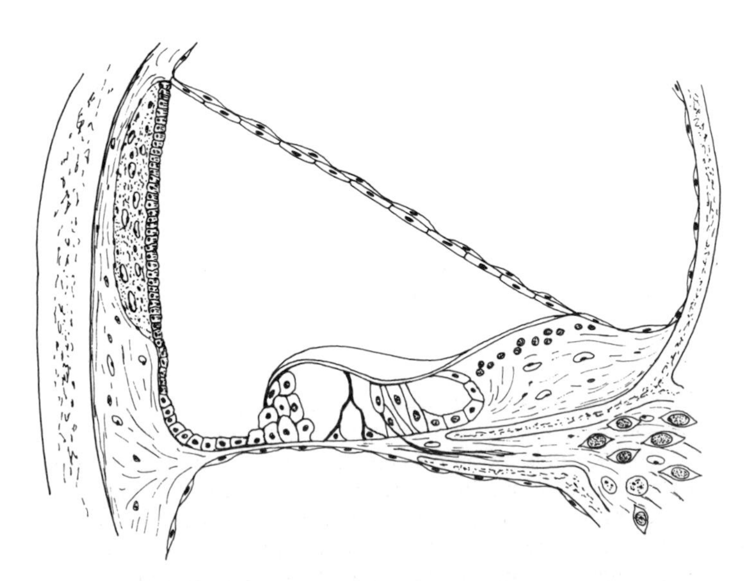

Fig. 3 Hair-cell loss in the basal cochlea.

of phonemic regression—loss of speech discrimination while pure-tone thresholds are relatively maintained (Gaeth 1948). Evidence of degenerative changes in the central nervous system, such as memory loss, intellectual decline, and motor inco-ordination, may be seen in association with particularly rapidly progressive neural presbycusis (Schuknecht 1974).

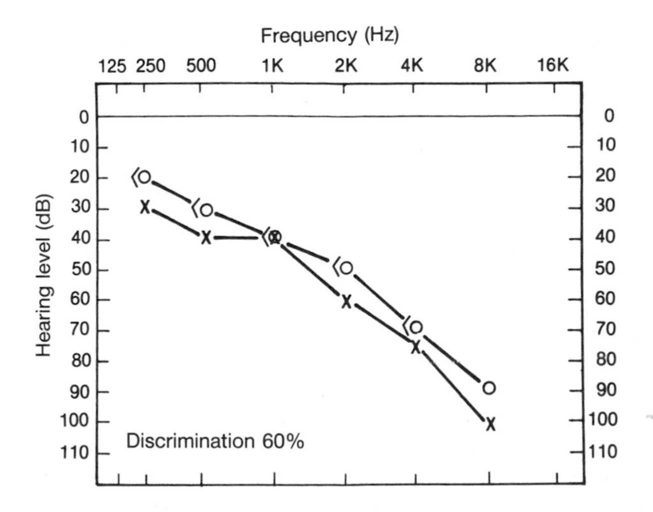

Fig. 4 Neural presbycusis, defined audiometrically.

The histopathological correlatives of neural presbycusis (Fig. 5) lie in the depletion of first-order neurones and their fibres out of proportion to the loss of the organ of Corti, generally most severe in the basal turn of the cochlea (Schuknecht 1974).

Ultrastructural studies have shown lipofuscin accumulation in the ganglion cells, disorganization of the myelin sheath of their dendrites and axons, and loss of synapses at the hair-cell bases (Nadol 1979). Presumably, the alterations in the myelin sheath are sufficient to disrupt normal saltatory conductive mechanisms, with consequent delay and energy loss as transmission of impulses occurs through the cell bodies.

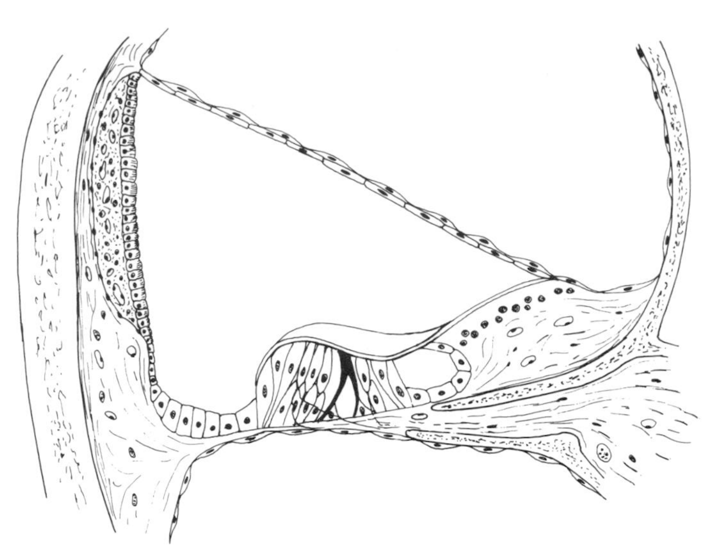

Fig. 5 Depletion of first-order neurones in neural presbycusis.

Metabolic presbycusis

This has its onset in the third to sixth decades of life, is slowly progressive, and appears to have a familial tendency (Schuknecht 1974). The audiogram typically has a 'flat' configuration, with speech discrimination scores remaining normal until the pure tone thresholds exceed 50 dB (Fig. 6).

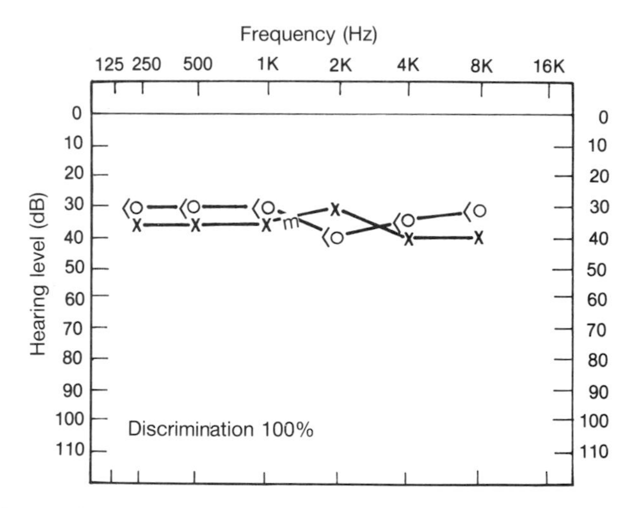

Fig. 6 Metabolic presbycusis, defined audiometrically.

Histopathological examination (Fig. 7) has correlated atrophy of the stria vascularis—the electrophysiological 'generator' of the cochlea—with this type of presbycusis (Schuknecht 1974). More recently, computer-aided morphometric techniques have been able to establish a statistically significant relationship between the degree of strial atrophy and the extent of hearing loss (Pauler *et al.* 1988). The degeneration of the stria has been variably attributed to vascular changes in the cochlea (Johnsson and Hawkins 1977) or to a genetically determined tendency for early cellular degeneration (Pauler *et al.* 1988). Although the marginal cells (those cells that face the fluid space of the cochlear duct) are most markedly affected, the entire stria may be reduced to a mere layer of basal cells (Kimura and Schuknecht 1970).

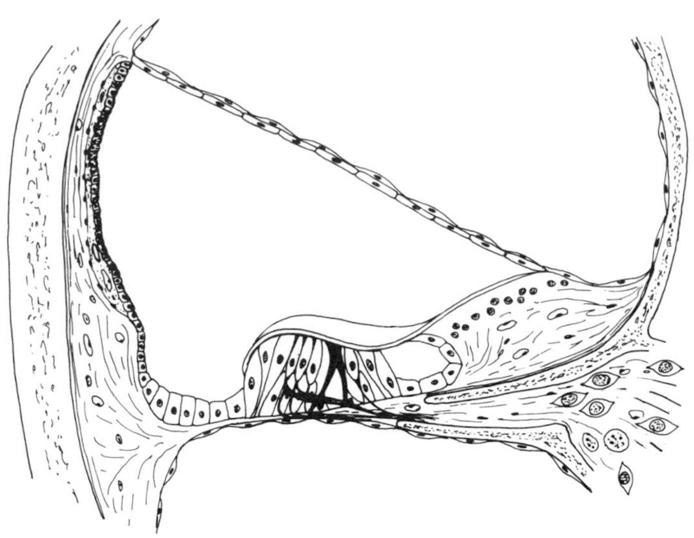

Fig. 7 Atrophy of the stria vascularis in metabolic presbycusis.

Cochlear conductive presbycusis

Cochlear conductive (mechanical) presbycusis describes a downward-sloping threshold curve (Fig. 8) associated with speech discrimination scores that are inversely proportional to the steepness of the slope of the curve (Schuknecht 1974).

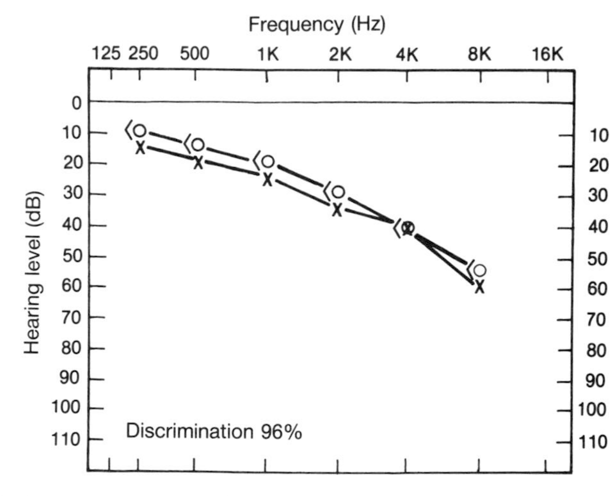

Fig. 8 Cochlear conductive presbycusis defined audiometrically.

Typically, histopathological examination of the cochlea is unable to reveal any alteration in the hair cells, neurones, or stria vascularis that could account for the hearing loss. It has been suggested that some alteration in the motion mechanics of the cochlea, centring particularly on the basilar membrane, underlies this type of presbycusis (Schuknecht 1974). Some support for this explanation comes from the light microscopical demonstration of hyalinization (Crowe *et al.* 1934), calcification (Mayer 1919–1920), and even a lipidosis of the basilar membrane (Nomura 1970). More convincingly, a marked thickening of the basilar membrane in the basal 10 mm of the cochlea has been found by electron microscopy in a patient with audiometric findings typical for cochlear conductive presbycusis (Nadol 1979).

Other pathological changes

Although 'pure forms of the four types of presbycusis are described above, various combinations can occur and result in differing audiometric patterns (Schuknecht 1974).

Alterations in the central auditory pathways have also long been suspected as being affected in ageing (Schuknecht 1974), but it has been difficult to obtain conclusive evidence. Findings which suggest that there is a loss of cells from the superior temporal gyrus (Brody 1955), ventral cochlear nucleus (Kirikae *et al.* 1964), dorsal cochlear nucleus (Hansen and Reske-Neilsen 1965), medial geniculate body (Kirikae *et al.* 1964), superior olivary nucleus (Kirikae *et al.* 1964), and inferior colliculus (Hansen and Reske-Neilsen 1965) with ageing have been contradicted (Konigsmark and Murphy 1972). Nonetheless, there is evidence that accumulation of lipofuscin and degeneration of myelin occur in the central auditory structures (Hansen and Reske-Neilsen 1965).

Correlation of specific functional deficits with specific central structural alterations is extremely difficult, particularly as peripheral auditory dysfunction/alteration nearly uniformly presents as an important confounding variable.

In a similar fashion, although vascular alterations, including atrophy with devascularization and an increased incidence of periodic acid—Schiff-positive thickening of capillary walls have been found in cochleae from elderly subjects (Jorgensen 1961), correlating such changes to degenerative changes in cochlear duct structures and hearing loss remains a formidable task.

The influence that a theoretical depletion of neural transmitters could have on age-associated hearing loss also warrants further investigation.

EVALUATION

Presbycusis may begin insidiously; usually the higher frequencies are involved first, but with progression, the frequencies of the upper range of human speech are affected, interfering particularly with consonant discrimination. Patients complain especially of difficulty with understanding high-pitched women's voices, children's voices, and conversations in crowded environments. Key factors to elicit in the history are the time course of the hearing loss, its progression, and any associated symptoms such as tinnitus, aural fullness, or fluctuation. Questioning should uncover any history of exposure to excessive noise or ototoxic drugs, as well as any family history of hearing loss or past history of ear infections or surgery. Asymmetrical, unilateral, sudden, or fluctuating hearing losses are not consistent with presbycusis and demand full evaluation with referral to an otolaryngologist/otologist.

The examination of any complaint of hearing loss comprises a pneumotoscopic examination of the ears with initial assessment of the hearing deficit by the Weber, Rinne, and whisper-threshold tests. Accumulated cerumen, which may interfere with hearing and/or auditory testing, is removed.

The complete audiogram, including pure-tone (air and bone conduction) threshold testing, determination of speech discrimination, and tympanometry is an integral part of evaluation and management. As discussed above, presbycusis is bilaterally symmetrical; hence, any significant asymmetry, as well as any suggestion of retrocochlear (e.g. acoustic neuroma) disease, demands further testing, including auditory brain-stem response (ABR) testing, and possibly a structural evaluation of the brain by computed tomography or magnetic resonance imaging. Otolaryngological/neurotological consultation may be indicated in such cases.

MANAGEMENT

The evaluation and fitting of hearing aids, speech reading, auditory training, and assistive listening devices can be used in various combinations to help the hearing-impaired person.

Hearing aids (Ruben and Kruger 1983)

Hearing aids are amplification devices that vary in size, power, and sophistication, yet have in common certain basic elements. A microphone is used to detect the incoming sound signals and convert them into electrical energy. An amplifier then boosts the energy of the signal by a factor, the 'gain' of the particular device. The output from the aid is then channelled to the ear canal by the ear-mould part of the device.

Individuals with either conductive or sensorineural hearing losses may benefit from appropriate evaluation and fitting of a hearing aid. In general, those hearing losses with relatively good discrimination scores are expected to perform better with hearing aids than those with poor speech discrimination; however, some individuals with poor discrimination do better than expected with a hearing aid, indicating that only a trial of properly fitted hearing aids can determine their potential benefit in the individual case.

Hearing aids range in size from tiny, 'in-the-canal' devices to the bulky 'body' aids. In general, the smaller the device, the less its power and the poorer its sound fidelity.

A modification of the mould called 'venting' is used for relatively selective amplification of high frequencies for those individuals whose low-frequency thresholds are less affected.

The contralateral routing of signals (CROS) aid is designed for individuals with a unilateral, profound hearing loss, but normal hearing in the remaining ear. A microphone worn on the 'bad' side transmits, by cord or FM signal (cordless), incoming signals to the normal, contralateral ear. The BICROS aid is a similar device, but appropriately amplifies the crossed signal to accommodate further for a hearing impairment in the 'good' ear.

Because of the limitations of fine-motor dexterity in some elderly people, it may be wise to recommend either a body aid with its more easily manipulated controls, or special 'geriatric' moulds. More recently, 'credit card' type, remote controls have been developed to simplify setting adjustments.

There are other modifications for hearing aids. Use of a telephone with the aid is facilitated by the 'T' switch. Automatic gain control (**AGC**) circuitry appropriately modifies incoming soft or loud stimuli. The AGC circuitry is especially helpful for the patient with a 'recruiting' ear—that is, an ear in which there is an abnormally rapid rise in the perceived loudness of, and discomfort associated with, an incoming sound signal.

Automatic and multiple signal processing describes modifications of the circuitry that are geared to improving the signal-to-noise ratio—i.e. improving human speech perception in the presence of competing environmental noise.

The oscillator of a bone-conduction aid is placed in direct contact with the skull, most commonly held at the mastoid by a head-set. Patients with uncontrollable otorrhoea, those with canal atresia, or other conditions that preclude the use of an air-conduction aid may gain significant benefit from a bone-conduction aid.

A new modification of the bone-conduction idea is that of an implantable bone-conduction device. A rare-earth magnet is implanted in the bone of the skull in a minor surgical procedure, and vibrates, with transmission of the vibration to the inner ear by means of a coupling with an external component that resembles an ordinary hearing aid. Currently, only candidates with a conductive hearing loss meeting certain criteria are eligible for this device, and evaluation by a qualified otologist/otolaryngologist is necessary.

The evaluation for a hearing aid is best conducted by a qualified audiologist who carefully matches the individual's frequency thresholds and speech discrimination scores to the recommended aid. In addition to the appropriateness of the selection and settings of the aid, other factors, such as motivation, the need to communicate in the patient's particular environment, and realistic expectations, all combine to determine whether the individual is a good candidate. Patients with recruiting ears tolerate amplification within a relatively narrow range before reaching uncomfortable levels of loudness, while others, with discrimination reduced out of proportion to pure-tone averages, may not find great benefit from a hearing aid.

Training in the proper cleaning, maintenance, and operation of the device is a necessary part of the evaluation and fitting. Often, it is helpful to have an interested family member participate in such training to help prompt the failing short-term memory of the senior citizen.

The cochlear implant

The cochlear implant (Corso 1985) is a step forward in the rehabilitation of bilaterally, profoundly deaf persons who derive no benefit from even the most powerful of hearing aids. The implanted electrode array conveys electrical impulses from an external receiver across the skin and directly stimulates the remaining neural population of the cochlea. Somewhat reminiscent of a hearing aid, signals are perceived by a microphone and specifically altered by a signal processor before being sent down into the cochlear array.

Candidacy is determined by careful audiological, radiological, and surgical evaluation; the implantation of the device by mastoidectomy requires a brief time in hospital. The cost of the device itself may be prohibitive. Nonetheless, despite the fact that only a select few can be helped by this device, the benefit perceived by trained recipients is not insignificant, and can help by diminishing the sense of isolation, improving speech-reading ability, and improving voice modulation.

Vibrotactile devices

Vibrotactile devices are alternative devices to help profoundly deaf persons perceive sound. Vibrators placed on the wrist, sternum, or around the waist transform environmental sound and speech into skin vibrations. After appropriate training, patients can use these devices to localize and identify sounds, in addition to improving their communication skills.

Speech reading

Speech reading (Senturia *et al.* 1983) is the term used today to describe the perception of linguistic information by observation of the speaker's facial expressions and gestures as well as lip movements. Speech reading may be used in conjunction with, or as an alternative to, a hearing aid. Poor visual acuity as well as failing short-term memory may impede the acquisition of this skill by some elderly people.

Auditory training

Auditory training (Senturia *et al.* 1983) attempts to educate the patient to discriminate differing sound stimuli, especially speech sounds, progressing to finer and finer distinctions. The awareness of subtle auditory clues thus developed is often used in conjunction with appropriate amplification.

As with hearing-aid care, the participation of an interested family member provides support for the elderly patient, as well as a memory aid.

Assistive listening devices

Assistive listening devices non-specifically help hearing-impaired persons in special circumstances, such as hearing in an auditorium or church, using the television or radio with people of normal hearing, or group conversations.

Many public facilities, such as churches and concert halls, have incorporated an assistive listening device, either by 'looping' the area of interest, or with an infrared transmission system. Hearing-impaired individuals can then 'tune their hearing aid in' to the looping system, or borrow the special receiver for the infrared system, to enjoy the presented programmes.

Television and/or radio sound signals can be amplified directly into an individual's hearing aid, while others present can continue to enjoy normal listening levels. Telecaptioning of television programmes is becoming more common, requiring a special decoder.

For conversations in small groups, small, easily portable systems consisting of a microphone, amplifier, and earphone are available at relatively small cost.

Alerting devices, either boosting the signal tone or using alternative stimuli such as flashing lights/vibration, are available to allow the perception of doorbells, telephone rings, etc.

GENERAL GUIDELINES

Communicating with any hearing-impaired individual can be a frustrating experience for both speaker and listener. A few simple guidelines can help alleviate some of the frustration (Matkin and Hodgson 1982). First, one must be sure that one has captured the listener's attention and that the face of the speaker is well illuminated. The elimination or reduction to an absolute minimum of any competing sound stimulus is essential. Shouting is not necessary; rather, one should speak slowly and clearly, favouring the better ear and maintaining an optimal distance of about 1 m. If a statement is not comprehended, try an alternative wording, rather than repeating an identical phrase.

CONCLUSION

Although hearing impairment, to a certain extent, is inevitable for many elderly people, there is much that can be done to reduce the degree of handicap it produces in the individual. It is the responsibility of every health care professional who works with elderly people to have a basic understanding of the psychosocial problems that hearing impairment presents and of how to initiate appropriate evaluation and treatment. By keeping elderly people in communication with their world, one would hope to make the later years of life less lonely and frustrating.

References

Brody, H. (1955). Organization of the cerebral cortex, III. A study of aging in the human cerebral cortex. *Journal of Comparative Neurology*, **102**, 511–56.

Corso, J.F. (1985). Communication, presbycusis, and technological aids. In *The aging brain: communication in the elderly* (ed. H.K. Ulatowska), pp. 33–51. College Hill Press, San Diego.

Crowe, S.J., Guild, S.T., and Polvogt, L.M. (1934). Observations on the pathology of high tone deafness. *Bulletin of the Johns Hopkins Hospital*, **54**, 315–79.

Etholm, B. and Belal, A., Jr. (1974). Senile changes in the middle ear joints. *Annals Otology, Rhinology and Laryngology*, **83**, 49–54.

Gaeth, J. (1948). Cited in Schuknecht, H.F. (1974).

Gleeson, M. and Felix, H. (1987). A comparative study of the effect of age on the human cochlear and vestibular neuroepithelia. *Acta Otolaryngologica* (Stockholm), **436** (Suppl.) 103–9.

Hansen, C.C. and Reske-Neilsen, E. (1965). Pathological studies in presbycusis: cochlear and central findings in 12 aged patients. *Archives of Otolaryngology*, **82**, 115–32.

Ishii, T., Murakami, Y., Kimura, R.S., and Balogh, K., Jr. (1967). Electron microscopic and histochemical identification of lipofuscin in the human inner ear. *Acta Otolaryngologica* (Stockholm), **64**, 17–29.

Johnsson, L.-G. and Hawkins, J.E., Jr. (1977). Age-related degeneration of the inner ear. In *Special senses in aging: a current biological assessment* (ed. S.S. Han and D.H. Coons), pp. 119–35. University of Michigan Press.

Jorgensen, M.B. (1961). Changes of aging in the inner ear. *Archives of Otolaryngology*, **74**, 164–70.

Kimura, R.S. and Schuknecht, H.F. (1970). The ultrastructure of the human stria vascularis. II. *Acta Otolargyngologica* (Stockholm), **70**, 301–18.

Kirikae, I., Sato, T., and Shitara, T. (1964). A study of hearing in advanced age. *Laryngoscope*, **74**, 205–20.

Konigsmark, B.W. and Murphy, E.A. (1972). Volume of the ventral cochlear nucleus in man: its relationship to neuronal population and age. *Journal of Neuropathology and Experimental Neurology*, **31**, 304–16.

Matkin, N.D. and Hodgson, W.R. (1982). Amplification and the elderly patient. *Otolaryngologic Clinics of North America*, **15**, 371–86.

Mayer, P. (1919–1920). Das anatomische Substrat der Altersschwerhorigkeit. *Archives Ohren Nasen Kehlkopfhielkunde*, **105**, 1–13.

Nadol, J.B., Jr. (1979). Electron microscopic findings in presbycusic degeneration of the basal turn of the human cochlea. *Otolaryngology—Head and Neck Surgery*, **87**, 818–36.

Nomura, Y. (1970). Lipidosis of the basilar membrane. *Acta Otolaryngologica* (Stockholm), **69**, 352–7.

Pauler, M., Schuknecht, H.F., and White, J.A. (1988). Atrophy of the stria vascularis as a cause of sensorineural hearing loss. *Laryngoscope*, **98**, 754–9.

Raafat, S.A., Linthicum, F.H., Jr., and Terr, L.I. (1987). Quantitative study of lipofuscin accumulation in ganglion cells of the cochlea. *Association for Research in Otolaryngology Abstracts*, **10**, 205.

Ruben, R.J. and Kruger, B. (1983). Hearing loss in the elderly. In *The neurology of aging* (ed. R. Katzman and R. Terry), pp. 123–47. F. A. Davis, Philadelphia.

Schuknecht, H.F. (1974). *Pathology of the ear*, pp. 388–403. Harvard University Press.

Senturia, B.H., Goldstein, R., and Hersperger, W.S. (1983). Otorhinolaryngologic aspects of geriatric care. In *Care of the geriatric patient in the tradition of E.V. Cowdry* (6th edn). Mosby, St. Louis.

Soucek, S., Michaels, L., and Frohlich, A. (1987). Pathological changes in the organ of Corti in presbyacusis as revealed by microslicing and staining. *Acta Otolaryngologica* (Stockholm), **436** (Suppl.), 93–102.

United States Department of Health and Human Services, Public Health Service, National Center for Health Statistics. (1985). *Vital and Health Statistics*. Series 10, No. 160.

18.15 Gait and mobility

LESLIE I. WOLFSON

Mobility is a basic human function, which is necessary for independence, social and intellectual interaction as well as the fundamental activities of daily living. Loss of mobility, along with the occurrence of falls, is a principal cause of a limited quality of life and increased dependence. Both are therefore important factors that lead to older persons becoming inhabitants of institutions. One-third of older persons living at home report a fall in the previous year (Baker and Harvey 1985), and falls occur considerably more often in long-term care facilities (0.67–2.0 falls per bed per year) (Gryfe *et al.* 1977; Rubenstein *et al.* 1988). This is not surprising in view of the prevalence of dementia and frailty in older people living in institutions, both of which are related to falls. Falls usually produce minor injury, although their high incidence in older persons make them responsible for one half of all accidental injuries in this age group (U.S. Bureau of the Census 1985). The result is 200 000 hip fractures plus numerous other major injuries (e.g. half of all head injuries in 55–65-year-olds) (Cooper *et al.* 1983) as well as 10 000 deaths per year in older Americans.

In addition to sensorimotor functions, mobility is dependent on joint function and overall fitness. Arthritis robs the joints of the legs of the ability to support mobility efficiently; deconditioned, frail older persons become short of breath after walking short distances and their impaired strength lends them only marginal support. Other factors, such as decreased hearing or vision, and orthostatic hypotension or cardiovascular disease producing syncope, are often related to the incidence of falls. Dysfunction of the sensorimotor control mechanism may involve the nervous system at any level, with each locus having characteristic signs and symptoms as well as different pathophysiological mechanisms.

This chapter will first briefly review the effects of age on sensorimotor function, balance, and the brain structures subserving them. Next, it will define the diseases that most frequently impair neuromuscular control, and then review 'idiopathic' dysfunction of gait and balance as well as the possible pathophysiology and strategies for treatment.

THE EFFECTS OF AGE ON SENSORIMOTOR FUNCTION

'Senile' is defined in one dictionary as 'showing characteristics of old age; weak of mind and body'. This definition demonstrates the misconception that age by itself is inevitably associated with a loss of cognitive and motor function. A vast published record now indicates that most cognitive loss occurs because of age-associated disease and that only mild memory changes occur in the absence of these diseases. The section that follows will review age-associated loss of sensorimotor function as well as the changes that occur in gait and balance to determine if falls and impaired mobility are primarily the result of age.

Balance

Maintaining an upright posture during standing and walking requires rapid responses to external forces. The inability to make these corrections may lead to falls. The postural reflexes that underlie these responses are mediated through neural structures and pathways.

Older persons have decrements in both static and dynamic balance when compared with younger subjects. The amplitude of swaying while standing on both feet (static balance) is increased in older people (Sheldon 1960); the time that they can stand on one leg is reduced (Bohannon *et al.* 1984). Many elderly people have electromyographic evidence of abnormal function in leg muscles during experiments with sudden, horizontal translations of a support platform (Woollacott *et al.* 1982): response latencies are prolonged, output amplitude is inconsistent, and the timing between distal and proximal muscles is reversed, particularly in perturbations causing posterior sway. Age-associated differences also occur during unexpected movements of the platform in other planes. Sudden upward or downward rotations of the support surface evoke destabilizing muscle synergies in normal older persons, causing 38 to 74 per cent of them to fall (Nashner 1976; Woollacott *et al.* 1982).

Abnormal vestibulospinal function has been implicated in balance disorders of ageing (Frol'Kis and Bezrukov 1978). The relative contribution of vestibular, visual, and somatosensory inputs to the maintenance of equilibrium can be assessed through the independent rotation of the support platform and a visual enclosure surrounding the subject. Inappropriate sensory (somatosensory and visual) input can be produced by programming the platform and its visual surround to rotate in proportion to the subject's angle of sway (sway referencing) (Mirka *et al.* 1988). When a subject is blindfolded and the support surface is sway-referenced, those in their 60s begin to show excessive amounts of sway (L. M. Nashner and J. Friedman, unpublished work), and 35 to 50 per cent of individuals in their 70s will fall (Woollacott *et al.* 1982; Mirka *et al.* 1988). Therefore, when visual and somatosensory inputs are experimentally eliminated, vestibular control alone may be inadequate to maintain balance. Balancing can be made even more difficult if both the visual surround and the support surface are sway-referenced. This produces false visual feedback that is in conflict with vestibular input. Younger subjects contend easily with this stress. In contrast, healthy older persons frequently fall under these conditions (Woollacott *et al.* 1982; Mirka *et al.* 1988), suggesting that processing delays entailed in resolving sensory conflict may lead to motor responses that arrive too late to maintain upright posture.

Profound deficits of the postural response were found in half of the elderly persons residing in a nursing home who had a history of falling (Wolfson *et al.* 1986). This suggests that impaired postural responses are a large factor underlying falls in elderly people. In contrast, the balance deficits found among community-living older people with a history of falls are subtle. However, these deficits can be elicited by the more difficult conditions made available through use of a balance platform (L. I. Wolfson *et al.*, unpublished observations).

The data suggest that the balance response of older persons with a history of falls has greater latency, with less use of ankle movement (the most efficient balance strategy), and that they have poorer balance overall than individuals without a history of falls. These are consistent with a defect within the central sensorimotor organization of an effective balance response (Woollacott *et al.* 1982) although it is likely that biomechanical factors (e.g. strength and joint function) are of great importance. (Research questions include the extent to which these changes simply reflect disease and to what extent they are modifiable by training.)

Gait

Walking slows with age, and the 'normal' gait of elderly is characterized by a decrease in stride and speed (Murray *et al.* 1969; Imms and Edholm 1979). These decrements represent a significant but modest decrease in these quantitative indices of gait. Not surprisingly, walking speed also varies with the level of activity (Imms and Edholm 1979). The gait of elderly individuals who fall is often more compromised than the gait of those who do not, and is characterized by a decreased speed and shorter stride (Imms and Edholm 1979; Guimaraes and Isaacs 1980; Wolfson *et al.* 1990). To assess gait we use simple measures of walking (e.g. velocity and stride length) as well as observer-rated analysis of videotaped gait (Wolfson *et al.* 1990). A study of nursing-home residents demonstrated a strong correlation between abnormal gait (stride length, walking speed, and qualitative assessment) and the occurrence of falls (Wolfson *et al.* 1990), but this remains to be reproduced in work on older people living at home.

In our experience there is as wide a range of gait characteristics in older persons, as there is in younger individuals (Wolfson *et al.* 1990). The gait of some older men has been described as small-stepped with anteroflexed posture (Murray *et al.* 1969), but this is by no means universal nor is it clear whether or not many of these individuals have diseases producing these changes.

Motor function

Strength increases through childhood and adolescence, peaks in the mid-20s, and then declines minimally until the age of 50, after which a greater decrement is seen (Larsson 1982). Cross-sectional isometric studies (power output with constant muscle length, i.e. no limb movement) have shown a 20 to 40 per cent decrease in strength from the mid-20s to the 70s (Moritani and de Vries 1980; Larsson 1980). Walking and balance, however, require power output during limb movement. Movement-associated strength during muscle shortening (i.e. limb movement) is most appropriately tested by isokinetic dynamometry. Cross-sectional isokinetic studies of leg strength demonstrate a 30 to 40 per cent average decrement from the third to seventh decades (Murray *et al.* 1980, 1985; Larsson 1982; Stalberg *et al.* 1989), with the decreases becoming more prominent at rapid rates of muscle contraction (Larsson 1982). This is consistent with studies of muscle morphology that show a diminished number of fast-twitch, type 2 muscle fibres in older persons (Tomonaga 1977; Grimby *et al.* 1982; Larsson 1982). The causes of the atrophy of type 2 muscle fibres are not clear, although there is evidence tying it to both disuse (related to diminished physical activity) (Engel 1970; Larsson 1978; Aniansson *et al.* 1980), and/or denervation–reinervation produced by neuropathic lesions (Tomonaga 1977; Grimby *et al.* 1982; Shields *et al.* 1984).

Isokinetic dynamometry has shown compromised motor function in specific leg muscles of nursing-home residents who fall (Whipple *et al.* 1987). In particular, the strength of ankle dorsiflexors was profoundly diminished at functional speed of contraction (10 per cent of control strength). The vulnerability of older subjects to defective control of posterior sway and hence backwards loss of balance may be associated with a loss of dorsiflexor strength.

Sensory function and electrophysiology

Abnormalities of proprioception occur in 15 to 20 per cent of elderly patients (Klawans *et al.* 1971). Quantitative studies of proprioception at the knee revealed a deterioration with age (Skinner *et al.* 1984), but a quantitative study of proprioception at the toe had equivocal findings (Kokmen *et al.* 1978). Vibration threshold increases significantly after the age of 50 (Steiness 1957), primarily in the lower parts of the legs (Klawans *et al.* 1971). Touch sensitivity also decreases with age (Axelrod and Cohen 1961; Dyck *et al.* 1972). In a study of nursing-home residents, clinically impaired proprioception was present in 40 per cent of elderly persons with a history of falling and in only 13 per cent of controls. A quantitative study of vibratory–tactile perception (large-fibre sensory function) also found that individuals with a history of falls have a higher threshold than controls (L. I. Wolfson and L. Nashner, unpublished observations). Thus, impaired sensory function may contribute to the poor postural response of fallers. The role of impaired vestibular and visual input in older individuals with a compromised balance response has not been studied, although an increased reliance on vision (Woollacott *et al.* 1982) in conjunction with impaired perception of visual orientation cues in fallers has been reported (Tobis *et al.* 1985).

Both the velocity and amplitude of sensory nerve action potentials decrease from the third to the eighth decades (Buchthal and Rosenfalck 1966; Buchthal *et al.* 1975; Schaumburg *et al.* 1983). In contrast, conduction velocity in the dorsal columns, as determined by somatosensory-evoked potentials, shows minimal decline until the age of 60, with a steady decrement thereafter (Dorfman and Bosley 1979). Abnormal conduction in the sural nerve and abnormal H-reflexes are significantly more common in individuals with either poor balance-testing scores or a history of falls (Pack *et al.* 1985). This supports the concept that a sensory neuropathy may be a factor in these problems, although there is little direct evidence of a widespread neuropathy underlying this functional impairment.

Structural changes

Neural structures serving motor function are affected by age. Cell counting in autopsy material shows a loss in neurones (20–50 per cent) in the motor cortex, substantia nigra, and cerebellar cortex, as well as a decreased dendritic tree in projection neurones of the motor cortex of older subjects compared with children (Shefer 1973; McGeer *et al.* 1977; Nakamura *et al.* 1985). No loss of peripheral nerve cells has been reported, although several studies suggest structural changes (Schaumburg *et al.* 1985). The most relevant to motor function of the numerous chemical changes within the brain is the decrease in striatal dopamine (Carlsson and Winblad 1976) along with a 50 per cent decline in tyrosine hydroxylase activity (rate-limiting in dopamine synthesis) (McGeer 1976). The significance of the structural and chemical changes with age are difficult to interpret, but the magnitude of the decreases are consistent with the decrements of sensorimotor function with age, suggesting the possibility of a relationship between the two. By comparison, autopsy studies of the brains of parkinsonian patients show an 80 per cent decrement in striatal dopamine due to severe loss of dopaminergic projection neurones in the substantia nigra, suggesting (Bernheimer *et al.* 1973) that a higher order of structural changes underlies disease states.

The losses of function related to disease are additive to those associated with age and therefore often lead to severe functional incapacity. By comparison, age-associated decrements may weaken and slow responses but reserves are adequate to support motor function. It is therefore a reasonable proposition that significant dysfunction of gait and balance is disease- rather than age-associated.

MEDICAL APPROACH TO AN OLDER PATIENT WITH A HISTORY OF FALLS

This chapter focuses on the motor dysfunction associated with falls, but falls have multiple causes. The details surrounding a fall are critical in determining its cause. The occurrence of altered consciousness, lightheadedness, or vertigo suggest transient ischaemic attack (**TIA**), seizure, or cardiovascular causes. Without historical evidence of symptoms suggestive of hemispheric or brain-stem ischaemia or the manifestations of epilepsy, one cannot make a diagnosis of TIA or seizure. Furthermore, both have rarely been reported to cause falls (Sheldon 1960; Overstall *et al.* 1977).

If there is no evidence of altered sensorium before a fall, the role of environmental hazards should be explored. In reported series these hazards are related to almost half of all falls (Sheldon 1960; Overstall *et al.* 1977). One must be cautious in evaluating the significance of these hazards. Often older individuals attribute a fall to insignificant hazards (e.g. a crack in the pavement or uneven flooring). The balance reflexes of normal (young and old) individuals are almost always capable of compensating for minor hazards. Thus, while one may stumble over a minor obstacle, a fall should not ensue. The occurrence of multiple falls suggests impaired balance. Usually, cognitively intact older individuals limit their activities to decrease the chance of falling. This fear of falling limits mobility, striking at the quality of life and functional independence. Thus it is important to determine the scope of mobility during the history. Many older persons shop daily, use public transport, and walk considerable distances, but others barely leave the safety of their homes. A mobility scale that we have adapted from other investigators is included for guidance in history taking (Table 1) (Isaacs 1985).

Table 1 Mobility scale

Life space	Frequency
1. Bedroom to bathroom	□
2. Rest of apartment (or floor)	□
3. Within building	□
4. Immediate exterior—25 yards (e.g. garden, porch)	□
5. Block (including crossing street)	□
6. Neighbourhood	□
7. Unrestricted local travel (i.e., public transportation and/or car)	□
Total score	□□

Scoring system:
Daily = 4; 3–4 times a week = 3; weekly = 2; monthly = 1; never = 0

Cardiovascular dysfunction may result in global cerebral ischaemia of variable magnitude, which may produce lightheadedness, black-out of vision, or syncope. The most common cause of these symptoms in older persons is orthostatic hypotension (see Chapter 18.11). The diagnosis is suspected if the symptoms occur on standing or after the individual is up for a while. Postprandial and night-time (on the way to the bathroom) symptoms are quite common. Lightheadedness relieved by sitting is the most common symptom but occasionally, even in the absence of syncope, a fall occurs. Reproducing the symptoms in association with a significant drop in blood pressure is evidence for orthostatic hypotension.

Orthostatic hypotension may be produced by autonomic dysfunction within the central (e.g. Shy-Drager syndrome) or peripheral (e.g. diabetic neuropathy) nervous system in elderly subjects. Its prevalence is also widespread in older individuals because of the use of medications that induce orthostatic hypotension. The diagnosis should be considered in all older persons with unexplained falls (especially if the gait and balance are satisfactory) and pursued with multiple determinations of blood pressure, lying and immediately after standing up if necessary. Although occasionally difficult to treat, often changing of medication produces a 'vast improvement in symptoms'. Often no identifiable cause is found.

Syncope associated with an arrhythmia or valvular heart disease is uncommon as a cause of falls. By contrast with orthostatic hypotension the symptoms frequently occur when lying or sitting, and loss of consciousness is common. Although it may be suspected clinically, the diagnosis is confirmed by contemporary cardiological techniques (e.g. ECG, Holter monitor, or echocardiography). Head turning (flexion–extension or rotation) in older individuals may produce lightheadedness, vertigo, or rarely loss of consciousness resulting in falls. Although reportedly common (5 per cent of falls) (Sheldon 1960; Overstall *et al.* 1977), in my experience head turning may sometimes produce lightheadedness or vertigo but rarely results in falls. No clear-cut cause for the symptoms is generally accepted, although carotid sinus hypersensitivity (Ritch 1975), labyrinthine dysfunction, and vertebrobasilar ischaemia due to bony compression have been suggested (Brain 1963). An attempt to reproduce the symptoms with head turning is worthwhile. Vertigo induced by head movement (postural vertigo), which is accompanied by nystagmus and nausea, is often produced by vestibular and/or labyrinthine dysfunction, usually lasting for several days. Evaluation of hearing and vestibular function as well as cranial nerve and cerebellar function is warranted, especially if the symptoms persist. In the absence of other symptoms suggestive of a TIA, a diagnosis of vertebrobasilar ischaemia is not warranted. Empirical daily treatment with a soft collar may, however, decrease the symptoms.

Drop attacks have been described as sudden falls produced by temporary inability of the limbs to support weight but not related to altered consciousness or other focal neurological symptoms. Generally the episodes are brief (lasting only 1 to 2 min) and primarily affect women (Stevens and Matthews 1973). Although reported as a frequent cause of repetitive falls in older subjects, our group have encountered it rarely as a cause of falls.

Lightheadedness, which is a common symptom among older individuals, is also related to the occurrence of falls. Lightheadedness may be episodic or continual and is associated with multiple sensory deficits, hyperventilation, and cardiovascular disease, although often a clear-cut cause is not established.

Neurological dysfunction that impairs gait and balance

The examination should determine the extent and type as well as localizing the site(s) of underlying neural dysfunction. Non-neural factors may affect function. These include arthritis, which produces pain and impairs joint function, frailty which may be associated with diminished capacity for aerobic work, generalized weakness, and medication, which may impair co-ordination (e.g. alcohol or sedative tranquillizers).

The examination should start with watching the patient arise from a chair, walk, turn, sit, and by testing balance. Arising from a low chair (without arm push-off) is an excellent test of the strength of proximal muscles (glutei and quadriceps) as well as of the ability to shift one's centre of gravity from the chair to a newly established base of support while standing. Standing should be accomplished with one fluid series of movements. Walking is also a series of synchronous automatic movements of the legs, torso, and arms, which have been divided into swing and stance phases. During the swing phase, one foot is lifted and moved forward while the foot in contact with the ground is rolling forward (from heel to toes) in preparation for the next step. During stance the heel of the swing foot strikes the ground while the other foot prepares for lift-off with a vigorous push from the toes and forefoot. The gait of healthy older individuals appears normal, although it may not be as rapid as in younger subjects. As in younger persons, turning in healthy older individuals should result from a smooth, pivoting movement rather than a series of small steps. Balance can be tested by creating minor, rapid shifts in the centre of gravity using small forward or backwards pushes. The healthy older person should easily correct for these without taking steps or losing balance. The examiner should protect the patient from a fall during this test by positioning himself/herself in the direction of the push.

An experienced clinician uses the type of gait and balance impairment to localize the site of dysfunction within the nervous system. This impression is confirmed by eliciting the appropriate signs during the neurological examination. The sites of dysfunction produce characteristic clinical syndromes, which will now be discussed.

Frontal lobe syndrome

This is produced by bilateral, frontal lobe lesions. The resultant gait is halting, with sliding of the feet along the floor as if they were magnetized (i.e. 'magnetic gait'). Initiation and changes of movement are laborious and the rapid postural corrections required in balance are impaired (Denny-Brown 1958; Meyer and Barron 1960; Barron 1967). The disability can be progressive and may become so severe that individuals can no longer correct for minor shifts of weight and so suffer spontaneous falls. Motor manifestations of the frontal lobe syndrome may be difficult to differentiate from Parkinson's disease, perhaps because the output from basal ganglia to the frontal lobes facilitates the same transcortical postural reflexes that are impaired directly by lesions of the frontal lobes.

In addition to the impaired motor function, there is a slowing of cognition. Responses, although delayed, are often correct, resulting in a mild dementia. Usually there is a blunting of affect, although occasionally inappropriate, labile responses are present. Often the syndrome includes urinary dysfunction with urgency and incontinence. In addition, examination often demonstrates frontal-lobe release signs, palmar and plantar grasps and evidence of dysfunction of the pyramidal tract (spasticity of the lower extremities and Babinski signs). The most common causes of this syndrome are multiple cerebral infarcts, tumours, the late stages of Alzheimer's disease, and normal-pressure hydrocephalus. Visualization of the brain by magnetic resonance imaging (**MRI**) allows differentation of small subcortical infarcts, mass lesions, and hydrocephalus. The diagnosis of normal-pressure hydrocephalus, currently felt to be an uncommon disease, was made frequently in the years immediately following the description of this syndrome (Adams *et al.* 1965). Patients with prominent dementia and cortical atrophy, in addition to hydrocephalus, were given shunts. The lack of response and the major complications of shunt surgery produced unsatisfactory results. This unusual diagnosis may be suspected in patients with a frontal lobe syndrome who, on imaging, have severe ventricular dilation and little or no cortical atrophy. The monitoring of intracranial pressure to record abnormal fluctuations (Launas and Lobata 1979), as well as the clinical response to removal of cerebrospinal fluid (Fisher 1982) are often used to predict the success of shunting, although many clinicians rely solely on clinical and imaging features. If appropriate patients are treated, the motor function often improves after the shunt.

A common cause of the bifrontal syndrome is infarction in the distribution of smaller end arteries feeding subcortical white matter. Autopsy studies of patients with bifrontal gait abnormalities have found a high frequency of lacunar infarcts (George *et al.* 1986; Steingart *et al.* 1987). Recently, MRI has made *ante-mortem* detection of small ischaemic lesions feasible. The role of subcortical ischaemic lesions in producing motor abnormalities will be discussed in more detail later.

Parkinson's disease

Parkinson's disease is a common, age-associated disease with increasing incidence in the sixth to eighth decades (Kurland *et al.* 1969). Because it is a syndrome as well as a specific disease state whose primary manifestations involve abnormalities of gait and balance, it is an important cause of impaired mobility and falls. The clinical features and management of Parkinson's disease in later life are discussed in Chapter 18.12.

Parkinson's disease progresses slowly over many years (5–20+) and if untreated usually produces severe disability. Treatment, however, provides significant improvement of symptoms at all stages. Walking becomes small-stepped and slow, with stooped posture and diminished associated arm and torso movement. Arising from a chair may require multiple efforts, with difficulty in establishing a stable standing posture. Turning becomes a series of small irregular steps rather than a fluid movement. A fundamental problem relates to balance: the patient is unable to keep the centre of gravity vertical to the base of support, and to make rapid postural corrections after encountering environmental hazards.

Other important signs of Parkinson's disease include a characteristic 3–8 c.p.s. resting tremor which is most prominent in the arms but may involve the legs or head. Motor activity is decreased, difficult to initiate and slow once underway. This bradykinesia is also manifest in an immobile (masked) facies, slowness in use of the hands, and in the gait/balance abnormalities previously described. There is also the ratchet-like resistance to passive stretching of cogwheel rigidity.

Parkinsonism has been used generically to denote patients with impaired gait and balance and motoric slowing. Therefore, patients with these features have been given this diagnosis when in reality they may have progressive supranuclear palsy (see below) or a bifrontal syndrome. Older patients with gait and balance dysfunction but none of the other features of the disease are often diagnosed as Parkinson's disease. In reality, these patients rarely have Parkinson's disease and usually do not respond to a therapeutic trial of L-dopa/carbidopa (Newman *et al.* 1985). On the other hand in patients with typical manifestations (three or more signs plus no features suggestive of an alternate diagnosis) Parkinson's disease is likely. In patients receiving phenothiazines or butyrophenones, the medication must be discontinued for several weeks before a diagnosis of Parkinson's disease can be made.

Disease impairing pyramidal tracts

Interruption of the descending corticospinal tract impairs voluntary motor control, resulting in weakness and clumsiness. Unilateral interruption of the tract results in a hemiparesis with a characteristic position of the legs and arms (circumduction at thigh and plantar flexion of the foot). The usual cause of hemiparesis in older patients is stroke, although defining the nature and extent of the infarction as well as evaluating the presence of other causes (e.g. subdural haematoma or tumour) warrants evaluation with computed tomography (**CT**) or MRI. Bilateral corticospinal lesions result in a stiff gait with hyperadduction and spasticity (scissoring) at the thighs and feet, which are everted and plantar flexed. Paraparesis results from multiple hemispheric lesions (often lacunar infarcts), spinal cord compression due to osteoarthritic overgrowth of bony elements within the canal, epidural metastatic lesions, and occasionally amyotrophic lateral sclerosis {or parasagittal lesions}. Ambulatory elderly patients who develop a progressive paraparesis should be evaluated and a diagnosis established. The possibility of significant benefit with little risk (at least during the investigations) dictates the approach. The spinal cord is well visualized by MRI, while CT allows delineation of bony elements within the canal. Myelography is occasionally required when questions remain. Patients with cervical spondylosis can be treated with a cervical collar. If progression continues, threatening function, healthy symptomatic elderly patients with one or two levels of compression often benefit greatly from surgical intervention. Similarly, patients with epidural spinal-cord compression due to metastases respond to radiation; occasionally a benign tumour (e.g. meningioma, neurofibroma) requires surgery.

Ataxia

A wide-based, unsteady gait with lurching steps and difficulty with postural transitions (turns, standing up) is most characteristic of cerebellar dysfunction. Sensory abnormalities produced by vestibular, visual, and tactile dysfunction may share common features although they are separable by neurological examination. Ataxia may be produced by mass lesions compressing the midline cerebellum or its output, but is more often produced by cerebral infarction in older individuals. Structural lesions within the posterior fossa are well demonstrated by MRI. Cerebellar ataxia may be part of an olivopontocerebellar or multiple-system atrophy as well as a manifestation of alcoholism, hypothyroidism, {a paraneoplastic syndrome}, or vitamin E deficiency.

Vestibular dysfunction due to end-organ problems (vestibular neuronitis) is accompanied by prominent positional vertigo, unsteadiness with nausea, and gaze-evoked nystagmus. Although the gait is ataxic, no other neurological signs are present unless the symptoms are a result of brain-stem ischaemia. Tactile proprioceptive loss results in an unsteady gait with difficulty on uneven surfaces compensated in part by visual cues. In older patients with accompanying visual impairment (e.g. glaucoma, cataracts) there may be mild ataxia and unsteadiness perceived as a lightheaded sensation.

Progressive supranuclear palsy and Alzheimer's disease

Progressive supranuclear palsy, often included within the symptomatic description of parkinsonism, has a prevalence of about 1 per cent of that of Parkinson's disease (Golbe *et al.* 1987). The disease is progressive (median survival, 6 years), with prominent impairment of gait and balance, but also includes dysarthria, dysphagia, and dementia (Golbe and Davis 1988). There is bradykinesia, a characteristic extensor rigidity of the neck and trunk muscles, and impaired voluntary eye movement particularly on down gaze (Golbe and Davis 1988). The dementia is often mild, with a characteristic slowness of processing, difficulty in using complex information, and an associated blunting of affect or depression (Golbe and Davis 1988). The pathology of progressive supranuclear palsy is distinct, with neuronal loss and the presence of neurofibrillary tangles (also present in Alzheimer's disease) in the basal ganglia as well as other deep neural structures (Golbe and Davis 1988). L-Dopa and synthetic dopamine agonists are ineffective (Golbe and Davis 1988). Parkinsonian signs and symptoms may be associated with a wide variety of neurological abnormalities (olivopontocerebellar and multiple-systems atrophy) as well as autonomic failure (Shy-Drager syndrome).

By contrast with progressive supranuclear palsy, Alzheimer's disease in its early stages has little obvious effect on motor function, although as the illness progresses motor signs become more prominent. In moderately advanced Alzheimer patients there are reports (Visser 1983) of diminished stride length, walking speed and balance (increased sway), and impaired postural reflexes. The motor dysfunction may be an important element in the disability in later stages of Alzheimer's disease, although it is overshadowed by the cognitive dysfunction.

Peripheral nerve and muscle dysfunction

As noted earlier, mild tactile proprioceptive loss in older subjects is presumably related to changes within receptors or sensory nerves. A significant neuropathy is not a part of healthy ageing and should be evaluated thoroughly. Neuropathy is often the result of systemic illness (e.g. diabetes, vasculitis) but may result from medication or toxic exposure. On the other hand, our group recently encountered two patients of over 70 years with progression of dysfunction in whom a hereditary sensorimotor neuropathy (Charcot–Marie–Tooth) was the most realistic diagnosis. The characteristic steppage gait with foot slapping is indicative of weakness of the foot dorsiflexors, often due to neuropathy. Myopathy results in proximal weakness with a characteristic waddling gait and difficulty in climbing stairs or arising from a chair. Inflammatory myopathies and hypothyroidism are potentially treatable, thereby making evaluation of weakness produced by nerve or muscle problems worthwhile. Muscle enzymes and electrodiagnostic evaluation of nerve and muscle function are essential in making the appropriate diagnosis.

IMPLICATIONS OF MOTOR DYSFUNCTION AND AGE

In recent years there has been widespread use of the term 'senile gait' to define changes in gait (and balance) that are associated with age. Earlier in the chapter, alterations in sensorimotor functions that some have associated with ageing were reviewed. Gait and balance slow modestly with age, and, although these age-associated changes require further definition, the functions remaining are capable of supporting a mobile active lifestyle.

Those patients with impaired functional gait and balance who do not get into the diagnostic categories of neurological diseases discussed earlier in the chapter are often diagnosed as having a 'senile gait disorder'. This diagnosis does not imply a specific type of gait or pathophysiology. In our experience, this represents the majority of the patients with failing mobility that a neurologist will encounter. Moreover, use of this definition implies an age-associated inevitability for motor dysfunction that the facts do not support. It remains for neurologists and geriatricians to broaden our understanding of age-associated diseases that produce this impairment so that a cause can be established for all patients. As an example, we recently reported a series of fallers with impaired gait and balance who, by comparison with unimpaired controls, had significant abnormalities within the subcortical white matter (Masdeu *et al.* 1989). Two of the subjects in this series who died had pathological evidence of widespread ischaemic lesions of subcortical white matter suggestive of small-vessel disease. The importance of arteriolar infarction as a mechanism for producing impairment of gait and balance by compromising connections in the white matter remains to be defined. By avoiding definitions that imply an inevitable deterioration, we may be able to define the underlying pathophysiological mechanisms that produce motor dysfunction so as to develop specific disease-oriented interventions.

In older patients without remediable specific diagnoses, treatment should be directed towards improving function. Experiments have shown that older patients improve their functional balance responses during a sequence of balance stresses on the balance platform (L. I. Wolfson *et al.*, unpublished observations). Long-term alteration of the functional balance response by balance-oriented interventions remains to be demonstrated, although it appears to be distinctly feasible at this time.

Another promising area of intervention relates to weak muscles in the lower legs, which are important for balance. Older subjects with a history of falls, poor balance, and impaired gait have severe weakness of the ankle muscles, which are critical for efficient balance reflexes (Whipple *et al.* 1987). Therefore, it is possible that a programme directed at judicious strengthening of the lower extremities might improve mobility. Empirical interventions in these disorders are necessary for validation of an effective, comprehensive strategy for rehabilitation. Even without such a defined rehabilitative approach, the practitioner can make useful suggestions. Activity should be encouraged, despite the subject's fear of falls. This can be accomplished by the family, friends, or even neighbours who are willing to accompany the patient on daily outings or in the safety of an indoor gym or old people's centre. A trial of effective strategies with a physiotherapist may be of value in dealing with remediable gait defects.

The correct choice of shoes may be of help in improving balance by moving the centre of gravity. Individuals who are unable to correct for backwards displacement of their body mass may be helped by heel lifts that move them forward; those who cannot resist forward-directed forces are helped by flat shoes. Shoes with non-slip (but not cleated) soles and a small amount of heel lift are suitable for the remainder of older persons with poor balance. A walking stick will broaden the base of support, thereby providing extra security to those with a fear of falling. Selection of an appropriate length (15° of elbow flexion when the stick is in contact with the floor), maximizes support while minimizing interference with gait. Walking frames provide more support but are disruptive to gait efficiency and speed and increase the effort required in walking. This trade-off may be necessary in very unstable older patients. A weighted shopping trolley is effective in providing stability in addition to aiding in important functions that support an independent lifestyle.

Vision is often impaired in older subjects. Visual input may be crucial in individuals with pre-existing tactile proprioceptive or vestibular dysfunction. Therefore, appropriate lenses for medium and far vision may significantly improve balance. Older individuals often live in suboptimal environments surrounded by potentially lethal hazards. Proper lighting, attention to uneven, loose, or stepped surfaces, along with the use of user-friendly environmental aids (grab bars or rails), can be provided with modest cost and effort. Older individuals and their family must also be aware of avoiding hazardous environments where possible (e.g. icy surfaces, tall, insecure step-stools, and steep, poorly lit stairs).

The realization that disorders of mobility are critical factors in the lives of our elderly populace is already widely accepted. Attention to defining the causes of this dysfunction as well as developing interventions is beginning to effect rapid changes in our approach to this widespread problem.

References

Adams, R.D., Fisher, C.M., Hakim, S., Ojeman, R.G., and Sweet, W.H. (1965). Symptomatic occult hydrocephalus with 'normal' cerebrospinal fluid pressure. *New England Journal of Medicine*, **273**, 117–26

Aniansson, A., Grimby, G., and Rundgren, A. (1980). Isometric and isokinetic quadriceps muscle strength in 70-year-old men and women. *Scandanavian Journal of Rehabilitation Medicine*, **12**, 161–8

Axelrod, S. and Cohen, L.D. (1961). Senescence and embedded figure performance in vision and touch. *Perception and Psychophysics*, **12**, 283–7

Baker, S.P. and Harvey, A.H. (1985). Fall injuries in the elderly. *Clinics in Geriatric Medicine*, **1**, 501–12.

Barron, R.E. (1967). Disorders of gait related to the aging nervous system. *Geriatrics*, **22**, 113.

Bernheimer, H., Birkmayer, W., Hornykiewicz, O., Jellinger, K., and Sertelberger F. (1973). Brain dopamine and the syndromes of Parkinson and Huntington. Clinical, morphological, and neurocorrelations. *Journal of Neurological Science*, **20**, 415–55.

Bohannon, R.W., Larkin, P.A., Cook, P.D., Gear, J., and Singer, J.

(1984). Decrease in timed balance test scores with aging. *Physical Therapy*, **64**, 1067–70.

Brain, R.W. (1963). Some unsolved problems of cervical spondylosis. *British Medical Journal*, **1**, 771.

Buchthal, F. and Rosenfalck, A. (1966). Evoked action potentials and conduction velocity in human sensory nerves. *Brain Research*, **3**, 1–22.

Buchthal, F., Rosenfalck, A., and Behse, F. (1975). Sensory potentials of normal and diseased nerve. In *Peripheral neuropathy* (ed P.J. Dyck, P.K. Thomas, and E.H. Lamber), pp. 442–64. W.B. Saunders, Philadephia.

Carlsson, A. and Winblad, B. (1976). Influence of age and time interval between death and autopsy on dopamine and 3-methoxytyramine levels in human basal ganglia. *Journal of Neural Transmission*, **38**, 271–6.

Cooper, K.D., Tabaddor, K., Hauser, W.A., Shulman, K., Feiner, C., and Factor, P.R. (1983). The epidemiology of head injury in the Bronx. *Neuroepidemiology*, **2**, 1–2; 70–88.

Denny-Brown, D. (1958). The nature of apraxia. *Journal of Neurologic Mental Disease*, **126**, 9.

Dorfman, L.J. and Bosley, T.M. (1979). Age-related changes in peripheral and central nerve conduction in man. *Neurology*, **29**, 38–44.

Dyck, P.J., Shultz, P.W., and O'Brien, P.C. (1972). Quantification of touch pressure sensation. *Archives of Neurology*, **26**, 465–73.

Engel, W.K. (1970). Selective and non-selective susceptibility of muscle fiber types. *Archives of Neurology* (Chicago), **22**, 97–117.

Fisher, C. M. (1982). Hydrocephalus as a cause of disturbances of gait in the elderly. *Neurology*, **32**, 1358–63.

Frol'Kis, V.V. and Bezrukov, V.V. (1978). Aging of the central nervous system. *Human Physiology*, **4**, 478–99.

George, A.E. *et al.* (1986). Leukoencephalopathy in normal and pathologic aging: CT of brain lucencies. *American Journal of Neuroradiology*, **7**, 561–6.

Golbe, L.I. and Davis, P.H. (1988). Progressive supranuclear palsy: recent advances. In *Parkinson's disease and movement disorders* (ed. J. Jankovic and E. Tolusa). Urban and Schwarzenberg, Baltimore.

Golbe, L.I., Davis, P.H., Schoenberg, B.S., and Pavoisin, R.C. (1987). The natural history and prevalence of progressive supranuclear palsy. *Neurology*, **37** (Suppl. 1), 121.

Grimby, G., Danneskiold-Sumsoe, B., Huid, K., and Saltin, B. (1982). Morphology and enzymatic capacity in arm and leg muscles in 78–81 year old men and women. *Acta Physiologica Scandinavica*, **115**, 125–34.

Gryfe, C.L., Amies, A., and Ashley, M.J. (1977). A longitudinal study of falls in an elderly population. I. Incidence and morbidity. *Age and Ageing*, **6**, 201–10.

Guimaraes, R.M. and Isaacs, B. (1980). Characteristics of the gait in old people who fall. *International Journal of Rehabilitative Medicine*, **2**, 177–80.

Hall, T.C., Mill, A.K.H., and Corsellis, J.H.N. (1975). Variations in the human Purkinje cell population according to age and sex. *Neuropathic Applied Neurobiology*, **1**, 267–92.

Horak, F. B., Shupert, C. L., and Mirka, A. (1989). Components of postural dyscontrol in the elderly. *Neurobiology of Aging*, **10**, 727.

Imms, F.J. and Edholm, O.G. (1979). Studies of gait and mobility in the elderly. *Age and Ageing*, **10**, 147–56.

Isaacs, B. (1985). Clinical and laboratory studies of falls in old people. *Clinics in Geriatric Medicine*, **1**, 513–25.

Klawans, H.L., Tufo, H.M., and Ostfeld, A.M. (1971). Neurologic examination in an elderly population. *Diseases of the Nervous System*, **32**, 274–9.

Kokmen, E., Bossemeyer, R.W., and Williams, W. (1978). Quantitative evaluation of joint motion sensation in an aging population. *Journal of Gerontology*, **33**, 62–7.

Kurland, L.T., Hauser, W.A., Okazake, H., and Nobrega, F.T. (1969). Epidemiologic studies of parkinsonism with special reference to the cohort hypothesis. In *Third Symposium on Parkinson's disease* (eds. F. J. Gillingham and I. M. L. Donaldson). F.S. Livingstone, Edinburgh.

Larsson, L. (1978). Morphological and functional characteristics of the aging skeletal muscle in man. A cross-sectional study. *Acta Physiologica Scandinavica*, **457** (Suppl.).

Larsson, L. (1982). Aging in mammalian skeletal muscle. In *The aging motor system* (eds. J.A. Mortimer, F.J. Pirazzolo, and G.J. Maletta), pp. 60–96. Praeger, New York.

Launas, C. and Lobata, R.D. (1979). Intraventricular pressure and CSF dynamics in chronic adult hydrocephalus. *Surgical Neurology*, **12**, 287.

Masdeu, J. C. *et al.* (1989). Brain white matter disease in elderly prone to falling. *Archives of Neurology*, **46**, 1292–6.

McGeer, E.G. (1976). Aging and neurotransmitter metabolism in the human brain. In *Alzheimer's disease, senile dementia and related disorders*, Ageing series, Vol. 7 (ed. R.K. Katzman, R.D. Terry and K.L. Byck), p. 427–40. Raven Press, New York.

McGeer, P.L., McGeer E.G., and Suzuki, J.S. (1977). Aging and extrapyramidal function. *Archives of Neurology*, **34**, 33–5.

Meyer, J.S. and Barron, D.W. (1960). Apraxia of gait: a clinicophysiologic study. *Brain*, **83**, 261–84.

Moritani, T. and de Vries, H.G. (1980). Potential for gross muscle hypertrophy in older men. *Journal of Gerontology*, **24**, 169–82.

Murray, M.P., Kory, R. C., and Clarkson, B.H. (1969). Walking patterns in healthy old men. *Journal of Gerontology*, **24**, 169–80.

Murray, M.P., Gardner, G.M., Mollinger, L.A., and Sepic, S.B. (1980). Strength of isometric and isokinetic contractions: knee muscle of men aged 20–86. *Physical Therapy*, **60**, 412–19.

Murray, M.R., Duthie, E.H., Gambert, S.R., Sepic, S.B., and Mollinger, T.A. (1985). Age-related changes in knee muscle strength in normal woman. *Journal of Gerontology*, **40**, 275–80.

Nakamura, S.I., Akiguchi, M., Kamegama, M., and Mizuno, W. (1985). Age-related changes to pyramidal cell basal dendrites in layers III and V of human motor cortex: a quantitative Golgi study. *Acta Neuropathologica* (Berlin), **65**, 281–4.

Nashner, L.M. (1976). Adapting reflexes controlling the human posture. *Experimental Brain Research*, **26**, 59–72.

National Center for Health Statistics (Hawlik, R.J., Liu, B.M., Suzman, R., Feldman, J.J., Harris, T., and Van Nostrand, J. (1987). *Health statistics on older persons, United States, 1986*, Vital and health statistics, Series 3, No. 25, DHHS Publication No. (PHS) 87-1409. Government Printing Office, Public Health Service, Washington.

Newman, R.P., LeWitt, P., Jaffe, M., Calne, D.B., and Larsen, T.A. (1985). Motor function in the normal aging population treatment with L-Dopa. *Neurology*, **35**, 571–3.

Overstall, P.W., Exton-Smith, A.N., Imms, F.J., and Johnson, A.L. (1977). Falls in the elderly related to postural imbalance. *British Medical Journal*, **i**, 261–4.

Pack, D., Wolfson, L.I., Amerman, P., Whipple, R., and Kaplan, J. (1985). Peripheral nerve abnormalities and falling in the elderly. *Neurology*, **356** (Suppl. 1), 79.

Pollock, M. and Hornabrook, R.W. (1966). The prevalence, natural history and dementia of Parkinson's disease. *Brain*, **89**, 429–88.

Ritch, A.E. (1975). The significance of carotid sinus hypersensitivity in the elderly. *Gerontologica Clinica*, **17**, 146.

Rubenstein, L.Z., Robbins, A.S., Schulman, B.C., Rosada, T., Osterweil, D., and Josephson, K.R. (1988). Falls and instability in the elderly. *Journal of American Geriatrics Society*, **36**, 266–8.

Sabin, T.D. (1982). Biologic aspects of falls and mobility in the elderly. *Journal of American Geriatrics Society*, **30**, 51–8.

Schaumburg, H.H., Spencer, P.S., and Ochoa, J. (1983). The aging human peripheral nervous system. In *The neurology of aging* (ed. R.K. Katzman and R. Terry), pp. 111–22. F.A. Davis, Philadelphia.

Shefer, U.F. (1973). Absolute number of neurons and thickness of the cerebral cortex during ageing, senile and vascular dementia and Pick's and Alzheimer's disease. *Neuroscience Behavioral Physiology*, **6**, 319–24.

Sheldon, J.H. (1960). On the natural occurrence of falls in old age. *British Medical Journal*, 1685.

Shields, R.W., Robbins, N., and Verrilli, A.A. (1984). The effects of chronic muscular activity of age-related changes in single fiber EMG. *Muscle and Nerve*, **7**, 273–7.

Skinner, H.B., Barrack, R.L., and Cook, S.D. (1984). Age-related decline in proprioception. *Clinical Orthopedics and Related Research*, **184**, 208–11.

Stalberg, E., *et al.* (1989). The quadriceps femoris muscle in 20–70 year old subjects: relationship between knee extension torque, electrophysiologic parameters and muscle fiber characteristics. *Muscle and Nerve*, **12**, 382–9.

Steiness, I. (1957). Vibratory perception in normal subjects. *Acta Medica Scandinavica*, **58**, 315–25.

Steingart, A. *et al.* (1987). Cognitive and neurologic findings in subjects with diffuse white matter lucencies on computed tomographic scan (leuko-araiosis). *Archives of Neurology*, **44**, 32–5.

Stevens, D.C. and Matthews, W.B. (1973). Cryptogenic drop attacks an affliction of women. *British Medical Journal*, **1**, 439.

Tobis, J.S., Reinsch, S., Swantston, J.M. Byrd, M., and Scharf, T. (1985). Visual perception dominance of fallers among community-dwelling older adults. *Journal of the American Geriatrics Society*, **33**, 330–3.

Tomonaga, M. (1977). Histochemical and ultrastructural changes in senile human skeletal muscle. *Journal of the American Geriatrics Society*, **25**, 125–31.

U.S. Bureau of the Census (1985). *Estimates of the population of the United States, by age, sex and race. 1980–1984.* Current population reports, Series p-25, No. 965. U.S. Government Printing Office, Washington, D.C.

Visser, H. (1983). Gait and balance in senile dementia of the Alzheimer type. *Age and Ageing*, **12**, 296.

Whipple, R., Wolfson, L.I., and Amerman, P. (1987). The relationship of knee and ankle weakness to falls in nursing home residents. An isokinetic study. *Journal of the American Geriatrics Society*, **35**, 13–20.

Wolfson, L.I., Whipple, R.H., Amerman, P., and Kleinberg, A. (1986). Stressing the postural response: a quantitative method for testing balance. *Journal of the American Geriatrics Society*, **34**, 845–50.

Wolfson, L.I., Whipple, R., Amerman, P., and Tobin, J.N. (1990). Gait assessment in the elderly: a gait abnormality rating scale and its relation to falls. *Journal of Gerontology*, **45**, M12–19.

Woollacott, M.H., Shumway-Cook, A., and Nashner, L. (1982). Postural reflexes and aging. In *The aging motor system* (ed. J.A. Mortimer, F.J. Pirozzolo, and G.J. Maletta), pp. 98–119. Praeger, New York.

SECTION 19
Voluntary muscle

19.1 Strength and power

ARCHIE YOUNG

STRENGTH

Advancing age brings a loss of muscle strength. Cross-sectional studies suggest that the deterioration begins around the age 45 years, with subsequent deterioration at about 1.5 per cent/year (Figs. 1 and 2). In the only longitudinal studies uncontaminated by a major change in life-style (Aniansson *et al.* 1983, 1986), the isometric strength of the quadriceps fell between the ages of 70 and 75 or 77 at 1.8 to 4.1 per cent/year in men. Isometric elbow flexion and extension strength were lost even faster, some 4.6 to 6.8 per cent/year. These dramatic figures require urgent confirmation and study. Do they indicate an acceleration of the loss of strength at around the age of 70 or do they simply point up the shortcomings of cross-sectional studies?

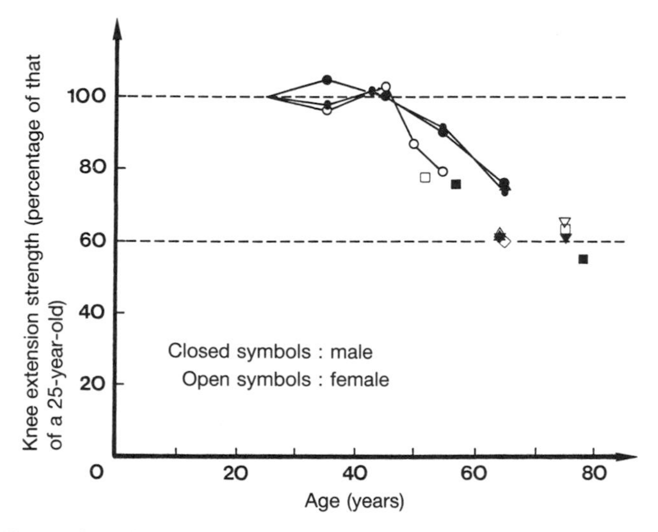

Fig. 1 Quadriceps strength in elderly subjects, expressed as percentages of the strength of subjects of the same sex in their mid-twenties (reproduced, with permission, from Young 1989, where the sources of data are referenced).

Men are consistently stronger than women of the same age (Fig. 2). The quadriceps strength of elderly men is approximately 50 per cent greater than that of elderly women (references in Young 1989) and the strength/body weight of weight-bearing muscles is approximately 10 to 20 per cent greater. For weight-bearing functional activities, therefore, elderly men have a significantly greater reserve capacity than women of the same age.

MUSCLE SIZE

Although some doubts remain about the details, it seems clear that both peripheral and proximal muscles of both sexes weaken steadily from, at the latest, the sixth decade onwards, and that much of this deterioration is due to a loss of muscle mass.

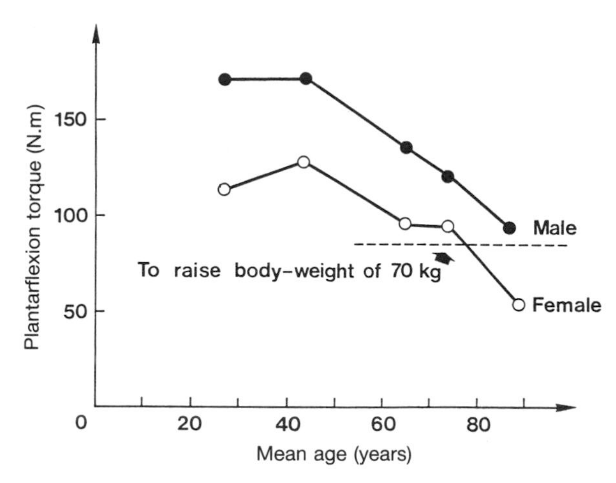

Fig. 2 Effect of age and sex on maximal, voluntary, isometric plantarflexion torque (drawn from the data of Vandervoort and McComas 1986). Also shown is the unilateral plantarflexion torque which must be exerted in order to raise a body weight of 70 kg, as in stepping up.

Fibre atrophy or fibre loss?

Do muscle fibres get smaller or fewer? Human data to answer this question are subject to several, unavoidable problems. First, there are the pitfalls in interpreting cross-sectional studies. In addition, biopsy specimens are imperfect indicators of the whole muscle. The sampling problem is less in autopsy studies but, even then, not all the fibres will be present in a single transverse section of a pennate muscle.

Another approach to the sampling problem has been to obtain more substantial specimens of muscle, as a byproduct of surgery. The results of such studies have been widely quoted, often with claims that they represent normal muscle. This is quite wrong; these studies have been based on patients with, for example, breast cancer, femoral fracture, hemiplegia, scleroderma, deforming arthritis, or gastric cancer.

Most studies of truly normal subjects have found the mean fibre area of the quadriceps (measured in transverse sections of biopsy specimens) well preserved at least until about 70 years of age. By age 78 there is a clear reduction in quadriceps mean fibre area but still no change in biceps brachii mean fibre area. The predominant picture, therefore, is of a relative preservation of fibre size in the presence of a declining muscle mass (see Fig. 2 in Grimby and Saltin 1983), implying that the muscle contains fewer fibres. This is borne out by an autopsy study in which fibres were counted across sections cut from whole human (male) vastus lateralis; a 25 per cent fall in fibre number was associated with an 18 per cent fall in the cross-sectional area of the whole muscle (**CSA**) from ages of 19 to 37 years to 70 to 73 years. (In contrast, the quadriceps wasting which follows knee injury and/or immobilization, and which may be at least as severe, is due to fibre atrophy, not fibre loss.)

The loss of muscle fibres with increasing age can probably be explained by the steady decline in the number of motor neurones in the lumbosacral expansion of the spinal cord as identified in cadavers of people aged 60 years and over. The loss of anterior horn cells is a sufficiently gradual process that, as Grimby and Saltin (1983) have argued, it is not surprising that it is not until about the age of 80 that denervation changes become apparent in biopsies from healthy subjects. That the underlying process is one of progressive denervation is further supported by the falling number of motor units electromyographically identifiable in extensor digitorum brevis from the age of 60. An increasing size of motor unit and the (later) appearance of fibre-type grouping imply terminal sprouting and reinnervation. The falling number of fibres, however, indicates that the process of denervation is only partially compensated.

MUSCLE SIZE AND STRENGTH

Only three studies have tested for the effects of both age and sex on the relationship between the strength of an individual human muscle and its directly measured CSA (Fig. 3). They do not agree on whether the strength/CSA of young women more resembles that of young men or those of elderly men and women. Nevertheless, three conclusions are possible: (1) a substantial part of the loss of strength with old age can be explained by the loss of muscle bulk; (2) in ageing men there is also a fall in strength/CSA; and (3) there is no sex difference in strength/CSA in old age.

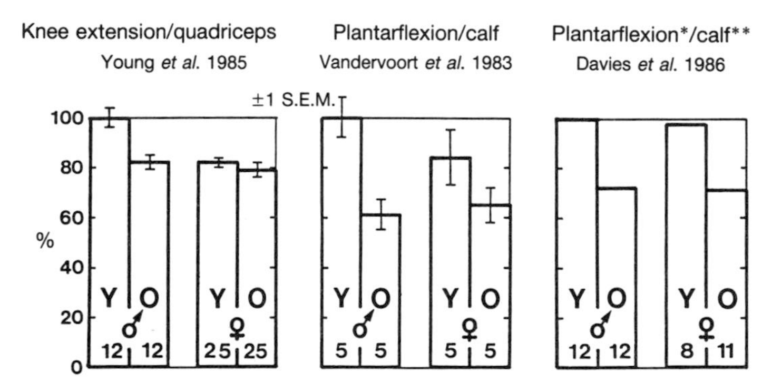

Fig. 3 Mean values (with number of subjects and ±1 SEM where available) for the ratio of a muscle's strength to its cross-sectional area in young (Y) and old (O) men and women (reproduced, with permission, from Young 1989, where the sources of data are referenced).
* Strength measurements potentially contaminated by hip flexion effort. ** Measurements of muscle cross-sectional areas contaminated by tibia and fibula.

It is interesting to speculate on the possible explanations for any age-difference in strength/CSA. Possibilities to be considered include incomplete voluntary activation of muscle by older subjects, replacement of contractile tissue by fat or fibrous tissue, and a reduced contribution of type II myofibres to the muscle mass. Studies with a twitch interpolation technique indicate that incomplete muscle activation by older subjects is unlikely to be an adequate explanation; complete or nearly complete activation was achieved with voluntary efforts, irrespective of age (Vandervoort and McComas 1986). (This may not be the case in the presence of joint pathology, see Chapter 19.4.)

There are conflicting views on how much the proportion of non-contractile tissue in human muscles increases in old age. For example, topical nuclear magnetic resonance spectroscopy showed no evidence of an increase in the fat content of the forearm muscles of 12 men and women in their 70s. In contrast, an early radiological study concluded that the amount of interstitial fat in the female quadriceps increased steadily with age (from 20 to 65 years) but, perhaps rather surprisingly, also reported no change in the total breadth of the muscle shadow. A more recent radiological study, using subjective evaluation of computed axial tomographs, demonstrated rather more intermuscular fat in the thighs (at least some of which would have been included in the measurements of 'quadriceps' cross-sectional area in other studies) and rather more intramuscular fat in both latissimus dorsi and the deep back muscles of men of mean age 69 than in men of mean age 46. Confirmatory studies would be valuable.

Influence of fibre-type composition

There is little or no change in the relative frequency of type I and type II fibres in human muscle, probably up to 100 years of age, but there does seem to be a rather sudden reduction in muscle fibre size at about the age of 70 and affecting type II fibres more severely (especially in men) (reviewed by Green 1986). The only important exception to this view is the report from Larsson *et al.* (1979) of a progressive reduction in the percentage of type II fibres in quadriceps biopsies from men aged between 22 and 65, and of fibre atrophy starting around age 50 (and affecting type II fibres rather more severely). As human type II fibres may be about twice as strong for their cross-sectional area as type I fibres (Young 1984), it seems possible that the fall in the strength/CSA of elderly male muscle could be explained by a change in fibre-type composition.

SPEED

Stimulated contractions

The speed with which muscle tension can be developed and released becomes slower in old age. In electrically stimulated contractions, twitch and tetanic contraction times and half-relaxation times are longer in people between 60 and 100 years old than in younger subjects.

Cooling to below 30°C increases the contraction times and half-relaxation times of the calf muscles of young men to values as great as those of elderly men (Davies *et al.* 1982).

Voluntary contractions

In isometric contractions of the quadriceps, older men take progressively longer to reach their maximal voluntary torque. They also achieve slower maximal extension velocities in voluntary isokinetic contractions of the quadriceps. In Aniansson's longitudinal study of the changes in quadriceps strength between 70 and 75 years of age, the loss of strength was more marked in faster isokinetic contractions than in slower or in isometric contractions (Aniansson *et al.* 1983).

Influence of fibre-type composition

As human type II fibres contract and relax only about twice as fast as type I fibres, the rather modest changes in fibre-type composition seem inadequate to explain fully the observed slowing of twitch contraction and relaxation in older muscle. Similarly, the decrease and increase in tetanic relaxation rate seen in hypo-and hyperthyroid disease are greater than can be explained by the, respectively, decreased and increased percentages of type II fibres in the quadriceps. It seems that ageing, like thyroid disease, may alter the speed of a muscle fibre without necessarily changing its myosin ATPase staining characteristics.

POWER

'Power' is defined as a rate of performing work. It may be calculated as the product of force and speed. The slowing of muscle with increasing age means that the fall in maximal power is greater than the fall in maximal strength. Shock and Norris (1970) compared a composite score for isometric strength around the shoulder with the power developed in a 15-s burst of cranking. The fall in strength was not apparent until the men were over 60 but power output, especially at the faster cranking speeds, faded from about 40 years of age. By age 80 there had been a 45 per cent fall in power for only a 27 per cent fall in the strength score (30 and 19 per cent, respectively, after adjustment for body weight). In a maximal vertical jump from a force plate, men and women in their early 70s produced an average power output of 70 to 75 per cent less than those in their early 20s, for a reduction in average downward force of about a half (Bosco and Komi 1980).

In jumping or climbing tests, the effect of a similar body weight combined with a reduced strength is that the older muscle must work at a slower point on its force/velocity curve. The older muscle's intrinsic slowness also impairs power development but for this to be seen in isolation, it is necessary to eliminate the effect of body weight. This can be achieved by testing power output on a cycle ergometer adapted to control pedal velocity and instrumented to record the forces exerted on the pedals. Men of mean age 69 years developed their peak power at a slower pedalling speed than men of mean age 22 but the difference between the loss of peak power (45 per cent) and the loss of maximum force (32 per cent) was less on the cycle than in a jump (51 and 16 per cent, respectively) (Davies *et al.* 1983).

FUNCTIONAL CONSEQUENCES

Thresholds

As one grows older, there comes a time when one's maximal strength (or power) in a particular action is the same as the minimum required to perform some everyday activity. For example, the threshold plantarflexion strength for raising the body weight on to the toes of one foot probably occurs, on average, at about age 80 in women and rather later in men (see Fig. 2). Once at a threshold, it needs only a small further decline in strength (or power) to go from being 'just able' to being 'just unable' to perform that activity (Fig. 2). If the activity is crucial to an independent life, that small change in muscle function may have catastrophic consequences for the quality of life.

Strength and everyday life

By combining information from several sources and assuming a rise time of less than 3 s (references in Young 1986), it would seem that a healthy 20-year-old female requires to make quadriceps contractions of 50 to 70 per cent of her maximum to rise from a low, armless chair but that a healthy 80-year-old female must make maximal quadriceps contractions (Fig. 4). That is, it seems likely that, on average, a healthy woman will reach the 'threshold' for this activity at about 80 years of age. Instead of a low, armless chair, consider a low, armless, toilet pedestal. If there is no grab-rail or other means for arm strength to supplement quadriceps strength, this could well be an important threshold. The predictions of Fig. 4, however, have yet to be verified by formal testing.

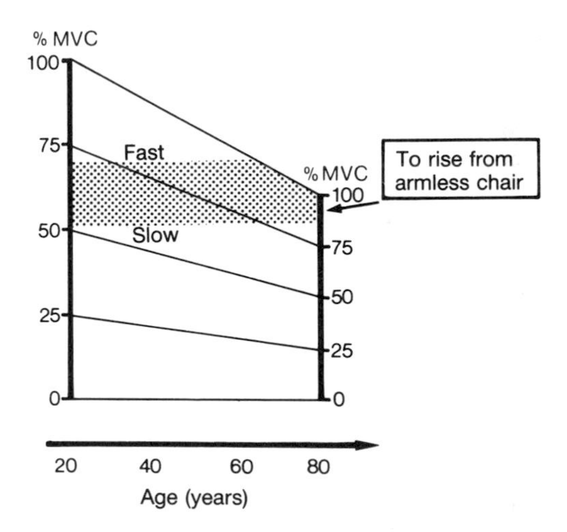

Fig. 4 Effect of age on the estimated percentage of a maximal voluntary contraction (% MVC) of the quadriceps muscles required to raise a healthy woman from a low, armless chair (reproduced, with permission, from Young 1986).

The ability of men and women to step up and down from steps of different heights has been tested for large groups of 70- and 79-year-olds in Göteborg (Aniansson *et al.* 1980; Lundgren-Lindquist *et al.* 1983, respectively) and a rather smaller group of 80-year-olds in Copenhagen (Danneskiold-Samsøe *et al.* 1984). Half of the older men and more than 85 per cent of the older women were unable to negotiate a 50-cm step without the help of a hand-rail. Even among the 70-year-old subjects, about 10 per cent of the men and 60 per cent of the women could not do this. Unfortunately, much more work is required before laboratory measurements can be used to predict which individuals are at, or near to, these and other, similar, thresholds.

Walking speed is linked with both calf muscle function and quadriceps function. A group of 79-year-old Swedish women had their walking speeds measured over a 30-m course, analogous to crossing a major road (Lundgren-Lindquist *et al.* 1983). For the whole group, the mean comfortable speed was only 0.9 m/s and the mean maximum speed 1.2 m/s. In Sweden, pedestrian crossings have their signals set for a minimum walking speed of 1.4 m/s. Even walking flat out, less than a third of the 31 'healthy' 79-year-old women within the larger group were fast enough to beat the traffic. Walking speeds for a similar group of men were rather faster but, even so, only 72 per cent of them were able to walk at 1.4 m/s or faster. This seems to be another example of a situation where the loss of muscle power or strength with advancing age puts large numbers of people on the wrong side of a functionally important threshold at an age which is far from exceptional.

Falls and fractures

Falls by people aged 65 and over account for 52 per cent of all deaths from domiciliary accidents (Ramprakash and Daly 1983). Discussions of possible explanations have tended to emphasize impairment of postural control. Preventing a stumble becoming a fall depends not only on postural control but also on the ability to develop a critical ('threshold') power output in the lower limb and trunk muscles. Muscle weakness and wasting may therefore be important contributory factors; muscle wasting is more common amongst fallers than controls, patients sustaining a femoral neck fracture are lighter than average for their height, and biopsies from their quadriceps have a reduced mean fibre area.

Figure 5 emphasizes the considerable reductions in the power-generating ability of the lower limbs which are associated with old age and which result from cooling. The very low level of power that can be developed by elderly women may help to explain the relative excess of females amongst fallers.

Cooling the calves of young men by 8.4 degrees centigrade reduced the peak power developed in a standing jump by 43 per cent (Davies and Young 1983). If cooling elderly muscle produces a similar, relative impairment of peak power, the power-generating ability of cold, elderly, female muscle will be very low indeed (Fig. 5); it is not surprising that an 'excess' of thin patients present in winter after a femoral fracture (usually sustained indoors) (Bastow *et al.* 1983). An additional contributory factor may be that the thinnest elderly patients have an impaired heat-production response to a lowered environmental temperature (see Chapter 19.3) rendering their muscles more susceptible to cooling.

IMPROVING PERFORMANCE

The strength of elderly muscles can be increased by appropriate physical training. The quadriceps strength of men in their early 70s was increased by 10 to 20 per cent by a strength-training programme, three times a week for 12 weeks, using only body weight as the training resistance (Aniansson and Gustafsson 1981).

As in younger age groups, strength-training by weight lifting produces an improvement in weight lifting performance which is much greater than the improvement in the strength of the muscle concerned. For example, after 12 weeks of weight training, men aged 60 to 72 had a 107 per cent mean increase in the maximum weight lifted by knee extension but only a 10 to 17 per cent mean increase in the isokinetic strength of the quadriceps (Frontera *et al.* 1988). Although simple compared with many daily activities, weight-training movements are more complex than measurements of isokinetic strength. They require the co-ordinated activation of other

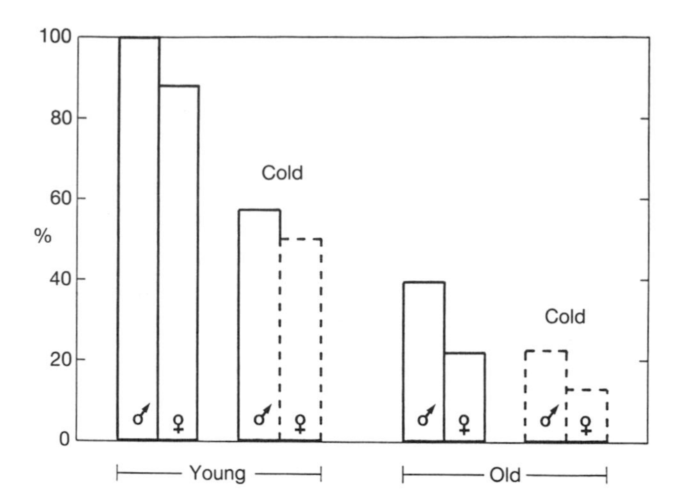

Fig. 5 Effect of age, sex, and cooling on the power-generating ability of the lower limb. Data expressed as percentages of the power-generating ability of a young adult male. Solid columns are based on data from Bosco and Komi (1980), Davies and Young (1983), and Davies *et al.* (1983). Broken columns are hypothetical extrapolations, as yet untested.

muscles, to stabilize the body. Weight lifting involves a large learning component to co-ordinate the muscles involved and this is highly specific to the movements practised. It seems possible, therefore, that an elderly person's functional ability might not be enhanced if she merely strengthens the relevant muscles but does not practise the specific movements. The restoration of strength might not be sufficient, of itself, to restore a functional ability to someone who had dropped below the corresponding 'threshold'.

A study of 58 men and 67 women, mostly aged 65 to 75 years, demonstrated statistical associations between calf muscle strength and chosen walking speed and, in the men, the number of steps taken per day (Bassey *et al.* 1988). This hints at the possibility that a high level of customary activity may help protect against some of the age-associated loss of muscle mass and performance. Further studies of the effect of strengthening exercises on the ability to perform everyday functional activities should be a priority.

References

Aniansson, A. and Gustafsson, E. (1981). Physical training in elderly men with special reference to quadriceps strength and morphology. *Clinical Physiology*, **1**, 87–98.

Aniansson, A., Rundgren, Å., and Sperling, L. (1980). Evaluation of functional capacity in activities of daily living in 70-year-old men and women. *Scandinavian Journal of Rehabilitation Medicine*, **12**, 145–54.

Aniansson, A., Sperling, L., Rundgren, Å., and Lehnberg, E. (1983). Muscle function in 75-year-old men and women: a longitudinal study. *Scandinavian Journal of Rehabilitation Medicine*, **suppl. 9**, 92–102.

Aniansson, A., Hedberg, M., Henning, G.B., and Grimby, G. (1986). Muscle morphology enzymatic activity, and muscle strength in elderly men: a follow-up study. *Muscle and Nerve*, **9**, 585–91.

Bassey, E.J., Bendall, M.J., and Pearson, M. (1988). Muscle strength in the triceps surae and objectively measured customary walking activity in men and women over 65 years of age. *Clinical Science*, **74**, 85–9.

Bastow, M.D., Rawlings, J., and Allison, S.P. (1983). Undernutrition, hypothermia, and injury in elderly women with fractured femur: an injury response to altered metabolism? *Lancet*, **i**, 143–6.

Bosco, C. and Komi, P.V. (1980). Influence of aging on the mechanical behavior of leg extensor muscles. *European Journal of Applied Physiology*, **45**, 209–19.

Danneskiold-Samsøe, B., Kofod, V., Munter, J., Grimby, G., Schnohr, P., and Jensen, G. (1984). Muscle strength and functional capacity in 78–81-year-old men and women. *European Journal of Applied Physiology*, **52**, 310–14.

Davies, C.T.M. and Young, K. (1983). Effect of temperature on the contractile properties and muscle power of triceps surae in humans. *Journal of Applied Physiology*, **55**, 191–5.

Davies, C.T.M., Mecrow, I.K., and White, M.J. (1982). Contractile properties of the human triceps surae with some observations on the effects of temperature and exercise. *European Journal of Applied Physiology*, **49**, 255–69.

Davies, C.T.M., White, M.J., and Young, K. (1983). Electrically evoked and voluntary maximal isometric tension in relation to dynamic muscle performance in elderly male subjects, aged 69 years. *European Journal of Applied Physiology*, **51**, 37–43.

Frontera, W.R., Meredith, C.N., O'Reilly, K.P., Knuttgen, H.G., and Evans, W.J. (1988). Strength conditioning in older men: skeletal muscle hypertrophy and improved function. *Journal of Applied Physiology*, **64**, 1038–44.

Green, H.J. (1986). Characteristics of aging human skeletal muscles. In *Sports medicine for the mature athlete* (ed. J.R. Sutton and R.M. Brock), pp. 17–26. Benchmark Press, Indianapolis.

Grimby, G. and Saltin, B. (1983). The ageing muscle. *Clinical Physiology*, **3**, 209–18.

Larsson, L., Grimby, G., and Karlsson, J. (1979). Muscle strength and speed of movement in relation to age and muscle morphology. *Journal of Applied Physiology*, **46**, 451–6.

Lundgren-Lindquist, B., Aniansson, A., and Rundgren, Å. (1983). Functional studies in 79-year-olds. III. Walking performance and climbing capacity. *Scandinavian Journal of Rehabilitation Medicine*, **15**, 125–31.

Ramprakash, D. and Daly, M. (1983). *Social Trends*, **13**, Table 7.16. HMSO, London.

Shock, N.W. and Norris, A.H. (1970). Neuromuscular coordination as a factor in age changes in muscular exercise. *Medicine and Sport*, **4**, 92–9.

Vandervoort, A.A. and McComas, A.J. (1986). Contractile changes in opposing muscles of the human ankle joint with aging. *Journal of Applied Physiology*, **61**, 361–7.

Young, A. (1984). The relative isometric strength of type I and type II muscle fibres in the human quadriceps. *Clinical Physiology*, **4**, 23–32.

Young, A. (1986). Exercise physiology in geriatric practice. *Acta Medica Scandinavica*, **suppl. 711**, 227–32.

Young, A. (1989). Muscle function in old age. *New Issues in Neuroscience*, **1**, 149–64 and 235–66.

19.2 Aerobic exercise

CAROLYN GREIG AND ARCHIE YOUNG

The ability to sustain muscular activity for more than just a minute or two depends on ensuring the delivery of an adequate supply of oxygen to the working muscles and its utilization in their mitochondria. It depends, therefore, not only upon central cardiorespiratory function, but also on the capacity of the peripheral muscles for aerobic metabolism. This chapter summarizes the changes in endurance-related, aerobic muscle function that are associated with increasing age and their possible causes (reviewed in detail by Saltin 1986*a*). It also examines their practical implications and the effects of physical training.

MAXIMAL AEROBIC POWER

Both cross-sectional and longitudinal studies have demonstrated a decline in maximal aerobic power (i.e. maximal oxygen uptake, $\dot{V}O_2$max) with advancing age. The underlying mechanisms are not clearly understood. Whether expressed in absolute units (1/min) or relative to body mass (ml/kg. min), the reduction is some 10 per cent per decade, comparable to the loss of muscle strength. Indeed, maximal oxygen uptake appears almost independent of age when expressed relative to total body muscle and it can be argued that the age-associated decline in $\dot{V}O_2$max may owe more to the shrinking total muscle mass than to any central, cardiac changes (Asmussen 1980; see also Section 9).

Another factor which has been implicated in the reduction of maximal oxygen uptake with age is the customary level of physical activity. Hagberg (1987) suggests that the rate of decline (in ml/kg.min) may be reduced by as much as half by a high level of customary physical activity. This is, however, a controversial issue and opinion remains divided as to whether the rate of decline of maximal oxygen uptake with age can be slowed by regular, vigorous, physical activity. Shephard (1987) has suggested that the rate of decline is similar to that observed in sedentary individuals and that apparent differences are probably due more to confounding variables, such as changes in body mass, rather than a genuine effect of maintained activity.

Elite, veteran sportsmen are an interesting and highly selected group of people. The top performers are highly motivated, have a very high level of customary physical activity, and are probably free of significant exercise-limiting disease. Studies of such individuals could be said to offer a glimpse of the effect of 'pure ageing' on maximal exercise performance. Saltin (1986*a*) reports that inactive and still active former élite orienteers differ in their $\dot{V}O_2$max values but not in the rates of decline in $\dot{V}O_2$max. The rate of decline in performance (about 8 per cent per decade) is broadly similar to the decline in $\dot{V}O_2$max. Similar rates of decline in prolonged exercise performance are seen in the marathon (13 per cent per decade (Åstrand 1986)) and prolonged cycling events (e.g. 11 per cent per decade for the distance covered in 12 h solo; Fig. 1).

It is important to stress how few of the published studies concern measurements of maximal oxygen uptake in people over 70 years of age.

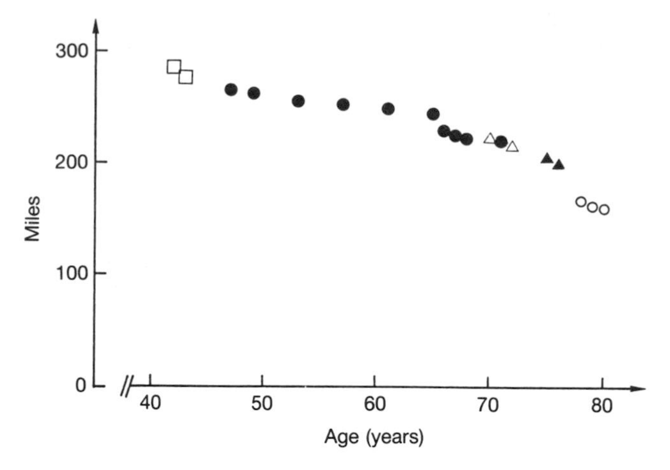

Fig. 1 British Veterans' records for cycling 12 h solo (as at 31 December 1980), including both longitudinal and cross-sectional data (*Veterans' Time Trials Association Handbook* 1981). ● Various different record holders; □ records both held by one individual; △ records both held by one individual; ▲ records both held by one individual; ○ records all held by one individual.

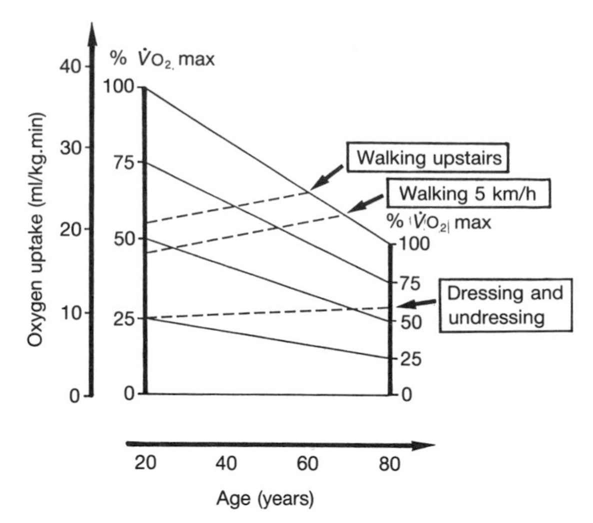

Fig. 2 Effect of age on the percentage of $\dot{V}O_2$max required for the performance of some everyday activities by an average, healthy woman (reproduced and translated, with permission, from Saltin 1980).

AEROBIC POTENTIAL OF ELDERLY MUSCLE

The potential of muscle for aerobic metabolism, that is its capillary density and oxidative enzyme activities, seems well maintained at least into the 70s (reviewed by Green 1986 and Saltin 1986*a*). A few studies, however, have reported decreased oxidative enzyme activities in older subjects and one has reported an increased plasma lactate concentration at a constant relative work rate (Strandell 1964). Interpretation of these studies is difficult as it is only with élite athletes that one can be reasonably confident of adequate matching of the habitual physical activity of the old and young subjects. Nevertheless, it is clear that any differences due to age are small when compared with those due to the customary level of physical activity.

FUNCTIONAL CONSEQUENCES

The reduction of maximal oxygen uptake with advancing age means that a given task or activity will require a greater proportion of maximal aerobic power. Thus, activities with a low absolute oxygen cost may yet demand a high proportion of an old person's maximal oxygen uptake (Fig. 2). This situation is made worse by the fact that not only the relative but also the absolute energy cost of some activities increases with age. The increasing absolute energy cost of walking (Bassey and Terry 1986), for example, can be explained by the shortening of stride length as age increases (Himann *et al.* 1988).

Many elderly people are in circumstances where it would need only a small, further reduction in maximal aerobic power to render some everyday activities either impossible or so dependent on anaerobic metabolism as to be unpleasant to perform.

Further reductions in maximal aerobic power, below the level expected for a person's age, may be a direct effect of disease or may be due to inactivity. Inactivity, whether chosen or enforced (perhaps as a result of illness), may reduce $\dot{V}O_2$max by a further 10 to 20 per cent. Surgery results in similar reductions in maximal aerobic power (evidence, mostly from younger age groups, reviewed by Young (1990)). Figure 3 illustrates the drastic consequences of a 15 per cent

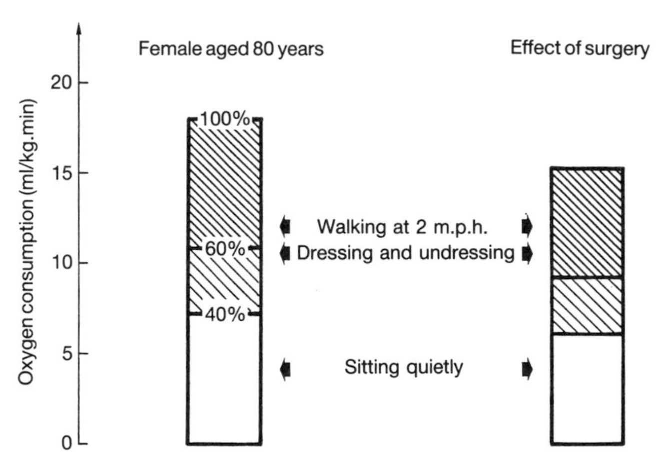

Fig. 3 Implications for an otherwise healthy 80-year-old woman of a 15 per cent fall in $\dot{V}O_2$max, such as might follow surgery.

reduction in $\dot{V}O_2$max for an average, healthy 80-year-old woman, with respect to 40 per cent of maximum (the greatest work rate that can be sustained for a full 8-hour working day) and 60 per cent of maximum (when the rate of lactate accumulation increases). Even just dressing or undressing would then entail a significant lactate accumulation, as the relative energy cost would be high. (The situation would be worse still if the absolute energy cost of dressing and undressing had been increased, perhaps as a result of joint stiffness, weakness, or dyspraxia.) Just sitting quietly for a full day would be difficult, requiring the same relative energy cost as the greatest that a younger person can sustain for a full working day. It seems likely that these changes will predispose to even greater immobility and its attendant clinical risks, e.g. deep-vein thrombosis and pulmonary embolism.

The good news is that even seemingly trivial activities may

be expected to make an important contribution to the restoration or maintenance of aerobic fitness in old age. Encouraging patients to dress in their own clothes each day is important, not only to practise the necessary skills and for self-respect, but perhaps also as a potent aerobic training stimulus.

There are several groups of elderly patients who suffer an accelerated decline in functional ability because an outcome of their condition is that the absolute energy cost of everyday tasks is increased. A very large proportion of maximal aerobic power must then be used for simple, everyday activities. For example, a stroke may cause spasticity and joint stiffness, which, combined with a change in the body's centre of gravity, will cause unproductive increases in energy expenditure. Osteoarthritis may cause an increase in the energy cost of walking and a large decrement in walking efficiency (Pugh 1973).

For elderly patients the oxygen cost of walking (in ml/kg.m) after an amputation may be elevated by some 30 per cent (Syme's amputation), 60 per cent (below knee), or 120 per cent (unilateral above knee, or bilateral below knee) (Waters *et al.* 1976; DuBow *et al.* 1983). Those with above-knee amputations have a slower cadence and shorter stride length than those with more distal amputations. Despite walking more slowly, they use a higher proportion of $\dot{V}o_2$max (Waters *et al.* 1976). The level of amputation is of extreme importance for the preservation of endurance capacity and the subsequent ability to perform everyday activities.

The majority of amputations are the result of peripheral vascular disease. Most of these patients are elderly and many also have evidence of coronary vascular disease. For some of them, particularly those who have had an above-knee amputation, the absolute and relative energy cost of ambulation is prohibitive and ambulation may be an unrealistic goal.

IMPROVING PERFORMANCE

In younger adults a marked increase in the ability to perform aerobic exercise can be achieved by physical training. The improvement occurs as a result of the enlargement of cardiorespiratory dimensions (central adaptation), and from an increased content of intramuscular oxidative enzymes, and from an increased arterio–venous oxygen difference (peripheral adaptation). It seems that the central adaptation allows a greater maximal oxygen uptake and the peripheral adaptation an increased capacity for prolonged aerobic exercise (Gollnick and Saltin 1982; Saltin 1986b).

The older individual also responds to endurance training. The published data are rather variable (Table 1), possibly reflecting not only differences in intensity or duration of training, but also differences in the stringency of the experimental protocols. Nevertheless, it seems that a 10 to 20 per cent improvement in maximal oxygen uptake can be expected, a similar percentage improvement to that seen with younger adults. Few of the subjects studied have been more than 70 years of age. It is worth noting, therefore, that in one study where all 13 subjects were over 70 (men, aged 70–81 years) there was no increase in maximal oxygen uptake, although there was a reduced heart-rate response to a constant submaximal work-rate and the training period was only 5 to 6 weeks long (Benestad 1966).

Table 1 Reported changes in the directly determined maximal oxygen uptake of healthy, elderly subjects undergoing endurance training

Age range* (years)	*N*	Duration (weeks)	$\dot{V}o_2$max (%)	Source
61–67	11	52	30	Seals *et al.* 1984*a*
57–69	88	52	12	Thomas *et al.* 1985
60–65	33	52	10	Cunningham *et al.* 1986
62–79	9	12	8	Tonino and Driscoll 1988
70–79	16	26	22	Hagberg *et al.* 1989

* Published range or calculated as mean ±2SD.
Studies have been included only if they included an adequate control group, control period, or stable pretraining baseline. Overall, ignoring differences in duration and intensity of training, the mean improvement is 13.6 per cent.

Training greatly increases the ability to sustain exercise at a fixed, submaximal, absolute level of energy expenditure but there is uncertainty about its effect upon the ability to sustain exercise at a constant relative energy expenditure. The question is not adequately settled for any age group. Some studies have shown no improvement (Saltin *et al.* 1969; Rumley *et al.* 1988), but one (Seals *et al.* 1984*b*) of men and women aged 61 to 67 years found that after 1 year of endurance training the plasma lactate at 67 per cent of $\dot{V}o_2$max was reduced by a quarter.

The central adaptations to aerobic training have not been well studied in old people but seem likely to be similar to those in younger subjects. The expected peripheral adaptation, increased oxidative enzyme activities, has been demonstrated in 70 to 75-year-old men (Örlander *et al.* 1980) and 69-year-old women (Suominen *et al.* 1977).

The functional implications of improvements in maximal aerobic power are clear: a given everyday task or activity will require a lower proportion of maximal power and may therefore be performed for much longer and with greater ease, thus alleviating functional limitation. If the rate of deterioration of $\dot{V}o_2$max can be slowed, so much the better. Even if this is not the case, a 10 to 20 per cent, training-induced improvement in $\dot{V}o_2$max may effectively postpone the crossing of functionally important thresholds for some 10 to 20 years.

References

Asmussen, E. (1980). Aging and exercise. In *Environmental physiology: aging, heat and altitude* (ed. S.M. Horvath and M.K. Yousef), pp. 419–28. Elsevier, North Holland.

Åstrand, P.-O. (1986). Exercise physiology of the mature athlete. In *Sports medicine for the mature athlete* (ed. J.R. Sutton and R.M. Brock), pp. 3–13. Benchmark Press, Indianapolis.

Bassey, E.J. and Terry, A. (1986). The oxygen cost of walking in the elderly. *Journal of Physiology*, **373**, 42P.

Benestad, A.M. (1966). Trainability of old men. *Acta Medica Scandinavica*, **178**, 321–7.

Cunningham, D.A., Rechnitzer, P.A., and Donner, A.P. (1986). Exercise training and the speed of self-selected walking pace in men at retirement. *Canadian Journal on Ageing*, **5**, 19–26.

DuBow, L.L., Witt, P.L., Kadaba, M.P., Reyes, R., and Cochran, G. Van B. (1983). Oxygen consumption of elderly persons with bilateral below knee amputations: ambulation vs wheelchair propulsion. *Archives of Physical Medicine and Rehabilitation*, **64**, 255–9.

Gollnick, P.D. and Saltin, B. (1982). Significance of skeletal muscle oxidative enzyme enhancement with endurance training. *Clinical Physiology*, **2**, 1–12.

Green, H.J. (1986). Characteristics of aging human skeletal muscles. In *Sports medicine for the mature athlete* (ed. J.R. Sutton and R.M. Brock), pp. 17–26. Benchmark Press, Indianapolis.

Hagberg, J.M. (1987). Effect of training on the decline of $\dot{V}O_2$max with aging. *Federation Proceedings*, **46**, 1830–3.

Hagberg, J.M. *et al.* (1989). Cardiovascular responses of 70- to 79-year-old men and women to exercise training. *Journal of Applied Physiology*, **66**, 2589–94.

Himann, J.E., Cunningham, D.A., Rechnitzer, P.A., and Patterson, D.H. (1988). Age-related changes in the speed of walking. *Medicine and Science in Sports and Exercise*, **20**, 161–6.

Örlander, J., Kiessling, K.H., and Ekblom, B. (1980). Time course of adaption to low intensity training in sedentary men: dissociation of central and local effects. *Acta Physiologica Scandinavica*, **108**, 85–90.

Pugh, L.G.C.E. (1973). The oxygen intake and energy cost of walking before and after unilateral hip replacement, with some observations on the use of crutches. *Journal of Bone and Joint Surgery*, **55B**, 742–5.

Rumley, A.G., Taylor, R., Grant, S., Pettigrew, A.R., Findlay, I., and Dargie, H. (1988). Effect of marathon training on the plasma lactate response to submaximal exercise in middle-aged men. *British Journal of Sports Medicine*, **22**, 31–4.

Saltin, B. (1980). Fysisk vedligholdelse hos ældre. I. Aerob arbejdsevne. *Månedsskrift for Praktisk Lægegerning* (Copenhagen), **58**, 193–216.

Saltin, B. (1986*a*). The aging endurance athlete. In *Sports medicine for the mature athlete* (ed. J.R. Sutton and R.M. Brock), pp. 59–80. Benchmark Press, Indianapolis.

Saltin, B. (1986*b*). Physiological adaptation to physical conditioning. In *Physical activity in health and disease* (ed. P.-O. Åstrand and G. Grimby), pp. 11–24. Almqvist and Wiksell International, Stockholm.

Saltin, B., Hartley, L.H., Kilbom, A., and Åstrand, I. (1969). Physical training in sedentary middle-aged and older men. *Scandinavian Journal of Clinical and Laboratory Investigation*, **24**, 323–34.

Seals, D.R., Hagberg, J.M., Hurley, B.F., Ehsani, A.A., and Holloszy, J.O. (1984*a*). Endurance training in older men and women. I. Cardiovascular responses to exercise. *Journal of Applied Physiology*, **57**, 1024–9.

Seals, D.R., Hurley, B.F., Schultz, J., and Hagberg, J.M. (1984*b*). Endurance training in older men and women. II. Blood lactate response to submaximal exercise. *Journal of Applied Physiology*, **57**, 1030–3.

Shephard, R.J. (1987). *Physical activity and ageing* (2nd edn). Croom Helm, London.

Strandell, T. (1964). Circulatory studies on healthy old men. *Acta Medica Scandinavica*, **175** (suppl. 414), 1–44.

Suominen, H., Heikkinen, L., and Parkatti, T. (1977). Effects of 8 weeks physical training on muscle and connective tissue of the M. vastus lateralis in 69-year old men and women. *Journal of Gerontology*, **32**, 33–7.

Thomas, S.G., Cunningham, D.A., Rechnitzer, P.A., Donner, A.P., and Howard, J.H. (1985). Determination of the training response in elderly men. *Medicine and Science in Sports and Exercise*, **17**, 667–72.

Tonino, R.P. and Driscoll, P.A. (1988). Reliability of maximal and submaximal parameters of treadmill testing for the measurement of physical training in older persons. *Journal of Gerontology*, **43**, M101–4.

Waters, R.L., Perry, J., Antonelli, D., and Hislop, H. (1976). Energy cost of walking of amputees. The influence of level of amputation. *Journal of Bone and Joint Surgery*, **58A**, 42–6.

Young, A. (1990). Exercise, fitness and recovery from surgery, disease, or infection. In *Exercise, Fitness, and Health: A Consensus of Current Knowledge* (eds. C. Bouchard *et al.*) pp. 589–600. Human Kinetics, Champaign.

19.3 The uptake, storage, and release of metabolites by muscle

MARK PARRY-BILLINGS, E. A. NEWSHOLME, AND ARCHIE YOUNG

INTRODUCTION

Muscle is the most plentiful tissue in the body. Its metabolic activity is not confined to the processes which fuel the making and breaking of actomyosin cross-bridges. It also plays a major role in the uptake, storage, and release of metabolites, in the generation of heat, and in the metabolic control of these processes. Its ability to perform these functions may be significantly altered in old age, as a result of both qualitative changes in the tissue and the considerable reduction in total muscle mass.

Even in health, quadriceps cross-sectional area falls on average by 25 to 35 per cent between 25 and 75 years of age (see Chapter 19.1). This is associated with a similar loss of muscle mass throughout the body. The progressive reduction in creatinine excretion described by Tzankoff and Norris (1977) corresponds to a 45 per cent reduction in total muscle mass between 23 and 90 years of age (Fig. 1). In frail and/or chronically malnourished elderly patients, the loss of muscle mass may be even more dramatic. The resulting changes in drug pharmacokinetics are well known, with a greatly prolonged half-life for fat-soluble drugs and an elevated peak plasma level following the administration of a water-soluble drug. The metabolic consequences of a reduced muscle mass are less well known although it seems likely that some may be at least as important, limiting the elderly patient's ability to mount an adequate metabolic response in times of stress due, for example, to cold, trauma, or infection.

Sensitivity in metabolic control

In order to understand metabolic changes that may depend on changes in elderly muscle, it is necessary to appreciate the mechanisms that may enhance sensitivity in the control of metabolic processes. Sensitivity in metabolic regulation can be defined as the quantitative relationship between the relative change in enzyme activity and the relative change in concentration of the regulator (see Newsholme and Leech

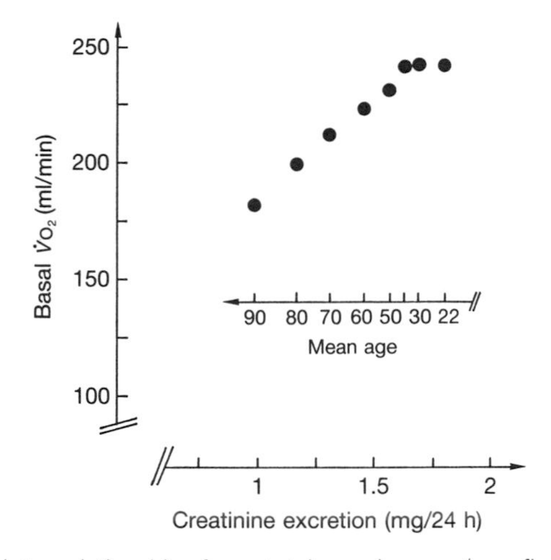

Fig. 1 Inter-relationship of age, total muscle mass (as reflected in 24-h creatinine excretion), and basal metabolic rate (basal oxygen consumption) in male volunteers (redrawn from Tzankoff and Norris 1977).

1983). The greater the response in enzyme activity to a given increase in regulator concentration, the greater is the sensitivity.

The rates of many processes change hugely in response to physiological stimuli (e.g. walking upstairs requires an increase in the rate of glycolysis of several hundredfold). Such large changes in flux may be mediated by mechanisms which increase sensitivity, e.g. 'substrate cycles' (Newsholme 1980) and 'branched point sensitivity' (Newsholme *et al.* 1988). These mechanisms may become less effective in old age.

Substrate cycles

A substrate cycle exists if a reaction which is non-equilibrium in the 'forward' direction of a pathway operates simultaneously with an opposing reaction which is non-equilibrium in the reverse direction of the pathway and which is catalysed by a different enzyme (see Fig. 2(a)). At least one of the reactions involves the hydrolysis of ATP. Thus, a substrate cycle is an energy-consuming and heat-producing process. As sensitivity is proportional to the ratio of cycling rate to flux, high sensitivity requires an increase in cycling rate and, as a result, increased heat production. Conditions which require an increase in sensitivity would be expected to increase cycling rates. This has been shown to be the case for stress, injury, fasting, and after exercise.

Catecholamines increase the rate of cycling and may be involved in all these circumstances. There are several examples of tissues and reactions where responsiveness to both adrenaline and noradrenaline is decreased in old age, even though the plasma levels may be elevated. If the responsiveness of substrate cycling to catecholamines is decreased in elderly human muscle, this might partly explain the decreased rates of thermogenesis seen in old age (see section on 'heat production' below).

Branched point sensitivity

The increased sensitivity in control provided by the branched point mechanism is similar in principle to that achieved in the substrate cycle; a continuous high flux in one branch of a pathway provides optimal conditions for the precise regulation of the (much smaller) flux in the other branch (see Fig. 2(b)).

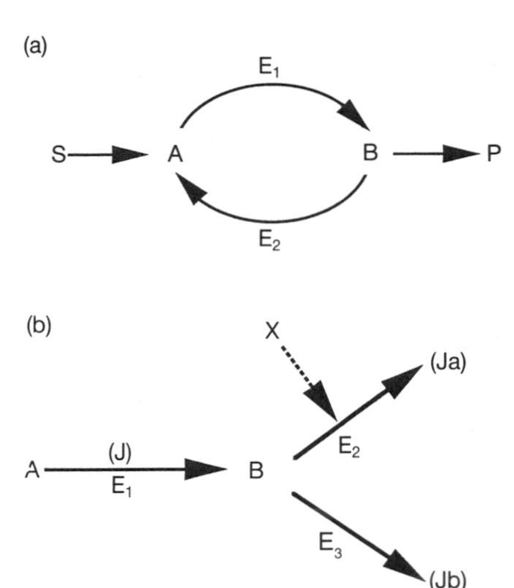

Fig. 2 (a) A hypothetical substrate cycle. A substrate cycle is produced when a non-equilibrium reaction in the forward direction ($A \rightarrow B$) catalysed by enzyme E_1 is opposed by another non-equilibrium reaction in the reverse direction ($B \rightarrow A$) catalysed by enzyme E_2. The highest sensitivity is achieved when the cycling rate is high compared to the flux through the pathway ($S \rightarrow P$). (b) The regulation of flux through a branch in a branched pathway. A branched pathway is produced when the flux (J) divides into two separate fluxes (Ja) and (Jb). The overall response of Ja to X is most effective when Jb is much greater than Ja: that is when Ja is much smaller than the total flux (J).

HEAT PRODUCTION

Even healthy elderly subjects seem to be less able than young adults to maintain their core temperature when exposed to a low ambient temperature (Collins *et al.* 1985). Amongst elderly subjects, those who are thinnest are most at risk, showing a fall in core temperature (and therefore in muscle temperature) after only a mild degree of cooling. This is not merely because of their poorer insulation but because they are less able to mount a protective increase in their metabolic rate (Fig. 3) (Fellows *et al.* 1985). This may be due, in part, to their smaller total muscle mass but it seems likely that there is also a qualitative change which results in a decreased ability to stimulate substrate cycles in muscle and other tissues (especially adipose tissue). Perhaps the catecholamine sensitivity of thermogenic cycles is even lower in cachectic than in well-nourished elderly people. This has yet to be confirmed, however.

GLUCOSE TOLERANCE

It is well known that old age is accompanied by impairment of the ability to correct the hyperglycaemia produced by an oral glucose load. As muscle is the major site of uptake, it is not surprising that a diminished total muscle mass should result in a decreased rate of glucose disposal. The decrements in insulin responsiveness and sensitivity, however, are too

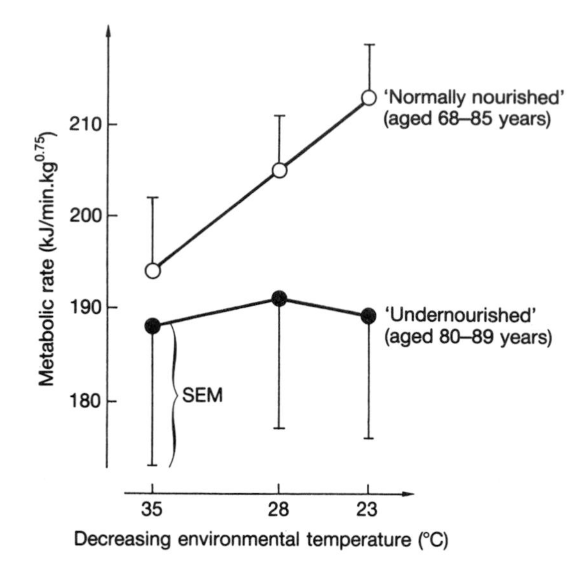

Fig. 3 Effect of nutritional status on the ability of elderly women to increase their metabolic rate in response to a cold challenge (drawn from the data of Fellows *et al.* 1985).

great for this to be an adequate explanation on its own; the sensitivity of muscle to insulin is also decreased (Leighton *et al.* 1989). There is evidence that chronic elevation of the plasma level of adrenaline increases the sensitivity of muscle to insulin. A decrease in the sensitivity of elderly muscle to the effects of catecholamines may thus contribute to its decreased insulin sensitivity and to impairment of thermogenesis (Newsholme 1989).

IMMUNE FUNCTION

Glutamine and glucose are important fuels for lymphocytes and macrophages. Both fuels are only partially oxidized and their rates of utilization are high, even in relatively 'quiescent' cells. These processes provide not only energy but also intermediates for biosynthesis, e.g. glutamine, ammonia, and aspartate (from glutaminolysis) for synthesis of purine and pyrimidine nucleotides, and glucose 6-phosphate (from glycolysis) for synthesis of ribose 5-phosphate. However, the rates of glutaminolysis and glycolysis in lymphocytes are hugely (more than 400-fold) in excess of the apparent maximal capacity of the biosynthetic processes (Szondy and Newholme 1989). This is an example of branched point sensitivity; the high rates of glutamine and glucose utilization in lymphocytes and macrophages (and, incidentally, in tumour cells) allow high precision in the control of the rates of synthesis of purine and pyrimidine nucleotides at specific times during the cell cycle. This allows a rapid rate of cell division. Conversely, lowering the concentration of glutamine decreases the rate of lymphocyte proliferation in culture (Szondy and Newsholme 1989).

What is the source of the glutamine used by macrophages and lymphocytes? Most dietary glutamine is used by intestinal cells and does not enter the bloodstream. Skeletal muscle appears to be the major tissue involved in the production of glutamine and its release into the blood. Synthesis of glutamine requires nitrogen and it is likely that branched-chain amino acids are the source of this in muscle. Hence, three processes in skeletal muscle, the transfer of nitrogen, the formation of glutamine, and its transport across the cell membrane, are considered to be important for the functioning of macrophages and lymphocytes. Therefore skeletal muscle, and especially its ability to control the rate of release of glutamine into the bloodstream, may be considered to be an important part of the immune system (Newsholme *et al.* 1988).

This concept raises the possibility that the reduced total muscle mass of old age, and perhaps also qualitative changes that occur in old age, may impair the ability of muscle to control the rate of glutamine release in relation to its requirement by the immune system (e.g. during critical illness). Indeed, it is known that the *in-vitro* rate of glutamine release from muscle is lower in the aged rat (Parry-Billings *et al.* 1991).

METABOLIC EFFECTS OF CRITICAL ILLNESS

The metabolic effects of the increasingly severe challenges imposed, respectively, by elective surgery, multiple trauma, sepsis, and burns are discussed in the *Oxford Textbook of Medicine* and in several reviews (e.g. Rennie 1985; Little and Frayn 1986; Wernerman and Vinnars 1987; see also Chapter 4.2).

Cuthbertson's hypermetabolic 'flow' phase after major trauma, sepsis, or burns starts with increased plasma concentrations of adrenaline, glucose, glucagon, and cortisol, followed later by increased levels of insulin and growth hormone. Despite prolonged insulin resistance, glucose turnover is increased, with increased hepatic gluconeogensis from lactate, from alanine, and, after its intestinal conversion to alanine, from glutamine. The rates of efflux of alanine and glutamine from skeletal muscle are increased and their levels in muscle are decreased. There is increased uptake of alanine by the liver, and of glutamine by the intestine and probably also by cells of the immune system. (Some of the increased glutamine uptake by the intestine may, in fact, be due to cells of the immune system; approximately a quarter of the cells of the intestine are immune cells and arteriovenous differences measured across the intestine may also reflect uptake by mesenteric lymph nodes.) Thus, there is an increase in the body flux of nitrogen and a net loss of muscle nitrogen. It is suggested that the increased breakdown of skeletal muscle protein provides amino acids for several functions but particularly provides nitrogen for an increase in the rate of formation and release of glutamine, an 'essential' amino acid for the function of macrophages, lymphocytes, and cells involved in repair (Newsholme *et al.* 1988).

Sympathetic stimulation is at least partially responsible for the elevation of metabolic rate and core temperature after injury. Until recently there had been no satisfactory explanation for this increase in thermogenesis. There is now evidence that it may result, in part, from stimulation of substrate cycles. For example, a 15-fold increase in the rate of the intracellular triacylglycerol/fatty acid cycle (which can be blocked by propranolol), and a doubling of the rate of glycolytic/gluconeogenic substrate cycles have been observed in severely burned patients (Wolfe *et al.* 1987). It is possible that these and other cycles in muscle may also be important in thermogenesis. The Cori cycle can be considered as a large substrate cycle (Fig. 4). Its function would be to act as a dynamic buffer to provide adequate amounts of circulating glucose and/or lactate for tissues which may need to use them at high rates during illness (e.g. lymphocytes, macrophages, fibroblasts, and endothelial cells). Failure to increase the rate of such cycles, that is, to maintain a high flux through such pathways, may lead to impaired control of the supply of essential nutrients and, coincidentally, to a smaller increase in thermogenesis in an elderly patient.

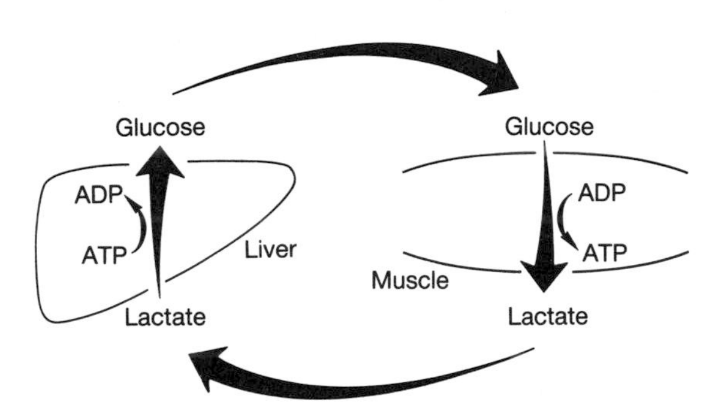

Fig. 4 The Cori cycle, viewed as a substrate cycle. More ATP is used to convert lactate to glucose than is synthesized in converting glucose to lactate; the result is that the overall cycle converts chemical energy into heat.

The role of skeletal muscle in illness has been seen as providing a pool of amino acids that can be mobilized. This is important not only for hepatic gluconeogensis but also as a source of the raw materials for hepatic synthesis of acute-phase plasma proteins, enzymes, and metal-binding proteins. As this requirement for amino acids will be small, we consider that the major role of muscle may be to provide glutamine at a high enough rate to allow branched point sensitivity to operate in cells of the immune system and cells involved in repair, so enabling them to respond with rapid proliferation and biosynthesis.

Beisel (1986) has concluded that 'the combined metabolic and physiological responses to severe surgical illness and sepsis appear to be purposeful and ultimately beneficial', but warns that a prolonged or particularly severe illness may severely deplete the labile store of nitrogen in muscle. This store may now be identified more specifically as the amount of glutamine that can be synthesized and/or released from muscle. During recovery, the replenishment of nitrogen stores by the regrowth of muscle may take many weeks. During that time not only does the patient remain severely weakened, but both tissue repair and immune competence also remain impaired. This may now be interpreted in terms of the impaired availability of glutamine.

Allison (1986) reviews some of the evidence that the smaller the muscle mass before the illness, the worse the impact of the illness on the patient. After correction for the severity of injury, the rate of production of urea (a crude index of protein catabolism) is proportional to the patient's initial body weight. In rats, the urinary loss of nitrogen is less after fracture in those previously given a low protein diet. If the mobilization of muscle protein is, indeed, beneficial, these observations imply that a small muscle mass or pre-existing malnutrition weaken the metabolic defence which can be mounted. This is borne out by the observations from Allison's own group of the effect of cachexia on mortality after femoral fracture (described below) and the much earlier observations that the greater the preoperative weight loss the greater the mortality associated with surgery for peptic ulcer (Studley 1936). Allison concludes: 'Perhaps we should give pre- and postoperative feeding to mimic the substrate mobilization of injury in the patients who have no substrate to mobilise.'

The reduced total muscle mass of old age may be a factor in reducing the reserve of amino acids that can be mobilized; this may significantly limit the metabolic and immunological defence mounted in the event of major illness. It is striking that age is a component of the APACHE II scoring system which, in turn, is a significant predictor of the likelihood of death in intensive care medicine. However, the validity of chronological age as an independent predictor has been challenged (Editorial 1991). Nevertheless, it is agreed that the effect of age on the response to injury is poorly documented. The plasma cortisol after inguinal herniorrhaphy is probably greater than in younger subjects (Blichert-Toft *et al.* 1979). After fractures, elderly patients may have a more prolonged elevation of plasma glucose and cortisol than younger subjects but this may be related more to their greater immobility than to any difference in response to the fracture (Frayn *et al.* 1983).

Metabolic interventions in critical illness

The loss of nitrogen can be inhibited by minimizing heat losses (see, for example, Carli *et al.* 1982). It can also be reduced or even reversed by nutritional measures, especially if these include a source of glutamine (Roth *et al.* 1988) or if they are combined with an anabolic agent such as growth hormone (Ponting *et al.* 1988), but corresponding clinical benefits have yet to be demonstrated. Manipulations to increase the plasma glutamine may be beneficial. These might involve the administration of glutamine or the enhancement of its release from muscle. Preliminary studies show that administration of growth hormone to rats increases the level of glutamine in muscles and its rate of release from muscle *in vitro* (M. Parry-Billings, unpublished work). Perhaps this helps explain the beneficial effect of growth hormone on wound healing in both normal and tumour-bearing rats (Pessa *et al.* 1985). As the rate of glutamine release from elderly rat muscle (*in vitro*) is less than that from the muscle of younger rats, it seems that this use of growth hormone might be of particular relevance to elderly patients but it has yet to be studied.

There is evidence that supplementary feeding postoperatively improves healing of experimental colonic anastomoses in moderately malnourished mice (Ward *et al.* 1982). The results of studies of supplementary nasogastric feeding after femoral fracture (below) are also strongly suggestive of clinical benefit.

Recovery after femoral fracture

Nutritional status, as judged from anthropometric indices, is closely linked with death rate after femoral fracture. In a Nottingham study of 744 elderly women with fractures of the neck of the femur, 47 per cent had triceps skin-fold thicknesses and mid-arm circumferences similar to a reference population and had a fatality of 4.4 per cent. Those with values between 1 and 2 standard deviations below the reference means (34 per cent of the total) had a fatality of 8 per cent. The remaining 19 per cent were even thinner and had a fatality of 18 per cent (Bastow *et al.* 1983*a*).

Supplementary nasogastric tube feeding of thin and very thin patients increased their caloric intake, improved their anthropometric indices, and reduced the time taken to achieve independent mobility. In addition, but only in the 'very thin' group, it hastened the recovery of normal serum levels of a short half-life protein (thyroid- binding prealbumin) and reduced the length of hospital stay. Fatality in the 'very thin' group was reduced from 22 per cent to 8 per cent by supplementary feeding, but this difference did not achieve conventional statistical significance (Bastow *et al.* 1983*b*). A Swiss study also suggests that supplementary nutrition is beneficial for elderly patients with femoral fracture (Delmi *et al.* 1990).

These reports indicate a major effect of nutritional status on the recovery and rehabilitation of elderly women who have suffered a femoral fracture. The possibility of benefit suggests that there is a role for supplementary feeding in helping to meet the heavy metabolic demands imposed on elderly muscle by trauma and surgery. Moreover, supplementary feeding may also reduce the amount of muscle tissue sacrificed to meet those metabolic requirements and/or lost as a result of immobilization.

References

Allison, S.P. (1986). Some metabolic aspects of injury. In *The scientific basis for the care of the critically ill* (ed. R.A. Little and K.N. Frayn), pp. 169–83. Manchester University Press.

Bastow, M.D., Rawlings, J., and Allison, S.P. (1983a). Undernutrition, hypothermia, and injury in elderly women with fractured femur: an injury response to altered metabolism? *Lancet*, **i**, 143–6.

Bastow, M.D., Rawlings, J., and Allison, S.P. (1983b). Benefits of supplementary tube feeding after fractured neck of femur: a randomised controlled trial. *British Medical Journal*, **287**, 1589–92.

Beisel, W.R. (1986). Sepsis and metabolism. In *The scientific basis for the care of the critically ill* (ed. R.A. Little and K.N. Frayn), pp. 103–22. Manchester University Press.

Blichert-Toft, M. *et al.* (1979). Influence of age on the endocrine–metabolic response to surgery. *Annals of Surgery*, **190**, 761–70.

Carli, F., Clark, M.M., and Woollen, J.W. (1982). Investigation of the relationship between heat loss and nitrogen excretion in elderly patients undergoing major abdominal surgery under general anaesthetic. *British Journal of Anaesthesia*, **54**, 1023–9.

Collins, K.J., Easton, J.C., Belfield-Smith, H., Exton-Smith, A.N., and Pluck, R.A. (1985). Effects of age on body temperature and blood pressure in cold environments. *Clinical Science*, **69**, 465–70.

Delmi, M., Rapin, C.-H., Bengoa, J.-M., Delmas, P.D., Vasey, H., and Bonjour, J.-P. (1990). Dietary supplementation in elderly patients with fractured neck of the femur. *Lancet*, **i**, 1013–16.

Editorial (1991). Intensive care for the elderly. *Lancet*, **i**, 209–10.

Fellows, I.W., Macdonald, I.A., Bennett, T., and Allison, S.P. (1985). The effect of undernutrition on thermoregulation in the elderly. *Clinical Science*, **69**, 525–32.

Frayn, K.N., Stoner, H.B., Barton, R.N., Heath, D.F., and Galasko, C.S.B. (1983). Persistence of high plasma glucose, insulin and cortisol concentrations in elderly patients with proximal femoral fractures. *Age and Ageing*, **12**, 70–6.

Leighton, B., Dimitriadis, G.D., Parry-Billings, M., Lozeman, F.J., and Newsholme, E.A. (1989). Effects of ageing on the sensitivity and responsiveness of insulin-stimulated glucose metabolism in skeletal muscle. *Biochemical Journal*, **261**, 383–7.

Little, R.A. and Frayn, K.N. (ed.) (1986). *The scientific basis for the care of the critically ill*. Manchester University Press.

Newsholme, E.A. (1980). A possible metabolic basis for the control of body weight. *New England Journal of Medicine*, **302**, 400–5.

Newsholme, E.A. (1989). A common mechanism to account for changes in thermogenesis and insulin sensitivity. In *Hormones, thermogenesis and obesity* (ed. H. Landy and F. Statman), pp. 47–58. Elsevier, New York.

Newsholme, E.A. and Leech, A.R. (1983). *Biochemistry for the medical sciences*. Wiley, Chichester.

Newsholme, E.A., Newsholme, P., Curi, R., Challoner, E., and Ardawi, M.S.M. (1988). A role for muscle in the immune system and its importance in surgery, trauma, sepsis and burns. *Nutrition*, **4**, 261–8.

Parry-Billings, M., Leighton, B., Dimitriadis, G.D., Bond, J., and Newsholme, E.A. (1991). The effects of ageing on skeletal muscle glutamine metabolism. *Proceedings of the Nutrition Society*, in press.

Pessa, M.E., Bland, K.I., Sitren, H.S., Miller, G.J., and Copeland, E.M. (1985). Improved wound healing in tumor-bearing rats treated with perioperative synthetic human growth hormone. *Surgical Forum*, **36**, 6–8.

Ponting, G.A., Halliday, D., Teale, J.D., and Sim, A.J.W. (1988). Postoperative positive nitrogen balance with intravenous hyponutrition and growth hormone. *Lancet*, **i**, 438–40.

Rennie, M. (1985). Muscle protein turnover and the wasting due to injury and disease. *British Medical Bulletin*, **41**, 257–64.

Roth, E. *et al.* (1988). Alanylglutamine reduces muscle loss of alanine and glutamine in post-operative anaesthetized dogs. *Clinical Science*, **75**, 641–8.

Studley, H.O. (1936). Percentage of weight loss. A basic indicator of surgical risk in patients with chronic peptic ulcer. *Journal of the American Medical Association*, **106**, 458–60.

Szondy, Z. and Newsholme, E.A. (1989). The effect of glutamine concentration on the activity of carbamoyl-phosphate synthase II and on the incorporation of [^{3}H] thymidine into DNA in rat mesenteric lymphocytes stimulated by phytohaemagglutinin. *Biochemical Journal*, **261**, 979–83.

Tzankoff, S.P. and Norris, A.H. (1977). Effect of muscle mass decrease on age-related BMR changes. *Journal of Applied Physiology*, **43**, 1001–6.

Ward, M.W.N., Danzi, M., Lewin, M.R., Rennie, M.J., and Clark, C.G. (1982). The effects of subclinical malnutrition and refeeding on the healing of experimental colonic anastomoses. *British Journal of Surgery*, **69**, 308–10.

Wernerman, J. and Vinnars, E. (1987). The effect of trauma and surgery on interorgan fluxes of amino acids in man. *Clinical Science*, **73**, 129–33.

Wolfe, R.R., Herndon, D.N., Jahoor, F., Miyoshi, H., and Wolfe, M. (1987). Effect of severe burn injury on substrate cycling by glucose and fatty acids. *New England Journal of Medicine*, **317**, 403–8.

19.4 Muscle disease in old age

ARCHIE YOUNG

INTRODUCTION

The neuromuscular diseases which present as muscle weakness in old age are well described in conventional accounts of muscle disease (see *Oxford Textbook of Medicine*; Engel and Banker 1986). The reader is referred to these texts for detailed accounts of polymyositis and dermatomyositis, cancer-related myositis, myopathy and neuropathy, metabolic and endocrine myopathies, alcohol-induced and drug-induced myopathies, motor neurone disease, myasthenia gravis, the limb girdle syndrome, facioscapulohumeral dystrophy, and oculopharyngeal dystrophy. Instead, this chapter will highlight selected aspects of particular conditions and draw attention to some of the potential pitfalls in the diagnosis of muscle pathology in elderly patients.

As always in geriatric medicine, it is crucial that a thorough search is made for any potentially reversible contributory factors. The most important of these are the metabolic and endocrine causes of muscle weakness, external agents such as drugs or alcohol, and the inflammatory myositides (despite the rising prevalence of associated carcinoma).

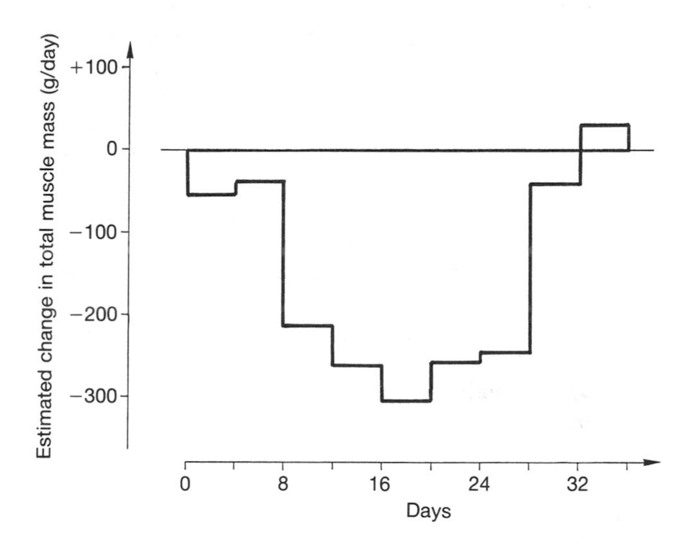

Fig. 1 Changes in total muscle mass (g/day), calculated from data on calcium, phosphorus, and nitrogen balance for successive 4-day periods. Patient: A 64-year-old man with chronic airflow obstruction, on a 1000 kcal reducing diet throughout and taking 60 mg prednisolone daily during periods 3 to 7 inclusive (patient of Professor R.H.T. Edwards).

DRUG-INDUCED MUSCLE WEAKNESS

Excessive exposure to diuretics (a common problem for elderly people) may produce a severe, generalized weakness and lassitude, probably related principally to the loss of potassium. This, of course, is readily reversible. So too is the ill understood but common complaint of fatigue experienced by patients taking a beta-blocker. The myasthenic syndrome occasionally resulting from treatment with D-penicillamine, on the other hand, may prove irreversible.

Treatment with steroids is probably the best known, and the most common, example of drug-induced weakness. Because the fluorinated steroids are more liable to cause weakness, they are usually avoided. Nevertheless, weakness from high-dose or prolonged treatment with non-fluorinated steroids may be severe. As in Cushing's syndrome, the weakness is due principally to muscle atrophy, possibly aggravated by hypokalaemia. The loss of muscle tissue on exposure to high-dose steroids may be profound. As Fig. 1 illustrates, a study of metabolic balance in a 64-year-old man starting a trial of 60 mg of prednisolone daily (to determine the reversibility of his chronic airflow obstruction) suggested a loss of some 250 g of muscle per day, starting within the first 4-day balance period and continuing at that rate of loss for a further four 4-day periods until the prednisolone was stopped.

OSTEOMALACIA

Osteomalacia is common in old age and muscle weakness is common in osteomalacia. The weakness has a predominantly proximal distribution and may be the presenting symptom of the disease. The severity of the weakness bears no relation to the degree of hypocalcaemia or hypophosphataemia. The presence of weakness is associated with reduced mean levels of muscle ATP and phosphocreatine, but its severity bears no relation to the muscle's concentration of these high-energy phosphates. Some have reported a high prevalence of 'myopathic' electromyograms—i.e., with short-duration, polyphasic potentials, often of low amplitude—but this has not been a universal finding and a normal electromyogram is by no means uncommon. The plasma concentration of creatine kinase is normal.

Needle biopsies of muscle from patients with osteomalacia (including five aged between 60 and 71 years) showed no evidence of replacement of muscle by fat or fibrous tissue, nor of denervation, inflammation, or necrosis (Young *et al.* 1981). Mean fibre size was reduced, its variability was increased, and there was often a severe degree of preferential type II fibre atrophy (Fig. 2). Relaxation of the quadriceps muscles from electrically stimulated contractions is slow but neither the cause nor the relevance of this slowing is known. It is not as pronounced as in hypothyroidism and is not clinically evident. There are some animal studies which suggest that it may be related to impairment of calcium uptake by the sarcoplasmic reticulum.

The anabolic influence of vitamin D on D-deficient muscle probably starts promptly on the initiation of treatment. Nevertheless, the recovery of strength is a slow process, measured in weeks or months (Fig. 3). It is associated with growth of muscle fibres, especially of the type II fibres, and the rate of recovery appears similar during treatment with vitamin

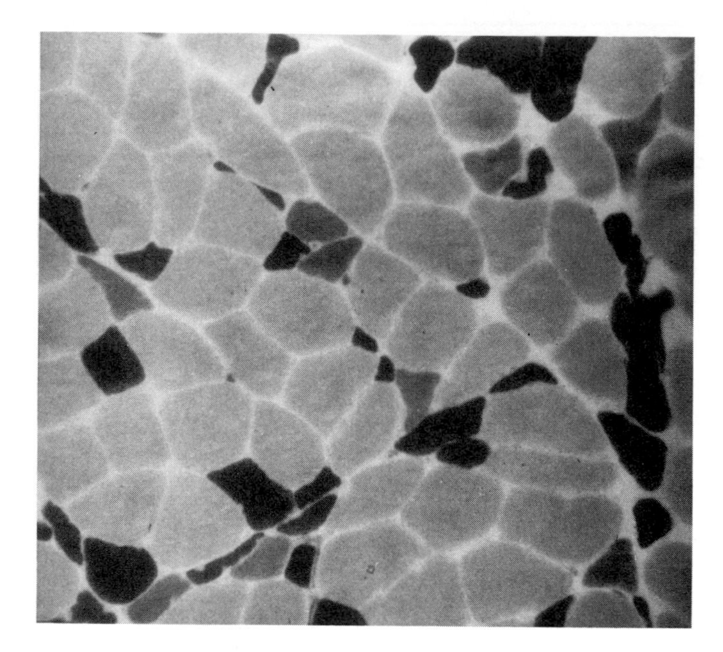

Fig. 2 Transverse section of a needle biopsy specimen from the quadriceps muscle, stained to show the activity of myosin ATPase (pH 9.4) and demonstrating marked preferential atrophy of the type II fibres. Patient: A 60-year-old, East African Hindu woman with nutritional osteomalacia, who presented with a 9-month history of being confined to bed, unable to walk (see also Fig. 3; patient of Dr L.C.A. Watson).

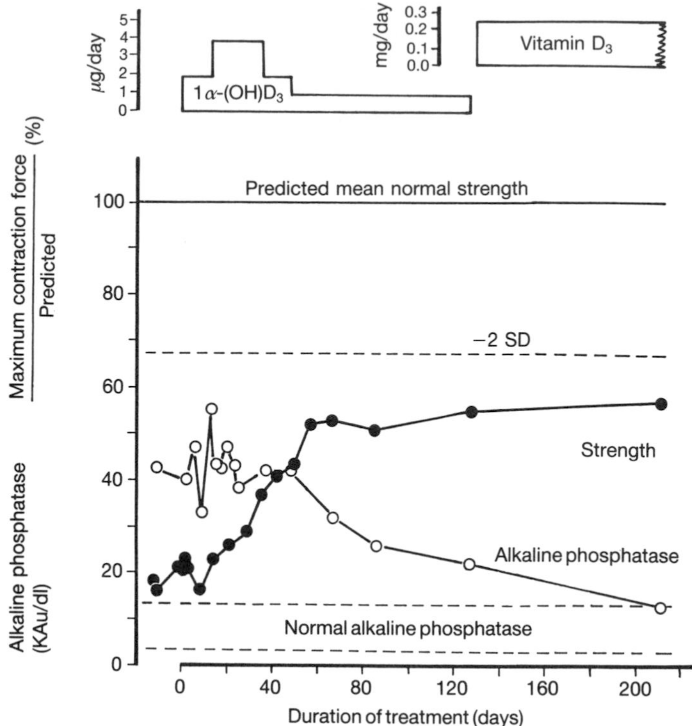

Fig. 3 Sequential measurements of isometric quadriceps strength and plasma alkaline phosphatase in a patient (as in Fig. 2) with nutritional osteomalacia treated with 1α-hydroxycholecalciferol and then vitamin D_3. (The normal range for quadriceps strength is age matched but not race matched.)

D_2 or D_3 and with 1α-hydroxycholecalciferol. After 4 weeks of treatment, the subject of Fig. 3, a 60-year-old Indian woman with nutritional osteomalacia, was showing a daily nitrogen balance equivalent to an increase in total body muscle of some 0.5 per cent per day, associated with a similar or slightly faster rate of increase in quadriceps strength.

A feature of patients receiving treatment for osteomalacia may be a disparity between the rates of recovery of objectively measured muscle strength and of subjective well-being and general physical ability. Even a small increase in strength may result in a large increase in a patient's functional ability. Changes in strength occur on a continuous scale but functional changes are quantal. It may need only a small gain in strength to take a patient from being 'just unable' to being 'just able' to perform some functionally important activity (see discussion of 'thresholds' in Chapter 19.1).

HYPERPARATHYROIDISM

Occasionally, muscle weakness is a prominent feature in patients with primary hyperparathyroidism. Plasma creatine kinase is normal, muscle biopsy may show type II atrophy, and a variety of moderate electromyographic abnormalities have been reported. The mechanism is unknown.

THYROID DISEASE

Muscle weakness and fatigue are common features of thyrotoxicosis. The weakness is associated with atrophy of muscle fibres of both fibre types. In hypothyroidism, muscle stiffness, aching, and mild weakness are common and a wide range of mildly myopathic microscopic features have been described. Muscle cross-sectional area may be increased without a corresponding increase in strength.

It is well known that hypothyroid muscle relaxes slowly. Thyrotoxic muscle relaxes rapidly but this is harder to demonstrate as there is a greater overlap with the normal range. These changes in muscle speed are associated with changes in fibre-type composition, with hypothyroid muscle commonly showing a predominance of type I fibres and thyrotoxic muscle showing a tendency to type II predominance. The changes in muscle speed, however, are much greater than can be explained merely on the basis of altered fibre-type composition.

Alterations in the relaxation rate of muscle are probably unimportant for the expression of maximal strength. In submaximal contractions, however, the increased relaxation rate of thyrotoxicosis means that a higher stimulus frequency is required to achieve tetanic fusion. Also, the increased rate of turnover of ATP means that the length of time for which a submaximal contraction may be sustained is reduced. Conversely, in hypothyroidism, the length of time for which a submaximal contraction may be sustained is greatly increased, reflecting the low rate of turnover of ATP in the slower relaxing muscle and the lower stimulus frequency necessary to achieve a fully fused tetanus.

Unlike the other metabolic and endocrine myopathies, the plasma level of creatine kinase is often abnormal in patients with thyroid disease. For reasons which are not understood, the plasma creatine kinase may be greatly elevated in hypothyroidism and tends to be towards or even below the lower limit of normal in thyrotoxicosis.

'LATE-ONSET' MUSCULAR DYSTROPHY

In 1885, Landouzy and Dejerine examined the relatives of their original case of facioscapulohumeral dystrophy. One was an 8-year-old girl with facial weakness. She lived until she was 86 or 87, although bed-bound from about 70 years of age (Justin-Besançon *et al.* 1964). Patients with limb girdle dystrophy (Shields 1986) or facioscapulohumeral dystrophy (Munsat 1986) may achieve a normal life-span. Rarely, they may even present to medical attention for the first time in old age. Perhaps this is because the steady loss of strength is not apparent until the patient crosses a functionally important 'threshold' and is suddenly unable to perform an important everyday activity (see 'thresholds' in Chapter 19.1). Alternatively, the patient who is losing strength may, unconsciously, develop compensatory trick movements which, for a time, maintain function and so conceal deterioration. Nevin (1936) describes a blacksmith with dystrophy mainly of the upper limb girdle who presented aged 60 with a history of symptoms for only 3 years. On questioning his family, however, there was evidence for weakness from the age of 27.

Patients with oculopharyngeal dystrophy (Tomé and Fardeau 1986) may present in old age with life-threatening dysphagia, but with a progressive ptosis dating back to their 40s. In its later stages, the disease may also involve other external ocular muscles and proximal limb muscles. Family details may reflect its autosomal-dominant inheritance.

Although the muscle biopsy and electromyographic appearances may be unmistakably myopathic, the plasma creatine kinase may be within the normal range (for example, Fig. 4). This is because the underlying dystrophy is relatively benign and, as the patient is elderly, the total muscle mass is small.

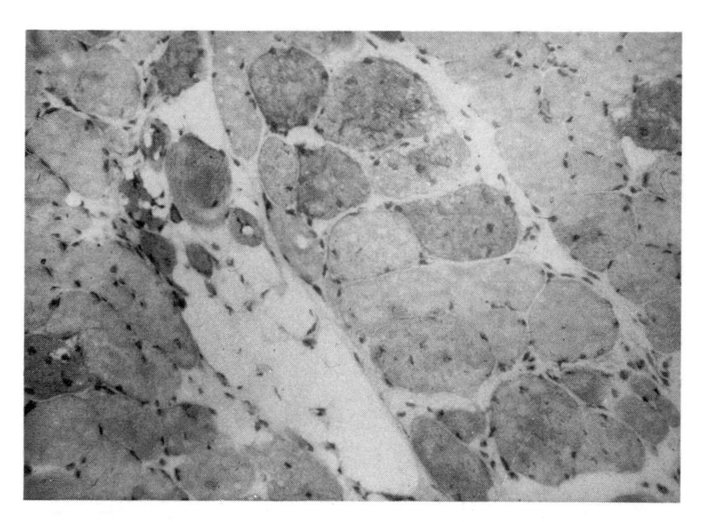

Fig. 4 Transverse section of a needle biopsy specimen from the quadriceps muscle, stained with haematoxylin and eosin and demonstrating evidence of muscular dystrophy, viz. increased variation in fibre size, central nuclei, moth-eaten and whorled fibres, ring fibres, increased endomysial connective tissue, and fat infiltration. Patient: a 78-year-old man who presented with an 18-month history of progressive muscle weakness, culminating in 5 weeks of difficulty in dressing and feeding, and 'bottom-shuffling' on stairs. He had a severe kyphoscoliosis, weakness, and wasting worse in the upper limbs and worse proximally, and was unable to whistle. Plasma creatine kinase was normal and electromyograms of an upper limb showed a population of brief, polyphasic, low-amplitude, single motor units, typical of a muscle fibre disease. When aged 33, he had had some spinal curvature but had been passed A1 for wartime military service (patient of the late Dr R.A. Griffiths).

RHABDOMYOLYSIS

Much the most common cause of rhabdomyolysis in old age is pressure necrosis of muscle in the patient who has lain for several hours after a fall. In most cases, the muscle damage is of little clinical significance except as a biochemical diagnostic 'red herring'. Elevated levels of plasma creatine kinase, aspartate transaminase, and hydroxybutyrate dehydrogenase do not necessarily imply damage to myocardial muscle (Fig. 5).

Not only are enzymes such as creatine kinase, aspartate transaminase, and hydroxybutyrate dehydrogenase released from damaged muscle; myoglobin is also released. Myoglobin is normally cleared rapidly from the circulation by the kidney but particularly high circulating levels may cause acute tubular necrosis and consequent acute renal failure. This is a life-threatening condition (Fig. 6). Urine, if there is any, will be positive for 'blood' on strip testing but microscopy will show no red cells. The damaged muscle tissue takes up calcium and releases potassium, phosphate, creatine, and purines. Therefore, for a given degree of renal failure, the plasma calcium is unusually low and plasma levels of potassium, phosphate, creatinine, and urate are unusually high. Treatment is by careful, forced, alkaline diuresis and close monitoring of plasma levels of calcium and potassium. A period of dialysis may be required (see *Oxford Textbook of Medicine*).

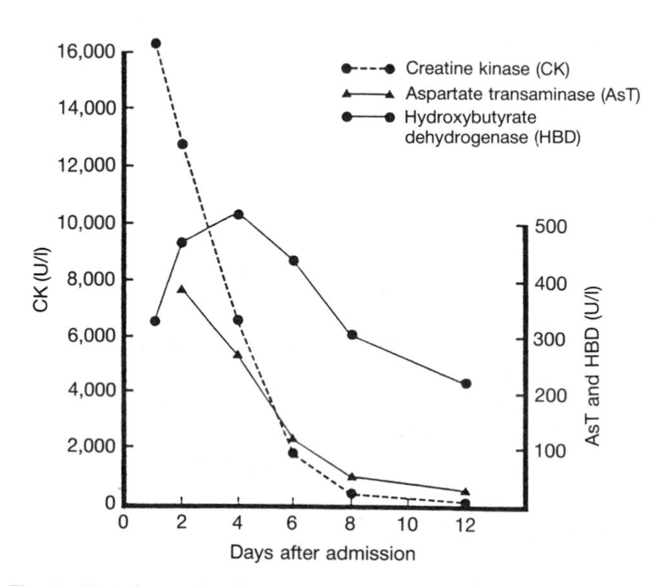

Fig. 5 Plasma creatine kinase, aspartate transaminase, and hydroxybutyrate dehydrogenase in an 81-year-old woman with parkinsonism and cardiac failure, admitted to hospital after lying undiscovered for some 7 h after a fall (reproduced, with permission, from Mallinson and Green 1985).

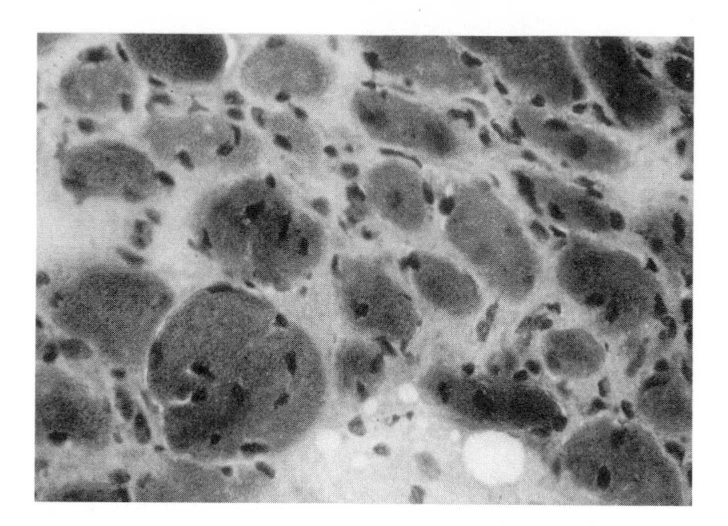

Fig. 6 Transverse section of a needle biopsy specimen from the left anterior tibial muscle, showing extensive muscle necrosis with macrophage infiltration and regeneration. Patient: An 82-year-old man, admitted as a medical emergency, confused, pyrexial, dehydrated, and with a swollen left calf. He had lain on a stone floor for several hours before discovery. Plasma aspartate transaminase >500 i.u./l (normal, 5–35); plasma creatine kinase >17 000 i.u./l (normal <100); peak plasma urea 68 mmol; peak plasma creatinine 1041 mmol/l. Treatment included haemodialysis. Good recovery and discharged home (patient of the late Dr R.A. Griffiths; see Ratcliffe *et al.* 1983).

POLYMYALGIA RHEUMATICA AND ARTERITIS

Polymyalgia rheumatica is three times as common in women over 80 as in those aged 60 to 69 years. It is also quite possible that many cases are not brought to medical attention. The predominant symptoms are pain, aching, and stiffness of the limb girdle musculature, especially in the neck and shoulders. Morning stiffness, generalized malaise, fatigue, and anorexia are common features. There may be a low-grade fever, mild anaemia, and elevation of the plasma alkaline phosphatase. The most striking laboratory abnormality is elevation of the erythrocyte sedimentation rate (**ESR**), sometimes to a considerable degree. (It is important not to be misled by this as only a minority of elderly patients with a greatly elevated ESR prove to have polymyalgia rheumatica.) Although the symptoms are predominantly myalgic, the true seat of the pathology remains obscure. Electromyography, muscle biopsy, and plasma creatine kinase are all normal. Diagnosis rests primarily on the history.

There is a considerable overlap of polymyalgia rheumatica with the arteritides, particularly temporal arteritis. Indeed, they may even be considered to be on the same spectrum of disease. This, and the treatment of the conditions, is discussed in greater detail elsewhere. A further potential source of diagnostic confusion is the fact that an elderly patient with a connective tissue disease, especially one producing inflammation of joints, may experience myalgia of the limb girdle and stiffness in the early morning for some months before the true nature of the underlying condition becomes apparent.

ARTHROGENOUS AMYOTROPHY

Pathology in a joint is often associated with weakness and wasting of muscles acting across the involved joint. Much of this is due to 'reflex inhibition', that is, a reflexly mediated inhibition of anterior horn cells by afferent stimuli arising in or around the joint (Stokes and Young 1984; Young *et al.* 1987). Reflex inhibition is well seen in the patient with knee pathology. In particular, an acute effusion of the knee may result in profound weakness of the quadriceps, even in the absence of perceived pain. Not only are voluntary contractions inhibited, so too is reflex activation of the muscle (as indicated by the reduced size of its H reflex). Long-standing reflex inhibition may result in severe atrophy, seemingly resistant to treatment, and microscopic changes indistinguishable from those of denervation.

Experimental evidence indicates that the severity of inhibition may be much greater when the knee is in full extension. It is tempting to speculate that a sudden increase in inhibitory afferent activity may explain some 'drop attacks', in which the patient's knees suddenly buckle without any alteration in the level of consciousness.

INTERPRETATION OF MUSCLE BIOPSIES

As discussed in Chapter 19.1, slowly progressive denervation is a feature of aged muscle. Clearly, therefore, evidence of denervation in muscle biopsies from elderly patients must be interpreted with great caution. There is a risk of overdiagnosing pathological denervation in elderly muscle. This risk may be even greater if the patient is not only elderly but also cachectic as a result of some other underlying disease (Tomlinson *et al.* 1969).

This may not be the only potential pitfall. It seems that biopsies from elderly patients without apparent neuromuscular disease may show other changes which might usually be interpreted as indicating neuromuscular pathology (Table 1).

Table 1 (a) Data from a review of muscle biopsies performed in Oxford between 1971 and 1985 as part of the investigation of neuromuscular symptoms in 63 patients aged 65 and over (mean age 73) (Squier 1987); (b) data from biopsies taken from the trunk and thigh muscles of 63 patients (mean age 75) undergoing surgery but without apparent neuromuscular disease (Tomonaga 1977)

Pathological 'diagnosis'	(a) Neuromuscular patients	(b) Other patients
Denervation, neuropathic	18	17
'Myopathy'	11	15
Type II atrophy	14	25
Myositis	17	0
Other	8	15
Normal	1	?

References

Engel, A.G. and Banker, B.Q. (ed.) (1986). *Myology*. McGraw-Hill, New York.

Justin-Besançon, L., Péquignot, H., Contamin, F., Delauvierre, P., and Rolland, P. (1964). Myopathie de type Landouzy–Dejerine. Présentation d'une observation historique. *Revue Neurologique*, **110**, 56–7.

Mallinson, W.J.W. and Green, M.F. (1985). Covert muscle injury in aged patients admitted to hospital following falls. *Age and Ageing*, **14**, 174–8.

Munsat, T.L. (1986). Facioscapulohumeral dystrophy and the scapuloperoneal syndrome. In *Myology* (ed. A.G. Engel and B.Q. Banker), pp. 1251–66. McGraw-Hill, New York.

Nevin, S. (1936). Two cases of muscular degeneration occurring in late adult life, with a review of the recorded cases of late progressive muscular dystrophy (late progressive myopathy). *Quarterly Journal of Medicine*, **5**, 51–68.

Ratcliffe, P.J., Berman, P., and Griffiths, R.A. (1983). Pressure induced rhabdomyolysis complicating an undiscovered fall. *Age and Ageing*, **12**, 245–8.

Shields, R.W. (1986). Limb girdle syndromes. In *Myology* (ed. A.G. Engel and B.Q. Banker), pp. 1349–65. McGraw-Hill, New York.

Squier, M.V. (1987). The pathology of neuromuscular disease in the

Squier, M.V. (1987). The pathology of neuromuscular disease in the elderly. In *Degenerative neurological disease in the elderly* (ed. R.A. Griffiths and S.T. McCarthy), pp. 119–29. Wright, Bristol.

Stokes, M. and Young, A. (1984). The contribution of reflex inhibition to arthrogenous muscle weakness. *Clinical Science*, **67**, 7–14.

Tomé, F.M.S. and Fardeau, M. (1986). Ocular myopathies. In *Myology* (ed. A.G. Engel and B.Q. Banker), pp. 1327–47. McGraw-Hill, New York.

Tomlinson, B.E., Walton, J.N., and Rebeiz, J.J. (1969). The effects of ageing and of cachexia upon skeletal muscle. A histopathological study. *Journal of the Neurological Sciences*, **9**, 321–46.

Tomonaga, M. (1977). Histochemical and ultrastructural changes in senile human skeletal muscle. *Journal of the American Geriatrics Society*, **25**, 125–31.

Young, A., Edwards, R.H.T., Jones, D.A., and Brenton, D.P. (1981). Quadriceps muscle strength and fibre size during the treatment of osteomalacia. In *Mechanical factors and the skeleton* (ed. I.A.F. Stokes), pp. 137–45. Libbey, London.

Young, A., Stokes, M., and Iles, J.F. (1987). Effects of joint pathology on muscle. *Clinical Orthopaedics and Related Research*, **219**, 21–7.

SECTION 20
Psychiatric aspects of the medicine of later life

20.1 The epidemiology of mental disorders in elderly people

A. S. HENDERSON

INTRODUCTION

Mental disorders as they present in hospitals or general practice are only a part of what occurs in the population (Goldberg and Huxley 1980; Henderson 1988). This is nowhere more true than with older persons, where there may be considerable impediments to professional recognition and treatment. It is knowledge from epidemiological studies, involving both treated and untreated cases, which provides a more complete picture of the main categories of disorder. The information generated from epidemiological research is of four categories: (i) knowing how frequently a specific disorder occurs among all the elderly people in a community, what groups are particularly affected, and what disability the disorder causes; (ii) determining the outcome or natural history of each disorder; (iii) determining what the risk factors may be for a disorder in the personal, social, and biological domains; and (iv) knowledge of the services used by those afflicted. This information is of fundamental value, both for understanding aetiology and for the administration of services. There is sometimes another purpose, which is the ultimate service of epidemiology when it can be achieved: this is the prevention of a disorder.

TYPES OF EPIDEMIOLOGICAL INFORMATION

It is convenient to define two sources of epidemiological information. The first is data from services. These refer to persons who have been in contact with hospitals, clinics, or general practitioners' surgeries. The data may commonly include diagnostic categories, as used by the medical staff. When such sets of data apply to a population of known size (the catchment area), and continue over several years, it is known as a 'case register'. The second type of data comes from field surveys. Here, an equal-probability sample is drawn from within a defined population (such as a town or city, or geographical area), and an attempt made to contact and examine each person in the sample. The examination may be by a trained interviewer or by a clinician, in either instance using a standardized interview to measure symptoms and signs (the dependent variable) and personal, demographic, or experiential factors (the independent variables). In this way, an accurate estimate is obtained of the prevalence of specified disorders in the total population, and factors associated with these disorders. The examination usually takes place in the person's own home, after an approach has been made by the investigators to seek co-operation. It should be noted here that, unlike medical practice, none of the respondents has sought an examination. The investigator is a guest. On the other hand, response rates of 85 per cent or more are commonly obtained, providing comprehensive coverage of morbidity in the whole sample. A cross-sectional survey yields information about the prevalence of specific disorders (or symptomatic states short of such disorders); and about their association with psychological, personal, or social factors. Causal effects, though, cannot usually be established in cross-sectional studies. Longitudinal surveys are highly labour-intensive, but yield information on incidence, and on the possible causal effects of the other factors on morbidity. They also throw light on natural history. Both cross-sectional and longitudinal studies provide information on what services are needed but not met; and who is using those services currently available. In this way, epidemiological studies show where limited resources can be best deployed.

The principal mental disorders of later life are the dementias and affective disorder. Others which are clinically important, but about which there is much less epidemiological information, are anxiety disorders, substance abuse, delirium, and chronic psychoses. There is reason to suspect that the prevalence of these, particularly in their milder forms, is much higher than appears from hospital-based practice. Certainly states of confusion, often lasting only a day or two, are likely to occur quite frequently, especially in the very old, but epidemiological data are lacking on this.

DEMENTIA

Cognitive decline and the diagnosis of dementia in community surveys

To consider a population as being sharply divided into cases of dementia and normal subjects is not realistic. In epidemiological data, such as scores on a test of memory or cognition, the distribution is continuous. The same holds in autopsy series when amyloid plaques, neurofibrillary tangles, or the deposition of A4 (β-amyloid) protein are measured. The implication is that individuals move progressively from an unaffected state towards increasing impairment. In Alzheimer's disease, whether of early or later onset, this progression is accelerated. Certainly in multi-infarct dementia the impairment is usually acute in onset, and unlike Alzheimer's disease such cases are categorically distinct from the healthy elderly individual. Persons encountered by clinicians tend to be already well down a scale of impairment. But persons in the general population present a very different pattern, with most being unimpaired or only mildly affected. The cutpoint, whereby clinicians define who is a case, is arbitrary and in no sense reflects a naturally occurring dichotomy. {This continuity of cognitive scores in community samples does not necessarily imply that there is a corresponding continuity in the underlying pathological processes in the brain. The effect of brain damage on cognitive and social functioning is modulated by environmental factors and by the previous intelligence and educational status of the sufferer. Highly educated patients can compensate for loss of brain tissue better than patients of lower intellectual achievement, and such influences will 'smooth out' the impact of pathology. The issue is still controversial, but many workers consider that in terms of numbers and distribution the changes of Alzheimer's disease are distinct from the superficially similar changes seen in some non-dementing older subjects.}

This dimensional view helps explain why states of impaired thinking and memory short of dementia have been reified with names such as 'benign senescent forgetfulness' and 'age-associated memory impairment'. In the former condition, it is implied that there is no progression. There is indeed some evidence that a few persons who develop impairment, but who do not have vascular disease, do not go on to have the full picture of Alzheimer's disease, but remain with only mild impairment.

Diagnostic criteria

In contemporary epidemiological studies, internationally established research criteria are now used for diagnosing cases. These are the *International Classification of Diseases*, draft 10th revision (World Health Organization 1990) (**ICD-10**) and the *Diagnostic and Statistical Manual of Mental Disorders*, 3rd edition revised (American Psychiatric Association, 1987) (**DSM-IIIR**) criteria, both for dementia in general and for specific dementias. The ICD-10 criteria are in two forms: one for clinical use, called the 'Clinical descriptions and diagnostic guidelines', and the other the 'Diagnostic criteria for research'. The latter have much in common with DSM-IIIR for both dementia and depression. What they do is to specify what symptoms, signs, or impairments have to be present for a person to be considered as a case of a specified disorder. In this way, they set a threshold. It should be borne in mind that both the content of phenomenology and the threshold are determined by consensus among experts, and not by empirical data. From ICD-10 and DSM-IIIR, algorithms can be written which state in a logical manner what ingredients must be present for the threshold to be reached and an individual thereby considered a case.

When it comes to diagnosing Alzheimer's disease, multi-infarct dementia, or mixed states in the setting of a community survey, there may be some difficulties in reaching a confident diagnosis, unless the elderly person has a good informant and is agreeable to having a computed tomography (**CT**) scan.

Standardized interviews

Epidemiological studies have been greatly enhanced by having not only these internationally accepted criteria but also interview instruments that yield those items of information needed for their diagnostic algorithms. Several such instruments have now been developed. They include the Geriatric Mental State Examination, the CAMDEX, the St Louis instrument, and the Canberra Interview for the Elderly. These can be used by non-clinicians after some training and are appropriate for clinical as well as epidemiological research.

The epidemiology of dementia

From the 50 or more field surveys of dementia in elderly people conducted throughout the world, four main findings have emerged (Mortimer and Schuman 1981; Jorm 1990). First, the prevalence increases exponentially with age, doubling every 5.1 years. This suggests that some process or processes related to the dementias are accelerated in late life. Second, the rates vary greatly between surveys, owing largely or perhaps entirely to differences in the methods for identifying and defining a case. Third, the overall prevalence of dementia is the same in men and in women, but Alzheimer's disease is more common in women. This is based on age and sex-specific denominators, so cannot be explained by the large numbers of women surviving to a late age. It suggests that some attribute of being female contributes to the onset of Alzheimer's disease. This could be an environmental exposure or some inherent biological attribute. Fourth, the prevalence rates of Alzheimer's disease and multi-infarct dementia are significantly different between world regions: the rates for multi-infarct dementia are said to be higher in Japan and the USSR. The rates in the USSR are probably due to diagnostic artefact, but the Japanese rates may be valid, for reasons which remain unexplained. Recent field studies from China have also found higher rates for multi-infarct dementia and a lower prevalence of Alzheimer's disease.

Attempts have been made to look for regional variation in rates for both early- and late-onset Alzheimer's disease within a country. The case-finding methods have included field surveys, hospital records, case registers, data from death certification, and the results of CT scans. A comparison of dementia in population samples from London and New York found a higher prevalence in the latter city (Gurland *et al.* 1983). This was based on the use of the same case-finding instrument in the hands of a jointly trained team of clinical interviewers. The observation remains unconfirmed, with no leads to its possible basis. A field survey in Finland found significantly higher rates in more rural areas. Hospital records and case registers have so far not provided satisfactory evidence of regional variation. Data from death certification have been found to be too inaccurate for reliable case-definition.

A secular trend in the prevalence of dementia was considered at one stage, based on incidence data over 25 years from the Lundby study in Sweden. It seemed possible that the rates were declining. Further analysis, though, showed that the rates for both multi-infarct dementia and Alzheimer's disease were constant over the period observed. {A study from the United States has suggested higher prevalence rates than found in earlier work but this finding has not yet been explained (Evans *et al.* 1989).}

Special populations

Interest has been shown in the prevalence of dementia in special populations, either for aetiological research or for administrative purposes. Attempts have been made to estimate prevalence in highly intelligent élites, or in persons sharing the same physical and social environment, including diet, over many years, such as nuns. No conclusive findings have so far emerged. It has been proposed that the highly intelligent have a greater cognitive reserve, and that they tend to continue mental activity to a late age, so that they may be partly protected from Alzheimer's disease. Elderly people living in nursing homes have been the subject of several studies of prevalence. Estimates vary considerably, from about 25 to 75 per cent, depending on the nursing home and the criteria used. Such estimates are useful in ensuring that appropriate staff are available and in planning appropriate environments and programmes for care.

Risk factors

For Alzheimer's disease, a sizeable number of risk factors have been proposed. The only ones that are well-established, though, are age, sex, having Down's syndrome and, for the familial form, a history of dementia in a first-degree relative. The others (Table 1), while most interesting, are only hypotheses.

Table 1 Some hypothetical risk factors for Alzheimer's disease

Education	Ethnicity
Down's syndrome in a relative	Vascular dementia
Diabetes	Thyroid disease
Head injury	Previous depressive illness
Urban versus rural living	Physical inactivity
Geographic variation	Smoking
Analgesics	Aluminium
Organic solvents	Vibrating tools

The present view is that genetic factors account, at most, for up to half of all cases of Alzheimer's disease. Some environmental exposures are therefore likely to be involved in the remaining cases, and are therefore well worth pursuing.

DEPRESSION

So much attention has been accorded recently to the epidemiology of dementia in elderly people that it has eclipsed depression as another major cause of disability in this age group. As with dementia, those persons with depression who present themselves to general practitioners or physicians are only a subset of all those so afflicted in the general population of elderly persons; and they are likely to differ systematically in their personal and social attributes from those who do not reach care.

There have been at least 25 population surveys of depression in elderly subjects, mainly in the United States, the United Kingdom, West Germany, Scandinavia, Australia, and Japan. The presence of a depressive disorder has been determined by very variable criteria and equally variable instruments. The latter have included depression rating scales, unstandardized clinical interviews and, more recently, standardized instruments such as the Geriatric Mental State, Comprehensive Assessment for Referral and Evaluation, Diagnostic Interview Schedule (**DIS**),and CAMDEX instruments. Not surprisingly, prevalence estimates had varied markedly until the introduction of these last methods.

On theoretical grounds, and according to popular stereotypes, it might be expected that elderly individuals should have higher rates for depressive disorder than younger adults. The psychological theories of Erikson predict either ego-integrity or despair in old age. Elderly people have major losses to endure, including bereavement, loss of status, income, mobility, and physical health. But on DSM-III criteria as assessed by the DIS standardized interview, the 1-month prevalence of affective disorder over five areas of the United States was 2.5 per cent for persons aged 65 and over. This was the lowest across all age groups, and contrasted markedly with a value of 6.4 per cent for persons aged 24 to 44 years. The present view is that elderly people do not have an excess of depressive disorder, but they may have more depressive symptoms. The latter, though, may be partly related to ageing itself rather than to depression, as they include sleep disturbance, loss of energy, and diminished appetite.

What seems quite likely from epidemiological studies is that while elderly persons probably have no higher a prevalence of depressive disorders than younger adults, there may be fewer cases who reach professional help and are also recognized as depressed. Some of the symptoms of depression may be attributed to old age and its 'inevitable' accompaniments both by the elderly person and his or her family or by physicians.

Epidemiological evidence about aetiology

Two sources of data are available on the causes of depression in elderly people: clinical services, where the individuals have reached treatment by a psychiatrist, usually in a teaching hospital setting; and persons found in the course of community surveys to have a depressive disorder. Depression in elderly persons, just as in younger adults, has been found to be closely associated with recent adverse life-events. Elderly people may actually have fewer of these, and be more resistant, on average, to adverse experiences, but the association with the onset of depression nevertheless holds. The absence of a close confiding relationship has been found to confer an increased risk of depression in elderly individuals.

Elderly persons living alone have been found in some studies not to be specially at risk for depression, probably because they are a selected group who are able to live independently. But where hostels or sheltered care are not readily available, this group may indeed have a higher prevalence. They do have a number of depressive symptoms, including the complaint of loneliness.

Physical illness has consistently been found to be associated with depression in elderly patients, and is probably causal. As every clinician would expect, the other main vulnerability factors are a previous history of depression, and a predisposing cognitive style or other personality attributes related to marked shifts in mood throughout earlier adult life.

The co-occurrence of cognitive decline and depression

It is familiar to clinicians that some elderly persons with cognitive decline (or actual dementia) also have depressive symptoms; and that persons with depressive illness may also have cognitive impairment. Evidence from epidemiological studies on community samples support this. It is likely that depressive symptoms are more likely to be present in the earlier stages of dementia, particularly when this is due to vascular causes. In the presence of depressive disorder, it is firmly established that attention, registration, and information-processing can be affected, leading to cognitive impairment. The significance of these findings for clinical practice is that depression should be looked for in elderly persons presenting with cognitive decline; and patients presenting with depression may be found also to be cognitively impaired.

Social supports and mortality

There is now persuasive evidence from several large-scale studies that elderly persons who have plentiful social relationships have a lower mortality rate than those with a sparse social network (House *et al.* 1988). This is a remarkable observation, suggesting that the social environment may influence physical health. The association seems to hold after allowing for health-related behaviours such as diet, smoking, alcohol use, and access to medical services. What is unclear is whether the association applies to close or more diffuse social relationships, and how it may be mediated through the brain.

Primary care and mental disorders in elderly people

As primary medical care is central to the health care of elderly people, epidemiological information can be put to good use in improving its efficiency. For this, we need to know the following. (1) Of all elderly persons in a community who have a clinically significant psychiatric disorder, what proportion go forward to seek help from their primary care physician (**PCP**). This is the so-called first filter of Goldberg and Huxley (1980). (2) What proportion of these elderly consulters are recognized as cases by the PCP (the second filter)? (3) Is the treatment given appropriate? (4) Is it principally by the use of supportive psychotherapy and counselling, by medication, or some combination? These questions provide a rich agenda for useful enquiry.

The present information is that elderly people tend to under-consult, attributing both physical and mental symptoms to their years and circumstances. Their relatives may share these beliefs. A minority of cases of dementia identified in the community have simultaneously been receiving treatment from their doctors. The performance of PCPs in recognizing, say, a case of depression or early dementia, has been the subject of several studies. In one well-known survey, as few as one-fifth of dementia and depression cases had passed both filters, so that they were known about by their doctors. More recent studies suggest this fraction has greatly increased. For dementia, recent estimates are particularly encouraging, and suggest that PCPs and practice nurses do know about the majority of cases, though they may identify as demented some persons who have a depressive disorder. PCPs are less likely to refer elderly persons than younger adults to a psychiatric clinic, for reasons that are not clear. As part of programmes of continuing education, some intervention trials have been established to improve the performance of PCPs in the recognition and management of depression and dementia. This is a key area in the health care of elderly persons, involving the application of epidemiological findings to practical issues.

References

American Psychiatric Association (1987). *Diagnostic and statistical manual of mental disorders* (3rd edn revised). APA, Washington.

Evans, D.A., *et al.* (1989). Prevalence of Alzheimer's disease in a community population of older persons. Higher than previously reported. *Journal of the American Medical Association*, **262**, 2551–6.

Goldberg, D.P. and Huxley, P. (1980). *Mental illness in the community: the pathway to psychiatric care*. Tavistock Publications, London.

Gurland, B., Copeland, J., Kuriansky, J., Kelleher, M., Sharpe, L., and Dean, L.L. (1983). *The mind and mood of aging, Mental health problems of the community elderly in New York and London*. Croom Helm, London.

Henderson, A.S. (1988). *An introduction to social psychiatry*. Oxford University Press.

House, J.S., Landis, K.R., and Umberson, D. (1988). Social relationships and health. *Science*, **241**, 540–5.

Jorm, A.F. (1990). *The epidemiology of Alzheimer's disease and related disorders*. Chapman and Hall, London.

Mortimer, J.A. and Schuman, L.M. (ed.) (1981). *Epidemiology of dementia*. Oxford University Press.

World Health Organization (1990). *The international classification of diseases* (draft 10th revision of Chapter V: categories F00–F99, mental and behavioural disorders). WHO, Division of Mental Health, Geneva.

20.2 Concepts of depression in old age

ELAINE MURPHY

INTRODUCTION

The majority of old people feel that life has turned out better for them than they expected. We also know from surveys of mental disorder in a number of different communities that the great majority of elderly people do not feel depressed, unhappy, or unfulfilled. The pessimism with which many young people regard their future old age is mainly the result of stereotyped misconceptions. Depression in old age has frequently been perceived as a predictable, understandable response to the losses and declines in the last season of life and the wintry themes of ageing and sorrow have been closely linked in literature. Those of us who make a career of working with elderly people with health problems tend to meet those at greatest risk of feeling burdensome, unhappy, dependent, and sad; and perhaps, therefore, professionals tend to maintain the stereotyped view of old age during the course of their daily work.

It is important not to exaggerate the extent of depression in the elderly population. Nevertheless, the minority of people who develop severe and significant depression generate a substantial demand for treatment and care. Even though health-care and social-service professionals see only the tip of the iceberg in terms of proportions of the total number of depressed people there are in the community, those who do reach professionals make a heavy demand on our services. Jolley and Arie (1976) reported that 40 per cent of all referrals to a comprehensive district psychiatric service for the elderly were for depression. Patients requiring inpatient treatment are heavy users of psychiatric beds. Over a 4-year period from 1978 to 1982, one-quarter of all the acute beds available in one large psychiatric hospital serving an East London catchment area was occupied by elderly patients with depression. The chronic and recurrent nature of severe depression in old age determines the extensive demand on services, both medical and psychiatric.

In spite of the impact of depression on the sufferer and his or her family, the condition is frequently overlooked or dismissed. The symptoms are easily confused with organic physical symptoms or regarded as an understandable and untreatable response to the inevitable stresses of later life. The importance of recognizing the symptoms of depression lies in the fact that the condition is treatable by a variety of medical and social measures; in general, an isolated episode carries a favourable prognosis for recovery.

HISTORICAL VIEW

Depression as a specific entity related to old age received little attention until the mid-nineteenth century, when Griesinger and his followers proposed a theory of the relationship of melancholia and dementia that has been surprisingly hard to shift from medical thinking. Griesinger proposed that melancholy was the forerunner of all mental illness, including dementia. Furthermore, the view was widely held at that time that dementia was the end-point for all psychiatric illness, providing the patient lived long enough. Thus it was a natural assumption that most melancholia in old age would progress to senile dementia. Observant psychiatrists, however, were in no doubt whatsoever that depressions in old age could and did occur in the setting of a well-preserved intellect and that some patients made a full recovery at a great age without deterioration into dementia. The question was finally settled by Roth (1955), who demonstrated beyond any doubt in his follow-up study of patients in hospital that affective illnesses had a markedly better prognosis than organic senile psychoses, and that affective patients did not in general develop organic psychoses.

CLASSIFICATION

There are now few advocates of the notion that the term 'depression' connotes an homogeneous entity, or that mere severity of the condition accounts for all the apparent clinical subgroups. There is as yet, however, no satisfactory scheme for the classification of depressive disorders at any age. This lack of an adequate nosology leads to even more confusion over depression in elderly people. We do not know whether to regard depressions occurring for the first time in 70- or 80-year-olds as similar or quite separate disorders from those in young or middle-aged people.

We are equally ignorant about what proportion of those who have a first episode in their teens or 20s will have further depressions in later life. It is likely that the answers to these questions will vary according to the type of depressive disorder, the subsequent life story of an individual, and his or her personal capacity to adapt.

Most classification systems aim to improve the distinction between subgroups, and so better to indicate prognosis and choice of therapy. But the first question to be asked is how to define the distinction between depression as 'depressive illness' and depression as a normal, understandable response to unhappy circumstances. This question has bedevilled much epidemiological work, particularly in surveys in which investigation of the extent of simple depressed mood and other minor symptoms unaccompanied by biological changes has found a very high prevalence of these in elderly compared with younger people. But where investigators have confined the concept of depression to a disorder lasting at least for some weeks, characterized by symptoms of mood disorder together with physiological and cognitive disturbances, far fewer elderly people are identified as cases in surveys, and the difference in prevalence between older and younger groups becomes much smaller.

The arbitrary line dividing a case of affective disorder from a normal but unhappy person is important clinically, especially in primary care and in hospital medicine, where the decision to treat an elderly person with psychotropic medication should not be taken lightly. The distinction is also important in research, where we are interested in delimiting disease that will have a characteristic aetiology, course, and prognosis, and we must be able to convey to colleagues the severity and characteristics of the disorder precisely. These issues have been partly resolved by emphasis on using specific diagnostic criteria to categorize symptoms elicited on standardized schedules by researchers who are trained to rate each symptom reliably. Schedules designed specifically for use with elderly populations include the **CARE** (Comprehensive Assessment for Referral and Evaluation; Gurland *et al.* 1983) and the **DIS** (Diagnostic Interview Schedule; Robins *et al.* 1981).

Some characteristic presentations of depression in elderly people are described more fully later in the chapter. One constellation of symptoms was at one time believed to represent a separate diagnostic entity, 'involutional melancholia'. After work showing that this condition was indistinguishable on clinical grounds from other depressions, it was omitted from the World Health Organization *International Classification of Diseases* (**ICD 9**) and from the *Diagnostic and Statistical Manual* (**DSM III**); American Psychiatric Association 1980).

The term involutional melancholia has now fallen out of favour but the issue of whether depression in old age is a separate category of illness has continued to be of interest. Many investigators have attempted to compare patients whose first illness began late in life with those with an earlier onset. Roth and Kay (1956) compared two groups of patients with affective disorder treated in hospital. Their work suggested that those who were older at onset were a less vulnerable and better integrated group, and that the appearance of affective disorder for the first time in old age was in part a measure of the resistance of this group to the ordinary stresses of life. They further suggested that physical illness was the main exogenous stress to provoke depression in the elderly.

Kay and Bergmann (1966) suggested that elderly patients with affective disorder fell into two groups. One group consisted of patients who were retarded, and had depressive, self-reproachful ideas, or hypochondriacal and nihilistic delusions; this group appeared to correspond to the major 'endogenous' disorders of earlier life, and serious physical illness was relatively rare among them. The second, 'late onset' group was predominantly male, anxious, irritable, 'attention seeking', and with many somatic complaints. Over half of this group had a serious physical illness and they often had a history of neurotic personality traits. However, the validity of separating these two groups has not been tested empirically and is not wholly supported by other research.

The predominant view at present is that depression is depression at any age and there is nothing that clearly distinguishes those whose first illness begins late in life from those whose first onset occurs in their younger days.

In clinical practice, and this is particularly the case in geriatric medical practice, depressions in many elderly people do not fall very easily into the traditional categories of 'endogenous' or 'neurotic' beloved of psychiatrists. 'Endogenous' depression has come to mean a characteristic syndrome of early-morning waking, diurnal variation of mood, weight loss, psychomotor retardation, loss of concentration, and a persistent, severe, mood disturbance. It is frequently associated with delusional ideas of guilt, hypochondriasis, and nihilistic delusions and is sometimes therefore referred to as psychotic depression. Factor analytic studies have repeatedly shown that this group of symptoms cluster together (Kendell 1976). Furthermore, clinical and biochemical findings in support of the concept are quite strong. For example, the syndrome appears to predict a good response to antidepressant drugs. The term 'endogenous' is an unfortunate one, however, as studies of both younger and older patients have demonstrated that so-called 'endogenous' depressions are as likely to be preceded by adverse life events as 'neurotic' and 'reactive' depressions (Brown and Harris 1978; Murphy 1982).

The 'endogenous' pattern of symptoms does, however, correlate quite closely with age, being more common in older age groups. Post (1972) attempted to differentiate three groups of older patients on the basis of their mental state. The three groups did not differ significantly in terms of predisposing factors or in the future course of the illness, and Post concluded there was no evidence among elderly people of separate syndromes.

Other classifications of depression have been less useful in studies of elderly persons. Assignment to a class using the unipolar–bipolar distinction, for example, can only be made after several episodes of illness and then not with certainty—six episodes of depression in a person's 30s, 40s, and 60s may be followed by one isolated episode of mania in their 70s. Genetic and family studies have consistently demonstrated an increased familial prevalence in families with bipolar manic–depressive disorder. Bipolar illness usually starts earlier in life; there are characteristically more frequent episodes and greater social impairment among patients with this condition, and they have a higher suicide rate. It is worth remembering, however, that both recurrent unipolar depressives and bipolar manic–depressives grow old in time, carrying their psychiatric burden into old age. Episodes of illness sometimes become more frequent and more prolonged with age, and it may be only in old age that episodes become frequent enough to persuade the patient and family of the wisdom of taking prophylactic medication. The specialist physician working with elderly people needs to be vigilant when a patient presents who is already taking lithium prophylactically. A sudden withdrawal of lithium can precipitate a sudden mood swing and should not be undertaken without consultation with the specialist services.

The classification of affective disorder in old age is unsatisfactory and confusing. It is likely to remain so until biochemical or genetic aetiological factors validate a classification system that can be confidently used to predict course, response to treatment, and prognosis in individual cases.

CHARACTERISTICS

Clinically, a major depressive disorder in old age is generally the same as in younger people. However, there are sometimes features that lead the diagnostician astray, especially in those with concomitant physical illness. Depressed mood is frequently not the main presenting symptom and may be overshadowed by somatic complaints, delusional beliefs, bizarre behaviour disturbances, or a picture resembling dementia.

Major depressive disorders in the elderly frequently present in florid form, with significant weight loss, sleep disturbance, and early-morning waking. Psychomotor agitation and retardation are common and often go together, seen as restless pacing, wringing of the hands, and clinging, importunate begging for help, which engenders a feeling of irritation in those around. At the same time the sufferer often feels slowed up, unable to think as fast as usual, and answers questions in a retarded, monosyllabic, and distracted fashion. Loss of concentration and muddled thinking, which are usual, can give the appearance of confusion. This, together with a lack of energy, prevents the sufferer from completing the simplest task effectively, often interpreted by the patient and sometimes by relatives or nursing staff as evidence of 'laziness', or as a sign of senility and dementia. Delusions of guilt, poverty and debt, and of severe illness—especially involving cancer, venereal disease (including now AIDS), punishment, and impending death—are often accompanied by somatic complaints of an inability to swallow, blocked bowels, and a feeling that the insides are rotting or diseased.

The classic presentation of an agitated depressive psychosis described here is characteristic of a severe illness seen in a cohort of elderly people born around the turn of the last century. The best description in the psychiatric literature of affective psychosis of this type was written by Aubrey Lewis, describing, in 1934, his young patients in their 20s, 30s, and 40s. We are now seeing the same cohort of patients in their old age. The psychopathology of future generations of elderly people may take a different form. Specialist physicians quite frequently encounter such patients, presenting with puzzling or bizarre physical complaints and weight loss.

Depressive stupor

Retardation may be so severe that the patient appears stuporose, immobile, and silent. This condition of akinetic mutism is rare but easily confused with neurological stupor. A clue to the diagnosis is the determined rejection of food and drink, the alert, sometimes frightened eyes, and the lack of neurological signs. The fatality rate of depressive stupor in old age is very high because of the rapid dehydration and risk of subsequent pneumonia. Rehydration and electroconvulsive therapy may not only be life-saving but can be curative. However, the situation is made more complex by the fact that severe depressive stupor quite frequently occurs in the setting of cerebrovascular disease or other major physical illness.

Milder forms of depressive illness

Not all depressions are severe and milder forms of major depressive disorders, presenting with querulous irritability, anxious clinging, and apprehensive pessimism, may be difficult to spot. Such elderly people are at risk of being labelled as having 'just personality problems' or 'getting crabby in old age'. Mistakes are easy to avoid if a good history is taken. Usually a point in time of change from the previous, normal, agreeable personality can be clearly established.

In the hospital ward or medical outpatient department, depression may present simply as withdrawal from the life of the ward or the immediate environment, an overcomplaining attitude, a preoccupation with poor health out of keeping with the severity of the physical illness, profound pessimism about the future, and so on. As all these symptoms may, on the other hand, be appropriate to the person's real problems, careful interview and assessment of the overall condition are essential.

Depression may also present as acute phobic anxiety, usually an intense fear of being alone, complaints of desperate loneliness and fear. Someone who may have adjusted to years of widowhood and living alone will suddenly become determined to stay close to relatives or friends, needing constant reassurance and continuous company.

Minor depression is much commoner than the florid major depressions, up to five times as common in some surveys (Blazer and Williams 1980; Gurland *et al.* 1983). Where depressed mood is not accompanied by biological symptoms, it has sometimes been referred to as 'dysphoria' to distinguish the condition from 'real' depressive disorders. However, there is no clear borderline between normal unhappiness, as a reaction to life's circumstances, and mild depression as an illness. It is probably more useful to regard depressions as falling along a spectrum of disorder with a hierarchy of symptoms that help to define the condition more clearly as its severity and their number increase (Gurland *et al.* 1983).

Distinguishing depression from dementia; 'pseudodementia'

The majority of elderly depressed patients, when specifically tested, show no evidence of impaired intellect or cognitive decline. Depressed old people frequently complain of poor memory and fears of intellectual decline but their performances in psychometric tests of immediate and delayed recall do not differ significantly from those of normal elderly people. The patient's perception is likely to be the direct consequence of difficulties in concentration. So common is the complaint of memory problems among depressed elderly people that 15 per cent of elderly people referred to a memory clinic for the investigation of early dementia were diagnosed as having depression. Indeed it has been said that an old person complaining of a failing mind is far more likely to be suffering from depression than dementia.

A small but important minority of elderly people with depression presents a confused, withdrawn picture that is very difficult to distinguish from dementia. Psychological testing reveals numerous patchy gaps and lowered performance scores, which are difficult to interpret. Computerized axial tomography (**CAT** scan) may be equally unhelpful, a finding of mild cortical atrophy being of little significance. The syndrome of 'pseudodementia' is overdiagnosed and tends to feature in referrals to psychiatrists from non-specialists who have not taken enough trouble to interview the patient at length to identify the depressive symptoms, which are clear to the person who takes time to draw out the patient's complaint. The rare cases that present real difficulty in diagnosis have been studied by a number of investigators (e.g., Rabins *et al.* 1984). Previous history of depression, rapid onset, and variable psychometric performance all point to a diagnosis of depression. It has been suggested that temporary impairment of cognitive functions may indicate some underlying, unspecified, cerebral organic disorder, consequent upon ageing, which perhaps affects the physiological arousal mechanisms, and further that this impairment might predispose the patient to depression and recurrences of depression in old age. However, the memory and learning impairments found in elderly people with depression are qualitatively different from those seen in the dementias of old age. In clinical practice, if the physician finds it difficult to distinguish depression from dementia after lengthy interview and observation, then the issue will not be solved by psychological testing but by a therapeutic trial of antidepressant medication.

The most common reason for a puzzling, mixed picture is that the patient has both depression and dementia at the same time. This is said to happen more frequently with multi-infarct dementia than with the Alzheimer type, but a review of the evidence that mood changes occur more frequently in multi-infarct dementia does not support this opinion.

Hypochondriasis and pain

Hypochondriasis affects approximately two-thirds of elderly depressed people (De Alarcon 1964) and somatic complaints in general are more commonly found than in younger patients (Gurland 1976). The predominance of somatic complaints, frequently compounded by the presence of real physical illness, can mislead the diagnostician when mood is not also obviously lowered. Pain of a persistent, unpleasant, unbearable quality that is difficult to pin down to a precise anatomical location is a common symptom of depression. It carries a rather poor prognosis for treatment if it has been going on for many months or years. Atypical facial pain is a variant that perhaps carries a better prognosis if vigorously treated. Because physical disability and pain are legitimate triggers for sympathy and extra help from others in our society, anyone who is felt to be 'exaggerating', 'inventing', or 'imagining' illness is given short shrift by relatives and professionals alike. The overcomplaining depressed person often feels ostracized and neglected for reasons that, to them, may be unfathomable. The management of hypochondriasis in the setting of depression therefore requires special understanding and skilled handling.

Depression presenting as behaviour disturbance

Behaviour problems of a wide variety of types are frequently symptoms of depression in elderly people who are heavily dependent on others for their day-to-day care. This happens, for example, in residential care, in long-stay hospitals, or where the elderly person is living with younger members of the family with whom they have a long-standing, difficult relationship. Depressive behavioural problems often occur in the setting of mild intellectual impairment and there is a danger then that the sufferer will be regarded as suffering from advanced dementia. Food refusal and wilful starvation, and inappropriate urinary and faecal incontinence with faecal smearing of walls and furniture, are reminiscent of the young child with a behaviour disorder. Persistent, intermittent, blood-curdling screaming, especially at night and frequently denied by the elderly person, seems usually to occur in response to anxious panic as a demand for instant help. In residential and nursing homes, one should suspect depressive disorder in the person who has recurrent, apparently wilful 'falls', throwing him- or herself on the floor in a theatrical fashion; the person who has recently 'fallen out' with all the other residents; and the person who has begun to bite and scratch the caring staff, or who has become 'a management problem'.

Depression, then, can be a rather chameleon-like disorder in elderly people, easily overlooked when the mood itself is not obviously sad or distressed. The key to diagnosis is a history of change in the person's mental state coming on over some days or weeks or a month or two rather than years. If there is any doubt about the diagnosis and the condition is severe enough to warrant pharmacological intervention, a good trial of antidepressant medication is justified and frequently rewarded by a remarkable reversal of bizarre symptoms.

EPIDEMIOLOGY

The epidemiology of psychiatric disorders in elderly people living at home and in residential institutions has received a good deal of attention over the past 15 years. A comprehensive and perhaps more optimistic picture is emerging of the distribution of depression among elderly people in general. A clinician tends to see patients who are not only at the most severe end of the range of depressive disorders but also at the height of the disorder. It is difficult for a specialist to gain a broad perspective throughout the course of the illness, and even the primary-care physician, who has a better opportunity to observe the waxing and waning of a disorder over some years, may only have regular contact with a patient during the 'down' times.

The concept of a subclinical iceberg of unrecognized depression among elderly people in the community has been supported by studies which suggest that general practitioners were unaware of more than three-quarters of cases of depression occurring in elderly people (Gruer 1975). The truth may be more complex as, more recently, studies of elderly people attending their own family doctors in both the United States and in Britain have suggested that depression is generally recognized by the practitioners but that very often no action is taken to investigate, treat, or modify it either socially or pharmacologically (see Macdonald 1986). It is possible, then, that much depression in old age is recognized but that no specific help is given to the sufferer. This may be a consequence of the nihilism that still bedevils the approach to treating all ailments in old age, or it may reflect the family doctor's assessment that he or she has little to offer therapeutically to the mildly and understandably depressed elderly person.

The major problem that has beset psychiatric epidemiologists has been the definition of what constitutes a 'case' of depression. Because depressed mood is an entirely appropriate response to unhappy circumstances, and older people may carry a large burden of losses, social difficulties, and health problems likely to cause sadness and unhappiness, studies in which a case has been identified merely by depressed mood may be expected to give a higher prevalence of the disorder among elderly than younger people. This has indeed been the outcome of studies with these low-threshold criteria, using self-rating scales or standard questionnaires (see Srole and Fischer 1980). Clinically this does not have much meaning because, for the majority of dysphoric elderly people, medical and/or social intervention would be considered inappropriate. However, society at large should perhaps be concerned that elderly people are often dissatisfied and unhappy.

Clinical epidemiologists who want to seek out cases similar in severity to those seen in clinical practice have used different approaches to psychiatric case definition. This has given rise to problems in comparing results between studies. On the one hand, a quantitative threshold for symptoms assigns caseness on the basis of a predefined, arbitrary, cut-off level. Most community-based surveys of elderly populations have used global rating scales of psychiatric impairment (e.g., Gurland *et al.* 1983). The alternative is to have clearly defined operational criteria for specified case diagnosis. In this, signs and symptoms are elicited by interviewers who are highly trained to rate reliably the answers to questions on a structured interview schedule. The ratings are then used to generate diagnoses for specific classification systems. For example, the DIS has been extensively used in the United State in a number of studies sponsored by the National Institute for Mental Health, the Epidemiological Catchment Area (ECA) programme; it generates diagnoses using DSM III criteria. This is an attractive method but many clinicians question the usefulness of schedules developed to increase reliability of diagnosis in elderly people that were originally derived from symptomatology of younger adults. Elderly people seen in clinical practice do not always fit conveniently into existing classifications, as has been noted above.

These two approaches both have the advantage that studies using these methods can be compared with one another. Urban/rural and international comparisons can be made, allowing us for the first time to examine and test theories of aetiological risk between different populations. What have we learned so far from studies with these standardized methods? The most surprising finding of all studies, contrasting with stereotyped views of old age, has been that the prevalence rates amongst elderly people for most non-organic mental disorders are in fact similar to or even slightly lower than those at other stages of the life cycle (Gurland 1976; Myers

Table 1 Rates of depressive disorders in community surveys

Reference	Population	Diagnostic method	Findings
Weissman and Myers (1978)	New Haven, CT (66+ years)	SADS[1] structured interview to elicit Research Diagnostic Criteria	5.4% 'major depression'; 2.7% 'minor depression'
Blazer and Williams (1980)	South-eastern US county (65+ years)	OARS[2] depression scale	14.7% with 'significant dysphoria'
Gurland *et al.* (1983)	Community sample, London and New York (65+ years)	Structured interview—CARE[3]	'Pervasive depression'; 13% New York; 12.4% London
Myers *et al.* (1984)	ECA[4] programme (65+ years): community + institutions—6-month prevalence, New Haven, Baltimore, St Louis, North Carolina	DIS interview DSM III criteria[5]	'Major depression', 1.9–3.5%; 'dysthymia, 2.1–3.8%
Morgan *et al.* (1987)	Home sample, Nottingham (65+ years)	SAD[6] scale: cut-off 6, depression subscale 4, validation by clinical interview	'Depression': 65–74 years 10%; 75–79 years 8.9% 80+ years 10.2%
Lindesay *et al.* (1989)	Home sample, South-east London	Structured interview, CARE[3]	'Major depression', 4.3%; 'pervasive depression', 13.5%

[1] SADS, Schedule for Affective Disorders and Schizophrenia.
[2] OARS, Older Americans Resources and Services.
[3] CARE, Comprehensive Assessment for Referal and Evaluation.
[4] ECA, Epidemiological Catchment Area.
[5] DSM, *Diagnostic and Statistical Manual* (see text).
[6] SAD, Symptoms of anxiety and depression scale.

et al. 1984). Overall, women do not have an increased prevalence of depression in old age; men have a small rise in prevalence through their 70s and 80s. Recent work in Nottingham (an English industrial city) suggests a steady prevalence of significant depression of about 10 per cent of all age groups among elderly people living at home (Morgan *et al.* 1987).

Epidemiological surveys of elderly people are made more difficult by the fact that psychiatric illness is an important reason for entering permanent residential care, so it is important to include an appropriate sample of those living in institutions of all kinds. Some surveys have included those in residential care, others have specifically excluded them, and findings should be interpreted accordingly.

Table 1 shows the prevalence rates of depressive disorders in some recent community surveys. The gender difference in prevalence is maintained throughout life, women having rates approximately 50 per cent above those of men.

Depression in residential homes and hospitals

Only a small minority of elderly people is permanently resident in hospitals and homes. The proportion of retired people in institutional care in the United Kingdom is approximately 6 per cent. Dementia is the major psychiatric problem found in elderly people in residential care and is the primary cause of admission in most cases. However, it is increasingly recognized that mentally alert residents and those with mild dementias have a markedly higher rate of depression than those living at home in the community. A survey of residents in old people's homes in one London borough found 38 per cent had the kind of pervasive depression that was found in only 13 per cent of the community-dwelling elderly people in the same city (Mann *et al.* 1984*a*). The rate in London homes was significantly higher than in similar institutions in New York and Mannheim, Germany, but even in those cities the rate was higher than might be expected (Mann *et al.* 1984*b*). It is possible that the drab quality of life provided in many residential homes is one reason for the high prevalence, but it is also possible that chronically depressed elderly people are preferentially selected into residential care as a result of their dependence on others and failure to cope adequately alone at home.

As physical illness is a common predisposing cause of depression, studies of elderly people in hospital might be expected to generate high prevalence rates and this has mostly been the case.

The picture overall in the general population of elderly is that moderate severities of depressive disorder are common, but no more common than in any other age group. First-admission rates to psychiatric hospitals for more severe depressions are also not markedly different in the older age groups. Recurrent admissions are more common among elderly people, however, and depression is common in institutions and in hospitals. Furthermore, there is a good deal of general unhappiness and 'low spirits' amongst elderly people in Western countries, which though not reaching a level of clinical significance, nevertheless draws attention to the less than ideal circumstances of many elderly people.

THE CONTRIBUTION OF BRAIN CHANGES

It has long been postulated that cerebral ageing may play a part in the aetiology of depression arising for the first time in old age, especially severe depression.

Neurotransmitter changes

Postmortem studies have focused mainly on the concentration of noradrenaline and serotonin in the brain, as these transmitters are implicated in depression. Normal people have an age-associated decrease in these concentrations in the hindbrain but 5-hydroxyindole acetic acid (**5-HIAA**) and monoamine oxidase increases with age, the increase in the latter being greater in women. It is tempting to suggest that these changes might predispose elderly people to depression or perhaps attenuate the healing effects of antidepressants and promote chronic illness.

Postmortem studies of the brains of depressed people have not revealed consistent changes in noradrenaline or its metabolites. However, decreased levels of 5-hydroxytryptamine and 5HIAA have been found in the raphe nuclei of suicide victims, and many studies have reported a lowered concentration of 5HIAA in the cerebrospinal fluid of depressed patients. It has been proposed that affective disorders result from an imbalance between the noradrenergic and the cholinergic systems, depression occurring when the cholinergic system is dominant and the noradrenergic system depleted.

Neuroendocrine changes; the dexamethasone suppression test

The possibility that changes in the concentrations of endocrine secretions might act as biological markers for depressive illness has long interested clinicians, and an extensive published record has grown up about the dexamethasone suppression test. There is an association between hypercortisaemia and depression, cortisol secretion being increased through its 24-h cycle in some depressed patients. Cortisol secretion by the adrenals is controlled by adrenocorticorticotrophic hormone, released from the pituitary under the control of corticotrophin-releasing factor from the hypothalamus. The hormone and the releasing factor are inhibited by high concentrations of circulating cortisol in the usual endocrine feedback system. The high concentrations of cortisol in depressed patients usually return to normal when the patient recovers and thus the concentrations are linked to the episode of depression itself.

Dexamethasone is a synthetic steroid that suppresses cortisol in the normal subject for up to 24 h. A positive or abnormal test is one in which the serum cortisol is not suppressed. However, several factors interfere with the suppression test to give positive results: for example, serious medical illness, especially diabetes, severe infections, significant weight loss, some drugs such as benzodiazepines, and withdrawal from alcohol and other psychotropic drugs. The test is therefore rarely of use in geriatric medical practice. Perhaps most unfortunately for the specialist working with the elderly, half of those suffering from dementia also have a positive test. Carroll (1982) has claimed that the dexamethasone suppression test provides a specific laboratory test for melancholia but others are less sanguine about its specifity or indeed whether it will prove to be useful at all as a diagnostic or prognostic aid.

Dexamethasone suppression is influenced by age, depressed elderly people being rather more likely to have abnormal results than younger people. However, dexamethasone may not be fully absorbed during the test in elderly people, leading to spuriously high results. The uses of the test in elderly people are few. Certainly it will not help in distinguishing depression from dementia. However, it may be helpful in monitoring the progress of treatment. Most patients with depression who have a positive dexamethasone test return to having a normal test after treatment, but a few do not. These 'continuing non-suppressors' are said to have a worse prognosis, even when they do make a clinical recovery.

Other endocrine abnormalities have been reported. The thyrotrophin-releasing hormone test is sometimes abnormally attenuated in elderly and some depressed patients but is not specific for depression. Release of growth hormone, which can be artifically stimulated in adults by some drugs, is also diminished in depression. However, at present these abnormalities are of no clinical significance.

Neuroradiological changes

The advent of CAT has brought a safe and non-invasive method of imaging the brain. There are changes in the brain with age in 15 per cent of elderly people who have no evidence of psychiatric disorder, cortical atrophy and ventricular enlargement being the most common changes reported. A longitudinal study of normal elderly people found 16 per cent had enlarged cerebral ventricles at first assessment, and a further 10 per cent had this enlargement at follow up, at an average 2.5 years later. Enlargement was correlated with reduced scores on cognitive tests, as one might expect. There was also a higher than expected number of subjects with enlarged ventricles in the 9 per cent of the total sample who developed depression during the follow-up study (Bird *et al.* 1986). This contrasts with the earlier work of Jacoby and Levy (1980), who found no difference in the proportion of depressed elderly patients with enlarged ventricles when compared with age-matched controls. Interestingly, the 9 out of 41 depressed elderly people who did have enlarged ventricles were described as clinically dissimilar from other depressed patients, being on the whole older, and having more features of endogenous, retarded depression. At follow up 2 years later, 5 of the 9 with the enlarged ventricles were dead compared with only 4 of the 31 depressed patients with normal ventricles, suggesting that venticular enlargement may influence the course of depression.

The same investigators found that the regional brain densities on CAT scans of depressed elderly patients were intermediate between those of normal people and those with dementia. At present the significance of this finding is unknown, but it may be linked to the presence of cerebrovascular disease. The nature of the relationship between vascular disease generally and depression is unclear but many have suspected a close link. Kay (1962) found a higher than expected rate of cerebrovascular disease as the attributed cause of death on death certificates in a sample of patients from mental hospital diagnosed as having 'functional psychosis'. The evidence from studies of depression after stroke are far less clear. Robinson and his colleagues in Baltimore have claimed that up to 60 per cent of patients may develop some form of depression after a stroke (Robinson *et al.* 1984). They also suggest that depression was more common in patients with lesions towards the left frontal pole than in those with right-sided lesions and posterior lesions on both sides. Other studies of stroke patients admitted to hospitals (Ebrahim *et al.* 1987) have not replicated these findings on side of lesion although they do support the very high rate of depression after stroke in this selected group. More recently, House (1987) has studied depression after stroke in the Oxford Community Stroke Survey. The sample included a more representative group of first-time stroke victims. His sample from general practice covered both those not admitted and the 40 per cent or so who were admitted to hospital. He found that although there was the expected increase in depression 1 month after the stroke, by the end of a year stroke victims were not significantly more depressed than age-matched controls. He too was not able to confirm Robinson's findings on the side of the lesion.

Neurophysiological studies

There are no characteristic abnormalities on the waking EEG of patients with depressive illness. However, there are changes reported in the sleep EEG: delayed onset of sleep, early-morning wakening, decreased latency of rapid eye-movement (**REM**) sleep, and increased REM activity. There is a close correlation between abnormal dexamethasone suppression and shortened REM latency. These findings are of little clinical importance at present. Further information has come from studying event-related potentials, i.e., the waveform generated by events in the brain-stem or sensory cortex by auditory, visual, and somatosensory stimuli. There is an age-associated increase in the latency of various potentials, which is increased markedly in patients with dementia. Both visual- and auditory-evoked potentials are delayed in elderly depressed people. These changes do not return to normal after clinical recovery, suggesting an enduring cerebral organic defect.

Abnormalities of circadian rhythm

The sleep–wake cycle is disturbed in depressive illness. Studies of 24-h secretion of neurotransmitter metabolites such as cortisol, growth hormone, and 3-methoxy-4-hydroxy-phenol ethylene glycol all show characteristic alterations from their normal circadian rhythms. Interest has been shown recently in melatonin secretion from the pineal gland, which normally occurs at night, but is known to decrease with age and is also inhibited in patients with evidence of increased cortisol secretion, such as those with severe depression. A specific abnormality of the beta-adrenergic system could lead to a decrease in melatonin production decrease and to the other biological abnormalities found in depression. Again, the relevance of these findings to clinical depressions is unknown.

Many measurements of biological factors are abnormal in depressed elderly people. Ageing *per se* produces some of these changes and it appears that ageing may itself produce changes to explain the onset of depression in old age in some individuals. We do not yet know, however, which individuals are at risk because of cerebral biological factors, or indeed if these changes are merely a reflection of changes that are a consequence of depression rather than a cause.

PHYSICAL ILLNESS, DISABILITY AND DEPRESSION

Many have commented on the close association between physical ill health and depression in old age. However, as both occur commonly in old age, this is not surprising. To examine whether there is a special relationship between the two, we need to estimate the probability of whether an association between physical and psychiatric illness in old age is based on chance or coincidence alone. In order to do this, we need to know the magnitude of the problems each presents in old age.

The increased prevalence of physical illness and disability associated with advancing age has been well documented. In surveys of old people living at home, up to 70 per cent describe themselves as in good physical health, yet fewer than 20 per cent of those who reach their seventh or eighth decade of life are disease-free, and over half have at least one activity-limiting disorder (Jarvik and Perl 1981). However, the evidence for an age-related increase in depression is more equivocal, as has been stated earlier in this chapter. Depressive symptoms, as identified by symptom check-lists and self-rating scales, are more common in older than in younger age groups, but depression as an illness is probably not more common with increasing age. The lack of clear evidence of an age-related increase in the prevalence of depressive illness lessens the probability that the observed relationship between depression and physical ill health is more frequent than would occur by chance alone. However, there is other evidence of more specific connections in three areas: specific illnesses predisposing to a high rate of depression; the evidence that depression and bereavement may cause increased fatality; and the influence on the cause and outcome of depression of physical ill health.

Physical disease predisposing to depression

At least five possible reasons could account for the association between depression and physical disorder: (i) depression could be the consequence of treatment for physical illness; (ii) depression may be a direct consequence of the cerebral organic effects of certain specific physical disorders; (iii) depression may result from the psychological reaction to physical illness and the process of adapting to a future life of handicap and disability; (iv) depression may predispose to the onset of physical disease; and lastly (v) the behavioural consequences of depressed mood may cause physical ill health through starvation, self-neglect, self-harm, and so on.

It is also possible that physical disorder and disability may increase the individual's vulnerability to other adverse life events that predispose to depression, and may inhibit recovery from depression.

Treatment factors are important because they are a potentially avoidable cause of depression. Treatment for known conditions is often fatiguing or painful and a patient may not share the physician's understanding of the mechanisms of treatment or prognosis. But even treatments that in themselves are painless may contribute to depression. For example, at least 23 medications can induce depression (Ouslander 1982). Such medications may act specifically by affecting neurotransmitter pathways, through a toxic action.

Many physical disorders have a specific association with depression where the order of presentation suggests a biogenic effect. One example of this is mood disorder after stroke or other localized brain injury. Robinson *et al.* (1984) reported a rate of 60 per cent for major affective disorder (DSM III criteria) in the first 6 months after a stroke; this has been commented on above. They also found that patients with lesions of the left frontal pole were more likely to develop depression than patients with lesions on the right side. They suggested that disruptions in catecholaminergic pathways may play a part in the aetiology of depression, for which there is some evidence from studies in the rats. Others, however, have not confirmed these findings, either reporting no differences in rates of depression between patients with lesions on different sides or finding that different sites were important. Overall, evidence is emerging that the side of the lesion is not especially important and that other items, such as residual physical disability, physical dependence on others for care, and social factors, may be important. At present, then, the mechanisms underlying the onset of depression after stroke are not straightforwardly attributable to biological cerebral change.

There is evidence that depression occurs more frequently in those neurological disorders with specific reductions in neurotransmitters that have been linked to the monoamine theory of depression. For example, depression is common in Parkinson's disease and in Huntington's chorea, a link between reductions in dopamine or γ-aminobutyric acid might predispose to imbalance of the neurotransmitter systems that maintain mood.

Many types of malignancies have been linked to severe depressions, especially carcinomas of the pancreas, stomach, and bronchus. Other commonly described associations of depression are cardiovascular diseases, and some metabolic and endocrine disorders (Ouslander 1982). The true incidence of depressive illness in elderly people after the onset of physical disorder is now known. Biochemical explanations of depression in ageing individuals are appealing in part because they suggest methods for intervention by biological treatment. However, physical disorder has a major impact on social well-being, dependence on others, the ability to maintain social activities, and interpersonal relationships. Illness brings home to the individual the closeness of death and, in influencing the course of one's life, the meaning an illness has cannot be easily separated out from the biological components of causality.

Physical illness caused by depression

Depression can be physically disabling. Fatigue, sleep disturbance, and loss of appetite may be compounded by self-neglect, inactivity, and a reduction in the patient's motivation to take treatment for physical health problems. All this is fairly straightforward. But can depression and psychological adversity make people physically ill? The published record on this topic is vast and much research has been methodologically unsound. However, the best evidence is from Parkes' studies of recently bereaved people, largely middle-aged and elderly. He found an increased number of deaths from cardiovascular disease among 4500 widowers in the 6 months after the death of the wife. The increase was confined to men and limited to these first 6 months. However, these men would have presumably had significant cardiovascular disease before their bereavement. The general public certainly believes that adverse life events can make people physically ill, particularly in the case of strokes and heart attacks, but the published accounts are unfortunately inconclusive on this point.

The influence on the course of depression

The course and outcome of depression does appear to be influenced by the presence of physical disorder. Murphy (1983) noticed that physical illness during a 1-year follow up was significantly more common among depressed patients who had a poor psychiatric outcome than among those who made a good recovery from depression. Shephard (1983) also showed that an improvement in physical condition correlated with a reduction in psychological illness. These findings are in accord with clinical commonsense and will surprise no one.

The implications of more severe physical illness in patients with depression

The majority of elderly people presenting to their family doctors or referred on to psychiatrists with depression have substantial physical health problems. They often have chronic, degenerative disorders in which the symptoms may be improved or controlled by the right medical treatment, yet illness will continue to be a major long-term problem for them to cope with. Vigilance is needed to spot depression accompanying chronic ill health; it is also needed by the psychiatrist so as to detect and treat relevant physical health problems. A close working relationship between psychiatrists and physicians with a special interest in the elderly is essential; joint assessment units staffed by teams of both types of specialist can make the treatment of patients with a complex mix of physical and psychiatric disorders considerably easier.

PSYCHOSOCIAL FACTORS

The health, vigour, and well-being of elderly people have more to do with economics and social organization than with the biological inevitability of the laws of nature. Many of the problems of the elderly are susceptible to change by social evolution and political intervention. The exploration of how and to what extent social factors are responsible for mental disorder in old age, especially depression, is therefore of great importance because it has implications for prevention.

Few concepts in psychiatry have been pursued with such enthusiasm over the last 20 years as the notion that adverse factors in the social environment lead to mental illness in general and to depression in particular. To the lay public, a causal relationship between these is a foregone conclusion, especially for elderly people. It is certainly true that in Western societies, older people are a socially underprivileged group. They are in general poorer, more likely to live alone, and to occupy the worst housing. They have more physical illness and are consequently less mobile, and are more likely to have poor sight and impaired hearing. Reduction of income, and loss of status, and sometimes of a useful role, are commonplace in old age. All this is not in doubt. But what empirical evidence is there that low social class, poverty, isolation, or the loss events of old age really contribute to the onset of serious depressive illness?

Social models of depression use psychological constructs to explain the origin of depressive disorder. The models postulate that interpersonal and social events external to the individual have enough effect on the emotions to lead to depression. The experience of depression can be traced to the perception of these events, often reflecting the emotional and cognitive reactions of the individual to events in the remote past, usually in early childhood. This model in no way excludes the biological perspective, and a comprehensive understanding of the aetiology of depression will need to incorporate both dimensions. At this point some of the empirical research evidence that social factors are relevant to the aetiology of depression in old age will be discussed.

Social class

The majority of studies have demonstrated that, at all ages, both severe depressive psychoses and milder neurotic illnesses are more common at the bottom of the social-class range. There could be several explanations for this. Probably of major importance for elderly people is the very wide variation in physical health status between the top and bottom social strata. Strokes, coronary artery disease, diabetes, and other major disabling conditions are more common among the lowest social classes. The close relationship between physical illness and depression has repeatedly cropped up throughout this chapter. There is also a difference in social isolation for elderly people in different classes. Lowenthal and Haven (1968) found an enormous class difference in the identity of close confidants reported by elderly people in San Francisco; more than three times as many of those with higher socioeconomic status reported a spouse as confidant, while those with lower status were more likely to report having no confidant of any kind. If the quality of close supporting relationships is related to vulnerability to depression, then it follows that we would expect a social-class difference in depression.

Life events

There are two quite separate disputes in recent publications concerned with research into the role of recently experienced life events. First, is there any causal link between these events and the onset of a depressive illness? Secondly, if we can satisfy ourselves that events do play a role, do they act merely to trigger off an episode that would have occurred sooner or later anyway in that particular individual, or can an event be the major formative factor, producing depression in someone who otherwise would have remained healthy?

There are important methodological problems in life-event research, but these cannot be gone into in great detail here. Recent methods devised by Brown and Harris (1978) provide a sophisticated way of overcoming the largest problems. Murphy (1982) used these methods to examine recent life events, chronic difficulties, and quality of confiding relationships in a study of two groups of depressed subjects who had experienced an onset of depression in the last year. The patients were compared with a group of normal elderly subjects in the general population. The depressed subjects were from two groups. The first consisted of 100 elderly patients referred to psychiatric services for the elderly in east London, a relatively socially deprived area of the inner city and its adjacent suburbs, and in addition to this 100 patients there were 19 subjects in a general population sample who had experienced an onset of depression in the past year. The comparison group was of 168 elderly subjects from the general population who were found to be free of psychiatric disorder at interview.

Murphy's findings were very similar to those that Brown and Harris made among younger subjects. Some 48 per cent of depressed patients and 68 per cent of depressed subjects in the community had experienced a severe life event in the year preceding onset, compared with only 23 per cent of the normal group. The severe events that were more common in depressed subjects were the death of a spouse or child, serious physical illness, life-threatening illness to someone close, severe financial loss, and enforced change of residence as a result of a demolition programme. Major social difficulties, lasting 2 years or more, were also significantly associated with depression. A notable difference between the earlier finding for younger subjects and Murphy's findings for elderly people was the predicted but conspicuous role of chronic poor physical health and grave personal health events in older depressed people. The overall risk of developing an onset of depression in the year for the total sample from the general population was approximately 10 per cent. For those in good health with no major social problems and no recent severe events, the risk was only 2.5 per cent, while the risk in the presence of one of these factors was 16 per cent. At this point it is worth recalling that the kind of depression that these subjects were suffering from was not simply 'dysphoric', general unhappiness, or 'the blues'. All had a major affective disorder as defined by DSM III; 24 per cent had a severe depressive psychosis; one-half were treated as inpatients at some time during the year. The rate of preceding severe events did not distinguish the 'psychotic' group from the rest, and the most severely ill were as likely to have predisposing social problems as the less severely ill.

Social circumstances and the events of a person's life do then seem to play an important role in depression. However, caution must be used in applying these research findings to individual clinical cases. As almost a quarter of the normal elderly population experienced an event that did not lead to depression, in a given clinical case it is impossible to be sure whether a reported event is causal or whether it is coincidental.

Critics of life-events research have pointed out that the magnitude of the effect of events on the causation of depression may be quite small. Paykel (1982) attempted a hypothetical calculation of the risk of developing depression after a severe event and estimated that it was of a similar order to the risk of developing clinical tuberculosis after exposure to the bacillus. Just as tuberculosis affects a vulnerable group among those who have contact with the bacillus, we must look for vulnerability factors that predispose individuals to depressive breakdown after life stress.

Social isolation

In Murphy's study (above) vulnerability to adverse life events, poor health, and major social problems were three times more common in those elderly who reported having no intimate, confiding relationship. The question of how the quantity and quality of social relationships predisposes to or protects from depression is a complex one in which research is fraught with methodological problems. Most are now agreed that simply living alone or having relatively few daily contacts is not especially disadvantageous for risk of developing depression. Furthermore, depressed people of all ages are more likely to report feeling lonely and unsupported but elderly people do not report more loneliness than younger age groups. The evidence emerging from most studies is that the perceived quality and adequacy of relationships are the key factors. It is likely that lifelong adjustment of personality and the capacity to form good social relationships are very important variables affecting vulnerability.

Social factors and biological factors tend to be studied separately by clinicians, psychologists, and social scientists, who pursue their own particular area of interest in isolation. The time has come to combine these two approaches if we are to make headway in developing a more comprehensive model of the aetiology of depression.

PHYSICAL METHODS OF TREATMENT

Drug therapy

Reviews by Veitch (1982) and Busse and Simpson (1983) concluded that there was little evidence to recommend any one drug as more effective than any other in the treatment of depression in elderly persons. The choice of medication is usually made from the physician's experience of a particular drug, the individual patient's ability to tolerate the side-effects of the traditional tricyclic antidepressants, and the physical health of the patient. The side-effects most frequently attributed to specific drugs are not always found in clinical practice. For example, amitriptyline and doxepin are not always sedating; imipramine is not necessarily 'activating'. Gerner *et al.* (1980) reported that, contrary to expectations, imipramine had a much better effect on agitation and anxiety than the serotonergic drug, trazodone. Thus the choice of antidepressants for elderly patients cannot at present be based on the rational options of blockade of serotonin or adrenaline.

That there is a positive correlation between age and the steady-state plasma concentrations of drugs has been firmly established. However, individual elderly people may have unexpectedly high plasma concentrations with relatively low doses of antidepressants. The concentration of antidepressant in plasma is not especially valuable as an indicator of therapeutic response but, in elderly people, may be a useful warning indicator of the likelihood of serious side-effects and toxicity. Dawling *et al.* (1981) studied 10 inpatients, mean age 82 years: they were given a single oral dose of 50 mg nortriptyline and its plasma concentrations were measured over the next 24 h; subsequent daily doses were calculated by predictions from Dawling's previous work on elderly depressives. Eight out of 10 patients required only 50 mg nortriptyline or less to maintain a therapeutic level: this suggests that in elderly patients it may well be possible to achieve a therapeutic response with lower doses of the traditional drugs but further studies of other tricyclics are required before any general guidelines can be formulated.

It is the anticholinergic effects of tricyclic antidepressants that are usually the most troublesome in elderly people. Postural hypotension leading to dizziness and falls is possibly the most important risk. The risk of cardiotoxicity has probably been overstressed. Glassman and Bigger (1981) concluded that tricyclic antidepressants are not cardiotoxic at therapeutic levels but that certain patients with abnormalities of intraventricular conduction are at increased risk because tricyclics prolong the PR interval and QRS complexes. Veitch (1982) found that tricyclics had no significant adverse effects on ventricular function in a group of patients with atherosclerotic or hypertensive heart disease.

There have been few double-blind trials of antidepressant medication for depression in elderly people. Those that have addressed the problem have had difficulty in recruiting enough patients without major physical disease and have found a high drop-out rate because of side-effects and intercurrent illness. Jarvik *et al.* (1982) studied elderly depressives (mean age 67 years), comparing imipramine and doxepin against cognitive and group therapy. Both drugs were significantly more effective than placebo, although at 26 weeks only 45 per cent of the patients were in remission, and 36 per cent had a poor outcome. Interestingly, good outcome was correlated with early response, within the first 2 weeks of treatment, and those who did respond were on only moderate doses of antidepressants, 25 to 50 mg/day. These findings call into question the usual clinical practice of doggedly increasing the dose up to tolerance level and waiting for at least 6 weeks before changing the medication.

The treatment of elderly patients with severe delusional depression has special problems. Should one treat with antidepressants alone or add neuroleptic medication? Spiker *et al.* (1985) investigated the pharmacological treatment of delusional depression by random, double-blind assignation of patients (all between the ages of 18 and 65 years) to amitriptyline alone, perphenazine alone, or a combination of the two. The combination treatment was clearly superior: 14 (78 per cent) of the 18 patients assigned to amitriptyline plus perphenazine were responders, compared with 7 (41 per cent) of 17 patients treated with amitriptyline alone, and 3 (19 per cent) of the 16 patients treated with perphenazine alone.

There has been renewed interest in the use of monoamine oxidase inhibitors in elderly patients, particularly since the finding that monoamine oxidase concentrations in the hindbrain are progressively depleted with age. Georgotas *et al.* (1983) treated a group of elderly depressed patients, mean age 68 years, with phenelzine up to 75 mg/day. These patients all had intractable depressions with a mean duration of 5 years and had received adequate regimens of tricyclic antidepressants before the trial; 37 per cent had also received courses of electroconvulsive therapy. As in previous studies, the drop-out rate was high, only 20 out of 30 completing 2 weeks of treatment and only 10 completing a full 7 weeks of treatment. However, 65 per cent of the 20 completing

2 weeks' treatment showed a significant improvement and one must conclude that monoamine oxidase inhibitors are useful ancillary weapons to have ready in the armoury for a few intractable cases.

The newer antidepressants such as trazodone have been shown to be as effective as imipramine in elderly patients, and are well-tolerated (Gerner *et al.* 1980). Similarly, mianserin, a tetracyclic antidepressant, is apparently as effective as amitriptyline. Nevertheless, many psychiatrists remain unconvinced of their efficacy and many severe depressions seem not to be influenced by them. They are consequently more popular with physicians than psychiatrists.

Drug manufacturers produce new antidepressants every year. There is currently a wave of new inhibitors of 5-hydroxytryptamine re-uptake inhibitors on the market that have yet to be adequately evaluated for elderly people.

Benzodiazepines have long been used to treat mild depressions and non-specific neurotic symptoms in primary care. The sedative anxiolytic effects are temporarily comforting but the long-term disadvantages far outweigh the advantages. The rapidly acquired tolerance leads to increasing doses and marked physical dependence. Psychiatric services for elderly patients are seeing increasing numbers who have developed serious dependence on diazepam, nitrazepam, lorazepam, and the shorter-acting hypnotics such as temazepam. These drugs are not antidepressant and should only be used sparingly and in very short courses as an adjunct to antidepressant therapy where severe anxiety and insomnia are not alleviated by a sedative antidepressant. Many psychiatrists and physicians consider that there are no indications at all for using benzodiazepines in elderly people.

The use of lithium as a prophylactic against recurrent unipolar depression has not been explored specifically for elderly people but those studies that have included some elderly patients suggest that it may be useful. Lithium is of proven value in the prophylaxis of bipolar manic–depressive illness and there is no evidence of a fall off in the efficacy of lithium with age. However, elderly people are more liable to lithium toxicity and plasma concentrations should probably be maintained at between 0.4 to 0.7 mcg/l to avoid toxicity (Jefferson 1983).

Electroconvulsive therapy

Elderly people with severe depressions have long been regarded as good candidates for electroconvulsive therapy. Fraser and Glass (1980) studied 29 inpatients with a mean age of 73 years: after 5 weeks a good outcome was reported in 12 patients and by 3 weeks after the last electroconvulsion, all but one had a satisfactory outcome. Unilateral electroconvulsion produced less postictal confusion than did bilateral, and more unilateral than bilateral treatments were required. However, not all have agreed with this finding, and it has been suggested that if a response is not forthcoming with unilateral treatment, a course of bilateral treatment should follow. Karlinsky and Shulman (1984) reported a slightly less optimistic response rate in a similar group of elderly inpatients: 42 per cent had a good immediate response, 36 per cent a moderate improvement, and 21 per cent a poor outcome. Electroconvulsive therapy can undoubtedly be dramatically effective for a severely retarded, depressed person who may be gradually starving to death and refusing fluid. The relapse rate of such illnesses is high but, if good prophylaxis is established with antidepressants of lithium, the risk of relapse may be diminished.

Psychological approaches to treatment

The provision of emotional support by professionals and relatives during treatment for depression is essential. Depressed elderly patients feel they are a burden to others, a failure, and a nuisance, and are highly self-critical. The continued interest of a doctor or a specially trained nurse helps to counteract the feelings of low self-esteem, if the support is given regularly and predictably at a time convenient to the patient. The patient's family also needs support. They need to be told what depression is, why their relative appears to be so anxious, irritable, clinging, 'impossible'. Relatives need to learn that depression is treatable and that recovery will eventually take place. Relatives who have unhelpful responses are sometimes amenable to education and regular reassurance.

Most elderly people referred to psychiatrists with depression are ill enough to warrant psychotropic medication. To embark on a psychotherapeutic approach without first establishing a foothold on the biological symptoms is doomed to failure. However, this is not the case for the physician in primary care, nor for the geriatrician faced with mild or fluctuating degrees of distress and depression accompanying physical ill health. Psychiatrists also see patients in the aftermath of bereavement or after a reluctant, enforced move into residential care, where a psychotherapeutic approach may be more appropriate.

There has been little exploration of suitable psychotherapeutic approaches for elderly people. In the United States, Butler's (1968) 'life review' technique of counselling elders has received a good deal of support. Butler's idea was that past experiences may be surveyed and reintegrated with one's present life, enhancing a sense of perspective and enabling people to let go of life, preparing for death with a feeling of leaving a legacy. Westcott's (1983) review of the published record on Butler's work suggests that the same kinds of technique help to prepare an elderly person for future life as well as for coming to terms with death.

More recently there has been interest in cognitive therapy for older depressed people. Cognitive therapy is a relatively short-term, problem-oriented therapy in which psychological disturbance is viewed as stemming from specific habitual errors or deficits in thinking. Therapy is directed at helping subjects to discover persistent illogicalities in their thought. Thompson and Gallagher and their colleagues in California have done a number of exploratory studies and have also attempted an evaluation trial of cognitive therapy for depressed elderly outpatients, with encouraging results. The same group of workers has also explored a group-training, 'self-help' approach based on the same principles of cognitive reshaping, which holds promise.

Psychotherapies for the elderly have not yet received the same research evaluation as the drug therapies and are likely to find little general favour until there is better evidence of their efficacy.

PROGNOSIS, COURSE, AND OUTCOME

The introduction of electroconvulsive therapy and specific antidepressant drugs has undoubtedly had a large effect on the short-term outcome of major depressive disorders in old age. In spite of the difficulties encountered in treatment, the majority of patients will respond well to treatment within a month (Jarvik *et al.* 1982), although a third will not respond very satisfactorily. However, the long-term course of depression in old age seems rather less optimistic than for younger people and there has been remarkably little improvement in that outcome over the past 15 years. The reasons for this, however, are not to do with the increased incidence of dementia in depressed elderly people. Cohorts of elderly depressives have been followed up for 8 years (Post 1962). These and later studies (Post 1972; Murphy 1983) have shown conclusively that the later appearance of dementia is no more frequent than in the age-matched general population. Suspicions (Kay 1962) that multi-infarct dementia may be more likely to occur with undue frequency as a late complication in late-onset depressives have not been confirmed.

Death

The death rate of patients with depression is consistently higher than expected when compared with the general population. Kay (1962) followed a cohort of elderly depressed patients admitted to the Stockholm Psychiatric Clinic between 1931 and 1937 until their death or up to mid-1956, and found their death rate was twice that expected. The increased rate was particularly marked in men (Kay and Bergmann 1966). The straightforward explanation for this excess of deaths is that elderly depressed patients have very poor physical health, but Murphy *et al* (1988) have shown that physical health problems alone do not satisfactorily explain it.

Quality of long-term outcome

There are difficulties in comparing studies of long-term outcome because of differences in judging what constitutes a satisfactory outcome, problems in classifying residual symptoms, and the choice of subjects studied. The quality of long-term outcome can be most clearly defined and compared in those patients in which recovery is described as a complete return to mental health. Post (1962) recorded a lasting recovery in 27 per cent of a group of in-patients who were admitted around 1950 and followed up to death or up to 6 years from index admission. In a later study (Post 1972), 26 per cent of a series of patients admitted in 1966 and followed for 3 years were similarly described as well and lastingly recovered. This similarity in the two outcomes was interesting because antidepressant drugs became available between the two series. Fourteen years later, in Murphy's 1979–80 series, 43 per cent had made a recovery in a 1-year follow up and at 4 years this proportion had fallen to a quarter, a figure very similar to Post's. More optimistic outcomes of 58 per cent recovered at 1 year have been reported, but with a similar result over the longer term, using a rather different method of investigation.

The other group in which comparison is fairly easy contains those who remain chronically and unremittingly ill with depression. Again, the proportions appear to have changed little over the years—17 per cent in 1950, 12 per cent, in 1966, and 14 per cent in 1979–80. Modern treatment does not appear to have reduced the hard core of persistently ill patients who make a very heavy demand on social and health services and pose a severe burden on the family.

There is a middle group of between a quarter and a third of patients who, while not remaining severely depressed, do not return to their former good mental health. While the biological symptoms remit, the person retains the cognitive and emotional changes of depressed mood. Post referred to this unsatisfactory outcome as 'residual depressive invalidism', a distressing, fluctuating condition that predisposes to further attacks of the full-blown disorder and that creates enormous social difficulties for the patient and family.

In conclusion, specific treatments appear to have shortened attacks of severe depression and far fewer patients now remain long term in hospital than in the earlier years of this century. However, a proportion remains seriously depressed for many years, many more have residual problems, and only a third have the kind of good recovery that we would like all our patients to achieve.

Predicting at the outset of an illness who will have a good recovery and who will not is extremely difficult. Illness of longer than a year before treatment is certainly one adverse factor. Not surprisingly, chronic physical health problems also militate against good recovery (Murphy 1983). Age at onset and the sex of the patient have not been found to be relevant. One study has shown that the more severely ill a patient when first seen, the less likely they will be to have recovered by the end of the first year (Murphy 1983) but it remains to be seen whether this holds for the longer term. In that study, only 10 per cent of those with severe psychotic illnesses had recovered in 1 year, and 23 per cent had died. In contrast, those that had 'classical' endogenous depressions with typical biological symptoms but no delusions or hallucinations did surprisingly well—70 per cent had recovered within the year. Type and severity of illness are likely, then, to have a very important influence on outcome.

Murphy also found that social factors played a part in influencing outcome in her series, with chronic social difficulties and intervening adverse life events affecting outcome negatively. However, Murphy's and Post's were small, local series of patients referred to specialist services. We need to know more about prognosis from large, prospective series of a broader range of categories and severities of depressive illness. It is possible that the kind of depression treated in primary-care settings has a more optimistic outlook.

The clinical importance of these studies in prognosis is that the practice of treating and managing patients and their families should be seen as a potential long-term commitment. Prophylactic medication to prevent relapse may be even more important in older than in younger subjects but, of course, will need close monitoring and supervision. Perhaps more importantly, the after-care arrangements to support those with residual disabilities need very careful thought. Surprisingly little is known about the best way to support such an individual to diminish the chances of relapse. Living accommodation, available social support, day care and respite care for families, and specific case work may all be of importance for certain individuals but evaluation of these various styles of management has so far been poor.

FUTURE DIRECTIONS

The term depression covers a wide range of types and severity of disorder. It is unlikely that one theoretical model will suffice as the explanation for the whole range of disorders and also that one treatment technique will prove to be universally effective. On the one hand there is burgeoning research about the ageing brain and cerebral organic factors, which might help us to explain the physiological changes of depression. The newer imaging techniques, which in the future may allow us to study the function of the brain rather than just its structure, may give us a clearer picture of these biochemical abnormalities. On the other hand, we need to understand more about individual vulnerability to depression, the role of that vague notion 'social support', and how we can manipulate the social environment professionally and, if necessary, politically, to create the circumstances that lessen vulnerability.

Social research has been carried out too often in isolation from the biological and yet clearly these two important aspects should be studied in tandem. It would also be helpful to have much more detailed data about the types and severities of disorder that respond or fail to respond to specific antidepressant drugs, and psychosocial approaches into the causes of these disabling disorders.

References

American Psychiatric Association (1980). *Diagnostic and statistical manual of mental disorders*, (3rd edn). American Psychiatric Association, Washington.

Bedford, A., Foulds, G.A., and Sheffield, B.F. (1976). A new personal disturbance scale. *British Journal of Social and Clinical Psychology*, **15**, 387–94.

Bird, J.M., Levy, R., and Jacoby, R.J. (1986). Computed tomography in the elderly: change over time in the normal population. *British Journal of Psychiatry*, **148**, 80–6.

Blazer, D.G. and Williams, C.D. (1980). The epidemiology of dysphoria and depression in an elderly population. *American Journal of Psychiatry*, **137**, 439–44.

Brown, G.W. and Harris, T.O. (1978). *Social origins of depression*. Tavistock, London

Busse, E. and Simpson, D. (1983). Depression and antidepressants and the elderly. *Journal of Clinical Psychiatry*, **44**, 35–9.

Butler, R.N. (1968). Towards a psychiatry of the life cycle. *Psychiatric Research Reports*, **23**, 233–48.

Carroll, B.J. (1982). The dexamethasone suppression test for melancholia. *British Journal of Psychiatry*, **140**, 292–304.

Dawling, S., Crome, P., Heyer, E.J., and Lewis, R.R. (1981). Nortriptyline therapy in elderly patients: dosage predictions from plasma concentration at 24 hours after a single 50 mg dose. *British Journal of Psychiatry*, **139**, 413–16.

De Alarcon, R.C. (1964). Hypochondriasis and depression in the aged. *Gerontologia Clinica*, **6**, 266–77.

Ebrahim, S., Barer, D., and Nouri, F. (1987). Affective illness after stroke. *British Journal of Psychiatry*, **151**, 52–6.

Fraser, R.M. and Glass, I.B. (1980). Unilateral and bilateral ECT in elderly patients. *Acta Psychiatrica Scandinavica*, **62**, 13–31.

Georgotas, A. *et al.* (1983). Resistant geriatric depressions and therapeutic response to monoamine oxidase inhibitors. *Biological Psychiatry*, **18**, 195–205.

Georgotas, A., Stokes, P., Krakowski, M., Farrelli, C., and Cooper, T. (1984). Hypothalamic-pituitary-adrenocortical function in geriatric depression. *Biological Psychiatry*, **19**, 685–93.

Gerner, R., Estabrook, W., Stener, J., and Jarvik, L. (1980). Treatment of geriatric depression with trazodone, imipramine and placebo: a double blind study. *Journal of Clinical Psychiatry*, **41**, 216–20.

Glassman, A.K. and Bigger, J.T. (1981). Cardiovascular effects of therapeutic doses of tricyclic antidepressants. *Archives of General Psychiatry*, **38**, 815–20.

Gruer, R. (1975). *Needs of the elderly in the Scottish borders*. Scottish Home and Health Department, Edinburgh.

Gurland, B. (1976). The comparative frequency of depression in various adult age groups. *Journal of Gerontology*, **31**, 283–92.

Gurland, B., Copeland, J., Kuriansky, J., Kelleher, M., Sharpe, L., and Dean, L.L. (1983). *The mind and mood of ageing*. Haworth, New York; Croom Helm, London.

House, A. (1987) Regular review: depression after stroke. *British Medical Journal*, **296**, 76–8.

Jacoby, R.J. and Levy, R. (1980). Computed tomography in the elderly. 3. Affective disorder. *British Journal of Psychiatry*, **136**, 270–5.

Jarvik, L.F. and Perl, M. (1981). Overview of physiologic dysfunctions related to psychiatric disorders in the elderly. In *Neuropsychiatric manifestations of physical disease in the elderly*, (ed. A.J. Levenson and R.C.W. Hall). Raven, New York.

Jarvik. L.F., Mintz, J., Stener, J., and Gerner, R. (1982). Treating geriatric depression: a 26 weeks interim analysis. *Journal of the American Geriatrics Society*, **30**, 713–17.

Jefferson, J.W. (1983). Lithium and affective disorders in the elderly. *Comprehensive Psychiatry*, **24**, 166–78.

Jolley, D. and Arie, T. (1976). Psychiatric services for the elderly: how many beds. *British Journal of Psychiatry*, **129**, 15–28.

Karlinsky, H. and Shulman, K. (1984). The clinical use of ECT in old age. *Journal of the American Geriatrics Society*, **32**, 183–6.

Kay, D.W.K. (1962). Outcome and cause of death in mental disorders of old age: a long term follow-up of functional and organic psychoses. *Acta Psychiatrica Scandinavica*, **38**, 249–76.

Kay, D.W.K. and Bergmann, K. (1966). Physical disability and mental health in old age. *Journal of Psychosomatic Research*, **10**, 3–12.

Kendell, R.E. (1976). The classification of depressions: a review of contemporary confusions. *British Journal of Psychiatry*, **129**, 15–28.

Lindesay, J., Briggs, K., and Murphy, E. (1989). The Guy's/Age Concern Survey: prevalence rates of cognitive impairment, depression and anxiety in an urban elderly community. *British Journal of Psychiatry*, **155**, 317–29.

Lowenthal, M.F. and Haven, C. (1968). Interaction and adaptation: intimacy as a critical variable. *American Sociological Review*, **33**, 20–30.

Macdonald, A.J.D. (1986). Do general practitioners 'miss' depression in elderly patients. *British Medical Journal*, **292**, 1365–7.

Mann, A.H., Graham, N., and Ashby, D. (1984a). Psychiatric illness in residential homes for the elderly: a survey in one London borough. *Age and Ageing*, **13**, 257–65.

Mann, A.H., Wood, K., Cross, P., Gurland, B., Schieber, P., and Hafner, H. (1984b). Institutional care of the elderly: a comparison of the cities of New York, London and Mannheim. *Social Psychiatry*, **19**, 97–102.

Morgan, K., Dallasso, H.M., Arie, T., Byrne, E.J., Jones, R., and Waite, J. (1987). Mental health and psychological well-being among the very old living at home. *British Journal of Psychiatry*, **150**, 801–7.

Murphy, E. (1982). Social origins of depression in old age. *British Journal of Psychiatry*, **141**, 135–42.

Murphy, E. (1983). The prognosis of depression in old age. *British Journal of Psychiatry*, **142**, 111–19.

Murphy, E., Smith, R., Lindesay, J., and Slattery, J. (1988). Increased mortality rates in late life depression. *British Journal of Psychiatry*, **152**, 347–53.

Myers, J.K. *et al.* (1984). Six month prevalence of psychiatric disorders in three communities. *Archives of General Psychiatry*, **41**, 959–67.

Ouslander, J.G. (1982). Physical illness and depression in the elderly. *Journal of the American Geriatrics Society*, **30**, 593–9.

Paykel, E.S. (1982). Life events and early environment. In *Handbook*

of Affective Disorders, (ed. E.S. Paykel). Churchill Livingstone, London.
Post, F. (1962). *The significance of affective symptoms in old age*, Maudsley Monographs 10. Oxford University Press.
Post, F. (1972). The management and nature of depressive illness in late life: a follow through study. *British Journal of Psychiatry*, **121**, 393–404.
Rabins, P., Merchant, A., and Nestradt, G. (1984). Criteria for diagnosing reversible dementia caused by depression: validation by two-year follow-up. *British Journal of Psychiatry*, **144**, 488–92.
Robins, L.N., Helzer, J., Croughan, J., and Ratcliff, K.S. (1981). National Institute of Mental Health Diagnostic Interview Schedule: its history, characteristics and validity. *Archives of General Psychiatry*, **38**, 381–9.
Robinson, R.G., Book Starr, L., and Price, T.R. (1984). A two year longitudinal study of mood disorders following stroke: a six month follow-up. *British Journal of Psychiatry*, **144**, 256–62.
Roth, M. (1955). The natural history of mental disorder in old age. *Journal of Mental Science*, **101**, 281–301.
Roth, M., and Kay, D.W.K. (1956). Affective disorders arising in the senium in physical disability as an aetiological factor. *Journal of Mental Science*, **102**, 141–48.
Shephard, R.J. (1983). Physical activity and the healthy mind. *Canadian Medical Association Journal*, **128**, 525.
Spiker, D.G. *et al.* (1985). The pharmacological treatment of delusional depression. *American Journal of Psychiatry*, **142**, 430–6.
Srole, L. and Fischer, A.K. (1980). The midtown Manhattan longitudinal study versus the 'Paradise Lost' doctrine. *Archives of General Psychiatry*, **37**, 209–221.
Veitch, R.C. (1982). Depression in the elderly: pharmacological considerations in treatment. *Journal of the American Geriatrics Society*, **30**, 581–6.
Weissman, M.M. and Myers, J.K. (1978). Rates and risks of depressive symptoms in a U.S. urban community. *Acta Psychiatrica Scandinavica*, **57**, 219–31.
Westcott, N.A. (1983). Application of the structured life review technique in counselling elders. *Personal and Guidance Journal*, **62**, 180–1.

20.3 Functional psychiatric disorders in old age

KLAUS BERGMANN

Functional psychiatric disorders in old age can be viewed in diametrically opposed ways: as an almost universal accompaniment of ageing, the reaction of a failing organism with a failing brain confronting an unfavourable period of relative or absolute deprivation; or as a period of life when all conflict and strife are over, psychosexual issues become less significant, and the social and economic career struggles are long past—tranquillity reigns!

Earlier work suggested that certainly for the functional disorders subsumed within the broad area of the neuroses the prevalence of both hospital and outpatient cases of neurosis fell markedly with increasing age. Later studies (Shepherd *et al.* 1966) showed evidence for decline of new 'cases' of neurosis in general practice, of hospital referrals for neurotic disorders, and of psychotherapy and counselling in general practice. On a more sinister note, the prescribing of sedatives and tranquillizers rose with increasing age. An informed medical approach to functional disorder in elderly patients will depend on an awareness of the size of the problem and of both its independence of organic cerebral and somatic diseases and, on the other hand, its interdependence.

EPIDEMIOLOGICAL CONSIDERATIONS

The major diagnostic entities within the group of functional disorders can be listed as follows:

1. neurotic and personality disorders;
2. major affective disorders (including both mania and depression;
3. schizophrenia-like illnesses and other persecutory disorders.

As major affective disorders are dealt with elsewhere, only items (1) and (3) will be considered here, but it should be emphasized that the boundary between major depressive disorders and those more likely to be diagnosed as neurotic is even more blurred in old age than in earlier life.

The outcome of studies of the natural history of hospital patients with psychiatric disorder and the demonstration of the absence of gross neuropathological changes in functional disorders (Blessed *et al.* 1968) allowed definitions of functional disorders to be used in epidemiological surveys with some degree of confidence.

A summary of some of the major surveys (Table 1) shows not only a range of prevalence figures for neurosis but also varying partitions between those abnormalities designated as neuroses or ascribed to the effects of personality disorder of some type, differences being more often due to nosological habits than to true variations of prevalence. A range of prevalences between 5 and 20 per cent for moderately severe neuroses and personality disorders might allow us to settle for an intermediate figure of 12 per cent, a figure consistent with a more specific enquiry into the neuroses of old age (Bergmann 1971).

Table 1 Prevalence of neurosis and personality disorder in seven community studies ($n = 3054$) (adapted from Simon 1980)

	Mean (%)	Range (%)
Neurosis	6.2	1.4–10.4
Personality disorder	5.8	2.2–12.6
Total	12.0	7.4–17.6

Why then is neurosis in old age so invisible: has it become mild and attenuated with advancing years? Certainly follow-up studies of neurosis from earlier life show amelioration of the disorder with the progression of the years. However, when the time of onset of neurotic disorder is examined in community samples, then onset after the age of 60 years occurs in 60 per cent of neurotic subjects (Bergmann 1971). Certain differences between late-onset neurosis and more long-standing, chronic neuroses can be demonstrated (Table 2). Chronic neurotics may have quite prominent symptoms, yet they seem to suffer from little except hypochondriasis, while experiencing no worse health than the population as a whole. Late-onset neurotics, however, have significantly worse physical health, especially cardiac disease, and experience considerable suffering and deprivation. In spite of this, late-onset neurotics seem invisible to general practitioners and, to some extent, hospital physicians (Bergmann and Eastham 1974), though a prevalence of around 17 per cent in series of acutely ill patients over the age of 65 years merits some consideration.

Table 2 A comparison of chronic and late-onset neurotic groups (from Jacoby and Bergmann 1986)

	Chronic	Late onset
Childhood		
History of neurotic traits	++	–
Poor relations with parents	++	–
Late in birth order	+	–
Problems in adult life	+	+
Stress in old age		
Physical disability	–	++
Poor mobility	–	++
Low income	–	+
Problems in old age		
Loneliness	–	++
No hobbies or holidays	–	++
Impaired self-care	–	++
Hypochondriasis	++	+

++, strong association; +, some association; –, not significantly different from normal population.

What inferences can be drawn from the epidemiology of neurotic disorders in later life? The prevalence of neurosis is little different from that found in younger people; it is neither omnipresent nor has it disappeared. Chronic neurosis probably ameliorates with the passing years and late-onset neurosis is relatively invisible to those caring for older people, but is associated with considerable suffering and linked to the vicissitudes of life in the senium. There is little valid research to explain why late-onset neurosis should be 'invisible' but one may speculate that the search for neurotic disorder often arises when a patient fails to function adequately in his or her role—at work, socially, or within personal relationships—and no physical cause for this failure is elucidated. For older patients there are two major barriers to this process. First, the role expectations demanded by society of most older people are so low that the disability engendered by neurotic disorder does not reach a threshold that commands attention. Secondly, physical ill health of a kind to provide a necessary but not necessarily a sufficient explanation is frequently found and the contribution of neurosis can readily be overlooked.

Neurotic syndromes and important groups of neurotic symptoms will be described, as will personality deviations, which have an important bearing on the medical management of the old person:

1. neurotic depression (entirely dealt with in Chapter 20.2);
2. anxiety states;
3. hypochondriasis;
4. hysterical symptoms
5. obsessional symptoms;
6. Diogenes syndrome.

ANXIETY STATES

Anxiety states in their pure form are rarely found in old age. In one survey that used a standardized rating, less than 1 per cent of a random sample could be said to be suffering from such a state. However, phobic, agoraphobic, and space-phobic states (Marks 1981) can be a prominent part of mixed states in which the picture also includes depressive symptoms. Agoraphobic anxiety symptoms are particularly important in the setting of the sequelae of an acute physical illness. Typically a patient may have had an acute myocardial infarction, have made a good functional recovery but yet be housebound and not able to function outside the home, no longer going shopping or carrying out social visits and engagements. It is too often assumed that this must either be a direct result of the illness or an inexplicably acute effect of normal ageing. Careful enquiry can often reveal an increasing feeling of tension and a mysterious feeling of impending doom or disaster, a fear of collapsing in public and being the focus of all eyes, being sick, losing consciousness or being shown up in some other way. The possibility of treating such disability by psychological or pharmacological means may be readily missed.

Acute anxiety and panic attacks with fear of imminent death are not uncommon within the broad range of emotional disorders. Such conditions may well be a prominent cause of the untimely removal of an elderly person to institutional care after the latest of a series of phone calls to their supporters, not least because these calls so often seem to be in the early hours. Psychological help with anxiety management and social support of an appropriate type may prevent an unnecessary disruption of the older person's life.

Treatment of anxiety

Benzodiazepines are very dangerous for older patients, where not only the risks of addiction and habituation have to be taken into account but the dangers of confusion, hypotension, and falls are greatly increased. There is no place for this group of drugs in the long-term management of anxiety in older people; an account of their psychopharmacological effects in old age is available in a review by Higgett (1988).

Elderly patients are amenable to psychological techniques of management and frequently benefit from referral to a clinical psychologist. They can receive help with relaxation techniques in an individual or group setting and in the case of phobic states, systematic desensitization. For a more detailed review, see Woods and Britton (1985).

If, for a limited period, psychopharmacological treatment is required, then it is best to consider very small doses of major tranquillizers (e.g. thioridazine, 10 mg, three times a day, or flupenthixol tablets, 0.5 to 1.0 mg twice a day). As so many anxiety states in old age have a mixed depressive picture the use of sedative antidepressants may be helpful.

HYPOCHONDRIASIS

Hypochondriasis resembles in some ways the proverbial elephant, which can be recognized by all but is nevertheless very difficult to define. The *Oxford English Dictionary* has an archaic definition, which includes the statement that the hypochondriac is 'chiefly characterized by the unfounded belief that he is suffering from some serious bodily disease'. Such a definition may be quite satisfactory when applied to younger people but suffers from defects when applied to the aged. Many older people suffer not from one but from several serious diseases; at this stage of the life-cycle, Occam's razor has become a blunt instrument!

Pilowsky's definition (1978, 1983) seems better to fit the concept of hypochondriasis in later life: 'Hypochondriasis is a form of illness behaviour in which the individual experiences and manifests a degree of concern over his state of health, which is out of proportion to the amount considered appropriate to the degree of objective evidence for the presence of disease.' This definition therefore requires an accurate evaluation of the health of the older person, and understanding of the usual reaction pattern of the older individual to various types of disability. It is unlikely that those without extensive experience of assessing the health of elderly people can feel safe in making the diagnosis of hypochondriasis.

Further classifications of hypochondriasis are probably helpful; these include the division into primary and secondary hypochondriasis. Secondary hypochondriasis consists of those hypochondriacal states arising in association with depressive illness, schizophrenic disorders, and organic states. Patients with primary hypochondriasis can also be divided into three states: those with bodily preoccupation; those with disease phobia; and those experiencing disease conviction. The first of these is self-evident but the other two merit some explanation. Disease phobia is the fear of catching some particular type of disease, and disease conviction is the profound conviction that the patient harbours some specific condition, sometimes held with an intensity bordering on the delusional.

What then is the significance of hypochondriasis in old age? Is it the central aspect of psychopathology in older people? The Duke University studies point strongly in this direction (Busse and Pfeiffer 1969). These studies, however, were carried out with patients and volunteers, and bias in the selection of populations may have influenced the clinical picture. A randomly selected community sample in Newcastle upon Tyne, in which 300 subjects aged 65 years to 80 years, without psychosis or dementia, were seen, yielded the prevalences of hypochondriacal preoccupation shown in Table 3. These findings suggest that hypochondriasis is not a common finding in normal old people but is a type of neurotic reaction more common in those with long-standing neurotic problems than in the physically more ill group of late-onset neurotic patients.

The practical implication of this observation is that before settling for the label of 'hypochondriasis' an enquiry into hypochondriacal tendencies in earlier life is of value. This enquiry may, in the absence of a positive history, lead the physician to search again for an occult illness or consider the possibility that the hypochondriacal picture may be a mask for a depressive illness with its own appreciable attendant risk of suicide.

Table 3 Prevalence of hypochondriacal preoccupation in a sample ($n = 300$) of community-living elderly people in Newcastle upon Tyne, England

	Prevalence (%)		
	None	Mild	Moderate/severe
Normal	93	7	<1
Recent neurosis	67	26	7
Chronic neurosis and personality disorder	64	19	17

Data from Bergmann, K. (1970). MD thesis, University of Sheffield, England.

The management of hypochondriasis

The following outline is based on my own views, the general overview of Pilowsky (1983), and the more specifically gerontological approach of Busse and Pfeiffer (1969). The general principles, which may be of some value to outline, can be listed as follows.

1. All hypochondriacal patients should receive a thorough physical assessment initially and whenever a change of circumstances justifies it.
2. Patients should be seen regularly and allowed, in the initial part of the interview, to air their physical concerns.
3. The doctor should be alert to the emergences of underlying worries and causes of grief, and should encourage their ventilation at the expense of the physical symptoms.
4. Appointments should be time-limited by prearrangement and manipulation of these arrangements should be resisted.
5. Always check for an underlying depression and check again!

IMPORTANT HYSTERICAL SYMPTOMS

Primary hysterical disorders do not arise *de novo* in later life and to label a patient 'hysterical' may lead to dangerous assumptions that hinder the search for underlying physical illness or functional or organic psychoses.

Nevertheless, prominent hysterical symptoms can be found in older people and these may take the form of conversion symptoms, dissociative phenomena, and abnormal personality developments of the type described as 'Briquet's syndrome'. Conversion symptoms are of the kind in which there are paralyses or ataxias, not corresponding to any recognizable pattern of organic neurological disease. Sensory disturbance of an hysterical type seems infrequent in later life. Dissociative phenomena manifested by disturbances of consciousness and atypical patterns of amnesia may be found not only in the absence of organic brain disease but also as an overlay to an early, subtle, organic psychosyndrome. Briquet's syndrome refers to the manifestation of histrionic acting-out of manipulative and demanding behaviour, which to many physicians represents what they mean when they refer to someone as 'hysterical'. Even these patients, especially when such a change is of relatively recent origin, may be indicating by their behaviour the presence of an underlying disorder.

It is wise always to assume that an underlying condition is responsible for the development of hysterical 'symptoms'. The underlying conditions may include functional psychoses, both depressive and schizophreniform, various forms of organic brain disease, and occult physical disease such as carcinoma.

OBSESSIONAL SYMPTOMS

Obsessional symptoms are commonly found in later life but the acute severe primary illnesses known as obsessional neuroses are no longer found in old age nor do they arise at this time of life (Pollitt 1957). Symptoms may take the form of obsessional thoughts, where the patient entertains an idea, recognized as irrational, often resisted, and having a recurring and compulsive quality. Ideas may include preoccupations with health, cleanliness, or financial matters. Obsessional acts includes rituals that have to be carried out, for example, hand washing, fixed ways of doing things. As with hysterical symptoms, underlying depressive illness has to be suspected and organic brain disease borne in mind.

DIOGENES SYNDROME

Self-neglect in old age is not uncommon and usually this can be adequately explained by severe physical inanition, psychosis, or dementia. However, sometimes scenes of appalling squalor, filth, and degradation can be encountered in the presence of a relatively normal mental state or, where some impairment is evident, it does not seem enough to account for the findings. Both psychiatric and geriatric descriptions of these cases have been made (Macmillan and Shaw 1966; Clark *et al.* 1975). These reports stress the large number of cases with no evidence of formal psychiatric disorder and suggest that this was a reaction of a certain type of personality to stress and isolation in later life. Other suggestions are that these patients are best seen as an end-stage of a personality disorder manifesting itself in the form of senile reclusiveness. A recent case study (Orrell *et al*, 1989) has suggested the possibility that frontal lobe impairment may play a significant part in accounting for some of the clinical features. The management of such cases is very difficult: constant supervision often in institutional care or at the very least with intensive day-care support has been found necessary to maintain these patients in a satisfactory state.

PERSECUTORY AND OTHER DELUSIONAL STATES IN LATER LIFE

These conditions, though relatively uncommon, are persistent. The prevalence of these disorders is about 1 per cent in community samples, and their incidence is 1 per 1000 per annum in those over 65 years (Kay and Bergmann, 1980; Holden 1987). They exhibit troublesome behaviour and have a higher profile than many other functional disorders. The persecutory states are also of interest because they bestride the borderlines between organic psychosis, functional psychosis, and personality disorder.

Paraphrenia as a diagnostic entity was distinguished from dementia praecox by the delimited nature of the delusion and the relatively good preservation of the patient's personality and capacities. This seemed particularly applicable to those patients whose delusional illness started in later life. These patients are conveniently grouped within the syndrome of late paraphrenia, though its heterogeneity has been well recognized and described by Post (1966).

A description of a typical paraphrenic patient would be of a single or late-married and childless woman, probably deaf for many years, of a rather prickly and withdrawn personality, who, in the absence of gross dementia and in clear consciousness, develops a delusional belief often accompanied by auditory hallucinations. Such a unitary picture is deceptive, though Graham (1984) has argued that late paraphrenia may be very similar to paranoid schizophrenia in earlier life. However, Holden (1987), in a retrospective case-note study and follow-up of a cohort could identify clinical subtypes. These include the following categories:

1. paranoid schizophrenia with first-rank diagnostic symptoms;
2. persecutory delusions, without the features of schizophrenia, arising often from pre-existing hallucinations or out of a long-standing abnormal personality state;
3. a mixed depressive and schizophrenic picture;
4. paranoid states proceeding fairly rapidly to dementia.

The aetiology of paraphrenia is multifactorial and only the major elements will be reviewed here.

Genetic

Genetic factors are not so prominent as in the schizophrenias of earlier life, though they still have some influence on the manifestation of the disease (Fundig 1961; Kay and Roth 1961). Also the female predominance would suggest some genetic contributions. More recently, the association of HLA antigens with late-onset paraphrenias has added to the evidence pointing in this direction (Naguib *et al.* 1987).

Sensory deprivation

Clinical reports of an association between paraphrenia and deafness have been extant for some time, but Cooper *et al.* (1976) demonstrated audiometrically the importance of long-standing conduction deafness which preceded the psychosis and related more significantly to paraphrenia than to a matched sample of the elderly with affective disorder. There have now been case reports of the development of paranoid states in association with the development of sensory impairment (Soni 1988).

Brain damage

A comparison of paraphrenics with an age- and sex-matched sample of normal subjects showed impaired performance in psychometric tests and ventricular enlargement on the computed tomographic scan of the brain (Naguib and Levy 1987). Follow-up of this sample (Hymas *et al.* 1989) showed that even though with treatment the psychotic symptoms got better, the psychometric and neuroradiological measures had deteriorated with the passage of time.

Treatment and prognosis

The treatment of paraphrenic disorders rests heavily on the use of phenothiazines; compliance with the prescribed drug regimen is the only predictive factor that distinguishes between a good or a bad outcome (Post 1966). The choice of oral medication should be preferred whenever possible to enable the minimum necessary dose to be titrated and action against side-effects to be taken promptly. As compliance is inversely proportional to the frequency of dosage, a phenothiazine such as trifluoperazine or pimozide may be worth choosing, as both can be given once daily. Dosage levels well below those used for young adults may be enough; trifluoperazine, 2 to 10 mg and pimozide, 2 to 4 mg, often suffice as a maintenance dose to keep the psychosis at bay. For very isolated, non-compliant, disturbed elderly patients, depot injections of major tranquillizers may have to be given. Half the young-adult dosage or less should be prescribed. Side-effects include extrapyramidal symptoms and tardive dyskinesia. Social factors are of importance and choices can be made between various forms of support including community psychiatric nurses, outpatient clinics or day hospitals, separately or in combination. The prognosis for complete cure with restoration of good insight is poor but most paraphrenics remain well preserved, coping with day-to-day life and managing their own affairs competently.

Functional disorder in old age is best seen as a response to the stresses of later life, both within the body and brain and from the outside world. The borderland between the organic and the functional is more blurred than in younger patients, and the assumption that clearly recognizable functional symptoms indicate the absence of any 'serious' disease becomes even less tenable than in earlier life.

References

Bergmann, K. (1971). The neuroses of old age. In *Recent developments in psychogeriatrics*, (ed. D.W.K. Kay and A. Walk), pp. 39–50. Headley Bros, Ashford.

Bergmann, K. and Eastham, E.J. (1974). Psychogeriatric ascertainment and assessment for treatment in an acute medical ward setting. *Age and Ageing*, **3**, 174–88.

Blessed, G., Tomlinson, B.E., and Roth, M. (1968). The association between quantitative measures of dementia and of senile change in the cerebral grey matter of elderly subjects. *British Journal of Psychiatry*, **114**, 797–811.

Busse, E.W. and Pfeiffer, E. (1969). Functional Psychiatric Disorders in Old Age: Hypochondriasis. In *Behaviour and Adaptation in later life*, pp. 202–9. Little, Brown and Co., Boston.

Clark, A.N.G., Mankikar, G.D., and Gray, I. (1975). Diogenes syndrome: a clinical study of gross neglect in old age. *Lancet*, **i**, 366–73.

Cooper, A.F., Garside, R.F., and Kay, D.W.K. (1976). A comparison of deaf and non deaf patients with paranoid and affective psychoses. *British Journal of Psychiatry*, **129**, 532–8.

Fundig, T. (1961). Genetics of paranoid psychosis in later life. *Acta Psychiatrica Scandinavica*, **37**, 267–82.

Graham, P.S. (1984). Schizophrenia in old age (late paraphrenia). *British Journal of Psychiatry*, **145**, 493–5.

Higgett, A. (1988). Editorial: indications for benzodiazapine prescriptions in the elderly. *International Journal of Geriatric Psychiatry*, **3**, 239–43.

Holden, N.L. (1987). Later paraphrenia or the paraphrenias? A descriptive study with a 10 year follow-up. *British Journal of Psychiatry*, **150**, 635–9.

Hymas, N., Naguib, M. Levy,R. (1989). Late paraphrenia—a follow up study. *International Journal of Geriatric Psychiatry*, **4**, 23–9.

Jacoby, R. and Bergmann, K. (1986). The psychiatry of old age. In *Essentials of postgraduate psychiatry*, (ed. R.M. Murray, A. Thorley, and P.Hill), p. 516. Grune and Stratton, London.

Kay, D.W.K. and Bergmann, K. (1980). Epidemiology of mental disorder among the aged in the community. In *Handbook of mental health and ageing*, (ed. J.E. Birren and R.B. Sloane), pp. 34–56. Prentice Hall, Englewood Cliffs NJ.

Macmillan, D. and Shaw, P. (1966). Senile breakdown of personal and environmental standards of cleanliness. *British Medical Journal*, **ii**, 1032–7.

Marks, I.M. (1981). Space 'phobia': a pseudo-agoraphobic syndrome. *Journal of Neurology, Neurosurgery and Psychiatry*, **44**, 387–91.

Naguib, M. and Levy, R. (1987). Late paraphrenia: neuropsychological impairment and structural brain abnormalities on computed tomography. *International Journal of Psychiatry*, **2**, 83–90.

Naguib, M., McGuffin, P., Levy, R., Festenshine, H., and Alonso, A.C. (1987). Genetic markers in late paraphrenia—a study of HLA antigens. *British Journal of Psychiatry*, **150**, 124–7.

Orrell, M.W., Sahakian, B. J., and Bergmann, K. (1989) Case reports: self-neglect and frontal lobe malfunction. *British Journal of Psychiatry*, **55,** 101–5.

Pilowsky, I. (1978). A general classification of abnormal illness behaviour. *British Journal of Medical Psychology*, **51**, 131–7.

Pilowsky, I. (1983). Hypochondriasis. In *Handbook of psychiatry 4, The neuroses and personality disorders*, (Ed. G.F.M. Russell and L.A. Hersov), pp. 319–23. Cambridge University Press.

Pollitt, J. (1957). Natural history of obsessional states. *British Medical Journal*, **i**, 194–8.

Post, F. (1966). *Persistent persecutory states of the elderly*. Pergamon, Oxford.

Shepherd, M., Cooper, B., Brown, A.C., and Kalton, G.W. (1966). *Psychiatric illness in general practice*. Oxford University Press.

Simon, A. (1980). The neuroses, personality disorders, alcoholism, drug use and misuse and crime in the aged. In *Handbook of mental health and ageing*, (ed. J.E. Birren and R.B. Sloane), p. 654. Prentice Hall, Englewood Cliffs, NJ.

Soni, S.D. (1988). Relationship between peripheral sensory disturbance and onset of symptoms in elderly paraphrenics. *International Journal of Geriatric Psychiatry*, **3**, 275–9.

Woods, R.T. and Britton, P.G. (1985). In *Clinical psychology with the elderly*, pp. 215–49. Treatment approaches for affective disorders. Croom Helm, London.

20.4 Alcoholism and the elderly patient

THOMAS P. BERESFORD AND ENOCH GORDIS

INTRODUCTION

Concern about covert alcoholism among elderly people has only recently begun to grow. With increasing numbers surviving into the seventh, eighth, and ninth decades of life, and the relative proportions of these elderly persons increasing every year, this concern appears warranted. When we recall the signs of healthy adjustment in this age group—active participation in work or play, and active social contact with long-time friends or acquaintances, or with family members of more than one generation—it is easy to see how problem drinking or alcohol dependence may threaten this adjustment. Alcohol use that is uncontrolled leads to isolation, loneliness, and often a sense of despair than can mimic depression. It is up to health-care professionals to recognize this impairment and to assist patients in taking up constructive social relationships once again. Social reintegration, being welcomed once again into the family or among elderly friends or acquaintances, is both a healthy and a reasonable goal.

It is important to recall that alcoholism is a chronic disease: it takes time to manifest itself, once drinking has become uncontrolled, and requires time for recovery. Family and friends of an older adult who is a 'problem drinker' will need time to rebuild their confidence in a person whom they may feel has betrayed their family's affection or friendship. Therapists must also approach this process prudently and patiently, avoiding the temptations of 'a quick cure' on the one hand, or a sense of therapeutic uselessness on the other. In this age group, as with others, recovery from alcohol abuse or dependence will generally result in fewer medical problems, fewer visits to emergency room and other primary medical-care centres, and a longer period of independent living. In short, there are many good reasons to diagnose and treat alcoholism when it occurs among elderly people and no good reasons not to.

EPIDEMIOLOGY

For the present discussion we have adopted the general definition of an aged person as someone who has passed their sixty-fifth birthday. According to estimates provided by the National Institute on Aging, there were approximately 25 million persons in the United States over 65 years of age in 1980. By the year 2030 this number will have grown to about 65 million. Estimates of the prevalence of alcoholism have ranged from 2 to as high as 10 per cent for this age group. Using a figure of 5 per cent, we estimate that, in the United States, approximately 1.25 million persons over the age of 65 suffered from alcohol abuse or dependence in 1980. By the year 2030 this number is likely to double to an estimated 3 million. This is probably a conservative figure because of cohort effects. Current estimates of prevalence may be artificially low because of commonly held social sanctions against alcohol use among those now 65 and older that come from their formative years. Social researchers point out that similar sanctions do not appear to apply for those born after the Second World War, who will be in their elderly years by the year 2030. Because of this, many expect the proportion of alcoholic elderly patients to increase more quickly than the proportion of elderly persons in the general population. It appears that the occurrence of ageing and alcoholism will likely represent a convergence of two public-health problems with large implications for health-care delivery.

In the recent published record an interesting clinical division has been made between problem drinking of 'early' and 'late' onset. A large number of elderly people began their heavy drinking within 5 years of their first contact with a hospital or clinic. Estimates put roughly one-third of alcoholic persons over the age of 65 in this category of 'late' onset. Some investigators have stated that the term 'late' onset may be a misnomer and that 'recent' onset would be a more apt description. They point out that a 70 year old who had begun drinking at the age of 55 would not necessarily be a drinker of recent onset but would qualify as one of 'late onset'. The importance of this distinction lies in the amount of damage, physical and social, linked to the absolute amount of alcohol drunk over time. The implication is that the problem drinker of 'recent' onset may have a better prognosis, and that intervention may be more effective when the drinking pattern is still being formed and before the illness has robbed the drinker of needed social and physical resources.

There have been considerable disparities between estimates of prevalence derived from clinical samples and those taken from surveys of drinking practices in large populations. Surveys of habits set the rate of problem drinking (that is, the frequency of social or medical problems caused by drinking) in the range of 1 to 3 per cent. But studies of this kind may miss the distinction between heavy, non-addicted drinkers and those who drink addictively while consuming similar amounts of alcohol. Hence, surveys based solely on the number of drinks per day or on the frequencies of drinking problems may miss the signs and symptoms of addictive drinking noted by clinical researchers. Another confounding variable comes from evidence suggesting that persons tolerate less alcohol as the ageing process continues: they appear to require less alcohol to achieve intoxicating effects (see below). As a result, surveys of elderly populations that merely quantify drinking may again miss the effects of amounts of drinking that would be unremarkable in younger populations. Finally, surveys portraying low prevalence of alcoholism in the latter years very often fail to take into account the effect of early deaths of long-term drinkers who may, for example, perish in their seventh decade despite a life expectancy that might have brought them into their ninth. Confounding variables such as these suggest that current notions of present and projected frequencies of alcoholism among elderly individuals are probably underestimates.

In addition to these variables, few studies have included racial, ethnic, or sex differences in drinking and alcohol dependence among elderly people. The current estimate of alcoholism among aged women, for example, is very low in large population surveys. Early data from prospective surveys done in clinical settings suggest that the rate may be significantly higher. {Clinical experience in the United Kingdom suggests an increasing prevalence of alcohol-related problems among elderly women. This is thought to be due to a number of factors including bereavement, loneliness, and poverty, which have always been with us. Social changes may also be important, including the recent general availability of alcoholic beverages in supermarkets rather than in the traditionally recognized—and traditionally stigmatized—'off-licence' drink shops. A further factor may be the changing image of the alcohol-consuming woman, who is depicted on television no longer as degenerate but as sophisticated, and even the drunken woman may be presented as humorous or pitiable rather than contemptible.}

The clinical setting itself may explain some of the discrepant findings: alcoholic persons of any age gravitate into medical and surgical wards and clinics. In one tertiary-care medical centre, for example, the rate of alcohol dependence among medical and surgical inpatients of 60 years and older was estimated at 15 per cent. At the same time, this age group accounted for 68 per cent of the total yearly admissions. The estimated absolute number of alcoholic elderly persons in the general hospital was high because of the high frequency of hospital use by this age group generally. Taken in their entirety, both population surveys and clinical studies of prevalence suggest that our knowledge of the epidemiology of alcoholism in the elderly age groups is in an early stage and will require considerably more refinement. For the clinician, it is important to know that alcoholism occurs significantly often in this population and should be part of a differential diagnosis when the impairments discussed below and elsewhere in this text present themselves in the office, clinic, or hospital.

AGEING, PHYSIOLOGY, AND ALCOHOL METABOLISM

The physiological changes that characterize ageing bear directly on the effects of alcohol among elderly people. For example, the proportion of body fat increases with age and that of body water decreases. As a result, the available volume of water into which alcohol may be absorbed becomes less. As alcohol is distributed freely in water, comparable doses will usually result in higher blood alcohol concentrations with increasing age. This relationship may explain in part the empirical observation of decreasing alcohol use with age, which is found both in cross-sectional studies of different age groups and in longitudinal follow-up studies of cohorts over time. Clinically, this can mean that patients who begin heavy drinking late in life may achieve a significant clinical tolerance using much less alcohol than would be necessary for a comparable effect in a younger person. The alteration in these physiological variables underscores the need to assess each patient's 'baseline' style and quantity of non-dependent drinking and to compare that against the history of increasing tolerance.

Other factors may contribute to the apparently decreased tolerance of alcohol among elderly people. Researchers have suggested that alcohol acts as a neurotoxin and has effects on the stability of either membranes or neurotransmitters in the central nervous system, or on both. None of these suggestions has completely accounted for the clinical phenomena of tolerance and withdrawal. Even further from present day understanding is the answer to why ageing systems and structures of the central nervous system appear to tolerate alcohol less well than younger ones. Some investigators suggest that the ageing brain itself may be more sensitive to the effects of alcohol because of variables such as the loss of brain mass or an impairment of blood flow in specific regions of the cortex. These hypotheses are being tested, but as yet they remain only interesting ideas. It is clear, by contrast, that the apparent decrease in tolerance to alcohol among aged people is not due to an impairment of its clearance from the bloodstream. Present data suggest that hepatic metabolism of alcohol is unimpaired in elderly patients who are free of the clinical signs and symptoms of hepatic failure. The zero-order kinetics that characterize alcohol metabolism in younger persons also obtain among older subjects.

IMPAIRMENT

Because ethyl alcohol is a very simple molecule that is miscible in all proportions with water, it rapidly diffuses through the body and has the potential to affect nearly every physiological system. Two systems have lately gained the attention of researchers who consider the problems of alcohol use among elderly people. In the first place, clinicians have long wondered whether the effects of alcohol either enhanced or mimicked the cognitive changes seen among persons who age normally. For example, recent memory loss in persons who drink heavily, with or without the Wernicke–Korsakoff syndrome, may appear to be very much like the mild forgetfulness that can accompany normal ageing. To date, both neuropsychological studies of cortical function and studies of brain structure using computed axial tomography (**CAT**) suggest that the two processes are very different. There is no consistent evidence that alcohol use *per se* hastens the processes of normal ageing in the central nervous system. Most neuropsychological studies indicate that chronic, heavy use of alcohol results in an inability to remember recent visual or verbal information and a reduced ability to carry out complex motor tasks. These are not concomitants of normal ageing and they are reversible on stopping drinking. Unless they are part of a pronounced confusional state as discussed below, these changes may not be evident during the examination in the survey or at the bedside. In the presence of a significant drinking history, however, they deserve evaluation. Whether and when CAT scanning is indicated in an elderly person who presents without confusion but with a history of significant alcohol use remains open to question. While early studies suggested that ageing and alcohol use might synergistically alter (CAT evidence of) brain volume, later more careful studies have described these as independent effects. There appears to be little association between CAT measures of loss of brain volume and concurrent neuropsychological measures of function. Most of the measurements of reduced brain volume reverse as a function of time free of alcohol use. At the same time, however, more acute pathological processes, such as subdural haematoma or stroke, with or without focal signs, occur more frequently among elderly people and perhaps even more often among elderly alcoholics. Heavy alcohol use has been identified as a risk factor for stroke although this may be partly mediated through a common association with cigarette smoking. The possible interaction between ageing and alcohol abuse or dependence has not been specifically studied. We suggest a clinical approach that errs on the patient's behalf until a better answer to this question appears.

Another area of recent inquiry has been the question of whether alcohol use diminishes the immune function of elderly people. This interest comes from epidemiological studies that suggest greater vulnerability to infections, especially respiratory infections, among elderly people who drink heavily than among those who drink little or not at all. *In-vitro* studies suggest that there may be a direct effect of alcohol on immune function, as measured by various T-cell and B-cell activities. Other findings suggest synergy between alcohol use and smoking that can incapacitate the immune system to a far greater extent than can either agent alone. In the presence of nicotine, concentrations of ethanol in the range of those seen in alcohol dependence can account for impaired immunity. This effect may explain the increased incidence of respiratory infections. Whether this synergy is more profound in elderly persons remains unclear as does the possibility of differential effects on antibacterial or antiviral immunity.

A second concern linking alcohol use and immune function relates to the incidence of cancers among elderly persons. There has long been a recognized association between cancers of the head and neck and the use of alcohol. Some postulate that this association has more to do with the convergence of alcohol and tobacco use rather than a direct effect of alcohol alone. Others have suggested that the effect of alcohol may be more pronounced among elderly people and may be linked to other cofactors such as impaired nutrition. These notions are interesting, but as yet there is no empirical support for them.

MISSED DIAGNOSIS

Alcohol dependence itself is an illness that be easily hidden until the late stages. By that time, the stereotypical 'skid row drunk' serves as the popular myth of what an alcoholic looks like. This stock character, often portrayed as an elderly person, represents a very small minority of those suffering from alcoholism. At the same time, this character's opposite—a well-situated elderly person who has begun filling free time by drinking—is rarely thought of, again because of the stereotype of persons reaching their 'golden years'. Drinking problems may go unnoticed in this age group because the thought of problem drinking is not consonant with this benign stereotypical view of an elderly person.

At the same time, questionnaires and screening tests that have traditionally been used to identify alcohol-dependent persons in their 20s, 30s, and 40s may not apply to persons at 65 and beyond. Questions about lost time at work because of alcohol use, for example, may have no relevance in a retirement community. The same may be true for the loss of friends because of alcohol use. During the latter decades of life, friendships may wane or end because of retirement, relocation, illness, or death. Trying to pin-point alcohol use as a cause of social isolation becomes more difficult in this age group.

Other areas of inquiry are likely to be more useful. For example, the complaints of a spouse or family member often signal an alcohol problem in younger age groups. In those elderly persons who have a spouse or who do not live alone, similar complaints may be relied upon simply because of the greater amount of time that spouses spend in each other's company after retirement. In another example, one study found that the threatened loss of a driving licence was an especially useful means of assuring treatment compliance among elderly persons with alcoholism. Independence of living and of movement takes on a special meaning in the elderly age group and the loss of these appears to carry much more weight than they might at an earlier age.

A variety of clinical features can lead an astute observer to the serendipitous identification of a previously unsuspected alcohol problem in an elderly patient. These include unexplained falls and fractures—especially rib fractures—macrocytosis on a routine blood film, hypertension, confusional episodes, late-onset epilepsy, sleep disturbance (see below), outbursts of ill-temper, non-compliance with medical advice, or simply a failure to maintain previous standards of dress and demeanour. For all of these, alcohol abuse should be specifically canvassed in the differential diagnosis.

A factor that bears on the use of screening questionnaires in this age group is the possibility of cognitive impairment that may render a questionnaire ineffective. One report suggests that alcohol dependence may be much more common than previously suspected among elderly patients who are brought to the family physician because of episodes of confusion. An impaired ability to attend to a task or to recall recent events can make it difficult for an elderly person to answer the 30 items on the Michigan Alcohol Screening Test accurately. The United States' Surgeon General's workshop on Health Promotion and Aging recently emphasized the task of developing age-specific screening examinations for alcohol dependence in this age group. Improved methods of screening for alcoholism, especially in medical settings, will provide an important tool for recognizing a treatable illness that often masquerades as other medical and psychiatric conditions.

MEDICAL MIMICRY

Heavy alcohol use or alcohol dependence can cause symptoms more commonly seen in other medical conditions that beset the elderly. As implied above, chronic use of alcohol in the later decades may result in confusion that can be mistaken as a dementia. Reversible states of confusion may occur as a result of intoxication or as a part of the withdrawal syndrome that follows cessation of drinking. Although there is little systematic information, clinical experience suggests that the mental confusion associated with either cause may last longer in elderly than in younger patients. This, in turn, can give the appearance of an irreversible impairment of cognition in the absence of the more acute peripheral nervous signs and symptoms of the alcohol withdrawal syndrome.

The differential diagnosis is further complicated by the occurrence of the two common dementing disorders due to alcohol: the Wernicke–Korsakoff syndrome and alcoholic dementia. The syndrome as classically described and well known to most practising physicians, involves a global confusional state in the presence of specific short-term memory deficits, an ophthalmoplegia, and a gait disorder. It can be reversed in more acute cases with prompt recognition and vitamin B replacement. In contrast, alcoholic dementia is poorly defined but characterized by an irreversible, global confusional state associated with a history of chronic drinking. It is generally used as a diagnosis of last resort. As in all the dementias, the clinician's task is to differentiate those that can be reversed from those, like Alzheimer's, that will continue to progress. The benefit of clinical doubt should be extended to the patients through (1) careful detoxification, (2) serial evaluation of cognitive function after detoxification, (3) attention to the nutritional status during detoxification and rehabilitation, and (4) long-term follow-up at regular intervals to chart the improvement in cognition.

Alcoholism can present as a major depressive disorder in older patients. Prolonged ingestion of a central depressant such as alcohol usually results in a depressed mood. It is not uncommon to see the 'hopeless–hapless–worthless' triad of a classical mood disorder among alcoholic persons. At the same time, chronic alcohol use can mimic the more 'biological' or 'vegetative' signs and symptoms of major depressive disease. Chronic use of alcohol, for example, alters the normal physiology of sleep such that affected individuals do not progress to very deep, restful, stage IV sleep. Because of this, they may wake early in a pattern that is indistinguishable from the 'terminal' insomnia of the depressive patient. At the same time, withdrawal symptoms after a rapid nocturnal drop in the blood alcohol may coincide with the terminal insomnia and create a clinical picture resembling the morning depression seen in major affective disorders. Morning drinking or daily use of a sedative tranquillizer relieves this pseudo-depression, mimicking the diurnal variation of mood seen in major depressive disorder. As in major depression so in alcoholics, loss of appetite and weight often ensue, because the prime source of calories is alcohol itself. While the body's mechanisms of satiety may perceive the intake of calories as sufficient, the calories are burned and not stored. The satiety leads to less eating followed by possible weight loss and malnutrition. Similarly, chronic alcohol use can account for other symptoms of major depression such as lack of energy, lack of interest in activities that were formerly found to be pleasurable, and an increasing number of physical complaints. In elderly drinkers, the vegetative and mood-related symptoms brought on by alcohol dependence generally lift after a month of abstinence from alcohol. In most cases, it is important to allow at least this period of time to pass before beginning antidepressant medicines or other therapy. Because alcoholics are no less prone to depression than the general population, however, the clinician can expect to see *bona fide* major depressive disorder in approximately 3 to 4 per cent of alcoholic patients. Characteristically, these mood-related and vegetative symptoms do not wane after a suitable period of abstinence and very often the clinician will find a history of prior depressive episodes, usually antedating the drinking, that have responded favourably to pharmacological treatment for depressive disorders.

The most catastrophic symptom of depression that can be mimicked or enhanced by alcohol dependence is the wish to commit suicide. Chronic alcohol dependence has been recognized as a significant cofactor in lethal suicide attempts. The triad of depressed mood, recent personal losses, and chronic alcohol use is an especially dangerous combination among patients contemplating suicide. This cluster of symptoms should be regarded as an emergency in any patient regardless of age, and may be especially relevant in the elderly population, which experiences depression and more significant losses. Prompt professional intervention and admission to hospital are generally indicated.

Active alcohol dependence can also mimic other psychiatric disorders. Profound agitation seen during the withdrawal syndrome can imitate the triad of symptoms seen in mania: hyperactivity, mood that is inappropriately euphoric or irritable, and speech characterized by flight of ideas. These symptoms depart with proper treatment within the first few days of the alcohol withdrawal syndrome, as noted below. The use of antimanic agents such as lithium carbonate should be reserved for patients with a prior history of mania or whose symptoms do not lift after proper treatment for alcohol withdrawal.

Similarly, perceptual disorders such as alcoholic hallucinosis or delirium tremens may be mistaken for the hallucinations of schizophrenia, mania, or the organic mental disorders. In alcoholic hallucinosis, the hallucinations characteristically occur in the presence of a clear sensorium, are generally auditory in nature, and usually respond very poorly to antipsychotic agents. In delirium tremens, the hallucinations are very often visual or tactile, and are generally quieted by rapid treatment with large doses of benzodiazepines {or chlormethiazole in the United Kingdom}, occasionally supplemented by small doses of antipsychotic agents. Long-term maintenance doses of antipsychotic agents have little role in either condition.

Alcoholic withdrawal has long been known to mimic seizure disorders by producing 'rum fits'. In most cases a seizure will appear on or about the first or second day following a rapid drop in the blood alcohol. While usually self-limiting, seizures can be obviated by rapid, adequate treatment of the underlying withdrawal disorder. Withdrawal seizures cannot be taken lightly because of their significant risk of death by asphyxiation. In the great majority of cases, treatment with a sedative agent is all that is required. Long-term antiepileptic treatment should be reserved for those with pre-existing seizures that are known to improve with therapy, or in patients continuing to suffer from seizures despite adequate treatment of the withdrawal state. Other causes of seizures should be carefully sought out in this latter group. Recent clinical data suggest that alcohol ingestion as well as withdrawal may precipitate fits.

Alcohol dependence can cause two other common medical conditions of elderly persons: peripheral neuropathy and hypertension. The neuropathy, classically described as a stocking glove sensory loss, can be mistaken for that seen in diabetes and other metabolic disorders. This is of special concern because of the high prevalence of diabetes mellitus in the latter decades of life. A history of alcohol use should be a routine part of the evaluation for elderly patients who present with a peripheral neuropathy. Similarly, hypertension has recently been linked to chronic alcohol use in large epidemiological studies. Hypertension is a characteristic symptom of the withdrawal syndrome, as noted below. In most cases, hypertension due to alcohol withdrawal returns to a normal state in 3 to 5 days after adequate treatment. Failing to recognize this clinical phenomenon may result in unneeded use of antihypertensive agents such as diuretics. For elderly patients, in whom relative dehydrating may result in cerebrovascular accident, unneeded antihypertensive medicines can produce an unnecessary risk to life and independence.

Finally, alcohol use adds a special risk to the most common cause of confusion among elderly adults, toxic states from medicines, whether these are prescribed or bought over the counter. As mentioned many times in this volume, elderly patients use more medicines and use them more frequently than younger persons. Many of these medicines, such as the commonly prescribed sedatives or over-the-counter antihistamines, act synergistically with alcohol, often resulting in an over-sedated or confused patient who is at risk for falls, accidents, dehydration, or aspiration pneumonia, to name only a few possibilities. The use of multiple medications by elderly persons has now gained more attention, but it is clear that concurrent use of alcohol should also be assessed routinely in the same context.

DIAGNOSIS

The general categories of signs and symptoms that are necessary for the diagnosis of alcohol dependence are the same for any age group, but their expression may be unique in such as the older age groups. There are four general symptom categories, of which the first is tolerance. This refers to the need of more alcohol to produce the same subjective effect as was originally achieved with very much less when the patient was either a beginning drinker or had an established pattern of alcohol use. In younger people, tolerance may be assessed by asking about early drinking experiences and charting the effects of known quantities, for example a 12-ounce can of beer, at that time. With older patients, a more useful approach may be to ask both the patient and a family member to describe the patient's typical drinking pattern prior to the time when alcohol use became identified as a problem. In one case report, for example, an elderly man described a drinking pattern that had been stable for nearly 50 years: he imbibed four mixed drinks daily, two at lunch, and two at dinner. He rarely varied from this routine and neither he nor his wife described an increase in his ability to tolerate alcohol. After a severe illness that robbed him of his mobility, he began drinking more frequently and more heavily and quickly developed a tolerance to as much as a fifth to a quarter gallon of spirits daily. The important clinical symptom of tolerance was his ability to drink three times his prior daily intake. For other elderly persons, whose ability to tolerate alcohol may be low, a shift from little or no alcohol use to daily consumption of significant amounts may indicate increased tolerance. There is wide variation among individuals, however, and much clinical research must be done to elucidate the factors involved both in initial sensitivity and in the growth of tolerance among different age groups.

The second category of symptoms may be grouped under the alcohol withdrawal syndrome. As tolerance to alcohol builds up, the body's reaction to a rapid drop in blood alcohol becomes ever more pronounced. For most patients, the withdrawal syndrome begins 6 to 12 h after the last drink or rapid fall in blood alcohol concentration. Many of the signs are those of a sympathetic discharge: tachycardia (greater than 110 beats per min), tachypnoea, hypertension, a low-grade fever, diaphoresis, and a profound sense of anxiety or impending disaster. Along with these the patient may experience nausea, vomiting, and a characteristic tremor of the hands. In severe cases, the patient may exhibit hyperreflexia or ankle clonus upon physical examination. These last two signs are especially worrying because they indicate a hyperactive central nervous system and often forbode an impending seizure. These physical signs call for rapid and aggressive treatment of the withdrawal syndrome.

In cases of uncomplicated alcohol withdrawal, the syndrome runs its course in 5 to 7 days. In elderly patients, however, clinical experience suggests that the syndrome may be associated with an increased frequency of cognitive impairment. When present, this impairment can persist well beyond the expected end of the syndrome itself and may continue for as long as a month or more. Studies of this phenomenon are badly needed, both to chart the course of the illness and to elucidate an approach to treatment. The primary factors associated with the severity of the withdrawal syndrome appear to be the total length of exposure to alcohol and the number of previous episodes of withdrawal that the patient has experienced. An intriguing factor of unknown import is the effect of age itself. We do not know whether those elderly persons who become dependent on alcohol very late in life are more or less likely than their younger counterparts to develop severe alcohol withdrawal. Much remains to be learned on this subject.

The third general category of symptoms can be considered as an impaired control phenomenon. As drinking increases in frequency and quantity, with tolerance increasing and the withdrawal symptoms becoming more pronounced, most alcohol-dependent patients experience a point of no return beyond which drinking becomes uncontrollable once it has begun. No one understands the pathophysiological mechanism of this, although it is generally regarded as the central factor in considering alcohol dependence as a disease. From the patient's point of view it means that he or she is unable to predict the consequences of their drinking once it has begun. As a result, patients spend more and more time and energy trying to compensate for behaviour that has become uncontrollable. Most dependent patients can describe attempts to 'go on the wagon' or to control their drinking by artifices such as switching beverages or brands of alcohol, or by drinking only after a certain hour. For a significant minority of patients, there is also an intense, if sporadic, craving for alcohol between drinking bouts. This may be set off by environmental cues and may be a cause for resumption of drinking for some patients. Little is known about the physiological concomitants of this symptom although some argue that it indicates a high affinity between the patient and alcohol transduced by an as yet unknown biological mechanism.

As the drinking continues to increase, driven by (1) the need for more alcohol to achieve a desired effect, (2) the necessity of frequent doses of this short-acting sedative in order to avoid the unpleasant withdrawal symptoms, and (3) the increasing inability to control the drinking once it has begun, the fourth major category of symptom obtains: social decline. As more of the patient's energies go to foster both the drinking behaviour and attempts to compensate for it socially, less energy remains to be devoted to the normal activities of life such as relationships with others and useful work or recreational activities. As a result, many patients become increasingly isolated from persons or activities outside the family, while becoming overly involved with specific family members. This latter is generally in response to the alcoholic person's enlisting the aid of family members in trying to control the use of alcohol. This activity usually leads to increased isolation when family members become frustrated and give up their efforts to 'cure' their alcoholic relative. Other symptoms may include difficulties with the law around issues such as drunk driving or, in some cases, shoplifting or other minor offences.

Over the past 40 years a series of investigators has taken the categories of symptoms and construed them in diagnostic schemes. From this has emerged a useful criterion published in the American Psychiatric Association's *Diagnostic and Statistical Manual* (third edition, revised). In this scheme, a patient may be considered to be alcohol dependent when tolerance or withdrawal syndrome can be demonstrated in the presence of either loss of control or social decline. This diagnosis is an important one to make because, when present, it forebodes a chronic, progressive, ultimately fatal illness. The diagnosis of alcohol dependence requires a careful interview with the patient, which may take an hour or more, along with a corroborating history from a family member.

To assist in the efficient use of this diagnosis, clinical groups have developed screening questions that allow the clinician to develop a high index of suspicion about the presence of alcohol dependence. One alternative is the Michigan Alcohol Screening Test (the MAST) or one of its derivatives. This is a 30-item pencil-and-paper questionnaire that can be quickly scored. For the busy clinician, the four-item CAGE questions may be easily woven into a standard, brief clinical history (see Table 1). Both screening devices were developed for specific populations: the MAST for relatively young drunk-driving offenders and the CAGE for patients on medical and surgical units of general hospitals. The extent to which either is especially sensitive or specific to alcohol problems among the elderly is unknown. Some research groups are developing age-specific, alcohol screening measures but these are not yet available. For now, either instrument appears to be useful, with the CAGE having specific application to general medical hospitals and clinics that are likely to see large numbers of elderly patients.

Table 1 CAGE questions

Question	
1. Have you ever felt you should cut down on your drinking?	C
2. Have people annoyed you by criticizing your drinking?	A
3. Have you ever felt bad or guilty about drinking?	G
4. Have you ever taken a drink first thing in the morning (eye opener) to steady your nerves or get rid of a hangover?	E

SHORT-TERM STABILIZATION AND MANAGEMENT

Faced with an elderly patient who is clearly dependent upon alcohol, the clinician must first attend to the dangers associated with intoxication and withdrawal. The first may be managed through careful monitoring at home or in hospital in an effort to avoid accidents or medical complications such as dehydration. Intoxicated elderly patients who threaten suicide require a prompt psychiatric evaluation and, probably, admission to hospital.

The dangers associated with alcohol-withdrawal syndrome are primarily those associated with seizures and with delirium tremens. For elderly alcoholics with a history of previous episodes of withdrawal, the risk is especially high. In one co-operative study done in the Veterans Administration, the risk of withdrawal seizure was as high as 7 per cent and the risk of delirium tremens as high as 5 per cent among patients presenting with the alcohol-withdrawal syndrome. Delirium tremens, characterized by disorientation and confusion, visual or tactile hallucinations, and autonomic hyperactivity, must be considered a medical emergency; the death rate is in the range of 10 to 15 per cent among untreated cases. Patients with this constellation of symptoms deserve serious consideration for admission to an intensive-care unit. Most of the serious sequelae of the alcohol-withdrawal syndrome, however, can be avoided with prompt, vigorous treatment of the syndrome itself.

In most instances, pharmacological treatment of the syndrome requires prescription of a centrally acting sedative in doses large enough to relieve the tachycardia, the tremor, the hyper-reflexia, or the ankle clonus. Not all of the symptoms of the withdrawal syndrome will be present in all patients. Each patient's target symptoms must be described and monitored carefully as the sedative dose is adjusted. This is especially important in elderly patients because of the danger of oversedation. Elderly patients with chronic pulmonary disease, for example, appear to have an increased risk of aspiration pneumonia or other pulmonary problems. At the same time, oversedation may worsen the cognitive capacity and may result in a confused, apparently delirious appearing patient. In general, the sedative should be used vigorously enough to relieve the withdrawal symptoms but not so vigorously as to result in somnolence.

For many elderly people, a long-acting benzodiazepine (such as chlordiazepoxide or diazepam) can be given in carefully adjusted doses during the first 24 h until the symptoms remit. The dose may then be reduced by one-third daily over the ensuing 2 or 3 days and then stopped completely. Because of their long half-lives, these agents will generally carry an elderly patient through the 5 to 7 days of the withdrawal syndrome. For patients in whom a long-acting benzodiazepine is contraindicated, for example, those with liver disease or compromised pulmonary function, a short-acting benzodiazepine (such as lorazepam or oxazepam) may be given. These agents must be given much more frequently, however, owing to their much shorter half-lives, and must be given throughout the entire course of the syndrome. Adrenergic blockers (such as clonidine) are probably not useful in this population because of the risk of hypotension that may result in a stroke. Agents such as chlormethiazole regarded as the drug of choice in several European countries may be used in place of the benzodiazepines, although this agent is not at present available in the United States.

Careful attention must be given to the patient's cognitive status during the withdrawal period. This is especially important in assisting patients in the transition from short-term, acute stabilization into long-term treatment for the alcohol dependence itself. Counselling, didactic sessions, and self-help groups may be appropriate but are of little use to the patient whose mental faculties are temporarily impaired. Similarly, sedation to the point of somnolence may render the patient unavailable to appropriate interventions at this particularly critical time in the course of treatment. As mentioned above, for most patients, other medications such as antidepressants or antipsychotics should be avoided unless there is a very clear indication for their use.

LONG-TERM MANAGEMENT

Once the dangers of the withdrawal syndrome have been negotiated, the hard work of approaching long-term treatment begins. At present, there is no specific pharmacological intervention that can relieve alcohol dependence. However, illness-specific behavioural interventions do exist and should be used in programmes tailored to the individual patient. The fundamental ingredient is the patient's own sense of recognition of the disease process and the employment of his sense of hope in stopping the progression of the dependence. The physician's first task is to present the patient with the diagnosis and to explain its basis in terms of symptoms and signs. The physician should make every attempt to involve other family members in an effort to focus upon the disease itself and its consequences. It is important to remember, however, that the relationship between the elderly person and alcohol is an ambivalent one. While the physician and the family may be all too aware of the negative consequences of the condition, they may present only those negative aspects to the patient, leaving him no alternative but to find positive reasons for continuing to drink. The wise clinician will present both sides of the story to the patient and focus the responsibility for the decision not to drink where it belongs: with the patient. Having done so, the physician is free to recommend any of a series of programmes as may be appropriate to each patient.

One scholar in the field identified four therapeutic areas that were associated with a likelihood of abstinence for more than 3 years. The first involved replacing the time spent drinking with substitute activities that involved other people and did not involve drinking. These may include participation in self-help groups such as Alcoholics Anonymous, therapy groups through an alcoholism treatment clinic, voluntary community interests whether for recreation or for work, or any combination of these. Secondly, patients who were able to maintain long-term abstinence were often those who found a source of increased self-esteem or hope in their lives. For many this referred to the 'higher power' and the spiritual concept of sobriety in Alcoholics Anonymous. For others it was a return to an organized religion. For still others it was merely coming to terms with a belief in a concept or an object with which they could share the responsibility for their daily decision not to drink. Thirdly, long-term abstainers frequently looked toward a sustaining, rehabilitative relationship with another person who viewed them as a valuable human being while at the same time drawing clear limits on the acceptability of their drinking behaviour. For elderly people this could be a spouse, a physician, a sponsor in Alcoholics Anonymous, or a psychotherapist. Experience to date in programmes specifically designed to treat elderly patients suggests that a supportive, low-key approach is the most effective intervention for patients in this age group. Aggressive, insight-oriented, 'How does that make you feel?' interventions are generally ineffective for patients in the older age range.

Fourthly and finally, there was a role for negative reinforcers of drinking in the effort to promote long-term abstinence. Such negative consequences must be certain, immediate, and perceived as noxious to the drinking person. In one group's experience with elderly drinkers, the threatened loss of a driving licence fulfilled all of these criteria. For other patients, episodes of pancreatic pain have the same effect. More subtle illnesses, such as heart, liver, or kidney failure, do not carry the same therapeutic impact because of their relatively subtle symptoms. For younger patients, the effectiveness of disulfiram therapy also depends upon a certain, immediate, noxious consequence to drinking. For elderly patients, with a possibility of a compromised cardiovascular system and a higher incidence of heart disease and stroke, prescription of disulfiram appears to take on a much higher risk and should probably be avoided.

As a general rule, the first three 'positive' interventions outlined above should bear most of the therapeutic load in an individual treatment plan. Alcoholics Anonymous appears to be effective as a treatment because it involves all three of those components. For many patients, however, Alcoholics Anonymous does not appear to 'work'. A careful treatment plan can be devised for these patients that includes the above ingredients in the service of enlisting the patient's own sense of hope in remission from this disease.

The physician should bear in mind that alcohol dependence remains a chronic, remitting/relapsing illness. Elderly patients may well have 'slips' and return to drinking. Such instances are often followed by shame and guilt that may prevent the elderly patient from returning to see their physician. Missed appointments, therefore, should be routinely followed up by an invitation to return to the clinic and resume treatment. Hope should be supported in these instances and the healing powers of the doctor–patient relationship should be offered.

CONCLUSION

It is important to remember that no medical specialty or discipline outside of medicine has a monopoly on the treatment of elderly patients suffering from alcohol dependence. We have much to learn from each other about this illness in this age group. Because of its high frequency as well as its chronic nature, we also have much to gain from each other in the way of morale, a very important ingredient in keeping up hope in the context of an illness that all too often is perceived nihilistically. If we are to treat an illness that responds only to hope and to do so in the context of advanced age, professionals must be especially attuned to the vitality of their own hope and the ways in which that is communicated to patients and their families.

It is important for physicians to understand, for example, that we do not bear the total responsibility for the care of the alcoholic patient. We do have a responsibility for diagnosis, for presenting the problem to patient and family, and for appropriate referral. By attending to these duties in a knowledgeable and consistent manner, patients can be returned to the health and vitality that should be theirs after a lifetime in the struggle with the human condition. Recovery from alcohol dependence is a hopeful process among the elderly that can achieve a substantial reduction in drinking and its consequences for most, and complete abstinence for some. For those with intact social resources that can be used in the services of health, there is a likelihood of high rates of successful remission. Until we can document these rates in this specific age group and further elucidate the factors that apply to successful treatment, it is fitting for physicians and patients alike to proceed knowledgeably in the service of true hope.

Bibliography

Beresford, R., Blow, F., Brower, K., and Singer, K. (1988). Screening for alcoholism. *Preventive Medicine*, **17**, 653–63.

Berglund, M., Hagstadius, S., Risberg, J., Johanson, T., Bliding, A., and Mubrin, Z. (1987). Normalization of regional cerebral blood flow in alcoholics during the first 7 weeks of abstinence. *Acta Psychiatrica Scandinavica*, **75**, 202–8.

Dorus, W. (1987). Symptoms and diagnosis of depression in alcoholism. *Alcoholism (NY)*, **11**, 150–4.

Grant, I., Reed, R., and Adams, K.M. (1987). Diagnosis of intermediate-duration and subacute organic mental disorders in abstinent alcoholics. *Journal of Clinical Psychiatry*, **48**, 319–23.

MacGregor, R.R. (1986). Alcohol and immune defense. *Journal of the American Medical Association*, **256**, 1474–9.

Vaillant, G.E. (1983). *The natural history of alcoholism*. Harvard University Press.

Vestal, R. *et al.* (1977). Aging and alcohol metabolism. *Clinical Pharmacology and Therapeutics*, **21**, 343–54.

20.5 Behaviour management

JEFFREY GARLAND

For this review, the term 'behaviour management' is used to cover a number of psychological interventions that have been used with older people to influence the thoughts, feelings, and behaviour of patients and carers. These are reported under various headings, including behaviour modification, behaviour therapy, behavioural psychotherapy, and cognitive–behavioural therapy.

While such approaches focus on the importance of manipulating environmental contingencies, practitioners acknowledge that disturbed behaviour also can be triggered by physical illness, the effects of treatment, or concurrent psychiatric illness (Gelder *et al.* 1989), and that such factors should be thoroughly explored before embarking on an individual behavioural programme.

Much of the language of behaviour management appears mechanistic, but in practice treatment programmes are developed within a framework that sees the patient as an individual, and with an appreciation that a behaviour which appears grossly inappropriate and dysfunctional, and therefore a prime target for modification, can from the patient's viewpoint appear essential to retain, because it helps that individual to make some kind of sense of his or her experience and to maintain precarious self-esteem. Thus intervention is aimed at building behaviours that strengthen the patient's sense of self, and reducing those behaviours that appear to be diminishing the patient's competence and individuality.

Behaviour management is not reserved solely for use by psychologists or even by mental health professionals in general. With supervision it can be used by others in health or social care, including informal carers and old people themselves, as many of these interventions prescribe comprehensive involvement of the patient's support network.

Three major categories of problem are addressed: absence of behaviour that would appear to be within a patient's presumed range of competence but is not being shown; behaviour that would be desirable in normal frequency, but is being produced excessively; and behaviour that is undesirable *per se*.

A central proposition is that many of our voluntary acts operate on our environment to produce consequences, and hence can be analysed as operant behaviours shaped and maintained by such consequences. If an individual's action is followed by consequences valued by that individual, these reinforce the action, making it more likely to recur. If an action is followed by consequences that an individual does not value, these do not reinforce the action, making it less likely to recur. Therefore, if carers systematically change the consequences of a problem behaviour, it may become possible progressively to increase desirable behaviour by positive reinforcement and decrease undesirable behaviour by extinction (withdrawal of reinforcement that may have been given inadvertently).

Behaviour may also be shaped by the 'push' of antecedent events (such as environmental cues or the occurrence of habitual thought patterns), as well as the 'pull' of consequences. Hence ABC analysis (recording of antecedent events, specific behaviour, and consequences), is fundamental for much behaviour management. For example, an elderly woman resident is referred by a home's care staff to a psychologist as 'aggressive', with a request that she be admitted to a psychiatric assessment unit 'to find out what's causing the aggression'. Asked to give a specific instance, the staff report that the resident suffers from chronic constipation, and that two of them had therefore approached her (antecedent) in the home's lounge, asking her to come to the clinical room for an enema to be given ('to be debunged', in language understood by staff and residents). The resident then 'shouted, and screamed, and waved her arms around' (behaviour). Alarmed by this reaction, and not wishing 'to make a scene', the staff retreated, and the resident did not have the enema on that occasion (consequences).

It can be seen from this brief example that ABC analysis can not only highlight a treatable physical factor contributing to the resident's irritability, but also give staff the scope to look at how their approach behaviour and the consequences for the resident in asserting control and avoidance can be understood and managed, in a way that would not be possible if the 'aggression' was simply transferred to an assessment unit to be dealt with.

There is evidence that a behavioural approach is of value with certain problems (Woods and Britton 1985). Both patients and carers can benefit from attributing problem behaviours to environmental contingencies, which can be modified, rather than to processes of senescence seen as essentially untreatable. Targeting specific thoughts or action is likely to be more productive than addressing global issues such as 'lacks motivation' or 'manipulative'.

Planning, monitoring, and review of behaviour management with older people needs to be thought through and documented with care, often within a goal-planning framework based on the principles set out by Houts and Scott (1976):

1. Involve the patient from the beginning;
2. Use the patient's strengths to set goals which help meet his or her needs;
3. Set reasonable goals;
4. Spell out the steps necessary to reach each goal;
5. State clearly what the patient will do, what the staff will do, and when you expect to reach each goal.

However, a number of qualifying clauses must be registered. The field is still in its early conceptual development, with research rarely definitive (Smyer *et al.* 1990). Trivialization—much time being spent on short-term demonstration projects from which only limited generalizations are possible—remains a cause for concern.

While much of the problem behaviour of older patients can be contained or even ameliorated, it is rarely realistic to expect 'cure'. Environmental constraints, such as lack of staff or rapid turnover, can frustrate even the most practical and relevant of interventions.

Because many behaviour problems of elderly patients tend to be attributed by patients and their carers to an inevitable process of ageing, intervention may be seen as irrelevant and the incentive to co-operate can be lacking.

Furthermore, it must be remembered that psychological intervention is just one component in an array of influences on the older patient, which include illnesses, the effects of treatment of illnesses, and admission to hospital with its attendant pressures. As yet, little is known with certainty about the complex linkages between elderly patients, their behaviour, and their environment. Prediction of the response to intervention is difficult, as it is necessary to consider individual characteristics such as cognitive functioning, diagnosis, lifelong adaptation, place of residence, social functioning, sex, health, and chronological age.

Nevertheless, some general issues of relevance for theory and practice with older patients can be described. The following broad categories of problem behaviour under which this description will be carried out are, in general, designated here by labels commonly used by referrers. The use of these headings does not necessarily imply that they can all be seen as sufficiently explicit, but with a few rare exceptions practitioners of behaviour management have not come up with better alternatives. It must also be pointed out that the categories can overlap, and that it is not infrequent to find a patient with problems in several of these areas.

ABSENCE OF PURPOSEFUL BEHAVIOUR

As used by Kennedy and Kennedy (1982), this term indicates near total lack of self-preserving or self-care activity. The patient is likely to show deficits in all areas of functioning, and to be classified as requiring total nursing care.

A prerequisite of management is to ensure that staff attitudes reflect positive expectations and belief in a worthwhile and dignified life for patients. Related to this is the choice of possible skills for the patient to develop, which Kennedy and Kennedy advise should be determined by examining their potential benefit to the patient's quality of life rather than by considerations of easing his or her management.

Intervention with such patients progresses through painstaking task analysis, breaking down into small steps the sequences of parallel actions for patient and carers, with highly selective use of reinforcement (non-verbal cues with gesture or touch are more efficient than a running commentary of verbal instructions and encouragement, which tends to distract from the task in hand).

However, while there are reports of such approaches being used successfully in patients with learning disabilities or in rehabilitation settings, little work has been done on the older patient with absence of purposeful behaviour. In clinical practice referral of such patients is extremely rare.

'AGGRESSION'

It is not uncommon for a patient who feels belittled and frustrated to react with violence and abuse: his or her complaints may be well-founded, based on real shortcomings of care, which sometimes can be rectified by the institution or individuals responsible without requiring further analysis of the patient's actions.

Other causes of 'aggression' include: the patient's perception that personal territory is being invaded by strangers (carers whom he or she has failed to recognize); another's obtrusive and embarrassing attempt to correct a mistake by the patient; frustration triggered by some failure in an activity of daily living, being restrained from other undesirable behaviour, or inability to make him/herself understood; misunderstanding of another's action; blaming others for the patient's own memory failure (accusing them of hiding or stealing possessions); and resentment of other patients' unpredictable actions or apparent competition for carers' attention.

Among preventive measures to reduce the frequency of 'aggressive' behaviour by older patients receiving institutional care are: careful preparation for admission and induction wherever possible; a policy for responding to aggression, with the means to inquire after each serious incident to discover causes, anticipate, and minimize the risk of a repeat; a routine that is clear, regular, and leisurely; and a range of opportunities for activity and stimulation at a pace and level to suit all patients.

As such a milieu is not often found in geriatric care, advice on the management of 'aggressive' activity is one of the more common reasons for referral. Advice generally given to carers includes: stay calm; remain detached; do not crowd the patient; direct others to draw back; ask the patient what is troubling them; listen to complaints; engage in distracting activity; physical (or pharmacological) restraint may be required, but should be a last resort.

More specific instruction may use video tapes of simulated incidents based on real-life events that staff need to learn from. For example, a carer would be shown noticing that the hair of a woman patient needs combing, picking up a comb, and approaching the patient. The film would then be stopped for the learners to comment on what was happening, suggest alternative action, and predict outcome. The film would then continue from the patient's viewpoint of a hand with a dimly discerned object approaching her head. Again the film would be stopped for a similar procedure. Finally, the outcome of the encounter would be shown for further discussion. After a series of such clips, learners would be invited to role-play incidents from their own experience that had added to their knowledge of the management of 'aggression'.

It is unfortunate that while such instruction has been used to good effect with other types of patient, it has not yet found wide acceptance in the care of older people.

ANXIETY

This is a fact of life for many older patients as Hussian (1981, p. 102) suggests:

> 'Age-related anxiety-producing stimuli . . . may include changes in residence, fears of losing control, financial problems, death and dying, being alone, stairways, steps, going outdoors alone, being too far away from a bathroom, inability to remember recent or remote events with any clarity, spending holidays alone, fear of post-holiday depression, losing one's sensory acuity, losing one's mind, choking, being robbed or accosted, being laughed at, various modes of transportation, going shopping, physical examinations, and minor surgery.'

Cognitive–behavioural therapy, which can be individual or group-based, proposes that the ways in which a person acts are determined by immediate situations and by the person's interpretation of those situations. It can be applied productively with older patients, as Woods and Britton (1985) show. This is a highly structured collaboration, focusing on self-help and aiming to reduce anxiety by teaching the patient to identify, evaluate, control, and modify negative thoughts and associated behaviour. The therapist, though, has an important role in promoting the patient's trust and in encouraging the use of strategies to manage anxiety. Among these are relaxation; becoming aware of, examining, and challenging negative thoughts and expectations; introducing positive coping strategies, problem-solving, 'homework' on difficult tasks; and both imagined and real-life exposure to anxiety-related stimuli in a series of graded exercises.

However, such approaches take time to set up and show results, and cognitive–behavioural therapists do not as yet participate widely in geriatric medicine. Use of tranquillizers for moderate to severe anxiety in older patients retains priority.

'ATTENTION SEEKING'

Attention from others is, of course, a basic human need, supporting our sense of identity, enabling us to feel secure and in control, bringing stimulation and variety, and helping us to feel valued. It is understandable therefore that some older patients threatened by illness (and often with sensory impairment and/or impaired mobility that restricts their ability to engage others) are seen by carers as seeking excessive attention. This may take the form of persistent calling, questioning, bids for reassurance, or complaints of pain (see 'Pain complaining', below) illness or incapacity. For some of the most demanding patients, hard-pressed carers may report episodes of incontinence, falls, refusal to eat, and a variety of other actions under the rubric of 'attention-seeking', not always with apparent justification.

A 'menu' of advice and procedures from which selections can be made in the management of excess noise-making is offered by Garland (1987).

1. Noise that is made because of delirium and/or terminal illness is unlikely to respond to behaviour management. However, if noise-making is clearly affected by events in the patient's surroundings, behaviour management can be applied. Share awareness of the problem with the patient. Attempt to get his or her views. Begin taping and/or charting the frequency and/or duration of noise, having consulted the patient.
2. Check possible rewards from noise: self-stimulation; tension release; feeling of control over others; feeling of security when attention comes; satisfaction from the stimulation attention brings; means of being sedated with consequent relief from distress.
3. Check for and correct immediate causes that may have been overlooked, e.g. physical discomfort, worry about toileting. Explore any underlying causes, e.g. concern about results of disability, fear of death.
4. If noise-making persists, tell the patient that changes in routine will be made 'because it's our job to see you get what you need without having to shout for it'. Repeat this at intervals.
5. Tape the noise and play it back to patient. Playback may need to be done several times. Be prepared for the possibility that the patient will disown the sounds, denying responsibility.
6. Often the noise-making is isolated by sensory impairment. This can be compounded if s/he is kept to their room to buffer others from the noise. Make the im-

mediate environment more stimulating. Gradually reintroduce increased contact.

7. Approach the patient if s/he is quiet and awake for a while; and respond with a verbal recognition that s/he is being quiet. Identify those reinforcers the patient is known to respond to (for example, a sweet, cigarette, drink, newspaper, back rub), and present one with your approach.
8. Do not respond to noise (unless an emergency). If approaching during the noise, use a minimum, standard response.
9. Prompt the patient's listening to radio/tape recorder with headphones. Use programmes or tapes known to be of individual appeal.
10. If noise-making persists and the patient is placed in seclusion at intervals, this should be done with a brief explanation and no further comment. Return him/her when quieter. Use seclusion for brief, set intervals only.

Persistent questioning or requests for reassurance that are relieved only temporarily by a matter-of-fact answer are possibly being reinforced inadvertently by carers' responses, which can feed a need for attention without satisfying it. It can be more productive to respond to the feeling behind a question rather than its content. For example, a 'confused' patient who asks a carer 'Where is my mother?' but is not satisfied for long with a factual answer may be feeling like a lost child and is communicating a need for affection and security that can be responded to directly.

For the most persistent questioners, the introduction of a substitute topic more acceptable to the carers can be successful. Since her arrival 2 years before, a resident of an old people's home had been asking carers 'Where is my home?' Two return visits to her former home had been made in an attempt to settle her mind,and carers had invested much time in trying to elicit how she felt about her move and encouraging her to feel at home. By serendipity, a successful behaviour-management programme was based on the discovery that the resident had just become attached to the 8-year-old son of a new carer who had recently visited the home. The resident was provided with a photo and ongoing information about the boy and, by arrangement, carers spontaneously opened conversation about him, making an excuse and leaving if the resident reverted to the topic of her former home. In time, conversation became generalized to the carers' own children or grandchildren, on a basis of reciprocity.

Patients may also be assisted to manage their own need for information directly, particularly if their attention-seeking is linked with forgetfulness. Cue cards carried in a handbag or wallet, or displayed on a bedside table or locker door, may be used to convey key information, with the patient being encouraged to 'look for yourself' when carers are approached.

'CONFUSION'

The psychological management of dementing patients aims at improvement in the quality of life for patients and carers, encouraging as much activity, independence,and socializing as possible.

Reality orientation, described by Holden and Woods (1988), deals with the introduction and maintenance of a range of environmental cues to orientate patients verbally and behaviourally, and so to increase their awareness of what is happening around them. Informal or 24-h reality orientation is used in carer–patient interactions, and patients are encouraged to refer to memory aids, signs, and notices to orientate themselves. In so-called classroom reality orientation, small groups, usually of about four to six patients, engage in supervised, structured activities and discussion with cues, prompts, immediate reinforcement of efforts to maintain orientation, and with minimal attention being given to disorientated behaviour.

The status of reality orientation as a philosophy of care that encourages communication with 'confused' patients and gives carers a rationale for doing so is widely accepted. It is not a treatment as such, in that there is no substantial evidence of generalization or maintenance of improved orientation outside a programme of reality orientation.

DEATH AND DYING

The development and application of knowledge and techniques relevant to the understanding and management of terminal illness, life-threatening behaviour, and grief is described by Sobel (1981) as 'behavioural thanatology'. A principal aim is the practice of behavioural self-control to help a patient or carers to monitor their thoughts and actions, and to plan and carry out their own adaptive responses, reinstating a sense of personal control and choice.

As grief is an intense personal experience embedded in a complex of biological and sociocultural factors, the application of behaviour management techniques may appear as intrusive and unnecessary. However, recovery from bereavement involves much relearning; and our understanding of bereavement can be enhanced by an appreciation that some of the signs of grief relate to the extinction of previously reinforced responses, while new maladaptive responses can be sustained inappropriately by the undiscriminating support that may be seen as the entitlement of a bereaved person.

A respondent model, which sees grieving as a conditioned emotional response, has also been advanced. It is suggested that death functions as a non-contingent aversive stimulus triggering distress and shock. In normal grieving the bereaved strives unsuccessfully to escape, eventually suppresses his or her responses, and finally develops a new repertoire. In pathological grieving an effective escape or avoidance response is achieved, so that suppression does not take place.

As yet the published record offers only a few studies of behavioural intervention with older patients who have been bereaved. This may reflect the effectiveness of normal grieving processes for most patients, and the readiness of competent counsellors to assist the remainder.

DEPRESSION

Complaints of being slowed down or fatigued, dissatisfied with sleep, or having no appetite are not infrequent in older patients who are not necessarily clinically depressed. However, clinical depression is quite common in later life, and physical treatment including antidepressant medication and electroconvulsive therapy is widely used in the psychiatry of old age.

Psychological approaches to depression in later life focus on loss of social reinforcement and on learned helplessness whereby a patient not only perceives that s/he has lost control over reinforcers (so that others' positive responses appear much less frequent and much more arbitrary than s/he would wish), but also tends to blame him/herself for this state of affairs, going into a downward spiral of lowered mood.

For the cognitive–behavioural therapist, depression in an older patient raises treatment issues that have a good deal in common with those arising in the management of anxiety. Cognitive–behavioural therapy emphasizes the importance of thought as mediators of mood state. Patients are taught to monitor mood and behaviour on a daily basis and to record the relationship between them; and to proceed to a systematic, progressive increase in pleasant activities and a decrease in unpleasant activities as part of a self-change plan formulated with the therapist.

As older patients can tend to frame problems in a global way and to honour the proverb 'you can't teach an old dog new tricks', the therapist generally needs to work hard on encouraging them to redefine depression in terms of small, specific units of thinking or action for which a behavioural skill will be relevant, and to see the intervention as an experiment 'worth a try' rather than as a task with criteria for success.

Ongoing support to help the patient keep on target, an explicit maintenance plan, and subsequent booster sessions are important components of this type of intervention. Throughout it is emphasized that it is more productive to view thoughts and behaviour as influencing mood state, rather than *vice versa*. The message is not 'when I feel better, I'll do it', but 'I do it, then I feel better'. Thoughts and feelings are presented as open to choice, and modifiable through self-instruction.

For example, a woman patient in a geriatric ward who encounters a series of undignified and uncomfortable experiences early in her stay begins to think 'nothing ever goes right for me here', and becomes increasingly unhappy. If she voices this spontaneously or can be persuaded to do so by a carer who has noted her low mood, systematic progress towards introducing positive self-statements can be made through a series of stages, assisting the patient to identify the source of worry, prepare for action, build confidence in handling her problems, cope with feelings of being overwhelmed, and give herself reinforcement for successful use of positive self-statements such as 'that went right', 'one day at a time'.

The published record on both individual and group cognitive–behavioural therapy with depressed older people is relatively substantial and generally positive. However, controlled studies indicate that in general such patients respond well to structured, short-term psychotherapy of any type, and further comparative study needs to be done.

INCONTINENCE

Psychological approaches to understanding and treating functional incontinence (arising from the patient's inability or unwillingness to use a toilet appropriately) are summarized by Smith and Smith (1987) and in the NIH Consensus Conference (1989). Thorough assessment of older patients before any intervention is crucial.

Among the most common causes and remedies for functional incontinence are: inadvertent reinforcement by a carer who responds when the patient is wet but who does not respond to the patient being dry or being toileted successfully (show the carer how to reinforce being dry and not to respond with unnecessary attention when patient is wet); the patient cannot recognize the whereabouts of the toilet (use distinctive colour- or sign-coding for the toilet with carer prompts); cannot reach the toilet in time because of poor mobility (improve on this situation, use a commode or other aids); cannot reliably find the toilet because of memory impairment (practise route, use cues and signs); cannot dress or undress efficiently and in correct sequence (practise dressing skills, provide adapted clothing); finds the toilet unattractive and uncomfortable (improve facilities); and does not ask for the toilet, perhaps because of expressive aphasia (increase the sensitivity of the carer to the patient's attempts to communicate need).

The published account of results is encouraging, although, as Smith and Smith suggest, many of the earlier studies were done in residential settings in which the baseline of care was so modest that any intervention would have looked good. The record covers contingency management, self-monitoring exercises, habit training, bladder drill (Fantl *et al.* 1991), and biofeedback to increase control of the bladder and the muscles of the pelvic floor. The most important component of effective intervention is the extent to which the patient stops seeing him/herself as victim of the problem and begins to believe that s/he can cope with it.

PAIN COMPLAINING

Pain is a significant component in the experience of chronic illness increasingly prevalent in late life. Beliefs, expectancies, mood state, personal interpretation of pain, and a history of reinforcement for pain complaining are among factors that help to account for differences between the experience of pain and actions by the patient that signal to others 'I am in pain'.

A vocal minority of older patients complain of pain (by facial expression, gait, medicine intake, as well as verbal complaints): to an extent that their carers deem excessive; in a manner observed as illogical (for example, where there is no apparent physical cause, or where the pain changes in intensity or location with undue speed, or where the description of the pain appears contradictory or muddled); or in circumstances where carers suspect secondary gain (for example, a patient who is reluctant to leave hospital, when his or her departure is due, may describe 'new pains I haven't told you about before').

Understandably, such complaining is rarely ignored completely. The interview and investigations that follow can unearth treatable physical or psychological causes such as underlying depression, or may lead to a temporary cessation or reduction in complaints. The patient reports 'they' have listened to him/her, his or her treatment has been adjusted, and anxiety is reduced through reassurance.

The change is seldom lasting. Complaints recur, further

investigations are ordered, more reassurance is given and accepted, and the cycle of pain complaining goes on. As Sturgis *et al.* (1987) point out, multimodal approaches recommended for pain complaining in older patients include: raising the patient's level of functional independence: improving self-control; decreasing disability secondary to pain; continuing management of pain behaviour; pacing of adaptive behaviour; and teaching of cognitive coping skills.

Even patients whose complaining is not judged as inappropriate may benefit from such training in self-help to describe how the experience of the pain varies, to test out reasons for this variation, to use relaxation, to have a range of distractors available, and to make changes in everyday activities.

PASSIVITY

Terms such as 'apathy', 'dependency', 'excess disabilities', or 'withdrawal' have been used to label the passive adjustment made by some older patients to residential care, whether in hospitals or old people's homes.

Baltes and Baltes (1986) have drawn attention to the psychological importance of an older patient's feeling that s/he is in control of the way s/he experiences and regulates his/her behaviour in terms of reasons and causes for action and their consequences.

Differential patterns of reinforcement have been demonstrated in the way care staff respond to dependent and independent behaviour. What older people in residential care do is not a major determinant of the subsequent behaviour of carers, with one exception: when the older person produces dependent behaviour during activities of daily living, social reinforcement in the form of attention tends to be provided by care staff. (In family care these contingencies generally are reversed—independent behaviour by the older person gets most attention from the informal carers).

A wide variety of attempts that have been made to reverse contingencies in residential care by encouraging residents to be more active include: making recreational materials more widely available; making the opportunity for a choice of activities available to residents; letter-writing programmes (one with the evocative title 'Project RSVP'); training in telephone skills; art, music, movement, occupational, or sensory-awareness therapies; use of animal pets in therapy; volunteers for one-to-one contact; use of older people themselves to work with mentally retarded adults or in teaching adopted 'grandchildren'; controlled use of alcohol as a social lubricant for small group discussions; and prompting, modelling, or reinforcement to increase interaction and maintain presence at leisure or social events.

A voluminous published account describes improvements in mood, morale, orientation, or participation in activity in many of these studies (which testify incidentally to the bleakness that can characterize such care settings.) As with reality orientation, such gains do not appear to generalize once the environmental enhancement is withdrawn.

SEXUALITY

A focus of concern is expression of sexual needs in a disinhibited way that offends carers or fellow-patients. Comfort (1980) reminds us that there is no reason why, give appropriate circumstances, sex cannot continue to be a rewarding part of life for the elderly person; but the carers of an older patient may have extreme difficulty in granting that patient permission to continue sexual activity, or support in developing new ways of meeting sexual needs where established behaviour has been disrupted by disability, lack of a partner, or the restrictions of institutional life.

Hussian (1981) describes, in a case study of a resident who masturbated publicly, how through behaviour management it is possible to teach an older person successfully to distinguish between expressing a sexual need acceptably in private, and acting unacceptably in public—but privacy itself can be difficult to ensure.

Initially, carers may be more in need of behaviour management than the patient, as their own attitudes to sexuality in older people, often equivocal and tinged with teasing and provocation that rapidly turn to angry rejection, can need resolution before productive work relating to the patient's sexuality can be undertaken.

SLEEP DISSATISFACTION

Rational, non-pharmacological approaches to improving older patients' satisfaction with sleep are described by Morgan (1987), who points out that successful sleepers tend to have developed effective stimulus control (a set of regular personal routines and habits constituting environmental signals making sleep more likely to occur predictably).

The onset of insomnia for a patient can be viewed as disruption of his or her pattern, which can generally be reinstated with information and support in stimulus-control procedures (but not with the agitated and 'confused' patient whose sleep/waking pattern is profoundly disturbed).

From the patient's viewpoint, stimulus control may be summarized as follows: the bed as a cue for sleep is strengthened, while being weakened as a cue for activities incompatible with sleep; prepare to sleep only when sleepy; use bed only for sleep or sex; if unable to sleep get up within 10 min and go to another room returning to your bed only when you are prepared to sleep; repeat the process again each time you are unable to sleep; get up at the same time each day; do not sleep during the day.

Self-monitoring and recording the sleep pattern in a sleep diary is an important component of this procedure, helping patients to appreciate the influence of exacerbation cycles (being awake at night worrying about sleep, which in itself can lead to sleeplessness, underestimation of time spent asleep, and overestimation of the latency of sleep onset). (See also Chapter 18.8.)

'WANDERING'

This tendency to move around, in an apparently aimless fashion, or in pursuit of a goal that appears unobtainable, may have a number of causes including: disorientation in new surroundings or familiar surroundings that are not recognized; onset of darkness experienced as disturbing and sometimes associated with frightening past events such as 'the Blitz' with its air raids by night; under-occupation; physical discomfort; a lifelong pattern of coping with stress ('going for a walk'); continuation of family or work role ('fetching the children from school', 'getting the cows in'); and a search for security (looking for spouse, friends, parent).

Unless the patient is at serious risk to self or others, such behaviour needs to be accepted as far as possible and redirected rather than restricted. Depending on the apparent causes, the environment, and the individual, management pointers embrace: use of reality orientation; mapping out the environment with cues and signs, and stopping-off points attractive to the patient to delay progress; if there are major no-go areas, conditioning the patient with warning signs at exits and other risk points that have been paired repeatedly with the presentation of an aversive stimulus such as a very loud noise; sheltered, protected space for strolling; a varied activity programme with visits and trips; increased use of supervised exercise (a floor-mounted exercise bicycle has been used effectively); developing awareness of signs of restlessness in the patient and to times when wandering is most common; intervening with distraction; reinforcing behaviour incompatible with wandering; unobtrusive tagging and security systems indicating an unauthorized exit (these have been used successfully); viewing physical restraint as a last resort.

CONCLUSIONS

In presenting some elements of the rationale for behaviour management with older patients, and outlining some considerations for its application across thirteen heterogeneous problem areas, this review has addressed clinical issues of everyday concern.

As Smyer *et al*. (1990) indicate, the scientific status of psychological interventions with such patients requires continued support from improved research into the effects of specific approaches on specific patterns of adjustment, together with a more thorough understanding of systems of care and how they operate.

Meanwhile, it can be anticipated that demand for behaviour management to be deployed in care of older people will grow. The abiding challenge for the discipline will be to keep progress towards scientific rigour in balance with the concurrent response to the urgent needs to service providers and patients.

References

Baltes, M.M. and Baltes, P.B. (1986). *The psychology of control and aging*. Erlbaum, Hillsdale, NJ.

Comfort, A. (1980). Sexuality in later life. In *Handbook of mental health and aging*, (ed. J.E. Birren and R.B. Sloane), pp. 885–92. Prentice-Hall, Englewood Cliffs, NJ.

Fantl, J.A., *et al*. (1991). Efficacy of bladder training in older women with urinary incontinence. *Journal of the American Medical Association*, **265**, 609–13.

Garland, J. (1987). Working with the elderly. In *What is clinical psychology?*, (ed. J.S. Marzillier and J. Hall), pp. 163–88. Oxford University Press.

Gelder, M., Gath, D., and Mayou, R. (1989). *Oxford Textbook of Psychiatry*. Oxford University Press.

Holden, U.P. and Woods, R.T. (1988). *Reality orientation: psychological approaches to the 'confused' elderly*. Churchill Livingstone, Edinburgh.

Houts, P.S. and Scott, R.A. (1976). *Individualized Goal Planning in Nursing Care Facilities*. Department of Behavioral Science, The Pennsylvania State University College of Medicine.

Hussian, R.A. (1981). *Geriatric psychology. A behavioural perspective*. Van Nostrand Reinhold, New York.

Kennedy, R.W. and Kennedy, A.B. (1982). Absence of purposeful behavior: issues in training the profoundly impaired elderly. In *Mental health interventions for the aging*, (ed. A.M. Horton, Jr), pp. 69–83. Praeger, New York.

Morgan, K. (1987). *Sleep and ageing*. Croom Helm, London.

NIH Consensus Conference on Urinary Incontinence in Adults. (1989). *Journal of the American Medical Association*, **261**, 2685–90.

Smith, P.S. and Smith, L.J. (1987). *Continence and incontinence. Psychological approaches to development and treatment*. Croom Helm, London,

Smyer, M.A., Zarit, S.H., and Qualls, S.H. (1990). Psychological intervention with the aging individual. In *Handbook of the Psychology of Aging*, (ed. J.E. Birren and K.W. Schaie), pp. 375–403. Academic Press, London.

Sobel, H.J. (1981). Toward a behavioral thanatology in clinical care. In *Behavior Therapy in Terminal Care. A Humanistic Approach* (ed. H.J. Sobel), pp. 3–38. Ballinger, Cambridge, Mass.

Sturgis, E.T., Dolce, J.J., and Dickerson, P.C. (1987). Pain management in the elderly. In *Handbook of clinical gerontology*, (ed. L.L. Carstensen and B.A. Edelstein), pp. 190–203. Pergamon, New York.

Woods, R.T. and Britton, P.G. (1985). *Clinical psychology with the elderly*. Croom Helm, London.

20.6.1 Problems of carers: the United States view

ELAINE M. BRODY

The family—forming much the greatest part of the 'informal system of care'—has been steadfast in helping its aged members by providing the vast majority of day-to-day health services required by those who are disabled. The capacities of the family to do so are being strained, however, owing to the confluence of certain broad socioeconomic and demographic trends. As a result, many family members suffer negative consequences to their health and well-being. Health professionals depend on family carers for referrals and for implementation of treatment and rehabilitation prescriptions as well as for day-to-day care of elderly patients. Achievement of health care goals therefore depends on viewing carers from two perspectives: as partners and allies to professionals, and as potential patients if their caring roles become unduly stressful.

This chapter will summarize the trends that have contributed to the vast increase in older people's needs for care, the scope of caring by family members, the effects the family experiences as a result of their efforts, and factors that are associated with their strains. Some sources of carers' strains that are amenable to amelioration will be identified—notably, inadequacies in public policy, organizational problems in the formal health systems, approaches and attitudes of professionals and other non-family helpers, and the role overload and psychological problems of some carers. Continuing and developing demographic and socioeconomic trends that have implications for family care-giving will be identified (see also Brody 1990).

DEMOGRAPHIC AND SOCIOECONOMIC TRENDS

The fundamental reason that caring has become more of a problem nowadays is, of course, the dramatic rise in the number and proportion of very old people in the population—those who are most vulnerable to the chronic diseases that result in disability.

Because of the increases in longevity, many contemporary families contain two generations of older people so that many parent-caring, adult children are in late middle age or early old age. It is estimated that 10 per cent of all people 65 or over have at least one adult child who also is over the age of 65. Two decades ago, Townsend (1968) predicted that the emphasis might shift from the problem of which of the children looks after a widowed parent to the problem of how a middle-aged couple can reconcile dependent relationships with both sets of parents. Now that the four-generation family is so common (about half of all older people have at least one great-grandchild), some adult grandchildren may find themselves helping both their parents and grandparents.

While the demography and the kinds of help needed by the old were changing, still another influential trend occurred: the lifestyles of the women who are the main group of carers were being transformed. One of the most visible of those changes has been the vastly increased number of women who are employed outside of the home. In addition to the changing values about women's roles, another major contributor to that trend has been economic considerations that compel many women to seek paid work. The proportion of women who work has quadrupled in the past half-century, rising from 24 per cent of working age women in 1930 to 51 per cent by 1979, and projected to rise to 80 per cent by 1995. The most rapid rate of increase has been among women between the ages of 45 and 64 (Lingg 1975)—those most likely to be in the parent-care years.

Part of the problem experienced by many women is the conflict in the different values they hold. Despite their changing values about working and about egalitarian gender roles, the value that care of their elderly relative is their responsibility has remained firm (Brody *et al.* 1983). That is, egalitarian attitudes are not reflected in actual behaviour; women still continue to provide the bulk of elder-care.

Yet another trend that bears on the capacity of women to provide care for older relatives has been the rapidly changing pattern of marriage and divorce. About 44 per cent of care-giving daughters are not married (Stone *et al.* 1987) (being widowed, divorced, never married), a marked contrast from 25 per cent a quarter of a century ago (Shanas 1961). Such carers do not have the emotional support of a spouse and often experience strains unique to the intersection of parent care with their particular marital status (Brody 1990).

Additional trends that hold the potential for increasing the problems of carers will be summarized at the end of this chapter.

THE SCOPE OF CARE

The kinds of help required by contemporary older people are determined by the nature of chronic diseases. Those diseases often result in disability—an inability to function in one's daily life without the assistance of another person. Chronic ailments often last for years, requiring a sustained input of services rather than time-limited episodic care.

As discussed in Chapters 1.1 and 1.2, almost one-fourth of all people 65 years of age and older are functionally disabled requiring assistance with Activities and Daily Living (**ADL**: bathing, dressing, eating, toileting, getting out of bed, and getting around indoors) or Instrumental Activities of Daily Living (**IADL**: money management, moving about outdoors, shopping, doing heavy housework, meal preparation, making phone calls, and taking medication). They may in addition require more technical nursing care such as injections and tube feeding. Four-fifths of disabled older persons receive home care, with about 85 per cent of that care being provided by the family; the extent of help needed varies greatly.

In addition to direct functional assistance, families provide other forms of help such as emotional support, mediation with formal system providers, financial assistance, and sharing their households when the older person can no longer live alone.

Emotional support

This is the most universal form of family care-giving, the one most wanted by older people from their children, and the one adult children deem the most important service they can give their disabled parents. It includes being the confidant, providing social contacts, help with decision-making, and giving the older person the sense of having someone who is interested and concerned and on whom to rely.

Mediation with organizations

To obtain and monitor available services and entitlements is a major function of the family. The importance of that function is now being recognized. As 'case management', 'care management', or 'resource coordination', it is usually thought of as a professional task, though families actually perform the role to a much greater extent than do professionals

Financial support

Financial support of older persons by their children to meet day-to-day living expenses and to pay the costs of acute medical care is now a much less important pattern in all industrialized countries than was true earlier in the century. Social Security and similar programmes have established an income floor for older persons (though it is often quite low). Pensions and savings have helped. In the United States the proportion of older persons who are totally dependent on their children for economic support dropped from about half in 1937 to 1.5 per cent in 1979 (Upp 1982); about double that proportion received some regular financial help from children in 1985 (US Department of Commerce 1988). Similarly, Medicare, despite its limitations, has in the main prevented aged people's catastrophic illnesses from wiping out their own and their adult children's savings.

Apart from direct dollar outlays, the average long-term care services adult children in the United States provide have a computed market value of $4529 annually, costed out on the basis of the minimum wage (Doty 1986). Invisible until recently are the opportunity costs incurred when people leave their jobs or reduce their working hours in order to take care of elderly family members (Brody 1985; Stone *et al.* 1987). There is virtually no information about the financial costs of health care (mental and physical) for those carers who suffer negative effects from helping the old. Comparable data from other countries are not readily available but it can be assumed that similar values would be found.

Sharing one's household

Sharing in this way with an elderly parent is a special form of care-giving. Elderly parents and adult children agree that this is not a desirable living arrangement; they join households primarily when the disabled old person can no longer manage independent living or cannot afford to live alone. The prevalence of shared households is low at any given time; in the United States in 1975 less than 18 per cent of elderly people lived with adult children (Mindel 1979). A significantly higher proportion do so at some time in their lives, however. Thus, in the recent national government survey, more than one-third of the severely disabled elderly (those in need of ADL assistance) lived with a child, and one-fifth of those households also contained at least one of the adult child's own children under the age of 18 (Stone *et al.* 1987).

Families perform other important services that are not usually counted by surveys. These include response and dependability in emergencies (such as acute illnesses and admissions to hospital), help with special needs such as moving from one residence to another, and convalescent and rehabilitative care.

FAMILY CARERS

There is almost invariably one person in the family who becomes the principal or primary carer (Brody 1990). Other family members may help out on a regular basis or from time to time, but play much smaller roles in terms of the intensity and amount of the help they give.

When the disabled older person is married, the spouse is the main provider of care. Spouses are enormously loyal in caring for each other. Elderly men are much more likely than elderly women to have a spouse on whom to rely, owing to differences between the sexes in life expectancy and the tendency of men to marry women younger than themselves. Most of the elderly widowed are women; at age 65 and over, most older women (52 per cent) are widowed and most older men (77 per cent) are married. Rates of widowhood rise sharply with advancing age so that widows outnumber widowers in a ratio of four to one (see Chapter 1.1).

Though older people exert extreme efforts in caring for an ill spouse, their capacities are limited by their own ages, reduced energy and strength, and age-associated ailments. As the most vulnerable group of care-givers, they often experience severe strains such as low morale, isolation, loneliness, economic hardship, and 'role overload' due to multiple responsibilities. Caring spouses therefore require attention to their own needs for respite, concrete helping services, and emotional support. The physical strain may be accompanied by tremendous anxiety and fear of losing one's partner in a marriage that may have endured half a century or more.

When an elderly couple has children, they assist the 'well' spouse in caring for the patient; when the patient is widowed, the bulk of care is given by adult children. A solid body of research has documented the strength of intergenerational ties and responsible filial behaviour, the frequency of contacts between generations, the strenuous efforts of family members to avoid institutional placement of the old, and their predominance over professionals in caring for the non-institutionalized impaired elderly. Most older people realize their preference to live near, but not with their children. About 84 per cent of those with children live with or less than an hour away from one of them.

Daughters outnumber sons about 4 to 1 as principal caregivers and in sharing their homes with elderly parents when need be (see Brody (1990), for review). Those women provide more than one-third of all long-term care services and more than half of all such services to those who are the most severely disabled (Stone *et al.* 1987). The power of gender in determining who in the family becomes the carer extends to other relatives as well. When there is no spouse or adult child, female relatives such as daughters-in-law, sisters and nieces predominate over their male counterparts. Little is known about patterns of care by such relatives or by grandchildren who may fill in the care-giving gap when adult children die. Friends and neighbours provide important forms of help, but by no means does that help approach family help in level or duration.

Demands for parent care often occur at a time of life when the adult children on whom the frail older person depends may be experiencing age-related interpersonal losses, the onset of chronic ailments, lower energy levels, and even retirement. As women advance from 40 years of age to their early 60s, those who have a surviving parent(s) are more and more likely to have that parent be dependent on them, to spend more and more time providing care, to do more difficult care-giving tasks, and to share their households with the parent (Lang and Brody 1983). Though most care-giving daughters are in their late 40s or early 50s, about one-third are either under 40 or over 60. Because of the wide variations in carers' ages, parent care is not a 'normal developmental stage' of life; it can overlay many different ages and stages in different people and different families.

Most daughters who help severely disabled parents (those in need of ADL help) devote an average of 4 h daily to doing so (Stone *et al.* 1987). The average duration of care-giving to a particular elderly person is 4 to 5 years. Many women have care-giving careers (Brody 1985), however, in that they help more than one older person simultaneously or sequentially. About three-fifths of adult daughters are in the workforce during their parent-care years.

In general, daughters who become the main carers receive relatively little regular practical help from other family members, though the latter share to a greater extent in providing emotional and financial support to the elderly parent.

The emotional support provided by husbands of married carers is considerable and is deeply appreciated by the wives; their not-married counterparts are acutely aware that they are lacking such support. Similarly, carers wish for emotional and instrumental support from their siblings, with intersibling problems arising when such help is not forthcoming or when the siblings have critical attitudes about the care. In some families, conflicts among the adult children concern such issues as fair sharing of care-giving responsibilities, with whom the older person should live, or who should help with money.

Expectedly, sibling relationships as they focus on parent care range from being cooperative and amicable to those characterized by considerable conflict. Notwithstanding the small amount of day-to-day assistance most carers have from other relatives, the latter do act as an invisible back-up system; they often help considerably in emergencies, at times of special need, and when important decisions must be made (such as nursing-home placement).

EFFECTS OF CARE-GIVING

The vast majority of families willingly provide care for their elderly family members and want the older people to continue to live on in a state of well-being. Carers derive many positive benefits such as satisfaction from fulfilling what they see as their responsibility, adhering to religious and cultural values, expressing their feelings of affection, and reciprocating for the help the older person had given them in the past. It is undeniable, however, that many carers do experience negative effects as a result of care-giving, though the nature and levels of strains they report vary widely. Different families react differently to the demands of parent care, but genuine concern and affection for the older person are generally at work. At the same time, there is concern about themselves, other family members, and the duration and intensity of the care-giving efforts they will need to exert in the future.

The most pervasive and severe effects of care-giving are in the realm of mental/emotional health. Research has found consistently that at least half of family carers experience moderate to severe stress effects in the form of anxiety, depression, anger, guilt, lowered morale, sleeplessness, frustration, and emotional exhaustion. Related to those symptoms are restrictions on time and freedom; relationship problems; conflict from competing demands of carers' various responsibilities; difficulties in setting priorities; and changes in family lifestyle, opportunities for socialization, and future plans. When the older person lives in the carer's household there is loss of privacy and the arena for potential conflict is widened. Smaller, but significant proportions of carers experience negative effects on their physical health (15–20 per cent).

Among the factors that have been identified as predictors of carers' strains are the older person's severe functional disability requiring 'heavy' care and the sharing of the care-giver's household that often becomes necessary. Characteristics of the older person that have been implicated as stressors are incontinence and the disordered behaviour and management difficulties that often characterize victims of Alzheimer's disease. Mental disabilities, whether functional or organic, are particularly stressful and difficult to deal with.

Information has been emerging recently about the effects of care-giving on people's work lives. A significant minority of care-givers (13.5 per cent of wives, 11.4 per cent of husbands, 11.6 per cent of daughters, and 5 per cent of sons) were found to have left the labour force when work and caring competed (Brody *et al.* 1987; Stone *et al.* 1987). About one-quarter of working care-givers rearrange their work schedules, reduce the number of their working hours, or experience problems on the job.

The personality of the disabled person is probably more important in producing carers' strain than has been emphasized. It is particularly upsetting, for example, when the recipient of care is critical of the carer and other family members, an issue that is more likely to arise when the older person lives in the carer's household and the latter also has a husband and children who live at home.

Depending on their own personalities as well as on the reality demands, carers may feel guilty about not doing enough for the older person. Some are angry (though they may not be aware of their anger or able to express it) at finding themselves in the predicament of needing to do more than they feel able. Unresolved relationship problems (and most families have at least some vestiges of these) may be reactivated so that the older person becomes the focus of difficulties between elderly spouses, among the adult children and their spouses, and across several generations.

Adult daughters are often under considerable stress from their multiple competing responsibilities, a situation that led to their characterization by the author as 'women-in-the-middle'. Most women try to meet all of the various demands on their time and energy, giving up only their own free time and opportunity for relaxation and socialization. Those pressures, compounded by emotional conflicts related to the need to set priorities, may place them at high risk of mental/physical problems. Many who seek health care for themselves may be doing so in the context of such role strains.

An intriguing research finding is that caring daughters consistently experience more strain than caring sons when in comparable positions. Daughters report significantly more strain among principal carers and among local and geographically distant siblings of the carer (see Brody (1990) for review). While adult sons are not remiss in affection for their parents or in a sense of responsibility, our culture has designated parent care as well as child care to women as gender-appropriate. When sons become principal care-givers (usually when a daughter is not available), they do less, receive more help from their spouses (the daughters-in-law), and experience less stress.

Though the mental-health literature is replete with information about the greater vulnerability of women to emotional strain, there are additional reasons specific to parent care. (see Brody (1990) for full discussion). Women are deeply and powerfully socialized to the expectation that caring for older (and other dependent) family members is their role. In addition to the fundamental acceptance of their role as nurturer is their feeling of responsibility for the emotional well-being (even happiness) of those for whom they care.

On some level, women feel that they must provide the disabled parent (who is usually the mother) with the total and complete care that the parent gave them as children.

Such 'role reversal' is, of course, a fallacy. A 60-year-old daughter cannot care for an 85-year-old incontinent parent as that parent had cared for the daughter when she was a baby. Apart from the physical differences, the carer's emotional reactions and commitment are different, the trajectory of change in the older person is usually downward rather than upward as with a child who is moving developmentally toward independence, and a young mother and a middle-aged women have different sets of responsibilities. When the unrealistic goal of total care is not achieved, women many experience a sense of failure.

In addition, the shift in the dependency needs of a disabled older person may set in motion a struggle for control of the caring situation. In some families, adaptations are relatively free of problems; in others the struggle may be bitter, with the carer unable to please the elderly parent. Service workers hear over and again that a service cannot be used because the older person wants care only from her daughter or the daughter feels that no one else can provide care as well.

NURSING-HOME CARE

Carers place their elderly relative in nursing homes as a last resort, in the main doing so after many arduous years of care. Those who are in nursing homes are severely disabled, in advanced old age, and have many fewer spouses and adult children than their counterparts who live in the community (Brody 1990). Patients with Alzheimer's disease are greatly over-represented in nursing homes because of the severe difficulties their care entails; 50 to 60 per cent of residents have Alzheimer's or a related dementia compared with 5 to 6 per cent of all older people.

Though it is often thought that placement of an aged person signals relief of the carers' strains, recent studies have shown that stress continues though it may flow from different sources. Many negative emotional symptoms are predicted, for example, by the quality of their visits to the older people, time pressure, and perceptions that nursing-home staff have poor attitudes towards the older people and their family members.

TOWARD AMELIORATING CARERS' PROBLEMS

There is no simple or total solution to all of the problems carers experience. Families will continue to care for and about their elderly relative, to be concerned and sad at the declines of those old people, to provide affection and support, to do what they are able, and to arrange (if possible) for the needed services that they cannot supply themselves. The strains of many carers are not completely preventable or remediable. There are, however, certain steps that can be taken to mitigate severe burdens and prevent family breakdown. These efforts are in several different, but interrelated areas, as follow.

Public policy

In the United States, policy has moved forward primarily in the areas of income maintenance and acute medical services for elderly people. Little progress has been made toward creating a supportive health-care system—the continuum of acute, transitory, and long-term care services and facilities dictated by chronicity. Income maintenance and payment of acute medical costs for older people have been of direct benefit to their families as well as to older people themselves. While there are still some poor older people, poverty no longer characterizes most of the elderly population.

Modest progress has been made in the provision of housing, which figures prominently in the emotional well-being of the old and their families. Thirty-five years ago, subsidized housing for older people that included services such as meals, housekeeping, and health care simply did not exist. Such housing remains scarce, but in the main, the increase in 'senior housing' and more importantly the improved income position of older people, have enabled many more to live as they wish—close to their children but in separate households (Shanas 1979). While retirement and life-care communities now offer more options for elderly residents, such arrangements are not subsidized by government and are therefore available only to those who can afford them.

In the United States the major agenda for social policy is to develop insurance for long-term care. This insurance should create a broad array of the kinds of day-to-day sustained services needed by functionally disabled older people and care-giving family members—respite care, day care, and in-home services such as personal care and homemaker services and transportation, for example. These could relieve some of the strains that family carers experience. At present such services are in very short supply and are uneven regionally. Insurance should also cover both short-term and long-term nursing-home residence and be oriented to functional status rather than medical needs alone.

Currently, community health and social services are limited and disproportionately underfinanced as compared to acute medical care services for the aged (Brody 1979). Inequities exist among the various states in their expenditures for Medicaid services, both for nursing-home care and community-based services.

Nursing homes are appropriate long-term living arrangements for the severely disabled older people who need them. In the United States the major public provision of nursing-home care is for Skilled Nursing Facilities (**SNF**) under the Medicaid programme; in many parts of the country, rates of government reimbursement are too low and standards are either inadequate or not enforced. While some facilities provide high-quality care, poor and even dangerous conditions continue in too many. In most states, elderly spouses have to impoverish themselves by 'spending down' so that the disabled older person becomes eligible for nursing-home care under Medicaid. The destitute 'intact' spouse then is at risk of economic dependency on children. When Medicaid beds are unavailable, some adult children purchase nursing-home care for their parent(s) and experience financial strain as a result. Among the remedies needed are adequate reimbursement, strict oversight and monitoring of facilities, and an emphasis on quality that includes attention to the social as well as medical aspects of care.

The policy trend in recent years has been to embrace community care as an 'alternative' to nursing-home care and to create barriers to institutional placement. The 'either/or' approach that places nursing homes in competition with community care has been proven to be inappropriate; different populations are served and community care is not cheaper for the severely disabled. Yet at the same time that the so-called 'alternative' of community care is encouraged, health and social services have been systematically reduced, and housing subsidies have been virtually eliminated, reinforcing the burden on care-giving families by closing avenues that might relieve them.

There is consensus that the long-term care system should focus on the needs of the entire family rather than solely on the needs of the older people. Part of the problem is the confusion about the meaning of words like 'dependency' and family 'reliability' and 'responsibility'. There are different kinds of dependency—financial, physical, and emotional—that have quite different meanings to those involved. Expectations of older persons for family support vary when anchored to specific forms of help such as emotional support, financial help, household tasks, or nursing care.

In the United States, financial assistance to care-giving families is virtually non-existent except for a token programme that permits a tax deduction for the adult child purchasing care for an economically dependent parent and a Veterans Administration allowance for caring for a spouse disabled from wartime injury. Economic supplements to caregiving families (such as attendance allowances or social security credits for care-givers remaining at home) are much more prevalent in other industrialized nations (Gibson 1984). Among the policy options that have been proposed are financial compensation for care-givers such as direct payments for care-giving, social security benefits for those who leave jobs for care-giving, and tax relief.

Providing and paying for long-term care is a major policy issue in many countries. In the United States the Pepper Commission, mandated by Congress, has just issued its recommendations for a combined public and private insurance approach, emphasizing home support services as the preferred first choice.

The organization of health and social services

Even if public policy should permit the creation of all of the services needed, it would be an exercise in futility unless those services were actually accessible to the families in need. Obtaining even those services that now exist is often a frustrating, defeating enterprise.

> Health services for the aged are multiple, parallel, overlapping, non-continuous and at the very least confusing ... Rarely do they meet the collective criteria of availability, accessibility, affordability or offer continuity of care in a holistically organized system. Virtually all the services are funded by differing public money streams and have varied administrative arrangements, widely ranging eligibility requirements and different benefits for the same or similar services (Brody (1979), p. 18).

The major public effort in the social service arena has been the mounting of demonstration projects to rationalize these intermittent efforts and establish a pattern of continuous care. If existing services are to reach those who need them, one of the major tasks is organizational and individual case management. This involves arranging the varied planning and service-delivery agencies so that they constitute a system in each community. Such a system would articulate the formal and informal, medical and social/health services and relate the acute care services to what have been called the short-term, long-term care (**STLTC**) and long-term, long-term care (**LTLTC**) subsystems (Brody and Magel 1990) so they can respond in a co-ordinated manner to each individual and family in need. Separating out these three systems is no longer relevant; continuity of care should follow the trajectory of chronic illness. Organizational arrangements should respond to the values and resources of each locality, with the underlying objective of providing a continuum of appropriate levels of care as the older person's needs change in kind, level, and intensity.

Formal intervention designed to meet these needs are variously called information and referral service, service management, outreach, case management, case management matrix management, or resources co-ordination. Though case management is usually thought of as a professional role, basically it is enacted by the family, which usually needs help in doing so. Where there is no available family, the need for professional help is obvious.

More subtle issues also impede service utilization and speak to the need for counselling or casework—specifically, psychological barriers that inhibit some families' use of needed services. In addition, the acceptability of services differs among older people and families with different socio-economic and ethnic backgrounds, and with diverse personalities and expectations. Therefore, whatever the label given to the process now most commonly referred to as case management, it should include the enabling process called counselling or casework.

Because of the complex nature of the problems with which care managers deal, effective care management is a highly skilled, knowledge- and values-based activity. It should not be seen solely as the mechanical manipulation or arrangement of services, but should include the offering of sensitive help with psychosocial issues. The blending of these activities cannot be carried out by untrained people, no matter how well-meaning they may be.

The need for training extends to the workers who actually deliver the direct, 'hands on' care. Families who try to use non-family help in caring for disabled older members are often discouraged by problems such as high turnover rates, unreliability, lack of skill, and inappropriate attitudes. Whether they are to be employed for in-home services or in nursing facilities, the recruitment and training of a sufficient cadre of such workers is an urgent and major task.

Professional approaches

Physicians occupy positions of enormous power in affecting the well-being of carers. They are looked to for help and guidance in decision-making and planning as well as for medical management of the patient. In fulfilling this major responsibility, it is necessary for the doctor to evaluate the situation of the carer and her family: their ages and stages of life, their health, other responsibilities, availability, the quality of family relationships, and the effects of care-giving they are experiencing. The doctor's sympathetic attitude, accompanied by a word or two conveying understanding and concern about the carer's problems, can be particularly meaningful. Moreover, the carer's guilt when nursing-home placement is inevitable can be relieved significantly by the physician's recommendation or 'permission' for such placement to take place.

Physicians cannot be expected to give in-depth counselling for emotional or relationship problems, to explore the family's specific service needs in great depth, to have detailed knowledge about the availability and eligibility criteria for each service, to make referrals to the various service sources, or to monitor the services over time. They can, however, act as the gatekeeper to introduce patient and family to the complex and confusing array of programmes that constitutes the 'formal' support system. To do so, they can refer their patients to the agencies in the community that can provide counselling and whose job it is to be informed about the complete range of available services and facilities, to connect older people and their families with those services, and to help to mobilize them. Examples of such organizations in the United States are Area Agencies on Aging, family counselling agencies, information and referral services, and (when the patient is in hospital) hospital social-service departments.

The physician's need to consider the needs of the total family as well as those of the elderly patient is a challenging task that can prevent breakdown in the physical and mental health of the carers. A first priority is close attention to their emotional strains.

Those emotional strains would not be eliminated even in the best of all possible worlds in which all needed services existed and were available, accessible, and affordable. Though understanding of the dynamic intra- and interpersonal processes at work is by no means complete, the information already at hand has not been fully integrated into the curricula of professional schools or applied in practice.

The non-use and underuse of services by some of the neediest care-givers is a constant frustration to service workers. Studies of subsidized respite services, for example, consistently report that they are grossly underutilized by carers despite concerted efforts to educate and counsel them (George 1988; Lawton *et al.* 1991). Among the barriers to service use are the powerful conviction that care of the elderly is a family responsibility; the belief held by particular ethnic, racial, or religious groups that they 'take care of their own'; the social norm of family independence; involvement in care-giving that becomes so consuming that the carers' own needs are ignored; reluctance to exhaust financial resources or to accept 'charity'; and demeaning means testing. When care-givers have tried using services, some are discouraged by quality considerations such as workers' unreliability, high turnover, and lack of motivation, knowledge, and training.

Recently there has been a flow of books of advice on parent care. Carers are being instructed in techniques of nursing care for older people and what symptoms to look for, methods of reducing stress (relaxation techniques, for example), and how to manage older people who present personality or behaviour problems. Support groups have proliferated and are helpful to many carers, though others need more or different forms of help than support groups can offer.

All of those approaches are good and necessary. However, the therapeutic professions have a responsibility to turn their attention to the profound subjective problems some care-givers experience that prevent them from using help to ease their distressing (even health-threatening) situations. A disturbing aspect of the care-giving drama is the resignation of some carers—their acceptance that this is the way it has to be. That acceptance is given such sanction and powerful reinforcement in our society that relatively few women can resist as individuals. Because of the intense pressure on women to embrace care-giving as their lot exclusively, no matter the effects on their lives, some have low expectations of help from others. The expectations of other people reinforce the women's expectations of themselves and loom large in producing the strains and symptoms to which women are so vulnerable.

A major imponderable in the provision of care to older persons is how the tension between changing and constant values will be resolved. While values about family care of their elderly member have remained firm across the generations, the new values about women's roles have taken hold unevenly. Continuing and newly emerging trends hold the potential for affecting patterns of care for elderly people and increasing the difficulties of family carers. Among them are the following:

1. The number and proportion of older people in the population continues to increase—particularly the very old who are most vulnerable to the need for care. Unless a major biomedical breakthrough should occur, for example, the frequency of the most burdensome ailments of all—Alzheimer's disease and related dementias—will increase rapidly.
2. The birthrate continues to fall so that more adult children (daughters in particular) will have multiple caring responsibilities for their older relations, not only in their role as daughters, but also as daughters-in-law.
3. The rise in the divorce rate and in the number of women who do not marry not only increases their labour-force participation, but means that these women (now 44 per cent of parent-caring daughters) do not have the emotional and instrumental support of husbands in their parent care efforts.
4. High rates of remarriage complicate the responsibilities of the adult children of remarried people. With two (and sometimes more) sets of parents, to whom does filial loyalty go when those parents become disabled?
5. The older ages at which some women are having their first child will result in more of them having young children during their parent-care years, a pattern called 'double dependency'.
6. Another form of double dependency results from the increased longevity of developmentally disabled people. More carers thus care for elderly parents and their own disabled children simultaneously (often well into the latters' adulthood).
7. The nests of middle-generation people are remaining filled with offspring for longer periods of time and are being refilled with young adults who return home because of the high divorce rate and the economic situation.
6. Increasing geographic mobility (combined with fewer children) creates special problems when distance separates children from their disabled parent(s).

Finally the emphasis in this chapter on the problems of family carers should not obscure the problems of older people who do not have close family members to fill the caring role. Such elderly people will be more numerous in the future. At present, about one-fifth of all older people are childless either because they never had children or because being in advanced old age increases the possibility of outliving one or more of their children. While other relatives—nieces, nephews, elderly siblings, and grandchildren—rise to the occasion to help the childless when need be, they do not provide care as intensive and extensive as that provided by spouses and children. Caring for older persons who are essentially alone presents a special challenge to society and health professionals.

References

Brody, E.M. (1985). Parent care as a normative family stress. The Donald P. Kent Memorial Lecture. *The Gerontologist*, **25**, 19–29.

Brody, E.M. (1990). *Women in the middle: their parent care years*. Springer, New York.

Brody, E.M., Johnsen, P.T., Fulcomer, M.C., and Lang, A.M. (1983). Women's changing roles and help to the elderly: attitudes of three generations of women. *Journal of Gerontology*, **38**, 597–607.

Brody, E.M., Kleban, M.H., Johnsen, P.T., Hoffman, C., and Schoonover, C.B. (1987). Work status and parent care; a comparison of four groups of women. *The Gerontologist*, **27**, 201–8.

Brody, S.J. (1979). The thirty-to-one paradox: health needs and medical solutions. In *Aging: agenda for the eighties*, National Journal Issues Book. Washington, DC.

Brody, S.J. and Magel, J.S. (1990). LTC: the long and short of it. In *Caring for the elderly: reshaping health policy* (ed. C. Eisdorfer). Johns Hopkins University Press.

Doty, P. (1986). Family care of the elderly: the role of public policy. *The Milbank Quarterly*, **64**, 34–71.

George L. K. (1988). Why won't caregivers use community services? Unexpected findings from a respite care demonstration/evaluation. Paper presented at 41st Annual Meeting of The Gerontological Society of America, San Francisco, CA.

Gibson, M.J. (1984). Women and aging. Paper presented at International Symposium on Aging. Georgian Court College, Lakewood, NJ.

Lang, A. and Brody, E.M. (1983). Characteristics of middle-aged daughters and help to their elderly mothers. *Journal of Marriage and the Family*, **45**, 193–202.

Lawton, M.P., Brody, E.M., and Saperstein, A.R. (1991). *Respite service for Alzheimer's caregivers*. Springer, New York.

Lingg, B. (1975). Women social security beneficiaries aged 62 and older, 1960–1974. *Research and Statistics Notes*, No. 13. 9/29.

Mindel, C.H. (1979). Multigenerational family households: Recent trends and implications for the future. *The Gerontologist*, **19**, 456–63.

Shanas, E. (1961). *Family relationships of older people*, Health Information Foundation, Research Series 20. University of Chicago.

Shanas, E. (1979). Social myth as hypothesis: the case of the family relations of old people. *The Gerontologist*, **19**, 3–9.

Stone, R., Cafferata, G.L., and Sangl, J. (1987). Caregivers of the frail elderly: a national profile. *The Gerontologist*, **27**, 616–26.

Townsend, P. (1968). The household and family relations of old people. In *Old people in three industrial societies*, (ed. E. Shanas, P. Townsend, D. Wedderburn, D. Friis, P. Milhoj, and J. Stehouwer). p. 178. Atherton Press, New York.

US Department of Commerce, Bureau of the Census. (1988). *Who's helping out? Support networks among American families*, Current Population Reports, Household Economic Studies. Series P-70. No. 13, October.

Upp, M., (1982). A look at the economic status of the aged then and now. *Social Security Bulletin*, **45**, 16–22.

20.6.2 Problems of carers: United Kingdom view

DEE JONES

INTRODUCTION

For many years it has been the philosophy and policy of the Department of Health and Social Security to maintain elderly people in the community rather than in residential care (Department of Health and Social Security 1981). The recent WHO Congress on ageing held in Vienna reinforced this by saying elderly people should be supported to live in the community and furthermore to live in the community with their families (United Nations 1982).

This policy has implications not only for the elderly population, community, social, and medical services, but also for those members of the community or their families who have to support these increasingly disabled and frail people. Any increase in the proportion of very elderly people in the population makes it important to know more about their needs, how they cope with their problems, and who their informal supporters are. These supporters play a crucial role in the quality of life for elderly persons and in their ability to remain out of hospitals or other institutions and in the community.

DEMOGRAPHY

In recent years, owing to reduced infant mortality, we have seen a major change in demographic structure in the United Kingdom and in the Western countries in general. There has been a dramatic increase in the proportion of elderly people and, more recently, an increase in the very elderly (over 75 years old) in particular. This latter trend will continue into the next century when we can expect 9 per cent of our population to be aged 75 years or over (OPCS and CSO 1984). This, whilst not necessarily being a problem, is an issue that needs to be addressed and planned for (see Chapter 1.1). At the same time there have also been other changes in our society, which have implications for the informal care system. Since 1971 the average number of children per family has declined from 2.2 to 1.8, and those couples having only one child have increased (OPCS 1987). This, together with a changing divorce rate, has altered the nature of 'the family' considerably.

Other social trends have also had an impact on the nature

of the informal care system and families. During the second half of this century there has been a consistent increase in the proportion of adult women who have an occupation. Currently 60 per cent of adult women are in paid employment (Central Statistical Office 1989). A large proportion of children migrate away from their parents, particularly among the higher social classes or the upwardly mobile.

All these demographic changes have had an impact on the availability of informal carers to support their elderly frail kin. In 1985 about one adult in eight in the United Kingdom was looking after an elderly or disabled person, representing about 6 million informal carers overall.

NEEDS OF ELDERLY PEOPLE

There are nearly 9 million people aged 65 years or over in the United Kingdom. A high proportion of elderly people in Britain live in the community, in their own homes, or with relatives (usually daughters). Only 4 per cent live in some form of institutional care, be it long-term hospital care, local authority homes, or private nursing and residential homes. This proportion is one of the lowest in the Western world. Of the 96 per cent who are currently living in the community, 43 per cent are living with their partners, 31 per cent live alone, and 26 per cent with other members of their families (OPCS and CSO 1984).

Mental and physical frailty increase with age among elderly people so that 8 per cent of those aged 60 to 69 years were scored as 5 or more on a 10-point disability score, 16 per cent of those aged 70 to 79 years, and 38 per cent of those aged 80 years or over (Martin *et al.* 1988). There are many tasks that present difficulties to functionally disabled elderly people; the most common are shopping, cleaning, and washing and ironing.

The needs of elderly people are many and varied. There is general agreement that approximately a fifth of people aged 70 years or more need help at least once a week from an informal carer, and obviously within that there is a great range of need for assistance. It must, however, be remembered that the majority (three-quarters) of elderly people enjoy good health, are independent, and need no regular assistance from informal carers or domiciliary services.

Of those elderly people who are supported regularly by informal carers, almost half are severely disabled and a third moderately so. Those who are living in the same home as their carer are more disabled than those who are not. A third of those cared for are significantly anxious and a fifth significantly depressed. Almost a quarter of those cared for suffer from memory impairment—a tenth severely so. Mental and physical disability are strongly associated with each other and the presence of both increases markedly with age.

Approximately one-third of elderly people with carers who help at least once a week need help every day, and a further quarter need constant help and attention. Elderly people coresident with their carers are considerably more dependent than elderly people with non-resident carers (Jones 1986).

CHARACTERISTICS OF INFORMAL CARERS

The main source of informal care in the community is 'the family'. Where there are elderly couples, the spouses are the main source of support but where elderly people are widowed, daughters are the main source of support (Table 1). As women outlive men by, on average, 5 years, it is mostly women who survive alone and are cared for by daughters.

Table 1 Relationship of carer to elderly persons, by residency (from Jones 1986)

	Resident		Non-resident		Both	
Relationship	No.	%	No.	%	No.	%
Spouse	66	40	—	—	66	26
Daughter	56	34	47	51	103	40
Daughter-in-law	5	3	8	9	13	5
Son	11	7	8	9	19	7
Son-in-law	1	<1	1	1	2	1
Sibling	9	6	2	2	11	4
Other relative	10	6	7	8	17	7
Friend	3	2	9	10	12	5
Neighbour	—	—	7	8	7	3
Other	2	1	4	4	6	2
Total	163	100	93	102	256	100

Spouses (more often wives) and daughters care for the more dependent and frail elderly people; those with weaker kinship ties and unrelated carers (i.e. friends and neighbours) care for less frail elderly people. Eighty per cent of informal carers are women and the majority are aged 45–54 years (Jones 1984). With changing demography, daughters and spouses are likely to be older as the age of frail elderly people increases, the 'old old' being cared for by the 'young old'.

It is well documented that most frail elderly people are cared for by one informal carer, on whom the main part of the burden falls; where there are secondary carers it is usually a daughter helping a parent to care for the dependent parent or another child helping the daughter to care for an elderly parent. Half of all carers report that they receive no help from family or domiciliary services to care for their dependants.

Two-thirds of carers share a household with their dependant: all spouses, half of daughters, and two-thirds of other related carers are resident carers. Unrelated carers are much less likely to be sharing a household with the person they assist.

Many (50 per cent) carers are themselves not in good health and inevitably many of the carers who are spouses and themselves elderly have disabilities; more than half of carers have health problems that make it difficult for them to care (Charlesworth *et al.* 1983; Levin *et al.* 1983; Jones 1984).

Some carers also have other dependants to care for and therefore other demands on their time and emotional support: children, disabled husbands, or other elderly relatives (aunts, in-laws, etc.). Many carers also have their own occupations: children are more likely to be in paid employment than are spouses.

THE ROLE OF DOMICILIARY SERVICES

For many years we have had, in Britain, a well-developed system of domiciliary care. Many of the services included within this were developed from charitable and voluntary organizations. Social services that are particularly relevant to elderly people living in the community are home helps or home care assistants, meals-on-wheels, and occupational therapists. Those employed by the health service are community nurses, health visitors, and physiotherapists.

Home helps and community nurses have been the mainstays of domiciliary care for the frail elderly. In recent years the role of home helps has changed considerably, even being replaced in some areas by care assistants. The role of the care assistant is more flexible and broader than that of home help, including as it does housework, shopping, help with mobility, and also personal care, that is, washing, bathing, and toileting where necessary. It is felt by many that the policy of social services is that home helps should, in preference, be provided to elderly people living alone with no informal support and not to those with informal carers, particularly if those carers are female. Seventeen per cent of elderly people with regular informal carers are assisted by home helps (Table 2) (Jones and Vetter 1985).

Table 2 Services received, by residency (from Jones 1986)

	Resident (n=163)		Non-resident (n=93)		Both (n=256)	
Services	No.	%	No.	%	No.	%
Home help	20	12	24	26	44	17
Meals-on-wheels	7	4	12	13	19	7
Social worker	8	5	5	5	13	5
Community nurse	40	25	10	11	50	20
Day hospital	17	10	11	12	28	11
Occupational therapist	1	<1	—	—	1	<1
Physiotherapist	2	1	—	—	2	<1
Voluntary worker	4	2	—	—	4	2

Fifty-six per cent received no services.

Social workers only visit a small minority (5 per cent) of frail elderly people or their informal carers. They usually visit as a response to crises or when institutional care has been requested, that is, when the informal system is breaking down.

Twenty per cent of elderly people with informal carers report receiving help from community nurses; this help most often comes in the form of nursing and personal care—bathing, washing, and dressing. Health visitors spend most of their time working with young families and, with some notable exceptions, rarely provide support to informal carers of elderly people.

Community-based therapists (physiotherapists and occupational therapists) are infrequent visitors to elderly people or their carers.

A number of elderly people attend day hospitals or day centres for a variety of reasons. In general, day hospitals are intended to perform a rehabilitative function for elderly people, but often patients continue to attend for respite care to provide relief for their carers (Brocklehurst and Tucker 1980). The receipt of regular, planned relief is all too rare and relief care is more often than not a response to crises rather than part of an overall plan to support informal carers.

The receipt of support from these domiciliary services varies between areas but also with the characteristics of both elderly clients and their carers.

Services have often been designed to satisfy the needs of professions and their organizations rather than addressing the requirements of elderly people and their informal carers and providing appropriate services to their needs and preferences. This has led to inflexibility in terms of job descriptions and hours at which the services are available—usually daytime and weekdays only.

In the past it has been repeatedly demonstrated that there is a paucity of systematic assessment of the needs of frail elderly people and their carers, whether this be undertaken by systematic screening, case-finding, or indeed opportunistic screening. This is despite the growing evidence of the benefit of such schemes for elderly people living in the community (Tulloch and Moore 1979; Vetter *et al.* 1983, 1984, 1986; Hendriksen *et al.* 1984). As a consequence, services to elderly people have not been allocated in a co-ordinated or planned way, but rather as a reaction to crises or serendipity. The benefits of undertaking full assessments and of designing care plans with the wishes of clients taken into account are increasingly reported. In many areas those plans are expedited and supervised by key workers from health or social services.

The fact that some domiciliary professionals are employed by and accountable to social services and others to health services has obstructed any attempt at co-ordination of domiciliary care.

Any real attempt to co-ordinate services must include the voluntary sector, which is more easily able to respond rapidly, flexibly, even idiosyncratically, to the needs of older people and their supporters. It is the voluntary sector that has so often been innovatory in its provision of community support.

As new situations arise, owing to the change in demography and policy, services need to be innovatory in their responses to these challenges: care assistants and 'Crossroads' care-attendant schemes are successful examples of such approaches.

That the provision of appropriate, good domiciliary care to support informal carers will displace carers has been a long-standing myth for which there is no evidence. Rather, many studies have concluded that the provision of appropriate services will prevent the breakdown of the informal care system and prevent unwanted institutionalization: most carers express a preference to continue caring for their elderly dependant provided that appropriate, planned, regular, dependable, and adequate support is provided for them (Levin *et al.* 1983; Jones 1986).

IMPACT ON CARERS

It is inevitable that caring for an elderly person regularly will have an impact on the life of an informal carer in terms of their occupational life, mental and physical well-being, social life, and family life. Many studies have now reported on the nature and extent of the deleterious effects experienced by informal carers (see also Chapter 20.6.1).

Economic

The effect of caring on employment will vary according to the age of the carer and also the degree of dependency of the dependant. Carers of those who become dependent earlier are more likely to be in the middle of their career. On the other hand, many carers of elderly dependants (particularly spouses, siblings, or friends) are themselves retired. The group therefore most likely to be affected are sons or daughters (many of whom are in mid-career or intending to start a new career having cared for children). Studies report that between a seventh and a quarter of carers ceased employment or reduced working hours as a direct consequence of their caring responsibilities; also about a tenth reported having to take time off work periodically. Daughters (and women generally) are more likely to reduce hours or give up employment completely than are other carers (Equal Opportunities Commission 1982; Wright 1983; Jones 1986).

For carers, particularly those of working age, employment outside the domestic environment can be beneficial, giving them a valued break from their caring responsibilities. Also, daughters who are main wage-earners dread the loss of their jobs if they stay off work (Isaacs *et al*. 1972).

Loss of employment may also have major economic consequences in terms of loss of income or opportunity costs (loss of promotion opportunities). It has been estimated that opportunity costs to families with wives unable to take on employment totalled £4500 per annum; for those in employment but having to work reduced hours the total opportunity costs have been estimated at £1900 per annum (1982 prices) (Nissel and Bonnerjea 1982).

In 1977, Hyman calculated that carers who had to cease employment altogether to care for an adult lost on average £120 per week (1980 prices) (cited by Equal Opportunities Commission 1982). The loss of earnings is but one aspect of the economic costs of caring. The expenditure on daily living expenses for a dependent person is likely to be higher than for an independent person, for example, in heating, washing, and special diets.

Health

The heavy nursing care required by some dependants such as assisting up and down stairs, lifting in and out of bed, in and out of baths and chairs, inevitably has an impact on the health of carers. In New Zealand a study reported that of those carers who were unable to continue caring for elderly dependants, two-thirds had experienced some effect on their health (Koopman-Boyden 1979). In a study of a general population of elderly dependants, more than a quarter of their carers perceived their health as being deleteriously affected by their caring role. As with other aspects of carers' quality of life, this varied with the characteristics of carers and dependants: daughters and wives reported this more often as did carers of the more frail elderly people (Jones 1986). A breakdown in health is often the cause of a reduced duration of informal care and is also often given as a future possible cause of carers seeking institutional care for their dependants (Levin *et al*. 1983; Jones 1986).

Social life

The frequent or constant care and attention provided by carers inevitably has a deleterious effect on their social life. A quarter of carers report that their social life (excluding family life) is affected negatively as a result of being carers (Equal Opportunities Commission 1982; Jones 1986). As with other aspects the extent varies with both the characteristics of carers and of the elderly dependants: resident carers are more likely to be affected in this way than non-resident carers and daughters more than other carers; and the more frail (mentally or physically) the dependant the more likely the carer is to be tied to the home and unable to leave it during the day or evenings apart from short periods, thus rendering any involvement in regular social activities very difficult (Jones 1986).

Family life

The lack of privacy and personal freedom necessarily has detrimental effects on carers, particularly those who share a home with their dependants; a quarter of carers report these effects. Detrimental effects on family life are more commonly reported among those caring for the more physically impaired. For carers who are not spouses, one of the main problems seems to be that of conflict of responsibilities and conflict of interests, which cause strain and tension to carers. It is probably for the latter reason that daughters more often report negative affects on their family life than do other carers (Jones 1986). Indeed, many resident daughters report not being able to spend an evening outside the home with their husbands.

Emotional stress

Long-term care and responsibility lead to emotional stress, which in turn influences the outcome of the duration of the caring relationship. It has been demonstrated with both a generally dependent population and an elderly, mentally infirm population that those dependants whose carers are experiencing a high level of stress are likely to move into long-term care sooner (Levin *et al*. 1983; Jones 1986).

A fifth of carers seem to report a large or an unbearable degree of stress. Consistent with the findings about family and social life, daughters perceive themselves as suffering more stress than do other carers, indeed a fifth of all resident daughters experience unbearable stress—twice as many as non-resident daughters. Research findings indicate that perceived stress is higher among carers of very frail dependants (Jones 1986).

Certain characteristics of elderly dependants that lead to stress of their carers have been identified: heavy incontinence, night disturbance, inability to communicate, and dangerous behaviour. Of all the aspects of quality of life upon which caring has an impact it seems to be found consistently that detrimental affects on social and family life are most likely to lead to perceived stress and also objectively measured anxiety (Gilleard *et al*. 1981; Levin *et al*. 1983; Gilhooly 1984; D. A. Jones, unpublished work).

Alleviation of stress

Time apart from dependants seems to be one of the most important factors in the alleviation of carer stress, whether this respite is a consequence of the carer being able to leave the home to participate in paid employment or leisure activities, or the dependant spending time away from home for a while. Support from other members of the wider family is associated with lower levels of stress (Nissel and Bonnerjea 1982). At least half of all carers receive no help from other members of the family—a situation which leads to feelings of resentment for the carers and guilt for the other family members.

Several studies have shown that carers who receive support and assistance from domiciliary services (in particular home helps and community nurses) show less stress than those who do not. Short-term hospital care as well as long-term hospital or residential care have also been shown to result in the reduction of stress among carers (Jones 1986; Levin *et al*. 1983). It must also be borne in mind that in certain conditions the stress can only be relieved by long-term residential care. Nevertheless, the key to community care must be in developing appropriate forms of support to informal carers. It seems that the provision of appropriate service support can alleviate the stress on carers and prolong the maintenance of elderly people in the community.

POLICY: PAST AND FUTURE

That the greater part of community care falls upon informal carers there can be no doubt. The policy that elderly people should remain in the community for as long as possible assumes that families and friends will support frail elderly people when they are dependent. Indeed, families who care for their elderly relatives mostly choose to do so, despite the costs, and would not consider institutional care unless the dependency became more than they could cope with or their own health failed (Jones and Salvage 1991). Also, there can be no doubt that families have cared and continue to care for dependent elderly people at great costs to themselves, in terms of finance and quality of life. With the increasing numbers of 'old old' we cannot continue to take such care for granted and assume that the next generation will be prepared to be similarly exploited.

That there has been little dialogue between elderly people and statutory services is well documented: there has been all too little involvement or indeed choice for elderly people in the design and running of services for their benefit. Fortunately this situation has begun to change; and the preferences and choices, in some areas of the United Kingdom at least, are beginning to be taken into account. This partnership is in its infancy.

As yet, however, there is little evidence of any involvement or consultation of informal carers in the development of health and social services. For example, there are no services that are currently designed for the benefit of informal carers, only services designed for the benefit of elderly people. Discharge from hospital, assessments for service, and financial benefits are decisions too often arrived at without consultation with informal carers.

There are some striking examples of consultation within the voluntary sector, for example, 'Crossroads' schemes, where the service is designed for the primary benefit of informal carers, who are involved in decisions about the nature and time of assistance provided. It is to be hoped that domiciliary services will emulate some of these examples of good practice.

Nature of the past

In the past, services have not been tailored to the needs of elderly people and their informal carers. The nature of the work that they could undertake and the times at which they could provide support were inflexible; home helps could only provide certain forms of household care, not personal care, and were only available during office hours. Partly because of the divide between health, social services, and family health service authorities, services provided to elderly people have not been well co-ordinated. Developments in the key-worker approach, which names one person (from either health or social services) as responsible and accountable for the appropriate package of care for elderly people, have to a degree reduced this problem, as indeed have some developments of primary health care teams. As general practitioners are the professionals in contact with most elderly people most often, it is unlikely that support to elderly people and their carers will be co-ordinated successfully if communication continues to be poor between social services, health services, and general practitioners.

Effective communication necessitates the breakdown of professional barriers and prejudices, and the sharing of information; the latter has obvious ethical and confidentiality implications. Voluntary agencies also have much to offer if they can be integrated into community care.

Any co-ordinated and targeted community care system depends on good assessment. It is all too rare that the needs of elderly people are properly assessed and reassessed; even more rare are assessments of the needs of informal carers. It has been demonstrated that general practitioners do not appear to perceive needs of informal carers even when home-visiting (Vetter *et al*. 1984). As with all community care services they are oriented towards the needs of elderly people in the community. Indeed, support is allocated according to criteria based on the disability of elderly people rather than the needs of carers.

Many carers are regularly undertaking heavy nursing tasks, either developed incrementally over many years or suddenly after a major event, for example, a stroke, for which they have never received training or instruction. Also, it is well documented that informal carers have, in the past, been uninformed about domiciliary services, voluntary organizations, and carers' associations. Together with elderly people, carers also need information about available and appropriate domiciliary services, voluntary organizations, and also relevant allowances and benefits. Many studies have reported on the under-utilization of benefits including attendance allowance and invalidity benefit.

Hope for the future

If appropriate sources of support are to be provided to informal carers in the future there must be systematic assessments and reassessments of the needs of informal carers as much as of elderly people. This should lead to referrals to appropriate agencies – be they health, social services, or voluntary organizations – and also to the provision of necessary information on available sources of support and finances.

If carers are to continue to be the main source of community care, including nursing care, they will need to be trained. This will include training in lifting techniques to avoid any injuries, nursing tasks, and, more generally, in developing skills to enable elderly people to care for themselves as far as possible rather than creating or encouraging dependency. At the point of discharge from hospital is one obvious time for contacting and involving informal carers in training programmes.

The amount of stress and consequent morbidity of carers is well documented and, one hopes, could be reduced by support and counselling in the early stages of their caring role, before crises develop. Carers need to learn how to cope with stress and deal with conflicting family interests. Feelings of guilt and resentment are common features among carers, as are resentment at their loss of independence and freedom, and reduction in their quality of life; and guilt because of their feelings of resentment towards their own husbands, mothers, or fathers. Carers of elderly, mentally infirm people experience particular problems, including feelings of bereavement because the 'known' personality has died despite the person still being alive.

If, in the future, informal carers are to continue to take on the lion's share of caring for an increasing number of elderly people, services must be oriented to the needs and wishes of informal carers and their elderly dependants. The needs of carers cannot be adjusted to fit into the structures and policies of professional organizations. Carers also need not only to be consulted about future policies but to be involved in this development. To meet the needs of carers, a more innovative approach to their ongoing support and respite will need to be taken. Much can be learnt in this area from voluntary organizations.

Any innovatory schemes need monitoring and evaluation: careful monitoring and rigorous evaluation of such innovations are invaluable to the development of effective and efficient community care.

References

Brocklehurst, J.C. and Tucker, J.S. (1980). *Progress in geriatric day care*. King Edward's Hospital Fund for London, London.

Central Statistical Office (1989). *Social Trends* **19**. HMSO, London.

Charlesworth, A., Wilkin, D., and Durie, A. (1983). *Carers and services: a comparison of men and women caring for dependent elderly people*. Department of Psychiatry and Community Medicine, University of Manchester.

Department of Health and Social Security (1981). *Growing older*, Cmnd. 8173. DHSS, London.

Equal Opportunities Commission (1982). *Caring for the elderly and handicapped: community care policies and women's lives*. EOC, Manchester.

Gilhooly, M.L. (1984). The impact of care-giving on care givers: factors associated with the psychological well-being of people supporting a dementing relative in the community. *British Journal of Medical Psychology*, **57**, 35–44.

Gilleard, C.J., Watt, G., and Boyd, W.D. (1981). *Problems of caring for the elderly mentally inform at home*. Paper presented at the 12th International Congress of Gerontology, 12–17 July, Hamburg, West Germany.

Hendriksen, C., Lund, E., and Stromgard, E. (1984). Consequences of assessment and intervention among elderly people: a three-year randomised controlled trial. *British Medical Journal*, **289**, 1522–4.

Isaacs, B., Livingstone, M., and Neville, Y. (1972). *Survival of the unfittest: a study of geriatric patients in Glasgow*. Routledge & Kegan Paul, London.

Jones, D.A. (1984). A survey of those who care for the elderly at home: their problems and their needs. *Social Science and Medicine*, **19**, 511–14.

Jones, D.A. (1986). *A survey of carers of elderly dependents living in the community: a report*. Research Team for the Care of the Elderly, Cardiff.

Jones, D.A. and Vetter, N.J. (1985). Formal and informal support received by carers of elderly dependents. *British Medical Journal*, **291**, 643–5.

Jones, D.A. and Salvage, A.V. (1991). *Attitudes to caring among a group of informal carers of elderly people*, in press.

Koopman-Boyden, P. (1979). Problems arising from supporting the elderly at home. *New Zealand Medical Journal*, **89**, 265–8.

Levin, E., Sinclair, I., and Gorbach, P. (1983). *The supporters of confused elderly people at home: extract from the main report*. National Institute for Social Work Research Unit, London.

Martin, J., Meltzer, H., and Elliot, D. (1988). *The prevalence of disability among adults*. HMSO, London.

Nissel, M. and Bonnerjea, L. (1982). *Family care of the handicapped elderly: who pays*? Policy Studies Institute, London.

OPCS and CSO. (1984). *Britain's elderly population*. HMSO, London.

OPCS Social Survey Division (1987). *General Household Survey 1985*. HMSO, London.

Tulloch, A.J. and Moore, V. (1979). A randomised controlled trial of geriatric screening and surveillance in general practice. *Journal of the Royal College of General Practitioners*, **29**, 733–5.

United Nations (1982). *Report on the World Assembly on Ageing*, Publication No. E.82.1.16. UN, New York and Geneva.

Vetter, N.J., Jones, D.A., and Victor, C.R. (1983). *The effectiveness of health visitors working with the elderly in general practice: final report*. Research Team for the Care of the Elderly, Cardiff.

Vetter, N.J., Jones, D.A., and Victor, C.R. (1984). The effectiveness of health visitors working with elderly patients in general practice: a randomised controlled trial. *British Medical Journal*, **288**, 369–72.

Vetter, N.J., Jones, D.A., and Victor, C.R. (1986). A health visitor affects the problems others do not reach. *Lancet*, **ii**, 30–2.

Wright, F. (1983). Single Carers: Employment, housework and caring. In *A labour of love: women, working and caring* (ed. J Finch and D. Groves). Routledge & Kegan Paul, London.

SECTION 21
Anaesthesiology and medical aspects of surgery

21.1 Risk prediction in medicine and surgery

D. GWYN SEYMOUR

One of the major challenges facing a clinician in day-to-day clinical practice is the need to make predictions, often on the basis of incomplete data (Feinstein 1983). The challenge is all the greater when the patient is elderly, as atypical presentation of disease and multiple medical problems become commoner with increasing age.

When predicting events that are yet to happen, it is usual to speak of 'prognosis'. When the aim is to evaluate the likelihood of a disease being present in a patient with a particular constellation of signs, symptoms, and tests, then the process is called 'diagnosis'. Mathematically there are many similarities between the processes needed to quantify prognosis and diagnosis, and numerical methods introduced for one purpose can often be adapted for the other (Spiegelhalter and Knill-Jones 1984) although this is not invariably the case.

In clinical medicine, risk prediction is rarely an end in itself, but is usually an intermediate step in the process of 'clinical decision-making'. The latter is a rapidly expanding field and it is impossible to do it justice here. The interested reader is referred to the introductory texts by Wulff (1981), Sackett *et al.* (1985), Knoebel (1986), and Sox *et al.* (1988), and the detailed account of clinical decision analysis of Weinstein and Fineberg (1980). Computers are also being used increasingly in the field of medical decision-making and the potential of 'probabilistic' and 'knowledge-based' systems is discussed by Spiegelhalter and Knill-Jones (1984), Chytil and Engelbrech (1987), and Gale (1986).

The present discussion confines itself to the more limited topic of risk prediction. It primarily considers how the results of one or more tests can be combined together in order to aid the process of prognosis or diagnosis. A useful starting point for examining the value of a single test or observation is the classical 2×2 ('2 by 2') table. The 2×2 table shown in Table 1 summarizes the relationship between a 'test' or observation (which can be either positive or negative) and a diagnosis or outcome (which can either be present or absent). The numerical example shown in Table 1 is drawn from a prospective study of patients aged 65 years and over, undergoing general surgery in Dundee (Seymour and Pringle 1983). In this example, the 'test' is whether or not a patient smoked in the 6 weeks prior to surgery, and the 'outcome' is whether or not a postoperative respiratory complication developed. The 2×2 table is laid out in the usual way with the test on the vertical axis, and the outcome on the horizontal axis. Following the convention of Sackett *et al.* (1985), the four cells are labelled A, B, C, D.

There are a bewildering number of ways that the four cells of a 2×2 table can be defined and manipulated in order to produce indices such as specificity, sensitivity, and relative risk. For reference purposes, some of the more important indices are listed in Table 1, and these are now briefly considered in turn.

The four cells A, B, C and D of Table 1 contain the number of true positives, false positives, false negatives, and true negatives, respectively. These terms are used predominantly in situations where it is desired to estimate the value of a test for screening.

The overall accuracy of the test is the second formula listed in Table 1. In the example given the test was 62 per cent accurate, comprising 48 true positives (smokers who developed a postoperative chest complication) and 111 true negatives (non-smokers who avoided a chest complication) out of a total of 257 patients. Superficially, it might appear that the accuracy of a test is all that a clinician requires in making a prediction, but this is far from true. This is because the formula by which accuracy is defined in Table 1 places equal weight on true positives and true negatives, which is rarely what the clinician demands from a test. For example, consider a situation where an outcome occurs in only one patient in a hundred (e.g. a postoperative stroke in elderly patients undergoing elective surgery). By simply predicting that none of the patients will develop a postoperative stroke it is possible to achieve 99 per cent accuracy. This example also illustrates a general principle that is relevant to many of the indices listed: it is difficult to produce a single index that adequately describes the interrelationships of a 2×2 table. The usual way around this problem is to quote more than one index, e.g. the specificity *and* the sensitivity.

Another general principle to be considered when evaluating the indices in Table 1 is that an index which is 'prevalence dependent' may be less useful than a prevalence-independent index. A prevalence-dependent index is one whose value is different in populations where the frequency of the outcome (or diagnosis) is different. (Strictly speaking, when the outcome is a postoperative respiratory complication we should be referring to incidence rather than prevalence, but 'prevalence dependent' now has widespread currency in the literature.) Of the indices listed in Table 1, positive predictive value, negative predictive value, and relative risk are all prevalence dependent (Sackett *et al.* 1985; Peters and Newcombe 1990), which diminishes their value in tests of comparisons between different populations. The 'prevalence-independent' indices in Table 1 are the sensitivity, specificity, odds ratio, and likelihood ratio. Note that formulae calculated 'horizontally' (i.e. with $A+B$ or $C+D$ as denominators) are prevalence dependent, while those calculated 'vertically' are not.

Despite its prevalence dependence the relative risk is still a useful index, and it has a central importance in prospective epidemiological studies in estimating the strengths of a relationship between a risk factor and an outcome. One of the great attractions of the relative risk is that it is a concept that is easy to grasp. For instance; using the numerical example given in Table 1 we could say that 'compared with non-smokers, elderly smokers have 1.6 times the risk of developing postoperative respiratory complications'. While it is true that the relative risk will be lower where the overall prevalence of chest complications is higher (Table 2), the statement still contains information that can be used as part of a bedside assessment.

Table 1 Terms used in assessing diagnostic tests and in risk prediction

Definition

		Diagnosis (or predicted outcome) +ve	Diagnosis (or predicted outcome) −ve
Test (or risk factor)	+ve	*A* (true positive)	*B* (false positive)
	−ve	*C* (false negative)	*D* (true negative)

Worked example

		Postoperative respiratory problem: Yes	Postoperative respiratory problem: No	
Current smoker ?	Yes	48	45	93
	No	53	111	164
		101	156	257

Term	Definition	Worked example
Prevalence of diagnosis (or incidence of outcome)	$=\frac{A+C}{A+B+C+D}$	$=\frac{48+53}{257}=0.39=39\%$
Accuracy of test	$=\frac{A+D}{A+B+C+D}$	$=\frac{48+111}{257}=0.62=62\%$
Positive predictive value (predictive value of a positive test)	$=\frac{A}{A+B}$	$=\frac{48}{93}=0.52=52\%$
Negative predictive value (predictive value of a negative test)	$=\frac{D}{C+D}$	$=\frac{111}{164}=0.68=68\%$
Sensitivity (or true positive rate)	$=\frac{A}{A+C}$	$=\frac{48}{101}=0.475=48\%$
Specificity (or true negative rate)	$=\frac{D}{B+D}$	$=\frac{111}{156}=0.712=71\%$
Relative risk (or risk ratio)	$=\frac{A}{A+B}\div\frac{C}{C+D}=\frac{A}{(A+B)}\times\frac{(C+D)}{C}$	$=\frac{48}{93}\times\frac{164}{53}=1.60$
Odds ratio (or cross-product ratio)	$=\frac{A}{B}\div\frac{C}{D}=\frac{A}{B}\times\frac{D}{C}$	$=\frac{48}{45}\times\frac{111}{53}=2.23$
Indices used when applying Bayes theorem		
Likelihood ratio of a positive test	$=\frac{A}{A+C}\div\frac{B}{B+D}=\frac{A}{(A+C)}\times\frac{(B+D)}{B}$	$=\frac{48}{101}\times\frac{156}{45}=1.65$
Likelihood ratio of a negative test	$=\frac{C}{A+C}\div\frac{D}{B+D}=\frac{C}{(A+C)}\times\frac{(B+D)}{D}$	$=\frac{53}{101}\times\frac{156}{111}=0.74$
Likelihood ratios can also be calculated as follows;		
Likelihood ratio of a positive test	$=\frac{\text{Sensitivity}}{1-\text{Specificity}}$	$=\frac{0.475}{1-0.712}=1.65$
Likelihood ratio of a negative test	$=\frac{1-\text{Sensitivity}}{\text{Specificity}}$	$=\frac{1-0.475}{0.712}=0.74$
Weight of evidence = 100 × ln likelihood ratio (where ln is the natural logarithm, $\log_e$)	For a likelihood ratio of 1.65, weight of evidence = 100 × ln 1.65 = 50.1 For a likelihood ratio of 0.74, weight of evidence = 100 × ln 0.74 = −30.1	

The odds ratio is prevalence independent, and the relative risk is numerically similar to the odds ratio when the prevalence of outcome is very low (Table 2). The odds ratio also has the useful property that it can be calculated retrospectively from a case-control study, whereas the relative risk can usually only be estimated from a prospective study (Fleiss 1981).

Table 2 The effect of prevalence on positive predictive value, negative predictive value, and relative risk of an excellent test*

Prevalence (%)	Positive predictive value(%)	Negative predictive value(%)	Relative risk	Odds ratio	Accuracy (%)
0.1	1.9	99.995	354.0	361	95
0.5	8.7	99.97	330.0	361	95
1.0	16.1	99.95	303.0	361	95
5.0	50.0	99.7	181.0	361	95
10.0	67.9	99.4	117.0	361	95
20.0	82.6	98.7	64.0	361	95
25.0	86.4	98.3	50.0	361	95
30.0	89.1	97.8	40.0	361	95
40.0	92.7	96.6	27.0	361	95
50.0	95.0	95.0	19.0	361	95
60.0	96.6	92.7	13.0	361	95
70.0	97.8	89.1	8.9	361	95
75.0	98.3	86.4	7.2	361	95
80.0	98.7	82.6	5.7	361	95
90.0	99.4	67.9	3.1	361	95
95.0	99.7	50.0	2.0	361	95
99.0	99.95	16.1	1.2	361	95

*Following Sackett *et al.* (1985), the test is assumed to have a sensitivity and specificity of 0.95.

The terms sensitivity and specificity, used in the context of evaluating a clinical test, will be familiar to most clinicians. These tests are prevalence independent and so it is possible to publish lists of sensitivity and specificity of different tests and outcomes with some confidence that they will be applicable in populations elsewhere. Both sensitivity and specificity need to be considered when a test is being evaluated; an excellent test would have both values close to 100 per cent, but 80 to 90 per cent is much commoner in clinical practice, even for widely accepted tests.

The final group of formulae in Table 1 relate to likelihood ratios. These are less familiar to clinicians, but they deserve to be better known as they allow Bayes' theorem (see below) to be applied to everyday clinical diagnosis (Sackett *et al.* 1985; Simel 1985). Like the sensitivity and specificity (to which they are related) likelihood ratios are prevalence independent. Sackett *et al.* (1985) foresee a time in the near future when lists of likelihood ratios will be commonplace in medical textbooks.

In some computer applications, the natural logarithms of likelihood ratios are used, and these when multiplied by 100 are referred to as the 'weights of evidence' (Card and Good 1974; Spiegelhalter 1985; Seymour *et al.* 1990*a*.). They have the useful property that they can be added when Bayes' theorem is applied, whereas the likelihood ratios need to be multiplied. This is considered below when the use of Bayes' theorem to weigh up several risk factors is described.

BAYES' THEOREM AND THE EVALUATION OF ONE TEST

Thomas Bayes (1702–1761) was a clergyman and mathematician. The theorem that bears his name was first proposed as an aid to medical diagnosis by Ledley and Lusted (1959) and has been repeatedly used in diagnosis and prognosis since that time. Examples of successful applications of Bayes' theorem include de Dombal's system for diagnosing surgical patients with an 'acute abdomen' (Adams *et al*, 1986), and two systems originating in Glasgow for determining prognosis after head injury (Barlow *et al.* 1987) and the differential diagnosis of gastroenterology patients (Spiegelhalter and Knill-Jones 1984; Spiegelhalter 1985).

Before explaining how to apply Bayes' theorem it is helpful to consider the two main reasons why it is useful.

1. In making a diagnosis or establishing a prognosis, the clinician is faced with one or more clinical findings (e.g. symptoms, signs, or test results) and wishes to estimate the chances of a disease being present or of a particular outcome occurring. Unfortunately, most medical textbooks and articles present their information in the opposite fashion: diseases are taken as the starting point and the clinical findings associated with those diseases are then described (Wulff 1981; Sackett *et al.* 1985; Knoebel 1986; Sox *et al.* 1988). Bayes' theorem allows the clinician to proceed from a clinical finding to a disease (or outcome) even though the available information is expressed the other way round.
2. In the terminology usually employed, Bayes' theorem allows a clinician to convert a 'prior probability' to a 'posterior probability' as the result of applying a 'test'. The prior probability (or pretest probability) is the estimate of risk before the test is carried out. In the case of a postoperative chest complication, a reasonable estimate of prior probability would be the rate of postoperative chest complications in previous studies of broadly similar patients. The posterior (or post-test) probability is the revised estimate of the risk of a postoperative chest complication once the result of the test is known. While the terminology might be unfamiliar, the general strategy of making a provisional assessment of a patient, applying a key test, then revising that assessment will be familiar to most clinicians as one of the major processes used in diagnosis and prognosis.

As usually written down, Bayes' theorem looks unappetizing to the non-mathematician (Chard 1988):

$$P(D:CF) = \frac{P(CF:D)\,P(D)}{P(CF:D)\,P(D) + P(CF:\text{not } D)\,P(\text{not } D)}$$

where P is probability, D is a diagnosis or outcome, CF is a set of one or more clinical features, and the expression $P(D:CF)$ is equivalent to saying 'the probability of diagnosis D given the presence of features CF'. '*not D*' designates the absence of D.

In formal diagnostic systems such as the de Dombal system referred to above, a microcomputer usually performs the calculations and so the mathematical details are relatively unimportant. However, if Bayes' theorem is being used in a less formal way in day-to-day diagnosis, a form of the equation which can be applied using mental arithmetic is preferable. This equation is achieved by using odds rather than probabilities, and it is expressed as follows (Simel 1985; Sackett *et al.* 1985; Sox *et al.* 1988):

$$\text{posterior odds} = \text{prior odds} \times \text{likelihood ratio}.$$

The likelihood ratios of positive and negative tests referred to in this equation can be easily calculated (as shown in Table 1) either by using raw data from a 2×2 table or through knowledge of the sensitivity and specificity of a test.

To convert probabilities to odds, and vice versa, the following formulae can be used:

$$\text{odds} = \frac{\text{probability}}{1 - \text{probability}}$$

$$\text{probability} = \frac{\text{odds}}{\text{odds} + 1}$$

This process can be best illustrated by a numerical example. From Table 1 the prior probability of postoperative respiratory complications can be estimated from the overall incidence of complications to be $101/257 = 0.39$. As

$$\text{odds} = \frac{\text{probability}}{(1 - \text{probability})}$$

the prior odds are $0.39/(1 - 0.39) = 0.64$.

From Table 1, it can be seen that the likelihood ratio for a positive test (i.e. smoking) is 1.65, and that of a negative test is 0.74. Thus, in a smoker, the equation:

$$\text{posterior odds} = \text{prior odds} \times \text{likelihood ratio}$$
$$= 0.64 \times 1.65 = 1.06,$$

while for a non-smoker the posterior odds are $0.64 \times 0.74 = 0.47$. Because probability = odds/(odds + 1), the posterior probabilities are $1.06/2.06 = 0.51$ for a smoker, and $0.47/1.47 = 0.32$ for a non-smoker.

To summarize, before applying the test, the best estimate of the probability of postoperative respiratory complications was 39 per cent. Once it was known the patient was a smoker, the estimated probability increased to 51 per cent, while in a non-smoker the estimate would have fallen to 32 per cent. A nomogram can be used to avoid calculations (Fagan 1975).

In a population where the average rate of postoperative respiratory complications is not 39 per cent but 25 per cent, then the likelihood ratios of Table 1 should still apply as they are prevalence independent. The prior odds in such a population would be $0.25/0.75 = 0.33$. Thus the posterior odds in a smoker would be $0.33 \times 1.65 = 0.54$, which corresponds to a posterior probability of 0.35 (or 35 per cent). In a non-smoker, the posterior odds would be $0.33 \times 0.74 = 0.24$, which corresponds to a posterior probability of 0.19 (or 19 per cent).

BAYES' THEOREM AND MORE THAN ONE TEST

In the clinical example given above it was shown how information about current smoking habits could be incorporated into an estimate of the risk of postoperative respiratory complications. However, in the process of diagnosis or prognosis we are rarely concerned with a single risk factor or the result of only one test. Could Bayes' theorem be used to look at the influence of more than one risk factor, say the effect of smoking and a history of chronic bronchitis, on postoperative respiratory complications? The simplest approach to the problem is to use the posterior probability from the initial analysis (with smoking as the risk factor) as the prior probability for the next analysis (with chronic bronchitis as the risk factor). To do this we need to know the likelihood ratios for chronic bronchitis and postoperative respiratory complications. Using data from the study quoted in Table 1, the likelihood ratio when there is a history of chronic bronchitis can be calculated to be 3.3, while the likelihood ratio when there is no history of chronic bronchitis is 0.70.

From the example previously presented it was found that the posterior probability of postoperative respiratory complications in a smoker was 0.51 (posterior odds = 1.06) and for a non-smoker was 0.32 (posterior odds = 0.47). Using these as prior odds for the analysis of the effect of chronic bronchitis gives the following results: as posterior odds = prior odds × likelihood ratio, then in a smoker with a history of chronic bronchitis, posterior odds = $1.06 \times 3.3 = 3.5$, which corresponds to a posterior probability of postoperative respiratory complications of $3.5/4.5 = 78$ per cent. (For a smoker with no chronic bronchitis the posterior probability can be calculated to be 43 per cent; for a non-smoker with chronic bronchitis the posterior probability is 61 per cent; for a non-smoker with no chronic bronchitis the posterior probability is 25 per cent.)

Fired by the apparent precision of these estimates we might proceed with further analyses based on likelihood ratios for the site of incision, sex, age, peak flow tests, etc. However, the longer this chain of calculations becomes, the more suspect are the predictions, as we are tacitly assuming that each of these risk factors is acting independently of all of the others in terms of predicting postoperative respiratory complications. If we return to the example of smoking and chronic bronchitis we can see that there is a strong chance that these two risk factors are not acting independently. As smoking is a major cause of chronic bronchitis there will be a tendency for bronchitics to be current smokers (although this might be partly offset if some severe bronchitics quit smoking for health reasons).

Applications of Bayes' theorem that combine several risk factors but assume they are acting independently are called 'independence Bayes' (or 'naive Bayes' 'or 'idiot Bayes'). While some authors have objections to this approach, others point out that, provided the risk factors are chosen with care, independence Bayes' systems are robust and work well in practice. Examples of successful systems using this approach are the acute abdomen diagnostic system and the head injury prognostic system referred to above.

Where independence Bayes' systems make errors they tend to be 'over-confident', i.e. their risk predictions tend to be too high in high-risk patients and too low in low-risk patients. Spiegelhalter and Knill-Jones (1984) have suggested a method that corrects for this overconfidence and that has several other advantages. Here regression methods are used to allow for lack of independence between risk factors. The practicalities involve working not with the likelihood ratios themselves but with multiples of their logarithms (i.e. the 'weights of evidence' referred to in Table 1). Details of this method can be found in Spiegelhalter (1985) and Seymour *et al.* (1990*a*).

While the newer techniques may overcome some of the objections to the use of Bayes' theorem, many statisticians and clinicians still have misgivings about its use in risk prediction. The concept of prior probabilities is a major stumbling block. In some cases, prior probabilities can be estimated with some precision on the basis of large surveys of similar patients. In other cases the prior probability is little more than an educated guess. As the final risk assessment (posterior probability) is based on a mixture of the prior probability and the clinical features in the individual patient (expressed as likelihood ratios), then it is theoretically possible for two individuals with identical clinical signs and symptoms to be given two completely different risk assessments on the basis of different prior-probability estimates. Another potential problem of Bayes' theorem is encountered in the diagnosis of a rare disease if the database is not large enough to provide several examples of that disease, although this problem is not confined to Bayesian techniques.

Wagner (1982) has referred to Bayes' theorem, in the context of diagnosis, as 'an idea whose time has come'. This is probably a fair assessment as long as Bayes' theorem is regarded as a potentially useful tool rather than a diagnostic or prognostic panacea (O'Connor and Sox 1991). The use of Bayes' theorem is not appropriate in all diagnostic and prognostic situations, and some clinical problems require a mixture of Bayesian and non-Bayesian approaches (Gale 1986). At present, there is the added problem that the raw data needed to apply Bayes' theorem 'at the bedside' are often not readily available (Harris 1981). One of the tasks of the interested geriatrician might be to assemble information of a type that would be useful in Bayesian analyses in elderly patients.

COMBINING INFORMATION FROM SEVERAL TESTS: NON-BAYESIAN PROBABILISTIC METHODS

As previous described, the methods of medical decision support can be broadly divided into 'knowledge-based' and 'probabilistic' categories. The former category falls under the general heading of Artificial Intelligence, the medical potentials of which are discussed by Gale (1986) and Chard (1988). The probabilistic or statistical category of techniques can be broadly subdivided into Bayesian and non-Bayesian methods. This section looks at the way that non-Bayesian statisticians approach the problem of predicting outcome on the basis of multiple risk factors. The two techniques discussed are multiple linear regression and logistic regression.

Multiple linear regression was the first of these two techniques to be widely employed in biology and medicine. It is used when the outcome variable is continuous (e.g. weight in kg; blood urea in mmol/l). The basic equation (Matthews and Farewell 1988) is:

$$Y = \mathrm{a} + \mathrm{b}_1X_1 + \mathrm{b}_2X_2 \ldots \mathrm{b}_kX_k,$$

where Y is the predicted outcome, and X_1, X_2, $X_3 \ldots X_k$ are the k parameters used to predict this outcome. The constants a, b_1, $\mathrm{b}_2 \ldots \mathrm{b}_k$ are calculated by computer methods beyond the scope of this discussion, but the aim is to produce a straight-line equation which 'best' fits the relationship between the explanatory variables X_1, X_2, $X_3 \ldots X_k$ and the outcome Y. As Matthews and Farewell (1988) point out, the computer analysis from which this equation is derived assumes that for specified values of the explanatory variables X_1, X_2, X_3, etc., the distribution of Y is Normal (with a mean value which can be calculated by fitting these variables into the equation $Y = \mathrm{a} + \mathrm{b}_1X_1 + \mathrm{b}_2X_2 \ldots \mathrm{b}_kX_k$).

The coefficients b_1, b_2, etc. represent the contribution that the explanatory variables make to predicting the value of Y, when all the other explanatory variables are taken into account. Thus, in theory at least, if two variables X_1 and X_2 are being used to predict Y, then the b_1 coefficient will reflect how much of the prediction is due to the effect of X_1 and the b_2 coefficient will reflect how much the prediction is due to X_2. The technique is thus more sophisticated than an approach such as 'Independence Bayes' where the variables X_1, X_2, X_3, etc. are assumed to be acting independently. In practice, however, it is prudent to avoid placing highly intercorrelated explanatory variables in the regression analysis (Armitage and Berry 1987).

Multiple linear regression equations are intended for situations where the outcome variable is continuous. Unfortunately, much of clinical medicine is concerned with outcomes that are binary (i.e. outcomes that take one of two values, e.g. dead or alive). In such circumstances we do not need an outcome measure (such as Y), which can take any value from minus infinity to plus infinity, but a measure that will estimate the *probability* of, say, a postoperative respiratory complication, on a scale ranging from 0 to 1 (as probabilities cannot be less than zero or greater than unity). The most widely used solution to this problem is to adopt the technique of logistic regression (Matthews and Farewell 1988).

The starting point of a logistic regression analysis is to construct an equation of the same form as that used in multiple linear regression (i.e. $Y = \mathrm{a} + \mathrm{b}_1X_1 + \mathrm{b}_2X_2 \ldots \mathrm{b}_kX_k$). In logistic regression, however, this is not the end of the analysis as Y is not the predicted outcome itself, although it is mathematically related to the predicted outcome by the equation:

$$Y = \log_e \text{(odds of the outcome)}$$

$$\text{As odds} = \frac{\text{probability}}{1 - \text{probability}}$$

it follows that:

$$Y = \log_e \frac{\text{probability}}{1 - \text{probability}}$$

which is equivalent to:

$$e^Y = \frac{\text{probability}}{1 - \text{probability}}.$$

A little algebra produces an equation that allows the probability to be calculated from Y:

$$\text{Probability} = \frac{e^Y}{1 + e^Y}$$

The non-statistician might want to take all of this on trust and simply note that the first step in a logistic regression is to calculate Y, and the second step is to calculate the pre-

Table 3 Prediction of postoperative respiratory complications

Use of predictive equation [derived from Dundee data using binary (logistic) regression]

Step A: calculate $Y = -1.79 + (\text{preoperative chest} \times 1.23) + (\text{smoking} \times 0.841) + (\text{incision site} \times 0.507) + (\text{volume depletion} \times 2.48)$

(Codings

For preoperative chest problems and smoking	— code 0 if absent, code 1 if present
For incision site	— code 0 if simple hernia or non-abdominal incision — code 1 if mid- or lower abdominal incision — code 2 if upper abdomen or thoracic incision
For volume depletion	— code 0 if nil or mild depletion — code 1 if otherwise).

Step B: to calculate predicted percentage rate of postoperative complications,

$$\text{compute } 100 \times \left(\frac{e^{Y}}{1+e^{Y}}\right)$$

or refer to Table 4

Example

In a patient with a history of chronic bronchitis, who did not smoke, undergoing upper abdominal surgery, with no evidence of volume depletion

Step A: $Y = -1.79 + (1 \times 1.23) + (0 \times 0.841) + (2 \times 0.507) + (0 \times 2.48) = 0.454$

Step B: predicted chances of postoperative respiratory complications

$$= 100 \times \left(\frac{e^{0.454}}{1+e^{0.454}}\right) = 100 \times \left(\frac{1.575}{2.575}\right) = 61\%$$

(alternatively consult Table 4 for $Y = 0.454$)

dicted probability either by using the above equation or by looking up a table.

An example of the use of a logistic regression equation for predicting postoperative respiratory complications in elderly, general surgical patients (based on the data set referred to in Table 1) is shown in Table 3.

Multiple linear regression and logistic regression equations are becoming ever more common in medical publications and useful insights into these (and other statistical topics related to medicine) are provided by Matthews and Farewell (1988). More detailed accounts of regression methods, written primarily for the non-statistician, are to be found in Edwards (1985) and Weisberg (1985). Table 4 shows how Y scores are converted to probabilities.

Table 4 Conversion of Y scores (logistic regression) to probabilities

Y value	Probability level (%)	Y value	Probability level (%)
−4.50	1.1	0.00	50.0
−4.00	1.8	0.20	55.0
−3.80	2.2	0.40	59.9
−3.60	2.7	0.60	64.6
−3.40	3.2	0.80	69.0
−3.20	3.9	1.00	73.1
−3.00	4.7	1.20	76.9
−2.80	5.7	1.40	80.2
−2.60	6.9	1.60	83.2
−2.40	8.3	1.80	85.8
−2.20	10.0	2.00	88.1
−2.00	11.9	2.20	90.0
−1.80	14.2	2.40	91.7
−1.60	16.8	2.60	93.1
−1.40	19.8	2.80	94.3
−1.20	23.1	3.00	95.3
−1.00	26.9	3.20	96.1
−0.80	31.0	3.40	96.8
−0.60	35.4	3.60	97.3
−0.40	40.1	3.80	97.8
−0.20	45.0	4.00	98.2
0.00	50.0	4.50	98.9

THE CONCEPT OF TRAINING AND TEST DATA SETS

A decision aid, whether it is in the form of a complex predictive equation or an apparently sophisticated computer system, may evoke a variety of responses in the medical reader. Two extreme responses are blind acceptance ('if it's complicated it must be correct') and total rejection ('if it involves mathematics and/or computers, then it's too much trouble and irrelevant to day-to-day medicine'). Rejection of (or indifference to) decision aids appears to be a more common reaction than blind acceptance, and part of the problem has been the poor presentation of some of the systems. Thus, when trouble has been taken to make systems 'user-friendly' and to provide facilities whereby the reasoning behind a system can be questioned, medical acceptance tends to be much higher (Spiegelhalter and Knill-Jones 1984; Adams *et al.* 1986).

A more fundamental question that needs to be asked of

any decision aid (or indeed of any new piece of complicated equipment) is 'is it likely to work in my own clinical practice'. A system that is not robust enough to produce accurate predictions in everyday clinical circumstances will be of no practical use. While the ultimate answer to this question can only be provided by a series of 'field trials' (such as those that recently showed the value of the de Dombal system for diagnosing the acute abdomen (Adams *et al.* 1986)), important insights can be gained by establishing whether the predictive system or decision aid was initially evaluated in both 'training' and 'test' data sets.

A training data set is that set of data from which the predictive system was initially derived. The system is essentially a model or summary of the training data set, and will thus have incorporated any 'unusual' features in the data that might have arisen from random variation or peculiarities of the sample population. The key test of a predictive system or decision aid is therefore not how well it fits the training data set from which it was derived, but how it performs in another 'test' data set (Wasson *et al.* 1985). It will be appreciated that performance in a test data set is unlikely to be as good as it was in the training data set, but a useful degree of predictive ability still needs to be demonstrated. In practice, it is best if several test data sets are used, some of which should ideally involve investigators and patients far removed from the setting of the original system. A study that simply reports the performance of a decision aid in a training data set is incomplete and the results can be regarded only as provisional (Seymour *et al.* 1990*b*).

The concept that test data sets should be used to validate predictive systems and decision aids is an important one that embraces systems of all complexity, from the humble sensitivity calculations of a 2×2 table, to the most sophisticated statistical or Artificial Intelligence systems. It should never be forgotten that such systems are ultimately, in the phrase of de Dombal 'just another test', and that the final decisions should be made by the clinician and the patient (de Dombal 1985, 1987) in the light of all the available evidence.

References

Adams I.D. *et al.* (1986). Computer aided diagnosis of acute abdominal pain: a multicentre study. *British Medical Journal*, **293**, 800–4.

Armitage, P. and Berry, G. (1987). *Statistical methods in medical research*, (2nd edn.). Blackwell Scientific, Oxford.

Barlow, P., Murray, G.D., and Teasdale, G. (1987). Outcome after severe head injury: the Glasgow model. In *Medical applications of microcomputers* (ed. W.A. Corbett), pp. 105–26. Wiley, New York.

Card, W.I. and Good, I J. (1974). A logical analysis of medicine. In *A companion to medical studies*, Vol. III (ed. R. Passmore and J. S. Robson), pp. 60.1–60.3. Blackwell Scientific, Oxford.

Chard, T. (1988). *Computing for clinicians*. Elmore-Chard, London.

Chytil, M.K., Engelbracht, R. (ed.) (1987). *Medical expert systems*. Sigma Press, Wilmslow.

de Dombal, F.T., (1985). Analysis of symptoms in the acute abdomen. *Clinics in Gastroenterology*, **14**, 531–43.

de Dombal, F.T. (1987). Ethical considerations concerning computers in medicine in the 1980s. *Journal of Medical Ethics*, **13**, 179–84.

Edwards, A.L. (1985). *Multiple regression and the analysis of variance and covariance*, (2nd edn). Freeman, New York.

Fagan, T.J. (1975). Nomogram for Bayes's theorem. *New England Journal of Medicine*, **293**, 257.

Feinstein, A.R. (1983). An additional basic science for clinical medicine. I. The constraining fundamental paradigms. *Annals of Internal Medicine*, **99**, 393–7.

Fleiss, J.L. (1981). *Statistical methods for rates and proportions*, (2nd edn). Wiley, New York.

Gale, W. A. (ed.) (1986). *Artificial intelligence and statistics*. Addison-Wesley, Reading, MA.

Harris, J.M. (1981). The hazards of bedside Bayes. *Journal of the American Medical Association*, **246**, 2602–5.

Knoebel, S.B. (1986). *Perspectives on clinical decision making*. Futura, New York.

Ledley, R.S. and Lusted, L.B. (1959). Reasoning foundations of medical diagnosis. *Science*, **130**, 9–21.

Matthews, D.E. and Farewell, V.T. (1988). *Using and understanding medical statistics*, (2nd edn). Karger, Basel.

O'Connor, G.T., and Sox, H.C. (1991). Bayesian reasoning in medicine: the contribution of Lee B. Lusted, M.D. *Medical Decision Making*, **11**, 107–11.

Peters, T.J. and Newcombe, R.S. (1991). Risk assessment in obstetrics; methodological and statistical issues. In *Reproductive and perinatal epidemiology* M. Kiely, (ed.) pp. 469–89. CRC Press, Boca Raton, FA.

Sackett, D.L., Haynes, R.B., and Tugwell, P. (1985). *Clinical epidemiology*. Little, Brown and Co., Boston.

Seymour, D.G. and Pringle, R. (1983). Post-operative complications in the elderly surgical patient. *Gerontology*, **29**, 262–70.

Seymour, D.G., Green, M., and Vaz, F.G. (1990*a*) Making better decisions: the construction of clinical scoring systems using the Spiegelhalter and Knill-Jones approach. *British Medical Journal*, **300**, 223–6.

Seymour, D.G., Green, M., Vaz, F.G., and Coles, E.C. (1990*b*). Risk prediction in medicine and surgery: ethical and practical considerations. *Journal of the Royal College of Physicians of London*, **24**, 173–7.

Simel, D.L. (1985). Playing the odds. *Lancet*, **i**, 329–30.

Sox, H.C., Blatt, M.A., Higgins, M.C., and Marten, K.I. (1988). *Medical Decision Making*. Butterworths, Boston.

Spiegelhalter, D.J. (1985). Statistical methodology for evaluating gastrointestinal symptoms. *Clinics in Gastroenterology*, **14**, 490–515.

Spiegelhalter, D.J. and Knill-Jones, R.P. (1984). Statistical and knowledge-based approaches to clinical decision-support systems, with an application in gastroenterology. *Journal of the Royal Statistical Society*, Series A **147**, 35–77.

Wagner, H.N. (1982). Bayes' theorem: an idea whose time has come? *American Journal of Cardiology*, **49**, 875–7.

Wasson, J.H., Sox, H.C., Neff, R.K., and Goldman, L. (1985). Clinical prediction rules. Applications and methodological standards. *New England Journal of Medicine*, **313**, 793–9.

Weinstein, M.C. and Fineberg, H.V. (1981). *Clinical decision analysis*. Saunders, Philadelphia.

Weisberg, S. (1985). *Applied linear regression*, (2nd edn). Wiley, New York.

Wulff, H.R. (1981). *Rational diagnosis and treatment: an introduction to clinical decision-making*, (2nd edn). Blackwell Scientific, Oxford.

21.2 Perioperative assessment and management in older patients

DIANE G. SNUSTAD AND RICHARD W. LINDSAY

It is estimated that one-half of all older persons will undergo surgery at some time during the remainder of their lives. Numbers of operations undertaken have been rising as the bias against operating on older patients has been slowly declining over the years. However, there is still evidence that some bias remains (Samet *et al.* 1986; Greenfield *et al.* 1987). There may be several reasons for this. First, many underestimate the expected lifespan of older patients, not realizing that the average 70-year-old man can expect to live to 81 years, and a woman aged 70 can expect to reach 84. Secondly, studies in the past have often implicated age as a risk factor for surgical fatality. However, these studies often failed to take into account the presence of comorbid conditions and the frequency of emergency operations, both of which are more prevalent in older patients and would be expected to increase fatality. Diminished homeostatic mechanisms and altered physiology may increase risk of any illness, but better monitoring and perioperative care are decreasing the excess morbidity and fatality associated with surgery at later ages. Carefully selected and monitored, even very elderly patients can face acceptable levels of risk of morbidity and fatality from surgery. Recent data indicate a fatality rate of 0.7 per cent within 48 h of elective surgery in those older than 90 years (Hosking *et al.* 1989).

This chapter will review preoperative assessment and care in older patients, as well as some aspects of perioperative and postoperative care, with special attention to some of the physiological changes of ageing that can affect management. It is primarily aimed at the medical practitioner in his or her role as consultant throughout the patient's time in hospital.

PREOPERATIVE ASSESSMENT

The role of the medical consultant who is asked to evaluate an elderly surgical patient should involve both preoperative assessment and assistance in postoperative care. The consultant should identify and assess the magnitude of any risk factors, as well as establish a baseline with which to compare any postoperative changes. He or she needs to establish a plan to decrease these risks, and to discuss the risks and prognosis with the patient, family, and surgeon. Postoperatively, the consultant can assist in the early identification of any complications, and aid in their management.

Preoperative assessment includes a history and physical examination, as well as certain laboratory investigations. In the history, special attention should be paid to any pulmonary or cardiac symptoms, such as dyspnoea, cough, sputum, chest pain, or syncope. One must also enquire about possible atypical symptoms of myocardial ischaemia or infarction such as dizziness or nausea, as well as atypical presentations of congestive heart failure, such as weakness or headache. A history of anorexia or recent weight loss can suggest the need for nutritional evaluation and possible supplementation before elective surgery.

A careful drug history, including non-prescribed medications and any allergies, should be taken, and degree of compliance should be ascertained. Not uncommonly, patients who may be only partially compliant with their medications at home develop complications when given full doses in hospital. Alcohol abuse is often a hidden problem, so relatives as well as the patient should be questioned about alcohol consumption. A smoking history is important in predicting pulmonary disease.

A complete functional assessment is invaluable for several reasons. Present functional status and potential changes anticipated as a result of surgery should be considered both in making the decision to operate, and in deciding how to manage the patient postoperatively. One needs to consider if the proposed surgery may cure the disease but leave the patient with diminished functional status and quality of life. Cognitive impairment and emotional problems such as depression may adversely affect the patient's willingness and ability to give consent or be cooperative with the procedures necessary for the operation and in the recovery period. A preoperative test of mental status is essential to help predict these problems as well as to serve as a baseline against which to judge any postoperative changes. It can also help predict the risk of delirium, as patients with dementia are more likely to develop this complication in the postoperative period. Testing mental status {for example, using the Mini-Mental State Examination (MMSE; Folstein *et al.* 1975)} often detects poor cognitive function even in patients who do not appear to be demented, and so should be done on most elderly patients preoperatively. Information from the family regarding any recent decline in cognitive function is also very valuable.

Social factors should be considered, such as where the patient will convalesce and what supports are available, factors that may influence how long the patient needs to stay in the hospital. If home supports seem inadequate for the convalescence, then early assistance from social workers can help in making arrangements for additional support in the home or temporary placement in a nursing home or other facility.

In the physical examination, particular attention should be paid to hydration and nutritional status. Pulmonary examination is especially important, as any abnormal findings should prompt consideration of pulmonary function tests. A careful search for signs of congestive heart failure, especially third heart sound or jugular venous distension, is essential, although one should remember that basal rales on chest examination may indicate atelectasis rather than pulmonary congestion, and that pedal oedema may be related to venous insufficiency rather than fluid overload. The assess-

Table 1 American Society of Anesthesiologists Physical Status Scale

Class	Description	Fatality within 1 month for patients over 80
I	Healthy patient	0.5 ($n = 87$)
II	Mild to moderate systemic disease	4 ($n = 256$)
III	Severe systemic disease	25 ($n = 56$)
IV	Severe systemic disease that is a threat to life	100 ($n = 1$)
V	Moribund patient unlikely to survive 24 h with or without operation	
E	Added to Class if emergency surgery	

ment of aortic stenosis, a predictor of operative risk, is often difficult without echocardiography because of the absence, on physical examination alone, of the classical findings.

Other parts of the physical examination that are especially pertinent in the older patient include a screening for hearing and visual problems, a careful skin examination for signs of impending or actual breakdown, a search for signs of Parkinson's disease or other neurological disorders that may predispose to immobility and its complications postoperatively, and a prostate examination (as well as any history of difficulty in voiding) to search for hypertrophy which may increase the risk of urinary retention or incontinence.

The value and content of preoperative laboratory testing is controversial. Preoperative blood tests should include electrolytes, glucose, blood urea nitrogen, serum creatinine and albumin, and a complete blood count. Prothrombin time and partial thromboplastin time need be ordered only if there are signs or symptoms of a bleeding diasthesis. An electrocardiogram is commonly recommended for all patients over the age of 55. The need for a preoperative chest radiograph remains the subject of much debate; it should certainly be taken in those who have pulmonary or cardiac signs or symptoms, or who are scheduled for thoracic or upper abdominal procedures. However, even among those elderly patients without a specific indication, many have unsuspected but significant findings on chest radiography and some may need a postoperative film, for which a baseline examination is invaluable. In some cases, further testing of cardiac or pulmonary function may be warranted (see below).

Surgical risk can further be assessed by using the American Society of Anaesthesiologists' physical status scale (Dripps *et al*. 1982) (Table 1). This has been shown to be well correlated with surgical fatality in the elderly patient (Djokovic and Hedley-Whyte 1979).

ASSESSMENT OF SPECIFIC ORGAN SYSTEMS AND STATES

Cardiovascular

Cardiac complications are one of the major risks to elderly patients undergoing surgery. Studies have indicated a cardiac fatality rate of between 0.5 and 11 per cent in elderly patients undergoing non-cardiac surgery. Cardiac morbidity is also high, with reports of recognized postoperative myocardial infarction of 1 to 4 per cent, and congestive heart failure in 4 to 10 per cent (Seymour 1986). Nearly half of postoperative deaths are cardiovascular.

All of the above complications have a higher rate in older than in young patients. Whether this is due to effects of age alone or to the increased incidence of cardiovascular disease is controversial. There are changes in the heart and vascular system with age that may in themselves increase risk. Although in an elderly person free of atherosclerotic disease, cardiac output is maintained at levels equivalent to those of a young person at any given level of exertion, the maximum exercise that can be accomplished is less in the older patient (Rodeheffer *et al*. 1984), and, as a group, elderly patients have, on average, a decrease in cardiac output and cardiac index. Baroreceptor responses are impaired, increasing the danger associated with hypotension. These factors make even a healthy elderly patient more susceptible to problems during and after surgery. This is most likely a minor problem, however, compared to the risk imposed by the atherosclerotic and other cardiac diseases that are present in many older surgical patients. Studies in which all patients have heart disease show no difference in the rate of cardiac complications between young and old.

The stresses imposed on the heart by surgery are multiple. Cardiac output is increased, elevating myocardial oxygen consumption. Anaesthetic agents can depress myocardial function, increase myocardial irritability, and cause vasodilatation. Intraoperative hypotension and hypertension can both occur, especially in hypertensive patients. The cardiac risks extend beyond the actual time of surgery, with the risk of congestive heart failure from mobilization of fluids given intraoperatively lasting at least 2 days postoperatively, and the risk of myocardial infarction being present for 5 or 6 days. The fact that up to a quarter of these infarctions are silent suggests the need for close monitoring with serial electrocardiography or cardiac isoenzymes.

Several methods have been proposed for assessing cardiac risk preoperatively. The Goldman cardiac risk index (Table 2) is the most commonly used. This index was developed for use in patients over the age of 40, and assigns a point value to various risk factors. The total number of points is then used to predict risk of cardiac morbidity and mortality (Table 3). More recent work by Goldman indicates that premature ventricular contractions in a healthy heart do not increase a patient's risk, and that elective abdominal aortic surgery actually carries a 40 per cent greater cardiac risk than indicated in the original index (Weitz and Goldman 1987). The estimate of the operative risk in patients with recent myocardial infarcts has also been revised. A patient with an uncomplicated infarction has been shown to have less risk of morbidity and fatality if surgery is undertaken within 6 months than would a patient with a complicated infarction. The risk of surgery can also be lessened by improving the cardiac status as far as is possible preoperatively, by use of invasive monitoring, and by active treatment of any haemodynamic abnormality. With this approach, fatality has been reduced to 5.7 per cent in those with a history of myocardial infarction in the previous 3 months (Rao *et al.* 1983).

Table 2 Cardiac Risk Index (adapted with permission from Goldman *et al.* (1977))

Criteria	Points
Historical	
Age over 70 years	5
Myocardial infarction in previous 6 months	10
Examination	
S3 gallop/jugular venous distension	11
Significant aortic valvular stenosis	3
ECG	
Premature atrial contractions or rhythm other than sinus	7
More than 5 premature ventricular contractions per minute	7
General status	3
P_{O_2} <60 or P_{CO_2} >50 mmHg (P_{O_2} <8 or P_{CO_2} >6.7 kPa) or K<3.0 or HCO_3 <20 mEq/l or BUN >50 or creatinine >3.0 mg/100 ml or liver disease or bedridden from non-cardiac causes	
Operation	
Emergency	1
Intraperitoneal; thoracic; aortic	3
Total possible	53

Table 3 Correlation of cardiac risk points and postoperative cardiac problems (adapted with permission from Goldman *et al.* (1977))

Point total	Life-threatening complications (%)	Cardiac deaths (%)
0–5	0.7	0.2
6–12	5.0	2.0
13–25	11.0	2.0
Greater than 26	22.0	56.0

Several other methods to assess cardiac risk in the elderly patient have been proposed. In a study comparing the usefulness of the American Society of Anaesthesiologists' class, the Goldman cardiac risk index, and rest and exercise radionuclide ventriculograms, the most sensitive indicator of cardiac risk was the inability to do 2 minutes' bicycle exercise in the supine position to a heart rate above 99 beats/min (Gerson *et al.* 1985). Those who were unable to do this had an 11-fold increase in perioperative cardiac complications. In patients undergoing vascular surgery, a 24-h electrocardiogram monitoring (Raby *et al.* 1990) and dipyridamole-thallium scanning (Boucher *et al.* 1985) have both been shown to be useful techniques to assess cardiac risk.

Other studies have shown additional possible cardiac risk factors, including a myocardial infarction occurring more than 6 months before surgery, lack of physical activity, pre-existing lung disease, and multiple surgical procedures being done at the same time. Angina is thought to be a significant risk factor only if unstable or of New York Heart Association Class III or IV, after controlling for other criteria in the Goldman index. Mild to moderate hypertension with a diastolic blood pressure less than 110 mmHg by itself is not a major risk factor (Goldman and Caldera 1979). Hypertension does, however, increase the risk of both hypertension and hypotension during surgery, each occurring in a quarter of hypertensive patients whether or not the hypertension is adequately treated preoperatively. Other studies indicate that intraoperative lability of blood pressure is decreased when hypertension is under good control preoperatively (Prys-Roberts *et al.* 1971). Any antihypertensive that the patient had been taking, with the possible exception of diuretics, should be continued up until the morning of surgery, and resumed immediately postoperatively.

The need for invasive monitoring of the elderly surgical patient has been much debated. Del Guerico and Cohn (1980) found that only 13.5 per cent of older patients cleared for surgery had normal results on invasive monitoring; many had abnormalities that affected preoperative care, and 23 per cent had findings that made them unacceptable risks for surgery. Because of this, they suggested that all elderly patients should undergo Swan–Ganz catheterization before surgery. The results of this invasive monitoring were not compared with a non-invasive evaluation of cardiac risk, other than the Dripps' classifications. Many physicians prefer to use this technique only for patients found to be at high risk by other methods or in those with obvious haemodynamic alterations (Tuman *et al.* 1989).

The risk that asymptomatic stenosis of a carotid artery will lead to an intraoperative stroke is a matter of some controversy. Some studies do not show an increased incidence of cerebrovascular accidents in the asymptomatic patient with a carotid bruit undergoing elective surgery, while others do, particularly during cardiac surgery. Current guidelines do not recommend further testing in asymptomatic patients regardless of the type of surgery planned (Feussner and Matchar 1988). However, patients with symptoms of transient ischaemic attacks should be fully evaluated before surgery.

Pulmonary

Respiratory complications cause between one-sixth and one-third of the deaths in elderly surgical patients; 12 to 46 per cent of older surgical patients suffer some postoperative pulmonary problem (Seymour 1986). As with cardiac risk, pulmonary risk is due to a combination of normal changes of ageing and the consequence of pre-existing diseases, with the latter being the most significant.

Those changes in pulmonary function that are common with age and that might increase surgical risk include a decrease in most static lung volumes, flow rate, pulmonary clearance mechanisms, compliance and P_{O_2}, and increased closing volume. These changes tend to promote atelectasis and decrease protective mechanisms. Elderly patients also have decreased sensitivity of the respiratory centre and a blunted response to hypercapnia and hypoxia. Impaired laryngeal reflexes contribute to the risk of aspiration pneumonia.

Abdominal (especially upper abdominal) and thoracic operations may cause further decreases in vital capacity, forced expiratory volume in 1 s ($FEV_{1.0}$), functional residual capacity, and protective reflexes, increasing the risk of respiratory complications. The incidence of these abnormalities is maximal on the first and second days after the surgery but can remain high for as long as 14 to 21 days. General anaesthesia also affects respiratory function by impairing oxygenation and elimination of CO_2, and by decreasing functional residual capacity, laryngeal reflexes, and hypoxic drive.

In general, elderly patients have the same risk factors for postoperative respiratory complications as do younger patients. The major factors are lung disease, smoking, and surgery on the thorax or upper abdomen. Other factors that may predict problems include obesity, decreased albumin, and longer operative procedures. Age is sometimes listed as an independent risk factor, but others have found that this is no longer significant when controlled for site and duration of surgery.

Preoperative assessment of pulmonary risk begins with the history and physical examination. A history of smoking or previous lung disease doubles the risk of respiratory complications. Such a history or the presence of abnormal findings on the lung examination should prompt further evaluation with pulmonary function tests. Preoperative pulmonary function tests have been found to predict respiratory complications in those undergoing lung resection and may be helpful in coronary artery bypass patients. Although some studies have indicated that pulmonary function tests are useful prior to upper abdominal procedures, this has not been firmly established. In other types of surgery the role of pulmonary function tests in non-smoking asymptomatic patients remains unclear (Zibrak 1990). When pulmonary function tests are indicated, simple spirometry, measurement of maximal ventilatory volume, and arterial blood gases constitute a reasonable assessment. An FEV_1 of less than 1 l, a maximal ventilatory volume below 50 per cent of predicted, or a $P\text{CO}_2$ above 45 mmHg indicate a significant risk.

Several steps should be taken to improve lung function before surgery in those at risk. Primary among these is the cessation of smoking for at least 2 months prior to surgery, although short periods of abstinence may be sufficient to improve cardiovascular parameters in patients with obstructive lung disease. Bronchodilators can be used in patients with obstructive lung disease, and any infection should be treated appropriately. Incentive spirometry, which promotes maximal inspiration, will help prevent postoperative atelectasis, and patients should be instructed in its use preoperatively.

After the operation, the patient should be encouraged to use the incentive spirometer frequently. Early mobilization should be a goal, as the supine position worsens the decrease in lung volumes that predispose to atelectasis and infection. In high-risk patients, the danger of adult respiratory distress syndrome can be lessened by use of a ventilator with continuous positive airways pressure or positive end-expiratory pressure for 24 h postoperatively.

Renal

Renal function in older patients is extremely variable, with some showing normal function as they age, and others having significant decline in creatinine clearance. It is important to remember that because of a concomitant decline in muscle mass, the serum creatinine is often not a good reflection of renal function, and nomograms or, ideally, determination of 24-h creatinine clearance are necessary. Fluid and electrolyte homeostasis may also be impaired, necessitating close attention to these aspects throughout the hospital stay. This is especially important because volume depletion, as well as overload, can significantly increase surgical risk. Older patients are more susceptible to urinary tract infections, especially if indwelling catheters are used. They are also at higher risk for the nephrotoxic effects of drugs.

Hepatic

Hepatic function is also variably affected in older persons. The most consistent change is a decrease in the mixed-function oxidase capacity, changing the metabolism of some medications, most notably some of the benzodiazepines such as diazepam and chlordiazepoxide. This change is not reflected in liver function tests, but should be assumed to be present, and drug dosages and choices should be adjusted accordingly. Many anaesthetic agents are metabolized and excreted by the liver, and can themselves affect hepatic function by decreasing hepatic blood flow.

The immune system—antimicrobial prophylaxis

Impaired immunological function with age causes the elderly patient to be more susceptible to infective complications from surgery. Antimicrobial prophylaxis has been shown to decrease the incidence of postoperative infection in a variety of surgical procedures.

One of the major uses of prophylactic antibiotics is in the prevention of endocarditis. The incidence of infective endocarditis in elderly people has been rising. The primary predisposing conditions include prosthetic heart valves, valvular degeneration, and mitral valve prolapse. Indications for prophylaxis and the antibiotics used are the same as in the younger patient (Tables 4, 5, 6, 7).

A variety of other procedures require prophylaxis against wound infection and sepsis, even in the absence of a cardiac abnormality (Kroenke 1987). Any elderly patient undergoing cholecystectomy should receive antibiotics as they are likely to have bacterial contamination. Patients undergoing other gastrointestinal procedures such as appendicectomy and colonic surgery also should be given antibiotics. Orthopaedic procedures that introduce a prosthesis or involve an open fracture are also associated with risk of infection, as are vascular procedures and hysterectomy. Urological surgery in the absence of infected urine does not usually require prophylaxis. {The use of prophylactic antibiotics and the agents to be used should ideally be subject to regular review as part of local policy for control of hospital infection.}

Nutrition

The state of nutrition is an important consideration both before and after surgery. Studies of elderly patients in hospital have indicated between 17 and 65 per cent are malnourished (Russell *et al.* 1986). Hypoalbuminaemia in the elderly surgical patient has been shown to double the risk of sepsis and increase fatality 10 fold (Seymour 1986). Wound healing is also impaired in these patients, and the risk of postoperative infections and decubitus ulcers increased. A weight of

Table 4 Dental or surgical procedures*

Endocarditis prophylaxis recommended
Dental procedures known to induce gingival or mucosal bleeding, including professional cleaning
Tonsillectomy and/or adenoidectomy
Surgical operations that involve intestinal or respiratory mucosa
Bronchoscopy with a rigid bronchoscope
Sclerotherapy for oesophageal varices
Oesophageal dilatation
Gallbladder surgery
Cystoscopy
Urethral dilatation
Urethral catheterization if urinary tract infection is present†
Urinary tract surgery if urinary tract infection is present†
Prostatic surgery
Incision and drainage of infected tissue†
Vaginal hysterectomy
Vaginal delivery in the presence of infection†
Endocarditis prophylaxis not recommended‡
Dental procedures not likely to induce gingival bleeding, such as simple adjustment of orthodontic appliances or fillings above the gum line
Injection of local intraoral anaesthetic (except intraligamentary injections)
Shedding of primary teeth
Tympanostomy tube insertion
Endotracheal intubation
Bronchoscopy with a flexible bronchoscope, with or without biopsy
Cardiac catheterization
Endoscopy with or without gastrointestinal biopsy
Caesarean section
In the absence of infection for urethral catheterization, dilatation and curettage, uncomplicated vaginal delivery, therapeutic abortion, sterilization procedures, or insertion or removal of intrauterine devices

*This table lists selected procedures but is not meant to be all-inclusive.
†In addition to prophylactic regimen for genitourinary procedures, antibiotic therapy should be directed against the most likely bacterial pathogen.
‡In patients who have prosthetic heart valves, a previous history of endocarditis, or surgically constructed systemic-pulmonary shunts or conduits, physicians may choose to administer prophylactic antibiotics even for low-risk procedures that involve the lower respiratory, genitourinary, or gastrointestinal tracts.

Table 5 Recommended standard prophylactic regimen for dental, oral, or upper respiratory tract procedures in patients who are at risk*

Drug	Dosing regimen
	Standard regimen
Amoxicillin	3.0 g orally 1 h before procedure; then 1.5 g 6 h after initial dose
	Amoxicillin/penicillin-allergic patients
Erythromycin or	Erythromycin ethylsuccinate, 800 mg, or erythromycin stearate, 1.0 g, orally 2 h before procedure; then half the dose 6 h after initial dose
Clindamycin	300 mg orally 1 h before procedure and 150 mg 6 h after initial dose

*Includes those with prosthetic heart valves and other high-risk patients.

Table 6 Alternate prophylactic regimens for dental, oral, or upper respiratory tract procedures in patients who are at risk

Drug	Dosing regimen
Patients unable to take oral medications	
Ampicillin	Intravenous or intramuscular administration of ampicillin, 2.0 g, 30 min before procedure; then intravenous or intramuscular administration of ampicillin, 1.0 g, or oral administration of amoxicillin, 1.5 g, 6 h after initial dose
Ampicillin/amoxicillin/penicillin-allergic patients unable to take oral medications	
Clindamycin	Intravenous administration of 300 mg 30 min before procedure and an intravenous or oral administration of 150 mg 6 h after initial dose
Patients considered high risk and not candidates for standard regimen	
Ampicillin, gentamicin, and amoxicillin	Intravenous or intramuscular administration of ampicillin, 2.0 g, plus gentamicin, 1.5 mg/kg (not to exceed 80 mg), 30 min before procedure; followed by amoxicillin, 1.5 g, orally 6 h after initial dose; alternatively, the parenteral regimen may be repeated 8 h after initial dose
Ampicillin/amoxicillin/penicillin-allergic patients considered high risk	
Vancomycin	Intravenous administration of 1.0 g over 1 h, starting 1 h before procedure; no repeated dose necessary

Table 7 Regimens for genitourinary/gastrointestinal procedures

Drug	Dosage regimen
Standard regimen	
Ampicillin, gentamicin, and amoxicillin	Intravenous or intramuscular administration of ampicillin, 2.0 g, plus gentamicin, 1.5 mg/kg (not to exceed 80 mg), 30 min before procedure; followed by amoxicillin, 1.5 g, orally 6 h after initial dose: alternatively, the parenteral regimen may be repeated once 8 h after initial dose
Ampicillin/amoxicillin/penicillin-allergic patient regimen	
Vancomycin and gentamicin	Intravenous administration of vancomycin, 1.0 g, over 1 h plus intravenous or intramuscular administration of gentamicin, 1.5 mg/kg (not to exceed 80 mg), 1 h before procedure; may be repeated once 8 h after initial dose
Alternate low-risk patient regimen	
Amoxicillin	3.0 g orally 1 h before procedure; then 1.5 g 6 h after initial dose

less than 80 per cent of the ideal for the patient's height, decreased albumin and transferrin, anergy to skin testing, and reduced triceps skin-fold thickness are indications of malnutrition and predictors of poor surgical outcome in studies done in younger patients. All of these variables may be affected by age and concomitant medical conditions, so malnutrition may be more difficult to diagnose accurately in the elderly patient. Despite this, the above indicators combined with a careful dietary history and clinical judgement can serve to alert one to the possibility of malnutrition.

If a patient is thought to be malnourished, and the surgery can be postponed, nutritional supplements can be provided either enterally or parenterally, depending on the function of the patient's gastrointestinal tract. Although intuitively it would seem that improving a patient's nutritional status would decrease postoperative complications, studies of this question have produced mixed results (Cooper 1987). The costs of delaying surgery, both in monetary terms and in terms of the possible progression of, or further complications from, the disease must also be weighed. In general, however, when time permits, raising the nutritional status to its optimum is advised, especially in those elderly patients who are severely malnourished.

Nutrition is also an important concern postoperatively and supplementation needs to be considered early in the patient who is not meeting his or her needs owing to anorexia, medical complications, or the inability to take food by mouth.

INTRAOPERATIVE CONCERNS

The choice of specific anaesthetic agents for the geriatric patient should be made by the anaesthetist. The medical consultant and surgeon may, however, assist in the decision to use local, regional, or general anaesthesia. The advantages and disadvantages must be weighed in each individual case. Local anaesthesia is generally safe, and may be used in procedures such as cataract or dental/oral surgery, or surgery involving the extremities. Regional anaesthesia is often used for hernia repair, hip surgery, vaginal hysterectomies, and transurethral resection of the prostate. Despite common belief, there is no evidence of a decrease in respiratory complications with regional as opposed to general anaesthesia. Regional anaesthesia can also be associated with significant haemodynamic effects, including hypotension and decreased cardiac output, which may be less controllable than with general anaesthesia. Studies comparing postoperative mental changes between patients after regional or general anaesthesia vary in their conclusions; some have found less confusion after regional anaesthesia, and others have shown no difference.

General anaesthesia should be used in major abdominal and thoracic procedures, when control of the airway is essential and when the patient cannot comply with instructions. Most general anaesthetics will depress myocardial function to a greater or lesser degree, with variable effects on heart rate and the peripheral vasculature. Some of these effects are influenced by medications, such as β-blockers and anti-arrhythmic agents, and the anaesthetist needs to be informed of all drugs the patient has been taking before surgery.

The effects of general anaesthesia may be prolonged if the patient becomes hypothermic in the operating room. This is a dangerous occurrence in elderly patients, as they are more susceptible both to the development of hypothermia and its consequences. Anaesthesetized patients of any age become poikilothermic and at risk for hypothermia in a cold operating room. The risk is heightened with increasing age as there is an associated decrease in basal metabolic rate, muscle mass, and vasomotor response to cold, all of which predispose to the development of hypothermia. Other risk factors for hypothermia include operations longer than 3 h, exposure of the major body cavities, or major vascular surgery. The hypothermic patient is more susceptible to cardiac arrhythmias, congestive heart failure, and pneumonia. Peripheral vasoconstriction while hypothermic can cause hypertension in the operating room, and the vasodilatation that occurs when the patient is rewarmed can cause significant hypotension. Shivering as the patient awakens from anaesthesia can also be dangerous, as it dramatically increases oxygen demand. Hypothermia can be avoided by careful monitoring of core temperatures, warmer operating rooms, warming blankets, and intravenous fluids given at body temperature.

Another risk to the elderly patient is development of decubitus ulcers while lying on a hard operating table for a long time without movement. This risk is particularly high in the presence of malnutrition, anaemia, and vascular disease. Proper padding of the table and intermittent relief of pressure on bony prominences can avoid this problem.

POSTOPERATIVE CONCERNS

Pain control

Adequate pain control should be a goal both for the comfort of the patient, and to prevent postoperative complications by facilitating early mobilization, coughing, and deep breathing. Assessing the need for medications against pain may be difficult in demented or delirious patients, as they may not remember or know how to ask for medication. Pain in these patients may be manifested by agitation or increased confusion rather than by complaints of discomfort. On the other hand, overmedication can also be manifested by increased confusion, and the physician and nurses must use judgment to determine which is actually the problem in the individual case.

There are many analgesics available for use in postoperative pain. Narcotics are very effective, but their prolonged half-lives in the elderly patient can lead to accumulation, and constipation and nausea are common side-effects. They should initially be given regularly rather than on demand as this will improve pain relief and reduce the total dose needed. The same results can be obtained in those who are mentally intact by patient-controlled analgesia, which allows them to administer their own medication through a bedside pump and has been shown to be useful even in the frail elderly patient (Egbert 1990). Epidural analgesia can be very effective, with fewer systemic effects. For less severe pain, non-steroidal anti-inflammatory agents either parenterally or orally or paracetamol (acetaminophen) can be helpful. Another option is a transcutaneous electrical stimulation unit, which has been used successfully on elderly patients and avoids the many untoward side-effects of medication.

Prophylaxis against venous thrombosis

Patients of any age undergoing surgery, particularly orthopaedic cases, are at risk of developing deep venous thrombosis. Fibrinogen scanning has detected thrombosis in the legs in a quarter of general surgical patients with 1.6 per cent developing clinically significant pulmonary embolism (Consensus Conference 1986). The risk is highest in hip surgery and knee reconstruction, with 45 to 70 per cent of these patients developing deep vein thrombosis, and 20 per cent of patients for hip surgery developing a pulmonary embolism. The incidence of both these conditions increases with age. Predisposing medical conditions include heart failure, acute myocardial infarction, cancer, obesity, and stroke with weakness of the legs.

There are various ways of preventing deep venous thrombosis and pulmonary embolism; their use should be routine in elderly patients in whom there are no contraindications. The incidence of both conditions in the general surgical patient has been significantly reduced by the use of low-dose subcutaneous heparin. Giving heparin 2 h before surgery and every 8 to 12 h after cuts the incidence of deep venous thrombosis to 10 per cent and of embolism to 0.8 per cent (Consensus Conference 1986). Heparin-induced thrombocytopenia is a rare complication, and there is an increased risk of post-operative haematoma, but in most patients, this regimen does not significantly increase serious bleeding complications.

Other regimens used to prevent venous thrombosis include dihydroergotamine/heparin combinations, and adjusted-dose heparin therapy. Dihydroergotamine is an α-adrenergic blocking agent that stimulates the vascular smooth muscle. Adding this agent to heparin augments the prophylactic effect, but the vasoconstriction it induces may cause complications in those patients with vascular disease. In patients undergoing hip surgery, adjusted-dose heparin has been shown to be superior to the standard heparin regimen. Adjusting the dose of heparin to keep the activated partial thromboplastic time between 31.5 and 36 s decreased the incidence of proximal deep-vein thrombosis from 32 per cent in those treated with fixed-dose heparin to 5 per cent (Leyvraz *et al.* 1983).

Warfarin can also be used, beginning 1 to 2 days before surgery and continuing until the patient is fully able to walk. In elective surgery, two-step warfarin therapy is preferred by some. In this treatment, the patient's prothrombin time is maintained at 2 to 3 s longer than the control for 10 to 14 days before surgery, and at 1.5 times the control after surgery.

Non-pharmacological methods of prophylaxis are also available. Properly fitted, graduated, elastic compression stockings improve venous return and can be used in combination with other therapies. The use of external pneumatic-compression stockings is a newer technique. It should be begun before surgery and continued at least 3 days or until the patient is fully ambulatory. This may be the preferred approach in patients who are at significant risk from bleeding with anticoagulation.

Delirium

Delirium is a common complication in surgical patients, occurring in 10 to 15 per cent of all ages, and often in older patients (Gamino *et al.* 1985). Studies of elderly patients with hip fractures have shown that up to 61 per cent develop post-operative confusion (Gustafson *et al.* 1988). The occurrence of delirium is associated with prolonged stays in hospital, more need for long-term care after discharge, a decreased ability to walk, falls, and a higher incidence of postoperative complications and death. Besides age, risk factors include dementia, depression, Parkinson's disease, impaired hearing or vision, anticholinergic drugs, emergency surgery, previous psychiatric conditions, and medical problems. A knowledge of preoperative mental status is important in order to detect subtle changes that may indicate the presence of delirium.

The causes of delirium are multiple (see Chapter 18.5). It can often be the presenting manifestation of medical illness such as pneumonia, urinary tract infection, or myocardial infarction. It can also be caused by metabolic abnormalities, sensory deprivation or overload, and a variety of medications, especially those with anticholinergic properties. Withdrawal from alcohol or psychoactive drugs can also be a cause.

The best approach to the problem of delirium is to try to prevent it. Keep the patient oriented by verbal and visual cues, by ensuring they have their glasses and or hearing aids, and by allowing them some familiar objects from home. Medications should be kept to a minimum and their side-effects carefully monitored. Finally, good medical care and constant vigilance will minimize the medical complications that can cause delirium.

If delirium does occur, management consists of the orientation techniques described above and discontinuation or a decrease in the dosage of any medications that might be implicated. If the patient is agitated, violent, or hallucinating, psychotropics, such as haloperidol at the lowest effective dose, can be helpful. If the agitation is related to alcohol withdrawal, it should be treated with a short-acting benzodiazepine {or chlormethiazole may be preferred in countries when it is available. Increasing numbers of cases of delirium in elderly patients are due to withdrawal of benzodiazepine hypnotics or tranquillizers}.

Other postoperative concerns

There are a variety of postoperative complications that may be less common and less dramatic than events in the cardiac, pulmonary, or central nervous systems, but that can nevertheless cause considerable discomfort and inconvenience. These include urinary incontinence or retention, constipation, and decubitus ulcers. Being aware of the risk factors and instituting preventive measures may decrease the frequency with which these occur.

Urinary problems are common in elderly patients in hospital, with a prevalence of 17 per cent on the general surgical ward in one study (Sullivan and Lindsay 1984). Medication, immobility, faecal impaction, and sedation, all common factors postoperatively, can each increase the incidence of incontinence. Adverse consequences include decubitus ulcers, falls, and the insertion of indwelling catheters with their attendant risk of infection. The chances of a patient remaining continent can be enhanced by the provision of a urinal and/or bedside commode, regular toileting, and avoidance of constipation, oversedation, and unnecessary use of drugs that might induce incontinence. Indwelling catheters should not be used for the management of incontinence unless there is skin breakdown or urine output needs to be monitored closely.

Postoperative urinary retention is also a common problem. Many anaesthetic agents can induce retention by altering the effect of the autonomic nervous system on the bladder and urethra. Older men are especially at risk owing to a high prevalence of urethral narrowing from prostatic hypertrophy. Indwelling catheters are commonly used intraoperatively, and when used for 24 h after orthopaedic surgery have been shown to decrease the incidence of postoperative retention as compared to intermittent catheterization (Michelson *et al.* 1988). Unless absolutely necessary for management of fluid status, they should be removed 24 to 48 h after surgery.

Both urinary incontinence and retention can be induced by faecal impaction. Constipation is extremely common after surgery, owing to direct effects of surgery on the bowel, drugs, dehydration, poor intake of dietary fibre, immobilization, pain, anxiety, bedpans, and lack of privacy in the hospital. Early mobilization, adequate hydration, and the judicious use of pain medications and prophylactic stool softeners and laxatives can all minimize this problem.

Decubitus ulcers can also complicate the postoperative course, especially if the patient is sedated, immobilized, undernourished, incontinent of bowel or bladder, hypoxic, or anaemic. Unrelieved pressure on a bony prominence can induce skin breakdown within 2 h. Pressure sores are associated with increased hospital stays, fatality, and risk of osteomyelitis and sepsis. Attention to the above risk factors, frequent turning, and the use of special mattresses, pads, and beds in high-risk patients can all be valuable in preventing pressure sores.

Early mobilization

Many of the above complications, including pulmonary infections, thrombophlebitis, urinary problems, constipation, pressure sores, and delirium, are associated with immobility, and can be prevented or treated with early mobilization. Attempts to mobilize an elderly patient may be affected by poor preoperative functional status, arthritis, orthostatic hypotension, confusion, and pain. These should be seen not as reasons to stay in bed, but as factors that must be addressed and included in the plan for mobilization and rehabilitation. Physical and occupational therapists, as well as the nursing staff, can be invaluable in developing and instituting a rehabilitation plan, which should include an early use of usual dress, frequent visits by family and friends, and avoidance of the use of bedside rails or other restraints.

CONCLUSION

As the elderly population continues to increase, and surgical techniques and management continue to improve, the number of elderly surgical patients will continue to grow. An understanding of the physiology of ageing and an awareness of concomitant diseases, in combination with careful selection, monitoring, and preoperative and postoperative assessment and care, can assure a satisfactory outcome for the vast majority of these patients.

References

Boucher, C.A., Brewster, D.C., Darling, R.C., Okada, R.D., Strauss, H.W., and Pohost, G.M. (1985). Determination of cardiac risk by dipyridamole-thallium imaging before peripheral vascular surgery. *New England Journal of Medicine*, **312**, 389–94.

Consensus Conference (1986). Prevention of venous thrombosis and pulmonary embolism, *Journal of the American Medical Association*, **256**, 744–9.

Cooper, J.K. (1987). Does nutrition affect surgical outcome? *Journal of the American Geriatrics Society*, **35**, 229–32.

Dajani, A.S., *et al.* (1990). Prevention of bacterial endocarditis: Recommendations by the American Heart Association. *Journal of the American Medical Association*, **264**, 2919–22.

Del Guercio, L.R. and Cohn, J.D. (1980). Monitoring operative risk in the elderly. *Journal of the American Medical Association*, **243**, 1350–5.

Djokovic, J. and Hedley-Whyte, J. (1979). Prediction of outcome of surgery and anesthesia in patients over 80. *Journal of the American Medical Association*, **242**, 2301–6.

Dripps, R., Eckenhoff, J., and Vandam, L. (1982). *Introduction to Anesthesia: the principles of safe practice*, (6th edn), pp. 17–18. Saunders, Philadelphia.

Egbert, A.M., Parks, L.H., Short, L.M., and Burnett, M.L. (1990). Randomized trial of postoperative patient-controlled analgesia vs. intramuscular narcotics in frail elderly men. *Archives of Internal Medicine*, **150**, 1987–2003.

Feussner, J.R. and Matchar, D.B. (1988) When and how to study the carotid arteries. *Annals of Internal Medicine*, **109**, 805–18.

Folstein, M.F., Folstein, S.E., and McHugh, P.R. (1975). Mini-Mental State: a practical method for grading the cognitive state of patients for the clinician. *Journal of Psychiatric Research*, **12**, 189–198.

Gamino, L.A., Hunter, R.B., and Brandon, R.A. (1985). Psychiatric complications associated with geriatric surgery. *Geriatric Clinics of North America*, **1**, 417–22.

Gerson, M.C. *et al.* (1985). Cardiac prognosis in noncardiac geriatric surgery. *Annals of Internal Medicine*, **103**, 832–7.

Goldman, L. and Caldera, D.L. (1979). Risks of general anaesthesia and elective operation in the hypertensive patient. *Anaesthesiology*, **50**, 285–92.

Goldman, L. *et al.* (1977) Multifactorial index of cardiac risk in noncardiac surgical procedures. *New England Journal of Medicine*, **297**, 845–50.

Greenfield, S., Blanco, D.M., Elashoff, R.M., and Ganz, P.A. (1987). Patterns of care related to age of breast cancer patients. *Journal of the American Medical Association*, **257**, 2766–70.

Gustafson, Y. *et al.* (1988). Acute confusional states in elderly patients treated for femoral neck fracture. *Journal of the American Geriatrics Society*, **36**, 525–30.

Hosking, M.P., Warner, M.A., Lobdell, C.M., Offord, K.P., and Melton, J. (1989). Outcomes of surgery in patients 90 years of age and older. *Journal of the American Medical Association*, **261**, 1909–15.

Kroenke, K. (1987). Clinical reviews: preoperative evaluation: the assessment and management of surgical risk. *Journal of General Internal Medicine*, **2**, 257–69.

Leyvraz, P.F. *et al.* (1983). Adjusted versus fixed-dose subcutaneous heparin in the prevention of deep-vein thrombosis after total hip replacement. *New England Journal of Medicine*, **309**, 954–7.

Michelson, J.D., Lotke, P.A., and Steinberg, M.E. (1988). Urinary-bladder management after total joint-replacement surgery. *New England Journal of Medicine*, **319**, 321–6.

Prys-Roberts, C., Meloche, R., and Foex, P. (1971). Studies of anesthesia in response to hypertension. *British Journal of Anaesthesia*, **43**, 112–37.

Raby, K.E., *et al.* (1989). Correlation between preoperative ischemia and major cardiac events after peripheral vascular surgery. *New England Journal of Medicine*, **321**, 1296–1300.

Rao, T.L.K., Jacobs, K.H., and El-Etr, A.A. (1983). Reinfarction following anesthesia in patients with myocardial infarction. *Anesthesiology*, **59**, 499–505.

Rodeheffer, R.J., Gerstenblith, G., Becker, L.C., Fleg, J.L., Weisfeldt, M.L., and Lakatta, E.G. (1984). Exercise cardiac output is maintained with advancing age in healthy human subjects: cardiac dilatation and increased stroke volume compensate for a diminished heart rate. *Circulation*, **69**, 203–13.

Russell, R., Sahyoun, N., and Whinston-Perry, R (1986). Nutritional assessment. In *The practice of geriatrics*, (1st edn) (ed. E. Calkins, P. Davis, and A. Ford), p. 135. Saunders, Philadelphia. PA.

Samet, J., Hunt, W.C., Key, C., Humble, C.G. and Goodwin, J. S. (1986). Choice of cancer therapy varies with age of patient. *Journal of the American Medical Association*, **255**, 3385–90.

Seymour, D.G. (1986). *Medical assessment of the elderly surgical patient*. Aspen Publications, Rockville, MD.

Sullivan, D. and Lindsay, R. (1984). Urinary incontinence in the geriatric population of an acute care hospital. *Journal of the American Geriatrics Society*, **32**, 646–50.

Tuman, K. *et al.* (1989). Effect of pulmonary artery catheterization on outcome in patients undergoing coronary artery surgery. *Anesthesiology*, **70**, 199–206.

Weitz, H.H. and Goldman, L. (1987). Noncardiac surgery in the patient with heart disease. *Medical Clinics of North America*, **71**, 413–32.

Zibrak, J.D., O'Donnell, C.R., and Marton, K. (1990). Indications for pulmonary function testing. *Annals of Internal Medicine*, **112**, 763–71.

SECTION 22
Symptom management and palliative care

22 Symptom management and palliative care

MARY J. BAINES

'I conceive it the office of the physician not only to restore the health but to mitigate pain and dolours; and not only when such mitigation may conduce to recovery but when it may serve to make a fair and easy passage.' Francis Bacon (1561–1626)

'The nurses battled on heroically. They emerged with far greater credit than we (the doctors) who are still capable of ignoring the conditions which make muted people suffer. The dissatisfied dead cannot noise abroad the negligence they have suffered.' (Hinton 1967)

The physician, in Bacon's time, could do very little to 'restore health'; not surprisingly he concentrated on mitigating 'pains and dolours'. But, with the advent of scientific medicine and the study of diseases rather than their symptoms, the situation radically changed. The successful treatment of tuberculosis relieved cough; vitamin B_{12} prevented the lethargy and dyspnoea of pernicious anaemia.

With the increased availability of diagnostic tools and therapeutic options there was the inevitable tendency to institute active therapy even in patients with very advanced or incurable disease, often associated with extreme old age. Such endeavour could easily mean that the relief of symptoms and the support of patients and family became low priorities. It was the lack of appropriate care for the dying that resulted, in the 1960s, in a number of influential reports (Hinton 1967) and the advent of the modern hospice movement.

THE DIAGNOSIS OF DYING

'There is a general understanding that terminal care refers to the management of patients in whom the advent of death is felt to be certain and not too far off and for whom medical effort has turned away from therapy and become concentrated on the relief of symptoms and the support of both patient and family.' (Holford 1973)

The decision that active treatment, aimed at prolongation of life, is no longer appropriate must always be difficult. The common respiratory or cardiovascular diseases may have fluctuating courses with unexpected but worthwhile improvements after treatment. In such diseases the diagnosis of dying is made by exclusion, after a failure to respond to standard therapeutic and rehabilitative endeavours. Experienced nursing staff are often the first to recognize this (Blackburn 1989).

In advanced malignant disease the decision to abandon active treatment is often easier, with the expectation of a steadily deterioration. It is partly for this reason that most hospices only accept cancer patients, and most writing, including this chapter, concentrates on the palliative care of those with malignant disease. However, many cancer patients die with cardiac failure, pneumonia, uraemia, paralysis, or organic brain disease, and it is to be expected that much of the symptom control practised in hospices will have wider application.

The support for an isolated, elderly woman soon to be bereaved, the question of how to break bad news, the misplaced anger directed against staff—all these and many other aspects of palliative care are just as relevant in the patient's home, the geriatric ward, and the nursing home. Perhaps the distinction between active treatment and palliative care should not be clearly defined for they can occur together. Symptom control and family support should be practised while active treatment is being undertaken. Many patients thought to be dying are able to resume active therapy once their symptoms and mental state improve with appropriate treatment.

PALLIATIVE SYMPTOM CONTROL

Palliation is often defined as 'alleviating without curing'. In recent years, the terms palliative medicine or palliative care have been specifically applied to the relief of suffering in the last weeks or months of life. The foundation of such treatment is meticulous attention to symptom control, for without this it is difficult, perhaps impossible, to give the necessary emotional support to the patient and family. The elderly patient with terminal illness usually presents a complex clinical picture. Even if there is advanced malignant disease, there are often coexisting cardiac, respiratory, neurological, or locomotor problems. Resistance to infection may be reduced, hepatic or renal function may be impaired, and there are often serious problems with communication. It is therefore not surprising that such patients can present a multiplicity of symptoms as death approaches.

PREVALENCE OF SYMPTOMS

A small number of studies have sought to determine the prevalence of symptoms in dying patients (Table 1). The hospice series in the table consists purely of patients with malignant disease whereas the other two series refer to patients with a variety of diagnoses.

Studies from geriatric units indicate that distressing physical symptoms are considerably less common in elderly patients than in younger ones. For example, two series showed a 14 per cent and 19 per cent prevalence of pain, but breathlessness, at 24 per cent, was more common (Exton-Smith 1961; Wilson *et al.* 1987). In contrast, confusion was often present, as were communication difficulties (Blackburn 1989) and a distressing but non-specific agitation (Wilson *et al.* 1987).

DIAGNOSIS OF SYMPTOMS

Palliative treatment involves making a diagnosis of the cause or causes of each symptom. Even in patients with advanced

Table 1 Prevalence of symptoms in terminal illness (percentage of patients)

Symptom	Hospice[1]	Hospital[2]	Hospital and home[3]
Weakness	92	88	52
Anorexia	75	92	32
Pain	69	69	52
Dyspnoea	50	69	42
Nausea/vomiting	49	54	20 (vomiting only)
Constipation	48	54	—
Cough	47	77	24
Insomnia	29	88	19
Confusion	27	35	23
Bedsores	25	61	9
Catheter/incontinence	20	31	35

[1] Symptoms on admission to St Christopher's Hospice, 69 per cent of patients aged 65 and over.
[2] Survey of terminally ill patients in a general hospital (Hockley *et al.* 1988).
[3] Retrospective study, interviewing general practitioner, nurse, and relative, Hospital death:home death = 2:1; 62 per cent patients over 70 years (Wilkes 1984).

and incurable disease it is often possible to reverse the immediate cause of a particular symptom—anorexia may be caused by constipation as well as by liver metastases.

Even if the cause of the symptom proves irreversible, a knowledge of the mechanism involved may point to the correct symptomatic treatment—uraemic vomiting requires different management from the vomiting of malignant gastroduodenal obstruction.

The diagnosis should include the anatomical site and pathological process involved, any contributing biochemical abnormality, and the psychological and social components. The main steps are as follows:

1. Believe the patient's complaint. In chronic or terminal illness, patients often minimize their symptoms.
2. Take a detailed history. A report of 'vomiting for 3 days' is inadequate. The patient or family should be asked about associated nausea and the size, nature, and timing of vomits.
3. Make a careful physical examination. This should record the patient's general condition and take especial notice of findings relevant to the symptoms.
4. Assess the psychological state of the patient.
5. Order appropriate investigations. These should not be done routinely but sometimes a biochemical profile, full blood count, or radiograph will help elucidate a distressing symptom.
6. Record the symptom and your diagnosis of it. This will assist other doctors who will be involved and help in your continuing treatment. For example: Mrs M.E. (78). Carcinoma of the cervix with ureteric involvement. Vomiting due to uraemia exacerbated by constipation, a recent increase in morphine dose, and considerable anxiety.
7. Explain the diagnosis to the patient and family, together with your planned treatment.

PRINCIPLES OF TREATMENT

Assessment of the cause of a particular symptom will often point to specific treatment. Most symptoms are caused by multiple factors and it is often possible to reverse one or more of these with considerable symptomatic benefit. For example, a patient with lung cancer may be severely dyspnoeic, not only due to the tumour but also to long-standing chronic obstructive airways disease and a recent chest infection. Specific treatment will be bronchodilators, antibiotics, and, perhaps, corticosteroids. However, as the disease progresses, specific treatments are less often effective, and appropriate, purely symptomatic treatment is required.

Drug therapy is the mainstay of treatment. Medication should be given regularly, 'by the clock' (World Health Organization 1986), to maintain control. The dose interval will depend on the plasma half-life of the drug used. For example, both morphine and diazepam can be used for malignant dyspnoea. Morphine, in solution, has a plasma half-life of about 3 h and therefore should be given 4-hourly. Diazepam, however, has a plasma half-life of over 30 h and can therefore be given in a single daily dose.

The oral route is preferred but rectal administration or injections are sometimes needed.

Non-drug methods should also be used as appropriate; diversional therapy, breathing exercises, and relaxation techniques may be helpful. The effect of every treatment is increased by the support given by members of the health-care team to each patient and family, enabling them to share their anxieties and fears.

Good symptom control requires regular review as symptoms change and new symptoms develop. In the sections which follow, many of the major symptoms of terminal illness are discussed, with their common causes and suggested treatment. Hospice experience has been with patients with advanced malignant disease and motor neurone disease but it is expected that much of the symptom control practised can be applied to patients dying from other diseases.

ANOREXIA

A diminished desire for food is common in a variety of systemic diseases, including infections, hepatic or renal failure, cancer, and depression. It is exacerbated by malnutrition, which in elderly people may be due to poor mobility or motivation.

There are three treatment options:

1. *Correction of causative factors*. Infections or a depressive illness should be treated appropriately. Pain, nausea, vomiting, and constipation require symptomatic treatment.

2. *Dietary measures*. Food should be attractively served, preferably a small amount on a small plate. Taste changes should be noted and catered for, often serving more strongly flavoured food to compensate for the increased taste threshold that can occur with age and illness. Special feeding equipment may be helpful; modified cutlery with large handles or a heating plate to keep food hot. As patients weaken, many find that they cannot cope with solid food but will accept oral nutritional supplements. These are made in various flavours and it is important to prevent boredom by varying those offered. Hyperalimentation, enteral or parenteral, is not appropriate in terminal illness.

3. *Drug treatment*. Corticosteroids provide the most effective drug treatment for anorexia and normally lead to an

improvement in appetite within a week (Willox *et al.* 1984). Prednisolone, 15 to 30 mg daily, or dexamethasone, 2 to 4 mg daily, is the recommended starting dose. In patients with advanced disease it can usually be continued without causing side-effects. Alcohol, before or with meals, can improve both appetite and mood.

ANXIETY

Anxiety may be primary, the generalized anxiety disorder, or secondary to physical disease, depression, or altered external circumstances. Anxiety may manifest itself in physical symptoms such as palpitations, dyspnoea, headache, and weakness.

Many elderly patients become anxious as their physical condition deteriorates and they approach death. Most are concerned about the process of dying, fearing pain or loss of dignity. They may be leaving an elderly spouse, or a splintered family, or major financial problems. Doctors and nurses need time to address these issues, remembering that the phrase 'Don't worry', which is 'reassurance without explanation' can actually increase anxiety. Anxiolytic drugs may be needed, for a short time, to allow patients to express their fears. Both benzodiazepines, such as diazepam (2 to 10 mg at night), and phenothiazines, such as chlorpromazine (25 mg thrice daily) are used. They can be withdrawn slowly as counselling and perhaps relaxation training prove effective.

BEDSORES

The breakdown of skin in pressure areas is a major problem both in terms of the suffering it causes and the cost of prevention and treatment. Poor general condition due to advanced disease, immobility, incontinence, inadequate nutrition, and impaired consciousness predispose to skin breakdown (Colburn 1987).

Pressure sores, especially superficial ones, are frequently exquisitely painful. Less often recognized are the psychological effects, on patient, family, and nurses, of discharging and offensive wounds. Traditionally, the care of the skin has been a nursing issue and it is only in more recent years that research has been done to assess the effectiveness of the treatment used. This has shown that poor equipment, lack of knowledge associated with an element of folklore, and delays in the process of care all contribute to the formation of pressure sores. (See also Chapter 3.2.)

Prevention of pressure sores

1. Assessment of risk factors (see above) so that vulnerable patients start treatment immediately.
2. Skin care, avoiding overfrequent washing (removes sebum) or massaging. An occlusive dressing will prevent friction and shearing movements.
3. Relieve pressure by rotating the patient through different positions.
4. Use appropriate appliances, if available. For example: pressure-relieving mattresses, sheepskins, bed cradles, and low-pressure profile beds.

Treatment of pressure sores

In addition to the measures used to prevent skin breakdown, it is necessary to apply dressings that provide optimum conditions for healing. The following factors are important:

1. Maintenance of warm, moist environment at wound/dressing interface.
2. Absorption of exudate.
3. Allowance of gaseous exchange.
4. Ease of removal without trauma.
5. Hypoxia under occlusive dressing stimulates healing and contributes to pain relief.

There are a growing number of proprietary preparations that meet many of these criteria. If the nurse has both time and enthusiasm it is usually possible to improve patient comfort and reduce odour, even if complete healing is not achieved.

CANCER PAIN

Pain is the symptom most expected and most feared by cancer patients and their families. Sadly, that fear has often been justified for, in the past, cancer pain has been poorly understood and inadequately treated. Fortunately, the situation is now improving and, in the developed world, most patients with cancer should be assured of good relief of pain with appropriate use of analgesics, adjuvant drugs, radiotherapy, and anaesthetic techniques.

Diagnosis of pain

This should include the site involved, the pathological process, and the patient's psychological state. This diagnosis is the key to providing the right treatment. Some of the common cancer pains are as follows:

1. Bone pain: metastases in bone are thought to cause pain by their production of prostaglandins, which sensitize free nerve endings.
2. Nerve involvement: there may be compression, infiltration or destruction of nerves, leading to deafferentation. Pain is felt in the corresponding dermatome(s) and is often associated with motor, sensory, or autonomic changes.
3. Visceral pain: this is due to tumour involving abdominal, pelvic, or intrathoracic organs.
4. Lymphoedema.
5. Intestinal colic from constipation or malignant obstruction.
6. Headaches from raised intracranial pressure.
7. Secondary infection.

Management of pain

Treatment should be based on an accurate diagnosis wherever possible. However, sometimes the patient is too confused or ill for a full assessment to be made and in this situation adequate analgesia must not be withheld.

For most patients, a combination of the following methods will be required: analgesic drugs; adjuvant analgesic drugs; psychological and emotional support; palliative radiotherapy; anaesthetic techniques. The last two methods should always be considered, but in the majority of terminally ill

Table 2 Alternatives to oral morphine

Name	Dose interval (h)	Tablet	Morphine equivalent (as solution 4- hourly)	Comment
Buprenorphine	8	0.2 mg 8-hourly	7 mg	Partial antagonist, therefore do not use with morphine; ceiling effect at 3 mg/day
Dextromoramide	2	5 mg (and 10 mg)	15 mg (peak effect)	Too short acting for regular use; good for 'breakthrough pain'
Diamorphine	4	10 mg (but usually prescribed in an elixir)	10–15 mg	Identical in use to morphine; more soluble
Dipipanone Co	4–6	10 mg (with cyclizine 30 mg)	5 mg	Presence of cyclizine causes marked sedation if more than one tablet required
Methadone	8–12	5 mg	7.5 mg (single dose)	Accumulation occurs with regular administration
Oxycodone	8	30 mg suppository 8-hourly	15 mg	Preferred analgesic suppository
Pethidine	2–3	50 mg	5 mg	Short acting; CNS side-effects
Phenazocine	8	5 mg 8-hourly	20 mg	Useful alternative to morphine; ceiling effect at 45–60 mg/day

patients the correct treatment is with skilled use of drugs and the support of patient and family.

Analgesic drugs

The criterion for giving analgesia, and especially opioids, is the presence of pain—not the expected length of life. Unfortunately, many patients and doctors still feel that if morphine is started early 'it will lose its effect'. The reverse is nearer the truth, and the really intractable pain, seen occasionally at the end of life, has often been preceded by months of inadequate control leading to depression, anger, and fear. Analgesics should be given regularly, if possible by mouth. The dose should be individually determined, being the lowest compatible with controlling pain. Experience has shown that tolerance is a minor problem; addiction does not occur and the side-effects of morphine, such as nausea or constipation, can be controlled.

Paracetamol (acetaminophen), 1 g 4-hourly, is recommended for the treatment of mild pain. If this proves ineffective, the choice is between adding a weak opioid, e.g., dextropropoxyphene with paracetamol (coproxamol), or a low dose of a strong opioid. With escalating pain or a very sick patient it is advisable to change directly from paracetamol to morphine, starting at 5 to 10 mg 4-hourly.

Morphine

Morphine is well absorbed when given by mouth, through the buccal mucosa, or rectally. It is rapidly distributed throughout the body. Metabolism mainly occurs in the liver, where it is broken down into glucuronides, and it is in this form that most renal excretion occurs (Hanks and Hoskin 1987).

Preparations of morphine include:

1. Morphine solution (morphine sulphate or hydrochloride in chloroform water); strengths from 5 to 200 mg, each in 10 ml; the dose is given 4-hourly.
2. Slow-release morphine sulphate tablets (MST), 10 mg, 30 mg, 60 mg, and 100 mg are available and are given 12-hourly.
3. Morphine suppositories, 15 mg or 30 mg: they need to be given 4-hourly so, in practice, oxycodone 8-hourly is preferred.
4. Morphine injection: where diamorphine is allowed (as in the United Kingdom) it is preferred because of its greater solubility; both morphine and diamorphine have a ratio of injected dose:oral dose of 1:2. Injections should be given 4-hourly. An alternative method, if pain is accompanied by severe vomiting or dysphagia, is to give the opioid by continuous subcutaneous infusion over 24 h using a portable syringe driver. Epidural morphine is given 12-hourly or by infusion. The ratio of epidural dose:oral dose is 1:10

Alternatives to morphine

Although morphine is the recommended strong analgesic, there are a few patients who complain of severe nausea or drowsiness, which they attribute to the drug, and others who refuse to have it. For such patients, alternatives to oral morphine are needed. Table 2 lists alternative strong analgesics, and can also be used in converting to morphine from other drugs.

Adjuvant analgesic drugs

The correct use of adjuvants for analgesia may mean that morphine is not required, or it may be possible to give it in a lower dose. Response to adjuvant treatment is variable, so it is recommended that, unless delayed response is likely, the drug is discontinued after a week if there is no improvement. Table 3 shows the indications for adjuvant drug treatment.

Other treatment

Radiotherapy may be helpful during the last weeks of life, provided that it is given without delay and with the minimum number of treatments. The most common indication is the development of a painful bony metastasis; worthwhile pain relief is achieved in 80 per cent of patients, often with a single fraction (Price *et al.* 1986).

The value of anaesthetic techniques will depend on the

Table 3 Adjuvant analgesic drugs

Type of pain	Drug
Bone pain	Naproxen 500 mg twice daily or other non-steroidal anti-inflammatory drug
Nerve compression or destruction (deafferentation)	Amitriptyline 25 mg at night, with increasing dose Carbamazepine 100–200 mg thrice daily Epidural steroids
Headache from raised intracranial pressure	Dexamethasone 16 mg daily
Intestinal colic	Hyoscine butylbromide (Buscopan) 40–80 mg daily
Pelvic or hepatic pain	Non-steroidal anti-inflammatory drug

availability of an anaesthetist specializing in pain control. Epidural steroids, morphine, and bupivacaine have proved valuable, as have some peripheral or autonomic nerve blocks. Transcutaneous electrical nerve stimulation (**TENS**) is occasionally useful for the control of cancer pain; acupuncture has rarely been found to be effective.

CHRONIC BENIGN PAIN

Chronic pain is unfortunately very common in elderly patients. There are many causes, which include osteoarthritis and rheumatoid arthritis, trigeminal and postherpetic neuralgia, ischaemic legs, back problems, bedsores, and the pain of immobility from stroke or severe disability. The management of many of these conditions is covered elsewhere. The principles of treatment for benign pain are the same as those for cancer pain. It is important to make a clear diagnosis, institute specific treatment if possible, use analgesic drugs regularly in adequate dose, and give appropriate psychological support. Non-opioids (paracetamol and aspirin), non-steroidal anti-inflammatory drugs, and weak opioids (codeine and coproxamol) are preferred. But with severe pain, for example from gangrene, morphine should be given in adequate dose.

Psychotropic drugs have a major role in the management of many types of chronic benign pain. Tricyclic antidepressants, sometimes combined with anticonvulsants, are used for postherpetic and trigeminal neuralgia, diabetic neuropathy, and phantom-limb pain.

Nerve blocks include the infiltration of trigger areas with local anaesthetic, epidural steroids for root irritation, and sympathetic blocks for ischaemia.

Physical methods, such as local anaesthetic sprays, TENS, and acupuncture, seem of greater value in chronic benign pain than in pain of malignant origin.

CONFUSION

The differential diagnosis and treatment of confusion are two of the most difficult problems facing a doctor who is caring for dying patients. Whereas pain, or perhaps incontinence, is the symptom most dreaded by the patient, confusion is probably the symptom that most distresses the family. Causes and management of delirium and dementia are discussed elsewhere (Chapters 18.4 and 18.5). In practice, confusion is usually of mixed aetiology. For example, an elderly cancer patient with a mild degree of dementia may be 'just about' coping at home. The development of a chest infection necessitating hospital admission tips the balance and precipitates an acute confusional state.

Reversible causes for confusion should be sought and treated appropriately. Non-drug methods of management are most important. These include the provision of a familiar routine, if possible in home surroundings. Hearing aids and glasses should be checked. Many patients value a simple explanation of the actual cause of their confusion and the steps planned to help with this.

The 'quietly muddled' patient with dementia requires no medication and sedative drugs may worsen the confusion. Psychotropic drugs are used to reduce the agitation sometimes associated with confusion. Thioridazine is valuable for the elderly patient but can be replaced by chlorpromazine if considerable sedation is required. Haloperidol gives a little sedation but extrapyramidal symptoms may cause problems.

CONSTIPATION

The great majority of patients with terminal illness become constipated. This may be due to a specific, potentially reversible condition such as hypercalcaemia, intestinal obstruction, hypothyroidism, or depression. Much more often the constipation is caused by a combination of the following: inactivity, a diet low in roughage, general weakness, confusion, and constipating drugs such as opioids, phenothiazines, and tricyclic antidepressants.

Unless the patient is very frail, an attempt should be made to increase activity, add fibre to the diet, and maintain a good oral fluid intake. Assistance with mobility should be available so that the patient can use the lavatory rather than a commode or bedpan. However, in spite of these measures, constipation often remains a problem and, in practice, laxatives and rectal procedures are required.

Laxatives are often divided into those that stimulate bowel peristalsis and those that soften and bulk the stool. This is not pharmacologically tenable, as increased stool size in itself causes increased peristalsis, but the division remains useful in the clinical choice of laxatives (Brunton 1985).

Faecal softeners retain water in the bowel by osmosis or increase water penetration of stool. These include magnesium hydroxide or sulphate, lactulose, and docusate. Stimulant laxatives stimulate colonic muscle activity. Examples are senna, bisacodyl, and cascara.

Persistent constipation in the terminally ill is best managed with a combination of both types of laxative. This will avoid either painful colic or a bowel loaded with soft faeces, problems that arise if a stimulant of softening laxative is given alone. Examples are:

- lactulose, 10 to 20 ml twice daily, with bisacodyl 5 mg twice daily;
- magnesium hydroxide and liquid paraffin emulsion, 10 to 20 ml twice daily, with senna tablets, 2 twice daily.

The dose should be gradually increased until a regular soft bowel action is obtained.

Suppositories or an enema or manual removal may be

needed if a patient has a loaded rectum or if the laxative regimen is ineffective. It is a good general rule to do a rectal examination on the third day if the bowels have not opened, inserting a glycerine or bisacodyl suppository if the rectum is loaded. 'High' impaction at thc rcctosigmoid junction may also occur in elderly patients.

COUGH

Cough can be caused by bacterial or viral infection, or tumours, involving the upper or lower respiratory tract. It may be associated with chronic obstructive airways disease, asthma, congestive cardiac failure, or smoking.

Specific treatment with antibiotics, bronchodilators, diuretics, or corticosteroids should be given unless life expectancy is very short. Advice to stop smoking will depend on the patient's prognosis and the degree of dependence on the habit.

If specific treatment is inappropriate or ineffective, symptomatic treatment should be offered. There are three types of cough:

1. productive—patient able to expectorate;
2. productive—patient too weak to expectorate adequately;
3. dry or irritant.

Centrally acting cough suppressants should not be given to the patient with a productive cough who is expectorating.

Antitussives with peripheral action (on the respiratory tract) include the inhalation of warm, moist air, which is often helpful, as are steam or benzoin inhalations. Simple linctus (cough syrup), which soothes pharyngeal irritation, is cheap and often helpful. Mucolytics and 'expectorants' are not of proven efficacy. Antitussives with central action (on the cough centre in the medulla) should be reserved for patients with a dry cough, those who are too weak to expectorate, or occasionally at night if cough disturbs sleep. Codeine, 30 to 60 mg 4-hourly (in linctus or tablet), is often effective, but with intractable cough, as is often associated with lung cancer, morphine is required. The starting dose is morphine 5 mg 4-hourly, or slow-release morphine sulphate tablets (MST), 10 mg twice daily, increasing slowly until relief is obtained.

Compound cough preparations should be avoided. They may contain codeine, antihistamines, sympathomimetics, or 'expectorants' that are usually in subtherapeutic doses and may have opposing effects.

DEPRESSION

Appropriate sadness is a natural reaction to declining strength or the approach of death; the diagnosis of clinical depression in elderly patients with terminal illness is not easy.

Physical symptoms, such as anorexia, insomnia, and weakness, cannot be used for diagnosis. The most helpful criteria are feelings of unworthiness and anhedonia—a lack of pleasure in things previously enjoyed.

Treatment is with a combination of counselling and medication. Tricyclic antidepressants need to be started in low dosage in elderly patients, especially if they are receiving other centrally acting drugs.

DYSPHAGIA

Difficulty in swallowing can be caused by obstruction of the lumen by pharyngeal or oesophageal tumours, or by benign oesophageal stricture. Mediastinal lymphadenopathy may lead to external compression. Neurological causes include cerebrovascular disease, cerebral tumours, and motor neurone disease. Dysphagia can be caused, or exacerbated, by pain on swallowing, a dry mouth, anorexia, and anxiety.

Having diagnosed the cause of dysphagia, specific treatment may be possible. Thrush infections of the mouth or oesophagus should be treated with nystatin or ketoconazole. Oesophageal cancer can be treated with external or intracavitary radiotherapy, palliative laser treatment, bouginage, or an indwelling oesophageal tube.

Radiotherapy should be considered for malignant mediastinal nodes but, if the patient is not fit enough, the dysphagia may respond, temporarily, to corticosteroids (dexamethasone, 8 mg/day, or prednisolone, 60 mg/day). These can cause shrinkage of peritumour inflammatory oedema.

If such specific treatment is inappropriate or ineffective, then dietary modification is required. In general, patients with obstructive lesions need a nourishing but fluid diet, whereas those with neurological problems prefer semisolid food. An oesophageal tube should be regularly flushed with carbonated drinks. Correct positioning and adequate time for feeding is important.

Drugs for symptom relief should be continued but may need to be given by suppository, subcutaneous infusion, or intramuscular injection. The use of a nasogastric tube, parenteral nutrition, or gastrostomy in the patient with advanced and irreversible disease is controversial. Their use may simply prolong the process of dying. Most patients with severe dysphagia sooner or later develop an aspiration pneumonia, which should usually be treated symptomatically.

DYSPNOEA

Dyspnoea is defined as a distressing difficulty in breathing. It is therefore a subjective syndrome that may be poorly related to tachypnoea or to pulmonary pathology. Breathing is under the control of the medullary respiratory complex. This can be stimulated via chemical changes in the blood ($\uparrow P_{CO_2}$, $\downarrow P_{O_2}$, $\uparrow H^+$), by afferent impulses from the respiratory muscles and lung sensory receptors, and from the cerebral cortex. Increase in any of these stimuli can lead to dyspnoea.

There are a great number of causes of dyspnoea in the terminally ill. The most common include chronic obstructive airways disease, pneumonia, lung tumour, cardiac failure, anaemia, and anxiety. Frequently these causes coexist.

Reversible causes of dyspnoea should be treated as long as worthwhile benefit results. But the time comes when, for example, a pleural effusion rapidly accumulates after aspiration or a pneumonia no longer responds to antibiotics. At this stage the aim must be to relieve the symptom rather than alter the progress of the disease

Explanation and adjustment

Severe dyspnoea is frightening to the patient and family and the resultant anxiety exacerbates the symptom. Fears such as 'Will I choke to death?' or 'Will I suffocate?' are common. Medical and nursing staff should encourage patients to voice their fears and give appropriate explanation and reassurance.

Many patients need help in adjusting their lifestyle. Simple measures such as avoiding stairs, a backrest in bed, or an open window or fan may prove beneficial. Breathing exercises or relaxation therapy are sometimes used.

Opioids

Morphine and other opioids reduce the sensitivity of the medullary respiratory centre to hypoxia or hypercapnia and thus alter the central perception of breathlessness. They lessen the heightened respiratory drive of the 'pink puffer' so that an improved exercise tolerance may be noted as well as relief of symptoms.

Many years of experience in hospices has shown that a low dose of oral morphine is the best treatment for intractable malignant dyspnoea. It does not lead to severe respiratory depression or even to an inevitable adverse change in the arterial blood gas tensions (Walsh *et al.* 1981). A suggested starting dose is morphine, 5 mg 4-hourly; this can be gradually increased as necessary. Dihydrocodeine, 15 mg thrice daily, has also been shown to lessen dyspnoea, though the side-effects of nausea and constipation may cause problems (Johnson *et al* 1983).

Anxiolytic drugs

Diazepam is used for its anxiolytic effect, sometimes in combination with morphine. A single night-time dose of 5 to 10 mg is suggested, though drug accumulation may occur.

Respiratory emergencies sometimes occur. These include acute tracheal compression, major haemorrhage, or pulmonary embolus. Intravenous midazolam (or diazepam), 10 to 20 mg, given slowly, or a rectal solution of diazepam, 10 to 20 mg, should be given immediately. An alternative treatment is with hyoscine and diamorphine (see below).

Hyoscine (scopolamine) and atropine

These drugs reduce exocrine secretions and relax smooth muscle. Atropine is a central stimulant while hyoscine is sedative (though can occasionally cause excitement). They are used to lessen bronchial secretions that accumulate in the last hours of life causing the 'death rattle'. The dose is 0.3 to 0.6 mg by injection every 4 h. Hyoscine is preferable because of its sedative effect.

Both drugs should be given with morphine or diamorphine to reduce dyspnoea further and to prevent any central stimulation. Hyoscine and diamorphine, by injection, can also be used in an emergency.

Oxygen

This is of value in acute dyspnoea and in the long-term management of some types of non-malignant dyspnoea. Drug methods give better control in the terminally ill and they avoid the use of the oxygen mask or nasal catheter, which may act as a distressing barrier between patient and family.

FISTULAE

A fistula is an abnormal connection between two hollow organs or between one or more of these and the skin. Fistulae are found in 3 per cent of patients with advanced malignant disease, most arising from the bowel. Patients find fistulae particularly distressing as they usually develop in a suture line where it is difficult to fit the usual stoma devices. Surgical intervention should always be considered but may be contraindicated in very frail patients with advanced disease. Management of some of the more common fistulae is as follows.

1. *Enterocutaneous fistulae* may develop after surgery for malignant intestinal obstruction. They form in the laparotomy scar or at a drainage site. It is sometimes possible to reduce the amount of discharge by bulking or firming intestinal contents by giving methylcellulose or codeine phosphate. Stomadhesive can be used to fill in skin creases so that a stoma bag can be fixed.
2. *Rectovaginal or rectovesical fistulae* cause major problems and palliative colostomy should always be considered. The severe dysuria caused by a rectovesical fistula can be eased by using a large catheter with regular wash-outs.
3. *Vesicovaginal fistulae* can often be managed with the use of tampons and a large-size catheter.
4. *Buccal fistulae* can sometimes be plugged with a silicone foam casting (Regnard and Meehan 1982).

HICCUP

Common causes include gastric distension, irritation of the diaphragm or phrenic nerve, uraemia, and cerebrovascular disease or cerebral tumour. An acute attack of hiccups can often be stopped by pharyngeal stimulation—sipping iced water or swallowing granulated sugar. Rebreathing into a paper bag, which increases the P_{CO_2}, is sometimes effective. Gastric distension will be reduced by peppermint water or metoclopramide, a drug which increases gastric emptying. For a severe and distressing attack of hiccup, intravenous chlorpromazine, 25 to 50 mg, may be necessary. It is less effective given by intramuscular injection.

Prevention of hiccup depends on the underlying cause. Metoclopramide, 10 mg four times daily, will reduce gastric distension; chloropromazine, 25 mg four times daily, is more effective for the hiccup of uraemia. Anticonvulsants, such as phenytoin or carbamazepine, are used when hiccup is caused by intracerebral disease (Williamson and MacIntyre 1977).

INSOMNIA

Many patients with advanced disease sleep poorly owing to unrelieved physical or mental distress. Before prescribing a hypnotic it is necessary to enquire if sleep is disturbed by pain, night sweats, fear of incontinence, or other physical symptoms for which appropriate relief can be given. More often anxiety, depression, or a fear of dying in the night prevent normal sleep. A careful enquiry will usually uncover such factors and they can be helped both by counselling and by psychotropic drugs. Hypnotics include:

1. Chlormethiazole, 1 to 2 caps. (192–384 mg). A useful hypnotic with very short half-life, for the elderly patient, even if confused. It may cause nasal irritation in younger patients.
2. Temazepam, 10 to 30 mg. A benzodiazepine with plasma half-life of 6 to 8 h, but which may have undesirably prolonged effects in some elderly patients.
3. Diazepam, 5 to 10 mg. This has a very long half-life with a risk of accumulation, especially in the elderly, and dependence. However, it is of value as a hypnotic when there is associated intractable dyspnoea or for the short-term management of severe anxiety.
4. Amitriptyline or other sedative tricylic antidepressants when insomnia is associated with clinical depression.

NAUSEA AND VOMITING

Vomiting results from the stimulation of the vomiting centre in the medullary reticular formation. This receives impulses from the chemoreceptor trigger zone, the vestibular apparatus, the cerebral cortex, and afferents innervating the gastrointestinal tract. Thus nausea and vomiting have a large range of potential causes originating from many parts of the body.

The recommended treatments for common causes of vomiting in the terminally ill are listed next. Details of the dosage and administration of antiemetic drugs are found in Table 4.

1. *Drug-induced vomiting*. Opioids should be started with metoclopramide or prochlorperazine. The antiemetic can usually be withdrawn after a week. Anti-inflammatory drugs should be given with food, but may have to be discontinued.

2. *Uraemia*. Haloperidol may be effective but often the more sedating drugs, chlorpromazine or methotrimeprazine, are required.

3. *Hypercalcaemia*. If active treatment is not indicated, then vomiting should be treated as for uraemia. Corticosteroids may be valuable in this situation.

4. *Raised intracranial pressure*. Dexamethasone, 16 mg daily, reducing as possible. Cyclizine if corticosteroids are contraindicated.

5. *Delayed gastric emptying*. Due to pressure from an enlarged liver or morphine-induced gastric stasis. Metoclopramide or domperidone are usually effective.

6. *Carcinoma of stomach*. The vomiting is difficult to control and antiemetics, except the most sedating ones, are disappointing. Sometimes corticosteroids (dexamethasone, 8 mg daily by injection) give temporary relief.

7. *Constipation*. (See above.)

8. *Inoperable intestinal obstruction*. Drugs for pain, colic and vomiting are given by subcutaneous infusion. Haloperidol is often effective, sometimes methotrimeprazine is required (Baines 1987).

9. *Motion sickness*. Cyclizine or hyoscine, 0.3 to 0.6 mg sublingually.

10. *Anxiety*. (See above.)

These causes frequently coexist and a combination of antiemetics may be needed.

SORE MOUTH

A dry or painful mouth is common in those with terminal disease. Causes include drugs with anticholinergic effect, oral thrush, poorly fitting dentures, oral tumours, and radiotherapy.

Good oral hygiene is most important. Teeth and dentures

Table 4 Antiemetic drugs

Name	Preparations: Tablet	Syrup	Injection	Suppository	Use subcutaneously in syringe driver	mg dose/ 24 h	Comments
Prochlorperazine	5 mg (25 mg)	5 mg/ 5 ml	12.5 mg	25 mg (5 mg)	No (skin reaction)	15–50	Effective in many types of vomiting. Minimal sedation
Chlorpromazine	10 mg 25 mg 50 mg 100 mg	25 mg and 100 mg in 5 ml	25 mg 50 mg	100 mg	No (skin reaction)	50–150	As prochlorperazine but more sedating
Methotrimeprazine	25 mg	—	25 mg	—	Yes (but some skin reaction)	50–150	Very potent antiemetic and analgesic Considerable sedation
Haloperidol	1.5 mg 5 mg 10 mg 20 mg	10 mg/ 5 ml	5 mg 10 mg	—	Yes	5–20	Minimal sedation. Occasional extrapyramidal side-effects
Cyclizine	50 mg	—	50 mg	—	Yes	150	Drug of choice in vestibular-induced vomiting. Dry mouth
Metoclopramide	10 mg	5 mg/ 5 ml	10 mg (100 mg)	—	Yes	30–60	Increases gastric emptying. Not sedative. Occasional extrapyramidal side-effects
Domperidone	10 mg	5 mg/ 5 ml	—	30 mg	—	30–60	Increases gastric emptying. Not sedative. No extrapyramidal side-effects

should be cleaned regularly. Mouthwashes with chlorhexidine 0.2 per cent or povidone iodine 1 per cent both moisten and cleanse the mouth. Chewing pineapple chunks, which contain the enzyme ananase, will clean a furred tongue.

Thrush infections may cause various clinical pictures. A smooth, sore, red tongue as well as the more common white plaques may be caused by thrush. Treatment is with nystatin suspension, 100000 ml, 4-hourly for 14 days. Dentures should be removed and treated, preferably overnight, as well. Nystatin pastilles are an alternative. Fungal resistance does not occur but in persistent cases of thrush, ketoconazole, 200 mg daily for 8 to 10 days, is usually effective.

Buccal analgesics include benzydamine, a non-steroidal anti-inflammatory agent absorbed through the mucosa, and choline salicylate, which can be applied to ulcerated areas before meals. Triamcinolone in orabase is used for painful aphthous ulcers.

URINARY INCONTINENCE

A degree of incontinence is not uncommon in the elderly population; it is a frequent complication of neurological or pelvic disease so, inevitably, it will often occur in those who are terminally ill. Patients find incontinence deeply humiliating, it is a major cause of breakdown in home care, and its management occupies a vast amount of nursing time on geriatric wards (Williams and Pannill 1982).

As with any symptom, a diagnosis of the cause should be made if possible. Unfortunately, in this group of patients, few causes are reversible, though occasionally the incontinence is due to urinary infection, faecal impaction, diuretics, or depression or simply inaccessibility of toilet facilities, all of which can be treated or ameliorated.

Physical methods of treatment, such as bladder training for detrusor instability, are obviously inappropriate. Imipramine or propantheline are occasionally helpful but the great majority of these patients require a urinary catheter or condom.

A latex or silicone catheter with a 5 to 10 ml balloon is used; bladder wash-outs with saline or chlorhexidine are given if there is a lot of sediment. Urinary infections are only treated if they cause symptoms. While it is acknowledged that long-term catheterization leads to many problems, they are irrelevant in this group of patients, for whom comfort and dignity are paramount.

THE LAST DAYS

The correct management of the last few days of life is extremely important. Those who visit the bereaved will be only too aware how the last hours become imprinted on the memory with unanswerable questions: 'I wonder if she was trying to say something to me?' 'Do you think he was in pain?'

As death approaches it is important to stop most drugs, for example diuretics and antibiotics, but to continue those needed for symptom control. These should be charted so that they can be given regularly by injection or suppository if swallowing becomes impossible.

Most patients become more drowsy and lapse into coma as death approaches but some become restless and agitated. This may be due to unrelieved pain or a distended bladder or rectum. Frequently no cause can be identified.

If a patient becomes restless, a sedating phenothiazine is the first choice—methotrimeprazine, 50 to 150 mg/24 h or chlorpromazine, 100 to 200 mg/24 h. Diazepam, 10 mg given rectally, can be used to control muscle twitching or to increase sedation. Hyoscine, 0.4 to 0.6 mg by injection, usually controls the 'death rattle' and it is also sedative.

Each of these drugs should be given with diamorphine if there is pain or dyspnoea. Medication, although important, is not the only way of managing a restless dying patient. Staff often notice that the presence of the family or nurse sitting by the bedside, holding the hand or speaking quietly to an apparently unconscious patient, has a remarkable calming effect.

ETHICAL ISSUES IN TERMINAL ILLNESS

'Must patients always be given food and water?' 'Will adequate sedation shorten his life?' 'Should antibiotics be withheld?' These, and many similar questions are heard frequently in hospitals, hospices, and homes where the old and terminally ill are cared for. They may come from families but more often are the concerns of caring staff. In the section which follows, the problem of dehydration will be considered in some detail, followed by a short reference to other ethical problems found in palliative care.

Dehydration and thirst

In the past, malnutrition and dehydration often accompanied deaths resulting from a prolonged illness, because sick and dying people could not consume adequate amounts of food and fluid. However, with the availability of intravenous fluids, tube feeding, and total parenteral nutrition, the situation has changed dramatically. With the recognition of the value of fluid and nutritional support in general medicine and surgery, the same methods have been applied to the terminally ill. The aims of such treatment are to extend life, relieve distressing symptoms, and '... an affirmation of the physician's role as a caring professional. A nurturing and symbolic act that avoids any appearance of abandoning the patient' (Siegler and Shiedermayer 1987). A survey of American physicians showed that three-quarters would routinely administer intravenous fluids to the terminally ill (Micetich *et al*. 1983).

Current reappraisal

Over the last few years there has been a reappraisal of the value of continuing hydration for the very ill patient with no prospect of recovery. Well publicized cases in the United States have brought into public awareness the issues involved. The 'Living Will' has been signed by 10 million Americans but it must be assumed that millions more agree with its tenets (Editorial 1987):

> If I should have an incurable or irreversible condition that would cause my death within a relatively short time, and am no longer able to make decisions regarding my medical treatment, I direct my attending physician ... to withhold or withdraw treatment that only prolongs the process of dying and is not necessary to my comfort or to alleviate the pain.

The British Medical Association Working Party on Euthanasia devoted a section to the management of hydration: 'The Working Party believes that feeding/gastrostomy tubes for nutrition and hydration are medical treatments and are warranted only when they make possible a decent life in which the patient can reasonably be thought to have a continued interest' (British Medical Association 1988).

At the same time as these ethical considerations have been widely debated, there has been a reappraisal of the presumed suffering caused by dehydration. Experienced nurses working with the terminally ill state that dehydration is not painful and that these patients rarely complain of thirst (Andrews and Levine 1989). The only distressing symptom is a dry mouth and this can be easily treated with local measures. Indeed, increasingly it is felt that continuing hydration actually increases the distress of dying. A fully hydrated patient will have more problems with pulmonary oedema, vomiting, urinary incontinence, and the 'death rattle', as well as any discomfort from the intravenous line or nasogastric tube. The fear that families will demand continued hydration is rarely justified. Indeed, most relatives consider that the patient should be allowed to die rather than have life prolonged by artificial means. Their only proviso is that suffering must be adequately relieved. Factors such as these have undoubtedly influenced medical practice so that, in some centres, long-stay geriatric patients do not receive intravenous fluids in the last week of life and fluids are discontinued in assessment-ward patients before death (Wilson *et al.* 1987)

Suggested management

A patient may suddenly start to refuse food and drink but, more often, the process is gradual, with a declining food intake that is inadequate to sustain life.

In either state, it is important to diagnose the cause of food refusal. This may be a physical symptom, such as persistent nausea or a painful mouth, or a psychological problem, such as the onset of clinical depression. There may be a deterioration in the underlying illness or an intercurrent problem such as a urinary tract infection. Some of these causes of food refusal are reversible with appropriate treatment, and it is reasonable to use intravenous fluids, if necessary, while investigations and treatment are proceeding.

However, if the cause of food refusal is irreversible, it is inappropriate to start or continue artificial feeding. A dry mouth, the only distressing symptom of dehydration, can be relieved by regular sips of water, crushed ice to suck, and meticulous attention to oral hygiene. Most medication should be discontinued but drugs required for symptom control, such as opioids or antiemetics, should be given rectally or by injection.

This conservative approach to the management of dehydration requires to be understood and accepted by both the professional team and the family. Care-givers have a deep instinctive desire to give food, and medical staff need to give time to explain what is happening. 'He won't eat and therefore he will die', needs to be turned into 'He is dying and therefore won't eat.'

The natural urge of families to 'do something' for someone who is dying can be harnessed by skilled nursing staff. Relatives can help with practical things such as mouth care, washing, and turning. They should be reminded that hearing and touch are the last senses to go, so their physical contact and speech can continue to bring comfort.

Other ethical issues

The care of dying patients and their families brings up many other ethical considerations, few of which have easy answers. A century ago, in *The Principles and Practice of Medicine*, Sir William Osler wrote: 'Pneumonia may well be called the friend of the aged. Taken off by it in an acute, short, not often painful illness, the old man escapes those cold degradations of decay so distressing to himself and to his friends.' But with modern antibiotics, pneumonia can usually be treated and 'the old man' foiled in his 'escape'.

In practice, each patient who develops a chest infection should be individually and carefully considered. It would be as inappropriate to give antibiotics by injection and physiotherapy to a virtually moribund patient, as it would be to withhold such treatment from one with a prognosis of some months who needed to settle business affairs, or, perhaps, to come to terms with dying. The doctor's judgement will be based on an imponderable factor—the quality of life offered. Guidelines in assessing this will be the patient's general physical and mental condition, with special notice being taken of respiratory function, realizing that each succeeding chest infection may reduce this. If active treatment is not given, it is essential to relieve dyspnoea, cough, and chest pain, usually with opioids (see above).

Similar considerations apply to the treatment of hypercalcaemia, the reduction of corticosteroids in patients with cerebral tumours, and the sedation of the agitated and confused.

SPEAKING OF DEATH

It has been shown that patients are less anxious or depressed in situations where open discussion about the illness is welcomed (Hinton 1979). Many patients will want to ask questions, not just about the diagnosis but about future symptoms they fear, chiefly pain, incontinence, and confusion.

Many elderly patients dread the prospect of the final stages of life being prolonged 'artificially', and are reassured that this will not be done. Some have major fears about the actual process of dying and the surprisingly common terror of being nailed into the coffin while still alive. A question such as 'What worries you most?' will often bring such fears to light. Talking about death cannot be hurried, it will not fit easily into the busy out-patient clinic or large ward-round. Medical and nursing staff need time to see the patient alone so that there is opportunity for these issues to be raised. They need to record a summary of what is said so that other members of the team can continue to help the patient in his or her individual journey of understanding and acceptance.

SUPPORT FOR THE FAMILY

Much of the distress of dying is due to the fear of separation from other family members and the anxiety about how they will manage in the future. Such fears will be increased if the spouse is very old or ill or isolated, or if there is a conspiracy of silence so that conversation avoids the main issues.

Unfortunately, doctors sometimes abet this by telling important facts about the illness to spouse or patient on their own.

Experience has shown that many couples and families are greatly relieved when encouraged slowly to share the truth (Earnshaw-Smith 1981). This may require the presence of a doctor, nurse, or social worker as a third party, helping the family to talk openly to each other.

Families may be reluctant to talk to the patient about important problems: where the surviving spouse will live, or how he or she will manage financially. They believe that this will add to the burden, whereas, in fact, the reverse is probably true. Discussing these practical issues reaffirms the importance of the patient in the family and brings them closer together.

There is often an irrational guilt felt by the relatives of the dying. This is sometimes because they feel themselves responsible for the illness: 'He was so busy looking after me he didn't go to the doctor in time.' More often they feel guilty about being unable to care at home until the end, especially if this was the patient's wish. These feelings need exploring gently, going over the history of the disease and emphasizing that the time often comes when full-time professional help is needed.

Nurses can often involve the spouse in simple practical care. The best preparation for the process of bereavement is done before death by involving the family in the physical and emotional care of the dying.

SPIRITUAL PAIN

Faced with serious illness or impending death, many patients will begin to think more deeply about their life and its meaning. Some will have a profound sense of guilt and failure about the past, things left undone or failed relationships. Others feel deeply the meaninglessness of life, that there is no point or purpose in it. A few are troubled by the fear of what happens after death.

These, and many other considerations, are aspects of spiritual pain and are very real to many terminally ill patients. Doctors working with the dying need increasingly to be aware of these issues. They should be informed of the patient's religious beliefs so that specific religious practices of the Jew, Moslem, Hindu, Buddhist, or Christian can be adhered to. If the patient has even a vague connection with a church he will probably appreciate the offer to contact the priest or minister.

However, the burning questions that start—'Why me?', 'If only', 'What is the point of it all?'—cannot wait until the chaplain is summoned. They arise during a physical examination or following an enquiry about last night's sleep. Sometimes all that the doctor can do is to listen and share. Surprisingly, quite often, he will be asked, 'What do you think?' and the opportunity comes to say something about a God who loves and cares.

STAFF PAIN

Doctors and nurses working with the elderly and dying need considerable support, especially when they start in this work. They may wonder if everything possible was done, they grieve with the families, and the work can be physically and emotionally exhausting.

Some form of group support needs to be planned. This will often occur in the context of the ward report, especially if the ward doctor and social worker are present. At least as important is the spontaneous conversation, in the dining room or on the stairs, when an individual's anxiety or anger or sadness can be expressed. But the greatest support comes from the work itself. Giving effective pain and symptom control is very satisfying. It is rewarding to see an anxious and denying patient come to terms with his or her illness and share this with the family. The dying have much to teach us about the meaning of life, for it is at this time that people are at their most mature and their most courageous.

References

Andrews, M.R. and Levine, A.M. (1989). Dehydration in the terminal patient: perception of hospice nurses. *American Journal of Hospice Care*, **1**, 31–4.

Baines, M.J. (1987). Medical management of intestinal obstruction. In *Contemporary palliation of difficult symptoms* (ed. T. Bates), Bailliere's Clinical Oncology, Vol. 1, pp. 357–71.

Blackburn, A.M. (1989). Problems of terminal care in elderly patients. *Palliative Medicine*, **3**, 203–6.

British Medical Association (1988). *Euthanasia*, Report of the Working Party to review the British Medical Association's Guidance on Euthanasia. BMA, London.

Brunton, L.L. (1985). Laxatives. In *The pharmacological basis of therapeutics* (7th edn) ed. A.G. Gilman, L.S. Goodman, T.W. Rall, and F. Murad, pp. 994–1003. Macmillan, New York.

Colburn, L. (1987). Pressure ulcer prevention for the hospice patient. *American Journal of Hospice Care*, **4**, 22–6.

Earnshaw-Smith, E. (1981). Dealing with dying patients and their relatives. *British Medical Journal*, **282**, 1779.

Editorial (1987). *British Medical Journal*, **295**, 1221–2.

Exton-Smith, A.N. (1961). Terminal illness in the aged. *Lancet*, **ii**, 305–8.

Hanks, G.W. and Hoskin, P.J. (1987). Opioid analgesics in the management of pain in patients with cancer. A review. *Palliative Medicine*, **1**, 1–25.

Hinton, J. (1967). *Dying*. Penguin, Harmondsworth.

Hinton, J. (1979). Comparison of places and policies for terminal care. *Lancet*, **i**, 29–32.

Hockley, J.M., Dunlop, R., and Davies, R.J. (1988). Survey of distressing symptoms in dying patients and their families in hospital and the response to a symptom control team. *British Medical Journal*, **296**, 1715–17.

Holford, J.M. (1973). Terminal care. In *Care of the Dying*, Proceedings of a National Symposium, November 29th, 1972. HMSO, London.

Johnson, M.A., Woodcock, A.A., and Geddes, D.M. (1983). Dihydrocodeine for breathlessness in 'pink puffers'. *British Medical Journal*, **286**, 675–7.

Micetich, K.C., Steinnecker, P.H., and Thomasma, D.C. (1983). Are intravenous fluids morally required for a dying patient? *Archives of Internal Medicine*, **143**, 975–8.

Price, P., Hoskin, P.J., Easton, D., Austin, S., Palmer, S.G., and Yarnold, J.R. (1986). Prospective randomised trial of single and multifraction radiotherapy schedules in the treatment of painful bony metastases. *Radiotherapy and Oncology*, **6**, 247–55.

Regnard, C. and Meehan, S. (1982). The use of silicone foam dressing in the management of malignant oral-cutaneous fistulae. *British Journal of Clinical Practice*, **36**, 243–5.

Siegler, M. and Schiedermayer, D.L. (1987). Should fluid and nutritional support be withheld from terminally ill patients? *American Journal of Hospice Care*, **4**, 32–5.

Walsh, T.D., Baxter, R., Bowman, K., and Leber B. (1981). High dose morphine and respiratory function in chronic cancer pain. *Pain*, **1**, 5.

Wilkes, E. (1984). Dying now. *Lancet*, **1**, 950–2.

Williams, M.E. and Pannill, F.C. (1982). Urinary incontinence in the elderly. *Annals of Internal Medicine*, **97**, 895–907.

Williamson, B.W.A. and MacIntyre, I.M.C. (1977). Management of intractable hiccup. *British Medical Journal*, **2**, 501–3.

Willox, J.C., Corr, J., Shaw, J., Richardson, M., and Calman, K.C. (1984). Prednisolone as an appetite stimulant in patients with cancer. *British Medical Journal*, **288**, 27.

Wilson, J.A., Lawson, P.M., and Smith, R.G., (1987). The treatment of terminally ill geriatric patients. *Palliative Medicine*, **1**, 149–53.

World Health Organization (1986). *Cancer pain relief*. WHO, Geneva.

SECTION 23
Services to older persons

23.1 Services for older persons: a North American perspective

MARGARET F. DIMOND

INTRODUCTION

In most developed countries there is a wide range of services available for the older adult. These services target both health and social needs of elderly people. The many varieties of services are both a benefit and a liability. The complexity of regulations that govern access to some services is sufficiently overwhelming to offset the value of the service. As Libow and Sherman (1981) comment, older persons are disadvantaged in the long-term care system of the United States; it requires mobility, familiarity with insurance codes and regulations, adequate income in some cases, social supports and access to transport. Thus, the system that is intended to be of service becomes a hindrance to service.

The physician who is the medical gatekeeper for many of these services plays a key role in assuring the older person's access to the appropriate type of service. It is incumbent on the physician, therefore, to know the types of services available, to know by whom and how the services will be paid for, and to select those services that offer the best 'fit' for the older patient. This responsibility requires the physician really to know the patient (not just the patient's illness or disability), including the patient's financial status. In addition, the physician needs to be aware of the patient's family members (if they exist) and their ability and willingness to participate in care. The determination of the services required should be a collaborative decision between patient, family, physician, and other health-care providers as necessary.

Discussed in this chapter are: various services for older persons, including long-term care services, with special attention to the nursing home; considerations in the selection of services for older people; the important role of the family in the provision of care for frail older family members, including some of the difficulties that occur because of this; mental health services, and selected special issues related to such care: attitudes, paying for care, and the problems and issues peculiar to long-term care services for older women.

LONG-TERM CARE

Most services for elderly people, with the possible exception of acute hospital care, can be described within the context of long-term care. The important, but often not achieved, goal of long-term care is to match the patient's needs with the most appropriate long-term care service and to do this in a way that achieves continuity of care. Unfortunately, the system of long-term care is frequently fraught with fragmentation and discontinuities. Physicians and other health-care providers can diminish this fragmentation by understanding the many services offered as a part of long-term care and by working together to ensure the most appropriate placement of elderly people in the system.

Several definitions of long-term care have appeared in the literature. In this chapter the preferred definition is 'a range of services that addresses the health, personal care, and social needs of individuals who lack some capacity for self-care' (Kane and Kane 1982). A comprehensive system of long-term care must include linkages to both inpatient service and community care, and must address biomedical, psychological, socioeconomic, and sociocultural needs. {In the United Kingdom the term 'long-term care' is often used to designate permanent institutional care rather than the whole care system.}

Services in the continuum of long-term care should have the following characteristics (Weiler and Rathbone-McCuan 1978):

1. Supports should be continuous or for specific time intervals that extend beyond the acute problem.
2. Access should be flexible and promptly available.
3. Mechanisms should exist to assess individual need at the community level.
4. The environmental setting should encourage the individual to make use of functional strengths.
5. There should be follow-up on the continuing needs of the individual.
6. Services should be appropriate to the target population.
7. Services should vary in their degree of complexity and avoid duplication of resources.

All recipients of long-term care services have some measure of functional impairment. It is generally agreed that the most common problems which occasion the need for such care are instability, immobility, incontinence, intellectual impairment, infections, iatrogenesis, isolation/depression, and impoverishment (Kane and Kane 1982). It is these conditions that give rise to the frustrations, fears, and fatigue of family care-givers, and require the physician to intervene and decide with the family and patient the appropriate placement for continuing care and service.

The dimensions of an integrated and comprehensive system of long-term care can be considered on the basis of the primary setting of the service, i.e. residential care, institutional care, and community-based services.

Residential care services include various types of homes and are distinguished by the fact that they do not provide nursing care. A variety of names have been used with this classification including group homes, family homes, personal care, boarding care, foster care, and congregate living. In most of these settings the older person lives relatively independently, but has the security of knowing that others are available should he or she need assistance.

Institutional care services are licensed health facilities that provide a wide range of medical, nursing, and other types of professional care. These facilities are most familiar to the lay public and offer skilled nursing services, intermediate care services, psychiatric services, and rehabilitation for a wide range of situations, e.g. stroke, fractured hip. Older persons who require institutional care services are generally not able to live independently, either for a short period of time, for example, during rehabilitation after a fractured hip, or for indefinite periods of time, as in the case of the later stages of Alzheimer's disease.

In the United States, community-based services cover a very wide range of care opportunities, many of which are regulated by federal, state, or local governments or other formal entities. Included in community-based services are hospice care for the terminally ill, respite care, which provides short-term care to relieve care-givers, day health care for health and rehabilitation services, day care, which offers social programmes without health services, and community mental-health centres. In addition, community-based services usually also provide legal assistance, protective services, information and referral, transportation, low-income housing, home health aides, homemaker chore services, and meal preparation. Any combination of these services can be arranged. Often with minimal assistance from community-based services the elderly person can manage to remain independent for a considerable period of time. It behoves the physician to be familiar with residential, institutional, and community-based services so that elderly people in his or her care can benefit from this wide array of opportunity.

While the terminology of these services is specific to the United States, most European and other developed countries provide similar services under varying nomenclature. In order to best serve elderly people and their families, physicians should have a good grasp of how these services are organized, their accessibility and eligibility requirements, and the methods of payment.

THE NURSING HOME

While the nursing home is only one aspect of the total complex of long-term care services, it is the service most feared by elderly people in the United States, creates the most guilt in relatives who must place their loved one in such a facility, and conjures up the most negative picture in the minds of the general public. It is, as many believe, the place of last resort, the place to be avoided at all cost, even in some cases at the cost of destroying families who try to care for their older relatives when they are no longer able to do so.

The physician, however, is not only responsible for the placement of older patients into nursing homes, but also has the power to change the negative image of the nursing home in the local community. He or she can influence the care through the role of medical director for one or more homes. In this position the physician can oversee the care provided in the home and can work to improve it, e.g. can influence the full and appropriate functioning of the interdisciplinary team in the facility. The physician can also exercise the prerogative of being an active citizen in a given community and work for improvement of nursing homes through legal and political action.

Although in the United States and other developed countries only 5 per cent or less of the older population are in a nursing home at any one time, fully 20 per cent will spend time in a nursing home at some point in their lives. This fact should motivate all health professionals to seek to improve care in nursing homes. Physicians are in a particularly strategic place to do this because of the respect that is accorded them by the lay public.

Persons entering nursing homes typically follow either of two paths: short stays or long stays. Those who are candidates for short stays usually come from hospitals and require intensive nursing and rehabilitative services, with the goal of discharge home or to some less intensive level of care. In this category are also those admitted for terminal care of a rapidly fatal illness. {In the United Kingdom these are functions usually provided by a hospital geriatric service.} Those who will have a long stay are those for whom return to home is no longer a possibility. They may reside in a nursing home for many months or even years before they die. For this latter group the nursing home is *home*. {In the United Kingdom until recently most care of this type has been provided within the hospital-based geriatric service.}

The physical, social, and psychological environment of the nursing home is critical to the well-being of the long-stay resident. The nursing staff is an integral aspect of the nursing-home environment, and in selecting a nursing home for an elderly client the physician should take into account the level of commitment, skill, knowledge, and enthusiasm of the nursing staff.

Consider the series of events that may precede admission to a nursing home. These may include loss of health (physical or emotional), loss of independence or diminished capacity to provide for oneself, loss of home, and of the material symbolic possessions associated with home. These losses often occur in rapid succession, leaving little time for adjustment. Losses overlap. Before one is resolved another occurs, resulting in what has been described as a 'piecemeal loss of self'. By the time the elderly person is admitted to a nursing home, he or she may be experiencing the most profound loss of all—loss of hope. The staff of the nursing home, beginning with the private physician or medical director, have a professional as well as a moral responsibility to do everything possible to restore the individual to a level of hope and self-worth that will enable him or her to achieve maximal functional ability.

MENTAL HEALTH SERVICES

Mental health services for older persons are generally poorly distributed. The availability of community-based services is often minimal while institutional care, though often inappropriate and very costly, is available.

As with all health-care services for older persons, the 'fit' between the mentally ill individual and the care environment is critical to the quality of life. What Lawton (1970) has described as the hypothesis of docility is particularly relevant for mentally ill elderly persons, i.e. the greater the functional disability, the more influence the environment will have on the individual. Thus, the physician must carefully consider the options for mental health care from the least to the most restrictive. Along this continuum the service settings for the elderly individual include family home care, day care, day hospital, foster home, nursing home, and hospital (for acute events requiring short-term intensive care). Families caring for mentally ill elderly people in non-institutional settings must have the support of the family physician. The partnership of physician and family will ensure that the context of care is the most appropriate for the well-being of the patient.

While older people are at risk of a wide range of mental disability, they are not consistent users of available mental health services. Factors associated with this low use include lack of finance, biases on the part of mental health professionals against elderly persons, lack of transport, and dependence on the family physician for all physical and emotional problems. There is some evidence, however, that elderly persons (and their families) lack accurate knowledge about the services that mental health professionals provide, and thus may not consider them as useful helpers (Woodruff *et al.* 1988). Thus, the role of physician as teacher becomes significant in the treatment of mental illness for elderly people. This role is part of the services provided to individual clients who visit the physician's office, but it is also important for the physician to be a teacher in the community, where larger numbers of elders and families can be informed of the role of the mental health professionals in treating mental illness.

THE FAMILY: THE BASIC UNIT OF SERVICE

Contrary to beliefs held by many, families have not abandoned their historical role of caring for elderly parents and other family members. The vast majority of long-term care is provided by spouses, children, and other relatives. Much of this care is directly responsible for allowing the older person to continue to live at home. (See detailed discussion of the family as care-givers in Chapter 20.6.1.)

Services by the family in the home differ from the formal services provided by agencies of various kinds. The objective of the latter is to provide for the medical, nursing, social, and related services, through co-ordinated planning, evaluation, and follow-up procedures (Kermis 1986), either as direct support to frail persons living alone or supplemental to what family members are already providing.

Unfortunately, in the United States and other developed countries that do not have some form of socialized medicine, financial barriers often preclude the attainment of objectives in the formal care system. The system of financial reimbursement is strongly biased in favour of institutional care, and even more specifically, acute institutional care. In addition, even when formal in-home services are reimbursable, the complexity of the reimbursement system is sometimes more than the average person can decipher.

While families are considered to be the 'best' care-givers for frail or dependent elders, not all families are equipped to handle this responsibility. Further, not all families want to provide care in the home for their older relatives. In these situations and in cases where there is a known history of family conflict or violence, it is the responsibility of the physician or other care provider to divert the client to other care-giver services to prevent the potential problem of abuse. In those cases where the older person is already known as abused, the physician may need to resort to adult protective services. Such services are defined as 'a system of preventive, supportive, and surrogate services for the elderly living in the community to enable them to maintain independent living and avoid abuse and exploitation' (Regan 1978). Although adult protective services have themselves been open to criticism from various civil rights groups, they do provide an avenue of help in an emergency.

SPECIAL CONSIDERATIONS IN THE SELECTION OF SERVICES FOR ELDERLY PEOPLE

Given the multiplicity of health-care services for older persons, their purposes, uses and limitations, how does the physician (or other health-care worker) decide from among so many options? The critical factor in selecting health services is the 'fit' between the needs of the client and what is offered by the service. A fundamentally sound and increasingly valued method for assessing needs begins with an assessment of the functional status of the client. Professional care-givers are becoming more aware of the primacy of functional status in elderly people over the mere assignment of a diagnostic label. It is the consequences of the medical diagnosis that determine the need for services, i.e. what do the findings from a physical examination imply for the older individual's ability to function independently?

But it is not physical findings alone that guide the practitioner in the selection of services. There is in the elderly population a greater interrelatedness of problems than in younger populations. That is, the physical, social, emotional, and environmental events experienced are in close association, so that a disturbance in any one dimension is likely to be reflected in one or more of the other dimensions. It takes a special kind of practitioner to recognize changes and intervene appropriately in this complex situation. Often when nothing more can be done about a disease process, a great deal more can be done to help the older person cope with the condition.

While the above comments are directed at a single health discipline, i.e. medicine, it is very important to recognize that no single discipline can provide the wide range of health services required by elderly people. The multidisciplinary team is the model that is most widely accepted as the appropriate vehicle for geriatric care. In the Institute of Medicine (1986) report *Improving the quality of care in nursing homes*, it is noted that the comprehensive assessment of a resident (older person) entails a team effort involving at a minimum, a nurse, a physician, social worker, and physiotherapist. The importance of involving physicians as active members (whether or not they function nominally as leader) of interprofessional health-care teams is widely accepted and endorsed (Williams 1986; Kapp 1987). The smooth functioning of the professional team requires respect and open acknowledgement of the contributions of the disciplines represented on the team. Not only must professional attitudes and conduct be re-evaluated, but patients, families, and society must be educated about the value of interprofessional geriatric care and encouraged to accept and, indeed, expect it (Kapp 1987). In considering health-care service options for the older client, the physician is well advised to give primacy to those in which the interdisciplinary team is an integral component of the service.

SPECIAL ISSUES

The provision of health services to older persons in such a way as to be both appropriate to the functional level of individuals and sensitive to their special vulnerabilities is a challenge for all health-care professionals. The topics already discussed delineate some of these challenges. In concluding this chapter, three special areas of concern are noted. While the physician may not be able to have a direct impact in these areas, he or she will certainly encounter them in the practice of geriatric medicine. First, the majority of patients seen by the physician will be women. Second, negative attitudes and stereotypical thinking about older people will be apparent both in medicine and in other service settings. Third, paying for health-care services for the older population will be a continuing issue in all countries.

Women: a population at risk

As discussed in Chapter 1.1, the world's ageing community is mostly female. Not only do demographic factors shape the context of older women's lives, but the problems of ageing are increasingly women's problems. Older women are more likely than older men to be poor; to have inadequate retirement income; to be widowed, divorced, and alone and to be care-givers to other relatives (Hooyman and Kiyak 1988).

In countries where paying for care is an issue, older women have less access to health insurance than men at the same time as they experience a higher prevalence of chronic health problems. This fact is an outcome of the employment history of most women, where health insurance is tied to employment.

Approximately 70 per cent of nursing-home residents are 75 years and older, and about the same percentage are women. Nursing-home residents are disproportionately single, widowed, childless, and they are poorer than the elderly population in general. Women over the age of 75 have few available resources for home-based care, and are often unable to afford home health services. Thus, women are more likely to be in institutions for social rather than medical reasons, and may be inappropriately placed when alternative community supports might have allowed more independent life-styles (Kahana and Kiyak 1980). How ironic that women who have been care-givers all their lives have no one to care for them and must depend on public institutional care in their final days.

Attitudes toward elderly people

It is commonly acknowledged among geriatric health-care professionals that a problem equally as challenging as the population forecasts is the fact that relatively few health-care professionals choose to care for older persons. This may be due in part to the limited clinical opportunities available in curricula of schools of medicine, nursing, and other related professions. However, the problem is much more basic than that. We all are products of our cultural and social heritage. To a greater or lesser extent, societies worldwide instil negative attitudes toward ageing and thus toward the aged.

These attitudes, unfortunately, also colour the way health professionals deal with older adults. 'Professional ageism' as Butler (1969) noted can affect the kinds of services recommended for elderly people by physicians and others. For example, if one does not believe that an old person can benefit from psychotherapy or from various rehabilitative services, then less therapeutic services, like nursing-home placement, might be recommended.

It is not unusual to hear old people report humiliating experiences as they try to apply for food stamps or similar services for which they are eligible. People who work in public assistance programmes for older and other disadvantaged persons need to be oriented and trained in ways that transform these services into humane encounters.

Paying for care

In the United States, long-term care services for older persons are financed through both public and private sources. There is a variety of different financing mechanisms, some at the federal level, some at state and local levels, and some that are a combination of several of these sources. This variety creates considerable difficulty for the older person who must ferret out the eligibility criteria and the scope of benefits (services) that will be reimbursed in order to know the extent of out-of-pocket funds that he or she will need in order to receive the services. This task may be so overwhelming and confusing that the older person gives up trying to understand it, and may therefore not receive all the financial assistance available. It is important that the physician understands, if not the intricacies of the payment system, at least the fact that the system is extraordinarily complex and where to obtain expert advice and assistance. Some physicians hire or use volunteer retired individuals to function as advocates to ensure appropriate access to financial assistance for patients.

In addition to the complexity and fragmentation of the payment system for long-term care services in the United States, there is concern on the part of physicians and other health-care providers that payment only covers medical problems and often does not include other health services that might prevent episodes of illness. For example, foot care (other than surgery) is not covered under most payment plans and yet healthy feet are necessary for mobility, and active, mobile elderly people are not as likely to have the major disabilities that are the sequelae of immobility. Similarly, healthy teeth or good fitting dentures have implications for nutritional status, yet in some instances dentures are not covered by the long-term care payment systems. The same is true for spectacles and hearing aids. The ability to walk, talk, eat, see and hear has enormous implications for the physical and mental health of elderly people.

SUMMARY AND CONCLUSIONS

The physician is the vital link in the continuum of services for older people. Matching the appropriate service to the needs of the older person is critical to well-being, particularly as this encompasses functional ability. Maintaining or enhancing functional well-being is the highest goal of geriatric medicine. It is the physician's responsibility, therefore, to be knowledgeable about the variety of geriatric-care services in the community, the scope of care provided by each, and the accessibility of the services (both physical and economic) to his–her patients.

Recognizing the integral role of the family in the provision of geriatric services, the physician incorporates them into the plan of care, relying on their knowledge of the client to inform decisions about geriatric services. The physician's familiarity with the family as care providers will function to prevent 'burning out' of care-givers. The physician's role in the provision of geriatric-care services is central. With involvement of the physician, services are appropriately used and continuity of care preserved. Without such involvement, services are fragmented and clients are poorly served.

Bibliography

Busse, E. and Blazer D. (ed.) (1980). *Handbook of geriatric psychiatry*. Van Nostrand Reinhold, New York.

Butler, R. (1969). Ageism: another form of bigotry. *The Gerontologist*, **9**, 243–6.

Hooyman, N. and Kiyak, A. (1988). *Social gerontology*. Allyn and Bacon, Boston.

Institute of Medicine (1986). *Improving the quality of care in nursing homes*. National Academy, Washington.

Kahana, E. and Kiyak, A. (1980). The older woman: impact of widowhood and living arrangements on service needs. *Journal of Gerontological Social Work*, **3**, 17–29.

Kane, R. and Kane, R. (1982). *Values and long-term care*. Heath, Lexington.

Kapp, M. (1987). Interprofessional relationships in geriatrics: ethical and legal considerations. *The Gerontologist*, **27**, 547–52.

Kermis, M. (1986). *Mental health in late life*. Jones and Bartlett, Boston.

Lawton, M.P. (1970). Ecology and ageing. In *Spatial behavior of older people*, (ed. L. Pastalan and D. Carson), pp. 40–67. University of Michigan and Wayne State University Press.

Libow, L. and Sherman, F. (1981). *The core of geriatric medicine*. Mosby, St. Louis.

Markson, E. (1983). *Older women*. Lexington Books, Lexington.

Regan, J. (1978). Intervention through adult protective services programs. *The Gerontologist*, **18**, 250–4.

Stone, R., Cafferata, G. and Sangl, J. (1987). Caregivers of the frail elderly: a national profile. *The Gerontologist*, **27**, 616–26.

Weiler, P. and Rathbone-McCuan, E. (1978). *Adult day care: community work with the elderly*. Springer, New York.

Williams, T.F. (1986). Geriatrics: the function of the clinician reconsidered. *The Gerontologist*, **26**, 345–9.

Woodruff, J., Donnan, H. and Halpin, G. (1988). Changing elderly persons' attitudes toward mental health professionals. *The Gerontologist*, **28**, 800–8.

23.2 Hospital services for elderly people. The United Kingdom experience

J. GRIMLEY EVANS

Modern health-care systems evolved when the chief problems afflicting populations were acute—particularly infective—diseases in younger people. These problems have receded in frequency and severity, and the numbers of elderly people have grown. Modern medicine is therefore dominated by needs generated by chronic disease in older people. This situation has required adaptations in the health-care system in the range of services on offer and the criteria for their deployment, but also changes in the recruitment, training, and attitudes of health-care staff. There are different ways of organizing health and social services for ageing populations, but the principles which must underlie any successful system should be identified so as to provide an aid to design and a standard for audit.

PRINCIPLES OF SERVICE DESIGN

Characteristics of need

Table 1 sets out the characteristics of illness in older people that must be planned for in the provision of medical services for them. They are essentially the consequences of the age-associated loss of biological adaptability that is the essential characteristic of senescence. Their crucial implication is that an old person who falls ill needs immediate access to the best of modern medicine since diagnosis will require more investigation and recovery more active care than would be required for a similar illness in a younger patient. Furthermore, a correct and full diagnosis has to be made and appropriate treatment instituted rapidly if the progress of the disease is to be interrupted early and prolonged illness prevented. In the early days of the National Health Service in the United Kingdom, old people were all too often admitted to second-rate and under-resourced geriatric or general practitioner hospitals. This was both inhumane and inefficient and it is now generally recognized that elderly people needing hospital admission should be assessed in the same fully equipped modern general hospitals considered appropriate for younger patients (Evans 1981).

Table 1 Characteristics of disease in old age

Multiple pathology
Non-specific or insidious presentation
Rapid deterioration if untreated
High incidence of secondary complications of disease and treatment
Need for rehabilitation to encourage recovery
Importance of environmental factors (including housing and social support) in determining recovery and return to the community

Table 2 outlines some other factors that are germane to the design of specialist geriatric services, and Table 3 some of the principles of good practice. While each Health District within the United Kingdom now has a specialist geriatric service, there is considerable variety in the ways in which these services embody the principles of care derived from Tables 1 and 2. The three main models of service will be

Table 2 Other factors to be considered in designing health services for an elderly population

Suitable management structure:
(a) to ensure economic efficiency as well as clinical effectiveness
(b) to deal knowledgeably with the administrative and financial complexities of community care
Need to provide a suitable ambience for research and teaching
Need to provide attractive working conditions and career structure to foster recruitment of staff

Table 3 Principles of good geriatric practice

An old person who falls ill should have immediate access to full diagnosis and treatment
No one should be denied the benefit of treatment on grounds of age
Care should have regard to the patient's situation and view of life
Any change of dependency of an old person should lead to assessment for a therapeutic intervention before a prosthetic service is imposed
No one should be permanently institutionalized until all possible alternatives have been explored with a specialist team
Old people should be cared for as far as possible where they wish
Relocation should be minimized
A single person should be responsible for co-ordinating services for and communicating with the patient and carers
Interventions should have clearly defined, realistic, and acceptable objectives
Progress towards objectives should be continuously reviewed
The planning of interventions should consider the well-being of carers as well as of the patient

described below but all have in common the basic structure of a providing in one way or another the three inpatient functions of acute assessment, rehabilitation, and long-term nursing care. An essential feature of a British geriatric service is that these three are managed and co-ordinated by the same medical personnel so that with the associated facilities of day hospital, outpatient clinic, home visiting programme, and within-hospital liaison they form a comprehensive service that is managed for the benefit of a geographically defined population of elderly people (Evans 1983*a*, *b*). This structure can ensure both clinical effectiveness and provide a basis for economical management.

Acute assessment

Acute assessment by a multidisciplinary geriatric team (Table 4) extends the traditional process of medical diagnosis into an appraisal of mental and physical function, family, social

Table 4 The members of the core geriatric team

Physician
Nurse
Physiotherapist
Occupational therapist
Social worker

and economic resources, together with enquiry into the patient's personality and wishes for the future. From this assessment a care plan is specified, which includes medical and nursing care, treatments by physiotherapist and occupational therapist, other therapies, such as speech therapy as appropriate, and plans for future placement. At regular intervals this care plan is formally reviewed by the team in the light of the patient's progress. These reviews may identify new problems that have arisen (for example, the stroke patient who is making poorer progress than expected because of an intervening depression) or modifications to the plan that have become necessary owing to the acquisition of new information. As many patients as possible should be managed in the ward to which they are first admitted; relocation can be a traumatic experience and is believed to add extra days to the total length of stay as rehabilitation is delayed while the patient adjusts to a new environment and new staff. Planning for discharge back to the community should begin as soon as the prognosis is established. Discharge may be facilitated by a preliminary home visit by the patient accompanied by an occupational therapist and other relevant staff to assess and arrange what forms of aftercare will be needed in the patient's home.

Rehabilitation

Approximately 70 per cent of patients aged 60 and over who are admitted to the medical services of an acute hospital can be discharged directly back into the community (Evans 1983*a*). A further 5 to 10 per cent or so require more prolonged specialist rehabilitation. In most, but not all, geriatric services this is provided in separate units as the nursing skills and ambience needed for rehabilitation differ from those appropriate to acute care. Here again the principle of a formal care plan agreed by the multidisciplinary team and reviewed formally at regular meetings is continued. In addition to formal rehabilitative therapy the regimen on a rehabilitation ward is founded on a return towards 'real' life and on meaningful activity. Patients are expected to dress and care for themselves with advice and 'hands-off' help from staff as far as possible. They are encouraged to walk to the toilet, with staff help if necessary rather than to use commodes or urine bottles by their beds, and social interaction is encouraged by patients being invited to take their meals at small tables with others. Except when illness intervenes, patients are all dressed and out of bed. Physical restraints, including cot-sides (bed-rails), are proscribed throughout a geriatric service and pharmacological restraint, including hypnotics, is used only on the specific instruction of senior medical staff. Difficult behaviour is managed as far as possible by behavioural approaches (see Chapter 20.5).

Long-stay nursing care

A small proportion of the rehabilitation patients (of the order of 1 to 2 per cent of those originally admitted to the acute assessment service) do not become sufficiently independent to be able to be discharged to care at home or in a non-nursing residential home. Traditionally these patients have been offered places in the long-term nursing wards of the geriatric service but as a result of government policy more of this care is now being provided in the profit-making and voluntary sectors. In the geriatric service, long-term care is usually provided in specialist units, again because of important requirements in the recruitment and training of their nursing staff, and the need for the units to be furnished and equipped in a more 'home-like' style than is appropriate for an acute or rehabilitative hospital. Furthermore it is important that an elderly patient should be placed for long-term institutional care near to where his or her friends and relatives live. A study showed that although visiting by relatives and friends is independent of distance for the first 6 weeks or so of an elderly patient's hospital admission, after that time visiting frequency falls off in proportion to the travelling time (Cross and Turner 1974). Long-term nursing units need therefore to be scattered throughout a community rather than localized on a central hospital site.

Placement of an elderly patient in a long-stay nursing unit should be seen as a positive decision with explicit care aims, not as a discarding of someone who has 'failed' rehabilitation. The care aims need to be discussed with relatives as well as the patient. Aims are defined mainly in objective terms such as maintenance of functional level, as measured perhaps by a Barthel score, and also in subjective terms specifying the patient's personal goals and desired state of mind. Aims for a particular patient might include fitness to attend a grandchild's wedding, or attaining psychological adjustment to disability and bereavement, for example. Nursing staff usually take the lead in defining care aims as they are much closer to the life of the patient and his or her family than are medical or other staff. In general, cardiopulmonary resuscitation is not provided in British long-term wards and referral back to an acute hospital is not normal except for treatment of fractures or for surgical care. Whether an intercurrent illness such as pneumonia is treated with antibiotics or with symptomatic relief only is decided by agreement between nursing and medical staff in the light of the patient's wishes and after discussion with relatives.

PRINCIPLES OF GOOD PRACTICE

Table 3 listed some of the aspects of good practice that should be built into a geriatric service. Some of these are self-evident, some have already been alluded to, while others will be discussed briefly here.

It is a fundamental principle of British geriatric practice that no elderly patient should be placed in any form of institutional care until all possible alternatives have been explored and this exploration includes a trial of rehabilitation. This approach is conceptually different from that of the geriatric evaluation units (Rubenstein *et al.* 1984) and geriatric consultation teams (Hogan and Fox 1990) of North America, in which a preliminary triage leads to only those patients deemed likely to respond to specialist care being accepted. The British approach arose because the teams who provided the rehabilitation services were also responsible for the long-term nursing units. It is probably more costly to the hospital budget than the North American system but undoubtedly a proportion of patients who in North America would be sent directly from acute hospitals to nursing homes return to live in the community in Britain. The economic benefit that this represents (to be offset against the hospital costs) has not been estimated. The viability of the system depends on the financing structure and unselective access of elderly patients in Britain to rehabilitation services may be threatened by impending changes in health and social services' administration.

The principle underlying attempts to avoid unnecessary institutionalization of older people has been broadened by geriatricians to that of ensuring that no older person is provided with a prosthetic service (home help or bath attendant, for example) until the potential benefit from a therapeutic service has been considered. Both types of service may help an old person to retain autonomy, the primary objective of care, but therapeutic services also help towards the secondary objective of independence (Evans 1984). An old person given a prosthetic service without associated therapeutic intervention, except as a temporary measure during convalescence from an acute illness, will nearly always become permanently dependent on it. Various forms of multidisciplinary assessment of elderly people living in the community have therefore become increasingly available in the United Kingdom in recent decades. These include geriatric out-patient clinics and day hospitals, and extending nuclear primary health-care teams of doctors and nurses to include therapists and social workers.

An important requirement in Table 3 is that there should be a single member of the team who takes personal responsibility for the standard of service provided for the patient and the carers and for communication with the family. This may or may not be the physician in the team, depending on the patient's main problems and the setting. In the rehabilitation and long-term care setting it is often the nurse who takes on this role, while at the time of discharge home the social worker or one of the therapists may be the key worker. It is necessary to be clear about who has this role for each patient, as one of the dangers of a team approach is that the individual professional responsibilities of its members become delegated and lost into the collective impersonality of a bureaucracy.

MODELS OF GERIATRIC SERVICE

There are three main models of geriatric service in the United Kingdom, although many local variants of each exist. These may be designated the traditional, the age-defined, and the integrated.

The traditional model

In this model of service, geriatricians provide a medical service in parallel to the general (internal) medicine **(G(I)M)** services offered to patients of all ages. Elderly patients are selected by referring doctors, general practitioners, or hospital consultants for referral to the geriatric service. The criteria for this selection may include positive identification of a need for specialist rehabilitation, but may also unfortunately include perception of the patients as not meriting normal hospital care because they are very old, do not have an 'interesting' disease, are mentally impaired or, in the worst case, because they are of low social class. (Middle-class families may object vociferously to having one of their members classified as 'geriatric'!) The problem with this type of service is that the referring doctors may not know enough about the geriatric service to which they are referring patients to make a proper judgement of whether the referral is appropriate or not, or more importantly at what stage of the patient's illness it should take place. This model is basically a relic of the early days of geriatric medicine when it began as a low-status specialty confined to the old workhouses that had been redesignated as geriatric hospitals after the advent of the National Health Service (Editorial 1985). For this reason it has come to be associated with the continuance of discriminatory attitudes towards older people and has become largely obsolescent.

The age-defined model

As noted above, modern geriatric medicine is based on the management of acute, rehabilitative, and long-stay care as a comprehensive system. Historically one of the main difficulties in achieving this has been the difficulty geriatricians have experienced in gaining control of the fully resourced acute facilities in hospital which traditionally had been the purlieu of G(I)M services. Where the definition of what constitutes a 'geriatric' patient has been left to the G(I)M and other referring agencies, geriatricians have experienced problems with planning for a predictable work-load. The first approach to solving these difficulties was to set up age-defined services in which geriatricians provided all acute medical care for patients above a certain age (Horrocks 1982). The rationale for this model is that age is used as a screening variable to distinguish those patients who will do best under geriatric care from those who will do best under G(I)M care. The specificity and sensitivity of different ages has never been estimated, however, and the wide range of defining ages in the United Kingdom, ranging from 65 to over 80, indicates that the real basis for the model is the resources available to the geriatric service rather than any rational consideration. The model has been criticized as providing poorer quality of care to older patients in that they may not have access to appropriate specialist care. For example, an elderly patient with an acute myocardial infarct in an age-defined system will be looked after by a geriatrician in a geriatric ward, while younger patients will be cared for by a cardiologist in a cardiology ward. The model is attractive to administrators, however, as it offers opportunity to provide cheaper care for a section of the population; it is an accepted tradition for geriatric services to be funded at a lower level than those for younger patients. Geriatricians also find the model attractive as it provides them with a definable territory, free from competition with staff in other specialties.

The integrated model

In this, the newest of the models of geriatric service, geriatric medicine is deployed like any other subspecialty of medicine (Evans 1983*a*). Geriatricians work as members of multispecialty teams sharing junior medical staff and joining in the assessment of elderly patients admitted to a single emergency unit. Subsequent allocation of patients to specialist geriatric facilities rather than those of another specialty is based on their assessed needs rather than their age. The pooling of emergency resources between the specialties leads to more efficient use of resources and the spread of expertise between the specialties; the co-operative use of junior medical staff and nursing teams provides an ambience particularly suited to teaching centres (Parkhouse and Campbell 1983). Successful implementation of this model requires a high degree of co-operation between the different specialties, which is not always easy to achieve, and the erasing of traditional negative attitudes towards older patients and their problems. An important benefit of this model is that old people have the same access to assessment for 'high technology', specialist medicine as do the young. This will be increasingly important as fiscal considerations lead to growing pressure from politicians to exclude elderly people from access to expensive medical or surgical care, even though they may be able to benefit from it.

Whatever model of service is deployed, the essential requirement is that all health-service staff agree on its ground rules, and that referring agencies (particularly in primary care) are kept informed about how the system works and its procedures for access.

References

Cross, K.W. and Turner, R.D. (1974) Factors affecting the visiting pattern of geriatric patients in a rural area. *British Journal of Preventive and Social Medicine*, **28**, 133–9.

Editorial (1985). Geriatrics for all? *Lancet*, **i**, 674–5.

Evans, J.G. (1981). Institutional care. In *Health care of the elderly*, (ed. T. Arie), pp. 176–93. Croom Helm, London.

Evans. J.G. (1983*a*). Integration of geriatric with general medical services in Newcastle. *Lancet*, **i**, 1430–3.

Evans, J.G. (1983*b*). The appraisal of hospital geriatric services. *Community Medicine*, **5**, 242–50.

Evans, J.G. (1984). Prevention of age-associated loss of autonomy: epidemiological approaches. *Journal of Chronic Diseases*, **37**, 353–63.

Hogan, D.B. and Fox, R.A. (1990). A prospective controlled trial of a geriatric consultation team in an acute-care hospital. *Age and Ageing*, **19**, 107–13.

Horrocks, P. (1982). The case for geriatrics as an age-related specialty. In *Recent advances in geriatric medicine 2*, (ed. B. Isaacs), pp.259–77. Churchill Livingstone, Edinburgh.

Parkhouse, J. and Campbell, M.G. (1983). Popularity of geriatrics among Newcastle qualifiers at preregistration stage. *Lancet*, **ii**, 221.

Rubenstein, L.Z., *et al.* (1984). Effectiveness of a geriatric evaluation unit: a randomized clinical trial. *New England Journal of Medicine*, **311**, 1664–70.

SECTION 24
Preventing disease and promoting health in old age

24 Preventing disease and promoting health in old age

J. A. MUIR GRAY

For some problems, prevention is not possible because the causal factor, or factors, are at present unknown although they are presumed to be extrinsic. In a second type of problem, for example osteoarthrosis of the hips, some of the causal factors have been identified but it is not yet possible to work out a preventive strategy. For a third group of problems preventive action must take place in youth and middle age, for example the prevention of chronic rheumatic heart disease is carried out by ensuring that the drugs and medical services are available to treat streptococcal infections in youth. There is, however, a fourth group of problems that can be prevented by action initiated after the age of 65, and the aim of this chapter is to discuss these problems and the steps that can be taken to prevent them. The objectives of this chapter are:

(1) to discuss the techniques that can be used to prevent disease and promote health in old age;
(2) to discuss not only the benefits but also the costs of disease prevention so that the reader may decide on the type of preventive services that will be most appropriate for his or her population.

CLASSIFYING PREVENTIVE ACTIVITIES

1. By the natural history of the disease

The most common means of classifying those activities that can prevent disease is to group them with respect to the stage of development of the relevant disease. Commonly, disease is described as having three stages: a stage at which only a precursor condition or risk factor is present, an asymptomatic stage, and a symptomatic stage. However, when considering health in old age it is common to divide the stage of symptomatic disease into two parts—the first when the disease is known only to the old person and has not been presented to the relevant health service, and the second being the stage at which the disease is known to the health service. These stages can be set out diagrammatically as shown in Fig. 1.

Preventive activities are often classified as either being primary, secondary, or tertiary, corresponding to the stages of the disease as shown in Fig. 1. The term 'tertiary prevention' is, however, clumsy and is hardly ever used, and there is also some confusion about the term 'secondary prevention', both because it can be used to describe the prevention of recurrence—for example, the prevention of a second myocardial infarction—and because it could be used to describe both the detection of asymptomatic disease and the identification of people with symptomatic health problems who have not presented those problems to the relevant health service. It is more common, therefore, to use the terms set out in Fig. 1.

No disease: risk factor present	Asymptomatic disease	Symptomatic disease	
		Not presented to any Health Service	Known to the relevant Health Service
Health education and health promotion	Screening	Case-finding	

Fig. 1 The three types of preventive activity.

2. By the type of intervention

This chapter is written in order to help those responsible for providing health services to individual old people or populations of older people. The main means doctors have of preventing disease is by offering or recommending preventive services such as counselling, screening, or immunization. However, preventive health services are only one of the three types of intervention that can prevent disease and promote health. The other two may be equally, if not more, effective but the physician has little influence over these other two types of intervention—changes in the physical environment and changes in the social environment.

Changes in the physical environment

Some of the problems that old people face are caused by the physical environment in which they lived when younger. Many old people with bronchitis and emphysema are disabled because of air pollution or industrial exposure in the past. However, the physical environment is also relevant to the development of health problems in old age, for example elderly people are at risk of death and injury on the roads when out walking. Similarly, poor housing contributes to ill health and the design of city public buildings increases the problems that old people have in obtaining their rights or getting access to facilities for recreation and leisure; and the inadequacy of public transport throughout the world compounds the problems of older people. The physical environment is, by and large, designed by young fit people who drive cars. The needs of older disabled people should be advocated frequently and assertively to create an environment that promotes health in old age.

Changes in the social environment

One of the major problems faced by mainly older people is poverty. Doctors can play a part in mitigating the effect of poverty, for example by ensuring their service is set up so as to provide older people with information about the financial benefits available and the opportunity of claiming those benefits if they so wish.

However, decisions about the entitlements of old people and about retirement age are political decisions made by politicians based partly on their assumptions about the prevailing views of society at large and partly on their own prejudices about old age. The individual physician is relatively powerless when it comes to influencing these decisions but the medical profession can make a contribution to them, and to the perception that old people have of themselves, by taking every opportunity to remind people how many of the problems of old age would be solved or mitigated if the prevalence of poverty were reduced.

RISK FACTOR MANAGEMENT

The term 'risk factor' is used to describe a characteristic of an old person or their circumstances that has been shown to be associated with an increased probability of disease. It is important to remember that the existence of the risk factor does not necessarily mean that modification of the risk factor will reduce the risk of disease. Ideally there should always be randomized trials to assess the effect that modification of the risk factor has on the probability of disease before action is taken, but such trials are difficult to organize in old age. One of the reasons for this is that more than one risk factor is often present; it is difficult to design studies with sufficient power to evaluate the effects of modifying one of these risk factors because the effect of modifying a risk factor may be small, though of clinical significance.

The objectives of health care for old people are not only to reduce the incidence of disease but also to improve quality of life, to maintain independence, and to allow people to stay in their own homes if that is what they wish. The following risk factors have been shown to be relevant to these objectives (Gray 1985).

Poverty
Bad housing
Physical disease or disability
Mental illness
Cigarette smoking
A diet deficient in essential vitamins or with too much energy for the person's requirements
Inactivity
Poor quality health care

Interventions to modify risk factors

There are three techniques that can be used to try to modify these risk factors, techniques which should not be considered in isolation but which should be integrated with one another as part of a comprehensive prevention programme.

Individual counselling

Many old people are isolated and a considerable proportion have no living relatives. Many decisions, such as the decision to stop smoking or change one's diet or buy an annuity with capital, are made by younger people after discussion with friends and family, but this type of decision-making is denied to many older people. To help older people make decisions it is therefore useful to try to provide them with an individual who can help them reach the decision. The individual may be someone with authority and knowledge—for example it has been shown that a general practitioner can give advice about lifestyle more effectively than other forms of intervention because of the respect that the average person has for the opinion of a doctor. However, much good advice can be given by people other than trained professionals, and an increasing number of schemes involve older people helping one another, either acting as their advocates or simply providing information and support.

Health education

The provision of information and support to individuals by counselling can be complemented by the provision of information through leaflets, newspapers, or advertising, or, perhaps more effectively, by linking with groups of older people, such as old people's clubs or networks of retired employees, to ensure that people are informed about health options and opportunities.

Health promotion

Health promotion is a broader concept than health education. It is broader in two ways: first, because it is designed to help older people feel better as well as helping them lower the risk of disease; secondly, health promotion also involves attempts to modify the environment in which the old person lives, for example, by trying to campaign for higher pensions as well as by ensuring that the old person claims the most benefits that are currently available.

SCREENING

Screening is the detection of asymptomatic disease, for example early breast cancer, or the precursor of disease, for example high blood pressure or carcinoma-in-situ. The screening test itself is usually not definitive, indicating only an increased probability of disease and requiring further investigations to confirm the diagnosis.

Markedly different approaches to screening can be found in the United Kingdom and the United States of America, differences which indicate different approaches to health care and the different culture in which each professional group has to work.

The United States approach

The United States Department of Health and Human Sciences set up a US Preventive Services Task Force which has produced and published a guide to clinical preventive services (US Preventive Services Task Force 1989). This guide reviews the scope for prevention both by health promotion and by screening for a number of different age groups and conditions, and its recommendations for people aged over 65 are reproduced in Table 1.

The United Kingdom approach

The United Kingdom's recommendations for preventive services in old age do not specify any particular laboratory or diagnostic procedures, such as those set out in the *Guide to Clinical Preventive Services*. However, the consensus of opinion in the United Kingdom is that only three screening tests can be supported as being appropriate for older people,

Table 1 Recommendations of the *Guide to Clinical Preventive Services* (US Preventive Task Forces 1989) for those aged 65 and over; procedures scheduled for every year*

Screening	Counselling	Immunizations
History	*Diet and exercise*	Tetanus–diphtheria (Td) booster[5]
Prior symptoms of transient ischaemic attack	Fat (especially saturated fat), cholesterol, complex carbohydrates, fibre, sodium, calcium[3]	Influenza vaccine[1]
Dietary intake	Caloric balance	Pneumococcal vaccine
Physical activity	Selection of exercise programme	**High-risk groups**
Tobacco/alcohol/drug use	*Substance use*	Hepatitis B vaccine (HR16)
Functional status at home	Tobacco cessation	
Physical examination	Alcohol and other drugs:	
Height and weight	Limiting alcohol consumption	
Blood pressure	Driving/other dangerous activities while under the influence	
Visual acuity	Treatment for abuse	
Hearing and hearing aids	*Injury prevention*	
Clinical breast exam[1]	Prevention of falls	
High-risk groups	Safety belts	
Auscultation for carotid bruits (HR1)	Smoke detector	
Complete skin exam (HR2)	Smoking near bedding or upholstery	
Complete examination of the mouth (HR3)	Hot water-heater temperature	
Palpation of thyroid nodules (HR4)	Safety helmets	
Laboratory/diagnostic procedures	**High-risk groups**	
Non-fasting total blood cholesterol	Prevention of childhood injuries (HR12)	
Dip-stick urinalysis	*Dental health*	
Mammogram[2]	Regular dental visits, tooth brushing, flossing	
High-risk groups	*Other primary preventive measures*	
Fasting plasma glucose (HR5)	Glaucoma testing by eye specialist	
Tuberculin skin test (PPD) (HR6)	**High-risk groups**	
Electrocardiogram (HR7)	Discussion of oestrogen replacement therapy (HR13)	
Papanicolaou smear[4] (HR8)	Discussion of aspirin therapy (HR14)	
Faecal occult blood/sigmoidoscopy (HR9)	Skin protection from ultraviolet light (HR15)	
Faecal occult blood/colonoscopy (HR10)		

*The recommended schedule applies only to the periodic visit itself. The frequency of the individual preventive services listed in this table is left to clinical discretion, except as indicated in other footnotes.

[1] Annually. [2] Every 1–2 years for women until age 75, unless pathology detected. [3] For women. [4] Every 1–3 years. [5] Every 10 years.

HR1 Persons with risk factors for cerebrovascular or cardiovascular disease (e.g. hypertension, smoking, coronary artery disease, atrial fibrillation, diabetes) or those with neurological symptoms (e.g. transient ischaemic attacks) or a history of cerebrovascular disease.

HR2 Persons with a family or personal history of skin cancer, or clinical evidence of precursory lesions (e.g. dysplastic naevi, certain congenital naevi), or those with increased occupational or recreational exposure to sunlight.

HR3 Persons with exposure to tobacco or excessive amounts of alcohol, or those with suspicious symptoms or lesions detected through self-examination.

HR4 Persons with a history of upper-body irradiation.

HR5 The markedly obese, persons with a family history of diabetes, or women with a history of gestational diabetes.

HR6 Household members of persons with tuberculosis or others at risk for close contact with the disease (e.g. staff of tuberculosis clinics, shelters for the homeless, nursing homes, substance-abuse treatment facilities, dialysis units, correctional institutions); recent immigrants or refugees from countries in which tuberculosis is common (e.g. Asia, Africa, Central and South America, Pacific Islands); migrant workers; residents of nursing homes, correctional institutions, or homeless shelters; or persons with certain underlying medical disorders (e.g. HIV infection).

HR7 Men with two or more cardiac risk factors (high blood cholesterol, hypertension, cigarette smoking, diabetes mellitus, family history of coronary artery disease); men who would endanger public safety were they to experience sudden cardiac events (e.g. commercial airline pilots); or sedentary or high-risk males planning to begin a vigorous exercise programme.

HR8 Women who have not had previous documented screening in which smears have been consistently negative.

HR9 Persons who have first-degree relatives with colorectal cancer; a personal history of endometrial, ovarian, or breast cancer; or a previous diagnosis of inflammatory bowel disease, adenomatous polyps, or colorectal cancer.

HR10 Persons with a family history of familiary polyposis coli or cancer-family syndrome.

HR11 Recent divorce, separation, unemployment, depression, alcohol or other drug abuse, serious medical illnesses, living alone, or recent bereavement.

HR12 Persons with children in the home or automobile.

HR13 Women at increased risk for osteoporosis (e.g. Caucasian, low bone mineral content, bilateral oopherectomy before menopause or early menopause, slender build) and who are without known contraindications (e.g. history of undiagnosed vaginal bleeding, active liver disease, thromboembolic disorders, hormone-dependent cancer).

HR14 Men who have risk factors for myocardial infarction (e.g. high blood cholesterol, smoking, diabetes mellitus, family history of early-onset coronary artery disease) and who lack a history of gastrointestinal or other bleeding problems, or other risk factors for bleeding or cerebral haemorrhage.

HR15 Persons with increased exposure to sunlight.

HR16 Homosexually active men, intravenous drug users, recipients of some blood products, or persons in health-related jobs with frequent exposure to blood or blood products.

This list of preventive services is not exhaustive. It reflects only those topics reviewed by the US Preventive Services Task Force. Clinicians may wish to add other preventive services on a routine basis, and after considering the patient's medical history and other individual circumstances. Examples of target conditions not specifically examined by the Task Force include: chronic obstructive pulmonary disease; hepatobiliary disease; bladder cancer; endometrial disease; travel-related illness; prescription drug abuse; occupational illness and injuries.

Remain alert for: depression symptoms; suicide risk factors (HR11); abnormal bereavement; changes in cognitive function; medications that increase risk of falls; signs of physical abuse or neglect; malignant skin lesions; peripheral arterial disease; tooth decay, gingivitis, loose teeth.

using the criteria for screening first formulated in a paper (Wilson and Jungner 1968) written for the World Health Organization (Table 2). The three tests to be discussed in detail are blood pressure measurement, cervical screening, and mammography.

Table 2 Criteria for screening (after Wilson and Jugner 1968)

The condition sought should pose an important health problem.
The natural history of the disease should be well understood.
There should be a recognizable early stage.
Treatment of the disease at an early stage should be of more benefit than treatment started at a later stage.
There should be a suitable test.
The test should be acceptable to the population.
There should be adequate facilities for the diagnosis and treatment of abnormalities detected.
For diseases of insidious onset, screening should be repeated at intervals determined by the natural history of the disease.
The chance of physical or psychological harm to those screened should be less than the change of benefit.
The cost of the screening programme should be balanced against the benefit it provides.

Screening for high blood pressure over the age of 65

Much of the early work on the effectiveness of reducing blood pressure was carried out on people under the age of 60. During the 1980s a number of trials of population screening and blood pressure reduction in people aged over 60 were reported. The National Institutes of Health produced a 'Statement on hypertension in the elderly', which stated that 'although a number of studies have demonstrated the benefits of antihypertensive therapy in elderly patients, many questions unique to this population remain unanswered' (Working Group on Hypertension in the Elderly 1986). However, it is accepted that there are benefits from reducing diastolic hypertension in people aged over 60, with three trials—in the United States, Australia, and Europe—showing a reduction in deaths from stroke and cardiac disease (National Heart Foundation of Australia 1981; Amery *et al.* 1985). The Systolic Hypertension in the Elderly Programme and a randomized trial of the treatment of hypertension in elderly patients in primary care considered the effect of treating people who had only systolic hypertension (Coope and Warrender 1986; SHEP Cooperative Research Group 1991). Both these studies showed that there were benefits from treating people who had only systolic hypertension. The reduction in cardiac mortality was not a constant feature in trials and may depend not only on the degree of blood-pressure reduction achieved but also on the drugs used, with β-blockers having a greater impact than other types of antihypertensive medication.

The precise level at which intervention is justified and the target blood pressure varies from trial to trial, with the recommendation from the United States proposing more vigorous management of blood pressure than European recommendations with, for example, the National Institutes of Health recommendation being that 'the goal of treatment is similar to the goal for younger patients, namely to reduce the diastolic blood pressure to below 90 mmHg or by at least 10 mmHg if pretreatment diastolic blood pressure is less than 100 mmHg'. In addition the doctor would wish to take into consideration the circumstances and characteristics of the individual with raised blood pressure before deciding on the use of drug treatment, weighing up the balance that has to be set in all preventive and therapeutic care between benefits and adverse effects.

Although it is widely believed that side-effects are more common in older people, this has not been found in all of the trials (Coope and Warrender 1986), but there are important studies that remind us of the adverse effects of hypertension treatment, for example the contribution that inappropriate treatment can make to the pathogenesis of stroke (Jansen *et al.* 1986). Adverse effects are always regrettable but particularly so when they are produced in a person who was not ill before treatment started.

Finally, it is important to remember that for an individual patient there will come a time when it may be appropriate to stop medication. This occurs in any population of people who had high blood pressure at the time when treatment was started but is particularly appropriate when managing high blood pressure in old age because there appears to be little evidence of benefit from lowering blood pressure over the age of 80 (Flamenbaum and Cohen 1985) (see also Chapter 11.6).

Mammographic screening

Mammographic screening for breast cancer is an effective means of reducing mortality in women over 65. However, because it was envisaged that the response to invitations for breast-cancer screening would be so low in this age group as to make the population impact of screening negligible, a decision was taken that women over the age of 65 should not be routinely invited for breast cancer screening in the United Kingdom (Department of Health 1986). The evidence on which this policy is based is the relatively low response rate to written invitations of women in Sweden. However, attitudes towards preventive activities are probably cohort-related, namely it cannot be assumed that women aged over 65 in future will have the same beliefs and attitudes as women aged over 65 at present. For this reason the screening programme in the United Kingdom will continue to invite women for mammographic screening after the age of 65 if they have been recruited into the programme before the age of 65.

Cervical screening

It has been said that cervical screening is unnecessary in women who have had one, or preferably more than one, negative smear test before the age of 65 because such women are at very low risk of developing cervical cancer (Intercollegiate Working Party 1987). There is no conclusive scientific evidence to support this belief or policy but it is one that is firmly established in screening programmes, in part perhaps because the cervical smear test becomes increasingly difficult to perform with advancing age as a result of the atrophy of tissues around the vaginal wall. Furthermore, cervical smear tests are often inconclusive in women over 65 because of atrophic changes in the epithelium.

Thus most screening programmes cease to invite women to continue attending if they have had one, or more than one, negative smear test at the age of 65. However, cervical cancer continues to be a disease of women over 65. It is therefore becoming increasingly common for screening programmes to try to reach women aged 65 and over who have never had a smear test and to offer them such a test, even though the test is difficult to do, will be unacceptable to many older women, and may provide unsatisfactory results even if it is done.

CASE FINDING—THE SEARCH FOR UNREPORTED PROBLEMS

For many reasons, old people often fail to report health and social problems to an appropriate source of help. A high prevalence of unreported problems has been shown in some but not all studies. Influenced by such findings a number of projects were developed in the United Kingdom to identify and solve the unmet needs of older people, but it was not until 1979 that the effectiveness of identifying and tackling unreported problems was evaluated by means of a randomized controlled trial. There have now been five such trials—three in England, one in Wales, and one in Denmark (Tulloch and Moore 1979; Hendriksen *et al.* 1984; Carpenter and Demopolus 1990; McKeown *et al.* 1990). One or more of these trials demonstrated the following benefits from early intervention:

(1) higher quality of life;
(2) reduction in the number of institutional-bed days used;
(3) increased take up of domiciliary services;
(4) decreased rate of admission to nursing homes;
(5) decreased mortality rate.

Few adverse effects result from case finding.

Dealing with the problems revealed by case finding

Although some early studies revealed a high prevalence of medical needs, more recent studies in the United Kingdom have shown that the principal problems faced by old people are social rather than medical, notably problems with income and housing. Of the health problems reported, difficulties with hearing, seeing, and feeding were the most common. Paradoxically, the unmet medical need revealed may be a need for the review of medical care currently being given, in particular the need to review medication.

Organizing case finding

The National Health Service in the United Kingdom decided in 1990 to make all general practitioners carry out an annual health check of people aged over 75 as part of their terms of service. Advice was given on the topics that should be covered in case finding and these are:

sensory functions
mobility
mental condition
physical condition, including continence
social environment
use of medicines.

It is possible to devise simple questionnaires to cover these topics, and simple short questionnaires are the only ones worth using if costs are to be kept to a minimum.

Those responsible for providing health services to older people can organize case finding in one of two ways. First, lists of addresses or registers may be used to identify older people and to invite them to attend for a health interview. The response rate to such invitations is usually low, in part because of the beliefs and attitudes of older people but principally because of the problems they have with transport. Secondly, use can be made of the fact that about 80 per cent of people aged over 75 initiate contact with a health professional working in primary care—a doctor or nurse—at some time during the course of any year. The opportunity offered by this contact can be used to carry out a health review, and if a record is kept of those who have had a health review carried out in this way it is comparatively simple for the primary-care team or provider of services to identify and make contact with that small proportion of people who have not themselves initiated contact. This approach is sometimes called the 'opportunistic approach'.

Weighing up the costs of case finding

Case finding has a number of costs. One cost is to the older people themselves who may have their hopes raised but not fulfilled. Most older people, however, have had so many disappointments in their lives that they can cope with further disappointments and usually welcome the fact that health professionals are trying to identify and solve their problems. They also appreciate that it is only by revealing problems that the necessary political action can be taken to solve them, if not for them at least for older people in cohorts to come.

Case finding does, however, have considerable costs for health services, costs that they have to compare with the costs of investing those same resources in treatment services. These costs can be reduced by using volunteers, who appear to be effective in carrying out the initial review of problems (Carpenter and Demopolus 1990).

Weighing up the opportunity costs of prevention

Much can be done to improve the health of older people, in part by health-service action, in part by action taken by other government agencies and voluntary societies. Those who provide and pay for health services have to consider not only the benefits of disease prevention and health promotion but also the costs, because the money invested in disease prevention and health promotion is money that could otherwise be invested in treatment services.

Some people are of the opinion that the over-riding priority of a health service is to provide for ill people and that deficiencies in treatment services, particularly for acutely ill people, should be made good before money is invested in preventive services. Others, however, believe that some money must be invested in disease prevention and health promotion, in part to prevent heavier costs falling upon the health service in years to come, in part because they believe that healthy older people have a right to preventive services in the same way that ill old people have a right to treatment services.

There is, unfortunately but not unexpectedly, no simple formula that can be used to compare the costs and benefits of prevention and treatment services. Attempts have been made to develop economic measures, for example QALY—the quality-adjusted life year—but such measures have many deficiencies (Harris 1987). What is important is to state clearly the benefits and adverse effects that result from investment of resources and to achieve, in discussion with older people or their representatives, some equitable balance between health promotion, disease prevention, and treatment services.

References

Amery, A. *et al.* (1985). Mortality and morbidity results from the European Working Party on High Blood Pressure in the Elderly trial. *Lancet*, **i**, 1349–54.

Carpenter, G.I., and Demopoulos, G.R. (1990). Screening the elderly in the community: control trial of dependency surveillance using a questionnaire administered by volunteers. *British Medical Journal*, **300**, 1253–6.

Coope, J. and Warrender, T.S. (1986). Randomised trial of treatment of hypertension in elderly patients in primary care. *British Medical Journal*, **293**, 1145–51.

Department of Health (1986). *Breast Cancer Screening*, Report of Working Party chaired by Sir Patrick Forrest. HMSO, London.

Flamenbaum, W. and Cohen, N.S. (1985). Editorial. The decision to 'unmedicate'. *Journal of the American Medical Association*, **253**, 687–8.

Gray, J.A.M. (1985). *Prevention of disease in the elderly*. Churchill Livingstone, Edinburgh.

Harris, J. (1987). QALYfying the value of life. *Journal of Medical Ethics*, **13**, 117–23.

Hendriksen, C., Lund, E., and Stromgard, E. (1984). Consequences of assessment and intervention among elderly people: a three-year randomised controlled trial. *British Medical Journal*, **289**, 1522–4.

Intercollegiate Working Party (1987). *Report on the Intercollegiate Working Party on Cervical Cytology Screening*. Royal College of Obstetricians and Gynaecologists, London.

Jansen, P.A.F., Gribnau, F.W.J., Schulte, B.P.M., and Poels, E.F.J. (1986). Contribution of inappropriate treatment for hypertension to pathogenesis of stroke in the elderly. *British Medical Journal*, **293**, 914–17.

McEwan, R.T., Davison, N., Foster, P.P., Pearson, T., and Stirling, E. (1990). Screening elderly people in primary care: a randomised control trial. *British Journal of General Practice*, **40**, 94–7.

National Heart Foundation of Australia (1981). Treatment of mild hypertension in the elderly. *Medical Journal of Australia*, October 17, 398–402.

SHEP Cooperative Research Group (1991). Prevention of stroke by antihypertensive treatment in older persons with isolated systolic hypertension. Final results of the Systolic Hypertension in the Elderly Program (SHEP). *Journal of the American Medical Association*, **265**, 3255–64.

Tulloch, A.J. and Moore, V. (1979). A randomised controlled trial of geriatric screening and surveillance in general practice. *Journal of the Royal College of General Practitioners*, **29**, 733–42.

US Preventive Services Task Force (1989). *Guide to clinical preventive services*. Williams & Wilkins, Baltimore.

Vetter, N.J., Jones, D.A., and Victor, C.R. (1984). Effect of health visitors working with elderly patients in general practice: a randomised controlled trial. *British Medical Journal*, **288**, 369–72.

Wilson, J.M.C. and Jungner, G. (1968). *Principles and practice of screening for disease*, Public Health Paper No. 34. WHO, Geneva.

SECTION 25
Ethical issues in the medicine of later life

25 Ethical issues in the medicine of later life

CHRISTINE K. CASSEL

The principles of ethical action are shaped by one's cultural context within the tradition of Western culture, including the traditions of major world religions and secular, political, and social philosophies. Nonetheless, there is a remarkably broad commonality of basic principles among these different cultural sources. Scholars examining ethical problems in medicine over the last two decades have established an analytical framework for problem solving which includes these widely held, fundamental principles. The cardinal principles of respect, beneficence, and justice are here considered first.

RESPECT FOR PEOPLE

Respect for people is a principle on which many rights and constitutional entitlements are based, and is essential for harmonious interactions in the human community. In medical care, this principle implies full disclosure of information to patients as a basis for informed consent. One should, in most cases, be frank with patients about their diagnosis, potential treatments, and prognosis. The physician should enlist the patient as a partner when decision making of any ambiguous nature is necessary. The exceptions to this are the few cases when it truly seems medically or psychiatrically dangerous to give the information to the patient or, as sometimes happens, when the patient requests not to be told and asks that the doctor make all the decisions. Even with patients who appear ambivalent, the effort must be made, and it is often surprising how much is understood. With people who have memory deficits, one must be prepared patiently to repeat the information, sometimes many times over. Informed consent is often an ongoing process of communication rather than a single encounter focused on the signing of a piece of paper.

Another aspect of this principle is self-determination, that the patient be allowed to decide for himself or herself what is the most favourable course, according to his or her own values. It is a common pitfall for health professionals to judge a patient's decisions by their own values rather than by those of the patient.

A third aspect of respect for people is constituted by respectful action. If a patient is not aware enough to engage in meaningful dialogue about the nature and prognosis of his or her disease, one must still take care to treat that patient respectfully. This includes how the patient is addressed, handled, clothed, and treated during the course of clinic visits or stays in hospital. Our actions towards those who are most vulnerable not only reflect our attitudes towards the human community, but also may strengthen or instil those attitudes in others.

One also shows respect for patients by ensuring that patients do not have a long or uncomfortable wait, that they have enough time with the physician, that there is acoustic and visual privacy for the interaction, and that the patient and the physician can openly discuss the patient's concerns and wishes about his or her care when death is near.

BENEFICENCE

Beneficence, or 'doing good', has a corollary in a widely quoted Hippocratic statement: 'Do no harm'. Often the two aspects of this principle come into conflict with each other, as when a very risky or painful therapy has a chance of benefiting a patient. The imperative is to do good, but in geriatrics one wonders if the decision to follow the more conservative route of 'doing no harm' would be the most ethical. In this kind of conflict, it is, whenever possible, the patient's own decision to make. In the case of an uncommunicative, comatose, or severely demented person, one is helped enormously by prior knowledge of the patient, his or her life's values and plans, and contributive information from family and friends.

It is important in understanding the principle of beneficence to distinguish doing what is best for the patient from paternalistic action. Paternalism is a stance in which one makes decisions on behalf of another, as a parent would for a child, when that person can either not decide for himself or herself or when it is believed that he or she is making the wrong decision. The history of medicine is largely one of paternalistic attitudes toward patients. It is only recently that health professionals have begun a serious attempt to share medical information with their patients. The image of the caring family physician who takes all the troubles of the patient on his or her shoulders, makes all the hard decisions, and simply tells the patient not to worry or that 'I did everything I could' is fading from view. However, we should not reject all the qualities of that era, for there is a great deal of caring concern that emerges from such an image. In the new model of enhanced autonomy of patients, it would be a mistake to lose the caring and beneficent aspect of health care practice.

In the name of beneficence, it is morally correct at times to let a patient die to put an end to needless suffering that cannot be relieved. Merciful and compassionate treatment of the dying is a very important part of any medical practice. The physician must have courage and sensitivity to make the ethical decisions involved, and these attributes ought to be emphasized in the training of every health care professional.

In ethical dilemmas concerning elderly patients, one often encounters the issue of 'quality of life'. Decisions made on considerations of 'quality of life' must be examined scrupulously because there is significant risk of generalizing one's own values in such an assessment. This caveat is clearly exemplified by consideration of a patient with dementia. Clinical decisions may be made on the assumption that a person who is demented, particularly one who is in a nursing home, has such a poor quality of life that it is not worth living. Although we may not explicitly acknowledge this, there may also be an assumption that such patients ought mercifully to be allowed to die.

Before such decisions are made and acted on, it is useful to examine the following issues. First, decisions that are made to relieve the suffering of the patient must take into account whether or not that patient is actually suffering. In fact, one of the most disconcerting aspects of dementing syndromes is that the patient loses insight about his or her own condition. If they are well cared for, patients with dementia often do not appear to suffer. The care-givers may suffer, seeing the reflection of a potential future loss of self. Physicians, in particular, find the prospect of developing a dementing illness fearful and abhorrent. But for some people, there is a significance to life that goes beyond cognitive ability. For example, it is not morally acceptable routinely to refuse medical treatment to retarded children, even though their cognitive function may not be any greater than that of a moderately demented elderly person. Secondly, part of our worry about their suffering is justified by the realities of acute and long-term care facilities. If we must place patients in institutions where they are given poor care, are neglected, and may even be mistreated, then concern for suffering is certainly warranted. But the moral obligation here is to take socially responsible action toward changing these conditions rather than to let the patient die because the conditions are so poor.

JUSTICE

The third major principle is that of justice. Simply stated, the resources (material and otherwise) of human society should be distributed as fairly as possible. This seems to be an ideal that is appealing to most people but not one that is easy to describe in practical terms, especially in its application to the allocation of scarce health care resources.

In the past decade there has been a great deal of discussion about the high cost of medical care and about the physician's role in keeping unnecessary costs down. Obviously one should not needlessly waste scarce resources in medicine any more than in any other sphere. Nor should one use expensive procedures on a patient who has asked to be allowed to die or for whom medical care cannot provide help. Except for those two situations, a physician who is making a decision for any patient on the basis of how much the treatment will cost is not acting strictly in the medical role. The physician should have a role in setting policies about what resources are available for medical care—setting priorities for spending within the health care budget and at times also arguing the value of spending on health care rather than in some other areas of public expenditures. This is a social perspective, and one in which our institutions should carefully weigh the fairness of distributive policies. However, the physician has not been trained as an arbiter of justice in the role of patient care, and he or she is the patient's advocate within the broad constraints set by policy makers. Except in wartime, overwhelming disasters, and other instances of dire scarcity, the rationing of health care resources must occur primarily at the level of social policy. The relationship between the physician and the patient is a fiduciary one, requiring faithfulness and implying trust that the physician is acting first and foremost as the advocate of the patient. Health professionals should participate more in the formation of social policy, but as individuals their primary concern should be that of loyalty towards each patient.

Older people are often spotlighted when concerns about rising medical costs are discussed. While it is true that the care of such people accounts for a greater share of health costs, is this necessarily wrong? Many older people are, in fact, in greater need of health care because they are more vulnerable to illness. Health care is a good of our society that should be distributed according to needs rather than to some other, more mathematical, distribution.

Life expectancy continues to increase, and thus there are times when many years of life might be gained by medical treatment of older persons. Whether the quality of those years is acceptable is a decision best made by the physician and patient, perhaps in conjunction with involved family members.

PATIENT AND FAMILY TRUST

The key to maintaining faithfulness, on which the fiduciary relationship is based, is to maintain the trust which can exist between the physician and the patient. One crucial component is to engage in honest and benevolently motivated information sharing. But that is only the beginning of what 'maintaining trust' requires. From this beginning the trust between a patient and the health professionals depends upon their collaborative effort. The collaboration entails providing the best treatment possible, whether oriented towards cure, maintenance of function, or palliation.

Throughout the patient's illness a dimension of treatment involves giving emotional support to persons closest to the patient. Sometimes relatives and close friends are frustrating to health professionals because they are experiencing anger, confusion, or intense sorrow about what is happening to their loved one, and they take it out on the professionals. Their demands, worries, questions, and interference with treatment can be disconcerting, especially when their actions call into question the physician's best judgement. Even in these situations, it must be remembered that any time the patient's most intimate sources of support are alienated or harmed the patient inevitably suffers deleterious consequences, too. At best, such persons are of great assistance to the health professional's efforts; at worst, they should not be unnecessarily excluded from the health professional's support and deprived of relevant information.

CAPACITY FOR DECISION MAKING

Despite the strong legal and ethical presumption toward respect for the individual resident's autonomous right to make decisions concerning his or her own life, including choices about medical treatment and financial management, for a significant proportion of frail, elderly people the capacity to make and express clear decisions has been compromised by biological factors (for example, Alzheimer's disease, other dementing illness, stroke, and depression). The combination of illness and institutionalization may further impair the resident's ability to make and communicate autonomous choices on important matters. Even the best nursing-care facility, where resident's rights are assiduously respected, may—simply because it is a total institution—exert a debilitating influence on the patient's sense of control.

Thus the physician is forced to consider the question of a patient's capacity to participate in decisions, or as it is often referred to in legal contexts, 'competence'. While courts generally grant petitions for substitute decision-makers for frail, elderly patients, in contested cases there should be a strong preference for letting older persons make, and live or die with, their own decisions. However, the great majority of such cases are quite properly—and without adverse legal consequences—managed by the physician or nursing home, in conjunction with the family whenever possible, without formal court involvement. In most circumstances, competency should be addressed as an ethical matter by those who are closest to the patient.

In some instances, such as the patient in a long-term coma, or a persistent vegetative state, or severe dementia, the determination of incompetence is fairly straightforward. In many circumstances, however, clinical conditions leading to potential incompetence are fluctuating or uncertain. Transient incapacity may be due to acute illness or the side-effects of medication, mental illness or emotional problems, or physical handicap.

In daily practice, it is frequently the attending physician acting alone, in his or her sole discretion, who decides when a person is not capable of making decisions and a substitute should be involved, without using any explicit standards for that determination. In fact, there exists no single, uniform standard of competence. In determining competency, most scholars urge that emphasis should not be placed on the objective nature of the patient's clinical diagnosis or on the specific choice he or she makes, but rather on the person's capacity and the subjective thought process followed in arriving at a decision. The focus is on functional ability.

Under a functional inquiry, the fundamental questions suggested are these:

1. Can the person make and communicate (by spoken words or otherwise) choices concerning his or her own life?
2. Can the person offer any reasons for the choices made?
3. Are the reasons underlying the choice rational? For instance, the person who declines amputation of a gangrenous leg because she does not wish to continue living with only one leg is acting more rationally than someone who cannot express that clear reason.
4. Is the person able to understand the likely risks and benefits of the alternatives presented and the fact that those implications apply to him or her?

Under this functional approach, the patient does not need to understand the scientific theory underlying the physician's recommendations as long as he or she comprehends the general nature and likely consequences of the choices presented. Also, under this approach, competency must be determined on a decision-specific basis; that is, a patient may be capable of rationally making certain sorts of decisions but not others.

Additionally, competency may wax or wane for a particular patient according to environmental factors, such as (1) time of day; (2) day of the week; (3) physical location; (4) acute, transient medical problems; (5) other persons involved in supporting or pressuring the resident's decision; and (6) reactions to medication. Care-givers should manipulate environmental barriers wherever possible in an attempt to maximize the decision-making capacity of a resident. Thus, if a decision can be delayed until a resident is in a more lucid phase, or medications can be altered to allow the resident a clearer head to contemplate choices, this is preferable to proceeding unnecessarily on the basis of substituted decision making. Also, many acute physical or mental problems of elderly patients that impinge on the decision-making capacity can be successfully treated medically, and that course should be vigorously pursued before considering the person incompetent.

ADVANCE DIRECTIVES

Because of the high value of self-determination, every attempt is made to make medical decisions consistent with the patient's own values. If the patient cannot decide, or cannot express a decision, it is very helpful to the physician to have evidence of 'prior expressed wishes' or 'advance directives'. In the United States, the 'living will' statutes exists in most states, which designate a legally valid statement to the effect that, in the event of critical or terminal illness where recovery is extremely unlikely, the patient does not wish to be kept alive by extraordinary measures. These statements inevitably are imprecise, but do provide evidence of the patient's own values. Recently, another form of advance directive—called Durable Power of Attorney for Health Care (**DPA-HC**)—has been legislated for in several States. This kind of advance directive authorizes a specific person to make proxy decisions in the event that one cannot speak for oneself. Most experts think the DPA-HC is better than the 'living will' because it does not have to specify definitions of 'terminal' or 'critical' illness and 'heroic' or 'extraordinary' measures, but allows a friend or family member to make decisions that reflect the patient's values in any kind of situation. While advance directives are not essential to ethical decision making, the concept is very useful in geriatric medicine. Especially in a long-standing physician/patient relationship, there is the opportunity to engage the patient in discussions about life, death, and personal decision making.

While some people will not have thought about these issues, there are many who have and who are eager to discuss their views with an interested physician. As people age, most will have experienced the critical illness or death of someone close to them. Such people are more likely to have considered the implications for themselves of modern medical advances, especially life-sustaining technology. When possible, it is always better to have the patient's own views rather than those of a family member or surrogate. Even in very close families, proxy decision-makers commonly make incorrect assumptions or misinterpreted judgements about what the patient would have wanted. Especially in critical illness, knowing the patient's own expressed wishes about extraordinary measures should be a major contribution of the geriatrician.

TO FOREGO LIFE-SUSTAINING TREATMENT: NUTRITION AND FLUID

There has been growing acceptance that certain life-sustaining interventions are sometimes no longer indicated for a dying patient. Case law and medical practice have established that mechanical ventilation, cardiopulmonary resuscitation, intensive care units, and major surgical procedures ethically may be omitted if the patient refuses the treatment or if there is no reasonable chance that it will help. Except in hospice care, life-sustaining food and fluid have not been seen in this category until the past few years. While there remains considerable controversy about this issue, many experts in law and ethics agree that in permanently comatose or severely impaired patients expected to survive less than a year, nutritional support can be viewed as 'extraordinary' treatment, and thus withheld, especially if there is evidence that the patient would have wished such support removed.

The health care team should be considered in any difficult ethical decision, and the complex decisions covering feeding in frail or terminally ill elderly patients is an excellent example of the need for a multidisciplinary approach. We usually consider the geriatric health care team to include physicians, nurses, and social workers, but other people may be appropriate in ethical decision making, for example the chaplain, legal experts, family members, and in some cases the patient. Including the feelings and ideas of several relevant care givers in the decision is not always simple, but has the advantage of requiring clarification of ethical reasoning.

Tube feeding is a common example of a practice in which the feelings of others are too easily overlooked. A physician may decide not to institute tube feeding in a severely demented and bedfast patient in a long-term care facility because of the potential discomfort to the patient and that even tube feeding is too aggressive a treatment to impose on a patient who can benefit so little from it. In such a patient who is not taking nourishment, the physician may write an order to the nursing staff to feed the patient by hand. The people who actually have to carry out this order are generally nurses' aides who may spend most of their day trying to feed patients who do not want to eat or who cannot eat. Often it seems to hurt the patient or causes choking. The patient may become combative, distressed, or frightened when food is placed inside an unwilling mouth. For care-givers who must go through this with several patients, day after day, distancing and dehumanization are probably the defences that make it endurable. These psychological consequences of the decision may therefore contribute to poor care of other and future patients. On the other hand, the act of feeding a patient can be a positive, caring act with positive effects on both patient and care giver, as long as the patient is allowed to determine the pace and amount of intake.

In any ethical decision, as in any medical decision, the physician must be clear and direct. If the prognosis is not clear, especially in acute illness, a therapeutic trial, which includes adequate (if possible, optimal) nutrition, is necessary if any accurate conclusions are to be drawn about the patient's potential for recovery. It is up to the physician to provide leadership in finding solutions to the problem of how to manage this feeding, including gastrostomy or jejunostomy, and to allow nurses, dietitians, psychologists, and rehabilitation therapists to perform their work in an humane context, and to allow frustrations and moral concerns to be expressed and discussed.

If a decision is then made that feeding is not of benefit to the patient, active and supportive involvement of the physician will help care-givers deal with inevitably difficult and painful feelings about the futility of treatment and the death of the patient. Once a decision is made, in particular, a decision to withdraw or not to institute an aggressive therapy, the ethical course requires sensitivity and responsibility. The less dramatic, more intimate details of caring for a patient can and should continue and perhaps even intensify in that situation.

ALLOWING DEATH TO OCCUR AND THE QUESTION OF EUTHANASIA

Currently, in the published record for medicine and ethics, there is spirited discussion about the morality of direct intervention mercifully to end a person's life. This is contrasted with the less direct practice of allowing a person to die by stopping treatment that supports life. Unfortunately, the important moral distinctions are sometimes clouded by the plethora of terms used.

Traditionally, the term 'euthanasia' has simply meant 'a good death', but eventually it became identified with the idea of direct medical intervention (i.e., administration of a drug) mercifully to end life in persons suffering from advanced stages of incurable illness.

In contrast, the term 'letting die' refers to the use of withdrawal which results in the hastened death of a person who has an incurable illness, but the procedure itself does not actively induce death; questions of 'letting die' involve mercifully withdrawing or withholding measures. {Some United Kingdom authors regard the distinction as casuistical.}

Obviously there are instances in which the boundary between the two types of acts seems blurred. Nonetheless, the distinction enables greater understanding of the vast majority of these difficult decisions. Ethicists make a distinction between acts of commission and acts of omission. Direct intervention would be treated as an act of commission, and withdrawal or withholding as acts of omission.

From an ethical point of view, some believe that the type of act itself is a primary distinguishing factor. The argument supporting the distinction is that the motive of mercy or compassion does not extend to the direct taking of human life. One might be able to cite a rare exception, but an exception does not negate it as a rule that can reliably guide practice within the medical domain. In other words, the duty to do no harm generally includes the practice of supporting human life, rather than taking it.

In such weighty matters, one must always consider consequences to society. In fact, those who argue against the legalization of active euthanasia, even for terminal patients in extreme suffering, assert that it would erode the moral fibre of the profession and of society to allow physicians actually to administer a lethal dose of a drug to a patient. This is the same argument that is used against having physicians involved in capital punishment, even though the form of death thereby may be more merciful. Some would counter this argument with the claim that compassionate care of the dying may have to include instances of assisted voluntary suicide, and that the influence on society could be positive because responsible caring for the dying deepens our sense of humanity. This debate continues, especially in The Netherlands, but there is no legal basis, even where the most liberal practices of mercy killing or euthanasia have been instituted. Such an important and difficult issue deserves the most thorough and serious examination by health care professionals and in public discourse.

CONCLUSION

Ethics has always been a discipline central to the sound practice of medicine, but has probably never been more important than at the present time. As with any aspect of patient care, certain general guidelines apply: continuity of care and consistency of care planning, regardless of the site of care; adequate communication between the team of health care providers, and between the health care providers and the patient and family; good faith and technical competence in the delivery of care; respect and compassion for the patient; an ability to identify and analyse ethical problems; and a willingness to seek consultation if the resolution to a dilemma is not clear. The many advances of modern medicine carry both a promise of improved quality of life and of increased longevity. These same advances may represent threats to the quality of life or autonomy of a patient if prudence and thoughtfulness are not included in clinical decision making. In addition, the pressures of cost containment and shifting incentives for reimbursement pose threats of both under- and overuse of health care services by elderly people. Physicians can maintain the integrity of their relationships with patients, and also deal with their social responsibilities by developing an ethically based framework for decision making.

Bibliography

Daniels, N. (ed.) (1988). Justice between generations and health care for the elderly. *The Journal of Medicine and Philosophy*.

Elford, J.R. (ed.) (1987). Medical ethics and elderly people. Churchill Livingstone, Edinburgh.

Pearlman, R. and Jonsen, A.R. (1985). Use of quality of life considerations in medical decision making. *Journal of the American Geriatrics Society*, **33**, 344–52.

Sachs, G.A. and Cassel, C.K. (1989). Ethical aspects of dementia. *Clinics in Neurology*, **7**, 845–58.

Society for the Right to Die (1985). *The physician and the hopelessly ill patient: legal, medical and ethical guidelines*. Society for the Right to Die, New York.

Wanzer, S.H. *et al.* (1989). The physician's responsibility toward hopelessly ill patients: A second look. *New England Journal of Medicine*, **320**, 844–9.

Zweibel, N.R. and Cassel, C.K. (1988). Clinical and policy issues in the care of the nursing home patient. *Clinics in Geriatrics* 4(3). Saunders, Philadelphia.

Zweibel, N.R. and Cassel, C.K. (eds.) (1989). Treatment choices at the end of life: A comparison of decisions by older patients and their physician selected proxies. *The Gerontologist*, **29**, 615–21.

SECTION 26
Reference values for biological data in older persons

26 Reference values for biological data in older persons

H. M. HODKINSON

Reference values for many biological measurements change with age but for others there may be no significant change. This chapter looks at a variety of mechanisms that may underlie age changes whilst the appendix lists examples of values unchanged with age and gives ranges for the more important tests that are different in older persons.

GROWTH AND INVOLUTION

Perhaps the most obvious group of age-associated changes are those that directly relate to growth, the achievement of full size, and then subsequent involution. An obvious example is height, where population data show progressive increases through childhood, a particularly fast increase in adolescence representing the adolescent growth spurt, and a peak in early adult life, all these changes being due to bone growth, particularly of the long bones, until the closure of epiphyses. Thereafter, all populations will show progressive declines in height, mainly ascribed to two processes, ageing of the intervertebral discs, which become less hydrated and lose volume and height, and, particularly in older age groups and especially in women beyond the menopause, osteoporosis of the vertebrae with consequent collapse and loss of vertebral height.

Many other biological measurements show the same pattern of increase, with growth to a peak in early adult life because they are also in some way dependent on body size or on the size of an organ whose size is correlated with total body size. However, the pattern of involution may be different and due to great variety of individual processes, just as changes in intervertebral discs and vertebral osteoporosis were specific to loss of height. Thus, similar patterns of rising to a peak in early adult life and subsequent decline with age may be seen for many physiological measurements that are to some way dependent on body size, but declines have no common explanation. For example, renal function declines because of actual loss of functional nephrons whilst decline in muscle strength is related rather to changes in habitual exercise levels and not loss of muscle cells. Furthermore, the involution changes may not always be those of a decline. Levels of serum alkaline phosphatase, a main contributor to which is the bone isoenzyme derived from osteoblasts, closely reflect skeletal growth in the young. Thus levels are relatively high in children, peak in adolescence during the growth spurt, and then decline to their lowest level in early adult life. Levels then slowly rise with age (Gillibrand *et al.* 1980). This is perhaps because of greater skeletal turnover as osteoporosis becomes more prevalent and because some individuals included in reference samples may have occult Paget's disease. So here we have a series of changes which, though related to skeletal growth and subsequent involution, show a quite different time trajectory. Not all organs or structures grow in concert with overall body size, moreover. For example, the lens grows continuously throughout life and this, together with progressive loss of elasticity of lens tissue, leads to progressive decline in accommodation and presbyopia.

CHANGES IN ADULT LIFE

In the remainder of this Section the emphasis will be on the changes that occur in adult life after the growth phase of development is complete, and will explore mechanisms that may operate and so result in altered values in older persons as contrasted with younger adults.

Organ involution

The loss of functioning nephrons with age has already been given as an example; this underlies such age-associated changes as the progressive fall in glomerular filtration rate with age, which results in a steady rise in reference ranges for blood urea, serum creatinine, and serum urate with age so that, for example, the range for urea is approximately 50 per cent higher in old age than in middle life. In general, reference data for static tests are less affected by organ changes than are results from dynamic tests of function, particularly those that relate to maximum capacity of the organ or system. Thus we may contrast the modest increases in fasting blood glucose with age with the far more striking changes in glucose tolerance tests (Jackson 1984) or the maintenance of unchanged levels of thyroxine in old age whilst dynamic tests such as the TSH response to TRH administration are considerably impaired (Rochman 1988).

Sex hormone changes

There are changes in the sex steroids and their trophic hormones in both sexes, although these changes are far more clear-cut and dramatic in women, in whom ovarian secretion of oestrogens stops at the menopause. In men there is a gradual decline in androgen production with age. In both sexes, gonadotrophins rise in older age groups. Peripheral conversion of the adrenal steroid androstenedione in adipose tissue becomes the main source of oestrogens after the menopause in women, and obesity may thus confer some protection against osteoporosis in postmenopausal women because of better oestrogen levels. These changes are reviewed by Kenny and Fotherby (1984).

The actions of oestrogens on skeletal status have influences on other laboratory analyses. Thus, a fall in oestrogens results in increased osteoclastic destruction of bone, which is manifested by abrupt rises in serum calcium, phosphate, and alkaline phosphatase (McPherson *et al.* 1978), and these changes persist into old age. In the case of phosphate, quite a large sex difference opens up in old age, with old men having phosphate values some 20 to 25 per cent lower than those of women of the same age. Sex steroids have a considerable influence on the synthesis of some plasma proteins, for example the specialized carrier proteins thyroxine-binding globulin, sex hormone-binding globulin, corticosteroid-binding globulin, and caeruloplasmin. Typically, oestrogens increase levels whilst androgens reduce them and, in keeping with expectation, sex hormone-binding globulin levels increase two-fold in elderly men with the fall in testosterone levels (Kenny and Fotherby 1984) and there is a modest rise in thyroxine-binding globulin also (Freeman and Cox 1984). However, levels of the other sex steroid-dependent proteins show no gross changes with age. The modest falls in serum albumin with age that are widely reported for both men and women may also be due to reduced anabolic drive as a consequence of lower levels of sex hormones. Changes in carrier proteins with age may also affect the total levels of substances bound to them and this accounts for the modest falls of serum calcium with age reported for men.

Lifestyle changes

Many changes in biological data may relate to changes in lifestyle as individuals grow older. Most strikingly, exercise levels tend to fall throughout adult life, especially in the developed countries, and total energy expenditure falls appreciably as a consequence. Food consumption tends to fall roughly in parallel with the fall in energy expenditure but, nonetheless, body fat tends to increase with age up to a peak in late middle life whilst muscle mass falls progressively in response to lower exercise levels. The proportion of total body mass due to fat thus tends to increase with age, whilst that due to lean tissue falls. Basal metabolism per unit of body weight thus falls with age but if expressed per unit of lean body mass would seem to be unchanged. These changes in body composition have effects on other biological measurements, for example serum creatinine; creatinine, being derived from muscle protein turnover, rises less steeply with age than urea because the effects of falling muscle mass partially cancel out those of diminishing renal function.

The fall in total food intake with age also means that intakes of specific nutrients tend to fall in proportion to the decline in energy expenditure. Ranges for levels of specific nutrients such as vitamins in the body fluids may thus fall with age, particularly as for many such substances homeostatic controls are few or absent. Thus levels for nutrients such as ascorbic acid, folic acid, and vitamin A, and for enzyme activities such as the red-cell transketolase, dependent upon thiamine status, and red-cell glutathione reductase, reflecting riboflavin status, are lower in old age. Levels of vitamin D may be particularly low in old age as exposure to sunlight falls with decreasing out-of-doors activity at the same time as dietary intakes are falling.

Selective effects of survival

Older people represent a selected group of individuals who have survived and so characteristics that confer poorer prognosis may be eliminated by earlier death and so be seen less often in old age. Changes in the serum cholesterol with age may be a case in point. Levels peak in late middle life in both sexes but then fall progressively in older age groups. This could be due to earlier death of subjects with higher cholesterol values, given that these represent an important risk factor for ischaemic heart disease and that this is a major cause of death in late middle age.

Unrecognized contribution of occult disease

The possibility that the elevation of the range of serum alkaline phosphatase in older subjects might be due, at least in part, to the inclusion of individuals with unrecognized and asymptomatic Paget's disease in the reference population has already been referred to. This could have a major effect, as the disease has a prevalence of the order of 5 per cent in old age (see Chapter 14.3). Similarly, elevation of reference ranges for globulins and for the erythrocyte sedimentation rate (which is affected by changes in globulins) may also be wholly or partly due to unrecognized occult disease. Elevations of gammaglobulins are perhaps most relevant. Some elevation may be due to unrecognized specific diseases, such as myeloma or chronic infections, but others may persist for long periods without deterioration in the individual's health status, the so-called benign gammopathies. If such values are included when ranges are established they may have a considerable effect and, unfortunately, published work is often evasive as to what policies have been adopted for the inclusion or exclusion of such 'outliers'. Similar problems arise in areas affected by changes in immune surveillance with age and the development of autoimmune phenomena. So, for example, many elevated values for gastrin are found in elderly subjects and are associated with autoimmune gastritis and the presence of gastric parietal-cell antibodies. Inclusion of such elevated values from an apparently well reference population of old people gives a very much elevated range for gastrin but if all those individuals having achlorhydria or antibodies are excluded, a far lower range results. Divergent philosophical approaches to such situations can greatly complicate considerations of age-related changes in biological data; at least investigators need to make their position explicit. The completely healthy old person would be difficult to define and is likely to be something of a philosophical abstraction!

Changes in homeostasis

Homeostasis may deteriorate with age and this would result in wider ranges for the less well-controlled metabolites. This can be seen, for example, for glucose, many studies having shown a rise in both the mean and the variance of blood glucose with age. Similarly, the elevation of means for urea and creatinine with age is also accompanied by greater variance and this can be ascribed to poorer homeostasis, i.e. the decline in renal function. This is by no means a universal phenomenon in old age, however. Serum sodium may be taken as a tightly controlled variable where neither mean nor variance show any significant change with age. However, though the homeostatic control in health is unchanged in older subjects, there is considerable evidence to suggest that homeostasis is more easily overcome by illness in elderly people. Hyponatraemia is commonly seen in ill old people. This may indeed be a common feature of homeostasis in old age and one may take as another example the regulation of thyroid hormone levels. These are virtually unchanged in well old people but disturbances in the ill old patient, e.g. euthyroid hyperthyroxinaemia, are relatively common.

Increased variability of values of biological data in old people may not be simply due to altered homeostatic mechanisms, however. The inclusion of individuals with occult disease will increase variance as well as alter the mean. Variance of a skew distribution will be higher with an increase in the mean if the data are not suitably transformed to normalize them, and so some increases in variance with age are merely manifestations of a rise in mean. Conversely, some skew data whose mean falls with age will have lower apparent variance, as for example for vitamins such as ascorbic acid and vitamin D. Increased variability cannot therefore be taken to be a general phenomenon with increasing age.

REVIEWS

Reviews giving further information on age changes in reference ranges include Gillibrand *et al.* (1980), Hodkinson (1984), Rochman (1988), and Dean (1988); this last author also reviews anthropomorphic, psychological, and physiological data as well as clinical laboratory data.

References

Dean, W. (1988). *Biological aging measurement*. The Center for Biogerontology, Los Angeles.

Freeman, H. and Cox, M.L. (1984). Plasma proteins. In *Clinical biochemistry of the elderly* (ed. H.M. Hodkinson), pp. 46–74. Churchill Livingstone, Edinburgh.

Gillibrand, D., Grewall, D., and Blattler, D.P. (1980). Chemistry reference values as a function of old age and sex, including paediatric and geriatric subjects. In *Aging—its chemistry* (ed. A.A. Dietz), pp. 366–89. American Association for Clinical Chemistry, Washington.

Hodkinson, H.M. (ed.) (1984). *Clinical biochemistry of the elderly*. Churchill Livingstone, Edinburgh.

Jackson, R.A. (1984). Blood sugar and diabetes. In *Clinical biochemistry of the elderly* (ed. H.M. Hodkinson), pp. 209–36. Churchill Livingstone, Edinburgh.

Kenny, R.A. and Fotherby, K. (1984). The sex steroids and trophic hormones. In *Clinical biochemistry of the elderly* (ed. H.M. Hodkinson), pp. 246–58. Churchill Livingstone, Edinburgh.

McPerson, K., Healy, M.J.R., Flynn, F.V., Piper, K.A.J., and Garcia-Webb, P. (1978). The effect of age, sex and other factors on blood chemistry in health. *Clinica Chimica Acta*, **84**, 373–97.

Rochman, H. (1988). *Clinical pathology in the elderly*. Karger, Basel.

Appendix

Table 1 Commonly used laboratory data for which reference ranges are essentially unchanged in old age

Serum sodium
Serum bicarbonate
Serum chloride
Serum magnesium
Serum aspartate transaminase
Serum lactate dehydrogenase
Serum alanine transferase
Serum amylase
Serum 5′-nucleotidase
Serum creatinine kinase
Serum bilirubin
Haemoglobin and red-cell indices
Coagulation tests
Thyroid hormones

Table 2 Commonly used laboratory data where reference ranges are altered in older persons

Test	Reference range in old age	Change in old age
Serum albumin	33–49 g/l	Lowered
Serum globulin	20–41 g/l	Raised
Serum potassium	3.6–5.2 mmol/l	Raised
Blood urea	3.9–9.9 mmol/l	Raised
Serum creatinine	52–159 μmol/l	Raised
Serum uric acid	148–461 μmol/l	Raised
Serum calcium (women)	2.18–2.68 mmol/l	Raised
Serum calcium (men)	2.19–2.59 mmol/l	Unchanged
Serum phosphate (women)	0.94–1.56 mmol/l	Raised
Serum phosphate (men)	0.66–1.27 mmol/l	Lowered
Serum alkaline phosphatase	22–82 i.u./l	Raised
Random blood glucose	3.4–9.3 mmol/l	Raised
Leucocyte count	3100–8900 mm^3	Lowered

Index

Note: Since the main subjects of this book are ageing and the elderly, index entries have been kept to a minimum under these keywords (i.e. entries under 'Ageing' have been restricted to those relating to mechanisms, origins and theories), and readers are advised to seek more specific references.

Page numbers in **bold** refer to principal discussions in the text. Page numbers in *italics* refer to pages on which tables appear.

'vs' denotes differential diagnosis.

Abbreviations used in subentries without clarification:

HDL high-density lipoprotein NSAIDs non-steroidal anti-inflammatory drugs SLE systemic lupus erythematsosus TIA transient ischaemic attack

Additional abbreviations appear within the index.